PHARMACEUTICAL MANUFACTURING HANDBOOK

Production and Processes

PHARMACEUTICAL MANUFACTURING HANDBOOK

Production and Processes

SHAYNE COX GAD, PH.D., D.A.B.T.
Gad Consulting Services
Cary, North Carolina

WILEY-INTERSCIENCE
A JOHN WILEY & SONS, INC., PUBLICATION

Published by John Wiley & Sons, Inc., Hoboken, New Jersey
Published simultaneously in Canada

For general information on our other products and services or for technical support, please contact our Customer Care Department within the United States at (800) 762-2974, outside the United States at (317) 572-3993 or fax (317) 572-4002.

Wiley also publishes its books in a variety of electronic formats. Some content that appears in print may not be available in electronic formats. For more information about Wiley products, visit our web site at www.wiley.com.

Library of Congress Cataloging-in-Publication Data is available.

ISBN: 978-0-470-25958-0

Printed in the United States of America

10 9 8 7 6 5 4 3 2 1

CONTRIBUTORS

Susanna Abrahmsén-Alami, AstraZeneca R&D Lund, Lund, Sweden, *Oral Extended-Release Formulations*

James Agalloco, Agalloco & Associates, Belle Mead, New Jersey, *Sterile Product Manufacturing*

Fakhrul Ahsan, Texas Tech University, Amarillo, Texas, *Nasal Delivery of Peptide and Nonpeptide Drugs*

James Akers, Akers Kennedy & Associates, Kansas City, Missouri, *Sterile Product Manufacturing*

Raid G. Alany, The University of Auckland, Auckland, New Zealand, *Ocular Drug Delivery; Microemulsions as Drug Delivery Systems*

Monique Alric, Université d'Auvergne, Clermont-Ferrand, France, *Recombinant* Saccharomyces Cerevisiae *as New Drug Delivery System to Gut: In Vitro Validation and Oral Formulation*

Sacide Alsoy Altinkaya, Izmir Institute of Technology, Urla-Izmir, Turkey, *Controlled Release of Drugs from Tablet Coatings*

Maria Helena Amaral, University of Porto, Porto, Portugal, *Vaginal Drug Delivery*

Anil Kumar Anal, Living Cell Technologies (Global) Limited, Auckland, New Zealand, *Controlled-Release Dosage Forms*

Gavin Andrews, Queen's University Belfast, Belfast, Northern Ireland, *Effects of Grinding in Pharmaceutical Tablet Production*

Sophia G. Antimisiaris, School of Pharmacy, University of Patras, Rio, Greece, *Liposomes and Drug Delivery*

Robert D. Arnold, The University of Georgia, Athens, Georgia, *Biotechnology-Derived Drug Product Development*

C. Scott Asbill, Samford University, Birmingham, Alabama, *Transdermal Drug Delivery*

Maria Fernanda Bahia, University of Porto, Porto, Portugal, *Vaginal Drug Delivery*

Bernard Bataille, University of Montpelier 1, Montpellier, France, *Tablet Design*

Gerald W. Becker, SSCI, West Lafayette, Indiana, *Biotechnology-Derived Drug Product Development; Regulatory Considerations in Approval of Follow-On Protein Drug Products*

B. Wayne Bequette, Rensselaer Polytechnic Institute, Troy, New York, *From Pilot Plant to Manufacturing: Effect of Scale-Up on Operation of Jacketed Reactors*

Erem Bilensoy, Hacettepe University Faculty of Pharmacy, Ankara, Turkey, *Cyclodextrin-Based Nanomaterials in Pharmaceutical Field*

Stéphanie Blanquet, Université d'Auvergne, Clermont-Ferrand, France, *Recombinant* Saccharomyces Cerevisiae *as New Drug Delivery System to Gut: In Vitro Validation and Oral Formulation*

Gary W. Bumgarner, Samford University, Birmingham, Alabama, *Transdermal Drug Delivery*

Isidoro Caraballo, University of Sevilla, Seville, Spain, *Tablet Design*

Stephen M. Carl, Purdue University, West Lafayette, Indiana, *Biotechnology-Derived Drug Product Development; Regulatory Considerations in Approval of Follow-On Protein Drug Products*

Sudhir S. Chakravarthi, University of Nebraska Medical Center, College of Pharmacy, Omaha, Nebraska, *Biodegradable Nanoparticles*

D.F. Chowdhury, University of Oxford, Oxford, United Kingdom, *Pharmaceutical Nanosystems: Manufacture, Characterization, and Safety*

Barbara R. Conway, Aston University, Birmingham, United Kingdom, *Solid Dosage Forms*

José das Neves, University of Porto, Porto, Portugal, *Vaginal Drug Delivery*

Osama Abu Diak, Queen's University Belfast, Belfast, Northern Ireland, *Effects of Grinding in Pharmaceutical Tablet Production*

Brit S. Farstad, Instititue for Energy Technology, Isotope Laboratories, Kjeller, Norway, *Radiopharmaceutical Manufacturing*

Dimitrios G. Fatouros, School of Pharmacy and Biomedical Sciences, Portsmouth, England, *Liposomes and Drug Delivery*

Jelena Filipoviç-Grčič, Faculty of Pharmacy and Biochemistry, University of Zagreb, Zagreb, Croatia, *Nasal Powder Drug Delivery*

Eddy Castellanos Gil, Center of Pharmaceutical Chemistry and University of Havana, Havana, Cuba; University of Sevilla, Seville, Spain; University of Montpelier 1, Montpellier, France, *Tablet Design*

Anita Hafner, Faculty of Pharmacy and Biochemistry, University of Zagreb, Zagreb, Croatia, *Nasal Powder Drug Delivery*

A. Atilla Hincal, Hacettepe University Faculty of Pharmacy, Ankara, Turkey, *Cyclodextrin-Based Nanomaterials in Pharmaceutical Field*

Michael Hindle, Virginia Commonwealth University, Richmond, Virginia, *Aerosol Drug Delviery*

Bhaskara R. Jasti, University of the Pacific, Stockton, California, *Semisolid Dosages: Ointments, Creams, and Gels*

Yiguang Jin, Beijing Institute of Radiation Medicine, Beijing, China, *Nanotechnology in Pharmaceutical Manufacturing*

David Jones, Queen's University Belfast, Belfast, Northern Ireland, *Effects of Grinding in Pharmaceutical Tablet Production*

Anne Juppo, University of Helsinki, Helsinki, Finland, *Oral Extended-Release Formulations*

Paraskevi Kallinteri, Medway School of Pharmacy, Universities of Greenwich and Kent, England, *Liposomes and Drug Delivery*

Gregory T. Knipp, Purdue University, West Lafayette, Indiana, *Biotechnology-Derived Drug Product Development; Regulatory Considerations in Approval of Follow-On Protein Drug Products*

Anette Larsson, Chalmers University of Technology, Göteborg, Sweden, *Oral Extended-Release Formulations*

Beom-Jin Lee, Kangwon National University, Chuncheon, Korea, *Pharmaceutical Preformulation: Physiochemical Properties of Excipients and Powders and Tablet Characterization*

Xiaoling Li, University of the Pacific, Stockton, California, *Semisolid Dosages: Ointments, Creams, and Gels*

David J. Lindley, Purdue University, West Lafayette, Indiana, *Biotechnology-Derived Drug Product Development*

Roberto Londono, Washington State University, Pullman, Washington, *Liquid Dosage Forms*

Ravichandran Mahalingam, University of the Pacific, Stockton, California, *Semisolid Dosages: Ointments, Creams, and Gels*

Kenneth R. Morris, Purdue University, West Lafayette, Indiana, *Biotechnology-Derived Drug Product Development; Regulatory Considerations in Approval of Follow-On Protein Drug Products*

Erin Oliver, Rutgers, The State University of New Jersey, Piscataway, New Jersey, *Biotechnology-Derived Drug Product Development; Regulatory Considerations in Approval of Follow-On Protein Drug Products*

Iván Peñuelas, University of Navarra, Pamplona, Spain, *Radiopharmaceutical Manufacturing*

Omanthanu P. Perumal, South Dakota State University, Brookings, South Dakota, *Role of Preformulation in Development of Solid Dosage Forms*

Katharina M. Picker-Freyer, Martin-Luther-University Halle-Wittenberg, Institute of Pharmaceutics and Biopharmaceutics, Halle/Saale, Germany, *Tablet Production Systems*

Satheesh K. Podaralla, South Dakota State University, Brookings, South Dakota, *Role of Preformulation in Development of Solid Dosage Forms*

Dennis H. Robinson, University of Nebraska Medical Center, College of Pharmacy, Omaha, Nebraska, *Biodegradable Nanoparticles*

Arcesio Rubio, Caracas, Venezuela, *Liquid Dosage Forms*

Maria V. Rubio-Bonilla, Research Associate, College of Pharmacy, Washington State University, Pullman, Washington, *Liquid Dosage Forms*

Ilva D. Rupenthal, The University of Auckland, Auckland, New Zealand, *Ocular Drug Delivery*

Maria Inês Rocha Miritello Santoro, Department of Pharmacy, Faculty of Pharmaceutical Sciences, University of São Paulo, São Paulo, Brazil, *Packaging and Labeling*

Helton M.M. Santos, University of Coimbra, Coimbra, Portugal, *Tablet Compression*

Raymond K. Schneider, Clemson University, Clemson, South Carolina, *Clean-Facility Design, Construction, and Maintenance Issues*

Anil Kumar Singh, Department of Pharmacy, Faculty of Pharmaceutical Sciences, University of São Paulo, São Paulo, Brazil, *Packaging and Labeling*

João J.M.S. Sousa, University of Coimbra, Coimbra, Portugal, *Tablet Compression*

Shunmugaperumal Tamilvanan, University of Antwerp, Antwerp, Belgium, *Progress in Design of Biodegradable Polymer-Based Microspheres for Parenteral Controlled Delivery of Therapeutic Peptide/Protein; Oil-in-Water Nanosized Emulsions: Medical Applications*

Chandan Thomas, Texas Tech University, Amarillo, Texas, *Nasal Delivery of Peptide and Nonpeptide Drugs*

Gavin Walker, Queen's University Belfast, Belfast, Northern Ireland, *Effects of Grinding in Pharmaceutical Tablet Production*

Jingyuan Wen, The University of Auckland, Auckland, New Zealand, *Microemulsions as Drug Delivery Systems*

Hui Zhai, Queen's University Belfast, Belfast, Northern Ireland, *Effects of Grinding in Pharmaceutical Tablet Production*

CONTENTS

PREFACE

This *Handbook of Manufacturing Techniques* focuses on a new aspect of the drug development challenge: producing and administering the physical drug products that we hope are going to provide valuable new pharmacotherapeutic tools in medicine. These 34 chapters cover the full range of approaches to developing and producing new formulations and new approaches to drug delivery. Also addressed are approaches to the issues of producing and packaging these drug products (that is, formulations). The area where the most progress is possible in improving therapeutic success with new drugs is that of better delivery of active drug molecules to the therapeutic target tissue. In this volume, we explore current and new approaches to just this issue across the full range of routes (oral, parenteral, topical, anal, nasal, aerosol. ocular, vaginal, and transdermal) using all sorts of forms of formulation. The current metrics for success of new drugs in development once they enter the clinic (estimated at ranging from as low as 2% for oncology drugs to as high as 10% for oral drugs) can likely be leveraged in the desired direction most readily by improvements in this area of drug delivery.

The *Handbook of Manufacturing Techniques* seeks to cover the entire range of available approaches to getting a pure drug (as opposed to a combination product) into the body and to its therapeutic tissue target. Thanks to the persistent efforts of Michael Leventhal, these 34 chapters, which are written by leading practitioners in each of these areas, provide coverage of the primary approaches to these fundamental problems that stand in the way of so many potentially successful pharmacotherapeutic interventions.

SECTION 1

MANUFACTURING SPECIALTIES

1.1

BIOTECHNOLOGY-DERIVED DRUG PRODUCT DEVELOPMENT

STEPHEN M. CARL,[1] DAVID J. LINDLEY,[1] GREGORY T. KNIPP,[1] KENNETH R. MORRIS,[1] ERIN OLIVER,[2] GERALD W. BECKER,[3] AND ROBERT D. ARNOLD[4]

[1]*Purdue University, West Lafayette, Indiana*
[2]*Rutgers, The State University of New Jersey, Piscataway, New Jersey*
[3]*SSCI, West Lafayette, Indiana*
[4]*The University of Georgia, Athens, Georgia*

Contents

Pharmaceutical Manufacturing Handbook: Production and Processes, edited by Shayne Cox Gad

1.1.1 INTRODUCTION

Although the origins of the first biological and/or protein therapeutics can be traced to insulin in 1922, the first biotechnology-derived pharmaceutical drug product approved in the United States was Humulin in 1982. In the early stages of pharmaceutical biotechnology, companies that specialized primarily in the development of biologicals were the greatest source of research and development in this area. Recent advances in molecular and cellular biological techniques and the potential clinical benefits of biotechnology drug products have led to a substantial increase in their development by biotechnology and traditional pharmaceutical companies. In terms of pharmaceuticals, the International Conference on Harmonization (ICH) loosely defines biotechnology-derived products with biological origin products as those that are "well-characterized proteins and polypeptides, their derivatives and products of which they are components, and which are isolated from tissues, body fluids, cell cultures, or produced using rDNA technology" [1]. In practical terms, biological and biotechnology-derived pharmaceutical agents encompass a number of therapeutic classes, including cytokines, erythropoietins, plasminogen activators, blood plasma factors, growth hormones and growth factors, insulins, monoclonal antibodies, and vaccines [1]. Additionally, short interfering and short hairpin ribonucleic acids (siRNA, shRNA) and antisense oligonucleotide therapies are generally characterized as biotechnology-derived products.

According to the biotechnology advocacy group, The Biotechnology Industry Organization (BIO), pharmaceutical-based biotechnology represents over a $30 billion dollar a year industry and is directly responsible for the production of greater than 160 drug therapeutics and vaccines [2]. Furthermore, there are more than 370 biotechnology-derived drug products and vaccines currently in clinical trials around the world, targeting more than 200 diseases, including various cancers, Alzheimer's disease, heart disease, diabetes, multiple sclerosis, acquired immunodeficiency syndrome (AIDS), and arthritis. While the clinical value of these products is well recognized, a far greater number of biotechnology-derived drug products with therapeutic potential for life-altering diseases have failed in development.

As the appreciation of the clinical importance and commercial potential for biological products grows, new challenges are arising based on the many technological limitations related to the development and marketing of these complex agents. Additionally, the intellectual property protection of an associated agent might not provide a sufficient window to market and regain the costs associated with the discovery, research, development, and scale-up of these products. Therefore, to properly estimate the potential return on investment, a clear assessment of potential therapeutic advantages and disadvantages, such as the technological limitations in the rigorous characterization required of these complex therapeutic agents to gain Food and Drug Administration (FDA) approval, is needed prior to initiating research. Clearly, research focused on developing methodologies to minimize these technological limitations is needed. In doing so we hypothesize the attrition rate can be reduced and the number of companies engaged in the development of biotechnology-derived products and diversity of products will continue to expand.

Technological limitations have limited the development of follow-on, or generic biopharmaceutical products that have lost patent protection. In fact, the potential pitfalls associated with developing these compounds are so diverse that regulatory guidance concerning follow-on biologics is relatively obscure and essentially notes that products will be assessed on a case-by-case basis. The reader is encouraged to see Chapter 1.2 for a more detailed discussion concerning regulatory perspectives pertaining to follow-on biologics.

Many of the greatest challenges in producing biotechnology-derived pharmaceuticals are encountered in evaluating and validating the chemical and physical nature of the host expression system and the subsequent active pharmaceutical ingredient (API) as they are transferred from discovery through to the development and marketing stages. Although this area is currently a hotbed of research and is progressing steadily, limitations in analytical technologies are responsible for a high degree of attrition of these compounds. The problem is primarily associated with limited resolution of the analytical technologies utilized for product characterization. For example, without the ability to resolve small differences in secondary or tertiary structure, linking changes to product performance or clinical response is impossible. The biological activity of traditional small molecules is related directly to their structure and can be determined readily by nuclear magnetic resonance (NMR), X-Ray crystallography (X-ray), mass spectrometry (MS), and/or a combination of other spectroscopic techniques. However, methodologies utilized for characterizing biological agents are limited by resolution and reproducibility. For instance, circular dichroism (CD) is generally considered a good method to determine secondary structural elements and provides some information on the folding patterns (tertiary structure) of proteins. However, CD suffers from several limitations, including a lower resolution that is due in part to the sequence libraries used to deconvolute the spectra. To improve the reliability of determining the secondary and tertiary structural elements, these databases need to be developed further. An additional example is the utility of two-dimensional NMR (2D-NMR) for structural determination. While combining homonuclear and heteronuclear experimental techniques can prove useful in structural determination, there are challenges in that 2D-NMR for a protein could potentially generate thousands of signals. The ability to assign specific signals to each atom and their respective interactions is a daunting task. Resolution between the different amino acids in the primary sequence and their positioning in the covalent and folded structures become limited with increasing molecular weight. Higher dimensional techniques can be used to improve resolution; however, the resolution of these methods remains limited as the number of amino acids is increased.

Understanding the limitations of the analytical methodologies utilized for product characterization has led to the development of new experimental techniques as well as the refined application of well-established techniques to this emerging field. Only through application of a number of complementary techniques will development scientists be able to accurately characterize and develop clinically useful products. Unfortunately, much of the technology is still in its infancy and does not allow for a more in-depth understanding of the subtleties of peptide and protein processing and manufacturing. For instance, many of the analytical techniques utilized for characterization will evaluate changes to product conformation on the macroscopic level, such as potential denaturation or changes in folding, as observed with CD. However, these techniques do not afford the resolution to identify subtle changes in conformation that may either induce chemical or physical instabilities or unmask antigenic epitopes.

Further limiting successful product development is a lack of basic understanding as to critical manufacturing processes that have the potential to affect the structural integrity and activity of biopharmaceuticals. As with traditional small molecules, stresses associated with the different unit operations may affect biopharmaceutical products differently. In contrast to traditional small molecules, there is considerable difficulty in identifying potentially adverse affects, if any, that a particular unit operation may have on the clinically critical structural elements of a drug. Considering that many proteins exhibit a greater potential for degradation from shear stress, it is particularly important to assess any negative effects of mixing, transport through tubing, filtration, and filling operations. Essentially all unit operations for a given manufacturing process could create enough shear stress to induce minor structural changes that could lead to product failure. The difficulty is establishing what degree of change will have an impact on the stability, bioactivity, or immunogenic potential of the compound. Unfortunately, unless exhaustive formulation development studies are conducted, coupled with a comprehensive spectrum of analytical methodologies, these effects may not be readily evident until after scale-up of the manufacturing process or, worse yet, in the clinical setting. Moreover, modeling shear and stress using fluid dynamic structurally diverse molecules is a foreboding task. Extending these models to validate process analytical technologies (PAT) and incorporate critical quality by design (QbD) elements in the development process for a collection of biopharmaceuticals would be largely hindered by the daunting nature of the task at hand.

The use of biological systems to produce these agents results in additional variability. Slight changes in nutrient profile could affect growth patterns and protein expression of cultured cells. Furthermore, microbial contamination in the form of viruses, bacteria, fungi, and mycoplasma can be introduced during establishment of cell lines, cell culture/fermentation, capture and downstream processing steps, formulation and filling operations, or drug delivery [3]. Therefore, establishing the useful life span of purification media and separation columns remains a critical issue for consistently producing intermediates and final products that meet the defined quality and safety attributes of the product [4]. In short, understanding the proper processability and manufacturing controls needed has been a major hurdle that has kept broader development of biopharmaceutical products relatively limited.

Notwithstanding the many technological hurdles to successfully develop a pharmaceutically active biotechnology product, they offer many advantages in terms of therapeutic potency, specificity, and target design (not generally limited to a particular class or series of compounds). This is an iterative approach, whereby every new approved compound, new lessons, and applications to ensure successful product development are realized, thereby adding to our knowledge base and facilitating the development of future products. This chapter will discuss some of the fundamental issues associated with successful biopharmaceutical drug product development and aims to provide an understanding of the subtleties associated with their characterization, processing, and manufacturing.

1.1.2 FORMULATION ASSESSMENT

In order to select the most appropriate formulation and route of administration for a drug product, one must first assess the properties of the API, the proposed therapeutic indication, and the requirements/limitations of the drug and the target patient population. Development teams are interdisciplinary comprised of individuals with broad expertise, for example, chemistry, biochemistry, bioengineering, and pharmaceutics, that can provide insight into the challenges facing successful product development. As such, knowledge gained through refinement of the API manufacturing process through to lead optimization is vital to providing an initial starting point for success. Information acquired, for example, in the way of analytical development and API characterization, during drug discovery or early preclinical development that can be applied to final drug product development may contribute to shorter development times of successful products.

The host system utilized for API production is critical to the production of the final product and will determine the basic and higher order physicochemical characteristics of the drug. Typically biopharmaceuticals are manufactured in *Escherichia coli* as prokaryotic and yeast and Chinese hamster ovary (CHO) cells as eukaryotic expression systems [5]. While general procedures for growth condition optimization and processing and purification paradigms have emerged, differences in posttranslational modifications and host–system related impurities can exist even with relatively minor processing changes within a single production cell line [5]. Such changes have the potential to alter the biopharmaceutical properties of the active compound, its bioactivity, and its potential to elicit adverse events such as immunogenic reactions. These properties will be a common theme as they could potentially play a major role in both analytical and formulation development activities.

During the process of lead optimization, characterization work is performed that would include a number of parameters that are critical to formulation and analytical development scientists. The following information is a minimalist look at what information should be available to support product development scientists:

- Color
- Particle size and morphology (for solid isolates)
- Thermoanalytical profile (e.g., T_g for lyophiles)
- Hygroscopicity
- Solubility with respect to pH
- Apparent solution pH

- Number and pK_a of ionizable groups
- Amino acid sequence
- Secondary and tertiary structural characteristics
- Some stability parameters with respect to
  - pH
  - Temperature
  - Humidity
  - Light
  - Mechanical stress
  - Oxygen sensitivity
- Impurity profile
  - Misfolded/misaligned active
  - Potential isoforms
  - Expression system impurities
- Potency [median inhibitory concentration (IC_{50})]
- Animal Pharmacokinetic/Pharmacodynamic (PK/PD) and T_{ox} profiles

All of the above information will prove invaluable in determining the potential methods for rational drug delivery. Particular attention should be paid to the relative hygroscopicity of the API, of course, any stability information, as well as the impurity profile and ADMET (absorption, distribution, metabolism, excretion, and toxicity) information. In short, the more information that is available when development activities are initiated, the easier it is to avoid common pitfalls and make development decisions more rationally.

1.1.2.1 Route of Administration and Dosage

Biologics are traditionally very potent molecules that may require only picomolar blood concentrations to elicit a therapeutic effect. Given that the amount of drug required per dosage will be commensurate with the relative potency of the molecule, small concentrations are generally required for any unit dose. Biopharmaceuticals typically have large molecular weights relative to conventional pharmaceutical agents, which may be increased further by posttranslational modifications. The pharmacokinetics (ADMET) of biotechnology products have been reviewed elsewhere [6], but generally they have short circulating half-lives [7]. As such, biological products are most often delivered parenterally and formulated as solutions, suspensions, or lyophilized products for reconstitution [8, 9]. However, one must first ascertain the potential physiological barriers to drug delivery and efficacy before assessing potential routes of administration. These barriers may include actual physical barriers, such as a cell membrane, that could restrict the drug from reaching its site of action or chemical barriers, including pH or enzymatic degradation. Based on current drug delivery approaches, the proteinaceous nature of biological products limits their peroral delivery due to their susceptibility to proteases and peptidases present in the gastrointestinal tract as well as size limitations for permeating through absorptive enterocytes [10].

Difficulties in peroral delivery have stimulated researchers to explore alternate delivery mechanisms for biologics, such as through the lungs or nasal mucosa [11, 12]. Further, advances in technology and our understanding of the mechanisms limiting oral delivery of biotechnology products have led to innovative drug delivery approaches to achieve sufficient oral bioavailability. However, no viable products have successfully reached the market [13]. As a result of the technological limitations inherent in biopharmaceutical delivery, these compounds are largely delivered parenterally through an injection or implant.

When assessing the potential routes of administration, one must consider the physicochemical properties of the drug, its ADMET properties, the therapeutic indication, and the patient population, some of which are discussed below. Table 1 provides a list of some of those factors that must be addressed when determining the most favorable route of administration and the subsequent formulation for delivery. Ideally the route of administration and subsequent formulation will be optimized after identifying critical design parameters to satisfy the needs of patients and health care professionals alike while maintaining the safety and efficacy of the product.

Parenteral administration is the primary route of delivering biopharmaceutical agents (e.g., insulin); however, issues associated with patient compliance with administration of short-acting molecules are a challenge. Yet, the risk-to-benefit ratio must be weighed when determining such fundamental characteristics of the final dosage form. For instance, a number of biopharmaceutical compounds are administered subcutaneously, but this route of parenteral administration exhibits the highest potential for immunogenic adverse events due to the presence of Langerhans cells [14]. A compound's immunogenic potential is related to a host of factors, both

TABLE 1 Factors That Determine Route of Administration

Site of action
Therapeutic indication
Dosage
Potency/biological activity
Pharmacokinetic profile
Absorption time from tissue vs. IV
Circulating half-life
Distribution and elimination kinetics
Toxicological profile
Immunogenic potential
Patient population characteristics
Disease state
Pathophysiology
Age
Pharmacodynamic profile
Onset and duration of action
Required clinical effect
Formulation considerations
Stability
Impurity profile

patient and treatment related, however, if an alternate, potentially safer route of administration is available, it may be prudent to consider it. Other factors, such as the frequency of dosing (especially into an immune organ such as the skin) and the duration of treatment, can also dramatically increase the potential for immunogenic reactions [14]. Many of the factors that contribute to the immunogenic potential of biopharmaceuticals, such as impurities, degradation products, and native antigenic epitopes, can be mitigated through altering the physicochemical properties of the drug (e.g., pegylation [15, 16], acylation [17, 18], increased glycosylation to mask epitopes [19]) or changing the characteristics of the formulation [20, 21]. In reality, the pharmaceutical industry has done a good job of recognizing the potential implications of immunogenic reactions and readily embraced technologies that can either mask or eliminate potential antigenic epitopes. However, additional research is needed to further identify and remove immunogenic epitopes.

1.1.2.2 Pharmacokinetic Implications to Dosage Form Design

Biological agents are generally eliminated by metabolism into di- and tripeptides, amino acids, and smaller components for subsequent absorption as nutrients or clearance by the kidney, liver, or other routes. Renal elimination of peptides and proteins occur primarily via three distinct mechanisms. The first involves the glomerular filtration of low-molecular-weight proteins followed by reabsorption into endocytic vesicles in the proximal tubule and subsequent hydroysis into small peptide fragments and amino acids [22]. Interleukin 11 (IL-11) [23], IL-2 [24], insulin [25], and growth hormone [26] have been shown to be eliminated by this method. The second involves hydrolysis of the compound at the brush border of the lumen and subsequent reabsorption of the resulting metabolites [6]. This route of elimination applies to small linear peptides such as angiotensin I and II, bradykinin, glucagons, and leutinizing hormone releasing hormone (LHRH) [6, 27, 28]. The third route of renal elimination involves peritubular extraction from postglomerular capillaries and intracellular metabolism [6]. Hepatic elimination may also play a major role in the metabolism of peptides and proteins; however, reticuloendothelial elimination is by far the primary elimination route for large macromolecular compounds [29].

Biopharmaceutical drug products are subject to the same principles of pharmacokinetics and exposure/response correlations as conventional small molecules [6]. However, these products are subject to numerous pitfalls due to their similarity to nutrients and endogenous proteins and the evolutionary mechanisms to break them down or prevent absorption. The types of pharmacokinetic-related problems that a biotechnology drug development team may encounter range from lack of specificity and sensitivity of bioanalytical assays to low bioavailability and rapid drug elimination from the system [6]. For example, most peptides have hormone activity and usually short elimination half-lives which can be desirable for close regulation of their endogenous levels and function. On the other hand, some proteins such as albumin or antibodies have half-lives of several days and formulation strategies must be designed to account for these extended elimination times [6]. For example, the reported terminal half-life for SB209763, a humanized monoclonal antibody against respiratory syncytial virus, was reported as 22–50 days [30]. Furthermore, some peptide and protein products that persist in the bloodstream exhibit the potential for idiosyncratic adverse affects as well as increased immunogenic potential. Therefore, the indication and formulation strategy can prove crucial design parameters simply based on clearance mechanisms.

1.1.2.3 Controlled-Release Delivery Systems

Given that the majority of biopharmaceutical products are indicated for chronic conditions and may require repeated administrations, products may be amenable to controlled-release drug delivery systems. Examples include Lupron Depot (leuprolide acetate), which is delivered subcutaneously in microspheres [31], and Viadur, which is implanted subcutaneously [32]. Various peptide/protein controlled delivery systems have been reviewed recently by Degim and Celebi and include biodegradable and nondegradable microspheres, microcapsules, nanocapsules, injectable implants, diffusion-controlled hydrogels and other hydrophilic systems, microemulsions and multiple emulsions, and the use of iontophoresis or electroporation [33]. These systems offer specific advantages over traditional delivery mechanisms when the drug is highly potent and if prolonged administration greater than one week is required [5, 33]. However, each of these systems has its own unique processing and manufacturing hurdles that must be addressed on a case-by-case basis. These factors, coupled with the difficulties of maintaining product stability, limit the widespread application of these technologies. However, the introduction of postapproval extended-release formulations may also provide the innovator company extended patent/commercial utility life and, as such, remains a viable option for postmarketing development. A current example of this is observed in the development of a long-acting release formulation of Amylin and Eli Lilly's co-marketed Byetta product.

1.1.3 ANALYTICAL METHOD DEVELOPMENT

The physical and chemical characterization of any pharmaceutical product is only as reliable as the quality of the analytical methodologies utilized to assess it. Without question, the role of analytical services to the overall drug product development process is invaluable. Good analytical testing with proper controls could mean the difference between a marketable product and one that is eliminated from development. Analytical methodologies intended for characterization and/or assessment of marketed pharmaceutical products must be relevant, validatable, and transferable to manufacturing/quality assurance laboratories.

1.1.3.1 Traditional and Biophysical Analytical Methodologies

Typically, there are a handful of traditional analytical methodologies that are utilized to assess the physical, chemical, and microbiological attributes of small-molecule pharmaceutical products. While many of these testing paradigms can still be utilized to assess biopharmaceuticals, these molecules require additional biophysical, microbiological, and immunogenic characterization as well. In brief, analytical methodologies should evaluate the purity and bioactivity of the product and must also be suitable to assess potential contaminants from expression systems as well as different isoforms and degradation products of the active. Biophysical methodologies allow for assessment of the structural elements of the product with respect to its activity. Such assessments include structural elements, such as the folding of the molecule, and also encompass potential posttranslational modifications and their impact on structure. A list of typical analytical parameters and methodologies utilized to assess those parameters can be found in Table 2.

The impact of a molecule's biophysical characteristics on its clinical efficacy should be readily quantifiable. With respect to rational drug design, it is also extremely important to minimize external factors that may influence the formation of any adverse response. One such factor is the presence of degradation products and drug-related impurities that may be responsible for an immune response. One such industrial example is granulocyte-macrophage colony-stimulating factor [GM-CSF, or Leukine (sargramostim), by Berlex Co.], which is produced as a recombinant protein synthesized and purified from a yeast culture, *Saccharomyces cerevisiae*. As expected, the expression system has an impact on the final product: sargramostim, manufactured from *S. cerevisiae*, yields an O-glycosylated protein, while molgramostim (Leucomax), synthesized using an *E. coli* expression system, is nonglycosylated [34]. The *E. coli*–derived product exhibited a higher incidence of adverse reactions in clinical trials and never made it to the market. With respect to the drug product, the immunogenic reactions included [34, 35]:

TABLE 2 Analytical Methodologies and Their Utility for API and Drug Product Characterization

Parameter Assessed	Methodologies	Utility
Appearance	Visual appearance, colorimetric assays, turbidity	Simple determination of physical stability, i.e., are there particles in solution, is the solution the correct color/turbidity? Is the container closure system seemingly intact?
Purity, degradation products and related substances	GPC/SEC-HPLC, RP-HPLC, gel electrophoresis, immunoassays, IEF, MS, CD, CE	Gives a general idea of the relative purity of the API and the drug product. Are there impurities related to the expression system? Are there alternate API isoforms present? Can degradation products be distinguished from the active component(s)?
Molecular weight determination	GPC/SEC-HPLC, gel electrophoresis, multiangle laser light scattering (MALLS), laser diffraction	Is the product a single molecular weight or polydisperse? Is the molecular weight dependent on posttranslational modifications?
Potency	Biological activity (direct or indirect)	Does the compound have reproducible in vitro activity and can this be correlated to in vivo?
pH	Potentiometric assays	Is the product pH labile or do pH changes affect potency is such ways that are not evident in other assays, i.e., minimal degradation and/or unfolding?

TABLE 2 ***Continued***

Parameter Assessed	Methodologies	Utility
Primary structural elements	Protein sequencing, *N*-term degradation (Edman degradation), peptide mapping, amino acid composition, 2D-NMR	Verifies primary amino acid sequence and gives preliminary insight into activity.
Secondary structural elements	CD, 2D-NMR, in silico modeling from AA sequence	Secondary structural elements result from the primary sequence and help define the overall conformation (3D folding) of the compound.
Tertiary structural elements	Disulfide content/position, CD	Determines correct folding and overall integrity of the 3D product. Qualitative determination for denaturation potential. Also correlates to immunogenic potential.
Agglomeration/ aggregation	Subvisual and visual Particle size analysis, immunogenicity	Indicator of physical instability. Also gives an indication of immunogenic potential.
Carbohydrate analysis	RP-HPLC, gel electrophoresis, AE-HPLC, CE, MALDI-MS, ES-MS, enzyme arrays	Ensures proper posttranslational modifications and carbohydrate content.
Water content (lyophilized products)	Karl Fischer, TGA, NIR	Indicator of hydrolytic potential and process efficiency.
Immunogenic potential	Surface plasmon resonance, ELISA, immunoprecipitation	Methodologies generally only give positive/negative indicators of immunogenic potential. In vitro methodologies do not always correlate to in vivo.
Sterility	Membrane filtration	Indicator of microbial contaminants from manufacturing operations.
Bacterial endotoxins	Limulus amebocyte lysate (LAL)	Gives an idea of processing contaminants and potentially host organism contaminants.
Container closure integrity	Dye immersion, NIR, microbial ingress/sterility	Demonstrates viability of container closure system over the life of the product.

Abbreviations: gel permeation chromatography (GPC), size exclusion chromatography (SEC), high-performance, or high-pressure, liquid chromatography (HPLC), reverse phase (RP), isoelectric focusing (IEF), mass spectrometry (MS), circular dichroism (CD), capillary electrophoresis (CE), nuclear magnetic resonance (NMR), anion exchange (AE), matrix-assisted laser desorption ionization (MALDI), electrospray ionization (ES), thermogravimetric analysis (TGA), near infrared (NIR), enzyme-linked immunosorbent assay (ELISA)

1. Formation of antibodies which bind and neutralize the GM-CSF
2. Formation of antibodies which bind but do not affect the efficacy of GM-CSF
3. Antibody formation against proteins not related to GM-CSF, but to proteins from the expression system (*E. coli*)
4. Antibodies formed against both product- and non-product-related proteins
5. No antibody formation

This example clearly illustrates not only the range of clinical manifestations with respect to antibody formation to drug therapy but also how the choice of an expression system can affect the final product. In this example, the expression system was responsible for the adverse events reported. This finding is certainly clinically relevant considering the homologous product, sargramostim, has been on the U.S. market for quite some time.

The above example also gives an indication of the relative importance of carbohydrate analysis. Without question, protein glycosylation is the most complex of all posttranslational modifications made in eukaryotic cells, the importance of which cannot be underestimated. For many compounds, glycosylation can readily affect protein solubility (as influenced by folding), protease resistance, immunogenicity, and pharmacokinetic/pharmacodynamic profiles (i.e., clearance and efficacy) [36]. Typical analytical methodologies used to assess carbohydrate content are also listed in Table 2.

1.1.3.2 Stability-Indicating Methodologies

Analytical methodologies that are specific to the major analyte that are also capable of separating and quantifying potential degradation products and impurities, while simultaneously maintaining specificity and accuracy, are deemed stability indicating. Traditional stability-indicating high-performance liquid chromatography (HPLC) methodologies for small molecules are developed and validated with relative ease. Typically, the stability-indicating nature of an analytical method can be demonstrated by subjecting the product to forced degradation in the presence of heat, acid, alkali, light, or peroxide [37]. If degradation products are sufficiently well resolved from the active while maintaining specificity and accuracy, the method is suitable. In contrast to small molecules, there is no one "gold standard" analytical methodology that can be utilized to determine the potential degradation products and impurities in the milieu that may constitute a biopharmaceutical drug product. Furthermore, a one-dimensional structure assessment (e.g., in terms of an absorption spectrum) does not give any indication of the overall activity of the product, as is the case with traditional small molecules. Thus, the stability assessment of biopharmaceuticals will typically comprise a multitude of methodologies that when taken together give an indication of the stability of the product. The overall goal is to assess the structural elements of the compound as well as attempt to determine the relative quantities of potential degradation products, as well as product isoforms and impurities, that are inherent to the expression systems utilized for API manufacture. However, it is still advised that bioactivity determinations are made at appropriate intervals throughout the stability program, as discussed below. Furthermore, any biopharmaceutical stability program should also minimally include an evaluation of the in vitro immunogenicity profile of the product with respect to time, temperature, and other potential degradative conditions.

1.1.3.3 Method Validation and Transfer

Analytical method validation is the process by which scientists prove that the analytical method is suitable for its intended use. Guidances available on validation procedures for some traditional analytical methodologies [38] can be adapted to nontraditional methodologies. The United States Pharmacopeia (USP) and National Formulary (NF) do provide some guidance on designing and assessing biological assays [39], as does the U.S. FDA [40]. Essentially, validation determines the acceptable working ranges of a method and the limitations of that method. At a minimum the robustness, precision, and accuracy of quantitative methodologies should be determined during support of API iteration and refinement, while at the very least the robustness of qualitative methodologies should be assessed. Of particular importance for successful analytical method validation is ensuring that the proper standards and system suitability compounds have been chosen and are representative or analogous to the compound to be analyzed and traceable to a known origin standard, such as the National Institute of Standards and Technology (NIST) or USP/NF. If a reference standard from an "official" source is not available, in-house standards may be used provided they are of the highest purity that can be reasonably obtained and are thoroughly characterized to ensure its identity, strength, quality, purity, and potency.

Methods developed and validated during the product development phase are routinely transferred to quality control or contract laboratories to facilitate release and in-process testing of production batches. Ensuring that method transfer is executed properly, with well-defined and reproducible system suitability and acceptance criteria, is the responsibility of both laboratories. Experiments should consist of all those parameters assessed during method validation and should include an evaluation of laboratory-to-laboratory variation. This information will give an idea of the reliability of the methodology and equipment used under the rigors of large-scale manufacturing.

1.1.4 FORMULATION DEVELOPMENT

The previous sections have highlighted some of the limitations and difficulties in developing biotechnology-derived pharmaceuticals. Although there are major technological limitations in working with these products, their synthesis and manufacturing are significantly more reproducible compared to naturally derived biologics. Determining the most appropriate route of administration and subsequent formulation is dependent on a number of factors, including the product's indication, duration of action, pharmacokinetic parameters, stability profile, and toxicity. As mentioned previously, biopharmaceuticals are typically delivered parenterally, and thus we will focus on those studies required to successfully develop a parenteral formulation of a biopharmaceutical agent. The goal of formulation development is to design a dosage form that ensures the safety and efficacy of the product throughout its shelf life while simultaneously addressing the clinical needs of both the patient and caregivers to ensure compliance. Formulation development is truly a balancing act, attempting to emphasize the benefits of the therapy and patient compliance while maximizing drug efficacy and minimizing toxicity. As such, a number of studies are required to properly design and develop a formulation, many of which are discussed below.

1.1.4.1 Processing Materials and Equipment

An important factor in the quality and reproducibility of any formulation development activity is the materials utilized for formulating and processing studies. In addition, the choice of container closure systems for the API and the formulation needs to be considered carefully to provide maximum product protection and optimal stability. Variability between small- and larger scale development stages may also be significant depending on the API and materials involved during process scale-up. It is important to conduct process development studies utilizing equipment representative of what will be used for large-scale production, if possible. Implementing this design approach will enable at least some limited dimensional analysis, allowing for early identification of critical design parameters, thereby facilitating scale-up or permitting earlier attrition decisions and cost savings. Regardless, it is important to consider the chemical composition and material properties of every manufacturing component that may contact the drug product. For instance, processing vessels may be made of glass, glass-lined steel, or bare steel, while stir paddles used to ensure homogeneity made be manufactured of a number of different materials. In short, any manufacturing unit that could potentially come into intimate contact with either the formulation or the API should be demonstrated to be compatible with the product, including sampling instruments, sample vials, analytical and processing tubing, and so forth. Material incompatibility could result in something as simple as unexplained analytical variability due to a loss of drug through adsorptive mechanisms to something as serious as a loss of bioactivity or an increase in immunogenic potential. Therefore, equipment design and materials would ideally be consistent from formulation development through to scale-up and process validation; however, this may not be readily feasible. As such, determining the chemical and physical compatibility of each piece of processing equipment with the API is critical to maintaining the physical and chemical attributes of the product. Furthermore, such studies help eliminate potential sources of experimental variability and give a better indicator as to the relative technological hurdles to successful product development.

Material compatibility protocols must be clearly defined and require that analytical methodologies be suitable for their intended use. Typically, product purity methods and cleaning methodologies utilized to determine residual contaminating product on processing equipment are used for compatibility studies as they are sufficiently sensitive and rugged to accurately determine product content in the presence of a multitude of potential confounding factors. This is particularly important when assessing potential metal, glass, and tubing compatibilities. Compatibility is a function not only of the product's intimate contact with surrounding materials but also of the contact time and surface area with these equipment. As such, protocols should be designed to incorporate expected real-world conditions the product will

One of the critical factors in excipient selection and concentration is the effect on preferential hydration of the biopharmaceutical product [53, 54]. Preferential hydration refers to the hydration layers on the outer surface of the protein and can be utilized to thermodynamically explain both stability enhancement and denaturation. Typical excipients used in protein formulations include albumin, amino acids, carbohydrates, chelating and reducing agents, cyclodextrins, polyhydric alcohols, polyethylene glycol, salts, and surfactants. Several of these excipients increase the preferential hydration of the protein and thus enhance its stability. Cosolvents need to be added in a concentration that will ensure their exclusion from the protein surface and enhance stability [54]. A more comprehensive review of excipients utilized for biopharmaceutical drug products is available elsewhere [48].

Buffer Selection In addition to maintaining solution pH, buffers serve a multitude of functions in pharmaceutical formulations, such as contributing toward overall isotonicity, preferential hydration of proteins and peptides, and serving as bulking agents in lyophilized formulations. The buffer system chosen is especially important for peptide and proteins that have sensitive secondary, tertiary, and quaternary structures, as the overall mechanisms contributing to conformational stabilization are extremely complex [48]. Furthermore, a protein's propensity for deamidation at a particular pH can be significant, as illustrated by Wakankar and Borchardt [55]. This study illustrated stability concerns with peptides and proteins at physiological pH in terms of asparagine (Asn) deamidation and aspartate (Asp) isomerization, which can be a major issue with respect to circulating half-life and potential in vivo degradation. This study and others also provide insight into predicting potential degradative mechanisms based on primary and secondary structural elements allowing for formulation design with these pathways in mind.

Selecting the appropriate buffer primarily depends on the desired pH range and buffer capacity required for the individual formulation; however, other factors, including concentration, effective range, chemical compatibility, and isotonicity contribution, should be considered [56]. Some acceptable buffers include phosphate (pH 6.2–8.2), acetate (pH 3.8–5.8), citrate (pH 2.1–6.2, p*K* 3.15, 4.8, and 6.4), succinate (pH 3.2–6.6, p*K* 4.2 and 5.6), histidine (p*K* 1.8, 6.0, and 9.0), glycine (p*K* 2.35 and 9.8), arginine (p*K* 2.18 and 9.1), triethanolamine (pH 7.0–9.0), trishydroxymethylaminomethane (THAM, p*K* 8.1), and maleate buffer [48]. Additionally, excipients utilized solely for tonicity adjustment, such as sodium chloride and glycerin, may not only differ in ionic strength but also could afford some buffering effects that should be considered [52].

Preservatives In addition to those processing controls mentioned above (Section 3.1.4.3), the sterility of a product may be maintained through the addition of antimicrobial preservatives. Preservation against microbial growth is an important aspect of multidose parenteral preparations as well as other formulations that require preservatives to minimize the risk of patient infection upon administration, such as infusion products [52]. Aqueous liquid products are prone to microbial contamination because water in combination with excipients derived from natural sources (e.g., polypeptides, carbohydrates) and proteinaceous active ingredients may serve as excellent media for the growth [57]. The major criteria for the selection of an appropriate preservative include efficiency against a wide spectrum of micro-

see when in contact with the material. For instance, temperature, light, and mechanical stimulation should mimic usage conditions, although study duration should include time intervals that surpass expectations to estimate a potential worst case. These factors should all be considered when examining potential process-related stability.

1.1.4.2 Container Closure Systems

The ICH guideline for pharmaceutical development outlines requirements for container closure systems for drugs and biologics [41]. The concept paper prepared for this guidance specifically states that "the choice of materials for primary packaging should be justified. The discussion should describe studies performed to demonstrate the integrity of the container and closure. A possible interaction between product and container or label should be considered" [42]. In essence, this indicates that the container closure system should maintain the integrity of the formulation throughout the shelf life of the product. In order to maintain integrity, the container closure system should be chosen to afford protection from degradation induced by external sources, such as light and oxygen. In addition to the primary container, the stability of the product should also be examined in the presence of IV administration components if the product could be exposed to these conditions (see Section 1.1.5.6). Understanding the potential impact of product-to-container interactions is integral to maintaining stability and ensuring a uniform dosage. For example, adsorption of insulin and some small molecules has been demonstrated to readily occur in polyvinyl chloride (PVC) bags and tubing when these drugs were present as additives in intravenous (IV) admixtures [43].

In addition to their use in large-volume parenterals and IV sets, thermoplastic polymers have also recently found utility as packaging materials for ophthalmic solutions and some small-volume parenterals [43]. However, there are many potential issues with using these polymers as primary packaging components that are not major concerns with traditional glass container closure systems, including [44]:

1. Permeation of vapors and other molecules in either direction through the wall of the plastic container
2. Leaching of constituents from the plastic into the product
3. Sorption (absorption and/or adsorption) or drug molecules or ions on the plastic material

These concerns largely preclude the utility of thermoplastic polymers as the primary choice of container closure system for protein and peptide therapeutics, although the formulation scientist should be aware of the potential advantages of these systems, such as the ease of manufacturability and their cost. These systems are also finding greater utility in intranasal and pulmonary delivery systems.

Parenterally formulated biopharmaceuticals are typically packaged in glass containers with rubber/synthetic elastomeric closures. Pharmaceutical glass is composed primarily of silicon dioxide tetrahedron which is modified with oxides such as sodium, potassium, calcium, magnesium, aluminum, boron, and iron [45]. The USP classifies glass formulations as follows:

Type I, a borosilicate glass
Type II, a soda–lime treated glass
Type III, a soda–lime glass
NP, a soda–lime glass not suitable for containers for parenterals

The tendency of peptides to adsorb onto glass surfaces is well known and a major concern in the pharmaceutical industry. This is especially important when the dose of the active ingredient is relatively small and a significant amount of drug is adsorbed to these surfaces. In addition, the leaching of atoms or elements in the glass's silicate network into solution is also a potential issue. This is especially important for terminally heat sterilized products where oxide additives included in the silicate network are relatively free to migrate/leach, resulting in increased solution pH, reaction catalysis, and so on [45]. As such, only type 1 treated glass is traditionally used for parenterally administered formulations, where these alkaline-rich phases in the glass have been eliminated, thus decreasing the potential for container closure system interactions. Additional approaches, including surface treatment with silicone (siliconization), have also been developed to minimize the interaction of biotechnology products with free silanols (Si–OH) [46].

Elastomeric closures are typically used for syringe and vial plungers and closures. For vials, elastomers provide a soft and elastic material that can permit the entry of a hypodermic needle without loss of the integrity [45]. For syringes, the closures not only provide a permeation barrier but also allow for a soft gliding surface facilitating plunger movement and drug delivery. Elastomeric polymers, however, are very complex materials composed of multiple ingredients in addition to the basic polymers, such as vulcanizing agents, accelerators, activators, antioxidants, fillers, lubricating agents, and pigments [45]. As leaching of these components into solution is a potential issue, the compatibility of the drug formulation with the closures must be studied early during the formulation development process. The choice and type of elastomeric closure depends on the pH and buffer, if any preservatives are present, the sterilization method, moisture vapor/gas protection, and active compatibility [47]. In addition, the problem of the additives in rubber leaching into the product can be reduced by the coating with specific polymers such as Teflon [48].

Container closure systems required for implantable devices are further restricted by the fact that they are required to be compatible with the formulation over the intended shelf life and therapeutic application time as well as being biocompatible. This means that the system not only must afford protection to and contain the formulation but also cannot cause any potential adverse effects, such as allergy. Typically, implantable systems are composed of biocompatible metals, such as titanium or polymers such as polyethylene glycol or polylactic-co-glycolic acid.

1.1.4.3 Sterility Assurance

Maintaining the sterility of biopharmaceutical products is especially important due to the relative potency and their innate potential for immunogenic reactions. Further, the biochemical nature of these compounds enables them to serve as potential nutrients for invading organisms. Methods for sterilizing small molecules include

heat terminal sterilization, terminal filtration coupled with aseptic processing t
niques, ultraviolet (UV) and gamma irradiation, ethylene oxide exposure (for c
tainers and packaging only), and electron beam irradiation. While terminal l
sterilization is by far the most common sterilization technique, it normally car
readily be utilized for peptide or protein formulations due to the potential eff
of heat and pressure on the compound's structure [48]. Furthermore, irradiation
affect protein stability by cross-linking the sulfur-containing and aromatic resid
resulting in protein aggregation [49].

To overcome these issues, sterile filtration coupled with aseptic processing
filling is the preferred manufacturing procedure for biopharmaceuticals. Garfir
et al. refer to aseptic processing as "those operations performed between the ste
ization of an object or preparation and the final sealing of its package. These ope
tions are, by definition, carried out in the complete absence of microorganisms" [
This highlights the importance of manufacturing controls and bioburden monitor
during aseptic processes. Newer technologies such as isolator technology have b
developed to reduce human intervention, thereby increasing the sterility assuran
These technologies have the added benefit of facilitating aseptic processing with
construction of large processing areas, sterile suites, or gowning areas [50].

Even the most robust monitoring programs do not ensure the sterility of the fi
formulation. As such, aseptically processed formulations are traditionally filte
through a retentive final filter, which ensures sterility. Coupled with proper com
nent sterilization, traditionally by autoclaving, these processes ensure product ste
ity. However, filtration is a complex unit operation that can adversely affect the d
product through increased pressure, shear, or material incompatibility. Therefo
filtration compatibility must be assessed thoroughly to demonstrate both prod
compatibility, and sufficient contaminant retention [51]. Parenteral Drug Assoc
tion (PDA) technical report 26 provides a thorough systematic approach to selecti
and validating the most appropriate filter for a sterilizing filtration applicati
[51].

1.1.4.4 Excipient Selection

Pharmaceutical products are typically formulated to contain selected nonacti
ingredients (excipients) whose function is to promote product stability and enab
delivery of the active pharmaceutical ingredient(s) to the target site. These su
stances include but are not limited to solubilizers, antioxidants, chelating agen
buffers, tonicity contributors, antibacterial agents, antifungal agents, hydroly
inhibitors, bulking agents, and antifoaming agents [45]. The ICH states that "t
excipients chosen, their concentration, and the characteristics that can influence t
drug product performance (e.g. stability, bioavailability) or manufacturability shou
be discussed relative to the respective function of each excipient" [42]. Excipien
must be nontoxic and compatible with the formulation while remaining stab
throughout the life of the product. Excipients require thorough evaluation an
optimization studies for compatibility with the other formulation constituents a
well as the container/closure system [52]. Furthermore, excipient purity may b
required to be greater than that listed in the pharmacopeial monograph if a specif
impurity is implicated in potential degradation reactions (e.g., presence of trac
metals) [48].

organisms, stability (shelf life), toxicity, sensitizing effects, and compatibility with other ingredients in the dosage form [57]. Typical antimicrobial preservatives include *m*-cresol, phenol, parabens, thimerosal, sorbic acid, potassium sorbate, benzoic acid, chlorocresol, and benzalkonium chloride. Cationic agents such as benzalkonium chloride are typically not utilized for peptide and protein formulations because they may be inactivated by other formulation components and their respective charges may induce conformational changes and lead to physical instability of the API. Further, excipients intended for other applications, such as chelating agents, may exhibit some antimicrobial activity. For instance, the chelating agent ethylenediaminetetraacetic acid (EDTA) may exhibit antimicrobial activity, as calcium is required for bacterial growth.

Identifying an optimal antimicrobial preservative is based largely on the effectiveness of that preservative at the concentration chosen. In short, it is not enough to assess the compatibility of the preservative of choice with the API and formulation and processing components. There also needs to be a determination of whether the preservative concentration is sufficient to kill certain standard test organisms. The USP presents standard protocols for assessing the relative efficacy of a preservative in a formulation using the antimicrobial effectiveness test (AET) [58]. Briefly, by comparing the relative kill efficiency of the formulation containing varying concentrations of the preservative, the formulator can determine the minimal concentration required for preservative efficacy and design the formulation accordingly.

1.1.5 DRUG PRODUCT STABILITY

1.1.5.1 Defining Drug Product Storage Conditions

From a regulatory standpoint, the primary objective of formulation development is to enable the delivery of a safe and efficacious drug product to treat and/or mitigate a disease state throughout its proposed shelf life. The efficacy and in many cases the safety of a product are directly related to the stability of the API, both neat and in the proposed formulation under processing, storage, and shipping conditions as well as during administration. As such, the concept of drug stability for biotechnology-derived products does not change substantially from that of small molecules, although the level of complexity increases commensurate with the increased complexity of the APIs in question and the formulation systems utilized for their delivery.

Stability study conditions for biotechnology-derived APIs and their respective drug products are not substantially different from those studies conducted for small molecules. Temperature and humidity conditions under which to conduct said studies are outlined in ICH Q1A(R2), which incorporates ICH Q1F, stability study conditions for zones III and IV climactic conditions [59]. Additional guidance specific to conducting stability studies on biopharmaceutical drug products is given in ICH Q5C [1]. However, the intention of ICH Q5C is not to outline alternate temperature and humidity conditions to conduct primary stability studies; rather it provides guidance with respect to the fact that the recommended storage conditions and expiration dating for biopharmaceutical products will be different from product to product and provides the necessary flexibility in letting the applicant determine

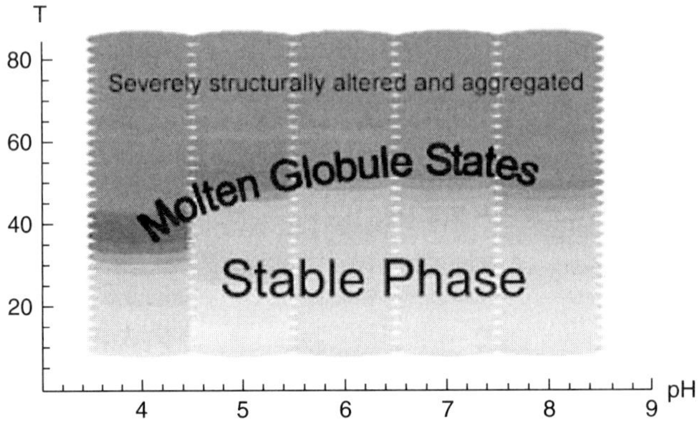

FIGURE 1 Empirical phase diagram for ricin toxin A-chain generated using CD molar ellipticity at 208 nm, ANS fluorescence, and intrinsic Trp fluorescence intensity data. Labels indicate the state of the protein within the same region of color based on evaluation of a compilation of data sets. (Reproduced with permission from ref. 62.)

the proper storage conditions for their respective product. Furthermore, this document provides general guidance in directing applicants in the types of analytical methodologies that may be used and direction on how to properly assess the stability of these complex molecules [1]:

> Assays for biological activity, where applicable, should be part of the pivotal stability studies. Appropriate physicochemical, biochemical and immunochemical methods for the analysis of the molecular entity and the quantitative detection of degradation products should also be part of the stability program whenever purity and molecular characteristics of the product permit use of these methodologies.

One recent approach to aid in defining the design space for protein and peptide therapeutics has been to create empirical phase diagrams indicating the relative stability of compounds based on altering conditions and assessing conformational changes via a compilation of analytical techniques (Figure 1) [60–62]. These empirical phase diagrams can be generated based on pH, temperature, salt concentration, and so on, and, although seemingly laborious at first glance, could provide invaluable information in defining the extremes to which a compound may be subjected without altering its conformation. For instance, if an empirical phase diagram determines the safe temperature range for a compound is up to 35 °C and an excursion occurs to 33 °C, this information would give the stability scientist a guideline as to the appropriate course of action. Under the traditional testing paradigm of ICH Q1A, where stability testing is limited to 25, 30, and 40 °C, one may not know the compound's upper transition temperature to induce conformational changes. If the information is not already available, then additional excursion studies may need to be conducted to assimilate this information and take the appropriate course of action.

1.1.5.2 Mechanisms of Protein and Peptide Degradation

The inherent heterogeneity of peptide and protein drug substances results in their relative sensitivity to processing, storage, and handling conditions as well as a mul-

TABLE 3 Potential Degradative Mechanisms of Peptides and Proteins

	Degradative Mechanism	Site of Occurrence
Chemical degradative mechanisms	Oxidation	Intrachain disulfide linkages Met, Trp, Tyr
	Peptide bond hydrolysis	AA backbone
	N-to-O migration	Ser and Thr
	α- to β-Carboxy migration	Asp and Asn
	Deamidation	Asn and Gln
	Acylation	α-Amino and ε-amino group
	Esterification/carboxylation	Glu, Asp, and C-term
Physical degradative mechanisms	Unfolding	Partial unfolding of tertiary structure
	Aggregation	Aggregation of subunits could result in precipitation
	Adsorption	Adsorption to processing equipment and container closure systems

Source: Modified from Crommelin et al. [5].

titude of other factors. Most importantly, this heterogeneity results in a whole host of potential degradative mechanisms, some of which are compiled in Table 3 and include chemical instability pathways such as oxidation, hydrolysis of side chains and potentially the peptide backbone, and deamidation of Asn and Gln side chains. Also, physical instability manifesting in the form of protein unfolding, formation of intermediate structures, aggregation, and adsorption to the surfaces of containers and other equipment can be a major technical hurdle in developing any biopharmaceutical and may or may not be related to chemical instability [63]. Further complicating matters is that instability can potentially manifest in various ways and may or may not be detectable by any one method. Taken together, however, the compilation of methodologies utilized for stability assessment should give a good approximation as to the degradative mechanisms of the compound in its respective formulation. Further, bioactivity and immunogenicity assays should play integral roles in assessing the relative stability of any biopharmaceutical compound. Briefly stated, the chemical and physical stability of products is extraordinarily difficult to assess and will not be belabored here as good reviews on this topic are readily available in the literature [63, 64].

1.1.5.3 Photostability

In certain cases, exposure of pharmaceutical compounds to UV and visible light could result in electronic excitation, termed vertical transition, that could ultimately result in light-induced degradation. The ICH guideline Q1B [65] is a reference on how to conduct photostability stress testing for pharmaceutical compounds. In brief, compounds are exposed to an overall illumination of not less than 1.2 million lux hours and an integrated near-UV energy of not less than 200 Wh/m^2 [65]. These requirements are in addition to normal stability stress testing and require the additional caveat that analytical methodologies are suitable to also detect photolytic degradation products, as discussed above. A comprehensive discussion of small-molecule photolytic degradative mechanisms is available for further review [66].

1.1.5.4 Mechanical Stress

Regulatory guidance on appropriate methods to evaluate the effect of shear stress and process-handling stability studies is not available. However, these studies are integral in determining the relative stability of the product with respect to mechanical stresses introduced during development and manufacturing. Although not typically recognized as a major degradative pathway for most small-molecule dosage forms, the introduction of mechanical stress is recognized as a major challenge in the formulation of semisolids and can potentially induce physical instability of biopharmaceuticals, although the extent of this effect is currently unknown. For example, processing shear may influence the protein's outer hydration shell, altering the stabilizing energy provided from preferential hydration and resulting in the exposure of internal, nonpolar residues. This may facilitate aggregation if enough shear force is provided. Alternately, the shear energy required to force unfolding has been studied but has not been related to the fluid dynamic shear experienced during processing. Therefore, stress studies should include meticulous controls in the form of temperature, light and humidity, and fluid dynamic shear as a function of time. Data from these studies could be incorporated into empirical phase diagrams, and/or response surfaces, to help further define the design space for the active and finished drug product. Understanding the effects of stress introduced during the manufacturing processing of biopharmaceutical products could facilitate the selection of appropriate PAT tools and QbD incorporation in the development of these products. Clearly, there is a considerable need for research in this area, and until the extent of the possible effects are understood, this lack of knowledge poses an unknown risk and prevents adequate risk assessment for biopharmaceutical development activities consistent with ICH Q9.

1.1.5.5 Freeze–Thaw Considerations and Cryopreservation

The rapid or continuous freezing and thawing of protein products could contribute significantly to instability of the API. Such studies are typically designed to assess the implications of potential transport and handling conditions. These conditions include not only the manufacturing processing, storage, and shipment to warehouses and pharmacies but also subsequent pharmacy storage and patient handling [52]. Unpredictable and somewhat modest temperature fluctuations could easily induce degradation or conformational changes that may reduce bioactivity or expose antigenic epitopes [5]. These effects could also be a result of altered preferential hydration at the surface of the peptide or protein through salting-out effects induced by rapid freezing, which could easily denature the product [67].

1.1.5.6 Use Studies

Stability of biopharmaceutical compounds should also be determined under conditions that mimic their normal usage. For instance, the stability of reconstituted lyophilized products should be assessed with respect to time and temperature and, if applicable, light and mechanical stimuli. Likewise, the stability of a compound included in implantable devices and controlled-release microsphere formulations should be determined over the course of its required use, under conditions which mimic the heat, moisture, light, and enzymatic physiological conditions to which it will be implanted. Such studies should also determine the release profile of the compound over these specified conditions.

Drug products intended for IV administration are generally dosed as an initial bolus followed by a slow infusion. Consequently, admixture studies of the compound in potential IV fluids, such as 0.9% (w/v) saline, 5% (w/v) dextrose, and Ringer's solution, should also be assessed to determine the relative stability of the compound in this new environment. These studies are critical as the formulation dynamic that protected and stabilized the compound has now been altered dramatically with dilution. This environmental change could potentially impact the preferential hydration of the compound as well as directly induce conformational changes based on the diluent chosen and the compound's potential degradative mechanism(s). Additional contributing factors to instability in admixture solutions could be due to changes in pH, mechanical mixing of the compound in the IV bag, adsorption of the compound to the bag itself (which is typically polymeric), or IV sets used for administration, as well as an increased potential for oxidative degradation. The suitability of analytical methodologies should also be determined in the presence of these additional analytes.

1.1.5.7 Container Closure Integrity and Microbiological Assessment

Ensuring that parenteral pharmaceuticals maintain their sterility over the course of their shelf life is an integral part of any stability assessment [68]. Parenteral dosage forms must be free from microbiological contamination, bacterial endotoxins, and foreign particulate matter. Selection of the adequate sterile manufacturing process has been briefly discussed above. Determining the microbiological integrity of the product over its shelf life also gives an indication of the relative quality of the container closure system chosen for the formulation. Compendial sterility and endotoxin testing are often used for this purpose; however, sampling is dependent on a statistical evaluation of the batch size, unit fill volume, and method of product sterilization [68]. Additionally, since these tests are destructive, it would be impossible to test an entire stability batch to ensure viability of a container closure system. Other nondestructive tests have been developed to determine the integrity of a container's closure system [69]. These tests could also serve as a surrogate indicator of product manufacturing quality over time.

1.1.5.8 Data Interpretation and Assessment

Interpretation of primary stability data for determining expiration dating and primary storage conditions has been outlined by ICH Q1E [70]. This guidance document delineates broad methodologies for interpreting primary and accelerated stability data and extrapolation of said data for determining expiry dating. Of course, expiry dating cannot be made without reference to specifications for those primary stability-indicating parameters assessed, which is discussed below. Traditionally, stability assessments performed during preformulation will give an indication of the potential storage conditions as well as allow for extrapolation of accelerated stability studies to kinetic degradation rates. Typically this is done through Arrhenius manipulations. However, as one would expect, these analyses are not readily useful for biopharmaceutical products, as there is rarely a linear correlation between temperature and the compound's degradative rate. This is primarily due to the complex and often competing degradative mechanisms as well as the potential for so-called molten globule intermediate phases. In spite of these limitations, ICH Q5C does provide relevant guidance in illustrating the flexibility required for determining storage conditions, as these products usually require a very narrow temperature condition to maintain optimal stability. Further guidances may be needed to enhance uniformity in testing methodology and enable the utilization of validated PAT methodologies.

1.1.6 QUALITY BY DESIGN AND SCALE-UP

1.1.6.1 Unit Operations

Unit operations are defined as the individual basic steps in a process that when linked together define the process train and result in the final product. In practical terms, a unit operation is often defined as an individual step that is carried out on one piece of equipment. Typical biopharmaceutical API unit operations may include fermentation or bioreactor processes, cell separation through centrifugation or microfiltration, virus removal or inactivation, cell lysis and inclusion body precipitation, product refolding, and purification steps [71]. Conversely, those unit operations for drug product manufacturing procedures would be similar to those seen in the manufacture of a small molecule of comparable dosage form, namely mixing, fluid transfer, sterile filtration, dose filling, lyophilization, and so on. Of course, unit operations will be dependent on the manufacturing process for the specific dosage form, but careful preformulation and characterization studies will enable relatively straightforward process design and ease subsequent scale-up activities. Modeling of unit operations for both small and large molecules is a recognized gap in our ability to achieve QbD. The application of accepted engineering methods to the problem is the subject of active research.

1.1.6.2 Bioburden Considerations

Bioburden refers to the amount of microbial flora that can be detected on an item, on a surface, or in a solution [68]. As mentioned previously, microbial contamination and bioburden are especially important for biotechnology-derived parenteral products since these products are typically capable of supporting microbial growth. Special care should be taken to ensure not only that the final packaged product does not contain microbial contamination but also that manufacturing equipment is also free from contamination. Monitoring bioburden and determining potential levels of microbial contamination on equipment surfaces are particularly important with respect to the material being evaluated.

In general, bioburden counts in parenteral solutions are obtained by conducting the total aerobic counts and total yeast and mold counts as specified in the USP microbial limits test (61) or an equivalent test [72]. In addition, membrane filtration of larger than specified volumes may also be used to detect any microbial contamination when sample results are expected to contain a negligible number of microbial flora or in the presence of potential confounding factors, such as antimicrobial preservatives [68, 72]. It is important to note that the presence of a high bioburden count can present an endotoxin contamination problem, as whole microbial cells and spores can be removed by sterilizing grade filtration (0.2 μm), while endotoxins are not [68]. These issues also underscore the importance of cleaning methods and their respective validation as well as assessing relevant product contamination on manufacturing equipment.

1.1.6.3 Scale-Up and Process Changes

The FDA defines process validation as "establishing documented evidence that provides a high degree of assurance that a specific process will consistently produce a product meeting its predetermined quality attributes" [73]. While validation studies are typically performed at full scale, in most cases scale-down or laboratory-scale models were used to initially develop the manufacturing process. Consequently, scale-down process precharacterization and characterization studies are considered crucial to successful process validation for both API and drug product manufacturing schemes [74]. Although they do require qualification work and a significant commitment of time and resources, characterization studies provide significant insight into the critical process and control parameters for each unit operation as well as improved success rates for process validation due to a better, more complete understanding of the process [74]. In engineering terms, characterization studies identify the critical parameters useful for dimensional analysis that enable successful process scale-up.

While the above explanation attempts to simplify the scale-up process, it is not meant to trivialize it. In fact, scale-up is probably the most difficult manufacturing challenge for traditional small molecules, let alone biopharmaceuticals. Issues such as homogeneous mixing, bulk product holding and transfer, and sterile filtration could all be potentially compounded due to the increased scale and introduced stress. However, a QbD approach to rational drug design should enable simplified process scale-up and validation. This is only true if experimental design approaches have been utilized to identify the design space for the processes involved in the production of the molecule. This is also where the greatest benefit of developing empirical phase diagrams early in development could materialize. Essentially, the QbD approach identifies the quality attributes of the product based on scientific rationale as opposed to attempting to fit the proverbial square peg into a round hole through a trial-and-error approach. This rational design approach goes further to identify the limiting factors of each unit operation and provides the means of attempting to correlate how each unit operation affects the final product quality attributes.

In order to initiate a successful QbD program, the first step is to identify those process parameters that are essential to product quality and develop well-validated analytical methodologies to monitor those parameters. In short, the process involves identification of the potential design space for production of the molecule and confirmation that design space through rational, deliberate experimentation. Ideally, process monitoring should be done in real time to minimize production time and if possible online; however, this may not always be the case or even necessary depending upon the relative duration of the process to the test. Recognizing potential quality metrics earlier in the development process could also potentially facilitate greater flexibility during product development and subsequent process characterization [74]. Certainly, manufacturing site-specific differences could also potentially introduce variability into processes. It is for this reason that site-specific personnel training, process/technology transfer and validation, and stability assessments are required to ensure product quality.

By definition, a process designed under the auspices of QbD should enable a degree of process knowledge that allows for controlled process changes without affecting the final product or requiring regulatory approval. For immediate- and controlled-release solid dosage products, SUPAC guidelines provide direction on the studies to conduct to determine the impact of a process change. Although there is some regulatory guidance available for biological products (e.g., "Changes to an Approved Application for Specified Biotechnology and Specified Synthetic Biological Products" or "FDA Guidance Concerning Demonstration of Comparability of Human Biological Products, Including Therapeutic Biotechnology-Derived Products"), process changes need to be evaluated on a case-by-case basis. The comparative analysis of process changes should also be evaluated with respect to defined product specifications. PAT will be invaluable in determining the potential impact of process changes. While stability is often the main metric for small-molecule drug product, bioactivity and immunogenicity will need to be added metrics for biopharmaceuticals. Therefore, any process change should be approached subjectively and care should be taken to validate the relative impact on the safety and efficacy of the product.

1.1.7 CONCLUDING REMARKS

Although the goals are the same, developing biotechnology molecules presents challenges that are unique compared to the development of conventional small molecules. The innate complexity of the molecular and macromolecular structures requires three dimensionally viable stability assays and understanding. The complexity of possible physiological responses and interactions requires an enhanced understanding of the formulation and processing stresses to identify the minor but critical changes that result in product unacceptability. A key to addressing these challenges is the development of analytical techniques with the sensitivity and reliability to detect and monitor such changes and to provide data to another gap-closing activity—modeling unit operations. Also the need to develop meaningful kinetic models is obvious to everyone involved in the development of both large and small molecules. Linking this type of information to the major efforts in the discovery arena is a necessary step to bringing the products of the future to market.

The use of biotechnology products is increasing exponentially and many opportunities exist to improve their development. The first step may be defining rational biotechnology-derived drug "developability" standards that can be assessed during preclinical/early development testing. Such a tiered approach based upon the potential risk, the confidence in methodology, and benefit has of course been a proven strategy for small molecules, and a preliminary version applicable to biotechnology drug products is likely possible today given the topics discussed in this chapter.

ACKNOWLEDGMENTS

The authors would like to thank The School of Pharmacy and Pharmaceutical Sciences, the Department of Industrial and Physical Pharmaceutics of Purdue University and the National Institutes of General Medical Sciences (R01-GM65448) for their financial support.

REFERENCES

1. International Conference on Harmonisation of Technical Requirements for Registration of Pharmaceuticals for Human Use, Q5C: Quality of biotechnological products: Stability testing of biotechnological/biological products, Nov. 30, 1995.
2. Biotechnology Industry Organization. Available: http://www.bio.org/, accessed Jan. 10, 2007.
3. Nims, R., Presene, E., Sofer, G., Phillips, C., and Chang, A. (2005), Adventitious agents: Concerns and testing for biopharmaceuticals, in Rathore, A. S., and Sofer, G., Eds., *Process Validation in Manufacturing of Biopharmaceuticals: Guidelines, Current Practices, and Industrial Case Studies*, Taylor and Francis, Boca Raton, FL.
4. Rathore, A. S., and Sofer, G. (2005), Life span studies for chromatography and filtration media, in Rathore, A. S., and Sofer, G., Eds., *Process Validation in Manufacturing of Biopharmaceuticals: Guidelines, Current Practices, and Industrial Case Studies*, Taylor and Francis, Boca Raton, FL.
5. Crommelin, D. J. A., Storm, G., Verrijk, R., de Leede, L., Jiskoot, W., and Hennink, W. E. (2003), Shifting paradigms: Biopharmaceuticals vs. low molecular weight drugs, *Int. J. Pharm.*, 266, 3–16.
6. Tang, L., Persky, A. M., Hochhaus, G., and Meibohm, B. (2004), Pharmacokinetic aspects of biotechnology products, *J. Pharm. Sci.*, 93(9), 2184–2204.
7. Roberts, M. J., Bentley, M. D., and Harris, J. M. (2002), Chemistry for peptide and protein pegylation, *Adv. Drug Deliv. Rev.*, 54, 459–476.
8. Frokjaer, S., and Otzen, D. (2005), Protein drug stability: A formulation challenge, *Nat. Rev.*, 4, 298–306.
9. Niu, C., and Chiu, Y. (1998), FDA perspective on peptide formulation and stability issues, *J. Pharm. Sci.*, 87, 1331–1334.
10. Washington, N., Washington, C., and Wilson, C. G. (2001), *Physiological Pharmaceutics: Barriers to Drug Absorption*, 2nd ed., Taylor and Francis, New York.
11. Hussain, A., Arnold, J. J., Khan, M. A., and Ashan, F. (2004), Absorption enhancers in pulmonary protein delivery, *J. Controlled Release*, 94, 15–24.
12. Alpar, H. O., Somavarapu, S., Atuah, K. N., and Bramwell, V. W. (2005), Biodegradable mucoadhesive particulates for nasal and pulmonary antigen and DNA delivery, *Adv. Drug Deliv. Rev.*, 57, 411–430.
13. Thanou, M., Verhoef, J. C., and Junginger, H. E. (2001), Chitosan and its derivatives as intestinal absorption enhancers, *Adv. Drug Deliv. Rev.*, 50, 91–101.
14. Schellekens, H. (2002), Immunogenicity of therapeutic proteins: Clinical implications and future prospects, *Clin. Ther.*, 24, 1720–1740.
15. Bhadra, D., Bhadra, S., Jain, P., and Jain, N. K. (2002), Pegnology: A review of PEG-ylated systems, *Pharmazie*, 57, 5–29.
16. Matthews, S. J., and McCoy, C. (2004), Peginteferon α2a: A review of approved and investigational uses, *Clin. Ther.*, 26, 991–1025.

17. Knudsen, L. B., Nielsen, P. F., Huusfeldt, P. O., Johansen, N. L., Madsen, K., Pedersen, F. Z., Thogersen, H., Wilken, M., and Agerso, H. (2000), Potent derivatives of glucagon-like peptide-1 with pharmacokinetic properties suitable for once daily administration, *J. Med. Chem.*, 43, 1664–1669.
18. Foldvari, M., Attah-Poku, S., Hu, J., Li, Q., Hughes, H., Babiuk, L. A., and Kruger, S. (1998), Palmitoyl derivatives of interferon α: Potential for cutaneous delivery, *J. Pharm. Sci.*, 87, 1203–1208.
19. Egrie, J.C., and Browne, J. K. (2001), Development and characterization of novel erythropoiesis stimulating protein (NESP), *Nephrol. Dial. Transplant.*, 16(Suppl 3), 3–13.
20. Haselbeck, A. (2003), Epoetins: Differences and their relevance to immunogenicity, *Curr. Med. Res. Opin.*, 19, 430–432.
21. Hochuli, E. (1997), Interferon immunogenicity: Technical evaluation of interferon-α2A, *J. Int. Cytokine Res.*, 17, S15–S21.
22. Maack, T., Johnson, V., Kau, S. T., Figueiredo, J., and Sigulem, D. (1979), Renal filtration, transport, and metabolism of low-molecular-weight proteins: A review, *Kidney Int.*, 16, 251–270.
23. Takagi, A., Masuda, H., Takakura, Y., and Hashida, M. (1995), Disposition characteristics of recombinant human interleukin-11 after a bolus intravenous administration in mice, *J. Pharmacol. Exp. Ther.*, 275, 537–543.
24. Anderson, P. M., and Sorenson, M. A. (1994), Effects of route and formulation on clinical pharmacokinetics of interleukin-2, *Clin. Pharmacokinet.*, 27, 19–31.
25. Rabkin, R., Ryan, M. P., and Duckworth, W. C. (1984), The renal metabolism of insulin, *Diabetologia*, 27, 351–357.
26. Johnson, V., and Maack, T. (1977), Renal extraction, filtration, absorption, and catabolism of growth hormone, *Am. J. Phsiol.*, 233, F185–F196.
27. Carone, F. A., and Peterson, D. R. (1980), Hydrolysis and transport of small peptides by the proximal tubule, *Am. J. Physiol.*, 238, F151–F158.
28. Carone, F. A., Peterson, D. R., and Flouret, G. (1982), Renal tubular processing of small peptide hormones, *J. Lab. Clin. Med.*, 100, 1–14.
29. Braeckman, R. (2000), Pharmacokinetics and pharmacodynamics of protein therapeutics, in Reid, R., Ed., *Peptide and Protein Drug Analysis*, Marcel Dekker, New York.
30. Meissner, H. C., Groothuis, J. R., Rodriguez, W. J., Welliver, R. C., Hogg, G., Gray, P. H., Loh, R., Simoes, E. A., Sly, P., Miller, A. K., Nichols, A. I., Jorkasky, D. K., Everitt, D. E., and Thompson, K. A. (1999), Safety and pharmacokinetics of an intramuscular monoclonal antibody (SB 209763) against respiratory syncytial virus (RSV) in infants and young children at risk for severe RSV disease, *Antimicrob. Agent Chemother.*, 43, 1183–1188.
31. Prescribing information for Lupron Depot®. Manufactured for TAP Pharmaceuticals, Inc., Lake Forest, IL 60045.
32. Prescribing information for Viadur®. Manufactured by Alza Corporation, Mountain View, CA 94043.
33. Degim, I. T., and Celebi, N. (2007), Controlled delivery of peptides and proteins, *Curr. Pharm. Des.*, 13, 99–117.
34. Sylvester, R. K. (2002), Clinical applications of colony-stimulating factors: A historical perspective, *Am. J. Health-Syst. Pharm.*, 59, s6–s12.
35. Dorr, R. T. (1993), Clinical properties of yeast-derived versus *Escherichia coli*-derived granulocyte-macrophage colony-stimulating factor, *Clin. Ther.*, 15, 19–29.
36. Jenkins, N., Shah, P. M., and Buckberry, L. D. (2000), Carbohydrate analysis of glycoproteins and glycopeptides, in Reid, R., Ed., *Protein and Peptide Drug Analysis*, Marcel Dekker, New York.

37. Berglund, M., Bystroem, K., and Persson, B. (1990), Screening chemical and physical stability of drug substances, *J. Pharm. Biomed. Anal.*, 8, 639–643.
38. International Conference on Harmonisation of Technical Requirements for Registration of Pharmaceuticals for Human Use, Q2(R1): Validation of analytical procedures: Text and methodology, Nov. 2005.
39. U.S. Pharmacopeia (USP), Chapter ⟨111⟩ Design and analysis of biological assays, USP 26, 2003, Rockville, MD.
40. Center for Drug Evaluation and Research, FDA (2001, May). Guidance for industry, bioanalytical method validation, FDA, Washington, DC.
41. International Conference on Harmonisation of Technical Requirements for Registration of Pharmaceuticals for Human Use, Q8: Pharmaceutical development, Nov. 10, 2005.
42. International Conference on Harmonisation of Technical Requirements for Registration of Pharmaceuticals for Human Use, Q8: Concept paper, available: http://www.ich.org/LOB/media/MEDIA3096.pdf, accessed Sept. 19, 2003.
43. Avis, K. E., and Levchuk, J. W. (2000), *Remington 20th Edition: Parenteral Preparations*, Lippincott, Williams & Wilkins, Philadelphia.
44. Autian, J. (1968), Interrelationship of the properties and uses of plastics for parenterals, *Bull. Parenteral Drug Assoc.*, 22, 276–288.
45. Avis, K. E. (1986), Sterile Products, in Lachman, L., Lieberman, H. A., and Kanig, J. L., Eds., *The Theory and Practice of Industrial Pharmacy,* 3rd ed., *Sterile Products*, 3rd Ed., Lippincott, Williams and Wilkins, reprint with permission by Stipes Publishing, Champaign, IL.
46. Gombotz, W., Pankey, S., Bouchard, L., Phan, D., and MacKenzie, A. (2002), Stability, characterization, formulation, and delivery system development for transforming growth factor-beta1, in Pearlman, R., and Yang, Y., Eds., *Pharmaceutical Biotechnology*, Vol. 9, *Formulation, Characterization, and Stability of Protein Drugs: Case Histories*, Springer, New York.
47. Bontempo, J. A. (1997), Considerations for elastomeric closures for parenteral biopharmaceutical drugs, in Bontempo, J. A. Ed., *Development of Biopharmaceutical Parenteral Dosage Forms*, Marcel Dekker, New York.
48. Banga, A. K. (2006), *Therapeutic Peptides and Proteins: Formulation, Processing and Delivery Systems*, 2nd ed., Taylor and Francis, Boca Raton, FL.
49. Yamamoto, O. (1992), Effect of radiation on protein stability, in Ahern, T. J., and Manning, M. C. Eds., *Pharmaceutical Biotechnology*, Vol. 2, *Stability of Protein Pharmaceuticals. Part A: Chemical and Physical Pathways of Protein Degradation*, Plenum, New York.
50. Garfinkle, B. D., and Henley, M. W. (2000), *Remington 20th Edition: Sterilization*, Lippincott, Williams & Wilkins, Philadelphia.
51. PDA Technical Report No. 26 (1998), Sterilizing filtration of liquids, *PDA J. Pharm. Sci. Tech.*, 52(3, Suppl).
52. Defelippis, M. R., and Akers, M. J. (2000), *Pharmaceutical Formulation Development of Peptides and Proteins: Peptides and Proteins as Parenteral Suspensions: An Overview of Design, Development, and Manufacturing Considerations*, Taylor and Francis, Philadelphia.
53. Timasheff, S. N. (1998), Control of protein stability and reactions by weakly interacting cosolvents: The simplicity of the complicated, *Adv. Protein Chem.*, 51, 355–432.
54. Shimizua, S., and Smith, D. J. (2004), Preferential hydration and the exclusion of cosolvents from protein surfaces, *J. Chem. Phys.*, 121, 1148–1154.
55. Wakankar, A. A., and Borchardt, R. T. (2006), Formulation considerations for proteins susceptible to asparagine deamidation and aspartate isomerization, *J. Pharm. Sci.*, 95, 2321–2336.

56. Windheuser, J. J. (1963), The effect of buffers on parenteral solutions, *Bull. Parenteral Drug Assoc.*, 17, 1–8.
57. Im-Emsap, W., Paeratakul, O., and Siepmann, J. (2002), Disperse Systems, in Banker, G. S., and Rhodes, C. T., Eds., *Modern Pharmaceutics*, 4th Ed., Marcel Dekker, New York.
58. U.S. Pharmacopeia (USP 26 2003), Chapter ⟨51⟩ Antimicrobial effectiveness testing, Rockville, MD.
59. International Conference on Harmonisation of Technical Requirements for Registration of Pharmaceuticals for Human Use, Q1A(R2): Stability testing of new drug substances and products, Feb. 6, 2003.
60. Fan, H., Li, H., Zhang, M., and Middaugh, C. R. (2007), Effects of solutes on empirical phase diagrams of human fibroblast growth factor 1, *J. Pharm. Sci.*, 96, 1490–1503.
61. Harn, N., Allan, C., Oliver, C., and Middaugh, C. R. (2006), Highly concentrated monoclonal antibody solutions: Direct analysis of physical structure and thermal stability, *J. Pharm. Sci.*, 96, 532–546.
62. Peek, L. J., Brey, R. N., and Middaugh, C. R. (2007), A rapid, three-step process for the preformulation of a recombinant ricin toxin A-chain vaccine, *J. Pharm. Sci.*, 96, 44–60.
63. Violand, B. N., and Siegel, N. R. (2000), Protein and peptide chemical and physical stability, in Reid, R., Ed., *Protein and Peptide Drug Analysis*, Marcel Dekker, New York.
64. Manning, M. C., Patel, K., and Borchardt, R. T. (1989), Stability of protein pharmaceuticals, *Pharm. Res.*, 6, 903–918.
65. International Conference on Harmonisation of Technical Requirements for Registration of Pharmaceuticals for Human Use, Q1B: Stability testing: Photostability testing of new drug substances and products, Nov. 6, 1996.
66. Fasani, E., and Albini, A. (2005), Photostability stress testing, in Baertschi, S. W., Ed., *Pharmaceutical Stress Testing: Predicting Drug Degradation*, Taylor and Francis, Boca Raton, FL.
67. Ghosh, T., Kalra, A., and Garde, S. (2005), On the salt-induced stabilization of pair and many-body hydrophobic interactions, *J. Phys. Chem. B. Condens. Matter Mater. Surf. Interfaces Biophys.*, 109, 642–651.
68. Birrer, G. A., Murthy, S. S., Liu, J., and Estrada, J. (2001), Parenteral dosage forms, in Ahuja, S., Ed., *Handbook of Modern Pharmaceutical Analysis*, Academic, New York.
69. Birrer, G. A., Liu, J., Halas, J. M., and Nucera, G. G. (2000), Evaluation of a container closure integrity test model using visual inspection with confirmation by near infra-red spectroscopic analysis, *PDA J. Pharm. Sci. Technol.*, 54, 373–382.
70. International Conference on Harmonisation of Technical Requirements for Registration of Pharmaceuticals for Human Use, Q1E: Evaluation for stability data, Feb. 6, 2003.
71. Godavarti, R., Petrone, J., Robinson, J., Wright, R., and Kelley, B. D. (2005), Scale-down models for purification processes: Approaches and applications, in Rathore, A. S., and Sofer, G. Eds., *Process Validation in Manufacturing of Biopharmaceuticals: Guidelines, Current Practices, and Industrial Case Studies*, Taylor and Francis, Boca Raton, FL.
72. U.S. Pharmacopeia (USP 26 2003), Chapter ⟨61⟩ Microbial limits test, USP, Rockville, MD.
73. Center for Drug Evaluation and Research, FDA (1987, May), Guideline on general principles of process validation.
74. Seely, J. E. (2005), Process characterization, in Rathore, A. S., and Sofer, G. Eds., *Process Validation in Manufacturing of Biopharmaceuticals: Guidelines, Current Practices, and Industrial Case Studies*, Taylor and Francis, Boca Raton, FL.

1.2

REGULATORY CONSIDERATIONS IN APPROVAL OF FOLLOW-ON PROTEIN DRUG PRODUCTS

Erin Oliver,[1] Stephen M. Carl,[2] Kenneth R. Morris,[2] Gerald W. Becker,[3] and Gregory T. Knipp[1]

[1]*Rutgers, The State University of New Jersey, Piscataway, New Jersey*
[2]*Purdue University, West Lafayette, Indiana*
[3]*SSCI, West Lafayette, Indiana*

Contents

Pharmaceutical Manufacturing Handbook: Production and Processes, edited by Shayne Cox Gad

1.2.1 INTRODUCTION

The ongoing need to provide the U.S. population with cost-effective pharmacological therapies has led to an emergent public health initiative in this country, namely for generic versions of therapeutic proteins. Greater access to generic drugs was made possible by the passage of the 1984 Drug Price Competition & Patent Term Restoration Act, commonly referred to as Hatch–Waxman. Generics have historically afforded considerable savings to the American consumer in need of prescription medication. Ten years after the Hatch–Waxman amendments, the Congressional Budget Office estimated that purchasers saved a total of $8–$10 billion on prescriptions at retail pharmacies by substituting generic drugs for their brand-name counterparts in 1994 [1]. To put those numbers in the context of today's pharmaceutical landscape, a recent report issued by the U.S. Department of Health and Human Services estimates that generic drugs constitute 63% of the total prescription medicines sold in the United States [2]. This same report suggests that generic drugs cost approximately 11% of the total cost of branded pharmaceuticals (on a per-dose basis).

At the same time, the development and use of therapeutic proteins have increased dramatically, with more than 850 biotechnology drug products and vaccines currently in trials [3]. Further, it is estimated that by the year 2010 nearly one-half of all newly approved medicines will be of biological origin [4]. The industrial financial incentives for the pursuit of follow-on biologics (heretofore termed biogenerics) are substantial with sales of biotechnology medicines in the United States rising 17% to approximately $30 billion in 2005 and growing at an annual rate of about 20% thereafter [3].

Not unexpectedly, the U.S. Food and Drug Administration (FDA) is experiencing mounting pressure to progress the cause of biogenerics. In a letter dated February 10, 2006, Senators Henry Waxman and Orrin Hatch (authors of the original "generic" legislation) urged the FDA to develop and implement clear guidelines for the approval of follow-on biological products for certain well-characterized proteins like insulin and human growth hormone (HGH) [5]. Additionally, recent litigation has compelled the FDA to take action on a pending drug application for a follow-on protein (FOP) drug product [Omnitrope, somatropin (recombinant DNA, rDNA origin)] [6].

A significant barrier to the emergence of "biogenerics" is the absence of a clear, efficient abbreviated pathway for approval. This hurdle is linked to significant scientific and legal issues in the United States in terms of how proteins are classified (drug vs. biologic) and subsequently regulated as well as how "generics" are traditionally defined in terms of equivalence and substitutability. However, an examination of the vast array of biologicals on the market today reveals that not all proteins are created equal. This range of complexity may provide an opportunity for stepwise progress on the regulatory front. This chapter presents the background to this multifaceted issue and examines the key regulatory challenges facing biogenerics today. An appropriate regulatory paradigm for the approval of FOPs is proposed and supported though a discussion of recent case studies.

1.2.1.1 Emergence of Biotechnology Industry

The explosion of scientific advances over the last quarter century has spawned the biotechnology industry and whole new classes of therapeutic agents for the treatment and prevention of disease. In October of 1982, the FDA approved the first protein-based therapeutic created by DNA technology in the form of Humulin (recombinant insulin). Developed by Eli Lilly & Co., with technical assistance from Genentech, Humulin is indicated for the treatment of diabetes. At the time, the use of recombinant technology was somewhat limited to the production of smaller, nonglycosylated proteins such as insulin (51 amino acids) and HGH (191 amino acids) using bacterial hosts. The seminal discovery by Columbia's Richard Axel of the process of cotransformation enabled complex protein production and glycosylation and thus spurred the emergence of the modern biotechnology industry [7].

The phenomenal growth observed in the biotechnology sector is notable in terms of the extraordinary number and diversity of therapeutic peptides and proteins that have been developed within a period of only about 20 years. Examples of therapeutic proteins in current use include cytokines, clotting factors, vaccines, and monoclonal antibodies, as illustrated in Table 1 [8].

As presented in Table 2, many of these "early" biotechnology products have reached the end of their period of patent exclusivity [4–9]. Thus, it is appropriate to now consider the next steps in the "life cycle" of these products as potential generic drugs.

1.2.1.2 Challenges Facing "Biogenerics"

The diversity and complexity of biologic molecules that drive their utility as therapeutic agents also contribute to the difficulty in classifying them as pharmacological entities, namely, whether they are drugs or biologics. This difficulty in classification is of profound importance since there are fundamental differences in how the FDA regulates drugs and biologics.

To appreciate the current challenges facing the pharmaceutical and biotechnology industry, it is informative to review the historical background associated with the classification and regulation of biologics in the United States, particularly in the context of the nation's evolving drug regulation system.

1.2.2 HISTORY OF BIOLOGICS REGULATION IN UNITED STATES

Due to the scientific limitations of the early to mid-1900s, significant differences existed between the approaches taken to manufacture and analyze biologics and

TABLE 1 Examples of Therapeutic Peptide and Protein Molecules Currently Marketed in United States

Peptides	***Antibiotics:*** bacitracin, bleomycin, gramicidine, capreomycin ***Hormones:*** corticotropin, glucagon, gonadrolein HCl, leuprolide acetate, histrelin acetate, oxytocin, secretin, goserelin acetate, vassopressin ***Others:*** polymixin B, eptifibatide, cyclosporine
Nonglycosylated proteins	***Interleukins:*** andresleukin (IL-1), denileukin diftitox (fusion, protein-IL-2+ DT), anakinra (IL-2) ***Interferons:*** interferon alpha-n1, interferon alpha-n3, interferon alpha-2a, peg interferon alfa-2b, interferon alfacon-1, Interferon alpha-2b, interferon beta-1b, interferon gamma-1b, ***Enzymes/inhibitors:*** anistreplase, asparaginase, lactase, trypsin, alpha-1 proteinase inhibitor, urokinase, deoxyribonuclease, fibrinolysin, chymotrypsin, pancreatin, papain, urokinase ***Growth factors/hormones:*** Filigrastim pegfilgrastim, somatropin, becaplermin, somatrem, menotropins ***Antithrombotic agents:*** thrombin, fibrinogen, hirudin, hirulog, fibrin ***Others:*** insulin, gelatin, prolactin, albumin (human), hemoglobin, collagen
Glycosylated proteins	Interferon beta-1a ***Antithrombotic agents:*** alteplase, drotrecogin alfa, antithrombin III ***Antianemic:*** darbopoetin alfa, erythropoietin ***Growth hormones:*** follitropin alpha, follitropin beta, chorionic gonadotropin (Human) ***Immuno globulins (IG):*** pertusssis IG, rabies IG, tetanus IG, hepatitis B IG, varicella zoster IG, rho(D) IG, normal immune globulin, lymphocyte anti-thymocyte, IB (equine) ***Coagulation factors:*** factor VII antihemophilic factor, factor IX (human, recombinant) ***Factor VIII (others):*** etanercept (CSF), sargramostim (TNF)
Monoclonal antiobodies	avciximab, alemtuzamub, basiliximab, gentuzumab, satumomab, inflixibam, palivizumab

drugs. This reality led to the creation of separate and distinct regulatory pathways for drugs and biologics. As noted earlier, the developments in analytical chemistry and improvements in process technologies have, in recent times, blurred the lines between drug and biologic drug development. In the current era of pharmaceutical development and standards harmonization, one might question the continued need for two distinct pathways. Recognizing the shifting paradigm of drug development, the history of biologics regulation is discussed below in two parts: early history and present day.

1.2.2.1 Early Biologics Regulation (1800s–1990s)

This country's earliest experience with biologics dates back to the infectious scourges of the late 1800s and early 1900s when epidemics of typhoid, yellow fever, smallpox, diphtheria, and tuberculosis were being battled by new advances in immunology. The discovery and development of vaccines and antitoxins led to the creation of a

TABLE 2 Patent Expiration Dates for U.S. Marketed Biologics

Brand Name	Generic Name	Indication	Company	Patent Expiry
Humulin	Recombinant insulin	Diabetes	Eli Lilly	Expired
Nutropin	Somatropin	Growth disorders	Genentech	Expired
Abbokinase	Eudurase urokinase	Ischaemic events	Abbott	Expired
Ceredase	Alglucerase	Gaucher disease	Genzyme	Expired
Cerezyme	Imiglucerase	Gaucher disease	Genzyme	Expired
Streptase	Streptokinase	Ischaemic events	AstraZeneca	Expired
Intron A	IFN-α-2b	Hepatitis B and C	Biogen/Roche	Expired
Serostim	Somatropin	AIDS wasting	Serono	Expired
Humatrope	Somatropin	Growth disorders	Eli Lilly	Expired
Geref	Sermorelin	Growth hormone deficiency	Serono	Expired (2004)
Synagis	Palivizumab	Respiratory syncytial virus	Abbott	Expired (2004)
Novolin	Human insulin	Diabetes	Novo Nordisk	2005
Protropin	Somatrem	Growth hormone deficiency	Genentech	2005
TNKase	Tenecteplase TNK-tPA	Acute myocardial infarction	Genentech	2005
Actimmmune	IFN-γ-1b	Chronic granulomatous disease; malignant osteoporosis	InterMune	2005, 2006, 2012
Activase, Alteplase	tPA	Acute myocardial infarction	Genentech	2005, 2010
Proleukin	IL-2	HIV	Chiron	2006, 2012
Epogen, Procrit, Eprex	Erythropoietin	Anemia	Amgen	2013
Neupogen	Filgrastim (G-CSF)	Anemia, leukemia, neutropenia	Amgen	2015

Note: Based on our search of available patent sites for only the reference product.
IFN-Interferon; tPA-Tissue Plasminogen Activator, IL-interleukin; HIV-Human Immunodeficiency Virus; G-CSF- Granulocyte-Colony Stimulating Factor; TNKase- Tenecteplase.

whole new "biopharmaceutical" industry. As demand increased, the pharmaceutical manufacturers responded and in turn supplanted the government's role in the public supply of vaccines (per Vaccine Act of 1813) [10]. Unfortunately, the commercialization of vaccines by smaller, less experienced, and likely less scrupulous manufacturers led to problems. Similar to the history of drug regulation, early advances in biologics regulation could be characterized as responsive rather than proactive. Change often occurred following tragedy and the result of government's attempt to respond. Some of the key milestones of early biologics regulation are summarized in Table 3. The following years saw many administrative changes in terms of the specific governmental agency responsible for regulating biologics, but with few substantive changes to the regulations themselves.

TABLE 3 Key Milestones in Early Biologics Regulation

1901	Ten children died in St. Louis from administration of tetanus-contaminated diphtheria antitoxin. In this case, no safety testing had been performed prior to use.
1902	Biologics Control Act (BCA) signed into law: • Authorizing the regulation of commercial viruses, serums, toxins, and analogous products • Requiring the licensure of biologics manufacturers and establishments • Providing governmental inspectional authority • Making it illegal for the commercial distribution of product not manufactured and labeled in accordance with the act
1906	Pure Food and Drug Act enacted (precursor of modern-day drug regulation). Lack of mention of biologics as a class effectively represents first distinction between drug and biologic regulation.
1919	BCA amended: • Required reporting of changes in equipment, manufacturing processes, personnel; establishment of formal quality control procedures; and submission of samples for regulatory inspection and approval for release • Recognized potential that slight changes to manufacturing conditions (raw materials, process, personnel, etc.) could have significant and adverse effect on product quality • Required strict control of input (environment and manufacturing conditions) rather than end-stage testing of quality attributes due to limitations in analytical methodology to detect these effects
1937	Elixir sulfanilamide, containing the poisonous solvent diethylene glycol, kills 107, many of whom are children.
1938	Food, Drug and Cosmetic Act (FDCA) enacted: • Established concept of "new drugs" requiring proof of safety prior to marketing • Required submission of an investigational new Drug (IND) application prior to clinical use of an experimental drug in humans • Required approval of a new drug application (NDA) prior to commercial sale of drugs • Granted federal government power of seizure of misbranded or adulterated drugs • Defined "drugs" comprehensively; not excluding potential of "biologics" to function as drugs
1941	• Approximately 300 deaths and injuries result from distribution of sulfathiazole tablets tainted with the sedative phenobarbital. • Insulin Amendment passed to require FDA testing/certification of purity and potency.
1944	Public Health Service (PHS) Act enacted to consolidate and codify previous biologics laws: • Outlined licensing requirements for biologics—for both product (product licensing application, or PLA) and establishment where the product was manufactured (establishment licensing application, or ELA) • Required submission of samples of each manufactured lot of all biologicals for government testing and certification prior to commercial release • Required sponsors to own all of manufacturing facilities, effectively eliminating multiparty or contract manufacturing

1.2.2.2 Modern Biologics Regulation (1990s–Today)

Whereas early biologics regulation was grounded by technical limitations, modern biologics regulation is driven by tremendous advances in scientific knowledge. Development of analytical tools and techniques has dramatically increased the ability to characterize proteins and substantiate the structure, composition, and function of the therapeutic molecule. These advances enable the detection of small differences in molecular weight; elucidation of primary, secondary, and tertiary protein structures; detection and quantification of posttranslational modifications (i.e., patterns of glycosylation); and improved understanding of structure–function relationships and potential immunogenic responses. Simultaneously, developments in the fields of pharmaceutical and biotechnological manufacturing have greatly improved process efficiency and control. This recent technological evolution has had a direct impact on biologics regulation as reflected below in several key events:

- In 1995, the FDA agreed to eliminate lot testing requirements for certain highly characterized products once the company's ability to consistently manufacture product of acceptable quality was established.
- In 1996, the FDA and Congress dismantled the dual-licensing process, requiring the submission of a single BLA (biologics license application), making the content and format of a biologics application similar to that required for new drug applications (NDAs).
- In 1996, the Center for Biologics Evaluation and Research (CBER) liberalized its definition of "legal manufacturer" and eliminated many of the barriers to cooperative, multiparty manufacturing arrangements.
- In 1997, Congress passed a noteworthy piece of legislation affecting modern pharmaceutical regulation in the Food and Drug Modernization Act (FDAMA). Among the many goals of the act was to harmonize the drug and biologic approval processes.

In fact, current pharmaceutical/regulatory initiatives appear to extract the best practices from biologic and drug approaches which can apply equally to both classes of products:

- The Quality Systems Approach and GMPs for the 21st Century, two initiatives being pursued by the FDA for drugs and devices, emphasize the utility of building quality into the process, consistent with the strict control of "input factors" seen in early biologic regulation.
- Initiatives such as Process Analytical Technologies build on the concept of conventional drug product testing using increasingly sophisticated analytical techniques to provide continuous process monitoring and finished-product quality assurance of multiple pharmaceutical dosage forms.
- The current global initiative to harmonize electronic submission format and content requirements effectively creates one standard data package for drugs or biologics. Thus, the eNDA (electronic new drug application) or eBLA (electronic biologics license application) will eventually be replaced by the eCTD (electronic common technical document).

1.2.3 REGULATORY CLASSIFICATION OF PROTEINS

Despite the blurring of lines between drugs and biologics, there remain two different mechanisms to bring protein drug products to the U.S. marketplace. The choice of approval framework is dependent on the protein's classification as a drug or biologic. The history of this regulatory distinction is rooted in the technical differences between small-molecule drugs and macromolecular biologics. Traditionally, drugs were characterized as having well-defined chemistry. Conversely, biologics were large, complex macromolecules whose active moiety defied characterization and quantitation. By necessity, different means of assuring the safety and efficacy of these therapeutic products were required at the time. The modern-day consequence is a legal system that distinguishes between proteins as drugs and proteins as biologics. The distinction is based on statutory definitions as well as historical precedent and has implications in terms of the approval pathways for original and follow-on products.

1.2.3.1 Definitions and Key Terminology

Drugs are defined by the U.S. Food and Drug Act [FD&C Act, 21 U.S.C. 321(g)(1)]*by function* as any article Federal Food, Drug and Cosmetic Act (a) intended for use in the diagnosis, cure, mitigation, treatment, or prevention of disease in man or animals and (b) intended to affect the structure or function of the body [11].

Biologics as a class may be regulated as drugs but are defined within the Public Health Service Act [PHSA, 42 U.S.C. 262(a)] *by category* as "a virus, therapeutic serum, toxin, antitoxin, vaccine, blood, blood component or derivative, allergenic product, or analogous product, or arsphenamine (or any other trivalent organic arsenic compound), applicable to the prevention, treatment, or cure of diseases or injuries of humans" [12].

A cursory examination of these definitions reveals that they are not mutually exclusive, leading to confusion about how to appropriately and consistently apply them. This point is illustrated when one reviews the history of how the FDA has categorized and subsequently regulated these drugs and biologics as shown below.

1.2.3.2 Application of Definitions to Proteins: Is It a Drug or a Biologic?

The answer to this fundamental question is not straightforward and has evolved over time. Historically, some natural-source-derived proteins such as insulin, hyaluronidase, menotropins, and Human Growth Hormone (HGH) have been regulated as drugs. While other natural-source-proteins such as blood factors were regulated as biologics. When recombinant proteins and monoclonal antibodies began development in the 1970s–1980s, these were regulated as follows:

1. By the Center for Drug Evaluation and Research (CDER) under the Food, Drug and Cosmetic Act (FDCA) as drugs when they were hormones such as insulin, HGH, and parathyroid hormone (PTH) derivatives
2. By the CBER under the PHSA as biologics when they were cytokines or blood factors such as factor VIII for hemophilia

TABLE 4 FDA Center Regulatory Responsibility for Therapeutic Biological Products

CDER	CBER
Monoclonal antibodies (in vivo use) Proteins intended for therapeutic use: Cytokines (e.g., interferons) Enzymes (e.g., thrombolytics) other novel proteins except those assigned to CBER Immunomodulators (nonvaccine, nonallergenic) Growth factors, cytokines, some hormones and monoclonal antibodies intended to mobilize, stimulate, decrease, or otherwise alter the production of hematopoietic cells in vivo	Cellular product, including products composed of human, bacterial, or animal cells Vaccines Allergenic extracts Antitoxins, antivenoms, venoms Blood, blood components, plasma-derived products (e.g., albumin, immunoglobulins, clotting factors, fibrin sealants, proteinase inhibitors), recombinant and transgenic versions of plasma derivatives

As other recombinant proteins and monoclonal antibodies came under development, the CBER held primary responsibility for this review, with the CDER retaining responsibility for hormones such as insulin and HGH. However, in 2003 all therapeutic proteins were transferred from the CBER to the CDER. This reassignment of review responsibility did not impact the legal classification of these protein products, such that the Center for *Drug* Evaluation and Research assumed responsibility for the review and approval of *biologics* approved under Section 351 of the PHSA.

The basic distribution of these therapeutic biologics to the respective FDA center is reflected in Table 4; however, many of the current complex biotechnology-derived products do not fit neatly into accepted definitions and require case-by-case classification [13].

1.2.3.3 Regulatory Approval Path for Proteins

The relevance of the preceding discussion becomes important with the understanding that therapeutic proteins classified as drugs are governed under a different set of laws than those classified as biologics. Drugs are approved via submission of NDAs under Section 505 of the FD&C Act, while biologics are supported by BLAs under the PHSA. These two approval paths are similar in terms of application content, that is, requirement of complete reports of clinical safety and efficacy data to support approval. However, only the drug regulation, that is, Section 505 of the FD&C Act, has been amended to outline an abbreviated approval mechanism for generic products.

1.2.4 REGULATION OF GENERIC DRUGS

1.2.4.1 History of Generic Drug Legislation in United States

In 1984, Congress responded to America's need for safe, affordable medicines by passing a pivotal piece of legislation, The Drug Price Competition and Patent Term

Restoration Act (Hatch–Waxman amendments). The intent of this act was to effectively balance the need to encourage pharmaceutical innovation with the desire to accelerate the availability of lower cost alternatives to approved drugs. The act also sought to eliminate unnecessary or redundant clinical testing to protect patients (reduce the number of patients in need receiving placebo in controlled clinical trials) and conserve industry and agency resources. To accomplish the goal of faster to market, cheaper alternatives, the amendments stipulated the following [14]:

- *For Innovator Companies* The act encouraged continued innovation, research, and development activities by providing manufacturers with meaningful incentives in the form of patent protection/restoration and marketing exclusivity, thus allowing them to recoup some of their investments.
- *For Generic Companies* The act provided access to certain innovator information without the threat of legal action via patent infringement suits (safe harbor provisions), allowing generics the opportunity to prepare for market introduction prior to the expiration of patent/exclusivity terms. This effectively limited the period of innovator exclusivity to the statutory timelines.

1.2.4.2 Approval Process for Generic Drugs

The act served as a boon to the generic industry by paving the path to abbreviated and accelerated drug approvals. From a legal perspective, the Hatch–Waxman amendments modified Section 505 of the FD&C Act to create two new abbreviated approval pathways (see Table 5) [14].

In essence, the abbreviated NDA (ANDA) and 505(b)(2) processes allow generic manufacturers the ability to rely on what is already known about the drug and refer to the agency's finding of safety and efficacy for the innovator. For an ANDA, the generic product must meet certain criteria related to bioequivalence and product sameness. However, a 505(b)(2) application often describes a drug with substantial differences to the innovator (which would seem more closely related to FOPs).

1.2.4.3 Application of Generic Regulations to Biologics

A central question is "Do biologics fall under the provisions of the Hatch–Waxman Act?" Since the Hatch–Waxman Act specifically amended the FD&C Act, biologics

TABLE 5 Description of NDA Approval Mechanisms

Traditional path	1. 505(b)(1)—Application that contains full reports of investigations of safety and effectiveness to which sponsor has right of reference (stand-alone NDA)
Abbreviated path	2. 505(b)(2)—Application that contains full reports of investigations of safety and effectiveness, where the sponsor relies on studies conducted by someone else to which the sponsor does not have right of reference
Abbreviated path	3. 505(j)(1)—Abbreviated new drug application (ANDA) containing information to show the product is a duplicate of an already approved drug product

approved via a BLA under the PHSA are not covered by this legislation nor does the PHSA have similar provisions for biogenerics. However, those few therapeutic proteins approved via Section 505 of the FDCA as NDAs are covered by the Hatch–Waxman amendments and thus are *legally* considered appropriate for filing a 505(b)(2) or 505(j)(1) application. For simple, well-characterized peptides and proteins regulated under Section 505 of the FD&C Act, mechanisms are already in place to bring FOPs to the market. In fact, several FOPs have already been approved by the FDA, including GlucaGen (glucagon recombinant for injection), Hylenex (hyaluronidase recombinant human), Hydase and Amphadase (hyaluronidase), Fortical (calcitonin salmon recombinant) Nasal Spray, and Omnitrope [somatropin (rDNA origin)] [15]. Further details related to the latter two are presented in the discussion of actual case studies.

1.2.5 LEGAL ARGUMENTS RELATED TO FOLLOW-ON PROTEINS

The legal arguments regarding the approval of biogenerics relate to several different aspects of drug/biologics law.

1.2.5.1 Constitutionality of 505(b)(2) Process for Drugs

The agency's authority to grant approval of drugs via the 505(b)(2) process has previously been challenged by several companies. The nature of these challenges has questioned the FDA's right to use proprietary information of the innovator in support of another company's drug approval. Recall that the 505(b)(2) process allows a company to use data for which it does not have right of reference (i.e., another company's safety and efficacy data) in support of its own application. The FDA's long-standing interpretation of the statute seems firm and well founded in precedent since over 80 applications for drugs have been approved via the 505(b)(2) route since its inception with indications ranging from cancer pain to Attention Deficit Disorder (ADD) [16].

1.2.5.2 Constitutionality of 505(b)(2) Process for Follow-On Proteins

The constitutionality issues related to FOPs are similar to those mentioned above for drugs, namely protection of proprietary information and intellectual property rights. Some critics opine that issues unique to FOPs create additional legal hurdles. For example, the rules pertaining to the disclosure of safety and effectiveness information are different for biologics licensed under the PHSA and drugs approved under the FDCA. When the rules were originally written (1974), it was thought that safety and effectiveness for one biologic would not support the licensure of another. So these data were deemed not to be protected trade secrets and could be publicly disclosed immediately after issuance of the biologic's license [see 21 CFR 601.51(e), 1974]. However, since this language applies strictly to the PHSA, it has no bearing on discussions related to the 505(b)(2) process.

In other public challenges, opponents argue that the unique and complex nature of biologics and the close relationship between their method of preparation and clinical attributes require that the FDA use and disclose the manufacturing methods and process information contained in an innovator's application. Further, this use and disclosure would violate Trade Secret and Constitutional Law (Fifth Amendment "taking clause") [17, 18].

The concept of "the product is the process" may have been applicable to early biologics, but current capabilities allow the chemical, biologic, and functional comparison of well-characterized protein drugs. The follow-on manufacturer need not necessarily utilize the identical method of manufacture or proprietary technology to reproduce a follow-on biologic with similar clinical safety and efficacy. Additionally, it is important to distinguish between the regulatory requirements for approval of an actual generic protein (duplicate of innovator; see discussion below) and those associated with a 505(b)(2), which requires a showing of similarity between two products. Any differences between the two would need to be adequately supported by bridging studies and appropriate clinical and/or nonclinical data.

The FDA has confirmed this interpretation in its response to petitions filed regarding FOPs (both in general and targeted to specific applications). The FDA has clearly said, "the use of the 505(b)(2) pathway does not entail disclosure of trade secret or confidential commercial information, nor does it involve unauthorized reliance on such data" [18].

1.2.5.3 Applicability of 505(j)(1) or ANDA Process to Biogenerics

Biogenerics per se, that is, protein drug products approved via 505(j)(1), would need to demonstrate their bioequivalence to the innovator protein. However, due to their complexity and heterogeneity, the classical biopharmaceutical principles upon which the current ratings of therapeutic equivalence are based do not apply in their current language to complex macromolecules. For example, due to the nature and complexity of an immunogenic response, one concern would be if traditional bioequivalence appropriately addresses the complex safety issues associated with biologics.

1.2.5.4 Current Rules Relating to Bioequivalence of Generic Drugs

The list of approved drug products with therapeutic equivalence (Orange Book) was originally intended as an information source to states seeking formulary guidance [19]. The list provides the FDA's recommendations as to which generic prescription drug products are acceptable substitutes for innovator drugs. The term *innovator* is used to describe the reference listed drug, or RLD [21 CFR 314.94(a)(3)], upon which an applicant (generic) relies in seeking approval of its ANDA. In layman's terms the RLD describes the original NDA-approved drug and is often referred to as the "pioneer" drug.

Under the Drug Price Competition and Patent Term Restoration Act of 1984, manufacturers seeking approval to market a generic drug need to submit data to the FDA demonstrating that their proposed drug product is bioequivalent to the pioneer (innovator) drug product. A major premise underlying the 1984 law is that bioequivalent drug products are therapeutically equivalent, will produce the same clinical effect and safety profile as the innovator product, and are therefore, interchangeable [19].

So how would FOPs be classified using conventional definitions of bioequivalence? To answer this question, it is necessary to review current legal definitions of bioequivalence terms [19]:

- Two products are bioequivalent in " the absence of a significant difference in the rate and extent to which the active ingredient or active moiety in *pharmaceutical equivalents* or *pharmaceutical alternatives* becomes available at the site of drug action when administered at the same molar dose under similar conditions in an appropriately designed study" [21 CFR 320.1(e)]. An appropriately designed comparison could include (1) pharmacokinetic (PK) studies, (2) pharmacodynamic (PD) studies, (3) comparative clinical trials, and/or (4) in vitro studies.
- *Pharmaceutical equivalents* are those drug products which are formulated to contain the same amount of active ingredient in the same dosage form to meet the same (compendial or other applicable) standards of quality.
- *Pharmaceutical alternatives* are drug products that contain the same therapeutic moiety, or its precursor, but not necessarily in the same amount or dosage form. Drug products are considered to be *therapeutic equivalents* only if they are pharmaceutical equivalents and if they can be expected to have the same clinical effect and safety profile when administered to patients under the conditions specified in the labeling. Although pharmaceutical alternatives may ultimately be proven bioequivalent, given their differences they are not automatically presumed to be.

Given these definitions, FOPs would likely be considered pharmaceutical alternatives if one presumes that pioneer and follow-on proteins are identical at a precursor stage, prior to potential post-translational modification. This presumption may also be consistent with the similarity standard the agency applies to ascertain orphan drug status (see discussion in Section 1.2.6). Follow-on proteins cannot be considered to be therapeutic equivalents since they are not pharmaceutical equivalents and cannot be expected to have the same clinical effect and safety profile in the absence of testing. This assertion is supported by the following:

- The potential impact of how posttranslational modifications, such as glycosylation, can directly impact protein conformation and subsequently affect biological activity, including the overall safety and efficacy of the drug product.
- An underlying premise of bioequivalence assessments is a clearly defined pharmacokinetic/pharmacodynamic relationship; however, the relation between blood levels and effect is less clearly established for proteins [20].

Consequently, within the current regulatory framework, FOPs are unique products that may be "similar" but are not the same as innovator proteins, consistent with their approval via a 505(b)(2) pathway. This interpretation is supported by the FDA's designation of Omnitrope as having a BX rating in the Orange Book. The code BX in the Orange Book refers to drug products for which the data are insufficient to determine therapeutic equivalence as compared to a therapeutic rating of A indicative of interchangeability. This concept of similarity is also consistent with the definitions proposed by the European Agency for the Evaluation of Medicinal Products (EMEA) for generic versions of proteins [21]:

> Bio-similar products: second and subsequent versions of biologics that are independently developed and approved after a pioneer has developed an original version. Bio-similar products may or may not be intended to be molecular copies of the innovator's product; however, they rely on the same mechanism of action and therapeutic indication.

1.2.5.5 Statutory Authority

Unlike the FDCA, which affords therapeutic protein drugs the legal pathway of abbreviated drug approval for a FOP, the PHSA currently has no similar provisions. Such a pathway for approval or licensure of FOP products under the PHSA would require new legislation and recent congressional developments suggest that work is underway to create this statutory pathway.

Legislation proposed on September 29, 2006, by U.S. Representative Henry Waxman (D-CA) and Senator Charles Schumer (D-NY) seeks to amend the PHSA to authorize the FDA to approve abbreviated applications for biologic products that are "comparable" to previously approved (brand name) biologic products. Entitled The Access to Life-Saving Medicine Act, this bill outlines a process by which the FDA could determine, on a product-by-product basis, the studies necessary to demonstrate comparability of a FOP product to a brand name product and assure its safety and effectiveness. The act allows for an applicant to seek interchangeability with a brand name product, recognizing that the extent of data to support such a designation must be discussed with the FDA. To encourage the development of interchangeable products, the bill would authorize tax incentives and periods of marketing exclusivity. The bill would also seek to create an improved process to facilitate early resolution of patent disputes which might otherwise delay competition [22].

1.2.6 SCIENTIFIC ISSUES RELATED TO FOLLOW-ON PROTEINS (DATA REQUIREMENTS)

The challenge of FOPs demonstrating similar quality, safety, and efficacy to the innovator product relates to the poor predictability of physicochemical characteristics and biologic activity. For example, there are several different interferon-α and erythropoietin α and β products currently on the market. These variants are characterized by differences in sequence, glycosylation pattern, and in vitro measures of specific activity; however, their clinical safety and efficacy profiles are considered similar [20].

In contrast, different formulations of insulin and growth hormone containing the same active ingredient exhibit significant differences in bioavailability [20]. Additionally, the inability to adequately predict immunogenic responses from in vitro data or animal studies remains a concern.

The answer to the challenge is that generic manufacturers must go through a similar process of in-depth characterization, including identification of critical structural elements of the product (structure/function) when developing a FOP. Although the regulatory standards for demonstrating similarity are currently undefined, some insight can be gleaned from consideration of FDA expectations in terms of granting orphan drug status to similar proteins and assessing postapproval Chemistry, Manufacturing and Controls (CMC) changes for innovator proteins.

1.2.6.1 "Sameness" as per Orphan Drugs Regulations

The Orphan Drug Act of 1983 was implemented in response to the government's concern that viable treatments for rare diseases were not being explored due to excessive costs of drug development in comparison to the relatively small population of potential users (and sales). Orphan drugs are (a) those used to treat rare diseases, defined by the act as affecting <200,000 persons in the United States, or (b) those drugs whose development costs would not be recovered through sales of the drug. To encourage development, the government authorized incentives in the form of marketing exclusivity (seven years), tax credits, protocol assistance, and grants/contracts, with the first being of primary importance to most drug sponsors.

Since exclusivity is awarded only to the first designated product to obtain approval for a given drug/indication, competition is fierce. No approval would be given to a subsequent sponsor's application for the same product/indication unless it was shown to be clinically superior (i.e., not the same). Thus, the agency needed to develop criteria upon which it would make these determinations.

In 1992, the FDA's orphan drug regulations first established the conditions under which the agency could determine product "sameness" of protein drugs and therefore take action to block the approval of a second orphan drug product: "two protein drugs would be considered the same if the only differences in structure between them were due to post-translational events, or infidelity of translation or transcription, or were minor differences in amino acid sequence; other potentially important differences, such as different glycosylation patterns or different tertiary structures, would not cause the drugs to be considered different unless the differences were shown to be clinically superior" [23]. It should be noted that there may exist exceptions to this rule that depend on the interpretation of each individual case. For example, Eli Lilly & Co. successfully received orphan drug status in the late 1980s for the naturally occurring HGH to compete with the previously marketed Met-HGH, which only differed in the N-terminal methionine.

The support for clinical superiority could be based on evidence of greater effectiveness and increased safety or represent a "major contribution to patient care." In short, orphan drug regulations utilize clinical data to demonstrate product differences. Examples include [23]:

- 1996: Biogen's Avonex (interferon β) was considered to be clinically superior to Berlex's Betaseron based on improved safety (fewer site injection reactions).
- 1999: In a law suit involving generic paclitaxel, Baker Norton, challenged the FDA's sameness determinations based on active moiety alone, arguing that factors such as formulation and labeling should be considered. The challenge was unsuccessful.
- 2002: Serono's Rebif (interferon β_{1a}) was awarded exclusivity based on the clinical demonstration of improved efficacy (reduced Multiple Sclerosis (MS) exacerbations).

Therefore, it would appear that the orphan drug regulations provide some flexibility to the sponsor (generic) in establishing product sameness but also reaffirm the important role of clinical data in supporting product safety and efficacy.

1.2.6.2 "Sameness" as per Postapproval Change Guidances

Guidelines for supporting postapproval changes to the chemistry, manufacturing, and controls of approved products (SUPAC guidances) take a somewhat different approach to establishing sameness. In essence, the SUPAC guidelines reflect risk management practices in evaluating the potential of certain CMC changes to impact the identity, strength, quality, purity, and potency of the product as they may relate to overall safety and efficacy.

A long-held contention within the biologics industry is that the product is the process and, by extension, change is strongly discouraged. Without qualification, this rather dated thinking is inconsistent with the flexibility required in managing change throughout the life cycle of a product. Further, this thinking may serve to discourage the implementation of advanced technologies designed to improve not only efficiency but also product quality. Even current biologics regulations recognize the need to accommodate change; 21 CFR 601.12 (for biologics) states that for changes in the product, production process, quality controls, equipment, facilities, and so on, an applicant must assess the effects of the change and demonstrate through appropriate validation and/or other clinical and/or nonclinical laboratory studies the lack of adverse effect of the change on the identity, strength, quality, purity, or potency of the product as they may relate to the safety or effectiveness of the product.

In fact, many of the challenges that generic manufacturers face in demonstrating sameness of FOPs to reference listed drugs are similar to those encountered by innovators in managing the dynamic CMC life cycle of a product. One of the tools available to assess the potential impact of product differences is a comparability protocol. The FDA described its expectations of the data requirements necessary to support postapproval CMC changes to protein drug product and biologic products in a Guidance to Industry on the use of comparability protocols for such products issued in 2003 [24]. Underpinning the successful application of a comparability protocol are extensive product development and characterization.

Initial Product Development Prior to undertaking any comparative analysis, a manufacturer must perform two critical steps. First, the manufacturer needs to conduct thorough process development and optimization of the therapeutic protein product. Second, the sponsor (generic or innovator) needs to prospectively examine the impact of changes to all critical processing parameters during the development phase and determine the minimum data requirements necessary to assure the absence of adverse impact to product quality, safety, or efficacy. The current state of technology provides us with better tools to more fully characterize the protein drug substance and drug product at all stages of production.

Physicochemical Characterization and Process Development Some of the key steps to process development and product characterization include:

- Production of a cell line/clone
- Identification and characterization of critical raw materials (media, resins, formulation excipients)
- Development of internal standards, in-process controls, product specifications
- Conduct of extensive pilot-scale manufacturing development: fermentation and downstream processing (separation and purification)
- Performance of process scale-up and optimization studies

TABLE 6 Analytical Techniques for Physicochemical Characterization of Proteins

Parameter	Test
Primary structure	Amino acid sequencing, N-terminal Edman sequencing, peptide mapping
Higher order structure	CD, NMR, FTIR, Raman
Mass	LC-ESI-MS, MALDI-TOF-MS
Size	SDS-PAGE, DLS, SEC-MALLS
Hydrophobicity	RP-HPLC
Binding	Immunological binding
Sulfhydryl groups/disulfide bridges	Peptide mapping (under reducing and nonreducing conditions)
Glycan analysis:	
Monosaccharide analysis	HPLC, MS
Sialic acid content	HPLC
Molecular weight	MALDI-MS, ESI-MS
Impurity profile	
Process-related impurities	Immunoassay, HPLC, SDS-PAGE, MS, CD, capillary gel electrophoresis, size exclusion chromatography
• Cell substrate derived	
• Cell culture derived	
• Downstream derived	
Product-related impurities	
• Truncated forms	
• Other modified forms (i.e., deamidated, isomerized)	
• Aggregates	
Evaluation of stability	HPLC

CD, Circular Dichroism; NMR, Nuclear Magnetic Resonance; FTIR, Fourier transform infrared spectroscopy; LC-ESI-MS, Liquid chromatography electrospray ionisation mass spectrometry; MALDI-TOF-MS, Matrix-assisted laser desorption ionization-time of flight-mass spectrometry; SDS-PAGE, Sodium dodecyl sulfate (SDS) polyacrylamide gel electrophoresis; DLS. Dynamic light scattering; SEC-MALLS, Size exclusion chromatography-multi-angle laser light scattering; RP-HPLC, Reversed phase-high performance liquid chromatography; HPLC, High performance liquid chromatography; MALDI-MS, Matrix-assisted laser desorption ionization mass spectrometry; ESI-MS, Electrospray ionisation mass spectrometry; MS, Mass spectrometry.

- Application of a comprehensive array of analytical techniques to fully characterize the drug product at each stage of development. Table 6 provides examples of methods to probe virtually every property of the protein and develop a fingerprint of the molecule.

Other Testing Requirements The need for additional supportive studies beyond physicochemical characterization will increase proportionately with the complexity of the protein drug. The entire battery of tests may not be required for each FOP but may include the following data, bioassay, preclinical (pharmacology/toxicology/pharmacokinetic/pharmacodynamic), clinical safety and efficacy, and immunogenicity. The nature, number, and size of the trials should relate directly to the particular drug/indication/patient population.

Bioassay A biological assay, or "bioassay," is an analytical procedure capable of measuring the biologic activity of a substance based on a specific functional, biologic response of the test system. Bioassays should be predictive of clinical effect and are therefore used as a means of quantifying activity (in nonclinical manner) and ensuring efficacy throughout development. They are informative in equivalence studies to the extent that a change affects a part of the molecule, which in turn impacts the molecule's biologic activity. Bioassays may be based on animal models, in vitro cell lines, cell-based biochemical assays (i.e., kinase receptor activity), receptor binding assays, or enzyme assays. The selection of an appropriate bioassay is driven in part by the ability to demonstrate a correlation to clinical effect. An example of a predictive bioassay is the measurement of the antiviral activity of interferon as a function of its cytopathic effect on host cells [25].

Nonclinical (Pharmtox, PK, PD) In the context of FOPs, the original sponsor will have demonstrated what the molecule per se does to the body; however, since the formulation is likely different, nonclinical studies are useful in demonstrating a lack of adverse impact due to dosage form, route of administration, excipient changes, manufacturing contaminants, and supporting sameness of the active moiety.

Appropriate toxicology studies would include acute or subchronic testing in at least one relevant small animal species. Pharmacokinetic studies are highly useful in assessing the impact of changes in the manufacture of natural-source- and recombinant-derived proteins. Standard approaches used in bioequivalence studies [measurement of the area under the curve (AUC), $C_{max,}$ t_{max}] can be used to make direct comparisons of innovator and follow-on profiles. Pharmacodynamic studies are similarly very informative. Direct comparison of innovator and follow-on products can be made by evaluating appropriate surrogate markers of efficacy (i.e., platelet aggregation following anticoagulation therapy).

Clinical (PK, PD, Safety and Efficacy) Human clinical studies can range in complexity from standard-design PK studies to complicated, long-term efficacy trials evaluating one or more indications in multiple populations. Human PK studies are used as the benchmark for establishing bioequivalence of conventional dosage forms. For traditional pharmaceuticals for which reliance on systemic exposure may not be suitable, PD or clinical safety and efficacy may be performed to show equivalence.

The appropriate clinical program is influenced by many factors, including the degree of molecular complexity of the particular protein and the extent of physicochemical characterization; the mode of action, indication(s), and use population(s); the presence of established structure–activity relationships and validated bioassays; and the results of preclinical testing.

As such, the nature and scope of each clinical support program need to be determined on a case-by-case basis in consultation between the sponsor and the regulatory agency.

Immunogenicity The observation of serious adverse events with the use of some recombinant and natural source proteins [i.e., pure red cell aplasia (PRCA) detected with erythropoietin use] has highlighted immunogenicity as a major issue for consideration when assessing within and between manufacturer changes [8]. Although the exact immunological mechanism responsible for the increased number of PRCA cases is unknown, it appears to be linked to a formulation change associated with Eprex, a European epoetin-α product. Replacement of the stabilizer human serum albumin with polysorbate 80 and glycine correlated with a surge in PRCA reported cases [4]. An immunogenic effect may have no clinical impact or it could have serious clinical consequences as seen above. The immune response of the therapeutic protein should be fully characterized using both immunoassays which detect antibodies that bind to the drug as well as bioassays which detect neutralizing antibodies that might block the protein's desired biological effect. Ultimately, this testing needs to be performed in humans, as animal testing is not truly predictive of human immune response. Antibody detection techniques include enzyme-linked immunosorbent assay radioimmunoassay, (ELISA), and surface plasmon resonance [8].

Comparability Testing to Demonstrate "Sameness" Following the developmental studies described above, comparative studies to directly evaluate pre- and post-change materials to one another and assess the impact of any process changes may be conducted. In a similar manner, comparative studies between pioneer drug and the follow-on can be used to systematically evaluate the impact of any differences between reference listed drug and proposed generic protein drug. When compiling information into an analytical characterization database, the data should be directly compared to the reference product and variation observed in multiple batches of test product (generic) should be similar to that of the reference innovator product.

The FDA's expectations in this regard are apparent in their description of the CMC data package supporting the comparability of Omnitrope to the innovator protein Genotropin. The FDA asserted [18]:

> Each biotechnology manufacturer, whether producing a new molecular entity or a follow-on product must independently develop its own cell expression, fermentation, isolation and purification systems for the active ingredient in its product. Thus, the manufacturing process for each active ingredient is unique to each manufacturer. Nevertheless, as Sandoz has demonstrated in its Omnitrope application, for this relatively simple recombinant protein, it is possible to determine that the end products of different manufacturing processes are highly similar, without having to compare or otherwise refer to the [proprietary] processes.

1.2.7 PROPOSED REGULATORY PARADIGM: CASE STUDIES

Based on the nature and complexity of therapeutic protein products, an approval pathway for follow-ons may require moving away from the traditional generic paradigm in place for small molecules and creating a biosimilar paradigm for complex molecules. The proposed regulatory paradigm for the approval of FOP products could be similar for protein drugs approved under Section 505 of the FDCA or licensed as biologics under the PHSA and mirror the current 505(b)(2) process. This pathway permits the sponsor and agency to determine exactly what studies are necessary to support the proposed differences (see 21 CFR 314.54(a) ["a 505(b)(2) application need contain only that information needed to support the modification(s) of the listed drug"]. Application of a 505(b)(2) paradigm removes the need to demonstrate bioequivalence per se and potentially reduces innovator intellectual property concerns that arise if a generic must "duplicate" the innovator. Guidance as to how similar a "biosimilar" needs to be exists in the form of current regulations related to orphan drugs and postapproval manufacturing changes.

Several recent drug approvals illustrate how this regulatory framework may be applied and are described in the sections to follow.

1.2.7.1 Case Study 1: Fortical [Calcitonin-Salmon (rDNA origin)]

On August 17, 1995, the FDA approved Novartis's NDA for Miacalcin (calcitonin-salmon) Nasal Spray (Miacalcin NS) for the treatment of postmenopausal osteoporosis in females greater than five years postmenopause with low bone mass relative to healthy premenopausal females. The active ingredient in Miacalcin NS is synthetic salmon calcitonin. On March 6, 2003, Unigene submitted a new drug application under Section 505(b)(2) for Fortical [calcitonin-salmon (rDNA origin)] Nasal Spray which relied in part on data submitted in the Miacalcin NS NDA.

Comparability Program Fortical and Miacalcin NS differed in certain aspects, such as the use of recombinant versus synthetic salmon calcitonin and the use of different types and amounts of excipients. Given these differences, Unigene was required to submit data to establish that the findings of safety and efficacy for Miacalcin were relevant to Fortical (i.e., contain the same active ingredient and have comparable bioavailability) and that the formulation differences did not impact previous clinical profile [26, 27].

Comparability Results

Physicochemical Analysis Salmon calcitonin is a 32-amino-acid, nonglycosylated peptide hormone. It is structurally simple, possessing limited secondary structure and a single disulfide bond. The physicochemical characterization studies demonstrated that the primary and secondary structure of Fortical's recombinant salmon calcitonin (sc) was identical to that of Miacalcin's synthetic sc or naturally occurring sc. Further, the tertiary structures of the three were indistinguishable.

Nonclinical PK/T_{ox} The pharmacokinetic profile of Fortical by different routes of administration was compared to Miacalcin, demonstrating similarity in PK profiles between the synthetic and recombinant peptides and toxicity results (28-day rat intranasal toxicity study) were acceptable, particularly in light of clinical safety data.

Clinical PK/PD Calcitonin has a well-established mechanism of action; published literature supports that salmon calcitonin, mediated through calcitonin receptors located on osteoclasts, inhibits bone resorption, thereby increasing bone mineral density. Since serum beta-CTx (C-telopeptides of type 1 collagen, corrected for creatinine) is a recognized marker of bone resorption, the effect of administered salmon calcitonin on serum beta-CTx is considered to be an adequate surrogate for pharmacodynamic comparisons.

Fortical's PD equivalence was shown in a double-blind, active-controlled, 24-week study in 134 postmenopausal women randomized to Fortical (200 IU per day) or Miacalcin (200 IU per day). The primary outcome measure was change in serum beta-CTx from baseline. The results fell within prespecified PD equivalence limits (−0.08 to 0.06 ng/mL; equivalence margin of ±0.2 ng.mL) and indicated Fortical was not inferior to Miacalcin.

Fortical's PK equivalence was assessed by comparing the relative bioavailability of Fortical to Miacalcin in a multidose, crossover study of 47 healthy female volunteers. Results indicated that Fortical was slightly more bioavailable than Miacalcin, but given the demonstration of similar PD activity, this difference were not considered to be clinically significant.

Immunogenicity Archived samples from the 24-week PD study were used to compare the immunogenicity potential of both products. The results indicated there was no difference in terms of total immune response and the response of neutralizing antibodies between the two drugs.

Conclusion to Case Study 1 On August 12, 2005, the FDA approved Unigene's 505(b)(2) application for Fortical for the same indication as Miacalcin NS [26, 27]. In the FDA's analysis no statistically and/or clinically significant differences were noted in any aspect of the comparability profile, including clinical performance, and Fortical was approved.

The basis of this comparison was strongly challenged in a citizen petition claiming that (1) recombinant salmon calcitonin is not the same as the synthetic version which could potentially cause differences in product efficacy, safety, or both and (2) only a long-term clinical study (actual bone fracture data) would provide adequate support of sameness [28]. The FDA responded to this citizen petition by asserting its decision that the comparability data presented above collectively constituted sufficient demonstration of sameness [27, 29].

1.2.7.2 Case Study 2: Omnitrope [Somatropin (rDNA origin)]

On August 24, 1995, the FDA approved NDA20-280 filed by the Pharmacia & Upjohn Company for Genotropin (somatropin) (rDNA origin) for injection. Since that time, Genotropin has been marketed as a safe and effective therapy for growth hormone deficiency (GHD) in children and adults.

On July 30, 2003, Sandoz submitted a 505(b)(2) application for the approval of its recombinant HGH product (recombinant somatropin) indicated for long-term treatment of pediatric patients who have growth failure due to an inadequate secretion of endogenous growth hormone and for long-term replacement therapy in adults with GHD of either childhood or adult onset. This application relied in part on data submitted in the Genotropin NDA.

Comparability Program As with the Fortical case study, Omnitrope and Genotropin differed in certain aspects. As such, Sandoz was required to submit substantial data to establish that Omnitrope was sufficiently similar to Genotropin to warrant reliance on FDA's finding of safety and effectiveness for Genotropin to support the approval of Omnitrope [18].

Comparability Results

Physicochemical Analysis In terms of complexity, HGH is fairly simple and well-characterized. Human growth hormone is a single-chain, 191-amino-acid, nonglycosylated protein with two intramolecular disulfide bonds. Sandoz used a variety of physicochemical tests and analytical methods to confirm the primary, secondary, and tertiary structures, molecular weight, and impurity profile. Characterization studies performed to verify somatropin as the active ingredient in Omnitrope included reverse-phase liquid chromatography/mass spetrometry (RP-HPLC/MS), DNA sequencing, N-terminal and C-terminal sequencing, peptide mapping, circular dichroism (CD) analysis, UV spectroscopy, one-dimensional nuclear magnetic resonance spectroscopy (1D NMR), two-dimensional (2D) NMR, size exclusion chromatography (SEC), isoelectric focusing (IEF), sodium dodecyl sulfate polyacrylamide gel electrophoresis (SDS–PAGE), and capillary zone electrophoresis.

Nonclinical PK/T_{ox} Minimal toxicity data were needed on recombinant HGH (rHGH) itself, since the clinical effects of HGH excess are well established and understood and are extensively documented in published literature. Sandoz performed toxicity studies to appropriately qualify impurities specific to Omnitrope, that is, a subacute 14-day rat study and a local (skin) tolerance study in rabbits. Further, the bioactivity of Omnitrope was assessed using a validated weight gain bioassay using a hypophysectomized (growth-hormone-deficient) rats.

Clinical PK/PD HGH has a well-established mechanism of action. Omnitrope was demonstrated to be pharmacokinetically and pharmacodynamically "highly similar" to Genotropin. The dataset comprised a total of three PK/PD studies, including a double-blind, randomized, two-way crossover study comparing Omnitrope and Genotropin. Additionally, Sandoz conducted three sequential, multicenter phase 3 pivotal trials in 89 pediatric patients with GHD providing data in some patients for up to 30 months. A fourth phase 3 trial (n = 51, 24 months) was submitted as part of its safety update. Collectively, these data in conjunction with the demonstrated comparability to the reference listed product provide substantial evidence of Omnitrope's safety and effectiveness.

Immunogenicity A significant number of patients who were administered an earlier version Omnitrope developed anti–growth hormone antibodies during the first and second phase 3 clinical trials. In response, Sandoz implemented changes to the drug product to address this immunogenicity and evaluated the impact of these changes clinically. Data from the 24-month clinical study demonstrated that Omnitrope has a low and acceptable level of immunogenicity (comparable to other rHGH products) as none of the patients developed anti–growth hormone antibodies during the duration of the study and only one patient developed anti–host cell protein antibodies, which were of no detectable clinical consequence.

Conclusion to Case Study 2 This case provoked significant challenges from interested parties voiced via several citizen petitions [18]. Furthermore, the FDA's delay in approval prompted Sandoz to file suit to compel the FDA to rule on its application. On April 10, 2006, the Washington, D.C., District Court ruled that the FDA must meet its statutory obligations and take action on Sandoz's outstanding NDA[6]. On May 30, 2006, the FDA approved Omnitrope [somatropin (rDNA origin)] as a "follow-on protein product" for use in the treatment of pediatric GHD. At the same time, the FDA responded to the related citizen petitions and defended its position that the data were adequate to demonstrate that Omnitorpe was sufficiently similar to Genotropin to enable reliance on the agency's previous findings of safety and efficacy for Genotropin. These data, in conjunction with the independent evidence of safety and efficacy provided by Sandoz, supported Omnitrope's approval.

1.2.7.3 Case Study 3: Generic Salmon Calcitonin

On February 17, 2004, Nastech Pharmaceutical Company announced its filing of an ANDA for a salmon calcitonin nasal spray drug product for the treatment of post-menopausal osteoporosis. As with Fortical, Novartis's Miacalcin was cited as the reference listed drug; however, Nastech chose to submit an ANDA via the 505(j)(1) route, rather than a 505(b)(2) application.

The distinction between the two regulatory routes has significant implications for FOPs. Whereas 505(b)(2) allows products to be "sufficiently similar," an ANDA requires the applicant establish "sameness" of the active ingredients. The scope of data necessary to demonstrate that the actives are the same is unclear. Additionally, use of the ANDA route is appropriate for circumstances in which "clinical studies are not necessary to show safety and effectiveness." If clinical data are required as proof of sameness, as in the previous example where clinical data were used to demonstrate comparable immunogenicity, then the ANDA route may not represent a viable regulatory path.

On July 10, 2006, Nastech was notified by the FDA that its ANDA for intranasal calcitonin salmon was not approvable at present based on concerns relating to the potential for immunogenicity that might result from a possible interaction between calcitonin salmon and chlorobutanol, the preservative in the formulation.

Nastech has indicated it will continue to work with the agency to understand the data requirements and regulatory options, but the final resolution remains presently unknown. This case study highlights the fact that demonstration of *sameness* of therapeutic proteins is more complex than for other drugs and that true "biogenerics" may be hard to come by due to the complexity in establishing sameness versus similarity.

1.2.8 SUMMARY AND CONCLUSIONS

This chapter provides an overview of the complex scientific, legal, and policy issues facing the development of biogenerics today. Given the rising cost of health care and prescription medications in this country and the pivotal and expanding role of biologically derived products within the pharmaceutical landscape, these issues present a challenge to industry, regulators, and legislators alike. Substantial progress has already been made and the regulatory climate continues to evolve in response to advancing science and technology. Recent FDA approvals provide insight into the technical requirements for approval of well-characterized FOP products. They also demonstrate the appropriate use of an abbreviated approval pathway, that is, the 505(b)(2) pathway in place for drugs approved under the FDCA. Importantly, recent legislative proposals seek to amend the PHSA to eliminate the current legal barriers which prohibit abbreviated approval of protein biologics. This legislation reaffirms the need for the FDA to determine on a case-by-case basis the nature and extent of supporting data required for a given product.

REFERENCES

1. Congressional Budget Office (1998), *How Increased Competition from Generic Drugs Has affected Prices and Returns in the Pharmaceutical Industry*, Congressional Budge Office, Washington, DC.
2. Crawford, L. M., Acting Commissioner of the Food and Drug Administration, in a Speech to the Generic Pharmaceutical Association on February 26, 2005, available: http://www.fda.gov/oc/speeches/2005/GPhA0301.html, accessed Apr. 23, 2005.
3. Comments of the Generic Pharmaceutical Association (GPhA) (Sept. 29, 2006), available: http://www.gphaonline.org/AM/Template.cfm?Section=Media&Template=/CM/HTMLDisplay.cfm&ContentID=2849, accessed Jan. 23, 2007.
4. Schellekens, H. (2005), Follow-on biologics: Challenges of the "next generation", *Nephrol. Dial. Transplant.* 20(Suppl. 4), iv31–iv36.
5. Congressional letter from Senators O. Hatch and H. Waxman to Andrew von Eschenbach, Acting Commissioner of the Food and Drug Administration (Feb. 10, 2006), available: http://www.henrywaxman.house.gov/news_letters_2006.htm, accessed Dec. 21, 2006.
6. Messplay, G. C., and Heisey, C. (2006), Follow-on biologics: The evolving regulatory landscape, *Bioexec Int.*, May, 42–45.
7. *Biotechnol. Law Rept.*, 2003, 22(5), 485–508.
8. Comments from R. Williams, U.S. Pharmacopoeia (USP), to FDA Docket No. 2004N-0355, Mar. 15, 2005.
9. Herrera, S. (2004), Biogenerics standoff, *Nat. Biotechnol.*, 22(11), 1343–1346.
10. Scott, S. R. (2004) What is a biologic?, Chapter 1 in Mathieu, M., Ed., *Biologics Development: A Regulatory Overview*, 3rd ed., Paraxel Intl., Waltham, MA, pp. 1–16.
11. Federal Food Drug and Cosmetic Act, available: http://www.fda.gov/opacom/laws/fdcact/fdctoc.htm, accessed Apr. 21, 2005.
12. Public Health Service Act, available: http://www.fda.gov/opacom/laws/phsvcact/phsvcact.htm, accessed Apr. 21, 2005.
13. U.S. Department of Health and Human Services, Food and Drug Administration Transfer of Therapeutic Products to the Center for Drug Evaluation and Research, available: http://www.fda.gov/cber/transfer/transfer.htm, accessed Apr. 23, 2005.
14. U.S. Department of Health and Human Services (DHHS) (1999, Oct.), Food and Drug Administration, Center for Drug Evaluation and Research, Guidance for Industry: Applications covered by Section 505(b)(2), DHHS, Washington, DC.
15. U.S. Department of Health and Human Services, Food and Drug Administration, Omnitrope (somatropin [rDNA origin]) questions and answers, available: http://www.fda.gov/cder/drug/infopage/somatropin/qa.htm, accessed Dec. 21, 2006.
16. Letter of J. Woodcock. (CDER, FDA) to Docket Nos. 2001P-0323/CP1, 2002P-0447/CP1, and 2003P-0408/CP1 (Oct. 14, 2003).
17. Glidden, S. (2001), The generic industry going biologic, *Biotechnol. Law Rept*, 20(2), 172–181.
18. Letter from S. Galson (CDER, FDA) in response to Docket Nos. 2004P-023 11CP1 and SUP 1,2003P-0 1 76lCP 1 and EMC 1, 2004P-0171lCP1 and 2004N-0355 (May 30, 2006).
19. U.S. Department of Health and Human Services, Food and Drug Administration, Center for Drug Evaluation and Research, Office of Pharmaceutical Science, Office of Generic Drugs, Electronic orange book: Approved drug products with therapeutic equivalence evaluations, available: http://www.fda.gov/cder/ob/default.htm, accessed Apr. 21, 2005.
20. Schellekens, H. (2004), How similar do "biosimilars" need to be? *Nat. Biotechnol.* 22(11), 1357–1359.

21. Webber, K. (2005), Relevant terminology. A presentation conducted at the Public Workshop on the Development of Follow-On Protein Products, Sept. 14, 2004, available: http://www.fda.gov/cder/meeting/followOn/followOnPresentations.htm, accessed Feb. 2, 2005.
22. Waxman, H., Schumer, C. E., and Clinton, H. R. (2006), Congress of the United States, H.R. 6257, Access to Life-Saving Medicine Act," available: http://www.waxman.house.gov/pdfs/bill_generic_biologics_9.29.06.pdf, accessed Sept. 29, 2006.
23. Mathieu, M., and Evans, A. G. (2005), The FDA's Orphan Drug Development Program, in Ed., *New Drug Development: A Regulatory Overview*, 7th ed., Paraxel Intl., Waltham, MA, pp. 307–317.
24. U.S. Department of Health and Human Services (DHHS) (2003, Sept.), Food and Drug Administration, Center for Drug Evaluation and Research, Guidance for industry: Comparability protocols—Protein drug products and biological products—Chemistry, manufacturing and controls information, DHHS, Washington, DC.
25. Beatrice, M. (2002), Regulatory considerations in the development of protein Pharmaceuticals, in Nail, S., and Akers, M., Eds., *Development and Manufacture of Protein Pharmaceuticals, Pharmaceutical Biotechnology*, Vol. 14, Kluwer Academic/Plenum, New York, pp. 405–457.
26. Letter from R. Levy (Unigene) to FDA Docket No. 2004P-0015 (Apr. 11, 2005).
27. Letter from S. Galson (CDER, FDA) in response to Docket No. 2004P-0015/CP1 (Aug. 12, 2005).
28. Letter from N. Buc to FDA Docket No. 2004P-0115/CP1 (Jan. 9, 2004).
29. *FDA Week*, 11(34), Aug. 26, 2005.

1.3

RADIOPHARMACEUTICAL MANUFACTURING

BRIT S. FARSTAD[1] AND IVÁN PEÑUELAS[2]

[1]*Institute for Energy Technology, Isotope Laboratories, Kjeller, Norway*
[2]*University of Navarra, Pamplona, Spain*

Contents

Pharmaceutical Manufacturing Handbook: Production and Processes, edited by Shayne Cox Gad

1.3.1 INTRODUCTION

1.3.1.1 Radiopharmacy

Radiopharmacy is a patient-oriented science that includes the scientific knowledge and professional judgment required to improve and promote health through assurance of the safe and efficacious use of radiopharmaceuticals. Radiopharmacy encompasses studies related to the pharmaceutical, chemical, physical, biochemical, and biological aspects of radiopharmaceuticals.

Radiopharmacy comprises a rational understanding of the design, preparation, and quality control of radiopharmaceuticals, the relationship between the physicochemical and biological properties of radiopharmaceuticals and their clinical applications, as well as radiopharmaceutical chemistry and issues related to the management, selection, storage, dispensing, and proper use of radiopharmaceuticals.

1.3.1.2 Characteristics of Radiopharmaceuticals

A radiopharmaceutical is any medicinal product which, when ready for use, contains one or more radionuclides (radioactive isotopes) included for a medicinal purpose. This generic definition of radiopharmaceutical thus includes both diagnostic and therapeutic radiopharmaceuticals.

A radiopharmaceutical can be as simple as a radioactive element such as ^{133}Xe, a simple salt such as ^{131}INa, a small labeled molecule such as L-(*S*-[^{11}C]methyl)methionine, or a protein labeled with a radionuclide such as ^{99m}Tc-labeled albumin or ^{90}Y-labeled monoclonal antibodies.

In clinical nuclear medicine, roughly 95% of radiopharmaceuticals are used with diagnostic purposes. Radiopharmaceuticals are administered to the patients only once, or a few times at most, in their lifetime. They contain minute amounts of active ingredients, with a radionuclide somehow linked to or being the active ingredient itself, with the main purpose of obtaining an image or a measure of their biodistribution. Radiopharmaceuticals do not usually show any measurable pharmacodynamic activity, as they are used in tracer quantities. Hence, there is no dose–response relationship in this case and thus differs significantly from conventional drugs.

Radiation is of course an inherent characteristic of all radiopharmaceuticals. Hence, patients always receive an unavoidable radiation dose. In the case of therapeutic radiopharmaceuticals, radiation is what produces the therapeutic effect.

The terms *tracer, radiotracer*, and *radiodiagnostic agent*, although long used as equivalent to radiopharmaceutical, should be avoided. The preferred and correct term is *radiopharmaceutical*, as the other names can be confusing or do not clearly show the nature of these compounds as pharmaceuticals.

The composition of radiopharmaceuticals is not constant as it varies with time as the radionuclide disintegrates. Very often, the half-life of the labeled molecule is so short that it must be readily prepared just before its administration to the patient. This implies in many cases the use of "semimanufactured", such as radionuclide generators, precursors, and cold kits that are also considered a medicinal product according to directive 2001/83/EC.

1.3.1.3 Ideal Characteristics of Radiopharmaceuticals

Radiopharmaceuticals should have several specific characteristics that are a combination of the properties of the radionuclide used as the label and of the final radiopharmaceutical molecule itself. The radiopharmaceutical should ideally be easily produced (both the radionuclide and the unlabeled molecule) and readily available. The half-life of the radionuclide should be adequate to the diagnostic or therapeutic purpose for which it is designed. It has to be considered that radiopharmaceuticals disappear from the organism by a combination of two different processes. The biological half-life (showing the disappearance of a radiopharmaceutical from the body due to biological processes such as metabolization, excretion, etc.) and the physical half-life (due to the radioactive decay of the radionuclide). The combination of both parameters gives the *effective half-life*:

$$T_e = \frac{T_p T_b}{T_p + T_b}$$

where T_e is the effective half-life, T_p the physical half-life, and T_b the biological half-life. Radiopharmaceuticals should have an effective half-life adequate to the use for which they are intended. It should be short (hours) for diagnostic radiopharmaceuticals (not longer than the time necessary to complete the study in question) and longer for therapeutic radiopharmaceuticals (most often days) as the intended effect should have a sufficient duration.

The type of decay of the radiopharmaceutical should also be adequate for its intended use. Diagnostic radiopharmaceuticals should decay by γ emission, electron capture, or positron emission, and never emit α or even β particles. On the contrary, therapeutic radiopharmaceuticals should decay by α or β emission because the intended effect is in fact radiation damage to specific cells.

Regarding the energy emission of diagnostic radiopharmaceuticals, the finally produced γ rays should be powerful enough to be detected from outside of the body of the patient. The ideal energy for nuclear medicine equipment is around 150 keV. γ rays should be monochromatic and photon abundance should be high to decrease the imaging time.

1.3.1.4 Radioactive Decay

Radionuclides are unstable nuclei that are stabilized upon radioactive decay. More than 2000 unstable nuclides have been described so far, most of them radioactive.

The stabilization process can proceed by several different processes, such as spontaneous fission, α-particle emission, β-particle emission, positron emission, γ-ray emission, or electron capture. In all decay processes the mass, energy, and charge of radionuclides must be conserved, and many nuclides can decay by a combination of any of the above-mentioned processes.

Fission is the process in which a nucleus breaks down into two fragments (thus leading to two different new nuclides) with an emission of two or three neutrons and a lot of energy. Spontaneous fission is a rare process that can only occur in heavy nuclei. Fission can also be produced by bombardment of certain nuclides with high-energy particles (such as neutrons) and is in fact the nuclear process used for the production of energy in nuclear energy plants by bombardment of highly enriched uranium with neutrons.

The usual decay process of heavy nuclei is α-particle emission. An α particle is a helium ion containing two protons and two neutrons. Alpha particles are heavy particles that have a very short range in matter due to their mass, and radiopharmaceuticals labeled with α emitters are used only with therapeutic purposes. Their clinical use is very limited, and they are mainly used for research purposes or in early phase clinical studies.

Radioactive nuclides that are neutron rich disintegrate by β decay. A β^- particle is originated by the conversion of a neutron into a proton, along with the emission of an antineutrino to conserve energy in the decay process. Beta-emitting radionuclides are also used in radiopharmaceuticals for therapeutic purposes.

Positron decay occurs in proton-rich nuclei. In this case, the positron (or β^+ particle) is originated by conversion of a proton into a neutron, along with the emission of a neutrino to conserve the energy. Positrons are the antiparticle of electrons. In a very fast process (10^{-12} s), emitted positrons collide with an electron of a nearby atom and both particles disappear in a process called annihilation. The necessary conservation of mass and energy accounts for the transformation of the mass of both particles into energy, which is characteristically emitted in the form of two 511-keV photons almost in opposite directions. Consequently, positron emitters are used to label radiopharmaceuticals produced with diagnostic purposes by imaging.

Proton-rich nuclei can also decay by electron capture. In this process, an electron from the innermost electron shell orbitals is captured into the nucleus and transforms a proton into a neutron (and a neutrino is emitted for conservation of energy). The vacancy created by the lost electron is filled by the transition of an electron from a higher level orbital, and the energy difference between the intervening orbitals is emitted as energy in the form of an X ray.

For any particular nucleus, several different energy states can be defined by quantum mechanics. All the excited energy states above the ground state are referred to as isomeric states and decay to the ground state by the so-called isomeric transition. In β, positron, or electron-capture decay processes, the parent nucleus may reach any of these isomeric states of the daughter nucleus. The energy difference between the nuclear energy states can be emitted as γ rays. A particular situation for isomeric transition is that in which the excited state is long lived and is then called the metastable state.

Radioactive Decay Equations, Magnitudes, and Units Radioactive decay is a random process, being impossible to tell which particular atom from a group of atoms will decay at a specific moment. It is then only possible to talk about the average number of atoms that disintegrate during a certain period of time, giving the disintegration rate ($-dN/dt$) of a particular radionuclide that is proportional to the total number of radioactive atoms present at that time. This magnitude is usually called the radioactivity (or mainly simply the activity) of a radionuclide and denoted by

$$A = -\frac{dN}{dt} = \lambda N$$

where λ is the decay constant and N the number of radioactive atoms. The previous differential equation is mathematically solved leading to the exponential equation

$$N_t = N_0 e^{-\lambda t}$$

where N_t and N are the number of radioactive atoms present at time $t = 0$ and $t = t$, respectively.

Radioactivity is expressed in becquerels (Bq), the Internationale System (SI) unit for the magnitude A. One Becquerel is defined as one disintegration per second (dps). Usual activities used in radiopharmacy are in the range of megabecquerels or gigabequerels. There is (as usual) a non-SI unit called the curie (Ci). It was initially defined in a trivial way as the disintegration rate of one gram of radium, which was considered to be 3.7×10^{10} dps. Thus the equivalence between the becquerel and the curie is as follows:

$$1\,\text{Bq} = 2.7 \times 10^{-11}\,\text{Ci} \qquad 1\,\text{Ci} = 37\,\text{GBq}$$

The decay constant λ is a specific characteristic of any single radionuclide, but being related to probability, it is difficult to understand its meaning. Thus, a new magnitude is defined: the half-life ($t_{1/2}$), which is the time required to reduce the initial activity of a radionuclide to one-half. In consequence, after one half-life the activity of a radionuclide would be $A/2$, after two half-lives $A/4$, after three half lives $A/8$, and so on.

The relationship between the decay constant λ and the half-life $t_{1/2}$ can be derived from the general radioactive decay equation

$$t_{1/2} = \frac{\ln 2}{\lambda}$$

An additional (and commonly misunderstood concept) is the mean life τ,which is the average life of a certain group of radioactive atoms that is mathematically also derived from the decay constant λ as

$$\tau = \frac{1}{\lambda} = \frac{t_{1/2}}{\ln 2} = 1.44 t_{1/2}$$

1.3.1.5 Principles of Radiation Protection

Production, transportation, and use of radiopharmaceuticals, as radioactive products, is governed by regulatory agencies dealing with radiation protection and nuclear safety.

In any case, and albeit the different regulation in different countries, as a general principle only licensed personnel working in an authorized facility are authorized to handle and use radiopharmaceuticals. Facilities and procedures are subject to periodic inspection by official radiation safety officers that control production and handling of radioactive material, its transportation, proper use, as well as personnel dosimetry and radioactive waste disposal.

The general principles of radiation protection are very simple:

Justification. All procedures involving radioactive material must be justified.

Optimization. The radiation exposure to any individual should be as low as reasonable achievable. This principle is the widely known ALARA concept, an acronym derived from as low as reasonable achievable.

Limitation. The radiation dose received by the personnel handling radioactive material will never exceed the legally established dose limits. It has to be taken into account that such limitations do not apply to patients receiving radiopharmaceuticals as either diagnostic or therapeutic agents. But nuclear medicine physicians, nuclear physicists, and radiopharmacists must ensure that the amount of radiopharmaceutical administered to a patient is adapted to his or her disease and optimized to obtained the intended result.

Operational Radiation Protection The fundamentals of operational radiation protection (i.e., how to proceed when working with radioactive products) are based on three factors: distance, time, and shielding. In any case, it is obvious that the radiation hazard is increased with the activity of the radiation source, as can be derived from the mathematical equation to calculate the exposure rate X given by

$$X = \frac{A\Gamma}{d^2}$$

where A is the activity of the radiation source, Γ a constant that is characteristic of every radionuclide, and d the distance to the source.

Distance should be increased as much as possible to decrease exposition and exposure time should be reduced to a minimum. Adequate shielding (depending on the radionuclide and its emission characteristics) should be used whenever possible and handling of high activities should only be carried out by either automated systems or proper manipulators.

1.3.1.6 Detection Devices for Clinical Nuclear Imaging

Diagnostic radiopharmaceuticals are mostly used for in vivo imaging of the biodistribution of the radiopharmaceutical. Depending on whether γ or positron emitters are used, different devices are employed for clinical imaging. In any case, imaging devices are based on detection of the high-energy photons coming from the body of the patient upon administration and specific uptake of a radiopharmaceutical. Advances in nuclear medicine imaging devices now permit in vivo noninvasive imaging of such biodistribution and to obtain tomographic (i.e., three-dimensional) images that can also give quantitative or semiquantitative information about the amount of radiopharmaceutical and even its kinetics.

1.3.2 PRODUCT DEVELOPMENT

1.3.2.1 Radionuclides

When designing a radiopharmaceutical one should have in mind the potential hazard the product may have to the patient. The goal must be to have maximum amounts of photons with a minimum radiation exposure of the patient. For use in therapy, β emitters and α emitters are particularly useful. For diagnostic purposes, γ emitters are most widely used. In general, those γ emitters with a short physical half-life and with a γ energy between 100 and 300 keV are most widely used in medical application, since these can easily be detected by standard γ cameras.

However, positron emission tomography (PET) radiopharmaceuticals involve short-lived radionuclides (positron emitters) giving a double set of photons at 511 keV each.

1.3.2.2 Carrier Molecules/Active Ingredients

The function of the carrier molecule is to carry the radioactivity to the target organ and to make sure the radioactivity stays there. The uptake of radioactivity should be as specific as possible in order to minimize irradiation of other organs and parts of the body. This is particularly important when using radiopharmaceuticals for therapy. But, also, for use in diagnostics, it is desirable that the radiopharmaceutical is localized preferentially in the organ under study since the activity from nontarget areas can obscure the structural details of the pictures of the target organ. It is therefore important to know the specific uptake in an organ for a potential chemical carrier and also the rate of leaking out of the organ/organ system. Thus, the target-to-background activity ratio should be large. There are several approaches to develop targeting radiopharmaceuticals. Radioimmunotargeting is one approach frequently used for radiopharmaceuticals, where monoclonal antibodies (MAbs) or fractions of MAbs are the carrier molecules for the radioactivity. These are binding specifically to receptors on cell surfaces in the target organs.

The target-binding surface of the cell has been well explored with a range of tumor-associated and other antigens, identified, and used for pathological tissue characterizations.

The active analog approach in general, whereby a set of compounds is synthesized so as to mimic features of a chosen natural compound, has been successful [1]. The active analog approach includes the pharmacophore. The concept of the pharmacophore is to look at features common to a set of drugs or compounds binding to and acting on the same receptors.

1.3.2.3 Radiolabeling Techniques

When a labeled compound is to be prepared, the first criterion to consider is whether the label can be incorporated into the molecule to be labeled [2]. This may be assessed from knowledge of the chemical properties of the two partners. Furthermore, one needs to know the amount of each component to be added. This is particularly important in tracer level chemistry and in ^{99m}Tc chemistry.

In a radiolabeled compound, atoms, or groups of atoms of a molecule, are substituted by similar or different radioactive atoms or groups of atoms. Saha [2] lists six major methods employed in the preparation of labeled compounds for clinical use: isotope exchange reactions, introduction of a foreign label, labeling with bifunctional chelating agents, biosynthesis, recoil labeling, and excitation labeling. Among these, three frequently used methods in radiopharmaceutical synthesis are briefly described below.

Isotope Exchange Reactions In isotope exchange reactions, isotopes of the same elements having different mass numbers replace one or more atoms in a molecule. Examples are labelling of iodide-containing material with iodine radioisotopes. Since the radiolabeled and parent molecules are identical except for the isotope effect, they are expected to have the same biological and chemical properties.

Introduction of a Foreign Label In this type of labelling, a radionuclide is incorporated into a molecule primary by the formation of covalent or coordinated covalent bonds. The tagging radionuclide is foreign to the molecule and does not label it by exchange of one of its isotopes. Examples are ^{99m}Tc–DTPA (Diethylenetriaminepentacetic acid), ^{51}Cr-labeled red blood cells, and many iodinated proteins and enzymes. In many compounds of this category, the chemical bond is formed by chelation. In chelation, one atom donates a pair of electrons to the foreign acceptor atom, which is usually a transition metal. Most of the ^{99m}Tc-labeled compounds used in nuclear medicine are formed by chelation.

Labeling by Bifunctional Chelating Agents In this approach, a bifunctional chelating agent is conjugated to a macromolecule (e.g., protein) on one side and to a metal ion by chelation on the other side. Examples of bifunctional chelating agents are DTPA, metallothionein, diamide dimercaptide (N_2S_2), and dithiosemicarbazone [2].

There are two methods: the preformed radiometal–chelate method and the indirect chelator—antibody method. Various antibodies are labelled by the latter, where the bifunctional chelating agent is initially conjugated to a macromolecule, which is then allowed to react with a metal ion, to form a metal–chelate–macromolecule complex. Due to the presence of the chelating agent, the biological properties of the labeled protein may be altered and must be assessed before clinical use.

1.3.2.4 Manufacturing Scale-Up

As the radiolabeled substances emerge from the laboratory to the clinics, there will be a need for scaling up the batch size of the product. This can be done by increasing either the total volume of the produced batches or the specific activity of the product or both. When doing this, the following aspects should be considered:

- The influence on the stability of the product itself due to possible radiolysis
- The need for additional operator protection due to handling of increased amounts of radioactivity

Product Stability The stability of a labeled compound is one of the major problems in labeling chemistry. It must be stable both in vitro and in vivo. Many labeled compounds are decomposed by radiation emitted by the radionuclides in them. This kind of decomposition is called radiolysis. Radiation may also decompose the solvent, producing free radicals that can break down the chemical bonds of the labeled compounds (indirect radiolysis). In general, the risk of radiolysis increases with higher specific activity of the product. In addition, the more energetic the radiation, the greater is the radiolysis. Alpha emitters, leaving most of its energy close by the molecules, and thus a high potential risk of radiolysis, give rise to major challenges when scaling up is necessary.

Operator Radiation Protection Even for the largest commercial manufacturer of radiopharmaceuticals, the batch volumes are small compared to nonradioactive pharmaceuticals. So even a scaled-up production batch can be contained within a limited space. When scaling up a radiopharmaceutical production, one always has to assure that the radiation outside the contained work unit is acceptable for the operator.

The production of a radiopharmaceutical will normally take place within a contained box unit. Depending on the kind of radionuclides used and the amount of radioactivity handled in the production process, the box units are shielded by lead walls, typically 5–15 cm in thickness. When the box is used for production of radiopharmaceuticals incorporating α- or β-emitting radionuclides, closed box units without any lead coating are sufficient. Working with these types of radionuclides or with smaller amounts of γ-emitting radioactivity, as in research scale, suitable glove boxes can be used. When working with larger quantities of γ-emitting radionuclides, the material must be handled by either remote control equipment or manipulator tongs incorporated in the wall.

1.3.2.5 Automation

Because of the unique operational and safety requirements of radiopharmaceutical synthesis, the motivation for the development of automated systems is clear. These unique constraints include short synthesis times and control from behind bulky shielding structures that make both access to and visibility of radiochemical processes and equipment difficult. The need for automated systems is particularly expressed for PET radiopharmaceutical synthesis, with the short-lived radionuclides emitting high-energy γ photons at 511 keV. Automated synthesis systems require no direct human participation. The short half-lives of the PET radionuclides may require repeated synthesis during the day, thus being a potential radiation burden for the operator when not using automated systems.

Furthermore, radiopharmaceutical synthesis must be reliable and efficient and result in pharmaceutical-quality products. In addition, the processes must be well documented and controlled. Automated systems may support all these challenges and requirements.

One must keep in mind, though, that success in synthesis automation requires first and foremost innovative chemistry. PET radiosynthesis draws from a broad chemistry knowledgebase rooted in synthetic organic chemistry [3].

1.3.3 MANUFACTURING ASPECTS

1.3.3.1 Design of Manufacturing Sites

The manufacturing of radiopharmaceuticals is potentially hazardous. Both small- and large-scale production must take place on premises designed, constructed, and maintained to suit the operations to be carried out. Radiation protection regulations stipulate that radionuclides must only be used in specially designed and approved "radioisotope laboratories." National regulations with regard to the design and classification of radioisotope laboratories must be fulfilled. Such laboratories are normally classified according to the amount of the various radionuclides to be handled at any time and the radiotoxicity grading given to each radionuclide. When planning the layout of the laboratory, it is recommended to allocate separate working areas or contained units for the various procedures to avoid possible cross-contamination of radionuclides [4, 5]. Premises must be designed with two important aspects in mind:

The product should not be contaminated by the operator.

The operator and the environment should be protected from contamination by the radioactive product.

This is the basic principle of good radiopharmaceutical practice (GRPP).

One of the most important factors in planning a radioisotope laboratory is the design of the ventilation system. Laboratories with medium and high grading must be designed with the purpose of protecting the personnel from inhaling radioactive gases or particles. The system should be designed to provide lower pressure at the actual working area compared to the surrounding environment. Furthermore, the system should have an appropriate number of air changes per hour and the replacement air should be filtered. Air extracted from the area where radioactive products are handled, though, should not be recirculated. Exhaust air to the environment should be monitored for radioactivity, and it may be necessary to install active charcoal filters to absorb radioactive gases and small particles [4].

Aseptic production of radiopharmaceuticals, that is, when the products cannot be terminally sterilized, will increase the requirements for the design and construction of the premises. Contained workstations and clean-room technology will be applied to a much higher degree. The general requirements for the design of such premises are the same as for nonradioactive pharmaceuticals, including entry of staff and the introduction of materials through air locks. The main difference is found in the planning and design of the ventilation system. Laboratories for aseptic work normally have a positive pressure relative to the surrounding areas. On the other hand, in laboratories for work with radioactivity, it is good practice to have a negative pressure to avoid the spread of radioactive material. In order to meet both pharmaceutical and radiation protection requirements, it is necessary to balance carefully the air pressures in the clean rooms, the air locks, and the surrounding areas. From a pharmaceutical point of view a negative pressure in the area designated for aseptic work can only be accepted in special cases. There are various ways to meet the required balance between these apparently contradictory principles. A frequently chosen solution is to use sealed production units or contained workstations supplied with unidirectional airflow (UDAF) and with a lower pressure compared to the aseptic laboratory. The laboratory itself may then have positive pressure in relation to the surrounding premises.

Waste management is an important aspect when planning a radiopharmaceutical manufacturing site. The key factor is to reduce the amount of radioactive waste to a minimum. There should be a system for dividing the waste according to physical half-life and radiotoxicity, both for solid and liquid waste. As an example, waste containing α emitters is normally kept separately, when possible. National legislation will vary considerably and influence the requirement that must be set for handling of radioactive waste material.

1.3.3.2 Design of Production Processes

The design of a radiopharmaceutical production process depends very much upon the kind of radiopharmaceutical to be made. Although most radiopharmaceuticals are intended for parenteral use, also oral radiopharmaceuticals in different forms are widely used. One must emphasize different factors when planning for production of parenteral radiopharmaceuticals compared to oral radiopharmaceuticals. Still, a common factor is the involvement of radioactive materials, and the radiation protection of the personnel must always be an integral part of the design.

The production of a radiopharmaceutical will normally take place within a contained box unit, consisting of either plastic walls or a combination of plastic and stainless steel. The latter is more optimal for clean-room work. The box units may be shielded by lead, either as large lead panels or as lead brick walls (see Figure 1). Depending on the kind of radionuclides used and the amount of radioactivity handled in the box, the walls are typically 5–15 cm in thickness. Shielded production units like these are often called "hotcells."

When the box is used for productions of radiopharmaceuticals incorporating α- or β-emitting radionuclides, closed box units without any lead coating may be sufficient. When handling radionuclides with mixed emitting properties, a possibility is to concentrate the shielding to critical parts of the process. This can be done by use of local shielding inside the production unit. However, for aseptic production, one must keep in mind a potential disturbance of the airflow inside the box.

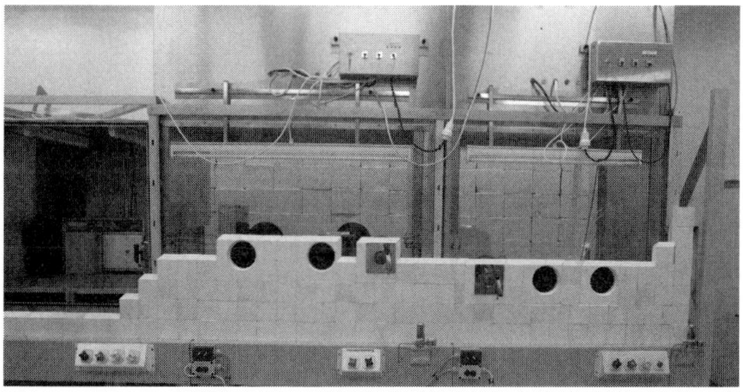

FIGURE 1 Shielding of box units (hot cells) with lead bricks. (Photo courtesy of Institute for Energy Technology.)

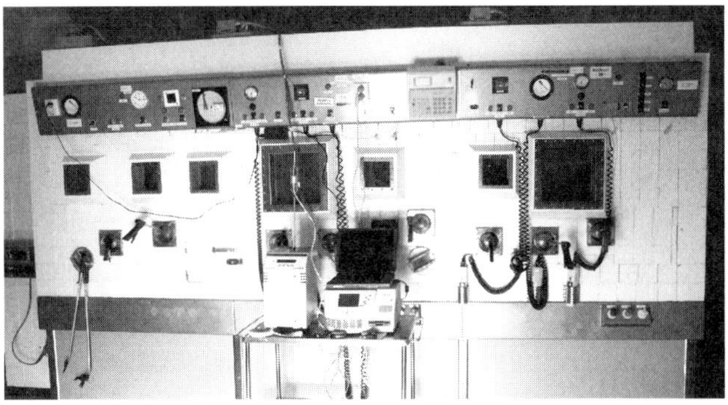

FIGURE 2 Hot cells for manufacturing of larger quantities of γ-emitting radionuclides. (Photo courtesy of Institute for Energy Technology.)

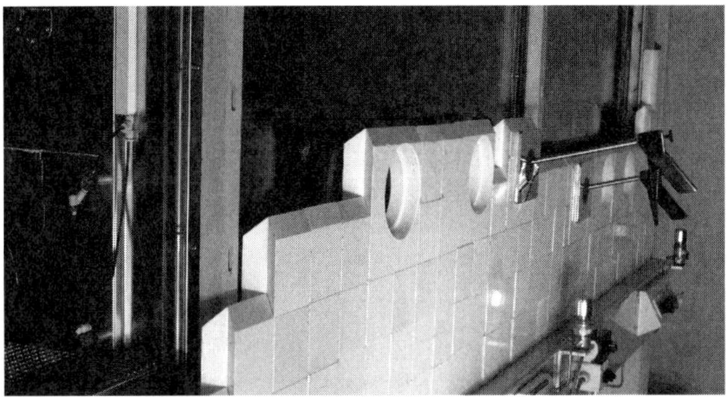

FIGURE 3 Lead bricks are interlocked when they are stacked on top of each other. (Photo courtesy of Institute for Energy Technology.)

Working with α and β radionuclides and also limited amounts of γ-emitting radioactivity, the boxes may be mounted with special protection gloves. When working with larger quantities of γ-emitting radionuclides, the material must be handled by either remote control equipment or manipulator tongs incorporated in the wall (see Figure 2).

The design of the elements and their assembly on the production unit should be such that there are no radiation leaks at the interface.

When using lead bricks to construct the wall of the production unit, they should have a special design. When they are stacked on top of each other, they should interlock (see Figure 3). This is important to avoid cracks in the wall through which radiation can escape.

Manipulator tongs are fitted into the wall as part of a large tungsten sphere which acts as a ball bearing and thereby allows more flexibility for the movement of the tong inside the box (see Figure 2). Lead glass windows, with good shielding properties, are fitted in the lead bricks to allow the operator to overlook the process.

When large lead panels are used, they should be reinforced with suitable steel structures [International Organization for Standardization (ISO) 10648-1: 1997].

The surfaces of the lead shielding must be smooth and easy to clean. This can be achieved by painting the surface of the wall.

In general, the manufacturing of most radiopharmaceuticals consists of the following:

Nuclear synthesis, synthesis of the radionuclide
Synthesis of the radiolabeled compound
Pharmaceutical formulation of the radiopharmaceutical

Nuclear Synthesis Except for radionuclides with ultrashort half-lives, like most PET radionuclides, the production of these is normally performed well in advance (see Section 1.3.4.1). Thus, the radionuclide is considered as a starting material and must undergo controls as a starting material.

Synthesis of Radiolabeled Compound The complexity of a radiopharmaceutical may differ greatly, with the radioactive element itself or simple salts as the less complex. Very often, though, the radiolabel is part of a larger molecule, and thus a radiolabeling procedure is required. This is part of the synthesis of the radiopharmaceuticals, which also may involve chemical alteration of a precursor of the active ingredient. Both labeling methods and synthesis may involve steps at elevated temperatures or even cooling steps. Thus, equipment for heating or cooling must be part of the production line. Furthermore, an important part of a synthesis is often the purification step, and equipment for this must be available. Typically, this is simple chromatographic or ion exchange columns.

Planning of the process very much depends on the complexity of the process. In general, keeping in mind the limited possibilities of direct handling of the materials, it is important to keep the processes as simple as possible. For more complex processes, automation may be the best solution, if available.

Pharmaceutical Formulation Even when the radiochemical part of a product is simple, the radiopharmaceutical may be a complex solution. A pharmaceutical formulation often contains additives in the form of buffers and preservatives: buffers to keep the solution at a pH suitable for injection and preservatives to preserve the integrity and efficiency of the radiopharmaceutical.

Ideally, a solution for injection should be an isotonic solution with a neutral (physiological) pH. However, the pH of a radiopharmaceutical is very important for its stability, and for labelled compounds, the pH for optimal stability is not always equivalent to physiological pH. For iodide solutions, the pH should be alkaline to prevent loss of radioiodine. Reducing agents, such as thiosulfate, are often added to radioiodide solutions to help this situation. A preservative can act as a stabilizer, an antioxidant, or a bactericidal agent.

Some additives, like benzyl alcohol, are added for a double action. Benzyl alcohol 0.9% is widely used as a bactericide. In addition, benzyl alcohol reduces radiolysis in radiopharmaceuticals and thus acts as a stabilizer.

1.3.3.3 Design of Production Equipment

The equipment used for manufacturing operation should be reserved exclusively for radiopharmaceuticals [6]. Furthermore, two principles are of utmost importance in the design of production equipment [4]:

The equipment must be easy to repair after it has been installed in the production unit.

The equipment must have a simple construction and be easy to assemble, so a substitution can be done quickly when total renovation of the equipment is necessary.

Glass is an important material in the construction of production equipment for radiopharmaceuticals. This material will become discolored and brittle when affected by radiation, and thus repair and/or change of parts of the equipment may be necessary. Due to radioactive contamination of the equipment, repair and maintenance can often be complicated, and time for decay must be included in the maintenance period. To secure the continuous supply of products, it may be necessary to construct two production lines in separate production units, where one is kept as a backup facility.

Sometimes it will be necessary to substitute not only parts of a production line but also the assembly of equipment as a whole. To facilitate this operation and thereby reduce time and radiation exposure, it can be advantageous to build the whole production line on a stainless steel support frame fitted with simple connections to electricity, water, and air supplies [4]. The complete withdrawal of a production line from a box and the introduction of a new one can then be performed in a very short time.

It is also important to keep in mind, when designing production equipment, that all sense of touch is lost when fingers are replaced by remote handling tongs.

The design of the equipment must therefore be as simple as possible. On the other hand, when using hot-cell units mounted with handling tongs, it may be favorable to use more automated systems in the production line. Systems like these can be run and controlled from steering panels outside the box unit.

Finally, equipment should be constructed so that surfaces that come in contact with the product are not reactive, additive, or absorptive so as to alter the quality of the radiopharmaceutical.

1.3.3.4 Cleaning and Sanitation of Production Equipment

Preparation equipment should be designed so it can be easily and thoroughly cleaned. Procedures for cleaning, sanitation, and storage of production equipment used in radiopharmaceutical production must be established. Special training is necessary for personnel involved in this kind of work with regard to both clean-room aspects and radiation protection aspects.

Before any equipment or materials used during production are removed from the production unit, a check for radioactive contamination must be performed. After removal, the equipment should be allowed to decay further in a special storage area before it is cleaned and made ready for assembly again.

Glass equipment will normally be sterilized by dry-heat sterilization. Smaller equipment, like plastic tubes and rubber stoppers, can be sterilized by autoclaving.

If available locally, also γ irradiation may be a suitable method for sterilization of equipment. One must keep in mind, though, that sterilization by irradiation may change the composition of plastic and rubber materials. In addition, glass materials may be discolored by γ irradiation.

Production equipment that cannot be sterilized must be sanitized and disinfected by an appropriate method. This can be done by use of biocides like alcohols (70%), hydrogen peroxide, or formaldehyde-based chemicals or a combination of these. These can either be used for surface disinfections by wiping or spraying or even better by use of gas or dry fog systems for application of the disinfectants. The effect of cleaning and sanitation should be monitored. Microbiological media contact plates can be used to test critical surfaces, as inside the hot cells or glove boxes. The test samples must then be handled and monitored as radioactive contaminated units.

A system must be established for sanitation of all equipment before these are transferred into clean areas.

1.3.3.5 Environmental Control

Workstations and their environment should be monitored with respect to radioactivity, particulate, and microbiological quality. Active air sampling from production units for radioactive products (hot cells or glove boxes) is subject to a safety consideration. There is always a risk of bringing radioactive contaminated air outside the workstation. To avoid the spread of radioactivity during the test, all possible exhaust from the test equipment must be sampled and/or controlled.

A possible approach for testing of particulate and microbiological quality of air inside the hot cells or glove boxes is to gain information about airborne particles during simulated operations (without radioactivity).

The use of settle plates is common practice for monitoring of the microbiological quality of air inside production units. These must then be placed as close a possible to critical parts of the production process in order to show the real microbiological burden to the product.

Warning systems must be installed to indicate failure in the filtered air supply to the laboratory. Recording instruments should monitor the pressure difference between areas where this difference is of importance.

1.3.3.6 Sterilization of Radiopharmaceuticals

Sterile radiopharmaceuticals may be divided into those which are manufactured aseptically and those which are terminally sterilized. In general, it is advisable to use a terminal sterilization whenever this is possible. Terminal sterilization is defined as a process that subjects the combined product/container/closure system to a sterilization process that results in a specified assurance of sterility [7]. Since sterilization of solutions normally means autoclaving (steam sterilization), one must assure that the radiopharmaceutical product does not decompose when it is heated to temperatures above 120°C. Many radiolabeled compounds are susceptible to decomposition at higher temperatures. Proteins, such as albumin, are good examples of this. Others, such as ^{18}F-fluodeoxyglucose (FDG), can be autoclaved in some formulation but not in others.

Furthermore, these processes take time, typically 20–30 min in total when heating up to 121°C. For very short-lived product with a half-life of only a few minutes, this is not an adequate method. On the other hand, these short-lived products are not subject to any storage, and thus the risk of microbiological growth is more limited.

Alternatively, a shorter cycle at a higher temperature might be used, assuming that the temperature does not decompose the radiopharmaceutical.

If terminal sterilization is not possible, aseptic processing must be performed. Aseptic processing is a process that combines presterilized materials and presterilized equipments in a clean area.

Heating of radioactive solutions, particularly under elevated pressure (e.g., steam sterilization), is also a matter of safety. In order to avoid any contaminated air to escape if a container or a seal is broken, autoclaves used for radioactive solutions should be placed inside negative-pressure sealed units. Autoclaves used for sterilizing high-energy γ-emitting radiopharmaceuticals should in addition be supplied with proper lead shielding.

1.3.3.7 Starting Materials

As for manufacturing of other pharmaceuticals, a system should be established to verify the quality of the starting materials used in manufacturing radiopharmaceuticals. This system must assure that no material is used for production until it has been released by a competent person [qualified person (QP) or others given this responsibility].

The starting materials as well as the packaging materials should be purchased from qualified vendors. It is recommended to use materials described in a pharmacopoeia, whenever this is available. Supplier approval should include an evaluation that provides adequate assurance that the material consistently meets specifications.

Radionuclides involved in manufacturing radiopharmaceuticals must be considered as starting materials. For very short-lived radionuclides, where batch analysis is not possible, the validation of the production process of the radionuclide is of utmost importance.

1.3.3.8 Labeling and Packaging

Packaging material should be purchased from qualified vendors. Primary containers and closures must be tested to verify that there are no interactions between the radiopharmaceutical and packaging material during storage of the product.

Due to the risk of radiation exposure, it is accepted that most of the labeling of the primary (direct) container is done prior to manufacturing. The empty vial can be prelabeled with partial information prior to filtration and filling [6]. This procedure should be designed so as to not compromise sterility or prevent visual inspection of the filled vial. After filling of radioactive products, the primary containers (vials) must be placed within a shielded container. These containers, which can be made of lead or tungsten, vary in size and thickness depending on the amount of radioactivity in the vial as well as the radiation properties of the radionuclide. Radiopharmaceuticals containing α or β emitters may be placed in thin lead pots, typically 2–4 mm in wall thickness. On the other hand, for vials containing regular doses of high-energy γ emitters, such as PET radionuclides, shielding with 3–5 cm lead/tungsten may be needed.

Necessary information about the product must be given on the label of the lead or tungsten container. Hence, there is no need to study the label on the direct container. The name of the radiopharmaceutical, including the radionuclide, together with the amount of radioactivity in the vial at a stated calibration time is part of the necessary information. So is the expiry date of the product. Furthermore, the symbol for radioactivity, designed as a black propeller, is obligatory on labels for radioactive solutions.

When the products are intended for distribution and transport, the packaging and labeling of the outer packages must be done according to the national regulation of the country from which the shipments will depart, transfer, and arrive. The outer packaging material must be properly tested in accordance with the type of shipment, most frequently type A packages for radiopharmaceuticals. Furthermore, the packages must be labeled with radionuclide data, such as type and amount of radioactivity, along with the transport index (TI), which indicates the radiation from the package at 1 m distance. While the information on the product itself (outside the lead pot) is intended for the physicians, the information outside the package is intended for the transport personnel.

1.3.4 PRODUCT MANUFACTURING

1.3.4.1 Production of Radionuclides

Radiopharmaceuticals are labeled with artificial radionuclides that are obtained by bombardment of stable nuclei with subatomic particles or photons. Nuclear reactions produced in such a way convert stable in unstable (radioactive nuclei). Several kind of devices are used for such purposes, including nuclear reactors, particle accelerators, and generators.

Various types of targets have been designed and used for both reactor and cyclotron irradiation. In the design of targets, primary consideration is given to heat deposition in the target by irradiation with neutrons in the reactors or charged particles in the cyclotrons [2]. As the temperature can rise to 1000°C during irradiation in both reactors and cyclotrons, the target needs proper cooling to avoid burning. Most often, the targets are designed in the form of a foil to maximize the heat dissipation. The target element should ideally be monoisotopic or an enriched isotope to avoid extraneous nuclear reactions.

Nuclear Reactors Nuclear reactors are highly complex systems in which two kinds of nuclear reactions are useful for the production of clinically useful radionuclides: Neutrons produced by the fission of heavy nuclides (such as ^{235}U or ^{239}Pu) are used in a neutron capture (n, γ) reaction to produce an isotope of the same element that is bombarded by the neutrons. Such reactions can be produced almost in all elements with different probability. Examples of useful nuclear reactions are ^{130}Te(n, γ)^{131}Te (which produces ^{131}I after emission of β particles with a half-life of 25 min), ^{50}Cr(n, γ)^{51}Cr, ^{58}Fe(n, γ)^{59}Fe, and ^{98}Mo(n, γ) ^{99}Mo. The second possibility for the use of nuclear reactors is to use fission reactions (n,f) in which a heavy nuclide is broken down into two fragments. Many clinically relevant radionuclides can be produced from thermal fission of ^{235}U, such as ^{131}I, ^{117}Pd, ^{133}Xe, and ^{137}Cs. The isotopes produced by this kind of fission reaction must be separated and purified by appropriate chemical procedures, but since the chemical behavior of many different heavy elements is similar, contamination can often become a problem in the isolation of the radionuclide of interest.

As an example, and due to the particular interest of ^{99}Mo in radiopharmacy (as it is the parent nuclide of ^{99m}Tc in the ^{99}Mo–^{99}mTc generator), the complex process used to produce and purify ^{99}Mo is described below.

Molybdenum-99 is produced by fission of ^{236}U as follows:

$$^{235}\mathrm{U} + {}^{1}\mathrm{n} \rightarrow {}^{236}\mathrm{U} \rightarrow {}^{99}\mathrm{Mo} + {}^{135}\mathrm{Sn} + 2{}^{1}\mathrm{n}$$

After irradiation of the uranium target, it is dissolved in nitric acid and the final solution adsorbed on an alumina column that is washed with nitric acid to remove uranium (and other fission products). Molybdenum is finally eluted with ammonium hydroxide and further purified by absorption on an anion exchange column from which ammonium molibdate is eluted with dilute hydrochloric acid after washing the resin with concentrated HCl. The ^{99}Mo is obtained in no-carrier-added conditions, and the most common contaminants can be ^{131}I and ^{103}Ru.

Particle Accelerators: Cyclotrons Both linear and circular particle accelerators (cyclotrons) can be used, but the latter have many advantages and are mainly used for the production of clinically relevant radionuclides.

A cyclotron is basically a cylinder-shaped high-vacuum chamber in which by means of a magnetic field and a radio-frequency system used to generate an alternating electric field, elemental particles can be accelerated to very high energies and used as projectiles. The bombardment of stable elements loaded in a properly designed target (either solid or filled with a liquid or a gas) induces different types of nuclear reactions that finally lead to the production of radioactive elements.

Most cyclotrons accelerate negative particles (such as ^{2}H, ^{1}H, or even heavier particles such as helium cations) that are stripped off the electrons in the stripping foils that are used also to focus the beam on the target. As the energy of the incident particle is increased, a much greater variety of nuclides can be produced.

When the nuclides produced have atomic numbers different from those of the target elements, such preparations have no stable isotope of the intended element and can be considered to be produced in no-carrier-added conditions.

The target material should ideally be monoisotopic to avoid the production of extraneous radionuclides. However, in many cases this is not possible and only isotopically enriched targets can be used, thus leading to the production of different radionuclides. In this case appropriate methods must be used to separate the different elements produced in the target.

An interesting concept that must always be taken into account in cyclotron-produced radionuclides is the saturation activity characteristic of each target and each nuclear reaction. The saturation activity is the activity of the radionuclide in which the secular equilibrium is obtained between the activity produced in the target and the disintegration of the radioisotope. The activity produced at a target can be calculated by the equation

$$A = A_S(1 - e^{-(\ln 2)t/T})(\mu\mathrm{A})$$

where A is the activity obtained for a radionuclide with a half-life of T after irradiation of the target during a time of t at a current of μA microamperes. From the practical point of view, almost 97% of the saturation activity value is reached after irradiation of the target for five half-lives of the radionuclide. Longer irradiation times do not produce significant increases in the activity obtained. Methods to obtain several cyclotron-produced radionuclides are described below.

Iodine-123 can be produced either directly or indirectly in a cyclotron. Direct reactions usually lead to ^{123}I contaminated with other iodine radioisotopes, such as ^{124}I or ^{125}I, due to side nuclear reactions. Using nuclear reactions such as $^{123}Te(p,n)^{123}I$, $^{122}Te(d,n)^{123}I$, or $^{124}(p,2n)^{123}I$ produces ^{123}I that is obtained after dissolving the target in hydrochloric acid by distillation into dilute NaOH.

In the indirect methods the radionuclide produced after bombardment of the target is not ^{123}I, but a radionuclide that decays to ^{123}I with a short half-life. The most widely used nuclear reactions produce ^{133}Xe (which decays to ^{123}I with a half-life of 2.1 h) by bombardment with high-energy ^{3}He or ^{4}He particles or ^{123}Cs (which decays to ^{123}Xe with a half-life of 5.9 min, and then ^{123}Xe decays to ^{123}I) after irradiation of ^{124}Xe with high-energy protons. Complex processing and purification processes must be used to obtain ^{123}I in any of these cases, and adequate design and composition of the target are critical to facilitate the process.

Thallium-201 is obtained using an indirect reaction such as $^{203}Tl(p,3n)^{201}Pb$ in which ^{201}Pb decays to ^{201}Tl with a half-life of 9.4 h. Thallium-201 can in this way be obtained pure and free from other contaminants after several purification steps and letting the target product decay for 35 h.

Indium-111 is produced by a direct nuclear reaction by irradiation of an ^{111}Cd target with 15-MeV protons. After irradiation the target is dissolved in HCl and purified in an anion exchange column.

Positron emission tomography has become a widely used diagnostic technique in nuclear medicine. Ultrashort half-live radionuclides are used in these cases, and such radionuclides are mostly obtained in small cyclotrons with high yields and short irradiation times. The overall process will be described further in this chapter when PET radiopharmaceuticals are described.

Generators A generator is constructed on the principle of the decay–growth relationship between a parent radionuclide with longer half-life that produces by disintegration a daughter radionuclide with shorter half-life. The parent and the daughter radionuclide must have sufficiently different chemical properties in order to be separated. The daughter radionuclide is then used either directly or to label different molecules to produce radiopharmaceutical molecules.

A typical radionuclide generator consists of a column filled with adsorbent material in which the parent radionuclide is fixed. The daughter radionuclide is eluted from the column once it has grown as a result of the decay of the parent radionuclide. The elution process consists of passing through the column a solvent that specifically dissolves the daughter radionuclide leaving the parent radionuclide adsorbed to the column matrix.

The main advantage of the generators is that they can serve as top-of-the-bench sources of short-lived radionuclides in places located far from the site of a cyclotron or nuclear reactor facilities.

A generator should ideally be simple to build, the parent radionuclide should have a relatively long half-life, and the daughter radionuclide should be obtained by a simple elution process with high yield and chemical and radiochemical purity. The generator must be properly shielded to allow its transport and manipulation.

Several different generators are used in radiopharmaceutical procedures, but the $^{99}Mo/^{99m}Tc$ is with great difference the most important generator of all of them and will be described in detail later on in this chapter.

1.3.4.2 Production of Radiopharmaceuticals

More than 90% of the radiopharmaceuticals used in nuclear medicine are for diagnostic use. PET radiopharmaceuticals, with their ultrashort half-lives, have become a significant part of this group of products. Hence PET investigation has been the fastest growing imaging modality worldwide the last few years [8].

Also for conventional radiopharmaceuticals used in diagnostic, it is favorable to use products with short half-lives. Radionuclide generator systems are widely used for supply of short-lived radionuclides/radiopharmaceuticals. Several generator systems are available and routinely in use within nuclear medicine. Some of these are listed in Table 1.

Because of the short half-life, the coupling of the radionuclide to the carrier molecule must be done immediately before the administration. Hence, there is a need to have a constant supply of carrier molecules that can be labeled efficiently on site. For this purpose, several preparation kits have been developed.

Ready-for-use diagnostic radiopharmaceuticals which are intended for transport over some distance typically include radionuclides with half-lives from 13 h and up.

Among these, products involving the radionuclide ^{131}I are used for both diagnostic and therapeutic indications. This is based upon the mixed emitting properties of the radionuclide, giving both β and γ emission. The availability, price, and half-life (8 days) of this radionuclide, together with the physical properties, have probably made it the most commonly used radionuclide in radiotherapy. Although ^{131}I also is frequently used for diagnostic purposes, the radiation characteristics of this radionuclide are not really ideal for use in conventional scintigraphy (SPECT) due to the high γ energies. In addition, the β emission from this radionuclide gives the patients an unnecessary radiation burden. Hence, other radionuclides are preferred for use in diagnostic nuclear medicine.

The radioiodine ^{123}I, on the other hand, is very useful in nuclear medicine because it has good radiation characteristics for scintigraphy, such as decay by electron capture, a half-life of 13 h, and γ emmision of 159 keV. However, the much shorter half-life, together with the more complex radionuclide production, makes this radionuclide less available and more expensive compared to ^{131}I.

There are several ^{131}I and ^{123}I radiopharmaceuticals on the market, for both oral and parenteral administration. Ready-for-use radiopharmaceuticals that contain

TABLE 1 Several Radionuclide Generator Systems Useful in Nuclear Medicine

Parent Nuclide	$t_{1/2}$		Daughter Nuclide	$t_{1/2}$
^{68}Ge	280 days	→	^{68}Ga	68 min
^{81}Rb	4.7 h	→	^{81m}Kr	13 s
^{99}Mo	66 h	→	^{99m}Tc	6 h
^{113}Sn	117 days	→	^{113m}In	100 min
^{188}W	69.4 days	→	^{188}Re	17 h

these radionuclides will normally be manufactured by radiopharmaceutical companies and distributed to the marked according to a marketing authorization (MA).

Although therapeutic application represents less than 10% of the nuclear medicine investigations, therapeutic radiopharmaceuticals are a very important group of radiopharmaceuticals. Hence, a brief description is outlined for production of therapeutic radiopharmaceuticals following some other selected groups of radiopharmaceuticals.

^{99}Mo/^{99m}Tc Generators The essential part of the most commonly available generator system is a simple chromatography column to which the mother radionuclide is absorbed on a suitable support material. The daughter radionuclide is a decay product of the mother nuclide. Since it is the daughter nuclide that is used to label the carrier molecules, it must be possible to separate this from the parent nuclide by a chemical separation.

In a ^{99}Mo/^{99m}Tc generator, the ^{99}Mo (molybdenum) is fixed as molybdate to aluminum oxide in the column. The daughter nuclide, ^{99m}Tc (technetium), is eluted from the column as pertechnetate when using saline solution. Molybdenum-99 has a half-life of 66 h, while ^{99m}Tc has a half-life of 6 h. This is an ideal combination of half-lives, giving a system where the daily supply of ^{99m}Tc can easily be calculated from the known amount of ^{99}Mo on the column. The half-life of ^{99m}Tc, along with the radiation characteristics of the nuclide, makes it excellent for use in nuclear medicine imaging. After reconstitution of kits and formation of various radiopharmaceuticals, this radionuclide is used in a major part of all nuclear medicine procedures.

Although the principle for the generators is similar, the design of ^{99}Mo/^{99m}Tc generators from different manufacturers can differ a lot. A drawing of a ^{99}Mo/^{99m}Tc generator is shown in Figure 4. In general, the generator consists of a column with adsorbent material where the radionuclide ^{99}Mo is applied. The column is combined with a needle system necessary for the elution process. A sterile filter is fitted on the air inlet side of the needles to keep an aseptic system during elution. The saline solution for elution may be supplied as a bulk solution sufficient for several elutions

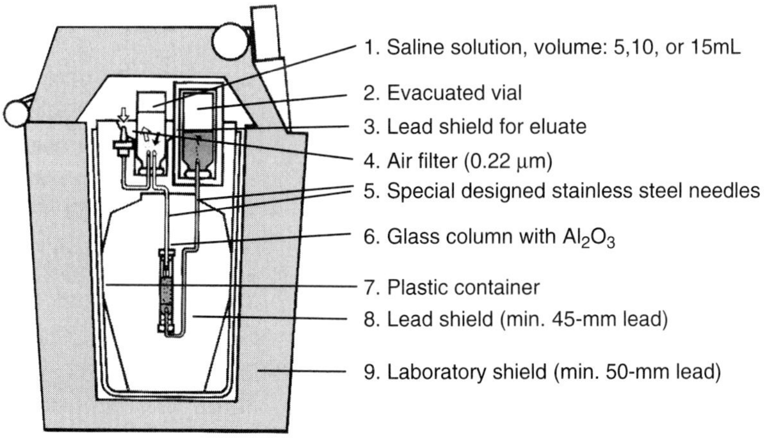

FIGURE 4 Typical radionuclide generator system (ISOTEC, GE Healthcare, AS).

or dispensed volumes sufficient for a single elution. For both, vacuum is normally used to run the elution of the column using sterile evacuated vials.

Finally, due to the relatively high radiation from ^{99}Mo, the system must be properly shielded by either lead or a combination of lead and tungsten.

Whether the column is designed to contain liquid after and between elutions, determine if this is a wet-column generator or a dry-column generator. When liquid is retained at the column (wet generator), radiolysis of water on the column may occur as a result β irradiation from ^{99}Mo. This may change the chemistry on the column and thus reduce the yield when eluting the generator. Most commonly, when manufacturing wet-column generators, oxidizing agents are added either to the saline or to the column itself to avoid reduction of pertechnetate on the column.

A radionuclide generator must be sterile and pyrogen free. Most commonly, the generator is sterilized by autoclaving the entire column after the molybdate has been bound to the aluminum oxide. Other critical procedures during the production and the assembly of the generator must be performed under aseptic conditions. Elution of the generator must also be carried out under aseptic conditions while using only sterile accessories.

Other Generators Of the generators listed in Table 1, two systems are of particular interest in nuclear medicine today along with the ^{99}Mo/^{99m}Tc generator, namely the ^{68}Ge/^{68}Ga generator and the ^{81}Rb/^{81m}Kr generator.

^{68}Ge/^{68}Ga Generator Germanium-68 has a half-life of 271 days, and ^{68}Ga (gallium) a half-life of 68 min. Gallium-68 is a PET emitter, and this generator system is a valuable source of a short-lived radionuclide in a radiopharmacy or nuclear medicine department. However, the system is not as easy or efficient as the ^{99}Mo/^{99m}Tc generator. On the other hand, the longer half-life of the mother nuclide allows use of the system for several months.

This generator can be made up of aluminum loaded on a plastic or glass column. Carrier-free ^{68}Ge in concentrated HCl is neutralized in ethylenediaminetetraacetic acid (EDTA) solution and adsorbed to the column. Then ^{68}Ga is eluted from the column with 0.005 *M* EDTA solution. Alternatively, ^{68}Ge is adsorbed on a stannous dioxide column and ^{68}Ga is eluted with 1 *N* HCl [2].

^{81}Rb/^{81m}Kr Generator Rubidium-81 has a half-life of 4.6 h and decays to ^{81m}Kr (krypton) by electron capture. Krypton-81m has a half-life of 13 s and decays by isomeric transition emitting γ rays of 190 keV. Being an inert gas ^{81m}Kr is used for lung ventilation study.

The parent ^{81}Rb is adsorbed on an ion exchange resin, and the daughter ^{81m}Kr is eluted with air. Because of the very short half-life of 13 s, the studies can be repeated every few minutes, and no radiation safety precaution for trapping ^{81m}Kr is needed [2].

Radiopharmaceutical Kits Radiopharmaceutical kits are nonradioactive ("cold") products containing the sterile ingredients needed to prepare the final radiopharmaceutical. Immediately before administration to the patient, the radionuclide is added. From the point of licensing, these semimanufactured products are defined as radiopharmaceuticals, as they have no other application in medicine [2].

Most of these preparation kits have been developed for labeling of various substances with ^{99m}Tc. Labeling is normally a single- or two-step procedure consisting of adding a solution of ^{99m}Tc-pertechnetate to the preparation kit. The preparation kit contains the ingredient necessary for labeling, such as the substance or ligand to be labeled, a reducing agent, buffers for pH adjustments, and various stabilizers. The reducing agent, very often a stannous salt, is added to bring the radionuclide into a valence state with high reactivity.

Most preparation kits are lyophilized, and the reason for this is to extend the shelf life of the products. Some preparation kits can in fact be stored for more than one year. Since these products are not radioactive, conventional clean rooms and clean-room technology can be applied for production of preparation kits. Most of these products have to be produced aseptically, as they cannot be sterilized with other methods. During lyophilization of the preparation kits used for ^{99m}Tc labeling, it is very important to remove all the oxygen from the kit vial. This is to ensure the right valence of the tin salt. Normally, the vials are filled with an inert gas, such as nitrogen, before the vials are closed completely. It is important, though, that the gas is dried. Some manufacturer chose to not completely replace the removed oxygen, giving a slightly negative pressure inside the kit vial. This may be favorable for the kit-labeling procedure.

Therapeutic Radiopharmaceuticals Radiopharmaceuticals used for therapy (radiotherapy) are designed such that, after administration, they act locally at a target by either damaging or killing cells by irradiation. One of the attractions of radionuclide therapy is the existence of radiation with quite different dimensions of effectiveness, ranging from subcellular (Auger electrons) to hundreds of cell diameters (β particles). In between, α emitters have a tissue range equivalent to only a few cell diameters [9]. Alpha emitters have a very high linear energy transfer (LET), being very potent at short distances.

Table 2 lists a selection of radionuclides and radiopharmaceuticals used in radiotheraphy.

TABLE 2 Selected Radionuclides and Radiopharmaceuticals Used for Radiotherapy in Routine Use or as Part of Clinical Investigations

Radionuclide	Mode of decay	$t_{1/2}$	Radiopharmaceuticals
^{131}I	β^-/γ	8.04 days	^{131}I-NaI, ^{131}I-MIBG, ^{131}I-mAbs
^{90}Y	β^-	2.7 days	^{90}Y-colloid, ^{90}Y-DOTATOC, ^{90}Y-mAbs
^{186}Re	β^-/γ	3.8 days	^{186}Re-sulfide, ^{186}Re-HEDP
^{188}Re	β^-/γ	17 h	^{188}Re-HEDP
^{177}Lu	β^-/γ	6.6 days	^{177}Lu-DOTA-Tyr3-octreotide
^{153}Sm	β^-/γ	1.9 days	^{153}Sm-EDTMP
^{89}Sr	β^-	50.6 days	^{89}Sr-chloride
^{223}Ra	α/γ	11.4 days	^{223}Ra-chloride
^{211}At	α	7.2 h	^{211}At-mAbs
^{213}Bi	α	46 min	^{213}Bi-mAbs
^{166}Ho	β^-/γ	26.8 days	^{166}Ho-colloid
^{169}Er	β^-	9.4 days	^{169}Er-citrate colloid
^{165}Dy	β^-/γ	2.3 h	^{165}Dy-ferric hydroxide macroaggregate
^{32}P	β–	14.3 days	^{32}P-*ortho*-phosphate

Pure α and β emitters are easy to shield, and thus production involving these can be performed in sealed production units with no lead protection. One must keep in mind, though, the potential hazard when inhaling some of these materials. Moreover, many radionuclides used for radiotherapy have an additional γ component. Hence, local lead shielding may be necessary. If the γ component is larger or represents very high energy emission, a total lead shielded unit may be necessary. The latter will be the case when manufacturing ^{131}I radiopharmaceuticals for therapy, since ^{131}I is a radionuclide consisting of a high-energy γ photon together with the β component.

Radiopharmaceuticals for therapeutic use must have a high target-to-background ratio. Targeted radiotherapy involves the use of molecular carrier such as a receptor-avid compound or an antibody to deliver a radionuclide to cell populations.

A challenge when performing radiolabeling of carrier molecules for targeted radiotherapy is the potential risk of radiolysis due to the radiation characteristics of the radionuclides involved. When increasing the specific activity, as part of the scaling up, the risk of radiolytic decomposition of the labeled compound also increases. This is particularly pronounced when using α emitters. The addition of stabilizers in the form of scavengers can reduce this risk. Benzyl alcohol is an example of a compound that acts as a scavenger by catching up with free radicals in the solution.

Another approach is to use kit formulations also for this kind of product. Therapeutic radiopharmaceuticals have been developed where the carrier molecule is formulated in a lyophilized kit and supplied together with the radionuclide. An example of this is the MAb ibritumomab tiuxetan formulated for labeling with the β-emitting radionuclide ^{90}Y. Yttrium-90 ibritumomab tiuxetan (Zevalin) is used in the treatment of non-Hodgkin's lymphoma (NHL).

The labeling is performed in a centralized radiopharmacy, hospital radiopharmacy, or nuclear medicine department immediately before use.

Radioactive Sanitary Products Radioactive sanitary products could be considered as radiopharmaceuticals according to the definition given in directive 2001/83/EC, although there are significant differences between radioactive sanitary products and classical radiopharmaceuticals. The former can in fact be considered as encapsulated radioactive sources, although with the use of microencapsulated sanitary products (such as micrometer-sized glass or polymer beads loaded with a radionuclide), the difference between both types is becoming more difficult to establish.

In any case, radioactive sanitary products are delivered locally (and not systemically or orally) for the local treatment of a disease. The idea is to give a high dose of radiation to a specific part of the body by the implantation of the corresponding sanitary product in the desired zone. The sanitary product must not be metabolized, destroyed, or removed from the place it has been located during a sufficiently long time as to give the desired high radiation dose.

The most commonly used radioactive sanitary products are millimeter-sized seeds or needles loaded with ^{103}P, ^{192}Ir, ^{90}Sr, or ^{125}I. Currently micrometer-sized or even nanometer-sized beads loaded with ^{90}Y are being used for the treatment of specific diseases.

PET Radiopharmaceuticals PET radiopharmaceuticals are labeled with short-lived positron-emitting radionuclides. Such radionuclides can either be produced in a cyclotron or obtained from an appropriate radionuclide generator.

General Considerations The synthesis of PET radiopharmaceuticals has several peculiarities substantially different from the procedures followed to prepare conventional γ-emitting radiopharmaceuticals. A very important issue that must be considered is the specific activity. For all radiopharmaceuticals it is usually very high and can be calculated from the formula

$$A_e = \frac{k}{AT_{1/2}}$$

where A_e is the specific activity, A the mass number of the radionuclide, and $T_{1/2}$ its half-life. It is then clear that the achievable specific activity is higher for radionuclides with shorter half-lives, as is the case for the most relevant PET radionuclides (^{18}F and ^{11}C). As an example, ^{18}F produced in no-carrier-added conditions can be obtained with specific activities of almost 10^{10} Ci/mmol, resulting in PET radiopharmaceuticals with extremely high specific activities.

For PET radiopharmaceuticals we must always consider that synthesis processes must be extremely fast. Consequently, synthesis schemes with as few steps as possible must be used, and each of the steps must proceed with high efficiency. The incorporation of the radionuclide to the molecule should ideally be done in the final steps of the synthesis. In this way two objectives can be achieved: reduce the overall synthesis time (thus increasing the yield) and reduce the number of side reactions and secondary undesired products obtained during the synthesis.

The synthesis of PET radiopharmaceuticals is always carried out at very small scale (only a few dozen micrograms of the radiopharmaceutical are obtained) and each batch can sometimes only be used for a single patient or a few patients at most. Consequently, there is always a big excess of the precursor in the reaction medium, and proper purification systems must be used to get rid of all the possible contaminants. Such systems must also be very efficient and fast, and the most usual is to apply either semipreparative high-performance liquid chromatography (HPLC) or solid-phase extraction-based procedures.

The position of the radionuclide in the molecule of interest is also critical as it will affect the biological behavior of the radiopharmaceutical. Chemical reactions must be designed to be stereospecific in many cases, as the production of a mixture of different stereoisomers complicates the purification of the final radiopharmaceutical. Synthesis procedures must also be easy to automate, as very elevated activities are used for the synthesis of PET radiopharmaceuticals (several curies usually) and appropriate radiation protection systems must be used.

PET Generators Table 3 summarizes the characteristics of some PET generators. So far, the most widely used system has been the ^{82}Sr/^{82}Rb generator, although due to the specific physical and chemical characteristics of the daughter radionuclide and the half-life of the parent radionuclide, the ^{68}Ge/^{68}Ga generator is probably one of the most interesting systems. Recent advances in gallium chemistry have permitted the development of ^{68}Ga radiopharmaceuticals of clinical interest making

TABLE 3 Selected of PET Generators

Generator	Parent	$T_{1/2}$	Daughter	$T_{1/2}$
Fe/Mn	^{52}Fe	8.27 h	^{52m}Mn	21.1 min
Zn/Cu	^{62}Zn	9.13 h	^{62}Cu	9.73 min
Ge/Ga	^{68}Ge	270.8 days	^{68}Ga	68.3 min
Sr/Rb	^{82}Sr	25.6 days	^{82}Rb	76.4 s

TABLE 4 Physical Characteristics of Some Positron Emitters of Clinical Use

Isotope	$T_{1/2}$ (min)	%	$E_{\beta+}$ (keV)
^{11}C	20.4	99.7	960
^{13}N	9.9	99.8	1198
^{15}O	2.0	99.9	1732
^{18}F	109.6	96.7	634

TABLE 5 Nuclear Reactions for Production of Most Widely Used Positron Emitters

^{11}C	^{13}N	^{15}O	^{18}F
^{14}N(p,α)^{11}C	**^{16}O(p,α)^{13}N**	**^{14}N(d,n)^{15}O**	**^{18}O(p,n)^{18}F**
^{10}B(d,n)^{11}C	^{13}C(p,n)^{13}N	^{15}N(p,n)^{15}O	^{20}Ne(d,α)^{18}F
^{11}B(d,2n)^{11}C	^{12}C(d,n)^{13}N		^{16}O(α,pn)^{18}F
^{11}B(p,n)^{11}C			^{19}F(p,pn)^{18}F
^{12}C(p,pn)^{11}C			

Note: Most common reactions used in small cyclotrons are bolded. Different energies of the incident particle are needed for the different nuclear reactions

available PET studies at stand-alone PET centers without a cyclotron with other compounds different from the classical ^{18}FDG.

PET Cyclotrons Cyclotrons used to produce positron emitters of clinical interest (see Tables 4 and 5), mainly ^{18}F and ^{11}C, do not need to be very big. In fact, small devices installed in hospital or academic institutions have long been used for such purposes (see Figure 5). These devices are easy to operate and maintain, and even with single-particle low-energy cyclotrons, it is possible to produce multicurie amounts of ^{18}F and ^{11}C.

Some Positron Emitters of Clinical Interest Fluorine-18 is undoubtedly the most widely used positron-emitting radionuclide. This is mainly due to the wide use of ^{18}FDG, the PET radiopharmaceutical that has permitted PET to become an everyday clinical tool. With the exception of ^{18}FDG and probably ^{18}FDOPA, the use of other ^{18}F-labeled radiopharmaceuticals is very limited. However, the chemical and physical characteristics of ^{18}F are excellent:

FIGURE 5 Small (less than 2 m in diameter) dual-beam negative ion cyclotron capable of easily producing multicurie amounts of ^{18}F and ^{11}C. (Photo courtesy of PET-CUN Center, University of Navarra.)

It can easily be produced in very high quantities (up to 7–9 Ci per batch) even in small cyclotrons with just a few hours irradiation time.

The mean positron emission energy of ^{18}F is just 0.64 MeV (the lowest of all positron emitters with clinical use) and this has several important consequences: The dose of radiation received by the patient will be lower and the distance between disintegration of the radionuclide and the annihilation site (after collision of the positron with an electron) is reduced, thus making PET images with higher resolution possible.

The half-life of ^{18}F (109 min) is sufficiently long to carry out complex synthesis procedures, apply long PET imaging protocols, and carry out metabolite analysis. Furthermore, it is possible to produce the radiopharmaceutical in a laboratory and transport it to a distant site only equipped with an imaging device. These kinds of "satellite PET centers" have boomed all around the world and permitted the fast expansion of PET as an everyday clinical tool in certain pathologies (mainly in oncological diseases).

Fluorine is not common in biological molecules, but many drugs contain this atom. Fluorine and hydrogen have quite similar radii, and changing a hydrogen to a fluorine atom in a molecule does not usually generate substantial steric differences between both molecules. Nonetheless, the electronegativity of fluorine is usually going to change substantially the physicochemical properties of the molecule (reactivity, hydrogen bonding, interactions with cognate receptors, metabolization, etc.). It is not possible to assume that the biological behavior of a molecule and its fluorinated analog is going to be similar. On the contrary, it is advisable to find substantial differences in lipophilicity, biodistribution, protein binding, affinity for receptors, and so on. However, such modifications are in many cases very useful to permit the use of a ^{18}F-fluorinated analog as a PET radiopharmaceutical. In fact, that is the case for the most widely use one: FDG. This compound, which accounts for probably more than 90% of the PET studies performed in the world every day, is a glucose analog that is taken up by the cells by GLUT transporters and metabolized just as glucose at the very first steps of glicolysis. But as a consequence of the change of the C_2 OH group in natural glucose by a ^{18}F atom in FDG, the latter cannot be isomerized (once phosphorilated) and suffers metabolic trapping being specifically accumulated in tumoral cells.

Carbon-11 has a very short half-life (just 20.4 min) but the chance to substitute a carbon atom in any biological molecule by a positron-emitting ^{11}C is a very interesting possibility. This has led to a substantial development of ^{11}C-labeled tracers. The short half-life conditions everything and only PET centers equipped with a cyclotron can have a clinical program with ^{11}C tracers. The production of the radiopharmaceutical must in these cases be performed just before the imaging study and is usually not started until the patient is already on the PET scanner.

The ^{12}C–^{11}C substitution will produce chemically identical molecules and give the chance to study many biological processes by this noninvasive methodology and can also be used in new-drug research and development (R&D).

Synthesis of PET Radiopharmaceuticals Albeit the requirements for the synthesis of PET radiopharmaceuticals previously described, the synthesis process could conceptually be reduced to a very simple scheme, as shown in Figure 6.

The concept is really simple, but there are considerable difficulties in each of the steps. In many cases it is difficult to synthesize a properly designed cold precursor that will permit a simple direct reaction with few secondary products. No modifica-

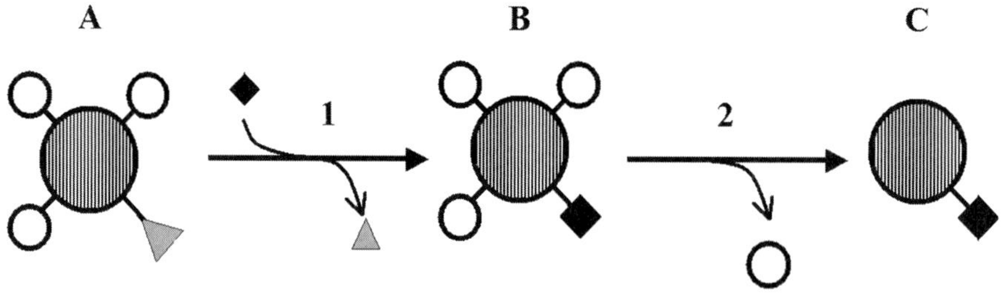

FIGURE 6 General reaction scheme for synthesis of PET radiopharmaceuticals. The precursor molecule (A) is designed with the adequate protecting groups (○) and a reactive leaving group (Δ). A reactive form of the radionuclide (■) is covalently joined to the precursor at the reaction site, while the leaving group is eliminated. An intermediate radioactive product (B) is obtained that is hence deprotected (2) to produce the final radiopharmaceutical (C). A fast and efficient purification process of C is needed to get read of unreacted cold precursor, radionuclide, and intermediate products.

tions in the configuration of the chiral centers should be produced during the overall process and a simple purification system able to purify the final product in a very short time should be found. Additionally, all the reactions should be very fast (just several minutes at most) and be easy to automate to be performed in a computer-controlled device placed in a shielded hot cell.

Production Process and Quality Control The production process includes the following:

- Production of the radionuclide in the cyclotron and sending it to the PET radiopharmaceutical laboratory
- Reaction of the radionuclide with an appropriate cold precursor, either in solution or in solid phase
- Purification of the radiopharmaceutical, usually by semipreparative radio HPLC or solid-phase extraction
- Formulation of the final product as an injectable solution (frequently including phase change in a rotary evaporator) and the adjustment of tonicity and pH
- Sterile filtration or autoclaving

The quality control of the final product must be carried out before release of the batch (except for the sterility and the endotoxin tests for extremely short-lived radionuclides). Consequently, all procedures must not only be very fast but also very accurate, and in all cases it is very important to have a properly established quality assurance system that might permit parametric release of the produced batches. The quality control assays that must be carried out in the radiopharmaceutical includ the following:

- Radionuclidic purity
- Radionuclidic identity
- Chemical purity
- Radiochemical purity
- Specific activity
- Residual solvents
- Visual inspection
- Tonicity
- pH
- Sterility
- Endotoxin

A PET radiopharmaceutical laboratory must include the cyclotron bunker (where positron-emitting radionuclides are produced), the production laboratory, the quality control laboratory, and several different ancillary areas.

In the production laboratory all synthesis and purification processes are carried out in remote-operated fully automated computer-controlled systems (synthesis modules, see Figure 7) located in heavily shielded hot cells (see Figure 8). Dispensing of individual doses is in many cases also carried out by automated systems.

FIGURE 7 Automated synthesis module for PET radiopharmaceutical synthesis located in a shielded hot cell. (Photo courtesy of PET-CUN Center, University of Navarra.)

FIGURE 8 Production laboratory for PET radiopharmaceuticals. The 10-cm lead shielded hot cells contain computer-controlled automated synthesis modules. (Photo courtesy of PET-CUN Center, University of Navarra.)

1.3.5 QUALITY CONSIDERATIONS

1.3.5.1 Documentation

Good documentation constitutes an essential part of the quality assurance system. As claimed in the European Community (EC) Guide to Good Manufacturing Practice (GMP), Chapter 4: "Clearly written documentation prevents errors from spoken communications and permits tracing of batch history." In general, the requirements for documentation related to manufacturing of pharmaceuticals, as set in the GMP regulations, are also valid for manufacturing of radiopharmaceuticals. A recent draft proposal of EC GMP Annex 3, "Manufacture of Radiopharmaceuticals," outlines the following regarding this issue:

> All documents related to the manufacture of radiopharmaceuticals should be prepared, reviewed, approved, and distributed according to written procedures.
>
> Specifications should be established and documented for raw materials, labeling and packaging materials, critical intermediates, the finished radiopharmaceutical, and any other critical material.
>
> Acceptance criteria should be established for the radiopharmaceutical, including criteria for release and shelf life specifications.
>
> Records of major equipment use, cleaning, sanitization or sterilization, and maintenance should show the product, batch number, date and time, and signatures of the persons involved.
>
> Records should be retained for at least three years unless another time frame is specified in national requirements.

It is of utmost importance to have a system for implementing such documents. Any new master document or a new version of such a document must be followed by a training process for relevant operators. This training must be recorded as well.

The recording of production data will make it necessary to bring batch documentation into the radioisotope laboratory. Hence, it is important to have routines that minimize the risk for radioactive contamination of the documents and to ensure that any contaminated documents will not leave the controlled area. Today, the use of computers instead of paper documents in the laboratory leaves most of the paperwork outside the controlled area.

1.3.5.2 Qualification of Personnel

As a general principle in GMP, there should be sufficient qualified personnel to carry out all the tasks that are the responsibility of the manufacturer. Furthermore, individual responsibilities should be clearly understood by the individuals and recorded.

For personnel working with radiopharmaceuticals, training and qualification should cover general principles of GMP and radiation protection. This includes also personnel in charge of cleaning premises and equipment used for this type of production. All manufacturing operations should be carried out under the responsibility of a QP with additional competence in radiation protection.

1.3.5.3 Quality Control

All quality control procedures that are applied to nonradioactive pharmaceuticals are in principle applicable to radiopharmaceuticals. In addition, tests for radionuclidic and radiochemical purity must be carried out. Furthermore, since radiopharmaceuticals are short-lived products, methods used for quality control should be fast and effective. Still, some radiopharmaceuticals with very short half-lives may have to be distributed and used after assessment of batch documentation even though all quality control tests have not been completed. It is acceptable, though, for these products to be released in a two-stage process, before and after full analytical testing. In this case there should be a written procedure detailing all production and quality control data that should be considered before the batch is dispatched. A procedure should also describe the measures to be taken by the QP if unsatisfactory test results are obtained after dispatch (GMP, Annex 3).

The quality control tests fall in two categories: biological tests and physiochemical tests. The biological tests establish the sterility and apyrogenicity, while the physiochemical tests include radionuclidic, chemical, and radiochemical purity tests along with determination of pH, osmotic pressure, and physical state of the sample (for colloids).

For lyophilized preparation kits containing reducing agents, such as ^{99m}Tc kits, a test for moisture content can be necessary. Residual water in the freeze-dried pellet may lead to oxidation of the reducing agent.

Radionuclidic Purity Radionuclidic purity is defined as the fraction of the total radioactivity in the form of the desired radionuclide present in a radiopharmaceutical. Radionuclide impurities may arise from impurities in the target material or from fission of heavy elements in the reactor [2]. In radionuclide generator systems, the appearance of the parent nuclide in the daughter nuclide product is a radionuclidic impurity. In a ^{99}Mo/^{99m}Tc generator, ^{99}Mo may be found in the ^{99m}Tc eluate due to breakthrough of ^{99}Mo on the aluminum column. The presence of these extraneous radionuclides increases the radiation dose to the patient and may also obscure the scintigraphic image.

Radionuclidic purity is determined by measuring the characteristic radiations emitted by individual radionuclides. Gamma emitters are distinguished from another by identification of their γ energies on the spectra obtained from a NaI crystal or a Ge (germanium) detector. This method is called γ spectroscopy.

Pure β emitters are not as easy to check as the γ emitters. However, they may be checked for purity with a β spectrometer or a liquid scintillation counter.

Radiochemical Purity The radiochemical purity (RCP) of a radiopharmaceutical is the fraction of the total radioactivity in the desired chemical form in the radiopharmaceutical. Radiochemical impurities arise from decomposition due to the action of solvent, change in temperature or pH, light, presence of oxidizing or reducing agents, and radiolysis [2]. Examples of radiochemical purity are free ^{99m}Tc-pertechenetate and hydrolyzed ^{99m}Tc in labeled ^{99m}Tc radiopharmaceuticals. The presence of radiochemical impurities in a radiopharmaceutical results in poor-quality images due to the high background from the surrounding tissues and blood. It also gives the patient unnecessary radiation doses.

A number of analytical methods are used to detect and determine the radiochemical impurities in a given radiopharmaceutical. Most commonly used are methods like paper (PC), thin-layer (TLC), and gel chromatography, paper and gel electrophoresis, HPLC, and precipitation. A common principle for the different methods is that they can chemically separate the different radiolabeled components in the radiopharmaceutical. It may sometimes be necessary to perform more than

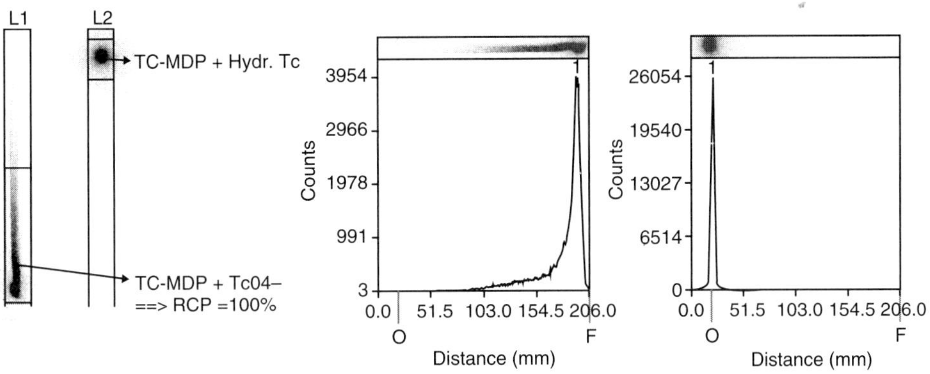

FIGURE 9 Typical chromatograms for ^{99m}Tc-MDP. The left strip and chromatogram are obtained with ITLC-SG in sodium acetate. The right strip and chromatogram are obtained in methyl ethyl ketone (MEK). When combining these, any free pertechnetate ($^{99m}TcO^{4-}$) and/or hydrolyzed ^{99m}Tc can be detected. Thus the fraction representing ^{99m}Tc-MDP (RCP) can be calculated.

one test method, for instance, TLC and HPLC, to get a complete picture of the different radiochemical impurities. Alternatively, one can use one chromatographic method consisting of a constant stationary phase but varying the mobile phase (solvent). An example is the radiochemical purity test of ^{99m}Tc-methylenediphosphate (MDP), a radiolabeled phosphate used in bone scintigraphy. When using two TLC systems, one with sodium acetate as a solvent and one with methyl ethyl ketone (MEK) as a solvent, the different ^{99m}Tc compunds in the product can be determined. A small aliquot of the radiopharmaceutical preparation is spotted on an instant thin-layer chromatography (ITLC) strip. The strip is dipped into the chromatography flask while keeping the spot above the solvent. During the chromatography process, the different components of the sample distribute differently in the ITLC strip, depending on the solubility and polarity of the components. In systems like this, each component is characterized by an R_f value, defined as the ratio of the distance traveled by the component to the distance the solvent front has advanced from the original point of application of the test material. The distribution of the radioactive components on the strips can be monitored by use of an appropriate device for measuring radioactivity and printed in a chromatogram. Figure 9 shows typical chromatograms for ^{99m}Tc-MDP in the TLC systems described above.

Chemical Purity The chemical purity of a radiopharmaceutical is the fraction of the material in the desired chemical form. Chemical impurities may arise from the breakdown of the material either before or after labeling. Chemical impurities may also arise from the manufacturing process, such as aluminum in a ^{99m}Tc eluate, coming from the aluminum column on the generator. Residuals of solvent from the radiopharmaceutical synthesis are also considered as chemical impurities. If the chemical impurity is present before labeling, the result may be undesirable labeled molecules. Furthermore, chemical impurities may cause a toxic effect. High-performance liquid chromatography and gas chromatography (GC) are important methods for determination of chemical impurities in a radiopharmaceutical.

Sterility and Pyrogen Testing Sterility indicates the absence of any viable bacteria or microorganisms in a radiopharmaceutical preparation. Hence, sterility testing is performed to prove that radiopharmaceuticals are essentially free of viable microorganism. The test for microbial contamination of these products is best carried out with filter methods. It is a great advantage to incubate only the filters instead of the radioactive solutions.

The test is performed according to the Ph.Eur/USP monograph on Sterility tests [13, 14], but with an important modification. Small batch sizes, typical for radiopharmaceuticals, make it necessary to use smaller test volumes than required in the monographs. Also the risk for radiation exposure supports this modification.

All radiopharmaceuticals for human administration are required to be pyrogen free. Also the tests for apyrogenicity must be modified when applied for these products. The classical rabbit test for pyrogens was never a convenient test for parenteral radiopharmaceuticals. Practical problems due to radioactive rabbits and the need for larger test volumes made this a difficult task. Today, the Limulus amebocyte test (LAL) is the method of choice and has been accepted by the Ph. monographs for many years. This test is normally done within an hour, compared to several days for the rabbit test.

However, even the LAL test may be too time consuming for the very short lived PET radiopharmaceuticals. Hence, less time consuming methods are in progress and will probably improve this situation. Meanwhile, it is accepted that the test for apyrogenicity, like the sterility test is for most radiopharmaceuticals, is finished after release of the most short lived radiopharmaceuticals.

Bubble Point Testing of Filters Parenteral radiopharmaceuticals that are not terminally sterilized must undergo a sterile filtration process as part of the aseptic production procedure. Although the supplier certifies the filters used, they must be checked for integrity after use to assure that there has been no leakage during the filtration. The integrity of the filter may be demonstrated by *bubble point testing*. In this test, the filter is placed and monitored under controlled pressure. When the test is done on wet filters, the pressure needed to push gas through the filter is defined as the bubble point. A filter with given pore width has a corresponding bubble point value. Most frequently, sterile filtration is performed by 0.22-μm filters; hence the bubble point is about 3–4 bars. However, the filter supplier should specify the bubble point valid for a specific filter.

Since this is an in-process test, special caution must be given to radiation protection. The test equipment should be placed within a closed and shielded unit and a system should be in place to collect any radioactive spill from the test.

When the filter integrity test fails, the sterile filtration process must be rejected.

Visual Inspection of Finished Product As part of the quality control, all parenterals will be subject to an inspection for the possible content of particles. Visual inspection of radiopharmaceuticals is more complicated than for other pharmaceuticals, as radiation protection guidelines strongly discourage any direct eye contact with radioactive sources. Normally, the visual inspection of a radiopharmaceutical is performed by placing the vial on a rotating station connected to a camera. The station is properly shielded, and the operators can study the solution on a distant screen.

1.3.5.4 Validation and Control of Equipment and Procedures

Preventive maintenance, calibration, and qualification programs should be operated to ensure that all facilities and equipment used in the manufacture of radiopharmaceuticals are suitable and qualified (GMP, Annex 3). Special emphasis should be put on critical equipment for handling of radiopharmaceuticals, such as dose calibrators that are used to check the accuracy of the dispensing of patient doses. Particular programs are outlined for checking the dose calibrator, including constancy, accuracy, linearity, and geometry. The general principles of validation outlined in the GMP regulations are valid for radiopharmaceuticals as well as for other pharmaceuticals. All validation activities should be planned and clearly defined and documented in a validation master plan (VMP). Special emphasis should be given on the validation of aseptic processes in the production of radiopharmaceuticals. Studies, including media fill tests, must be performed and recorded to demonstrate maintenance of sterility throughout the production process. This is particularly important since most radiopharmaceuticals are dispatched and used before the sterility test is finished.

1.3.5.5 Stability Aspects of Radiopharmaceuticals

As discussed already, radiopharmaceuticals are exposed to stability problems, particularly when radiolabeled compounds are involved. Decomposition of labeled compounds by radiolysis depends on the specific activity of the radioactive material, the energy of the emitted radiation, and the half-life of the radionuclide. Particles, such as α and β radiation, are more damaging than γ rays, due to their short range and local absorption in matter. The stability of a compound is time dependent on exposure to light, change in temperature, and radiolysis. The longer a compound is exposed to these conditions, the more it will tend to break down.

Stabilizers such as ascorbic acid and benzyl alcohol may be added to inhibit or delay the decompostion. Many preparations are stored in the dark under refrigeration to slow down the degradation of the material [2]. The expiry date of a radiopharmaceutical is based upon data from stability studies designed to demonstrate the described effects on the product after storage.

Hence, for most stability studies on radiolabeled compounds, the radiochemical purity and pH are the most important physiochemical parameters to study. Moreover, for parenteral radiopharmaceuticals, a stability study also has to demonstrate the maintenance of sterility and apyrogenicity after storage.

1.3.6 EXTEMPORANEOUS PREPARATION OF RADIOPHARMACEUTICALS

An extemporaneous preparation is defined as a product which is dispensed immediately after preparation and not kept in stock [10]. Hence, many radiopharmaceuticals could fall into this category due to their limited shelf life.

The use of extemporaneous preparation should be limited to situations where there is no product with marketing authorization (MA) available. This could be prepared based upon a prescription for a named patient (magistral preparation) or a production based upon a formula and prepared on a regular basis. The latter is a common situation for many radiopharmaceuticals. For radiopharmaceuticals with short half-lives or rare indications, no sizable commercial market exists. Consequently, no pharmaceutical company will be prepared to obtain a MA for a product that will not yield a profit due to these limitations. Still, there is a need from a medical point of view to have such products available. For radiopharmaceuticals incorporating radionuclides with a physical half-life of only a few minutes, only local production is feasible. They are therefore prepared in hospital pharmacies or laboratories and supplied for individual or small numbers of patients on a daily basis.

The extemporaneous preparation of radiopharmaceuticals is regulated on a national level, and hence this regulation may differ from country to country. The Pharmaceutical Inspection Convention (PIC/S) has drafted a guide to good practices for preparations of medicinal products in pharmacies [10], valid for medicinal products that do not have a MA, prepared extemporaneously or for stock. For medicinal products prepared to a larger extent or for use in clinical trials, industrial GMPs are applicable. Although the suggested guide outlines a general principle according to GMP, different requirements are particularly evident when it comes to documentation and quality control testing. There is also a discussion about the grades of background environment needed for production, with a differentiation between products with shelf lives less than or longer than 24h [10]. While aseptic manufacturing according to industrial GMP has to be performed in grade A with a grade B background, this proposed guide opens for a relaxation to this. For an aseptic preparation of a product with a shelf life of less than 24h, using a biohazard safety cabinet (BSC), the background environment may be grade D. Even for products with a shelf life longer than 24h, an extensively documented procedure may allow grade C in background, as long as grade B clothing is worn. In general, the referred draft guide is based much upon a risk related approach and is graduated, depending on the size and type of prepared medicinal products.

As to documentation for extemporaneous prepared products, the proposed guide set as a minimum requirement to specify the name, strength, and expiry date of the product. If a product is prepared for a single patient (magistral production), it is assumed that no end product testing will be required. For radiopharmaceuticals, though, the activity in each dose must be measured before administration. Chemical and microbiological quality control is not required for products that have a shelf life of 24h or less, provided that frequent process validation is performed. In addition, chemical and microbiological information must be available to justify the shelf life for the product.

For products that are prepared extemporaneously at a regular basis or even for a limited stock, a product specific documentation (product file) is needed. This will include specifications, instructions, and records but also a pharmaceutical assessment of safety data, toxicity, biopharmaceutical aspects, stability, and product design. The product file should also include a product review as soon as a product is used repeatedly or over longer periods.

Furthermore, the drafted guide suggests that the level of end-product testing for those products will depend on the associated risk connected to the scale of operation, shelf life of the product, frequency of preparation, as well as type of product (parenterals, orals) and type of facility where the product has been prepared.

Independent of which regulation applies at a national level to extemporaneous or magistral preparation of radiopharmaceuticals, the patients should be entitled to expect that these products are prepared accurately, are suitable for use, and will meet the expected standards for quality assurance. Pharmacists involved in this kind of production must ensure that they and any other staff involved are competent to undertake the tasks to be performed and that the requisite facilities and equipment are available [11]. As for other radiopharmaceutical production, systems must be in place to ensure the operator safety due to handling of radioactive materials. All involved staff must have sufficient training in radiation safety issues, in addition to training in GMP.

REFERENCES

1. Britton, K. (1996), Radiopharmaceuticals for the future, *Curr. Dir. Radiopharma. Res. Dev.* (Ed. by Stephen Mather), viii. Developments in Nuclear medicine, Vol XXX. London, UK.
2. Saha, G. B. (1998), *Fundamentals of Nuclear Pharmacy*, 4th ed., Springer, Heidelberg, Germany.
3. Alexoff, D. L. Automation for the synthesis and application of PET radiopharmaceuticals, BNL-68614 Officinal File Copy.
4. Bremer, P. O. (1995), Aseptic production of radiopharmaceuticals, in *Aseptic Pharmaceutical Manufacturing*, Vol. II, *Application for the 1990s*, Interpharm, Michael J. Groves and Ram Murty, pp. 153–180.
5. Nordic Council on Medicines. (1989), *Radiopharmacy: Preparation and Control of Radiopharmaceuticals in Hospitals*, NLN Publications No. 26, Uppsala, Sweden.
6. European Commision (2003), *EU Guide to Good Manufacturing Practice*, Annex 1 and 3, Brussels, Belgium, October 8.
7. Dabbah, R. (1995), Controlled environments in the pharmaceutical and medical products industry: A global review from regulatory, compendial, and industrial perspectives, in *Aseptic Pharmaceutical Manufacturing,* Vol. II, *Application for the 1990s*, Interpharm, Michael J. Groves and Ram Murty, pp. 11–40.
8. Lee, M. C., PET and PET/CT are the fastest growing imaging modalities worldwide, paper presented at the 5th International Conference on Isotopes (5ICI), Brussels, Belgium, Apr. 25–29, 2005.
9. Zalutsky, M. R., Pozzi, O., and Vaidyanatha, G., Targeted radiotherapy with alpha particle emitting radionuclides, paper presented at the International Symposium on Trends in Radiopharmaceuticals (ISTR-2005), Vienna, Austria, Nov. 14–18, 2005.
10. Pharmaceutical Inspection Convention (2006, Aug.), PIC/S guide to good practices for preparation of medicinal products in pharmacies, PE 010-1 (Draft 2), Geneva, Switzerland.
11. Standards for good professional practice (2000), *Pharm. J.* 265(7109), 233.
12. Kowalsky, R. J., and Falen, S. W. (2004), *Radiopharmaceuticals in Nuclear Pharmacy and Nuclear Medicine*, American Pharmacists Association, Forrester Center, WV.

FURTHER READINGS

European Commision. (2006), *EU Guide to Good Manufacturing Practice*, Annex 3; draft proposal, Brussels, Belgium, Apr. 12.

Rootwelt, K. (2005), *Nukleærmedisin*, 2nd ed. Gyldendal Norsk Forlag AS, Oslo, Norway.

Schwochau, K. (2000), *Technetium: Chemistry and Radiopharmaceutical Applications*, VCH Verlagsgesellschaft Mbh, Weinheim, Germany.

Welch, M. J., and Redvanly, C. S., Eds. (2002), *Handbook of Radiopharmaceuticals*, Wiley, Hoboken, NJ.

SECTION 2

ASEPTIC PROCESSING

2.1

STERILE PRODUCT MANUFACTURING

JAMES AGALLOCO[1] AND JAMES AKERS[2]

[1]*Agalloco & Associates, Belle Mead, New Jersey*
[2]*Akers Kennedy & Associates, Kansas City, Missouri*

Contents

Pharmaceutical Manufacturing Handbook: Production and Processes, edited by Shayne Cox Gad

2.1.1 INTRODUCTION

The manufacture of sterile products is universally acknowledged to be the most difficult of all pharmaceutical production activities to execute. When these products are manufactured using aseptic processing, poorly controlled processes can expose the patient to an unacceptable level of contamination. In rare instances contaminated products can lead to microbial infection resulting from products intended to hasten the patient's recovery. The production of sterile products requires fastidious design, operation, and maintenance of facilities and equipment. It also requires attention to detail in process development and validation to ensure success. This chapter will review the salient elements of sterile manufacturing necessary to provide acceptable levels of risk regarding sterility assurance.

Commensurate with the criticality associated with sterile products, the global regulatory community has established a substantial number of the basic requirements that firms are expected to adhere to in the manufacture of sterile products. The most extensive of these are those defined by the Food and Drug Administration

(FDA) in its 2004 Guideline on Sterile Drug Products Produced by Aseptic Processing and the European Agency for the Evaluation of Medicinal Products (EMEA) Annex 1 on Sterile Medicinal Products [1, 2]. Substantial additional information is available from the International Organization for Standardization (ISO), the Parenteral Drug Association (PDA), and the International Society for Pharmaceutical Engineering (ISPE) (see Appendix) [3]. The organizations have provided a level of practical, experience-based detail not found in the regulatory documents, thereby better defining practices that are both compliant with regulatory expectations and based upon rational, evidence-based science and engineering.

Consideration of patient risk associated with pharmaceutical production emerged largely from regulatory impetus, by which the regulatory community stated its intended goal to structure its inspectional process using patient safety as a major focus in determining where to allocate their inspectional and review resources. Emanating from the International Conference on Harmonization (ICH) efforts to produce a harmonized approach to pharmaceutical regulation, risk-based compliance has been adopted in Europe, Japan, and the United States [4, 5]. Sterile products, especially those made by aseptic processing, have been properly identified as a high priority by the global regulatory community. Several risk analysis approaches have been developed that can help the practitioner review practices with the goal of minimizing risk to the patient [6–8].

2.1.2 PROCESS SELECTION AND DESIGN

The production of sterile products is profoundly impacted both by formulation and the selection of primary packaging components. Design parameters for a facility and selection of appropriate manufacturing technologies for the product require that the formulation process and packaging components be chosen and evaluated in advance.

2.1.2.1 Formulation and Compounding

The vast majority of parenteral formulations are solutions requiring a variety of tankage, piping, and ancillary equipment for liquid mixing (or powder blending), filtration, transfer, and related activities. Suspensions, ointments, and other similar products, including the preparation of the solutions for lyophilized products, can be manufactured in the same or very similar equipment. The scale of manufacturing can vary substantially, with the largest batches being well in excess of 5000 L (typically for large-volume parenteral production), down to less than 50 mL for radiopharmaceuticals or biologicals customized for a particular patient.

The majority of this equipment is composed of 300 series austenitic stainless steel, with tantalum or glass-lined vessels employed for preparation of formulations sensitive to iron and other metal ions. The vessels can be equipped with external jackets for heating and/or cooling and various types of agitators, depending upon the mixing requirements of the individual formulation. In many facilities, a variety of tank sizes are available for use. Larger facilities may have the high-capacity tanks permanently installed and permanently connected to process utilities. Smaller vessels are generally mobile and positioned in individual processing booths or rooms as needed.

After sterilizing filtration (or sterilization by heat or other means), comparably sized vessels are sometimes utilized to contain the product prior to and during the filling process. These holding vessels are often steam sterilized along with the connecting piping prior to use. There are a number of firms that fill directly from the compounding vessel using in-line filtration eliminating the intermediate vessel. When this approach is used, a small moist-heat-sterilized surge tank or reservoir tank may be required, particularly with modern time–pressure filling systems. This practice may reduce initial facility and equipment cost but places additional constraints on operational flexibility. The use of disposable equipment for compounding and holding of sterile formulations is coming into greater use. This eliminates the cleaning of vessels prior to reuse, but confirmation of material compatibility is required. Disposable equipment is often used with products manufactured in small to moderate volumes, and while reducing initial equipment expenses disposable equipment also results in contaminated waste, which cannot be recycled or reused and must be treated appropriately.

Aseptic compounding as required for suspensions and other formulations in which open-vessel processes are required mandate an ISO 5 environment providing ideally >400 air changes/hour in which these steps can be performed with minimal opportunity for adventitious contamination. This could be accomplished using a protective curtain and a unidirectional flow hood (UFH) or other more evolved designs such as a restricted access barrier (RABs) system or an isolator (technologies that provide a higher level of employee separation from the area in which materials are handled can get by with lower air exchange rates). All activities requiring opening of processing lines such as sampling or filter integrity testing should be performed using similar protective measures. The preparation of sterile suspensions requires a facility/equipment design capable of safe addition of sterile solids to a liquid vehicle and is conventionally performed using a specifically designed processing area to minimize contamination potential. Comparable and greater complexity is generally required for creams, ointments, emulsions, and the increasingly common liposome formulations.

Some sterile powder formulations (these are predominantly, but not exclusively, antibiotics) may require sampling, mixing, milling, and subdivision activities similar to those found in oral powder manufacturing. The facilities and equipment utilized for these products is substantially different from that used for liquids, and the production area bears little resemblance to that utilized for liquids. These materials are received sterile and must be processed through sterilized equipment specifically intended for powder handling in a fully aseptic environment with ISO 5 protection over all open container activities.

2.1.2.2 Primary Packaging

The primary package for parenteral formulations provides protection to the sterile materials throughout the shelf life. The components of the primary package are every bit as important to contamination control and hence safety of the finished product as the formulation itself, and their preparation must be given a comparable level of consideration. The most commonly used container is glass; vials are still the most common, although increasingly prefilled syringes are chosen. Glass ampoules are still seen. However, although convenient from a manufacturing perspective, the difficulty involved in opening ampoules while at the same time avoiding problems with glass particulate or microbial contamination has reduced their popularity. The use of plastic containers (as vials, ampoules, or syringes) is increasingly common given their reduced weight and resistance to breakage. Blow-fill seal (BFS) and form-fill seal (FFS) are utilized for the filling of numerous ophthalmic and other noninjectable formulations in predominantly low-density polyethylene (LDPE) containers. With the exception of ampoules and BFS/FFS, an elastomeric closure system is also necessary to seal the containers. Some delivery systems (i.e., prefilled syringes, multichamber vials, and others may require more than one elastomeric component to operate properly. In the case of vials, an aluminum crimp is applied to secure the closure to the vial. Prefilled syringes may require the preparation and assembly of additional components such as needles, needle guards, stoppers, diaphragms, or plungers, depending on the specifics of the design. Lyophilization is required to ensure the stability of some formulations and requires the use of closures that allow venting of the container during the freeze-drying process. Full seating of the closure is accomplished within the lyophilizer using moving shelves to seat the closure.

Glass is ordinarily washed prior to sterilization/depyrogenation to reduce contamination with foreign material prior to filling. In aseptic fill processes, the glass is then depyrogenated using dry heat. This can be accomplished using either a continuous tunnel (common for larger volumes and high-speed lines) or a dry heat oven (predominantly for small batches). The depyrogenation process serves to sterilize the glass at the same time, and thus the glass components must be protected post-processing. This is generally accomplished by short-term storage in an ISO 5 environment often accompanied by covering within a lidded tray. There are suppliers that offer depyrogenated glass vials and partially assembled syringes in sealed packages for filling at a customer's site. In this instance, the supplier assumes responsibility for the preparation, depyrogenation, and aseptic packaging. Glass ampoules are available presealed and depyrogenated; the end user has merely to open, fill, and reseal the syringe under appropriate conditions.

Plastic components (whether container or closure) can be sterilized using steam, ethylene oxide, hydrogen peroxide, or ionizing radiation. The γ irradiation is accomplished off-site by a subcontractor with appropriate expertise as these methods are considered the province of specialists because of the extreme health hazards directly related to the sterilization method. Electron beam sterilization may also be done by a contractor, although compact lower energy electron beam systems have been introduced that allow sterilization in-house. Steam sterilization is ordinarily performed in house, though many common components are becoming available presterilized by the supplier. Preparation steps prior to sterilization vary with the component and the methods used to produce the component. Rubber components are washed to reduce particles, while this is less common with plastic materials.

Syringes vary substantially in design details and can be aseptically assembled from individual components. However, increasingly, these are supplied as presterilized partial assemblies in sealed containers.

The BFS and FFS are unique systems in that the final container is formed as a sterile container just prior to the aseptic filling step. The BFS requires careful control over the endotoxin content of the LDPE (and other polymeric materials) beads used to create the containers as well as the melting conditions utilized to form them.

The FFS utilizes in-line sterilization/drying of the film prior to shaping of the containers.

2.1.2.3 Process Objectives

The production of parenteral products requires near absolute control over microorganisms. Endotoxin contamination is a serious health concern, particularly among neonates and infants and also requires a high level of control and validation. Additionally, the control of foreign matter, including particles and fibers of various types, is also vitally important to end-user safety. Assuring appropriate control over these potential contaminants requires careful attention to several factors: facility design, equipment selection, sterilization procedures, cleaning regimens, management of personnel, and the process details associated with compounding, filling, and sealing of product containers. Each of these will be discussed in detail.

2.1.3 FACILITY DESIGN

To provide control of microbial, pyrogen, and particles controls over the production environment are essential. The facility concerns encompass the entire building, but the most relevant components are those in which production materials are exposed to the environment.

2.1.3.1 Warehousing

Environmental protection of materials commences upon receipt where samples for release are taken from the bulk containers. Protection of the bulk materials is accomplished by the use of ISO 7 classified environments for sampling. All samples should be taken aseptically, which mandates unidirectional airflow and full operator gowning. This practice is mandated by current good manufacturing practice (CGMP) and assures that sampling does not introduce contaminants to the materials that will be used in the production. Where central weighing/subdivision of active ingredients and excipients are performed, similar protection is provided for identical reasons. The expectation is that these measures reduce the potential for contamination ingress into materials that have yet to receive any processing at the site. Materials and components that are supplied sterile are received in this area, but samples are often packaged separately by the supplier to eliminate the need for potentially invasive sampling of the bulk containers. Where so-called delivery samples are used, it is critical that these samples are known to be fully representative of the production process. Additionally, where sterility or bioburden control of sampled materials is critical, thought must be given to the methods used to reseal the containers to ensure that moisture levels, bioburden levels, or in the case of sterile products sterility assurance are not compromised.

2.1.3.2 Preparation Area

The materials utilized for production of sterile processes move toward the filling area through a series of progressively cleaner environments. Typically, the first step is transfer into an ISO 8 [Class 100,000, European Union (EU) Grade D] environment in which the presterilization preparation steps are performed. Wooden pallets and corrugated materials should always be excluded from this zone (and any classified environment), and transfers of materials are performed in air locks designed to reduce the potential for particle ingress and to a lesser extent microbial ingress. Preparation areas provide protection to materials and components for a variety of activities: component washing (glass, rubber, and other package components), cleaning of equipment (product contact fill parts, process tools, etc.), and preassembly/wrapping for sterilization. In some facilities, this area is also utilized to support compounding operations in which case process utensils, small containers, and even portable equipment will be cleaned and prepared for sterilization.

Careful attention must be given to material flow patterns for clean and dirty equipment to prevent cross contamination. In larger facilities, the equipment wash room may be a separate room proximate to the preparations area with defined flows for materials and personnel. Ideally, materials should move through the facility in a unidirectional fashion, with no cross over of any kind.

The preparations area typically includes storage areas where clean and wrapped change parts, components, and vessels can be held until required for use in the fill or compounding areas. (Just-in-time practices are desirable for all parenteral operations to avoid extensive and extended storage of materials in the higher classified fill or compounding areas.) The preparations area is ordinarily located between the warehouse and the filling/compounding areas and connected to each of those by material/equipment air locks.

Preparation areas are supplied with high-efficiency particulate air (HEPA) filters (remote-mounted HEPAs are commonplace). The common design requirement is more than 20 air changes per hour, turbulent airflow (see below), and temperature and relative humidity controlled for personnel comfort. As in any clean room area designed for total particulate control, the air returns should be low mounted. Wall and ceiling surfaces should be smooth, easily cleaned, and tolerant of localized high humidity. Floors should be typically monolithic with integral drains to prevent standing water. Common utilities are water for injection, deionized water, compressed air, and clean/plant steam. Clean-in-place (CIP) and sterilize-in-place (SIP) connections may be present if the prep area supports compounding as well.

Ordinarily, present within the preparation area are localized areas of ISO 5 unidirectional airflow (Class 100) utilized to protect washed components prior to sterilization and/or depyrogenation. These areas are not aseptic and should not be subjected to the more rigorous microbial expectations of aseptic processing. They are designed to reduce/eliminate the potential for particle contamination of unwrapped washed materials. Operators accessing these protective zones wear gloves at all times when handling materials.

Operators in the preparations area are typically garbed in low particle uniforms (or suits) with shoe, hair, and beard covers. The use of latex or other gloves is required when contacting washed components. Sterilized gowns and three-stage gowning facilities are not required to enter or work in this ISO 8 environment. Gowns are generally donned within a single-stage airlock, which is maintained at a pressure slightly negative to the ISO 8 working environment. Separate personnel entry/exit are not typically necessary for this lower classified environment.

Equipment within the preparations area varies with the practices of the firm and can include manual or ultrasonic wash/rinse sinks; single or double door automated parts washers; batch or continuous glass washers; stopper washers for closure components; CIP/SIP stations; equipment wrap areas (as described above); and staging areas for incoming (prewash) components, dirty equipment, and cleaned components/equipment. An adjacent classified storage area(s) may be present in larger facilities to accommodate the full variety of change parts and equipment that is not in immediate use. Where the preparations area also supports compounding, it may include additional equipment such as pH meters, filter integrity apparatus, and the like in support of those operations. (*Note:* Where compounding requires aseptic conditions for rigorous control of bioburden, as is the case for unpreserved biologics and other contamination-sensitive products, it is best to provide separate entry for compounding. The moisture level and hence contamination potential in a typical preparation area is unsuitable for entry into an aseptic compounding area).

Depending on the scale of the operation, the preparations area may include the loading areas for both sterilizers and ovens. In high-throughput operations where the use of tunnels for glass depyrogenation is more prevalent, glass washers and tunnels for each filling line may be in separate ISO 8 rooms accessed from the preparations area.

2.1.3.3 Compounding Area

The manufacture of parenteral solutions is ordinarily performed in ISO 7 (Class 10,000, EU Grade C) controlled environments in which localized ISO 5 unidirectional flow hoods are utilized to provide greater environmental control during material addition. These areas are designed to minimize the microbial, pyrogen, and particle contributions to the formulation prior to sterilization. Depending upon the scale of manufacture, this can range from small containers (up to 200 L) (disposable containers are coming into use for these applications), to portable tanks (up to 600 L) to large fixed vessel (10,000 L or more have been used) in which the ingredients are formulated using mixing, heating, cooling, or other unit operations. Smaller vessels are placed or rolled onto scales, while fixed vessels are ordinarily mounted on weigh cells. The vessels may be equipped for temperature and pressure measurement instruments, as mandated by process requirements. Compounding areas often include equipment for measuring mass and volume of liquid and solid materials including, for example, graduated cylinders, and scales of various ranges, transfer and metering pumps, homogenizers, prefilters, and a variety of other liquid/powder handling equipment. Liquid handling may be accomplished by single-use flexible hose, assemblies of sanitary fittings, or some combination thereof. A range of smaller vessels to be used for the addition of formulation subcomponents or excipients to the primary compounding tank may be required as well. Because parenteral formulations can include aqueous and nonaqueous vehicles, suspensions, emulsions, and other liquids, the capabilities of the compounding area may vary. Agitators can be propeller, turbine, high shear, or anchor designs depending upon the requirements of the products being manufactured, and it is not uncommon to find examples of each in larger facilities. It is preferable to perform as much of the process as possible while the formulated liquid is nonsterile to ease sterilization requirements, although precautions to prevent microbial and endotoxin contamination are important risk abatement features.

The formulation area is customarily a combination of open floor space, adjacent to three-sided booths and individual processing rooms in which the ingredients are handled and individual batches are produced. Walls and ceiling materials are selected to be impervious to liquids and chemical spills and are easy to clean. Floors in these areas are monolithic and should be sloped (at 1–3:100) to drains with appropriate design elements and control procedures to eliminate backflow potential (regulatory bans on drains in classified areas are focused on protecting aseptic environments and are inappropriate for nonsterile compounding areas). Pit scales should be avoided in new installations; floor-mounted scales intended for cleaning underneath the base are preferable.

Compounding areas are supplied with HEPA filters (ceiling-mounted terminal HEPAs are more common, though central supply is possible in areas of low contamination risk). The common design requirement is more than 50–60 air changes per hour, turbulent airflow (see below), with temperature and relative humidity for personnel comfort. Air returns may be at or near floor level, with localized extraction provided as necessary to minimize dusting of powder materials. Where substantial heat is generated from processing or sterilization, a ceiling or high wall return may be more appropriate. Wall and ceiling surfaces are smooth, easy to clean, and tolerant of localized high humidity. Floors are typically monolithic with integral drains to prevent standing water. Common utilities are water for injection, deionized water, nitrogen, compressed air, clean/plant steam, and heating and cooling media for the fixed and portable tanks. Water for injection use points are often equipped with sanitizable heat exchangers for operator safety.

Cleaning of the fixed vessels and portable tanks is accomplished using either manual sequenced cleaning procedures or more commonly with a CIP system. Cleaning of other items can be accomplished in a wash area accessed from the compounding area or in a common wash room incorporating both filling and compounding equipment. Sterilization of the nonsterile processing equipment and vessel is often provided for as an option, even where it is not routinely required to control product bioburden. Where production volumes or physical location dictate, the compounding area may have a separate preparations area from that utilized to support filling operations.

Personnel working in the compounding area typically wear a coverall (which may be sterilized for contamination control as required), with head/beard covers, as well as dust masks and sterile gloves. Additional personnel protective equipment may be necessary for some of the materials being processed. A fresh gown should be donned upon each entry into the compounding area. Separate gowning/degowning rooms should be provided to minimize cross-contamination potential for personnel working with different materials. As nonsterile compounding areas are often ISO 6–7 environments but are not aseptic, the more rigorous contamination controlling designs required of aseptic gowning areas (see below) are somewhat reduced.

2.1.3.4 Aseptic Compounding Area (If Present)

Where products are filled using in-line filtration direct to the filling machine, an aseptic compounding area may not be present. In those instances the final sterilizing filter will be located in the fill room.

Products that are held/processed in sterilized vessels prior to filling require an aseptic compounding area. This is typically an ISO 7 in environment with localized

ISO 5 unidirectional flow present where open-product containers or aseptic operations are conducted. Some products may require larger ISO 5 suites with full HEPA coverage rather than the more common ISO 5 clean booth design. Fixed vessels in this area are cleaned and sterilized in situ, while portable vessels are typically relocated to the wash area for cleaning. Sterilization of portable vessels may be accomplished at an SIP station in the aseptic core, compounding, or preparations areas. When accomplished outside the aseptic processing area, resterilization of the connecting lines may be appropriate. Filters for sterilization of solutions from compounding to holding vessels are typically located in this environment as well, with sterilization by either SIP or sterilization in an autoclave. The use of integrated, programmable logic controlled (PLC) filter skids with automatic CIP/SIP and filter integrity testing is frequently seen for contamination sensitive products.

Depending upon the formulations being produced, additional sterilized processing equipment may be present in this area for use in the process. This can include in-line homogenizers, static mixers, and colloid mills. Where sterile powders are produced, the aseptic compounding processes can include blending, milling, and subdivision equipment.

Aseptic compounding areas typically require a means to introduce sterile equipment, tubing, and other items, so access to a sterilizer is desirable. The aseptic compounding area may be contiguous to the aseptic filling suites. If it is not, separate gowning areas must be provided for personnel as well as separate air locks/pass-throughs (see below).

Personnel working in aseptic compounding wear full aseptic garb: sterile gown, hood, face mask, goggles, foot covers, and gloves. Adaptations may be necessary for potent/toxic compounds to assure operators are properly protected from hazardous materials. Gowning areas are ordinarily shared with aseptic filling, but where they are not shared a comparable design, albeit on a smaller scale, is appropriate.

The facility design features match that of the aseptic filling room/aseptic processing areas described in greater detail below. Utility services would mimic those utilized in the nonsterile compounding area that is usually adjacent (next to or above) to the aseptic compounding area. Temperature and humidity should be controlled to similar levels as those required for aseptic filling. Since CIP/SIP systems tend to generate heat and humidity, sufficient capacity must be available to control temperatures to approximately 18–20°C and <50% relative humidity (RH).

2.1.3.5 Aseptic Filling Rooms and Aseptic Processing Area*

The filling of aseptic formulations (and many terminally sterilized products as well, by reason of their lesser number) is performed in an ISO 5 (Class 100) environment, which is accessed from an ISO 6/7 background environment in which personnel are present. Some measure of physical separation is provided between the ISO 5 and ISO 6/7 environments as a means of environmental protection as well as a reminder to personnel to restrict their exposure to ISO 5.

*This section describes the conventional manned clean room; a later section in this chapter will address alternative aseptic processing environmental control designs with somewhat different features and control measures.

In large operations an aseptic filling room is generally one of a multiple suite of aseptic rooms which allow simultaneous production of multiple products. The filling rooms are independent of each other; however, sharing the supporting rooms is common. Sterilizer unload rooms, corridors, air locks, storage rooms, lyophilizer loading rooms, and gowning rooms (each will be briefly described as well) may all be present, and their arrangement must suit production volumes. Where shared common areas are required, the design should feature unidirectional materials flow to prevent cross-contamination and to minimize the potential for mix-ups. In the smallest facilities, only the gowning area might be separate from the fill room, and all of the supportive activities could be inclusive in a single room (however, unloading activities should not occur during filling operations). All of these aseptic processing areas (APAs) are built to the same design standards: smooth, impervious ceilings, walls and floors, flush-mounted windows, clean room door designs, coved corners, finishes capable of withstanding the aggressive chemicals utilized for cleaning and sanitization. Air returns throughout the APA are located at or near floor level. Unidirectional airflow is provided over all exposed sterile materials, that is, fill zone, sterilizer/oven/tunnel unload areas, and anywhere else sterile materials are exposed to the environment. Air changes in these ISO 5 environments can approach 600 per hour, though lesser values have proven successful. Air changes in the background environment vary from 60 to 120 per hour.

The glass container fill rooms filling machines are connected to depyrogenating tunnels and exit ports leading to capping stations. Batch handling of glass is discouraged unless isolator systems are employed. In some operations, the in-feed and discharge of containers/components may utilize trays, tubs, or bag systems for material feed/discharge. Wherever possible, automation of component feeding should be considered to reduce contamination risk. Supportive equipment present might include carts, weigh stations, stoppering, crimping, sealing, and other fill system related machinery depending upon requirements.

The product contact surfaces in this environment are typically removed for cleaning; however, in some installations, the sterilization, transfer, and reinstallation of the component feed hoppers present such difficulty that these systems are decontaminated in situ with a sporicidal agent, rather than removed after each use. These units should still be removed for cleaning and sterilization on a validated periodic basis to prevent the buildup of residues that might impact their in-situ decontamination or create particle control problems. All other product contact surfaces should be sterilized prior to each use. Nonsterilized items should not be allowed to enter the ISO 5 portion of the fill zone, and sanitization is essential for all nonproduct surfaces in the fill zone, as well as the surrounding background environment.

Discharge of sealed containers can be accomplished via a exit port or "mouse hole" that allows for the passage of the containers from the APA to the surrounding environment. Proper design of the mouse hole system ensures protection of the classified fill area from contamination flowing against the flow of the containers. In many instances the discharge is into a nonclassified inspection area that may lead directly to the secondary labeling/packaging area.

Personnel working in aseptic compounding wear full aseptic garb: sterile gown, hood, face mask, goggles, foot covers, and gloves. Adaptations may be necessary for potent/toxic compounds to assure operators are properly protected from hazardous materials.

2.1.3.6 Capping and Crimp Sealing Areas

The application of aluminum seals over rubber stoppers is essential to secure them properly. In many older facilities this was accomplished outside the aseptic processing area in an unclassified environment. Current practice requires that air supplied to this activity meet ISO 5 under static conditions. The protection of crimping has resulted in a variety of designs to meet the requirement: Sterile crimps can be applied with the aseptic core on the filling line; sterile crimps can be applied in a separate crimping room accessible from the filling room. If the crimpling operation is located within the APA, it should be in a separate room maintained at a negative pressure differential relative to the filling environment. Crimping may alternatively be performed in a classified room accessed from a controlled but unclassified environment. In this case it is imperative to verify that the environmental controls satisfy regulatory expectations for all relevant markets.

2.1.3.7 Sterilizer Unload (Cooldown) Rooms

Sterilizers/ovens are unloaded and items staged prior to transfer to the individual fill rooms. ISO 5 air is provided over the discharge area of ovens (and autoclaves if items are sterilized unwrapped) to provide protection until the items are ready for transfer. The heat loads in this room may be such that special high-temperature sprinkler heads may be necessary to avoid unintentional discharge when unloading hot materials. This room may not be separate from the corridor used to connect the fill rooms. It is ordinarily adjacent to any aseptic storage area.

2.1.3.8 Corridors

Corridors serve to interconnect the various rooms that comprise the APA. Fill rooms, air locks, and gowning rooms are accessed from the corridor. They can also be utilized for modest storage as well.

2.1.3.9 Aseptic Storage Rooms

In general, extensive use of in-process storage areas should be avoided. It is best to operate the aseptic facility in a just-in-time mode in which components and equipment are sterilized shortly before they are required for use in the filling or compounding areas. Some limited storage is necessary for nonproduct contact materials such as sanitizing agents, environmental supplies and equipment, and other items.

2.1.3.10 Lyophilizer Loading and Unloading Rooms

The loading of lyophilizers is accomplished under ISO 5 environmental conditions within the aseptic processing area. Several possible locations are possible: within the aseptic fill room itself, in a separate room adjacent to the fill room, or in a separate room remote from the fill room. There are pros and cons with each of these selections which should be carefully considered in the facility design. There is a universal expectation that filled containers of product should be maintained under ISO 5 conditions during transfer and lyophilizer loading. Many modern facilities incorporate automatic lyophilizer loading and unloading. Automation of loading, unloading, and in the case of vials transfer to the crimping station greatly reduces contamination risk and is highly recommended.

If manual transfer is unavoidable, location of the lyophilizer relatively close to the filling line enables protected transfer to be accomplished rather easily. Remote locations may require transfer of product in carts capable of providing ISO 5 quality air. These carts will generally require battery power in order to run the necessary air blowers and control systems. Alternatively, product trays could be placed in airtight carriers; this activity and the sealing of the carriers would have to be accomplished under ISO 5 conditions. Locating the lyophilizer in the fill room may restrict the ability to unload the dryer while the filling line is in use, particularly if the lyophilizer is loaded and unloaded manually, which would increase the clean room personnel load and potentially increase contamination risk.

The use of trays during lyophilization is less common, nevertheless, ring trays with removable bottoms are sometimes used to transfer vials to/from the lyophilizer. Where trays are used, they must be cleaned and sterilized prior to each batch. Large lyophilization facilities will sometimes use an automated loading/unloading system in which all shelves or a shelf at a time are processed. Regardless of the practice, ISO 5 conditions are required for all areas of the facility in which partially stoppered containers are transferred or handled. As previously mentioned, it may be possible in some operations to transfer containers in a manner that they are not exposed to the environment during transfer.

Upon completion of the drying process, the containers will ordinarily have their stoppers fully seated on the container within the freeze dryer. The stoppered containers are then passed through a sealing station in which aluminum crimps are applied. This may be accomplished on the fill line, or using a separate crimping machine. Precautions will need to be taken to ensure that only fully stoppered vials are transferred to the crimping station. This can be accomplished by automatic inspection systems of various designs. It is increasingly common for product transfer to crimping and crimping itself to be done under unidirectional airflow. It should be noted that a crimpling station will generally not meet ISO 5 particulate air quality requirements when the crimper is operating since the generation of relatively high levels of particulate is an inherent feature of this process.

2.1.3.11 Air Locks and Pass-Throughs

Air locks serve as transition points between one environment and another. Ordinarily, they are designed to separate environments of different classification: that is, ISO 6 from ISO 7. When this is the case, they are designed to achieve the higher of the two air quality levels in operation. If they are utilized for decontamination purposes for materials/equipment that cannot be sterilized, but must be introduced into the higher air quality environment, they may be fitted with ultraviolet (UV) lights, spray systems, vapor phase hydrogen peroxide generators, or other devices that may be effectively utilized for decontamination of materials. Regardless of the design or the decontamination method employed, the process should be validated to ensure consistent efficacy. The doors at each end can be automatically interlocked or managed by standard operating procedure. In some instances a demarcation line is used to delineate the extent to which individuals from one side should access the air lock. It is good practice to carefully control and to minimize the time that any operator spends accessing an air lock, therefore transfer of materials should be carefully planned to minimize frequent and spontaneous access. Additionally, the capacity of the air lock should be carefully considered relative to the actual production requirements. Air locks that lack sufficient capacity and that cannot provide sufficient air exchange will be less suited to the control of contamination into more critical areas of the aseptic processing environment.

A smaller scale system with comparable capabilities is the pass-through. This differs from the air lock primarily in dimension, as items are typically placed into the pass-through by personnel, whereas the air lock is customary for pallet, portable tanks, and larger items that are either rolled or mechanically lifted into position. The operation of the pass-through can be either manual or automatic with similar capabilities to that of the air lock described above. In general pass-throughs should be supplied with HEPA filters and should be designed to meet the air quality level of the higher air quality classification room served. Pass-throughs should also be interlocked and provide adequate facilities for decontamination of materials being transferred.

Air locks and pass-throughs are bidirectional and can be used for movement in either direction. When used as an exit route, the decontamination procedure can be omitted. Where production volumes warrant separate entry and exit, air locks may be necessary to maintain both adequate capacity and separation between clean and used items. In an emergency, airlocks can serve as emergency exits for personnel, in which case the interlocks can be overridden.

2.1.3.12 Gowning Rooms

The gowning area used for personnel entry/exit presents some unique problems. Gowning facilities must be designed to the standards of the aseptic processing area, yet personnel upon entry are certainly not gowned. Because ungowned staff will release higher concentrations of contaminants into the environment, gown rooms must be designed with sufficient air exchange so that this contamination is effectively and promptly removed. In general, the contamination load within a gowning environment will require air exchange rates at the high end of recommended levels for a given ISO 14644 air quality classification. Gowning areas are separated into well-defined zones where personnel can progress through the various stages of the gowning process.

The most common approach in industry is a three-stage gowning area design in which three linked rooms with increasing air quality levels are utilized to efficiently and safely affect clothing change. Staff should enter the first state of the gowning room wearing plant uniforms. No articles of outerwear worn outside the facility should be worn to the gowning area. Therefore, a pregowning room equipped with lockers is required so that operators can change into dedicated plant clothing prior to moving to the gowning area. Generally, the pregowning locker area is not classified, although entry is controlled and temperature and humidity are maintained at 20–24°C and 50% ± 10%. The pregown area should have extensive hand-washing facilities equipped with antibacterial soap, warm water, and brushes for cleaning finger nails. Soap and water dispensing should be automatic and hands should be air rather than towel dried. The pregown area should have typical clean room wall and floor finishes along for frequent and rigorous cleaning and sanitization. The pregown area is bidirectional as it is used as both an entry and exit point. Separate pregown areas are required for female and male personnel. A typical complement of garments for exit of the pregown area includes surgical scrubs or other nonparticulate shedding plant uniform. Ideally, the uniform should have a high neck and sleeves which extend to the lower wrist. Hair covers and beard covers are donned in the pregown area.

Upon entry into the first-stage gowning room, which is generally designed to an ISO 7 air quality level, the operators often don a second hair cover, sterilized gloves, and a sterilized surgical mask. In the second and third stages of the gowning area room classification is typically ISO 6 or ISO 6 followed by ISO 5 at the exit point. Different firms have different gowning sequences. However, in every case the flow of personnel and arrangement of gowning materials should be such that personnel flow is in one direction. In the last of the three gowning stages, secondary protective equipment can be donned, including sleeve covers and a second set of gloves. Some firms will use tape to secure the gloves to the sleeves to prevent separation. A dry glove decontamination point utilizing disinfectant foam is generally provided prior to exiting the gowning area; this should be a hands-free operation. In some facilities air showers, which provide a high-intensity blast of HEPA air for a predetermined length of time, are employed after gowning is completed. Side-by-side gowning of personnel should be avoided to preclude adventitious contamination. Similarly, personnel exiting the aseptic area should use a separate degowning area. These design practices are appropriate in all but the very smallest facilities where only a single aseptic operator is present.

2.1.3.13 Terminal Sterilization Area

The terminal sterilization of finished product containers may be performed in the same sterilizers utilized to supply the aseptic processing operations. The differing process needs of terminal sterilization will sometimes dictate the use of sterilizers specifically designed for terminal sterilization incorporating air-over pressure systems, internal fans, and spray cooling. Where this is the case, the terminal sterilizer is located proximate to the crimping/sealing areas. A double-door sterilizer design is preferred with staging areas for filled containers to be sterilized and a separate area for containers that have completed the process. Classification of these areas is not required as the containers are closed throughout the sterilization process. The flooring materials in this area should be monolithic to allow for easy cleanup in the event of container breakage.

2.1.3.14 Inspection, Labeling, and Packaging

These activities are performed on finished product containers in unclassified environments. The primary design requirements are straightforward: separation of products to prevent mix-up, adequate lighting for the processes, and control over labeling materials.

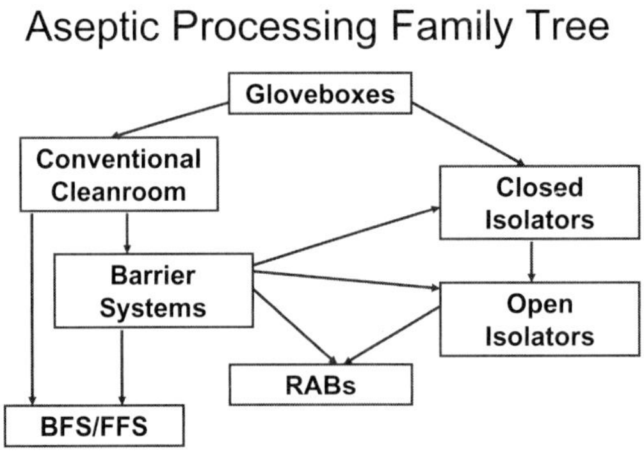

FIGURE 1 Aseptic processing family tree.

2.1.4 ASEPTIC PROCESSING FACILITY ALTERNATIVES

The successful production of parenteral drugs by aseptic processing requires an environment in which microorganisms and particles are very well controlled. The means to accomplish this has undergone substantial change over the last 50 years (see Figure 1) with continuing refinement. The earliest aseptic processing systems used glove boxes with minimal (if any) airflow and manual disinfection in which manual processes were performed. The availability of HEPA filters in the late 1950s led to human-scale clean rooms in which processing equipment could be installed. Aseptic processing changed radically once entire clean rooms became feasible.

As it had always been recognized that personnel were the dominant source of contamination, the majority of designs utilized some measure of physical separation between the operator and the critical zone (sterile field) in which the aseptic processing activities were performed. Separative devices (a term that is now embodied in ISO 14644-7 Separative Enclosures) of different design and varying capability have been successfully employed including flexible curtains and fixed plastic shields with or without integrated gloves/sleeves [9]. In the most evolved designs operation of the equipment is interlocked with the surrounding enclosure, such that equipment stops running when the doors are opened. These latter designs represented the pinnacle of clean room-based aseptic processing into the early 1990s.

Isolators represent a return to operator separation principles utilized during the glove box era, albeit with substantial improvements in the form of rapid transfer ports for material transfer, air-handling systems utilizing modern HEPA filters, and reliable decontamination systems. The salient element of all isolator designs is the completeness of separation between the internal and external environments. This single feature affords vastly superior performance relative to manned clean rooms in excluding personnel-derived contamination and has comparable advantages for the containment of potent compounds. While initial adoption of the technology was slowed by the novelty that isolators presented to users, much of the initial reluctance has been overcome [10, 11]. Isolators for aseptic processing vary in complexity, size, and amount of processing equipment. They can be utilized for processing ranging from manual compounding of small batches to high-speed filling of final product containers. Depending upon the process requirements, isolators can be utilized for containment of potent compounds (under negative pressure while still nonsterile) during the compounding, aseptic operation (under positive pressure) for preparation and transfer of components and aseptic containment (also under positive pressure) for aseptic filling of the potent drug solution.

Firms that were intimidated by or unconvinced of the superiority of isolators developed the restricted-access barrier (RAB) system as a potentially less complex and less costly alternative [12]. The real-world utility of RABs systems is unknown; there are still relatively few installations; thus, the experience base is still emerging. Also unconfirmed at this point are the actual validation and ongoing process control requirements which make direct comparison of project time lines and overall costs with isolators somewhat speculative.

There are specialized technologies such as BFS and FFS that are appropriate for aseptic processing, but these are restricted to filling processes only. A number of other new technologies are being developed for use in aseptic processing, including vial isolators and closed vial filling [13–15]. All of these have the objective of reducing contamination through reduction in human involvement or increased protection of the container. Further advances in processing including gloveless isolator designs, robotics, and others are already under active development to further improve the safety of parenteral products.

2.1.4.1 Expandability

Large facilities often include design elements that facilitate later expansion of the facility to add additional capacity. The most common of these is extension of an aseptic corridor to additional filling suites; reservation of space for additional sterilizers; and allocating space for additional or oversizing initial utility systems. Obviously, these types of changes require careful design and must be properly managed during execution to avoid impact on existing operations.

Isolation technology changes this dynamic significantly by eliminating most of the disruption on current activities, as fabrication of the isolator occurs off-site, and installation can be minimally disruptive compared to what is required with a cleanroom design. Isolators are generally installed in ISO 8 space; therefore, it is possible to build a rather large ISO 8 facility in which equipment can be moved, replaced, or reconfigured quite easily compared to conventional human-scale zoned aseptic processing areas.

2.1.5 UTILITY REQUIREMENTS

Any utility in direct product contact is subject to formal qualification through confirmation of the quality of the delivered material at each use point. Water-for-injection (WFI) systems are considered the most critical of all, and the qualification period for WFI is the longest and may be as long as 3 months. The remaining product contact utilities can be qualified more rapidly. Nonproduct utilities requirements can be satisfied by commissioning.

2.1.5.1 Water for Injection

The most important utility in sterile manufacturing is WFI. Not only is it a major component in many formulations, it is also utilized as a final rinse of process equipment, product contact parts, utensils, and components. In some facilities it may be the only grade of water available and is used for initial cleaning of items as well. The WFI may be produced by either distillation (multiple effect or vapor compression) or reverse osmosis (generally in conjunction with deionization) and is ordinarily stored and recirculated at an elevated temperature greater than 70°C to prevent microbial growth [16, 17]. Where cold water is required, it may be supplied by use point heat exchangers or using a separate cold loop (usually without a storage capability). Point-of-use cool water drops and reduced temperature circulation loops are generally sterilized or high-temperature sanitized at defined and validated intervals. The design details of the WFI system varies with the incoming water quality, local utility costs, and operational demands. Very small operations may not have a WFI system and will utilize larger (5 L or larger) packages of WFI for formulation and cleaning.

Other grades of water may be present in parenteral facilities for use as initial rinses and detergent cleaning. The water utilized for these purposes is generally of relatively low bioburden and is often deionized, softened, ultra-filtered, or in some instances prepared by distillation or reverse osmosis, resulting in chemical purity similar to, if not identical to, WFI. Systems for the preparation of this water are subject to qualification, validation, and routine analysis to assure consistent quality.

2.1.5.2 Clean (Pure) Steam

Sterilizers and SIP systems in the facility are supplied with steam which upon condensation meets WFI quality requirements (testing steam condensate for microbial content is not fruitful). The steam can be produced directly from the water of sufficient purity to meet the input requirements of the steam generator. Steam generators are phase transition technologies that operate like a still, so it is no more necessary to provide these devices with WFI feed water than it would be to double distill WFI. (Production from WFI is certainly possible, but that is both expensive and an unnecessary precaution.) Modest quantities of steam can be produced from the first effect of a multiple effect WFI still, however, with a resultant loss of WFI output [18].

2.1.5.3 Process Gases

Air or nitrogen used in product contact is often supplied in stainless steel piping and ordinarily equipped with point-of-use filters; quite often an additional filter is placed within the distribution loop or at the entry point into a room resulting in a form of redundant filtration. Compressed air is typically provided by oil-free compressors to minimize potential contaminants and is often treated with a drier to obviate the possibility of condensation within the lines which could be a source of contamination. Nitrogen is supplied as a bulk cryogenic liquid. Argon and carbon dioxide have also been utilized as inerting gases, while propane or natural gas may be needed for sealing of ampoules.

2.1.5.4 Other Utilities

The operation of a parenteral facility often entails other utilities for the operation of the equipment. These include plant steam, jacket cooling water, and instrument air.

2.1.6 STERILIZATION AND DEPYROGENATION

The preparation of the drug formulation, components, and equipment entails the use of various sterilization/depyrogenation treatments to control bioburden, avoid excessive pyrogens, and to sterilize. The selection of the specific process must always fully consider the impact of the treatment on the items being sterilized/depyrogenated. Sterilization and heat depyrogenation processes must balance the effect of the treatment on the microorganism with the effect of that same treatment on the materials being processed. The choice of one method over another is often based upon achieving the desired sterilization/depyrogenation effect with minimal impact on the items critical quality attributes.

2.1.6.1 Steam Sterilization

The method of choice in nearly every instance is moist heat due to its lethality, simplicity, speed, and general ease of process development and validation. For the majority of items, this is accomplished in a double-door steam sterilizer, which is conventionally located between the preparations and aseptic processing (filling or compounding) areas. Steam sterilizers are routinely utilized for items such as elastomeric closures, process and vent filters, product contact parts, heat stabile environmental monitoring equipment, tools and utensils, hoses, sample containers, and other items unaffected by contact with saturated steam at commonly used sterilizing temperature and pressure [19]. Similar items utilized in the nonsterile compounding area would be processed in a similar manner. Regardless of their final destination or usage, items for steam sterilization should be protected from post-sterilization contamination by materials that are permeable to steam, air, or condensate but impenetrable by microorganisms. The wrapping materials would be maintained on the sterilized items until just prior to use. There are numerous publications that provide additional details on steam sterilization procedures [19–21].

Sealed containers of aqueous solutions, suspensions, and other liquids can be processed through steam sterilizers as well. These liquids might be used in formulation or cleaning procedures, and sterilization in this manner may be more efficient and more reliable than sterilizing filtration. Larger volumes of aqueous liquids are often sterilized in bulk using a jacketed and agitated pressure vessel (the vessel is usually rated for full vacuum as well).

Steam SIP is a widely used practice for the sterilization of equipment prior to the introduction of process materials and is the method of choice for holding tanks, process transfer lines, lyophilizers, and other large items. Conceptually, it has many similarities to sterilization in autoclaves but differs markedly due to the often custom designs of process equipment requiring SIP. Systems must be designed with careful consideration given to air removal and condensate draining, process sequencing, and poststerilization integrity to assure success [22]. Terminal sterilization of finished product containers is addressed later in this chapter.

2.1.6.2 Dry-Heat Sterilization and Depyrogenation

The use of dry heat for depyrogenation (and sterilization) is almost universal for glass containers. Temperatures of 250°C or higher are utilized to render the glass endotoxin free. The depyrogenation is necessary because the washing of glass to reduce particles can introduce unacceptable levels of gram-negative microorganisms whose presence could result in pyrogen formation. The depyrogenation process can assist in component surface treatment (siliconization is required for some formulations) and will also render the glass sterile as well (depyrogenation temperature conditions far exceed those needed for sterilization [23]).

Sterilization by dry heat is only infrequently used, preference being given to the use of steam (due to its higher speed) or dry-heat depyrogenation (affording an added measure of safety using the same equipment). Where it is employed temperatures in the range of 170–180°C are employed, and a batch oven is customarily used.

Dry-heat processes are conducted in either batch ovens or continuous tunnels, which are also installed between preparations and aseptic processing areas. Ovens have lower capacity and are typically found in smaller facilities. They offer the ability to handle items other than final product containers and thus can replace autoclaves in facilities where filling parts, feed hoppers, tools, and other items that must be extremely dry. Ovens should be equipped with internal HEPA filters, recirculating fans, heating/cooling coils, and a sophisticated control system [24]. Items prepared for dry-heat treatment in ovens are inverted or covered to protect them after exiting from the oven as there are no sealed protective systems suitable for the higher temperatures necessary for dry-heat depyrogenation or sterilization. Oven discharge is typically into a cool-down area (usually the same as that used for the sterilizer), though in small facilities it might discharge directly into the fill room. Unless ovens are used in conjunction with isolators, they require direct operator intervention to transfer containers to the filling line and to charge the line with depyrogenated glass. This constitutes a risky intervention which should be avoided. For this reason, batch glass processing is rare in all but the lowest throughput facilities.

Dry-heat tunnels are typically utilized where the production volumes are higher and allow for continuous supply of depyrogenated glass to the aseptic fill room. Tunnels are operated at high temperatures (>300°C) to increase processing speed and include a cooling zone that facilities discharge at or near room temperature. Typically, heating of the glass to 300°C or more for 3 or more minutes will result in much greater than the three-log endotoxin reduction required in current industry standards. The air inside the tunnel is HEPA filtered, and newer designs allow for dry-heat sterilization of the cooling zone as an added protective measure. Tunnels must be positioned with some care as they ordinarily will terminate into a fill room. A pressure differential between the cooling zone of the tunnel and the fill room is critical for proper operation of the tunnel. The pressure differential must conform to the requirements stipulated by the tunnel manufacturer. It is not necessary to have a >12.5 PA (particulate air) differential between the in-feed side of the heating zone of the tunnel and the exit side of the cooling zone. It has been suggested by some that, since the in-feed side of the tunnel is typically in ISO 7 or 8 space, a greater differential is required; however, this is not true since the cooling zone is ISO 5, and the heating zone is certain to be sterile and is also ISO 5 in terms of particulate air quality. Their in-feed is often direct from a glass washer, which may be remote from the main preparations area utilized for washing, wrapping, and sterilizer loading.

2.1.6.3 Gas and Vapor Sterilization

The sterilization of materials using noncondensing gases (ethylene oxide, chlorine dioxide, or ozone) or condensing vapors such as hydrogen peroxide is a supplementary process intended for items that cannot be exposed to heat. The utilization of gas/vapor designs is coming into increased use as a supportive technology for isolation technology for presterilized items such as syringes and stoppers that must be introduced into the isolators aseptic zone. Air locks using these agents can be utilized in similar fashion for the supply of materials to manned clean rooms. Control over agent concentration or injection mass, relative humidity, and temperature may be required for these systems. There are different types of vapor processes available, and users should generally follow the cycle development strategy suggested by the manufacturer of the equipment they have chosen. Specific temperature and humidity ranges may be required for some vapor processes to assure appropriate efficacy [25, 26].

2.1.6.4 Radiation Sterilization

The use of radiation within a parenteral facility would have been considered unthinkable prior to the start of the twenty-first century. While γ irradiation is typically a contracted service provided off-site, electron beam sterilization advances can make the installation of an in-house (and generally an in-line) system a real possibility. An in-line system would be utilized similarly to the gas/vapor systems described above for treatment of external surfaces for entry into either a clean room or isolator-based aseptic processing facility. The use of this same technology for terminal sterilization is also possible [1]. Association for the Advancement of Medical Instrumentation (AAMI)/ISO 11137 provides widely accepted guidance on the development and validation of radiation sterilization processes.

2.1.6.5 Sterilization by Filtration

Filters are utilized to sterilize liquids and gases by passage through membranes that retain microorganisms by a combination of sieve retention, impaction, and attractive mechanisms [27]. In contrast with the other forms of sterilization that are destructive of the microorganisms, filters rely on separation of the undesirable items (microorganisms as well as nonviable particles) from the fluid. Because filtration requires passage of the fluid from the "dirty" (upstream) side of filter to the clean (downstream) side of the filter, the downstream piping and equipment must be both "clean" and sterile prior to the start of the filtration process. This will ordinarily require the use of SIP procedures or sterilization followed by aseptic assembly.

Sterilizing filtration of parenterals is a complex and often inadequately considered subject, and numerous controls are required on the filter, fluid, and sterilizing/

operating practices employed. PDA Technical Reports 26 and 40 can be instructive in understanding the relevant concerns [28, 29].

2.1.7 FACILITY AND SYSTEM: QUALIFICATION AND VALIDATION

Facilities for the manufacture of sterile products require the qualification/validation of the systems/equipment and procedures utilized for that production. Each system described above and others with a direct/indirect impact on the quality of the products being produced should be placed into operation using a defined set of practices. The general approach is described below, and best practices include the development of traceable documentation from project onset. The preferred approach begins during a project's conceptual design phase where provisions for meeting the CGMP expectations and user requirement specifications establishing the technical basis for the processes are first defined. This is commonly followed by the validation master planning exercise in which the user requirement specifications are used as a basis for the development of acceptance criteria for process control studies. This effort should be accompanied by an analysis of risk that considers product attributes, target patient population, as well as technical and compliance requirements. Detailed design follows in which the specifics of the various systems are refined. Construction of the facility and fabrication of the process equipment follows and a variety of controls are necessary during these activities to satisfy user requirements for compliance of the various elements of the facility. Typically, factory acceptance testing (FAT) will be done on all key process equipment, usually at the manufacturer's plant site; much of the information gathered during FAT can be referenced in the qualification activities to follow. Physical completion is followed by a well-defined step termed commissioning in which construction and fabrication errors and omissions are addressed. Site acceptance testing of installed process equipment may be done in parallel with facility commissioning. Formal qualification of the facility ensues in which the installed systems and equipment are evaluated for their conformance to the design expectations. The very last steps in this process are variously termed performance qualification. Detailed discussion of these subjects is not possible within the constraints of this chapter, however the qualification/validation of equipment, systems, and processes has been extensively addressed in the literature [30].

2.1.8 ENVIRONMENTAL CONTROL AND MONITORING

Confirmation of appropriate conditions for aseptic processing and its supportive activities is required by regulation. In the highest air quality environment utilized for aseptic processing, ISO 5, there is a general expectation that the air and surfaces be largely free of microbial contamination and the number of particles be within defined limits (less than 3500 particles greater than 0.5 $\mu m/m^3$). Proving the complete absence of something is an impossible requirement, so the usual expectation is that 99+% of all samples taken from this most critical environment be free of detectable microorganisms. The minimum monitoring expectations for these environments as defined by the regulators are consistently attainable in nearly all instances, especially those with lesser expectations. This is accomplished by proper design, periodic facility disinfection, and measures to control the ingress of microorganisms and particles for materials entering each environment from adjacent less clean areas [31].

2.1.8.1 Sanitization and Disinfection

Disinfection is customarily performed by gowned personnel during nonoperating periods using such agents as phenolics, quaternary ammonium compounds, aldehydes, and other nonsporicidal agents. The frequency of treatment varies with the ability of the facility to maintain the desired conditions between disinfection. Sporicidal agents such as dilute hydrogen peroxide or bleach are reserved for those occasional periods when control over the spore population warrants and is often employed after lengthy maintenance shutdowns or at the end of construction. Isolation technology replaces the manual disinfection with reproducible decontamination with a sporicidal agent and thus assures a superior level of environmental control as compared to manned environments. The manual treatments fall short of this level of control due to the uncertainties of the manual procedure and recontamination of the environment as a consequence of the very personnel and activities utilized to disinfect it. To mitigate these weaknesses, automatic sporicidal disinfection of manned clean spaces has been developed by multiple vendors. Disinfection of the less critical environments is accomplished in the same manner albeit on a less frequent interval befitting their higher allowable levels of microorganisms.

2.1.8.2 Monitoring

Aseptic environments are subject to a variety of monitoring systems including air, surface, and personnel monitoring for viable microorganisms and for nonviable particles. Environmental monitoring programs are often developed during the qualification of a new facility using a multiphase approach. Methods for the monitoring and expectations for performance have been extensively discussed in the literature and will only be addressed briefly in the context of this chapter [1, 2, 31, 32]. In general, the frequency and intensity of monitoring and concern for cleanliness increases as the product progresses from preparation steps (typically in ISO 7/8 environments) to more important activities (nonsterile compounding in ISO 6) and ultimately into the aseptic core (aseptic compounding and filling in ISO 5). Sampling site and time selection should be a balance between the need to collect meaningful data and avoidance of sampling interventions that could adversely (and inadvertently) impact product quality. Microbiological sampling must always be done by well-trained staff utilizing careful aseptic technique. This will both minimize risk to the product and also improve the reliability of the data by reducing the likelihood of false-positive results.

Air Sampling The relative cleanliness of air in the most critical environment is assessed using passive sampling systems such as settle plates or estimated volumetrically using active air samplers. Active air samplers should be designed to be isokinetic in operation to avoid disruptions to unidirectional airflow. Considerable variability has been reported among the several sampling methods employed for active air sampling, and there are also reports that active air sampling may have advantages in terms of sensitivity. Passive sampling using settle plates can be a useful adjunct in critical areas with limited access and where an active sampler might interfere with airflow or entail a worrisome intervention risk. It must be recognized that attempts to support the "sterility" of the cleanest aseptic environments (those in ISO 5) by aggressive sampling may have exactly the opposite effect. Sampling too frequently will increase process contamination risk by causing critical interventions that are best avoided within these very clean environments. As personnel are the greatest single source of microbial contamination and conduct the sampling, sampling intensity should be carefully considered. There is no value to taking air samples beyond those required to assess the relative cleanliness level within the environment.

Surface Sampling Surfaces in the classified environments are monitored using a variety of methods but most commonly with contact plates (on smooth surfaces) or swabs (for irregular surfaces). Surface sampling in aseptic environments (ISO 5/6) is typically performed after the completion of the process to avoid the potential for adventitious contamination of the production materials as a consequence of sampling activities during the process. Fortunately, studies indicate that contamination does not build up during typical processing operations in modern clean rooms. Sampling with these materials may leave a trace of media or water on the sampled surface, and cleaning of the surface immediately after sampling is commonplace. Sampling of product contact surfaces (i.e., fill needles, feeder bowls, etc.) should only be performed after completion of the process, and the results of this testing should **not** be considered as an additional sterility test on the products. As in any form of manual environmental sampling, the risk of contamination by samplers during the processing of a sample makes the data less than completely reliable. Sampling of surfaces such as walls and floors should not be overdone because with good attention to aseptic technique they should be of little concern relative to actual process risk. Sampling on these surfaces is probably most useful in assessing ongoing changes in microflora and to confirm the adequacy of the disinfection program.

Personnel Sampling The monitoring of personnel gown surfaces is an adaptation of surface sampling in which samples are taken from surfaces on the operator. In ISO 5 environments, this ordinarily entails the gloved hands and perhaps forearms. As with any other sampling of a critical surface (the gloved hand is often in closest proximity to sterile product contact surfaces and sterilized components), the sampling should be performed at the conclusion of the aseptic activity. Sampling during the midst of the process risks contamination of the product and should be avoided. Sampling of other aseptic gown surfaces is ordinarily restricted to gowning certification or postmedia fill testing, where more aggressive sampling can sometimes be informative. Whenever a gowned individual is sampled, the sample should be taken in the background environment (not ISO 5), and the individual should immediately exit and regown before continuing any further activity in the aseptic core area. Sampling of personnel in less critical environments can be useful; however, meeting regulatory expectations in these areas is ordinarily straightforward. Recommended contamination levels often distinguish among the different room classification levels found within clean rooms. While this may seem reasonable, it is not completely logical since operators often move frequently between these different levels of classification during the conduct of their work.

Total Particulate Monitoring Confirming the ability of the facility's heating, ventilation, and air-conditioning (HVAC) system to maintain the appropriate conditions throughout (to the extent practical) the classified environments is most easily accomplished using electronic total particle counters that can provide near immediate feedback on conditions during production operations. Total particle samples can be taken automatically, using permanently installed probes oriented into the unidirectional airflow. As such, they can be positioned proximate to critical activities to reaffirm the continued quality of the air in the vicinity of the sterile materials and surfaces. Manual total particulate air sampling can be a dangerous intervention and therefore if required should be timed so as to minimize risk to product. Attempts to correlate total particle counts with microbial counts have proven difficult. Correlations are only meaningful when the source of foreign material is personnel since people are the only source of airborne contamination within an aseptic processing area. When personnel are the only source of particulate, the ratio between viable and nonviable particles have been consistently found to be >1000:1, which means that in ISO 5 environments even relatively large total particulate count excursions would typically contribute microbial contamination that fell far below the limit of detection. Process equipment can and often does contribute airborne particulate matter but not detectable levels of microbial contamination. Also, microbial sampling is highly variable with respect to sensitivity, accuracy, precision, and limit of detection making correlations, particularly in rooms of highest air quality. So, it might seem logical to think that particle excursions are indicative of coincident microbial excursions especially in the cleaner environments (ISO 5) where the aseptic process takes place.

It is common practice for firms to interrupt their aseptic processes when atypical total particulate excursions are observed so that the scientists and engineers can determine the source of the foreign material. Monitoring frequency and expectations in the less critical environments is always reduced relative to the critical aseptic environments.

Where firms have introduced unidirectional air systems in preparations and compounding areas for particle control, there is often the temptation to expect these areas to meet the same microbial limits that these locations might attain in the aseptic core. This temptation should be resisted to avoid unnecessary sampling and deviations associated with expecting these environs to meet the conditions of aseptic areas where sanitization frequency, background environment, and most importantly personnel gowning are far superior to that found in the less clean locales [33].

Housekeeping An important component of environmental control are the housekeeping activities utilized to clean the facility external to the controlled environments. Aseptic operations utilize a series of protective environments to protect the sterile field. Controls on the surrounding unclassified areas are an important part of the overall control scheme for sterile manufacturing. These unclassified areas support sterile operations in a variety of ways, and it is important to conduct activities therein that assist in the environmental control. Routine housekeeping, periodic sanitization, and even occasional environmental monitoring may be appropriate to assure that microbial and particle loads on items, equipment, and personnel entering the classified environments is appropriately controlled.

2.1.9 PRODUCTION ACTIVITIES

The preparation of sterile materials requires execution of a number of supportive processes that together constitute the manufacturing process. They are intended to control bioburden, reduce particle levels, remove contaminants, sterilize, and/or depyrogenate. Nearly all of these activities occur within the controlled environments and are subject to qualification/validation.

2.1.9.1 Material and Component Entry

Prior to the start of any production activity, materials and components must be transferred from a warehouse environment into a classified environment. For most items this will necessitate removal from boxes or cartons, transfer to a nonwooden pallet, and passage through an air lock which serves as the transfer system between the controlled and uncontrolled environments. Often components are contained within plastic bags within a box or carton, and in some cases there are multiple bag layers to facilitate disinfection and passage through air locks into different zones of operation within the aseptic area. The firm may utilize an external disinfection of the materials in conjunction with this transfer. The concern is for minimization of particles and bioburden on these as yet unprocessed items in order to protect the controlled environment.

Raw materials may be weighed in a weigh area in which they are transferred to plastic bags and/or noncorrugate containers prior to the transfer. The weighing area provides ISO 7 or better conditions, and may be a dedicated portion of the warehouse proper; in a central weighing/dispensing area; or in a location contiguous to the compounding area. Sterile ingredients are never opened anywhere other than an aseptic environment and must be handled aseptically at all times including sampling and processing of samples.

2.1.9.2 Cleaning and Preparation

Once the container component items have been introduced into the preparations area, they must be readied for sterilization/depyrogenation. For many items this consists of washing/rinsing processes designed to remove particles and reduce bioburden and endotoxin levels. The application of silicone suspensions for glass or closure materials is sometimes employed to provide lubrication allowing smoother feeding of components or dispensing (elimination of product accumulation on vial). Following the cleaning, items for sterilization are dried, wrapped, and staged/stored for steam sterilization. Washed containers are either placed in trays or boxes for depyrogenation in ovens or are directly loaded into dry-heat tunnels. It is common practice to protect all washed items with ISO 5 air from the completion of washing, through either wrapping or placement into a sterilizer or oven for passage into the aseptic area. The intention is to avoid foreign matter that could result in contamination of product.

It is increasingly common for components to be supplied by the vendor in a ready-to-sterilize condition (washed and pretreated as necessary). Some items are available in a ready-to-use configuration with the supplier providing sterile and pyrogen-free components. The use of supplier-prepared items eliminates the need for preparation activities at the fill site and requires modification of material in-feed practices relative to on-site prepared items.

The process equipment (portable tanks, valves, fill needles, etc.) and consumable materials (filters, hoses, gaskets, etc.) are prepared using a variety of methods. Portable tanks are subjected to CIP (and perhaps SIP as well) in the preparation area. Smaller items are disassembled (if necessary) and cleaned either manually or in a cabinet washer. After cleaning they are wrapped and staged/stored prior to sterilization. Tubing should not be reused; its preparation typically consists of flushing with WFI followed by cutting to the required length. It is best to preassemble fill sets with tubing, filters, and fill needles/pumps and then wrap them in preparation for sterilization. This process obviates poststerilization assembly steps and therefore mitigates contamination risk. These steps may be performed in ISO 5 environments to reduce total particulate contamination on the items.

There are items that must be transferred into the aseptic processing area that cannot be treated within a sterilizer/oven. These include portable tanks, electronic equipment, and containers of sterile materials (ready-to-use items, sterile powders, environmental monitoring media, etc.). Air locks, pass-throughs, and similar designs are employed in which the exterior surfaces of the items are disinfected. The disinfection process may be completed by personnel outside and/or inside the aseptic area depending upon the specifics of the design.

At the completion of the cleaning process, the items should be free of contaminating residues including traces of prior products, free of endotoxin, and well-controlled in terms of total particulate and microbial levels. This level of control would be appropriate regardless of whether the items, equipment, or components are to be sterilized or not. Sterilization, other than by relatively high temperature dry heat, has only a modest impact on endotoxin levels; cleaning provides the only means to control endotoxin for materials and equipment that is sterilized by other means.

2.1.9.3 Compounding

Fixed equipment in the compounding area (nonaseptic or aseptic) is cleaned in place. This eliminates traces of prior products, particles, and pyrogens. Sterilization in place is required for the aseptic fixed equipment and is sometimes employed for the nonaseptic equipment as well as a bioburden control measure. Fixed transfer lines must be cleaned and sterilized as well, and this is accomplished independently or in conjunction with the vessels. The reuse of hoses and tubing is discouraged as cleaning and extractables cannot be confirmed beyond a single use.

The preparation of the product is performed within a classified environment with careful attention to the batch record, especially for time limits and appropriate protection of materials during handling to guard against all forms of contamination. This is proper for nonsterile compounding to minimize contamination prior to filtration/sterilization and is required for aseptic compounding activities. Barrier designs and other means of physically separating the worker from the product are recommended as a minimum even in nonaseptic compounding. As compounding may expose the worker to a variety of potent/toxic materials, the use of personnel protective equipment may be required. In extreme cases, the use of containment system may be required to protect the compounding operator.

Where the compounding is nonaseptic, careful control over the environment, materials, and equipment is still appropriate to reduce viable/nonviable levels and to reduce the potential for endotoxin. Time limits should be imposed on manufacturing operations for additional control over microorganisms and thus microbial toxins.

Once the materials have been sterilized, interventions near either the formulation or product contact surfaces/parts should be minimized. Direct handling of these materials should only be done with sterilized tools or implements; nonsterile objects, such as operator gloves, should never directly contact a sterilized surface. Sampling, filter integrity testing, process connection, and other activities should all be designed to eliminate the need for personnel exposure to sterile items.

Aseptic compounding is often a required activity for sterile products that cannot be filter sterilized. The preparation of the sterile solids for use in these formulations is outside the scope of this chapter, but it is often acknowledged as the most difficult of all pharmaceutical processes to properly execute. Handling these materials at the fill site is performed using ISO 5 environments, and the use of closed systems is preferred [34].

2.1.9.4 Filling

Aseptic filling is performed in ISO 5 environments, and a variety of approaches are utilized with the technology choice largely dependent upon the facility design, batch size, and package design. Older plants utilize manned clean rooms in which aseptically gowned personnel operate the filling equipment: performing the setup, supplying components, making any required adjustments, and conducting the environmental monitoring. As human operators are directly or indirectly responsible for essentially all microbial contamination, aseptic filling operations are increasingly designed to minimize the potential for operator contamination to enter the critical environment. Barriers of various sophistication and effectiveness are employed to increase the protection afforded to sterile materials. The most evolved of the cleanroom designs are RAB systems in which personnel interventions are restricted to defined locations. Many newer facilities utilize isolation technology in which the filling environment is fully enclosed and personnel contamination is completely avoided.

Filling designs for syringes and ampoules differ only with respect to the details of component handling and closure design. However, it is wise not to underestimate the influence of both component quality and component handling reliability on contamination control in aseptic processing. Components that minimize the need for intervention and equipment that is rather tolerant of component variability will result in better contamination control performance. Aside from these distinctions, the range of filling technologies previously described is also possible.

The filling of plastic containers is accomplished using two very different approaches. Pre-formed containers can be sterilized in bulk, introduced into the aseptic suite via air locks, oriented (unscrambled), and filled. Blow-fill-seal prepares sterile bottles (most often LDPE) on line just prior to filling and sealing.

Filling of suspensions, emulsions, and other liquids may require slightly different filling designs to assure uniformity of dose in each container. Ointments and creams are sometimes filled at elevated temperatures to improve their flow properties through the delivery and filling equipment. These are ordinarily filled into presterilized plastic tubes that have largely replaced aluminum tubes for these formulations. Powders are typically filled in vials using equipment specifically engineered for that purpose.

An inerting gas (typically nitrogen, but other gases can be utilized) may be added to the headspace of the container to protect formulations that are oxygen sensitive. If the product is particularly sensitive to oxygen, purging may be done in the empty container prior to filling and again immediately after filling. Products may also be filled in an isolator under a nitrogen atmosphere if required. Products that require inert gas purging will also generally require inert gas for pressurization of tanks to provide motive force to drive the product through the filter(s) and into the filling reservoir.

2.1.9.5 Stoppering and Crimping

If the product is not freeze dried, the primary closure or "stopper" is applied shortly after completion of the filling process to better assure the sterility of the contents. When the product is to be lyophilized, the stopper may be partially inserted after filling and be fully seated after completion of the lyophilization cycle. Alternatively, the container could be left open and a stopper applied after completion of the drying.

Crimping is the act of securing the closure to the vial. It must be performed with sufficient uniform downward force to assure the container is properly secured. Too little downward force results in inadequately secured closures, while excessive force can result in container breakage. The force contributed by the crimp roller may be controllable as well.

Applying the closure to syringes, ampoules, and other containers usually differs in methodology from the approaches used for vials, but the objective is identical to secure the container's contents fully assuring the product's critical quality attributes (especially sterility) are maintained throughout its shelf life.

2.1.9.6 Lyophilization

Lyophilization (or freeze-drying) is a process utilized to convert a water-soluble material filled into a container to a solid state by removal of the liquid while frozen. The process requires the use of deep vacuums and careful control of temperatures. By conducting the process under reduced pressure, the water in the container converts from ice directly to vapor as heat is applied and is removed from the container by the vacuum. The dissolved solids in the formulation cannot undergo this phase change and remain in the container. At the completion of the cycle, the container will be returned to near atmospheric pressure; stoppers are applied or fully seated and crimped as described above. Lyophilization is particularly common with biological materials whose stability in aqueous solution may be relatively poor. The time period in solution and the temperature of the solution are kept at a specified low temperature to prevent product degradation [35].

As partially stoppered but unsealed containers must be transferred to the lyophilizer from the fill line, various designs have been utilized to protect the containers during this transit. Among the common alternatives utilized are the following:

- Placement of the lyophilization in the wall of the fill room to allow for direct loading
- Battery-operated unidirectional airflow carts to a remote lyophilizer
- ISO 5–protected conveyors with single shelf loading
- Transfer utilizing isolator technology

The use of trays for supporting the containers during the transfer, loading, lyophilization, and unloading steps was at one time common. The major problem with the use of trays for this purpose was the heat/handling-related distortion of the tray bottom that impacted the uniformity of the heating process in the freeze dryer. This was overcome by the use of trays with bottoms that were removed after loading and reinserted after completion of the drying. The current preference is for the placement of the containers directly on the shelf eliminating the trays entirely. This is accomplished by single height loading/unloading of the individual shelves with various pusher designs.

The use of thermocouples to monitor product temperature inside selected vials with the lyophilizer is still the prevalent practice. The utility of this data is questionable and the current trend is to eliminate this "requirement" as soon as possible to better assure sterility of the unsealed vials by eliminating placement of the thermocouples.

The lyophilizer chamber and condenser should be cleaned with a CIP system after each batch to prevent cross-contamination and, after cleaning, both should be sterilized. If a slot door loading system is utilized, periodic opening of a full door in the lyophilizer may be required to remove stoppers and glass that may have fallen.

2.1.10 PERSONNEL

Aseptic processing in the pharmaceutical industry is almost entirely dependent upon the proficiency of the personnel assigned to this most critical of all activities. The operators must be able to consistently aseptically transfer sterile equipment and materials in a manner that avoids contamination of those materials [1]. This is no mean feat given the contamination continuously released by personnel and the prevailing need for personnel for execution of the process activities.

Personnel proficiency in aseptic operations must be firmly established before they are allowed to conduct critical aseptic process steps. Operators must master a number of relevant skills in order to be declared competent. The usual progression is from classroom training (CGMP, microbiology, sterilization, etc.) to relevant practical exercises (aseptic media transfers, aseptic gowning rehearsals) and ultimately to the core aseptic skills required (aseptic gowning certification, aseptic assembly/technique) using a growth medium. Through this approach the operator gradually acquires the necessary skills to be a fully qualified member of the production staff. Training/qualification of personnel is an ongoing requirement and must be repeated periodically to assure the skills are maintained. Continuing evaluation of operator qualification is accomplished using written examinations, practical challenges, documented observation, and participation in process simulation trials.

There is general acknowledgment of the risk associated with heavy reliance on personnel for aseptic processing. This has fostered much of the innovative designs for aseptic filling such as RABS and isolators where personnel are largely removed from the critical environment. The future will undoubtedly witness aseptic technologies where human interaction with sterile materials has been eliminated.

2.1.11 ASEPTIC PROCESSING CONTROL AND EVALUATION

The preparation of any pharmaceutical product requires controls over the production operations to assure the end result is a product that meets the required quality attributes. The methods utilized for this control are supported by formalized validation studies in which proof of consistency is demonstrated by appropriately designed experiments. The definition of appropriate operating parameters is the primary objective of the development activities and is further confirmed during scale-up to commercial operations. The validation supports that the routine controls applied to the process are appropriate to assure product quality [36]. This is typically accomplished in formalized validation activities in which expanded sampling/testing of the product materials is performed to substantiate their uniformity and suitability for use [30].

2.1.11.1 In-Process Testing

The sampling and testing of in-process materials during the course of the manufacturing process can confirm that essential conditions have been provided. This is appropriate in preparation, compounding, and filling activities. Sampling in preparation processes can confirm the absence of particles, proper siliconization levels, and cleanliness of equipment to assure that production items and equipment are suitable for use. Samples for microbiological quality, must, as previously mentioned, always be done by fully gowned staff under ISO 5 conditions using excellent aseptic techniques. During compounding, in-process testing can confirm proper pH, dissolution of materials, bioburden, and potency prior to filling. Filling operations can be monitored for fill volume (weight), headspace oxygen, and particles. These activities can all be automated to reduce interventions. These are typical examples of in-process controls utilized to assure acceptability of the process while it is underway. In the event of an abnormal result, corrective measures could be applied before further processing. The validation effort supports that these control measures are sufficient to assure product quality, when met during production operations. The sample intervals, sizes, and locations for in-process testing are chosen to enhance the validation. The tolerance limits are usually tightened relative to the release requirements to further assure that no out-of-tolerance materials are produced.

2.1.11.2 End-Product Testing

Upon completion of the process, samples are taken to establish that the batch meets the final product specifications defined for release. Predefined sampling plans are utilized to obtain representative samples of the entire batch, the prior validation effort having assured through an expanded sampling effort that the process provides a uniform product. End-product sampling often suffers from the inability to link an anomalous result with a specific portion/segment of the batch. If the validation is insufficiently rigorous, an out-of-specification result will ordinarily result in rejection of the batch and little opportunity to take effective corrective action.

The FDA has been supportive of the use of process analytical technologies (PATs) as an improvement on end-product testing [37]. These are intended to act as on-line indicators of critical product attributes enabling immediate corrective action and preventing the production of off-specification materials. This approach is common in the continuous process industries where feedforward controls are often employed. Their application to the more batch-oriented pharmaceutical/biotechnology industry is an acknowledgment that this approach can assure product quality more fully than a sampling-based approach. The PAT applications are still relatively few in number, but their utility in lieu of traditional quality methods is certainly promising.

The preceding relates solely to product quality attributes, based upon chemical or physical requirements. Assurance of sterility, the most critical of all the quality components for an aseptically filled sterile product relies on the following:

- The validation of the various sterilization processes for preparation of materials, equipment, and formulations
- The design of the aseptic manufacturing process and facility
- The establishment and maintenance of a proper processing environment
- Most importantly, the proficiency of the operating personnel directly involved with the aseptic process

There is no direct means to evaluate the cumulative capability of these measures. We infer success in aseptic processing through the evaluation of indirect measures of performance: air pressure differentials, total particle counts, viable monitoring results, and end-product sterility testing. The enormous challenge of aseptic processing is that none of the in-process or end-product testing results can prove that the attribute of sterility is attained with a high degree of certainty. Therefore, we rely on validation and the demonstration of a validated state of control to infer the adequacy of our contamination control efforts.

2.1.11.3 Process Simulations

An indirect means of assessing a facility's aseptic processing performance is the process simulation (or media fill) test [38]. This test substitutes a growth medium for the product in the process from the point of sterilization through to closure of the product container. The expectation is that successful handling of the growth media through the operating steps provides assurance that product formulations handled in a similar fashion would also be successful [39]. Process simulations culminate in the incubation of the media-filled containers with success defined as a limited number of contaminated units in a larger number of filled units. The result is a contamination rate for the media fill, and not a direct indication of the level of sterility assurance afforded to aseptically processed materials using the same procedures. At the present time, the level of sterility provided to aseptically processed materials cannot be measured. The FDA and EMEA have harmonized their expectations relative to process simulation performance, but they have also asserted that the goal in every process simulation is zero contamination [1, 2]. This formalized expectation and recognition that patient safety should always be preeminent have resulted in substantial improvements in aseptic processing technology over the last 20 years.

2.1.12 TERMINAL STERILIZATION

Terminal sterilization is a process by which product is sterilized in its final container. Terminal sterilization is the method of choice for products that are sufficiently stabile when subjected to a compatible lethal treatment. Because the process utilized is expected to be lethal to the microorganisms present, is highly reproducible, and generally readily validated, there is a clear preference for its use [1, 40, 41].

The predominant method for terminal sterilization is moist heat, and a substantial percentage of sterile products are processed in this manner. (Estimates range from 5 to 15% of all sterile products are terminally sterilized.) The sterilization often requires the attainment of a balance between sterility assurance and degradation of the material's essential properties [42]. The overkill sterilization method is preferred for heat-resistant materials, and may be usable for terminal sterilization where the formulation can tolerate substantial heat input. The bioburden/biological indicator approach uses less heat input but requires increased control over the titer and resistance of the bioburden organisms present.

The large-volume parenteral (LVP) industry sometimes uses dedicated nonaseptic filling systems for its containers prior to subjecting them to terminal treatments. These LVP systems may approach the aseptic designs described earlier, but they are not supported by the same levels of environmental monitoring nor process simulation. Application of terminal sterilization at small volume parenteral producers may be done after the product is aseptically filled, although this practice is usual only where the firm produces predominantly aseptically filled products and would not have a filling system dedicated to terminally sterilized formulations. Product that will be subject to terminal sterilization may be filled under clean conditions with reduced environmental monitoring and control. However, control of total particulate levels requires unidirectional airflow for critical filling or assembly processes.

Terminal sterilization is most commonly accomplished by moist heat. Terminal sterilization by other means is certainly possible, and a very limited number of parenteral drugs are treated with dry heat or radiation after filling. There is growing interest in the use of radiation, including low-energy E-beam, as a terminal treatment suggesting more products will be processed in this manner.

Although there are numerous advantages to terminal sterilization, there can be very good reasons for aseptically filling products that are stabile enough to be compatible with a sterilization process. For example, multichamber containers that cannot withstand terminal sterilization may provide a very important safety benefit to the patient by reducing aseptic admixture or reconstitution in the clinic. These aseptic activities when conducted in clinics are generally not able to be done within anything like the controls required in industrial aseptic processing. It is often beneficial to discuss processing technology choices with regulatory authorities early in the development of a new product.

2.1.13 CONCLUSION

The manufacture of parenteral drugs by aseptic processing has long been considered a difficult technical challenge. These products require careful control and stringent attention to detail to assure their safety. Aseptic processing done with discipline and taking advantage of the numerous technical developments that have occurred over the years results in sterile products that can be administered with complete confidence. The wider adaptation of advanced aseptic processing will result in further evolutionary improvements in aseptic processing. The industry is at the beginning of an era in which human-scale aseptic processing will be completely replaced by separative technologies and process automation. Additionally, improved in-process controls are likely to be implemented making validation easier and easing the compliance burden.

APPENDIX

Parenteral Drug Association, Bethesda, Maryland

TM 1: Validation of Steam Sterilization Cycles, 1978

TR 3: Validation of Dry Heat Processes used for Sterilization & Depyrogenation, 1981

TR 7: Depyrogenation, 1985

TR 11: Sterilization of Parenterals by Gamma Irradiation, 1988

TR 13: Fundamentals of an Environmental Monitoring Program, 2001

TR 22: Process Simulation Testing for Aseptically Filled Products, 1996

TR 26: Sterilizing Filtration of Liquids, 1998

TR 28: Process Simulation Testing for Sterile Bulk Pharmaceutical Chemicals, 2006

TR 34: Design & Validation of Isolator Systems for the Manufacture & Testing of Health Care Products, 2001

TR 36: Current Practices in the validation of Aseptic Processing, 2002

TR 40: Sterilizing Filtration of Gases, 2005

International Society For Pharmaceutical Engineering, Tampa, Florida

Baseline Guide, Vol. 3: Sterile Manufacturing Facilities, 1999

Baseline Guide, Vol. 4: Water and Steam Systems, 2001

Baseline Guide, Vol. 5: Commissioning and Qualification, 2001

REFERENCES

1. U.S. Food and Drug Administration (FDA) (2004), Guideline on sterile drug products produced by aseptic processing, FDA, Washington, DC.
2. European Union (EU) (2006), Annex 1—Sterile medicinal products—draft revision.
3. International Organization for Standardization (ISO), international standard 14644 1-3.
4. U.S. Food and Drug Administration (FDA) (2004), Pharmaceutical CGMPs for the twenty-first century—A risk-based approach, FDA, Washington, DC.
5. International Conference on Organization (ICH) (2005), Draft consensus guideline quality risk management Q9, draft.
6. Whyte, W., and Eaton, T. (2004), Microbiological contamination models for use in risk assessment during pharmaceutical production, *Eur J Parenteral Pharm Sci*, 9(1).
7. Whyte, W., and Eaton, T. (2004), Microbial risk assessment in pharmaceutical cleanrooms, *Eur J Parenteral Pharm Sci*, 9(1).
8. Agalloco, J., and Akers, J. (2006), Simplified risk analysis for aseptic processing: The Akers-Agalloco method, *Pharm Technol*, 30(7), 60–76.
9. International Organization for Standardization (ISO) (2004), Cleanrooms and associated controlled environments—Part 7: Separative devices (clean air hoods, gloveboxes, isolators and mini-environments), ISO 14644-7.
10. Agalloco, J. (2006), Thinking inside the box: The application of isolation technology for aseptic processing, *Pharm Technol.*, p. S8–11.
11. Lysford, J., and Porter, M. (2003), Barrier isolators history and trends, *Pharm Eng*, 23(2), 58–64.
12. ISPE (2005), Restricted access barrier systems (RABS) for aseptic processing, ISPE definition, Aug. 16.
13. Wikol, M. (2004), GoreTM vial isolator, ISPE presentation, Feb. 12.
14. Py, D. (2004), Development challenges for intact sterile filling, PDA presentation, Mar. 9.
15. Thilly, J. (2004), CVFL technology from lab scale to industry, PDA presentation, Mar. 8.
16. ISPE (2001), Water and Steam Systems Baseline® guide.
17. ISPE (1999), Sterile Manufacturing Facilities Baseline® guide.
18. ISPE (2001), Water and Steam Systems Baseline® guide.
19. PDA (2006), Technical Monograph 1, Industrial moist heat sterilization in autoclaves, draft 17.
20. Perkins, J. (1969), *Principles and Methods of Sterilization in Health Sciences*, Charles Thomas, Springfield, IL.
21. Phillips, G. B., and Morrissey, R. F. (1993), *Sterilization Technology: A Practical Guide for Manufacturers and Users of Health Care Products*, Van Nostrand Reinhold, New York.
22. Agalloco, J. (1998), Sterilization in place technology and validation, in Agalloco, J., and Carleton, F. J., Eds., *Validation of Pharmaceutical Processes: Sterile Products*, Marcel Dekker, New York.
23. PDA (1981), Technical Report 3, Validation of dry heat processes used for sterilization and depyrogenation.
24. Case, L., and Heffernan, G. (1998), Dry heat sterilization and depyrogenation: Validation and monitoring, in Agalloco, J., and Carleton, F. J., Eds., *Validation of Pharmaceutical Processes: Sterile Products*, Marcel Dekker, New York.
25. Burgess, D., and Reich, R. (1993), Industrial ethylene oxide sterilization, in Phillips, G. B., and Morrissey, R. F. Eds., *Sterilization Technology: A Practical Guide for Manufacturers and Users of Health Care Products*, Van Nostrand Reinhold, New York.

26. Sintim-Damao, K. (1993), Other gaseous sterilization methods, in Phillips, G. B., and Morrissey, R. F. Eds., *Sterilization Technology: A Practical Guide for Manufacturers and Users of Health Care Products*, Van Nostrand Reinhold, New Youk.
27. Meltzer, T., Agalloco, J., et al. (2001), Filter integrity testing in liquid applications; Revisited, Part 1, *Pharm Technol*, 25(10), and Part 2, *Pharm Technol*, 25(11).
28. PDA (1998), Technical Report 26, Sterilizing filtration of liquids.
29. PDA (2005), Technical Report 40, Sterilizing filtration of gases.
30. Agalloco, J., and Carleton, F. J., Eds. (1998), *Validation of Pharmaceutical Processes: Sterile Products*, Marcel Dekker, New York.
31. PDA (2001), Technical Report 13, Fundamentals of an environmental control program.
32. USP ⟨1116⟩ (2005), Microbiological control and monitoring environments used for the manufacture of healthcare products, *Pharm Forum*, 31(2), Mar.–Apr.
33. Agalloco, J. (1996), Qualification and validation of environmental control systems, *PDA J Pharm Sci Technol*, 50(5), 280–289.
34. PDA (2006), Technical Report 28, Process simulation testing for sterile bulk pharmaceutical chemicals.
35. Trappler, E. (1998), Validation of lyophilization, in Agalloco, J., and Carleton, F. J., Eds., *Validation of Pharmaceutical Processes: Sterile Products*, Marcel Dekker, New York.
36. Chapman, K. G. (1984), The PAR approach to process validation, *Pharm Technol*, 8(12), 22–36.
37. Food and Drug Administration (FDA) (2004), PAT guidance for industry—A framework for innovative pharmaceutical development, manufacturing, and quality assurance, FDA, Washington, DC.
38. PDA (1998), Technical Report 22, Process simulation testing for aseptically filled products.
39. Agalloco, J., and Akers, J. (2006), Aseptic processing for dosage form manufacture: Organization & validation, in Carleton, F. J., and Agalloco, J. P., Eds., *Validation of Pharmaceutical Processes: Sterile Products*, Marcel Dekker, New York.
40. Food and Drug Administration (FDA) (1991), Use of aseptic processing and terminal sterilization in the preparation of sterile pharmaceuticals, FR 56, 354–358.
41. PIC/S41. (1999), Decision trees for the selection of sterilisation methods (CPMP/QWP/054/98).
42. PDA (2006), Technical Monograph 1, Industrial moist heat sterilization in autoclaves, draft 17.

ADDITIONAL READINGS

Akers, J. (2001), An overview of facilities for the control of microbial agents, in Block, S. S., Ed., *Disinfection, Sterilization and Preservation*, 5th ed, Lippincott, Williams and Wilkins, Philadelphia, pp. 1123–1138.

Akers, J., and Agalloco, J. (1997), Sterility and sterility assurance, *J Pharm Sci Technol* 51, 72–77.

Cole, J. C. (1990), *Pharmaceutical Production Facilities—Design and Application*, Ellis Norwood, Chicester.

Institute of Environmental Science and Technology (IEST) (1995), Compendium of standards, practices, and similar documents relating to contamination control, CC009/IESCC009.2, IEST, Mt. Prospect, IL.

Ljungvist, B., and Reinmueller, B. (1995), *Ventilation and Airborne Contamination in Clean Rooms*, Pharmacia A/B, Stockholm.

Reinmuller, B. (2000), Microbiological risk assessment of airborne contaminants in clean zones, Bulletin No. 52, Royal Institute of Technology/Building Services and Engineering, Stockholm.

United States Pharmacopoeia/National Formulary (2006), 29, Chapter 1116, Microbial evaluation of clean rooms, Rockville, Maryland, pp. 2969–2976.

SECTION 3

FACILITY

3.1

FROM PILOT PLANT TO MANUFACTURING: EFFECT OF SCALE-UP ON OPERATION OF JACKETED REACTORS

B. Wayne Bequette
Rensselaer Polytechnic Institute, Troy, New York

Contents

Pharmaceutical Manufacturing Handbook: Production and Processes, edited by Shayne Cox Gad

3.1.1 MOTIVATION

There are many phases of process development between the discovery of an active pharmaceutical ingredient and the design, construction, and operation of a manufacturing process to produce a drug. A sequence of reactions and separations that is successful at the bench scale may lead to a process that is unsafe, is difficult to operate, or produces unsatisfactory product at the manufacturing scale. A manufacturing process typically has a large sequence of steps, involving several different unit operations (heat exchangers, reactors, separators, etc.), and a complete review of the design and scale-up of these unit operations would constitute a chemical engineering curriculum; thus, the focus of this chapter is the scale-up of jacketed batch chemical reactors from the laboratory to the pilot plant and manufacturing. These reaction vessels often serve many functions, including mixing, heating, cooling, distillation, and crystallization.

Temperature control for laboratory reactors is typically easy because of high heat transfer area–reactor volume ratios, which do not require large driving forces (temperature differences) for heat transfer from the reactor to the jacket. Pilot- and full-scale reactors, however, often have a limited heat transfer capability. A process development engineer will usually have a choice of reactors when moving from the laboratory to the pilot plant. Kinetic and heat of reaction parameters obtained from the laboratory reactor, in conjunction with information on the heat transfer characteristics of each pilot plant vessel, can be used to select the proper pilot plant reactor.

Similarly, when moving from the pilot plant to manufacturing, a process engineer will either choose an existing vessel or specify the design criteria for a new reactor. A necessary condition for operation with a specified reactor temperature profile is that the required jacket temperature is feasible. We have therefore chosen to focus on heat transfer–related issues in scale-up. Clearly there are other scale-up issues, such as mixing sensitive reactions. See Paul [1] for several examples of mixing scale-up in the pharmaceutical industry.

In this chapter we discuss important issues as we move from laboratory to pilot plant and manufacturing. A review of batch process operation and pharmaceutical research is covered in Section 3.1.2, followed by laboratory vessels and reaction calorimetry in Section 3.1.3. In Section 3.1.4 heat transfer in process vessels is presented, including the effect of reactor type and heat transfer fluid on the vessel heat transfer capability. In Section 3.1.5 dynamic behavior based on simulation studies is discussed.

3.1.2 BACKGROUND

3.1.2.1 Pharmaceutical Process Development

Anderson [2] presents a wide range of topics on pharmaceutical process development, including a number of different problems related to process scale-up, such as solvent and reagent selection, purification, and limitations to various operations. He notes that most reactors used for scale-up operations are selected for flexibility in running many different processes, especially for pilot plants and multiproduct manufacturing plants.

Pisano [3] discusses the management of process development projects in the pharmaceutical industry. Case studies are used to illustrate the effect of resource allocation decisions at different stages of a project. While there has been a focus on product development in the pharmaceutical industry, clearly process development plays an important role in getting a product to market and lowering the long-term product manufacturing costs.

3.1.2.2 Batch Reactors

Batch processes present challenging control problems due to the time-varying nature of operation. Chylla and Haase [4] present a detailed example of a batch reactor problem in the polymer products industry. This reactor has an overall heat transfer coefficient that decreases from batch to batch due to fouling of the heat transfer surface inside the reactor. Bonvin [5] discusses a number of important topics in batch processing, including safety, product quality, and scale-up. He notes that the frequent repetition of batch runs enables the results from previous runs to be used to optimize the operation of subsequent ones.

LeLann et al. [6] discuss tendency modeling (using approximate stoichiometric and kinetic models for a reaction) and the use of model predictive control (linear and nonlinear) in batch reactor operation. Studies of a hybrid heating–cooling system on a 16-L pilot plant are presented.

Various aspects of the effect of process scale-up on the safety of batch reactors have been discussed by Gygax [7], who presents methods to assess thermal runaway. Shukla and Pushpavanam [8] present parametric sensitivy and safety results for three exothermic systems modeled using pseudohomogenous rate expressions from the literature. Caygill et al. [9] identify the common factors that cause a reduction in performance on scale-up. They present results of a survey of pharmaceutical and fine chemicals companies indicating that problems with mixing and heat transfer are commonly experienced with large-scale reactors.

3.1.2.3 Reaction Calorimetry

The microanalytical methods of differential thermal analysis, differential scanning calorimetry, accelerating rate calorimetry, and thermomechanical analysis provide important information about chemical kinetics and thermodynamics but do not provide information about large-scale effects. Although a number of techniques are available for kinetics and heat-of-reaction analysis, a major advantage to heat flow calorimetry is that it better simulates the effects of real process conditions, such as degree of mixing or heat transfer coefficients.

Regenass [10] reviews a number of uses for heat flow calorimetry, particularly process development. The hydrolysis of acetic anhydride and the isomerization of trimethyl phosphite are used to illustrate how the technique can be used for process development. Kaarlsen and Villadsen [11, 12] provide reviews of isothermal reaction calorimeters that have a sample volume of at least 0.1 L and are used to measure the rate of evolution of heat at a constant reaction temperature. Bourne et al. [13] show that the plant-scale heat transfer coefficient can be estimated rapidly and accurately from a few runs in a heat flow calorimeter.

Landau et al. [14] use a heat flow calorimeter to investigate feasible pilot plant operating conditions for the production of a pharmaceutical intermediate. They determine kinetic and heat flow parameters using the calorimeter and further estimate heat transfer parameters for a pilot-scale reactor. Simulation studies are used to find the required jacket temperature for desired batch reactor temperature profiles. Semibatch operation is shown to be safer than normal batch operation. Landau [15] provides a detailed review of reaction calorimetry, including mathematical expressions for energy balances, and a number of application examples.

3.1.3 LABORATORY VESSELS AND REACTION CALORIMETERS

As reviewed in Section 3.1.2.3, reaction calorimeters can be used to better understand and characterize scale-related process phenomena, such as mixing and heat transfer. A heat flow calorimeter, the Mettler RC1e, is shown in Figure 1. A schematic of a similar calorimeter system is shown in Figure 2 [16]. A heat flow calorimeter can be used to estimate:

- Physical parameters (heat capacity)
- Reaction rate constants
- Heat transfer coefficients (overall, U or, or film, h_i)

3.1.3.1 Material and Energy Balances

The overall energy balance for a process with no reaction has the form

Energy accumulation = energy in heat transfer from jacket
+ energy in by calibration probe − energy lost by ambient heat transfer

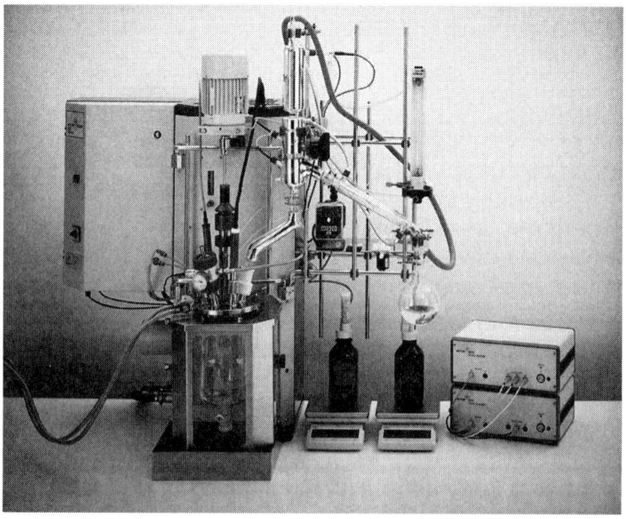

FIGURE 1 Mettler RC1e heat flow calorimeter system (www.mt.com).

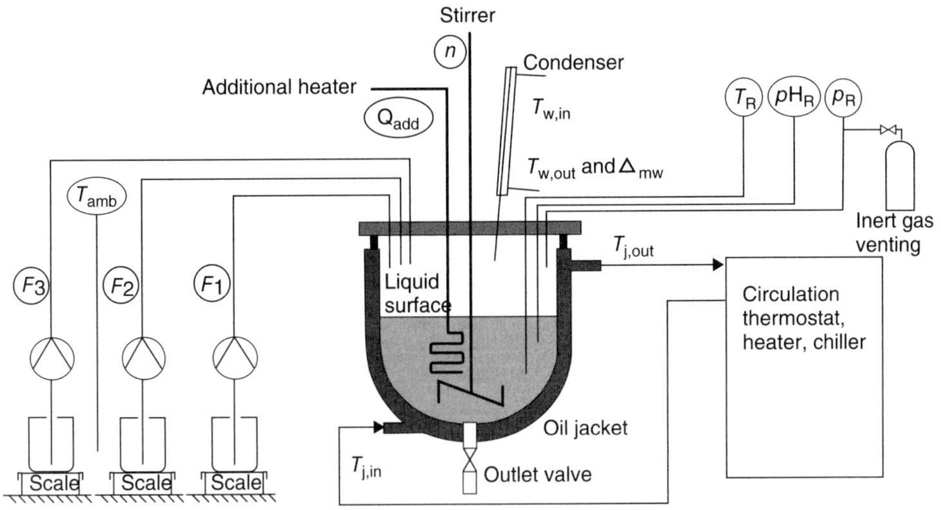

FIGURE 2 Schematic of HEL SIMULAR reaction calorimeter. From ref. 16.

which is shown mathematically as

$$(mc_p)_r \frac{dT}{dt} = -UA(T - T_j) + q_{\text{cal}} - k_{\text{loss}}(T - T_{\text{amb}}) \qquad (1)$$

where $(mc_p)_r$ is the reactor thermal capacitance, T is the reactor temperature, T_j is the jacket temperature, U is the overall heat transfer coefficient, A is the area for heat transfer, q_{cal} is the heat flow from the calibration probe, and the final term accounts for heat loss from the reactor system. The thermal capacitance is composed of the fluid in the vessel as well as the inert components in contact with the fluid, including the vessel wall, agitator (stirrer), and sensors (e.g., thermocouple), as shown in the equation

$$(mc_p)_r = V\rho c_p + m_v c_{pv} \qquad (2)$$

where V is the volume of liquid, ρ is the liquid density, c_p is the liquid heat capacity, m_v is the mass of the vessel wall and other inerts, and c_{pv} is the average heat capacity of the vessel wall and inerts. The inert contributions and heat transfer to the ambient can be found from extensive calibration studies. For small-scale reactors, such as reaction calorimeters, the thermal mass of the inerts can be significant. The thermal capacitance ratio, sometimes called the Lewis number, is given as

$$\phi = \frac{(mc_p)_r}{V\rho c_p} = 1 + \frac{m_v c_{pv}}{V\rho c_p} \qquad (3)$$

which can be on the order of 1.5–2 for a small-scale reactors and adiabatic calorimeters but is often 1.05–1.10 for small pilot plant reactors and less than 1.02 for manufacturing-scale reactors.

3.1.3.2 Estimating Fluid Properties and Heat Transfer Coefficients from Calorimeter Data

In a heat flow calorimeter, a feedback controller is used to maintain a constant desired reactor temperature by adjusting the jacket temperature. From (1), with a constant calibration probe heat flow, at steady state ($dT/dt = 0$), the overall heat transfer coefficient can be found from

$$UA = \frac{q_{\mathrm{cal}} - k_{\mathrm{loss}}(T - T_{\mathrm{amb}})}{T - T_j} \tag{4}$$

Also, the fluid heat capacity can be found by ramping up the reactor temperature and using

$$(mc_p)_r = \frac{-UA(T - T_j) + q_{\mathrm{cal}} - k_{\mathrm{loss}}(T - T_{\mathrm{amb}})}{dT/dt} \tag{5}$$

and solving for c_p from (2), assuming that the reactor inert component contributions are known from previous studies. An example calibration study is shown in Figure 3, where a constant heat flow is applied from 35 to 42 min, enabling the heat transfer coefficient to be estimated from the temperature difference using Equation (4). Then, the heat capacity is estimated from the temperature ramp applied between 5 and 20 min. It should be noted that the heat transfer coefficient and heat capacity of the fluid may vary with concentration and temperature. Typically, calibration experiments are performed before and after the reaction; then the heat transfer coefficient and heat capacity are assumed to vary linearly with conversion or batch time. For polymerization reactions in particular, the viscosity can increase tremendously with conversion, causing a substantial decrease in the heat transfer coeffi-

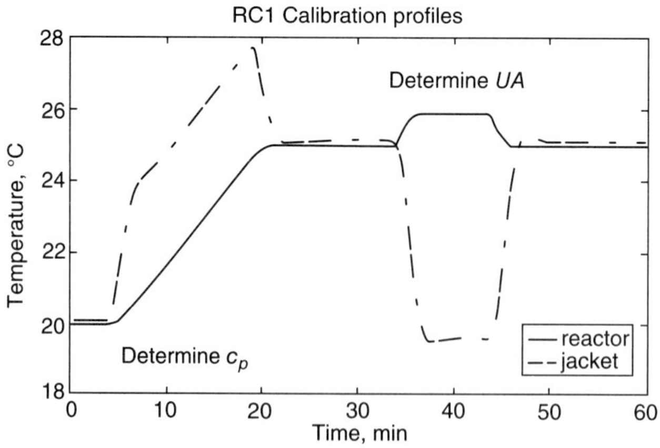

FIGURE 3 Example reaction calorimetry study without reaction. The overall heat transfer coefficient area can be found during the steady-state temperature difference and known calibration probe heat flow, between 35 and 42 min. The heat capacity can then be found from the temperature ramp between 5 and 20 min.

cient. Reaction experiments can be run at several temperatures to find the functional relationship with temperature.

Since the heat transfer area as a function of liquid volume is known, the overall heat transfer coefficient U can be calculated from (4). The overall heat transfer coefficient is calculated as

$$\frac{1}{U} = \frac{1}{h_i} + \frac{x_g}{k_g} \tag{6}$$

where the jacket side resistance is negligible. The glass vessel heat transfer resistance (x_g/k_g, thickness/thermal conductivity) can be used to find the reactor fluid heat transfer coefficient (h_i).

3.1.3.3 Estimating Heat Flows

The reaction heat flow can be found by rearranging (1), with the calibration heat probe replaced by the reaction heat flow, to find

$$q_r = (mc_p)_r \frac{dT}{dt} + UA(T - T_j) + k_{\text{loss}}(T - T_{\text{amb}}) \tag{7}$$

The total heat released during the reaction can be found by integrating (7),

$$Q_{\text{tot}} = \int_0^{t_f} q_r dt \tag{8}$$

or, represented as a scaled (per-unit mass) total heat release,

$$\bar{Q}_{\text{tot}} = \frac{Q_{\text{tot}}}{V\rho} = \frac{Q_{\text{tot}}}{m} \tag{9}$$

The molar heat of reaction can be found from

$$\Delta H_{\text{rxn}} = \frac{-Q_{\text{tot}}}{n_{\text{rxn}}} \tag{10}$$

where n_{rxn} is the molar amount reacted.

As a “first-pass” calculation, if it is assumed that the dominant heat transfer resistance is on the reactor side, then the overall heat transfer coefficient (U) from (4) can be used for scale-up.

3.1.3.4 Relating Heat Flows and Conversion

The reaction heat flows are directly related to the conversion of reactants [14]. Consider a first-order reaction of a limiting reactant, with the rate expression

$$\frac{dC}{dt} = -kC \tag{11}$$

where C is the molar concentration of the reactant. The heat flow is

$$q_r = \left(\frac{dC}{dt}\right)\Delta H_{\text{rxn}} V = -kC\Delta H_{\text{rxn}} V \tag{12}$$

with an initial heat flow of

$$q_{r0} = -kC_0\Delta H_{\text{rxn}} V \tag{13}$$

dividing (12) by (13), we find the relationship between concentration and heat flow:

$$\frac{C}{C_0} = \frac{q_r}{q_{r0}} \tag{14}$$

For an isothermal reaction, the solution to (11) is

$$\frac{C}{C_0} = e^{-kt} \tag{15}$$

so, the heat flow for an isothermal reaction is

$$\frac{q_r}{q_{r0}} = e^{-kt} \tag{16}$$

Thus, the reaction rate constant k can be estimated from the reaction heat flow without making any concentration measurements. Assuming an Arrhenius rate expression

$$k = A_0 e^{-E/RT} \tag{17}$$

the rate constant at several temperatures can be used to estimate the frequency factor (A_0) and activation energy (E). (See ref. 14 for an example application.)

3.1.3.5 Semibatch Reactions

For extremely exothermic reactions it is necessary to slowly add the feed over time, that is, operate in a semibatch fashion. The heat flow for a semibatch reaction can be found from

$$q_r = UA(T - T_j) + (mc_p)_r \frac{dT}{dt} + \dot{m}_f c_{pf}(T - T_f) \tag{18}$$

where $\dot{m}_f$ is the mass flow rate of the feed stream. If the reactor temperature is maintained constant, this reduces to

$$q_r = UA(T - T_j) + \dot{m}_f c_{pf}(T - T_f) \tag{19}$$

For reactions with essentially instantaneous kinetics, the reaction rate is limited by the feed addition rate. For other reactions, particularly if the reactor is operated at too low of a temperature, a reactant concentration can "build up," eventually reaching an unsafe level that could lead to a rapid temperature rise and explosion. It is important for these reactions to monitor the heat flow to confirm that the reactant concentration is not increasing to unacceptable levels.

3.1.3.6 Rapid Scale-Up Relationships

Lacking knowledge of the larger scale reactor, it is tempting to simply assume that only the area for heat transfer varies upon scale-up. A natural parameter is the *cooling time*,[1] defined as

$$\tau_{co} = \frac{(mc_p)_r}{UA} = \frac{V\rho c_p \phi}{UA} \tag{20}$$

The heat transfer area varies with the square of the vessel diameter, and the volume varies with the cube of the vessel diameter. Thus the area–volume ratio (A/V) varies with volume as

$$\frac{A}{V} \sim \frac{1}{V^{1/3}} \tag{21}$$

The inverse cooling time relationship for scale-up from volume V_1 to V_2 is

$$\left[\frac{UA}{V\rho c_p \phi}\right]_2 = \left[\frac{UA}{V\rho c_p \phi}\right]_1 \left(\frac{V_1}{V_2}\right)^{1/3} \tag{22}$$

The required reactor-jacket temperature difference on scale-up, with a constant Lewis number, is

$$[T - T_j]_2 = [T - T_j]_1 \left(\frac{V_2}{V_1}\right)^{1/3} \tag{23}$$

so the temperature difference can increase dramatically when a process is scaled up several orders of magnitude. Reactor-jacket temperature difference constraints can be particularly important for glass-lined vessels, where the limit is often 75 °C.

3.1.3.7 Safety under a Cooling System Failure

In the event of a cooling system failure it can be assumed that the reactor operates adiabatically. The adiabatic temperature rise can be found from

[1]The notion of cooling time can be understood by writing (1) and assuming no calibration energy or heat loss. Then (1) becomes $\tau_{co}(dT/dt) = -(T - T_j)$. If a constant temperature difference $T - T_j$ is applied, it will take τ_{co} time units for the reactor temperature to change by the temperature difference.

$$\Delta T_{\mathrm{ad}} = \frac{Q_{\mathrm{tot}}}{(mc_p)_r} \tag{24}$$

and the final temperature is

$$T_{\mathrm{final}} = T_{\mathrm{initial}} + \Delta T_{\mathrm{ad}} \tag{25}$$

As long as the final temperature is less than some critical "onset" temperature where a secondary decomposition reaction occurs, then the process can safely handle a cooling system failure. If a batch reactor temperature cannot be assured to remain less than the onset temperature after a cooling system failure, then a semibatch operation should be used. As noted in Section 3.1.3.5, it is necessary to assure that reactant concentration is not increasing above an onset concentration where a similar decomposition could occur with a cooling system failure.

3.1.4 HEAT TRANSFER IN PROCESS VESSELS

Based on initial heat flow calorimetry studies, a process development engineer must choose the appropriate reactor vessels for pilot plant studies. A pilot plant typically has vessels that range from 80 to 5000 L, some constructed of alloy and others that are glass lined. In addition some vessels may have half-pipe coils for heat transfer, while others have jackets with agitation nozzles. A process drawing for a typical glass-lined vessel is shown in Figure 4. In Sections 3.1.4.1 and 3.1.4.2 we review fundamental heat transfer relationships in order to predict overall heat transfer coefficients. In Section 3.1.4.3 we review experimental techniques to estimate heat transfer coefficients in process vessels.

3.1.4.1 Heat Transfer Relationships

Reactor-Side Coefficient The reactor-side heat transfer coefficient is calculated as

$$h_i = a\frac{k_i}{D_i}\mathrm{Re}_i^{0.67}\,\mathrm{Pr}_i^{0.33} \tag{26}$$

where a is the agitation constant (0.33), k_i is the fluid thermal conductivity, Re_i is the Reynolds number, and Pr_i is the Prandtl number,

$$\mathrm{Re}_i = \frac{D_{ag}^2 N \rho_i}{\mu_i} \tag{27}$$

$$\mathrm{Pr}_i = \frac{\mu_i c_{pi}}{k_i} \tag{28}$$

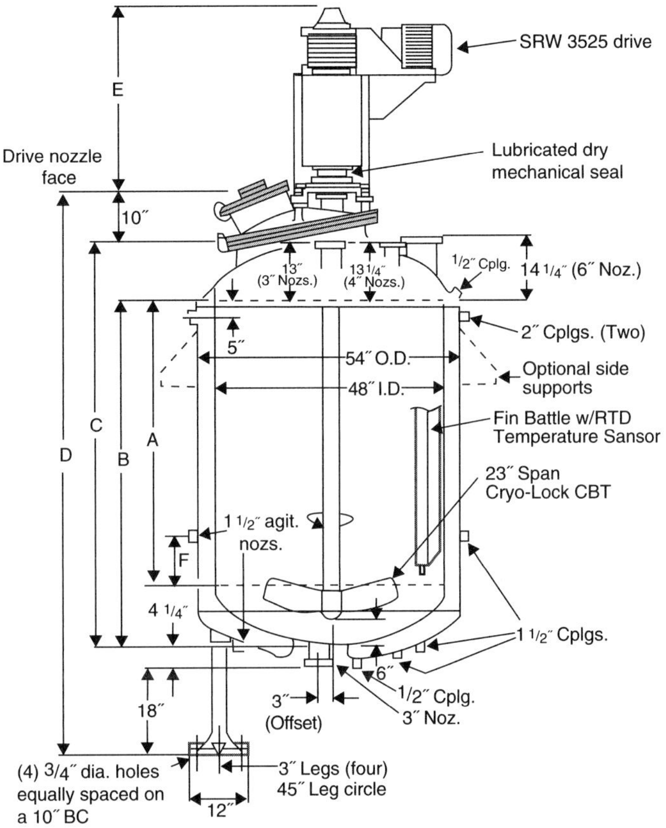

FIGURE 4 Typical 300- or 500-gal jacketed vessel (www.pfaudler.com).

and N is the agitator rotation rate. It should be noted that the film heat transfer coefficient varies inversely with the viscosity, that is,

$$h_i \sim \frac{1}{\mu_i^{0.33}} \tag{29}$$

Reactions where the viscosity increases substantially with conversion, such as some polymerization reactions, can be particularly difficult to control upon scale-up.

Jacket-Side Coefficient Here the calculations are shown for a jacket equipped with agitation nozzles that greatly increase the jacket fluid velocity. The jacket "swirl velocity" v_j is calculated (iteratively) from the nonlinear algebraic relationship [17]

$$\dot{m}_n(v_n - v_j) = \left(\frac{4fL}{D_e}\right)\left(\frac{v_j^2}{2}\right)\rho A_f \tag{30}$$

where $\dot{m}_n$ the is the nozzle mass flow rate, v_n is the nozzle velocity, the friction factor is

$$f = \frac{2 \times 0.023}{\mathrm{Re}^{0.2}} \tag{31}$$

the jacket-side film coefficient is

$$h_j = \frac{k_j}{D_e} 0.027\,\mathrm{Re}_j^{0.8}\,\mathrm{Pr}_j^{0.33} \tag{32}$$

and the Reynolds and Prandtl numbers are

$$\mathrm{Re}_j = \frac{D_e v_j \rho_j}{\mu_j} \tag{33}$$

$$\mathrm{Pr}_j = \frac{\mu_j c_{pj}}{k_j} \tag{34}$$

Overall Coefficient The overall heat transfer coefficient is found from the sum of the resistances,

$$\frac{1}{U} = \frac{1}{h_i} + \frac{1}{h_j} + \frac{x_m}{k_m} + \frac{x_g}{k_g} + ff_i + ff_j \tag{35}$$

which includes reactor film, jacket film, vessel metal, vessel glass, and fouling factors for both the reactor and jacket sides.

3.1.4.2 Effect of Reactor Type, Jacket Heat Transfer Fluid, and Reactor Fluid Viscosity

Here we present examples of how the reactor type and heat transfer fluid affect the heat transfer coefficient. When the reactor fluid has a low viscosity, the dominant heat transfer resistance tends to be on the jacket side. When the reactor fluid has a high viscosity, however, the dominant resistance is typically on the reactor side. Parameter values for the studies are presented in Figures 5–7 and are given in the literature [18].

The overall heat transfer coefficient is much higher for an alloy reactor/half-pipe jacket than for a glass-lined carbon steel reactor/agitation nozzle jacket, as shown in Figure 5, where Syltherm is the heat transfer fluid. Syltherm has a significantly lower heat transfer coefficient than an ethylene glycol mixture, as shown in Figure 6, but is capable of operating over a wider range of temperatures. The reactor fluid viscosity has a tremendous effect on the overall heat transfer coefficient, as shown in Figure 7. This can be particularly important in polymerization reactions where viscosity increases with conversion.

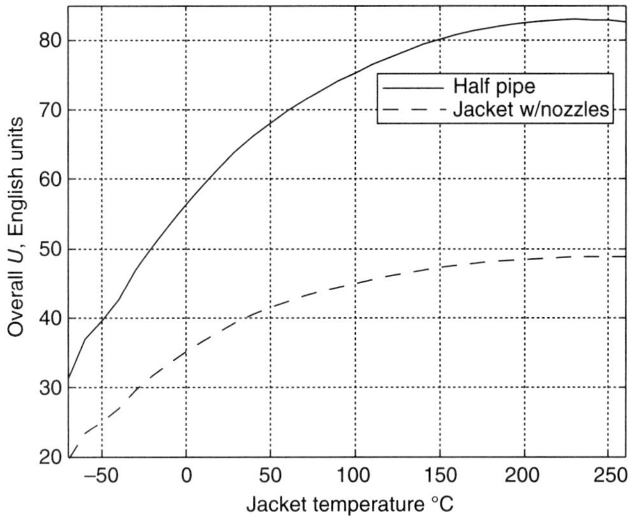

FIGURE 5 Overall heat transfer coefficient for 500-gal reactors. Comparison of alloy half pipe with glass-lined carbon steel (GLCS). Syltherm is the heat transfer fluid. (From ref. 18, with permission.)

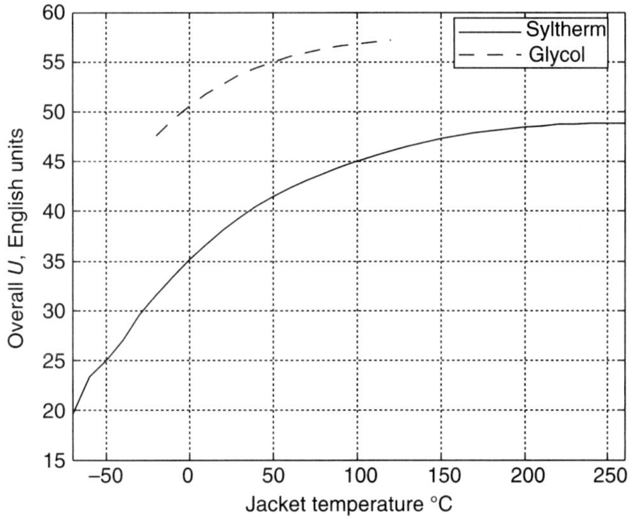

FIGURE 6 Overall heat transfer coefficient for 500-gal GLCS reactor. Comparison of Syltherm with Glycol. (From ref. 18, with permission.)

3.1.4.3 Pilot- and Production-Scale Experiments

The relationships shown in Section 3.1.3 are also pertinent to large-scale reactors. By using different solvents and volumes of solvent, pilot and production reactor heat transfer characteristics can be determined from a series of experiments. A primary limitation, compared to reaction calorimeter characterization, is that a calibration probe is rarely available. Thus, heat-up and cool-down studies, performed

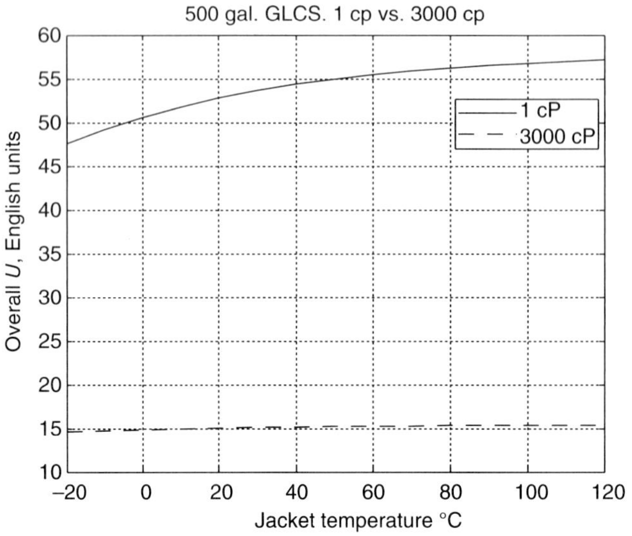

FIGURE 7 Overall heat transfer coefficient for 500-gal GLCS reactor with glycol heat transfer fluid. Comparison of effect of reactor-side viscosity.

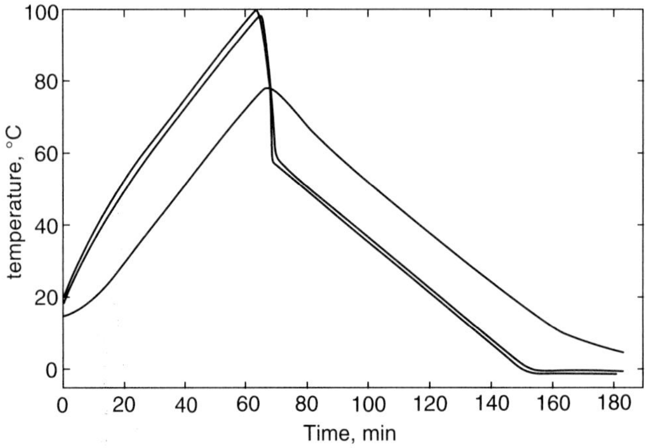

FIGURE 8 Temperature profiles (jacket inlet, jacket outlet, and reactor) for a pilot plant reactor. (From ref. 19.)

by varying the jacket temperature and observing the changes in the reactor temperature (for solvents with known heat capacity), are used to characterize the reactor. The inverse cooling time,

$$\frac{UA}{(mc_p)_r} = \frac{dT/dt}{T_j - T} \tag{36}$$

can be estimated from the temperature data collected from a heat-up/cool-down study. A characteristic example for a pilot-scale reactor is shown in Figure 8. The

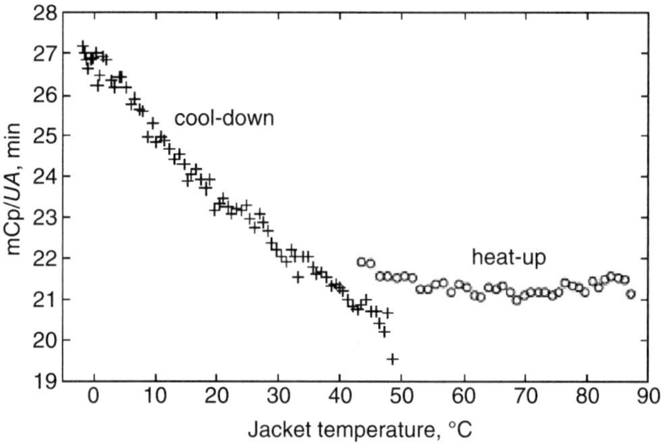

FIGURE 9 Cooling time estimates based on data presented in Figure 8. (From ref. 19.)

resulting cooling time estimates are shown in Figure 9. Notice that the overall heat transfer coefficient is clearly a function of the jacket temperature. The reduced heat transfer at the lower jacket temperatures is due to the strong relationship between viscosity and temperature for the 40% glycol solution used in the jacket. The discontinuity in the cooling time estimate at around 45 °C may be due to two factors. One factor is the assumption of no heat loss from the vessel, which would tend to lower the *UA* estimates during the heat-up phase. Another factor is the assumption that the metal and glass inerts in the reactor are at the temperature of the reactor; in practice it might be a better assumption that the reactor wall in particular is at a temperature that is intermediate between the jacket and reactor temperatures.

The fluid and inert thermal masses can be independently estimated by conducting experiments with a number of different solvent amounts. From the cooling time expression

$$\frac{(mc_p)_r}{UA} = \frac{m_v c_{pv}}{UA} + \frac{V\rho c_p}{UA} \tag{37}$$

writing this as a function of the reactor fluid volume,

$$\frac{(mc_p)_r}{UA} = \frac{m_v c_{pv}}{UA} + \frac{\rho c_p}{UA} \cdot V \tag{38}$$

and conducting experiments at a number of different fluid volumes or, equivalently, masses ($V\rho$),

$$\frac{(mc_p)_r}{UA} = \frac{m_v c_{pv}}{UA} + \frac{c_p}{UA} \cdot V\rho \tag{39}$$

the linear regression can be used to find the slope and intercept and thus estimate the *UA* and $m_v c_{pv}$ terms [19]. This approach is shown in Figure 10 for a jacket tem-

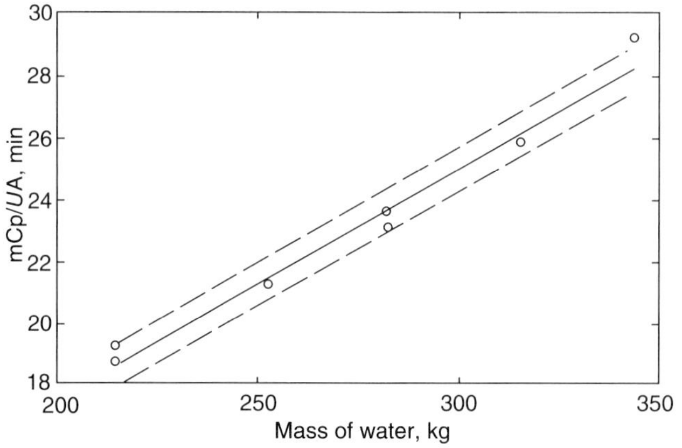

FIGURE 10 Linear regression to estimate thermal mass and *UA*. (From ref. 19.)

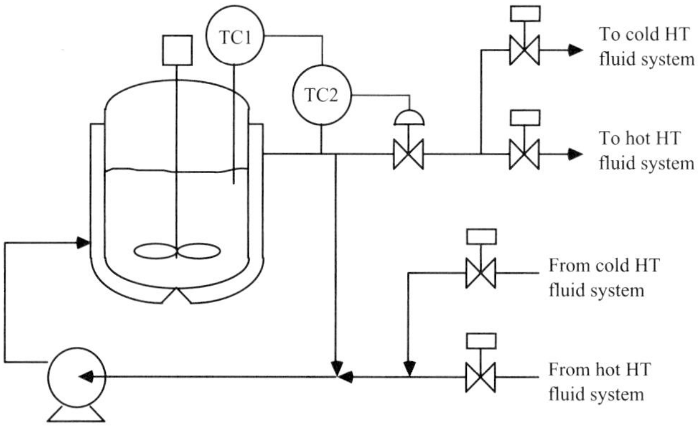

FIGURE 11 Characteristic pilot plant vessel control strategy. Slave (secondary) controller based on jacket outlet temperature is shown. The control valve is on the outlet stream to minimize temperature gradients (when switching from hot to cold fluids) that would be imposed if the valve was on the inlet. (From ref. 18, with permission.)

perature of 60 °C (based on a total of eight experiments at five different reactor fluid volumes).

3.1.5 DYNAMIC SIMULATION STUDIES

Older pilot plant and manufacturing processes often used steam for heating and water for cooling, with a switch-and-purge strategy between the two modes. Recent process designs have two heat transfer fluid systems (hot and cold heat transfer fluids) that are used for most of the heating and cooling needs. In addition, some vessels may have nitrogen coolers for cryogenic operation.

A simplified schematic for a jacket heat transfer service is shown in Figure 11 [18]. Here, two separate heat transfer fluid headers are used, and the control valve is on the outlet stream to reduce the temperature shocks that might occur if a single control valve was on the inlet stream. Depending on the range of temperatures, either ethylene glycol or a proprietary fluid such as Syltherm is used. Depending on whether heating or cooling is needed, either the hot or cold process control valve is open. Similarly, on–off valves return fluid to the appropriate distribution system.

Although the heat transfer fluid can be used over a wide range of temperatures, the film heat transfer coefficient is a strong function of temperature due to viscosity effects. The "cooling time" of a large reactor operating at a low temperature can be substantially longer than that of a small reactor operating at a high temperature due to this strong temperature effect. Simulation studies can be used to:

- Understand the effect of heat transfer fluid
- Understand possible performance limitations due to scale and operating conditions
- Test the effect of specified temperature gradient constraints
- Assist with controller design and selection of tuning parameters for system start-up

Various levels of models can be used to describe the behavior of pilot-scale jacketed batch reactors. For online reaction calorimetry and for rapid scale-up, a simple model characterizing the heat transfer from the reactor to the jacket can be used. Another level of modeling detail includes both the jacket and reactor dynamics. Finally, the complete set of equations simultaneously describing the integrated reactor/jacket and recirculating system dynamics can be used for feedback control system design and simulation. The complete model can more accurately assess the operability and safety of the pilot-scale system and can be used for more accurate process scale-up.

In the simulation studies that follow, it is assumed that the reactor and jacket are well mixed, resulting in differential equations for the material and energy balances [18]. The reactor shell (including a glass lining, if used) and reactor internals (agitator and baffles) are at the same temperature as the reactor, so their "thermal mass" is included in the reactor energy balance. Similarly, the jacket shell is at the jacket temperature, with an associated thermal mass. The heat transfer area A is proportional to the reactor liquid level (between volumes associated with the minimum and maximum heat transfer area); also, the reactor shell thermal mass varies linearly with the liquid level. Heat transfer coefficients are calculated using the relationships presented in Section 3.1.4; see Garvin [20] or Dream [21] for detailed examples. Parameters, viscosity in particular, are a function of temperature.

We focus on the effect of reactor size and material of construction on the expected dynamic behavior of the reactors. Details on the model development and simulation environment are presented elsewhere [18]. Figure 12 illustrates that a vessel can have significantly different dynamic behavior depending on whether it is being heated or cooled (for illustrative purposes, the freezing point of water is neglected in this simulation). The increase in reactor temperature results in a much faster response than a decrease for two reasons: (i) the jacket heat transfer fluid has a much higher viscosity (resulting in a lower overall heat transfer coefficient) at low temperatures and (ii) the fluid flow rate/jacket temperature gain is proportional to

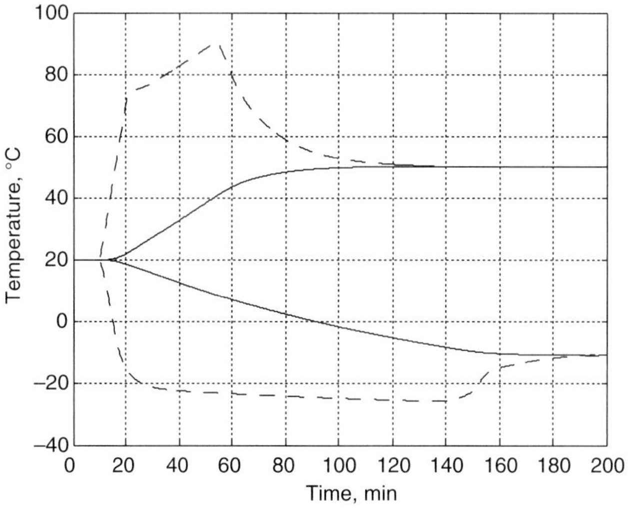

FIGURE 12 Comparison of responses for ±30 °C reactor temperature setpoint changes at t = 10 min; 500-gal GLCS filled with water (1925 kg).

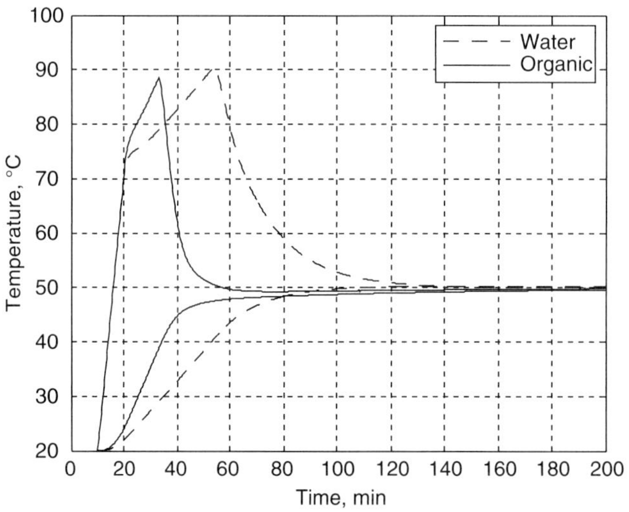

FIGURE 13 Comparison of temperature responses for 30 °C batch setpoint change; 500-gal GLCS, water (1925 L) vs. organic (1700 L).

the difference between the jacket temperature and make-up fluid temperature, which becomes small at low jacket temperatures. Notice that the initial response for the temperature increase is constrained by the ramp limit of 5 °C/min on the jacket temperature. The temperature response of an organic solvent is much faster than water because of the heat capacity difference, as shown in Figure 13. The previous plots were for simple heating/cooling applications (ref. 18 presents further studies for cryogenic and semibatch systems).

3.1.6 SUMMARY

In this chapter we have presented an overview of scale-up considerations involved as one moves from bench-scale reaction calorimetry to larger scale pilot plant and production reactors. Our focus has been on heat transfer and single-phase processes, addressing primarily the problem that the heat transfer area per unit reactor volume decreases with scale. Clearly, there are many challenging problems associated with multiphase vessels, with evaporation/distillation and crystallization as obvious examples, but these topics are beyond the scope of this chapter.

REFERENCES

1. Paul, E. L. (1988), Design of reaction systems for specialty organic chemicals, *Chem. Eng. Sci.*, 43(8), 1773–1782.
2. Anderson, N. G. (2000), *Practical Process Research and Development*, Academic, New York.
3. Pisano, G. P. (1997), *The Development Factory*, Harvard Business School, Boston.
4. Chylla, R. W., and Hasse, D. R. (1993), Temperature control of semi-batch polymerization reactors, *Comp. Chem. Eng.*, 17(3), 257–264.
5. Bonvin, D. (1998), Optimal operation of batch reactors—A personal view, *J. Proc. Cont.*, 8(5–6), 355–368.
6. LeLann, M. V., Cabassud, M., and Casamatta, G. (1999), Modeling, optimization and control of batch chemical reactors in fine chemical production, *Annu. Rev. Control*, 23, 25–34.
7. Gygax, R. W. (1990, Feb.), Scale-up principles for assessing thermal runaway risks, *Chem. Eng. Prog.*, 86(2), 53–60.
8. Shukla, P. K., and Pushpavanam, S. (1994), Parametric sensitivity, runaway, and safety in batch reactors: Experiments and models, *Ind. Eng. Chem. Res.*, 33(12), 3202–3208.
9. Caygill, G., Zanfir, M., and Gavrildis, A. (2006), Scalable reactor design for pharmaceuticals and fine chemicals production. 1: Potential scale-up obstacles, *Org. Proc. Res. Dev.*, 10(3), 539–552.
10. Regenass, W. (1985), Calorimetric monitoring of industrial chemical processes, *Thermochim. Acta*, 95, 351–369.
11. Kaarlsen, L. G., and Villadsen, J. (1987), Isothermal reaction calorimeters—I. A literature review, *Chem. Eng. Sci.*, 42(5), 1153–1164.
12. Kaarlsen, L. G., and Villadsen, J. (1987), Isothermal reaction calorimeters—II. Data treatment, *Chem. Eng. Sci.*, 42(5), 1165–1173.
13. Bourne, J. R., Buerli, M., and Regenass, W. (1981), Heat transfer and power measurements in stirred tanks using heat flow calorimetry, *Chem. Eng. Sci.*, 36, 347–354.
14. Landau, R. N., Blackmond, D. G., and Tung, H.-H. (1994), Calorimetric investigation of an exothermic reaction: Kinetic and heat flow modeling, *Ind. Eng. Chem. Res.*, 33, 814–820.
15. Landau, R. N. (1996), Expanding the role of reaction calorimetry, *Thermochim. Acta*, 289, 101–126.
16. Obenndip, D. A., and Sharratt, P. N. (2006), Towards an information-rich process development. Part I: Interfacing experimentation with qualitatitive/semiquantitative modeling, *Org. Proc. Res. Dev.*, 10(3), 430–440.

17. Bolliger, D. H. (1982), Assessing heat transfer in process-vessel jackets, *Chem. Eng.*, Sept. 20, 95–100.
18. Bequette, B. W., Holihan, S., and Bacher, S. (2004), Automation and control issues in the design of a pharmaceutical pilot plant, *Control Eng. Practice*, 12, 901–908.
19. Zima, A., Spencer, G., and Bequette, B. W. (1996), Model development for batch reactor calorimetry and control, Preprint, presented at the AIChE Annual Meeting, Chicago, IL, Nov. 1996.
20. Garvin, J. (1999), Understand the thermal design of jacketed vessels, *Chem. Eng. Prog.*, 95(6), 61–68.
21. Dream, R. F. (1999), Heat transfer in agitated jacketed vessels, *Chem. Eng.*, Jan., 90–96.

3.2

PACKAGING AND LABELING

MARIA INÊS ROCHA MIRITELLO SANTORO AND ANIL KUMAR SINGH
University of São Paulo, São Paulo, Brazil

Contents

Pharmaceutical Manufacturing Handbook: Production and Processes, edited by Shayne Cox Gad

3.2.1 INTRODUCTION

The packaging of a pharmaceutical product fulfils a variety of roles, such as product presentation, identification, convenience, and protection until administration or use. Selection of packaging requires a basic knowledge of packaging materials, the environmental conditions to which the product will be exposed, and the characteristics of the formulation. Several types of packaging are used to contain and protect the pharmaceutical preparations, such as the primary packaging around the product and secondary packaging such as a carton and subsequent transit cases [1].

The principal objective of the modern pharmaceutical industry is to manufacture pharmaceutical preparations presenting high quality, identity, purity, effectiveness, and innocuity in order to guarantee the satisfaction and safety of patients. The development of a new drug must involve the synthesis of a molecule, determination of its pharmacological activity, industrial-scale production, and its commercialization to guarantee quality of the final product.

Packaging system development, including primary and secondary packaging components, is of critical importance. The material should be selected based on the characteristics of pharmaceutical product and dosage form. After the production phase, packaging must be planned according to regulatory requirements and its quality should be controlled according to the specifications.

Commercially, the packaging material is used as a barrier to protect the pharmaceutical preparations against external factors that can degrade them and consequently decrease their effectiveness and increase toxic effects.

Once the type of packaging material is decided based on such factors as size, shape, capacity, and physicochemical properties, all these data, including quality control tests, should be included in the specification of the products in order to assure the therapeutic effectiveness during its shelf life.

Several types of materials are in use in the preparation of containers and closure systems: glass, plastics, metals, and combinations of these materials. However, care should be taken in the selection of appropriate material. These materials should not present any physical or chemical reactivity that could modify drug activity, quality, purity, or physical characteristics of the drug and pharmaceutical preparations. Any minor modification in the pharmacopeial specification is acceptable if it does not present a threat to patient's health.

The aim of this chapter is to discuss the importance of the packaging and labeling of pharmaceutical preparations. The role of packaging and labeling in the pharmaceutical industry has grown substantially over the past decade. The total packaging operation is part of any drug development program. Pharmaceutical products generally require a standard of packaging which is superior to that of most other products in order to support and comply with their main requirements, such as efficacy, integrity, purity, safety, and stability.

For these reasons packaging technology should be based on the understanding of pharmaceutical products, characteristics of formulations, and dosage forms, including the physical and chemical properties of the drug substance.

In the past, packaging concerns often arose only during the later steps of product development. Today, packaging is integrated with the development step and is among the earliest considerations of new pharmaceutical preparations being studied. Labels of products can vary from the simple to the extremely complex. But, even at the most basic level, product identification should meet regulatory requirements. More complex are the labels that make use of bar code technologies. New components such as microchips, biosensors, and deoxyribonucleic acid (DNA) arrays are making possible the development of new technologies leading to finished products individually packed that require specialized packaging materials and design expertise. The challenge now is to maintain low packaging cost, that is, always integrated into the cost of the product itself.

Packaging in the post–World War II period benefited immensely from the commercialization of plastics, which were little known or used in prior years. Since then, the packaging industry has openly adopted plastics as a powerful new tool in the development of new packaging forms and functions.

Quality control of a packaging component starts at the design stage. All aspects of package development that may give rise to quality problems must be identified and minimized by good design. Identifying and correcting mistakes in packaging will avoid product recall and rejection of pharmaceutical preparations [2, 3].

3.2.2 PACKAGING MATERIALS

3.2.2.1 General Considerations

Packaging refers to all the operations, including filling and labeling, through which a bulk product should pass to become a finished product. Usually, sterile filling is not considered part of the packing process, although the bulk product is contained in a primary container.

A packaging component means any single part of a container closure system. Typical components are containers (e.g., ampules, vials, bottles), container liners (e.g., tube liners), closures (e.g., screw caps, stoppers), closure liners, stopper overseals, container inner seals, administration ports [e.g., on large-volume parenterals (LVPs)], overwraps, administration accessories, and container labels [4].

A primary packaging component is one that is or may be in direct contact with the dosage form. A secondary packaging component is one that is not and will not be in direct contact with the dosage form [4].

A container closure system refers to the sum of packaging components that together contain and protect the dosage form. This includes primary packaging components and secondary packaging components, if the latter are intended to provide additional protection to the drug product. A packaging system is equivalent to a container closure system [4].

The role of packaging material on the overall perceived and actual stability of the dosage form is well established. Packaging plays an important role in quality maintenance, and the resistance of packaging materials to moisture and light can significantly affect the stability of drugs and their dosage forms. It is crucial that stability testing of dosage forms in their final packaging be performed. The primary role of packaging, other than its esthetic one, is to protect the dosage forms from moisture and oxygen present in the atmosphere, light, and other types of exposure, especially if these factors affect the overall quality of the product on long-term storage [5].

The compliance packaging such as for fixed-dose combination pills and unit dosage form packaging is a therapy-related intervention and is designed to facilitate medication regimens and so potentially improve adherence. Compliance packaging can be defined as a prepackaged unit that provides one treatment cycle of the medication, to both the pharmacist and the patient, in a ready-to-use package. This innovation type of packaging is usually based on blister packaging that contain unit therapeutic dose for one time use. The separate dosage units and separate days are usually indicated on the dosage cards to help remind the patient when and how much of the medication to take, for example, blister packed oral dosage forms with drug information leaflets and contraceptive pills [6, 7].

The selection of packaging material for any pharmaceutical product is as important as proper pharmaceutical dosage form. To guarantee the safe and adequate delivery of drug product to the patient and improve patient compliance, the manufacturer should consider the following factors:

1. Compatibility and safety concerns raised by the route of administration of the drug product and the nature of the dosage form (e.g., solid or liquid based)
2. Kinds of protection the container closure system should provide to the dosage form (e.g., photosensitive, hygroscopic, easily oxidized drug products)
3. Potential effect of any treatment or handling that may be unique to the drug product in the packaging system
4. Patient compliance to the treatment and ease of drug administration
5. Safety, efficacy, and quality of drug product throughout its shelf-life

The acquisition, handling, and quality control of primary and secondary packaging materials and of printed materials should be accomplished in the same way as that for the raw materials. The printed materials should be stocked in a reserved place so the possibility of unauthorized access is avoided. The labels and other rejected printed materials should be stored and transported with proper identification before being destroyed. There should be a destruction record of the printed materials. Each batch of printed material and packaging material should receive a specific reference number for identification.

The identification affixed on the containers, on the equipment, in the facilities, and on the product containers should be clear, without ambiguity, and in a format approved by the company and contain the necessary data. Besides the text, differentiated colors indicating its condition could be used (e.g., in quarantine, approved, rejected, and cleaned).

The packing materials should attend to the specifications, giving emphasis to the compatibility of the same with the pharmaceutical product that it contains. The material should be examined with relation to visible physical and critical defects as well as the required specifications.

3.2.2.2 Glass as Packaging Material

A packaging system found acceptable for one drug product may not be appropriate for another. Each application should contain enough information to show that each proposed container closure system and its components are suitable for the intended use.

Nonsterile Products

Solids Some topical drug products such as powders may be considered for marketing in glass bottles with appropriate dispenser. These topical drug products may be sterile and could be subject to microbial limits.

The most common glass-packed solid oral dosage forms are oral powders and granules for reconstitution. A typical solid oral dosage form container closure system is a glass bottle (although plastic bottles are also used) with a screw-on or snap-off closure. A typical closure consists of a metal cap, often with a liner and frequently with an inner seal.

The dry powders that are reconstituted in their marketed container need not be sterile; however, the possibility of an interaction between the packaging components and the reconstituting fluid can't be discarded. Although the contact time will be relatively short when compared to the component/dosage form contact time for liquid-based oral dosage forms, it should still be taken into consideration when the compatibility and safety of the container closure system are being evaluated.

Powders for oral administration that are reconstituted in their market container, however, have an additional possibility of interaction between the packaging components and the reconstituting fluid. Although the contact time will be relatively short when compared to the component/dosage form contact time for liquid-based oral dosage forms, it should still be taken into consideration when the compatibility and safety of the container closure system are being evaluated.

Nonsolids For nonsterile products the preservative provides some protection, but continual microbial challenge will diminish the efficacy of the preservative, and spoilage or disease transmission may occur [8].

Antimicrobial preservatives such as phenylmercuric acetate are known to partition into rubbers during storage, thus reducing the formulation concentration below effective antimicrobial levels [9]. A complication of modern packaging is the need for the application of security seals to protect against deliberate adulteration and maintain consumer confidence.

Sterile Products The sterile dosage forms share the common attributes that they are generally solutions, emulsions, or suspensions and are all required to be sterile. Injectable dosage forms represent one of the highest risk drug products (Table 1). Any contaminants present (as a result of contact with a packaging component or due to the packaging system's failure to provide adequate protection) can be rapidly and completely introduced into the patient's general circulation. Injectable drug products may be liquids in the form of solutions, emulsions, or suspensions or dry solids that are to be combined with an appropriate vehicle to yield a solution or suspension.

Although ophthalmic drug products can be considered topical products, they have been grouped here with injectables because they are required to be sterile and the descriptive, suitability, and quality control information is typically the same as that for an injectable drug product.

The potential effects of packaging component/dosage form interactions are numerous. Hemolytic effects may result from a decrease in tonicity and pyrogenic effects may result from the presence of impurities. The potency of the drug product or concentration of the antimicrobial preservatives may decrease due to adsorption or absorption.

A cosolvent system essential to the solubilization of a poorly soluble drug can also serve as a potent extractant of plastic additives.

A disposable syringe may be made of plastic, glass, rubber, and metal components, and such multicomponent construction provides a potential for interaction that is greater than when a container consists of a single material.

Injectable drug products require protection from microbial contamination (loss of sterility or added bioburden) and may also need to be protected from light or exposure to gases (e.g., oxygen).

Performance of a syringe is usually addressed by establishing the force to initiate and maintain plunger movement down the barrel and the capability of the syringe to deliver the labeled amount of the drug product.

Solids For solids that must be dissolved or dispersed in an appropriate diluent before being injected, the diluent may be in the same container closure system (e. g., a two-part vial) or be part of the same market package (e.g., a kit containing a vial of diluent).

Sterile powders or powders for injection may need to be protected from exposure to water vapor. For elastomeric components, data showing that a component meets the requirements of U.S. Pharmacopeia (USP) elastomeric closures for injections will typically be considered sufficient evidence of safety.

Nonsolids The package must prevent the entry of organisms; for example, packaging of sterile products must be absolutely microorganism proof—hence the continued use of glass ampules. Liquid injections are classified as small-volume parenterals (SVPs), if they have a solution volume of 100 mL or less, or as LVPs, if the solution volume exceeds 100 mL [10]. Liquid-based injectables may need to be protected from solvent loss.

An SVP may be packaged in a vial or an ampule. An LVP may be packaged in a vial, a glass bottle or, in some cases, as a disposable syringe. Packaging material for vials, and ampules are usually composed of type I or II glass. Stoppers and septa in cartridges, and vials are typically composed of elastomeric materials.

Pharmaceuticals may interact with packaging and containers, resulting in the loss of drug substances by adsorption onto and absorption into container components and the incorporation of container components into pharmaceuticals. Diazepam in intravenous fluid containers and administration sets exhibited a loss during storage due to adsorption onto glass [11, 12].

Glass surfaces are also known to adsorb drug substances. Chloroquine solutions in glass containers decreased in concentration owing to adsorption of the drug onto the glass [13].

Rubber closures are also known to absorb materials, including drugs. Absorption of preservatives such as chlorocresol into the rubber closures of injectable formulations has been studied extensively [13].

The water permeability of rubber closures used in injection vials is considered an important parameter in assessing the closures, but quantitative prediction of water permeability through rubber closures is difficult because the diffusion coefficient of water is dependent on relative humidity [14].

Liquid-based oral drug products are usually dispensed in glass bottles (sometimes in plastic), often with a screw cap with a liner, and possibly with a tamper-resistant seal or an overcap that is welded onto the bottle. The same cap liners and inner seals are sometimes used with solid oral dosage forms. A laminated material can be used to overwrap glass bottles for extra safety.

The USP-grade glass packaging components are chemically resistant and can be considered sufficient evidence of safety and compatibility. In some cases (e.g., for some chelating agents), a glass packaging component may need to meet additional criteria to ensure the absence of significant interactions between the packaging component and the dosage form.

Several ophthalmic preparations are commercialized in glass containers. Although the risk factors associated with ophthalmic preparations are generally considered to be lower than for injectables, any potential for causing harm to the eyes demands caution.

A large-volume intraocular solution (for irrigation) may be packaged in a polyolefin (polyethylene and/or polypropylene) container.

The liquid-based oral dosage forms may be marketed in multiple-unit bottles. The dosage form may be used as is or admixed first with a compatible diluent or dispersant. Liquid-based oral drug products in glass container must meet the requirements for USP containers. Glass containers are accepted as sufficient evidence of safety and compatibility. Performance is typically not a factor for liquid-based oral drug products but should be considered while treating pressurized liquid-based oral drug products (e.g., elixir spray).

Topical dosage forms such as unpressurized sprays, lotions, ointments, solutions, and suspensions may be considered for marketing in glass bottles with appropriate dispenser. Some topical drug products, especially ophthalmic, are sterile or may be subject to microbial limits. In these cases, packaging material and handling should be done as those for injectables.

3.2.2.3 Plastic as Packaging Material

For plastic components, data from USP biological reactivity tests will typically be considered sufficient evidence of safety. Whenever possible, extraction studies should be performed using the drug product. If the extraction properties of the drug product vehicle may reasonably be expected to differ from that of water (e.g., due to high or low pH or to a solubilizing excipient), then drug product should be used as the extracting medium. If the drug substance significantly affects extraction characteristics, it may be necessary to perform the extractions using the drug product vehicle. If the total extract significantly exceeds the amount obtained from water extraction, then an extraction profile should be obtained. It may be advisable to obtain a quantitative extraction profile of an elastomeric or plastic packaging component and to compare this periodically to the profile from a new batch of the packaging component. Extractables should be identified whenever possible.

Nonsterile Products

Solids The most common solid oral dosage forms are capsules and tablets. A typical solid oral dosage forms container closure system is a plastic, usually high-density polyethylene (HDPE), bottle with a screw-on or snap-off closure and a flexible packaging system such as a pouch or a blister package. A typical closure consists of a cap, often with a liner, frequently with an inner seal. If used, fillers, desiccants, and other absorbent materials are considered primary packaging components.

A change in the selection of packing materials combined with a change in storage conditions or conditions during administration of the drug products may provoke stability problems.

Many studies have been conducted on predicting the role of packaging in moisture adsorption by dosage forms. Adsorption of moisture by tablets contained in polypropylene films was successfully modeled from storage temperature and the difference in water vapor pressure between the inside and outside of the packaging [15].

Chemical and physical degradation of packaged dosage forms caused by moisture adsorption has been predicted from the moisture permeability of the packaging. For example, strength changes of lactose–corn starch tablets in strip packaging [16] and discoloration of sugar-coated tablets of ascorbic acid [17, 18] were predicted using the moisture permeability coefficient of the packaging.

Typical flexible forms of packaging containing solid oral dosage forms are the blister package and the pouch. A blister package usually consists of a lidding material and a forming film. The lidding material is usually a laminate which includes a barrier layer (e.g., aluminum foil) with a print primer on one side and a sealing agent (e.g., a heat-sealing lacquer) on the other side.

The sealing agent contacts the dosage form and the forming film. The forming film may be a single film, a coated film, or a laminate. A pouch typically consists of film or laminate which is sealed at the edges by heat or adhesive.

Solid oral dosage forms generally need to be protected from the potential adverse effects of the following:

1. Water vapor (e.g., moisture may affect the decomposition rate of the active drug substance or the dissolution rate of the dosage form)
2. Incident light (e.g., in case of photosensitive products)
3. Reactive gases (e.g., oxygen could provoke oxidative reactions)

Carefully selected packaging material may help protect drug products. For example, a blister or pouch and use of secondary packing may be used to protect pack photosensitive material, especially when a dark polymeric film with a covering lid made of aluminum is used for blister packing. Blister packaging using multilayer HDPE material and selection of an adequate sealing technique may help prevent moisture in the blister system. However, plastics and glass for packaging of solid oral dosage forms and for powders for reconstitution should meet the requirements of the USP container test. Incorporating oxygen adsorbents such as iron powder into packaging units can reduce the effect of oxygen. Protection from light can be achieved using primary packaging (packaging that is in direct contact with the dosage forms) and secondary packaging made of light-resistant materials. May be involved in the photolytic degradation kinetics. The velocity of the photochemical reaction may be affected not only by the light source, intensity, and wavelength of the light but also by the size, shape, composition, and color of the container.

Great effort should be taken to stabilize a formulation in such a way that the shelf life becomes independent of the storage conditions. The photostability of drugs and excipients should be evaluated at the formulation development stage in order to assess the effects of packaging on the stability of the final product. Molsidomine tablet preparations in inadequate primary containers (blister) without secondary containers when exposed to irradiation may produce morpholine. These results illustrate the importance of packaging for the stability of molsidomine [19].

Three standard tests for water vapor permeation have been established by the USP for use with solid oral dosage forms.

1. Polyethylene containers (USP ⟨661⟩) [10]
2. Single-unit containers and unit-dose containers for capsules and tablets (USP ⟨671⟩)
3. Multiple-unit containers for capsules and tablets (USP ⟨671⟩) [10]

The cotton and rayon used as fillers in solid oral dosage form containers may not meet pharmacopeial standards, but through appropriate tests and acceptance criteria for identification and moisture content, their adequacy should be shown. For example, rayon has been found to be a potential source of dissolution problems for gelatin capsules and gelatin-coated tablets.

Desiccants are often used to eliminate moisture in packaging when the moisture resistance of the packaging is not sufficient to prevent exposure. The utility of desiccants has been assessed based on a sorption–desorption moisture transfer model [20].

Desiccants or other absorbent materials are primary packaging component. The components should differ in shape and/or size from the tablets or capsules with which they are packaged. Their composition should be provided and their inertness should be proved through appropriate tests, and acceptance criteria should be established.

A topical powder product may be marketed in a sifter-top container made of flexible plastic tubes or as part of a sterile dressing (e.g., antibacterial product). The topical formulations in a collapsible tube can be constructed from low-density polyethylene (LDPE), with or without a laminated material. Normally, there is no product contact with the cap during storage. Thus usually there is no cap liner, especially in collapsible polypropylene screw caps. Normally separate applicator devices are made from LDPE. Product contact is possible if the applicator is part of the closure, and therefore an applicator's compatibility with the drug product should be established, as appropriate (e.g., vaginal applicators).

Nonsolids Typical liquid-based oral dosage forms are elixirs, emulsions, extracts, fluid extracts, solutions, gels, syrups, spirits, tinctures, aromatic waters, and suspensions. These products are usually nonsterile but typically need to be protected from solvent loss, microbial contamination, and sometimes exposure to light or reactive gases (e.g., oxygen).

The presence of a liquid phase implies a significant potential for the transfer of materials from a packaging component into the dosage form.

The higher viscosity of semisolid dosage forms and transdermal systems may cause the rate of migration of leachable substances into these dosage forms to be slower than for aqueous solutions. Due to extended contact, the amount of leachables in these drug products may depend more on a leachable material's affinity for the liquid/semisolid phase than on the rate of migration.

In addition to absorption onto and absorption into containers, transfer of container components into pharmaceuticals may affect the perceived stability/quality of drug dosage forms. Adsorption of volatile components from rubber closures onto freeze-dried parenterals during both dosage form processing and storage brought about haze formation upon reconstitution [21–23]. Leaching of dioctyl phthalate, a plasticizer used especially in polyvingl chloride (PVC) plastics, into intravenous solutions containing surfactants was observed [24, 25]. Plastics contain additives to enhance polymer performance. PVC may contain phthalate diester plasticizer, which can leach into infusion fluids from packaging [26].

The liquid-based oral dosage forms may be marketed in multiple-unit bottles or in unit-dose or single-use pouches or cups. The dosage form may be used as is or admixed first with a compatible diluent or dispersant. A liquid-based oral drug pouch may be a single-layer plastic or a laminated material. The pouches may use an overwrap, which is usually a laminated material.

For LDPE components, data from USP container tests are typically considered sufficient evidence of compatibility. The USP general chapters do not specifically address safety for polyethylene (HDPE or LDPE), polypropylene (PP), or laminate components.

In such cases, an appropriate reference to the indirect food additive regulations [27] is typically considered sufficient. This reference is considered valid for liquid-based oral dosage forms which the patient will take only for a relatively short time.

For liquid-based oral drug products which the patient will continue to take for an extended period, that is, months or years, and is expected to extract greater amounts of substances from plastic packaging components than from water (presence of cosolvents), additional extractable information may be needed to address safety issues.

Topical dosage forms such as creams, emulsions, gels, lotions, ointments, pastes, and powders may be marketed in plastic materials. Topical dosage formulations are for local (not systemic) effect and are generally applied to the skin or oral mucosal surfaces. Some vaginal and rectal creams and nasal, otic, and ophthalmic solutions may be considered for topical drug products.

A rigid bottle, a collapsible tube, or a flexible pouch made of plastic material may be used for liquid-based topical product. These preparations are marketed in a single- or multiple-unit container. For example, dissolved drug (or any substance, e.g., benzocaine) may diffuse in the suppository base and can, for instance, partition into polyethylene linings of the suppository wrap.

Topical delivery systems are self-contained, discrete dosage forms that are designed to deliver drug for an extended period via intact skin or body surface, for example, transdermal, ocular, and intrauterine.

These systems may be constructed of a plastic or polymeric material loaded with active ingredients or a coated metal. Each of these systems is generally marketed in a single-unit soft blister pack or a preformed tray with a preformed cover or overwrap. The compatibility and safety for topical delivery systems are addressed in the same manner as for topical drug products. Performance and quality control should be addressed for the rate-controlling membrane.

Sterile Products

Nonsolids An SVP may be packaged in a disposable cartridge, a disposable syringe, or a flexible bag made of polymeric plastic. Flexible bags are typically constructed with multilayered plastic (Table 2).

An LVP may be packaged in a vial, a flexible bag, or, in some cases, a disposable syringe. Packaging material for cartridges, syringes, vials, and ampules are usually composed of polypropylene (Table 2).

Stoppers and septa in cartridges and syringes are typically composed of elastomeric materials. An overwrap may be used with flexible bags to retard solvent loss and to protect the flexible packaging system from rough handling.

Diazepam in intravenous fluid containers and administration sets exhibited a loss during storage due to adsorption onto and absorption into plastics [11, 12].

Absorption of clomethiazole edisylate and thiopental sodium into PVC infusion bags was observed [28].

The pH dependence of adsorption/absorption of acidic drug substances such as warfarin and thiopental and basic drug substances such as chlorpromazine and diltiazem indicates that only the un-ionized form of the drug substance is adsorbed onto or absorbed into PVC infusion bags [29].

The absorption was correlated to the octanol–water partition coefficients of the drugs, suggesting that prediction of absorption from partition data is possible [30, 31]. Polymers such as Nylon-6 (polycaprolactam) are known to adsorb drug substances such as benzocaine [32].

The ophthalmic drug products are usually solutions marketed in a LDPE bottle with a dropper built into the neck. A few solution products use a glass container due to stability concerns regarding plastic packaging components.

3.2.2.4 Metal as Packaging Material

Metal tubes constructed of a single material are the packaging material of choice for topical dosage forms and may be tested readily for stability with a product. Tubes with a coating, however, present additional problems. The inertness of coating material must be established through adequate tests and guarantee that it completely covers underlying material. The coating material must be resistant to creaking, leaking, leaching, and solvent erosion. For example, frequently used aluminum tubes have demonstrated reactivity with fatty alcohol emulsions, mercurial compounds, and preparations with pH below 6.5 and above 8.0. Nonreactive, epoxy linings have been found to make aluminum tubes resistant to attack [6].

TABLE 1 Plastic Additives

Type	Purpose	Examples
Lubricants	Improve processability	Stearic acid paraffin waxes, polyethylene (PE) waxes
Stabilizers	Retard degradation	Epoxy compounds, organotins, mixed metals
Plasticizers	Enhance flexibility, resiliency, melt flow	Phthalates
Antioxidants	Prevent oxidative degradation	Hindered phenolics (BHT), aromatic amines, thioesters, phosphites
Antistatic agents	Minimize surface static charge	Quaternary ammonium compounds
Slip agents	Minimize coefficient of friction, especially polyolefins	
Dyes, pigments	Color additives	

Source: From ref. 6.

TABLE 2 Parenteral Drug Administration Devices

Sterile Device	Plastic Material
Containers for blood products	Polyvinyl chloride
Disposable syringes	Polycarbonate, polyethylene, polypropylene
Irrigating solution containers	Polyethylene, polypropylene, polyvinyl chloride
Intravenous infusion fluid containers	Polyethylene, polypropylene, polyvinyl chloride
Administration sets	Nylon (spike), polyvinyl chloride (tubing), polymethylmethacrylate (needle adapter), polypropylene (clamp)
Catheter	Teflon, polypropylene, thermoplastic elastomers

Source: From ref. 6.

Some examples of plastic additives and parenteral drug administration devices used as packaging materials for sterile products can be seen in Tables 1 and 2.

Ophthalmic ointments are marketed in a metal tube with an ophthalmic tip. Ophthalmic ointments that are reactive toward metal may be packaged in a tube lined with an epoxy or vinyl plastic coating.

Metal containers, pressurized or not, may also be used for topical drug products. Topical dosage forms include aerosols, emulsions, gels, powders, and solutions and may be marketed in metallic flasks, pressurized or not. Topical dosage formulations are for local (not systemic) effect and are generally applied to the skin or oral mucosal surfaces. Some vaginal and rectal creams and nasal and otic spray drug products may be considered for marketing in metallic containers for topical use.

A number of topical products marketed as a pressurized aerosol may be dispensed in a metallic bottle with a screw cap. Topical dosage forms in aluminum tubes usually include a liner. A tube liner is frequently a lacquer or shellac whose composition should be stated. A metallic pressurized packaging system for a liquid-based topical product may deter solvent loss and may provide protection from light when appropriate.

The droplet size of topical aerosol sprays does not need to be carefully controlled, and the dose usually is not metered as in inhalers. The spray may be used to apply the drug to the skin (topical aerosol) or mouth (lingual aerosol) and the functionality of the sprayer should be addressed. The drug product has no contact with the cap and short-term contact with the nozzle. A topical aerosol may be sterile or may conform to acceptance criteria for microbial limits. However, the physical stability of aerosols can lead to changes in total drug delivered per dose and total number of doses that may be obtained from the container.

3.2.2.5 Applications: Some Examples

Many research papers in the scientific literature present studies showing the importance of the effect of packaging materials on the stability of pharmaceutical and cosmetic preparations:

1. Santoro and co-workers [33] presented results of the stability of oral rehydration salts (ORSs) in different types of packaging materials. The objective of the research was to give guidance on the adequate choice of packaging material presenting the indispensable characteristics in order to protect ORS preparation. This pharmaceutical preparation is essential to children living in developing countries with tropical climate and its distribution is one of the programs of the World Health Organization (WHO) [34].

It has been proved in several research papers that water is the most important factor in the component's degradation of ORSs. To proceed with the study, the pharmaceutical formulation was prepared by a pharmaceutical manufacture. The batch was packed in six types of packaging material. After storage of samples for 36 weeks maintained at ambient temperature, at ambient temperature and 76% relative humidity, and at 40°C with 80% relative humidity, analyses of water determination were made at different intervals of time. Water determination was performed by loss on drying at 50°C and Karl Fisher methods.

The studied ORS preparation contained anhydrous glucose (20 g), sodium chloride (3.5 g), trisodium dehydrate citrate (2.9 g), and potassium chloride (1.5 g)

According to the results, the packaging material that better protected the ORS preparation is the one constituted of polyester (18 g), aluminum (35%), and polyethylene (50 g).

2. The effect of packaging materials on the stability of sunscreen emulsions was also studied by Santoro and co-workers [35, 36]. The purpose of the research was to study the stability of an emulsion containing UVA, UVB, and infrared sunscreens after storage in different types of packaging materials (glass and plastic flasks, plastic and metallic tubes). The samples, emulsions containing benzophenone-3 (B-3), octyl methoxycinnamate (OM), and Phycocorail , were stored at 10, 25, 35, and 45°C and representative samples were analyzed after 2, 7, 30, 60, and 90 days. Stability studies were conducted by analyzing samples at predetermined intervals by high-performance liquid chromatography (HPLC) along with periodic rheological measurements.

The proposed HPLC method enabled the separation and quantitative determination of B-3 and OM present in sunscreens. The method was successfully applied in the stability studies of the emulsions. The method is simple, precise, and accurate; there was no interference from formulation components. The sample emulsions stored at different temperatures presented similar rheological behavior, at least during the period of the study (three months). Most of the samples showed a pseudoplastic non-Newtonian thixotropic profile. There were no significant changes in the physical and chemical stability of emulsions stored in different packaging material. The studied glass and plastic packaging materials were found adequate for storing referred solar protector emulsions.

3. Sarbach and co-workers [48], studied the effect of plastics packaging materials on parenteral pharmaceutical products. Compatibility studies of these containers with different contents are required for drug registration. The authors demonstrated the migration phenomena which occurred between a trilaminated film and a parenteral solution of metronidazole at 0.5%. The main migration products found in the solution were e-caprolactam and a phthalic derivative. The authors also separated several unidentified compounds probably coming from the polyurethane adhesive.

4. Molsidomine is sensitive to light and shows a fast decomposition in solutions and in tablets. Thoma and co-workers [37] showed the importance of light-resistant packaging material for photolabile pharmaceuticals. They irradiated molsidomine preparations over a period of 72 h in a light cabinet according to storage at daylight for about 4–6 weeks. Losses of 23–90% in tablets and 43–60% in solutions were found. The photodegradation could be overcome by selection of suitable packaging materials, colorants or vanillin. The degradation product morpholine after dansylation was determined by HPLC and showed contents of 0.10–0.67 mg in tablets and 0.10–0.38 mg/mL in solution after irradiation.

These examples, among many others described in the scientific literature, illustrate the importance of proper selection of packaging material for the stability and effectiveness of pharmaceutical dosage forms.

3.2.3 QUALITY CONTROL OF PACKAGING MATERIAL

3.2.3.1 General Considerations

Several regulatory agencies as well as private agencies [Food and Drug Administration (FDA), British Pharmacopoeia, WHO, USP] [4, 10, 34, 38] have issued guidelines on the safety evaluation of materials and container closure systems. However, the ultimate proof of the safety and suitability of a container closure system and the packaging process is established by full shelf life stability studies. An important step in such evaluations is characterization of the packaging materials and the chemicals that can migrate or extract from container closure system components to the drug product. This extractable material belongs to diverse chemical classes that can migrate from polymeric materials, such as antioxidants, contaminants, lubricants, monomers, plasticizers, and preservatives. Such basic information is critical to understanding the biological safety and suitability of a container.

Establishing the safety of container closure systems is of key importance to the medical and pharmaceutical industries (Table 3). It is no less important than the contents themselves. The FDA's document "Guidance on Container Closure Systems for Packaging Human Drugs and Biologics" makes this point clear [4].

TABLE 3 Examples of Packaging Concerns for Common Classes of Drug Products

Degree of Concern Associated with Route of Administration	Likelihood of Packaging Component–Dosage Form Interaction		
	High	Medium	Low
Highest	Inhalation aerosols and solutions; injection; injectable suspensions	Sterile powders and powders for injections and inhalation powders	
High	Ophthalmic solutions and suspensions; transdermal ointments and patches; nasal aerosols and sprays		
Low	Topical solutions and suspensions; topical and lingual aerosols; oral solutions and suspensions	Topical powders; oral powders	Oral tablets and oral (hard and soft; gelatin) capsules

The FDA's guidance document requires the evaluation of four attributes to establish suitability: protection, compatibility, safety, and performance/drug delivery. The document also provides a structured approach to ranking packaging concerns according to the route of drug administration and likelihood of packaging component–dosage form interaction. A container closure system acceptable for one drug product cannot be assumed to be appropriate for another. Each product should have sufficient information to establish that a container and its components are right for their intended use [4].

To establish suitability, all four attributes must be evaluated and be shown to pose no concern to the drug product or to product performance. Suitability refers to the tests used for the initial qualification of the container closure system with regard to its intended use. The guidance defines what tests must be done to evaluate each of the attributes of suitability.

While the tests and methods described in Table 4 allow one to provide data that the container closure system is suitable for its intended use, an application must also describe the quality control (QC) measures that will be used to ensure consistency in the packaging components. The principal considerations for the QC measures are the physical characteristics and the chemical composition. By choosing two or three of the tests done in the initial suitability study, a QC program can be established that will ensure the consistency of the container closure system (Table 4).

Protection A container closure system should provide the dosage form with adequate protection from factors (e.g., temperature, light) that can cause a degradation in the quality of that dosage form over its shelf life. Common causes of such degradation are exposure to light, loss of solvent, exposure to reactive gases (e.g., oxygen), absorption of water vapor, and microbial contamination.

A container intended to provide protection from light or offered as a light-resistant container must meet the requirements of the USP ⟨661⟩ light transmission test. The procedure requires the use of a spectrophotometer, with the required sensitivity and accuracy, adapted for measuring the amount of light transmitted by the plastic material used for the container.

The ability of a container closure system to protect against moisture can be ascertained by performing the USP ⟨661⟩ water vapor permeation test. The USP sets limits on the amount of moisture that can penetrate based upon the size and composition of the plastic components [HDPE, LDPE, or polyethylene terephthalate (PET)].

Evaluating the integrity of the container can be done in several ways. Two of the most common tests are dye penetration and microbial ingress. Container closure systems stored in a dye solution and exposed to pressure and vacuum cycles are examined for dye leakage into the container. The microbial ingress is similar in fashion but determines the microbial contamination of the contents when soaked in a media contaminated with bacteria. Other quantitative tests that can be run are vacuum/pressure decay, helium mass spectrometry, and gas detection.

Compatibility Packaging components that are compatible with a dosage form will not interact sufficiently to cause unacceptable changes in the quality of either the dosage form or the packaging component. A leachability study designed to evaluate the amount and/or nature of any chemical migrating from the plastic material to the drug product should be considered. The study should evaluate substances that migrate into the drug product vehicle for the length of shelf life. The drug product should be evaluated at regular intervals, such as at one, three, or six months or one or two years, until the length of the shelf-life claim has been met.

Analytical techniques such as liquid chromatography/mass spectrometry (LC/MS) to evaluate nonvolatile organics, gas chromatography/mass spectrometry (GC/MS) to evaluate semivolatile organics, and inductively coupled plasma (ICP) spectroscopy to detect and quantitate inorganic elements should be a part of this study.

Unknown impurities and degradation products can be identified using liquid or gas chromatography along with MS. Information or substances identified from extractable chemical evaluation can be used to help prepare standards specific for the plastic container being studied during leachability studies. Development and validation of the selective analytical method should be thoroughly studied before its application in the detection of leachable chemicals in active drug substance and drug product.

Organoleptic and chemical changes such as precipitates, discoloration, strange odor, and pH modification are signs of degradation of drug product. Changes in the physical characteristics of the container, such as brittleness, should be evaluated using thermal analysis and hardness testing. An infrared spectroscopic scan can fingerprint the materials and also provide proof of identity. Spectrophotometry and LC with ultraviolet detection can be used for the analysis of drug product stored at different stress conditions. These tests can be used for the quality control of drug product as well as for conducting stability studies on different products stored in the same container material.

Safety Packaging components should be constructed of materials that will not leach harmful or undesirable amounts of substances to which a patient will be exposed when being treated with the drug product. This consideration is especially

TABLE 4 Properties of Suitability Concerns and Interactions

Attributes	Concerns and Interactions	Proposed Methods
Protection	Exposure to light, moisture, microbial ingress, and oxidation from presence of oxygen	USP⟨661⟩ light transmission and water vapor permeation, container integrity (microbial ingress, dye penetration, helium leak)
Compatibility	Leachable induced degradation, absorption or adsorption of drug, precipitation, change in pH, discoloration, brittleness of packaging materials	Leachability study (migration of chemicals into drug product) using LC/MS, GC/MS, ICP/AA, pH, appearance of drug and container, thermal analysis (DSC, TGA), and infrared (IR)
Safety	No leached harmful or undesirable amounts of substances to expose patients treated with drug	Extraction study (USP physicochemical tests–plastics), USP elastomeric closures for injections, toxicological evaluation, USP biological reactivity and complies with CFR, additives and purity
Performance	Container closure system functionality, drug delivery	Functionality (improved patient compliance or use), delivery (transfer dose in right amount or rate)

Abbreviations: DSC, differential scanning calorimetry; ICP, Inductively coupled plasma spectrometer; AA, Atomic absorption.
Source: From ref. 39.

important for those packaging components which may be in direct contact with the dosage form, but it is also applicable to any component from which substances may migrate into the dosage form (e.g., an ink or adhesive). Determining the safety of a packaging component is not a simple process, and a standardized approach has not been established. However, an extraction study should be one of the first considerations.

A good knowledge regarding possible extractable material could help analysts develop specific and selective methods to identify extractables from container closure components under various control extraction study conditions.

Precise information on the synthesis of the polymer and descriptions of the monomers used in the polymerization, the solvents used in the synthesis, and the special additives that have been added during material production as well as knowledge of degradation products that may be released into the drug product are also important.

Some potential extractable chemicals from packaging materials are water soluble, while others are soluble only in nonpolar environments. The USP includes physicochemical tests for plastics based on water extracts, while water, alcohol, and hexane extracts are required for polyethylene containers under controlled temperature and time parameters (70°C for 24 h for water and alcohol and 50°C for 24 h for hexane).

The USP physicochemical tests for extractables should be a part of all suitability programs, regardless of the criticality of the drug dosage form. USP elastomeric closures for injections should also be a part of the extractables study to establish safety. These USP tests, which have evolved over many years, are relevant, sensitive, rapid, and inexpensive. They help establish material safety.

The safety of material can be guaranteed by using appropriate analytical methods and instrumentation to identify and quantitate extracted chemicals. Liquid and gas chromatography and MS are powerful analytical tools that can separate and quantitate volatile and nonvolatile chemicals along with useful structural information. The mass spectrum or fragmentation pattern acquired for each molecule makes these excellent and effective tools for identifying unknown impurities or degradation products.

Toxicological evaluation of identified and unidentified impurities from a container can help improve the safety index of drug products. The toxicological evaluation should take into consideration container closure system properties, drug product formulation, dosage form, route of administration, and dose regimen. A close correlation between chemical and toxicological information can provide better control on safety and compatibility of containers and closures.

Performance The fourth attribute of the suitability of the container closure system, performance and drug delivery, refers to its ability to function in the manner for which it was designed. There are two major considerations when evaluating performance. The first consideration is functionality that may improve patient compliance, [e.g., a two-chamber vial or intravenous (IV) bag], or improve ease of use (e.g., a cap that contains a counter, a prefilled syringe). The second consideration is drug delivery, which is the ability of the packaging system to deliver the right amount or rate (e.g., a prefilled syringe, a transdermal patch, a metered tube, a dropper or spray bottle, a dry-powder inhaler, and a metered-dose inhaler).

3.2.3.2 Packaging Components

Quality control refers to the tests typically used and accepted to establish that, after the application is approved, the components and the container closure system continue to possess the characteristics established in the suitability studies.

To ensure consistency, protection, compatibility, safety, and performance of the packaging components, it is necessary to define QC measures that will be used to ensure consistency in the packaging components. These controls are intended to limit unintended postapproval variations in the manufacturing procedures or materials of construction for a packaging component and to prevent adverse effects on the quality of a dosage form.

The USP tests and studies for establishing suitability and QC of container closure system and for associated component materials are summarized in Table 5.

Hydrolysis and oxidation are the two main routes of degradation for the majority of drugs. To gain more information, the drug could be subjected to a range of temperature and relative humidity conditions. In addition, photostability studies could be conducted by exposure to artificial or natural light conditions. Elevated temperature, humidity, and light stress the drug and induce rapid degradation. Harmonized guidelines are available for new drug substances and products and may provide useful information to characterize degradation processes and selection of appropriate packaging material.

The primary packaging must physically protect the product from the mechanical stresses of warehousing, handling, and distribution. Mechanical stress may take a

TABLE 5 U.S. Pharmacopeia General Tests and Assays

Chapter	Topic
⟨1⟩	Injections
⟨51⟩	Antimicrobial preservatives—effectiveness
⟨61⟩	Microbial limit tests
⟨71⟩	Sterility tests
⟨87⟩	Biological reactivity tests, in vitro
⟨88⟩	Biological reactivity tests, in vivo
⟨161⟩	Transfusion and infusion assemblies
⟨381⟩	Elastomeric closures for injections, biological test procedures, physicochemical test procedures
⟨601⟩	Aerosols
⟨661⟩	Containers: light transmission; chemical resistance—glass containers; biological tests—plastics and other polymers; physicochemical tests—plastics; containers for ophthalmics—plastics; polyethylene containers; polyethylene terephthalate bottles and polyethylene; terephthalate G bottles; single-unit containers and unit-dose containers for nonsterile; solid and liquid dosage forms; customized patient medication packages
⟨671⟩	Containers—permeation: multiple-unit containers for capsules and tablets; single-unit containers and unit-dose containers for capsules and tablets
⟨691⟩	Cotton (or the monograph for purified rayon USP)
⟨771⟩	Ophthalmic ointments
⟨1041⟩	Biologics
⟨1151⟩	Pharmaceutical dosage forms

Source: From ref. 10.

variety of forms, from impact through vibration in transit and compression forces on stacking.

The demands for mechanical protection will vary with product type: Glass ampules will require greater protection than plastic eye drop bottles, for example. Other protection is required from environmental factors such as moisture, temperature changes, light, gases, and biological agents such as microorganisms and, importantly, humans.

The global market for medicinal products requires that the products are stable over a wide range of temperatures ranging from subzero in the polar region, 15°C in temperate zones, up to 32°C in the tropics. Along with this temperature variation, relative humidity can vary from below 50% to up to 90%, a feature that the packaging should be able to resist if necessary. The majority of packaging materials (including plastics) are to some degree permeable to moisture and the type of closure employed, such as screw fittings, may also permit ingress of moisture. The susceptibility of the product to moisture and its hygroscopicity will have to be considered and may require packaging with a desiccant or the use of specialized strip packs using low-permeability materials such as foil.

Temperature fluctuations can lead to condensation of moisture on the product and, with liquids, formation of a condensate layer on top of the product. This latter problem is well known and can lead to microbiological spoilage as the condensate is preservative free.

If the product is sensitive to photolysis, then opaque materials may be required. Most secondary packaging materials (e.g., cartons) do not transmit light, but in some cases specialized primary packaging designed to limit light transmission is employed.

The package must also prevent the entry of organisms; for example, packaging of sterile products must be microorganism proof—hence the continued use of glass ampules. For nonsterile products the preservative provides some protection, but continual microbial challenge will diminish the efficacy of the preservative, and spoilage or disease transmission may occur.

The packaging material must not interact with the product either to adsorb substances from the product or to leach chemicals into the product. Plastics contain additives to enhance polymer performance. PVC may contain phthalate diester plasticizer, which can leach into infusion fluids from packaging. Antimicrobial preservatives such as phenylmercuric acetate are known to partition into rubbers and plastics during storage, thus reducing the formulation concentration below effective antimicrobial levels.

A complication of modern packaging is the need for the application of security seals to protect against deliberate adulteration and maintain consumer confidence. The active products used must also be stability tested in the proposed packaging material.

The FDA guidance for industry suggests considering consistency in physical and chemical composition. Using a few simple tests, the quality of components and ultimately the container closure system can be monitored.

Physical Characteristics The physical characteristics of interest include dimensional criteria (e.g., shape, neck finish, wall thickness, design tolerances), physical parameters critical to the consistent manufacture of a packaging component (e.g., unit weight), and performance characteristics (e.g., metering valve delivery volume or the ease of movement of syringe plungers). Unintended variations in dimensional parameters, if undetected, may affect package permeability, drug delivery performance, or the adequacy of the seal between the container and the closure. Variation in any physical parameter is considered important if it can affect the quality of a dosage form.

Physical considerations such as water vapor transmission to evaluate seal integrity, thermal analysis such as DSC to monitor melting point and glass transitions of plastics, and IR scanning to prove identity should be part of an ongoing quality control monitoring program.

Chemical Composition The chemical composition of the materials of construction may affect the safety of a packaging component. New materials may result in new substances being extracted into the dosage form or a change in the amount of known extractables. The chemical composition may also affect the compatibility, functional characteristics, or protective properties of packaging components by changing rheological or other physical properties (e.g., elasticity, resistance to solvents, or gas permeability).

The chemical composition should also be evaluated by performing the simple but informative USP physicochemical tests using water, drug product vehicle, and alcohol extractions of plastic components. Specifications should be set for nonvolatile residue (total extractables) during the initial suitability tests and then used to monitor the level of polar and nonpolar extractables as part of a quality control plan.

A change in the composition of packaging raw material or a change in formulation is considered a change in the specifications. Due care must be taken to guarantee the safety, compatibility, and performance of a new dosage form in a new packaging system.

The use of stability studies for monitoring the consistency of a container closure system in terms of compatibility with the dosage form and the degree of protection provided to the dosage form is essential. Except for inhalation drug products, for which batch-to-batch monitoring of the extraction profile for the polymeric and elastomeric components is routine, no general policy concerning the monitoring of a packaging system and components with regard to safety is available.

Secondary packaging components are not intended to make contact with the dosage form. Examples are cartons, which are generally constructed of paper or plastic, and overwraps, fabricated from a single layer of plastic or from a laminate made of metal foil, plastic, and/or paper. In special cases, secondary packaging components provide some additional measure of protection to the drug product. In such cases it could be considered a potential source of contamination and the safety of the raw materials should be taken into consideration.

3.2.3.3 Inhalation Drug Products

Inhalation drug products include inhalation aerosols (metered-dose inhalers); inhalation solutions, suspensions, and sprays (administered via nebulizers); inhalation powders (dry-powder inhalers); and nasal sprays. The carboxymethylcellulose (CMC) and preclinical considerations for inhalation drug products are unique in that these drug products are intended for respiratory tract compromised patients. This is reflected in the level of concern given to the nature of the packaging components that may come in contact with the dosage form or the patient (Table 4).

In October 1998, the FDA issued guidance for industry regarding container closure systems such as metered-dose inhaler (MDI) and dry-powder Inhaler (DPI) drug products.

3.2.3.4 Drug Products for Injection and Ophthalmic Drug Products

Injectable dosage forms are sterile and represent one of the highest risk drug products. Injectable drug products may be liquids in the form of solutions, emulsions, and suspensions or dry solids that are to be combined with an appropriate vehicle to yield a solution or suspension.

Cartridges, syringes, vials, and ampules are usually composed of type I or II glass or polypropylene frequently used to deliver SVP and LVPs. Flexible bags are typically constructed with multilayered plastic. Stoppers and septa in cartridges, syringes, and vials are typically composed of elastomeric materials. An overwrap may be used with flexible bags to retard solvent loss and to protect the flexible packaging system from rough handling.

Injectable drug products require protection from microbial contamination (loss of sterility or added bioburden) and may also need to be protected from light or exposure to gases (e.g., oxygen). Liquid-based injectables may need to be protected from solvent loss, while sterile powders or powders for injection may need to be protected from exposure to water vapor.

For elastomeric components, data showing that a component meets the requirements of USP elastomeric closures for injections should typically be performed to assure safety. For plastic components, USP biological reactivity tests are recommended to assure evidence of safety. Whenever possible, the extraction studies described in USP should be performed using the drug product. Extractables should be identified whenever possible. For a glass packaging component, data from USP "Containers: Chemical resistance—Glass containers" will typically be considered sufficient evidence of safety and compatibility. In some cases (e.g., for some chelating agents), a glass packaging component may need to meet additional criteria to ensure the absence of significant interactions between the packaging component and the dosage form.

The performance of a syringe is usually addressed by establishing the force to initiate and maintain plunger movement down the barrel and the capability of the syringe to deliver the labeled amount of the drug product.

Ophthalmic drug products are usually solutions marketed in a LDPE bottle with a dropper built into the neck or ointments marketed in a metal tube lined with an epoxy or vinyl plastic coating with an ophthalmic tip.

Since ophthalmic drug products are applied to the eye, compatibility and safety concerns should also address the container closure system's potential to form substances which irritate the eye or introduce particulate matter into the product (USP ⟨771⟩, ophthalmic ointments).

3.2.3.5 Liquid-Based Oral Products, Topical Drug Products, and Topical Delivery Systems

The presence of a liquid phase implies a significant potential for the transfer of materials from a packaging component into the dosage form.

Liquid-Based Oral Drug Products Typical liquid-based oral dosage forms are elixirs, emulsions, extracts, fluid extracts, solutions, gels, syrups, spirits, tinctures, aromatic waters, and suspensions. These products are usually nonsterile but may be monitored for changes in bioburden or for the presence of specific microbes.

A liquid-based oral drug product typically needs to be protected from solvent loss, microbial contamination, and sometimes exposure to light or reactive gases (e.g., oxygen). For glass components, data showing that a component meets the requirements of USP "Containers: Glass containers" are accepted as sufficient evidence of safety and compatibility. For LDPE components, data from USP container tests are typically considered sufficient evidence of compatibility.

The USP general chapters do not specifically address safety for polyethylene (HDPE or LDPE), PP, or laminate components. A patient's exposure to substances extracted from a plastic packaging component (e.g., HDPE, LDPE, PP, laminated components) into a liquid-based oral dosage form is expected to be comparable to a patient's exposure to the same substances through the use of the same material when used to package food [27].

Topical Drug Products Topical dosage forms include aerosols, creams, emulsions, gels, lotions, ointments, pastes, powders, solutions, and suspensions. These dosage forms are generally intended for local (not systemic) effect and are generally applied to the skin or oral mucosal surfaces. Topical products also include some nasal and otic preparations as well as some ophthalmic drug products. Some topical drug products are sterile and should be subject to microbial limits.

A rigid bottle or jar is usually made of glass or polypropylene with a screw cap. The same cap liners and inner seals are sometimes used as with solid oral dosage forms. A collapsible tube is usually constructed from metal or is metal lined from LDPE or from a laminated material.

Topical Delivery Systems Topical delivery systems are self-contained, discrete dosage forms that are designed to deliver drug via intact skin or body surface, namely transdermal, ocular, and intrauterine.

Each of these systems is generally marketed in a single-unit soft blister pack or a preformed tray with a preformed cover or overwrap. Compatibility and safety for topical delivery systems are addressed in the same manner as for topical drug products. Performance and quality control should be addressed for the rate-controlling membrane. Appropriate microbial limits should be established and justified for each delivery system.

3.2.3.6 Solid Oral Dosage Forms and Powders for Reconstitution

The most common solid oral dosage forms are capsules and tablets. A typical container closure system is a plastic (usually HDPE) or a glass bottle with a screw-on or snap-off closure and a flexible packaging system, such as a pouch or a blister package. If used, fillers, desiccants, and other absorbent materials are considered primary packaging components.

Solid oral dosage forms generally need to be protected from the potential adverse effects of water vapor, light, and reactive gases. For example, the presence of moisture may affect the decomposition rate of the active drug substance or the dissolution rate of a dosage form. The potential adverse effects of water vapor can be determined with leak testing on a flexible package system (pouch or blister package). Three standard tests for water vapor permeation have been established by the USP, namely polyethylene containers (USP ⟨661⟩), single-unit containers and unit-dose containers for capsules and tablets (USP ⟨671⟩), and multiple-unit containers for capsules and tablets (USP ⟨671⟩).

3.2.4 IMPORTANCE OF PROPER PACKAGING AND LABELING

The Poison Prevention Packaging Act (www.cpsc.gov/businfo/pppa.html) requires special packaging of most human oral prescription drugs, oral controlled drugs, certain normal prescription drugs, certain dietary supplements, and many over-the-counter (OTC) drug preparations in order to protect the public from personal injury or illness from misuse of these preparations.

In many countries there are very strict regulations for packaging of many drug substances. Nevertheless, special packaging is not required for drugs dispensed within a hospital setting for inpatient administration. Manufacturers and packagers of bulk-packaged prescription drugs do not have to use special packaging if the drug will be repackaged by the pharmacist.

Various types of child-resistant packages are covered in ASTM International standard D-3475. Medication errors linked to poor labeling and packaging can be controlled through the use of error potential analysis. The recognition that a drug name, label, or package may constitute a hazard to safety typically occurs after the drug has been approved for use and is being marketed. Calls for change almost always result from accumulating reports of serious injuries associated with the use of a drug.

Numerous reports of medication errors are being reported, some of which have resulted in patient injury or death. In a number of these reports, a medication was mistakenly administered either because the drug container (bag, ampule, prefilled syringe and bottle) was similar in appearance to the intended medication's container or because the packages had similar labeling. Obviously, the severity of such errors depends largely on the medication administered.

The problem of medical errors associated with the naming, labeling, and packaging of pharmaceuticals is being very much discussed. Sound-alike and look-alike drug names and packages can lead pharmacists and nurses to unintended interchanges of drugs that can result in patient injury or death. Simplicity, standardization, differentiation, lack of duplication, and unambiguous communication are human factors that are relevant to the medication use process. These factors have often been ignored in drug naming, labeling, and packaging.

The process for naming a marketable drug is always lengthy and complex and involves submission of a new entity and patent application, generic naming, brand naming, FDA—or other corresponding organization all over the world—review, and final approval. Drug companies seek the fastest possible approval and may believe that the incremental benefit of human factor evaluation is small. Very often, the drug companies are resistant to changing, for example, brand names. Although a variety of private-sector organizations in many countries have called for reforms in drug naming, labeling, and packaging standards, the problem remains.

Drug names, labels, and packages are not selected and designed in accordance with human factor principles. FDA standards or other corresponding organizations in other countries do not require application of these principles, the drug industry has struggled with change, and private-sector initiatives have had only limited success.

A number of factors can contribute to the mistaking of one medication for another. Failure to read the package label is one cause. Another if a medication is stored in the wrong location or if clinicians select the medication based solely on the appearance of its package. Also, confusion can occur between medications with names that look alike or sound alike or between premixed medications packaged in similar-looking containers. Another potential source for confusion with premixed medications is the presence of different concentrations of the same medication in a particular location (e.g., a package with 100 mg/mL concentration of a drug could be mistaken for one with 10 mg/mL concentration).

Daily, physicians, nurses, and pharmacists base medical decisions on the information provided by a drug product's labeling and packaging. Unfortunately, poor labeling and packaging have been linked all too often to medication errors. To help practitioners avoid errors, drug manufacturers should present information in a clear manner that can be grasped quickly and easily.

To determine what presentation is most clear, manufacturers should invite and consider the input of physicians, nurses, and pharmacists, because they work with these products every day and are more likely than label and package designers to discover potential problems. Such input provides the basis for failure and effect analysis (FMEA), also known as error potential analysis or error prevention analysis. FMEA is a systematic process that can predict how and where systems might fail. Using FMEA, health care practitioners examine a product's packaging or labeling in order to identify the ways in which it might fail. A number of steps to reduce confusion and improve the readability of a drug product's label have already been determined through the use of FMEA.

The first step is to reduce label clutter. Only essential information, such as the brand and generic names, strength or concentration, and warnings, should appear prominently on the front label. Numerous deaths have been prevented through the addition of a warning to concentrated vials of injectable potassium chloride, for example. Another step includes the use of typeface to enhance distinctive portions of look-alike drug names on look-alike packaging.

Medication errors are also associated with poor product packaging design. Unfortunately, medication errors linked to poor labeling and packaging are sometimes used in the health care environment to justify the damage. Participation of an expertise from health care practitioners, during labeling and packaging design phase, might have prevented several errors.

Whether for established drugs or new entities going through the approval process, the principles of safe practice in naming, labeling, and packaging are the same and must be very well controlled. Safety experts may differ about specific details, but there is little disagreement about the fundamental principles that should be incorporated into the drug approval process.

Based on reports of errors associated with packaging and labeling, many recommendations have been proposed. Some of them are:

1. Avoid storing medications with similar packaging in the same location or in close proximity.
2. Follow the American Society of Health System Pharmacists (ASHP) guidelines or other legislation of a specific country for preventing hospital medication errors [40, 41]. The ASHP's recommendations include the following:

   Fully document all medication prescription and deliveries and instruct staff that discrepancy or misunderstanding about prescription or patient information should be verified with the prescribing physician. Staff members should be told that all caregivers (regardless of level) have the duty to question the prescribing physician (regardless of the physician's relative position in the hospital hierarchy) if they have concerns about a drug, dose, or patient.

   Periodically train staffs in practices that will help avoid medication errors.

   Ensure that the medication storage and distribution to hospital locations outside the pharmacy are supervised by hospital pharmacy staff only.

Nonpharmacists should not be allowed to enter the pharmacy if it is closed.

3. Perform failure mode and effects analysis. This is a technique used to identify all medication errors that could occur, determine how they occur, and estimate what their consequences would be. Steps then should be taken to prevent errors from occurring, when possible, and to minimize the effects of any errors that do occur.
4. Report any information relating to medication errors to the Medication Errors Reporting Program operated by USP convention [10] and the Institute for Safe Medication Practice (ISMP) or other corresponding institutions in the different countries. The program shares information on medication errors with health care professionals to prevent similar errors from recurring.
5. Hospitals should report incidents in which a device caused or helped cause a medication error.
6. Urge suppliers to provide clear and unique labels and packages for their various individual medications.

Some other considerations relating to standards for drug names, labeling standards, and packaging standards are as follows:

1. *Standard for Drug Names.* The most critical issue in drug name selection is that one name should not be easily confused with another. This applies to both generic and brand names. A name must neither sound like that of another drug nor look like another drug name when it is written out by hand. From the industry's standpoint, the challenge is to find a name that is easy to recollect and appropriate for the connotation desired, do not lead astray (safe), and not already a trade name.

Nowadays, increasing sophisticated and effective methods are available for determining the likelihood of confusion by sound or sight.

2. *Labeling Standards.* To minimize the possibility of error, labels should be easy to read and avoid nonessential material. The name of the drug, and not the name of the manufacturer, should be the most prominent feature and should be in at least 12-point type. The use of color is very controversial; some believe that all colors should be prohibited to force personnel to read the labels.

In the 1990s, a Washington State legislator proposed that every drug product entering the state must have a color-coded label. There was concern on the part of many that the state legislature would turn this idea into law. The prospect of having to color-code all the drugs entering a single state galvanized a response by industry, regulators, practitioners, and safe experts who agreed to revise pharmaceutical labeling. A Committee to Reduce Medication Errors was formed to study the problem. The effort eventually satisfied the color coders and the proposed legislation was dropped.

The committee made several recommendations for standardizing and simplifying labels:

1. Eliminate unnecessary words from the label, such as "sterile," "nonpyrogenic," and "may be habit forming."
2. Allow some abbreviations such as "HCl" and "Inj."
3. Make label information consistent.

4. Require that vials containing medication that must be diluted bear the words "Concentrated, must be diluted" in a box on the label, that the vial have a black flip-top with those words on it, and that the ampules carry a black band.

3. *Packaging Standards.* While there is no evidence that trademark colors and logos on boxes pose a problem, the use of color on bottle tops and labels creates many difficulties. There are dozens of drugs whose names are quite different but whose packages look alike. This creates the potential for error when people "see" what they expect to see on the label.

Standards need to be set for color on both caps and labels. Some believe that prohibiting all color would be safest—in effect, taking away a cue that could divert someone from reading the label.

3.2.5 REGULATORY ASPECTS

3.2.5.1 General Considerations

Once the finished dosage form is made, the product should be packed into the primary container and labeled. Additional packaging and labeling are also included. Because of the many products and labeling materials, personnel in this area must be alert to prevent mix-ups. Controls and in-process checks should be carried out throughout the packaging/labeling operation to ensure proper labeling.

Some examples of good manufacturing practices (GMP) requirements specific to packaging and labeling in different countries are as follows:

In the United States the requirements should be written procedures designed to assure that correct labels, labeling, and packaging materials are used for drug products; such written procedures should be followed. These procedures should incorporate features such as prevention of mix-ups and cross-contamination by physical or spatial separation from operations on other drug products.

In Canada, packaging operations are performed according to comprehensive and detailed written operating procedures or specifications, which include identification of equipment and packaging lines used to package the drug, adequate separation, and, if necessary, the dedication of packaging lines packaging different drugs and disposal procedures for unused printed packaging materials. Packaging orders are individually numbered.

In the European Union, the requirements should be formally authorized in the "packaging instructions" for each product containing pack size and type. They are normally included in process controls with instructions for sampling and acceptance limits [42].

3.2.5.2 Food, Drug and Cosmetic Act

About 100 years after its foundation, the Congress of the United States recognized that subjects related to safety and public health could not exclusively be state dependent and measures should be taken to protect the population in vital areas. Therefore, the federal government became interested in regulating products for consumption.

In 1906, the Congress approved the Wiley Law to avoid the production, sale, or transport of food, medications, and alcoholic beverages that were inadequate or falsified, poisonous, or harmful. It was the first food and medication regulation adopted in interstate commerce. The Congress was given power to regulate commerce between foreign nations and several U.S. states.

In 1912 a civil code law was enacted prohibiting any false affirmation of curing or therapeutic effect on medication labels. The current law was enacted on June 27, 1938, and regulates food, medications, medical devices, and diagnostic and cosmetic products. The law of 1938 stopped regulating the trade of alcoholic beverages. This law stated, among other recommendations, the following:

1. The label of each medication had to give the name of each active component and the quantity of some specific substances, active or not.
2. Cosmetics had to be inoffensive and be properly labeled and packaged.

The 1938 law states that the label of a medication should contain adequate information regarding its use. However, in practice, it became evident that some pharmaceuticals and medications had to be administered by or under the orientation of a medical practitioner, due to the inability of a layman to diagnose a disease, choose an effective treatment, and recognize the cure or the symptoms. Several products were thus classified, but "the prescription concept of a medication" was introduced only after Alteration in the Law of Durham-Humphrey's in 1951. Since then, a label had to carry the warring "Caution, the Federal Law prohibits dispensation without medical prescription." The use of these medications had to be restricted to prescription by a practitioner and the packing or printed material inside had to contain adequate information so that the practitioner could prescribe them safely.

Alterations in 1962 of the 1938 Law constituted an attempt to establish rigid controls on the research, production, divulging, promotion, sale, and use of medications as well as to assure its quality, efficiency, and effectiveness [43].

3.2.5.3 New Drugs

Before starting clinical trials in humans, an authorization should be obtained from the FDA. This is known as a clinical trial authorization request for a new medication (AEM), on which it is necessary to establish the following:

1. The name that best describes the medication, including the chemical name and the structure of any new molecule
2. A complete list of medication components.
3. A quantitative composition of the medication.
4. The name and address of the vendor and an acquired description of the new drug
5. The methods, facilities, and controls used for the production, processing, and packing of a new medication
6. All available results available from preclinical and clinical trials
7. Copies of medication labels and the informative material that will be supplied to the researchers

8. A description of the scientific training and the appropriate experience considered by the proponent to qualify a researcher as an adequate expert to investigate the medication
9. The names and "curricula vitae" of all researchers
10. An investigation layout planned for test accomplishment in humans

Solicitations for release of new medications are generally very extensive, sometimes thousands of pages. The information has to be enough to justify the affirmations contained in the label of the proposed medication with respect to effectiveness, dosage, and safety. The exact composition of the content on the medication label is usually decided by consensus between the proponent and the FDA.

The requisites for solicitation of new medications, whether by prescription or not, are identical. The instructions contained in the medication labels for use without prescription should demonstrate that the medication can be used safely without medical supervision.

Once the medications are perfected, the publicity related to them has to be routinely presented to the FDA.

The rules of 1985 also changed the requisites regarding addendums that are necessary when alterations are proposed in the medication or in its labeling, for example.

In regulations promulgated by the FDA on February 12, 1972, a clinic should be called upon regarding the effectiveness of a medication. After that the information may be included in the label or in the drug informative leaflet with eligible sentence and defined by dark lines that contour it [43].

Other dispositions contained in the alterations to the 1962 law are as follows:

1. Immediate registration with the FDA before starting the production, repacking, or relabeling of medications and later annual registration, with inspections to be made at least once every two years.
2. Supportive inspections in the factory, particularly where prescription medications are produced.
3. The procedures used by the manufacturers should be in conformity with the good manufacturing practices, which permits the government to better inspect of all the operations.
4. The common name should be presented on the label.
5. The publicity of a prescription medication should present a brief summary mentioning the secondary effects, the contraindications, and the medication effectiveness.
6. All antibiotics are subject for certification procedures.

3.2.5.4 Labeling Requisites

According to a 1962 law, the main requisites for labeling are as described below. The labeling of over-the-counter medications is regulated by the Food, Drug and Cosmetic Act, which states:

> A medication should be considered falsified unless the label contains: 1. Indications of adequate use and 2. Adequate warnings regarding the pathological indications in those it should not be used or not for children use, when its use can be dangerous for health, of dosages, methods or interval of administration, or unsafe application, of mode and in necessary form for patients' protection.

"Indications of use" were defined in the regulations as information with which even a layman can use the medication safely and for the purpose to which it is designated.

The label of an over-the-counter medication must refer to the active substances, but it is not necessary to indicate its relative quantity, except where the ingredient leads to habituation. In this case the warning "Can lead to habituation" should appear on the label.

A drug can be considered falsified if it does not provide, besides indications of adequate use, warnings against its use in some pathological conditions (or for children) in which the medication can constitute a health risk. Regulations have suggested warnings that can be used for most well-known dangerous substances.

3.2.5.5 Prescription Drugs

Specific requisites for labeling of ethical medications or of prescription medications are also found in the Food, Drug and Cosmetic Act. These need not to contain "adequate indications of use"; however, they must contain indications for the practitioner, inside or outside the package in which the medication is going to be dispensed, with adequate information for its use. This information may include indications, effects, dosages, route of administration, methods, frequency and duration of administration, important dangers, contraindications, secondary effects, and cautions "according to which the practitioners can prescribe the medication assuredly and for the desirable effects, including those for which it is proclaimed."

Regarding all medications, the act requires that the label present a precise affirmation on the weight of the content, measure or counting, as well as the name and manufacturer's address, packer or distributor.

The label of a prescription medication destined for oral administration has to contain the quantity or proportion of each active substance.

If the medication is for parenteral administration, the quantity or proportion of all the excipients have also to be mentioned on the label, except for those that are added to adjust pH or make it isotonic, in which case only the name and its effect are needed. However, if the vehicle for injection is water, this does not need to be mentioned.

If the medication is not to be administered by any of the routes mentioned above, for example, a pomade or a suppository, all excipients must to be mentioned, except for perfuming agents. Perfumes can be designated as such without the need to mention the specific components.

Coloring agents can be assigned without being specified individually, unless this is required in a separate section for regulation of coloring agents, and inoffensive substances added exclusively for individual identification of each product need not be mentioned.

The only warning that is necessary, "Attention: the Law prohibits the dispensing without prescription," should be on the label of a prescription medication or in its secondary packing if the label is too little to contain it.

3.2.5.6 Drug Information Leaflet

The inclusion of a drug information leaflet is not compulsory whatever the medication. However, all medications, whether prescription or of over the counter, have to contain a label with adequate indications for use. If the medication label does not have enough space to contain all the information, the drug information leaflet has to be included with necessary information. The drug information leaflet and labels containing indication information must include the date when the text was last revised.

To satisfy the act, the drug information leaflet usually included in the prescription medication packaging should contain "adequate information on usage, including indications, effects, dosages, methods, route, frequency and duration of administration. Any important dangers, contraindications, secondary effects and cautions, based on which the practitioner can prescribe the medication safely and for desirable effects, including those for which a clam is made." To present the information in a uniform manner, the FDA issued labeling policies describing its format and the order and headings for the drug information leaflet description, action, indication, contraindications, alerts, cautions, adverse reactions, dosage and administration, overdose (when applicable), and as it is supplied.

The drug information leaflet can contain the following optional information:

Animal pharmacology and toxicology
Clinical studies
References

Other specific cautions on medication have to appear in a visible manner at the beginning of the drug information leaflet so that practitioners, pharmacists, and patients can easily see them.

According to GMP, an inspector should be cautious with several aspects of drug production, including the following:

1. Product containers and other components have to be tested and be considered adequate for their intended use only if they are not reactive, departure by-products, or even have absorption capacity; so that they do not affect the safety, identity, potency, quality, or purity of the medication or its components.
2. Packing and labeling operations should be adequately controlled to (1) guarantee that only those medications that own quality standards and attain established specifications in their production and control be distributed, (2) avoid mix-ups during the filling operations, packing, and labeling, (3) assure that the labels and labeling used are correct for the medication, and (4) identify the finished product with a batch or a control number that allows determination of the batch production and control history.

Application of the federal law on food, drug, and cosmetics is the FDA's responsibility, which is a subdivision of the Department of Health and Human Services.

The institution is managed by a Commissaries and is subdivided into several departments: Food safety and applied nutrition (CFSAN), Drug evaluation and research (CDER), Biologics evaluation and research (CBER), Devices and radiological health (CDRH), Veterinary medicine (CVM), Toxicological research (NCTR), Regulatory affairs (ORA) and the office of the commissioner (OC) [4, 44].

3.2.5.7 Other Regulatory Federal Laws

There are other federal laws with which a pharmacist should be familiar. Perhaps the most important are laws on packing and labeling, operations that are regulated by the FDA and the Federal Communications Commission (FCC). The law on packing and labeling is targeted mostly to protecting the consumer. In the case of liquid the ingredients should be on the visible part of the package. The law presents specific requisites concerning the location and size of the type. Violation of this law can lead to apprehension by the FDA or a withdrawal order from the FCC.

Many times a pharmacist involved in developing a product is called upon in the publicity of the medication. For this, he or she must understand the politics of the regulatory agency involved. The FCC, according to the Federal Law of Commerce, has jurisdiction over the announcement and promotion of all consumables, including medications and cosmetics.

This law extends to all publicity and has to do with practices of fraudulent publicity and with promotion that is understood to be false and fraudulent. In general, the FCC controls the publicity of nonprescription drugs and cosmetics with respect to false or fraudulent affirmations, and the FDA is responsible for labeling of medications and for all publicity related to prescription medications. The principal objective of this is to avoid unnecessary duplication of procedures while enforcing the law. The agencies work closely together and the FCC relies strongly on the FDA due to its scientific knowledge. Any government has the right to approve laws for its citizens' protection. This right constitutes the base on which laws regulate the drug substance, the drug product, and its production, distribution, and sale. It is common that these laws exist at a district level, state level, and national level and deal with falsification and adulteration, fraudulent publicity, and maintenance of appropriated sanitary conditions.

Most U.S. states specify the purity requisites, labeling, and applicable packaging of a medication that are generally defined in identical language in federal law. Almost all states, prohibit the commercialization of a new medication until an authorization request for commercialization of a new medication has been submitted to the FDA and has been approved. Medication labeling requisites in each state are established, just as the local laws are defined, taking into consideration arguments and information, such as name and place of activity (production), content quantity, drug name, name of ingredients, quantity or proportion of some ingredients, usage indications, warning regarding dependence, caution against deterioration (degradation), warning about situations in which the use can be dangerous, and special requisites for labeling of official drugs [43].

3.2.5.8 Fair Packaging and Labeling Act [44]

The FDA through Fair Packaging and Labeling Act regulates the labels on many consumer products, including health products. Title 15: Commerce and Trade

Chapter 39: Fair Packaging and Labeling Program [44]

Section 1451. Congressional Delegation of Policy Informed consumers are essential to the fair and efficient functioning of a free market economy. Packages and their labels should enable consumers to obtain accurate information as to the quantity of the contents and should facilitate value comparisons. Therefore, it is hereby declared to be the policy of the Congress to assist consumers and manufacturers in reaching these goals in the marketing of consumer goods [44].

Section 1452. Unfair and Deceptive Packaging and Labeling: Scope of Prohibition

(a) Nonconforming Labels It shall be unlawful for any person engaged in the packaging or labeling of any consumer commodity (as defined in this chapter) for distribution in commerce, or for any person (other than a common carrier for hire, a contract carrier for hire, or a freight forwarder for hire) engaged in the distribution in commerce of any packaged or labeled consumer commodity, to distribute or to cause to be distributed in commerce any such commodity if such commodity is contained in a package, or if there is affixed to that commodity a label, which does not conform to the provisions of this chapter and of regulations promulgated under the authority of this chapter.

(b) Exemptions The prohibition contained in subsection (a) of this section shall not apply to persons engaged in business as wholesale or retail distributors of consumer commodities except to the extent that such persons (1) are engaged in the packaging or labeling of such commodities, or (2) prescribe or specify by any means the manner in which such commodities are packaged or labeled.

Section 1453. Requirements of Labeling; Placement, Form, and Contents of Statement of Quantity; Supplemental Statement of Quantity

(a) Contents of Label No person subject to the prohibition contained in section 1452 of this title shall distribute or cause to be distributed in commerce any packaged consumer commodity unless in conformity with regulations which shall be established by the promulgating authority pursuant to section 1455 of this title which shall provide that:

- (1) The commodity shall bear a label specifying the identity of the commodity and the name and place of business of the manufacturer, packer, or distributor;
- (2) The net quantity of contents (in terms of weight or mass, measure, or numerical count) shall be separately and accurately stated in a uniform location upon the principal display panel of that label, using the most appropriate units of both the customary inch/pound system of measure, as provided in paragraph (3) of this subsection, and, except as provided in paragraph (3)(A)(ii) or paragraph (6) of this subsection, the SI metric system;
- (3) The separate label statement of net quantity of contents appearing upon or affixed to any package:
  - (A)
    - (i) if on a package labeled in terms of weight, shall be expressed in pounds, with any remainder in terms of ounces or common or decimal fractions of

the pound; or in the case of liquid measure, in the largest whole unit (quart, quarts and pint, or pints, as appropriate) with any remainder in terms of fluid ounces or common or decimal fractions of the pint or quart;

- (ii) if on a random package, may be expressed in terms of pounds and decimal fractions of the pound carried out to not more than three decimal places and is not required to, but may, include a statement in terms of the SI metric system carried out to not more than three decimal places;
- (iii) if on a package labeled in terms of linear measure, shall be expressed in terms of the largest whole unit (yards, yards and feet, or feet, as appropriate) with any remainder in terms of inches or common or decimal fractions of the foot or yard;
- (iv) if on a package labeled in terms of measure of area, shall be expressed in terms of the largest whole square unit (square yards, square yards and square feet, or square feet, as appropriate) with any remainder in terms of square inches or common or decimal fractions of the square foot or square yard;

- (B) shall appear in conspicuous and easily legible type in distinct contrast (by topography, layout, color, embossing, or molding) with other matter on the package;
- (C) shall contain letters or numerals in a type size which shall be
  - (i) established in relationship to the area of the principal display panel of the package, and
  - (ii) uniform for all packages of substantially the same size; and
- (D) shall be so placed that the lines of printed matter included in that statement are generally parallel to the base on which the package rests as it is designed to be displayed; and

- (4) The label of any package of a consumer commodity which bears a representation as to the number of servings of such commodity contained in such package shall bear a statement of the net quantity (in terms of weight or mass, measure, or numerical count) of each such serving.
- (5) For purposes of paragraph (3)(A)(ii) of this subsection the term "random package" means a package which is one of a lot, shipment, or delivery of packages of the same consumer commodity with varying weights or masses, that is, packages with no fixed weight or mass pattern.
- (6) The requirement of paragraph (2) that the statement of net quantity of contents include a statement in terms of the SI metric system shall not apply to foods that are packaged at the retail store level.

(b) Supplemental Statements No person subject to the prohibition contained in section 1452 of this title shall distribute or cause to be distributed in commerce any packaged consumer commodity if any qualifying words or phrases appear in conjunction with the separate statement of the net quantity of contents required by subsection (a) of this section, but nothing in this subsection or in paragraph (2) of subsection (a) of this section shall prohibit supplemental statements, at other places on the package, describing in nondeceptive terms the net quantity of contents: *Provided*, That such supplemental statements of net quantity of contents shall not

include any term qualifying a unit of weight or mass, measure, or count that tends to exaggerate the amount of the commodity contained in the package.

Section 1454. Rules and Regulations

(a) Promulgating Authority The authority to promulgate regulations under this chapter is vested in (A) the Secretary of Health and Human Services (referred to hereinafter as the "Secretary") with respect to any consumer commodity which is a food, drug, device, or cosmetic, as each such term is defined by section 321 of title 21; and (B) the Federal Trade Commission (referred to hereinafter as the "Commission") with respect to any other consumer commodity.

(b) Exemption of Commodities from Regulations If the promulgating authority specified in this section finds that, because of the nature, form, or quantity of a particular consumer commodity, or for other good and sufficient reasons, full compliance with all the requirements otherwise applicable under section 1453 of this title is impracticable or is not necessary for the adequate protection of consumers, the Secretary or the Commission (whichever the case may be) shall promulgate regulations exempting such commodity from those requirements to the extent and under such conditions as the promulgating authority determines to be consistent with section 1451 of this title:

(c) Scope of Additional Regulations Whenever the promulgating authority determines that regulations containing prohibitions or requirements other than those prescribed by section 1453 of this title are necessary to prevent the deception of consumers or to facilitate value comparisons as to any consumer commodity, such authority shall promulgate with respect to that commodity regulations effective to:

- (1) establish and define standards for characterization of the size of a package enclosing any consumer commodity, which may be used to supplement the label statement of net quantity of contents of packages containing such commodity, but this paragraph shall not be construed as authorizing any limitation on the size, shape, weight or mass, dimensions, or number of packages which may be used to enclose any commodity;
- (2) regulate the placement upon any package containing any commodity, or upon any label affixed to such commodity, of any printed matter stating or representing by implication that such commodity is offered for retail sale at a price lower than the ordinary and customary retail sale price or that a retail sale price advantage is accorded to purchasers thereof by reason of the size of that package or the quantity of its contents;
- (3) require that the label on each package of a consumer commodity (other than one which is a food within the meaning of section 321(f) of title 21) bear (A) the common or usual name of such consumer commodity, if any, and (B) in case such consumer commodity consists of two or more ingredients, the common or usual name of each such ingredient listed in order of decreasing predominance, but nothing in this paragraph shall be deemed to require that any trade secret be divulged; or

- (4) prevent the nonfunctional-slack-fill of packages containing consumer commodities. For purposes of paragraph (4) of this subsection, a package shall be deemed to be nonfunctionally slack-filled if it is filled to substantially less than its capacity for reasons other than (A) protection of the contents of such package or (B) the requirements of machines used for enclosing the contents in such package.

(d) Development by Manufacturers, Packers, and Distributors of Voluntary Product Standards Whenever the Secretary of Commerce determines that there is undue proliferation of the weights or masses, measures, or quantities in which any consumer commodity or reasonably comparable consumer commodities are being distributed in packages for sale at retail and such undue proliferation impairs the reasonable ability of consumers to make value comparisons with respect to such consumer commodity or commodities, he shall request manufacturers, packers, and distributors of the commodity or commodities to participate in the development of a voluntary product standard for such commodity or commodities under the procedures for the development of voluntary products standards established by the Secretary pursuant to section 272 of this title. Such procedures shall provide adequate manufacturer, packer, distributor, and consumer representation.

(e) Report and Recommendations to Congress upon Industry Failure to Develop or Abide by Voluntary Product Standards If (1) after one year after the date on which the Secretary of Commerce first makes the request of manufacturers, packers, and distributors to participate in the development of a voluntary product standard as provided in subsection (d) of this section, he determines that such a standard will not be published pursuant to the provisions of such subsection (d), or (2) if such a standard is published and the Secretary of Commerce determines that it has not been observed, he shall promptly report such determination to the Congress with a statement of the efforts that have been made under the voluntary standards program and his recommendation as to whether Congress should enact legislation providing regulatory authority to deal with the situation in question.

Section 1455. Procedures for Promulgation of Regulations

(a) Hearings by Secretary of Health and Human Services Regulations promulgated by the Secretary under section 1453 or 1454 of this title shall be promulgated, and shall be subject to judicial review, pursuant to the provisions of subsections (e), (f), and (g) of section 371 of title 21. Hearings authorized or required for the promulgation of any such regulations by the Secretary shall be conducted by the Secretary or by such officer or employees of the Department of Health and Human Services as he may designate for that purpose.

(b) Judicial Review; Hearings by Federal Trade Commission Regulations promulgated by the Commission under section 1453 or 1454 of this title shall be promulgated, and shall be subject to judicial review, by proceedings taken in conformity with the provisions of subsections (e), (f), and (g) of section 371 of title 21 in the same manner, and with the same effect, as if such proceedings were taken by the Secretary pursuant to subsection (a) of this section. Hearings authorized or required

for the promulgation of any such regulations by the Commission shall be conducted by the Commission or by such officer or employee of the Commission as the Commission may designate for that purpose.

(c) Cooperation with Other Departments and Agencies In carrying into effect the provisions of this chapter, the Secretary and the Commission are authorized to cooperate with any department or agency of the United States, with any State, Commonwealth, or possession of the United States, and with any department, agency, or political subdivision of any such State, Commonwealth, or possession.

(d) Returnable or Reusable Glass Containers for Beverages No regulation adopted under this chapter shall preclude the continued use of returnable or reusable glass containers for beverages in inventory or with the trade as of the effective date of this Act, nor shall any regulation under this chapter preclude the orderly disposal of packages in inventory or with the trade as of the effective date of such regulation.

3.2.5.9 United States Pharmacopeia Center for the Advancement of Patient Safety [45]

For nearly 33 years, the USP has been reporting programs for health care professionals to share experiences and observations about the quality and safe use of medications. This year, the USP Center for the Advancement of Patient Safety publishes its sixth annual report to the nation on medication errors reported to MEDMARX (Table 6). It was observed that drug product packaging/labeling is one of the main courses of medication errors in hospitals.

3.2.5.10 National Agency of Sanitary Vigilance (ANVISA, Brazil)

ANVISA is a federal organization linked to Brazil's Health Ministry, which has the incumbency of looking after medication quality and other health products aimed at patients' safety. Several documents regarding GMP and quality control are easily accessed. The agency is also responsible for establishing enforcing the rules and can take corrective measures and punish the offenders [46].

Product stability and compatibility with the conditioning material are distinct, separate, and complementary concepts which should be applied to the pharmaceutical product before being made available for health care.

TABLE 6 Selected Causes of Error Related to Equipment, Product Packaging/Labeling, and Communication in ICUs

Cause of Error	*N* (Nonharmful + Harmful)	Percent Harmful
Label (the facility's) design	1,236	6,9
Similar packaging/labeling		
Packaging/container design		
Label (manufacturer's) design		
Brand/generic names look-alike		

Source: MEDMARX Data Report: A Chartbook of 2000–2004 Findings from Intensive Care Units (ICUs) and Radiological Services.

In the compatibility test between formulation and the conditioning material, several options of conditioning materials are evaluated to determine the most adequate for the product.

The environmental conditions and periodicity analyses can be the same as those mentioned for the stability studies for the formulation. In this phase, the possible interactions between the product and the conditioning material which is in direct contact with the medication are verified. Phenomena such as absorption, migration, corrosion, and others that compromise integrity can be observed. Considering that these types of tests are generally destructive, it is necessary to define the number of samples to be tested.

In ANVISA's documents, different types of tests are established that should be carried out with different types of available materials and employed for conditioning medications and cosmetics (cellulose packagings, metallic, plastic, pressurized, etc.) [46].

3.2.5.11 International Committee on Harmonization (ICH)

In the document "Good Manufacturing Practice Guide for Active Pharmaceutical Ingredients (APIs)" of the ICH Harmonized Tripartite Guideline, the following instructions are given for packaging and identification labeling of APIs and intermediates [47].

General

- There should be written procedures describing the receipt, identification, quarantine, sampling, examination and/or testing and release, and handling of packaging and labeling materials.
- Packaging and labeling materials should conform to established specifications. Those that do not comply with such specifications should be rejected to prevent their use in operations for which they are unsuitable.
- Records should be maintained for each shipment of labels and packaging materials showing receipt, examination, or testing, and whether accepted or reject.

Packaging Materials

- Containers should provide adequate protection against deterioration or contamination of the intermediate or API that may occur during transportation and recommended storage.
- Containers should be clean and, where indicated by the nature of the intermediate or API, sanitized to ensure that they are suitable for their intended use. These containers should not be reactive, addictive, or absorptive so that the quality of the intermediate or API complies with the specifications.
- If containers are reused, they should be cleaned in accordance with documented procedures and all previous labels should be removed or defaced.

Label Issuance and Control

- Access to the label storage areas should be limited to authorized personnel.
- Procedures should be used to reconcile the quantities of labels issued, used, and returned and to evaluate discrepancies found between the number of containers labeled and the number of labels issued. Such discrepancies should be investigated and the investigation should be approved by the quality unit(s).
- All excess labels bearing batch numbers or other batch-related printing should be destroyed. Returned labels should be maintained and stored in a manner that prevents mix-ups and provides proper identification.
- Obsolete and outdated labels should be destroyed.
- Printing devices used to print labels for packaging operations should be controlled to ensure that all imprinting conforms to the print specified in the batch production record.
- Printed labels issued for a batch should be carefully examined for proper identity and conformity to specifications in the master production record. The results of this examination should be documented.
- A printed label representative of those used should be included in the batch production record.

Packaging and Labeling Operations

- There should be documented procedures designed to ensure that correct packaging materials and labels are used.
- Labeling operations should be designed to prevent mix-ups. There should be physical or spatial separation from operations involving other intermediates or APIs.
- Labels used on containers of intermediates or APIs should indicate the name or identifying code, the batch number of the product, and storage conditions, when such information is critical to assure the quality of intermediate API.
- If the intermediate or API is intended to be transferred outside the control of the manufacturer's material management system, the name and address of the manufacturer, quantity of contents and special transport conditions, and any special legal requirements should also be included on the label. For intermediates or APIs with an expiry date, the expiry date should be indicated on the label and certificate of analysis. For intermediates or APIs with a retest date, the retest date should be indicated on the label and/or certificate of analysis.
- Packaging and labeling facilities should be inspected immediately before use to ensure that all materials not needed for the next packaging operation have been removed. This examination should be documented in the batch production records, the facility log, or other documentation system.
- Packaged and labeled intermediates or APIs should be examined to ensure that containers and packages in the batch have the correct label. This examination should be part of the packaging operation. Results of these examinations should be recorded in the batch production or control records.

- Intermediate or API containers that are transported outside of the manufacturer's control should be sealed in a manner such that, if the seal is breached or missing, the recipient will be alerted to the possibility that the contents may have been altered.

3.2.5.12 European Union Regulatory Bodies

European regulatory requirements say little to date about container closure integrity of parenteral or sterile pharmaceutical products. Regulations provide for package integrity verification of parenteral vials to be supported by the performance of sterility tests as part of the stability program. More specific information is described in the European Union (EU) 1998 "Rules Governing Medical Products in the European Union, Pharmaceutical Legislation." These GMP regulations require that the sealing or closure process be validated. Packages sealed by fusion (e.g., ampules) should be 100% integrity tested. Other packages should be sampled and checked appropriately. Packages sealed under vacuum should be checked for the presence of vacuum. While not as detailed as the FDA guidances, it is evident that the EU rules also require the verification of parenteral product package seal integrity. It is important to note that the EU rules specifically require 100% product testing for fusion-sealed packages, sampling and testing of all other packages, and vacuum verification for packages sealed under partial pressure [42].

The vacuum/pressure decay test is performed by placing the package in a tightly closed test chamber, a pressure or vacuum is applied inside the chamber, and then the rate of pressure/vacuum change in the chamber over time is monitored. The rate or extent of change is compared to that previously exhibited by a control, nonleaking package. Significantly greater change for a test package is indicative of a leak.

REFERENCES

1. Griffin, J. P. Ed. (2002), *The Textbook of Pharmaceutical Medicine*, 4th ed., BMJ Publishing, London.
2. Harburn, K. (1990), *Quality Control of Packaging Materials in the Pharmaceutical Industry*, Marcel Dekker, New York.
3. O'Brien, J. D. (1990), *Medical Device Packaging Handbook*, Marcel Dekker, New York.
4. U.S. Food and Drug Administration (FDA) (1999, May) Guidance on container closure systems for packaging human drugs and biologics, U.S. Department of Health and Human Services, FDA, Washington, DC.
5. Yoshioka, S. (2000), *Stability of Drugs and Dosage Forms*, Kluwer Academic Publishers: New York, NY, USA, p 272.
6. Banker, G. S., and Rhodes, C. T. (2002), *Modern Pharmaceutics*, 4th ed., rev. and expanded, Marcel Dekker, New York.
7. Connor, J., Rafter, N., and Rodgers, A. (2004), Do fixed-dose combination pills or unit-of-use packaging improve adherence? A systematic review. *Br. World Health Org.*, 82, 935–939.
8. Bloomfield, S. F. (1990), Microbial contamination: Spoilage and hazard, in Denyer, S., and Baird, R., Eds., *Guide to Microbiological Control in Pharmaceuticals*, Ellis Horwood, Chichester, England, pp 29–52.

9. Aspinall, J. A., Duffy, T. D., Saunders, M. B., and Taylor, C. G. (1980), The effect of low density polyethylene containers on some hospital-manufactured eye drop formulations. 1. Sorption of phenyl mercuric acetate, *J. Clin. Hosp. Pharm.*, 5, 21–29.
10. *United States Pharmacopeia* (2006), 29th ed., United States Pharmacopeial Convention, Rockville, MD.
11. Parker, W. A., and MacCara, M. E. (1980), Compatibility of diazepam with intravenous fluid containers and administration sets, *Am. J. Hosp. Pharm.*, 37, 496–500.
12. Mizutani, T., Wagi, K., and Terai, Y. (1981), Estimation of diazepam adsorbed on glass surfaces and silicone-coated surfaces as models of surfaces of containers, *Chem. Pharm. Bull.*, 29, 1182–1183.
13. Yahya, A. M., McElnay, J. C., and D'Arcy, P. F. (1985), Binding of chloroquine to glass, *Int. J. Pharm.*, 25, 217–223.
14. Vromans, H., and Van Laarhoven, J. A. H. (1992), A study on water permeation through rubber closures of injection vials, *Int. J. Pharm.*, 79, 301–308.
15. Matsuura, I., and Kawamata, M. (1978), Studies on the prediction of shelf life. III. Moisture sorption of pharmaceutical preparation under the shelf condition, *Yakugaku Zusshi*, 98, 986–996.
16. Nakabayashi, K., Tuchida, T., and Mima, H. (1980), Stability of packaged solid dosage forms. I. Shelf-life prediction of packaged tablets liable to moisture damage, *Chem. Pharm. Bull.*, 28, 1090–1098.
17. Nakabayashi, K., Shimamoto, T., and Mima, H. (1980), Stability of packaged solid dosage forms. II. Shelf-life prediction for packaged sugar-coated tablets liable to moisture and heat damage, *Chem. Pharm. Bull.*, 28, 1099–1106.
18. Nakabayashi, K., Shimamoto, T., and Mima, H. (1980), Stability of packaged solid dosage forms. III. Kinetic studies by differential analysis on the deterioration of sugar-coated tablets under the influence of moisture and heat, *Chem. Pharm. Bull.*, 28, 1107–1111.
19. Tonnesen, H. H. (1996), *Photostability of Drugs and Drug Formulations*, CRC Press, London.
20. Kontny, M. J., Koppenol, S., and Graham, E. T. (1992), Use of the sorption–desorption moisture transfer model to assess the utility of a desiccant in a solid product, *Int. J. Pharm.*, 84, 261–271.
21. Pikal, M. J., and Lang, J. E. (1978), Rubber closures as a source of haze in freeze dried parenterals: Test methodology for closure evaluation, *J. Parenteral drug Assoc.*, 32, 162–173.
22. Jaehnke, R. W. O., Kreuter, J., and Ross, G. (1990), Interaction of rubber closures with powders for parenteral administration, *J. Parenteral sci. Tech.*, 44, 282–288.
23. Jaehnke, R. W. O., Kreuter, J., and Ross, G. (1991), Content/container interactions: The phenomenon of haze formation on reconstitution of solids for parenteral use, *Int. J. Pharm.*, 77, 4755.
24. Moorhatch, P., and Chiou, W. L. (1974), Interactions between drugs and plastic intravenous fluid bags. II: Leaching of chemicals from bags containing various solvent media, *Am. J. Hosp. Pharm.*, 31, 149–152.
25. Venkataramanan, R., Burckart, G. J., Ptachcinski, R. J., Blaha, R., Logue, L. W., Bahnson, A. C., and Brady, G. J. E. (1986), Leaching of diethylhexyl phthalate from polyvinyl chloride bags into intravenous cyclosporine solution, *Am. J. Hosp. Pharm.*, 43, 2800–2802.
26. Boruchoff, S. A. (1987), Hypotension and cardiac arrest in rats after infusion of mono (2ethylhexyl) phthalate (MEHP), a contaminant of stored blood, *N. Engl. J. Med.*, 316, 1218–1219.

27. U.S. Food and Drug Administration, *Code of Federal Regulations* (CFR)—Title 21, Food and drugs, Chapters 174–186, available: http://www.access.gpo.gov/nara/cfr/index.html, accessed Mar. 11, 2005.
28. Kowaluk, E. A., Roberts, M. S., Blackburn, H. D., and Polack, A. E. (1981), Interactions between drugs and polyvinyl chloride infusion bags, *Am. J. Hosp. Pharm.*, 38, 1308–1314.
29. Illum, L., and Bundgaard, H. (1982), Sorption of drugs by plastic infusion bags, *Int. J. Pharm.*, 10, 339–351.
30. Illum, L., Bundgaard, H., and Davis, S. S. (1983), A constant partition model for examining the sorption of drugs by plastic infusion bags, *Int. J. Pharm.*, 17, 183–192.
31. Atkinson, H. C., and Duffull, S. B. (1990), Prediction of drug loss from PVC infusion bags, *J. Pharm. Pharmacol.*, 43, 374–376.
32. Richardson, N. E., and Meakin, B. J. (1974), The sorption of benzocaine from aqueous solution by nylon 6 powder, *J. Pharm. Phamacol.*, 26, 166–174.
33. Santoro, M. I. R. M., Kedor-Hackmann, E. R. M., and Moudatsos, K. M. (1993), Estabilidade de sais de reidratação oral em diferentes tipos de embalagem. *Bol. Sanit. Panam.*, 115, 310–315.
34. World Health Organization (WHO) (2003), *The International Pharmacopoeia, Tests and General Requirements for Dosage Forms: Quality Specifications for Pharmaceutical Substances and Tablets*, 3rd ed., Vol. 5, WHO, Geneva.
35. Santoro, M. I. R. M., Oliveira, D. A. G. C., Kedor-Hackmann, E. R. M., and Singh, A. K. (2004), Quantifying benzophenone-3 and octyl methoxycinnamate in sunscreen emulsions, *Cosm. & Toil.*, 119, 77–82.
36. Santoro, M. I. R. M., Oliveira, D. A. G. C., Kedor-Hackmann, E. R., and Singh, A. K. (2005), The effect of packaging materials on the stability of sunscreen emulsions, *Int. J. Pharm.*, 13, 197–203.
37. Thoma, K., and Kerker, R. (1992), Photoinstability of drugs. 6. Investigations on the photosansibility of molsidomine, *Pharm. Ind.*, 54, 630–638.
38. *British Pharmacopoeia* (2002), Her Majesty's Stationary Office, London, pp A144, 135–136, 196, 671–673, 778–780, 976–978, 1145–1146.
39. Albert, D. E. (2004), Evaluating pharmaceutical container closure systems, *Pharm. & Med. Packaging News*, 3, 76–78.
40. ASHP Council on Professional Affairs (1993), ASHP Guidelines on preventing medication errors in hospital, *Am. J. Hosp. Pharm.*, 50, 305–314.
41. ASHP Council on Professional Affairs (2001), ASHP guidelines on preventing medication errors in hospital, *Am. J. Hosp. Pharm.*, 58, 3033–3041.
42. *European Pharmacopoeia* (2001), 4th ed., Council of Europe, Strasbourg.
43. Lachman, L., Lieberman, H. A., and Kanig, J. L. (2001), *Teoria e prática na indústria farmacêutica*, Fundação Calouste Gulbenkian, Lisboa.
44. U.S. Food and Drug Administration, Fair Packaging and Labeling Act. Title 15—Commerce and Trade, Chapter 39—Fair Packaging and Labeling Program, available: http://www.fda.gov/opacom/laws/fplact.htm accessed Mar. 11, 2005.
45. Santell, J. P., Hicks, R. W., and Cousins, D. D. (2005), *MEDMARX Data Report: A Chartbook of 2000–2004 Findings from Intensive Care Units and Radiological Services*, USP Center for Advancement of Patient Safety, Rockville, MD.
46. Agência Nacional de Vigilância Sanitária (ANVISA) (2004), *Guia de Estabilidade de Produtos Cosméticos*, ANVISA, Brasília.
47. International Organization on Harmonisation (2000), ICH harmonized tripartite guideline: Good manufacturing practice guide for active pharmaceutical ingredients, available: http://www.ICH.org, accessed June 23, 2005.
48. Sarbach, C., Yagoubi, N., Sauzieres, J., Renaux, C., Ferrier, D., and Postaire, E. (1996), Migration of impurities from a multi-layer plastics container into a parenteral infusion solution, *Int. J. Pharm.*, 140, 169–174.

3.3

CLEAN-FACILITY DESIGN, CONSTRUCTION, AND MAINTENANCE ISSUES

RAYMOND K. SCHNEIDER
Clemson University, Clemson, South Carolina

Contents

Pharmaceutical Manufacturing Handbook: Production and Processes, edited by Shayne Cox Gad

3.3.1 INTRODUCTION

While there are discrete steps in the design and construction of a pharmaceutical manufacturing plant project, those projects deemed successful incorporate certain practices that promote flow of the construction process toward completion on time and within budget. Proper front-end planning is not completed until it results in appropriate values for design parameters, "buy-in" at all levels of management, and clear direction for the design phase. Engineering the clean room in accordance with recognized industry practice would produce construction documents that facilitate clear procurement and construction planning as well as a focused, efficient, construction effort. A full return on the energy expended through the construction phase cannot be realized without a well-executed start-up and validation process that provides baseline data for effective ongoing operation and maintenance.

The steps in the clean-room construction project include:

Needs assessment
Front-end planning
Preliminary design
Construction document development
Procurement
Construction
Start-up and validation

One of the truisms of the construction industry is that the greatest impact on the cost of a facility can be made at the earliest stages of the process. The construction process can be likened to a snowball rolling down a snow-covered hill. It grows and gains momentum, seemingly taking on a life of its own, until it can only be brought under control with a major effort. So too with manufacturing plant projects. Careful work during the first three stages will ensure that the project begins on a well-directed course and moves to a successful conclusion.

Sometimes the special nature of pharmaceutical manufacturing plant projects clouds the fact that building such a plant is in fact a construction project. The facility engineering team of a small to medium company may be tempted to turn away from such projects due to the projects' perceived uniqueness and leave the key decision making to others. In fact, it is the construction experience of that team that is most required to keep the project costs under control. The way to accomplish this is for the team to be involved in the process from its earliest stages.

Let us review the steps in such a project and identify what should occur at each step and the potential for trouble.

3.3.2 PLANNING FOR PROJECT SUCCESS

3.3.2.1 Needs Assessment

It is during this early stage that a requirement for a clean manufacturing facility is perceived. The need for the facility may be precipitated by a new product, an improved product, an improved manufacturing methodology, new or more stringent regulation requirements, or perhaps a change in marketing strategy.

At this point a study should be undertaken to determine the benefits to be realized by the new facility as well as the costs to be incurred. Costs arise from not only construction but also ongoing operation and maintenance. These costs are affected by the plant location and the availability of a trained or trainable workforce. Does the day-to-day operation of the facility generally require that special attire be worn? Are special procedures, possibly more time consuming than those presently used, required? It is important that this study is complete and accurate in order to prevent any unrealistic expectations on the part of management and plant operations and to permit advanced planning for revised procedures once the facility is in use.

The study should describe the goals of the project, its impact on present operations, budget restraints, tentative schedule, and path forward. It will serve as the basis for front-end planning and will provide the standard against which the success of the program is measured.

3.3.2.2 Front-End Planning

While the needs assessment study may be conducted by a limited number of people, the front-end planning process should be open to all. Plant facilities people will be bearing the brunt of the responsibility for bringing the facility online, on schedule, and within budget. Process people are responsible for ensuring that the facility will adequately house the process equipment and that the facility incorporates sufficient space, utilities, process flow considerations, and provision for flow of people and material to support the goals of the building program. Human resources people have to staff the facility, either out of the present employee pool or from the general local labor market. They must know the requirements of potential employees as well as the conditions under which they will be working.

Procurement people will be purchasing furnishings and process equipment for the plant as well as overseeing the contracts let to the design and construction professionals. Operations people should have input regarding design parameters such as temperature, humidity, lighting, vibration, cleanliness class, and energy needs. Materials handling people should participate in order to understand the requirements for storing and transporting raw materials as well as retrieving, storing, and shipping finished goods from the plant.

An integral part of the front-end planning team should be the design professionals charged with developing the plant design based on client input in such a way as to satisfy as many requirements developed in needs assessment as possible. This team may be assembled internally but frequently is drawn from specialty builders, architectural and engineering (A&E) firms, and design/build firms active in the pharmaceutical industry. The team of design professionals should have pharmaceutical experience on facilities comparable in size and complexity to that being planned as well as extensive experience in construction projects of all types. The design team may offer design only, design/build, procurement, construction management, or combinations of these services. This design team should be considered a resource during the front-end planning phase. It is the wise client who takes advantage of the experience of the design team, permitting them a large role as facilitators of the planning sessions.

An appropriate design team will demonstrate expertise in contamination control philosophies, space planning, code compliance, and mechanical and electrical design and will be familiar with materials of construction currently being used in pharmaceutical projects. It is frequently helpful to include a member of the construction team in the front-end planning effort to advise on constructibility of the facility being planned. Unrealistic construction schedules will be avoided and field rework will be minimized if appropriate attention is paid to the construction phase early in the planning process.

3.3.2.3 Preliminary Design

Front-end planning typically utilizes the expertise of client process people to convey the requirements of the pharmaceutical facility to the design team. With this information in hand the design team begins the facility design incorporating process needs, code requirements, safety issues, material and personnel flow, work-in-process storage, utility needs, and so on, into a first-cut approach.

Client representatives have an opportunity to review the effort and begin fine tuning the design to incorporate late-breaking process changes. The preliminary design is a target that helps both the design team and the client solidify design goals. Change is inexpensive, and therefore encouraged, at this stage and buy-in by all concerned is a major objective of this phase of the design effort.

A budget based on the agreed-upon preliminary design should be developed to make sure that the overall project is on course. This will minimize surprises further along in the design/build process. Ideally the design will be "cast in stone" at the end of the preliminary phase. This permits the production work on the design documents to proceed unhindered. The more unknowns left at the end of the preliminary phase, the more difficult it will be to complete design documents in a timely fashion.

Construction Document Development The construction documents should convey the intent of the design team and client to the construction team. A good set of construction documents should result in a tight spread of construction bids as there should be little room for varying interpretation on the part of the potential construction contractors. The drawings should have sufficient notes to convey the design intent without creating a cluttered appearance. The written specifications should be as brief as possible consistent with clarity.

Complicated documents create the impression that a project may be more involved, and therefore more costly, than it should be. Cautious contractors may unnecessarily inflate their bid to cover perceived contingencies. Specifications that are too wordy may be difficult to follow and similarly result in higher prices as bidders make sure all bases are covered. No one likes surprises.

The development of construction documents should be a straightforward process with little involvement by the client except to monitor the process and ensure that the original design intent is followed. While changes will always occur during this phase ("cast in stone" is a euphemism for "let's keep the changes under control"), they are certainly less costly at this point than during the construction phase. It is desirable to minimize such changes. A continuous sequence of changes suggests that the preliminary design phase was not entered into seriously. It demonstrates a lack of preparedness on the part of the client and a lack of ability to communicate and draw out the client's needs on the part of the design team. A sense of clarity of purpose slips away with ongoing change and the possibility for errors in construction documents, which eventually surface as costly construction changes, increases.

3.3.2.4 Procurement

A detailed scope of work describing the materials and services required is a vital part of the procurement process. There is no purpose to keeping the project bidders in the dark regarding what is required of them. The role of the procurement function is to obtain maximum value, that is, the best quality and schedule at the lowest price. The clearer the scope of work and construction documents, the better will be the chance of this happening. A low price is not a good value if the schedule slips by several months as a result. A marginal plant that does not maintain design conditions or meet production goals is a poor value even if it was delivered within schedule.

The procurement process should qualify potential bidders by ensuring that similar pharmaceutical projects have been delivered on time, within budget, and on schedule. References should be checked. It is expected that references offered by a potential bidder would have good things to say about that bidder, but this is not a certainty and pointed questioning about personnel, schedule, quality, change orders, follow-up, and so on, can help develop a warm feeling or an uncertain feeling about potential bidders. If bids are in fact quite close, it is the quality of references that might suggest a particular bidder be given preference.

There are a number of ways in which the project can be procured. Use of in-house engineering and construction expertise may work in special situations or on smaller projects. Typically problems arise when facilities departments, stretched to their limit with ongoing plant requirements, must lower the priority of the new facility to meet other commitments. Schedules may stretch out unacceptably.

A number of specialty contractors have proven over the years to be adept at installing small turnkey facilities of limited complexity in a timely and economical fashion. If extensive engineering is required, if local code compliance becomes an issue, if complex process requirements must be met, or if the client requirements exceed the experience of the supplier there could be cause for concern.

Design/build is a popular approach in that it suggests a single source of responsibility for all phases of the project. Frequently firms billing themselves as "design/build" are strong in either design *or* build, but not both. The strong design firm can put the essentials on paper but the final price and schedule may suffer. The strong construction firm may lack the expertise to create an appropriate manufacturing environment, particularly where clean-room expertise is required. The project may be outstanding in all respects except performance. A good review of references is essential before selecting a design/build firm.

Construction management has been increasingly used on larger projects. A good construction management firm will work closely with the client-selected design company to review constructability and adequacy of construction documents. It will assist to qualify bidders, maintain schedule, track costs, administer and oversee, and generally ensure that a team incorporating the strongest skills is assembled to complete the project. Pharmaceutical experience is essential.

3.3.2.5 Construction

The construction process should proceed smoothly if the remarks presented above are followed. Cost can increase during this phase if changes must be implemented. While change is inevitable, a construction change procedure negotiated during the bidding phase and in place during construction will keep such change from getting out of control.

The requirement for "building clean" has arisen in recent years as more stringent clean rooms have become more popular. Imposing a clean construction protocol on contractors can lengthen the schedule and increase cost. The protocol should be developed during the construction document phase and be an integral part of the bid documents. Once the decision is made to work clean, protocols developed should be followed by *everyone* on the jobsite associated with the clean areas. A poorly conceived and enforced protocol will be a costly and futile exercise. The tendency to build clean on every new or retrofit project should be carefully evaluated and a practical protocol should be developed consistent with the needs of the project.

Client end users should be encouraged to observe construction as it progresses. They will be more intelligent about how their facility was built and therefore more attuned to maintaining the facility once it is completed and in operation. While suggestions should be welcomed as construction progresses, it is important that a chain of command be enforced. Any questions or suggestions or concerns should not be expressed to workers on the site but rather through project management channels. In this way good ideas can be implemented and bad ideas shelved without impacting the construction effort in a negative manner. Note the one exception to this practice is in regard to safety. Everyone on the site has safety responsibility. Any unsafe acts should be questioned and supervisors consulted immediately.

3.3.2.6 Start-Up and Validation

Subcontractors on the jobsite should be responsible for start-up as well as installation of equipment. Equipment manufacturers typically have personnel available to ensure appropriate start-up procedures are followed. If several trades are involved in the installation of a particular piece of equipment, then one trade should be assigned, by contract, as having coordinating responsibility for that piece of equipment. This will minimize "finger pointing" when equipment does not start or operate properly. This can be a sensitive issue and a construction manager can set the tone for cooperation in this area.

An independent contractor responsible to the construction manager or owner should do testing and balancing (TAB) of mechanical systems. All start-up should be complete and initial valve or damper settings made (and recorded) by the subcontractor before testing and balancing begins. The TAB contractor should not have to repair equipment or troubleshoot inoperative equipment but rather only adjust and verify performance of equipment.

A separate contractor should certify clean-room areas. This might be the TAB contractor if that firm is suitably qualified. There should be no question of equipment being operative at this stage of the project since start-up and testing and balancing are complete. Certification is the verification of facility compliance with clean-room specifications. If the facility design is well conceived and the construction team has installed a quality project, any certification test failure will most likely be corrected through fairly minor adjustments. Failure of the clean room to pass certification tests might require redesign but more frequently requires some equipment adjustment or perhaps a filter repair and then a retest. It is important that a clear understanding of responsibility be communicated before problems are encountered. Failure to plan for potential problems could result in extending the schedule and incurring unforeseen costs at a crucial point in the project.

3.3.2.7 Summary

Recognizing the step-by-step process involved in even the smallest pharmaceutical project can help focus attention in a manner that will result in a successful project. The formal schedule of a well-conceived project will include needs assessment, front-end planning, and preliminary design. It is important that project progress is measured against such a schedule and not just by the visual impact caused by bricks and mortar being installed.

3.3.3 DESIGN OPTIONS

3.3.3.1 Clean-Facility Scope

The purpose of this section is to identify design and construction options for those parts of a pharmaceutical facility intended to house process equipment. These suggestions are intended to assure that the facilities, when used as designed, will meet the requirements of current good manufacturing practices (cGMPs). Air cleanliness within the facility may range from International Organization for Standardization (ISO) 5 (Class 100) through ISO 8 (Class 100,000). In addition, areas may be considered clean or labeled as "controlled environment" without having a cleanliness class assigned to the space. Note that throughout this chapter cleanliness class will be described using the designation presented in the new ISO 14644 (e.g., ISO 5, ISO 8) and parenthetically as presented in the currently obsolete (but widely understood and quoted) U.S. Federal Standard 209 (e.g., Class 100, Class 100,000).

A cleanliness classification in accordance with the latest revision of ISO 14644 is generally inadequate by itself to describe a facility used for pharmaceutical processes. The presence of viable particles (living organisms) within the particle count achieved by applying methods described in the standard may affect the product within the facility. A measure of both viable and nonviable particles is required to provide sufficient information upon which to base a decision regarding the suitability of the clean room for its intended purpose.

The options presented herein are intended to provide facilities that will effectively restrict both viable and nonviable particles from entering the clean areas, minimize contamination introduced by the facility itself, and continuously remove contaminants generated during normal operations.

Measurement of total particle count in the clean room is described in ISO 14644. This count may be composed of viable, nonviable, or nonviable host particles with a viable traveler. There is no generally accepted relationship between total particle count and viable particle count. While maintaining appropriate particle counts is important in clean-room design and operation, a protocol designed to identify viable particles should be inherent in the certification/validation testing of a pharmaceutical clean room.

No facility design can compensate for excessive contamination generated within it. In addition to effective facility design, the user must also institute a routine maintenance program as well as maintain personnel and operational disciplines that limit particles both entering and being generated within the facility.

While this section identifies options for contamination control in facility design, any such options must be implemented in accordance with all appropriate government and regulatory building and safety codes. The design guideline is nonspecific as regards biological or chemical materials that may be used within the facility but generally addresses bulk pharmaceutical chemical plants (BPCs), secondary manufacturing chemical plants, bulk biopharmaceutical plants, and plants used for fill and finish operations. Good practice as well as any regulations governing biological and pharmaceutical processes conducted within the facility must be adhered to as required and could modify some of the suggestions contained herein.

3.3.3.2 Design Parameters

The design of the facility is based upon specification of certain design parameters. These in turn are used to calculate building system equipment capacities and aid in the selection of the appropriate types of equipment that are required. Design parameters that may be critical are discussed below.

Cleanliness Classification The classification of the clean areas is determined by the using organization consistent with the level of nonviable and viable particulate contamination acceptable to the process conducted within the facility. This may be governed by regulatory agencies, client organizations, or company protocols. Target goals are set for nonviable particle count in accordance with the ISO. Viable particle target goals should be stated in colony-forming units (CFU) per square centimeter. In accordance with ISO 14644, particle goals will typically be identified for "at rest" and "operational" modes.

In the absence of other guidance governing the cleanliness classification and acceptable levels of microbial contamination of the clean room, the values presented in Table 1 may be used. The room grades presented are from most critical (A) to least critical (E). The definition of criticality is left to the clean-room user organization.

Other Design Parameters Facility design parameters that support the process within the clean room should be established by the user organization. Parameters such as temperature, humidity, lighting requirements, sound level, and/or vibration may be process driven or comfort driven and therefore are selected to accommodate specific process or comfort requirements as determined by the end user.

Local Control Under some circumstances, cleanliness requirements can be achieved through the use of localized controls such as clean tents, glove boxes, minienvironments, or isolators. These provide unidirectional filtered airflow within

TABLE 1 An Example of Cleanliness Classification Goals

		Particle Counts[e]		Microbial Contamination			
Room Grade	Cleanliness Class[a]	At Rest	Operational	Air Sample (cfu/m^3)	Settle Plates (cfu/4 h)[b]	Contact Plates (cfu/plate)[c]	Glove Print (cfu/glove)[d]
A[f]	M3.5 (100)	3,500	3,500	<1	<1	<1	<1
B[g]	M3.5 (100)	3,500	35,000	10	5	5	5
C	M5.5 (10000)	350,000	3,500,000	100	50	25	—
D	M6.5 (100000)	3,500,000	N/A	200	100	50	—
E	Uncontrolled	N/A	N/A	N/A	N/A	N/A	N/A

[a]In accordance with U.S. Federal Standard 209E.
[b]90-mm-diameter settling plate. These are average values and individual plates may have <4 h of exposure.
[c]55-mm contact plates.
[d]Five-fingered glove.
[e]Maximum particle counts per cubic meter >0.5 μm.
[f]Unidirectional airflow at 90 ft/min.
[g]Non-unidirectional airflow.

a limited area. They may be located within a facility that provides the necessary temperature and humidity conditions or they may be provided with integral environmental control equipment designed to maintain necessary conditions.

Air Change Rate The airflow pattern and air change rate in a clean room largely determines the class of cleanliness that can be maintained during a given operation. Non-unidirectional flow clean rooms rely on air dilution as well as a general ceiling-to-floor airflow pattern to continuously remove contaminants generated within the room. Unidirectional flow is more effective in continuously sweeping particles from the air due to the piston effect created by the uniform air velocity. The desired air change rate is determined based on the cleanliness class of the room and the density of operations expected in the room. An air change rate of 10–25 per hour is common for a large, low-density ISO 8 (Class 100,000) clean room. ISO 7 (Class 10,000) clean rooms typically require 40–60 air changes per hour. In unidirectional flow clean rooms, the air change rate is generally not used as the measure of airflow but rather the average clean-room air velocity is the specified criterion. The average velocity in a typical ISO 5 (Class 100) clean room will be 70–90 ft/min. A tolerance of plus or minus 20% of design airflow is usually acceptable in the clean room. The foregoing values have been found to be appropriate in many facilities. Generally air change rate or air velocity is not a part of regulations. It is left to the user to demonstrate that the selected design parameter is appropriate for the products being manufactured. An exception to this may be in the case of filling operations where a unidirectional flow velocity of 90 ± 20 ft/min may be required.

Pressurization A pressure differential should be maintained between adjacent areas, with the cleaner area having the higher pressure. This will minimize infiltration of external contamination through leaks and during the opening and closing of personnel doors. A minimum overpressure between clean areas of 5 Pa [0.02 in. of water column (in. WC)] is recommended. The pressure between a clean area and an adjacent unclean area should be 12–14 Pa (0.05 in. WC). Where several clean rooms of varying levels of cleanliness are joined as one complex, a positive-pressure hierarchy of cleanliness levels should be maintained, including air locks and gowning rooms. Note that for certain processes and products it may be desirable to have a negative pressure relative to the surrounding ambient in one or more rooms when containment is a major concern. A "room within a room" may have to be designed to achieve this negative pressure yet still meet the needs of clean operation.

Temperature Control Where occupant comfort is the main concern, a temperature of 68–70°F ± 2°F will usually provide a comfortable environment for people wearing a typical lab coat. Where a full "bunny suit" or protective attire is to be worn, room temperature as low as 66°F may be required. If the temperature is to be controlled in response to process concerns, the value and tolerance should be specified early in the design phase to ensure that system selection is appropriate and that budgeting is accurate. Note that a tight tolerance (e.g., ±1°F or less) will typically be more costly to maintain than a less stringent tolerance.

Humidity Control The humidity requirement for comfort is in the range of 30–60% relative humidity (RH). If process concerns suggest another value, it should be specified as soon as possible in the design process. Biopharmaceutical materials sensitive to humidity variations or excessively high or low values may require stringent controls.

3.3.3.3 Architectural Design Issues

Facility Layout The facility layout should support the process contained within the clean room. While a rectangular shape is easiest to accommodate, other shapes may be incorporated into the facility as long as appropriate attention is paid to airflow patterns. The facility should be able to accommodate movement of equipment, material, and personnel into and out of the clean room. The layout of the clean suite should facilitate maintaining cleanliness class, pressure differentials, and temperature/humidity conditions by isolating critical spaces and by excluding nonclean operations. See Figure 1. The potential for cross-contamination is addressed as both an architectural and a mechanical issue. Generally, in a facility where multiple products are to be processed, each product has a dedicated space, isolated physically from adjacent spaces, and each has its own air conditioning system, independent of adjacent systems.

Air Locks or Anteroom This is a room between the clean room and an unrated or less clean area surrounding the clean room or between two rooms of differing cleanliness class. The purpose of the room is to maintain pressurization differentials between spaces of different cleanliness class while still permitting movement between the spaces. An air lock can serve as a gowning area. Certain air locks may be designated as an equipment or material air lock and provide a space to remove packaging material and/or to clean equipment or materials before they are introduced into the clean room. Interlocks are recommended for air lock door sets to prevent opening of both doors simultaneously. The air lock is intended to separate the clean from the unclean areas.

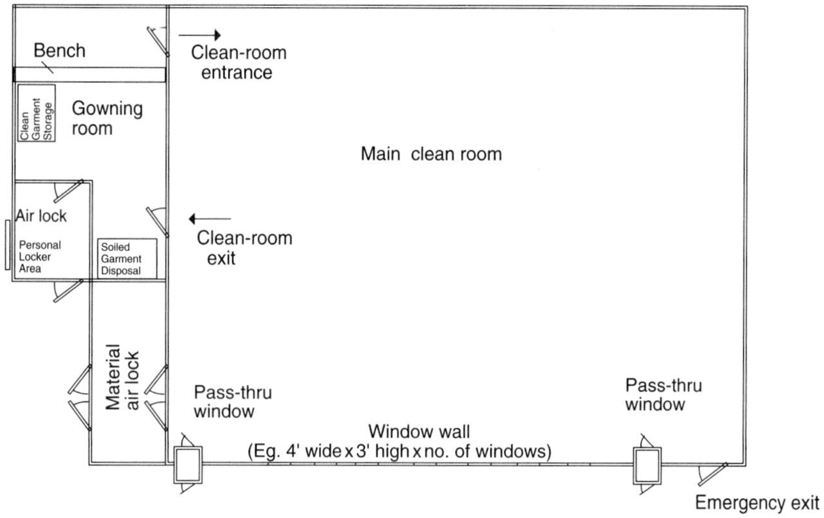

FIGURE 1 Sample clean-room lay-out.

Prior to equipment or raw materials being introduced into the clean room, they should be prepared. This may mean removing an outer package wrap or perhaps surface cleaning of the object. Material handling equipment used within the clean room should be dedicated to the clean room. Physical barriers may be integrated into the material air lock design to prevent material handling equipment from leaving the clean room or outside equipment from passing into the clean room.

Windows Windows are recommended in interior clean-room walls to facilitate supervision and for safety, unless prohibited by the facility protocol for visual security reasons. Windows in exterior building walls adjacent to a clean space are problematic. Windows can be a source of leakage and can result in contaminants entering the space. Windows should be placed to permit viewing of operations in order to minimize the need for non-clean-room personnel to enter the clean room. Windows should be impact-resistant glass or acrylic, fully glazed, installed in a manner that eliminates or minimizes a ledge within the clean space. Double glazing is frequently used to provide a flush surface on both sides of the wall containing the window. Windows may be included if there is a public relations requirement for visitors to view the operations. Speaking diaphragms or flush, wall-mounted, intercom systems are recommended near all windows to facilitate communication with occupants of the clean room.

Pass-Through A pass-through air lock should be provided for the transfer of product or materials from uncontrolled areas into the clean room or between areas of different cleanliness class (Figure 2). The pass-through may include a speaking diaphragm, intercom, or telephone for communication when items are transferred and interlocks to prevent both doors from being opened at the same time. A cart-size pass-through installed at floor level can be used to simplify the movement of carts between clean areas. Stainless steel is typically the material of choice (Figure 3).

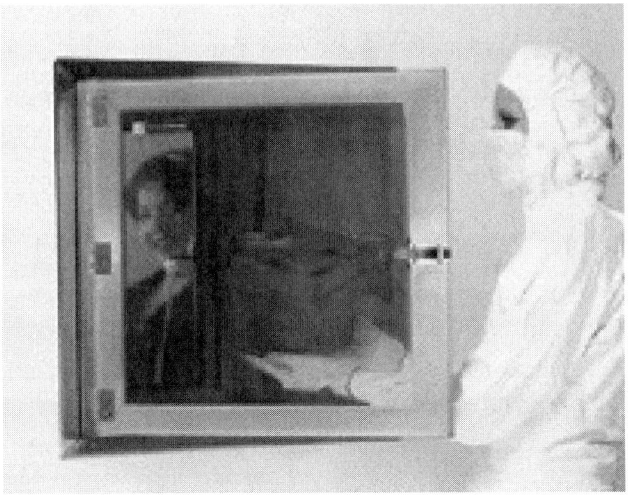

FIGURE 2 Stainless steel pass-through with interlock designed to permit safe passage of small items between spaces of differing cleanliness. (Courtesy of Terra Universal.)

FIGURE 3 Cart pass-through enabling larger amounts and sizes of items to be transported. Note that the cart shown is not to be taken from the clean room. Typically a physical barrier is incorporated into the cart pass-through design. (Courtesy of Terra Universal.)

Gowning Room Gowning rooms should be designed to support the garment protocol established for the facility. A typical gowning room may have a wall- or floor-mounted coat rack for clean garment storage (Figure 4); a bench specifically designed for clean-room use (Figure 5); a full-length mirror installed near the door for gowning self-inspection; storage for new packaged garments; and bins for disposal of soiled garments.

Personal lockers and coat racks for the storage of notebooks, coats, and personal items should be located outside the gowning room or in an anteroom separate from

FIGURE 4 Furnishings in the gowning room are typically of a nonshedding material such as the stainless steel designs shown. The gown rack will generally have a ceiling-mounted HEPA filter above it to continually bathe the garments in clean air. (Courtesy of Terra Universal.)

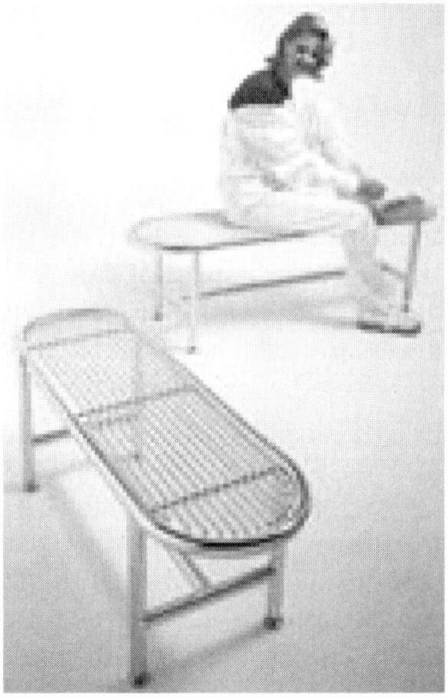

FIGURE 5 The stainless steel clean benches have has a perforated seat to permit airflow from ceiling to floor essentially unobstructed. (Courtesy of Terra Universal.)

the clean gowning area. Restroom facilities may also be located outside the gowning room or in an anteroom adjacent to the clean gowning area. A common gowning room design has two areas divided by a bench. The "unclean" area is used to remove and store outer garments. Stepping over the bench as the clean-room footwear is being put on ensures that the "clean" side of the gowning room will remain that way. Final donning of the clean-room garb is then accomplished.

Male and female gowning rooms may be required depending on the make-up of the work force and the type of garments being used.

Siting A clean room that serves as an element of a larger process line should be integrated into the line to permit movement of personnel and materials in and out of the room. A free-standing clean room may be located in any convenient site; however, certain conditions adjacent to the facility may degrade its performance. Vibration sources inside or near a clean room will encourage particle release within the room and under severe conditions may cause leaks in filters and ductwork. Heavy equipment, including the heating, ventilation, and air conditioning (HVAC) system components, pumps, house and vacuum system, ought to be vibration isolated. Location of a clean room directly adjacent to heavy equipment or loading docks that see heavy truck traffic and other sources of vibration, shock, and noise may be problematic. The outdoor air intake for the clean-room makeup air must be carefully located to prevent overloading of filters or entrance of contaminating gases that the filter will not remove. Clean-room air intakes should not be located near loading docks, traffic lanes, or other areas where vehicles may drive through or idle. These intakes should not be located near the exhaust locations of other processing facilities. Use of gas-phase filtration may be required if the quality of make-up air is not acceptable.

3.3.3.4 Materials of Construction

Walls Generally wall material selection should be based on the operations and material handling equipment to be used within the space. The walls should be strong enough to withstand repeated impact of carts or other equipment without deterioration. The materials should also be selected with the sanitizing protocol in mind. Chemicals, high-pressure wash, and steam can cause reduced wall life if proper materials are not selected. Seamless walls, to the extent possible, are desirable.

Basic steel stud construction with gypsum board paneling can be used in biopharmaceutical clean rooms when appropriately coated with a nonshedding finish. Modular wall systems utilizing coated steel or aluminum panel construction are growing in popularity due to the ability to easily retrofit a lab or production space at a later date with minimal disruption and construction debris. Stainless steel may be appropriate but costly. Modular systems have been developed that address the concerns of the biopharmaceutical clean-room user relative to surface finish integrity and smooth surfaces. The joint between adjacent modular panels is commonly treated with a gunnable sealant to provide a smooth, cleanable joint that will not hold contaminants.

Concrete masonry unit (CMU) construction is widely used (Figure 6). It can prevent buildup of contaminants when finished with an epoxy or other smooth, chemical-resistant coating. Where retrofit is not a regular practice, the strength of concrete block and its long life recommend it.

Rounded, easy-to-clean corners and smooth transitions between architectural features such as windows and walls (Figure 7) should be featured in all wall system designs, whether modular or "stick built."

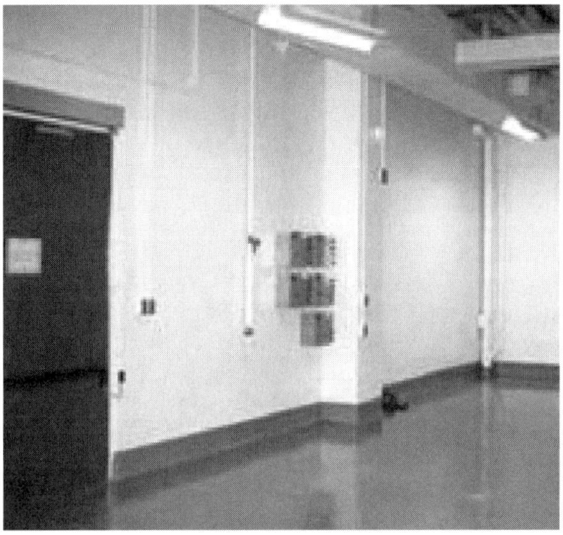

FIGURE 6 A CMU wall treated with block filler and epoxy finish to provide a smooth, cleanable wall surface. (Courtesy of Niagara Walls.)

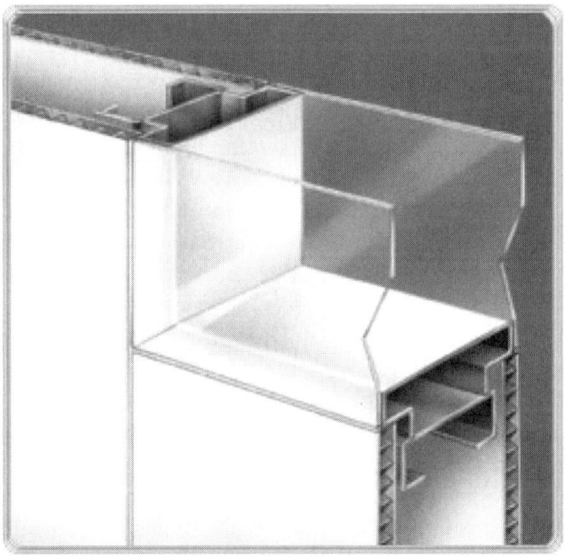

FIGURE 7 A window detail that provides a smooth, easy-to-clean surface on both wall faces. (Courtesy of Portafab.)

Wall Finishes Inexpensive latex wall paints will deteriorate over time and are unacceptable in clean rooms. Acceptable wall finishes include epoxy paint, polyurethane, or baked enamel of a semigloss or gloss type. These may be applied in the factory to metal wall system panels. Field application of epoxy to gypsum board or CMU should be done to ensure a smooth, nonporous, monolithic surface that will not provide a breeding site for organisms. Exposed outside corners in high traffic areas as well as on lower wall surfaces may have stainless steel facings or guards to prevent impact

FIGURE 8 A modular wall system has been installed in a manner that provides a smooth surface for cleaning. The fit of the components and the method of sealing are important when a modular wall is selected. (Courtesy of Portafab.)

damage to the wall. This is particularly true when gypsum board construction is used. Corner and wall guards should extend from the floor to at least the 4-ft height. Traditionally the clean room has been white throughout as an indication of the clean nature of the facility and to identify it as a special work space. Other colors may be used in the clean room to provide an interesting environment as long as the materials of construction do not contribute particles to the air stream and will withstand the sanitizing agents and procedures used in the facility (Figure 8).

Doors Entry should be through air locks to maintain clean-room pressure differentials. Emergency exit doors should incorporate a panic-bar mechanism (or a similar emergency opening device) with alarms for exit only. Emergency exit doors must be secured in a manner that prevents entry from the outside yet permits exiting from within. All doors should include essentially air-tight seals. Neoprene seals are generally acceptable. Brush-type door seals are not recommended. Foam rubber door seals are not recommended as these have been found to quickly deteriorate and shed particles. All personnel doors and swinging equipment doors should include self-closing mechanisms. Manual and automatic sliding doors may be useful when space is an issue or to facilitate movement between spaces of similar cleanliness class for personnel whose hands are otherwise engaged. As the mechanism of such doors can generate particles, a design specifically intended for clean-room application should be selected.

Ceilings The ceiling finish should be similar to that used on the walls. The requirements for sanitizing typically address the ceiling as well as the walls and ceiling material and finish selection should reflect this. Suspended ceilings using an inverted-T grid and lay-in panels may have a place in that part of the clean-room suite not subjected to the rigors of frequent sanitizing and where the possibility of trapped spaces to support organism growth is not considered an issue (Figure 9). When suspended panel ceilings are used, the panels must be securely clipped or sealed in place to prevent movement due to air pressure changes.

Modular wall systems designed for biopharmaceutical applications frequently have a "walk-on" ceiling designed using materials and finish similar to the wall. A rounded, easy-to-clean intersection between ceiling and walls should be a feature of the clean-room ceiling design, whether modular or stick built. Monolithic (seamless) ceilings can be installed using inverted-T grid supports and gypsum panels (Figure 10). This design permits incorporation of filtration and lighting into what is essentially a monolithic ceiling.

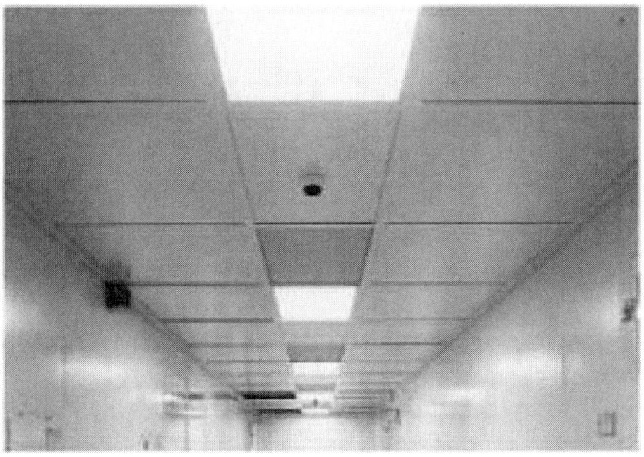

FIGURE 9 A suspended ceiling utilizing lay-in panels and lay-in lighting troffers. A variety of cleanable materials can be used for the panels. The lay-in lights should be of a design that will provide appropriate service based on the cleaning protocol to be used. (Courtesy of CleanTek.)

FIGURE 10 An area of HEPA filters is installed above a process machining. Tear drop lighting is used to permit maximum filter coverage. A monolithic ceiling construction of gypsum panels suspended from a framework. The panels are finished with an epoxy coating compatible with cleaning/sterilization procedures. (Courtesy of CleanTek.)

FIGURE 11 The process area is subjected to substantial chemical action due to the sterilizing protocol. It has a troweled epoxy, easy-to-clean finish. (Courtesy of Dex-O-Tex.)

Floors Commonly used floor finishes for biopharmaceutical clean rooms include sheet vinyl installed using heat-welded or chemically fused seams to provide a seamless surface. Troweled epoxy and epoxy paint (Figure 11) have also found wide use. Compatibility of the floor material with solvents, chemicals, and cleaning agents to be used in the room must be considered. A minimum 4-in. cove at the junction of floor and walls is recommended to facilitate cleaning. Some modular wall systems have a recess or offset that permits sheet vinyl to be installed in a manner that creates a seamless junction between floor and wall. When a stick-built approach is used, care should be taken to design cleanable intersections of walls and floors (Figure 12).

3.3.3.5 HVAC System

Air Side The clean-room HVAC system must be designed to maintain the required particulate cleanliness, temperature, humidity, and positive pressure at the expected outside environmental extremes and during the expected worst-case use operations. Rapid recovery from upset conditions such as door openings and contaminant-generating events is also a consideration. The high cost of conditioning outside air suggests that as much air as possible be recirculated. Recirculated air should be high-efficiency particulate air (HEPA) filtered in those spaces requiring a cleanliness classification in accordance with ISO 14644. Air that may be hazardous to health, even after HEPA filtration, should be exhausted after appropriate treatment. The required quantity of make-up air is calculated based on process exhaust plus air leakage from the clean room. A rate of two air changes per hour for clean room pressurization may be used in the absence of a more detailed calculation of air

FIGURE 12 The wall system used in the facility incorporates a monolithic sheet vinyl flooring junction between floor and wall face. Note the coving run up the wall around the edges to provide a smooth surface for cleaning. (Courtesy of Portafab.)

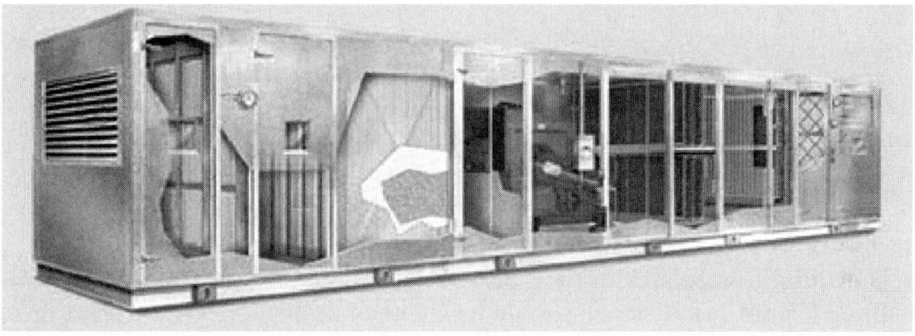

FIGURE 13 The air handler has several stages of filtration combined with heating, cooling, humidification, and dehumidification capability. (Courtesy of Air Enterprises.)

leakage. Make-up air should be drawn from the outdoors, conditioned, and filtered as necessary before being introduced into the clean-room recirculation air stream. Care should be taken to ensure that make-up air intakes are not drawing in contaminated air.

The potential for cross-contamination is an issue that should be addressed. A flexible manufacturing facility is one in which a variety of products can be manufactured simultaneously. If the facility has a single air-handling system, the likelihood of materials from one space intruding into an adjacent space is high. For this reason each filling or compounding operation, or operation where noncompatible product can be expected to be picked up by the air stream, should be served by its own air-handling system (Figure 13). Isolated systems will minimize the possibility of cross-contamination. This can be a costly option and should not be undertaken

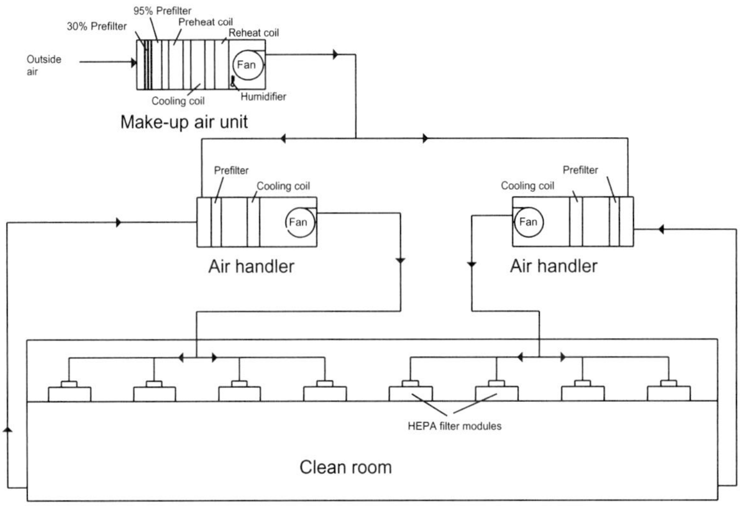

FIGURE 14 Non-unidirectional clean-room with lay-in HEPA filter modules.

lightly. The current use of the plant and the anticipated future use should be assessed before a blanket decision that may lead to costly duplicated systems is made.

Filtration The filtration system for a biopharmaceutical clean room typically consists of several stages of filters. Prefilters are selected, sized, and installed to maximize the life of the final HEPA filters. With proper selection of prefilters, the final HEPA filters should not require replacement within the life of the filter media and seal materials, a period of several years (perhaps as long as 10–15 years). Make-up air is commonly filtered by a low-efficiency [30% as set by the American Society of Heating, Refrigerating, and Air Conditioning Engineers (ASHRAE)] prefilter followed by an intermediate- (60% ASHRAE) or high-efficiency (95% ASHRAE) final filter (Figure 14). A screen should be included at the make-up air inlet to keep out pests and large debris. The make-up air is then directed to the recirculating air handler which also may have a low-efficiency prefilter, although prefiltration of recirculated clean-room air is often omitted because of its high cleanliness level even after having passed through the clean room. The air is then directed through HEPA filters into the clean room. HEPA filters must be a minimum of 99.97% efficient on 0.3-μm particles in accordance with military standard Mil-F-51068 or the Institute of Environmental Science and Technology IEST-RP-CC001. Note that the filtration system for an unrated "controlled area" is the same, except that the HEPA filter stage may be omitted. Refer to Figure 15.

Filter Location HEPA filters may be installed in a facility either within an air handler or at the inlet to a plenum above the clean room or in the clean room ceiling. High-velocity HEPA filters, that is, filters with a face velocity up to 500 ft/min, are frequently installed in air handlers serving Class 100,000 clean rooms and are also used in make-up air handlers. Where hazardous materials may be trapped by the filters a "bag-in–bag-out" filter arrangement, such as that depicted in Figure 16, may

FIGURE 15 Several panel-type filters commonly used as prefilters in air handlers. Second from left is a high-dust-loading filter available in ASHRAE efficiency as high as 95% frequently used in make-up air handlers. If HEPA filtration of the make-up air is required, the high-velocity duct-mounted HEPA filter third from the left is appropriate. It can tolerate face velocities up to 500 fpm, compared to the standard HEPA, which is usually designed for face velocity on the order of 90–100 fpm. A standard HEPA designed for bio-pharma facility ceiling installation is shown at right. (Courtesy of of CamFil.)

FIGURE 16 The "bag-in–bag-out" filter unit contains a HEPA filter and permits personnel to change the filter without coming into contact with possibly hazardous materials that may have been filtered from the air. (Courtesy of Flanders Filters Inc.)

be employed. Figure 17 shows a schematic arrangement with HEPA filters installed in the air handler. During the design phase care should be taken to provide access to both the upstream and downstream face of these filters to permit periodic challenging and leak testing.

To provide HEPA filtered air over a limited area within a larger controlled space, a ceiling-mounted pressure plenum may be used. This plenum has an air distribution means at its lower face that permits air to be introduced in a unidirectional manner over the critical process area. Refer to Figure 18.

HEPA filters are installed at the upper face of the pressure plenum and the plenum is pressurized with filtered air. The ceiling-mounted HEPA filters have a face velocity up to 100–120 ft/min. This is somewhat higher than the HEPA filters serving the rest of the clean room. The filters are commonly supplied with air by a

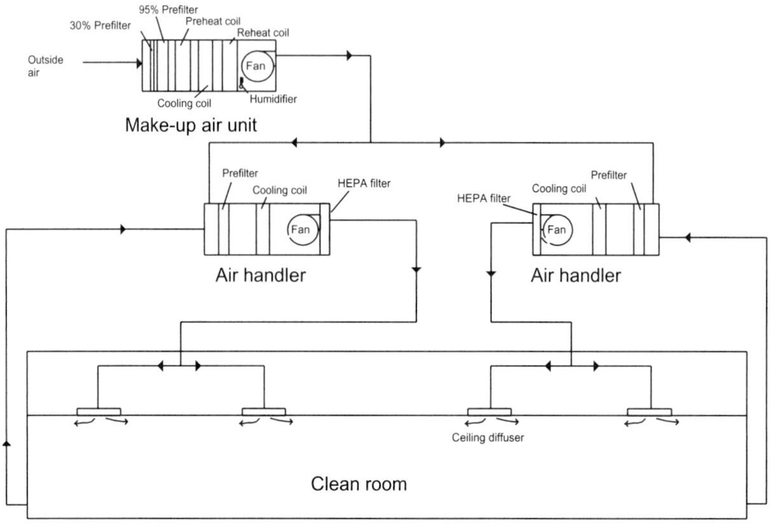

FIGURE 17 Non-unidirectional clean-room with air handler mounted HEPA filters.

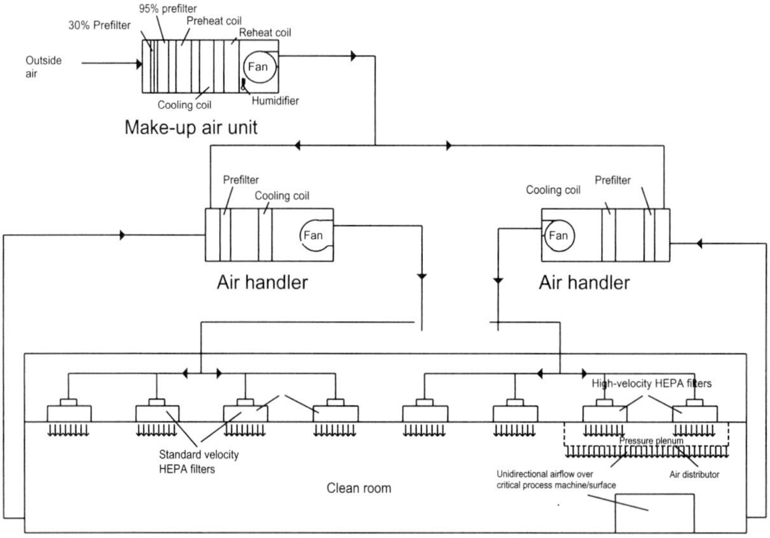

FIGURE 18 Non-unidirectional clean-room with critical area unidirectional flow plenum.

duct distribution network consisting of rectangular or round trunk ducts and flexible or rigid round branch ducts. Full coverage, typical for ISO 5 (Class 100) clean rooms, or partial coverage, for higher class (less stringent) clean rooms, can be accomplished using 2 × 4-ft lay-in HEPA filter modules installed in the ceiling.

3.3.3.6 Clean-Room Testing

ISO 14644 describes methodology and instrumentation for particle counting in the clean room. The tests described there are the basis for assigning a cleanliness rating to the facility. IEST-RP-CC006 similarly provides a procedure for particle counting but goes beyond that to a full series of tests that can be conducted to determine the effectiveness of the clean-room design and operability. The determination of which tests should be run is up to the clean-room end user. As a minimum, particle counting, room pressurization, and filter leakage tests should be run. Other tests dealing with airflow patterns, temperature, humidity, lighting, and sound levels are available. The array of tests selected is determined by the owner based on the effect the various design parameters will have on the product. The data obtained in acceptance tests become baseline data against which future testing is compared to determine if clean-room performance is changing over time. Ongoing periodic monitoring of the facility will ensure that clean-room performance degradation is identified as it occurs. Pass–fail criteria are not part of the ISO standards but are to be developed on a case-by-case basis by the end user of the facility. These standards become part of the operational protocol of the facility.

The clean-room testing described here is part of the commissioning or validation process wherein all equipment in the facility is run, tested, and observed to ensure it is working as designed.

3.3.3.7 Utilities

Biopharmaceutical clean-rooms typically house process equipment requiring utilities such as pure water, electricity, vacuum, and clean compressed air. The source of these utilities is usually outside the clean room. During the design phase a utility matrix is developed, in conjunction with end users and equipment manufacturers, identifying all equipment and the utilities required. This is the basis for determining the capacity of the utility systems as well as the point-of-use location of specific utilities.

When bringing the utilities to the point of use, care should be taken to ensure that the clean room is not compromised. A clean construction protocol should be implemented and wall, ceiling, and floor penetrations, if needed, should be flashed and sealed in such a manner as to prevent contaminants from entering the clean room. Such entry points should also be smoothly sealed to ensure that there are no crevices to harbor organisms. Drains should be avoided in the clean room wherever possible. When this is not possible, the drains should be covered when not in use with a means specifically designed for biopharmaceutical clean-room application. Such means are tight, smooth, cleanable, and corrosion resistant.

In small facilities an individual pipeline may be run from outside the facility to the point of use. In large facilities a utility chase (Figure 19) that enables major utility lines to be brought to the vicinity of process tools may be provided. Final hook-up between the chase and point of use then becomes a relatively simple, minimally intrusive procedure. The utility chase concept is also beneficial in facilities that undergo frequent retrofit or upgrade.

3.3.4 CONSTRUCTION PHASE: CLEAN BUILD PROTOCOL

Ongoing experience has demonstrated that an aggressive clean construction protocol program is generally not required for biopharmaceutical facilities that do not

FIGURE 19 The utility chase is located between two clean rooms. Major utility lines are installed within the chase and hook-up lines for local pieces of process equipment are connected through the clean-room walls. A major benefit of this arrangement is that installers need not be fully garbed in clean-room attire to access the utility lines. (Courtesy of CleanTek.)

carry a cleanliness rating. Where cleanliness classifications less stringent than Class 10,000 are used, standard construction techniques followed by careful cleanup and wipe-down within the clean space have proven quite acceptable. Cleanliness levels of Class 1000 or Class 10,000 are achievable shortly after startup and maintainable thereafter. For cleanliness levels of Class 100 a somewhat more restrictive protocol is required. Once a facility is up and running, any intrusion into clean areas for retrofit work should be done in conjunction with some level of clean build protocol in place, dependent on the rating of the facility and the degree of disruption encountered during the retrofit project.

The levels of clean construction described herein can provide a practical means of meeting operational cleanliness goals in a cost-effective fashion. Each project, whether new construction or retrofit of an existing process, should have as part of it an evaluation of the required elements of the build clean protocol to be employed. The information provided below is broad and can act as a template for the protocol put in place for a specific project.

A key to successful clean construction is the appointment of an individual as a clean-room monitor who is well versed in the clean-room construction protocol. That person is charged with maintaining a clean environment and monitoring the activities of all personnel within the clean area during the construction phase *and* is concerned with maintaining budget and schedule goals. The clean-room monitor should have the confidence to make "real-world" decisions supporting the "spirit" of the protocol as well as the "letter."

3.3.4.1 General

All clean-facility construction, while employing standard construction techniques, should be accomplished in a manner that does not create excessive particulate contamination. A temporary lay-down area within the building adjacent to the clean area should be set aside for storage of clean construction components. All tools used for clean construction should be in an "as-new" condition and be cleaned and inspected prior to use. The pass–fail criteria for tool and material inspection is "no visible dirt."

Cleanup within the clean area at the end of each shift should consist of broom cleaning *and* vacuum cleaning the floor with a clean vacuum, that is, a vacuum with a HEPA filter (99.97% efficient on 0.3-μm particles).

Clean-facility construction materials should be left in an outer shipping wrap until moved to the temporary lay-down area, where they should then be unwrapped and wiped down before being moved into the clean space.

Adherence to these guidelines will make final clean-up faster and acceptable start-up and certification/validation more certain. While a goal of clean construction is rapid start-up and certification/validation, a long-range goal is the maintenance of the facility cleanliness without intrusion, over an extended period of time, of contaminants deposited during construction due to a poor protocol or improper implementation of the protocol.

Appendix A and Appendix B offer a template for working in a clean environment as well as clean-room cleaning procedures. Procedures should be modified with caution to suit a particular project.

3.3.4.2 Level I Clean Construction

Level I clean construction is used for all areas with a cleanliness rating of Class 1000 (ISO 6) or higher (less stringent), including those spaces within which clean processes are conducted in minienvironments/isolators and those unrated areas identified as being "controlled environments."

Standard construction techniques are used until the clean-room envelope is completed, HEPA filters with protective film in place are installed and air handlers are ready to start. The clean envelope consists of clean-room walls, ceiling, and floor. Prior to starting the air handlers, a thorough clean-up of the space within the clean envelope is accomplished as described in Appendix B, 2A–2J. Following clean-up and start-up of the clean-space air-handling system, particle counts should quickly drop to well within operational requirements.

Once the clean room is operational, as described above, additional construction related to process equipment installation and facility modification within the clean room can be done in compliance with the Guidelines in Appendix A.

3.3.4.3 Level II Clean Construction

This is used for construction of those areas rated at Class 100 (ISO 5) employing a 100% HEPA filter ceiling. Generally standard construction techniques should be used. The clean-room envelope includes walls, ceiling, floor, return ductwork, supply fans, and supply ductwork.

All ductwork sections should be cleaned and sealed with plastic wrap at the time of fabrication until just prior to installation or start-up to prevent contaminants from accumulating inside air-handling passageways. The sections of ductwork should be unsealed only as required for installation. Open ends of ducts and fans should remain sealed until connecting duct is about to be installed. A final isopropyl alcohol (IPA) wipe-down of all interior duct sections and fan surfaces should be done immediately prior to installation.

When general construction of the clean room is completed, steps 2A–2J of Appendix B describing coarse cleaning can be implemented. Following coarse cleaning the protective film can be removed from the HEPA filters and the air-handling system started.

Successful completion of the cleaning process described above will indicate that installation of process equipment may begin. Note that procedures described in Appendix A should be followed. After installation of all process equipment or when the clean room is to be prepared for certification, steps 2K–2O of Appendix B for final wipe-down can be followed.

A black-and-white felt rub-down test is performed to demonstrate adequate cleanliness of the interior clean-envelope surfaces. This test consists of both black-and-white felt being wiped over any surface for 1 m linear distance with a firm hand pressure. No residue should be visible on the cloth. Each cloth should be 60 cm square black or white static-free natural fiber felt folded with cut edges inside to a 25-cm square. The cut edges should be sealed with an approved latex sealant.

3.3.5 MAINTENANCE

To maximize the life and effectiveness of the facility, it must be maintainable. The facility should be designed to permit ongoing day-to-day preventive maintenance of the mechanical systems and, should a failure occur, permit needed repairs to be made in an expeditious manner. Perhaps of equal importance is the janitorial maintenance required to keep the facility suitable for pharmaceutical manufacturing. Proper janitorial maintenance begins with the design of the facility and evolves into an operational protocol, personnel training, and effective implementation.

In the design phase it is important to provide sufficient access to mechanical and process equipment to enable preventive maintenance procedures to be carried out with minimum effort. Typically manufacturer's installation instructions offer guidelines as to how much space should be left open around equipment to permit removal of critical components. One driver of construction cost is floor space. Making a space as small as possible to house an operation presumably will result in first-cost savings. If the space does not provide sufficient access for lubrication, filter changes, belt adjustments, and the like, there is a strong possibility that this preventive maintenance will be ignored. A predictable result is shortened equipment life and the disappearance of any first-cost savings that may have been realized. If there is a major equipment failure that requires replacement of an inaccessible component, the cost associated with knocking down a wall to gain equipment access will very likely negate first-cost savings.

Storage of maintenance items should be identified early in the design process. Spare-parts storage, janitorial supply storage, janitors' sink closets, repair work shops, and storage space for consumable maintenance items (e.g. air filters) will require floor space in the facility design. Frequently tools are dedicated to the clean facility or are required specifically for unique process equipment and must also have a storage area. Accommodation of these items is an important part of the planning process.

A requirement of a clean facility is that the cleaning materials should be specifically intended for use in a "clean" operation, should be kept in good ("like new") repair, and should not be used in other, nonclean, areas of the facility. Using general cleaning materials manned by the "house" janitorial staff will invariable introduce more contamination into the clean portions of the facility than it removes.

A central housekeeping vacuum is very useful in keeping contamination under control. While "wet-and-dry" versions of the central vacuum are available, the manner in which each is to be used should be carefully reviewed to ensure that it is in keeping with the sanitary requirements of the facility. A common housekeeping procedure addresses spills with local clean-up and uses a dry-type central vacuum for dry particulate contaminants.

APPENDIX A: GUIDELINES FOR CONSTRUCTION PERSONNEL AND WORK TOOLS IN A CLEAN ROOM

1.0 General requirements

A. Makeup will not be allowed inside the clean room.
B. Smoking will not be allowed in or around the clean room.
C. Tobacco chewing will not be allowed in the clean room.
D. Paper or paper by-products will not be allowed in the clean room except clean-room approved paper and pens.
E. Prints or papers will be allowed only if totally laminated in plastic and cleaned with isopropyl alcohol prior to entry.
F. Lead pencils will not be allowed in the clean room. Ball point pens only.
G. Clean-room garments, to include shoe covers, coveralls, and head cover, will be worn within the clean room.
H. Clean-room garments will not be unfastened or unzipped while inside the clean room.
I. No writing will be allowed on the clean-room garments.
J. Food and drink will not be allowed in the clean room.
K. Combing of hair will not be allowed in the clean room or gowning area.
L. Stepping on chairs, work benches, test equipment or process equipment is not allowed.
M. Damaged garments (rips, worn booties, torn gloves) will be replaced immediately. Do not wait for a convenient time. DO IT NOW!
N. Tool pouches are not allowed in a clean room.
O. All work areas and adjacent areas will be vacuumed after completion of work and prior to leaving the clean room.
P. **ALWAYS** wash hands before entering the clean room to remove residues from food, smoke, and/or other sources.

2.0 Personnel

A. All personnel working inside a clean room will be required to follow all dress codes associated with the particular clean space.

B. Street clothes or company uniforms will be allowed as standard undergarments provided they are well maintained and clean. No such garments will be allowed that are soiled with grease, dirt, or any detectable stains.
C. Any garments producing excessive fibers (such as fuzzy sweaters) will not be allowed as an undergarment.
D. Standard safety shoes (or other specified footware) will be required. Shoe covers must be worn.
E. Bare feet, socks, and stockings are not allowed inside booties.
F. Coats, lunches, and private items will not be allowed inside the clean room.

3.0 Gowning procedure

A. Each individual is responsible for knowing and using the correct method of gowning prior to entering the clean room. (See Figure 20.)
   1. Clean shoes prior to entering the gowning room.
   2. The order of dress should be as follows:
      a. Shoe covers
      b. Hairnet/beard cover (required after final cleaning)
      c. Hood
      d. Coveralls
      e. Face cover (required after final cleaning)
      f. Gloves (required after final cleaning)

B. Ensure that hoods are tucked inside neck opening of coveralls and pants legs are tucked and snapped inside booties. Garments are to be snapped

FIGURE 20 Clean-room garments are intended to keep contaminants from entering the clean room. In a critical environment the "bunny suit" shown may be required. In a less critical environment a lab coat may suffice. The clean-room construction protocol should identify the type of garment that will be required at various stages of construction and for process equipment installation. (Courtesy of Terra Universal.)

closed at the neck, wrist and ankle opening and sleeves tucked inside gloves.

C. All head hair must be covered at all times.

D. Do not allow garments to touch the floor while dressing or undressing.

E. Avoid leaning on walls, lockers, or other personnel at all times. **DO NOT** place feet on benches.

F. The order of undress should be as follows:

1. Gloves
2. Coveralls
3. Face cover and hood
4. Shoe covers

G. If you will be reentering the clean room, unsoiled garments may be hung for reuse; gloves are not to be reused.

4.0 Work tools, parts, and equipment

A. All tools and equipment used in a clean room should be in like-new condition.

B. All parts will be removed from their shipping container prior to cleaning and introduction into the clean room. **NO PAPER PRODUCTS** will be allowed inside the clean room.

C. All tools, parts, and equipment will be properly cleaned prior to entering the clean room. Minimum cleaning should be a total wipe-down with isopropyl alcohol, using certified clean-room wipes, to assure that the last wipe does not leave visible residue on the wipe. Parts should be blown off *outside* the clean room using filtered nitrogen when available.

D. All parts and equipment should be sent through the equipment wipe-down area (material air lock) and not carried through the gowning area.

5.0 Working in a Clean Room

A. A major concern when working in a clean room is the generation of particles of the size that cannot be seen and spreading these particles throughout the clean room. Every possible precaution must be taken to contain these contaminants and protect the clean-room environment. Everything that is done as a standard operation must be analyzed to determine if it will adversely affect the cleanroom. *If you have any concerns, ask the clean-room monitor before you damage the environment and incur unnecessary clean-up cost.*

B. *All procedures must be reviewed with the clean-room monitor to ensure compliance with clean-room operation practices.* All procedures that can generate particles should be done outside the clean-room whenever possible. In the listing below all prohibited procedures are subject to review by the clean-room monitor. The intent is to get the job done; however, some preplanning with the clean-room monitor can result in a positive result and a clean facility.

1. Drilling:
   a. All power drills will be wrapped to encapsulate any contamination generated during operation.
   b. Drills may be operated in sealed enclosures equipped with an exhaust vacuum.

c. Surface to be drilled will be tented or vacuumed to prevent the spread of contamination.
2. Grinding: NO GRINDING will be allowed in the clean room.
3. Welding: NO WELDING will be allowed inside the clean room.
4. Soldering: May be allowed after total review.
5. Painting: NO PAINTING will be allowed after the start-up of the clean room.
6. Sanding/filing: Will be allowed only within a properly tented space.
7. Cutting: Will be allowed with clean-room approved vacuums removing particles created.

APPENDIX B: CLEANING THE CLEANROOM

1.0 Final cleaning: During this cleaning phase, the clean-room proper should be prepared for certification to the appropriate cleanliness level. Extreme care must be exercised by all those involved in this procedure to minimize the potential for contamination.

2.0 Procedure

A. Secure the entrance to the clean space "envelope," the gowning area, and the entrance from the gowning room to the clean room with locks or limited access via card keys. Post a restriction notice: YOU ARE ENTERING A CONTROLLED, PARTICULATE-FREE ENVIRONMENT, CONTACT __________ FOR PERMISSION TO ENTER.
B. Perform two coarse cleanings of the clean room. Each cleaning should include the following:
   1. Wipe-down of the HEPA filter grid, all lights, walls, floors, windows, and all exposed interior surfaces. This will include any outlet boxes or floor/wall recesses.
   2. Wipe-down should be by clean potable water and mild nonphosphate detergent using clean, lint-free cloths approved by the clean-room monitor.
   3. A second wash-down should commence using clean potable water in the same manner.
   4. Floors should be scrubbed and polished using a floor-polishing machine. No wax is to be used.
C. Partitions and floors should be washed to maintain a dust-free condition.
D. Access into the clean room should be restricted to discourage infiltration of outside particulates.
E. Tacky mats 3 ft by 6 ft should be installed inside the entrance of the gowning area as well as at the entrance to the clean room.
F. Foot covers should be worn by all personnel entering the clean room after the second coarse cleaning is complete.
G. Caulking crew should be assigned and commence work after the second coarse cleaning is completed. They should complete all caulking as required by specification.
H. Simultaneously with the caulking procedure, air handlers and associated ducts and plenums should be checked for cleanliness.

I. The HEPA filter protective film should be removed. Air-handling equipment should be activated and the clean space pressurized to maintain a positive static pressure.
J. From this point forward, clean-room garments and head covers should be worn by all personnel entering the clean space.
K. Commence with the first of two final wipe-downs. Nonshedding clean-room wipes (saturated with isopropyl alcohol) or tacky wipes should be used. *All exposed surfaces should be wiped.*
L. A particle counter should be installed in the clean space and samples taken at several control points over the next 48 hours. A steady decrease in the particle count over time should be achieved.
M. If particle counts stabilize at a level above that desired, a search for filter leakage will be required.
N. If the search for filter leakage fails to find a leak, the entire area should be recleaned as described for the final wipe-down.
O. Once the air standards are achieved, final air balance can begin followed by clean-room certification testing.

BIBLIOGRAPHY

U.S. Food and Drug Administration, Washington, DC

21 CFR Part 210, Current good manufacturing practice in manufacturing, processing, packing, or holding of drugs.

21 CFR Part 211, Current good manufacturing practice for finished pharmaceuticals.

Institute of Environmental Sciences and Technology, Rolling Meadows, IL

IEST-RP-CC001.4: *HEPA and ULPA Filters*, Nov. 7, 2005.

IEST-RP-CC002.2: *Unidirectional Flow Clean-Air Devices*, Jan. 19, 1999.

IEST-RP-CC003.3: *Garments Systems Considerations for Cleanrooms and Other Controlled Environments*, Aug. 11, 2003.

IEST-RP-CC004.3: *Evaluating Wiping Materials Used in Cleanrooms and Other Controlled Environments*, Aug. 23, 2004.

IEST-RP-CC005.3: *Gloves and Finger Cots Used in Cleanrooms and Other Controlled Environments*, May 1, 2003.

IEST-RP-CC006.3: *Testing Cleanrooms*, Aug. 30, 2004.

IEST-RP-CC008-84: *Gas-Phase Adsorber Cells*, Nov. 1, 1984.

IEST-RP-CC012.1: *Considerations in Cleanroom Design*, Mar. 1, 1998.

IEST-RP-CC013-86-T: *Equipment Calibration or Validation Procedures*, Nov. 1, 1986.

IEST-RP-CC016.2: *The Rate of Deposition of Nonvolatile Residue in Cleanrooms*, Nov. 15, 2002.

IEST-RP-CC018.3: *Cleanroom Housekeeping: Operating and Monitoring Procedures*, Jan. 1, 2002.

IEST-RP-CC019.1: *Qualifications for Organizations Engaged in the Testing and Certification of Cleanrooms and Clean-Air Devices*, Jan. 23, 2006.

IEST-RP-CC023.2: *Microorganisms in Cleanrooms*, Jan. 31, 2006.

IEST-RP-CC026.2: *Cleanroom Operations*, July 21, 2004.

IEST-RP-CC027.1: *Personnel Practices and Procedures in Cleanrooms and Controlled Environments*, Apr. 1, 1999.

IEST-RP-CC028.1: *Minienvironments*, Sept. 1, 2002.

IEST-RP-CC034.2: *Hepa and ULPA Filter Leak Tests*, June 23, 2005.

IEST-STD-CC1246D: *Product Cleanliness Levels and Contamination Control Program*, Jan. 1, 2002.

International Organization for Standardization (ISO) Standards

ISO 14644-1: Classification of air cleanliness.

ISO 14644-2: Specifications for testing and monitoring to prove continued compliance.

ISO 14644-3: Test methods.

ISO 14644-4: Design, construction and start-up.

ISO 14644-5: Operations.

ISO 14644-6: Terms and definitions.

ISO 14644-7: Separative devices (clean air hoods, gloveboxes, isolators, and minienvironments).

ISO 14644-8: Classification of airborne molecular contamination.

ISO 14698-1: Bicontamination control—General principles.

ISO 14698-2: Biocontamination control—Evaluation and interpretation of biocontamination data.

ISO 14698-3: Biocontamination control—Methodology for measuring the efficiency of processes of cleaning and/or disinfection of inert surfaces bearing biocontaminated wet soiling or biofilms.

SECTION 4

NORMAL DOSAGE FORMS

4.1

SOLID DOSAGE FORMS

BARBARA R. CONWAY
Aston University, Birmingham, United Kingdom

Contents

Pharmaceutical Manufacturing Handbook: Production and Processes, edited by Shayne Cox Gad

Drug substances are most frequently administered as solid dosage formulations, mainly by the oral route. The drug substance's physicochemical characteristics, as well the excipients added to the formulations, all contribute to ensuring the desired therapeutic activity. Tablets and capsules are the most frequently used solid dosage forms, have been in existence since the nineteenth century, and are unit dosage forms, comprising a mixture of ingredients presented in a single rigid entity, usually containing an accurate dose of a drug. There are other types of solid dosage forms designed to fulfill specific delivery requirements, but they are generally intended for oral administration and for systemic delivery. The major solid oral dosage form is the tablet, and these can range from relatively simple, single, immediate-release dosage forms to complex modified-release systems. Tablets offer advantages for both patients and manufacturers (Table 1). Most tablets are intended to be swallowed whole and to rapidly disintegrate and release drug in the gastrointestinal tract. Tablets are classified by their route of administration or their function, form, or manufacturing process. For example, some tablets are designed to be placed in the oral cavity and to dissolve there or to be chewed before swallowing, and there are many kinds of formulation designed for sustained or controlled release (Table 2).

Solid dose formulations, including tablets, must have the desired release properties coupled with manufacturability and aesthetics and must involve rational formulation design. The dose of the drug and its solubility are important considerations

TABLE 1 Advantages of Tablets as a Dosage Form

Easy to handle
Variety of manufacturing methods
Can be mass produced at low cost
Consistent quality and dosing precision
Can be self-administered
Enhanced mechanical, chemical, and microbiological stability compared to liquid dosage forms
Tamperproof
Lend themselves to adaptation for other profiles, e.g., coating for sustained release

TABLE 2 Types of Solid Dosage Form

Formulation type	Description
Immediate-release tablet/capsule	Intended to release the drug immediately after administration
Delayed-release tablet/capsule	Drug is not released until a physical event has occurred, e.g., change in pH
Sustained-release tablet/capsule	Drug is released slowly over extended time
Soluble tablets	Tablet is dissolved in water prior to administration
Dispersible tablet	Tablet is added to water to form a suspension prior to administration
Effervescent tablet	Tablet is added to water, releasing carbon dioxide to form a effervescent solution
Chewable tablet	Tablet is chewed and swallowed
Chewable gum	Formulation is chewed and removed from the mouth after a directed time
Buccal and sublingual tablets	Tablet is placed in the oral cavity for local or systemic action
Orally disintegrating tablet	Tablet dissolves or disintegrates in the mouth without the need for water
Lozenge	Slowly dissolving tablet designed to be sucked
Pastille	Tablet comprising gelatin and glycerine designed to dissolve slowly in the mouth
Hard gelatin capsule	Two-piece capsule shell that can be filled with powder, granulate, semisolid or liquid
Soft gelatin capsule (softgel)	One-piece capsule containing a liquid or semisolid fill

in the design of the formulation as are the type of dosage form and its method of preparation.

4.1.1 BIOPHARMACEUTICS CLASSIFICATION SYSTEM

Dissolution of the drug must occur before or on reaching the absorption site before absorption can occur, and generally water-soluble drugs do not exhibit formulation difficulties. For poorly water-soluble drugs, the absorption rate may be dictated by the dissolution rate, and, if dissolution is slow, bioavailability may be compromised. The solubility of a drug should, therefore, be considered along with its dose when designing formulations, and unsuitable biopharmaceutical properties is the major reason for the failure of new drugs.

In 1995, the Biopharmaceutics Classification System (BCS) was devised to classify drugs based on their aqueous solubility and intestinal permeability [1]. According to the BCS, drug substances are classified as follows [2]:

Class I: high permeability, high solubility
Class II: high permeability, low solubility
Class III: low permeability, high solubility
Class IV: low permeability, low solubility

A dose solubility volume can be defined for all drugs (i.e., the volume required to dissolve the dose). A drug substance is considered highly soluble when the highest dose strength is soluble in ≤250 mL water over a pH range of 1–7.5. A drug substance is considered highly permeable when the extent of absorption in humans is determined to be ≥90% of an administered dose, based on mass balance or in comparison to an intravenous reference dose. A drug product is considered to be rapidly dissolving when ≥85% of the labeled amount of drug substance dissolves within 30 min using U.S. Pharmacopeia (USP) apparatus I or II in a volume of ≤900 mL buffer solutions.

It was recognized that dissolution rate has a negligible impact on bioavailability of highly soluble and highly permeable (BCS class I) drugs when dissolution of their formulation is sufficiently rapid. As a result, various regulatory agencies including the U.S. Food and Drug Administration (FDA) now allow bioequivalence of formulations of BCS class I drugs to be demonstrated by in vitro dissolution (often called a biowaiver) [3]. Therefore, one of the goals of the BCS is to recommend a class of immediate-release (IR) solid oral dosage forms for which bioequivalence may be assessed based on in vitro dissolution tests.

4.1.2 SYSTEMATIC FORMULATION DEVELOPMENT

Systematic development approaches are needed to gather a full and detailed understanding of marketable formulations in order to satisfy the requirements of regulatory bodies and to provide a research database. Efficient experimental design using in-house or commercial software packages can ensure quality while avoiding expensive mistakes and lost time. Information from various categories such as the properties of the drug substance and excipients, interactions between materials, unit operations, and equipment are required [4]. Design of experiments (DOE) and statistical analysis have been applied widely to formulation development. Using DOE facilitates evaluation of all formulation factors in a systematic and timely manner to optimize the formulation and manufacturing process. Abbreviated excipient evaluation techniques such as Plackett–Burman design can be applied to minimize the number of experiments and identify critical components or processes. Optimization processes can then be applied. When the formulation and manufacturing processes of a pharmaceutical product are optimized by a systematic approach, the scale-up and processes validation can be very efficient because of the robustness of the formulation and manufacturing process.

Innovations in statistical tools such as multivariate analysis, artificial intelligence, and response surface methodology have enabled rational development of formulations, and such methods allow formulators to identify critical variables without having to test each combination.

4.1.3 STANDARD AND COMPRESSED TABLETS

The simplest tablet formulations are uncoated products that are made by direct compression or compression following wet or dry granulation. They are a versatile drug delivery system and can be intended for local action in the gastrointestinal

(GI) tract or for systemic effects. General design criteria for tablets are accuracy and uniformity of drug content, stability of the drug candidate and the formulation, optimal dissolution and availability for absorption (whether immediate or extended release), and patient acceptability in terms of organoleptic properties and appearance. Flocculant, low-density drugs can be difficult to compress and formulate into tablets. This is a particular issue with drugs of low potency. Also some poorly water-soluble, poorly permeable drugs or highly metabolized drugs cannot be given this way. Additionally, local irritant effects can be harmful to the mucosa of the GI tract.

Tablets are a popular dosage form due to their simplicity and economy of manufacture, relative stability, and convenience in packaging, shipping, and storage. For the patient, uniformity of dose, blandness of taste, and ease of administration ensure their popularity. Thus, the purpose of the formulation and the identification of suitable excipients are of primary importance in the development of a successful formulation. A well-designed formulation should contain, within limits, the stated quantity of active ingredient, and it should be capable of releasing that amount of drug at the intended rate and site. Tablets need to be strong enough to withstand the rigors of manufacture, transport, and handling, and they need to be of acceptable size, taste, and appearance. A typical manufacturing process for a tablet product includes weighing, milling, granulation and drying, blending and lubrication, compression, and coating. Each processing step involves several process parameters. For a given formulation, all processing steps should be thoroughly evaluated so that a robust manufacturing process can be defined, and DOE can be applied effectively to optimize this process.

Direct compression is a simple process being more economical and less stressful to ingredients in terms of heat and moisture, However, there are limitations governed by the physical properties of the ingredients, and raw materials must be carefully controlled. It is difficult to form directly compressed tablets containing high-dose and poorly compactible drugs. Granulation can be employed to improve the compaction characteristics of the powder. Granulation can also improve flow properties and reduce the tendency for segregation of the mix due to a more even particle size and bulk density. Granules can be produced by either wet or dry methods based on the stability of the drug and excipients.

Although the basic mechanical process of producing tablets by compression has not changed, there has been much work on improving tableting technology [5]. Understanding of the physical and mechanical properties of powders and the compaction process has improved and will continue to improve product design while increases in the speed and uniformity of action of tableting presses improve the process.

4.1.4 EXCIPIENTS IN SOLID DOSE FORMULATIONS

In addition to the active ingredients, solid oral dosage forms will also contain a range of substances called excipients. The role of excipients is essential in ensuring that the manufacturing process is successful and that the quality of the resultant formulation can be guaranteed. The appropriate selection of excipients and their relative concentrations in the formulation is critical in development of a successful product.

TABLE 3 Excipients Used in Solid Dose Formulations

Excipient Category	Examples
Fillers/diluents	Lactose, sucrose, glucose, microcrystalline cellulose
Binders	Polyvinyl pyrrolidone, starch, gelatin, cellulose derivatives
Lubricants	Magnesium stearate, stearic acid, polyethylene glycol, sodium chloride
Glidants	Fine silica, talc, magnesium stearate
Antiadherents	Talc, cornstarch, sodium dodecylsulfate
Disintegrants and superdisintegrants	Starch, sodium starch glycollate, cross-linked polyvinyl pyrrolidone
Colorants	Iron oxide, natural pigments
Flavor modifiers	Mannitol, aspartame

Although they are often categorized as inert, preformulation studies can determine the influence of excipients on stability, bioavailability, and processability. Excipients are categorized into groups according to their main function, although some may be multifunctional, and examples of common excipients used in the manufacture of tablets and capsule are detailed in Table 3.

4.1.4.1 Diluents

An inert substance is frequently added to increase the bulk of a tablet for processing and handling. The lower weight limit for formulation of a tablet is usually 50mg. Ideally, diluents should be chemically inert, nonhygroscopic, and hydrophilic. Having an acceptable taste is important for oral formulations, and cost is always a significant factor in excipient selection.

Lactose is a common diluent in both tablets and capsules, and it fulfils most of these criteria but is unsuitable for those who are lactose intolerant. Various lactose grades are commercially available which have different physical properties such as particle size distribution and flow characteristics. This permits the selection of the most suitable material for a particular application. Usually, fine grades of lactose are used for preparation of tablets by wet granulation or when milling during processing is carried out, since the fine size permits better mixing with other formulation ingredients and facilitates more effective action of the binder [6].

Diluents for direct compression formulations are often subject to prior processing to improve flowability and compression, for example, amorphous lactose, but this can contribute to reduced stability especially under high-humidity conditions when reversion to the crystalline form is more likely [6].

Microcrystalline cellulose (Avicel) is purified partially depolymerized cellulose, prepared by treating α-cellulose with mineral acids. In addition to being used as a filler, it is also used as dry binder and disintegrant in tablet formulations. Depending on the preparation conditions, it can be produced with a variety of technical specifications depending on particle size and crystallinity. It is often used as an excipient in direct compression formulations but can also be incorporated as a diluent for tablets prepared by wet granulation, as a filler for capsules and for the production of spheres.

Diluents, although commonly presumed inert, do have the ability to influence the stability or bioavailability of the dosage form. For example, dibasic calcium phosphate (both anhydrous and dihydrate forms) is the most common inorganic salt used as a filler–binder for direct compression. It is particularly useful in vitamin products as a source of both calcium and phosphorous. Milled material is typically used in wet-granulated or roller-compacted formulations. The coarse-grade material is typically used in direct compression formulations. It is insoluble in water, but its surface is alkaline and it is therefore incompatible with drugs sensitive to alkaline pH. Additionally, it may interfere with the absorption of tetracyclines [7].

4.1.4.2 Binders

Binders (or adhesives) are added to formulations to promote cohesiveness within powders, thereby ensuring that the tablet remains intact after compression as well as improving the flow by forming granules. A binder should impart adequate cohesion without retarding disintegration or dissolution. Binders can be added either as a solution or as a dry powder. Binders added as dry powders are mixed with other powders prior to agglomeration, dissolving in water or solvent added during granulation, or added prior to compaction. Solution binders can be sprayed, poured, or mixed with the powder blend for agglomeration and are generally more effective, but further dry binder can be added prior to tableting. Starch, gelatin, and sugars are used along with gums, such as acacia and sodium alginate, and are used at concentrations between 2 and 10% w/w. Celluloses and polyvinyl pyrrolidone (PVP) are also utilized, often as dry binders.

4.1.4.3 Lubricants

Lubricants can reduce friction between the tablet and the die wall during compression and ejection by interposing an intermediate film of low shear strength at the interface between the tablet and the die wall. The best lubricants are those with low shear strength but strong cohesive tendencies perpendicular to the line of shear [8]. The hydrophobic stearic acid and stearic acid salts, primarily magnesium stearate, are the most widely used and are included at concentrations less than 1% w/w in order to minimize any deleterious effects on disintegration or dissolution. They should be added after the disintegrant to avoid coating it and preferably at the final stage prior to compression to ensure mixing time is kept to a minimum. Hydrophilic lubricants such as polyethylene glycols (PEGs) and lauryl sulfates can be used to redress the issues with dissolution but may not be as efficient as their hydrophobic counterparts.

4.1.4.4 Glidants and Antiadherents

Like lubricants, glidants are fine powders and may be required for tablet compression at high production speeds to improve the flow properties of the material into the die or during initial compression stages. They are added in the dry state immediately prior to compression and, by virtue of their low adhesive potential, reduce the friction between particles. Colloidal silica is popular, as are starches and talc.

Antiadherents can also be added to a formulation that is especially prone to sticking to the die surface (or picking). Water-insoluble lubricants such as magnesium stearate can be used as antiadherents, as can talc and starch.

4.1.4.5 Disintegrants

Disintegrants are added to a formulation to overcome the cohesive strength imparted during compression, thus facilitating break up of the formulation in the body and increasing the surface area for dissolution. They can be either intragranular, extragranular, or both, and there is still a lack of understanding concerning their precise mechanism of action. On contact, disintegrants can draw water into the tablet, swelling and forcing the tablet apart. Starch, a traditional and still widely used disintegrant, will swell when wet, although it has been reported that its disintegrant action could be due to capillary action [6]. Levels can be increased beyond the normal 5% w/w to 15–20% w/w if a rapid disintegration is required. Surfactants can also act as disintegrants promoting wetting of the formulation, and sodium lauryl sulfate can be combined with starch to increase effectiveness.

Tablet disruption following production of carbon dioxide is another mechanism used to enhance disintegration. This uses a mixture of sodium bicarbonate and a weak acid such as citric acid or tartaric acid and is exploited for effervescent formulations.

4.1.4.6 Superdisintegrants

Compared to the more traditional starch, newer disintegrants are effective at much lower levels and comprise three groups: modified starches, modified cellulose, and cross-linked povidone. Their likely mechanism of action is a combination of proposed theories including water wicking, swelling, deformation recovery, repulsion, and heat of wetting [9]. Superdisintegrants are so called because of the relatively low levels required (2–4% w/w). Sodium starch glycollate (Primojel, Explotab) is made by cross-linking potato starch and can swell up to 12-fold in less than 30s. Crospovidone is completely insoluble in water, although it rapidly disperses and swells in water, but does not gel even after prolonged exposure. It rapidly exhibits high capillary activity and pronounced hydration capacity with little tendency to form gels and has a greater surface area–volume ratio compared to other disintegrants. Micronized versions are available to improve uniformity of mix. Croscarmellose sodium, a cross-linked polymer of carboxymethyl cellulose sodium is also insoluble in water, although it rapidly swells to 4–8 times its original volume on contact with water [6].

4.1.4.7 Added Functionality Excipients

Adverse physiochemical and mechanical properties of new chemical entities prove challenging for formulation development. There is an increasing demand for faster and more efficient production processes. Also, biotechnological developments and various emerging protein-based therapies are broadening the definition for excipient products. Although the description of excipients from inactive ingredients is shifting toward functionally active materials and will continue to grow in this area, the introduction of improved versions of long-existing excipients is probably the more successful development. New single-component and coprocessed products have been introduced, for example, filler–binders. In addition, there have been advances in the understanding of how such substances act and hence how they can be optimally designed. Excipients for use in direct compression product forms or physically or chemically modified excipients used in relatively new drug delivery systems, such as patches or inhalation systems, are examples of these developments.

4.1.4.8 Colorants

Colorants are frequently used in uncoated tablets, coated tablets, and hard and soft gelatin capsules. They can mask color changes in the formulation and are used to provide uniqueness and identity to a commercial product. Concerns over the safety of coloring agents in formulations generally arise from adverse effects in food substances. Colorants are therefore subject to regulations not associated with other pharmaceutical excipients. Legislation specifies which colorants may be used in medicinal products and also provides for purity specifications. The number of permitted colors has decreased in recent years, and a list of approved colorants allowed by regulatory bodies can vary from country to country.

Colorants can be divided into water-soluble dyes and water-insoluble pigments. Some of the insoluble colors or pigments can also provide opacity to tablet coatings or gelatin shells, which can promote stability of light-sensitive active materials. Pigments such as the iron oxides, titanium dioxide, and some of the aluminum lakes are especially useful for this purpose.

Water-soluble dyes are usually incorporated within the granulation process to ensure even distribution throughout the formulation, but there can be an uneven distribution due to migration of the dye during drying processes. Therefore, water-soluble dyes can also be adsorbed into a carrier such as starch or lactose and dry blended prior to the final mix. Water-insoluble pigments are more popular in direct compression and are dry blended with the other ingredients.

Lakes are largely water-insoluble forms of common synthetic water-soluble dyes and are prepared by adsorbing the sodium or potassium salt of a dye onto a very fine substrate of hydrated alumina, followed by treatment with a further soluble aluminum salt. The lake is then purified and dried. Lakes are frequently used in coloring tablet coatings since they are more stable and have greater opacity than a water-soluble dye [6].

4.1.4.9 Interactions and Safety of Excipients

Because there is such a wide selection available, rational choice of the necessary excipients and their concentration is required. Consideration must also be given to cost, reliability, availability, and international acceptability. Although generally considered inert, formulation incompatibility of excipients is also necessary. Lactose, for example, can react with primary and secondary amines via its aldehyde group by Maillaird condensation reaction [6], and calcium carbonate is incompatible with acids due to acid–base chemical reaction and with tetracyclines due to complexation. Additionally, excipients can contribute to the instability of the active substance through moisture distribution.

Despite the importance of drug–excipient compatibility testing, there is no generally accepted method available for this purpose. After identification of any major known incompatibilities, a compatibility screen needs to be proposed. Issues such as sample preparation, storage conditions, and methods of analysis should be addressed and factorial design applied to reduce the number if tests required. Drug–excipient compatibility studies can be performed with minimal amounts of materials. Usually, small amounts of each material are weighed into a glass vial, in a ratio representative of the expected ratio in the formulation. The vials can be sealed as is or with additional water, either in an air environment or oxygen-free (nitrogen head space) environment, and stored in the presence or absence of ambient light, at various temperatures. Factorial or partial factorial design experiments can be set up to determine important binary and multiple component interaction factors. This information helps determine which excipients should be avoided and whether oxidation or light instability in the formulation is a consideration. Controls consisting of the active pharmaceutical ingredient (API) alone in the various conditions also should be run to determine whether the API is susceptible alone or must have the mediating excipient or water additives for instability.

4.1.5 COATED TABLETS

Tablets are often coated to protect the drug from the external environment, to mask bitter tastes, add mechanical strength, or to enhance ease of swallowing. A coating can also be used for aesthetic or commercial purposes, improving product appearance and identity.

4.1.5.1 Sugar-Coated Tablets

Sugar coating can be beneficial in masking taste, odors, and colors. It is useful in protecting against oxidation, and sugar coating was once very common due to its aesthetic results and cheapness of materials. Use has declined in recent years due to the complexity of the process and skills required, but advances in technology have led to a resurgence in popularity. Typical excipients used are sucrose (although this can be substituted with low-calorie alternatives), fillers, flavors, film formers, colorants, and surfactants. It is usually carried out in tumbling coating pans and comprises several stages.

The first sealing stage uses shellac or cellulose acetate phthalate, for example, to prevent moisture from reaching the tablet core. This has to be kept to a minimum to prevent impairment of drug release. The subcoating is an adhesive coat of gum (such as acacia or gelatin) and sucrose used to round off the edges, and the tablets can be dusted with substances such as kaolin or calcium carbonate to harden the coating. A smoothing coat is built up in layers using 70% v/v sucrose syrup and often opacifiers such as titanium dioxide, and the tablets are dried between each application. A colorant is added to the final few layers and followed with a final polishing step which can make further embossing difficult. The coating is relatively brittle, prone to chipping or cracking, and there is a substantial increase in weight, up to 50%, and size of the product.

4.1.5.2 Compression Coating and Layered Tablets

A coating can be applied by compression using specially designed tablet presses. The same process can be used to produce layered tablets which can comprise two or even three layers if complete separation of the ingredients is required. This process is used when physical separation of ingredients is desired due to incompatibility or to produce a repeat-action product. The formulation can also be designed to provide an immediate and a slow-release component. Release rates can be controlled by modification of the geometry, the composition of the core, and the inclusion of a membrane layer.

The technique involves using a preliminary compression step to produce a relatively soft tablet core which is then placed in a large die containing coating material. Further coating material is added and the content compressed. A similar light compression is used for the production of layers and a final main compression step used to bind the layers together.

4.1.5.3 Film-Coated Tablets

Film coating, although most often applied to tablets, can also be used to coat other formulations including capsules. Film coating imparts the same general characteristics as sugar coating but weight gain is significantly less (typically up to 5%), it is easier to automate, and it has capacity to include organic solvents if required. The main methods involved are modified conventional coating pans, side-vented pans, and fluid-bed coating. Celluloses are often used as film-forming polymers, as detailed in Table 4, and usually require addition of a compatible plasticizer as glass transition temperatures are higher than the temperatures used in the process. Polyethylene glycol, propylene glycol, and glycerol are commonly used, and colorants and opacifiers can also be added to the coating solution. Specialist coatings such as Opadry fx and Opaglos 2 can be used to give a high gloss finish to improve brand identity and consumer recognition.

4.1.5.4 Tablet Wrapping or Enrobing[1]

Recent innovations in tablet coating include the use of gelatin and non-animal-derived coatings for tablets that require formulation of a pre-formed film that is then used to encapsulate the product (e.g., Banner's Soflet Gelcaps or Bioprogress' Nrobe technology). The coated formulations are tamper evident and can be designed with different colors for branding purposes. They are reported to be preferred by patients due to their ease of swallowing and superior taste- and odor-masking properties. An alternative is the Press-fit Geltabs system, which uses a high-gloss gelatin capsule shell to encapsulate a denser caplet formulation.

4.1.6 HARD AND SOFT GELATIN CAPSULES

Capsules are solid oral dosage forms in which the drug is enclosed within a hard or soft shell. The shell is normally made from gelatin and results in a simple, easy-to-swallow formulation with no requirement for a further coating step. They can be

[1]See http://www.banpharm.com/technologiesSofletGelcap.cfm and http://www.fmcmagenta.com/NRobe/tabid/145/Default.aspx.

TABLE 4 Polymers Commonly Used in Film Coating of Tablets

Polymer	Soluble in	Description
Methylcellulose (MC)	Cold water, GI fluids, and organic solvents	Low-viscosity grades best for aqueous films
Ethylcellulose (EC)	Organic solvents and GI fluids (insoluble in water)	Used in combination with water-soluble agents for immediate release
Hydroxyethyl cellulose (HEC)	Water and GI fluids	Similar to MC with clear solutions
Hydroxypropyl cellulose (HPC)	Cold water, GI fluids, and polar solvents	Results in a tacky coat and used in combination to promote adhesion
Hydroxypropylmethyl cellulose (HPMC)	Cold water, GI fluids, and alcohols	Excellent film former, low-viscosity grades best
Sodium carboxymethyl cellulose	Water and polar solvents	Cannot be used if presence of moisture is a problem
Methylhydroxyethyl cellulose (MHEC)	Water and GI fluids	Similar to HPMC but less soluble in organic solvents
Povidone (PVP)	Water, GI fluids, alcohol, and isoproplyl alcohol (IPA)	Can lead to tackiness during drying, often brittle and hygroscopic
PEGs	Water, GI fluids, some organic solvents	High molecular weights best for film forming and low molecular weights used as plasticizer; can be waxy
Enteric coatings such as poly(methacrylates) or cellulose acetate phthalate	Soluble at elevated pHs	Used for delayed-release formulations

Source: Adapted from refs. 5 and 10.

either hard or soft depending on the nature of the capsule shell, with soft capsules possessing a flexible, plasticized gelatin film. Hard gelatin capsules are usually rigid two-piece capsules that are manufactured in one procedure and packed in another totally separate operation, whereas the formulation of soft gelatin capsules is more complex but all steps are integrated.

There is a growing interest in using non-animal-derived products for formulation of the capsule shells to address cultural, religious, and dietary requirements. HPMC (e.g., V-caps, Quali-VC, Vegicaps) and pullulan shells (NPCaps) and starch are alternatives.

4.1.6.1 Hard-Shell Gelatin Capsules

Although the challenges of powder blending, homogeneity, and lubcricity exist for capsules as for tablets, they are generally perceived to be a more flexible formulation as there is no requirement for the powders to form a robust compact. This means that they may also be more suitable for delivery of granular and beadlike formulations, fragile formulations that could be crushed by the normal compaction step. They are commonly employed in clinical trials due to the relative ease of blinding and are useful for taste masking.

Capsules are usually more expensive dosage forms than an equivalent tablet formulation due to the increased cost of the shells and the slower production rates. Even with modern filling equipment, the filling speeds of capsule machines are much slower than tablet presses. However, increased costs can be offset by avoiding a granulation step. Capsules, although smoother and easy to swallow, also tend to be larger than corresponding tablet formulations, potentially leading to retention in the esophagus. Humidity needs to be considered during manufacture and storage, with moisture leading to stickiness and desiccation causing brittleness. Cross-linking of gelatin in the formulation can also lead to dissolution and bioavailability concerns.

Capsule excipients are similar to those required for formulation of tablets and include diluents, binders, disintegrants, surfactants, glidants, lubricants, and dyes or colorants. The development of a capsule formulation follows the same principles as tablet development, and consideration should be given to the same BCS issues. The powder for encapsulation can comprise simple blends of excipients or granules prepared by dry granulation or wet granulation. There is a reduced requirement for compressibility, and often the flow properties are not as critical as in an equivalent tablet formulation. The degree of compressibility required is the major difference, and capsules can therefore be employed when the active ingredient does not possess suitable compression characteristics.

4.1.6.2 Manufacture of Hard Gelatin Shells

Gelatin is a generic term for a mixture of purified protein fractions obtained either by partial acid hydrolysis (type A gelatin) or by partial alkaline hydrolysis (type B gelatin) of animal collagen. Type A normally originates from porcine skin while B is usually derived from animal bones, and they have different isoelectric points (7.0–9.0 and 4.8–5.0, respectively) [6]. The protein fractions consist almost entirely of amino acids joined together by amide linkages to form linear polymers, varying in molecular weight from 15,000 to 250,000. Gelatin can comprise a mixture of both types in order to optimize desired characteristics, with bone gelatin imparting firmness while porcine skin gelatin provides plasticity. Gelatin Bloom strength is measured in a Bloom gelometer, which determines the weight in grams required to depress a standard plunger in a 6.67% w/w gel under standard conditions. Bloom strength and viscosity are the major properties of interest for formulation of capsules, and Bloom strength of 215–280 is used in capsule manufacture.

Gelatin is commonly used in foods and has global regulatory acceptability, is a good film former, is water soluble, and generally dissolves rapidly within the body without imparting any lag effect on dissolution. Gelatin capsules are strong and robust enough to withstand the mechanical stresses involved in the automated filling and packaging procedures.

In addition to gelatin, the shells may contain colorants, opacifiers, and preservatives (often parabens esters). There are eight standard capsule sizes, and the largest capsule size considered suitable for oral use is size 0 (Table 5).

To manufacture the shells, pairs of molds, for the body and the cap, are dipped into an aqueous gelatin solution (25–30% w/w), which is maintained at about 50 °C in a jacketed heating pan. As the pins are withdrawn, they are rotated to distribute the gelatin evenly and blasted with cool air to set the film. Drying is carried out by

TABLE 5 Capsule Size and Corresponding Volume or Weight of Fill

Size	Volume (mL)	Fill weight[a] (g)
000	1.37	1.096
00	0.95	0.760
0	0.68	0.544
1	0.50	0.400
2	0.37	0.296
3	0.30	0.240
4	0.21	0.168
5	0.13	0.104

Source: Adapted from http://capsugel.onlinemore.info/download/BAS192-2002.pdf.

[a]Assumes a powder density of $0.8\,g/cm^3$.

passing dry air over the shell as heating temperatures are limited due to the low melting point of gelatin. The two parts are removed from the pins, trimmed, and joined using a prelock mechanism. The external diameter of the body is usually wider at the open end than the internal diameter of the cap to ensure a tight fit. They can be made self-locking by forming indentations or grooves on the inside of both parts so that when they are engaged, a positive interlock is formed (e.g., Posilok, Conicap, Loxit).

Alternatively, they may be hermetically sealed using a band of gelatin around the seam between the body and the cap (Qualicaps). This can be applied without the application of heat and provide a tamper-evident seal. LEMS (liquid encapsulation microspray sealing) used in Licaps is a more elegant seal in which sealing fluid (water and ethanol) is sprayed onto the joint between the cap and body of the capsule. This lowers the melting point of gelatin in the wetted area. Gentle heat is then applied which fuses the cap to the body of the Licaps capsule. The moisture content of manufactured shells is 15–18% w/w and levels below 13% will result in problems with the capsule filling machinery. Therefore, capsules are stored and filled in areas where relative humidity is controlled to between 30 and 50%.

4.1.6.3 Hard-Gelatin Capsule Filling

The filling material must be compatible with the gelatin shell and, therefore, deliquescent or hygroscopic materials cannot be used. Conversely, due the moisture content in the capsule shells, they cannot be used for moisture-sensitive drugs. All ingredients need to be free of even trace amounts of formaldehyde to minimize cross-linking of gelatin.

Powders and granules are the most common filling materials for hard-shell gelatin capsules, although pellets, tablets, pastes, oily liquids, and nonaqueous solutions and suspensions have been used. Filling machines are differentiated by the way they measure the dose of material and range in capacity from bench-top to high-output, industrial, fully automated machines. Those that rely on the volume of the shell are known as capsule dependent, whereas capsule-independent forms measure the quantity to be filled in a separate operation. The simplest dependent method of filling is leveling where powder is transferred directly from a hopper to the capsule

TABLE 6 Liquid Excipients Compatible with Hard Gelatin Capsules

Peanut oil	Paraffin oil
Hydrogenated peanut oil	Cetyl alchohol
Castor oil	Cetostearyl alcohol
Hydrogenated castor oil	Stearyl alcohol
Fractionated coconut oil	Stearic acid
Corn oil	Beeswax
Olive oil	Silica dioxide
Hydrogenated vegetable oil	Polyethylene glycols
Silicone oil	Macrogol glycerides
Soya oil	Poloxamers

Source: Adapted from http://www.capsugel.com/products/licaps_oil_chart.php.

body, aided by a revolving auger or vibration. Additional powder can be added to fill the space arising, and the fill weight depends on the bulk density of the powder and the degree of tamping applied.

Most automated machinery is of the independent type and compresses a controlled amount of powder using a low compression force (typically 50–200 *N*) to form a plug. Most are piston-tamp fillers and are dosator or dosing disk machines. The powder is passed over a dosing plate containing cavities slightly smaller than the capsule diameter, and powder that falls into the holes is tamped by a pin to form a plug. This can be repeated until the cavity is full and the plugs (or slugs) are ejected into the capsule shells. The minimum force required to form a plug should be used to reduce slowing of subsequent dissolution.

In the dosator method, the plug is formed within a tube with a movable piston that controls the dosing volume and applies the force to form the plug. The dose is controlled by the dimensions of the dosator, the position of the dosator in the powder bed, and the height of the powder bed. Fundamental powder properties to ensure even filling are good powder flow, lubricity, and compressibility. The auger or screw method, now largely surpassed, uses a revolving archimedian screw to feed powder into the capsule shell.

A liquid fill can be useful when manufacturing small batches if limited quantities of API are available. Liquid fills also offer improved content uniformity for potent, low-dose compounds and can reduce dust-related problems arising with toxic compounds. Two types of liquid can be filled into hard gelatin capsules: nonaqueous solutions and suspensions or formulations that become liquid on application of heat or shear stress. These require hoppers with heating or stirring systems. For those formulations that are liquid at room temperature, the capsule shells need to be sealed after filling to prevent leakage of the contents and sticking of the shells. It is essential to ensure the liquid is compatible with the shell (Table 6).

4.1.6.4 Soft-Gelatin Capsules

Soft gelatin capsules are hermetically sealed one-piece capsules containing a liquid or a semisolid fill. Like liquid-filled hard capsules, although the drug is presented in a liquid formulation, it is enclosed within a solid, thus combining the attributes of both. Soft gelatin capsules (softgels) offer a number of advantages including improved bioavailability, as the drug is presented in a solubilized form, and enhanced drug stability. Consumer preference regarding ease of swallowing, convenience, and taste can improve compliance, and they offer opportunities for product differentiation via color, shape, and size and product line extension. The softgels can be enteric coated for delayed release. They are popular for pharmaceuticals, cosmetics, and nutritional products, but highly water-soluble drugs and aldehydes are not suitable for encapsulation in softgels. Formulations are tamper evident and can be used for highly potent or toxic drugs. However, they do require specialist manufacture and incur high production costs.

4.1.6.5 Manufacture of Soft Gelatin Capsules

The shell is primarily composed of gelatin, plasticizer, and water (30–40% wet gel), and the fill can be a solution or suspension, liquid, or semisolid. The size of a softgel represents its nominal capacity in minims, for example, a 30 oval softgel can accommodate 30 minims (or $1.848\,cm^3$). Glycerol is the major plasticizer used, although sorbitol and propylene glycol can also be used. Other excipients are dyes, pigments, preservatives, and flavors. Up to 5% sugar can be added to give a chewable quality. Most softgels are manufactured by the process developed by Scherer [11]. The glycerol–gelatin solution is heated and pumped onto two chilled drums to form two separate ribbons (usually 0.02–0.04 in. thick) which form each half of the softgel. The ribbons are lubricated and fed into the filling machine, forcing the gelatin to adopt the contours of the die. The fill is manufactured in a separate process and pumped in, and the softgels are sealed by the application of heat and pressure. Once cut from the ribbon, they are tumble-dried and conditioned at 20% relative humidity.

Fill solvents are selected based on a balance between adequate solubility of the drug and physical stability. Water-miscible solvents such as low-molecular-weight PEGs, polysorbates, and small amounts of propylene glycol, ethanol, and glycerin can be used. Water-immiscible solvents include vegetable and aromatic oils, aliphatic, aromatic, and chlorinated hydrocarbons, ethers, esters, and some alcohols. Emulsions, liquids with extremes of pH (<2.5 and >7.5), and volatile components can cause problems with stability, and drugs that do not have adequate stability in the solvents can be formulated as suspensions. In these instances, the particle size needs to be carefully controlled and surfactants can be added to promote wetting.

Vegicaps soft capsules from Cardinal Health are an alternative to traditional softgels, containing carageenan and hydroxyproyl starch. As with traditional soft gelatin capsules, the most important packaging and storage criterion is for adequate protection against extremes of relative humidity. The extent of protection required also depends on the fill formulation and on the anticipated storage conditions.

4.1.6.6 Dissolution Testing of Capsules

In general, capsule dosage forms tend to float during dissolution testing with the paddle method. In such cases, it is recommended that a few turns of a wire helix around the capsule be used [12]. Inclusion of enzymes in the dissolution media must be considered on case-by-case basis. A Gelatin Capsule Working Group (including participants from the FDA, industry, and the USP) was formed to assess the noncompliance of certain gelatin capsule products with the required dissolution specifications and the potential implications on bioavailability [13]. The working group recommended the addition of a second tier to the standard USP and new drug and abbreviated new drug applications (NDA/ANDA) dissolution tests: the incorporation of enzyme (pepsin with simulated gastric fluid and pancreatin with simulated intestinal fluid) into the dissolution medium. If the drug product fails the first tier but passes the second tier, the product's performance is acceptable. The two-tier dissolution test is appropriate for all gelatin capsule and gelatin-coated tablets and the phenomenon may have little significance in vivo.

4.1.7 EFFERVESCENT TABLETS

Effervescence is the reaction in water of acids and bases to produce carbon dioxide, and effervescent tablets are dissolved or dispersed in water before administration. Advantages of effervescent formulations over conventional formulations are that the drug is usually already in solution prior to ingestion and can therefore have a faster onset of action. Although the solution may become diluted in the GI tract, any precipitation should be as fine particles that can be readily redissolved. Variability in absorption can also be reduced. Formulations can be made more palatable and there can be improved tolerance after ingestion. Thus, the types of drugs suited to this formulation method are those that are difficult to digest or are irritant to mucosa. Analgesics such as paracetamol and aspirin and vitamins are common effervescent formulations. The inclusion of buffering agents can aid stability of pH-sensitive drugs. There is also the opportunity to extend market share and to deliver large doses of medication.

Effervescents comprise a soluble organic acid and an alkali metal carbonate salt. Citric acid is most commonly used for its flavor-enhancing properties. Malic acid imparts a smoother after taste and fumaric, ascorbic, adipic, and tartaric acids are less commonly used [14]. Sodium bicarbonate is the most common alkali, but potassium bicarbonate can be used if sodium levels are a potential issue with the formulation. Both sodium and potassium carbonate can also be employed. Other excipients include water-soluble binders such as dextrose or lactose, and binder levels are kept to a minimum to avoid retardation of disintegration. All ingredients must be anhydrous to prevent the components within the formulation reacting with each other during storage.

Lubricants such as magnesium stearate are not used as their aqueous insolubility leads to cloudy solutions and extended disintegration times. Spray-dried leucine and PEG are water-soluble alternatives [15, 16]. Both artificial and natural sweeteners are used and an additional water-soluble flavoring agent may also be required. If a surfactant is added to enhance wetting and dissolution, the addition of an antifoaming agent may also be considered [17].

4.1.7.1 Manufacture of Effervescent Tablets

Essentially, effervescent formulations are produced in the same way as conventional tablets, although due to the hygroscopicity and potential onset of the effervescence reaction in the presence of water, environmental control of relative humidity and water levels is of major importance during manufacture. A maximum of 25% relative humidity (RH) at 25°C is required. Closed material-handling systems can be used or open systems with minimum moisture content in the ventilating air.

A dry method of granulation is preferred as no liquid is involved but may not always be possible. Wet granulation can be carried out under carefully controlled conditions using two separate granulators for the alkaline and acid components. Water can be added at 0.1–1.0% w/w, and it initiates a preeffervescent reaction. The cycle is stopped by drying, usually by transfer into a preheated fluidized-bed dryer. Fluid-bed spray granulation is a process wherein granulation and drying are simultaneous and can be useful for effervescent formulations. Water (or a binder solution) is sprayed onto the mixture, which is suspended in a stream of hot, dry air. Organic solvents can also be employed for granulation avoiding the need for water and are useful for heat-labile formulations, although complex handling equipment is required.

Effervescent formulations must contain less than 0.3% w/w water and are often quite large. Sticking due to insufficient lubrication can be overcome by adaptation of punches for external lubrication or using flat-faced punches with disks of poly (tetrafluoroethylene) (PTFE). Poor lubrication can also be the cause of poor flow characteristics, and this can be addressed by using a constant level powder feed system. The tablets should be stored in tightly closed containers or moisture-proof packs. In tube arrangements, dry air is added prior to sealing and desiccants to reduce enclosed moisture levels once the pack has been opened. Foil packaging should be heavy gauge to minimize risk of holes, and the surrounding pocket should be large enough to hold the tablets but minimize inclusion of air.

In-process quality control is of major importance for these formulations as are stability testing and stress testing of packaging. Tablet disintegration and dissolution are of prime importance, and disintegration should be carried out using representative conditions. Hardness and friability are also important as these large tablets tend to chip easily. Common areas for problems are that the packaging permits entry of water, the seal is compromised or that the excipients can react with each other.

4.1.8 LOZENGES

Lozenges are tablets that dissolve or disintegrate slowly in the mouth to release drug into the saliva. They are easy to administer to pediatric and geriatric patients and are useful for extending drug form retention within the oral cavity. They usually contain one or more ingredient in a sweetened flavored base. Drug delivery can be either for local administration in the mouth, such as anaesthetics, antiseptics, and antimicrobials or for systemic effects if the drug is well absorbed through the buccal lining or is swallowed. More traditional drugs used in this dosage form include phenol, sodium phenolate, benzocaine, and cetylpyridinium chloride. Decongestants and antitussives are in many over-the-counter (OTC) lozenge formulations, and there are also lozenges that contain nicotine (as bitartrate salt or as nicotine polacrilex resin), flurbiprofen (Strefen), or mucin for treatment of dry mouth (A.S Saliva Orthana).

Lozenges can be made by molding or by compression at high pressures, often following wet granulation, resulting in a mechanically strong tablet that can dissolve in the mouth. Compressed lozenges (or troches) differ from conventional tablets in that they are nonporous and do not contain disintegrant. As the formulation is designed to release drug slowly in the mouth, it must have a pleasant taste, smoothness, and mouth feel. The choice of binder, filler, color, and flavor is therefore most important. The binder is particularly important in ensuring retardation of dissolution and pleasant mouth feel. Suitable binders include gelatin, guar gum, and acacia gum. Sugars such as sucrose, dextrose, and mannitol are preferred to lactose, and xylitol is often included in sugar-free formulations. In order to ensure adequate sweetness and taste masking, artificial sweeteners including aspartame, saccharin, and sucralose are also included subject to regulatory guidelines.

Other variations include hard-candy-type and soft or chewable lozenges. Most hard-candy-type lozenges contain sugar, corn syrup, acidulant, colorant, and flavors. They are made by heating sugars and other ingredients together and then pouring the mixture into a mold. Corn syrup combined with sucrose and dextrose can form an amorphous glass suitable for such formulations [18]. Colorants can be added to enhance product appearance or to mask products of degradation. Stability and compatibility with the drug must be established along with the other excipients. Flavors tend to be complex entities, and stability or compatibility can pose major formulation challenges. Acidulants such as citric and tartaric acids are often added to enhance flavors, thus lowering pH of the formulation as low as 2.5–3.0. Addition of bases such as calcium carbonate, sodium bicarbonate, and magnesium trisilicate is common to increase pH and enhance drug stability. For example, in vivo and in vitro studies confirmed that the pH of the dissolved lozenge solution was the single most influential, readily adjustable formulation parameter influencing the activity of cetylpyridinium chloride activity in candy-based lozenges [19]. The dosage form needs a low moisture content (0.5–1.5% w/w), so water is evaporated off by boiling the sugar mixture during the compounding process, thus limiting the process to nonlabile drugs, and the manufacture requires specialized candy processing facilities. Packaging also needs to protect the formulated product from moisture and ranges from individual bunch wrapping to foil wraps.

4.1.8.1 Chewable Lozenges

Chewable lozenges are popular with the pediatric population since they are "gummy-type" lozenges. Most formulations are based on a modified suppository formula consisting of glycerin, gelatin, and water. These lozenges are often highly fruit-flavored and may have a slightly acidic taste to cover the acrid taste associated with glycerin. Soft lozenges typically comprise ingredients such as PEG 1000 or 1450, or a sugar–acacia base. Silica gel can be added to prevent sedimentation, and again this dosage form requires flavors and sweeteners to aid compliance. Soft lozenges tend to dissolve faster than gelatin bases and can be used if taste masking is not effective.

4.1.9 CHEWABLE TABLETS

Chewable tablets are designed to be mechanically disintegrated in the mouth. Potential advantages of chewable tablets are mainly concerning patient convenience and acceptance, although enhanced bioavailability is also claimed. This can be due to a rapid onset of action as disintegrate is more rapid and complete compared to standard formulations that must disintegrate in the GI tract. The dosage form is an appealing alternative for pediatric and geriatric consumers. Chewable tablets also offer convenience for consumers, avoiding the necessity of coadministration with water, and creation of palatable formulations can increase compliance. Antacids and pediatric vitamins are often formulated as chewable tablets, but other formulations include antihistamines (Zyrtec), antimotility agents (Imodium Plus) and antiepileptic agents (Epanutin Infatabs), antibiotics (Augmentin Chewable), asthma treatments (Singulair), and analgesics (Motrin).

Constraints with these systems are that many pharmaceutical actives have an unpleasant bitter taste that can actually reduce compliance among patients. Iron salts, for example, can impart a rusty taste, and some antihistamines such as promethazine HCl can have a bitter aftertaste. As such, active formulations require very effective taste-masking strategies to provide acceptable patient tolerance and to ensure patient adherence to their pharmaceutical regimen.

Formulation factors governing design are similar to standard formulations (e.g., compactability, flow, etc.), but disintegrants are not included. Organoleptic properties are a major concern, especially in the design of products for children, and usage has been limited as formulators have encountered difficulties in achieving satisfactory sensory characteristics.

Certain diluents are beneficial in the formulation of chewable tablets by compression such as mannitol, lactose, sucrose, and sorbitol. They can aid disintegration upon chewing and can help with acceptable taste and mouth feel. Mannitol, for example, can impart a cooling or soothing sensation. Specialist excipients with improved sensory components such as mouth feel and lack of grittiness have been developed for formulation of chewable tablets. For example, Avicel CE-15 [a combination of microcrystalline cellulose (MCC) and guar gum] can reduce grittiness, leading to a creamier mouth feel and improved overall compatibility.

Citric acid, grape, raspberry, lemon, and cherry flavors are often used in chewable tablets and lozenges (Table 7). Flavoring agents are commonly volatile oils, and they can be dissolved in alcohol and then sprayed onto another excipient or granules. They are usually added immediately prior to compression to avoid loss due to their volatile nature. Dry flavors have advantages in terms of stability and ease of handling and are formed by emulsification of the flavor into an aqueous solution of a carrier followed by drying, encapsulating the flavor within the carrier. This is useful if the agent is prone to oxidation. Common carrier substances are acacia gum, starch, and maltodextrin. Sweeteners such as aspartame can also be added. Low-calorie and non-sugar-based excipients may present a marketing advantage.

Issues of taste masking for chewable formulations may be addressed by coating in wet granulation. The granulating/coating agent should form a flexible rather than

TABLE 7 Flavor Groups for Taste Types

Sweet	Vanilla, grape, maple, honey
Sour	Citrus, raspberry, anise
Salty	Mixed fruit, mixed citrus, butterscotch, maple
Bitter	Licorice, coffee, mint, cherry, grapefruit
Metallic	Grape, lemon, lime

Source: From ref. 18.

brittle film, have no unpleasant taste of its own, not interfere with dissolution, and be insoluble in saliva. Microencapsulation for taste masking can be achieved by phase separation or coacervation and may also impart stability. The same taste-masking technologies may be used to encapsulate drugs for formulation into chewable, softchew, and fast dissolving dosage forms. Coating materials include carboxymethylcellulose, polyvinyl alcohol (PVA), and ethylcellulose. Xylitol is the sweetest sugar alcohol, and it has a high negative heat of solution, making it a good candidate as an excipient for chewable tablets. There are many types of compressible sugars today, and most of them are composed of sucrose granulated with small amounts of modified dextrins in order to make the sucrose more compressible [20]. Modifications to sugar-based excipients such as spray-dried crystalline maltose and directly compressible sucrose (95% sucrose and 5% sorbitol) to facilitate direct compression are also aiding development in this area [21].

4.1.9.1 Testing of Chewable Tablets

Dissolution testing for chewable tablets should be the same as that used for regular tablets [22]. This is because patients could swallow the dosage form without adequate chewing, in which case the drug would still need to be released to ensure the desired pharmacological action. However, as outlined, chewable tablets will typically have different excipients than standard formulations, including agents to either mask or add flavor, and may undergo a different manufacturing process. Where applicable, test conditions would preferably be the same as used for non-chewable tablets of the same active pharmaceutical ingredient, but because of the nondisintegrating nature of the dosage form, it may be necessary to alter test conditions (e.g., increase the agitation rate) and specifications (e.g., increase the test duration). The reciprocating cylinder (USP apparatus 3) with the addition of glass beads may also provide more intensive agitation for in vitro dissolution testing of chewable tablets. As another option, mechanical breaking of chewable tablets prior to exposing the specimen to dissolution testing could be considered. Chewable tablets should also be evaluated for in vivo bioavailability and/or bioequivalence.

Additional concerns in the testing of chewable tablets are organoleptic, chemical, and physical stability. As it is a critical factor in the design of such formulations, taste masking should be incorporated into excipient testing during preformulation studies. Technologies like the "electronic tongue" can be used to match desirable taste characteristics [23, 24].

4.1.10 CHEWING GUMS

4.1.10.1 Composition of Chewing Gum

Medicated chewing gums are gums made with a tasteless masticatory gum base that consists of natural or synthetic elastomers [25]. They include excipients such as fillers, softeners, and sweetening and flavoring agents. Natural gum bases include chicle and smoked natural rubber and are permitted in formulations by the FDA, but modern gum bases are mostly synthetic in origin and approved bases include

TABLE 8 Typical Formulation of Gum Base

Ingredient	Weight (%)	Example
Elastomer	10	Styrene–butadiene rubber
Plasticizer	30	Rosin esters
Texture agent/filler	35	Calcium carbonate
Wax	15	Paraffin wax
Lipid	7	Soya oil
Emulsifier	3	Lecithin
Miscellaneous	1	Colorant, antioxidant

Source: From ref. 26.

TABLE 9 Example Chewing Gum Formulations

Ingredient (%)	Sugar Gum	Sugar-Free Gum
Gum base	19.4	25.0
Corn syrup	19.8	—
Sorbitol, 70%	—	15.0
Sugar	59.7	—
Glycerin	0.5	6.5
Sorbitol	—	52.3
Flavor	0.6	1.2

Source: From ref. 26.

styrene–butadiene rubber, polyethylene, and polyvinylacetate. Gum base usually forms about 40% of the gum, but can comprise up to 65%, and is a complex mixture, insoluble in saliva, comprising mainly of elastomer, plasticizers, waxes, lipids, and emulsifiers (see Table 8). It will also contain an adjuvant such as talc to modify the texture of the gum and low quantities of additional excipients including colorants and antioxidants such as butylated hydroxyanisole. Elastomers control the gummy texture while the plasticizers and texture agents regulate the cohesiveness of the product. The lipid and waxes melt in the mouth to provide a cooling, lubricating feeling while the juicy feel of the gum texture is from the emulsifiers. The choice and formulation of gum base will affect the release of active ingredient, and the texture, stability, and method of manufacture of the product.

The remaining ingredients in the chewing gum itself include drug, sweeteners, softeners, and flavoring and coloring agents. A typical chewing gum formulation is shown in Table 9. The sugar is for sweetening the product while the corn syrup keeps the gum fresh and flexible. Softeners or fillers are included to help blend the ingredients and retain moisture. Sugar-free gum has sorbitol, mannitol, aspartame, or saccharin instead of sugar. Optimized chewing gum formulations will require tailoring for each individual product. For example, nicotine-containing gums are formulated with the nicotine within an ion exchange resin and pH-modifying carbonates and/or bicarbonates to increase the percentage of the drug in its free base form in saliva.

4.1.10.2 Manufacture of Chewing Gum

The majority of chewing gum delivery systems today are manufactured using conventional gum processes. The gum base is softened or melted and placed in a kettle mixer where sweeteners, syrups, active ingredients, and other excipients are added at a defined time. The gum is then sent to a series of rollers that form it into a thin, wide ribbon. During this process, a light coating of an antisticking agent can be added (e.g., magnesium stearate, calcium carbonate, or finely powdered sugar or sugar substitute). Finally, the gum is cut to the desired size and cooled at a carefully controlled temperature and humidity.

As the heating process involved in conventional methods may limit the applicability of the process for formulation of thermally labile drugs, directly compressible, free-flowing powdered gums such as Pharmagum (SPI Pharma) and MedGumBase (Gumbase Co) have been proposed to simplify the process. These formulations can be compacted into a gum tablet using a conventional tablet press and have the potential to simplify the manufacture, facilitating inclusion of a wider range of drugs.

4.1.10.3 Drug Release from Chewing Gums

Until recently, the release of substances from chewing gums during mastication was studied using a panel of tasters and chew-out studies. During the mastication process, the medication contained within the gum product should be released into the saliva and is either absorbed through the buccal mucosa or swallowed and absorbed via the GI tract. The need for, and value of, in vitro drug release testing is well established for a range of dosage forms, however, standard dissolution apparatus is not suitable for monitoring release of drug from chewing gums as mastication is essential in order to provide a renewable surface for drug release after chew action. A number of devices to mimic the chewing action have been reported [26–28]. In 2000, the European Pharmacopoeia produced a monograph describing a suitable apparatus for studying the in vitro release of drug substances from chewing gums [25]. The chewing machine consists of a temperature-controlled chewing chamber in which the gum piece is chewed by two electronically controlled horizontal pistons driven by compressed air. The two pistons transmit twisting and pressing forces to the gum while a third vertical piston operates alternately to the two horizontal pistons to ensure that the gum stays in the right place (see Figure 1). The temperature of the chamber can be maintained at 37°C ± 0.5°C and the chew rate varied. Other adjustable settings include the volume of the medium, distance between the jaws and the twisting movement. The European Pharmacopoeia recommends using 20 mL of unspecified buffer in a chewing chamber of 40 mL and a chew rate of 60 strokes per minute. This apparatus has been used to study release of nicotine from commercial gums and directly compressible gums [26].

Factors affecting the release of medicament from chewing gum can be divided into three groups: the physicochemical properties of the drug, the gum properties, and chew-related factors, including rate and frequency. Drugs can be incorporated into gums as solids or liquids. For most pharmaceuticals, aqueous solubility of the drug will be a major factor affecting the release rate. In order for drugs to be

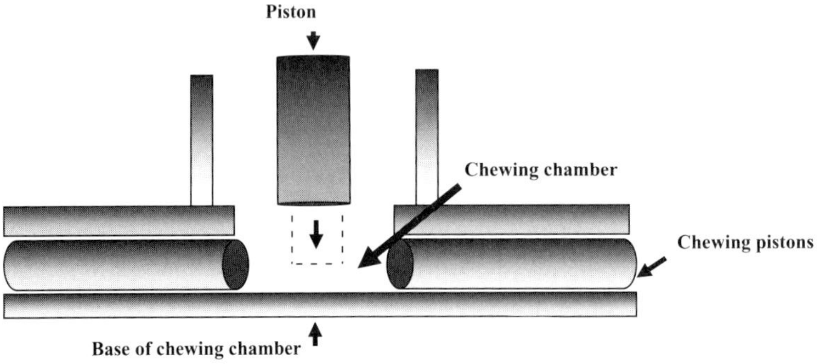

FIGURE 1 Schematic of chewing chamber of in vitro chewing apparatus [26].

released, the gum would need to become hydrated; the drugs can then dissolve and diffuse through the gum base under the action of chewing.

For treatment of local conditions, a release period less than 1 h may be desirable, but a faster release may be required if a rapid onset of action is required for a systemically absorbed formulation. There are a number of strategies that can be undertaken in order to achieve the desired release rate. Decreasing the amount of the gum base will enhance the release of lipophilic drugs and addition of excipients designed to promote release can also be considered. Release can be sustained using, for example, ion exchange resins as described for nicotine-containing gums. Changes in gum texture as a consequence of changes in excipient levels provide a further challenge to controlling the release of drugs. A quantitative measure of gum texture during the process is possible using texture analysis techniques [26].

4.1.10.4 Applications for Chewing Gums

The promotion of sugar-free gums to counteract dental caries by stimulation of saliva secretion has led to a more widespread use and acceptance of gums. Medicated gums for delivery of dental products to the oral cavity are marketed in a number of countries, for example, fluoride-containing gums as an alternative to mouthwashes and tablets or chlorhexidine gum for treatment of gingivitis. The potential use of medicated chewing gums in the treatment of oral infections has also been reported. Gums have been prepared containing antifungal agents such as nystatin [29] and miconazole [30] or antibiotics, such as penicillin and metronidazole for the treatment of oral gingivitis [31].

Chewing gum is also useful as a delivery system for agents intended for systemic delivery. Drug that is released from the gum within the oral cavity can act locally, be absorbed via the buccal mucosa, or swallowed with the saliva. The buccal mucosa is well vascularized, and if a drug is absorbed by this route, then first-pass metabolism could be avoided. Associated increases in bioavailability can permit the use of lower dosages. Like orally disintegrating tablets, chewing gum is a convenient dosage form; it can be administered without water and to those who have difficulty swallowing. Although medicated gums are generally intended to be chewed for 10–30 min and can therefore be designed for sustained release, a fast onset of action can result either from buccal absorption or as a consequence of the active being already dissolved in the saliva prior to swallowing. Guidance can be given regarding chewing conditions (e.g., time, frequency), but factors such as the force of chewing and salivary flow will impact on drug release and the fraction of drug absorbed via the oral mucosa. Released drug can be swallowed with the saliva, therefore leading to multiple absorption sites, which can result in variable pharmacokinetics.

Along with nicotine replacement patches, nicotine chewing gum for smoking cessation therapy has met with major sales success. The principal active ingredient of currently marketed nicotine chewing gums is nicotine polacrilex USP. The nicotine is loaded at approximately 18% w/w on an ion exchange resin (Amberlite IRP64). Recent product variations have been launched with improved flavors such as mint and fruit, rather than the original peppery flavoring, designed to reduce the unpleasant taste and burning sensation arising from nicotine itself and flavored coated gums that are sweeter and easier to chew.

Other applications for chewing gum formulations include delivery of antacids such as calcium carbonate, antiemetics for travel sickness, and vitamins and minerals. However, the potential for a buccal delivery, a fast onset of action, and the opportunity for product line extension makes it an attractive alternative delivery form for other applications.

4.1.11 ORALLY DISINTEGRATING TABLETS

The demand for fast-dissolving/disintegrating tablets or fast-melting tablets that can dissolve or disintegrate in the mouth has been growing particularly for those with difficulty swallowing tablets such as the elderly and children. They are referred to using a range of terminologies: fast dissolving, orodispersible, and fast melting and the FDA has adopted the term orally disintegrating tablets (ODTs). Patients with persistent nausea or those who have little or no access to water could also benefit from ODTs. Other advantages include product differentiation and market expansion, and applications exist in the veterinary market for oral administration to animals.

Orally disintegrating tablets disintegrate and/or dissolve rapidly in the saliva without the need for water, within seconds to minutes. Some tablets are designed to dissolve rapidly in saliva, within a few seconds, and are true fast-dissolving tablets. Others contain agents to enhance the rate of tablet disintegration in the oral cavity and are more appropriately termed fast-disintegrating tablets, as they may take up to a minute to completely disintegrate. Increased bioavailability using such formulations is sometimes possible if there is sufficient absorption via the oral cavity prior to swallowing [32]. However, if the amount of swallowed drug varies, there is the potential for inconsistent bioavailability. Patented orally disintegrating tablet technologies include OraSolv, DuraSolv, Zydis, FlashTab, WOWTAB, and others. They are generally prepared using freeze drying, compaction, or molding. Examples of marketed products, excipients, and technologies used are given in Table 10.

Platform technologies based on freeze drying include Zydis (Cardinal Health) and Quicksolv (Janssen Pharmaceutica). Zydis was the first ODT to be successfully launched, and it is ideal for poorly soluble drugs. It can incorporate doses up to 400 mg, but high loadings can extend disintegration time. The porous matrix consists of a network of water-soluble carriers and active ingredient. The maximum dose for

TABLE 10 Examples of Marketed ODT Products and Technologies

Name (Company)	Examples	Ingredients[a]	Technology
Zydis (Cardinal Health)	Claritin Reditab	*Micronized loratadine (10 mg)*, citric acid, gelatin, mannitol, mint flavor	Freeze drying
Zydis (Cardinal Health)	Zofran ODT	*Ondansetron (4 or 8 mg)*, aspartame, gelatin, mannitol, methylparaben sodium, propylparaben sodium, strawberry flavor	Freeze drying
Zydis (Cardinal Health)	Zyprexa Zydis	*Olanzapine (5, 10, 15, or 20 mg)*, gelatin, mannitol, aspartame, methylparaben sodium, propylparaben sodium	Freeze drying
Oralsolv (CIMA Labs Inc.)	Remeron Soltab	*Mirtazepine (15, 30, or 45 mg)*, aspartame, citric acid, crospovidone, hydroxypropyl methylcellulose, magnesium stearate, mannitol, microcrystalline cellulose, polymethacrylate, povidone, sodium bicarbonate, starch, sucrose, orange flavor	Compression
Durasolv (CIMA Labs Inc.)	Zomig ZMT	*Zolmitriptan (2.5 mg)*, mannitol, microcrystalline cellulose, crospovidone, aspartame, sodium bicarbonate, citric acid, anhydrous, colloidal silicon dioxide, magnesium stearate, orange flavor	Compression
WOWTAB (Yamanouchi Pharma Technologies, Inc.)	Benadryl Allergy & Sinus Fastmelt	*Diphenhydramine citrate (19 mg), pseudoephedrine HCl (30 mg)*, aspartame, citric acid, D&C red no. 7 calcium lake, ethylcellulose, flavor, lactitol, magnesium stearate, mannitol, and stearic acid	Compression molded tablet
Flashtab (Prographarm/ Ethypharm)	Excedrin Quicktabs	*Acetaminophen (500 mg), caffeine (65 mg)*, aminoalkyl methacrylate copolymers, citric acid, colloidal silicon dioxide, crospovidone, distilled acetylated monoglycerides, ethylcellulose, flavors, magnesium stearate, mannitol, methacrylester copolymer, polyvinyl acetate, povidone, propylene glycol, propyl gallate, silica gel, sodium lauryl sulfate, sucralose, talc	Compression

[a]Active ingredients appear in italics.

water-soluble drugs is 60 mg, and particle sizes of drug and excipients should be below 50 μm.

Excipients used in the formulation usually include a mixture of a water-soluble polymer and a crystalline sugar. Mannitol and natural polysaccharides such as gelatin and alginates are used. Microencapsulation and complexation with ion exchange resins can be combined with additional flavors and sweeteners for taste masking of bitter drugs. The fairly complex nature of manufacture and scale-up contributes to a relatively high manufacturing cost. Manufacture comprises three stages. The production sequence begins with the bulk preparation of an aqueous drug solution or suspension and subsequent precise dosing into preformed blisters. It is the blister that actually forms the tablet shape and is, therefore, an integral component of the total product package. The second phase of manufacturing entails passing the filled blisters through a cryogenic freezing process to control the ultimate size of the ice crystals. This aids in ensuring porosity and the product is freeze dried. The final phase of production involves sealing the open blisters via a heat-seal process to ensure stability and protect the fragile tablet during removal by the patient.

The manufacture of Flashdose (Fuisz Technologies/Biovail) is patented as Shearform process and utilizes a unique spinning mechanism to produce a flosslike or shear-form crystalline structure, much like cotton candy. The matrix comprises saccharides or polysaccharides which are subjected to simultaneous melting and centrifugal force and then partially recrystallized [33]. High temperatures are involved so the technology is only suitable for thermostable agents. Drug can then be incorporated, either as coated or uncoated microspheres, into the sugar and the formulation is compressed into a tablet. Manufacture of the microspheres is patented as Ceform and will help with taste masking. The final product has a very high surface area for dissolution and it disperses and dissolves quickly once placed onto the tongue. Like freeze-drying processes, the manufacture is expensive and resultant formulations are friable and moisture sensitive, therefore requiring specialized packaging.

Most commercial ODTs have been developed using mannitol as the bulk excipient of choice because of its extremely low hygroscopicity, excellent compatibility, good compressibility, better sweetness, and relatively slower dissolution kinetics. Although lactose also has a relatively low aqueous solubility compared with other excipients that have acceptable palatabilities, the dispersibility of a bulk excipient is more important than its aqueous solubility for a successful ODT formulation. Many of the initially marketed ODTs were prepared by the wet granulation of mannitol followed by direct compression. However, added functionality mannitols are now available to simplify the process of ODT manufacturing by direct compression.

Direct compression is, as for normal tablets, the most straightforward process for manufacturing ODTs. Conventional equipment can be used and high doses can be incorporated. The excipients play a major role in the successful formulation and superdisintegrants, hydrophilic polymers, and effervescent compounds are included. Patented technologies include Orasolv and Durasolv (Cima Labs) and Ziplets (Eurand). The OraSolv technology is best described as a fast disintegrating, slightly effervescing tablet; the tablet matrix dissolves in less than one minute, leaving coated drug powder. Both the coating and the effervescence contribute to taste masking in OraSolv. The tablet is prepared by direct compression but at a low pressure, yielding a weaker and more brittle tablet in comparison with conventional tablets. For that reason, Cima developed a special handling and packaging system for OraSolv called Packsolv. Acidic compounds such as citric or fumaric acid are included in the formulation together with a carbonate or bicarbonate. An advantage that goes along with the low degree of compaction of OraSolv is that the particle coating used for taste masking is not compromised by fracture during processing.

DuraSolv is Cima's second-generation fast-dissolving/disintegrating tablet formulation and is also produced using direct compression but using higher compaction pressures during tableting, resulting in a stronger product. It is thus produced in a faster and more cost-effective manner and may not require specialized packaging. Large amounts of finely milled conventional fillers are used (mannitol, lactose) while the effervescing agents are reduced. It is best suited to potent drugs, requiring only low doses, and the taste-masking coating can be disturbed following compaction. DuraSolv is currently available in two products: NuLev and Zomig ZMT.

Compression following wet or dry granulation is also employed in the manufacture of ODTs. Patented formulations include WOWTAB and Flashtab. WOWTAB relies on a combination of low moldable sahharides (mannitol, glucose, sucrose) with a highly moldable saccharide (malitol, sorbitol, maltose) using conventional granulation and tableting techniques to form a tablet of suitable mechanical properties with desired disintegration. It is manufactured by compression of molded granules, can accommodate a high level of drug loading (up to 50% in some cases), and can be packed using conventional methodology.

Flashtab (Ethypharm) is the technology behind Exedrin QuickTabs and uses swellable agents and disintegrants along with sugars and polyalcohols to achieve a fast dispersible formulation. The manufacture involves either wet or dry granulation of the excipients, blending with the active followed by direct compression.

4.1.11.1 Dissolution Testing of ODTs

Taste masking (drug coating) is very often an essential feature of ODTs and thus can also be the rate-determining mechanism for dissolution/release. If taste masking is not an issue, then the development of dissolution methods is comparable to the approach taken for conventional tablets and pharmacopeial conditions should be used [34]. Due to the nature of the product, the dissolution of orally disintegrating tablets is very fast when using USP monograph conditions, and slower paddle speeds can be used to obtain a profile. Other media such as 0.1 N HCl can also be used. USP 2 paddle apparatus is the most suitable and common choice for orally disintegrating tablets, with a paddle speed of 50 rpm commonly used [34]. Faster agitation rates may be necessary in the case of sample mounding. The method can be applied to the ODTs (finished product) as well as to the bulk intermediate (in the case of coated drug powder/granulate). A potential difficulty for in vitro dissolution testing may arise from floating particles [35]. Similarly, difficulties can arise using USP I due to trapping of disintegrated fragments.

A single-point specification is considered appropriate for ODTs with fast dissolution properties. For ODTs that dissolve very quickly, a disintegration test may be used in lieu of a dissolution test if it is shown to be a good discriminating method. If taste masking (using a polymer coating) is a key aspect of the dosage form, a multipoint profile in a neutral pH medium with early points of analysis (e.g., $\leq$5 min) may be recommended [34].

4.1.12 SOLID DOSAGE FORMS FOR NONORAL ROUTES

Although the majority of tablets and capsules are intended for oral delivery, there are a number of other delivery routes suitable for drug delivery by these formulations. Some buccal formulations have been discussed above, and tablets can also be administered via the rectal and vaginal routes for local and systemic treatment.

Many types of product have been designed for vaginal administration with creams, gels, and pessaries being most popular, although powders and tablets have also been used. Despite the effectiveness of systemic vaginal absorption, the majority of products administered by this route are for the treatment of localized infections, especially *Candida albicans*, (e.g., Canestan vaginal tablets). Estradiol tablets (Vagifem) were also designed for delivery via vaginal route to address patient preference issues with vaginal creams. The formulations are administered with an applicator and are designed to dissolve or erode slowly in the vaginal secretions [36]. Bioadhesion as a means of retaining the formulation at the site of delivery is widely accepted to retain formulations in the buccal cavity [37] and has also been reported for the vaginal route [38]. An increased residence time may improve drug absorption by these routes.

REFERENCES

1. Amidon, G. L., Lennernas, H., Shah, V. P., and Crison, J. R. (1995), A theoretical basis for a biopharmaceutic drug classification: The correlation of *in vitro* drug product dissolution and *in vivo* bioavailability, *Pharm. Res.*, 12, 413–420.
2. Waiver of *in-vivo* bioavailability and bioequivalence studies for immediate release solid oral dosage forms based on a biopharmaceutics classification system, available: http://www.fda.gov/cder/OPS/BCS_guidance.html.
3. Waiver of in vivo bioequivalence studies for immediate release solid oral dosage forms based on a biopharmaceutics classification system, available: http://www.fda.gov/cder/guidance/index.html.
4. Zhao, C., Jain, A., Hailemariam, L., Suresh, P., Akkisetty, P., Joglekar, G., Venkatasubramanian, V., Reklaitis, G.V., Morris, K., and Basu, P. (2006), Toward intelligent decision support for pharmaceutical product development, *J. Pharm. Innovation*, 1, 23–35.
5. Banker, G. S., and Rhodes, C. T. (2002), *Modern Pharmaceutics*, 4th ed., Drugs and the Pharmaceutical Sciences 121, Marcel Dekker, New York.
6. Rowe, R. C., Sheskey, P. J., and Weller, P. J. (2001), *Handbook of Pharmaceutical Excipients*, 4th ed., Pharmaceutical Press, London.
7. Weiner, M., and Bernstein I. L. (1989), *Adverse Reactions to Drug Formulation Agents: A Handbook of Excipients*, Marcel Dekker, New York, pp. 93–94.
8. Lachman, L., Lieberman, H. A., and Kanig, J. L. (1986), *The Theory and Practice of Industrial Pharmacy*, 3rd ed., Lea & Febiger, Philadelphia.
9. Augsburger, L. L., Hahm, H. A., Brzeczko, A. W., and Shah, U. (2002), Superdisintegrants: Characterization and function, in Swarbrick, J., and Boylan, J. V., Eds., *Encyclopedia of Pharmaceutical Technology*, Vol. 3, 2nd ed., Marcel Dekker, New York.
10. Gibson, M. (2001), Pharmaceutical preformulation and formulation; a practical guide from candidate drug selection to commercial dosage form, IHS Health Group, Englewood, Colorado.
11. Stanley, J. P. (1986), Soft gelatin capsules, in Lachman, L., Lieberman, H. A., and Kanig, J. L., Eds., *The Theory and Practice of Industrial Pharmacy*, 3rd ed., Lea & Febiger, Philadelphia.

12. *The United States Pharmacopeia*, 26th revision, United States Pharmacopeial Convention, Rockville, MD, 2003.
13. Gelatin Capsule Working Group, Collaborative development of two-tier dissolution testing for gelatin capsules and gelatin-coated tablets using enzyme-containing media, *Pharmacop. Forum*, 24(5), Sept./Oct. 1998.
14. Lee, R. E., Effervescent tablets. Key facts about a unique effective dosage form, *Tablets Capsules*, available: http://www.amerilabtech.com/EffervescentTablets&KeyFacts.pdf, accessed Aug. 6, 2004.
15. Rotthauser, B., Kraus, G., and Schmidt, P. C. (1998), Optimization of an effervescent tablet formulations containing spray-dried L-leucine and polyethylene glycol 6000 as lubricants using a central composite design, *Eur. J. Pharm. Biopharm.*, 46, 85–94.
16. Stahl, H. (2003), Effervescent dosage manufacturing, *Pharm. Technol. Eur.*, 4, 25–28.
17. Lindberg, N.-O., and Hansson, H. (2002), Effervescent pharmaceuticals, in Swarbrick, J., and Boylan, J. V., Eds., *Encyclopedia of Pharmaceutical Technology*, Vol. 2, 2nd ed., Marcel Dekker, New York.
18. Mendes, R. W., and Bhargava, H. (2002), Lozenges, in Swarbrick J., and Boylan, J. V., Eds., *Encyclopedia of Pharmaceutical Technology*, Vol. 2, 2nd ed., Marcel Dekker, New York.
19. Richards, R. M. E., Xing, J. Z., and Weir, L. F. C. (1996), The effect of formulation on the antimicrobial activity of cetylpyridinium chloride in candy based lozenges, *Pharm. Res.*, 13, 583–587.
20. Bolhuis, G., and Armstrong, A. N. (2006), Excipients for direct compaction—An update, *Pharm. Dev. Technol.*, 11, 111–124.
21. Bowe, K. E. (1998), Recent advances in sugar-based excipients, *Pharm. Sci. Technol. Today*, 1(4), 166–173.
22. FDA guidance for industry: Bioavailability and bioequivalence studies for orally administered drug products—General considerations, Oct. 2000.
23. Murray, O. J., Dang, W., and Bergstrom, D. (2004), Using an electronic tongue to optimize taste-masking in a lyophilized orally disintegrating tablet formulation, *Pharm. Technol. Outsourcing Res.*, 42–52.
24. Zheng, J. Y., and Keeney, M. P. (2004), Taste masking analysis in pharmaceutical formulation development using an electronic tongue, *Int. J. Pharm.*, 310, 118–124.
25. *European Pharmacopoeia*, 4th ed., European Directorate for the Quality of Medicines, Strasbourg, 2001.
26. Morjaria, Y., Irwin, W. J., Barnett, P. X., Chan, R. S., and Conway, B. R. (2004), *In vitro* release of nicotine from chewing gum formulations, *Dissolution Technol.*, 11, 12–15.
27. Rider, J. N., Brunson, E. L., Chambliss, W. G., Cleary, R. W., Hikal, A. H., Rider, P. H., Walker, L. A., Wyandt, C. M., and Jones, A. B. (1992), Development and evaluation of a novel dissolution apparatus for medicated chewing gum products, *Pharm. Res.*, 9, 255–260.
28. Kvist, C., Andersson, S.-B., Fors, S., Wennergren, B., and Berglund, J. (1999), Apparatus for studying *in vitro* release from medicated chewing gums, *Int. J. Pharm.*, 89, 57–65.
29. Andersen, T., Gram-Hansen, M., Pedersen, M., and Rassing, M. R. (1990), Chewing gum as a drug delivery system for nystatin. Influence of solubilising agents on the release of water-soluble drugs, *Drug Dev. Ind. Pharm.*, 16, 1985–1994.
30. Pedersen, M., and Rassing, M. R. (1991), Miconazole chewing gum as a drug delivery system test of release promoting additives, *Drug Dev. Ind. Pharm.*, 17, 411–420.
31. Emslie, R. D. (1967), Treatment of acute ulcerative gingivitis. A clinical trial using chewing gums containing metronidazole or penicillin, *Br. Dent. J.*, 122, 307–308.

32. Habib, W., Khankari, R., and Hontz, J. (2002), Fast-dissolve drug delivery system, *Crit. Rev. Therap. Drug Carrier Syst.*, 17, 61–72.

33. Dobetti, L. (2001), Fast-melting tablets: Developments and technologies, *Pharm. Technol. Drug Deliv. Suppl.*, 44–50.

34. Klanke, J. (2003), Dissolution testing of orally disintegrating tablets, *Dissolution Technol.*, 10(2), 6–8.

35. Siewert, M., Dressman, J., Brown, C., and Shah, V. P. (2003), FIP/AAPS. Guidelines for dissolution/*in vitro* release testing of novel/special dosage forms, *Dissolution Technol.*, 10(1), 6–15.

36. Conine, J. W., and Pikal, M. J. (1989), Special tablets, in Lieberman, H. A., Lachman, L., and Schwartz, J. B. Eds., *Pharmaceutical Dosage Forms: Tablets*, Vol. 1, 2nd ed., Marcel Dekker, New York.

37. Sudhakar, Y., Kuotsu, K., and Bandyopadhyay, A. K. (2006), Buccal bioadhesive drug delivery—A promising option for orally less efficient drugs, *J. Controlled Release*, 114, 15–40.

38. Hussain, A., and Ahsan, F. (2005), The vagina as a route for systemic drug delivery, *J. Controlled Release*, 103, 301–313.

4.2

SEMISOLID DOSAGES: OINTMENTS, CREAMS, AND GELS

RAVICHANDRAN MAHALINGAM, XIAOLING LI, AND BHASKARA R. JASTI
University of the Pacific, Stockton, California

Contents

4.2.1 INTRODUCTION

Semisolid dosage forms are traditionally used for treating topical ailments. The vast majority of them are meant for skin applications. They are also used for treating ophthalmic, nasal, buccal, rectal, and vaginal ailments. Various categories of drugs

Pharmaceutical Manufacturing Handbook: Production and Processes, edited by Shayne Cox Gad

such as antibacterials, antifungals, antivirals, antipruritics, local anesthetics, anti-inflammatories, analgesics, keratolytics, astringents, and mydriatic agents are incorporated into these products. Drugs incorporated into semisolids either show their activity on the surface layers of tissues or penetrate into internal layers to reach the site of action. For example, an antiseptic ointment should be able to penetrate the skin layers and reach the deep-seated infections in order to prevent the growth of microbes and heal the wound.

Systemic entry of drugs from these products is limited due to various physicochemical properties of dosage forms and biological factors. The barrier nature of most surface biological layers such as skin, cornea and conjunctiva of the eye, and mucosa of nose, mouth, rectum, and vagina greatly limits their entry into the systemic circulation. Systemic delivery of drugs from topical dosages is however feasible by suitable formulation modifications. Semisolid dosage forms are also used in nontherapeutic conditions for providing protective and lubricating functions. They protect the skin against external environments such as air, moisture, and sun rays and hence their components do not necessarily penetrate the skin layers. Cold creams and vanishing creams are classic examples of such semisolid preparations.

The formulation, evaluation, and regulatory feature of the three most commonly used semisolid dosage forms, ointments, creams, and gels, are described in this chapter.

4.2.2 OINTMENTS AND CREAMS

4.2.2.1 Definition

Ointments are semisolid preparations intended for topical application. They are used to provide protective and emollient effects on the skin or carry medicaments for treating certain topical ailments. They are also used to deliver drugs into eye, nose, vagina, and rectum. Ointments intended for ophthalmic purposes are required to be sterile. When applied to the eyes, they reside in the conjunctival sac for prolonged periods compared to solutions and suspensions and improve the fraction of drug absorbed across ocular tissues. Ophthalmic ointments are preferred for nighttime applications as they spread over the entire corneal and conjunctival surface and cause blurred vision.

Creams are basically ointments which are made less greasy by incorporation of water. Presence of water in creams makes them act as emulsions and therefore are sometimes referred as semisolid emulsions. Hydrophilic creams contain large amounts of water in their external phase (e.g., vanishing cream) and hydrophobic creams contain water in the internal phase (e.g., cold cream). An emulsifying agent is used to disperse the aqueous phase in the oily phase or vice versa. As with ointments, creams are formulated to provide protective, emollient actions or deliver drugs to surface or interior layers of skin, rectum, and vagina. Creams are softer than ointments and are preferred because of their easy removal from containers and good spreadability over the absorption site.

4.2.2.2 Bases

Bases are classified based on their composition and physical characteristics. The U.S. Pharmacopeia (USP) classifies ointment bases as hydrocarbon bases (oleaginous bases), absorption bases, water-removable bases, and water-soluble bases (water-miscible bases) [1].

Hydrocarbon bases are made of oleaginous materials. They provide emollient and protective properties and remain in the skin for prolonged periods. It is difficult to incorporate aqueous phases into hydrocarbon bases. However, powders can be incorporated into these bases with the aid of liquid petrolatum. Removal of hydrocarbon bases from the skin is difficult due to their oily nature. Petrolatum USP, white petrolatum USP, yellow ointment USP, and white ointment USP are examples of hydrocarbon bases.

Absorption bases contain small amounts of water. They provide relatively less emollient properties than hydrocarbon bases. Similar to hydrocarbon bases, absorption bases are also difficult to remove from the skin due to their hydrophobic nature. Hydrophilic petrolatum USP and lanolin USP are examples of absorption bases.

Water-removable bases are basically oil-in-water emulsions. Unlike hydrocarbon and absorption bases, a large proportion of aqueous phase can be incorporated into water-removable bases with the aid of suitable emulsifying agents. It is easy to remove these bases from the skin due to their hydrophilic nature. Hydrophilic ointment USP is an example of a water-removable ointment base.

Water-soluble bases do not contain any oily or oleaginous phase. Solids can be easily incorporated into these bases. They may be completely removed from the skin due to their water solubility. Polyethylene glycol (PEG) ointment National Formulary (NF) is an example of a water-soluble base.

Selection of an appropriate base for an ointment or cream formulation depends on the type of activity desired (e.g., topical or percutaneous absorption), compatibility with other components, physicochemical and microbial stability of the product, ease of manufacture, pourability and spreadability of the formulation, duration of contact, chances of hypersensitivity reactions, and ease of washing from the site of application. In addition, bases that are used in ophthalmic preparations should be nonirritating and should soften at body temperatures. White petrolatum and liquid petrolatum are generally used in ophthalmic preparations. Table 1 summarizes

TABLE 1 Some Compendial Bases Used in Ointments and Creams

Name	Synonyms	Official Compendia	Specifications
Carnauba wax	Caranda wax, Brazil wax	BP, JP, PhEur, USPNF	Melting range 80–88 °C[a]; iodine value 5–14[b]; acid value 2–7; saponification value 78–95; total ash ≤0.25%
Cetyl alcohol	Cetanol, Avol, Lipocol C	BP, JP, PhEur, USPNF	Melting range 47–53 °C[b]; residue on ignition ≤0.05%[b]; iodine value ≤5.0; acid value ≤2.0; saponification value ≤2.0[a]
Cetyl ester wax	Crodamol SS, Ritachol SS, Starfol wax	USPNF	Melting range 43–47 °C; acid value ≤5.0; saponification value 109–120; iodine value ≤1.0
Emulsifying wax	Collone HV, Crodex A, Lipowax PA	BP	Saponification value ≤2.0; iodine value ≤3.0[c]

TABLE 1 ***Continued***

Name	Synonyms	Official Compendia	Specifications
Hydrous lanolin	Hydrous wool fat, Lipolan	BP, JP, PhEur	Melting range 38–44 °C; acid value ≤0.8; saponification value 67–79; nonvolatile matter 72.5–77.5%; iodine value 18–36[b]
Lanolin	Wool fat, purified wool fat, Corona	BP, JP, PhEur, USPNF	Melting range 38–44 °C; loss on drying ≤0.25%; residue on ignition ≤0.1%; iodine value 18–36; acid value ≤1.0[b]
Lanolin alcohols	Argowax, Ritawax, wool wax alcohol	BP, PhEur, USPNF	Melting range ≥56 °C; loss on drying ≤0.50%; residue on ignition ≤0.15%; acid value ≤2.0; saponification value ≤12
Microcrystalline wax	Petroleum ceresin	USPNF	Melting range 54–102 °C; residue on ignition ≤0.10%
Paraffin	Paraffin wax, hard wax, hard paraffin	BP, JP, PhEur, USPNF	Melting range 47–65 °C
Petrolatum	Yellow soft paraffin, yellow petroleum jelly	BP, JP, PhEur, USPNF	Melting range 38–60 °C; residue on ignition ≤0.1%
Poloxamer	Polyethylene–propylene glycol, Lutrol, Pluronic	BP, PhEur, USPNF	Melting point ≈50 °C
Polyethylene glycol (PEG)	Macrogol, Carbowax, PEG, Lutrol	BP, JP, PhEur, USPNF	Melting range of PEG 1000, 37–40 °C; melting range of PEG 8000, 60–63 °C; residue on ignition ≤0.1%
Stearic acid	Emersol, Hystrene	BP, JP, PhEur, USPNF	Melting range ≥54 °C; iodine value ≤4.0
Stearyl alcohol	Lipocol S, Cachalot, Rita SA	BP, JP, PhEur, USPNF	Melting range 55–60 °C; residue on ignition 0.05%[b]; iodine value ≤2.0; acid value ≤2.0; saponification value ≤2.0[a]
White wax	Bleached wax	BP, JP, PhEur, USPNF	Melting range 62–65 °C; acid value 17–24; saponification value 87–104[a]
Yellow wax	Refined wax	BP, JP, PhEur, USPNF	Acid value 17–22[a]; saponification value 87–102[a]

Note: BP, British Pharmacopoeia; JP, Japanese Pharmacopoeia; PhEur, European Pharmacopoeia; USPNF, U.S. Pharmacopeia/National Formulary. All are USPNF specifications, except as indicated below.

[a]European Pharmacopoeia.

[b]Japanese Pharmacopoeia.

[c]British Pharmacopoeia.

compendial status, synonym, and specifications of some of the bases used in ointments and creams.

The following sections describe the source, physicochemical properties, formulation considerations, stability, incompatibility, storage, and hypersensitivity reactions of some of these bases.

Lanolin Lanolin is a refined, decolorized, and deodorized material obtained from sheep wool. It is available as a pale yellow, waxy material with a characteristic odor. It is extensively used in the preparation of hydrophobic ointments and water-in-oil creams. As lanolin is prone to oxidation, antioxidants such as butylated hydroxytoluene are generally included. Although lanolin is insoluble in water, it is miscible with water up to 1:2 ratio. This property favors in preparing physically stable creams. Addition of soft paraffin or vegetable oil improves the emollient property of lanolin preparations. Exposure of lanolin to higher temperature usually leads to discoloration and rancidlike odor, and hence prolonged heating is avoided during the preparation and preservation of lanolin-containing preparations. Gamma sterilization or filtration sterilization is usually employed for sterilizing ophthalmic ointments containing lanolin. Lanolin and some of its derivatives are reported to cause hypersensitivity reactions and therefore are avoided in patients with known hypersensitivity. One of the reasons for hypersensitivity reactions is free fatty alcohols. Modified lanolins containing reduced levels of free fatty alcohols are commercially available [2, 3].

Hydrous Lanolin Incorporation of about 25–30% of water into lanolin gives hydrous lanolin. Gradual addition of water into molten lanolin with constant stirring helps in water incorporation. It is available as a pale yellow, oily material with a characteristic odor. The water uptake capacity of hydrous lanolin is higher than lanolin, and it is used for preparing topical hydrophobic ointments or water-in-oil creams with larger aqueous phase. Exposure of these preparations to higher temperatures results in separations of oily and aqueous layers. Addition of antioxidants and preservation in well-filled, airtight, light-resistant containers in a cool and dry place improve the stability of lanolin products. Well-preserved preparations can be stored up to two years. Hydrous lanolin that contains free fatty alcohols is avoided in hypersensitive patients [2, 3].

Lanolin Alcohols Lanolin alcohol is prepared from lanolin by the saponification process and is used as a hydrophobic vehicle in pharmaceutical ointments and creams. It is composed of steroidal and triterpene alcohols and is available as a brittle solid material pale yellow in color with a faint characteristic odor. The brittle powder becomes plastic under warm conditions. It is practically insoluble in water and soluble in boiling ethanol. Lanolin alcohol possesses emollient properties, which makes it suitable for preparing dry-skin ointments, eye ointments, and water-in-oil creams. Creams containing lanolin alcohols do not show surface darkening and do not produce objectionable odor compared to lanolin-containing preparations. Inclusion of about 0.1% antioxidant, however, minimizes the oxidation on storage. Preparations containing lanolin alcohols can be stored up to two years if preserved in well-filled, well-closed, light-resistant containers in a cool and dry place. As with other lanolin bases, hypersensitivity reactions may occur in some individuals while using preparations containing lanolin alcohols [2, 3].

Petrolatum Petrolatum is also known as yellow soft paraffin. It is an inert material obtained from petroleum, which contains branched and unbranched hydrocarbons. It is available as soft oily material and appears pale yellow to yellow in color. Various grades of petrolatum are commercially available with varying physical properties. All these grades are generally insoluble in water and possess emollient properties. Concentrations up to 30% are used in creams. Petrolatum shows phase transitions on heating to about 35 °C. As it possesses a higher coefficient of thermal expansion, prolonged heating is avoided during processing. The presence of minor impurities can oxidize petrolatum and discolor the product. Antioxidants are therefore added to prevent such physical changes in preparations during storage. Butylated hydroxyanisole, butylated hydroxytoluene, or α-tocopherol is generally incorporated as an antioxidant in petrolatum products. In addition, use of well-closed, airtight, light-resistant containers and storage in a cool and dry place improve stability of preparations. Minor quantities of polycyclic aromatic hydrocarbon impurities in petrolatum sometimes cause hypersensitivity reactions. Substituting yellow soft paraffin with white soft paraffin reduces such reactions [4].

Petrolatum and Lanolin Alcohols Various quantities of lanolin alcohols are mixed with petrolatum to form these mixtures. Wool ointment British Pharmacopoeia (BP) 2001 contains 6% lanolin alcohols and 10% petrolatum. These proportions can be varied to alter physical properties such as consistency and melting range. They are available as soft solids pale ivory in color and possess a characteristic odor. These mixtures are insoluble in water, and concentrations ranging 5–50% are used for preparing hydrophobic ointments. They are also used for preparing water-in-oil emollient creams. Preparations containing petrolatum and lanolin alcohols need to be preserved in airtight, well-closed, light-resistant containers in a cool and dry place to avoid oxidation of impurities and discoloration. Antioxidants improve the stability of these products. Although these mixtures are safe for topical applications, hypersensitivity reactions may occur in some individuals due to the presence of lanolin alcohol [5].

Paraffin Paraffin is obtained by distillation of crude petroleum followed by purification processes. The purified fraction contains saturated hydrocarbons. Paraffin is available as a white color solid and does not possess any specific odor or taste. Different purity grades are available. Use of highly purified grades can avoid batch-to-batch variations in formulations, especially the hardness, melting behavior, and malleability. Paraffin is insoluble in water and is generally used to prepare hydrophobic topical ointments and water-in-oil creams. Repeated heating and congealing are avoided during formulation as they change the physical properties of paraffin. These preparations need to be preserved in well-closed container at room temperature. Synthetic paraffins, which melt between 96 and 105 °C, are sometimes used to increase the melting point and stiffness of formulations [6].

Polyethylene Glycol Also known as macrogol, PEG is synthesized by condensation of ethylene oxide and water under suitable reaction conditions. Based on the

TABLE 2 Properties of Different Grades of PEG

	By Grade							
Property	200	400	600	1000	2000	3000	4000	8000
Physical state	Liquid	Liquid	Liquid	Solid	Solid	Solid	Solid	Solid
Average molecular weight	190–210	380–420	570–613	950–1050	1800–2200	2700–3300	3000–4800	7000–9000
Melting (°C)	—	—	—	37–40	45–50	48–54	50–58	60–63
Density (g/cm^3)	1.11–1.14	1.11–1.14	1.11–1.14	1.15–1.21	1.15–1.21	1.15–1.21	1.15–1.21	1.15–1.21
Kinematic viscosity[a] (cS)	3.9–4.8	6.8–8.0	9.9–11.3	16.0–19.0	38–49	67–93	110–158	470–900

[a]At 98.9°C.

number of oxyethylene groups present, their molecular weights vary from few hundreds to several thousands. Usually the number that follows PEG represents their average molecular weight. They are available as liquids or solids based on molecular weight. PEGs 600 or less are liquids, whereas PEGs above 1000 are solids. PEG liquids are usually clear or pale yellow in color. Their viscosity increases with increase in molecular weight. Solid PEGs are usually white in color and available as pastes, waxy flakes, or free-flowing solids based on their molecular weight. Table 2 shows the physicochemical properties of some PEGs.

PEGs are hydrophilic materials and are extensively used in the preparation of hydrophilic ointments and creams. They are nonirritants and are easily washed from skin surfaces. Products with varying consistency are prepared by mixing different grades of PEGs. Excessive heating is avoided while melting PEGs. This will prevent oxidation and discoloration of products. In addition, use of purified grades that are free from peroxide impurities, inclusion of suitable antioxidants, and heating under nitrogen atmosphere can minimize the oxidation. PEGs are prone to etherification or esterfication reactions due to the presence of two terminal hydroxyl groups. They are incompatible with some antibiotics, antimicrobial preservatives, iron, tannic acid, and salicylic acid and also interact with plastic containers made of polyvinyl chloride and polyethylene. PEG-containing products are usually packed in aluminum, glass, or stainless steel containers to avoid such interactions. Although low-molecular-weight PEGs are hygroscopic, they do not promote microbial growth. PEG-containing products are generally stored in well-closed containers in a cool, dry place. These products can cause stinging sensation on mucus and some hypersensitivity reactions, especially when applied onto open wounds [7, 8].

Stearic Acid Stearic acid is obtained by hydrolysis of fat or hydrogenation of vegetable oils. Compendial stearic acid contains a mixture of stearic acid and palmitic acids. It is available as powder or crystalline solid which is white to yellowish white in color and possesses a characteristic odor. Although stearic acid is insoluble in water, partially neutralized grades form a cream base when combined with about 10 times its weight of aqueous solvents. The appearance and consistency of these grades are based on the proportion of alkali or triethanolamine used for neutralization. Concentrations up to 20% are used for formulating creams and ointments. Different grades of stearic acids are commercially available with varying stearic acid content, melting temperature, and other physical properties. A suitable antioxidant is included in formulations containing stearic acid. As stearic acid interacts with metals, it is avoided in preparations which contain salts, especially divalent metals such as calcium and zinc. It also reacts with metal hydroxides and some drugs. Compatibility evaluation between stearic acid and other formulation components is therefore essential when formulating newer products with stearic acid [9].

Carnauba Wax Carnauba wax contains a mixture of esters of acids and hydroxyacids isolated from Brazilian carnauba palm. It also contains various resins, hydrocarbons, acids, polyhydric alcohols, and water. It is available as lumps, powder, or flakes which are brown to pale yellow in color and possesses a characteristic odor. Carnauba wax is practically insoluble in water and melts at 80–88 °C. Being a hard material, it improves the stiffness of topical preparations [6].

Cetyl Alcohol Cetyl alcohol is obtained by hydrogenolysis or esterfication of fatty acids and contains not less than 90% cetyl alcohol along with other aliphatic alcohols. It is available as flakes or granules white in color and possesses a characteristic odor. Different grades are commercially available with varying proportions of cetyl alcohol, stearyl alcohol, and related alcohols. Although insoluble in water, cetyl alcohol has good water-absorptive and emulsifying properties. This property makes it suitable for preparing emollient ointments and creams. Its viscosity-enhancing properties reduce coalescence of dispersed phase and improves the physical stability of creams. Concentrations ranging from 2 to 10% are used in topical preparations to impart emollient, emulsifying, water-absorptive, and stiffening properties. Mixtures of petrolatum and cetyl alcohol are sometimes used for preparing creams. Such mixtures minimize the quantity of additional emulsifying agents in preparations. Although cetyl alcohol forms stable preparations, it is incompatible with strong oxidizing materials and some drugs. Compatibility studies are therefore conducted when including cetyl alcohol into formulations. Highly purified grades are free from hypersensitivity reactions [3, 10].

Emulsifying Wax Emulsifying wax, also known as anionic emulsifying wax, is a mixture of cetostearyl alcohol, sodium lauryl sulfate, and purified water. Emulsifying wax BP contains about 90% cetostearyl alcohol, 10% sodium lauryl sulfate, and 4% purified water. Emulsifying wax USP contains nonionic surfactants. It is available as flakes or solids which are white to pale yellow in color and possesses a characteristic odor. Although emulsifying wax is insoluble in water, its emulsifying properties help in preparing hydrophilic oil-in-water emulsions. Ointment bases are prepared by mixing up to 50% emulsifying wax with liquid or soft paraffins. At concentrations up to 10%, it forms creams. Although emulsifying wax is compatible with many acids and alkalis, it is incompatible with many cationic materials and polyvalent metal salts. Stainless steel vessels are preferred for mixing operations.

Preparations containing emulsifying wax are preserved in well-closed container in a cool, dry place [11].

Cetyl Esters Wax Cetyl esters wax is obtained by esterification of some fatty alcohols and fatty acids. It is available as crystalline flakes which are white to off-white in color and possesses a characteristic aromatic odor. It is insoluble in water and has emollient and stiffening properties. About 10% of cetyl ester wax is used for preparing hydrophobic creams and about 20% is used for preparing topical ointments. Various grades of cetyl esters wax are available commercially and vary in their fatty alcohol and fatty acids content and melting range. As this wax is incompatible with strong acids and bases, it should be avoided in certain formulations. Cetyl ester wax–containing products are stored in well-closed containers in a cool, dry place [6].

Hydrogenated Castor Oil It is used as stiffening agent in hydrophobic ointments and creams due to its higher melting point. Hydrogenated castor oil contains triglyceride of hydroxystearic acid and is available as white color flakes or powder. It is insoluble in water and melts at 85–88 °C. Different grades with varying compositions and physical properties are commercially available. Products can be prepared at higher temperatures, as hydrogenated castor oil is stable up to 150 °C. It is compatible with other waxes obtained from vegetable and animal sources. Preparations containing hydrogenated castor oil need to be preserved in well-closed containers in a cool and dry place [12].

Microcrystalline Wax Microcrystalline wax is obtained from petroleum by solvent fractionation and dewaxing procedures. It contains many straight-chain and branched-chain alkanes, with carbon chain lengths ranging from 41 to 57. It is available as fine flakes or crystals which are white or yellow in color. Microcrystalline wax is insoluble in water and possesses a wide melting range (54–102 °C). High-melting and stiffening properties of microcrystalline wax make it suitable for preparing ointments and cream with higher consistency. Acids, alkalis, oxygen, and light do not affect its stability [6].

Stearyl Alcohol Reduction of ethyl stearate in the presence of lithium aluminum hydride yields stearyl alcohol, which contains not less than 90% of 1-octadecanol. It is available as flakes or granules which are white in color and possesses a characteristic odor. It is insoluble in water and melts at 55–60 °C. Stearyl alcohol has stiffening, viscosity-enhancing, and emollient properties and hence is used in the preparation of hydrophobic ointments and creams. Its weak emulsifying properties help in improving the water-holding capacity of ointments. Hypersensitivity reactions are sometimes observed due to the presence of some minor impurities. Stearyl alcohol preparations are compatible with acids and alkalis and are preserved in well-closed containers in a cool and dry place [6].

White Wax White wax is a bleached form of yellow wax which is usually obtained from the honeycomb of bees and hence is known as *bleached wax* or *white bees wax*. It contains about 70% esters of straight-chain monohydric alcohols, 15% free acids,

12% carbohydrates, and 1% free wax alcohols and stearic esters of fatty acids. It is available as granules or sheets which are white in color and possesses a characteristic odor. White wax is insoluble in water and melts between 61 and 65 °C. It has stiffening and viscosity-enhancing properties and therefore is used in hydrophobic ointments and oil-in-water creams. Although it is thermally stable, heating to above 150 °C results in reduction of its acid value. White wax is incompatible with oxidizing agents. The presence of small quantities of impurities results in hypersensitivity reactions in rare occasions. Preparations are stored in well-closed, light-resistant containers in a cool, dry place [13].

Yellow Wax Yellow wax, also known as yellow beeswax, is obtained from honey combs. It contains about 70% esters of straight-chain monohydric alcohols, 15% free acids, 12% carbohydrates, and 1% free wax alcohols and stearic esters of fatty acids. It is available as noncrystalline pieces which are yellow in color and possesses a characteristic odor. It is practically insoluble in water and melts at 61–65 °C. It is used in the preparation of hydrophobic ointments and water-in-oil creams because of its viscosity-enhancing properties. Concentrations up to 20% are used for producing ointments and creams. It is incompatible with oxidizing agents. Esterification occurs while heating to 150 °C and hence should be avoided during preparation. Hypersensitivity reactions sometimes occur on topical application of yellow wax–containing ointments and creams due to the presence of some minor impurities. These products are preserved in well-closed, light-resistant containers [13].

Combinations of bases are sometimes used to acquire better stability. Gelling agents such as carbomers and PEG are also included in some ointment and cream preparations. Table 3 shows examples of cream bases used in some commercial cream preparations.

4.2.2.3 Preparation and Packaging

In addition to the base and drug, ointments and creams may also contain other components such as stabilizers, preservatives, and levigating agents. Usually levigation and fusion methods are employed for incorporating these components into the base. Levigation involves simple mixing of base and other components over an ointment slab using a stainless steel ointment spatula. A fusion process is employed only when the components are stable at fusion temperatures. Ointments and creams containing white wax, yellow wax, paraffin, stearyl alcohol, and high-molecular-weight PEGs are generally prepared by the fusion process. Selection of levigation or the fusion method depends on the type base, the quantity of other components, and their solubility and stability characteristics.

Oleaginous ointments are prepared by both levigation and fusion processes. Small quantities of powders are incorporated into hydrocarbon bases with the aid of a levigating agent such as liquid petrolatum, which helps in wetting of powders. The powder component is mixed with the levigating agent by trituration and is then incorporated into the base by spatulation. All solid components are milled to finer size and screened before incorporating into the base to avoid gritty sensation of the final product. Roller mills are used for producing large quantities of ointments in pharmaceutical industries. Uniform mixing can be obtained by the geometric dilution procedure, which usually involves stepwise dilution of solids into the ointment

TABLE 3 Cream Bases Present in Some Commercial Creams

Commercial Name	Drug	Cream Base (s) Used
Dritho-Calp, Psoriatec	Anthralin, 0.5%, 1.0%	White petrolatum, cetostearyl alcohol
Temovate E	Clobetasol propionate, 0.05%	Propylene glycol, glyceryl monostearate, cetostearyl alcohol, glyceryl stearate, PEG 100 stearate, white wax
Eurax	Crotamiton, 10%	Petrolatum, propylene glycol, cetyl alcohol, carbomer-934
Topicort	Desoximetasone, 0.25%	White petrolatum USP, isopropyl myristate NF, lanolin alcohols NF, mineral oil USP, cetostearyl alcohol NF
Apexicon, Maxiflor, Psorcon	Diflorasone diacetate, 0.05%	Hydrophilic vanishing cream base of propylene glycol, stearyl alcohol, cetyl alcohol
Lidex Cream, Vanos	Fluocinonide, 0.05%, 0.10%	Polyethylene glycol 8000, propylene glycol, stearyl alcohol
Carac	Fluorouracil, 0.5%, 1.0%, 5.0%	Carbomer-940, PEG 400, propylene glycol, stearic acid
Halog	Halcinonide, 0.1%	Polyethylene and mineral oil gel base with PEG 400, PEG 6000, PEG 300, PEG 1450
Cortaid, Anusol-Hc, Proctosol HC	Hydrocortisone, 2.5% water washable	Petrolatum, stearyl alcohol, propylene glycol, carbomer-934
Monistat-Derm	Miconazole nitrate, 2%	Water-miscible base consisting of pegoxol 7 stearate, peglicol 5 oleate, mineral oil, butylated hydroxyanisole

base. The fusion method is followed when the drugs and other solids are soluble in the ointment bases. The base is liquefied, and the soluble components are dissolved in the molten base. The mixture is then allowed to congeal by cooling. Fusion is performed using steam-jacketed vessels or a porcelain dish. The congealed mixture is then spatulated or triturated to obtain a smooth texture. Care is taken to avoid thermal degradation of the base or other components during the fusion process.

Absorption-type ointments and creams are prepared by incorporating large quantities of water into hydrocarbon bases with the aid of a hydrophobic emulsifying agent. Water-insoluble drugs are added by mechanical addition or fusion methods. As with oleaginous ointments, levigating agents are also included to improve wetting of solids. Water-soluble or water-miscible agents such as alcohol, glycerin, or propylene glycol are used if the drug needs to be incorporated into the internal aqueous phase. If the drug needs to be incorporated into the external oily phase, mineral oils are used as the levigating agent. Incorporation of water-soluble components is achieved by slowly adding the aqueous drug solution to the hydrophobic base using pill tile and spatula. If the proportion of aqueous phase is larger, inclusion of additional quantities of emulsifier and application of heat may be needed to achieve uniform dispersion. Care must be taken to avoid excessive heating as it can result in evaporation aqueous phase and precipitation of water-soluble components and formation of stiff and waxy product.

Water-removable ointments and creams are basically hydrophilic-type emulsions. They are prepared by fusion followed by mechanical addition approach. Hydrocarbon components are melted together and added to the aqueous phase that contains water-soluble components with constant stirring until the mixture congeals. A hydrophilic emulsifying agent is included in the aqueous phase in order to obtain stable oil-in-water dispersion. Sodium lauryl sulfate is used in the preparation of hydrophilic ointment USP.

Water-soluble ointments and creams do not contain any oily phase. Both water-soluble and water-insoluble components are incorporated into water-soluble bases by both levigation and fusion methods. If the drug and other components are water soluble, they are dissolved in a small quantity of water and incorporated into the base by simple mixing over an ointment slab. If the components are insoluble in water, aqueous levigating agents such as glycerin, propylene glycol, or a liquid PEG are used. The hydrophobic components are mixed with the levigating agent and then incorporated into the base. Heat aids incorporation of a large quantity of hydrophobic components.

A wide range of machines are available for the large-scale production of ointments and creams. Each of these machines is designed to perform certain unit operations, such as milling, separation, mixing, emulsification, and deaeration. Milling is performed to reduce the size of actives and other additives. Various fluid energy mills, impact mills, cutter mills, compression mills, screening mills, and tumbling mills are used for this purpose. Alpine, Bepex, Fluid Air, and Sturtevant are some of the manufacturers of these mills. Separators are employed for separating materials of different size, shape, and densities. Either centrifugal separators or vibratory shakers are used for separation. Mixing of the actives and other formulation components with the ointment or cream base is performed using various types of low-shear mixers, high-shear mixers, roller mills, and static mixers. Mixers with heating provisions are also used to aid in the melting of bases and mixing of components. Chemineer, Fryma, Gate, IKA, Koruma (Romaco), Moorhouse-Cowles, Ross, and Stokes Merrill are some of the manufacturers of semisolids mixers.

Creams are produced with the help of low-shear and high-shear emulsifiers. These emulsifiers are used to disperse the hydrophilic components in the hydrophobic dispersion phase (e.g., water-in-oil creams) or oleaginous materials in aqueous dispersion medium (oil-in-water creams). Bematek, Fryma, Koruma (Romaco), Lightnin, Moorhouse, and Ross supply various types of emulsifiers. Entrapment of air into the final product due to mixing processes is a common issue in the large-scale manufacturing of semisolid dosage forms. Various offline and in-line deaeration procedures are adopted to minimize this issue. Effective deaeration is generally achieved by using vacuum vessel deaerators. Some of the recent large-scale machines are designed to perform heating, high-shear mixing, scrapping, and deaeration processes in a single vessel. Figure 1 shows the design feature of a semisolid production machine manufactured by Ross.

Various low- and high-shear shifters are used to transfer materials from the production vessel to the packaging machines. In the packaging area, various types of holders (e.g., pneumatic, gravity, and auger holders), fillers (e.g., piston, peristaltic pump, gear pump, orifice, and auger fillers), and sealers (e.g., heat, torque, microwave, indication, and mechanical crimping sealers) are used to complete the unit

FIGURE 1 Semisolid production machine with heat jacketed vessel, high-shear mixer, scrapper, vacuum attachments, and control station. (Courtesy of Ross, Inc.)

operations. These equipments are supplied by various manufacturers, namely Bosch, Bonafacci, Erweka, Fryma-Maschinenbau, IWKA, Kalish, and Norden.

Sterility of ointments, especially those intended for ophthalmic use, is achieved by aseptic handling and processing. Improper processing, handling, packing, or use of ophthalmic ointments lead to microbial contaminations and eventually result in ocular infections. In general, the empty containers are separately sterilized and filled under aseptic condition. Final product sterilization by moist heat sterilization or gaseous sterilization is ineffective because of product viscosity. Dry-heat sterilization is associated with stability issues. Strict aseptic procedures are therefore practiced when processing ophthalmic preparations. Antimicrobial preservatives such as benzalkonium chloride, phenyl mercuric acetate, chlorobutanol, or a combination of methyl paraben and propyl paraben are included in ophthalmic ointments to retain microbial stability.

Packaging An ideal container should protect the product from the external atmosphere such as heat, humidity, and particulates, be nonreactive with the product components, and be easy to use, light in weight, and economic [14]. As tubes made of aluminum and plastic meet most of these qualities, they are extensively used for packaging semisolids. Aluminum tubes with special internal epoxy coatings are commercially available for improving the compatibility and stability of products. Various modified plastic materials are used for making ointment tubes. Tubes made

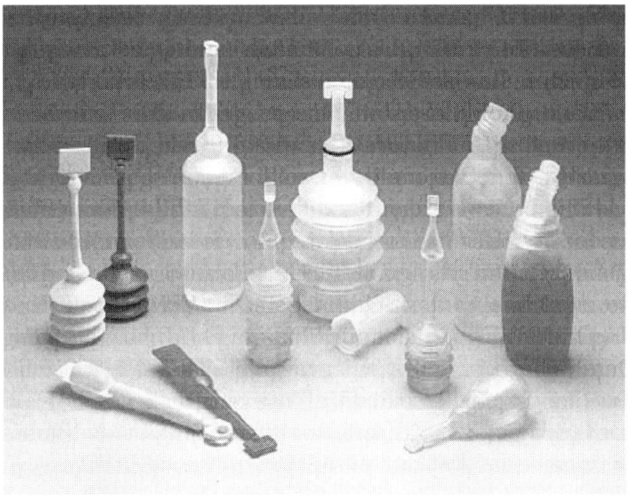

FIGURE 2 Custom-designed LDPE containers made by BFS process for packaging topical products. (Courtesy of Rommelag USA, Inc.)

of low-density polyethylene (LDPE) are generally soft and flexible and offer good moisture protection. Tubes made of high-density polyethylene (HDPE) are relatively harder but offer high moisture protection. Polypropylene containers offer high heat resistance. Plastic containers made of polyethylene terephthalate (PET) are transparent and provide superior chemical compatibility. Ointments meant for ophthalmic, nasal, rectal, and vaginal applications are supplied with special application tips for the ease of product administration.

A recent method known as blow fill sealing (BFS) performs fabrication of container, filling of product, and sealing operations in a single stage and hence is gaining greater attention. The products can be sterile filled, which makes BFS a cost-effective alternative for aseptic filling. All plastic materials are suitable for BFS processing. In most cases, monolayered LDPE materials are used for making small-size containers. If the product is not compatible with the LDPE or sensitive to oxygen, barrier layers are added to the container wall by coextrusion methods. As the container is formed inside the BFS machine, upstream handling problems are avoided. The BFS machine can hand the container off to any secondary packaging operation that needs to be performed. Typically a secondary overwrap is added to the containers prior to cartooning. An additional advantage of BFS containers is the integrated design of the applicator into the product container. Figure 2 shows some of the custom-designed BFS containers for topical products.

4.2.2.4 Evaluation

Ointments and creams are evaluated for various pharmacopeial and nonpharmacopeial tests to ascertain their physicochemical, microbial, in vitro, and in vivo characteristics. These tests help in retaining their quality and minimizing the batch-to-batch variations. The USP recommends storage and labeling, microbial screening, minimum fill, and assays for most ointments and creams. Tables 4 and 5 summarize the compendial requirements for some pharmacopeial ointments and creams.

TABLE 4 USP Specifications for Some Official Ointments

Drug	Quality Control Tests	Packaging and Storage Requirements
Acyclovir	*Staphylococcus aureus, Pseudomonas aeruginosa,* minimum fill, limit of guanine, and assay	Tight containers; store between 15 and 25 °C in a dry place
Alclometasone dipropionate	*S. aureus, P. aeruginosa,* minimum fill, and assay	Collapsible tubes or tight containers, store at controlled room temperature
Amphotericin B	Minimum fill, water, and assay	Collapsible tubes or other well-closed containers
Anthralin	Assay	Tight containers; in a cool place; protect from light
Bacitracin	Minimum fill, water, and assay	Well-closed containers containing not more than 60 g; controlled temperature
Benzocaine	*S. aureus, P. aeruginosa,* minimum fill, and assay	Tight containers; protect from light; avoid prolonged exposure to temperatures exceeding 30 °C
Betamethasone valerate	*S. aureus, P. aeruginosa,* minimum fill, and assay	Collapsible tubes or tight containers; avoid exposure to excessive heat.
Clioquinol	Assay	Collapsible tubes or tight, light-resistant containers
Clobetasol propionate	*S. aureus, P. aeruginosa, Escherichia coli, Salmonella* species, total aerobic microbial count, minimum fill, and assay	Collapsible tubes or in tight containers; store at controlled room temperature; do not refrigerate
Erythromycin	Minimum fill, water, and assay	Collapsible tubes or in tight containers at controlled room temperature
Fluocinolone acetonide	*S. aureus, P. aeruginosa,* and assay	Tight containers
Gentamycin sulfate	Minimum fill, water, and assay	Collapsible tubes or in tight containers; avoid exposure to excessive heat
Hydrocortisone valerate	*S. aureus, P. aeruginosa,* total microbial count, minimum fill, and assay	Tight container; store at room temperature
Ichthammol	Assay	Collapsible tubes or in tight containers; avoid prolonged exposure to temperatures exceeding 30 °C
Lidocaine	*S. aureus, P. aeruginosa,* minimum fill, and assay	Tight containers
Mometasone furoate	*S. aureus, P. aeruginosa, E. coli, Salmonella* species, minimum fill, and assay	Well-closed containers
Nitrofurazone	Completeness of solution and assay	Tight, light-resistant containers; avoid exposure to direct sunlight, strong fluorescent lighting, and excessive heat
Nitroglycerine	Minimum fill, homogeneity, and assay	Tight containers
Nystatin	Minimum fill, water, and assay	Well-closed containers at controlled room temperature
Tetracycline hydrochloride	Minimum fill, water, and assay	Well-closed containers at controlled room temperature
Zinc oxide	Minimum fill, calcium, magnesium, other foreign substances, and assay	Tight containers; avoid prolonged exposure to temperatures exceeding 30 °C

TABLE 5 USP Specifications for Some Official Creams

Cream	Quality Control Tests	Packaging and Storage Requirements
Alclometasone dipropionate	Microbial limits, minimum fill, and assay	Collapsible tubes or tight containers; store at controlled room temperature
Amphotericin B	Minimum fill and assay	Collapsible tubes or other well-closed containers
Benzocaine	Microbial limits, minimum fill, and assay	Tight containers, protected from light, and avoid prolonged exposure to temperatures exceeding 30 °C
Betamethasone dipropionate	Minimum fill and assay	Collapsible tubes or tight containers; store at 25 °C; excursions permitted between 15 and 30; protect from freezing
Ciclopirox olamine	Minimum fill, pH, content of benzyl alcohol, and assay	Collapsible tubes at controlled room temperature
Clobetasol propionate	Microbial limits, minimum fill, pH, and assay	Collapsible tubes or tight containers; store at controlled room temperature; do not refrigerate
Clotrimazole	Assay	Collapsible tubes or tight containers at a temperature between 2 and 30 °C
Desoximetasone	Minimum fill, pH, and assay	Collapsible tubes at controlled room temperature
Dibucaine	Microbial limits, minimum fill, and assay	Collapsible tubes or in tight, light-resistant containers
Dienestrol	Minimum fill and assay	Collapsible tubes or in tight containers
Diflorasone diacetate	Microbial limits, minimum fill, and assay	Collapsible tubes, preferably at controlled room temperature
Fluocinolone acetonide	Microbial limits, minimum fill, and assay	Collapsible tubes or in tight containers
Fluorouracil	Microbial limits, minimum fill, and assay	Tight containers and stored at controlled room temperature
Gentamycin sulfate	Minimum fill and assay	Collapsible tubes or in other tight containers; avoid exposure to excessive heat
Hydrocortisone butyrate	Microbial limits, minimum fill, pH, and assay	Well-closed containers
Hydroquinone	Minimum fill and assay	Well-closed, light-resistant containers
Lindane	pH and assay	Tight containers
Meclocycline sulfosalicylate	Minimum fill and assay	Tight containers, protected from light
Miconazole nitrate	Minimum fill and assay	Collapsible tubes or tight containers; store at controlled room temperature
Monobenzone	Assay	Tight containers; avoid exposure to temperatures higher than 30 °C
Nystatin	Minimum fill and assay	Collapsible tubes or in other tight containers; avoid exposure to excessive heat
Prednisolone	Minimum fill and assay	Collapsible tubes or in tight containers
Tetracaine hydrochloride	Microbial limits, minimum fill, pH between 3.2 and 3.8, and assay	Collapsible, lined metal tubes
Triamcinolone acetonide	Microbial limits, minimum fill, and assay	Tight containers

Packaging and Storage The USP recommends packaging and storage requirements for each official ointment and cream. Generally collapsible tubes, tight containers, or other well-closed containers are recommended for packing. They are stored in either a cool place or at controlled room temperatures. In some cases, special storage conditions are recommended: for example, protect from light, avoid exposure to excessive heat, avoid exposure to direct sunlight, avoid strong fluorescent lighting, do not refrigerate, and avoid prolonged exposure to temperatures exceeding 30 °C.

Minimum Fill This test is performed to compare the weight or volume of product filled into each container with their labeled weight or volume. It helps in assessing the content uniformity of product. A minimum-fill test is applied only to those containers that contain not more than 150 g or mL of preparation. It is performed in two steps. Initially, labels from the product containers are removed. After washing and drying the surface, their weights are recorded (W_1). In the second step, the entire product from each container is removed. After cleaning and drying, the weight of empty containers is recorded (W_2). The difference between total weight (W_1) and empty-container weight (W_2) gives the weight of product. The USP recommends that the average net content of 10 containers should not be less than the labeled amount. If the product weight is less than 60 g or mL, the net content of any single container should not be less than 90% of the labeled amount. If the product weight is between 60 and 150 g or mL, the net content of any single container should not be less than 95% of the labeled amount. If these limits are not met, the test is repeated with an additional 20 containers. All semisolid topical preparations should meet these specifications [15].

Water Content The presence of minor quantities of water may alter the microbial, physical, and chemical stability of ointments and creams. Titrimetric methods (method I) are usually performed for determining the water content in these preparations. These methods are based on the quantitative reaction between water and anhydrous solution of sulfur and iodine in the presence of a buffer that can react with hydrogen ions. Special titration setups and reagents (Karl Fischer, KF) are used in these determinations. In the direct method (method Ia), about 35 mL of methanol is titrated with sufficient quantity of KF reagent to the electrometric or visual endpoint (color change from canary yellow to amber). This blank titration helps to consume any moisture that may be present in the reaction medium. A known quantity of test material (ointment or cream) is added to the reaction medium, mixed, and again titrated with KF reagent to the reaction endpoint. The water content is determined by considering the volume of KF reagent consumed and its water equivalence factor. In the residual titration method (method Ib), a known excess quantity of KF reagent is added to the titration vessel, which is then back titrated with standardized water to the electrometric or visual endpoint. In the coulometric titration method (method Ic), the sample is dissolved in anhydrous methanol and injected into the reaction vessel that contains the anolyte, and the coulometric reaction is performed until the reaction endpoint. In some cases, methanol is replaced with other solvents. The maximum allowable limit of water in ointment preparations varies between 0.5 and 1.0%. The limit of water in bacitracin, chlortetracycline hydrochloride, and nystatin ointments is not more than 0.5%, whereas amphotericin

B, erythromycin, gentamycin sulfate, neomycin sulfate, and tetracycline hydrochloride ointments may contain up to 1% moisture [15].

Metal Particles This test is required only for ophthalmic ointments. The presence of metal particles will irritate the corneal or conjunctival surfaces of the eye. It is performed using 10 ointment tubes. The content from each tube is completely removed onto a clean 60-mm-diameter petridish which possesses a flat bottom. The lid is closed and the product is heated at 85 °C for 2 h. Once the product is melted and distributed uniformly, it is cooled to room temperature. The lid is removed after solidification. The bottom surface is then viewed through an optical microscope at 30× magnification. The viewing surface is illuminated using an external light source positioned at 45 ° on the top. The entire bottom surface of the ointment is examined, and the number of particles 50 μm or above are counted using a calibrated eyepiece micrometer. The USP recommends that the number of such particles in 10 tubes should not exceed 50, with not more than 8 particles in any individual tube. If these limits are not met, the test is repeated with an additional 20 tubes. In this case, the total number of particles in 30 tubes should not exceed 150, and not more than 3 tubes are allowed to contain more than 8 particles [15].

Leakage Test This test is mandatory for ophthalmic ointments, which evaluates the intactness of the ointment tube and its seal. Ten sealed containers are selected, and their exterior surfaces are cleaned. They are horizontally placed over absorbent blotting paper and maintained at 60 ± 3 °C for 8 h. The test passes if leakage is not observed from any tube. If leakage is observed, the test is repeated with an additional 20 tubes. The test passes if not more than 1 tube shows leakage out of 30 tubes [15].

Sterility Tests Ophthalmic semisolids should be free from anaerobic and aerobic bacteria and fungi. Sterility tests are therefore performed by the membrane filtration technique or direct-inoculation techniques. In the membrane filtration method, a solution of test product (1%) is prepared in isopropyl myristate and allowed to penetrate through cellulose nitrate filter with pore size less than 0.45 μm. If necessary, gradual suction or pressure is applied to aid filtration. The membrane is then washed three times with 100-mL quantities of sterile diluting and rinsing fluid and transferred aseptically into fluid thioglycolate (FTG) and soybean–casein digest (SBCD) medium. The membrane is finally incubated for 14 days. Growth on FTG medium indicates the presence of anaerobic and aerobic bacteria, and SBCD medium indicates fungi and aerobic bacteria. Absence of any growth in both these media establishes the sterility of the product. In the direct-inoculation technique, 1 part of the product is diluted with 10 parts of sterile diluting and rinsing fluid with the help of an emulsifying agent and incubated in FTG and SBCD media for 14 days. In both techniques, the number of test articles is based on the batch size of the product. If the batch size is less than 200 the containers, either 5% of the containers or 2 containers (whichever is greater) are used. If the batch size is more than 200, 10 containers are used for sterility testing [15].

Microbial Screening Semisolid preparations are required to be free from any microbial contamination. Hence, most of the topical ointments are screened for the presence of *Staphylococcus aureus* and *Pseudomonas aeruginosa*. In some cases, screening for *Escherichia coli*, *Salmonella* species, and total aerobic microbial counts is recommended by the USP. For instance, clobetasol propionate ointment USP and mometasone furoate ointment USP are screened for all these organisms. In addition, preparations meant for rectal, vaginal, and urethral applications are tested for yeasts and molds [15].

Test for S. aureus *and* P. aeruginosa The test sample is mixed with about 100 mL of fluid soybean–casein digest (FSBCD) medium and incubated. If microbial growth is observed, it is inoculated in agar medium by the streaking technique. Vogel–Johnson agar (VJA) medium is used for *S. aureus* screening, and cetrimide agar (CA) medium is used for screening *P. aeruginosa*. The petridishes are then closed, inverted, and incubated under appropriate conditions. The appearance of black colonies surrounded by a yellow zone over VJA medium and greenish colonies in CA medium indicates the presence of *S. aureus* and *P. aeruginosa*, respectively. Various other agar media are also available for screening these organisms. A coagulase test is then performed for confirming the presence of *S. aureus* and oxidase and pigment tests for confirming *P. aeruginosa*.

Test for Salmonella *Species and* E. coli The test sample is mixed with about 100 mL of fluid lactose (FL) medium and incubated. If microbial growth is observed, the contents are mixed and 1 mL is transferred to vessels containing 10 mL of fluid selinite cystine (FSC) medium and fluid tetrathionate (FT) medium and incubated for 12–24 h under appropriate conditions. To identify the presence of *Salmonella*, samples from the above two media are streaked over brilliant green agar (BGA) medium, xylose lysine desoxycholate agar (XLDA) medium, and bismuth sulfite agar (BSA) medium and incubated. The appearance of small, transparent or pink-to-white opaque colonies over BGA medium, red colonies with or without black centers over XLDA medium, and black or green colonies over BSA medium indicates the presence of *Salmonella*. It is further confirmed in triple sugar iron agar medium. The presence of *E. coli* is screened by streaking the samples from FL medium over MacConkey agar medium. The appearance of brick red colonies indicates the presence of *E. coli*. It is further confirmed using Levine eosin methylene blue agar medium.

TOTAL AEROBIC MICROBIAL COUNTS The plate method or multiple-tube method is performed to estimate the total count. About 10 g or 10 mL of the test sample is dissolved or suspended in sufficient volume of phosphate buffer (pH 7.2), fluid soybean casein digest (FSBCD) medium, or fluid casein digest–soy lecithin–polysorbate 20 medium to make the final volume 100 mL. In the plate method, about 1 mL of this diluted sample is mixed with molten soybean–casein digest agar (SBCDA) medium and solidified at room temperature. The plates are inverted and incubated for two to three days. The number of colonies that are on the surface of nutrient media are counted. The multiple tube method is performed using sterile fluid SBCD medium. The number of colonies formed should not exceed the limits specified in an individual monograph. For example, clobetasol propionate ointment USP and hydrocortisone valerate ointment USP contains less than 100 colony-forming units (CFU) per gram of sample.

Test for Yeasts and Molds The plate method is used for testing molds and yeast in semisolids. The procedure is similar to that of the total count test. Instead of SBCDA medium, Sabouraud dextrose agar (SDA) medium or potato dextrose agar (PDA) medium is used. Samples are incubated for five to seven days at 20–25 °C to identify the presence of yeasts and molds.

Assay The quantity of drug present in a unit weight or volume of ointment or cream is determined by various methods. Spectrophotometric, titrimetric, chromatographic, and in some cases microbial assays are performed. Selection of a particular method is based on the nature of drug, its concentration in the product, interference between the drug and other formulation components, and official requirements. Although spectrophotometric methods are accurate and easy to perform, the complexity of ointment matrix sometimes reduced the specificity of analysis compare to liquid chromatographic methods. The USP prescribes high-performance liquid chromatographic (HPLC) assays for many official ointments due to its specificity, accuracy, and precision. For example, amcinonide, anthralin, betamethasone dipropionate, clobetasol propionate, dibucaine, nitroglycerine, hydrocortisone, and triamcinolone acetonide are assayed by HPLC methods. These methods involve extraction of drug from the formulation matrix using suitable solvents followed by chromatographic separation using suitable reversed-phase columns followed by ultraviolet (UV) detection. Clioquinol preparation is assayed by gas chromatography. The USP also recommends potentiometric titrations (benzocaine, lidocaine, and ichthammol) and complexometric titrations (zinc oxide) for some semisolid preparations.

Microbial assays are recommended for certain preparations containing antibiotics such as amphotericin B, bacitracin, chlortetracycline hydrochloride, gentamycin sulfate, neomycin sulfate, and nystatin. These tests evaluate the potency of an antibiotic by means of its inhibitory effects on specific microorganism. Two types of microbial assays are performed to determine the antibiotic potency. They are known as cylindrical plate or plate assays and turbidimetric or tube assays. The plate method measures the extent of growth inhibition of a particular microorganism in solidified agar medium in the presence of the test antibiotic (commonly known as *zone of inhibition*). The tube method measures the turbidity of a liquid medium that contains a particular organism in the presence and absence of the test antibiotic. These methods involve extracting drug from the formulation matrix, diluting the drug to a known concentration, and measuring the zone of inhibition or turbidity.

In Vitro Drug Release Studies These studies are conducted to ascertain release of drug from the formulation matrix. Open-chamber diffusion cells such as Franz cells are used for performing in vitro studies. These cells consist of a donor side and a receiver side separated by a synthetic membrane such as cellulose acetate/nitrate mixed ester, polysulfone, or polytetrafluoroethylene. The membranes are usually pretreated with the receiver fluid to avoid any lag phase in drug release. The receiver side is filled with a known volume of release medium and is heated to 32 ± 0.5 °C by circulating warm water through an outer jacket. Aqueous buffers are used for water-soluble drugs. Phosphate buffer solution of pH 5.4 is considered most appropriate for dermatological products as it mimics the pH of skin. Hydroalcoholic or other suitable medium may also be used for sparingly water soluble drugs. A known quantity of the test product is applied uniformly over the membrane on the donor side and samples are withdrawn from the receiver side at different time intervals. After each sampling, an equal volume of fresh medium is replaced to the receiver side. The sampling time points are different for different formulations; however, at least five samples are withdrawn during the study period for determining the release rate. A typical sample time sequence for a 6-h study is 0.5, 1.0, 2.0, 4.0, and 6.0h. The receiver samples are analyzed by a suitable analytical method to quantify the amount of drug released from the formulation at different time intervals. The slope of the straight line which is obtained from a plot of cumulative amount drug release across 1-cm^2 membrane versus the square root of time represents the release rate. Experiments are conducted in hexaplicate to obtain statistically significant results [16].

In Vivo Bioequivalence Studies In vivo studies are conducted to establish the biological availability or activity of the drug from a topically applied semisolid formulation. Dermatopharmacokinetic studies, pharmacodynamic studies, or comparative clinical trials are generally conducted to assess the bioequivalence of topical products [16, 17].

Dermatopharmacokinetic (DPK) studies are applicable for topical semisolid products that contain antifungals, antivirals, corticosteroids, and antibiotics and vaginally applied products. They are not applicable for ophthalmic, otic, and other products that damage stratum corneum. DPK studies involve measurement of drug concentrations in stratum corneum, drug uptake, apparent steady state, and elimination after application of the test product onto skin. These studies are conducted in healthy human subjects adopting crossover design. The test and the reference products are applied onto eight to nine sites in the forearm. The surface area of each site is based on the strength of drug, extent of drug diffusion, exposure time, and sensitivity of the analytical technique. The application site is washed and allowed to normalize for at least 2h prior to drug application. A known amount of product is applied onto these selected sites. At appropriate time intervals, the excess of drug from each area is removed using cotton swabs or soft tissue papers. Care is taken to avoid stratum corneum damage during sample collection. Stripping of stratum corneum is performed using adhesive tape-strips (e.g., D-Squame, Transpore). In the elimination phase, the excess drug is removed at the steady-state time point, and the stratum corneum is harvested at succeeding times over 24h. The drug content from strips from each time point are extracted using suitable solvents and quantified by a validated analytical method. A stratum corneum drug levels–time curve is developed, and pharmacokinetic parameters such as maximum concentration at steady state ($C_{\text{max-ss}}$), time to reach $C_{\text{max-ss}}$ ($T_{\text{max-ss}}$), and areas under the curve for the test and standard (AUC_{test} and $AUC_{\text{reference}}$) are computed. DPK studies are performed in either one or two occasions. If performed in one occasion, both arms of a single subject are used to compare the test and reference products. If performed in two occasions, a wash-out period of at least 28 days is allowed to rejuvenate the harvested stratum corneum.

Pharmacodynamic (PD) studies are also performed to estimate the bioavailability and bioequivalence of drugs from topically applied semisolids. In this case, known therapeutic responses from drug products such as skin blanching due to vasoconstrictor effects caused by corticosteroids and transepidermal water loss caused by retinoids are measured and compared between the test and reference products. Comparative clinical studies are rarely conducted due to the difficulties involved in performing the study, variability in study results, and their poor sensitivity.

4.2.2.5 Typical Pharmacopeial/Commercial Examples

The vast majority of topical ointments and creams are meant for dermatological applications. They are used to treat various skin conditions such as eczema, dermatitis, allergies, inflammatory and pruritic manifestations, minor skin wounds, pain, insect bites, psoriasis, herpes and other infections of the skin (e.g., impetigo), acne, and precancerous and cancerous skin growths. Similarly, ophthalmic conditions such as infections, inflammation, allergy, and dry-eye symptoms are treated with semisolid preparations. Products are also available for certain eye examinations. Vaginal preparations are available for treating genital herpes, yeast infections, and vaginosis caused by bacteria and to reduce menopausal symptoms (e.g., vaginal dryness), and rectal preparations are available for treating minor pain, itching, swelling, and discomfort caused by hemorrhoids and other problems of the anal area. Tables 6 and 7 show some of the commercially available compendial ointment and cream preparations used for treating various topical ailments.

4.2.3 GELS

4.2.3.1 Definition

Gels are semisolid preparations that contain small inorganic particles or large organic molecules interpenetrated by a liquid. Gels made of inorganic materials are usually two-phase systems where small discrete particles are dispersed throughout the dispersion medium. When the particle size of the dispersed phase is larger, they are referred to as magmas. Gels made of organic molecules are single-phase systems, where no apparent physical boundary is seen between the dispersed phase and the dispersion medium. In most cases, the dispersion medium is aqueous. Hydroalcoholic or oleaginous dispersion media are also used in some cases. Unlike dispersed systems like suspensions and emulsions, movement of the dispersed phase is restricted in gels because of the solvated organic macromolecules or interconnecting three-dimensional networks of particles.

Gels are attractive delivery systems as they are simple to manufacture and suitable for administering drugs through skin, oral, buccal, ophthalmic, nasal, otic, and vaginal routes. They also provide intimate contact between the drug and the site of action or absorption. With the advancement in polymer science, gel-based systems that respond to specific biological or external stimuli like pH, temperature, ionic strength, enzymes, antigens, light, magnetic field, ultrasound, and electric current are being designed and evaluated as smart delivery systems for various applications.

4.2.3.2 Characteristics

Gels may appear transparent or turbid based on the type of gelling agent used. They exhibit different physical properties, namely, imbibition, swelling, syneresis, and

TABLE 6 Examples of Compendial/Commercial Ointments

Drug[a]	Category	Indication	Commercial Names	Strength(s) Available (%)
Acyclovir	Antiviral	Genital herpes, herpes infections of the skin, and oral herpes	Zovirax	5
Atropine sulfate (oph)	Mydriatic	Relax muscles in the eye, treat inflammation of certain parts of the eye (uveal tract), and used for certain eye exams	Atropisol, Isopto Atropine	0.5, 1
Bacitracin	First-aid antibiotic	Treat or prevent skin infections	Baciguent Oint, Bacitracin Top	500 units/g
Bacitracin (oph)	Antibiotic	Treat or prevent eye infections	Ak-Tracin, Bacticin	500 units/g
Benzocaine	Antipruritic and local anesthetic	Itching, minor skin wound pain, and insect bites	Americaine	20
Ciprofloxacin (oph)	Antibiotic	Eye infections	Ciloxan	0.3
Clobetasol propionate	Anti-inflammatory agent	Relieve inflammatory and pruritic manifestations of corticosteroid-responsive dermatoses	Temovate Ointment	0.05
Erythromycin	Antibiotic	Treatment of acne	Akne-Mycin	2.0
Erythromycin (oph)	Antibiotic	Infections of eye or ear	Erythromycin Ophthalmic	0.5
Gentamicin sulfate (oph)	Antibacterial	Infections of eye or ear	Gentamicin Sulfate	0.3
Hydrocortisone	Anti-inflammatory agent	Minor pain, itching, swelling, and discomfort caused by hemorrhoids and other problems of anal area	Cortaid, Anusol-HC, Proctosol HC	2.5
Mupirocin	Antibiotic	Treat certain skin infections (e.g., impetigo)	Bactroban	2.0
Sodium chloride (oph)	Miscellaneous	Treat fluid accumulation in cornea of eye causing swelling	Muro-128, Sochlor	2.0

[a]oph: ophthalmic ointment.

TABLE 7 Examples of Compendial/Commercial Creams

Drug[a]	Category	Indication	Commercial Products	Strength(s) Available (%)
Alclometasone dipropionate	Anti-inflammatory	Eczema, dermatitis, allergies, and rash	Aclovate	0.05
Amcinonide	Anti-inflammatory	Lymphomas of the skin, atopic dermatitis, contact dermatitis, and skin rash	Cyclocort	0.1
Amphotericin B	Antifungal antibiotic	Treat skin infection due to a *Candida* yeast and diaper rash	Fungizone	3.0
Anthralin	Keratolytic	Long-term psoriasis	Dritho-Calp, Psoriatec	0.5, 1.0
Betamethasone dipropionate	Anti-inflammatory	Eczema, dermatitis, allergies, and rash	Diprolene AF	0.05
Butoconazole nitrate (vag)	Antifungal	Vaginal yeast infections	Gynazole-1	2.0
Clindamycin phosphate (vag)	Antibiotic	Vaginosis caused by bacteria	Clindesse	2.0
Clobetasol propionate	Anti-inflammatory	Inflammatory and pruritic manifestations of corticosteroid-responsive dermatoses	Temovate E Cream	0.05
Clotrimazole	Antifungal	Ringworm of groin area, athlete's foot, ringworm of the body, fungal infection of the skin with yellow patches, skin infection due to a candida yeast, and diaper rash	Lotrimin	1.0
Crotamiton	Scabicidal and antipruritic	Scabies infection and itching	Eurax	10.0
Desoximetasone	Anti-inflammatory	Eczema, dermatitis, allergies, and rash	Topicort	0.25
Dienestrol	Estrogen	Reduce menopause symptoms such as vaginal dryness	Ortho-Dienestrol	0.01
Diflorasone diacetate	Anti-inflammatory	Eczema, dermatitis, allergies, and rash	Apexicon, Maxiflor	0.05
Fluocinonide	Anti-inflammatory and antipruritic	Psoriasis, eczema, dermatitis, allergies, and rash	Lidex, Vanos	0.05, 0.10
Fluorouracil	Anticancer	Precancerous and cancerous skin growths	Fluoroplex, Carac, Efudex	0.5, 1.0, 5.0

TABLE 7 *Continued*

Drug[a]	Category	Indication	Commercial Products	Strength(s) Available (%)
Halcinonide	Anti-inflammatory and antipruritic	Eczema, dermatitis, allergies, and rash	Halog	0.1
Mometasone furoate	Anti-inflammatory	Eczema, dermatitis, allergies, and rash	Elocon	0.1
Naftifine hydrochloride	Antifungal	Jock itch, athlete's feet, or ringworm	Naftin	1
Nystatin	Antifungal	Fungal skin infections	Mycostatin	100,000 units/g

[a]vag, vaginal cream.

thixotropy. *Imbibition* refers to the uptake of water or other liquids by gels without any considerable increase in its volume. *Swelling* refers to the increase in the volume of gel by uptake of water or other liquids. This property of most gels is influenced by temperature, pH, presence of electrolytes, and other formulation ingredients. *Syneresis* refers to the contraction or shrinkage of gels as a result of squeezing out of dispersion medium from the gel matrix. It is due to the excessive stretching of macromolecules and expansion of elastic forces during swelling. At equilibrium, the system still maintains its physical stability because the osmotic forces of swelling balance the expanded elastic forces of macromolecules. On cooling, the osmotic pressure of the system decreases and therefore the expanded elastic forces return to normal. This results in shrinkage of the stretched molecules and squeezing of dispersion medium from the gel matrix. Addition of osmotic agents such as sucrose, glucose, and other electrolytes helps in retaining higher osmotic pressure even at lower temperatures and avoids syneresis of gels. *Thixotropy* refers to the non-Newtonian flow nature of gels, which is characterized by a reversible gel-to-sol formation with no change in volume or temperature [18].

4.2.3.3 Classification

Gels are classified as hydrogels and organogels based on the physical state of the gelling agent in the dispersion. Hydrogels are prepared with water-soluble materials or water-dispersible colloids. Organogels are prepared using water-insoluble oleaginous materials.

Hydrogels Natural and synthetic gums such as tragacanth, sodium alginate, and pectin, inorganic materials such as alumina, bentonite, silica, and veegum, and organic materials such as cellulose polymers form hydrogels in water. They may either be dispersed as fine colloidal particles in aqueous phase or completely dissolve in water to gain gel structure. Gums and inorganic gelling agents form gel structure due to their viscosity-increasing nature. Organic gelling agents which are generally high-molecular-weight cellulose polymer derivatives produce gel structure because of their swelling and chain entanglement properties. The swollen molecular chains are held together by secondary valence forces, which help in retaining their gel structure. The physical strength of the gel structure is based on the quantity of gelling agent, nature and molecular weight of gelling agent, product pH, and gelling temperature. Generally high-molecular-weight polymers at higher concentrations produce thick gels. The gel-forming temperature (*gel point*) varies with different polymers. Generally natural gums form gel at lower temperatures. Gelatin, a natural protein polymer, forms gel at about 30 °C. If the temperature is increased, gel consistency is not obtained even at higher concentrations of gelatin. On the other hand, polymers such as methylcellulose gain gel structure only when the temperature is above 50 °C due to its decreased solubility and precipitation. Knowledge of the gel point for each gelling agent is therefore essential for preparing physically stable hydrogels.

Organogels Organogels are also known as oleaginous gels. They are prepared using water-insoluble lipids such as glycerol esters of fatty acids, which swell in water and form different types of lyotropic liquid crystals. Widely used glycerol esters of fatty acids include glycerol monooleate, glycerol monopalmito stearete, and glycerol monolinoleate. They generally exist as waxes at room temperature and form cubic liquid crystals in water and increase the viscosity of dispersion. Waxes such as carnauba wax, esparto wax, wool wax, and spermaceti are used in cosmetic organogel preparations. A large quantity of water is entrapped between the three-dimensional lipid bilayers. The equilibrium water content in organogels is about 35%. The structural properties of the lipid, quantity of water in the system, solubility of drug incorporated, and external temperature influence the nature of the liquid crystalline phase. The bipolar nature of organogels allows incorporation of both hydrophilic and lipophilic drugs. Release rates can be controlled by altering the hydrophilic and hydrophobic components. Biodegradability of these waxes by the lipase enzyme in the body makes organogels suitable for parenteral administration.

The water present in the gel framework can be completely removed with some gelling agents. Gelatin sheets, acacia tears, and tragacanth ribbons are generally prepared by removal of water from their respective gel matrix. These dehydrated gel frameworks are called as xerogels.

4.2.3.4 Stimuli-Responsive Hydrogels

The three-dimensional networks of hydrophilic polymers absorb a large quantity of water and form soft structures which resemble biological tissues. Swelling properties of these hydrogels can be altered by various physicochemical parameters. Physical factors such as temperature, pH, and ionic strength of the swelling medium and chemical factors such as the structure of polymer (linear/branched) and chemical modifications (cross-linking) can be altered to tailor their swelling rate. This feature makes them very attractive for drug delivery and biomedical applications [19–23].

pH-Responsive Hydrogels Some polymers show pH-dependent swelling and gelling characteristics in aqueous media. A polymer that exhibits such phase transition properties is very useful from the point of drug delivery. Methacrylic acids (e.g., carbomers) that contain many carboxylic acid groups exist as solution at lower pH conditions. When the pH is increased, they undergo a sol-to-gel transition. This is because of the increase in the degree of ionization of acidic carboxylic groups at higher pH conditions, which in turn results in electrostatic repulsions between chains and, increased hydrophilicity and swelling. Conversely, polymers that contain amine-pendant groups swell at lower pH environment due to ionization and repulsion between polymer chains. The ionic strength of surrounding fluids significantly influences the equilibrium swelling of these pH-responsive polymers. Higher ionic strength favors gel–counter ionic interactions and reduces the osmotic swelling forces.

Thermoresponsive Hydrogels A dispersion which exists as solution at room temperature and transforms into gel on instillation into a body cavity can improve the administration mode and help in modulating the drug release. Many polymers with thermoresponsive gelling properties are currently being synthesized and evaluated. A triblock copolymer that consists of polyethylene glycol–polylactic acid, glycolic acid–polyethylene glycerol (PEG–PLGA–PEG) is solution at room temperature and gels at body temperature. Poloxamers, which are made of triblock poly(ethylene oxide)–poly(propylene oxide)–poly(ethylene oxide), exhibit gelatin properties at body temperatures. Similarly, xyloglucan and xanthan gum aqueous dispersions are solutions at room temperature and become gel at body temperature. These are considered convenient alternatives for rectal suppository formulations which usually cause mucosal irritations due to their physical state. The physicochemical properties of these chemically modified thermoresponsive hydrogels are altered by changing the ratio of hydrophilic and hydrophobic segments, block length, and polydispersity. ReGel by MacroMed contains a triblock copolymer PLGA–PEG–PLGA, undergoes sol-to-gel transition on intratumoral injection, and releases paclitaxel for six weeks.

Ionic-Responsive Hydrogels Administration of sodium alginate aqueous drops into the eye results in alginate gelation due to its interaction with calcium ions in the tear fluid. Alginate with high guluronic acid and deacetylated gellan gum (Gelrite) show sol-to-gel conversions in the eye due to their interaction with cations in the tear fluid. Timolol maleate sterile ophthalmic gel-forming solution (Timoptic-XE) that contains Gelrite gellan gum is commercially available.

4.2.3.5 Gelling Agents

A large number of gelling agents are commercially available for the preparation of pharmaceutical gels. In general, these materials are high-molecular-weight compounds obtained from either natural sources or synthetic pathways. They are water dispersible, possess swelling properties, and improve the viscosity of dispersions. An ideal gelling agent should not interact with other formulation components and should be free from microbial attack. Changes in the temperature and pH during preparation and preservation should not alter its rheological properties. In addition, it should be economic, readily available, form colorless gels, provide cooling sensation on the site of application, and possess a pleasant odor. Based on these factors, gelling agents are selected for different formulations. Table 8 summarizes the

TABLE 8 Some Compendial Gelling Agents Used in Gels

Name	Molecular Weight	Gelling Strength (%)	Synonyms	Official Compendia
Bentonite	359.16	10–20	Magnabrite, mineral soap, Polargel, Veegum HS	BP, JP, PhEur, USPNF
Carbomer	7×10^5–4×10^9	0.5–2.0	Acritamer, Carbopol, polyacrylic acid	BP, PhEur, USPNF
Carboxymethyl cellulose sodium	90,000–700,000	3.0–6.0	Akucell, Aquasorb, Sodium CMC, Tylose CB	BP, JP, PhEur, USPNF
Carrageenan	≈1,000,000	0.3–2.0	Gelcarin, Genu, Hygum Marine colloids	USPNF
Colloidal silicon dioxide	60.08	2.0–10.0	Aerosil, colloidal silica, fumed silica	BP, PhEur, USPNF
Gelatin	15,000–25,0000	10.0–20.0	Cryogel, Solugel	BP, JP, PhEur, USPNF
Glyceryl behenate	1059.8	1.0–15.0	Docosanoic acid, glycerol behenate	BP, PhEur, USPNF
Guar gum	≈220,000	1.0–5.0	Galactosal, Guar flour, Jaguar gum, Meyproguar	BP, PhEur, USPNF
Hydroxypropyl cellulose	50,000–1,250,000	2.0–5.0	Hyprolose, klucel, Methocel	BP, JP, PhEur, USPNF
Hydroxypropylmethyl cellulose	10,000–1,500,000	1.0–10.0	Hypromellose	BP, JP, PhEur, USPNF
Magnesium aluminum silicate	—	5.0–15.0	Veegum, aluminosilicic acid, Carrisorb, Magnabite	BP, PhEur, USPNF
Methylcellulose	10,000–220,000	1.0–5.0	Benecel, Methocel, Metolose	BP, JP, PhEur, USPNF
Poloxamer	2090–17,400	15.0–20.0	Lutrol, Monolan, Pluronic, Supronic	BP, phEur, USPNF
Polyvinyl alcohol	~20,000–200,000	2.5–10.0	Airvol, Elvanol, PVA, vinyl alcohol	USP
Povidone	2500–3,000,000	2.0–20.0	Kollidon, Plasdone, Polyvidone, PVP	BP, JP, PhEur, USPNF
Sodium alginate	20,000–240,000	10.0–20.0	Algin, alginic acid, sodium salt, Protanal	BP, PhEur, USPNF
Tragacanth	840,000	1.0–8.0	Gum Benjamin, Gum dragon, Trag, Tragant	BP, JP, PhEur, USPNF

Note: BP, British Pharmacopoeia; JP, Japanese Pharmacopoeia; PhEur, European Pharmacopoeia; USPNF, U.S. Pharmacopeia/National Formulary molecular weight, gelling strength, synonyms, and compendial status of some of these agents.

The following sections briefly describe the source, physicochemical properties, formulation, and preservation of some pharmacopeial gelling agents.

Alginic Acid Alginic acid is tasteless and odorless and occurs as a yellowish white fibrous powder. The main source for this naturally occurring hydrophilic colloidal polysaccharide is different species of brown sea weed, known as Phaeophyceae. It consists of a mixture of D-mannuronic acid and L-glucuronic acids. It is used in gels due to its thickening and swelling properties. Alginic acid is insoluble in water; however, it absorbs 200–300 times its own weight of water and swells. The viscosity of alginic acid gels can be altered by changing the molecular weight and concentration. Addition of calcium salts increases the viscosity of alginic acid gels. Its viscosity decreases at higher temperature. Depolymerization due to microbial attack also results in viscosity reduction. Inclusion of an antimicrobial preservative avoids depolymerization and viscosity reduction during storage [6].

Bentonite Bentonite is a naturally occurring colloidal hydrated aluminum silicate and contains traces of calcium, magnesium, and iron. It is odorless, available as fine crystalline powder, and is cream to grayish in color. The particles are negatively charged. Its high water uptake and swelling and thickening properties make it suitable for preparing gels. It swells to about 12-fold when it comes in contact with water. The viscosity of bentonite dispersion increases with increase in concentration. The gel-forming properties increase with addition of alkaline materials such as magnesium oxide and decrease with addition of alcohol or electrolytes. Use of hot water and stirring improve wetting and dispersion of bentonite particles in the preparation of the gel. Mixing with magnesium oxide or zinc oxide prior to addition helps in dispersion of bentonite in water. Prior trituration of bentonite with glycerin also helps in easy dispersibility in water. These dispersions are generally left for about 24 h to complete the swelling process. At lower concentration (10%) bentonite suspension exhibits the properties of shear thinning systems and at high concentrations (about 50–60%) it forms gel with finite yield strength [24].

Carbomer Carbomers are one of the widely used gelling agents in topical preparations due to their extensive swelling properties. They are obtained by cross-linking acrylic acid with allyl sucrose or allyl pentaerythritol. Various grades with varying degree of cross-linking and molecular weight are commercially available. Carbomers are generally available as hygroscopic powders, are white in color, and possess a characteristic odor. Presence of about 60% carboxylic acid in its composition makes them acidic. Carbomer 934P, 971P, 974P, and so on, are used for preparing clear gels. Aqueous dispersions of carbomers are usually low viscous, and on neutralization they form high-viscous gels. Basic materials such as sodium hydroxide, potassium hydroxide, sodium bicarbonate, and borax are being used for neutralizing carbomer dispersions. About 0.4 g of sodium hydroxide is used to neutralize 1 g of carbomer dispersion. The viscosity of gels depends on the molecular weight of carbomer and its degree of cross-linking. Inclusion of antioxidants, protection from light, and preservation at room temperature help in retaining their viscosity for prolonged periods. Microbial stability of carbopol gels can be improved by adding antimicrobial preservatives. These gels are prone to discoloration in the presence of large amounts of electrolytes, strong acids, and cationic polymers. Glass, plastic, and resin-lined containers which possess good corrosion-resistant properties are used for packing carbomer gels [6].

Carboxymethylcellulose Calcium (Calcium CMC) A calcium salt of polycarboxymethyl ether of cellulose, calcium CMC is obtained by carboxymethylation of cellulose and conversion into calcium salt. Different molecular grades are prepared by changing the degree of carboxymethylation. It is available as a fine powder, white to yellowish white in color, and hygroscopic in nature. Calcium CMC has swelling and viscosity-enhancing properties in water. It can swell twice its volume in water [25].

Carboxymethylcellulose Sodium (Sodium CMC) A sodium salt of polycarboxymethyl ether of cellulose, sodium CMC is obtained by treating alkaline cellulose with sodium monochloroacetate. It is available as white-colored granular powder. Various viscosity grades of sodium CMC commercially available basically differ in their degree of substitution. The degree of substitution represents the average number of hydroxyl groups that are substituted per anhydroglucose unit. It is readily dispersible in water and forms clear gels. The aqueous solubility of CMC sodium is governed by the degree of substitution. Higher concentrations generally yield thicker gels. Although the viscosity of gels is stable over a wide range of pH (4–10), a fall in pH below 2 or a rise to above 10 results in physical instability and viscosity reduction. Higher viscosity is obtained at neutral pH conditions. Exposure of gels to higher temperature also results in viscosity reduction. Preservation at optimum temperature and inclusion of an antimicrobial preservative improve the physical and microbial stability of CMC sodium gels [25].

Carrageenan Extraction of some red seaweed belonging to the Rhodophyceae class with water or aqueous alkali yields carrageenan. It is a hydrocolloid and mainly contains sodium, potassium, calcium, magnesium, and ammonium sulfate esters of galactose and copolymers of 3, 6-anhydrogalactose. They differ in their ester sulfate and anhydrogalactose content. It is available as a coarse to fine powder which is yellow to brown in color. It is odorless and tasteless. Carrageenan is soluble in hot water and forms gels at 0.3–2.0%. ι-Carrageenan and κ-carrageenan show good gelling properties [26].

Colloidal Silicon Dioxide Colloidal silicon dioxide is a fumed silica obtained by vapor hydrolysis of chlorosilanes. It is available as nongritty amorphous powder which is bluish white in color. It is tasteless and odorless and possesses low tapped density. Although insoluble in water, it readily forms a colloidal dispersion due to its fine particle size, higher surface area, and water-adsorbing properties. The bulk density and particle size of colloidal silicon dioxide can be altered by changing the method of manufacture. Transparent gels can be formed by mixing with other materials that possess similar refractive index. Under acidic and neutral pH conditions, it shows viscosity-increasing properties. This property is lost at higher pH conditions because of its dissolution. Viscosity of gels is not generally affected by temperature [27].

Ethylcellulose Ethylcellulose is a synthetic polymer made of β-anhydroglucose units connected by acetyl linkages. It is obtained by ethylating alkaline cellulose solution with chloroethane. Ethylcellulose is available as a free-flowing powder which is tasteless and white in color. Although it is insoluble in water, it is incorporated into topical preparations due to its viscosity-enhancing properties. Ethanol or a mixture of ethanol and toluene (2:8) is used as a solvent. A decrease in the ratio of alcohol increases the viscosity. The viscosity of the dispersion is increased by increasing the concentration of ethylcellulose or by using a high-molecular-weight material. As ethylcellulose is prone to photo-oxidation at higher temperature, and gels are prepared and preserved at room temperature and dispensed in airtight containers [28].

Gelatin Gelatin is a protein obtained by acid or alkali hydrolysis of animal tissues that contain large amounts of collagen. Based on the method of manufacture, it is named type A or type B gelatin. Type A is obtained by partial acid hydrolysis and type B is obtained by partial alkaline hydrolysis. They differ in their pH, density, and isoelectric point. Gelatin is available as yellow-colored powder or granules. It swells in water and improves the viscosity of dispersions. Different molecular weights and particle size grades are commercially available. Gels can be prepared by dissolving gelatin in hot water and cooling to 35 °C. Temperature greatly influences the viscosity and stability of gelatin dispersions. It transforms to a gel at temperatures above 40 °C and undergoes depolymerization above 50 °C. The viscosity of gelatin gel is also affected by microbes [29].

Guar Gum Guar gum is a high-molecular-weight polysaccharide obtained from the endosperms of guar plant. It mainly contains D-galactan and D-mannan. It is available as powder which is odorless and white to yellowish white in color. It readily disperses in water and forms viscous gels. The viscosity of gel is influenced by the particle size of material, pH of the dispersion, rate of agitation, swelling time, and temperature. Viscosity reduces on long-time heating. Maximum viscosity can be achieved within 2–4 h. Gels are stable at pH between 7 and 9 and show liquification below pH 7. Addition of antimicrobial preservatives improves the microbial stability of guar gum gels. Rheological properties of these gels can be modified by incorporating other plant hydrocolloids such as tragacanth and xanthan gum [30].

Hydroxyethyl Cellulose (HEC) HEC is a partially substituted poly(hydroxyethyl) ether of cellulose. It is obtained by treating alkali cellulose with ethylene oxide. HEC is available as a powder and appears light tan to white in color. Different viscosity grades of HEC are commercially available which differ in their molecular weights. Clear gels are prepared by dissolving HEC in hot or cold water. Dispersions can be prepared quickly by altering the stirring rate of dispersion, temperature, and pH. Slow stirring at room temperature during the initial stages favors wetting. Increasing the temperature at this stage increases the rate of dispersion. Although HEC dispersions are stable over a wide pH range, maintaining basic pH improves the dispersion. The preservation temperature, formulation pH, and microbial attack influence the rheological properties of HEC dispersions. Viscosity reduces at higher temperature, but reverts to the original value on returning to room temperature. Lower and higher pH of the vehicle usually results in hydrolysis or oxidation of HEC, respectively. Some of the enzymes secreted by microbes decrease the viscosity of HEC dispersions. The presence of higher levels of electrolytes may also destabilize HEC dispersions. Inclusion of a suitable antimicrobial preservative is essential to retain the viscosity of HEC gels [31].

Hydroxyethylmethyl Cellulose (HEMC) HEMC is a partially *o*-methylated and *o*-(2-hydroxyethylated) cellulose. It is available as powder or granules which are white, grayish white, or yellowish white in color. Various viscosity grades of HEMC are commercially available, and form viscous colloidal dispersions or gels in cold water which has a pH in the range of 5.5–8 [6].

Hydroxypropyl Cellulose (HPC) HPC is a partially substituted poly(hydroxypropyl) ether of cellulose. It is obtained by treating alkali cellulose with propylene oxide at higher temperatures. It is available as tasteless and odorless powder which is yellowish or white in color. Different viscosity grades are commercially available. Gradual addition of HPC powder into vigorously stirred water yields clear viscous dispersions or gels below 38 °C. Increase in temperature destabilizes the dispersion and leads to precipitation. The viscosity of dispersions can be increased by increasing the concentration of HPC or by using high-molecular-weight grades. Inclusion of a cosolvent such as dichloromethane or methane produces viscous dispersion or gels with modified texture. The viscosity of HPC dispersions can be increased by mixing with an anionic polymer. High concentrations of electrolytes destabilize HPC dispersions. HPC dispersions are neutral in pH (6–8). They undergo acid hydrolysis at lower pH and oxidation at higher pH. Both processes decrease the dispersion viscosity. In addition, certain enzymes produced by microbes degrade HPC and reduce its viscosity. Addition of an antimicrobial preservative is therefore recommended for HPC gels. Preservation of these gels from light can further improve its physical stability [25].

Hydroxypropylmethyl Cellulose (HPMC) HPMC is a partly *o*-methylated and *o*-(2-hydroxypropylated) cellulose obtained by treating alkali cellulose with chloromethane and propylene oxide. It is available as odorless and tasteless granular or fibrous powder which is creamy white or white in color. HPMC is soluble in cold water. Aqueous dispersions are prepared by dispersing material in about 25% hot water (80 °C) under vigorous stirring. On complete hydration of HPMC, a sufficient quantity of cold water is added and mixed. The gel point of HPMC dispersions varies from 50 to 90 °C. Gels are stable over a wide pH range (3–11). The viscosity HPMC dispersions depends on the concentration of material used, its molecular weight, vehicle composition, presence of preservatives, and so on. Viscous gels can be prepared using high concentrations of high-molecular-weight grades. Inclusion of organic solvents such as ethanol or dichloromethane improves the viscosity. Addition of an antimicrobial preservative (e.g., benzalkonium chloride) minimizes microbial spoilage of HPMC gels [25].

Glyceryl Behenate Glyceryl behenate is a mixture of glycerides of fatty acids which is obtained by esterification of glycerin with behenic acid. It may also contain arachidic acid, stearic acid, erucic acid, lignoceric acid, and palmitic acid. It is available as a waxy mass or powder, possesses a faint odor, and is white in color. It is practically insoluble in water and soluble in dichloromethane and chloroform. It is used as a viscosity-increasing agent in silicon gels [6].

Glyceryl Monooleate (GMO) GMO is a mixture of glycerides of fatty acids obtained by esterification of glycerol with oleic acid. It may also contain linoleic acid, palmitic acid, stearic acid, linolenic acid, arachidic acid, and eicosenoic acid. It is available as a partially solidified or oily liquid. GMO is insoluble in water. Self-emulsifying grades that contain an anionic surfactant disperse easily and swell in water. The nonemulsifying grades are used as emollients in topical preparations and self-emulsifying grades are used as emulsifiers in aqueous emulsions [6].

Magnesium Aluminum Silicate (MAS) MAS is also known as veegum. It is a polymeric complex of magnesium, aluminum, silicon, oxygen, and water and is obtained from silicate ores. Based on the ratio of aluminum and magnesium and viscosity, it is classified as types IA, IB, IC, and IIA. It is available as fine powder that is odorless, tasteless, and off-white to creamy white in color. Although MAS is insoluble in water, it swells to a large extent and produces viscous colloidal dispersions. Use of higher concentration, addition of electrolytes, and heating of dispersion usually improve the viscosity [32].

Methylcellulose (MC) MC is a long-chain cellulose polymer with methoxyl substitutions at positions 2, 3, and 6 of the anhydroglucose ring. It is synthesized by methylating alkali cellulose with methyl chloride. The degree of substitution of methoxy groups influences the molecular weight, viscosity, and solubility characteristics of MC. It is available as powder or granules and is odorless, tasteless, and white to yellowish white in color. MC is insoluble in hot water but slowly swells and forms viscous colloidal dispersions in cold water. Gels can be prepared by initially mixing the methylcellulose with half the volume of hot water ($\approx 70\,^{\circ}C$) followed by addition of the remaining volume of cold water. Viscosity of these dispersions can be increased by using high-concentration or high-molecular-weight grades of methylcellulose. Higher processing or preservation temperatures reduce the viscosity of formulations, which regain their original state on cooling to room temperature. MC aqueous dispersions show pH values of 5–8. Reduction in pH to less than 3 leads to acid-catalyzed hydrolysis of glucose–glucose linkages and results in low viscosity. Antimicrobial preservatives are generally included to enhance the microbial stability of dispersions. Salting out is observed when high concentrations of electrolytes are added to methylcellulose dispersions. The viscosity of methylcellulose dispersions is also influenced by the presence of formulation excipients and drugs [25].

Poloxamer Poloxamers are copolymers of ethylene oxide and propylene oxide. Different molecular weight grades that are different in physical form, solubility, and melting point are available. Poloxamer 124 is a colorless liquid, whereas poloxamers 188, 237, 338, and 407 are solids at room temperature. All poloxamer grades are freely soluble in water and form gels at higher concentrations. The pH of aqueous liquids ranges between 5 and 7.5 [33].

Polyethylene Oxide Polyethylene oxide is a nonionic homopolymer of ethylene oxide synthesized by polymerization of ethylene oxide. It is available as a free-flowing powder white to off-white in color with a slight ammonia odor. Various molecular weight grades of polyethylene oxide are commercially available. They swell in water and form viscous dispersions or gels based on the concentration and grade used. Inclusion of alcohol improves the rheological stability of polyethylene oxide dispersions [6].

Polyvinyl Alcohol (PVA) PVA is a synthetic polymer prepared by hydrolysis of polyvinyl acetate. It is available as a granular powder which is odorless and white in color. Mixing with water at room temperature, heating for about 5 min at 90 °C, followed by cooling with constant mixing yield aqueous dispersions or gels. Higher viscosities can be obtained by using high-viscosity grades. Addition of borax improves the gelling properties of PVA, whereas inorganic salts destabilize these dispersions. The pH of PVA dispersions ranges between 5 and 8. Physical and chemical decompositions occur at lower and higher pH conditions. Incorporation of an antimicrobial preservative and storage at room temperature improve its stability [6].

Povidone Povidone is a synthetic polymer consisting of 1-vinyl-2-pyrrolidinone units. It is available as a fine powder and appears white to creamy-white in color. Various molecular weight grades of povidone are available which differ in their degree of polymerization. Povidone is soluble in water and forms viscous solutions and gels based on the concentration and viscosity grade used. Decomposition occurs when dispersions are heated to about 150 °C. The pH of aqueous dispersions ranges from 3 to 7. The microbial stability of povidone aqueous dispersions can be increased by adding preservatives [6].

Propylene Carbonate (PC) PC is prepared by reacting propylene chlorohydrin with sodium bicarbonate. It is available as a clear liquid with a faint odor. Mixtures of PC and propylene glycol are good solvents for corticosteroids in topical preparations. It is incompatible with strong acids, bases, and amines. The pH of 10% aqueous dispersion is 6.0–7.5 [34].

Propylene Glycol Alginate (PGA) PGA is a propylene glycol ester of alginic acid obtained by treating alginic acid with propylene oxide. It is available as granular or fibrous powder which is odorless, tasteless, and white to yellowish-white in color. PGA is soluble in water and forms viscous colloidal dispersions. The viscosity of these dispersions is based on the concentration of PGA, temperature, and pH. Its aqueous solubility decreases at higher temperatures. The aqueous dispersions are acidic in nature and more stable at pH 3–6. Higher pH leads to saponification. As these dispersions are prone to microbial spoilage, antimicrobial preservatives are generally included [6].

Sodium Alginate Sodium alginate is obtained by extraction of alginic acid from brown seaweed followed by neutralization with sodium bicarbonate. Alginic acid is composed of D-mannuronic acid and L-guluronic acid. It is available as a powder which is tasteless, odorless, and white to yellowish-brown in color. Sodium alginate forms viscous gels in water. Dispersing agents such as glycerol, propylene glycol, sucrose, and alcohol are added to improve dispersion. The presence of low concentration of electrolytes improves the viscosity, whereas at high concentrations salting out takes place. The viscosity of gel is based on the concentration of sodium alginate, temperature, pH, and other additives. Various viscosity grades of sodium alginate are commercially available. Aqueous dispersions are stable at pH 4–10. Precipitation or decrease in viscosity is observed when the pH is below or above these values. Autoclaving or heating above 70 °C results in depolymerization and decrease in viscosity. Inclusion of preservative is essential to maintain the microbial stability of sodium alginate topical gels [35].

Tragacanth Tragacanth is a polysaccharide polymer obtained from some *Astragalus* species. It is composed of two polysaccharides: bassorin (water insoluble) and tragacanthin (water soluble). It is available as odorless powder white to yellowish in color and possesses mucilaginous taste. Tragacanth swells about 10 times its weight in water and forms viscous solutions or gels. Tragacanth is usually added with vigorous stirring to avoid lump formation. Wetting agents such as glycerin, propylene glycol, and 95% ethanol are used in initial stages to improve wetting and dispersion of tragacanth in water. The viscosity of tragacanth dispersions is influenced by the processing temperature and formulation pH. High temperature usually increases the viscosity of gels. Tragacanth dispersions show higher viscosity at pH 8 and starts decreasing at higher pH. These gels usually contain preservatives such as benzoic acid or a combination of methyl and propyl parabens for effective preservation from microbial attack. The viscosity of tragacanth dispersions reduces in the presence of strong mineral and organic acids and sodium chloride [6].

4.2.3.6 Preparation and Packaging

Gels are relatively easier to prepare compare to ointments and creams. In addition to the gelling agent, medicated gels contain drug, antimicrobial preservatives, stabilizers, dispersing agents, and permeation enhancers. Some of the factors discussed below are essential to obtain a uniform gel preparation.

Order of Mixing The order of mixing of these ingredients with the gelling agent is based on their influence on the gelling process. If they are likely to influence the rate and extent of swelling of the gelling agent, they are mixed after the formation of gel. In the absence of such interference, the drug and other additives are mixed prior to the swelling process. In this case, the effects of mixing temperature, swelling duration, and other processing conditions on the physicochemical stability of the drug and additives are also considered. Ideally the drug and other additives are dissolved in the swelling solvent, and the swelling agent is added to this solution and allowed to swell.

Gelling Medium Purified water is the most widely used dispersion medium in the preparation of gels. Under certain circumstances, gels may also contain cosolvents or dispersing agents. A mixture of ethanol and toluene improves the dispersion of ethylcellulose, dichloromethane and methanol increase the viscosity of hydroxypropyl cellulose dispersions, alcohol improves the rheological stability of polyethylene oxide gels, and inclusion of glycerin, propylene glycol, sucrose, and alcohol improves the dispersion of sodium alginate dispersions. Borax is included in polyvinyl alcohol gels and magnesium oxide, zinc oxide, and glycerin are included in bentonite gel as dispersing agents. Care should be taken to avoid the evaporation or degradation of these cosolvents and dispersing agents during the preparation of gels.

Processing Conditions and Duration of Swelling The processing temperature, pH of the dispersion, and duration of swelling are critical parameters in the preparation of gels. These conditions vary with each gelling agent. For instance, hot water is preferred for gelatin and polyvinyl alcohol, and cold water is preferred for methylcellulose dispersions. Carbomers, guar gum, hydroxypropyl cellulose, poloxamer, and tragacanth form gels at weakly acidic or near-neutral pH conditions (pH 5–8). Gelling agents such as carboxymethyl cellulose sodium, hydroxypropylmethyl cellulose, and sodium alginate form gels over a wide pH range (4–10). Hydroxyethyl cellulose forms gel at alkaline pH condition. A swelling duration of about 24–48 h generally helps in obtaining homogeneous gels. Natural gums need about 24 h and cellulose polymers require about 48 h for complete hydration.

Removal of Entrapped Air Entrapment of air bubbles in the gel matrix is a common issue, especially when the swelling process involves a mixing procedure or the drug and other additives are added after the swelling process. Positioning the propeller at the bottom of the mixing container minimizes this issue to a larger extent. Further removal of air bubbles can be achieved by long-term standing, low-temperature storage, sonication, or inclusion of silicon antifoaming agents. In large-scale production, vacuum vessel deaerators are used to remove the entrapped air.

Packaging Being viscous and non-Newtonian systems, gels need high attention during packing into containers. Usually they are packed into squeeze tubes or jars made of plastic materials. Aluminum containers are also used when the product pH is slightly acidic. Pump dispensers and prefilled syringes are sometimes used for packing gels. As most of the gels contain an aqueous phase, preservation in airtight containers helps in protecting them from microbial attack. Usually they are preserved at room temperature and protected from direct sunlight and moisture.

In large-scale production, different mills, separators, mixers, deaerators, shifters, and packaging machines are used. Most of this equipment is similar to those discussed under ointments and creams. Figure 3 shows a "one-bowl" vacuum processing machine manufactured by FrymaKoruma-Rheinfelden (Romaco) for the preparation gels. Batch sizes ranging from 15 to 160 L are processed using this machine. It uses an extremely efficient high-shear rotor/stator system for homogenizing and a counterrotating mixing system for macromixing. The raw materials are drawn into the multichamber system of the homogenizer by vacuum and then mixed and pumped into the homogenizing zone. The product which enters the vessel is mixed, sheared, and recirculated. All the entrapped air is removed during the recirculation. The machine also has an insulated jacket for controlling the processing temperature.

4.2.3.7 Evaluation

Various pharmacopeial and nonpharmacopeial tests are carried out to evaluate the physicochemical, microbial, in vitro, and in vivo characteristics of gels. These tests are meant for assessing the quality of gel formulations and minimizing the batch-

FIGURE 3 Vacuum processing machine used for preparation of gels. (Courtesy of FrymaKoruma-Rheinfelden, Switzerland.)

to-batch variations. Some of the tests recommended by the USP for gels include minimum fill, pH, viscosity, microbial screening, and assay. In some cases sterility and alcohol content are also specified. The USP also recommends storage for each compendial gel formulation. Table 9 shows the quality control tests and storage requirements that are specified for some pharmacopeial gels. The procedures for minimum fill, microbial screening, sterility, assay, in vitro drug release, and in vivo bioequivalence are similar to those of ointments and creams. The procedures for additional tests such as homogeneity, surface morphology, pH, alcohol content, rheological properties, bioadhesion, stability, and ex vivo penetration are described below.

Homogeneity and Surface Morphology The homogeneity of gel formulations is usually assessed by visual inspection and the surface morphology by using scanning electron microscopy. Generally, the swollen gel is allowed to freeze in liquid nitrogen and then lyophilized by freeze drying. It is assumed that the morphologies of the swollen samples are retained during this process. The lyophilized hydrogel is gold sputter coated and viewed under an electron microscope.

pH Many gelling agents show pH-dependent gelling behavior. They show highest viscosity at their gel point. Determination of pH is therefore important to maintain consistent quality. As conventional pH measurements are difficult and often give erratic results, special pH electrodes are used for viscous gels. Flat-surface digital

TABLE 9 USP Specifications for Some Official Gels

Drug	Quality Control Tests	Packaging and Storage Requirements
Aminobenzoic acid	Minimum fill, pH (4.0–6.0), alcohol content, and assay	Tight, light-resistant containers
Benzocaine	Microbial limits, minimum fill, and assay	Well-closed containers
Benzoyl peroxide	pH (2.8–6.6), and assay	Tight containers
Betamethasone benzoate	Microbial limits, minimum fill, and assay	Tight containers; store at 25 °C, excursions permitted between 15 and 30 °C; protect from freezing
Clindamycin phosphate	Minimum fill, pH (4.5 and 6.5), and assay	Tight containers
Desoximetasone	Minimum fill, alcohol content, and assay	Collapsible tubes at controlled room temperature
Dexamethasone	Minimum fill and assay	Collapsible tubes; keep tightly closed; avoid exposure to temperatures exceeding 30 °C
Dyclonine hydrochloride	pH (2.0 and 4.0), and assay	Collapsible, opaque plastic tubes or in tight, light-resistant glass containers
Erythromycin	Minimum fill and assay	Tight containers
Fluocinonide	Minimum fill and assay	Collapsible tubes or tight containers
Hydrocortisone	Minimum fill and assay	Tight containers
Lidocaine hydrochloride	Sterility, minimum fill, pH (7.0–7.4), and assay	Tight containers
Metronidazole	Minimum fill, pH (4.0 and 6.5), and assay	Laminated collapsible tubes at controlled room temperature
Naftifine hydrochloride	Microbial limits, minimum fill, pH (5.5–7.5), content of alcohol, and assay	Tight containers
Salicylic acid	Alcohol content and assay	Collapsible tubes or tight containers, preferably at controlled room temperature
Sodium sulfide	pH (11.5–13.5) and assay	Tight containers at controlled room temperature or in a cool place
Stannous fluoride	Viscosity, pH (2.8–4.0), stannous ion content, total tin content, and assay	Well-closed containers
Tolnaftate	Minimum fill and assay	Tight containers

pH electrodes from Crison, combination electrodes that contain a built-in temperature probe, a bridge electrolyte chamber and movable sleeve junction from Mettler, and combination pH puncture electrodes with spear-shaped tip from Mettler are some commercially available pH measurement systems for semisolid formulations.

Alcohol Content Alcohol levels in some gel preparations are determined by gas chromatographic (GC) methods. Desoxymetasone gel USP and naftifine hydrochloride gel USP contain 18–24% and 40–45% (w/w) of ethyl alcohol, respectively. In a desoxymetasone gel, the sample is dissolved in methanol and injected into a gas chramatograph for quantitative analysis. Isopropyl alcohol is used as an internal standard. In naftifine hydrochloride gel, *n*-propyl alcohol is used as an internal standard and water is used as the diluting solvent [15].

Rheological Studies Viscosity measurement is often the quickest, most accurate, reliable method to charactreize gels. It gives an idea about the ease with which gels can be processed, handled, or used. Some of the commonly used tests for characterizing rheology of gels are yield stress, critical strain, and creep. *Yield stress* refers to the stress that must be exceeded to induce flow. This helps in characterizing the flow nature of non-Newtonian systems. *Critical strain* or *gel strength* refers to the minimum energy needed to disrupt the gel structure. When samples are subjected to increasing stress, viscosity is maintained as long as the gel structure is maintained. When the gel's intermolecular forces are overcome by the oscillation stress, the sample breaks down and the viscosity falls. The higher the critical strain, the better the physical integrity of gel systems. *Creep* or *recovery* helps in assessing the strength of bonds in a gel structure. This is assessed by determining relaxation times, zero-shear viscosity, and viscoelastic properties.

Based on the nature of the test material, different techniques are employed to measure the rheological parametrs of gels. Very sophisticated automatic equipment is commercially available for measurements. Cup-and-bob viscometers and cone-and-plate viscometers are widely used for viscous liquids and gels. They measure the frictional force that is created when gels start flowing. These viscometers are usually calibrated with certified viscosity standards before each measurement. General-purpose silicone fluids which are less sensitive to temperature change are used as standards. The viscosity of gels is affected by the experimental temperature and shear rate and the gels exhibit liquid- or solidlike properties. Hence the viscosity of these non-Newtonian systems are recorded at several shear rates under controlled temperatures. The USP specifies the operating conditions for each gel formulation. Commercially available viscometers include Brookfield rotational viscometers, Haake rheometers, Schott viscoeasy rotational viscometers, Malvern viscometers, and Ferranti-Shirley cone-and-plate viscometers.

Bioadhesion This test is performed to assess the force of adhesion of a gel with biological membranes. The bioadhesive property is preferred for ophthalmic, nasal buccal, and gastroretentive gel formulations. Drugs applied as solutions, viscous solutions, and suspensions drain out from these biological locations within a short time and only a limited fraction of drug elicits the pharmacological activity. Products with higher bioadhesion thus help in increasing the contact time between drugs and absorbing surface and improve their availability. The bioadhesive properties of gels are measured using various custom-designed equipment. All the equipment, however, measures the force required to detach the gel from a biological surface under controlled experimental conditions (e.g., temperature, wetting level, contact time, contact force, surface area of tissue). A typical bioadhesion measurement system consists of a moving platform and a static platform. A tissue from a particular biological region is fixed onto these platforms and a known quantity of the test product is uniformly applied to the tissue surface of the lower static platform. The upper moving platform is allowed to contact with the product surface with a known contact force. After allowing for a short contact time, the moving platform is separated from the product with a constant rate. The force required to detach the mucosal surface from the product is recorded. The analog signals generated by precision load cells are then converted to digital signals through data acquisition systems and processed using specific software programs.

Stability Studies Being dispersed systems containing water in their matrix, gels are prone to physical, chemical, and microbial stability issues. Syneresis is a commonly observed physical stability problem with gels. It involves squeezing out dispersion medium due to elastic contraction of polymeric gelling agents. This results in shrinkage of gels. Syneresis can be determined by heating the gels to a higher temperature followed by rapid cooling using an ice water bath at room temperature. The sample is preserved at 4 °C for about a week, and water loss from the gel matrix is measured. Water loss is measured by weighing the mass of the gel matrix after centrifugation. Absence of syneresis indicates higher physical stability of gels. The chemical stability of drugs in the gel matrix is determined using stability-indicating analytical methods. Studies are conducted at accelerated temperature, moisture, and light conditions to determine the possible degradation of drug in the gel.

Ex Vivo Penetration Ex vivo studies are carried out to examine the permeation of drug from gels through the skin or any other biological membrane. As with in vitro release studies, ex vivo penetration is conducted using vertical diffusion cells or modified cells with flow-through design. In this case, the receiver side is filled with phosphate buffer solution of pH 7.4 to simulate the biological pH of human blood. Skin samples from different animal sources such as rats, rabbits, pigs, and human cadavers are used for screening dermatological products. The stratum corneum layer of the skin is separated from the dermis before mounting onto the diffusion cells. The epidermis is separated by immersing the skin sample in normal saline or purified water which is maintained at 60 °C for 2 min followed by immersion into cold water for 30 s. Careful peeling helps in the separation of the epidermis layer from the dermis. This layer is mounted between the donor and receiver sides and studies are conducted after application of test gel over the surface of the stratum corneum in the donor side. Samples are withdrawn at different time intervals and analyzed for drug permeation by suitable analytical techniques.

4.2.3.8 Typical Pharmacopeial and Commercial Examples

Gels are becoming popular dosage forms for delivering various categories of drugs for treating dermatological, oral, ophthalmic, vaginal, and other conditions. Many dermatological gels are used for treating mild to moderate acne, eczema, dermatitis, allergies, rash, and psoriasis and for removal of common warts. Oral gels are available for relieving painful mouth sores, treating tooth decay, preventing tooth plaque, and relieving inflammation of the gums, and vaginal gels are available for treating certain type of vaginal infections (e.g., bacterial vaginosis). Some special types of gels are available for preventing or controlling pain during certain medical procedures, numbing and treating urinary tract inflammation (urethritis), and numbing mucous membranes. Table 10 shows some of the commercially available compendial gels.

TABLE 10 Examples of Compendial/Commercial Gels

Drug[a]	Category	Indication	Commercial Products	Strength (%)
Benzocaine	Local anesthetic	In mouth to relieve pain or irritation caused by many conditions	Oratect Gel, Num Zit Gel	7.5, 10
Benzoyl peroxide	Keratolytic	Mild to moderate acne	Persa-Gel, 5 Benzagel 10	5.0, 10
Betamethasone	Anti-inflammatory	Eczema, dermatitis, allergies, and rash	Diprolene	0.05
Clindamycin phosphate (vag)	Antibiotic	Certain types of vaginal infection (e.g., bacterial vaginosis)	Cleocin T	1.0
Desoximetasone	Anti-inflammatory	Eczema, dermatitis, allergies, and rash.	Topicort	0.05
Dyclonine hydrochloride	Antipruritic and local anesthetic	Relieve painful mouth sores	Dyclone	0.5, 1.0
Erythromycin	Antibiotic	Acne and skin infection due to bacteria	Erygel	2.0
Fluocinonide	Anti-inflammatory	Psoriasis, eczema, dermatitis, allergies, and rash	Lidex	0.05
Lidocaine hydrochloride	Local anesthetic	Prevent and control pain during certain medical procedures, numb and treat urinary tract inflammation (urethritis), and numb mucous membranes	Xylocaine, Anestacon	2.0
Metronidazole (vag)	Antifungal	Certain types of bacterial infections in the vagina	Metrogel	0.75, 1.0
Naftifine hydrochloride	Antifungal	Fungal infections of skin such as jock itch, athlete's feet, or ringworm	Naftin	1.0
Salicylic acid	Keratolytic	Removal of common warts	Sal-Plant Gel	17.0
Stannous fluoride	Fluoride	Treat tooth decay, prevent tooth plaque and inflammation of gums	Flo-Gel, Gel-Kam	0.4
Tolnaftate	Antifungal	Skin infections such as athlete's foot, jock itch, and ringworm	Tolnaftate	1.0

[a]vag: vaginal gel.

4.2.4 REGULATORY REQUIREMENTS FOR SEMISOLIDS

Regulatory agencies such as the Center for Drug Evaluation and Research (CDER) have at the Food and Drug Administration (FDA) have set forth guidelines for various pharmaceutical activities to ensure the identity, strength, quality, safety, and efficacy of semisolid drug products. A manufacturer of semisolid formulations needs to fulfill these requirements at the time of filing for investigational new drug (IND), abbreviated new drug application (ANDA), or abbreviated antibiotic drug application (AADA). Standard chemistry, manufacturing, and control (CMC) tests are necessitated for all dermatological drug products. Additional information on polymorphic form, particle size distribution, and other characteristics is needed for submitting an NDA. When an ANDA for a semisolid product is filed, the manufacturer should meet the standards of compendial requirements if available and match the important in vitro and in vivo characteristics of the reference listed drug (RLD). If such information is not available, appropriate in vitro release methods are submitted to ensure batch-to-batch consistency. Even at later stages, if changes are made for an approved semisolid product with respect to its components, composition, equipment, process, batch size, and manufacturing site, the formulator should submit necessary details to the regulatory agency.

A typical guideline that defines the types and levels of scale-up and postapproval changes (SUPAC) is outlined in Table 11. Based on the type and level of change, the manufacturer needs to submit application and compendial product release requirements, executed batch records, accelerated and long-term stability data, identification and assay for new preservative, preservative effectiveness test at lowest specified level, validation methods to support absence of interference of preservative with other tests, in vitro release test, and in vivo bioequivalence data to the FDA. When changes are made with respect to the quality and quantity of excipients or crystallinity of drug, especially if the drug is in suspension, in vivo bioequivalence studies are recommended. As routine pharmacokinetic studies do not produce measurable quantities of drug in blood, plasma, urine, and other extracutaneous biological fluids, dermatopharmacokinetic (DPK) studies and pharmacodynamic or comparative clinical trials are recommended to establish bioequivalence of topical products. Table 12 shows specific requirements for various SUPAC levels. If bioavailability or bioequivalence data of a highest strength product are already available,

TABLE 11 Description on SUPAC Guidelines for Nonsterile Semisolid Dosage Forms

Type of Change	Level	Description
Change in components and composition	1	Partial deletion or deletion of color, fragrance, or flavor; up to 5% excipient change in approved amount; change in supplier for structure forming or technical-grade excipient
	2	Excipient changes from 5 to 10%; change in supplier for structure forming excipient, which is not covered under level 1; change in technical grade of structure forming excipient; change in particle size of drug if drug is in suspension
	3	Qualitative and quantitative changes in excipient not covered under levels 1 and 2; any change in crystallinity of drug if drug is in suspension

TABLE 11 ***Continued***

Type of Change	Level	Description
Change in preservative components and composition	1	Less than 10% quantitative change in preservative
	2	10–20% quantitative change in preservative
	3	Deletion or more than 20% quantitative change in preservative; inclusion of a different preservative
Change in manufacturing equipment	1	Change to automated or mechanical equipment for transfer of ingredients; use of alternative equipment of same design and operating principles
	2	Use of alternative equipment of different design and operating principles; change in type of mixing equipment
Change in manufacturing process	1	Changes in process within approved application ranges; addition of formulation additives
	2	Changes in process outside approved application ranges; process of combining phases
Change in batch size	1	Batch size changes upto 10 times of pivotal clinical trial or biobatch
	2	Batch size changes above 10 times of pivotal clinical trial or biobatch
Change in manufacturing site	1	Changes within existing facility
	2	Changes within same campus or facilities in adjacent city blocks
	3	Change to different campus; change to a contract manufacturer

TABLE 12 SUPAC Requirements for Nonsterile Semisolid Dosage Forms

		Requirements[a]										
Parameter	Change Level	A	B	C	D	E	F	G	H	I	J	K
Change in components and composition	1	•		•								
	2	•	•		•			•				
	3	•	•			•		•	•			
Change in preservative components and composition	1	•								•		
	2	•								•		
	3	•	•		•					•	•	
Change in manufacturing equipment	1	•		•								
	2	•	•				•	•				
Change in manufacturing process	1	•										
	2	•	•				•	•				
Change in batch size	1	•	•	•								
	2	•	•		•							

TABLE 12 ***Continued***

Parameter	Change Level	Requirements[a] A	B	C	D	E	F	G	H	I	J	K
Change in manufacturing site	1	•										
	2	•	•	•								•
	3	•	•			•		•				•

[a]Only those highlighted with black circles:

A: Application/compendial product release requirement.
B: Executed batch records.
C: long-term stability data for first production batch.
D: 3-month accelerated stability data for 1 batch and long-term data for first production batch.
E: 3-month accelerated stability data for 1 batch and long-term data for first 3 production batches if significant information is available or 3-month accelerated stability data for 3 batches and long-term data for first 3 production batches if significant information is not available.
F: 3-month accelerated stability data for 1 batch and long-term data for first production batch if significant information is available or 3-month accelerated stability data for 3 batches and long-term data for first 3 production batches if significant information is not available.
G: In vitro release test.
H: In vivo bioequivalence.
I: Preservative effectiveness test at lowest specified preservative level.
J: Identification and assay for new preservative; validation methods to support absence of interference with other tests.
K: Location of new site.

in vitro release data are used to evaluate the in vivo bioequivalence of a lower strength product [16].

REFERENCES

1. Ansel, H. C., Allen, L. V., and Popovich, N. G. (1999), *Pharmaceutical Dosage Forms and Drug Delivery Systems*, 7th ed., Lippincott Williams and Wilkins, Philadelphia, pp. 245–250.
2. Breit, J., and Bandmann, H.J. (1973), Dermatitis from lanolin, *Br. J. Dermatol.*, 88, 414–416.
3. Smolinske, S. C. (1992), *Handbook of Food, Drug, and Cosmetic Excipients*, CRC Press, Boca Raton, FL, pp. 225–229.
4. Barker, G. (1977), New trends in formulating with mineral oil and petrolatum. *Cosmet. Toilet.* 92(1), 43–46.
5. Davis, S. S. (1969), Viscoelastic properties of pharmaceutical semisolids I: Ointment bases. *J. Pharm. Sci.*, 58, 412–418.
6. Rowe, R. C., Sheskey, P. J., and Weller, P. J. (2003), *Handbook of Pharmaceutical Excipients*, 4 ed., Pharmaceutical Press and American Pharmaceutical Association, IL, pp. 16–18, 89–92, 260–263, 287–288, 417–418, 491–492, 508–513, 524–525, 618–619, 654–656, 679–684.
7. Hadia, I. A., Ugrine, H. E., Farouk, A. M., and Shayoub, M. (1989), Formulation of polyethylene glycol ointment bases suitable for tropical and subtropical climates I. *Acta Pharm. Hung.*, 59, 137–142.

8. Fisher, A. A. (1978), Immediate and delayed allergic contact reactions to polyethylene glycol, *Contact Dermatitis*, 4, 135–138.
9. Mores, L. R. (1980), Application of stearates in cosmetic creams and lotions, *Cosmet. Toilet.* 95(3), 79–84.
10. Mapstone, G. E. (1974), Crystalization of cetyl alcohol from cosmetic emulsions, *Cosmet. Perfum.*, 89(11), 31–33.
11. Eccleston, G. M. (1984), Properties of fatty alcohol mixed emulsifiers and emulsifying waxes, in Florence, A. T., Ed., *Materials Used in Pharmaceutical Formulations: Critical Reports on Applied Chemistry*, Vol. 6, Blackwell Scientific, Oxford, UK, pp. 124–156.
12. Kline, C. H. (1964), Thixcin, R-Thixotrope, *Drug Cosmet. Ind.*, 95(6), 895–897.
13. Cronin, E. (1967), Contact dermatitis from cosmetics, *J. Soc. Cosmet. Chem.*, 18, 681–691.
14. Forcinio, H. (1998), Tubes: The ideal packaging for semisolid products, *Pharm. Tech.*, 22, 32–36.
15. *The United States Pharmacopeia/The National Formulary*, 28th/23rd ed., U.S. Pharmacopeial Convention, Rockville, HD, 2005, pp. 2246–2255, 2378, 2434–2435, 2441, 2510 2512.
16. Guidance for industry, Nonsterile semisolid dosage forms: Scale-up and postapproval changes, U.S. Food and Drug Administration, May 1997, pp. 1–37.
17. Shah, V. P., Glynn, G. L., and Yacobi, A. (1998), Bioequivalence of topical dosage forms—methods of evaluation of bioequivalence, *Pharm. Res.*, 15, 167–171.
18. Allen, L. V. (2002), *The Art, Science, and Technology of Pharmaceutical Compounding*, American Pharmaceutical Association, Washington, DC, pp. 301–312.
19. Jeong, B., Bae, Y. H., and Kim, S. W. (1999), Thermoreversible gelation of PEG-PLGA-PEG triblock copolymers aqueous solutions, *Macromolecules*, 32, 7064–7069.
20. Miyazaki, S., Suisha, F., Kawasaki, N., Shirakawa, M., Yamatoya, K., and Attwood, D. (1998), Thermally reversible xyloglucan gels as vehicles for rectal drug delivery, *J. Controlled Release*, 56, 75–83.
21. Watanabe, K., Yakou, S., Takayama, K., Isowa, K., and Nagai, T. (1996), Rectal absorption and mucosal irritation of rectal gels containing buprenorphine hydrochloride prepared with water-soluble dietary fibers, xanthan gum and locust bean gum, *J. Controlled Release*, 38, 29–37.
22. Cohen, S., Lobel, E., Trevgoda, A., and Peled, T. (1997), A novel in situ—forming ophthalmic drug delivery system from alginates undergoing gelation in the eye, *J. Controlled Release*, 44, 201–208.
23. Carlfors, J., Edsman, K., Petersson, R., and Jornving, K. (1998), Rheological evaluation of Gelrite in situ gels for ophthalmic use, *Eur. J. Pharm. Sci.*, 6, 113–119.
24. Altagracia, M., Ford, I., Garzon, M. L., and Kravzov, J. (1987), A comparative mineralogical and physico-chemical study of some crude Mexican and pharmaceutical grade montmorillonites, *Drug Dev. Ind. Pharm.*, 13, 2249–2262.
25. Doelker, E. (1993), Cellulose derivatives, *Adv. Polym. Sci.*, 107, 199–265.
26. Lev, R., Long, R., and Mallonga, L. (1997), Evaluation of carrageenan as a base for topical gels, *Pharm. Res.*, 14(11), 42.
27. Daniels, R., Kerstiens, B., Tishinger-Wagner, H., and Rupprecht, H. (1986), The stability of drug absorbates on silica, *Drug Dev. Ind. Pharm.*, 12, 2127–2156.
28. Ruiz-Martinez, A., Zouaki, Y., and Gallard-Lara, V. (2001), In vitro evaluation of benzylsalicylate polymer interaction in topical formulation, *Pharm. Ind.*, 63, 985–988.
29. Ling, W. C. (1978), Thermal degradation of gelatin as applied to processing of gel mass, *J. Pharm. Sci.*, 67, 218–223.

30. Goldstein, A. M., Alter, E. N., and Seaman, J. K. (1973), Guar gum, in Whistler, R. L., Ed., *Industrial Gums*, 2nd ed., Academic, New York, pp. 303–321.
31. Gauger, L. J. (1984), Hydroxyethylcellulose gel as a dinaprostone vehicle, *Am. J. Hosp. Pharm.*, 41, 1761–1762.
32. Farley, C. A., and Lund, W. (1976), Suspending agents for extemporaneous dispensing: Evaluation of alternatives to tragacanth, *Pharm. J.*, 216, 562–566.
33. Cabana, A., Ait-Kadi, A., and Juhasz, J. (1997), Study of the gelation process of polyethylene oxide copolymer (poloxamer 407) aqueous solutions, *J. Colloid Interface Sci.*, 190, 307–312.
34. Burdick, K. H., Haleblian, J. K., Poulsen, B. J., and Cobner, S. E., (1973), Corticosteroid ointments: Comparison by two human bioassays, *Curr. Ther. Res.*, 15, 233–242.
35. Pavics, L. (1970), Comparison of rheological properties of mucilages, *Acta Pharm. Hung.*, 40, 52–59.

4.3

LIQUID DOSAGE FORMS

Maria V. Rubio-Bonilla[1], Roberto Londono[1], and Arcesio Rubio[2]

[1]*Washington State University, Pullman, Washington*

[2]*Caracas, Venezuela*

Contents

Pharmaceutical Manufacturing Handbook: Production and Processes, edited by Shayne Cox Gad

4.3.1 INTRODUCTION

Liquid dosage forms are designed to provide the maximum therapeutic response in a target population with difficulty swallowing tablets and capsules and/or to produce rapid therapeutic effects. The major ingredient in most liquid dosage forms is water. While it is the safest and most palatable solvent option, water quality is significant for the stability of pharmaceutical dosage forms. Furthermore, the Food and Drug Administration (FDA) "Guide to Inspections of Dosage Form Drug Manufacturer's—CGMPRs" considers microbial contamination due to inappropriate design and control of purified water systems as the most common problem of liquid dosage forms. Solutions and dispersions studied in this chapter are chemically, microbiologically, and/or physically unstable systems that require a high level of organizational management of manufacturing processes in order to maintain a state of apparent stability, at least until the expiration date [1].

The pharmaceutical industry manufactures dosage forms in large-scale formulations. The decision to scale up is based on the economics of the production process related to costs of materials, personnel, equipment, and control [2]. To reduce costs of wastes and to obtain high-quality and efficacious drug products, the strategic plan to be applied during the process has to be developed carefully. In fact, the variables that affect product quality are identified and understood in the process instead of tested into the final product [3].

Commercial liquid dosage forms reach large-scale production after being preformulated at the laboratory level followed by formulation at the small scale and then at the pilot plant scale. Due to the complexity of the manufacturing process, scale-up from pilot to commercial production is not a simple extrapolation. The approaches to the four levels of production are different. Most of the formulation ingredients are analyzed, studied, and selected at the laboratory scale. While small-scale production is more focused on the liquid preparation procedure with higher amounts of ingredients, the main issues at the pilot plant scale are the design of infrastructure and reduction of costs. Commercial production introduces problems that are not a major issue on a small scale, for instance, materials handling and storage, pulverizing, mixing, dissipation of the generated heat during production, time control, personnel administration, and bottle-filling capabilities. Furthermore, purified water is essential for the manufacturing of these products as well as on-site packing capabilities [2].

The organization and advance of the pharmaceutical industry should be focused on three main points of project development based on quality by design (QbD): product objective (design of experiment, DoE), production resources (process analytical technology, PAT), and product acceptability (quality system) [3]. Manufacturers of liquid dosage forms must ensure safety, efficacy, stability, elegance, and acceptability of the final drug product while achieving development and clinical milestones [1]. Achieving the desirable clinical attributes of the product efficacy means ensuring potency stability by confirming the functionability of the manufacturing process and quality system. The stability and safety of the product are goals to be reached through chemical and microbiological stability by establishing and updating manufacturing and quality control protocols of product development. Time control and aesthetic considerations reflect the physical stability through product elegance and acceptability. Flavoring, sweetening, coloring, and texturing are both challenges and opportunities. They are challenges because "no single correct method exists to solve significant problems of elegance"; they are opportunities because "they enable a pharmacist to prepare a product more easily accepted by the patient" [4]. Although the most important characteristic of a dosage form is efficacy, there are other characteristics that remain important subjects for the manufacturing of liquid dosage forms such as safety as well as chemical, physical, and microbiological stabilities. From a pharmaceutical point of view, stability problems are the main causes of safety complaints. Despite its significance, some companies decide to outsource stability services [5]. To solve or minimize stability problems in drug products, it is necessary to analyze and enhance the development of critical control points in each operation of the full manufacturing process as well as expected variances and tolerance limits.

Except some aqueous acids, water in aqueous solutions is an excellent media for microbiological growth, such as molds, yeast, and bacteria. Typical microorganisms affecting drug microbiological stability are *Pseudomonas, Escherichia coli Salmonella*, and *Staphylococcus* [1]. Deficient methods or an insufficient preservative system may be the principal causes of microbiological contamination in the pharmaceutical industry of liquid manufacturing [6].

Chemical instability reactions appear with or without microbiological contribution through reactions such as hydrolysis, oxidation, isomerization, and epimerization. Interactions between ingredients and ingredients with container closure materials are established as the principal causes of these reactions [1], for instance, the hydrolysis of cefotaxime sodium, the oxidation of vitamin C, the isomerization of epinephrine, and the epimerization of tetracycline [7].

In most cases, physical instabilities are consequences of previous chemical instabilities. Physical instabilities can arise principally from changes in uniformity of suspensions or emulsions, difficulties related to dissolution of ingredients, and volume changes [6]. For instance, some cases where physical stability has been affected are cloudiness, flocculence, film formation, separation of phases, precipitation, crystal formation, droplets of fog forming on the inside of container, and swelling of the container [8].

Although commercial oral solution and emulsion dosage forms rarely present bioequivalence issues, some bioequivalence problems have been reported for oral suspensions such as phenytoin [9]. The possibility of microbiological contamination and physicochemical instabilities during the manufacturing process also needs to be carefully considered. To approach the stability problems of liquid dosage forms, in this chapter, the main critical aspects during the manufacturing process are based on FDA inspection. From physical plant systems to batching management and packing, the potential sources of microbiological, chemical, and physical instabilities will be analyzed using definitions, case-by-case explanations, and practical examples.

Final product stability, which determines the therapeutic activity and uniformity among other characteristics of the final product, reflects the dynamic of the production process. Conceptualization of stability issues is important to determine the changes to enhance the design space as well as protocols of manufacturing and quality control [3]. An information technology (IT) solution like Enterprice Resource Planning (ERP) may support the pharmaceutical industry's current challenges of organization [10].

4.3.2 GENERALITIES

4.3.2.1 Dosage Form

According to the FDA: "A dosage form is the physical form in which a drug is produced and dispensed. In determining dosage form, FDA examines such factors as (1) Elegance: physical appearance of the drug product, (2) Stability: physical form of the drug product prior to dispensing to the patient, (3) Acceptability: the way the product is administered, (4) Efficacy: frequency of dosing, and (5) Safety: how pharmacists and other health professionals might recognize and handle the product" [11]. The term *dosage form* is different from "dose," which is defined as a specific amount of a therapeutic agent that can be taken at one time or at intervals.

4.3.2.2 Liquid Dosage Form

The physical form of a drug product that is pourable displays Newtonian or pseudoplastic flow behavior and conforms to its container at room temperature. In contrast, a semisolid is not pourable and does not flow at low shear stress or conform to its container at room temperature [12]. According to its physical characteristics, liquid dosage forms may be dispersed systems or solutions.

4.3.2.3 Dispersed Systems

Dispersed systems are dosage forms composed of two or more phases, where one phase is distributed in another [2]. If a dispersed system is formed by liquid phases, then it is known as an "emulsion." In contrast, the dispersed system is named a "suspension" when the liquid dosage form is accomplished by the distribution of a solid phase suspended in a liquid matrix. The solid phase of a suspension is usually the drug substance, which is insoluble or very poorly soluble in the matrix [12].

4.3.2.4 Solutions

A solution refers two or more substances mixed homogeneously [2]. Although solubility refers to the concentration of a solute in a saturated solution at a specific temperature, in pharmacy, solution liquid dosage forms are unsatured to avoid crystallization of the drug by seeding of particles or changes of pH or temperature [13]. The precipitation of drug crystals is one of the most important physical instabilities of solutions that may affect its performance [14]. Water is the most used solvent in solutions manufacturing; however, there are also some commercial nonaqueous solutions in the pharmaceutical market [1].

4.3.2.5 Manufacturing of Nonparenteral Liquid Dosage Forms

The manufacturing of liquid dosage forms with market-oriented planning includes the following stages with respect to special good manufacturing practice (GMP) requirements: planning of material requirements, liquid preparation, filling and packing, sales of drug products, vendor handling, and customer service [15]. From the viewpoint of product stability, each stage of the process includes critical batches that are more decisive than others. Also, each decisive batch contains one or several unit operations that are more critical than others. The FDA inspection focuses on those critical unit operations to ensure the safety and stability of the liquid dosage forms [6].

4.3.2.6 Optimizing Drug Development Strategies

> According to Sokoll [16]: The phases of drug development include discovery, preclinical development, clinical development, filing for licensure, approval/licensure and post-approval. Discovery typically includes basic research, drug identification and early-stage process and analytical method development. . . . Emerging companies that review their pipeline objectively and strike a balance between properly resourcing and developing their lead candidates in the clinic while nurturing their next generation of drug candidates will have the best chance for success and sustainability.

4.3.2.7 Unit Operation or Batch

A "batch" job or operation is defined as a unit of work. Raw materials, semifinished drug products (bulk), and finished drug products are handled in batches. Each different type of material used during the process, such as product packing, should be managed by batches. This applies also to process aids and operation facilities [15].

4.3.2.8 Batch Management

The batch management of production simplifies the process and makes it easier to control the status of transformation between raw and final products [2]. Some of the data used to follow the material performance around and out of the product manufacturing process are batch-where-used-list, initial status, batch determinations, master data, and expiration date check [15].

The functionality of the overall process to manufacture liquid dosage forms depends on the successful linkage of one unit operation to another. To use mathematical formulations to scale up the manufacturing process, it is necessary to divide the process into stages, batches, and unit operations. Each single unit operation is scalable, but the composite manufacturing process is not. Production problems result from attempts to follow a process scale-up instead of a unit operation scale-up. By using mathematical formulations, it is possible to understand the level of similarity between two scale sizes. In addition, nonlinear similarities between two scale sizes might require the use of conversion factors to achieve an extrapolation point for the scale [2].

4.3.2.9 Steps of Liquids Manufacturing Process

Establishing short-term goals makes it easier to measure efficiency as well as evaluate the difficulties [2]. Based on these concepts, the problems of manufacturing liquid dosage forms can be approached as problems in one or more batches of the following process steps [6, 15]:

Planning of Material Requirements Research and development of protocols and selection of materials; acquisition and analysis of raw materials; physical plant design, building, and installation; equipment selection and acquisition; personnel selection and initial training; and monitoring information system.

Liquid Preparation Research and development of protocols concerning liquid compounding; scale-up of the bulk product compounding; physical plant control and maintenance; equipment maintenance and renovation; continuous training of personnel and personnel compensation plan; and supervision of system reports.

Filling and Packing Research and development of protocols concerning filling and packing; scale-up of the finished drug product filling and packing; physical plant control and maintenance; equipment maintenance and renovation; continuous training of personnel and personnel compensation plan; and supervision of system reports.

Sales of Drug Products Research and development of protocols concerning product storage; distribution process; continuous training of personnel and personnel compensation plan; and supervision of system reports.

Vendor Handling Research and development protocols concerning precautions to maintain product stability; control of vendor stock; and sales system reports.

Customer Service Research and development of protocols concerning home storage and handling to maintain product stability; relations with health insurance companies and health care professionals; educational materials for patient counseling; and customer service system reports.

4.3.2.10 Protocols

Protocols are patterns developed by repeating procedures and fixing the identified problems each time that the procedure is followed. Therefore, protocols are dynamic entities that originally can be developed at a laboratory level but must be adjusted in every new step of the scal -up process. When the manufacturing process moves up in scale, the number of people affected by the protocol increases geometrically. Initially, the information can be obtained from library references, personal tests, interpersonal training, and previous laboratory protocols. However, when the production is scaled up, the information required to fine tune the process comes from monitoring the process itself [2].

4.3.3 APPROACHES

Quality by Design is a systemic approach that applies the scientific method to the process. QbD theory contains components of management, statistics, psychology, and sociology. The FDA's new century has identified the QbD approach as its "key component" based on process quality control before industry end results [3, 17].

The cooperation between industry members and regulators is increased when the industry explains clearly what it is doing and the agency can understand the formulation and production process. In these cases, regulatory relief appears when industry explores its issues and receives active guidance and programs from the FDA. The agency takes the role of facilitator, or even partner of the industry, in order to improve the strength of the process and formulation [3, 17].

To apply QbD as a systemic approach, the company starts by understanding, step by step, the space design, the design of the dosage form, the manufacturing process, and the critical process parameters to be controlled in order to reach the new building block which is the expectation of variances within those critical process parameters that can be accepted. This approach allows the establishment of priorities and flexible boundaries in the process [3]. Inflexible specifications allow uncontrolled small variances that can follow the butterfly effect of the theory of chaos by producing unpredictable large variations in the long-term behavior of the product shelf-life [18, 19]. In contrast, flexibility, with knowledge of potential variances, reduces changes in the approved spaces and manufacturing protocols [3, 17].

According to the FDA [6], critical parameters during the manufacturing process of nonparenteral liquid dosage forms may appear in the design of physical plant systems, equipment, protocols of usage and maintenance, raw materials, compounding, microbiological quality control, uniformity of suspensions and emulsions, and filling and packing [6].

Process isolation and installation of an appropriate air filtration system in the physical plant may reduce product exposition to chemical and microbiological contaminations. In addition, the use of a suitable dust removal system as well as a heating, ventilation, and air conditioning system (HVACS) may help to repress product chemical instabilities [6].

The equipment of sanitary design, including transfer lines, as well as appropriate cleaning and sanitization protocols may reduce chemical and microbiological contaminations in the final product. Chemical instabilities may be reduced by weighting the right amount of liquids instead of using a volumetric measurement, avoiding the common use of connections between processes, and using appropriate batching equipment [6].

Particle sizes of raw materials are critical to control dissolution in solutions as well as uniformity in suspensions and emulsions. Temperature control during compounding is important since heat helps to support mixing and/or filling operations, but, in contrast, high-energy mixers may produce adverse levels of heat that affect product stability. Too much heat may cause chemical and physical instabilities such as change of particle size or crystallization of drugs in suspensions, dissolution and potency loss of drugs in suspensions, oxidation of components, and activation of microbiological growth after degradation of compounds as well as precipitation of dissolved compounds in solution [20]. In addition, uniformity of suspensions depends on viscosity and segregation factors while solubility, particle size, and crystalline form determine uniformity of emulsions. Application of pharmaceutical GMP for product processes and storage assures microbiological quality. A deficient deionizer water-monitoring program and product preservative system facilitate microbial contamination. Filling uniformity is indispensable for potency uniformity of unit-dose products and depends on the mixing operation. Calibration of provided measuring devices and the use of clean containers will allow administering the right amount of the expected components in the liquid dosage form [6].

Principal product specifications are microbial limits and testing methods, particle size, viscosity, pH, and dissolution of components. Process validation requires control of critical parameters observed during compounding and scale-up. Product stability examination is based on chemical degradation of the active components and interactions with closure systems, physical consequences of moisture loss, and microbial contamination control [6].

4.3.4 CRITICAL ASPECTS OF LIQUIDS MANUFACTURING PROCESS

4.3.4.1 Physical Plant

Heating, Ventilation, and Air Conditioning System The manufacturer has to warrant adequate heating, ventilation, and air conditioning in places where labile drugs are processed [6].

The effect of long processing times at suboptimal temperatures should be considered at the production scale in terms of the consequences on the physical or chemical stability of individual ingredients and product. A pilot plant or production scale differs from laboratory scale in that their volume-to-surface-area ratio is relatively large. Thus, for prolonged suboptimal temperatures, jacketed vessels or immersion heaters or cooling units with rapid circulation times are absolutely necessary [2].

For heat-labile drugs, uncontrolled temperature increments can activate auto-oxidation chains when the drug product ingredients react with oxygen and generate free radicals but without drastic external interference. Vitamins, essential oils, and almost all fats and oils can be oxidized. A good example of a heat-labile drug solution is clindamycin, which has to be stored at room temperature and away from excess heat and moisture [19]. Auto-oxidation chains are finished when free radicals react with each other or with antioxidant molecules (quenching). The tocopherols, some esters of gallic acid, as well as BHA and BHT (butylated hydroxyanisole and butylated hydroxytoluene) are common antioxidants used in the pharmaceutical industry [1].

Isolation of Processes To minimize cross-contamination and microbiological contamination, the manufacturer may develop special procedures for the isolation of processes. The level of facilities isolation depends on the types of products to be manufactured. For instance, steroids and sulfas require more isolation than over-the-counter (OTC) oral products [6]. To minimize exposure of personnel to drug aerosols and loss of product, a sealed pressure vessel must be used to compound aerosol suspensions and emulsions [21]. An example of cross-contamination with steroids was the controversial case of a topical drug manufactured for the treatment of skin diseases. High-performance liquid chromatography/ultraviolet and mass spectrometry (HPLC/UV, HPLC/MS) techniques were used by the FDA for the detection of clobetasol propionate, a class 1 superpotent steroid, as an undeclared steroid in zinc pyrithione formulations. The product was forbidden and a warning was widely published [22].

Dust Removal System The efficiency of the dust removal system depends on the amount and characteristics of dust generated during the addition of drug substance and powdered excipients to manufacturing vessels [6]. Pharmaceutical industries usually generate some type of dust or fume during processing. Important factors for selecting dust collectors are maintenance, surrogate test, economics, and containment. In addition, reentrainment of the fine particles, vertical or horizontal position, efficiency, pressure resistant, service life time, as well as sealing capacity to work through the bag are significant factors concerning filter selection of dust removal systems. Some examples of dust collection applications in the manufacture of liquid dosage forms are handling and pulverization of raw materials, spray dryers, and general room ventilation [23].

Air Filtration System The efficiency of the air filtration system has to be demonstrated by surface or air-sampling data where the air is recirculated [6]. To monitor the levels of contamination in the air, there are commercial automatic samplers for microbiological contamination or gas presence. Air trace environmental samplers for pharmaceutical industries are based on the slit-to-agar impaction technique for the presence of viable microorganisms. Automatic samplers for compressed gas analyze the presence of a specified gas in 1 m^3 by absorbing air at a fixed flow rate for a sampling period of 1 h or a different adjusted time. These solutions to the sampling needs of the pharmaceutical industry are robust, require low maintenance, and are easy to use. This allows for validation of sampling data at the moment of application filling to support the process control. Sampling time and selection of microbiological growth media or analysis technique are important components to consider when developing a sampling plan [24].

4.3.4.2 Equipment

Sanitary Design Pumps, valves, flowmeters, and other equipment should be easily sanitized. Some examples of identified sources of contamination are ball valves, packing in pumps, and pockets in flowmeters [6].

The sanitary design and performance of equipment make it accessible for inspection, cleaning, and maintenance. It has to be cleanable at a microbiological level and its performance during normal operations should contribute to sanitary conditions. The materials used in the design have to assure hygienic compatibility with other equipment, the product, the environment, other systems such as electrical, hydraulics, steam, air, and water, as well as the method and products used for cleaning and sanitation. The equipment should be self-draining to assure product or liquid collection. Small niches, for example, pits, cracks, corrosion, recesses, open seams, gaps, lap seams, protruding ledges, inside threads, bolt rivets, and dead ends, as well as inaccessible cavities of equipment such as entrap and curlers must be eliminated whenever possible; otherwise they have to be permanently sealed. Enclosures, for example, push buttons, valve handles, switches, and touch screens, should be prepared for a hygienic design of maintenance. Standards have been developed by the American Meat Institute [25].

Standard Operating Procedures for Cleaning Production Equipments Current GMPs are defined as the basic principles, procedures, and resources required to guarantee an environment appropriate for manufacturing products of adequate quality [26]. To minimize cross-contamination and microbiological contamination, it is GMP for a manufacturer to create and pursue written standard operating procedures (SOPs) to clean and sanitize production equipment in a way that avoids contamination of in-progress and upcoming batches. When the drug is known as a potent generator of allergic reactions, such as steroids, antibiotics, or sulfas, cross-contamination becomes an issue of safety [20]. In addition, validation and data analysis procedures,

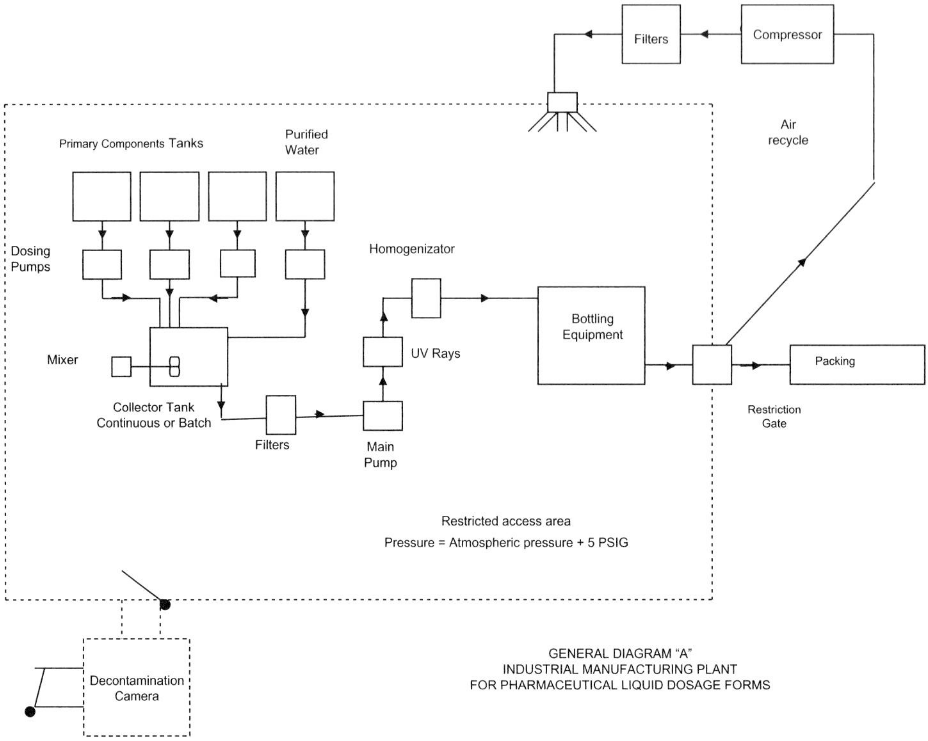

FIGURE 1 Mixing and filling lines for pharmaceutical dosage forms. Positive indoor pressure of 5 psi over outdoor pressure assures constant airflow from inside to outside in order to reduce entrance of contaminating agents.

including drawings of the manufacturing and filling lines [6], are especially important for clean-in-place (CIP) systems, as indicated in Figures 1 and 2.

Many companies have problems with standardizing operating procedures for cleaning steps and materials used [6]. Appropriate SOPs are necessary to determine the scope of the problem in investigations about possible cross-contaminations or mix-ups. The best approach to validate a SOP is to test it, use it as a training tool, and observe the results obtained by different persons. This includes the worst-case situation in order to enhance the step-by-step writing methodology as well as standardizing the materials used. A typical SOP contains a header to present the SOP title, date of issue, date of last review, total number of pages, responsible person, and approval signature. Typically, a SOP includes position of responsible person, SOP purpose and scope, definitions, equipment and materials, safety concerns, step-by-step procedure, explanation of critical steps, tables to keep data, copies of forms to fill, and references [26]. The forms to keep the records must show the date, time, product, and lot number of each batch processed. However, the most important points of the SOP are equipment identity, cleaning method(s) with documentation of critical cleaning steps, materials approved for cleaning that have to be easily removable, names and position of persons responsible for cleaning and inspection, inspection methods, and maintenance and cleaning history of the equipment [20].

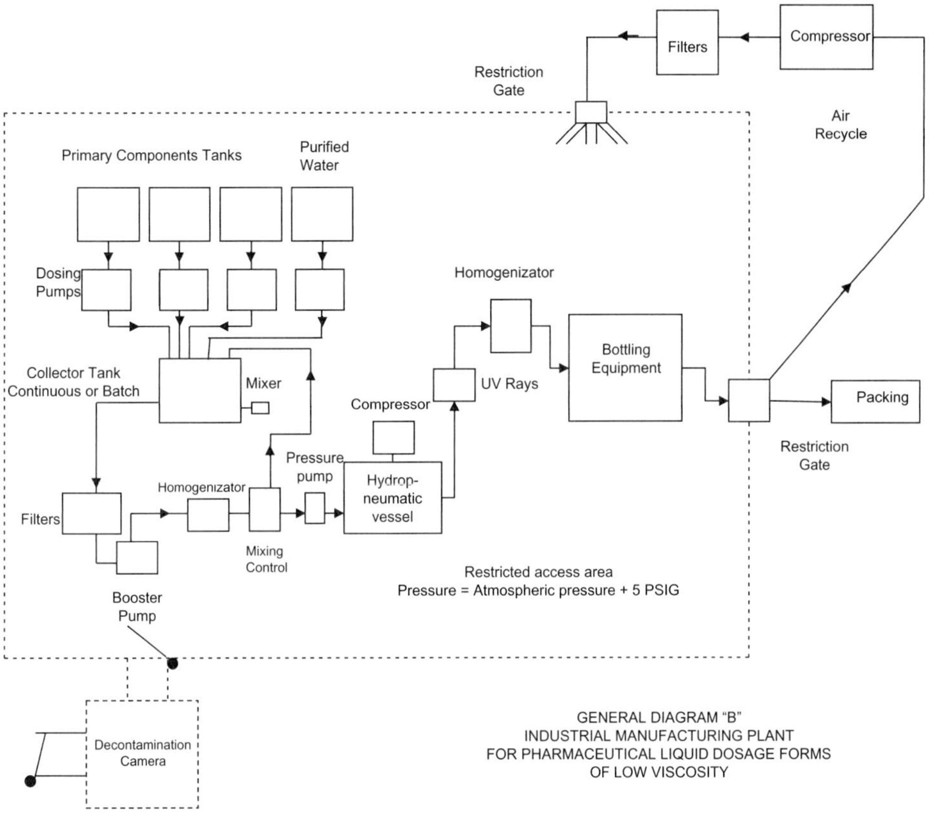

FIGURE 2 Mixing and filling lines for pharmaceutical dosage forms. Using this hydro-pneumatic system, instead of the mechanical system in Figure 1, the liquid moves by the pressure generated in a compressed air tank.

Cleaning and Sanitizing Transfer Lines Pipes should be hard, easily cleaned, and sanitized. To avoid moisture collection and microbiological contamination, hoses should be stored in a way that allows them to drain rather than be looped. For example, transfer lines are an important source of contamination when flexible hoses are handled by operators, lying on the floor, and after they are placed in transfer or batching tanks [6].

Heat is considered one of the most efficient physical treatments for sanitizing pharmaceutical equipment and could be used for sanitizing hoses that have already been cleaned. The recirculation of hot water at a temperature of 95°C for at least 100 min allows bacteria elimination [14].

Due to the important amount of insoluble residues left on piping and transfer lines after emulsion manufacturing, such as topical creams and ointments, equipment cleaning becomes difficult to address. To avoid cross-contamination, some manufacturers have decided to dedicate lines and hoses to specific products.

However, these decisions have to appear in the written production protocols and SOPs [20].

Sampling Cleaned Surfaces for Presence of Residues The cleaning method is validated by sampling the cleaned surfaces of the equipment for the existence of residues. The equipment characteristics and residue solubility are factors to support the selection of the sampling method to be used [6]. There are two acceptable general types of sampling methods: direct surface sampling by swabbing of surfaces and rinse sampling with a routine production in-process control. Although surface residues will not be identical on each part of the surface, statistically the most advantageous is direct surface sampling because it allows evaluation of the hardest areas to clean as well as insoluble or "dried-out" residues by physical removal. The type of sampling material and solvent used for extraction from the sampling material should be validated in order to determine their impact on the test data. The second method, rinse sampling, is used for larger surfaces or inaccessible systems. Contaminants that are physically occluded and insoluble residues are disadvantages of the rinse sampling method. To validate this cleaning process, direct measurement of the contaminant in the rinse water has to be tested instead of a simple test for water quality. Routine production in-process control is used as indirect testing for large equipment that has to be cleaned by the rinse sampling method. The uncleaned equipment has to give an unacceptable result for the indirect test [27].

Establishing Appropriate Limits on Levels of Postequipment Cleaning Residues Very low levels of residue are possible to be determined since technological advances offer more sensitive analytical methods. The manufacturer should know the toxicological information of the materials used and potential amounts of residues after exposure to the equipment surface. Accordingly, the manufacturer has to establish proper limits of residues after equipment cleaning and scientifically justify these limits. The established limits must be clinically and pharmaceutically safe, realistic, viable, and verifiable [20]. The sensitivity of the analytical method will determine the logic of the established limits since absence of residues could indicate a low sensitivity of the analytical method or a poor sampling procedure. Sometimes thin-layer chromatography (TLC) screening must be used in addition to chemical analyses. Some practical levels established by manufacturers include 10 ppm of chemicals, 1/1000 of the biological activity levels met on a normal therapeutic dose, and no visible residues of particles determined organoleptically [27].

Connections Connectors and manifolds should not be for common use. For example, sharing connectors in a water supply, premix, or raw material supply tanks may be a source of cross-contamination [6].

Time between Completion of Manufacturing and Initiation of Cleaning The time that may elapse from completion of a manufacturing operation to initiation of equipment cleaning should also be stated where excessive delay may affect the adequacy of the established cleaning procedure. For example, residual product may dry and become more difficult to clean [20].

SOPs are an example of deficiency in many manufacturers regarding time limitations between batch cleaning and sanitization [6]. Lack of communication between departments responsible for the production at different levels is the main cause of time control problems. Typically each department, from human resources to finances, manufacturing, and warehouse, has its own computer system optimized for the particular ways that the department does its work. Therefore, time control becomes a primordial issue when labile materials are transferred from one department to another [28].

To facilitate communication between different departments, some useful softwares have been developed. For example, ERP is an integrated approach which may have positive payback if the manufacturer installs it correctly. An ERP is a type of software that can improve communication between planning and resources. The software attempts to integrate all departments and functions in a company onto a single computer system that can serve each particular need, such as finance, human resources, manufacturing management, process manufacturing management, inventory management, purchasing management, quality management, and sales management. Each department has its own software, except now the software is linked together, so that, for example, someone in manufacturing can look into the maintenance software to see if specific batch cleaning and sanitization have been scheduled or realized and someone in finance can review the warehouse software to see if a specific order has been shipped. The information is online and not in someone's heads or on papers that can be misplaced. People in different departments may see the same information, update it if they are allowed to do, and make right decisions faster. However, the software is less important than the changes companies make in the ways they work. Reorganization and training are the keys of ERP's success to fix integration problems. There are three different ways to install an ERP: big bang, franchising strategy, and slam dunk. Big bang is the most ambitious way whereby companies install a single ERP across the entire company. By the franchising strategy, departments do not share many common processes across, whereas slam dunk is focused on just a few key processes [28].

Weight in Formulations Flow properties of liquids rarely vary due to their constant density at a constant temperature. Oral solutions and suspensions are formulated on a weight basis (gravimetry) in order to be able to measure the final volume by weight before filling and packing. Volumetric measurements of liquid amounts to be used for manufacturing liquid dosage forms have shown greater variability than weighted liquids. For instance, the inaccurate measurement of the final volume by using dip sticks or a line on a tank may cause further analytical errors and potency changes [6].

The importance of selecting gravimetry instead of volumetry to measure liquid amounts in the pharmaceutical industry of liquid dosage forms is well illustrated by the volume contraction of water–ethanol and volume expansion of ethyl acetate–carbon disulfide liquid mixtures as well as a CS2–ethyl acetate system. The National Formulary (NF) diluted alcohol is a typical example of the volume nonadditivity of liquid mixtures [29]. This solution is prepared by mixing equal volumes of alcohol [U.S. Pharmacopeia (USP)] *USP* and purified water (USP). The final volume of this solution is about 3% less than the sum of the individual volumes because of the contraction due to the mixing phenomenon [1]. In addition, molecular interactions of surfactants in mixed monolayers at the air–aqueous solution interface and in mixed micelles in aqueous media also cause some contraction of volume upon mixing [30].

Location of Bottom Discharge Valve in Batching Tank The bottom discharge valve should be located exactly at the bottom of the tank. In some cases valves have been found to be several inches to a foot above the bottom of the tank [6].

For a tank suspected of having substantial deposits at the bottom, a fiber-optic camera can be inserted in the tank to provide a view and positive confirmation of the tank bottom condition. These camera and light vision systems are sanitary equipment able to provide a computational real-time visual inspection of the inside tank under process conditions or pressure vessel. In addition, they are used to control several parameters during the manufacturing process, such as product level and thickness, solids level, uniformity of suspensions, foam, and interface and/or cake detection [31].

Batching Equipment to Mix Solution Ingredients of solutions have to be completely dissolved. For instance, it has been observed that some low-solubility drugs or preservatives can be kept in the "dead leg" below the tank, and the initial samples have reduced potency [6]. When there is inadequate solubility of the drug in the chosen vehicle, the dose is unable to contain the correct amount of drug in a manageable size unit, that is, one teaspoonful or one tablespoonful. Thus, ingredients as well as handling and storage conditions should be chosen to manage the problem [14].

In solutions, the most important physical factors that influence the solubility of ingredients are type of fluid, mixing equipment, and mixing operations. Generalized Newtonian fluids are ideal fluids for which the ratio of the shear rate to the shear stress is constant at a particular time. Unfortunately, in practice, usually liquid dosage forms and their ingredients are non-Newtonian fluids in which the ratio of the shear rate to the shear stress varies. As a result, non-Newtonian fluids may not have a well-defined viscosity [32].

When all the ingredients are miscible liquids, the combination and distribution of these components to obtain a homogeneous mixture are called blending. Whenever possible, ingredients should be added together and the impeller mixer often is located near the bottom of the vessel [21]. Mixing of high-viscosity materials requires higher velocity gradients in the mixing zone than regular blending operations. In fact, the fundamental laws of physics regarding the performance of Newtonian fluids in the production process may be studied using computational tools. For example, VisiMix is a software that is routinely used to calculate shear rates [2].

Finally, if it is determined that there is a bigger problem of insolubility coming from the formulation, then addition of cosolvents, surfactants, as well as the preparation of the ionized form of an acid or base, drug derivatization, and solid-state manipulation are approaches to manipulating the solubility of the drug [14].

Batching Equipment to Mix Suspension In the case of suspensions, the flow necessary to overcome settling in a satisfactory suspension depends on the mixing equipment and is predicted by Stokes's law. Thus, to use the Stokes's law, suspensions are considered as Newtonian fluids if the percentage of solids is below 50%. Mixing equipment uses a mechanical device that moves through the liquid at a given velocity. Dispersing and emulsifying equipment is categorized as "high-shear" mixing equipment. The maximum shear rate with such equipment occurs very close to the mixing impeller. Therefore, the diameter of the impeller and the impeller speed directly influence the power applied by the mixer to the liquid [21].

Batching Equipment to Mix Emulsion The most common problems of mixing emulsions are removing "dead spots" of the mixture and scrapping internal walls of the mixer. Dead spots are quantities of ingredients that are not mixed and become immobile. Where dead spots are present, that quantity of the formula has to be recirculated or removed and not used. If the inside walls of the mixer keep residual material, operators should use hard spatulas to scrape the walls; otherwise the residual material will become part of the next batch. In both cases, the result may be nonuniformity. Stainless steel mixers have to include blades made of hard plastic, such as Teflon, to facilitate the scrapping of the mixer walls without damaging the mixer. Scrapper blades should be flexible enough to remove internal material but not too rigid to avoid damaging the mixer [20]. The mixing will be successful if the macroscale mixing offers sufficient flow of components in all areas in the mixing tank and the microscopic examination shows a correct particle size distribution [33].

4.3.4.3 Particle Size of Raw Materials

Raw materials in Solution The types of raw materials used to be part of solutions are presented in Table 1. They have different purposes and can be cosolvents, electrolytes, buffers, antioxidants, preservatives, coloring, flavoring and sweetener agents, among others.

Particle Size of Raw Materials in Solution Particle size is affected by the breaking process of the particle, crystal form, and/or salt form of the drug. The particle size can affect the rate of dissolution of raw materials in the manufacturing process. Raw materials of a finer particle size may dissolve faster because they have a larger surface area in contact with the solvent than those of a larger particle size when the product is compounded [6]. Mixing faster causes the particle to break down and dissolve more quickly. In addition, hydrated particles are less soluble than their anhydrous partners [37].

TABLE 1 Solutions: pharmaceutical excipients

Purpose	Agent
Protecting the active product ingredients	- Buffers - Antioxidants - Preservatives
Maintaining the appearance	- Colorings - Stabilizers - Cosolvents - Antimicrobial preservatives - Electrolytes
Taste/Small Masking	- Sweeteners - Flavorings

Source: From ref. 4, 34, 35, 36

Solid drugs may occur as pure crystalline substances of definite identifiable shape or as amorphous particles without definite structure. In addition, when a drug particle is broken up, the total surface area is increased as well as its rate of dissolution. The amorphous form of a chemical is usually more soluble than the crystalline form while the crystalline form usually is more stable than the amorphous form [37]. Processing conditions used for providers to obtain raw materials can dramatically impact their quality and stability; for instance, the presence of different polymorphs may depend on the thermal history of freezing, concentration of solvents, and drying conditions [38]. The polymorphism of a crystalline form is the capacity of a chemical to form different types of crystals, depending on the conditions of temperature, solvents, and time followed for its crystallization. Among different polymorphs, only one crystalline form is stable at a given temperature and pressure. Over time, the other crystalline forms, called metastable forms, will be transformed into stable forms. Transformations longer than the shelf-life of metastable forms into stable forms of a drug are very common in final products and compromise its stability and efficacy to different extents depending on quality control [37].

While the metastable forms offer higher dissolution rates, many manufacturers use a particular amorphous, crystalline, salt, or ester form of a drug with the solubility needed to be dissolved in the established conditions, for instance, to prepare a chloramphenicol ophthalmic solution [39]. Thus, the selection of amorphous or crystalline form of a drug may be of considerable importance to facilitate the formulation, handling, and stability [37].

However, the dissolution rate of an equal sample of a slowly soluble raw material usually will increase with increasing temperature or rate of agitation as well as with reduce viscosity, changes of pH or nature of the solvent. In addition, other alternative mechanisms to enhance the solubility of insoluble drugs are: 1) hydrophilization: the reduction in contact angle or angle between the liquid and solid surface [40], which can be accessed by intensive mixing of the hydrophobic drug with a small amount of methylcellulose solution [41]; 2) the formation of microemulsions: by covering small particles with surfactants to obtain micromicelles that are visible only in the form of an opalescence; and, 3) the formation of complexing compounds: by adding a soluble substance to form soluble reversible complexes. However, the last method is used with some restrictions [42].

Raw Materials in Suspension The types of raw materials used to be part of suspensions are presented in Table 2. They have different purposes and can be wetting agents, salt formation ingredients, buffers, polymers, suspending agents, flocculating agents, electrolytes, antioxidants, poorly soluble Active Product Ingredients, preservatives, coloring, flavoring and sweetener agents, among others.

Particle Size of Drug in Suspension The physical stability of a suspension can be enhanced by controlling the particle size distribution [43]. Uncontrolled changes of drug particle size in a suspension affect the dissolution and absorption of the drug in the patient. Drug substances of finer particle size may be absorbed faster and bigger particles may not be absorbed. Aggregation or crystal growth is evaluated by particle size measurements using microscopy and a Coulter counter [21] or preferably techniques that allow samples to be investigated in the natural state. Allen [44] offers an academic and industrial discussion about particle characterization.

TABLE 2 Suspensions: pharmaceutical excipients

Purpose	Agent
Facilitating the connection between Active Product Ingredient and vehicle	- Wetting agents particle size (>0.1 μm) - Salt formation ingredients - Sugars
Protecting the Active Product Ingredients	- Buffering–systems - Polymers - Antioxidants - Poorly soluble drugs
Maintaining the suspension appearance	- Colorings - Suspending agent - Flocculating agent - Antimicrobial preservatives - Electrolytes
Masking the unpleasant taste/smell	- Sweeteners - Flavorings - Poorly soluble Active Product Ingredient

Source: From ref. 4, 34, 35, 36

Powder properties and behavior, sampling, numerous potential particle size measuring devices, available equipment as well as surface and pore size are his principal themes.

Particles are usually very fine (1–50 μm). For instance, topical suspensions use less than 25 μm particle size [6]. The particle size of the drug is the most important consideration in the formulation of a suspension, since the sedimentation rate of disperse systems is affected by changes in particle size. Finer particles become interconnected and produce particle aggregation followed by the formation of nonresuspendable sediment, known as caking of the product. The two main causes of aggregation and caking are energetic bonding and bonding through shared material. A statistical wide distribution of particle sizes gives more compact packing and energetic bonding than narrower distributions. It has been observed that heat treatments can cause agglomeration of particles, not only due to energetic bonding but also by formation of crystal bridges. Also, when the application of shear forces to mix and homogenize the suspension uses too high energy inputs, then the probability for aggregation increases [43].

Examples of oral suspensions in which a specific and well-defined particle size specification for the drug substance is important are phenytoin suspension, carbamazepine suspension, trimethoprim and sulfamethoxazole suspension, and hydrocortisone suspension [6].

There are some useful methods to improve the physical stability of a suspension, such as decreasing the salt concentration, addition of additives to regulate the osmolarity, as well as changes in excipient concentrations, unit operations in the process, origin and synthesis of the drug substance, polymorphic behavior of the drug substance crystals, and other particle characteristics. However, methods based on changes of the particle properties and the surfactants used are the most successful [43].

TABLE 3 Emulsions: pharmaceutical excipients

Purpose	Agent
Particle Size	- Solid particles (10 nanometers to 5 micrometers size) - Droplet particles (0.1–1.0 micrometers size)
Protecting the Active Product Ingredients	- Buffering-Systems - Polymers - Antioxidants - Distribution pattern (O/W, W/O)
Maintaining the appearance	- Colorings - Emulsifying agents - Penetration enhancers - Gelling agents - Stabilizers - Antimicrobial preservatives
Taste/smell Masking	- Sweeteners - Flavorings - Relation oil vs. water

Source: From ref. 4, 34, 35, 36

To approach physical stability problems of suspensions, effectiveness and stability of surfactants as well as salt concentrations must be checked with accelerated aging. In addition, unit operations affecting particle size distribution, surface area, and surfactant effectiveness should be approached, taking into account that different types of distributions, for instance, volume or number weighted, give a different average diameter for an equal sample [43].

Raw Materials in Emulsions The types of raw materials used to be part of emulsions are presented in table 3. They have different purposes and can be buffers, polymers, emulsifying agents, penetration enhancers, gelling agents, stabilizers, antioxidants, preservatives, coloring, flavoring and sweetener agents, among others.

Particle Size in Emulsions When a solid drug is suspended in an emulsion, the liquid dosage form is known as a coarse dispersion. In addition, a colloidal dispersion has solid particles as small as 10 nm–5 μm and is considered a liquid between a true solution and a coarse dispersion [44].

4.3.4.4 Compounding: Effects of Heat and Process Time

Oxygen Oxygen removal for processing materials that require oxygen to degrade is possible by methods such as nitrogen purging, storage in sealed tanks, as well as special instructions for manufacturing operations [6]. For instance, sealing glass ampules containing a liquid dosage form with heat under an inert atmosphere is a packing mechanism used to prevent oxidation. Some aspects of oxygen sensitiveness that should be taken into account are the necessity of water and headspace deoxygenation in ampules before sealing, the avoidance of multidose vials that facilitate oxygen contact with the product after opened, and rubber stoppers for vial sealing that are permeable to oxygen as well as release additives to catalyze oxidative reactions. Rubber stoppers soften and get sticky over time because all rubber products degrade as sulfur bonds induced during vulcanization revert. Connors et al. [45] present the oxygen content of water at different temperatures and an interesting discussion of calculations for the case of captopril as an oxygen-sensitive drug.

Dissolution of Drugs in Solutions Although some compounds, such as poloxamers, decrease their aqueous solubility with an increase in temperature [46], usually, drugs dissolve more quickly when the temperature increases because particle vibration is augmented and the molecules move apart to form a liquid. Chemical instabilities by oxidation due to high temperature or prolonged periods of heat exposure can occur when trying to increase the dissolution of poorly soluble raw materials. To control such instabilities, charts of time and amount of temperature treatments to dissolve materials as well as tests of dissolution are required [6]. In addition, precipitations and other reactions may occur between salts in solution and can be anticipated by using heat-of-mixing data and activation energy calculations for decomposition reactions. Connors et al. [45] provide examples of calculations about effects of temperature on chemical stability of pharmaceuticals in solution. Regarding the instability of the product, the reasons to limit temperature amounts can go from controlling final concentration changes to controlling burn-on/fouling when too-high temperatures are applied [45]. Usually salts are more soluble in water and alcohol than weak acids or bases. The reason salts are not always the best choice to increase the solubility of a drug is its permeability. Oral drug absorption depends not only on solubility and dissolution but also on permeability through the cellular membrane. Drugs have to be able to dissolve not only in the aqueous fluids of the body before reaching the intestinal wall but also in the lipophilic environment of the cellular membrane in order to reach the internal part of the cell and interfere with its functionability. Therefore, the cosolvent approach is essential if the drug presents problems in dissolving in the media. The dielectric constant of a solvent is a relative measure of its polarity. Comparing the hydroxyl–carbon ratio of the solvent molecule allows establishing the relative polarity of the cosolvent as determined by its dielectric constant [47]. Remington describes the formulations of some solutions, such as the ferrous sulfate syrup, amantadine hydrochloride syrup, phenobarbital elixir, and theophylline elixir [1].

Potency of Drugs in Suspension To avoid degradation of the suspended drug substance by high temperature or prolonged periods of heat exposure, it is necessary to record the time and amount of temperature treatments on charts [6]. The rate of dissolution of a suspended drug increases with the increase in temperature. The potency stability of a suspended drug depends on the concentration of the dissolved drug since drug decomposition occurs only in solution [48]. The goal is to avoid the dissolution of suspensions. Changing the pH of the vehicle or replacing the drug with a less soluble molecule may result in enhanced potency stability of the suspended drug [48].

For instance, when the chemical stability of a suspension of ibuprofen powder and other ibuprofen–wax microspheres was studied with a modified HPLC procedure for three months, the amount of drug released from the microspheres was affected by the medium pH, type of suspending agent, and storage temperature without observing chemical degradation of the drug [49].

Temperature Uniformity in Emulsions During the preparation of emulsions, heat may be increased as part of the manufacturing protocol or mixing operation system. Temperature measurements should be monitored and documented continuously using a recording thermometer if the temperature control is critical or using a handheld thermometer if it is not a critical factor. Temperature may be critical in the manufacturing process depending on the thermosensitivity of the drug product and excipients as well as the type of mixer used. To guarantee the temperature uniformity during the mixing operation, manufacturers may consider the relation between the container size, mixer speed, blade design, viscosity of the contents, and rate of heat transfer [20].

Fong-Spaven and Hollenbeck [50] studied the apparent viscosity as a function of the temperature from 25 to 75 °C of an oil–water emulsion stabilized with 5% triethanolamine stearate (TEAS) using a Brookfield digital viscometer. They observed that the viscosity decreased when the temperature reached about 48 °C, but surprisingly viscosity increased to a small peak at 54 °C and then continued decreasing after that peak. The viscosity peak was attributed to a transitional gellike arrangement molecular structure of TEAS that is destroyed as soon as the temperature continues increasing, the TEAS crystalline form reappears, and viscosity again decreases [36].

Microbiological Control To avoid chemical instabilities that yield microbiological and physical instabilities, as a result of high temperature or prolonged periods of exposure, it is necessary to record the time and amount of temperature treatments on charts [6].

Product Uniformity Charts of storage and transfer operation times for the bulk product are required to control the risk of segregation. Transfers to the filling line and during the filling operation are the most critical moments to keep the suspension uniformity [6]. The implementation of an ERP for time scheduling is the best solution for time control and organization of resources. However, it could be difficult due to the reluctance of people to change [10]. The constant flow of the liquid through the piping, the constant mixing of the bulk product in the tank, as well as the transfer of small amounts near the end of the filling process to a smaller tank during the filling process may minimize segregation risks [6].

Final Volume Excess heating produces variations of the final volume over time [6]. Although increasing solute concentration can elevate the boiling point and reduce evaporation of water, changes in drug concentration are undesirable because they yield different final products. Regarding the instability of the product, the reasons to limit temperature amounts can go from controlling final concentration changes to controlling burn-on/fouling when too-high temperatures are applied [36].

A solution is a liquid at room temperature that passes into the gaseous state when heated at very high temperature, forming a vapor with determined vapor pressure, through a process called vaporization. The kinetic energy is not evenly distributed between the molecules of the liquid. When the liquid is in a closed container at a constant temperature, the molecules with the highest kinetic energy leave the surface of the liquid and become gas molecules. Some of the gas molecules remain as gas and others condense and return to the liquid. When, at a determined temperature, the rate of condensation equals the rate of vaporization, the equilibrium vapor pressure is reached. However, vapor pressure increases with increases in liquid temperature, resulting in more molecules leaving the liquid surface and becoming gas molecules [51].

Storage Charts of time and temperature of storage are important to control the increased levels of degradedness [6]. Shelf life is defined as the amount of time in storage that a product can maintain quality and is equivalent to the time taken to reach 90% of the composition claim or have 10% degradation. The availability of an expiration date is assumed under specified conditions of temperature. Based on zero- and first-order reaction calculations, Connors et al. [45] show the estimation methods to determine the shelf life of a drug product at temperatures different from the one specified under standard conditions.

4.3.4.5 Uniformity of Oral Suspensions

Keeping the particles uniformly distributed throughout the dispersion is an important aspect of physical stability in suspensions. Based on Stokes's law for dilute suspensions where the particles do not interfere with one another, there are different factors that control the velocity of particle sedimentation in a suspension, for instance, particle diameter, densities of the dispersed phase and the dispersion medium, as well as viscosity of the dispersion medium [36]. Remington describes the formulation of trisulfapyrimidines oral suspension [1]. In addition, Lieberman et al. [42, 48] are also good sources of typical formulations for suspensions.

Viscosity Depending upon the viscosity, many suspensions require continuous or periodic agitation during the filling process [6].

Segregation in Transfer Lines When the stored bulk of a nonviscous product is transferred to filling equipment through delivery lines, some level of segregation is expected. The manufacturer has to write the procedures and diagrams for line setup prior to filling the product [6]. Delivery lines of suspensions increase the tendency of particles of the same size to assemble together. However, slightly increasing the global mixing in the lines can easily reverse the segregation without enhancing the global mixing [52]. Shear stress versus rate of shear can be plotted to determine the flow pattern of a specific suspension as pseudoplastic, Newtonian, or dilatant. The type of flow is determined by the slope of the plot. While shaking increases the yield stress and causes particles flow, the cessation of shear and rest rebuilds the order of the system. A good-quality suspension is known as a thixotrophic system and is obtained when the particles at rest avoid or show reduced sedimentation. The rheogram of a thixotrope system presents a typical hysteresis or curve representing different shear stresses over time [33].

Quality Control The GMPs for suspensions include testing samples at different checkpoints in the procedure, at the beginning, middle, and end, as well as samples from the bulk tank. The uniformity will be successful only if, on microscopic analysis, the components are dispersed to the expected particle size distribution established by product development. Visual and microscopic examinations should consist of looking for verification of foam formation, segregation, and settling, although testing for viscosity is important to determine agitation during the filling process. Samples used for tests should not be combined again with the lot [6, 33].

4.3.4.6 Uniformity of Emulsions

Remington describes the following three typical formulas of emulsions: type A gelatin, mineral oil emulsion (USP), and oral emulsion (O/W) containing an insoluble drug [1]. In addition, Lieberman et al. [42, 50] are also good sources of typical formulations for emulsions. The components of the emulsion system may present physical and chemical instabilities reflected on the distribution of an active ingredient, component migration from one phase to another, polymorphic changes in components, and chemical degradation of components [33].

Solubility The soluble active ingredient should be added to the liquid phase that will be its carrier vehicle. Data of solubility have to be determined as part of the process validation. Potency uniformity has to be tested by demonstrating satisfactory distribution in the emulsified mix [20].

Particle Size Regarding globule diameter in emulsions, the size–frequency distribution of particles in an emulsion over time may be the only method for determining stability [36]. Drug activity and potency uniformity of insoluble active ingredients depend upon control of particle size and distribution in the mix [6]. In addition, aggregation of the internal phase droplets, formation of larger droplets, and phase separation are categorized as emulsion system instabilities that are reflected in the particle size distribution of the emulsion. The measurement of particle size distribution over time allows the characterization of the emulsion stability and determines the rheological behavior of the emulsion. Well-accepted approaches to determine particle size distribution include microscopy, sedimentation, chromatography, and spectroscopy. However, these analyses are problematic in a multiphase emulsion [33].

Crystalline Form Uncontrolled temperature or shear can induce changes in component crystallinity or solubility. For this reason, analytes originally present in each phase of the product should be counted as well possible interactions with the container or closure and the processing equipment analyzed. Some techniques used to obtain information about the emulsion system and its components are microscopic examination, macro- and microlaser Raman, and rheological studies [33]. The FDA guidance offers the following example: "in one instance, residual water remaining in the manufacturing vessel, used to produce an ophthalmic ointment, resulted in partial solubilization and subsequent recrystallization of the drug substance; the substance recrystallized in a larger particle size than expected and thereby raised questions about the product efficacy" [20].

4.3.4.7 Microbiological Quality

Microbial Specifications These specifications are determined by the manufacturer. The USP Chapters 61, 62, and 1111 present the microbial limits to assess the significance of microbial contamination in a dosage form [53]. However, the USP does not determine specific methods for water-insoluble topical products. The microbial specifications are presented as a manufacturer's document that details the methods to isolate and identify the organisms as well as the number of organisms permitted and action levels to be taken when limits are exceed and the potential causes are investigated [6]. The *Pharmaceutical Microbiology Newsletter* (PMF) presents several articles to discuss topics such as microbial identification, methods, data analysis, and preservation as well as topics related to USP and FDA regulations [54]. To minimize the differences about microbial limits and test methods, the USP is trying to harmonize the standards with the European Pharmacopoeia (EP) [55].

Microbial Test Methods The selected microbial test methods determine specific sampling and analytical procedures. When the product has a potential antimicrobial effect and/or preservative, the spread technique on microbial test plates must be validated. In addition, the personnel performing the analytical techniques have to be qualified and adequately trained for this purpose [6].

Usually, total aerobic bacteria, molds, and yeasts are counted by using a standard plate count in order to test the microbial limits. The microbial limit test may be customized by performing a screening for the occurrence of *Staphylococcus aureus, Pseudomonas aeruginosa, Pseudomonas cepacia, Escherichia coli, and Salmonella* sp. [56].

Investigation of Exceeded Microbiological Limits A high number of organisms may indicate deficiencies in the manufacturing process, such as excessive high temperature, component quality, inadequate preservative system, and/or container integrity. Information about the health hazards of all organisms isolated from the product has different meanings depending on the type of dosage form and group of patients to be treated. For instance, in oral liquids, pseudomonads are usually a high-risk contamination. Examples presented by the FDA are Nystatin antifungal suspension, used as prophylaxis in AIDS patients [57]; antacids, with which *P. aeruginosa* contamination can promote gastric ulceration [58]; and the presence of *Pseudomona putida*, which could indicate the presence of other significant contaminants such as *P. aeruginosa* [6].

Deionizer Water-Monitoring Program Deionizing systems must be controlled in order to produce purified water, required for liquid dosage forms and USP tests and assays [1]. The monitoring program has to include the manufacturer's documentation about time between recharging and sanitizing, microbial quality and chlorine levels of feed water, establishment of water microbial quality specifications, conductivity monitoring intervals, methods of microbial testing, action levels when microbial limits are exceeded, description of sanitization and sterilization procedures for deionizer parts, and processing conditions such as temperature, flow rates, use and sanitization frequency, and regenerant chemicals for ion exchange resin beds [6, 59].

Effectiveness of Preservative Manufacturing controls and shelf life must ensure that the specified preservative level is present and effective as part of the stability program [6]. Depending on the type of product, the selection of the preservative system is based on different considerations, such as site of use, interactions, spectrum, stability, toxicity, cost, taste, odor, solubility, pH, and comfort. The USP and other organizations describe methods to validate the preservative system used in the dosage form. Compounds used as preservatives are alcohols, acids, esters, and quaternary ammonium compounds, among others. For instance, to preserve ophthalmic liquid dosage forms, these products are autoclaved or filtrated and require an antimicrobial preservative to resist contamination throughout their shelf life, such as chlorobutanol, benzalkonium chloride, or phenylmercuric nitrate [1].

4.3.4.8 Filling and Packing

Constant Mixing during Filling Process Due to the tendency of suspensions to segregate during transport through transfer lines, special attention is required on suspension uniformity during the filling process. Appropriate constant mixing of the bulk to keep homogeneity during the filling process and sampling of finished products and other critical points are indispensable conditions to assure an acceptable quality level during the filling and packing process [20].

Mixing Low Levels of Bulk Near End of Filling Process Constant mixing during the filling process includes mixing low levels of bulk near the end of the filling process. Large-size batches of bulk suspension require the transfer of the residual material to a smaller tank in order to assure appropriate mixing of components before filling and packing the containers [20].

Potency Uniformity of Unit-Dose Products Products manufactured have to be of quality at least as good as the established acceptable quality level (AQL). The quality level should be based on the limits specified by the USP. However, when the bulk product is not properly mixed during filling and packing processes, liquid dosage forms, and specially suspensions, are not homogeneous and unit-dose products contain very different amounts of the active component and potency. For these reasons, finished products have to be tested to assure that the final volume and/or weight as well as the amount of active ingredient are within the specified limits [6].

Calibration of Provided Measuring Devices Measuring devices consist of droppers, spoons for liquid dosage forms, and cups labeled with both tsp and mL. Measuring devices have to be properly calibrated in order to assure the right amount of ingredients per volume to be administered [6].

Container Cleanliness of Marketing Product The previous cleanliness of containers filled with the product will depend on their transportation exposure, composition, and storage conditions. Glass containers usually carry at least mold spores of different microorganisms, especially if they are transported in cardboard boxes. Other containers and closures made with aluminum, Teflon, metal, or plastic usually have smooth surfaces and are free from microbial contamination but may contain fibers or insects [45]. Some manufacturers receive containers individually wrapped to reduce contamination risks and others use compressed air to clean them. However, the cleanliness of wrapped containers will depend on the provider's guarantee of the manufacturing process and compressed-air equipment may release vapors or oils that have to be tested and validated [6].

4.3.4.9 Stability

The typical stability problems are color change, loss of active component, and clarity changes for solutions; inability to resuspend the particles and loss of significant amounts of the active component for suspensions; and creaming and breaking (or coalescence) for emulsions [1]. These instabilities are usually related to the following:

Active and Primary Degradant. A liquid dosage form is stable while it remains within its product specifications. When chemical degradation products are known, for stability study and expiration dating, the regulatory requirements for the primary degradant of a active component are chemical structure, biological effect and significance at the concentrations to be determined, mechanism of formation and order of reaction, physical and chemical properties, limits and methods for quantitating the active component and its degradant molecule at the levels expected to be present, and pharmacological action or inaction [45]. Examples of drugs in liquid dosage forms that are easily degraded are vitamins and phenothiazines [6].

Interactions with Closure Systems. Elastomeric and plastic container and closure systems release leachable compounds into the liquid dosage form, such as nitrosamines, monomers, plasticizers, accelerators, antioxidants, and vulcanizing agents [44]. Each type of container and closure with different composition and/or design proposed for marketing the drug or physician's samples has to be tested and stability data should be developed. Containers should be stored upright, on their side, and inverted in order to determine if container–closure interactions affect product stability [6, 45].

Moisture Loss. When the containers are inappropriately closed, part of the vaporized solvent is released and the concentration and potency of the active component may be increased [6].

Microbiological Contamination. Inappropriate closure systems also increase the possibilities of microbial contamination when opening and closing containers [6].

4.3.4.10 Process Validation

Objective Process validation has the objectives of identifying and controlling critical points that may vary product specifications through the manufacturing process [6].

Amount of Data To validate the manufacturing process, the manufacturer has to design and specify in the protocol the use of data sheets to keep information about the control of product specifications from each batch in-process as well as finished-product tests. Some formats are common to different products, though each type of product has some specific information to be kept on special sheets. Thus, the amount of data varies from one type of product to another [6].

Scale-Up Process Data obtained using special batches for the validation of the scale-up process are compared with data from full-scale batches and batches used for clinical essays [6].

Product Specifications The most important specifications or established limits for liquid dosage forms are microbial limits and test methods, medium pH, dissolution of components, viscosity, as well as particle size uniformity of suspended components and emulsified droplets. Effectiveness of the preservative system depends on the dissolution of preservative components and may be affected by the medium pH and viscosity. In addition, dissolved oxygen levels are important for components sensitive to oxygen and/or light [6].

Bioequivalence or Clinical Study In the patient, the general or systemic circulation is responsible for carrying molecules to different tissues of the body. To assure the expected bioactivity of a product, the amount of drug that reaches the systemic circulation per unit of time is analyzed and is known as bioavailability. Bioequivalence is the comparison of the bioavailability of a product with a reference product. While oral solutions may not always need bioequivalence studies because they are considered self-evidente, suspensions usually require bioequivalence or clinical studies in order to demonstrate effectiveness. However, OTC suspension products such as antacids are exempt from these studies [6].

Control of Changes to Approved Protocol The manufacturing process of a specific product is validated and approved internally by the quality control unit and externally by the FDA. Any change in the approved protocol has to be documented to explain the purpose and demonstrate that the change will not unfavorably affect product safety and efficacy. Factors include potency and/or bioactivity as well as product specifications. However, the therapeutic activity and uniformity of the product are the main concerns after formulation and process changes [20].

4.3.5 LIQUID DOSAGE FORMS*

Douche A liquid preparation, intended for the irrigative cleansing of the vagina, that is prepared from powders, liquid solutions, or liquid concentrates and contains one or more chemical substances dissolved in a suitable solvent or mutually miscible solvents.

Elixir A clear, pleasantly flavored, sweetened hydroalcoholic liquid containing dissolved medicinal agents; it is intended for oral use.

Emulsion A dosage form consisting of a two-phase system comprised of at least two immiscible liquids, one of which is dispersed as droplets (internal or dispersed phase) within the other liquid (external or continuous phase), generally stabilized with one or more emulsifying agents. (Note: Emulsion is used as a dosage form term unless a more specific term is applicable, e.g. cream, lotion, ointment.).

Enema A rectal preparation for therapeutic, diagnostic, or nutritive purposes.

Extract A concentrated preparation of vegetable or animal drugs obtained by removal of the active constituents of the respective drugs with a suitable menstrua, evaporation of all or nearly all of the solvent, and adjustment of the residual masses or powders to the prescribed standards.

*The definitions in this section are from ref. 11.

For Solution A product, usually a solid, intended for solution prior to administration.

For Suspension A product, usually a solid, intended for suspension prior to administration.

For Suspension, Extended Release A product, usually a solid, intended for suspension prior to administration; once the suspension is administered, the drug will be released at a constant rate over a specified period.

Granule, Effervescent A small particle or grain containing a medicinal agent in a dry mixture usually composed of sodium bicarbonate, citric acid, and tartaric acid which, when in contact with water, has the capability to release gas, resulting in effervescence.

Inhalant A special class of inhalations consisting of a drug or combination of drugs, that by virtue of their high vapor pressure can be carried by an air current into the nasal passage where they exert their effect; the container from which the inhalant generally is administered is known as an inhaler.

Injection A sterile preparation intended for parenteral use; five distinct classes of injections exist as defined by the USP.

Injection, Emulsion An emulsion consisting of a sterile, pyrogen-free preparation intended to be administered parenterally.

Injection, Solution A liquid preparation containing one or more drug substances dissolved in a suitable solvent or mixture of mutually miscible solvents that is suitable for injection.

Injection, Solution, Concentrate A sterile preparation for parenteral use which, upon the addition of suitable solvents, yields a solution conforming in all respects to the requirements for injections.

Injection, Suspension A liquid preparation, suitable for injection, which consists of solid particles dispersed throughout a liquid phase in which the particles are not soluble. It can also consist of an oil phase dispersed throughout an aqueous phase, or vice-versa.

Injection, Suspension, Liposomal A liquid preparation, suitable for injection, which consists of an oil phase dispersed throughout an aqueous phase in such a manner that liposomes (a lipid bilayer vesicle usually composed of phospholipids which is used to encapsulate an active drug substance, either within a lipid bilayer or in an aqueous space) are formed.

Injection, Suspension, Sonicated A liquid preparation, suitable for injection, which consists of solid particles dispersed throughout a liquid phase in which the particles are not soluble. In addition, the product is sonicated while a gas is bubbled through the suspension and these result in the formation of microspheres by the solid particles.

Irrigant A sterile solution intended to bathe or flush open wounds or body cavities; they're used topically, never parenterally.

Linament A solution or mixture of various substances in oil, alcoholic solutions of soap, or emulsions intended for external application.

Liquid A dosage form consisting of a pure chemical in its liquid state. This dosage form term should not be applied to solutions.

Liquid, Extended Release A liquid that delivers a drug in such a manner to allow a reduction in dosing frequency as compared to that drug (or drugs) presented as a conventional dosage form.

Lotion An emulsion, liquid dosage form. This dosage form is generally for external application to the skin.

Lotion/Shampoo A lotion dosage form which has a soap or detergent that is usually used to clean the hair and scalp; it is often used as a vehicle for dermatologic agents.

Mouthwash An aqueous solution which is most often used for its deodorant, refreshing, or antiseptic effect.

Oil An unctuous, combustible substance which is liquid, or easily liquefiable, on warming, and is soluble in ether but insoluble in water. Such substances, depending on their origin, are classified as animal, mineral, or vegetable oils.

Rinse A liquid used to cleanse by flushing.

Soap Any compound of one or more fatty acids, or their equivalents, with an alkali; soap is detergent and is much employed in liniments, enemas, and in making pills. It is also a mild aperient, antacid and antiseptic.

Solution A clear, homogeneous liquid dosage form that contains one or more chemical substances dissolved in a solvent or mixture of mutually miscible solvents.

Solution, Concentrate A liquid preparation (i.e., a substance that flows readily in its natural state) that contains a drug dissolved in a suitable solvent or mixture of mutually miscible solvents; the drug has been strengthened by the evaporation of its nonactive parts.

Solution, for Slush A solution for the preparation of an iced saline slush, which is administered by irrigation and used to induce regional hypothermia (in conditions such as certain open heart and kidney surgical procedures) by its direct application.

Solution, Gel Forming/Drops A solution, which after usually being administered in a drop-wise fashion, forms a gel.

Solution, Gel Forming, Extended Release A solution that forms a gel when it comes in contact with ocular fluid, and which allows at least a reduction in dosing frequency.

Solution/Drops A solution which is usually administered in a drop-wise fashion.

Spray A liquid minutely divided as by a jet of air or steam.

Spray, Metered A non-pressurized dosage form consisting of valves which allow the dispensing of a specified quantity of spray upon each activation.

Spray, Suspension A liquid preparation containing solid particles dispersed in a liquid vehicle and in the form of coarse droplets or as finely divided solids to be applied locally, most usually to the nasal-pharyngeal tract, or topically to the skin.

Suspension A liquid dosage form that contains solid particles dispersed in a liquid vehicle.

Suspension, Extended Release A liquid preparation consisting of solid particles dispersed throughout a liquid phase in which the particles are not soluble; the suspension has been formulated in a manner to allow at least a reduction in dosing frequency as compared to that drug presented as a conventional dosage form (e.g., as a solution or a prompt drug-releasing, conventional solid dosage form).

Suspension/Drops A suspension which is usually administered in a dropwise fashion.

Syrup An oral solution containing high concentrations of sucrose or other sugars; the term has also been used to include any other liquid dosage form prepared in a sweet and viscid vehicle, including oral suspensions.

Tincture An alcoholic or hydroalcoholic solution prepared from vegetable materials or from chemical substances.

Notes:

1. A liquid is pourable; it flows and conforms to its container at room temperature. It displays Newtonian or pseudoplastic flow behavior.
2. Previously the definition of a lotion was "The term lotion has been used to categorize many topical suspensions, solutions, and emulsions intended for application to the skin." The current definition of a lotion is restricted to an emulsion.
3. A semisolid is not pourable; it does not flow or conform to its container at room temperature. It does not flow at low shear stress and generally exhibits plastic flow behavior.
4. A colloidal dispersion is a system in which particles of colloidal dimension (i.e., typically between 1 nm and 1 μm) are distributed uniformly throughout a liquid.

REFERENCES

1. Crowley, M. M. (2005), Solutions, emulsions, suspensions, and extracts, in USIP, *Remington: The Science and Practice of Pharmacy*, Lippincott Williams & Wilkins, Philadelphia, pp. 745–774.
2. Block, L. H. (2002), Nonparenteral liquids and semisolids, in Levin, M., Ed., *Pharmaceutical Process Scale-Up*, Marcel Dekker, New York, pp. 57–94.
3. Spurgeon, T. (2007), Quality by design in solid dosage processes: How will QbD impact manufacturing? *Contract Pharm.*, 3–1.
4. Allen, L. V. *The art, Science, and Technology of Pharmaceutical Compounding*. American Pharmaceutical Association (APhA). Washington D.C., 2002. Pages 93, 95.
5. Qu, A., and Maglayo, A. (2003), Outsourcing stability services. *Contract Pharma.* 11–12. Available at: http://www.contractpharma.com/articles/2003/11/outsourcing-stability-services
6. *USA Department of Health and Human Services: Food and Drug Administration (FDA).* Guides to inspections. Available at http://www.fda.gov/ora/Inspect_ref/igs/iglist.html

7. Kourounakis, P. N. (1994), *Advanced Drug Design and Development*, Taylor and Francis, p. 141.
8. Papich, M. G. (2005), Drug compounding for veterinary patients, *AAPS J.*, 7(2), E281–E287.
9. Wiberg, C. C., Leppik, I. E., and Cloyd, J. C. (2005), Lower phenytoin serum levels in persons switched from brand to generic phenytoin, *Neurol. J.*, 64(8), 1485–1486.
10. Brooks, K. (2007), Pharma IT outsourcing: ERP consolidates IT infrastructure, *Contract Pharm*, March 2007, available: http://www.contractpharma.com/articles/2007/03/pharma-it-outsourcing
11. *USA Department of Health and Human Services, Food and Drug Administration (FDA).* Center for Drug Evaluation and Research (CDER) Data Standards Manual: Dosage Form (4/92) available at: http://www.fda.gov/cder/dsm/DRG/drg00201.htm.
12. Drugs@FDA glossary of terms, U.S. Department of Health and Human Services, Food and Drug Administration (FDA), available: http://www.fda.gov/cder/drugsatfda/glossary.htm#L.
13. DeLuca, P. P. (1992), Formulation of small volume parenterals, in Avis, K. E., Lieberman, H. A., and Lachman, L., Eds., *Pharmaceutical Dosage Forms: Parenteral Medications*, Vol. 1, Marcel Dekker, New York.
14. Allen, L. V., Popovich, N. G., and Ansel, H. C., *Ansel's Pharmaceutical Dosage Forms and Drug Delivery Systems*, Lippincott William & Wilkins, Baltimore, 2004, Chapters 2 and 5.
15. SAP Information Center, Best practices for automotive: Pharmaceutical industry, Liquid dosage forms, available: http://help.sap.com/.
16. Sokoll, K. (2007), Optimizing drug development strategies: Exploring methods to improve the process, *Contract Pharm.* Marzo 2007. Available: http://www.contractpharma.com/articles/2007/03/optimizing-drug-development-strategies
17. Carleton, F. J., and Agalloco, J. P. (1999), *Validation of Pharmaceutical Processes: Sterile Products*, Marcel Dekker, New York.
18. Stacey, R. D., Griffin, D., and Shaw, P. (2000), *Complexity and Management: Fad or Radical Challenge to Systems Thinking?* Taylor and Francis Group, Boca Raton, Fl.
19. Hardman, J. G., Limbrid, L. E., Molinoff, P. B., and Ruddon, R. W. *Goodman and Gilman's The Pharmacological Basis of Therapeutics*, McGraw-Hill, New York, 2006.
20. Scott, R. R. (1989), A practical guide to equipment selection and operating techniques, in Lieberman, H. A., Rieger, M. M., and Banker, G. S., Eds., *Pharmaceutical Dosage Forms: Disperse Systems*, Vol. 2, Marcel Dekker, New York, pp 1–71.
21. Reepmeyer, J. C., Revelle, L. K., and Vidavsky, I. (1998), Detection of clobetasol propionate as an undeclared steroid in zinc pyrithione formulations by high-performance liquid chromatography with rapid-scanning ultraviolet spectroscopy and mass spectrometry. *Journal of Chromatography A*, Vol. 828, Number 1, 18 December 1998, pp. 239–246(8).
22. FARR Air Pollution Control (2005), FARR dust collectors, HEPA BIBO filtration systems, Pharmaceutical Industry Standard On-line, available: http://www.pharmaceuticalonline.com/downloads/detail.aspx?docid=74B77B1C-0ED9-4E7B-A317-5DD0D23124B3&VNETCOOKIE=NO.
23. Biotrace International (2004), Air samplers. LIT055/002/29.06.04, available: http://www.biotrace.co.uk/uploads/documents/103.pdf.
24. 3-A Sanitary Standards, Technical Resource Center, resource papers, available: http://www.3-a.org/techresource/papers.htm.
25. Cook, Jr., J. L. (1998), *Standard Operating Procedures and Guidelines*, PennWell Books, Tulsa, OK, p. 318.

26. Guides to inspections: Emulsions, U.S. Department of Health and Human Services, Food and Drug Administration (FDA), available: http://www.fda.gov/ora/Inspect_ref/igs/iglist.html.
27. Guides to inspections: Validation of cleaning processes, U.S. Department of Health and Human Services, Food and Drug Administration (FDA), available: http://www.fda.gov/ora/Inspect_ref/igs/valid.html.
28. Thierauf, R. J., and Hoctor, J. J. (2006), *Optimal Knowledge Management: Wisdom Management Systems Concepts and Applications*, Idea Group, Hershey, PA.
29. Petruševski, V. M., and Metodija, Z. N. (2001), Volume non-additivity of liquid mixtures: Modifications to classical demonstrations, *Chem. Ed. J.*, 6, 3.
30. Zhou, Q., and Milton, R. (2003), Molecular interactions of surfactants in mixed monolayers at the air/aqueous solution interface and in mixed micelles in aqueous media: The regular solution approach, *Am. Chem. Soc.*, 19(11), 4555–4562.
31. CANTY Process Technology (2007), CANTY Lighting Systems, available: http://www.jmcanty.com/images/overview/16pagermaster.pdf.
32. Wikipedia, Non-Newtonian fluid definition, available: http://en.wikipedia.org/wiki/Non-Newtonian_fluid.
33. Lieberman, H. A., Rieger, M. M., and Banker, G. S. (1988), *Pharmaceutical Dosage Forms: Disperse Systems*, Vol. 1, Marcel Dekker, New York.
34. Gibson, M., *Pharmaceutical Preformulation and Formulation: A Practical Guide from Candidate Drug Selection to Commercial Dosage Form*, Interpharm, Boca Raton, FL, 2004, pp. 581.
35. Davidow, L. W., Davidow, L., and Thompson, J. E., *A Practical Guide to Contemporary Pharmacy Practice*, Lippincott Williams & Wilkins, 2003. pp. 704.
36. Sinko, P. J. *Martin's Physical Pharmacy and Pharmaceutical Sciences*, Lippincott Williams & Wilkins, Baltimore, MD., 2005, pp. 795. 47.
37. Nusim, S. (2005), *Active Pharmaceutical Ingredients: Development, Manufacturing, and Regulation*, Taylor and Francis Group, Boca Raton, FL.
38. Chongprasert, S., and Nail S. L. (1998), Influence of drug polymorphism on the physical chemistry of freeze-drying, Digital dissertations and theses, Purdue e-pubs, *ETD Collection for Purdue University*, available: http://docs.lib.purdue.edu/dissertations/AAI9939331/.
39. Borka, L. (1971), The stability of chloramphenicol palmitate polymorphs. Solid and solution phase transformations, *Acta Pharm. Suec.*, 9–8(4), 365–372.
40. Rosenholm, J. B. (2007), Wetting of surfaces and interfaces: A conceptual equilibrium thermodynamic approach, in *Colloid Stability: The Role of Surface Forces, Part II*, Tharwat F. Tadros Colloids and Interface Science Series, Vol. 2, Wiley-VCH, Weinheim.
41. Lerk, C. F., Lagas, M., Lie-a-Huen, L., Broersma, P., and Zuurman, K. (1979), In vitro and in vivo availability of hydrophilized phenytoin from capsules, *J. Pharm. Sci.*, 5–68(5), 634–638.
42. Rosoff, M. (1988), Specialized pharmaceutical emulsions, in Lieberman, H. A., Rieger, M. M., and Banker, G. S., Eds., *Pharmaceutical Dosage Forms: Disperse Systems*, Vol. 1, Marcel Dekker, New York, pp. 245–283.
43. Moorthaemer, B., Sprakel, J. (2006) Improving the stability of a suspension. Pharmaceutical Technology Europe, O1 February 2006. Available: http://www.ptemag.com/pharmtecheurope/article/articleDetail.jsp?id=306687
44. Allen, T. (1997), *Particle Size Measurement*, Chapmann and Hall, London.
45. Connors, K. A., Amidon G. L., and Stella V. J. (1986), *Chemical Stability of Pharmaceuticals: A Handbook for Pharmacists*, Wiley-Interscience, New York.

46. Miller, S. C., and Drabik, B. R. (1984), Rheological properties of poloxamer vehicles, *Int. J. Pharm.*, 18, 269.
47. Millard, J. W., Alvarez-Nunez, F. A., and Yalkowsky, S. H. (2002), Solubilization by cosolvents—Establishing useful constants for the log-linear model, *Int. J. Pharm.*, 245(1), 153–166 (14).
48. Nash, R. (1996), Pharmaceutical suspensions, in Lieberman, H. A., Rieger, M. M., and Banker, G. S., Eds., *Pharmaceutical Dosage Forms: Disperse Systems*, Vol. 2, Marcel Dekker, New York, pp. 1–46.
49. Adeyeye, C. M., and Price, J. C. (1997), Chemical dissolution stability and microscopic evaluation of suspensions of ibuprofen and sustained release ibuprofen-wax microspheres, *J. Microencapsul.*, 14(3), 357–377.
50. Fong-Spaven, F., and Hollenbeck, R. G. (1986), Thermal rheological analysis of triethanolamine-stearate stabilized mineral oil in water emulsions, *Drug Dev. Ind. Pharm.*, 12, 289.
51. Moelwyn-Hughes, E. A. (1961), *Physical Pharmacy*, Pergamon, New York, pp. 297.
52. Lan, C. W. (2005), Flow and segregation control by accelerated rotation for vertical Bridgman growth of cadmium zinc telluride: ACRT versus vibration, *J. Crystal Growth*, 274 (3–4), 379–386.
53. USP 29: U.S. Pharmacopeia and the National Formulary (USP 29/NF 24), United States Pharmacopeial Convention, Rockville, MD. 2006.
54. PMF Pharmaceutical Microbiology Forum Newsletters, available: http://www.microbiologyforum.org/news.htm.
55. Sutton, S. (2006), The harmonization of the microbial limits test—Enumeration, *Microbiol. Net.—PMF*, 12(3), 2–3.
56. Carstensen, J. T., and Rhodes, C. T. (2000), *Drug Stability: Principles and Practices*, Macel Dekker, New York, pp. 773.
57. Robinson, E. P. (2006), Pseudomonas aeruginosa contamination of liquid antacids: A survey, *J. Pharm. Sci.*, 60(4), 604–605.
58. Steer, H. W., and Colin-Jones, D. G. (1975), Mucosal changes in gastric ulceration and their response to carbenoxolone sodium. *Gut BMJ*, 16, 590–597.
59. Dreeszen, P. (2003), *Biofilm: Key to Understanding and Controlling Bacterial Growth in Automated Drinking Water Systems*, Edstrom Industries, Waterford, WI, pp. 19.

SECTION 5

SPECIAL/NEW DOSAGE FORMS

5.1

CONTROLLED-RELEASE DOSAGE FORMS

ANIL KUMAR ANAL
Living Cell Technologies (Global) Limited, Auckland, New Zealand

Contents

Pharmaceutical Manufacturing Handbook: Production and Processes, edited by Shayne Cox Gad

5.1.1 INTRODUCTION

Therapeutic value and pharmaeconomic value have in recent years become major issues in defining health care priorities under the pressure of cost containment [1]. The improvement in drug therapy is a consequence of not only the development of new chemical entities but also the combination of active substances and a suitable delivery system. The treatment of an acute disease or chronic illness is mostly accomplished by delivery of one or more drugs to the patient using various pharmaceutical dosage forms. Tablets, pills, capsules, suppositories, creams, ointments, liquids, aerosols, and injections are in use as drug carriers for many decades. These conventional types of drug delivery systems are known to provide a prompt release of the drug. Therefore, to achieve as well as to maintain the drug concentration within the therapeutically effective range needed for treatment, it is often necessary to take this type of drug several times a day, resulting in the significant fluctuation in drug levels [2]. For all categories of treatment, a major challenge is to define the optimal dose, time, rate, and site of delivery. Recent developments in drug delivery techniques make it possible to control the rate of drug delivery to sustain the duration of therapeutic activity and/or target the delivery of drug to a specific organ or tissue. Many investigations are still going on to apply the concepts of controlled delivery for a wide variety of drugs [3].

5.1.2 RATIONALE

The basic rationale for controlled drug delivery is to alter the pharmacokinetics and pharmacodynamics of pharmacologically active moieties by using novel drug delivery systems or by modifying the molecular structure and/or physiological parameters inherent in a selected route of administration. It is desirable that the duration of drug action become more a design property of a rate-controlled dosage form and less, or not at all, a property of the drug molecules' inherent kinetic properties.

The rationale for development and use of controlled dosage forms may include one or more of the following arguments [4]:

- Decrease the toxicity and occurrence of adverse drug reactions by controlling the level of drug and/or metabolites in the blood at the target sites.
- Improve drug utilization by applying a smaller drug dose in a controlled-release form to produce the same clinical effect as a larger dose in a conventional dosage form.
- Control the rate and site of release of a drug that acts locally so that the drug is released where the activity is needed rather than at other sites where it may cause adverse reactions.
- Provide a uniform blood concentration and/or provide a more predictable drug delivery.
- Provide greater patient convenience and better patient compliance by significantly prolonging the interval between administrations.

However, there are also disadvantages attached to the use of controlled-release dosage forms. These include higher cost of manufacturing, unpredictability, poor in vitro/in vivo correlation, reduced potential, and poor systemic availability in general and the effective release period is influenced and limited by the gastrointestinal (GI) residence time [5]. The transit time of a dosage form through the GI tract is dependent on the physical characteristics of the formulation as well as on physiological factors such as stomach emptying time and effect of food on the absorption process.

Only drugs with certain properties are suitable for controlled-release dosing. Characteristics that may make a drug unsuitable for controlled-release dosing include a long or short elimination half-life, a narrow therapeutic index, a large dose, low/slow solubility, extensive first-pass clearance, and time course of circulating drug levels different from that of the pharmacological effect. The ideal drug delivery should be inert, biocompatible, mechanically strong, comfortable for the patient, capable of achieving high drug loading, simple to administer, and easy to fabricate and sterilize [6]. A range of materials have been employed to control the release of drugs and other active substances. Controlled-release dosage forms have been developed for over four decades. One of the first practically used controlled-release oral dosage forms was the Spansule capsule, which was introduced in the 1950s. Spansule capsules were manufactured by coating a drug onto nonpareil particles and further coating with glyceryl stearate and wax. Subsequently, ion exchange resins were proposed for application as sustained-release delivery systems of accessible drug. Since then numerous products have been introduced and commercialized.

5.1.3 GENERAL DESIGN PRINCIPLE FOR CONTROLLED-RELEASE DRUG DELIVERY SYSTEMS

In the drug delivery system, the pharmacodynamics of active molecules becomes more a function of design and less one of inherent kinetic properties. Therefore, a deep understanding of the design of controlled-release systems of the pharmacokinetics and pharmacodynamics of the drug is required [7]. The conventional tablet or capsule provides only a single and transient burst of drug. A modification introduced to the molecular structure of the drug (often used to decrease the elimination rate) or a system for modified release rate is the common approaches used to increase the interval between two doses. The objective of both these approaches is to decrease the fluctuations in plasma levels during multiple dosing. This allows the dosing interval to increase without compromising the required dosage levels. If the half-life of a drug is less than 6h or the passage time in the smaller intestinal track is decreased, there might not be enough time to allow proper absorption, thus making frequent dosing compulsory. For other routes, where the residence time is not a constraint, dosing intervals can be as long as months or even years.

A controlled-release drug delivery system serves primarily two functions [8]. First, it involves the transport of the drug to a particular part of the body. This may be accomplished in two ways, parenterally and nonparenterally. Second, the release of active ingredients occurs in a controlled manner, depending on the preparation of dosage forms. This determines the rate at which a drug is made available to the body once it has been delivered. Controlled drug delivery occurs when a biomaterial, either natural or synthetic, is judiciously combined with a drug or other active agent in such a way that the active agent is released from the material in a predesigned manner. To be successfully used in controlled drug delivery formulations, a material must be chemically inert and free of leachable impurities. It must also have an appropriate physical structure, with minimal undesired aging, and be readily processable.

Controlled-release systems provide numerous benefits over conventional dosage forms. Conventional dosage forms are not able to control either the rate of drug delivery or the target area of administration and provide an immediate or rapid drug release. This necessitates frequent administration in order to maintain a therapeutic level. As a result, as shown in Figure 1, drug concentrations in the blood fluctuate widely. The concentrations of drug remain at a maximum value, which may represent a toxic level, or a level at which undersized side effects might occur, and a minimum value, below which the drug is no longer effective. The duration of therapeutic efficacy is dependent upon the frequency of administration, the half-life of the drug, and the release rate of dosage forms. In contrast, controlled-release dosage forms not only are able to maintain therapeutic levels of drug with narrow fluctuations but also make it possible to reduce the frequency of drug administration. The drug concentrations, as shown in Figure 1, released from controlled-release dosage forms fluctuate within the therapeutic range over a longer period of time. The plasma concentration profile depends on the preparation technology, which may generate different release kinetics, resulting in different pharmacological and pharmacokinetic responses in the blood or tissues.

The primary objectives of controlled drug delivery are to ensure safety and to improve efficacy of drugs as well as patient compliance. This is achieved by better control of plasma drug levels and less frequent dosing. For conventional dosage forms, only the dose (D) and dosing interval (τ) can vary above which undesirable or side effects are elicited. As an index of this window, the therapeutic index (TI) can be used. This is often defined as the ratio of lethal dose (LD_{50}) to median effective dose (ED_{50}). Alternatively, it can be defined as the ratio of maximum drug concentration (C_{max}) in blood that can be tolerated to the minimum concentration (C_{min}) needed to produce an acceptable therapeutic response.

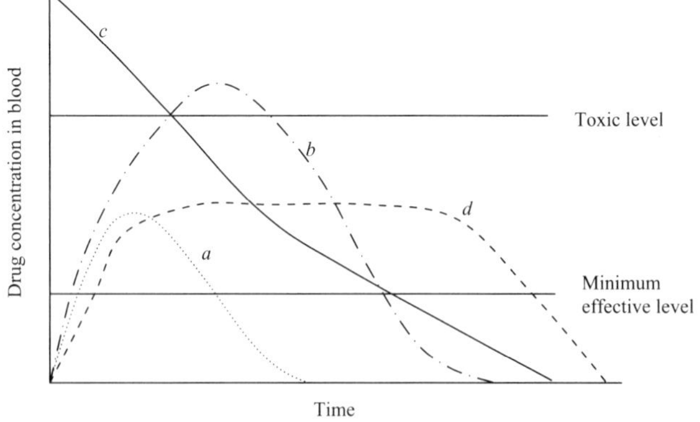

FIGURE 1 Theoretical plasma concentration after administration of various dosage forms: (*a*) standard oral dose; (*b*) oral overdose; (*c*) IV injection; (*d*) controlled-release system.

Different types of modified release systems can be defined [4, 8]:

- Sustained release (extended release) that permits a reduction in dosing frequency as compared to the situation in which the drug is presented as a conventional form
- Delayed release when the release of the active ingredient comes sometimes other than promptly after administration
- Pulsatile release when the device actively controls the dosage released following predefined parameters

In general, the sustained-release dosage form is designed to maintain therapeutic blood or tissue levels of the drug for an extended period of time. This is accomplished by attempting to obtain zero-order release from the designed dosage form. Zero-order release constitutes drug release from the dosage form that is independent of the amount of drug in the delivery system at a constant release rate. Systems that are designed for prolonged release can also be attributed as achieving sustained-release delivery systems. Repeat-action tablets are an alternative method of sustained release in which multiple doses of drugs are contained within a dosage form and each dose is released at a periodic interval, while delayed-release systems may not be sustaining, since often the function of these dosage forms is to maintain the drug within the dosage form.

5.1.4 PHYSICOCHEMICAL AND BIOLOGICAL FACTORS INFLUENCING DESIGN AND PERFORMANCE OF CONTROLLED-RELEASE FORMULATIONS

A number of variables, such as drug properties including stability, solubility, partitioning characteristics. charge and protein binding behavior, routes of drug delivery, target sites, acute or chronic therapy, the disease, and the patient, must be considered to establish the criteria for designing controlled-release products [9]. The performance of a drug in its release pattern from the dosage form as well as in the body proper is a function of its properties. These properties can at times prohibit placement if the drug is in a controlled-release form, restrict the route of drug administration, and significantly modify performance for one reason or another. There is no clear distinction between physicochemical and biological factors since the biological properties of a drug are a function of its physicochemical properties while biological properties result from typical pharmacokinetic studies on the absorption, distribution, metabolism, and excretion (ADME) characteristics of a drug as well as those resulting from pharmacological studies.

5.1.4.1 Physicochemical Factors

Physicochemcial properties, such as aqueous solubility, partition coefficient and molecular size, drug stability, and protein binding, are those that can be determined from in vitro experiments.

Ionization, pK$_a$, and Aqueous Solubility Most drugs are weak acids or bases. It is important to note the relationship between the pK_a of the compound and the absorptive environment. Delivery systems that are dependent on diffusion or dissolution will likewise be dependent on the solubility of drug in the aqueous media. Since drugs must be in solution before they can be absorbed, compounds with very low aqueous solubility usually have the oral bioavailability problems because of limited GI transit time of the undissolved drug particles and they are limited at the absorption site. Unfortunately, for many of the drugs and bioactive compounds, the site of maximum absorption occurs at the site where solubility of these compounds is least.

The drug (e.g., tetracycline) for which the maximum solubility is in the stomach but high absorption takes place in the intestinal region may be poor candidates for controlled-release systems, unless the system is capable of retaining the drug in the stomach and gradually releasing it to the small intestine or unless the solubility is made higher and independent of the external environment by encapsulating those compounds in a membrane system. Other compounds, such as digoxin [10], with very low solubility, are inherently sustained, since their release over the time course of a dosage form in the gastrointestinal tract is limited by dissolution of the drug. Although the action of a drug can be prolonged by making it less soluble, this may occur at the expense of consistent and incomplete bioavailability.

The choice of mechanism for oral sustained/controlled-release systems is limited by the aqueous solubility of the drug. Thus, diffusional systems are poor choices for low aqueous-soluble drugs since the driving force for diffusion, the concentration in aqueous solution, will be low. The lower limit for the solubility of a drug to be formulated in a controlled-release system has been reported to be 0.1 mg/mL [11].

Partition Coefficient and Molecular Size Following administration, drugs and other bioactive compounds must traverse a variety of membranes to gain access to the target area. The partition coefficient and molecular size influence not only the permeation of drug across biological membranes but also diffusion across or through a rate-controlling membrane or matrix. The partition coefficient is generally defined as the ratio of the fraction of drug in an oil phase to that of an adjacent aqueous phase. Drugs with extremely high partition coefficient (i.e., those that are highly oil soluble) readily penetrate the membranes but are unable to proceed further, while the excessive high aqueous-soluble compounds, having low oil/water partition coefficients, cannot penetrate the membranes. A balance in the partition coefficient is needed to give an optimum flux for permeation through the biological and rate-controlling membranes. The ability of drugs to diffuse through membranes, also known as diffusivity, is related to its molecular size by the following equation:

$$\log D = -s_V \log V + k_V = -s_M \log M + k_M$$

where D is the diffusivity, M is the molecular weight, V is the molecular volume, and s_V, s_M, k_V, and k_M are constants in a particular medium. Generally, there is smaller diffusivity with the denser medium.

Drug Stability The stability of drug in the environment where it is to be exposed is an essential physicochemical factor to be considered before designing controlled dosage forms [12]. For example, orally administered drugs are subjected to both acid–base hydrolysis and enzymatic degradation [13]. For drugs that are unstable in the stomach, the dosage forms can be designed in so that they can be placed in a slowly soluble form or have their release delayed until they reach the intestine. This type of approach can be ineffective and the drug may be unstable in the small intestine or undergo extensive gut-wall metabolism. To obtain better bioavailability for such types of drugs, which are unstable even in the intestine, a different route of administration (e.g., transdermal with controlled-release dosage forms) can be a better option [14]. A transdermal patch of nitroglycerin is a good example. The details for transdermal dosage forms will be described later in this chapter.

5.1.4.2 Biological Factors

A drug, being a chemical/biological agent or a mixture of chemical and biological agents, is recognized as a xenobiotic by the human body. Subsequently, the drug will be prevented from entering the body and/or eliminated after its entry. As a result, the defense mechanisms of the human body become barriers to the delivery of drugs. A drug may encounter physical, physiological, enzymatic, or immunological barriers on its way to the site of action. Hence, the design of controlled-release product should be based on a comprehensive picture of drug disposition. This would entail a complex examination of the ADME characteristics of the drug. The details of these effects on various controlled-release dosage systems will be given in the following sections of this chapter.

5.1.5 CONTROLLED-RELEASE ORAL DOSAGE FORMS

Oral drug delivery is the preferred route for drug administration because of its convenience, economy, and high patient compliance compared with several other routes. About 90% of the drugs are administered via the oral route [15, 16]. For the oral controlled administration of drugs, several research and development activities have shown encouraging signs of progress in the development of programmable controlled-release dosage forms as well as in the search for new approaches to overcome the potential problems associated with oral drug administration. Many oral drugs are perceived as "patient friendly" for compliance, often requiring that the medication only needs to be taken once a day. The most prescribed drugs in the United States that use oral drug delivery technologies include Lipitor (atorvastatin calcium), manufactured by Pfizer, and AstraZeneca's Toprol-XL (metoprolo succinate). These potential developments and recently developed approaches are discussed here along with an overview of GI physiology.

5.1.5.1 Anatomical and Physiological Considerations

Anatomically, the alimentary canal can be divided into a conduit region and digestive and absorptive regions. The conduit region includes the mouth, pharynx, esophagus, and lower rectum. The digestive and absorptive regions include the stomach, small intestine, and all parts of the large intestine except the very distal region.

TABLE 1 Gastrointestinal Tract: Physical Dimensions and Dynamics

			Transit Time	
Region	Surface area (m^2)	pH	Fluid	Solid
Stomach	0.1–0.2	1.2	50 min	8 h
Small intestine	100	6.8	2–6 h	4–9 h
Large intestine	0.5–1	7.5	2–6 h	3–72 h

The role of the stomach in drug and nutrition absorption is very limited, and it acts primarily as a reception area for oral dosage forms. Nonionic, lipophilic molecules of moderate size can be absorbed through the stomach only to a limited extent owing to the small epithelial surface area and the short duration of contact with the stomach epithelium in comparison with the intestine [17]. The transit time in the GI tract varies from one person to another and also depends upon the physical properties of the object ingested and the physiological conditions of the alimentary canal (Table 1). After passing through the stomach, the next organ that a drug or bioactive compound encounters is the small intestine. The intestinal epithelium is composed of absorptive cells (enterocytes) interspersed with goblet cells (specialized for mucus secretion) and a few enteroendocrine cells (that release hormones). The enterocytes of intestinal epithelium are the most important cells in view of the absorption of drugs and nutrients [18]. Histologically, colonic mucosa resembles the small intestinal mucosa, the absence of villi being the major difference [19]. The microvilli of the large intestine enterocytes are less organized than those of the small intestine. The resulting decrease in the surface area of the colon leads to a low absorption potential in comparison with the small intestine. However, the colonic residence time is longer than that for the small intestine, providing extended periods of time for the slow absorption of drugs and nutrients [20]. Figure 2 shows the various physiological processes encountered by an orally administered drug during the course of GI transit.

5.1.5.2 Fundamentals of Controlled-Release Oral Dosage Forms

Oral controlled drug delivery is a system that provides the continuous delivery of drugs at predictable and reproducible kinetics for a predetermined period throughout the course of GI transit [21]. Also included are systems that target the delivery of a drug to a specific region within the gastrointestinal tract (GIT) for either local or a systemic action. All the oral controlled drug delivery systems have limited utilization in the GI controlled administration of drugs if the systems cannot remain in the vicinity of the absorption site for the lifetime of the drug delivery. In the exploration of oral controlled-release dosage forms, one encounters three areas of potential challenges [22]:

1. *Drug Delivery System* To develop a viable oral controlled-release drug delivery system capable of delivering a drug at a therapeutically effective rate to a desirable site for the duration required for optimal efficacy.
2. *Modulation of GI Transit Time* To modulate the GI transit time so that the drug delivery system developed can be transported to a target site or to the

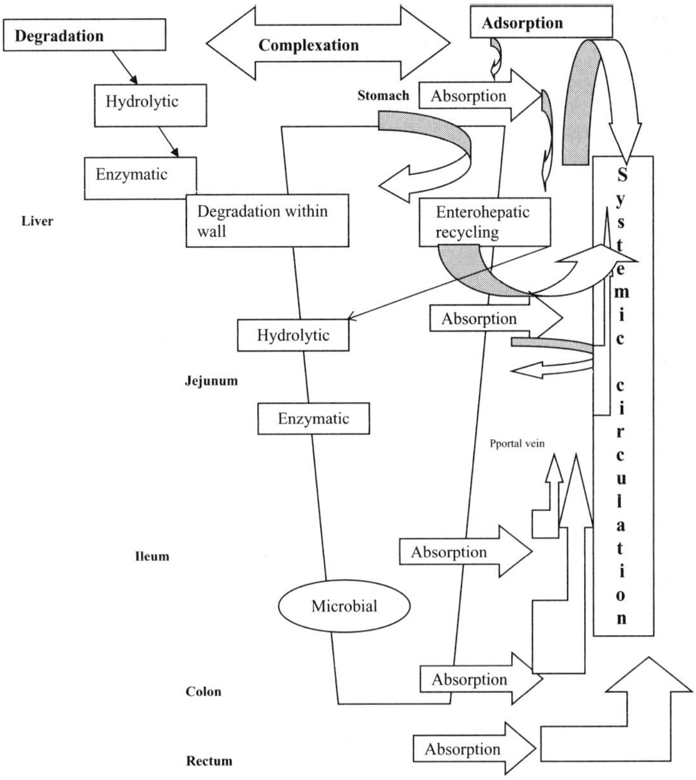

FIGURE 2 Physical model illustrating various physiological processes during gastrointestinal transit.

vicinity of an absorption site and reside there for a prolonged period of time to maximize the delivery of a drug dose.

3. *Minimization of Hepatic First-Pass Elimination* If the drug to be delivered is subjected to extensive hepatic first-pass elimination, preventive measures should be devised to either bypass or minimize the extent of hepatic metabolic effect.

With most orally administered drugs, targeting is not a primary concern, and it is usually intended for drugs to permeate to the general circulation and perfuse to other body tissues, except it is medicated intentionally for localized effects in the GIT. There is a general assumption that increasing concentration at the absorption site will increase the rate of absorption and, therefore, increase circulating blood levels, which in turn promotes greater concentrations of drug at the site of action. If toxicity is not an issue, therapeutic levels can thus be extended as shown in Figure 3. In essence, "drug delivery" by these systems usually depends on release from specific types of dosage forms and permeation through an epithelial membrane to the blood. Still, the biological and physicochemical factors that come across play important roles in the design of such systems. The physicochemical properties have been described earlier in this chapter while biological factors involved with oral dosage forms will be described below.

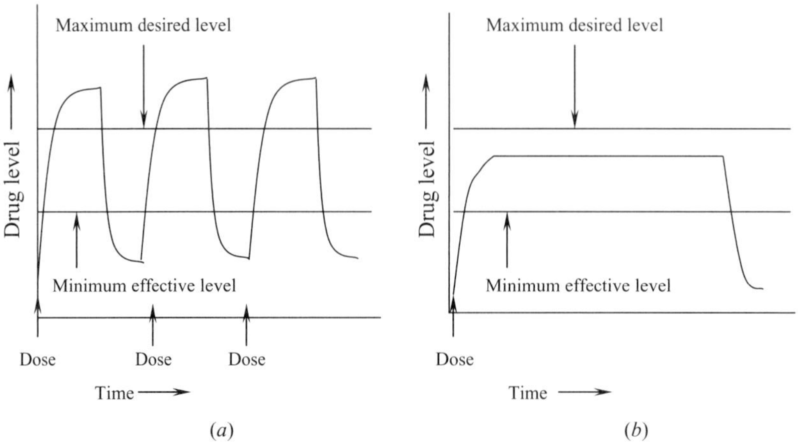

FIGURE 3 Profile of drug level in blood: (*a*) traditional dosing of tablets; (*b*) controlled drug delivery dose.

The degree to which a delivery system can achieve standard release profiles for a variety of chemically and physically diverse, pharmaceutically active molecules is a measure of a delivery system's efficacy and flexibility (Figure 3). Among the most challenging profiles, linear, zero-order release of highly soluble actives over a 12–24-h period could be considered a reasonable performance standard against which delivery systems may be judged.

5.1.5.3 Factors Influencing Oral Controlled-Release Dosage Forms

Biological Half-Life The usual goal of an oral controlled-release dosage form is to maintain therapeutic blood levels, as shown in Figure 3, over an extended period of time. A drug must be absorbed and enter the circulation at approximately the same rate at which it is eliminated. The elimination rate is quantitatively described by the half-life ($t_{1/2}$). Each drug has its own characteristic elimination rate, which is the sum of all elimination process, including metabolism, urinary excretion, and all other processes that permanently remove drug from the bloodstream.

Therapeutic compounds with short half-lives are excellent candidates for controlled/sustained-release preparations, since this can reduce dosing frequency [23]. In general, drugs with half-lives shorter than 2 h, such as furosemide or levodopa, are poor candidates for controlled-release preparations. Compounds with longer half-lives, such more than 8 h, also do not need to be in the form of controlled release, since their effect is already sustained. Digoxin, warfarin and phenytoin are some examples [24–26]. However, drugs having even longer half-lives can be used in other forms of modified release, such as pulsatile release.

Gastrointestinal Tract and Absorption The design of a controlled-release dosage form should be based on a comprehensive picture of drug disposition. Both the pharmacokinetic property and biological response parameter have a useful range for the design of sustained- and controlled-release products. The potential problems inherent in oral controlled-release oral dosage forms generally relate to (i) interactions between the rate, extent, and location that the dosage form releases the drug and (ii) the regional differences in GI physiology [27].

Total GI transit time in the normal population varies from 5 to 36h, with an average total transit time of approximately 24h [28]. There is still much more to explore about the physiological processes involved and factors that influence gastric emptying, intestinal transit, and colon residence. One of the major factors is food administration, which delays the "housekeeper wave," causing delay in gastric emptying. A high-fat meal may delay gastric emptying from 3 to 5h, and the total GI transit delay is largely a function of delay in gastric emptying in this case. If we presume that the transit time of most drugs and devices in the absorptive areas of the GI tract is about 8–12h, the maximum half-life for absorption should be approximately 3–4h; otherwise, the device will pass out of the potential absorptive regions before drug release is complete. That many controlled-release products are having somewhat lesser bioavailability than their conventional dosage forms may be due to incomplete release of the dosage form or release at such a slow rate that the drug has passed the actual site of absorption. Compounds that demonstrate a true lower absorption rate constant will probably be poor candidates for controlled-release dosage forms [28].

An understanding of the behavior of dosage forms in the stomach has been gained largely from scintographic studies in which phases of a meal and formulations are labeled with different nucleotides, particularly technetium-99 and indium-111 [29]. Such studies have demonstrated that retention times of formulations in the stomach are dependent on the size of formulations and whether or not the formulation is taken with a meal. Enteric-coated or enteric matrix tablets may be retained for a considerable time if dosed with heavy meals or breakfast. Multiparticulate dosage forms will empty more slowly in the presence of food, and because the dosage forms mix evenly with the food, the entry into the small intestine will be strongly influenced by the caloric density and bulk of the ingested meal. The rate of gastric emptying, therefore, predicts the absorption behavior [29, 30]. However, the absorption of drugs from small, soft gelatin capsules is sometimes less predictable [31].

In the small intestine, contact time with the absorptive epithelium is limited, and the small intestinel transit time is 3.5–4.5h [32]. The scintographic data show that many prolonged release products, particularly those intended for twice- or once-daily administration, actually release some of their drug contents in the colon where it may be absorbed into the systemic circulation for higher bioavailability. It is anticipated that conditions of dissolution, absorption, and metabolism in the distal portions of the intestine are different than in the proximal regions, due to differences in pH, lumen fluid, mucosal morphology, and motility.

For most formulations, colonic absorption represents the only real opportunity to increase the interval between doses. Transit through the lower part of the gut is quoted at about 24h, but in reality only the ascending colonic environment has sufficient fluid to facilitate dissolution. In the cecum, the fermentation of soluble fiber produces fatty acids and gas [33]. The gas rises into the transverse colon and can form temporary pockets, restricting access of water to the formulation. Consequently, distal release of drug is associated with poor spreading, reduced surface area, and restricted absorption. In the colon, water availability is also low past the hepatic flexure, as the ascending colon is extremely efficient at water absorption [34].

5.1.6 DESIGN AND FABRICATION OF CONTROLLED-RELEASE DOSAGE FORMS

5.1.6.1 Microencapsulation

Microencapsulation has been the subject of massive research efforts since its inception around 1950. Today, it is the mechanism utilized by approximately 65% of all sustained-release systems [35]. The technique's popularity can be attributed mainly to its wide variety of applications. Hundreds of drugs have been microencapsulated and used as controlled-release systems. Some examples are Arthritis Bayer, Dexatrim Capsules, and Dimetapp Elixir.

Microencapsulation provides more efficient drug delivery because it increases the ability of the drug to interact with the body. The active ingredient of a drug is encapsulated into a particle that may be as small as 1 μm. The greatest feature of microencapsulation is the control provided by the choice of coating [36–38]. This control allows microencapsulation to be a controlled-release device. Microcapsules can be engineered to gradually release drugs to the body. To achieve this type of delivery, equilibrium is established which will monitor the liberation of medicine from those microcapsules. A microcapsule may be opened by many different means. Release mechanisms include fracture by heat, solvation, diffusion, and pressure [39]. A coating may also be designed to open specific areas of the body. A microcapsule containing drugs that will be consumed by GI fluids must not be fractured until after it passes through the stomach [36–39]. A coating can therefore be used that is able to withstand stomach acids and allow the drug to pass through the stomach.

Methods of Microencapsulation

1. *Air Suspension* This method, known as the Wurster process or fluidized-bed coating, involves dispersing solid particulate core materials in a supporting air stream and the spray coating of the suspended materials [40]. The design of the chamber and its operating parameters effect a recirculating flow of the particles through the coating zone of the chamber, where a coating material, usually a polymer solution, is sprayed onto the fluidized particles. The cyclic process is repeated until the desired coat thickness is obtained.

2. *Pan Coating* This process has been around for many decades and is commonly associated with sugar coating. It is essential that the particles be greater than 600 μm for effective coating. The process has been extensively utilized for the preparation of controlled-release beads.

The coating by this method is applied as a solution or as an atomized spray to the desired solid core material in the coating pan [41]. Warm air is passed over the coated materials as the coatings are applied in the coating pans to remove the coating solvent. Sometimes the final solvent removal is carried out in a drying oven.

3. *Multiorifice Centrifugal Process* This is a mechanical process involving the use of centrifugal forces to hurl a material particle through an enveloping microencapsulation membrane to effect mechanical encapsulation [42]. The microcapsules are then hardened and congealed. This method is capable of microencapsulating liquids and solids dispersed in a liquid.

4. *Coacervation Phase* Encapsulation by coacervation is the one of the more popular methods commonly studied. The process consists of three steps carried out under continuous agitation [43]:

Step 1 The core material is dispersed in a solution of coating polymer, the solvent for the polymer being the liquid manufacturing vehicle phase.

Step 2 Deposition of the coating, accomplished by controlled, physical mixing of the coating material and the core material in the manufacturing vehicle.

Step 3 Rigidization of the coating by thermal, cross-linking, or desolvation techniques to form self-sustaining microcapsules.

Since the core materials are microencapsulated while being dispersed in some liquid manufacturing vehicle, subsequent drying operations are usually required.

5. *Solvent Evaporation Technique [39, 44, 45]* The process is carried out in a liquid manufacturing vehicle. The core material to be encapsulated is dissolved or dispersed in the coating polymer solution. With agitation, the core coating material mixture is dispersed in the liquid manufacturing vehicle phase to obtain the appropriate microcapsule size. The microcapsules can be used in suspension form, coated onto substrates or isolated as powders. A schematic of the emulsification/solvent evaporation technique to prepare drug-loaded microparticles is shown in Figure 4.

6. *Spray Drying* Spray drying, by definition, is the transformation of feed from a fluid state to a dried particulate form by spraying the feed into a hot drying medium [44, 46]. The feed can be a solution, suspension, or paste. A schematic is shown in Figure 5.

Spray drying consists of four process stages:

(i) Atomization of feed into spray
(ii) Spray–air contact (mixing and flow)
(iii) Drying of spray (moisture evaporation)
(iv) Separation of dried product from the air

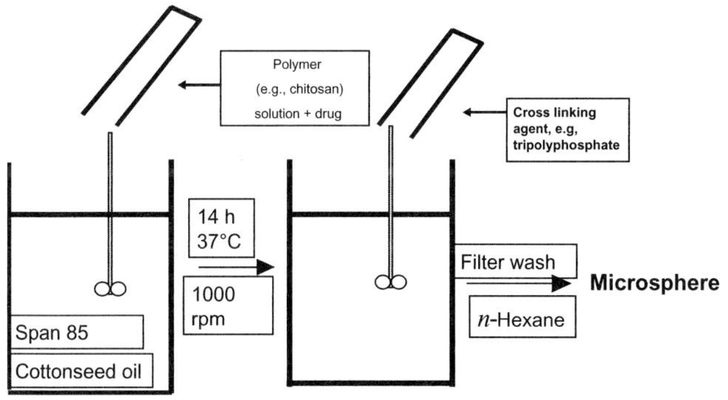

FIGURE 4 Schematic of microspheres prepared by emulsification/solvent evaporation method.

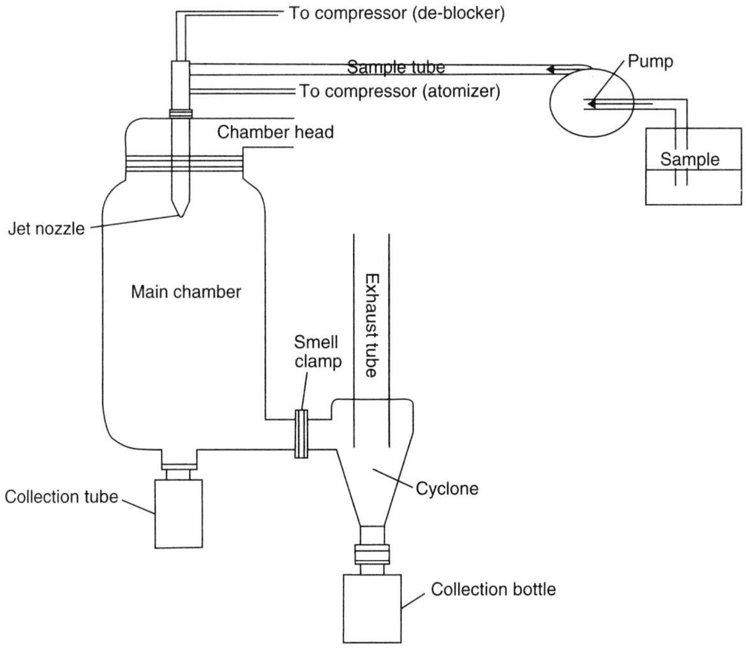

FIGURE 5 Schematic of spray drying method.

7. *Spray Congealing [47]* Spray congealing is similar to spray drying in that it involves dispersing the core material in a liquefied coating substance. Coat solidification is accomplished by congealing the molten coating material or by solidifying a dissolved coating material by introducing the coat–core material mixture into a nonsolvent. Removal of the nonsolvent is then achieved by sorption, extraction, or evaporation techniques. Waxes, fatty acids and alcohol, and polymers and sugars, which are solids at room temperature but melt at high temperatures, are applicable to spray congealing methods.

5.1.6.2 Nanostructure-Mediated Controlled-Release Dosage Forms

The efficiency of drug delivery to various parts of the body is directly affected by particle size. Nanostructure-mediated drug delivery, a key technology for the realization of nanomedicine, has the potential to enhance drug bioavailability, improve the timed release of drug molecules, and enable precision drug targeting [48]. Nanoscale drug delivery systems can be implemented within pulmonary therapies, as gene delivery vectors, and in stabilization of drug molecules that would otherwise degrade too rapidly. Additional benefits of using targeted nanoscale drug carriers are reduced drug toxicity and efficient drug distribution.

Anatomic features such as the blood–brain barrier, the branching pathways of the pulmonary system, and the tight epithelial junctions of the skin make it difficult for drugs to reach many desired physiological targets. Nanostructured drug carriers will help to penetrate or overcome these barriers to drug delivery [49]. Greater uptake efficiency has also been shown for GI absorption and transcutaneous permeation, with particles around 100 and 50 nm in size, respectively. However, such small particles traveling in the pulmonary tract may help with delivery to the pulmonary extremities. For instance, the outer layers of the carrier architecture may be formulated to biodegradable as the carrier travels through the pulmonary tract. As the drug carrier penetrates further into the lung, additional shedding will allow the encapsulated drug to be released. Biodegradable nanoparticles of gelatin and human serum albumin (HSA) show promise as drug carriers.

Advantages of nanostructure-mediated drug delivery include the ability to deliver drug molecules directly into cells and the capacity to target tumors within healthy tissue [50]. The mechanisms of cellular uptake of external particulates include calthrin- and caveoli-mediated endocytosis, pinocytosis, and phagocytosis. However, phagocytosis may not play a role in the uptake of nanoscale particles because of the small size of such particles.

Nanoscale drug delivery architectures are able to penetrate tumors due to the discontinuous, or "leaky," nature of the tumor microvasculature, which typically contains pores ranging from 100 to 1000 nm in diameter. The microvasculature of healthy tissue varies by tissue type, but in most tissues, including the heart, brain, and lung, there are tight intercellular junctions less than 10 nm. Therefore, tumors within these tissue types can be selectively targeted by creating drug delivery nanostructures greater than the intercellular gap of the healthy tissue but smaller than the pores found within the tumor vasculature.

Various nanoscale architecture can be designed, including solid spheres, hollow spheres, tubes, porous particles, solid particles, and branched structures (Table 2). To achieve such nanostructures, different fabrication methods are used depending on the types of material. The methods used for nanoscale assembly include molecular self-assembly, bioaggregation, nanomanipulation, photochemical patterning, molecular imprinting, layer-by-layer electrsostatic deposition, and vapor deposition.

TABLE 2 Nanoscale Drug Delivery Technologies

Drug Delivery Technology	Materials	Nanostructure Forms
Biologicals	Lipids, peptides, nucleic acids, polysaccharides, viruses	Vesicles, nanotubes, rings, nanoparticles, nanocapsules, nanospheres
Polymeric	poly(lactic acid), poly(glycolic acid), poly(alkylcyanoacrylate), poly(3-hydroxybutanoic acid), Poly(organophosphazene), poly(ethylene glycol), poly(caprolactone), poly(ethylene oxide), poly(amidoamine), poly(L-glutamic acid), poly(propylene imine)	Vesicles, nanospheres, nanoparticles, micelles, dendrimers
Silicon based	Silicon, silicon dioxide	Porous nanoparticles, nanoneedles
Carbon based	Carbon	Nanotubes
Metallic	Gold, silver, palladium, platinum	Nanoparticles, nanoshells

Manufacturing and Characterization of Nanoparticles/Nanocapsules/ Nanospheres Production of nanoparticles of soft materials is much more difficult and challenging than that of hard materials because of the high stickiness of the former. The bulk pharmaceuticals are available in solids of large sizes, which often can be easily solubilized in solvent to obtain particular sizes. Hence, there are two extremes of sizes: molecular size (each particle containing one molecule) and larger sizes (e.g., each particle containing on the order of 10^{18} molecules). To obtain nanoparticles in the range of 50–300 nm of drug delivery, one requires on the order of 10^4–10^8 molecules in each particle. This size has to be achieved from either solution-phase (single-molecule) or millimeter-size particles (10^{18} molecules).

Pearl/Ball Milling Technology for Production of Drug Nanocrystals There are two different drug nanocrystal products, prepared by using pearl/ball milling technology. The Rapamune coated tablet is the more convenient formulation, introduced by Wyeth Pharmaceuticals in 2002. The Emend, introduced in 2003 by MSD, Sharp and Dohme Gmbh, is a capsule composed of sucrose, microcrystalline cellulose, hyperlose, and sodium dodecylsulfate [51].

Traditional equipment used for micronization of drug powders such as rotor-stator colloid mills or jet mills are of limited use for the production of nanocrystals. For example, jet milling leads to a drug powder with a size range of roughly 0.1–20 μm, containing only a very small fraction of about 10% in the nanometer range. However, it has been shown when running a pearl mill over a sufficiently long milling time that drug nanosuspensions can be obtained [52–54]. These mills consist of a milling container filled with fine milling pearls or larger sized balls. The container can be static and the milling material is moved by a stirrer; alternatively, the complete container is moved in a complex movement leading consequently to movement of the milling pearls.

The different milling materials available include traditional steel, glass, and zircon dioxide as well as new special polymers such as hard polystyrene. A general problem associated with this technology is the erosion from the milling materials during manufacturing [55]. Surfactants and stabilizers have to be added for the physical stability of the produced nanosuspensions. In the production process, the coarse drug powder is dispersed by high-speed stirring or homogenization in a surfactant/stabilizer solution to yield macro- and nanosuspensions. The choice of surfactant or stabilizers depends not only on the properties of the particles to be suspended (e.g., affinity of surfactant/stabilizer to the crystal surface) but also on the physical principles (e.g., electrostatic or steric stabilization) and the route of administration. In general, steric stabilization is recommended as it is less susceptible to electrolytes in the gut and blood.

There are number of pearl mills available on the market, ranging from laboratory-scale to industrial-scale volumes. The ability of large-scale production is an essential prerequisite for the introduction of a product to the market. One advantage of pearl mills, apart from their low cost, is their ability in scaling up.

Nanoparticles/Nanoemulsions/Nanospheres Prepared by High-pressure Homogenization High-pressure homogenization is a technology that has been applied for many years in various areas of the production of emulsions and suspensions. A distinct advantage of this technology is its ease in scaling up, even to very large volumes.

In the pharmaceutical industry, parenteral emulsions such as Intralipid and Lipofundin (mean droplet diameter 200–400 nm) are generally produced by this technology [56]. Typical pressures for the production of drug nanosuspensions are 1000–1500 bars (corresponding to 100–150 MPa); the number of required homogenization cycles varies from 10 to 20 depending on the properties of the drug. Most of the homogenizers used are based on the piston gap principle; an alternative can be jet stream technology [57].

In the piston gap homogenizer, the liquid is forced through a tiny homogenization gap, typically in the range of 5–20 μm (depending on the pressure applied and the viscosity of the dispersion medium). Using a Micron Lab 40, the suspension is supplied from a metal cylinder by a piston, and the cylinder diameter is approximately 3 cm. The suspension is moved by the piston having an applied pressure between 100 and 1500 bars. The piston gap homogenizer corresponds to a tube system in which the tube diameter narrows from 3 to 20 μm. The Microfluidizer (Microfluidics Inc.) is based on the jet stream principle [58]. Two streams of liquid collide, diminution of droplets or crystals is achieved mainly by particle collision, but occurrence of cavitation is achieved mainly by particle collision.

Nanoparticles/Nanocapsules Obtained by Interfacial Polymerization Nanoparticles/nanocapsules can be obtained by fast polymerization of a monomer at the interface between the organic and the aqueous phase of an emulsion. Alkylcyanoacrylates have been proposed for the preparation of both oil- and water-containing nanocapsules [59]. These monomers polymerize within a few seconds, initiated by hydroxyl ions from equilibrium dissociation of water or by nucleophilic groups of any compound of the polymerization medium.

1. *Formation of Nanocapsules/Nanospheres Containing Oil Core* This type of nanocapsule is preferred for the encapsulation of lipophilic and oil-soluble compounds. The general procedure for the preparation of oil-containing nanocapsules by interfacial polymerization of alkylcyanoacrylates consists of preparing a very fine oil-in-water (O/W) emulsion with an additional water-miscible organic solvent such as ethanol or acetone [60, 61]. These solvents are used to disperse the oil as very small droplets in the aqueous phase, which contains a hydrophilic surfactant. The solvent also serves as a vehicle for the monomer. Gallardo et al. [62] proposed a mechanism to explain nanocapsule formation. An important factor is that the organic solvent must be completely water miscible so that the formation of small enough oil droplets occurs spontaneously while the solvent is diffusing toward the aqueous phase and the water is diffusing toward the organic phase. The polymerization of monomer is also induced by contact with hydroxyl ions from the water phase, which should be very fast to allow efficient formation of a thin layer of coating around the oil droplet and thus achieve effective encapsulation of drugs.

The organic phase containing the oil, the monomer, and the bioactive compounds, dissolved in the water-miscible organic solvent, is injected into the aqueous phase containing water and a hydrophilic surfactant under strong magnetic stirring. The nanocapsules/nanospheres are formed immediately to give a milky suspension. The organic phase is then removed under reduced pressure using a rotary evaporator. The most commonly used materials for the preparation of oil-loaded nanocapsules are given in Table 3.

TABLE 3 Main Components Used to Prepare Oil-Containing Nanocapsules/Nanospheres by Interfacial Polymerization

Components	Examples
Oil	Miglyol, Lipiodol, benzylbenzoate
Monomers	Ethylcyanoacrylate, isobutylcyanoacrylate, *n*-butylcyanoacrylate, isohexylcyanoacrylate
Organic solvents	Ethanol, acetone, acetonitrile, *n*-butanol, isopropanol
Surfactants	Poloxamer 188, poloxamer 238, poloxamer 407, Triton X100, Tween 80

Nanocapsule/nanosphere size ranges between 200 and 350 nm were observed to be affected by both the oil–ethanol ratio and the oil–monomer ratio [63, 64]. It is also influenced by the particular oil, water-miscible organic solvent, and nonionic surfactant in the aqueous phase. The pH of the aqueous phase and the temperature also affect the size distribution.

2. *Nanocapsules/Nanospheres Containing Aqueous Core Obtained by Interfacial Polymerization* Nanocapsules/nanospheres with an aqueous core are a recent technology developed for the efficient encapsulation of water-soluble compounds, which are generally very difficult to include within carrier systems. This type of nanocapsule/nanosphere may also be obtained by interfacial polymerization, but in this case monomers are added to a water-in-oil (W/O) emulsion. Anionic polymerization of the cyanoacrylates in the water phase is initiated at the interface by nucleophiles such as hydroxyl ions in the aqueous phase, leading to the formation of nanocapsules/nanospheres with an aqueous core. A typical procedure as described by Lambert et al.[65] consists of preparing an aqueous phase composed of ethanol (20% v/v) in water (pH 7.4) which is emulsified in an organic phase containing Miglyol oil and Montane 80. Slow addition of cyanoacrylic monomer under mechanical stirring (about 4 h) allows the polymerization to occur. Thus the water droplets are surrounded by the polymer-forming nanocapsules with an aqueous core dispersed in an oily phase.

Polymeric Nanocapsules/Nanospheres/Nanoparticles

1. *Manufacturing by Interfacial Nanodeposition/Nanoprecipitation* The nanoprecipitation procedure generally consists of a water-miscible organic phase such as an alcohol or a ketone containing oil (with or without lipophilic surfactant) with an aqueous phase containing a hydrophilic surfactant. The polymer, which may be preformed synthetic, semisynthetic, or natural, is solubilized in the organic phase (or in a phase in which the polymer is soluble). After addition of the organic phase to the aqueous phase, the polymer diffuses with the organic solvent and is stranded at the interface between oil and water. The driving force for the formation of nanocapsules/nanospheres is the rapid diffusion of the organic solvent in the aqueous phase inducing interfacial nanoprecipitation of the polymer around droplets of the oily phase. The polymers which may be used to manufacture nanoparticles by this method include natural polymers such as gum arabic, chitosan, alginate, gelatin, ethylcellulose, hydroxypropyl methylcellulose (HPMC), and hydroxypropyl methylcellulose phthalate (HPMCP); semisynthetic polymers such as diacyl β-cyclodextrin; and synthetic polymers such as poly(D,L-lactide), poly(ε-caprolactone),

and poly(alkylcyanoacrylate), which are most commonly employed [65]. Similarly, a broad range of oils is suitable for the preparation of nanocapsules/nanospheres, including vegetable or mineral oils and pure compounds such as ethyl oleate and benzyl benzoate. The criteria for the selection of these compounds are the nontoxicity and the low solubility of the oil in the polymers and vice versa.

Both hydrophilic and lipophilic surfactants can be used to stabilize the polymeric nanoparticles. Generally the lipophilic surfactant is a natural lecithin of relatively low phosphotidylcholine content, whereas the hydrophilic one is synthetic: anionic (lauryl sulfate), cationic (quaternary ammonium), or more commonly nonionic [poly(oxyethylene)-poly(propylene)glycol]. Nanoparticles can be prepared in the absence of surfactants, but there are lots more chances to get aggregated during storage.

2. *Nanoparticles Obtained by Multiple Emulsion/Solvent Evaporation Method* The multiple emulsion/solvent evaporation method was initially developed for the preparation of microcapsules. This method consists in first dissolving the drug in an aqueous solution with or without a surfactant and the polymer in a volatile organic solvent that is not miscible to water. Polymers used for the formation of such types of particles have been mainly poly(lactide-*co*-glycolide) and poly(ε-caprolactone) [66, 67]. The inner water phase is then poured into the organic phase. This mixture is generally emulsified forming the first inner emulsion or the primary W/O emulsion, which is then mixed vigorously into an aqueous phase (outer water phase) that contains an emulsifier forming the water-in-oil-in-water (W/O/W) multiple emulsion. The resulting multiple nanoemulsion is continuously stirred and the solvent is allowed to evaporate, inducing precipitation of polymer and, thereby, the formation of solid drug loaded nanoparticles.

5.1.6.3 Liposomes

Liposomes were discovered in the mid-1960s [68] and were originally studied as cell membrane models. They have since gained recognition in the field of drug delivery. Liposomes are formed by the self-assembly of phospholipid molecules in an aqueous environment. The amphophilic phospholipid molecules form a closed bilayer sphere in an attempt to shield their hydrophobic groups from the aqueous environment while still maintaining contact with aqueous phase via the hydrophilic head group. The resulting closed sphere may encapsulate aqueous soluble drugs within the central aqueous compartment or lipid-soluble drugs within the bilayer membrane. Alternatively, lipid-soluble drugs may be complexed with other polymers (e.g., cyclodextrin) and subsequently encapsulated within the liposome aqueous compartment. The encapsulation within/association of drugs with liposomes alters the drug pharmacokinetics.

Attractive Biological Properties of Liposomes [69]

- Liposomes are biocompatible.
- Liposomes can entrap hydrophilic bioactive compounds in their internal compartment and hydrophobic into the membrane.
- Liposome-incorporated bioactives are protected from the inactivating effect of external conditions yet do not cause undesirable side reactions.

TABLE 4 Liposomal Drugs Approved for Clinical Application or Undergoing Clinical Evaluation

Active Drug	Product Name	Applications
Daunorubicin	DaunoXome	Sarcoma
Doxorubicin	Mycet	Breast cancer
	Doxil/Caelyx	Sarcoma, ovarian cancer, breast cancer
Amphotericin B	AmBisome	Fungal infections
Cytarabine	DepoCyt	Lymphomatous meningitis
Vincristine	Onco TCS	Non-Hodgkin's lymphoma
Lurtotecan	NX211	Ovarian cancer
Nystatin	Nyotran	Topical antifungal agent
All-*trans* retinoic acid	Altragen	Leukemia, carcinomas
DNA plasmid encoding HLA-B7 and α_2-microglobulin	Allovectin-7	Metastatic melanoma

- Liposomes provide a unique opportunity to deliver pharmaceuticals into cells or even inside individual cellular compartments.
- The size, charge, and surface properties of liposomes can be easily changed by adding new ingredients to the lipid mixture before liposome preparation and/or by variation of preparation methods.

The clinical applications of liposomes are well known (Table 4). The initial success achieved with many liposome-based drugs has fueled further clinical investigations. One of the drawbacks of the use of liposomes is the fast elimination from the blood and capture of liposomal preparations by the cells of the reticuloendothelial system (RES), primarily in the liver.

There are a number of different types of liposomal vesicles [69]:

- *Multilamellar Vesicles* These range in size from 500 to 5000 nm and consist of several concentric bilayers.
- *Small Unilamellar Vesicles* These are around 100 nm in size and are formed by a single bilayer.
- *Large Unilamellar Vesicles* These range in size from 200 to 800 nm.
- *Long Circulating Liposomes* Different methods have been suggested to achieve long circulation of liposomes in vivo, including coating the liposome surface with inert, biocompatible polymers, such as polyethylene glycol (PEG), which form a protective layer over the liposome surface and slow down its recognition by opsonins and therefore subsequent clearance of liposomes. An important feature of protective polymers is their flexibility, which allows a relatively smaller number of surface-grafted polymer molecules to create an impermeable layer over the liposome surface. These types of modified liposomes demonstrate dose-dependent, nonsaturable, long-linear kinetics, and increased bioavailability.
- *Immunoliposomes* To increase liposomal drug accumulation in the desired tissues and organs, the use of targeted liposomes with surface-attached ligands capable of recognizing and binding to cells of interest has been suggested [70]. Immunoglobulins (Ig) of the IgG class and their fragments are the most widely

used targeting moieties for liposomes, which can be attached, without affecting liposomal integrity or the antibody properties, by covalent bonding to the liposome surface or by hydrophobic insertion into the liposomal membrane after modification with hydrophobic residues.

5.1.6.4 Niosomes

The success achieved with liposomal formulations stimulated the search for other vesicle-forming amphiphiles. Nonionic surfactants were among the first alternative materials studied and a large number of surfactants have since been found to self-assemble into closed bilayer vesicles which may be used for drug delivery [71]. Anticancer niosomes are expected to accumulate within tumors. The niosomal encapsulation of methotrexate and doxorubicin increases drug delivery to the tumor and tumoricidal activity. Unlike nonstealth liposomes, doxorubicin niosomes (size 800 nm) possessing a triglycerol or doxorubicin niosomes (size 200 nm) possessing a muramic acid surface are not taken up significantly by the liver. As such, these triglycerol niosomes accumulate in the tumor. However, muramic acid vesicles do accumulate in the spleen. Uptake by the liver and spleen make niosomes ideal for targeting diseases manifesting in these organs. One such condition is leishmaniasis, and number of studies have shown that niosomal formulations of sodium stibogluconate improve parasite suppression in the liver, spleen, and bone marrow. Niosomes may also be used as depot systems for short-acting peptide drugs on intramuscular administration [72].

Niosomal antigens are potent stimulators of the cellular and humoral immune response. The formulation of antigens as a niosome in W/O emulsions further increases the activity of antigens. The controlled-release property of these types of emulsion formulations is responsible for enhancing the immunological responses.

5.1.7 TECHNOLOGIES FOR DEVELOPING TRANSDERMAL DOSAGE FORMS

Continuous intravenous infusion at a programmed rate has been recognized as a superior model of drug delivery not only to bypass the hepatic first-pass elimination but also to maintain a constant, prolonged, and therapeutically effective drug level in the body. A closely monitored intravenous infusion can provide both the advantages of direct entry of drugs into the systemic circulation and control of circulating drug levels. However, such a mode of drug delivery entails certain risks and therefore necessitates hospitalization of patients and close medical supervision of the medication. Recently there has been an increasing awareness that the benefits of intravenous drug infusion can be easily duplicated, without its potential hazards, by continuos transdermal drug administration through intact skin [73].

Advances in transdermal delivery systems (TDSs) and the technology involved have been rapid because of the sophistication of polymer science, which now allows incorporation of polymeric additives in TDSs in adequate quantity. Drugs with which transdermal therapy was pioneered include scopolamine, nitroglycerine, isosorbide dinitrite, clonidine, estradiol, nicotine, and testosterone [74].

Advantages of Transdermal Drug Delivery System [75]

- Avoids GI absorption (pH effects, enzymatic activity, drug interactions)
- Substitute for oral route
- Avoids first-pass effect (drug deactivation by digestive and liver enzymes)
- Multiday therapy with a single application
- Extends the activity of the drugs with short half-lives
- Provides capacity to terminate drug effects rapidly
- Rapid identification of medication in emergency

Limitations of Transdermal Drug Delivery [75]

- Not for all drugs
- Limited time that the patch can remain affixed
- Variable intra- and interindividual percutaneous efficiency absorption efficiency
- Variable adhesion to different skin types
- Skin rashes and sensitization
- Bacterial and enzymatic drug metabolism under the patch
- Complex technology/high cost

Skin Site for Transdermal Drug Administration The skin is one of the most extensive and readily accessible organs of the human body. The skin of an average adult body covers a surface area of approximately $2\,m^2$ and receives about one-third of the blood circulating through the body. It is elastic, rugged, and, under normal physiological conditions, self-regenerating. It serves as a barrier against physical and chemical attacks and shields the body from invasion by microorganisms.

Microscopically the skin is a multilayered organ composed of, anatomically, many histological layers, but it is generally described in terms of three tissue layers: the epidermis, the dermis, and the subcutaneous fat tissue.

Microscopic sections of the epidermis show two main parts: the stratum corneum and the stratum germinativum. The stratum corneum forms the outermost layer of the epidermis and consists of many layers of compacted, flattened, dehydrated, keratinized cells in stratified layers. In normal stratum corneum, the cells have a water content of only approximately 20% compared to the normal physiological level of 70% in the physiologically active 10% (w/w) to maintain flexibility and softness. It becomes rough and brittle, resulting in so-called dry skin, when its moisture content decreases at a rate faster than can be resupplied from the underlying tissues. The stratum corneum is responsible for the barrier function of the skin. It also behaves as the primary barrier to percutaneous absorption. The thickness of this layer is mainly determined by the extent of stimulation of the skin surface by abrasion and weight bearing; hence thick palms and soles develop.

Several technologies have been successfully developed to provide rate control over the release and skin permeation of drugs. These technologies can be classified into four basic approaches which are described below [73, 75].

Polymer Matrix Diffusion-Controlled Transdermal Drug Delivery (TDD) System In this approach, the drug reservoir is formed by homogeneously dispersing the drug solids in a hydrophilic or lipophilic polymer matrix, and the medicated polymer formed is then molded into medicated disks with a defined surface area and controlled thickness. This drug-reservoir-containing polymer disk is then mounted onto an occlusive baseplate in a compartment fabricated from a drug-impermeable plastic baking (Figure 6*a*).

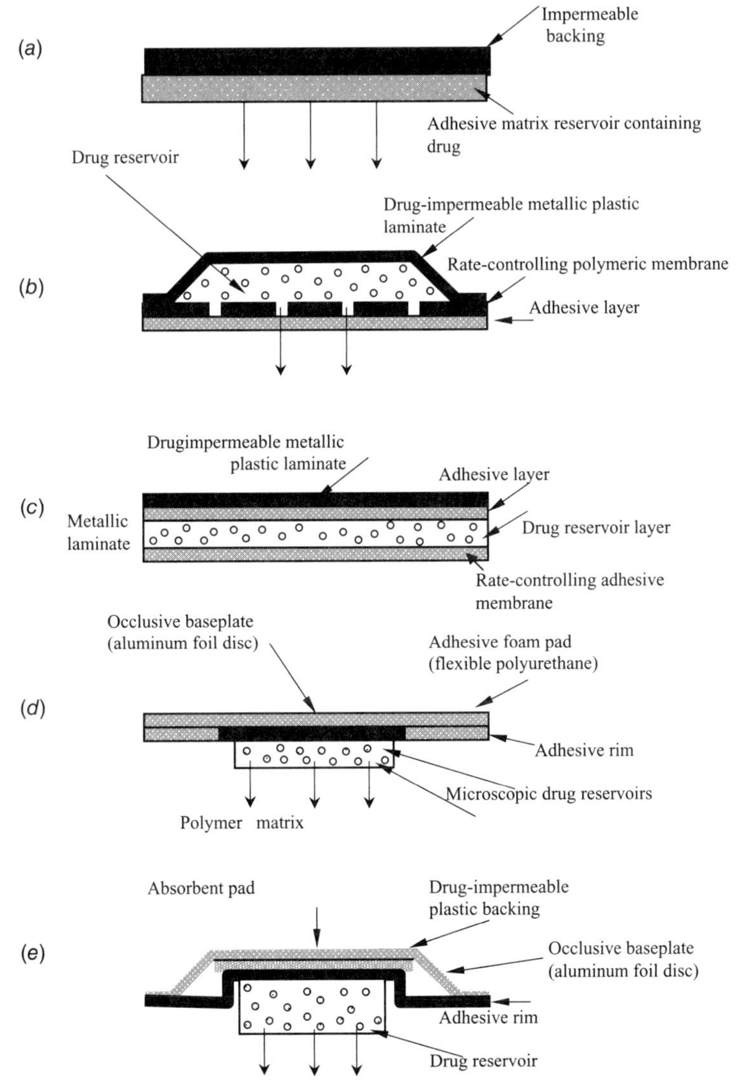

FIGURE 6 Cross-sectional view of several TDSs: (*a*) poly(sebacic anhydride) (PSA) matrix device; (*b*) membrane-moderated TDS; (*c*) adhesive-controlled TDS; (*d*) microreservoir-type TDS; (*e*) matrix dispersion–type TDS.

Polymer Membrane Permeation-Controlled TDD System In this system, the drug reservoir is sandwiched between a drug-impermeable backing laminate and a rate-controlling polymeric membrane (Figure 6*b*). The drug molecules are permitted to release only through the rate-controlling polymeric membrane.

Drug Reservoir Gradient-Controlled TDD System The rate of drug release from this type of drug reservoir type is gradient controlled. In this system the thickness of the diffusion path through which the drug molecule diffuses increases with time (Figure 6*c*).

Microreservoir Dissolution-Controlled TDD System This type of drug delivery system can be considered a hybrid of the reservoir and matrix dispersion-type drug delivery systems. In this approach, the drug reservoir is formed by first suspending the drug solids in an aqueous solution of water–miscible solubilizer (e.g., PEG), and then homogeneously dispersing the drug suspension with controlled aqueous solubility, in a lipophilic polymer, by high-shear mechanical force, to form thousands of unleachable microscopic drug reservoirs. This thermodynamically unstable dispersion is quickly stabilized by immediately cross-linking the polymer chains in situ, which produces a medicated polymer disk with a constant surface area and a fixed thickness. Mounting the medicated disk at the center of an adhesive pad then produces a TDD system (Figure 6*d* and *e*).

5.1.8 OCULAR CONTROLLED-RELEASE DOSAGE FORMS

The eye is unique in its therapeutic challenges. The eye drop dosage form is easy to instill but suffers from the inherent drawback that the majority of the medication it contains is immediately diluted in the tear as soon as the eye drop solution is instilled. Usually less than 10% of a topically applied dose is absorbed into the eye, leaving the rest of the dose to potentially absorb into the bloodstream, resulting in unwanted side effects [76]. The objectives of most of the controlled delivery system are to maintain the drug in the precorneal area and allow its diffusion across the cornea. Polymeric matrices seem to reduce the drainage significantly, but other newer methods of controlled-release dosage forms can also be used.

The sustained release of artificial tears has been achieved by a hydroxypropylcellulose polymer insert [77]. However, the best known application of diffusional therapy in the eye, Ocusert-Pilo, as shown in Figure 7, is a relatively simple structure with two rate-controlling membranes surrounding the drug reservoir containing

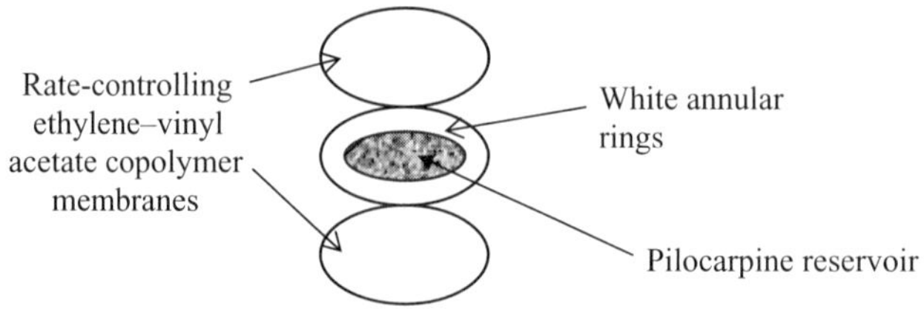

FIGURE 7 Schematic of Ocusert intraocular device for controlled release of pilocarpine.

pilocarpine. The unit is placed in the eye and resides in the lowe cul-de-sac, just below the cornea. Since the device itself remains in the eye, the drug is released into the tear film. The advantage of such a device is that it can control intraocular pressure for up to a week. Controlled release is achieved with less drug and fewer side effects, since the release of drug is zero order. However, it is difficult to keep it in the eye for a longer time and can cause discomfort.

The prodrug administration is also getting attention as ocular controlled-release dosage forms. Since the corneal surface presents an effective lipoidal barrier, especially to hydrophilic compounds, it seems reasonable that a prodrug that is more lipophilic than the parent drug will be more successful in penetrating this barrier. Recently, dipivalyl epinephrine (Dividephrine), a dipivalyl ester of epinephrine, has been formulated [78]. Epinephrine itself is poorly absorbed owing to its polar characteristics and is highly metabolized. The prodrug form is 10 times as effective at crossing the cornea and produces substantially higher aqueous humor levels.

New sustained technologies are also gaining much interest in ocular delivery, as in other routes. Liposomes as drug carriers have achieved enhanced ocular delivery of certain drugs, antibiotics, and peptides. Prolonged delivery of pilocarpine can be achieved with a polymeric dispersion or submicrometer emulsions [79].

5.1.9 VAGINAL AND UTERINE CONTROLLED-RELEASE DOSAGE FORMS

Controlled-release devices for vaginal and uterine areas are most often for the delivery of contraceptive steroid hormones. The advantages are prolonged release, minimal side effects, and increased bioavailability. First-pass metabolism that inactivates many steroid hormones can also be avoided.

Therapeutic levels of the medroxyprogestrone vaginal ring have been achieved at a total dose that is one-sixth the required oral dose [80]. The sustained release of progesterone from various polymers given vaginally has also been found useful in cervical ripening and the induction of labor. A possible new use of the vaginal route is for long-term delivery of antibodies. When various antibodies, including monoclonal IgG, were administered from polymer vaginal rings in test animals, antibody concentrations remained high over a month in vaginal secretions and detectable in blood serum [81].

The hormone-releasing devices in uterus have a closer resemblance to controlled release because they involve the release of a steroid compound by diffusion [82, 83]. Progesterone, the active ingredient, is dispersed in the inner reservoir, surrounded by ethlene/vinyl acetate copolymer membrane. The release of progesterone from this system is maintained almost constant for about a year [84–86].

5.1.10 RELEASE OF DRUGS FROM CONTROLLED-RELEASE DOSAGE FORMS

There are three primary mechanisms by which active agents can be released from a delivery system: diffusion, degradation, and swelling followed by diffusion. Any or all of these mechanisms may occur in a given release system. Probable

TABLE 5 Probable Mechanism of Drug Delivery from Hydrogels with Certain Environmental Conditions

Stimulus	Hydrogel	Mechanism
pH	Acidic or basic hydrogel	Change in pH, swelling–diffusion, erosion or burst release of drug
Ionic strength	Ionic hydrogel	Change in ionic strength, change in concentration of ions inside gel, change in swelling, release of drug
Chemical species	Hydrogel containing electron-accepting groups	Electron-donating compounds, formation of charge transfer complex, change in swelling, release of drug
Enzyme–substrate	Hydrogel containing immobilized enzymes	Substrate present, enzymatic conversion, product changes, swelling of gel, release of drug
Magnetic	Magnetic particles dispersed in alginate microspheres	Applied magnetic field, change in pores in gel, change in swelling, release of drug
Thermal	Thermoresponsive hydrogel	Change in temperature, change in polymer–polymer and water–polymer interactions, change in swelling, release of drug
Electrical	Polyelectrolyte hydrogel	Apply electric field, membrane charging, electrophoresis of charged drug, change in swelling, release of drug
Ultrasound irradiation	Ethylene–vinyl alcohol hydrogel	Ultrasound irradiation, temperature increase, release of drug

mechanisms of drug release from controlled-release dosage forms are briefly described in Table 5. Diffusion occurs when a drug or other active agent passes through the polymer that forms the controlled-release device. The diffusion can occur on a macroscopic scale—as through pores in the polymer matrix—or on a molecular level—by passing between polymer chains [4, 87].

Table 5 describes the probable mechanisms of drug delivery from controlled-release dosage forms under contain environmental conditions. A polymer and active agent have generally been mixed to form a homogeneous system, also referred to as a matrix system. Diffusion occurs when the drug passes from the polymer matrix into the external environment. As the release continues, its rate normally decrease with this type of system, since the active agent has a progressively longer distance to travel and therefore requires a longer diffusion time to release. The drug release is accomplished only when the polymer swells. Because many of the potentially most useful pH-sensitive polymers swell at high pH values and collapse at low pH values, the triggered drug delivery occurs upon an increase in the pH of the environment. Such materials are ideal for systems such as oral delivery, in which the drug is not released at low pH values in the stomach but rather at high pH in the upper small intestine.

In reservoir systems, a reservoir—whether solid drug, dilute solution, or highly concentrated drug solution with in polymer matrix—is surrouned by a film or membrane of a rate-controlling material [5]. The only structure effectively limiting the release of the drug is the polymer layer surrounding the reservoir. Since this polymer coating is essentially uniform and of a nonchanging thickness, the diffusion rate of the active agent can be kept fairly stable throughout the lifetime of the delivery system.

For the diffusion-controlled systems ddescribed thus far, the drug delivery device is fundamentally stable in the biological environment and does not change its size through either swelling or degradation [4]. It is also possible for a drug delivery system to be designed so that it is incapable of releasing its agent or agents until it is placed in an appropriate biological environment. Swelling-controlled-release systems are initially dry and, when placed in the body, will absorb water or other body fluids and swell. The swelling increases the aquenous solvent content with the formulation as well as the polymer mesh size, enabling the drug to diffuse through the swollen network into the external environment. Most of the materials used in swelling-controlled-release systems are based on hydrogels, which are polymers that well without dissolving when placed in water or other biological fluids.

5.1.10.1 Time-Controlled-Release Dosage Forms

To achieve a drug release which is independent of the environment (e.g., pH, enzymatic activity, intestinal motility), the lag time prior to the drug release has to be controlled primarily by the delivery system. The release mechanisms employed include bulk erosion of polymers in which drug release by diffusion is restricted, surface erosion of layered devices composed of alternation drug-containing and drug-free layers, osmotically controlled rupture, and enzymatic degradation of liposomes. The device environment may modulate the release profile of any of these systems and may depend on factors such as the amount of free moisture, regional blood flow, and various cellular activities at the site [88, 89].

Systmes with Eroding or Soluble Barrier Coatings These types of delivery systems comprise reservoir devices coated with a barrier layer. the barrier dissolves or erodes after a specified lag period, after which the drug is released rapidly from the reservoir core. In general, the lag time prior to drug release from a reservoir-type device can be controlled by the thickness of the coating layer, for example, the Chronotropic systems, which consists of a drug-containing core layered with hydroxy propyl methyl cellulose (HPMC), optionally coated with an outer enteric coating. The lag time prior to drug release is controlled by the thickness and the viscosity grade of the HPMC layer. After erosion or dissolution of the HPMC layer, a distinct pulse was observed. The Chronotropic system [90, 91] is an oral dosage form designed to achieve time-controlled delivery. This system has been developed keeping in mind the interaction between GI fluid and coating polymer, which causes time- or site-controlled release. The reaction causes the liberation of drugs by the mechanism of swelling of polymer, increased permeability, and dissolution/erosion phenomena. This system probably works better for poorly water soluble drug because highly water soluble drugs could diffuse through the swollen HPMC layer prior to complete erosion.

The TIME CLOCK system for the oral dosage should enable fast and complete release of drug after a predetermined lag time [92]. A tablet was made containing the drug molecule and bulking agents (lactose, polyvinlpyrrolidine, corn starch, and

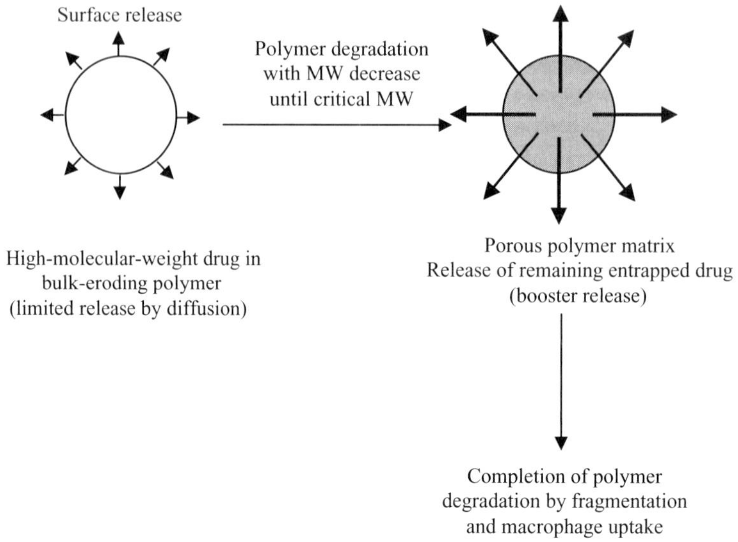

FIGURE 8 Theoretical controlled release from a surface-eroding polymeric system. (Adapted from ref. 93 with permission of Elsevier Copyright 1999.)

magnesium stearate). This core was coated with a hydrophobic dispersion of carnauba wax, bees' wax, poly(oxyethylene) sorgitan monooleate, and HPMC in water. By altering the coating thickness, the lag time could be proportionally modulated. In vitro results indicated that the drug core was dissolved immediately after direct immersion in water and release was completed within 30 min, while a rapid release was observed after a certain lag time for the TIME CLOCK system with the hydrophobic coating. In vivo results revealed that drug disintergration was modulated by the coating thickness of the drug core as well as the food intake before drug administration. This approach may also be used to control the release onset time. Since the drug core is formulated with soluble ingredients, shell dissolution/distintegration becomes the key factor to control the lag time. Furthermore, drug release is independent of normal physiological conditions, such as pH, digestive state, and anatomical position at the time of release. This approach could be applicable for oral as well as for implant systems. Figure 8 illustrates the theoretical description of drug release from surface-eroding polymeric controlled-release dosage forms [93].

Systems with Rupturable Coatings This class of reservoir-type pulsatile release system is based on rupturable coatings. The drug is released from a core after rupturing of a surrounding polymer layer, caused by a pressure buildup within the system, as shown in Figure 9 [93]. The pressure necessary to rupture the coating can be achieved with gas-producing effervescent excipients, inner osmotic pressure, or swelling agents.

An effervescent mixture of citric acid and sodium bicarbonate was incorporated in a tablet core coated with ethyl cellulose. The carbon dioxide development after water penetration into the core resulted in a pulsatile release after rupture of the coating, which was strongly dependent on the mechanical properties of the coating layer: The weak and nonflexible ethyl cellulose film ruptured sufficiently when

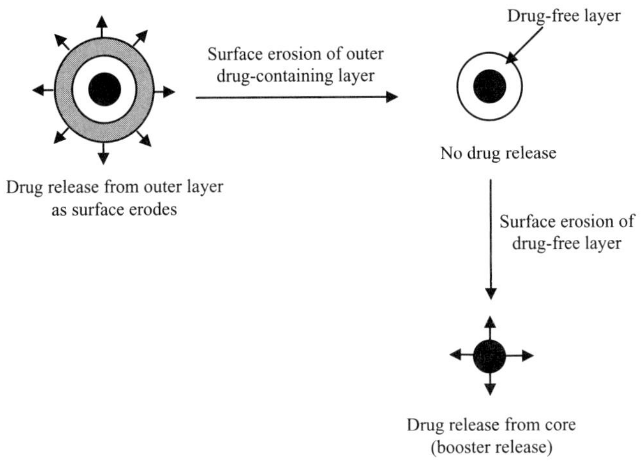

FIGURE 9 Theoretical controlled release from an osmotically driven system. (Adapted from ref. 93 with permission of Elsevier Copyright 1999.)

compared with more flexible films. The lag time before release increased with increasing coating level and increasing hardness of the core tablet. The effectiveness of so-called superdistintegrants, which are highly swellable agents, was demonstrated for a capsule-based system consisting of a drug containing a core capsule, and swelling layer, and a rupturable polymeric layer. Croscarmellose, sodium starch glycolate, or low-substituted hydroxypropyl cellulose (L-HPC) were used as swelling substances, which resulted in a complete film rupture followed by a rapid drug release. The lag time is controlled by the composition of the outer polymer layer: Water-soluble polymers such as HPMC increase the permeability and therefore reduce the lag time. The swelling energy of several excipients decreased in the following order: croscamellose sodium > L-HPC > crospovidone > HPMC. Bothe solid and liquid drug formulations could be delivered with this system [94–96].

A novel capsule was made from ethyl cellulose for the time-controlled release of drugs in the colon [97]. Initially the ethyl cellulose capsule was prepared using a gelatin capsule with ethyl cellulose, followed by dissolution of the gelatin in water. The thickness of the ethyl cellulose capsule body was varied and the effect of wall thickness on the release of drugs in the capsules was investigated. Ethyl cellulose capsules contained a large number of mechanically made micropores (400 μm) at the bottom. Also located in the bottom of the capsule body was a swellable layer consisting of L-HPC. Above the swellable layer was the drug reservoir, which contained a mixture of the model drug, fluorescein, and a bulking agent, such as lactose or starch. The capsule was thus capped and sealed with a concentrated ethyl cellulose solution. After administration of drug-containing capsule, water molecules penetrated the capsule through the micropores in the bottom of the capsule body. Hydration and swelling of HPC induced an increase in the internal osmotic pressure, which resulted in the "explosion" of the capsule and a burstlike drug release was observed. By altering the thickness of the capsule, the lag time of the drug release could be altered. A similar approach for the pulsatile release of drug was reported in which a hydrostatic pressure was generated inside the capsules.

Systems with Capsular Structure Several single-unit pulsatile dosage forms with a capsular design have been developed. Most of them consist of an insoluble capsular body which contains the drug and a plug which is removed after a predetermined lag time because of swelling, erosion, or dissolution.

The Pulsincap system consists of a water-insoluble capsule body filled with the drug formulation. The capsule half is closed at the open end with a swellable hydrogel plug. The dimension and the position of the plug can control the lag time prior to the release. In order to assure a rapid release of the drug content, effervescent agents or disintegrants can be included in the drug formulation, in particular with water-insoluble drugs. The system is coated with an enteric layer which dissolves upon reaching the higher pH region of the small intestine. This system comprises insoluble capsules and plugs. The plugs consist either of swellable materials, which are coated with insoluble but permeable polymers (e.g., polymethacrylates), or of erodible substances, which are compressed (e.g., HPMC, polyvinyl alcohol, polyethylene oxide) or prepared by congealing of melted polymers (saturated polyglycolated glycerides of glyceryl monooleate). The erosion of the plug can also be controlled enzymatically: A pectin plug can be degraded by pectinolytic enzymes being directly incorporated into the plug [98–100].

Linkwitz et al. [101] described the delivery of agents from osmotic systems based on the technology of an expandable orifice. The system is in the form of a capsule from which the drug is delivered by the capsule's osmotic infusion of moisture from the body. The delivery orifice opens intermittently to achieve a pulsatile delivery effect. The orifice forms in the capsule wall, which is constructed of an elastic material, preferably elastomer (e.g., styrene–butadiene copolymer), which strectches under apressure dufferential caused by the pressure rise inside the capsule as the osmotic infusion progessses. The orifice is small enough that when the elastic wall is relaxed, the flow rate of drug through the orifice is substantially zero, but when the elastic wall is stretched due to the pressure differential across the wall exceeding a threshold, the orifice expands sufficiently to allow the release of the drug at a phsiologically required rate. This osmotically driven delivery device as an implant can used in the anal–rectal passageway, in the cervical canal, as an artificial gland, in the vagina, as ruminal bolus, and the like.

A core-shelled cylindrical dosage form is available comprising a hydrophobic polycarbonate coating and a cylindrical core of alternating polyanhydride isolating layer and drug-loaded poly[ethyl glycinate) (benzyl amino acethydroxamate) phosphazene] (PEBP) layer for a programmable drug delivery system for single-dose vaccine and other related applications [102]. The pulsatile release of model compounds [fluorescein isothiocyanate (FITC)–dextran and myoglobin] whith a certain lag time (18–118 h) was achieved on the basis of the pH-sensitive degradation of PEBP and its cooperative interaction with polyanhydrides. In another experiment, Jiang and Zhu [103, 104] designed laminated devices comprising of polyanhydrides as isolating layers and pH-sensitive complexes of poly(sebacic anhydride)-*b*-polyethylene glycol (PSA-*b*-PEG) and poly(trimellitytylimdoglycine-*co*-sebacic anhydride)-*b*-polyethylene glycol [P(TMA-gly-*co*-SA)-*b*-PEG] as protein-loaded layers. The release of model proteins [bovine serum albumin (BSA) and myoglobin] showed a typical pulsatile fashion. The lag time prior to the release correlated with the hydrolytic druation of polyanhydrides, which varied from 30 to 165 h depending on polymer type and isolating layer thickness.

5.1.10.2 Stimuli-Induced Controlled-Release Systems

Several polymeric delivery systems undergo phase transitions and demonstrate marked swelling–deswelling change in reponse to environmental changes, including solvent composition ionic strength, temperature, electric fields, and light [105]. Responsive drug release from those systems results from the stimuli-induced changes in the gels or micelles, which maydeswell, swell, or erode in response to the respecive stimuli. The mechanisms of drug release include ejection of the drug from the gel as the fluid phase synerses out and drug diffusion along a concentration gradient.

pH-Responsive Drug Release Dosage Forms pH-sensitive enteric coatings have been used routinely to deliver drug to the small intestine. These polymer coatings are insensitive to the acidic conditions of the stomach yet dissolve at the higher pH environment of the small intestine. This pH differential has also been attempted for colonic delivery purposes, although the polymers used for colonic targeting tend to have a threshold pH for dissolution that is higher than for those used in conventional enteric coating applications [104–106].

The synthesis and characterization of series of novel azo hydrogels for colon-targeting drug delivery have been described. The colon specificity is achieved dure to the presence of pH-sensitive monomers and azo cross-linking agents in the hydrogel structures. Most commonly, copolymers of methacrylic acid and methylmethacrylate that dissolve at pH 6 (Eudragit L) and pH 7 (Eudragit S) have been extensively investigated [106, 107]. This approach is based on the assumption that gastrointestinal pH increases progressively from the small intestine to the colon. The pH in the distal small intestine is usually around 7.5, while the pH in the proximal colon is closer to 6.

To overcome the premature release of drugs, a copolymer of methacrylic acid, methylmethacrylate, and ethyl acetate (Eudragit ES), which dissolves at a slower rate and at a higher threshold pH (7–7.5), has been developed [108]. The trn A series of in vitro dissolution studies with this polymer have highlighted clear benefits over the Eudragit S polymer for colon targeting. A gamma scintigraphy study comparing the in vivo performance of these various polymers revealed that Eudragit S (coating over the tablets) was superior to the older polymers in terms of retarding drug release in the small intestine, although, in some cases, the coated tablets did not break up at all. pH-sensitive delivery systems are commercially available for mesalazine (5-iminosalicylic cid) (Asacol and Salofalk) and budesonide (Budenofalk and Entocort) for the treatment of ulcerative colitis and Crohn's disease, respectively [109].

Natural polysaccharides are being used for the development of solid dosage forms for pH-dependent delivery and for targeting the release of drugs in colon [110]. Various major approaches utilizing polysaccharides are fermentable coating of the drug core, embedding of the drug in biodegradable matrix, and formation of drug–saccharide conjugate (prodrugs). A large number of polysaccharides have already been studied for their potential in these types of delivery systems, such as chitosan, alginate, pectin, chondroitin sulfate, cyclodextrin, dextrans, guar gum, inulin, amylose, and locust bean gum [111].

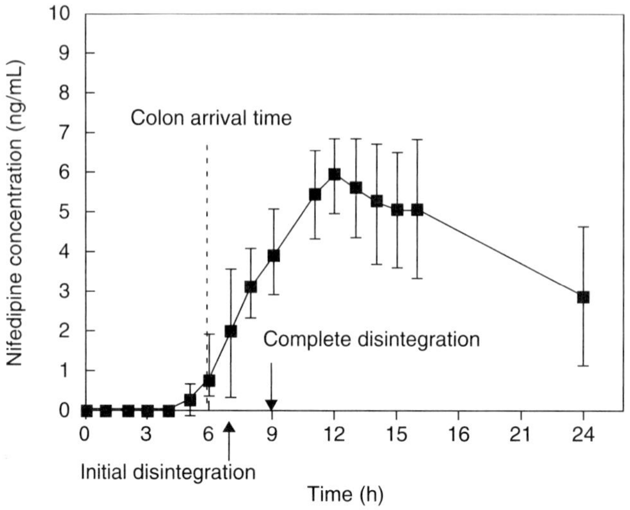

FIGURE 10 Nifedipine plasma concentration–time profile from pectin–galactomannan-coated tablets. (Adapted from ref. 110 with permission of Elsevier Copyright 2002.)

A pectin-and-galactomannan coating was developed by Lee et al. [112]. It consists of a conventional tablet coated with pectin and galactomannan. The coating from aqueous solutions of pectin and galactomannan was shown to be strong, elastic, and insoluble in gastric fluid. Figure 10 shows the plasma concentration profile of nifedifine from pectin–galactomannan-coated tablets and associated in vivo transit and disintegration characteristics. The mean plasma concentration of nifedifine was negligible for more than 5 h postdose and then increased rapidly.

CODES Technology CODES is a unique colon-specific drug delivery technology that was designed to avoid the inherent problems associated with pH- or time-dependent systems [113, 114]. The design of CODES exploited the advantages of certain polysaccharides that are only degraded by microorganisms available in the colon [115]. This is coupled with a pH-sensitive polymer coating. Since the degradation of polysaccharides occurred only in the colon, this system exhibited the capability to achieve colon delivery consistently and reliably. As schematically presented in Figure 11, one typical configuration consists of a core table coated with three layers of polymer coatings. The first coating (next to the core tablet) is an acid-soluble polymer (e.g., Eudragit E) and an outer coating is enteric with a HPMC barrier layer in between the oppositely charged polymers. The polysaccharides, degradable by enteroorganisms, generate organic acid, including mannitol, maltose, lactulose, and fructooligosaccharides. During the transit through the GI tract, CODES remains intact in the stomach, but the enteric and barrier coatings disolve in the intestines. In vivo performance of CODES in beagle dogs was studied using acetaminophen as the model drug and lactulose as the matrix-forming excipient in the core tablet. Compared with enteric-coated tablet, the onset of acetaminophen release form CODES was delayed more than 3 h, as shown in Figure 12.

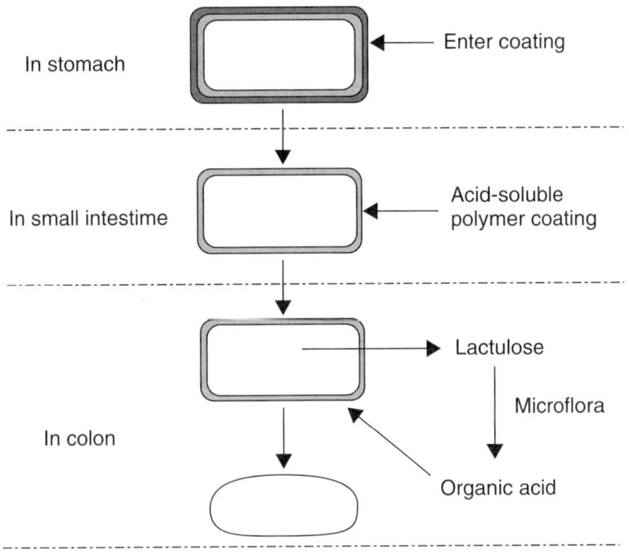

FIGURE 11 Conceptual design of CODES technology. (Adapted from ref. 114 with permission of Elsevier Copyright 2002.)

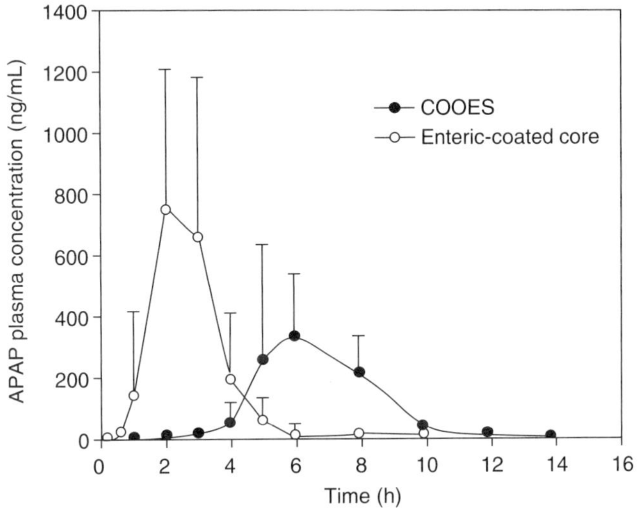

FIGURE 12 Percentage of acetaminophen released from CODES and enteric-coated core tablets in beagle dogs. (Adapted from ref. 114 with permission of Elsevier Copyright 2002.)

Thermoresponsive Drug Release Dosage Forms Temperature is the most widely utilized triggering signal for a variety of modulated or pulsatile drug delivery systems. The use of temperature as a signal has been justified by the fact that the body temperature often deviates from the physiological temperature (37°C) in the presence of pathogens or pyrogens. This deviation sometimes can be a useful stimulus that achivates the release of therapeutic agents from various temperature-responsive drug delivery systems for diseases accompanying fever. The drug delivery systems that are responsive to temperature utilize various polymer properties, including the thermally reversible coil/globule transition of polymer molecules, swelling change of networks, glass transition, and crystalline melting.

Thermoresponsive hydrogels have been investigated as possible drug delivery carriers for stimuli-responsive drug delivery systems [116–118]. The common characteristics of temperature-sensitive polymers are the presence of hydrophobic groups, such as methyl, ethyl, and propyl groups. Of the many temperature-sensitive polymers, poly(*N*-isopropylacrylamide) (PIPPAm) is probably the most extensively used. PIPPAm cross-linked gels have shown thermoresponsive, discontinuous swelling/deswelling phases—swelling, for example, at temperatures below 32°C while shrinking above this temperature. A sudden temperature increase above the transition temperature of these gels resulted in the formation of a dense, shrunken layer on the gel surface, which hindered water permeation from inside the gel into the environment. Drug release from the PIPPAm hydrogels at temperatures below 32°C was governed by diffusion, while above this temperature drug release was stopped completely, due to the "skin layer" formation on the gel surface (on–off drug release regulation).

Swelling–deswelling kinetics of conventional cross-linked hydrogels are normally the reciprocal of the square of the gel dimension. This mobility of the cross-linked chains in the gel is affected by the surrounding chains and the swelling–deswelling phases of the gel are governed by the collective diffusions of the network chains. Thus, to accelerate structural changes of the gel in response to external stimuli, several approaches have been developed which form porous structure within the gel and decrease gel size. Kaneko et al. [119, 120] introduced a method to accelerate gel swelling–deswelling kinetics based on the molecular design of the gel structure by grafting the free mobile linear PIPPAm chains within the cross-linked PIPPAm hydrogels. These novel graft-type PIPPAm gels had the same transition temperature as conventional cross-linked PIPPAm gels and existed in the swollen state below the transition temperature, while above this temperature, they shrank. Adense skin layer formed on the conventional PIPPAm gels upon temperature change above the transition temperature, which limited the complete shrinkage of the gel. In contrast, the PIPPAm-grafted gels showed rapid deswelling kinetics without the formation of a skin layer on the gel surface. This is probably due to the rapid dehydration of the graft chains formed by hydrophobic aggregation on the three-dimensional cross-linked gel chains. The low-molecular-weight compounds released immediately from conventional PIPPAm gels after a temperature increase, after which the release was teminated due to the formation of a dense impermeable skin layer on the surface. In comparison, 65% of the drug was released in one burst from free PIPPAm-grafted hydrogels with a graft molecular weight (MW) of 9000 following the temperature increase. Graft-type gels with a molecular weight of 4000 showed oscillating drug release profiles. The release of high-molecular-weight compound (e.g., dextran, MW 9300) from PIPPAm graft-type gels was shown to burst after a temperature increase of 40°C. The difference in drug release profiles for two graft-type gels is probably due to the different strengths of aggregation forces between the formed hydrophobic cores within the graft-type gels. That is, the high-molecular-weight graft chains formed more hydrophobic cores within the gels upon the temperature increase, which induced rapid gel deswelling.

Temperature-sensitive hydrogels can also be placed inside a rigid capsule containing holes or apertures. The on–off release is achieved by the reversible volume change of temperature-sensitive hydrogels. Such a device is called a squeezing hydrogel device because the drug release is affected by the hydrogel dimension. In addition to temperature, hydrogels can be made to respond to other stimuli, such as pH. In this type of system, the drug release rate was found to be proportional to the rate of squeezing of the drug-loaded polymer.

Clinical applications of thermosensitive hydrogels based on NIPAAm and its derivatives have limitations [121]. The monomers and cross-linkers used in the synthesis of the hydrogels are still not known to be biocompatible and biodegradable. The observation that acrylamide-based polymers activate platelets upon contact with blood, together with the unclear metabolism of poly(NIPAAm), requires extensive toxicity studies before clinical applications can merge.

Recently some studies have been conducted on anocomposite hydrogels for photothermally modulated drug delivery. Gold nanoshells can be designed to absorb light strongly at desired wavelengths, in particular, in the near infrared between 800 and 1200 nm, where tissue is relatively transparent [122]. When optically absorbing gold nanoshells are embedded in a matrix material, illuminating them at their resonance wavelength causes the nanoshells to transfer heat to their local environment. This photothermal effect can be used to optically modulate drug release from a nonshell polymer composite drug delivery system. To accomplish photothermally modulated release, the matrix polymer material must be thermally responsive. The authors observed the pulsatile release of insulin and other proteins in response to near-infrared irradiation when gold nanoshells were embedded in NIPAAm-*co*-acrylamide hydrogels [122].

Electroresponsive Release An electric field as an external stimulus has advantages, such as the availability of equipment, which allow precise control with regard to the magnitude of current, duration of electric pulses, interval between pulses, and so on. Electrically responsive delivery systems are prepared from polyelectrolytes (polymers which contain relatively high concentration of ionizable groups along the backbone chain) and are thus pH responsive as well as electroresponsive. Under the influence of an electric field, electroresponsive hydrogels generally deswell or bend, depending on the shape of the gel that lies parallel to the electrodes, whereas deswelling occurs when the hydrogel lies perpendicular to the electrodes. Synthetic (e.g., acrylate and methacrylae derivatives) as well as naturally occurring polymers (including hyaluronic acid, chondroitin sulfate, agarose, carbomer, xanthan gum, and calcium alginate separately or in combination) have been used. Complex multicomponent gels or interpenetrating networks have been prepared in order to enhance the gels or interpenetrating networks have been prepared in order to enhance the gel's electroresponsiveness [123]. Electrically enhanced transdermal delivery provides the time-dependent delivery. Ionotophoresis, the electromigrational movement of charged molecules through the skin under a low-voltage and continuous driving force, has been reported for a number of bioactive compounds, such as leutinizaing hormone relesing hormone (LHRH), testosterone, and buserelin.

Electronic Microelectromechanical System for Controlled Release Electronic microelectromechanical devices are manufactured using standard microfabrication techniques that are used to create silicon chips for computers, and they often have moving parts or components that enable some physical or analytical function to be performed by the device. Microfabrication techniques, the same processing techniques used to make microprocessors for computers and other microelectronic devices, have been used increasingly to produce microscale devices whose primary functions are mechanical, chemical and optical in nature. Such devices are commonly referred to as *microelectromechanical systems* (MEMS) and are found in ink-jet printers, automotive applications, and microtube engines in the aerospace industry. MEMS for biological applications are classified as either microfluidic devices or nonmicrofluidic devices. The ultimate goal of MEMS is to develop a microfabricated device with the ability to store and release multiple chemical substances on demand by a mechanism devoid of moving its parts [124, 125]. A wide variety of microreservoirs, micropumps, cantilevers, rotors, channels, valves, sensors, and other structures have been fabricated, typically from the materials that have been demonstrated to be biocompatible and can be sterilely fabricated and hermetically sealed. The digital capabilities of MEMS may allow greater temporal control over drug release compared to traditional polymer-based systems, while the batch-processing techniques used in the microelectronics industry can lead to greater device uniformity and reproducibility than is currently available to the pharmaceutical industry. The use of MEMS for drug delivery necessitates the existence of drug depot or supply within or on the device. One straightforward approach to achieve this drug reservoir is the fabrication of silicon microparticles that contain an internal reservoir loaded with drug. These devices could be used for oral drug delivery, with release of the drug triggered by binding of a surface-functionalized molecule to cells in the digestive tract.

The completely implantable minipump made by Minimed has a pulsatile, radio-controlled injection rate through a catheter into the intraperitoneal region [126]. One study found that patients with the implantable pump did not differ from control subjects on any meansure of psychosocial function but that puump users monitored their blood glucose levels more frequently and had lower average blood glucose levels. Even though this type of device may improve patient's mbility and reduce infections by eliminating transcutaneous catheters, they may still be hampered by their size, cost, ability to deliver only drugs in solution, and limited stability of some drugs in solution at 37°C. Ikemoto and Sharpe [127] have developed a stepmotor micropump for the injection of nanoliter volumes of D-amphetamine solution into discrete brain regions of freely moving rats that was well tolerated. This micropump delivered a reliable volume of 50 nL per infusion over an hour at a rate of one infusion per minute.

Another development in MEMS technology is the microchip. The microchip consists of an array of reservoirs that extend through an electrolyte-impermeable substrate. The prototype microchip is made of silicon and contains a number of drug reservoirs. Each reservoir is sealed at one end by a thin gold membrane of material that serves as an anode in an electrochemical reaction and dissolves when an electric potential is applied to it in an electrolyte solution. The reservoirs are filled with any combination of drug or drug mixtures in any form (i.e., solid, liquid, or gel) through the opening opposite the anode membrane by ink-jet printing or microinjection and are then sealed with a waterproof material. A cathode is also required for the electrochemical reaction to take place, and the cathode is usually made of the same conductive material as the anode to simplify the fabrication procedure. The device is submerged in an electrolyte solution containing ions and upon electric stimulation forms a soluble complex with the anode in its ionic form. When release is desired, an electric potential is applied between an anode membrane and a cathode, and the gold membrane anode is dissolved within 10–20 s and allows the drug in the reservoir to be released. This electric potential causes oxidation of the anode material to form a soluble complex with the electrolytes which when dissolves allowing release of the drug. Complex releasc patterns (such as simultaneous constant and pulsatile release) can be achieved from the microchips. The microchip has the ability to control both release time and release rate. The rate of release from a reservoir is a function of the dissolution rate of the materials in the reservoir, the diffusion rate of these materials out of the reservoir, or both. Therefore, the release rate from an individual reservoir can be tailored to a particular application by proper selection of the materials placed inside the reservoir [e.g., pure drug(s), drugs with polymers] [124, 125].

A microchip with insulin-filled reservoirs could eventually provide a better alternative for the treatment of insulin-dependent diabetes mellitus (IDDM) [125]. Because the microchip is capable of being programmed as well as integrated with other electronic devices, it is supposable that the microchip could be incorporated into a closed-loop biofeedback system. An electronic apparatus that continuously measures the blood glucose levels could provide the stimulus to the microchip and result in release of insulin into the bloodstream. Although such a system could still not perfectly mimic an endogenous system of healthy person, it could practically meet the needs of IDDM patients. Pulsatile release of synthetic gonadotropin-releasing hormone (GnRH) can be achieved with a programmed microchip. A subcutaneous implanted microchip containing 1000 drug reservoirs would be adequate to administer a month's worth of drug therapy. The implanted microchip would be a convenient means to achieve the desired pharmacotherapeutic outcome of ovulation without interfering with the patient's daily activities or causing phlebitis.

While microchip drug delivery would be the most technologically advanced delivery system, it has itself limited storage capacity for therapeutic drugs [125]. Because most applications of this technology require implantation within bodily tissues, the question arises, "What would be done when the chip runs out of drug?" Some sort of procedure would be required to retrieve the empty chip cartridge once it has emptied. Due to the limited quantity of drug that can be stored on one chip, this technology is only ideal for potent drugs. If a larger dose of a medication is required, the chip would not be adequate for dispensing larger quantities of drug.

Magnetically Induced Release Magnetic carriers receive their magnetic response to a magnetic field from incorporated materials such as magnetite, iron, nickel, and cobalt. For biomedical applications, magnetic carriers must be water based, biocompatible, nontoxic, and nonimmunogenic. Earlier, Langer et al. [128] embedded magnetite or iron beads into a drug-filled polymer matrix and then showed that they could activate or increase the release of the drug from the polymer by moving a magnet over it or by applying an oscillating magnetic field. When the frequency of the applied field was increased from 5 to 11 Hz, the release of BSA from ethylenevinylacetate copolymer (EVAc) matrices slowed in a linear fashion. The rate of release could be modulated by altering the position, orientation, and magnetic strength of the embedded materials as well as by changing the amplitude of frequency of the magnetic field. The micromovement within the polymer produced microcracks in the matrix and thus made the influx of liquid, dissolution, and efflux of the drug. Done repeatedly, this would allow the pulsatile delivery of insulin.

Another mechanistic approach based on magnetic attraction is the slowing down of oral drugs in the gastrointestinal system. This is possible by filling an additional magnetic component into capsules or tablets. The speed of travel through the stomach and intestines can then be slowed down at specific positions by an external magnet, thus changing the timing and/or extent of drug absorption into stomach or intestines. Slowing down the passage of magnetic liposomes with a magnet actually increased the blood levels of drug. Babincova et al. [129] developed magnetoliposomes for triggered release of drug. In their delivery systems, they entrapped dextran–megnetite and model drug 6-carboxyfluorescein in the liposomes and used laser to trigger the release of drug. The magnetite absorbs the laser light energy to heat the lipid bilayer above the gel–liquid crystal-phase transsition temperature T_c, which is 41°C for dipalmitoyl-phosphatidylcholine. Liposomes made from this lipid release their content as soon as the temperature reaches this level. They have also suggested that the absorption of laser energy by magnetite particles provides a means for localized heating and controlled release of liposome with a single laser pulse. This may have potential applications for selective drug delivery, especially to the eyes and skin. Even though the magnetic-modulated therapeutic approach is promising, it still needs very careful attention for a number of physical and magnetism-related properties. The magnetic force, which is defined by its field and field gradient, needs to be large and carefully shaped to activate the delivery system within the target area. The magnetic materials should be tissue stable and compatible.

Chemically Induced Release

Gluose-Responsive Insulin Release Device A decrease in or the absence of insulin secretion from pancreatic islets is the cause of diabetes mellitus. An effective glucose-responsive insulin delivery system should be composed of a glucose-sensing component and an insulin-releasing component. The sensing component detects a change in the glucose level and produces a signal that affects the releasing component. The magnitude of the signal increases with increasing glucose concentration, and so does the rate of insulin release. Based on this principle, various polymer-based glucose-responsive delivery systems have been designed, most of which are hydrogels that can alter their volume and degree of hydration in response to glucose concentration. Several systems have already been developed which are able to respond to glucose concentration changes, such as glucose oxidase (GOD), which catalyzes glucose oxidation [130]. Glucosylated insulin bound to concanavalin (Con) A was released through exchange with external glucose, due to the difference in their binding constants. This system needs direct injection of microcapsules into the peritoneal cavity of patients, which may cause undesirable side effects arising from the immune response to Con A if Con A was directly exposed to immune systems after breakage of microcapsules. Obaidat and Park [131] prepared a copolymer of acrylamide and allyl glucose. The side-chain glucose units in the copolymer were bound to Con A. These hydrogels showed a glucose-responsive, sol–gel phase transition dependent upon the external glucose concentration. The nonlinear dependence of this sol–gel phase transition with regard to the glucose concentration was due not only to the increased binding affinity of allyl glucose to Con A compared to native glucose but also to the cooperative interaction between glucose-containing copolymer and Con A. Kataoka et al. [132] developed glucose and thermoresponsive hydrogels using acrylamidophenylboronic acid and *N*-isopropylacrylamide (IPAAm). The obtained gels, containing 10 mol % phenylboronic acid moieties, showed a transition temperature of 22°C in the absence of glucose. Below this temperature, the gels existed in a swollen state. The introduction of glucose to the medium altered the transition temperature of the gels in such a way that the transition temperature increased with increasing glucose concentration to reach 36°C at 5 g/L glucose concentration. Boronic acid was in equilibrium between the undissociated and dissociated forms. With increasing glucose concentration, the equilibrium shifted to increase the amount of dissociated boronate groups and gels became more hydrophilic. Although all of the glucose-sensitive insulin delivery systems are elegant and highly promising, many improvements need to be made for them to become clinically useful. First of all, the response of these systems upon changes in the environment occurs too slowly. In clinical situations, these systems need to respond to ever-changing glucose concentrations at all times, requiring hydrogels that can respond reproducibility and with rapid-response onset times on a long-term basis. An additional constant is that all the components used in the systems must be biocompatible.

Chemotactic Factor-Induced Controlled-Release Systems With physical or chemical stress such as injury and broken bones, an inflammation reaction takes place at the injured site. At the inflammatory sites, inflammation-responsive phagocytic cells such as macrophages and polymorphonuclear cells play a role in healing the injury. During inflammation, hydroxyl radicals ($OH^{\bullet}$) are produced from the cells. Yui and co-workers [133, 134] developed inflammatory-induced hydroxyl radicals and designed drug delivery system which responded to the hydroxyl radicals and degraded in a limited manner. They used hyaluronic acid (HA), a linear aminopolysaccharide composed of repeating disaccharide subunits of *N*-acetyl-D-glucosamine and D-guluronic acid. In the body, HA is mainly degraded by hyaluronidase, or hydroxyl radicals. Degradation of HA via the enzyme is very low in a normal state of health. Degradation via hydroxyl radicals, however, is usually dominant and rapid when HA is injected at inflammatory sites. These authors prepared cross-linked HA with ethyleneglycol diglydylether or polyglycerol polygluycidalether. These HA gels degraded only when the hydroxyl radicals were generated through the reaction between the iron (Fe^{2+}) ions and the hydrogen peroxide in vitro. Thus, a surface erosion type of degradation was achieved. When microspheres were incorporated in the HA hydrogels as a model drug, these microspheres were released when hydroxyl radicals induced HA gel degradation. Furthermore, degradation of HA in vivo tests showed that HA gels are degraded only when inflammation was induced by surgical incision. Control HA gels were stable over 100 days. It is possible to treat locally in inflammatory diseases such as rheumatoid arthritis using anti-inflammatory drug incorporated in HA gels [135].

5.1.11 SUMMARY

As pharmaceutical scientists have increased their knowledge of pharmacokinetics and pharmacodynamics, it has become apparent that these factors can result in more efficacious drugs. The number of new drug entities appearing on the market yearly has declined and pharmaceutical manufacturers have shown a renewed interest in improving existing dosage forms and developing more sophisticated drug delivery systems, including those employing the principles of controlled drug release.

Current research in this area involves numerous new and novel systems, many of which have strong therapeutic potential. In this chapter, we have tried to emphasize the importance of oral routes as well as others, such as ocular, transdermal, intrauterine, and vaginal. The various microencapsulation, nanoencapsulation, and liposome technologies and the release of drugs and bioactive compounds from such products have been described.

REFERENCES

1. Bhalla, H. L. (1999), Drug delivery research in India: A challenge and an opportunity, *J. Controlled Release*, 62, 65–68.
2. Breimer, D. D. (1999), Future challenges for drug delivery, *J. Controlled Release*, 62, 3–6.
3. Chien, Y. W. (1992), *Novel Drug Delivery Systems*, Marcel Dekker, New York, pp. 1, 2.
4. Brannon-Peppas, L. (1997), Polymers in controlled drug delivery, *Biomaterials*, 11, 1–14.
5. Kydoneieus, A. F. (1980), *Controlled Release Technologies: Methods, Theory, and Applications*, CRC Press, Boca Raton, FL.
6. Murano, E. (1998), Use of natural polysaccharides in the microencapsulation techniques, *J. Appl. Ichthyol.* 14, 245–249.
7. Junginger, H. E., and Verhoef, J. C. (1998), Macromolecules as safe penetration enhancers for hydrophilic drugs: A fiction, *Pharm. Sci. Today*, 1, 375–376.
8. Robinson, J. R., and Lee, V. H. L. (1987), *Controlled Drug Delivery: Fundamentals and Applications*, Marcel Dekker, New York.
9. Hofman, A. F., Pressman, J. H., Code, C. F., and Witztum, K. F. (1983), Controlled entry of orally-administered drugs: Physiological considerations. *Drug Dev. Ind. Pharm.*, 9, 1077.
10. Manninen, V., Melin, J., and Reissel, P. (1972), Tablet disintegration: Possible link with biological availability of digoxin, *Lancet*, 26, 490–491.
11. Fincher, J. H. (1968), Particle size of drugs and its relationship to absorption and activity, *J. Pharm. Sci.*, 57, 1825–1835.
12. Lee, V. H., and Robinson, J. R. (1987), *Sustained and Controlled Release Drug Delivery Systems*, Marcel Dekker, New York, p. 71.
13. Fix, J. A. (1996), Oral controlled release technology for peptides: Status and future prospectus. *Pharm. Res.*, 13, 1760–1763.
14. Chien, Y. W. (1983), Logics for transdermal controlled drug administration, *Drug Dev. Ind. Pharm.*, 9, 497–520.
15. Gandhi, R., Kaul, C. L., and Panchangula, R. (1999), Extrusion and spheronization in the development of oral controlled-release dosage forms, *Pharmaceutical Science and Technology Today*, 2, 160–170.

16. Anal, A. K. (2007), Time-controlled pulsatile delivery systems for bioactive compounds, Recent Patents Drug Deliv. *Formulat.*, 1, 73–79.
17. Chan, O. H., and Stewart, B. H. (1996), Physico-chemical and drug delivery considerations for oral drug bioavaibility. *Drug Discovery Today*, 1, 461–472.
18. Wilson, C. G. (2000), Gastrointestinal transit and drug absorption, in *Oral Drug Absorption: Prediction and Assessment*, Marcel Dekker, New York, pp. 1–9.
19. Knutson, L., Knutson, F., and Knutson, T. (2000), Permeability in the gastrointestinal tract, in *Oral Drug Absorption: Prediction and Assessment*, Marcel Dekker, New York, pp. 11–16.
20. Waterbeemd, H. V. (2000), Intestinal permeability: Prediction from theory, in *Oral Drug Absorption: Prediction and Assessment*, Marcel Dekker, New York, pp. 31–49.
21. Chien, Y. W. (1992), *Novel Drug Delivery Systems*, Vol. I, Marcel Dekker, New York, pp. 139–196.
22. Jantzen, G. M., and Robinson, J. R. (2000), Sustained and controlled release drug delivery systems, in *Modern Pharmaceutics*, Marcel Dekker, New York, pp. 501–528.
23. Welling, P. G. (1983), Oral controlled drug adminstration: Pharmacokinetic considerations. *Drug Dev. Ind. Pharm.*, 9, 1185–1125.
24. Hofman, A. F., Dressman, J. H., Code, C. F., and Witztum, K. F. (1983), Controlled entry of oral administered drugs: Physiological considerations, *Drug Dev. Ind. Pharm.*, 9, 1077–1083.
25. Fara, J. W. (1983), Gastrointestinal transit of solid dosage forms. *Pharm. Technol.*, 7, 23–26.
26. Meyer, J. H., Ohashi, H., Jehn, D., and Thomson, J. B. (1981), Size of particles emptied from the human stomach, *Gastroenterology*, 80, 1489–1496.
27. Welling, P. G., and Dobrinska, M. R. (1987), Dosing considerations and bioavailability assessment of controlled drug delivery systems, in *Controlled Drug Delivery: Fundamentals and Applications*, Marcel Dekker, New York, pp. 253–289.
28. Hinton, J. M., Lennard-Jones, J. E., and Young, A. C. (1969), Alimentary canal transit time for an inert object in humans, *Gut*, 10, 842.
29. Frier, M., and Perkins, A. C. (1994), Radiopharmaceuticals and the gastrointestinal tract, *Eur. J. Nucl. Med. Mol. Imaging*, 21, 1234–1242.
30. Basit, A. W., Podczeck, F., Newton, M., Waddington, W. A., Ell, P. J., and Lacey, L. F. (2004), The use of formulation technology to assess regional gastrointestinal drug absorption in humans, *Eur. J. Pharm. Sci.*, 21, 179–189.
31. Delie, F., and Blanco-Príeto, M. J. (2005), Polymeric particulares to improve oral bioavailability of peptide drugs, *Molecules*, 10, 65–80.
32. Walter, E., Kissel, T., and Amidon, G. E. (1996), The intestinal peptide carrier: A potential transport system for small peptide-derived drugs, *Adv Drug Deliv. Rev.*, 20, 33–58.
33. Scheline, R. R. (1973), Metabolism of foreign compounds by gastrointestinal microorganisms, *Pharmacol. Rev.*, 25, 451–523.
34. Edwards, C. (1997), Physiology of colorectal barrier. *Adv. Drug Deliv. Rev.*, 28, 173–190.
35. Whateley, T. L. (1992), *Microencapsulation of Drugs*, Harwood Academic, Geneva, Switzerland.
36. Anal, A. K., and Stevens, W. F. (2005), Chitosan-alginate multilayer beads for controlled release of ampicillin, *Int. J. Pharm.*, 290, 45–54.
37. Anal, A. K., Bhopatkar, D., Tokura, S., Tamura, H., and Stevens, W. F. (2003), Chitosan-alginate multilayer beads for gastric passage and controlled intestinal release of protein, *Drug Dev. Ind. Pharm.*, 29, 713–724.

38. Hui, H. W., Lee, V. H. L., and Robinson, J. R. (1987), Design and fabrication of oral controlled release drug delivery systems, in *Controlled Drug Delivery: Fundamentals and Applications*, Marcel Dekker, New York, pp. 373–421.
39. Anal, A. K., and Singh, H. (2007), Recent advances in microencapsulation of probioitcs for industrial applications and targeted delivery, *Trends Food Sci. Technol.*, 18, 240–251.
40. Mathir, Z. M., Dangor, C. M., Govender, T., and Chetty, D. J. (1997), In vitro characterization of a controlled-release chlorpheniramine maleate delivery system prepared by the air-suspension technique, *J. Microencapsul.*, 14, 743–751.
41. Deasy, P. B. (1991), Microencapsulation of drugs by pan and air suspension techniques, *Crit. Rev. Ther. Drug Carrier Syst.*, 8, 39–89.
42. O'Connor, R. E., and Schwartz, J. B. (1989), Extrusion and spheronization technology, in *Pharmaceutical Pelletization Technology*, Marcel Dekker, New York, pp. 187–215.
43. Gopferich, A., Alonso, M. J., and Langer, R. (1994), Development and characterization of microencapsulated microspheres, *Pharm. Res.*, 11, 1568–1574.
44. Anal, A. K., Stevens, W. F., and López, C. R. (2006), Ionotropic cross-linked chitosan microspheres for controlled release of ampicillin, *Int. J. Pharm.*, 312, 166–173.
45. Weidenauer, U., Bodmeier, D., and Kissel, T. (2003), Microencapsulation of hydrophilic drug substances using biodegradable polyesters. Part I: Evaluation of different techniques for the encapsulation of pamidronate di-sodium salt, *J. Microencapsul.*, 20, 509–524.
46. Cui, J. H., Goh, J. S., Park, S. Y., Kim, P. H., and Le, B. J. (2001), Preparation and physical characterization of alginate microparticles using air atomization method, *Drug Dev. Ind. Pharm.*, 27, 309–319.
47. Purvis, T., Vaughan, J. M., Rogers, T. L., Chen X., Overhoff, K. A., Sinswat, P., Hu, J., McConville, J. T., Johnston, K. P., and Williams, R. O. (2006), Cryogenic liquids, nanoparticles, and microencapsulation, *Int. J. Pharm.*, 4, 43–50.
48. Charman, W. N., Chan, H. K., Finnin, B. C., and Charman, S. A. (1999), Drug delivery: A key factor in realizing the full therapeutic potential of drugs, *Drug Dev. Res.*, 46, 316–327.
49. Kayser, O., Lemke, A., and Hernández-Trejo, N. (2005), The impact of nanobiotechnolgy on the development of new drug delivery systems, *Curr. Pharm. Biotechnol.*, 6, 3–5.
50. Roco, M. C. (2003), Nanotechnology: Convergence with modern biology and medicine, *Curr. Opin. Biotechnol.*, 14, 337–346.
51. Merck & Co. (2004), Drug Information: Emend, capsules, available: www.merck.com.
52. Liversidge, G. G., Cundy, K. C., Bishop, J. F., and Czekai, D. A. (1992), Surface modified drug nanoparticles, Un. St. Patent 5,145,684, Sterling Drug, New York.
53. Merisko-Liversidge, E., Liversidge, G. G., and Cooper, E. R. (2003), Nanosizing: A formulation approach for poorly water-soluble compounds, *Eur. J. Pharm. Sci.*, 18, 113–120.
54. Merisko-Liversidge, E., Sparpotdar, P., and Bruno, J. (1996), Formulation and antitumor activity evaluation of nanocrystalline suspension of poorly soluble anticancer drugs, *Pharm. Res.*, 13, 272–278.
55. Buchmann, S., Fischli, W., Thiel, F. P., and Alex, R. (1996), Aqueous microsuspension, an alternative intravenous formulation for animal studies, paper presented at the 42nd Annual Congress of the International Association for Pharmaceutical Technology, Mainz, p. 124.
56. Müller, R. H., and Heinemann, S. (1989), Surface modeling of microparticles as parenteral systems with high tissue affinity, in *Bioadhesion-Possibilities and Future Trends*, Wissenschaftliche Verlagsgesellschaft, Stuttgart, pp. 202–214.

57. Müller, R. H., and Peters, K. (1998), Nanosuspensions for the formulation of poorly soluble drugs: I: Preparation by a size-reduction technique, *Int. J. Pharm.*, 160, 229–237.
58. Maa, Y. F., and Hsu, C. C. (1998), Performance of sonication and microfluidisation for liquid-liquid emulsification, *Pharm. Dev. Technol.*, 4, 233–240.
59. Roller, J. M., Covereur, P., Robolt-Treupel, L., and Puisieux, F. (1986), Physicochemical and morphological characterization of polyisobutyl cyanoacrylates nanocapsules, *J. Pharm. Sci.*, 75, 361.
60. Chouinard, F., Kan, F. W., Leroux, J. C., Foucher, C., and Lenaerts, V. (1991), Preparation and purification of polyisohexylcyanoacrylate nanocapsules, *Int. J. Pharm.*, 72, 211.
61. Ammoury, N., Fessi, H., Devissaguet, J. P., Ouisieux, F., and Benita, S. (1989), Physicochemical characterization of polymeric nanocapsules and in vitro release evaluation of indomethacin as a drug model, *STP Pharma*, 5, 642.
62. Gallardo, M. M., Courraze, G., Denizor, B., Treupel, L., Couvreur, P., and Puisieux, F. (1993), Preparation and purification of isohexylcyanoacrylate nanocapsules, *Int. J. Pharm.*, 100, 55–64.
63. Chouinard, F., Buczkowski, S., and Lenaerts, V. (1994), Poly(alkylcyanoacrylate) nanocapsules: Physico-chemical characterization and mechanism of formation, *Pharm. Res.*, 11, 869.
64. Wohlgemuth, M., Machtle, W., and Mayer, C. (2000), Improved preparation and physical studies of polybutylcyanoacrylate nanocapsules, *J. Microencapsul.*, 17, 437.
65. Lambert, G., Fattal, E., Pinto-Alphandary, H., Gulik, A., and Couvereur, P. (2000), Polyisobutylcyanoacrylate nanocapsules containing an aqueous core the delivery of oligonucleotides, *Int. J. Pharm.*, 214, 13.
66. Couvreur, P., Barratt, G., Fattal, E., Legrand, P., and Vauthier, C. (2002), Nanocapsule technology: A review. *Crit. Rev. Therap. Drug Carrier Syst.*, 19, 99–134.
67. Quintanar-Guerrero, D., Allemann, E., Doelker, E., and Fessi, H. (1998), Preparation and characterization of nanocapsules from performed polymers by a new process based on emulsification-diffusion technique, *Pharm. Res.*, 15, 1056.
68. Ravi, K. M. N. (2000), Nano and microparticles as controlled drug delivery devices, *J. Pharm. Sci.*, 3, 234–258.
69. Torchilin, V. P. (2005), Recent advances with liposomes as pharmaceutical carriers, *Nature Rev.: Drug Discovery*, 4, 145–159.
70. Leserman, L. (2004), Liposomes as protein carriers in immunology, *J. Liposome Res.*, 14, 175–189.
71. Uchegbu, I. F., and Vyas, S. P. (1998), Non-ionic surfactant based vesicles (niosomes) in drug delivery, *Int. J. Pharm.*, 172, 33–70.
72. Uchegbu, I. F. (2000), Synthetic surfactant vesicles: Niosomes and other non-phospholipid vesicular systems, in *Drug Targeting and Delivery*, Vol. 11, Harwood Academic, Amsterdam.
73. Bodor, N., and Loftsson, T. (1987), Novel chemical approaches for sustained drug delivery, in *Controlled Drug Delivery: Fundamentals and Applications*, Marcel Dekker, New York, pp. 337–369.
74. Sugibayashi, K., and Morimoto, Y. (1994), Polymers for transdermal drug delivery systems, *J. Controlled Release*, 29, 177–185.
75. Chien, Y. W. (1987), Transdermal therapeutic systems, in *Controlled Drug Delivery: Fundamentals and Applications*, Marcel Dekker, New York, pp. 524–549.
76. Chien, Y. W. (1982), Ocular controlled release drug administration, in *Novel Drug Delivery Systems*, Marcel Dekker, New York, pp. 13–48.

77. Gelatt, K. N., Gum, G. G., williams, L. W., and Peiffer, R. L. (1979), Evaluation of a soluble sustained-release ophthalmic delivery unit in the dog, *Am. J. Vet. Res.*, 40, 702–704.
78. Chien, D. S., and Schoenwald, R. D. (1990), Ocular pharmacokinetics and pharmacodynamics of phenylephrine and phenylephrine oxazolidine in rabbit eyes, *Pharm. Res.*, 7, 476–483.
79. Vyas, S. P., Ramchandraiah, S., Jain, C. P., and Jain, S. K. (1992), Polymeric pseudolatices bearing pilocarpine for controlled ocular delivery, *J. Microencapsul.*, 9, 347–355.
80. Lyrenäs, S., Clason, I., and Ulmsten, U. (2001), In vivo controlled release of PGE2 from a vaginal insert (0.8 mm, 10 mg) during induction of labor, *Br. J. Obstet. Gynaecol.*, 108, 169–178.
81. Sherwood, J. K., Zeitlin, L., Whaley, K. J., Richard, A. C., and Saltzman, M. (1996), Controlled release of antibodies for long-term topical passive immunoprotection of female mice against genital herpes, *Nat. Biotechnol.*, 14, 468–471.
82. Nilsson, C., Lachteenmaki, P., and Luukkainen, T. (1980), Patterns of ovulation and bleeding with a low levonregesterol-releasing device, *Contraception*, 21, 225–233.
83. van Laarhoven, J., Kruft, M., and Vromans, H. (2002), In vitro release properties of etonogestrel and ethynylestradiol from a contraceptive vaginal ring, *Int. J. Pharm.*, 232, 163–173.
84. Rathbone, M. J., Macmillan, K. L., JöChle, W., Boland, M., and Inskeep, E. K. (1998), Controlled release products for the control of the estrous cycle in cattle, sheep, goats, deer, pigs, and horses, *Crit. Rev. Ther. Drug Carrier Syst.*, 15, 285–380.
85. Strix, J. (1999), German Patent Application DE98-19809243.
86. Duncan, R., and Seymour, L. (1989), Controlled Release Technolgies, Elsevier Advanced Technology, Amsterdam, 11.
87. Lee, E. S., Kim, S. W., Kim, S. H., Cardinal, J. R., and Jacobs, H. (1980), Drug release from hydrogel devices with rate-controlling barriers, *J. Membr. Sci.*, 7, 293–303.
88. El-Nokaly, M. A., Piatt, D. M., and Charpentier, B. A. (1993), *Polymeric Delivery Systems: Properties and Applications*, ACS Symposium Series, Washington, DC.
89. Hui, H. W., Lee, V. H. L., and Robinson, J. R. (1987), Design and fabrication of oral controlled release drug delivery systems, in *Controlled Drug Delivery: Fundamentals and Applications*, Marcel Dekker, New York, pp. 373–421.
90. Sangalli, M. E., Maroni, A., Zema, L., Busseli, C., Giordano, F., and Gazzaniga, A. (2001), *In vitro* and *in vivo* evaluation of oral system for time and/or site specific drug delivery, *J. Controlled Release*, 73, 103–110.
91. Sangalli, M. E., Maroni, A., Foppoli, A., Zema, L., Giordano, F., and Gazzaniga, A. (2004), Different HPMC viscosity grades as coating agents for an oral time and/or site-controlled delivery systems: A study on process parameters and in vitro performances, *Eur. J. Pharm. Sci.*, 22, 469–476.
92. Pozzi, F., Furlani, P., Gazzaniga, A., Davis, S. S., and Wilding, I. R. (1994), The TIME CLOCK® system: A new oral dosage form for fast and complete of drug after a predetermined lag time, *J. Controlled Release*, 31, 99–108.
93. Medlicott, N. J., and Tucker, I. J. (1999), Pulsatile release from subcutaneous implants, *Adv. Drug Deliv. Rev.*, 38, 139–149.
94. Bussemer, T., Peppas, N. A., and Bodmeier, R. (2003), Evaluation of the swelling, hydration and rupturing properties of the swelling layer of a rupturable pulsatile drug delivery system, *Eur. J. Pharm. Biopharm.*, 56, 261–270.
95. Bussemer, T., Peppas, N. A., and Bodmeier, R. (2003), Time-dependent mechanical properties of polymers coating used in rupturable pulsatile release dosage forms, *Drug Dev. Ind. Pharm.*, 29, 623–630.

96. Sungthongjeen, S., Puttipipatkhachorn, S., Paeratakul, O., Dashevsky, A., and Bodmeier, R. (2004), Development of pulsatile release tablets with swelling and rupturable layers, *J. Controlled Release*, 95, 147–159.
97. Niwa, K., Takaya, T., Morimoto, T., and Takada, K. (1995), Preparation and evaluation of a time-controlled release capsule made of ethylcellulose for colon delivery of drugs, *J. Drug Target.*, 3, 83–89.
98. Krogel, I., and Bodmeier, R. (1997), Pulsatile drug release from an insoluble capsule body controlled by an erodible plug, *Pharm. Res.*, 15, 474–481.
99. Krogel, I., and Bodmeier, R. (1999), Evaluation of an enzyme-containing capsular shaped pulsatile drug delivery system, *Pharm. Res.*, 16, 1424–1429.
100. Gohel, M. C., and Sumitra, M. (2002), Modulation of active pharmaceutical material release from a novel "tablet in capsule system" containing an effervescent blend, *J. Controlled Release*, 79, 157–164.
101. Linkwitz, A., Magruder, J. A., and Merrill, S. (1994), Osmotically driven delivery device with expandable orifice for pulsatile delivery effect, U.S. Patent 5,318,558.
102. Qui, L. Y., and Zhu, K. J. (2001), Design of core-shelled polymer cylinder for potential programmable drug delivery, *Int. J. Pharm.*, 219, 151–160.
103. Jiang, H. L., and Zhu, K. J. (1999), Preparation, characterization and degradation characteristics of polyanhydrides containing polyethlene glycol, *Poly. Int.*, 48, 47–52.
104. Jiang, H. L., and Zhu, K. J. (2000), Pulsatile protein release from a laminated device comprising of polyanhydride and pH-sensitive complexes, *Int. J. Pharm.*, 194, 51–60.
105. Anal, A. K. (2007), Stimuli-induced pulsatile or triggered release of bioactive compounds, Recent Patents Endocrine, *Metabolic Immune Drug Discovery*, 1, 83–90.
106. Rubinstein, A. (2005), Colonic drug delivery, *Drug Discovery Today: Technol.*, 2, 33–37.
107. Deshpande, A. A., Rhodes, C. T., Shah, N. H., and Mallick, A. W. (1996), Controlled-release drug delivery systems for prolonged gastric residence: An overview, *Drug Dev. Ind. Pharm.*, 22, 531–539.
108. Rubinstein, A. (1995), Approaches and opportunities in colon-specific drug delivery. *Curr. Rev. Ther. Drug Carrier Syst.*, 12, 101–149.
109. Basit, A., and Bloor, J. (2003), Perspectives on colonic drug delivery, *Pharmatech*, 185–190.
110. Vandamme, T. F., Lenoury, A., Charrueau, C., and Chaumeil, J.-C. (2002), The use of polysaccharides to target drugs to the colon. *Carbohydr. Poly.*, 48, 219–231.
111. Sinha, V. R., and Kumria, R. (2001), Polysaccharides in colon-specific drug delivery, *Int. J. Pharm.*, 224, 19–38.
112. Lee, S., Lim, C. B., Lee, S., and Pai, C. M. (1999), Colon selective drug delivery composition, WO 99/01115.
113. Watanabe, S., Kawai, H., Katsuma, M., and Fukul, M. (2002), Colon-specific drug release system, U.S. Patent 6,368,629.
114. Takemura, S., Watanabe, S., Katsuma, M., and Fukul, M. (2002), Gastrointestinal transit study of a novel colon delivery system (CODES™) using gamma scintigraphy, *Proc. Int. Symp. Controlled Release Bioactive Mater.*, 27, 445–446.
115. Li, J., Yang, L., Ferguson, S. M., Hudson, T. J., Watanabe, S., Katsuma, M., and Fix, J. A. (2002), In vitro evaluation of dissolution behaviour for a colon-specifc drug delivery system (CODESTM) in multi-pH media using USP apparatus II and III, *AAPS Pharm. Sci. Technol.*, 3, article 33, available: www.aapspharmascitech.org.
116. Bae, Y. H., Okano, T., and Kim, S. W. (1991), "On-off" thermocontrol of solute transport. I. Temperature dependence of swelling of *N*-isopropylacrylamide networks modified with hydrophobic components in water, *Pharm. Res.*, 8, 531–537.

117. Bae, Y. H., Okano, T., and Kim, S. W. (1991), "On-off" thermocontrol of solute transport. II. Temperature dependence of swelling of *N*-isopropylacrylamide networks modified with hydrophobic components in water, *Pharm. Res.*, 8, 624–628.

118. Okano, T., Bae, Y. H., Jacobs, H., and Kim, S. W. (1990), Thermally on-off switching polymers for drug permeation and release, *J. Controlled Release*, 11, 255–265.

119. Kaneko, Y., Sakai, K., Kikuchi, A., Yoshida, R., Sakurai, Y., and Okano, T. (1995), Influence of freely mobile grafted chain length on dynamic properties of comb-type grafted poly(*N*-isopropylacrylamide) hydrogels, *Macromolecules*, 28, 7717–7723.

120. Kaneko, Y., Sakai, K., Kikuchi, A., Sakurai, Y., and Okano, T. (1996), Fast swelling/deswelling kinetics of comb-type grafted poly(*N*-isopropylacrylamide hydrogels). *Macromol. Chem.*, 109, 41–53.

121. Bromberg, L. E., and Ren, E. S. (1998), Thepertaure-responsive gels and thermo gelling polymer matrices for protein and peptide delivery, *Adv. Drug Deliv. Rev.*, 31, 197–221.

122. Sershen, S. R., Westcott, S. L., Halas, N. J., and West, N. J. (2000), Temperature-sensitive polymer-nanoshell composites for photothermally modulated drug delivery, *J. Biomed. Mater. Res.*, 51, 293–298.

123. Yuk, S. H., Cho, S. H., and Lee, H. B. (1992), Electric current-sensitive drug delivery systems using sodium-alginate/polyacrylic acid composites, *Pharm. Res.*, 9, 955–957.

124. Santini, J. T., Cima, M. J., and Langer, R. (1999), A controlled-release microchip, *Nature*, 397, 335–338.

125. Santini, J. T., Richards, A. C., Schiedt, R., Cima, M. J., and Langer, R. (2000), Microchips as controlled-drug delivery devices, *Chem. Int. Ed.*, 39, 2396–2407.

126. Barrett, D. H., Davidson, P. C., Steed, L. J., Abel, G. G., Loman, K. E., and Saudek, C. D. (1995), Evaluation of the psychosocial impact of the minimed variable-rate implanatable insulin pump, *Med. J.*, 88, 1226–1230.

127. Ikemoto, S., and Sharpe, L. G. (2001), A head-attachable device for injecting anolitre volumes of drug solutions into brain sites of freely moving rats, *J. Neurosci. Meth.*, 110, 135–140.

128. Edelman, E., Kost, J., Bobeck, H., and Langer, R. (1985), Regulation of drug release from polymer matrices by oscillating magnetic fields, *J. Biomed. Mater. Res.*, 19, 67.

129. Babincová, M., Sourivong, P., Chorvát, D., and Babinec, P. (1999), Laser triggered drug release from magnetoliposomes, *J. Magnetism Magnetic Mater.*, 194, 163–166.

130. Kim, S. W., Pai, C. M., Makino, K., Seminoff, L. A., Holmberg, D. L., Gleeson, J. M., Wilson, D. E., Mack, E. J. (1990), Self-regulated glycosylated insulin delivery, *J. Controlled Release*, 11, 193–201.

131. Obaidat, A. A., and Park K. (1997), Characterization of protein release through glucose-sensitive hydrogel membranes, *Biomaterials*, 18, 801–806.

132. Kataoka, K., Miyazaki, H., Bunya, M., Okano, T., and Sakurai, Y. (1998), Totally synthetic polymer gels responding to external glucose concentration: Their preparation and application to on-off regulation of insulin release, *J. Am. Chem. Soc.*, 120, 12694–12695.

133. Yui, N., Okano, T., and Sakurai, Y. (1992), Inflammation responsive degradation of crosslinked hyaluronic acid gels, *J. Controlled Release*, 22, 105–116.

134. Yui, N., Nihira, J., Okano, T., and Sakurai, Y. (1993), Regulated release of drug microspheres from inflammation responsive degradable matrices of crosslinked hyaluronic acid, *J. Controlled Release*, 25, 133–143.

135. Kikuchi, A., and Okano, T. (2002), Pulsatile drug release control using hydrogels, *Adv. Drug Deliv. Rev.*, 54, 53–77.

5.2

PROGRESS IN DESIGN OF BIODEGRADABLE POLYMER-BASED MICROSPHERES FOR PARENTERAL CONTROLLED DELIVERY OF THERAPEUTIC PEPTIDE/PROTEIN

SHUNMUGAPERUMAL TAMILVANAN*
University of Antwerp, Antwerp, Belgium

Contents

*Current address: Department of Pharmaceutics, Arulmigu Kalasalingam College of Pharmacy, Anand Nagar, Krishnankoil, India

Pharmaceutical Manufacturing Handbook: Production and Processes, edited by Shayne Cox Gad

5.2.1 INTRODUCTION

At the cellular level, deoxyribonucleic and ribonucleic acids (DNA and RNA, respectively) serve as an endogenious vehicle not only to store genetic information but also to transfer genetic information from one generation to their offsprings of all known living organisms. In addition, utilizing the rule of complementary base pairing, the DNA undergoes replication and transcription processes to produce respectively a new double-stranded DNA molecule and a complementary single-stranded RNA molecule. Following the translation process, peptide and protein are synthesized/constructed in ribosomal subunits through peptidic linkages between available 20 amino acids. The peptide and protein thus constructed perform a wide variety of functions and each cell contains several thousands of different proteins. Peptide- and protein-mediated, important physiological and biological processes of the human body include ligands/hormones for signaling, enzymes for biotransformation reactions, receptors for pharmacological response elucidation, antibodies in immune system interactions, transcription, and translation. Hence these molecules play a vital role to ensure proper development and functioning of entire organs of the human body.

Webster's New World Dictionary defines a drug as "any substances used as a medicine or as an ingredient in a medicine." Indeed, peptides and recombinant proteins are highly potent, relatively macromolecular and promising therapeutic agents that emerged out from the significant development of biotechnic and biogenetic engineering technologies. Peptide and protein therapeutics include semisynthetic vaccines, monoclonal antibodies, growth factors, cytokines, soluble receptors, hormones, and enzymes. The advent of recombinant DNA technology allowed the possibility of the commercial production of proteins for pharmaceutical applications from the early 1980s and, in fact, manufacture of therapeutic proteins represented the first true industrial application of this technology [1]. During the 1980s the term *biopharmaceutical* became synonymous with *therapeutic protein produced by recombinant DNA technology* (or, in the case of a small number of therapeutic monoclonal antibodies, *by hybridoma technology*). Clinical evaluation of nucleic acid–based drugs used for the purposes of gene therapy and antisense technology commenced in the 1990s, and today the term biopharmaceutical also encompasses such (as-yet-experimental) drugs [2]. The first such recombinant therapeutic protein (insulin) was approved for general medical use only 24 years ago. Today there are in excess of 100 such products approved in some world regions at least, with 88 having received approval within the European Union (EU). This represents 36% of all new drug approvals since the introduction of the new centralized European drug approval system in 1995 [3]. Over the coming decade, therefore, in the region of a dozen new therapeutic proteins should, on average, gain regulatory approval each year. While EU figures are difficult to locate, the American Association of Pharmaceutical Researchers and Manufacturers (PhRMA) estimates that there are currently some 371 biotechnology medicines in development [4]. Out of these 371 biotechnology medicines, as estimated by PhRMA, more than 300 are protein based, with recombinant vaccines and monoclonal/engineered antibodies representing the two most promising categories. Incidentally, all 88 biopharmaceutical products currently approved within the EU are protein based. Of the proteins thus far approved, hormones and cytokines represent the largest product categories (23 and 18 products, respectively). Hormones approved include several recombinant insulins, displaying both native and modified amino acid sequences. In addition, several recombinant gonadotrophins [follicle-stimulating hormone (FSH), luteinizing hormone (LH), and human chorionic gonadotrophins (hCG)] have been approved for the treatment of various forms of subfertility/infertility. Cytokines approved include a range of recombinant hematopoietic factors, including multiple erythropoietin-based products used for the treatment of anemia as well as a colony-stimulating factor aimed at treating neutropenia. Additional approved cytokines include a range of interferon-based products, mainly used to treat cancer and various viral infections, most notably hepatitis B and C, and a recombinant tumor necrosis factor (TNF) used as an adjuvant therapy in the treatment of some soft tissue cancers. Blood-related approved therapeutic proteins include a range of recombinant blood coagulation factors used to treat hemophilia, recombinant thrombolytics, and recombinant anticoagulants. Additional product categories include a range of subunit vaccines containing at least one recombinant component [mainly hepatitis B surface antigens (HBsAg)] and a variety of monoclonal antibody–based products indicated for the treatment/detection of various cancers or the prevention of organ transplant rejection. In summary, ailments that can be treated more effectively by this new class of therapeutic agents include cancers, autoimmune disease, memory impairment, mental disorders, hypertension, and certain cardiovascular and metabolic diseases [5, 6].

Poor absorption and easy degradation by endogenous proteolytic enzymes present in eye tissues, nasal mucosa, and gastro intestinal tract and low transdermal bioavailabilities due to relatively large size make the peptide/protein molecules to be administered only through parenteral routes either by multiple injections or infusion therapy in order to achieve desired therapeutic plasma levels for prolonged periods of time. Nevertheless, because of remarkably short half-lives within the in vivo arena, the therapeutic usuage of most of the peptide/protein is practically possible only through daily multiple injections under close medical supervision. Hence, the commercial success of peptides/proteins as therapeutic agents depends mainly on development of novel drug delivery systems which could potentially reduce the injection frequencies and thus eliminate the accompanying serious problem of patient compliance.

Among the several technologies that have been suggested for reducing injection frequencies of therapeutic peptide/protein, microspheres prepared from biodegradable polymers are widely recognized for controlled drug delivery following parenteral administration. Polyester polymers such as poly(lactic acid) (PLA), poly(glycolic acid) (PGA), and their copolymer poly(lactic acid-*co*-glycolic acid) (PLGA) are used routinely for the preparation of injectable microspheres after taking into consideration their well-known biocompatibility, controlled biodegradability, absorbability, and no toxicity of degradation products [7]. Furthermore, the PLGA types and related poly(hydroxyalkanonates) have a long history of medical and pharmaceutical use in fields as diverse as sutures, bone fixatives, artificial skins and cartilages, dental materials, materials for bone generation, drug delivery, and many others, as reviewed by Ueda and Tabata [8]. In conjunction with a long safety record of PLGA polymers, at least 12 different peptide/protein-loaded PLGA microsphere products are available in the market from nine different companies worldwide for the treatment of some life-threatening diseases (Table 1). In recent years, poly(ε-caprolactone) (PCL) has been investigated as an alternative to PLGA to make microspheres [9, 10]. A glimpse of ongoing research activities utilizing biodegradable polymer-based microspheres for various peptide/protein is shown in

TABLE 1 Currently Marketed Preparations (Injectable Microspheres) Containing Peptide/Protein Molecules

Commercial Name	API	Polymer	Company	Indication
Lupron Depot	Leuprolide	PLGA or PLA	TAP	Prostate cancer, endometriosis
Enantone Depot	Leuprolide	PLGA or PLA	Takeda	Prostate cancer, endometriosis
Trenantone	Leuprolide	PLGA or PLA	Takeda	Prostate cancer, endometriosis
Enantone Gyn	Leuprolide	PLGA or PLA	Takeda	Prostate cancer, endometriosis
Sandostatin LAR	Octreotide	PLGA-glucose	Novartis	Acromegaly
Nutropin	Somatropin	PLGA	Genentech	Growth deficiencies
Trelstar Depot	Triptorelin	PLGA	Pfizer	Prostate cancer
Decapeptyl SR	Triptorelin	PLGA or PLA	Ipsen-Beaufour	Prostate cancer
Decapeptyl	Triptorelin	PLGA	Ferring	Prostate cancer
Suprecur MP	Buserelin	PLGA	Aventis	Endometriosis
Somatuline LA	Lanreotide	PLGA	Ipsen-Beaufour	Acromegaly
Parlodel LAR	Bromocriptine	PLGA-Glu	Novartis	Parkinsonism

Abbreviations: PLA: polylactide; PLGA: poly(lactide-*co*-glycolide); API: active pharmaceutical ingredient; PLGA-Glu: poly(D,L-lactide-*co*-glycolide-D-glucose).

Table 2 (incorporating refs. 12–35). However, overcoming the propensity for peptides/proteins to undergo degradation processes during incorporation into the biodegradable microspheres or after injection into the body awaiting release is one of the key hurdles in bringing microencapsulated systems for these drugs to market. This partially explains the limited and only a countable number of formulations available on the market. Furthermore, irrespective of the various microencapsulation techniques adopted to prepare peptide/protein-loaded microspheres, several transfer-required processes such as filtration, centrifugation, and vacuum or freeze drying are necessary to obtain a final product, and these processes might be obstacles when scaling up the manufacturing technique to produce sufficient quantities of sterile material for clinical trial and, ultimately, commercialization [11].

This chapter encompasses investigations made progressively on the design of injectable peptide/protein-loaded PLGA microspheres. It covers an update on the state of art of the manufacturing of peptide/protein-loaded microspheres through both conventional and newer microencapsulation techniques, different analytical methods used for microsphere characterization, immune system interaction with microspheres following parenteral administration, and potential application of microspheres having therapeutic peptides/proteins. Special emphasis is given particularly on various instability problems and investigated mechanistic ways to obviate the possible instability problems of peptide/protein drug during microsphere preparation as well as its release from the microspheres. It should be added that although the chapter focuses mainly on PLGA microspheres, many of the destabilization mechanisms and stabilization approaches described herein can be valid to some extent for other polymeric delivery systems, too.

TABLE 2 Injectable Peptide/Proteins/Vaccines Encapsulated in Biodegradable Microspheres

Peptides, Protein, Vaccine	Technique	Polymer	Reference
	Vaccine		
SPf 66 malaria vaccine	Double emulsion	PLGA	28
Multivalent vaccines of *Haemophilus influenzae* type b (Hib), diphtheria toxoid (DT), tetanus toxoid (TT), pertussis toxin (PT)	Spray drying	PLGA	29
Rotavirus	Double emulsion	PLG	30
	Polypeptides and Proteins		
Insulin	Double emulsion	PLA polyethylene glycol (PEG)	12
Recombinant human epidermal growth factor (rhEGF)	Double emulsion	PLA	13
Ribozyme	Double emulsion	PLA, PLGA	14
Vapreotide (somatostatin analogue)	Spray drying	PLGA	15
Insulinlike growth factor-1 (IGF-1)	Double emulsion	PLGA-PEG	16
Ornitide acetate leuteinizing hormone releasing hormone [(LHRH) antagonist]	Dispersion/solvent extraction/evaporation	PLA, PLGA	17
Vascular endothelial growth factor (VEGF)	Single emulsion	PLGA/PEG	18
Human chorionic gonadotropin (hCG)	Double emulsion	PLA, PLGA	19
Calcitonin	Double emulsion	PLGA	20
FITC-bovine serum albumin (BSA)	Double emulsion	Poly(ε-caprolactone)	9
Levonorgestrel and ethinylestradiol	Double emulsion	Poly(ε-caprolactone)	10
Recombinant human bone morphogenetic protein	Double emulsion	PLGA	21–24
Transforming growth factor beta	Double emulsion	PLGA or PLGA-PEG	25–27
Recombinant human erythropoietin (rhEPO)	Modified double emulsion	LPLG-PEO-LPLG	31
Protein-C	Double emulsion	PLA	32
Ovalbumin	Double emulsion	PLGA	33
Human serum albumin	Double emulsion	PLA	34
Bovine serum albumin	Nonaqueous oil-in-oil (o/o) emulsion	PLG	35

Abbreviations: FITC: fluroscein isothiocyanate; PLG: poly(lactide-co-glycolide); LPLG-PEO-LPLG: copoly(l-lactic-co-glycolic acid-b-oxyethylene-b-l-lactic-co-glycolic adic); PEO: polyethylenenoxide.

5.2.2 PEPTIDE/PROTEIN-LOADED MICROSPHERE PRODUCTION METHODS

The development of delivery systems for therapeutic peptides/proteins depends on biophysical, biochemical, and physiological characteristics of these molecules, including molecular size, biological half-life, immunogenicity, conformational stability, dose requirement, site and rate of administration, pharmacokinetics, and pharmacodynamics [36]. Unlike conventional drug molecules, the unique conformational structure of peptidic/proteinic therapeutic agents poses a great challenge right from the beginning of the selection of suitable microencapsulation techniques to make microspheres. Table 3 lists the considerations to be taken before choosing a particular encapsulation technique. Apart from the traditional phase separation technique, other techniques suitable for peptide/protein-loaded microsphere production can be divided into two main categories: during microsphere preparation, those involved in utilizing a hydrous environment such as emulsion-based methods and those based on an anhydrous environment such as spray freeze drying, spray drying, freeze drying, grinding, jet milling, liquid-phase antisolvent precipitation, and supercritical CO_2-based methods [37–40]. In the following sections, the various production techniques to make injectable peptide/protein-loaded microspheres are briefly introduced; however, a detailed discussion is beyond the scope of this chapter.

5.2.2.1 Phase Separation (A Traditional Technique)

Polymer phase separation or coacervation is an excellent technique for the encapsulation of water-soluble drugs including peptide/protein into a final microsphere product [41]. The peptide/protein molecule is dispersed in solid form into solution containing dichloromethane and PLGA. Silicone oil is added to this dispersion at a defined rate, reducing the solubility of polymer in its solvent. The polymer-rich liquid phase (coacervate) encapsulates the dispersed peptide/protein molecules and embryonic microspheres are subjected to hardening and washing using heptane. The process is quite sensitive to polymer properties, and residual solvent is also an important issue. Decapeptyl [triptorelin, a luteinizing hormone releasing hormone (LHRH) analogue] [42] and Somatuline LA (lanreotide, a somatostatin analogue) [43] are microsphere commercial products developed by this technique (Table 1).

TABLE 3 Factors in Selection of Microencapsulation Method to Prepare Peptide/Protein-Loaded Microspheres

Optimal peptide loading
High microsphere yield
Batch content uniformity
Interbatch reproducibility
Peptide stability during preparation and release
Size uniformity
Adjustable release profile
Low burst release
Flowability of final product
Residual solvent and polymer monomer control
Sterilization (both aseptic and terminal)

5.2.2.2 Double Emulsion (A Hydrous Technique)

Oil-in-water (o/w) and water-in-oil-water (w/o/w) are the two hydrous techniques representing respectively the single- and double-emulsion formation during microsphere preparation. However, the w/o/w technique is most commonly employed [44]. In this process, peptides/proteins in aqueous solution are emulsified with nonmiscible organic solution of polymer to form a w/o emulsion. Dichloromethane serves as organic solvent and the o/w primary emulsion is formed using either high-speed homogenization or ultrasonication. This primary emulsion is then rapidly transferred to an excess of aqueous medium containing a stabilizer, usually polyvinyl alcohol. Again homogenization or intensive stirring is necessary to initially form a double emulsion of w/o/w. Subsequent removal (by evaporation) of organic solvent by heat, vacuum, or both results in phase separation of polymer and core to produce microspheres. Instead of solvent evaporation, solvent extraction with a large quantity of water with or without a stabilizer can also be undertaken to yield microspheres containing peptide/protein. Although the w/o/w microencapsulation technique seems to be conceptually simple to carry out, the particle formation process is quite complicated, and a host of process parameters influence the properties of peptide/protein-loaded PLGA microspheres [45]. In spite of that, different peptides and proteins such as bovine serum albumin (BSA) or ovalbumin (OVA), insulin, recombinant human insulinlike growth factor-1 (rhIGF-1), recombinant human epidermal growth factor (rhEGF), human chorionic gonadotropin (hCG), protein C, recombinant human bone morphogenetic protein (rhBMP), and calcitonin, along with antigens and other therapeutically relevant proteins such as recombinant human erythropoietin (rhEPO), have been successfully encapsulated (see Table 2) by the w/o/w double-emulsion technique. Lupron Depot/Enantone Depot/Trenantone/Enantone Gyn (all having leuprolide acetate, a LHRH analogue) are very popular commercial microsphere products produced by this technique [46, 47], available both in the EU and United States, for the treatment of either prostate cancer of man or infertility (endometriosis) of women (Table 1).

5.2.2.3 Spray Drying (An Anhydrous Technique)

Spray drying is a rapid, convenient technique which can be conducted under aseptic conditions. First, a polymer—prevalently PLGA is applied—is dissolved in a volatile organic solvent such as dichloromethane or acetone. The protein is suspended as a solid or emulsified as aqueous solution in this organic solution by homogenization. After that, the resulting dispersion is atomized through a (heated) nozzle into a heated airflow. The organic solvent evaporates, thereby forming microspheres with dimensions of typically 1–100 μm. The microspheres are collected in a cyclone separator. For the complete removal of the organic solvent, a vacuum drying or lyophilization step can follow downstream.

The internal structure of the resulting polymeric microspheres depends on the solubility of the peptide/protein in the polymer before being spray dried leading to the formation of reservoir- or matrix-type products (see Figure 1). When the initial dispersion is solution, the final product obtained following spray drying is matrix or monolithic type, that is, polymer particles with a dissolved or dispersed nature of the active ingredient (defined as microspheres). Conversely, when the initial

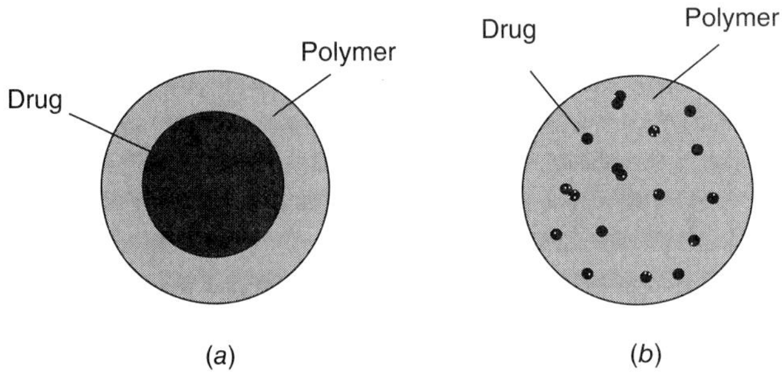

FIGURE 1 Polymeric delivery systems: (a) reservoir systems (microcapsules); (b) matrix systems (microspheres).

dispersion is in suspension, the product obtained is reservoir type, that is, a distinct polymeric envelope/shell encirculating a liquid core of dissolved active ingredient (defined as microcapsules). Recombinant human erythropoietin [48] and bromocriptine mesylate, Parlodel Depot [49], are examples of microspheres (matrix type) obtained by the spray drying technique.

5.2.2.4 New Trends in Production Methods

Several issues such as reducing cost, reducing scale-up difficulties, improving protein stability, allowing for terminal sterilization, and eliminating the need for organic solvents during addition of the peptide/protein motivate the development of new methods to manufacture microspheres. Moreover conventional microencapsulation methods involve relatively harsh conditions that are not generally tolerated by peptide/protein molecules without stabilization. Therefore, new and improved processes shielding the peptide/protein from deleterious conditions have been proposed and evaluated.

Modified Conventional Methods The w/o/w solvent evaporation or extraction is probably one of the most widely used methods for peptide and protein microencapsulation [44], despite its many drawbacks. Improvements and alternatives have therefore been proposed such as oil in water (o/w), *o/w (the asterisk including cosolvent) and oil in oil (o/o) [50].

Utilising a modified w/o/w method, the rhIGF-1 was encapsulated into PLGA microspheres after increasing the pH of the protein solution from 4.5 to 5.5–6.0, where rhIGF-1 formed a viscous gel [51]. High entrapment efficiency of fully bioactive protein was achieved, and 92–100% of pure, monomeric, and bioactive rhIGF-1 was released in vitro over 21 days. The lowering of the rhIGF-1 solubility at pH 5.5–6.0 probably restricted its conformational flexibility and changes upon exposure to the polymer solvent. Without pH adjustment, approximately 10–32% of rhIGF-1 was lost upon solvent exposure, due to degradation and aggregation. Elsewhere, a $w/o_1/o_2$ system was investigated for encapsulating different proteins and peptides, with the o_1 and o_2 phases consisting of acetonitrile/dichloromethane and liquid paraffin/Span 80, respectively [52]. The acetonitrile mediated the partial mixing of the w and o_1 phases and subsequent protein/peptide precipitation, which was a prerequisite for microencapsulation. The proteins BSA, tetanus toxoid (TT), and lysozyme did precipitate at low acetonitrile concentration, resulting in efficient microencapsulation (more than 90%), while a decapeptide and a linear gelatine did not precipitate so rapidly, resulting in poor entrapment. TT and lysozyme released during the burst phase (15%) maintained their bioactivity, although lack of further release suggested aggregation within the microspheres.

Another approach consisted of dispersing the protein antigen in a mineral oil before encapsulation into PLGA microspheres by a $o_1/o_2/w$ method [53]. The mineral oil (o_1) was intended as a barrier to protect the antigen during emulsification with the polymer solution and from exposure to moisture during release. Over 92% of enzyme-linked imunosorbert assay (ELISA) reactive TT was released from the reservoir-type microspheres in a pulsatile pattern, proceeding with an initial burst and followed by a second release pulse between 14 and 35 or 35 and 63 days, depending on the polymer type used. The latter stage of release was ascribed to TT diffusion through the oily phase, once an appreciable loss of polymer mass had occurred. The authors claimed the mineral oil was the key to protecting the solid antigen during polymer erosion, where acidic degradants and moisture would otherwise have led to antigen inactivation.

To improve solvent extraction, a novel method using a static micromixer was presented where a w_1/o dispersion (aqueous BSA in organic PLGA solution) is fed into an array of microchannels and the extraction fluid (w_2) into a second array of interdigitated channels [54]. The two fluids, transported separately through the channels, are discharged through an outlet slit where alternating fluid lamellae are formed with the w_1/o fluid lamella disintegrating into microdroplets, which harden quickly to form microspheres. This process offers easy scale-up, methodological robustness, continuous production, and a simple setup, making it ideally suited for aseptic production, a strongly needed feature for microsphere vaccine formulations.

ProLease Technology (Cryogenic Spray Drying) A variation of the conventional spray drying method is a cryogenic method which will described below. A novel low-temperature spraying technique (called ProLease technology) for preparing PLA and PLGA microspheres has been reported by Khan et al. [55] and the group at Alkermes [56, 57]. The method relies on the use of stabilizing and release controlling agents, low processing temperature, and nonaqueous microencapsulation. Typically, a protein powder is micronized, possibly with a stabilizer, by spray freeze drying and then suspended in an organic polymer solution. The suspension is atomized into a vessel containing liquid N_2 underlaid by frozen ethanol (extraction solvent). The atomized droplets freeze in the liquid N_2 and deposit on the surface of the frozen ethanol. As liquid N_2 evaporates, the frozen ethanol liquefies (T_m approximately −110°C) so that the frozen polymeric droplets will transfer into the ethanol where the polymer solvent is extracted, yielding solid microspheres [58, 59]. To date, the ProLease system has been effectively applied to the encapsulation of zinc-complexed human growth hormone in PLGA microspheres, resulting in a one-month effect after one single injection [37, 57, 60]. As a reference, the recombinant human growth hormone (rhGH) was unstable in contact with ethyl acetate or dichloromethane [61]. The only protein-containing PLGA microspheres, Nutropin Depot, is produced by this novel technique. However, this product containing rhGH marketed initially in the United States in 1999 was pulled from the market voluntarily by the manufacturer in June 2004 because of high costs of production and commercialization (http://www.gene.com/gene/news/press-releases/, accessed May 25, 2006).

ProLease technology was also used for encapsulating recombinant human vascular endothelial growth factor (rhVEGF) and rhIGF-1 [62, 63]. Both proteins were stabilized in aqueous solution, prior to spray freeze drying, and encapsulated (9–20% w/w) into PLGA microspheres. The microspheres also contained $ZnCO_3$ (3–6% w/w) as release modifier. The resistance of rhIGF-1 to aggregation and oxidation, determined from in vitro release studies, hardly changed. Protein, released in an almost pulsatile fashion over 21 days, was composed of predominantly monomeric rhIGF-1 with only minor amounts (~6%) of degradants forming toward day 21. Similarly, the integrity of rhVEGF dimer released over 21 days was good and its bioactivity remained largely unaffected, regardless of the extent of aggregation and degradation. In view of these studies, ProLease technology appears to have potential for sustaining antigen stability and release from microspheres.

Techniques Using Supercritical Fluids Generally, the application of supercritical (SC) fluids for the encapsulation of peptides and proteins has been fueled by the recognition that the established methods implicate some drawbacks. The application of supercritical fluids, especially of supercritical carbon dioxide (CO_2), can minimize or even eliminate the use of organic solvents and renders work at moderate temperatures possible [64]. The term *supercritical* defines the area above the critical point, which specifies the final point of the liquid–gas phase transition curve. Beyond that critical point, isobar/isotherm alterations of pressure or temperature alter the density of the critical phase but do not lead to a separation into two phases. A density change is directly associated with a change of the solvent power, and thus the method features a high variability. Usually CO_2 is used as supercritical fluid due to its critical point (T_c = 31.1°C, P_c = 73.8 bars), which can be easily reached. That allows a moderate working temperature and leaves no toxic residues since it returns to the gas phase at ambient conditions. Two SC CO_2-based processes have been reported for the preparation of drug-loaded polymeric microspheres: first, the rapid expansion from supercritical solutions (RESS) process, whereby a SC CO_2 solution of an active agent and a polymeric carrier is rapidly expanded. This quickly transforms the SC CO_2 into a liquid that is a much poorer solvent, thereby precipitating the active agent–carrier mixture as small particles [65]. Second is the aerosol solvent extraction system (ASES), also referred as the gas antisolvent spray precipitation (GAS) process [66]. Here, a solution of the active agent and the polymeric carrier is sprayed into a chamber loaded with SC CO_2. The SC CO_2 extracts the solvent from the spray droplets and induces coprecipitation of the active agent and the polymeric carrier in the form of small, solvent-free particles [67, 68]. However, the use of organic solvents cannot be avoided, which is to be deemed a major disadvantage of both techniques.

In peptide/protein pharmaceuticals, the GAS process is predominantly applied for the preparation of microparticulate protein powders as an alternative to common drying processes. However, Winters et al. [69] reported an increase of β-sheet aggregates during the precipitation of lysozyme, trypsin, and insulin as a consequence of stress parameters such as organic solvent, pressure, and shear forces. One reason why these methods were not credited as encapsulation techniques for protein within PLGA may be the tendency of several polymers to rapidly precipitate and agglomerate during the process [70].

ASES has been compared with conventional spray drying in terms of effects on the stability of the peptide tetracosactide [71]. Almost no intact peptide was recovered from spray-dried PLA particles, whereas the tetracosactide was well protected against oxidation during ASES (~94% unmodified peptide). In general, the particle formation step seems to be less detrimental to proteins than the loading step. For example, emulsification in an aqueous phase or spray drying of rhEPO/PLGA emulsions was mild compared to the first emulsification step [72]. Also, variation of the particle formation step (spray drying or coacervation) had a minor impact on diphtheria toxoid (DTd) antigenicity when compared to other process variables [73].

A serious limitation of GAS, ASES, and RESS for producing microspheres is the need of polymer types that form discrete crystalline domains upon solidification, such as l-PLA [74, 75]. The advantages of these methods offer (e.g., over spray drying) are the low critical temperatures for processing (34°C) and the avoidance of oxygen exposure during atomization, with both parameters being potentially important to peptide/protein stability.

Ultrasonic Atomization Ultrasonic atomization of w/o dispersions is presently under investigation for preparing especially protein antigen containing microspheres. In one setup, the atomized antigen/polymer dispersion was sprayed into a nonsolvent where the polymer solvent was extracted, resulting in microspheres [76]. A comparable technique was proposed where the antigen or polymer dispersion was atomized into a reduced pressure atmosphere and the preformed microspheres hardened in a collection liquid [77]. Similarly, PLGA solutions were also atomized by acoustical excitation and the atomized droplets transported by an annular stream of a nonsolvent phase [aqueous polyvinyl alcohol (PVA)] into a vessel containing aqueous PVA [78]. Solvent evaporation and microsphere hardening occurred in the vessel over several hours. The main advantages of these atomization techniques encompass the possibility of easy particle size control and scale-up, processing at ambient or reduced temperature, and the suitability for aseptic manufacturing in a small containment chamber such as an isolator.

In Situ Formed Injectable Microspheres All the encapsulation techniques discussed so far rely on the preparation of solid microspheres. However, a method for preparing a stable dispersion of protein containing semisolid PLGA microglobules has been reported [79]. Here, a protein dissolved in PEG 400 was added to a solution of PLGA in triacetin or triethyl citrate. This mixture, stabilized by Tween 80, was added dropwise and under stirring to a solution of Miglyol 812 or soybean oil, containing Span 80, resulting in a stable dispersion of protein inside semisolid PLGA microglobules. The microglobules remained in an embryonic state until mixed with an aqueous medium, so that the water-miscible components were extracted and protein containing matrix-type microspheres formed. Myoglobin was encapsulated and found to remain physically unchanged (circular dichroism analysis) after the process and during storage of the microglobular dispersion (15 days, 4°C).

Preformed Porous Microspheres A new approach for attaining sustained release of protein is introduced involving a pore-closing process of preformed porous PLGA microspheres [80]. Highly porous biodegradable PLGA microspheres were fabricated by a single w/o emulsion solvent evaporation technique using Pluronic F127 as an extractable porogen. The rhGH was incorporated into porous microspheres by a simple solution dipping method. For its controlled release, porous microspheres containing rhGH were treated with water-miscible solvents in the aqueous phase for production of pore-closed microspheres. These microspheres showed sustained-release patterns over an extended period; however, the drug loading efficiency was extremely low. To overcome the drug loading problem, the pore-closing process was performed in an ethanol vapor phase using a fluidized-bed reactor. The resultant pore-closed microspheres exhibited high protein loading amount as well as sustained rhGH release profiles. Also, the released rhGH exhibited structural integrity after the treatment.

Charged (Anionic and Cationic) PLGA Microspheres PLGA or any other type of microspheres can be readily decorated with positive or negative surface charges by simply preparing the particles by a w/o/w solvent evaporation/extraction process where the second water phase contains a cationic emulsion stabilizer [hexadecyltrimethylammonium bromide; poly(ethyleneimine); stearylamine] or an anionic emulsifier [sodium dioctyl-sulfosuccintate; sodium dodecyl sulfate (SDS)]. Such compounds attach tightly to PLGA surfaces during preparation and provide the necessary surface charge for ionic adsorption of counterions. It is known that a protein's surface charge depends on its isoelectric point (pI) and the pH of the medium in which it is dispersed. The use of particles with ionic surface charge offers several advantages over classical microencapsulation, among which the mild conditions for loading are probably the most attractive. PLGA microspheres with surface-adsorbed protein antigens and DNA have been highly efficient in inducing strong immune responses, as reviewed by Singh et al. [81] and Jilek et al. [82]. Nonetheless, it remains to be shown whether such particles are also suitable for eliciting long-term immunity after one or two injections.

Jabbal-Gill et al. [83, 84] noted the tendency for microencapsulated protein antigens to distribute heavily at the surface of PLGA microspheres and developed polymeric lamellar substrate particles (PLSP) by precipitating a highly crystalline poly(*l*-lactic acid)/organic solvent solution with water, followed by removal of remaining organic solvent with nitrogen purge. The particles, which can be sterilized by gamma irradiation and stored as a suspension for several months without changes to antigen absorption [84], possessed a large lamellar surface area and highly negative zeta potential (~−35 to −42 mV) and could adsorb significant amounts of antigen (up to 50 μg/mg microspheres) depending on pH, ionic strength, antigen–polymer ratio, and other factors. Release of protein antigen (TT) could be extended to over 1 month with minimal antigenic losses in released antigen, although most of the antigen was lost to the initial burst or to apparent irreversible adsorption (as indicated by the absence of reaching 100% release). Elevated antibody responses in mice were elicited using PLSP similar to one dose of aluminum adjuvant following subcutaneous administration of OVA at elevated doses (100 or 300 μg). Both immunoglobulin IgG1 and IgG2a antibody subtypes were of similar magnitude over 28 days in the PLSP/OVA groups, and cellular immunity was also observed following immunization with a 38-kDa protein antigen against tuberculosis [85]. Similarly, Kazzaz et al. [86] created anionic PLGA microparticles by substituting the standard nonionic emulsifier PVA with anionic SDS during microsphere preparation. In addition to eliciting elevated antibody responses in mice relative to the soluble antigen, the adsorbed antigen elicited a potent cytotoxic T-cell (CTL) response, similar to that observed after infection from virus expressing the p55 gag and polymerase proteins. Moreover, the CTLs were formed from the more challenging intramuscular route but not significantly by the soluble antigen, even at elevated doses. The SDS-PLGA particles could also be gamma irradiated before adsorption and were shown to effectively boost antigen in nonhuman primates [87].

5.2.3 ANALYTICAL CHARACTERIZATION OF PEPTIDE/PROTEIN-LOADED MICROSPHERES

An area requiring additional efforts is analytical characterization of peptides and proteins encapsulated in PLGA microspheres. The high complexity of the therapeutic peptides and proteins requires not only physicochemical methodologies but also immunochemical and biological techniques for the characterization and quality control of these substances. In general, the analytical methods can be broadly viewed from the following study perspectives: methods meant for microsphere product quality checking, methods used for peptide/protein stability identification inside the microspheres, and methods called for peptide integrity detection following liberation from the microspheres immediately upon placement in release medium either in vitro or in vivo. Therefore, in most cases, a combination of several analytical methods is necessary for a comprehensive characterization of the peptide/protein under investigation and for appropriate quality control of the product concerning identity, purity, and potency. However, some of the analytical methods have potentially appealing applications to interplay among the mentioned perspectives. In Table 4, a selection of widely used analytical methods is given, showing which technology is applicable for the testing of identity, purity, and potency of peptides and proteins. In addition, peptide/protein integrity evaluation is indeed likely to be affected by artefacts during the sample preparation before analysis and during the analysis itself. Therefore, artefacts might prevent the scientist from critically ascribing detected protein denaturation to manufacturing conditions [88].

In order to measure the extent of peptide/protein degradation within the carriers and during release, the encapsulated molecule has to be removed from the polymeric matrix. Moreover, for avoiding artefacts such as underestimation of drug content, recovery methods need to be tried by an empirical trial-and-error approach as each peptide/protein is different one from the other. Recovery methods so far reported include extraction-based method with the help of potentially deleterious organic solvents, hydrolysis of the polymer matrix with alkaline medium, dissolution of polymer matrix in an organic solvent, recovery of suspended insoluble protein by filtration [89], total protein quantification after complete digestion of carriers followed by amino acid analysis [90, 91], electrophoretic extraction of the protein using sodium dodecyl sulfate–polyacrylamide gel electrophoresis (SDS–PAGE) [92, 93], and direct dissolution of both the polymer and the protein drug in a single liquid

TABLE 4 Analytical Methods for Characterization and Quality Control of Pharmaceutical Peptides and Proteins

	Indicated Usage/Checking		
Methods	Identity	Purity	Potency
Physicochemical			
Chromatography			
Reversed-phase high-performance liquid chromatography (HPLC, RP-1)	+	+	−
Ion exchange	−	+	−
Affinity	−	+	−
Size exclusion chromatography (SEC)	−	+	−
Spectroscopy			
Infrared spectroscopy	+	+	−
Raman spectroscopy	+	+	−
Fluorescence spectroscopy	−	+	−
Ultraviolet/visible (UV/VIS) spectroscopy	+	+	−
NMR spectroscopy	+	+	−
Mass spectrometry	+	+	−
Circular dichroism (CD)	+	+	−
Matrix-assisted light desorption ionization–time-of-flight (MALDI-TOF) mass spectrometry	+	+	−
Electrophoresis			
Capillary electrophoresis	−	+	−
SDS–polyacrylamide gel electrophoresis (PAGE)	+	+	−
Isoelectric focusing	−	+	−
Immunochemical			
Radioimmunoassay (RIA)	−	−	+
ELISA	−	−	+
Western blot	+	−	−
Biological			
In vivo assays	−	−	+
In vitro (cell culture) assays	−	−	+

Abbreviation: SDS: sodium do decyl sulfate.

phase containing water-miscible organic solvents such as acetonitrile or dimethylsulfoxide (DMSO) [94, 95].

Following successful recovery of peptide/protein molecule from the microspheres, a simple spectrophotometric method does not always allow discrimination between the monomeric protein form and its aggregates. However, HPLC might separate these species and thus provides more accurate qualitative data [96]. But HPLC cannot quantify exclusively the amount of active protein antigen, as is the case with ELISA techniques [97]. Nowadays, Fourier transform infrared (FTIR) spectroscopy has become a popular, noninvasive method, as it is able to characterize the secondary structure of entrapped proteins [26, 95, 98–101]. Only recently, the integrity of their primary structure was evaluated, thanks to a new matrix-assisted laser desorption ionization–time-of-flight (MALDI-TOF) mass spectrometry method [94, 102]. The method was shown to require little sample material and only a simple dissolution of the carrier was needed prior to the analysis. The MALDI-TOF allowed elucidation of a new degradation pathway, that is, peptide acylation within PLGA carriers resulting from a chemical interaction between peptide and degraded polymer [102]. Moreover, the method was also useful for quantification, and it should be underlined that no interference from PLGA was detected during the measurements. For all the reasons cited above, mass spectrometry should be considered one of the most promising methods for protein analysis inside polymeric carriers including microspheres. Using erythropoietin as an example, an exploratory and elaborative discussion was made on the analytical techniques used for the characterization and quality control of pharmaceutical peptides and proteins [103]. A similar discussion was also done on the analytical techniques critical to (as a part of) the quality assurance after process changes of the production of a therapeutic antibody [104].

With an increasing level of sophistication in the design of new protein antigens and adjuvants (including polymer controlled-release systems), efforts both in the United States and the EU are underway to respond with more appropriate regulations [105–107]. For example, the Committee for Proprietary Medicinal Products (CPMP), the primary scientific body in EU regulatory matters, is currently updating its "notes for guidance," which guide/direct industry and regulatory authorities on content and evaluation of marketing authorization applications for vaccines [105]. Early drafts of these updates include more rigorous guidelines for new non-aluminum-based adjuvants, including antigen stability requirements (see Sesardic and Dobbelaer [105] for a discussion). Similar discussions ongoing in the United States have attempted to standardize requirements of controlled-release parenterals [106, 107], including specifics regarding in vitro release assays and the need to account for 80% or more of the encapsulating agent during the release period.

5.2.4 IMMUNE SYSTEM INTERACTION WITH INJECTABLE MICROSPHERES

Since microspheres are capable of forming a drug depot, the encapsulated peptide or protein is being slowly released over days or months at the injection site. Interestingly, the size of microspheres plays an important role in immune response. Microspheres with sizes smaller than 10 μm can be directly taken up macrophages (and dendritic cells) through a phagocytosis mechanism while sizes greater than 10 μm need to undergo biodegradation before phagocytosis can occur [108]. It was shown that within a few days of intramuscular injection PLGA microspheres less than 10 μm are completely engulfed in a thin layer of connective tissue and thus evidenced infiltration by macrophages as a consequence of wound-healing response to injected particles [109]. It is feasible that the influx of these macrophages may cause degradation of the encapsulated protein and available protein released in the vicinity of the microspheres. Furthermore, it has been suggested that these macrophages are capable of producing proteolytic enzymes [110], which may result in the release and circulation of altered, inactive, or immunogenic forms of the encapsulated peptide or protein.

On the other hand, degradation, protein antigen release, location, and antigen presentation of microspheres larger than 10 μm are expected to be different from smaller ones. Larger microspheres can provide an extracellular depot for secondary immune responses by way of B-cell stimulation [111–113]. In both cases, upon administration of microspheres, a foreign-body response occurs resulting in an acute initial inflammation despite the excellent tissue compatibility and biodegradability properties of polymers such as PLGA. This initial inflammation is followed by the infiltration of small foreign-body giant cells and neutrophils [114]. These immune cells could consume the released peptide or protein and produce an immune response. However, if released protein is recognized as a self-protein (e.g., homologus), the probability of an immune response by these cells is reduced. It is therefore always essential to release the protein in its native conformation. The release of aggregated or denatured protein from the microspheres may, in fact, result in an unwanted immune response [115]. It should be added that systematic studies to explore the effects of tissue response on the bioavailability of incorporated peptide or protein drug have not appeared extensively in the literature, with a few exceptions as described below. Using a light microscopic technique, bumps containing residual amounts of microspheres were observed at the injection site two weeks after administration of TT-encapsulated PLGA microspheres to mice and guinea pigs [116]. These bump formations may be due to chronic reactions, long-term immunogenicity, and immunological priming of mice and guinea pigs against the injected polymeric microspheres. The immunogenicity of microsphere-encapsulated vaccines can be varied to some extent by changing the physicochemical properties of the microspheres, for example, size, surface properties, and release kinetics of the antigen from the microspheres [111]. An interesting review by Jiang et al. [117] details the various reports on the relationship between in vitro protein antigen stability and immunogenicity, modulation of cell-mediated immune responses, and different formulation approaches to achieve the appropriate immune response with microencapsulated vaccine antigens. There has been some debate, arising from some animal experiments, that the antigenicity does not directly correlate with immunogenicity. However, the stability of protein antigens is considered to play a significant role in the quality and magnitude of immune response for the controlled-release single-dose vaccines as degraded or nonantigenic proteins may not be able to provide a continuous boost for generation of protective levels of high-affinity antibodies.

5.2.5 EXCIPIENT INCLUSION: INJECTABLE PEPTIDE/PROTEIN-LOADED MICROSPHERES

Peptide and protein molecules are highly prone to degradation mechanisms that can be divided into two classes: physical and chemical [118]. Whereas chemical degradation leads to the loss of the protein's primary structure through oxidation, deamidation, peptide bond hydrolysis, isomerization, disulfide exchange, and covalent aggregation, physical degradation refers to the changes in the higher order structure (secondary and above) mainly by noncovalent aggregation and precipitation. In particular, aggregates formation during the encapsulation process must be avoided because these aggregates always represent loss of therapeutic efficacy and increased immunogenicity which can endanger the patient's health [119, 120]. The following few examples indicate the fragility of peptide and protein molecules due to physical or chemical degradation: Aggregation of insulin has been well characterized and depends on unfolding of the insulin molecules [121]; aggregation of lyophilized formulations of BSA, β-galactoglobulin, and glucose oxidase are attributed to disulfide interchange [118]; deamidation contributes to reduction in catalytic activity of lysozyme [122] and ribonuclease at high temperatures [123]; and peptide bond hydrolysis results in loss of activity of lysozyme when heated to 90–100°C [122]. A recent introduction to this list is formaldehyde-mediated aggregation pathway (FMAP) unique to formaldehyde-treated protein antigens such as TT [117, 124, 125].

The formulator of injectable microspheres for peptide and protein faces multiple challenges: (i) to maximize physical and chemical stability, (ii) to prolong biological half-life, (iii) to increase absorption, (iv) to decrease antigenicity, and (v) to minimize metabolism. Thus, it is quite obvious that the fabrication of peptide- and protein-loaded microspheres requires several kinds of excipients for effective stabilization or immobilization of encapsulated therapeutic molecules. Excipients of choice are included specifically for controlling protein degradation in microspheres due to (a) external and internal environmental changes, (b) manipulating the initial burst release, (c) preventing protein adsorption onto delivery devices, and (d) neutralizing the causative acidic microclimate formation due to the acids liberated by the biodegradable lactic/glycolic-based polymers. Therefore, it is generally best to find conditions to stabilize the protein before other aspects of the formulation, such as controlled-release characteristics, are optimized. Typically, the appropriate excipients for the protein under investigation are experimentally selected among various substances by screening. This tedious experimental screening is partly necessary due to the present inability to predict protein stability after addition of such excipients. Moreover, since individual entrapped peptides and proteins differ in terms of physicochemical properties and chemical/therapeutic function, each species is expected to demonstrate a different degree of sensitivity to stress and react differently to the same stabilization strategy. For example, a sugar, amino acid, or antacid excipient may be required to stabilize protein, each of which can increase water uptake in the polymer matrix leading to an increase in release rate. In the scenario in which controlled-release conditions are optimized before such a stabilizer has been identified, it is likely that upon addition of the new stabilizer the release kinetics may change enough to require reformulation. Certainly, there is a sharp contrast between encapsulating a highly water soluble protein [126] or a poorly soluble zinc–protein complex [37, 57]. Switching between these two cases would be expected to alter the requirements in the formulation necessary to attain the controlled-release function (e.g., low versus high polymer matrix permeability for the protein, respectively) because protein solubility in water may be important for any diffusion component of release. The principal stresses causing instability of encapsulated peptide/proteins in PLGA microspheres are elaborated in a book chapter [127] and in a jounal publication [128]. This subject was again reexamined in a review based on new findings since the previous book chapter by the same author [129]. An interesting review from the same research group was published on the biodegradable PLGA microparticles for injectable delivery of vaccine antigens [117], where they focused on mechanistic approaches to improve the stability of PLGA-encapsulated protein antigens.

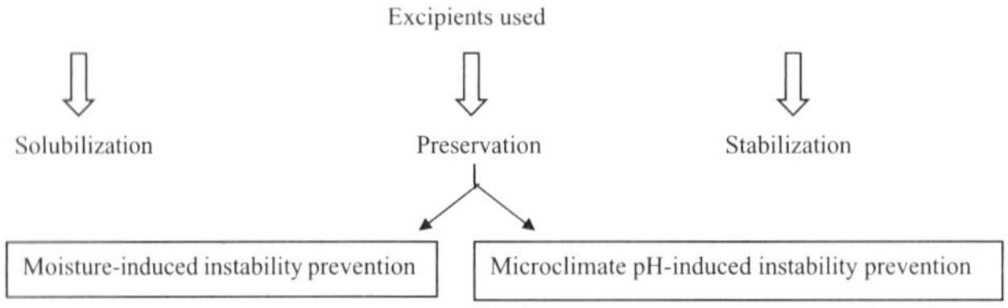

FIGURE 2 Flow chart of excipients used to prevent/minimize protein instability problems.

Another review by Bilati et al. [130] also envisioned the strategic approaches for overcoming peptide and protein instability within biodegradable nano- and microparticles. The reader is also referred to related publications edited by Sanders and Hendren [131] and Senior and Radomsky [132] for information on excipients used in injectable peptide/protein-loaded formulations including microspheres. This section will not cover all excipients used in parenteral protein formulations because the aforementioned publications already do so. Rather this section highlights examples of synergestic and antagonistic interactions that have been reported mainly between the excipients and the peptide/protein drugs, especially before microsphere preparation, followed by a brief discussion of a major instability problem of proteins/peptides inside the microspheres. The published research paper is being organized according to major functions of parenteral excipients, namely, solubilization, stabilization, and preservation (see Figure 2) [133].

5.2.5.1 Solubility- and Stability-Increasing Excipients

The traditional approach is to solubilize directly the peptide/protein in organic solvents. This can be achieved by different means. Cleland and Jones [61] assumed that native protein conformation could be maintained by precipitating the protein at its pI. The molecule is then free of charge and can be readily solubilized in organic solvents. Conversely, an alternative concept is based on the freeze drying of the protein at a pH away from its pI value before formulating it. It was thought that this strategy could increase protein solubility and stability in various polar and water-miscible organic solvents such as DMSO [134, 135]. It should be noted that a preformulation procedure consisting of using spray freeze drying with a suitable excipient was able to stabilize BSA before encapsulation by a nonaqueous method [35]. Using the dissolution approach, lysozyme was successfully formulated but incomplete lysozyme release from microspheres was observed and ascribed to aggregation [136]. Protein solubility can also be increased via an ion-pairing mechanism. The protein is modified by adding an oppositely charged surfactant that binds to the protein, so as to obtain a neutral hydrophobic entity and thus reduce direct contact between the protein and the organic solvent. Positively charged proteins and negatively charged surfactants should be employed, since cationic surfactants might have toxic side effects. This technique was shown to improve lysozyme conformational stability after a hydrophobic complex between lysozyme and oleic acid [137, 138]. A new interesting concept is to encapsulate an aggregated protein in a reversibly dissociable form in order to avoid the formation of irreversible aggregates during processing and to promote the sustained release of the native monomeric form. Growth hormone was successfully formulated with this approach [139].

Cyclodextrins (CD) have emerged as very effective additive compounds for solubilizing hydrophobic drugs. In the parenteral dosage form area, modified cyclodextrins such as hydroxylpropyl-β-cyclodextrin and sulfobutylether-β-cyclodextrin have been reported to solubilize and stabilize many injectable drugs, including dexamethasone, estradiol, interleukin-2, and other proteins and peptides [140] without apparent compatibility problems [141]. In addition, CD-containing formulations (either 0.1 *M* sulfobutylether-β-cyclodextrin or 0.1 *M* hydroxylpropyl-β-cyclodextrin) were shown to cause less damage to venous epithelial cells at the site of injection compared with formulations containing organic cosolvents [142]. When CD were coentrapped in the internal aqueous phase, erythropoietin (EPO) covalent aggregate formation was significantly reduced during microsphere preparation by the double-emulsion method [72] and lysozyme stability was improved [88]. Although the precise mechanism is unclear, interactions between amino acids and the hydrophobic inner cavity of CD may play a role [143]. However, CD showed no protecting effect on insulinlike growth factor-1 (IGF-1) [144] and hepatitis B core antigen (HBcAg) [145] and even promoted the loss of superoxide dismutase activity at high CD concentrations [146]. By contrast, carboxymethylcellulose (CMC) did not efficiently stabilize HBcAg and GH against dichloromethane-induced denaturation [61, 145]. Various types (α, β, and γ) of CD were examined for encapsulating TT in PLGA microspheres [147], with γ-hydroxypropyl-cyclodextrin effectively increasing TT encapsulation. However, CD also showed low efficiency in retaining spray-dried TT antigenicity, probably due to antigenic epitopes being buried inside the molecular CD core [147].

Surfactants have the ability to lower surface tension of protein solutions and prevent protein adsorption and/or aggregation at hydrophobic surfaces such as PLGA. Among them, nonionic surfactants are generally preferred as ionic surfactants might bind to groups in proteins and cause denaturation. Tween 20 was shown to greatly reduce the rate of formation of insoluble aggregates of recombinant human factor XIII caused by both freeze thawing and agitation stresses [148]. Maximum protection occurs at concentrations close to the critical micelle concentration of Tween 20, independent of initial protein concentration. In another report, Tween 20 at a 1% (w/v) concentration caused precipitation of a relatively hydrophobic protein (*Humicola lanuginosa* lipase) by inducing nonnative aggregates [149]. Similarly, nonionic surfactants such as Tween 20 or 80 were not good stabilizers for lysozyme and rhGH against the unfolding effect of the water–dichloromethane interface. It has been assumed that both the hydrophilic (PEG chains) and hydrophobic (fatty acid chain) parts of the polysorbate molecules were preferentially partitioned in the dichloromethane phase, leading to low protection efficacy [61, 98]. Exchange of Tween 20 for a less hydrophobic surfactant, PEG 3350, provided almost complete rhGH recovery irrespective of protein concentration. However, an opposing trend was seen with EPO encapsulation in PLGA microspheres [72]. Encapsulated protein aggregates increased (~15%) with different PEG types codissolved in the w_1 phase. Conversely, when three nonionic surfactants of different hydrophilic–lipophilic balances (HLBs) were coencapsulated with insulin by the w/o/w double-emulsion method, only Tween 20 was able to improve insulin stability within particles and to limit formation of high-molecular-weight products during the sustained-release period [150].

Tween 80 is well known to protect proteins against surface-induced denaturation [151]. Tween 80 was demonstrated to reduce hemoglobin aggregation in solution by preventing the protein from reaching the air–liquid interface or the liquid–surface interfaces [152]. Polyoxyethylene surfactants such as Tween 80 can form peroxide impurities after long-term storage. Knepp et al. [153] concluded that Tween 80 and other nonionic polyether surfactants undergo oxidation during bulk material storage and subsequent use and the resultant alkyl hydroperoxides formed can contribute to the degradation of proteins. In such formulations, they further reported that thiols such as cysteine, glutathione, and thioglycerol were most effective in stabilizing protein formulations containing peroxide-forming nonionic surfactants.

The Pluronics, also known as poloxamers (e.g., poloxamer 188, British Pharmacopoeia standard) are a well-studied series of commercially available, nonionic, triblock copolymers with a central block composed of the relatively hydrophobic poly(propylene oxide) flanked on both sides by blocks of the relatively hydrophilic poly(ethylene oxide) [154, 155]. The Pluronics possess an impressive safety profile and are approved selectively by the Food and Drug Administration (FDA) for pharmaceutical and medical applications, including parenteral administration [156]. The strong safety profile, commercial availability, ease of preparation, and well-studied physical properties make the Pluronics particularly appealing for drug delivery purposes. They have been used in several patented protein formulations as stabilizers and sustained-release injectables in development as solubilizing and stabilizing agents [157]. However, poloxamers, like Tweens, can form peroxide impurities over time. Poloxamer 188 was successfully used when mixed with PLGA for prolonged release of active interferon-α (INF-α) [158], but such a formulation had no effect on BSA secondary structure compared to PLGA alone [35]. Poloxamer 188 was not effective in preventing nerve growth factor (NGF) aggregation during in vitro release from microspheres generated by spray drying [159]. Complex interactions between poloxamer, BSA, and PLGA were believed to have influenced BSA microencapsulation [160]. The gelling property of the amphiphilic poloxamer 407 was successfully employed for urease encapsulation. The protein was likely protected during the microsphere preparation by a hydrated gelled structure due to the hydrophilic polyoxyethylene chains [161]. EPO aggregates in PLGA microspheres decreased when poloxamer 407 was incorporated at a level of 10% (w/w) [72].

Interleukin-1α (IL-1α) was protected by phosphatidylcholine from damage during the double-emulsion process but underwent inactivation during microsphere incubation [162]. Sodium dodecyl sulfate significantly reduced insulin aggregation at the dichloromethane–water interface, whereas dodecyl maltoside did not, this surfactant being more efficient at air–water or solid–water interfaces [163]. It should be mentioned that surfactants are used along with sugars, proteins, and polymers effectively for solubilization and stabilization purposes of peptide/protein in microspheres. Bilati et al. [130] give an overview on various proteins and polymers that act as stabilizing excipients during the development of peptide/protein-loaded microspheres.

5.2.5.2 Preservation-Imparting Excipients

Prevention/Minimization of Moisture-Induced Instability Moisture- and microclimate acid pH–induced instability (typically the aggregation) of the peptide/protein encapsulated in PLGA microspheres has been monitored. Even several formulation strategies to inhibit these instability problems are being actively investigated. If the protein is expected to exist in the solid state within the PLGA polymer, the protein is remarkably prone to aggregation when formulated under conditions that allow moisture- and microclimate acid pH–induced instability. The two covalent aggregation mechanisms commonly described during exposure of the solid protein to moisture are the disulfide interchange/exchange [164] and the FMAP, which is operative for protein antigens that have been detoxified with formaldehyde exposure [124]. In the former pathway, the reaction is typically initiated by a thiolate ion on the protein or free thiolate ions that accompany β elimination of an intact disulfide [165]. Decreasing the concentration of the reactive species (e.g., lowering pH to favor the nonionized thiol, covalently blocking the thiol group, or oxidizing free thiols as they appear with divalent copper ion) has been shown to block this mechanism [165]. To inhibit the FMAP, strongly formaldehyde-interacting amino acids such as histidine and lysine [166] have been colyophilized with the formalinized protein antigen. On exposure to moisture, the amino acids appear to bind with the reactive Schiff base or equivalent electrophile [167] in the protein before a neighboring protein nucleophile can react to form an intermolecular cross-link [124]. Sorbitol has also been identified to inhibit the FMAP of TT at the maximal aggregating water content of the antigen, about 30 g H_2O/g protein [168], although whether this is a humectant effect [169] or a possible covalent reaction with the highly reactive electrophile in the antigen has not been determined. Several techniques have been developed to successfully bypass the destabilizing stress either by altering the role of water in the solid or immobilizing the protein or, alternatively, by directly inhibiting the aggregation. Clearly, one of the most significant findings in the field of peptide/protein stability in polymers is the success of the immobilization strategy of Zn^{2+} precipitation, as performed with human growth hormone [37, 57, 170, 171]. The 2:1 mole ratio Zn–protein complex, which immobilizes the rhGH as a solid precipitate in a near-native state [99], has been shown to confer superior stability on the protein encapsulated in PLGA for a one-month release incubation. Since then, other proteins such as INF-α [172] and NGF [173] were also stabilized in PLGA microspheres by this approach. Another interesting approach originating in the patent literature is the precipitation of erythropoietin with salting-in salt, ammonium sulfate [174], which is a technique commonly used in protein processing. Other methods to alter the role of water in the reaction involve the addition of agents that alter the amount of water sorbed in the polymer and/or the activity of the water present. For example, both water-soluble salts (NaCl) and antacid excipients ($Mg(OH)_2$) are known to dramatically increase the amount of water sorbed in PLGAs, with the former due to osmosis and the latter to a complex effect of neutralizing acidic degradation products and end groups of the polymer (which also involves an osmotic component) [126]. In contrast, for a given moisture content, humectants such as sorbitol, which dissolve in water bound to the protein, reduce the available free water necessary to mobilize the protein or perform other roles in deleterious reactions [169].

The alternative to bypassing the deleterious role of moisture is to inhibit the aggregation mechanism directly. Several ways to accomplish this have been reported, particularly in the solid state and in the absence of the polymer. Well-referenced and useful book chapters by Johnson [175] and Carpenter and Chang [176] are available to thoroughly focus on the importance of making a lyophilized powder before loading the peptide or protein into an injectable microspheres. It has been stated that, in comparison to protein solution, the protein in the solid state would be less susceptible to shear forces that occur during an emulsification procedure or denaturation at oil–water interfaces. However, special precautions should be taken during freeze drying because the drying process itself will expose the protein to destabilizing stresses. To circumvent this problem, cryo- and lyoprotectants and bulking agents are usually included along with a peptide or protein solution while it undergoes the drying stages of the lyophilization process.

Cryo- and Lyoprotectants and Bulking Agents Various mechanisms are proposed to explain why excipients serve as cryo- or lyoprotectants. The most widely accepted mechanism to explain the action of cryoprotection is the preferential exclusion mechanism [177]. Excipients that will stabilize proteins against the effects of freezing do so by not associating with the surface of the protein. Such excipients actually increase the surface tension of water and induce preferential hydration of the protein. Examples of solutes that serve as cryoprotectants by this mechanism include amino acids, polyols, sugars, and polyethylene glycol.

For lyoprotection, that is, stabilization of proteins during the drying stages of freeze drying and during storage in the dry state, two mechanisms are generally accepted. One is the water substitute hypothesis [178] and the other is the vitrification hypothesis [179]. Both are legitimate theories, but both also have exceptions; that is, neither fully explain the stabilization of proteins by excipients during dehydration and dry storage [180]. The water substitute hypothesis states that a good stabilizer is one that hydrogen bonds to the protein just as water would do where it presents and, therefore, serves as a water substitute. Sugars are good water substitutes. (It may at first appear contradictory that sugars can serve both as cryoprotectants because of being excluded from the surface of the protein and as lyoprotectants that hydrogen bond to the protein. However, keep in mind that the excluded solute concept involves a frozen aqueous system whereas the water substitute concept occurs in a dry system.) This is why many freeze-dried protein formulations contain sucrose or trehalose. Nevertheless, during a w/o/w procedure to prepare peptide/protein-loaded PLGA microspheres, sugars are often added to the inner aqueous phase. Trehalose was shown to partially improve the BSA secondary-structure protection within PLGA microspheres and to facilitate BSA monomer release [26]. Trehalose and mannitol had a significant effect on the recovery of soluble nonaggregated interferon-γ (INF-γ) and rhGH after emulsification and ultrasonication [61], whereas no or very little protecting effect on IGF-1 against these stress factors was observed [144]. No effect of trehalose, mannitol, and sucrose was observed against o/w interface-induced degradation of lysozyme, whereas lactose and lactulose significantly improved its structural stability and activity, mostly if these additives were also added to the second aqueous phase [88, 100]. Lysozyme and trypsin activity was not improved by addition of sucrose, which was unable to protect them from an emulsion-induced denaturation and from sonication [98, 181]. Mannitol and sucrose dissolved together in the inner aqueous phase had slight effect on NGF activity [182] and neither mannitol nor lactose improved HBcAg immunogenicity during dichloromethane/water emulsification [145]. Surprisingly, sucrose and trehalose even decreased urease bioactivity, showing the opposite effect to that expected [161]. Coencapsulation of maltose reduced α-chymotrypsin aggregation [183]. With respect to microspheres generated by spray drying, trehalose was effective in retaining TT antigenicity [147] and in preventing BSA secondary-structure degradation [35]. Trehalose protected efficiently NGF during the processing but did not prevent its aggregation during in vitro release [159].

The vitrification hypothesis states that excipients that remain amorphous (glass formers) form a glassy matrix with the protein with the matrix serving as a stabilizer. Acceptance of this hypothesis requires formulators to determine glass transition temperatures of formulations to be freeze dried and to develop freeze-dry cycles that maintain drying temperatures below the glass transition temperature. Reports are available to indicate that excipient stabilizers, which are capable of undergoing crystallization during storage, caused degradation (typically aggregation and loss of potency) of the protein [176, 184, 185].

Freeze-dried formulations typically contain one or more of the following bulking agents: mannitol, lactose, sucrose, trehalose, dextran 40, and povidone. These excipients may also serve as cryo- and/or lyoprotectants in protein formulations. Fakes et al. [186] studied these bulking agents for moisture sorption behavior before and after freeze drying. Moisture uptake certainly can affect drug stability in the freeze-dried state, particularly with peptides and proteins. When selecting a bulking agent, these properties, particularly the tendency for moisture uptake, must be considered by the formulation scientist in developing an optimally stable freeze-dried formulation. Several excipients can serve as stabilizers for proteins that are unstable during the drying phases of freeze drying and/or during long-term storage in the dry state. Typically, additives that will crystallize during lyophilization (e.g., mannitol) or will remain amorphous but unable to hydrogen bond to the dried protein (e.g., dextran) are not effective lyoprotectants for proteins. Excipients that will crystallize during freeze drying will also be relatively ineffective, as was shown with sucrose in *H. lanuginosa* lipase formulations [149]. However, these authors also reported that sucrose crystallization could be inhibited by decreasing the mass ratio of sucrose to protein and by minimizing the moisture content that serves to decrease the glass transition temperature during storage. The reverse can also be true for certain small molecules. For example, excipients (mannitol or sodium bicarbonate) that promoted the crystallization of cyclophosphamide during freeze drying stabilized the final product whereas excipients that did not allow crystallization (e.g., lactose) destabilized the final product [187]. Costantino et al. [188] studied the effects of a variety of parenteral excipients on stabilizing human growth hormone in the lyophilized state. Mannitol, sorbitol, methyl *a*-D-mannopyranoside, lactose, trehalose, and cellobiose all provided significant protection of the protein against aggregation, particularly at levels (131:1 excipient-to-protein molar ratio) to potentially satisfy water binding sites on the protein in the dried state. At higher excipient-to-protein ratios, mannitol and sorbitol crystallized and were not as effective in stabilizing the protein compared with low levels in which they remained in the amorphous, protein-containing phase.

Reducing sugars may not be as effective as other bulking agents, cryoprotectants, or lyoprotectants because they may potentially react with proteins via the Maillard reaction. For example, glucose will form covalent adducts with side-chain amino acids lysine and arginine of human relaxin [189]. In addition, a significant amount of serine cleavage from the C terminal of the B chain of relaxin was formed when glucose was used as the excipient. These reactions did not occur if mannitol and trehalose replaced glucose in the lyophilized formulation. Lactose will react with primary amines in the well-known Maillard-type condensation reaction to form brown-colored degradation products [190]. Thus, lactose is known to be incompatible with amine-containing compounds such as aminophylline, amphetamines, and amino acids/peptides. This reaction occurs more readily with amorphous lactose than crystalline lactose.

Hydrophilic additives such as glucose are known to increase the porosity of microspheres, causing an increase in permeability to mass transport and a higher burst. However, a significant reduction in initial burst release of a highly water-soluble model peptide, octreotide acetate, from poly(D,L-lactide-*co*-glycolide) microspheres by the coencapsulation of a small amount of glucose (e.g., 0.2% w/w) was reported [191]. Using the double emulsion–solvent evaporation method of encapsulation, the effect of glucose on initial burst in an acetate buffer pH 4 was found to depend on polymer concentration, discontinuous phase/continuous phase ratio, and glucose content. Extensive characterization studies were performed on two microsphere batches, ±0.2% glucose, to elucidate the mechanism of this effect. However, no significant difference was observed with respect to specific surface area, porosity, internal and external morphology, and drug distribution. Continuous monitoring of the first 24-h release of octreotide acetate from these two batches disclosed that, even though their starting release rates were close, the microspheres plus glucose exhibited a much lower release rate between 0.2 and 24 h compared to those without glucose. The microspheres plus glucose showed a denser periphery and a reduced water uptake at the end of the 24-h release, indicating decreased permeability. However, this effect at times was offset as glucose content was further increased to 1%, causing an increase in surface area and porosity. In summary, these authors concluded that the effects of glucose on initial burst are determined by two factors: (1) increased initial burst due to increased osmotic pressure during encapsulation and drug release and (2) decreased initial burst due to decreased permeability of microspheres [191].

Mannitol is probably the most widely used bulking agent in lyophilized formulations because of its many positive properties with respect to crystallinity, high eutectic temperature, and matrix properties. However, some lots of mannitol can contain reducing sugar impurities that were implicated in the oxidative degradation of a peptide in a lyophilized formulation [192]. Mannitol at or above certain concentrations and volumes in glass vials is well known to cause vial breakage because of the unique crystallization properties of mannitol-ice during the primary drying states of freeze drying [193, 194].

Other Freeze-Dry Excipients High-molecular-weight carbohydrates such as dextran have higher glass transition temperatures than peptides/proteins. Therefore, when mixed with proteins, the overall glass transition temperature presumably can be increased with resultant increases in protein storage stability. Typically, carbohydrates (sucrose, trehalose, or dextran) alone do not result in appreciable increases in the storage stability of proteins. However, combinations of disaccharide and polymeric carbohydrates do tend to improve protein storage stability [195]. However, singular carbohydrates (sucrose or trehalose at 60 m*M*) were also just as effective in stabilizing a model recombinant humanized monoclonal antibody as combinations of sucrose and mannitol or trehalose and mannitol. Interestingly, with this model monoclonal antibody, mannitol alone at 60 mM provided less protection during storage than sucrose or trehalose alone. A specific sugar/protein molar ratio was sufficient to provide storage stability for this particular monoclonal antibody [196, 197].

Low-molecular-weight additives such as osmolytes (*N,N*-dimethylglycine, trehalose, and sucrose) or salts (sodium chloride, sodium phosphate, ammonium sulfate, and sodium citrate) were found to be highly effective in stabilizing keratinocyte growth factor, both against thermal denaturation and enhancing long-term storage stability [198]. Nevertheless, the stabilizing properties of osmolytes appear to be balanced between their binding to (deteriorating effect) and exclusion from (stabilizing effect) the peptide/protein surface. As binding or exclusion predominantly results from hydrophobic interactions, hydrogen bonding, and electrostatic interactions, the sum of the various interaction parameters are dissimilar for different proteins. Therefore, it becomes crucial to examine the individual nature of the additive toward each individual protein and to assess whether it will offer a stabilizing or destabilizing effect [199, 200].

Polyvinyl pyrrolidone (PVP) and glycine were found to stabilize lyophilized sodium prasterone sulfate whereas dextran 40 or mannitol did not [201]. PVP and glycine stabilized the pH of the reconstituted solution by neutralizing the acidic degradation product, sodium bisulfate, formed by the hydrolysis of prasterone sulfate. Dextran 40 or mannitol was ineffective because of no buffer capacity. Buffering agents, such as phosphate–citrate buffer and some neutral and basic amino acids (L-arginine, L-lysine, and L-histidine), also stabilized prasterone sulfate. L-Cysteine is an example of an amino acid that did not stabilize the drug, presumably because of its weak buffer capacity. Although the efficiency of proteinic additives for protein stabilization has been clearly demonstrated in several occasions even during encapsulation processes [31, 72, 98, 144], their use in pharmaceuticals is at present not desirable from a strictly regulatory point of view. Additionally, such agents might contribute to complicate all subsequent protein characterization within the formulation. Among these additives, albumins and gelatins are those mainly used for protection purposes. The protective effect of albumins against protein unfolding and aggregation has been extensively documented and is likely due to their surface-active properties (see Bilati et al. [130] for details).

Prevention/Minimization of Microclimate pH-Induced Instability Evidence for acidification within degrading microspheres is investigated and local pH values between 1.5 and 4.7 are being reported [202–204]. Methods to measure microclimate pH in PLGA microspheres include (i) ensemble average measurements using electron paramagnetic resonance (EPR) [203, 204], nuclear magnetic resonance (NMR) [205], and potentiometry and (ii) direct visualization techniques such as confocal imaging of pH-sensitive dyes [206, 207]. In the EPR method, the constant of hyperfine splitting, $2aN$, was used to determine an average pH inside PLGA microspheres. Because the experiments relied on the mobility of spin-labeled protein, with an increase of the microviscosity in the later hours of the experiments, the spectra of EPR was changed and the signal-to-noise ratio decreased to prevent the measurement of pH throughout the release period [203]. The potentiometric measurements can give rapid values of pH for thin polymer films, and the pH of the thin water film between the electrode and polymer mimics the microclimate pH of aqueous pores inside the polymer-based drug delivery system. However, it is difficult to mimic microclimate pH of a small-scale system, such as microspheres or nanospheres, which may have unique microstructures, excipient/drug distributions, and transport characteristics. Overall, the ensemble average measurements described above could give a general picture of microclimate pH at the macroscopic level. However, the microscopic level of the detection can only be achieved through direct visualization techniques, such as microscopic imaging. Shenderova et al. [207] first developed the confocal microscope imaging method to relate the microclimate pH with the fluorescent intensity. Because of the difficulty of controlling and predicting the fluorescein concentration in the aqueous pore inside the microsphere, the method was only semiquantitative. Fu et al. [206] improved the confocal microscopic imaging method by coencapsulating two dextran fluorescent dye (NERF and SNARF-1) conjugates inside microspheres and related the ratio of the two dye images with microclimate pH in order to eliminate the poorly controlled effects of dye concentration and pore distribution. However, both of the dyes emit in the green range (535 nm for NERF and 580 nm of SNARF), giving rise to poor resolution without a narrow-bandwidth detector. Because of the high noise-to-signal ratio from the ratio images, the prediction of pH is also expected to be semiquantitative. In order to overcome the aforementioned drawbacks in microclimate pH measurement, a new quantitative ratiometric method based on laser scanning confocal microscopic imaging was developed to create a pixel-by-pixel neutral range microclimate pH map inside PLGA microspheres [208]. This method was then applied to both acid-neutralized and nonneutralized PLGA microspheres during extended incubation in physiological buffer. In another study, the PLGA water-soluble acid distribution has been measured with prederivatization HPLC [209].

Ongoing acidification of the microsphere interior was shown to induce deamidation and covalent dimerization of nonreleased insulin [202]. Despite the evidence of acidification mentioned above, there is controversy on this subject. It has been pointed out that the sampling scheme has a significant impact on the degree of acidification; frequent replenishment of the release medium or the use of a dialysis bag can effectively prevent the acidification of the medium with subsequent reduced protein degradation [93, 210]. It is unsure, however, whether this also reflects the situation in vivo, in which the PLGA microspheres are often surrounded by a fibrous capsule that may reduce efflux of acidic degradation products from the PLGA matrix [93]. On the other hand, studies on rhGH-loaded PLGA microspheres showed a reasonable in vitro–in vivo correlation (IVIVC) only when a strong high-capacity buffer [200 m*M* *N*-(2-hydroxyethyl)piperazine-*N*′-2-ethanesulfonic acid (HEPES), pH 7.4] was used, which effectively minimized the pH drop [211].

As indicated by the prevention of acid-induced physical aggregation of BSA in an abstract [212], three principal ways have been identified thus far to avoid the formation of highly acidic microclimate regions in the PLGAs during protein release:

(i) Increasing the permeability of the polymer to facilitate escape of the water-soluble hydrolytic products of the PLGA polyester [125]
(ii) Decreasing the degradation rate of the polyester [213]
(iii) Coencapsulating additives to neutralize the weak acids formed by PLGA hydrolysis [126]

In addition, two more ways that are likely to favor a lowering of microclimate pH are elevated initial acid content in the polymer [214] and low-frequency release media exchange [206].

The concept of controlling polymer permeability is difficult because attempts to increase permeability can spoil the controlled-release function of the polymer and cause the encapsulated protein to be released too rapidly. For example, Jiang and Schwendeman [213] increased the permeability of slow-degrading PLA (molecular weight (MW) 145 kDa) by blending in PEG (MW 10 or 30 kDa) at 0, 10, 20, and 30%. Insoluble BSA aggregation in the PLA microspheres containing 4.5–5% w/w BSA was found in 0 and 10% PEG after a one-month incubation, but not in those preparations containing 20 or 30% PEG. The structural integrity of BSA was also intact in the stabilized formulations. However, between 10 and 30% PEG, the release rate of BSA increased rapidly and by 30% PEG, 60% of the protein encapsulated was released in only three days [213]. In contrast, an abstract [212] implied that 5% BSA encapsulated in a more permeable PLA (MW 77 kDa), the BSA formed <2% insoluble aggregates over one month. This result suggested strongly that in some instances the slow degradation rate of the non-glycolic-acid-containing PLA is sufficient to inhibit acid formation in the microclimate.

In instances in which it is desirable to increase permeability and/or decrease the hydrolytic rate of PLGAs, that is, where a highly water-soluble protein requires release for one month or longer, it becomes necessary to coencapsulate a basic additive. Antacids such as $MgCO_3$, $Mg(OH)_2$ or $ZnCO_3$ have been found to be particularly potent in preventing instabilty of acid-labile proteins [126, 215]. By means of thin films coating pH glass electrodes to measure directly the microclimate acidic environment in PLGA microspheres, the stabilization against insoluble acid-induced noncovalent BSA aggregation afforded by a series of antacid excipients has been correlated with the ability of the antacid to neutralize acidic pores in films of the same lot of PLGA coating pH glass electrodes [215]. Though much of the physical chemistry of microclimate pH adjustment with antacid additives is currently unclear, the strength of the base, the base solubility, and the association of the divalent cation with the carboxylate of the degradation products and/or polymer end groups appear to be important. For instance, Shenderova [216] has shown that from microclimate pH measurements in PLGA films coating pH glass electrodes, $MgCO_3$ and $Mg(OH)_2$ were found to be very similar under conditions which favor homogenous neutralization (i.e., high protein loading sufficient to make pores for the base to diffuse all regions of the polymer matrix), but $MgCO_3$ was found to increase microclimate pH higher than $Mg(OH)_2$. This result was consistent with the improved BSA stability in PLGA 50/50 microspheres when $MgCO_3$ was used in place of $Mg(OH)_2$ [126].

5.2.6 PEPTIDE/PROTEIN ENCAPSULATED INTO BIODEGRADABLE MICROSPHERES: CASE STUDY

Selected examples of therapeutic peptide and protein including vaccines which have been encapsulated into biodegradable polymer-based microspheres are discussed in this section. Besides what is mentioned below, many other proteins and vaccines have been encapsulated in biodegradable polymers, so a glimpse of ongoing research on microsphere delivery systems using biodegradable polymers is shown in Table 2.

5.2.6.1 Vaccines

Group B Streptococcus Vaccine Group B streptococcus (GBS) is the leading bacterial cause of neonatal sepsis and meningitis. Although antibiotic prophylaxis has decreased the infection rate, the best long-term solution lies in the development of effective vaccines. The GBS capsular polysaccharide (CPS) is a major target of antibody-mediated immunity. The feasibility of producing a GBS having the ability to produce both a local IgA immune response at the mucosal surface and humoral IgG response having capability of transplacental passive immunization was investigated [217]. Inactivated GBS antigen was encapsulated in PLGA by a w/o/w multiple-emulsion technique along with immunostimulatory synthetic oligodeoxynucleotides containing cytosine phosphate guanosine (CpG) as potent adjuvant [217]. Immunization of female mice with normal immune systems was done with these PLGA microspheres containing GBS type III polysaccharide and CpG adjuvant (PLGA/GBS/CpG) and results indicated a significantly higher GBS antibody response as compared to nonencapsulated GBS antigen or PLG-encapsulated GBS PS vaccine without the addition of the CpG.

Diphtheria Toxoid (DT) Diphtheria is a communicable disease caused by *Corynebacterium diphtheriae* which colonizes and forms a pseudomembrane at the infection site. This pathogen produces a potent protein toxin, diphtheria toxin, which is responsible for the typical systemic toxemia. DT is required for active immunization against diphtheria. DT was encapsulated in different types of PLA and PLGA microspheres by spray drying and coacervation. Immunization of guinea pigs with DT microspheres made with relatively hydrophilic PLGA 50:50 resulted in specific and sustained antibody responses to alum adjuvanted toxoid in contrast to microspheres made with hydrophobic polymers where very low antibody responses were determined confirming the feasibility of microsphere vaccines to induce strong, long-lasting protective antibody responses after single immunization [218].

In an endeavor toward development of multivalent vaccines based on biodegradable microspheres, Peyre et al. [219] tested the immunological performance of several divalent microsphere formulations against tetanus and diphtheria. Microspheres were made by separate microencapsulation of tetanus and diphtheria toxoid in PLGA by either spray drying or coacervation. Guinea pigs were subcutaneously immunized by a single injection of the divalent vaccines or, for control, an equivalent dose of a licensed vaccine containing both antigens adsorbed on aluminum hydroxide. All microsphere formulations were strongly immunogenic, irrespective of particle size and hydrophobicity. Endpoint titers of ELISA antibodies, mainly of the IgG1 subtype, were comparable to those obtained after immunization with the licensed vaccine. The microspheres provided increasing levels of antibodies, during the 16 weeks of testing, and the antibodies were weakly polarized toward tetanus. The induced antibodies were also toxin neutralizing, as determined for both diphtheria (1–4 IU/mL) and tetanus (5–9 IU/mL) eight weeks after immunization. These neutralization levels were several orders of magnitude above the level considered minimum for protection (0.01 IU/mL). When the animals were challenged with tetanus or diphtheria toxins six weeks after immunization, microsphere vaccines produced protective immunity that was comparable to or better than that induced by the licensed divalent vaccine. In conclusion, this study showed that a single administration of biodegradable microsphere vaccines provided protective immunity against diphtheria and tetanus and that this immunization approach might be feasible for multivalent vaccines. In a separate study, the same group have studied for the first time the fate of immunogenic fluorescent-labeled PLGA microspheres loaded with DT in vivo following a subcutaneous injection route [220].

A unique instability problem of DT is being foreseen when the DT would be encapsulated in PLGA microspheres along with a preservative such as thiomersal [221]. Thimerosal (TM)—also known as thiomersal, Merthiolate, or sodium ethylmercuri-thiosalicylate—is a water-soluble derivative complex of thiosalicylic acid (TSA) that has been used as bactericide in parenteral vaccines and ophthalmic products for decades. It has been reported that this preservative can be decomposed by oxidation to 2,2-dithiosalicylic acid, ethyl mercuric ion, 2-sulfenobenzoic acid, 2-sulfobenzoic acid, and 2-sulfinobenzoic acid [222]. Namura et al. [221] demonstrated in vitro that the TSA, produced after the reduction of TM by lactic acid, reduces the S–S bridge of the previously incubated DT. This reduction is immediately followed by blocking the two SH groups formed by the same TSA molecules. In light of these conclusions, it is necessary now to reinterpret the in vitro protein degradation–stabilization data in the presence of PLGA microsphere, mainly for those proteins which contain S–S. The authors propose that all the PLGA microsphere microencapsulation studies and protein structural considerations should be done in the absence of TM as preservative.

Tetanus Toxoid (TT) Tetanus is considered a major health problem in developing and underdeveloped countries, with approximately one million new cases occurring each year. Tetanus is an intoxication manifested primarily by neuromuscular dysfunction. So vaccination is required for prevention of this disease. TT was encapsulated using PLGA with different molar compositions (50:50, 75:25) by the w/o/w multiple-emulsion technique and protein integrity was evaluated during antigen release in vitro in comparison to alum-adsorbed TT for in vivo induction of tetanus-specific antibodies [223]. TT microspheres elicited antibody titers as high as conventional alum-adsorbed TT, which lasted for 29 weeks, leading to the conclusion that TT microspheres can act as potential candidates for single-shot vaccine delivery systems.

The study by Determan et al. [224] focuses on the effects of polymer degradation products on the primary, secondary, and tertiary structure of TT, OVA, and lysozyme after incubation for 0 or 20 days in the presence of ester (lactic acid and glycolic acid) and anhydride [sebacic acid and 1,6-bis(*p*-carboxyphenoxy)hexane] monomers. The structure and antigenicity or enzymatic activity of each protein in the presence of each monomer was quantified. SDS-PAGE, circular dichroism, and fluorescence spectroscopy were used to assess/evaluate the primary, secondary, and tertiary structures of the proteins, respectively. ELISA was used to measure changes in the antigenicity of TT and OVA and a fluorescence-based assay was used to determine the enzymatic activity of lysozyme. TT toxoid was found to be the most stable in the presence of anhydride monomers, while OVA was most stable in the presence of sebacic acid, and lysozyme was stable when incubated with all of the monomers studied.

Jaganathan et al. [225] compared the efficiency of microspheres produced from PLGA and chitosan polymers by using protein stabilizer (trehalose) and acid-neutralizing base [$Mg(OH)_2$]. The immunogenicity of PLGA- and chitosan microsphere–based single-dose vaccine was evaluated in guinea pigs and compared with multiple doses of alum-adsorbed TT. Results indicated that a single injection of PLGA and chitosan microspheres containing TT could maintain the antibody response at a level comparable to the booster injections of conventional alum-adsorbed vaccines. Both the PLGA- and chitosan-based stable vaccine formulations produced an equal immune response. Hence chitosan can be used to replace the expensive polymer PLGA. This approach should have potential application in the field of vaccine delivery.

The study by Kipper et al. [226] focuses on the development of single-dose vaccines based on biodegradable polyanhydride microspheres that have the unique capability to modulate the immune response mechanism. The polymer system employed consists of copolymers of 1,6-bis(*p*-carboxyphenoxy)hexane and sebacic acid. Two copolymer formulations that have been shown to provide extended-release kinetics and protein stability were investigated. Using TT as a model antigen, in vivo studies in C3H/HeOuJ mice demonstrated that the encapsulation procedure preserves the immunogenicity of the TT. The polymer itself exhibited an adjuvant effect, enhancing the immune response to a small dose of TT. The microspheres provided a prolonged exposure to TT sufficient to induce both a primary and a secondary immune response (i.e., high antibody titers) with high-avidity antibody production, without requiring an additional administration. Antigen-specific proliferation 28 weeks after a single immunization indicated that immunization with the polyanhydride microspheres generated long-lived memory cells and plasma cells (antibody-secreting B cells) that generally do not occur without maturation signals from T helper cells. Furthermore, by altering the vaccine formulation, the overall strength of the T-helper type 2 immune response was selectively diminished, resulting in a balanced immune response, without reducing the overall titer. This result is striking, considering free TT induces a T-helper type 2 immune response and has important implications for developing vaccines to intracellular pathogens. The ability to selectively tune the immune response without the administration of additional cytokines or noxious adjuvants is a unique feature of this delivery vehicle that may make it an excellent candidate for vaccine development.

Polylactide (PLA) polymer particles entrapping TT were evaluated in terms of particle size, antigen load, dose, and additional adjuvant for achieving high and sustained anti-TT antibody titers from single-point intramuscular immunization [227]. Admixture of polymer-entrapped TT and alum improved the immune response in comparison to particle-based immunization. High and long-lasting antibody titer was achieved upon immunization with 2–8-μm size microparticles. Microspheres within the size range 50–150 μm elicited very low serum antibody response. Immunization with very small particles (<2 μm) and with intermediate-size-range particles (10–70 μm) elicited comparable antibody response from single-point immunization but lower in comparison to that achieved while immunizing with 2–8-μm particles. Potentiation of antibody response on immunization of admixture of microspheres and alum was also dependent on particle size. These results indicate the need of optimal particle sizes in micrometer ranges for improved humoral response from single-point immunization. Increasing antigen load on polymer particles was found to have a positive influence on the generation of antibody titers from particle-based immunization. Maximum peak antibody titer of ~300 μg/mL was achieved on day 50 upon immunization with particles having the highest load of antigen (94 μg/mg of polymer). Increase in dose of polymer-entrapped antigen resulted in concomitant increase in peak antibody titers, indicating the importance of antigen stability, particle size, and load on generating a reproducible immune response. Optimization of particle size, antigen load, dose, and use of additional adjuvant resulted in high and sustained anti-TT antibody titers over a period of more than 250 days from single-point immunization. Serum anti-TT antibody titers from single-point immunization of admixture of PLA particles and alum were comparable with immunization from two divided doses of alum-adsorbed TT.

Vibrio Cholerae *(VC) Whole-Cell Vaccine* Cholera, an acute intestinal infection caused by the bacterium *Vibrio cholerae*, produces an enterotoxin that causes a copious, painless, watery diarrhea that can quickly lead to severe dehydration and death if treatment is not promptly given. For prevention of cholera, cholera vaccine is usually given. VC was successfully entrapped in the PLGA microspheres by a double-emulsion method with trapping efficiencies up to 98%. The immnunogenic potential of VC-loaded microspheres physically mixed with or without amphotericin B was evaluated in adult mice by oral immunization in comparison to VC solution. The immunogenicity of VC-loaded microparticles mixed with amphotericin B in evoking *Vibrio*-specific serum IgG and IgM responses was higher than that of VC-loaded microparticles only [228, 229]. However, VC was loaded in different polymer compositions (50:50 PLGA, 75:25 PLGA, and PLA/PEG blended), the higher antibody responses and serum IgG, IgA, and IgM responses were obtained when sera from both VC-loaded 75:25 PLGA and PLA/PEG-blended microparticles immunized mice were titrated against VC solution [230].

Japanese Encephalitis Virus (JEV) Japanese encephalitis is a disease that is spread to humans by infected mosquitoes in Asia. It is one of a group of mosquito-borne viral diseases that can affect the central nervous system and cause severe complications and even death. Vaccination is one of the ways of treating it. JEV vaccine was encapsulated in PLGA microspheres by a double-emulsion technique and influences of various process variables such as stirring rate, types and concentration of emulsifier, and polymer concentration were studied on size, size distribution, and biodegradation. The mean size of microspheres decreased with increasing speed, increasing concentration of emulsifier, and decreasing polymer concentration. Rate of biodegradation of nonporous microspheres was slower than that of porous microspheres, leading to the conclusion that PLGA microspheres can be used to apply oral vaccination through Peyers patches across the gastrointestinal tract (GIT) [231].

Several approaches to develop an improved JEV vaccine are in progress in various laboratories. Of these, immunization of mice with plasmid DNA encoding JEV envelope (E) protein has shown great promise. The technology, developed by Kaur et al. [232], involved the adsorption of DNA onto cetyltrimethyl-ammonium bromide (CTAB) containing cationic poly(lactide-*co*-glycolide) (PLG) microspheres.

The microsphere-adsorbed DNA induced a mixed Th1–Th2 immune response as opposed to Th1 immune responses elicited by the naked DNA.

JEV-loaded poly(lactide) (PLA) lamellar and PLG microspheres were successfully prepared with low-molecular-weight PLA by the precipitate method and with 6% w/v PLG in the organic phase, 10% w/v PVP, and 5% w/v NaCl in the continuous phase by using a w/o/w emulsion/solvent extraction technique, respectively [233]. The JEV incorporation, physicochemical characterization data, and animal results obtained in this study may be relevant in optimizing the vaccine incorporation and delivery properties of these potential vaccine targeting carriers.

Hepatitis B Virus Hepatitis B is one of the most important infectious diseases in the world. Approximately 350 million people worldwide are chronic carriers of the hepatitis B virus (HBV), which accounts for approximately one million deaths annually. PLGA microspheres loaded with recombinant HBsAg were formulated using a double-emulsion technique. The pharmaceutical characteristics of size, surface morphology, protein loading efficiency, antigen integrity, release of HBsAg-loaded PLGA microspheres, and degradation of the polymer in vitro were evaluated [234–237]. Based on these findings in vitro and in vivo, it was concluded that HBsAg was successfully loaded into the PLGA microspheres, which can autoboost an immune response, and the HBsAg-loaded PLGA microsphere is a promising candidate for the controlled delivery of a vaccine.

5.2.6.2 Proteins

Prolidase Deficiency of this enzyme results in chronic intractable ulcerations of the skin, particularly of lower limbs, since it is involved in the final stages of protein catabolism. To counteract the problem, the enzyme was encapsulated in PLGA microspheres by a double- or multiple-emulsion technique, in vitro and ex vivo evaluations were done, and the results indicated that microencapsulation stabilizes the enzymatic activity inside the PLGA microspheres resulting in both in vitro and ex vivo active enzyme release, hence opening the doors for the possibility of enzyme replacement therapy through microencapsulation [238]. Further evaluation from the same research group for prolidase-loaded PLGA microspheres is reported elsewhere [239, 240].

Insulin Insulin is the most important regulatory hormone in the control of glucose homeostasis. The World Health Organization (WHO) has indicated that more than 50 million people around the world suffer from diabetes and require daily parenteral injections of insulin to stay healthy and live normally. For the treatment of type I diabetes insulin still is number one, with three subcutaneous injections to be taken per day. A controlled-release system for a long-term therapy of this disease is the need of the hour, as this can obviate the need for painful injection given a number of times to the diabetes patients. Insulin was encapsulated in blends of poly(ethylene glycol) with PLA homopolymer and PLGA copolymer by a w/o/w multiple-emulsion technique with entrapment efficiencies up to 56 and 48% for PLGA/ PEG and PLA/ PEG, respectively [12]. Insulin-loaded microspheres were capable of controlling the release of insulin for 28 days with in vitro delivery rates of 0.94 and 0.65 μg insulin/mg per particle per day in the first 4 days and steady release with a rate of 0.4 and 0.43 μg insulin/mg per particle per day over the following 4 weeks, respectively, along with the extensive degradation of PLGA/ PEG microspheres as compared to PLA/ PEG blends which resulted in a stable particle morphology along with reduced fragmentation and aggregation of associated insulin.

Two types of injectable cationized microspheres were prepared based on a native gelatin (NGMS) and aminated gelatin with ethylenediamine (CGMS) to prolong the action of insulin [241]. Release of rhodamin B isothiocyanate insulin from CGMS was compared with that from NGMS under in vitro and in vivo conditions. Lower release of insulin from CGMS compared with that from NGMS was caused by the suppression of initial release. The disappearance of ^{125}I-insulin from the injection site after intramuscular administration by NGMS and CGMS had a biphasic profile in mice. Almost all the ^{125}I-insulin had disappeared from the injection site one day after administration by NGMS. The remaining insulin at the injection site after administration by CGMS was prolonged, with approximately 59% remaining after 1 day and 16% after 14 days. The disappearance of CGMS from the injection site was lower than that of NGMS. However, the difference in these disappearance rates was not great compared with those of ^{125}I-insulin from the injection site by NGMS and CGMS. The time course of disappearance of ^{125}I-CGMS from the injection site was similar to that of ^{125}I-insulin by CGMS. The initial hypoglycemic effect was observed 1 h after administration of insulin by NGMS, and thereafter its effect rapidly disappeared. The hypoglycemic effect was observed 2–4 h after administration by CGMS and continued to be exhibited for 7 days. The prolonged hypoglycemic action by CGMS depended on the time profiles of the disappearance of insulin from muscular tissues, which occurs due to the enzymatic degradation of CGMS.

A novel controlled-release formulation was developed with PEGylated human insulin encapsulated in PLGA microspheres that produces multiday release in vivo [242]. The insulin is specifically PEGylated at the amino terminus of the B chain with a relatively low molecular weight PEG (5000 Da). Insulin with this modification retains full biological activity but has a limited serum half-life, making microencapsulation necessary for sustained release beyond a few hours. PEGylated insulin can be codissolved with PLGA in methylene chloride and microspheres made by a single o/w emulsion process. Insulin conformation and biological activity are preserved after PEGylation and PLGA encapsulation. The monolithic microspheres have inherently low burst release, an important safety feature for an extended-release injectable insulin product. In PBS at 37°C, formulations with a drug content of approximately 14% show very low (<1%) initial release of insulin over one day and near-zero-order drug release after a lag of three to four days. In animal studies, PEG-insulin microspheres administered subcutaneously as a single injection produced <1% release of insulin in the first day but then lowered the serum glucose levels of diabetic rats to values <200 mg/dL for approximately nine days. When the doses were given at seven-day intervals, steady-state drug levels were achieved after only two doses. PEG-insulin PLGA microparticles show promise as a once-weekly dosed, sustained-release insulin formulation.

Shenoy et al. [243] developed an injectable, depot-forming drug delivery system for insulin based on microparticles technology to maintain constant plasma drug concentrations over a prolonged period of time for the effective control of blood sugar levels. Formulations were optimized with two well-characterized biodegradable polymers, namely PLGA and poly-ε-caprolactone, and evaluated in vitro for physicochemical characteristics, drug release in phosphate-buffered saline (pH 7.4), and evaluated in vivo in streptozotocin-induced hypoglycemic rats. With a large volume of internal aqueous phase during a w/o/w double-emulsion solvent evaporation process and high molecular weight of the polymers used, they could not achieve high drug capture and precise control over subsequent release within the study period of 60 days. However, this investigation revealed that upon subcutaneous injection the biodegradable depot-forming polymeric microspheres controlled the drug release and plasma sugar levels more efficiently than plain insulin injection. Preliminary pharmacokinetic evaluation exhibited steady plasma insulin concentration during the study period. These formulations, with their reduced frequency of administration and better control over drug disposition, may provide an economic benefit to the user compared with products currently available for diabetes control.

Interferon α_{2a} (IFN α_{2a}) Interferon α_{2a} is indicated for the treatment of adults with chronic hepatitis C virus infection who have compensated liver disease and have not been previously treated with interferon α. To improve the stability and loading efficiency of protein drugs, a new microsphere delivery system comprises calcium alginate cores surrounded by PELA [poly-*D,L*-lactide-poly-(ethylene glycol)]. Recombinant IFNα_{2a} as a model drug was entrapped within calcium alginate cores surrounded by PELA by a w/o/w multiple-emulsion technique [244]. Core-coated microspheres stabilized the IFN in the PELA matrix. The core-coated microspheres indicated high encapsulation efficiency and biological retention as compared to conventional PLGA microspheres. The extent of burst release reduced to 14% in core-coated microspheres from 31% in conventional microspheres, indicating a new approach for water-soluble macromolecular drug delivery.

5.2.7 CONCLUSION

From this chapter, it has become apparent that a number microencapsulation methods are available today for the preparation of microspheres on an industrial scale. In fact, parenteral drug delivery systems based upon biodegradable microspheres are a true success story for the concept of drug delivery. However, the production of biodegradable microspheres containing a stable therapeutic peptide or protein still remains a major challenge in terms of technical obstacles. Ideally, peptides/proteins of therapeutic interest should be studied case by case, so as to bring to the fore processing steps and stress factors which damage them. Continued efforts to establish methods for stable protein, especially antigen, delivery from microspheres may hopefully pave the way for future microsphere-based vaccines. Areas of further research should focus on the performance of peptide/protein-loaded microspheres under in vitro and in vivo conditions. Interestingly, the addition of medium-chain triglycerides (MCT) modifies/shifts the triphasic release pattern of leuprolide acetate-loaded PLGA microspheres to a more continuous release in vitro [245]. Alternatively, BSA-loaded PLGA microspheres were coated with a thermosensitive gel, Pluronic F127 (PF127) [246]. The results demonstrated that PF127, which gelled at 37°C, inhibited the initial burst release of BSA from microspheres effectively. It is anticipated that more efforts will be invested in the future to develop novel ways to reduce the initial burst release of entrapped peptide/protein and to attain a more continuous release. In addition, an in vitro release model mimicking the fate of biodegradable microspheres applied through the parenteral route would be highly desirable. Also, new strategies to stabilize proteins in microspheres during manufacturing, shelf life, or in vivo could be of general interest. Moreover, the use of analytical techniques such as FTIR or MALDI-TOF mass spectrometry certainly constitutes a step forward for protein analysis in more appropriate conditions.

REFERENCES

1. Brooks, G. (1998), *Biotechnology in Healthcare*, Pharmaceutical Press, London.
2. Walsh, G. (2002), Biopharmaceuticals and biotechnology medicines: An issue of nomenclature, *Eur. J. Pharm. Sci.*, 15, 135–138.
3. Walsh, G. (2003), Pharmaceutical biotechnology products approved within the European Union, *Eur. J. Pharm. Biopharm.*, 55, 3–10.
4. New biotechnology medicines in development report (2002), available: http://pharma.org, accessed May 25, 2006.
5. Banga, A. K., and Chien, Y. W. (1988), Systemic delivery of therapeutic peptides and proteins, *Int. J. Pharm.*, 48, 15–50.
6. Lee, V. H. L. (1987), Ophthalmic delivery of peptides and proteins, *Pharm. Technol.*, 11, 26–38.
7. Athanasiou, K. A., Niederauer, G. G., and Agrawal, C. M. (1996), Sterilization, toxicity, biocompatibility and clinical applications of polylactic acid/polyglycolic acid copolymers. *Biomaterials*, 17, 93–102.
8. Ueda, H., and Tabata, Y. (2003), Polyhydroxyalkanonate derivatives in current clinical applications and trials, *Adv. Drug Deliv. Rev.*, 55, 501–518.
9. Ramesh, V. D., Medlicott, N., Razzak, M., and Tucker, I. G. (2002), Microencapsulation of FITC-BSA into poly(e-caprolactone) by a water-in-oil-in-oil solvent evaporation technique. *Trends Biomater. Artif. Organs*, 15, 31–36.
10. Dhanaraju, M. D., Vema, K., Jayakumar, R., and Vamsadhara, C. (2003), Preparation and haracterization of injectable microspheres of contraceptive hormones, *Int. J. Pharm.*, 268, 23–29.
11. Tracy, M. A. (1998), Development and scale-up of a microsphere protein delivery system, *Biotechnol. Prog.*, 14, 108–115.
12. Yeh, M. K. (2000), The stability of insulin in biodegradable microparticles based on blends of lactide polymers and polyethylene glycol, *J. Microencapsul.*, 17, 743–756.
13. Han, K., Lee, K. D., Gao, Z. G., and Park, J. J. (2001), Preparation and evaluation of poly(L-lactic acid) microspheres containing rhEGF for chronic gastric ulcer healing, *J. Controlled Release*, 75, 259–269.
14. Jackson, J., Liang, L., Hunter, W., Reynolds, M., Sandberg, J., Springate, C., and Burt, H. (2002), The encapsulation of ribozymes in biodegradable polymeric matrices, *Int. J. Pharm.*, 243, 43–47.
15. Blanco-Prieto, M. J., Besseghir, K., Zerbe, O., andris, D., Orsolini, P., Heimgartner, F., Merkle, H. P., and Gander, B. (2000), In vitro and in vivo evaluation of a somatostatin analogue released from PLGA microspheres, *J. Controlled Release*, 67, 19–28.
16. Yuksel, E., Weinfeld, A. B., Cleek, R., Waugh, J. M., Jensen, J., Boutros, S., Shenaq, S. M., and Spira, M. (2000), De novo adipose tissue generation through long-term, local delivery of insulin and insulin like growth factor-1 by PLGA/PEG microspheres in an in vivo rat midel: A novel concept and capability, *Plastic Reconstruct. Surg.*, 105, 1721–1729.

17. Kostanski, J. W., Thanoo, B. C., and Deluca, P. P. (2000), Preparation, characterization and in vitro evaluation of 1- and 4-month controlled release ornitide PLA and PLGA microspheres, *Pharm. Dev. Technol.*, 5, 585–596.

18. King, T. W., and Patrick, Jr., C. W. (2000), Development and in vitro characterization of vascular endothelial growth factor (VEGF) loaded poly(DL-lactic-*co*-glycolicacid)/poly(ethylene glycol) microspheres using a solid encapsulation/single emulsion/solvent extraction technique, *J. Biomed. Mater. Res.*, 51, 383–390.

19. Zhu, K. J., Jiang, H. L., Du, X. Y., Wang, J., Xu, W. X., and Liu, S. F. (2001), Preparation and characterization of hCG-loaded polylactide or poly(lactide-*co*-glycolide) microspheres using a modified water-in-oil-in-water (w/o/w) emulsion solvent evaporation technique, *J. Microencapsul.*, 18, 247–260.

20. Prabhu, S., Sullivan, J. L., and Betageri, G. V. (2002), Comparative assessment of in vitro release kinetics of calcitonin polypeptide from biodegradable microspheres, *Drug Deliv.*, 9, 195–198.

21. Oldham, J. B., Lu, L., Zhu, X., Porter, B. D., Hefferan, T. E., Larson. D. R., Currier, B. L., Mikos, A. G., and Yaszemski, M. J. (2000), Biological activity of rhBMP-2 released from PLGA microspheres, *J. Biomech. Eng.*, 122, 289–292.

22. Woo, B. H., Fink, B. F., Page, R., Schrier, J. A., Jo, Y. W., Jiang, G., DeLuca, M. Vasconez, H. C., and DeLuca, P. P. (2001), Enhancement of bone growth by sustained delivery of recombinant human bone morphogenetic protein-2 in a polymeric matrix, *Pharm. Res.*, 18, 1747–1753.

23. Weber, F. E., Eyrich, G., Gratz, K. W., Maly, F. E., and Sailer, H. F. (2002), Slow and continuous application of human recombinant bone morphogenetic protein via biodegradable poly(lactide-*co*-glycolide) foamspheres, *Int. J. of Oral Maxillofacial Surg.*, 31, 60–65.

24. Bordem, M. D., Khan, Y., Attawia, M., and Laurencin, C. T. (2001), Tissue engineered microsphere-based matrices for bone repair: Design, evaluation and optimisation, *Biomaterials*, 23, 551–559.

25. Lu, L., Stamatas, G. N., and Mikos, A. G. (2000), Controlled release of transforming growth factor beta 1 from biodegradable polymer microparticles, *J. Biomed. Mater. Res.*, 50, 440–451.

26. Fu, K., Griebenow, K., Hsieh, L., Klibanov, A. M., and Langer, R. (1999), FTIR characterization of the secondary structure of proteins encapsulated within PLGA microspheres, *J. Controlled Release*, 58, 357–366.

27. Bezemer, J. M., Radersma, R., Grijpma, D. W., Dijkstra, P. J., van Blitterswijk, C. A., and Feijen, J. (2000), Microspheres for protein delivery prepared from amphiphilic multiblock copolymers. I. Influence of preparation techniques on particle characteristics and protein delivery, *J. Controlled Release*, 67, 233–248.

28. Rosas, J. E., Pedraz, J. L., Hernandez, R. M., Gascon, A. R., Igartua, M., Guzman, F., Rodrignuez, R., Cortes, J., and Patarroyo, M. E. (2002), Remarkably high antibody levels and protection against *P. falciparum* malaria in Aotus monkey after a single immunization of SPf66 encapsulated in PLGA microspheres, *Vaccine*, 20, 1707–1710.

29. Boehm, G., Peyre, M., Sesardie, D., Huskisson, R. J., Mawas, F., Douglas, A., Xing, D., Merkle, H. P., Gander, B., and Johansen, P. (2002), On technological and immunological benefits of multivalent single-injection microspheres vaccines. *Pharm. Res.*, 19, 1330–1336.

30. Sturesson, C., Artursson, P., Ghaderi, R., Johansen, K., Mirazimi, A., Uhnoo, I., Svensson, L., Alberstsson, A. C., and Carlfors, J. (1999), Encapsulation of rotavirus into poly(lactide-*co*-glycolide) microspheres, *J. Controlled Release*, 59, 377–389.

31. Morlock, M., Kissel, T., Li, Y. X., Koll, H., and Winter, G. (1998), Erythropoietin loaded microspheres prepared from biodegradable LPLG-PEO-LPLG triblock copolymers: Protein stabilization and in-vitro release properties, *J. Controlled Release*, 56, 105–115.

32. Zambaux, M. F., Bonneaux, F., Gref, R., Dellacherie, E., and Vigneron, C. (1999), Preparation and characterization of protein C loaded PLA nanoparticles, *J. Controlled Release*, 60, 179–188.

33. Cho, S. W., Song, S. H., and Choi, Y. W. (2000), Effects of solvent selection and fabrication on the characteristics of biodegradable poly(lactide-*co*-glycolide) microspheres containing ovalbumin, *Arch. Pharm. Res.*, 23, 385–390.

34. Li, X., Deng, X., and Huang, Z. (2001), In vitro protein release and degradation of poly-DL-lactide-poly(ethylene glycol) microspheres with entrapped human serum albumin: Quantitative evaluation of the factors involved in protein release phase, *Pharm. Res.*, 18, 117–124.

35. Carrasquillo, K. G., Stanley, A. M., Aponte-Carro, J. C., DeJesus, P., Costantino, H. R., Bosques, C. J., and Griebenow, K. (2001), Non-aqueous encapsulation of excipient stabilized spray freeze dried BSA into poly(lactide-*co*-glycolide) microspheres results in release of native protein, *J. Controlled Release*, 76, 199–208.

36. Lee, V. H. L. (1986), Peptide and protein drug delivery: Opportunities and challenges, *Pharm. Int.*, 7, 208–212.

37. Johnson, O. L., Cleland, J. L., Lee, H. J., Charnis, M., Duenas, E., Jaworowicz, W., Shepard, D., Shihzamani, A., Jones, A. J. S., and Putney, S. D. (1996), A month-long effect from a single injection of microencapsulated human growth hormone, *Nat. Med.*, 2, 795–799.

38. Costantino, H. R., Firouzabadian, L., Hogeland, K., Wu, C., Beganski, C., Carrasquillo, K. G., Cordova, M., Griebenow, K., Zale, S. E., and Tracy, M. A. (2000), Protein spray-freeze drying: Effect of atomization conditions on particle size and stability, *Pharm. Res.*, 17, 1374–1383.

39. Bustami, R. T., Chan, H. K., Dehghani, F., and Foster, N. R. (2000), Generation of microparticles of proteins for aerosol delivery using high pressure modified carbon dioxide, *Pharm. Res.*, 17, 1360–1366.

40. Alder, M., Unger, M., and Lee, G. (2000), Surface composition of spray-dried particles of bovine serum albumin/trehalose/surfactant, *Pharm. Res.*, 17, 863–870.

41. Ruiz, J. M., Tissier, B., and Benoit, J. P. (1989), Microencapsulation of peptides: A study of the phase separation of poly(D,L-lactic acid-*co*-glycolic acid) copolymers 50/50 by silicone oil, *Int. J. Pharm.*, 49, 69–77.

42. Redding, T. W., Schally, A. V., Tice, T. R., and Meyers, W. E. (1984), Long acting delivery systems for peptides: Inhibition of rat prostate tumors by controlled release of DTrp6 leutinizing hormone releasing hormone injectable microcapsules, *Proc. Nat. Acad. Sci. USA*, 81, 5845.

43. Heron, I., Thomas, F., Dero, M., Poutrain, J. R., Henane, S., Catus, F., and Kuhn, J. M. (1993), Traitement de l'acromegalie par une forme a liberation prolongee du lanreotide, un nouvel analogue de la somatostatine, *Presse Med.*, 22, 526–531.

44. O'Donnell, P. B., and McGinity, J. W. (1997), Preparation of microspheres by the solvent evaporation technique, *Adv. Drug Deliv. Rev.*, 28, 25–42.

45. Cleland, J. L. (1998), Solvent evaporation processes for the production of controlled release biodegradable microsphere formulations for therapeutics and vaccines, *Biotechnol. Prog.*, 14, 102–107.

46. Ogawa, Y., Yamamoto, M., Takada, S., and Shimamoto, T. (1988), Controlled release of leuprolide acetate from polylactic acid or copolymer ratio of polymer, *Chem. Pharm. Bull.*, 36, 1502–1507.

47. Ogawa, Y., Yamamoto, M., Okada, H., Yashiki, T., and Shimamoto, T. (1988), A new technique to efficiently entrap leuprolide acetate into microcapsules of polylactic acid or copoly(lactic/glycolic) acid, *Chem. Pharm. Bull.*, 36, 1095–1103.

48. Bittner, B., Morlock, M., Koll, H., Winter, G., and Kissel, T. (1998), Recombinant human erythropoietin (rHEPO) loaded poly(lactide-*co*-glycolide) microspheres: Influence of the encapsulation technique and polymer purity on microspheres characteristics, *Eur. J. Pharm. Biopharm.*, 45, 295–305.

49. Kissel, T., Brich, Z., Bantle, S., Lancranjam, I., NimmerFall, F., and Vit, P. (1991), Parentral depot systems as the basis of biodegradable polyesters, *J. Controlled Release*, 6, 27–34.

50. Herrmann, J., and Bodmeier, R. (1998), Biodegradable, somatostatin acetate containing microspheres prepared by various aqueous and non-aqueous solvent evaporation methods. *Eur. J. Pharm. Biopharm.*, 45, 75–82.

51. Singh, M., Shirley, B., Bajwa, K., Samara, E., Hora, M., and O'Hagan, D. (2001), Controlled release of recombinant insulin-like growth factor from a novel formulation of polylactide-coglycolide microparticles, *J. Controlled Release*, 70, 21–28.

52. Viswanathan, N. B., Thomas, P. A., Pandit, J. K., Kulkarni, M. G., and Mashelkar, R. A. (1999), Preparation of non-porous microspheres with high entrapment efficiency of proteins by a (water-in-oil)-in-oil emulsion technique, *J. Controlled Release*, 58, 9–20.

53. Sanchez, A., Gupta, R. K., Alonso, M. J., Siber, G. R., and Langer, R. (1996), Pulsed controlled-release system for potential use in vaccine delivery, *J. Pharm. Sci.*, 85, 547–552.

54. Freitas, S., Walz, A., Merkle, H. P., and Gander, B. (2003), Solvent extraction employing a static micromixer: A simple, robust and versatile technology for the microencapsulation of proteins, *J. Microencapsul.*, 20, 67–85.

55. Khan, M. A., Healy, M. S., and Bernstein, H. (1992), Low temperature fabrication of protein loaded microspheres, *Proc. Int. Symp. Controlled Release Bioactive Mater.*, 19, 518–519.

56. Herberger, J. D., Wu, C., Dong, N., and Tracy, M. A. (1996), Characterization of Prolease® human growth hormone PLGA microspheres produced using different solvents, *Proc. Int. Symp. Controlled Release Bioactive Mater.*, 23, 835–836.

57. Johnson, O. L., Jaworowicz, W., Cleland, J. L., Bailey, L., Charnis, M., Duenas, E., Wu, C., Shepard, D., Magil, S., Last, T., Jones, A. J. S., and Putney, S. D. (1997), The stabilization and encapsulation of human growth hormone into biodegradable microspheres, *Pharm. Res.*, 14, 730–735.

58. Gombotz, W. R., Healy, M., Brown, L. R., and Auer, H. E. (1990), Process for producing small particles of biologically active molecules, Pabst Patrea, Patent No. WO 90/13285.

59. Gombotz, W. R., Healy, M. S., and Brown, L. R. (1989), Very low temperature casting of controlled release microspheres, Enzytech Inc., Application No. 89-346143, U.S. Patent 5,019,400.

60. Costantino, H. R., Johnson, O. L., and Zale, S. F. (2004), Relationship between encapsulated drug particle size and initial release of recombinant human growth hormone from biodegradable microspheres, *J. Pharm. Sci.*, 93, 2624–2634.

61. Cleland, J. L., and Jones, A. J. S. (1996), Stable formulations of recombinant human growth hormone and interferon-g for microencapsulation in biodegradable microspheres, *Pharm. Res.*, 13, 1464–1475.

62. Cleland, J. L., Duenas, E. T., Park, A., Daugherty, A., Kahn, J., Kowalski, J., and Cuthbertson, A. (2000), Development of poly-(D,L-lactide-*co*-glycolide) microsphere formulations containing recombinant human vascular growth factor to promote local angiogenesis, *J. Controlled Release*, 72, 13–24.

63. Lam, X. M., Duenas, E. T., Daugherty, A. L., Levin, N., and Cleland, J. L. (2000), Sustained release of recombinant human insulin-like growth factor-I for treatment of diabetes. *J. Controlled Release*, 67, 281–292.

64. Ribeiro Dos Santos, I., Richard, J., Pech, B., Thies, C., and Benoit, J. P. (2002), Microencapsulation of protein particles within lipids using a novel supercritical fluid process, *Int. J. Pharm.*, 242, 69–78.

65. Debenedetti, P. G., Tom, J. W., Yeo, S., Lim, G. B. (1993), Application of supercritical fluids for the production of sustained delivery devices, *J. Controlled Release*, 24, 27–44.

66. Randolph, T. W., RAndolph, A. D., Mebes, M., and Yeung, S. (1993), Submicrometer-sized biodegradable particles of poly(L-lactic acid) via the gas antisolvent spray precipitation process, *Biotechnol. Prog.*, 9, 429–435.

67. Waßnus, W., Bleich, J., and Müller, B. W. (1991), Microparticle production by using supercritical gases, *Clin. Pharmacol.*, 13, 367.

68. Bleich, J., Mueller, B. W., and Wassmus, W. (1993), Aerosol solvent extraction system. A new microparticle production technique, *Int. J. Pharm.*, 97, 111–117.

69. Winters, M. A., Debenedetti, P. G., Carey, J., Sparks, H. G., Sane, S. U., and Przybycien, T. M. (1997), Long-term and high-temperature storage of supercritically-processed microparticulate protein powders, *Pharm. Res.*, 14, 1370–1378.

70. Bodmeier, R., Wang, H., Dixon, D. J., Mawson, S., and Johnston, K. P. (1995), Polymeric microspheres prepared by spraying into compressed carbon dioxide, *Pharm. Res.*, 12, 1211–1217.

71. Witschi, C., and Doelker, E. (1998), Peptide degradation during preparation and in vitro release testing of poly(L-lactic acid) and poly(DL-lactic-*co*-glycolic acid) microparticles, *Int. J. Pharm.*, 171, 1–18.

72. Morlock, M., Koll, H., Winter, G., and Kissel, T. (1997), Microencapsulation of rh-erythropoietin using biodegradable poly(D,L-lactide-*co*-glycolide): Protein stability and the effects of stabilizing excipients, *Eur. J. Pharm. Biopharm.*, 43, 29–36.

73. Bittner, B., and Kissel, T. (1999), Ultrasonic atomization for spray drying: A versatile technique for the preparation of protein loaded bio-degradable microspheres, *J. Microencapsul.*, 16, 325–341.

74. Breitenbach, A., Mohr, D., Kissel, T. (2000), Biodegradable semicrystalline comb polymers influence the microsphere production by means of a supercritical fluid extraction technique, *J. Controlled Release*, 63, 53–68.

75. Engwicht, A., Girreser, U., and Muller, B. W. (1999), Critical properties of lactide-*co*-glycolide polymers for the use of microparticle preparation by the aerosol solvent extraction system, *Int. J. Pharm.*, 185, 61–72.

76. Felder, C. B., Blanco-Príeto, M. J., Heizmann, J., Merkle, H. B., and Gander, B. (2003), Ultrasonic atomization and subsequent polymer desolvation for peptide and protein microencapsulation into biodegradable polyesters, *J. Microencapsul.*, 20, 553–567.

77. Freitas, S., Merkle, H. P., and Gander, B. (2004), Ultrasonic atomisation into reduced pressure atmosphere—envisaging aseptic spray drying for microencapsulation, *J. Controlled Release*, 95, 185–195.

78. Berkland, C., Kim, K., and Pack, D. W. (2001), Fabrication of PLG microspheres with precisely controlled and monodispersed size distributions, *J. Controlled Release*, 73, 59–74.

79. Jain, R. A., Rhodes, C. T., Railkar, A. M., Malik, A. W., and Shah, N. H. (2000), Controlled release of drugs from injectable in situ formed biodegradable PLGA microspheres: Effect of various formulation variables, *Eur. J. Pharm. Biopharm.*, 50, 257–262.

80. Kim, H. K., Chung, H. J., and Park, T. G. (2006), Biodegradable polymeric microspheres with "open/closed" pores for sustained release of human growth hormone, *J. Controlled Release*, 112, 167–174.
81. Singh, M., Kazzaz, J., Ugozzoli, M., Chesko, J., and O'Hagan, D. (2004), Charged poly(lactide-*co*-glycolide) microparticles as novel antigen delivery systems, *Expert Opin. Biolog. Ther.*, 4, 483–491.
82. Jilek, S., Merkle, H. P., and Walter, E. (2005), DNA-loaded biodegradable microparticles as vaccine delivery systems and their interaction with dendritic cells, *Adv. Drug Deliv. Rev.*, 57, 377–390.
83. Jabbal-Gill, I., Lin, W., Jenkins, P., Watts, P., Jimenez, M., Illum, L., Davis, S. S., Wood, J. M., Major, D., Minor, P. D., Li, X. W., Lavelle, E. C., and Coombes, A. G. A. (1999), Potential of polymeric lamellar substrate particles (PLSP) as adjuvants for vaccines, *Vaccine*, 18, 238–250.
84. Jabbal-Gill, I., Lin, W., Kistner, O., Davis, S. S., and Illum, L. (2001), Polymeric lamellar substrate particles for intranasal vaccination, *Adv. Drug Deliv. Rev.*, 51, 97–111.
85. Venkataprasad, N., Coombes, A. G. A., Singh, M., Rohde, M., Wilkinson, K., Hudecz, F., Davis, S. S., and Vordermeier, H. M. (1999), Induction of cellular immunity to a mycobacterial antigen adsorbed on lamellar particles of lactide polymers, *Vaccine*, 17, 1814–1819.
86. Kazzaz, J., Neidleman, J., Singh, M., Ott, G., and O'Hagan, D. T. (2000), Novel anionic microparticles are a potent adjuvant for the induction of cytotoxic T lymphocytes against recombinant p55 gag from HIV-1, *J. Controlled Release*, 67, 347–356.
87. Otten, G., Schaefer, M., Greer, C., Calderon-Cacia, M., Coit, D., Kazzaz, J., Medina-Selby, A., Selby, M., Singh, M., Ugozzoli, M., zur Megede, J., Barnett, S. W., O'Hagan, D., Donnelly, J., and Ulmer, J. (2003), Induction of broad and potent antihuman immunodeficiency virus immune responses in rhesus macaques by priming with a DNA vaccine and boosting with protein-adsorbed polylactide coglycolide microparticles, *J. Virol.*, 77, 6087–6092.
88. Kang, F., Jiang, G., Hinderliter, A., DeLuca, P. P., and Singh, J. (2002), Lysozyme stability in primary emulsion for PLGA microsphere preparation: Effect of recovery methods and stabilizing excipients, *Pharm. Res.*, 19, 629–633.
89. Sanchez, A., Villamayor, B., Guo, Y., McIver, J., and Alonso, M. J. (1999), Formulation strategies for the stabilization of tetanus toxoid in poly(lactide-*co*-glycolide) microspheres, *Int. J. Pharm.*, 185, 255–266.
90. Sharif, S., and O'Hagan, D. T. (1995), A comparison of alternative methods for determination of the levels of proteins entrapped in poly(lactide-*co*-glycolide) microparticles, *Int. J. Pharm.*, 115, 259–263.
91. Gupta, R. K., Chang, A. C., Griffin, P., Rivera, R., Guo, Y. Y., and Siber, G. R. (1997), Determination of protein loading in biodegradable polymer microspheres containing tetanus toxoid, *Vaccine*, 15, 672–678.
92. Lu, W., and Park, T. G. (1995), Protein release from poly(lactic-*co*-glycolic acid) microspheres: Protein stability problems, *PDA J. Pharm. Sci. Technol.*, 49, 13–19.
93. Park, T. G., Lu, W., and Crotts, G. (1995), Importance of in vitro experimental conditions on protein release kinetics, stability and polymer degradation in protein encapsulated poly(D,L-lactic acid-*co*-glycolic acid) microspheres, *J. Controlled Release*, 33, 211–222.
94. Bilati, U., Pasquarello, C., Corthals, C. L., Hochstrasser, D. F., Allémann, E., and Doelker, E. (2005), Matrix-assisted laser desorption/ionization time-of-flight mass spectrometry for quantitation and molecular stability assessment of insulin entrapped within PLGA nanoparticles, *J. Pharm. Sci.*, 94, 1–7.

95. van de Weert, M., Van't Hof, R., Van der Weerd, J., Heeren, R. M., Posthuma, G., Hennink, W.E., and Crommelin, D. J. (2000), Lysozyme distribution and conformation in a biodegradable polymer matrix as determined by FTIR techniques, *J. Controlled Release*, 68, 31–40.
96. Blanco, M. D., and Alonso, M. J. (1997), Development and characterization of protein-loaded poly(lactide-*co*-glycolide) nanospheres, *Eur. J. Pharm. Biopharm.*, 43, 287–294.
97. Johansen, P., Tamber, H., Merkle, H. P., and Gander, B. (1999), Diphtheria and tetanus toxoid microencapsulation into conventional and end-group alkylated PLA/PLGAs, *Eur. J. Pharm. Biopharm.*, 47, 193–201.
98. van de Weert, M., Hoechstetter, J., Hennink, W. E., and Crommelin, D. J. (2000), The effect of a water/organic solvent interface on the structural stability of lysozyme, *J. Controlled Release*, 68, 351–359.
99. Yang, T. H., Dong, A., Meyer, J., Johnson, O. L., Cleland, J. L., and Carpenter, J. F. (1999), Use of infrared spectroscopy to assess secondary structure of human growth hormone within biodegradable microspheres, *J. Pharm. Sci.*, 88, 161–165.
100. Perez, C., De Jesus, P., and Griebenow, K. (2002), Preservation of lysozyme structure and function upon encapsulation and release from poly(lactic-*co*-glycolic) acid microspheres prepared by the water-in-oil-in-water method, *Int. J. Pharm.*, 248, 193–206.
101. Jorgensen, L., Vermehren, C., Bjerregaard, S., and Froekjaer, S. (2003), Secondary structure alterations in insulin and growth hormone water-in-oil emulsions, *Int. J. Pharm.*, 254, 7–10.
102. Na, D. H., Youn, Y. S., Lee, S. D., Son, M. W., Kim, W. B., DeLuca, P. P., and Lee, K. C. (2003), Monitoring of peptide acylation inside degrading PLGA microspheres by capillary electrophoresis and MALDI-TOF mass spectrometry, *J. Controlled Release*, 92, 291–299.
103. Gilg, D., Riedl, B., Zier, A., and Zimmermann, M. F. (1996), Analytical methods for the characterization and quality control of pharmaceutical peptides and proteins, *Pharm. Acta Helv.*, 71, 383–394.
104. Brass, J. M., Krummen, K., and Moll-Kaufmann, C. (1996), Quality assurance after process changes of the production of a therapeutic antibody, *Pharm. Acta Helv.*, 71, 395–403.
105. Sesardic, D., and Dobbelaer, R. (2004), European union regulatory developments for new vaccine adjuvants and delivery systems, *Vaccine*, 22, 2452–2456.
106. Burgess, D. J., Crommelin, D. J. A., Hussain, A. S., and Chen, M. L. (2004), Assuring quality and performance of sustained and controlled release parenterals: EUFEPS workshop report, *AAPS PharmSciTech*, 6, 1–12.
107. Burgess, D. J., Hussain, A. S., Ingallinera, T. S., and Chen, M. L. (2002), Assuring quality and performance of sustained and controlled release parenterals: Workshop report, *AAPS PharmSciTech*, 4, (article 7).
108. Gupta, R. K., Chang, A. C., and Siber, G. R. (1998), Biodegradable polymer microspheres as vaccine adjuvants and delivery systems, *Dev. Biol. Stand.*, 92, 63–78.
109. Visscher, G. E., Robison, R. L., Maulding, H. V., Fong, J. W., Pearson, J. E., and Argentieri, G. J. (1985), Biodegradation of and tissue reaction to 50:50 poly(DL-lactide-*co*-glycolide) microcapsules, *J. Biomed. Mater. Res.*, 19, 349–365.
110. Schakenraad, J. M., Hardonk, M. J., Feijen, J., Molenaar, I., and Nieuwenhuis, P. (1990), Enzymatic activity toward poly(L-lactic) acid implants, *J. Biomed. Mater. Res.*, 24, 529–545.
111. Johansen, P., Men, Y., Merkle, H. P., and Gander, B. (2000), Revisiting PLG/PLGA microspheres: An analysis of their potential in parenteral vaccination, *Eur. J. Pharm. Biopharm.*, 50, 129–146.

112. Men, Y., Audran, R., Thomasin, C., Eberl, G., Demotz, S., Merkle, H. P., Gander, B., and Corradin, G. (1999), MHC class I-and class II-restricted processing and presentation of microencapsulated antigens, *Vaccine*, 17, 1047–1056.
113. Raychaudhuri, S., and Rock, K. L. (1998), Fully mobilizing host defense: Building better vaccines, *Nat. Biotechnol.*, 16, 1025–1031.
114. Visscher, G. E., Robinson, R. L., and Argentieri, G. J. (1987), Tissue response t biodegradable injectable microcapsules, *J. Biomater. Applic.*, 2, 118–131.
115. Cleland, J. L., Powell, M. F., and Shire, S. J. (1993), The development of stable protein formulations: A close look at protein aggregation, deamidation, and oxidation, *Crit. Rev. Therap. Drug Carrier Syst.*, 10, 307–377.
116. Gupta, R. K., Alroy, J., Alonso, M. J., Langer, R., and Siber, G. R. (1997), Chronic tissue reactions, long term immunogenicity and immunological priming of mice and guinea pigs to tetanus toxoids encapsulated in biodegradable polymer microspheres composed of polylactide-*co*-glycolide polymers, *Vaccine*, 15, 1716–1723.
117. Jiang, W., Gupta, R. K., Deshpande, M. C., and Schwendeman, S. P. (2005), Biodegradable poly(lactic-*co*-glycolic acid) microparticles for injectable delivery of vaccine antigens, *Adv. Drug Deliv. Rev.*, 57, 391–410.
118. Johnson, O. L., and Tracy, M. A. (1999), Peptide and protein drug delivery, in Mathiowitz, E., Ed., *Encyclopedia of Controlled Drug Delivery*, Vol. 2, Wiley, New York, pp. 816–833.
119. Braun, A., and Alsenz, J. (1997), Development and use of enzyme-linked immunosorbent assays (ELISA) for the detection of protein aggregates in interferon-alpha (IFN-alpha) formulations, *Pharm. Res.*, 14, 1394–1400.
120. Hochuli, E. (1997), Interferon immunogenicity: Technical evaluation of interferon-alpha 2a, *J. Interferon Cytokine Res.*, 17 (Suppl 1), 15–21.
121. Volkin, D. B., and Middaugh, C. R. (1992), in: Ahern, T. J., Manning, M. C., Eds., *Stability of Protein Pharmaceuticals Part A: Chemical and Physical Pathways of Protein Degradation*, Plenum, New York, pp. 109–134.
122. Manning, M. C., Patel, K., and Borchardt, R. T. (1989), Stability of protein pharmaceuticals, *Pharm. Res.*, 6, 903–918.
123. Zale, S. E., and Klibanov, A. M. (1986), Why does ribonuclease irreversibly inactivate at high temperatures? *Biochemistry*, 25, 5432–5443.
124. Jiang, W., and Schwendeman, S. P. (2000), Formaldehyde-mediated aggregation of protein antigens: Comparison of untreated and formalinized model antigens, *Biotechnol. Bioeng.*, 70, 507–517.
125. Jiang, W., and Schwendeman, S. P. (2001), Stabilization of a model formalinized protein antigen encapsulated in poly(lactide-*co*-glycolide)-based microspheres, *J. Pharm. Sci.*, 90, 1558–1569.
126. Zhu, G., Mallery, S. R., and Schwendeman, S. P. (2000), Stabilization of proteins encapsulated in injectable poly(lactide-*co*-glycolide), *Nat. Biotechnol.*, 18, 52–57.
127. Schwendeman, S. P., Cardamone, M., Klibanov, A., and Langer, R. (1996), Stability of proteins and their delivery from biodegradable polymer microspheres, in Cohen, S., and Bernstein, H., Eds., *Microparticulates Systems for the Delivery of Proteins and Vaccines*, Marcel Dekker, New York, pp. 1–49.
128. van de Weert, M., Hennink, W. E., and Jiskoot, W. (2000), Protein instability in poly(lactic-*co*-glycolic acid) microparticles, *Pharm. Res.*, 17, 1159–1167.
129. Schwendeman, S. P. (2002), Recent advances in the stabilization of proteins encapsulated in injectable PLGA delivery systems, *Crit. Rev. Therap. Drug Carrier Syst.*, 19, 73–98.

130. Bilati, U., Allémann, E., and Doelker, E. (2005), Strategic approaches for overcoming peptide and protein instability within biodegradable nano- and microparticles, *Eur. J. Pharm. Biopharm.*, 59, 375–388.

131. Sanders, L., and Hendren. W. (1998), *Protein Delivery: Physical Systems*, Plenum, New York.

132. Senior, J., and Radomsky, M. (2000), *Sustained-Release Injectable Products*, Interpharm, Denver.

133. Akers, M. J. (2002), Excipient-drug interactions in parenteral formulations, *J. Pharm. Sci.*, 91, 2283–2300.

134. Stevenson, C. L. (2000), Characterization of protein and peptide stability and solubility in non-aqueous solvents, *Curr. Pharm. Biotechnol.*, 1, 165–182.

135. Chin, J. T., Wheeler, S. L., and Klibanov, A. M. (1994), On protein solubility in organic solvents, *Biotechnol. Bioeng.*, 44, 140–145.

136. Park, T. G., Lee, H. Y., and Nam, Y. S. (1998), A new preparation method for protein loaded poly(D,L-lactic-*co*-glycolic acid) microspheres and protein release mechanism study, *J. Controlled Release*, 55, 181–191.

137. Yoo, H. S., Choi, H. K., and Park, T.G. (2001), Protein–fatty acid complex for enhanced loading and stability within biodegradable nanoparticles, *J. Pharm. Sci.*, 90, 194–201.

138. Quintanar-Guerrero, D., Allémann, E., Fessi, H., and Doelker, E. (1997), Applications of the ion-pair concept to hydrophilic substances with special emphasis on peptides, *Pharm. Res.*, 14, 119–127.

139. Kim, H. K., and Park, T. G. (2001), Microencapsulation of dissociable human growth hormone aggregates within poly(D,L-lactic-*co*-glycolic acid) microparticles for sustained release, *Int. J. Pharm.*, 229, 107–116.

140. Johnson, M. D., Hoesterey, B. L., and Anderson, B. D. (1995), Solubilization of a tripeptide HIV protease inhibitor using a combination of ionization and complexation with chemically modified cyclodextrins, *J. Pharm. Sci.*, 83, 1142–1146.

141. Brewster, M. E., Simpkins, J. W., Hora, M. S., Stern, W. C., and Bodor, N. (1989), The potential use of cyclodextrins in parenteral formulations, *J. Parenteral Sci. Technol.*, 43, 231–240.

142. Medlicott, N. J., Foster, K. A., Audus, K. L., Gupta, S., and Stella, V. J. (1998), Comparison of the effects of potential parenteral vehicles for poorly water soluble anticancer drugs (organic cosolvents and cyclodextrin solutions) on cultured endothelial cells (HUV-EC), *J. Pharm. Sci.*, 87, 1138–1143.

143. Matsuyama, K., El-Gizway, S., and Perrin, J. H. (1987), Thermodynamics of binding of aromatic amino acids to α, β, and γ cyclodextrins, *Drug Dev. Ind. Pharm.*, 13, 2687–2691.

144. Meinel, L., Illi, O. E., Zapf, J., Malfanti, M., Peter, M. H., and Gander, B. (2001), Stabilizing insulin-like growth factor-I in poly(D,L-lactide-*co*-glycolide) microspheres, *J. Controlled Release*, 70, 193–202.

145. Uchida, T., Shiosaki, K., Nakada, Y., Fukada, K., Eda, Y., Tokiyoshi, S., Nagareya, N., and Matsuyama, K. (1998), Microencapsulation of hepatitis B core antigen for vaccine preparation, *Pharm. Res.*, 15, 1708–1713.

146. Youan, B. B. C., Gillard, J., and Rollmann, B. (1999), Protein-loaded poly(ε-caprolactone) microparticles. III. Entrapment of superoxide dismutase by the (water-in-oil)-in water solvent evaporation method, *STP Pharma Sci.*, 9, 175–181.

147. Johansen, P., Men, Y., Audran, R., Corradin, G., Merkle, H. P., and Gander, B. (1998), Improving stability and release kinetics of microencapsulated tetanus toxoid by co-encapsulation of additives, *Pharm. Res.*, 15, 1103–1110.

148. Krielgaard, L., Jones, L. S., Randolph, T. W., Frokjaer, S., Flink, J. M., Manning, M. C., Carpenter, J. F. (1998), Effect of Tween 20 on freeze-thawing- and agitation-induced aggregation of recombinant human factor XIII, *J. Pharm Sci.*, 87, 1593–1603.

149. Krielgaard, L., Frokjaer, S., Flink, J .M., Randolph, T. W., and Carpenter, J. F. (1999), Effects of additives on the stability of Humicola lanuginosa lipase during freeze-drying and storage in the dried solid, *J. Pharm. Sci.*, 88, 281–290.

150. De Rosa, G., Iommelli, R., La Rotonda, M. I., Miro, A., and Quaglia, F. (2000), Influence of the co-encapsulation of different non-ionic surfactants on the properties of PLGA insulin-loaded microspheres, *J. Controlled Release*, 69, 283–295.

151. Chang, B. S., Kendrick, B. S., and Carpenter, J. F. (1996), Surface-induced denaturation of proteins during freezing and its inhibition by surfactants, *J. Pharm. Sci.*, 85, 1325–1330.

152. Kerwin, B. A., Heller, M. C., Levin, S. H., and Randolph, T. W. (1998), Effects of Tween 80 and sucrose on the acute short term stability and long term stability storage at –20 degrees of a recombinant hemoglobin, *J. Pharm. Sci.*, 87, 1062–1068.

153. Knepp, V. M., Whatley, J. L., Muchnik, A., and Calderwood, T. S. (1996), Identification of antioxidants for prevention of peroxide-mediated oxidation of recombinant human ciliary neurotrophic factor and recombinant human nerve growth factor, *J. Parenteral Sci. Technol.*, 50, 163–171.

154. Nace, V. (1996), *Nonionic Surfactants: Polyoxyalkylene Block Copolymers*, Marcel Dekker, New York.

155. Kabanov, A., Batrakova, E., and Alakhov, V. (2002), Pluronic block copolymers as novel polymer therapeutics for drug and gene delivery, *J. Controlled Release*, 82, 189–212.

156. U.S. Food and Drug Administration (2003, Nov. 7), Inactive ingredient search for approved drug products, "poloxamer," available http:/www.accessdata.fda.gov/scripts/cder/iig/index.cfm, accessed May 21, 2006.

157. Johnston, T. P., and Miller, S. C. (1985), Toxicological evaluation of poloxamer vehicles for intramuscular use, *J. Parenteral Sci. Technol.*, 39, 83–88.

158. Sanchez, A., Tobio, M., Gonzalez, L., Fabra, A., and Alonso, M. J. (2003), Biodegradable micro- and nanoparticles as long-term delivery vehicles for interferon-α, *Eur. J. Pharm. Sci.*, 18, 221–229.

159. Lam, X. M., Duenas, E. T., and Cleland, J. L. (2001), Encapsulation and stabilization of nerve growth factor into poly(lactic-co-glycolic) acid microspheres, *J. Pharm. Sci.*, 90, 1356–1365.

160. Nihant, N., Schugens, C., Grandfils, C., Jerôme, R., and Teyssie, P. (1994), Polylactide microparticles prepared by double emulsion/evaporation technique: I. Effect of primary emulsion stability, *Pharm. Res.*, 11, 1479–1484.

161. Sturesson, C., and Carlfors, J. (2000), Incorporation of protein in PLG-microspheres with retention of bioactivity, *J. Controlled Release*, 67, 171–178.

162. Chen, L., Apte, R. N., and Cohen, S. (1997), Characterization of PLGA microspheres for the controlled delivery of IL-1a for tumor immunotherapy, *J. Controlled Release*, 43, 261–272.

163. Kwon, Y. M., Baudys, M., Knutson, K., and Kim, S. W. (2001), In situ study of insulin aggregation induced by water–organic solvent interface, *Pharm. Res.*, 18, 1754–1759.

164. Costantino, H. R., Langer, R., and Klibanov, A. (1995), Aggregation of a lyophilized pharmaceutical protein, recombinant human albumin: effect of moisture and stabilization excipients, *BioTechnology*, 13, 493–496.

165. Costantino, H. R., Langer, R., and Klibanov, A. M. (1994), Solid-phase aggregation of proteins under pharmaceutically relevant conditions, *J. Pharm. Sci.*, 83, 1662–1669.

166. Means, G. E., and Feeney, R. E. (1971), *Chemical Modification of Proteins*, Holden-Day, San Francisco.
167. Schwendeman, S. P., Costantino, H. R., Gupta, R. K., Siber, G. R., Klibanov, A. M., and Langer, R. (1995), Stabilization of tetanus and diphtheria toxoids against moisture-induced aggregation, *Proc. Nat. Acad. Sci. USA*, 92, 11234–11238.
168. Costantino, H. R., Schwendeman, S. P., Griebenow, K., Klibanov, A. M., and Langer, R. (1996), The secondary structure and aggregation of lyophilized tetanus toxoid, *J Pharm. Sci.*, 85, 1290–1293.
169. Hageman, M. J. (1988), The role of moisture in protein stability, *Drug Dev. Ind. Pharm.*, 14, 2047–2070.
170. Lee, H. J., Riley, G., Johnson, O., Cleland, J. L., Kim, N., and Charnis, M. (1997), In vivo characterization of sustained release formulations of human growth hormone, *J. Pharmacol. Exp. Ther.*, 281, 1431–1439.
171. Herbert, P., Murphy, K., Johnson, O., Dong, N., Jaworowicz, W., and Tracy, M. A. (1998), A large-scale process to produce microencapsulated proteins, *Pharm. Res.*, 15, 357–361.
172. Tracy, M. A., Bernstein, H., and Kahn, M. A. (2000), Controlled release of metal cation-stabilized interferon, U.S. Patent 6,165,508.
173. Lam, X. M., Duenas, E. T., and Cleland, J. L. (1998), Stabilization of nerve growth factor during microencapsulation and release from microspheres [abstract], *Proc. Int. Symp. Controlled Release Bioactive Mater.*, 25, 491.
174. Zale, S. E., Burke, P. A., Bernstein, H., and Brickner, A. (1997), Composition for sustained release of non-aggregated erythropoietin, U.S. Patent 5,674,534.
175. Johnson, O. L. (2000), Stabilization of proteins in solutions by lyophilization, in Wise, D. E., *Handbook of Pharmaceutical Controlled Release Technology*, Marcel Dekker, New York, pp. 693–749.
176. Carpenter, J. F., and Chang, B. S. (1996), Lyophilization of protein pharmaceuticals, in Avis, K. E., and Wu, V. L., Eds., *Biotechnology and Biopharmaceutical Manufacturing, Processing, and Preservation*, Interpharm, Buffalo Grove, IL, pp. 199–264.
177. Carpenter, J. F., and Crowe, J. H. (1988), The mechanism of cryoprotection of proteins by solutes, *Cryobiology*, 25, 244–255.
178. Arakawa, T., Prestrelski, S., Kinney, W., and Carpenter, J. F. (1993), Factors affecting short-term and long-term stabilities of proteins, *Adv. Drug Deliv. Rew.*, 10, 1–28.
179. Franks, F. (1990), Freeze drying: From empiricism to predictability, *Cryoletters* 11, 93–110.
180. Prestrelski, S. J., Tedeschi, N., Arakawa, T., and Carpenter, J. F. (1993), Dehydration-induced conformational changes in proteins and their inhibition by stabilizers, *Biophys. J.*, 65, 661–671.
181. Tabata, Y., Gutta, S., and Langer, R. (1993), Controlled delivery systems for proteins using polyanhydride microspheres, *Pharm. Res.*, 10, 487–496.
182. Péan, J. M., Boury, F., Venier-Julienne, M. C., Menei, P., Proust, J. E., and Benoit, J. P. (1999), Why does PEG 400 co-encapsulation improve NGF stability and release from PLGA biodegradable microspheres? *Pharm. Res.*, 16, 1294–1299.
183. Perez-Rodriguez, C., Montano, N., Gonzalez, K., and Griebenow, K. (2003), Stabilization of a-chymotrypsin at the CH_2Cl_2/water interface and upon water-in-oil-in-water encapsulation in PLGA microspheres, *J. Controlled Release*, 89, 71–85.
184. Izutsu, K., Yoshioka, S., and Teroa, T. (1993), Decreased protein-stabilizing effects of cryoprotectants due to crystallization, *Pharm. Res.*, 10, 1232–1237.
185. Izutsu, K., Yoshioka, S., and Teroa, T. (1994), Effect of mannitol crystallinity on the stabilization of enzymes during freeze drying, *Chem. Pharm. Bull.*, 42, 5–8.

186. Fakes, M. G., Dali, M. V., Haby, T. A., Morris, K. R., Varia, S. A., and Serajuddin, A. T. M. (2000), Moisture sorption behavior of selected bulking agents used in lyophilized products, *PDA J. Pharm. Sci. Technol.*, 54, 144–149.

187. Kovalcik, T. R., and Guillory, J. K. (1988), The stability of cyclophosphamide in lyophilized cakes. Part I. Mannitol, lactose, and sodium bicarbonate as excipients, *J. Parenterol Sci. Technol.*, 43, 80–83.

188. Costantino, H. R., Carrasquillo, K. G., Cordero, R. A., Mumenthaler, M., Hsu, C. C., and Griebenow, K. (1998), Effect of excipients on the stability and structure of lyophilized recombinant human growth hormone, *J. Pharm. Sci.*, 87, 1412–1420.

189. Li, S., Patapoff, T. W., Overcashier, D., Hsu, C., Nguyen, T. H., and Borchardt, R. T. (1996), Effects of reducing sugars on the chemical stability of human relaxin in the lyophilized state, *J. Pharm. Sci.*, 85, 873–877.

190. Hartauer, K. J., and Guillory, J. K. (1991), A comparison of diffuse reflectance FT-IR spectroscopy and DSC in the characterization of a drug-excipient interaction, *Drug Dev. Ind. Pharm.*, 17, 617–630.

191. Wang, J., Wang, B. M., and Schwendeman, S. P. (2004), Mechanistic evaluation of the glucose-induced reduction in initial burst release of octreotide acetate from poly(D,L-lactide-*co*-glycolide) microspheres, *Biomaterials*, 25, 1919–1927.

192. Dubost, D. C., Kaufman, M. J., Zimmerman, J. A., Bogusky, M. J., Coddington, A. B., and Pitzenberger, S. M. (1996), Characterization of a solid state reaction product from a lyophilized formulation of a cyclic heptapeptide: A novel example of an excipient-induced oxidation, *Pharm. Res.*, 13, 1811–1814.

193. Williams, N. A., Lee, Y., Polli, G. P., and Jennings, T. A. (1986), The effects of cooling rate on solid phase transitions and associated vial breakage occurring in frozen mannitol solutions, *J. Parenteral Sci. Technol.*, 40, 135–141.

194. Williams, N. A., and Dean, T. (1991), Vial breakage by frozen mannitol solutions: Correlation with thermal characteristics and effect of stereoisomerism, additives, and vial configuration, *J. Parenteral Sci. Technol.*, 45, 94–100.

195. Allison, S. D., Manning, M. C., Randolph, T. W., Middleton, K., Davis, A., and Carpenter, J. F. (2000), Optimization of storage stability of lyophilized actin using combinations of disaccharides and dextran, *J. Pharm. Sci.*, 89, 199–214.

196. Cleland, J. L., Lam, X., Kendrick, B., Yang, J., Yang, T., Overcashier, D., Brooks, D., Hsu, C., and Carpenter, J. F. (2001), A specific molar ratio of stabilizer to protein is required for storage stability of a lyophilized monoclonal antibody, *J. Pharm. Sci.*, 90, 310–321.

197. Lam, X. M., Patapoff, T. W., and Nguyen, T. H. (1997), Effect of benzyl alcohol on stability of recombinant human interferon gamma, *Pharm. Res.*, 14, 725–729.

198. Chen, B. L., and Arakawa, T. (1996), Stabilization of recombinant human keratinocyte growth factor by osmolytes and salts, *J. Pharm. Sci.*, 85, 419–422.

199. Arakawa, T., Kita, Y., and Carpenter, J. F. (1991), Protein–solvent interactions in pharmaceutical formulations, *Pharm. Res.*, 8, 285–291.

200. Zhou, Y., and Hall, C. K. (1996), Solute excluded-volume effects on the stability of globular proteins: A statistical thermodynamic theory, *Biopolymers*, 38, 273–284.

201. Sugimoto, L., Ishihara, T., Habata, H., and Nakagawa, H. (1981), Stability of lyophilized sodium prasterone sulfate, *J. Parenteral Sci. Techn.*, 35, 88–92.

202. Shao, P. G., and Bailey, L. C. (2000), Porcine insulin biodegradable polyester microspheres: Stability and in vitro release characteristics, *Pharm. Dev. Technol.*, 5, 1–9.

203. Mäder, K., Bittner, B., Li, Y., Wohlauf, W., and Kissel, T. (1998), Monitoring microviscosity and microacidity of the albumin microenvironment inside degrading microparticles from poly(lactide-coglycolide) (PLG) or ABA-triblock polymers containing hydropho-

bic poly(lactide-*co*-glycolide) A blocks and hydrophilic poly(ethyleneoxide) B blocks, *Pharm. Res.*, 15, 787–793.

204. Brunner, A., Mader, K., and Gopferich, A. (1999), pH and osmotic pressure inside biodegradable microspheres during erosion, *Pharm. Res.*, 16, 847–853.

205. Burke, P. A. (1996), Determination of internal pH in PLGA microspheres using ^{31}P NMR spectroscopy, *Proc. Inter. Symp. Controlled Release Bioactive Mater.*, 23, 133–134.

206. Fu, K., Pack, D. W., Klibanov, A. M., and Langer, R. (2000), Visual evidence of acidic environment within degrading poly(lactic-*co*-glycolic acid) (PLGA) microspheres, *Pharm. Res.*, 17, 100–106.

207. Shenderova, A., Burke, T. G., and Schwendeman, S. P. (1999), The acidic microclimate in poly(lactide-*co*-glycolide) microspheres stabilizes camptothecins, *Pharm. Res.*, 16, 241–248.

208. Li, L., and Schwendeman, S. P. (2005), Mapping neutral microclimate pH in PLGA microspheres, *J. Controlled Release*, 101, 163–173.

209. Ding, A. G., and Schwendeman, S. P. (2004), Determination of water-soluble acid distribution in poly(lactide-*co*-glycolide), *J. Pharm. Sci.*, 93, 322–331.

210. Cleland, J. L., Mac, A., Boyd, B., Yang, J., Hsu, C., Chu, H., Mukku, V., and Jones, A. J. S. (1997), The stability of recombinant human growth hormone in poly(lactic-*co*-glycolic acid) (PLGA) microspheres, *Pharm. Res.*, 14, 420–425.

211. Cleland, J. L., Duenas, E., Daugherty, A., Marian, M., Yang, J., Wilson, M., Celniker, A. C., Shahzamani, A., Quarmby, V., Chu, H., Mukku, V., Mac, A., Roussakis, M., Gillette, N., Boyd, B., Yeung, D., Brooks, D., Maa, Y. F., Hsu, C., and Jones, A. J. S. (1997), Recombinant human growth hormone poly(lactic-*co*-glycolic acid) (PLGA) microspheres provide a long lasting effect, *J. Controlled Release*, 49, 193–205.

212. Schwendeman, S. P. (2001), Stabilization of vaccine antigens encapsulated in PLGA microspheres [abstract 320], *Proc. Int. Symp. Controlled Release Bioactive Mater.*, 28.

213. Jiang, W., and Schwendeman, S. P. (2001), Stabilization and controlled release of bovine serum albumin encapsulated in poly(D,L-lactide) and poly(ethylene glycol) microsphere blends, *Pharm. Res.*, 18, 878–885.

214. Bittner, B., Witt, C., Mäder, K., and Kissel, T. (1999), Degradation and protein release properties of microspheres prepared from biodegradable poly(lactide-*co*-glycolide) and ABA triblock copolymers: Influence of buffer media on polymer erosion and bovine serum albumin release, *J. Controlled Release*, 60, 297–309.

215. Shenderova, A., Zhu, G., and Schwendeman, S. P. (2000), Correlation of measured microclimate pH with the stability of BSA encapsulated in PLGA microspheres abstract 0413], *Proc. Int. Symp. Controlled Release Bioactive Mater.*, 27.

216. Shenderova, A. (2000), The microclimate in poly(lactide-*co*-glycolide) microspheres and its effect on the stability of encapsulated substances, Ph.D thesis, The Ohio State University.

217. Hunter, S. K., andracki, M. E., and Kreig, A. M. (2001), Biodegradable microspheres containing group B Streptococcus vaccine: Immune response in mice, *Am. J. Obstet. Gynecol.*, 185, 1174–1179.

218. Johansen, P., Moon, L., Tamber, H., Merkle, H. P., Gander, B., and Sesardic, D. (1999), Immunogenicity of single-dose diphtheria vaccines based on PLA/PLGA microspheres in guinea pigs, *Vaccine*, 18, 209–215.

219. Peyre, M., Sesardic, D., Merkle, H. P., Gander, B., and Johansen, P. (2003), An experimental divalent vaccine based on biodegradable microspheres induces protective immunity against tetanus and diphtheria, *J. Pharm. Sci.*, 92, 957–966.

220. Peyre, M., Fleck, R., Hockley, D., Gander, B., and Sesardic, D. (2004), In vivo uptake of an experimental microencapsulated diphtheria vaccine following sub-cutaneous immunisation, *Vaccine*, 22, 2430–2437.
221. Namura, J. A. M., Takata, C. S., Moroc, A. M., Politi, M. J., de Araujo, S., Cuccovia, I. M., and Bueno da Costa, M. H. (2004), Lactic acid triggers, in vitro, thiomersal to degrade protein in the presence of PLGA microspheres, *Int. J. Pharm.*, 273, 1–8.
222. Tan, M., and Parkin, J. E. (2000), Route of decomposition of thiomersal (Thimerosal®), *Int. J. Pharm.*, 208, 23–34.
223. Jung, T., Koneberg, R., Hungerer, K. D., and Kissel, T. (2002), Tetanus toxoid microspheres consisting of biodegradable poly(lactide-*co*-glycolide)- and ABA-triblock-copolymers: Immune response in mice, *Int. J. Pharm.*, 234, 75–90.
224. Determan, A. S., Wilson, J. H., Kipper, M. J., Wannemuehler, M. J., and Narasimhan, B. (2006), Protein stability in the presence of polymer degradation products: Consequences for controlled release formulations, *Biomaterials*, 27, 3312–3320.
225. Jaganathan, K. S., Rao, Y. U. B., Singh, P., Prabakaran, D., Gupta, S., Jain, A., and Vyas, S. P. (2005), Development of a single dose tetanus toxoid formulation based on polymeric microspheres: a comparative study of poly(D,L-lactic-*co*-glycolic acid) versus chitosan microspheres, *Int. J. Pharm.*, 294, 23–32.
226. Kipper, M. J., Wilson, J. H., Wannemuehler, M. J., and Narasimhan, B. (2006), Single dose vaccine based on biodegradable polyanhydride microspheres can modulate immune response mechanism, *J. Biomed. Mater. Res. Part A*, 76, 798–810.
227. Katare, Y. K., Muthukumaran, T., and Panda, A. K. (2005), Influence of particle size, antigen load, dose and additional adjuvant on the immune response from antigen loaded PLA microparticles, *Int. J. Pharm.*, 301, 149–160.
228. Yeh, M. K., Liu, Y. T., Chen, J. L., and Chiang, C. H. (2002), Oral immunogenicity of the inactivated *Vibrio cholerae* whole cell vaccine encapsulated in biodegradable microparticles, *J. Controlled Release*, 82, 237–247.
229. Yeh, M. K., Chen, J. L., and Chiang, C. H. (2002), Vibrio cholerae-loaded poly(DL-lactide-*co*-glycolide) microparticles, *J. Microencapsul.*, 19, 203–212.
230. Yeh, M. K., and Chiang, C. H. (2004), Inactive *Vibrio cholerae* whole-cell vaccine-loaded biodegradable microparticles: In vitro release and oral vaccination, *J. Microencapsul.*, 21, 91–106.
231. Khang, G., Cho, J. C., Lee, J. W., Rhee, J. M., and Lee, H. B. (1999), Preparation and characterization of Japanese encephalitis virus vaccine loaded poly(L-lactide-*co*-glycolide) microspheres for oral immunization, *Biomed. Mater. Eng.*, 9, 49–59.
232. Kaur, R., Rauthan, M., and Vrati, S. (2004), Immunogenicity in mice of a cationic microparticle-adsorbed plasmid DNA encoding Japanese encephalitis virus envelope protein, *Vaccine*, 22, 2776–2782.
233. Yeh, M. K., Coombes, A. G. A., Chen, J. L., and Chiang, C. H. (2002), Japanese encephalitis virus vaccine formulations using PLA lamellar and PLG microparticles, *J. Microencapsul.*, 19, 671–682.
234. Feng, L., Qi, X. R., Zhou, X. J., Maitani, Y., Wang, S. C., Jiang, Y., and Nagai, T. (2006), Pharmaceutical and immunological evaluation of a single-dose hepatitis B vaccine using PLGA microspheres, *J. Controlled Release*, 112, 35–42.
235. Jaganathan, K. S., Singh, P., and Prabakaran, D. (2004), Development of a single-dose stabilized poly(D,L-lactide-*co*-glycolide) microspheres-based vaccine against hepatitis B, *J. Pharm. Pharmacol.*, 56, 1243–1250.
236. Shi, L., Caulfield, M. J., Chern, R. T., Wilson, R. A., Sanyal, G., and Volkin, D. B. (2002), Pharmaceutical and immunological evaluation of a single-shot hepatitis B vaccine formulated with PLGA microspheres, *J. Pharm. Sci.*, 91, 1020–1035.

237. Singh, M., Li, X. M., McGee, J. P., Zamb, T., Koff, W., Wang, C. Y., and O'Hagan, D. T. (1997), Controlled release microparticles as a single dose hepatitis B vaccine: Evaluation of immunogenicity in mice, *Vaccine*, 15, 475–481.

238. Genta, I., Perugini, P., Pavanetto, F., Maculotti, K., Modena, T., Casado, B., Lupi, A., Iadarola, P., and Conti, B. (2001), Enzyme loaded biodegradable microspheres in vitro ex vivo evaluation, *J. Controlled Release*, 77, 287–295.

239. Perugini, P., Genta, I., Pavanetto, F., Modena, T., Maculotti, K., and Conti, B. (2002), Evaluation of enzyme stability during preparation of polylactide-*co*-glycolide microspheres, *J. Microencapsul.*, 19, 591–602.

240. Lupi, A., Perugini, P., Genta, I., Modena, T., Conti, B., Casado, B., Cetta, G., Pavanetto, F., and Iadarola, P. (2004), Biodegradable microspheres for prolidase delivery to human cultured fibroblasts, *J. Pharm. Pharmacol.*, 56, 597–603.

241. Morimoto, K., Chono, S., Kosai, T., Seki, T., and Tabata, Y. (2005), Design of novel injectable cationic microspheres based on aminated gelatin for prolonged insulin action, *J. Pharm. Pharmacol.*, 57, 839–844.

242. Hinds, K. D., Campbell, K. M., Holland, K. M., Lewis, D. H., Piché, C. A., and Schmidt, P. G. (2005), PEGylated insulin in PLGA microparticles. In vivo and in vitro analysis, *J. Controlled Release*, 104, 447–460.

243. Shenoy, D. B., D'Souza, R. J., Tiwari, S. B., and Udupa, N. (2003), Potential applications of polymeric microsphere suspension as subcutaneous depot for insulin, *Drug Dev. Ind. Pharm.*, 29, 555–563.

244. Zhou, S., Deng, X., He, S., Li, X., Jin, W., Wei, D., Zhang, Z., and Ma, J. (2002), Study on biodegradable microspheres containing recombinant interferon-alpha-2a, *J. Pharm. Pharmacol.*, 54, 1287–1292.

245. Luan, X., and Bodmeier, R. (2006), Modification of the tri-phasic drug release pattern of leuprolide acetate-loaded poly(lactide-*co*-glycolide) microparticles, *Eur. J. Pharm. Biopharm.*, 63, 205–214.

246. Wang, Y., Gao, J. Q., Chen, H. L., Zheng, C. H., and Liang, W. Q. (2006), Pluronic F127 gel effectively controls the burst release of drug from PLGA microspheres, *Die Pharma.*, 61, 367–368.

5.3

LIPOSOMES AND DRUG DELIVERY

Sophia G. Antimisiaris,[1] Paraskevi Kallinteri,[2] and
Dimitrios G. Fatouros[3]

[1]*School of Pharmacy, University of Patras, Rio, Greece*
[2]*Medway School of Pharmacy, Universities of Greenwich and Kent, England*
[3]*School of Pharmacy and Biomedical Sciences, Portsmouth, England*

Contents

5.3.1 INTRODUCTION

Liposomes are vesicles in which an aqueous volume is entirely surrounded by a phospholipid membrane and their size can range between 30 and 50 nm up to several micrometers. They can consist of one (unilamellar) or more (multilamellar) homocentric bilayers of amphipathic lipids (mainly phospholipids). Based on their lamellarity (number of lamellae)—and size—they are characterized as SUVs/LUVs

Pharmaceutical Manufacturing Handbook: Production and Processes, edited by Shayne Cox Gad

(small or large unilamellar vesicles) or MLVs (multilamellar vesicles). MLV liposomes are always large (at least cannot be considered small) and aqueous spaces exist in their center and also between their bilayers.

Liposomes were initially invented by Alec Bangham [1] to serve as a model for cell membranes in biophysical studies. In the 1970s they started to be investigated as promising drug carriers [2, 3]. The main advantages of liposomes as a drug delivery system are the following: (i) They have a very versatile structure which can be easily tairored in order to bear the properties needed for each specific application. (ii) They can accommodate any type of drug molecules either in their aqueous compartments (hydrophilic drugs) or in their bilayers (lipophilic drugs) or both (amphiphilic drugs). (iii) Last, but not least, they are nontoxic, nonimmunogenic, and fully biodegradable. Early attempts to use liposomes as a drug delivery system revealed the main limitation of the system and resulted in disappointment due to the fast nonspecifc clearance of liposomes from circulation by reticuloendothelial system (RES) cells [4]. However, it was realized that recognition by the RES macrophage system could be useful in antigen presentation, macrophage activation or killing, and elimination of parasitic infections (what is called passive targeting). From then on many different types of liposomes or methods to obtain longer circulation half-lives have been successfully invented by controlling the physicochemical characteristics of lipid bilayers and their interaction with the biological environment. Some of the basic methods to control the properties of liposomes are presented in Table 1. Today (Figure 1), several different liposome types are available, mainly conventional, long circulating (or "stealth," sterically stabilized, or PEGylated), targeted [or ligand bearing or immunoliposomes (when antibodies are used as targeting ligands)], cationic (for genetic material delivery), and deformable or elastic (or "transferosomes") [will be discussed in the paragraph about of skin delivery in Section 5.3.3.3]. As a consequence of these advances in liposome technology, a very broad range of liposome applications for drug delivery are being explored, some of which have resulted in life-saving products on the market or under late-stage clinical testing.

In this chapter we will provide information about the basic characteristics of liposomes staring from their building blocks, that is, phospholipids. After this, liposome structure, physicochemical properties, and stability, which are most important for their in vivo performance, will be discussed as well as methods used for liposome preparation, characterization, and stabilization. Following this first part which is more technological, we will move into the biological part and talk about the fate of conventional liposomes and sterically stabilized liposomes, as well as liposomal drugs, after in vivo administration by different routes [mainly intravenous (i.v.), intraperitoneal (i.p.), or subcutaneous (s.c.)] and also give some information about other possible routes for in vivo administration of liposomes. Finally, specific applications of liposomes in therapeutics will be presented, some in more detail, mainly for the therapy of different types of cancer.

5.3.2 LIPOSOME STRUCTURE AND CHARACTERISTICS

The main building blocks of most liposomal drug formulations are phospholipids. This chapter will start with an introduction of phospholipid structure and briefly

TABLE 1 Methods and Results of Modifying Physicochemical Properties of Liposomes

Properties	Method	Result
Physicochemical properties		
Size distribution	Sonication, extrusion, microfluidization	Control of circulation time, Increased extravasation
Membrane permeability	Lipid composition modification (cholesterol, thermotropic transitions)	Increased liposome stability, proton or temperature-induced sensitivity
Tendency for aggregation or fusion	Lipid composition modification, addition of cations	Formation of structures (as cochleates) that can carry antigens, deoxyribonucleic acid (DNA), vaccine formulations
Surface hydrophilicity	Steric stabilization by grafting hydrophilic molecules on liposome surface (linear dextrans [5], sialic acid–containing gangliosides [6], lipid derivatives of hydrophilic polymers [as PEG [7, 8], poly-*N*-vinylpyrrolidones [9], and polyvinyl alcohol [10])	Increase of circulation time, modification of pharmacokinetics and tissue dispotition of liposomes and encapsulated drugs
Drug encapsulation efficiency	Active or remote loading [11]	Stable liposome encapsulation of drugs at high drug-to-lipid ratios, Many applications in drug delivery
Elasticity, rigidity	Introduction of detergents or edge activators and "skin lipids" (as ceramides) in liposome preparations [12]	Increased skin penetration and retention in skin which results in increased transdermal absortption of liposomal drugs
Surface properties, liposome stability, and other properties	Addition of spefific ligand on liposome surface, together with steric stabilization	Increased potential for efficient targeting

describe the most important (for liposomes) physichochemical characteristics of lipids and lipid membranes. Afterward properties of liposomes and finally methods used for the preparation of liposomal drug formulations will be discussed.

5.3.2.1 Phospholipids: Structure Stability and Characterization of Lipid Membranes

Phospholipids are naturally occurring biomacromolecules that play an important role in the physiology of humans as they serve as structural components of biological membranes and support organisms with energy [13]. They are amphiphilic molecules with poor aqueous solubility and typically consist of two parts: a water-soluble group, the so called polar head, and an insoluble one, the backbone (Figure 2). The polar head group contains hydroxyl groups which are responsible for the surface

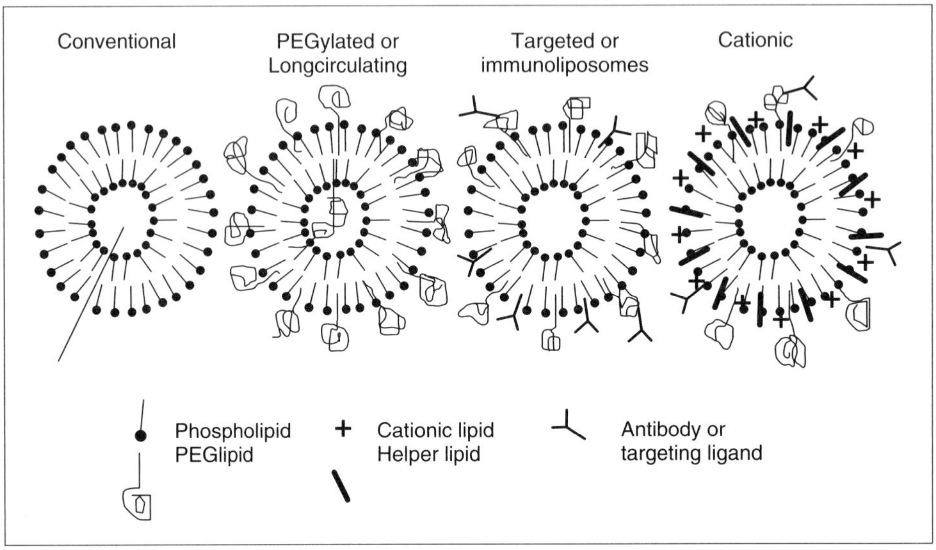

FIGURE 1 Liposome types. *Conventional liposomes* are composed of phospholipids that form bilayers enclosing an aqueous compartment. Cholesterol may be included in the bilayer to increase membrane rigidity. Hydrophilic drugs can be encapsulated in the aqueous interior of the vesicles and lipophilic drugs can be included in their membranes. *Pegylated or long-circulating liposomes* have a surface coating of polyethylene glycol (PEG) molecules that permits liposomes to escape opsonization (coating with plasma proteins—opsonins—that make liposomes visible by RES macrophages). PEG-conjugated lipids are used for the preparation of this type of vesicle. *Targeted liposomes* or immunoliposomes are liposomes that in addition to a PEG coating (in most cases) have targeting moiety on their surface that directs them to the preferred target. This targeting moiety may be a sugar (i.e., galactose, to target cells with galactose receptors on their membranes) or other type of molecule or an antibody (usually monoclonal antibody), in which case the liposomes are characterized as *immunoliposomes. Cationic liposomes* are vesicles that consist of positively charged lipids (cationic lipids) which may form complexes with negatively charged DNA molecules and thus are used for gene delivery or targeting applications. For their preparation an additional lipid (helper lipid) is usually required. Cationic liposomes can also have PEG molecules on their surface (for longer circulation in the bloodstream) and/or targeting moieties. The last type of liposome is the so called *transformable liposome* or *elastic liposome*. (structure is presented and explained in Figure 10).

charge of the lipids that can be positively or negatively charged, zwitterionic, or noncharged. During liposome formation, these molecules arrange themselves by exposing their polar parts toward the water phase, while the hydrocarbon moieties (hydrophobic) adhere together in the bilayer. Two classes of lipids are mainly used for liposome preparation: double-chain polar lipids and sterols (mainly cholesterol). Such lipids form bilayers, in contrast to single-chain lipids (e.g., short-chain phosphatidylcholine) that form micelles, upon their dispersion in water. [14].

Double-chain lipids are either naturally occurring or synthesized in the laboratory. They consist of glycerol or sphingosine and a polar head containing a phosphor- or glyco-group. Glycerophopsholipids or phospholipids are the most popular, among the other lipids, for the preparation of liposomal dispersions [15]. Phosphatidylcho-

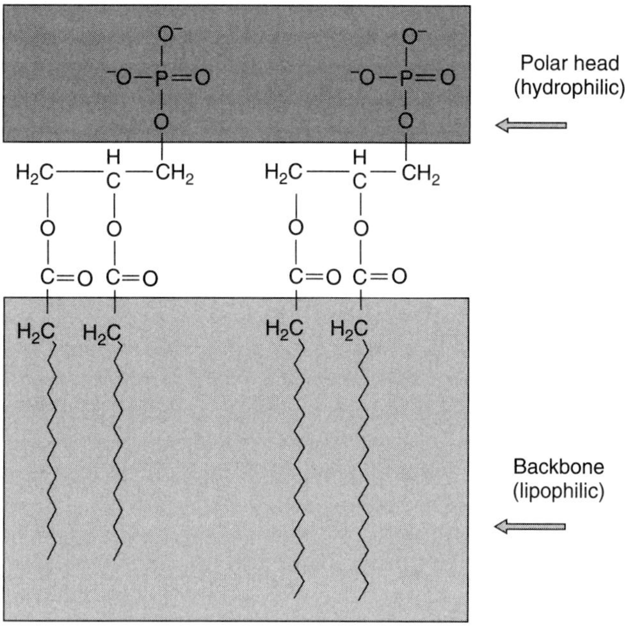

FIGURE 2 Typical phospholipid structure.

line (PC) or lecithin, one of the main components of the liposomal bilayer, belongs to this group [1]. Phosphatidylcholine is zwitterionic in the range of physiological pH [16] and can be found in egg yolk. Other naturally occurring glycerophospholipids are the following: phosphatidylethanolamine (PE), which is isolated from brain lipids and is zwitterionic; phosphatidylserine (PS), which is found in bovine brains and posseses a negative charge; cardiolipin (CL), which is isolated from heart tissue or mitochondrial membranes and is negatively charged; phosphatidylglycerol (PG), which can be found in mitochondria or chloroplasts of mammalian cells and is negatively charged; and finally phosphatidylinositol (PI), a negatively charged lipid found in mammalian tissues [13].

Another group of naturally occurring lipids with applications in liposome technology is comprised of the sphingophospholipids (mainly sphingomyelin) which are derivatives of ceramides [17]. Sphingomyelin (SM) is found in the outer leaflet of plasma membranes [17] and has many similarities with PC since they both have the same zwitterionic polar group and two hydrophobic acyl chains.

Custom-made lipids can be produced by de- or reacylation of natural lipids. Commonly used phospholipids with polar heads containing myristoyl (14:0), palmitoyl (16:0), stearoyl (18:0) fatty acids are all classified by four-letter abbreviations, for example, DMPC, where DM stands for the number and type of fatty acids (di-myristoyl) and PC for the type of polar head (phosphatidylglycero-choline), and similarly DPPC and DSPC (Figure 2).

Positively charged lipids are capable of making complexes with deoxyribonucleic acid (DNA) (since it is negatively charged) and are currently very popular. Examples of such lipids are: *N*-[1-(2,3-dioleyloxy)propyl]-*N*,*N*,*N*-trimethylammonium chloride (DOTMA), 1,2-dioleoyl-3-trimethylammoniopropane (DOTAP), and dicetyl phosphate (DCP), and dioctadecyldimethylammonium bromide (DODAB).

The last group of amphiphiles contains sterols that are present in the membranes of cells. The most popular among them is cholesterol (Chol), which can be easily incorporated in lipid bilayers, increasing their rigidity and making them less permeable, due to the interactions taking place with phospholipids in lipid membranes which result in modification of the lipid acyl-chain conformation.

Recently polyethyleneglycol (PEG, of varying molecular weight)–lipid conjugates have become commercially available and are frequently used in liposome applications. Aditionally, functionalized phospholipids exist for the covalent or noncovalent attachment of proteins, peptides, or drugs to the liposome surface. Most of these lipids fall into three major classes of functionality: Conjugation through amide bond formation, disulfide or thioether formation, or biotin/streptavidin binding.

Active lipids—mostly with anticancer activity—have also been added in liposome membrane for production of active liposomes. Examples of such lipids are ether lipids [18] and arsonolipids [19].

Several techniques are employed for the physicochemical characterization of lipid membranes, as summarized in Table 2. Thermal analysis, mainly differential scanning calorimetry (DSC), has been used extensively, offering information on the thermodynamics of various types of liposomes. The phase behavior of lipid components of membranes determines membrane fluidity. Each lipid has a characteristic lipid chain transition temperature, T_m. Changes in the structure of lipids occur below and above this temperature [20]. The temperature at which these changes occur depends on the head group, the chain length, and the degree and type of unsaturation of each lipid [21]. Using DSC studies in has been demonstrated that heat capacity curves are affected by the size of vesicles [20, 22], and can be modified by introduction of drugs [23, 24] or peptides [25, 26] in the lipid membranes (due to interactions between incorporated molecules and lipids).

TABLE 2 Methods for Physicochemical Characterization of Lipid Membranes

Method	Information	References
Thermal analysis, mainly Differential scanning calorimetry (DSC)	Membrane fluidity Lipid chain transition temperature, T_m	20–26
Fluorescence spectroscopy	Phase transitions	27, 28
	Membrane dynamics	29, 30
Nuclear magnetic resonance (NMR)	Polymorphism	31, 32
	Lamellarity	33
	Membrane dynamics	34, 35
Electron paramagnetic resonance (EPR)	Fluidity of membranes	36
	Liposomal internal volume	37
	Membrane dynamics	38
	Membrane–drug interactions	39
Fluorescence quenching	Fusion processes	40, 41
X-ray diffraction	Structural information; thickness of the membrane and water layers	42–45

5.3.2.2 Physicochemical Properties of Liposomes

The in vitro and in vivo performance of liposomes is highly dependent on their structural and surface properties. Liposome size and size distribution, surface charge (zeta potential), and trapping efficiency of the drug incorporated in the liposomes are important parameters that should be measured when developing a liposomal drug formulation. To obtain optimum performance of a liposomal preparation, parameters influencing both the liposome and the drug need to be carefully considered during early stages of development. In this chapter the most important physicochemical properties of liposomes will be discussed in terms of the way they may affect liposomal drug performance as well as the techniques used for their measurement.

Liposome Size Distribution Liposomes have to be smaller than the vascular pore cutoff (380–780 nm) to extravasate and reach solid tumors [46]. Liposome size also plays an important role in complement activation and RES clearance of liposomes [47] and [48]. In general, vesicles that are larger than 100 nm require additional strategies for preventing surface opsonization and prolonging their circulation half-life (Table 1). Light scattering, field flow fractionation, microscopy, size exclusion chromatography, and turbidity are commonly used techniques for the physicochemical characterization of liposomal dispersions.

Quasi-elastic light-scattering or photon correlation spectroscopy is the most popular light-scattering technique. The Brownian motion of the particles causes fluctuations in the light intensity versus time. The hydrodynamic radius and the polydispersity index of liposomes can be easily determined from these studies [49–51]. Field flow fractionation is another approach to measure the particle size and the surface charge of liposomes [52–54]. The liposomes are exposed to a perpendicular field under laminal flow and their size and mass distribution can be determined. Size exclusion chromatography is a simple method to determine the size distribution of liposome dispersions [13, 55], especially on a routine basis, if other sophisticated equipment is not available. Samples with high heterogeneity are suitable to be analyzed with this method [55]. The selection of the proper gel, pretreatment of the column with sonicated vesicles to avoid any loss of material by adsorption, and use of isotonic buffer to avoid osmotic shock and frequent column calibration can secure the reproducibility of the method [13, 55].

Turbidity is a spectroscopic technique determining the optical density of colloidal particles. A wavelength between 350 and 500 nm is the first choice for such studies [56]. Turbidity measurements can offer important information on the kinetics of membrane–surfactant interactions since membrane solubilization changes reflect changes to the optical density of the dispersion [57–60]. It is also widely used as a technique to investigate liposome aggregation and fusion [61]. However the exact particle size of liposomes cannot be determined using turbidity techniques.

Optical microscopy is applicable for LUVs, MLVs, and especially giant liposomes [62, 63]. For instance, mechanical properties of the liposomal membrane can be studied by combining optical microscopy and a micropipette technique [64]. Electron microscopy can give information about both the morphology and homogeneity of liposomes. Quantification in terms of number can be carried out measuring at least 300 particles from different images [65]. Transmission electron microscopy (TEM) is a powerful technique since magnifications up to 200,000 and a resolution of approximately 1 nm can be achieved. Negative staining is a quite common technique for visualization studies. A small amount of the sample is dried on a grid coated with carbon film. Then the film is coated with an electron-dense solution (e.g., tungsten molybdate). However the technique suffers from some drawbacks, such as artifacts due the fixation process or the extraction of lipid material by embedments [66, 67]. With freeze-fracture electron microscopy (FFEM) samples are quickly frozen and fractured. Compared with negative staining technique, FFEM has the advantage of preserving water-dependent lipid phases because no dehydration steps are involved. Therefore phase transitions, lipid polymorphism, or fusion processes can be visualized with this approach [68–71]. Recently cryogenic transmission electron microscopy (cryo-TEM) has been employed for visualization studies. With this technique very precise morphological assessment of liposome interior

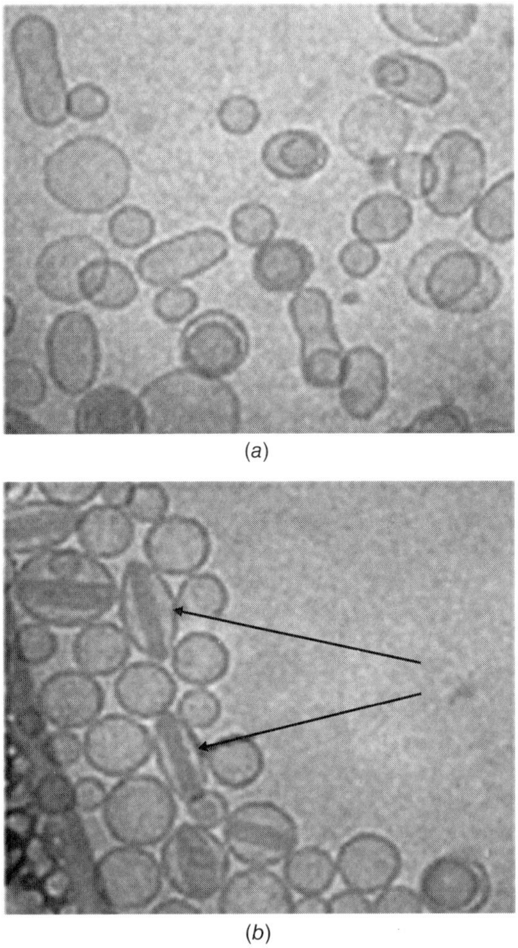

FIGURE 3 Morphology of liposome of liposome by cryo-TEM (transmission electron microscopy). (*a*) Empty liposomes (liposomes with entrapped unbuffered $CuSO_4$ in the absence of drug). (*b*) Topotecan-encapsulating liposomes (drug was added to the empty liposomes of (*a*) to achieve a final drug-to-lipid ratio of 0.2 mol/mol, and the system was incubated at 20 °C). (Reproduced from ref. 72 with permission of Elsevier.)

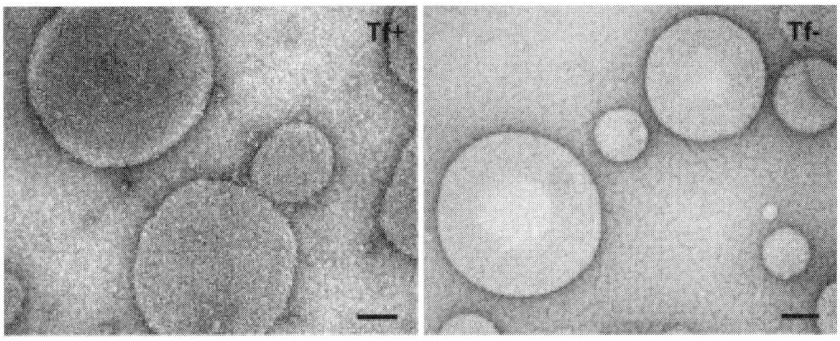

(*a*) (*b*)

FIGURE 4 Morphological assessment of liposome surface coating using cryo-TEM. Liposomes that have been conjugated with transferrin display small particles on their surface (*a*), while control preparations appear to be smooth and undecorated (*b*). (Reproduced from ref. 73 with permission of Elsevier.)

(Figure 3) as well as liposome surface (Figure 4) can be carried out as recently demonstrated for Copper-Topexan encapsulating [72] and transferring-coated liposomes [73], respectively. Compared with the previous techniques the main advantage of cryo-TEM is avoidance of any fixation of the grid, which can create artifacts induced by staining and thus keeps the sample close to the original state [74, 75]. The samples for the cryo-TEM studies are prepared in a controlled environment vitrification system (CEVS). A small amount of the sample is placed on a grid-supported film. The grid is quenched in liquid ethane and it is vitrified. Then the samples are characterized with a TEM microscope. In a manner similar to the previous discussion, lipid polymorphism or fusion can be investigated [61, 76–79]. Finally scanning probe microscopy (SPM) including atomic force microscopy (AFM) has been recently applied in the liposome field [73, 80]. Some of the advantages of the technique are the high resolution in atomic dimensions (Figure 5 [73]), the production of three-dimensional images with high resolution, and the versatility of the operation conditions (vacuum air liquid). Aditionally, sample preparation does not involve any staining, freezing, embedding, or fracturing procedures.

Surface Charge of Liposomes The electrical properties of liposomal surfaces can influence the physical stability of liposomal dispersions during storage as well as the behavior of liposomes in the biological milieu and their interaction with cells [47, 81].

Microelectrophoresis is used to measure the electrophoretic mobility or, in other words, the movement of liposomes under the influence of an electric field. From the electrophoretic mobility the electrical potential at the plane of shear or ζ (zeta) potential can be determined (by the Helmoholtz–Smoluchowski equation). From the zeta potential values the surface charge density (σ) can be calculated.

Aggregation of liposomes both in vitro and in vivo is one of their main stability problems. According to the Derjaguin-Landau-Verwey-Overbeek (DLVO) theory, or theory of colloidal stability, a colloidal system is stable if the electrostatic repulsion forces between two particles are larger than the attraction van der Waals forces. Therefore charged liposomal formulations are highly desirable. Manipulation of

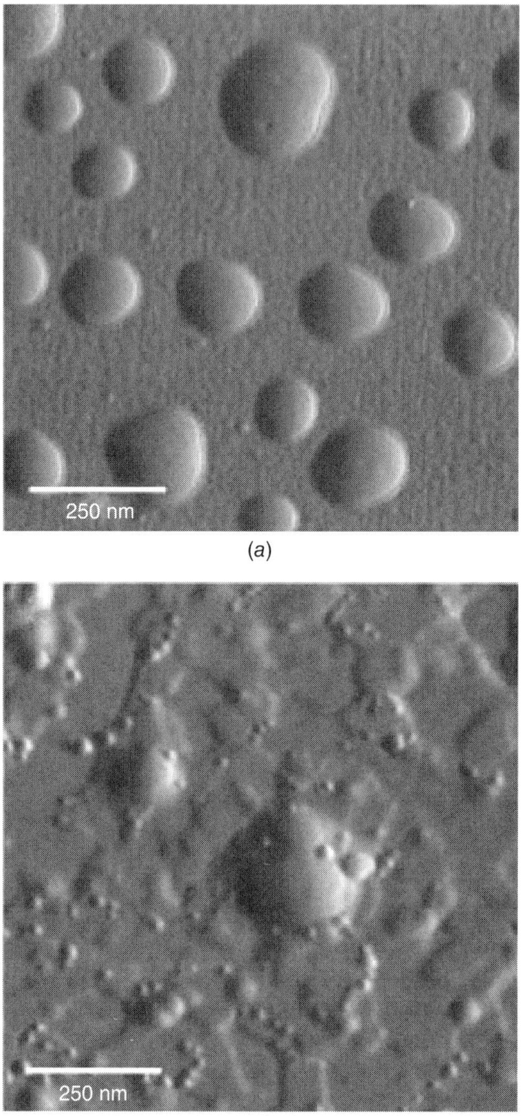

FIGURE 5 Morphological assessment of liposome surface coating using atomic force microscopy (AFM). AFM images of liposomes prepared with DSPE–PEG 200–COOH. (*a*) Plain liposomal formulations. The liposomes show a smooth surface morphology. (*b*) Liposomes covalently modified with transferring. Small globular structures are visible at the surface. (Reproduced from ref. 73 with permission of Elsevier.)

liposome surface renders them stable against self-aggregation or nonspecific interactions, and this can be achieved by introducing other molecules on the surface, which may be natural molecules such as glycolipids [82, 83] or antibodies [84] or lectins [85] (that are usually grafted on the liposome surface by chemical linking).

As also mentioned in the introduction, the RES has been the "Achilles heel" to the delivery of liposomes by injection to the blood stream because of their rapid uptake by the macrophages of the RES. Grafting of liposome surface with hydrophilic polymers has been used successfully as a method to protect the liposomes with a steric barrier which inhibits the adsorption of blood components [7, 86, 87]. It has been shown that *sterically stabilized* liposomes have significantly lower zeta potential values compared with conventional liposomes [87], indicating that the presence of PEG might shift the plane of shear away from the phosphate moieties. Thereby, zeta-potential modification can serve as proof of successful coating of liposomes by polymers.

Drug Loading Efficiency and Techniques The amount of drug incorporated (for amphiphilic or lipophilic drugs) or encapsulated (for hydrophilic drugs) is a very important parameter that determines largely the selection of liposome type and components. Terms used to quantitate drug loading are drug–lipid (mol/mol) ratio (or D/L) and trapping or encapsulation efficiency (which is the percent D/L in the final liposomal formulation compared to the initial D/L used for liposome preparation). In many cases, the percent encapsulation is mentioned (percentage of drug encapsulated in relation with the amount of drug offered for encapsulation during liposome preparation); however, this number is practically useless and misdirecting, since it highly depends on the initial amount of drug offered to the specific lipid quantity, while the amount of lipid is not quantified.

Several parameters influence the encapsulation of drugs into liposomes. The types of vesicle and drug used play significant roles in the percentage of encapsulation. Hydrophobic drugs have higher loading efficiency in MLVs since they consist of high numbers of bilayers and low aqueous volumes [88–91]. In contrast, hydrophilic drugs have higher encapsulation in LUVs [91, 92]. Generally, for hydrophilic drugs, the percentage of encapsulation increases in the order of SUV $\leq$ MLV $<$ LUV [92].

The method used for liposome preparation has a significant impact on drug loading as well. Larger surface areas for the formation of thin films are preferable since they facilitate the hydration process of the bilayer, as demonstrated for doxorubicin (DOX) using five different hydration protocols [93]. The highest encapsulation was achieved when a thin film with large surface area was formed. Glass beads can be used for this purpose. From 8 to 10 times higher values for hydrophobic drug encapsulation were obtained after the addition of such beads [94]. The type of hydration (conditions and media used) also influences drug loading.

Drug encapsulation and stability of liposomes are also affected by the length of the acyl chain and the degree of saturation of the lipids used for their formation. As the acyl chain length of the lipid increases, so does the partitioning of hydrophobic drugs in the lipid membrane [95], as demonstrated for atenolol and propranolol in MLVs and SUVs [96]. The impact of the head group of different lipids used for the preparation of liposomes on encapsulation of citicoline was investigated by Puglisi et al. [97]. The fluidity of the membrane has also been demonstrated to influence encapsulation of different compounds [98–100].

When galactocerebroside was incorporated into liposomes, the percentage of encapsulation of mintoxantrone increased proportionally to the amount of glycolipid present in the membrane [101]. Numerous studies have demonstrated the impact of vesicle surface charge on drug loading efficiency, which is important for charge-bearing molecules. Encapsulation of hydroxycobalamin [102] and doxorubicin [103] was higher in negatively charged liposomes compared to neutral ones. In a similar manner, higher amounts of calcitonin were encapsulated in positively charged compared to neutral liposomes [104] while the highest loading of indomethacin in liposomes was obtained in the following order: positively charged > negatively charged > neutral liposomes [105].

Cholesterol and α-tocopherol are used quite often to increase the rigidity and stability of liposomal membranes [88, 106–108]. In most cases cholesterol appears to improve the encapsulation of both hydrophilic and hydrophobic compounds. However, if the drug is lipophilic and partitions in the liposome membrane, there is a good chance that it might be displaced by adding increasing amounts of cholesterol in the bilayer (as observed in the case of dexamethasone encapsulating liposomes in our laboratory).

A novel system for enhancing the drug loading of lipophilic drugs combining liposomes and cyclodextrin–drug complexes by forming drug-in-cyclodextrins-in-liposome preparations has been proposed [109–111]. Cyclodextrins (CDs) are hydrophobic, cavity-forming, water-soluble oligosaccharides that can accommodate water-insoluble drugs in their cavities, increasing their water solubility. The basic intention is to encapsulate a stable water-soluble drug–cyclodextrin complex in the aqueous compartments of liposomes (Figure 6). As will be discussed below, this system can also serve as a method to increase the retention of lipophilic drugs in diluted liposome dispersions.

Finally, remote loading and active loading [11] are other methods used to achieve high trapping efficiency in liposome formulations, but unfortunately they can only be applied to a small number of drugs with specific physicochemical properties. This technique will be discussed below in Section 5.3.2.3.

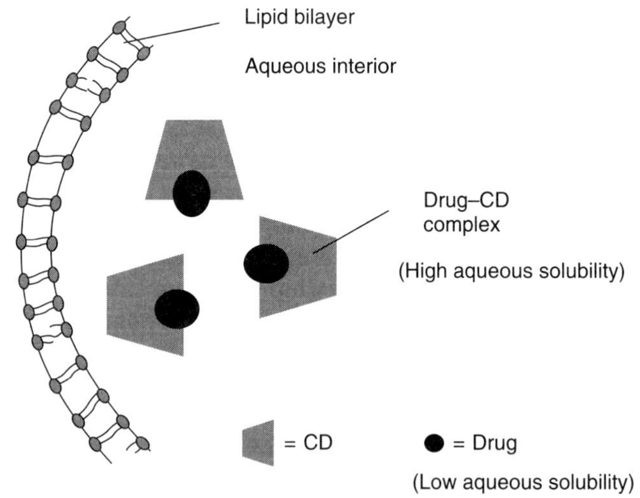

FIGURE 6 Representation of drug in–cyclodextrin in liposome technique for encapsulation of lipophilic drugs in aqueous interior of vesicles. Drug molecules have low aqueous solubility and thus cannot be encapsulated in the aqueous compartment of the vesicle. However, the drug–cyclodextrin complex has high aqueous solubility and can thus be encapsulated in high concentrations in the vesicles.

Stability of Liposomes A shelf life of at least two years is requested for pharmaceutical products. Therefore, chemical stability and physical stability are important parameters for the overall performance of liposomal formulations. Additionaly, another very important factor is the retention of encapsulated drug.

Several studies have sufficiently addressed the chemical degradation of liposomes during storage [112–115]. This is due to the hydrolysis of the phospholipids to fatty acids and 1- and 2-acyl-lysophospholipids. Further hydrolysis leads to the production of glycerol phospho compounds. Antioxidants (α-tocopherol), complexing agents [e.g., ethylenediaminetetraacetic acid (EDTA)], and inert atmosphere (e.g., nitrogen) are most commonly used to overcome this problem. Moreover, the presence of α-tocopherol can reduce the auto-oxidation of lipids, which is usually induced by light, metal ions, and temperature. Prevention of the chemical decomposition of the lipids by adding α-tocopherol in liposomal dispersions can increase the shelf life of liposomes [116]. Furthermore, the coexistence of cholesterol and α-tocopherol in the lipid bilayer can improve the antioxidant activity of tocopherol [117].

An alternative to circumvent problems related to the chemical decomposition of liposomes is their storage in the dry state (freeze dried). However, the protection against damage by freezing (cryoprotection) [118–120] and the protection against damage by dehydration (lyoprotection) [121, 122] require special attention for the proper storage of liposomes. Depending on the drug encapsulated in the liposomes and possible interactions between drug molecules and components of the lipid membrane, initial studies should be carried out in order to find the proper cryo- and lyoprotectant which preserves the integrity of the specific liposomal formulation.

The physical stability of liposome dispersions is mainly related with possible aggregation and leakage of the liposomal membrane. The size and surface charge play a significant role for the stability of liposomal dispersions, as has been discussed in the previous paragraph.

In addition to physical stability, the retention of drug in liposomal formulations is particularly important, not only during storage, but also during in vivo administration. Especially when targeted or long-circulating liposomes are used, association of the drug in the liposome carrier until the carrier reaches its biological target is a prerequisite for achieving any therepeutic benefit. However, in many cases, although the amount of drug loaded in liposomes is initially high, for different reasons most of the liposome-associated drug is rapidly released from the vesicles. The main causes for such behavior are different according to the physicochemical properties of the drug. *For hydrophilic drugs:* (i) The low integrity of liposome membranes after in vivo administration and contact with blood compomnents, which results in removal of some lipid molecules and concurrent opening of pores in the liposomal bialyer through which the loaded drug molecules may leak out, and (ii) the physical instability of liposomes that results in aggregation and fusion. and finally release of drug from liposomes. *For amphiphilic and lipophilic drugs*, the main problem is caused by the dilution of liposome dispersions, which usually occurs immediately after in vivo administration [by most routes, especially i.v. (diluted in 4 L of blood)]. This results in release of drug (since the drug can permeate membranes) until the drug saturates the full aqueous volume in which the liposome vesicles are diluted.

For hydrophilic drugs, the problem can be solved by increasing the rigidity of the liposome membrane while concurrently decreasing their tendency for aggregation, after proberly selecting the liposome lipid components. However, for membrane-permeable drugs, drug leaking upon dilution cannot be easily, if at all, confronted. Theoretically, one method may be to increase the affinity of the drug molecule with the lipid membrane (and at the same time decrease its aqueus solubility) by chemical modification (conjugation to a lipid or fatty acid). However, this is neither easy nor generally applicable. The association of liposomes with cyclodextrins has been recently proposed as a method to ensure high and stable entrapment of lipophilic drugs in aqueous compartments of liposomes [109–111, 123] (Figure 6), providing that the drug has high affinity for the cyclodextrin molecule and is not displaced from the cyclodextrin cavity by components of the lipid membrane (mostly cholesterol). In such cases, this approach may not increase drug retention, but encapsulation of drug may be substantially improved [111].

5.3.2.3 Preparation of Liposomes

Liposome Preparation Techniques In most cases, liposomes are named by the preparation method used for their formation, Such as sonicated, dehydrated–rehydrated vesicle (DRV), reverse-phase evaporation (REV), one step, and extruded. Several reviews have summarized available liposome preparation methods [91, 124, 125]. Liposome formation happens spontaneously when phospholipids are dispersed in water. However, the preparation of drug-encapsulating liposomes with high drug encapsulation and specific size and lamellarity is not always an easy task. The most important methods are highlighted below.

Thin Film Method It was in 1964 that Alec Bangham introduced the "thin-film method" for liposome preparation [1, 126]. Lipids are dissolved in organic solvents (chloroform or mixtures with methanol) and the solvent is removed under a high-vacuum rotor evaporator forming a thin film on the walls of round-bottomed flask. Depending on the phase transitions of the lipids used for the preparation of liposomes, the aqueous phase for the rehydration should be prewarmed at temperatures above the phase transition of the lipid. After addition of the aqueous phase the thin film is detached from the flask walls by agitation and a highly heterogeneous population of MLVs is produced. Depending on the physicochemical properties of drugs, they can be introduced either in the thin film together with the lipids (lipophilic compound) or in the rehydration solution (hydrophilic compound).

Sonication This approach uses energy (ultrasound) and can be applied to a dispersion of MLVs [127] or to solid lipids mixed with aqueous solution. The flask with the liposome dispersion is placed in a bath sonicator or a probe sonicator (tip) is immersed in the tube containing the liposome dispersion. With the first setup it is difficult to reduce the size to the nanometer level since the energy produced by the bath sonicators is rather low. However, it has the advantage that there is no contact with the liposome dispersion. The position of the flask in the sonicator is equally important. It is easy to understand if it is in the right place from the noise produced by ultrasound waves. For instance, if foams are produced or there is no noise at all, that implies the sample is misplaced and finally the size of vesicles will not be reduced.

In the second approach (probe sonicator), the size of the liposomes can be reduced to nanometers and SUVs can be produced. As previously, the position of the probe plays an important role on the ability to minimize vesicle size. Because the energy produced from the transducer is high, overheating of the system is quite common; therefore, a water bath filled with ice is recommended. After sonication, fragments of Ti originated from the probe are scattered in the dispersion. Centrifugation for 4–5 min at 10.000 rpm will cause sedimentation of these fragments, giving clear liposome dispersions.

Injection Methods

Ethanol Injection Small unilamellar vesicles (with diameter of 30 nm) can be prepared with the ethanol injection technique [128]. Lipids are dissolved in ethanol and injected rapidly in the aqueous solution under stirring (final concentrations up to 7.5% (v/v) ethanol can be applied). The method is very easy, having the advantage of avoiding chemical or physical treatment of lipids. However, there is an extra step to remove ethanol and the concentration of vesicles produced is rather low. Also encapsulation of hydrophilic drugs is also low, due to the high volumes used.

Ether Injection The general principle of this method is the same as ethanol injection. The only difference is that the lipid is injected slowly in the aqueous solution that is warm [129]. Furthermore, the concentrations used in this case are somewhat higher (up to 10 m*M*) compared to the ethanol injection approach.

Extrusion (Extruded Vesicles) The extrusion method, which today is very popular for the production of homogenous vesicle samples of a predetermined size, was introduced by the group of D. Papahadjopoulos [65]. Multilamellar vesicles are extruded through filters with well-defined pores under pressure. Polycarbonate is the most commonly used material for these membranes, which have pore sizes from 30 nm up to several micrometers. For lipids with high melting point, the extrusion should be carried out above their phase transition temperature. The operating volumes are from 1 to 50 mL with a liposome dispersion concentration up to 150 m*M*. Generally, repeated extrusions reduce the number of lamellae and the produced liposomes are mainly unilamellar. High pressure can cause disruption of membranes. The reproducibility of the method is good; however, it is quite time consuming and membrane rupture problems occur quite often.

French Press With this approach MLV liposomes are introduced in a cell and a piston presses the dispersion [130]. Pressures up to 25,000 psi can be achieved and SUVs are produced. The main disadvantage of this technique is that it is not applicable for lipids with phase transition temperatures lower than 20 °C and the concentration of liposomes that can be used is relatively low (maximum 20 m*M*).

Microfluidization During microfluidization, MLVs are circulated with a pneumatic pump operating under high pressure through a prefilter and then to the interaction chamber [131]. From there, they are separated into two streams and they pass through defined microchannels under high velocity to a heat exchanger which is connected with a water bath. This is repeated many times until the size of the liposomes is significantly reduced. They can operate with volumes from 0.1 up to 10 L of liposomes and with concentrations up to 300 m*M*, which is by far the highest capacity from all other methods. Small unilamellar vesicles less than 100 nm can be produced, but the population is not completely homogeneous.

Reverse-Phase Evaporation The REV method was developed by Szoka and Papahadjopoulos [132]. Lipids are dissolved in organic solvent and the solvent is removed with evaporation. The thin film is resuspended in diethyl ether (1 mL solvent/mL liposomes) followed by the addition of one-third of water and sonication in a bath sonicator for 1 min. This water-in-oil (w/o)-emulsion is evaporated until a dry gel is formed, and finaly the gel is broken by agitation and water addition. Sometimes this step is quite difficult. The remnants of the organic solvent are removed by evacuation and the resulting dispersion is REV liposomes.

Dehydrated–Rehydrated Vesicles DRV liposomes were developed by Kirby and Gregoriadis in 1984 [133] and are capable of encapsulating high amounts of aqueous soluble molecules under mild conditions (conditions that do not cause decomposition or loss of activity). The high entrapment ability of this type of liposomes is due to the fact that preformed, "empty" SUVs are disrupted during a freeze-drying step in the presence of the solute destined for entrapment. Subsequently, during controlled rehydration that is carried out in the presence of concentrated solution of the solute (to be encapsulated), the vesicles fuse into large oligolamellar vesicles entrapping high amounts of solute. The produced liposomes are multilamellar and their size is between 200 and 400 nm up to a few micrometers. Recently, with the addition of certain amounts of sucrose, DRV liposomes with diameter between 90 and 200 nm were obtained entrapping considerable proportions (up to 87) of the solute [134].

Giant Vesicles Large or giant liposomes have been developed by Reeves and Dowben [135]. Briefly, lipids are dissolved in chloroform/methanol 2:1 and dispersed on a piece of glass. Water is added for their rehydration; however, their population is quite heterogeneous. Other types of particle-encapsulating giant liposomes [136] can be prepared by applying a double-emulsion technique followed by a freeze–drying step.

Detergent Depletion With this method phospholipid–detergent mixed micelles are initially produced, and during controlled-rate detergent removal, liposomes are formed. The rate and method of detergent removal determine the size and size homogeneity of the liposomes produced. Gel filtration and dialysis are the most popular approaches [137, 138]. Although liposomes are produced under mild conditions (low temperature and low shear mechanical forces applied), this method suffers from low encapsulation efficiency of hydrophilic drugs.

Large Unilamellar Vesicles from Cochleates Large unilamellar vesicles can be produced with the "cochleate" approach [139]. Small unilamellar vesicles consisted from phosphatidylserine adopt a cochleate shape after addition of calcium. Addition of EDTA creates complexes with calcium, turning the cochleates to LUVs.

One-Step Method The "one-step method" has been introduced by Talsma et al. [140]. Lipid dispersions are hydrated at high temperatures in the presence of a steam of N_2. Liposomes between 200 and 500 nm can be prepared with this approach.

Large-Scale Manufacturing Despite the fact that there are many methods to prepare liposomal dispersions not all of them are applicable for scaling up. In fact, scaling up to larger batches could be a monumental task. Among all the methods reviewed so far, microfluidization and homogenization are the most powerful methods to produce large quantities of liposomes. New homogenizers have a capacity of 1000 L/h and require a minimal sample volume of 2 L [141]. The fact that lipid concentrations up to 300 m*M* can be used secures high encapsulation capacity.

Remode and Active Drug Loading Techniques The main advantages of these approaches are the high encapsulation efficiency and low leakage of the encapsulated material.

An in situ method for "remote" drug loading based on the development of a pH gradient across the internal and external water phases of the membrane has been established. A transmembrane pH gradient induces the uptake of charged drugs into liposomes. Drug encapsulation is based on its partitioning between the lipid and aqueous parts. This process is governed mainly by pH and to some extent by the ionic strength of the medium. Drugs that are weak bases can diffuse through the lipid membrane as unprotonated species. The presence of a proton gradient makes them more hydrophilic, allowing them to accumulate in the intraliposomal aqueous phase. Encapsulations up to 90% have been reported for doxorubicin [142–144] and vincristine [145].

An alternative but similar technique is based on an ammonium sulfate gradient used to obtain "active" loading of amphipathic weak bases into the aqueous compartment of liposomes. This has been used for active loading of anthracyclines, acridine orange, epirubicin, and doxorubicin [11, 146, 147] at very high efficiency (>90%). In the case of doxobubicide most of the intraliposomal drug is present in the aggregated state. Additionally, antracycline accumulation in liposomes is stabilized for prolonged periods of storage due to aggregation and gelation of antracycline sulfate salt.

Active entrapment and loading stability are dependent on liposome lipid composition, lipid quality, medium composition, and temperature as well as on the pK_a and hydrophobicity of the base. The ammonium sulfate gradient approach differs from most other chemical approaches used for remote loading of liposomes, since it does not require liposomes with acidic pH interior or an alkaline extraliposomal phase.

In addition to the remote or active loading techniques mentioned above, metal complexation reactions have been demonstrated to achieve accumulation of doxorubicin in liposomes [148, 149]. Furthermore, copper–topotecan complexation has been recently seen to mediate drug accumulation into liposomes and is proposed as a methodology to prepare liposomal camptothecin formulations [72].

5.3.2.4 Functionalization of Liposomes

"Active targeting" is used to describe the specific liposomal drug localization achieved by grafting various moieties (antibodies, lectins, polymers, etc.) on the carrier surface.

There are a number of techniques available to attach the suitable ligand on the liposome surface, either by covalent or noncovalent coupling [150, 151].

These techniques should be fast, efficient, and reproducible, yielding stable nontoxic bonds, while the conjugated ligands should maintain the ability to recognize the target site with high binding affinity. Also, the coupling method should not affect the blood clearance of the formulation, colloidal stability, drug incorporation, and release in a negative way. For example, when antibodies were attached on the liposome surface where PEG molecules were grafted as well to ensure prolonged carrier retention in the blood, it was shown that the ligand binding efficiency on the bilayer was low as well as the binding ability to the target [152, 153]. The latter problem was opposed by attaching the ligand at the distal end of the PEG molecules already grafted on the liposomal bilayer [154].

Covalent Binding of Ligands The majority of ligand coupling is achieved by covalent reactions with hydrophobic anchors. The procedure could be carried out in two patterns: Either the hydrophobic anchor is included already in the liposomal bilayer and the ligand interacts with the anchor on preformed liposomes [155] or the ligand–hydrophobic anchor conjugate in the form of micelles is mixed with the liposomes [156]. In the first instance, where a hydrophobic anchor is mediated between the liposomal surface and ligand, covalent attachment can occur via a thioether bond [157–161], via a disulfide bridge, between carboxylic acid and the primary amine group [162], via hydrazone, or via cross-linking between two primary amine groups (Figure 7).

The formation of the stable thioether bond is a reaction between thiol moieties of proteins mainly with maleimide groups. Usually, PE or PEG–PE or PEG–DSPE (distearoyl-phosphatidylethanolamine) has been functionalized with maleimido (Mal-), maleimido-phenylpropionate (MP), or pyridil-dithio-propionylamino (PDP) groups, which eventually will react with the thiol groups. Also, sometimes the ligand does not carry enough thiol groups or those are completely absent, so they have to be introduced using a heterobifunctional cross-linker, such as SPDP [*N*-hydroxysuccinimidyl 3-(2-pyridyldithio) propionate] or SATA (succinimidyl-*S*-acetylthioacetate), which introduce one amine group. Still, in the case of SATA, deacetylation using hydroxylamine is necessary to uncover the thiol group, while if SPDP is used, the produced disulfide bond has to be reduced to thiol groups with dithiothreitol (DTT). The former cross-linker is more preferable as only mild conditions are used to make the thiol group available.

It has been shown that attaching the ligand at the distal end of PEG molecules combines the advantages of a specific drug delivery system with steric stabilization for higher stability in the blood [163, 164]. However, Longmuir et al. showed that introducing a peptide from *Plasmodium* at the free end of PEG_{3400}–AP (aminopropane) was not capable of retaining the liposome stability, so PEG_{5000}–PE molecules were added to the liposomal composition resulting in PC/PE–PEG_{5000}/AP–PEG_{3400}-peptide liposomes with molar ratio of 86:10:4 [165]. The coupling efficiency of ligand attached on PEG was higher (60–70% or even 100% in some cases) compared to that achieved using, for example, *N*-(4′-4″-maleimidophenyl)butyrol)-dioleoylphosphatidylethanolamine (MPB–DOPE) (only 10%) [164, 166].

Antibodies, whole or fragments, have been attached successfully on liposomal surfaces and more commonly at the free end of PEG molecules, as has been reported

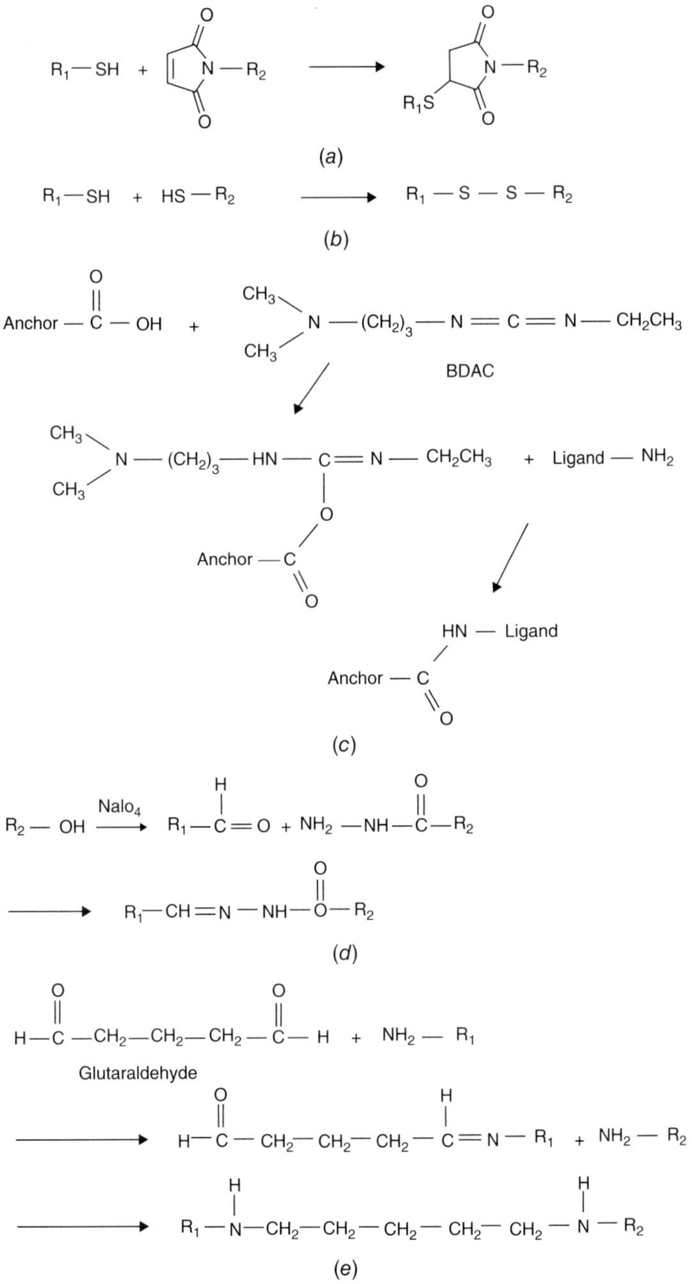

FIGURE 7 Schematic of different coupling methods used: (*a*) reaction between meleimide and thiol functions; (*b*) formation of disulfide bond; (*c*) reaction between carboxylic acid and primary amine group; (*d*) reaction between hydrazide and aldehyde functions; (*e*) cross-linking between two primary amine functions. (Reproduced from ref. 150 with permission of Elsevier.)

in a number of studies since the interaction with the cells is much more favored [161]. Using Fab′ fragments is more advantageous as the Fc part which mediates MPS activation through a receptor, is omitted [157, 160]. Moreover, the distance of the antibody fragment from the liposomal surface is another important factor which determines the drug delivery system uptake by the cells, as reported by Mamot et al. [167]. Besides the latter has been reported elsewhere even if the coupling reaction is different from thioether formation [168].

Examples of peptides attached on liposomal surface via thioether bonds were TAT-peptide and antagonist-G on Mal–PEG2000–DSPE and PDP–PEG–DSPE, respectively [169, 170]. Both ligands exhibited significant increase in cell uptake.

Another possible way for ligand conjugation on liposomes is the formation of a disulfide bridge, which is quite unstable in serum and thus it is not used as much [171].

However, an amide bond formed between the carboxylic acid group on the liposome surface (DSPE–PEG–COOH) and the primary amine of the ligand is favored as the ligand modification is not necessary. According to this method, an acyl amino ether is produced in the presence of 1-ethyl-3-(dimethylaminopropyl) carbodiimide (EDAC) and *N*-hydroxysulfosuccinimide (NHS), which eventually will react with the primary amine of the ligand [150]. For example, Wartchow et al. improved the efficacy of a small integrin antagonist of the extracellular domain of the αvβ3 integrin by grafting it on dextran-coated liposomes [172]. DPPE-succinate was included in the liposomal bilayer and a 3-amino-2-hydroxypropyl ether derivative of dextran was added to preformed liposomes in the presence of EDAC, while unreacted succinyl groups were converted to amides. The amino groups of dextran were succinylated and integrin antagonist (IA) was attached on the succinamidodextran liposomes in the presence of EDAC. The final IA–dextran liposomes had a size of 110 nm, which is attributed to dextran coating, as the liposomes without dextran were 60 nm in diameter. The antiangiogenic mechanism of IA–dextran liposomes as well as the apoptotic potency was proved after a series of studies.

Moreover, Voinea et al. attached antibodies against vascular cell adhesion molecule-1 (VCAM-1) overexpressed on activated human endothelial cells on liposomes with the intention of using them as drug carriers [162]. *N*-glurary1-PE was used as membrane anchor for the antibody coupling via its free amino groups after its activation with carbodiimide. There is no necessity of antibody modification before the coupling reaction.

Also, transferrin was grafted onto liposomes containing *N*-glutaryl-PE activated with carboxidiimide with the final plan to use those as carriers for inhalation therapy for lung cancer [173]. Tfr liposome uptake was significantly higher from immortalized or cancer cell lines, but to reduce uptake from alveolar macrophages, PEG molecules are attached on the liposome surface. In general, transferrin is a glycoprotein that consists of a single chain of amino acids which has been coupled on the liposome surface for a number of applications because its receptor is overexpressed at malignant cells so a higher amount of transferrin binds on the cell membrane [174]. Its covalent attachment on liposomes takes place either by conjugation between transferrin-lipid and insertion of it on the preformed vesicles [174, 175] or preparation of liposomes with activated lipids and reaction with activated transferrin [176].

In addition, Torchilin et al. synthesized pNP (*p*-nitrophenylcarbonyl)–PEG–DOPE to enable protein coupling via its amino group at the distal end of PEG molecules on liposome surface in a quantitative manner at pH around 8.0 [177]. The only disadvantage mentioned is the hydrolysis of pNP groups from PEG–DOPE at pH higher than neutral (complete hydrolysis occurs in 1–2 h at pH around 8.0). Therefore, the coupling reaction between protein and pNP has to take place faster or at least at the same ratio. It was shown that 65% of pNP binds to ligand at pH 9, so the binding efficiency and time are adequate at the conditions studied. Also, the amount of pNP–PEG–DOPE was critical for successful protein binding; it was shown that 1 mol % of pNP–PEG–DOPE was enough to bind approximately 100 protein molecules. Incorporation of antibodies, lectins, avidin, and nucleosomes did not seem to alter the activity of those molecules even at high concentrations of pNP–PEG–DOPE. However, Savva et al. conjugated a genetically modified tumor necrosis factor (TNF) at the free end of the PE_{3500} molecule by reacting the latter with NHS and DCC (*N,N′*-dicyclohexyl carbodiimide) to introduce the succinyl groups which will react with the phospholipid to produce DOPE–PEG–COOH [178]. Then, the carboxyl group of the derivatized PEGylated molecule located on the liposomal bilayer was activated with EDAC and NHS. Consequently, recombinant TNF was added in the liposome suspension for the final conjugation to occur at 4 °C where the overall coupling efficiency was approximately 55%. However, it was shown that the biological activity of TNF was reduced when attached on the liposomes as the degree of PEG modification increased irrespective of the PEG molecular weight (MW). Also, the formulation did not show the prolonged blood circulation expected due to the PEG presence. Those results were attributed to a number of reasons, including either damage of the protein during the coupling reaction or possible dissociation of the trimeric form of rKRKTNF to a monomeric less active form or cross-linking between PEG and rKRKTNF during the coupling reaction.

A hydrazone bond is an alternative way of antibody coupling onto liposomal surface so as to avoid the use of essential (for the recognition) amino groups at the coupling reaction (by using the maleimido method) or the risk of rapid clearance due to Fc moiety/segment [150, 179]. According to this method, the carboxylic group of the heavy chain of the antibody undergoes mild oxidation by sodium periodate or galactose oxidase to aldehyde groups. The oxidized product can be coupled either on the hydrazide-hydrophobic anchor inserted in the lipid bilayer [180] or at the free end of PEG molecules [181, 182]. Hydrazide cross-linkers used are, for example, *S*-(2-thiopyridyl)-L-cysteine hydrazide (TCPH), *N*-acetylmercaptobutyric hydrazide (AMBH), and 3-(2-pyridyldithio)propionic acid hydrazide (PDPH). TCPH is structurally closely related to PDPH and could be expected to behave in a similar manner [183]. Unprotected mercaptohydrazides such as AMBH are unsuitable since the free thiol function is susceptible to oxidation and may also reduce the disulfide bonds in immunoglobulin IgG at the concentrations required for conjugation. The disadvantage of this method is the low coupling efficiency (17% only) [166]. According to Ansell et al. [179], the drawbacks reported was the possible damage of some amino acid residues, such as methionine, tyrosine, and tryptophan, due to periodate. Therefore, antibodies sensitive to periodate treatment would be unsuitable candidates for the PDPH protocol. Also, the hydrazone bond might undergo hydrolysis after six weeks of storage, which would be a potential problem if the conjugate would not be used soon after preparation. However, it is possible to stabilize the bond using sodium cyanoborohydride to reduce the hydrazone linkage if long-term storage is required [179].

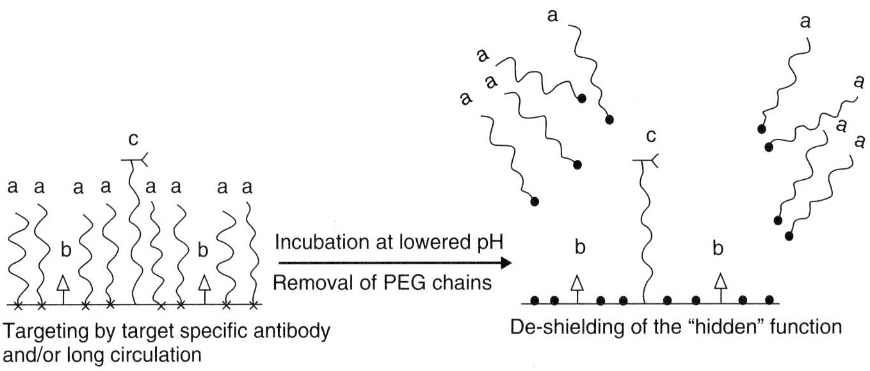

FIGURE 8 Schematic for design of multifunctional drug delivery system (DDS) that includes pH-cleavable PEG-Hz-PE (a), temporarily "shielded" biotin or TATp (b), and monoclonal antibody (c) attached to surface of DDS via pH-noncleavable spacer. (Reprinted with permission from ref. 183. Copyright 2006 by the Americam Chemical Society.)

However, the labile nature of the hydrazone bond is used to formulate "smart" drug delivery systems where the basic idea is to introduce in parallel a pH-responsive ligand spacer in the lipid bilayer (which is PEG_{5000}–Hz–PE), a temporarily shielded biotin or TATp and mAb (monoclonal antibody) attached to the surface of the drug delivery system via a noncleavable bond (TATp–PEG_{2000}–PE) [183]. Such a system will be able to respond to environment stimuli such as pH changes, where, for example, at acidic pH (5.0–6.0) PEG5000 molecules will be detached from the carrier surface and biotin or TATp will be available to either bind to avidin or be internalized by the cells, respectively (Figure 8). The monoclonal antibody and biotin or TAT has been attached on pNP–PEG–PE. The produced mAb DDS (drug delivery system) demonstrated clear immunoreactivity toward the antigen. However, some affinity decrease was observed for the antibodies modified with the pNP–PEG–PE anchor and incorporated onto the immunoliposomes. Biotin binding to avidin was pH dependent with higher retention (75%) at pH 5.0, where the shielding PEG molecules were cleaved away. Significant increase of DDS uptake by the cells was achieved when TAT was incorporated on the liposome surface at pH 5.0.

An avidin–biotin system has been used to attach antibodies in the bilayer of DDSs. Xiao et al. developed a three-step strategy to improve the tumor-to-tissue ratio of anticancer agents [184]. Two antibodies specific for the CA-125 antigen that is highly expressed on NIH:OVCAR-3 cells were used. These cells were prelabeled with biotinylated anti-CA-125 antibody and fluoroscein isothiocyanate (FITC)–labeled streptavidin (SAv) prior to administration of biotinylated liposomes. Both antibodies were specifically bound to the cell surface of OVCAR-3 cells but not to SK-OV-3 cells, which do not express the specific antibody. Antibody biotinylation did not affect its immunoreactivity.

Schnyder et al. explored the targetability of biotinylated immunoliposomes to skeletal muscle cell line in vitro [185]. OX26 mAb binds to transferrin receptor and is covalently attached to streptavidin by introducing thiol groups using 2-iminothiolane (Traut's reagent). Immunoliposomes consisted of DSPC (5.2 μmol), cholesterol

(4.5 µmol), PEG–DSPE, (0.3 µmol), and linker lipid (bio-PEG–DSPE; 0.015 µmol). OX26 mAb–streptavidin was added to preformed liposomes in a 1:1 ratio. According to estimations, the average number of bio-PEG–DSPE molecules was 30, assuming that one 100-nm liposome contains 100,000 phospholipid molecules. Uptake experiments with muscle cell line using the OX26 mAb, fluorescence-labeled OX26–streptavidin, or fluorescent OX26–immunoliposomes demonstrated cellular uptake and accumulation within an intracellular compartment of the OX26 mAb and its conjugates.

All the methods described earlier consider that coupling of the ligand on the anchor already existed on the liposome surface. Another option is the ligand–anchor conjugate in the form of micelles to mix with the liposomes. According to that, anti-CD19 mAb was thiolated using Traut's reagent and reacted with Mal-PEG–DSPE in a micellar form with PEG–DSPE and molar ratio 4:1 [186]. Antibodies were coupled at the end of the PEG–DSPE. Consequently, micelles were incubated with preformed liposomes at molar ratio 0.05:1, respectively, for 1 h at 60 °C.

In another study, mAb 2C5 with nucleosome-restricted specificity, which recognizes specifically human brain tumor cells, was tested as a potential ligand candidate for liposome targeting to brain tumor cells [187]. The mAb was attached to pNP–PEG–DSPE and the formed micelles incubated with preformed liposomes. The 100–200 mAbs bound per single liposome of approximately 200 nm in diameter. A slight reduction in immunoreactivity was observed for a single antibody molecule for a number of reasons; the overall evaluation was sufficient target recognition and affinity due to multipoint attachment of immunoliposomes to the target via several antibody molecules. Indeed, the immunoliposomes showed threefold higher accumulation in the tumors compared to nonspecific carriers.

At this point it has to be emphasized that this method seems to be the most advantageous one because damaging chemical reactions are excluded as they happen at a different stage. Also, this method provides the flexibility of attaching a large variety of ligands on liposomes of various compositions loaded with different drugs. Apparently, targeted liposomes produced with this last technique have shown similarities in the in vitro drug leakage, cell association, and therapeutic efficacies to liposomes made by conventional coupling procedures.

Noncovalent Binding of Ligands According to this procedure, the ligand is added to the lipid mixture during liposome preparation. Small molecules such as sugars have been attached on the liposome or lipoplex surface in this way. At first galactose, mannose, and fucose were modified to Gal–C4–Chol, Man–C4–Chol, and Fuc–C4–Chol and they were added in the lipid mixture of DSPC/Chol/Sugar–C4–Chol with a ratio 60:35:5. Chol was chosen due to the stability in the liposomal membrane while only one sugar was conjugated so the lipophilicity of the final glycolipid would not be altered considerably, and thus the stability of the latter in the liposomal membrane would be more secure [188]. After in vivo administration of 0.5% Gal, Man, and Fuc liposomes it was found that the ratio of their uptake from parenchymal/nonparenchymal liver cells was 15.1, 0.6, and 0.2, respectively. Also, in high doses, 5% Gal liposomes are taken up by nonparenchymal liver cells as well as the parenchymal ones, while they are capable of inhibiting the uptake of Fuc liposomes by nonparenchymal cells.

Even if this is a simple and mild technique, there is always a concern about ligand orientation, very low attachment efficiency achieved (4–40%), and the liposome aggregation often observed.

5.3.3 IN VIVO DISTRIBUTION

Successful treatment depends not only on the formulation characteristics but also on the route of administration. For example, the schistosomicidal drug tartar emetic incorporated in PEGylated liposomes was delivered either intraperitoneally or subcutaneously (27 mg Sb/kg) to mice infected with *Schisostoma mansoni* [189]. Indeed, 82 and 67% reduction levels of worm were obtained, respectively. However, the efficacy of the formulation given by either administrative route was not significantly different. The only difference was the slower liposome absorption by the subcutaneous route.

Also, the therapeutic effect of liposomal adriamycin (PC/Chol, 120 nm) was enhanced significantly after concurrent i.v. and s.c. administration to rabbits bearing VX2 tumors in the mammary gland [190]. The i.v. route significantly inhibits breast tumor and metastasis, while the s.c. route acts on local-regional lymph nodes. That was proved by slowed growth rates, decreased messenger ribonucleic acid (mRNA) expression of proliferating cell nuclear antigen, and extensive necrosis and apoptosis of tumor cells. Even if allergic reactions have not been reported after s.c. injection of liposomes, there is more to be done on systemic toxicity.

The therapeutic efficacy of paclitaxel is stronger after drug incorporation in magnetoliposomes injected either i.v. or s.c to an EMT-6 breast cancer mouse model [191]. The carrier manipulation due to the application of a magnetic field led to their increased accumulation to the tumor site. However, paclitaxel accumulation is slightly lower after s.c. administration, probably attributed to time delay during the drug transportation process to the circulation.

In another study by Wang et al., i.v. injection of liposomes carrying rat insulin promoter (RIK)–thymidine kinase (TK) was found to be less toxic to the liver than the i.p. injection of the same formulation to severe combined immunodeficient mice (SCID) [192]. The direct injection of the liposomes to abdominal cavity probably leads to higher local absorption and, thus, higher liver toxicity. In contrast, the i.v. injected volume is smaller (70 μL to 100 μL of i.p. injection), which is diluted fast as soon as it enters the body.

5.3.3.1 Conventional Liposomes

Conventional liposomes are those that do not carry any sterically stabilizing or targeting moieties on their surface. Their biodistribution depends strongly on their physicochemical properties (size, ζ potential, composition) and physiological and pathological conditions of the body [193]. Thus, conventional liposomes comprise the passive targeting of drug molecules.

Intravenous Administration Liposomes administered intravenously face barriers such as the endothelial lining of the vasculature and the blood–brain barrier. Extravasation of the liposomes occurs only in organs such as liver, spleen, and bone marrow (due to leaky fenestrae and loose junctions between the endothelial cells) and under certain pathological conditions (presence of tumors, inflammation, infection). Besides, neutral (uncharged) liposomes of size smaller than 100 nm show slow blood clearance compared to others of larger size and/or positive or negative charge (due to presence of opsonins) [194, 195]. Also, lipid exchange between liposomal carriers and plasma lipoproteins contributes to liposomal membrane rupture and consequently loss of the therapeutic substance [196].

Therefore, conventional liposomes are used mostly in treating the RES system or to mask the toxic side effects of anticancer drugs. Many anticancer drugs entrapped in liposomes have shown altered biodistribution and reduced toxicity. Plain SUV liposomes consisting of DSPC/Chol (known as "Stealth") in a molar ratio 2:1 or 1:1 have shown particularly promising vehicles as reported in a number of studies of animal models. They have undergone preclinical and clinical studies due to relatively low levels of RES uptake and the high level of tumor targeting exhibited [197–200]. Indeed, ^{111}In-labeled DSPC/Chol liposomes have proven capable of targeting a number of tumors [201–206]. Although the significant RES uptake was a fact, about 45–50% of liposomes remained in the blood circulation 4 h after i.v. injection. Positive images on gamma camera were reported. In patients with recurrent high-grade gliomas, 1% of the injected liposomal dose was taken up by the tumor [205].

Moreover, anthracyclines have been formulated in conventional liposomes and are commercially available. Doxorubicin for i.v. use is commercially available in the form of Myocet, which consists of egg PC and cholesterol [207]. It is recommended in combination with cyclophosphamide for metastatic breast cancer. Thus, drug entrapment in liposomal vesicles may reduce the incidence of cardiotoxicity and lower the potential for local necrosis, but infusion reactions, sometimes severe, may occur. Hand–foot syndrome (painful, macular reddening skin eruptions) occurs commonly with liposomal doxorubicin and may be dose limiting. Daunorubicin also has general properties similar to those of doxorubicin. A liposomal formulation (DSPC/Chol, 2:1, size 45 nm) for i.v. use is licensed for patients with AIDS-related Kaposi's sarcoma [207, 208]. The plasma pharmacokinetics of DaunoXome differs significantly from the results reported for free daunorubicin hydrochloride. DaunoXome has a small steady-state volume of distribution of 6.4 L (probably because it is confined to vascular fluid volume) and clearance of 17 mL/min. These differences in the volume of distribution and clearance result in a higher daunorubicin exposure [in terms of plasma "Area on the curve" or "bioavailabity" (AUC)] from DaunoXome than with free daunorubicin hydrochloride. The apparent elimination half-life of DaunoXome (daunorubicin citrate liposome injection) is 4.4 h, far shorter than that of daunorubicin, and probably represents a distribution half-life. Although preclinical biodistribution data in animals suggest that DaunoXome crosses the normal blood–brain barrier, it is unknown whether DaunoXome crosses the blood–brain barrier in humans [207].

In addition, TAS-103 (a novel quinoline derivative, topoisomerase inhibitor) incorporated in DPPC/Chol (2:1) liposomes (size < 80 nm) enhanced the survival time of mice with Lewis lung carcinoma to 42 days in comparison to the 38.6 days of those treated with free TAS-103 [209]. The increases in lifetime were 45 and 58% for the free TAS-103 and liposomal TAS-103, respectively.

In another study, the antibiotic cefoxitine was incorporated in DMPC/Chol (2:1) liposomes prepared using the reverse-phase evaporation technique in order to increase the efficacy of the drug characterized by a short half-life and poor intracellular diffusion [210]. It was shown that the cefoxitin levels achieved in liver and spleen 5 h postinjection were 6- and 16-fold higher than those observed after administration of free antibiotic. Also, the elimination rate through the kidney was slower.

Intraperitoneal Administration Intraperitoneal administration has the biological and pharmacological advantage of creating direct exposure of the tumor, infection, or inflammation to the therapeutic agent. This drug delivery method increases the dose intensity within the peritoneal cavity [211–213]. Intraperitoneal administration of liposomal formulations of anticancer drugs is preferred to the i.v. one, due to the higher drug accumulation in the tumors and lower drug plasma concentration minimizing drug toxicity [214–216].

Size, liposomal composition, charge, drug density in the liposomal membrane, and preparation method are some of the important parameters which need to be considered carefully to design an efficient DDS. Sadzuka et al. assessed DOX-incorporating liposomes made by either DMPC or DSPC of a variety of sizes (150, 600, and 4000 nm) and surface charge (positive and negative) on the therapy of solid tumors and peritoneal dissemination in Ehrilch ascites carcinoma-bearing mice [214]. When using small negative liposomes, lipid composition did not affect the clearance or stability of liposomes in the abdominal cavity. However, for neutral liposomes, DSPC ones were found more effective for the treatment of the solid tumor due to the higher stability of those liposomes in comparison to DMPC ones. Thus DSPC exhibited longer plasma circulation. As for the effect of surface charge, the positive vesicles were cleared faster from the abdominal cavity until 1 h postinjection and then showed a slower clearance rate until 48 h, in opposition to the negative ones. Larger particles were found in abundance in the peritoneal cavity and stayed longer there, inducing toxicity due to liposomal membrane disruption and release of the anticancer drug. Overall, it was concluded that the larger liposomes were effective against peritoneal dissemination and the smaller ones against the solid tumor.

The same author evaluated the method of preparation by using DOX-encapsulating liposomes on the peritoneal dissemination of tumor in Ehrlich ascites carcinoma-bearing mice [217]. The liposomal carriers were made either with the method of Bangham et al. [1] (BLDOX), the pH gradient [144] (PLDOX), or the gelation method (GLDOX) [147]. It was shown that survival in the BLDOX group was significantly prolonged compared to that in the DOXsol (DOX as solution) group, while there was no effect on survival of the GLDOX group. BLDOX liposomes appeared to be less stable and released DOX in a higher degree than the other formulations. The latter seems to be of high importance due to increased DOX level in the abdominal cavity and enhancement of drug efficiency for the local therapy.

Also, positive outcome was achieved after i.p. administration of the liposomal formulation of an L-dopa prodrug derivative to rats [218]. It was shown that the level of dopamine in rat striatum was 2.5-fold higher to what was obtained after i.p. administration of L-dopa or the free prodrug itself.

Subcutaneous Administration Liposomes given s.c. aim to target the lymphatic system for imaging, distribution of therapeutic agents, or vaccination [219, 220]. According to Oussoren and Storm, the determining factors influencing lymphatic absorption are liposome size and site of injection [219]. Liposome charge, composition, or PEG coating does not have a significant effect on the fate of the liposome trip in the lymphatic system.

Liposomes injected s.c. that do not enter the bloodstream either enter the lymphatic capillaries or stay at the site of injection. In the first case, 1–2% of the injected liposomal formulation is captured by the lymphatic nodes 12h postinjection. However, this depends on liposome size. Neutral vesicles smaller than 100nm pass through the interstitium and then to the lymphatics a lot easier than the larger particles. Drug carriers remaining at the site of injection will release the entrapped molecule. Often, 40% of the injected dose of small liposomes (about 70nm) is retained at the injection site. Therefore, liposome surface modification was attempted using nonspecific human antibodies and saccharides. Only saccharide-modified liposomes enhanced absorption from the injection site and enhanced lymph node uptake was in comparison to control liposomes [221]. Also, the specific site of injection is very important and species dependent. Taking the rat as animal model, s.c. injection in the dorsal foot or the footpad results in higher liposome uptake by the lymph nodes, in contrast to the flank as an injection point.

As mentioned earlier, the s.c. route for delivery of anticancer agents could prevent the metastatic spread of tumors that occurs often through the lymphatic system. However, a number of limiting factors, such as incomplete absorption of drug-loaded liposomes, which would increase, for example, the toxicity of the released drug at the surrounding tissue and the development of tumors in the regional lymph nodes could limit the therapeutic potential of liposomes.

In addition, for imaging studies, only liposome-encapsulated gadolinium was used successfully [222, 223]. In a more recent study, electron spin resonance (ESR) was applied successfully to investigate the integrity of MLV and the possibility of a depot effect after the s.c. injection in mice [224].

Also, Gregoriadis et al. evaluated the type and degree of immune response after s.c. injection of ovalbumin (OVA)–encoding plasmid DNA (2.5 or 10μg) either alone or in liposomes, in mice [220]. Anti-OVA serum antibody titers were detected in animals immunized with 10μg of liposomal DNA (after a single dose of antigen) and with both 2.5 and 10μg of liposomal DNA (after two doses of antigen) [225]. However, the anti-OVA response was not detected using the DNA alone.

Similarly, significantly higher humoral responses were obtained after s.c. administration of either a lipid and/or a nonionic-based vesicle-entrapped plasmid for the nucleoprotein of H_3N_2 influenza virus in comparison to the naked DNA alone [226].

5.3.3.2 Long-Circulating or PEGylated Liposomes

The liposome biodistribution profile changes significantly when the vesicle surface [227–229] is coated with polymers, usually PEG. Longer blood circulation, lower liver uptake, and higher accumulation in tumors have been reported. The presence of the hydrophilic groups of PEG on the liposome surface provides electrostatic and steric repulsion between PEG-grafted liposomes. PEG molecules neutralize the surface charge of vesicles and thus prevent their opsonization. Also, liposome opsonization is reduced due to inability of opsonins to bind to hydrophilic surfaces. Moreover, the thickness of the PEG layer influences the interaction of the liposomes with the macrophages. The thickness of the PEG layer depends on the PEG molecular weight and the amount (%) incorporated in the liposomal composition. For example (Figure 9), the fast clearance of positively charged stearylamine liposomes can be reversed by attachment of 6% mol of PEG with molecular weight of 750 or 5000 kDa [230]. In the case of phosphatidic acid–containing liposomes, only PEG 5000 can prolong blood circulation while phosphatidylserine-containing liposomes are eliminated fast due to the insufficient effect of 6% PEG 750 or 5000.

However, an optimum level of PEGylation (PEG 2000 kDa) was estimated for PC (1.85 mol %)/Chol (1 mol %) liposomes as to the effect on blood circulation [231]. As reported, after 5 mol % of PEG incorporation the accumulation in the liver was significantly decreased, while the minimum uptake by the spleen was achieved with 9.6 mol % of PEG insertion. The same authors showed that as the PEG amount grafted increased, the liposome accumulation in the heart increased. But above a 9.6 mol % of PEG, the circulation time was slightly decreased in blood and was increased in the liver and spleen. The uptake of liposomes by RES was even higher when 13.7% of PEG was present on the vesicle surface. As shown by another group, the optimum PEG amount required for liposome stabilization and prolonged half-life was 5–10 mol % DSPE–PEG 2000 [232]. Higher amounts than that led to disruption of the vesicles.

Intravenous Administration The effect of PEG on liposome biodistribution presented in the previous paragraph is for i.v. administration of liposomes. Several recent examples for the effect of PEG coating of liposomal drug formulations on the biodistribution (and thus pharmacological outcome) for different drugs follow. DOX concentrations were estimated in various organs after i.v. administration of DOX-loaded liposomes (PC/Chol/PEG2000, molar ratio 55:40:5) in xenograft tumor-bearing mice [233]. Obviously, the encapsulation of DOX in conventional or PEGylated liposomes reduced the drug concentration in liver, heart, kidney, and stomach compared to the drug solution and prolonged the circulation half-life to 46.09 h, in contrast to 26.04 and 23.72 of conventional and free DOX, respectively. However, PEGylated DOX showed higher antitumor activity in comparison to that entrapped in conventional liposomes. In comparison with free DOX, the inhibition rate of both liposomal formulations was higher. Doxil is a commercially available formulation of DOX entrapped in HSPC/Chol/mPEG liposomes [234]. Thereafter, PEGylated liposomes are in common use in many applications. Covalent attachment of specific molecules (folate) at the free end of PEG molecules results in liposomes able to be recognized by specific receptors (folate receptors on cancer cells). There are a huge number of research papers on the active targeting of modified PEGylated liposomes [235–237] with very promising results.

Intraperitoneal Administration The potential of PEGylated liposomes administered intraperitoneally has been evaluated for cancers located in the peritoneal cavity. For example, Syrigos et al. [238] studied the biodistribution of indium (^{111}In)–labeled PEGylated liposomes [Hydrogenated soya PC (HSPC)/Chol/PEG–DSPE] compared to free ^{111}In via either i.p. or i.v. route to non-tumor bearing mice. The AUC of In-PEG liposomes was 74-fold higher than that of free indium. The relative ratio of the AUCs (RR-AUCs) for i.p. versus i.v. administration for peritoneum was 1.36 [423.6 vs. 311.3% individual dose (ID) g/h]. The blood AUC values for i.p. and

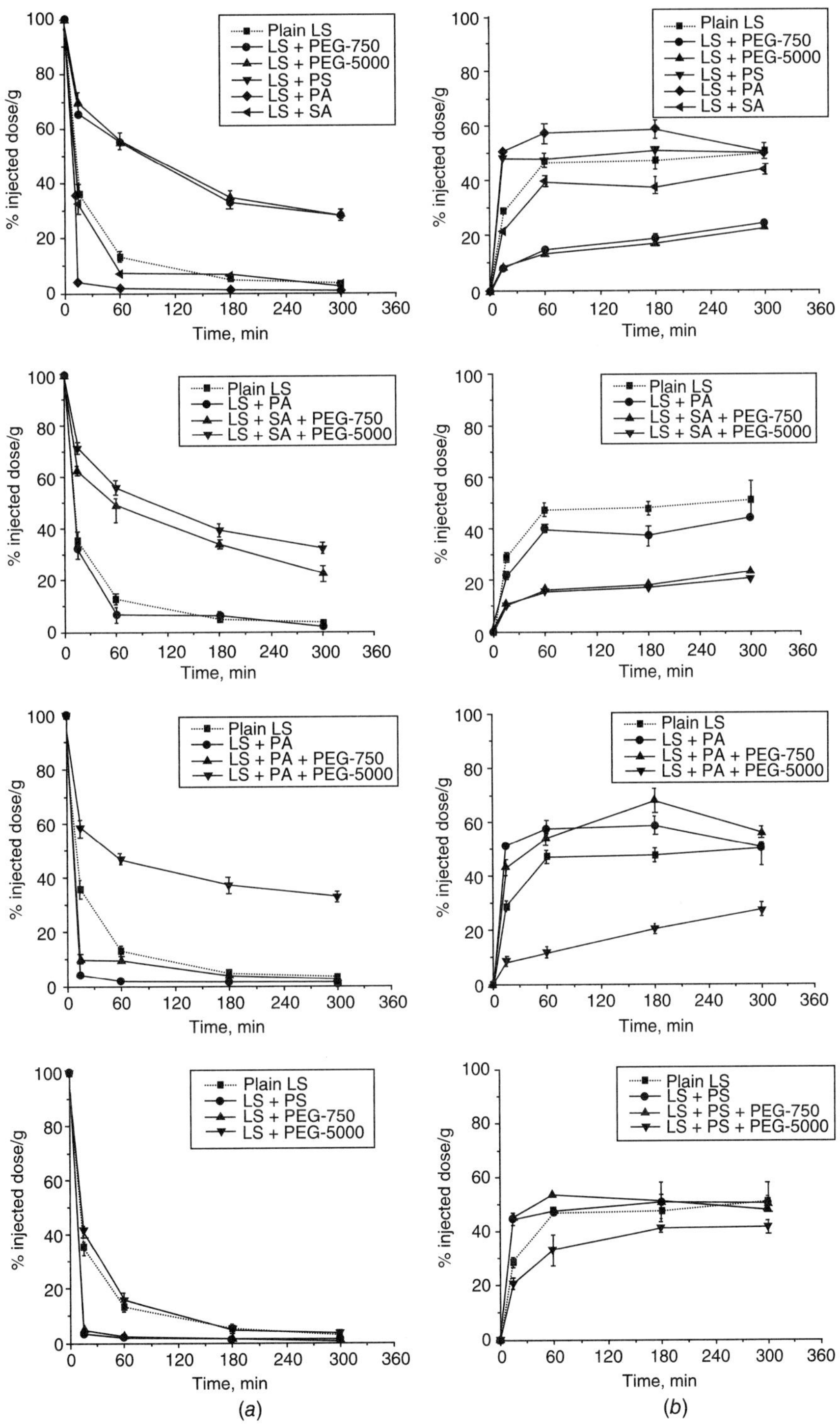

FIGURE 9 (*a*) Clearance from circulation and (*b*) accumulation in liver and spleen of liposomes composed of lecithin (LS) mixed or not with charged lipids (PS = phosphadityl serine, PA = phosphatidic acid, SA = stearylamine) and bearing or not, a surface coating with PEG molecules after IV administration (PEG-750 and PEG-5000 = polyethylene glycol with molecular weights 750 and 5000). (Reproduced from ref. 230 with permission of Elsevier.)

i.v. administration were essentially the same (RR-AUC 1.03; 453.7 vs. 439.2% ID g/h) 18h postinjection despite the delayed absorption of the liposomes from the peritoneal cavity. However, the relevant values for organs in the peritoneum were higher in case of i.p. administration. An increase in the range 1.2–5.1 was seen for organs such as stomach, pancreas, ileum, colon, gallbladder, ovary, and adrenal glands. This is an advantage compared to the i.v. administration because the drug can target both the primary site and any peritoneal deposits. The encapsulation of doxorubicin and cis-platin (small molecules with high toxicity and short half-lives) in PEGylated liposomes might increase the retention from the peritoneal cavity and reduce the drug toxicity.

Another study points out the vesicle size rather than the presence of PEG as a more determining factor to successfully tackle peritoneal cancers [239]. According to this study, the synergistic effect of Doxil after coadministration of PEG–SUV–interleukin-2 (IL-2) or MLV–IL-2 via either the i.p. or i.v. route to mice bearing M109 lung adenocarcinoma was studied. The cancer was inoculated i.p., resulting in multiple i.p. masses. Small PEGylated liposomes as vehicles for IL-2 for systemic treatment of metastatic lung cancer boosted the antitumor effect of Doxil to the same level achieved with soluble IL-2. In contrast, in the regional model, the most effective combination was Doxil with MLV–IL-2 liposomes. This is attributed to the retention and slow release of IL-2 in the peritoneal cavity due to the inability of MLVs to enter the circulatory system or the draining lymph vessels, whereas IL-2 in small liposomes or in soluble form escapes rapidly from the peritoneal cavity. Another possibility could be the enhanced immunostimulation results from the uptake of MLV–IL-2 by peritoneal macrophages as opposed to the stealth properties of PEGylated SUV–IL-2.

Subcutaneous Administration As for the use of PEGylated liposomes subcutaneously, there is not much reported. In one paper the influence of the administration either s.c. or i.m. of mitoxantrone-loaded liposomes was studied. It was reported that mitoxantrone showed reduced irritation when the formulation was administered i.m. rather than s.c. However, when PEG was incorporated on the liposome surface, there was no apparent protective effect of the liposomes [240].

5.3.3.3 Other Routes of Administration

In general, the administration route plays an important role on the impact of the therapeutic treatment, and it is chosen according to the kind or purpose of the treatment (local or systemic), toxicity, and accessibility of the diseased area. In this section specific characteristics of other routes of administration that are currently receiving attention will be emphasized and the most recent developments with respect to liposomal drug applications will be presented.

Pulmonary Drug Delivery Pulmonary epithelium offers many advantages for drug delivery due to easy access and the large surface area provided by alveoli [241]. Also, macromolecules can penetrate the lungs much easier and faster than other noninvasive routes avoiding the first-pass meabolism; therefore, many promising applications are being considered for the delivery of proteins and peptides. For a drug or drug delivery system to reach the lungs successfully, it has to be aerosolized with optimum aerodynamic particle diameter between 1 and 3 μm. The latter, in combination with how the patient inhales determines if the drug particles deposit primarily in the conducting ways or in the alveoli.

Pulmonary delivery of liposomes has focused on the treatment of asthma, infectious diseases, genetic diseases (cystic fibrosis), and lung injury and lately on gene therapy.

Special attention has been paid on the physical characterization of liposome aerosols [242–244], including dry powder formulations [245], and cationic liposome DNA complexes [246, 247]. Corticosteroid therapy using liposomal formulations has focused on the development of aerosols containing beclomethasone dipropionate, triamcinolone acetonide, and triamcinolone acetonide phosphates [248–251]. Aminoglycosides have been considered as good candidates for pulmonary delivery because of their potency and their ability to directly target the lungs. When amikacin was encapsulated in liposomes [fluid or rigid state (chol containing)] and administered in sheep, the drug mean residence time (MRT) increased 5 times compared to the instilled solution (rigid liposomes gave 2 times higher MRT compared to fluid ones) [252]. In a similar study, prolonged retention of liposome-encapsulated tobramycin was reported after administration in *Pseudomonas aeruginosa*–infected rat lungs [253]. Liposomal formulation of antioxidants has also been investigated for pulmonary delivery [254]. Liposomes containing α-tocopherol prolonged the residence time of radioactively labeled glutathione [255]. Paraquat poisoning, which causes extended damage to lung tissues, was attenuated after pretreatment with α-tocopherol-containing liposomes [256]. Liposomes containing CuZn/superoxide dismutase and catalase were found to protect the lungs of premature rabbits when exposed to hyperoxia [257].

Drugs for the treatment of infections such as aspergillosis, tuberculosis, and anticancer therapy have been formulated in liposomes and tested in vivo by administration via the respiratory system [258–260]. More specifically, rifampicin used to treat pulmonary tuberculosis associated with AIDS has been incorporated in MLV liposomes (PC/Chol, 7:3 molar ratio) with the aim of increasiing its efficacy to macrophages and reduce side effects [258]. The rifampicin retention was found higher in PC/Chol/DCP (7:3:0.1) liposomes tested due to electrostatic interaction between DCP and the drug. Both liposomal formulations showed greater accumulation in the lungs when compared to the controls. In the case of free drug and after 0.5 h postadministration, only 39.12% of the administered dose was retained in the lungs and 29.84% of it was found in the serum. No drug was estimated in the lungs 24 h later. PC/Chol MLVs demonstrated higher lung accumulation (49.03%) after 0.5 h, but the overall distribution pattern was not much different to that of the free drug in solution with no drug estimation at the 24-h time point. It seems that they were rapidly passed to the systemic circulation and to the RES organs (liver and spleen). The negatively charged MLVs showed even higher drug accumulation (53.86% of the administered dose) while 4.14% of it was still present after 24 h. This is attributed to the interaction of the negatively charged liposomes with the scavenger receptors on alveolar macrophages [261]. Also, it has to be mentioned here that the ligands maleylated bovine serum albumin (MBSA) and *O*-steroyl amylopectin (O-SAP) were incorporated on the liposome surface because they are recognized by the scavenger receptors of the macrophages. The lung accumulation levels were 61.49 and 65.14% of the administered dose, respectively, after 0.5 h, while after 24 h they were 8.12 and 10.75%, respectively. The relative lung retention of the various formulations after 6 h of administration was 1.3 times for PC/Chol MLVs, 3.4 times for PC/Chol/DCP MLVs, 4.53 times for MBSA-PC/Chol, and 4.76 for O-SAP–PC/Chol in comparison to plain drug solution administered by aerosolization.

The same group (Vyas et al.) studied the impact of amphotericin B entrapped in liposomes (MLVs: PC/Chol, 7:3 molar ratio) in the absence or presence of ligands [*O*-palmitoylated mannan (OPM) or *O*-palmitoylated pullulan (OPP)] on the vesicle surface as potential use for the treatment of aspergillosis [259]. Optimized formulation was the one of PC/Chol 7:3 molar ratio with the production of spherical MLVs of approximate mean vesicle size 2.56 μm and maximum entrapment efficiency 78.2%. Again the lung uptake of the ligand-appended liposomes was higher compared to the plain liposomes. The lung accumulation levels of OPM- and OPP-coated liposomes were 58.12 and 55.02% of the administered dose 5 h postadministration, respectively, while 24 h later the relevant lung retentions were 11.23 and 9.86%, respectively.

Methotrexate (MTX) (a folic acid antagonist) was entrapped into liposomes (PC/PI/Chol 2:1:1 molar ratio) so as to reduce nephrotoxicity and investigate the pharmacokinetics of the liposomal MTX [260]. Indeed, the latter showed increased MTX retention in the lungs while the biodistribution in spleen and kidney was less than that obtained with free MTX. Similar results were obtained by liposomalization of various anticancer drugs elsewhere [262–264]. For example, cytosine arabinoside was administered intratracheally to rats in the free or the liposomal form [262, 263]. Liposomal drug was effective into the lung but not other tissues, contrary to the free drug. Moreover, the antitumor properties of the anticancer 9-nitrocamptothecin (9NC) after its liposomalization were tested in three different human cancers xenografted s.c. in mice as well as murine melanoma and human osteosarcoma pulmonary metastases [264]. The liposomal form of anticancer drug inhibited the growth of subcutaneous tumors and metastatic pulmonary cancers given via the respiratory system. Intramuscularly administered liposomal drug exhibited some anticancer activity, but that achieved using the aerosol was superior. Thus, the liposomal 9NC aerosol was proved to be of high potential for the treatment of cancers throughout the body.

Recently the gene therapy of pulmonary diseases using liposomal formulations has attracted a lot of attention. Transfection of the lungs of animals with aerosolized cationic liposome–DNA complexes has been attempted [265, 266]. However, the transfection efficiency was rather low despite the large amounts of DNA used in these studies. Cationic liposomes from 2, 3-dioleyloxy-N-[2(sperminecarboxamido)ethyl]-N, N-dimethyl-1-propanaminium trifluoracetate (DOSPA), (±) N-(2-hydroxyethyl)-N, N-dimethyl-2, 3-bis(tetradecyloxy)-1-propanaminium bromide (DMRIE) mixed with DOPE, and DNA at fixed ratios of DNA/lipid 1:4 for DMRIE and 1:3 for DOSPA were tested with two different types of jet aerosols (Aerotech II and Puritan-Bennett 1600) [246]. The decrease in transfection activity was gradual with Puritan-Bennett 1600, in contrast with Aerotech II, which rapidly lost transfection efficiency. That was attributed to the increased throughput of the Aerotech II, resulting in more frequent cycling and therefore damage of the complex.

The impact of the zeta potential of the formulation was emphasized by Eastman et al. [247]. The maximal aerosol transfer efficiency of cationic lipid/DNA complexes was achieved in the presence of a salt concentration of 25 m*M*. The authors attributed that to the fact that the formulation kept its zeta potential between –40 and –50 mV. As a closing remark, one of the major problems and challenges of aerosol delivery is the duration to deliver therapeutic doses of DNA to the lungs. This possibly could be overcome with dry powders avoiding volume limitations of aqueous dispersions.

Oral Delivery The destructive effects of the conditions in the gastrointestinal tract (GIT), especially due to the presence of bile salts, are known and well established [267, 268]. The lipid composition of liposomes determines to a large extent the possibility of remaining intact under such conditions [58]. Quadachi et al. attempted to provoke IgA response from the M cells of Peyer's patches after oral administration of OVA-loaded MLV liposomes made by either soya PC or DSPC to a model of hypersensitivity to OVA Balb/C mouse [269]. Clearly, DSPC MLVs were much more stable as demonstrated in in vitro stability studies in simulating GIT media. However, liposome incorporation of OVA did not cause any significant impact on the reduction of hypersensitivity to that obtained with the free allergen. Surprisingly, the empty liposomes show some immunoresponse which was attributed to nonspecific stimulation.

The biodistribution of novel liposomal-like spherical carriers called Spherulites, consisting of PC/Chol/polyoxyethylene alcohol (43:4:3 w/w/w), was shown to be more promising [270]. These were prepared by shearing the phospholipid bilayer and labeling with ^{111}In. Their integrity was demonstrated by an increase in radioactivity in the blood 1 h after oral administration to fasted rats, while no increase was seen for free label.

Skin Delivery Dermal delivery of phospholipid-based vesicles first appeared in the literature in the early 1980s [271, 272]. Liposomes can play a dual role after their application to the skin: retention (and perhaps protection) of the active compound across the stratum corneum and acting as a penetration enhancer. The composition, size, and vesicle surface charge are parameters that can influence the transport rate of drugs contained in liposomal formulations across the skin [273]. Regarding their composition, liquid-state liposomes resulted in higher skin permeation rates compared with gel–liquid liposomes for progesterone and Triamcinolone (TRMA) [274, 275]. Moreover skin lipid liposomes have provided higher drug disposition in the deeper layers of the skin for corticosteroids and acyclovir [276, 277]. Neverthless, in another study [278] higher amounts of acyclovir in the skin were delivered from conventional lipid liposomes compared with liposomes containing skin-based lipids. These studies emphasize that a careful design is needed to define optimal compositions, depending also on the specific objective which may be either dermal (topical) delivery of the liposome-encapsulated drug (which is easier since only increased retention of the drug in the skin is desired) or transdermal delivery (systemic absorption of the drug), which is more complicated, and in general deeper penetration of the liposome carrier in the skin is required. What is also important as regards the final outcome (especially when transdermal delivery is the objective) is the type of final formulation with respect to the conditions applying, occlusive or nonocclusive.

Amphotericin B encapsulated in charged liposomes demonstrated 10-fold higher transport rates across the skin compared with neutral liposomes [279]. In a similar manner, the retention of acyclovir from positively charged vesicles was much higher compared with other formulations [277]. This could be attributed to the attraction between the positively charged liposomes and the negative charge of the skin [277]. The pore size of the skin is approximately 0.3 nm; however it can be opened up to 40 nm without significant damage to its structure. To some extent, vesicles can transport across the skin via the follicular and transcellular routes. Therefore the size of the vesicles used for transdermal delivery is a crucial parameter for the overall performance of the formulation [280–282]. Several studies have also emphasized the impact of the size of the liposomes on the transport rate of active compounds across the skin [277, 280, 283, 284]. Biologically active macromolecules, including superoxide dismutase [285] and interferon-γ [286], have been successfully applied to skin in liposomal formulations. Generally it is believed that the main pathway for transdermal delivery of active compounds is either intercellular or paracellular. However, the appendage transport (e.g., follicular route) has attracted a lot of interest lately. The pilocebaceous units (hair follicle, hair shaft, sebaceous glands) can be used for drug targeting. Combinations of liposomes with DNA [287, 288] and monoclonal antibodies [289] have demonstrated that liposome composition, hair structure, and hair cycle play significant roles in the transfection of human hair follicles. Two different mechanisms have been suggested for the incorporation of active compounds to the hair shaft: (1) direct permeation of the vesicular formulations with the active compound to the hair shaft and (2) incorporation of the active compounds in the follicular matrix cells and then into the hair shaft as the matrix cells develop and differentiate into new hair shafts [290].

Recently, for the transdermal delivery of drugs using carrier systems, attention has been focused on the development of *transformable* [284, 285] or *elastic* vesicles [12]. These vesicles are liposomes that contain surfactants or in general "edge activators" in addition to phospholipids in their lipid membranes (Figure 10), a fact that

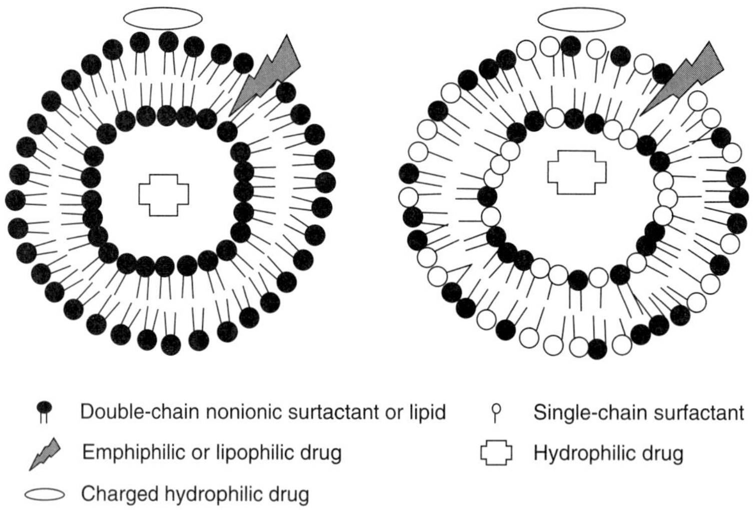

FIGURE 10 Conventional liposomes/elastic vesicles. Charged hydrophilic, amphiphilic, and lipophilic drug molecules can be associated with the bilayers of the vesicles, whereas hydrophilic substances can also be entrapped in the vesicles. Rigid vesicles consist of double-chain nonionic surfactants or lipids in the presence or absence of cholesterol (left image). Elastic vesicles consist of double-chain surfactants or lipids and an edge activator. The edge is often a single-chain surfactant (right image). (Reprinted from ref. 273 with permission of Elsevier.)

increases their elasticity (ability to be deformed, without being disrupted, after applying pressure on them). Modified liposomes called ethosomes (containing alcohol) have shown increased skin permeability. Because of their structure, ethosomes are able to encapsulate and deliver through the skin highly lipophilic molecules such as cannabinoids, testosterone, and minoxidil as well as cationic drugs such as propranolol and trihexyphenidil or even plasmids and insulin [291]. Although the mechanism of the increased transdermal delivery of drug molecules still remains a controversial issue, mainly with respect to the depth at which intact vesicles can travel in the skin, many interesting results are being generated. However, due to the limited permeability of the skin membrane, physical enhancement mechanisms, including iontophoresis, electroporation, and ultrasound, have been used in combination with chemical enhancers (liposomes) to increase skin permeability. One of the methods, application of an electric current to the skin, has been shown to promote the transdermal transport of drugs by an additional driving force, namely, an electrical potential gradient across the skin [292]. Transdermal iontophoretic transport of a liposomal formulation across human cadaver skin was first reported for [Leu5] enkephalin [293]. Liposomes could penetrate into the skin. Enkephalin, when delivered iontophoretically from liposomes carrying positive or negative charge on their surface, was found to be the same or less than that of the controls; however, the degradation of enkephalin was less in liposome formulations as compared to controls, demonstrating that liposomes can protect peptides from the proteolytic environment of skin. However, when enoxacin was encapsulated in different liposome formulations and was transported (electrically assisted) across the skin, the drug transport results showed that the permeability of enoxacin released from liposomes was higher compared to that of free drug [294]. In vitro transdermal iontophoretic delivery of estradiol from ultradeformable liposomes, saturated aqueous solution [295], and conventional liposomes [296] has demonstrated the superiority of ultradeformable liposomes (Transfersomes) due to their lipid composition. Liposomal formulations of β blockers were iontophoresced in vivo to hairless rats [297]. Skin irritation was significantly reduced when a liposomal formulation of the propranolol base was used rather than the base itself, emphasizing another important role liposomes could play. When adriamycin was delivered via the hair follicles using various liposomes and iontophoresis combined with application of ionic liposomes, higher transport rates were obtained with the latter, emphasizing their synergistic effect [298]. Moreover skin electroporation was applied to enhance gene transfer into subcutaneous MC2 murine breast tumor skin in combination with cationic liposomes demonstrating significant transfection improvement [299]. However, when electoporation combined with estradiol-loaded liposomes were applied to skin, the estradiol skin penetration was not affected significantly. That was attributed to the antienhancer or retardant effect of liposomes [300]. In a mechanistic study, anionic phospholipids were found to enhance the transdermal transport of molecules by electroporation compared to cationic or neutral phospholipids, offering new insights to design better enhancers for transdermal drug and vaccine delivery [301]. Finally higher transport across the skin obtained after combined application of ultrasound and liposomal formulations of diclophenac demonstrated a synergistic effect [302].

Ocular Delivery Ocular drug delivery has evolved into a great challenge and a subject of interest for many scientists with different backgrounds, including medical, clinical, pharmaceutical, physical, chemical, biochemical, and toxicological sciences. For ailment of eye diseases, topical administration is preferred over systemic in order to avoid systemic toxicity, for rapid onset of action, and for decreasing the required dose.

The main route for intraocular absorption is across the cornea [303]. In terms of drug delivery, the cornea presents an effective barrier to the absorption of both hydrophilic and lipophilic compounds. Actually, the main constraints in topical ocular delivery are (i) poor ocular retention of conventional dosage forms [304] and (ii) poor corneal absorption. Various approaches have been developed to increase the bioavailability and duration of therapeutic action of ocular drugs. One such approach is based on the use of drug delivery systems [305, 306], which provide controlled and continuous delivery of drugs and can also provide improved (increased) residence time of the drug at the delivery site. Recently, intravitreal drug injection has evolved into a preferred administration method for therapy of disorders in the posterior segment of the eye [305]. The procedure is associated with a high risk of complications, particularly when frequent, repeated injections are required. Thus, sustained-release technologies are being proposed, and the benefits of using colloidal carriers in intravitreal injections are currently under investigation for posterior drug delivery.

Between the different types of particulate drug delivery systems, liposomes offer additional advantages for ophthalmic delivery, since they are completely biodegradable and relatively nontoxic and thus are well tolerated by the eye [305, 306]. Indeed, when using other types of colloidal systems, for example, nanoparticles consisting of polyalkyl cyanoacrylate, inflammation and damage of the corneal epithelium have been reported [307–309]. Another potential advantage of liposomes is their ability to come in intimate contact with the corneal and conjunctival surfaces. This results in increased probability of ocular drug absorption [310, 311].

The potential of liposomes in topical ocular drug delivery was first exploited in the 1980s by a number of research groups [303, 310–315]. As an example, higher levels of inulin were found in the cornea when it was encapsulated in liposomes as compared to its aqueous solution [301, 314], and this was attributed to the physical adsorption of lipid vesicles onto the epithelial surface of the membrane [315, 316]. More recently a number of liposomal applications for ocular delivery have been under investigation [305, 306] for anterior as well as posterior segment administration, as outlined in Table 3. Indeed, a large number of ophthalmic drugs used in cases of ocular surface disorders (such as dry eye syndrome) [317], keratitis and uveitis [318–327], and keratoplasty [328–331] have been studied in liposomal form, and in most cases the results were promising in terms of drug penetration and retention in the various ocular tissues (cornea, sclera, retina, and choroids), following subconjunctival administration. In some cases, detectable levels of drugs were found in ocular tissues up to 7 days after administration [305, 306].

As mentioned above, the ability to adsorb to the cornea and an optimal drug release rate have been defined as the two liposomal attributes most responsible for increasing bioavailability after topical ocular administration. A number of factors, including drug encapsulation efficiency, liposome size and charge, distribution of the drug within liposomes, stability of liposomes in the conjunctival sac and ocular tissues, their retention in the conjunctival sac, and most importantly their affinity toward the corneal surface and the rate of release of the encapsulated drug, have

TABLE 3 Experimental Liposomal Preparations of Drugs for Anterior and Posterior Segment Administration

Drug Class	Anterior Segment Drugs	Posterior Segment Drugs
Antibiotics	Gentamicin Norfloxacin Tobramycin	Clindamycin Gentamicin Penicillin
Antifungals	Amphotericin B	Amphotericin B
Antivirals	Acyclovir Idoxuridine	Ganciclovir Trifluorothymidine (Trifluridine)
Steroids	Dexamethasone	
Immunosuppressives	Cyclosporine FK506 (Tacrolimus)	Cyclosporine
Antimetabolites	5-Fluorouracil (5-FU) 5-Fluoroorotate	5-Fluorouracil (5-FU) 5-Fluorouridine (5-FUR) Bleomycin Cytosine arabinoside (Cytarabine) Daunomycin Daunorubicin Etoposide (VP-16)
Platelet-aggregating agents	Adenosine diphosphate	Adenosine diphosphate
Photosensitive cytotoxic agents		Verteporfin (BPD-MA)
Miscellaneous	Dichloromethylene diphosphonate (Clodronate) Disulfiram Immunoglobulins	Dichloromethylene diphosphonate (Clodronate)

Source: S. Ebrahim, G. Peyman, and P. J. Lee, *Survey of Ophthalmology*, 50, 167–182, 2005.

been found to influence the effectiveness of liposomes in topical ocular drug delivery [310, 323, 332–335, 337]. Indeed, liposomal manipulation to increase corneal adherence has met with some success [310].

Positively charged liposomes seem to be preferentially captured at the negatively charged corneal surface as compared with neutral or negatively charged liposomes. Aditionally, cationic vehicles are expected to slow down drug elimination by the lacrymal flow both by increasing solution viscosity and by interacting with the negative charges of the mucus [334, 335]. Indeed, positively charged phospholipids yielded superior retention of liposomes at the corneal surface in rabbits [336]. Schaeffer et al. [310] worked with indoxole and penicillin G and reported that liposome uptake by the cornea is greatest for positively charged liposomes, less for negatively charged liposomes, and least for neutral liposomes. Positively charged unilamellar liposomes enhanced transcorneal flux of penicillin G across isolated rabbit cornea more than fourfold. Similar results were also obtained by others [336, 323]. By observing the morphology of corneal surface treated with liposomes, it was suggested that positively charged liposomes formed a completely coated layer on the corneal surface [323]. These liposomes bind intimately on the corneal surface, leading to an increase in residence time and therefore to an increase in corneal absorption time.

With respect to vesicle size, larger particles are more likely to be entrapped under the eyelids or in the inner canthus and can thus remain in contact with the corneal and conjunctival epithelia for extended periods. Indeed, larger liposomes have been found to resist drainage at the inner canthus and are more bioavailable at the ocular surface [337]. However, for patient comfort, it is considered that solid particles intended for ophthalmic use should not exceed 5–10 μm diameter [338].

Assessment of ocular irritability of neutral or positively charged liposomes by the Draize test, histological examination, and the rabbit blinking test has also been reported in the literature [339]. The mean total score (MTS) of the Draize test was found to show a slight increase immediately following instillation of liposome preparations. However, it did not exceed the "practically nonirritating level," and no corneal histological alteration was observed by optical microscopy. Neutral liposome preparations were confirmed to be nonirritating; however, positively charged liposomes may cause initial pain or unpleasantness following instillation. Thus, althouth positive charge helps in improving the contact time with the cornea, at the same time it can lead to irritation. Additionally, the release rate of the drug is found to be more in neutral liposomes, while increased liposome size restricts solution drainage, thus prolonging contact time of the drug, but it can be increased within the limits of not inducing any irritancy. From all the above it is understood that in each case liposome properties have to be adjusted for best in vivo therapeutics results. As mentioned above, liposome stability is another important factor. Barber and Shek reported that increasing the cholesterol content of the liposomal membrane decreased the rate of tear-induced release of its contents [340].

The use of bioadhesive polymers (e.g., a polyacrylic acid, chitosan, hyaluronic acid) to prolong the residence time of an ocular preparation in the precorneal region, due to increased formulation viscosity, is another approach which can further improve liposomal drug delivery. In this respect "collasomes" (liposomes coupled to collagen matrices) increased the bioadhesive ability of liposomes, were well tolerated, and could be instilled safely and effectively by patients in the same fashion as ointments or drops [341]. It was also demonstrated that liposomes coated with collagen layer bound to cell monolayer with higher affinity [342]. Other approaches were used to increase the contact time of ocular liposomes, such as the case in which prolonged retention of liposomal suspension of oligonucleotide was achieved by dispersing liposomes within an in situ gel-forming medium [343–345]. Novel measures to further enhance adsorption of liposomes and increase penetration of the cornea included application of a natural lectin which promoted binding of ganglioside-containing liposomes to the cornea [346].

Liposomes can also be used as promising dosage forms for topical administration of immunosuppressive compounds for the treatment of ocular immune-mediated diseases [327]. Indeed, it was found that liposomes containing immunosuppressive compound FK506 were effective in delivering significantly higher drug concentrations to all ocular tissues and particularly aqueous humor and vitreous humor as compared to the oil formulation of the agent. Further, liposomes can be used to protect drug molecules from attack of metabolic enzymes present at the tear/corneal epithelium interface, as demonstrated for the *O*-palmitoyl prodrug of tilisolol [347].

As mentioned above, intravitreal injection of drugs should be used in many cases to achieve therapeutic intravitreal drug levels. This is especially true for cases of viral retinitis, such as cytomegalovirus (CMV) retinitis and acute retinal necrosis (ARN) which require intravitreal injection of antivirals, or for the treatment of bacterial and fungal endophthalmitis or proliferative vitreoretinopathy [305]. It still remains a controversial issue whether liposomes can reach the retina after intravitreal injections and which vesicle physicochemical characteristics should be preferred for such formulations.

In addition to conventional drugs, liposomes were also used for intravitreal administration of oligonucleotides in order to treat ocular viral infections such as herpes simplex virus or CMV [344]. Antisense oligonucleotides are poorly stable in biological fluids and their intracellular penetration is limited. Hence a system that is able to permit a protection of oligonucleotides against degradation and their slow delivery into the vitreous should be favorable for improving the therapeutic outcome in addition to patient compliance. It was found that lipid vesicles are able to protect oligonucleotides against degradation by nucleases [344, 345]. Furthermore, they increase the retention time of many drugs in the vitreous. Thereby, the use of liposomes for intravitreal administration is a very promising approach.

Nasal Delivery Liposomes are able to decrease mucociliary clearance in the case of nasal administration due to their low viscosity. This is attributed to the incorporation of liposomal lipids in the membranes of the nasal epithelial cells, which results in the opening of pores in the paracellular tight junctions [348]. MLVs containing nifedipine administered via the nasal route could attain a constant plasma level [349]. Liposomal formulations of levonorgestrel containing carbopol or chitosan demonstrated prolonged contact time with the absorptive surfaces, resulting in increased bioavailability of the intranasally administered drug [350].

The nasal route may be highly promising for candidate vaccines against potential pathogens (and infection-related diseases such as cancers) that utilize this route of infection [351, 352]. In addition to the likelihood of increased patient compliance, immune responses elicited by nasal administration may be more predictable when compared with the vaginal route due to immunological changes in the female reproductive tract during the menstrual cycle [352]. Previous work with liposome-containing vaccines for nasal delivery demonstrated that liposomes can confer adjuvancy to the subunit influenza vaccine, but also empty liposomes administered 48h prior to immunization resulted in immune stimulation, emphasizing that the properties and composition of the liposomes play a significant role facilitating the transport of the antigen across the membrane [353]. More recently, several promising liposome-based vaccines are being designed and investigated for delivery by the nasal route, such as the liposome-encapsulated plasmid DNA-encoding influenza virus hemagglutinin, which has been reported to elicit mucosal, cellular, and humoral immune responses after intranasal administration in Balb/C mice [354]. Additionally, nasal immunization studies using liposomes loaded with tetanus toxoid were perfomed, and it was found that intranasal administration of liposome-encapsulated vaccines can be an effective way for inducing mucosal immune responses [355]. Furthermore, intranasal immunization studies have been carried out with liposomes containing recombinant meningococcal opacity proteins [356] and with anthrax-protective antigen protein incorporated in liposome–protamine particles [357], both with promising results.

Vaginal Delivery The vaginal route has been under investigation in the last years, especially for the topical delivery of drugs that are intended to act in the vagina, as contraceptives, microbicides and antibiotics, as recently reviewed [358]. In such cases the main advantage of using liposomes would be the controlled and sustained release of the drug at the site, which would result in a less frequent drug administration and improved patient compliance.

However, the major limitation of using liposomes topically and vaginally is the liquid nature of preparation. Nevertheless, several formulation characteristics should be optimized in order to achieve the needed rheological and mucoadhesice properties for maximum retention of the delivery system in the vagina. Research for the development of liposomal gels (gels that contain liposomes) with the required properties is currently ongoing [358–360].

It has been demonstrated that, by their incorporation in an adequate vehicle, such as carbopol resins, the original structure of vesicles is preserved [359], while liposomes are fairly compatible with gels made from polymers derived from such resins. A previous study has suggested application of liposomes containing antimicrobial drugs for the local therapy of vaginitis [359], while recently the design and in vitro evaluation of bioadhesive liposome gels containing clotrimazole and metronidazole, or acyclovir, were carried out [359, 360].

Some other applications are arising lately concerning liposomes and vaginal administration. As mentioned above (in the nasal administration section), mucosal surfaces serve as a gateway to disease. It was recently demonstrated that RNA interference can be used to manipulate mucosal gene expression in vivo. Using a murine model, it was shown by Zhang et al. [361] that direct application of liposome-complexed small interfering RNA (siRNA) mediates gene-specific silencing in cervicovaginal and rectal mucosa. A single vaginal or rectal administration of siRNA targeting hematopoietic or somatic cell gene products reduced corresponding messenger RNA (mRNA) levels by up to 90%. Additionaly, liposomal siRNA formulations proved nontoxic, did not elicit a nonspecific interferon response, and provided a means for genetic engineering of mucosal surfaces in vivo.

In addition, it was recently found that when human immunodeficiency virus type 2 (HIV-2) DNA vaccine were formulated with cationic liposomes [362], stronger immune responses in mice were observed compared with naked DNA alone. Using this knowledge, very recently a vaccine consisting of some HIV-2 genes (*tat*, *nef*, *gag*, and *env*) was formulated within cationic liposomes by Lochera et al. [363]. Baboons (*Papio cynocephalus hamadryas*) that were immunized by the intramuscular, intradermal, and intranasal routes with these expression constructs were challenged with HIV-2_{UC2} by the intravaginal route, and the results of this study demonstrate that partial protection against HIV-2 vaginal challenge, as measured by reduced viral load, can be achieved using only a DNA vaccine formulation.

5.3.4 APPLICATIONS OF LIPOSOMES IN THERAPEUTICS

It is well known that liposomes have many applications in drug delivery. Initially, after the limitations of conventional liposomes were noticed, great effort was given

TABLE 4 Currently Marketed Liposome-Based Products

Active Agent[a]	Application
Daunorubicin (DaunoXome, Gilead Sciences, Inc.)	Kaposi's sarcoma
Doxorubicin (Doxil/Caelyx, Ortho Biotech ProductsLP/Sequus Pharmaceuticals)	Kaposi's sarcoma
Amphotericin B (Ambisome/Abelcet, Fujisawa Healthcare, Wyeth Pharmaceuticals)	Fungal infections in immunocompromised patients
Doxorubicin (Myocet/Evacet, Sopherion/ Liposome Company)	Metastatic breast cancer
Hepatitis A virus envelope proteins (Epaxal, Berna Biotech)	Hepatitis A
Influenza virus (Inflexal V, Berna Biotech)	Influenza
Verteporfin (Visudyne, Novartis Ophthalmics)	Age-related macular degeneration

Source: D. Felnerova, J. F. Viret, R. Gluck, and C. Moser, *Current Opinion in Biotechnology*, 15, 518–529, 2004.

[a]Product names and companies given in parentheses.

toward the therapy of parasitic diseases, due to the fact that the targeting of RES macrophages, the place were parasites are mainly located, was considered to be very easy and fast (usually mentioned as passive targeting). Indeed, even now, research is ongoing, such as the treatment of drug-resistant visceral leishmaniasis with liposomal amphotericin-B [364] or with sterically stabilized liposomes containing camptothecin [371]. Nevertheless, possibly due to the high cost of liposome manufacturing, in relation to other drug formulations, and the fact that the main need for such medicaments would be for third world countries, such products have not been marketed for these diseases, despite the fact that they offer therapeutics advantages. In addition, again due to the same etiology, a very small part of recent research efforts and money are devoted to such diseases.

A list of the marketed liposomal products is presented in Table 4 [364–366]. In addition, liposomes are currently being investigated for a variety of conventional and novel drugs: therapeutic agents, including antibiotics (as amikacin [367], vancomycin, and ciprofloxacin [368]); anticancer agents (e.g., paclitaxel [369] and cisplatin [370]; camptothecin and analogs [371–373]), biologics such as antisense oligonucleotide [374], DNA, and siRNA [375]; and muramyl tripeptide [376]. A list of most liposome-based products currently under clinical investigation is presented in Table 5. Additionally many products are currently being evaluated in preclinical studies. In many of the latest studies, the liposomes used have been surface modified with active targeting ligands to improve delivery of therapeutics to target cells [238, 377–379].

After investigating the recent literature, we have seen that most recent efforts connected with the use of liposome in therapeutics are mainly connected with the treatment of cancer. This is the reason why here we deal with the most common cancer types (brain, breast, lung, and ovarian cancer), and ongoing research and clinical treatments are discussed in more detail. However, we do not want to imply that the future of liposomes in drug delivery is limited to cancer therapy. Indeed, liposome structure, characteristics, and versatility are sure to find, in

TABLE 5 Liposome-Based Products Currently under Clinical Testing

Active Agent or Product[a]	Application	Company and Trial Phase/Reference
Drug delivery		
Caelyx	Bladder cancer	Schering Plough; approved EU
Doxil	Multiple myeloma	Schering Plough; III ALZA Pharm; III
	Bladder, liver cancer	ALZA Pharm; II, III
	Pancreatic cancer	ALZA Pharm; II, Sequuz; II
Doxorubicin combined with ATB	Prostate cancer	Pharmacia; III, Neopharm; III
	Bladder cancer	Neopharm; II, III Pharmacia; II, III
Myocet combined with ATB	Bladder cancer	Liposome; III
Liposomal ether lipid (TLC ELL12)	Bladder cancer	Liposome; I
	Lung, prostate, skin cancer	Elan Pharm; I
Platinum (Aroplatin)	Cervical, ovarian, kidney	Aronex Pharm; II
	Lung, pancreatic cancer	Aronex Pharm; I, II
Annamycin	Leukemia	
Paclitaxel	Head and neck cancer	Pharmacia; II, III
	Lung cancer	Neopharm; II, III
Vincristine (Onco-TCS)	Lung cancer	Inex Pharm; II
Topoisomerase inhibitor (OSI 211)	Lung, ovarian cancer	OSI Pharm; II
All-trans retinoic acid (ATRA-IV)	Lung cancer	Antigenics; II
Mitoxanthrone	Other cancers	Neopharm; II
Nystatin (Nyotran)	Leukemia (antifungal)	Pharmacia; II
	Lung (antifungal)	Aronex Pharm; II
	Prostate cancer (antifungal)	Abbott Laboratory; II
DNA delivery		
Human leucocyte antigen (HLA) B and β_2 microglobulin plasmid DNA (Allovectin)	Head and neck cancer	Vical; II
Interleukin-2 plasmid DNA	Kidney, prostate cancer	Vical; II
Antisense toraf-1 (LerafON)	Leukemia	Neopharm; I
Antigen delivery:		
MUC-1 peptide: BLP25 (human epithelial mucin peptide)	Lung cancer	Biomira; II Merck; III

Source: D. Felnerova, J. F. Viret, R. Gluck, and C. Moser, *Current Opinion in Biotechnology*, 15, 518–529, 2004.

[a]Commercial names are given in parentheses. ATB, antibiotika.

addition to those existing already, numberous applications in drug delivery in the future.

Recently, a multicomponent liposomal drug delivery system consisting of doxorubicin and antisense oligonucleotides targeted to MRP1 mRNA and BCL2 mRNA to suppress pump resistance and non–pump resistance, respectively, has been developed [379]. This liposomal system successfully delivered the antisense oligonucleotides and doxorubicin to cell nuclei, inhibited MRP1 and BCL2 protein synthesis, and substantially potentiated the anticancer action of doxorubicin by stimulating the caspase-dependent pathway of apoptosis in multidrug resistant human lung cancer cells.

5.3.4.1 Anticancer Drug Delivery

Brain Tumors Brain tumors are classified as gliomas (astrocytomas, oligodendrogliomas, ependymomas) and primitive neuroectodermal tumors (PNET) (medulloblastoma and supratentorial PNETs). Approximately half of all primary brain tumors are gliomas, while 80% of those are astrocytomas and glioblastomas. Most chemotherapeutic drugs are toxic to the healthy tissue and have damaging side effects due to their nonspecific nature. Incorporation of those drugs in liposomes can enhance the therapeutic efficacy and reduce the toxicity. Drug delivery to the brain via the intravenous route has been a very challenging task due to the strict selectivity of the blood–brain barrier (BBB) as to the number and kind of molecules able to pass through.

The BBB is the tight junction formed between the cerebral endothelial cells (Figure 11). These cells are in close contact with astrocytes and pericytes connected over a basal membrane. Only small (MW < 400–600) lipophilic molecules can diffuse

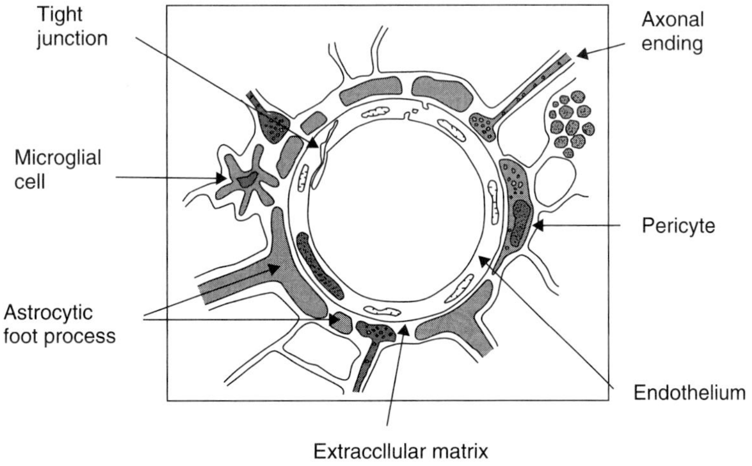

FIGURE 11 Schematic of neurovascular unit/cell forming BBB (brain–blood barrier). (Reproduced from D. J. Begley, *Pharmacology & Therapeutics*, 104, 29–45, 2004, with permission by Elsevier.)

through the BBB, while the majority of the circulating drugs cannot access the brain. Nutrients and peptides pass through the BBB via receptor-mediated or carrier-mediated transport systems. These mechanisms are exploited in an attempt to deliver drug-loaded liposomes into the central nervous system (CNS) [380]. The most common are low-density lipoprotein (LDL) receptors, insulin receptors, and transferrin receptors.

Thus, mAb against the transferrin receptor OX26 mAb has been conjugated via a stable thioether bond to the end of the PEG chain inserted on the liposome surface. Tritiated daunomycin was incorporated in OX26 mAb–PEG liposomes and the formulation was given intravenously to rats. The brain volume of distribution of daunomycin increased with time and exceeded 200 µL/g 24 h after injection [381]. In contrast, the pharmacokinetics of the free drug and the drug loaded in conventional liposomes was much lower [382].

In another study, the OX26 mAb has been grafted on the PEG chains by the biotin streptavidin coupling [383]. Brain tissue distribution obtained using biotinylated immunoliposomes was the same with that reported in the previous work where the mAb was chemically linked on the distal end of PEG. Therefore, the coupling method has not had a great impact on the brain accumulation of immunoliposomes. However, accumulation of the biotinylated PEG immunoliposomes was quite high in tissues such as liver, spleen, heart, muscle, and kidney. The latter was attributed to either the fact that the OX26–biotinylated PEG immunoliposomes could pass through the BBB by an active transport system or the biotinidase activity, which could mediate cleavage of the targeting antibody from PEG and interfere with the tissue distribution of the formulation.

An increase in therapeutic efficacy and lower toxicity was reported with liposomes where the bradykinin analogue RMP-7 was chemically attached at the end of PEG molecules of PEGylated liposomes (approximate size 70 nm) [384]. RMP-7 exhibits high selectivity for the B2 receptor of the BBB endothelial cells, which "shrunk" and let the RMP-7–PEG liposomes to pass into the brain. Actually the mechanism used in that study was based on opening the tight junctions of the BBB. Liposome-incorporated nerve growth factor (NGF) concentration increased 10 times in comparison to free NGF, while they accumulated mainly in striatum, hippocampus, and cortex.

A different type of immunoliposome was developed using antinuclear autoantibodies with nucleosome (NS)–restricted specificity [187]. Anti-NS mAb 2C5 specifically recognizes human brain tumor cells. Evaluation of immunoliposomes 2C5–PEG–PC/Chol was carried out in nude mice bearing subcutaneous brain tumor (U-87 astrocytoma) and exhibited a threefold higher accumulation in the tumor in comparison to control (IgG–PEG liposomes).

Moreover, disialoganglioside (GD_2) is expressed in abundance by neuroectodermal cancer cells and in low levels by normal cells located mainly in cerebellum and peripheral nerves. The Fab′ fragment of the monoclonal antibody anti-GD_2 was grafted on PEG chains of sterically stabilized liposomes loaded with doxorubicin and their potential in treating neuroblastoma was assessed in nude mice with HTLA-230 xenografts [160]. Mice receiving intravenously the immunoliposomes showed significant improvement in long-term survival compared with other mice that received free DOX, freeGD_2 Fab′, Fab′–PEG liposomes, PEG liposomes–DOX. The control mice died from metastatic disease, while the immunoliposome-treated

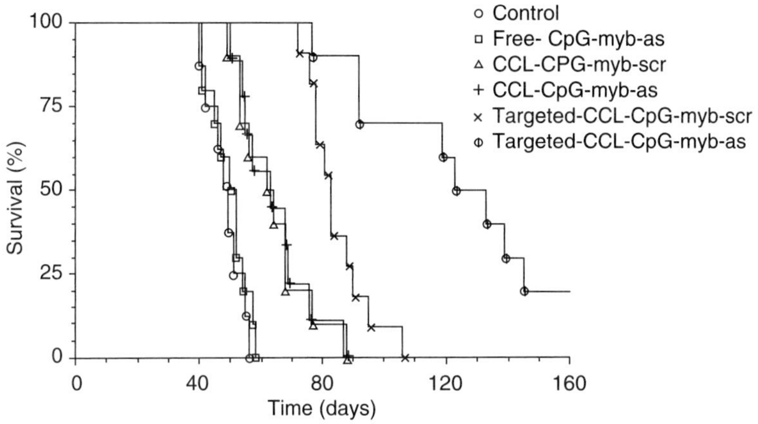

FIGURE 12 Survival of NB-bearing nude mice after injection of oligonucleotides (free or encapsulated within liposomal formulations). Nude mice were injected intravenously with 3.5×10^6 HTLA-230 neuroblastoma cells. After 4h each mouse received 50μg of oligonucleotides either free or encapsulated in targeted or nontargeted liposomes. Control mice received HEPES-buffered saline. (Reprinted from ref. 385 with permission of Elsevier.)

group lived 4 months longer. The same ligand, anti-GD_2–Fab′, was used with liposomes carrying the antisense oligonucleotide c-myb [385]. After intravenous injection of the targeted liposomes into nude mice with HTLA-230 xenografts, significant prolonged survival times were obtained in comparison to controls (Figure 12). The suggested mechanism was downregulation of c-myb proto-oncogene expression.

Although all the active targeting liposomes mentioned earlier have not left the laboratory, nonspecific sterically stabilized liposomes are being tested in clinical trials. Doxorubicin is the anticancer agent which is used as standard therapy, and it has the most serious side effects (mucositis, cardiotoxicity), so its incorporation in liposomes and bioavailability enhancement are under scrutiny [386–388].

It has also been demonstrated that PEGylated liposomal doxorubicin, Caelyx, can cross the BBB with a consequent accumulation in primary and secondary brain lesions [389]. In 10 patients with metastatic brain tumors treated with radiolabeled Caelyx concurrent with radiotherapy, the accumulation of the liposomal doxorubicin was 7–13 times higher in the metastatic lesions compared to the normal brain.

Liposomal formulations, type of tumor, anticancer agent, delivery pathway, day of treatment, and general conclusions are given in Table 6. DaunoXome (liposomal daunorubicin) and Doxil (liposomal doxorubicin) have been proved to have good response in clinical trials, in the range of approximately 40%.

Also, according to Arnold et al. [390], the dose scheme of doxorubicin-loaded liposomes affects the drug accumulation in various tissues as well as the tumor. The authors reported that after repetitive doses of sterically stabilized liposomes (SSL)-DOX every week, the plasma half-life of the drug increased, the deposition in liver and spleen decreased, and peak concentrations of DOX in the heart were threefold

TABLE 6 Recently Completed and Ongoing Clinical Trials of Low-Molecular-Weight Drug-Carrying Liposomes for Brain Tumors

Tumor	Number of Patients	Treatment	Delivery	Day of Treatment	General Conclusions
Recurrent tumor	14	Daunorubicin	IV	Once every 4 weeks	6/14: Patients showed positive response
GBM	8	Daunorubicin	IV	24 h before surgery	Concentration similar in tumor mass and peripheral regions
Pediatric glioma	7	Daunorubicin + free carboplatin and etoposide	IV	Daunorubicin: day 1 and 2; carboplatin and etoposide: day 1	5/7: Showed positive response with monthly treatment
GBM	15	Doxorubicin + radiation	IV	Doxorubicin: days 1, 21; radiation: days 1–21, 21–23	4/10: Patients completely responded
Solid tumor	22	Doxorubicin	IV	Once every 4 weeks	Phase I: dose-limiting toxicity: 70 mg/m^2
Glioma	40	Daunorubicin + free tamoxifen	IV	PEG-dox: day 4 every 14 days	Phase II: response (including stabilization) 40%; tamoxifen: day 1 every 4 days
Glioma	8	^{111}In labeled	IV	Contrast only	Tumor uptake: 1.1% max tumor/brain contrast 7:5
Glioma	3	Bleomycin	IT	Twice weekly for up to 6 weeks	All patients deteriorated, no toxicity
Recurrent meningeal malignancies	15	Cytarabine	IT	Once every 2 weeks for 2 courses; in positive response, second induction in 2 weeks after first dose	
Meningeal neoplasms	100	Cytarabine against free methotrexate	IT	Once every 2 weeks	Ongoing

Source: From G. H. Huynh, D. F. Deen, and F. C. Szoka, Jr., *Journal of Controlled Release*, 110, 236–259, 2006. Reprinted with permission of Elsevier.

Note: GBM: glioblastoma; IV, intravenous; IT: intratumor; PEG: poly(ethylene glycol). See http://www.clinicaltrials.gov.

lower. These results were not obtained using free DOX. In addition, a significant increase in survival was achieved in animals treated weekly with SSL-DOX, while animals treated with free drug did not survive longer than the untreated controls.

In phase I clinical trials children with recurrent or refractory tumors previously treated with free doxorubicin were administered i.v. Doxil in various doses (40–70 mg/m^2) in order to determine the best tolerated one [391]. It was concluded that the maximum tolerated dose was 60 mg/m^2 because at the highest, 70 mg/m^2, some patients developed mucositis, so dose adjustment was necessary.

Administration of a combination of liposomal anthracyclines in parallel with other anticancer agents has been found of great advantage as they prolong the patients' survival. Caraglia et al. evaluated in a phase II study the use of combination Doxil with temozolomide in the treatment of brain metastases from brain tumors [392].

It is worth mentioning that temozolomide accumulates significantly in the brain after oral administration. It is well tolerated and therefore is considered a potential candidate for combination chemotherapy. Administration of 200 mg/m^2 of temozolomide for 5 days and liposomal doxorubicin 35 mg/m^2 on day 1 was performed on 19 patients. The overall response rate was 36.8% and the median overall survival was 10.0 months.

Moreover, Fiorillo et al. [393] studied the effect of a combination of liposomal daunorubicin, etoposide, and carboplatin administered to seven children with recurrent malignant supratentorial brain tumors as a second-line therapy. Chemotherapy consisted of infusion of liposomal daunorubicin on days 1 and 2 and infusion of etoposide and carboplatin on day 1 whereas courses were repeated every 3–4 weeks. After a total of eight courses, five of seven children evaluated were alive 12–64 months after diagnosis and 8–29 months from the start of the second-line chemotherapy. Of the seven children, three showed complete response, two partial responses, one stable disease, and one progressive disease. The time to the best response was 3–10 months, while the median time to progression was 23 months. The toxicity observed was minimum.

Boron neutron capture therapy (BNCT) is also of high interest in treating brain tumors, especially glioblastoma multiforme, due to the high degree of normal brain infiltration, the high histological complexity, and the heterogeneity of the cellular composition of the latter. This method is based on the nuclear reaction which occurs when boron-10 is irradiated with low-energy thermal neutrons, producing high linear energy transfer of α particles and lithium-7 nuclei [394]. Development of BNCT has been ongoing over the last 50 years and the greatest challenge is to achieve selective tumor targeting at a sufficient therapeutic dose with minimal toxicity. Various compounds are currently being used or have been used in BNCT (shown in Figure 1 of [394]), such as BPA [(L)-4-dihydroxy-borylphenylalanine] and BSH (sodium mercaptoundecahydro-*closo*-dodecaborate), which are the first most successfully used chemical compounds (so-called second-generation compounds) due to low toxicity, longer retention to the tumor site and tumor/brain, and tumor/blood ratios higher than 1. However, these drugs are not ideal, but they are safe after i.v. administration, so they are being used in clinical trials in Europe, the United States, Japan, and Argentina. The next group of advanced boron molecules (third generation) consists of stable boron or a cluster attached to a tumor-targeting moiety, such as monoclonal antibodies or low-MW biomolecules with amphiphilic properties. Other anionic compounds show little specificity, so their potency increases when they are incorporated in either targeted or nontargeted liposomes [395, 396].

According to Feakes et al. [396], boron-loaded DSPC/Chol liposomes of 40nm average size were prepared and injected i.v. in murine mice carrying EMT6 tumors. Those liposomes showed high tumor retention, while boron amount was at therapeutic levels through the entire course of the experiment (more than 15μg B/g tumor). However, targeted liposomes would, in theory, assure higher boron accumulation than the nontargeted ones [397].

Due to the overexpression by glioma cells, the most potent ligands for glioma treatment are endothelial growth factor receptor (EGFR) [398], the *vIII* mutation of EGFR [399], platelet-derived growth factor (PDGFR) [400], and tenascin epitopes [401].

Liposomes are also used as carriers for gene delivery to gliomas while the cationic ones have demonstrated better interaction with cells in comparison to other types of liposomes [402, 403]. However, cationic liposomes suffer from toxicity, which varies according to cell type, duration of exposure, and density of cell culture. Antisense genes have been incorporated into liposomal carriers. For example, the EGFR antisense gene was packaged in PEGylated immunoliposomes [carrying human insulin receptor antibody (HIR)] [404]. It was reported [405] that the liposomes could cause 70–80% inhibition in human glioma cell growth. The same authors reported a 100% increase in survival time of mice with intracranial human brain cancer with weekly i.v. injections of antisense gene therapy directed at the human EGFR [405].

Moreover, double immunoliposomes were developed in order to treat intracranial human brain cancer in mice [406]. The mAbs used were the rat 8D3 mAb to the mouse transferrin receptor and the mAb against the HIR. RNAi (intereference RNA) is a new antisense gene therapy, where an expression plasmid encodes for a shRNA (short hairpin RNA). The shRNA is processed in the cell to an RNA duplex. The latter mediated RNAi. Indeed, weekly i.v. RNAi gene therapy reduced tumor expression of immunoreactive EGFR and caused an 88% increase in survival time of mice with advanced intracranial brain cancer.

In another study, the herpes simplex virus thymidine kinase (HSVtk) gene was evaluated as to its potency to increase the sensitization of ganciclovir (GCV) to glioma cell lines when that gene was incorporated in liposomes [407].

Although the efficiency of transfection was 18.6% in vivo after intratumoral injection of DNA liposomal complexes, the sensitivity to ganciclovir was improved as tumor weight induction was observed. In 2001, the FDA approved a clinical protocol relevant to liposomal gene therapy with the HSVtk/GCV system for the treatment of glioblastoma multiforme [408].

Hybrid vectors consisting of adeno-associated virus (AAV) vectors enclosed in liposomes lead to a10-fold increase in transduction efficiency compared to liposomes containing plasmid DNA and 6-fold increase compared to AAV vector alone [409].

However, alternative routes are currently used in the clinic according to the site/location and type of tumor to achieve higher therapeutic efficiency [410] because systemically administred drugs are not able to pass to cerebrospinal fluid (CSF), whereas there is direct tumor exposure to the drug, characterized with increased drug concentration and half-life [410]. Thus, drugs can be administered in CNS via intrathecal, intraventricular, and intraparenchymal routes. Intrathecal and intraventricular routes result in a high drug concentration in the bulk CSF but with limited penetration to parenchyma. Thus, they are suited for treating meningeal and ventricular diseases. The intraparenchymal route assures the delivery in a local region in the parenchyma and so is useful for solid tumors and degenerative diseases that are surgically accessible.

DepoCyt, a slow-release liposomal cytosine–arabinoside, undergoes clinical trials, as in some previous preclinical and clinical tests it was shown of good potency for the treatment of meningeal malignancies. For example, DepoCyt was administered to children with refractory neoplastic meningitis via lumbar puncture using an Ommaya reservoir or intraventricularly [411]. That study demonstrated the safety and feasibility of using liposomal cytarabine in children older than three years at the recommended dose of 35 mg with concomitant administration of dexamethasone. The latter is given to reduce the side effects of liposomal cytarabine, which is mild headache and back or neck pain. In a relevant study on adults with lymphomatous meningitis, DepoCyt given once every two weeks at a dose of 50 mg yielded a response rate of 71%, whereas free cytarabine at a dose of 50 mg twice a week produced a response rate of only 15% [412]. In patients with solid tumor neoplastic meningitis, 50 mg DepoCyt given once every two weeks yielded a response rate in evaluable patients of 36%, whereas free methotrexate 10 mg given twice a week produced a response rate of 21%. However, a problem accompanies the direct delivery to the brain parenchyma: the limited diffusion coefficient of particles in general (liposomes, nanoparticles, viral vectors) from the injection site in the brain tissue [410]. Thus, intraparenchymal delivery of liposomal carriers is facilitated using convection-enhanced delivery (CED) to distribute the drug through a larger region in the tissue. CED utilizes a bulk flow mechanism to deliver and distribute macromolecules to clinically significant volumes of solid tissues providing a larger volume of distribution. A range of parameters affecting CED are connected with the physicochemical properties of liposomes, that is, size, surface charge, and steric stabilization [413].

According to investigations by MacKay et al. [413], the ideal liposome (or in general nanoparticle) for CED will be PEGylated and of less than 100 nm in diameter, have neutral or negative surface charge, and need a targeting ligand to direct the particle to the target cell. For example, mannosylated liposomes containing clodronate infused directly to the fourth ventricle of the rat brain were selectively taken up by macrophages but not from microglia [414]. Also, the lipid dose is important as it was shown that infusing a high total lipid concentration reduced the fraction of the dose taken up by perivascular cells in the brain.

However, MacKay and co-workers [413] carried out the previously mentioned study using healthy animals. Mamot et al. administered liposomes of either 40 or 90 nm with 25% mannitol in U-87 glioma xenograft animals [415]. In contrast to MacKay et al. [413], the small 40-nm liposomes infused via CED without mannitol achieved distribution over nearly all the tumor tissue. The addition of mannitol to CED further enhanced distribution, as liposomes accumulated throughout all sections of tumor and further penetrated beyond the tumor boundary into adjacent normal brain tissue. For the 90-nm liposomes, CED with mannitol resulted in more than 50% tumor penetration of the total area evaluated, which also included extension into surrounding normal brain. Distribution was somewhat less extensive than with the 40-nm liposomes, though.

The efficacy of liposome administration with CED was demonstrated using real-time magnetic resonance imaging (MRI) of rat brain tumors [416]. Two types of liposomes were used: doxorubicin-loaded liposomes and gadolinium (Gd)–loaded (MRI contrast agent) liposomes. According to the authors, MRI facilitated the distinction of distribution between different tumor models, such as C6 gliomas and 9L-2 gliomas. Also, after CED of Doxil, the drug presence was identified in the tissue several weeks after a single administration, while the therapeutic response achieved was greater compared to systemic administration of Doxil. The fact that therapeutic liposomes can be coinfused with liposomal Gd to successfully monitor the distribution of the drug carrier in the nonhuman primate brain was shown by Saito and co-workers [417]. A 2:1 ratio between the volume of distribution and the volume of liposomes infused was found while liposomal Gd was still detectable 48 h postinfusion, confirming the previous finding of prolonged retention of liposomal Gd in the tissues and negligible toxicity in rat. With this study, the authors showed that CED is a technique enabling safe and extensive liposome distribution. Also, real-time MRI is a potentially useful tool to estimate the concentration and tissue half-lives of liposome-loaded drugs within target tissues.

However, CED might not always be successful as good technical skills are required (if the catheter is not placed properly, the liposomes will escape within the CSF). Also CED is an invasive technique and could cause inflammation and neurotoxicity and it is determined by many formulation characteristics and diffusional properties of the latter in brain tissue [418].

Breast Cancer Breast cancer is the second leading cause of cancer deaths in women today (after lung cancer) and is the most common cancer among women, excluding nonmelanoma skin cancers [419]. According to the World Health Organization, more than 1.2 million people will be diagnosed with breast cancer this year worldwide. Breast cancer stages range from stage 0 (very early form of cancer) to stage IV (advanced, metastatic breast cancer). Each patient's individual tumor characteristics, state of health, genetic background, and so on, will impact her survival. In addition, levels of stress, immune function, will to live, and other unmeasurable factors also play a significant role in a patient's survival. The majority of women with breast cancer will undergo surgery as part of their cancer treatment (lumpectomy and mastectomy). In addition to surgery, some women will receive adjuvant (additional) treatment (chemotherapy, radiation therapy, and drug treatments) to stop cancer growth, spread, or recurrence. Occasionally women may be treated with chemotherapy, radiation, or drugs without having breast surgery.

From a clinical point of view and with relevance to liposomes, administration of either PEGylated (Doxil) or non-PEGylated liposomal doxorubicin (Myocet) has improved the safety and tolerance of patients with breast cancer by significantly reducing the main side effect of those drugs, cardiotoxicity. However, drug efficacy remains the same either incorporated or not in the liposomes as this is reported after conducting various clinical studies [420, 421]. For example, Myocet (M) received European approval for use in patients with metastatic breast cancer at a dose of

60 mg/m^2 in combination with cyclophosphamide (C) having shown equivalent efficacy at a phase III study [422]. Also, doxorubicin is given at a dose of 60 or 75 mg/m^2 [423]. Chan et al. [424] studied the efficacy and tolerability of Myocet in a dose of 75 mg/m^2 in combination with cyclophosphamide in 160 patients. A high incidence of neutropenia was obtained. Thus, that group confirmed that the use of 60 mg/m^2 is an appropriate choice. However, administering a combination of either Myocet and cyclophosphamide or epirubicin and cyclophosphamide to patients with metastatic breast cancer, the response rate was not significantly different, which showed that the use of the liposomal formulation just reduced the cardiotoxicity of doxorubicin.

In a phase I study, Myocet was administered in combination with docetaxel in 21 metastatic breast cancer patients with the aim of estimating the safety and maximum tolerated dose of Myocet in parallel with docetaxel [425]. The latter is proved highly active in metastatic breast cancer as well as doxorubicin. The maximum tolerated dose was 50 mg/m^2 of Myocet and 25 mg/m^2 of docetaxel. As reported in the previous study, neutropenia was the most common toxicity effect while some incidents of congestive heart failure were observed after a total doxorubicin dose of 540 mg/m^2. Moreover, a combination of Myocet (75 mg/m^2) with gemcitabine (350 mg/m^2) and docetaxel (75 mg/m^2) was injected intravenously in 44 patients with early breast cancer every three weeks for six cycles [426]. The overall clinical response rate was 80% and the pathological complete response was 17.5%. The toxicity of the regimen was moderate and, as expected, neutropenia and leukopenia were the most prominent side effects. The latter effects were well managed and treatment discontinuation was not required.

Using PEGylated liposomal doxorubicin (Caelyx), Keller et al. [427] compared the efficacy of the liposomal formulation with that of a common regimen in patients with taxane-refractory advanced breast cancer. The regimen scheme was Caelyx (50 mg/m^2 every 28 weeks) or vinorelbine (30 mg/m^2) or mitomycin C (10 mg/m^2 every 28 days) plus vinblastine (5 mg/m^2 at days 1, 14, 28, 42). Finally, progression free survival and overall survival were similar for Caelyx and the comparative regimen.

The efficacy and toxicity of Caelyx in combination with paclitaxel (Taxol) were investigated as a first-line therapy in 34 patients with advanced breast cancer in a multicentric phase II study [428]. Paclitaxel at a dose of 175 mg/m^2 and Caelyx (30 mg/m^2) were administered intravenously every 3 weeks. It was shown that the response rates of the combination were over 70% while the median time to treatment failure was 45 weeks. No significant clinical cardiotoxicity was observed and the usual side effects (mucositis, stomatitis, hand–foot syndrome) were treated accordingly.

In another phase II study, Caelyx was used in combination with cyclophosphamide (CP) as a first-line therapy in patients with metastatic or recurrent breast cancer [429]. Three different schemes were given in groups of patients: Caelyx 50 mg/m^2 intavenously on day 1 and CP 100 mg/m^2 (orally on days 1–14) every 28 days, 30 mg/m^2 Caelyx and 600 mg/m^2 i.v. on day 1 every 21 days, and 35 mg/m^2 Caelyx plus 600 mg/m^2 CP i.v on day 1 every 21 days. The responses were similar among the different groups (51%) of patients, but less side effects were observed in group 2, making that regimen the one of choice as a first-line therapy for patients with metastatic or recurrent breast cancer. The median duration of response was 35.1 weeks.

Coleman et al. [430] demonstrated that using Caelyx in a dose format of 50 mg/m^2 every four weeks in patients with metastatic breast cancer is quite effective (objective partial responses 31%) while the main side effect was hand–foot syndrome.

From a research point of view, liposomes have been used in order to facilitate or alter the pharmacokinetics/bioavailability of various molecules. For example, ceramide is an antiproliferative and proapoptotic molecule produced after sphingomyelin metabolism [431]. Ceramide C6–loaded PEG(750)-C8 liposomes were injected intravenously in mice bearing syngeneic or human xenografts of breast adenocarcinoma. Administration of 36 mg/kg of liposomal C6 over a three-week period caused a sixfold decrease of tumor size in the case of syngeneic tumor–bearing mice. Liposomal C6 accumulated in caveolae and mitochondria, while a marked increase of apoptotic cells was observed. The PEGylated liposomal ceramide followed first-order kinetics in the blood and a steady-state concentration was achieved in tumor tissue (Figure 13). Also, a decrease in tumor size was obtained in the human xenograft model. Minimal toxicities were observed in tumor-bearing mice, suggesting that the bioactive concentration of C6 achieved in the tumor tissue is not active in normal tissues. Besides, contortrostatin (CN), a molecule isolated from snake venom, has been proved to interact with integrins on tumor cells and with newly growing vascular endothelial cells via its two Arg–Gly–Asp sites [432]. Thus, interactions between CN and integrins disrupt several steps critical to tumor growth: angiogenesis and metastasis. CN was incorporated in PEGylated liposomes with an encapsulation efficiency of approximately 80% while retaining full biological activity. Intravenous injection of approximately 100 μg of liposomal CN twice a week caused a significant reduction of 94% of microvascular density and hindered the tumor growth in the MDA-MB-435 xenograft model. Also, liposomal CN demonstrated prolonged circulation ($t_{1/2}$ values were 1.9 and 0.5 h for liposomal and nonencapsulated CN, respectively).

In another study, the authors modified the liposome composition by including polyoxy-ethylenedodecyl ether [C_{12}(EO)n] in the DMPC-made liposomes [433]. These vesicles caused activation of caspases 3, 8, 9 and eventually induction of apoptosis in MDA-MB-453 cells. Another idea is based on using methods to direct the drug loading carriers to the cancer site. More specifically, the paclitaxel is successfully used for the treatment of breast cancer among other types of cancers but has side effects such as neutropenia, peripheral neuropathy, and hypersensitivity reactions. Also, it is formulated with Cremophor EL due to its high lipophilicity. In order to decrease the drug toxicity and enhance the therapeutic potential of that drug, paclitaxel was incorporated in negatively charged magnetic liposomes and its efficacy was evaluated in vivo [191]. A magnetic field of suitable strength was used to direct the magnetoliposomes to the desired site. Indeed, after intravenous injection the AUC obtained for the magnetic carriers and the Cremophor EL/ethanol were 20.7 and 6.8 h·μg/mL, respectively. The liposomal paclitaxel concentration was much higher in the tumor in comparison to other organs while the concentration peak was reached sooner (19.85 μg/g at 0.25 h). Even after 8 h, paclitaxel concentration in the plasma was 2.71–0.33 μg/mL for the Cremophor EL formulation. The antitumor efficacy of magnetoliposomes

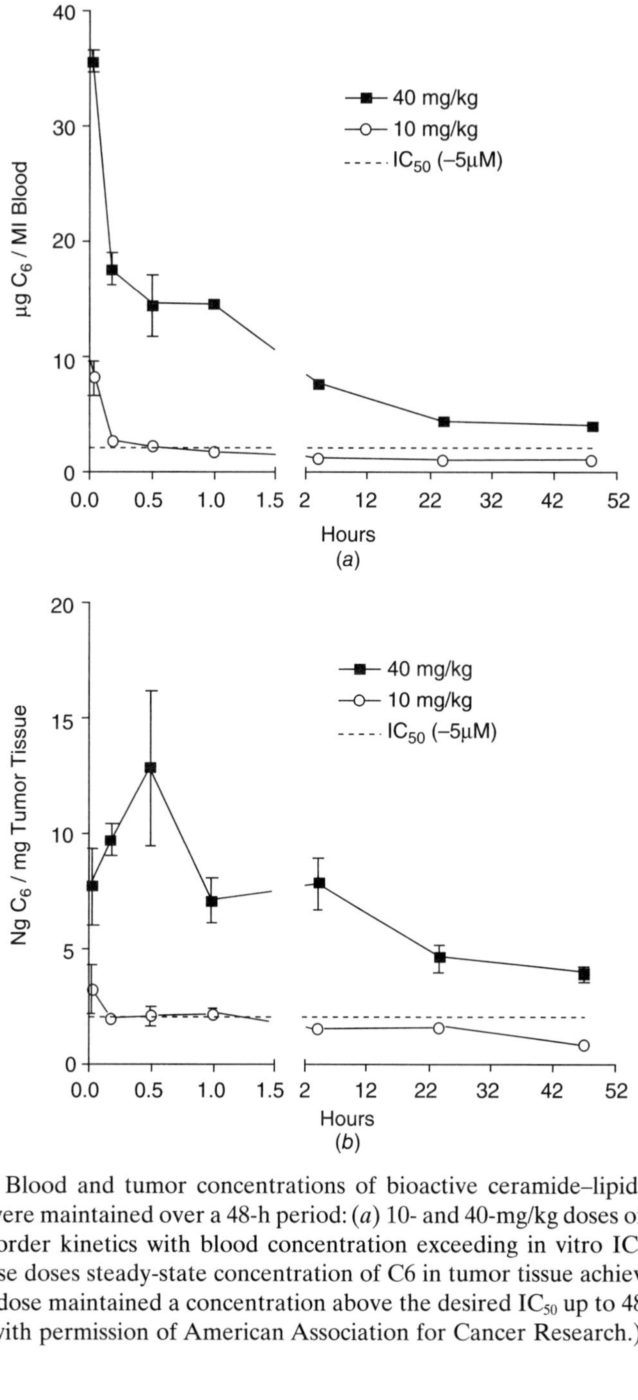

FIGURE 13 Blood and tumor concentrations of bioactive ceramide–lipid C6 in tumor-bearing mice were maintained over a 48-h period: (*a*) 10- and 40-mg/kg doses of liposomal-C6 followed first-order kinetics with blood concentration exceeding in vitro IC_{50} sustaibed at 48 h; (*b*) at these doses steady-state concentration of C6 in tumor tissue achieved at ~60 min. The 40-mg/kg dose maintained a concentration above the desired IC_{50} up to 48 h. (Reprinted from ref. 431 with permission of American Association for Cancer Research.)

(expressed as change rate of tumor weight) after subcutaneous injection of 10 mg/kg drug was equivalent to that achieved with Cremophor EL 50.4 and 51.9%. When the administered dose was 20 mg/kg, the change rate of tumor weight was the highest, 60.5%.

Pulsed high-intensity focused ultrasound (HIFU) is another suggested method to enhance the delivery and therapeutic effect of Doxil in a murine breast cancer tumor model with the final aim of reducing the Doxil dose given during therapy [434]. However, the results were not particularly encouraging as there was not any significant difference between free and liposomal Doxil distribution in the tumor after HIFU exposure. The latter was possibly attributed to the ability of liposomes to extravasate easily through the leaky vasculature of the tumor, so the use of HIFU did not add any therapeutic or other advantage.

In an attempt to achieve active targeting, sigma receptor ligands were incorporated on to the liposomal surface. Sigma receptors are overexpressed in various human tumors, including breast cancer cells [435]. Haloperidol has shown high affinity for sigma receptors; thus it has been attached at the end of PEG molecules, which are protruding from the surface of cationic lipoplexes. Indeed, haloperidol–PEG lipoplexes showed more than 10-fold higher gene expression in MCF-7 (breast carcinoma) cells than control lipoplexes, while the presence of haloperidol or 1,3-ditolylguanidine suppressed the expression of the reporter gene [435].

Transferrin is another ligand attached on the lipoplex surface so to cause higher transfection efficiency, as has been reported [436]. Thus, Basma et al. [437] studied the effect of *cis*-diaminedichloro platinum (CDDP) in combination with bcl-2 antisense treatment on p53(+)MCF-7 and p53(−)MCF-7/E6 breast cancer cells using transferrin bcl-2 lipoplexes. The median inhibitory concentration (IC_{50}) for bcl-2 antisense delivered with lipoplexes plus transferrin was approximately 1.4 μM for MCF-7 and 1.2 μM for MCF-7/E6 cells (Figure 2 of ref. 437). In CDDP-treated cells, the IC_{50} was approximately 5 μM for both cell lines. In general, bcl-2 antisense delivered in the form of transferrin bcl-2 lipoplexes in combination with cisplatin induced cell death and apoptosis in a higher degree in MCF-7/E6 cells rather than MCF-7 cells. Cisplatin demonstrated higher caspase-8 activation compared to the targeted lipoplexes, which suggested that possibly caspase-8 is the major pathway for cancer cell death. G3139 is another oligonucleotide (ODN) capable of down-regulating bcl-2 and, its efficient delivery to breast cancer cells would potentially make it a successful candidate for antisense therapy [438]. Thus, the efficiency of different liposomal formulations for the lipid composition (DOTAP, DC-Chol, CCS)[1] with or without helper lipids DOPC, DOPE, and Chol and liposomal size [approximately 100 nm LUV or unsized heterolamellar vesicles (UHV)] on MCF-7 breast cancer cells was examined [438]. Out of 18 tested formulations, only the CCS-bcl-2 lipoplexes (UHV derived) have been effective, causing a larger than 50% decrease of cell growth in comparison to free ODN. The possible mechanism of action of the particular formulation is attributed to the presence of one primary and two secondary amines on the spermine moiety of the CCS molecule. Because of this, CCS lipoplexes are taken up by cells via adsorptive endocytosis and the secondary amines cause a "proton sponge effect" due to which ODN escapes from the endosomes and so has the ability to interact with the bcl-2 neutralizing it. Of course, the effect of the L^+/DNA^- ratio is a determining factor along with the lipid composition and size.

[1]DOTAP: *N*-(1-(2,3-dioleoyloxy)propyl)-*N*,*N*,*N*-trimethylammonium chloride; DC-Chol: 3β[*N*-(*N′N′*-di methylaminoethane)carbamoyl]-cholesterol; CCS: ceramide carbomoyl spermine.

Diagnosis is always the first aim in effectively treating breast cancer, so the need to develop or reveal more tumor markers at an early stage of the cancer is absolutely critical [439]. Such are the circulating epithelial cells, the cyclins, and the urokinase-type plasminogen activator and plasminogen activator inhibitor, which indicate breast cancer or metastatic spread apart from the already existing markers estrogen receptor, progesterone receptor, and human epidermal growth factor receptor-2 (HER-2). Liposomal formulations have facilitated anticancer activity of highly toxic anticancer drugs as well as altered their biodistribution, while targeted drug-loaded liposomes have shown some promising results in the laboratory. Development of new drugs and use of particulate carriers to increase bioavailability could be one way forward in the battle for improvement of quality of life for patients suffering from breast cancer.

Lung Cancer Lung cancer might be the most common form of cancer and the most common cause of death in both men and women, although it affects more men than women [440]. There are three main types of lung cancer, based on their appearance when examined by a pathologist: small cell carcinoma, squamous cell carcinoma, and adenocarcinoma. The latter two consist of non–small cell lung cancer. It is important to know which type of cancer a patient has because small cell cancers respond best to chemotherapy (anticancer medicines) whereas the other types (often referred to collectively as non–small cell cancer) are better treated with surgery or radiotherapy (X-ray treatment).

Surgery can cure lung cancer, but only one in five patients are suitable for this treatment. If the tumor has not spread outside the chest and does not involve vital structures such as the liver, then surgical removal may be possible, but only if the patient does not also have severe bronchitis, heart disease, or other illnesses. Small cell lung cancer is treated with chemotherapy. Non–small cell cancer may be treated with radiotherapy and chemotherapy (as part of a research trial) or with supportive care. Radiotherapy is either "radical" or "palliative."

Regarding drug treatment of lung cancer, Merck KGaA and Biomira will soon start phase III trial on their BLP-25 liposomal vaccine for patients with non–small cell lung carcinoma [441]. L-BLP25 is a synthetic MUC1 peptide vaccine. MUC1 is a mucinous protein expressed on the apical borders of normal epithelial cells. It is overexpressed and glycosylated on tumor cells, where it appears to be antigenically distinct from the normal protein. The liposomal vaccine can induce a MUC1-specific T-cell response. Median survival for patients with stage IIIB locoregional non–small cell lung cancer who received L-BLP25 in a phase IIb study was 30.6 months in the vaccinated group compared with 13.3 months for the unvaccinated group.

Caelyx is liposomal doxorubicin very well used as a treatment of choice for a number of cancers with good tolerability and antitumor activity, as has been demonstrated in many phase I or II clinical trials. One such example is the conduct of phase I study of Caelyx (PEGylated liposomal doxorubicin, 25–40 mg/m^2) in combination with cyclophosphamide (750–1000 mg/m^2) and vincristine (1.2 mg/m^2) every 21 days in patients with relapsed or refractory small cell lung cancer [442]. The suggested doses were CaelyxTM 35 mg/m^2, cyclophosphamide 750 mg/m^2, and vincristine 1.2 mg/m^2 intravenously every 21 days. This combination was well tolerated while antitumor activity was observed for patients with relapsed small cell lung carcinoma with survival duration of 5 months. The latter is similar to what is achieved using only camptothecin analogues taxanes or vinorelbine (survival duration ≤6 months).

Also, paclitaxel is the most widely used anticancer agent for non–small cell lung cancer [420]. Liposomal encapsulated paclitaxel faces the problem of formulation due to the drug's high hydrophobicity. However, a phase I trial with liposomal paclitaxel reported dose-limiting toxicity at the dose of 150 mg/m^2/week. Besides, the whole blood clearance of paclitaxel was similar for liposomal and free paclitaxel.

Promising results that were referred to liposomal encapsulated paclitaxel easy to use (LEP-ETU; NeoPharm) consisted of DSPC/Chol/cardiolipin molar ratio 90:5:5, lipid/drug molar ratio 33:1, and paclitaxel concentration 2 mg/mL (mean liposome size 150 nm). In a phase I study of increasing doses of liposomal paclitaxel (135–375 mg/m^2) in 25 patients, the enhanced effectiveness was proved with much better tolerability and with 3 partial remissions and 11 patients with stable disease. Lurtotecan, a camptothecin analogue, was incorporated in PC/Chol (2:1 molar ratio) liposomes (size 50–100 nm) with a lipid/drug molar ratio 20:1. A remarkable 1500-fold AUC increase was obtained after administration in nude mice and several xenograft models.

9-Nitro-20(*S*)-camptothecin (9NC), another lipophilic camptothecin analogue, showed antitumor effects in mice and milder effect in humans after oral administration. Thus, the potency of 9NC-loaded di-lauryl-PC (DLPC) liposomes was investigated in patients with primary or metastatic lung cancer after aerosol administration (aerosol droplet size 1–3 μm) for five consecutive days per week [373]. Indeed, the most serious side effect of 9NC, hematological toxicity, did not appear in any of the patients, while the dose-limiting toxicity, chemical pharyngitis, was observed at a dose of 26.6 μg/kg/day. 9NC plasma levels, after aerosol administration (13.3 μg/kg/day), were similar to that given orally (2 mg/m^2), despite the lower dose administered in the first case. More specifically, C_{max} and AUC were 76.7 ± 39.1 ng/mL and 275 ± 149 ng-h/mL, respectively, after aerosol administration and 111 ng/mL and 194.4 ± 108.4 ng-h/mL, respectively, after oral administration. The recommended dose for phase II studies is 13.3 μg/kg/day on a 1-h exposure for five consecutive days per week and for eight weeks, with 0.4 mg/mL 9NC concentration in the nebulizer.

Moreover, targeting moieties have been grafted on the vesicle surface, exploiting the fact that antigen/receptor overexpression on cancer cell membranes increases the specificity of the active substance. For example, antagonist-G is a hexapeptide which blocks the action of multiple mitogenic neuropeptides by interacting with their receptors [170]. Antagonist-G was chemically attached at the distal end of PEG molecules of stealth liposomes (HSPC/Chol/mPEG–DSPE/PDP–PEG–DSPE, 2:1:0.08:0.02 molar ratio) (SLG) loaded with doxorubicin. The antiproliferative activity of doxorubicin was estimated on the human variant small cell lung carcinoma (SCLC) H82 cell line. Indeed, 20- to 30-fold increase of both, binding, and internalization took place after SLG incubation on cells in comparison to stealth liposomes only, or PEGylated liposomes without the antagonist-G. The 0.03 μg of antagonist-G on the liposome surface was enough to cause 50% cell liposome association. Doxorubicin accumulation in the whole cell was 20-fold higher with SLG. Eventually, there was the suggestion that the main antagonist-G mechanism of

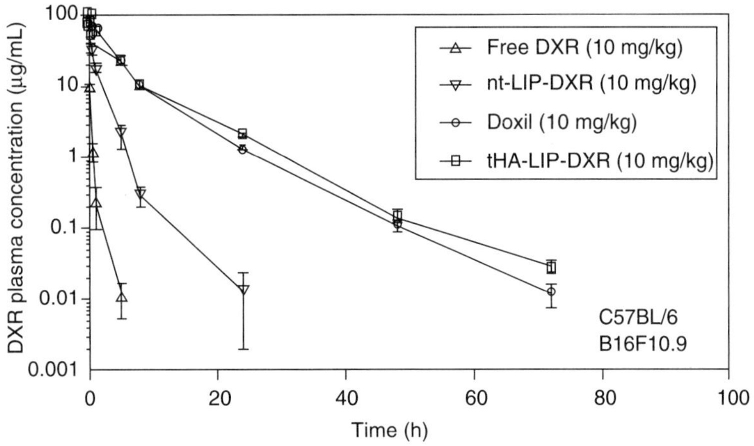

(*a*)

FIGURE 14 Doxorubicin (DXR) plasma concentration (μg/mL) as function of time from dosing: (*a*) C57BL/6 mice inoculated (by intraveneous injection) with B16F10.9 cells; (*b*) healthy C57BL/6 mice; (*c*) BALB/c mice inoculated with C-26 cells (injected into right-hind footpad). A single dose of the selected formulation was injected into the tail vein. DXR formulations and doses are specified. (Reprinted from ref. 443 with permission of Neoplasia Press, Inc.)

action is to bind to vasopressin receptor, which is expressed abundantly on the SCLC cells.

Hyaluronan (HA) is another potential candidate for tumor targeting because it has been proven that hyaluronan receptors CD44 and RHAMM are overexpressed on several tumor types [443]. In this case, hyaluronan is used not only as a targeting moiety but also to prolong the circulation half-life of the vesicles in question; in other words, it replaces PEG molecules. HA was chemically attached on the vehicle surface (57 μg/μmol lipid), which was comprised of PC/PE/Chol (3:1:1) (HA–LIP). Doxorubicin was entrapped in 78 ± 5% encapsulation efficiency, while the ζ potential and size of the targeted liposomes were −13.1 ± 3.9 mV and 81 ± 13. Pharmacokinetic/biodistribution studies were carried out with Doxil (PEGylated liposomal doxorubicin) as a comparison to the new hyaluronan liposomes. Those studies plus therapeutic responses were tested on three mice models, one of which was C57Bl/6-bearing B16F10.9 lung metastasis. Clearly, ha-liposomal (HA–LIP)–DXR exhibited prolonged blood circulation similar to Doxil (approved formulation in the market), which indicated that the amount of attached HA on the vehicle surface is enough to stabilize and offer the steric stabilization required (Figure 14). Also, accumulation of HA–LIP–DXR was much more reduced in liver of C57Bl/6-bearing B16F10.9 lung metastasis mice, while DXR accumulation in the tumor site was threefold higher to that achieved using Doxil. Significant improvement was obtained in both metastatic burden and survival with Doxil and HA–LIP–DXR (Figure 15). Besides, the HA–LIP–DXR had positive results on all tumor types tested.

Manipulating the genetic material by injecting tumor-suppressing genes has been an alternative way in cancer treatment. The 3p *FUS1* gene is a tumor suppressor

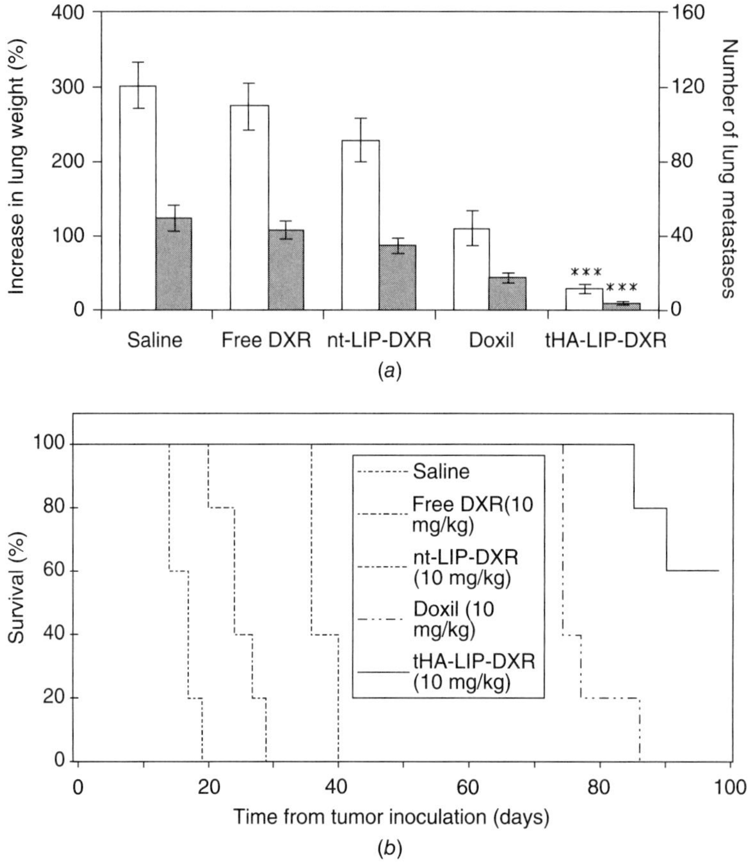

FIGURE 15 Therapeutic responses of mice bearing B16F10.9-originating lung metastatic disease. Doxobubicin (DXR) formulations and doses are specified. Treatments were on days 1, 5, and 9 by injection of selected formulation to the tail vein. (*a*) Lung metastatic burden. Light-shaded bars are data for increase in lung weight; dark-shaded bars are data for number of lung metastasis: (∗∗∗) $P < 0.001$ compared with free drug. (*b*) Survival ($n = 5$). Each line connects the symbols representing the daily survival state of the group. (Reprinted from ref. 443 with permission of Neoplasia Press, Inc.)

gene which belongs to a 120-kb region of the 3p chromosome on the region 3p21.3. That 120-kb region is missing so injection of DOTAP/Chol lipoplexes with the *FUS1* gene could induce apoptosis and inhibit tumor growth [444]. A significant inhibition of lung metastatic tumor was obtained in animals (bearing subcutaneous lung tumor xenografts) treated with liposome–FUS1 DNA complex (total of six doses). Also, intravenous treatment of lung tumor-bearing animals with DOTAP/Chol–FUS1 complex prolonged the animal survival; 40% of the animals were still alive after approximately 100 days. Also, the antitumor potency of FUS1 was demonstrated to be superior to the one obtained with p53 as the same therapeutic effect was achieved by using less amount of *FUS1* gene in the lipoplexes in comparison to p53. This would be an advantage because lowering the DNA dose would result in much lower toxicity of DNA due to inflammatory response.

Moreover, the efficiency of gene therapy depends very much on the vector used to achieve the gene delivery, the properties and stability of the final lipoplexes in the presence or absence of serum, and the pharmacokinetics and biodistribution. On this basis, Li et al. studied the effect of lipoplex size on the lipofection efficiency of TFL-3/pDNA (plasmid-encoding luciferase) liposomes in the absence and presence of serum and investigated the correlation between in vitro and in vivo results using either B16BL6 murine melanoma cell line or the same cells injected i.v. in C57BL/6 mice in order to produce pulmonary metastases [445]. The authors incubated a range of different ratios of pDNA/TFL-3 (P/L) ranging from 8 to 120 P/L on the previously mentioned cell line in the absence and presence of serum. Serum had a dramatic effect on lipoplex size; in the absence of serum, the size increased as the pDNA amount increased and it reached a maximum value at the ratio 80 g/mol P/L, while the ζ potential decreased as the P/L ratio increased and at 80 g/mol P/L was approximately +2 mV. Maximum luciferase activity was observed with lipoplexes of 80 g/mol. In contrast, in the presence of serum, the size increased from 8 to 20 g/mol and decreased from 40 to 120 g/mol as the P/L increased. Maximum luciferase activity was obtained with 40 g/mol in the presence of serum. However, the in vitro transfection efficiency of the lipoplexes in the presence of serum at the highest point of 40 g/mol was twofold less than the one achieved at the peak of transfection efficiency at 80 g/mol in the absence of serum. However, the plasmid expression was higher with 8 and 80 g/mol lipoplexes in tumor-bearing animals in contrast to the in vitro findings in the presence of serum. This was attributed to less aggregate formation in the blood due to the lipoplex/serum ratio (50–100 μL serum/200 μL lipoplex dispersion) so there is no complete interaction between lipoplexes and serum proteins. Also, pulmonary gene expression was dependent on the time after cell inoculation. It was shown that gene expression takes place at different parts of the lung; for example, 8 g/mol P/L lipoplexes expressed luciferase in the cells surrounding the tumor, while 80 g/mol lipoplexes expressed the gene in the entire lung without any specificity. Possible explanations for the latter are the extended lung capillary bed and increased lipoplex uptake by tumor cells compared to the normal cells. Of course many other factors could contribute to the different patterns of gene expression, such as endosomal release, nuclear uptake, increased translation, and transcription.

All-trans retinoic acid (ATRA) has shown anticancer activity in a number of types of cancer cells [446]. However, some non–small cell lung carcinoma is resistant to ATRA probably due to the high lipophilicity of the molecule, which makes it unable to pass through the cellular membrane. Therefore, cationic liposomes (DOTAP/Chol 1:1 molar ratio) were used to incorporate and facilitate ATRA's action in the resistant NSCLC in A549 human lung cancer cells. The produced ATRA–DOTAP/Chol lipoplexes were 125 nm in diameter with 50 mV of ζ potential, while for the DSPC/Chol liposomes used as control these were 110 nm and −3 mV, respectively. The former lipoplexes exhibited higher uptake by the cells in comparison to DSPC/Chol due probably to the positive surface charge, which interacts to a higher degree with the negatively charged cell membrane. Thus, the apoptosis induced on these cells in the presence of ATRA lipoplexes was much higher than the one caused by ATRA only or ATRA–DSPC/Chol (Figure 16).

Lastly, although antisense therapy is considered a simple and efficient procedure, its products have never reached the market. This is due to the instability of phos-

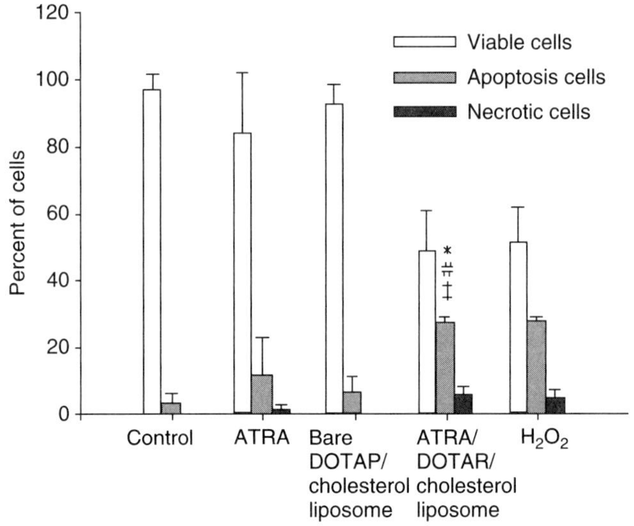

FIGURE 16 Flow cytometric analysis of A549 cells treated with 1.0 μ*M* ATRA, bare DOTAP/cholesterol liposomes, or ATRA incorporated in DOTAP/cholesterol liposomes for 48 h in A549 cells. As a control. Cells were incubated with 1% DMSO and 400 μM H_2O_2. Significant differences: (*) $P < 0.05$ vs. control; (#) $P < 0.05$ vs. ATRA; (‡) $P < 0.05$ vs. DOTAP/cholesterol liposomes. (Reprinted from ref. 446 after permission of Elsevier.)

phodiester oligonucleotides (PO-ONs) in the cytoplasm, while the phosphothioate ONs demonstrate unwanted side effects [447]. In order to design better delivery systems for PO-ONs, the intracellular fate of the PO-ON and PS-ON lipoplexes with DOTAP/DOPE was investigated after their application on A549 cells. After endocytosis, endosomal localization, and endosomal escape of the lipoplexes, the ONs were localized in the nuclei. However, PO-ONs were degraded in the nucleus (degradation products diffuse out of the cells after 2 h) whereas PS-ONs were still intact and remained in the nucleus for 8 h. It is worth mentioning that PS-ONs were eliminated from the nucleus and were found in cytoplasmic granules, which indicated that the cell has a mechanism of elimination of intact PS-ONs. Also, it seemed that the amount of PO-ONs was quite important for the ONs' stability as the degradation was reduced after injecting 22 μ*M*, in contrast to complete degradation with injection of 2 μ*M*. From the study it was concluded that ON degradation happened after their release from the lipoplexes. Thus, for the successful delivery of ONs, they have to remain complexed with their carrier. Polyethylenimine (PEI) was suggested as a possible carrier due to its proton sponge effect, leading to endosomal rupture without releasing the ONs as well as graft-pDMAEMA–PEG [poly(2-dimethylamino)ethyl methacrylate-*co*-aminoethyl methacrylate–bearing polyethylene glycol chains]. According to the authors, probably cationic polymers are more efficient than cationic lipids, but lots of work still needs to be done on the matter.

Ovarian Cancer Ovarian cancers start at the ovaries [448]. They can be either benign, and so never spread from the ovary, or malignant, in which case they can metastasize to other parts of the body. Malignant cancers are classified into germ cell, stromal, and epithelial tumors. Germ cells produce eggs, stromal cells produce progesterone and estrogen, and epithelial cells cover the ovary. The majority of the malignant cases are epithelial cancer (almost 90%). As with most tumors, the usual treatment is surgery, chemotherapy, and radiotherapy.

Chemotherapy refers to drug administration with highly serious side effects, such as nausea, hand and foot rashes, mouth sores, and increased risk of infection, easy bruising, and so on. Therefore, liposomal carriers have been used in order to improve the drug's biodistribution and protect the patient from those side effects. The main anticancer drugs used to treat ovarian cancer are carboplatin and cisplatin, paclitaxel, topotecan, and lurtotecan. PEGylated liposomal doxorubicin has been approved as a regimen for patients with metastatic ovarian cancer refractory to both paclitaxel and platinum based-therapy [449].

A number of clinical trials have been reported and many refer to a combined therapeutic regimen of liposomes with other anticancer drugs. PEGylated liposomal doxorubicin was used in a phase II trial on patients with platinum/paclitaxel pretreated ovarian cancer with or without topotecan. Approximately 36% of patients showed stable disease [420]. A phase III study followed up the one already mentioned in 474 patients with pretreated epithelial ovarian cancer that failed or recurred upon platinum-based combination chemotherapy. The patients received either PEGylated liposomal doxorubicin (50 mg/m^2 for 1 day every 4 weeks) or topotecan (1.5 mg/m^2 for 1–5 days every 3 weeks). The response rates and the overall progression-free survival (PFS) were similar for the two formulations. The platinum-sensitive patients showed longer PFS and overall survival of 108 weeks against the 72 weeks obtained using topotecan. The interesting result of the study was the reduced cytotoxicity recorded for PEG–liposomal DXR compared to topotecan, proving that the former treatment improves patient quality of life.

A phase III study was conducted to compare PEG–liposomal DXR (50 mg/m^2 every 4 weeks) with paclitaxel (175 mg/m^2 every 3 weeks) using 214 patients with relapse after first-line platinum-based chemotherapy [420]. As previously, the response rates and PFS were not significantly different, but again the liposomal formulation was notably less toxic as fewer patients recorded with grade 4 adverse effects (17% compared to 71% for topotecan) and thus was more tolerable.

The most common toxic effects associated with liposomal DXR treatment were hand-and-foot syndrome and stomatitis, which can be handled by modifying the dose so there is no need for the regimen to cease. For example, administration of 40 mg/m^2 liposomal DXR every 4 weeks reduces the incidences of hand-and-foot syndrome and stomatitis compared to 50 mg/m^2 liposomal DXR every 4 weeks without loss of drug potency [450]. A series of clinical trials using patients with a variety of ovarian cancers in terms of characteristics (relapsed, refractory, platinum, paclitaxel resistant) were carried out. The optimized dose regimen for PEGylated liposomal doxorubicin was 10–12.5 mg/m^2 per week when it is used as a single therapy. Combining liposomal DXR with other anticancer drugs is an equally or more effective strategy due to the lower dosage regimen of the two formulations avoiding relevant toxicities.

While platinum compounds and paclitaxel comprise the first-line treatment in combination with PEGylated liposomal doxorubicin for patients with ovarian cancer, vinorelbine has been used in combination with liposomes as a palliative second-line therapy for patients with refractory/resistant ovarian cancer to platinum–paclitaxel [451]. The best MTD (mean therapeutic dose) was 25 mg/m^2 at days 1 and 8 for vinorelbine and 30 mg/m^2 at day 1 for liposomal DXR of every 21 days, which was well tolerated with moderate hematologic and mild nonhematologic toxicities. A phase II study relevant to the clinical efficacy, toxicity, and pharmacokinetics of that combined therapeutic regimen was carried out by Katsaros et al. in 30 patients with platinum–paclitaxel pretreated recurrent ovarian cancer [452]. Caelyx (30 mg/m^2) and vinorelbine (30 mg/m^2) were administered every 3 weeks for six cycles. The regimen was proved of significant activity for patients pretreated with paclitaxel–platinum first-line therapy. The overall response rate was 37% with 10% of patients demonstrating stable disease. Vinorelbine bioavailability was higher under the current regimen. The overall survival was 9 months while the toxicity was mild and reversible. There were no treatment-related deaths and there were only 2 patients, one reported with grade 4 and the other with grade 3 hand–foot syndrome. Also, the toxicity due to liposomal formulation was much lower compared to that reported from a phase III study with the drug given as a single agent [453]. In another phase II study, the combination of liposomal doxorubicin and infused topotecan was studied in 27 patients with platinum-resistant ovarian cancer in two cohorts [454]. Liposomal DXR (30 mg/m^2 at day 1) and topotecan (1 mg/m^2 for 5 days) were infused and the cycle was repeated every 21 days. The overall response rate of the regimen was 28% and the median overall survival was 40 weeks. However, neutropenia and thrombopenia were observed at 70 and 41% of the patients. Therefore, the topotecan dose was reduced to 0.75 mg/m^2 and liposomal DXR was increased to 40 mg/m^2 in order to reduce the toxicity. After that, the cytotoxic effect due to liposomal doxorubicin increased and the bone marrow cytotoxicity remained the same despite the effectiveness of the regimen.

Lurtotecan is a more advanced camptothecin analogue with probably greater potency with regard to topotecan and therefore was encapsulated in liposomes to investigate the toxicity and pharmacokinetics [455]. In a multi-institutional open-label phase II study, 22 patients with topotecan-resistant ovarian cancer were administered liposomal lurtotecan in a dose of 2.4 mg/m^2 on days 1 and 8 every 21 days. Although the toxicity profile of the drug was lower, no response was observed. Others used a regimen of liposomal delivery at days 1 and 3 with more promising therapeutic results, but higher toxicity, too [456].

A variety of new molecules either in combination with liposomal doxorubicin or not are in development at the moment [457]. For example, a phase III study will be conducted to test the efficacy and safety of pattupilone versus PEG–liposomal DXR in taxane/platinum refractory/resistant patients with recurrent epithelial ovarian, primary fallopian, or primary peritoneal cancer. A phase III randomized study of Telcyta with Doxil/Caelyx versus Doxil/Caelyx has been planned in patients with platinum-refractory or platinum-resistant ovarian cancer. A phase II study relevant to side effects and best dose of ixabepilone combined with liposomal DXR will be assessed in patients with advanced ovarian epithelial, peritoneal cavity, or fallopian tube cancer or metastatic breast cancer.

Gemcitabine is a clinically active antineoplastic drug in platinum-refractory ovarian cancer. The efficacy and tolerability of the particular drug in combination with liposomal DXR were investigated in athymic mice bearing cisplatin-resistant human ovarian carcinoma [458].

Using two therapeutic regimens, either 80 mg/kg of gemcitabine and 15 mg/kg of liposomes or 20 mg/kg of gemcitabine and 6 mg/kg of liposomes, the same trend of response was observed, with some of the animals having complete tumor regression at the end of the study. The lack of toxicity and therapeutic efficacy observed makes that regimen promising for clinical trials.

In terms of drug development, a novel analogue of vitamin E assembled into liposomes was evaluated as a potent anticancer agent in combination with cisplatin in mice bearing human ovarian cancer xenografts [459]. The analogue is the 2,5,7,8-tetramethyl-2*R*-(4*R*, 8*R*, 12-trimethyltridecyl)chimoran-6-yloxyacetic acid (α-TEA), which has apoptotic properties to cancer and not the normal cells in a dose- and time-dependent manner. Liposomes were administered in the form of aerosol while cisplatin was injected intraperitoneally to mice with either small or large tumor volume. It was shown that the combined therapeutic scheme demonstrated the highest antitumor activity compared to α-TEA itself in both cases. Also, the current regimen significantly reduced micrometastases observed in lungs and lymph nodes, while the ratio of proliferating to apoptotic cells in the tumor was decreased due to induction of apoptosis. However, the specific apoptotic mechanism of the particular molecule needs to be elucidated.

By grafting folate molecules on the liposome surface, their accumulation is highly increased in macrophages located in tumor ascites fluid but not in solid tumors, which have been tested and the existence of functional folate receptors has been confirmed [460]. Folate receptors are overexpressed on epithelial tumor cells, and thus folate was attached on PEG molecules via cysteine (folate–cysteine–PEG3400–PE), which, consequently, was one of the components of the liposomal bilayer (PC/Chol/PEG–PE). The ligand-bearing liposomes (diameter of 65–90 nm) were injected i.p. in tumor-bearing mice. Significant folate-bearing liposome accumulation was obtained in the ascites fluid and more specifically by macrophages, which indicates macrophage activation. A reduced targeted liposome accumulation in the tumor could occur for a number of reasons, such as poor liposome penetration into solid tumor mass. Activated macrophages secrete immunosuppressive cytokines and angiogenic factors, so liposome targeting could possibly eliminate them during malignancy therapy.

Gene delivery is another approach trying to tackle the problem of cancer. Mutations of p53 tumor suppressor gene contribute to genetic abnormalities in ovarian cancer. Kim et al. developed a nonviral vector for delivery of p53 in ovarian cancer cells (OVCAR-3) [461]. The nonviral (liposomal) vector consisted of DOTAP, DOPE, and Chol in the molar ratio 1:0.7:0.3. High expression of p53 mRNA and proteins in OVCAR-3 cells indicated successful transfection of the lipoplexes to the cells used. The latter was indicated with the inhibition of cell growth obtained in OVCAR-3 cells due to apoptosis caused by that protein. Intratumoral injection of DDC/pp53-GFAP lipoplexes into mice bearing the OVCAR-3 cells clearly showed the tumor growth inhibition, which suggests the therapeutic efficiency of the particular lipoplexes.

The neutral 1,2-dioleoyl-*sn*-glycerol-3-phosphatidylcholine (DOPC) was used to make complexes with the siRNA targeting the oncoprotein EphA2 and then injected i.v. into ovarian tumor-bearing mice [462]. According to the authors, the significance of the particular study is the formulation of siRNA in liposomes with successful outcome after delivery of the loaded carrier, which showed significant reduced protein expression and tumor growth (Figure 17).

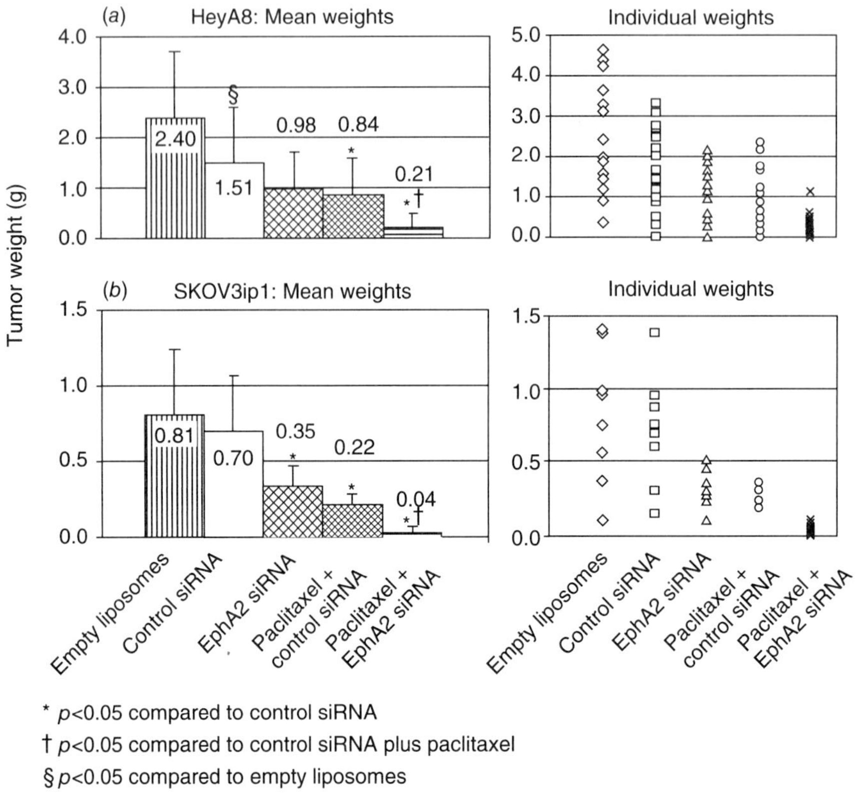

FIGURE 17 Therapeutic efficacy of siRNA-mediated EphA2 down regulation. A and B nude mice were injected i.p. with (*a*) 2.5 × 105 HeyA8 cells or (*b*) 1.0 × 106 SKOV3ip1 cells and randomly allocated to one of five groups, with therapy beginning 1 week after cell injection. (a) Empty DOPC liposomes, (b) control siRNA in DOPC, (c) EphA2-targeting siRNA in DOPC, (d) paclitaxel + control siRNA in DOPC, or (e) paclitaxel + EphA2 siRNA in DOPC. siRNA liposomes were injected twice weekly at a dose of 150 μg/kg siRNA. Paclitaxel (100 μg) or vehicle (first three groups) was injected i.p. once weekly. (Reprinted from ref. 462 and with permission of the American Association for Cancer Research.)

REFERENCES

1. Bangham, A. D., Standish, M. M., and Watkins, J. C. (1965), Diffusion of univalent ions across the lamellae of swollen phospholipids, *J. Mol. Biol.*, 13, 238–252.
2. Papahadjopoulos, D. (1978), Liposomes and their uses in biology and medicine, *Ann. N.Y. Acad. Sci.*, 408, 1.
3. Gregoriadis, G. (1978), Liposomes in therapeutic and preventive medicine: The development of the drug-carrier concept, *Ann. N.Y. Acad. Sci.*, 308, 343–370.
4. Kirby, C., Clarke, J., and Gregoriadis, G. (1980), Cholesterol content of small unilamellar liposomes controls phospholipid loss to high density lipoproteins in the presence of serum, *FEBS Lett.*, 111, 324–328.
5. Pain, D., Das, P. K., Ghosh, P. C., and Bachhawat, B. K. (1984), Increased circulatory half-life of liposomes after conjugation with dextran, *J. Biosci.*, 6, 811–816.

6. Allen, T. M., and Chonn, A. (1987), Large unilamellar liposomes with low uptake into the reticuloendothelial system, *FEBS Lett.*, 223, 42–46.
7. Papahadjopoulos, D., Allen, T. M., Gabizon, A., Mayhew, E., Matthay, K., Huang, S. K., et al. (1991), Sterically stabilized liposomes: Improvements in pharmacokinetics and antitumor therapeutic efficacy, *Proc. Natl. Acad. Sci. USA*, 88, 11460–11464.
8. Lasic, D. D., Martin, F. J., Gabizon, A., Huang, S. K., and Papahadjopoulos, D. (1991), Sterically stabilized liposomes: A hypothesis on the molecular origin of the extended circulation times, *Biochim. Biophys. Acta*, 1070, 187–192.
9. Torchilin, V. P., Levchenko, T. S., Whiteman, K. R., Yaroslavov, A. A., Tsatsakis, A. M., Rizos, A. K., et al. (2001), Amphiphilic poly-*N*-vinylpyrrolidones: Synthesis, properties and liposome surface modification, *Biomaterials*, 22, 3035–3044.
10. Takeuchi, H., Kojima, H., Yamamoto, H., and Kawashima, Y. (2001), Evaluation of circulation profiles of liposomes coated with hydrophilic polymers having different molecular weights in rats, *J. Controlled Release*, 75, 83–91.
11. Haran, G., Cohen, R., Bar, L., and Barenholz, Y. (1993), Transmembrane ammonium sulphate gradients in liposomes produce efficient and stable entrapment of amphipathis weak bases, *Biochim. Biophys. Acta*, 1151, 201–215.
12. Loan Honeywell-Nguyen, P., Wouter, H. W., Groenink, A., de Graaff, M., and Bouwstra, J. A. (2003), The in vivo transport of elastic vesicles into human skin: Effects of occlusion, volume and duration of application, *J. Controlled Release*, 90, 243–255.
13. Lasic, D. (1993), *Liposomes from Physics to Applications*, Elsevier, London.
14. Hauser, H. (2000), Short-chain phospholipids as detergents, *Biochim. Biophys. Acta*, 1508, 164–181.
15. Eibl, H. (1980), Synthesis of glycerophospholipids, *Chem. Phys. Lipids*, 26, 405–429.
16. Barenholz, Y., Thompson, T. E. (1999), Sphingomyelin: Biophysical aspects, *Chem. Phys. Lipids*, 102, 29–34.
17. Barenholz, Y. (2004), Sphingomyelin and cholesterol: From membrane biophysics and rafts to potential medical applications, *Subcell. Biochem.*, 37, 167–215.
18. Patel, G. B., and Sprott, G. D. (1999), Archaeobacterial ether lipid liposomes (archaeosomes) as novel vaccine and drug delivery systems, *Crit. Rev. Biotechnol.*, 19, 317–357.
19. Fatouros, D., Ioannou, P. V., and Antimisiaris, S. G. (2006), Novel nanosized arsenic containing vesicles for drug delivery: Arsonoliposomes, *J. Nanosci. Nanotechnol.*, 6, 2618–2687.
20. Biltonen, R. L., and Lichtenberg, D. (1993), The use of differential scanning calorimetry as a tool to characterize liposome preparations, *Chem. Phys. Lipids*, 64, 129–142.
21. Champan, D., Urbina, J., and Keough, K. M. (1974), Studies of lipid water systems using differential scanning calorimetry, *J. Biol. Chem.*, 249(8), 2512–2521.
22. van Osdol, W. W., Johnson, M. L., Ye, Q., and Biltonen M. (1991), Relaxation dynamics of the gel to liquid-crystalline transition of phosphatidylcholine bilayers. Effects of chain length and vesicle size, *Biophys. J.*, 59, 775–785.
23. van Osdol, W. W., Ye, Q., Johnson, M. L., and Biltonen, M. (1992), Effects of the anesthetic dibucaine on the kinetics of the gel-liquid crystalline transition of dipalmitoylphosphatidylcholine multilamellar vesicles, *Biophys. J.*, 63, 1011–1017.
24. Momo, F., Fabris, S., Bindoli, A., Scutari, G., and Stevanato, R. (2002), Different effects of propofol and nitrosopropofol on DMPC multilamellar liposomes, *Biophys. Chem.*, 95, 145–155.

25. Zhang, F., and Rowe, E. S. (1994), Calorimetric studies of the interactions of cytochrome c with dioleoylphosphatidylglycerol extruded vesicles: Ionic strength effects, *Biochim. Biophys. Acta*, 1193, 219–225.
26. Lo, Y. L., and Rahman, Y. E. (1995), Protein location in liposomes, a drug carrier: A prediction by differential scanning calorimetry, *J. Pharm. Sci.*, 84, 805–814.
27. Parente, R. A., and Lentz, B. R. (1985), Advantages and limitations of 1-palmitoyl-2-[[2-[4-(6-phenyl-trans-1,3,5-hexatrienyl)phenyl]ethyl]carbonyl]-3-sn-phosphatidylcholine as a fluorescent membrane probe, *Biochemistry*, 24, 6178–6185.
28. Metso, A. J., Zhao, H., Tuunainen, I., and Kinnnunen, P. K. (2004), Characterization of the main transition of dinervonoylphosphocholine liposomes by fluorescence spectroscopy, *Biochim. Biophys. Acta*, 1663, 222–231.
29. van Langen, H., Van Ginkel, G., Shaw, D., and Levine, Y. K. (1989), The fidelity of response by 1-[4-(trimethylammonio)phenyl]-6-phenyl-1,3,5-hexatriene in time-resolved fluorescence anisotropy measurements on lipid vesicles. Effects of unsaturation, head-group and cholesterol on orientational order and reorientational dynamics, *Eur. Biophys. J.*, 17, 37–48.
30. Davenport, L., and Targowski, P. (1996), Submicrosecond phospholipid dynamics using a long-lived fluorescence emission anisotropy probe, *Biophys. J.*, 71, 1837–1852.
31. de Kruijff, B., and Cullis, P. R. (1980), The influence of poly(L-lysine) on phospholipid polymorphism. Evidence that electrostatic polypeptide-phospholipid interactions can modulate bilayer/non-bilayer transitions, *Biochim. Biophys. Acta*, 601, 235–240.
32. Vasilenko, I., de Kruijff, B., and Verkleij, A. J. (1982), The synthesis and use of thionphospholipids in 31P-NRM studies of lipid polymorphism, *Biochim. Biophys. Acta*, 685, 144–152.
33. Perkins, W. R., Mincey, S. R., Ostro, M. J., Taraschi, T. F., and Janoff, A. S. (1988), The captured volume of multilamellar vesicles, *Biochim. Biophys. Acta*, 943, 103–107.
34. Davis, J. H. (1983), The description of membrane lipid conformation, order and dynamics by 2H-NMR, *Biochim. Biophys. Acta*, 737, 117–171.
35. Wu, W. G., Dowd, S. R., Simplaceanu, V., Peng, Z. Y., and Ho, C. (1985), 19F NMR investigation of molecular motion and packing in sonicated phospholipid vesicles, *Biochemistry*, 24, 7153–7161.
36. Bertoli, E., Masserini, M., Sonnino, S., Ghidoni, R., Cestaro, B., and Tettamanti, G. (1981), Electron paramagnetic resonance studies on the fluidity and surface dynamics of egg phosphatidylcholine vesicles containing gangliosides, *Biochim. Biophys. Acta*, 647, 196–202.
37. Shin, Y. K., and Freed, J. H. (1989), Dynamic imaging of lateral diffusion by electron spin resonance and study of rotational dynamics in model membranes. Effect of cholesterol, *Biophys. J.*, 55, 537–550.
38. Lai, C. S., Joseph, J., and Shih, C. C. (1989), Molecular dynamics of antitumor ether-linked phospholipids in model membranes: A spin-label study, *Biochem. Biophys. Res. Commun.*, 160, 1189–1195.
39. Ondrias, K., Stasko, A., Marko, V., and Nosal, R. (1989), Influence of beta-adrenoceptor blocking drugs on lipid-protein interaction in synaptosomal membranes. An ESR study, *Chem. Biol. Interact.*, 69, 87–97.
40. Hoekstra, D. (1982), Fluorescence method for measuring the kinetics of Ca^{2+}-induced phase separations in phosphatidylserine-containing lipid vesicles, *Biochemistry*, 21, 1055–1061.
41. Hoekstra, D. (1990), Fluorescence assays to monitor membrane fusion: Potential application in biliary lipid secretion and vesicle interactions, *Hepatology*, 12, 61S–66S.

42. LeNeveu, D. M., Rand, R. P., Parsegian, V. A., and Gingel, D. (1976), Apparent modification of forces between lecithin bilayers, *Science*, 191, 399–400.

43. Franks, N. P. (1976), Structural analysis of hydrated egg lecithin and cholesterol bilayers. I. X-ray diffraction, *J. Mol. Biol.*, 100, 345–358.

44. Worcester, D. L., and Franks, N. P. (1976), Structural analysis of hydrated egg lecithin and cholesterol bilayers. II. Neutron diffraction, *J. Mol. Biol.*, 100, 359–378.

45. LeNeveu, D. M., and Rand, R. P. (1977), Measurement and modification of forces between lecithin bilayers, *Biophys. J.*, 18, 209–230.

46. Hobbs, S. K., Monsky, W. L., Yuan, F., Roberts, W. G., Griffith, L., Torchillin, V. P., et al. (1998), Regulation of transport pathways in tumor vessels: Role of tumor type and microenvironment, *Proc. Natl. Acad. Sci. USA*, 95, 4607–4612.

47. Drummond, D. C., Meyer, O., Hong, K., Kirpotin, D. B., and Papahadjopoulos, D. (1999), Optimizing liposomes for delivery of chemotherapeutic agents to solid tumour, *Pharmacol. Rev.*, 51, 691–743.

48. Devine, D. V., Wong, K., Serrano, K., Chonn A., and Cullis, P. R. (1994), Liposome-complement interactions in rat serum: Implications for liposome survival studies, *Biochim. Biophys. Acta*, 1191, 43–51.

49. Kolchens, S., Ramaswami, V., Birgenheier, J., Nett, L., and O'Brien, D. F. (1993), Quasi-elastic light scattering determination of the size distribution of extruded vesicles, *Chem. Phys. Lipids*, 65, 1–10.

50. Lesieur, S., Grabielle-Madelmont, C., Paternostre, M. T., and Ollivon, M. (1991), Size analysis and stability study of lipid vesicles by high-performance gel exclusion chromatography, turbidity, and dynamic light scattering, *Anal. Biochem.*, 1, 192, 334–343.

51. Matsuzaki, K., Murase, O., Sugishita, K., Yoneyama, S., Akada, K., Ueha, M., Nakamura, A., and Kobayashi, S. (2000), Optical characterization of liposomes by right angle light scattering and turbidity measurement, *Biochim. Biophys. Acta*, 31, 219–226.

52. Moon, M. H., and Giddings, J. C. (1993), Size distribution of liposomes by flow field-flow fractionation, *J. Pharm. Biomed. Anal.*, 11, 911–920.

53. Korgel, B. A., van Zanten, J. H., and Monbouquette, H. G. (1998), Vesicle size distributions measured by flow field-flow fractionation coupled with multiangle light scattering, *Biophys. J.*, 74, 3264–3272.

54. Lee, H., Williams, S. K., Allison, S. D., and Anchordoquy, T. J. (2001), Analysis of self-assembled cationic lipid-DNA gene carrier complexes using flow field-flow fractionation and light scattering, *Anal. Chem.*, 73, 837–843.

55. Grabielle-Madelmont, C., Lesieur, S., and Ollivon, M. (2003), Characterization of loaded liposomes by size exclusion chromatography, *J. Biochem. Biophys. Methods*, 56, 189–217.

56. Goni, F. M., and Alonso, A. (2000), Spectroscopic techniques in the study of membrane solubilization, reconstitution and permeabilization by detergents, *Biochim. Biophys. Acta*, 1508, 51–68.

57. Alonso, A., Villena, A., and Goñi, F. M. (1981), Lysis and reassembly of sonicated lecithin vesicles in the presence of Triton X-100, *FEBS Lett.*, 123, 200–204.

58. Kokona, M., Kallinteri, P., Fatouros, D., and Antimisiaris, S. G. (2000), Stability of SUV liposomes in the presence of cholate salts and pancreatic lipases: Effect of lipid composition, *Eur. J. Pharm. Sci.*, 9, 245–252.

59. Almog, S., Litman, B. J., Wimley, W., Cohen, J., Wachtel, E. J., Barenholz, Y., Ben-Saul, A., and Lichtenberg, D. (1990), States of aggregation and phase transformations in mixtures of phosphatidylcholine and octyl glucoside, *Biochemistry*, 29, 4582–4592.

60. Lopez, O., de la Maza, A., Coderch, L., Lopez-Iglesias, C., Wehrli E., and Parra, J. L. (1998), Direct formation of mixed micelles in the solubilization of phospholipid liposomes by Triton X-100, *FEBS Lett.*, 426, 314–318.
61. Fatouros, D. G., Piperoudi, S., Gortzi, O., Ioannou, P. V., and Antimisiaris, S. G. (2005), Physical stability of sonicated arsonoliposomes: Effect of calcium ions, *J. Pharm. Sci. US*, 94, 46–55.
62. Akashi, K., Miyata, H., Itoh. H., and Kinoshita, K., Jr. (1996), Preparation of giant liposomes in physiological conditions and their characterization under an optical microscope, *Biophys. J.*, 71, 3242–3250.
63. Shohda, K., Toyota, T., Yomo, T., and Saguwara, T. (2003), Direct visualization of DNA duplex formation on the surface of a giant liposome, *Chembiochem.*, 4, 778–781.
64. Henriksen, J. R., and Ipsen, J. H. (2004), Measurement of membrane elasticity by micropipette aspiration, *Eur. Phys. J. E. Soft Matter*, 14, 149–167.
65. Olson, F., Hunt, C. A., Szoka, F. C., Vail, W. J., and Papahadjopoulos, D. (1979), Preparation of liposomes of defined size distribution by extrusion through polycarbonate membranes, *Biochim. Biophys. Acta*, 557, 9–23.
66. Weibull, C., Christiansson, A., and Carlemalm, E. (1983), Extraction of membrane lipids during fixation, dehydration and embedding of Acholeplasma laidlawii-cells for electron microscopy, *J. Microsc.*, 129, 201–207.
67. Dermer, G. B. (1968), An autoradiographic and biochemical study of oleic acid absorption by intestinal slices including determinations of lipid loss during preparation of electron microscopy, *J. Ultrastruct. Res.*, 22, 312–325.
68. Hope, M. J., Walker, D. C., and Cullis, P. R. (1983), Ca^{2+} and pH induced fusion of small unilamellar vesicles consisting of phosphatidylethanolamine and negatively charged phospholipids: A freeze fracture study, *Biochem. Biophys. Res. Commun.*, 110, 15–22.
69. Hope, M. J., Wong, K. F., and Cullis, P. R. (1989), Freeze-fracture of lipids and model membrane systems, *J. Electron. Microsc. Tech.*, 13, 277–287.
70. Burger, K. N., Nieva, J. L., Alonso, A., and Verkleij, A. J. (1991), Phospholipase C activity-induced fusion of pure lipid model membranes. A freeze fracture study, *Biochim. Biophys. Acta*, 1068, 249–253.
71. Meyer, H. W., and Richter, W. (2001), Freeze-fracture studies on lipids and membranes, *Micron.*, 32, 615–644.
72. Taggar, A. S., Alnajim, J., Anantha, M., Thomas, A., Webb, M., Ramsay, E., and Bally, M. B. (2006), Copper-topotecan complexation mediates drug accumulation into liposomes, *J. Controlled Release*, 114, 78–88.
73. Anabousi, S., Laue, M., Lehr, C. M., Bakowsky, U., and Ehrhardt. C. (2005), Assessing transferring modification of liposomes by atomic force microscopy and tramsmission electron microscopy, *Eur. J. Pharm. Biopharm*, 60, 295–303.
74. Bellare, J. R., Davis, H. T., Scriven, L. E., and Talmon, Y. (1988), Controlled environment vitrification system. An improved sample preparation technique, *J. Elec. Micr. Tech.*, 10, 87–111.
75. Dubochet, J., Adrian, M., Chang, J., Homo, J. C., Lepault, J., McDowell, A. W., and Schultz, P. (1988), Cryo-electron microscopy of vitrified specimens, *Q. Rev. Biophys.*, 21, 129–228.
76. Frederik, P. M., Stuart, M. C., and Verklej, A. J. (1989), Intermediary structures during membrane fusion as observed by cryo-electron microscopy, *Biochim. Biophys. Acta*, 979, 275–278.
77. Siegel, D. P., Green, W., and Talmon, J. (1989), Intermediates in membrane fusion and bilayer/nonbilayer phase transitions imaged by time-resolved cryo-transmission electron microscopy, *Biophys J.*, 56, 161–169.

78. Frederik, P. M., Burger, K. N., Stuart, M. C., and Verklej, A. J. (1991), Lipid polymorphism as observed by cryo-electron microscopy, *Biochim. Biophys. Acta*, 1062, 133–141.

79. Siegel, D. P., Green, W., and Talmon, J. (1994), The mechanism of lamellar-to-inverted hexagonal phase transitions: A study using temperature-jump cryo-electron microscopy, *Biophys. J.*, 66, 402–414.

80. Mozafari, M. R., Reed, C. J., Rostron, C., and Hasirci, V. (2005), A review of scanning probe microscopy investigations of liposome-DNA complexes, *J. Liposome Res.*, 15, 93–107.

81. Papadimitriou, E., and Antimisiaris, S. G. (2000), Interactions of PC/Chol and PS/Chol liposomes with human cells in vitro, *J. Drug Target.*, 5, 335–351.

82. Haywood, A. M., and Boyer, B. P. (1986), Ficoll and dextran enhance adhesion of Sendai virus to liposomes containing receptor (ganglioside GD1a), *Biochemistry*, 25, 3925–3929.

83. Shichijo, S., and Alving, C. R. (1986), Inhibitory effects of gangliosides on immune reactions of antibodies to neutral glycolipids in liposomes, *Biochim. Biophys. Acta*, 858, 118–124.

84. Ho, R. J., Rouse, B. T., and Huang, L. (1986), Target-sensitive immunoliposomes: Preparation and characterization, *Biochemistry*, 25, 5500–5506.

85. Bogdanov, A. A., Gordeeva, L. V., Torchilin, V. P., and Margolis, L. B. (1989), Lectin-bearing liposomes: Differential binding to normal and to transformed mouse fibroblasts, *Exp. Cell Res.*, 181, 362–374.

86. Woodle, M. C., and Lasic, D. D. (1992), Sterically stabilized liposomes, *Biochim. Biophys. Acta*, 1113, 171–199.

87. Woodle, M. C., Collins, L. R., Sponsler, E., Kossovsky, N., Papahadjopoulos, D., and Martin, F. J. (1992), Sterically stabilized liposomes. Reduction in electrophoretic mobility but not electrostatic surface potential, *Biophys. J.*, 61, 902–910.

88. Mayhew, E., Lazo, R., and Vail, W. J. (1989), Preparation of liposomes entrapping cancer chemotherapeutic agents for experimental in vivo and in vitro studies, in Gregoriadis, G., Ed., *Liposome Technology*, Vol. II, CRC Press, Boca Raton, FL, Chapter 2.

89. Fatouros, D. G., and Antimisiaris, S. G. (2001), Physicochemical properties of liposomes incorporating hydrochlorothiazide and chlorothiazide, *J. Drug Target.*, 9, 61–74.

90. Fatouros, D. G., and Antimisiaris, S. G. (2002), Effect of amphiphilic drugs on the stability and zeta-potential of their liposome formulations: A study with prednisolone, diazepam and griseofulvin, *J. Coll. Interf. Sci.*, 251, 271–277.

91. Szoka, F. C., and Papahadjopoulos, D. (1980), Comparative properties and methods of preparation of lipid vesicles (liposomes), *Ann. Rev. Biophys. Bioeng.*, 75, 4194–4199.

92. Kulkarni, S. B., Betageri, G. V., and Singh, M. (1995), Factors affecting microencapsulation of drugs in liposomes, *J. Microencapsul.*, 12, 229–246.

93. Amselem, S., Gabizon, A., and Barenholz, Y. (1990), Optimization and up scaling of doxorubicin-containing liposomes for clinical use, *J. Pharm. Sci.*, 79, 1045–1052.

94. Mezei, M., and Nugent, F. J. (1984), Method of encapsulating biologically active materials in multilmellar lipid vesicles, U.S. Patent 4, 485,054.

95. Ma, L., Ramachandran, C., and Weiner, N. D. (1991), Partitioning of a homologous series of alkyl *p*-amino benzoates in dipalmitoyl phosphatidyl choline liposomes: Effect of liposomes type, *Int. J. Pharm.*, 70, 209–218.

96. Betageri, G. V., and Parsons, D. L. (1992), Drug encapsulation and release from multilamellar and unilamellar liposomes, *Int. J. Pharm.*, 81, 235–241.

97. Puglisi, G., Fresta, M., La Rosa, C., Ventura, C. A., Panic, A. M., and Mazzonne, A. (1992), Liposomes as potential drug carrier for citicoline (CDP-choline) and the effect of formulation conditions on encapsulation efficiency, *Pharmazie*, 47, 211–215.

98. Betageri, G. V. (1993), Liposomal encapsulation and stability of dideoxynosine triphosphate, *Drug Dev. Ind. Pharm.*, 19, 531–539.

99. Fresta, M., Villari, A., Puglisi, G., and Cavallaro, G. (1993), 5-Fluorouracil: Various kinds of loaded liposomes: Encapsulation efficiency, storage stability and fusogenic properties, *Int. J. Pharm.*, 99, 145–156.

100. Elorza, B., Elorza, M. A., Frutos, G., and Chantres, J. R. (1993), Characterization of 5-flourouracil loaded liposomes prepared by reverse phases evaporation or freezing thawing extrusion methods: Study of release, *Biochim. Biophys. Acta*, 1153, 135–142.

101. Law, S. L., Chang, P., and Lin, C. H. (1991), Characteristics of mitoxantrone loading on liposomes, *Int. J. Pharm.*, 70, 1–7.

102. Alpar, O. H., Bamford, J. B., and Walters, V. (1981), In vitro incorporation and release of hydroxycobalamin by liposomes, *Int. J. Pharm.*, 7, 349–351.

103. Crommelin, D. J. A., and Van Bloois, L. (1983), Preparation and characterization of doxorubicin containing liposomes. Part 2. Loading capacity, long-term stability and doxorubicin bilayer interaction mechanism, *Int. J. Pharm.*, 17, 135–144.

104. Arien, A., Coigoux, C., Baquey, C., and Dupuy, B. (1993), Study of in vitro and in vivo stability of liposomes loading with calcitonin or indium in GIT, *Life Sci.*, 53, 1279–1290.

105. D'Silva, J. B., and Notari, R. E. (1982), Drug stability in liposomal suspensions: Hydrolysis of indomethacin, cyclocytidine and *p*-nitrophenyl acetate, *J. Pharm. Sci.*, 71, 1394–1398.

106. Kirby, C., Clarke, J., and Gregoriadis, G. (1980), Effect of the cholesterol content of small unilamellar liposomes on their stability in vivo and in vitro, *Biochem. J.*, 186, 591–595.

107. Kirby, C., and Gregoriadis, G. (1980), The effect of the cholesterol content of small unillamellar liposomes on the fate of their lipid components in vivo, *Life Sci.*, 27, 2223–2230.

108. Layton, D., and Trouet, A. (1980), A comparison of the therapeutic effects of free and liposomally encapsulated vincristine in leucemic mice, *Eur. J. Cancer*, 16, 945–951.

109. McCormack, B., and Gregoriadis, G. (1994), Drugs-in-cylcodextrins-liposomes: A novel concept in drug delivery, *Int. J. Pharm.*, 112, 249–258.

110. McCormack, B., and Gregoriadis, G. (1996), Comparative studies of the fate of free and liposome-entrapped hydroxypropyl-β-cyclodextrin-drug complexes after intravenous injection into rats: Implications in drug delivery, *Biochim. Biophys. Acta*, 1291, 237–244.

111. Fatouros, D. G., Hatzidimitirou, K., and Antimisiaris, S. G. (2001), Liposomes encapsulating prednisolone and prednisolone-cyclodextrin complexes: Comparison of membrane integrity and drug release, *Eur. J. Pharm. Sci.*, 13, 287–296.

112. Grit, M., and Crommelin, D. J. A. (1993), Analysis and hydrolysis kinetics of phospholipids in aqueous liposome dispersions, in Gregoriadis G., Ed., *Liposome Technology*, Vol. I, CRC Press, Boca Raton, FL, pp. 455–487.

113. Grit, M., and Crommelin, D. J. A. (1993), Chemical stability of liposomes: Implications for their stability, *Chem. Phys. Lipids*, 64, 3–18.

114. Zuidam, N. J., and Crommelin, D. J. A. (1995), Chemical hydrolysis of phospholipids, *J. Pharm. Sci.*, 84, 1113–1119.

115. Grit, M., de Smidt, J., Struijke, H. A., and Crommelin, D. J. A. (1989), Hydrolysis of natural soybean phosphatidylcholine in aqueous liposome dispersions, *Int. J. Pharm.*, 50, 1–6.

116. Hunt, C. A., and Tsang, S. (1981), α-Tocopherol retards autoxidation and prolongs the shelf life of liposomes, *Int. J. Pharm.*, 8, 101–110.

117. Fukuzawa, K., Chida, H., Akira, T., and Tsukatani, H. (1981), Autooxidative effect of α-tocopherol incorporation into lecithin liposomes on ascorbic acid Fe^{++}-induced lipid peroxidation, *Arch. Biochem. Biophys.*, 206, 173–180.

118. Talsma, H., van Steenbergen, M. J., and Crommelin, D. J. A. (1992), The cryopreservation of liposomes. 2. Effect of particle size on crystallization behavior and marker retention, *Cryobiology*, 29, 80–86.

119. Talsma, H., van Steenbergen, M. J., and Crommelin, D. J. A. (1991), The cryopreservation of liposomes. 3. Almost complete retention of a water-soluble marker in small liposomes in a cryoprotectant containing dispersion after a freezing/thawing cycle, *Int. J. Pharm.*, 77, 119–126.

120. Kristiansen, J. (1992), Leakage of a trapped fluorescent marker from liposomes: Effects of eutectic crystallization of NaCl and internal freezing, *Cryobiology*, 29, 575–584.

121. Crowe, L. M., and Crowe, J. H. (1988), Trehalose and dry dipalmitoloylphosphatidylcholine revisited, *Biochim. Biophys. Acta*, 946, 193–201.

122. Crowe, J. H., Leslie, S. B., and Crowe, L. M. (1994), Is vitrification sufficient to preserve liposomes during freeze-drying? *Cryobiology*, 31, 355–366.

123. Piel, G., Piette, M., Barillaro, V., Castagne, D., Evrard, B., and Delattre, L. (2006), Betamethasone-in-cyclodextrin-in-liposome: The effect of cyclodextrins on encapsulation efficiency and release kinetics, *Int. J. Pharm.*, 312, 75–82.

124. Woodle, M. C., and Papahadjopoulos, D. (1989), Liposome preparation and size characterization, *Meth. Enzym.*, 171, 193–217.

125. New, R. R. C., Ed. (1989), *Liposomes. A Practical Approach*, IRL Press, Oxford. UK.

126. Bangham, A. D., and Horne, R. W. (1964), Negative staining of phospholipids and their structured modification by surface active agents as observed in the electron microscope. *J. Mol. Biol.*, 8, 660–668.

127. Papahadjopoulos, D., and Watkins, J. C. (1967), Phospholipid model membrane 2: Permeability properties of hydrated liquid crystals, *Biochim. Biophys. Acta*, 135, 639–652.

128. Batzri, S., and Korn, E. D. (1973), Single bilayer vesicles prepared without sonication, *Biochm. Biophys. Acta*, 298, 1015–1019.

129. Deamer, D., and Bangham, A. D. (1976), Large volume liposomes by an ether vaporization method, *Biochm. Biophys. Acta*, 443, 629–634.

130. Barenholz, Y., Amselem, S., and Lichtenberg, D. (1979), A new method for preparation of phospholipid vesicles, *FEBS Lett.*, 99, 210–214.

131. Mayhew, E., Lazo, R., Vail, W. J., King, J., and Green, A. M. (1984), Characterization of liposomes prepared using microfluidizer, *Biochim. Biophys. Acta*, 775, 169–174.

132. Szoka, F. C., and Papahadjopoulos, D. (1978), Procedure for preparation of liposomes with large internal aqueous space and high capture by reverse phase evaporation, *Proc. Natl. Acad. Sci. USA*, 75, 4194–4198.

133. Gregoriadis, G., Da Silva, H., and Florence A. T. (1990), A procedure for the efficient entrapment of drugs in dehydration-rehydration liposomes, *Int. J. Pharm.*, 65, 235–242.

134. Zadi, B., and Gregoriadis, G. (2000), A novel method for high-yield entrapment of solutes into small liposomes, *J. Lipos. Res.*, 10, 73–80.

135. Reeves, J. P., and Dowben R. M. (1969), Formation and properties of thin phospholipids vesicles, *J. Cell. Phys.*, 73, 49–60.

136. Antimisiaris, S. G., Jayesekera, P., and Gregoriadis, G. (1993), Liposomes as vaccine carriers: Incorporation of soluble and particulate antigens in giant vesicles, *J. Immunol. Methods*, 166, 271–280.
137. Mimms, L. T., Zampighi, G., Nozaki, Y., Tanford, C., and Reynolds, J. A. (1981), Phospholipid vesicles formation and transmembrane protein incorporation using octyl glucoside, *Biochemistry*, 20, 833–840.
138. Hauser, H., Brunner, J., and Skrabal, P. (1976), Single bilayer vesicles prepared without sonication, *Biochim. Biophys. Acta*, 455, 322–331.
139. Papahadjopoulos, D., Vail, W. J., Jacobson, K., and Poste, G. (1975), Cochleate lipid cylinders formation by fusion of unilamellar lipid vesicles, *Biochim. Biophys. Acta*, 394, 483–491.
140. Talsma, H., Van Steenbergen, M. J., Borchert, J. C. H., Crommelin, D. J. A. (1994), A novel technique for the one-step preparation of liposomes and nonionic surfactant vesicles without the use of organic solvents. Liposome formation in a continuous gas stream: The "bubble" method, *J. Pharm. Sci.*, 83, 276–280.
141. Avestin, Inc., Ottawa, Canada, http://www.avestin.com/products.html, accessed Jan. 1, 2004.
142. Mayer, L. D., Bally, M. B., and Cullis, P. R. (1990), Characterization of liposomal systems containing doxorubicin entrapped in response to pH gradients, *Biochim. Biophys. Acta*, 1025, 143–151.
143. Nichols, J. W., and Deamer, D. W. (1990), Catecholamine uptake and concentration of liposomes maintaining pH gradients, *Biochim. Biophys. Acta*, 455, 269–278.
144. Harrigan, P. R., Wong, K. F., Redelmeier, T. E., Wheller, J. J., and Cullis, P. R. (1993), Accumulation of doxorubicin and other lipophilic amines into LUVs in response to transmembrane pH gradient, *Biochim. Biophys. Acta*, 1149, 237–244.
145. Boman, N. L., Mayer, L. D., and Cullis, P. R. (1993), Optimization of the retention and properties of vincristine in liposomal systems, *Biochim. Biophys. Acta*, 1152, 253–258.
146. Gabizon, A., Shiota, R., and Papahadjopoulos. D. (1989), Pharmacokinetics and tissue distribution of doxorubicin encapsulated in stable liposomes with long circulation times, *J. Nat. Can. Inst.*, 81, 1484–1488.
147. Gabizon, A., Barenholz, Y., and Bialer, M. (1993), Prolongation of the circulation time of doxorubicin encapsulated in liposomes containing a polyethylene glycol derivatized phospholipid: Pharmacokinetic studies in rodents and dogs, *Pharm. Res.*, 10, 703–708.
148. Cheung, B. C., Sun, T. H., Leenhouts, J. M., and Cullis, P. R. (1998), Loading of doxorubicin into liposomes by forming Mn^{2+}–drug complexes, *Biochim. Biophys. Acta*, 1414, 205–216.
149. Abraham, S. A., Edwards, K., Karlsson, G., MacIntosh, S., Mayer, L. D., McKenzie, C., and Bally, M. B. (2002), Formation of transition metal–doxorubicin complexes inside liposomes, *Biochim. Biophys. Acta*, 1565, 41–54.
150. Nobs, L., Buchegger, F., Gurny, R., and Allemann, E. (2004), Current methods for attaching targeting ligands to liposomes and nanoparticles, *J. Pharm. Sci.*, 93, 1980–1991.
151. Sapra, P., and Allen, T. M. (2003), Ligand-targeted liposomal anticancer drugs, *Prog. Lipid Res.*, 42, 439–462.
152. Klibanov, A. L., Maruyama, K., Beckerleg, A. M., Torchilin, V. P., and Huang, L. (1991), Activity of amphipathic poly(ethylene glycol) 5000 to prolong the circulation time of liposomes depends on the liposome size and is unfavorable for immunoliposome binding to target, *Biochim. Biophys. Acta*, 1062, 142–148.
153. Maruyama, K., Takizawa, T., Takahashi, N., Nagaike, K., and Iwatsuru, M. (1997), Targeting efficiency of PEG-immunoliposome-conjugated antibodies at PEG terminals, *Adv. Drug Deliv. Rev.*, 24, 235–242.

154. Maruyama, K., Takizawa, T., Yuda, T., Kennel, S. J., Huang, L., and Iwatsuru, M. (1995), Targetability of novel immunoliposomes modified with amphipathic poly(ethylene glycol)s conjugated at their distal terminals to monoclonal antibodies, *Biochim. Biophys. Acta*, 1234, 74–80.

155. Park, J. W., Benz, C. C., and Martin, F. J. (2004), Future directions of liposome- and immunoliposome-based cancer therapeutics, *Semin. Oncol.*, 31(6 Suppl 13), 196–205.

156. Ishida, T., Iden, D. L., and Allen, T. M. (1999), A combinatorial approach to producing sterically stabilized (Stealth) immunoliposomal drugs, *FEBS Lett.*, 460, 129–133.

157. Park, C. G., Thiex, N. W., Lee, K. M., Szot, G. L., Bluestone, J. A., and Lee, K. D. (2003), Targeting and blocking B7 costimulatory molecules on antigen-presenting cells using CTLA4Ig-conjugated liposomes: In vitro characterization and in vivo factors affecting biodistribuiton, *Pharm. Res.*, 20, 1239–1248.

158. Bartsch, M., Weeke-Klimp, A. H., Meijer, D. K. F., Scherphof, G. L., and Kamps, J. A. A. (2002), Massive and selective delivery of lipid-coated cationic lipoplexes of oligonucleotides targeted in vivo to hepatic endothelial cells, *Pharm. Res.*, 19, 676–680.

159. Chekhonin, V. P., Zhirkov, Y. A., Gurina, O. I., Ryabukhin, I. A., Lebedev, S. V., Kashparov, I. A., and Dmitriyeva, T. B. (2005), PEGylated Immunoliposomes directed against brain astrocytes, *Drug Deliv.*, 12, 1–6.

160. Brignole, C., Marimpietri, D., Gambini, C., Allen, T. M., Ponzoni, M., and Pastorino, F. (2003), Development of Fab′ fragments of anti-GD_2 immunoliposomes entrapping doxorubicin for experimental therapy of human neuroblastoma, *Cancer Lett.*, 197, 199–204.

161. Demirovic, A. M., Marty, C., Console, S., Zeisberger, S. M., Ruch, C., Jaussi, R., Schwendener, R. A., and Ballmer-Hofer, K. (2005), Targeting human cancer cells with VEGF receptor-2-directed liposomes, *Oncol. Rept.*, 13, 319–324.

162. Voinea, M., Manduteanu, I., Dragomir, E., Capraru, M., and Simionescu, M. (2005), Immunoliposomes directed toward VCAM-1 interact specifically with activated endothelial cells—a potential tool for specific drug delivery, *Pharm. Res.*, 22, 1906–1917.

163. Park, J. W., Hong, K., Kiprotin, D. B., Meyer, O., Papahadjopoulos, D., and Benz, C. C. (1997), Anti-HER2 immunoliposomes for targeted therapy of human tumors, *Cancer Lett.*, 118, 153–160.

164. Mercadal, M., Domingo, J. C., Petriz, J., Garcia, J., and de Madariaga, M. A. (1999), A novel strategy affords high-yield coupling of antibody to extremities of liposomal surface-grafted PEG chains, *Biochim. Biophys. Acta*, 1418, 232–238.

165. Longmuir, K. J., Robertson, R. T., Haynes, S. M., Baratta, J. L., and Waring, A. J. (2006), Effective targeting of liposomes to liver and hepatocytes in vivo by incorporation of a *Plasmodium* amino acid sequence, *Pharm. Res.*, 23, 759–769.

166. Hansen, C. B., Kao, G. Y., Moase, E. H., Zalipsky, S., and Allen, T. M. (1995), Attachment of antibodies to sterically stabilized liposomes: Evaluation, comparison and optimization of coupling procedures, *Biochim. Biophys. Acta*, 1239, 133–144.

167. Mamot, C., Ritschard, R., Kung, W., Park, J. W., Herrmann, R., and Rochlitz, C. F. (2006), EGFR-targeted immunoliposomes derived from the monoclonal antibody EMD72000 mediate specific and efficient drug delivery to a variety of colorectal cancer cells, *J. Drug Targe.*, 14, 215–223.

168. Kallinteri, P., Papadimitriou, E., and Antimisiaris, S. G., (2001), Uptake of liposomes which incorporate a glycopeptide fraction of asialofetuin by HepG2 cells, *J. Lipos. Res.*, 11, 175–193.

169. Fretz, M. M., Koning, G. A., Mastrobattista, E., Jiskoot, W., and Storm, G. (2004), OVCAR-3 cells internalize TAT-peptide modified liposomes by endocytosis, *Biochim. Biophys. Acta*, 1665, 48–56.

170. Moreira, J. N., and Gaspar, R. (2004), Antagonist G-mediated targeting and cytotoxicity of liposomal doxorubicin in NCI-H82 variant small cell lung cancer, *Braz. J. Med. Biol. Res.*, 37, 1185–1192.

171. Martin, F. J., Hubbell, W. L., and Papahadjopoulos, D. (1981), Immunospecific targeting of liposomes to cells: A novel and efficient method for covalent attachment of Fab′ fragments via disulfide bonds, *Biochemistry*, 20, 4229–4238.

172. Wartchow, C. A., Alters, S. E., Garzone, P. D., Li, L., Choi, S., DeChene, N. E., Doede, T., Huang, L., Pease, J. S., Shen, Z., Knox, S. J., and Cleland, J. L. (2004), Enhancement of the efficacy of an antagonist of an extracellular receptor by attachment to the surface of a biocompatible carrier, *Pharm. Res.*, 21, 1880–1885.

173. Anabousi, S., Bakowsky, U., Schneider, M., Huwer, H., Lehr, C. M., and Ehrhardt, C. (2006), In vitro assessment of transferrin-conjugated liposomes as drug delivery systems for inhalation therapy of lung cancer, *Eur. J. Pharm. Sci., Jul 22*, 29, 367–374.

174. Lopez-Barcons, L. A., Polo, D., Llorens, A., Reig, F., and Fabra, A. (2005), Targeted adriamycin delivery to MXT-B2 metastatic mammary carcinoma cells by transferring liposomes: Effect of adriamycin ADR-to-lipid ratio, *Oncol. Rept.*, 14, 1337–1343.

175. Afzelius, P., Demant, E. J. F., Hansen, G. H., and Jensen, P. B. (1989), Covalent modification of serum transferrin with phospholipids and incorporation into liposomal membranes, *Biochim. Biophys. Acta*, 979, 231–238.

176. Stavridis, J. C., Deliconstantinos, G., Psallidopoulos, M. C., Armenakas, N. A., Hadjiminas, D. J., and Hadjiminas, J. (1986), Construction of transferrin-coated liposomes for on vivo transport of exogenous DNA to bone marrow erythroblasts in rabbits, *Exp. Cell Res.*, 164, 568–572.

177. Torchilin, V. P., Levchenko, T. S., Lukyanov, A. N., Khaw, B. A., Klibanov, A. L., Rammohan, R., Samokhin, G. P., and Whiteman, K. R. (2001), *p*-Nitrophenylcarbonyl-PEG-PE-liposomes: Fast and simple attachmentof specific ligands, including monoclonal antibodies, to distal ends of PEG chains via *p*-nitrophenylcarbonyl groups, *Biochim. Biophys. Acta*, 1511, 397–411.

178. Savva, M., Duda, E., and Huang, L. (1999), A genetically modified recombinant tumor necrosis factor-α conjugated to the distal terminals of liposomal surface grafted polyethyleneglycol chains, *Int. J. Pharm.*, 184, 45–51.

179. Ansell, S. M., Tardi, P. G., and Buchkowsky, S. S. (1996), 3-(2-Pyridyldithio)propionic acid hydrazide as a cross-linker in the formation of liposome-antibody conjugates, *Bioconjugate Chem.*, 7, 490–496.

180. Chua, M. M., Fan, S. T., and Karush, F. (1984), Attachment of immunoglobulin to liposomal membrane via protein carbohydrate, *Biochim. Biophys. Acta*, 800, 291–300.

181. Koning, G. A., Morselt, H. W., Velinova, M. J., Donga, J., Gorter, A., Allen, T. M., Zalipsky, S., Kamps, J. A., and Scherphof, G. L. (1999), Selective transfer of a lipophilic prodrug of 5-fluorodeoxyuridine from immunoliposomes to colon cancer, *Biochim. Biophys. Acta*, 1420, 153–167.

182. Harding, J. A., Engbers, C. M., Newman, M. S., Goldstein, N. I., and Zalipsky, S. (1997), Immunogenicity and pharmacokinetic attributes of poly(ethylene glycol)-grafted immunoliposomes, *Biochim. Biophys. Acta*, 1327, 181–192.

183. Sawant, R. M., Hurley, J. P., Salmaso, S., Kale, A., Tolcheva, E., Levchenko, T. S., and Torchilin, V. P. (2006), "SMART" drug delivery systems: Double-targeted pH-responsive. Pharmaceutical nanocarriers, *Bioconjugate Chem.*, 17, 943–949.

184. Xiao, Z., McQuarrie, S. A., Suresh, M. R., Mercer, J. R., Gupta, S., and Miller, G. G. (2002), A three-step strategy for targeting drug carriers to human ovarian carcinoma cells in vitro, *J. Biotechnol.*, 94, 171–184.

185. Schnyder, A., Krahenbuhl, S., Torok, M., Drewe, J., and Huwyler, J. (2004), Targeting of skeletal muscle in vitro using biotinylated immunoliposomes, *Biochem. J.*, 377, 61–67.

186. Allen, T. M., Mumbengegwi, D. R., and Charrois, J. R. (2005), Anti-CD19-targeted doxorubicin improves the therapeutic efficacy in murine B-cell lymphoma and ameliorates the toxicity of liposomes with varying drug release rates, *Clin. Cancer Res.*, 11, 3567–3573.

187. Gupta, B., Levchenko, T. S., Mongayt, D. A., and Torchilin, V. P. (2005), Monoclonal antibody 2C5-mediated binding of liposomes to brain tumor cells in vitro and in subcutaneous tumor model in vivo, *J. Drug Target.*, 13, 337–343.

188. Kawakami, S., Wong, J., Sato, A., Hattori, Y., Yamashita, F., and Hashida, M. (2000), Biodistribution characteristics of mannosylated, fucosylated, and galactosylated liposomes in mice, *Biochim. Biophys. Acta*, 1524, 258–265.

189. de Melo, A. L., Silva-Barcellos, N. M., Demicheli, C., and Frezard, F. (2003), Enhanced schistosomicidal efficacy of tartar emetic encapsulated in pegylated liposomes, *Int. J. Pharm.*, 255, 227–230.

190. Chen, J-H., Ling, R., Yao, Q., Li, Y., Chen, T. Wang, Z., and Li, K-Z. (2005), Effect of small-sized liposomal adriamycin administered by various routes on a metastatic breast cancer model, *Endocr.-Related Cancer*, 12, 93–100.

191. Zhang, J. Q., Zhang, Z. R., Yang, Q. Y., Qin, S. R., and Qiu, X. L. (2005), Lyophilized paclitaxel magnetoliposomes as a potential drug delivery system for breast carcinoma via parenteral administration: In vitro and in vivo, *Pharm. Res.*, 22, 573–583.

192. Wang, X. P., Yazawa, K., Templeton, N. S., Yang, J., Liu, S., Li, Z., Li, M., Yao, Q., Chen, C., and Brunicardi, F. C. (2005), Intravenous delivery of liposome-mediated nonviral DNA is less toxic than intraperitoneal delivery in mice, *World J. Surg.*, 29, 339–343.

193. Kamps, J. A. A. M., and Scherphof, G. L. (2004), Biodistribution and uptake of liposomes in vivo, *Methods Enzymol.*, 387, 257–266.

194. Moghimi, S. M., and Hunter, A. S. (2001), Recognition by macrophages and liver cells of opsonized phospholipid vesicles and phospholipid headgroups, *Pharm. Res.*, 18, 1–8.

195. Chiu, G. N., Bally, M. B., and Mayer, L. D. (2001), Selective protein interactions with phosphatidylserine containing liposomes alter the steric stabilization properties of poly(ethylene glycol), *Biochim. Biophys. Acta*, 1510, 56–69.

196. Harrington, K. J., Syrigos, K. N., and Vile, R. G. (2002), Liposomally targeted cytotoxic drugs for the treatment of cancer. *J. Pharm. Pharmacol.*, 54, 1573–1600.

197. Forssen, E. A., Coulter, D. M., and Proffitt, R. T. (1992), Selective in vivo localization of daunorubicin small unilamellar vesicles in solid tumours, *Cancer Res.*, 52, 3255–3261.

198. Forssen, E. A., Male-Brune, R., Adler-Moore, J. P., Lee, M. J., Schmidt, P. G., Krasieva, T. B., Shimizu, S., and Tromberg, B. J. (1996), Fluorescence imaging studies for the disposition of daunorubicin liposomes (DaunoXome) within tumour tissue, *Cancer Res.*, 56, 2066–2075.

199. Ogihara-Umeda, I., Sasaki, T., Kojima, S., and Nishigori, H. (1996), Optimal radiolabelled liposomes for tumour imaging, *J. Nucl. Med.*, 37, 326–332.

200. Sadzuka, Y., and Hirota, S. (1998), Does the amount of an antitumor agent entrapped in liposomes influence its tissue distribution and cell uptake? *Cancer Lett.*, 131, 163–170.

201. Turner, A. F., Presant, C. A., Proffitt, R. T., Williams, L. E., Winsor, D. W., and Werner, J. L. (1988), In-111-labelled liposomes: Dosimetry and tumour depiction, *Rdiology*, 166, 761–765.

202. Presant, C. A., Proffitt, R. T., Turner, A. F., Williams, L. E., Winsor, D., Werner, J. L., Kennedy, P., Wiseman, C., Gala, K., and McKenna, R. J. (1988), Successful imaging of human cancer with In-111-labelled phospholipid vesicles, *Cancer*, 62, 905–911.

203. Presant, C. A., Blayney, D., Proffitt, R. D., Turner, A. F., Williams, L. E., Nadel, H. I., Kennedy, P., Wiseman, C., Gala, K., and Crossley, R. J. (1990), Preliminary report: Imaging of Kaposi sarcoma and lymphoma in AIDS with indium-111-labelled liposomes, *Lancet*, 335, 1307–1309.

204. Kubo, A., Nakamura, K., Sammiya, T., Katayama, M., Hashimoto, T., Hashimoto, S., Kobayashi, H., and Teramoto, T. (1993), Indium-111-labelled liposomes: Dosimetry and tumour depiction in patients with cancer, *Eur. J. Nucl. Med.*, 20, 107–113.

205. Khalifa, A., Dodds, D., Rampling, R., Paterson, J., and Murray, T., (1997), Liposomal distribution in malignant glioma: Possibilities for therapy, *Nucl. Med. Commun.*, 18, 17–23.

206. Zucchetti, M., Boiardi, A., Silvani, A., Parisi, I., Piccolrovazzi, S., and D'Incalci, M. (1999), Distribution of daunorubicin and daunorubicinol in human glioma tumors after administration of liposomal daunorubicin, *Cancer Chemother. Pharmacol.*, 44, 173–176.

207. British National Formulary (BNF) Edition 51, 2006, Pharmaceutical Press, London.

208. http://www.gilead.com/pdf/daxpius.pdf#search='DaunoXome'. Gilead Sciences, Inc., 2007.

209. Shimizu, K., Takada, M., Asai, T., Kuromi, K., Baba, K., and Oku, N. (2002), Cancer chemotherapy by liposomal 6-[[2-(dimethylamino)ethyl]amino]-3-hydroxy-7H-indeno[2,1-*c*]quinolin-7-one dihydrochloride (TAS-103), a novel anticancer agent, *Biol. Pharm. Bull.*, 25, 1385–1387.

210. Wu, P. C., Tsai, Y. H., Liao, C. C., Chang, J. S., and Huang, Y. B. (2004), Tha characterization and biodistribution of cefoxitin-loaded liposomes, *Int. J. Pharm.*, 271, 31–39.

211. Alberts, D. S., Markman, M., Armstrong, D., Rothenberg, M. L., Muggia, F., and Howell, S. B. (2002), Intraperitoneal therapy for stage III ovarian cancer: A therapy whose time has come! *J. Clin. Oncol.*, 20, 3944.

212. Koga, S., Hamazoe, R., Maeta, M., Shimizu, N., Murakami, A., and Wakatsuki, T. (1988), Prophylactic therapy for peritoneal recurrence of gastric cancer by continuous hyperthermic peritoneal perfusion with mitomycin C, *Cancer*, 61, 232.

213. Speyer, J. L. (1985), The rationale behind intraperitoneal chemotherapy in gastrointestinal malignancies, *Semin. Oncol.*, 12, 23.

214. Sadzuka, Y., Hirota, S., and Sonobe, T. (2000), Intraperitoneal administration of doxorubicin encapsulating liposomes against peritoneal dissemination, *Toxicol. Lett.*, 116, 51–59.

215. Sadzuka, Y., Nakai, S., Miyagishima, A., Nozawa, Y., and Hirota, S. (1997), Effects of administered route on tissue distribution and antitumour activity of polyethyleneglycol-coated liposomes containing adriamycin, *Cancer Lett.*, 111, 77–86.

216. Marchettini, P., Stuart, A., Mohamed, F., Yoo, D., and Sugarbaker, P. H. (2002), Docetaxel: Pharmacokinetics and tissue levels after intraperitoneal and intravenous administration in a rat model, *Cancer Chemother. Pharmacol.*, 49, 499–503.

217. Sadzuka, Y., Hirama, R., and Sonobe, T. (2002), Effects of intraperitoneal administration of liposomes and methods of preparing liposomes for local therapy, *Toxicol. Lett.*, 126, 83–90.

218. Di Stefano, A., Carafa, M., Sozio, P., Pinnen, F., Braghiroli, D., Orlando, G., Cannazza, G., Ricciutelli, M., Marianecci, C., and Santucci, E. (2004), Evaluation of rat striatal L-dopa and DA concentration after intraperitoneal administration of L-dopa prodrugs in liposomal formulations, *J. Controlled Release*, 99, 293–300.

219. Oussoren, C., and Storm, G. (2001), Liposomes to target the lymphatics by subcutaneous administration, *Adv. Drug Deliv. Rev.*, 50, 143–156.

220. Gregoriadis, G., Bacon, A., Caparros-Wanderley, W., and McCormack, B. (2002), A role for liposomes in genetic vaccination, *Vaccine* 20, B1–B9.

221. Wu, M. S., Robbins, J. C., Bugianesi, R. L., Ponpipom, M. M., and Shen, T. Y. (1981), Modified in vivo behaviour of liposomes containing synthetic glycolipids, *Biochim. Biophys. Acta*, 674, 19–29.

222. Fujimoto, Y., Okuhata, Y., Tyngi, S., Namba, Y., and Oku, N. (2000), Magnetic resonance lymphography of profunded lymphnodes with liposomal gadolinium-diethylenetriamine penta-acetic acid, *Biol. Pharm. Bull.*, 23, 97–100.

223. Misselwitz, B., and Sachse, A. (1997), Interstitial MR lymphographyusing GD-carrying liposomes, *Acta Radiol. Suppl.*, 412, 51–55.

224. Moll, K. P., Stober, R., Hermann, W., Borchert, H. H., and Utsumi, H. (2004), In vivo ESR studies on subcutaneously injected multilamellar liposomes in living mice, *Pharm. Res.*, 21, 2017–2024.

225. Bacon, A., Caparrós-Wanderley, W., Zadi, B., and Gregoriadis, G. (2002), Induction of a cytotoxic T lymphocyte (CTL) response to plasmid DNA delivered by Lipodine™, *J. Liposome Res.*, 12, 173–183.

226. Perrie, Y., Barralet, J. E., McNeil, S., and Vangala, A. (2004), Surfactant vesicle-mediated delivery of DNA vaccines via the subcutaneous route, *Int. J. Pharm.*, 284, 31–41.

227. Klibanov, A. L., Maruyama, K., Torchilin, V. P., and Huang, L. (1990), Amphipathic polyethyleneglycols effectively prolong the circulation time of liposomes, *FEBS Lett.*, 268, 235–237.

228. Gabizon, A., and Papahadjopoulos, D. (1992), The role of surface charge and hydrophilic groups on liposome clearance in vivo, *Biochim. Biophys. Acta*, 1103, 94–100.

229. Ishida, O., Maruyama, K., Sasaki, K., and Iwatsuru, M. (1999), Size-dependent extravasation and interstitial localization of polyethyleneglycol liposomes in solid tumour-bearing mice, *Int. J. Pharm.*, 190, 49–56.

230. Levchenko, T. S., Rammohan, R., Lukyanov, A. N., Whiteman, K. R., and Torchilin, V. P. (2002), Liposome clearance in mice: The effect of a separate and combined presence of surface charge and polymer coating, *Int. J. Pharm.*, 240, 95–102.

231. Lee, C. M., Choi, Y., Huh, E. J., Lee, K. Y., Song, H. C., Sun, M. J., Jeong, H. J., Cho, C. S., and Bom, H. S. (2005), Polyethylene glycol (PEG) modified ^{99m}Tc-HMPAO-liposome for improving blood circulation and biodistribution: The effect of the extent of PEG ylation, *Cancer Biother. Radiopharm.*, 20, 620–628.

232. Bradley, A. J., Devine, D. V., and Ansell, S. M. (1998), Inhibition of liposome-induced complement activation by incorporated poly(ethylene glycol)-lipids, *Arch. Biochem. Biophys.*, 357, 185–192.

233. Lu, W. L., Qi, X. R., Zhang, Q., Li, R. Y., Zhang, R. J., and Wei, S. L. (2004), A PEGylated liposomal platform: pharmacokinetics, pharmacodynamics and toxicity in mice using doxorubicin as a model drug, *J. Pharmacol. Sci.*, 95, 381–389.

234. Ortho Biotech Products, L. P., http://www.doxil.com/common/prescribing_information/DOXIL/PDF/DOXIL_PI_Booklet.pdf#search='DOXIL', 2003–2007.

235. Blume, G., Cevc, G., Crommelin, M. D., Bakker-Woudenberg, I. A., Kluft, C., and Storm, G. (1993), Specific targeting with poly(ethylene glycol)-modified liposomes: Coupling of homing devices to the ends of the polymeric chains combines effective target binding with long circulation times, *Biochim. Biophys. Acta*, 1149, 180–184.

236. Gabizon, A., Horowitz, A. T., Goren, D., Tzemach, D., Shmeeda, H., and Zalipsky, S. (2003), In vivo fate of folate-targeted polyethylene-glycol liposomes in tumor-bearing mice, *Clin. Cancer Res.*, 9, 6551–6559.

237. Maeda, N., Takeuchi, Y., Takada, M., Sadzuka, Y., Namba, Y., and Oku, N. (2004), Antineovascular therapy by use of tumour neovasculature-targeted long-circulating liposome, *J. Controlled Release*, 100, 41–52.

238. Syrigos, K. N., Vile, R. G., Peters, M., and Harrington, K. J. (2003), Biodistribution and pharmacokinetics of 111In-DTPA-labelled pegylated liposomes after intraperitoneal injection, *Acta Oncol.*, 42, 147–153.

239. Cabanes, A., Even-Chen, S., Zimberoff, J., Barenholz, Y., Kedar, E., and Gabizon, A. (1999), Enhancement of antitumor activity of polyethylene glycol-coated liposomal doxorubicin with soluble and liposomal interleukin 2, *Clini. Cancer Res.*, 5, 687–693.

240. Oussoren, C., Eling, W. M., Crommelin, D. J., Storm, G., and Zuidema, J. (1998), The influence of the route of administration and liposome composition on the potential of liposomes to protect tissue against local toxicity of two antitumour drugs, *Biochim. Biophys. Acta*, 1369, 159–172.

241. Patton, J. S., Fishburn, C. S., and Weers, J. G. (2004), The lungs as a portal of entry for systemic drug delivery, *Proc. Am. Thorac. Soc.*, 1, 338–344.

242. Niven, R. W., and Scheier, H. (1990), Nebulization of liposomes. I. Effects of lipid composition, *Pharm. Res.*, 7, 1127–1133.

243. Niven, R. W., Speer, M., and Scheier, H. (1991), Nebulization of liposomes. II. The effects of size and modeling of solute release profiles, *Pharm. Res.*, 8, 217–221.

244. Niven, R. W., Carvajal, M. T., and Scheier, H. (1992), Nebulization of liposomes. III. Effect of operating conditions, *Pharm. Res.*, 9, 515–520.

245. Taylor, K. M. G., Taylor, G., Kellaway, I. W., and Stevens, J. (1990), The stability of liposomes to nebulization, *Inter. J. Pharm.*, 58, 57–61.

246. Schwarz, L. A., Johnson, J. L., Black, M., Cheng, S. H., Hogan, M. E., and Waldrep, J. C. (1996), Delivery of DNA-cationic liposome complexes by small particle aerosol, *Hum. Gene Ther.*, 24, 35–36.

247. Eastman, S. J., Tousignant, J. D., Lukason, M. J., Murray, H., Siegel, C. S., Constantino, P., Harris, D. J., Cheng, S. H., and Scheule, R. K. (1997), Optimization of formulations and conditions for the aerosol delivery of functional cationic lipid: DNA complexes, *Hum. Gen. Ther.*, 8, 313–322.

248. Gonzalez-Rothi, R. J., and Schreier, H. (1993), Pulmonary delivery of liposomes, *J. Controlled Release*, 5, 149–161.

249. Waldrep, J. C., Gilbert, B. E., Black, M., and Knight, V. (1997), Operating characteristics of 18 different continuous-flow jet nebulizers with beclomethasone dipropionate liposome aerosol, *Chest*, 111, 316–323.

250. Hochhaus, G., Gonzalez-Rothi, R. J., Lukyanov, A., Derendorf, H., and Dallas Costa, T. (1995), Assessment of glycocorticoid lung targeting by ex vivo receptor binding studies in rats, *Pharm. Res.*, 12, 134–137.

251. Gonzalez-Rothi, R. J., Suarez, S., Hochhaus, G., Schreier, H., Lukyanov, A., Derendorf, H., and Dallas Costa, T. (1996), Pulmonary targeting of liposomal triamcinolone acetonide, *Pharm. Res.*, 13, 1699–1703.

252. Schreier, H., McNicol, K. J., Ausborn, M., and Soucy, D. W., Derendorf, H., Stecenko, A. A., and Gonzalez-Rothi, R. J. (1992), Pulmonary delivery of amikacin liposomes and acute liposomes toxicity in the sheep, *Int. J. Pharm.*, 87, 183–193.

253. Omri, A., Bealuc, C., Bouhajib, M., Montplaisir, S., Sharkawi, M., and Lagace, J. (1994), Pulmonary retention of free and liposome-encapsulated tobramycin after intratracheal administration in uninfected rats and rats infected with Pseudomonas aeruginosa, *Antimicr. Agents Chemother.*, 38, 1090–1095.

254. Shek, P. N., Suntres, Z. E., and Brooks, J. I. (1994), Liposomes in pulmonary applications: Physicochemical considerations, pulmonary distribution and antioxidant delivery, *J. Drug Targe.*, 2, 431–442.

255. Suntres, Z. E., and Shek, P. N. (1994), Incorporation of α-tocopherol in liposomes promotes the retention of liposome-encapsulated glutathione in the rat lung, *J. Pharm. Pharmacol.*, 46, 23–28.

256. Suntres, Z. E., and Shek, P. N. (1995), Intratracheally administered liposomal alpha-tocopherol protects the lung against long-term toxic effects of paraquat, *Biomed. Environ. Sci.*, 8, 289–300.

257. Walther, F. J., David-Cu, R., and Lopez, S. L. (1995), Antioxidant-surfactant liposomes mitigate hyperoxic lung injury in premature rabbits, *Am. J. Physiol.*, 269, L613–L617.

258. Vyas, S. P., Kannan, M. P., Jain, S., Mishra, V., and Singh, P. (2004), Design of liposomal aerosols for improved delivery of rifampicin to alveolar macrophages, *Int. J. Pharm.*, 269, 37–49.

259. Vyas, S. P., Quraishi, S., Gupta, S., and Jaganathan, K. S. (2005), Aerosolized liposome-based delivery of amphotericin B to alveolar macrophages, *Int. J. Pharm.*, 296, 12–25.

260. Doddoli, C., Ghez, O., Barlesi, F., D'Journo, B., Robitail, S., Thomas, P., and Clerc, T. (2005), In vitro and in vivo methotrexate disposition in alveolar macrophages: Comparison of pharmacokinetic parameters of two formulations, *Int. J. Pharm.*, 297, 180–189.

261. Fidler, I. J., Raz, A., Fogler, W. E., Kirsh, R., Bugleski, P., and Poste, G. (1980), Design of liposomes to improve delivery of macrophage-augmenting agents to alveolar macrophages, *Cancer Res.*, 40, 4460–4466.

262. Juliano, R. L., and McCullough, H. N. (1980), Controlled delivery of an antitumor drug: Localized action of liposome encapsulated cytosine arabinoside administered via the respiratory system, *Am. Soc. Pharm. Exp. Ther.*, 214, 381–387.

263. McCullough, H. N., and Juliano, R. L. (1979), Organ-selective action of an antitumor drug: Pharmacologic studies of liposome-encapsulated beta-cytosine arabinoside administered via the respiratory system of the rat, *J. Natl. Cancer Inst.*, 63, 727–731.

264. Knight, V., Kleinerman, E. S., Waldrep, J. C., Giovanella, B. C., Gilbert, B. E., and Koshkina, N. V. (2000), 9-Nitrocamptothecin liposome aerosol treatment of human cancer subcutaneous xenografts and pulmonary cancer metastases in mice, *Ann. N.Y. Acad. Sci.*, 922, 151–163.

265. Canonico, A. E., Conary, J. T, Meyrick, B. O., and Brigham, K. L. (1994), Aerosol and intravenous transfection of human a1-antitrypsin gene to lungs of rabbits, *Am. J. Respir. Cell Mol. Biol.*, 10, 24–29.

266. Stribling, R., Brunette, E., Liggitt, D., Gaensler, K., and Debs, R. (1992), Aerosol gene delivery in vivo, *Proc. Natl. Acad. Sci. USA*, 89, 11277–11281.

267. Rowland, R. N., and Woodley, J. F. (1980), The stability of liposomes in vitro to pH, bile salts and pancreatic lipase, *Biochim. Biophys. Acta*, 620, 400–409.

268. Nagata, M., Yotsuyanagi, T., and Ikeda, M. (1988), A two step model of disintegration kinetics of liposomes in bile salts, *Chem. Pharm. Bull.*, 36, 1508–1513.

269. Quadachi, S., Paternostre, M., andre, C., Genin, I., Thao, T. X., Puisieux, F., Devissaguet, J., and Barratt, G. (1998), Liposomal formulations for oral immunotherapy: In-vitro stability in synthetic intestinal media and in-vivo efficacy in the mouse, *J. Drug Targe.*, 5, 365–378.

270. Freund, O. (2001), Biodistribution and gastrointestinal drug delivery of new lipidic multilamellar vesicles, *Drug Deliv.*, 8, 239–244.

271. Mezei, M., and Gulasekharam, V. (1981), Liposomes: A selective drug delivery system for topical route of administration-gel dosage form, *J. Pharm. Pharmacol.*, 34, 473–474.

272. Mezei, M., and Gulasekharam, V. (1980), Liposomes: A selective drug delivery system for topical route of administration. I-lotion dosage form, *Life Sci.*, 26, 1473–14777.

273. Honeywell-Nguyen, P. L., and Bouwstra, J. A. (2005), Vesicles as a tool for transdermal and dermal delivery, *Drug Discovery Today: Technol.*, 2, 67–74.

274. Knepp, V. M., Hinz, R. S., Szoka, F. C., and Guy, R. H. (1988), Controlled drug release from a novel liposomal delivery system. I. Investigation of transdermal potential, *J. Controlled Release*, 5, 211–221.

275. Yu, H. Y., and Liao, H. M. (1996), Triamcinolone permeation from different liposome formulations through rat skin in vitro, *Int. J. Pharm.*, 127, 1–7.

276. Fresta, M., and Puglisi, G. (1997), Corticosteroid dermal delivery with skin lipid liposomes, *J. Controlled Release*, 44, 141–151.

277. Liu, H., Pan, W. S., Tang, R., and Luo, S. D. (2004), Topical delivery of acyclovir palmitate liposome formulations through rat skin in vitro, *Pharmazie*, 59, 203–206.

278. Egbaria, K., Ramanchandran, C., and Weiner, N. (1991), Topical application of liposomally entrapped cyclosporin evaluated by in vitro diffusion studies with human skin, *Skin Pharmacol.*, 4, 21–28.

279. Manosroi, A., Konganeramit, L., and Manosroi, J. (2004), Stability and transdermal absorption of topical amphotericin B liposome formulations, *Int. J. Pharm.*, 270, 279–286.

280. Aguillela, V., Kontturi, K., Murtomaki, L., and Ramirez, P. (1994), Estimation of the pore size and charge density in human cadaver skin, *J. Controlled Release*, 32, 249–257.

281. Cevc, G., Schatzlein, A., and Richardsen, H. (2002), Ultradeformable lipid vesicles can penetrate the skin and other semi-permeable barriers unfragmented. Evidence from double label CLSM experiments and direct size measurements, *Biochim. Biophys. Acta*, 564, 21–30.

282. Cevc, G. (1996), Transferosomes, liposomes and other lipid suspensions on the skin: Permeation enhancement, vesicle penetration and transdermal drug delivery, *Crit. Rev. Ther. Drug Carrier Syst.*, 13, 257–388.

283. de Plessis, J., Ramachandran, C., Weiner, N., and Muller, D. G. (1994), The influence of particle size of liposomes on the deposition of the drug into skin, *Int. J. Pharm.*, 103, 277–282.

284. Verma, D. D., Verma, S., Blume, G., and Fahr, A. (2003), Particle size of liposomes influences dermal delivery of substances into skin, *Int. J. Pharm.*, 258, 141–151.

285. Miyachi, Y., Imamura, S., and Niwas, Y. (1987), Decreased skin superoxide dismutase activity by a single exposure of ultraviolet radiation is reduced by liposomal superoxide dismutase pretreatment, *J. Invest. Dermatal.*, 89, 111–112.

286. Short, S. M., Rubas, W., Paasch, B. D., Mrsny, R. (1995), Transport of biologically active interferon-gamma across human skin in vitro, *Pharm. Res.*, 12, 1140–1145.

287. Cotsarelis, G. (2002), The hair follicle as a target for gene therapy, *Ann. Dermatal Venereol.*, 129, 841–844.

288. Raghavachari, N., and Fahl, W. E. (2002), Targeted gene delivery to skin cells in vivo: A comparative study of liposomes and polymers as delivery vehicles, *J. Pharm. Sci.*, 91, 615–622.

289. Balsari, A. L., Morelli, D., Menard, S., Veronesi, U., and Colnaghi, M. I. (1994), Protection against doxorubicin-induced alopecia in rats by liposome-entrapped monoclonal antibodies, *Res. Commun.*, 8, 226–230.

290. Weiner, N., Williams, N., Birch, G., Ramachandran, R., Shipman, C., Jr., and Flynn, G. (1989), Topical delivery of liposomally encapsulated interferon evaluated in a cutaneous herpes guinea pig model, *Antimicrob. Agents Chemoter.*, 33, 1217–1221.

291. Godin, B., and Touitou, E. (2003), Ethosomes: New prospects in transdermal delivery, *Crit. Rev. Ther. Drug Carrier Syst.*, 20, 63–102.

292. Kalia, Y. N., Naik, A., Garrison, J., Guy, R. H. (2004), Iontophoretic drug delivery, *Adv. Drug Deliv. Rev.*, 56, 619–658.

293. Vulta, N. B., Betageri, G. V., and Banga, A. K. (1996), Transdermal iontophoretic delivery of enkephalin formulated in liposomes, *J. Pharm. Sci.*, 85, 5–8.

294. Fang, J. Y., Sung, K. C., Lin, H. H., and Fang, C. L. (1999), Transdermal iontophoretic delivery of enoxacin from various liposome-encapsulated formulations, *J. Controlled Release*, 60, 1–10.

295. Essa, E. A., Bonner, M. C., and Barry, B. W. (2002), Iontophoretic estradiol skin delivery and tritium exchange in ultradeformable liposomes, *Int. J. Pharm.*, 240, 55–66.

296. Essa, E. A., Bonner, M. C., and Barry, B. W. (2004), Electrically assisted skin delivery of liposomal estradiol; phospholipid as damage retardant, *J. Controlled Release*, 95, 535–546.

297. Conjeevaram, R., Chaturvedala, A., Betageri, G. V., Sunkara, G., and Banga, A. K. (2003), Iontophoretic in vivo transdermal delivery of beta-blockers in hairless rats and reduced skin irritation by liposomal formulation, *Pharm. Res.*, 9, 1496–1501.

298. Han, I., Kim, M., and Kim, J. (2004), Enhanced transfollicular delivery of adriamycin with a liposome and iontophoresis, *Exp. Dermatol.*, 13, 86–92.

299. Wells, J. M., Li, L. H., Sen, A., Jahreis, G. P., and Hui, S. W. (2000), Electroporation-enhanced gene delivery in mammary tumors, *Gene Ther.*, 7, 541–547.

300. Essa, E. A., Bonner, M. C., and Barry, B. W. (2003), Electroporation and ultradeformable liposomes; human skin barrier repair by phospholipids, *J. Controlled Release*, 92, 163–172.

301. Sen, A., Zhao, Y. L., and Hui, S. W. (2002), Saturated anionic phospholipids enhance transdermal transport by electroporation, *Biophys. J.*, 83, 2064–2073.

302. Vyas, S. P., Singh, R., Asati, R. K. (1995), Liposomally encapsulated diclofenac for sonophoresis induced systemic delivery, *J. Microencapsul.*, 12, 149–154.

303. Ahmed, I., and Patton, T. F. (1987), Disposition of timolol and inulin in the rabbit eye following corneal versus noncorneal absorption, *Int. J. Pharm.*, 38, 9–21.

304. Mishima, S., Gasset, A. Klyce, S. D., and Baum, J. L. (1966), Determination of tear volume and tear flow, *Invest. Ophthalmol.*, 5, 264–276.

305. Ebrahim, S., Peyman, G., and Lee, P. J. (2005), Applications of liposomes in ophtalmology, *Sur. Opthalmol.*, 50, 167–182.

306. Kaur, I. P., Garg, A., Singla, A. K., and Aggarwal, D. (2004), Vesicular systems in ocular drug delivery: An overview, *Int. J. Pharm.*, 269, 1–14.

307. Zimmer, A. K., and Kreuter, J. (1991), Studies on the transport pathway of PBCA nanoparticles in ocular tissues, *J. Microencapsul.*, 8, 497–504.

308. Marchal-Heussler, Sirbat D., and Hoffman, M. (1993), Poly-E-caprolactone nanocapsules in carteolol ophthalmic delivery, *Pharm. Res.*, 10, 386–390.

309. Calvo, P., Thomas, C., and Alonso, M. J. (1994), Study of the mechanism of interaction of poly-ξ-caprolactone nanocapsules with the cornea by confocal laser scanning microscopy, *Int. J. Pharm.*, 103, 283–291.

310. Schaeffer, H. E., Brietfelter, J. M., and Krohn, D. L. (1982), Lectin-mediated attachment of liposomes to cornea: Influence on transcorneal drug flux, *Invest. Ophthalmol. Vis. Sci.*, 23, 530–533.

311. Dharma, S. K., Fishman, P. H., and Peyman, G. A. (1986), A preliminary study of corneal penetration of 125I-labelled iodoxuridine liposome, *Acta Ophthalmol. (Copenh.)*, 64, 298–301.

312. Smolin, G., Okumoto, M., Feiler, S., and Condon, D. (1981), Iodoxuridine-liposome therapy for herpes simplex keratitis, *Am. J. Ophthalmol.*, 91, 220–226.

313. Schaeffer, H. E., and Krohn, D. L. (1982), Liposomes in topical drug delivery, *Invest. Ophthalmol. Vis. Sci.*, 22, 220–227.

314. Stratford, R. E. J., Yang, D. C., Redell, M. A., and Lee, V. H. L. (1983), Effects of topically applied liposomes and disposition of epinephrine and inulin in albino rabbit eye, *Int. J. Pharm.*, 13, 263–272.

315. Stratford, R. E. J., Yang, D. C., Redell, M. A., and Lee, V. H. L. (1983), Ocular distribution of liposome encapsulated epinephrine and inulin in the albino rabbit, *Curr. Eye Res.*, 2, 377–386.

316. Lee, V. H. L., Takemoto, K. A., and Iimoto, D. S. (1984), Precorneal factors influencing the ocular distribution of topically applied liposomal inulin, *Curr. Eye Res.*, 3, 585–591.

317. Angelucci, D. (2001, Nov.), New solutions for dry eye, EyeWorld, available: http://eyeworld.org.pastissue.php.

318. Lin, H. H., Ko, S. M., Hsu, L. R., and Tsai, Y. H. (1996), The preparation of norfloxacin-loaded liposomes and their in-vitro evaluation in pig's eye, *J. Pharm. Pharmacol.*, 48, 801–805.

319. Frucht-Perry, J., Assil, K. K., Ziegler, E., et al. (1992), Fibrin-enmeshed tobramycin liposomes: Single application topical therapy of Pseudomonas keratitis, *Cornea*, 11, 393–397.

320. Assil, K. K., Frucht-Perry, J., Ziegler, E., et al. (1991), Tobramycin liposomes. Single subconjunctival therapy of pseudomonal keratitis, *Invest. Ophthalmol. Vis. Sci.*, 32, 3216 3220.

321. Barza, M., Doft, B., and Lynch, E. (1993), Ocular penetration of ceftriaxone, ceftazidime, and vancomycin after subconjunctival injection in humans, *Arch. Ophthalmol.*, 111, 492–494.

322. Pleyer, U., Grammar, J., Pleyer, J. H., et al. (1995), Amphotericin B—bioavailability in the cornea. Studies with local administration of liposome incorporated amphotericin B, *Ophthalmologe*, 92, 469–475.

323. Law, S. L., Huang, K. J., and Chiang, C. H. (2000), Acyclovir-containing liposomes for potential ocular delivery. Corneal penetration and absorption, *J. Controlled Release*, 63(1–2), 135–140.

324. Fresta, M., Panico, A. M., Bucolo, C., et al. (1999), Characterization and in-vivo ocular absorption of liposome-encapsulated acyclovir, *J. Pharm. Pharmacol.*, 51, 565–576.

325. Al-Muhammed, J., Ozer, A. Y., Ercan, M. T., and Hincal, A. A. (1996), In-vivo studies on dexamethasone sodium phosphate liposomes, *J. Microencapsul.*, 13, 293–306.

326. Pleyer, U., Elkins, B., Ruckert, D., et al. (1994), Ocular absorption of cyclosporine A from liposomes incorporated into collagen shields, *Curr. Eye Res.*, 13, 177–181.

327. Pleyer, U., Lutz, S., Jusko, W. J., et al. (1993), Ocular absorption of topically applied FK506 from liposomal and oil formulations in the rabbit eye, *Invest. Ophthalmol. Vis. Sci.*, 34, 2737–2742 [erratum appears *Invest. Ophthalmol. Vis. Sci.*, 34, 3481, 1993].

328. Torres, P. F., Slegers, T. P., Peek, R., et al. (1999), Changes in cytokine mRNA levels in experimental corneal allografts after local clodronate-liposome treatment, *Invest. Ophthalmol. Vis. Sci.*, 40, 3194–3201.
329. Slegers, T. P., van Rooijen, N., van Rij, G., and van der Gaag, R. (2000), Delayed graft rejection in pre-vascularised corneas after subconjunctival injection of clodronate liposomes, *Curr. Eye Res.*, 20, 322–324.
330. Van der Veen, G., Broersma, L., Dijkstra, C. D., et al. (1994), Prevention of corneal allograft rejection in rats treated with subconjunctival injections of liposomes containing dichloromethylene diphosphonate, *Invest. Ophthalmol. Vis. Sci.*, 35, 3505–3515.
331. Van der Veen, G., Broersma, L., Van Rooijen, N., et al. (1998), Cytotoxic T lymphocytes and antibodies after orthotropic penetrating keratoplasty in rats treated with dichloromethylene diphosphonate encapsulated liposomes, *Curr. Eye Res.*, 17, 1018–1026.
332. Lee, V. H., and Carson, L. W. (1986), Ocular disposition of inulin from single and multiple doses of positively charged multilamellar liposomes: Evidence for alterations in tear dynamics and ocular surface characteristics, *J. Ocul. Pharmacol.*, 2, 353–364.
333. Elorza, B., Elorza, M. A., Sainz, M. C., and Chantres, J. R. (1993), Comparison of particle size and encapsulation parameters of three liposomal preparations, *J. Microencapsul.*, 10, 237–248.
334. Felt, O., Furrer, P., Mayer, J. M., Plazonnet, B., Buri, P., and Gurny, R. (1999), Topical use of chitosan in ophthalmology: Tolerance assessment and evaluation of precorneal retention, *Int. J. Pharm.*, 180, 185–193.
335. Felt, O., Baeyens, V., Zignani, M., Buri, P., and Gurny, R., (1999), Mucosal drug delivery, in Mathiowitz, E., Ed., *The Encyclopedia of Controlled Drug Delivery*, Wiley, New York, pp. 605–626.
336. McCalden, T. A., and Levy, M. (1990), Retention of topical liposomal formulations on the cornea, *Experientia*, 46, 713–715.
337. Grass, G. M., and Robinson, J. R. (1988), Mechanisms of corneal drug penetration. I: In vivo and in vitro kinetics, *J. Pharm. Sci.*, 77, 3–14.
338. Burrow, J., Tsibouklis, J., and Smart, J. D., (2002), Drug delivery to the eye, *The Drug Delivery Company Report*, p. 4.
339. Taniguchi, K., Yamamoto, Y., Itakura, K., Miichi, H., and Hayashi, S. (1988), Assessment of ocular irritability of liposome preparations, *J. Pharmacobiodyn.*, 11, 607–611.
340. Barber, R. F., and Shek, P. N. (1990), Tear induced release of liposome entrapped agents, *Int. J. Pharm.*, 60, 219–227.
341. Kaufman, H. E., Steinemann, T. L., Lehman, E., Thompson, H. W., Varnell, E. D., Jacob-LaBarre, J. T., and Gebhardt, B. M. (1994), Collagen-based drug delivery and artificial tears, *J. Ocul. Pharmacol.*, 10, 17–27.
342. Yerushalmi, N., and Margalit, R. (1994), Bioadhesive, collagen-modified liposomes: Molecular and cellular level studies on the kinetics of drug release and on binding to cell monolayers, *Biochim. Biophys. Acta*, 1189, 13–20.
343. Bochot, A., Fattal, E., Gulek, A., and Aonarraze, G. (1998), Liposome dispersed within thermosensitive gel. A new dosage form for ocular delivery of oligonucleotide, *Pharm. Res.*, 15, 1364–1369.
344. Bochot, A., Couvreur, P., and Fattal, E. (2000), Intravitreal administration of antisense oligonucleotides: Potential of liposomal delivery, *Prog. Retin. Eye Res.*, 19, 131–147.
345. Bochot, A., Fattal, E., Boutet, V., Deverre, J. R., Jeamy, J. C., Chacun, H., and Couvreur, P. (2002), Intravitreal delivery of oligonucleotides by sterically stabilized liposomes, *Invest. Opthalmol. Vis. Sci.*, 43, 253–259.

346. Schmidt-Erfurth, U., Flotte, T. J., Gragoudas, E. S., et al. (1996), Benzoporphyrin-lipoprotein-mediated photodestruction of intraocular tumors, *Exp. Eye Res.*, 62, 1–10.

347. Kawakami, S., Yamamura, K., Mukai, T., Nishida, K., Nakamura, J., Sakaeda, T., Nakashima, M., and Sasaki, H. (2001), Sustained ocular delivery of tilisolol to rabbits after topical administration or intravitreal injection of lipophilic prodrug incorporated in liposomes, *J. Pharm. Pharmacol.*, 53, 157–161.

348. Harris, A. S., Svensson, E., Wagner, Z. G., Lethagen, S., and Nilsson, I. M. (1988), Effect of viscosity on particle-size, deposition, and clearance of nasal delivery systems containing desmopressin, *J. Pharm. Sci.*, 77, 405–408.

349. Vyas, S. P., Goswami, S. K., and Singh, R. (1995), Liposomes based nasal delivery system of nifedipine—development ad characterization, *Int. J. Pharm.*, 118, 23–30.

350. Shahiwala A, and Misra, A. (2004), Nasal delivery of levonorgestrel for contraception: An experimental study in rats, *Fertil. Steril.* 81(Suppl. 1), 893–898.

351. Kuper, H., Adami, H. O., and Trichopoulos, D. (2000), Infection as a major preventable cause of human cancer, *J. Intern. Med.*, 248, 171–183.

352. Alpar, H. O., Somavarapu, S., Atuah, K. N., and Bramwell, V. W. (2005), Biodegradable mucoadhesive particulates for nasal and pulmonary antigen and DNA delivery, *Adv. Drug Deliv. Rev.*, 57, 411–430.

353. De Haan, A., Renegar, K. B., Small, P. A., Wilschut, J., (1995), *Vaccine*, 13, 613–616.

354. Wang, D., Christopher, M. E., Nagata, L. P., Zabielski, M. A., Li, H., Wong, J. P., and Samuel, J. (2004), Intranasal immunization with liposome-encapsulated plasmid DNA encoding influenza virus hemagglutinin elicits mucosal, cellular and humoral immune responses, *J. Clin. Virol.*, 31(Suppl. 1), 99–106.

355. Tafaghodi, M., Jaafari, M.-R., and Abolghasem, S., and Tabassi, S. (2006), Nasal immunization studies using liposomes loaded with tetanus toxoid and CpG-ODN, *Eur. J. Pharm. Biopharm.*, 64, 138–145.

356. de Jonge, M. I., Hamstra, H. J., Jiskoot, W., Roholl, P., Williams, N. A., Dankert, J., van Alphen, L., and Van der Ley, P. (2004), Intranasal immunisation of mice with liposomes containing recombinant meningococcal OpaB and OpaJ proteins, *Vaccine*, 22, 4021–4028.

357. Sloat, B. R., and Zhengrong, C. (2006), Strong mucosal and systemic immunities induced by nasal immunization with anthrax protective antigen protein incorporated in liposome-protamine-DNA particles, *Pharm. Res.*, 23, 262–269.

358. Paveli , Z., Škalko-Basnet, N., and Schubert, R. (2001), Liposomal gels for vaginal drug delivery, *Int. J. Pharm.*, 219, 139–149.

359. Paveli , Z., Škalko-Basnet, N., and Jalšenjak, I. (2005), Characterisation and in vitro evaluation of bioadhesive liposome gels for local therapy of vaginitis, *Int. J. Pharm.*, 301, 140–148.

360. Paveli , Z., Škalko-Basnet, N., Filipovi -Gr i , J., Martinac, A., and Jalšenjak, I. (2005), Development and in vitro evaluation of a liposomal vaginal delivery system for acyclovir, *J. Controlled Release*, 106, 34–43.

361. Zhang, Y., Cristofaro, P., Silbermann, R., Pusch, O., Boden, D., Konkin, T., Hovanesian, V., Monfils, P. R., Resnick, M., Moss S. F., and Ramratnam, B. (2006), Engineering mucosal RNA interference in vivo, *Mol. Ther.*, 14, 336–342.

362. Hartikka, J., Bozoukova, V., Ferrari, M., Sukhu, L., Enas, J., Sawdey, M., et al. (2001), Vaxfectin enhances the humoral immune response to plasmid DNA-encoded antigens, *Vaccine*, 19, 1911–1923.

363. Lochera, C. P., Witta, S. A., Ashlocka, B. M., Polacinob, P., Hub, S. L., Shiboskic, S., Schmidtb, A. M., Agyb, M. B., anderson, D. M., Stapransd, S. I., Megedee, J., and Levy, J.

A. (2004), Human immunodeficiency virus type 2 DNA vaccine provides partial protection from acute baboon infection, *Vaccine*, 22, 2261–2272.

364. Davidson, R. N., Croft, S. L., Scott, A., Maini, M., Moody, A. H., and Bryceson, A. D. (1991), Liposomal amphotericin B in drug-resistant visceral leishmaniasis, *Lancet*, 337, 1061–1062.

365. Guaglianone, P., Chan, K., DelaFlor-Weiss, E., Hanisch, R., Jeffers, S., Sharma, D., et al. (1994), Phase I and pharmacologic study of liposomal daunorubicin (DaunoXome), *Invest. New Drugs*, 12, 103–110.

366. Gabizon, A., Peretz, T., Sulkes, A., Amselem, S., Ben-Yosef, R., Ben-Baruch, N., et al. (1989), Systemic administration of doxorubicin-containing liposomes in cancer patients: A phase I study, *Eur. J. Cancer Clin. Oncol.*, 25, 1795–1803.

367. Donald, P. R., Sirgel, F. A., Venter, A., Smit, E., Parkin, D. P., VandeWal, B. W., et al. (2001), The early bactericidal activity of a low-clearance liposomal amikacin in pulmonary tuberculosis, *J. Antimicrob. Chemother.*, 48, 877–880.

368. Kadry, A. A., Al-Suwayeh, S. A., Abd-Allah, A. R., and Bayomi, M. A. (2004), Treatment of experimental osteomyelitis by liposomal antibiotics, *J. Antimicrob. Chemother.*, 54, 1103–1108.

369. Nishiyama, N., and Kataoka, K. (2006), Current state, achievements, and future prospects of polymeric micelles as nanocarriers for drug and gene delivery, *Pharmacol. Therap.*, 112, 630–648.

370. Zamboni, W. C., Gervais, A. C., Egorin, M. J., Schellens, J. H., Zuhowski, E. G., Pluim, D., et al. (2004), Systemic and tumor disposition of platinum after administration of cisplatin or STEALTH liposomal-cisplatin formulations (SPI-077 and SPI-077 B103) in a preclinical tumor model of melanoma, *Cancer Chemother. Pharmacol.*, 53, 329–336.

371. Proulx, M. E., Desormeaux, A., Marquis, J. F., Olivier, M., and Bergeron, M. G. (2001), Treatment of visceral leishmaniasis with sterically stabilized liposomes containing camptothecin, *Antimicrob. Agents Chemother.*, 45, 2623–2627.

372. Giles, F. J., Tallman, M. S., Garcia-Manero, G., Cortes, J. E., Thomas, D. A., Wierda, W. G., et al. (2004), Phase I and pharmacokinetic study of a low-clearance, unilamellar liposomal formulation of lurtotecan, a topoisomerase 1 inhibitor, in patients with advanced leukemia, *Cancer*, 100, 1449–1458.

373. Verschraegen, C. F., Gilbert, B. E., Loyer, E., Huaringa, A., Walsh, G., Newman, R. A., et al. (2004), Clinical evaluation of the delivery and safety of aerosolized liposomal 9-nitro-20(s)-camptothecin in patients with advanced pulmonary malignancies, *Clin. Cancer Res.*, 10, 2319–2326.

374. Rudin, C. M., Marshall, J. L., Huang, C. H., Kindler, H. L., Zhang, C., Kumar, D., et al. (2004), Delivery of a liposomal c-raf-1 antisense oligonucleotide by weekly bolus dosing in patients with advanced solid tumors: A phase I study, *Clin. Cancer Res.*, 10, 7244–7251.

375. Chien, P. Y., Wang, J., Carbonaro, D., Lei, S., Miller, B., Sheikh, S., et al. (2004), Novel cationic cardiolipin analogue-based liposome for efficient DNA and small interfering RNA delivery in vitro and in vivo, *Cancer Gene Ther.*, 12, 321–328.

376. Asano, T., and Kleinerman, E. S. (1993), Liposome-encapsulated MTP-PE: A novel biologic agent for cancer therapy, *J. Immunother.*, 14, 286–292.

377. Dagar, S., Krishnadas, A., Rubinstein, I., Blend, M. J., and Onyuksel, H. (2003), VIP grafted sterically stabilized liposomes for targeted imaging of breast cancer: In vivo studies, *J. Controlled Release*, 91, 123–133.

378. Park, J. W., Hong, K., Kirpotin, D. B., Colbern, G., Shalaby, R., Baselga, J., et al. (2002), Anti-HER2 immunoliposomes: Enhanced efficacy attributable to targeted delivery, *Clin. Cancer Res.*, 8, 1172–1181.

379. Pakunlu, R. I., Wang, Y., Tsao, W., Pozharov, V., Cook, T. J., and Minko, T. (2004), Enhancement of the efficacy of chemotherapy for lung cancer by simultaneous suppression of multidrug resistance and antiapoptotic cellular defense: Novel multicomponent delivery system, *Cancer Res.*, 64, 6214–6224.

380. Pardridge, W. M. (1996), Vector-mediated drug delivery to the brain, *Adv. Drug Deliv.*, 36, 299–321.

381. Huwyler, J., Wu, D., and Pardridge, W. M. (1996), Brain drug delivery of small molecules using immunoliposomes, *Proc. Natl. Acad. Sci. USA*, 93, 14164–14169.

382. Huwyler, J., Yang, J., and Pardridge, W. M. (1997), Targeted delivery of daunomycin using immunoliposomes: Pharmacokinetics and tissue distribution in the rat, *J. Pharmacol. Exp. Ther.*, 282, 1541–1546.

383. Schnyder, A., Krahenbuhl, S., Drewe, J., and Huwyler, J. (2005), Targeting of daunomycin using biotinylated immunoliposomes: Pharmacokinetics, tissue distribution and in vitro pharmacological effects, *J. Drug Targeting.*, 13, 325–335.

384. Xie, Y., Ye, L., Zhang, X., Cui, W., Lou, J., Nagai, T., and Hou, X. (2005), Transport of nerve growth factor encapsulated into liposomes across the blood-brain barrier: In vitro and in vivo studies, *J. Controlled Release*, 105, 106–119.

385. Brignole, C., Marimpietri, D., Pagnan, G., Di Paolo, D., Zancolli, M., Pistoia, V., Ponzoni, M., and Pastorino, F. (2005), Neuroblastoma targeting by c-myb-selective antisense oligonucleotides entrapped in anti-GD2 immunoliposome: Immune cell-mediated antitumor activities, *Cancer Lett.*, 228, 181–186.

386. Siegal, T., Horowitz, A., and Gabizon, A. (1995), Doxorubicin encapsulated in sterically stabilized liposomes for the treatment of a brain tumor model: Biodistribution and therapeutic efficacy, *J. Neurosurg.*, 83, 1029–1037.

387. Sharma, U. S., Sharma, A., Chau, R. I., and Straubinger, R. M. (1997), Liposome-mediated therapy of intracranial brain tumors in a rat model, *Pharm. Res.*, 14, 992–998.

388. Aoki, H., Kakinuma, K., Morita, K., Kato, M., Uzuka, T., Igor, G., Takahashi, H., and Tanaka, R. (2004), Therapeutic efficacy of targeting chemotherapy using local hyperthermia and thermosensitive liposome: Evaluation of drug distribution in a rat glioma model, *Int. J. Hypertherm.*, 20, 595–605.

389. Koukourakis, M. I., Koukouraki, S., Fezoulidis, I., Kelekis, N., Kyrias, G., Archimandritis, S., and Karkavitsas, N. (2000), High intratumoural accumulation of stealth liposomal doxorubicin (Caelyx) in glioblastomas and in metastatic brain tumors, *Br. J. Cancer*, 83, 1281–1286.

390. Arnold, R. D., Mager, D. E., Slack, J. E., and Straubinger, R. M. (2005), Effect of repetitive administration of Doxorubicin-containing liposomes on plasma pharmacokinetics and drug biodistribution in a rat brain tumor model, *Clin. Cancer Res.*, 11, 8856–8865.

391. Marina, N. M., Cochrane, D., Harney, E., Zomorodi, K., Blaney, S., Winick, N., Bernstein, M., and Link, M. P. (2002), Dose escalation and pharmacokinetics of pegylated liposomal doxorubicin (Doxil) in children with solid tumors: A pediatric oncology group study, *Clin. Cancer Res.*, 8, 413–418.

392. Caraglia, M., Addeo, R., Costanzo, R., Montella, L., Faiola, V., Marra, M., Abruzzese, A., Palmieri, G., Budillon, A., Grillone, F., Venuta, S., Tagliaferri, P., and Del Prete, S. (2006), Phase II study of temozolomide plus pegylated liposomal doxorubicin in the treatment of brain metastases from solid tumors, *Cancer Chemother. Pharmacol.*, 57, 34–39.

393. Fiorillo, A., Maggi, G., Greco, N., Migliorati, R., D'Amico, A., Del Basso De Caro, M., Sabbatino, M. S., and Buffardi, F. (2004), Second-line chemotherapy with the association of liposomal daunorubicin, carboplatin and etoposide in children with recurrent malignant brain tumors, *J. Neuro-Oncol.*, 66, 179–185.

394. Barth, R. F., Coderre, J. A., Graca, M., Vicente, H., and Blue, T. E. (2005), Boron neutron capture therapy of cancer: Current status and future prospects, *Clin. Cancer Res.*, 11, 3987–4002.

395. Hawthorne, M. F., Feakes, D. A., and Shelly, K. (1996), Recent results with liposomes as boron delivery vehicles from boron neutron capture therapy, in Mishima, Y., Ed., *Cancer Neutron Capture Therapy*, Plenum, New York, pp 27–36.

396. Feakes, D. A., Waller, R. C., Hathaway, D. K., and Morton, V. S. (1999), Synthesis and in vivo murine evaluation of Na4[1-(1′-B10H9)-6-SHB10H8] as a potential agent for boron neutron capture therapy, *Proc. Natl. Acad. Sci. USA*, 96, 6406–6410.

397. Carlsson, J., Kullberg, E. B., Capala, J., Sjoberg, S., Edwards, K., and Gedda, L. (2003), Ligand liposomes and boron neutron capture theory, *J. Neuro-Oncol.*, 62, 47–59.

398. Barth, R. F., Yang, W., Adams, D. M., Rotaru, J. H., Shukla, S., Sekido, M., Tjarks, W., Fenstermaker, R. A., Ciesielski, M., Nawrocky, M. M., and Coderre, J. A. (2002), Molecular targeting of the epidermal growth factor receptor for neutron capture therapy of gliomas, *Cancer Res.*, 62, 3159–3166.

399. Wikstrand, C. J., Cokgor, I., Sampson, J. H., and Bigner, D. D. (1999), Monoclonal antibody therapy of human gliomas: Current status and future approaches, *Cancer Metastasis Rev.*, 18, 451–464.

400. Hermanson, M., Funa, K., Koopmann, J., Maintz, D., Waha, A., Westermark, B., Heldin, C. H., Wiestler, O. D., Louis, D. N., von Deimling, A., and Nister, M. (1996), Association of loss of heterozygosity on chromosome 17p with high platelet derived growth factor alpha receptor expression in human malignant gliomas, *Cancer Res.*, 56, 164–171.

401. Akabani, G., Cokgor, I., Coleman, R. E., Gonzalez Trotter, D., Wong, T. Z., Friedman, H. S., Friedman, A. H., Garcia-Turner, A., Herndon, J. E., DeLong, D., McLendon, R. E., Zhao, X. G., Pegram, C. N., Provenzale, J. M., Bigner, D. D., and Zalutsky, M. R. (2000), Dosimetry and dose–response relationships in newly diagnosed patients with malignant gliomas treated with iodine-131-labeled anti-tenascin monoclonal antibody 81C6 therapy, *Int. J. Radiat. Oncol. Biol. Phys.*, 46, 947–995.

402. Yoshida, J., and Mizuno, M. (2003), Clinical gene therapy for brain tumors. Liposomal delivery of anticancer molecule to glioma, *J. Neuro-Oncol.*, 65, 261–267.

403. Felgner, P. L., Gadek, T. R., Holm, M., Roman, R., Chan, H. W., Wenz, M., Northrop, J. R., Ringold, G. M., and Danielson, M. (1987), Lipofection: A highly efficient, lipid-mediated DNA-transfection procedure, *Proc. Natl. Acad. Sci. USA*, 84, 7413–7414.

404. Zhang, Y., Jeong, Lee, H., Boado, R. J., and Pardridge, W. M. (2002), Receptro-mediated delivery of an antisense gene to human brain cancer cells, *J. Gene Med.*, 4, 183–194.

405. Zhang, Y., Zhu, C., and Pardridge, W. M. (2002), Antisense gene therapy of brain cancer with an artificial virus gene delivery system, *Mol. Ther.*, 6, 67–72.

406. Zhang, Y., Zhang, Y. F., Bryant, J., Charles, A., Boado, R. J., and Pardridge, W. M. (2004), Intravenous RNA interference gene therapy targeting the human epidermal growth factor receptor prolongs survival in intracranial brain cancer, *Clin. Cancer Res.*, 10, 3667–3677.

407. Zerrouqi, A., Rixe, O., Ghoumari, A. M., Yarovoi, S. V., Mouawad, R., Khayat, D., and Soubrane, C. (1996), Liposomal delivery of the herpes simplex virus thymidine kinase gene in glioma: Improvement of cell sensitization to ganciclovir, *Cancer Gene Ther.*, 3, 385–392.

408. Voges, J., Weber, F., Reszka, R., Sturm, V., Jacobs, A., Heiss, W. D., Wiestler, O., and Kapp, J. F. (2002), Clinical protocol; Liposomal gene therapy with the herpes simplex thymidine kinase gene/ganciclovir system for the treatment of glioblastoma multiforme, *Hum. Gene Ther.*, 13, 675–685.

409. Mizuno, M., and Yoshida, J. (1998), Improvement of transduction efficiency of recombinant adeno-associated virus vector by entrapment in multilamellar liposomes, *Jpn. J. Cancer Res.*, 89, 352–354.

410. Huynh, G. H., Deen, D. F., and Szoka, F. C., Jr. (2006), Barriers to carrier mediated drug and gene delivery to brain tumours, *J. Controlled Release*, 110, 236–259.

411. Bomgaars, L., Geyer, J. R., Franklin, J., Dahl, G., Park, J., Winick, N. J., Klenke, R., Berg, S. L., and Blaney, S. M. (2004), Phase I trial of intrathecal liposomal cytarabine in children with neoplastic meningitis, *J. Clin. Oncol.*, 22, 3916–3921.

412. Glantz, M., LaFollette, S., Jaeckle, K., Shapiro, W., Swinnen, L., Rozental, J., Phuphanich, S., Rogers, L., Gutheil, J., Batchelor, T., Lyter, D., Chamberlain, M., Maria, B., Schiffer, C., Bashir, R., Thomas, D., Cowens, W., and Howell, S. B. (1999), A randomized trial of a slow-release versus a standard formulation of cytarabine for the intrathecal treatment of lymphomatous meningitis, *J. Clin. Oncol.* 17, 3110–3116.

413. MakCay, A. J., Deen, D. F., and Szoka, F. C., Jr. (2005), Distribution in brain of liposomes after convection enhanced delivery; modulation by particle charge, particle diameter, and presence of steric coating, *Brain Res.*, 1035, 139–153.

414. Polfliet, M. M., Goede, P. H., van Kesteren-Hendrikx, E. M., van Rooijen, N., Dijkstra, C. D., and van De Berg, T. K. (2001), Amethod for the selective depletion of perivascular and meningeal macrophages in the central nervous system, *J. Neuroimmunol.*, 116, 188–195.

415. Mamot, C., Nguyen, J. B., Pourdehnad, M., Hadaczek, P., Saito, R., Bringas, J. R., Drummond, D. C., Park, J. W., and Bankiewicz, K. S. (2004), Extensive distribution of liposomes in rodent brains and brain tumors following convection-enhanced delivery, *J. Neuro-Oncol.*, 68, 1–9.

416. Saito, R., Bringas, J. R., McKnight, T. R., Wendland, M. F., Mamot, C., Drummond, D. C., Kiprotin, D. B., Park, J. W., Berger, M. S., and Bankiewicz, K. S. (2004), Distribution of liposomes into brain and rat brain tumor models by convection-enhanced delivery monitored with magnetic resonance imaging, *Cancer Res.*, 64, 2572–2579.

417. Saito, R., Krauze, M. T., Bringas, J. R., Noble, C., McKnight, T. R., Jackson, P., Wendland, M. F., Mamot, C., Drummond, D. C., Kiprotin, D. B., Hong, K., Berger, M. S., Park, J. W., and Bankiewicz, K. S. (2005), Gadolinium-loaded liposomes allow for real-time magnetic resonance imaging of convection-enhaced delivery in the primate brain, *Exp. Neurol.*, 196, 381–389.

418. Groothuis, D. R. (2000), The blood-brain and blood-tumor barriers: A review of strategies for increasing drug delivery, *Neuro-Oncol.*, 2, 45–59.

419. Imaginis Corp. (The Women's Health Resource), http://www.imaginis.com/breasthealth/treatment.asp, 1997–2007.

420. Hofheinz, R. D., Gnad-Vogt, S. U., Beyer, U., and Hochhaus, A. (2005), Liposomal encapsulated anti-cancer drugs, *Anti-Cancer Drugs*, 16, 691–707.

421. Gradishar, W. J. (2005), The future of breast cancer: The role of prognostic factors, *Breast Cancer Res. Treat.*, 89, S17–S26.

422. Batist, G., Ramakrishnan, G., and Rao, C. S. (2001), Reduced cardiotoxicity and preserved antitumor efficacy of liposome-encapsulated doxorubicin and cyclophosphamide compared with conventional doxorubicin and cyclophosphamide in a randomized, multicenter trial of metastatic breast cancer, *J. Clin. Oncol.*, 19, 1444–1454.

423. Paridaens, R., Van Aaelst, F., and Georgoulias, V. (2003), A randomized phase II study of alternating and sequential regimens of docetaxel and doxorubicin as first line chemotherapy for metastatic breast cancer, *Ann. Oncol.*, 14, 433–440.

424. Chan, S., Davidson, N., Juozaityte, E., Erdkamp, F., Pluzanska, A., Azarnia, N., and Lee, L. W. (2004), Phase III trial of liposomal doxorubicin and cyclophosphamide compared with epirubicin and cyclophosphamide as first line therapy for metastatic breast cancer, *Ann. Oncol.*, 15, 1527–1534.

425. Mrozek, E., Rhoades, C. A., Allen, J., Hade, E. M., and Shapiro, C. L. (2005), Phase I trial of liposomal encapsulated doxorubicin (Myocet™; D-99) and weekly docetaxel in advanced breast cancer patients, *Ann. Oncol.*, 16, 1087–1093.

426. Schmid, P., Krocker, J., Jehn, C., Michniewicz, K., Lehenbauer-Dehm, S., Eggemann, H., Heilmann, V., Kummel, S., Schulz, C. O., Dieing, A., Wischnewsky, M. B., Hauptmann, S., Elling, D., Possinger, K., and Flath, B. (2005), Primary chemotherapy with gemcitabine as prolonged infusion, non-pegylated liposomal doxorubicin and docetaxel in patients with early breast cancer: Final results of a phase II trial, *Ann. Oncol.*, 16, 1624–1631.

427. Keller, A. M., Mennel, R. G., Georgoulias, V. A., Nabholtz, J. M., Erazo, A., Lluch, A., Vogel, C. L., Kaufmann, M., Minckwitz, G., Henderson, G., Mellars, L., Alland, L., and Tendler, G. (2004), Randomized phase III trial of pegylated liposomal doxorubicin versus vinorelbine or mitomycin C plus vinblastine in women with taxane-refractory advanced breast cancer. *J. Clin. Oncol.*, 22, 3893–3901.

428. Vorobiof, D. A., Rapoport, B. L., Chasen, M. R., Slabber, C., McMichael, G., Eek, R., and Mohammed, C. (2004), First in line therapy with paclitaxel (Taxol®) and pegylated liposomal doxorubicin (Caelyx®) in patients with metstatic breast cancer: A multicentre phase II study, *Breast*, 13, 219–226.

429. Overmoyer, B., Silvermann, P. Holder, L. W., Tripathy, D., and Henderson, I. C. (2005), Pegylated liposomal doxorubicin and cyclophosphamide as first-line therapy for patients with metastatic or recurrent breast cancer, *Clin. Breast Cancer*, 6, 150–157.

430. Coleman, R. E., Biganzoli, L., Canney, P., Dirix, L., Mauriac, L., Chollet, P., Batter, V., Ngalula-Kabanga, E., Dittrich, C., and Piccart, M. (2006), A randomized phase II study of two different schedules of pegylated liposomal doxorubicin in metastatic breast cancer (EORTC-10993), *Eur. J. Cancer*, 42, 882–887.

431. Stover, T. C., Sharma, A., Robertson, G. P., and Kester, M. (2005), Systemic delivery of liposomal short-chain ceramide limits solid tumor growth in murine models of breast adenocarcinoma, *Clin. Cancer Ther.*, 11, 3465–3474.

432. Swenson, S., Costa, F., Minea, R., Sherwin, R. P., Ernst, W., Fujii, G., Yang, D., and Markland, F. S., Jr. (2004), Intravenous liposomal delivery of the snake venom disintegrin contortrostatin limits breast cancer progression, *Mol. Cancer Ther.*, 3, 499–511.

433. Nagami, H., Matsumoto, Y., and Ueoka, R. (2006), Induction of apoptosis by hybrid liposomes for human breast tumor calls along with activation caspases, *Biol. Pharm. Bull.*, 29, 380–381.

434. Frenkel, V., Etherington, A., Greene, M., Quijano, J., Xie, J., Hunter, F., Dromi, S., and Li, K. C. P. (2006), Delivery of liposomal doxorubicin (Doxil) in a breast cancer tumor model: Investigation of potential enhancement by pulsed-high intensity focused ultrasound exposure, *Acad. Radiol.*, 13, 467–479.

435. Mukherjee, A., Prassad, T. K., Rao, N. M., and Banerjee, R. (2005), Haloperidol-associated stealth liposomes, *J. Biol. Chem.*, 280, 16(22), 1561–1562.

436. Seki, M., Iwakawa, J., Cheng, H., and Cheng, P. W. (2002), p53 and PTEN/MMAC1/TEP1 gene therapy of human prostate PC-3 carcinoma xenograft, using transferrin-facilitated lipofection gene delivery strategy, *Hum. Gene Ther.*, 13, 761–773.

437. Basma, H., El-Refaey, H., Sgagias, M. K., Cowan, K. H., Luo, X., and Cheng, P. W. (2005), Bcl-2 antisense and cisplatin combination treatment of MCF-7 breast cancer cells with or without functional p53, *J. Biom. Sci.*, 12, 999–1011.
438. Meidan, V. M., Glezer, J., Salomon, S., Sidi, Y., Barenholz, Y., Cohen, J. S., and Lilling G. (2006), Specific lipoplex-mediated antisense against bcl-2 in breast cancer cells: A comparison between different formulations, *J. Liposome Res.*, 16, 27–43.
439. Gradishar, W. J. (2005), The future of breast cancer: The role of prognostic factors, *Breast Cancer Res. Treat.*, 89, S17–S26.
440. NetDoctor.co.uk, http://www.netdoctor.co.uk/diseases/facts/lungcancer.htm, 2007.
441. http://www.merck.de/servlet/PB/menu/1508710/index.html.
442. Leighl, N., Burkes, R. L., Dancey, J. E., Lopez, P. G., Higgins, B. P., Walde, P. L. D., Rudinskas, L. C., Rahim, Y. H., Rodgers, A., Pond, G. R., and Shepherd, F. A. (2003), A phase I study of pegylated liposomal doxorubicin (Caelyx™) in combination with cyclophosphamide and vincristine as second-line treatment of patients with small-cell lung cancer, *Clin. Lung Cancer*, 5, 107–112.
443. Peer, D., and Margalit, R. (2004), Tumor-targeted hyaluronan nanoliposomes increase the antitumor activity of liposomal doxorubicin in syngeneic and human xenograft mouse tumor models, *Neoplasia*, 6, 343–353.
444. Ito, I., Ji, L., Tanaka, F., Saito, Y., Gopalan, B., Branch, C. D., Xu, K., Atkinson, E. N., Bekele, B. N., Stephens, L. C., Minna, J. D., Roth, J. A., and Ramesh, R. (2004), Liposomal vector mediated delivery of the 3p FUS1 gene demonstrate potent antitumor activity against human lung cancer in vivo, *Cancer Gene Ther.*, 11, 733–739.
445. Li, W., Ishida, T., Okada, N., and Kiwada, H. (2005), Increased gene expression by cationic liposomes (TFL-3) in lung metastases following intravenous injection, *Biol. Pharm. Bull.*, 28, 701–706.
446. Kawakami, A., Suzuki, S., Yamashita, F., and Hashida, M. (2006), Induction of apoptosis in A549 human lung cancer cells by all-trans retinoic acid incorporated in DOTAP/cholesterol liposomes, *J. Controlled Release*, 110, 514–521.
447. Remaut, K., Lucas, B., Braeckmans, K., Sanders, N. N., Demeester, J., and De Smedt, S. C. (2006), Delivery of phosphodiester oligonucleotides: Can DOTAP/DOPE liposomes do the trick? *Biochemistry*, 45, 1755–1764.
448. American Cancer Society, Inc., http://www.cancer.org, 2007.
449. Johnston, S. (2004), Ovarian cancer: Review of the national institute for clinical excellence (NICE) guidance recommendations, *Cancer Inv.*, 22, 730–742.
450. Rose, P. (2005), Pegylated liposomal doxorubicin: Optimizing the dosing schedule in ovarian cancer, *Oncologist*, 10, 205–214.
451. Tambaro, R., Greggi, S., Iaffaioli, R. V., Rossi, A., Pisano, C., Manzione, L., Ferrari, E., Di Maio, M., Iodice, F., Casella, G., Laurelli, G., and Pignata, S. (2003), An escalating dose finding study of liposomal doxorubicin and vinorelbine for the treatment of refractory or resistant epithelial ovarian cancer, *Ann. Oncol.*, 14, 1406–1411.
452. Katsaros, D., Oletti, M. V., Rigault de la Longrais, I. A., Ferrero, A., Celano, A., Fracchioli, S., Donadio, M., Passera, R., Cattel, L., and Bumma, C. (2005), Clinical and pharmacokinetic phase II study of pegylated liposomal doxorubicin and vinorelbine in heavily pretreated recurrent ovarian carcinoma, *Ann Oncol.*, 16, 300–306.
453. Gordon, A. N., Fleagle, J. T., Guthrie, D., et al. (2001), Recurrent epithelial ovarian carcinoma: A randomized phase III study of pegylated liposomal doxorubicin versus topotecan, *J. Clin. Oncol.*, 19, 3312–3322.
454. Verhaar-Langereis, M., Karakus, A., van Eijkeren, M., Voest, E., and Witteveen, E. (2006), Phase II study of the combination of pegylated liposomal doxorubicin and topotecan in platinum-resistant ovarian cancer, *Int. J. Gynecol. Cancer*, 16, 65–70.

455. Seiden, M. V., Muggia, F., Astrow, A., Matulonis, U., Campos, S., Roche, M., Sivret, J., Rusk, J., and Barrett, E. (2004), A phase II study of liposomal lurtotecan (OSI-211) in patients with topotecan resistant ovarian cancer, *Gynecol. Oncol.*, 93, 229–232.

456. Calvert, A. H., Grimshaw, R., Poole, C., Dark, G., Swnerton, K., Gore, M., et al. (2002), Randomized phase II trial of two intravenous schedules of the liposomal topoisomerase I inhibitor, NX211, in women with relapsed epithelial ovarian cancer (OVCA): An NCIC CTG study, *Proc. Am. Soc. Clin. Oncol.*, 21, 208a.

457. National Cancer Institute, U.S. National Institutes of Health, http://www.cancer.gov/search/ResultsClinicalTrials, 2007.

458. Gallo, D., Fruscella, E., Ferlini, C., Apollonio, P., Mancuso, S., and Scambia, G. (2006), Preclinical in vivo activity of a combination gemcitabine/liposomal doxorubicin against cisplatin-resistant human ovarian cancer (A2780/CDDP), *Int. J. Gynecol. Cancer*, 16, 222–230.

459. Anderson, K., Lawson, K. A., Simmons-Menchaca, M., Sun, L., SAnders, B. G., and Kline, K. (2004), α-TEA plus cisplatin reduces human cisplatin-resistant ovarian cancer cell tumor burden and metastasis, *Exp. Biol. Med.*, 229, 1169–1176.

460. Turk, M. J., Waters, D. J., and Low, P. S. (2004), Folate-conjugated liposomes preferentially target macrophages associated with ovarian carcinoma, *Cancer Lett.*, 213, 165–172.

461. Kim, C. K., Choi, E. J., Choi, S. H., Park, J. S., Haider, K. H., and Ahn, W. S. (2003), Enhanced p53 gene transfer to human ovarian cancer cells using the cationic nonviral vector, DDC, *Gynecol. Oncol.*, 90, 265–272.

462. Landen, C. N., Jr., Chavez-Reyes, A., Bucana, C., SchmAndt, R., Deavers, M. T., Lopez-Berestein, G., and Sood, A. K. (2005), Therapeutic EphA2 gene targeting in vivo using neutral liposomal small interfering RNA delivery, *Cancer Res.*, 65, 6910–6918.

5.4

BIODEGRADABLE NANOPARTICLES

SUDHIR S. CHAKRAVARTHI AND DENNIS H. ROBINSON

University of Nebraska Medical Center, College of Pharmacy, Omaha, Nebraska

Contents

Pharmaceutical Manufacturing Handbook: Production and Processes, edited by Shayne Cox Gad

5.4.1 INTRODUCTION

Interest in nanotechnology has increased exponentially in many scientific areas, including drug delivery, nanoimaging, and other medical-related applications. Nanoparticles can be fabricated in many different shapes and sizes using a wide range of organic and inorganic materials. However, by definition, these particles must be within the size range of 1–1000 nm. Because the use of nanoparticles in drug delivery and nanomedicine invariably requires parenteral administration, there has been, and continues to be, a major need for the use of polymeric carriers that are both biocompatible and biodegradable. This review will focus on the application of nanotechnology to deliver therapeutic or diagnostic agents using biodegradable polymeric nanoparticles for systemic, localized, or targeted delivery.

5.4.1.1 Classification of Nanoparticles

Depending on the method of preparation and resulting structure, nanoparticles are broadly classified as either matrix-type or encapsulated particles. Hence, a drug is either homogenously dispersed in the polymeric matrix or encapsulated within the core of the particle. In drug delivery applications, biodegradable polymers used to prepare nanoparticles are either natural or synthetic in origin (Table 1). Natural polymers, or biopolymers, include alginates, chitosan, cellulose, gelatin, gliadin, and pullulan. A recent review describes the applications of these polymers in gene delivery and tissue engineering [1]. Natural polymers may vary widely in their composition and therefore physicochemical properties. Such variability in properties may result in poor reproducibility in delivery characteristics, such as drug loading and release kinetics. Further, their purification from natural sources may be difficult. In contrast, synthetic polymers can be prepared with relatively precise properties such as molecular weight, solubility, and permeability characteristics. Examples of synthetic polymers used to make biodegradable nanoparticles include polylactide, poly(lactide-*co*-glycolide) (PLGA), polyanhydride, polycyanoacrylate, poly(ε-caprolactone), and polyphosphoester. As liposomes are more commonly prepared in the micrometer-size range, they were considered out of the scope of this review. Because there are numerous reviews of these polymers in the literature, their physicochemical properties, methods of synthesis, applications, and biological fate of each of these polymers are only briefly described in this chapter.

5.4.2 NATURAL BIODEGRADABLE POLYMERIC NANOPARTICLES

Natural polymers extracted and purified from plant and animal sources often vary significantly in their purity. For example, alginate is available in over 200 different

TABLE 1 Classification of Natural and Synthetic Polymers and Their Methods of Purification or Synthesis

Polymer	Source	Method of Purification or Synthesis	Solubility
Sodium alginate	Natural (seaweed)	Alkali-based extraction	Water soluble (pH > 3), insoluble in organic solvents
Chitosan	Natural (crab shells)	Deacetylation of chitin	Soluble in aqueous solutions (low pH), insoluble in organic solvents
Gelatin	Natural (collagen)	Hydrolysis	Soluble in hot water (>34 °C), acetic acid, forms insoluble gel with water at room temperature, insoluble in organic solvents
Polysaccharides	Natural	Enzymatic reactions	Pullulans (soluble in water), dextrans (soluble in water)
Albumin	Natural (plants, animals)	Separation techniques (chromatography)	Soluble in water
Gliadin	Natural (wheat)	Alcohol extraction	Insoluble in water, soluble in ethanol
Poly(lactide) and Poly(lactide-*co*-glycolide)	Synthetic	Ring-opening polymerization	Insoluble in water, soluble in organic solvents
Poly(ε-caprolactone)	Synthetic	Anionic, cationic, free-radical, ring-opening polymerization	Soluble in select organic solvents such as chloroform, dichloromethane
Polyanhydrides	Synthetic	Melt condensation, ring-opening polymerization	Most polyanhydrides soluble in organic solvents, insoluble in water
Poly-alkylcyanoacrylates	Synthetic	Emulsion and interfacial polymerization	Soluble in organic solvents
Polyphosphoesters	Synthetic	Polyaddition, ring-opening polymerization	Available as water-soluble and water-insoluble types

grades and is extracted from various sources that differ in molecular weight and the percentage and arrangement of guluronic and mannuronic acid blocks. Further, chitosan, poly-β(1-4-D-glucosamine), is available in grades varying in molecular weight, degree of deacetylation (from parent compound, chitin), and viscosity [2].

Some natural polymers may be chemically modified to tailor solubility properties. An example is the reaction of the free amino groups of chitosan to form the more water soluble derivative, methoxy-polyethylene glycol (PEG) chitosan [3]. Collagen is marketed as six different types, I–VI, depending on its source and physiological applicability. Similarly, the properties of gelatin are dependent on the method of preparation using acid- or base-catalyzed hydrolysis from collagen.

5.4.2.1 Physical Properties of Natural Polymers and Methods Used to Prepare Nanoparticles

Sodium Alginate Alginates are primarily derived from the algae *Macrocystis pyrifera* and *Laminaria hyperborea*. These are linear, unbranched polymers containing β-(1–4)-linked mannuronic acid and α-(1–4)-linked guluronic acid residues that are either arranged in blocks, commonly called G blocks and M blocks, or alternate with each other. Alginates are hydrophilic, anionic polymers that vary in molecular weight, depending primarily on the G and M blocks. They are characterized by the ratio of guluronic and mannuronic acids, which can be quantified by ultraviolet (UV) spectrophotometry, gas chromatography, and high-performance liquid chromatography (HPLC) [4]. For example, the polymer obtained from *M. pyrifera* has an M/G ratio of 1.6. Alginate obtained from seaweed must be purified by one of several applicable alkali and acid treatment protocols [5, 6].

Alginate nanoparticles can be prepared using ionotropic gelation, emulsification/internal gelation, and emulsification/solidification methods. Ionotropic gelation results when the anionic alginate reacts with cationic ions or molecules such as calcium or poly-*l*-lysine. Gelation occurs when cations chelate the guluronic and mannuronic acid groups to produce an "egg-box" structure that encapsulates the drug. The size of the alginate particles is determined by the molar concentration of calcium or poly-*l*-lysine and the method of addition of these counterions to alginate [7]. In the emulsification/internal gelation method, the sodium alginate and an insoluble calcium salt are dispersed in a vegetable oil and the calcium ions are liberated to form an alginate gel when the pH of the dispersion is lowered [8, 9]. An advantage of the use of alginate polymers to deliver drugs is that nanoparticles are prepared in aqueous media and may be more suitable to formulate compounds that are unstable in organic solvents such as proteins and peptides. However, since the chelation to form the gel is reversible, a disadvantage of unmodified, alginate-based delivery systems is rapid drug release due to the collapse of the egg-box structure when exposed to monovalent ions in physiological media.

Chitosan Chitosan is a nontoxic, biodegradable polymer obtained by hydrolysis of chitin, a natural polysaccharide that is a chief component of the crustacean exoskeleton. Unmodified chitosan is soluble in acidic media and has significant mucoadhesive properties.

Chitosan nanoparticles may be prepared using various methods, including emulsion cross-linking, coacervation–precipitation, spray drying, emulsion droplet coalescence, ionic gelation, reverse-micellar method, and sieving. A relatively recent review describes the methods and applications of chitosan nanoparticles in drug delivery [10]. Chitosan nanoparticles have also been prepared using water-soluble cross-linking agents such as carbodiimide with the size being controlled by changing pH [11]. More monodisperse nanoparticles may be prepared using fractionated and deacetylated chitosan [12]. In general, the size of nanoparticles will depend on the molecular weight of chitosan, its concentration, and its surface charge [13]. The physicochemical properties of chitosan are determined by the solution pH and ionic strength [14].

Gelatin Gelatin is obtained by either alkaline or acidic hydrolysis of collagen. It has a triple helical structure with a high content of glycine, proline, and hydroxyproline residues. Gelatin that is formed from alkaline treatment of collagen has more carboxyl groups and a lower isoelectric point than that derived from acidic hydrolysis [15]. The physicochemical properties of gelatin depend on the method of extraction and the extent of thermal denaturation that occurs during the purification.

Gelatin nanoparticles can be prepared by various methods, including chemical cross-linking, water-in-oil (w/o) emulsification, and desolvation. Gelatin is cross-linked with agents such as glutaraldehyde. Efficient cross-linking usually results in decreased rate of drug release. The w/o emulsification involves extruding a pre-heated, aqueous solution of gelatin into vegetable oils, such as corn or olive oils [15]. The two-step desolvation method involves the dropwise addition of a water-miscible nonsolvent such as acetone and ethanol [16]. While the use of collagen, the parent compound of gelatin, in drug delivery is rare, collagen nanoparticles have been used to deliver genes by exploiting the electrostatic interaction between the positively charged polymer and negatively charged deoxyribonucleic acid (DNA) [17].

Polysaccharides The macromolecular polysaccharides that include pullulan, mannan, and dextran are the main constituents of the cellular glycocalyx and play an important role in cell–cell adhesion and the cell–cell recognition process [18]. Pullulan is a nonimmunogenic, nontoxic, water-soluble, linear, nonionic polysaccharide with $\alpha(1\text{–}4)$ and $\alpha(1\text{–}6)$ linkages with free hydroxyl groups for drug conjugation [19]. Pullulans are intracellularly synthesized and secreted by a fungus, *Aureobasidium pullulans* [20]. On the other hand, dextrans are anionic glucose polymers derived from sucrose with $\alpha(1\text{–}6)$ glucosidic linkage. A class of enzymes, glucansucrases, produced by two genera of lactic acid bacteria, namely, *Leuconostoc* and *Streptococcus,* catalyze the synthesis of dextrans from sucrose. The extraction of dextrans as well as their physical properties and drug delivery aspects has been reviewed [21, 22]. When coated with mannans, the biological response of both natural and synthetic polymeric nanoparticles may be changed [23].

To make pullulan nanoparticles, the polymer must first be made hydrophobic, typically by conjugating alkyl groups, cholesterol groups, or succinyl groups. Hydrophobic pullulans self-assemble to form stable hydrogel nanoparticles [24]. Alternately, pullulan nanoparticles can be formed by cross-linking reverse micelles of the polymer with glutaraldehyde [25]. Nanoparticles form when a solution of pullulan acetate in *N,N*-dimethyl acetamide is used and dialyzed with borate buffer [26]. Dextran is commonly conjugated to other polymers, such as PLGA, PEG, polystyrene, and poly(methyl methacrylate) for preparation of nanoparticles. Complex coacervation has been used to prepare dextran nanoparticles using oppositely charged polymers such as polyethyleneimine [27]. Although nanoparticle formulation using mannans has not been reported, mannan-coated nanoparticles increased cell binding and macrophage uptake [28].

Albumin Human serum albumin, the most abundant plasma protein, is a positively charged, multifunctional protein and is involved in transport, ligand binding, and enzymatic activities. Albumin is a globular protein containing approximately 585 amino acids in an α-helical tertiary structure. Several exogenous and endogenous compounds can covalently or reversibly bind to albumin, and because of the excellent adsorptive properties of human serum albumin, this polypeptide can be adsorbed onto the surface of polymeric nanoparticles [29]. In addition, as it is amphoteric, albumin can be used as a surfactant during the preparation of nanoparticles where it is irreversibly adsorbed onto the surface of biodegradable polymers such as poly(lactic acid) (PLA) and PLGA [30]. These protein–particle interactions are mainly driven by electrostatic forces further stabilized by hydrophobic forces [31].

Albumin nanoparticles can be prepared by controlled desolvation, pH-induced coacervation, and/or chemical cross-linking with glutaraldehyde. Briefly, the pH of an aqueous solution of albumin is raised to about 9.0 and nanoparticles are precipitated by adding a miscible cosolvent such as acetone [32]. A study reports attempts to optimize the desolvation method to prepare albumin nanoparticles of a more controlled particle size and narrower particle size distribution [33].

Gliadin Gliadin, a glycoprotein derived from gluten, is extracted from wheat and separated by capillary electrophoresis [34, 35]. Gliadin is classified as ω-5, ω-1,2, α, and γ- type based on its structure and electrophoretic mobility [36]. The glycoprotein is water insoluble due to the presence of interpolypeptide disulfide bonds and hydrophobic interactions. A limitation on the use of gliadins is that patient sensitivity causes an autoimmune disorder called celiac disease.

Gliadin nanoparticles are prepared by the desolvation method by first pouring an organic solution of polymer into an aqueous phase such as physiological saline containing a surfactant stabilizer (e.g., Pluronic). The nanoparticles are formed by evaporating the organic solvent.

5.4.2.2 Drug Delivery Applications and Biological Fate of Natural Polymeric Nanoparticles

Alginate Nanoparticles Alginate nanoparticles have been used to formulate a wide range of drugs. Because they are prepared in an aqueous environment under mild conditions, alginate nanoparticles are particularly suitable for formulating proteins, peptides, and oligonucleotides [37]. Further, in addition to being biodegradable, alginates are nonimmunogenic. To decrease the rate of exchange of cations such as Ca^{2+} with monovalent ions in the dissolution medium, the anionic alginates are often treated with cationic molecules such as chitosan, poly-*l*-lysine, or tripolyphosphate. Some examples of the wide range of applications of alginate-based nanoparticles are described. Alginate nanoparticles prepared with tripolyphosphate were used in oral delivery [38]. A study of physical properties demonstrated that alginate–chitosan nanoparticles are suitable for the delivery of DNA [39]. Alginate-coated chitosan nanoparticles increased stability and decreased the burst release of ovalbumin [40]. A study reported that chitosan-stabilized alginate nanoparticles increased bioavailability and sustained release of antifungal drugs compared to PLGA nanoparticles [41]. Although predominantly used for oral administration, inhaled alginate nanoparticles improved the bioavailability of antitubercular drugs [42]. In vivo, alginate nanoparticles accumulate in the Kupffer cells, parenchymal cells of liver, and phagocytes of spleen and lungs [43, 44]. Alginate nanoparticles have also been reported to be absorbed into Peyer's patches, suggesting that they may enhance targeting to the intestinal mucosa [45]. In the body, the alginates degrade by acidic hydrolysis of the guluronic and mannuronic segments [46].

Chitosan Nanoparticles In addition to low-molecular-weight drugs and nutraceuticals, chitosan nanoparticles are widely used in the delivery of macromolecules such as DNA and small interfering ribonucleic acid (siRNA) [47]. Apart from sustaining the release of macromolecules, chitosan nanoparticles protect them from nucleases. Placebo chitosan nanoparticles exhibited antibacterial activity against several microbes, including *Escherichia coli* [48]. The surface of chitosan nanoparticles was hydrophobically modified with linoleic acid for delivery of trypsin [49]. Other applications of chitosan nanoparticles include lung [50] and ocular delivery [51]. The primary amine group at the 2 position can be modified to tailor chitosan for specific applications. For example, chemical conjugation of these amine groups to methoxy-PEG groups increased water solubility [52]. Thiolation of chitosan enhanced the mucosal permeation of the nanoparticles [53]. Hydrophobically modified glycol chitosans that self-assemble into nanoparticles have been used to deliver doxorubicin [54]. Targeting chitosan nanoparticles to folate receptors on the surface of cells enhanced the transfection efficiency of DNA [55]. *N*-Succinyl chitosan nanoparticles containing 5-fluorouracil demonstrated excellent activity against sarcoma tumors [56]. Self-assembled *N*-acetyl histidine–conjugated glycol chitosan nanoparticles were efficiently internalized into cells by adsorptive endocytosis [57]. No toxic effects have been observed with chitosan nanoparticles. Upon intravenous (i.v.) administration, chitosan nanoparticles accumulated in the liver with minimal concentrations in the heart and lung [54].

Gelatin Nanoparticles Gelatin nanoparticles have been used as a delivery system for several drugs, including pilocarpine, hydrocortisone [58], methotrexate [59], paclitaxel [60], and chloroquine [61]. High protein loading and sustained release were achieved using composite gelatin and PLGA nanoparticles [62]. Surprisingly, placebo gelatin nanoparticles exhibited antimelanoma activity in vivo [63]. Primary amine groups of gelatin molecule can be chemically conjugated or cross-linked using bifunctional cross-linkers. This is demonstrated in the delivery of biotinylated peptide nucleic acid using avidin-cross-linked gelatin nanoparticles [64]. PEGylation of gelatin nanoparticles containing hydrophilic drugs prolonged circulation time in the body [65]. Thiolated gelatin nanoparticles produced effective transfection of plasmid DNA encoding the green fluorescent protein [66]. Following endocytic uptake by the cells, gelatin nanoparticles concentrate in the perinuclear region [65]. In vitro, the gelatin nanoparticles are efficiently internalized into macrophages and monocytes [67]. In tumor-bearing mice, PEGylated gelatin nanoparticles predominantly accumulated in the liver and tumor [68] while in dendritic cells they are primarily localized in the lysosomes [69]. In vivo, gelatin is degraded by proteases to amino acids. Although cardiotoxicity and mild immunogenicity were reported with gelatin nanoparticles covalently coupled to doxorubicin, this was attributed to the coupling reagent glutaraldehyde [70].

Pullulan and Dextran Nanoparticles Pullulan nanoparticles successfully delivered HER2 oncoprotein to induce humoral and cellular immune responses against HER2-expressing murine sarcomas [71]. Hydrophobic polysaccharides such as pullulan and mannan enable soluble proteins to induce cellular immunity and therefore may be a potential delivery vehicle for vaccines [72]. pH-sensitive pullunan nanoparticles prepared by conjugating sulfonamides are stable at physiological pH but aggregate and release the encapsulated drug when exposed to the lower tumor pH [73]. Coating of magnetic nanoparticles with pullulan enhanced their cellular uptake by endocytosis [74]. Amphotericin-loaded, dextran–polyethyleneimine nanoparticles were active against *Candida albicans* [27]. Similarly, insulin-containing dextran–polyethyleneimine nanoparticles prolonged the hypoglycemic effect in diabetic rats [75]. Hydrogels prepared from blends of polyvinyl alcohol and dextrans have been used as matrices to entrap PLGA nanoparticles [76]. In vitro, immunofluorescent staining illustrated that pullulan nanoparticles are internalized by active endocytosis [25]. However, other studies suggest that both absorption and internalization of pullulan nanoparticles are inhibited by coating them with dextran [77].

Albumin Nanoparticles A significant development in the drug delivery of albumin nanoparticles has been the recent marketing of the commercial product Abraxane® for chemotherapy of breast cancer. This delivery system is prepared by nab™ technology, which involves noncovalent complexation of albumin with paclitaxel. Albumin nanoparticles have been used to deliver antisense oligonucleotides, interferon γ, and anticytomegaloviral drugs [78, 79]. Intravitreal injection of ganciclovir-loaded albumin nanoparticles was attempted to prolong residence time in the eye [79]. PLA nanoparticles coated with albumin degrade more rapidly in the gastrointestinal (GI) region, resulting in efficient delivery of water-soluble drugs across the GI tract [80]. In addition, intra-arterial chemotherapy with paclitaxel-containing albumin nanoparticles effectively treated squamous cell carcinoma [81]. Human serum albumin–polyethyleneimine nanoparticles optimized transfection of the luciferase gene in human, embryonic, and epithelial kidney cells [82]. Conjugation of the cellular targeting agent, folic acid, resulted in increased cellular uptake of albumin nanoparticles compared to unmodified particles [83, 84]. Transferrin-conjugated PEGylated albumin nanoparticles demonstrated enhanced uptake into the brain tissues [85]. Glycyrrhizin was conjugated to the amine groups of albumin nanoparticles targeting hepatocytes [86]. After i.v. administration, albumin-coated PLA nanoparticles were distributed in the liver, bone marrow, lymph nodes, spleen, and peritoneal macrophages [87]. In the body, albumin nanoparticles are actively taken up by macrophages.

Gliadin Nanoparticles The hydrophobic nature of gliadin makes this polymer ideal for delivery of hydrophobic compounds such as all-trans retinoic acid and vitamin E. Gliadin nanoparticles adhered to the stomach mucosa and significantly increased the bioavailability of carbazole [88]. Typically, encapsulation efficiency and drug loading of gliadin nanoparticles are higher for hydrophobic drugs [89]. Gliadin nanoparticles were targeted to *Helicobacter pylori* by chemical conjugation of lectin glycoproteins to their surface, resulting in a twofold increase in inhibition of bacterial activity compared to unmodified nanoparticles [90]. Although gliadins adhere to the mucosa, internalization of gliadin nanoparticles into cells has not been reported. More definitive studies are required to fully understand the biological fate of these particles.

5.4.3 SYNTHETIC BIODEGRADABLE POLYMERIC NANOPARTICLES

A detailed description of the methods of polymerization and variables employed in polymer synthesis are beyond the scope of this review and can be found in many texts and review articles. The focus of this section is to provide an overview of the properties and methods used to prepare nanoparticles from each class of synthetic polymer.

5.4.3.1 Synthetic Polymers: Physical Properties and Methods of Preparation of Nanoparticles

Poly(lactic acid) and Poly(lactide-*co*-glycolide) These poly-hydroxy acids are approved for human use by the Food and Drug Administration (FDA) and have been widely used to prepare biodegradable nanoparticles. PLA exists in optically active and inactive forms and is a semicrystalline, hydrophobic molecule that degrades in the body over a period of months. Conversely, poly(glycolic acid) is amorphous and hydrophilic and degrades more rapidly than PLA. In aqueous media, these polymers degrade by random hydrolysis of ester bonds that is autocatalyzed in acidic media to form lactic and glycolic acids [91]. The factors that affect the rate of hydrolytic degradation include type and composition of the polymer backbone, nature of pendent groups, molecular weight, pH, enzymes, and geometry of the delivery device.

The preparation and characterization of PLA and PLGA nanoparticles have been extensively reviewed elsewhere [92, 93]. Various techniques may be used to prepare PLA and PLGA nanoparticles, including simple and multiple emulsions, nanoprecipitation, gas antisolvent method, supercritical fluid technology, coacervation/phase separation, and spray drying [91]. Briefly, in the single-emulsion method, an organic solution of the polymer and drug is emulsified with an aqueous solution of surfactant such as polyvinyl alcohol (PVA). While PLA and PLGA nanoparticles containing hydrophobic drugs are prepared by the two-phase emulsion method, a w/o/w multiple-emulsion method is needed to encapsulate hydrophilic drugs. In the phase separation method, the addition of a nonsolvent precipitates or coacervates the polymer from solution to encapsulate the drug. The experimental variables for each protocol can be altered to influence the physicochemical properties, such as particle size, particle size distribution, morphology, and zeta potential [93]. The release of encapsulated drug from PLA and PLGA nanoparticles may occur by a combination of diffusion and polymer degradation at a rate that is influenced by properties of the polymer and nanoparticles and the environment. The surface of both PLGA and PLA nanoparticles can be modified to target cells and organs by conjugation with ligands such as folates, transferrin, HIV-TAT, aptamers, heparin, and lectins. The negative zeta potential of PLA and PLGA nanoparticles can be altered by coating with cationic polymers such as chitosan and polyethyleneimine, which promote nanoparticle–cell interaction. As with liposomes, PEGylation of PLGA nanoparticles prolongs circulation times in the body.

Poly(ε-caprolactone) Poly(ε-caprolactone) is a semicrystalline polymer synthesized by anionic, cationic, free-radical, or ring-opening polymerization [94]. It is available in a range of molecular weights and degrades by bulk hydrolysis autocatalyzed by the carboxylic acid end groups. The presence of enzymes such as protease, amylase, and pancreatic lipase accelerates polymer degradation [95]. The various methods of preparation of poly(ε-caprolactone) nanoparticles include emulsion polymerization, interfacial deposition, emulsion–solvent evaporation, desolvation, and dialysis. These methods and various applications are extensively reviewed [94].

Polyanhydrides Polyanhydrides have a hydrophobic backbone with a hydrolytically labile anhydride linkage. These polymers widely vary in chemical composition and include aliphatic, aromatic, and fatty acid–based polyanhydrides. The rate of degradation depends on the chemical composition of the polymer. In general, aliphatic polyanhydrides degrade more rapidly than the aromatic polymer. Hence, copolymer blends with varying ratios of aliphatic-to-aromatic polyanhydrides can be synthesized to suit specific applications.

The synthesis and physical properties of polyanhydrides have been reviewed [96, 97]. Polyanhydride nanospheres are commonly prepared by the emulsion–solvent evaporation method using PVA as a stabilizer. However, as polyanhydrides are hydrolabile, they need to be flash frozen in liquid nitrogen and lyophilized immediately [98]. An example of their use to deliver drugs is entrapment of bovine zinc insulin by phase inversion nanoencapsulation [99]. Although not used to formulate nanoparticles, polyanhydride microspheres have been prepared using alternate techniques involving nonaqueous solvents, such as solid/oil/oil double emulsion and cryogenic atomization techniques [100]. The surface of polymeric nanoparticles can be modified for targeted delivery by reaction of the anhydride with an amino group to form an amide linkage with a ligand [101]. However, application of the ligand-conjugated nanoparticles for drug delivery is yet to be explored extensively.

Poly(alkyl-cyanoacrylates) As poly(alkyl-cyanoacrylates) form strong bonds with polar substrates including the skin and living tissues, they exhibit bioadhesive properties. These polymers are synthesized by free-radical, anionic, or zwitterionic polymerization. As detailed in a recent review, poly(alkyl-cyanoacrylate) nanoparticles are prepared by emulsion polymerization, interfacial polymerization, nanoprecipitation, and emulsion–solvent evaporation methods [102].

Solid–lipid Nanoparticles Solid–lipid nanoparticles (SLNs) are obtained by high-pressure homogenization of molten lipids in the presence of surfactant. The major advantage of using SLN is the ability to have high drug loading and prolonged stability with lipophilic compounds. In addition to drug delivery, SLNs have been used in dermatological and cosmetic preparations. The most common carriers used to prepare SLNs are triglycerides, glycerides, fatty acids, steroids, and waxes and may contain a wide range of emulsifiers. Methods of preparation of SLNs include high speed hot and cold homogenization, ultrasound, emulsion–solvent evaporation, and microemulsion. The various parameters involved in the preparation of SLNs have been optimized and thoroughly reviewed and their physicochemical properties elucidated [103, 104].

Other Synthetic Biodegradable Polymers Although well investigated for drug delivery, polyorthoesters, polyurethanes, and polyamides have found limited application as nanoparticles. A report documents the synthesis and characterization of polyorthoester nanoparticles [105].

5.4.3.2 Drug Delivery Applications and Biological Fate of Synthetic Biodegradable Polymers

PLA/PLGA ***Nanoparticles*** A wide range of hydrophilic and hydrophobic drugs, including low- and high-molecular-weight compounds, have been encapsulated into PLGA/PLA nanoparticles for a wide range of therapeutic applications and routes of administration, including oral, intravenous, intra-arterial, nasal, and inhalation delivery [92, 106]. Extensive reviews describing the application of PLGA nanoparticles in drug therapy are available [92, 107, 108].

After i.v. administration, the PLGA nanoparticles are removed from systemic circulation by the mononuclear phagocytic system in the liver [109]. PLGA nanoparticles enter cells by absorptive endocytosis and may escape the lysosomes to accumulate in cytoplasm [110, 111]. In the body, PLA and PLGA degrade into the monomers lactic and glycolic acids, which enter the citric acid cycle, where they are metabolized and eliminated as CO_2 and H_2O. Glycolic acid may also be excreted through the kidney [91]. Humoral response to these results in mild, acute, and chronic inflammation [112].

Poly(ε-caprolactone) Nanoparticles As important applications of poly(ε-caprolactone) nanoparticles have been reviewed previously, only representative examples will be given [94]. Decreased cardiovascular adverse effects of cartelol was observed upon ophthalmic administration of poly(ε-caprolactone) nanocapsules [113]. Poly(ε-caprolactone) nanoparticles, nanocapsules, or nanoemulsions increased the ocular uptake of indomethacin [114]. The cytotoxicity of retinoic acid was enhanced when delivered in core-shell-type nanoparticles formed from poly(ε-caprolactone–polyethylene glycol) blends [115]. Alternately, these nanoparticles were also chemically modified with folic acid to target the folate receptors for enhanced cellular uptake [116]. Coating poly(ε-caprolactone) nanoparticles with polysaccharides such as galactose resulted in lectin-dependent aggregation, demonstrating the potential as a targeted delivery system to hepatocytes [117]. Stable complexes were formed between anionic DNA and chitosan-modified poly(ε-caprolactone) nanoparticles, demonstrating high transfection efficiency [118]. After i.v. administration, these particles are eliminated by macrophages of the reticuloendothelial system and biodegradation occurs by bulk scission of polymer chains [94]. However, dextran-coated poly(ε-caprolactone) nanoparticles lowered their uptake into macrophages [119].

Polyanhydride Nanoparticles Polyanhydrides have been more commonly used to prepare microparticles than nanoparticles. However, the technology is adaptable for nanoparticles. The transfection efficiency of firefly luciferase DNA was enhanced when delivered in nanoparticles prepared from polyanhydride–lactic acid blends, demonstrating the potential application in gene delivery [120]. The degradation and elimination of polyanhydrides have been reviewed [97]. In vivo, the anhydride bond degrades to form diacid monomers that are eliminated from the body.

Poly(alkyl-cyanoacrylate) Nanoparticles The applications of poly(alkyl-cyanoacrylate) nanoparticles have been reviewed elsewhere and therefore only representative examples are described [102]. Because of their adhesive properties, nanoparticles have the potential to prophylactically treat candidiasis of the oral cavity [121]. Not surprisingly, poly(alkyl-cyanoacrylate) nanoparticles have been used to deliver drugs to tumors [122]. Enhanced absorption and prolonged hypoglycemic effect were observed when insulin was delivered in poly(alkyl-cyanoacrylate) nanoparticles [121]. Nuclear accumulation of antisense oligonucleotides into vascular smooth muscle cells was increased when delivered using poly(alkyl-cyanoacrylate) nanoparticles [123]. Dextran-coated poly(alkyl-cyanoacrylate) nanoparticles lowered protein adsorption in the blood [124].

Poly(alkyl-cyanoacrylates) degrade by hydrolysis of the ester bond of the alkyl side chains to form water-soluble alkyl alcohol and poly(cyanoacrylic acid). After in vivo administration, poly(alkyl-cyanoacrylate) nanoparticles are predominantly distributed in the liver, spleen, and bone marrow where they are endocytosed into the cells to become localized in the lysosomes. However, the mechanisms of lysosomal escape have not been identified [102].

Solid–Lipid Nanoparticles SLNs have been used to deliver small molecules and macromolecules such as DNA and peptides. The in vitro and in vivo applications of SLNs are reviewed elsewhere [125, 126]. The stability and oral bioavailability of insulin were enhanced when administered in wheat germ agglutinin–conjugated nanoparticles [127]. A polyoxyethylene stearate coat on the SLN confers stealth properties [128].

5.4.4 THERMOSENSITIVE AND pH-SENSITIVE NANOPARTICLES

5.4.4.1 Physical Properties and Methods of Preparation

Thermosensitive Polymeric Nanoparticles Thermoresponsive "smart" polymers that change their physical characteristics, such as shape, surface properties, or solubility, in response to changes in temperature have been developed to target drugs [129]. The encapsulated drug is released when the nanoparticles are exposed to changes in temperature, such as body temperature or external heat source. For example, poly(*N*-isopropyl acrylamide), or poly(NIPAAm), is water soluble at room temperature but aggregates and is insoluble above its lower critical solution temperature (LCST), which typically ranges between 37 and 42 °C [129]. The methods used to prepare thermosensitive nanoparticles have been thoroughly reviewed [127,

129–131]. Interestingly, when heated above the LCST, poly(NIPAAm)–PEG block copolymers spontaneously self-assemble into nanoparticles whose size is controlled by the rate of heating [132].

pH-Sensitive Polymeric Nanoparticles Enteric-coated polymers have long been used to protect drugs from the acidic pH in the stomach. Chitosan is a pH-sensitive polymer that is soluble only in acidic media. Similarly, the solubility of sulfonamide-modified pullulans is dependent on pH [133]. Chitosan–insulin nanoparticles are stable at low pH but dissociate at physiological pH, releasing insulin [134]. The pH of tumor interstitium is lower than the normal tissue. The pH-sensitive polyethylene oxide–poly(β-amino ester) microparticles containing paclitaxel significantly reduced tumor burden [135, 136]. Polyketals, a new generation of acid-sensitive polymers, degrade by acidic hydrolysis [137]. Low pH inside the endosomes facilitated the escape of pH-responsive plasmid–lipid nanoparticles, resulting in enhanced transfection efficiency [138].

5.4.4.2 Drug Delivery Applications and Biological Fate of Thermosensitive and pH-Sensitive Nanoparticles

Thermosensitive block copolymer nanoparticles containing doxorubicin increased cytotoxicity against Lewis lung carcinoma cells when activated by heating above the LCST [139]. Chitosan was chemically conjugated to NIPAAm/vinyl laurate copolymer to enhance gene transfection in mouse myoblast cells [140]. Upon i.v. administration, poly(NIPAAm) nanoparticles are taken up by the reticuloendothelial cells of the liver and mild inflammatory and fibrotic responses are observed [141].

After internalization into SKOV-3 (ovarian adenocarcinoma) cells, polyethylene oxide–modified poly(β-amino ester) nanoparticles rapidly disintegrated and released the drug in the low pH of the endosomes [142]. Intravenous administration of polyethylene oxide–modified poly(β-amino ester) nanoparticles containing paclitaxel significantly reduced tumor burden in mice with ovarian cancer [142]. *N*-Acetyl histidine–conjugated glycol chitosan nanoparticles were used to deliver drugs into the cytoplasm. These pH-responsive nanoparticles are endocytosed where their structural integrity is lost due to protonation of imidazoles, resulting in their endolysosomal escape [57]. As most of the pH-sensitive biodegradable polymers are blends of natural and synthetic polymers, they are degraded by mechanisms specific to individual polymers.

5.4.5 APPLICATIONS OF BIODEGRADABLE NANOPARTICLES OTHER THAN DRUG DELIVERY

Diagnosis and imaging are important applications of nanoparticles that are briefly described. The ability to encapsulate or conjugate fluorescent compounds into or onto biodegradable nanoparticles has been used extensively in imaging. Compounds that have been encapsulated into nanoparticles for imaging include gadolinium, fluorescein isothiocyanate (FITC)–dextrans, Bodipy, and the autofluorescent anticancer drug doxorubicin. Nanoparticles encapsulating radioactive ligands, such as

^{99m}Tc-labeled colloids and ^{111}In, have been used in scintigraphic imaging [143]. In addition, fluorescent or radioactive moieties can be targeted by noncovalently or covalently tagging the nanoparticles through avidin–biotin conjugation and thiol formation [144]. In vitro imaging enables the dynamics of cellular internalization and localization of nanoparticles to be studied. For example, Bodipy—loaded PLGA nanoparticles have been used to study their cellular disposition in vitro [145] as well as the effect of storage temperature on their physical properties [146]. The biotinylated antibody, specific to the CD3 antigen on lymphocytes, was chemically conjugated to nanoparticles and their binding to leukemic and primary T lymphocytes investigated [143]. The instrumentation that facilitates imaging of nanoparticles includes confocal laser scanning microscopy, liquid scintigraphy, and flow cytometry.

5.4.6 PHYSICOCHEMICAL CHARACTERIZATION OF POLYMERIC NANOPARTICLES

The selection of polymer is critical to the performance, properties, and application of nanoparticles. Further, the physicochemical properties of the polymer will determine the surface properties of nanoparticles with polymer molecular weight, hydrophobicity, and glass transition temperature being particularly important. The surface properties that influence their biodistribution and cellular response include particle size, zeta potential, and surface hydrophilicity.

5.4.6.1 Molecular Weight

Many reviews and textbooks document the experimental methods available to determine number-average, weight-average, viscosity-average, and z-average molecular weights. The molecular weight of the polymer will influence many parameters during the preparation of nanoparticles as well as their properties such as drug loading and rate and extent of drug release.

5.4.6.2 Hydrophobicity

The experimental methods of determining hydrophobicity include interaction chromatography and two-phase partition using fluorescent or radiolabeled hydrophobic probes [147]. Generally, a more hydrophobic polymer degrades more slowly and releases drugs at a decreased rate. Hence, a blend of hydrophobic and hydrophilic polymers can be used to tailor drug release kinetics. Depending on the conditions of polymerization, mixtures of two or more monomers of different types yield block, alternate, or random cocopolymers, each of which will possess different hydrophobicities and consequently drug release kinetics. Additionally, the degree of hydrophobicity can change the mechanism of degradation. For example, by inserting hydrophilic ethylene glycol moieties within the hydrophobic backbone, polyanhydrides have been tailored to specifically degrade by bulk or surface erosion [148]. Incorporation of aromatic side chains generally increases hydrophobicity. A difference in hydrophobicity of individual polymers results in microphase separation of copolymers followed by a thermodynamic partition and altered release profile of encapsulated drugs [149].

5.4.6.3 Glass Transition Temperature

The morphology and physical properties of nanoparticles are affected by the glass transition temperature and physical state of the polymer or polymer blends. The glass transition temperature (T_g) is the temperature at which polymers undergo a change in heat capacity and transform their physical arrangement. The crystallinity or amorphous nature of the polymers can be altered by synthesizing polymer blends of varying ratios. The T_g of a polymer is experimentally determined by differential scanning calorimetry (DSC). The T_g is mathematically calculated using multidimensional lattice representations and statistical methods [150]. The various factors that influence the T_g of polymers include molecular weight, composition and stereochemistry of the polymer backbone, type and length of pendent groups, additives such as copolymers, and plasticizers. Additives and copolymers may also be used to alter the T_g. For example, incorporation of mPEG significantly lowered the T_g of PLA, resulting in rapid release of drug from the nanoparticles [151].

5.4.6.4 Particle Size and Particle Size Distribution Instrumentation

Particle size is a critical characteristic of nanoparticles and by definition differentiates them from microparticles. Particle size and particle size distribution play an important role in the biological performance of the nanoparticles. Important techniques for measuring particle size are photon correlation spectroscopy (PCS), electron microscopy, and atomic force microscopy, which have been comprehensively described [147]. Particle size and particle size distribution are determined by the method of preparation and experimental variables during manufacture. For example, in the emulsion–solvent evaporation method, the particle size is determined by controlling the energy of emulsification and the resulting droplet size of the internal phase. Control of particle size is also possible by altering experimental variables such as the volume and phase ratio of the internal and external phases and the concentration, type, and viscosity of the emulsifying agent. As an example, the experimental parameters that can influence the particle size and size distribution of PLGA nanoparticles include the method used, polymer concentration, surfactant concentration, stirring speed, ratio of aqueous and organic phases, and concentration of the emulsifier [152]. Methods of separation such as filtration and centrifugation can also influence particle size distribution. When chitosan nanoparticles are prepared using ionic gelation, the critical parameters for a narrow particle size distribution are molecular weight, degree of deacetylation, concentration and molar ratio of chitosan, and the presence and concentrations of counterions (e.g., tripolyphosphate) [12].

5.4.6.5 Surface Charge and Zeta Potential

The surface charge of nanoparticles is important because it determines the nature and extent of aggregation of colloids and their interaction with cells and other biological components within the body. The zeta potential is the potential at the solid–liquid interface and is commonly determined using light scattering [153]. Decreasing the zeta potential of nanoparticles below a critical value increases the rate and extent of their aggregation, resulting in conglomerates with different physical properties. The surface charge can also be altered by chemical conjugation of ligands to nanoparticles. As the following examples demonstrate, surface charge may be modified to facilitate targeting. The degree of positive charge on chitosan–tripolyphosphate nanoparticles was increased by controlling the processing parameters, resulting in a stronger electrostatic interaction with negatively charged cell surfaces [13]. An increase in surface charge increased the adsorption of nanoparticles to plasma proteins [154]. Positively charged ligands such as chitosan as well as peptides neutralize the surface charge of negatively charged particles made using polymers such as PLGA [155]. Positively charged tripalmitin nanoparticles increased circulation time and higher blood concentrations of etoposide compared to negatively charged particles [156]. On the other hand, neutrally charged particles may be protected from opsonization [157].

5.4.6.6 Surface Hydrophilicity

The surface hydrophilicity of nanoparticles also influences the nature and extent of their interaction with cells and their behavior in the biological environment. The techniques for determining surface hydrophilicity have been described previously [147]. As hydrophobic nanoparticles are opsonized and eliminated by the mononuclear phagocytic system, surface hydrophilicity is an important parameter to ensure longer circulation times. The hydrophilicity to the nanoparticles may be modified using several methods, including adsorption of nonionic surfactants such as Poloxamer as well as adsorption or conjugation of hydrophilic polymers such as polysaccharides, polyacrylamides, PVA, poly(*N*-vinyl-2-pyrrolidone), PEG, polyoxamines, and polysorbates [157]. Stealth nanoparticles can be prepared by adsorbing or conjugating PEG to the particle surface, which protects them from opsonization [157]. The biological properties of stealth nanoparticles have been reviewed elsewhere [158].

5.4.6.7 Drug Loading and Encapsulation Efficiency

Drug loading or the weight of the drug encapsulated in the polymeric carrier is expressed as a percentage (w/w) of the delivery system. Encapsulation efficiency is the difference between the amount of the drug encapsulated into nanoparticles compared to the total amount added during preparation. The encapsulation efficiency is dependent on the properties of the polymer and excipients and the method of preparation of nanoparticles. For example, increased miscibility between the nanoparticle and water at lower temperatures and high rate of solvent evaporation at high temperatures result in the formation of an outer sphere wall, increasing encapsulation efficiency [159]. Efficient encapsulation depends on (1) physicochemical properties of the drug, such as solubility, hydrophilicity, and crystallinity; (2) physicochemical properties of the polymer, including molecular weight, hydrophobicity, drug–polymer interactions, and solubility parameter; and (3) variables involved in the preparation of the nanoparticles, such as drug–polymer weight ratio, solvents, method of preparation of particles, solvent evaporation rate, type as well as volume, and concentration of the surface-active agent used.

5.4.6.8 Drug Release

Many cumulative and differential dissolution methods are used to monitor the rate and extent of drug release. The rate of drug release from nanoparticles depends on the chemical properties of the polymer, properties of particles such as hydrophobicity and surface area, and environmental factors such as pH. Drug release may occur by one or a combination of the following: diffusion, dissolution, degradation, or swelling. Hence, drug release normally follows first- rather than zero-order kinetics [160–162]. A disadvantage of nanoparticles is that they may have a significant burst release (~40%) due to their high surface area–mass ratio. Surface treatment of nanoparticles can reduce the burst release of encapsulated drug [163, 164].

5.4.6.9 Physical Stability of Polymeric Nanoparticles

The main indication of physical instability of nanoparticles is irreversible aggregation. The principal factors that affect the extent of aggregation are type of polymer, zeta potential, duration and temperature of storage, and presence of electrolytes. Generally, homopolymers such as PLA and poly(ε-caprolactone) are more stable than copolymers. However, the stability of PLGA copolymers can be prolonged by appropriate storage [165]. The duration of storage can also be important. For example, poly(σ-caprolactone) nanoparticles aggregated after four months [166]. Another study compared the physical stability of PLGA nano- and microspheres after incubation at different temperatures and demonstrated that, while microspheres did not aggregate, the extent of aggregation of nanospheres increased as storage temperature increased [146]. Ideally, PLGA nanoparticles should be stored desiccated at 4 °C. Aqueous dispersions of SLN are stable for three years [167]. It is important to note that electrolytes can accelerate the aggregation and instability of nanoparticles. For example, multivalent ions caused solid–lipid nanoparticles to gel [168]. Physical stability can be improved by coating the surface of nanoparticles with hydrophilic polymers or surfactants [169]. As discussed previously, a reduction in the zeta potential below critical values causes flocculation of colloidal systems. For example, neutralization of the surface charge of colloidal systems by addition of cationic oligonucleotides resulted in aggregation [170].

5.4.7 TARGETING NANOPARTICLES BY SURFACE CONJUGATION WITH LIGANDS

Nanoparticles can be targeted to cells using specific or nonspecific ligands. Specific ligands can be covalently conjugated to the nanoparticles to target them to a selected site of action. Some examples of targeting ligands are folic acid [171], transferrin [172], lectin [173], and epidermal growth factor [174]. Nonspecific targeting is achieved by attaching ligands that alter the biodistribution of the nanoparticles. Examples of nonspecific targeting include the use of PEG for imparting stealth properties or hydrolytically cleavable linkers that protect the drug from degradation. Common methods of conjugation of ligands include carbodiimide coupling, glutaraldehyde conjugation, peptide bond formation, disulfide, and thiol linkages [175–178].

5.4.8 CELLULAR TRAFFICKING OF BIODEGRADABLE NANOPARTICLES

Biodegradable nanoparticles are internalized by one or more of the following mechanisms: phagocytosis, macropinocytosis, and clathrin- and caveolin-mediated endocytosis (Figure 1). While phagocytosis by macrophages eliminates nanoparticles from the body, efficient cellular uptake occurs when high-affinity receptors capture the nanoparticles through receptor-mediated endocytosis [179]. On the cell surface, nanoparticles activate caveolin, a dimeric protein, resulting in their internalization through caveolae. Clathrin-mediated endocytosis occurs when nanoparticles accumulate on the plasma membrane and clathrin-coated pits are formed to transport the nanoparticles into the cell, resulting in the formation of endosomes. Macropinocytosis is restricted to larger particles, such as nanoparticles typically greater than 800 nm. The mechanism by which particles enter cells depends on the composition of the nanoparticles, type of the cell, and particle size. For example, the uptake of chitosan nanoparticles into lung epithelial and Caco-2 cells was mediated, in part, by the clathrin-mediated pathway [180, 181]. However, PLGA nanoparticles of size 100 nm are internalized by the clathrin- and caveolin-independent pathway [110]. There are comprehensive reviews of the mechanisms responsible for the uptake of nanoparticles [182–184].

Nanoparticles that are internalized into cells by these mechanisms first enter the primary endosomes of the cell and are then transported into sorting endosomes. While some nanoparticles in the sorting endosomes are transported out of the cell by recycling endosomes, the remaining nanoparticles are transported into secondary endosomes that fuse with the lysosomes [107]. The surface charge of PLGA nanoparticles is reversed in the acidic lysosome, resulting in their escape into the cytoplasm [111]. A high external concentration of nanoparticles outside the cell prolongs their intracellular concentration within the cytoplasm [107].

Ligand-mediated endocytosis targets specific cell surface receptors and these nanoparticles are internalized by a receptor-mediated endocytic pathway. Examples of targeting transferrin, folate, lectins, and epidermal growth factor receptors are contained in the literature [185–187].

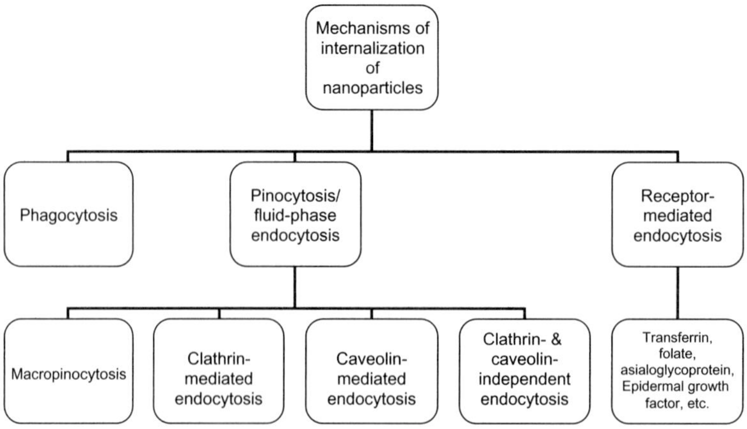

FIGURE 1 Mechanisms of cellular internalization of biodegradable nanoparticles.

5.4.9 CONCLUSIONS

Biodegradable nanoparticles are a very active area of research in drug delivery, imaging, and diagnostics. This review has primarily focused on drug delivery applications. Natural and synthetic polymers used to prepare nanoparticles were discussed as well as their physicochemical properties that influence the biological performance of particles. Methods used to prepare and characterize the properties of nanoparticles have also been reviewed. Specific and nonspecific methods used to target nanoparticles to cells were mentioned as well as the mechanism for cellular and intracellular transport. As research into the various uses of biodegradable nanoparticles increases, so will our knowledge to further optimize their preparation and formulation and hence improve drug therapy and diagnosis.

REFERENCES

1. Dang, J. M., and Leong, K. W. (2006), Natural polymers for gene delivery and tissue engineering, *Adv. Drug Deliv. Rev.*, 58(4), 487–499.
2. Singla, A. K., and Chawla, M. (2001), Chitosan: Some pharmaceutical and biological aspects—an update, *J. Pharm. Pharmacol.*, 53(8), 1047–1067.
3. Kulkarni, A. R., Lin, Y. H., Liang, H. F., Chang, W. C., Hsiao, W. W., and Sung, H. W. (2006), A novel method for the preparation of nanoaggregates of methoxy polyethyleneglycol linked chitosan, *J. Nanosci. Nanotechnol.*, 6(9–10), 2867–2873.
4. Sanchez-Machado, D. I., Lopez-Cervantes, J., Lopez-Hernandez, J., Paseiro-Losada, P., and Simal-Lozano, J. (2004), Determination of the uronic acid composition of seaweed dietary fibre by HPLC, *Biomed. Chromatogr.*, 18(2), 90–97.
5. Dusseault, J., Tam, S. K., Menard, M., Polizu, S., Jourdan, G., Yahia, L., and Halle, J. P. (2006), Evaluation of alginate purification methods: Effect on polyphenol, endotoxin, and protein contamination, *J. Biomed. Mater. Res. A.*, 76(2), 243–251.
6. Tonnesen, H. H., and Karlsen, J. (2002), Alginate in drug delivery systems, *Drug Dev. Ind. Pharm.*, 28(6), 621–630.
7. Rajaonarivony, M., Vauthier, C., Couarraze, G., Puisieux, F., and Couvreur, P. (1993), Development of a new drug carrier made from alginate, *J. Pharm. Sci.*, 82(9), 912–917.
8. Poncelet, D., Lencki, R., Beaulieu, C., Halle, J. P., Neufeld, R. J., and Fournier, A. (1992), Production of alginate beads by emulsification/internal gelation. I. Methodology, *Appl. Microbiol. Biotechnol.*, 38(1), 39–45.
9. Reis, C. P., Neufeld, R. J., Vilela, S., Ribeiro, A. J., and Veiga, F. (2006), Review and current status of emulsion/dispersion technology using an internal gelation process for the design of alginate particles, *J. Microencapsul.*, 23(3), 245–257.
10. Agnihotri, S. A., Mallikarjuna, N. N., and Aminabhavi, T. M. (2004), Recent advances on chitosan-based micro- and nanoparticles in drug delivery, *J. Controlled Release*, 100(1), 5–28.
11. Bodnar, M., Hartmann, J. F., and Borbely, J. (2005), Preparation and characterization of chitosan-based nanoparticles, *Biomacromolecules*, 6(5), 2521–2527.
12. Zhang, H., Oh, M., Allen, C., and Kumacheva, E. (2004), Monodisperse chitosan nanoparticles for mucosal drug delivery, *Biomacromolecules*, 5(6), 2461–2468.
13. Gan, Q., Wang, T., Cochrane, C., and McCarron, P. (2005), Modulation of surface charge, particle size and morphological properties of chitosan-TPP nanoparticles intended for gene delivery, *Colloids Surf. B. Biointerfaces*, 44(2–3), 65–73.

14. Lopez-Leon, T., Carvalho, E. L., Seijo, B., Ortega-Vinuesa, J. L., and Bastos-Gonzalez, D. (2005), Physicochemical characterization of chitosan nanoparticles: Electrokinetic and stability behavior, *J. Colloid Interface Sci.*, 283(2), 344–351.
15. Young, S., Wong, M., Tabata, Y., and Mikos, A. G. (2005), Gelatin as a delivery vehicle for the controlled release of bioactive molecules, *J. Controlled Release*, 109(1–3), 256–274.
16. Azarmi, S., Huang, Y., Chen, H., McQuarrie, S., Abrams, D., Roa, W., Finlay, W. H., Miller, G. G., and Lobenberg, R. (2006), Optimization of a two-step desolvation method for preparing gelatin nanoparticles and cell uptake studies in 143B osteosarcoma cancer cells, *J. Pharm. Pharmacal. Sci.*, 9(1), 124–132.
17. Lee, C. H., Singla, A., and Lee, Y. (2001), Biomedical applications of collagen, *Int. J. Pharm.*, 221(1–2), 1–22.
18. Sihorkar, V., and Vyas, S. P. (2001), Potential of polysaccharide anchored liposomes in drug delivery, targeting and immunization, *J. Pharm. Pharmacal. Sci.*, 4(2), 138–158.
19. Jeong, Y. I., Na, H. S., Oh, J. S., Choi, K. C., Song, C. E., and Lee, H. C. (2006), Adriamycin release from self-assembling nanospheres of poly(DL-lactide-*co*-glycolide)-grafted pullulan, *Int. J. Pharm.*, 322(1–2), 154–160.
20. Leathers, T. D. (2005), in Steinbuchel, A. R., (Ed.), *Polysaccharides and Polyamides in the Food Industry. Properties, Production, and Patents*, Vol. 1, Wiley-VCH, Weinheim, pp. 387–421.
21. Garach, V. (1975), The synthesis of dextrans and levans—a review of the literature, *Diastema*, 4(3), 25–28.
22. Mehvar, R. (2000), Dextrans for targeted and sustained delivery of therapeutic and imaging agents, *J. Controlled Release*, 69(1), 1–25.
23. Tizard, I. R., Carpenter, R. H., McAnalley, B. H., and Kemp, M. C. (1989), The biological activities of mannans and related complex carbohydrates, *Mol. Biother.*, 1(6), 290–296.
24. Hasegawa, U., Nomura, S. M., Kaul, S. C., Hirano, T., and Akiyoshi, K. (2005), Nanogel-quantum dot hybrid nanoparticles for live cell imaging, *Biochem. Biophys. Res. Commun.*, 331(4), 917–921.
25. Gupta, M., and Gupta, A. K. (2004), Hydrogel pullulan nanoparticles encapsulating pBUDLacZ plasmid as an efficient gene delivery carrier, *J. Controlled Release*, 99(1), 157–166.
26. Na, K., Lee, E. S., and Bae, Y. H. (2003), Adriamycin loaded pullulan acetate/sulfonamide conjugate nanoparticles responding to tumor pH: pH-dependent cell interaction, internalization and cytotoxicity in vitro, *J. Controlled Release*, 87(1–3), 3–13.
27. Tiyaboonchai, W., Woiszwillo, J., and Middaugh, C. R. (2001), Formulation and characterization of amphotericin B-polyethylenimine-dextran sulfate nanoparticles, *J. Pharm. Sci.*, 90(7), 902–914.
28. Cui, Z., Hsu, C. H., and Mumper, R. J. (2003), Physical characterization and macrophage cell uptake of mannan-coated nanoparticles, *Drug Dev. Ind. Pharm.*, 29(6), 689–700.
29. Quinlan, G. J., Martin, G. S., and Evans, T. W. (2005), Albumin: Biochemical properties and therapeutic potential, *Hepatology*, 41(6), 1211–1219.
30. Verrecchia, T., Huve, P., Bazile, D., Veillard, M., Spenlehauer, G., and Couvreur, P. (1993), Adsorption/desorption of human serum albumin at the surface of poly(lactic acid) nanoparticles prepared by a solvent evaporation process, *J. Biomed. Mater. Res.*, 27(8), 1019–1028.
31. Bousquet, Y., Swart, P. J., Schmitt-Colin, N., Velge-Roussel, F., Kuipers, M. E., Meijer, D. K., Bru, N., Hoebeke, J., and Breton, P. (1999), Molecular mechanisms of the adsorption

of a model protein (human serum albumin) on poly(methylidene malonate 2.1.2) nanoparticles, *Pharm. Res.*, 16(1), 141–147.

32. Lin, W., Garnett, M. C., Davis, S. S., Schacht, E., Ferruti, P., and Illum, L. (2001), Preparation and characterisation of rose Bengal-loaded surface-modified albumin nanoparticles, *J. Controlled Release*, 71(1), 117–126.
33. Langer, K., Balthasar, S., Vogel, V., Dinauer, N., von Briesen, H., and Schubert, D. (2003), Optimization of the preparation process for human serum albumin (HSA) nanoparticles, *Int. J. Pharm.*, 257(1–2), 169–180.
34. Mamone, G., Addeo, F., Chianese, L., Di Luccia, A., De Martino, A., Nappo, A., Formisano, A., De Vivo, P., and Ferranti, P. (2005), Characterization of wheat gliadin proteins by combined two-dimensional gel electrophoresis and tandem mass spectrometry, *Proteomics*, 5(11), 2859–2865.
35. Ezpeleta, I., Irache, J. M., Stainmesse, S., Chabenat, C., Gueguen, J., Popineau, Y., and Orecchioni, A. (1996), Gliadin nanoparticles for the controlled release of all-trans-retinoic acid, *Int. J. Pharm.*, 131(2), 191–200.
36. Wieser, H. (1996), Relation between gliadin structure and coeliac toxicity, *Acta Paediatr. Suppl.*, 412, 3–9.
37. Lambert, G., Fattal, E., and Couvreur, P. (2001), Nanoparticulate systems for the delivery of antisense oligonucleotides, *Adv. Drug Deliv. Rev.*, 47(1), 99–112.
38. Bodmeier, R., Chen, H. G., and Paeratakul, O. (1989), A novel approach to the oral delivery of micro- or nanoparticles, *Pharm. Res.*, 6(5), 413–417.
39. Douglas, K. L., and Tabrizian, M. (2005), Effect of experimental parameters on the formation of alginate-chitosan nanoparticles and evaluation of their potential application as DNA carrier, *J. Biomater. Sci. Polym. Ed.*, 16(1), 43–56.
40. Borges, O., Borchard, G., Verhoef, J. C., de Sousa, A., and Junginger, H. E. (2005), Preparation of coated nanoparticles for a new mucosal vaccine delivery system, *Int. J. Pharm.*, 299(1–2), 155–166.
41. Pandey, R., Ahmad, Z., Sharma, S., and Khuller, G. K. (2005), Nano-encapsulation of azole antifungals: Potential applications to improve oral drug delivery, *Int. J. Pharm.*, 301(1–2), 268–276.
42. Zahoor, A., Sharma, S., and Khuller, G. K. (2005), Inhalable alginate nanoparticles as antitubercular drug carriers against experimental tuberculosis, *Int. J. Antimicrob. Agents*, 26(4), 298–303.
43. Yi, Y. M., Yang, T. Y., and Pan, W. M. (1999), Preparation and distribution of 5-fluorouracil (125)I sodium alginate-bovine serum albumin nanoparticles, *World J. Gastroenterol.* 5(1), 57–60.
44. Ahmad, Z., Pandey, R., Sharma, S., and Khuller, G. K. (2006), Pharmacokinetic and pharmacodynamic behaviour of antitubercular drugs encapsulated in alginate nanoparticles at two doses, *Int. J. Antimicrob. Agents*, 27(5), 409–416.
45. Borges, O., Cordeiro-da-Silva, A., Romeijn, S. G., Amidi, M., de Sousa, A., Borchard, G., and Junginger, H. E. (2006), Uptake studies in rat Peyer's patches, cytotoxicity and release studies of alginate coated chitosan nanoparticles for mucosal vaccination, *J. Controlled Release*, 114(3), 348–358.
46. Holtan, S., Zhang, Q., Strand, W. I., and Skjak-Braek, G. (2006), Characterization of the hydrolysis mechanism of polyalternating alginate in weak acid and assignment of the resulting MG-oligosaccharides by NMR spectroscopy and ESI-mass spectrometry, *Biomacromolecules*, 7(7), 2108–2121.
47. Chen, L., and Subirade, M. (2005), Chitosan/beta-lactoglobulin core-shell nanoparticles as nutraceutical carriers, *Biomaterials*, 26(30), 6041–6053.

48. Qi, L., Xu, Z., Jiang, X., Hu, C., and Zou, X. (2004), Preparation and antibacterial activity of chitosan nanoparticles, *Carbohydr. Res.*, 339(16), 2693–2700.

49. Liu, C. G., Desai, K. G., Chen, X. G., and Park, H. J. (2005), Preparation and characterization of nanoparticles containing trypsin based on hydrophobically modified chitosan, *J. Agric. Food Chem.*, 53(5), 1728–1733.

50. Grenha, A., Seijo, B., and Remunan-Lopez, C. (2005), Microencapsulated chitosan nanoparticles for lung protein delivery, *Eur. J. Pharm. Sci.*, 25(4–5), 427–437.

51. Enriquez de Salamanca, A., Diebold, Y., Calonge, M., Garcia-Vazquez, C., Callejo, S., Vila, A., and Alonso, M. J. (2006), Chitosan nanoparticles as a potential drug delivery system for the ocular surface: Toxicity, uptake mechanism and in vivo tolerance, *Invest. Ophthalmol. Vis. Sci.*, 47(4), 1416–1425.

52. Saito, H., Wu, X., Harris, M., and Hoffman, A. (2003), Graft copolymers of poly(ethylene glycol) (PEG) and chitosan, *Macromol. Rapid Commun.*, 18(7), 547–550.

53. Bernkop-Schnurch, A. (2000), Chitosan and its derivatives: Potential excipients for peroral peptide delivery systems, *Int. J. Pharm.*, 194(1), 1–13.

54. Hyung Park, J., Kwon, S., Lee, M., Chung, H., Kim, J. H., Kim, Y. S., Park, R. W., Kim, I. S., Bong Seo, S., Kwon, I. C., and Young Jeong, S. (2006), Self-assembled nanoparticles based on glycol chitosan bearing hydrophobic moieties as carriers for doxorubicin: In vivo biodistribution and anti-tumor activity, *Biomaterials*, 27(1), 119–126.

55. Mansouri, S., Cuie, Y., Winnik, F., Shi, Q., Lavigne, P., Benderdour, M., Beaumont, E., and Fernandes, J. C. (2006), Characterization of folate-chitosan-DNA nanoparticles for gene therapy, *Biomaterials*, 27(9), 2060–2065.

56. Yan, C., Chen, D., Gu, J., and Qin, J. (2006), Nanoparticles of 5-fluorouracil (5-FU) loaded *N*-succinyl-chitosan (Suc-Chi) for cancer chemotherapy: Preparation, characterization—in-vitro drug release and anti-tumour activity, *J. Pharm. Pharmacol.*, 58(9), 1177–1181.

57. Park, J. S., Han, T. H., Lee, K. Y., Han, S. S., Hwang, J. J., Moon, D. H., Kim, S. Y., and Cho, Y. W. (2006), *N*-acetyl histidine-conjugated glycol chitosan self-assembled nanoparticles for intracytoplasmic delivery of drugs: Endocytosis, exocytosis and drug release, *J. Controlled Release*, 115(1), 37–45.

58. Vandervoort, J., and Ludwig, A. (2004), Preparation and evaluation of drug-loaded gelatin nanoparticles for topical ophthalmic use, *Eur. J. Pharm. Biopharm.*, 57(2), 251–261.

59. Cascone, M. G., Lazzeri, L., Carmignani, C., and Zhu, Z. (2002), Gelatin nanoparticles produced by a simple W/O emulsion as delivery system for methotrexate, *J. Mater. Sci. Mater. Med.*, 13(5), 523–526.

60. Lu, Z., Yeh, T. K., Tsai, M., Au, J. L., and Wientjes, M. G. (2004), Paclitaxel-loaded gelatin nanoparticles for intravesical bladder cancer therapy, *Clin. Cancer Res.*, 10(22), 7677–7684.

61. Bajpai, A. K., and Choubey, J. (2006), Design of gelatin nanoparticles as swelling controlled delivery system for chloroquine phosphate, *J. Mater. Sci. Mater. Med.*, 17(4), 345–358.

62. Li, J. K., Wang, N., and Wu, X. S. (1997), A novel biodegradable system based on gelatin nanoparticles and poly(lactic-*co*-glycolic acid) microspheres for protein and peptide drug delivery, *J. Pharm. Sci.*, 86(8), 891–895.

63. Farrugia, C. A., and Groves, M. J. (1999), The activity of unloaded gelatin nanoparticles on murine melanoma B16-F0 growth in vivo, *Anticancer Res.*, 19(2A), 1027–1031.

64. Coester, C., Kreuter, J., von Briesen, H., and Langer, K. (2000), Preparation of avidin-labelled gelatin nanoparticles as carriers for biotinylated peptide nucleic acid (PNA), *Int. J. Pharm.*, 196(2), 147–149.

65. Kaul, G., and Amiji, M. (2002), Long-circulating poly(ethylene glycol)-modified gelatin nanoparticles for intracellular delivery, *Pharm. Res.*, 19(7), 1061–1067.
66. Kommareddy, S., and Amiji, M. (2005), Preparation and evaluation of thiol-modified gelatin nanoparticles for intracellular DNA delivery in response to glutathione, *Bioconjug. Chem.*, 16(6), 1423–1432.
67. Coester, C. J., Langer, K., van Briesen, H., and Kreuter, J. (2000), Gelatin nanoparticles by two step desolvation—a new preparation method, surface modifications and cell uptake, *J. Microencapsul.*, 17(2), 187–193.
68. Kaul, G., and Amiji, M. (2004), Biodistribution and targeting potential of poly(ethylene glycol)-modified gelatin nanoparticles in subcutaneous murine tumor model, *J. Drug Target.*, 12(9–10), 585–591.
69. Coester, C., Nayyar, P., and Samuel, J. (2006), In vitro uptake of gelatin nanoparticles by murine dendritic cells and their intracellular localisation, *Eur. J. Pharm. Biopharm.*, 62(3), 306–314.
70. Leo, E., Arletti, R., Forni, F., and Cameroni, R. (1997), General and cardiac toxicity of doxorubicin-loaded gelatin nanoparticles, *Farmaco*, 52(6–7), 385–388.
71. Gu, X. G., Schmitt, M., Hiasa, A., Nagata, Y., Ikeda, H., Sasaki, Y., Akiyoshi, K., Sunamoto, J., Nakamura, H., Kuribayashi, K., and Shiku, H. (1998), A novel hydrophobized polysaccharide/oncoprotein complex vaccine induces in vitro and in vivo cellular and humoral immune responses against HER2-expressing murine sarcomas, *Cancer Res.*, 58(15), 3385–3390.
72. Shiku, H., Wang, L., Ikuta, Y., Okugawa, T., Schmitt, M., Gu, X., Akiyoshi, K., Sunamoto, J., and Nakamura, H. (2000), Development of a cancer vaccine: Peptides, proteins, and DNA, *Cancer Chemother. Pharmacol.*, 46 (Suppl), S77–82.
73. Na, K., and Bae, Y. H. (2002), Self-assembled hydrogel nanoparticles responsive to tumor extracellular pH from pullulan derivative/sulfonamide conjugate: Characterization, aggregation, and adriamycin release in vitro, *Pharm. Res.*, 19(5), 681–688.
74. Gupta, A. K., and Gupta, M. (2005), Cytotoxicity suppression and cellular uptake enhancement of surface modified magnetic nanoparticles, *Biomaterials*, 26(13), 1565–1573.
75. Tiyaboonchai, W., Woiszwillo, J., Sims, R. C., and Middaugh, C. R. (2003), Insulin containing polyethylenimine-dextran sulfate nanoparticles, *Int. J. Pharm.*, 255(1–2), 139–151.
76. Cascone, M. G., Pot, P. M., Lazzeri, L., and Zhu, Z. (2002), Release of dexamethasone from PLGA nanoparticles entrapped into dextran/poly(vinyl alcohol) hydrogels, *J. Mater. Sci. Mater. Med.*, 13(3), 265–269.
77. Jaulin, N., Appel, M., Passirani, C., Barratt, G., and Labarre, D. (2000), Reduction of the uptake by a macrophagic cell line of nanoparticles bearing heparin or dextran covalently bound to poly(methyl methacrylate), *J. Drug Target.*, 8(3), 165–172.
78. Segura, S., Gamazo, C., Irache, J. M., and Espuelas, S. (2007), Interferon-γ loaded onto albumin nanoparticles: In vitro and in vivo activity against brucella abortus, *Antimicrob. Agents Chemother.*, 51(4), 1310–1314.
79. Irache, J. M., Merodio, M., Arnedo, A., Camapanero, M. A., Mirshahi, M., and Espuelas, S. (2005), Albumin nanoparticles for the intravitreal delivery of anticytomegaloviral drugs, *Mini. Rev. Med. Chem.*, 5(3), 293–305.
80. Landry, F. B., Bazile, D. V., Spenlehauer, G., Veillard, M., and Kreuter, J. (1998), Peroral administration of 14C-poly(D,L-lactic acid) nanoparticles coated with human serum albumin or polyvinyl alcohol to guinea pigs, *J. Drug Target.*, 6(4), 293–307.
81. Damascelli, B., Patelli, G. L., Lanocita, R., Di Tolla, G., Frigerio, L. F., Marchiano, A., Garbagnati, F., Spreafico, C., Ticha, V., Gladin, C. R., Palazzi, M., Crippa, F., Oldini, C.,

Calo, S., Bonaccorsi, A., Mattavelli, F., Costa, L., Mariani, L., and Cantu, G. (2003), A novel intraarterial chemotherapy using paclitaxel in albumin nanoparticles to treat advanced squamous cell carcinoma of the tongue: Preliminary findings, *Am. J. Roentgenol.*, 181(1), 253–260.

82. Rhaese, S., von Briesen, H., Rubsamen-Waigmann, H., Kreuter, J., and Langer, K. (2003), Human serum albumin-polyethylenimine nanoparticles for gene delivery, *J. Controlled Release*, 92(1–2), 199–208.
83. Zhang, L. K., Hou, S. X., Mao, S. J., Wei, D. P., Song, X. R., and Qiao, X. R. (2004), [Study on the tumor cell targetability of folate-conjugated albumin nanoparticles] *Sichuan Da Xue Xue Bao Yi Xue Ban*, 35(2), 165–168.
84. Zhang, L., Hou, S., Mao, S., Wei, D., Song, X., and Lu, Y. (2004), Uptake of folate-conjugated albumin nanoparticles to the SKOV3 cells, *Int. J. Pharm.*, 287(1–2), 155–162.
85. Mishra, V., Mahor, S., Rawat, A., Gupta, P. N., Dubey, P., Khatri, K., and Vyas, S. P. (2006), Targeted brain delivery of AZT via transferrin anchored pegylated albumin nanoparticles, *J. Drug Target.*, 14(1), 45–53.
86. Mao, S. J., Hou, S. X., Zhang, L. K., Jin, H., Bi, Y. Q., and Jiang, B. (2003), [Preparation of bovine serum albumin nanoparticles surface-modified with glycyrrhizin], *Yao Xue Xue Bao*, 38(10), 787–790.
87. Bazile, D. V., Ropert, C., Huve, P., Verrecchia, T., Marlard, M., Frydman, A., Veillard, M., and Spenlehauer, G. (1992), Body distribution of fully biodegradable [^{14}C]-poly(lactic acid) nanoparticles coated with albumin after parenteral administration to rats, *Biomaterials*, 13(15), 1093–1102.
88. Arangoa, M. A., Campanero, M. A., Renedo, M. J., Ponchel, G., and Irache, J. M. (2001), Gliadin nanoparticles as carriers for the oral administration of lipophilic drugs. Relationships between bioadhesion and pharmacokinetics, *Pharm. Res.*, 18(11), 1521–1527.
89. Duclairoir, C., Orecchioni, A. M., Depraetere, P., Osterstock, F., and Nakache, E. (2003), Evaluation of gliadins nanoparticles as drug delivery systems: A study of three different drugs, *Int. J. Pharm.*, 253(1–2), 133–144.
90. Umamaheshwari, R. B., and Jain, N. K. (2003), Receptor mediated targeting of lectin conjugated gliadin nanoparticles in the treatment of *Helicobacter pylori*, *J. Drug Target.*, 11(7), 415–423; discussion 423–424.
91. Jain, R. A. (2000), The manufacturing techniques of various drug loaded biodegradable poly(lactide-*co*-glycolide) (PLGA) devices, *Biomaterials*, 21(23), 2475–2490.
92. Bala, I., Hariharan, S., and Kumar, M. N. (2004), PLGA nanoparticles in drug delivery: The state of the art, *Crit. Rev. Ther. Drug Carrier Syst.*, 21(5), 387–422.
93. Astete, C. E., and Sabliov, C. M. (2006), Synthesis and characterization of PLGA nanoparticles, *J. Biomater. Sci. Polym. Ed.*, 17(3), 247–289.
94. Sinha, V. R., Bansal, K., Kaushik, R., Kumria, R., and Trehan, A. (2004), Poly-epsilon-caprolactone microspheres and nanospheres: An overview, *Int. J. Pharm.*, 278(1), 1–23.
95. Seppala, J. V., Helminen, A. O., and Korhonen, H. (2004), Degradable polyesters through chain linking for packaging and biomedical applications, *Macromol. Biosci.*, 4(3), 208–217.
96. Kumar, N., Langer, R. S., and Domb, A. J. (2002), Polyanhydrides: An overview, *Adv. Drug Deliv. Rev.*, 54(7), 889–910.
97. Jain, J. P., Modi, S., Domb, A. J., and Kumar, N. (2005), Role of polyanhydrides as localized drug carriers, *J. Controlled Release.*, 103(3), 541–563.
98. Pfeifer, B. A., Burdick, J. A., and Langer, R. (2005), Formulation and surface modification of poly(ester-anhydride) micro- and nanospheres, *Biomaterials*, 26(2), 117–124.

99. Carino, G. P., Jacob, J. S., and Mathiowitz, E. (2000), Nanosphere based oral insulin delivery, *J. Controlled Release*, 65(1–2), 261–269.

100. Torres, M. P., Determan, A. S., and Anderson, G. L., Mallapragada, S. K., and Narasimhan, B. (2007), Amphiphilic polyanhydrides for protein stabilization and release, *Biomaterials*, 28(1), 108–116.

101. Gao, J., Niklason, L., Zhao, X. M., and Langer, R. (1998), Surface modification of polyanhydride microspheres, *J. Pharm. Sci.*, 87(2), 246–248.

102. Vauthier, C., Dubernet, C., Fattal, E., Pinto-Alphandary, H., and Couvreur, P. (2003), Poly (alkylcyanoacrylates) as biodegradable materials for biomedical applications, *Adv. Drug Deliv. Rev.*, 55(4), 519–548.

103. Mehnert, W., and Mader, K. (2001), Solid lipid nanoparticles: Production, characterization and applications, *Adv. Drug Deliv. Rev.*, 47(2–3), 165–196.

104. Uner, M. (2006), Preparation, characterization and physico-chemical properties of solid lipid nanoparticles (SLN) and nanostructured lipid carriers (NLC): Their benefits as colloidal drug carrier systems, *Pharmazie*, 61(5), 375–386.

105. Heller, J., Barr, J., Ng, S. Y., Abdellauoi, K. S., and Gurny, R. (2002), Poly(ortho esters): Synthesis, characterization, properties and uses, *Adv. Drug Deliv. Rev.*, 54(7), 1015–1039.

106. Barichello, J. M., Morishita, M., Takayama, K., and Nagai, T. (1999), Encapsulation of hydrophilic and lipophilic drugs in PLGA nanoparticles by the nanoprecipitation method, *Drug Dev. Ind. Pharm.*, 25(4), 471–476.

107. Panyam, J., and Labhasetwar, V. (2003), Biodegradable nanoparticles for drug and gene delivery to cells and tissue, *Adv. Drug Deliv. Rev.*, 55(3), 329–347.

108. Avgoustakis, K. (2004), Pegylated poly(lactide) and poly(lactide-*co*-glycolide) nanoparticles: Preparation, properties and possible applications in drug delivery, *Curr. Drug Deliv.*, 1(4), 321–333.

109. Panagi, Z., Beletsi, A., Evangelatos, G., Livaniou, E., Ithakissios, D. S., and Avgoustakis, K. (2001), Effect of dose on the biodistribution and pharmacokinetics of PLGA and PLGA-mPEG nanoparticles, *Int. J. Pharm.*, 221(1–2), 143–152.

110. Qaddoumi, M. G., Gukasyan, H. J., Davda, J., Labhasetwar, V., Kim, K. J., and Lee, V. H. (2003), Clathrin and caveolin-1 expression in primary pigmented rabbit conjunctival epithelial cells: Role in PLGA nanoparticle endocytosis, *Mol. Vis.*, 9, 559–568.

111. Panyam, J., Zhou, W. Z., Prabha, S., Sahoo, S. K., and Labhasetwar, V. (2002), Rapid endo-lysosomal escape of poly(DL-lactide-*co*-glycolide) nanoparticles: Implications for drug and gene delivery, *FASEB J.*, 16(10), 1217–1226.

112. Shive, M. S., and Anderson, J. M. (1997), Biodegradation and biocompatibility of PLA and PLGA microspheres, *Adv. Drug Deliv. Rev.*, 28(1), 5–24.

113. Marchal-Heussler, L., Sirbat, D., Hoffman, M., and Maincent, P. (1993), Poly(epsilon-caprolactone) nanocapsules in carteolol ophthalmic delivery, *Pharm. Res.*, 10(3), 386–390.

114. Calvo, P., Vila-Jato, J. L., and Alonso, M. J. (1996), Comparative in vitro evaluation of several colloidal systems, nanoparticles, nanocapsules, and nanoemulsions, as ocular drug carriers, *J. Pharm. Sci.*, 85(5), 530–536.

115. Jeong, Y. I., Kang, M. K., Sun, H. S., Kang, S. S., Kim, H. W., Moon, K. S., Lee, K. J., Kim, S. H., and Jung, S. (2004), All-trans-retinoic acid release from core-shell type nanoparticles of poly(epsilon-caprolactone)/poly(ethylene glycol) diblock copolymer, *Int. J. Pharm.*, 273(1–2), 95–107.

116. Park, E. K., Lee, S. B., and Lee, Y. M. (2005), Preparation and characterization of methoxy poly(ethylene glycol)/poly(epsilon-caprolactone) amphiphilic block

copolymeric nanospheres for tumor-specific folate-mediated targeting of anticancer drugs, *Biomaterials*, 26(9), 1053–1061.

117. Cade, D., Ramus, E., Rinaudo, M., Auzely-Velty, R., Delair, T., and Hamaide, T. (2004), Tailoring of bioresorbable polymers for elaboration of sugar-functionalized nanoparticles, *Biomacromolecules*, 5(3), 922–927.
118. Haas, J., Ravi Kumar, M. N., Borchard, G., Bakowsky, U., and Lehr, C. M. (2005), Preparation and characterization of chitosan and trimethyl-chitosan-modified poly-(epsilon-caprolactone) nanoparticles as DNA carriers, *AAPS Pharm. Sci. Tech.*, 6(1), E22–30.
119. Lemarchand, C., Gref, R., Passirani, C., Garcion, E., Petri, B., Muller, R., Costantini, D., and Couvreur, P. (2006), Influence of polysaccharide coating on the interactions of nanoparticles with biological systems, *Biomaterials*, 27(1), 108–118.
120. Pfeifer, B. A., Burdick, J. A., Little, S. R., and Langer, R. (2005), Poly(ester-anhydride): poly(beta-amino ester) micro- and nanospheres: DNA encapsulation and cellular transfection, *Int. J. Pharm.*, 304(1–2), 210–219.
121. McCarron, P. A., Donnelly, R. F., Canning, P. E., McGovern, J. G., and Jones, D. S. (2004), Bioadhesive, non-drug-loaded nanoparticles as modulators of candidal adherence to buccal epithelial cells: A potentially novel prophylaxis for candidosis, *Biomaterials*, 25(12), 2399–2407.
122. Vauthier, C., Dubernet, C., Chauvierre, C., Brigger, I., and Couvreur, P. (2003), Drug delivery to resistant tumors: The potential of poly(alkyl cyanoacrylate) nanoparticles, *J. Controlled Release*, 93(2), 151–160.
123. Toub, N., Angiari, C., Eboue, D., Fattal, E., Tenu, J. P., Le Doan, T., and Couvreur, P. (2005), Cellular fate of oligonucleotides when delivered by nanocapsules of poly(isobu tylcyanoacrylate), *J. Controlled Release*, 106(1–2), 209–213.
124. Labarre, D., Vauthier, C., Chauvierre, C., Petri, B., Muller, R., and Chehimi, M. M. (2005), Interactions of blood proteins with poly(isobutylcyanoacrylate) nanoparticles decorated with a polysaccharidic brush, *Biomaterials*, 26(24), 5075–5084.
125. Wissing, S. A., Kayser, O., and Muller, R. H. (2004), Solid lipid nanoparticles for parenteral drug delivery, *Adv. Drug Deliv. Rev.*, 56(9), 1257–1272.
126. Manjunath, K., Reddy, J. S., and Venkateswarlu, V. (2005), Solid lipid nanoparticles as drug delivery systems, *Methods Find. Exp. Clin. Pharmacol.*, 27(2), 127–144.
127. Zhang, N., Ping, Q., Huang, G., Xu, W., Cheng, Y., and Han, X. (2006), Lectin-modified solid lipid nanoparticles as carriers for oral administration of insulin, *Int. J. Pharm.*, 327(1–2), 153–159.
128. Wang, Y., and Wu, W. (2006), In situ evading of phagocytic uptake of stealth solid lipid nanoparticles by mouse peritoneal macrophages, *Drug Deliv.*, 13(3), 189–192.
129. Chilkoti, A., Dreher, M. R., Meyer, D. E., and Raucher, D. (2002), Targeted drug delivery by thermally responsive polymers, *Adv. Drug Deliv. Rev.*, 54(5), 613–630.
130. Sakuma, S., Suzuki, N., Sudo, R., Hiwatari, K., Kishida, A., and Akashi, M. (2002), Optimized chemical structure of nanoparticles as carriers for oral delivery of salmon calcitonin, *Int. J. Pharm.*, 239(1–2), 185–195.
131. Piskin, E. (2004), Molecularly designed water soluble, intelligent, nanosize polymeric carriers, *Int. J. Pharm.*, 277(1–2), 105–118.
132. Neradovic, D., Soga, O., Van Nostrum, C. F., and Hennink, W. E. (2004), The effect of the processing and formulation parameters on the size of nanoparticles based on block copolymers of poly(ethylene glycol) and poly(*N*-isopropylacrylamide) with and without hydrolytically sensitive groups, *Biomaterials*, 25(12), 2409–2418.

133. Na, K., Lee, K. H., and Bae, Y. H. (2004), pH-sensitivity and pH-dependent interior structural change of self-assembled hydrogel nanoparticles of pullulan acetate/oligo-sulfonamide conjugate, *J. Controlled Release*, 97(3), 513–525.
134. Ma, Z., Yeoh, H. H., and Lim, L. Y. (2002), Formulation pH modulates the interaction of insulin with chitosan nanoparticles, *J. Pharm. Sci.*, 91(6), 1396–1404.
135. Lynn, D. M., Amiji, M. M., and Langer, R. (2001), pH-responsive polymer microspheres: Rapid release of encapsulated material within the range of intracellular pH, *Angew Chem. Int. Ed. Engl.*, 40(9), 1707–1710.
136. Potineni, A., Lynn, D. M., Langer, R., and Amiji, M. M. (2003), Poly(ethylene oxide)-modified poly(beta-amino ester) nanoparticles as a pH-sensitive biodegradable system for paclitaxel delivery, *J. Controlled Release*, 86(2–3), 223–234.
137. Heffernan, M. J., and Murthy, N. (2005), Polyketal nanoparticles: A new pH-sensitive biodegradable drug delivery vehicle, *Bioconjug. Chem.*, 16(6), 1340–1342.
138. Choi, J. S., MacKay, J. A., and Szoka, F. C. Jr. (2003), Low-pH-sensitive PEG-stabilized plasmid-lipid nanoparticles: Preparation and characterization, *Bioconjug. Chem.*, 14(2), 420–429.
139. Na, K., Lee, K. H., Lee, D. H., and Bae, Y. H. (2006), Biodegradable thermo-sensitive nanoparticles from poly(L-lactic acid)/poly(ethylene glycol) alternating multi-block copolymer for potential anti-cancer drug carrier, *Eur. J. Pharm. Sci.*, 27(2–3), 115–122.
140. Sun, S., Liu, W., Cheng, N., Zhang, B., Cao, Z., Yao, K., Liang, D., Zuo, A., Guo, G., and Zhang, J. (2005), A thermoresponsive chitosan-NIPAAm/vinyl laurate copolymer vector for gene transfection, *Bioconjug. Chem.*, 16(4), 972–980.
141. Weng, H., Zhou, J., Tang, L., and Hu, Z. (2004), Tissue responses to thermally-responsive hydrogel nanoparticles, *J. Biomater. Sci. Polym. Ed.*, 15(9), 1167–1180.
142. Devalapally, H., Shenoy, D., Little, S., Langer, R., and Amiji, M. (2007), Poly(ethylene oxide)-modified poly(beta-amino ester) nanoparticles as a pH-sensitive system for tumor-targeted delivery of hydrophobic drugs: Part 3. Therapeutic efficacy and safety studies in ovarian cancer xenograft model, *Cancer Chemother. Pharmacol.*, 59(4), 477–484.
143. Brigger, I., Dubernet, C., and Couvreur, P. (2002), Nanoparticles in cancer therapy and diagnosis, *Adv. Drug Deliv. Rev.*, 54(5), 631–651.
144. Mulder, W. J., Strijkers, G. J., van Tilborg, G. A., Griffioen, A. W., and Nicolay, K. (2006), Lipid-based nanoparticles for contrast-enhanced MRI and molecular imaging, *NMR Biomed.*, 19(1), 142–164.
145. De, S., Miller, D. W., and Robinson, D. H. (2005), Effect of particle size of nanospheres and microspheres on the cellular-association and cytotoxicity of paclitaxel in 4T1 cells, *Pharm. Res.*, 22(5), 766–775.
146. De, S., and Robinson, D. H. (2004), Particle size and temperature effect on the physical stability of PLGA nanospheres and microspheres containing Bodipy, *AAPS Pharm. Sci. Tech.*, 5(4), e53.
147. Alonso, M. J. (1996), Nanoparticulate drug carrier technology, Cohen, S., and Bennstein, H. Eds. *Microparticulate Systems for the Delivery of Proteins and Vaccines*, Marcel Dekkes, New York, pp. 203–242.
148. Torres, M. P., Vogel, B. M., Narasimhan, B., and Mallapragada, S. K. (2006), Synthesis and characterization of novel polyanhydrides with tailored erosion mechanisms, *J. Biomed. Mater. Res. A.*, 76(1), 102–110.
149. Shen, E., Pizsczek, R., Dziadul, B., and Narasimhan, B. (2001), Microphase separation in bioerodible copolymers for drug delivery, *Biomaterials*, 22(3), 201–210.

150. DiBenedetto, A. K. (2003), Prediction of the glass transition temperature of polymers: A model based on the principle of corresponding states, *J. Polym. Sci. Part B: Polym. Phys.*, 25(9), 1949–1969.

151. Dong, Y., and Feng, S. S. (2006), Nanoparticles of poly(D,L-lactide)/methoxy poly(ethylene glycol)-poly(D,L-lactide) blends for controlled release of paclitaxel, *J. Biomed. Mater. Res. A*, 78(1), 12–19.

152. Zweers, M. L., Grijpma, D. W., Engbers, G. H., and Feijen, J. (2003), The preparation of monodisperse biodegradable polyester nanoparticles with a controlled size, *J. Biomed. Mater. Res. B Appl. Biomater.*, 66(2), 559–566.

153. Kirby, B. J., and Hasselbrink, E. F. Jr. (2004), Zeta potential of microfluidic substrates: 2. Data for polymers, *Electrophoresis*, 25(2), 203–213.

154. Gessner, A., Lieske, A., Paulke, B., and Muller, R. (2002), Influence of surface charge density on protein adsorption on polymeric nanoparticles: Analysis by two-dimensional electrophoresis, *Eur. J. Pharm. Biopharm.*, 54(2), 165–170.

155. Kumar, M. N., Mohapatra, S. S., Kong, X., Jena, P. K., Bakowsky, U., and Lehr, C. M. (2004), Cationic poly(lactide-*co*-glycolide) nanoparticles as efficient in vivo gene transfection agents, *J. Nanosci. Nanotechnol.*, 4(8), 990–994.

156. Reddy, L. H., Sharma, R. K., Chuttani, K., Mishra, A. K., and Murthy, R. R. (2004), Etoposide-incorporated tripalmitin nanoparticles with different surface charge: Formulation, characterization, radiolabeling, and biodistribution studies, *AAPS J.*, 6(3), e23.

157. Owens, D. E., 3rd, and Peppas, N. A. (2006), Opsonization, biodistribution, and pharmacokinetics of polymeric nanoparticles, *Int. J. Pharm.*, 307(1), 93–102.

158. Moghimi, S. M., and Szebeni, J. (2003), Stealth liposomes and long circulating nanoparticles: Critical issues in pharmacokinetics, opsonization and protein-binding properties, *Prog. Lipid Res.*, 42(6), 463–478.

159. Freiberg, S., and Zhu, X. X. (2004), Polymer microspheres for controlled drug release, *Int. J. Pharm.*, 282(1–2), 1–18.

160. Siepmann, J., and Gopferich, A. (2001), Mathematical modeling of bioerodible, polymeric drug delivery systems, *Adv. Drug Deliv. Rev.*, 48(2–3), 229–247.

161. Grassi, M., and Grassi, G. (2005), Mathematical modelling and controlled drug delivery: Matrix systems, *Curr. Drug Deliv.*, 2(1), 97–116.

162. Kanjickal, D. G., and Lopina, S. T. (2004), Modeling of drug release from polymeric delivery systems—a review, *Crit. Rev. Ther. Drug Carrier Syst.*, 21(5), 345–386.

163. Fu, K., Harrell, R., Zinski, K., Um, C., Jaklenec, A., Frazier, J., Lotan, N., Burke, P., Klibanov, A. M., and Langer, R. (2003), A potential approach for decreasing the burst effect of protein from PLGA microspheres, *J. Pharm. Sci.*, 92(8), 1582–1591.

164. Dhoot, N. O., and Wheatley, M. A. (2003), Microencapsulated liposomes in controlled drug delivery: Strategies to modulate drug release and eliminate the burst effect, *J. Pharm. Sci.*, 92(3), 679–689.

165. Lemoine, D., Francois, C., Kedzierewicz, F., Preat, V., Hoffman, M., and Maincent, P. (1996), Stability study of nanoparticles of poly(epsilon-caprolactone), poly(D,L-lactide) and poly(D,L-lactide-*co*-glycolide), *Biomaterials*, 17(22), 2191–2197.

166. Molpeceres, J., Aberturas, M. R., Chacon, M., Berges, L., and Guzman, M. (1997), Stability of cyclosporine-loaded poly-sigma-caprolactone nanoparticles, *J. Microencapsul.*, 14(6), 777–787.

167. Freitas, C., and Muller, R. H. (1999), Correlation between long-term stability of solid lipid nanoparticles (SLN) and crystallinity of the lipid phase, *Eur. J. Pharm. Biopharm.*, 47(2), 125–132.

168. Freitas, C., and Muller, R. H. (1999), Stability determination of solid lipid nanoparticles (SLN) in aqueous dispersion after addition of electrolyte, *J. Microencapsul.*, 16(1), 59–71.

169. Santander-Ortega, M. J., Jodar-Reyes, A. B., Csaba, N., Bastos-Gonzalez, D., and Ortega-Vinuesa, J. L. (2006), Colloidal stability of Pluronic F68-coated PLGA nanoparticles: A variety of stabilisation mechanisms, *J. Colloid Interface Sci.*, 302(2), 522–529.

170. Zobel, H. P., Werner, D., Gilbert, M., Noe, C. R., Stieneker, F., Kreuter, J., and Zimmer, A. (1999), Effect of ultrasonication on the stability of oligonucleotides adsorbed on nanoparticles and liposomes, *J. Microencapsul.*, 16(4), 501–509.

171. Hattori, Y., Sakaguchi, M., and Maitani, Y. (2006), Folate-linked lipid-based nanoparticles deliver a NFkappaB decoy into activated murine macrophage-like RAW264.7 cells, *Biol. Pharm. Bull.*, 29(7), 1516–1520.

172. Sahoo, S. K., Ma, W., and Labhasetwar, V. (2004), Efficacy of transferrin-conjugated paclitaxel-loaded nanoparticles in a murine model of prostate cancer, *Int. J. Cancer*, 112(2), 335–340.

173. Gao, X., Tao, W., Lu, W., Zhang, Q., Zhang, Y., Jiang, X., and Fu, S. (2006), Lectin-conjugated PEG-PLA nanoparticles: Preparation and brain delivery after intranasal administration, *Biomaterials*, 27(18), 3482–3490.

174. Sokolov, K., Follen, M., Aaron, J., Pavlova, I., Malpica, A., Lotan, R., and Richards-Kortum, R. (2003), Real-time vital optical imaging of precancer using anti-epidermal growth factor receptor antibodies conjugated to gold nanoparticles, *Cancer Res.*, 63(9), 1999–2004.

175. Garnett, M. C. (2001), Targeted drug conjugates: Principles and progress, *Adv. Drug Deliv. Rev.*, 53(2), 171–216.

176. Nobs, L., Buchegger, F., Gurny, R., and Allemann, E. (2004), Current methods for attaching targeting ligands to liposomes and nanoparticles, *J. Pharm. Sci.*, 93(8), 1980–1992.

177. Kreuter, J. (1994), Drug targeting with nanoparticles, *Eur. J. Drug Metab. Pharmacokinet.*, 19(3), 253–256.

178. Jain, K. K. (2006), Nanoparticles as targeting ligands, *Trends Biotechnol.*, 24(4), 143–145.

179. Conner, S. D., and Schmid, S. L. (2003), Regulated portals of entry into the cell, *Nature*, 422(6927), 37–44.

180. Huang, M., Ma, Z., Khor, E., and Lim, L. Y. (2002), Uptake of FITC-chitosan nanoparticles by A549 cells, *Pharm. Res.*, 19(10), 1488–1494.

181. Ma, Z., and Lim, L. Y. (2003), Uptake of chitosan and associated insulin in Caco-2 cell monolayers: A comparison between chitosan molecules and chitosan nanoparticles, *Pharm. Res.*, 20(11), 1812–1819.

182. Liu, J., and Shapiro, J. I. (2003), Endocytosis and signal transduction: Basic science update, *Biol. Res. Nurs.*, 5(2), 117–128.

183. Steinman, R. M., Mellman, I. S., Muller, W. A., and Cohn, Z. A. (1983), Endocytosis and the recycling of plasma membrane, *J. Cell Biol.*, 96(1), 1–27.

184. Okamoto, C. T. (1998), Endocytosis and transcytosis, *Adv. Drug Deliv. Rev.*, 29(3), 215–228.

185. Hilgenbrink, A. R., and Low, P. S. (2005), Folate receptor-mediated drug targeting: From therapeutics to diagnostics, *J. Pharm. Sci.*, 94(10), 2135–2146.

186. Park, I. K., Seo, S. J., Akashi, M., Akaike, T., and Cho, C. S. (2003), Controlled release of epidermal growth factor (EGF) from EGF-loaded polymeric nanoparticles composed

of polystyrene as core and poly(methacrylic acid) as corona in vitro, *Arch. Pharm. Res.*, 26(8), 649–652.

187. Douglas, S. J., Davis, S. S., and Illum, L. (1987), Nanoparticles in drug delivery, *Crit. Rev. Ther. Drug Carrier Syst.*, 3(3), 233–261.

5.5

RECOMBINANT *SACCHAROMYCES CEREVISIAE* AS NEW DRUG DELIVERY SYSTEM TO GUT: IN VITRO VALIDATION AND ORAL FORMULATION

STÉPHANIE BLANQUET AND MONIQUE ALRIC
Université d'Auvergne, Clermont-Ferrand, France

Contents

5.5.1 ENGINEREED MICROORGANISMS AS DELIVERY VECTORS TO HUMAN GASTROINTESTINAL TRACT

5.5.1.1 What Is the "Biodrug" Concept?

The development of recombinant deoxyribonucleic acid (DNA) technology has allowed the emergence of novel applications such as drug production directly in the

Pharmaceutical Manufacturing Handbook: Production and Processes, edited by Shayne Cox Gad

human digestive environment ("in situ") by ingested living recombinant microorganisms [1–3].

This new kind of vector offers several advantages over classical dosage forms. First, the microorganisms, by protecting the active compounds, can allow the administration of drugs known to be sensitive to digestive conditions when given in classical pharmaceutical formulations. Second, the regulation of gene expression (e.g., using an inducible promoter) makes it possible to target specific sites throughout the digestive tract (i.e., the absorption or reaction site of the drug) and to control drug release. Thus, similar therapeutic effects can potentially be obtained at lower doses [4, 5] and the degradation of the active compound by acid or proteases should be avoided upstream from its absorption or reaction site.

In the digestive tract, genetically modified microorganisms can either carry out a reaction of bioconversion or produce compounds of interest. The bioconversion reaction can lead either to the production of an active product or to the removal of undesirable compounds. The active compound produced in situ can be secreted in the digestive medium [5], be bound to the cells [6, 7], or accumulate inside the cells and be released in the digestive environment by cell lysis [8].

5.5.1.2 Medical Applications

The biodrug concept involves the use of orally administered recombinant microorganisms as a new drug delivery route to prevent or treat various diseases. The potential medical applications are numerous and can be classified in terms of bioconversion or production of active compounds. Validation studies have yet been conducted in animals (and even sometimes in human being), as described below and summarized in Table 1.

Bioconversion Three main types of medical applications have been considered. First, recombinant microorganisms could be administered to perform "biodetoxication" in the gut [9]. Here, the objective is to increase the body's protection against environmental xeniobiotics, mainly those ingested with food (e.g., pesticides, procarcinogens, or chemical additives), by ingesting microorganisms expressing enzymes that play a major role in the human detoxication system [e.g., phase I xenobiotic metabolizing enzymes such as cytochrome P450 or phase II–like glutathione *S*-transferase (GST)]. Therefore, recombinant microorganisms could be used to prevent multifactorial diseases that have been associated with anomalies in human detoxication processes. For instance, a deficiency in GST-M1 has been correlated with an increased susceptibility to different cancers, endometriosis, and chronic bronchitis [10].

Second, modified microorganisms could correct errors of metabolism resulting from either gastric or intestinal enzyme deficiencies (e.g., lipase or lactase) [11] or organ failure (by removing urea in the case of kidney failure or ammonia in the case of liver failure) [12, 13]. This could constitute an alternative to current therapy such as renal dialysis, which is time consuming and uncomfortable for the patient.

The third potential application is the use of recombinant cells to control the activation of prodrug into drug directly in the digestive tract. This is of interest when the drug, but not the prodrug, is either toxic at high concentrations or damaged by digestive secretions [1].

TABLE 1 Potential Medical Applications of Biodrug Concept and Their Validation in Animals

Biodrug Concept	Applications	Examples	Validation: Recombinant Microorganisms (Heterologous Gene)	Validation: Experimental Models	Validation: Effect
Bioconversion	Biodetoxication	Removal of benzo(*a*)pyrene [9]	*Escherichia coli* (P450 1A1 and glutathione *S*-transferase)	Stomach and duodenum of a mouse combined with mutagenesis assay (Ames assay)	Decrease of the mutagenic potential of B(a)P
	Correction of errors of metabolism	Correction of lipase deficiency [11]	*Lactococcus lactis* (lipase)	Pig with pancreatic deficiency, oral administration	Increase of lipid digestibility
		Correction of urease deficiency [13]	*Escherichia coli* DH5 (urease)	Rat with renal failure, oral administration	Decrease of plasma uric acid
In situ production of active compounds	Synthesis of biological mediators	Secretion of Interleukine [5]	*Lactococcus lactis* (IL-10)	Mouse treated with dextran sulfate sodium, oral administration	Reduction of colitis symptoms
	Oral vaccines	Vaccination against bacteria [17]	*Lactobacillus plantarum* (*Helicobacter pylori* urease B subunit)	Mouse, intragastric immunization	Induction of immune response (anti-UreB Ig), partial protection against *H. felis*
		Vaccination against virus [19]	Attenuated *Salmonella typhimurium* (SIV capsid antigen p27)	Macaques, oral administration	Induction of immune response (anti-p27 Ig)
		Control of food allergy [20]	*Lactococcus lactis* (bovine β-lactoglobulin)	Model of mouse allergy, oral administration	Induction of BLG-specific Th1 response

In Situ Production of Active Compounds The first (and main) medical application derived from the biodrug concept is the development of oral vaccines [14]. In that case, the microorganisms will locally deliver the antigens to the digestive mucosa in order to stimulate an immune response (production of immunoglobulins) and ensure a protection against bacterial [7, 15–17] or viral [6, 18, 19] diseases or being used in the management of food allergy [4, 20]. For several immunological and practical reasons, these new vaccines represent a promising alternative to the traditional injectable ones [21]. In particular, oral immunization is the most efficient way to induce a protective local immune response at the site of pathogen contact. Recently, clinical trials have shown the vaccinal efficiency of different recombinant strains of attenuated *Salmonella typhi* in humans, but the patients were presenting undesirable side effects (diarrhea) [22].

Other medical applications involve the direct production in the digestive medium of various biological mediators, such as insulin, cytokines [23], or growth factors. In particular, new approaches to treating inflammatory bowel diseases (IBD) such as Crohn's disease, celiac disease, or ulcerative colitis have been considered [24]. Anti-inflammatory and immunosuppressive therapies are commonly used for the treatment of IBD. However, patients are often subject to unpleasant side effects owing to the high level of the drug in the body (systemic administration) and some of them remain refractory to such treatments. Steidler et al. [5] have investigated the potential of alternative therapeutics and shown the interest of delivering interleukin-10 (IL-10), a strong anti-inflammatory cytokine, in a localized manner by the action of recombinant *Lactococcus lactis*. Oral administration of the strain producing mouse IL-10 led to 50% curing of dextran sulfate sodium–induced colitis and prevented the onset of the pathology in mice [5]. The amount of IL-10 required to achieve the healing effect was 10,000-fold lower when the cytokine was in situ delivered by *L. lactis* compared to systemic treatment with anti-inflammatory drugs (e.g., dexamethasone). A recent study of the same author has reported that the treatment of Crohn's disease patients with *L. lactis* secreting human IL-10 was safe and allowed a decrease in disease activity [25].

5.5.1.3 Choice of Candidate Host Microorganisms

Recombinant bacteria, particularly lactic acid bacteria, have been mostly suggested as potential hosts for this new kind of drug delivery system [2, 3]. However, yeasts can offer advantages, especially when a eukaryotic environment is required for the functional expression of human genes. Moreover, the absence of bacterial sequences liable to promote gene transfer to host bacteria can be ensured using the efficient site-targeted homologous recombination machinery of yeasts for introduction of the heterologous gene into the yeast genome. Lastly, yeasts are not sensitive to antibacterial agents, allowing the simultaneous administration of the recombinant microorganisms and antibiotics. In this study, the common baker's yeast *Saccharomyces cerevisiae* was chosen owing to its "generally recognized as safe" (GRAS) status and its easy culture and genetic engineering. *Saccharomyces* spp. have already been used in humans, mainly in the treatment of intestinal functional disorders such as colitis [26] or antibiotic-associated diarrhea [27].

5.5.2 EVALUATION OF SCIENTIFIC FEASIBILITY OF BIODRUG CONCEPT USING YEAST AS VECTOR

5.5.2.1 Approach

The scientific feasibility of our approach was recently evaluated using recombinant *S. cerevisiae* expressing model genes and an original artificial digestive system simulating human gastrointestinal conditions. The survival rate and heterologous activity of the recombinant model yeasts were followed in this in vitro system.

Recombinant Model Yeasts Three different yeast strains, all derived from the haploid strain W303-1B, were used to evaluate the scientific feasibility of the biodrug concept using yeast as vector.

The first one, WRP45073A1 (provided by Denis Pompon, CNRS, Gif-sur-Yvette, France), expresses the plant P45073A1 when grown in the presence of galactose [28]. P45073A1 was chosen as a model for a reaction catalyzed by a P450 owing to the nontoxicity and easy quantification of both substrate and product [29]. It catalyzes the 4-hydroxylation of *trans*-cinnamic acid into *p*-coumaric acid [cinnamate 4-hydroxylase (CA4H) activity]. In this model, the heterologous protein is an intracellular enzyme, and it was synthesized (induction of CYP73A1 by galactose during the last 12 h of culture) before yeast introduction into the artificial digestive system. The in situ CA4H activity of yeasts was quantified following the simultaneous introduction of recombinant yeasts and *trans*-cinnamic acid into the in vitro system.

The two other strains—WppGSTV$_5$H$_6$ and WppV$_5$H$_6$—were genetically engineered [30] to secrete (i) a model protein derived from the commonly used reporter protein GST, named GST–V$_5$H$_6$ [molecular weight (MW) 31.5 kDa], and (ii) a model peptide, peptide–V$_5$H$_6$ (MW 5.6 kDa). The recombinant protein and peptide were expressed in fusion with the V$_5$ epitope (V$_5$) and the polyhistidine (H$_6$) tag to allow their immunological detection and make easier their purification, respectively. A leader sequence derived from that of the α-factor precursor was used to direct the secretion of the heterologous protein compounds into the extracellular medium [30]. In that case, recombinant yeasts and galactose, the inductor of the heterologous genes, were simultaneously introduced into the artificial digestive system to evaluate the yeast ability to initiate the synthesis and secrete protein compounds of various sizes, directly in the digestive environment.

Artificial Digestive System TIM: Powerful In Vitro Tool The system TIM (TNO gastrointestinal tract model) is the in vitro model that at present time allows the closest simulation of in vivo dynamic physiological processes occurring within the lumen of the stomach and small intestine of humans [31]. It is composed of four successive compartments reproducing the stomach and the three parts of the small intestine: the duodenum, jejunum, and ileum (Figure 1). This dynamic, computer-controlled system has been designed to accept parameters and data from in vivo studies on human volunteers. The main parameters of digestion, such as pH, body temperature, peristaltic mixing and transport, gastric, biliary, and pancreatic secretions, and passive absorption of small molecules (e.g., nutrients, drugs) and water,

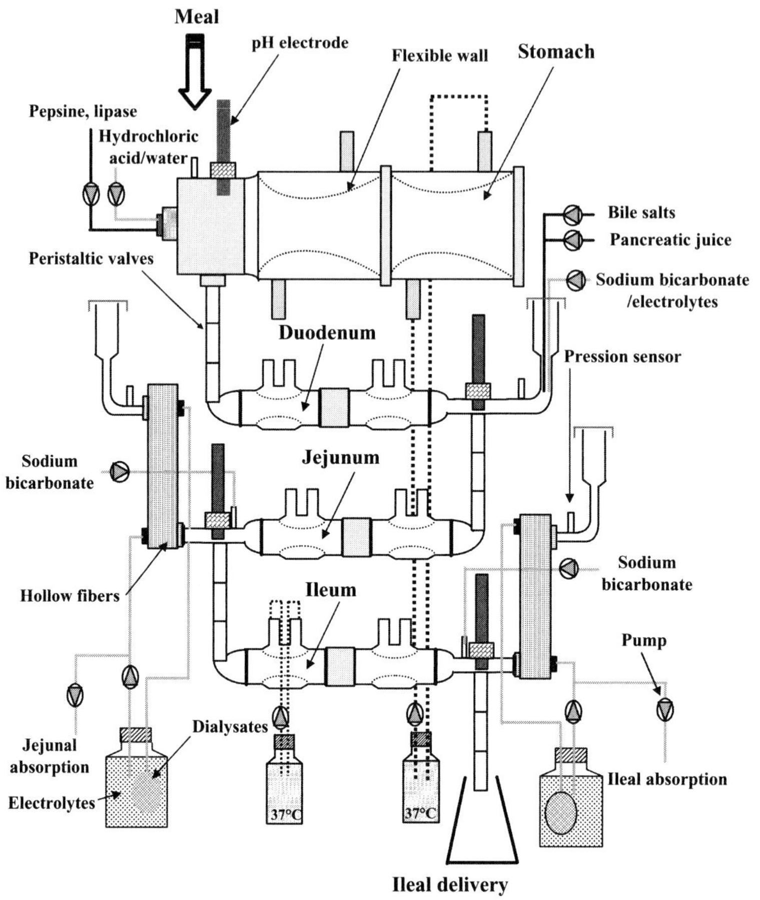

FIGURE 1 Gastric and small intestinal system TIM.

are reproduced as accurately as possible (Table 2). This system has been previously described in detail [1, 29–31].

Compared with animal experiments, this in vitro system offers accuracy, reproducibility, easy manipulation, and the possibility of collecting samples at any level of the digestive tract and at any time during digestion with no ethical constraint. It has been validated by microbial, nutritional, and pharmaceutical studies. For instance, validation experiments demonstrate the predictive value of the TIM with regard to the survival rate of probiotic bacteria [32, 33], the digestibility of nutrients [31, 34], and the availability for absorption of minerals [35], vitamins [36], food mutagens [37], and drugs such as paracetamol [33, 38].

In the present study, the TIM was programmed to reproduce gastrointestinal conditions of the adult after the intake of a liquid meal, according to in vivo data [31, 39–41]. The initial "meal" (introduced into the artificial stomach at the beginning of digestion) was composed of (i) 10^{10} WRP45073A1 cells and 200 µmol of *trans*-cinnamic acid or (ii) 3×10^{10} WppGSTV$_5$H$_6$ or WppV$_5$H$_6$ cells and galactose (40 g/L)

TABLE 2 Digestive Parameters Reproduced in Gastrointestinal Model TIM and Their Simulation

pH	The pH is computer monitored and continuously controlled in each digestive compartment. The fall of gastric pH is reproduced by adding hydrochloric acid. The pH is kept to 6.5, 6.8, and 7.2 in the duodenum, jejunum, and ileum, respectively, by secreting sodium bicarbonate.
Temperature	The compartments are surrounded by water at body temperature (37 °C).
Peristaltic mixing	Peristaltic mixing is mimicked by alternate compression and relaxation of the flexible walls containing the chyme, following changes in the water pressure.
Dynamic of chyme transit	A mathematical model using power exponential equations [39] is used to reproduce gastric and ileal deliveries ($f = 1 - 2^{-(t/t1/2)\beta}$, where f represents the fraction of meal delivered, t the time of delivery, $t_{1/2}$ the half-time of delivery, and β a coefficient describing the shape of the curve). Chyme transit is regulated by opening or closing the peristaltic valves that connect the compartments.
Volume	The volume in each compartment is monitored with a pressure sensor connected to the computer.
Digestive secretions	Simulated gastric, biliary, and pancreatic secretions are introduced into the corresponding compartments by computer-controlled pumps.
Absorption of small molecules and water	Semipermeable membrane units are connected to the jejunum and ileum to remove the products of digestion as well as water.

in suspension in 300 mL of yeast culture medium. The parameters of in vitro digestion are summarized in Table 3.

5.5.2.2 Yeast Survival Rate in Simulated Gastrointestinal Conditions

At the end of in vitro digestion, yeast survival rate was evaluated by comparing the total ingested yeasts with the living yeasts recovered in both the ileal effluents of the TIM and the residual digestive content. After 240 or 270 min digestion (depending on the strain), 79.5 ± 12.1% (n = 3), 63.9 ± 2.4% (n = 3) and 75.5 ± 25.3% (n = 4) of the ingested WRP45073A1, $WppV_5H_6$, and $WppGSTV_5H_6$, respectively, were recovered in the ileal effluents (Figure 2). When the yeasts remaining in the residual chyme were added (t = Tf), 95.6 ± 10.1% (n = 3), 83.1 ± 9.6% (n = 3), and 95.3 ± 22.7% (n = 4) survival percentages, respectively, were found, showing the high resistance of recombinant yeasts to gastric (pepsin and lipase) and small intestinal (bile salts and pancreatic juice) secretions and low gastric pH [29, 30]. This high survival rate was confirmed (Figure 2), no significant difference ($p < 0.05$) being observed during digestion between the ileal recovery profiles of recombinant yeasts (except for $WppV_5H_6$) and that of a nonabsorbable marker, blue dextran, added in the artificial stomach at the beginning of digestion, as previously described [31].

The survival rate of other microorganisms, such as lactic acid bacteria, has also been studied in the TIM. At the end of digestion, Marteau et al. [32] found a bacte-

TABLE 3 Parameters of In Vitro Digestion in TIM When Simulating Gastrointestinal Conditions of Adult After Intake of Liquid Meal

Gastric compartment	Initial volume	300 mL
	Time (min)/pH	0/6
		20/4.2
		40/2.8
		60/2.1
		90/1.8
		120/1.7
	Secretions	0.25 mL/min pepsin (590 IU/mL)
		0.25 mL/min lipase (37.5 IU/mL)
		0.25 mL/min HCl 0.5 *M* if necessary
	Time of half emptying $t_{1/2}$	30 min
	ß coefficient	1
Duodenal compartment	Volume	30 mL
	pH	Maintained at 6.5
	Secretion	0.5 mL/min bile salts (4% during first 30 min of digestion, then 2%)
		0.25 mL/min pancreatic juice (10^3 USP/mL)
		0.25 mL/min intestinal electrolyte solution
		0.25 mL/min $NaHCO_3$ 1 *M* if necessary
Jejunal compartment	Volume	70 mL
	pH	Maintained at 6.8
	Secretion	0.25 mL/min $NaHCO_3$ 1 *M* if necessary
	Dialysis	10 mL/min of jejunal fluid solution
Ileal compartment	Volume	70 mL
	pH	Maintained at 7.2
	Secretion	0.25 mL/min $NaHCO_3$ 1 *M* if necessary
	Dialysis	10 mL/min ileal fluid solution
	Time of half emptying $t_{1/2}$	160 min
	ß coefficient	1.6

rial cumulative delivery from the ileum between 0 and 25% (depending on the tested strain). In this work, under similar experimental conditions, about 75% of ingested yeasts were recovered. Until now, the feasibility of the biodrug concept had been mainly evaluated with lactic acid bacteria (see Section 5.5.1.2). The high viability of *S. cerevisiae* in the digestive tract might favor the choice of yeasts over lactic acid bacteria as hosts for the development of biodrugs, particularly if the viability of the microorganisms is required for their in situ activity.

5.5.2.3 Yeast Heterologous Activity in Simulated Gastrointestinal Conditions

Bioconversion The CA4H activity of WRP45073A1 yeasts was quantified measuring *p*-coumaric acid production by high-performance liquid chromatography. Control experiments showed that both *trans*-cinnamic and *p*-coumaric acids were stable under digestive conditions when no yeast was introduced into the TIM. In the presence of yeasts with no CA4H gene in their plasmid, no *p*-coumaric acid was produced, showing the specificity of the enzymatic reaction catalyzed by the recombinant model yeasts.

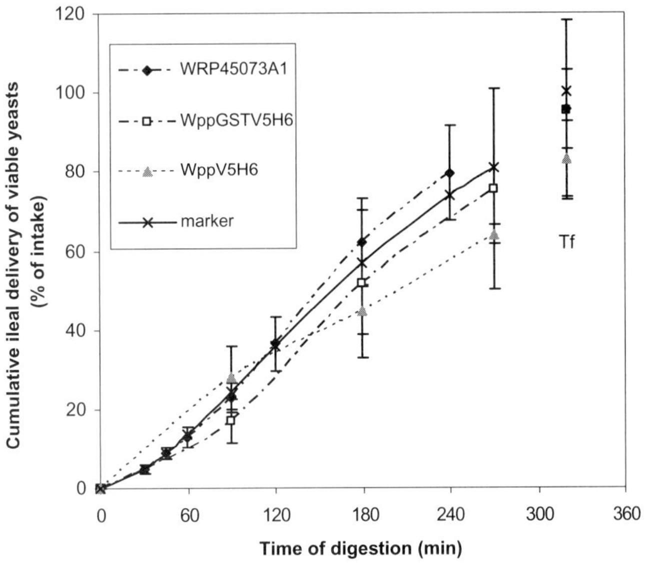

FIGURE 2 Survival rate of three recombinant model yeasts in TIM. The cumulative ileal deliveries of viable yeasts and that of a nonabsorbable marker, bleu dextran, are represented. At the end of digestion, the percentages obtained in the cumulative ileal effluents (0–240 min or 0–270 min depending on strain) and in the residual digestive content are added (t = Tf). Results are expressed as mean percentages ± SD (n = 3 for WRP45073A1 and $WppV_5H_6$, n = 4 for $WppGSTV_5H_6$) of intake.

At the end of digestion (240 min), 41.0 ± 5.8% (n = 3) of initial *trans*-cinnamic acid was converted into *p*-coumaric acid (Figure 3*a*) [29]. By means of a computer simulation [29], in each compartment of the in vitro system, the amount of *p*-coumaric acid resulting from the CA4H activity of yeasts could be dissociated from that delivered by the previous compartment. After calculation, *trans*-cinnamic acid conversion rates of 8.9 ± 1.6%, 13.8 ± 3.3%, 11.8 ± 3.4%, and 6.5 ± 1.0% (n = 3) were found in the stomach, duodenum, jejunum, and ileum, respectively (Figure 3*b*). The enzymatic reaction occurred throughout the artificial gastrointestinal tract, but mostly in the duodenum and jejunum. This could be explained by the fact that yeasts were no longer stressed by the acid pH of the stomach and could metabolize the *trans*-cinnamic acid that had previously easily entered the cells, owing to the low pH (*trans*-cinnamic acid is essentially in a cationic form which easily diffuses through the yeast membrane [42]). Also, previous studies have demonstrated that bile salts can favor enzymatic reactions [43]. The lower activity in the ileum might result from a decrease in the availability of *trans*-cinnamic acid owing to its previous conversion into *p*-coumaric acid in the upper digestive compartments. The computer simulation that was developed here should prove useful in future stages of the development of biodrugs, especially if a specific level of the digestive tract has to be targeted for drug action.

Further calculations were performed to quantify the specific enzymatic activity of the WRP45073A1 yeasts. Yeast specific activity in the TIM (from $0.05 \pm 0.04 \times 10^{-10}$ to $3.36 \pm 0.86 \times 10^{-10}$ µmol/cell/min, depending on the digestive compartment and the sampling time [29]) was close to that observed in classical batch cultures,

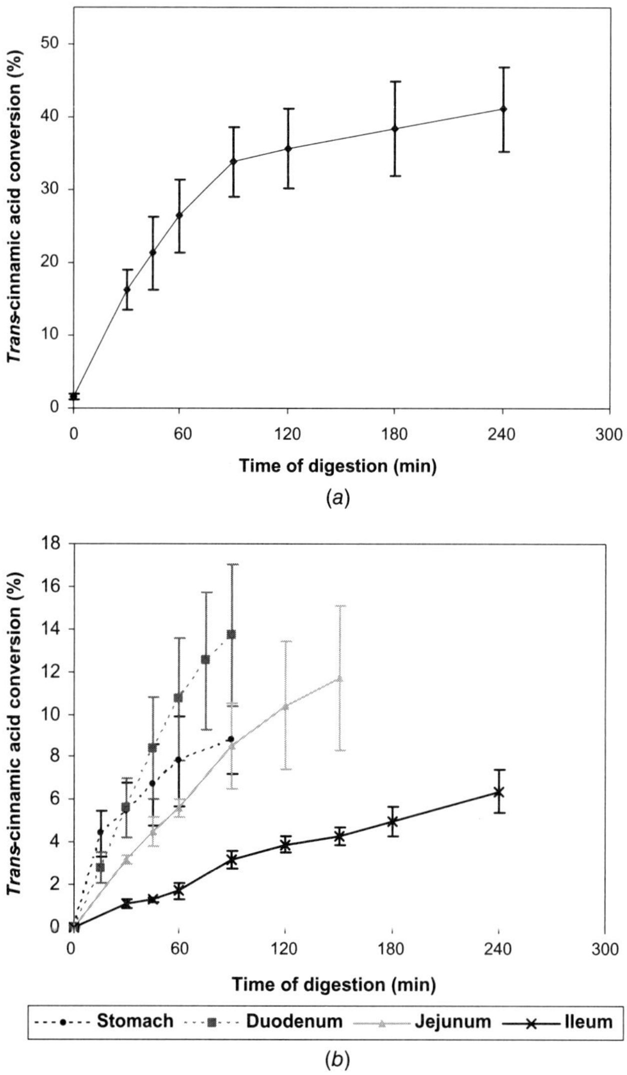

FIGURE 3 CA4H activity of WRP45073A1 in TIM. The *trans*-cinnamic acid conversion rate was evaluated in (*a*) overall TIM and (*b*) each compartment of TIM. Results are expressed as mean percentages ± SD ($n = 3$) of ingested *trans*-cinnamic acid converted into *p*-coumaric acid. (Reprinted with permission from Blanquet et al., *Applied and Environmental Microbiology*, 69, 2889.)

which is very encouraging for a potential use of *S. cerevisiae* as a biodetoxication system to the gut.

Secretion of Peptides or Proteins The production of the GST–V_5H_6 and peptide–V_5H_6 in the TIM was examined by Western blotting (data not shown). No signal was detectable during control digestions without yeast or with yeasts with no heterologous gene in their plasmid. The model protein and peptide were detected as early as 90 min after the yeast intake/gene induction in each compartment of the in vitro system and remained until 270 min of digestion in the lower part of the small intestine [30]. No signal was detectable in the "meal" before introduction into the artificial stomach, showing the efficient initiation of protein compound synthesis and secretion by galactose in the digestive environment.

The amount of GST–V_5H_6 produced in each compartment of the TIM was quantified by enzyme-linked immunosorbent assay (ELISA). The protein concentrations in the digestive medium reached 15 ng/mL, the highest values being found in the jejunum and ileum from 150 min to the end of digestion (Figure 4*a*). The GST–V_5H_6 concentrations in the digestive environment were close to those measured in stan-

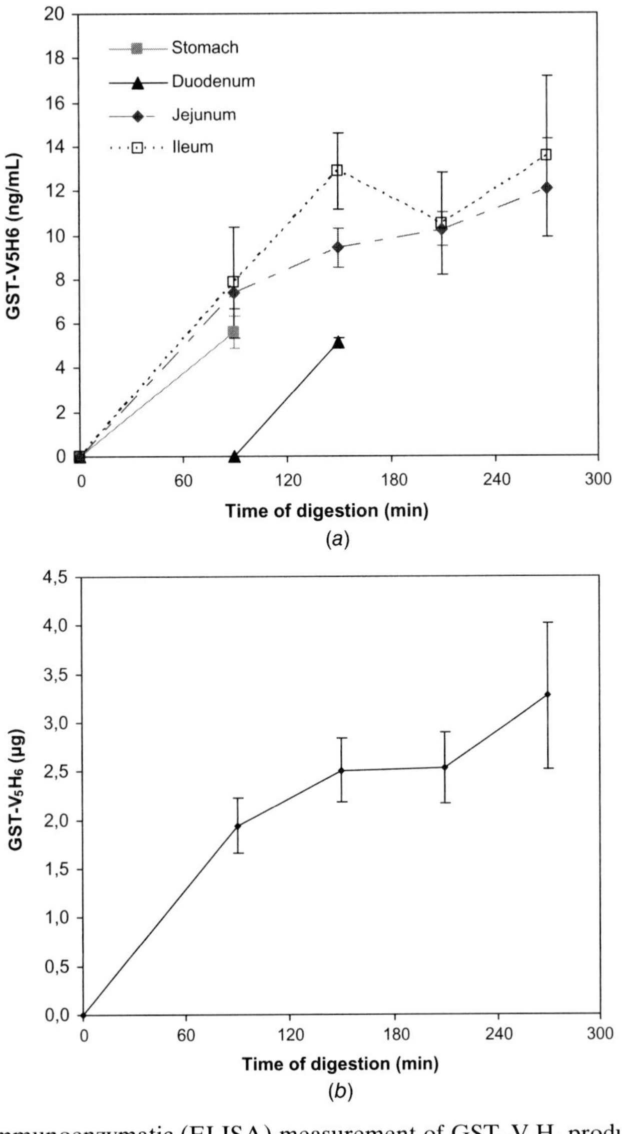

FIGURE 4 Immunoenzymatic (ELISA) measurement of GST–V_5H_6 produced by WppG-STV$_5$H$_6$ in (*a*) different compartments of TIM (ng/mL) and (*b*) the overall TIM (μg). Error bars represent standard deviations (n = 4). (Reprinted with permission from 110, Blanquet et al., *Journal of Biotechnology*, 45, 2006. Copyright 2006 by Elsevier.)

dard batch cultures. Therefore, the low secretion levels of GST–V_5H_6 may be only imputed to the large size of the protein and/or the genetic construction, but not to the particular digestive conditions. This hypothesis is consistent with the results of Zsebo et al. [44], who showed that *S. cerevisiae* is an efficient host for small polypeptide secretion, but not for larger proteins, which accumulate in the periplasmic place and cell wall of the yeast. Improved secretion levels might be obtained with another heterologous protein or a different expression vector. In the overall in vitro system, the total amount of GST–V_5H_6 regularly increased during digestion to reach $3.3 \pm 0.7\,\mu g$ ($n = 4$) after 270 min digestion (Figure 4*b*).

To check that the GST–V_5H_6 recovered in the TIM was truly secreted by living recombinant yeasts and did not result from cell lysis, a control strain producing an intracellular form of the model protein was contructed (without the leader sequence) and tested in similar experimental conditions. Some GST–V_5H_6 was found in the ileum showing that cell lysis occurred in this compartment. At the end of digestion, the total amount of GST–V_5H_6 released by the control strain represented 30% ($1 \pm 0.2\,\mu g$, $n = 2$) of the protein produced by WppGSTV$_5$H$_6$, showing a significant contribution of cell lysis to the release of the protein in the TIM.

For the first time, the amount of heterologous proteins secreted by recombinant *S. cerevisiae* was evaluated throughout the upper digestive tract. Until now, the secretion efficiency of recombinant microorganisms had never been directly quantified throughout the digestive tract. The ability of recombinant strains to produce heterologous proteins in the digestive environment had been mainly demonstrated indirectly, following the biological effect of the protein: immune response (antibody production) [7, 15–17, 19, 20], growth improvement [45], or reduction in colitis symptoms [5]. Nevertheless, the ability of recombinant bacteria to initiate protein synthesis in situ has been reported in a few studies. Oozeer et al. [46] have shown that engineered *Lactobacillus casei* was able to initiate the synthesis of luciferase during its transit in the digestive tract of a human flora-associated mouse model. Steidler et al. [5] have demonstrated in mice the in situ synthesis of mouse IL-10 by recombinant *L. lactis*, the viability of these microorganisms being required to achieve their therapeutic effect (see Section 5.5.1.2). They have further documented this result by showing the de novo synthesis of IL-10 in the colon of IL-10 –/– mice. Moreover, these authors have quantified the amount of IL-10 produced by the engineered *L. lactis* in two animal models, but only in a limited part of their digestive tract : (i) 7 ng of mouse IL-10 was recovered in the colon of IL-10 –/– mice, but the interleukin was not detectable in other areas of the gastrointestinal tract [5] and (ii) about 470 pg/mL of human IL-10 was found in an ileal loop of a pig 4 h after injection of the recombinant bacteria [47]. Unlike what was previously obtained in the TIM with GST–V_5H_6, the concentrations of IL-10 found in the digestive tract of the animals were much lower than that recovered in batch cultures [5, 47].

5.5.2.4 Conclusion

For the first time, the ability of engineered *S. cerevisiae* to carry out a bioconversion reaction [29] and initiate the synthesis and secrete protein compounds of various sizes [30] was shown throughout the upper gastrointestinal tract in human simulated digestive conditions. The CA4H specific activity of WRP45073A1 and the secretion level of GST–V_5H_6 were surprisingly similar to that obtained in classical batch cultures. This is particularly remarkable as the expression strategy of the model genes had not yet been adapted to the particular constraints of the digestive environment and promising for a future use of recombinant *S. cerevisiae* as host for biodrug development.

5.5.3 ORAL FORMULATION OF RECOMBINANT YEASTS

Once the scientific feasibility of the new drug delivery system was established, the development of pharmaceutical formulations allowing the oral administration of the genetically modified *S. cerevisiae* was considered. Ideally, these oral drug dosage forms would improve both the survival and the heterologous activity of yeasts in the digestive environment. The following works were carried out only with the strain expressing the model P450. In a preliminary step, the effect of a preservation technique (lyophilization) and an immobilization procedure (entrapment in whey protein beads) on the survival rate and heterologous activity of the model strain WRP45073A1 was assessed in simulated digestive conditions.

5.5.3.1 Freeze Drying of Recombinant Model Yeasts

Freeze-Drying Conditions Freeze drying is a technique of dehydration commonly used for the formulation of drugs containing nonrecombinant *Saccharomyces* spp. [48–50]. Standard freeze-drying conditions derived from the literature [51–54] and our own experiments were applied for the lyophilization of the genetically modified model yeasts. The effect of cryoprotectants was further investigated because it appears as one of the most important parameters during lyophilization [51, 52]. Following galactose induction of the heterologous CYP73A1, yeasts in the beginning of their stationary growth phase (10^9 cells/mL) were lyophilized in suspensions of trehalose 10% w/v, maltose 10% w/v, lactose 10% w/v, or a mixture of 5% w/v milk proteins and 10% w/v trehalose. The parameters of lyophilization are summarized in Table 4 [55].

Saccharomyces cerevisiae WRP45073A1 survives freeze drying and yeast survival rates were dependent on the nature of the cryoprotectants: 13.1 ± 1.8%, 9.5 ± 6.0%,

TABLE 4 Parameters of Lyophilization Used for Recombinant Model Yeast WRP45073A1

Recombinant yeasts	Growth phase: beginning of stationnary growth phase Cell concentration in freeze-drying flasks: 10^9 cells/mL
Cryoprotectants	10% w/v trehalose 10% w/v maltose 10% w/v lactose 5% w/v milk proteins and 10% w/v trehalose Control: physiological water
Freeze-drying conditions	Volume of sample: 5 mL Cooling rate: 1 °C/min Condenser plate temperature: –40 °C Time of lyophilization: 24 h Heating temperature (secondary drying): 23 °C

7.7 ± 4.6%, and 7.1 ± 4.0% ($n = 5$) for the milk protein–trehalose mix, lactose, maltose, and trehalose, respectively [55]. The protective effect of trehalose [51–53] and maltose [52] (but not that of lactose) compared to physiological water (survival rate 0.3 ± 0.2%, $n = 5$) had already been shown in non genetically modified *S. cerevisiae*. Several mechanisms have been proposed to explain this protective effect. One hypothesis is related to the ability of these carbohydrates to form a glassy structure during drying, responsible for the long stability of biological materials [56]. The milk protein–trehalose mix led to a higher survival compared with trehalose alone ($p < 0.05$). A similar result had already been observed by Abadias et al. [54], who showed an improvement in the viability of another yeast, *Candida sake*, from 7 to 29% when 5% skim milk is used in combination with 10% trehalose.

Influence of Cryoprotectants on Viability and Heterologous Activity of Lyophilized Yeasts in Simulated Gastrointestinal Conditions To evaluate the influence of cryoprotectants on both the survival rate and CA4H activity of WRP45073A1 in simulated gastrointestinal conditions, 10^{10} viable freeze-dried yeasts and 200 μmol of *trans*-cinnamic acid were simultaneously introduced into the TIM. Yeast cells were lyophilized in the presence of the milk protein–trehalose mix, trehalose, lactose, or maltose, as previously explained (see above). The freeze-dried samples were introduced into the artificial stomach suspended in 300 mL of yeast culture medium without any storage period. The number of viable cells introduced into the TIM was determined from previously obtained survival rates (cf. Section 5.5.3.1).

Viability of Freeze-Dried Yeasts in TIM Freeze-dried yeasts showed a high tolerance to gastric and small intestinal conditions. At the end of digestion (240 min), 61.5 ± 0.7%, 59.9 ± 3.8%, 56.3 ± 5.4%, and 55.6 ± 6.0% ($n = 3$) of the ingested cells were recovered in the ileal effluents of the TIM, following freeze drying in the presence of the milk protein–trehalose mix, maltose, lactose, and trehalose, respectively (Figure 5). When the yeasts remaining in the residual chyme were added (t = Tf), 84.7 ± 3.5%, 87.0 ± 6.4%, 83.7 ± 6.2%, and 70.7 ± 9.2% ($n = 3$) survival percentages were found, respectively. Nevertheless, whatever the protective agent, the cumulative ileal delivery of freeze-dried yeasts remained significantly lower ($p < 0.05$) than that of the nonabsorbable marker (see Section 5.5.2.2) and that of nondried ("native") yeasts. The higher sensitivity to digestive conditions of freeze-dried cells compared with native ones may be linked to the damage caused to cells during drying and rehydration, resulting in increased membrane permeability [57].

In addition, no significant difference was observed between the ileal recovery profiles of yeasts with the various cryoprotectants, showing their lack of influence on the survival of freeze-dried WRP45073A1 in the TIM.

A few studies have evaluated the survival rate of freeze-dried *S. cerevisiae* spp. in human volunteers following their oral administration. Nevertheless, comparison between in vitro results in the TIM and these in vivo data is hampered by the fact that yeast survival had been evaluated only in feces (and not at the end of the ileum) after a single or multiple oral administration of the microorganisms. Klein et al. [49] found a fecal recovery of 0.12 ± 0.04% ($n = 8$) after a single dose of 1 g of *S. boulardii* [$10^{10.4}$ colony-forming units (CFU)] to healthy volunteers, and Blehaut et al. [50] measured a steady-state fecal recovery of 0.36 ± 0.31% ($n = 8$) after oral administra-

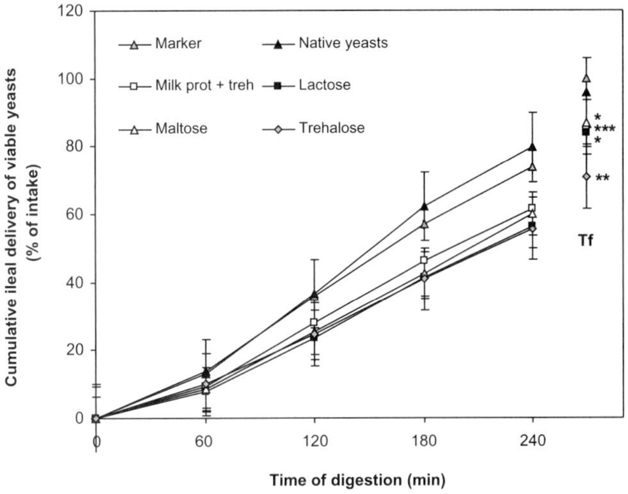

FIGURE 5 Effect of cryoprotectants on survival rate of WRP45073A1 in TIM. The cumulative ileal deliveries of viable freeze-dried and native yeasts and that of the nonabsorbable marker are represented. At the end of digestion, the percentages obtained in the cumulative ileal effluents (0–240 min) and the residual digestive content are added (t = Tf). Results are expressed as mean percentages ± SD ($n = 3$) of intake. Significantly different from the marker (t = Tf) at $p < 0.05$ (*), $p < 0.01$ (**) and $p < 0.001$ (***). (Reprinted with permission from Blanquet et al., *European Journal of Pharmaceutics and Biopharmaceutics*, 61, 37, 2006. Copyright 2006 by Elsevier.)

tion of 1 g of *S. boulardii* for 15 days. These survival rates in feces are much lower than those obtained in the ileal effluents of the TIM. However, a recent work [29] has shown that native WRP45073A1 yeasts are very sensitive to colonic conditions: Yeast viability in an artificial digestive system reproducing the human large intestine [58] was only 1.2 ± 0.4% ($n = 3$) after 12 h and no more yeast could be detected following 24 h fermentation.

Heterologous Activity of Freeze-Dried Yeasts in TIM The ability of freeze-dried recombinant yeasts to catalyze the bioconversion of *trans*-cinnamic acid into *p*-coumaric acid in the in vitro system was shown whatever the tested protectant. At the end of the experiment, conversion rates of 24.2 ± 1.0%, 17.7 ± 2.2%, 15.1 ± 3.3%, and 16.5 ± 0.7% ($n = 3$) were found for the yeasts lyophilized in the presence of milk proteins plus trehalose, maltose, lactose, and trehalose, respectively (Figure 6). During all the in vitro digestion, the CA4H activity of freeze dried cells remained significantly lower ($p < 0.01$) than that of native ones. This lower activity could result from the P45073A1 damage during freeze drying, probably aggravated by the membranous location of the enzyme. Among those tested, the cryoprotectant that allowed the highest CA4H activity was the milk protein–trehalose mix ($p < 0.05$).

Conclusion Although the impact of freeze drying on both the survival rate and heterologous activity of yeasts in the artificial digestive system was found to be adverse, lyophilization appears to be a convenient technique for the dehydration of recombinant *S. cerevisiae*. Among the tested cryoprotectants, the association of milk proteins and trehalose was the most efficient to maintain the CA4H activity of

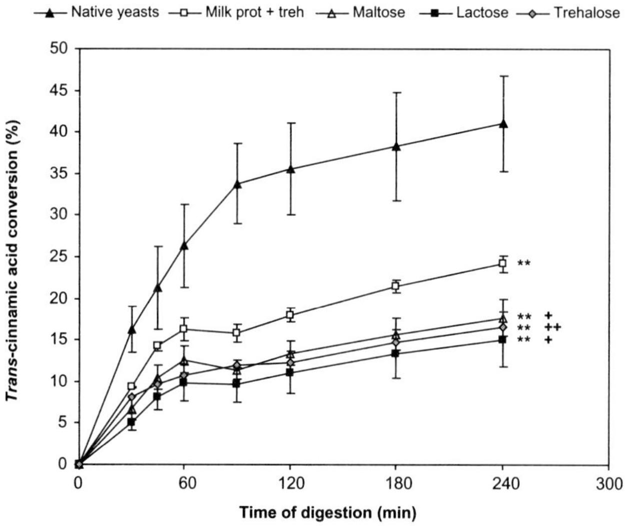

FIGURE 6 Effect of cryoprotectants on CA4H activity of WRP45073A1 in TIM. The CA4H activity of freeze-dried and native yeasts was evaluated in the overall TIM. Values are expressed as mean percentages ± SD (n = 3) of initial *trans*-cinnamic acid converted into *p*-coumaric acid. Significantly different from native yeasts (t = 240 min) at $p < 0.01$ (**). Significantly different from milk proteins/trehalose group (t = 240 min) at $p < 0.05$ (+) and $p < 0.01$ (++). (Reprinted with permission from Blanquet et al., *European Journal of Pharmaceutics and Biopharmaceutics*, 61, 38, 2006. Copyright 2006, by Elsevier.)

recombinant model yeasts in the stringent digestive conditions. This study also gives an example of the usefulness of the TIM in the prescreening of pharmaceutical excipients, such as cryoprotectants.

5.5.3.2 Immobilization of Recombinant Model Yeasts in Whey Protein Beads

Cell Immobilization Among the available techniques, the entrapment in gel beads is frequently used for the immobilization of living cells in food sciences [59, 60] because of its simplicity and low cost. This technique has been recently extended to microorganisms with probiotic activity with the aim of increasing their survival in the human digestive environment and particularly in the stomach [61, 62].

Whey proteins have been recently considered a potential alternative to the commonly used alginate [59, 61–63] for the production of gel beads. They have been used to entrap drugs such as retinol [64] and living microorganisms such as bifidobacteria [65], but until now not yeasts. A new immobilization system using whey proteins was then developed for entrapping the recombinant model yeasts WRP45073A1 in order to ensure their oral administration [66].

The formation of beads is a two-step process based on the cold gelation of whey proteins in the presence of divalent cations, such as Ca^{2+} [67]. Briefly, the whey protein isolate (WPI) solution (10% w/v in deionized water) was (i) adjusted at pH 7 to favor the apparition of negative charges implied in ionic bounds with Ca^{2+} ions and (ii) heated (80 °C, 45 min) to denaturate the proteins. Recombinant cells in the beginning of their stationnary growth phase were suspended in a sterile solution of

10% w/v lactose (final concentration 10^9 cells/mL) and added to denaturated WPI solution (7% v/v). The extrusion of the mixture through a needle led to the production of droplets forming gel beads in a calcium bath (0.1 *M* $CaCl_2$). This protocol allows the obtaining of spherical beads (diameter 2 605 ± 18 μm, $n = 3$) with an homogeneous distribution of yeasts through the matrix (1.15×10^6 viable cells per bead) [66]. The lack of influence of the immobilizing procedure on the viability of yeasts was also shown [66].

Gastric Digestion Protocol A most sought-after property of gel beads is their potential resistance to gastric conditions. Authors have already shown in vitro that beads resulting from the cold-induced gelation of a whey protein–oil emulsion [64] or a whey protein–polysaccharide mix [65] were gastroresistant. To further investigate the involvement of whey proteins in the gastroresistance of beads, the behavior of entrapped WRP45073A1 yeasts was followed in simulated gastric conditions.

The human gastric environment was reproduced using a simple in vitro model adapted from that initially developped by Yvon et al. [68]. The main parameters of gastric digestion are reproduced according to in vivo data: decrease of pH, pepsin supply, body temperature, mixing, and gastric emptying (Figure 7). This system was validated by studies on the digestability of milk proteins (unpublished data). In the present work, it was programmed to reproduce gastric conditions of the adult after the intake of a glass of milk. Initially, 10^{10} viable entrapped WRP45073A1 cells and 200 μmol of *trans*-cinnamic acid were simultaneously introduced into the artificial stomach, suspended in 300 mL of yeast culture medium. Table 5 summarizes the parameters of in vitro gastric digestion.

Release of Yeasts from Beads in Simulated Gastric Conditions The release of entrapped WRP45073A1 cells from whey protein beads was followed in the artificial

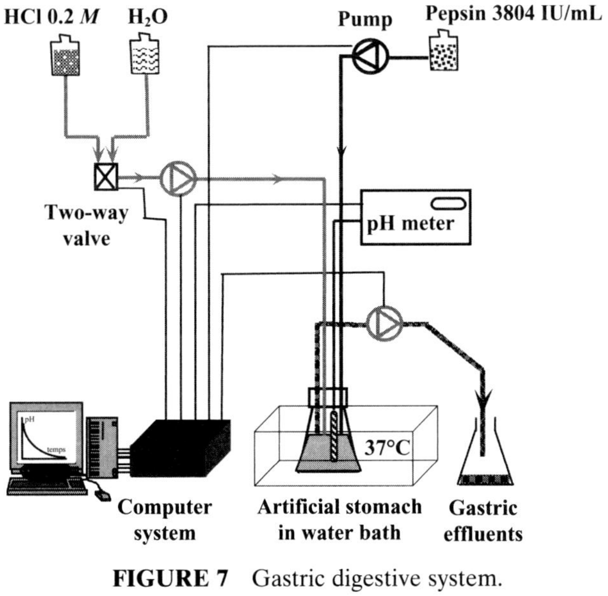

FIGURE 7 Gastric digestive system.

TABLE 5 Parameters of In Vitro Digestion in Artificial Stomach When Simulating Gastric Conditions of Adult after Intake of Glass of Milk

Initial "meal"	
Volume	300 mL (constant during all digestion)
Pepsin	75 IU/mL
Acidification	
Time (min)/pH	0/6.5, 60/2.1 15/4.4, 75/1.9 30/3.2, 90/1.7 45/2.5, 120/1.6
Flow rate	2 mL/min HCl 0.2 *M* if necessary
Exponential base	$e = 1.04$
Pepsin supply	
Time (min)/pepsine (IU/mL)	0/77.6, 60/139.2 15/99.7, 75/146.7 30/117.1, 90/152.8 45/129.7, 120/163.0
Flow rate	From 13 ($t = 0$) to 1 ($t = 120$ min) mL/min pepsin 3804 IU/mL
Exponential base	$e = 1.03$
Gastric emptying	
Flow rate	From 10 ($t = 0$) to 3 ($t = 120$ min) mL/min
Exponential base	$e = 1.03$
Time of digestion	120 min

stomach. During the first 60 min of gastric digestion, a few percentage points (2.2 ± 0.9%, $n = 3$) of initial entrapped yeasts was recovered in the gastric medium (Figure 8). This low percentage cannot be explained by cell death since control experiments showed the high survival rate of free yeasts (about 90%) during all the digestion [66]. These results are in agreement with those of Beaulieu et al. [64], who have observed that only 5–10% of the incorporated retinol was released from the whey protein–oil matrix following 30 min incubation in HCl 0.1 *M* and pepsin 24 mg/L. The low release of yeasts in the first hour of digestion indicates that the beads are resistant to acidification until pH 2 (cf. Table 5) and pepsin attack, which implies that they might cross the gastric barrier in humans.

From 60 min digestion, the percentage of released yeasts increased regularly to reach 39 ± 5% ($n = 3$) at 120 min. This phenomenon might be explained by two hypotheses: (i) a swelling of beads due to an increase in the acidity of the medium or (ii) a degradation of the matrix resulting from a raise in pepsin concentration. Complementary studies conducted to further evaluate the effect of pH and pepsin on beads have shown that pepsin has no effect on the protein matrix whatever the tested pH [66]. As already suggested by other authors [69], this resistance to enzyme attack might be due to the formation of hydrophobic interactions between aromatic amino acids of ß-lactoglobulin, the major whey protein. On the contrary, incubation at pH 2 led to an increase in the diameter of beads, certainly due to high electrostatic repulsive forces (between positive charges of protoned amino acids and Ca^{2+}), which induced a raise in the pore size. In conclusion, the release of yeasts observed from

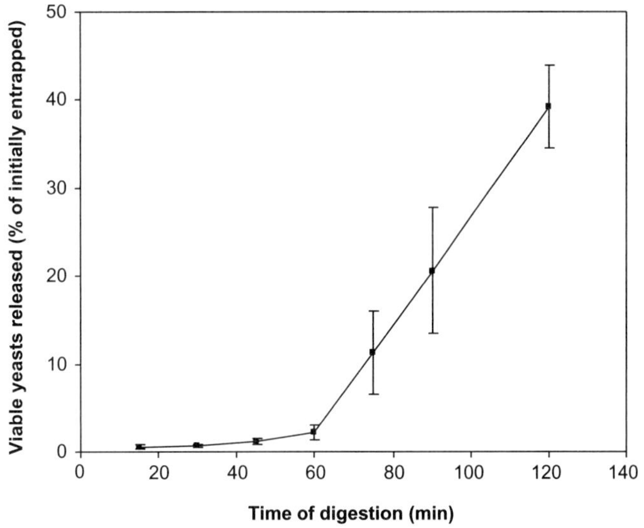

FIGURE 8 Release of WRP45073A1 cells from whey protein beads in simulated gastric conditions. Results are expressed as mean percentages ± SD ($n = 3$) of initial entrapped yeasts. (Reprinted with permission from Hebrard et al., *Journal of Biotechnology*, in press. Copyright 2006 by Elsevier.)

60 min digestion is provoked by acidic conditions rather than by enzymatic degradation of beads.

Influence of Entrapment on Heterologous Activity of Yeasts in Simulated Gastric Conditions In order to evaluate the influence of the entrapment process of the CA4H activity of WRP45073A1, the heterologous activity of free and entrapped yeasts was followed in the gastric system under similar experimental conditions. In both cases, *p*-coumaric acid was detected in the gastric medium as soon as 15 min after the beginning of the experiment (Figure 9). This implies that both *trans*-cinnamic and *p*-coumaric acid could diffuse through bead pores.

During all the digestion, the CA4H activity of entrapped yeasts was significantly ($p < 0.05$) higher than that of free ones (expected for $t = 15$ min and $t = 75$ min). At 120 min, 63.4 ± 1.6% ($n = 3$) of initial *trans*-cinnamic acid was converted into *p*-coumaric acid for immobilized yeasts versus 51.5 ± 1.8% ($n = 3$) for control yeasts. This phenomenon was particularly marked from 30 to 60 min of digestion when a very low amount of recombinant yeasts was released from beads. As suggested by Bienaimé et al. [70], beads might create a microenvironment (e.g., a buffer effect toward low pH or enzyme attack) favoring the heterologous activity of yeasts. In a similar way, the microenvironment resulting from the presence of an alginate matrix improves the invertase activity of recombinant *S. cerevisiae* in batch cultures [71].

Conclusion This preliminary work reveals whey proteins as a convenient material for immobilizing recombinant yeasts. Gel beads were resistant to acidification until pH 2 and pepsin attack, suggesting that they should cross the gastric barrier in humans. Moreover, the presence of the protein matrix seemed to create "microconditions" that favor the heterologous activity of entrapped yeasts.

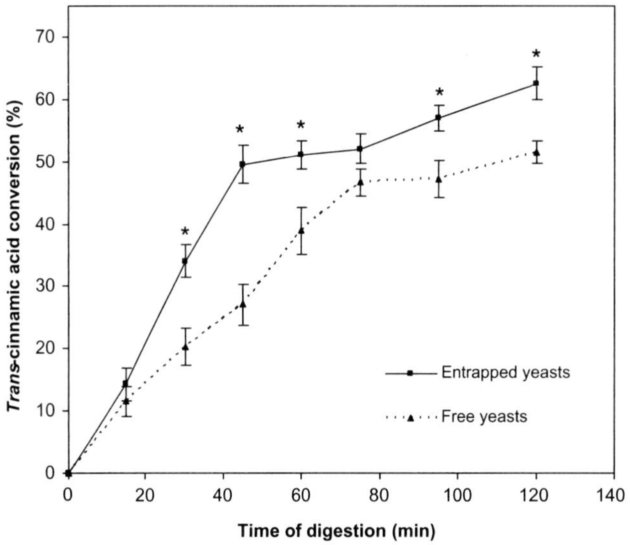

FIGURE 9 Influence of immobilization on CA4H activity of WRP45073A1 in simulated gastric conditions. The CA4H activity of entrapped and free yeasts was evaluated in the artificial gastric system. Results are expressed as mean percentages ± SD ($n = 3$) of *trans*-cinnamic acid converted into *p*-coumaric acid. Significantly different from free yeasts at $p < 0.05$ (*). (Reprinted with permission from Hebrard et al., *Journal of Biotechnology*, in press. Copyright 2006 by Elsevier.)

5.5.4 GENERAL CONCLUSION AND FUTURE DEVELOPMENTS

Using genetically engineered microorganisms as new delivery vehicles to the gut is an important challenge for the development of innovative drugs. A potential application directly issued from the present work is the development of drug delivery systems based on orally administered yeasts carrying out a bioconversion reaction or secreting compounds directly in the human digestive tract.

Soon, the choice of candidate genes as well as the most appropriate dosage forms will be made according to the therapeutic target. Oral formulations will be optimized in order to (i) control the release of yeasts according to their action site in the gastrointestinal tract, (ii) maximize the heterologous activity of yeasts (by addition of the appropriate substrate and/or inductor), and (iii) ensure a stability of both yeasts and pharmaceutical dosage forms before administration to the patient. Of course, heterologous gene expression strategies have to be tailored for a safe use in humans, the presence of mobilizable vectors, antibiotic selection markers, and bacterial sequences liable to promote gene transfer to host microflora being prohibited. In addition, environmental confinement of recombinant cells has to be achieved by introducing a suicide process that triggers the elimination of the microorganisms upon leaving the digestive tract. Two types of biological containment systems may be considered [1, 3]: (i) the active system, which should provide control of the recombinant microorganism dissemination through the conditional production of a toxic protein [72–74], and (ii) the passive system, which could render the cell growth dependent on the complementation of an auxotrophy or other gene defects [75, 76].

Steidler et al. [47] have already developed and validated in pigs a passive containment system for the *L. lactis* expressing human IL-10 by deleting the thymidylate synthase gene which is essential for their growth (the resulting strain being dependent on thymidine or thymine).

The present study also shows the particular interest of the TIM in drug development and testing. This artificial system will constitute a powerful alternative to animal experimentation during all preclinical phases of biodrug development. The in vitro model can aid in the selection of pharmaceutical formulations, ensuring both the release of yeasts directly at their action site and their optimal activity. The efficiency of newly developed molecular tools (e.g., promoters, selection markers, and vectors) can also be evaluated to optimize the functionality of recombinant strains in the digestive environment. As the mucosal layer is not involved in the actual configuration of the TIM, this system may be used in combination with intestinal cells in culture (e.g., Caco-2) to study the mucosal transport and metabolism of the active compounds produced by recombinant yeasts. Moreover, experiments in a large intestinal model [58], complementary of the gastric and small intestinal system, could provide necessary data on the biological safety of engineered microorganisms. For example, the potential gene transfer to the human flora can be studied and the cell death outside of the digestive tract can be checked to ensure there is no dissemination in the environment.

This study opens up new opportunities in the development of new drug delivery vectors based on engineered living yeasts for the prevention or treatment of various diseases in human.

REFERENCES

1. Blanquet, S., Marol-Bonnin, S., Beyssac, E., Pompon, D., Renaud, M., and Alric, M. (2001), The Biodrug concept: An innovative approach to therapy, *Trends Biotechnol.*, 19, 393–400.
2. Corthier, G., and Renault, P. (1999), Future directions for research on biotherapeutic agents: Contribution of genetic approaches on lactic acid bacteria, in Elmer, G. W., McFarland, L., Surawicz, C., Eds., *Biotherapeutic Agents and Infection Diseases*, Humana, Totowa, NJ, pp. 269–304.
3. Steidler, L. (2003), Genetically engineered probiotics, *Best Practise Res. Clin. Gastroenterol.*, 17, 861–876.
4. Chatel, J. M., Langella, P., Adel-Patient, K., Commissaire, J., Wal, J. M., and Corthier, G. (2001), Induction of mucosal immune response after intranasal or oral inoculation of mice with *Lactococcus lactis* producing bovine beta-lactoglobulin, *Clin. Diag. Lab. Immunol.*, 8, 545–551.
5. Steidler, L., Hans, W., Schotte, L., Neirynck, S., Obermeier, F., Falk, W., Fiers, W., and Remaut, E. (2000), Treatment of murine colitis by *Lactococcus lactis* secreting Interleukin-10, *Science*, 289, 1352–1355.
6. Schreuder, M. P., Deen, C., Boersma, W. J. A., Pouwels, P. H., and Klis, F. M. (1996), Yeast expressing hepatitis B virus surface antigen determinants on its surface: Implications for a possible oral vaccine, *Vaccine*, 14, 383–388.
7. Reveneau, N., Goeffroy, M. C., Locht, C., Chagnaud, P., and Mercenier, A. (2002), Comparison of the immune responses induced by local immunizations with recombinant

Lactobacillus plantarum producing tetanus toxin fragment C in different cellular locations, *Vaccine*, 20, 1769–1777.

8. Drouault, S., Corthier, G., Ehrlich, S. D., and Renault, P. (1999), Survival, physiology, and lysis of *Lactococcus lactis* in the digestive tract, *Appl. Environ. Microbiol.*, 65, 4881–4886.
9. Fahl, W. E., Loo, D., and Manoharan, H. (1999), Chemoprotective bacterial strains, International Patent WO 99/27953.
10. Baranov, V. S., Ivaschenko, T., Bakay, B., Aseev, M., Belotserkovskaya, R., Baranova, H., Malet, P., Perriot, J., Mouraire, P., Baskakov, V. N., Savitskyi, G. A., Gorbushin, S., Deyneka, S. I., Michnin, E., Barchuck, A., Vakharlovsky, V., Pavlov, G., Shilko, V. I., Guembitzkaya, T., and Kovaleva, L. (1996), Proportion of the GSTM1 0/0 genotype in some Slavic populations and its correlation with cystic fibrosis and some multifactorial diseases, *Hum. Genet.*, 97, 516–520.
11. Drouault, S., Juste, C., Marteau, P., Renault, P., and Corthier, G. (2002), Oral treatment with *Lactococcus lactis* expressing *Staphylococcus hyicus* lipase enhances lipid digestion in pigs with induced pancreatic insufficiency, *Appl. Environ. Microbiol.*, 68, 3166–3168.
12. Chang, T. M., and Prakash, S. (1998), Therapeutic uses of microencapsulated genetically engineered cells, *Mol. Med. Today*, 4, 221–227.
13. Prakash, S., and Chang T. M. S. (2000), In vitro and in vivo uric acid lowering by artificial cells containing microencapsulated genetically engineered *E. coli* DH5 cells, *Int. J. Artif. Organs*, 23, 429–435.
14. Pouwels, P. H., Leer, R. J., Shaw, M., Bak-Glashouwer, M. J. H. D., Tielen, F. D., Smit, E., Martinez, B., Jore, J., and Conway, P. L. (1998), Lactic acid bacteria as antigen delivery vehicles for oral immunization purposes, *Int. J. Food Microbiol.*, 41, 155–167.
15. Robinson, K., Chamberlain, L. M., Schofield, K. M., Wells, J. M., and Le Page, R. W. F. (1997), Oral vaccination of mice against tetanus with recombinant *Lactococcus lactis*, *Nat. Biotechnol.*, 15, 653–657.
16. Nayak, A. R., Tinge, S. A., Tart, R. C., McDaniel, L. S., and Briles, D. E., 3rd, and Curtiss, R. (1998), A live recombinant avirulent oral *Salmonella* vaccine expressing pneumococcal surface protein A induces protective responses against Streptococcus pneumoniae, *Infect. Immun.*, 66, 3744–3751.
17. Corthesy, B., Boris, S., Isler, P., Grangette, C., and Mercenier, A. (2005), Oral imunization of mice with lactic acid bacteria producing *Helicobacter pylori* urease B subunit partially protects against challenge with Helicobacter felis, *J. Infect. Dis.*, 192, 1441–1449.
18. Barron, M. A., Blyveis, N., Pan, S. C., and Wilson, C. C. (2006), Human dendritic cell interactions with whole recombinant yeasts: Implication for HIV-1 vaccine development, *J. Clin. Immunol.*, 26, 251–264.
19. Steger, K. K., Valentine, P. J., Heffron, F., So, M., and Pauza, C. D. (1999), Recombinant, attenuated *Salmonella typhimurium* stimulate lymphoproliferative responses to SIV capsid antigen in rhesus macaques, *Vaccine*, 17, 923–932.
20. Adel-patient, K., Ah-Leung, S., Creminon, C., Nouaille, S., Chatel, J. M., Langella, P., and Wal, J. M. (2005), Oral administration of recombinant *Lactococcus lactis* expressing bovine beta-lactoglobulin partially prevents mice from sensitisation, *Clin. Exp. Allergy*, 35, 539–546.
21. Mercenier, A., Müller-Alouf, H., and Grangette, C. (2000), Lactic acid bacteria as live vaccines, *Curr. Issues Mol. Biol.*, 2, 17–25.
22. Bumann, D., Hueck, C., Aebisher, T., and Meyer, T. F. (2000), Recombinant live *Salmonella spp.* for human vaccination against pathogens, *FEMS Immunol. Med. Microbiol.*, 27, 357–364.

23. Steidler, L. (2002), In situ delivery of cytokines by genetically engineered *Lactococcus lactis*, *Ant. Leeuwenh.*, 82, 323–331.

24. Steidler, L. (2001), Microbiological and immunological strategies for treatment of inflammatory bowel disease, *Microbes Infect.*, 3, 1157–1166.

25. Braat, H., Rottiers, P., Hommes, D. W., Huyghebaert, N., Remaut, E., Remon, J. P., Van Deventer, S. J., Neirynck, S., Peppelenbosch, M. P., and Steidler, L. (2006), A phase I trial with transgenic bacteria expressing interleukin-10 in Crohn's disease, *Clin. Gastroenterol. Hematol.*, 4, 754–759.

26. Lewis, S. J., and Freedman, A. R. (1998), Review article: The use of biotherapeutic agents in the prevention and treatment of gastrointestinal disease, *Aliment. Pharmacol. Ther.*, 12, 807–822.

27. Bergogne-Berezin, E. (2000), Treatment and prevention of antibiotic-associated diarrhea, *Int. J. Antimicrobial. Agents*, 16, 521–526.

28. Urban, P., Werck-Reichhart, D. H., Teutsch, G., Durst, F., Regnier, S., Kazmaier, M., and Pompon, D. (1994), Characterization of recombinant plant cinnamate 4-hydroxylase produced in yeast. Kinetic and spectral properties of the major plant P450 of the phenylpropanoid pathway, *Eur. J. Biochem.*, 222, 843–850.

29. Blanquet, S., Meunier, J. P., Minekus, M., Marol-Bonnin, S., and Alric, M. (2003), Recombinant *Saccharomyces cerevisiae* expressing a P450 in artificial digestive systems: A model for biodetoxication in the human digestive environment, *Appl. Environ. Microbiol.*, 69, 2884–2892.

30. Blanquet, S., Antonelli, R., Laforet, L., Denis, S., Marol-Bonnin, S., and Alric, M. (2004), Living recombinant *Saccharomyces cerevisiae* secreting proteins or peptides as a new drug delivery system in the gut, *J. Biotechnol.*, 110, 37–49.

31. Minekus, M., Marteau, P., Havenaar, R., and Huis in't Veld, J. H. J. (1995), A multicompartmental dynamic computer-controlled model simulating the stomach and small intestine, *ALTA*, 23, 197–209.

32. Marteau, P., Minekus, M., Havenaar, R., and Huis in't Veld, J. H. J. (1997), Survival of lactic acid bacteria in a dynamic model of the stomach and small intestine: Validation and the effects of the bile, *J. Dairy Sci.*, 80, 1031–1037.

33. Blanquet, S., Zeijdner, E., Beyssac, E., Meunier, J. P., Denis, S., Havenaar, R., and Alric, M. (2004), A dynamic gastrointestinal system for studying the behavior of orally administered drug dosage forms under various physiological conditions, *Pharm. Res.*, 21, 585–591.

34. Minekus, M., Jelier, M., Xiao, J. Z., Kondo, S., Iwatsuki, K., Kokubo, S., Bos, M., Dunnewind, B., and Havenaar, R. (2005), Effect of partially hydrolyzed guar gum (PHGG) on the bioaccessibility of fat and cholesterol, *Biosci., Biotechnol. Biochem.*, 69, 932–938.

35. Larsson, M., Minekus, M., and Havenaar, R. (1997), Estimation of the bioavailability of iron and phosphorus in cereals using a dynamic in vitro gastro-intestinal model, *J. Sci. Food Agric.*, 74, 99–106.

36. Verwei, M., Arkbage, K., Havenaar, R., Van den Berg, H., Witthöft, C., and Schaafsma, G. (2003), Folic acid and 5-methyl-tetrahydrofolate in fortified milk are bioaccessible as determined in a dynamic in vitro gastrointestinal model, *J. Nutr.*, 133, 2377–2383.

37. Krul, C. A. M., Luiten-Schuite, A., Baan, R., Verhagen, H., Mohn, G., Feron, V., and Havenaar, R. (2000), Application of a dynamic in vitro gastrointestinal tract model to study the availability of food mutagens, using heterocyclic aromatic amines as model compounds, *Food Chem. Toxicol.*, 38, 783–792.

38. Souliman, S., Blanquet, S., Beyssac, E., and Cardot, J. M. (2006), A level A in vitro/in vivo correlation in fasted and fed states using different methods: applied to solid immediate release oral dosage form, *Eur. J. Pharm. Sci.*, 27, 72–79.

39. Elashoff, J. D., Reedy, T. J., and Meyer, J. M. (1982), Analysis of gastric emptying data, *Gastroenterology*, 83, 1306–1312.
40. Heading, R. C., Tothill, P., McLoughlin, G. P., and Shearman, D. J. C. (1976), Gastric emptying rate measurement in man. A double isotope scanning technique for simultaneous study of liquid and solid components of a meal, *Gastroenterology*, 71, 45–50.
41. Chung, Y. C., Kim, Y. S., Shadchehr, A., Garrido, A., Macgregor, I. L., and Sleisenger, M. H. (1979), Protein digestion and absorption in human small intestine, *Gastroenterology*, 76, 1415–1421.
42. Castelli, F., Uccella, N., Trombetta, D., and Saija, A. (1999), Differences between coumaric and cinnamic acids in membrane permeation as evidenced by time-dependent calorimetry, *J. Agric. Food Chem.*, 47, 991–995.
43. Zarate, G., Chaia, A. P., Gonzalez, S., and Oliver, G. (2000), Viability and beta-galactosidase activity of diary propionibacteria subjected to digestion by artificial intestinal fluids, *J. Food Protection*, 63, 1214–1221.
44. Zsebo, K. M., Lu, H. S., Fieschko, J. C., Goldstein, L., Davis, J., Duker, K., Suggs, S. V., Lai, P. H., and Bitter, G. A. (1986), Protein secretion from *Saccharomyces cerevisiae* directed by the prepro-a-factor leader region, *J. Biolog. Chem.*, 261, 5858–5865.
45. Chen, C. M., Cheng, W. T. K., Chang, Y. C., Chang, T. J., and Chen H. L. (2000), Growth enhancement of fowls by dietary administration of recombinant yeast cultures containing enriched growth hormones, *Life Sci.*, 67, 2102–2115.
46. Oozeer, R., Goupil-Feuillerat, N., Alpert, C. A., Vande Guchte, M., Anba, J., Mengaud, J., and Corthier, G. (2002), *Lactobacillus casei* is able to survive and initiate protein synthesis during its transit in the digestive tract of human flora-associated mice, *Appl. Environ. Microbiol.*, 68, 3570–3574.
47. Steidler, L., Neirynck, S., Huyghebaert, N., Snoeck, V., Vermeire, A., Goddeeris, B., Cox, E., Remon, J. P., and Remaut, E. (2003), Biological containment of genetically modified *Lactococcus lactis* for intestinal delivery of human interleukin-10, *Nat. Biotechnol.*, 21, 785–789.
48. Harms, H. K., Bertele-Harms, R. M., and Bruer-Kleis, D. (1987), Enzyme-substitution therapy with the yeast *Saccharomyces cerevisiae* in congenital sucrase-isomaltase deficiency, *N. Engl. J. Med.*, 316, 1306–1309.
49. Klein, S. M., Elmer, G. W., McFarland, L. V., Surawicz, C. M., and Levy, R. H. (1993), Recovery and elimination of the biotherapeutic agent, *Saccharomyces cerevisiae*, in healthy human volunteers, *Pharm. Res.*, 10, 1615–1619.
50. Blehaut, H., Massot, J., Elmer, G. W., and Levy, R. (1989), Disposition kinetics of *Saccharomyces boulardii* in man and rat, *Biopharm. Drug Dispos.*, 10, 353–364.
51. Berny, J. F., and Hennebert, G. L. (1991), Viability and stability of yeast cells and filamentous fungus spores during freeze-drying: Effects of protectants and cooling rates, *Mycologia*, 83, 805–815.
52. Lodato, P., Segovia de Huergo, M., and Buera, M. P. (1999), Viability and thermal stability of a strain of *Saccharomyces cerevisiae* freeze-dried in different sugars and polymer matrices, *Appl. Microbiol. Biotechnol.*, 52, 215–220.
53. Diniz-Mendes, L., Bernardes, E., De Araujo, P. S., Panek, A. D., and Paschoalin, V. M. F. (1999), Preservation of frozen yeast cells by trehalose, *Biotechnol. Bioeng.*, 65, 572–578.
54. Abadias, M., Benabarre, A., Teixido, N., Usall, J., and Vias I. (2001), Effect of freeze-drying and protectants on viability of the biocontrol yeast Candida sake, *Int. J. Food Microbiol.*, 65, 173–182.
55. Blanquet, S., Garrait, G., Beyssac, E., Perrier, C., Denis, S., Hébrard, G, and Alric, M. (2005), Effects of cryoprotectants on the viability and activity of freeze-dried recombi-

nant yeasts as novel oral drug delivery systems assessed by an artificial digestive system, *Eur. J. Pharm. Biopharm.*, 61, 32–39.

56. Aquilera, J. M., and Karel, M. (1997), Preservation of biological materials under desiccation, *Crit. Rev. Food Sci. Nutr.*, 37, 287–309.
57. Beker, M. J., and Rapoport, A. I. (1987), *Advances in Biochemical Engineering/Biotechnology*, Springer, Berlin, pp. 128–171.
58. Minekus, M., Smeets-Peter, M., Bernalier, A., Marol-BonnIn, S., Havenaar, R., Marteau, P., Alric, M., Fonty, G., and Huis in't Veld, J. H. J. (1999), A computer-controlled system to simulate conditions of the large intestine with peristaltic mixing, water absorption and absorption of fermentation products, *Appl. Microbiol. Biotechnol.*, 53, 108–114.
59. Yadav, B. S., Rani, U., Dhamija, S. S., Nigam, P., and Singh, D. (1996), Process optimization for continuous ethanol fermentation by alginate-immobilized cells of *Saccharomyces cerevisiae* HAU-1, *J. Basic Microbiol.*, 36, 205–210.
60. Tsen, J. H., Lin, Y. P., and King, V. A. E. (2004), Fermentation of Banana media by using κ-carrageenan immobilized *Lactobacillus acidophilus*, *Int. J. Food Microbiol.*, 91, 215–222.
61. Lee, K. Y., and Heo, T. R. (2000), Survival of *Bifidobacterium longum* immobilized in calcium alginate beads in simulated gastric juices and bile salt solution, *Appl. Environ. Microbiol.*, 66, 869–873.
62. Chandramouli, V., Kailasapathy, K., Peiris, P., and Jones, M. (2004), An improved method of microencapsulation and its evaluation to protect *Lactobacillus* spp. in simulated gastric conditions, *J. Microbiol. Methods*, 56, 27–35.
63. Lamas, C. M., Bregni, C., D'Aquino, M., Degrossi, J., and Firenstein, R. (2001), Calcium alginate microspheres of Bacillus subtilis, *Drug Dev. Ind. Pharm.*, 27, 825–829.
64. Beaulieu, L., Savoie, L., Paquin, P., and Subirade, M. (2002), Elaboration and characterization of whey protein beads by an emulsification/cold gelation process: Application for the protection of retinol, *Biomacromolecules*, 3, 239–248.
65. Guérin, D., Vuillemard, J. C., and Subirade, M. (2003), Protection of bifidobacteria encapsulated in polysaccharide-protein gel beads against gastric juice and bile, *J. Food Prot.*, 66, 2076–2084.
66. Hébrard, G., Blanquet, S., Beyssac, E., Remondetto, G., Subirade, M., and Alric, M. (2006), Use of whey protein beads as a new carrier system for recombinant yeasts in human digestive tract. *J. Biotechnology*, 127, 151–160.
67. Hongsprabhas, P., and Barbut, S. (1998), Ca^{2+}-induced cold gelation of whey protein isolate: Effect of two-stage gelation, *Food Res. Int.*, 30, 523–527.
68. Yvon, M., Beucher, S., Scanff, P., Thirouin, S., and Pelissier, J. P. (1992), In vitro simulation of gastric digestion of milk proteins: Comparison between in vitro and in vivo data, *J. Agric. Food Chem.*, 40, 239–244.
69. Remondetto, G. E., and Subirade, M. (2003), Molecular mechanisms of Fe^{2+}-induced-β-lactoglobuline cold gelation, *Biopolymers*, 69, 461–469.
70. Bienaimé, C., Barbotin, J. N., and Nava-Saucedo, J. E. (2003), How to build an adapted and bioactive cell microenvironment? A chemical interaction study of the structure of Ca-alginate matrices and their repercussion on confined cells, *J. Biomed. Mater. Res., A*, 67, 376–388.
71. Rossi-Alva, J. C., and Miguez Rocha-Leaõ, M. H. (2003), A strategic study using mutant-strain entrapment in calcium alginate for the production of *Saccharomyces cerevisiae* cells with high invertase activity, *Biotechnol. Appl. Biochem.*, 38, 43–51.
72. Molin, S. (1993), Environmental potential of suicide genes, *Curr. Opin. Biotechnol.*, 4, 299–305.

73. Diaz, E., Munthali, M., De Lorenzo, V., and Timmis, K. N. (1994), Universal barrier to lateral spread of specific genes among microorganisms, *Mol. Microbiol.*, 13, 855–861.

74. Kaplan, D. L., Mello, C., Sano, T., Cantor, C., and Smith, C. (1999), Steptavidin-based containment systems for genetically engineered microorganisms, *Biomol. Eng.*, 16, 135–140.

75. Hols, P., Defrenne, C., Ferain, T., Derzelle, S., Delplace, B., and Delcour, J. (1997), The alanine racemase gene is essential for growth of *Lactobacillus plantarum*, *J. Bacteriol.*, 179, 3804–3807.

76. Fu, X., and Xu, J. G. (2000), Development of a chromosome-plasmid balanced lethal system for *Lactobacillus acidophilus* with thyA gene as selective marker, *Microbiol. Immunol.*, 44, 551–556.

5.6

NASAL DELIVERY OF PEPTIDE AND NONPEPTIDE DRUGS

CHANDAN THOMAS AND FAKHRUL AHSAN
Texas Tech University Health Sciences Center, Amarillo, Texas

Contents

Pharmaceutical Manufacturing Handbook: Production and Processes, edited by Shayne Cox Gad

5.6.1 INTRODUCTION

The delivery of drugs via the nasal route has been practiced since ancient times; for example, psychotropic and hallucinogenic agents have been used as snuff in many parts of the world for hundreds and possibly thousands of years. More recently, especially over the past two decades, intranasal drug delivery has shown great promise in various fields of medical practice. At present, a number of conditions (e.g., rhinitis, migraine, nasal congestion, and osteoporosis) are being treated successfully with nasal formulations. In addition, an ever-increasing number of nasally delivered, systemically acting drugs are in the pipeline. Recently this form of therapy received encouragement with the approval of FluMist (MedImmuneVaccines, Gaithersburg, MD), an intranasal vaccine against *Haemophilus influenzae,* the influenza virus. This vaccine is the first to be given by the nasal route as a mist rather than by injection. With its approval, many of the pharmaceutical companies, including some giants of the pharmaceutical industry, are increasingly looking toward the area of nasal drug delivery. The market for such therapy in 2005 is reported to have reached $2.4 billion. With 16 of the 20 major pharmaceutical companies conducting active programs in this area, the field of nasal drug delivery is expected to grow at an estimated 33% annually [1]. As reported by Koch [2], the U.S. drug delivery market in 2005 was somewhere in excess of $50 billion and has been predicted to be around $67 billion by the year 2009, whereas the nasal drug delivery market is expected to be valued at $9 billion by 2008 [1–3]. Nasal administration—as compared with injection and oral administration—is more feasible and convenient, especially in view of the rising number of peptide and protein therapeutics that are rapidly being developed. In particular, the possibility of delivering drugs to the brain by the nasal route is eliciting increased interest, especially due to the possibility of accessing or targeting the local receptors and also of circumventing the blood–brain barrier.

Drug delivery via the nasal route offers a number of advantages, the most important of which is the possibility of needle-free treatment. It also means that—in addition to the newly developed peptide- and protein-based drugs—this method is also suitable for a wide variety and perhaps most of the drugs that are currently in use. However, it is not only convenience that sets nasal drug delivery apart: This method also provides a rapid onset of action and high bioavailability.

Because of its rich vasculature and highly permeable structure, the nasal route can be used as an alternative to parenteral routes of delivery. It circumvents hepatic first-pass metabolism and gut-wall enzyme-mediated degradation. It is also easily accessible for self-administration without the help of a health professional and there are no associated needle-stick hazards. Other advantages of nasal drug delivery systems include a rapid onset of action, reduced risk of overdose, and improved patient compliance. However, there are also several disadvantages, including the impermeability of the nasal mucosa to lipophilic and high-molecular-weight drugs, mucotoxicity associated with long-term use of some formulations, the requirement for an expensive delivery device, and possible dose inaccuracy.

In order to understand the delivery and absorption of drugs by the nasal route and appreciate the factors that may affect it, one must begin with a clear picture of the anatomy and physiology of the nose.

5.6.2 NASAL ANATOMY AND PHYSIOLOGY

5.6.2.1 Structure of Nasal and Olfactory Mucosa

The human nasal passage is about 12 cm long and runs from the nostrils to the nasopharynx (Figure 1). The nasal cavity is divided into right and left halves by a midline septum, or cartilaginous wall, that extends posteriorly into the nasopharynx (E). Each half of the nasal cavity consists of three well-separated regions: (A) the vestibule, (B) the olfactory region, and (C) the respiratory region. The vestibule is the most anterior part of the nasal cavity, which opens to the face through the

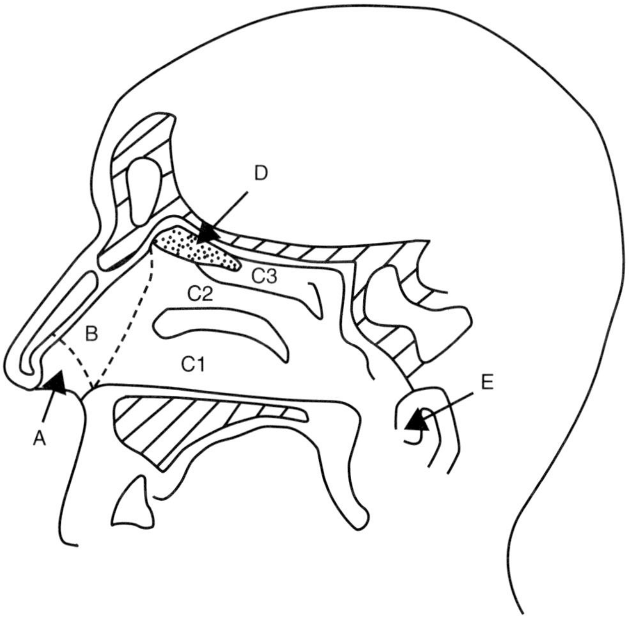

FIGURE 1 Sagittal section of nasal cavity showing nasal vestibule (A), atrium (B), respiratory area and inferior turbinate (C1), middle turbinate (C2) and superior turbinate (C3), olfactory region (D), and nasopharynx (E). (Reproduced from ref. 5 with permission of Pharmaceutical Press.)

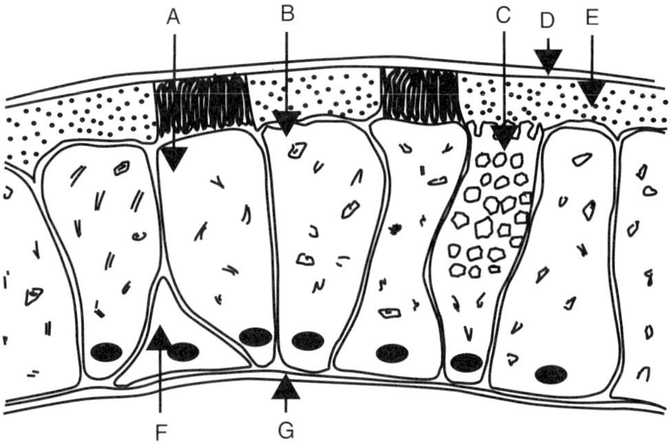

FIGURE 2 Cell types of nasal epithelium showing ciliated cell (A), nonciliated cell (B), goblet cell (C), gel mucous layer (D), sol layer (E), basal cell (F), and basement membrane (G). (Reproduced from ref. 5 with permission of Pharmaceutical Press.)

nostrils. The respiratory region and the turbinates make up most of the nasal cavity; this region has lateral walls that divide it into three chambers, including the superior nasal turbinate (C3) at the top, the middle nasal turbinate (C2) below, and the inferior nasal turbinate (C1) at the bottom (Figure 1). The total volume of nasal cavity is 15 mL; its surface area is 150 cm^2 [4, 5]. Air generally travels into the nose in an approximately parabolic pattern from the vestibule to the nasopharynx. The hard palatine bone and soft palate form the floor of the nose and the roof of the mouth. The human nasal epithelial surface is covered mainly by stratified squamous, olfactory, and respiratory epithelia. The stratified epithelium lies in the anterior part of the nose and becomes pseudostratified columnar epithelium at the posterior part, which constitutes the respiratory epithelium (Figure 2). The olfactory epithelium covers the olfactory region in the upper part of the nasal cavity. The nasal epithelial surface is covered by a continuous layer of mucus secreted by various mucosal and submucosal glands. The mucous layer comprises the gel layer (D) and sol layer (E) as shown in Figure 2. The sol layer is a low-viscosity fluid surrounding the cilia, and the viscous gel layer covers the tips of the cilia—fine, hairlike structures that move in an organized fashion to ease the flow of mucus across the epithelial surface. There are also microvilli, and every ciliated cell carries some 100 cilia. Each ciliated and nonciliated cell has about 300 microvilli. Figure 2 also shows both nonciliated cells and basal cells [4–6].

The olfactory mucosa is discussed further on in relation to the delivery of drugs via the nose to the brain.

5.6.2.2 Nasal Vasculature

The nasal surface is supplied with a dense network of blood vessels by the external and internal carotid arteries. Blood from the anterior part of the nose is drained through the facial vein, but the nose's main blood supply drains through the sphenopalatine foramen into the pterygoid plexus or via the superior ophthalmic vein

[6]. The nasal blood vessels can be greatly dilated with blood to facilitate warming and humidification of inspired air in response to prevailing conditions. Nasal blood flow is very sensitive to a variety of agents applied topically and systemically.

5.6.2.3 Enzymes and pH

Nasal secretions contain a mixture of secretory materials from the goblet cells, nasal glands, and lacrimal glands. The main constituents of nasal secretions are water, with 2–3% mucin and 1–2% electrolytes. Nasal secretions also contain several enzymes, including lysozyme, cytochrome P450–dependent monooxygenases, steroid hydroxylases, proteases such as neutral endopeptidase, leucine aminopeptidase, aminopeptidase peroxidase, carboxypeptidase N, and protease inhibitors [7]. However, because most studies of peptide and protein degradation are carried out in homogenates of nasal tissue, the peptides and proteins can be exposed to both intracellular and extracellular enzymes; therefore, the data regarding peptide stability must be interpreted with caution. The normal pH of nasal secretions in adults ranges from about 5.5 to 6.5; in young children, it ranges from 5.0 to 6.7. The nasal pH can vary depending on pathological conditions such as allergic rhinitis and environmental conditions such as cold and heat [8].

5.6.3 FACTORS INFLUENCING NASAL DRUG ABSORPTION

The nasal absorption of drugs is influenced by a multitude of factors, including nasal physiology, nasal pathology, physicochemical properties of drugs, dosage forms, and delivery method. The presence of pathological conditions such as allergic rhinitis and the common cold further complicates nasal drug delivery. Moreover, the intimate contact between the nasal mucosa and the atmosphere leads to variability in absorption with changes in temperature and humidity.

5.6.3.1 Physiological Factors

Nasal Blood Flow The nasal vasculature differs from the tracheobronchial tree due to the presence of (a) venous sinusoids, (b) arteriovenous anastosomes, and (c) the nasal vasculature, which shows cyclical changes of congestion, hence giving rise to the nasal cycle. In the nasal vasculature, the arterioles lack an internal elastic membrane, making the endothelial basement membrane continuous with the basement membrane of the smooth muscle cells. Also present are the fenestrated type of capillaries lying just below the surface epithelium and surrounding the glands. Because of this, the capillaries facilitate rapid movement of fluid through the vascular wall, allowing water to escape into the airway lumen. The conditioning of the inhaled air is greatly influenced by the nasal blood vessels. In essence, air is heated and humidified by the flow of nasal blood in the opposite direction to the incoming airflow. The nasal blood flow also controls the size of the nasal passage's lumen. Changes in ambient temperature and humidity, nasal administration of vasoactive drugs, nasal trauma, and compression of large veins in the neck may adversely affect blood flow in the nose [9, 10]. Other factors such as mood changes, hyperventilation, and even exercise can have an effect on the nasal blood flow and hence the nasal passages [10–12]. In conclusion, any change in blood flow can alter the absorption of a nasally administered drug.

Nasal Enzymes The enzymes present in the nasal cavity may be involved in the extensive enzymatic degradation of drugs administered nasally. The presence of the enzymes in the nasal epithelium acts as a defensive mechanism or a barrier against the entry of xenobiotics [9]. Examples include nasal decongestants, essences, anesthetics, nicotine, and cocaine, which are metabolized by the nose's P450-dependent monooxygenase system [13, 14]. The nasal mucosa also includes oxidative phase I and conjugative phase II enzymes. The phase I enzymes include aldehyde dehydrogenase, carboxyl esterase, and carbonic anhydrases; phase II enzymes include glucuronyl, sulfate, and glutathione transferases. 17β-Estradiol has shown significantly more conjugation when administered via the nasal route as compared to the intravenous route. A variety of other drugs have been shown to be metabolized by nasal enzymes, including progesterone, testosterone, and insulin [15, 16].

5.6.3.2 Nasal Mucociliary Clearance

Nasal mucociliary clearance is the transport of the mucous layer covering the nasal epithelium toward the nasopharynx by ciliary beating for its eventual discharge into the gastrointestinal tract. Nasal mucociliary clearance plays a very important role in the upper respiratory tract in preventing various noxious agents such as allergens, bacteria, viruses, and toxins from reaching the lungs. The ciliated cells of the nasal mucosa drive the movement of the mucus, and hence the physiological control of the ciliated cells and the rheological properties of the mucus determine the efficiency of the nasal mucociliary clearance system. In humans the normal mucociliary transit time is reported to be 12–15 min, and transit times of 30 min or more are likely to be an indication of impaired mucociliary clearance. Impairment of mucociliary clearance has been associated with longer contact times of various noxious agents as well as drugs with the nasal mucosa. On the other hand, increases in the mucociliary clearance rate decrease the contact between drug and the epithelium and consequently reduce drug absorption. Therefore nasal drug absorption can be augmented by the use of bioadhesive polymers or microspheres or by increasing the viscosity of the drug formulation [5, 17]. Hydroxypropyl methylcellulose, polyacrylic acid, and hyaluronan all enhance absorption by increasing nasal residence time [18–20]. The effect of mucociliary clearance may vary depending on the site of drug deposition. Ciliated epithelium is present in the middle and posterior parts of the turbinates, but there is little or no ciliary epithelium in the anterior regions of the nasal cavity [21, 22]. This is one of the reasons why a drug deposited in the posterior part of the nose is washed away more quickly than a drug deposited in the anterior site of the nasal cavity [23].

5.6.3.3 Pathological Condition of Nose

As mentioned earlier, the presence of nasal pathological conditions—such as allergic rhinitis, polyposis, and common colds—influences nasal drug absorption to a great extent. The majority of pathological conditions of the nose show bleeding, excessive secretion of mucus, nasal blockage, and crusting. It has been reported that excessive nasal secretion may wash away a nasally administered drug before it can be absorbed [24]. Nasal drug absorption and distribution are also influenced by the presence of nasal polyps and blockage. Several studies however have suggested that the presence of nasal pathological conditions do not affect nasally administered peptide drugs. For example, buserelin and desmopressin absorption studies have shown similar nasal absorption profiles in normal subjects and in those suffering from colds or rhinitis [25, 26].

5.6.3.4 Dose Volume and Site of Deposition

A dose of 25–200 μL per nostril is what can be maximally accommodated by the human nose. A dose higher than this will be drained off and hence shows lower absorption. Some studies have reported that a 100-μL volume resulted in a larger deposition area. Hence, taking into account the volume of administration becomes very important for manufacturers of nasal drug delivery systems. The site of deposition of the nasal formulation may also affect the nasal absorption of drugs since the anterior part of the nose provides greater contact between the nasal epithelium and drug, but the mucociliary clearance mechanism of the posterior tends to remove drug more rapidly [27]. It has been found that the permeability of the posterior area is greater than that of the anterior portion, and hence, based on the formulation, drugs may be administered in either the anterior or posterior parts of the nose. The nasal adapter's spray-cone angle defines the width of the nasal spray pattern and thus plays an important role in determining the site of deposition in the nasal cavity. Changes in the cone angle of the adapter from 60° to 35° or 30° can produce a larger and more posterior deposition and therefore higher drug deposition in the ciliated area [10].

5.6.3.5 Physicochemical Properties of Drugs

The absorption of a drug across the nasal mucosa is a function of its physicochemical properties, such as molecular weight, lipophilicity, and water solubility, as seen with most of the mucosal routes of delivery. The majority of studies on the effects of drug lipophilicity on nasal absorption are rather conflicting. The effect of lipophilicity on the nasal absorption of barbituric acids has been investigated. It was found that drug absorption through the nasal mucosa increases with an increase in the partition coefficient. Interestingly, there was only a fourfold increase in absorption between phenobarbital and barbital despite the fact that the partition coefficient of phenobarbital was 40-fold higher than that of barbital [28]. Similarly, increases in nasal absorption have been seen for hydrocortisone, testosterone, and progesterone with increases in the partition coefficient. However, a hyperbolic—rather than a linear—relationship was observed between the in vivo nasal bioavailability of a series of progesterone derivatives and their octanol–water partition coefficients [29]. In contrast to this, Kimura et al. [30] showed that, for a series of quaternary ammonium compounds structurally related to tetraethylammonium chloride, nasal absorption was inversely related to the partition coefficient. All these studies suggest that a drug's lipophilicity may not be an appropriate indicator of the extent of its nasal absorption. Besides the drug lipophilicity another important factor most studied for its influence on nasal absorption is the aqueous solubility of a drug. This is because the nasal mucosa is constantly kept moist by the nasal secretions and is well perfused with blood vessels. In addition to limiting drug absorption, it also is a limiting factor for the formulation of the drug. Further it is also important to understand the relation between the saturation solubility of the drug and drug absorption. However, the influence of drug solubility on absorption of drugs via the nasal route has not been significantly explored and needs much more attention. The relative effectiveness of nasal atropine and hyoscine was compared by Tonndorf et al. [31] by measuring each drug's capacity to arrest salivary secretion; they found that 0.65 mg of hyoscine, 40 times more soluble in water than atropine, is equivalent to 2 mg of atropine. Some authors have suggested that the aqueous pores in the nasal mucosa play a major role in absorption of hydrophilic drugs [32]. When nasal formulations are administered as inhaled powders or suspensions, drug dissolution rate becomes an important factor. Formulations deposited in the nostrils require proper dissolution for better absorption. Nasal mucociliary clearance may remove the drug if it remains undissolved in the nostrils. The size and shape of a drug molecule can affect its nasal absorption. Molecules with an average molecular weight less than 1000 Da are better absorbed nasally than higher molecular weight drugs whereas linear molecules are less effectively absorbed than compact ones [33]. The nasal and oral absorption of polyethylene glycol 600, 1000, and 2000 have been studied by Donovan et al. [34] in relation to molecular weight; they found that the greater the molecular weight, the less effective the absorption. This pattern of absorption was seen in the case of both the nasal and oral routes. The effect of water-soluble compounds such as 4-oxo-4*H*-1-benzopyran-2-carboxylic acid, *p*-aminohippuric acid, sodium cromoglycate, inulin, and dextran showing different molecular weights on the nasal absorption was studied by Fisher et al. [35]. A 43-fold decrease in the nasal absorption of the these compounds was observed with a 368-fold increase in the molecular weight [9]. Similarly, studies with 13 di-iodo-L-tyrosine-labeled dextran showed an inverse relationship between the percentage absorbed after nasal administration and the weight of the molecule; in fact, a 36 fold increase in molecular weight produced an 88-fold reduction in nasal absorption [36]. The effect of molecular weight on the nasal absorption of fluorescein isothiocyanate and diethylaminoethyl dextrans has been studied by Maitani et al. [37]. It was found that an inverse relationship between absorption and molecular weight existed for these compounds. However, since the nasal absorption of these compounds was low, enhancers were used; hence it is difficult to rule out the influence of the enhancers used on the extent of absorption obtained. The absorption of a drug after nasal administration is also influenced by the pH of a drug formulation as well as that of the nasal cavity—along with the pK_a of the drug substance. Biological membranes form a major barrier to the transport of drugs into the bloodstream. There are a number of transport mechanisms by which drugs are transported across the biological membranes. These include transcellular, paracellular, and carrier-mediated transport mechanisms. The most important factors that influence the above-mentioned mechanisms are the pH, pK_a, and partition coefficient of the drug. The pH of a nasal formulation should be in the range of 4.5–6.5 in order to minimize nasal irritation. However, the drug's pK_a must also be taken into account so as to maximize the drug's concentration in un-ionized form. The effect of pH on the nasal absorption of benzoic acid was studied by Hussain et al. [28], who showed that the absorption of benzoic acid is pH dependent.

5.6.3.6 Type of Delivery Device

Both the type of drug delivery system and the specific type of delivery device can affect drug absorption via the nasal route. The choice of delivery system depends mainly on the physiochemical properties of the drug, its desired site of action, and, more importantly, patient compliance and marketing aspects. The formulations most commonly used in nasal delivery are solutions, suspensions, gels, dry powders, and, most recently, nanoparticulate formulations.

Solutions are most commonly used for intranasal drug delivery. Such solutions may be used when the active ingredient is soluble in water or in some other vehicle approved by the Food and Drug Administration. At present, nasal solutions are available in the form of drops and sprays. Drops are the simplest and the most convenient nasal dosage form, and they are also easy to manufacture. However, their major drawback is that exact dosages cannot be administered with them. Another disadvantage is that they—like most solution-based medications—are vulnerable to microbial contamination; therefore, preservatives must often be added. These, in turn, have further disadvantages, as they may both cause irritation and hamper mucociliary clearance, thus decreasing compliance. Chemical stability is also often an issue with nasal drops.

Since the introduction of metered-dose inhalers, nasal solutions have increasingly been formulated as nasal sprays. Initially, aerosol-based systems containing chlorofluorocarbons were employed; however, the Montreal Protocol put an end to this. Thereafter, mechanical pumps or actuators were employed to deliver nasal formulations as sprays. These devices, using actuators, can precisely deliver as little as 25 μL and as much as 200 μL of a formulation. However, various factors must be considered in formulating the spray; these include viscosity, particle size, and surface tension, all of which may affect the accuracy of the dose administered.

Suspensions may also be used to deliver nasal formulations, though only rarely, since a number of complicating factors (e.g., particle size and morphology) must be considered. Suspensions offer the advantage of increasing residence time in the nasal cavity, thus possibly augmenting nasal bioavailability.

Gels are thickened solutions that may sometimes be used to deliver drugs via the nose, since they offer a number of advantages, such as reducing postnasal drip into the back of the throat and hence reducing the loss of the drug from the nasal cavity, anterior leakage, and the associated irritation. The use of gels is also reported to improve absorption and to mask the irritation associated with some ingredients by the addition of soothing agents and emollients. A vitamin B_{12} (cyanocobalamin) nasal gel, Nascobal (Nastech Pharmaceutical, Kirkland, WA), is available in a metered-dose formulation. Several other drugs, such as insulin, are being studied with a view to formulating them as nasal gels [9, 27, 38]. Although nasal powders are more stable than other formulations, they are rarely used because they tend to irritate the nasal tissue. However, a powder form may be useful when the active ingredient cannot be formulated as a solution or suspension. With the development of refinements in technology, many researchers are exploring the use of nanoparticle-based formulations to deliver drugs nasally. The main advantage of these state-of-the-art formulations is that they ensure increased absorption as well as better compliance. Microsphere- and liposome-based formulations are being increasingly tested. Some of these studies are discussed in Sections 5.6.6 and 5.6.7.

The type of nasal device employed in delivering a drug formulation plays a major role in the efficacy of the treatment. In general, two types of delivery systems are used: mechanical pumps and pressurized aerosol containers. The properties of the drug to be used influence the selection of the system. The various types of delivery devices are described in the following sections.

Unit-Dose Containers The unit-dose container offers a number of advantages. It is easy and convenient to carry and also does away with the need for preservatives, thus greatly increasing patient compliance. The unit-dose container is more accurate than the multidose container; metered-dose nasal sprays are still more accurate. The volume of drug held by such a container is usually determined by its filling volume, which greatly influences its accuracy. Another type of unit-dose container is based on an actuator and consists of a nasal adapter, or a small chamber that contains a piston [9]. Unit-dose containers are used mainly in emergencies (e.g., to manage pain), although they are not restricted to such use and may also be employed in instances where a single administration is required (e.g., vaccination). An example of a product that employs unit-dose containers is Imitirex (sumatriptan) (GlaxoSmithKline, Research Triangle Park, NC).

Squeeze Bottles This is a smooth plastic bottle with a jet outlet. When the bottle is pressed, a certain volume of its contents is atomized as the air inside the bottle is pushed out. This type of device is vulnerable to contamination as ambient air rushes into it following the release of pressure. The squeeze bottle is used mainly to deliver decongestants and not vasoconstrictors, as in the latter case the dose administered would be difficult to control [9].

Metered-Dose Nasal Pump Sprays Metered-dose pumps are the most widely used devices for the delivery of formulations via the nose. A number of commercially available products use this technology. The accuracy of the delivered dose is fairly high and makes it possible to administer dose volumes ranging from 25 to 200 μL. A metered-dose pump is made up of a container as well as a pump, valve, and actuator. The characteristics of the spray delivered will differ depending on the properties of the drug, the precompression mechanism, and the valve and pump selected. The length of the actuator is an important factor determining the deposition of the drug in the nose; the collection of residual drops on its tip will affect correct dosing.

Airless and Preservative-Free Sprays There are now pumps that prevent the entry of air into a dispensing device after use, thus increasing the stability of numerous compounds that are vulnerable to oxidation; this innovation has also minimized the use of preservatives. The working principle of these pumps is operation against a vacuum using a collapsible bag and a sliding piston. This is possible because the vacuum created when a dose is dispensed is accompanied by a reduction in the volume of the container, either by deforming the container itself or by dragging the sliding piston out of it. These maneuvers have no influence on the system's efficiency and, in fact, provide an advantage in that the container can be held in any position without significantly compromising the accuracy of the dose dispensed. This system is particularly suitable for use with children and bedridden hospitalized patients.

Whenever systems without preservatives are used for single or double doses, they pose little risk of contamination. However, multidose systems are generally used over a longer period of time; therefore, unless their formulations also include preservatives, the chances of contamination are increased. Scientists from Erich Pfeiffer and Qualis Laboratorium (both in Germany) have reviewed the latest trends in preservative-free nasal sprays and report that it is possible to prevent microbial contamination via the orifice in two ways. The first is by introducing a chemical additive, such as a bacteriostatic agent, into the nasal actuator so that it comes into contact with both the medication and the environment. However, in the case of an open system, the formulation within the actuator can still be contaminated. The second approach is the use of a mechanism whereby the system is sealed behind the orifice, thus preventing microorganisms gaining access [9, 39].

Some innovative technologies are being developed by a variety of pharmaceutical firms. The following section touches briefly on the latest of these.

Kurve Technology Kurve Technology has developed a unique system of controlled particle dispersion (CPD) by which it is possible to deliver drugs to the entire nasal cavity as well as the olfactory region and the paranasal sinuses. It uses the principle of *vortical* flow, by which inherent airflows of the nasal cavity are disrupted (Figure 3*a* and *b*). Its advantages include optimization of the size and trajectory of droplets, which makes it possible to saturate the nasal cavity. CPD also increases nasal residence time and reduces the deposition of compounds in the lungs and stomach, thus making the treatment more effective and efficient. ViaNase ID (Kurve Technology, Bothell, WA) is a CPD-powered electronic atomizer also developed by Kurve. Its advantages include generation of narrow droplet distribution between 3 and 50 μm and control of the atomization rate (i.e., the rate at which the droplets are generated and how rapidly they exit the device). CPD technology can be used to deliver both solutions and suspensions; currently, work is in progress to apply this principle to dry powders. Testing is also under way for the delivery of small and large molecules as well as peptides and proteins. Finally, CPD technology makes it possible to provide preservative-free packaging; unit-dose ampules; targeted deposition, as mentioned earlier; and monitoring of doses and compliance [40–42].

OptiNose OptiNose AS (Oslo, Norway) has introduced the novel idea of bidirectional intranasal drug delivery, which delivers a drug while the patient exhales and thus is said to prevent lung deposition. It utilizes the concept that exhalation against resistance leads to closure of the soft palate, thus separating the nasal cavity from the mouth as well as cutting off communication between the cranial surface of the soft palate and the posterior margin of the nasal septum. When this occurs, the air can enter one nostril through the sealing nozzle, turn 180°, and finally exit through the other nostril in the reverse direction. This concept is utilized in breath-actuated bidirectional delivery; that is, the air is blown out of the container and the sealing nozzle is used to direct its flow of air into the nose. When this approach was compared with conventional nasal drug delivery in 16 healthy subjects using ^{99m}Tc-labeled nebulized particles, it was found that bidirectional nasal delivery did, in fact, prevent deposition in the lungs. The single-use device is already developed and undergoing clinical testing for the delivery of a variety of compounds; a multidose liquid reservoir and powder delivery device are also being developed. The technol-

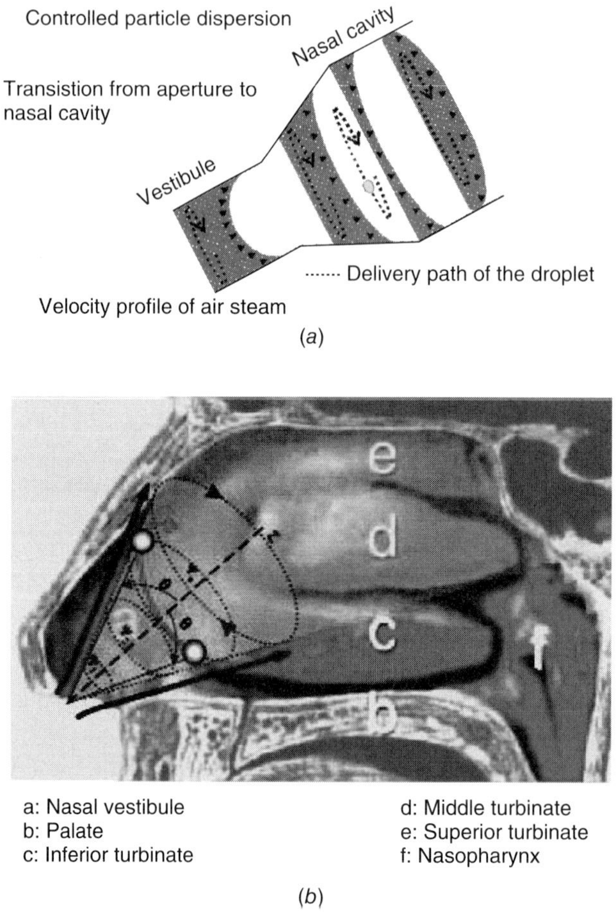

FIGURE 3 (*a*) Vertical droplet flow created by controlled particle dispersion used in ViaNase ID (Kurve Technology, Bothell, WA). (*b*) Deposition pattern produced by controlled particle dispersion. (Reproduced from ref. 42 with permission from *Drug Delivery Technology*.)

ogy is also being tested in connection with vaccination by the nasal route, and vaccines against diphtheria and influenza have already been shown to improve local and systemic immune responses. Another area being explored is the delivery of drugs to the brain via the nose. A phase I clinical trial of midazolam has been carried out and has shown an onset and level of sedation comparable to that of intravenous administration. The duration of sedation was found to be longer with the OptiNose technology, but the bioavailability of the drug was found to be only 68%, as compared with 100% with intravenous administration. Figures 4*a* and *b* show the Optinose multidose liquid device and multiuse powder device being developed respectively [43, 44].

DirectHaler The DirectHaler (Direct-Haler A/S, Copenhagen, Denmark) nasal delivery device takes advantage of the nasal anatomy and an innovation in device technology in order to improve nasal drug delivery and patient compliance. Similar

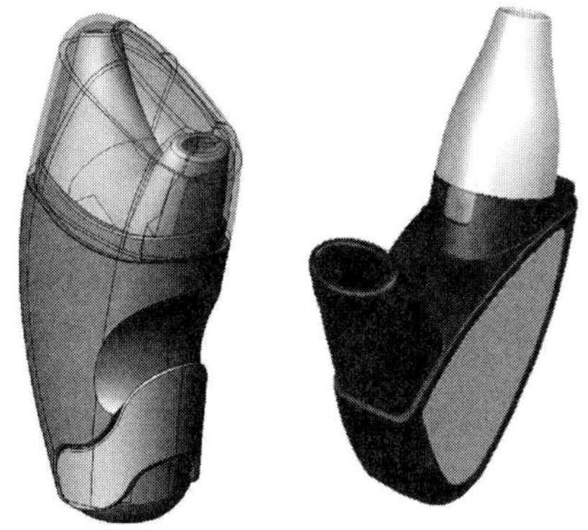

FIGURE 4 (*a*) Optinose multidose liquid device and (*b*) Optinose multiuse powder device. (Reproduced with permission of Per Gisle Djupesland by personal communication.)

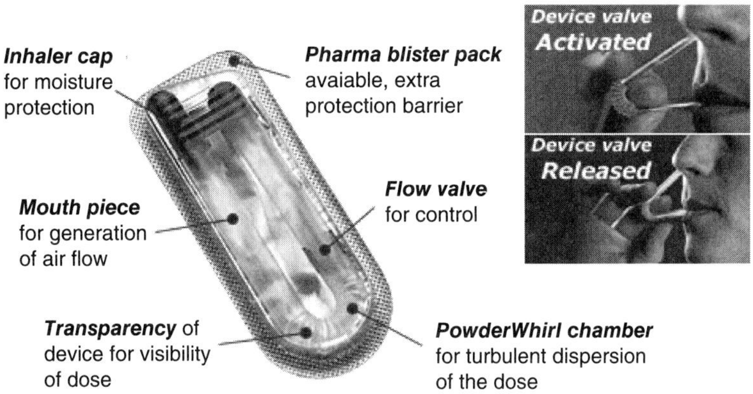

FIGURE 5 DirectHaler Nasal delivery device. (Reproduced with permission of Troels Keldmann by personal communication.)

to OptiNose, this novel drug delivery device avoids lung deposition. It takes advantage of the fact that, when air is blown out of the mouth against a particular resistance, the oral and nasal cavity airway passage closes on its own. Hence when a patient blows air into the DirectHaler Nasal device, the nasal dry powder dose is delivered into the nostril (Figure 5). The DirectHaler is reported to solve most of the problems that are associated with existing drug delivery devices, including dripping of the liquid dose out of the nostril following its delivery, swallowing of the dose after the administration and hence low absorption, and other problems of contamination associated with the liquid and multidose formulations. The device is quite easy and cost effective to manufacture, fill, and assemble using the latest high-speed technology. The tube of the device is made by using extrusion and roll forming while the device cap is manufactured by injection molding. A modified high-speed capsule-filling machine is used to carry out the powder dose filling. DirectHaler has also been developed for the combination of oral and nasal drug delivery. For pulmonary delivery DirectHaler Pulmonary has been developed. Further, for the complete targeting of the whole respiratory system, Novel DirectHaler Compliance system has been designed. The device has successfully undergone clinical trials for its use in nasal drug delivery and has shown patient compliance [45].

5.6.4 ANIMAL MODELS FOR NASAL ABSORPTION STUDIES

Nasal absorption studies can be carried out in mainly two types of animal models: the whole-animal model and the isolated organ perfusion model.

In vivo nasal absorption has been studied in a variety of mammals, including the rat, rabbit, dog, sheep, and monkey. Rats are the most widely used animals for testing intranasal drug delivery. They are easy to handle and simple and inexpensive to maintain. Use of the rat model for the nasal delivery of insulin was first described by Hirai and co-workers more than 25 years ago [46, 47]. The rat has been used in two basic models, in vivo and in situ—the latter also referred to as the ex vivo nasal perfusion model. Both models require anesthesia, though an in vivo model not requiring surgery has also been reported and is discussed later. In early studies of nasal absorption in rats, the animals were anesthetized and the passage of the nasopalatine tract was then sealed with acrylic glue to prevent the drainage of drug solution from the nasal cavity. The trachea was cannulated with a polyethylene tube, with another tube being inserted through the esophagus toward the posterior part of the nasal cavity. The drug solution was then delivered to the nose through either the nasal cavity or the esophageal tubing [8, 47]. Blood samples were then collected from the femoral vein.

However, a less traumatic and much more feasible rat model has recently been proposed by Pillion et al. and other groups [48, 49]. In this model, the rats are anesthetized and kept on their backs; then the drug solution is administered directly into the nose by inserting a pipette about 3–5 mm into each nostril. In some cases the drug is administered into only one nostril so as to prevent blockage. The rats then remain on their backs long enough for the formulation to come into contact with the nasal mucosa. Blood samples are then collected from the tail vein.

In support of the earlier models, it was argued that, by sealing the nasopalatine tract, the drug would necessarily be fully absorbed and transported into the circulation via permeation through the nasal mucosa. However, such nasal absorption with blockage of drainage from the nose to the mouth is not the normal physiological condition. Although the rat model has been used extensively by investigators throughout the world, application of the results of such studies to humans is very limited because of the small body size of these animals and significant interspecies differences. In fact, significant variability between the rat and the human was observed in studies of the bioavailability of insulin administered intranasally [50]. Furthermore, the use of anesthesia has raised concerns because of its potential to confound the test results.

In the rabbit model, drug solution is delivered by spray instillation into a nostril, keeping the rabbit's head in an upright position and allowing the rabbit to breathe normally. Blood samples are then collected from the ear vein. New Zealand White and Japanese White rabbits are most commonly used in such research. One of the advantages of the rabbit model is that the blood volume of these animals is large enough for multiple sampling and pharmacokinetic analysis [51].

Dogs, sheep, and monkeys can be kept conscious during nasal delivery to mimic the human [51]. Sheep, because of their large nostrils and docile nature, serve as excellent models for studies of this kind.

5.6.5 ENHANCEMENT OF INTRANASAL DRUG ABSORPTION

Lipophilic drugs or compounds have consistently been shown to be completely absorbed across the nasal mucosa; frequently, nasal absorption of these compounds is identical to that obtained with intravenous administration. In some reports, bioavailability after nasal administration reached almost 100% and $T_{\max}$ was similar to that obtained with intravenous administration. For example, lipophilic drugs such as propranolol, naloxone, buprenorphine, testosterone, and 17α-ethynylestradiol have been reported to be completely or almost completely absorbed after nasal administration in animal models. However, the same is not true of polar molecules and certain low-molecular-weight drugs and also high-molecular-weight peptides and proteins. In these instances absorption enhancers have been employed, and this is the method most widely used to improve nasal drug absorption [52–54]. Such absorption enhancers belong to a variety of different chemical groups and may have one or multiple ways of enhancing the absorption. Absorption enhancers work by (1) altering the mucous layer by decreasing its viscosity or disrupting it; (2) altering the tight junctions by sequestering extracellular calcium ions, which are reported to be essential in maintaining the integrity of these junctions; (3) inhibiting mucosal enzymatic degradation; (4) reverse-phase micelle formation—in certain cases, reverse-phase micelles may be formed within the cell membranes, thus creating an aqueous pore through which the drug can pass; and (5) altering membrane fluidity, which can be achieved when there is disorder in the membrane phospholipid component or leaching of proteins from the membrane or by a combination of these mechanisms [52–56]. Table 1 lists some of the selected absorption enhancers based on the mechanisms mentioned above.

5.6.6 NASAL DELIVERY OF PEPTIDE AND HIGH-MOLECULAR-WEIGHT DRUGS

Protein and peptide delivery by means other than injection is currently receiving enormous attention due to the increasing number of biotechnology-based products being developed. There have been numerous attempts to design systems for oral peptide, protein, and gene delivery, but these have unfortunately met with limited success, thus providing an impetus for exploring alternative noninvasive delivery methods. As mentioned earlier, more and more research has been directed to nasal drug delivery because of the numerous advantages it offers. The following section deals with the delivery of peptide and protein drugs.

5.6.6.1 Insulin

Insulin is produced by the β cells of the islets of Langerhans in the pancreas. It is made up of two peptide chains, which have 21 and 30 amino acid residues,

TABLE 1 Enhancement of Nasal Absorption

Type of Compound (*Absorption Enhancers*)	Examples	Mechanism of Absorption Enhancement
Bile salts and derivatives	Sodium deoxycholate, sodium glycocholate, sodium taurocholate, sodium taurodihydrofusidate, sodium glycodihydrofusidate	Alteration of tight junctions, membrane disruption, inhibition of mucosal enzymatic degradation, mucolytic activity
Chelating agents	Citric acid, EDTA, enamines, *N*-acyl derivatives of collagen, salicylates	Alteration of tight junctions
Enzyme Inhibitors	Amastatin, bestatin, camostat mesylate, boroleucine	Enzyme inhibition
Fatty acids and derivatives	Acylcarnitines, acylcholine, caprylic acid, capric acid, oleic acid, phospholipids, mono- and diglycerides, sodium laurate	Membrane disruption
Surfactants	Sodium lauryl sulfate, saponin, polyoxyethylene-9-lauryl ether, polyoxyethylene-20-lauryl ether, alkylmaltosides such as tetradecylmaltoside, dodecylmaltoside, and decylmaltoside	Membrane disruption

respectively. These two chains are held together by disulfide linkages between cysteine residues. Insulin is an anabolic and anticatabolic peptide hormone with a molecular weight of about 6000. It was first used in the successful treatment of diabetes mellitus in 1922. Today insulin is widely used in the treatment of both insulin-dependent and non-insulin-dependent diabetes mellitus [57–59]. A nonparenteral formulation of insulin has been approved only recently; before this development, insulin treatment required the use of painful injections. Exubera (Pfizer Labs, New York, NY) is the first insulin and first biotechnology-based medicine for the treatment of a systemic disorder that can be administered without an injection. It was developed by Nektar (San Carlos, CA) and is now a registered trademark of Pfizer (New York, NY). It still poses a few concerns, especially its effect on the lungs of patients with asthma or chronic obstructive pulmonary disease. Because of the problems associated with the parenteral injection of insulin, many diabetic patients flatly refuse to accept insulin therapy. As to subcutaneous injections, these fail to attain a physiological pattern of insulin owing to their adverse pharmacokinetics, and normoglycemia is often not achieved [47]. Of all the routes so far studied apart from delivery via the lungs, the nasal route would appear to be the most advantageous for the delivery of insulin.

The nasal delivery of insulin was demonstrated as early as 1922 by Woodyatt [59]. Since then, numerous studies have focused on this methodology. Some of the early studies included absorption of insulin from the nasal mucosa in human diabetics, the use of an insulin sprayer that contained saponin, and insulin in ethylene glycol or trimethylene glycol applied to the nose in the form of drops or sprays; the last of these methods demonstrated a significant fall in blood glucose levels in normal rabbits, dogs, and diabetic humans [60–62].

The enhancement of nasal absorption of insulin by hydrophobic bile salts has also been investigated. It was found that minor differences in the number, position, and orientation of the nuclear hydroxyl groups as well as alterations to side-chain conjugation can improve the adjuvant potency of bile salts. Moreover, the absorption of insulin positively correlated with an increase in the hydrophilicity of the steroid nucleus of the bile salts. In the presence of bile salts, nasal absorption of insulin reached peak levels within about 10 min, and some 10–20% of the dose was found to have been absorbed into the circulation. Marked increases in serum insulin levels were seen with sodium deoxycholate, the most lipophilic of the bile salts, whereas the least elevation—as well as least lowering of blood glucose levels—was seen with the most hydrophobic bile salt, sodium ursodeoxycholate [63].

Morimoto et al. [64] studied the nasal absorption of insulin using polyacrylic acid gel. When insulin was formulated with 0.1% w/v polyacrylic acid gel base (pH 6.5), the maximum hypoglycemic effect was seen 30 min following intranasal administration; in 1% w/v gel base, however, it took 1 h to reach the maximum effect. There was no effect of the pH (4.5, 6.5, and 7.5) of 0.1% w/v polyacrylic acid gel on the extent of nasal absorption.

Pillion and his group studied alkylmaltosides differing in chain length for their abilities to lower blood glucose levels when formulated with insulin [65]. Tetradecylmaltoside (TDM) was the most effective agent in producing the hypoglycemic effect, followed by dodecylmaltoside (DDM) and decylmaltoside (DM), all at concentrations of 0.060%. The onset of hypoglycemic action using these nasal drops was seen within 30 min and the maximum effect was obtained within 60–120 min. It was also demonstrated that insulin plus TDM at concentrations as low as 0.03% induced a hypoglycemic effect; however, insulin plus octylmaltoside (OM) failed to produce any hypoglycemic effect even at OM concentrations as high as 0.50%. Dodecylsucrose, which differs from DDM by only one carbohydrate residue, had a similar effect on blood glucose; however, decylsucrose was found to be less potent, and nonglucosides were able to enhance the nasal absorption of insulin only at concentrations ≥0.50% [65].

Insulin formulated with 0.06 or 0.125% hexadecylmaltoside produced a pronounced and rapid dose-dependent decrease in blood glucose levels after nasal administration. The effects of seven different alkylmaltosides were studied, and all the reagents (Figure 6) showed a similar maximal enhancement of insulin uptake when a concentration of 0.125% was employed. The figure demonstrates that TDM showed the greatest effect when concentrations of 0.03 and 0.06% were used.

Similar experiments were carried out using sucrose esters in nasal insulin formulations (Figure 7). It was observed that tetradecanoylsucrose and tridecanoylsucrose were more effective in stimulating insulin absorption as compared with decanoylsucrose and dodecanoylsucrose. But—compared with TDM at concentrations of 0.03%—the sucrose esters were less effective in promoting nasal absorption [66]. Sucrose cocoate (SL-40) is produced by the chemical esterification of coconut oil with sucrose; it has frequently been used in cosmetic and dental preparations as an excipient. When this excipient was formulated with insulin at 0.125, 0.25, and 0.5% concentrations, the associated plasma levels of insulin increased rapidly; whereas there was no enhancement of insulin plasma levels when insulin in saline was admin-

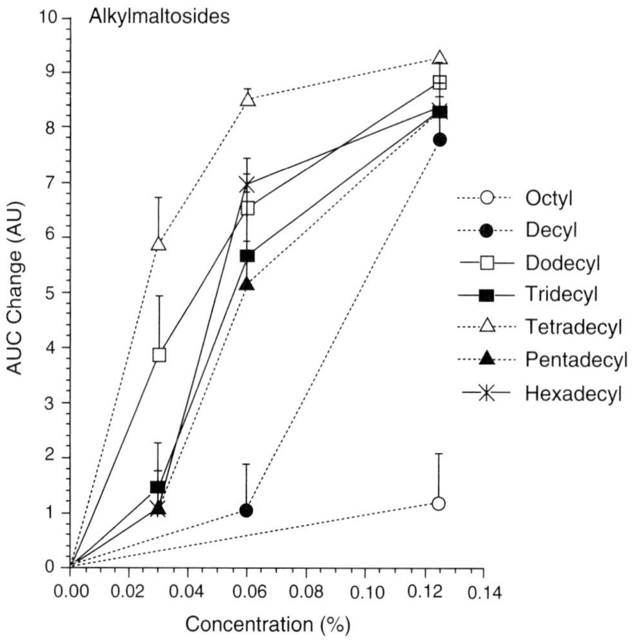

FIGURE 6 Changes in area under the curve (AUC_{0-120}) for blood glucose values in rats that received 2U insulin in presence of alkylmaltosides. Data represent mean change in AUC_{0-120} in arbitrary units (AU) ± standard error of the mean (SEM) compared with rats that received insulin formulated without alkylmaltoside ($n = 3, 4$). (Reproduced from ref. 66 with permission of John Wiley & Sons.)

istered. The levels of insulin in plasma increased from a baseline value of 10 µU/mL to a level of 200 µU/mL after the nasal administration of sucrose cocoate-containing formulations; the T_{max} was found to be 10 min (Figure 8) [67].

Chitosan is a linear cationic polysaccharide made up of copolymers of glucosamine and *N*-acetylglucosamine. It is commercially obtained by alkaline deacetylation of chitin [53, 68] and has been used for the nasal delivery of a number of drugs. The usefulness of chitosan in the enhancement of nasal absorption was reported first by Illum [69]. Later, Illum and his group also published experimental results indicating that solution formulations with 0.5% chitosan promoted the absorption of nasally administered insulin in rat and sheep [70].

The use of chitosan nanoparticles in the enhancement of the nasal absorption of insulin has also been investigated in rabbits. Chitosan nanoparticles were prepared by ionotropic gelation of chitosan and pentasodium tripolyphosphate (TPP). Two types of chitosan were used in the hydrochloride salt form (Seacure® 210 Cl and Protasan® 110 Cl). Insulin loaded in chitosan 210 Cl produced a significant increase in systemic absorption and also the greatest decline in the level of blood glucose, as much as 60% of basal levels; this result was found to be significantly different from that obtained with the insulin control solution as well as the insulin–chitosan solution [71]. A novel chitosan nanoparticle formulation was prepared by again employing ionotropic gelation of TPP and chitosan glutamate (A1), and postloaded insulin–chitosan nanoparticles (A2) were also prepared. Both these

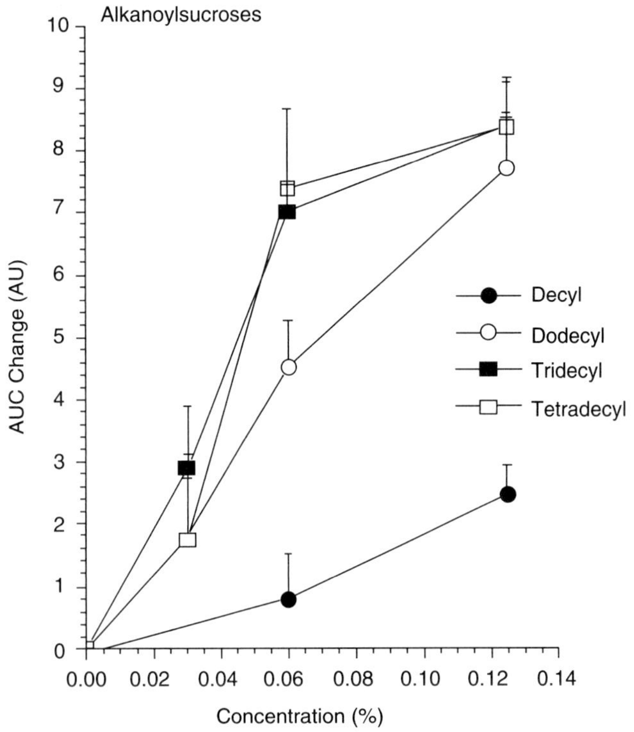

FIGURE 7 Changes in area under the curve (AUC_{0-120}) for blood glucose values in rats that received 2U insulin in presence of alkanoylsucroses. Data represent mean change in AUC_{0-120} in arbitrary units (AU) ± SEM compared to rats that received insulin formulated without alkanoylsucroses ($n = 3, 4$). (Reproduced from ref. 66 with permission of John Wiley & Sons.)

novel formulations were tested in rat and sheep models and also compared with the insulin–chitosan solution formulation (A3) and subcutaneous injection of insulin. In the rat model, it was found that A3 performed better than either A1 or A2. A1 showed a minimum blood glucose concentration (C_{min}) and time to reach C_{min} (T_{min}) of 40% and 90 min, respectively. The F_{dyn}—calculated as individual area over the curve ($AOC_{IN\ or\ SC} \times dose_{SC}/mean\ AOC_{SC} \times dose_{IN\ or\ SC}) \times 100$—was found to be around 48%, whereas the insulin–chitosan nanoparticles showed F_{dyn} values of 38 and 37 for the A1 and A2 formulations, respectively.

Since the concentrations of insulin to be administered in the sheep model would have been large, the insulin-loaded chitosan nanoparticles were not investigated in that model. However, the pharmacodynamics and pharmacokinetics of various insulin–chitosan preparations were compared with postloaded insulin–chitosan nanoparticles. It was found that chitosan solution and chitosan powder formulations were far better, with the chitosan powder formulation showing a bioavailability of 17% as against 1.3 and 3.6% for the chitosan nanoparticles and chitosan solution [72]. The effects of the concentration and osmolarity of chitosan and the presence of absorption enhancers in the chitosan solution on the permeation of insulin across the rabbit nasal mucosa in vitro and in vivo were investigated, and the same

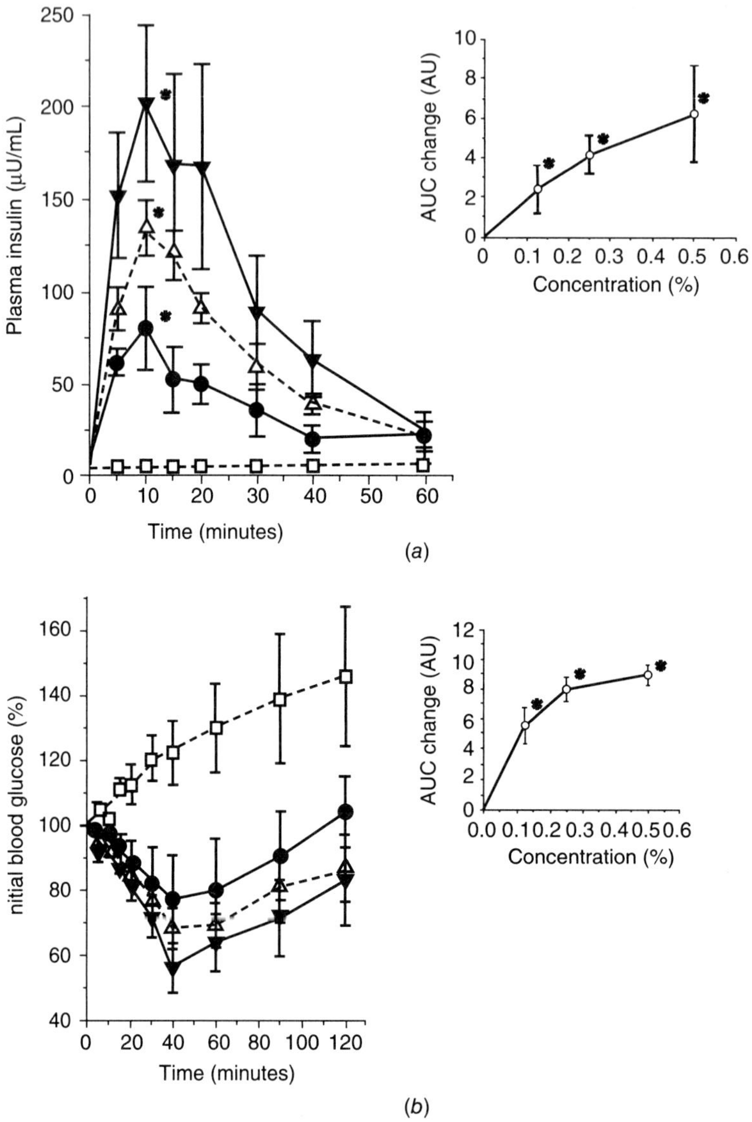

FIGURE 8 Changes in (*a*) plasma insulin levels and (*b*) blood glucose levels after nasal administration of 0.5 U insulin formulated in saline (□) or 0.125% (•), 0.25% (▼), and 0.5% (%) sucrose cocoate. Blood glucose concentrations at time 0 (250–350 mg/dL) were normalized to a value of 100% in each animal. Data represent mean ± SEM, $n = 3$. Inserts represent changes in plasma insulin AUC_{0-60} (*a*) and changes in blood glucose AUC_{0-120} (*b*). (Reproduced from ref. 67 with permission of Elsevier.)

investigation was also carried out in rats. The results suggested that, by increasing the concentrations of chitosan from 0 to 1.5%, insulin permeation in vitro could be increased. As compared with insulin without chitosan, there was a 25-fold increase in the permeability coefficient of insulin with 1.5% chitosan. A similar increase in permeability was seen in the case of a hyperosmotic solution, and there was also an

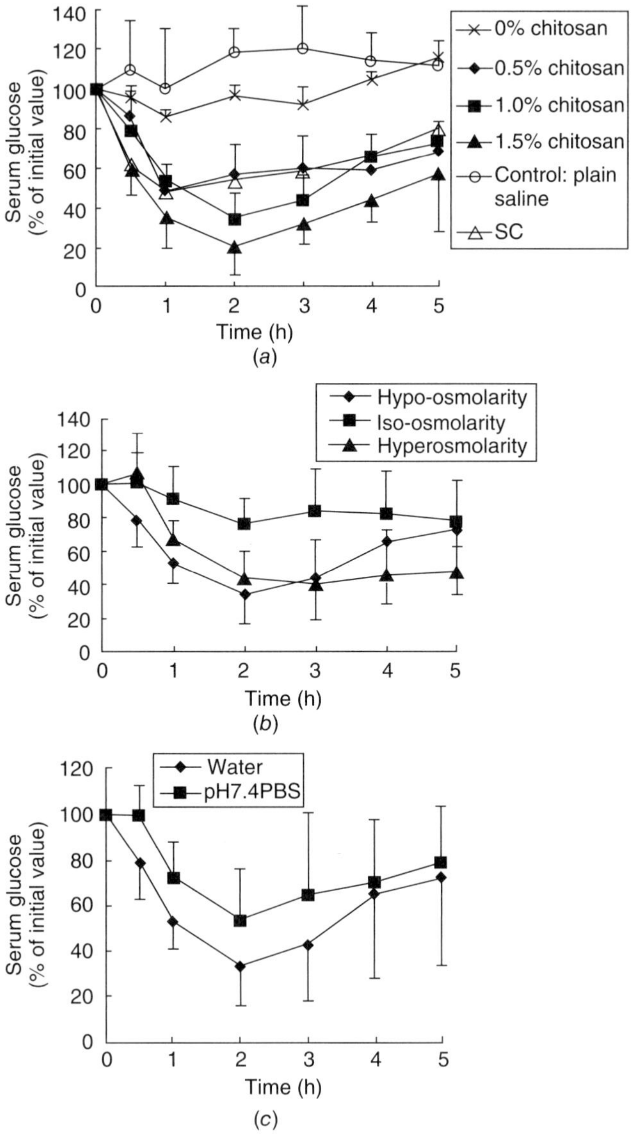

FIGURE 9 Effect of (*a*) concentrations, (*b*) osmolarity, and (*c*) medium of chitosan solution on mean serum glucose concentrations after nasal administration of 10 IU/kg insulin to rats. Bars represent the standard deviation (SD) of five experiment. (Reproduced from ref. 73 with permission of Elsevier.)

increase in permeability when insulin was formulated in deionized water as compared with phosphate buffer 7.4 (Figure 9). An increase in permeability was also seen when 5% hydroxypropyl β-cyclodextrin (HPβ-CD) plus 1% chitosan was included with insulin as compared with chitosan alone. However, there was no statistical difference in permeability when insulin was formulated with 0.1% ethylenediaminetetraacetic acid (EDTA) and 1% chitosan. In the in vivo studies in rats, insulin without chitosan did not show any reduction in blood glucose levels, but as the concentrations of chitosan were increased, a hypoglycemic effect was seen. With a formulation of 0.5% chitosan, the nadir in glucose levels was observed within 1 h after administration; with 1 and 1.5% chitosan–insulin formulations, the lowest levels were reached in about 2 h. In the case of isoosmotic formulations, the lowering of blood glucose levels was weaker as compared with the hypo- or hyperosmolar solutions; with deionized water and EDTA, effects similar to those of the permeability studies were seen in terms of lower serum glucose levels. When insulin was formulated with 1% chitosan, there was also a decrease in serum glucose levels, and a similar effect was seen with both 1% chitosan and 0.1% EDTA. A similar effect was also seen in the case of 5% Tween 80. Insulin formulated with 5% HPβ-CD and 1% chitosan was more effective in reducing serum glucose levels than when 5% HPβ-CD or 1% chitosan was used alone. These studies suggested that the concentrations, osmolarity, medium, and inclusion of absorption enhancers in chitosan solution influence absorption following the nasal delivery of insulin [73].

Varshosaz and co-workers recently explored the use of chitosan microspheres and chitosan gels in the nasal delivery of insulin [74]. They prepared microspheres of chitosan, and insulin was loaded into the microspheres. The formulations were administered through the nasal cavity by the method described earlier by Hussain et al. [47]. Varshosaz and colleagues found marked differences in the AUC of blood glucose reduction and the AUC of insulin concentration between the untreated controls and those animals that were treated either by intravenous administration and insulin-loaded chitosan (using ascorbyl palmitate as the cross-linking agent) or with microspheres. Serum glucose reduction in diabetic rats with nasal insulin–chitosan microspheres was around 67% in the intravenous group, and absolute bioavailability (F_{abs}) was around 44% [74]. Furthermore, the same researchers prepared insulin–chitosan gels containing different enhancers and investigated their nasal absorption. As seen in Figures 10 and 11, when EDTA was employed as an absorption enhancer in the chitosan gels and administered nasally, a significant increase in insulin absorption and a decrease in serum glucose levels by as much as

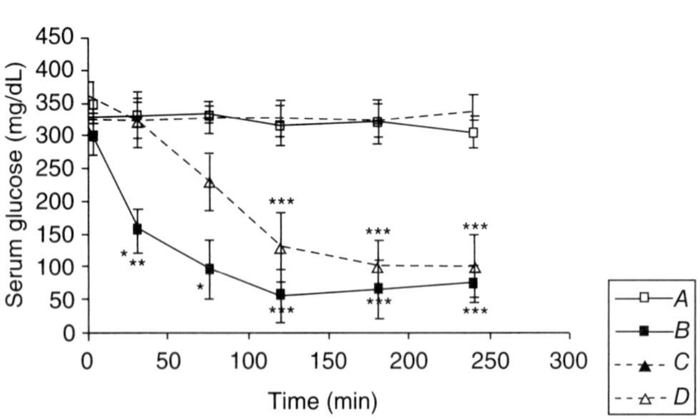

FIGURE 10 Serum glucose level in four groups of diabetic rats ($n = 6$): *A*, untreated control group; *B*, intravenous administration of 4 IU/kg insulin; *C*, nasal administration of blank gel base; *D*, nasal administration of 100 μL/kg chitosan gel containing 4000 IU/dL insulin. (Reproduced from ref. 75 with permission of Taylor & Francis.)

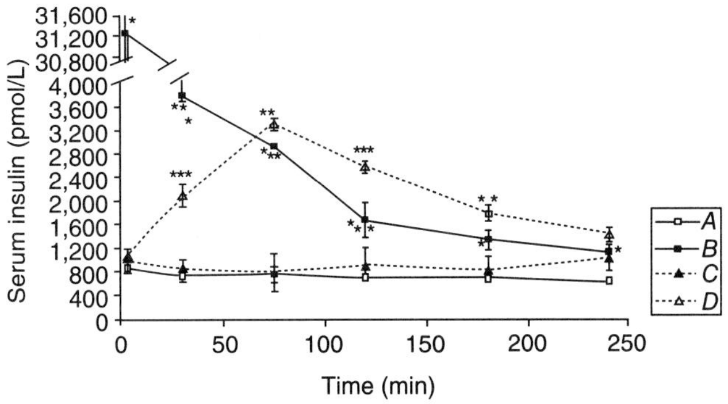

FIGURE 11 Serum insulin levels of four groups of diabetic rats ($n = 6$): *A*, untreated control group; *B*, intravenous administration of 4 IU/kg insulin; *C*, nasal administration of blank gel base; *D*, nasal administration of 100 μL/kg chitosan gel containing 4000 μg/dL insulin. (Reproduced from ref. 75 with permission of Taylor & Francis.)

46% compared to the intravenous route of administration were obtained. The authors suggested that this formulation would be beneficial in the controlled delivery of insulin by the nasal route [75].

5.6.6.2 Calcitonin

Calcitonin is a 32–amino-acid peptide with a molecular weight of 3418 that is cleaved from a larger prohormone. It has a single disulfide bond, which causes the amino terminus to assume a ring shape. Calcitonin is a hormone that participates in calcium and phosphorus metabolism. The major source of calcitonin in mammals is the parafollicular or C cells in the thyroid gland; it is also synthesized in other tissues, including the lungs and intestinal tract. When serum calcium levels are elevated, calcitonin is released from the thyroid gland. Salmon calcitonin (sCT) is more potent and longer lasting than the mammalian form and hence is used clinically. Calcitonin's main action is to reduce the plasma concentration of calcium. At pharmacological doses, calcitonin brings about reduction in bone resorption. It is indicated for the treatment of postmenopausal osteoporosis in women with low bone mass relative to healthy premenopausal women. The marketed version of intranasal salmon calcitonin is Miacalcin (calcitonin-salmon) Nasal Spray (Novartis Pharma AS, Huningue, France). Up to now, this calcitonin treatment has been approved only for treatment in women. It has been reserved as a second-line treatment, since it reduces fracture risk less than do other available treatments for osteoporosis. Miacalcin Nasal Spray, first manufactured by Sandoz Pharmaceuticals, was approved by the FDA in 1995 and is now distributed by Novartis Pharmaceuticals in the United States.

Pontiroli et al. [76] looked at the intranasal absorption of calcitonin in normal subjects. Their study included six healthy volunteers who had no family history of endocrine or metabolic diseases. Human calcitonin (Cibacalcin; Ciba-Geigy) was administered intravenously or mixed with sodium glycocholate, a surfactant, in distilled water and instilled as nose drops. Plasma concentrations of calcitonin were found to be consistently higher when compared with intranasal administration of the same dose of calcitonin. However, it was found that intranasal administration effected a reduction in the plasma concentrations of calcium to a similar extent as that seen when a similar dose of intravenous calcitonin was given. The data also showed that although a higher dose of calcitonin by the intranasal route brought about a higher plasma calcitonin concentration, there was no difference in the decrease in plasma concentrations when calcitonin was given at doses of 500 and 1000 μg.

Polyacrylic acid aqueous gel enhances the absorption of calcitonin after nasal as well as rectal administration. When [Asu1,7]-eel calcitonin (10 U/kg) was administered nasally in polyacrylic acid gel at a concentration of 0.1% w/v, a prominent hypocalcemic effect was seen in the first 30 min. Nasal administration of [Asu1,7]-eel calcitonin in saline had no hypocalcemic effect at the same dose when given by the nasal route. In addition to this, the effect of [Asu1,7]-eel calcitonin in the dose range of 1–10 U/kg has also been studied. The resulting data showed that a rapid reduction in plasma calcium concentrations can be achieved at doses of 5 and 10 U/kg; however, at doses of 1 U/kg only a small reduction in the plasma calcium concentration was observed, suggesting that polyacrylic acid gel can be used for the intranasal administration of peptides such as calcitonin. The possible side effects, however, were not known at the time the study was performed [76–78].

Commercially available sCT has been used mainly in the treatment of bone-related diseases such as Paget's disease, hypercalcemia, and osteoporosis. It can be used to alleviate pain due to its analgesic properties; hence there may be a need for its frequent administration. As reported in the literature, owing to chemical and enzymatic degradation, the polypeptide has a short half-life of only about 14 min. Hence its use calls for measures to improve its in vivo stability and overcoming the problem of rapid clearance. Polyethylene glycol (PEG) has been extensively used for this purpose in association with a variety of agents. Lee at al. [79] have prepared and characterized PEG-modified sCT and studied how blood clearance is affected with and without PEGylation. Succinimidyl carbonate mono-methoxy-polyethylene glycol (SC-mPEG) was prepared using mPEG, which has a molecular weight of 12,000, and bis(*N*-succinimidyl) carbonate as per the procedure of Miron and Wilcheck [80]. There are three main reactive sites for the activated PEG in sCT: the N-terminal amino group (Cys^{1}) and the ε-amino groups of two lysine residues (Lys^{11} and Lys^{18}), as shown in Figure 12. The site of conjugation of the PEG on sCT determines the stability of the sCT in the face of enzymatic attacks; hence the three possible mono-PEG-sCTs—Cys^{1}-PEG-sCT, Lys^{11}-PEG-sCT, and Lys^{18}-PEG-sCT—can withstand the effects of proteolytic enzymes. It was found that, on PEGylation, the plasma half-life improved to 11.2 and 54.0 min for the mono-PEG-sCT and the di-PEG-sCT as compared with the non-PEGylated sCT, which showed a plasma half-life of about 4.7 min (Figure 13).

Cys^{1}-Ser-Asn-Leu-Ser-Thr-Cys-Val-Leu-Gly-Lys^{11}-Leu-Ser-Gln-Glu-Leu-His-Lys^{18}-Leu-Gln-Thr-Tyr-Pro-Arg-Thr-Asn-Thr-Gly-Ser-Gly-Thr-Pro-NH_2

FIGURE 12 Primary structure of salmon calcitonin. Possible PEGylated sites are Cys^{1}, Lys^{11}, and Lys^{18}. (Reproduced from ref. 79 with permission of Taylor & Francis.)

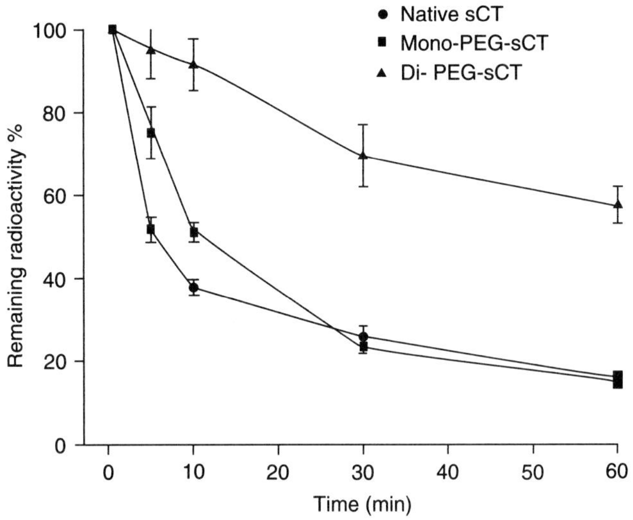

FIGURE 13 Blood clearance of PEGylated salmon calcitonins in rat. (Reproduced from ref. 79 with permission of Taylor & Francis.)

The same group—Shin et al. [81] —did nasal absorption studies of a low-molecular-weight PEG (2000) instead of the 5000 and 12,000 they tried previously. They also used commercially available succinimidyl-propionated monomethoxy-poly(ethylene glycol)-2000 (SP-mPEG) for the chemical modification instead of synthesizing it, as in their previous studies. The PEGylation of sCT was done by mixing SC-mPEG and sCT, and this mixture was shaken at 25°C. The reaction mixture was stopped by using an excess of 1.0 *M* glycine solution. This conjugated mixture was then subjected to size exclusion chromatography. Radioiodination of the sCT and the PEG-sCT were carried out and the radiolabeled ^{125}I-sCT and ^{125}I-mono-PEG2000-sCT were then used for the nasal absorption studies.

Tissue distribution studies were also done in rats after nasal administration. As seen in Figure 14 and Table 2, it was found that the elimination half-life of the unmodified sCT was 199 min, whereas the SP-mPEG2000-modified sCT showed an increased terminal elimination with a half-life of 923 min. It was also found that the SP-mPEG2000-modified sCT took a significantly longer time to reach its maximum concentration, 520 min, as compared with the 77 min for the unmodified sCT, and the AUC was found to be 20,638 μg/min/mL, which is much higher than the 3650 μg/min/mL for the unmodified sCT. The authors reported that the increase in the terminal half-life observed could be due to a flip-flop phenomenon. Also, when the tissue distribution of the formulation was examined 12 h after administration, the highest radioactivity was found in the liver. The details of the biodistribution studies are as shown in Table 3.

The same group [82] further studied the stability of these mono-PEG2000-sCT and the unmodified sCT in the rat nasal mucosa. It was found that PEGylated sCT exhibited significant resistance against trypticlike and nonspecific enzymatic degradation. Ahsan et al. [49] showed that when sCT was formulated with alkylglycosides, bioavailability was enhanced following both nasal and ocular administration. Miacalcin (Novartis Pharma AS, Huningue, France) was used to prepare the formulation

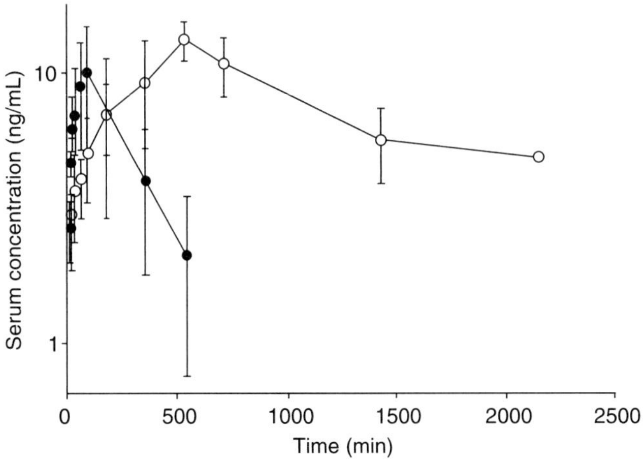

FIGURE 14 Average serum concentration of intact sCT–time curves following nasal administration of unmodified sCT (•) and Mono-PEG2K-sCT (○) in rats (n = 9 each). (Reproduced from ref. 81 with permission of The Pharmaceutical Society of Japan.)

TABLE 2 Pharmacokinetic Parameters (Mean ± SD) of Unmodified sCT and Mono-PEG2k-sCT obtained after nasal administration to rats (n = 9 each)

Parameter	Unmodified sCT	Mono-PEG$_{2k}$-sCT
C_{max} (ng/mL)	10.5 ± 4.7	12.9 ± 3.0
t_{max} (min)	77 ± 22	520 ± 167[a]
$t_{1/2,\lambda z}$ (min)	199 ± 97	923 ± 389[a]
Cl/F (mL/min)	7.4 ± 5.2	1.3 ± 1.0[a]
Vss/F (mL)	1802 ± 811	1392 ± 450
AUC (ng·min/mL)	3650 ± 1894	20638 ± 9486[a]
AUC/D (ng·min/mL/ng)	0.18 ± 0.09	1.03 ± 0.48[a]
MRT$_{\mu}$a (min)	314 ± 131	1505 ± 560[a]

Source: From ref. 74.

[a]Significantly different from unmodified sCT ($p < 0.05$).

TABLE 3 Extent of Total Radioactivity (Mean ± SD) in Various Body Organs after Nasal Administration of Unmodified sCT and Mono-PEG2k-sCT to Rats (n = 9 each)

	Radioactivity (%) in Whole Organ[a]	
Tissue	Unmodified SCT	Mono-PEG$_{2k}$-sCT
Liver	0.80 ± 0.41	1.03 ± 0.65
Kidney	0.30 ± 0.14	0.52 ± 0.24
Lung	0.18 ± 0.12	0.20 ± 0.12
Heart	0.06 ± 0.03	0.13 ± 0.05[b]
Spleen	0.04 ± 0.02	0.10 ± 0.05
Thyroid	0.04 ± 0.03	0.05 ± 0.02

Source: From ref. 74.

[a]Determined 12 h after administration.

[b]Significantly different from unmodified sCT ($p < 0.05$).

in solutions containing different concentrations of tetradecylmaltoside and octylmaltodise. These formulations were administered as described by Ahsan et al. [49]. It was found that when calcitonin was formulated in saline, absorption after administration by the nasal route was negligible; a similar result was seen with 0.125% OM. However, when calcitonin was formulated with 0.125% TDM, absorption was found to be increased, and maximal absorption (T_{max}) was achieved after 10 min. The AUC_{0-40} was found to be fourfold higher than with saline and OM formulations. The bioavailability was found to be 53% as compared with intravenously administered calcitonin at the same dose of 2.2 U. The AUC_{0-40} was found to be 6250 pg/mL·min as compared with 3500 pg/mL·min when the concentration of TDM was increased to 0.25% from 0.125%. It was also found that in the absence of the absorption enhancer increasing the amount of calcitonin from 2.2 to 22 U did not increase absorption by the nasal route. However, when formulated in the presence of 0.25% TDM, there was a 23-fold increase in the relative bioavailability of calcitonin. Also, when calcitonin was formulated with 0.25% TDM and given as nose drops, there was a significant reduction in the plasma calcium concentration as compared with the saline formulation.

5.6.6.3 Low-Molecular-Weight Heparins

Low-molecular-weight heparins (LMWHs) are fragments of natural heparin that are obtained by either enzymatic degradation or chemical depolymerization of unfractionated heparins (UFHs). The molecular weight of LMWHs, a heterogeneous mixture of sulfated glycosaminoglycans, is about one-third that of UFHs. Owing to the variations in the distribution of molecular weights, they show differences in their affinity for plasma proteins, activity against factor Xa, and thrombin, as well as duration of activity. They have been proven to be useful in the treatment and prevention of deep vein thrombosis. Of late, LMWHs have been preferred to therapy with conventional heparin. However, one of the main disadvantages of the use of LMWHs on a regular, noninstitutional basis is that they must be delivered by the subcutaneous or intravenous route. There have been concerns regarding patient compliance, longer hospital stays, and the need for skilled health professionals for therapeutic drug monitoring and administration. This has prompted a number of researchers to seek alternative forms of delivery.

The nasal administration of LMWHs was investigated in Sprague-Dawley rats using TDM as an absorption enhancer [83]. TDM was used at concentrations of 0.125 and 0.25%. Lovenox (Aventis Pharmaceuticals, Bridgewater, NJ) (enoxaparin sodium injection), a commercially available LMWH, was prepared with 0.25% TDM and the nasal absorption was compared with and without TDM. It had been previously reported that a plasma anti–factor Xa level of 0.20 U/mL or higher is required for an antithrombotic effect to be considered therapeutic [83]. It was found that enoxaparin when formulated only in saline did not produce a therapeutic anti–factor Xa level. However, when enoxaparin was formulated with 0.25% TDM, a significant increase in the AUC and C_{max} for the anti–factor Xa level was observed. Dalterparin, another commercially available LMWH, showed a similar response. When UFH was formulated with TDM, it produced an increase in the anti–factor Xa levels as compared with saline, but it was not in the therapeutic range (Figure 15).

The bioavailability (F_{abs}) of enoxaparin achieved by the subcutaneous route was found to be 83% of that achieved by the intravenous route. In the absence of TDM,

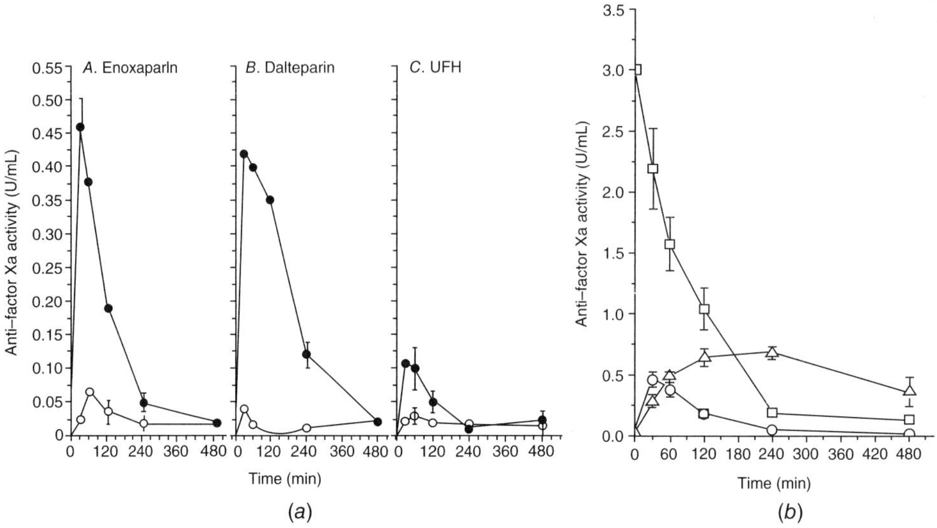

FIGURE 15 (*a*) Nasal administration of 100U of (*A*) enoxaparin, (*B*) dalteparin, or (*C*) UFH formulated with (•) and without (○) 0.25% tetradecylmaltoside. Data represent mean ± SEM, $n = 3$. Asterisks indicate results that are significantly different from those obtained with the drug formulated with saline, $P < 0.05$. (*b*) Administration of 100U of enoxaparin via the subcutaneous (▲), intravenous (□), and nasal (○) routes. Nasal administration was performed with a formulation that included 0.25% TDM. Data represent mean ± SEM, $n = 3$. (Reproduced from ref. 83 with permission of John Wiley & Sons.)

the nasal F_{abs} was 4.0%, but in the presence of 0.25% TDM, the F_{abs} was 19% compared with the intravenous route. The relative bioavailability (F_{rel}) of nasal enoxaparin plus 0.25% TDM was found to be 23% compared with the subcutaneous route.

The use of alkanoylsucroses in the enhancement of nasal absorption of LMWHs was investigated by our group [84]. As seen in Figure 16*a*, enoxaparin plus 0.125% octanoylsucrose showed no significant increase in anti–factor Xa levels; even when the concentration was increased to 0.5%, it barely reached the therapeutic level of 0.2U/mL. Similar results were reported for 0.125% decanoylsucrose; however, when the concentration was increased to 0.25 or 0.5%, there was an appreciable and rapid rise in anti–factor Xa levels (Figure 16*b*). The inability of octanoylsucroses and 0.125% decanoylsucrose to increase anti–factor Xa levels is attributed to the critical micellar concentration as reported by the authors [84]. In the case of dodecanoylsucroses, all the concentrations produced anti–factor Xa levels well above the therapeutic level required for an antithrombotic effect (Figure 16*c*).

Our group also compared the efficacy and potency of alkanoylsucroses with those of the well-known absorption enhancer 1% sodium glycocholate (Figure 16*d*); the results were similar to those seen with 0.5% dodecanoylsucrose. Also, 0.5% dodecanoylsucrose showed the highest increase in C_{max}, and it was found that an increase in the concentration of alkanoylsucroses led to a subsequent increase in the C_{max}. When the absolute and relative bioavailabilities of nasal LMWH plus 0.5% dodecanoylsucrose were compared with those of 1% sodium glycocholate, similar profiles were found [84].

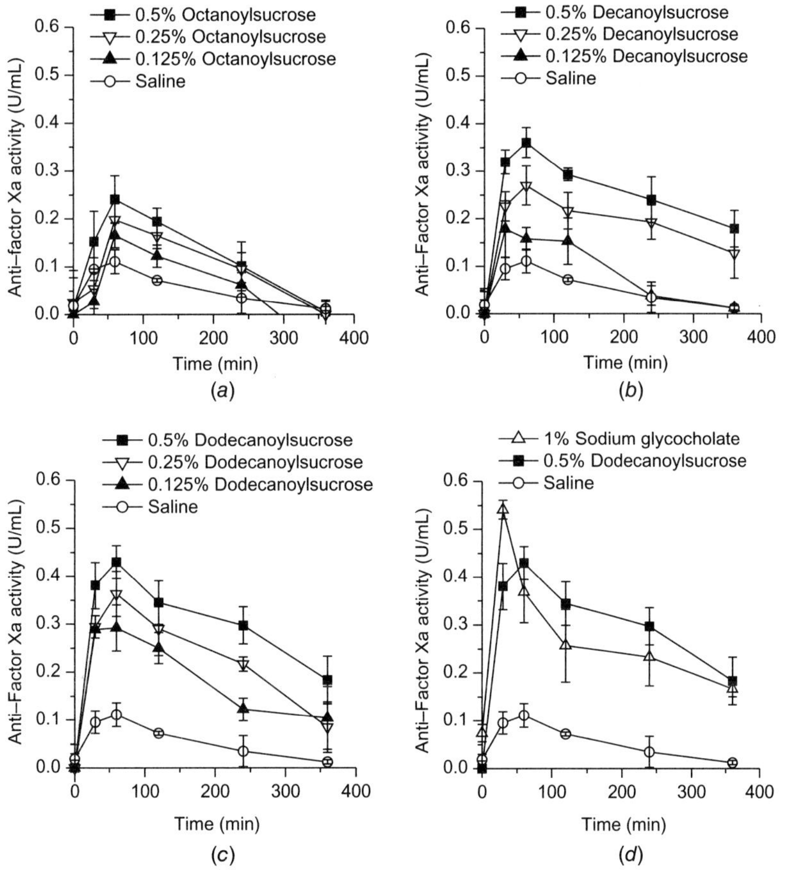

FIGURE 16 Changes in anti–factor Xa activity after nasal administration of enoxaparin formulated in saline or in presence of different concentrations of the following: (*a*) octanoylsucrose; (*b*) decanoylsucrose; (*c*) dodecanoylsucrose; (*d*) sodium glycocholate and dodecanoylsucrose to anesthetized rats (enoxaparin dose, 330 U/kg). Data represent mean ± SEM, $n = 3, 5$. (Reproduced from ref. 84 with permission of Pharmaceutical Press.)

The authors also studied the influence of the chain length of alkylmaltosides on the nasal absorption of enoxaparin. The results indicated that increases in the concentration of alkylmaltosides increased the AUC for plasma anti–factor Xa; it was found that the absolute and relative bioavailabilities of enoxaparin increased by twofold with an increase in alkyl chain length from 8 to 14 carbons. Of all the alkylmaltosides, TDM was found to be the most potent absorption enhancer [85].

Furthermore, we have also shown the efficacy of cyclodextrins in enhancing absorption following the nasal delivery of LMWHs. Three different cyclodextrins were employed: β-cyclodextrins (β-CD), hydroxypropyl β-CD (HPβ-CD), and dimethyl β-CD (DMβ-CD). β-CD showed therapeutic levels of anti–factor Xa only at 2.5 and 5% β-CD, but there was no significant difference between the two concentrations, which was attributed to their solubility limit. In the case of HPβ-CD, neither 1.25 nor 2.5% produced an appreciable increase in anti–factor Xa levels;

only 5% HPβ-CD showed levels above 0.2 U/mL, which, however, was not significant. Unlike the other two cyclodextrins, 1.25 and 2.5% DMβ-CD showed a fourfold increase in AUC profiles. The studies reported that 5% DMβ-CD produced the greatest increase in the bioavailability of LMWHs, with an eightfold increase in the AUC profile. It was also found that the reduction in transepithelial electrical resistance (TEER) and changes in tight junction protein ZO-1 distribution facilitated by 5% DMβ-CD were much greater than with β-CD or HPβ-CD [86]. Recently our group has shown that positively charged polyethylenimines (PEI) increases the nasal absorption of LMWHs by reducing the surface negative charge of the drug. When PEI 1000 kDa was employed at a concentration of 0.25%, a four-fold increase in the absolute and relative bioavailabilities was observed in comparison with the control formulation of LMWH plus saline [87].

5.6.6.4 Azetirelin

Azetirelin is a novel analog of the tripeptide thyrotropin-releasing hormone (TRH). It was discovered in 1969 when two different groups of researchers, led by Guillemin and Schally, showed that the hypothalamic substance that causes the anterior pituitary gland to release thyrotropin (thyroid-stimulating hormone, or TSH) is L-pyroglutamyl-L-histidyl-L-prolineamide (L-pGlu-L-His-L-ProNH2), now called TRH. In azetirelin the pyroglutamyl moiety of the TRH is replaced by an (oxo-azetidinyl) carbonyl moiety. It has been reported that the inhibition of pentobarbital-induced sleep and reserpine-induced hypothermia due to azetirelin in mice, as opposed to TRH, are about 10–100 times more effective as well as 8–36 times longer lasting. Azetirelin is stable in plasma and degrades much more slowly than TRH in brain homogenates, thus showing improved pharmacological potency as well as efficacy over TRH. It is highly potent when given intravenously; however, when administered by the oral route, it shows very low bioavailability of only 2% [88–91].

Kagatani et al. [90] studied the effect of acylcarnitines as drug absorption enhancers for the nasal delivery of azetirelin in a rat model. A buffered azetirelin sample solution was administered intranasally, as described previously [47]. The nasal and oral absorptions of azetirelin were then compared. The F_{abs} after nasal absorption was found to be 17.1%, which was 21 times greater than the 0.8% after oral administration. As reported above, a pilot study of oral azetirelin showed a bioavailability of about 2%. A bioavailability of about 20% was seen in the case of nasally administered TRH in humans as well as rats. The authors predicted that since azetirelin is an analog of TRH, its pharmacokinetic properties after nasal delivery in humans could also be about 20% [90, 91].

5.6.6.5 Growth Hormones

Recombinant human growth hormone (hGH) is a 22-kDa protein drug having 191 amino acids. It has been used to treat a number of conditions, including short stature in children, Turner syndrome, and chronic renal failure. It is said to play an important role in the metabolism of proteins, carbohydrates, and fats as well as electrolytes and hence influences weight and height. It has been reported that hGH secretion in humans is pulsatile, showing low basal serum levels in between peaks. It has been suggested that this secretory pattern of hGH can be mimicked by the nasal route as opposed to painful subcutaneous injections [8, 92, 93].

In a pharmacokinetic-based study by Hedin et al. [94], hGH was administered with a nasal permeation enhancer, sodium tauro-24, 25-dihydrofusidate (STDHF), in patients deficient in growth hormone (GH) using a reprocessed lyophilized form of hGH. The lyophilized material was formulated with STDHF and all the subjects received the formulation by both the nasal and subcutaneous routes. The dose given by the subcutaneous route was a standard dose of 0.1 IU/kg body weight (BW), whereas three different doses (of 0.2, 0.4, and 0.8 IU/kg BW) of the nasal formulation were given. As compared with the subcutaneous route, all three nasal formulations showed a rapid increase in the plasma levels of hGH, with T_{max} being reached 15–25 min after administration, as compared with 3–4 h in the case of the subcutaneous route. However, the C_{max} was higher in the case of the latter route, and the nasal formulations touched baseline after 3–4 h, as compared with 14–18 h after subcutaneous delivery.

Several studies have shown that frequent doses of hGH are more beneficial than the total amount given as a single dose and that higher peaks of plasma hGH with low troughs are found in taller children; hence the nasal therapy for deficiencies in hGH would not only be more convenient but also offer advantages, including a rapid decrease of the peaks to zero levels and a mimicking of the pulsatile pattern of the endogenous hormone [94–97]. Nasal irritation studies also were carried out, indicating that the nasal formulations showed only local short-term irritation.

Didecanoylphosphatidylcholine (DDPC) and α-cyclodextrin (α-CD) were used as enhancers and reversibility studies were carried out in vivo in rabbits. Three different combinations were used: DDPC, α-CD, and DDPC plus α-CD for the nasal administration of hGH. Vermehren et al. [98] used intravenous hGH as the reference. When hGH was administered with α-CD as a powder, 23.6% bioavailability was seen, as compared with 18.1% F_{rel} when given at the same time as two powders. When hGH only was given, a bioavailability of 8.3% was attained. DDPC plus hGH together showed a F_{rel} of 22.3 and 21.5% when given as two powders, while simultaneous administration of DDPC plus α-CD and hGH as two powders showed a F_{rel} of 14.3%. When dosed as one single powder, it showed a F_{rel} of 31.9%. Reversibility of the enhancer effect was seen when a dose of hGH in enhancer-free formulation was given 30 min after dosing of the test formulations; this resulted in reduction in the AUC and C_{max} by half.

Another group studied the enhancer effect of DDPC on the pharmacokinetics and the biological activity of nasally administered hGH in GH-deficient patients. Three different doses—0.05, 0.10, and 0.20 IU/kg with DDPC—were given by the nasal route and 0.10 and 0.015 IU/kg were given by the subcutaneous and intravenous routes. A short-lived serum GH peak was seen in the intravenous treatment, showing a peak value of around 128 μg/L, whereas the subcutaneous route showed a peak level of 13.98 μg/L and nasal doses showed peaks of about 3.26, 7.07, and 8.37 μg/L for the three treatment doses. The bioavailabilities of the nasal doses were found to be 7.8, 8.9, and 3.8%, respectively, as compared with the F_{abs} of 49.5% for the subcutaneous dose. It was also found that the serum insulinlike growth factor 1 (IGF-1) increased only upon subcutaneous administration There was no change in the serum IGF protein binding protein 3 levels in any of the nasal doses or in the subcutaneous or intravenous doses [92].

Leitner et al. [93] tried to overcome the limitations posed by the low bioavailability of nasally delivered hGH due to the drug's high molecular weight and hydrophilicity. The strategy employed was to use dry polymer particles to enhance absorption, taking advantage of the fact that these polymers would form a gellike layer and hence be cleared slowly, giving a longer circulating drug. Due to the presence of covalent immobilization of sulfhydryl groups on the backbone of the thiolated polymers or thiomers, the permeation-enhancing properties of these thiomers were improved. They have also been reported to have mucoadhesive and enzyme-inhibitory properties, and the addition of glutathione (GSH) was found to increase their permeation-enhancing properties further [93, 99–103]. A polycarbophil–cysteine (PCP–Cys) microparticulate system was prepared using GSH and hGH, and this was added to the formulation. The in vitro release profiles of the PCP–Cys and unmodified PCP containing hGH proved to be similar. Three different formulations were tested for in vivo nasal absorption: PCP–Cys/GSH/hGH microparticles, PCP/hGH microparticles, and mannitol/hGH powder against a subcutaneous hGH. The F_{rel} for mannitol/hGH powder was around 2.40%, the PCP–Cys/GSH/hGH microparticles showed a F_{rel} of 8.11%, and PCP/hGH showed 2.70%, which represents a threefold increase in nasal uptake when thiomer/GSH is used in the formulation. The microparticulate formulation of PCP–Cys/GSH/hGH showed a sixfold higher plasma concentration when compared to the PCP–Cys/GSH/hGH gel formulation.

5.6.7 NASAL DELIVERY OF NONPEPTIDE MOLECULES

5.6.7.1 Morphine

Morphine is a potent narcotic analgesic used preoperatively and as an anxiolytic agent in pediatric patients; it is also used in the management of postoperative pain as well in moderate to severe pain in cancer. Oral morphine in solution, immediate-release, or controlled-release formulations shows a bioavailability of only about 20%. Morphine absorption in humans is poor, and only 10% bioavailability is obtained when it is given as a solution rather than intravenously. Chitosan has been reported to be a potent absorption enhancer and greatly improves the absorption of small polar molecules and peptides. Intranasal administration was carried out in a sheep model; various formulations were also tested in humans and comparisons versus the intravenous route were made. The sheep model was used for nasal delivery because it has been reported that testing in sheep is highly predictive of results in humans [104, 105]. In sheep studies, when morphine HCl solution was given nasally (control), the C_{max} obtained was limited: 151 nM with a F_{rel} of 10%, with a T_{max} of about 20 min, suggesting a slow rate of absorption. When 0.5% chitosan was formulated with morphine as a solution, the C_{max} increased to 657 nM, and F_{rel} was 26.6%. The rate of absorption was also increased, as evidenced by a T_{max} of 14 min. With the formulation of chitosan into microspheres, the C_{max} was 1010 nM, T_{max} about 8 min, and F_{rel} about 54.6%. A further increase in nasal absorption was seen when morphine was formulated as a powder consisting of starch microspheres and L-α-lysophosphatidylcholine (LPC), with C_{max} being 1875 nM, T_{max} about 10 min, and F_{rel} about 75% (Figure 17). In the case of human phase I clinical trials, a dose of

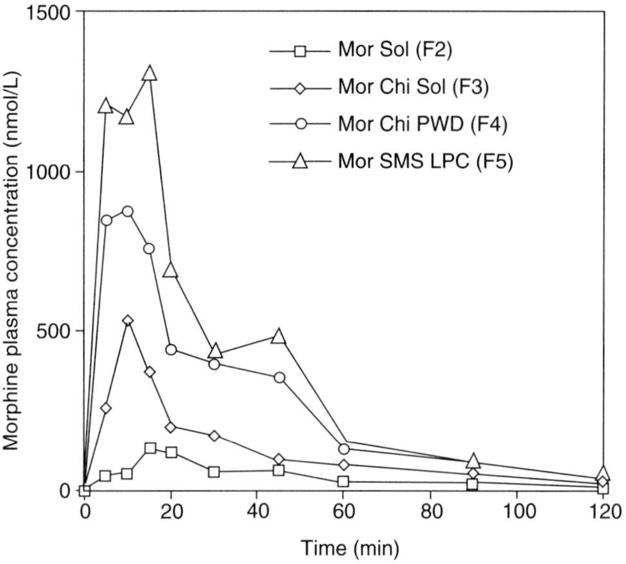

FIGURE 17 Morphine plasma concentration after nasal administration of morphine formulations in sheep: Mor Sol, morphine solution; Mor Chi Sol, morphine solution containing chitosan; Mor Chi PWD, morphine chitosan powder; Mor SMS LPC, starch microspheres with lysophosphatidylcholine and morphine as a freeze-dried powder. (Reproduced from ref. 105 with permission of the American Society for Pharmacology and Experimental Therapeutics.)

10 mg morphine sulfate led to a mean C_{max} of 336 nM following 30 min of intravenous administration. A 0.5% chitosan and morphine HCl solution led to a C_{max} of 98 nM and a T_{max} of about 16 min. The plasma half-life ($t_{1/2}$) in the case of nasal administration was found to be 2.98 h, as compared with 1.67 h via the intravenous route. The mean bioavailability with the nasal solution of morphine plus chitosan was 56%. Furthermore, the powder formulation comprising chitosan and morphine HCl showed a C_{max} of 92 nM, T_{max} of 21 min, $t_{1/2}$ of 2.72 h, and F_{rel} of about 56% (Figure 18). The nasal formulations were reported to be well tolerated.

5.6.7.2 Benzodiazepines

Diazepam has been the standard or preferred option for the treatment of all types of seizures in both children and adults. However, it has disadvantages, including a short duration of action, so that in some cases diazepam must be given rectally in order to manage prolonged seizures. Moreover, its use in the community is restricted because of the need for privacy, especially in the case of adult patients. Finally, the intravenous route is also reported to be inconvenient, since nonprofessional caregivers may not be comfortable enough to administer the drug in this way. Midazolam has been reported to be clinically effective with both intravenous and oral administration for the induction of sedation and reduction of anxiety. Owing to the drawbacks mentioned above, intranasal formulations of benzodiazepines could be highly beneficial [106, 107].

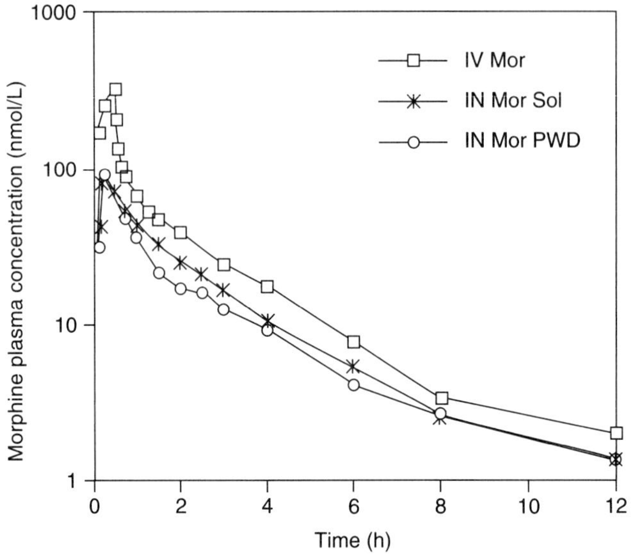

FIGURE 18 Morphine plasma concentration in human volunteers after intravenous administration of morphine and after nasal administration of morphine as chitosan solution and powder formulations: Mor Chi Sol, morphine solution containing chitosan; Mor Chi PWD, morphine–chitosan powder; IN, intranasal. (Reproduced from ref. 105 with permission of the American Society for Pharmacology and Experimental Therapeutics.)

Midazolam, Triazolam, and Flurazepam The feasibility of intranasal administration of midazolam, flurazepam, and triazolam has been studied and compared with oral absorption in dogs. There was a 3.4-fold increase in the C_{max} after nasal administration, from 5.5–8.7 ng/mL to 17.4–30.0 ng/mL. The mean $t_{1/2}$ showed comparable values for both routes. The T_{max} obtained after nasal administration of midazolam was found to be 15 min, as compared with the 15–45 min observed for oral dosing, while the C_{max} after nasal administration was 6.5–20.3 ng/mL, as compared with 3.0–8.6 ng/mL observed for the oral route. Like midazolam and triazolam, flurazepam also showed a shorter half-life, 15 min, as compared with 15–45 min with oral administration. The C_{max} for oral administration was 0.14–0.59 ng/mL; after nasal administration it was in the range of 2.6–11.1 ng/mL, a 16.4-fold increase. Since the gastrointestinal tract at bedtime is likely to be in the fed state, causing a twofold decrease in the absorption of midazolam and triazolam, the nasal route may be a better option for the treatment of amnesia, since these drugs cross the nasal mucosa effectively without the use of an absorption enhancer, as shown in these studies [108].

In situ nasal absorption studies of midazolam were carried out in rats. The effects of solution concentration, osmolality, and pH on nasal absorption were studied using the in situ perfusion technique. The absorption of midazolam was reported to be prevented at osmolalities in the range of 142–450 mOsm/kg; however, a hypoosmotic 3-mOsm/kg solution resulted in significant absorption, where the pH rose from 3.3 to 6.5. No lag time in absorption was observed when the solutions were buffered at a pH of either 5.5 or 7.4; however, at pH 3.3, no absorption was seen, suggesting

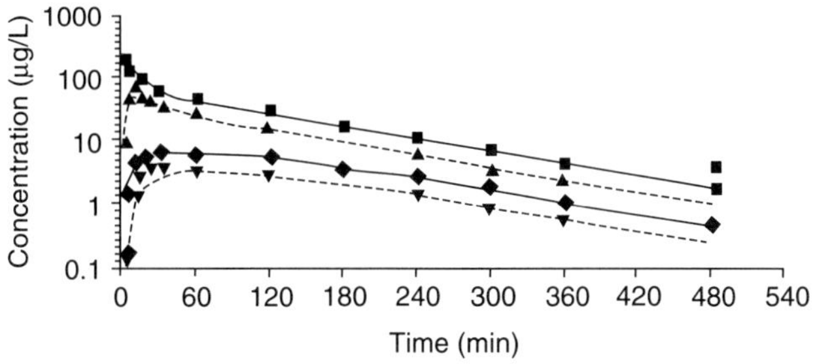

FIGURE 19 Fit of composite model to concentration–time data for midazolam and 1-hydroxymidazolam in one volunteer. Solid lines indicate the time course of midazolam concentrations (■) and 1-hydroxymidazolam concentrations (◆) after intravenous administration. Dotted lines indicate the time course of midazolam concentrations (▲) and 1-hydroxymidazolam concentrations (▼) after intranasal administration. (Reproduced from ref. 111 with permission of Blackwell Publishing.)

that pH was the main factor determining the absorption of midazolam (Figure 19) [109].

The pharmacokinetics and pharmacodynamics of midazolam after nasal administration were investigated in healthy volunteers in two different studies in comparison with the intravenous route [110, 111]. Studies reported in 1997 demonstrated that intranasal midazolam was rapidly absorbed, with maximal concentration attained in the range of 10–48 min, with a mean of 25 min. These results were only the maximum concentration achieved and the time taken to reach this maximum. The maximum concentration reported after intranasal administration was in the range of 91.0 to 224.3 ng/mL, with a mean of 147 ng/mL. Bioavailability reported in this study [110] was about 50%, in line with the results obtained with oral administration, as reported in an earlier study [112]. Knoester et al. [111] also carried out similar studies using a concentrated intranasal spray in healthy volunteers. The concentrated preparation was prepared by mixing midazolam HCl in a mixture of water and propylene glycol, pH 4. A Spruyt Hillen (IJsselstein, Netherlands) intranasal device was used to deliver the dose. Besides nasal irritation lasting 10 min and teary eyes, no other discomfort was reported. Midazolam was rapidly absorbed on nasal administration, showing a maximum concentration of 72 ng/mL within 14 min. A mean bioavailability of 0.89 was obtained. It was reported that intraindividual basal electroencephalogram (EEG) activity after intranasal administration was comparable with that after intravenous administration (Figures 19 and 20).

Diazepam As mentioned earlier, because of shortcomings of rectal administration, the nasal delivery of diazepam has gained interest. The nasal bioavailability of diazepam in sheep was estimated and further compared with results obtained earlier in humans and rabbits [106]; in this study, human and rabbit nasal bioavailability for the first 30 min was reported to be 37 and 54%, respectively [113]. Diazepam solubilized in PEG 300 was used for nasal administration via a modified nasal device, a Pfeiffer unit dose (Princeton, NJ). The sheep received the nasal formulations in a fixed standing position such that the head was slightly tilted back. It was found that the serum concentration after administration of a 7-mg solution of diazepam was

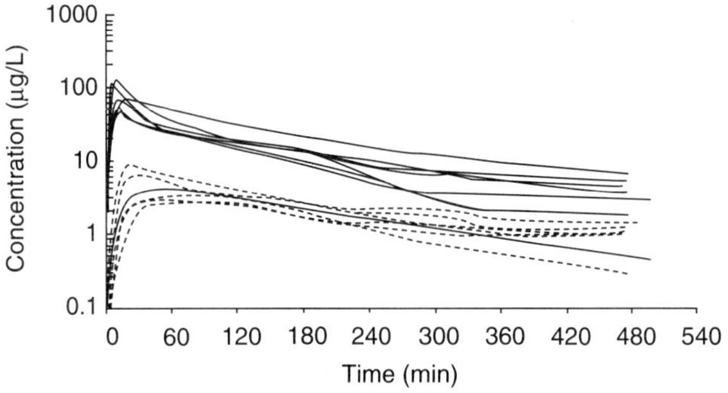

FIGURE 20 Individual plasma concentration–time curves for midazolam (solid lines) and 1-hydroxymidazolam (broken lines) after intranasal administration of 5 mg midazolam. The bold curves represent the mean pharmacokinetic model fit to the data. (Reproduced from ref. 111 with permission of Blackwell Publishing.)

lower than that obtained with a 3-mg solution, suggesting a low nasal bioavailability. The bioavailability after the initial 30 min was found to be 15%, as compared to the earlier mentioned bioavailability in human and rabbit, and a C_{max} of 934 ng/mL with a T_{max} of 5 min. The difference in bioavailability between animal and human was larger when periods shorter than 0–60 min were used in the calculations. When the results among the three species were compared, it was found that the bioavailability in sheep was higher than that in humans during the early or initial phases, after which the reverse was observed. Lindhardt et al. [113] used a profile that took into account the observation period with respect to the rate of bioavailability and found that similar profiles with respect to rate were observable in relation to all of the three nasal formulations given in humans; moreover, an optimal correlation between sheep and rabbit was observed. The authors suggested that use of the jugular vein in sampling blood from sheep could have resulted in the low bioavailability.

5.6.7.3 Buprenorphine

Buprenorphine is a derivative of the morphine alkaloid thebaine and is a partial opioid antagonist. It exerts an agonistic effect on the μ-muscarinic receptors and an antagonistic effect on the κ type. Buprenorphine at lower doses produces sufficient agonist effect to enable opioid-addicted individuals to discontinue the use of opioids without experiencing withdrawal symptoms. It has been reported that an intravenous dose of 0.3 mg of buprenorphine is equivalent to 10 mg of morphine and that oral delivery of buprenorphine results in a bioavailability of only about 15% due to first-pass metabolism. In addition to the intravenous formulation, there is a sublingual formulation offering the advantage of avoiding the first-pass metabolism effect. A clinical trial of nasally administered buprenorphine was reported in 1989 [114]. Buprenorphine is highly lipophilic and hence easily absorbed across the nasal epithelium. The buprenorphine formulation used in the clinical trial was prepared in 5% dextrose solution and a Pfeiffer atomizing pump operated manually was used to deliver it. The mean T_{max} and C_{max} for the intranasal dose were 30.6 min and 1.77 ng/mL, respectively. A relative nasal bioavailability of about 48.2% was attained.

Butorphanol, an analog of buprenorphine, showed a nasal bioavailability of 70% and also a much lower T_{max} after nasal absorption as compared with the sublingual and buccal routes [115]. Lindhardt et al. [106] compared buprenorphine formulated in 30% PEG-300 in sheep with that of the 5% dextrose formulation mentioned earlier. A unit-dose Pfeiffer device was again used to administer the formulation. It was found that nasal bioavailability in sheep was about 70% when buprenorphine was formulated in PEG-300 and approximately 89% when it was formulated with 5% dextrose. The rate of absorption was reported to be very fast, with a T_{max} of 10 min; the C_{max} was found to be 37 and 48 ng/mL for PEG-300 and dextrose, respectively. In sheep, the pharmacokinetics of buprenorphine showed a two-compartment model as compared to a three-compartment model in humans.

5.6.7.4 Hydralazine

Hydralazine is a vasodilator used in the treatment of malignant hypertension and hypertensive emergencies and is generally used in conjunction with other antihypertensive agents. Although the oral absorption is good, there is low oral bioavailability due to first-pass metabolism. Nasal absorption of hydralazine has been studied in rats using both in vivo nasal absorption and in situ nasal perfusion methods; the effect of surfactant and solution pH has also been reported. It was found that the nasal absorption of hydralazine was increased in the presence of surfactants such as sodium glycocholate and polyoxyethylene-9-lauryl ether. The nasal absorption of hydralazine was reported to be a pH-dependent passive process, with the absorption increasing as the pH was increased from 3.0 to 6.6. In nasal absorption studies in rats, peak levels of hydralazine were reached in 30 min at pH 3.0. The in situ absorption of hydralazine as a function of perfusion pH was also evaluated. The results of the in situ nasal perfusion studies demonstrated that hydralazine is eliminated from the nasal cavity and the perfusate by first-order kinetics. Even in the ionized form, the drug was well absorbed, and it was suggested that the aqueous channels in the nasal mucosa played an important role in the transport of hydralazine [116, 117].

5.6.7.5 Nitroglycerin

Nitroglycerin is delivered across the mucosal membranes in the management of acute ischemic conditions. Nitroglycerin carries out this function by arterial vasodilatation and venodilation, which leads to a decrease in both the preload and afterload and also improved coronary blood flow. The intranasal action of nitroglycerin, also called glyceryl trinitrate, appears to be similar [118] and brings about a reduction in myocardial oxygen consumption. Like that of hydralazine, the oral bioavailability of nitroglycerin is low; hence alternative routes of delivery and innovative delivery systems have been preferred, such as sublingual patches, ointments, or transdermal patches. The intravenous route ensures a rapid onset of action, but its preparation and standardization procedures make it costly. Intranasal nitroglycerin in various operative experiences has been found to have a rapid onset of action with predictable and consistent therapeutic effects. A peak level of nitroglycerin is reached 1–2 min after intranasal administration; it is barely detectable after 16 min. These studies were carried out in five patients who were undergoing elective coronary artery bypass surgery. The plasma levels were reported to be similar with intravenous administration and better than with sublingual administration [118–120]. The pressor response to endotracheal intubation in both normotensive and hypertensive patients can be attenuated by intranasal nitroglycerin in operative settings. It has been reported that when nitroglycerin was given 30s before the induction sequence in 40 hypertensive patients treated with β blockers, there was a blunted pressor response to intubation. In comparison to the placebo control group, the group that received nitroglycerin had a lower mean arterial pressure at 1, 3, and 5 min after induction. Thus intranasal nitroglycerin can be employed in the selective control of hypertension [118, 121, 122].

5.6.7.6 Propranolol and Other β-Adrenergic Blocking Agents

Propranolol is a nonselective β-adrenergic receptor blocking agent. It is clinically used in the management of hypertension and the treatment of angina pectoris. When given orally in humans, it has led to considerable variation in plasma drug levels. This, as well as its subsequently low bioavailability, is believed to be due to its extensive metabolism in the gut and in the liver. A study in the late 1970s showed that nasal absorption of propranolol at a dose of 1 mg in rats produced blood levels similar to those achieved with intravenous administration; however, the same dose administered orally resulted in very low blood levels [47, 123]. The feasibility of nasal absorption of propranolol in solutions and sustained-release formulations in rats and dogs was studied. The procedure described by Hirai et al. [46] was used to carry out the surgical operation in rats. In dogs, the formulations were given intranasally using a micropipette and syringe. The mean blood levels of propranolol by the nasal route were compared with those of the oral and intravenous routes in rats and dogs. As in the case of the results obtained in humans, oral administration of propranolol solutions resulted in low and variable drug levels in rats and dogs, whereas the nasal administrations of propranolol solution showed plasma drug levels that were similar to those achieved with intravenous administration. In the case of sustained-release formulations, it was found that there was an initial low level of drug; however, these levels were maintained for a longer time. The bioavailabilities obtained from the AUC were found to be identical, although the maximum blood levels in the case of sustained-release formulations were found to be much lower than with the propranolol solutions. A propranolol formulation containing 2% methylcellulose gels in humans was studied by the same group. Identical serum drug profiles were obtained after nasal administration as with intravenous administration [124].

The effect of intranasal propranolol on exercise parameters with the Bruce protocol [118, 125] was studied in 16 patients with chronic, stable, effort-induced angina pectoris. Propranolol was given as a single 5-mg/puff nasal spray to the patients. A mean plasma level of 20 ng/mL was obtained, and a significant increase in total exercise time was seen, from 460 to 530 s. This led to an increase in the time to 1 mm ECG as well as to the onset of angina. Both maximum heart rate and systolic blood pressure were lower than with placebo. This study demonstrated that propranolol in the form of a nasal spray elicited immediate β blockade and was useful in treating patients with angina pectoris, who showed improvement in exercise tolerance after receiving the drug [118, 125].

Although no adverse reactions have been reported with intranasal administration of propranolol, complications may occur, as ocular administration has produced some systemic side effects [118]. The influence of substrate lipophilicity on drug uptake by the nasal route was reported in humans. Alprenolol and metoprolol, β blockers with varying degrees of lipophilicity, were used. The findings from these studies demonstrate that the more hydrophilic drugs showed a lower bioavailability. Alprenolol showed rapid uptake into the systemic circulation by the nasal route and also a higher bioavailability [126, 127].

5.6.7.7 Sex Hormones

The low oral bioavailability of hormone-replacement drugs due to intestinal and first-pass metabolism requires the use of higher doses of these drugs, which are associated with many side effects. Parenteral administration of sex steroids as well as use of the transdermal route has been viewed as an alternative. However, the transdermal route has certain limitations, such as the visibility and palpability of the patch as well as possible skin irritation. These drawbacks have limited the use of this route. The intranasal route has therefore been considered as an alternative [128].

5.6.7.8 17β-Estradiol (E_2)

The most common form of estrogen in clinical use is 17β-estradiol. It is reported to reduce bone turnover, prevent postmenopausal bone loss, and decrease the risk of fracture in both early and late postmenopausal women. Lipophilic drugs such as sex steroids pose the problem of going into solution, thus leading to low estrogen levels. Hence a number of studies have attempted to solubilize 17β-estradiol using DMβ-CD for its intranasal delivery [129, 130]. An E_2 spray has been developed (S21400 or Aerodiol, Institut de Recherches Internationales Servier, France), which is estradiol that has been solubilized in water with randomly methylated β-CD. Estradiol delivered intranasally is rapidly absorbed by the nasal mucosa and shows maximum plasma levels within 10–30 min. Plasma levels return to 10% of the maximum plasma concentration within 2 h of administration and to untreated postmenopausal levels within 12 h. Hence after intranasal administration, estradiol has a pulsatile profile as compared to the more sustained plasma profile seen with the oral and transdermal routes. However, whether sustained levels are required for efficacy has not been determined, although intranasally delivered estradiol does increase serum estradiol to the same extent as is seen with oral administration [128, 130–134].

The short- and long-term effects of intranasal 17β-estradiol on bone marrow turnover and serum IGF-1 were studied in a double-blind placebo-controlled clinical trial and compared with oral 17β-estradiol. Some 425 Caucasian postmenopausal women with climacteric symptoms were studied. The nasal efficacy of estradiol was assessed using the Kupperman index (KI), which is a weighted evaluation of the incidence and severity of 11 menopausal symptoms summarized in a menopausal index as follows: hot flushes, the most heavily weighted (×4); night sweats (×2); and sleep disturbances and nervousness (each ×2). The lower weighted symptoms include depression, irritability, vertigo, fatigue, arthralgia, headache, tachycardia, and vaginal dryness (each ×1). The highest possible score is 51 [128, 135–137]. Various markers of bone resorption—such as urinary type I collagen telopeptides, the formation of serum osteocalcin, serum type I collagen, N-terminal extension propeptide (PINP), and serum bone marrow alkaline phosphatase (BAP)—were determined at baseline and after 1, 3, and 15 months. Urinary-type collagen C telopeptides were considerably reduced in all the treated groups within 1 month, and this reduction continued even at 3 months. Neither serum osteocalcin nor PINP showed any change at 1 month; however, they were reduced at 3 months with oral dosage. There was an increase in the bone formation parameters at 1 month for the higher doses of intranasal estradiol, but no reduction was seen at 3 months. No significant change from placebo-treated groups was observed at the end of 3 months in the case of circulating IGF-I after intranasal estradiol, but a significant decrease was seen with oral estradiol. After a year of treatment with intranasal estradiol at a dose of 300 μg/day, resorption and formation markers decreased to premenopausal levels. This study concluded that normalization of bone turnover to premenopausal levels can be achieved following 1 year of treatment with intranasal 17β-estradiol [131].

5.6.7.9 Testosterone

Testosterone, the most potent natural male sex hormone, is generally given intramuscularly. It is absorbed well orally but is extensively metabolized in the liver and the gastrointestinal tract. Owing to fluctuations in serum levels of testosterone esters, other viable routes of delivery are being explored. Two transdermal patches have become available commercially. However, due to the limitations associated with this route, it is not preferred by patients, particularly because the site of application is the scrotum [8, 138]. In order to improve systemic bioavailability, the nasal absorption of testosterone has been evaluated versus intravenous and intraduodenal administration in rats. When given nasally, the concentration of testosterone in the circulation increased and peak levels were reached within 2 min; blood levels were similar to those seen with intravenous administration. Intraduodenal administration produced low blood levels. A bioavailability of 99% at 25 μg/dose was seen and 90% at 50 μg, but the intraduodenal route showed a very low bioavailability of 1% [8]. Hussain et al. [139] showed in rats that testosterone can be absorbed intranasally, and an elimination half-life of about 40 min was obtained. However, since the problem of solubility is an obstacle to preparing the formulation for nasal administration, the use of a prodrug was evaluated by the same group. A water-soluble ester of testosterone, testosterone 17β-*N*, *N*-dimethylglycinate hydrochloride (TSDG), is completely absorbed when given intranasally at much lower doses as compared with testosterone itself. After absorption, conversion of the prodrug to testosterone begins almost immediately, and the terminal elimination half-life of testosterone was found to be 55 min, which is similar to that obtained after intravenous administration. Peak plasma concentration was reached within 12 min for the lower dose (equivalent to 25 mg/kg of TS) and 20 min for the higher dose (50 mg/kg) by both routes. The AUC also showed similarity, suggesting the complete absorption of the nasally administered prodrug.

5.6.8 THE NOSE: OPTION FOR DELIVERY OF DRUGS TO CENTRAL NERVOUS SYSTEM

The possibility of delivering agents to the central nervous system (CNS) via the nose has long been known, an example being the sniffing of cocaine in order to produce a sense of euphoria, which is attained within 3–5 min [53]. In short, nasal administration is not only an exciting possibility in the field of drug delivery but may also be the means of solving delivery problems for the innumerable agents that cannot cross the blood–brain barrier (BBB). Such agents are therefore being developed by nanoparticle-based systems or by formulating them as prodrugs. The literature offers numerous examples demonstrating ways of delivering such drugs to the brain. In particular, use of a direct pathway, as in the case of cocaine, from the nasal cavity to the CNS has been suggested. Illum [140] and Chow et al. [141] have shown in animal models that, in the early time points after nasal administration, the brain concentration of cocaine was higher than when the drug was given by the intravenous route. The specific site through which nose-to-brain delivery is believed to take place is the olfactory region. In the early 1900s it was shown that the olfactory region was responsible for the uptake (or rather entry) of viruses, in particular the poliomyelitis virus, into the brain [142–146]. Further work in support of this theory then demonstrated the presence of the poliomyelitis virus in the cerebrospinal fluid (CSF) and also in the systemic lymphatics of the olfactory mucosa [147–150]. In the period from 1970 to 1990, there were many reports of nose-to-brain delivery across the olfactory epithelium for a number of different agents, including metals and tracer materials such as colloidal gold, cadmium, potassium ferricyanide, and iron ammonium citrate [151–154].

There are certain aspects of drug delivery that must be clearly understood in designing a nose-to-brain drug delivery system. For example, lipophilic drugs are absorbed across the nasal epithelium almost immediately and efficiently to enter the systemic circulation. Therefore such drugs will show little sign of direct nose-to-brain delivery. Drugs that are on the hydrophilic side or polar molecules will not be readily absorbed across the nasal mucosa; these molecules generally undergo paracellular transport as compared to transcellular transport in the case of lipophilic drugs. Such molecules have a higher chance of being taken up by the olfactory mucosa for delivery to the brain. In general, drugs travel from the nose to the brain via (1) drug internalization into the olfactory epithelium's primary neurons followed by the intracellular axonal transport to the olfactory bulb or (2) drug absorption by paracellular or transcellular pathways across the olfactory sustentacular epithelial cells, following which the drug enters the CSF or CNS. In the human nervous system, the olfactory region is the only site in direct contact with the surrounding environment. It has been reported that the intracellular axonal pathway takes a longer time to deliver drugs to the brain [140, 155–157]. Table 4 gives an account of some of the drugs/molecules that have been delivered from the nose to the brain in human and animal models. The direct delivery of drugs to the brain or CSF via the olfactory epithelium is discussed in the following paragraph, briefly describing the olfactory mucosa.

The olfactory region is mainly involved in the detection of smell, and the makeup and organization of the epithelial layer enhance the accessibility of air to the

TABLE 4 Nose-to-Brain Delivery of Agents in Different Species

Drug/Molecule	Path Followed	Reference
	Humans	
Arginine-vasopressin	—[a]	158
Adrenocorticotrophin	—[a]	159
Cholecystokinin	—[a]	160
Diazepam	—[a]	161
Insulin	—[a]	158, 162
technetium-99m-diethylenetriaminepentacetic acid	Direct from nose to brain	163
Apomorphine	Nasal cavity → CSF	164
Melatonin/hydroxycobalamin	Nasal cavity → CSF	165
Melanocortin, vasopressin, and insulin	Nasal cavity → CSF	166
	Rats	
Zidovudine	Nasal cavity → CSF, nasal cavity → systemic circulation	167
Dextromethorphan HCl	Direct from nose to brain	168
Cephalexin	Nasal cavity → CSF, nasal cavity → systemic circulation	169
Sulfonamides	Nasal cavity → CSF, nasal cavity → systemic circulation	170, 171
Dextran	Nasal cavity → olfactory mucosa → CNS and also nasal cavity → systemic circulation	172
Dopamine	Direct from nose to brain	173
Cocaine	Direct from nose to brain	141
Dihydroergotamine	Direct from nose to brain	174
Insulin	Direct from nose to brain	175
Dopamine	Nasal cavity → CSF	173
Methotrexate	Nasal cavity → CSF	176
Nerve growth factor	Direct from nose to brain via olfactory pathway	177
WGA–HRP (wheat germ agglutinin–horseradish peroxidase)	Direct from nose to brain via olfactory nerve and bulb	178
Zolmitriptan	Direct from nose to brain	179
Leptin	Direct from nose to brain	180
Morphine	Direct nose to brain via olfactory pathway	181
Nimodipine	Direct nose to brain via olfactory bulb and nasal cavity → CSF	182
Meptazinol hydrochloride	Nasal cavity → CSF	183
Estradiol	Nasal cavity → CSF via olfactory neurons	184

[a]Facilitated transport to brain based on functional evidence in humans.

neuronal components comprising the odorant detectors. The olfactory region is mainly located on the nasal septum and partly on the superior and middle turbinates. It occupies only a small region in humans of about 10 cm^2 in the roof of the nasal cavity, as compared to around 150 cm^2 in dogs. The olfactory epithelium is a pseudostratified epithelium comprising three types of cells: olfactory receptor cells, supporting cells, and basal cells. The receptor cells are elongated bipolar neurons located in the middle stratum of the epithelium, interspersed among the sustentacular cells; the microvilli cover the supporting cells, which extend from the mucosal surface to the basal membrane; while the basal cells are located in the basal surface of the neuroepithelium. These basal cells go on to differentiate, becoming new receptor cells [140, 185, 186]. The surface of the nasal cavity measures about 180 cm^2 in humans as compared to about 10 cm^2 in rats, and the olfactory region is reported to constitute about 3% of the nasal cavity in humans and 50% in rats. Some other differences include the fact that in adult humans the volume of CSF is about 160 mL while it is only about 150 μL in rats; also, the CSF is replaced about three times daily in humans, whereas in rats it is replaced hourly. Hence, though there is sufficient evidence regarding nose-to-brain delivery, especially in rats and in some cases in humans, the impact of these factors on the interpretation of the results between the two species could be significant [155, 187].

5.6.9 NASAL DELIVERY OF VACCINES

The discovery of vaccines for smallpox, cholera, and typhoid and the variety of vaccines now available have led to a significant reduction in the mortality and morbidity due to many diseases, with smallpox being the first to have been completely eradicated and poliomyelitis targeted to be the next. At present, the World Health Organization is working toward the complete elimination of poliomyelitis throughout the world [188, 189]. However, since Jenner discovered the vaccine for smallpox more than two centuries ago [190], only some 50 vaccines have been approved for use, and few additional vaccines have been discovered. Most of those in current use are administered parenterally; they can induce only a systemic immune response, not mucosal immunity. Obviously the latter is very important in the prevention and treatment of infectious diseases, be they due to viral, bacterial, or parasitic pathogens that attack via the mucosal surfaces [190].

The criteria to be met in designing a vaccine formulation include the following: The vaccine should have the capacity to produce lifelong immunity, be able to act against the different strains and variants or the subtypes of the organisms, be effective in all age groups, be able to act quickly and also to induce immunity in the fetus when the mother is treated, be able to act effectively after a single treatment, and ideally be administered noninvasively. Finally, such a vaccine must be relatively inexpensive and remain active under a variety of conditions, especially not requiring the cold chain [191].

The following section addresses the need for needle-free vaccines with formulations based on safer adjuvant and delivery systems.

5.6.9.1 Nasal Vaccines: Ideal Noninvasive Route

When we talk about targeting the pathogens entering through the mucosal surfaces, the route that usually comes to mind is the oral route. However, this route has its drawbacks for several reasons, such as the fact that the antigen used is degraded along with the gastric contents; furthermore, there is also the difficulty of reaching the antigen-presenting cells [192]. In comparison to the oral route, nasal vaccination has been shown to require a lower antigen dose, which is essential considering the cost of the recombinant agents that may be used as antigens. Nasal administration has been shown to induce immune responses in the respiratory and genital surfaces. When compared with the other mucosal routes such as vaginal or rectal administration, the nasal route is much more acceptable in terms of both accessibility and overall convenience. Hence the nasal route is increasingly being seen as ideal for the administration of vaccines.

The nasal route has traditionally been used as an effective route in the treatment of respiratory infections, the rationale having been to target the infectious agents at their port of entry. Since most infectious disease pathogens enter at various mucosal sites, the nasal route has attracted increased attention as an alternative route for the delivery of vaccines. A further advantage offered by the nasal route in that it is capable of inducing both systemic and mucosal immunity as compared to the parenteral route, which brings about only systemic immunity [190, 193, 194]. Nasal immunization can result in distant as well as local mucosal immunity because of the mucosal immune system's common properties. This means that it is possible, via nasal immunization, to attain adequate immunity at other mucosal sites such as the respiratory, intestinal, and genital; hence vaccines administered nasally will have an important role in the prevention of respiratory infections and, more importantly, in the treatment of sexually transmitted diseases [190–196].

The earliest vaccines were live attenuated, inactivated toxins, or inactivated toxoids. But with advances in molecular biotechnology, it has become possible to produce extremely pure vaccines. However, the main drawbacks of these vaccines are their poor immunogenicity, so that the use of adjuvants is often required to attain the necessary immunogenicity. A vaccine adjuvant is especially important in subunit vaccines, which is how most of the vaccines available today are supplied. Adjuvants can be best defined as particular agents that increase the immunity produced when they are coformulated and delivered with the vaccine antigen. The adjuvants are formulated in the vaccines so as to produce a longer lasting immunity. The more time the adjuvant takes to be eliminated from the system, the longer it is able to induce the required or intended lasting immune response.

Vaccine adjuvants may be classified operationally as delivery or immunostimulating adjuvants. Vaccine delivery adjuvants simply act as delivery agents, that is, they present the vaccine antigens to the antigen-presenting cells (APCs), whereas the immune-stimulating agents act by stimulating the APCs to elicit an appropriate immune response. This results from the stimulation of the Toll-like receptors, present on macrophages and dendritic cells, the activation of which shows increased antigen presentation and cytokine release. The combination of both types of adjuvants may lead to better formulations for nasal vaccines [196–199].

5.6.9.2 Immunity after Intranasal Immunization

An understanding of the respiratory tract and the immune response following nasal vaccination is necessary to understand how the antigen used in the vaccine interacts with the surfaces of the human body and how the different adjuvants may interact, modify, and aid in generating an immune response. The nose is a component of the upper respiratory tract, which is composed of the mouth, nasopharynx, and larynx. The nasal passages have an extensive surface area which is richly vascularized.

However, the nasal epithelium has little ability to break down drugs. The extensive mucosal surface of the nose has a lining of pseudostratified epithelium as well as cilia and the goblet cells involved in the secretion of mucus. The lymphoid tissue primarily involved in the mucosal immune responses is the mucosal-associated lymphoid tissue (MALT). The different regions of the respiratory tract that play an influencing role in the immune system are as follows:

- The epithelial surface, which comprises immunocompetent cells in the connective tissue
- The lymphoid tissues linked to the respiratory tract, which are categorized into three parts: larynx-associated lymphoid tissue (LALT), nose-associated lymphoid tissue (NALT), and bronchus-associated lymphoid tissue
- The lymph nodes that drain the respiratory tract

The NALT, which is the organized lymphoid structure in the nasal passages and occurs in abundance in the nasal mucosa, plays a significant role in the mucosal surface's defense against invading pathogens. The NALT is equivalent to Waldeyer's ring, which is made up of the adenoids or tonsils situated in the roof of the nasopharynx, bilateral lymphoid bands, the palatine tonsils, and the tonsil at the base of the tongue (the lingual tonsil). The NALT is the main target site for vaccine antigens in humans [193, 194, 197, 200, 201]. As the mucosal immune system develops, the NALT is involved in a variety of important functions in relation to the host defense mechanism, from being an impediment to drug absorption, serving as a guard against attacking pathogens and antigens, facilitating the uptake of antigens, eliciting the secretory antibody response, and inducing the immune response in the other distant mucosal surfaces due to the function of the common mucosal immune system. Owing to mucosal tolerance, it is also involved in preventing serious allergic responses to inhaled antigens [194, 201, 202].

On administration, the antigen interacts first with the inductive tissue of the MALT, thus initiating a primary response. The IgA serum cells are found in the effector sites of the MALT, and local immunity results from the production of the secretory IgA (s-IgA) response. As mentioned earlier, the NALT and the tonsils form the main inductive sites in rodents and humans. These are composed of M cells involved in the uptake of antigen and its presentation to the underlying lymphoid tissues, the antigen being taken up by the M cells and the APCs, consisting of dendritic cells, macrophages, and B cells; all these cells together with the T cells produce the cellular and humoral immune responses [194].

Vaccination by the nasal route produces a mucosal protection using mucus, the epithelial surface, and both innate and acquired immune responses. The innate defense mechanism plays a very important role in that it influences the type of acquired immune mechanism, which mainly responds on the basis of "immune memory." The ability to attain these responses is the main principle of attaining protection from infection.

5.6.9.3 Need for Adjuvants

As mentioned earlier, the subunit vaccines in particular require the use of adjuvants in order to initiate an immune response leading to protection. The subunit vaccines

TABLE 5 Nasal Vaccination Delivery Systems Studied

Delivery System/Adjuvant	Antigen Employed	Remarks	Reference
Chitosan	Diphtheria toxin	Increased local as well as systemic effects	208
	Influenza virus	Immune response comparable to that of intramuscular administration	209
	Filamentous hemagglutinin and recombinant pertussis toxin (single or bivalent vaccine)	Chitosan stimulates mucosal and systemic effects	209, 210
IL-12	Tetanus toxoid	IL-12-induced IgA response	211
	Influenza hemagglutinin and neuraminidase	Induces protective immunity	212
CpG motifs	Hepatitis B surface antigen (HBsAg)	Potent enhancement of systemic and mucosal immune responses	213
	Formalin-inactivated influenza virus	Enhances the serum IgG and s-IgA responses	214
Liposomes	Influenza subunit vaccine	Induction of systemic IgG and s-IgA responses	215
	Inactivated measles virus	Stimulation of mucosal and systemic responses	216
	Bovine serum albumin (BSA)	—	217
	Streptococcus mutans	—	218
	Influenza hemagglutinin and neuraminidase (when used in combination with heat-labile toxin (HLT)	Good response in presence of HLT	219
	Influenza, hepatitis B, tetanus toxoid	—	220
	Yersinia pestis	—	221
ISCOMs (immune stimulating complex)	Influenza subunit	Protective immunity to challenge	222
	Measles nucleoprotein	Induces cytotoxic T-cell response	223
	Echinococcus surface antigen	—	224
	Respiratory syncytial virus (RSV) envelope antigen		225
Poly(lactic-*co*-glycolic acid (PLGA) microparticles	HBsAg	Strong systemic and mucosal immune responses	226
Cationic nanoparticles (SMBV)	HBsAg and β-galactosidase	Strong mucosal as well as systemic antibody and CTL responses	227

TABLE 5 ***Continued***

Delivery System/Adjuvant	Antigen Employed	Remarks	Reference
HBcAg	HBsAg	Stimulates strong Th1 response	228
Cholera toxin	Group B streptococci	High levels of IgA in cervicovaginal secretions	229
	Haemophilus influenzae	Effective nasal vaccination	230
	Influenza virus	—	231
	Synthetic peptide of RSV	Complete protection	232

available today, which are administered intramuscularly or subcutaneously, involve alum as an adjuvant. The drive for initiating further research into vaccine adjuvants has been stimulated by many factors, among the chief of which is that the aluminum-based adjuvants currently available have failed in many candidate vaccines or have not achieved the necessary immunity or induced a cytotoxic T-cell response. Nasally formulated vaccines, mainly subunit vaccines, currently being purified are less immunogenic and also cannot elicit the necessary T-cell response. As a result, research is now focusing on finding newer adjuvants for nasal DNA and subunit vaccines in order to attain specific immune responses as well as the necessary antibody subtype response. In addition to this, an adjuvant can help to reduce the dose of antigen required and also the number of doses needed to achieve mucosal immunity [203–207]. Despite the extensive research going on in the field of vaccine adjuvants, the only FDA-approved adjuvant for human use is alum. There are several other adjuvants, such as monophosphoryl lipid A (MPL), that have been approved in the European market; another, Corixa, is used as an adjuvant in Fendrix, the hepatitis B vaccine of GlaxoSmithKline Biologicals. The main hindrance to the approval of many adjuvants that reach clinical trials is their potential to elicit toxic side effects in clinical use. It has been reported that aluminum salts do induce some allergies in humans. As more purified and target-oriented or specific vaccines obtained by recombinant technology are being launched, it becomes more difficult for vaccine antigens alone to induce the necessary immune responses, as these recombinant antigens or synthetic peptides cannot jump start an immune response.

A number of adjuvants are awaiting approval for human use. The main impediment to the successful development of vaccine adjuvants is that their mechanism of action is not clearly understood. Table 5 offers a list of available nasal drug delivery systems and the various adjuvants that have been used in the development of nasal vaccines.

REFERENCES

1. Southall, J., and Ellis, C. (2000), Developments in nasal drug delivery, *Innovat. Pharm. Technol.*, 110–115.

2. Koch, M. (2003), The growing market for nasal drug delivery, *Pharmatech*, 1–3.
3. Ruppar, D. (2006), Intranasal delivery: Sniffing out new sources for growth, *Drug Deliv. Technol.*, 6, 0–33.
4. Illum, L., and Fisher, A. N. (1997), Intranasal delivery of peptides and proteins, in Adjei, A. L. and Gupta, P. K., Eds., *Inhalation Delivery of Therapeutic Peptides and Proteins*, Marcel Dekker, New York, pp. 135–184.
5. Ugwoke, M. I., Verbeke, N., and Kinget, R. (2001), The biopharmaceutical aspects of nasal mucoahdesive drug delivery, *J. Pharm. Pharmacol.*, 53, 3–21.
6. Illum, L. (1999), Bioadhesive formulations for nasal peptide delivery, in Mathiowitz, E., Chickering, D. E., and Lehr, C.-M., Eds., *Bioadhesive Drug Delivery Systems, Fundamentals, Novel Approaches and Development*, Marcel Dekker, New York, pp. 507–539.
7. Washington, N., Washington, C., and Wilson, C. G. (2001), *Physiological Pharmaceutics, Barriers to Drug Absorption*, Taylor and Francis, London.
8. Chien, Y. W., Su, K. S. E., and Chang, S.-F. (1989), *Nasal Systemic Drug Delivery*, Marcel Dekker, New York.
9. Behl, C. R., Pimplaskar, H. K., Sileno, A. P., deMeireles, J., and Romeo, V. D. (1998), Effects of physicochemical properties and other factors on systemic nasal drug delivery, *Adv. Drug Deliv. Rev.*, 29, 89–116.
10. Mygind, N., and Dahl, R. (1998), Anatomy, physiology and function of the nasal cavities in health and disease, *Adv. Drug Deliv. Rev.*, 29, 3–12.
11. Olson, P., and Bende, M. (1985), Influence of environmental temperature on human nasal mucosa, *Ann. Otol. Rhinol. Lryngol.*, 94, 153–157.
12. Paulsson, B., Bende, M., and Ohlin, P. (1985), Nasal mucosal blood flow at rest and during exercise, *Acta Otolaryngol. (Stockh.)*, 99, 140–144.
13. Hadley, W. M., and Dahl, A. R. (1983), Cytochrome P450-dependent mono-oxygenase in nasal membrane of six species, *Drug Metab. Dispos.*, 11, 275–279.
14. Longo, V., Pacifici, G. M., Panattoni, G., Ursino, F., and Gervasi, P. G. (1989), Metabolism of diethylnitrosamine by microsomes of human respiratory nasal mucosa and liver, *Biochem. Pharmacol.*, 38, 1867–1869.
15. Brittebo, E. G., and Rafter, J. J. (1984), Steroid metabolism by rat nasal mucosa: Studies on progesterone and testosterone, *Steroid. Biochem.*, 20, 1147.
16. Smith, E. L., Hill, R. L., and Borman, A. (1958), Activity of insulin degraded by leucine aminopeptidase, *Biochem. Biophys. Acta.*, 29, 207–214.
17. Marttin, E., Schipper, N. G. M., Verhoef, J. C., and Merkus, F. W. H. M. (1998), Nasal mucosal clearance as a factor in nasal drug delivery, *Adv. Drug Deliv. Rev.*, 29, 13–38.
18. Nagai, T., Nishimoto, Y., Nambu, N., Suzuki, Y., and Sekine, K. (1984), Powder dosage form of insulin for nasal administration, *J. Controlled Release*, 1, 15–22.
19. Nagai, T., and Machida, Y. (1990), Bioadhesive dosage forms for nasal administration, in Lenaerts, V., and Gurny, R., Eds., *Bioadhesive Drug Delivery Systems*, CRC Press, Boca Raton, FL, pp. 169–178.
20. Pritchard, K., Lansely, A. B., Martin, G. P., Helliwell, M., Marriott, C., and Benedetti, L. M. (1996), Evaluation of bioadhesive properties of hyaluronan derivatives, detachment weight and mucociliary transport rate studies, *Int. J. Pharm.*, 129, 137–145.
21. Marttin, E., Verhoef, J. C., Romeijn, S. G., Zwart, P., and Merkus, F. W. H. M. (1996), Acute histopathological effects of bezalkonium chloride and absorption enhancers on rat nasal epithelium in vivo, *Int. J. Pharm.*, 141, 151–160.
22. Ennis, R. D., Borden, L., and Lee, W. A. (1990), The effects of permeation enhancers on the surface morphology of rat nasal mucosa: A scanning electron microscopy study, *Pharm. Res.*, 7, 468–475.

23. Hardy, J. G., Lee, S. W., and Wilson, C. G. (1985), Intranasal drug delivery by spray and drops, *J. Pharm. Pharmacol.*, 37, 294–297.

24. Proctor, D. F. (1985), Nasal Physiology in intranasal drug administrations, in Chien, Y. W., Ed., *Transnasal Systemic Medications*, Elsevier, Amsterdam, pp. 101–106.

25. Greiff, L., Venge, P., Andersson, M., Enander, I., Linden, M., Myint, S., and Persson, C. G. (2002), Effects of rhinovirus-induced common colds on granulocyte activity in allergic rhinitis, *J. Infect.*, 45, 227–232.

26. Larsen, C., Niebuhr, J. M., Tommerup, B., Mygind, N., Dagrosa, E. E., Grigoleit, H. G., and Malerczyk, V. (1987), Influence of experimental rhinitis on the gonadotropin response to intranasal administration of buserelin, *Eur. J. Clin. Pharmacol.*, 33, 155–159.

27. Kublik, H., and Vidgren, M. T. (1998), Nasal delivery systems and their effect on deposition and absorption, *Adv. Drug Deliv. Rev.*, 29, 157–177.

28. Hussain, A. A., Bawarshi-Nassar, R., and Huang, C. H. (1985), Physicochemical considerations in intranasal drug administration, in Chien, Y. W., Ed., *Transnasal Systemic Medication*, Elsevier, Amsterdam, pp. 121–138.

29. Corbo, D. C., Huang, Y. C., and Chien, Y. W. (1989), Nasal delivery of progestational steroids in ovarectomized rabbits. II. Effect of penetrant hydrophilicity, *Int. J. Pharm.*, 50, 253–260.

30. Kimura, R., Miwa, M., Kato, Y., Sato, M., and Yamada, S. (1991), Relationship between nasal absorption and physicochemical properties of quaternary ammonium compounds, *Arch. Int. Pharmacodyn. Ther.*, 310, 13–21.

31. Tonndorf, J., Chinn, H. I., and Lett, J. E. (1953), Absorption from nasal mucous membrane: Systemic effect of hyoscine following intranasal administration, *Ann. Otol. Rhinol. Laryngol.*, 62, 630–634.

32. Rogerson, A., and Parr, G. D. (1990), Nasal drug delivery, in Florence, A. T., and Salole, E. G., Eds., *Topics in Pharmacy, Routes of Drug Administration*, Northants, England, pp. 1–29.

33. McMartin, C., Hutchinson, E. F., Hyde, R., and Peters, G. E. (1987), Analysis of structural requirements for the absorption of drugs and macromolecules from the nasal cavity, *J. Pharm. Sci.*, 76, 535–540.

34. Donovan, M. D., Flynn, G., and Amidon, G. (1990), Absorption of polyglycols 600 through 2000: The molecular weight dependence of gastrointestinal and nasal absorption, *Pharm. Res.*, 7, 863–868.

35. Fisher, A., Brown, K., Davis, S., Parr, G., and Smith, D. A. (1987), The effect of molecular size on the nasal absorption of water-soluble compounds in the albino rat, *J. Pharm. Pharmacol.*, 39, 357–362.

36. Fisher, A., Illum, L., Davis, S., and Schacht, E. (1992), Di-iodo-L-tyrosine labeled dextrans as molecular size markers for nasal absorption in the rat, *J. Pharm. Pharmacol.*, 44, 550–554.

37. Maitani, Y., Machida, Y., and Nagai, T. (1989), Influence of molecular weight and charge on nasal absorption of water soluble compounds in the albino rat, *Int. J. Pharm.*, 40, 23–27.

38. D'Souza, R., Mutalik, S., Venkatesh, M., Vidyasagar, S., and Udupa, N. (2005), Nasal insulin gel as an alternative to parenteral insulin: Formulation, preclinical and clinical studies, *AAPS PharmSciTech.*, 6, E184–E189.

39. Bommer, R., Kern, J., Hennes, K., and Zwisler, W. (2005), Preservative-free nasal drug delivery systems, *Drug Deliv. Technol.*, 5.

40. Giroux, M. (2005), Controlled Particle Dispersion™: Effective nasal delivery from a versatile, flexible technology platform, in *Nasal Drug Delivery: Rapid Onset via Convenient Route*, ONdrugdelivery, pp. 13–15.
41. Hwang, P., Woo, E., and Fong, K., Intranasal deposition of nebulized saline: A radionuclide distribution study, paper presented at the 50th annual meeting of the American Rhinologic Society, New York, NY, Sept. 2004.
42. Giroux, M., Hwang, P., and Prasad, A. (2005), Controlled Particle Dispersion™: Applying vortical flow to optimize nasal drug deposition, *Drug Deliv. Technol.*, 5, 44–49.
43. Djupesland, P. G., and Watts, J. (2005), Breath-actuated bidirectional delivery sets the nasal market on a new course, ONdrugdelivery, pp. 20–23.
44. Djupesland, P. G., Skretting, A., Winderen, M., and Holand, T. (2004), Bi-directional nasal delivery of aerosols can prevent lung deposition, *J. Aerosol. Med.*, 17, 249–259.
45. Keldmann, T. (2005), Advanced simplification of nasal delivery technology: Anatomy + innovative device = added value opportunity, ONdrugdelivery, pp. 4–7.
46. Hirai, S., Yashiki, T., and Matsuzawa, T., Nasal absorption of drugs. Effect of surfactants on the nasal absorption of insulin in rats, paper presented at the 98th annual meeting of the Pharmaceutical Society of Japan, Apr. 1978.
47. Hussain, A., Hirai, S., and Bawarshi, R. (1980), Nasal absorption of propranolol from different dosage forms by rats and dogs, *J. Pharm. Sci.*, 69, 1411–1413.
48. Pillion, D. J., Ahsan, F., Arnold, J. J., Balusubramanian, B. M., Piraner, O., and Meezan, E. (2002), Synthetic long-chain alkyl maltosides and alkyl sucrose esters as enhancers of nasal insulin absorption, *J. Pharm. Sci.*, 91, 1456–1462.
49. Ahsan, F., Arnold, J. J., Meezan, E., and Pillion, D. J. (2001), Enhanced bioavailability of calcitonin formulated with alkylglycosides following nasal and ocular administration in rats, *Pharm. Res.*, 18, 1742–1746.
50. Merkus, F. W. H. M., Verhoef, J. C., Marttin, E., Romeijn, S. G., van der Kuy, P. H. M., Hermens, W. A. J. J., and Schipper, N. G. M. (1999), Cyclodextrins in nasal drug delivery, *Adv. Drug Deliv. Rev.*, 6, 41–57.
51. Gizurarson, S. (1990), Animal models for intranasal drug delivery, *Acta. Pharm. Nord.*, 2, 105–122.
52. Behl, C. R., Pimplaskar, H. K., Sileno, A. P., Xia, W. J., Gries, W. J., de Meireles, J. C., and Romeo, V. D. (1998), Optimization of systemic nasal drug delivery with pharmaceutical ecipients, *Adv. Drug Deliv. Rev.*, 29, 117–133.
53. Illum, L. (2002), Nasal drug delivery: New developments and strategies, *Drug Discovery Today*, 7, 1184–1189.
54. Hinchcliffe, M., and Illum, L. (1999), Intranasal insulin delivery and therapy, *Adv. Drug Deliv. Rev.*, 35, 199–234.
55. Illum, L., and Fisher, A. N. (1997), Intranasal delivery of peptides and proteins, in Adjei, A. L., and Gupta, P. K., Eds., *Inhalation Delivery of Therapeutic Peptides and Proteins*, Marcel Dekker, New York, p. 135.
56. Lansley, A. B., and Martin, G. P. (2001), Nasal drug delivery, in Hillery, A. M., Lloyd, A. W., and Swarbrick, J., Eds., *Drug Delivery and Targeting*, 2nd ed., Taylor & Francis, New York, p. 237.
57. Triplitt, C. L., Reasner, C. A., and Isley, W. (2005), Diabetes mellitus, in Dipiro, J. T., Talbert, R. L., Yee, G. C., Matzke, G. R., Wells, B. G., and Posey, L. M. Eds., *Pharmacotherapy: A Pathophysiological Approach*, 6th ed., McGraw-Hill, New York, p. 1333.
58. Bliss, M. (1933), The history of insulin, *Diabetes Care*, 16, 4–7.
59. Woodyatt, R. T. (1922), The clinical use of insulin, *J. Metab. Res.*, 2, 793–801.

60. Collens, W. S., and Goldzeiher, M. A. (1932), Absorption of insulin by nasal mucous membranes, *Proc. Soc. Exp. Biol. Med.*, 29, 756.

61. Major, R. H. (1935), The intranasal application of insulin, *J. Lab. Clin. Med.*, 21, 278.

62. Major, R. H. (1936), The intranasal application of insulin, *Am. J. Med. Sci.*, 192, 257.

63. Gordon, G. S., Moses, A. C., Silver, R. D., Flier, J. S., and Carey, M. C. (1985), Nasal absorption of insulin: Enhancement by hydrophobic bile salts, *Proc. Natl. Acad. Sci. USA*, 82, 7419–7423.

64. Morimoto, K., Morisaka, K., and Kamada, A. (1985), Enhancement of nasal absorption of insulin and calcitonin using polyacrylic acid gel, *J. Pharm. Pharmacol.*, 37, 134–136.

65. Pillion, D. J., Atchison, J. A., Gargiulo, C., Wang, R. X., Wang, P., and Meezan, E. (1994), Insulin delivery in nosedrops: New formulations containing alkylglycosides, *Endocrinology*, 135, 2386–2391.

66. Pillion, D. J., Ahsan, F., Arnold, J. J., Balusubramanian, B. M., Piraner, O., and Meezan, E. (2002), Synthetic long-chain alkyl maltosides and alkyl sucrose esters as enhancers of nasal insulin absorption, *J. Pharm. Sci.*, 91, 1456–1462.

67. Ahsan, F., Arnold, J. J., Meezan, E., and Pillion, D. J. (2003), Sucrose cocoate, a component of cosmetic preparations, enhances nasal and ocular peptide absorption, *Int. J. Pharm.*, 251, 195–203.

68. Illum, L., Jabbal-Gill, I., Hinchcliffe, M., Fisher, A. N., and Davis, S. S. (2001), Chitosan as a novel nasal delivery system for vaccines, *Adv. Drug Deliv. Rev.*, 51, 81–96.

69. Illum, L. (1992), Nasal delivery of peptides, factors affecting nasal absorption, in *Topics in Pharmaceutical Sciences*, Medpharm Scientific, Stuttgart, p. 71.

70. Illum, L., Farraj, N. F., and Davis, S. S. (1994), Chitosan as a novel nasal delivery system for peptide drugs, *Pharm. Res.*, 11, 1186–1189.

71. Fernández-Urrusuno, R., Calvo, P., Remuñán-López, C., Vila-Jato, J. L., and Alonso, M. J. (1999), Enhancement of nasal absorption of insulin using chitosan nanoparticles, *Pharm. Res.*, 16, 1576–1581.

72. Dyer, A. M., Hinchcliffe, M., Watts, P., Castile, J., Jabbal-Gill, I., Nankervis, R., Smith, A., and Illum, L. (2002), Nasal delivery of insulin using novel chitosan based formulations: A comparative study in two animal models between simple chitosan formulations and chitosan nanoparticles, *Pharm. Res.*, 19, 998–1008.

73. Yu, S., Zhao, Y., Wu, F., Zhang, X., Lü, W., Zhang, H., and Zhang, Q. (2004), Nasal insulin delivery in the chitosan solution: In vitro and in vivo studies, *Int. J. Pharm.*, 281, 11–23.

74. Varshosaz, J., Sadrai, H., and Alingari, R. (2004), Nasal delivery of insulin using chitosan microspheres, *J. Microencapsul.*, 21, 761–774.

75. Varshosaz, J., Sadrai, H., and Heidari, A. (2006), Nasal delivery of insulin using bioadhesive chitosan gels, *Drug Deliv.*, 13, 31–38.

76. Pontiroli, A. E., Alberetto, M., and Pozza, G. (1985), Intranasal calcitonin and plasma calcium concentrations in normal subjects, *Br. Med. J.*, 290, 1390–1391.

77. Morimoto, K., Morisaka, K., and Kamada, A. (1985), Enhancement of nasal absorption of insulin and calcitonin using polyacrylic acid gel, *J. Pharm. Pharmacol.*, 37, 134–136.

78. Morimoto, K., Akatsuchi, H., Aikawa, R., Morishita, M., and Morisaka, K. (1984), Enhanced rectal absorption of [Asu1, 7]-eel calcitonin in rats using polyacrylic acid aqueous gel base, *J. Pharm Sci.*, 73, 1366–1368.

79. Lee, K. C., Tak, K. K., Park, M. O., Lee, J. T., Woo, B. H., Yoo, S. D., Lee, H. S., and DeLuca, P. P. (1999), Preparation and characterization of polyethylene-glycol-modified salmon calcitonins, *Pharm. Dev. Tech.*, 4, 269–275.

80. Miron, T., and Wilcheck, M. (1993), A simplified method for the preparation of succinimidyl carbonate polyethylene glycol for coupling to proteins, *Biconj. Chem.*, 4, 568–569.

81. Shin, B. S., Jung, J. H., Lee, K. C., and Yoo, S. D. (2004), Nasal absorption and pharmacokinetic disposition of salmon calcitonin modified with low molecular weight polyethylene glycol, *Chem. Pharm. Bull.*, 52, 957–960.

82. Na, D. H., Youn, Y. S., Park, E. J., Lee, J. M., Cho, O. R., Lee, K. R., Lee, S. D., Yoo, S. D., Deluca, P. P., and Lee, K. C. (2004), Stability of PEGylated salmon calcitonin in nasal mucosa, *J. Pharm. Sci.*, 93, 256–261.

83. Arnold, J. J., Ahsan, F., Meezan, E., and Pillion, D. J. (2002), Nasal administration of low molecular weight heparin, *J. Pharm. Sci.*, 91, 1707–1714.

84. Yang, T., Mustafa, F., and Ahsan, F. (2004), Alkanoylsucroses in nasal delivery of low molecular weight heparins: In-vivo absorption and reversibility studies in rats, *J. Pharm. Pharmacol.*, 56, 53–60.

85. Mustafa, F., Yang, T., Khan, M. A., and Ahsan, F. (2004), Chain length-dependent effects of alkylmaltosides on nasal absorption of enoxaparin, *J. Pharm. Sci.*, 93, 675–683.

86. Yang, T., Hussain, A., Paulson, J., Abbruscato, T. J., and Ahsan, F. (2004), Cyclodextrins in nasal delivery of low-molecular-weight heparins: In vivo and in vitro studies, *Pharm. Res.*, 21, 1127–1136.

87. Yang, T., Hussain, A., Bai, S., Khalil, I. A., Harashima, H., and Ahsan, F. (2006), Positively charged polyethylenimines enhance nasal absorption of the negatively charged drug, low molecular weight heparin, *J. Controlled Release*, 115, 289–297.

88. Mason, G. A., Garbutt, J. C., and Prange Jr., A. J. (2000), Thyrotropin releasing hormone: Focus On neurobiology, available: http://www.acnp.org/g4/GN401000048/CH048.html.

89. Yamamoto, M., and Shimizu, M. (1987), Effects of new TRH analogue, YM-14673 on the central nervous system, *Naunyn. Schmiedebergs. Arch. Pharmacol.*, 336, 561–565.

90. Kagatani, S., Inaba, N., Fukui, M., and Sonobe, T. (1998), Nasal absorption kinetic behavior of Azetirelin and its enhancement by acylcarnitines in rats, *Pharm. Res.*, 15, 77–81.

91. Mitsuma, T., and Nogimori, T. (1984), Changes in plasma thyrotrophin-releasing hormone, thyrotrophin, prolactin and thyroid hormone levels after intravenous, intranasal or rectal administration of synthetic thyrotrophin-releasing hormone in man, *Acta Endocrinol.*, 107, 207–212.

92. Laursen, T., Grandjean, T., Jorgensen, J. O., and Christiansen, J. S. (1996), Bioavailability and bioactivity of three different doses of nasal growth hormone (GH) administered to GH-deficient patients: Comparison with intravenous and subcutaneous administration, *Eur. J. Endocrinol.*, 135, 309–315.

93. Leitner, V. M., Guggi, D., Krauland A. H., and Bernkop-Schnurch, A. (2004), Nasal delivery of human growth hormone: In vitro and in vivo evaluation of a thiomer/glutathione microparticulate delivery system, *J. Controlled. Release*, 100, 87–95.

94. Hedin, L., Olsson, B., Diczfalusy, M., Flyg, C., Petersson, A. S., Rosberg, S., and Albertsson-Wikland, K. (1993), Intranasal administration of human growth hormone (hGH) in combination with a membrane permeation enhancer in patients with GH deficiency: A pharmacokinetic study, *J. Clin. Endocrinol. Metab.*, 76, 962–967.

95. Albertsson-Wikland, K., and Rosberg, S. (1988), Analyses of 24-hour growth hormone (GH) profiles in children, *J. Clin. Endocrinol. Metab.*, 67, 493–500.

96. Clark, R. G., Jansson, J. O., Isaksson, O., and Robinson, I. C. A. F. (1985), Intravenous growth hormone: Growth responses to patterned infusions in hypophysectomized rats, *J. Endocrinol.*, 104, 53–61.

97. Eden, S. (1979), Age and sex related differences in episodic growth hormone secretion in the rat, *Endocrinology*, 105, 555–560.
98. Vermehren, C., Hansen, H. S., and Thomsen, M. K. (1996), Time dependent effects of two absorption enhancers on the nasal absorption of growth hormone in rabbits, *Int. J. Pharm.*, 128, 239–250.
99. Clausen, A. E., and Schnürch-Bernkop, A. (2000), In vitro evaluation of permeation enhancing effect of thiolated polycarbophil, *J. Pharm. Sci.*, 89, 1253–1261.
100. Schnürch-Bernkop, A., and Steininger, S. (2000), Synthesis and characterization of mucoadhesive thiolated polymers, *Int. J. Pharm.*, 194, 239–247.
101. Schnürch-Bernkop, A., Zarti, H., and Walker, G. F. (2001), Thiolation of polycarbophil enhances its inhibition of soluble and intestinal brush border membrane bound aminopeptidase N, *J. Pharm. Sci.*, 90, 1907–1914.
102. Clausen, A. E., Kast, C. E., and Schnürch-Bernkop, A. (2002), The role of glutathione in the permeation enhancing effect of thiolated polymers, *Pharm. Res.*, 19, 602–608.
103. Schnürch-Bernkop, A., Guggi, D., and Pinter, Y. (2004), Thiolated chitosans: Development and in vitro evaluation of a mucoadhesive, permeation enhancing oral drug delivery system, *J. Controlled Release*, 94, 177–186.
104. Illum, L. (1996), Animal models for nasal delivery, *J. Drug Target.*, 3, 717–724.
105. Illum, L., Watts, P., Fisher, A. N., Hinchcliffe, M., Norbury, H., Jabbal-Gill, I., Nankervis, R., and Davis, S. S. (2002), Intranasal delivery of morphine, *J. Pharmcol. Exp. Ther.*, 301, 391–400.
106. Lindhardt, K., Ólafsson, D. R., Gizurarson, S., and Bechgaard, E. (2002), Intranasal bioavailability of diazepam in sheep correlated to rabbit and man, *Int. J. Pharm.*, 231, 67–72.
107. Kyrkou, M., Harbord, M., Kyrkou, N., Kay, D., and Coulthard, K. (2006), Community use of intranasal midazolam for managing prolonged seizures, *J. Intellect. Dev. Disabil.*, 31, 131–138.
108. Lui, C. Y., Amidon, G. L., and Goldberg, A. (1991), Intranasal absorption of flurazepam, midazolam and triazolam in dogs, *J. Pharm. Sci.*, 80, 1125–1129.
109. Olivier, J. C., Djilani, M., Fahmy, S., and Couet, W. (2001), In situ nasal absorption of midazolam in rats, *Int. J. Pharm.*, 213, 187–192.
110. Burstein, A. H., Modica, R., Hatton, M., Forest, A., and Gengo, F. M. (1997), Pharmcokinetics and pharmacodynamics of midazolam after intranasal administration, *J. Clin. Pharmacol.*, 37, 711–718.
111. Knoester, P. D., Jonker, D. M., van der Hoeven, R. T. M., Vermeij, A. C., Edelbroek, P. M., Brekelmans, G. J., and de Haan, G. J. (2002), Pharmacokinetics and pharmacodynamics of midazolam administered as a concentrated intranasal spray. A study in healthy volunteers, *Br. J. Clin. Pharmcol.*, 53, 501–507.
112. Allonen, H., Ziegler, G., and Klotz, U. (1981), Midazolam kinetics, *Clin. Pharmacol. Ther.*, 30, 653–661.
113. Lindhardt, K., Gizurarson, S., Stefansson, S., Olafsson, D. R., and Bechgaard, E. (2001), Electroencephalographic effects and blood levels after administration of diazepam to humans, *Br. J. Clin. Pharmcol.*, 52, 1–12.
114. Eriksen, J., Jensen, N. H., Kamp-Jensen, M., Bjarno, H., Friis, P., and Brewster, D. (1989), The systemic availability of buprenorphine administered by nasal spray, J. Pharm. Pharmacol., 41, 803–805.
115. Shyu, W. C., Mayol, R. F., Pfeffer, M., Pittman, K. A., and Barbhiya, R. H. (1993), Biopharmaceutical evaluation of transnasal, sublingual and buccal disk dosage forms of butorphanol, *Biopharm. Drug Dispos.*, 14, 371–379.

116. Kaneo, Y. (1983), Absorption from the nasal mucous membrane: I. Nasal absorption of hydralazine in rats, *Acta. Pharm. Suec.*, 20, 379–388.
117. Hirai, S., Yashiki, T., Matsuzawa, T., and Mima, H. (1981), Absorption of drugs from the nasal mucosa of rats, *Int. J. Pharm.*, 7, 317.
118. Landau, A. J., Eberhardt, R. T., and Frishman, W. H. (1994), Intranasal delivery of cardiovascular agents: An innovative approach to cardiovascular pharmacotherapy, *Am. Heart J.*, 127, 1594–1599.
119. Hill, A. B., Bowley, C. J., Nahrwold, M. L., Knight, P. R., Krish, M. M., and Denlinger, J. K. (1981), Intranasal administration of nitroglycerin, *Anethesiology*, 54, 346–348.
120. Krishnamoorthy, R., and Mitra, A. K. (1998), Prodrugs for nasal drug delivery, *Adv. Drug Deliv. Rev.*, 29, 135–146.
121. Fossoulaki, A., and Kaniaris, P. (1983), Intranasal administration of nitroglycerin attenuates the pressor response to laryngoscopy and intubation of the trachea, *Br. J. Anaesth.*, 55, 49–51.
122. Grover, V. K., Sharma, S., Mahajan, R. P., and Singh, H. (1987), Intranasal nitroglycerin attenuates pressor response to tracheal intubation in beta-blocker treated hypertensive patients, *Anesthesia*, 42, 884–887.
123. Hussain, A., Hirai, S., and Bawarshi, R. (1979), Nasal absorption of propranolol in rats, *J. Pharm. Sci.*, 68, 1196.
124. Hussain, A. A., Foster, T., Hirai, S., Kashihara, T., Batenhorst, R., and Jones, M. (1980), Nasal absorption of propranolol in humans, *J. Pharm. Sci.*, 69, 1240.
125. Landau, A. J., Frishman, W. H., Alturk, N., Adjei-Poku, M., Fornasier-Bongo, M., and Furia, S. (1993), Immediate beta-adrenergic blockade and improved exercise tolerance with intranasal propranolol in patients with angina pectoris, *Am. J. Cardiol.*, 72, 995–998.
126. Duchateau, G. S. M. J. E., Zuidema, J., Albers, W. M., and Merkus, F. W. H. M. (1986), Nasal absorption of alprenolol and metoprolol, *Int. J. Pharm.*, 34, 131–136.
127. Duchateau, G. S. M. J. E., Zuidema, J., and Merkus, F. W. H. M. (1986), Bioavailability of propranolol after oral, sublingual and intranasal administration, *Pharm. Res.*, 3, 108–111.
128. Wattanakumtornkul, S., Pinto, A. B., and Williams, D. B. (2003), Intranasal hormone replacement therapy, *Menopause.*, 10, 88–98.
129. Hermens, W. A. J. J., Belder, C. W., Merkus, J. M., Hooymans, P. M., Verhoef, J., and Merkus, F. W. (1991), Intranasal estradiol administration to oophorectomized women, *Eur. J. Obstet. Gynecol. Reprod. Biol.*, 40, 35–41.
130. Hermens, W. A., Belder, C. W., Merkus, J. M., Hooymans, P. M., Verhoef, J., and Merkus, F. W. (1992), Intranasal administration of estradiol in combination with progesterone to oophorectomized women, *Eur. J. Obstet. Gynecol. Reprod. Biol.*, 43, 65–70.
131. Garnero, P., Tsouderos, Y., Marton, I., Pelissier, C., Varin, C., and Delmas, P. D. (1999), Effects of intranasal 17 β-estradiol on bone marrow turnover and serum insulin-like growth factor I in postmenopausal women, *J. Clin. Endocrinol. Metab.*, 84, 2390–2397.
132. Lievertz, R. W. (1987), Pharmacology and pharmacokinetics of estrogens, *Am. J. Obstet. Gynecol.*, 156, 1289–1293.
133. Studd, J., Pornel, B., Marton, I., Bringer, J., Varin, C., Tsouderos, Y., and Christansen, C. (1999), Efficacy and acceptability of intranasal 17 β-estradiol on menopausal symptoms: A randomized dose-response study, *Lancet.*, 353, 1574–1578.
134. Devissaguet, J. P., Brion, N., Lhote, O., and Deloffre, P. (1999), Pulsed estrogen therapy: Pharmacokinetics of intranasal 17 β-estradiol (S21400) in postmenopausal women and

comparison with oral and transdermal formulations, *Eur. J. Drug Metab. Pharmacokinet.*, 24, 265–271.

135. Wilkund, I., Karlberg, J., and Mattson, L. A. (1993), Quality of life of postmenopausal women on a regimen of transdermal estradiol therapy: A double blind placebo-controlled study, *Am. J. Obstet. Gynecol.*, 168, 824–830.
136. Wilkund, I., Holst, J., Karlberg, J., et al. (1992), A new methodological approach to the evaluation of quality of life in postmenopausal women, *Maturitas*, 14, 211–224.
137. Kupperman, H. S., Blatt, M. H. G., Weisbader, H., and Filler, W. (1953), Comparative clinical evaluation of estrogenic preparations by the menopausal and amenorrheal indices, *Endocrinology*, 13, 688–703.
138. Hussain, A. A., Al-bayati, A. A., Dakkuri, A., Okochi, K., and Hussain, M. A. (2002), Testosterone 17 β-*N,N*-dimethylglycinate hydrochloride: A prodrug with a potential for nasal delivery of testosterone, *J. Pharm. Sci.*, 91, 785–789.
139. Hussain, A. A., Kimura, R., and Huang, C. H. (1984), Nasal absorption of testosterone in rats, *J. Pharm. Sci.*, 73, 1300–1301.
140. Illum, L. (2000), Transport of drugs from the nasal cavity to the central nervous system, *Eur. J. Pharm. Sci.*, 11, 1–18.
141. Chow, H. S., Chen, Z., and Matsuura, G. T. (1999), Direct transport of cocaine from the nasal cavity to the brain following intranasal cocaine administration in rats, *J. Pharm. Sci.*, 88, 754–758.
142. Landsteiner, K., and Levaditi, C. (1910), Etude experimentale de la poliomyelite aigue (Maladiede Heine Medin), *Ann. Inst. Pasteur.*, 24, 833–878.
143. Leiner, C., and von Weisner, R. (1910), Experimentale untersuchungen uber poliomyelitis acuta, *Wien. Med. Wchnschr.*, 60, 2482.
144. Flexner, S. (1912), The mode of action of infection epidemic poliomyelitis, *JAMA*, 59, 1371.
145. Flexner, S., and Clark, P. F. (1912), A note on the mode of infection in epidemic poliomyelitis, *Proc. Soc. Exp. Biol. Med.*, 10, 1.
146. Clark, P. F., Fraser, F. R., and Amoss, H. L. (1917), The relation to the blood of the virus of epidemic poliomyelitis, *J. Exp. Med.*, 19, 223–233.
147. Fairbrother, R. W., and Hurst, E. W. (1930), The pathogenesis of the propagation of the virus in experimental poliomyelitis, *J. Path. Bact.*, 33, 17–45.
148. Faber, H. K., and Gebhardt, L. D. (1938), Localization of poliomyelitic virus during incubation period after intranasal instillation in monkeys, *Proc. Soc. Exper. Biol. Med.*, 30, 879–880.
149. Faber, H. K., and Gebhardt, L. D. (1938), Localizations of the virus of poliomyelitis. In the central nervous system during the pre-paralytic period, after intranasal inoculation, *J. Exp. Med.*, 57, 933–954.
150. Sabin, A. B., and Olitsky, P. K. (1936), Influence of pathway of infection on pathology of olfactory bulbs in experimental poliomyelitis, *Proc. Soc. Exp. Biol. Med.*, 35, 300–301.
151. Bodian, D., Morgan, I., and Schwerdt, C. E. (1950), Virus and host factors influencing the titer of Lansing poliomyelitis virus in monkeys, cotton rats and mice, *Am. J. Hyg.*, 51, 126–133.
152. Gopinath, P. G., Gopinath, G., and Kumar, T. C. A. (1978), Target site of intranasally sprayed substances and their transport across the nasal mucosa. A new insight into the intranasal route of drug delivery, *Curr. Ther. Res.*, 23, 596–607.
153. Hastings, L. (1990), Sensory neurotoxicology: Use of the olfactory system in the assessment of toxicity, *Neurotoxicol. Teratol.*, 12, 455–459.

154. Hastings, L., and Evans, J. E. (1991), Olfactory primary neuron as a route of entry for toxic agents into the CNS, *Neurotoxicology*, 12, 707–714.

155. Illum, L. (2004), Is nose-to-brain transport of drugs in man a reality? *J. Pharm. Pharmacol.*, 56, 3–17.

156. Mathison, S., Nagilla, R., and Kompella, U. B. (1998), Nasal route for direct delivery of solutes to the central nervous system: Fact or fiction? *J. Drug Target.*, 5, 415–441.

157. Thorne, R. G., and Frey, W. H. (2001), Delivery of neurotropic factors to the central nervous system, *Brain. Res.*, 692, 278–282.

158. Pietrowsky, R., Struben, C., Molle, M., Fehm, H. L., and Born, J. (1966), Brain potential changes after intranasal vs. intravenous administration of vasopressin: Evidence for a direct nose-to-brain pathway for peptide effects in humans, *Biol. Psychiatry*, 39, 332–340.

159. Derad, I., Willeke, K., Pietrowsky, R., Born, J., and Fehm, H. L. (1998), Intranasal angiotensin II directly influences central nervous regulation of blood pressure, *Am. J. Hypertens.*, 11, 971–977.

160. Pietrowsky, R., Thiemann, A., Kern, W., Fehm, H. L., and Born, J. (1966), A nose-to-brain pathway for psychotropic peptides: Evidence from a brain evoked potential study with cholecystokinin, *Psychoneuroendocrinology*, 21, 559–572.

161. Lindhardt, K., Gizurarson, S., Stefansson, S. B., Olafsson, D. R., and Bechgaard, E. (2001), Electroencephalographic effects and serum concentration after intranasal and intravenous administration of diazepam to healthy volunteers, *Br. J. Clin. Pharmacol.*, 52, 521–527.

162. Kern, W., Peters, A., Fruehwald-Schultes, B., Deininger, E., Born, J., and Fehm, H. L. (2001), Improving influence of insulin on cognitive functions in humans, *Neuroendocrinology*, 74, 270–280.

163. Okuyama, S. (1997), The first attempt at radioisotopic evaluation of the integrity of the nose-brain barrier, *Life Sci.*, 60, 1881–1884.

164. Quay, S. C. (2001), Successful delivery of apomorphine to the brain following intranasal administration demonstrated in clinical study, PRNewswire, July 18.

165. Merkus, P. (2003), Transport of non-peptide drugs from the nose to the CSF, paper presented at the Nasal Drug Delivery Meeting, Management Forum, London, Mar. 24–25.

166. Born, J., Lange, T., Kern, W., McGregor, G. P., Bickel, U., and Fehm, H. L. (2002), Sniffing neuropeptides: A transnasal approach to the human brain, *Nat. Neurosci.*, 5, 514–516.

167. Seki, T., Sato, N., Hasegawa, T., Kawaguchi, T., and Juni, K. (1994), Nasal absorption of zidovudine and its transport to cerebrospinal fluid in rats, *Biol. Pharm. Bull.*, 17, 1135–1137.

168. Char, H., Kumar, S., Patel, S., Piemontese, D., Iqbal, K., Waseem Malick, A., Salvador, R. A., and Behl, C. R. (1992), Nasal delivery of [14C] dextromethorphan hydrochloride in rats: Levels in plasma and brain, *J. Pharm. Sci.*, 81, 750–752.

169. Sakane, T., Akizuki, M., Yoshida, M., Yamashita, S., Nadai, T., Hashida, M., and Sezaki, H. (1991), Transport of cephalexin to the cerebrospinal fluid directly from the nasal cavity, *J. Pharm. Pharmacol.*, 43, 449–451.

170. Sakane, T., Akizuki, M., Yamashita, S., Sezaki, H., and Nadai, T. (1994), Direct drug transport from the nasal cavity to the cerebrospinal fluid: The relation to the dissociation of the drug, *J. Pharm. Pharmacol.*, 46, 378–379.

171. Sakane, T., Akizuki, M., Yamashita, S., Nadai, T., Hashida, M., and Sezaki, H. (1991), The transport of a drug to the cerebrospinal fluid directly from the nasal cavity: The relation to the lipophilicity of the drug, *Chem. Pharm. Bull.*, 39, 2456–2458.

172. Sakane, T., Akizuki, M., Taki, Y., Yamashita, S., Sezaki, H., and Nadai, T. (1995), Direct drug transport from the nasal cavity to the cerebrospinal fluid: The relation to the molecular weight of the drugs, *J. Pharm. Pharmacol.*, 47, 379–381.

173. Dahlin, M., Jansson, B., and Bjork, E. (2001), Levels of dopamine in the blood and brain following nasal administration to rats, *Eur. J. Pharm. Sci.*, 14, 75–80.

174. Wang, Y., Aun, R., and Tse, F. L. S. (1998), Brain uptake of dihydroergotamine after intravenous and nasal administration in the rat, *Biopharm. Drug Dispos.*, 19, 571–575.

175. Gizurarson, S., Trovaldsson, T., Sigurdsson, P., and Gunnarsson, E. (1996), Selective delivery of insulin into the brain: Introlfatory absorption, *Int. J. Pharm.*, 140, 77–83.

176. Wang, F., Jiang, X., and Lu, W. (2003), Profiles of methotrexate in blood and CSF following intranasal and intravenous administration to rats, *Int. J. Pharm.*, 263, 1–7.

177. Frey, W. H., Liu, J., Thorne, R. G., and Rahman, Y. E. (1995), Intranasal delivery of ^{125}I-labelled nerve growth factor to the brain via the olfactory route, in Iqbal, K., Mortimer, J. A., Winbald, B., and Wisniewski, H. M., Eds., *Research Advances in Alzheimer's Disease and Related Disorders*, Wiley, New York, pp. 329–335.

178. Thorne, R. G., Emroy, C. R., Ala, T. A., and Frey, W. H. (1995), Quantitative analysis of the olfactory pathway for drug delivery to the brain, *Brain. Res.*, 692, 278–282.

179. Vyas, T. K., Babbar, A. K., Sharma, R. K., and Misra, A. (2005), Intranasal mucoadhesive microemulsions of zolmitriptan: Preliminary studies on brain-targeting, *J. Drug Target.*, 13, 317–324.

180. Fliedner, S., Schulz, C., and Lehnert, H. (2006), Brain uptake of intranasally applied radioiodinated leptin in wistar rats, *Endocrinology*, 147, 2088–2094.

181. Westin, U. E., Boström, E., Gråsjö, J., Hammarlund-Udenaes, M., and Björk, E. (2006), Direct nose-to-brain transfer of morphine after nasal administration to rats, *Pharm. Res.*, 23, 565–572.

182. Zhang, Q. Z., Zha, L. S., Zhang, Y., Jiang, W. M., Lu, W., Shi, Z. Q., Jhang, X. G., and Fu, S. K. (2006), The brain targeting efficiency following nasally applied MPEG-PLA nanoparticles in rats, *J. Drug Target.*, 14, 281–290.

183. Shi, Z., Zhang, Q., and Jiang, X. (2005), Pharmacokinetic behavior in plasma, cerebrospinal fluid and cerebral cortex after intranasal administration of hydrochloride meptazinol, *Life Sci.*, 77, 2574–2583.

184. Wang, X., Haibing, H., Leng, W., and Tang, X. (2006), Evaluation of brain-targeting for the nasal delivery of estradiol by the microdialysis method, *Int. J. Pharm.*, 317, 40–46.

185. Jones, N. (2001), The nose and paranasal sinuses physiology and anatomy, *Adv. Drug Deliv. Rev.*, 51, 5–19.

186. Graff, C. L., and Pollack, G. M. (2005), Nasal drug administration: Potential for targeted central nervous system delivery, *J. Pharm. Sci.*, 94, 1187–1195.

187. Sherwood, L. (1989), *Human Physiology. From Cells to Systems*, West Publishing, St. Paul, MN, Chapter 5.

188. Stanley, P. A., and Walter, O. A. (2004), in Plotkin, S. A., Orenstein, W. A., and Offit, P. A., Eds., *A Short History of Vaccination. Vaccines*, 4th ed., Saunders, Philadelphia, pp. 1–15.

189. Ogra, P. L., Faden, H., and Welliver, R. C. (2001, Apr.), Vaccination strategies for mucosal immune responses, *Clin. Microbiol. Rev.*, 14(2), 430–445.

190. Cui, Z. R., and Mumper, R. J. (2002), Intranasal administration of plasmid DNA-coated nanoparticles results in enhanced immune responses, *J. Pharm. Pharmacol.*, 54, 1195–1203.

191. Beverley, L. C. P. (2002), Immunology of vaccination, *Br. Med. Bull.*, 62, 15–28.

192. Illum, L., and Davis, S. S. (2001), Nasal vaccination: A non-invasive vaccine delivery method that holds great promise for the future, *Adv. Drug Deliv. Rev.*, 51, 1–3.

193. Davis, S. S. (2001), Nasal vaccines, *Adv. Drug Deliv. Rev.*, 51, 21–42.

194. Partidos, C. D. (2000), Intranasal vaccines: Forthcoming challenges, *PSTT*, 3, 273–281.

195. Mestecky, J., and McGhee, J. R. (1987), Immunoglobulin A: Molecular and cellular interactions in IgA biosynthesis and immune response, *Adv. Immunol.*, 40, 153–245.

196. Gill, J. I., Lin, W., Kistner, O., Davis, S. S., and Illum, L. (2001), Polymeric lamellar substrate particles for intranasal vaccination, *Adv. Drug Deliv. Rev.*, 51, 97–111.

197. Biesterveld, N. (2002), Nasal vaccine delivery, vaccine delivery adjuvants, Merck & Company, West Point, PA, a Literature review submitted to the Department of Chemistry, Lehigh University, Aug. 2002.

198. Gander, B. (2005), Trends in particulate antigen and DNA delivery systems for vaccines, *Adv. Drug Deliv. Rev.*, 57, 321–323.

199. Friede, M., and Aguado, T. M. (2005), Need for new vaccine formulation and potential of particulate antigen and DNA delivery system, *Adv. Drug Deliv. Rev.*, 57, 325–331.

200. Illum, L. (2003), Nasal drug delivery–possibilities, problems and solutions, *J. Controlled Release*, 87, 187–198.

201. Alpar, O. H., Somavarapu, S., Atuah, N. K., and Bramwell, W. V. (2005), Biodegradable mucoadhesive particulates for nasal and pulmonary antigen and DNA delivery. *Adv. Drug Deliv. Rev.*, 57, 411–430.

202. Mackay, M., Williamson, I., and Hastewell, J. (1991), Cell biology of Epithelia, *Adv. Drug Deliv. Rev.*, 7, 313–338.

203. Storni, T., Kundig, M. T., Senti, G., and Johansen, P. (2005), Immunity in response to particulate antigen-delivery systems, *Adv. Drug Deliv. Rev.*, 57, 333–355.

204. O'Hagan, T. D., and Rappuoli, R. (2004), Novel approaches to vaccine delivery, *Pharm. Res.*, 21(9), 1519–1530.

205. McNeela, A. E., and Mills, G. H. K. (2001), Manipulating the immune response: Humoral, versus, cell, mediated, immunity, *Adv. Drug Deliv. Rev.*, 51, 43–54.

206. Edelman, R. (2002), The development and use of vaccine adjuvants, *Mol. Biotechnol.*, 21, 129–148.

207. Singh, M., and O'Hagan, D. (1999), Advances in vaccine adjuvants, *Nat. Biotechnol.*, 17, 1075–1081.

208. McNeela, E. A., O'Connor, D., Gill, J. I., Illum, L., Davis, S. S., Pizza, M., Peppoloni, S., Rappuoli, R., and Mills, G. H. K. (2000), A mucosal vaccine against diphtheria: Formulation of cross reacting material (CRM_{197}) of diphtheria toxin with chitosan enhances local and systemic antibodies and Th2 responses following nasal delivery, *Vaccine.*, 19, 1188–1198.

209. Illum, L., Gill, J. I., Hinchcliffe, M., Fisher, N. A., and Davis, S. S. (2001), Chitosan as a novel nasal delivery system for vaccines, *Adv. Drug Deliv. Rev.*, 51, 81–96.

210. Jabbal-Gill, I., Fisher, A. N., Rappuoli, R., Davis, S. S., and Illum, L. (1998), Stimulation of mucosal and systemic antibody responses against Bordetella *pertussis* filamentous haemagglutinin and recombinant *pertussis* toxin after nasal administration with chitosan in mice, *Vaccine*, 16, 2039–2046.

211. Boyaka, P. N., Marinaro, M., Jackson, R. J., Menon, S., Kiyona, H., Jirillo, E., and McGhee, J. R. (1999), IL-12 is an effective adjuvant for induction of mucosal immunity, *J. Immunol.*, 162, 122–128.

212. Arulanandan, B. P., O'Toole, M., and Metzger, D. W. (1999), Intranasal interleukin-12 is a powerful adjuvant for protective mucosal immunity, *J. Infect. Dis.*, 180, 940–949.

213. McCluskie, M. J., and Davis, H. L. (2001), Oral, intrarectal and intranasal immunizations using CpG and non-CpG oligodeoxynucleotides as adjuvants, *Vaccine.*, 19, 413–422.

214. Moldoveanu, Z., Love-Homan, L., Huang, W. Q., and Krieg, A. M. (1998), CpG DNA, a novel immune enhancer for systemic and mucosal immunization with influenza virus, *Vaccine*, 16, 1216–1224.

215. de Haan, A., Geerligs, H. J., Huchshorn, J. P., van Scharrenburg, G. J., Palache, A. M., and Wilschut, J. (1995), Mucosal immunoadjuvant activity of liposomes: Induction of systemic IgG and secretory IgA responses in mice by intranasal immunization with an influenza subunit vaccine and coadministered liposomes, *Vaccine*, 13, 155–162.

216. de Haan, A., Tomee, J. F. C., Huchshorn, J. P., and Wilschut, J. (1995), Liposomes as an immunoadjuvant system for stimulation of mucosal and systemic antibody response against inactivated measles virus administered intranasally to mice, *Vaccine*, 13, 1320–1324.

217. Aramaki, Y., Fujii, Y., Yachi, K., Kikuchi, H., and Tsuchiya, S. (1994), Activation of systemic and mucosal immune response following nasal administration of liposomes, *Vaccine*, 12, 1241–1245.

218. Childers, N. K., Tong, G., Mitchell, S., Kirk, K., Russel, M. W., and Michalek, S. M. (1999), A controlled clinical study of the effect of nasal immunization with a *Streptococcus mutans* antigen alone or incorporated into liposomes on induction of immune responses, *Infect. Immun.*, 67, 618–623.

219. Gluck, U., Gebbers, J. O., and Gluck, R. (1999), Phase I evaluation of intranasal virosomal influenza vaccine with and without *Escherichia coli* heat-labile toxin in adult volunteers, *J. Virol.*, 73, 7780–7786.

220. Glück, R. (1999), Adjuvant activity of immunopotentiating reconstituted influenza virosomes (IRIVs), *Vaccine*, 17, 1782–1787.

221. Baca-Estrada, M. E., Foldvari, M., Snider, M., Harding, K., Kournikakis, B., Babiuk, L. A., and Griebel, P. (2000), Intranasal immunization with liposome-formulated *Yersinia pestis* vaccine enhances mucosal immune responses, *Vaccine*, 18, 2203–2211.

222. Lövgren, K., Kaberg, H., and Morein, B. (1990), An experimental influenza subunit vaccine (iscom): Induction of protective immunity to challenge infection in mice after intranasal or subcutaneous administration, *Clin. Exp. Immunol.*, 82, 435–439.

223. Hsu, S. C., Schadeck, E. B., Delmas, A., Shaw, M., and Steward, M. W. (1996), Linkage of a fusion peptide to a CTL epitope from the nucleoprotein of measles virus enables incorporation into ISCOMS and induction of CTL responses following intranasal immunization, *Vaccine*, 14, 1159–1166.

224. Carol, H., Nieto, A., Villacres-Eriksson, M., and Morein, B. (1997), Intranasal immunization of mice with *Echinococcus granulosus* surface antigens iscoms evokes a strong immune response, biased towards glucidic epitopes, *Parasite Immunol.*, 19, 197–205.

225. Hu, K. F., Elvander, M., Merza, M., Akerblom, L., BrAndenburg, A., and Morein, B. (1998), The immunostimulatory complex (ISCOM) is an efficient mucosal delivery system for respiratory syncytial virus (RSV) envelope antigens inducing high local and systemic antibody responses, *Clin. Exp. Immunol.*, 113, 235–243.

226. Jaganathan, K. S., and Vyas, S. P. (2006), Strong systemic and mucosal immune responses to surface-modified PLGA microspheres containing recombinant hepatitis B antigen administered intranasally, *Vaccine*, 24, 4201–4211.

227. Debin, A., Kravtzoff, R., Vaz Santiago, J., Cazales, L., Speranido, S., Melber, K., Janowicz, Z., Betbeder, D., and Moynier, M. (2002), Intranasal immunization with recombinant antigens associated with new cationic particles induces strong mucosal as well as systemic antibody and CTL responses, *Vaccine*, 20, 2752–2763.

228. Aguilar, J. C., Lobaina, Y., Muzio, V., Garcia, D., Penton, E., Iglesias, E., Pichardo, D., Urquiza, D., Rodeiguez, D., Silva, D., Petrovsky, N., and Guillen, G. (2004), Development of a nasal vaccine for chronic hepatitis B infection that uses the ability of hepatitis B core antigen to stimulate a strong Th1 response against hepatitis B surface antigen, *Immnol. Cell. Biol.*, 82, 539–546.

229. Hordness, K., Tynning, T., Brown, T. A., Haneberg, B., and Jonsson, R. (1997), Nasal immunization with group B streptococci can induce high levels of specific IgA antibodies in cervicovaginal secretions of mice, *Vaccine*, 15, 1244–1251.

230. Kurono, Y., Yamamoto, M., Fujihashi, K., Kodama, S., Suzuki, M., Mogi, G., McGhee, J. R., and Kiyono, H. (1991), Nasal immunization induces *Haemophilis influenzae*-specific Th1 and Th2 responses with mucosal IgA and systemic IgG antibodies for protective immunity, *J. Infect. Dis.*, 180, 122–132.

231. Hagiwara, Y., Komase, K., Chen, Z., Matsuo, K., Suzuki, Y., Aizawa, C., Kurata, T., and Tamura, S. (1999), Mutants of cholera toxin as an effective and safe adjuvant for nasal influenza vaccine, *Vaccine*, 17, 2918–2926.

232. Bastein, N., Trudel, M., and Simard, C. (1999), Complete protection of mice from respiratory syncytial virus infection following mucosal delivery of synthetic peptide vaccines, *Vaccine*, 17, 832–836.

5.7

NASAL POWDER DRUG DELIVERY

JELENA FILIPOVIĆ-GRČIĆ AND ANITA HAFNER
University of Zagreb, Zagreb, Croatia

Contents

5.7.1 INTRODUCTION

Intranasal drug administration has been practiced since ancient times. In Tibet extracts of sandalwood and aloewood were inhaled to treat emesis. Egyptians treated epistaxis and rhinitis using intranasal medication. North American Indians relieved headaches inhaling crushed leaves of *Ranunculus acris* [1]. Due to the rich vasculature and high permeability of nasal mucosa, the absorption rate and pharmacokinetics of nasally administrated drug are comparable to that obtained by intravenous drug delivery, while noninvasive nasal drug administration is more convenient to patients. As nasally administrated drugs avoid first-pass hepatic metabolism, improved bioavailability can be expected. However, rapid mucociliary

Pharmaceutical Manufacturing Handbook: Production and Processes, edited by Shayne Cox Gad

clearance reduces the residence time of nasal drug delivery system at the site of absorption. Dry powders have been shown to delay mucociliary clearance, thus prolonging the contact time between the drug delivery system and mucosa compared to liquid formulations. Most of the dry powder investigations are based on mucoadhesive swellable polymers as they can additionally improve drug absorption and bioavailability. Dry powder delivery systems such as microspheres are of special interest, offering the possibility of predictable and controlled drug release from the polymeric device [2, 3].

5.7.2 NASAL DRY POWDER FORMULATIONS

Liquid preparations are most frequently used nasal dosage forms at present. However, such preparations are characterized with short residence time in the nasal cavity, low drug concentration at the site of absorption, and problems linked to the chemical stability of the drug and the stability of the preparation. In the case of liquid formulations, drugs must be administered in small volumes. The maximum volume of a single dose in one nostril is about 200 μL. The volume of therapeutic dose should not exceed the capacity of the nasal cavity, as it would drain out of the nose. Thus, only low-dose or highly soluble drugs can be administered nasally in the form of a simple liquid formulation [4]. Dry powder formulations have been recognized as efficient nasal delivery systems offering numerous advantages over liquid formulations, such as avoidance of preservatives, improved formulation stability, and prolonged contact with the mucosa. For a powder formulation, the maximum quantity is approximately 50 mg, depending upon the bulk density of the material [5].

5.7.2.1 Benefits Associated with Nasal Powder Drug Delivery

A powder form was found to be more effective than liquid formulations in a number of investigations described in the literature [6–11]. Dry powders are characterized by prolonged residence time and higher drug concentration at the site of deposition as well as improved formulation stability with no requirement for preservatives [12]. Prolonged residence time of the powder delivery systems at the absorption site results in enhanced systemic bioavailability, compared to the liquid formulations. In the case of powders, higher drug concentration at the site of absorption causes rapid transmucosal diffusion and faster onset of action [13]. Most of the dry powder formulations are based on mucoadhesive swellable polymers (e.g., starch, dextran, chitosan) as they can additionally improve drug bioavailability, prolonging the residence time in the nasal cavity or even promoting drug absorption. Powder formulation with a water-insoluble and nonswellable drug carrier may also improve nasal bioavailability of the polar drugs. Ishikawa et al. [14] found that a nasal powder delivery system of elcatonin based on $CaCO_3$ significantly increased the systemic elcatonin bioavailability in rats and rabbits compared to the liquid formulation. Enhanced bioavailability has been primarily ascribed to the retardation of the clearance of the drug powder delivery system from the nasal cavity.

The use of dry powder formulations in nasal vaccine delivery has been extensively reviewed elsewhere [15–18]. The association of vaccines to some of the particulate systems has been proved to enhance the systemic and mucosal immune responses against the antigens [19–21]. Dry powder formulations for nasal vaccine delivery may also provide significant advantages with respect to stability shortcomings compared to conventional liquid intranasal and intramuscular formulations, which require frozen storage or refrigeration [22].

5.7.2.2 Drug Powder or Drug/Polymer Powder Formulation for Nasal Drug Delivery?

The drug candidate for nasal administration should possess a number of attributes, such as appropriate aqueous solubility and nasal absorption characteristics, minimal nasal irritation, low dose, no offensive odor or aroma, and suitable stability characteristics [23]. In the case of drug powder formulations it is possible to hide or alter the unfavorable characteristics of a drug using suitable polymers as drug carriers. Thus, improvement of the dissolution behavior of drugs of low aqueous solubility after incorporation in polymeric powder devices such as microspheres has been reported in the literature [24, 25]. The improvement of the drug dissolution rate from the microspheres has been ascribed to several factors, such as high microsphere surface–volume ratio, the hydrophilic nature of the polymer, and drug amorphization due to drug–polymer interaction and/or the microsphere preparation method [25, 26].

Nasally administrated polymer–drug powders were also characterized by improved drug absorption compared to pure drug powders [25, 27]. Teshima et al. [27] monitored changes in the plasma glucagons and glucose concentrations after nasal administration of the powder form of glucagon alone and glucagon mixed with the carrier, microcrystalline cellulose (MCC). Glucagon and glucose plasma concentrations remained unchanged after nasal administration of the powder form of glucagon alone while it increased after glucagon–MCC administration in an MCC content–dependent manner. Results of in vivo nasal administration of carbamazepine-loaded chitosan microspheres revealed an increase in carbamazepine concentration in serum compared to the pure carbamazepine powder [25]. Such an increase in drug absorption has been ascribed to both improved dissolution of carbamazepine and adhesion of the chitosan microspheres to the mucosal surface.

The influence of polymers on the drug stability in the powder formulation has also been reported [8]. Green coloration of polyacrylic acid powder dosage forms loaded with 60% apomorphine due to atmospheric drug oxidation upon storage has been observed. Dosage forms with lower drug loadings (and higher polymer content) showed no coloration, indicating the protective role of polymers against drug oxidation in powder formulations.

Powders intended for nasal administration have to be optimized in terms of particle size and morphology as these properties are related to potential irritation in the nasal cavity [23]. Certain procedures (e.g., spray drying process) can modify the particle size of the drug powder raw material, but in order to optimize the morphology and flowability properties of some pure drug powders, excipients need to be used. Sacchetti et al. [28] reported that the use of mannitol as a filler and hydroxypropylmethyl cellulose (HPMC) as a shaper of spray-dried caffeine microparticles modified the typical needle shape of spray-dried caffeine to a more convenient roundish shape. Further addition of polyethylene glycol (PEG) resulted in increased cohesiveness between particles, producing agglomerates considered as beneficial for nasal deposition.

5.7.2.3 Powder Properties Affecting Nasal Deposition and Drug Delivery

Powder formulation efficiency is strongly dependent on two technological aspects of delivery: first, the complete release of the dose through the powder delivery device, that is, the quantitative aspect, and, second, the distribution of the powder on the nasal mucosa, that is, the qualitative aspect. For example, rhinitis treatment requires drug deposition over the entire mucosal area, while brain targeting requires deposition strictly at the roof of the nose. The qualitative aspect is defined by the aerodynamic properties of the powder cloud produced (puff) and is assessed by a photographic technique providing information on the spray pattern and cloud geometry. The parameters involved are shape, height, area, density, particle size, and velocity. For nasal impacting the powder cloud should remain as compact as possible to achieve an efficient shot of powder to the nasal mucosa, whereas for sedimentation a larger cloud would be preferred [12].

The quantitative aspect of nasal powder delivery is largely dependent on the type of delivery device. Devices for powder dosage forms, including insufflators, monodose and multidose powder inhalers, and pressurized metered-dose inhalers, are extensively reviewed elsewhere [29, 30]. Particles intended for nasal delivery should have good flow properties to be reproducibly filled in the dose reservoir and easily insufflated to obtain appropriate nasal deposition [28]. Since the particle size of the applied powder formulation has a major impact on its nasal deposition, the characterization of this parameter is very important. Only particles over 5 μm are deposited in the nostrils while smaller particles can be inhaled into the lower parts of the respiratory system [23]. For that reason, it is necessary to determine particle sizes not only prior to filling the nasal delivery device but also after its actuation, since the particle size of the formulation leaving the device depends on the device-disaggregating properties [29]. Furthermore, it is known that powder deposition in the anterior part of the nasal epithelium contributes to nasal absorption more than powder deposition in the posterior part, since ciliated cells cover mainly the posterior part of the nasal epithelium, providing a faster clearance of particles [31, 32]. Thus, deposition studies must include both the amount of deposited particles and the location of the particle deposition [29]. Distribution of the formulation in the nasal cavity should be evaluated considering two different deposition types: the initial deposition immediately after application and the secondary deposition, owing to translocation by mucociliary clearance [29]. Cornaz et al. [33] investigated the ability of microsphere delivery, studying the shape and area of the clouds of microspheres emitted from a nasal device. The puff areas of loaded particles were smaller compared to unloaded particles due to different particle porosities. Pringels et al. [31] investigated the influence of the deposition pattern and spray characteristics of nasal powder formulations on insulin bioavailability. It has been shown that posterior deposition of the powder formulation in the nasal cavity lowered insulin bioavailability. To study the spray characteristics, the shape and cross section of the emitted powder cloud (puff) were analyzed. It has been concluded that the powder bulk density of the formulation influenced the spray pattern. Comparing two powders with different bulk densities and particle sizes, it has been revealed that the powder with higher bulk density and smaller particle size was more compact and was characterized by higher resistance to airflow, resulting in a slower spray time and larger spray pattern [31]. No influence of the powder bulk density and spray pattern on insulin bioavailability has been observed.

5.7.3 POLYMERS IN NASAL POWDER DELIVERY SYSTEM

It has been demonstrated that low absorption of drugs can be improved by using absorption enhancers or prolonging contact between drug and absorptive sites in the nasal cavity by delaying mucociliary clearance of the formulation. Some mucoadhesive polymers can serve both functions. They are typically high-molecular-weight polymers with flexible chains which can interact with mucin through hydrogen bonding, electrostatic, hydrophobic or van der Waals interactions [18, 34, 35].

The mucoadhesive polymers are often hydrophilic and swellable, containing numerous hydrogen bond–forming groups such as hydroxyl, carboxyl, or amine, which favors adhesion. When used in a dry form they attract water from the mucosal surface and swell, leading to polymer–mucus interaction, increased viscosity of polymer–mucus mixture, and reduced mucociliary clearance [34]. Beside the type of polymer functional groups, the mucoadhesive force of a polymer material is dependent on the polymer molecular weight, concentration, flexibility of the polymer chain, spatial conformation, contact time, environmental pH, and physiological factors such as mucin turnover and disease state [3]. There is a critical polymer molecular weight for each polymer type below or above which there is reduced adhesive power [34]. The mucoadhesive properties can also be affected by the degree of cross-linking of the polymer since mucoadhesion requires an adequate free chain length for interpenetration to occur. Hence, the more cross-linked the polymer, the less strong the mucoadhesive interaction [36]. Hydration and swelling present both polymer- and environment-related factors. Overhydration causes extended swelling, resulting in slippery mucilage formation [30]. The polymer concentration that is required for optimum mucoadhesion is different between gels and solid mucoadhesives. In the liquid state, an optimum concentration exists for each polymer for which best adhesion can occur, while with solid dosage forms, increased polymer concentration leads to increased mucoadhesive power [35]. Studies have shown that polymers with charge density can serve as good mucoadhesive agents, although their mucoadhesive properties are affected by the pH of the surrounding media. The presence of metal ions, which can interact with charged polymers, may also affect the adhesion process [35, 36]. It has also been reported that polyanion polymers are more effective bioadhesives than polycation polymers or nonionic polymers [37].

Research on nasal powder drug delivery has employed polymers such as starch, dextrans, polyacrylic acid derivatives (e.g., carbopol, polycarbophil), cellulose derivatives (microcrystalline cellulose, semicrystalline cellulose, hydroxypropylmethyl cellulose, hydroxypropyl cellulose, carboxymethyl cellulose), chitosan, sodium alginate, hyaluronans, and polyanhydrides such as poly(methyl vinyl ether-*co*-maleic anhydride) (PVM/MA). Many of these polymers have already been used as excipients in pharmaceutical formulations and are often referred to as first-generation bioadhesives [38–45]. In nasal dry powder a single bioadhesive polymer or a combination of two or more polymers has been formulated as freeze-dried or spray-dried particles or micropheres.

Crystalline cellulose, hydroxypropyl cellulose, and Carbopol 934 have been studied in combination with lyophilized insulin as bioadhesive powder dosage forms for nasal delivery. Each formulation tested resulted in an decrease in plasma glucose level after nasal administration in dog and rabbit models. The most effective formulation, crystalline cellulose blended with insulin, decreased the plasma glucose level to 49% of the control value. In ternary systems the lyophilized Carbopol 934 and insulin blend with crystalline cellulose powder has been the most effective, leading to a hypoglycemia on the order of one-third of the effect obtained after intravenous injection of the same dose of insulin. The plasma glucose levels obtained in the volunteers after administration of the insulin–Carbopol–crystalline cellulose powder formulation were quite variable [38].

The various powder formulations were prepared by dry blending of octreotide with microcrystalline cellulose, semicrystalline cellulose, hydroxyethyl starch, cross-linked dextran, microcrystalline chitosan, pectin, and alginic acid [40]. Their potential to enhance the nasal absorption of the somatostatin analogue peptide octreotide was studied in vivo in the rat model. The powder mixtures were also characterized in vitro regarding calcium binding, water uptake, and drug release. The bioavailabilities obtained for all of the powder formulations were low, with the highest values for alginic acid and cross-linked dextran powder formulations (4.1 and 5.56%, respectively).

Callens and Remon [46] have shown improved nasal absorption of insulin in rabbits by using a bioadhesive powder formulation containing drum-dried waxy maize starch (mainly amylopectin) and Carbopol 974P. The bioavailability of 14% has been obtained. They have shown [47] that the initial advantage of a longer residence time of the powder formulation in the nasal cavity might turn into a disadvantage after multiple administration and impact bioavailability. They investigated the influence of eight daily administrations of two powder formulations to rabbits on the bioavailability and therapeutic effect of the insulin [47]. The first powder formulation consisted of a co-spray-dried mixture of Amioca starch and Carbopol 974P and the second one has been a physical mixture of drum-dried waxy maize starch and Carbopol 974P. By a single nasal administration to rabbits, absolute bioavailabilities of 17.8 and 13.4% have been obtained, respectively. The lower insulin bioavailabilities (4.4 and 3.6%, respectively) after multiple administrations were observed with both formulations, mainly due to the high viscosity of the bioadhesive powders in the nasal mucus, causing a physical barrier toward absorption and a strongly decelerated mucociliary clearance. Long residence times of the powder formulations were also reported by Ugwoke et al. [42, 48], who noticed nasal residence times of more than 24 h using powder formulations containing Carbopol 971P and carboxymethyl cellulose.

Rhinocort is a commercially available mucoadhesive transnasal powder preparation of beclomethasone dipropionate with hydroxypropyl cellulose (HPC) as a gel-forming drug carrier developed by Suzuki and Makino [49]. The HPC has been shown to promote the absorption of low-molecular-weight drugs, but it was not that effective with a peptide drug salmon calcitonin. Microcrystalline cellulose has been shown to be effective for the promotion of absorption of calcitonin in humans, producing about 10% bioavailability with rapid absorption onset [49]. In the study of the effect of an HPC and MCC combination on the development of nasal powder preparations for peptide delivery, significant absorption enhancement of leuprolide, calcitonin, and fluorescein isothiocyanate (FITC)–dextran in rabbits has been obtained by the addition of 10–20% HPC to MCC. It has been suggested that MCC works as an absorption enhancer by causing a locally high concentration of drugs in the vicinity of the mucosa surface while HPC works to increase retention of drugs on the nasal mucosa due to its gel-forming property. In a comparative study [50] of a series of MCC nasal sprays and lyophilized powder formulations of ketorolac, the spray formulations have been shown to be better absorbed than powder formulations. The absolute bioavailability of ketorolac from a powder formulation has been 38%, and no significant differences in absorption between different powder formulations have been observed.

Lim et al. [51, 52] compared a number of mucoadhesive microspheres prepared by solvent evaporation composed of hyaluronic acid (HA), chitosan glutamate (CH), and a combination of the two with microcapsules of HA and gelatin prepared by complex coacervation. Some other polymers—such as alginates [53, 54], a natural polymer of low toxicity, irritability, and immunogenicity; epichlorohydrine cross-linked starch (Spherex) [55–61] and dextran (Sephadex) [62–65]; poly(lactide-*co*-glycolide) (PLGA) [66]; and the biocompatible and biodegradable copolymer of lactic and glycolic acids, which have also been approved by the Food and Drug Administration (FDA) [67]—have mainly been used in microspheres for nasal dry powder delivery and are referred to in more detail in the next section.

Recently, thiolated polymers or thiomers, a new generation of permeation-enhancing agents, have been introduced in the pharmaceutical literature. Thiomers are characterized by covalent immobilization of sulfhydryl groups on their polymeric backbone, which are responsible for improved permeation-enhancing properties combined with mucoadhesive and enzyme-inhibitory properties [68]. A further improvement of the permeation-enhancing effect of thiomers has been achieved by the addition of the permeation mediator glutathione [69]. The improvement of human growth hormone (hGH) bioavailability (8.11%) by intranasal administration of the microparticulate formulation composed of thiomer polycarbophil-cysteine (PCP-Cys) and permeation mediator glutathione has been shown. Evaluation of the effect of PCP-Cys on the ciliary beat frequency (CBF) of human nasal epithelial cells in vitro has shown no ciliotoxic effect [70].

Chitosan is a hydrophilic, biocompatible, and biodegradable polymer of low toxicity, and it has been extensively investigated for pharmaceutical and medical purposes. It has been included in the European Pharmacopoeia since 2002. Chitosan is a polysaccharide composed of *N*-acetyl-D-glucosamine (approximately 20%) and glucosamine (approximately 80%). It is derived by deacetylation of chitin, which after cellulose is the most abundant polymer found in nature. It is a polycation at acidic pH values where most of the amino groups are protonated and has an apparent pK_a of 5.5 [71–74]. In the context of drug delivery, chitosan has been used for the preparation of microcapsules and microspheres with encapsulated small polar molecules, proteins, enzymes, DNA, and cells, as a nasal delivery system for insulin [75], as a system for oral vaccination, and as a stabilizing constituent of liposomes. Several studies have highlighted the potential use of chitosan as an absorption-enhancing agent due to its mucoadhesive properties and ability to open the tight junctions in the mucosal cell membrane [72].

Its biodegradability and low toxicity in humans have aided the recent increased interest in chitosan as an immunopotentiating agent. In vivo studies have demonstrated that chitosan powder and solution formulations are able to enhance the systemic and mucosal immune responses after nasal vaccine delivery [19, 22, 76].

The nasal absorption of insulin after administration in chitosan powder was the most effective formulation for nasal delivery of insulin in sheep compared to chitosan nanoparticles and chitosan solution [11]. Similarly, chitosan powder formulations have been shown to enable an efficient nasal absorption of goserelin in a sheep model where bioavailabilities of 20–40% were obtained depending on the nature of the formulation [9].

There has been a report on chitosan utility in improving the intranasal absorption of high-molecular-weight (>10-kDa) therapeutic protein. Chitosan glutamate powder blend or granules with recombinant hGH have been evaluated for intranasal administration in sheep. Relative to subcutaneous injection the nasal formulations produced bioavailabilities of 14 and 15%, respectively [77].

Various chitosan derivatives of enhanced solubility, mucoadhesive, and permeation properties were developed. *N*-Trimethyl chitosan chloride (TMC) is a quaternized derivative of chitosan with superior aqueous solubility over a broader pH range and penetration-enhancing properties under physiological conditions [78]. Carboxymethylated chitosan (CMChi) is a polyampholytic polymer able to form viscoelastic gels in aqueous environments. CMChi appears to be less potent compared with the quaternized derivative. Neither TMC nor CMChi have been found to provoke damage of the cell membrane, and therefore, they should not alter the viability of nasal epithelial cells [79].

Thiolated chitosans, chitosan thioglycolic acid conjugates, chitosan–cysteine conjugates, and chitosan-4-thio-butyl-amidine conjugates are new-generation polymers that are pH sensitive. The pH range at which the gelation and mucoadhesion of these polymers are optimal is within the physiological range (pH 5–6.5) of the nasal mucosa, but these polymers have been primarily investigated for oral drug delivery [80].

5.7.4 MICROSPHERES AS NASAL DRUG DELIVERY DEVICES

Developing an appropriate drug delivery system for a given drug can completely alter the drug's unfavorable properties, such as improve its effectiveness or reduce its side effects. Dry powder delivery systems such as microspheres are of special interest. In the last two decades they have been extensively studied with respect to nasal delivery and a considerable number of studies have been reported on that subject [3, 23].

In general, microspheres as specialized drug delivery systems represent spherical polymeric devices that are small in size (from 1 to 1000 μm), are characterized by high surface-to-volume ratio, and are able to provide targeted and predictable controlled release of the drug [3]. In the scope of nasal delivery, except for controlled drug release rate, microspheres are beneficial due to their broad surface area, which can provide extensive interaction with the mucin layer and protection of incorporated drug from enzymatic degradation in the nasal cavity [10, 41].

Microspheres prepared with bioadhesive polymers have some additional advantages; they assure much more intimate and prolonged contact with the mucous layer and improved drug absorption due to additional delay in mucociliary clearance. Bioadhesive microspheres can significantly improve patient compliance as all the advantages described lead to reduction in the frequency of drug administration [3, 74].

Bioadhesive microspheres that have been extensively studied for nasal drug delivery are water insoluble but they swell in contact with the mucosa. Swollen microspheres form a gellike system that adheres onto the mucus, retaining drug at the absorption site for prolonged periods [2, 81]. Swelling of the microspheres causes mucosal dehydration and reversible shrinkage of the cells, resulting in the temporary widening of the tight junctions and increased permeability of hydrophilic compounds, or more precisely, paracellular absorption of the drug [62]. Oechslein et al. [40] suggested that the opening of tight junctions could be related to the local decrease in Ca^{2+} concentration as well, since the absorption-promoting effect of investigated particulate drug delivery systems correlated directly with their capability to bind Ca^{2+}. In order to additionally improve nasal drug absorption, bioadhesive particulate systems have been combined with biological absorption enhancers [55–57].

A number of studies of the nasal mucociliary clearance rate of microspheres confirmed their potential to retain drug at the absorption site longer than liquid formulations [58, 67, 81, 82, 83]. Methods to measure formulation clearance rates from the nasal cavity can be divided into three groups, differing in the detected substance [58]. The most exact method is gamma scintigraphy, which monitors the deposition and clearance of radiolabeled drug delivery systems. The second method involves mixing of a fluorescent dye with the formulation and monitoring the cumulative tracer amount in the pharynx. The third method is the saccharin test, in which saccharin is mixed with the formulation and the clearance rate is determined by the first perception of sweet taste [84]. The first study of the mucociliary clearance of microspheres using gamma scintigraphy was reported by Illum et al. [81]. They evaluated clearance rates of starch, dextran, and albumin microspheres: Three hours after nasal administration about 50% of albumin and starch microspheres and 60% of dextran microspheres were still detected at the site of deposition. Nasal clearance study of melatonin starch microspheres and the melatonin solution applied revealed that more than 80% of the starch microspheres remained in the nasal mucosa 2 h after administration, compared to only 30% for the melatonin solution [58]. The study of the clearance rate of alginate, PLGA, and Sephadex microspheres revealed that alginate and PLGA were suitable for nasal delivery as they had the best mucoadhesive properties [67]. It has also been shown that the limiting step of the mucociliary clearance of nasally administrated microspheres was their clearance from the initial deposition site. The same conclusion has been drawn for the Carbopol 971P and carboxymethyl cellulose microspheres [82]. Soane et al. [83] evaluated the clearance rate of chitosan microspheres and chitosan solutions, compared to the control solution from the nasal cavity in sheep, by gamma scintigraphy. They found that both chitosan systems had higher retention times compared to the control. Also, chitosan microspheres were cleared at a slower rate than the chitosan solution, with half times of clearance of 115 and 43 min, respectively. The nasal clearance rates found in the sheep model were similar to the clearance rates found in their previous study carried out on human subjects [85], indicating that the sheep could be a suitable model for in vivo nasal clearance studies.

Starch Microspheres Bioadhesive starch microspheres in the context of the nasal delivery system were first introduced by Illum et al. [81] in a study that examined human nasal mucociliary clearance. Since then starch microspheres have been shown to promote the nasal absorption of a number of drugs. Animal studies using the sheep model showed greatly improved absorption of gentamicin [59] and human growth hormone [56] when administered in combination with starch microspheres in a freeze-dried formulation. Similar findings have been reported for insulin [55, 60] and desmopressin [61] loaded starch microspheres compared to simple drug solutions administered to animal models. Biodegradable starch microspheres have also been investigated for nasal delivery of metoclopramide whereas enhanced bioavailability was achieved compared to nasal spray [86]. The bioadhesive starch microspheres were shown to act synergistically with the absorption enhancers improving the transport of insulin across the nasal membrane [57]. Recently, the potential of starch microspheres for the nasal delivery of melatonin was investigated [58]. An in vitro release study revealed a sustained drug release profile. Melatonin bioavailability after nasal administration of starch microspheres was high, 84%. A good correlation between the in vitro release profile and in vivo absorption has been observed.

The use of degradable starch microspheres has proved to be well tolerated in both experimental animals and humans. No alterations of nasal mucosa were detected after eight weeks of nasal administration of starch microspheres to rabbits. Additionally, a preliminary test on healthy volunteers also showed good acceptability [62]. Another study of healthy volunteers revealed no changes in mucociliary clearance or in the geometry of the nasal cavities after eight days of nasal administration of dry starch microspheres [87].

Dextran Microspheres Similar to starch microspheres, Illum et al. [81] introduced dextran microspheres as a bioadhesive drug delivery system able to prolong the residence time in the nasal cavity. However, Rydén and Edman [65] reported dextran microspheres were not shown to improve nasal absorption of insulin in rats as insulin was too strongly bound to the Diethylaminoethyl (DEAE) groups to be released by a solution with an ionic strength corresponding to physiological conditions. In a later study it has been shown that the localization of insulin influenced the in vivo behavior of dextran microspheres [63]. The distribution of insulin at the surface or inside the dextran microspheres after the lyophilization loading process was determined by the cut-off limit of the microspheres. Microspheres with insulin left at the surface showed higher insulin absorption–enhancing effect than the microspheres with insulin inside the dextran matrix. Dextran microspheres have also been evaluated in vivo as a delivery system for octreotide [40] and in vitro for nicotine [33]. Ciliotoxicity studies performed in vitro on explants from rat trachea showed that dextran microspheres had no effect on the ciliary beat frequency [64]. The immediate recovery of the ciliary movement after dextran microspheres washing off indicated that the cilia were not damaged by dextran microspheres.

Gelatin Microspheres Several studies characterizing gelatin microspheres as a nasal drug delivery system have been reported. Gelatin microspheres were shown to swell readily in contact with nasal mucosa and to have good bioadhesive properties [65]. An in vitro release study using a Franz diffusion cell on levodopa-loaded gelatin microspheres showed prolonged drug release as compared to drug alone [88]. Negatively and positively charged gelatin microspheres intended for nasal and intramuscular delivery of salmon calcitonin were prepared by Morimoto et al. [10]. Both types of microspheres enhanced nasal absorption of salmon calcitonin compared to the solution. Positively charged gelatin microspheres seamed to exhibit greater enhancing effect on nasal absorption than negatively charged gelatin microspheres. Recently gelatin and gelatin–poly(acrylic acid) microspheres were studied with respect to oral and nasal delivery of oxprenolol [89]. Combining the gelatin with poly(acrylic acid) resulted in microspheres with improved bioadhesive properties. Also, nasal administration of gelatin–poly(acrylic acid) microspheres resulted in improved bioavailability of the drug compared to nasal administration of the drug solution.

Polyacrylate Microspheres Cross-linked polyacrylate microspheres as nasal powder delivery systems have been investigated in several studies [39, 41, 90]. Microspheres were produced by spray drying and emulsification methods and their nasal drug delivery potential has been evaluated only in vitro. Carbopol 934P microspheres were shown to have the best bioadhesive properties compared to other hydrophilic microspheres prepared with polyvinyl alchohol, chitosan, and hydroxypropylmethyl cellulose [41]. Improved permeation-enhancing effect of polycarbophil microparticles was obtained when microparticles were prepared with the thiolated polycarbophil and the permeation mediator glutathione [68].

Chitosan Microspheres As a specific chitosan-based delivery system, chitosan microspheres have been extensively studied and number of reports has verified their potential regarding nasal drug delivery [9, 21, 25, 41, 53, 73, 91, 92]. Chitosan microspheres have been prepared by the emulsification solvent evaporation method [51, 52, 93], emulsification cross-linking process [91], spray drying method [25, 53], and ionic gelation process [94]. They have been shown to significantly reduce mucociliary clearance from the nasal cavity of sheep and humans compared to solutions [83, 85]. The bioadhesive properties of chitosan microspheres were shown to be inversely proportional to particle size: Among chitosan microspheres in the size class between 50 and 200 μm, smaller microspheres appeared to swell faster than large microspheres, providing a more powerful mucoadhesive system [41].

A modulated release rate of drug from the swellable chitosan microspheres has been achieved with cross-linking agents such as glutaraldehyde [95], citric acid [92], ascorbic acid, or ascorbyl palmitate [91] that reacted with chitosan forming covalent bonds with chitosan amino groups. However, to maintain the bioadhesive properties of cross-linked chitosan microspheres, the amount of cross-linking agent should be optimized [95].

Chitosan molecular weight has also been reported to influence drug release. Jiang et al. [94] studied *Bordetella bronchiseptica* dermonecrotoxin (BBD) release from chitosan microspheres prepared by tripolyphosphate ionic gelation. It has been shown that the BBD release rate increased with chitosan molecular weight decrease. It has been explained by the weaker BBD interaction with chitosan of lower molecular weight and lower content of free amine groups, responsible for their interaction.

Chitosan microspheres were shown to enhance nasal bioavailability of several peptide drugs such as insulin and goserelin. A simple chitosan–insulin powder formulation provided about 20% of absolute insulin bioavailability in sheep [96]. Improved bioavailability (of 44%, in rats) was obtained when insulin was loaded into chitosan microspheres prepared with ascorbyl palmitate as cross-linking agent [91]. Chitosan microspheres have also been shown to improve nasal goserelin absorption providing about 40% bioavailability relative to goserelin intravenous application [9].

Krauland et al. [93] prepared the microparticles with thiolated chitosan (chitosan-TBA; chitosan–4-thiobutylamidine conjugate) intended for nasal peptide delivery. During the preparation process microparticles were stabilized by the formation of inter- and intramolecular cross-linking via disulfide bonds. Chitosan–TBA microparticles were characterized by improved swelling ability and displayed 3.5-fold higher insulin bioavailability compared to unmodified chitosan microparticles.

Besides the polymer derivatization, combining the polymers in microsphere preparations can result in improved drug delivery and absorption characteristics. Hyaluronic acid–chitosan microspheres appeared to improve the absorption of incorporated gentamicin compared to the individual polymers, assembling the mucoadhesive potential of both polymers and the penetration-enhancing effect of chitosan [51, 52].

Chitosan microparticulate systems have also been investigated for vaccine nasal delivery and have proven to induce strong systemic and mucosal immune responses [18, 21, 76].

5.7.4.1 Preparation Methods

The design of bioadhesive microspheres includes selection of the most suitable preparation method, considering the nature of the drugs and polymers used as well as the route of administration. A number of methods for the preparation of microspheres have been described in the literature [3, 97]. In the scope of nasal delivery, the first microspheres in use were starch and dextran microspheres, prepared by an emulsion polymerization technique employing epichlorohydrine as a cross-linking agent [55, 56, 59, 60, 62, 81]. Currently techniques based on solvent removal, such as solvent evaporation [41, 51, 66, 93, 98] and solvent extraction [88, 99], are most frequently in use. There are tree processes involved in such microensapsulation procedures: the preparation of emulsion, solvent removal, and separation of the particles obtained. Selection of the type of (oil-and-water) emulsion system (O/W, W/O, W/O/W, W/O/O, etc.) depends on the physicochemical properties of the drug and polymer used. After the preparation of stable emulsion, solvent is removed from the system at high or low temperature, at low pressure, or by addition of another solvent that enables the extraction of polymer solvent to the continuous phase. Hardened microspheres are then washed, centrifuged, and lyophilized.

Emulsion techniques are suitable for the preparation of microspheres intended for nasal delivery since they allow controlling the size of the particles. Freiberg and Zhu [97] reviewed solvent evaporation process parameters (e.g., polymer concentration, viscosity, stirring rate, temperature and percentage of emulsifying agent) affecting microsphere size. It can be assumed that the particle size is directly proportional to polymer concentration and inversely proportional to stirring rate and percentage of emulsifying agent [41, 88], while there is a nonlinear correlation between particle size and process temperature. Yang et al. [100] reported that larger microspheres were produced at lower temperatures due to the higher viscosity of solution and at higher temperatures due to the higher solvent flow pressure moving more material from the microsphere center outward. In the same work encapsulation efficiency was also correlated with the temperature of solvent evaporation in the process of microsphere preparation. It was found that the highest encapsulation efficiencies occurred at the lowest and highest temperatures tested.

Recently, the spray drying method has been extensively used for the preparation of microspheres intended for nasal delivery [3, 25, 39, 44, 45, 53, 54, 101, 102]. Spray drying is a single-step procedure transforming liquid into dry particulate form (e.g., microparticles, microspheres, microcapsules) applicable to drugs and polymers of various solubility characteristics. It is a fast, simple, and reproducible technique that is easy to scale up [3]. Spray drying can be described as follows: The liquid is fed to the nozzle with a peristaltic pump, atomized by the force of the compressed air, and blown together with hot air into the chamber, where the solvent in the droplets is evaporated. The dry product is then collected in a collection bottle. Spray-dried microspheres are reported to have relatively low production yields, rarely higher than 50%. The loss of material during spay drying has been explained by the powder adhering to the cyclone walls, small amounts of materials processed in each batch, and loss of the smallest and lightest particles through the exhaust of the spray dryer apparatus, which lacks a trap to recover the lighter and smaller particles [25, 53, 103, 104].

Spray drying offers the possibility to control the particle size of the product. Microparticles of desirable size can be obtained by optimizing the spray drying process parameters, such as size of the nozzle, feeding pump rate, inlet temperature, and compressed airflow rate. In accordance with this, He et al. [105] reported that larger particles were formed at a larger size of nozzle and faster feeding pump rate, while smaller particles were formed at a greater volume of air input [105]. The size and other properties of spray-dried microspheres (e.g., morphology, density, shape, porosity, and flowability) can also be affected by the qualitative and quantitative composition of the liquid feed [28]. Thus, feed concentration has been reported to influence the particle size distribution, as spray drying of more concentrated liquid feeds resulted in the formation of larger particles [104, 106, 107].

5.7.4.2 Microsphere Characterization

Microspheres intended for nasal administration need to be well characterized in terms of particle size distribution, since intranasal deposition of powder delivery systems is mostly determined by their aerodynamic properties and particle sizes. Commonly used methods for particle size determinations described in the literature are sieving methods [108], light microscopy [58], photon correlation spectroscopy [66], and laser diffractometry [25, 41, 53, 93]. The morphology of the microparticles (shape and surface) has been evaluated by optical, scanning, and transmission electron microscopy [66, 95].

Determination of the zeta potential is an important part of microsphere characterization, as the zeta potential has a substantial influence on the adhesion of drug delivery systems onto biological surfaces [109]. For example, Jaganathan and Vyas [110] reported the reduction in the nasal clearance rate of PLGA microspheres modified with chitosan compared to unmodified PLGA microspheres due to the change in zeta potential from negative for PLGA microspheres to positive for surface-modified

PLGA microspheres. Methods to measure the zeta potential of microspheres are laser doppler anemometry [105] and photon correlation spectroscopy [110].

The physical state of the drug incorporated in a powder drug delivery system (e.g., degree of crystallinity and possible interactions with the polymer) is assessed by differential scanning calorimetry (DSC) or Fourier transform infrared (FTIR) spectroscopy. These observations can clarify the results of other parameter investigations, especially the results of in vitro drug release studies.

To predict microsphere performance in vivo, the swelling properties of nasal powder delivery systems need to be evaluated. Methods described in the literature are mostly based on the weight difference measurements between the dry and swollen powder [40]. Swelling properties of nasal powders such as water-absorbing capacity can be evaluated using a Franz diffusion cell [43, 107]. The swelling capacity may also be expressed as the volume expansion of the microspheres that is determined at equilibrium after placing the microspheres in water using a graduated cylinder [108]. Gavini et al. [53] determined the swelling properties of microspheres in vitro by laser diffractometry. That method allows us to evaluate the variation of particle size versus time.

In vitro evaluation of mucoadhesive properties is essential in the development of a nasal drug powder delivery system, since mucoadhesion is of great importance for the in vivo performance of formulation. A large number of in vitro and in vivo methods used to assess mucoadhesive properties of microspheres have been extensively described in the literature [3, 111, 112]. Many in vitro methods are based on the interaction of microspheres with mucin. Evaluation of that interaction can be performed using scanning and transmission electron microscopy [95] or photon correlation spectroscopy [45]. Scanning electron microscopy (SEM) provides the information on morphological changes on the microsphere surface in contact with mucin, while transmission electron microscopy confirms SEM results and reveals the ultrastructural features of the surface interactions between microspheres and mucin chains [95]. He et al. [102] evaluated the mucoadhesive properties of chitosan microspheres by measuring the amount of mucin adsorbed on the microspheres. Gavini et al. [53] evaluated the mucoadhesive properties of metoclopramide-loaded microspheres by determining the amount of microspheres that stuck to a filter paper saturated with mucin after exposure to the air stream. Vidgren et al. [39] used a tensiometer to measure the force required for the separation of two filter paper discs saturated with mucin and with the examined microspheres placed between them.

In the work reported by Witschi and Mrsny [54] mucoadhesion of dry powder microparticles was investigated using Callu-3 cells as a surrogate for human nasal epithelia: Microparticles were applied to the apical surface of cell sheets and at certain time points were washed with phosphate-buffered saline (PBS) to remove poorly adhering microparticles. Rango Rao and Buri [113] developed an in situ method to evaluate the bioadhesive properties of polymers and microparticles, based on washing off a mucous membrane covered with the formulation to be tested by simulated biological flow. The mucoadhesion of gelatin microspheres [10] was measured by an in situ nasal perfusion experiment. In the work reported by Lim et al. [51] the mucoadhesive properties of microspheres were evaluated by determining the mucociliary transport rate of the microparticles across an isolated frog palate. The boiadhesive properties of microspheres can be evaluated by the everted sac technique using a section of everted intestinal tissue or the CAHN dynamic contact angle analyzer [3].

Santos et al. [112] correlated these two methods and concluded that each method could be used alone as the relevant indicator of microsphere bioadhesion.

In vitro drug release experiments can be performed in order to characterize the release behavior of microparticles in general. For that purpose microparticles can be dispersed directly in the dissolution medium [51, 91] or a dynamic dialysis technique can be employed [58]. However, to obtain results comparable with the in vivo situation of nasal administration, it is necessary to provide experimental conditions similar to those encountered in the nasal cavity as nasally administrated powders are not being dispersed directly in the large quantity of liquid [8]. Such in vitro drug release experiments can be performed by a modified U.S. Pharmacopeia (USP) XXII rotating basket [8]. The drug-loaded powder formulation is weighed on a membrane filter placed between the filter holder and the cup and then immersed in the released medium. Thus, the membrane filter separates the donor and acceptor compartment but at the same time allows the powders to hydrate and to form a gel. Drug is released to the release medium after diffusion through the swollen gel of known surface area. Cornaz et al. [33] developed a special diffusion chamber that simulated the hydration conditions of the nasal mucosa to study the in vitro release of nicotine from dextran microspheres. A number of authors have used Franz diffusion cells for in vitro release experiments since that model provides conditions similar to those encountered in the nasal cavity and slow hydration of the microspheres [41, 43–45, 107].

5.7.4.3 Chitosan-Formulated Spray-Dried Microspheres

Chitosan, a biocompatible and biodegradable polycationic polymer with low toxicity, is known for its swelling ability and permeation-enhancing properties and represents a polymer of choice for the preparation of microspheres intended for nasal administration [74].

Spray drying has proved to be a suitable and simple technique for the preparation of chitosan microspheres with preserved chitosan properties, offering numerous advantages over other microencapsulation methods. It has been successfully used to entrap both hydrophilic and lipophilic drugs into the chitosan matrix, since a variety of colloidal systems (e.g., polymer solutions, emulsions, dispersions, suspensions) can be subjected to spray drying. Chitosan-based spray-dried microspheres have been prepared with chitosan alone (resulting in conventional microspheres) or in combination with another polymer (resulting in composed microspheres). Combining the polymers has been reported to result in microspheres with improved properties regarding surface characteristics, entrapment efficiency, or control over the drug release rate. Recently, several drugs, such as carbamazepine [25], propranolol hydrochloride [45], metoclopramide hydrochloride [53], and loratadine [103, 107], have been incorporated into chitosan-based nasal powder formulations produced by the spray drying-method. They were characterized in terms of encapsulation efficiency, morphology, size distribution, zeta potential, physical state of the drug, in vitro drug release behavior, and swelling and bioadhesive properties. Chitosan–ethyl cellulose composed microspheres improved loratadine entrapment compared to conventional chitosan microspheres, which influenced directly the microsphere surface characteristics: Loratadine was less present at the surface of the microspheres and consequently had less influence on their bioadhesive properties. Thus, although showing moderate swelling ability, loratadine-loaded composed microspheres were more bioadhesive than conventional chitosan microspheres. Composed microspheres showed good loratadine-sustained release potential in vitro, depending on the polymeric weight ratio and concentration of the spray-dried system [103, 107]. Gavini et al. [53] produced metoclopramide-loaded alginate and/or chitosan microspheres by the spray drying method. The results obtained revealed that complexation of chitosan with alginate in the microsphere preparation provided improved control of the drug release in vitro compared to chitosan alone. Despite the chitosan complexation, composed microspheres showed good bioadhesive properties. Ex vivo drug permeation tests carried out using sheep nasal mucosa showed higher drug permeation from chitosan-based microspheres than from alginate microspheres, confirming the well-known chitosan permeation-enhancing properties. Cerchiara et al. [45] developed spray-dried chitosan–poly(methyl vinyl ether-*co*-maleic anhydride) microparticles for nasal delivery of propranolol hydrochloride. Chitosan was combined with polyanhydride, able to enhance the formation of hydrogen bonds between the polymers and mucosal components through carboxylic acid groups generated after polyanhydride hydrolytical degradation. The swelling and bioadhesive properties of chitosan–polyanhydride microparticles increased in a pH-dependent manner. Both chitosan and chitosan–polyanhydride microparticles provided sustained propranolol hydrochloride release.

Microparticulate spray-dried delivery systems have shown great potential for nasal delivery of drugs characterized by poor water solubility. According to DSC analysis, the spray drying method together with the carriers seemed to promote the amorphization of loratadine [107] and carbamazepine [25]. Carbamazepine incorporated into chitosan microspheres was characterized by increased dissolution rate compared to carbamazepine raw material. It was explained not only by the promoted drug amorphization but also by the chitosan well-known dissolution rate enhancer properties and by the small size of microspheres (or high surface-to-volume ratio). Results of in vivo nasal administration of carbamazepine loaded chitosan microspheres revealed a remarkable increase in carbamazepine concentration in serum compared to the pure carbamazepine powder [25]. Such an increase in drug absorption has been ascribed to both improved dissolution of carbamazepine and adhesion of the chitosan microspheres to the mucosal surface.

The mucoadhesive function of chitosan has also been employed in vaccine dry powder delivery. Alpar et al. [18] produced bovine serum albumin (BSA)–loaded chitosan microspheres using the spray drying method. It has been shown that the stability of encapsulated BSA was preserved in the microspheres prepared, indicating that spray drying was appropriate even for the preparation of antigen-loaded microspheres. BSA-loaded chitosan microspheres generated higher immune response than the free BSA, thus proving to be a suitable system for nasal antigen delivery.

5.7.5 TOXICOLOGICAL CONSIDERATIONS

The possible toxicological effects of dry powder formulation on the nasal mucosa, including local irritation, effect on mucociliary clearance, and epithelial damage and recovery rate [114], should be investigated in an early stage of its development. There were some attempts to define the categories of toxic effects as well of the constituents of nasal formulations according to their toxic potential. Thus, Hvidberg et al. [115] introduced a scale of irritation in the nose (0, no irritation; 1, slight irritation; 2, acceptable; 3, unwilling to accept the treatment again) that was later used to evaluate the degree of nasal powder irritation after administration in human volunteers [27]. Ugwoke et al. [82] classified the degree of ciliary beat frequency change caused by nasally applied liquid formulation as follows: no effect (less than 10%), mild (10–20%), moderate (20–50%), and severe (more than 50%). Soon after, that classification was applied to a powder ciliotoxicity study [116]. Reversibility of ciliotoxic effect after washing out the tested compound was classified as well, resulting in three categories: reversible, partially reversible, and irreversible effects [82]. Merkus et al. [117] proposed the three categories of constituents of nasal formulations based on the recovery of ciliary beat frequency after the tested compound was washed out. The first category is cilio friendly, with ciliary beat frequency recovery of 75% or more; the second is cilio inhibiting, with recovery between 25 and 75%; and the third category is ciliostatic, with recovery of 25% or less.

Methods to evaluate the possible toxicological effect of formulations on the nasal mucosa described in the literature mainly refer to histopathological evaluation as a standard method for cytotoxicity evaluation [64] and a study on the release of marker enzymes [118]. In the work reported by Callens et al. [118] the possible toxicological effects of multiple starch and carbopol powder nasal administration were evaluated using rabbits by measuring the proteins and lactate dehyrogenase (LDH) release from the nasal mucosa. Contrary to the invasive in situ perfusion method, performable with anaesthetized animals [119], this method has been shown to be noninvasive, applicable to nonanaesthetized and nonsedated animals, and suitable for repeated measurement of the marker protein release on the same animal during a long-term administration study. A histopathological study has also been performed. In agreement with attempting to replace the use of vertebrates in scientific experiments with lower organisms such as invertebrates, plants, and microorganisms, Adriaens and Remon [120] developed a new mucosal toxicity screening method using the slug *Arion lusitanicus* as the model organism. The body wall of the slug resembles the nasal mucosa since it consists of a single-layered epithelium containing both ciliated and mucus-secreting cells. Callens et al. [118] successfully used that method for screening the mucosal irritation potential of bioadhesive starch and carbopol powder formulations. The possible toxicological effects of the powder formulations were evaluated by measuring the proteins LDH and alkaline phosphatase (ALP) release from the body wall of the slugs as well as the amount of mucus produced.

Among various nasal toxicity studies, ciliotoxicity studies are of special interest due to the importance of maintaining optimal ciliary beating to protect the lower respiratory system from infections. If the drug formulation inhibits the ciliary beating, such inhibition needs to be completely reversible upon formulation removal [116].

Methods to determine ciliary beat frequency and mucociliary transport in vitro and in vivo have been extensively reviewed elsewhere [84, 121]. In general, in vivo methods are more reliable for ciliotoxicity studies than in vitro methods and are essential to confirm the safety of nasal drug formulations. However, in vitro methods are more suitable for the large number of screening studies required during formulation development [122].

In vitro methods to measure ciliary beat frequency can be performed on explants of ciliated mucosa [64] or on different types of ciliated cells, such as ciliated chicken embryo trachea cells [123], nasal cell lines derived from carcinomas of epithelial origin (RPMI 2650, BT, NAS 2BL), or human lung adenocarcinoma cell line Calu-3

[124]. In vitro ciliotoxicity of dextran microspheres was evaluated on explants from rat trachea [64].

Human nasal epithelial cell cultures can serve as a relevant screening tool for prediction of nasal formulation toxicity in humans as long as the cells maintain differentiated morphological and biochemical characteristics of the original tissue [124]. Epithelial cells intended for initiation of primary cell culture should be taken from the regions of the nasal cavity where formulations are supposed to be deposited [125]. Considering toxicological investigations, human nasal epithelial cells cultured in an air–liquid interface system are the most promising ones at the moment, as in these culturing conditions cell differentiation closely resembles cell differentiation in vivo [124]. Monolayer immersion feeding and air–liquid interface cultures have already been used for ciliary beat frequency measurements. However, unstable ciliary activity and their short life span in the culture have proved to be the main shortcomings of these systems. Jorissen et al. [126] developed a suspension culture system of human nasal epithelium with actively beating cilia for several months. That system has been validated for ciliotoxicity investigations [122] and later applied in other studies [116, 127]. At this time it is the most successful model of human nasal epithelium culture useful for ciliary beat frequency determination [124].

In vivo methods to determine the toxicological effects of drug formulations on the nasal mucosa mainly refer to mucociliary clearance rate studies. Methods described in the literature are mostly based on the gamma scintigraphic technique, in which cleared formulations are labeled with the radiotracer [82]. In vivo scintigraphic evaluation of the nasal clearance of drug delivery systems requires a nondiffusible and stable radiotracer to prevent its absorption and decomposition. Another method described in the literature is the saccharin test [87], in which saccharin is mixed with the formulation and the clearance rate is determined by the first perception of sweet taste [84]. Gamma scintigraphy [31] is more relevant for the ciliary function monitoring than the saccharin test, since it investigates the whole mucosal surface, while the saccharin test investigates only the fastest flow rate [82]

A number of ciliotoxicity studies have pointed out a low correlation between the results obtained using different in vitro and in vivo methods [121]. The effects of nasal formulations on the ciliary beat frequency in vitro are usually more expressed than in vivo, since in vivo, cilia are partially protected by the mucous layer and investigated formulation is eventually cleared from the nasal cavity due to the mucociliary clearance mechanism. Also, toxic effects of the formulations on the cilia in vivo may be reversible due to the constant nasal mucosal cell turnover [121].

In conclusion, in order to make predictions regarding the safety of the nasal formulation on mucociliary clearance, both in vitro and in vivo studies have to be performed. It is also essential to determine long-term use effects in animals and in humans if the nasal formulation is intended for subchronic or chronic administration.

Dry powder formulations for nasal delivery of peptides and proteins have been investigated for the first time by Nagai and others [38]. Since then, much research work has been done on dry powders containing bioadhesive polymers for nasal drug administration. The bioavailability and duration of action of drugs administered by the nasal route are increased by the use of the principle of mucoadhesion and dry powder formulations. Research work on dry powder formulation containing bioadhesive polymers is summarized in Table 1.

TABLE 1 Summary of Research Work on Nasal Dry Powder Formulations

Powder Formulation	Preparation Method	Drug	Polymer	Studies	Comments	Reference
Microparticles	Spray drying	BSA	Starch, alginate, chitosan, carbopol	In vitro	Chitosan microparticles provided most desirable characteristics for protein delivery	54
Microparticles		Budesonide	Polymethacrylic acid–polyethylene glycol [P(MAA–PEG)]	In rabbits	Continuous drug release for at least 8h High bioavailability	133
Microparticles	Solvent evaporation	Gentamicin	Hyaluronic acid–chitosan	In rabbits	Polymers improved gentamicin absorption synergistically	52
Microparticles	Lyophilization	hGH	Polycarbophil–cysteine	In rats	Improved bioavailability	68
Microparticles	W/O emulsification solvent evaporation	Insulin	Chitosan–TBA	In rats	Improved bioavailability	93
Microparticles	Spray drying	Propranolol hydrochloride	CH-PVM/MA	In vitro	Sustained drug release	45
Microparticles, nanoparticles	W/O/W solvent evaporation	Model protein, tetanus toxoid	PLA-PEG	In rats	Size-dependent mucosal uptake	66
Microspheres	Ionic gelation with tripolyphosphate	Bordetella, bronchiseptica, dermonecrotoxin	Chitosan	In mice	Systemic and mucosal immune responses induced	21
Microspheres	Ionic gelation process with tripolyphosphate	Bordetella, bronchiseptica, dermonecrotoxin	Chitosan	In vitro	Chitosan molecular weight–related drug release profile	94
Microspheres	Spray drying	Carbamazepine	Chitosan hydrochloride, chitosan glutamate	In sheep	Increased drug dissolution rate and absorption	25

TABLE 1 ***Continued***

Powder Formulation	Preparation Method	Drug	Polymer	Studies	Comments	Reference
Microspheres	W/O/O emulsion solvent evaporation	α-Cobrotoxin	PLGA/P(CPP: CEFB)	In rats	Increased strength and duration of antinociceptive effect	98
Microspheres	Emulsion polymerization	Desmopressin	Starch	In rats, in sheep	Increased bioavailability with LPC	61
Microspheres	W/O emulsification solvent evaporation	FITC–dextran	Carbopol 394P Chitosan HPMC PVA	In vitro	Initial release at a constant rate Chitosan microspheres exhibited size-dependent release effect	41
Microspheres		FITC–dextran	Chitosan, Carbopol 934P	In rabbits	Improved bioavailability	131
Microspheres	Cross-linking with epichlorohydrin	Gabexate mesylate	Starch cyclodextrin	In vitro	Fast release rate	108
Microspheres	Solvent evaporation	Gentamicin	Chitosan, hyaluronan, gelatine	In vitro	Prolonged release, improved mucoadhesive properties	51
Microspheres	Emulsion polymerization	Gentamicin	Starch	In sheep, in rats	Improved bioavailability with microsphere/enhancer (LPC) system	59
Microspheres	Spray drying	Gentamicin sulfate	HPMC	In vitro	Modified drug release	44
Microspheres	Solvent evaporation	Heparin	Poly(lactic acid)	In rats	Sustained-release effect	134
Microspheres	Emulsion polymerization	hGH	Starch	In sheep	Enhanced nasal absorption with microspheres, rapid and higher absorption with microsphere/enhancer (LPC) system	56
Microspheres	Emulsion polymerization	Insulin	Starch, dextran	In rats	Rapid absorption	62
Microspheres	Emulsion polymerization	Insulin	Starch	In sheep	Improved bioavailability with microsphere/enhancer (LPC) system	55

Microspheres	Emulsion polymerization	Insulin	Starch	In rats	Improved nasal absorption	60
Microspheres	Emulsion polymerization	Insulin	Dextran	In rats	Promoted absorption	63, 65
Microspheres	Emulsification–cross-linking	Insulin	Chitosan	In rats	Promising absolute bioavailability of 44%	91
Microspheres	Emulsification–solvent evaporation	Insulin	Hyaluronic acid ester	In sheep	Increased nasal absorption	128
Microspheres	W/O emulsification solvent extraction	Levodopa	Gelatin	In vitro	Initial fast release rate, followed by a second slower release rate	88
Microspheres	Spray drying	Loratadine	Chitosan, chitosan/ethylcellulose	In vitro	Moderate swelling behavior, sustained drug release	107
Microspheres	Emulsification–cross-linking	Melatonin	Starch	In rabbits	Increased residence time, rapid absorption rate, high absolute bioavailability	58
Microspheres	Spray drying	Metoclopramide	Sodium alginate, chitosan, sodium alginate/chitosan	In vitro	Controlled drug release, promising properties as nasal drug carriers	53
Microspheres	Emulsification–cross-linking	Metoprolol tartrate	Alginate	In rabbits	Improved therapeutic efficacy	132
Microspheres	W/O emulsification	Morphine	Chitosan, starch	In sheep	High bioavailability	73
Microspheres	Emulsion polymerization	Nicotine	Dextran	In vitro	Rapid release, good dispersion ability	33
Microspheres	W/O emulsification solvent extraction	Oxprenolol	Gelatin–poly(acrylic acid)	In rats	Slow-release drug delivery system with good adhesive characteristics	89
Microspheres	W/O emulsion cross-linking	Pentazocine	Chitosan	In rabbits	Matrix diffusion controlled delivery, improved bioavailability, in vitro/in vivo correlation	129
Microspheres	Emulsion solvent evaporation	Salbutamol	Chitosan	In rabbits	Prolonged and controlled release	92

TABLE 1 ***Continued***

Powder Formulation	Preparation Method	Drug	Polymer	Studies	Comments	Reference
Microspheres	W/O emulsion cross-linking	Salmon calcitonin	Gelatin	In rats	Enhanced nasal absorption	10
Microsphers	Emulsion polymerization	Insulin	Starch	In sheep	Microspheres and absorption enhancers (LPC, GDC, and STDHF) acted synergistically to enhance absorption	57
Powder	Lyophilization	Apomorphine	Carbopol 971P, polycarbophil	In rabbits	Sustained release, improved bioavailability	8
Powder	Lyophilization	Apomorphine	Carboxymethyl cellulose	In rabbits	Sustained plasma level	42
Powder	Lyophilization	Apomorphine	Carbopol 971P, carboxymethyl cellulose	In rabbits	Increased residence time	48
Powder	Lyophilization	Apomorphine HCl	Carbopol 971P, Carbopol 974P, polycarbophil	In rabbits	In vitro release but not in vivo absorption has been influenced by drug loading	130
Powder	Spray drying	Cyanocobalamin	Microcrystalline cellulose, dextran, crospovidone	In rabbits	Improved bioavailability	43
Powder	Press-on force method	Glucagon	Microcrystalline cellulose	In human volunteers	Increased formulation stability, decreased irritability	27
Powder	—	Goserelin	Chitosan	In sheep	Improved bioavailability	9
Powder	Material mixing with a pestle	Insulin	Chitosan	In rats, in sheep	Improved bioavailability	11
Powder	Blending/ lyophilization	Insulin	Microcrystalline cellulose, hydroxypropyl cellulose, Carbopo 934	In dogs	Decreased plasma glucose level	38

Powder	Lyophilization	Insulin	Starch–Carbopol 974P, maltodextrin–Carbopol 974P	In rabbits	The highest absolute bioavailability obtained was 14.4%	46
Powder	Lyophilization	Ketorolac	Microcrystalline cellulose	In rabbits	Significantly lower bioavailability of drug from powders compared to spray formulation	50
Powder	—	Leuprolide, calcitonin, FITC–dextran	Hydroxypropyl cellulose, microcrystalline cellulose	In rabbits	Enhanced absorption	49
Powder	Manual blending using mortar and pestle	Morphine	Chitosan	In human volunteers	Rapid onset of pain relief, formulations well tolerated by patients	73
Powder	Dry blending	Octreotide	Dextran, microcrystalline cellulose, semicrystalline cellulose, hydoxyethyl starch, microcrystalline chitosan, pectin, alginic acid	In rats	Correlation between carrier calcium binding properties and their potential as nasal absorption enhancers for peptides	40
Powder, granules	Powder: manual mixing using mortar and pestle	Recombinant hGH	Chitosan	In sheep	Relative bioavailability of hGH from powder and granules have been 14 and 15%, respectively	77

Abbreviations: PLGA: poly(lactide-co-glycolide); P(CPP:CEFB): poly[1,3-bis(p-carboxy-phenoxy) propane-co-p-(carboxyethylformamido) benzoic anhydride]; chitosan–TBA: chitosan–4-thiobutylamidine conjugate; LPC: lysophosphatidyl choline; hGH: human growth hormone; GDC: glycodeoxycholate sodium; STDHF: sodium taurodihydroxyfusidate

REFERENCES

1. Quraishi, M. S., Jones, N. S., and Tomason, J. D. T. (1997), The nasal delivery of drugs, *Clin. Otolaryngol.*, 22, 289–301.
2. Pereswetoff-Morath, L. (1998), Microspheres as nasal drug delivery systems, *Adv. Drug Deliv. Rev.*, 29, 185–194.
3. Vasir, J. K., Tambwekar, K., and Garg, S. (2003), Bioadhesive microspheres as a controlled drug delivery system, *Int. J. Pharm.*, 255, 13–32.
4. Tirucherai, G. S., Pezron, I., and Mitra, A. K. (2002), Novel approaches to nasal delivery of peptides and proteins, *S. T. P. Pharm. Sci.*, 12, 3–12.
5. Davis, S. S. (1999), Delivery of peptide and non-peptide drugs through the respiratory tract, *Pharm. Sci. Technol. Today*, 2, 450–456.
6. Resta, O., Barbaro, M., and Carnimeo, N. (1992), A comparison of sodium cromoglycate nasal solution and powder in the treatment of allergic rhinitis, *Br. J. Clin. Practice*, 46, 94–98.
7. Schipper, N., Romeijn, S., Verhoef, J., and Merkus, F. (1993), Nasal insulin delivery with dimethyl beta cyclodextrin as an absorption enhancer in rabbits: Powder more effective than liquid formulations, *Pharm. Res.*, 10, 682–686.
8. Ugwoke, M. I., Exaud, S., Van Den Mooter, G., Verbeke, N., and Kinget, R. (1999), Bioavailability of apomorphine following intranasal administration of mucoadhesive drug delivery systems in rabbits, *Eur. J. Pharm. Sci.*, 9, 213–219.
9. Illum, L., Watts, P., Fisher, A. N., Jabbal-Gill, I., and Davis, S. S. (2000), Novel chitosan-based delivery systems for the nasal administration of a LHRH-analogue, *S. T. P. Pharm. Sci.*, 10, 89–94.
10. Morimoto, K., Katsumata, H., Yabut, T., Iwanaga, K., Kakemi, M., Tabata, Y., and Ikada, Y. (2001), Evaluation of gelatin microspheres for nasal and intramuscular administrations of salmon calcitonin, *Eur. J. Pharm. Sci.*, 13, 179–185.
11. Dyer, A. M., Hinchcliffe, M., Watts, P., Castile, J., Jabbal-Gill, I., Nankervis, R., Smith, A., and Illum, L. (2002), Nasal delivery of insulin using novel chitosan based formulations: A comparative study in two animal models between simple chitosan formulations and chitosan nanoparticles, *Pharm. Res.*, 19, 998–1008.
12. Colombo, P. (1999), Mucosal drug delivery, nasal, in Mathiowitz, E., Ed., *Encyclopedia of Controlled Drug Delivery*, J Wiley, New York, pp. 593–605.
13. Ishikawa, F., Murano, M., Hiraishi, M., Yamaguchi, T., Tamai, I., and Tsuji, A. (2002), Insoluble powder formulation as an effective nasal drug delivery system, *Pharm. Res.*, 19, 1097–1104.
14. Ishikawa, F., Katsura, M., Tamai, I., and Tsuji, A. (2001), Improved nasal bioavailability of elcatonin by insoluble powder formulation, *Int. J. Pharm.*, 224, 105–114.
15. Alpar, H. O., Eyles, J. E., Williamson, E. D., and Somavarapu, S. (2001), Intranasal vaccination against plaque, tetanus and diphteria, *Adv. Drug Deliv. Rev.*, 51, 173–201.
16. Vajdy, M., and O'Hagan, D. T. (2001), Microparticles for intranasal immunisation, *Adv. Drug Deliv. Rev.*, 51, 127–141.
17. Davis, S. S. (2005), The use of soluble polymers and polymer microparticles to provide improved vaccine responses after parenteral and mucosal delivery, *Vaccine*, 24, Suppl. 2, S7–S10.
18. Alpar, H. O., Somavarapu, S., Atuah, K. N., and Bramwell, V. W. (2005), Biodegradable mucoadhesive particulates for nasal and pulmonary antigen and DNA delivery, *Adv. Drug Deliv. Rev.*, 57, 411–430.

19. Van der Lubben, I. M., Verhoef, J. C., Borchard, G., and Junginger, H. E. (2001), Chitosan and its derivatives in mucosal drug and vaccine delivery, *Eur. J. Pharm. Sci.*, 14, 201–207.
20. Van der Lubben, I. M., Kersten, G., Fretz, M. M., Beuvery, C., Verhoef, J. C., and Junginger, H. E. (2003), Chitosan microparticles for mucosal vaccination against diphteria: Oral and nasal efficacy studies in mice, *Vaccine*, 21, 1400–1408.
21. Kang, M. L., Kang, S. G., Jiang, H.-L., Shin, S. W., Lee, D. Y., Ahn, J.-M., Rayamahji, N., Park, I.-K., Shin, S. J., Cho, C.-S., and Yoo, H. S. (2006), In vivo induction of mucosal immune responses by intranasal administration of chitosan microspheres containing *Bordetella bronchiseptica* DNT. *Eur. J. Pharm. Biopharm.*, 63, 215–220.
22. Huang, J., Garmise, R. J., Crowder, T. M., Mar, K., Hwang, C. R., Hickey, A. J., Mikszta, J. A., and Sullivan, V. J. (2004), A novel dry powder influenza vaccine and intranasal delivery technology: Induction of systemic and mucosal immune responses in rats, *Vaccine*, 23, 794–801.
23. Behl, C. R., Pimplaskar, H. K., Sileno, A. P., de Meireles, J., and Romeo, V. D. (1998), Effects of physicochemical properties and other factors on systemic nasal drug delivery, *Adv. Drug Deliv. Rev.*, 29, 89–116.
24. Genta, I., Pavanetto, F., Conti, B., Giunchedi, P., and Conte, U. (1995), Improvement of dexamethasone dissolution rate from spray-dried chitosan microspheres, *S. T. P. Pharm. Sci.*, 5, 202–207.
25. Gavini, E., Hegge, A. B., Rassu, G., Sanna, V., Testa, C., Pirisino, G., Karlsen, J., and Giunchedi, P. (2006), Nasal administration of carbamazepine using chitosan microspheres: In vitro/in vivo studies, *Int. J. Pharm.*, 307, 9–15.
26. Filipović-Grčić, J., Perissutti, B., Moneghini, M., Voinovich, D., Martinac, A., and Jalšenjak, I. (2003), Spray-dried carbamazepine-loaded chitosan and HPMC microspheres: Preparation and characterisation, *J. Pharm. Pharmacol.*, 55, 921–931.
27. Teshima, D., Yamauchi, A., Makino, K., Kataoka, Y., Arita, Y., Nawata, H., and Oishi, R. (2002), Nasal glucagon delivery using microcrystalline cellulose in healthy volunteers, *Int. J. Pharm.*, 233, 61–66.
28. Sacchetti, C., Artusi, M., Santi, P., and Colombo, P. (2002), Caffeine microparticles for nasal administration obtained by spray drying, *Int. J. Pharm.*, 242, 335–339.
29. Kublik, H., and Vidgren, M. T. (1998), Nasal delivery systems and their effect on deposition and absorption, *Adv. Drug Deliv. Rev.*, 29, 157–177.
30. Ugwoke, M. I., Verbeke, N., and Kinget, R. (2001), The biopharmaceutical aspects of nasal mucoadhesive drug delivery, *J. Pharm. Pharmacol.*, 53, 3–22.
31. Pringels, E., Callens, C., Vervaet, C., Dumont, F., Slegers, G., Foreman, P., and Remon, J. P. (2006), Influence of deposition and spray pattern of nasal powders on insulin bioavailability, *Int. J. Pharm.*, 310, 1–7.
32. Mygind, N., and Dahl, R. (1998), Anatomy, physiology and function of the nasal cavities in health and disease, *Adv. Drug Deliv. Rev.*, 29, 3–12.
33. Cornaz, A.-L., De Ascentis, A., Colombo, P., and Buri, P. (1996), In vitro characteristics of nicotine microspheres for transnasal delivery, *Int. J. Pharm.*, 129, 175–183.
34. Ugwoke, M. I., Agu, R. U., Verbeke, N., and Kinget, R. (2005), Nasal mucoadhesive drug delivery: Background, applications, trends and future perspectives, *Adv. Drug Deliv. Rev.*, 57, 1640–1665.
35. Smart, J. D. (2005), The basics and underlying mechanisms of mucosdhesion, *Adv. Drug Deliv. Rev.*, 57, 1556–1568.
36. Mathiowitz, E., and Chickering, D. E. (1999), Definitions, mechanisms and theories of bioadhesion, in Mathiowitz, E., Chickering, D. E., and Lehr, C.-M., Eds., *Bioadhesive*

Drug Delivery Systems: Fundamentals, Novel Approaches and Development, Marcel Decker, New York, pp. 1–10.

37. Lee, J. W., Park, J. H., and Robinson, J. R. (2000), Bioadhesive-based dosage forms: The next generation, *J. Pharm. Sci.*, 89, 850–866.
38. Nagai, T., Nishimoto, Y., Nambu, N., Suzuki, Y., and Sekine, K. (1984), Powder dosage form of insulin for nasal administration, *J. Controlled Release*, 1, 15–22.
39. Vidgren, P., Vidgren, M., Arppe, J., Hakuli, T., Laine, E., and Paronen, P. (1992), In vitro evaluation of spray-dried mucoadhesive microspheres for nasal administration, *Drug Dev. Ind. Pharm.*, 18, 581–597.
40. Oechslein, C. R., Fricker, G., and Kissel, T. (1996), Nasal delivery of octreotide: Absorption enhancement by particulate carrier systems, *Int. J. Pharm.*, 139, 25–32.
41. Abd El-Hameed, M. D., and Kellaway, I. W. (1997), Preparation and in vitro characterisation of mucoadhesive polymeric microspheres as intra-nasal delivery systems, *Eur. J. Pharm. Biopharm.*, 44, 53–60.
42. Ugwoke, M. I., Kaufmann, G., Verbeke, N., and Kinget, R. (2000), Intranasal bioavailability of apomorphine from carboxymethylcellulose-based drug delivery systems, *Int. J. Pharm.*, 202, 125–131.
43. García-Arieta, A., Torrado-Santiago, S., Goya, L., and Torrado, J. J. (2001), Spray-dried powders as nasal absorption enhancers of cyanocobalamin, *Biol. Pharm. Bull.*, 24, 1411–1416.
44. Hasçiçek, C., Gönül, N., and Erk, N. (2003), Mucoadhesive microspheres containing gentamicin sulfate for nasal administration: Preparation and in vitro characterisation, *Il Farmaco*, 58, 11–16.
45. Cerchiara, T., Luppi, B., Chidichimo, G., Bigucci, F., and Zecchi, V. (2005), Chitosan and poly(methyl vynil ether-co-maleic anhydride) microparticles as nasal sustained delivery systems, *Eur. J. Pharm. Biopharm.*, 61, 195–200.
46. Callens, C., and Remon, J. P. (2000), Evaluation of starch-maltodextrin-Carbopol 974 P mixtures for the nasal delivery of insulin in rabbits, *J. Controlled Release*, 66, 215–220.
47. Callens, C., Pringels, E., and Remon, J. P. (2003), Influence of multiple nasal administrations of bioadhesive powders on the insulin bioavailability, *Int. J. Pharm.*, 250, 415–422.
48. Ugwoke, M. I., Agu, R. U., Vanbilloen, H., Baetens, J., Augustijns, P., Verbeke, N., Mortelmans, L., Verbruggen, A., Kinget, R., and Bormans, G. (2000), Scintigraphic evaluation in rabbits of nasal drug delivery systems based on carbopol 971p and carboxymethylcellulose, *J. Controlled Release*, 68, 207–214.
49. Suzuki, Y., and Makino, Y. (1999), Mucosal drug delivery using cellulose derivatives as a functional polymer, *J. Controlled Release*, 62, 101–107.
50. Quadir, M., Zia, H., and Needham, T. E. (2000), Development and evaluation of nasal formulations of ketorolac, *Drug Deliv.*, 7, 223–229.
51. Lim, S. T., Martin, G. P., Berry, D. J., and Brown, M. B. (2000), Preparation and evaluation of the in vitro drug release properties and mucoadhesion of novel microspheres of hyaluronic acid and chitosan, *J. Controlled Release*, 66, 281–292.
52. Lim, S. T., Forbes, B., Berry, D. J., Martin, G. P., and Brown, M. B. (2002), In vivo evaluation of novel hyaluronan/chitosan microparticulate delivery systems for the nasal delivery of gentamicin in rabbits, *Int. J. Pharm.*, 231, 73–82.
53. Gavini, E.,Rassu, G., Sanna, V., Cossu, M., and Giunchedi, P. (2005), Mucoadhesive microspheres for nasal administration of an antiemetic drug, metoclopramide: In-vitro/ex-vivo studies, *J. Pharm. Pharmacol.*, 57, 287–294.

54. Witschi, C., and Mrsny, R. J. (1999), In vitro evaluation of microparticles and polymer gels for use as nasal platforms for protein delivery. *Pharm. Res.*, 16, 382–390.
55. Farraj, N. F., Johansen, B. R., Davis, S. S., and Illum, L. (1990), Nasal administration of insulin using bioadhesive microspheres as a delivery system, *J. Controlled Release*, 13, 253–261.
56. Illum, L., Faraj, N., Davis, S., Johansen, B., and O'Hagan, D. (1990), Investigation of the nasal absorption of biosynthetic human growth hormone in sheep: Use of a bioadhesive microsphere delivery system, *Int. J. Pharm.*, 63, 207–211.
57. Illum, L., Fisher, A. N., Jabbal-Gill, I., and Davis, S. S. (2001), Bioadhesive starch microspheres and absorption enhancing agents act synergistically to enhance the nasal absorption of polypeptides, *Int. J. Pharm.*, 222, 109–119.
58. Mao, S., Chen, J., Wei, Z., Liu, H., and Bi, D. (2004), Intranasal administration of melatonin starch microspheres, *Int. J. Pharm.*, 272, 37–43.
59. Illum, L., Farraj, N. F., Critchley, H., and Davis, S. S. (1988), Nasal administration of gentamicin using a novel microsphere delivery system, *Int. J. Pharm.*, 46, 261–265.
60. Björk, E., and Edman, P. (1988), Degradable starch microspheres as a nasal delivery system for insulin, *Int. J. Pharm.*, 47, 233–238.
61. Critchley, H., Davis, S. S., Farraj, N. F., and Illum, L. (1994), Nasal absorption of desmopressin in rats and sheep. Effect of a bioadhesive microsphere delivery system, *J. Pharm. Pharmacol.*, 46, 651–656.
62. Edman, P., Björk, E., and Rydén, L. (1992), Microspheres as a nasal delivery system for peptide drugs, *J. Controlled Release*, 21, 165–172.
63. Pereswetoff-Morath, L., and Edman, P. (1995), Dextran microspheres as a potential nasal drug delivery system for insulin—in vitro and in vivo properties, *Int. J. Pharm.*, 124, 37–44.
64. Pereswetoff-Morath, L., Bjurström, S., Khan, R., Dahlin, M., and Edman, P. (1995), Toxicological aspects of the use of dextran microspheres and thermogelling ethyl(hydroxyethyl) cellulose (EHEC) as nasal drug delivery systems, *Int. J. Pharm.*, 128, 9–21.
65. Rydén, L., and Edman, P. (1992), Effect of polymers and microspheres on the nasal absorption of insulin in rats, *Int. J. Pharm.*, 83, 1–10.
66. Vila, A., Sánchez, A., Évora, C., Soriano, I., McCallion, O., and Alonso, M. J. (2005), PLA-PEG particles as nasal protein carriers: The influence of the particle size, *Int. J. Pharm.*, 292, 43–52.
67. Tafaghodi, M., Tabassi, S. A. S., Jaafari, M. R., Zakavi, S. R., and Momen-nejad, M. (2004), Evaluation of the clearance characteristics of various microspheres in the human nose by gamma-scintigraphy, *Int. J. Pharm.*, 280, 125–135.
68. Leitner, V. M., Guggi, D., Krauland, A. H., and Bernkop-Schnürch, A. (2004), Nasal delivery of human growth hormone: In vitro and in vivo evaluation of a thiomer/glutathione microparticulate delivery system, *J. Controlled Release*, 100, 87–95.
69. Bernkop-Schnürch, A., Guggi, D., and Pinter, Y. (2004), Thiolated chitosans: Development and in vivo evaluation of a mucoadhesive permeation enhancing oral drug delivery system, *J. Controlled Release*, 94, 177–186.
70. Greimel, A., Dorly del Curto, M., D'Antonio, M., Palmberger, T., Sprinzl, G. M., and Bernkop-Schnürch, A. (2006), *In vitro* evaluation of thiomer microparticles for nasal drug delivery, *J. Drug Deliv. Sci. Technol.*, 16, 103–108.
71. Dodane, V., Khan, M. A., and Merwin, J. R. (1999), Effect of chitosan on epithelial permeability and structure, *Int. J. Pharm.*, 182, 21–32.

72. Singla, A. K., and Chawla, M. (2001), Chitosan: Some pharmaceutical and biological aspects—An update, *J. Pharm. Pharmacol.*, 53, 1047–1067.
73. Illum, L., Watts, P., Fisher, A. N., Hinchcliffe, M., Norbury, H., Jabbal-Gill, I., Nankervis, R., and Davis, S. S. (2002), Intranasal delivery of morphine, *J. Pharm. Exp. Ther.*, 301, 391–400.
74. Illum, L. (2003), Nasal drug delivery—Possibilities, problems and solutions, *J. Controlled Release*, 87, 187–198.
75. Illum, L., Farraj, N. F., and Davis, S. S. (1994), Chitosan as a novel nasal delivery system for peptide drugs, *Pharm. Res.*, 8, 1186–1189.
76. Illum, L., Jabbal-Gill, I., Hinchcliffe, M., Fisher, A. N., and Davis, S. S. (2001), Chitosan as a novel nasal delivery system for vaccines, *Adv. Drug Deliv. Rev.*, 51, 81–96.
77. Cheng, Y.-H., Dyer, A. M., Jabbal-Gill, I., Hinchcliffe, M., Nankervis, R., Smith, A., and Watts, P. (2005), Intranasal delivery of recombinant human growth hormone (somatropin) in sheep using chitosan-based powder formulations, *Eur. J. Pharm. Sci.*, 26, 9–15.
78. Thanou, M. M., Verhoef, J. C., Romeijn, S. G., Nagelkerke, J. F., Merkus, F. W. H. M., and Junginger, H. E. (1999), Effects of *N*-trimethyl chitosan chloride, a novel absorption enhancer, on Caco-2 intestinal epithelia and the ciliary beat frequency of chicken embryo trachea, *Int. J. Pharm.*, 185, 73–82.
79. Kotze, A. F., Lueßen, H. L., de Boer, A. G., Verhoef, J. C., and Junginger, H. E. (1998), Chitosan for enhanced intestinal permeability: Prospects for derivatives soluble in neutral and basic environments, *Eur. J. Pharm. Sci.*, 7, 145–151.
80. Bernkop-Schnürch, A., Hornof, M., and Guggi, D. (2004), Thiolated chitosans, *Eur. J. Pharm. Biopharm.*, 57, 9–17.
81. Illum, L., Jorgensen, H., Bisgaard, H., Krogsgaard, O., and Rossing, N. (1987), Bioadhesive microspheres as a potential nasal drug delivery system, *Int. J. Pharm.*, 39, 189–199.
82. Ugwoke, M. I., Agu, R. U., Jorissen, M., Augustijns, P., Sciot, R., Verbeke, N., and Kinget, R. (2000), Nasal toxicological investigations of Carbopol 971P formulation of apomorphine: Effects on ciliary beat frequency of human nasal primary cell culture and in vivo on rabbit nasal mucosa, *Eur. J. Pharm. Sci.*, 9, 387–396
83. Soane, R. J., Hinchcliffe, M., Davis, S. S., and Illum, L. (2001), Clearence characteristics of chitosan based formulations in the sheep nasal cavity, *Int. J. Pharm.*, 217, 183–191.
84. Jones, N. (2001), The nose and paranasal sinuses physiology and anatomy, *Adv. Drug Deliv. Rev.*, 51, 5–19.
85. Soane, R. J., Frier, M., Perkins, A. C., Jones, N. S., Davis, S. S., and Illum, L. (1999), Evaluation of the clearence characteristics of bioadhesive systems in humans, *Int. J. Pharm.*, 178, 55–65.
86. Vivien, N., Buri, P., Balant, L., and Lacroix, S. (1994), Nasal absorption of metoclopramide administrated to man, *Eur. J. Pharm. Biopharm.*, 40, 228–231.
87. Holmberg, K., Björk, E., Bake, B., and Edman, P. (1994), Influence of degradable starch microspheres on human nasal mucosa, *Rhinology*, 32, 74–77.
88. Brime, B., Ballesteros, M. P., and Frutos, P. (2000), Preparation and in vitro characterization of gelatin microspheres containing Levodopa for nasal administration, *J. Microencapsul.*, 17, 777–784.
89. Preda, M., and Leucuta, E. (2003), Oxprenolol-loaded bioadhesive microspheres: Preparation and *in vitro/in vivo* characterization, *J. Microencapsul.*, 20, 777–789.
90. Kriwet, B., and Kissel, T. (1996), Poly(acrylic acid) microparticles widen intercellular spaces of Caco-2 cell monolayers: An examination by confocal laser scanning microscopy, *Eur. J. Pharm. Biopharm.*, 42, 233–240.

91. Varshosaz, J., Sadrai, H., and Alinagari, R. (2004), Nasal delivery of insulin using chitosan microspheres, *J. Microencapsul.*, 21, 761–774.

92. Jain, S. K., Chourasia, M. K., Jain, A. K., Jain, R. K., and Shrivastava, A. K. (2004), Development and characterization of mucoadhesive microspheres bearing salbutamol for nasal delivery, *Drug Deliv.*, 11, 113–122.

93. Krauland, A. H., Guggi, D., and Bernkop-Schnürch, A. (2006), Thiolated chitosan microparticles: A vehicle for nasal peptide drug delivery, *Int. J. Pharm.*, 307, 270–227.

94. Jiang, H.-L., Park, I.-K., Shin, N.-R., Kang, S.-G., Yoo, H.-S., Kim, S.-I., Suh, S.-B., Akaike, T., and Cho, C.-S. (2004), In vitro study of the immune stimulating activity of an athrophic rhinitis vaccine associated to chitosan microspheres, *Eur. J. Pharm. Biopharm.*, 58, 471–476.

95. Genta, I., Costantini, M., Asti, A., Conti, B., and Montanari, L. (1998), Influence of glutaraldehyde on drug release and mucoadhesive properties of chitosan microspheres, *Carbohydr. Polym.*, 36, 81–88.

96. Hinchcliffe, M., and Illum, L. (1999), Intranasal insulin delivery and therapy, *Adv. Drug Deliv. Rev.*, 35, 199–234.

97. Freiberg, S., and Zhu, X. X. (2004), Polymer microspheres for controlled drug release, *Int. J. Pharm.*, 282, 1–18.

98. Li, Y., Jiang, H. L., Zhu, K. J., Liu, J. H., and Hao, Y. L. (2005), Preparation, characterization and nasal delivery of a-cobrotoxin-loaded poly(lactide-*co*-glycolide)/polyanhydride microspheres, *J. Controlled Release*, 108, 10–20.

99. Ugwoke, M. I., Verbeke, N., and Kinget, R. (1997), Microencapsulation of apomorphine HCI with gelatin, *Int. J. Pharm.*, 148, 23–32.

100. Yang, Y.-Y., Chung, T.-S., Bai, X.-L., and Chan, W.-K. (2000), Effect of preparation conditions on morphology and release profiles of biodegradable polymeric microspheres containing protein fabricated by double-emulsion method, *Chem. Eng. Sci.*, 55, 2223–2236.

101. Conte, U., Giunchedi, P., Maggi, L., and Torre, M. L. (1994), Spray dried albumin microspheres containing nicardipine, *Eur. J. Pharm. Biopharm.*, 40, 203–208.

102. He, P., Davis, S. S., and Illum, L. (1998), In vitro evaluation of the mucoadhesive properties of chitosan microspheres, *Int. J. Pharm.*, 166, 75–68.

103. Martinac, A., Filipovic-Grčic, J., Voinovich, D., Perissutti, B., and Franceschinis, E. (2005), Development and bioadhesive properties of chitosan-ethylcellulose microspheres for nasal delivery, *Int. J. Pharm.*, 291, 69–77.

104. Giunchedi, P., Juliano, C., Gavini, E., Cossu, M., and Sorrenti, M. (2002), Formulation and in vivo evaluation of chlorhexidine buccal tablets prepared using drug-loaded chitosan microspheres, *Eur. J. Pharm. Biopharm.*, 53, 233–239.

105. He, P., Davis, S. S., and Illum, L. (1999), Chitosan microspheres prepared by spray drying, *Int. J. Pharm.*, 187, 53–65.

106. Pavanetto, F., Genta, I., Giunchedi, P., Conti, B., and Conte, U. (1994), Spray dried albumin microspheres for the intra-articular delivery of dexamethazone, *J. Microencapsul.*, 11, 445–454.

107. Martinac, A., Filipovic-Grčic, J., Perissutti, B., Voinovich, D., and Pavelic, Ž. (2005), Spray-dried chitosan/ethylcellulose microspheres for nasal drug delivery: Swelling study and evaluation of *in vitro* drug release properties, *J. Microencapsul.*, 22, 549–561.

108. Fundueanu, G., Constantin, M., Dalpiaz, A., Bortolotti, F., Cortesi, R., Ascenzi, P., and Menegatti, E. (2004), Preparation and characterization of starch/cyclodextrin bioadhesive microspheres as platform for nasal administration of Gabexate Mesylate (Foy(r)) in allergic rhinitis treatment, *Biomaterials*, 25, 159–170.

109. Berthold, A., Cremer, K., and Kreuter, J. (1996), Preparation and characterization of chitosan microspheres as drug carrier for prednisolone sodium phosphate as model for anti-inflammatory drugs, *J. Controlled Release*, 39, 17–25.

110. Jaganathan, K. S., and Vyas, S. P. (2006), Strong systemic and mucosal immune responses to surface-modified PLGA microspheres containing recombinant Hepatitis B antigen administrated intranasally, *Vaccine*, 24, 4201–4211.

111. Callens, C., Ceulemans, J., Ludwig, A., Foreman, P., and Remon, J. P. (2003), Rheological study on mucoadhesivity of some nasal powder formulations. *Eur. J. Pharm. Biopharm.*, 55, 323–328.

112. Santos, C. A., Jacob, J. S., Hertzog, B. A., Freedman, B. D., Press, D. L., Harnpicharnchai, P., and Mathiowitz, E. (1999), Correlations of two bioadhesion assays: The everted sac technique and the CAHN microbalance, *J. Controlled Release*, 61, 113–122.

113. Rango Rao, K. V., and Buri, P. (1989), A novel in situ method to test polymers and coated microparticles for bioadhesion, *Int. J. Pharm.*, 52, 265–270.

114. Agu, R. U., Jorissen, M., Kinget, R., Verbeke, N., and Augustijns, P. (2002), Alternatives to *in vivo* nasal toxicological screening for nasally-administrated drugs, *S. T. P. Pharm. Sci.*, 12, 13–22.

115. Hvidberg, A., Djurup, R., and Hilsted, J. (1994), Glucose recovery after intranasal glucagon during hypoglycaemia in man, *Eur. J. Clin. Pharmacol.*, 46, 15–17.

116. Ugwoke, M. I., Agu, R. U., Jorissen, M., Augustijns, P., Sciot, R., Verbeke, N., and Kinget, R. (2000), Toxicological investigations of the effects carboxymethylcellulose on ciliary beat frequency of human nasal epithelial cells in primary suspension culture and in vivo on rabbit nasal mucosa, *Int. J. Pharm.*, 205, 43–51.

117. Merkus, P., Romeijn, S. G., Verhoef, J. C., Merkus, F. W., and Schouwenburg, P. F. (2001), Classification of cilio-inhibiting effects of nasal drugs, *Laryngoscope*, 111, 595–602.

118. Callens, C., Adriaens, E., Dierckens, K., and Remon, J. P. (2001), Toxicological evaluation of a bioadhesive nasal powder containing a starch and Carbopol 974 P on rabbit nasal mucosa and slug mucosa, *J. Controlled Release*, 76, 81–91.

119. Hirai, S., Yashiki, T., Matsuzawa, T., and Mima, H. (1981), Absorption of drugs from the nasal mucosa of rat, *Int. J. Pharm.*, 7, 317–325.

120. Adriaens, E., and Remon, J. P. (1999), Gastropods as an evaluation tool for screening the irritating potency of absorption enhancers and drugs, *Pharm. Res.*, 16, 1239–1243.

121. Marttin, E., Schipper, N. G. M., Verhoef, J. C., and Merkus, F. W. H. M. (1998), Nasal mucociliary clearance as a factor in nasal drug delivery, *Adv. Drug Deliv. Rev.*, 29, 13–38.

122. Agu, R. U., Jorissen, M., Willems, T., Van Den Mooter, G., Kinget, R., and Augustijns, P. (1999), Effects of pharmaceutical compounds on ciliary beating in human nasal epithelial cells: A comparative study of cell culture models, *Pharm. Res.*, 16, 1380–1385.

123. Merkus, F. W. H. M., Schipper, N. G. M., Hermens, W. A. J. J., Romeijn, S. G., and Verhoef, J. C. (1993), Absorption enhancers in nasal drug delivery: Efficacy and safety, *J. Controlled Release*, 24, 201–208.

124. Dimova, S., Brewster, M. E., Noppe, M., Jorissen, M., and Augustijns, P. (2005), The use of human nasal in vitro cell systems during drug discovery and development, *Toxicol. in Vitro*, 19, 107–122.

125. Schmidt, M. C., Peter, H., Lang, S. R., Ditzinger, G., and Merkle, H. P. (1998), In vitro cell models to study nasal mucosal permeability and metabolism, *Adv. Drug Deliv. Rev.*, 29, 51–79.

126. Jorissen, M., Van der Schueren, B., Tyberghein, J., Van Der Berghe, H., and Cassiman, J. J. (1989), Ciliogenesis and coordinated ciliary beating in human nasal epithelial cells cultured in vitro, *Acta Oto-Rhino-Laryngol. Begica*, 43, 67–73.

127. Agu, R. U., Jorissen, M., Willems, T., Van Den Mooter, G., Kinget, R., Verbeke, N., and Augustijns, P. (2000), Safety assesment of selected cyclodextrins: Effect on ciliary activity using a human cell suspension model exhibiting in vitro ciliogenesis, *Int. J. Pharm.*, 193, 219–226.

128. Illum, L., Farraj, N. F., Fisher, A. N., Gill, I., Miglietta, M., and Benedetti, L. M. (1994), Hyaluronic acid ester microspheres as a nasal delivery system for insulin, *J. Controlled Release*, 29, 133–141.

129. Sankar, C., Rani, M., Srivastava, A. K., and Mishra, B. (2001), Chitosan based pentazocine microspheres for intranasal systemic delivery: Development and biopharmaceutical evaluation, *Pharmazie*, 56, 223–226.

130. Ugwoke, M. I., Sam, E., Van Den Mooter, G., Verbeke, N., and Kinget, R. (1999), Nasal mucoadhesive delivery systems of the anti-parkinsonian drug, apomorphine: Influence of drug-loading on in vitro and in vivo release in rabbits, *Int. J. Pharm.*, 181, 125–138.

131. El-Shafy, M. A., Kellaway, I. W., Taylor, G., and Dickinson, P. A. (2000), Improved nasal bioavailability of FITC-dextran (Mw 4300) from mucoadhesive microspheres in rabbits, *J. Drug Target.*, 7, 355–361.

132. Rajinikanth, P. S., Sankar, C., and Mishra, B. (2003), Sodium alginate microspheres of metoprolol tartarate for intranasal systemic delivery: Development and evaluation, *Drug Deliv.*, 10, 21–28.

133. Nakamura, K., Maitani, Y., Lowman, A. M., Takayama, K., Peppas, N. A., and Nagai, T. (1999), Uptake and release of budesonide from mucoadhesive, pH-sensitive copolymers and their application to nasal delivery, *J. Controlled Release*, 61, 329–335.

134. Yildiz, A., Okyar, A., Baktir, G., Araman, A., and Özsoy, Y. (2005), Nasal administration of heparin-loaded microspheres based on poly(lactic acid), *Il Farmaco*, 60, 919–924.

5.8

AEROSOL DRUG DELIVERY

Michael Hindle
Virginia Commonwealth University, Richmond, Virginia

Contents

Pharmaceutical Manufacturing Handbook: Production and Processes, edited by Shayne Cox Gad

5.8.1 INTRODUCTION

Aerosol drug delivery to the lungs has long been the route of choice for the treatment of respiratory diseases, including asthma and chronic obstructive airway disease. Metered dose inhalers (MDIs), dry powder inhalers (DPIs), and nebulizers have been employed to successfully deliver a wide range of pharmaceuticals principally to the lungs for local action. However, with their unique characteristics, the lungs have now begun to be targeted as a means of noninvasive delivery of systemically acting compounds, including genes, proteins, peptides, antibiotics, and other small molecules [1, 2]. The primary function of the respiratory tract is gaseous exchange, transferring oxygen from the inspired air to the blood and removing carbon dioxide from the circulation. This pulmonary circulation offers rapid absorption and systemic distribution of suitable drugs deposited in the airways. Due to its anatomical structure, however, an important secondary role is the protection of the body from inhalation of foreign particles (including aerosol drug particles). The challenge of aerosol drug delivery is to overcome this highly effective barrier and accurately and reproducibly deliver aerosol drug particles in sufficient doses to their targeted sites within the lungs for either local action or systemic absorption. Effective aerosol drug delivery is tied to the aerosol inhaler that generates and delivers the respirable aerosol. This chapter will primarily focus on aerosol drug delivery devices, their development, and future prospects for pulmonary administration.

5.8.2 HUMAN RESPIRATORY TRACT AND AEROSOL PARTICLE DEPOSITION

5.8.2.1 Human Respiratory Tract

The human respiratory tract can be divided into three main regions: first, the upper airways, including the nose, mouth and throat (oropharynx), and the larynx [3]. The conducting airways consist of the regions from the trachea to the respiratory bronchioles and have airway diameters between 0.6 and 20 mm. The alveolar region consists of respiratory bronchioles and alveolar sacs and has airway diameters between 0.2 and 0.6 mm. The lungs are a branching system which commences asymmetrically, dividing first at the base of the trachea. The left and right bronchi branch dichotomously into the conducting airways. There are approximately 23 generations before the respiratory bronchioles give way to the alveoli, the site of gaseous exchange [4]. This branching produces a progressive reduction in airway diameter and also significantly increases the total surface area of the lower airways [3]. Another important characteristic with respect to drug delivery is the extensive vascular circulation. The blood vessels supplying the conducting airways are part of the systemic circulation. In contrast, the alveolar region is connected to the pulmonary circulatory pathway; drugs absorbed into this circulation will avoid first-pass hepatic metabolism effects.

5.8.2.2 Mechanisms of Particle Deposition

Aerosol particles are deposited in the lungs by three main mechanisms, and the site of deposition is dependent upon the predominating mechanism. Inertial impaction occurs because a particle traveling in an air stream has its own momentum (the product of its mass and velocity). As the direction of the airflow changes due to a bend or obstacle, the particle will continue in its original direction for a certain distance because of its inertia. Particles with a high momentum, due to high velocity or large size, are often unable to change direction before they impact on the surface in front of them [5]. Impaction of particles entering the mouth with a high velocity occurs either at the back of the mouth or at the bend where the pharynx leads to the trachea. Only a small fraction of particles greater than 15 μm will reach the trachea following mouth breathing. The majority, due to their size, will impact in the oropharyngeal region. Deposition by impaction will also occur as the trachea splits into the left and right bronchus. As the velocity of the particles decreases, inertial impaction becomes a less important mechanism of deposition in the smaller airways. Following the removal of larger particles in the upper airways by inertial impaction, gravitational sedimentation is the mechanism by which smaller particles (2–5 μm) are deposited in the respiratory bronchioles and alveoli. These particles settle under gravity and accelerate to a steady terminal velocity when the gravitational force is balanced by the resistance of the air through which it is traveling [6]. It is a time-dependent process which is aided by breath holding [7]. Brownian motion or diffusion is a mechanism which significantly affects only particles less than 0.5 μm in diameter. These particles are subjected to bombardment by surrounding gas molecules causing random movement of the particles. In this situation, the diffusivity of a particle is inversely proportional to its diameter. For an extensive mechanistic review of the area of particle deposition readers should consult Finlay (2001) [8].

Aerosol particle size and polydispersity are major determinants of the site and mechanism of pulmonary deposition. Fundamental deposition studies using monodisperse aerosols together with mathematical models have established the optimum aerosol particle size for lung deposition [9–12]. Aerosols larger than 10 μm will deposit in the oropharyngeal region and will not be inhaled. Particles less than 3 μm will be capable of penetrating into the alveolar region. Aerosols in the size range 3–10 μm will be distributed in the central and conducting airways [13]. A polydisperse aerosol containing a range of these particle sizes will allow deposition throughout the lungs. In theory, lung site deposition targeting should be possible by controlling the particle size of the inhalation aerosol [14]. However, a number of other significant variables can affect deposition within the respiratory tract and these often confound any efforts at targeting [15]. The patient's respiratory cycle, both the rate and depth of breathing, will affect aerosol deposition, and this is also the source of large intersubject variability in deposition [16]. Slow and deep inhalations are required for deposition in the peripheral airways, and this is the technique often recommended for inhalation with the MDI [17]. A different technique may be required for DPIs, where the patient's inspiratory effort is often the powder dispersion and delivery force. Flow rates greater than 60 L/min are commonly employed for powder inhalers [18, 19]. A final respiratory maneuver can be employed to promote deposition; breath holding up to 10 s is generally recommended to enhance deposition by sedimentation [17]. Other parameters that will affect lung deposition are the disease state within the lungs and its effect on airway caliber together with the patient's age and airway morphology [20–27].

5.8.2.3 Pharmacokinetics

Once deposited on the surface of the airways, the particle is subject to absorption and clearance processes depending upon its physical properties and the site of deposition [28–30]. For example, a lipophilic small molecule deposited in the central airways would have a different pharmacokinetic profile than a 50-kDa macromolecule deposited in the alveolar region. The former may undergo mucociliary clearance following deposition on a ciliated epithelial cell. Following dissolution, lipophilic drugs may be transported across the epithelium by passive transcytosis, while hydrophilic compounds are taken up by other pathways such as via tight junctions and endocytosis. Having overcome the barrier of the epithelial layer, the drug is available for distribution into the systemic circulation or to its site of action. Finally, the drug may also be subject to metabolism within the airways. For the macromolecule deposited in the peripheral airways, the absorption rate has been shown to be dependent upon molecular size. Larger molecules are subject to active processes such as caveolae or vesicular transport across the cell. Diffusion remains the predominant mechanism for smaller lipophilic macromolecules. Insoluble molecules can be phagocytosed by alveolar macrophages and removed via the lymphatic system or the mucociliary escalator. The pharmacokinetics of inhaled drugs is complicated by the fact that a significant fraction of the delivered dose is deposited in the oropharnyx or removed from the lungs via mucociliary clearance and in both cases subsequently swallowed [31]. An often desired goal for a pulmonary formulation is prolonged action within the lung. Rapid clearance or metabolism results in short duration of action for most inhaled drugs. A number of approaches using formulation excipient additives have been investigated to increase the residency or prolong release of drug at its site of action within the lungs [32, 33]. Microspheres containing nanoparticles have been formulated as dry powders for inhalation offering sustained-release properties [34]. In addition, prodrugs which are activated locally within the lungs have been used in an alternative approach [35–37].

The pharmacokinetic process of absorption, distribution, metabolism, and excretion within the lungs is an enormous subject area and readers are referred to specific reviews for further details [38–43]. Of particular interest may be the subject of absorption enhancer methodologies for lung delivery, which is beyond the scope of this chapter [44].

5.8.3 THERAPEUTIC INDICATIONS FOR AEROSOL DELIVERY

5.8.3.1 Current Applications

Aerosolized drug delivery is currently used to deliver a limited range of therapeutic classes of compounds. These are mainly for asthma and chronic obstructive airway disease. These classes of compounds include short- and long-acting β-adrenoceptor agonist, corticosteroids, mast cell stabilizers, and muscarinic antagonists. Of recent note is the popularity of combination products. These have obvious advantages from a patient compliance perspective. In addition, certain combinations of drugs have shown synergistic therapeutics benefits when compared to the drugs given by separate inhalers [45]. Long-acting β-adrenoceptor agonists and corticosteroids formulated as combination products are available as both MDIs and DPIs [46]. Also recently introduced was a MDI formulation, the R enantiomer of albuterol, which is believed to be mainly responsible for bronchodilation in the racemic mixture [47]. Zanamivir is licensed in the United States as an inhaled antiviral agent for the treatment of influenza [48]. Recombinant human deoxyribonuclease (rhDNAase) is available as a nebulizer product for the treatment of cystic fibrosis, in which it acts to liquefy viscous lung secretions [49]. And recently, insulin was approved as an inhaled powder for glycemic control in type I and II diabetes (see Section 5.8.6.5) [50].

5.8.3.2 Future Applications

Research and development are presently underway covering a vast array of novel applications. Clark (2004) provides an extensive list of products and their current state of development [51]. A significant future advance will be the development of inexpensive, noninvasive, stable, single-dose vaccine delivery via the lungs [52]. Efforts in this area are being led by the World Health Organization in the Measles Aerosol Project, and in a separate project, the Grand Challenges in Global Health initiative has funded a program to further develop an inhalation aerosol measles vaccine. Delivery of the measles vaccine via the lungs has been demonstrated to be both safe and effective [53–58]. Now the challenge of each of these projects is to produce stable inhalation vaccine formulations to be delivered via inexpensive inhalers while maintaining both safety and efficacy [59]. The use of inhaled vaccinations in the event of a bioterrorism attack is also a potential application [60, 61].

The use of the inhalation route for the delivery of gene therapy is also an area of significant interest [60, 62]. Cationic liposomes and polymers together with adenoviral vectors containing the reporting genes have been aerosolized using nebulizers for the majority of clinical studies. However, there are a significant number of challenges that must be overcome before pulmonary gene delivery is deemed completely successful, the most important being low gene transfer efficiency at the cellular level. This problem is not unique to inhalation therapy. Inhalation of a recombinant adenovirus containing the cystic fibrosis transmembrane regulator (Ad2/CFTR) demonstrated the feasibility of this approach for the treatment of cystic fibrosis [63, 64]. However, the limited duration of transfection and low cellular uptake efficiency still remain a barrier to full utilization of this route [60, 65]. There are a number of reviews that provide updates as to recent developments in this area [60, 66–70].

Given the success of delivering insulin, other peptides and proteins are being considered for pulmonary applications [71, 72]. Leuprolide is a nonapeptide which has been investigated as both an MDI and DPI formulation for the treatment of prostrate cancer [73–76]. Other hormones being investigated include calcitonin for the treatment of Paget disease and osteoporosis, parathyroid hormone to treat osteoporosis, growth hormone releasing factor for the treatment of pituitary dwarfism, and vasoactive intestinal peptide (VIP) for the treatment of pulmonary diseases [60, 77–81].

Other potential inhalation applications include drugs for both local and systemic delivery. Inhaled tobramycin is being investigated for the treatment of *Pseudomonas aeruginosa* exacerbations in cystic fibrosis [82, 83]. Liposomal ciprofloxacin is being developed as a first-line defense against biowarfare agents (e.g., anthrax) [61]. Inhaled cyclosporine has been shown to improve survival rates and extend periods of chronic rejection-free survival in lung transplant patients [84]. Apomorphine has been proposed as an inhalation formulation for the treatment of erectile dysfunction [85]. Aerosol delivery of chemotherapeutic drugs has been advocated for the treatment of lung cancer [86]. Morphine and fentanyl have been investigated for alternative routes of administering analgesics [87–90]. Heparin and low-molecular-weight heparins have been aerosolized and advocated for the treatment of emphysema and thrombosis [91–93]. Iloprost, a stable prostacyclin analog, has been aerosolized by nebulization for use in the treatment of pulmonary hypertension [94]. This list of potential new treatments approached via the inhalation route is not exhaustive; among the other compounds under investigation are α_1-antitrypsin, sumatriptan, ergotamine, nicotine as replacement therapy, pentamidine, and ribavirin. Readers should be aware that a large number of these examples are proof-of-concept studies that may not get beyond in vitro experiments and animal studies.

5.8.4 AEROSOL DRUG DELIVERY DEVICES

5.8.4.1 Introduction

As can be seen from the previous section, aerosol drug delivery continues to be an area of intensive research and development for the pharmaceutical industry. Not only are new applications for the pulmonary route being investigated, but also new delivery technologies are under development. The reformulation of MDIs with hydrofluoroalkane (HFA) propellants together with the potential of using the inhalation route as a means of systemic administration has led to significant technological advances in delivery devices. In parallel to MDI research DPIs have been developed from breath-actuated single-dose devices to both multiple-dose inhalers and active-dispersion DPIs. There is an extensive literature detailing the fundamental mechanisms of powder dispersion aimed at improving pulmonary deposition from powder inhalers. In addition, novel particle production technologies have been developed that provide alternatives to the traditional micronized powder for formulation in both MDIs and DPIs. Nebulizer technology has evolved from previously nonportable devices into high-efficiency, hand-held nebulizers that offer alternatives to the MDI and DPI for certain treatment regimes. Finally, novel soft mist inhalers that generate aerosols by solution atomization have emerged on the inhaler landscape. All this research has focused on improving aerosol deposition efficiency and reproducibility within the lungs, together with targeting the peripheral lungs for systemic absorption. The efforts of the last decade culminated in two significant events. First, the regulatory approval of Proventil HFA and QVAR, the first suspension and solution HFA MDIs, respectively. Second, in 2006, the U.S. and European regulatory authorities approved Exubera, an insulin DPI for the systemic treatment of type I and II diabetes. Exubera offered a noninvasive alternative to subcutaneous injections of insulin.

5.8.4.2 Characteristics of Ideal Delivery Device

With these developments, innovation continues toward development of the ideal inhaler. A number of authors have compiled lists of desired characteristics for an aerosol inhaler [95–97]. These can be grouped into patient-desired or industry-driven properties. From the patient's perspective the overriding requirement is a device that is simple to operate. This is becoming increasing difficult to achieve as evidenced by the intensive patient education initiative that is being planned for the launch of the Exubera insulin inhaler. Poor compliance and adherence to prescribed therapy may be related to patients' failure to use the inhaler correctly [98]. Inhalers should be portable and contain a large number of doses. The device should also give some indication to the patient when it is empty. The inhaler should be suitable for use by all of the population, especially children and the elderly. Ganderton (1999) cited that from a device perspective aerosol generation should be independent of the patient's inhalation and should continue for a substantial portion of the inspiratory cycle. This would minimize the reliance on coordinating inhalation and actuation of the device [99]. Breath-actuated devices have been developed to address this issue. In order to achieve lung deposition targeting, the particle size distribution of the aerosol should be capable of being altered depending upon the specific target region. For example, the central airways may be targeted with a 3–5 μm aerosol for the treatment of acute bronchoconstriction, while a smaller aerosol (1–3 μm) might be used for deep lung deposition and subsequent systemic absorption [99]. In addition, the dose should be delivered reproducibly with minimal oropharyngeal deposition, perhaps as a low-velocity aerosol. There should be a minimal number of small parts in the inhaler, and it should be robust and reliable when placed "in use." The manufacturer has the option of producing a disposable or refillable unit; however, the inhaler should protect the formulation from environment and not affect its stability.

Dolovich et al. (2005) have provided an extensive evidence-based evaluation of aerosol drug devices. They concluded that when selecting an inhalation delivery system the following should be considered: device and drug combination availability, clinical setting, patient age, the ability of the patient to use the device correctly, device use with multiple medications, cost and reimbursement, drug administration time, convenience in outpatient and inpatient settings, and patient and physician preference [100]. Other reviews have compared the benefits and disadvantages of inhalers from clinical and patients' perspectives [101, 102].

5.8.5 METERED DOSE INHALERS

5.8.5.1 Introduction

Since their development, MDIs have been widely used for pulmonary aerosol drug delivery [103]. Despite their recognized limitations, they remain the device of choice for many physicians around the globe. From a patient's perspective, they are light, portable, and robust and contain multiple doses of medication. They are also relatively simple to operate (press and fire); however, significant numbers of patients experience difficulties correctly using the MDI due to coordination problems [104]. To maximize lung drug deposition, actuation (pressing the MDI canister) by the patient must be coordinated with a slow, deep inhalation. Studies have reported that 51% of patients fail to operate the MDI correctly [104]. This leads to low lung deposition, high oropharyngeal deposition, and ultimately perhaps therapeutic failure. From the pharmaceutical industry perspective, the components are relatively inexpensive; however, the formulation and manufacturing have become increasingly complex. There are numerous studies describing the multifaceted and interactive effects of propellant [105–110], excipient [111–115], metering valve [110, 116], and actuator [116–119] on the aerosol particle size characteristics of the MDI [120, 121]. To date, the success of the MDI has relied in part on the potency and relative safety of the bronchodilators and corticosteroids commonly used for the treatment of respiratory disorders rather than its delivery efficiency. The relatively low and often variable aerosol deposition efficiency, only around 10–20% of the nominal dose being delivered to the lungs, is the challenge that is beginning to be addressed as the MDI looks to enter the next 50 years of aerosol drug delivery.

5.8.5.2 Metered Dose Inhaler and HFA Reformulation

The basic design and operation of the MDI has changed little over its lifetime. Aerosols are generated from a formulation of drug (0.1–1% w/w) either suspended or in solution in the liquefied propellant. The formulation is held under pressure in a canister.

Figure 1 shows the basic components of the MDI, consisting of a canister sealed with a metering valve which is inserted into a plastic actuator containing the mouthpiece. Aerosol generation takes place when the canister is pressed against the actuation sump by the patient. Actuation causes the outlet valve to open and the liquefied propellant formulation is released through the actuator nozzle and subsequently through the mouthpiece to the patient. Metered volumes between 20 and 100 μL are dispensed, and as the pressurized propellant is released, it forms small liquid droplets traveling at high velocity. These droplets evaporate to leave drug particles for inhalation [117]. Purewal and Grant (1998) have assembled a definitive reference source for issues relating to the design, manufacturing, and performance of MDIs [122].

The currently marketed MDIs may look similar to the devices that were first developed by Riker in 1950. However, due to the replacement of the ozone-depleting chlorofluorocarbon (CFC) propellants with HFA propellants, virtually all of the components of the MDI have been altered. In 1987, the Montreal Protocol was drawn up, leading to the eventual phase-out of CFC propellants. MDIs contain-

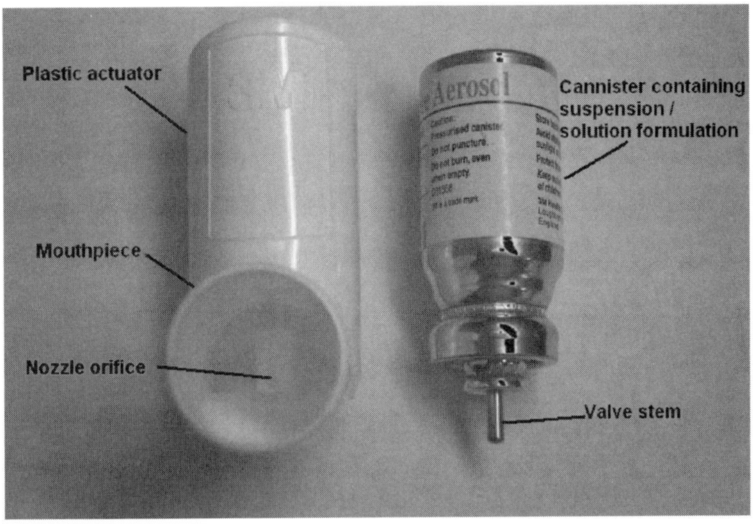

FIGURE 1 Schematic of MDI.

ing CFC propellants were granted essential-use exemptions until viable alternatives became available. Therefore, with this impending withdrawal, a consortium of pharmaceutical companies (IPACT-I and IPACT-II) worked to identify and toxicologically test alternative propellants for MDIs. HFA 134a and HFA 227 were identified as viable alternatives and the task of reformulation began. At first look, it appeared that the most expeditious route to replacing a CFC product would be to produce a suspension HFA MDI with exactly the same in vitro characteristics as the CFC MDI. This would prove to be a time-consuming route [123, 124]. While some manufacturers focused on producing HFA products with identical characteristics to the current CFC versions to accelerate the pathway through clinical testing to market. Others undertook extensive research and development in the area of HFA formulation options, and this has led to the possibility of utilizing the MDI for both local and systemic administration. During this reformulation effort, the industry has taken the opportunity to address some of the other shortcomings of the MDI [125]. Among these issues were poor peripheral lung delivery, variable dose delivery, and limitations as to the dose capable of being delivered to the lung (typically about 1 mg) [126].

The replacement of CFC MDIs with inhalers formulated with the HFA propellants is now well underway in Europe. Although progress in the United States has been slower, with the introduction of suitable alternatives for albuterol inhalers, the FDA has ordered that CFC albuterol MDIs be withdrawn from the market by the end of 2008 [127]. Examples of reformulated products available in the United States include Ventolin HFA, which is a suspension albuterol sulfate formulation using HFA 134a alone. ProAir is an alternative albuterol sulfate product manufactured by Ivax which contains ethanol and HFA 134a. Xopenex HFA has recently been approved for marketing in the United States [128]. This product contains levalbuterol tartrate (R-albuterol enantiomer) together with HFA 134a, dehydrated alcohol, and oleic acid as a suspension formulation. Table 1 summarizes the HFA

TABLE 1 Summary of HFA Metered Dose Inhaler Products Available in United States, June 2006

Product Name	Drug	Approval Date	Excipients	Type
Proventil HFA (3M)	Albuterol sulfate	August 1996	HFA 134a, ethanol, oleic acid	Suspension
Ventolin HFA (GSK)	Albuterol sulfate	April 2001	HFA 134a	Suspension
Proair HFA (IVAX)	Albuterol sulfate	October 2004	HFA 134a, ethanol	Suspension
QVAR (3M)	Beclomethasone dipropionate	September 2000	HFA 134a, ethanol, oleic acid	Solution
Flovent HFA (GSK)	Fluticasone propionate	May 2004	HFA 134a, ethanol	Suspension
Atrovent HFA (BI)	Ipratropium bromide	November 2004	HFA 134a, purified water, dehydrated alcohol, anhydrous citric acid	Solution
Xopenex (Sepracor)	Levalbuterol tartrate	March 2005	HFA 134a, dehydrated alcohol, oleic acid	Suspension

products currently available in the United States and their excipients. The following section will focus on the current options for formulation of drugs in HFA propellant systems and the challenges that are encountered as products are reformulated as HFA formulations.

5.8.5.3 Propellants

The CFC propellants primarily used in MDI formulations were CFC 11, 12, and 114. Blends of these propellants were held liquefied under pressures of 50–80 psig within the canister. Flocculated drug suspensions in CFC propellants were formulated using a surfactant (e.g., oleic acid and lecithin). In a suspension formulation, the aerosol particle size is dependent upon the initial micronized drug particle size (typically between 2 and 5 μm) and the evaporation of the propellant droplets. It has long been recognized that changes in CFC propellant vapor pressure result in changes in droplet size and velocity of the aerosol. Newman et al. (1982) showed that increasing the vapor pressure of the propellant blend in the MDI significantly increased whole-lung deposition and reduced oropharyngeal deposition [110].

The first challenge encountered during the reformulation with HFA propellants was the altered physicochemical properties of HFA 134a and HFA 227 compared to the CFC propellants [124]. Table 2 compares the physicochemical properties of the CFC and HFA propellants [124]. The increased polarity of HFA 134a and HFA 227 is illustrated by the increased dipole moments and dielectric constant. From a practical point of view, the altered solvency properties of the HFA propellant for the drug, excipient, water, and, surprisingly, components of the MDI have been the

TABLE 2 Comparison of Physicochemical Characteristics of CFC and HFA Propellants

	CFC			HFA	
Property	11	12	114	134a	227
Thermodynamic					
RRM	137	121	171	102	170
Boiling point, °C	24	−30	4	−26	−16
Vapor pressure, 20°C, kPa	89	566	182	572	390
Enthalpy vap., 20°C, kJ/mol (J/g)	25.1 (183)	17.2 (142)	22.1 (130)	18.6 (182)	19.6 (115)
C_p liquid, 20°C, J/mol·K (J/g·K)	120 (0.88)	118 (0.98)	168 (0.98)	143 (1.41)	210 (1.24)
Polarity					
Dielectric constant	2.3	2.1	2.2	9.5	4.1
Dipole moment (*D*)	0.45	0.51	0.58	2.1	1.2
Induced polarization	2.8	2.3	3.2	6.1	6.1
Solubility parameter (Hild. units)	7.5	6.1	6.4	6.6	6.2
Kauri-butanol value	60	18	12	9	13
log *P* octanol/water	2.0	2.2	2.8	1.1	2.1
Water solubility (ppm)	130[a]	120[a]	110[a]	2200[b]	610[b]
Liquid Phase					
Density (g/cm^3)	1.49	1.33	1.47	1.23	1.42
Viscosity (mPa·s)	0.43	0.20	0.30	0.21	0.27
Surface tension (mN/m)	18	9	11	8	7

Source: From ref. 124. [a]30° [b]25°.

major issues during reformulation. Suspension formulations of micronized drug in the liquefied CFC propellant blends with surfactants were replaced with either suspension or solution HFA formulations, depending upon the solubility of the individual drug in the HFA propellants. For suspension HFA formulations, however, it was observed that the conventional surfactants used in the CFC products were insoluble in the HFA propellants without the addition of a cosolvent (e.g., ethanol) [129, 130]. An alternative approach to produce suspension formulations was to develop a new class of surfactants suitable for use in the HFA systems [131]. It was also observed that some of the commonly used inhalation drugs were slightly soluble in the new propellants and therefore precluded their formulation as a suspension. The potential to formulate as a solution offered a number of advantages together with significant problems. Perhaps most importantly, changing from a suspension to a solution formulation altered the mechanism of aerosol particle formation. In the case of solution formulations, drug is dissolved in the liquefied propellant and a suitable cosolvent (if necessary) and particle formation takes place during evaporation of the propellant. This leads to much smaller particles being formed when propellant evaporation is complete. Stein and Myrdal (2006) recently described the MDI aerosol generation for solution formulations as a two-step process [132]. Droplet formation takes place as millions of atomized droplets are produced after the formulation exits the metering valve through the actuator sump. Initial droplet size is dependent upon the vapor pressure and surface tension of the formulation, valve size, and actuator orifice diameter. The second step is an evaporative or "aerosol maturation" phase, as the propellant and cosolvents (e.g., ethanol) rapidly evaporate leaving inhalable drug particles. The final size of these particles is dependent upon the initial droplet size, the vapor pressure of the formulation mixture, and the proportion of nonvolatiles in the formulation [132].

Leach et al. (1998) compared the pulmonary deposition of a suspension CFC formulation of beclomethasone diproprionate with a solution HFA formulation [133]. The marketed CFC product had a mass median aerodynamic diameter (MMAD) of 3.5 μm compared to 1.1 μm for the solution formulation, reflecting the altered aerosol formation mechanism. The gamma scintigraphy profile for the solution formulation showed the drug and label to be diffusely deposited throughout the airways with approximately 55–60% deposited in the lungs. In contrast, the CFC product was deposited mainly in the mouth and throat (90–94%), with only 4–6% being deposited in the airways. In many ways, this study summarized the deficiencies of the suspension formulation CFC MDI and offered the alternative of improved delivery efficiencies with the solution HFA formulation. A significant conclusion from this and other studies supported the hypothesis that improved pulmonary deposition and reduced oropharyngeal losses of aerosols would allow reduction in the dose required by the patient to achieve the same therapeutic effect [108, 133, 134].

The altered solubility profile of the HFA propellant, while providing attractive characteristics for solution formulations, also provide significant challenges with respect to their interactions with the basic MDI components. Leachables are compounds that can be transferred from MDI component parts to the formulation during the shelf life of the product. Berry et al. (2003) postulated that MDI orientation could affect the amount of leachables that entered a formulation and affect the particle size distribution of aerosol [135]. Extensive efforts are now required for extractable and leachable testing of the component materials prior to formulation of an MDI. Another by-product of replacing the CFC propellants was to tighten the impurity specifications required for the new propellants. A proposed U.S. Pharmacopeia (USP) monograph for HFA 134a has now been published detailing the impurity profile [136].

Manufacturing processes for MDIs have also required adapting for the use with the new propellant system [137]. There are two main manufacturing processes used for MDIs: cold filling and pressure filling [138]. Cold filling requires cooling the propellants to below −50°F and filling at that temperature prior to crimping the valve onto the canister. Pressure-filling techniques for MDIs are most commonly employed. These can be accomplished in either a one- or two-step process. In the single-step process, the formulation is placed in a pressurized mixing vessel. The empty canister is purged with propellant to remove the air. The valve is then crimped onto the canister and the formulation is metered through the valve. The absence of a HFA propellant that was liquid at room temperature was a major difference compared to the process employed for CFC manufacturing. In the two-step process, the formulation (excluding the propellant) are mixed together to form a concentrate. Previously, liquefied CFC 11 was used in this step of the process. However, there is no suitable HFA propellant that is liquid at room temperature. Therefore, cosolvents such as ethanol and glyercol are employed during this step to form the product concentrate. The concentrate is metered into the empty canister. The valve is then crimped onto the canister and the propellant is filled through the valve. Wilkinson (1998) provides an extensive history and review of the manufacturing procedures for MDIs [138].

5.8.5.4 Excipients

A number of excipients have been included in MDI formulations; however, the nature of the excipients has changed with the introduction of the HFA propellant aerosols. Oleic acid and sorbitan trioleate (SPAN 85) and lecithins were used in CFC suspension MDIs as suspending agents and valve lubricants [112]. Typical concentrations ranged from 0.1 to 2.0% w/w. Ethanol is now being used in HFA formulations as a cosolvent for suspension and solution formulations. The addition of ethanol to the formulation has a number of effects [121, 139]. Increasing the ethanol concentration has been shown to increase the initial droplet size [121]. In addition, ethanol can increase the hydrophilicity of the formulation and increase moisture uptake. Glycerol and polyethyleneglycol have also been added as cosolvents but also have the effect of increasing the residual droplet particle size due to their lower volatility [111]. In general, a relationship can be observed between the fraction of nonvolatile components (drug and nonvolatile excipients) in a solution HFA formulation and the final particle size of the aerosol. The MMAD was observed to be linearly proportional to the cube root of the nonvolatile concentration [119, 121]. Oligolactic acids (OLAs) have been investigated for their use in a variety of functions in HFA formulations. OLAs with repeating units of 6–15 units have been proposed as suspending agents [131]. They are readily soluble in both HFA 134a and 227. These molecules have also been shown to act as ion pair solubilizers for certain drugs (e.g., albuterol). The addition of ethanol to these OLA formulations synergizes the solubilizing effect [131].

A word of caution is required when considering introduction of novel excipients into any inhalation drug product formulation. Due to the unique toxicological challenges associated with administration and clearance from the lung, the qualification of novel excipients for inhalation has proven to be an expensive and time-consuming challenge. This has led to a limited number of compounds with an extensive "in-use" profile being commonly employed.

5.8.5.5 Valves

Metering valves are required to accurately meter and dispense the formulation upon MDI actuation. In addition, they perform an important contact closure role preventing moisture ingress and minimizing propellant evaporation. Figure 2 and Table 3 show the basic components of the metering valve. Currently, the most common valve type is the retention valve, consisting of a plastic metering chamber and two rubber gaskets. The remaining valve components are manufactured from plastic, metal, and elastomeric materials.

Material component evaluation and selection are critical steps in the development of a MDI formulation [140, 141]. The materials must be chemically resistant and compatible with all components of the formulation. Gaskets must have appropriate mechanical properties and work effectively as a seal, preventing leakage of the formulation and moisture ingress. While the basic components themselves have

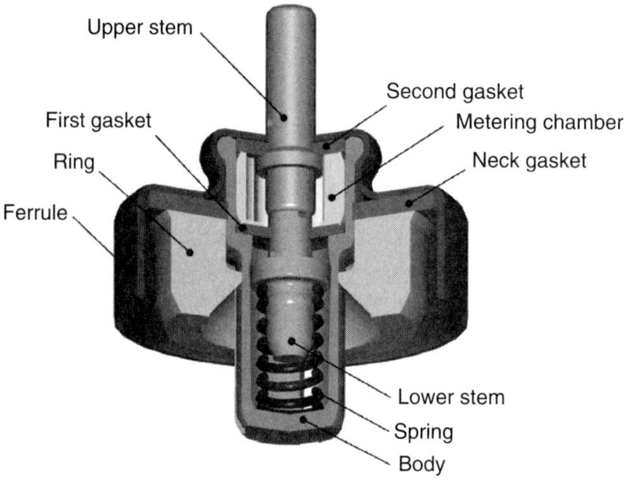

FIGURE 2 Schematic of components of metering valve. (Courtesy of Valois Pharm.)

TABLE 3 Summary of Components and Materials Used in Metered Dose Inhaler Valves

Component	Material
Metering chamber	Polyester
Core	Polyester/acetal
Core extension	Polyester/acetal
Body	Polyester/nylon
Seats/gaskets	EPDM/nitrile/butyl/chloroprene/ bromobutyl
Spring	Stainless steel
Ferrule	Aluminium

remained unchanged during the introduction of the HFA propellants, the materials used to manufacture the components have required significant adaptation. Nitrile was the most commonly employed elastomer in CFC MDIs; it has good mechanical and elastic properties. However, it has been shown to swell when in contact with HFA propellants and ethanol. Newer elastomers such as ethylene propylene diene monomer (EPDM), chloroprene, and bromobutyl are now used in HFA MDIs [142]. The ideal universal elastomer has yet to be developed and the newer materials must be assessed on a case-by-case basis for formulation compatibility and the desired moisture ingress characteristics. Among the many issues to be considered when screening materials are formulation–material compatibility, extractable profiles, and mechanical resistance. Manufacturers such as Valois, Solvay, and Bespak have extensive knowledge of drug/excipient/material component compatibility and should be used as the first point of reference when considering a MDI formulation project.

Another concern to formulators is the ingress of moisture into HFA-formulated MDIs [143]. HFA propellants have a higher moisture affinity compared to the CFC propellants, especially HFA 134a. In addition, the inclusion of ethanol in some formulations increases its hydrophilicity. Moisture entering the canister can have several effects; it may alter the physical or chemical stability of the formulation and aerosolization performance of HFA MDIs. Due to its lower volatility compared to the other components, water may affect aerosol generation and alter aerosol particle size [144]. The increased water content may increase the solubility of suspended polar drug particles or decrease the solubility of hydrophobic compounds [145]. Corrosion in aluminum canisters may also increase over the shelf life of the product. Williams and Hu (2000) reported that HFA 134a had a greater tendency to take up moisture during storage than did HFA 227 [144]. The issue of moisture ingress during storage has led to certain HFA MDIs being stored in moisture-protecting pouches prior to initial use (e.g., Ventolin HFA). An alternative approach to minimize the effects of moisture ingress has been taken by SkyePharma, which has incorporated subtherapeutic doses of cromolyn sodium into its HFA MDI formulations. Cromolyn sodium is used as a hygroscopic excipient to scavenge any moisture that penetrates into the formulation. Cromolyn sodium has been used widely by inhalation over the past 30 years and has an excellent safety profile via the inhalation route. Burel et al. (2004) reported that for a HFA 134a MDI formulation the inclusion of a polyamide (nylon 66) molded ring around the valve body reduced both the initial water content and the final water content (6 months) when stored under stress conditions [40°C and 75% relative humidity (RH)]. A combination of a thermoplastic elastomer sealing gasket in the MDI valve and a polyamide ring produced the lowest water ingress under these stress conditions [143]. For formulations that might be susceptible to water-induced stability issues, HFA 227 may be considered a more suitable propellant than HFA 134a. Given the possibility of moisture ingress, there is also the issue of propellant leakage. Leak testing is among the array of in-process quality assurance tests that are required. These include assay of the suspension or solution, moisture level, consistency of filling of both the concentrate and the propellant, valve crimp measurements, quality of sealing, in-line leak testing under stress conditions, and performance of the valve.

Another significant issue encountered during use of MDIs was related to loss of prime and dose reproducibility [129, 146]. Loss of prime relates to the fact that in conventional capillary retention metering valves the dose is filled into the valve immediately following the last actuation. Capillary retention valves require priming with one or two sprays prior to their first use. In addition, if there is a significant interval between the actuations and the inhaler is stored upside down or on its side or shaken, then the metering valve may actually partially empty, resulting in a low and variable dose being delivered to the patient. A review of patient information leaflets indicated varying instructions on priming MDIs. This ranged from Atrovent CFC and Combivent CFC requiring priming with 3 sprays "after 24 hours of nonuse." Ventolin HFA and Proventil HFA both required priming with 4 sprays after "2 weeks of nonuse." Flovent CFC required priming with 4 sprays after "4 weeks of nonuse." Clearly, such instructions add to the complexity for patients using MDIs and also contribute to drug waste issues. Loss of prime is also a significant issue for breath-actuated MDIs, where the opportunity to prime the inhaler is not readily possible. A number of new valve designs have been developed to address this issue. The fast-fill, fast-empty valves offer a solution to the priming and loss of prime issues. In these valves [(e.g., 3M Shuttle valve (3M), 3M Face Seal valve (3M), ACT (Valois), and Easifill valve (Bespak)], the metering valve is only isolated from the formulation canister reservoir immediately prior to dose actuation. Therefore, the metering chamber can be emptied and refilled with a fresh dose from the reservoir simply by shaking the canister prior to use.

5.8.5.6 Actuators

Nonvolatile component concentration has previously been described as one of the primary determinants of the initial droplet size for HFA solution formulations [119]. Perhaps, equally important is the MDI actuator [147]. The actuator consists of the sump block into which the metered dose is immediately delivered during MDI actuation. As expansion and vaporization of the propellant take place, the aerosol exits the sump via the actuator nozzle and then is inhaled through the actuator mouthpiece. From a practical perspective, in general, reducing the size of the orifice diameter for HFA solution formulations produced a relatively slower spray emitted with less force compared to marketed CFC products [118]. The nozzle orifice diameter has been considered to be the most important, although not the only, actuator variable determining the particle size distribution of HFA solution formulations [118, 147, 148]. Recently, Smyth et al. (2006) described three critical components of the actuator that could affect the aerosol performance of a solution HFA formulation. In addition to the orifice diameter, sump depth (and hence the expansion chamber volume) together with orifice length was observed to have significant effects on the aerosol particle size distribution and should be considered for optimization with an HFA formulation [147]. It has also been recognized that the electrostatic charge of all components of the MDI and its formulation may affect the aerosolization properties of the aerosol spray [149, 150].

5.8.5.7 Canisters

Aluminum canisters are widely used in commercial MDI products mainly due to their inert characteristics. Other materials, including stainless steel and glass, can be employed depending upon the particular formulation characteristics. These canisters were usually uncoated. Changes to the canister may be required when the formulation interacts with the interior surface of the canister altering the chemical stability of the formulation. The presence of ethanol in HFA formulations has also increased the risk of metal corrosion. Drug migration or absorption to the metal components of the canister and also the metal valve components has also been reported [141]. The loss of drug to the walls of the canister will result in variability in the delivered dose from the MDI during the shelf life of the inhaler. The use of canister coating and anodized canisters has been advocated to mitigate this problem [141].

5.8.5.8 Breath Actuation

In order to overcome the problems associated with many patients' inability to coordinate actuating the MDI and inhaling, breath-actuated MDIs were developed [107]. These devices allow the MDI to be automatically actuated only when the patient commences inhaling through the mouthpiece. Of critical importance here is ensuring that the patient has sufficient inspiratory flow rate to trigger actuation. While these devices offer little improvement for patients with a good inhaler technique, it has been shown that patients with poor coordination did have significantly greater lung drug deposition when inhaling using a breath-actuated MDI [151]. The 3M Autohaler was the first device marketed using this technology [151, 152]. In Europe, the Easibreathe and Autohaler breath-actuated MDIs are used to deliver β agonists and corticosteroids for the treatment of obstructive airway. Recently, in the United States, the MD Turbo has been launched, a device that allows patients to take their regular MDI canister and actuator and insert it into the MD Turbo. The MD Turbo acts as a generic breath actuator for a number of marketed MDIs and also incorporates a dose counter.

5.8.5.9 Spacers

Spacer devices have been developed as another alternative to overcome the problems associated with patients coordinating the beginning of their inspiratory effort with actuation of the MDI [153]. This problem is extenuated by the fact that the MDI emits a high-velocity, short-duration aerosol cloud. On actuation, the propellant spray is delivered into the spacer that often incorporates a one-way inhalation valve. The patient is now able to inhale the aerosol cloud. The large-volume spacers have an additional effect in that they allow evaporation of large propellant droplets prior to inhalation. These high-velocity droplets would previously have had a high probability of impacting in the patient's throat. Figure 3 shows the large number of spacer chambers that are available [154]. Spacers are advocated for use by children and elderly patients and people who experience difficulty coordinating actuation of the MDI. The use of spacers for the delivery of corticosteroids also minimizes oral deposition of the inhaled dose and therefore reduces the incidence of steroid-related side effects [155]. Both in vitro and clinical studies have shown the effectiveness of spacers with CFC MDIs [156–158]. It has been shown that electrostatic charge can have a significant effect on the performance of a spacer chamber and where possible the charge should be minimized to maximize drug delivery [159–161]. Finally, it should be noted that the use of any particular spacer–MDI combination should be evaluated at least in vitro to confirm the beneficial effect, especially when employed with solution-based HFA MDI formulations [162, 163].

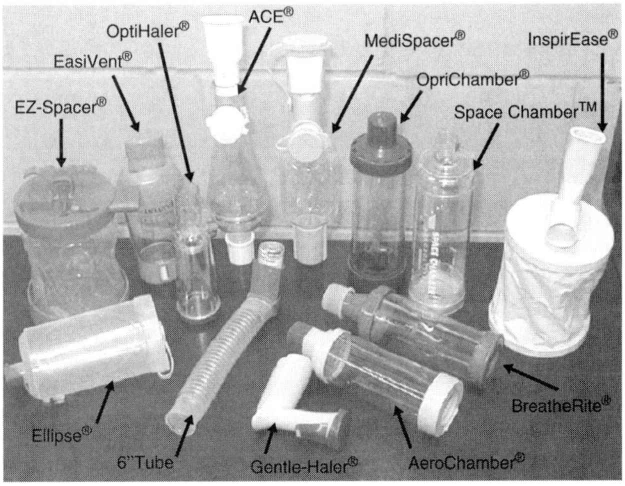

FIGURE 3 Example spacer chambers available for use with MDIs. (Reproduced from ref. 154 with permission of *Pharmacotherapy*.)

5.8.5.10 Dose Counters

A guidance document from the Food and Drug Administration (FDA) recommends the addition of a dose counter to the MDI. This would overcome a long-standing problem with the MDI, the inability of a patient to accurately know the number of doses remaining in the canister [164]. Dose counters have been incorporated successfully into multiple-dose DPIs. In general, the counter should give a clear indication of when approaching end of life and the actual end of life. It should be either numeric or color coded. If numeric, it should count downward and should be 100% reliable and avoid undercounting [165].

5.8.6 DRY POWDER INHALERS

5.8.6.1 Introduction

Dry power inhalers have been in use for over 40 years. They were developed as an environmentally friendly alternative to the MDI. The early DPIs were simple in design, portable, but again, a relatively inefficient means of delivering drugs to the lungs for local action [166]. The Spinhaler, the first DPI, has been prescribed in Europe since the late 1960s. In general, the acceptance and use of DPIs is much greater in Europe than in the United States. However, with the reformulation efforts for MDIs, there are an increasing number of DPIs becoming available in the United States (Figure 4).

Research and development for dry powder inhalers have two main focuses: the optimization of the powder formulation for use in these inhalers and investigations of novel DPI device designs and technology. An enormous literature now exists in each of these areas; for more extensive reviews readers should consult refs. 33, 167, or 168.

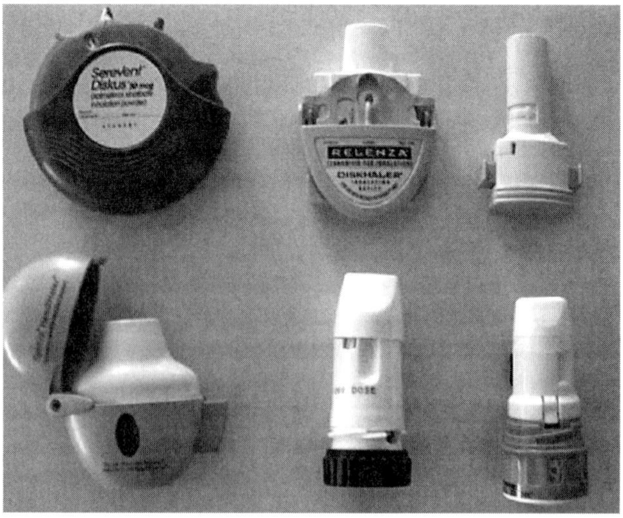

FIGURE 4 Example DPIs available in United States.

5.8.6.2 Size Reduction and Particle Formation Technologies

Dry powder inhaler formulations consist usually of either a drug-only formulation or an ordered mixture of drug and excipient, most commonly lactose monohydrate. In both cases, the first challenge is the production of drug particles with suitable size characteristics for inhalation (i.e., 1–5 μm). Traditionally, micronization or jet-milling methods have been employed as the method of choice for conventional small molecules. This method is identical to that employed for the production of fine particles for suspension MDIs. Using this method it is possible to produce primary particles between 1–5 μm. However, as a consequence of the particle size reduction there are a number of undesirable effects with respect to the powder properties. Micronized powders possess high intramolecular forces and are cohesive. They readily form aggregates that are difficult to disperse to the primary particles. Dispersion to its primary particle is essential for successful pulmonary deposition. In addition, they often possess high inherent electrostatic charges which cause particle adhesion to the components of the dry powder inhaler [169]. The high-energy micronization process also causes disruption of the crystal lattice and results in the formation of amorphous regions which may affect the long-term stability of the formulation [170]. Finally, it is not possible to control the drug particle morphology. Despite all of these problems, micronization remains the most common technique employed for respirable particle formation. Modifications to conventional micronization techniques have been investigated as alternative methods of particle size reduction [171–173].

A number of novel particle formation technologies now exist that are able to produce respirable drug particles for formulation in both DPIs and MDIs. Depending on the method of preparation, these particles offer unique and potentially advantageous physical and aerodynamic properties compared to conventional crystallization and micronization techniques. Some investigators have advocated that major improvements in aerosol particle performance may be achieved by lowering particle density and increasing particle size, as large, porous particles display less tendency to agglomerate than (conventional) small and nonporous particles. Also, large, porous particles inhaled into the lungs can potentially release therapeutic substances for long periods of time by escaping phagocytic clearance from the lung periphery, thus enabling therapeutic action for periods ranging from hours to many days [174].

Many of these techniques involve particle formation from solution formulations that contain novel excipients. Spray drying is the most advanced of these technologies and has been used to produce the powder formulation in the Exubera inhaler [175]. Various modifications of this basic technique, including co–spray drying with novel excipients, have been employed.

AIR particles are low-density lipid-based particles that are produced by spray drying lipid–albumin–drug solutions. These particles are characterized by their porous surface characteristics and large geometric diameter while having a low aerodynamic diameter [176, 177]. This technology has been used to produce porous particle powder formulations of L-dopa that have been investigated for the treatment of Parkinson's disease [178].

Pulmospheres are produced using a proprietary spray drying technique, with phosphatidylcholine as an excipient to produce hollow and porous particles with low interparticulate forces. These particles have been formulated as suspended particles in HFA MDIs. In comparison with conventional suspension MDIs, the Pulmosphere MDI exhibited significantly higher fine particle fractions. This technology has been used to produce cromolyn sodium, albuterol sulfate and formoterol fumarate microspheres [179]. Pulmospheres powder formulations containing tobramycin and budesonide have also been tested clinically [83, 180].

Technosphere technology has been developed as an alternative porous particle for pulmonary delivery [181]. These porous microspheres are formed by precipitating a drug-diketopiperazine derivative from an acidic solution. Para-thryroid hormone (PTH) Technospheres have been investigated for the treatment of osteoporosis following aerosol delivery [182].

The use of supercritical fluid processing technology has also been widely used for its application in controlled microparticle formation. Conventional small molecules and proteins for inhalation have been generated and formulated as powders for inhalation. [183–186].

The application of pulmonary delivery of nanoparticles (<1 um) for pharmaceuticals remains to be developed [187–189].

5.8.6.3 Drug–Lactose Formulations

The most common means of overcoming cohesion problems is by incorporation of a carrier excipient. Lactose monohydrate is used most often; it is inert, cheap, widely available, and a GRAS (generally regarded as safe) non-toxic excipient. A significant area of research has been undertaken to optimize the critical parameters involved in the formulation of drug–lactose blends. Micronized drug is typically blended with lactose (50–100 µm) to produce an ordered mix. The blend ratio is fixed depending upon the dose of drug to be delivered and the mass of powder blend in each dosage unit (typically between 5 and 25 mg). The aerosolization properties of the blend are related to the adhesive forces between the drug and lactose together with the cohesive forces between the drug particles. Reproducible dispersion of the blend either by the dry powder inhaler (active DPI) or by the patients' inspiratory effort (passive DPI) is required. This allows the detached micronized drug to be inhaled and deposited in the respiratory tract while the larger lactose particles are deposited by inertial impaction in the oropharnyx.

Formulators have become increasingly aware of the criticality of the drug and lactose powder surface characteristics and their relationship to the aerosolization performance in a DPI [190, 191]. A number of investigators have shown in vitro the importance of controlling the size of the lactose and the amount of "fines" (lactose particles less than 5 µm in size) in the drug–lactose blend [192–194]. Inherent fines are present in all lactose powders, and the fines are usually adhered to the surface of the larger lactose particle. These fines are believed to occupy "active" or high-energy sites on the lactose particle surface. Occupation of these sites by the lactose fines prevents the micronized drug from adhering to these positions. This allows the drug to adhere to less active sites and become detached easier from the lactose surface during inhalation. Obviously any significant change in the quantity of fines present in the lactose may alter the distribution of the micronized drug on the lactose particle and therefore the aerosolization characteristics of the powder blend [193–195]. Batch-to-batch control of the fines content of inhalation lactose has been recognized as critical to ensuring reproducible in vitro emitted and fine particle doses. Jones and Price (2006) have recently surveyed the literature in this area and provided a comprehensive review [196]. Modification of the surface characteristics of the lactose particle has been used as an alternative approach to control the adherence of drug particles to the lactose surface [197–200]. Alternative sugar carriers have also been investigated; these appear to possess many of the same performance-limiting characteristics as lactose [201]. Finally, tertiary additives have also been used to improve the aerosolization properties of DPI formulations [202]. The majority of the studies described above relate to in vitro testing of DPI formulation performance, and little is known about the clinical significance of these studies.

Moisture ingress into a powder formulation is a particular concern as it may significantly decrease the aerosolization performance of the formulation [203, 204]. Increased adhesion of particles is often seen following exposure to high-RH environments [205, 206]. Moisture ingress has also been shown to affect drug stability [170]. The pharmaceutical industry has used a number of approaches to protect powder formulations from the ingress of moisture during storage and for their "in-use" life. The Turbuhaler incorporates a desiccant in the base of the inhaler to keep the power reservoir free from moisture [207]. Unit-dose blisters used in the Diskus are sealed in a foil strip pack to protect each individual dose prior to inhalation [208]. It is also essential that the patient not exhale into the DPI immediately prior to inhaling the dose.

Electrostatic charge can also influence the performance of DPI formulations. A number of studies have investigated the interactions of drug and lactose particle charge with respect to aerosolization properties and drug retention by the plastic components within the inhaler [209–212].

5.8.6.4 Dry Powder Inhaler Design

Inhalation Flow Rate The main function of a DPI is to facilitate dispersion and delivery of inhalable drug particles. An extensive patent and scientific literature exists describing the ever-increasing number of DPI device designs [33]. Powder dispersion in the early passive DPIs was provided in part by the inspiratory effort of the patient. This removed the necessity to coordinate patient inhalation with actuation and delivery of the dose (in contrast to MDIs). These passive DPIs were "breath actuated," with the patients' inspiratory effort dispersing, aerosolizing, and delivering the powder during the inhalation cycle. The airflow rate through the inhaler was determined by the inherent device resistance and the inspiratory force exerted by the patient [18]. Devices such as the Spinhaler, Rotahaler, and Diskhaler are low-resistance devices requiring relatively high inspiratory flow rates to disperse the powder formulations by turbulent deaggregation. These simple devices have low aerosolization efficiencies with only 5–20% of the dose being delivered to the lungs [166] The inhalation flow rate dependence of passive DPIs has been cited as a potential problem in their use, especially given the large intersubject flow rate variability within the patient population (especially for the young and older patients). In vitro testing revealed that for certain DPIs there was large variability in both the delivered dose and the aerodynamic particle size distribution as a function of the inhalation flow rate [203, 213–216]. Similar clinical studies also revealed a flow rate dependence for certain DPIs while others were observed to perform with a degree

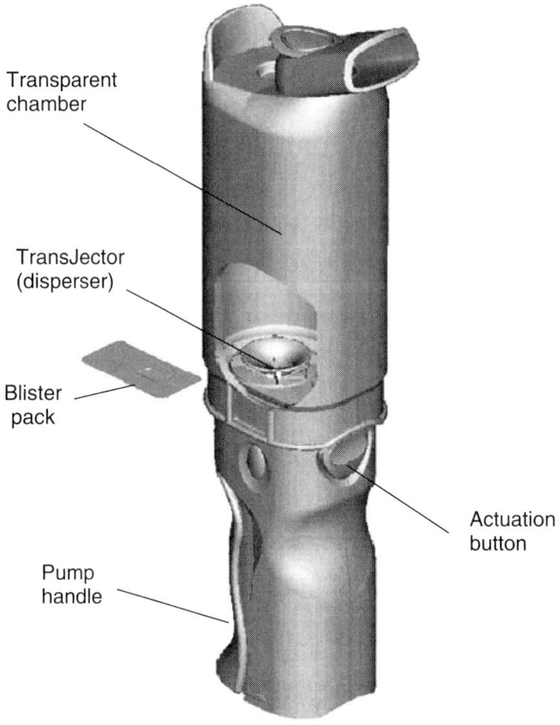

FIGURE 5 Exubera Inhaler. (Reprinted from ref. 175. Courtesy of Mary Ann Liebert, Inc.)

of flow rate independence [214, 217–221]. When choosing a DPI, the effect of inhalation flow rate should be assessed on a case by case basis for each individual DPI, and readers should be aware of contradictory studies, especially when comparing in vitro and clinical performance. The Turbuhaler is one such example, where some in vitro studies show high variability; however, this is not reflected in clinical studies [215, 220, 222].

From these and many other studies it can be concluded that a desirable characteristic for any DPI is that its dose delivery performance is independent of inhalation flow rate. A second generation of DPIs have been developed that incorporate a combination of improved powder formulations, more effective turbulent dispersion within the inhaler, and in some cases an active dispersion mechanism. The Exubera inhaler releases a bolus of compressed air through the formulation and actively generates an aerosol cloud from the powder (Figure 5). The cloud is held within a reservoir chamber from which the patient then inhales the insulin dose [175]. Active dispersion improves device aerosolization efficiency, with greater than 50% of the dose being deposited in the lungs, while minimizing the reliance on the patients' inspiratory effort.

Single- and Multiple-Dose DPIs Inhalation powder dose metering is one of the problems encountered by DPI formulators. The powder dose can range from 250 μg in the drug-only Pulmicort Turbuhaler formulation to 25 mg in the lactose-blended

Spinhaler formulations. In each case, accurate and reproducible metering of the powder is required for regulatory approval and therapeutic efficacy. This proved to be a technological challenge that was solved in a number of ways. Single-unit-dose inhalers were the first generation of DPIs, the unit dose being metered in the factory and subsequently loaded into the inhaler by the patient immediately prior to each dosing. Because metering takes place prior to batch release by the manufacturer, this allows for quality control and release testing, ensuring that dosage units were within acceptable criteria. Procedures such as capsule filling were common for early devices such as the Spinhaler and Rotahaler. This approach is still used by some of the newer devices being developed (e.g., Aerohaler and Cyclohaler) [185, 223]. While popular with the pharmaceutical industry, the single-unit-dose device required significant patient handling to load and empty the inhaler for each inhalation (unlike the MDI, which often contained up to 200 doses available for inhalation on demand). Two approaches were taken toward the design of multiple-dose DPIs; the multiple-unit-dose DPI (e.g., Diskhaler and Diskus) and the powder reservoir multidose DPI (e.g., Turbuhaler) [207]. For the multiple-unit-dose DPI, manufacturers sought to address the requirement for multiple doses while retaining the control of factory premetering. Perhaps the most successful DPI in this respect is the Diskus, in which the dose is premetered into a coiled foil covered strip containing individually sealed blister reservoirs [208]. Each blister is opened immediately prior to inhalation and up to 60 doses can be help in each foil strip. For the powder reservoir multidose DPI, volumetric dose metering of the powder takes place within the DPI immediately prior to inhalation in a manner analogous to MDIs. Among the devices that use this approach are the Turbuhaler, Clickhaler, Pulvinal, and Easyhaler [220, 224–226]. The Turbuhaler is used with a drug-only formulation (although lactose blends have also been used) that employs a proprietary powder agglomeration process to produce loosely bound aggregates that are easily dispersed by the patient's inhalation and by the turbulent flow path encountered in the DPI [227]. Besides the Diskus and Turbuhaler, there are four other devices currently available in the United States, the Asmanex Twisthaler, the Foradil Aerolizer, the Relenza Diskhaler, and the Spiriva Handihaler. Other devices in development include the Novolizer, a multidose, refillable, breath-actuated DPI that delivers up to 200 metered doses of drug from a single cartridge [228, 229]. The Ultrahaler offers yet another alternative DPI [230]. The Taifun inhaler, the JAGO inhaler, and the Airmax are other multidose DPIs [231–234].

5.8.6.5 Exubera

Systemic delivery of drugs via the lungs offers a noninvasive route of administration. Perhaps the most important and widely investigated molecule considered for this route has been insulin [235, 236]. Following over a decade of development, in January 2006, Pfizer and its partner Nektar received marketing approval for Exubera, their insulin DPI. This offered diabetics a noninvasive route of insulin administration rather than repeated subcutaneous injections [237]. Exubera has been indicated for the treatment of adult patients with diabetes mellitus for the control of hyperglycemia [238]. It has an onset of action similar to rapid-acting insulin analogs and has a duration of glucose-lowering activity comparable to subcutaneously administered regular human insulin [239]. Patton et al. (2004) provided an extensive review of the clinical pharmacokinetics and pharmacodynamics of inhaled insulin [240]. In patients with type I diabetes, Exubera should be used in regimens that include a longer acting insulin [241]. In patients with type II diabetes, Exubera can be used as monotherapy or in combination with oral agents or longer acting insulins [242–244]. Studies revealed that the same level of blood sugar control was achieved following inhalation compared to subcutaneous injection, although different nominal doses were required due to lung bioavailability issues [245]. The therapeutic efficacy and safety of inhaled insulin appears to have been proven, although there are a significant number of issues with its administration via this route [246]. It has been noted that asthmatics absorb less insulin from the lungs than nonasthmatics. In addition, smokers absorb more insulin than nonsmokers. Small and reversible changes in pulmonary lung function have been observed in some studies with inhaled insulin. Each of these issues has led to the development of specific prescribing guidelines and an intensive physician/patient education program for the inhaled insulin product. The Exubera insulin formulation is a spray-dried, amorphous insulin powder containing 60% insulin in a buffered, sugar-based matrix [175].

Other pharmaceutical companies are also continuing to develop their own inhalation insulin products. Aradigm and NovoNordisk are using a liquid insulin formulation in combination with the AERx IDMS inhaler [247–251]. Alkermes and Lily are developing an insulin product derived from their research on geometrically large, low-density particles that are formed by a spray drying process incorporating a natural phospholipid. MannKind is using its Technosphere technology to produce low-density porous insulin particles. This formulation is delivered using the MedTone inhaler. Other companies working in this area include Kos Pharmaceuticals, Mircodose Technologies, Coremed, and Biosante.

5.8.7 NEBULIZERS

Nebulization of liquid formulations has long been established as an effective, if not efficient, means of pulmonary drug delivery. The basic principle of nebulizer aerosol generation has remained unchanged; however, a number of technological advances have been made which have improved efficiency and reduced variability. Aerosols that were previously delivered in a continuous inhalation mode over 5–15 min are now delivered only during the inspiratory cycle, thus reducing drug waste. In general, nebulizers convert a liquid into a fine droplet mist, either by means of a compressed gas (jet nebulizer) or by high-frequency sound (ultrasonic nebulizer) [252]. Ultrasonic nebulizers use a piezoelectric source within the formulation reservoir to induce waves at the surface of the nebulizer formulation. Interference of these waves induces the formation of droplets which are then carried in a flowing air stream that is passed over the formulation. These devices are not suitable for the nebulization of suspension formulations [253]. Rau (2002) also observed that ultrasonic nebulizers can increase the solution reservoir temperature and may cause drug degradation [254]. In the case of the jet nebulizer, an aerosol is produced by forcing compressed air through a narrow orifice which is positioned at the end of a capillary tube. The negative pressure created by the expanding jet causes formulation to be drawn up to the capillary tube from the reservoir in which it is immersed. As the liquid emerges from the tip of the capillary, it is drawn into the air stream and broken up into droplets by the jet to produce an aerosol. Baffle structures within the nebulizer filter the large droplets from the aerosol by impaction and the deposited drug solution is recycled back into the drug reservoir [255]. Only the small aerosol droplets evade impaction on the baffles and are delivered to the patient for inhalation. Jet nebulizers can be categorized by function, for example, the DeVilbiss 646 is a conventional jet nebulizer with continuous drug output resulting in significant waste during exhalation. The Pari LC Plus system incorporated a valve system and operates as an active venturi jet nebulizer; although drug output is continuous, there is an increased output during inhalation. The patient's inspiratory effort increases the nebulizer airflow, thus increasing drug output for these breath-enhanced nebulizers. Finally, dosimetric jet nebulizers such as the Ventstream use a one-way valve system to emit aerosol only during inspiration and are also breath-enhanced nebulizers [256, 257]. It is this last type of nebulizer that offers the most significant advances in technology [258].

Jet nebulizers are commonly used in nonambulatory settings such as hospitals or the patient's home. In vitro studies comparing the performance of commercial nebulizers have concluded that there were large differences in drug delivery between nebulizers of different classes and even between nebulizers of apparently the same class [259–261]. The aerosolization performance of different nebulizers has been found to be dependent upon a number of factors, including the drug being aerosolized, the formulation fill volume, the compressed airflow rate, and breathing pattern [254, 260, 262, 263]. These parameters ultimately control the aerosol droplet size and rate of drug output [264]. However, probably the size of the conventional nebulizer, the duration of the treatment cycle (5–15 min), and the cost of the nebulizers are the main reasons that they are usually reserved for nonambulatory settings and remain less popular than the MDI and DPI.

Solutions or suspensions are available as nebulizer formulations. Due to the relative simplicity in formulating a liquid nebulizer formulation and because of the relatively large range of doses available for delivery, the nebulization method is often chosen as the aerosol method for proof-of-concept investigational studies.

Nebulizer technology continues to be developed to miniaturize and lower the cost of the devices while maintaining the quality of the aerosols generated. The Halolite incorporates adaptive aerosol delivery which monitors patients inspiratory cycle and delivers drug to patients during the first 50% of their inspiratory cycle [265–267]. The Pari eFlow is a hand-held device that uses a vibrating membrane nebulizer to generate a respirable aerosol [268]. Aerogen (now part of Nektar) has a range of nebulizer-based technologies, including the Aeroneb and Aerodose devices. Aerosols are generated as a liquid formulation passes through vibrating apertures [269, 270].

5.8.8 EMERGING TECHNOLOGIES

5.8.8.1 Soft Mist Aerosols

In recent years research has focused on a new method of pharmaceutical aerosol generation that involves passing a solution formulation through a nozzle or series of nozzles to generate a "soft mist" aerosol as a bolus dose [271]. Aerosol generation is achieved by mechanical, thermomechanical or electromechanical processes depending upon the particular technology employed [272]. It is worth noting that these devices are bolus dose delivery inhalers, rather than the new continuous-generation nebulizers which generate aerosols by vibrating porous membranes at ultrasonic frequencies. Such devices include the eFlow and Aerodose, which were described earlier.

While the precise mechanism of soft mist aerosol generation may differ between inhalers, a number of common characteristics can be observed. They are propellant free and produce slow-moving aerosols over an extended duration with high in vitro fine-particle fractions compared to MDIs and DPIs. The aerosols are often generated from simple solution formulations containing pharmaceutically acceptable excipients. Water and ethanol are the most commonly employed vehicles for soft mist aerosols [273]. Perhaps the most simple and advantageous vehicle is water. There is often a well-known and established stability profile of many pharmaceuticals in aqueous solutions, accelerating the route to the clinic in any development program. Drug solubility can be manipulated by choice of water, ethanol, or mixtures of the two to increase formulation options and doses. In multidose reservoir-type devices, a preservative would be required to prevent microbial contamination. This is in addition to the current federal regulations that all aqueous-based drug products for oral inhalation must be manufactured to be sterile.

5.8.8.2 Respimat

The Respimat inhaler was recently launched in Germany as a combination product of fenoterol and ipratropium hydrobromide (Berodual) and was licensed for the treatment of chronic obstructive airway disease. A large body of literature now exists documenting the aerosol characteristics and clinical performance of the Respimat inhaler with a number of different drugs [274, 275]. Aerosolized formulations include the steroids budesonide and flunisolide in addition to the β agonist fenoterol as well as the commercially available combination product of fenoterol and ipratropium bromide [276–281].

The Respimat device is a multidose reservoir system that is primed by twisting the device base (Figure 6). This compresses a spring and transfers a metered volume of formulation from the drug cartridge to the dosing chamber. The metered volume is between 11 and 15 μl depending upon the drug formulation. When the device is actuated (in coordination with the patient's inspiration), the spring is released. This forces a micropiston into the dosing chamber and pushes the solution through the uniblock. The uniblock is the heart of the aerosol generation system and consists of a filter structure with two fine outlet nozzle channels. The uniblock produces two fine jets of liquid that converge at a precisely set angle and then collide. This collision aerosolizes the liquid to form an aerosol [282].

Aerosols generated from the Respimat inhaler have been characterized as having a prolonged aerosol cloud duration compared to MDIs and have a slower cloud velocity as measured using video camera imaging. Hochrainer et al. (2005) measured the cloud duration of the Respimat aerosol to be 0.2–1.6 s compared to less than 0.2 s for HFA and CFC MDIs. Aerosol velocities have been reported as less than 1 m/s for the Respimat, compared to 6–8 m/s for CFC MDI inhalers [283]. While a degree of patient coordination is required to actuate the Respimat and to inhale,

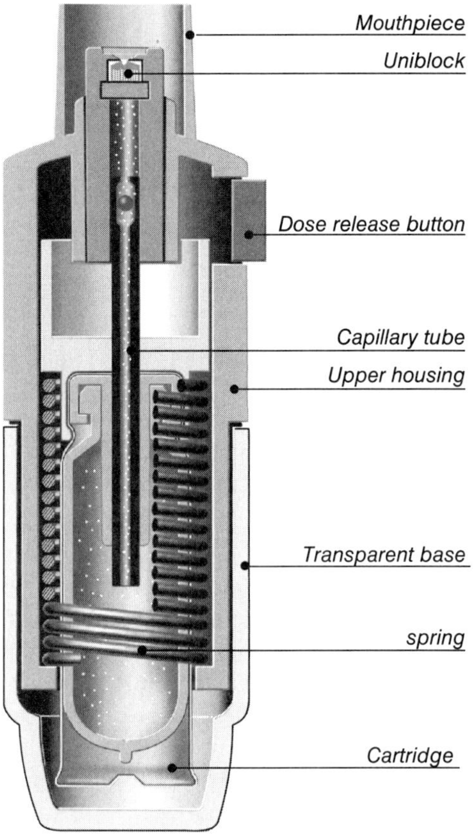

FIGURE 6 Respimat Inhaler. (Courtesy of Boehringer Ingelheim.)

the longer duration of aerosol cloud generation makes this maneuver less critical than with MDIs.

Aqueous and ethanolic formulations have been employed with the Respimat and the in vitro aerosol performance determined. Zierenberg (1999) reported fine-particle fractions of 66% for an aqueous fenoterol formulation and 81% for an ethanolic flunisolide formulation. The respective MMADs were $2.0 \pm 0.4\,\mu m$ for the aqueous formulation and $1.0 \pm 0.3\,\mu m$ for the ethanolic formulation [284].

5.8.8.3 AERx

The AERx system was developed for the systemic delivery of insulin. Unit-dose aqueous solution formulations were produced in a blister strip design. The first-generation AERx device is a battery-operated device that guides the patient through the inhalation technique required to successfully deliver a dose. It can also monitor dose times and frequency together with the facility to download dosing data in the clinic. A number of macromolecules, including insulin, and traditional small molecules (e.g., morphine) have been investigated using the AERx technology [89, 251].

Aerosol generation using the AERx system is achieved by mechanically forcing a dose of the liquid formulation though a nozzle array in its disposable unit-dose blisters. The electronic version of the AERx inhaler guides the patient to inhale at the required flow rate. A cam-operated piston mechanism is actuated to compress the blister and extrude the dose as an aerosol through the nozzle array into warmed flowing air. The nozzle array consists of a number of laser-drilled holes. Nozzle design characteristics can be altered depending upon the formulation characteristics and the desired droplet particle size. The single-use nature of the blister avoids potential problems such as microbial contamination from a dosing solution reservoir and nozzle-clogging issues.

A number of prototype versions of the AERx system have been investigated. In general, the in vitro aerosol characteristics revealed that about 50–60% of the loaded dose was emitted from the device, of which over 90% was respirable. MMADs ranged from 1 to 3 μm depending upon the formulation and nozzle array [285]. In a scintigraphic study, lung deposition following inhalation from the AERx was 53.3% (expressed as a percentage of the radioactivity in the AERx blister) compared to 21.7% for an MDI [285].

A number of clinical studies delivering insulin to diabetic patients using the AERx system are currently ongoing. Hermansen et al. (2004) concluded that in type II diabetics, preprandial inhaled insulin via the AERx was as effective as preprandial subcutaneous insulin in achieving glycemic control [286]. Clinical studies with morphine revealed comparable analgesic efficacy for a matched dose of inhaled and intravenous morphine in a postsurgical pain model [251]. In addition, the AERx inhaler has been employed for the topical delivery of rhDNase to cystic fibrosis patients. A mean relative increase in forced expiratory volume in 1 s (FEV1) of 7.8% was observed after 15 days treatment compared to control [287].

5.8.8.4 Mystic

The Mystic inhaler offers a soft mist aerosol generated from solution or suspension formulations. Unlike the previously described soft mist inhalers which use purely mechanical forces to generate the aerosol, the Mystic inhaler applies an electric field to the formulation within the spray nozzle [288]. An electric charge builds on the fluid surface and, as the droplets exit the nozzle, the repelling force of the surface charge overcomes the surface tension of the droplets to form a soft mist droplet aerosol. This process is known as electrohydrodynamic aerosolization or electrospray. The particle size characteristics of the aerosol can be controlled by adjusting the physical and chemical characteristics of the formulation together with the formulation flow rate and electrical field properties. The inhaler consists of a number of components, a drug containment system, metering system, aerosol nozzle, power supply, and microprocessor, all enclosed in a housing. To date, Ventaira reports that the inhaler has been successfully employed to generate aerosols from small-molecule formulations (albuterol, triamcinolone, cromolyn, budesonide, and terbutaline) and macromolecules, including insulin [288].

5.8.8.5 Capillary Aerosol Generator

In the capillary aerosol generator (CAG) system, the aerosol is formed by pumping the drug formulation through a small, electrically heated capillary. Upon exiting the capillary, the formulation is rapidly cooled by ambient air to produce an aerosol. The generated aerosol characteristics are dependent upon the formulation employed. Using propylene glycol as a condensing vehicle, drug containing condensation aerosols are generated [289]. When using water, ethanol, or combinations of both as noncondensing excipients, a stream of solid particles is delivered as a soft mist aerosol. In vitro studies using budesonide, cromolyn sodium, buprenophine, albuterol, and insulin have been performed to demonstrate various applications of the CAG technology. These studies are characterized by high emitted doses and high fine-particle fractions. Using noncondensing excipients, it is possible to produce aerosols with vastly different size characteristics, depending upon the required application.

5.8.8.6 Staccato

This technology utilizes a rapid heating technique to vaporize a thin film of drug. Following vaporization, the drug particles condense in the inhalation flow stream to form a respirable aerosol and are inhaled. Single- and multiple-dose breath-actuated inhalers are currently in development. As with any method involving heating of a formulation, drug degradation must be minimal. Rabinowitz et al. (2006) described the absorption of prochlorperazine from human lungs as similar to the pharmacokinetic profiles observed following intravenous administration [290, 291].

5.8.9 CONCLUSIONS

Pharmaceutical aerosol drug delivery has been established for over 50 years. Pulmonary administration remains the route of choice for local treatment of respiratory diseases. Over the past decade there have been changes in both the diseases treated by this route and the devices used for aerosol generation. Future advances will see pulmonary delivery of gene therapy and vaccines, together with improved drug targeting within the respiratory tract using novel inhalers.

ACKNOWLEDGMENTS

The author would like to thank Suparna Das Choudhuri and Deepika Arora for their assistance and discussions during the preparation of this chapter. In addition, he is grateful to Guillaume Brouet (Valois Pharm), Michael Spallek (Boehringer Ingelheim), and Joanne Peart (RDD) for their help in obtaining figures and tables used in this chapter. Finally, the review of soft mist inhalers has previously been published in the Drug Delivery Company Report (Autumn/Winter 2004), and the author acknowledges PharmaVentures and the Drug Delivery Company Report, which allowed reproduction of an abridged form of this paper.

The author received a research grant from Chrysalis Technologies, a division of Philip Morris USA, for the development of the CAG technology.

REFERENCES

1. Byron, P. R., and Patton, J. S. (1994), Drug delivery via the respiratory tract, *J. Aerosol Med.*, 7, 49–75.
2. Gonda, I. (2006), Systemic delivery of drugs to humans via inhalation, *J. Aerosol Med.*, 19, 47–53.
3. Douglas, R. B. (1987), The physiology of the lung, in Ganderton, D. and Jones, T. M., Eds., *Drug Delivery to the Respiratory Tract*, Ellis Horwood, Chichester, pp 13–26.
4. Weibel, E. R. (1963), *Morphometry of the Human Lung*, Academic, New York.
5. Gonda, I. (1990), Aerosols for delivery of therapeutic and diagnostic agents to the respiratory tract, *Crit. Rev. Ther. Drug Carrier Syst.*, 6, 273–313.
6. Stuart, B. O. (1973), Deposition of inhaled aerosols, *Arch. Intern. Med.*, 131, 60–73.
7. Palmes, E. D. (1973), Measurement of pulmonary air spaces using aerosols, *Arch. Intern. Med.*, 131, 76–79.
8. Finlay, W. H. (2001), Particle deposition in the respiratory tract, in Finlay, W. H., Ed., *The Mechanisms of Inhaled Pharmaceutical Aerosols*, Academic, London, pp 119–174.
9. Lippmann, M., and Albert, R. E. (1969), The effect of particle size on the regional deposition of inhaled aerosols in the human respiratory tract, *Am. Ind. Hyg. Assoc. J.*, 30, 257–275.
10. Foord, N., Black, A., and Walsh, M. (1975), Pulmonary deposition of inhaled particles with diameters in the range 2.5 to 7.5 micron, *Inhaled Part.*, 4(Pt. 1), 137–149.
11. Taulbee, D. B., and Yu, C. P. (1975), A theory of aerosol deposition in the human respiratory tract, *J. Appl. Physiol.*, 38, 77–85.
12. Yu, C. P., and Taulbee, D. B. (1975), A theory of predicting respiratory tract deposition of inhaled particles in man, *Inhaled Part.*, 4(Pt. 1), 35–47.
13. Pavia, D., and Thomson, M. L. (1976), The fractional deposition of inhaled 2 and 5 mum particles in the alveolar and tracheobronchial regions of the healthy human lung, *Ann. Occup. Hyg.*, 19, 109–114.
14. Bennett, W. D., Brown, J. S., Zeman, K. L., Hu, S. C., Scheuch, G., and Sommerer, K. (2002), Targeting delivery of aerosols to different lung regions, *J. Aerosol Med.*, 15, 179–188.
15. Dolovich, M. A. (2000), Influence of inspiratory flow rate, particle size, and airway caliber on aerosolized drug delivery to the lung, *Respir. Care*, 45, 597–608.
16. Bennett, W. D. (1988), Human variation in spontaneous breathing deposition fraction: A review, *J. Aerosol Med.*, 1, 67–80.
17. Hindle, M., Newton, D. A., and Chrystyn, H. (1993), Investigations of an optimal inhaler technique with the use of urinary salbutamol excretion as a measure of relative bioavailability to the lung, *Thorax*, 48, 607–610.
18. Clark, A. R., and Hollingworth, A. M. (1993), The relationship between powder inhaler resistance and peak inspiratory conditions in healthy volunteers-implications for in vitro testing, *J. Aerosol Med.*, 3, 99–110.
19. Broeders, M. E. A. C., Molema, J., Vermue, N. A., and Folgering, H. T. M. (2001), Peak inspiratory flow rate and slope of the inhalation profiles in dry powder inhalers, *Eur. Respir. J.*, 18, 780–783.
20. Phalen, R. F., and Oldham, M. J. (2001), Methods for modeling particle deposition as a function of age, *Respir. Physiol.*, 128, 119–130.
21. Segal, R. A., Martonen, T. B., Kim, C. S., and Shearer, M. (2002), Computer simulations of particle deposition in the lungs of chronic obstructive pulmonary disease patients, *Inhal. Toxicol.*, 14, 705–720.

22. Kim, C. S., and Hu, S. C. (1998), Regional deposition of inhaled particles in human lungs: Comparison between men and women, *J. Appl. Physiol.*, 84, 1834–1844.
23. Kim, C. S., and Kang, T. C. (1997), Comparative measurement of lung deposition of inhaled fine particles in normal subjects and patients with obstructive airway disease, *Am. J. Respir. Crit. Care Med.*, 155, 899–905.
24. Bennett, W. D., Zeman, K. L., and Kim, C. (1996), Variability of fine particle deposition in healthy adults: Effect of age and gender, *Am. J. Respir. Crit. Care Med.*, 153, 1641–1647.
25. Kim, C. S., Abraham, W. M., Garcia, L., and Sackner, M. A. (1989), Enhanced aerosol deposition in the lung with mild airways obstruction, *Am. Rev. Respir. Dis.*, 139, 422–426.
26. Kim, C. S., Lewars, G. A., and Sackner, M. A. (1988), Measurement of total lung aerosol deposition as an index of lung abnormality, *J. Appl. Physiol.*, 64, 1527–1536.
27. Palmes, E. D., Goldring, R. M., Wang, C., and Altshuler, B. (1970), Effect of chronic obstructive pulmonary disease on rate of deposition of aerosols in the lung during breath holding, *Inhaled Part.*, 1, 123–130.
28. Suarez, S., and Hickey, A. J. (2000), Drug properties affecting aerosol behavior, *Respir. Care*, 45, 652–666.
29. Groneberg, D. A., Witt, C., Wagner, U., Chung, K. F., and Fischer, A. (2003), Fundamentals of pulmonary drug delivery, *Respir. Med.*, 97, 382–387.
30. Labiris, N. R., and Dolovich, M. B. (2003), Pulmonary drug delivery. part I: Physiological factors affecting therapeutic effectiveness of aerosolized medications, *Br. J. Clin. Pharmacol.*, 56, 588–599.
31. Hindle, M., and Chrystyn, H. (1992), Determination of the relative bioavailability of salbutamol to the lung following inhalation, *Br. J. Clin. Pharmacol.*, 34, 311–315.
32. Niven, R. W. (2004), Modulated drug therapy with inhalation aerosols: Revisited, in Hickey, A. J., Ed., *Pharmaceutical Inhalation Aerosol Technology*, 2nd ed., Marcel Dekker, New York, pp 551–570.
33. Niven, R. W. (2002), Powders and processing: Deagglomerating a dose of patents and publications, in Dalby, R. N., Byron, P.R., Peart, J., and Farr, S. J., Eds., *Respiratory Drug Delivery VIII*, Davis Horwood International Publishing, Raleigh, NC, pp 257–266.
34. Cook, R. O., Pannu, R. K., and Kellaway, I. W. (2005), Novel sustained release microspheres for pulmonary drug delivery, *J. Controlled Release*, 104, 79–90.
35. Crooks, P. A., and Al-Ghananeem, A. M. (2004), Drug targeting to the lung: Chemical and biological considerations, in Hickey, A. J., Ed., *Pharmaceutical Inhalation Aerosol Technology*, 2nd ed., Marcel Dekker, New York, pp 89–154.
36. Sitar, D. S. (1996), Clinical pharmacokinetics of bambuterol, *Clin. Pharmacokinet.*, 31, 246–256.
37. Svensson, L. A. (1991), Mechanism of action of bambuterol: A beta-agonist prodrug with sustained lung affinity, *Agents Actions Suppl.*, 34, 71–78.
38. Taburet, A. M., and Schmit, B. (1994), Pharmacokinetic optimisation of asthma treatment, *Clin. Pharmacokinet.*, 26, 396–418.
39. Lipworth, B. J. (1996), Pharmacokinetics of inhaled drugs, *Br. J. Clin. Pharmacol.*, 42, 697–705.
40. Hochhaus, G., and Mollmann, H. (1992), Pharmacokinetic/pharmacodynamic characteristics of the beta-2-agonists terbutaline, salbutamol and fenoterol, *Int. J. Clin. Pharmacol. Ther. Toxicol.*, 30, 342–362.
41. Dahl, A. R., and Lewis, J. L. (1993), Respiratory tract uptake of inhalants and metabolism of xenobiotics, *Annu. Rev. Pharmacol. Toxicol.*, 33, 383–407.

42. Witek, T. J., Jr. (2000), The fate of inhaled drugs: The pharmacokinetics and pharmacodynamics of drugs administered by aerosol, *Respir. Care*, 45, 826–830.

43. Issar, M., Mobley, C., Khan, P., and Hochhaus, G. (2004), Pharmacokinetics and pharmacodynamics of drugs delivered to the lungs, in Hickey, A. J., Ed., *Pharmaceutical Inhalation Aerosol Technology*, 2nd ed., Marcel Dekker, New York, pp 215–252.

44. Hussain, A., Arnold, J. J., Khan, M. A., and Ahsan, F. (2004), Absorption enhancers in pulmonary protein delivery, *J. Controlled Release*, 94, 15–24.

45. Nelson, H. S., Chapman, K. R., Pyke, S. D., Johnson, M., and Pritchard, J. N. (2003), Enhanced synergy between fluticasone propionate and salmeterol inhaled from a single inhaler versus separate inhalers, *J. Allergy Clin. Immunol.*, 112, 29–36.

46. Dhillon, S., and Keating, G. M. (2006), Beclometasone dipropionate/formoterol: In an HFA-propelled pressurised metered-dose inhaler, *Drugs*, 66, 1475–1483.

47. Nowak, R., Emerman, C., Hanrahan, J. P., et al. (2006), A comparison of levalbuterol with racemic albuterol in the treatment of acute severe asthma exacerbations in adults, *Am. J. Emerg. Med.*, 24, 259–267.

48. Cass, L. M., Brown, J., Pickford, M., et al. (1999), Pharmacoscintigraphic evaluation of lung deposition of inhaled zanamivir in healthy volunteers, *Clin. Pharmacokinet.*, 36(Suppl. 1), 21–31.

49. Diot, P., Vecellio-None, L., Varaigne, F., Marchand, S., and Lemarie, E. (2003), Role of rhDNase in cystic fibrosis, *Rev. Mal. Respir.*, 20, S171–S175.

50. Anonymous (2006), First inhaled insulin product approved, *FDA Consum.*, 40, 28–29.

51. Clark, A. R. (2004), Pulmonary delivery technology: Recent advances and potential for the new millennium, in Hickey, A. J., Ed., *Pharmaceutical Inhalation Aerosol Technology*, 2nd ed., Marcel Dekker, New York, pp 571–591.

52. Roth, Y., Chapnik, J. S., and Cole, P. (2003), Feasibility of aerosol vaccination in humans, *Ann. Otol. Rhinol. Laryngol.*, 112, 264–270.

53. Sabin, A. B., Fernandez de Castro, J., Flores Arechiga, A., Sever, J. L., Madden, D. L., and Shekarchi, I. (1982), Clinical trials of inhaled aerosol of human diploid and chick embryo measles vaccine, *Lancet.*, 2, 604.

54. Sabin, A. B. (1983), Immunization against measles by aerosol, *Rev. Infect. Dis.*, 5, 514–523.

55. Sabin, A. B., Albrecht, P., Takeda, A. K., Ribeiro, E. M., and Veronesi, R. (1985), High effectiveness of aerosolized chick embryo fibroblast measles vaccine in seven-month-old and older infants, *J. Infect. Dis.*, 152, 1231–1237.

56. Sabin, A. B., Flores Arechiga, A., Fernandez de Castro, J., Albrecht, P., Sever, J. L., and Shekarchi, I. (1984), Successful immunization of infants with and without maternal antibody by aerosolized measles vaccine. II. Vaccine comparisons and evidence for multiple antibody response, *JAMA*, 251, 2363–2371.

57. Fernandez-de Castro, J., Kumate-Rodriguez, J., Sepulveda, J., Ramirez-Isunza, J. M., and Valdespino-Gomez, J. L. (1997), Measles vaccination by the aerosol method in Mexico, *Salud Publica Mex.*, 39, 53–60.

58. Bennett, J. V., Fernandez de Castro, J., Valdespino-Gomez, J. L., et al. (2002), Aerosolized measles and measles-rubella vaccines induce better measles antibody booster responses than injected vaccines: Randomized trials in Mexican schoolchildren, *Bull. World Health Organ.*, 80, 806–812.

59. LiCalsi, C., Maniaci, M. J., Christensen, T., Phillips, E., Ward, G. H., and Witham, C. (2001), A powder formulation of measles vaccine for aerosol delivery, *Vaccine*, 19, 2629–2636.

60. Laube, B. L. (2005), The expanding role of aerosols in systemic drug delivery, gene therapy, and vaccination, *Respir. Care*, 50, 1161–1176.

61. Blanchard, J. D. (2004), Pulmonary drug delivery as a first response to bioterrorism, in Dalby, R. N., Byron, P. R., Peart, J., Suman, J. D., and Farr, S. J., Eds., *Respiratory Drug Delivery IX*, Davis Healthcare International, River Grove, IL, pp 73–82.
62. Li, H. Y., Seville, P. C., Williamson, I. J., and Birchall, J. C. (2005), The use of amino acids to enhance the aerosolisation of spray-dried powders for pulmonary gene therapy, *J. Gene Med.*, 7, 343–353.
63. Joseph, P. M., O'Sullivan, B. P., Lapey, A., et al. (2001), Aerosol and lobar administration of a recombinant adenovirus to individuals with cystic fibrosis. I. Methods, safety, and clinical implications, *Hum. Gene Ther.*, 12, 1369–1382.
64. Perricone, M. A., Morris, J. E., Pavelka, K., et al. (2001), Aerosol and lobar administration of a recombinant adenovirus to individuals with cystic fibrosis. II. Transfection efficiency in airway epithelium, *Hum. Gene Ther.*, 12, 1383–1394.
65. Rochat, T., and Morris, M. A. (2002), Gene therapy for cystic fibrosis by means of aerosol, *J. Aerosol Med.*, 15, 229–235.
66. Garcia-Contreras, L., and Hickey, A. J. (2003), Aerosol treatment of cystic fibrosis, *Crit. Rev. Ther. Drug Carrier Syst.*, 20, 317–356.
67. Gautam, A., Waldrep, J. C., and Densmore, C. L. (2003), Aerosol gene therapy, *Mol. Biotechnol.*, 23, 51–60.
68. Anson, D. S., Smith, G. J., and Parsons, D. W. (2006), Gene therapy for cystic fibrosis airway disease—is clinical success imminent? *Curr. Gene Ther.*, 6, 161–179.
69. Griesenbach, U., Geddes, D. M., and Alton, E. W. (2006), Gene therapy progress and prospects: Cystic fibrosis, *Gene Ther.*, 13, 1061–1067.
70. Ziady, A. G., and Davis, P. B. (2006), Current prospects for gene therapy of cystic fibrosis, *Curr. Opin. Pharmacol*, 6, 515–521.
71. Cryan, S. A. (2005), Carrier-based strategies for targeting protein and peptide drugs to the lungs, *AAPS J.*, 7, E20–41.
72. Shoyele, S. A., and Slowey, A. (2006), Prospects of formulating proteins/peptides as aerosols for pulmonary drug delivery, *Int. J. Pharm.*, 314, 1–8.
73. Adjei, A., and Garren, J. (1990), Pulmonary delivery of peptide drugs: Effect of particle size on bioavailability of leuprolide acetate in healthy male volunteers, *Pharm. Res.*, 7, 565–569.
74. Adjei, A., Sundberg, D., Miller, J., and Chun, A. (1992), Bioavailability of leuprolide acetate following nasal and inhalation delivery to rats and healthy humans, *Pharm. Res.*, 9, 244–249.
75. Zheng, J. Y., Fulu, M. Y., Lee, D. Y., Barber, T. E., and Adjei, A. L. (2001), Pulmonary peptide delivery: Effect of taste-masking excipients on leuprolide suspension metered-dose inhalers, *Pharm. Dev. Technol.*, 6, 521–530.
76. Shahiwala, A., and Misra, A. (2005), A preliminary pharmacokinetic study of liposomal leuprolide dry powder inhaler: A technical note, *AAPS PharmSciTech.*, 6, E482–486.
77. Deftos, L. J., Nolan, J. J., Seely, B. L., et al. (1997), Intrapulmonary drug delivery of salmon calcitonin, *Calcif. Tissue Int.*, 61, 345–347.
78. Chan, H. K., Clark, A. R., Feeley, J. C., et al. (2004), Physical stability of salmon calcitonin spray-dried powders for inhalation, *J. Pharm. Sci.*, 93, 792–804.
79. Colthorpe, P., Farr, S. J., Smith, I. J., Wyatt, D., and Taylor, G. (1995), The influence of regional deposition on the pharmacokinetics of pulmonary-delivered human growth hormone in rabbits, *Pharm. Res.*, 12, 356–359.
80. Bosquillon, C., Preat, V., and Vanbever, R. (2004), Pulmonary delivery of growth hormone using dry powders and visualization of its local fate in rats, *J. Controlled Release*, 96, 233–244.

81. Ohmori, Y., Onoue, S., Endo, K., Matsumoto, A., Uchida, S., and Yamada, S. (2006), Development of dry powder inhalation system of novel vasoactive intestinal peptide (VIP) analogue for pulmonary administration, *Life Sci.*, 79, 138–143.
82. Ramsey, B. W., Pepe, M. S., Quan, J. M., et al. (1999), Intermittent administration of inhaled tobramycin in patients with cystic fibrosis. cystic fibrosis inhaled tobramycin study group, *N. Engl. J. Med.*, 340, 23–30.
83. Newhouse, M. T., Hirst, P. H., Duddu, S. P., et al. (2003), Inhalation of a dry powder tobramycin PulmoSphere formulation in healthy volunteers, *Chest*, 124, 360–366.
84. Iacono, A. T., Johnson, B. A., Grgurich, W. F., et al. (2006), A randomized trial of inhaled cyclosporine in lung-transplant recipients, *N. Engl. J. Med.*, 354, 141–150.
85. Staniforth, J. N. (2006), Nasal and pulmonary powder opportunities: New drugs and formulations for rapid systemic onset, in: Dalby, R. N., Byron, P. R., Peart, J., Suman, J. D., and Farr, S. J., Eds., *Respiratory Drug Delivery*, 2006, Davis Healthcare International, River Grove, IL, pp. 249–256.
86. Azarmi, S., Tao, X., Chen, H., et al. (2006), Formulation and cytotoxicity of doxorubicin nanoparticles carried by dry powder aerosol particles, *Int. J. Pharm.*, 319, 155–161.
87. Ward, M. E., Woodhouse, A., Mather, L. E., et al. (1997), Morphine pharmacokinetics after pulmonary administration from a novel aerosol delivery system, *Clin. Pharmacol. Ther.*, 62, 596–609.
88. Mather, L. E., Woodhouse, A., Ward, M. E., Farr, S. J., Rubsamen, R. A., and Eltherington, L. G. (1998), Pulmonary administration of aerosolised fentanyl: Pharmacokinetic analysis of systemic delivery, *Br. J. Clin. Pharmacol.*, 46, 37–43.
89. Otulana, B., Okikawa, J., Linn, L., Morishige, R., and Thipphawong, J. (2004), Safety and pharmacokinetics of inhaled morphine delivered using the AERx system in patients with moderate-to-severe asthma, *Int. J. Clin. Pharmacol. Ther.*, 42, 456–462.
90. Fulda, G. J., Giberson, F., and Fagraeus, L. (2005), A prospective randomized trial of nebulized morphine compared with patient-controlled analgesia morphine in the management of acute thoracic pain, *J. Trauma*, 59, 383–388; discussion 389–390.
91. Kohler, D. (1994), Aerosolized heparin, *J. Aerosol Med.*, 7, 307–311.
92. Bendstrup, K. E., Chambers, C. B., Jensen, J. I., and Newhouse, M. T. (1999), Lung deposition and clearance of inhaled (99m)Tc-heparin in healthy volunteers, *Am. J. Respir. Crit. Care Med.*, 160, 1653–1658.
93. Qi, Y., Zhao, G., Liu, D., et al. (2004), Delivery of therapeutic levels of heparin and low-molecular-weight heparin through a pulmonary route, *Proc. Natl. Acad. Sci. U S A*, 101, 9867–9872.
94. Olschewski, H., Rohde, B., Behr, J., et al. (2003), Pharmacodynamics and pharmacokinetics of inhaled iloprost, aerosolized by three different devices, in severe pulmonary hypertension, *Chest*, 124, 1294–1304.
95. Ganderton, D. (1997), General factors influencing drug delivery to the lung, *Respir. Med.*, 91(Suppl. A), 13–16.
96. Buck, H. (2001), The ideal inhaler for asthma therapy, *Med. Device Technol.*, 12, 24–27.
97. Virchow, J. C. (2005), What plays a role in the choice of inhaler device for asthma therapy? *Curr. Med. Res. Opin.*, 21(Suppl. 4), S19–S25.
98. Rau, J. L. (2005), Determinants of patient adherence to an aerosol regimen, *Respir. Care*, 50, 1346–1356; discussion 1357–1359.
99. Ganderton, D. (1999), Targeted delivery of inhaled drugs: Current challenges and future goals, *J. Aerosol Med.*, 12 (Suppl. 1), S3–S8.

100. Dolovich, M. B., Ahrens, R. C., Hess, D. R., et al. (2005), Device selection and outcomes of aerosol therapy: Evidence-based guidelines: American College of Chest Physicians/ American College of Asthma, Allergy, and Immunology, *Chest*, 127, 335–371.

101. O'Byrne, P. M. (1995), Clinical comparisons of inhaler systems: What are the important aspects? *J. Aerosol Med.*, 8(Suppl. 3), S39–46; discussion S47.

102. O'Callaghan, C., and Barry, P. W. (2000), How to choose delivery devices for asthma, *Arch. Dis. Child.*, 82, 185–187.

103. Newhouse, M. (1991), Advantages of pressurized canister metered dose inhalers, *J. Aerosol Med.*, 4, 139–150.

104. Crompton, G. K. (1990), The adult patient's difficulties with inhalers, *Lung*, 168(Suppl)., 658–662.

105. Harnor, K. J., Perkins, A. C., Wastie, M., et al. (1993), Effect of vapor pressure on the deposition pattern from solution phase metered dose inhalers, *Int. J. Pharm.*, 95, 111–116.

106. Leach, C. L., Davidson, P. J., Hasselquist, B. E., and Boudreau, R. J. (2002), Lung deposition of hydrofluoroalkane-134a beclomethasone is greater than that of chlorofluorocarbon fluticasone and chlorofluorocarbon beclomethasone: A cross-over study in healthy volunteers, *Chest*, 122, 510–516.

107. Leach, C. L., Davidson, P. J., Hasselquist, B. E., and Boudreau, R. J. (2005), Influence of particle size and patient dosing technique on lung deposition of HFA-beclomethasone from a metered dose inhaler, *J. Aerosol Med.*, 18, 379–385.

108. Leach, C. L. (1998), Improved delivery of inhaled steroids to the large and small airways, *Respir. Med.*, 92(Suppl. A), 3–8.

109. Williams, R. O., III, and Liu, J. (1998), Influence of formulation additives on the vapor pressure of hydrofluoroalkane propellants, *Int. J. Pharm.*, 166, 99–103.

110. Newman, S. P., Moren, F., Pavia, D., Corrado, O., and Clarke, S. W. (1982), The effects of changes in metered volume and propellant vapor pressure on the deposition of pressurized inhalation aerosols, *Int. J. Pharm.*, 11, 337–344.

111. Brambilla, G., Ganderton, D., Garzia, R., Lewis, D., Meakin, B., and Ventura, P. (1999), Modulation of aerosol clouds produced by pressurised inhalation aerosols, *Int. J. Pharm.*, 186, 53–61.

112. Vervaet, C., and Byron, P. R. (1999), Drug-surfactant-propellant interactions in HFA-formulations, *Int. J. Pharm.*, 186, 13–30.

113. Steckel, H., and Wehle, S. (2004), A novel formulation technique for metered dose inhaler (MDI) suspensions, *Int. J. Pharm.*, 284, 75–82.

114. Berry, J., Kline, L. C., Sherwood, J. K., et al. (2004), Influence of the size of micronized active pharmaceutical ingredient on the aerodynamic particle size and stability of a metered dose inhaler, *Drug Dev. Ind. Pharm.*, 30, 705–714.

115. Berry, J., Kline, L., Naini, V., Chaudhry, S., Hart, J., and Sequeira, J. (2004), Influence of the valve lubricant on the aerodynamic particle size of a metered dose inhaler, *Drug. Dev. Ind. Pharm.*, 30, 267–275.

116. Berry, J., Heimbecher, S., Hart, J. L., and Sequeira, J. (2003), Influence of the metering chamber volume and actuator design on the aerodynamic particle size of a metered dose inhaler, *Drug. Dev. Ind. Pharm.*, 29, 865–876.

117. Dunbar, C. A. (1998), Atomization mechanisms of the pressurized metered dose inhaler, *Part. Sci. Technol.*, 15, 253–271.

118. Gabrio, B. J., Stein, S. W., and Velasquez, D. J. (1999), A new method to evaluate plume characteristics of hydrofluoroalkane and chlorofluorocarbon metered dose inhalers, *Int. J. Pharm.*, 186, 3–12.

119. Lewis, D. A., Ganderton, D., Meakin, B. J., and Brambilla, G. (2004), Theory and practice with solution systems, in: Dalby, R. N., Byron, P. R., Peart, J., Suman, J.D., and Farr, S. J., eds., *Respiratory Drug Delivery IX*, Davis Healthcare International, River Grove, IL, pp. 109–115.
120. Polli, G. P., Grim, W. M., Bacher, F. A., and Yunker, M. H. (1969), Influence of formulation on aerosol particle size, *J. Pharm. Sci.*, 58, 484–486.
121. Stein, S. W., and Myrdal, P. B. (2004), A theoretical and experimental analysis of formulation and device parameters affecting solution MDI size distributions, *J. Pharm. Sci.*, 93, 2158–2175.
122. Purewal, T. S., and Grant, D. J. W., Eds. (1998), *Metered Dose Inhaler Technology*, CRC Press, Boca Raton.
123. Rogers, D. F., and Ganderton, D. (1995), Determining equivalence of inhaled medications, consensus statement from a workshop of the British association for lung research, held at Royal Brompton National Heart & Lung Institute, London 24 June 1994, *Respir. Med.*, 89, 253–261.
124. Meakin, B. J., Lewis, D. A., Ganderton, D., and Brambilla, G. (2000), Countering challenges posed by mimicry of CFC performance using HFA systems, in: Dalby, R. N., Byron, P. R., Peart, J., and Farr, S. J., Eds., *Respiratory Drug Delivery VII*, Serentec, Raleigh, NC, pp. 99–107.
125. Smyth, H. D., and Leach, C. L. (2005), Alternative propellant aerosol delivery systems, *Crit. Rev. Ther. Drug Carrier Syst.*, 22, 493–534.
126. Pritchard, J. N., and Genova, P. (2006), Adapting the pMDI to deliver novel drugs: Insulin and beyond, in: Dalby, R. N., Byron, P. R., Peart, J., Suman, J. D., and Farr, S. J., Eds., *Respiratory Drug Delivery*, 2006, Davis Healthcare International, River Grove, IL, pp. 133–141.
127. Jannick, P. (2006), CFC phase-out scenarios of pressurized metered dose inhalers: Current status, in: Dalby, R. N., Byron, P. R., Peart, J., Suman, J. D., and Farr, S. J., Eds., *Respiratory Drug Delivery*, 2006, Davis Healthcare International, River Grove, IL, pp. 789–792.
128. Anonymous (2006), A levalbuterol metered-dose inhaler (xopenex HFA) for asthma, *Med. Lett. Drugs Ther.*, 48, 21–22, 24.
129. Byron, P. R. (1994), Dosing reproducibility from experimental albuterol suspension metered-dose inhalers, *Pharm. Res.*, 11, 580–584.
130. Tansey, I. (1997), The technical transition to CFC-free inhalers, *Br. J. Clin. Pract. Suppl.*, 89, 22–27.
131. Stefely, J. D., Duan, D. C., Myrdal, P. B., Ross, D. L., Schultz, D. W., and Leach, C. L. (2000), Design and utility of a novel class of biocompatible excipients for HFA-based MDIs, in: Dalby, R. N., Byron, P. R., Peart, J., and Farr, S. J., Eds., *Respiratory Drug Delivery VII*, Serentec Press, Raleigh, NC, pp. 83–90.
132. Stein, S. W., and Myrdal, P. B. (2006), The relative influence of atomization and evaporation on metered dose inhaler drug delivery efficiency, *Aerosol Sci. Tech.*, 40, 335–347.
133. Leach, C. L., Davidson, P. J., and Boudreau, R. J. (1998), Improved airway targeting with the CFC-free HFA-beclomethasone metered-dose inhaler compared with CFC-beclomethasone, *Eur. Respir. J.*, 12, 1346–1353.
134. Seale, J. P., and Harrison, L. I. (1998), Effect of changing the fine particle mass of inhaled beclomethasone dipropionate on intrapulmonary deposition and pharmacokinetics, *Respir. Med.*, 92(Suppl. A), 9–15.
135. Berry, J., Kline, L. C., Hart, J. L., and Sequeira, J. (2003), Influence of the storage orientation on the aerodynamic particle size of a suspension metered dose inhaler containing propellant HFA-227, *Drug Dev. Ind. Pharm.*, 29, 631–639.

136. Anonymous (2005), Tetrafluoroethane, *Pharm. Forum.*, 31, 1672–1679.

137. Ashurst, I. C., Schultz, R. D., and Shurkus, D. D. (1996), Optimization and scale-up of a new pressure filling manufacturing process for albuterol/HFA 134a inhaler, in: Byron, P. R., Dalby, R. N., and Farr, S. J., Eds., *Respiratory Drug Delivery V*, Interpharm, Buffalo Grove, IL, pp. 294–296.

138. Wilkinson, A. (1998), The manufacture of metered dose inhalers, in: Purewal, T. S., and Grant, D. J. W., Eds., *Metered Dose Inhaler Technology*, CRC Press, Boca Raton, F.L., pp. 69–116.

139. Gupta, A., Stein, S. W., and Myrdal, P. B. (2003), Balancing ethanol cosolvent concentration with product performance in 134a-based pressurized metered dose inhalers, *J. Aerosol Med.*, 16, 167–174.

140. Tiwari, D., Goldman, D., Dixit, S., Malick, W. A., and Madan, P. L. (1998), Compatibility evaluation of metered-dose inhaler valve elastomers with tetrafluoroethane (P134a), a non-CFC propellant, *Drug Dev. Ind. Pharm.*, 24, 345–352.

141. Brouet, G., Robins, E., Hall, S., Butterworth, G., Hemy, J., and Turner, R. (2006), Developing new container closure options: A suppliers perpective, in: Dalby, R. N., Byron, P. R., Peart, J., Suman, J. D., and Farr, S. J., Eds., *Respiratory Drug Delivery 2006*, Davis Healthcare International, River Grove, IL, pp. 111–120.

142. Leone, P. (2006), Review of elastomer materials used in metering valves for pharmaceutical applications, paper presented at the International Rubber Conference IRC, Lyon, pp. 1–14.

143. Burel, S., Brouet, G., and Grandsire, T. (2004), Moisture uptake in pMDIs: The effect on valve component materials, paper presented at Drug Delivery to the Lungs XV, London, pp. 297–300.

144. Williams, R. O. III, and Hu, C. (2000), Moisture uptake and its influence on pressurized metered-dose inhalers, *Pharm. Dev. Technol.*, 5, 153–162.

145. Williams, R. O. III, and Hu, C. (2001), Influence of water on the solubility of two steroid drugs in hydrofluoroalkane (HFA) propellants, *Drug Dev. Ind. Pharm.*, 27, 71–79.

146. Ross, R. N. (1997), Loss of bronchodilator medication in priming a conventional metered dose inhaler: A cost of treating asthma, *Med. Interface*, 10, 141–146.

147. Smyth, H., Brace, G., Barbour, T., Gallion, J., Grove, J., and Hickey, A. J. (2006), Spray pattern analysis for metered dose inhalers: Effect of actuator design, *Pharm. Res.*, 23, 1591–1596.

148. Smyth, H. D., Brace, G., Barbour, T., Gallion, J., Grove, J., and Hickey, A. (2006), Actuator design variables for particle size modulation, in: Dalby, R. N., Byron, P. R., Peart, J., Suman, J. D., and Farr, S. J., Eds., *Respiratory Drug Delivery 2006*, Davis Healthcare International, River Grove, IL, pp. 857–860.

149. Peart, J., Orban, J. C., McGlynn, P., Redmon, M., Sargeant, C. M., and Byron, P. R. (2002), MDI electrostatics: Valve and formulation interactions which really make a difference, in: Dalby, R. N., Byron, P. R., Peart, J., and Farr, S. J., Eds., *Respiratory Drug Delivery XIII*, Davis Horwood International, Raleigh, NC, pp. 223–230.

150. Kwok, P. C., Glover, W., and Chan, H. K. (2005), Electrostatic charge characteristics of aerosols produced from metered dose inhalers, *J. Pharm. Sci.*, 94, 2789–2799.

151. Newman, S. P., Weisz, A. W., Talaee, N., and Clarke, S. W. (1991), Improvement of drug delivery with a breath actuated pressurised aerosol for patients with poor inhaler technique, *Thorax*, 46, 712–716.

152. Woodman, K., Bremner, P., Burgess, C., Crane, J., Pearce, N., and Beasley, R. (1993), A comparative study of the efficacy of beclomethasone dipropionate delivered from a breath activated and conventional metered dose inhaler in asthmatic patients, *Curr. Med. Res. Opin.*, 13, 61–69.

153. Terzano, C. (2001), Pressurized metered dose inhalers and add-on devices, *Pulm. Pharmacol. Ther.*, 14, 351–366.
154. Asmus, M. J., Liang, J., Coowanitwong, I., and Hochhaus, G. (2004), In vitro performance characteristics of valved holding chamber and spacer devices with a fluticasone metered-dose inhaler, *Pharmacotherapy*, 24, 159–166.
155. Goldberg, S., Einot, T., Algur, N., et al. (2002), Adrenal suppression in asthmatic children receiving low-dose inhaled budesonide: Comparison between dry powder inhaler and pressurized metered-dose inhaler attached to a spacer, *Ann. Allergy Asthma Immunol.*, 89, 566–571.
156. Dalby, R. N., Somaraju, S., Chavan, V. S., and Jarvis, D. (1998), Evaluation of aerosol drug output from the OptiChamber and AeroChamber spacers in a model system, *J. Asthma.*, 35, 173–177.
157. Newman, S. P., Millar, A. B., Lennard-Jones, T. R., Moren, F., and Clarke, S. W. (1984), Improvement of pressurised aerosol deposition with nebuhaler spacer device, *Thorax*, 39, 935–941.
158. Hindle, M., and Chrystyn, H. (1994), Relative bioavailability of salbutamol to the lung following inhalation using metered dose inhalation methods and spacer devices, *Thorax*, 49, 549–553.
159. Wildhaber, J. H., Waterer, G. W., Hall, G. L., and Summers, Q. A. (2000), Reducing electrostatic charge on spacer devices and bronchodilator response, *Br. J. Clin. Pharmacol.*, 50, 277–280.
160. Pierart, F., Wildhaber, J. H., Vrancken, I., Devadason, S. G., and Le Souef, P. N. (1999), Washing plastic spacers in household detergent reduces electrostatic charge and greatly improves delivery, *Eur. Respir. J.*, 13, 673–678.
161. Coppolo, D. P., Mitchell, J. P., and Nagel, M. W. (2006), Levalbuterol aerosol delivery with a nonelectrostatic versus a nonconducting valved holding chamber, *Respir. Care*, 51, 511–514.
162. Smyth, H. D., Beck, V. P., Williams, D., and Hickey, A. J. (2004), The influence of formulation and spacer device on the in vitro performance of solution chlorofluorocarbon-free propellant-driven metered dose inhalers, *AAPS PharmSciTech*, 5, E7.
163. Rahmatalla, M. F., Zuberbuhler, P. C., Lange, C. F., and Finlay, W. H. (2002), In vitro effect of a holding chamber on the mouth-throat deposition of QVAR (hydrofluoroalkane-beclomethasone dipropionate), *J. Aerosol Med.*, 15, 379–385.
164. Sander, N., Fusco-Walkert, S. J., Harder, J. M., and Chipps, B. E. (2006), Dose counting and the use of pressurized metered-dose inhalers: Running on empty, *Ann. Allergy Asthma Immunol.*, 97, 34–38.
165. Bradshaw, D. R. S. (2006), Developing dose counters: An appraisal based on regulator, pharma, and user needs, in: Dalby, R. N., Byron, P. R., Peart, J., Suman, J. D., and Farr, S. J., Eds., *Respiratory Drug Delivery 2006*, Davis Healthcare International, River Grove, IL, pp. 121–131.
166. Vidgren, M., Paronen, P., Vidgren, P., Vainio, P., and Nuutinen, J. (1990), In vivo evaluation of the new multiple dose powder inhaler and the rotahaler using the gamma scintigraphy, *Acta Pharm. Nord.*, 2, 3–10.
167. Smith, I. J., and Parry-Billings, M. (2003/2004), The inhalers of the future? A review of dry powder devices on the market today, *Pulm. Pharmacol. Ther.*, 16, 79–95.
168. Telko, M. J., and Hickey, A. J. (2005), Dry powder inhaler formulation, *Respir. Care*, 50, 1209–1227.
169. Clarke, M. J., Peart, J., Cagnani, S., and Byron, P. R. (2002), Adhesion of powders for inhalation: An evaluation of drug detachment from surfaces following deposition from aerosol streams, *Pharm. Res.*, 19, 322–329.

170. Ward, G. H., and Schultz, R. K. (1995), Process-induced crystallinity changes in albuterol sulfate and its effect on powder physical stability, *Pharm. Res.*, 12, 773–779.

171. Giry, K., Pean, J. M., Giraud, L., Marsas, S., Rolland, H., and Wuthrich, P. (2006), Drug/lactose co-micronization by jet milling to improve aerosolization properties of a powder for inhalation, *Int. J. Pharm*, 321, 162–166.

172. Rasenack, N., and Muller, B. W. (2004), Micron-size drug particles: Common and novel micronization techniques, *Pharm. Dev. Technol.*, 9, 1–13.

173. Steckel, H., Rasenack, N., Villax, P., and Muller, B. W. (2003), In vitro characterization of jet-milled and in-situ-micronized fluticasone-17-propionate, *Int. J. Pharm.*, 258, 65–75.

174. Edwards, D. A., Ben-Jebria, A., and Langer, R. (1998), Recent advances in pulmonary drug delivery using large, porous inhaled particles, *J. Appl. Physiol.*, 85, 379–385.

175. White, S., Bennett, D. B., and Cheu, S., et al. (2005), EXUBERA: Pharmaceutical development of a novel product for pulmonary delivery of insulin, *Diabetes Technol. Ther.*, 7, 896–906.

176. Vanbever, R., Mintzes, J. D., Wang, J., et al. (1999), Formulation and physical characterization of large porous particles for inhalation, *Pharm. Res.*, 16, 1735–1742.

177. Dunbar, C., Scheuch, G., Sommerer, K., DeLong, M., Verma, A., and Batycky, R. (2002), In vitro and in vivo dose delivery characteristics of large porous particles for inhalation, *Int. J. Pharm.*, 245, 179–189.

178. Bartus, R. T., Emerich, D., Snodgrass-Belt, P., et al. (2004), A pulmonary formulation of L-dopa enhances its effectiveness in a rat model of parkinson's disease, *J. Pharmacol. Exp. Ther.*, 310, 828–835.

179. Dellamary, L. A., Tarara, T. E., Smith, D. J., et al. (2000), Hollow porous particles in metered dose inhalers, *Pharm. Res.*, 17, 168–174.

180. Duddu, S. P., Sisk, S. A., Walter, Y. H., et al. (2002), Improved lung delivery from a passive dry powder inhaler using an engineered PulmoSphere powder, *Pharm. Res.*, 19, 689–695.

181. Pfutzner, A., Mann, A. E., and Steiner, S. S. (2002), Technosphere/Insulin—A new approach for effective delivery of human insulin via the pulmonary route, *Diabetes Technol. Ther.*, 4, 589–594.

182. Pfutzner, A., Flacke, F., Pohl, R., et al. (2003), Pilot study with technosphere/PTH (1-34)—A new approach for effective pulmonary delivery of parathyroid hormone (1-34), *Horm. Metab. Res.*, 35, 319–323.

183. Young, P. M., and Price, R. (2004), The influence of humidity on the aerosolisation of micronised and SEDS produced salbutamol sulphate, *Eur. J. Pharm. Sci.*, 22, 235–240.

184. Rehman, M., Shekunov, B. Y., York, P., et al. Optimisation of powders for pulmonary delivery using supercritical fluid technology, *Eur. J. Pharm. Sci.*, 22, 1–17.

185. Lobo, J. M., Schiavone, H., Palakodaty, S., York, P., Clark, A., and Tzannis, S. T. (2005), SCF-engineered powders for delivery of budesonide from passive DPI devices, *J. Pharm. Sci.*, 94, 2276–2288.

186. Schiavone, H., Palakodaty, S., Clark, A., York, P., and Tzannis, S. T. (2004), Evaluation of SCF-engineered particle-based lactose blends in passive dry powder inhalers, *Int. J. Pharm.*, 281, 55–66.

187. Pandey, R., Sharma, A., Zahoor, A., Sharma, S., Khuller, G. K., and Prasad, B. (2003), Poly (DL-lactide-*co*-glycolide) nanoparticle-based inhalable sustained drug delivery system for experimental tuberculosis, *J. Antimicrob. Chemother.*, 52, 981–986.

188. Dickinson, P. A., Howells, S. W., and Kellaway, I. W. (2001), Novel nanoparticles for pulmonary drug administration, *J. Drug Target.*, 9, 295–302.

189. Sham, J. O., Zhang, Y., Finlay, W. H., Roa, W. H., and Lobenberg, R. (2004), Formulation and characterization of spray-dried powders containing nanoparticles for aerosol delivery to the lung, *Int. J. Pharm.*, 269, 457–467.
190. Begat, P., Morton, D. A., Staniforth, J. N., and Price, R. (2004), The cohesive-adhesive balances in dry powder inhaler formulations I: Direct quantification by atomic force microscopy, *Pharm. Res.*, 21, 1591–1597.
191. Davies, M., Brindley, A., Chen, X., et al. (2005), Characterization of drug particle surface energetics and Young's modulus by atomic force microscopy and inverse gas chromatography, *Pharm. Res.*, 22, 1158–1166.
192. Larhrib, H., Zeng, X. M., Martin, G. P., Marriott, C., and Pritchard, J. (1999), The use of different grades of lactose as a carrier for aerosolised salbutamol sulphate, *Int. J. Pharm.*, 191, 1–14.
193. Zeng, X. M., Pandhal, K. H., and Martin, G. P. (2000), The influence of lactose carrier on the content homogeneity and dispersibility of beclomethasone dipropionate from dry powder aerosols, *Int. J. Pharm.*, 197, 41–52.
194. Lucas, P., Anderson, K., and Staniforth, J. N. (1998), Protein deposition from dry powder inhalers: Fine particle multiplets as performance modifiers, *Pharm. Res.*, 15, 562–569.
195. Begat, P., Morton, D. A., Staniforth, J. N., and Price, R. (2004), The cohesive-adhesive balances in dry powder inhaler formulations II: Influence on fine particle delivery characteristics, *Pharm. Res.*, 21, 1826–1833.
196. Jones, M., and Price, R. (2006), The influence of fine excipient particles on the performance of carrier-based dry powder inhalation formulations, *Pharm. Res.*, 23, 1665–1674.
197. Ming Zeng, X., Martin, G. P., Marriott, C., and Pritchard, J. (2001), The use of lactose recrystallised from carbopol gels as a carrier for aerosolised salbutamol sulphate, *Eur. J. Pharm. Biopharm.*, 51, 55–62.
198. Flament, M. P., Leterme, P., and Gayot, A. (2004), The influence of carrier roughness on adhesion, content uniformity and the in vitro deposition of terbutaline sulphate from dry powder inhalers, *Int. J. Pharm.*, 275, 201–209.
199. Zeng, X. M., Martin, G. P., Marriott, C., and Pritchard, J. (2001), Lactose as a carrier in dry powder formulations: The influence of surface characteristics on drug delivery, *J. Pharm. Sci.*, 90, 1424–1434.
200. El-Sabawi, D., Price, R., Edge, S., and Young, P. M. (2006), Novel temperature controlled surface dissolution of excipient particles for carrier based dry powder inhaler formulations, *Drug Dev. Ind. Pharm.*, 32, 243–251.
201. Steckel, H., and Bolzen, N. (2004), Alternative sugars as potential carriers for dry powder inhalations, *Int. J. Pharm.*, 270, 297–306.
202. Lucas, P., Anderson, K., Potter, U. J., and Staniforth, J. N. (1999), Enhancement of small particle size dry powder aerosol formulations using an ultra low density additive, *Pharm. Res.*, 16, 1643–1647.
203. Hindle, M., Jashnani, R. N., and Byron, P. R. (1994), Dose emissions from marketed dry powder inhalers: Influence of flow, volume and environment, in: Byron, P. R., Dalby, R. N., and Farr, S. J., Eds., *Respiratory Drug Delivery IV*, Interpharm, Buffalo Grove, IL, pp. 137–142.
204. Borgstrom, L., Asking, L., and Lipniunas, P. (2005), An in vivo and in vitro comparison of two powder inhalers following storage at hot/humid conditions, *J. Aerosol Med.*, 18, 304–310.
205. Price, R., Young, P. M., Edge, S., and Staniforth, J. N. (2002), The influence of relative humidity on particulate interactions in carrier-based dry powder inhaler formulations, *Int. J. Pharm.*, 246, 47–59.

206. Young, P. M., Price, R., Tobyn, M. J., Buttrum, M., and Dey, F. (2003), Investigation into the effect of humidity on drug-drug interactions using the atomic force microscope, *J. Pharm. Sci.*, 92, 815–822.
207. Borgstrom, L., Asking, L., and Thorsson, L. (2005), Idealhalers or realhalers? A comparison of diskus and turbuhaler, *Int. J. Clin. Pract.*, 59, 1488–1495.
208. Fuller, R. (1995), The diskus: A new multi-dose powder device—Efficacy and comparison with turbuhaler, *J. Aerosol Med.*, 8(Suppl. 2), S11–17.
209. Bennett, F. S., Carter, P. A., Rowley, G., and Dandiker, Y. (1999), Modification of electrostatic charge on inhaled carrier lactose particles by addition of fine particles, *Drug Dev. Ind. Pharm.*, 25, 99–103.
210. Byron, P. R., Peart, J., and Staniforth, J. N. (1997), Aerosol electrostatics. I: Properties of fine powders before and after aerosolization by dry powder inhalers, *Pharm. Res.*, 14, 698–705.
211. Carter, P. A., Rowley, G., Fletcher, E. J., and Stylianopoulos, V. (1998), Measurement of electrostatic charge decay in pharmaceutical powders and polymer materials used in dry powder inhaler devices, *Drug Dev. Ind. Pharm.*, 24, 1083–1088.
212. Murtomaa, M., Mellin, V., Harjunen, P., Lankinen, T., Laine, E., and Lehto, V. P. (2004), Effect of particle morphology on the triboelectrification in dry powder inhalers, *Int. J. Pharm.*, 282, 107–114.
213. Taylor, A., and Gustafsson, P. (2005), Do all dry powder inhalers show the same pharmaceutical performance? *Int. J. Clin. Pract. Suppl.*, 59, 7–12.
214. Cegla, U. H. (2004), Pressure and inspiratory flow characteristics of dry powder inhalers, *Respir. Med.*, 98(Suppl. A), S22–28.
215. Hindle, M., and Byron, P. R. (1995), Dose emissions from marketed dry powder inhalers, *Int. J. Pharm.*, 116, 169–177.
216. Hindle, M., and Byron, P. R. (1996), Impaction and impingement techniques for powder inhalers—Comparisons, problems and validation, in: Byron, P. R., Dalby, R. N., and Farr, S. J., Eds., *Respiratory Drug Delivery V*, Interpharm, Buffalo Grove, IL, pp. 263–272.
217. Pitcairn, G. R., Lim, J., Hollingworth, A., and Newman, S. P. (1997), Scintigraphic assessment of drug delivery from the ultrahaler dry powder inhaler, *J. Aerosol Med.*, 10, 295–306.
218. Pedersen, S., Hansen, O. R., and Fuglsang, G. (1990), Influence of inspiratory flow rate upon the effect of a turbuhaler, *Arch. Dis. Child.*, 65, 308–310.
219. Tukiainen, H., and Terho, E. O. (1985), Comparison of inhaled salbutamol powder and aerosol in asthmatic patients with low peak expiratory flow level, *Eur. J. Clin. Pharmacol.*, 27, 645–647.
220. Borgstrom, L. (1994), Deposition patterns with turbuhaler, *J. Aerosol Med.*, 7, S49–53.
221. Finlay, W. H., and Gehmlich, M. G. (2000), Inertial sizing of aerosol inhaled from two dry powder inhalers with realistic breath patterns versus constant flow rates, *Int. J. Pharm.*, 210, 83–95.
222. Borgstrom, L., Bengtsson, T., Derom, E., and Pauwels, R. (2000), Variability in lung deposition of inhaled drug, within and between asthmatic patients, with a pMDI and a dry powder inhaler, turbuhaler, *Int. J. Pharm.*, 193, 227–230.
223. Nielsen, K. G., Skov, M., Klug, B., Ifversen, M., and Bisgaard, H. (1997), Flow-dependent effect of formoterol dry-powder inhaled from the aerolizer, *Eur. Respir. J.*, 10, 2105–2109.
224. Koskela, T., Malmstrom, K., Sairanen, U., Peltola, S., Keski-Karhu, J., and Silvasti, M. (2000), Efficacy of salbutamol via easyhaler unaffected by low inspiratory flow, *Respir. Med.*, 94, 1229–1233.

225. Newhouse, M. T., Nantel, N. P., Chambers, C. B., Pratt, B., and Parry-Billings, M. (1999), Clickhaler (a novel dry powder inhaler) provides similar bronchodilation to pressurized metered-dose inhaler, even at low flow rates, *Chest*, 115, 952–956.

226. Meakin, B. J., Ganderton, D., Panza, I., and Ventura, P. (1998), The effect of flow rate on drug delivery from the pulvinal, a high-resistance dry powder inhaler, *J. Aerosol Med.*, 11, 143–152.

227. Newman, S. P., Moren, F., Trofast, E., Talaee, N., and Clarke, S. W. (1989), Deposition and clinical efficacy of terbutaline sulphate from turbuhaler, a new multi-dose powder inhaler, *Eur. Respir. J.*, 2, 247–252.

228. Kohler, D. (2004), The novolizer: Overcoming inherent problems of dry powder inhalers, *Respir. Med.*, 98(Suppl. A), S17–21.

229. O'Connor, B. J. (2004), The ideal inhaler: Design and characteristics to improve outcomes, *Respir. Med.*, 98(Suppl. A), S10–16.

230. Lim, J. G., Shah, B., Rohatagi, S., and Bell, A. (2006), Development of a dry powder inhaler, the ultrahaler, containing triamcinolone acetonide using in vitro–in vivo relationships, *Am. J. Ther.*, 13, 32–42.

231. McCormack, P. L., and Plosker, G. L. (2006), Inhaled mometasone furoate: A review of its use in persistent asthma in adults and adolescents, *Drugs*, 66, 1151–1168.

232. Pitcairn, G. R., Lankinen, T., Seppala, O. P., and Newman, S. P. (2000), Pulmonary drug delivery from the taifun dry powder inhaler is relatively independent of the patient's inspiratory effort, *J. Aerosol Med.*, 13, 97–104.

233. Keating, G. M., and Faulds, D. (2002), Airmax: A multi-dose dry powder inhaler, *Drugs*, 62, 1887–1895; discussion 1896–1897.

234. Iida, K., Leuenberger, H., Fueg, L. M., Muller-Walz, R., Okamoto, H., and Danjo, K. (2000), Effect of mixing of fine carrier particles on dry powder inhalation property of salbutamol sulfate (SS), *Yakugaku Zasshi*, 120, 113–119.

235. Patton, J. S., Fishburn, C. S., and Weers, J. G. (2004), The lungs as a portal of entry for systemic drug delivery, *Proc. Am. Thorac. Soc.*, 1, 338–344.

236. Patton, J. S. (2005), Unlocking the opportunity of tight glycaemic control: Innovative delivery of insulin via the lung, *Diabetes Obes. Metab.*, 7(Suppl. 1), S5–8.

237. Rosenstock, J., Cappelleri, J. C., Bolinder, B., and Gerber, R. A. (2004), Patient satisfaction and glycemic control after 1 year with inhaled insulin (exubera) in patients with type 1 or type 2 diabetes, *Diabetes Care*, 27, 1318–1323.

238. Bellary, S., and Barnett, A. H. (2006), Inhaled insulin: New technology, new possibilities, *Int. J. Clin. Pract.*, 60, 728–734.

239. Dunn, C., and Curran, M. P. (2006), Inhaled human insulin (exubera): A review of its use in adult patients with diabetes mellitus, *Drugs*, 66, 1013–1032.

240. Patton, J. S., Bukar, J. G., and Eldon, M. A. (2004), Clinical pharmacokinetics and pharmacodynamics of inhaled insulin, *Clin. Pharmacokinet.*, 43, 781–801.

241. Quattrin, T., Belanger, A., Bohannon, N. J., Schwartz, S. L., and Exubera Phase III Study Group (2004), Efficacy and safety of inhaled insulin (exubera) compared with subcutaneous insulin therapy in patients with type 1 diabetes: Results of a 6-month, randomized, comparative trial, *Diabetes Care*, 27, 2622–2627.

242. Barnett, A. H., Dreyer, M., Lange, P., and Serdarevic-Pehar, M. (2006), An open, randomized, parallel-group study to compare the efficacy and safety profile of inhaled human insulin (exubera) with glibenclamide as adjunctive therapy in patients with type 2 diabetes poorly controlled on metformin, *Diabetes Care*, 29, 1818–1825.

243. Rosenstock, J., Zinman, B., Murphy, L. J., et al. (2005), Inhaled insulin improves glycemic control when substituted for or added to oral combination therapy in type 2 diabetes: A randomized, controlled trial, *Ann. Intern. Med.*, 143, 549–558.

244. DeFronzo, R. A., Bergenstal, R. M., Cefalu, W. T., et al. (2005), Efficacy of inhaled insulin in patients with type 2 diabetes not controlled with diet and exercise: A 12-week, randomized, comparative trial, *Diabetes Care*, 28, 1922–1928.

245. Rave, K., Bott, S., Heinemann, L., et al. (2005), Time-action profile of inhaled insulin in comparison with subcutaneously injected insulin lispro and regular human insulin, *Diabetes Care*, 28, 1077–1082.

246. Davidson, M. B., Mehta, A. E., and Siraj, E. S. (2006), Inhaled human insulin: An inspiration for patients with diabetes mellitus? *Cleve. Clin. J. Med.*, 73, 569–578.

247. An, B., and Reinhardt, R. R. (2003), Effects of different durations of breath holding after inhalation of insulin using the AERx insulin diabetes management system, *Clin. Ther.*, 25, 2233–2244.

248. Boyd, B., Noymer, P., Liu, K., et al. (2004), Effect of gender and device mouthpiece shape on bolus insulin aerosol delivery using the AERx pulmonary delivery system, *Pharm. Res.*, 21, 1776–1782.

249. Cramer, J. A., Okikawa, J., Bellaire, S., and Clauson, P. (2004), Compliance with inhaled insulin treatment using the AERx iDMS insulin diabetes management system, *Diabetes Technol. Ther.*, 6, 800–807.

250. Farr, S. J., McElduff, A., Mather, L. E., et al. (2000), Pulmonary insulin administration using the AERx system: Physiological and physicochemical factors influencing insulin effectiveness in healthy fasting subjects, *Diabetes Technol. Ther.*, 2, 185–197.

251. Thipphawong, J., Otulana, B., Clauson, P., Okikawa, J., and Farr, S. J. (2002), Pulmonary insulin administration using the AERx insulin diabetes system, *Diabetes Technol. Ther.*, 4, 499–504.

252. Ferron, G. A., Kerrebijn, K. F., and Weber, J. (1976), Properties of aerosols produced with three nebulizers, *Am. Rev. Respir. Dis.*, 114, 899–908.

253. Nikander, K., Turpeinen, M., and Wollmer, P. (1999), The conventional ultrasonic nebulizer proved inefficient in nebulizing a suspension, *J. Aerosol Med.*, 12, 47–53.

254. Rau, J. L. (2002), Design principles of liquid nebulization devices currently in use, *Respir. Care*, 47, 1257–1275; discussion 1275–1278.

255. Nerbrink, O., and Dahlback, M. (1994), Basic nebulizer function, *J. Aerosol Med.*, 7, S7–11.

256. Nerbrink, O., Dahlback, M., and Hansson, H. C. (1994), Why do medical nebulizers differ in their output and particle size characteristics? *J. Aerosol Med.*, 7, 259–276.

257. Newnham, D. M., and Lipworth, B. J. (1994), Nebuliser performance, pharmacokinetics, airways and systemic effects of salbutamol given via a novel nebuliser delivery system ("ventstream"), *Thorax*, 49, 762–770.

258. Leung, K., Louca, E., and Coates, A. L. (2004), Comparison of breath-enhanced to breath-actuated nebulizers for rate, consistency, and efficiency, *Chest*, 126, 1619–1627.

259. Barry, P. W., and O'Callaghan, C. (1999), An in vitro analysis of the output of salbutamol from different nebulizers, *Eur. Respir. J.*, 13, 1164–1169.

260. O'Callaghan, C., White, J., Jackson, J., Barry, P. W., and Kantar, A. (2005), Delivery of nebulized budesonide is affected by nebulizer type and breathing pattern, *J. Pharm. Pharmacol.*, 57, 787–790.

261. Rau, J. L., Ari, A., and Restrepo, R. D. (2004), Performance comparison of nebulizer designs: Constant-output, breath-enhanced, and dosimetric, *Respir. Care*, 49, 174–179.

262. Raabe, O. G., Wong, T. M., Wong, G. B., Roxburgh, J. W., Piper, S. D., and Lee, J. I. (1998), Continuous nebulization therapy for asthma with aerosols of beta2 agonists, *Ann. Allergy Asthma Immunol.*, 80, 499–508.

263. Dolovich, M. B. (2002), Assessing nebulizer performance, *Respir. Care*, 47, 1290–1301; discussion 1301–1304.

264. Le Brun, P. P., de Boer, A. H., Heijerman, H. G., and Frijlink, H. W. (2000), A review of the technical aspects of drug nebulization, *Pharm. World Sci.*, 22, 75–81.

265. Byrne, N. M., Keavey, P. M., Perry, J. D., Gould, F. K., and Spencer, D. A. (2003), Comparison of lung deposition of colomycin using the HaloLite and the pari LC plus nebulisers in patients with cystic fibrosis, *Arch. Dis. Child.*, 88, 715–718.

266. Denyer, J., Nikander, K., and Smith N. J. (2004), Adaptive aerosol delivery (AAD) technology, *Expert Opin. Drug Deliv.*, 1, 165–176.

267. Nikander, K., Arheden, L., Denyer, J., and Cobos, N. (2003), Parents' adherence with nebulizer treatment of their children when using an adaptive aerosol delivery (AAD) system, *J. Aerosol Med.*, 16, 273–281.

268. Knoch, M., and Keller, M. (2005), The customised electronic nebuliser: A new category of liquid aerosol drug delivery systems, *Expert Opin. Drug Deliv.*, 2, 377–390.

269. Kim, D., Mudaliar, S., Chinnapongse, S., et al. (2003), Dose-response relationships of inhaled insulin delivered via the aerodose insulin inhaler and subcutaneously injected insulin in patients with type 2 diabetes, *Diabetes Care*, 26, 2842–2847.

270. Lipworth, B. J., Sims, E. J., Taylor, K., Cockburn, W., and Fishman, R. (2005), Dose-response to salbutamol via a novel palm sized nebuliser (aerodose inhaler), conventional nebuliser (pari LC plus) and metered dose inhaler (ventolin evohaler) in moderate to severe asthmatics, *Br. J. Clin. Pharmacol.*, 59, 5–13.

271. Hindle, M. (2004), Soft mist inhalers: A review of current technology, *The Drug Delivery Companies Report*, 13, 31–34.

272. Geller, D. E. (2002), New liquid aerosol generation devices: Systems that force pressurized liquids through nozzles, *Respir. Care*, 47, 1392–1404; discussion 1404–1405.

273. Patel, K. R., Pavia, D., Lowe, L., and Spiteri, M. (2006), Inhaled ethanolic and aqueous solutions via respimat soft mist inhaler are well-tolerated in asthma patients, *Respiration*, 73, 434–440.

274. Voshaar, T., Hausen, T., Kardos, P., et al. (2005), Inhalation therapy with respimat soft inhaler in patients with COPD and asthma, *Pneumologie*, 59, 25–32.

275. Kassner, F., Hodder, R., and Bateman, E. D. (2004), A review of ipratropium bromide/fenoterol hydrobromide (berodual) delivered via respimat soft mist inhaler in patients with asthma and chronic obstructive pulmonary disease, *Drugs*, 64, 1671–1682.

276. Pitcairn, G., Reader, S., Pavia, D., and Newman, S. (2005), Deposition of corticosteroid aerosol in the human lung by respimat soft mist inhaler compared to deposition by metered dose inhaler or by turbuhaler dry powder inhaler, *J. Aerosol Med.*, 18, 264–272.

277. Vincken, W., Dewberry, H., and Moonen, D. (2003), Fenoterol delivery by respimat soft mist inhaler versus CFC metered dose inhaler: Cumulative dose-response study in asthma patients, *J. Asthma*, 40, 721–730.

278. Newman, S. P., Brown, J., Steed, K. P., Reader, S. J., and Kladders, H. (1998), Lung deposition of fenoterol and flunisolide delivered using a novel device for inhaled medicines: Comparison of RESPIMAT with conventional metered-dose inhalers with and without spacer devices, *Chest*, 113, 957–963.

279. Newman, S. P., Steed, K. P., Reader, S. J., Hooper, G., and Zierenberg, B. (1996), Efficient delivery to the lungs of flunisolide aerosol from a new portable hand-held multidose nebulizer, *J. Pharm. Sci.*, 85, 960–964.

280. Schurmann, W., Schmidtmann, S., Moroni, P., Massey, D., and Qidan, M. (2005), Respimat soft mist inhaler versus hydrofluoroalkane metered dose inhaler: Patient preference and satisfaction, *Treat. Respir. Med.*, 4, 53–61.

281. Pavia, D., and Moonen, D. (1999), Preliminary data from phase II studies with respimat, a propellant-free soft mist inhaler, *J. Aerosol Med.*, 12(Suppl. 1), S33–39.

282. Dalby, R., Spallek, M., and Voshaar, T. (2004), A review of the development of respimat soft mist inhaler, *Int. J. Pharm.*, 283, 1–9.

283. Hochrainer, D., Holz, H., Kreher, C., Scaffidi, L., Spallek, M., and Wachtel, H. (2005), Comparison of the aerosol velocity and spray duration of respimat soft mist inhaler and pressurized metered dose inhalers, *J. Aerosol Med.*, 18, 273–282.

284. Zierenberg, B. (1999), Optimizing the in vitro performance of respimat, *J. Aerosol Med.*, 12(Suppl. 1), S19–24.

285. Farr, S. J., Warren, S. J., Lloyd, P., et al. (2000), Comparison of in vitro and in vivo efficiencies of a novel unit-dose liquid aerosol generator and a pressurized metered dose inhaler, *Int. J. Pharm.*, 198, 63–70.

286. Hermansen, K., Ronnemaa, T., Petersen, A. H., Bellaire, S., and Adamson, U. (2004), Intensive therapy with inhaled insulin via the AERx insulin diabetes management system: A 12-week proof-of-concept trial in patients with type 2 diabetes, *Diabetes Care*, 27, 162–167.

287. Geller, D., Thipphawong, J., Otulana, B., et al. (2003), Bolus inhalation of rhDNase with the AERx system in subjects with cystic fibrosis, *J. Aerosol Med.*, 16, 175–182.

288. Zimlich, W. C., Ding, J. Y., Busick, D. R., et al. (2000), The development of a novel electrohydrodynamic pulmonary drug delivery device, in Dalby, R. N., Byron, P. R., Peart, J., and Farr, S. J., Eds., *Respiratory Drug Delivery VII*, Serentec, Raleigh, NC, pp. 241–246.

289. Shen, X., Hindle, M., and Byron P. R. (2004), Effect of energy on propylene glycol aerosols using the capillary aerosol generator, *Int. J. Pharm.*, 275, 249–258.

290. Rabinowitz, J. D., Lloyd, P. M., Munzar, P., et al. (2006), Ultra-fast absorption of amorphous pure drug aerosols via deep lung inhalation, *J. Pharm. Sci*, 95, 2438–2451.

291. Rabinowitz, J. D., Wensley, M., Lloyd, P., et al. (2004), Fast onset medications through thermally generated aerosols, *J. Pharmacol. Exp. Ther.*, 309, 769–775.

5.9

OCULAR DRUG DELIVERY

ILVA D. RUPENTHAL AND RAID G. ALANY
The University of Auckland, Auckland, New Zealand

Contents

5.9.1 INTRODUCTION

Due to the accessibility of the eye surface, topical administration of ophthalmic medications is the most common method for treating conditions affecting the exterior eye surface. However, the unique anatomy and physiology of the eye renders it difficult to achieve an effective drug concentration at the target site. Therefore, efficient delivery of a drug past the protective ocular barriers accompanied with minimization of its systemic side effects remains a major challenge.

Conventional eye drops currently account for more than 90% of the marketed ophthalmic formulations [1]. However, after instillation of an eye drop, typically less than 5% of the applied drug penetrates the cornea and reaches the intraocular

Pharmaceutical Manufacturing Handbook: Production and Processes, edited by Shayne Cox Gad

tissues. This is due to the rapid and extensive precorneal loss caused by drainage and high tear fluid turnover.

As a consequence, the typical corneal contact time is limited to 1–2 min and the ocular bioavailability is usually less than 10% [2]. Furthermore, there is an initial peak dose of the drug, which is usually higher than that needed for a therapeutic effect, followed by a sharp drop-off in concentration to subtherapeutic levels.

Various ocular delivery systems, such a ointments, suspensions, micro- and nanocarriers, and liposomes, have been investigated during the past two decades pursuing two main strategies: to increase the corneal permeability and to prolong the contact time on the ocular surface [3].

On the other hand, the most efficient method for drug delivery to the posterior chamber of the eye so far has been intravitreal injection. This chapter focuses on the topical application of drugs to the surface of the eye and discusses the most recent formulation approaches in this area.

5.9.2 CHALLENGES IN OCULAR DRUG DELIVERY

5.9.2.1 Anatomical and Physiological Considerations

In order to research and develop an effective ophthalmic delivery system, a good understanding of the anatomy and physiology of the eye (the globe) is necessary. Figure 1 shows a cross section through the human eye. This chapter will mainly focus on the precorneal area and the transport barriers present in the eye.

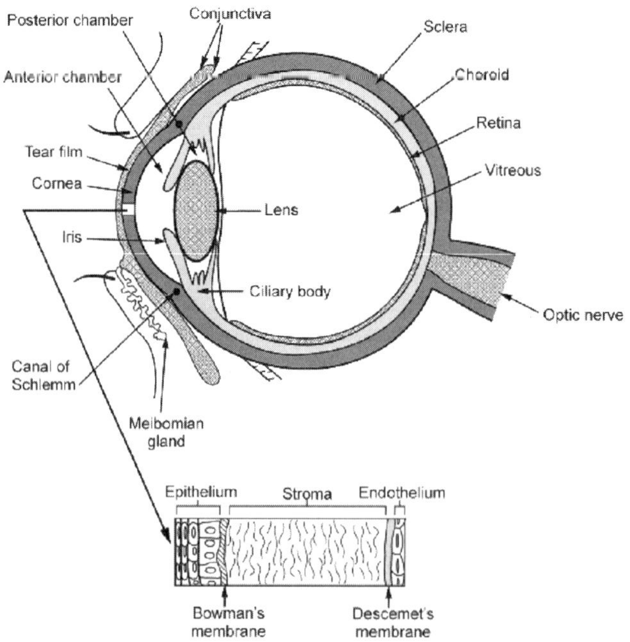

FIGURE 1 Cross section of the human eye and cornea.

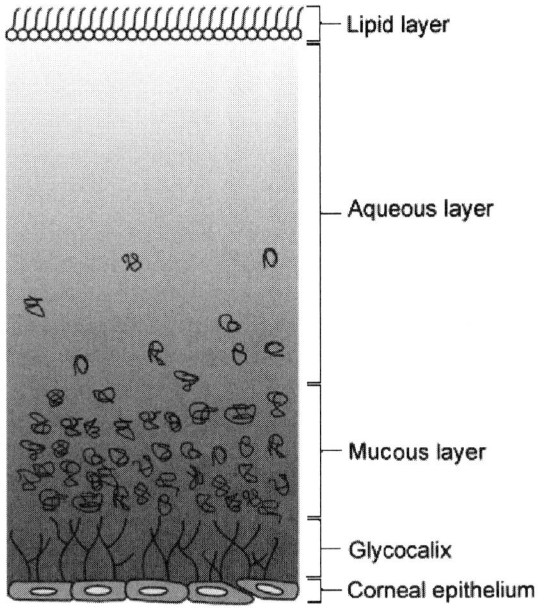

FIGURE 2 Structure of the precorneal tear film.

Precorneal Area

Precorneal Tear Film Corneal transparency and good visual function require a uniform eye surface. This is achieved by the tear film, which covers and lubricates the cornea and the external globe. It is about 7–8 μm thick and is the first structure encountered by topically applied drugs. The trilaminar structure of the tear film is shown in Figure 2.

Attached to the glycocalix of the corneal/conjunctival surface is a mucous layer, which consists mainly of glycoproteins. This layer is produced by the conjunctival goblet cells and the lacrimal gland. It plays an important role in the stability of the tear film as well as in the wetting of the corneal and conjunctival epithelium. The middle aqueous layer constitutes about 98% of the tear film and is mostly secreted by the main and accessory lacrimal glands [4]. It is composed of water, electrolytes, and various proteins such as lipocalin, lysozyme, and lactoferrin [5–8]. The outermost lipid layer is derived from the Meibomian and sebaceous Zeiss glands and prevents the evaporation of the tear fluid. It consists of sterol esters, triacylglycerols, and phospholipids and is spread over the aqueous layer during blinking.

Nasolacrimal Drainage System Figure 3 illustrates the nasolacrimal drainage system. The lacrimal gland, which is situated in the superior temporal angle of the orbit, is responsible for most of the tear fluid secretion. Secreted fluid is spread over the surface of the cornea during blinking and ends up in the puncta when the upper eye lid approaches the lower lid. The blinking process creates a suction mechanism which results in tears flowing through the lacrimal canaliculi into the lacrimal sac. Fluid from the lacrimal sac then drains into the 12-mm-long nasolacrimal duct, which empties into the inferior nasal passage. This passage is a highly vascular area

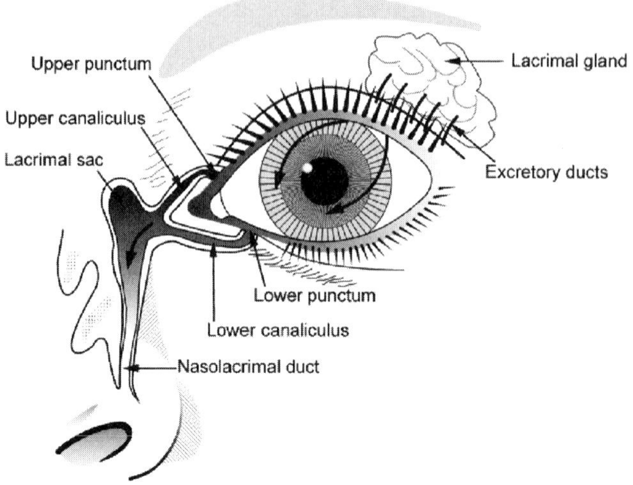

FIGURE 3 Schematic of the nasolacrimal drainage system.

and is responsible for most of the systemic drug absorption and subsequent systemic side effects.

The cul-de-sac normally holds 7–9 μL of tear fluid, with the normal tear flow rate being 1.2–1.5 μL/min [4]. The loss from the precorneal area by drainage, tear fluid turnover, and noncorneal absorption plays an important role in determining the ocular bioavailability of a drug. As the drainage rate is much faster than the ocular absorption rate, most of the topically applied drug is eliminated from the precorneal area within the first minute [4].

Tear production can be divided into basal, reflex, and emotional tearing [9]. Reflex tearing can be induced by many pharmaceutical/formulation factors, including the drug itself as well as pH and tonicity of the ocular dosage form.

Transport Barriers in the Eye Topical administration is the most common route for ocular drug delivery. Consequently, the cornea, conjunctiva, and sclera form the most essential barriers for drug penetration into the intraocular tissues.

Cornea The cornea is an important mechanical barrier protecting the intraocular tissues. It is considered to be the main pathway for ocular penetration of topically applied drugs. However, due to its unique structure, with the hydrophilic stroma sandwiched between the highly lipophilic epithelium and the less lipophilic endothelium, the penetration of compounds through the cornea depends on their *n*-octanol–water partition coefficient. Only drugs with a partition coefficient between 10 and 100 that show both lipid- and water-soluble properties can readily pass through the cornea.

The cornea is composed of five layers (see Figure 1): epithelium, Bowman's membrane stroma, Descemet's membrane, and endothelium.

The epithelium is the outermost layer and consists of five to six cell layers. These can be subdivided into one to two outermost layers of flattened superficial cells with microvilli on their anterior surface enhancing the cohesion and stability of the tear film [10], two to three layers of polygonal wing cells, and a single layer of basal columnar cells, allowing for minimal paracellular transport. The superficial cells adhere to one another via desmosomes and the cells are encircled by tight junctions [11]. Due to the nature of the tight junctions, the epithelium represents the rate-limiting barrier for hydrophilic compounds.

Bowman's membrane is composed of a layer of collagen fibers which form a relatively tight and impermeable barrier against microorganisms and therefore protect the stroma.

The stroma makes up approximately 90% of the corneal thickness and is mainly composed of hydrated collagen fibrils. It is highly hydrophilic, porous, and can be considered as an open structure, as it allows free passage of hydrophilic substances with a molecular weight below 500,000 Da but acts as a diffusion barrier to all lipophilic drugs [4, 12].

Descemet's membrane is the basement membrane of the endothelial cells. It is comprised of collagen fibers arranged in a hexagonal pattern and embedded in a matrix.

The endothelium is a single layer of flattened polygonal cells with microvilli which increase the surface area for removal of waste and absorption of nutrients [12]. It plays an important role in the maintenance of corneal hydration and transparency via active ion and fluid transport mechanisms.

Since the cornea exhibits hydrophilic as well as lipophilic characteristics, it represents an effective barrier for diffusion of both hydrophilic and lipophilic substances [3, 13].

Conjunctiva The conjunctiva is a thin and vascular mucous membrane consisting of two to three layers of epithelial cells overlying a loose, highly vascular connective tissue. The tight junctions present on the apical surface of the epithelium act as the main barrier for drug penetration (molecules > 20,000 Da) across the tissue, although not as tight as the corneal epithelium, which is impermeable to molecules larger than 5000 Da [14, 15]. The conjunctiva covers the anterior surface of the globe (bulbar conjunctiva), with the exception of the cornea, and is folded at the fornix (fornix conjunctiva) to form the palpebral conjunctiva, which lines the inner surface of the eyelids. The bulbar conjunctiva represents the first barrier for permeation of topically applied drugs via the noncorneal route [16].

Sclera The sclera is the outermost firm coat of the eye that serves as a protective barrier for the sensitive inner parts. It is composed of the same type of collagen fibers as the corneal stroma. However, the fibers are arranged in an irregular network rather than a lattice pattern, which makes the tissue appear opaque compared to the transparent cornea. The white sclera constitutes the posterior five-sixths of the globe, whereas the transparent cornea comprises the anterior one-sixth [17].

Iris–Ciliary Body The ciliary body comprises the ciliary muscle, which mainly enables accommodation, and the ciliary processes, which produce the aqueous humor.

The iris separates the anterior and posterior chambers and consists of the pigmented epithelial cell layer, the iridial sphincter and dilator muscles, and the stroma. The amount of melanin present in the stroma determines the color of the iris: few melanocytes exhibit blue, grey, or green, while many melanocytes are responsible for the brown appearance of the iris. There is often a considerable quantitative difference in drug response between light and heavily pigmented eyes [18]. The binding of drugs with melanin can decrease the aqueous humor concentration of free drug and is therefore likely to reduce the pharmacological response [19].

Lens The lens is the transparent biconvex structure situated behind the iris and in front of the vitreous. It plays an important role in the visual function of the eye and also enables accommodation together with the ciliary muscle. The lens is made up of slightly more than 30% protein (water-soluble crystallins) and therefore has the highest protein content of all tissues in the body [20]. The lens receives its nutrients from the aqueous humor and its transparency depends on the geometry of the lens fibres.

Blood–Ocular Barriers The blood–ocular barriers can be divided into the blood–aqueous barrier and the blood–retinal barrier.

The blood–aqueous barrier is located in the anterior part of the eye and is formed by the endothelial cells of the blood vessels in the iris and the nonpigmented cell layer of the ciliary epithelium [21]. It regulates the solute exchange between the blood and the intraocular fluid, preventing unspecific passage of solutes that could influence the transparency of the ocular tissues. The outward movement into the systemic blood circulation is less restricted, allowing especially small and lipophilic drug molecules to enter the uveal blood circulation [22]. These molecules are consequently removed more rapidly from the anterior chamber than larger, hydrophilic molecules, which are eliminated by the aqueous humor turnover only [23].

The blood–retinal barrier can be found in the posterior part of the eye. It prevents toxic molecules, plasma components, and water from entering the retina. It also forms a barrier for passage of systemically administered drugs into the vitreous, typically resulting in only 1–2% of the drug's plasma concentration in the intraocular tissues [24].

5.9.2.2 Pharmacokinetic Considerations

After topical application of an ophthalmic solution, the solution is instantly mixed with the tear fluid and then spread over the eye surface. However, various precorneal factors such as the drainage of the instilled solution, induced lacrimation, normal tear turnover, noncorneal absorption, drug metabolism, and enzymatic degradation limit the ocular absorption by shortening the contact time of the applied drug with the corneal surface [25]. As a result, typically less than 10% of the instilled dose is delivered into the intraocular tissues, whereas the rest is absorbed into the systemic circulation, leading to various side effects [3, 25, 26]. A summary of the drug deposition model in the eye after topical application as described by Lee and Robinson [27] is given below.

Upon instillation, the topically applied drug solution is instantly diluted by the resident tears, resulting in a significant decrease in the concentration gradient (driving force) and hence in the reduction of the transcorneal flux. Drainage of lacrimal fluid towards the nasolacrimal sac during blinking leads to a rapid elimination of the ocular solution via the canaliculi.

Any drug remaining on the ocular surface for a sufficient period of time can be absorbed into the anterior chamber via either the corneal or the conjunctival–scleral route.

Corneal absorption is considered to be the major penetration pathway for topically applied drugs. There are two mechanisms for absorption across the corneal epithelium, namely transcellular and paracellular diffusion [28]. Lipophilic drugs prefer the transcellular route while hydrophilic drugs penetrate primarily via the paracellular route. Transcorneal transport includes simple diffusion, facilitated diffusion, active transport, and endocytosis. Transport along the paracellular route is passive and is only limited by the pore size and the charge of the intracellular spaces. Only relatively small molecules can permeate through the pores. Negatively charged molecules permeate at a slower rate than positively charged and neutral ones [29]. In addition, a positive charge may also decrease the permeation due to the possible ionic interaction with the negatively charged carboxylic acid residues of the tight junction proteins [30].

The noncorneal route of absorption via the conjunctiva and the sclera is usually nonproductive, as most of the drug reaches the systemic circulation before gaining access to the intraocular tissues. As the surface area of the conjunctiva is much larger than that of the cornea and with the highly vascularized conjunctiva being more permeable, especially to larger hydrophilic molecules, drug loss through this route of absorption may be significant. Both transconjunctival and transnasal absorption via the nasolacrimal duct are generally undesirable, not only because of the loss of drug into the systemic circulation, but also because of the possible side effects [1].

5.9.2.3 Formulation Considerations

Irritation of the eye following the use of an ocular delivery system can be induced by a number of factors, including the instilled volume, the pH, and the osmolality of the formulation, as well as by the drug itself [31]. All these factors may induce reflex tearing and as a result increase lacrimal drainage. This is likely to reduce the ocular bioavailability of the drug and thus needs to be considered during the formulation process. In addition, general safety considerations such as sterility, ocular toxicity and irritation, and the amount of preservative used, need to be taken into account when formulating an ocular dosage form.

Physicochemical Drug Properties On the one hand, factors such as the chemical nature or the concentration of a drug can cause irritation of the ocular tissues, inducing reflex tearing and therefore reducing the retention time of the formulation in front of the eye.

However, more important physicochemical properties in terms of ocular bioavailability are the ones that affect the corneal permeability of the active compound. These include the lipophilicity of the drug as reflected by its *n*-octanol–water partition coefficient [32], the molecular size and shape [33], the charge [34], and the acid–base properties as determined by its pK_a [35].

According to Kaur and Smitha [36], the optimum lipophilicity for corneal absorption is found in drugs with an *n*-octanol–water partition coefficient between 10 and 100. For drugs with smaller partition coefficients (highly hydrophilic drugs), the lipophilic epithelium forms the rate-limiting barrier, whereas the hydrophilic stroma represents the primary barrier for transcorneal diffusion of highly lipophilic drugs [1, 3, 25]. In general, more lipophilic drugs penetrate via the transcellular pathway while the more hydrophilic drugs enter the cornea via the paracellular route.

In the case of ionized compounds (weak acids and bases), drug permeation depends on the chemical equilibrium between the ionized and the un-ionized form, both in the delivery system itself and in the lacrimal fluid. In general, un-ionized molecules penetrate lipid membranes more readily than ionized ones.

Besides lipophilicity and degree of ionization, the charge of the drug molecule may have an effect on its penetration. Cationic drugs permeate the cornea more easily than anionic compounds, which are repelled by the negative charge of the mucin layer on the ocular surface as well as the negatively charged pores present in the corneal epithelium [1,3,25,29]. However, a positive charge may also decrease permeation in some cases, due to possible ionic interactions between the positively charged molecules and the negatively charged carboxylic acid groups of the tight junction proteins [21, 30].

Finally, the molecular size of the drug has an effect on the corneal absorption. The cornea is impermeable to molecules larger than 5000 Da, whereas the conjunctival tissues allow compounds of up to 20,000 Da to penetrate [14, 15].

Buffer Capacity and pH The normal pH of the tear fluid is 7.4. Ocular formulations should ideally be formulated between pH 7.0 and 7.7 to avoid irritation of the eye [31]. However, in most cases the pH necessary for maximal solubility or stability of the drug is well outside this range. The tear fluid has only a limited buffering capacity, which is mainly due to the dissolved carbon dioxide and bicarbonate. It is therefore recommended to formulate using buffers with a low buffering capacity to allow the tears to regain their normal pH more rapidly [31].

Instillation Volume The cul-de-sac normally holds 7–9 μL [37], but can momentarily accommodate up to 30 μL without overflowing. Most commercial eye droppers however, deliver a volume of approximately 50 μL. The excess volume is rapidly removed, either by spillage from the conjunctival sac or through the puncta to the nasolacrimal drainage system, until the tears return to their normal volume [38]. Chrai et al. [39, 40] determined the influence of drop size on the rate of drainage of a solution instilled into the conjunctival sac of rabbits. The authors reported that the drainage process followed first-order kinetics and found that the rate of solution drainage from the conjunctival sac (as reflected by the elimination rate constant) was directly proportional to the instilled volume [40]. Similar observations were reported with other dosage forms such as suspensions [41] and liposomes [42]. Therefore, keeping the applied dose constant while decreasing the instilled volume substantially increases ocular bioavailability and decreases systemic absorption [3, 26, 43].

Osmotic Pressure The osmolality of the lacrimal fluid is mainly dependent on the number of ions dissolved in the aqueous layer of the tear film and normally ranges between 310 and 350 mOsm/kg [44]. When an ophthalmic solution is instilled into the eye, it mixes with the tear fluid, resulting in an osmotic pressure that is dependent on the osmolality of the tears as well as that of the formulation and the amount of the formulation instilled. In general, hypotonic solutions are better tolerated by the ocular tissues than hypertonic ones, which lead to increased lacrimation. If the tonicity of the formulation is lower than 260 mOsm/kg or higher than 480 mOsm/kg, the formulation becomes irritant [45], induces reflex tearing and blinking, and is therefore likely to reduce the bioavailability of topically instilled drugs.

5.9.3 FORMULATION APPROACHES TO IMPROVE OCULAR BIOAVAILABILITY

One of the major problems encountered with topical administration is the rapid precorneal loss caused by nasolacrimal drainage and high tear fluid turnover, which leads to drug concentrations of typically less than 10% of the applied drug. Approaches to improve the ocular bioavailability have been attempted in two directions: to increase the corneal permeability by using penetrations enhancers or prodrugs and to prolong the contact time with the ocular surface by using viscosity-enhancing or in situ gelling polymers. Table 1 summarizes conventional and novel ocular drug delivery approaches. Marketed ophthalmic delivery systems based on recent formulation approaches are listed in Table 2. An optimal ocular delivery systems would be administered in the form of an eye drop, causing neither blurred vision nor ocular irritancy, and would only need to be instilled once a day [1].

5.9.3.1 Conventional Dosage Forms

Conventional dosage forms such as solutions, suspensions, and ointments account for almost 90% of the currently accessible ophthalmic formulations on the market [1, 169]. They offer some advantages such as their ease of administration by the patient, ease of preparation, and the low production costs. However, there are also significant disadvantages associated with the use of conventional solutions in particular, including the very short contact time with the ocular surface and the fast nasolacrimal drainage, both leading to a poor bioavailability of the drug. Various ophthalmic delivery systems have been investigated to increase the corneal permeability and prolong the contact time with the ocular surface. However, conventional eye drops prepared and administered as aqueous solutions remain the most commonly used dosage form in ocular disease management.

Solutions The reasons behind choosing solutions over other dosage forms are their favorable cost advantage, the simplicity of formulation development and production, and the high acceptance by patients [170]. However, there are also a few drawbacks, such as rapid and extensive precorneal loss, high absorption via the conjunctiva and the nasolacrimal duct leading to systemic side effects, as well as the increased installation frequency resulting in low patient compliance.

Some of these problems have been encountered by addition of viscosity-enhancing agents such as cellulose derivates, which are believed to increase the viscosity of the preparation and consequently reduce the drainage rate. The use of viscosity enhancers will be discussed later in this section.

Suspensions Suspensions of the micronized drug (<10 μm) in a suitable aqueous vehicle are formulated, where the active compound is water insoluble. This is the

TABLE 1 Summary of Conventional and Novel Ocular Drug Delivery Approaches

Drug	Formulation Approach	Polymers/Bases	References
Pilocarpine	Viscosity enhancer	Methylcellulose	[40]
Fluorescein	Viscosity enhancer + collagen shields	Collagen shields in methylcellulose vehicle	[46]
Ciprofloxacin	Viscosity enhancers + penetration enhancer	Carbopol 934P/HPMC + dodecylmaltoside	[47]
Pilocarpine	Viscosity enhancers, mucoadhesives, in situ gelling systems	Gellan gum, xanthan gum, HEC, HPMC, PVA	[48]
Progesterone	Mucoadhesive	Cross-linked acrylic acid	[49]
Levobetaxolol	Mucoadhesive	Polyacrylic acid	[50]
Tropicamide	Mucoadhesive	CMC, HPCL, HPCM, PVP, PVA	[51]
		CMC-Na, HA-Na, PAA	[52]
Pilocarpine, tropicamide	Mucoadhesive	Hyaluronic acid, polyacrylic acid	[53]
Tertrahydrozoline	Mucoadhesive	Hyaluronic acid, polyacrylic acid, chitosan, gelatin	[54]
Tobramycin	Mucoadhesive	Chitosan	[55]
Tobramycin, ofloxacin	Mucoadhesive	Chitosan	[56]
Ofloxacin	Mucoadhesive	*N*-Trimethyl and *N*-carboxy-methyl chitosan	[57, 58]
Pilocarpine	Mucoadhesive	Carbopol 934P	[59]
		Carbomer 974P and 1342	[60]
		PVA, PVP, dextran, HPMC, HEC, MC, PAA, Na-hyaluronate, Na-alginate, gellan gum, chitosan	[61]
Timolol	Mucoadhesive	Xyloglucan	[62]
		Xanthan gum	[63]
		Carrageenan, gellan gum	[64]
		Carrageenan, locust bean gum, guar gum, xanthan gum, scleroglucan, xanthan gum, sodium alginate, β-cyclodextrin	[65]
		Hyaluronic acid	[66]
		HPMC, PVA, hyaluronic acid	[67]
		Carbomer, hyaluronic acid	[68]
		PAA, PVP	[69]
	In situ gelling system	Gellan gum	[70–78]
Indomethacine	In situ gelling system	Gellan gum	[79]
Pefloxacin mesylate	In situ gelling system	Gellan gum	[80]
Gatifloxacin	In situ gelling system	Alginate/HPMC	[81]
Ofloxacin	In situ gelling system	Carbopol 940/HPMC	[82]
Pilocarpine	In situ gelling system	Alginate	[83]
Carteolol	In situ gelling system	Alginate	[84]
Ciprofloxacin	In situ gelling system	Poloxamer/hyaluronic acid	[85]

TABLE 1 ***Continued***

Drug	Formulation Approach	Polymers/Bases	References
Pilocarpine	In situ gelling system	Pluronic F127, MC, HPMC	[86]
		Pluronic F127/carbopol	[87]
		Pluronic F127, xyloglycan	[88]
Timolol	In situ gelling system	Pluronic F127, MC, HPMC, CMC	[89]
Doxorubicin	In situ gelling system	Chitosan/glycerophosphate	[90]
		Pluronic F127 and F68, sodium hyaluronate	[91]
Pilocarpine, hydrocortisone	Nanoparticles	Gelatin	[92]
Hydrocortisone	Nanoparticles	Albumin	[93]
Pilocarpine	Nanoparticles	Albumin	[94, 95]
		Poly(methyl)methacrylate–acrylic acid copolymer (Piloplex)	[96–98]
		Cellulose acetate hydrogen phthalate (CAP)	[99–101]
Betaxolol	Nanoparticles	Poly-ε-caprolactone (PECL), poly (isobutyl)cyanoacrylate,polylac tic-*co*-glycolic acid (PLGA)	[102]
Carteolol	Nanoparticles	Poly-ε-caprolactone (PECL)	[103]
Pilocarpine	Nanoparticles	Poly(butyl)cyanoacrylate (PBCA)	[104–107]
		Poly(hexyl)cyanoacrylate (PHCA)	[104, 106, 107]
Ciclosporine	Nanoparticles	Chitosan	[108]
Indomethacine	Nanoparticles	Chitosan- and poly-L-lysine-coated poly-ε-caprolactone nanocapsules	[109]
Rhodamine	Nanoparticles	Chitosan- and PEG-coated poly-ε-caprolactone nanocapsules	[110]
Aciclovir	Nanoparticles	PEG-coated PLA nanospheres	[111]
Epinephrine	Nanoparticles	Poly-*N*-isopropylacrylamide	[112]
Ibuprofen	Nanoparticles	Eudragit RS100	[113, 114]
Flurbiprofen	Nanoparticles	Eudragit RS100 and RL100	[115]
Diclofenac	Nanoparticles	Eudragit RLPM and RSPM	[116]
Tobramycin	Nanoparticles	Various lipids [solid lipid nanoparticles (SLN)]	[117]
Gentamicin	Microspheres	Eudragit RS100 and RL100	[118]
Piroxicam	Microspheres	Pectin	[119]
		Albumin	[120]
Pilocarpine	Liposomes + mucoadhesive coating	Carbopol-coated liposomes	[121, 122]
Radioactive-labeled BSA	Liposomes + mucoadhesive coating	Chitosan-coated liposomes	[123]
Oligonucleotides (pdT16)	Liposomes + in situ gelling system	Liposomes in poloxamer 407 gel	[124–126]
Tropicamide	Liposomes + in situ gelling system	Liposomes in polycarbophil gel	[127]

TABLE 1 ***Continued***

Drug	Formulation Approach	Polymers/Bases	References
Timolol	Niosomes	Chitosan- and carbopol-coated niosomes	[128]
Indomethacine	Submicrometer emulsions	Phospholipids, lauroamphodiacetate	[129]
Pilocarpine	Submicrometer emulsions	Mono-dodecylphosphoric acid	[130]
	Microemulsions	Lecithin-based microemulsions	[131]
Retinol	Microemulsions	Tween 60 and 80, soy bean lecithin, *n*-butanol, triacetin, PG	[132]
		Crodamol EO, Crill 1 and Crillet 4	[133]
Epinephrine	Prodrug	Dipivalyl epinephrine	[134, 135]
Pilocarpine	Prodrug	*O, O′*(Xylylene)bispilocarpic acid esters	[136]
Ganciclovir	Prodrug	Ganciclovir acyl ester	[137]
Atenolol, betaxolol, Timolol	Penetration enhancers	Polyoxyethylene alkyl ethers (Brij), bile salts	[138]
Cromoclycin	Penetration enhancers	EDTA	[33]
Carbonic anhydrase inhibitors	Cyclodextrins	β-Cyclodextrin, hydroxypropyl-β-cyclodextrin	[139]
Ciprofloxacin	Cyclodextrins	Hydroxypropyl-β-cyclodextrin	[140]
Pilocarpine	Cyclodextrins	Hydroxypropyl-β-cyclodextrin	[141]
	Ocular films	Hydroxypropyl cellulose	[142]
		Poly(2-hydroxypropyl-methacrylate)	[143]
Pefloxacin mesylate	Ocular films	HPC, HPMC, PVP, PVA	[144]
	Ocular inserts	PVP, Eudragit RS and RL	[145]
Mitomycin C	Ocular inserts	Collagen implant	[146]
Pilocarpine	Ocular inserts	Collagen shield	[147]
		PAA, PVP, HPMC	[148]
		PVA, glyceryl behenate, xanthan gum, carrageenan, HPMC, HA; coated with Eudragit RS and RL	[149]
Timolol	Ocular inserts	HPC, coated with Eudragit RS and RL	[150]
Tilisolol	Ocular inserts	Poly(hydroxypropyl-methacrylate)	[151]
		Poly(2-hydroxypropyl-methacrylate), polypropylene tape	[152]
Ciprofloxacin	Ocular inserts	Sodium alginate, Eudragit, polyvinyl acetate	[153]
Pradofloxacin	Ocular inserts	Hydrogel coating on thin metallic wire (OpthaCoil)	[154]

TABLE 1 ***Continued***

Drug	Formulation Approach	Polymers/Bases	References
Fluorescein	Ocular inserts	HPMC lyophilisate on carrier strip	[155]
		HPMC lyophilisate on poly(tetra fluoroethylene) carrier strip	[156]
		Carbopol 974P, maize starch	[157, 158]
Gentamicin	Ocular inserts	CAP, carbomer, HPMC, HPC, EC	[159, 160]
Oxytetracycline	Ocular inserts	Silicone, PAA, PMA	[161]
Oxytetracycline, prednisolone, gentamicin	Ocular inserts	PEO 400, PEO 900	[162]
Ofloxacin	Ocular inserts	PEO 200, 400, 900, and 2000	[163]
		PEO 400, Eudragit L100	[164]
		PEO 900, PEO 2000, chitosan-thiolated PAA	[165]
Fluorescein, diclofenac	Ocular inserts		[166]
Gentamicin	Iontophoresis	Hydroxyethyl methacrylate	[167]
Pilocarpine, tropicamide	Dendrimers	Various poly(amidoamine) (PAMAM) dendrimers	[168]

case for most of the steroids. It is assumed that the drug particles remain in the conjunctival sac, thus promoting a sustained release effect [171]. There have been many studies trying to prove this assumption, but none of them has revealed a pronounced prolonged release profile [35, 172, 173].

According to Davies [174], topical ophthalmic suspensions have a number of limitations compared to solutions. They need to be adequately shaken before use to ensure correct dosing, a process which can result in poor patient compliance. In addition, they need to be sterilized, which may cause physical instability of the formulation. Furthermore, the amount of drug required to achieve only a moderate increase in bioavailability is very high, rendering suspensions expensive in terms of their production costs [175].

The drug particle size plays the most important role in the formulation process of suspensions. Particles greater than 10 μm cause patient discomfort. As they are perceived as foreign substances, they cause reflex tearing in order to eliminate the particles from the ocular surface [176]. A study by Schoenwald and Stewart [177] showed the influence of the particle size of dexamethasone on its bioavailability. The in vivo dissolution rate decreased with increasing particle size to the point when particles were removed from the conjunctival sac before the dissolution was complete.

As a result, achieving a near-solution state with small particles that are easy to resuspend and show minimal sedimentation remains the goal when formulation of a suspension is unavoidable [176].

TABLE 2 Marketed Ophthalmic Delivery Systems Based on Recent Formulation Approaches

Formulation Approach	Polymer/Base	Product	Company
Suspensions/ microparticulates	Carbomer ion exchange resin	Betoptic S	Alcon
Ointments	Wool fat, paraffin	Polyvisc	Alcon
	Liquid paraffin, white soft paraffin	LacriLube	Allergan
Viscosity enhancers/ mucoadhesives	Polyethylene glycol (PEG), propylene glycol (PG), HP-guar	Systane	Alcon
	Dextran, HPMC	Bion Tears	Alcon
	Carboxymethylcellulose sodium (CMC-Na)	Refresh Celluvisc	Allergan
		Refresh Liquigel	Allergan
	Polyvinyl alcohol (PVA)	Liquifilm Tears	Allergan
	HPMC	Lacrigel	Sunways
	Carbomer	Viscotears	Novartis
	Carbomer, polyvinyl alcohol (PVA)	Nyogel	Novartis
	Hyaluronic acid (HA)	Hy-Drop	Bausch&Lomb Fidia Oftal
	Sodium hyaluronate	Vismed	TRB Chemedica
	Carbomer	Pilopine HS	Alcon
	Polyacrylic acid (PAA)	Fucithalmic	Leo Pharma
In situ gelling systems	Gellan gum	Timoptic XE	Merck
	Polycarbophil	DuraSite	InSite Vision
	Polyacrylic acid, poloxamer	Smart Hydrogel	Advanced Medical Solutions
Prodrugs	Dipivefrin hydrochloride (epinephrin prodrug)	Propine	Allergan
Ocular inserts	Alginic acid	Ocusert	Alza
	Hydroxylpropyl cellulose	Lacrisert	Merck
	Silicone elastomer	Ocufit SR	Escalon Medical
	Collagen shield	MediLens	Chiron
		ProShield	Alcon

Ointments Ointments generally consist of a dissolved or dispersed drug in an appropriate vehicle base. They are the most commonly used semisolid preparations as they are well tolerated and fairly safe and increase the ocular bioavailability of the drug. The instilled ointment breaks up into small oily droplets that remain in the cul-de-sac as a drug depot. The drug eventually gets to the ointment–tear interface due to the shearing action of the eyelids [178].

Sieg and Robinson [35] compared the bioavailability of fluorometholone in a solution, a suspension, and an ointment. They found that the peak concentration

(c_{max}) of the drug in the aqueous humor of rabbits was comparable with all three formulations, whereas the time to peak concentration (t_{max}) occurred much later with the ointment, leading to a significantly greater total bioavailability of the drug.

Overall, ophthalmic ointments offer the following advantages: reduced dilution of the medication via the tear film, resistance to nasolacrimal drainage, and an increased precorneal contact time [179, 180]. However, oily viscous preparations for ophthalmic use (such as ointments) can cause blurred vision and matting of the eyelids and may also be associated with discomfort by the patient as well as occasional ocular mucosal irritation. Ointments are therefore generally used in combination with eye drops, which can be administered during the day, while the ointment is applied at night, when clear vision is not required.

5.9.3.2 Polymeric Delivery Systems

Polymeric systems used for ocular drug delivery can be divided into three groups: viscosity-enhancing polymers, which simply increase the formulation viscosity, resulting in decreased lacrimal drainage and enhanced bioavailability; mucoadhesive polymers, which interact with the ocular mucin, therefore increasing the contact time with the ocular tissues; and in situ gelling polymers, which undergo sol-to-gel phase transition upon exposure to the physiological conditions present in the eye. However, there are no defined boundaries between the different groups and most polymers exhibit more than one of these properties.

Viscosity-Enhancing Polymers In order to reduce the lacrimal clearance (drainage) of ophthalmic solutions, various polymers have been added to increase the viscosity of conventional eye drops, prolong precorneal contact time, and subsequently improve ocular bioavailability of the drug [40, 51, 181–184]. Among the range of hydrophilic polymers investigated in the area of ocular drug delivery are polyvinyl alcohol (PVA) and polyvinyl pyrrolidone (PVP), cellulose derivates such as methylcellulose (MC), and polyacrylic acids (carbopols).

Chrai and Robinson [40] evaluated the use of an MC solution of pilocarpine in albino rabbits and found a decrease in the drainage rate with increasing viscosity. Patton and Robinson [185] investigated the relationship between the viscosity and the contact time or drainage rate and demonstrated an optimum viscosity of 12–15 cps for an MC solution in rabbits. The influence of different polymers on the activity of pilocarpine in rabbits and human was reported by Saettone et al. [182]. Trueblood et al. [183] used lacrimal microscintigraphy to evaluate the corneal contact time for saline, PVA, and hydroxpropyl methylcellulose (HPMC) and observed the longest contact time for the formulation with HPMC as a viscosity-enhancing agent.

The ocular shear rate ranges from $0.03\,s^{-1}$ during interblinking periods to 4250–$28{,}500\,s^{-1}$ during blinking [186]. It has a great influence on the rheological properties of viscous ocular dosage forms and consequently the bioavailability of the incorporated drug [187]. Newtonian systems do not show any real improvement of bioavailability below a certain viscosity and blinking becomes painful, followed by reflex tearing, if the viscosity is too high [188]. While the viscosity of Newtonian systems is independent from the shear rate, non-Newtonian pseudoplastic or so-called shear-thinning systems exhibit a decrease in viscosity with increasing shear rates. This pseudoplastic behavior is favorable for ocular drug delivery systems as it offers less resistance to blinking and therefore shows greater acceptance by patients than Newtonian systems of the same viscosity [189].

Mucoadhesive Polymers Bioadhesion refers to the attachment of a drug molecule or a delivery system to a specific biological tissue by means of interfacial forces. If the surface of the tissue is covered by a mucin film, as is the case for the external globe, it is more commonly referred to as mucoadhesion.

In order to be an effective mucoadhesive excipient, polymers must show one or more of the following features [190]: strong hydrogen binding group, strong anionic charge, high molecular weight, sufficient chain flexibility, surface energy properties favoring spreading onto the mucus, and near-zero contact angle to allow maximum contact with the mucin coat.

The most commonly used bioadhesives are macromolecular hydrocolloids with numerous hydrophilic functional groups capable of forming hydrogen bonds (such as carboxyl, hydroxyl, amide, and sulfate groups) [191]. Hui and Robinson [49] were the first to demonstrate the usefulness of bioadhesive polymers in improving the ocular bioavailability of progesterone. Saettone et al. [53] evaluated a series of bioadhesive dosage forms for ocular delivery of pilocarpine and tropicamide and found hyaluronic acid to be the most promising mucoadhesive polymer. Lehr et al. [192] suggested that cationic polymers, which are able to interact with the negative sialic acid residues of the mucin, would probably show better mucoadhesive properties than anionic or neutral polymers. They investigated the polycationic polymer chitosan and demonstrated that the mucoadhesive performance of chitosan was significantly higher in neutral or slightly alkaline pH as it is present in the tear fluid.

However, according to Park and Robinson [193], polyanions are better than polycations in terms of binding and potential toxicity. In general, both anionic and cationic charged polymers demonstrate better mucoadhesive properties than nonionic polymer, such as cellulose derivates or PVA [194, 195].

The mechanism of mucoadhesion involves a series of different steps. First, the mucoadhesive formulation needs to establish an intimate contact with the corneal surface. Prerequisites are either good wetting or swelling of the mucoadhesive polymer as well as sufficient spreading across the cornea. The second stage involves the penetration of the mucoadhesive polymer chains into the crevices of the tissue surface and also the entanglement with the mucous chains [196]. On a molecular level, mucoadhesion is a results of van-der-Waals forces, electrostatic attractions, hydrogen bonding, and hydrophobic interactions [36].

Mucoadhesive polymers increase the contact time of a formulation with the tear film and simulate the continuous delivery of tears due to a high water-restraining capacity. As such, they allow a decrease in the instillation frequency compared to common eye drops and are therefore useful as artificial tear products [1, 197, 198].

In Situ Gelling Systems In situ gelling systems are viscous polymer-based liquids that exhibit sol-to-gel phase transition on the ocular surface due to change in a specific physicochemical parameter (ionic strength, temperature, pH, or solvent exchange). They are highly advantageous over preformed gels as they can easily be instilled in liquid form but are capable of prolonging the residence time of the formulation on the surface of the eye due to gelling [199].

The principal advantage of in situ gelling systems is the easy, accurate, and reproducible administration of a dose compared to the application of preformed gels [198].

The concept of forming gels in situ (e.g., in the cul-de-sac of the eye) was first suggested in the early 1980s, and ever since then various triggers of in situ gelling have been further investigated.

Polymers that may undergo sol-to-gel transition triggered by a change in pH are cellulose acetate phthalate (CAP) and cross-linked polyacrylic acid derivates such as carbopols, methacrylates and polycarbophils. CAP latex is a free-running solution at pH 4.4 which undergoes sol-to-gel transition when the pH is raised to that of the tear fluid. This is due to neutralization of the acid groups contained in the polymer chains, which leads to a massive swelling of the particles. The use of pH-sensitive latex nanoparticles has been described by Gurny et al. [100, 200]. Carbopols have apparent pK_a values in the range of 4–5 resulting in rapid gelation due to rise in pH after ocular administration.

Gellan gum is an anionic polysaccharide which undergoes phase transition under the influence of an increased ionic strength. In fact, the gel strength increases proportionally with the amount of mono- or divalent cations present in the tear fluid. As a consequence, the usual reflex tearing, which leads to a dilution of common viscous solutions, further enhances the viscosity of gellan gum formulations due to the increased amount of tear fluid and thus higher cation concentration [201]. Several studies have been performed comparing Gelrite formulations (low acetyl gellan gum) to conventional ophthalmic solutions of the same active compound. Shedden et al. [76] compared the plasma concentrations of timolol following multiple applications of Timoptic-XE and a timolol maleate solution. They found that a once-daily application of the in situ gelling formulation was sufficient to reduce the intraocular pressure to levels comparable to a twice-daily application of the solution, leading to better patient compliance as well as a reduction in systemic side effects.

Poloxamers or pluronics are block copolymers consisting of poly(oxyethylene) and poly(oxypropylene) units. They rapidly undergo thermal gelation when the temperature is raised to that of the ocular surface (32 °C), while they remain liquid at refrigerator temperature. Poloxamers exhibit surface active properties, but even if used in high concentrations (usually between 20 and 30%), Pluronic F127 was found no more damaging to the cornea than a physiological saline solution [202]. In order to reduce the total polymer concentration and achieve better gelling properties, several poloxamer combinations have been tested. Wei et al. [91] used a mixture of Pluronic F127 and F68 resulting in a more suitable phase transition temperature with a free-flowing liquid under 25 °C.

Combining thermal- with pH-dependent gelation, Kumar et al. [86] developed a combination of methylcellulose 15% and carbopol 0.3%. This composition exhibited a sol-to-gel transition between 25 and 37 °C with a pH increase from 4 to 7.4 [203]. A possible mechanism for the thermal effect could be the decrease in the degree of the methylcellulose hydration, while the polyacrylic acid can transform into a gel upon an increase in pH due to the buffering properties of the tear fluid [1].

5.9.3.3 Colloidal Delivery Systems

Colloidal carriers have been investigated as drug delivery systems for the past 30 years in order to achieve specific drug targeting, facilitate the bioavailability of drugs through biological membranes, and protect the drug against enzymatic degradation. Their use in topical administration and especially in ocular delivery however has only been studied for the last two decades [1, 3].

Colloidal carriers are small particulate systems ranging in size from 100 to 400 nm. As they are usually suspended in an aqueous solution, they can easily be administered as eye drops, thus avoiding the potential discomfort resulting from bigger particles present in ocular suspensions or from viscous or sticky preparations [38].

Most efforts in ophthalmic drug delivery have been made with the aim of increasing the corneal penetration of the drug. Calvo et al. [204] have shown that colloidal particles are preferably taken up by the corneal epithelium via endocytosis. It has also been stated by Lallemand and co-workers [205], that the cornea acts as a drug reservoir, slowly releasing the active compound present in the colloidal delivery system to the surrounding ocular tissues.

Nanoparticles Nanoparticles have been among the most widely studied particulate delivery systems over the past three decades. They are defined as submicrometer-sized polymeric colloidal particles ranging from 10 to 1000 nm in which the drug can be dissolved, entrapped, encapsulated, or adsorbed [206]. Depending on the preparation process, nanospheres or nanocapsules can be obtained. Nanospheres have a matrixlike structure where the drug can either be firmly adsorbed at the surface of the particle or be dispersed/dissolved in the matrix. Nanocapsules, on the other hand, consist of a polymer shell and a core, where the drug can either be dissolved in the inner core or be adsorbed onto the surface [207].

The first nanoparticulate delivery system studied was Piloplex, consisting of pilocarpine ionically bound to poly(methyl)methacrylate–acrylic acid copolymer nanoparticles [44]. Klein et al. [1, 98] found that a twice-daily application of Piloplex in glaucoma patients was just as effective as three to six instillations of conventional pilocarpine eye drops. However, the formulation was never accepted for commercialization due to various formulation-related problems, including the nonbiodegradability, local toxicity, and difficulty of preparing a sterile formulation [208].

Another early attempt to formulate a nanoparticulate system for the delivery of pilocarpine was made by Gurny [99]. This formulation was based on pilocarpine dispersed in a hydrogen CAP pseudolatex formulation. Gurny and co-workers [101] compared the formed nanoparticles to a 0.125% solution of hyaluronic acid some years after their first investigation and found that the viscous hyaluronic acid system showed a significantly longer retention time in front of the eye than the pseudolatex formulation.

The most commonly used biodegradable polymers in the preparation of nanoparticulate systems for ocular drug delivery are poly-alkylcyanoacrylates, poly-ε-caprolactone, and polylactic-*co*-glycolic acid copolymers. Marchal-Heussler et al. [102, 103] compared the three particulate delivery systems using antiglaucoma drugs including betaxolol and cartechol. Results showed that poly-ε-caprolactone (nanospheres and nanocapsules) exhibited the highest pharmacological activity when loaded with betaxolol. It seemed that the higher ocular activity was related to the hydrophobic nature of the carrier and that the mechanism of action seemed to be directly linked to the agglomeration of the particles in the conjunctival sac [1]. In general, nanocapsules displayed a much better effect than nanospheres probably due to the fact that the active compound was in its un-ionized form in the oily core and could diffuse faster into the cornea. Diffusion of the drug from the oily core of the nanocapsule to the corneal epithelium seems to be more effective than diffusion from the internal, more hydrophilic matrix of the nanospheres [1, 209].

In order to achieve a sustained drug release and a prolonged therapeutic activity, nanoparticles must be retained in the cul-de-sac and the entrapped drug must be released from the particles at a certain rate. If the release is too fast, there is no sustained release effect. If it is too slow, the concentration of the drug in the tears might be too low to achieve penetration into the ocular tissues [208]. The major limiting issues for the development of nanoparticles include the control of particle size and drug release rate as well as the formulation stability.

So far, there is only one microparticulate ocular delivery system on the market. Betoptic S is obtained by binding of betaxolol to ion exchange resin particles. Betoptic S 0.25% was found to be bioequivalent to the Betoptic 0.5% solution in lowering the intraocular pressure [208].

Liposomes Liposomes were first reported by Bangham in the 1960s and have been investigated as drug delivery systems for various routes ever since then. They offer some promising features for ophthalmic drug delivery as they can be administered as eye drops but will localize and maintain the pharmacological activity of the drug at its site of action [1]. Due to the nature of the lipids used, conventional liposomes are completely biodegradable, biocompatible, and relatively nontoxic [1].

A liposome or so-called vesicle consists of one or more concentric spheres of lipid bilayers separated by water compartments with a diameter ranging from 80 nm to 100 μm. Owing to their amphiphilic nature, liposomes can accommodate both lipohilic (in the lipid bilayer) and hydrophilic (encapsulated in the central aqueous compartment) drugs [207]. According to their size, liposomes are classified as either small unilamellar vesicles (SUVs) (10–100 nm) or large unilamellar vesicles (LUVs) (100–300 nm). If more than one bilayer is present, they are referred to as multilamellar vesicles (MLVs). Depending on their lipid composition, they can have a positive, negative, or neutral surface charge.

Liposomes are potentially valuable as ocular drug delivery systems due to their simplicity of preparation and versatility in physical characteristics. However, their use is limited by instability (due to hydrolysis of the phospholipids), limited drug-loading capacity, technical difficulties in obtaining sterile preparations, and blurred vision due to their size and opacity [42].

In addition, liposomes are subject to the same rapid precorneal clearance as conventional ocular solutions, especially the ones with a negative or no surface charge [127]. Positively charged liposomes, on the other hand, were reported to exhibit a prolonged precorneal retention due to electrostatic interactions with the negative sialic acid residues of the mucin layer [2, 127, 208, 210–213].

There have been several attempts to use liposomes in combination with other newer formulation approaches, such as incorporating them into mucoadhesive gels or coating them with mucoadhesive polymers [210]. Mucoadhesive polymers investigated in this regard were poly(acrylic acid) (PAA), hyaluronic acid (HA), chitosan, and poloxamer [36, 43, 121–122, 214].

Durrani and co-workers [122] reported on the influence of a carbopol coating on the corneal retention of pilocarpine-loaded liposomes, and demonstrated a biphasic response with an initial low intensity followed by a sustained reaction.

Bochot et al. [124, 125] developed a novel delivery system for oligonucleotides by incorporating them into liposomes and then dispersing them into a thermosensitive gel composed of poloxamer 407. They compared the in vitro release of the model oligonucleotides pdT16 from simple poloxamer gels (20 and 27% poloxamer) with the ones where pdT16 was encapsulated into liposomes and then dispersed within the gels. They found that the release of the oligonucleotides from the gels was controlled by the poloxamer dissolution, whereas the dispersion of liposomes within 27% poloxamer gel was shown to slow down the diffusion of pdT16 out of the gel.

Niosomes In order to circumvent some of the limitations encountered with liposomes, such as their chemical instability, the cost and purity of the natural phospholipids, and oxidative degradation of the phospholipids, niosomes have been developed. Niosomes are nonionic surfactant vesicles which exhibit the same bilayered structures as liposomes. Their advantages over liposomes include improved chemical stability and low production costs. Moreover, niosomes are biocompatible, biodegradable, and nonimmunogenic [215]. They were also shown to increase the ocular bioavailability of hydrophilic drugs significantly more than liposomes. This is due to the fact that the surfactants in the niosomes act as penetrations enhancers and remove the mucous layer from the ocular surface [209].

A modified version of niosomes are the so-called discomes, which vary from the conventional niosomes in size and shape. The larger size of the vesicles (12–60 μm) prevents their drainage into the nasolacrimal drainage system. Furthermore, their disclike shape provides them with a better fit in the cul-de-sac of the eye [26].

Vyas et al. [216] demonstrated that discomes entrapped higher amounts of timolol maleate than niosomes and that both niosomes and discomes significantly increased the bioavailability of timolol maleate when compared to a conventional timolol maleate solution.

Microemulsions Microemulsions (MEs) are colloidal dispersions composed of an oil phase, an aqueous phase, and one or more surfactants. They are optically isotropic and thermodynamically stable and appear as transparent liquids as the droplet size of the dispersed phase is less than 150 nm. One of their main advantages is their ability to increase the solubilization of lipophilic and hydrophilic drugs accompanied by a decrease in systemic absorption [217]. Moreover, MEs are transparent systems thus enable monitoring of phase separation and/or precipitation. In addition, MEs possess low surface tension and therefore exhibit good wetting and spreading properties.

While the presence of surfactants is advantageous due to an increase in cellular membrane permeability, which facilitates drug absorption and bioavailability [218], caution needs to be taken in relation to the amount of surfactant incorporated, as high concentrations can lead to ocular toxicity. In general, nonionic surfactants are preferred over ionic ones, which are generally too toxic to be used in ophthalmic formulations. Surfactants most frequently utilized for the preparation of MEs are poloxamers, polysorbates, and polyethylene glycol derivatives [219].

Similar to all other colloidal delivery systems discussed above it was hypothesized by numerous research teams that a positive charge (provided by cationic surfactants [220]) would increase the ocular residence time of the formulation due to electrostatic interactions with the negatively charged mucin residues. However, toxicological studies contradicted this assumption regarding the ocular effects, and so far there has been no publication demonstrating a distinct beneficial effect of charged surfactants incorporated into MEs.

Microemulsions can be classified into three different types depending on their microstructure: oil-in-water (o/w ME), water-in-oil (w/o ME), and bicontinuous ME. They have been investigated by physical chemists since the 1940s but have only gained attention as potential ocular drug delivery carriers within the last two decades.

Gasco and co-workers [221] investigated the potential application of o/w lecithin MEs for ocular administration of timolol, in which the drug was present as an ion pair with octanoate. The ocular bioavailability of the timolol ion pair incorporated into the ME was compared to that of an ion pair solution as well as a simple timolol solution. Areas under the curve for the ME and the ion pair solution respectively were 3.5 and 4.2 times higher than that of the simple timolol solution. A prolonged absorption was achieved using the ME with detectable amounts of the drug still present 120 min after instillation.

Various lecithin-based MEs were also characterized by Hasse and Keipert [131]. The formulations were tested in terms of their physicochemical parameters (pH, refractive index, osmolality, viscosity, and surface tension) and physiological compatibility (HET-CAM and Draize test). In addition, in vitro and in vivo evaluations were performed. The tested MEs showed favorable physicochemical parameters and no ocular irritation as well as a prolonged pilocarpine release in vitro and in vivo.

Muchtar and co-workers [129] prepared MEs with poloxamer 188 and soybean lecithin to deliver indometacin to the ocular tissues. They found a threefold increased indomethacin concentration in the cornea and aqueous humor 1 h post-instillation.

Beilin et al. [222] demonstrated a prolonged ocular retention of a submicrometer emulsion (SME) in the conjunctival sac using a fluorescent marker (0.01% calcein) as well as the miotic response of New Zealand Albino rabbits to pilocarpine. They found that the fluorescence intensity of calcein in SME was significantly higher than that of a calcein solution at all time points. Moreover, the pilocarpine SME exhibited a longer duration of miosis than the simple pilocarpine solution. It should be mentioned that SMEs are true emulsions, being different from MEs. They do not form spontaneously and are kinetically rather than thermodynamically stable. They generally require high-shear homogenization to form and are more susceptible to phase inversion. Furthermore, they are neither transparent nor translucent but rather turbid due their larger droplet size compared to MEs. While the two terms are used interchangeably in the the scientific literature, they actually refer to two distinct categories of dispersed systems and should be differentiated from each other.

The w/o MEs composed of water, Crodamol EO, Crill 1, and Crillet 4 were investigated as potential ocular delivery systems by Alany et al. [133]. It was hypothesized that w/o MEs undergo phase transition into lamellar liquid crystals (LCs) upon aqueous dilution by the tears, prolonging the precorneal retention time due to an increase in the formulation's viscosity. HET-CAM studies revealed no ocular irritancy by the excipients used. Ocular drainage was evaluated via γ-scintigraphy and demonstrated a significantly higher precorneal retention of the tested microemulsions compared to an aqueous solution.

The use of MEs in ocular delivery is very attractive due to all the advantages offered by these formulations. They are thermodynamically stable and transparent, possess low viscosity, and thus are easy to instill, formulate, and sterilize (via filtration). Moreover, they offer the possibility of reservoir and/or localizer effects. All these factors, in addition to the ones previously mentioned, render MEs promising ocular delivery systems.

5.9.3.4 Other Delivery Approaches

Many other ocular delivery approaches have been investigated over the past decades, including the use of prodrugs, penetration enhancers, cyclodextrins, as well as different types of ocular inserts. In addition, iontophoresis, which is an active drug delivery approach utilizing electrical current of only 1–2 mA to transport ionized drugs across the cornea, offers an effective, noninvasive method for ocular delivery. Another more recent approach is the use of dendrimers in ocular therapy. Dendrimers are synthetic spherical molecules named after their characteristic treelike branching around a central core with a size ranging from 2 to 10 nm in diameter [223]. So far, PAMAM (polyamidoamine) has been the most commonly studied dendrimer system for ocular use [224, 225].

Prodrugs Prodrugs are pharmacologically inactive derivates of drug molecules that require a chemical or enzymatic transformation into their active parent drug [226]. To be effective, an ocular prodrug should show an appropriate lipophilicity to facilitate corneal absorption, posses sufficient aqueous solubility and stability to be formulated as an eye drop, and demonstrate the ability to be converted to the active parent drug at a rate that meets therapeutic needs [227].

When considering ophthalmic drug molecules as prodrug candidates, the following factors need to be considered: the pathway and mechanism of ocular drug penetration, the functional groups of the drug candidate amenable to prodrug derivatization, and the enzymes present in the ocular tissues, which are necessary for prodrug activation [28].

The majority of ophthalmic prodrugs developed so far are chemically classified as ester. They are derived from the esterification of the hydroxyl or carboxylic acid groups present in the parent molecule. Of all enzymes participating in the activation of prodrugs, esterases, which are present in all anterior segment tissues except the tear film, have received the most attention [228, 229].

Prodrugs were introduced into the area of ocular drug delivery about 25 years ago [230], and steroids were probably the first ones to be utilized as prodrugs. However, the concept of prodrugs was not fully exploited until the introduction of dipivefrin (epinephrine prodrug) in the late 1970s. Kaback and co-workers [134] found that a 0.1% dipivalyl epinephrine solution lowered the intraocular pressure as effective as a 2% epinephrine solution, while significantly lowering the systemic side effects. Wei et al. [135] compared the ocular penetration, distribution, and metabolism of epinephrine and dipivalyl epinephrine and found the partition coefficient of the later to be 100–600 times higher than that of epinephrine, therefore leading to a 10-times faster absorption into the rabbit eye.

Dipivefrin was the only commercially available ophthalmic prodrug at that time. However, numerous prodrug derivates have been designed to improve the efficacy of ophthalmic drugs ever since.

Jarvinen and co-workers [136] synthesized unique *O, O′*-(xylylene)bispilocarpic acid esters containing two pilocarpic acid monoesters linked with one moiety. The found that prodrug showed a two- to seven-fold higher corneal permeability than pilocarpine itself despite the high molecular weight.

Tirucherai et al. [137] formulated an acyl ester prodrug of ganciclovir. The increased permeability was associated with a linear increased susceptibility of the ganciclovir esters to the esterases present in the cornea.

So far, aims that have been achieved by using prodrugs include the modification of the drug's duration of action, reduction of the systemic absorption, and reduction of ocular and systemic side effects. Although prodrugs are commonly used to treat diseases of the anterior segment, there have also been attempts to treat conditions associated with the posterior segment of the eye.

Penetration Enhancers The transport process across the corneal tissue is the rate-limiting step in ocular drug absorption. Increasing the permeability of the corneal epithelium by penetration enhancers is likely to enhance the drug transport across the corneal tissues and therefore improve ocular bioavailability of the drug.

Penetration enhancers act by increasing the permeability of the corneal cell membrane and/or loosening the tight junctions between the epithelial cells, which primarily restrict the entry of molecules via the paracellular pathway. Classes of penetration enhancers include surfactants, bile salts, calcium chelators, preservatives, fatty acids, and some glycosides such a saponin.

Surfactants are perceived to enhance drug absorption by disturbing the integrity of the plasma membranes. When present at low concentrations, surfactants are incorporated into the lipid bilayer, leading to polar defects in the membrane, which change the membrane's physical properties. When the lipid bilayer is saturated, micelles start to form, enclosing phospholipids from the membranes, hence leading to membrane solubilization [36]. Saettone et al. [138] found an increased corneal permeability for atenolol, timolol, and betaxolol by including 0.05% Brij 35, Brij 78, and Brij 98 into their formulations.

Bile salts are amphiphilic molecules that are surface active and self-associate to form micelles in aqueous solution. They increase corneal permeability by changing the rheological properties of the bilayer [231]. A number of bile salts such as deoxycholate, taurodeoxycholate, and glycocholate have been tested so far, and it was suggested, that a difference in their physicochemical properties (solubilizing activity, lipophilicity, Ca^{2+} sequestration capacity) is probably related to their performance as permeability-enhancing agents [36].

Another class of penetration enhancers includes calcium chelators such as ethylenediaminetetraacetic acid (EDTA). These molecules induce Ca^{2+} depletion in the cells. This leads to a global change within the cell and as a result loosens the tight junctions between superficial epithelial cells, thus facilitating paracellular transport

[138, 232]. Grass et al. [33] were among the first to emphasize the enhancing effects of chelating agents for ocular drug delivery. They found that 0.5% EDTA doubled the corneal absorption of topically applied glycerol and cromoclycin sodium.

Large numbers of penetration enhancers have been investigated to date. However, the unique structure of the corneal/conjunctival tissues requires caution. When selecting penetration enhancers for ocular delivery, their capacity to affect the integrity of the epithelial surfaces needs to be considered. Studies have shown that penetration enhancers themselves can penetrate ocular tissues, which could lead to potential toxicity. EDTA concentrations in the iris–ciliary body, for example, were found to be high enough to alter the permeability of the blood vessels in the uveal tract, therefore indirectly accelerating the drug's removal from the aqueous humor [233]. Similarly, benzalkonium chloride (BAC), a cationic surfactant which shows the highest penetration-enhancing effect among the currently used preservatives, was found to accumulate in the cornea for several days.

Cyclodextrins Cyclodextrins were introduced into the area of ocular drug delivery in the early 1990s. They are a group of homologous cyclic oligosaccharides with a hydrophilic outer surface and a lipophilic cavity in the center. Their initial aim was to increase the solubility of lipophilic drugs by forming inclusion complexes. Cyclodextrin complexation generally results in improved wettability, dissolution, solubility, and stability in solution as well as reduced side effects [234].

It is assumed that cyclodextrins are too large and hydrophilic to penetrate biological membranes. However, they act as penetration enhancers by assuring a high drug concentration at the corneal surface, from where the drug then partitions into the ocular tissues [235].

Even though cyclodextrins drug complexes seem to decrease ocular toxicity of irritant drugs, cyclodextrins themselves may exhibit ocular toxicity and should therefore be screened by performing corneal sensitivity studies. Among all cyclodextrin derivates investigated, hydroxy-propyl-β-cyclodextrin showed the most favorable properties in terms of toxicity [1].

Nijhawan and Agarwal [140] investigated inclusion complexes of ciprofloxacin hydrochloride and hydroxy-propyl-β-cyclodextrin and found that the complexes exhibited better stability, biological activity, and ocular tolerance than the uncomplexed drug in solution.

Aktas et al. [141] showed an increased permeation of pilocarpine nitrate complexed with hydroxy-propyl-β-cyclodextrin using isolated rabbit cornea. They found a significant reduction in the pupil diameter compared to a simple aqueous solution of the same active compound.

Cyclodextrins improve chemical stability, increase the drug's bioavailability, and decrease local irritation. However, the improvement of ocular bioavailability seems to be limited by the very slow dissociation of the complexes in the precorneal tear fluid.

Several studies have shown that combinations of cyclodextrins drug complexes and viscosity enhancers can significantly improved ocular absorption [141, 236–237] and should therefore be further investigated.

Ocular Inserts Solid ocular dosage forms such as films, erodible and nonerodible inserts, rods, and shields have been developed to overcome the typical pulse-entry-type drug release associated with conventional ocular dosage forms. This pulse entry is characterized by a transient overdose, a relatively short period of appropriate dosing, followed by a prolonged period of underdosing. Ocular inserts were developed in order to overcome these disadvantages by providing a more controlled, sustained, and continuous drug delivery by maintaining an effective drug concentration in the target tissues and yet minimizing the number of applications [238].

Ocular inserts probably represent one of the oldest ophthalmic formulation approaches. In 1948 the British Pharmacopoeia described an atropine-in-gelatin wafer and ever since then numerous systems have been developed applying various polymers and different release principals. However, the difficulty of insertion by the patient, foreign-body sensation, and inadvertent loss of inserts from the eye make these systems less popular, especially among the elderly. Furthermore, the high cost involved in manufacture prevented the insert market from taking off [197].

Two products, Alza Ocusert and Merck Lacrisert, have been marketed, although Ocusert is no longer available. Ocusert is a membrane-controlled reservoir system for the treatment of glaucoma. It contains pilocarpine and alginic acid in the core reservoir, sandwiched between two transparent, lipophilic ethylenevinyl acetate (EVA) rate-controlling membranes, which allow the drug to diffuse from the reservoir at a precisely determined rate for a period of seven days. This system is nonbiodegradable and must therefore be removed after use. Lacrisert, on the other hand, is a soluble minirod of hydroxypropylmethyl cellulose without any active ingredient. The system is placed in the conjunctival sac, where it softens within an hour and completely dissolves within 14–18 h. Lacrisert stabilizes and thickens the precorneal tear film and prolongs the tear film break-up time, which is usually accelerated in patients with dry-eye syndrome (keratoconjunctivitis sicca) [239].

A number of ocular inserts using different techniques, namely soluble, erodible, nonerodible, and hydrogel inserts with polymers such as cellulose derivates, acrylates, and poly(ethylene oxide), have been investigated over the last few decades.

An example of a degradable matrix system is the pilocarpine-containing inserts formulated by Saettone et al. [148]. Pilocarpine nitrate and polyacrylic acid were incorporated into a matrix containing polyvinyl alcohol and two types of hydroxypropyl methylcellulose. It was shown that all inserts significantly increased the pharmacological effect (miotic response) compared to a solution of pilocarpine nitrate.

Sasaki et al. [151] prepared nondegradable disc-type ophthalmic inserts of β- blockers using different polymers. They found that inserts made from poly(hydroxypropyl methacrylate) were able to control the release of tilisolol hydrochloride.

Numerous studies have also been performed on soluble collagen shields. Collagen shields are fabricated from porcine scleral tissue, which has a similar collagen composition to that of the human cornea. Drug loading is typically achieved by soaking the collagen shield in the drug solution prior to application. Designed to slowly dissolve within 12, 24, or 72 h, collagen shields have attracted much interest as potential sustained ocular drug delivery systems over the last years [240].

5.9.4 CONCLUSION

Although conventional eye drops still represent about 90% of all marketed ophthalmic dosage forms, there have been significant efforts towards the development of new drug delivery systems.

Only a few of these new ophthalmic drug delivery systems have been commercialized over the past decades, but research in the different areas of ocular drug delivery has provided important impetus and dynamism, with the promise of some new and exciting developments in the field.

An ideal ophthalmic delivery system should be able to achieve an effective drug concentration at the target site for an extended period of time while minimizing systemic side effects. In addition, the system should be comfortable and easy to use, as the patient's acceptance will continue to play an important role in the design of future ocular formulations.

All delivery technologies mentioned in this chapter hold unlimited potential for clinical ophthalmology. However, each of them still bears its own drawbacks. To circumvent these, newer trends are directed toward combinations of the different drug delivery approaches. Examples for this include the incorporation of particulates into in situ gelling systems or coating of nanoparticles with mucoadhesive polymers.

These combinations will open new directions for the improvement of ocular bioavailability, but they will also increase the complexity of the formulations, thus increasing the difficulties in understanding the mechanism of action of the drug delivery systems.

Many interesting delivery approaches have been investigated during the past decades in order to optimize ocular bioavailability, but much remains to be learned before the perfect ocular drug delivery system can be developed.

REFERENCES

1. Le Bourlais, C., Acar, L., Zia, H., Sado, P. A., Needham, T., and Leverge, R. (1998), Ophthalmic drug delivery systems—Recent advances, *Prog. Ret. Eye Res.*, 17(1), 33–58.
2. Le Bourlais, C. A., Treupel-Acar, L., Rhodes, C. T., Sado, P. A., and Leverge, R. (1995), New ophthalmic drug delivery systems, *Drug Dev. Ind. Pharm.*, 21(1), 19–59.
3. Jarvinen, K., Jarvinen, T., and Urtti, A. (1995), Ocular absorption following topical delivery, *Adv. Drug Deliv. Rev.*, 16(1), 3–19.
4. Sasaki, H., Yamamura, K., Mukai, T., Nishida, K., Nakamura, J., Nakashima, M., et al. (1999), Enhancement of ocular drug penetration, *Crit. Rev. Ther. Drug Carrier Syst.*, 16(1), 85–146.
5. Baeyens, V., and Gurny, R. (1997), Chemical and physical parameters of tears relevant for the design of ocular drug delivery formulations, *Pharm. Acta Helv.*, 72(4), 191–202.
6. Van Haeringen, N. J. (1981), Clinical biochemistry of tears, *Surv. Ophthalmol.*, 26(2), 84–96.
7. Nagyova, B., and Tiffany, J. M. (1999), Components responsible for the surface tension of human tears, *Curr. Eye Res.*, 19(1), 4–11.
8. Tiffany, J. M. (2003), Tears in health and disease, *Eye*, 17(8), 923–926.
9. Murube, J., Murube, L., and Murube, A. (1999), Origin and types of emotional tearing, *Eur. J. Ophthalmol.*, 9(2), 77–84.
10. Versura, P., Bonvicini, F., Caramazza, R., and Laschi, R. (1985), Scanning electron microscopy of normal human corneal epithelium obtained by scraping-off in vivo, *Acta Ophthalmol.*, 63(3), 361–365.

11. Klyce, S. D., and Crosson, C. E. (1985), Transport processes across the rabbit corneal epithelium: A review, *Curr. Eye Res.*, 4(4), 323–331.

12. Lens, A. (1999), *Ocular Anatomy and Physiology*, Slack, New York.

13. Ahmed, I. (2003), The noncorneal route in ocular drug delivery, in Mitra, A. K., Ed., *Ophthalmic Drug Delivery Systems*, 2nd ed., Marcel Dekker, New York, pp. 335–363.

14. Greaves, J. L., and Wilson, C. G. (1993), Treatment of diseases of the eye with mucoadhesive delivery systems, *Adv. Drug Deliv. Rev.*, 11(3), 349–383.

15. Huang, A. J. W., Tseng, S. C. G., and Kenyon, K. R. (1989), Paracellular permeability of corneal and conjunctival epithelia, *Invest. Ophthalmol. Vis. Sci.*, 30(4), 684–689.

16. Ahmed, I., and Patton, T. F. (1985), Importance of the noncorneal absorption route in topical ophthalmic drug delivery, *Invest. Ophthalmol. Vis. Sci.*, 26(4), 584–587.

17. Macha, S., Hughes, P. M., and Mitra, A. K. (2003), Overview of ocular drug delivery, in Mitra, A. K., Ed., *Ophthalmic Drug Delivery Systems*, 2nd ed., Marcel Dekker, New York, pp. 1–11.

18. Robinson, J. C. (1993), Ocular anatomy and physiology relevant to ocular drug delivery, in Mitra, A. K., Ed., *Ophthalmic Drug Delivery Systems*, Marcel Dekker, New York, pp. 29–58.

19. Mikkelson, T. J., Chrai, S. S., and Robinson, J. R. (1973), Altered bioavailability of drugs in the eye due to drug-protein interaction, *J. Pharm. Sci.*, 62(10), 1648–1653.

20. Stjernschantz, J., and Astin, M. (1993), Anatomy and physiology of the eye. Physiological aspects of ocular drug therapy, in Edman, P., Ed., *Biopharmaceutics of Ocular Drug Delivery*, CRC Press, Boca Raton, FL, pp. 1–25.

21. Hornof, M., Toropainen, E., and Urtti, A. (2005), Cell culture models of the ocular barriers, *Eur. J. Pharm. Biopharm.*, 60(2), 207–225.

22. Jumbe, N. L., and Miller, M. H. (2003), Ocular drug transfer following systemic drug administration, in Mitra, A. K., Ed., *Ophthalmic Drug Delivery Systems*, 2nd ed., Marcel Dekker, New York, pp. 109–133.

23. Urtti, A., and Salminen, L. (1993), Animal pharmacokinetic studies, in Mitra, A. K., Ed., *Ophthalmic Drug Delivery Systems*, Marcel Dekker, New York, pp. 121–136.

24. Duvvuri, S., Majumdar, S., and Mitra, A. K. (2003), Drug delivery to the retina: Challenges and opportunities, *Expert Opin. Biol. Ther.*, 3(1), 45–56.

25. Loftssona, T., and Jarvinen, T. (1999), Cyclodextrins in ophthalmic drug delivery, *Adv. Drug Deliv. Rev.*, 36(1), 59–79.

26. Kaur, I. P., Garg, A., Singla, A. K., and Aggarwal, D. (2004), Vesicular systems in ocular drug delivery: An overview, *Int. J. Pharm.*, 269(1), 1–14.

27. Lee, V. H., and Robinson, J. R. (1979), Mechanistic and quantitative evaluation of precorneal pilocarpine disposition in albino rabbits, *J. Pharm. Sci.*, 68(6), 673–684.

28. Lee, V. H. L., and Li, V. H. K. (1989), Prodrugs for improved ocular drug delivery, *Adv. Drug Deliv. Rev.*, 3(1), 1–38.

29. Rojanasakul, Y., and Robinson, J. R. (1989), Transport mechanisms of the cornea: Characterization of barrier permselectivity, *Int. J. Pharm.*, 55(2–3), 237–246.

30. Palmgren, J. J., Toropainen, E., Auriola, S., and Urtti, A. (2002), Liquid chromatographic-electrospray ionization mass spectrometric analysis of neutral and charged polyethylene glycols, *J. Chromatogr. A*, 976(1–2), 165–170.

31. Van Ooteghem, M. M. M. (1993), Formulation of ophthalmic solutions and suspensions. Problems and Advantages, in Edman, P., Ed., *Biopharmaceutics of Ocular Drug Delivery*, CRC Press, Boca Raton, FL, pp. 27–42.

32. Schoenwald, R. D., and Huang, H. S. (1983), Corneal penetration behavior of beta-blocking agents I: Physiochemical factors, *J. Pharm. Sci.*, 72(11), 1266–1272.
33. Grass, G. M., Wood, R. W., and Robinson, J. R. (1985), Effects of calcium chelating agents on corneal permeability, *Invest. Ophthalmol. Vis. Sci.*, 26(1), 110–113.
34. Liaw, J., Rojanasakul, Y., and Robinson, J. R. (1992), The effect of drug charge type and charge density on corneal transport, *Int. J. Pharm.*, 88(1–3), 111–124.
35. Sieg, J. W., and Robinson, J. R. (1975), Vehicle effects on ocular drug bioavailability: Evaluation of fluorometholone, *J. Pharm. Sci.*, 64(6), 931–936.
36. Kaur, I. P., and Smitha, R. (2002), Penetration enhancers and ocular bioadhesives: Two new avenues for ophthalmic drug delivery, *Drug Dev. Ind. Pharm.*, 28(4), 353–369.
37. Mishima, S., Gasset, A., Klyce, S. D., Jr, and Baum, J. L. (1966), Determination of tear volume and tear flow, *Invest. Ophthalmol.*, 5(3), 264–276.
38. Mainardes, R. M., Urban, M. C. C., Cinto, P. O., Khalil, N. M., Chaud, M. V., Evangelista, R. C., et al. (2005), Colloidal carriers for ophthalmic drug delivery, *Curr. Drug Targets*, 6(3), 363–371.
39. Chrai, S. S., Patton, T. F., Mehta, A., and Robinson, J. R. (1973), Lacrimal and instilled fluid dynamics in rabbit eyes, *J. Pharm. Sci.*, 62(7), 1112–1121.
40. Chrai, S. S., and Robinson, J. R. (1974), Ocular evaluation of methylcellulose vehicle in albino rabbits, *J. Pharm. Sci.*, 63(8), 1218–1223.
41. Sieg, J. W., and Triplett, J. W. (1980), Precorneal retention of topically instilled micronized particles, *J. Pharm. Sci.*, 69(7), 863–864.
42. Lee, V. H. L., Urrea, P. T., Smith, R. E., and Schanzlin, D. J. (1985), Ocular drug bioavailability from topically applied liposomes, *Surv. Ophthalmol.*, 29(5), 335–348.
43. Zimmer, A., and Kreuter, J. (1995), Microspheres and nanoparticles used in ocular delivery systems, *Adv. Drug Deliv. Rev.*, 16(1), 61–73.
44. Ludwig, A. (2005), The use of mucoadhesive polymers in ocular drug delivery, *Adv. Drug Deliv. Rev.*, 57(11), 1595–1639.
45. Ludwig, A., and Van Ooteghem, M. (1987), The influence of the osmolality on the precorneal retention of ophthalmic solutions, *J. Pharm. Belg.*, 42(4), 259–266.
46. Kaufman, H. E., Steinemann, T. L., Lehman, E., Thompson, H. W., Varnell, E. D., Jacob-LaBarre, J. T., et al. (1994), Collagen-based drug delivery and artificial tears, *J. Ocul. Pharmacol.*, 10(1), 17–27.
47. Ke, T.-L., Cagle, G., Schlech, B., Lorenzetti, O. J., and Mattern, J. (2001), Ocular bioavailability of ciprofloxacin in sustained release formulations, *J. Ocul. Pharmacol. Ther.*, 17(6), 555–563.
48. Meseguer, G., Buri, P., Plazonnet, B., Rozier, A., and Gurny, R. (1996), Gamma scintigraphic comparison of eyedrops containing pilocarpine in healthy volunteers, *J. Ocul. Pharmacol. Ther.*, 12(4), 481–488.
49. Hui, H.-W., and Robinson, J. R. (1985), Ocular delivery of progesterone using a bioadhesive polymer, *Int. J. Pharm.*, 26(3), 203–213.
50. Lele, B. S., and Hoffman, A. S. (2000), Insoluble ionic complexes of polyacrylic acid with a cationic drug for use as a mucoadhesive, ophthalmic drug delivery system, *J. Biomater. Sci. Polym. Ed.*, 11(12), 1319–1331.
51. Saettone, M. F., Giannaccini, B., Ravecca, S., La Marca, F., and Tota, G. (1984), Polymer effects on ocular bioavailability—The influence of different liquid vehicles on the mydriatic response of tropicamide in humans and in rabbits, *Int. J. Pharm.*, 20(1–2), 187–202.
52. Herrero-Vanrell, R., Fernandez-Carballido, A., Frutos, G., and Cadorniga, R. (2000), Enhancement of the mydriatic response to tropicamide by bioadhesive polymers, *J. Ocul. Pharmacol. Ther.*, 16(5), 419–428.

53. Saettone, M. F., Chetoni, P., Tilde Torracca, M., Burgalassi, S., and Giannaccini, B. (1989), Evaluation of muco-adhesive properties and in vivo activity of ophthalmic vehicles based on hyaluronic acid, *Int. J. Pharm.*, 51(3), 203–212.

54. Sandri, G., Bonferoni, M. C., Chetoni, P., Rossi, S., Ferrari, F., Ronchi, C., et al. (2006), Ophthalmic delivery systems based on drug-polymer-polymer ionic ternary interaction: In vitro and in vivo characterization, *Eur. J. Pharm. Biopharm.*, 62(1), 59–69.

55. Felt, O., Furrer, P., Mayer, J. M., Plazonnet, B., Buri, P., and Gurny, R. (1999), Topical use of chitosan in ophthalmology: Tolerance assessment and evaluation of precorneal retention, *Int. J. Pharm.*, 180(2), 185–193.

56. Felt, O., Baeyens, V., Buri, P., and Gurny, R. (2001), Delivery of antibiotics to the eye using a positively charged polysaccharide as vehicle, *Aaps Pharmsci*, 3(4), E34.

57. Di Colo, G., Zambito, Y., Burgalassi, S., Nardini, I., and Saettone, M. F. (2004), Effect of chitosan and of N-carboxymethylchitosan on intraocular penetration of topically applied ofloxacin, *Int. J. Pharm.*, 273(1–2), 37–44.

58. Di Colo, G., Zambito, Y., Burgalassi, S., and Saettone, M. F. (2003), Effects of chitosan and of its N-trimethyl and N-carboxymethyl derivatives on the ocular pharmacokinetics of ofloxacin in rabbits, paper presented at the 30th International Symposium on Controlled Release of Bioactive Materials, Glasgow, Scotland.

59. Davies, N. M., Farr, S. J., Hadgraft, J., and Kellaway, I. W. (1991), Evaluation of mucoadhesive polymers in ocular drug delivery. I. Viscous solutions, *Pharm. Res.*, 8(8), 1039–1043.

60. Edsman, K., Carlfors, J., and Harju, K. (1996), Rheological evaluation and ocular contact time of some carbomer gels for ophthalmic use, *Int. J. Pharm.*, 137(2), 233–241.

61. Hartmann, V., and Keipert, S. (2000), Physico-chemical, in vitro and in vivo characterisation of polymers for ocular use, *Pharmazie*, 55(6), 440–443.

62. Burgalassi, S., Chetoni, P., Panichi, L., Boldrini, E., and Saettone, M. F. (2000), Xyloglucan as a novel vehicle for timolol: Pharmacokinetics and pressure lowering activity in rabbits, *J. Ocul. Pharmacol. Ther.*, 16(6), 497–509.

63. Ceulemans, J., Vinckier, I., and Ludwig, A. (2002), The use of xanthan gum in an ophthalmic liquid dosage form: Rheological characterization of the interaction with mucin, *J. Pharm. Sci.*, 91(4), 1117–1127.

64. Verschueren, E., Van Santvliet, L., and Ludwig, A. (1996), Evaluation of various carrageenans as ophthalmic viscolysers, STP Pharma *Sci.*, 6(3), 203–210.

65. Albasini, M., and Ludwig, A. (1995), Evaluation of polysaccharides intended for ophthalmic use in ocular dosage forms, *Farmaco*, 50(Sept.), 633–642.

66. Snibson, G. R., Greaves, J. L., Soper, N. D., Prydal, J. I., Wilson, C. G., and Bron, A. J. (1990), Precorneal residence times of sodium hyaluronate solutions studied by quantitative gamma scintigraphy, *Eye*, 4(4), 594–602.

67. Snibson, G. R., Greaves, J. L., Soper, N. D., Tiffany, J. M., Wilson, C. G., and Bron, A. J. (1992), Ocular surface residence times of artificial tear solutions, *Cornea*, 11(4), 288–293.

68. Debbasch, C., De La Salle, S. B., Brignole, F., Rat, P., Warnet, J. M., and Baudouin, C. (2002), Cytoprotective effects of hyaluronic acid and Carbomer 934P in ocular surface epithelial cells, *Invest. Ophthalmol. Vis. Sci.*, 43(11), 3409–3415.

69. Oechsner, M., and Keipert, S. (1999), Polyacrylic acid/polyvinylpyrrolidone bipolymeric systems. I. Rheological and mucoadhesive properties of formulations potentially useful for the treatment of dry-eye-syndrome, *Eur. J. Pharm. Biopharm.*, 47(2), 113–118.

70. Shibuya, T., Kashiwagi, K., and Tsukahara, S. (2003), Comparison of efficacy and tolerability between two gel-forming timolol maleate ophthalmic solutions in patients with glaucoma or ocular hypertension, *Ophthalmologica*, 217(1), 31–38.
71. Rosenlund, E. F. (1996), The intraocular pressure lowering effect of timolol in gel-forming solution, *Acta Ophthalmol. Scand.*, 74(2), 160–162.
72. Stewart, W. C., Leland, T. M., Cate, E. A., and Stewart, J. A. (1998), Efficacy and safety of timolol solution once daily versus timolol gel in treating elevated intraocular pressure, *J. Glaucoma*, 7(6), 402–407.
73. Shedden, A., Laurence, J., and Tipping, R. (2001), Efficacy and tolerability of timolol maleate ophthalmic gel-forming solution versus timolol ophthalmic solution in adults with open-angle glaucoma or ocular hypertension: A six-month, double-masked, multicenter study, *Clin. Ther.*, 23(3), 440–450.
74. Carlfors, J., Edsman, K., Petersson, R., and Jornving, K. (1998), Rheological evaluation of Gelrite(R) in situ gels for ophthalmic use, *Eur. J. Pharm. Sci.*, 6(2), 113–119.
75. Schenker, H. I., and Silver, L. H. (2000), Long-term intraocular pressure-lowering efficacy and safety of timolol maleate gel-forming solution 0.5% compared with timoptic XE 0.5% in a 12-month study, *Am. J. Ophthalmol.*, 130(2), 145–150.
76. Shedden, A. H., Laurence, J., Barrish, A., and Olah, T. V. (2001), Plasma timolol concentrations of timolol maleate: Timolol gel-forming solution (TIMOPTIC-XE) once daily versus timolol maleate ophthalmic solution twice daily, *Doc. Ophthalmol.*, 103(1), 73–79.
77. Lindell, K., and Engstrom, S. (1993), In vitro release of timolol maleate from an in situ gelling polymer system, *Int. J. Pharm.*, 95(1–3), 219–228.
78. Rozier, A., Mazuel, C., Grove, J., and Plazonnet, B. (1989), Gelrite(R): A novel, ion-activated, in-situ gelling polymer for ophthalmic vehicles. Effect on bioavailability of timolol, *Int. J. Pharm.*, 57(2), 163–168.
79. Balasubramaniam, J., Kant, S., and Pandit, J. K. (2003), In vitro and in vivo evaluation of the Gelrite gellan gum-based ocular delivery system for indomethacin, *Acta Pharm.*, 53(4), 251–261.
80. Sultana, Y., Aqil, M., and Ali, A. (2006), Ion activated, Gelrite based in situ ophthalmic gels of pefloxacin mesylate: Comparison with conventional eye drops, *Drug Deliv.*, 13(3), 215–219.
81. Liu, Z., Li, J., Nie, S., Liu, H., Ding, P., and Pan, W. (2006), Study of an alginate/HPMC-based in situ gelling ophthalmic delivery system for gatifloxacin, *Int. J. Pharm.*, 315(1–2), 12–17.
82. Srividya, B., Cardoza, R. M., and Amin, P. D. (2001), Sustained ophthalmic delivery of ofloxacin from a pH triggered in situ gelling system, *J. Controlled Release*, 73(2–3), 205–211.
83. Cohen, S., Lobel, E., Trevgoda, A., and Peled, Y. (1997), A novel in situ-forming ophthalmic drug delivery system from alginates undergoing gelation in the eye, *J. Controlled Release*, 44(2–3), 201–208.
84. Demailly, P., Allaire, C., and Trinquand, C. (2001), Ocular hypotensive efficacy and safety of once daily carteolol alginate, *Br. J. Ophthalmol.*, 85(8), 921–924.
85. Cho, K. Y., Chung, T. W., Kim, B. C., Kim, M. K., Lee, J. H., Wee, W. R., et al. (2003), Release of ciprofloxacin from poloxamer-graft-hyaluronic acid hydrogels in vitro, *Int. J. Pharm.*, 260(1), 83–91.
86. Desai, S. D., and Blanchard, J. (1998), Evaluation of pluronic F127-based sustained-release ocular delivery systems for pilocarpine using the albino rabbit eye model, *J. Pharm. Sci.*, 87(10), 1190–1195.

87. Lin, H. R., and Sung, K. C. (2000), Carbopol/pluronic phase change solutions for ophthalmic drug delivery, *J. Controlled Release*, 69(3), 379–388.

88. Miyazaki, S., Suzuki, S., Kawasaki, N., Endo, K., Takahashi, A., and Attwood, D. (2001), In situ gelling xyloglucan formulations for sustained release ocular delivery of pilocarpine hydrochloride, *Int. J. Pharm.*, 229(1–2), 29–36.

89. El-Kamel, A. H. (2002), In vitro and in vivo evaluation of Pluronic F127-based ocular delivery system for timolol maleate, *Int. J. Pharm.*, 241(1), 47–55.

90. Wu, J., Su, Z.-G., and Ma, G.-H. (2006), A thermo- and pH-sensitive hydrogel composed of quaternized chitosan/glycerophosphate, *Int. J. Pharm.*, 315(1–2), 1–11.

91. Wei, G., Xu, H., Ding, P. T., Li, S. M., and Zheng, J. M. (2002), Thermosetting gels with modulated gelation temperature for ophthalmic use: The rheological and gamma scintigraphic studies, *J. Controlled Release*, 83(1), 65–74.

92. Vandervoort, J., and Ludwig, A. (2004), Preparation and evaluation of drug-loaded gelatin nanoparticles for topical ophthalmic use, *Eur. J. Pharm. Biopharm.*, 57(2), 251–261.

93. Zimmer, A. K., Maincent, P., Thouvenot, P., and Kreuter, J. (1994), Hydrocortisone delivery to healthy and inflamed eyes using a micellar polysorbate 80 solution or albumin nanoparticles, *Int. J. Pharm.*, 110(3), 211–222.

94. Zimmer, A. K., Zerbe, H., and Kreuter, J. (1994), Evaluation of pilocarpine-loaded albumin particles as drug delivery systems for controlled delivery in the eye I. In vitro and in vivo characterisation, *J. Controlled Release*, 32(1), 57–70.

95. Zimmer, A. K., Chetoni, P., Saettone, M. F., Zerbe, H., and Kreuter, J. (1995), Evaluation of pilocarpine-loaded albumin particles as controlled drug delivery systems for the eye. II. Coadministration with bioadhesive and viscous polymers, *J. Controlled Release*, 33(1), 31–46.

96. Mazor, Z., Ticho, U., Rehany, U., and Rose, L. (1979), Piloplex, a new long-acting pilocarpine polymer salt. B: Comparative study of the visual effects of pilocarpine and Piloplex eye drops, *Br. J. Ophthalmol.*, 63(1), 48–51.

97. Ticho, U., Blementhal, M., Zonis, S., Gal, A., Blank, I., and Mazor, Z. W. (1979), Piloplex, a new long-acting pilocarpine polymer salt. A: Long-term study, *Br. J. Ophthalmol.*, 63(1), 45–47.

98. Klein, H. Z., Lugo, M., Shields, M. B., Leon, J., and Duzman, E. (1985), A dose-response study of piloplex for duration of action, *Am. J. Ophthalmol.*, 99(1), 23–26.

99. Gurny, R. (1981), Preliminary study of prolonged acting drug delivery system for the treatment of glaucoma, *Pharm. Acta Helv.*, 56(4–5), 130–132.

100. Gurny, R., Boye, T., and Ibrahim, H. (1985), Ocular therapy with nanoparticulate systems for controlled drug delivery, *J. Controlled Release*, 2, 353–361.

101. Gurny, R., Ibrahim, H., Aebi, A., Buri, P., Wilson, C. G., Washington, N., et al. (1987), Design and evaluation of controlled release systems for the eye, *J. Controlled Release*, 6(1), 367–373.

102. Marchal-Heussler, L., Orallo, F., Vila Jato, J. L., and Alonso, M. J. (1992), Colloidal drug delivery systems for the eye: A comparison of the efficacy of three different polymers: Polyisobutylcyanoacrylate, polylacticcoglycolic acid, poly-epsilon-caprolactone, *STP Pharma Sci.*, 2, 98–104.

103. Marchal-Heussler, L., Sirbat, D., Hoffman, M., and Maincent, P. (1993), Poly(epsilon-caprolactone) nanocapsules in carteolol ophthalmic delivery, *Pharm. Res.*, 10(3), 386–390.

104. Harmia, T., Speiser, P., and Kreuter, J. (1986), Optimization of pilocarpine loading onto nanoparticles by sorption procedures, *Int. J. Pharm.*, 33(1–3), 45–54.

105. Diepold, R., Kreuter, J., Himber, J., Gurny, R., Lee, V. H., Robinson, J. R., et al. (1989), Comparison of different models for the testing of pilocarpine eyedrops using conventional eyedrops and a novel depot formulation (nanoparticles), *Graefes Arch. Clin. Exp. Ophthalmol.*, 227(2), 188–193.

106. Harmia, T., Speiser, P., and Kreuter, J. (1987), Nanoparticles as drug carriers in ophthalmology, *Pharm. Acta Helv.*, 62(12), 322–331.

107. Zimmer, A., Mutschler, E., Lambrecht, G., Mayer, D., and Kreuter, J. (1994), Pharmacokinetic and pharmacodynamic aspects of an ophthalmic pilocarpine nanoparticle-delivery-system, *Pharm. Res.*, 11(10), 1435–1442.

108. De Campos, A. M., Sanchez, A., and Alonso, M. J. (2001), Chitosan nanoparticles: A new vehicle for the improvement of the delivery of drugs to the ocular surface. Application to cyclosporin A, *Int. J. Pharm.*, 224(1–2), 159–168.

109. Calvo, P., Vila-Jato, J. L., and Alonso, M. J. (1997), Evaluation of cationic polymer-coated nanocapsules as ocular drug carriers, *Int. J. Pharm.*, 153(1), 41–50.

110. De Campos, A. M., Sanchez, A., Gref, R., Calvo, P., and Alonso, M. J. (2003), The effect of a PEG versus a chitosan coating on the interaction of drug colloidal carriers with the ocular mucosa, *Eur. J. Pharm. Sci.*, 20(1), 73–81.

111. Giannavola, C., Bucolo, C., Maltese, A., Paolino, D., Vandelli, M. A., Puglisi, G., et al. (2003), Influence of preparation conditions on acyclovir-loaded poly-*d,l*-lactic acid nanospheres and effect of PEG coating on ocular drug bioavailability, *Pharma. Res.*, 20(4), 584–590.

112. Hsiue, G.-H., Hsu, S.-h., Yang, C.-C., Lee, S.-H., and Yang, I.-K. (2002), Preparation of controlled release ophthalmic drops, for glaucoma therapy using thermosensitive poly-*N*-isopropylacrylamide, *Biomaterials*, 23(2), 457–462.

113. Pignatello, R., Bucolo, C., Ferrara, P., Maltese, A., Puleo, A., and Puglisi, G. (2002), Eudragit RS100(R) nanosuspensions for the ophthalmic controlled delivery of ibuprofen, *Eur. J. Pharm. Sci.*, 16(1–2), 53–61.

114. Bucolo, C., Maltese, A., Puglisi, G., and Pignatello, R. (2002), Enhanced ocular anti-inflammatory activity of ibuprofen carried by an eudragit RS100(R) nanoparticle suspension, *Ophthal. Res.*, 34(5), 319–323.

115. Pignatello, R., Bucolo, C., Spedalieri, G., Maltese, A., and Puglisi, G. (2002), Flurbiprofen-loaded acrylate polymer nanosuspensions for ophthalmic application, *Biomaterials*, 23(15), 3247–3255.

116. Khopade, A. J., and Jain, N. K. (1995), Self assembling nanostructures for sustained ophthalmic drug delivery, *Pharmazie*, 50(12), 812–814.

117. Cavalli, R., Gasco, M. R., Chetoni, P., Burgalassi, S., and Saettone, M. F. (2002), Solid lipid nanoparticles (SLN) as ocular delivery system for tobramycin, *Int. J. Pharm.*, 238(1–2), 241–245.

118. Safwat, S. M., and Al-Kassas, R. S. (2002), Evaluation of gentamicin-Eudragit microspheres as ophthalmic delivery systems in inflamed rabbit's eyes, STP Pharma *Sci.*, 12(6), 357–361.

119. Giunchedi, P., Conte, U., Chetoni, P., and Saettone, M. F. (1999), Pectin microspheres as ophthalmic carriers for piroxicam: Evaluation in vitro and in vivo in albino rabbits, *Eur. J. Pharm. Sci.*, 9(1), 1–7.

120. Giunchedi, P., Chetoni, P., Conte, U., and Saettone, M. F. (2000), Albumin microspheres for ocular delivery of piroxicam, *Pharm. Pharmacol. Commun.*, 6(4), 149–153.

121. Davies, N. M., Farr, S. J., Hadgraft, J., and Kellaway, I. W. (1992), Evaluation of mucoadhesive polymers in ocular drug delivery. II. Polymer-coated vesicles, *Pharm. Res.*, 9(9), 1137–1144.

122. Durrani, A. M., Davies, N. M., Thomas, M., and Kellaway, I. W. (1992), Pilocarpine bioavailability from a mucoadhesive liposomal ophthalmic drug delivery system, *Int. J. Pharm.*, 88(1–3), 409–415.
123. Henriksen, I., Green, K. L., Smart, J. D., Smistad, G., and Karlsen, J. (1996), Bioadhesion of hydrated chitosans: An in vitro and in vivo study, *Int. J. Pharm.*, 145(1–2), 231–240.
124. Bochot, A., Fattal, E., Gulik, A., Couarraze, G., and Couvreur, P. (1998), Liposomes dispersed within a thermosensitive gel: A new dosage form for ocular delivery of oligonucleotides, *Pharm. Res.*, 15(9), 1364–1369.
125. Bochot, A., Fattal, E., Grossiord, J. L., Puisieux, F., and Couvreur, P. (1998), Characterization of a new ocular delivery system based on a dispersion of liposomes in a thermosensitive gel, *Int. J. Pharm.*, 162(1–2), 119–127.
126. Bochot, A., Mashhour, B., Puisieux, F., Couvreur, P., and Fattal, E. (1998), Comparison of the ocular distribution of a model oligonucleotide after topical instillation in rabbits of conventional and new dosage forms, *J. Drug Target.*, 6(4), 309–313.
127. Nagarsenker, M. S., Londhe, V. Y., and Nadkarni, G. D. (1999), Preparation and evaluation of liposomal formulations of tropicamide for ocular delivery, *Int. J. Pharm.*, 190(1), 63–71.
128. Aggarwal, D., and Kaur, I. P. (2005), Improved pharmacodynamics of timolol maleate from a mucoadhesive niosomal ophthalmic drug delivery system, *Int. J. Pharm.*, 290(1–2), 155–159.
129. Muchtar, S., Abdulrazik, M., Frucht-Pery, J., and Benita, S. (1997), Ex-vivo permeation study of indomethacin from a submicron emulsion through albino rabbit cornea, *J. Controlled Release*, 44(1), 55–64.
130. Sznitowska, M., Zurowska-Pryczkowska, K., Dabrowska, E., and Janicki, S. (2000), Increased partitioning of pilocarpine to the oily phase of submicron emulsion does not result in improved ocular bioavailability, *Int. J. Pharm.*, 202(1–2), 161–164.
131. Haße, A., and Keipert, S. (1997), Development and characterisation of microemulsions for ocular application, *Eur. J. Pham. Biopharm.*, 43(2), 179–183.
132. Radomska, A., and Dobrucki, R. (2000), The use of some ingredients for microemulsion preparation containing retinol and its esters, *Int. J. Pharm.*, 196(2), 131–134.
133. Alany, R. G., Rades, T., Nicoll, J., Tucker, I. G., and Davies, N. M. (2006), W/O microemulsions for ocular delivery: Evaluation of ocular irritation and precorneal retention, *J. Controlled Release*, 111(1–2), 145–152.
134. Kaback, M. B., Podos, S. M., Harbin, T. S.Jr, ., Mandell, A., and Becker, B. (1976), The effects of dipivalyl epinephrine on the eye, *Am. J. Ophthalmol.*, 81(6), 768–772.
135. Wei, C. P., Anderson, J. A., and Leopold, I. (1978), Ocular absorption and metabolism of topically applied epinephrine and a dipivalyl ester of epinephrine, *Invest. Ophthalmol. Vis. Sci.*, 17(4), 315–321.
136. Jarvinen, T., Suhonen, P., Auriola, S., Vepsalainen, J., Peura, P., et al. (1991), *O,O′*-(1,4-Xylylene)bispilocarpic acid esters as new potential double prodrugs of pilocarpine for improved ocular delivery. Part 1. Synthesis and analysis, *Int. J. Pharm.*, 75(2–3), 249–258.
137. Tirucherai, G. S., Dias, C., and Mitra, A. K. (2002), Corneal permeation of ganciclovir: Mechanism of ganciclovir permeation enhancement by acyl ester prodrug design, *J. Ocul. Pharmacol. Ther.*, 18(6), 535–548.
138. Saettone, M. F., Chetoni, P., Cerbai, R., Mazzanti, G., and Braghiroli, L. (1996), Evaluation of ocular permeation enhancers: In vitro effects on corneal transport of four beta-blockers, and in vitro/in vivo toxic activity, *Int. J. Pharm.*, 142(1), 103–113.

139. Maestrelli, F., Mura, P., Casini, A., Mincione, F., Scozzafava, A., and Supuran, C. T. (2002), Cyclodextrin complexes of sulfonamide carbonic anhydrase inhibitors as long-lasting topically acting antiglaucoma agents, *J. Pharm. Sci.*, 91(10), 2211–2219.
140. Nijhawan, R., and Agarwal, S. P. (2003), Development of an ophthalmic formulation containing ciprofloxacin-hydroxypropyl-β-cyclodextrin complex, *Boll. Chim. Farm.*, 142(5), 214–219.
141. Aktas, Y., Unlu, N., Orhan, M., Irkec, M., and Hincal, A. A. (2003), Influence of hydroxypropyl beta-cyclodextrin on the corneal permeation of pilocarpine, *Drug Dev. Ind. Pharm.*, 29(2), 223–230.
142. Harwood, R. J., and Schwartz, J. B. (1982), Drug release from compression molded films: Preliminary studies with pilocarpine, *Drug Dev. Ind. Pharm.*, 8(5), 663–682.
143. Hsiue, G. H., Guu, J. A., and Cheng, C. C. (2001), Poly(2-hydroxyethyl methacrylate) film as a drug delivery system for pilocarpine, *Biomaterials*, 22(13), 1763–1769.
144. Bharath, S., and Hiremath, S. R. (1999), Ocular delivery systems of pefloxacin mesylate, *Pharmazie*, 54(1), 55–58.
145. Sultana, Y., Aqil, M., and Ali, A. (2005), Ocular inserts for controlled delivery of pefloxacin mesylate: Preparation and evaluation, *Acta Pharma.*, 55(3), 305–314.
146. Zimmermann, C., Drewe, J., Flammer, J., and Shaarawy, T. (2004), In vitro release of mitomycin C from collagen implants, *Curr. Eye Res.*, 28(1), 1–4.
147. Vasantha, V., Sehgal, P. K., and Rao, K. P. (1988), Collagen ophthalmic inserts for pilocarpine drug delivery system, *Int. J. Pharm.*, 47(1–3), 95–102.
148. Saettone, M. F., Giannaccini, B., Chetoni, P., Galli, G., and Chiellini, E. (1984), Vehicle effects in ophthalmic bioavailability: An evaluation of polymeric inserts containing pilocarpine, *J. Pharm. Pharmacol.*, 36(4), 229–234.
149. Saettone, M. F., Torracca, M. T., Pagano, A., Giannaccini, B., Rodriguez, L., and Cini, M. (1992), Controlled release of pilocarpine from coated polymeric ophthalmic inserts prepared by extrusion, *Int. J. Pharm.*, 86(2–3), 159–166.
150. Saettone, M. F., Chetoni, P., Bianchi, L. M., Giannaccini, B., Conte, U., and Sangalli, M. E. (1995), Controlled release of timolol maleate from coated ophthalmic mini-tablets prepared by compression, *Int. J. Pharm.*, 126(1–2), 79–82.
151. Sasaki, H., Tei, C., Nishida, K., and Nakamura, J. (1993), Drug release from an ophthalmic insert of a beta-blocker as an ocular drug delivery system, *J. Controlled Release*, 27(2), 127–137.
152. Sasaki, H., Nagano, T., Sakanaka, K., Kawakami, S., Nishida, K., Nakamura, J., et al. (2003), One-side-coated insert as a unique ophthalmic drug delivery system, *J. Controlled Release*, 92(3), 241–247.
153. Charoo, N. A., Kohli, K., Ali, A., and Anwer, A. (2003), Ophthalmic delivery of ciprofloxacin hydrochloride from different polymer formulations: In vitro and in vivo studies, *Drug Dev. Ind. Pharm.*, 29(2), 215–221.
154. Pijls, R. T., Sonderkamp, T., Daube, G. W., Krebber, R., Hanssen, H. H. L., Nuijts, R. M. M. A., et al. (2005), Studies on a new device for drug delivery to the eye, *Eur. J. Pharm. Biopharm.*, 59(2), 283–288.
155. Lux, A., Maier, S., Dinslage, S., Suverkrup, R., and Diestelhorst, M. (2003), A comparative bioavailability study of three conventional eye drops versus a single lyophilisate, *Br. J. Ophthalmol.*, 87(4), 436–440.
156. Dinslage, S., Diestelhorst, M., Weichselbaum, A., and Suverkrup, R. (2002), Lyophilisates for drug delivery in Ophthalmology: Pharmacokinetics of fluorescein in the human anterior segment, *Br. J. Ophthalmol.*, 86(10), 1114–1117.
157. Ceulemans, J., Vermeire, A., Adriaens, E., Remon, J. P., and Ludwig, A. (2001), Evaluation of a mucoadhesive tablet for ocular use, *J. Controlled Release*, 77(3), 333–344.

158. Weyenberg, W., Vermeire, A., Remon, J. P., and Ludwig, A. (2003), Characterization and in vivo evaluation of ocular bioadhesive minitablets compressed at different forces, *J. Controlled Release*, 89(2), 329–340.

159. Baeyens, V., Kaltsatos, V., Boisrame, B., Fathi, M., and Gurny, R. (1998), Evaluation of soluble Bioadhesive Ophthalmic Drug Inserts (BODI) for prolonged release of gentamicin: Lachrymal pharmacokinetics and ocular tolerance, *J. Ocul. Pharmacol. Ther.*, 14(3), 263–272.

160. Baeyens, V., Felt-Baeyens, O., Rougier, S., Pheulpin, S., Boisrame, B., and Gurny, R. (2002), Clinical evaluation of bioadhesive ophthalmic drug inserts (BODI(R)) for the treatment of external ocular infections in dogs, *J. Controlled Release*, 85(1–3), 163–168.

161. Chetoni, P., Di Colo, G., Grandi, M., Morelli, M., Saettone, M. F., and Darougar, S. (1998), Silicone rubber/hydrogel composite ophthalmic inserts: Preparation and preliminary in vitro/in vivo evaluation, *Eur. J. Pharm. Biopharm.*, 46(1), 125–132.

162. Di Colo, G., and Zambito, Y. (2002), A study of release mechanisms of different ophthalmic drugs from erodible ocular inserts based on poly(ethylene oxide), *Eur. J. Pharm. Biopharm.*, 54(2), 193–199.

163. Di Colo, G., Burgalassi, S., Chetoni, P., Fiaschi, M. P., Zambito, Y., and Saettone, M. F. (2001), Relevance of polymer molecular weight to the in vitro/in vivo performances of ocular inserts based on poly(ethylene oxide), *Int. J. Pharm.*, 220(1–2), 169–177.

164. Di Colo, G., Burgalassi, S., Chetoni, P., Fiaschi, M. P., Zambito, Y., and Saettone, M. F. (2001), Gel-forming erodible inserts for ocular controlled delivery of ofloxacin, *Int. J. Pharm.*, 215(1–2), 101–111.

165. Di Colo, G., Zambito, Y., Burgalassi, S., Serafini, A., and Saettone, M. F. (2002), Effect of chitosan on in vitro release and ocular delivery of ofloxacin from erodible inserts based on poly(ethylene oxide), *Int. J. Pharm.*, 248(1–2), 115–122.

166. Hornof, M., Weyenberg, W., Ludwig, A., and Bernkop-Schnurch, A. (2003), Mucoadhesive ocular insert based on thiolated poly(acrylic acid): Development and in vivo evaluation in humans, *J. Controlled Release*, 89(3), 419–428.

167. Frucht-Pery, J., Mechoulam, H., Siganos, C. S., Ever-Hadani, P., Shapiro, M., and Domb, A. (2004), Iontophoresis-gentamicin delivery into the rabbit cornea, using a hydrogel delivery probe, *Exper. Eye Res.*, 78(3), 745–749.

168. Vandamme, T. F., and Brobeck, L. (2005), Poly(amidoamine) dendrimers as ophthalmic vehicles for ocular delivery of pilocarpine nitrate and tropicamide, *J. Controlled Release*, 102(1), 23–38.

169. Lang, J. C. (1995), Ocular drug delivery conventional ocular formulations, *Adv. Drug Deliv. Rev.*, 16(1), 39–43.

170. Fitzgerald, P., and Wilson, C. G. (1994), Polymeric systems for ophthalmic drug delivery, in Dimitriuitra S., Ed., *Polymeric Biomaterials*, Marcel Dekker, New York, pp. 373–398.

171. Kupferman, A., Pratt, M. V., Suckewer, K., and Leibowitz, H. M. (1974), Topically applied steroids in corneal disease. 3. The role of drug derivative in stromal absorption of dexamethasone, *Arch. Ophthalmol.*, 91(5), 373–376.

172. Wilson, C. G., Olejnik, O., and Hardy, J. G. (1983), Precorneal drainage of polyvinyl alcohol solutions in the rabbit assessed by gamma scintigraphy, *J. Pharm. Pharmacol.*, 35(7), 451–454.

173. Olejnik, O., and Weisbecker, C. A. (1990), Ocular bioavailability of topical prednisolone preparations, *Clin. Ther.*, 12(1), 2–11.

174. Davies, N. M. (2000), Biopharmaceutical considerations in topical ocular drug delivery, *Clin. Exp. Pharmacol. Physiol.*, 27(7), 558–562.

175. Davies, N. M., Wang, G., and Tucker, I. G. (1997), Evaluation of a hydrocortisone/hydroxypropyl-[beta]-cyclodextrin solution for ocular drug delivery, *Int. J. Pharm.*, 156(2), 201–209.

176. Olejnik, O. (1993), Conventional systems in ophthalmic drug delivery, in Mitra, A. K., Ed., *Ophthalmic Drug Delivery Systems*, Marcel Dekker, New York, pp. 177–198.

177. Schoenwald, R. D., and Stewart, P. (1980), Effect of particle size on ophthalmic bioavailability of dexamethasone suspensions in rabbits, *J. Pharm. Sci.*, 69(4), 391–394.

178. MacKeen, D. L. (1980), Aqueous formulations and ointments, *Int. Ophthalmol. Clin.*, 20(3), 79–92.

179. Sieg, J. W., and Robinson, J. R. (1977), Vehicle effects on ocular drug bioavailability II: Evaluation of pilocarpine, *J. Pharm. Sci.*, 66(9), 1222–1228.

180. Greaves, J. L., Wilson, C. G., and Birmingham, A. T. (1993), Assessment of the precorneal residence of an ophthalmic ointment in healthy subjects, *Br. J. Clin. Pharmacol.*, 35(2), 188–192.

181. Lee, V. H., and Robinson, J. R. (1986), Topical ocular drug delivery: Recent developments and future challenges, *J. Ocul. Pharmacol.*, 2(1), 67–108.

182. Saettone, M. F., Giannaccini, B., Teneggi, A., Savigni, P., and Tellini, N. (1982), Vehicle effects on ophthalmic bioavailability: The influence of different polymers on the activity of pilocarpine in rabbit and man, *J. Pharm. Pharmacol.*, 34(7), 464–466.

183. Trueblood, J. H., Rossomondo, R. M., Wilson, L. A., and Carlton, W. H. (1975), Corneal contact times of ophthalmic vehicles. Evaluation by microscintigraphy, *Arch. Ophthalmol.*, 93(2), 127–130.

184. Patton, T. F., and Robinson, J. R. (1976), Quantitative precorneal disposition of topically applied pilocarpine nitrate in rabbit eyes, *J. Pharm. Sci.*, 65(9), 1295–1301.

185. Patton, T. F., and Robinson, J. R. (1975), Ocular evaluation of polyvinyl alcohol vehicle in rabbits, *J. Pharm. Sci.*, 64(8), 1312–1316.

186. Tiffany, J. M. (1991), The viscosity of human tears, *Int. Ophthalmol.*, 15(6), 371–376.

187. Van Ooteghem, M. (1987), Factors influencing the retention of ophthalmic solutions on the eye surface, in Saettone, M. F., Bucci, M., and Speiser, P., Eds., *Ophthalmic Drug Delivery: Biopharmaceutical, Technological and Clinical Aspects*, Vol. 11, Livinia, Padova, pp. 7–17

188. Ludwig, A., van Haeringen, N. J., Bodelier, V. M., and Van Ooteghem, M. (1992), Relationship between precorneal retention of viscous eye drops and tear fluid composition, *Int. Ophthalmol.*, 16(1), 23–26.

189. Greaves, J. L., Olejnik, O., and Wilson, C. G. (1992), Polymers and the precorneal tear film, *STP Pharma Sci.*, 2, 13–33.

190. Park, H., and Robinson, J. R. (1987), Mechanisms of mucoadhesion of poly(acrylic acid) hydrogels, *Pharm. Res.*, 4(6), 457–464.

191. Robinson, J. R., and Mlynek, G. M. (1995), Bioadhesive and phase-change polymers for ocular drug delivery, *Adv. Drug Deliv. Rev.*, 16(1), 45–50.

192. Lehr, C.-M., Bouwstra, J. A., Schacht, E. H., and Junginger, H. E. (1992), In vitro evaluation of mucoadhesive properties of chitosan and some other natural polymers, *Int. J. Pharm.*, 78(1–3), 43–48.

193. Park, K., and Robinson, J. R. (1984), Bioadhesive polymers as platforms for oral-controlled drug Delivery: Method to study bioadhesion, *Int. J. Pharm.*, 19(2), 107–127.

194. Meseguer, G., Gurny, R., Buri, P., Rozier, A., and Plazonnet, B. (1993), Gamma scintigraphic study of precorneal drainage and assessment of miotic response in rabbits

of various ophthalmic formulations containing pilocarpine, *Int. J. Pharm.*, 95(1–3), 229–234.

195. Thermes, F., Rozier, A., Plazonnet, B., and Grove, J. (1992), Bioadhesion: The effect of polyacrylic acid on the ocular bioavailability of timolol, *Int. J. Pharm.*, 81(1), 59–65.
196. Duchene, D., Touchard, F., and Peppas, N. A. (1988), Pharmaceutical and medical aspects of bioadhesive systems for drug administration, *Drug Dev. Ind. Pharm.*, 14(2–3), 283–318.
197. Calonge, M. (2001), The treatment of dry eye, *Surv. Ophthalmol.*, 45(Suppl. 2), S227–S239.
198. Zignani, M., Tabatabay, C., and Gurny, R. (1995), Topical semi-solid drug delivery: Kinetics and tolerance of ophthalmic hydrogels, *Adv. Drug Deliv. Rev.*, 16(1), 51–60.
199. Krauland, A. H., Leitner, V. M., and Bernkop-Schnurch, A. (2003), Improvement in the in situ gelling properties of deacetylated gellan gum by the immobilization of thiol groups, *J. Pharm. Sci.*, 92(6), 1234–1241.
200. Gurny, R., Ibrahim, H., and Buri, P. (1993), The development and use of in situ formed gels, triggered by pH, in Edman, P., Ed., *Biopharmaceutics of Ocular drug Delivery*, CRC Press, Boca Raton, FL, pp. 81–90.
201. Greaves, J. L., Wilson, C. G., Rozier, A., Grove, J., and Plazonnet, B. (1990), Scintigraphic assessment of an ophthalmic gelling vehicle in man and rabbit, *Curr. Eye Res.*, 9(5), 415–420.
202. Furrer, P., Plazonnet, B., Mayer, J. M., and Gurny, R. (2000), Application of in vivo confocal microscopy to the objective evaluation of ocular irritation induced by surfactants, *Int. J. Pharm.*, 207(1–2), 89–98.
203. Kumar, S., Haglund, B. O., and Himmelstein, K. J. (1994), In situ-forming gels for ophthalmic drug delivery, *J. Ocul. Pharmacol.*, 10(1), 47–56.
204. Calvo, P., Alonso, M. J., Vila-Jato, J. L., and Robinson, J. R. (1996), Improved ocular bioavailability of indomethacin by novel ocular drug carriers, *J. Pharm. Pharmacol.*, 48(11), 1147–1152.
205. Lallemand, F., Felt-Baeyens, O., Besseghir, K., Behar-Cohen, F., and Gurny, R. (2003), Cyclosporine A delivery to the eye: A pharmaceutical challenge, *Eur. J. Pharm. Biopharm.*, 56(3), 307–318.
206. Kreuter, J. (1990), Nanoparticles as bioadhesive ocular drug delivery systems, in Lenaerts, V., and Gurny, R., Eds., *Bioadhesive Drug Delivery Systems*, CRC Press, Boca Raton, FL, pp. 203–212.
207. Mainardes, R. M., and Silva, L. P. (2004), Drug delivery systems: Past, present, and future, *Curr. Drug Targets*, 5(5), 449–455.
208. Ding, S. (1998), Recent developments in ophthalmic drug delivery, *Pharm. Sci. Technol. Today*, 1(8), 328–335.
209. Kaur, I. P., and Kanwar, M. (2002), Ocular preparations: The formulation approach, *Drug Dev. Ind. Pharm.*, 28(5), 473–493.
210. Meisner, D., and Mezei, M. (1995), Liposome ocular delivery systems, *Adv. Drug Deliv. Rev.*, 16(1), 75–93.
211. El-Gazayerly, O. N., and Hikal, A. H. (1997), Preparation and evaluation of acetazolamide liposomes as an ocular delivery system, *Int. J. Pharm.*, 158(2), 121–127.
212. Fresta, M., Panico, A. M., Bucolo, C., Giannavola, C., and Puglisi, G. (1999), Characterization and in-vivo ocular absorption of liposome-encapsulated acyclovir, *J. Pharm. Pharmacol.*, 51(5), 565–576.

213. Law, S. L., Huang, K. J., and Chiang, C. H. (2000), Acyclovir-containing liposomes for potential ocular delivery: Corneal penetration and absorption, *J. Controlled Release*, 63(1–2), 135–140.

214. Alonso, M. J., and Sanchez, A. (2003), The potential of chitosan in ocular drug delivery, *J. Pharm. Pharmacol.*, 55(11), 1451–1463.

215. Carafa, M., Santucci, E., and Lucania, G. (2002), Lidocaine-loaded non-ionic surfactant vesicles: Characterization and in vitro permeation studies, *Int. J. Pharm.*, 231(1), 21–32.

216. Vyas, S. P., Mysore, N., Jaitely, V., and Venkatesan, N. (1998), Discoidal niosome based controlled ocular delivery of timolol maleate, *Pharmazie*, 53(7), 466–469.

217. Schmalfuss, U., Neubert, R., and Wohlrab, W. (1997), Modification of drug penetration into human skin using microemulsions, *J. Controlled Release*, 46(3), 279–285.

218. Bagwe, R. P., Kanicky, J. R., Palla, B. J., Patanjali, P. K., and Shah, D. O. (2001), Improved drug delivery using microemulsions: Rationale, recent progress, and new horizons, *Crit. Rev. Ther. Drug Carrier Syst.*, 18(1), 77–140.

219. Attwood, D. (1994), Microemulsions, in Kreuter, J., Ed., *Colloidal Drug Delivery Systems*, Marcel Dekker, New York, pp. 31–71.

220. Benita, S., and Levy, M. Y. (1993), Submicron emulsions as colloidal drug carriers for intravenous administration: Comprehensive physicochemical characterization, *J. Pharm. Sci.*, 82(11), 1069–1079.

221. Gasco, M. R., Gallarate, M., Trotta, M., Bauchiero, L., Chiappero, O., et al. (1989), Microemulsions as topical delivery vehicles: Ocular administration of timolol, *J. Pharm. Biomed. Anal.*, 7(4), 433–439.

222. Beilin, M., Bar-Ilan, A., Amselem, S., Schwarz, J., Yogev, A., and Neumann, R. (1995), Ocular retention time of submicron emulsion (SME) and the miotic response to pilocarpine delivered in SME, *Invest. Ophthalmol. Vis. Sci.*, 36(4), 166.

223. Esfand, R., and Tomalia, D. A. (2001), Poly(amidoamine) (PAMAM) dendrimers: From biomimicry to drug delivery and biomedical applications, *Drug Discov. Today*, 6(8), 427–436.

224. Cloninger, M. J. (2002), Biological applications of dendrimers, *Curr. Opin. Chem. Biol.*, 6(6), 742–748.

225. Patri, A. K., Majoros, I. J., Baker, J., and James R. (2002), Dendritic polymer macromolecular carriers for drug delivery, *Curr. Opin. Chem. Biol.*, 6(4), 466–471.

226. Stella, V. J., and Himmelstein, K. J. (1980), Prodrugs and site-specific drug delivery, *J. Med. Chem.*, 23(12), 1275–1282.

227. Lee, V. H. (1993), Improved ocular drug delivery by use of chemical modifications (prodrugs), in Edman, P., Ed., *Biopharmaceutics of Ocular Drug Delivery*, CRC Press, Boca Raton, FL, pp. 121–143.

228. Lee, V. H. (1983), Esterase activities in adult rabbit eyes, *J. Pharm. Sci.*, 72(3), 239–244.

229. Redell, M. A., Yang, D. C., and Lee, V. H. L. (1983), The role of esterase activity in the ocular disposition of dipivalyl epinephrine in rabbits, *Int. J. Pharm.*, 17(2–3), 299–312.

230. Jarvinen, T., and Jarvinen, K. (1996), Prodrugs for improved ocular drug delivery, *Adv. Drug Deliv. Rev.*, 19(2), 203–224.

231. Sasaki, H., Igarashi, Y., Nagano, T., Nishida, K., and Nakamura, J. (1995), Different effects of absorption promoters on corneal and conjunctival penetration of ophthalmic beta-blockers, *Pharm. Res.*, 12(8), 1146–1150.

232. Hochman, J., and Artursson, P. (1994), Mechanisms of absorption enhancement and tight junction regulation, *J. Controlled Release*, 29(Mar.), 253–267.

233. Grass, G. M., and Robinson, J. R. (1984), Relationship of chemical structure to corneal penetration and influence of low-viscosity solution on ocular bioavailability, *J. Pharm. Sci.*, 73(8), 1021–1027.

234. Szejtli, J. (1994), Medicinal applications of cyclodextrins, *Med. Res. Rev.*, 14(3), 353–386.

235. Sasaki, H., Yamamura, K., Nishida, K., Nakamura, J., and Ichikawa, M. (1996), Delivery of drugs to the eye by topical application, *Prog. Retin. Eye Res.*, 15(2), 583–620.

236. Jarvinen, K., Jarvinen, T., Thompson, D. O., and Stella, V. J. (1994), The effect of a modified beta-cyclodextrin, SBE4-beta-CD, on the aqueous stability and ocular absorption of pilocarpine, *Curr. Eye Res.*, 13(12), 897–905.

237. Jarho, P., Jarvinen, K., Urtti, A., Stella, V. J., and Jarvinen, T. (1996), Modified beta-cyclodextrin (SBE7-beta-CyD) with viscous vehicle improves the ocular delivery and tolerability of pilocarpine prodrug in rabbits, *J. Pharm. Pharmacol.*, 48(3), 263–269.

238. Sultana, Y., Jain, R., Aqil, M., and Ali, A. (2006), Review of ocular drug delivery, *Curr. Drug Deliv.*, 3(2), 207–217.

239. Ranade, V. V., and Hollinger, M. A. (2003), *Intranasal and ocular drug delivery. in Drug Delivery Systems*, 2nd ed., CRC Press, Boca Raton, FL, pp. 249–287.

240. Lee, V. H. (1990), New directions in the optimization of ocular drug delivery, *J. Ocul. Pharmacol.*, 6(2), 157–164.

5.10

MICROEMULSIONS AS DRUG DELIVERY SYSTEMS

RAID G. ALANY AND JINGYUAN WEN
The University of Auckland Auckland, New Zealand

Contents

5.10.1 HISTORICAL BACKGROUND, TERMINOLOGY, AND DEFINITION

Hoar and Schulman coined the term *microemulsion* (ME) in 1943 to define a transparent system obtained by titrating a turbid oil-in-water (o/w) emulsion with a

Pharmaceutical Manufacturing Handbook: Production and Processes, edited by Shayne Cox Gad

TABLE 1 Comparison of Emulsions and Microemulsions

Property	Emulsion	Microemulsion
Disperse-phase droplet size	0.2–10 μm	Less than 0.2 μm
Visual appearance	Turbid to milky	Transparent to translucent
Stability	Thermodynamically unstable	Thermodynamically stable
Formation	Requires energy input	Spontaneous

Source: From ref. 2.

medium-chain alcohol, namely hexanol. Since then the term has been used to describe systems comprising a nonpolar component, an aqueous component, a surfactant, and a cosurfactant. While a cosurfactant is usually present, a ME can be formulated without a cosurfactant, that is, using a single surfactant. It is important to point out that the term ME was (and occasionally is) used in the literature to describe various liquid crystalline systems (lamellar, hexagonal, and cubic), surfactant systems (micelles and reverse micelles), and even coarse emulsions that are micronized using external energy (submicrometer emulsions). To avoid such confusion, Danielsson and Lindmann [1] proposed the following definition: " Microemulsion is defined as a system of water, oil and amphiphile which is optically isotropic and a thermodynamically stable liquid solution." By this definition the following systems were excluded:

- Aqueous solutions of surfactants (micellar and nonmicellar) without additives or with water soluble nonelectrolytes as additives
- Liquid crystalline phases (mesophases)
- Coarse emulsions, including micronized coarse emulsions
- Systems that are surfactant free

The term ME is often incorrectly used in the literature to describe oil and water dispersions of small droplet size produced by prolonged ultrasound mixing, high-shear homogenization, and microfluidisation, that is, submicrometer emulsions. The major differences between a microemulsion and a coarse emulsion are shown in Table 1.

5.10.2 STRUCTURE AND FORMATION OF MICROEMULSION SYSTEMS

A ME system can be one of three types depending on the composition: oil in water (o/w ME), in which water is the continuous medium; water in oil (w/o ME), in which oil is the continuous medium, and water-and-oil bicontinuous ME, in which almost equal amounts of water and oil exist [3]. While the three types are quite different in terms of microstructure, they all have an interfacial amphiphile monolayer separating the oil and water domains.

The formation of a ME system can be explained using a simplified thermodynamic approach and with reference to the equation

$$\Delta G = \gamma \Delta A - T \Delta S \tag{1}$$

where ΔG-free energy of ME formation
γ-interfacial tension at oil–water interface
ΔA-change in interfacial area (associated with reducing droplet size)
S-system entropy
T-absolute temperature

The process of ME formation is associated with a reduction in droplet size, which results in an increase in the value of ΔA due to an overall increase in surface area that is associated with droplet size reduction. This is compensated by a very low interfacial tension that is normally achieved by using relatively high amphiphile concentrations. Furthermore, the process of ME formation is accompanied by a favorable entropy contribution (increased value of ΔS) that is due to the mixing of the two immiscible phases, surfactant molecules partitioning in favor of the interface rather than the bulk and monomer–micelle surfactant exchange. The net outcome is a negative value for ΔG which translates into a spontaneous ME formation [4].

The simplest representation of the ME microstructure is with reference to the droplet model in which an interfacial film consisting of amphiphile (surfactant/cosurfactant) molecules surrounds the dispersed droplets (Figures 1*a* and *b*). The orientation of the amphiphile at the interface differs depending on the type of the ME. Whether an o/w or w/o ME forms is dependent to a great extent on the volume fraction of oil and water as well as the nature of the interfacial film as reflected by the geometry of the amphiphile molecules forming the film. It follows that the

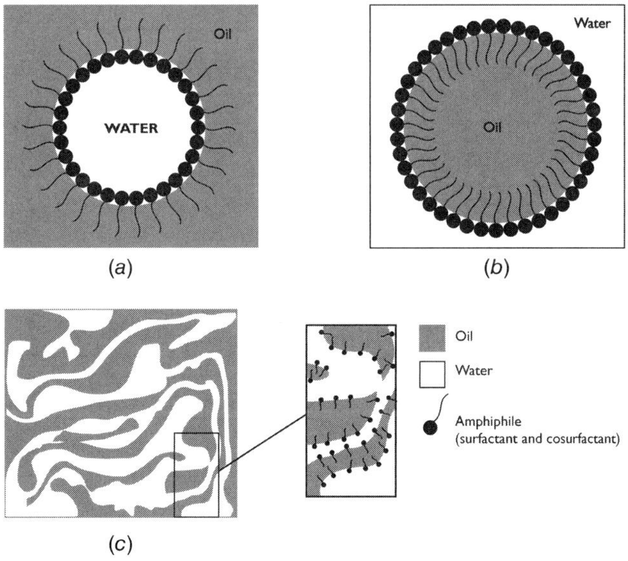

FIGURE 1 Diagrammatic representation of different types of ME systems: (*a*) w/o ME; (*b*) o/w ME; (*c*) water-and-oil bicontinuous ME. Droplet diameter for (*a*) and (*b*) is typically less than 140 nm.

presence of o/w ME droplets is more likely to happen in systems where the oil volume fraction is low, whereas w/o ME droplets form when the water volume fraction is low and oil is present in abundance. Interestingly, in systems containing comparable amounts of water and oil, a bicontinuous ME may exist (Figure 1*c*). In such systems both oil and water exist as microdomains that are separated by an amphiphile-stabilized interface with a zero net curvature.

Mitchell and Ninham [5] extended the theory of self-assembly of surfactant molecules forming micelles and bilayers [6] to ME systems. Accordingly, if the volume of the surfactant is v, its head group surface area a, and its length l, it follows that when the critical packing parameter (CPP = v/al) has values between 0 and 1, o/w MEs are likely to be formed. On the other hand, when CPP is greater than 1, w/o MEs are favored. When using surfactants with critical packing parameters close to unity (CPP $\approx$ 1) and at approximately equal volumes of water and oil, the mean curvature of the interfacial film approaches zero and droplets may merge into a bicontinuous structure (Figure 1*c*). It should be noted, however, that this approach is based solely on geometric considerations and does not account for penetration of oil and cosurfactant molecules into the interface and the hydration of surfactant head groups.

The ratio of the hydrophilic and the hydrophobic groups of the surfactant molecules, that is, their hydrophile–lipophile balance (HLB), is also important in determining interfacial film curvature and consequently the structure of the ME. The HLB system has been used for the selection of surfactants to formulate MEs and accordingly the HLB of the candidate surfactant blend should match the required HLB of the oily component for a particular system; furthermore a match in the lipophilic part of the surfactant used with the oily component is favorable [7].

Shinoda and Kuineda [8] highlighted the effect of temperature on the phase behavior of systems formulated with two surfactants and introduced the concept of the phase inversion temperature (PIT) or the so-called HLB temperature. They described the recommended formulation conditions to produce MEs with surfactant concentration of about 5–10% w/w being (a) the optimum HLB or PIT of a surfactant; (b) the optimum mixing ratio of surfactants, that is, the HLB or PIT of the mixture; and (c) the optimum temperature for a given nonionic surfactant. They concluded that (a) the closer the HLBs of the two surfactants, the larger the cosolubilization of the two immiscible phases; (b) the larger the size of the solubilizer, the more efficient the solubilisation process; and (c) mixtures of ionic and nonionic surfactants are more resistant to temperature changes than nonionic surfactants alone.

5.10.3 ROLE OF COSURFACTANTS/COSOLVENTS IN FORMATION AND STABILIZATION OF MICROEMULSIONS

Cosurfactants are molecules with weak amphiphilic properties that are mixed with the surfactant(s) to enhance their ability to reduce the interfacial tension of a system and promote the formation of a ME [3]. Cosolvents have also been described as weak amphiphilic molecules that tend to distribute between the aqueous phase, the oily phase, and the interfacial layer and act by making the aqueous phase less hydrophilic, the oily phase less hydrophobic, and the interfacial film more flexible and less condensed [9, 10].

Most single-chain surfactants do not sufficiently lower the oil–water interfacial tension to form MEs, nor are they of the right molecular structure (i.e., HLB) to act as cosolvents. To overcome such a barrier, cosurfactant/cosolvent molecules are added to further lower the interfacial tension between oil and water, fluidize the hydrocarbon region of the interfacial film, and influence the curvature of the film. Typically small molecules (C3–C8) with a polar head (hydroxyl or amine) group that can diffuse between the bulk oil and water phase and the interfacial film are suitable candidates [11].

All the abovementioned mechanisms are expected to facilitate the formation and stabilization of ME systems.

5.10.4 PHARMACEUTICAL FORMULATION OF MICROEMULSIONS

5.10.4.1 Selection of Microemulsion Ingredients

Pharmaceutically acceptable ME systems are formulated using at least GRAS- (generally regarded as safe) and preferably pharmaceutical-grade ingredients, that is, ones already in use in pharmaceutical formulation and devoid of serious side effects and toxicity in humans [12]. Nonionic and zwitterionic surfactants are among the most commonly used ingredients to formulate pharmaceutical MEs while vegetable oils, medium- and long-chain triglycerides, and esters of fatty acids are the most commonly used oils [2]. Among the range of nonionic surfactants used are sucrose esters [13], polyoxyethylene alkyl ethers [14], polyglycerol fatty acid esters [15], polyoxyethylene hydrogenated castor oil [16], and sorbitan esters [17]. Furthermore, systems formulated with zwitterionic phospholipids, particularly lecithin, have been widely investigated because of their biocompatible nature [9, 18–22].

The effect of the oily component on the phase behavior of o/w ME-forming systems formulated with nonionic surfactants was reported [23]. The authors showed that it is possible to formulate cosurfactant-free o/w ME systems suitable for use as drug delivery vehicles using either polyoxyethylene surfactants or amine-*N*-oxide surfactants. The major advantage of these ME systems is their ability to be diluted without destroying their integrity; however both classes of surfactants were shown to be sensitive to electrolytes.

The choice of cosurfactants to formulate pharmaceutically acceptable MEs is challenging as most of the cosurfactants investigated in fundamental ME research cannot be used for the development of pharmaceutically acceptable systems due to biocompatibility considerations. Among the pharmaceutically acceptable cosurfactants are ethanol [24], medium-chain mono- and diglycerides [25–28], 1,2-alkanediols [29, 30] and sucrose–ethanol combinations [31], alkyl monoglucosides, and geraniol [32]. Kahlweit et al. [29, 30] reported on the usefulness of certain 1,2-alkanediol cosurfactants as nontoxic substitutes to the physiologically incompatible short- and medium-chain alcohols. They suggested the possible use of these components for the formulation of nontoxic ME systems.

5.10.4.2 Phase Behavior Studies

Before a ME can be used as a drug delivery vehicle, the phase behavior of the particular combination of the candidate ingredients should be established. This is necessary due to the diverse range of colloidal and coarse dispersions that could be obtained when oil, water, and an amphiphile blend are mixed. Coarse emulsions, vesicles, lyotropic liquid crystals, and micellar systems are some examples. A variety of multiphase systems may coexist and the demarcations of the regional boundaries become important. One of the most suitable methods to study the phase behavior of such systems is to construct a ternary phase diagram using a Gibbs triangle (Figure 2). A ternary phase diagram can be constructed by two methods [33]:

- Titrating a mixture of two components with the third component
- Preparing a large number of samples of different composition

If all mixtures reach equilibrium rapidly, both methods give identical results. For mixtures that do not reach equilibrium quickly, the second method is recommended, as with the titration method the change in the ratio of components during titration may occur too fast, not allowing sufficient time to visually recognize phase changes [33].

As the formulation may contain more than three components, the complete phase behavior cannot be fully represented using a triangular diagram. However,

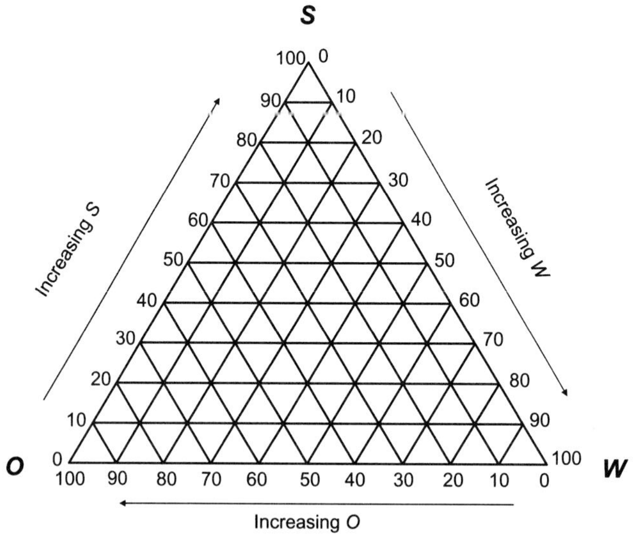

FIGURE 2 Ternary phase diagram used to elucidate ME formation regions. Each of the three corners represents 100% of the individual components. Apex S = 100% w/w surfactant (0% oil and water), apex W = 100% w/w water (0% oil and surfactant), and apex O = 100% w/w oil (0% water and surfactant). The three lines joining the corner points represent two-component systems. The area within the triangle represents all possible combinations of the three components.

the phase behavior of a four-component mixture at fixed pressure and temperature can be represented using a tetrahedron. Full characterization of such systems is a tedious task requiring a large number of experiments [34]. One acceptable approach for representing such systems is by fixing the mass ratio of two components (such as the two amphiphiles) and as such considered a single component. Such an approach, although regarded by many as an oversimplification of the systems, is yet acceptable for the purpose of phase behavior studies. Such systems are described as "pseudoternary" as they comprise more than three components (four or possibly five) yet are represented using a Gibbs triangle, which is a used to describe the phase behavior of a three-component system.

A novel approach to reduce the experimental effort associated with constructing pseudoternary phase diagrams is by using expert systems to predict the phase behavior of multicomponent ME-forming systems. Artificial neural networks have been investigated and were shown to be promising in phase behavior studies [17, 35, 36] as well as in the process of ingredient selection [37].

5.10.5 TECHNIQUES USED TO CHARACTERIZE MICROEMULSIONS AND RELATED SYSTEMS

The physicochemical and analytical techniques used to characterize MEs and related systems could be categorized into those used to:

- Elucidate the microstructure and monitor phase behavior changes
- Determine the droplet size of the disperse phase

The choice of a particular technique is limited by factors such as availability, feasibility, and the nature of the information sought. Pharmaceutical scientists are more focused on the usefulness of a particular ME system for a drug delivery application and the influence of the microstructure on that, rather than on the fundamental understanding of aspects such as microstructure and phase behavior. Polarized light microscopy is a readily available technique that could be used at the early formulation development stage to differentiate between isotropic and anisotropic systems. Transmission electron microscopy (TEM) is another available technique that has been shown to provide microstructural as well as size-related information on droplet and bicontinuous ME systems.

The main disadvantages of TEM applications (such as freeze fracture and cryo-TEM) are the lengthy and sophisticated experimental procedures associated with sample preparation and the possibility of creating artefacts during sample preparation. Other readily available and more user friendly techniques are electrical conductivity and viscosity measurements. Electrical conductivity measurements are widely used for their simplicity, feasibility, and sensitivity to structural changes in systems with increasing water content, particularly systems undergoing percolation transitions. Viscosity measurements, on the other hand, require more sophisticated instrumentation yet provide useful information on changes in the flow properties associated with structural changes of the systems. A brief overview of the key characterization techniques will follow.

5.10.5.1 Polarized Light Microscopy

When a mixture of oil, water, and surfactant(s) is examined under a polarized light microscope, the textures observed depend on the nature of the surfactant aggregate formed and the relative ratio of the comprising constituents. If the resulting aggregates are anisotropic, they tend to show strong birefringence and characteristic textures could be viewed when examined using a polarized light microscope. On the other hand, if the resulting aggregates are isotropic (as with ME systems or coarse emulsions), then polarized light microscopy would be of less value in disclosing structural information. Thermotropic and lyotropic liquid crystalline (LC) systems such as lamellar, hexagonal, and reverse hexagonal mesophases are anisotropic and exhibit characteristic textures when viewed under a polarizing light microscope. The only type of LC systems that is isotropic and would not display birefringence when viewed under a cross-polarizer is the cubic mesophase.

Polarized light microscopy is a simple technique to learn and use, readily available, and of great value to differentiate between various anisotropic LC systems. It is also of value to formulation scientists investigating amphiphile–oil–water mixtures with emphasis on colloidal systems in general and MEs in particular. This is mainly due to the fact that many LC systems may appear transparent to the naked eye and can be easily misinterpreted as isotropic ME systems. Thus it becomes essential when investigating systems of amphiphile–oil–water to confirm findings based on visual appearance with polarized light microscopic examination.

5.10.5.2 Transmission Electron Microscopy

Transmission electron microscopy was one of the earliest techniques used to investigate MEs [38, 39]. Freeze fracture along with replication is a sophisticated sample preparation method for TEM that requires careful attention to a variety of details to avoid formation of sample artefacts. The technique involves rapidly freezing the sample by immersing it in a cryogen (slush nitrogen, propane, ethane, and freons). For systems with volatile ingredients the freezing must be achieved rapidly to avoid phase separation or crystallization. Thus, high cooling rate and adequate environmental control of the samples before freezing are critical to prevent loss of volatile components. The frozen sample is then transferred to a vacuum chamber and split under vacuum by a fracturing device. The fractured surface is then shadowed with metal (usually platinum) deposited from one side. The shadowed surface is then coated with a layer of carbon that is directly deposited from above the specimen. The carbon layer is transparent when examined and forms a supportive backing for the shadowing metal deposited on the fractured surface. The specimen is then removed from the vacuum and treated with solvents of different polarities, leaving the metal carbon film as a replica of the fracture surface.

Although freeze-fracture TEM provides direct visualization of ME structures, it is not currently in wide use probably due to the experimental difficulties associated with the technique. The points to consider when preparing conventional TEM replicas are the physical and chemical sample properties, freezing, cleaving, etching, replication, cleaning, and mounting steps of the procedure.

Jahn and Strey [40] investigated systems with varying water-to-oil ratios at constant amphiphile concentration. The TEM images support the notion of a bicontinuous network for systems containing comparable amounts of the aqueous and oil components. The author also showed images of w/o droplet ME systems which showed a reduction in number densities of the dispersed phase droplets upon dilution with the organic phase.

Freeze-fracture TEM combined with nuclear magnetic resonance and quasi-elastic light scattering was used to study the microstructure of surfactant–water systems and dynamics of o/w and bicontinuous ME systems [41]. The authors reported a rather abrupt transition from a discontinuous droplet (o/w) to bicontinuous (oil-and-water) microstructure occurring at low surfactant concentration, close to a three-phase region in the constructed phase diagram of pentaethylene glycol dodecyl ether, water, and octane [41].

Direct imaging of the ME microstructure using cryo-TEM involves directly investigating a thin proportion of the specimen in the frozen hydrated stage by using a cryo stage in the transmission electron microscope. Cryo-TEM was used in combination with nuclear magnetic resonance (NMR), small-angle X-ray diffraction, and small-angle neutron scattering to investigate four-component nonionic systems composed of 1-dodecane, octa-ethylene glycol mono dodecyl ether, *n*-pentanol, and water [42]. These authors reported on the existence of at least two different colloidal microstructures, swollen spherical micelles with a diameter of around 8 nm and lamellar structures. Both o/w and w/o MEs were also visualized using cryo-TEM [43].

In the pharmaceutical field, very little has been done to elucidate the microstructure of ME systems using electron microscopy. Bolzinger et al. [44] reported on bicontinuous sucrose ester-based ME systems for transdermal drug delivery. The microstructure of the investigated ME systems was viewed by freeze-fracture TEM. The authors showed images of a bicontinuous structure and reported that incorporating the anti-inflammatory drug niflumic acid into the system did not alter the ME microstructure. Alany et al. (2001) reported on the microstructure of ME systems formulated using a blend of two nonionic surfactants, ethyl oleate and water with and without 1-butanol. They described two distinct microstructures, namely droplet w/o and bicontinuous MEs [35]. Their TEM observations were complemented by electrical conductivity and viscosity measurements.

5.10.5.3 Electrical Conductivity Measurements

Electrical conductivity measurements can provide valuable information concerning the structure and phase behavior of ME systems [45]. Schulman et al. [38] measured the conductivity of ME systems, but only in a qualitative way, as they did not monitor changes in conductivity that are associated with phase changes.

Shah and Hamlin [46] studied the changes in electrical resistance associated with change in the water-to-oil ratio in an ME system during inversion into various LC systems and coarse dispersions. In such systems the electrical conductivity may exhibit both maxima and/or minima, reflecting changes in ion mobility caused by variation in viscosity. The most important feature of systems undergoing ME-to-LC transition is the gradual change in electrical conductivity with changing composition.

On the other hand, a large electrical conductivity transition has been observed in several w/o ME systems [47]. A well-known feature of w/o ME systems is the steep rise in electrical conductivity as the water concentration increases [48]. This sudden change in electrical conductivity has been attributed to the percolation of spherical droplets (water droplets surrounded by an amphiphile shell, that is, water-swollen reverse micelles) in the oil phase [49]. The conductivity remains low up to a certain water volume fraction due to the nonconducting nature of the continuous phase of the w/o system. However, as the volume fraction of water reaches and exceeds the percolation threshold (φ_p), some of these conductive droplets begin to contact each other and form clusters which are sufficiently close to each other. This causes an efficient transfer of charge carriers between the dispersed droplets by charge hopping or transient merging of connected droplets resulting in an exponential increase of conductance from an almost zero value to much higher levels. It is reasonable to imagine a continuous pathway of water extending through the ME system, which by some authors is recognized as a sign of emergence of a bicontinuous structure [50]. Different methods were used to estimate the percolation threshold (φ_p) from the conductivity data by various investigators [35, 48, 51–54].

5.10.5.4 Viscosity Measurements

The viscosity of ME systems is also sensitive to structural changes and Newtonian flow is usually observed. The low viscosity of ME reflects the fluid character of the overall structure, which is a favorable feature for most ME applications. The pioneering work of Attwood et al. [55] on ME systems formulated using liquid paraffin, water, Span 60, and Tween 80 made reference to the equation

$$\eta_{rel} = a^{\phi/(1-\kappa\phi)} \tag{2}$$

where η_{rel}-relative viscosity
a-viscosity constant with theoretical value of 2.5 for solid spheres
φ-volume fraction of disperse phase
κ-hydrodynamic interaction coefficient

In the same study Attwood et al. [55] investigated the effect of increasing the surfactant concentration on the overall viscosity of an o/w ME system and obtained values for the viscosity constant a of 3.19–4.17. The authors concluded that allowance for the hydration of the polyoxyethylene chain of the used surfactant reduced the value of the viscosity constant a toward the theoretical value of 2.5 for a solid sphere. They also concluded that changing the ratio of the nonionic surfactants did not significantly affect the viscosity of the system.

Baker et al. [56] studied the viscosity of w/o ME systems containing water, xylene, sodium alkylbenzenesulfonate, and hexanol using Equation (1). They reported values of the viscosity constant a of 3.3–6.0, which is above the theoretical value of 2.5 for a sphere with an increase in the surfactant concentration. This finding was attributed to the increase in the ratio of surfactant layer thickness to droplet core radius as the surfactant concentration was increased. However, deviation in values of the viscosity constant a from the solid-sphere theoretical value of 2.5 could also be attributed to changes in the droplet shape or symmetry [57].

Viscosity studies have also been carried out to investigate the effect of the surfactant and cosurfactant concentrations as well as the surfactant–cosurfactant mass ratio on the hydration of the disperse-phase droplets for o/w ME systems [58]. A study was conducted on systems composed of isopropyl myristate, water, polysorbate 80, and sorbitol. The results showed an increase in the viscosity constant and a decrease in the hydrodynamic interaction coefficient with decreasing surfactant–cosurfactant mass ratio. The increase in the viscosity constant resulted from greater hydrodynamic volume of droplets as well as the associated increase in the bound solvent layer of the droplet core radius from 7 to 22%.

Kaler et al. [50] reported on the viscosity changes in association with a percolative phenomenon for systems containing the commercial surfactant TRS 10–80, octane, tertiary amyl alcohol, and various brines. Their viscosity results were interpreted as evidence for a smooth transition from an oil-continuous to a bicontinuous one in which both oil and water span the sample. A second transition was observed and was attributed to a transition from a bicontinuous to a water-continuous system.

Borkovec et al. [59] also reported on a two-stage percolation process for the ME AOT (Aerosol OT, bis(2-ethylhexyl)sodium sulfosuccinate) system AOT–decane–water. The structural inversions were investigated using viscosity, conductivity, and electro-optical effect measurements. The viscosity results showed a characteristic profile with two maxima, which was interpreted as evidence for two symmetrical percolation processes: an oil percolation on the water-rich side of the phase diagram and a water percolation process on the oil-rich side.

Alany et al. [11, 35] reported on the phase behavior of two pharmaceutical ME systems showing interesting viscosity changes. The viscosity of both systems increased with increasing volume fraction of the dispersed phase to 0.15 and flow was Newtonian. However, formation of LC in one of the two systems, namely the cosurfactant-free system, resulted in a dramatic increase in viscosity that was dependent on the volume fraction of the internal phase and a change to pseudoplastic flow. In contrast, the viscosity of the bicontinuous ME was independent of water volume fraction. The authors used two different mathematical models to explain the viscosity results and related those to the different colloidal microstructures described.

5.10.5.5 Other Characterization Techniques

Among the other techniques used to characterize ME systems with emphasis on droplet size determination are optical techniques such as static and dynamic light scattering and nonoptical techniques such as small-angle X-ray scattering and small-angle neutron scattering [22, 45, 60–67], pulsed field gradient NMR [21, 42, 61, 68–71], and dielectric measurements [72–74]. One limitation of dynamic light scattering or so-called photon correlation spectroscopy is the need to either dilute or heat the ME system to overcome droplet–droplet interactions as well as microstructure changes that are associated with increased viscosity. Such techniques are valuable to obtain useful information regarding such interactions, but the downside would be the inevitable phase behavior changes that would render these techniques of some limited value in determining the original droplet size [75].

5.10.6 MICROEMULSIONS AS DRUG DELIVERY SYSTEMS

ME systems have been attracting increasing interest as vehicles for drug delivery via the various routes. Particular emphasis has been put on the oral, transdermal, ocular, and parenteral routes. Moreover, these systems have been investigated for other applications that are relevant to the pharmaceutical, chemical, and biological sciences. The most significant development from a pharmaceutical perspective to this date would be undoubtedly the launch of the first oral cyclosporin A ME formulation, namely Sandimmnue Neoral. Other breakthroughs are likely to follow. Some of the key advantages related to ME systems include their thermodynamic stability, transparency, ease of preparation, low viscosity, and ultralow interfacial tension, to mention a few.

Furthermore, the presence of nanodomains of different polarity along with an interfacial surfactant/cosurfactnat film within the same systems allow for hydrophilic, lipophilic, and amphiphilic drugs to be accommodated together if needed. For o/w ME systems, their main advantage would be to improve the oral bioavailability of class two (II) drugs. These are drugs that display a dissolution-dependent bioabsorption [76]. The rationale would be to dissolve the poorly water soluble drug in the oil phase, thereby rendering it available in a molecular form ready for dissolution. However, it is important to remember that most drugs are insoluble in hydrocarbon and to some extent in vegetable oils, rather they would require more polar oils such as mixtures of mono- and diglycerides along the lines of Miglyol 840. Using such oils results in systems where the poorly soluble drug is dispersed as molecules in nanometer-sized oil droplets and as such are readily available for dissolution. Moreover, it is worth noting that the use of o/w MEs for delivery routes where extensive aqueous dilution is likely is less problematic than using w/o ME systems. This is because the external phase of an o/w ME will be diluted by water (being the main constituent of biological fluids) and will therefore retain its microstructure. Conversely, the potential use of w/o ME systems in the area of drug delivery where there is extensive aqueous dilution is rather complicated by the fact that such systems will be diluted by the aqueous biological fluids. This will result in droplet growth and subsequent phase changes with potential dose dumping. Despite the aforementioned disadvantages w/o MEs offer an option for the formulation of class III drugs. These drugs show permeability-limited bioabsorption in vivo [76]. Peptides/proteins, antisense oligonucleotides, deoxyribozymes, and small interfering ribonucleic acids (siRNAs) are good examples on class II drugs. Such molecules display little or no activity when delivered orally and are highly susceptible to the harsh gastrointestinal (GI) tract conditions and to their degrading enzymes in vivo. Water-in-oil MEs offer an exciting opportunity for optimizing the delivery of such molecules as they can be successfully incorporated in the internal (aqueous) phase and therefore are denied access to the harsh external conditions. Moreover the internal water droplets are likely to act as a nanoreservoir to control the mass transfer process of the loaded drug, that is, offer a mechanism for controlled release. In addition, the presence of surfactants/cosurfactants as constituents of the formulation can serve to increase membrane permeability, thereby improving drug absorption and possibly bioavailability. Bicontinuous MEs are the least investigated as drug delivery vehicles. They are highly fluid with low viscosity and possess ultralow interfacial tension. These properties render them potential candidates for topical and ocular drug delivery where their wetting and spreading properties come as an advantage.

The following section highlights some of the main drug delivery areas where ME systems have been researched as potential drug carriers.

5.10.6.1 Oral Drug Delivery

The most common method for drug delivery is through the oral route as it offers convenience and high patient compliance. However, recent advances in combinatorial chemistry is resulting in new molecules with very low water solubility. This leads to poor dissolution in the GI tract and subsequent erratic and unpredictable bioabsorption post–oral administration. Self-emulsifying drug delivery systems (SEDDSs) and self-micro-emulsifying drug delivery systems (SMEDDSs) offer an interesting option to optimize the delivery of such problematic drug molecules. SMEDDSs can be defined as isotropic, anhydrous systems comprising oil and surfactant that form o/w MEs upon mild agitation in the presence of water [77, 78]. The usefulness of SEDDSs in the area of oral drug delivery has been previously reported by Charman and co-workers, who showed improved pharmacodynamic properties of the investigational lipophilic drug WIN 54954 [79]. Systems comprising the medium-chain triglycerides Captex 355 and 800 in combination with Capmul MCM (medium chain mono- and di-glyceride mixture) and polyoxyethylene 20 sorbitan mono-oleate (Tween 80) were formulated by Constantinides et al. [26, 28, 80]. The authors reported on the improved bioavailability of calcein, RGD peptide, and a water-soluble marker when incorporated using a ME preconcentrate and w/o ME in comparison with an aqueous solution serving as a control. The same group also reported on the use of ME systems to increase oral absorption of poorly water soluble drugs [25]. The authors reported on the ability of the investigated systems to increase drug aqueous solubility and improve dissolution and oral bioavailability. Surprisingly, the formation of lamellar LC systems upon aqueous dilution of SEDDSs was reported to be a characteristic feature of the most effective SEDDSs [81]. Tenjarla summarized the factors that are likely to have an effect on the in vivo absorption of drugs from ME systems as being phase volume ratios, in vivo droplet size, partition coefficient of the drug between the two immiscible phases, the presence of a drug in an emulsified form or dispersed in oil, site or path of absorption, metabolism of the oil in the formulation, excipients that may act as absorption promoters, gastric emptying, and drug solubility in the ME excipients [2]. It was also reported that peptide uptake from MEs in the GI tract is dependent on particle size, type of ME lipid phase, digestability of lipid used, presence of bile slats, lipase, type of surfactants in ME, pH, and shedding of enterocytes [82].

The most remarkable story of success with ME research and development has to do with the oral delivery of cyclosporin A, marketed nowadays under the commercial name Sandimmune Neoral. Cyclosporin A is a cyclic peptide used post-transplantation surgery as an immune-suppressing agent. Unlike most peptides, cyclosporin A is hydrophobic and possesses very limited water solubility. The conventional cyclosporin oral formulation (Sandimmune) is in the form of a drug solution in olive oil along with ethanol and polyethoxylated oleic acid glycerides. Once given orally, the oily solution forms a coarse emulsion and as such behaves as a SEDDS with a bioabsorption process that is slow and incomplete. The net outcome is fluctuating drug plasma levels, poor and variable bioavailability, and pronounced inter- and intrapatient variability [83]. In an attempt to overcome some of these problems, Tarr and Yalkowsky [84] demonstrated that particle size reduction using high-shear homogenization can enhance absorption in rats. This improvement in oral absorption was attributed to the increased dosage form surface area that is associated with droplet size reduction. The Neoral formulation is composed of a concentrated blend of two surfactants based on medium-chain partial glycerides along with an equivalent chain length triglyceride serving as oil, a cosolvent, and the drug and is described by the manufacturer as "microemulsion preconcentrate." Exposure of this "preconcentrate" to water results in the formation of a w/o ME that upon further dilution undergoes a phase change into an o/w ME. The improved in vivo performance of Neoral over Sandimmune has been demonstrated on multiple occasions [83, 85, 86]. Furthermore, the available pharmacokinetic data have been reviewed and it was concluded that Neoral offers better predictable and more extensive drug absorption than Sandimmune. Other poorly water soluble molecules that have been recently formulated in the form of SMEDDSs with the aim of improving bioavailability include simvastatin [87] and paclitaxel [88].

Water-in-oil microemulsions, on the other hand, offer an exciting opportunity to enhance the oral bioavailability of water-soluble peptide drugs. Because of their low oral bioavailability, peptide drugs are mostly available as parenteral formulations. However, parenteral peptides have an extremely short biological half-life and would therefore require multiple daily injections. This is likely to be problematic in chronic conditions (insulin for diabetes management is a good example) where patient compliance is likely to be an issue. Hydrophilic peptide drugs of this nature can be successfully accommodated into the internal aqueous phase of w/o ME systems where they are provided with protection from enzymatic degradation post–oral administration [89]. Furthermore, the presence of surfactant and some cosurfactants (such as medium-chain glycerides) can act to increase GI membrane permeability through interacting with the cell membrane bilayer and as such improve oral bioavaialibility [25, 26, 80, 89–92].

One major concern regarding the safety profile of ME systems intended for oral administration is the comparatively high amphiphile content. Both o/w and w/o ME systems are amphiphile-rich systems compared to conventional emulsions and would contain in the most conservative case up to 15–20% w/w surfactant–cosurfactant. This is further complicated by the limited models available to evaluate chronic toxicology in comparison to conventional oral dosage forms such as tablets [91].

5.10.6.2 Transdermal Drug Delivery

Transdermal drug delivery to the systemic circulation is one of the oldest routes that have been exploited using ME systems. This route offers distinct advantages compared to traditional routes by avoidance of first-pass metabolism, potential of controlled release, ease of administration, and possibility of immediate withdrawal of treatment when necessary [93]. Transdermal drug delivery aims at maximizing drug flux into the systemic circulation through the skin, whereas dermal drug delivery aims at targeting either the epidermis or the dermis of the skin. The key challenge in both cases is to provide sufficient increase in drug flux with minimal or no significant irreversible alteration to the skin barrier function [93].

Several studies have reported on the enhanced bioavailability of cutaneous drugs using o/w and w/o MEs compared to conventional emulsions, gels or solutions, mesophases, micellar and inverse micellar systems, and vesicles [93]. Moreover, a diverse range of drug molecules such as ketoprofen, apomorphine, estradiol, lidocaine [94–97], indomethacin and diclofenac [98], prostaglandin E_1 [99], aceclofenac

[100], vinpocetine [101], azelaic acid [102], methotrexate [103], piroxicam [104], triptolide [105], fluconazol [106], and ascorbyl palmitate [107] were incorporated into different ME systems.

Kantaria et al. [108, 109] reported on gelatin ME-based organogels (MBGs) as potential iontophoretic systems for the transdermal delivery of drugs. The microstructure of the proposed MBG was elucidated using small-angle neutron scattering where w/o ME droplets were entrapped in an extensive network of gelatin–water percolative channels. Theses MBGs were found to be electrically conducting and were shown to successfully deliver a model drug (sodium salicylate). Theses systems were formulated using pharmaceutically acceptable ingredients, including Tween 80 as a surfactant and isopropyl myristate as the oily component [108].

Transdermal delivery of proteins and/or DNA vaccines for needle-free immunization has been attracting increasing interest. Cui et al. [110] reported on ethanol-in-fluorocarbone (E/F) MEs for topical immunisation. The authors showed that plasmid DNA incorporated into E/F MEs was found to be stable. Furthermore, after topical application to the skin, significant enhancements in luciferase expression, antibody production, and T-helper type 1 based immune response compared to an aqueous or ethanolic solutions of DNA were observed [110].

One major concern with the topical application of ME systems is their biocompatibility and toxicity potential, mostly due to their high surfactant–cosurfactant content. Fundamental ME research utilizes ionic surfactants and medium-chain alcohols. While these ingredients are interesting from a physicochemical perspective, they pose serious biocompatibility and toxicity concerns [75]. Nonionic and zwitter ion–based surfactants (such as certain phospholipids) offer a more pharmaceutically acceptable alternative. Several research groups have been focusing on formulating ME systems using a single surfactant such as lecithin [98, 111, 112] or *n*-alkyl POE (polyoxyethylene ethers) [14, 112, 113]. This approach tends to compromise the phase behavior of ME-forming systems. This is usually seen when the ME region in the constructed ternary phase diagrams tends to become smaller in size. This translates to less choice in terms of ME composition and possibly stability. Alternative cosurfactants with improved biocompatibility and lower skin irritation potential have been recently introduced. Plurol isostearique has been shown to be compatible with a range of surfactants and oils and was capable of providing sizable ME regions.

In conclusion, topically applied MEs have been shown to significantly increase the cutaneous uptake of both lipophilic and hydrophilic drugs. The favorable properties of ME systems include the large concentration gradient (between vehicle and skin) due to the high drug solubilization power of ME systems without increasing drug affinity to the vehicle compared to conventional topical delivery systems [93]. Moreover, the penetration-enhancing properties of the individual surfactant/cosurfactant ingredient, ease of preparation and "infinite" physical stability, and good wetting and spreading properties make ME promising for future topical applications.

5.10.6.3 Parenteral Drug Delivery

Flubiprofen o/w ME systems were prepared and evaluated as vehicles for parenteral drug delivery [114]. These systems were formulated using POE 20 sorbitan monolaurate (Tween 20) as the surfactant and ethyl oleate as the oil phase. Flubiprofen solubility in the o/w ME systems was eight times higher than that in an isotonic buffer; however, there was no significant differences in the pharmacokinetic parameters in rats between the ME formulation and the buffer [114]. Bicontinuous MEs designed for intravenous (i.v.) administration have been prepared and characterized [71]. The bicontinuous ME system underwent a phase change into an o/w emulsion upon aqueous dilution. In vitro investigations revealed small droplets with mean size radii of 60–200 nm. While solubilization studies were conducted using two drugs, namely felodipine and an antioxidant experimental drug (H 290/58), in vivo evaluations were conducted using the drug-free formulation. Doses of up to 0.5 mL/kg given i.v. to rats did not show any undesirable effects and had no significant effects on acid–base balance, blood gases, plasma electrolytes, arterial blood pressure, or heart rate [71].

Lecithin-based o/w MEs for parenteral use were formulated using polysorbate 80, IPM (Isopropyl myristate), lecithin, and water at different lecithin–polysorbate 80 weight ratios [115]. The formulated systems were shown to be highly stable and of minimal toxicity when evaluated in vitro. Phospholipid-based ME formulations of all-trans retinoic acid (ATRA) for parenteral administration were prepared and tested in vitro [116]. ATRA is effective against acute promyelocytic leukemia with highly variable oral bioavailability. Parenteral ME of ATRA was prepared using pharmaceutically acceptable ingredients, namely phospholipids and soybean oil. The inhibitory effect of ATRA on two human cancer cell lines (HL-60 and MCF-7) was not affected by incorporation into a ME formulation.

ME systems intended for parenteral application have to be formulated using nontoxic and biocompatible ingredients. The o/w ME systems would be suitable to improve the solubility of poorly water soluble drug molecules whereas w/o ME systems would be best suited for optimizing the delivery of hydrophilic drug molecules that are susceptible to the harsh GI conditions. Moreover, w/o systems can serve to prolong the release and mask any potential tissue irritation and site toxicity that are caused by intramuscular (i.m.) administration of hydrophilic drug molecules.

5.10.6.4 Ocular Drug Delivery

Aqueous solutions account for around 90% of the available ophthalmic formulations, mainly due to their simplicity and convenience [117]. However, extensive loss caused by rapid precorneal drainage and high tear turnover are among the main drawbacks associated with topical ocular drug delivery. Only 1–5% of the topically applied drug reaches the intraocular tissue with the remainder of the instilled dose undergoing nonproductive absorption via the conjunctiva or drainage via the nasolacrimal duct. This results in drug loss into the systemic circulation and provides undesirable systemic side effects [118]. Many strategies have been implemented to overcome such delivery challenges. These include the use of thermosetting in situ gelling polymer-based systems [119], nanoparticles, liposomes, and niosomes [120–123]. However, MEs offer a promising alternative as they comprise aqueous and oily components and can therefore accommodate both hydrophilic and lipophilic drugs. Moreover, they are transparent and thermodynamically stable and possess ultralow interfacial tension and therefore offer excellent wetting and spreading properties. Further advantages result form possible improvement of solubility and stability of incorporated drugs with potential increase in bioavailability; hence these systems could be a suitable alternative to conventional ocular formulations. So far, only few investigators [124–127] have considered the use of MEs for ocular drug delivery. Their work was solely focused on o/w MEs as ocular delivery carriers. Recently Alany et al. [128] reported on w/o MEs formulated using a blend of two nonionic surfactants (Crillet 4 and Crill 1), an oily component (Crodamol EO), and water. These systems were shown to be capable of undergoing a phase change to lamellar liquid crystals upon aqueous dilution. The ocular irritation potential of the individual components and final formulations was assessed using a modified hen's egg chorioallantoic membrane test (HET-CAM), and the preocular retention was investigated in the rabbit eye using gamma scintigraphy. The authors demonstrated that the retention of ME systems was significantly greater than an aqueous solution. The rapid clearance of the w/o ME formulated with 10% water compared to the LC system indicated that phase change is less likely to take place in the rabbit eye [128]. It was also concluded that w/o MEs may be of value as vehicles for the ocular drug delivery of irritant hydrophilic drugs as they appear to have a protective effect when evaluated using a modified HET-CAM test [128]. The potential of bicontinuous ME systems as vehicles for ocular drug delivery is yet to be investigated. A recent review reported on the potential of MEs as ocular drug delivery systems; however, submicrometer emulsions and systems requiring energy input to prepare were also covered and classified as MEs [129].

5.10.7 CONCLUDING REMARKS

Microemulsions represent an exciting opportunity for pharmaceutical formulators and drug delivery scientists. They are easy to prepare and thermodynamically stable. Moreover, they can accommodate drugs of different physicochemical properties and protect those that are labile. They have the potential to increase the solubility of poorly water soluble drugs, enhance the bioavailability of problematic drugs, reduce patient variability, and offer an option for controlled drug release. A critical look at the current literature shows that exciting and promising research is taking place. It is only a matter of time before new ME-based products will find their way to the market following the successful introduction of Sandimmune Neoral.

REFERENCES

1. Danielsson, I., and Lindman, B. (1981), The definition of microemulsion, *Colloid Surfaces*, 3, 391.
2. Tenjarla, S. (1999), Microemulsions: An overview and pharmaceutical applications [Review], *Crit. Rev. Ther. Drug Carrier Syst.*, 16(5), 461–521.
3. Attwood, D. (1994), Microemulsions, in Kreuter, J., Ed., *Colloidal Drug Delivery Systems*, Marcel Dekker, New York, pp 31–71.
4. Lawrence, M. J., and Rees, G. D. (2000), Microemulsion-based media as novel drug delivery systems, *Adv. Drug Del. Rev.*, 45(1), 89–121.
5. Mitchell, D., and Ninham, W. (1981), Micelles vesicles and microemulsions, *J. Chem. Soc. Faraday Trans.*, 77(2), 601–629.

6. Israelachvili, J. N., Mitchell, D., and Ninham, W. (1976), Theory of self-assembly of hydrocarbon amphiphiles into micelles and bilayers, *J. Chem. Soc. Faraday Trans. II*, 72, 1525–1568.
7. Prince, L. M. (1977), Formulation, in Prince, L. M., Ed., *Microemulsions: Theory and Practice*, Academic, New York, pp 33–49.
8. Shinoda, K., and Kuineda, H. (1973), Condition to produce so-called microemulsions: Factors to increase mutual solubility of oil and water solubilisers, *J. Colloid Interface Sci.*, 42, 381–387.
9. Aboofazeli, R., Lawrence, C. B., Wicks, S. R., and Lawrence, M. J. (1994), Investigations into the formation and characterization of phospholipid microemulsions. Part 3. Pseudo-ternary phase diagrams of systems containing water-lecithin-isopropyl myristate and either an alkanoic acid, amine, alkanediol, polyethylene glycol alkyl ether or alcohol as cosurfactant, *Int. J. Pharm.*, 111, 63–72.
10. Kahlweit, M., Busse, G., and Faulhaber, B. (1995), Preparing microemulsions with alkyl monoglucosides and the role of *N*-alkanols, *Langmuir*, 11(9), 3382–3387.
11. Alany, R. G., Rades, T., Agatonovic-Kustrin, S., Davies, N. M., and Tucker, I. G. (2000), Effects of alcohols and diols on the phase behaviour of quaternary systems, *Int. J. Pharm.*, 196(2), 141–145.
12. FDA (rev. Apr. 1, 2001), Food and drugs, Chapter I—Food and Drug Administration, in Department of Health and Human Services, Ed., *Code of Federal Regulations, Title 21* (Vol. 3), U.S. Government Printing Office, Washington, DC.
13. Thevenin, M. A., Grossiord, J. L., and Poelman, M. C. (1996), Sucrose esters cosurfactant microemulsion systems for transdermal delivery—assessment of bicontinuous structures, *Int. J. Pharm.*, 137(2), 177–186.
14. Malcolmson, C., Satra, C., Kantaria, S., Sidhu, A., and Lawrence, M. J. (1998), Effect of oil on the level of solubilization of testosterone propionate into nonionic oil-in-water microemulsions, *J. Pharm. Sci.*, 87(1), 109–116.
15. Ho, H. O., Hsiao, C. C., and Sheu, M. T. (1996), Preparation of microemulsions using polyglycerol fatty acid esters as surfactant for the delivery of protein drugs, *J. Pharm. Sci.*, 85(2), 138–143.
16. Kuineda, H., Hasegawa, Y., John, A. C., Naito, M., and Muto, M. (1996), Phase behaviour of polyoxyethylene hydrogenated caster oil in oil / water systems, *Colloid Surfaces*, 209–216.
17. Agatonovic-Kustrin, S., Glass, B. D., Wisch, M. H., and Alany, R. G. (2003), Prediction of a stable microemulsion formulation for the oral delivery of a combination of antitubercular drugs using ANN methodology, *Pharm. Res.*, 20(11), 1760–1765.
18. Aboofazeli, R., and Lawrence, M. J. (1993), Investigations into the formation and characterization of phospholipid microemulsions. Part 1. Pseudo-ternary phase diagrams of systems containing water-lecithin-alcohol-isopropyl myristate, *Int. J. Pharm.*, 93(May 31), 161–175.
19. Aboofazeli, R., and Lawrence, M. J. (1994), Investigations into the formation and characterization of phospholipid microemulsions. Part 2. Pseudo-ternary phase diagrams of systems containing water-lecithin-isopropyl myristate and alcohol: Influence of purity of lecithin, *Int. J. Pharm.*, 106, 51–61.
20. Aboofazeli, R., Patel, N., Thomas, M., and Lawrence, M. J. (1995), Investigations into the formation and characterization of phospholipid microemulsions. 4. Pseudo-ternary phase diagrams of systems containing water-lecithin-alcohol and oil—the influence of oil, *Int. J. Pharm.*, 125(1), 107–116.
21. Shinoda, K., Araki, M., Sadaghiani, A., Khan, A., and Lindman, B. (1991), Lecithin-based microemulsions; phase behaviour and microstructure, *J. Phys. Chem.*, 95, 989–993.

22. Saint-Ruth, H., Attwood, D., Ktistis, G., and Taylor, C. J. (1995), Phase studies and particle size analysis of oil-in-water phospholipid microemulsions, *Int. J. Pharm.*, 116, 253–261.
23. Warisnoicharoen, W., Lansley, A. B., and Lawrence, M. J. (2000), Nonionic oil-in-water microemulsions: The effect of oil type on phase behaviour, *Int. J. Pharm.*, 198(1), 7–27.
24. Park, K. M., Lee, M. K., Hwang, K. J., and Kim, C. K. (1999), Phospholipid-based microemulsions of flurbiprofen by the spontaneous emulsification process, *Int. J. Pharm.*, 183(2), 145–154.
25. Constantinides, P. P. (1995), Lipid microemulsions for improving drug dissolution and oral absorption: physical and biopharmaceutical aspects, *Pharm. Res.*, 12(Nov.), 1561–1572.
26. Constantinides, P. P., Lancaster, C. M., Marcello, J., Chiossone, D. C., Orner, D., Hidalgo, I., et al. (1995), Enhanced intestinal absorption of an RGD peptide from water-in-oil microemulsions of different composition and particle size, *J. Controlled Release*, 34, 109–116.
27. Constantinides, P. P., and Yiv, S. H. (1995), Particle size determination of phase-inverted water-in-oil microemulsions under different dilution and storage conditions, *Int. J. Pharm.*, 115(Mar. 7), 225–234.
28. Constantinides, P. P., Welzel, G., Ellens, H., Smith, P. L., Sturgis, S., Yiv, S. H., et al. (1996), Water-in-oil microemulsions containing medium-chain fatty acids/salts: Formulation and intestinal absorption enhancement evaluation, *Pharm. Res.*, 13(2), 210–215.
29. Kahlweit, M., Busse, G., Faulhaber, B., and Eibl, H. (1995), Preparing nontoxic microemulsions, *Langmuir*, 11(11), 4185–4187.
30. Kahlweit, M., Busse, G., and Faulhaber, B. (1996), Preparing microemulsions with alkyl monoglucosides and the role of alkanediols as cosolvents, *Langmuir*, 12, 861–862.
31. Joubran, R., Parris, N., Lu, D., and Trevino, S. (1994), Synergetic effect of sucrose and ethanol on formation of triglyceride microemulsions, *J. Disper. Sci. Technol.*, 15(6), 687–704.
32. Stubenrauch, C., Paeplow, B., and Findenegg, G. H. (1997), Microemulsions supported by octyl monoglucoside and geraniol. 1. The role of alcohol in the interfacial layer, *Langmuir*, 13, 3652–3658.
33. Bhargava, H. N., Narurkar, A., and Lieb, L. M. (1987), Using microemulsions for drug delivery, *Pharm. Technol.*, 11(Mar.), 46.
34. Bourrel, M., and Schechter, R. S. (1988), The R-ratio, in Bourrel, M., and Schechter, R. S. Eds., *Microemulsions and Related Systems*, Marcel Dekker, New York, pp 1–30.
35. Alany, R. G., Davies, N. M., Tucker, I. G., and Rades, T. (2001), Characterising colloidal structures of pseudoternary phase diagrams formed by oil/water/amphiphile systems, *Drug Dev. Ind. Pharm.*, 27(1), 33–41.
36. Alany, R. G., Agatonovic-Kustrin, S., Rades, T., and Tucker, I. G. (1999), Use of artificial neural networks to predict quaternery phase systems from limited experimental data, *J. Pharm. Biomed. Anal.*, 19(3–4), 443–452.
37. Richardson, C. J., Mbanefo, A., Aboofazeli, R., Lawrence, M. J., and Barlow, D. J. (1997), Prediction of phase behavior in microemulsion systems using artificial neural networks, *J. Colloid Interface Sci.*, 187(2), 296–303.
38. Schulman, J. H., Stoeckenius, W., and Prince, L. M. (1959), Mechanism of formaution and structure of microemulsions by electron microscopy, *J. Phys. Chem.*, 63, 1677–1685.
39. Schulman, J. H., Stoeckenius, W., and Prince, L. M. (1960), The structure of meylin figures and microemulsions as observed with the electron microscope, *Kolloid-Z*, 169, 170–179.

40. Jahn, W., and Strey, R. (1988), Microstructure of microemulsions by freeze fracture electron microscopy, *J. Phys. Chem.*, 92, 2294–2301.
41. Bodet, J. F., Bellare, J. R., Davis, H. T., Scriven, L. E., and Miller, W. G. (1988), Fluid microstructure transition from globular to bicontinuous midrange microemulsion, *J. Phys. Chem.*, 92, 1898–1902.
42. Regev, O., Ezrahi, S., Aserin, A., Garti, N., Wachtel, E., and Kaler, E. W., et al. (1996), A study of microstructure of a four component nonionic microemulsion by cryo-TEM, NMR, SAXS and SANS, *Langmuir*, 12, 668–674.
43. Krauel, K., Davies, N. M., Hook, S., and Rades, T. (2005), Using different structure types of microemulsions for the preparation of poly(alkylcyanoacrylate) nanoparticles by interfacial polymerization, *J. Controlled Release*, 106(1–2), 76–87.
44. Bolzinger, M. A., Thevenin, Carduner, T. C., and Poelman, M. C. (1998), Bicontinuous sucrose ester microemulsion: A new vehicle for topical delivery of niflumic acid, *Int. J. Pharm.*, 176(1), 39–45.
45. Kahlweit, M., Strey, R., Hasse, D., Kuineda, H., Schmeling, T., and Faulhaber, B., et al. (1987), How to study microemulsions, *J. Colloid Interface Sci.*, 118, 436–451.
46. Shah, D. O., and Hamlin, R. M. J. (1971), Structure of water in microemulsions: Electrical birefringence and nuclear magnetic resonance studies, *Sciences*, 171, 483–485.
47. Clausse, M., Nicholas-Morgantini, L., Zarbda, A., and Touraud, D. (1987), Water/sodium dodecylsulfate/1-pentanol/*n*-dodecanemicroemulsions realms of existence and transport properties, in Rosano, H. L., and Clausse, M., Eds., *Microemulsion Systems*, Marcel Dekker, New York, pp 387–425.
48. Lagues, M., and Sauterey, C. (1980), Percolation transition in water in oil microemulsions. Electrical conductivity measurements, *J. Phys. Chem.*, 84, 3503–3508.
49. Lagourette, B., Peyrelasse, J., Boned, C., and Clausse, M. (1979), Percolative conduction in microemulsion type system, *Nature*, 281, 60–62.
50. Kaler, E. W., Bennett, K. E., Davis, H. T., and Scriven, L. V. (1983), Toward understanding microemulsion microstructure I: A small-angle x-ray scattering study, *J. Phys. Chem.*, 79, 5673–5684.
51. Fang, J., and Venable, R. L. (1987), Conductivity study of the microemulsions system sodium dodecyl sulfate-hexylamine-heptane-water, *J. Colloid Interface Sci.*, 116, 269–277.
52. Gu, G., Wang, W., and Yan, H. (1996), Electric percolation of water-in-oil microemulsions: The application of effective medium theory to system sodium dodecylbenzenesulfonate (DDBS)/*n*-pentanol/*n*-heptane/water, *J. Colloid Interface Sci.*, 178(1), 358–360.
53. Mehta, S. K., and Bala, K. (1995), Volumetric and transport properties in microemulsions and the point of view of percolation theory, *Phys. Rev. E*, 51(6), 5732–5737.
54. Ray, S., Paul, S., and Moulik, S. P. (1996), Physicochemical studies on microemulsions: V. Additive effects on the performance of scaling equations and activation energy for percolation of conductance of water/AOT/heptane microemulsion, *J. Colloid Interface Sci.*, 183(1), 6–12.
55. Attwood, D., Currie, L. R. J., and Elworthy, P. H. (1974), Studies of solubilised micellar solutions, *J. Colloid Interface Sci.*, 46, 261–265.
56. Baker, R. C., Florence, A. T., Ottewill, R. H., and Tadros, T. F. (1984), Investigations into the formation and characterization of microemulsions. II. Light scattering conductivity and viscosity studies of microemulsions, *J. Colloid Interface Sci.*, 100(2), 332–349.
57. Florence, A. T., and Attwood, D. (1998), *Physicochemical Principles of Pharmacy*, 3rd ed., Macmillan, Basingstoke.

58. Ktistis, G. (1990), Viscosity study on oil-in-water microemulsions, *Int. J. Pharm.*, 61(June 30), 213–218.

59. Borkovec, M., Eicke, H. F., Hammerich, H., and Das Gupta, B. (1988), Two percolation process in microemulsions, *J. Phys. Chem.*, 92, 206–211.

60. Patel, N., Marlow, M., and Lawrence, M. J. (1998), Microemulsions: A novel pMD1 formulation, in *Drug Delivery to the Lungs IX*, AerosolSociety, London, pp 160–163.

61. Giustini, M., Palazzo, G., Colafemmina, G., Dealla Monica, M., Giomini, M., and Ceglie, A. (1996), Microstructure and dynamics of the water-in-oil CTAB/*n*-pentanol/*m*-hexane/water microemulsions: spectroscopic and conductivity study, *J. Phys. Chem.*, 100, 3190–3198.

62. Schurtenberger, P., Peng, Q., Leser, M. E., and Luisi, P. L. (1993), Structure and phase behaviour of lecithin-based microemulsions: A study of chanin length dependence, *J. Colloid Interface Sci.*, 156, 43–51.

63. Shioi, A., Harada, M., and Tanabe, M. (1995), Static light scattering from oil-rich microemulsions containing polydispersed cylindrical aggregates in sodium bis(2-ethylhexyl) phosphate system, *J. Phys. Chem.*, 99, 4750–4756.

64. Ktistis, G. (1997), Effect of polysorbate 80 and sorbitol concentration on in vitro release of indomethacin from microemulsions, *J. Disper. Sci. Technol.*, 18(1), 49–61.

65. Eugster, C., Rivara, G., Forni, G., and Vai, S. (1996), Marigenol-concentrates comprising Taxol and/or Taxan esters as active substances, *Panminerva Med.*, 38(4), 234–242.

66. Constantinides, P. P., and Scalart, J. P. (1997), Formulation and physical characterisation of water-in-oil microemulsions containing long versus medium chain length glycerides, *Int. J. Pharm.*, 58, 57–68.

67. Hantzschel, D., Enders, S., Kahl, H., and Quitzsch, K. (1999), Phase behaviour of quaternary systems containing carbohydrate surfactants-water-oil-cosurfactant, *Phys. Chem. Chem. Phys.*, 1, 5703–5701.

68. Olla, M., Monduzzi, M., and Ambrosone, L. (1999), Microemulsions and emulsions in DDAB/water/oil systems, *Colloids Surfaces A: Physicochem. Eng. Aspects*, 160, 392–401.

69. Carlfors, J., Blunte, I., and Schmidt, V. (1991), Lidocaine in microemulsion—a dermal-delivery system, *J. Disper. Sci. Technol.*, 12, 467–482.

70. Angelico, R., Palazzo, G., Colafemmina, G., Cirkel, P. A., Giustini, M., and Ceglie, A. (1998), Water diffusion and head group mobility in polymer-like reverse micelle: Evidence of a sphere-to-rod-sphere transition, *J. Phys. Chem. B*, 102, 2883–2889.

71. von Corswant, C., Thoren, P., and Engstrom, S. (1998), Triglyceride-based microemulsion for intravenous administration of sparingly soluble substances, *J. Pharm. Sci.*, 87(2), 200–208.

72. D'Angelo, M., Fioretto, D., Onori, G., Palmieri, L., and Santucci, A. (1996), Dynamics of water-containing sodium bis(2-ethylhexyl)sulfosuccinate (AOT) reverse micelles: A high-frequency dielectric study, *Phys. Rev. E*, 54(1), 993–996.

73. Feldman, Y., Kozlovich, N., Nir, I., and Garti, N. (1997), Dielectric spectroscopy of microemulsions, *Colloid Surfaces*, 128, 47–61.

74. Cirkel, P. A., van der Ploeg, J. P. M., and Koper, G. J. M. (1998), Branching and percolation in lecithin wormlike micelles studied by dielectric spectroscopy, *Phys. Rev. E*, 57, 6875–6883.

75. Alany, R. G. (2001), Microemulsions as vehicles for ocular drug delivery: Formulation, Physical-chemical characterisation and biological evaluation, Ph. D. thesis, University of Otago, Otago.

76. Amidon, G. L., Lennernas, H., Shah, V. P., and Crison, J. R. (1995), A theoretical basis for a biopharmaceutic drug classification: The correlation of in vitro drug product dissolution and in vivo bioavailability, *Pharm. Res.*, 12(3), 413–420.
77. Greiner, R. W., and Evans, D. F. (1990), Spontaneous formation of a water-continuous emulsion from water-in-oil microemulsion, *Langmuir*, 6, 1793–1796.
78. Shah, N. H., Carvajal, M. T., Patel, C. I., Infeld, M. H., and Malick, A. W. (1994), Self-emulsifying drug delivery systems with poly-glycolyzed glycerides for improving in vitro dissolution and oral absorption of lipophilic drugs, *Int. J. Pharm.*, 106, 15–23.
79. Charman, S. A., Charman, W. N., Rogge, M. C., Wilson, T. D., Dutko, F. J., and Pouton, C. W. (1992), Self-emulsifying drug delivery systems: Formulation and biopharmaceutic evaluation of an investigational lipophilic compound, *Pharm. Res.*, 9(1), 87–93.
80. Constantinides, P. P., Scalart, J. P., Lancaster, C., Marcello, J., and Smith, P. L., et al. (1994), Formulation and intestinal absorption enhancement evaluation of water-in-oil microemulsions incorporating medium-chain glycerides, *Pharm. Res.*, 11, 1385–1390.
81. Craig, D. Q. M., Barker, S. A., Banning, D., and Booth, S. W. (1995), An investigation into the mechanisms of self-emulsification using particle size analysis and low frequency dielectric spectroscopy, *Int. J. Pharm.*, 114(1), 103–110.
82. Ritschel, W. A. (1991), Microemulsions for improved peptide absorption from the gastrointestinal tract, *Methods Findings Exp. Clin. Pharm.*, 13(3), 205–220.
83. Holt, D. W., Mueller, E. A., Kovarik, J. M., van Bree, J. B., and Kutz, K. (1994), The pharmacokinetics of Sandimmun Neoral: a new oral formulation of cyclosporine, *Transplant. Proc.*, 26(5), 2935–2939.
84. Tarr, B. D., and Yalkowsky, S. H. (1989), Enhanced intestinal absorption of cyclosporine in rats through the reduction of emulsion droplet size, *Pharm. Res.*, 6(1), 40–43.
85. Kovarik, J. M., Mueller, E. A., van Bree, J. B., Tetzloff, W., and Kutz, K. (1994), Reduced inter- and intraindividual variability in cyclosporine pharmacokinetics from a microemulsion formulation, *J. Pharm. Sci.*, 83(3), 444–446.
86. Mueller, E. A., Kovarik, J. M., van Bree, J. B., Tetzloff, W., Grevel, J., and Kutz, K. (1994), Improved dose linearity of cyclosporine pharmacokinetics from a microemulsion formulation, *Pharm. Res.*, 11(2), 301–304.
87. Kang, B. K., Lee, J. S., Chon, S. K., Jeong, S. Y., Yuk, S. H., and Khang, G., et al. (2004), Development of self-microemulsifying drug delivery systems (SMEDDS) for oral bioavailability enhancement of simvastatin in beagle dogs, *Int. J. Pharm.*, 274(1–2), 65–73.
88. Yang, S., Gursoy, R. N., Lambert, G., and Benita, S. (2004), Enhanced oral absorption of paclitaxel in a novel self-microemulsifying drug delivery system with or without concomitant use of P-glycoprotein inhibitors, *Pharm. Res.*, 21(2), 261–270.
89. Sarciaux, J. M., Acar, L., and Sado, P. A. (1995), Using microemulsion formulations for drug delivery of therapeutic peptides, *Int. J. Pharm.*, 120, 127–136.
90. Swenson, E. S., and Curatolo, W. (1992), Intestinal permeability enhancement for proteins, peptides and other polar drugs: Mechanism and potential toxicity, *Adv. Drug Del. Rev.*, 8, 39–92.
91. Swenson, E. S., Milisen, W. B., and Curatolo, W. (1994), Intestinal permeability enhancement: Efficacy, acute local toxicity, and reversibility, *Pharm. Res.*, 11(8), 1132–1142.
92. Pouton, C. W. (1997), Formulation of self-emulsifying drug delivery systems for peptides: reduced plasma testosterone levels in male rats after a single injection, *Int. J. Pharm.*, 25, 47–58.
93. Kreilgaard, M. (2002), Influence of microemulsions on cutaneous drug delivery, *Adv. Drug Del. Rev.*, 54 Suppl 1, S77–S98.

94. Rhee, Y. S., Choi, J. G., Park, E. S., and Chi, S. C. (2001), Transdermal delivery of ketoprofen using microemulsions, *Int. J. Pharm.*, 228(1–2), 161–170.

95. Peira, E., Scolari, P., and Gasco, M. R. (2001), Transdermal permeation of apomorphine through hairless mouse skin from microemulsions, *Int. J. Pharm.*, 226(1–2), 47–51.

96. Peltola, S., Saarinen-Savolainen, P., Kiesvaara, J., Suhonen, T. M., and Urtti, A. (2003), Microemulsions for topical delivery of estradiol, *Int. J. Pharm.*, 254(2), 99–107.

97. Sintov, A. C., and Shapiro, L. (2004), New microemulsion vehicle facilitates percutaneous penetration in vitro and cutaneous drug bioavailability in vivo, *J. Controlled Release*, 95(2), 173–183.

98. Dreher, F., Walde, P., Walther, P., and Wehrli, E. (1997), Interaction of a lecithin microemulsion gel with human stratum corneum and its effect on transdermal transport, *J. Controlled Release*, 45(2), 131–140.

99. Ho, H. O., Huang, M. C., Chen, L. C., Hsia, A., Chen, K. T., and Chiang, H. S., et al. (1998), The percutaneous delivery of prostaglandin E1 and its alkyl esters by microemulsions, *Clin. Pharm. J.*, 50, 257–266.

100. Yang, J. H., Kim, Y. I., and Kim, K. M. (2002), Preparation and evaluation of aceclofenac microemulsion for transdermal delivery system, *Arch. Pharm. Res.*, 25(4), 534–540.

101. Hua, L., Weisan, P., Jiayu, L., and Hongfei, L. (2004), Preparation and evaluation of microemulsion of vinpocetine for transdermal delivery, *Pharmazie*, 59(4), 274–278.

102. Gasco, M. R., Gallarate, M., and Pattarino, F. (1991), In vitro permeation of azelaic acid from viscosized microemulsions, *Int. J. Pharm.*, 69(3), 193–196.

103. Alvarez-Figueroa, M. J., and Blanco-Mendez, J. (2001), Transdermal delivery of methotrexate: Iontophoretic delivery from hydrogels and passive delivery from microemulsions, *Int. J. Pharm.*, 215(1–2), 57–65.

104. Park, E. S., Cui, Y., Yun, B. J., Ko, I. J., and Chi, S. C. (2005), Transdermal delivery of piroxicam using microemulsions, *Arch. Pharm. Res.*, 28(2), 243–248.

105. Mei, Z., Chen, H., Weng, T., Yang, Y., and Yang, X. (2003), Solid lipid nanoparticle and microemulsion for topical delivery of triptolide, *Eur. J. Pharm. Biopharm.*, 56(2), 189–196.

106. El Laithy, H. M., and El-Shaboury, K. M. (2002), The development of Cutina lipogels and gel microemulsion for topical administration of fluconazole, *AAPS Pharm. Sci. Tech.*, 3(4), E35.

107. Jurkovic, P., Sentjurc, M., Gasperlin, M., Kristl, J., and Pecar, S. (2003), Skin protection against ultraviolet induced free radicals with ascorbyl palmitate in microemulsions, *Eur. J. Pharm. Biopharm.*, 56(1), 59–66.

108. Kantaria, S., Rees, G. D., and Lawrence, M. J. (1999), Gelatin-stabilised microemulsion-based organogels: Rheology and application in iontophoretic transdermal drug delivery, *J. Controlled Release*, 60(2–3), 355–365.

109. Kantaria, S., Rees, G. D., and Lawrence, M. J. (2003), Formulation of electrically conducting microemulsion-based organogels, *Int. J. Pharm.*, 250(1), 65–83.

110. Cui, Z., Fountain, W., Clark, M., Jay, M., and Mumper, R. J. (2003), Novel ethanol-in-fluorocarbon microemulsions for topical genetic immunization, *Pharm. Res.*, 20(1), 16–23.

111. Bonina, F. P., Montenegro, L., Scrofani, N., Esposito, E., Cortesi, R., and Menegatti, E., et al. (1995), Effects of phospholipid based formulations on in vitro and in vivo percutaneous absorption of methyl nicotinate, *J. Controlled Release*, 34(1), 53–63.

112. Malcolmson, C., and Lawrence, M. J. (1990), A comparison between nonionic micelles and microemulsions as a means of incorporating the poorly water soluble drug diazepam, *J. Pharm. Pharmcol.*, Suppl. 42, 6P.

113. Muller, M., Mascher, H., Kikuta, C., Schafer, S., Brunner, M., and Dorner, G., et al. (1997), Diclofenac concentrations in defined tissue layers after topical administration, *Clin. Pharmacol. Therap.*, 62(3), 293–299.

114. Park, K.-M., and Kim, C.-K. (1999), Preparation and evaluation of flurbiprofen-loaded microemulsion for parenteral delivery, *Int. J. Pharm.*, 181(2), 173–179.

115. Moreno, M. A., Ballesteros, M. P., and Frutos, P. (2003), Lecithin-based oil-in-water microemulsions for parenteral use: pseudoternary phase diagrams, characterization and toxicity studies, *J. Pharm. Sci.*, 92(7), 1428–1437.

116. Hwang, S. R., Lim, S.-J., Park, J.-S., and Kim, C.-K. (2004), Phospholipid-based microemulsion formulation of all-trans-retinoic acid for parenteral administration, *Int. J. Pharm.*, 276(1–2), 175–183.

117. Bourlais, C. L., Acar, L., Zia, H., Sado, P. A., Needham, T., and Leverge, R. (1998), Ophthalmic drug delivery systems—recent advances, *Prog. Retinal Eye Res.*, 17(1), 33–58.

118. Lang, J. C. (1995), Ocular drug delivery conventional ocular formulations, *Adv. Drug Del. Rev. Ocular Drug Del.*, 16(1), 39–43.

119. Miller, S. C., and Donovan, M. D. (1982), Effect of poloxamer 407 gel on the miotic activity of pilocarpine nitrate in rabbits, *Int. J. Pharm.*, 12(2–3), 147–152.

120. Fitzgerald, P., Hadgraft, J., Kreuter, J., and Wilson, C. G. (1987), A γ-scintigraphic evaluation of microparticulate ophthalmic delivery systems: Liposomes and nanoparticles, *Int. J. Pharm.*, 40(1–2), 81–84.

121. Calvo, P., Vila-Jato, J. L., and Alonso, M. J. (1997), Evaluation of cationic polymer-coated nanocapsules as ocular drug carriers, *Int. J. Pharm.*, 153(1), 41–50.

122. Pignatello, R., Bucolo, C., Ferrara, P., Maltese, A., Puleo, A., and Puglisi, G. (2002), Eudragit RS100 nanosuspensions for the ophthalmic controlled delivery of ibuprofen, *Eur. J. Pharm. Sci.*, 16(1–2), 53–61.

123. Vyas, S. P., Mysore, N., Jaitely, V., and Venkatesan, N. (1998), Discoidal niosome based controlled ocular delivery of timolol maleate, *Pharmazie*, 53(7), 466–469.

124. Gasco, M. R., Gallarate, M., Trotta, M., Bauchiero, L., Gremmo, E., and Chiappero, O. (1989), Microemulsions as topical delivery vehicles: Ocular administration of timolol, *J. Pharm. Biomed. Anal.*, 7(4), 433–439.

125. Gallarate, M., Gasco, M. R., Trotta, M., Chetoni, P., and Saettone, M. F. (1993), Preparation and evaluation in vitro of solutions and o/w microemulsions containing levobunolol as ion-pair, *Int. J. Pharm.*, 100(1–3), 219–225.

126. Hasse, A., and Keipert, S. (1997), Development and characterization of microemulsions for ocular application, *Eur. J. Biopharm.*, 43, 179–183.

127. Siebenbrodt, I., and Keipert, S. (1993), Poloxamer-Systems as potential ophthalmics II. Microemulsions, *Eur. J. Biopharm.*, 39, 25–30.

128. Alany, R. G., Rades, T., Nicoll, J., Tucker, I. G., and Davies, N. M. (2006), W/O microemulsions for ocular delivery: Evaluation of ocular irritation and precorneal retention, *J. Controlled Release*, 111(1–2), 145–152.

129. Vandamme, T. F. (2002), Microemulsions as ocular drug delivery systems: Recent developments and future challenges, *Prog. Retinal Eye Res.*, 21(1), 15–34.

5.11

TRANSDERMAL DRUG DELIVERY

C. Scott Asbill and Gary W. Bumgarner
Samford Unversity, Birmingham, Alabama

Contents

Pharmaceutical Manufacturing Handbook: Production and Processes, edited by Shayne Cox Gad

5.11.1 INTRODUCTION

During the last decade there has been an explosion of drug candidates in clinical trials and a high number of dosage forms approved by the Food and Drug Administration (FDA) that utilize the transdermal route of drug delivery (Table 1). The interest in transdermals can be attributed to the many advantages offered by this route of delivery [1]. These advantages include maintaining steady-state drug concentrations (Figure 1), bypassing first-pass metabolism, mimicking an intravenous (IV) infusion, less frequent dosing, and increased patient compliance (Table 2).

The disadvantages of transdermal patches are patient allergies to adhesives found in transdermal patches, patch excipients that sometimes produce local irritation, potential for abuse or misuse, temperature that can affect the delivery of drugs from certain patches, and the inability of most drugs to be delivered transdermally (Table 3) [2].

In the 1960s a proposal was made that suggested the transdermal exposure of nitroglycerin to munitions workers resulted in tolerance of the workers to nitroglycerin. Long-term exposure of nitroglycerin in high doses can lead to tolerance. Workers who were abruptly removed from exposure to the nitroglycerin often experienced cardiac arrest and cases of sudden death were sometimes reported [3]. Soon after, a correlation was made between the physical and chemical properties of nitroglycerin and its significant transdermal penetration [4].

TABLE 1 Selected FDA-Approved Transdermal Patches

Trade Name	Generic Name	Indication	Company
Androderm	Testosterone	Testosterone deficiency	Watson Pharma
Catapres TTS	Clonidine	Hypertension	Boehringer Ingelheim
Climara	Estradiol	Hormone replacement	Berlex
CombiPatch	Estradiol/norethindrone acetate	Hormone replacment	Novartis
Daytrana	Methylphenidate	Attention-deficit hyperactivity disorder	Shire Pharmaceuticals
Duragesic	Fentanyl	Pain managment	Janssen
Emsam	Selegiline	Depression	Somerset Pharmaceuticals
Estraderm	Estradiol	Hormone replacement	Novartis
Nicoderm CQ	Nicotine	Smoking cessation	GlaxoSmithKline
Nicotrol	Nicotine	Smoking cessation	Pfizer
Ortho Evra	Ethinyl estradiol/norelgestromin	Contraception	Ortho-Mcneil
Oxytrol	Oxybutynin	Urinary incontinence	Watson Pharma
Transderm Nitro	Nitroglycerin	Angina	Novartis
Transdrerm Scop	Scopolamine	Motion sickness	Novartis
Vivelle	Estradiol	Hormone replacement	Noven
Vivelle Dot	Estradiol	Hormone replacement	Noven

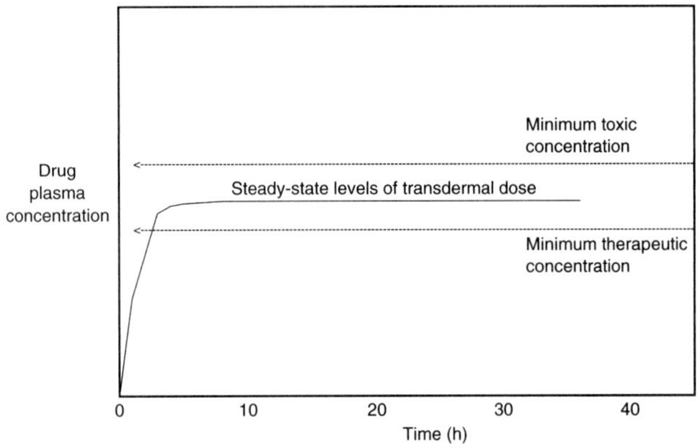

FIGURE 1 Steady-state drug levels achieved by transdermal delivery.

TABLE 2 Advantages of Transdermal Drug Delivery

Excellent for drugs with short half-lives
Analogous to IV infusion
Route bypasses first-pass metabolism
Reduced side effects
Decreased dosing
Zero-order kinetics
Self-administration
Increased patient compliance

TABLE 3 Disadvantages of Transdermal Drug Delivery

Allergies to adhesives
Product of local irritation
Potential for abuse or misuse
Temperature affects delivery
Drug needs to be potent and have desired physical and chemical properties

Another historical perspective on transdermal drug delivery involved tobacco farmers receiving transdermal doses of nicotine from handling tobacco leaves. A paper published by Gehlback et al. in 1974 introduced the concept of transdermal nicotine [5]. It was common for farm workers who had direct contact with tobacco plants during rainy and humid conditions to exhibit symptoms often associated with nicotine poisoning. This condition is called green tobacco sickness and has affected many farm workers, particularly in states such as Kentucky and North Carolina, where the farming of tobacco is significant. Moisture on tobacco leaves from rain or dew may contain significant amounts of nicotine due to its high water solubility. Farm workers directly handling tobacco leaves often retained the moisture-containing nicotine on their clothes or skin. Nicotine, because of its physical and chemical properties, is low molecular weight and has excellent transdermal penetration, and significant blood levels can lead to many adverse reactions, such as nausea and vomiting [6].

5.11.2 PHYSIOLOGY AND CHARACTERISTICS OF HUMAN SKIN

The skin is the largest organ in the human body and has many physiological functions. The skin serves to regulate overall body homeostasis, protect the body from external pathogens and chemicals, as well as control water loss from the body. The skin has three main layers. The epidermis, which is the outermost layer, is the thinnest layer of the skin and provides the most significant barrier function [7]. Beneath the epidermis, the dermis provides mechanical support to the skin and the third layer, immediately under the dermis, is a layer of subcutaneous fat called the hypodermis.

The epidermis consists of five principal layers and is an area of both intense biochemical activity and differentiation. These layers are the stratum corneum, stratum lucidum, stratum granulosum, stratum spinosum, and stratum basale. The stratum corneum (horny layer) is the uppermost layer of the epidermis and the skin. The stratum corneum is composed of dead keratinocytes, which are called corneocytes, and has an abundance of keratin and lipid structures [8]. The stratum corneum is considered the rate-limiting barrier for the diffusion of chemical compounds across the skin. The stratum lucidum (clear layer) is composed of two to three layers of dead flattened keratinocytes which appear translucent under a microscope and are present only in thick glabrous skin.

Stratum granulosum (granular layer) is an epidermal layer than consists of three to five layers of keratinocytes which have started to flatten out, and their nuclei and organelles have begun to disintegrate. Also, lipid and keratin granules have started to form inside the keratinocytes. Eventually the granules will give rise to the brick-and-mortar structure of the stratum corneum.

The stratum spinosum (prickly layer) is composed of several layers of keratinocytes which are starting to exhibit histological and biochemical changes that mark the beginning of the differentiation process. The shape of the keratinocytes has become irregular and enzymes responsible for lipid synthesis are present.

The stratum basale is the deepest layer of the epidermis and is composed mainly of keratinocytes with melanocytes making up approximately 10% of the cell population. The stratum basale is one cell layer thick and is a layer of rapid cell division where keratinocytes are rapidly dividing and giving rise to the uppermost layers of the epidermis.

The dermis is the largest layer of the skin. It is a region of strong and flexible connective tissue. The dermis consists of two primary layers, the papillary layer and the reticular layer. The papillary layer is the smallest layer of the dermis and is composed mainly of collagen and elastin fibers. The reticular layer is the largest layer of the dermis and is composed of mainly dense connective tissue. The layer of subcutaneous fat found directly beneath the dermis provides insulation and additional mechanical support to the skin.

5.11.2.1 Transepidermal Water Loss and Occlusion

Transepidermal water loss (TEWL) is a natural occurrence that takes place in the skin layers. TEWL is the result of movement of water from the deep skin layers across the epidermis into the outside atmosphere. It is a tightly regulated process that is controlled by the stratum corneum [9]. Occlusive topical bases and devices, such as transdermal patches, block TEWL and cause increased hydration of the skin. Hydration of the skin increases the permeation rates of compounds transdermally. The occluding effect of transdermal patches is an important mechanism that promotes increased diffusion of the compound across the skin into the systemic circulation [10].

5.11.2.2 Skin Lipids

As keratinocytes differentiate and move toward the stratum corneum, they synthesize an abundance of lipids, much of which are packaged into small organelles called lamellar granules [11]. When reaching the stratum corneum, the corneocytes eject these lipid granules forming a principal component of the brick-and-mortar structure of the stratum corneum. Lipids synthesized in the skin layers are thought to arrive from carbon sources derived from acetate obtained from the systemic circulation. The role of epidermal lipids in the brick-and-mortar structure of the stratum corneum and the barrier function of the skin is well established in the literature [12]. Many studies have suggested that when organic solvents such as dimethyl sulfoxide (DMSO) are used to dissolve epidermal lipids, an increase in skin permeability is found [13].

The major lipids found in the stratum corneum are ceramides, fatty acids, and cholesterol. Free fatty acids make up 10–15% of the lipid mass of the stratum corneum and predominantly consist of straight-chain saturated species ranging from 14 to 28 carbons in length. Cholesterol, a major lipid found in the stratum corneum, represents approximately 25% of the total stratum corneum lipid while cholesterol sulfate accounts for another 5%.

Ceramides are considered to be the largest group of lipids found in the stratum corneum, representing 50% of the total lipid weight. Six distinct ceramide fractions have been isolated and characterized [14]. Ceramide 1 is derived from the linoleate-rich acylglucosylceramide. Ceramide 2 consists of straight-chain saturated fatty acids amide linked to sphingosine and dihydrosphingosine bases. Ceramide 3 consists of saturated fatty acids amide linked to phytosphingosine, which has an additional hydroxyl group on carbon 4. Ceramides 4 and 5 both consist of α-hydroxyacids amide linked to sphingosines and dihydrosphingosines. Ceramide 6 contains α-hydroxyacids amide linked to phytosphingosines.

It has been suggested that these ceramides form a gel-phase membrane domain within the skin. Straight fatty acid chains as well as the small polar head groups on the ceramides are thought to produce a tightly packed domain which is less fluid and thereby less permeable than other liquid crystalline domains which are also present. Recent evidence using differential scanning calorimetery (DSC) and infrared absorption spectroscopy analyses verifies the presence of gel phases within the stratum corneum.

5.11.3 DIFFUSION

The diffusion of an active ingredient from a transdermal patch through the skin layers and into the systemic circulation can be attributed to Fick's law. Fick's law is best defined as a linear relationship between the flux of a chemical and the chemical's concentration gradient. The concentration gradient is the engine which triggers drug diffusion in all directions [15]. The chemicals move from a region of higher concentration to a region of lower concentration. Transdermal patches are typically loaded with large amounts of active ingredients in order to maximize diffusion. Once the drug is absorbed systemically the concentration gradient is maintained with "sink conditions" existing between the dosage form and the systemic circulation.

5.11.4 DRUG CANDIDATES FOR TRANSDERMAL DELIVERY

Several characteristics are important in determining which drugs are candidates for transdermal drug delivery. These are half-life, molecular weight, lipophilicity, and potency. Some pharmaceutical active ingredients have short biological half-lives whereas other have very long half-lives. For drugs such as estradiol with half-lives of less than 2 h, for the drug to be delivered orally, a frequent dosing schedule must be followed due to the rapid clearance of the drug from the systemic circulation. This high dosing frequency may be inconvenient and may lead to poor patient compliance. However, drugs with short half-lives are excellent candidates for transdermal drug delivery because the rate of drug delivery is constant and mimics the kinetics offered by an IV infusion. The steady-state drug delivery allows for less frequent dosing, which may result in better patient compliance.

Molecular weight is another property that must be considered when selecting potential candidates for transdermal drug delivery. Molecular weight is a major determinant in whether a molecule may pass through the restrictive barrier of the stratum corneum. Drugs that have low molecular weights have a better chance of penetrating the stratum corneum compared to high-molecular-weight compounds such as proteins and oligonucleotides, which are too large to passively diffuse across the stratum corneum. It has been proposed in the literature that compounds with a molecular weight of less than 400 daltons are potential candidates for transdermal drug delivery (Table 4) [16]. Lipophilicity of the drug is also an important factor concerning the chemical's ability to undergo transdermal delivery. Pharmaceutical active ingredients must have sufficient solubility in both the lipid portion and hydrophilic regions of the skin in order to have significant permeation. Drugs with a $\log P$ between –1.0 and 4.0 can potentially traverse the brick-and-mortar structure of the stratum corneum.

TABLE 4 Physico Chemical Factors That Affect Permeation of Compounds

Small molecular weight
Melting point <200°C
Suitable partition coefficient > 0.5×10^{-3} cm/h

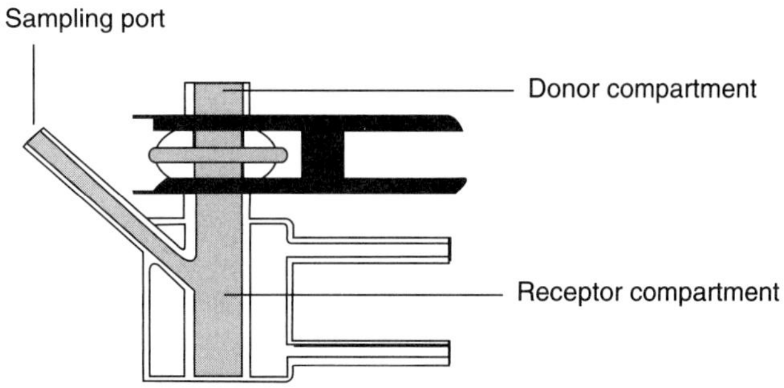

FIGURE 2 Typical in vitro permeation cell.

Another important parameter when selecting transdermal candidates is the drug's potency. Only drugs that provide therapeutic effects at low steady-state plasma levels are viable candidates. The input rate needed for a transdermal patch flux can be determined from multiplying the clearance of the drug with the desired plasma concentration of the drug at steady state. This is a useful predictor when selecting the feasibility of using an active ingredient in a transdermal patch.

5.11.5 IN VITRO TESTING OF TRANSDERMAL DEVICES AND DRUG CANDIDATES

The ability to test the transdermal penetration of compounds in vitro has been investigated extensively over the last decade. Hundreds of studies have shown that the permeation rate of compounds can be accurately measured using several available in vitro diffusion cells and various types of skin models [17].

A typical in vitro permeation cell is composed of both a donor compartment and a receptor compartment (Figure 2). The model skin or membrane for permeation testing is placed between the donor and receptor compartments. The donor compartment is where the drug solution, ointment, or transdermal patch is applied. The compound then permeates across the membrane into the receptor compartment. The receptor compartment typically contains a buffered solution maintained at 37°C and is continuously stirred [18]. Often antibiotics and preservatives are added to the receptor compartment to restrict growth of microorganisms and to maintain the integrity of the skin model that is being utilized for the permeation study [19]. Samples are withdrawn from the receptor compartment at predetermined intervals for analysis. Following the withdrawal of samples from the receptor compartment, an equal amount of diffusion buffer is added to the volume.

5.11.6 TRANSDERMAL PATCH DESIGN

5.11.6.1 Membrane-Moderated Patches

Membrane-moderated patches have been utilized in many FDA-approved transdermal patches such as Duragesic, Transderm Scop, and Catapress TTS. This type of patch utilizes a rate-controlling membrane to precisely control the release of

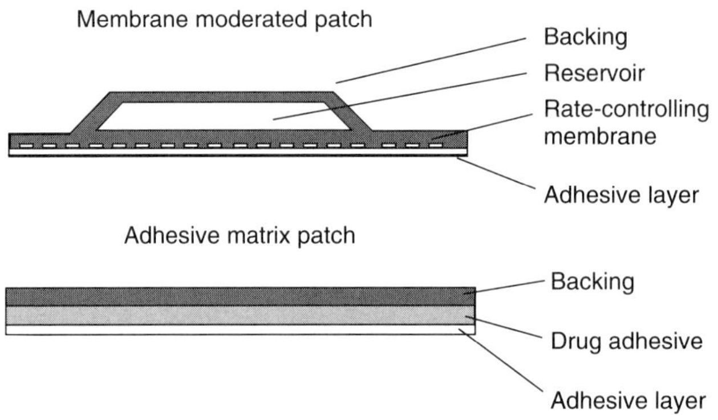

FIGURE 3 Diagram of membrane-moderated and adhesive matrix transdermal patches.

active ingredient from the transdermal patch and provides release profiles of active ingredients that exhibit zero-order kinetics [20]. The layers of a membrane-moderated patch are as follows: The uppermost layer is an occlusive backing that is impermeable; next is the reservoir layer, which consists of active ingredient dissolved in an appropriate solvent such as mineral oil or ethyl alcohol. The next layer is a rate-controlling membrane that is typically composed of ethylene vinyl acetate copolymer. Underlying the rate-controlling membrane is a layer of adhesive and then finally the release liner (Figure 3). Commonly burst doses of the active ingredient are placed in the adhesive layer to saturate the skin layers and reduce the lag time to steady-state blood levels [21] (Figure 1).

5.11.6.2 Adhesive Matrix

Adhesive matrix patches are increasingly common, and most of the recently approved transdermal drug delivery devices utilize this type of technology [22–24]. Advantages of this type of system are easier to manufacture than membrane-moderated patches, smaller in size, and resistant to manufacturing defects which could lead to dose dumping. The layers of an adhesive matrix patch consists of the backing followed by a layer containing the active ingredient dissolved and mixed with adhesive and then the release liner [25] (Figure 3).

5.11.7 COMMERCIALLY AVAILABLE PATCHES

5.11.7.1 Transderm Scop

In 1978 Tranderm Scop was the first transdermal patch to receive FDA approval. Scopolamine, the active ingredient, is a belladonna alkaloid that is frequently used to treat motion sickness and nausea resulting from anesthetics and analgesics. Transderm Scop is a membrane-moderated patch that has a three-day life span. It has a circular shape with an area (2.5 cm^2) approximately the size of a quarter [26].

5.11.7.2 Catapres TTS

Catapres TTS is the first and only transdermal patch approved by the FDA for the treatment of hypertension. Catapres TTS contains the active ingredient clondidine and was approved in 1985 [27]. Catapres TTS is a seven-day patch and is a membrane-moderated transdermal patch. The patch comes in three sizes, 0.1, 0.2, and 0.3 mg/day. In addition there is also a burst dose of clonidine in the adhesive layer. The presence of the burst doses provides an immediate-release dose of clonidine which promotes rapid systemic levels of the drug.

5.11.7.3 Androderm

Androderm is an FDA-approved membrane-moderated transdermal patch that delivers testosterone. Androderm is manufactured in two sizes and both are 24-h patches. The round Androderm patch contains 12.2 mg of testosterone and delivers a 2.5-mg dose over a 24-h period. The larger oval patch contains 24.3 mg of testosterone and delivers 5 mg over a 24-h period. The reservoir of Androderm contains testosterone gelled with alcohol and glycerin.

5.11.7.4 Estradiol Transdermal Systems

There are several FDA-approved estradiol patches currently on the market. These patches are three- to four-day patches used to treat symptoms associated with menopause. Estraderm is the only membrane-moderated estradiol system that is on the market. Estraderm is available in two sizes: 0.05 and 0.1-mg/day patches. Vivelle, Vivelle Dots, Alora, and Climara are examples of commercially available estradiol patches that utilize adhesive matrix patch design.

5.11.7.5 CombiPatch

CombiPatch is a three- or four-day patch that delivers both estradiol and norethindrone acetate. CombiPatch is available in two sizes: a 9-cm^2 patch that delivers 0.05 mg of estradiol per day and 0.14 mg of norethindrone acetate per day and a 16-cm^2 patch that delivers 0.05 mg of estradiol per day and 0.25 mg of norethindrone acetate per day. Estradiol is a lipophilic compound with a molecular weight of 272. The molecular weight of norethindrone acetate is 340. The design of the patch is considered an adhesive-matrix-type patch that consists of three layers. The backing is comprised of polyolefin and the adhesive layer contains a silicone adhesive, acrylate adhesive, estradiol, norethindrone acetate, oleic acid, and oleyl alcohol.

5.11.7.6 Duragesic

Duragesic is a transdermal patch that delivers the potent opioid analgesic fentanyl. The life span of the patch is three days and it is manufactured in five sizes: 12, 25, 50, 75, and 100 μg/h. Duragesic is a membraned-moderated patch and consists of four patch layers: a backing layer of polyester film, a drug reservoir that contains fentanyl, and U.S. Pharmacopeia (USP) alcohol gelled with hydroxyethyl cellulose, a rate-controlling membrane made of ethylene-vinyl acetate copolymer, and a layer of silicone adhesive with a burst dose of fentanyl. Fentanyl has a molecular weight of 336.5 and is a lipophilic drug with significant transdermal permeation.

5.11.7.7 Ortho Evra

Ortho Evra was approved by the FDA in 2002 as the world's first transdermal contraceptive patch. Ortho Evra contains both ethinyl estradiol and norelgestromin as active ingredient. Ortho Evra is a three-layered patch that is a matrix-type design. The molecular weight of ethinyl estradiol is 296.41 and the molecular weight of norelgestromin is 327.47. The adhesives used in this system are polyisobutylene and polybutene. Ortho Evra is a seven-day patch which is cycled three weeks on and one week patch free.

5.11.7.8 Oxytrol

Oxytrol is a three- or four-day patch used for the treatment of urinary incontinence that was recently approved by the FDA. It delivers the active ingredient oxybutynin. Oxybutynin is an antispasmotic and anticholinergic agent with a molecular weight of 357. Oxybutynin is considered to be a lipophilic drug. Oxytrol is a matrix-type patch with a surface area of 39 cm^2 and contains 36 mg of oxybutynin. Oxytrol has an in vivo delivery rate of approximately 3.9 mg/day.

5.11.7.9 Emsam

Emsam is a once-a-day patch that delivers an active ingredient selegiline that is used to treat depression. Selegiline is an irreversible monoamine oxidase inhibitor and has a molecular weight of 187.30. Emsam is a matrix-type patch that has three layers and is available in three size. The 6-mg/day patch has a surface area of 20 mg/20 cm^2, the 9-mg/day patch has a surface area of 30 mg/30 cm^2, and the 12-mg/day patch has a surface area of 40 mg/40 cm^2.

5.11.7.10 Daytrana

Daytrana is a newly approved methylphenidate transdermal system. It is a 9-h adhesive matrix patch and comes in four sizes: a 12.5-cm^2 patch that has a delivery rate of 1.1 mg/h, an 18.75-cm^2 patch that has a delivery rate of 1.6 mg/h, a 25-cm^2 patch that delivers 2.2 mg/h, and a 37.5-cm^2 patch that delivers 3.3 mg/h.

5.11.8 CHEMICAL AND PHYSICAL APPROACHES TO TRANSERMAL DELIVERY

Due to the brick-and-mortar structure of the stratum corneum, the skin is a difficult layer to permeate across for most active pharmaceutical ingredients. Because of this diffusional barrier, new strategies have been developed to allow compounds to better penetrate the stratum corneum [28]. These strategies can be defined as either chemical or physical approaches to disrupting the barrier function of the skin.

TABLE 5 Desired Properties of Chemical Permeation Enhancers

Provide reversible effects in skin layers
Low systemic bioavailability
High stability and compatibility with formulations
Possess no pharmacological activity
Should be nontoxic and nonirritating

TABLE 6 Investigated Chemical Permeation Enhancers

Azone	Oleic acid
Ethano	Oxazolidnesl
Dimethyl sulfoxide	Propylene glycol
Fatty acids	Sodium lauryl sulfate
Lecithin	Terpenes

5.11.8.1 Chemical Penetration Enhancers

Many approaches that have been investigated over the last several decades enhance the permeation of compounds across the skin using novel chemical compounds [29]. These permeation enhancers are compounds which partition into the stratum corneum and promote the passage of topically applied compounds across the skin layers by using three possible mechanisms of action (Table 5). The mechanisms by which these compounds enhance permeability of the skin have been previously described by Williams and Barry and are often referred to as the lipid–protein partitioning theory [30]. One proposed mechanism of action of permeation enhancers and the most common method by which chemical enhancers increase permeation is by fluidizing the intercellular lipid structures that are found within the stratum corneum. By interacting with and disorganizing these lipid structures, channels can be formed which allow the compound to better diffuse across the rate-limiting barrier of the stratum corneum.

Another method is based on the ability of some permeation enhancers to interact with intracellular proteins such as keratins inside corneocytes. The disassembly of these proteins structures within the corneocytes allows some compounds to transcellularly penetrate the stratum corneum. Also, some enhancers act as vehicles and cosolvents, increasing and promoting the partitioning of compounds into the stratum corneum (Table 6).

5.11.8.2 Physical Enhancement Methods

Microneedles Microneedles are a new type of transdermal device that is receiving tremendous attention by pharmaceutical companies because of the potential for the transdermal delivery of high-molecular-weight compounds. This type of transdermal device contains microscopic needles or projections that can be loaded with active ingredient [31]. The projections when placed on the skin penetrate beyond the stratum corneum into the living epidermis. This allows the compound to bypass passage directly through the stratum corneum, which is the rate-limiting barrier for passive diffusion. The drug can be loaded into the microneedles and then released into the lower skin levels or the microneedle patch may contain a reservoir and the drug will penetrate through channels in the stratum corneum that have been created by the microneedles. This allows compounds that normally would not passively diffuse across the stratum corneum, such as therapeutic proteins, to be delivered transdermally. Currently various formulations of microneedle patches are being investigated in clinical trials in the United States [32–34].

Iontophoresis and Sonophoresis In the early 1900s it was discovered that some chemical compounds could be delivered into the systemic circulation across the skin using an electric current. This phenomenon was later described as iontophoresis. Iontophoresis occurs when an electric potential difference is created across the skin layers by an electric current and this gradient drives the penetration of both charged and uncharged drugs across the skin [35].

One of the earliest FDA-approved products that utilized iontophoresis was the iontophoretic delivery of pilocaripne as a method for diagnosing cystic fibrosis [36]. The sweat of individuals with cystic fibrosis contains large amounts of both sodium and chloride ions. The pilocarpine that is delivered into the skin promotes increased sweating, which can be easily collected and analyzed.

Another product that uses iontophoresis has been recently approved by the FDA. This device is called Ionsys and is an iontophoretic system that delivers fentanyl hydrochloride transdermally [37, 38]. This is a patient-controlled device that provides on-demand delivery of fentanyl for up to 24 h or 80 doses. This device contains 10.8 mg of fentanyl hydrochloride and is designed to deliver a 40-μg dose of fentanyl over a 10-min period upon activation of the dose button by the patient.

Another popular physical enhancement method that has routes in physical therapy and sports rehabilitation clinics is sonophoresis. Sonophoresis involves the use of ultrasound as a source of disrupting intercellular lipid structures in the stratum corneum [39, 40]. The sound waves produced by the device induce cavitation of the lipids found within the stratum corneum, which then opens channels and allows the chemical compound to easily penetrate the skin. This is a safe and reversible process that has received much attention in the literature and by pharmaceutical companies.

5.11.9 FUTURE OF TRANSDERMAL DRUG DELIVERY DEVICES

Through practical application of recombinant deoxyribonucleic acid (DNA) technology there has been a recent explosion of biotech drugs that are in clinical trials, and the FDA has approved many. Unfortunately, due to the high molecular weights and other physical and chemical characteristics of these macromolecules, many delivery obstacles exist that prevent these compounds from being delivered transdermally. Many physical and chemical approaches are being investigated that enhance the delivery of the biotech agents across the skin.

TransPharma-Medical, an Israeli-based pharmaceutical company, is investigating the transdermal delivery of human parathyroid hormone fragment for the treatment of osteoporosis in addition to the delivery of human growth hormone [41]. This technology utilizes a 1-cm^2 patch that creates small channels or holes in the stratum corneum much like the microneedle technology that was previously discussed. The delivery of insulin has been studied for over a decade using transdermal systems. Unfortunately, due to the high molecular weight and dosing considerations with insulin, it has been difficult. However, insulin delivery transdermally has recent successes clinically using sonophoresis, permeation enhancers, iontophoresis, and microneedles [42, 43].

REFERENCES

1. Robinson, D. H., and Mauger, J. W. (1991), Drug delivery systems, *Am. J. Hosp. Pharm.*, 48(10 Suppl. 1), S14–23.
2. Murphy, M., and Carmichael, A. J. (2000), Transdermal drug delivery systems and skin sensitivity reactions. Incidence and management, *Am. J. Clin. Dermatol.*, 1(6), 361–368.
3. Forman, S. A., Helmkamp, J. C., and Bone, C. M. (1987), Cardiac morbidity and mortality associated with occupational exposure to 1,2 propylene glycol dinitrate, *J. Occup. Med.*, 29(5), 445–450.
4. Ben-David, A. (1989), Cardiac arrest in an explosives factory worker due to withdrawal from nitroglycerin exposure, *Am. J. Ind. Med.*, 15(6), 719–722.
5. Gehlbach, S. H., et al. (1974), Green-tobacco sickness. An illness of tobacco harvesters, *JAMA*, 229(14), 1880–1883.
6. Curwin, B. D., et al. (2005), Nicotine exposure and decontamination on tobacco harvesters' hands, *Ann. Occup. Hyg.*, 49(5), 407–413.
7. Xhauflaire-Uhoda, E., et al. (2006), Dynamics of skin barrier repair following topical applications of miconazole nitrate, *Skin Pharmacol. Physiol.*, 19(5), 290–294.
8. Norlen, L., and Al-Amoudi, A. (2004), Stratum corneum keratin structure, function, and formation: The cubic rod-packing and membrane templating model, *J. Invest. Dermatol.*, 123(4), 715–732.
9. Fluhr, J. W., Feingold, K. R., and Elias, P. M. (2006), Transepidermal water loss reflects permeability barrier status: Validation in human and rodent in vivo and ex vivo models, *Exp. Dermatol.*, 15(7), 483–492.
10. Casiraghi, A., et al. (2002), Occlusive properties of monolayer patches: In vitro and in vivo evaluation, *Pharm. Res.*, 19(4), 423–426.
11. Wertz, P. W. (2000), Lipids and barrier function of the skin, *Acta Derm. Venereol. Suppl. (Stockh.)*, 208, 7–11.
12. Wertz, P. W. (1997), Integral lipids of hair and stratum corneum, *Exs*, 78, 227–237.
13. Astley, J. P., and Levine, M. (1976), Effect of dimethyl sulfoxide on permeability of human skin in vitro, *J. Pharm. Sci.*, 65(2), 210–215.
14. Wertz, P. W., et al. (1985), The composition of the ceramides from human stratum corneum and from comedones, *J. Invest. Dermatol.*, 84(5), 410–412.
15. Barry, B. W. (2002), Drug delivery routes in skin: A novel approach, *Adv. Drug Deliv. Rev.*, 54 (Suppl 1), S31–40.
16. Buchwald, P., and Bodor, N. (2001), A simple, predictive, structure-based skin permeability model, *J. Pharm. Pharmacol.*, 53(8), 1087–1098.
17. El-Kattan, A., Asbill, C. S., and Haidar, S. (2000), Transdermal testing: Practical aspects and methods, *Pharm. Sci. Technol. Today*, 3(12), 426–430.
18. Asbill, C., et al. (2000), Evaluation of a human bio-engineered skin equivalent for drug permeation studies, *Pharm. Res.*, 17(9), 1092–1097.

19. El-Kattan, A. F., Asbill, C. S., and Michniak, B. B. (2000), The effect of terpene enhancer lipophilicity on the percutaneous permeation of hydrocortisone formulated in HPMC gel systems, *Int. J. Pharm.*, 198(2), 179–189.

20. Krishnaiah, Y. S., and Bhaskar, P. (2004), Studies on the transdermal delivery of nimodipine from a menthol-based TTS in human volunteers, *Curr. Drug Deliv.*, 1(2), 93–102.

21. Misra, A., et al. (1997), Biphasic testosterone delivery profile observed with two different transdermal formulations, *Pharm. Res.*, 14(9), 1264–1268.

22. Burkman, R. T. (2004), The transdermal contraceptive system, *Am. J. Obstet. Gynecol.*, 190(4, Suppl), S49–53.

23. Sane, N., and McGough, J. J. (2002), MethyPatch Noven, *Curr. Opin. Invest. Drugs*, 3(8), 1222–1224.

24. Amsterdam, J. D. (2003), A double-blind, placebo-controlled trial of the safety and efficacy of selegiline transdermal system without dietary restrictions in patients with major depressive disorder, *J. Clin. Psychiatry*, 64(2), 208–214.

25. Chien, Y. W. (1985), Microsealed drug delivery systems: Fabrications and performance, *Methods Enzymol.*, 112, 461–470.

26. Parrott, A. C. (1989), Transdermal scopolamine: A review of its effects upon motion sickness, psychological performance, and physiological functioning, *Aviat. Space Environ. Med.*, 60(1), 1–9.

27. Burris, J. F., and Mroczek, W. J. (1986), Transdermal administration of clonidine: A new approach to antihypertensive therapy, *Pharmacotherapy*, 6(1), 30–34.

28. Trommer, H., and Neubert, R. H. (2006), Overcoming the stratum corneum: The modulation of skin penetration. A review, *Skin Pharmacol. Physiol.*, 19(2), 106–121.

29. Xu, P., and Chien, Y. W. (1991), Enhanced skin permeability for transdermal drug delivery: Physiopathological and physicochemical considerations, *Crit. Rev. Ther. Drug Carrier Syst.*, 8(3), 211–236.

30. Williams, A. C., and Barry, B. W. (1991), Terpenes and the lipid-protein-partitioning theory of skin penetration enhancement, *Pharm. Res.*, 8(1), 17–24.

31. Henry, S., et al. (1998), Microfabricated microneedles: A novel approach to transdermal drug delivery, *J. Pharm. Sci.*, 87(8), 922–925.

32. Giudice, E. L., and Campbell, J. D. (2006), Needle-free vaccine delivery, *Adv. Drug Del. Rev.*, 58(1), 68–89.

33. Aqil, M., Sultana, Y., and Ali, A. (2006), Transdermal delivery of beta-blockers, *Expert Opin. Drug Deliv*, 3(3), 405–418.

34. Cormier, M., et al. (2004), Transdermal delivery of desmopressin using a coated microneedle array patch system, *J. Controlled Release*, 97(3), 503–511.

35. Fischer, G. A. (2005), Iontophoretic drug delivery using the IOMED Phoresor system, *Expert Opin. Drug Deliv.*, 2(2), 391–403.

36. Parad, R. B., et al. (2005), Sweat testing infants detected by cystic fibrosis newborn screening, *J. Pediatr.*, 147(3, Suppl), S69–72.

37. Miaskowski, C. (2005), Patient-controlled modalities for acute postoperative pain management, *J. Perianesth. Nurs.*, 20(4), 255–267.

38. Sinatra, R. (2005), The fentanyl HCl patient-controlled transdermal system (PCTS): An alternative to intravenous patient-controlled analgesia in the postoperative setting, *Clin. Pharmacokinet.*, 44(Suppl 1), 1–6.

39. Mitragotri, S., and Kost, J. (2000), Low-frequency sonophoresis: A noninvasive method of drug delivery and diagnostics, *Biotechnol. Prog.*, 16(3), 488–492.

40. Mitragotri, S., Blankschtein, D., and Langer, R. (1996), Transdermal drug delivery using low-frequency sonophoresis, *Pharm. Res*, 13(3), 411–420.
41. Levin, G., et al. (2005), Transdermal delivery of human growth hormone through RF-microchannels, *Pharm. Res.*, 22(4), 550–555.
42. Rastogi, S. K., and Singh, J. (2005), Effect of chemical penetration enhancer and iontophoresis on the in vitro percutaneous absorption enhancement of insulin through porcine epidermis, *Pharm. Dev. Technol.*, 10(1), 97–104.
43. King, M. J., et al. (2002), Transdermal delivery of insulin from a novel biphasic lipid system in diabetic rats, *Diabetes Technol. Ther.*, 4(4), 479–488.

5.12

VAGINAL DRUG DELIVERY

José das Neves, Maria Helena Amaral, and
Maria Fernanda Bahia
University of Porto, Porto, Portugal

Contents

Pharmaceutical Manufacturing Handbook: Production and Processes, edited by Shayne Cox Gad

5.12.1 INTRODUCTION

The vagina has been used for a long time as a route for drug administration, being as old as medicine and pharmacy themselves. Throughout the history of human civilization, vaginal administration of drugs has been practiced and recorded until the modern era. Some of the first records were found in Egypt, where the Kahun Papyrus, the oldest of the surviving medical papyri (ca. 1850 B.C.), included references to vaginal "preparations" containing substances such as mud, frankincense, oil, malachite, ass urine, myrrh, crocodile dung, honey, and sour milk, normally used in female genitalia–related conditions and contraception [1]. Latter papyri such as the Ramesseum Papyrus (ca. 1700 B.C.), the Ebers Papyrus (ca. 1550 B.C.), and the Greater Berlin Papyrus (ca. 1300 B.C.) also contained drug formulations to be administered in the vagina. Vaginal administration of drugs continued to be carried out by other civilizations from ancient Greece and Rome to the Middle Ages, in the Arabic and Oriental cultures, passing through the Renaissance, until our days [1, 2].

Although traditionally used for local action, some drugs can permeate the vaginal mucosa and reach the bloodstream in sufficient concentrations to have systemic effects. Current understanding of the vaginal anatomy, physiology, and pathophysiology is very considerable, contrasting with the still limited knowledge of the possibilities of vaginal drug delivery. Nonetheless, interest and contribution of pharmaceutical scientists toward vaginal drug delivery development have increased in the last years in response to the specific needs of this route of drug administration. Indeed, vaginal drug delivery has seen recent advances that make it very promising, in particular therapeutic fields such as the prevention of human immunodeficiency virus (HIV) and other sexually transmitted infections, contraception, hormone replacement therapy in menopausal women, and labor induction.

Therefore, this chapter discusses the main features of vaginal anatomy, physiology, and histology related to drug delivery as well as vaginal drug delivery systems and their evaluation. Also, past and current usage of the vagina as a route of drug administration, ongoing investigation, and promising strategies in this field are addressed.

5.12.2 THE HUMAN VAGINA

The knowledge of the human vagina's anatomy, histology, and physiology is fundamental in the design of drug delivery systems, helping researchers optimize vaginal products. Pharmaceutical scientists must comprehend the particularities of this organ in order to overcome its natural limitations and enhance its advantages over other drug delivery routes.

5.12.2.1 Anatomy

The organs of the female reproductive tract are classically divided into the internal and external genitalia [3]. The internal genital organs include the vagina, cervix, uterus, oviducts, ovaries, and surrounding supporting structures, as seen in Figure 1. The external genital organs, also known as the vulva or pudendum, are composed by the structures that surround the vaginal entrance, which are visible in the perineal area. These include the mons pubis, labia majora, labia minora, hymen, clitoris, vestibule, urethra, vestibular glands (Skene glands and Bartholin glands), and vestibular bulbs.

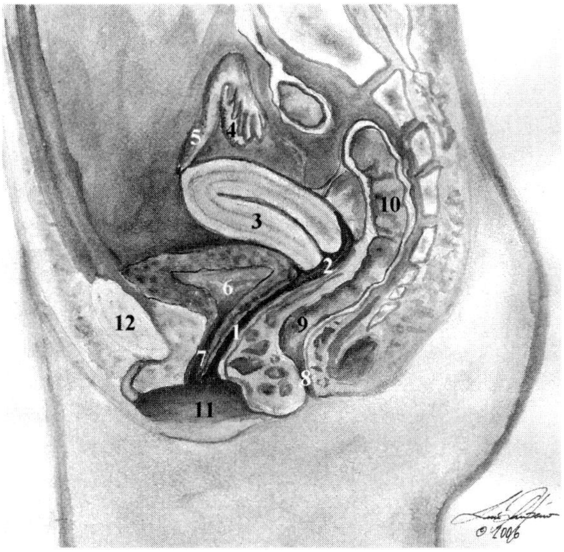

FIGURE 1 Female reproductive system and related structures: vagina (1), cervix (2), uterus (3), ovary (4), fallopian tube (5), urinary bladder (6), urethra (7), anus (8), rectum (9), colon (10), vestibule (11), and pubic symphysis (12). (Courtesy of Luís Paupério.)

The vagina is a tubular, fibromuscular organ approximately 9 cm long, extending from the cervix to the vestibule, being positioned between the urinary bladder and the rectum (Figure 1). Because of the protrusion of the cervix into the upper part of the vagina, two deep recessions (anterior and posterior) are formed, called fornices, the posterior fornix being considerably larger than the anterior one. Also, two lateral fornices can be considered [3, 4]. In the adult woman, the anterior wall of the vagina is approximately 7.5 cm long, while the posterior wall extends by approximately 9 cm. The width of the vagina varies throughout its length, the transverse diameter being higher at the level of the fornices, decreasing progressively toward the vaginal entrance. The vagina is commonly misunderstood as a straight tube, descending from the cervix to the introitus. The axes of the upper and lower vagina are different, forming a slight S-shaped curve: The axis of the lower vagina in relation to a standing woman is vertical and posterior, while the upper part (from the pelvic diaphragm to the cervix) becomes more horizontal, with the final portion of the vagina curving toward the hollow of the sacrum [3, 5]. The angle between these two portions is approximately 130°C. This fact is important in drug delivery, taking into consideration that pharmaceutical formulations must be retained in the vagina, resisting to gravity forces. The walls of the vagina are covered by a mucosal tissue that forms a series of transverse folds denominated rugae (more prominent in the lower third of the vagina), thus increasing the available surface for absorption and allowing considerable distension of this organ during penile penetration and childbirth [4, 6]. Indeed, the surface area of the vaginal mucosa is an important parameter concerning its coverage by vaginal products or drug absorption. Results by Pendergrass et al. varied through a wide range of values (65.73–107.07 mm^2), with a mean surface area of 87.46 mm^2, although these results can be an understatement because of limitations in the determination method [7]. In a more recent study, Barnhart et al. found higher surface values which ranged from 103.9 to 165.0 mm^2. These variable results suggest that one volume of vaginal product may not be appropriate for all women [5]. The walls of the vagina are normally in apposition and flattened in the anteroposterior diameter. Thus the vagina has the appearance of the letter H in cross section, although a W shape may also be observed [8].

The vagina has a high blood supply, which is primarily provided by the vaginal branches of the internal iliac artery. To a minor extension, blood is also supplied through branches of the uterine, middle rectal, and internal pudendal arteries. Blood drainage from the vagina returns through vaginal veins (the perineum venous plexus), running parallel to the course of the arteries, which flow into the pudendal vein, internal iliac vein, and then the vena cava, thus bypassing the portal circulation. Blood supply increases with sexual excitement, particularly to the clitoris and vestibular bulbs, causing the erection of these structures along with the expansion and elongation of the vagina, allowing the accommodation of the male penis [4]. The lymphatic drainage is characterized by its wide distribution and frequent crossovers between both sides of the pelvis. Generally, the upper third of the vagina drains to the external iliac nodes, the middle third to the common and internal iliac nodes, and the lower third to the common iliac, superficial inguinal, and perirectal nodes [3].

Sensory innervation of the vagina is provided by the pudendal plexus and is particularly developed near the vaginal orifice. The nervous sympathetic innervation of the vagina is provided by the hypogastric plexus, while the nervous parasympathetic innervation derives from the second and third sacral nerves. There is a lack of free nerve endings in the upper two-thirds of the vagina [3,4]. Thus, objects placed in this area, such as vaginal drug delivery systems, are likely to be unperceived by women.

5.12.2.2 Histology

The vaginal tissue is composed of four distinct layers: stratified squamous epithelium, lamina propria, muscular layer, and tunica adventitia, the first two usually referred to as vaginal mucosa (Figure 2). The epithelial layer is about 150–200 μm thick, corresponding approximately to 30–45 layers of cells, being nonkeratinized. The epithelium presents different types of cells, making it possible to identify five different layers, namely the basal, parabasal, intermediate, transitional, and superficial layers. Epithelial cells are closely joined by numerous desmosomes and occasionally tight junctions, the last being particularly abundant in the basal layer. Deposits of a lipid material are found near the surface of the epithelium, acting as a permeability barrier to large water-soluble molecules. The lamina propria, or tunica, is composed of fibrous connective tissue richly supplied by small blood and lymphatic vessels. Although being a mucosal surface, the vaginal mucosa does not

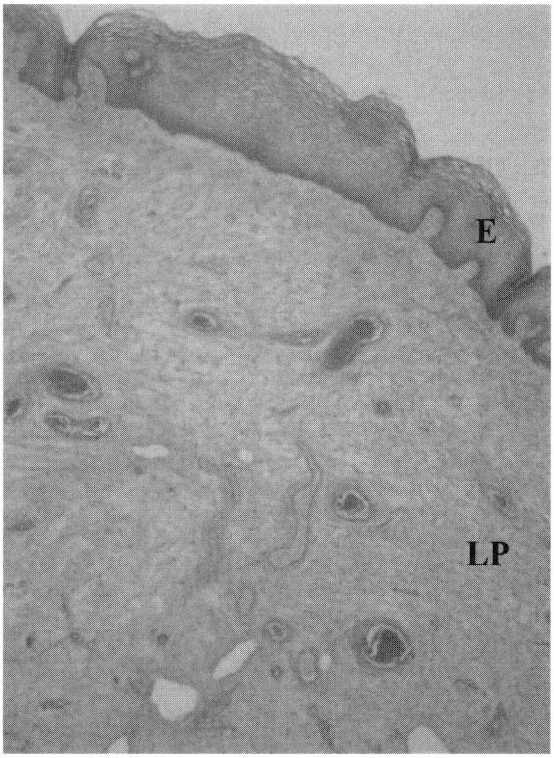

FIGURE 2 Detail of human vaginal mucosa of fertile adult (H&E, ×40). Image shows stratified squamous epithelium (E), which is nonkeratinized, and lamina propria (LP), richly supplied by small blood vessels and deprived of glands (Courtesy of Manuel Dias.)

have mucus-secreting glands (Figure 2). The muscular layer has a large number of interlacing fibers, making it possible to distinguish an inner circular layer and an outer longitudinal layer. These muscle fibers allow substantial elongation during childbirth, increasing approximately five-fold. The tunica adventitia consists of connective tissue containing a large plexus of blood vessels, lymphatic vessels, and nerves [9].

The characteristics of the vaginal mucosa depend on sexual hormones undergoing changes throughout the menstrual cycle. During the proliferative phase, estrogens increase vaginal blood flow and the integrity of the mucosa, inducing the proliferation of the vaginal epithelium, peaking at approximately midcycle. Nonetheless, this change in epithelial thickness is small and probably clinically insignificant [10, 11]. In the secretory phase progesterone opposes those actions, although without significantly influencing the epithelium thickness [12].

5.12.2.3 Physiology

Functions of the vagina include receiving the erect penis and semen during coitus and ejaculation and serving as a passageway for fetus and menses to the outside of the body [4].

The vaginal milieu is typically acidic in healthy women during their fertile years, playing an important role in preventing the proliferation of pathogenic microorganisms. Vaginal pH is around 3.5–4.5, changing during the menstrual cycle being lower in the middle cycle and higher around menses [13]. Also, vaginal pH may be altered by several conditions, including the presence of semen (pH 7.2–8.0) or bacterial infections. In this last case vaginal pH usually increases to values of 5.0–6.5 [14]. These and other variations of vaginal pH may alter the efficacy of administered drugs through quite a few mechanisms, including variable release from the delivery system, alteration in drug absorption and/or metabolism, and modulation of drug activity in the cervicovaginal milieu [15]. *Lactobacillus* species are present in the vagina of healthy women, being responsible for its acidity. These bacteria use glycogen, which is synthesized by epithelial cells under estrogen's influence in order to produce lactic acid, thus reducing vaginal pH [16]. Besides their contribution to vaginal pH, lactobacilli also have the ability to regulate the proliferation of microorganisms, namely pathogenic species, by other mechanisms: adherence to the mucus, forming a barrier which prevents colonization by pathogens or competition for the receptors of the epithelial cells, and the production of antimicrobial compounds such as hydrogen peroxide, lactic acid, bacteriocin-like substances, and possibly biosurfactants [17]. Although lactobacilli are the main organisms responsible for vaginal acidity, recent research suggests that other acid-producing microorganisms, such as *Atopobium* sp., *Megasphaera* sp., and *Leptotrichia* sp., may also contribute to the acidity of the vaginal tract [18]. The composition of vaginal flora is complex, being influenced by hormonal changes, although lactobacilli levels of colonization seem to remain relatively constant during the menstrual cycle [19]. Also, other factors such as glycogen content, pH, sexual intercourse, medication, and immunity status are known to influence the vaginal ecosystem [20].

Although deprived of secreting glandules, the vaginal mucosa is covered with a watery acidic fluid. This fluid includes contributions from vaginal transudation, Bartholin and Skene glands, exfoliated epithelial cells, residual urine, and fluids from the upper reproductive tract such as cervical mucus and endometrial and tube fluids. Vaginal fluid consists of 90–95% water, inorganic and organic salts, urea, carbohydrates, glycerol, mucins, fatty acids, albumins, immunoglobulins, enzymes, leukocytes, and epithelial debris. Although quite variable, normal daily production of vaginal fluid is estimated at around 6 mL, increasing at midcycle and decreasing around the menstruation period [21]. In fact, the menstrual cycle plays an important role in vaginal fluid characteristics, particularly on pH values, rheological properties, color (from milky white to transparent), and antimicrobial activity. Estrogens induce the production of vaginal fluid, leading to the lubrication of the mucosa. Decline of serum estrogen levels, as during prepubertal or postmenopausal periods, results in a reduction of vaginal moisture [22, 23]. This variability may influence drug release, dissolution, absorption, and removal and thus its activity. Also, vaginal fluid is selectively antimicrobial, lactic acid and to a lesser extent antimicrobial peptides and proteins (calprotectin, lysozyme, histones, and others) being partially responsible by the resistance of the normal vagina to colonization by exogenous microorganisms [24].

Cervical mucus is an important component of the vaginal fluid, although it presents substantial differences when compared with whole vaginal fluid, mainly in pH value (approximately 7.0). It is produced in the cervix, leaking down into the vagina, particularly during the three- to five-day interval that precedes ovulation. Its properties are also influenced by the menstrual cycle, particularly pH (range of 5.4–8.0) and antimicrobial activity [25]. Also, estrogens stimulate the secretion of abundant and fluid cervical mucus, while progesterone induces the formation of thick cervical mucus. These changes in viscosity influence the capacity of sperm and other substances, such as drugs, to pass the cervix and migrate to the uterus [22, 23].

Interaction between vaginal fluids and drug delivery systems is an important aspect that has to be managed during drug design, as it may influence the flow, retention, drug delivery kinetics, and bioactivity of vaginal formulations [21]. Since fluids present in the vaginal environment are difficult to obtain, simulated fluids have been developed in order to emulate their physical and chemical properties. Recently, a vaginal fluid simulant was proposed by Owen and Katz [21], whether Burruano et al. [26] developed a synthetic cervical mucus formulation. These fluids have been successfully used to evaluate vaginal formulations in vitro, being able to mimic with considerable accuracy the physiological fluids. Also, the standardization of a single composition provides the possibility of comparing results obtained by different investigators. Table 1 presents the main features of these two simulants.

Enzymes present in the vagina may influence drug delivery, as they can degrade and influence the permeability of the administered drugs. Although the enzymatic activity in the vagina is not as high as in other drug delivery sites, several enzymes can be found in the vaginal fluid and in different vaginal epithelium cells, namely succinic and lactic dehydrogenase, acid and alkaline phosphatases, β-glucuronidase, phosphoamidase, lactate dehydrogenase, aminopeptidase, and esterases [27, 28]. This enzymatic activity can limit drug bioavailability and decrease the stability of prolonged delivery formulations.

Understanding the immune mechanisms responsible for the defense of the female genital tract is of extreme importance concerning the development of vaccines that are effective against pathogens. The female genital tract has several mechanisms of defense against infectious agents, which appear complementary, additive, and even

TABLE 1 Main Features of Vaginal Fluid Simulant[a] and Synthetic Cervical Mucus[b]

	Vaginal Fluid Simulant	Synthetic Cervical Mucus
Composition	NaCl (3.51 g), KOH (1.40 g), $Ca(OH)_2$ (0.222 g), bovine serum albumin (0.018 g), lactic acid (2.00 g), acetic acid (1.00 g), glycerol (0.16 g), urea (0.4 g), glucose (5.0 g), HCl (to adjust pH), and water (e.q. 1 L)	Guar gum (1.00%), dried porcine gastric mucine (type III) (0.50%), imidurea, (0.30%), methylparaben (0.15%), propylparaben (0.02%), dibasic potassium phosphate (0.26%), monobasic potassium phosphate (1.57%), and water (96.20%)
pH	4.2	7.4

[a]From ref. 21.
[b]From ref. 26.

TABLE 2 Immune Response of Female Genital Tract

Nonimmune	Passive
	Synthesis of protective mucus
	pH
	Epithelial barrier
	Active
	Inflammatory reaction
	Secretion of humoral soluble factors
Preimmune	Humoral response, cellular response
Immune	Humoral response, cellular response

Source: From ref. 29.

synergistic. The immune response can be classified in three levels: nonimmune, preimmune, and acquired or specific (Table 2) [29]. Nonimmune response is very effective in limiting the infectious inoculum, while preimmune mechanisms hold up the infection long enough so that the immune response can be activated. These defense strategies are largely influenced by pathogenic agents presented in the vagina and the hormonal milieu [29, 30].

The genital tract is part of the common mucosal immune system, which comprises all mucosal tissues of the body, sharing similarities and common immunization mechanisms, being able to disseminate acquired immunity between them. Nonetheless, it has become clear that the genital tract presents several unique features that differentiate it from other mucosal sites [31]. One of the most important is the ability to only induce local immune responses. In fact, most of the antibodies are produced locally at the mucosa, while those derived from the circulation represent only a small fraction. Also, contrasting with other mucosal tissues, vaginal secretions contain more immunoglobulin (Ig) G than secretory IgA [30, 32]. Hence, immunity of the genital tract is conferred by local production of IgA and, to a less extent, by transudation of serum IgG, although cellular immune response should also be considered [33]. Local immune cell population in the vaginal mucosa includes Langerhans cells, macrophages, T cells, and neutrophils. Systemic immune response is also important in the reinforcement of the acquired mucosal immunity or to take over when this one has faded, although they are independent from one another [29, 30].

Sexual hormones, particularly estrogens, also influence the regulation of local immunity, as IgA levels in genital secretions, antigen presentation by vaginal cells, and lymphocyte proliferation vary throughout the menstrual cycle [34, 35]. Estrogens can decrease the concentrations of immunoglobulins, while progestogens can increase their levels [36].

Alterations in vaginal physiology before, during, and after sexual intercourse are important factors in the performance of those formulations intended to be used in this period. During sexual arousal, genital vasocongestion occurs, leading the clitoris and the labia minora to become enlarged with blood and the vagina to increase in length and diameter as a result of relaxation of the smooth muscular wall. The vaginal canal is lubricated by secretions from the uterus, Bartholin and Skene glands, and a fluid that transudates from the subepithelial vascular tissues, being passively transported through the intercellular spaces. Engorgement of the vaginal wall raises the pressure inside the capillaries, thus increasing the transudation of plasma through the vaginal epithelium [6]. The resulting vaginal fluid is increased in quantity, less acidic, and more diluted, allowing penile penetration and thus preventing the male and female genitalia from becoming irritated [3, 4].

5.12.2.4 Childhood, Pregnancy, and Menopause

The vagina undergoes several lifetime changes that may influence vaginal formulation performance. These changes are particularly important when developing vaginal products that are intended to be used in specific situations, such as labor inducers in pregnant women or hormonal supplements in postmenopausal women.

The vagina of the newborn exhibits the influence of residual maternal estrogens, presenting a stratified squamous epithelium rich in glycogen and becoming colonized with lactic acid–producing microorganisms shortly after birth. By the fourth postnatal week these estrogenic effects disappear, and the vaginal epithelium loses its stratification and glycogen content, becoming much thinner and exhibiting alkaline or neutral pH because of acid-producing microorganism depletion. These characteristics remain throughout childhood, until puberty. By this time, the vagina experiences changes due to adrenal and gonadal maturation. This organ increases in size, vaginal fornices develop, cervicovaginal secretions start being produced, and vaginal milieu becomes acidic. Also, vaginal epithelium thickens and intracellular glycogen production increases [37].

When the fertile woman becomes pregnant, the connective tissue of the vulva, vagina, and perineum relaxes, and the muscle fibers of the vaginal wall increases in size. These alterations prepare the vagina for childbirth. During delivery, perineal and vaginal musculature relaxes and the vaginal rugae flatten, allowing full expansion of the vaginal tract, accommodating the passage of the newborn. Normal morphology and dimensions of the vagina are recovered after 6–12 weeks [37]. The vaginal blood supply is substantially increased during pregnancy, which can enhance systemic absorption of drugs. High levels of estrogens during pregnancy lead to thickening of vaginal epithelium and stimulation of glycogen production. This increased glycogen content promotes lactobacilli growth, consequently decreasing vaginal pH by the enhancement of lactic acid synthesis.

Menopausal women experience a decline in estrogens, which leads to several changes in the genital organs, namely in the vagina. Such changes include atrophy of the labia majora and shortening and loss of elasticity of the vaginal barrel. The number of epithelial cell layers decreases (from 8–10 in premenopausal to 3–4 in postmenopausal), leading to alterations in the barrier capacity and potential increase of mucosal damage and pain and burning sensation during sexual intercourse. Vaginal fluids decrease approximately 50% because of the Bartholin glands atrophy and a decrease in the number and maturity of vaginal cells. A decrease in the colonization by lactobacilli species is observed in menopausal women as a result of the reduction of vaginal glycogen levels, leading to a low production of lactic acid and consequently to increased vaginal pH. In postmenopausal women without estrogen treatment the vaginal pH is estimated to be 5.5–6.8 or even higher [37, 38].

5.12.3 GENERAL FEATURES OF VAGINAL DRUG DELIVERY

Vaginal drug delivery is mostly used in gender-specific conditions, although it can be a viable alternative for drugs usually administered by other routes. Also, traditionally problematic drugs from a delivery point of view (e.g., peptides) may find in the vaginal route an interesting and promising way for nonparental administration. Limitations and potentialities of vaginal drug administration are intimately connected to this route's idiosyncrasy. Hence, acquaintance of particular features of vaginal drug delivery is required.

5.12.3.1 Advantages and Disadvantages of Vaginal Drug Delivery

The administration of drugs through the vagina, and eventually their absorption, is a function for which this organ is not physiologically conceived. Nonetheless, the vaginal route of administration presents some advantages. Substances absorbed through the vaginal mucosa bypass the liver before entering systemic circulation, avoiding hepatic first-pass metabolism. Thus, drugs that undergo extensive hepatic first-pass metabolism can benefit from vaginal administration, usually requiring less amount of drug to achieve the same biological effects. Steroids used in hormone replacement therapy or contraception are a good example of molecules that are largely metabolized in the liver, with approximately 95% of orally administered estrogens undergoing hepatic metabolism. Also, these molecules are able to damage the liver when administered by the oral route, an event that can be minimized with vaginal administration. Gastrointestinal side effects are common for many oral administered drugs; the vaginal route may be an alternative to their administration, with the benefit of increased patient compliance. Additionally, vaginal enzymatic activity is lower when compared with gastrointestinal activity, lacking even some important enzymes enrolled in drug metabolism, such as trypsin and chymotrypsin. The vaginal route offers women the possibility of easy self-insertion and removal of drug delivery systems as well as avoidance of the pain, tissue damage, and eventual infection often associated with parenteral routes. Ocular and buccal administration sites frequently become irritated after prolonged contact with drug delivery systems; conversely, the vagina presents less sensitivity, allowing the presence of drug formulations for long periods of time. Although absorption of substances is not a function of the vagina, features such as its relatively large surface area and rich blood supply contribute to this organ's high permeability to several drugs, allowing higher bioavailability of some active substances when compared to other routes [39–41]. On the other hand, the vaginal drug delivery route has some limitations. Gender specificity is the most important, as it restricts its use to females only. Others, such as misperceptions and cultural issues about genital manipulation and insertion of objects in the vagina, personal hygiene, influence with sexual intercourse, and variability in drug absorption related with menstrual cycle, menopause, and pregnancy, can also limit vaginal drug delivery route usage [41, 42].

5.12.3.2 Permeability and Drug Absorption

Although some substances are not desired to be absorbed, such as those targeted for local action, permeation of drugs through the vaginal wall into the bloodstream must occur if one seeks to obtain a systemic effect. Although vaginal administration of drugs has been performed for a long time, the capability of systemic drug absorption through this organ was not clarified until the early twentieth century by investigators such as Macht or Robinson [43,44]. Both conducted independent experiments with substances such as potassium iodide, morphine, atropine, sodium salicylate, quinine, sucrose, and phenol red, which evidenced the permeability of this mucosal tissue, thus opening the possibility of systemic drug delivery through the vagina. Nonetheless, the currently acknowledged variability of vaginal drug absorption with the hormonal status of women, which can limit the potential for systemic drug delivery, was documented in the 1940s by authors such as Rock et al. These researchers reported that although the vaginal administration of drugs, namely penicillin, could provide therapeutic blood levels, the extension of their absorption was highly variable [45].

Recently, several in vitro experiments substantiated the potential of the human vaginal mucosa as a good administration route relating to the degree of permeation when compared with other mucosal surfaces. In fact, the vagina can be more permeable to some commonly used model substances, such as water, 17β-estradiol (Figure 3), arecoline, arecaidine, and vasopressin, than colonic or small intestinal mucosa, or at least as permeable as when compared to human buccal mucosa [46, 47].

In general, systemic drug absorption requires three steps: drug release from the delivery system, drug dissolution in the vaginal fluid, and permeation of the vaginal mucosa. Knowledge of the permeability characteristics of the vaginal mucosa is an important step when developing a pharmaceutical product, with both the physicochemical properties of chemical substances (e.g., chemical nature, degree of ionization, molecular weight and size, conformation, and oil/water partition coefficient) and the biophysicochemical nature of the tissue influencing absorption. The epithelial layer of the vaginal mucosa presents itself as the main permeability barrier for drug absorption. As the epithelium is hormonally dependent, its permeability also changes, usually decreasing with higher estrogen levels because of the induced membrane thickening. However, contradictory findings of enhanced vaginal absorption in postmenopausal women treated with estrogen have been reported [48]. These results can be explained by the increment of vaginal mucosa blood flow that is also induced by estrogen. The transport mechanism of most vaginal absorbed substances is simple diffusion. Lipophilic substances are absorbed through the

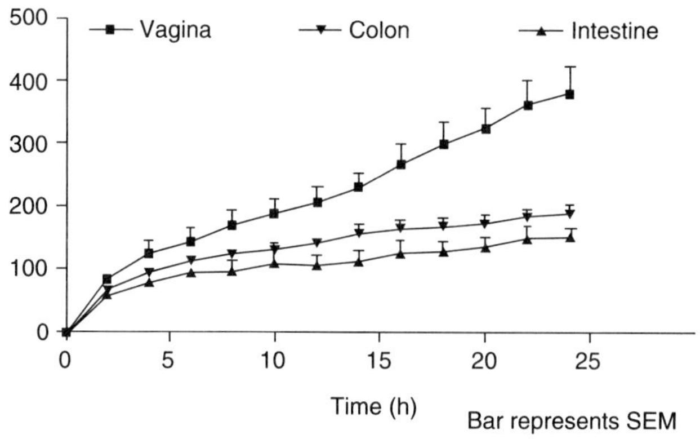

FIGURE 3 Overall mean flux values of 17β-estradiol across human vaginal, colonic, and small intestinal mucosa, as determined by flow-through diffusion cells. Flux values (J) were calculated as $J = Q/(At)$, where Q is quantity of 17β-estradiol crossing mucosa (in dpm), A is mucosa area exposed (in cm^2), and t is time of exposure (in min). SEM: standard error of the mean. (Reprinted with permission from P. van der Bijl and A. D. van Eyk, *International Journal of Pharmaceutics*, 261, 147–152, 2003. Copyright 2003 by Elsevier.)

intracellular (or transcellular) pathway, whereas hydrophilic substances are absorbed through the intercellular (or paracellular) pathway or across aqueous pores present in the vaginal mucosa [49]. Also, receptor-mediated transport mechanisms can be involved in the absorption of some substances.

The magnitude of the flux rate across the vaginal mucosa is mainly related to the molecular size and hydrophobicity of the permeating substances. In general, compounds with molecular weight over 300 Da have decreased flux rates, while hydrophobic properties usually increase permeation [50–52]. The influence of the penetrant hydrophobicity/hidrophilicity in the rate and extent of absorption through the vaginal mucosa was demonstrated by Corbo et al. [53, 54]. Experiments performed in rabbits showed that hydrophilicity influenced mucosal permeability of drugs such as progesterone, with increasing hidrophilicity leading to decrease in rate and extent of vaginal absorption. Nonetheless, one should keep in mind that a minimum degree of aqueous solubility is always required in order to ensure that the drug dissolves in the vaginal fluid. Also, pH, viscosity, and volume of this fluid change throughout the menstrual cycle, potentially influencing the extension of drug dissolution: Different pH values influence the drug degree of ionization and thus its solubility; increased viscosity of the vaginal fluid may enhance drug retention, although it may also present a barrier to drug absorption; and higher volumes of fluid benefit drug dissolution but also increase its clearance from the vagina.

However, permeability-enhancing strategies may be required, as many substances cannot permeate the vaginal mucosa in significant pharmacological levels. An interesting and helpful option is the use of several permeation enhancers which can increase drug absorption by interacting with epithelial tight junctions, providing a new intercellular penetration pathway [55]. Some of these substances, such as citric acid, benzalkonium chloride, laureth-9, lysophosphatidylglycerol, sodium taurodihydrofusidate, lysophosphatidylcholine, palmitoylcarnitine chloride, lysophosphatidylglycerol, and sodium glycodihydrofusidate, have been successfully tested [56–60].

Although this strategy may enhance permeability, it presents some disadvantages, particularly the possibility of mucosal damage. These unwanted effects are variable and not always observed; however, they may often be severe [61]. Also, some drugs achieve poor systemic levels after being delivered by the vaginal route, not because of poor permeation through the mucosa, but due to their fast inactivation by local enzymes, particularly when therapeutic peptides and proteins are considered. In these cases, enzymatic inhibitors may be a helpful solution. Vaginal peptidase inhibitors, such as ethylene diamine tetraacetic acid (EDTA), thimerosal, amastatin, bestatin, leuptin, and pepstatin A, were shown to be useful, promoting peptide absorption in rats, as they prevent drug degradation [60, 62]. At this point it is also important to notice that some substances can reduce the permeability of the vaginal mucosa [63], this possibility always being important when designing a drug delivery system. In addition, other approaches, such as the use of mucoadhesive polymers, in situ gelling formulations, or solubility enhancers, have been shown to be useful in improving vaginal permeability of several drugs [40].

5.12.3.3 First-Uterine-Pass Effect

The first-uterine-pass effect can be defined as a preferential transfer of vaginally administered drugs to the uterus. This effect is due to a countercurrent mode of exchange, with an upward vagina-to-uterus transport of substances absorbed in vaginal and lymphatic vessels and diffusing to nearby arteries [64]. Evidences of higher than expected uterine concentration after vaginal administration of drugs, namely progesterone, terbutaline, or danazol, led to the postulation and verification of this hypothesis [65, 66]. This effect can be of the utmost importance when the uterus is the desired locale for a drug to exert its effects, opening new therapeutic options for uterus-related conditions.

Further investigations showed that the placement of a formulation in different areas of the vagina dramatically influences the observation of the first-uterine pass-effect. Experimental findings suggest that this preferential transfer to the uterus is only observed when absorption occurs in the outer one-third of the vagina [67]. Thus, drugs intended to exert their effects in the uterus should not be inserted deeply in the vagina, as it is often recommended, instead they should be placed near the vaginal entrance.

5.12.4 VAGINAL DRUG DELIVERY SYSTEMS

5.12.4.1 Overview

A wide range of drug delivery systems have been used, although many of them are not specifically designed for intravaginal administration. Traditionally used vaginal drug delivery systems include solutions, ointments, creams, vaginal suppositories, and tablets. Recently, others, such as vaginal rings or vaginal films, have been developed. Also, several strategies and improvements have been tested in order to overcome natural limitations of drug delivery through this route, particularly low retention, limited absorption, and cyclic variations.

Most of the currently available vaginal formulations, particularly those that have been marketed for a longer period of time, have serious limitations such as poor spreadability, messiness, and small capacity of retention in the vagina. Nonetheless, recent advances allowed circumventing some of the major difficulties that hold back the use of this route of drug delivery as a serious alternative to the most traditional ones, with the consequent increase of commercially available drug delivery systems [39]. Drug release of most traditional formulations is rapid, needing frequent administrations to sustain therapeutic drug concentrations. Thus, in recent years, sustained release has been a new approach to deliver several active substances through the vaginal route. Also, vaginal drug delivery systems should ensure either an adequate penetration of the drug within the mucosa, in order to enhance the local effects and reduce systemic absorption, or an ideal permeation of the active substances into the bloodstream in order to assure an effective systemic response.

Before formulators choose a delivery system for a selected drug, several issues should be taken into consideration: physicochemical properties of the active substance, intended effect of the active substance, required drug release profile, excipients to be used and their compatibility with the active substance and vaginal mucosa, women's preferences, and economical implications.

5.12.4.2 Excipients

When vaginal drug delivery is considered, formulators must select a number of suitable excipients in order to design a drug delivery system able to ensure the therapeutic success of the active substance(s). In fact, it is known that excipients used in vaginal formulations can influence the pharmacological performance of active substances, being able to improve or diminish their activity [68, 69]. The decision of which excipients to use depends to a great extent on the final dosage form and desired characteristics of the drug delivery system. Some excipients can influence drug delivery system performance by changing some properties, such as viscosity, mucoadhesion, and distribution [70]. Although these variations do not interfere directly with the pharmacological effect of the active substances, their availability and thus the formulation clinical outcome can be compromised. Thus, excipient selection must be performed with utmost caution, taking into consideration the quality, safety, and functionality aspects of these materials. Indeed, Garg et al. recently compiled a list of excipients that are currently approved or have already been investigated for vaginal administration [71].

Although by definition excipients are deprived of pharmacological effects, some have showed that this is not always true. For instance, chitosan, an excipient that has attracted a lot of interest in the formulation of vaginal drug delivery systems, exhibits antimycotic effects, particularly against the common vaginal pathogen *Candida albicans* [72]. Also, other polymers commonly used in tablets and capsules, such as cellulose acetate phthalate, have been investigated in the formulation of vaginal microbicides, due to their antiviral effects against HIV [73].

Some commonly used excipients can interact with vaginal and cervical fluids, altering their properties. These interactions should be taken into consideration when designing a drug delivery system, as they can influence in vivo performance. For example, small amounts of nonionic (e.g., polyethylene glycol) and cationic (e.g., polyvinylpyridine) polymers are able to modify the gel structure of the cervical mucus, altering its barrier properties, while like-charged molecules (e.g., polyacrylic acid) interact little with this biological fluid. This approach has been taken into account, particularly as a new prevention strategy for pathogens that infect via the mucosa, as a new treatment option for diseases that affect the mucous layer itself or even as a strategy for systemic drug delivery routes [74, 75].

5.12.4.3 Solid Systems

Solid systems commonly administered by the vaginal route include tablets, capsules, and vaginal suppositories.

Tablets are frequently used as vaginal drug delivery systems, being inexpensive and easy to manufacture. They are also easily administered in the vagina, allowing a "clean" insertion that contrasts with the typical messiness of semisolid drug delivery systems. Although very similar to oral tablets, these systems have some particularities, such as being round or oval shaped and devoid of sharp edges, in order to avoid damage of mucosal tissue. These drug delivery systems are usually designed to rapidly release their active substances after being placed in the vagina. In fact, disintegration or dissolution problems, mainly due to the scarce amount of vaginal fluid, are important issues to be managed by formulators. This rapid release and solubility enhancement of the active substances can be important because of the rapid vaginal wash-off and low in situ retention. Increased and faster release of drug content has been achieved using effervescent tablets [76] or including specific excipients that can enhance its disintegration in vaginal fluids [77]. For instance, Karasulu et al. proposed an effervescent tablet made of a mixture of mucoadhesive microcapsules loaded with ketoconazole and effervescent granules [76]. This combination showed ability to improve retention with rapid onset of action. Additionally, other strategies have been used, such as inclusion complexes of poorly soluble drugs with cyclodextrins, in order to improve drug solubility, allowing a rapid onset of the pharmacological effect.

Although fast release of the active substances is a frequent goal, controlled-release tablets can be used in order to enhance their efficacy, because of their prolonged release, and prevent the irritation of the vaginal mucosa that may be caused by some drugs [78]. Nonoxynol-9, a commonly used microbicide and spermicide, known for its irritability when administered in the vagina, is a good example of a drug that can benefit from controlled release. Formulation of double-layer tablets (fast-release outer layer and slow-release core) obtained from coprecipitates of nonoxynol-9 with polyvinylpyrrolidone, can provide extended drug release, allowing a more prolonged spermicidal effect while reducing its irritating effect [79]. Also, controlled release prevents peaks in serum concentration of absorbable drugs, limiting possible systemic effects [80].

Vaginal tablets containing lactobacilli have been used in order to restore the normal vaginal flora. Formulation of these delivery systems requires specific proceedings in order to provide viability of lactobacilli and stability of the final product. Freeze drying of bacterial suspensions has been tested to obtain lyophilized powders for tablet production [81]. These powders were shown to be processable and tablet production was easy and reproducible. Also, the use of double-layer tablets (fast-release layer and slow-release layer) seems to be an interesting approach to lactobacilli administration.

It is common to use tablets designed for the oral route in order to deliver drugs through the vagina. Nonetheless, issues such as delivery system retention and distribution and drug release can influence the final performance of the formulation, being preferable to use specifically vaginal designed drug delivery systems, or at least study the pharmacokinetics of oral tablets after vaginal administration [82].

Capsules, particularly soft capsules, have been used as vaginal drug delivery systems, but with modest popularity. These systems are relatively stable, particularly when compared with semisolid formulations or vaginal suppositories, being an adequate way to deliver liquid drugs within a solid dosage form.

Vaginal suppositories, also referred as ovules or pessaries, are ovoid-shaped, solid (but generally malleable) dosage forms specifically designed for vaginal administration. These systems usually weigh 2–3 g, although formulations with up to 16 g have been used in the past [83]. Vaginal suppositories have a long history of use as vaginal drug delivery systems, mainly in the management of local conditions. Major advantages are their reduced price and ease of manufacture. However, they present some inconveniences, such as messiness, low retention in the vagina, and poor stability, the last feature due to their temperature and moisture sensibility.

Vaginal suppositories are very close to rectal suppositories in terms of excipient nature and manufacturing process. Thus, they are usually prepared by fusion of the excipient(s) (referred as "base") and incorporation of the active substance(s), this mixture being subsequently poured into molds and allowed to solidify. Other methods, such as by compression, can also be used. Several substances have been utilized as bases for the formulation of vaginal suppositories: gelatin and glycerin, cocoa butter, semisynthetic glycerides, and polyethylene glycol, among others [83]. Composition of vaginal suppositories is importantly related to their melting or dissolution, thus influencing drug release profile. Generally, it can be stated that drug release rate increases as the melting temperature of a suppository decreases or as its dissolution time in vaginal fluids increases. Also, affinity of the drug for the base influences its release from vaginal suppositories: Greater release of drug is expected when there is less affinity between the active substance(s) and the base [84]. The melting temperature and melting process of vaginal suppositories can be characterized by several techniques, such as differential scanning calorimetry and viscosity and dilatometry methods, among others [85]. Specific pharmaceutical characterization of vaginal suppositories includes the determination of disintegration time and breaking hardness. Also, other standard quality control tests include appearance description, surface texture evaluation, pH determination, uniformity of content, and microbial limit testing [86, 87].

Recently, sustained-release vaginal suppositories have been developed in order to attain drug delivery systems with improved performance. Sustained release can reduce the number of administrations, thus improving patient compliance. A base composition consisting of a polymeric gum (carboxymethylcellulose and xanthan gum), a dispersing agent (colloidal silicone dioxide), and polyethylene glycol, referred as long acting, sustained release of spermicide (LASRS), has recently been studied by Zaneveld et al. in order to deliver contraceptives and microbicides [88]. Results showed that a LASRS base is able to spread quickly and evenly over the mucosa, being retained in place for prolonged periods of time and allowing long-lasting efficacy for several active drugs. Preliminary human trials have confirmed these results [89]. In another study, Mandal developed hydrophilic vaginal suppositories comprising mixtures of miconazole cross-linked with poly(vinyl alcohol) by freeze thawing and different polyethylene glycols that were able to sustain release this antifungal drug for up to 108 h [90].

5.12.4.4 Semisolid Systems

Semisolid systems present several advantages over other drug delivery systems: They are easy to use and generally inexpensive and have good acceptability. Among their disadvantages, leakage has been one of the most disturbing, mainly because many conventional formulations are not mucoadhesive. The simplest way of dealing with this problem has been the recommendation for night administration, as the supine position diminishes leakage. Also, messiness and discomfort upon application and difficulties in dispensing an accurate dose are important limitations.

Once widely used, ointments have been largely substituted by creams and gels. Nonetheless, some of these drug delivery systems may still be encountered, particularly as hydrophilic bases.

Creams have been used for quite some time as vaginal drug delivery systems, particularly for the administration of sexual hormones and antimicrobials. The main advantage of creams over other semisolid systems is their ability to easily dissolve both hydrophobic and hydrophilic drugs. As most conventional creams do not possess bioadhesive properties, incorporation of bioadhesive polymers is an effective approach to improve their retention in the vagina. Recently, a new approach to vaginal drug delivery was developed using Site Release (SR) technology (KV Pharmaceutical, St. Louis, MO). This technology is based on bioadhesive controlled-release water-in-oil emulsions, being formulated as a vaginal cream (SR cream). The outer oily phase repels moisture (thereby resisting dilution) and retains the dispersed water phase containing the drug (allowing controlled release) [91]. The SR cream allows minimizing leakage and enhancing clinical outcome, requiring less total drug exposure per course of therapy. Site Release technology is currently available in the United States in two commercial products: one containing butoconazole nitrate 2% (Gynazole-1, Ther-Rx Co., St. Louis, MO) and the other containing clindamycin phosphate 2% (Clindesse, Ther-Rx Co.). Clinical findings demonstrated that a single application of Gynazole-1 makes it possible to achieve more rapid relief of vaginal candidiasis symptoms than standard oral therapy with fluconazole [92]. Similarly, Clindesse was show to be able to achieve prolonged local effective concentrations while presenting lower systemic bioavailability and thus less systemic adverse effects when compared with conventional formulations in the treatment of bacterial vaginitis [93]. Also, other drugs have been studied in order to further evaluate the potential of this versatile technology [94].

Since the pioneer work by Wichterle and Lim in the 1960s [95], gels have evolved greatly from simple formulations to advanced drug delivery systems. These systems were soon demonstrated to be good candidates to deliver drugs in the vagina, particularly because of their high bioavailability (mainly because of mucoadhesive properties), biocompatibility, spreadability, ease of usage, and economical savings [96]. Gels are extremely versatile, being used to deliver most of the currently used drugs through the vaginal route.

Recent advances in gel and polymer technology boosted research, opening new possibilities for vaginal drug delivery [97]. Indeed, the development of new and improved gelling agents, particularly concerning to their mucoadhesive properties, has been of great importance. For example, polycarbophil (Noveon AA-1, Noveon, Cleveland, OH), a mucoadhesive polyacrilic acid polymer, has been widely used as a gelling agent in vaginal gel formulations. This polymer is acidic in nature, which can be useful in reducing the elevated pH associated with bacterial vaginosis [98]. Additionally, acidic polycarbophil gels may be used in the treatment of dry vagina and menopause-related stress incontinence [99].

Also, gel microemulsions have been recently reported as safe and devoid of mucosal toxicity drug delivery systems, presenting intrinsic spermicide activity and the possibility of improving vaginal bioavailability of poorly soluble antimicrobial agents [100].

Although most currently available vaginal gels rapidly release their active substance(s), they can also be formulated to achieve modified drug release profiles [101]. It is not clear how controlled release is achieved, but the analysis of most formulations that claim to possess this feature suggests a combination of dissolution and diffusion control [102]. Gels are also known to be promising drug delivery systems in protein and peptide administration through the vagina, proving to be adequate to accommodate and stabilize sensible molecules such as leuprolide [103].

5.12.4.5 Liquid Systems

Vaginal douching with liquids containing antimicrobial drugs such as povidone-iodine has been a common practice among women, with the intention of improving personal hygiene and treat vaginitis [104]. These liquids are almost immediately removed from the vagina after administration, thus being inadequate for controlled release. Although vaginal washing is frequently performed by women all over the world, its practice is discouraged, as it is associated with increased risk of acquiring HIV, particularly when soap or other substances rather than water are used [105, 106]. Also, bacterial vaginitis and other adverse reproductive health effects are possible when vaginal douching is a frequent practice [107, 108].

In addition, several solutions are utilized by gynecologists in their office practice. For example, glacial acetic acid solutions (3–5%) are used to identify cervical dysplasia during colposcopy, and Lugol solution is employed to perform Schiller's test (diagnosis of cervix cancer).

5.12.4.6 Vaginal Rings

Vaginal rings are doughnut-shaped drug delivery systems designed to provide controlled release of drugs. Developed systems are made of flexible, inert, and nonirritating polymeric materials, presenting different dimensions, usually 54–58 mm in diameter and 4–9.5 mm in cross-sectional diameter [109, 110]. Vaginal rings present several advantages particularly important for hormonal contraceptives delivery: (1) They do not require daily attention, allowing higher compliance than with daily dosage forms; (2) flexibility of current rings allow them to be easily inserted and removed by the woman herself, not requiring medical assistance as in the case of subcutaneous or intrauterine devices; (3) the continuous and prolonged delivery (three weeks to one year) of hormones avoids the high peak concentrations and fluctuations seen with daily oral administration; (4) rings are not associated with adverse local effects, including cytological and normal flora changes; and (5) contraceptive rings may be removed from the vagina during sexual intercourse and up to 2h, without compromising their pharmacological effect [110–112]. Although vaginal rings have been essentially investigated and used for the delivery of sexual hormones with contraceptive purposes or as hormone replacement therapy, these drug delivery systems can be also useful for the administration of other drugs such as bromocriptine mesylate, danazol, oxybutynin, antigens, and microbicides [113–117].

In the 1960s, first reports that implants made of polysiloxane, containing sexual steroids, could release their content at constant rates in saline solutions provided early information that led to the development of the first vaginal rings [118]. Vaginal rings were initially developed in the 1970s as contraceptives. The first system was composed of a silicone rubber ring containing medroxyprogesterone acetate as the active substance [119]. Nonetheless, the first vaginal rings have just recently reached the market, due to several unpredictable obstacles such as formulation difficulties, safety issues, and poor ovulation suppression [120, 121]. Table 3 presents some vaginal rings currently available in the market.

Vaginal rings may present several designs, as seen in Figure 4. The first vaginal rings were made of a homogeneous matrix containing the mixture of poly(dimethylsiloxane) (matrix-forming polymer) and the active drug, usually referred as a matrix design. Unfortunately, these rings showed an initial burst effect due to rapid release of the drug contained in the system's surface followed by persistent linear decrease of the drug release rate. This later phenomenon is related to the gradually thickening of a drug-depleted boundary between the inner drug-loaded region and the release surface, which is created by continuous drug release from the outer layers. Thus, their use in clinical practice was compromised, namely

TABLE 3 Selected Vaginal Rings Currently Available

Commercial Name	Active Substance(s)	Clinical Indications	Availability	Company
Nuvaring	Etonogestrel and ethinyl estradiol	Contraception	United States, Europe, Brazil, Chile	Organon
Progering	Progesterone	Contraception in lactating women	Chile	Laboratorios Silesia
Femring	Estradiol acetate	Relief of vaginal and urogenital symptoms in menopausal women	United States, Netherlands	Warner Chilcott Laboratories
Estring	Estradiol	Relief of vaginal and urogenital symptoms in menopausal women	United States, Canada, Europe, South Africa	Pfizer

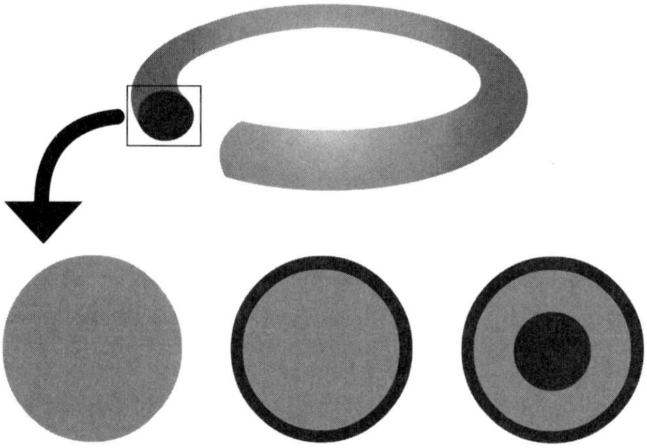

FIGURE 4 Cross sections of three vaginal rings presenting different designs: matrix design (left), core or reservoir design (center), and sandwich or shell design (right). Light gray represents drug–polymer mixture and dark gray represents polymer only.

as contraceptive devices. Later, and in order to improve control of the drug release, a layer of poly(dimethylsiloxane) without active substance was applied over the core containing the drug, acting as a drug release–limiting sheath (core or reservoir design). This strategy allowed achieving a near zero-order drug release profile. The diffusion rate of reservoir design rings is dependent on the drug concentration in the core, its partition coefficient between the core and membrane, the thickness and surface area of the membrane, and the diffusion coefficient of the drug in the membrane. In order to achieve a constant release rate, the drug should be much more permeable through the core than through the membrane [122]. Rings with several independent reservoirs containing different drugs have been obtained, thereby allowing the administration of two or more active substances from the same device. Also, another design has been developed in order to overcome drug release drawbacks, comprising a core of poly(dimethylsiloxane), an intermediate layer of poly(dimethylsiloxane) containing the active substance, and an outer drug release–limiting membrane of poly(dimethylsiloxane) (sandwich or shell design). As the drug is closer to the releasing surface, this strategy is particularly suited for substances presenting poor polymer diffusion characteristics [123]. As with reservoir design rings, a near zero-order drug release profile is obtained.

Besides poly(dimethylsiloxane), other elastomeric polymers have been employed in the manufacturing of vaginal rings, such as poly(dimethylsiloxane/vinylmethylsiloxane), styrene–butadiene–styrene block copolymer, and poly(ethylene-*co*-vinyl acetate) [123–125]. In fact, poly(ethylene-*co*-vinyl acetate) (commonly referred as EVA) appeared in the mid 1990s as an alternative to poly(dimethylsiloxane), when the manufacturer of this last material stopped supplying it for human use, demonstrating it to be very suitable for the production of controlled-release systems.

At the laboratory scale, silicone vaginal rings are usually obtained by injection molding, where poly(dimethylsiloxane) is mixed with a polymerization catalyst and the drug, being subsequently injected in ring-shaped molds. The mixture is allowed to cure for a period of time at a preestablished temperature, which can range from

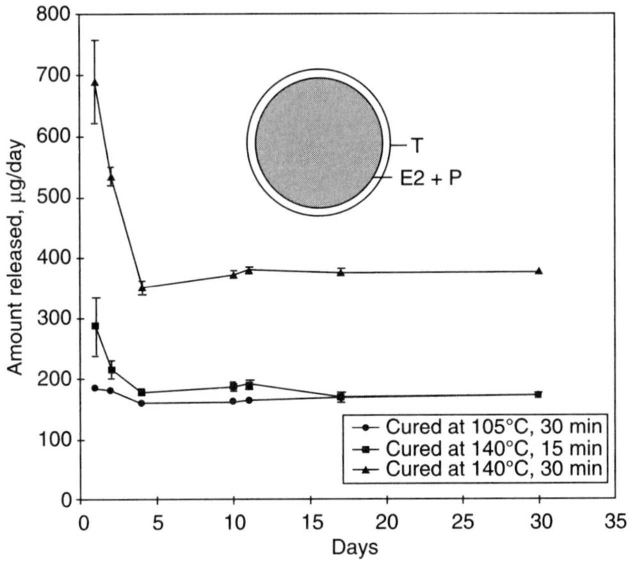

FIGURE 5 Effect of curing conditions on in vitro release profile of estradiol (μg) from core design rings containing estradiol and progesterone: T: polymer-only outer sheath [poly(dimethylsiloxane/vinylmethylsiloxane)]; E2 + P: core sheath, comprising mixture of polymer [poly(dimethylsiloxane/vinylmethylsiloxane)], estradiol, and progesterone. (Reprinted with permission from S. I. Saleh et al., *Journal of Pharmaceutical Sciences*, 92, 258–265, 2003. Copyright 2003 by Wiley-Liss.)

room temperature to 150°C and over (higher temperatures allow increasing the speed of the ring curing). In fact, curing time and temperature should be optimized as they influence the final performance of the ring, particularly drug release (Figure 5). This step leads to the formation of a three-dimensional network by means of a cross-linking reaction between polymer chains [126]. Afterward, other layers can be added, a step that is usually performed by injection molding or a dipping process. Although vaginal rings produced at the laboratory scale are useful during preclinical and clinical experimentation, the pharmaceutical industry needs other manufacture solutions to allow large-scale and financially viable production of these drug delivery systems. This process scale-up requires proof of bioequivalence between rings obtained by both processes [109].

The most common process of obtaining vaginal rings in the pharmaceutical industry is hot-melt extrusion (or hot-melt spinning), where the polymer, either alone or mixed with drugs or other additives, is melted (usually between 105 and 120°C) and forced by single- or twin-extrusion screws to pass through a die. After leaving the die, the obtained coaxial fiber is cooled and cut, the obtained fragments being shaped as rings by gluing both ends with an adequate pharmaceutical adhesive. Figure 6 presents a simplified scheme of the manufacturing process of a reservoir design ring similar to one used to produce Nuvaring (NV Organon, Oss, The Netherlands), an EVA vaginal ring containing etonogestrel and ethiny lestradiol. The core [polymer and active substance(s)] and the surrounding membrane polymer mixtures are extruded separately through two single-screw extruders (coextrusion)

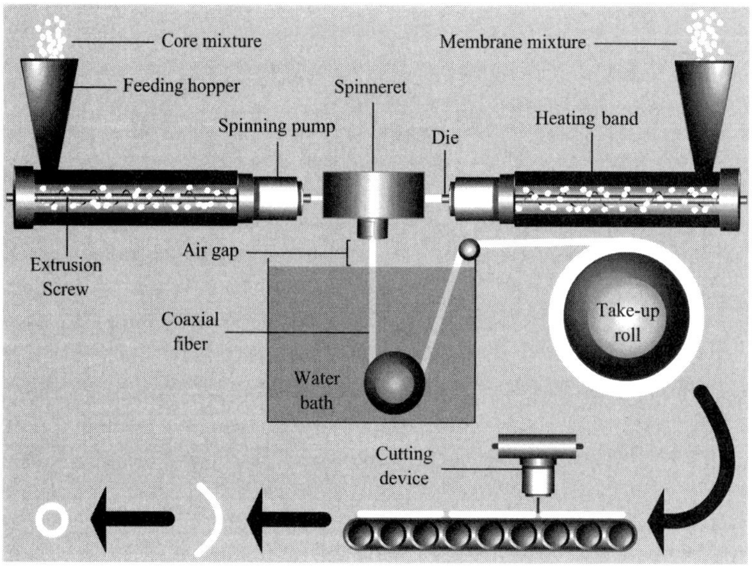

FIGURE 6 Schematic of manufacturing process of reservoir design ring by hot-melt extrusion.

that are connected to a spinning pump. In these cylindrical pipelines the polymers are melted and extruded through a die at an accurate flow rate. Then, the core and membrane polymers are combined in a spinneret, forming a coaxial fiber. The ratio between flow rates of both spinning pumps determines the thickness of the membrane. After leaving the spinneret, this fiber is cooled, first by air exposure and then by immersion in a water bath. At this stage, the fiber diameter is adjusted to the desired value by elongation with take-up rolls. In fact, after leaving the die, the obtained fiber expands its diameter as a result of the viscoelastic behavior of the polymers used [122, 127].

The drug release profile is conditioned by the polymeric structure of the systems, which is influenced by several parameters such as polymer composition, melt spinning process variables (namely feeding of polymer mixture and spinning velocity, extrusion temperature, spinline stress, cooling rate, and drawing back elongating force), and storage conditions [122, 128]. Also, drug release characteristics are largely influenced by the active substances' molecular weight and diffusion coefficient and solubility in the polymer. For example, extremely hydrophobic drugs and molecules with molecular weight above 450 Da are poorly released from silicone. Modulation of the drug solubility by adding various excipients (e.g., propylene glycol, polyethylene glycol, gelatin, and fluid silicone) can be used to change the release profile [129, 130]. The physical state of the drug is also an important parameter related to drug release. Taking the example of steroids, these compounds can be either in a solid crystalline state or in a molecularly dissolved state. In the first case, the concentration of the drug is fixed by its saturation solubility, making it possible to control the release rate by the thickness and the permeability of an outer membrane. When two drugs are present, this concept cannot be used: Both drugs need to be completely dissolved, their release rates being controlled by their concentrations [123].

5.12.4.7 Vaginal Films

Vaginal films are polymeric drug delivery systems shaped as thin sheets, usually ranging from 220 to 240 μm in thickness. These systems are often square (approximately 5 cm × 5 cm), colorless, and soft, presenting a homogenous surface. Vaginal films have some advantages, such as portability, ease of application, long time of retention in the vagina, and good drug stability [131]. Once placed in the vagina, the fluid present in the mucosa hydrates the polymer, covering the mucosa with the active substance. This coating may also be helped during sexual intercourse due to the spreading motion of the male penis.

Vaginal films are produced with polymers such as polyacrylates, polyethylene glycol, polyvinyl alcohol, and cellulose derivatives. Proper combination of these polymer is essential to achieve adequate mucoadhesion and optimal drug release profiles. Vaginal films can be produced by casting [132], in which polymer solutions containing the active substance(s) are poured into adequate molds and dried until a thin, solid, and flexible polymeric sheet is formed. Afterward, the sheet is cut in small pieces (individual films) and peeled off.

Vaginal films have been mostly used as spermicides, although they present inferior contraceptive success than hormonal methods, condoms, and intrauterine devices. A contraceptive film containing 28% nonoxynol-9 in a polyvinyl alcohol base (VCF, Apothecus Pharmaceutical) is currently available in the United States. Alternative contraceptive films containing different drugs and film-forming polymers have also been investigated, in order to obtain more effective and acceptable formulations [133].

5.12.4.8 Medicated Vaginal Tampons

Vaginal tampons have been studied for their feasibility as vaginal drug delivery systems. First experiments used commercially available tampons that were impregnated with active substances by a simple dipping process [134]. Currently, a medicated vaginal tampon, approved as a medical device by the Food and Drug Administration (FDA), is available (Ela Tampon, Rostam, Israel). This bifunctional tampon contains a polymeric delivery system (strips) that absorbs menstrual fluid while gradually releasing lactic acid and citric acid. These two drugs act by preserving the acid vaginal milieu, preventing the proliferation of potential pathogenic bacteria [135].

5.12.4.9 Vaginal Foams

Vaginal foams have been tested to deliver drugs in the vagina, mainly microbicides or spermicides [136, 137]. The efficacy of these systems was shows to be limited when compared to other available options, leading to a decline of their use. Nonetheless, foams are easy to use, providing a good coverage of the vaginal mucosa with minimal leaking. Also, most foam bases are nonirritating, unlike some other conventional formulations which are reported to cause burning and itching. Thus, improving their formulation can be an interesting approach to obtaining new vaginal drug delivery systems. In fact, foams containing antimicrobials, local anesthetics, and hormones have the potential to gradually substitute several currently available dosage forms, namely creams and ovules.

5.12.4.10 Vaginal Sponges

Vaginal sponges were once widely used as vaginal contraceptives, the oldest reference to their use in the Talmud (c. 500 B.C.), where a sponge soaked in vinegar was recommended in order to prevent pregnancy [138]. Although these devices have been disapproved by some researchers, since their use encouraged colonization and proliferation of bacteria in the vagina, predisposing women to vaginitis and other genital disorders [139], sponges containing spermicides are still used as contraceptives. These contraceptive devices have the ability to deliver the active drugs while absorbing semen and blocking the cervical canal. Also, vaginal sponges are inexpensive devices; less messy than creams, gels, and foams; and easier to insert and remove than diaphragms [140]. Nonetheless, these systems have demonstrated less efficacy than other contraceptive methods, such as the diaphragm [141].

A vaginal sponge (Today, Whitehall Robins) made of soft polyurethane foam (diameter of 5.5 cm and 2.5 cm thick) and saturated with 1 g of nonoxynol-9 was authorized as a spermicide in the United States in 1983 [140]. Advantages of this formulation include the possibility of being used up to 24 h without requiring additional application of spermicide. Although in 1995 the manufacturer discontinued its production because of increased costs related to FDA guideline compliance, Allendale Pharmaceuticals purchased the rights to the sponge, reintroducing the product in U.S. and Canadian markets [142]. Similar sponges are available outside the United States containing benzalkonium chloride alone (Pharmatex, Innotech International) or a combination of nonoxynol-9 and sodium cholate (Protectaid, Pirri Pharma).

5.12.4.11 Other Strategies and Vaginal Drug Delivery Systems

Bioadhesion may be defined as the state in which two materials, at least one of which is biological in nature, are held together for extended periods of time by interfacial forces. Mucoadhesion is a particular case of bioadhesion where one of these materials is a mucous membrane or the mucus [143]. Mucoadhesive drug delivery systems can circumvent the poor retention that most traditional vaginal formulations present, improving residence time, and even their specific location in the mucosa. Prolonged time of contact and intimate interaction with the vaginal mucosa are able to increase drug absorption and bioavailability and thus its therapeutic effect. Currently used mucoadhesive materials are polymeric in nature, many of them previously used for other specific purposes (e.g., as gelling agents in vaginal gels). Several polymers, alone or in combination, have been used or investigated for both systemic and local vaginal drug delivery in order to obtain pharmaceutical systems with mucoadhesive properties (see Table 4) [144]. Virtually almost all drug delivery systems may benefit from these mucoadhesives, particularly gels, creams, tablets, vaginal suppositories, and films. To date, polyacrylates are the most used and explored mucoadhesive polymers, although others such as chitosan, carrageenan, or sodium alginate have proved to be advantageous. Also, some widely used mucoadhesives in vaginal formulations, such as polycarbophil (Noveon AA-1, Noveon, Cleveland, OH) present the advantage of being useful as controlled- and sustained-release matrices [145]. Recently, thiolated polymers, usually referred as thiomers, have been studied in vaginal mucoadhesive systems, exhibiting higher mucoadhesion than nonmodified

TABLE 4 Examples of Mucoadhesive Polymers Used or Investigated in Formulation of Vaginal Drug Delivery Systems

Polymer Classes	Examples
Carrageenan	Iota-carrageenan
Cellulose derivatives	Sodium carboxymethylcellulose (NaCMC)
	Methylcellulose (MC)
	Hydroxypropyl methylcellulose (HPMC)
	Hydroxypropyl cellulose (HPC)
	Hydroxyethyl cellulose (HEC)
Chitosans	
Gelatin	
Gums	Pectin
	Tragacanth
	Dextran
	Xanthan
Hyaluronic acid and derivatives	Sodium hyaluronate
Polyacrylates	Carbopol 974P
	Polycarbophil (Noveon® AA-1)
Polyethylene glycols (PEGs)	PEG 600
Povidone	
Alginates	Sodium alginate
Starch	
Sulfated polysaccharides	Cellulose sulfate
Thiomers	Carbopol® 974P–cysteine
	Chitosan–thioglycolic acid (TGA)
	Chitosan–4-thio-butylamidin-conjugates (TBA)

Note: As reviewed in ref. 144.

polymers. Although thiomers can improve the mucoadhesion of the original polymers, they do not interfere with other advantageous characteristics such as the ability of controlled release [146, 147].

Environmentally sensitive drug delivery systems are characterized by their ability to respond to changes in pH, temperature, ionic strength, solvent composition, magnetic fields, light, and electric current, among others, most commonly used for the first two [148]. Thermoreversible systems are fluids that can be introduced into the body in a minimally invasive manner prior to solidifying or gelling within the local of administration because of temperature increase from room temperature to body temperature [149]. This innovative approach is particularly interesting in vaginal drug delivery, as a liquid system is easily introduced in the vaginal cavity, while the in situ formed gel facilitates retention. Indeed, some thermosensitive formulations have already been suggested to obtain vaginal drug delivery systems, poloxamer gels being the most extensively studied. A clotrimazole-containing vaginal thermosensitive gel (poloxamer 188 and polycarbophil) was demonstrated to be useful for the effective and convenient treatment of vaginal candidiasis in female rats, with the advantages of facilitating administration and reducing dosing interval when compared with conventional therapy [150].

The incorporation of drug-loaded liposomes in adequate formulations can improve their stability, allowing their applicability in vaginal drug delivery [151]. Additionally, entrapment of drugs in liposomes may improve their solubility and availability at the site of administration, reducing the administered dose and systemic effects. In order to administer liposomes, adequate dosage forms must be selected. Polyacrylate gels have emerged as good candidates, showing the potential to accommodate liposomes for vaginal drug delivery [151, 152]. Incorporation of liposomes in these gel bases can be achieved by gently mixing both components, which is an advantage over other dosage forms. Also, these novel drug delivery formulations enable sustained drug release, combining both the liposome limiting release effect and the bioadhesive properties of these gels.

When liposomes are considered for vaginal administration, their stability should be assessed under in situ conditions presented by both pre- and postmenopausal women. Drug formulation can help improve their stability. Several gels made of Carbopol 974P NF were shown to be suitable in improving the stability of liposomes containing clotrimazole, metronidazole, or acyclovir in a vaginal environment simulating media when compared to liposomal dispersions [153, 154]. Also, compatibility between liposomes and vaginal mucosa (animal or human) should be checked [155]. Nonetheless, liposomes still have some stability-related problems that limit their shelf life. Formulation of liposomes as proliposomes can prevent these stability issues. Proliposomes are dry, free-flowing products which on addition of water disperse to form liposomal suspensions [156]. As proliposomes are dry powders, they can be formulated as solid dosage forms, which can be more convenient to administer in the vagina. Recent work by Ning et al. demonstrated that clotrimazole-containing proliposomes can be suitable for vaginal administration, allowing a rapid conversion to sustained-release liposomes upon contact with physiological fluids [157].

Niosomes (nonionic surfactant vesicles) have been experimented as controlled- and prolonged-release drug delivery systems to administer insulin through the vagina. These particles might be good carriers for protein delivery, showing the potential to enhance the effects of these drugs when compared to the vaginal administration of free insulin. Results obtained in rats showed that niosomes containing insulin can achieve comparable bioavailability with subcutaneous administration of this protein [158]. Also, experiments performed with vaginal gels containing both niosomes and liposomes have been shown to be a promising strategy for the prolonged release and safe vaginal administration of clotrimazole [159].

Microparticles and nanoparticles present some advantageous features, namely mucoadhesive properties. They have demonstrated some potential in vaginal drug delivery, particularly in the formulation of delivery systems for vaccines or peptides and proteins [160, 161]. Nonetheless, these particles have to be incorporated in adequate carrier systems in order to be delivered. This task has been shown to be complex, it being hard to achieve controlled-release and steady-release profiles.

Cyclodextrins are commonly used in pharmaceutics, their applicability being no exception in vaginal drug delivery. Drug/cyclodextrin complexes allowed enhancement of solubility and achievement of prolonged-release properties of drugs such as clotrimazole and itraconazole when included in suitable vaginal drug delivery systems such as gels, creams, or tablets [162–164].

Cervical barrier devices such as diaphragms can be modified to include a reservoir that releases active substances such as spermicides or microbicides. Although these devices have limited applicability, they can be particularly helpful in preventing pregnancy, as they combine a physical barrier device with a chemical spermicide that faces the vagina, after correct placement in the cervix [165]. Lee et al. presented a diaphragm made from silicone containing 35% nonoxynol-9 that was able to achieve controlled drug release depending on the device design and size [166]. The diaphragm was prepared by using compression molding in a single-cavity aluminum mold.

Patches can provide delivery of very toxic drugs in very limited areas of the vaginal mucosa without leaking to the circumventing tissue. McCarron et al. described a bioadhesive patch that delivered 5-aminovulinic acid to intraepithelial neoplasia lesions. The proposed delivery system is based on a poly(methylvinylether/maleic anhydride) matrix that contains the active drug, providing local retention of up to 4 h and enhancing efficacy without damaging healthy epithelium [167]. Patchs with bilayer design may also be an option for the treatment of these neoplasic lesions, where a bioadhesive drug-loaded matrix bonded to a drug-impermeable backing layer is able to prevent drug spillage to healthy tissues, promoting unidirectional and deep drug penetration. Woolfson et al. presented a 5-fluorouracil vaginal patch with a bilayer design, comprising a flexible polyvinyl chloride (PVC) emulsion as a backing layer and a drug-loaded bioadhesive film made of 2% Carbopol 981 and 1% glycerin as a plasticizer [168].

Propess (Ferring Pharmaceuticals) is a commercially available controlled-release vaginal insert presented as a thin macrogol 8000 matrix (rectangular in shape with radiused corners) containing 10 mg of dinoprostone being used for labor inducement. This drug-loaded matrix is included within a knitted polyester retrieval system that ends in a long tail to help retrieval at the end of the dosing interval. This insert is capable of releasing the active substance at a rate of 0.8 mg/h over 12 h, after being exposed to vaginal moisture [169]. Clinical studies indicate that Propess is as equally effective and safe in achieving cervical ripening as other commercially available vaginal drug delivery systems containing dinoprostone [170].

Multiple emulsions have been studied as drug vehicles for vaginal administration that are able to include several active substances in the different phases. An antimicrobial water–oil–water (W/O/W) multiple emulsion containing lactic acid in the internal aqueous phase, octadecylamine in the oily phase, and benzalkonium chloride in the external aqueous phase was tested, proving to stabilize the included active substances while providing adequate viscosity properties to vaginal administration [171]. Nonetheless, the structure of these formulations is destroyed at high shear rates (e.g., shear rates observed during coitus), losing or diminishing their intended activity and thus limiting their applicability.

Recently, an interesting approach for protein delivery using genetically engineered normal vaginal flora as delivery systems has been proposed. This strategy is based on the natural affinity that these microorganisms have to adhere tightly to the epithelial surface, providing a direct delivery of the drugs to the mucosa and thus minimizing enzymatic and bacterial degradation. Lactic acid bacteria, transformed with plasmids that contained a gene encoding for the therapeutic protein to be administered, have been used as delivery systems. Obtained results showed enhancement of protein delivery when compared with a conventional solution containing the same molecule. This strategy looks particularly appealing for the development of vaccines that induce mucosal immunization [172, 173].

5.12.4.12 Packaging and Vaginal Applicators

Vaginal packaging and applicators are an integral part of vaginal products. Packaging is designed to accommodate and protect formulations, while applicators should allow their convenient administration in the vagina. Several materials have been used to manufacture these devices, such as plastics (e.g., polypropylene and polyethylene) and nonlatex rubber. Besides compatibility, stability, and suitability issues, these materials should be selected regarding the final cost of packaging and applicators.

Applicators are intended to be introduced in the vagina, adequately deliver the product, and then be removed. Their design relates to safety (e.g., relationship with product purity and stability, avoidance of local trauma associated with insertion or use), efficacy (e.g., consistent delivery of the required amount of product in the intended location), and acceptability (comfort, ease of use, convenience, aesthetic appeal) [174]. In general, they can be divided as single-use or reusable applicators. Single-use applicators are usually prefilled, while reusable applicators are filled by women prior to vaginal insertion. Several applicator designs have been used, such as barrel-and-plunger and squeeze tube, but they all should be easy to insert, comfortable, and deprived of cutting edges. Also, some specific formulations, such as those intended for vaginal douching, require other types of applicators. Typically, squeeze plastic bottles with variable volumes (approximately 100–200 mL) are used.

5.12.5 PHARMACEUTICAL EVALUATION OF VAGINAL DRUG DELIVERY SYSTEMS

Evaluation of pharmaceutical systems is consensually recognized as an important component of their development and quality control. Although evaluation of general features (e.g., drug content) is also required for vaginal formulations, this section focuses only upon some of the most important parameters that are intimately related to drug delivery systems specifically designed to be administered by this route.

5.12.5.1 Legal and Official Compendia Requirements

There is a lack of well-defined guidelines and regulations for vaginal products in most countries as well as official compendia information and requirements on quality control and other important aspects of vaginal drug delivery systems [39]. In fact, the latest editions of the U.S. Pharmacopeia, European Pharmacopoeia, and Japanese Pharmacopeia include limited or even no information related to the quality control and evaluation of vaginal drug delivery systems. Thus, most of the currently used evaluation procedures require standardization, making it difficult to compare results obtained by different research groups.

5.12.5.2 Drug Release and Permeability

When formulating vaginal drug delivery systems, it is important to consider the release of the active substances, as different formulations can greatly affect the release rate and ultimately their pharmacological effects. The choice of the dissolution method should be done on a case-by-case basis, while dissolution profiles can then be fitted to commonly used mathematical models, as reviewed by Costa and Sousa Lobo [175]. Conventional procedures and apparatus have been adapted taking into consideration vaginal physiological specifications such as pH, fluid volume, and temperature, among others. The experimental method may not inevitably imitate the vaginal environment, but it should test the main key performance indicators of the formulation. Although official methods are not available, some have been recommended for specific vaginal drug dosage forms [176]. For semisolid vaginal formulations, the Franz diffusion cell is considered the most promising apparatus for drug release investigation. In the case of hydrophilic vaginal suppositories, a basket apparatus, a paddle apparatus, or flow-through cells can generally be considered as suitable; for hydrophobic vaginal suppositories, modified flow-through cells would be preferable. Dissolution methods that use a basket instead of a paddle can be advantageous for vaginal solid formulations that tend to float, particularly those that may include a modification that prevents the formulations to form a cake inside the basket, limiting their dissolution [177, 178]. Drug release from vaginal rings is usually determined by placing these systems in conical flasks containing an adequate dissolution medium. Flasks are then placed in a water bath controlled at 37°C and shaken during the time of the assay, with samples being taken and release medium being replaced typically every 24 h [113, 129]. Correlation of these in vitro release tests with in vivo results proved to be satisfactory for the majority of the evaluated vaginal rings [125].

Absorption of drugs through the vagina is an important parameter to be evaluated, particularly when a systemic effect is required. Also, assessment of the absorption potential of drugs intended to locally exert their effects needs to be evaluated, as this event can lead to unwanted systemic effects. The evaluation of both formulated and unformulated drugs can be performed by in vitro or in vivo methods.

In vitro permeability studies have been performed in flow-through diffusion cells using either animal or human vaginal mucosa [179, 180]. After isolation and adequate treatment of vaginal mucosa specimens, small tissue disks are mounted in the apparatus in order to perform the permeation experiments. At the end of these procedures, routine histological examination of the used tissues can be performed in order to identify any changes in the normal structure of the vaginal mucosa. Several alterations can suggest potential mechanisms of permeation of tested drugs [181, 182]. Animal mucosa used in these experiments can be obtained from several species, namely rabbits and pigs. Also, porcine vaginal mucosa was demonstrated to be a good in vitro permeability model of human vaginal mucosa, particularly when hydrophobic substances are tested [50]. This feature can be explained because both mucosal tissues are very similar in many aspects, namely their lipid composition and histological structure. On the other hand, high molecular weight or charged molecules, such as oxytocin, may show different permeability profiles when tested with either porcine or human vaginal mucosa. Thus, researchers must be careful when interpreting and extrapolating results to human tissues because unexpected differences often occur. When human vaginal mucosa is used, samples are usually obtained from excess tissue removed from postmenopausal women after vaginal hysterectomy [181, 182]. Use of postmenopausal vaginal mucosa is advantageous as it is less altered (particularly in thickness) by hormonal stimulation, leading to more uniform results. Nonetheless, these features do not reflect the normal histological architecture of fertile women, being able to significantly alter the permeability profiles for many drugs. Along with vaginal mucosal tissue, other model membranes, such as vaginal and cervical cell monolayer membranes, have been suggested in recent years in order to predict in vivo absorption [172].

It is also noteworthy that many of the in vitro results of vaginal permeability studies cited in the literature have limitations related to the experimental conditions that were used, particularly pH values at which they have been performed. Differences in permeability values are particularly expected when ionization characteristics change between experimental pH and vaginal pH [183]. However, in vitro results should only be considered as evidence that the vaginal mucosa is able to be permeated.

In vivo studies performed in animals are an important step before considering human experimentation. Animal species commonly used in vaginal permeability studies include rabbits, rats, and mice [53, 158, 184]. Although potentially more accurate in predicting human vaginal absorption of drugs, animal experimentation have some limitations. A major problem is the variability of the vaginal epithelium properties throughout the estrous cycle, thus influencing drug absorption [184, 185]. In order to minimize this variability and standardize the thickness of the epithelium, animals are usually ovariectomized. Also, vaginal enzymatic activity is an important parameter in choosing animal models. It is recommended that the enzymatic profile of such animal should be comparable to that of the woman. Taking this into consideration, rats and rabbits seem to be good models for vaginal permeability studies, particularly when protein and peptide drugs are considered [27].

5.12.5.3 pH and Acid-Buffering Capacity

As already referred, pH is an important parameter concerning the health and normal physiology of the vagina, being important that vaginal formulations do not interfere with its normal value. Also, the pH of the vagina can be elevated due to changes in its normal physiology (e.g., bacterial vaginitis) or the presence of semen. Vaginal formulations presenting good acid-buffering capacity have the potential to reestablish normal pH or to prevent it from rising.

The acid-buffering capacity of a vaginal formulation can be measured by simple titration with an inorganic alkali, such as sodium hydroxide. The physiologically relevant acid-buffering capacity can be defined as the amount of alkali required to elevate the pH from its initial value to the maximum desirable value when considering the healthy vagina [186]. Also, mixtures of vaginal formulations with semen may be useful in determining their buffering capacity, this proceeding being particularly helpful when testing products used during sexual intercourse.

Variations in vaginal pH of the vagina can influence drug stability, particularly when extreme values are observed. Thus, the adjustment of the formulation pH can also be important in order to assure maximum stability or pharmacological activity of the active substance(s). As an example, the administration of antibodies in the vaginal milieu can compromise their activity because of the acidic pH. Generally monoclonal human antibodies are more stable at pH 4–7, losing binding and neutralizing activity below pH 4 [187]. These findings underline the importance of pH buffering when delivering pH-sensible molecules such as antibodies.

5.12.5.4 Rheological Studies

Rheological properties of semisolid vaginal formulations are crucial to their suitability as drug delivery systems. They experience in vivo a wide range of shear rates (from less than $0.1\,s^{-1}$ to about $1000\,s^{-1}$), both steady and transient, while being diluted with vaginal fluids, which influences their rheological properties and hence their spreading and retention. Events such as passive seeping, sliding, squeezing between vaginal walls, and coitus, among others, influence the rheological performance of vaginal formulations [188]. Thus, knowledge of rheological properties of semisolid vaginal drug formulations may assist in improving their design, being helpful in the process of predicting which formulations can retain their structural stability over time, particularly in the physiological environment [189, 190].

Qualitative and quantitative composition of a semisolid vaginal formulation can strongly influence its rheological properties [191]. This fact is particularly important when considering the optimization of a drug-containing formulation and its placebo formulation. These two systems should only differ in the absence of the active substance(s). However, this small difference can sometimes originate different rheological properties that can greatly influence the formulation's performance and even the results of clinical trials [192].

Rheological properties of a formulation can be studied either by simple flow measurements or by dynamic oscillatory measurements, although the latter are preferable as they allow a complete characterization of both elastic and viscous components. Also, they are nondestructive and, if the strain is not too high, the sample is not disturbed [193]. As already noticed, in vivo conditions, particularly temperature and fluids that may be present in the vagina (vaginal fluid, cervical mucus, and semen), can influence the rheology of pharmaceutical formulations. Indeed, these factors have to be taken into account when formulating a vaginal drug delivery system. Thus, optimization should not only focus upon the rheology of undiluted material but also include mixtures of formulations and fluids that may be present in the vagina [194]. However, these biological fluids are not always available and considerable differences between individuals limit their use. In order to abbreviate these limitations, some simulants of vaginal fluid [21], cervical mucus [26], and semen [195] have been used.

5.12.5.5 Textural Studies

Textural profile analysis is a widely used analytical method based on the measurement of the forces involved during the compression/decompression of a probe in a sample of the product to be tested. From the obtained results, important parameters can be calculated, including hardness (force required to attain a given deformation), compressibility/spreadability (the work required to deform the product during the first compression cycle of the probe), and adhesiveness of the product. Besides their influence in the ease of removal from a container (e.g., vaginal applicators), the ease of application, or retention, among others, it is consensual that textural properties of formulations will influence their clinical performance. Therefore, it is important to fully characterize these properties during the formulation process [196]. Also, these textural parameters can be converted into rheological properties, such as shearing stress, shear rate, and viscosity, using dimensional analysis, allowing the comparison of results generated by both techniques [197].

5.12.5.6 Mucoadhesion

As previously discussed, development of mucoadhesive drug delivery systems is a promising strategy in order to enhance vaginal drug administration. Several methods for the in vitro evaluation of mucoadhesive properties of vaginal formulations may be found in the literature. Although these tests present some limitations, they can offer easy and valuable tools in the initial formulation and evaluation of vaginal drug delivery systems. Most methods can be categorized as tensile strength or shear stress tests, where the force needed to separate a model membrane attached to the formulation measures mucoadhesion. Different results can be obtained for the same sample because of the different types of forces involved in these two methods [198]. For this reason, both types of tests should be performed in order to achieve a more complete characterization of the mucoadhesive potential of formulations.

Alternative methods have also been shown to be useful. For example, mucoadhesive properties of semisolid vaginal formulations can be assessed by the rheological characterization of the mucoadhesive interface, based on the assumption that the interpenetration extension between polymer gels and their mixtures with mucin can be detected by measuring differences in rheological parameters. Also, the texture analysis of these formulations/mucin interfaces was demonstrated to be a useful technique in measuring bioadhesion, with results in the same rank as the ones obtained with the rheological technique [199]. Kast et al. tested vaginal tablets for their mucoadhesiveness by a simple method, comprising the use of a dissolution apparatus [146]. In this procedure, the formulation to be tested is attached to a vaginal mucosa that is fixed on a cylinder; the cylinder is then immersed in the dissolution vessel and rotated in an adequate testing solution at approximately 37°C. The mucoadhesion is evaluated by the detachment time of the formulation. An alternative method based on a modified balance, in which the tablet is attached to a vaginal mucosa fixed to one side plate, has also been proposed [200]. In this case, mucoadhesion is measured by the weight required to detach the formulation.

5.12.5.7 Vaginal Distribution and Retention

Vaginal mucosa coating by a formulation is an important parameter to be determined, since its action may depend on the effectiveness of this phenomenon. Nonetheless, knowledge about the distribution and retention of commercially available products is limited. After being administered, liquid, semisolid, or solid (after liquefaction) formulations can spread throughout the vagina, with part of the material being able to exit this tube, either to the exterior or to the upper genital tract [201]. This distribution is governed by physical forces that include gravity, normal forces from contacting tissues, surface tension, and shearing. Formulation flow due to these forces is affected by many factors, including formulation physical properties, amount of formulation applied, surface interactions, surrounding tissue properties, vaginal secretions, baseline dimensions of the vagina, ambulation and posture changes of the user, and sexual intercourse [202, 203]. Drug delivery systems intended for local effect should ideally spread evenly throughout the vaginal mucosa. Also, evaluation of the possible erosion of a product coating during its residence in the vagina, particularly during sexual intercourse, is extremely important. Thus, vaginal distribution and retention studies may clarify some questions, such as the amount of product to cover evenly the vaginal surface, the time required to distribute to all areas and to be removed from the vagina, and the effect of daily activities such as ambulation and sexual activity in these phenomena.

Simple in vitro tests can be used to quantify the vaginal coating of a vaginal formulation. It is important that these tests model the natural history of a product in the vagina, from initial application and contact with fluids that may be present in the vagina to the period during and following sexual intercourse [204]. The distribution and retention of solid drug delivery systems can be evaluated by a simple method proposed by Ceschel et al. [205], where the formulations are placed in a vertical thermostated cellophane tube, the discharged liquid collected and measured throughout the time of the experiment. The amount of discharge liquid is related to the retention while the distribution of formulations can be assessed by dosing the amount of active substance(s) in different sections of the tube at the end of the experiment. Also, the contribution of gravity to the vaginal coating flows of vaginal semisolid drug delivery systems can be evaluated by a simple and objective method proposed by Kieweg et al. [203]. The proposed technique is based on the measurement of the flow behavior of a formulation sample after being placed in an inclined plane surface. Obtained results, together with mathematical models, can help formulators to select primary candidate formulations before in vivo tests commence.

In vivo assessment provides more reliable and complete information about the vaginal distribution and retention of drug delivery systems, although ethical and economical issues limit its applicability to routine evaluation. In recent years, imaging techniques, which have been used for other purposes for a long time, emerged as valuable tools for the evaluation of vaginal distribution and retention. Magnetic resonance imaging proved to be a helpful method, providing cross-sectional images of drug delivery systems that are administered in the vagina, both in animals and in humans, allowing precise reproducible data regarding the spread of vaginal formulations to be achieved [202]. As magnetic resonance imaging reflects the images of a chemical label, such as gadolinium, and not necessarily that of the drug delivery systems or carried drugs, it is necessary to perform the association of the chemical label with vaginal products and its validation [206]. Another interesting imaging technique that has been used in this type of assessment is gamma scintigraphy, shown to be a useful tool for evaluating and comparing the distribution, spreading, and clearance of vaginal delivery systems [207, 208].

Imaging methods, such as gamma scintigraphy and magnetic resonance imaging, are useful but have some limitations concerning their resolution, being unable to quantify or even identify the presence of vaginal coating layers of just a few hundred micrometers. In order to overcome such limitations an optical instrument capable of detecting coating layers as thick as 50 μm has been developed by Henderson and co-workers [209]. The device is inserted and remains stationary in the vagina, where both local video images and fluorescence intensity measurements of fluorescein-labeled formulations are obtained. Since the tube that is inserted in the vagina is shaped and sized like a phallus, the vaginal coating measured is analogous to that observed during sexual intercourse. Nonetheless, in vivo fluorescence-based methods present limitations, particularly because of the diffusion of the dye out of the formulation and photobleaching, limiting the interval over which measurements can be performed accurately. A new technique based on low-coherence interferometry that can overcome these difficulties is being developed, allowing extended time studies to be performed [210]. This easy, label-free, high-resolution method uses broadband light in an interferometry scheme to achieve depth-resolved reflection measurements. Future studies focus on the development of an easy-to-use endoscopic device that may be used in clinical studies.

5.12.5.8 Safety and Toxicology

It is accepted that a new pharmaceutical product should be assessed for its effects on the vaginal mucosa before being approved by drug-licensing agencies. Nonetheless, many of the older products that have been used for a long time by women all over the world do not have this type of information available. Safety issues are particularly important when a vaginal drug formulation is used repeatedly, as in the case of microbicides, spermicides, and contraceptive vaginal rings. Also, it is important to test vaginal applicators for their safety as they are considered an integral part of vaginal products, being able to induce alterations in the mucosa [211].

Local effects assessment should include not only short-term but also long-term protocols, as some formulations are intended to be used for large periods, in order to assess their real impact on vaginal health. Although systemic exposure to drugs intended for topical action is expected to be minor, vaginal formulations should also be assessed for systemic effects due to possible absorption. Alterations in blood parameters and liver and renal function should be investigated [212]. Also, drugs and formulations to be administered through the vaginal route must be assessed for fertility and teratogenic effects in animal models, before entering clinical trialing and human use [213].

In vitro testing helps formulation scientists understand and predict the potential harmful effects of formulations to the vaginal mucosa. Although animal testing still needs to be performed, these in vitro methods can be of great interest in initial screening of new products and formulations, reducing the amount of animal testing required. Also and unlike animal testing, in vitro testing can often differentiate products that are very mild in terms of toxicity potential. Thus, toxicity of vaginal products can be assessed using simple tests with epithelial cell monolayers, where maintenance of membrane integrity in the presence of testing formulation indicates potential safety [214, 215]. Nonetheless, monolayer cell cultures lack histological and functional resemblance with native ectocervical and vaginal tissues, which limits the interpretation of the obtained results. In order to respond to this and other problems, Ayehunie et al. recently proposed a fast and highly reproducible three-dimensional organotypic vaginal–ectocervical tissue model that simulates the structure of the vaginal epithelium [216].

The standard preclinical test of local vaginal irritation and toxicity of pharmaceutical products, and the only one recommended by the FDA, is the rabbit vaginal irritation test [217, 218]. However, reproducibility problems and differences in vaginal physiology when compared to women limit the interpretation of results. Other animal models have also been used, namely primates, dogs, guinea pigs, pigs, mice, and rats [219–223]. Classic animal testing is limited because of the number of animals required, which makes testing burdensome, expensive, and ethically questionable, and because of differences between species, which may jeopardize the extrapolation of results to humans [224]. Thus, simpler toxicity tests performed in nonvertebrate organisms, such as gastropods, are interesting alternatives to vertebrate animal testing. These tests also proved to be superior than in vitro testing, mainly because of the limitations that are intrinsic to simple cell culture models [225]. Recently, a simple in vitro test using slugs (*Arion lusitanus*) has been proposed as an alternative to vertebrates in order to screen new vaginal semisolid formulations for local tolerance early in the development process. The irritation potential is evaluated by mucus production, and protein and enzyme release (lactate dehydrogenase and alkaline phosphatase). Experimental results showed that the slug mucosa irritation test performance is comparable to the classically used rabbit vaginal irritation test [226].

When initiating human testing, symptoms and signs of genital irritation must be assessed [227]. These investigations should be performed comparing results between the formulation to be tested (vehicle plus active substance(s) and vehicle only) and formulations that are well known for their irritative effects. In the late 1980s and early 1990s colposcopy of the vagina and cervix began to be used in the in vivo safety assessment of vaginal products, becoming a standard technique [228]. The objective of this procedure is to detect epithelial changes, such as breaks in the epithelium, inflammation, or other not well characterized, that may be a consequence of vaginal products usage. Although very important, it presents several limitations such as costs, specific personnel training, and difficulty in understanding which colposcopic findings indicate risk [229]. Also, other techniques, such as Papanicolau stained smears or automated cytomorphometric analysis, have been used in order to assess the effect of formulations on the vaginal mucosa [230].

5.12.5.9 Other Characteristics

In addition to the discussed evaluation tests and methodologies, other characteristics of vaginal formulations may be assessed according to their individual specificities. For example, the compatibility of vaginal formulations with condoms is an important parameter to be determined, particularly when they are used during sexual intercourse. These studies are usually performed according to the American Society for Testing and Materials (ASTM) norm D3492-89 (Standard Specification for Rubber Contraceptives), where condoms are tested by accelerated testing for their tensile strength and elongation on break point after being exposed to vaginal formulations [231, 232]. Also, it is important to consider the effects of vaginal formulations that are used during sexual intercourse in the penis and the possibility of drug penetration through this organ. Although the human penis is covered with keratinized stratified epithelium, and thus the expected absorption should be less than that of the vaginal epithelium, it is always a possibility to be taken into account [233].

5.12.6 CLINICAL USAGE AND POTENTIAL OF VAGINAL DRUG DELIVERY

5.12.6.1 Microbicides

Microbicides (initially termed “virucides”) are anti-infective drugs formulated for topical self-administration in the vagina before sexual intercourse in order to protect against HIV and other sexually transmitted pathogens [234]. Once a neglected subject in the war against HIV and other sexually transmitted diseases, in the last decade microbicide investigations have gained important boosting and interest by the scientific community, being considered as a new approach for prevention [235]. Despite the fact that several microbicides are already in clinical trialing (see Table 5), currently there are no available products on the market. Also, 100% effective microbicides are not likely to be achieved, even though only partially effective, microbicides can be a big help in reducing the spread of HIV infection. Investigators estimate that the use of a 60% effective microbicide in only 20% of all coital acts could prevent approximately 2.5 million infections over a period of 3 years [236]. While waiting for the first-generation microbicides, preclinical research in new and improved microbicides is already in progress. Several candidates, such as PSC-RANTES [237], antimicrobial peptides [238], monoclonal antibodies [239], inhibitors of virus–cell fusion [240], and natural products [241], are currently being developed. Also, another strategy that seems to be gaining consensus among the

TABLE 5 Microbicides Currently Undergoing Clinical Trialing

Mechanism of Action	Active Substance(s)	Candidate Product	Drug Delivery Systems	Developers	Clinical Trial Status
Vaginal defense enhancer	—	ACIDFORM	Gel	CONRAD/Instead	Phase I
	—	BufferGel	Gel	Reprotect	Phases II/III
Membrane disruptive agent/ surfactant	C31G	Savvy	Gel	CONRAD	Phase III
Entry/fusion inhibitor	Carrageenan (PC-515)	Carraguard	Gel	Population Council	Phase III
	Cellulose acetate 1,2-benzenedicarboxylate	Cellacefate/ CAP	Gel	Lindsey F. Kimball Research Institute / Dow Pharmaceuticals, Inc.	Phase I
	Cellulose sulfate	Cellulose Sulfate Vaginal Gel	Gel	CONRAD	Phase III
	Sodium lauryl sulfate	Invisible Condom	Gel	Laval University	Phases I/II
	Naphthalene 2-sulfonate polymer	PRO 2000/5	Gel	Indevus Pharmaceuticals	Phase III
	SPL7013	VivaGel	Gel	Starpharma	Phase I
Replication inhibitor	Tenofovir	PMPA Gel	Gel	Gilead Sciences	Phases II/IIb
	Dapivirine (TMC120)	TMC120 Vaginal Gel	Gel and vaginal ring	International Partnership for Microbicides	Phases I/II
	UC-781	UC-781Gel	Gel	CONRAD	Phase I
Unknown mechanism	Extracts of *Azadirachta indica, Sapindus mukerossi*, and *Mentha citrata*	Praneem polyherbal	Vaginal tablet	Talwar Research Foundation	Phases II/IIb

scientific community is the synergistic association of microbicides with different action mechanisms in order to improve protection [242].

Early expectations created around the possibility of using nonoxynol-9 (widely used as a vaginal contraceptive) as an effective microbicide, based on its in vitro efficacy against HIV, were frustrated in clinical trials [243]. Several hypotheses for this failure were brought up, an inadequate choice of drug delivery system being one of them. These results confirm that preformulation and formulation studies play an important role in microbicide rational design and development, being a big challenge to overcome. Although clinical development recommendations have been extensively reviewed [244, 245], pharmaceutical development algorithms for microbicides are yet to be defined. Nonetheless, it is known that preformulation parameters such as organoleptic characteristics, stability, permeability, inherent bioadhesion and retention features, and compatibility with excipients and condoms of candidate drugs, among others, play a crucial role when developing drug delivery systems containing microbicides [246]. After collection of this information, formulation studies are necessary in order to obtain a final product that fulfills microbicide objectives and requirements, namely safety, efficacy, acceptability, affordability, and regulatory duties [247]. As well as providing effective protection, microbicide formulations must also be safe on multiple exposures over time, chemically and physically stable, compatible with latex and other materials used in barrier devices, and affordable and acceptable to the end user. Ideally, they should be colorless, odorless, tasteless, and nonmessy [234, 248]. Tested microbicides have been formulated mostly as gels, although creams, vaginal rings, foams, sponges, vaginal suppositories, and films may also be considered. Nonetheless, alternative innovative options for the delivery of microbicides have been developed in recent years. For instance, the formulation of a safe and inexpensive "universal" drug carrier for microbicidal substances is an interesting strategy that may potentially ensure the efficacy of most currently researched molecules [249]. Another interesting approach for the delivery of microbicides was proposed by Chang et al. These researchers studied the possibility of using genetically modified comensal vaginal bacteria (*Lactobacillus jensenii*) to produce anti-HIV proteins [250]. In vitro experiments showed that this strategy can be a new step toward an effective microbicide formulation.

5.12.6.2 Antimicrobials

Vaginitis is a common condition in women which can be caused by bacteria, yeasts, or protozoa. Treatment of vaginitis has been achieved by oral or vaginal administration of antimicrobials, often with similar efficacy rates [251]. Several drugs are currently available for intravaginal treatment of bacterial (e.g., metronidazole, clindamycin), fungal (e.g., azoles, boric acid, nystatin), and protozoal (e.g., metronidazole) vaginitis [252]. Also, alternative vaginal therapies have been investigated in order to treat vaginitis. For instance, herbal formulations, particularly those containing essential oils, have been referred as potential antimicrobials for the treatment of both fungal and bacterial infections [253, 254]. Several vaginal drug delivery systems have been used in order to administer antimicrobial substances, particularly gels, creams, tablets, and vaginal suppositories. Current research in antimicrobial vaginal drug delivery systems is focused on more convenient single-dose formulations that can achieve clinical cure while improving patient compliance [255–257].

Vaginal treatment may be advantageous when compared with oral treatment, as systemic adverse effects are less likely to occur [258]. For example, Cunningham and co-workers conducted a study where systemic levels of metronidazole after vaginal administration (5 g of a 0.75% metronidazole gel) were residual when compared to the oral administration of a 500-mg standard dose despite comparable clinical outcome [259]. This low level of vaginal absorption may be attributed to metronidazole's poor lipid solubility. Similar results have been obtained for clindamycin when comparing intravaginal and intravenous routes [260]. Also, contraindication during pregnancy and possible interference with oral contraceptives are situations that probably recommend vaginal administration of these therapeutic agents over oral administration in women requiring treatment [261].

As previously discussed, vaginal pH plays an important role in the normal physiology of the vagina, being elevated in bacterial vaginitis. Vaginal acidifiers, such as vitamin C in the form of tablets, can be effective in the treatment of vaginitis, with the advantage of maintaining or even improving the normal vaginal flora equilibrium, particularly in diabetic and pregnant women or those with recurrent bacterial vaginitis episodes [78, 262]. Also, other formulations such as acid-buffering gels were demonstrated to be potential helpers in the maintenance of a healthy vaginal milieu and as adjuvant of antimicrobials in the treatment of bacterial vaginosis [263, 264]. In fact, some gels may even be used in the treatment of bacterial vaginosis. For example, a mucoadhesive gel containing two polymers, polycarbophil and Carbopol 974P, was demonstrated to be effective in the treatment of bacterial vaginosis, even when compared with the clinical cure rate of vaginal metronidazole or tinidazole [265].

5.12.6.3 Hormonal Contraceptives and Hormonal Replacement Therapy

The vaginal route is one of the many available routes for estrogen and progestogen delivery. When compared to the oral and transdermal route, vaginal administration of these hormones presents the advantage of needing a smaller quantity of drug to achieve similar systemic effects with less relative variability [266, 267]. Also, the inconveniences associated with subcutaneous implants are circumvented. Thus, vaginal hormonal contraception and replacement therapy were soon considered as a possibility.

Hormonal contraception requires drug delivery systems that are able to achieve sustained blood levels of these substances, either by multiple administrations, as in the case of oral contraceptives, or by sustained drug release. Since daily vaginal administration would not be feasible and acceptable by most women, sustained-release systems were developed. This need led to the development of several devices, including vaginal rings [109]. Progestogen-only rings were initially developed, but menstrual bleeding problems and inefficacious control of ovulation led to discontinuation in several studies. Thus, combined estrogen–progestogen rings were the next natural step toward effective and acceptable formulations. Various combined estrogen–progestogen rings have proven to be highly effective as contraceptives, providing excellent inhibition of ovulation [41]. Also, serious vaginal lesions were shown to be unlikely to occur with short- and medium-term exposure

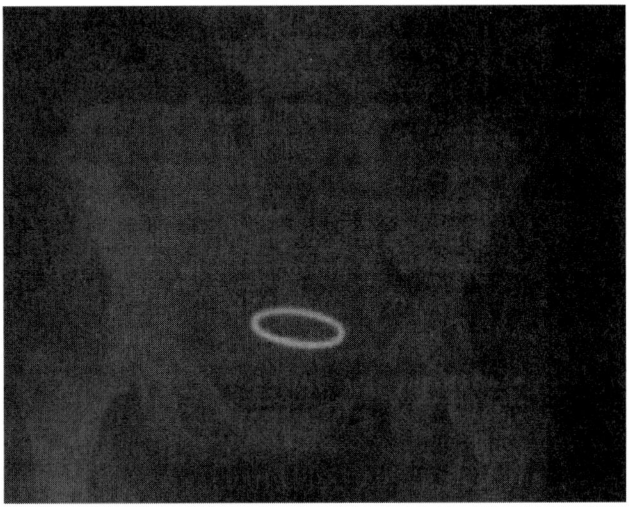

FIGURE 7 X-ray image of vaginal ring after placement in human vagina. (Reprinted with permission from K. Malcolm et al., *Journal of Controlled Release*, 90, 217–225, 2003. Copyright 2003 by Elsevier.)

to currently available rings, mainly due to their improved flexibility and small dimensions [268]. Although other alternatives have also been studied, the most popular schedule for contraceptive vaginal rings is three weeks in, one week out. With this regimen the ring is inserted in the vagina on day 5 of the menstrual cycle and left in place for three weeks. On week 4 the ring is removed, allowing menstrual bleeding. A new ring (one-month ring) or the same one (over-one-month ring) is inserted after one week ring free [112]. Vaginal rings are easy to use, being self-administered by women. Once inserted in the vagina, the ring fits in the upper vagina and delivers the active substance(s) by contact with the vaginal mucosa (see Figure 7).

Despite all the investigational work already performed, only one contraceptive vaginal ring, Nuvaring (Organon), is available in the United States and Europe. This three weeks in–one week out vaginal ring was the first approved by the FDA, being available in the market since 2002. It is a flexible and transparent ring containing etonogestrel and ethinyl estradiol in a poly(ethylene-*co*-vinylacetate) matrix, with an outer diameter of 54 mm and a cross section of 4 mm [269]. When placed in the vagina it releases 120 µg/day of etonogestrel and 15 µg/day of ethinyl estradiol. Clinical studies showed that Nuvaring is as effective and reliable as commonly used oral contraceptives, being well tolerated, convenient, and highly acceptable to most women and their partners [270–272]. Moreover, this contraceptive ring is able to provide lower and more stable systemic exposure to estrogens than other contraceptive options, namely combined oral contraceptives or transdermal patches, thereby reducing drug-related side effects (Figure 8) [267]. Other contraceptive vaginal rings that are still in premarket research include a one-year contraceptive ring being developed by the Population Council, New York. This nestorone and ethinyl estradiol containing ring has been shown to be as efficacious as oral contraceptives when used up to 12 months on a 3 weeks in–1 week out regimen [273, 274].

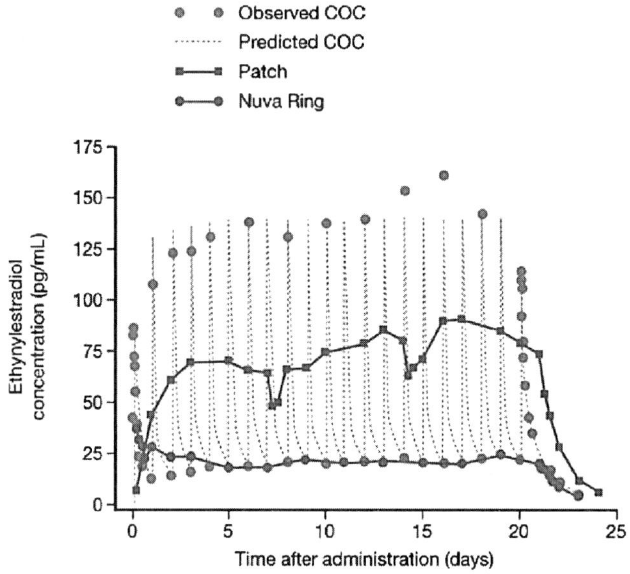

FIGURE 8 Mean serum ethinyl estradiol concentration versus time for subjects treated during 21 days with Nuvaring ($n = 8$), transdermal contraceptive patch (Evra, OrthoMcNeil Pharmaceutical; releases 20 µg ethinyl estradiol and 150 µg norelgestromin daily; $n = 6$), and combined oral contraceptive (COC) (Microgynon, Schering AG; contains 30 µg ethinyl estradiol and 150 µg levonorgestrel; $n = 8$). Subject exposure to ethinyl estradiol in Nuvaring group was on average 3.4 and 2 times lower than for subjects in patch and combined oral contraceptive groups, respectively. (Reprinted with permission from M. W. van den Heuvel et al., *Contraception*, 72, 168–174, 2005. Copyright 2005 by Elsevier.)

Although conventional emergency contraception usually comprises the use of orally administered levonorgestrel (either alone or in combination with ethinyl estradiol), the vaginal route has been shown to be an efficacious alternative. The obtained hormonal plasma levels for the same oral dose are lower, but presumably high enough to prevent pregnancy [275]. Nonetheless, increases in the administered dose by the vaginal route have been suggested [276, 277].

Hormone replacement therapy has been a common practice for a long time in order to improve the quality of life of women suffering from acute symptoms related to menopause. Several formulations for oral, buccal, subcutaneous, parenteral, intrauterine, nasal, transdermal, or vaginal administration have been used for this purpose [278]. Although in recent years hormone replacement therapy has been associated with an increased risk of fatal breast cancer [279], some postmenopausal women may still benefit from this treatment. Estrogens have been used through the vaginal route for the treatment of vaginal symptoms associated with hormone decline in menopausal woman, such as dryness, dyspareunia, pruritus, irritation, discomfort, and atrophy and has been shown to be as effective as systemic therapy [280, 281]. Creams and vaginal suppositories were the first vaginal drug delivery systems to be used to deliver estrogens. Recently, estradiol rings for urogenital symptoms therapy (low-dose rings) or vasomotor plus vaginal symptoms relief (usually higher dose rings) were approved in some countries, providing a new option for the administration of these drugs [282]. For instance, Estring (Pharmacia & Upjohn), a currently marketed vaginal ring containing low-dose estradiol (releases 7.5 μg/day when placed in the vagina), proved to be efficacious, well tolerated, and safe when used for up to a year in the treatment of urogenital symptoms in postmenopausal women [283, 284]. This silicone polymer ring has a diameter of 55 mm and a cross-sectional diameter of 9 mm and is used up to three months. Another similar ring containing estradiol acetate (Femring, Warner Chilcott), with a diameter of 56 mm and a cross-sectional diameter of 7.6 mm, is also currently available.

Estrogens and progestogens have also been administered through the vagina in order to manage other conditions. In fact, vaginally administered progesterone is commonly used for luteal-phase support in women undergoing assisted reproduction treatment, allowing optimal uterine concentrations without the high serum levels observed by other routes (oral and intramuscular), possibly due to the first-uterine-pass effect [285]. Several drug delivery systems such as capsules, tablets, vaginal suppositories, or gels have been used and have been shown to be equally efficacious in increasing the chance of becoming pregnant. Nonetheless, sustained-release formulations allow fewer administrations per day with lower doses [286]. Also, progesterone vaginal rings have been successfully used for luteal-phase support [287].

5.12.6.4 Spermicides

The vaginal use of spermicidal substances during sexual intercourse is perhaps the oldest method of contraception. However, the introduction of oral contraceptives and the intrauterine device in the 1960s led to the decay in their use. Since many of these substances also offer protection against sexually transmitted diseases, interest and investigation in this field have recently increased [288]. Also, the development of new potential spermicides, namely antibodies [289], contributed to further awareness. Currently used spermicides include nonoxynol-9, octoxynol, benzalkonium chloride, and chlorhexidine.

Nonoxynol-9 has been used for more than 30 years as a spermicide in over-the-counter vaginal products, such as semisolid formulations, sponges, foams, films, and others, in order to prevent pregnancy. Although relatively safe and effective, nonoxynol-9 formulations are still not able to achieve the same decrease in pregnancy risk obtained with hormonal methods [290]. Thus, some strategies have been used to enhance nonoxynol-9 effects while reducing its toxic effects. For instance, some synergistic associations with chelating agents that have themselves little spermicide activity, such as EDTA and ethylene glycol tetraacetic acid (EGTA), have been shown to be promising in this matter [291]. Also, coprecipitation of nonoxynol-9 with polyvinylpyrrolidone by a freeze-drying method can be useful, particularly when the formulation of a solid system is desirable, as this process is necessary to alter the chemical state (liquid to solid) of nonoxynol-9 [292, 293]. Using these coprecipitates, a tablet with an inner core that provides slow release of nonoxynol-9 after its fast release of the outer core was shown to be an efficient and safer way of delivering this spermicide in rabbits [294].

Semisolid formulations are often used as contraceptives, particularly gels. These drug delivery systems were demonstrated to be useful, namely in reducing the toxicity of nonoxynol-9 [295]. Adequate formulation of these products has also been shown to be essential for their efficacy. For example, the contraceptive effect of spermicides can be enhanced by adequate consistency of the formulation. When the viscosity of a formulation increases, contraceptive efficacy may increase as a result of becoming more tenacious and more resistant to sperm migration, consequently decreasing its capacity of reaching the site of fertilization [296]. Other gel properties, such as pH and osmolarity, may also influence its spermicidal effects [297]. Moreover, vaginal rings and inserts have been proposed as adequate vehicles for the delivery of nonhormonal contraceptives [298, 299]. These systems may be advantageous due to their controlled and prolonged release properties.

5.12.6.5 Labor Inducers and Abortifacients

Cervical ripening for the induction of labor has been a common practice in modern obstetrics. Several drugs have been tested with this purpose, misoprostol, dinoprostone, and oxytocin the three most frequently used. These labor inducers are normally administered by the vaginal route as gels, tablets, suppositories, or inserts, all demonstrated to be effective and safe. Nonetheless, drug formulation and choice of drug delivery are important factors to be considered. In fact, different times to achieve cervical ripening have been observed in several studies due to different drug release profiles [300, 301].

Vaginal administration of prostaglandins can be useful in the termination of pregnancy. Misoprostol has been used to terminate unwanted pregnancies, demonstrating a relatively high efficacy, even when different regimens have been tested [302–304]. Indeed, improved results with vaginal administered misoprostol may be expected, because of higher and prolonged serum drug concentrations obtained, when compared with the oral route [305]. It is also noteworthy that since vaginal formulations are usually not available, oral tablets containing misoprostol (Cytotec, Pfizer) have been routinely used to administer this drug by the vaginal route. As these tablets are not specifically designed to be administered in the vagina, suboptimal clinical results may occur.

5.12.6.6 Proteins and Peptides

A few proteins and peptides, such as insulin, leuprolide, and salmon calcitonin, have been tested for their possible administration by the intravaginal route. In spite of the rich blood supply, relatively large surface area, and good permeability, vaginal absorption of peptides and proteins is influenced by hormonal-induced changes in the mucosa histology and enzymatic activity, thus limiting their administration through this route [306]. Nonetheless, rational design of adequate and innovative drug delivery carriers allowed considerable progress in protein and peptide vaginal delivery. For example, microspheres have shown good potential to deliver peptides and proteins, such as calcitonin, being able to increase drug stability and absorption. This enhancement of absorption is thought to be related with the intimate contact between the microspheres and the mucosa, resulting in high local concentrations at the site of absorption [307, 308]. Another tested approach to vaginal administration of proteins and peptides has been polymeric matrices, identical to the ones used in the design of vaginal rings. For instance, antibodies have been successfully administered in mice using poly(ethylene-*co*-vinyl acetate) disks [309]. Also, these and other polymeric matrices can provide long-term (up to several years), controlled, and high-dose topical delivery of antibodies [310, 311].

Although considered a route that offers little or no real opportunities for insulin administration, mainly because of low and variable levels of absorption, efforts have been made to systemically deliver insulin through the vaginal mucosa. In the early 1980s, studies by Morimoto et al. showed that polyacrylic acid gels containing insulin were able to induce and maintain (up to 30 min) hypoglycemia when administered to rats and rabbits. However, sustained release was necessary to achieve longer time of hypoglycemia [312]. Later, Richardson et al. observed that insulin is almost not absorbed by the vaginal mucosa of ovariectomized rats in the absence of permeation enhancers, but the coadministration of substances such as sodium taurodihydrofusidate, polyoxyethylene-9-laurylether, lysophosphatidyl choline, palmitoylcarnitine chloride, and lysophosphatidyl glycerol significantly increased its absorption and consequently hypoglycemia [59]. Also, the use of mucoadhesive microspheres as delivery systems for insulin improved the absorption rate of this drug in sheep, particularly when associated with permeation enhancers [55]. Recently, Degim et al. developed vaginal chitosan gels as carriers for insulin. Studies performed in rabbits showed that chitosan gels containing 5% dimethyl-β-cyclodextrin as a penetration enhancer may provide longer insulin release, offering a potential alternative to the parenteral route [313]. Also, Ning et al. investigated the suitability of niosomes as insulin carriers for vaginal administration in rats. Results demonstrated that the bioavailability of insulin when administered through the vaginal route was comparable to that of the subcutaneous route [158].

5.12.6.7 Vaccines

Although once considered not to be a very promising approach, intravaginal vaccines have emerged in recent years as a potential noninvasive immunization strategy, particularly for the prevention of HIV transmission [314]. In fact, some animal and human experiments suggest that the obtained female genital tract immunization can be superior with vaginal administration of vaccines when compared to other routes, such as oral, nasal, or rectal [315–317]. Also, it is known that effective immunization against sexually transmitted diseases will require strong local genital tract as well as strong systemic antibody responses [318]. Mucosal vaccination presents several advantages over systemic immunization: improved safety profile, minimization of adverse effects, ease of administration, and potentially lower costs. However, limitations such as epithelium changes with menstrual cycle, which leads to reduced mucosa permeability and poor antigen presentation during certain stages, and inactivation of antigens by exposure to the vaginal environment can modulate the magnitude of immune response inducement [319].

Phosphate-buffered saline has been conveniently used as a vehicle for the delivery of antigens. Although these liquid formulations facilitate vaginal administration, they allow poor retention, controlled release, and protection of the delivered antigens. Thus, development of adequate vaginal drug delivery systems plays an important role in the success of vaginal immunization. Thermosensitive delivery systems with adequate mucoadhesiveness may be useful in enhancing the mucosal and systemic immune responses, as they can increase the exposure and contact of antigens to the vaginal mucosa [320]. The combination of thermosensitive polymers such as poloxamers and mucoadhesive polymers such as polycarbophil were shown to be a good strategy in the development of mucoadhesive, thermosensitive, and controlled-release formulations for vaginal delivery of vaccines [321]. Also, antigen susceptibility to the vaginal environment can be circumvented by administering these molecules in protective systems, such as the one described by Shen et al., which comprises poly(ethylene-*co*-vinyl acetate) matrices containing plasmid DNA [322]. These disk-shaped devices, produced by a solvent evaporation technique described by Luo et al. [323], were shown to be effective in inducing local immunity in mice, protecting and providing controlled and sustained delivery of plasmid DNA. In addition, other strategies for vaginal antigen administration have also been proposed, such as the use of nonpathogenic bacterial vectors [324–326] and microspheres [327].

5.12.6.8 Other Uses

Many other drugs have been administered in the vagina for the management of either local or systemic conditions. Table 6 summarizes some of these reports.

5.12.7 ACCEPTABILITY AND PREFERENCES OF WOMEN RELATED TO VAGINAL DRUG DELIVERY

Usage of vaginal products is intimately related to taboos and presumptions associated with the knowledge and handling of genitalia. Results from a recent international survey showed that, despite improvements in recent decades, society's attitude toward the vagina and its use as a drug delivery route is not very open and not as open as women would like it to be. Also, this study revealed that the vagina is not commonly recognized as a possible drug delivery route, with only approximately 35% of women acknowledging this fact [342]. Although frequently overlooked, acceptability studies of vaginal products are important in predicting women's compliance and thus their effectiveness. Dedicated studies to women's acceptability and preferences toward vaginal drug delivery are not common, most information available resulting from parallel research to clinical trials. Also, most of these studies have focused on microbicides and spermicides, since their acceptance is decisive for their consistent use [343–345]. In fact, there can be a sense of guilt or negative feeling with this kind of vaginal product, contrasting with others that are prescribed by a physician for a specific gynecological problem, which seem to be deprived of moral issues due to the "legitimacy" of their use [346]. Acceptability studies should also be extended to sexual partners, particularly when evaluating products meant to be used during sexual intercourse [347]. Data collected from previous acceptability studies must be considered during the development of formulations in order to improve their suitability. For example, lubrication provided by a vaginal product seems to be important in determining its acceptability, as this feature may be regarded advantageous or not by different users [348]. Also, negative perceptions regarding product characteristics should be identified during clinical trials in order to improve formulations.

Although results found in the literature related to women's preferences toward the vaginal route of drug administration may vary significantly, particularly when different locations and cultures are considered, some general statements can be

TABLE 6 Selected Drugs Administered in the Vagina

Drugs	Intended Use	Drug Delivery Systems	Comments	References
5-Fluorouracil	Treatment of intravaginal warts	Gel	Demonstrated to be effective, safe, and tolerable	328
	Maintenance therapy of cervical dysplasias after standard excisional or ablative therapy	Cream	Reduction of recurrence was achieved	329
Bromocriptine	Therapy of hyperprolactinemia	Oral tablet and vaginal suppository	Proved to be effective and safe, without the adverse effects of oral administration; vaginal suppository obtained higher reduction in serum prolactin	330
Cabergoline	Therapy of hyperprolactinemia	Oral tablet	Proved to be effective and safe, without adverse effects of oral administration	331
Chlorhexidine	Prevention of peripartum infections of newborn	Aqueous solution	Proved to be useful	332
Danazol	Treatment of pelvic endometriosis	Ring and vaginal suppository	Effective without increased serum concentrations observed during oral therapy	114
Etoposide	Management of cervical dysplasias associated with human papilloma virus (HPV)	Vaginal suppository	Demonstrated to be safe and tolerable	333
Imiquimod	Treatment of high-grade vaginal intraepithelial neoplasia	Cream	Results suggest it can be an alternative conservative therapy	334
Indomethacin	Tocolysis in preterm labor	Rectal suppository	Vaginal administration proved to be more effective than conventional rectal plus oral administration	335
Lignocaine	Cervical anesthesia during insertion of tenaculum	Spray and gel	No difference in pain management between spray and gel	336
Morphine	Pain control as alternative to parenteral administration	Tablet and vaginal suppository	Requires close monitoring due to unpredictable bioavailability	337
Oxybutinin	Treatment of urge urinary incontinence	Insert	Proved to be effective and safe in rabbits	338
Propranolol	Control of tachycardia	Tablet	Obtained serum levels were within β-blocking range and comparable to those achieved by oral route	339
Sildenafil	Endometrial development for embryo implantation	Vaginal suppository	Enhanced endometrial development was achieved	340
Trichloroacetic acid	Treatment of low-grade vaginal intraepithelial neoplasia	Solution	Proved to be effective as well as inexpensive and easy to perform	341

made. Ideally, formulations should be easy and comfortable to use, colorless, odorless, and messiness free. Also, products that are better retained in the vagina seem to be favored by women, since leakage is one of the most undesired feature of vaginal formulations [349]. Concerning preferred vaginal dosage forms, gels and creams seem to be the most popular among women. On the other hand, vaginal suppositories and tablets are among the most disliked dosage forms. Others, such as films, present ambiguous results, mostly related with difficulties during insertion [350, 351]. Vaginal rings have been shown to be highly acceptable by both women and their sexual partners, even during sexual intercourse. A trial conducted in North America and Europe showed that couples rarely felt the device during penile penetration, and when the ring was noticed, almost none of the partners seemed to mind [352]. Furthermore, insertion and removal of vaginal rings are judged to be easy by users [353].

Packaging and applicators may also influence women's choice. Products that are placed in the vagina by means of an applicator seem to be preferred because it facilitates administration and avoids direct touching of genitalia during insertion. Applicator characteristics, namely length, width, color, filling features (single-use or reusable applicators), and ease of usage, are also known to influence women's acceptability [351, 354].

5.12.8 VETERINARY VAGINAL DRUG DELIVERY

As with humans, veterinary vaginal drug administration has been performed for a long time, particularly for the treatment of local infections, traditionally involving the use of vaginal suppositories, liquid formulations, or gels. Advantages of this route include the avoidance of damage to the skin or to tissue that is associated with injections, minor stress inflicted to the animals, and possibility of ceasing drug delivery at will [355]. Nonetheless, the first major studies on veterinary drug delivery have been performed by Robinson in the 1960s with progestogen-impregnated polyurethane sponges [356]. Since then, the major use of the vaginal route has been the control of the estrus cycle in livestock by delivering progestogens and estrogens in a controlled fashion. Also, the administration of these hormones showed good results in treating reproductive disorders, such as ovarian quiescence, cystic ovary or cystic corpus luteum [357]. Estrus synchronicity is advantageous as it allows insemination of all or selected females in a herd or flock to occur during a single period of several hours or days [358]. With this purpose, several drug delivery systems have been developed, being generally based on polymeric matrices that are able to control release of drug content. These devices also present two common features: Retention is guaranteed by means of a gentle pressure applied to the mucosa and the existence of a mechanism (e.g., an attached string) that allows their easy removal at the end of the treatment. Table 7 presents a synopsis of these systems, as reviewed by Rathbone et al. [358, 359]. Also, some of these devices are shown in Figure 9. In addition to the vaginal administration of progestogens and estrogens, delivery of other therapeutic agents through this route for both local and systemic effects has been investigated, namely 1,25-dihydroxy vitamin D3 [360], lactic acid–producing lactobacilli [361], antimicrobials [362], local anesthetics [363], and vaccines [364].

TABLE 7 Selected Veterinary Vaginal Drug Delivery Systems for Control of Estrus Cycle in Livestock

Drug Delivery Systems	Active Substance(s)	Matrix-Forming Polymers	Brief Description	Comments
Sponges	Progesterone, estradiol, and several progestins	Polyurethane	Polymeric devices impregnated with active substance(s), usually cylindrical shaped; alternative designs allow achieving zero-order release profile	Variable vaginal retention; inexpensive and simple to prepare
PRID (Progesterone Releasing Intravaginal Device, InterAg, Hamilton, NZ)	Progesterone	Poly(dimethylsiloxane)	Spiral-shaped device obtained by molding and curing polymer (containing active substance) onto stainless steel spiral by high-temperature (~190°C) injection molding; also may enclose hard gelatin capsule (containing estradiol) glued to device	Excellent vaginal retention
CIDR (Controlled Internal Drug Release, CEVA, Libourne, France)	Progesterone	Poly(dimethylsiloxane)	T-shaped nylon spine over which the polymer (containing active substance) is molded and cured by high-temperature injection molding; several types developed: CIDR-S (rabbit eared in shape), CIDR-G (slimmer, straight T-shaped), CIDR-B (similar to CIDR-G but with increased dimensions)	Excellent vaginal retention; only types G and B currently available
INVAS (Intravaginal Application System)	Progesterone	Poly(dimethylsiloxane)	Similar to CIDR but obtained by lower temperature (<120°C) method	Not commercially available
Rings	—	Poly(dimethylsiloxane)	Similar to human vaginal rings	Not commercially available; abandoned because of poor retention properties
Rajamehendran intravaginal device	17β-estradiol/ progesterone	Poly(dimethylsiloxane)	Two C-shaped polymer tubes (impregnated with active substances) tied together, forming "umbrella-shaped" device	Excellent vaginal retention; not commercially available
IBD (Intelligent Breeding Device, Advanced Animal Technology, Hamilton, NZ)	Progesterone/ estradiol/ prostaglandin	Poly(dimethylsiloxane)	Composed by outer plastic sheath [designed to protect circuit board (which controls drug release) and two batteries, four polymeric drug reservoirs (a large one at base and three small ones at head of device)], retention mechanism, and tail	Allows release of active substances at different rates and at specific times

Source: From refs. 358 and 359.

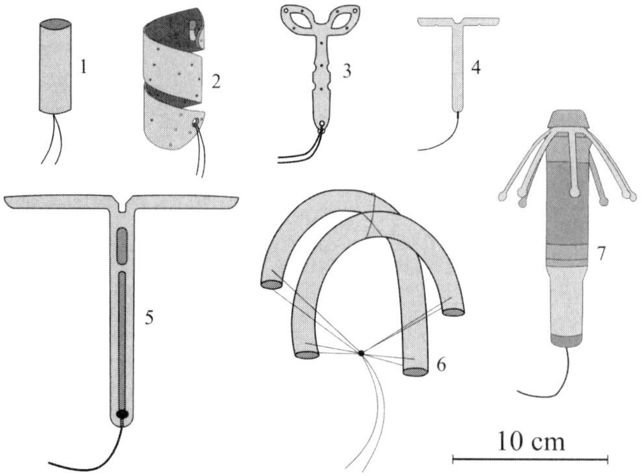

FIGURE 9 Some vaginal veterinary drug delivery systems used or investigated to control estrus cycle in livestock: sponge (1), PRID (2), CIDR-S (3), CIDR-G (4), CIDR-B (5), Rajamehendran intravaginal device (6), and IBD (7).

In order to improve the performance of drug delivery systems, several strategies have been tested. For example, improvement of currently available progestogen- and estrogen-delivering devices and development of new ones have been a common gold of vaginal drug delivery investigators [365]. In other therapeutic field, Gavini et al. used bioadhesive chitosan microspheres compressed into tablets with several excipients to deliver acriflavine [362]. In vitro results demonstrated that these systems have good mucoadhesive properties, allowing increased residence time in the vagina. Recently, Cross et al. proposed an interesting electronic device that allows controlling drug release in response to in loco stimuli (e.g., vaginal temperature) or external commanding via radio wireless link [366]. This device is placed inside and behind the piston of a modified syringe, being able to control the production of hydrogen from a gas cell. The increased gas pressure behind the piston propels a viscous pharmaceutical vehicle that fills the syringe. This drug release and monitoring unit (DMU) have been shown to be a promising strategy in controlling the delivery of active substances to cows while simultaneously collecting physiological data in the vaginal environment.

5.12.9 CONCLUSIONS AND FUTURE TRENDS

As other routes of drug administration presented serious difficulties to deliver some active substances, the vagina emerged as a feasible alternative. Undoubtedly, the vaginal route of drug delivery has attracted the interest of the scientific community, particularly in the last few decades. The latest developments in this field, namely mucoadhesive formulations, vaginal rings, and other controlled-release drug delivery systems, have boosted research and clinical use of this once-neglected route of drug administration. Also, women's emancipation in the last century has slowly led to higher acceptability of vaginal formulations, as old fears and preconceived ideas are being demystified.

Much work remains to be done, particularly in specific fields, such as the delivery of macromolecular drugs (e.g., proteins and peptides) and other substances that are poorly absorbed through the vaginal mucosa. Another promising area that needs further investigation is vaginal administration of vaccines and microbicides. Indeed, in an era where HIV and other sexually transmitted diseases are an increasing concern, vaginal preventive strategies are required. Formulating scientists can contribute decisively to these objectives, as optimization of drug delivery systems seems to be essential. Issues such as poor vaginal distribution and retention, inadequate drug release, limited drug protection from vaginal "aggressors," and adverse effects of currently available drug delivery systems still need to be solved.

ACKNOWLEDGMENTS

The authors would like to express their gratitude to Bruno Sarmento and Cláudia Carneiro for their kind review and useful comments on the manuscript.

REFERENCES

1. O'Dowd, M. J. (2001), *The History of Medications for Women: Materia Medica Woman*, 1st ed., Taylor & Francis, London, pp 53–58.
2. O'Dowd, M. J., and Philipp, E. E. (2000), *The History of Obstetrics and Gynecology*, 1st ed., Taylor & Francis, London, pp 1–40.
3. Stenchever, M. A., Droegemueller, W., and Herbst, A. L. (2002), *Comprehensive Gynecology*, 4th ed., Mosby, St. Louis, MO, pp 40–44.
4. Van De Graff, K. (2001), *Human Anatomy*, 6th ed., McGraw-Hill, London, pp 725–753.
5. Barnhart, K. T., Izquierdo, A., Pretorius, E. S., Shera, D. M., Shabbout, M., and Shaunik, A. (2006), Baseline dimensions of the human vagina, *Hum. Reprod.*, 21, 1618–1622.
6. Berman, J. R., and Bassuk, J. (2002), Physiology and pathophysiology of female sexual function and dysfunction, *World J. Urol.*, 20, 111–118.
7. Pendergrass, P. B., Belovicz, M. W., and Reeves, C. A. (2003), Surface area of the human vagina as measured from vinyl polysiloxane casts, *Gynecol. Obstet. Invest.*, 55, 110–113.
8. Barnhart, K. T., Pretorius, E. S., and Malamud, D. (2004), Lesson learned and dispelled myths: Three-dimensional imaging of the human vagina, *Fertil. Steril.*, 81, 1383–1384.
9. Fawcett, D. W., and Raviola, E. (1994), *Bloom and Fawcett, A Textbook of Histology*, 12th ed., Chapman & Hall, New York, pp 857–858.
10. Patton, D. L., Thwin, S. S., Meier, A., Hooton, T. M., Stapleton, A. E., and Eschenbach, D. A. (2000), Epithelial cell layer thickness and immune cell populations in the normal human vagina at different stages of the menstrual cycle, *Am. J. Obstet. Gynecol.*, 183, 967–973.
11. Ildgruben, A. K., Sjoberg, I. M., and Hammarstrom, M. L. (2003), Influence of hormonal contraceptives on the immune cells and thickness of human vaginal epithelium, *Obstet. Gynecol.*, 102, 571–582.
12. Mauck, C. K., Callahan, M. M., Baker, J., Arbogast, K., Veazey, R., Stock, R., Pan, Z., Morrison, C. S., Chen-Mok, M., Archer, D. F., and Gabelnick, H. L. (1999), The effect of one injection of Depo-Provera on the human vaginal epithelium and cervical ectopy, *Contraception*, 60, 15–24.

13. Wagner, G., and Ottesen, B. (1982), Vaginal physiology during menstruation, *Ann. Intern. Med.*, 96, 921–923.
14. Caillouette, J. C., Sharp, C. F., Jr., Zimmerman, G. J., and Roy, S. (1997), Vaginal pH as a marker for bacterial pathogens and menopausal status, *Am. J. Obstet. Gynecol.*, 176, 1270–1275.
15. Ramsey, P. S., Ogburn, P. L., Jr., Harris, D. Y., Heise, R. H., and Ramin, K. D. (2002), Effect of vaginal pH on efficacy of the dinoprostone gel for cervical ripening/labor induction, *Am. J. Obstet. Gynecol.*, 187, 843–846.
16. Boskey, E. R., Cone, R. A., Whaley, K. J., and Moench, T. R. (2001), Origins of vaginal acidity: High D/L lactate ratio is consistent with bacteria being the primary source, *Hum. Reprod.*, 16, 1809–1813.
17. Boris, S., and Barbés, C. (2000), Role played by lactobacilli in controlling the population of vaginal pathogens, *Microbes Infect.*, 2, 543–546.
18. Zhou, X., Bent, S. J., Schneider, M. G., Davis, C. C., Islam, M. R., and Forney, L. J. (2004), Characterization of vaginal microbial communities in adult healthy women using cultivation-independent methods, *Microbiology*, 150, 2565–2573.
19. Larsen, B., and Galask, R. P. (1982), Vaginal microbial flora: Composition and influences of host physiology, *Ann. Intern. Med.*, 96, 926–930.
20. Woolfson, A. D., Malcolm, R. K., and Gallagher, R. (2000), Drug delivery by the intravaginal route, *Crit. Rev. Ther. Drug Carrier Syst.*, 17, 509–555.
21. Owen, D. H., and Katz, D. F. (1999), A vaginal fluid simulant, *Contraception*, 59, 91–95.
22. Widmaier, E. P., Raff, H., and Strang, K. T. (2004), *Vander, Sherman, Luciano's Human Physiology: The Mechanisms of Body Function*, 9th ed., McGraw-Hill, New York, pp 658–688.
23. Terranova, P. F. (2003), The female reproductive system, in Rhoades, R. A., and Tanner, G. A., Eds. *Medical Physiology*, 2nd ed., Lippincott, Williams & Wilkins, Philadelphia, pp 667–683.
24. Valore, E. V., Park, C. H., Igreti, S. L., and Ganz, T. (2002), Antimicrobial components of vaginal fluid, *Am. J. Obstet. Gynecol.*, 187, 561–568.
25. Eggert-Kruse, W., Botz, I., Pohl, S., Rohr, G., and Strowitzki, T. (2000), Antimicrobial activity of human cervical mucus, *Hum. Reprod.*, 15, 778–784.
26. Burruano, B. T., Schnaare, R. L., and Malamud, D. (2002), Synthetic cervical mucus formulation, *Contraception*, 66, 137–140.
27. Acarturk, F., Parlatan, Z. I., and Saracoglu, O. F. (2001), Comparison of vaginal aminopeptidase enzymatic activities in various animals and in humans, *J. Pharm. Pharmacol.*, 53, 1499–1504.
28. Gibbs, D. F., Labrum, A. H., and Stagg, B. H. (1968), Vaginal fluid enzymology. A new assay method with enzyme-potassium ratios, *Am. J. Obstet. Gynecol.*, 102, 982–988.
29. Belec, L. (2002), Défenses non immunes, pré-immunes et immunes du tractus génital féminin contre les infections, *J. Gynecol. Obstet. Biol. Reprod. (Paris)*, 31, 034302.1–034302.7.
30. Mestecky, J., and Fultz, P. N. (1999), Mucosal immune system of the human genital tract, *J. Infect. Dis.*, 179(Suppl. 3), S470–474.
31. Russell, M. W. (2002), Immunization for protection of the reproductive tract: A review, *Am. J. Reprod. Immunol.*, 47, 265–268.
32. Mestecky, J., Moldoveanu, Z., and Russell, M. W. (2005), Immunologic uniqueness of the genital tract: Challenge for vaccine development, *Am. J. Reprod. Immunol.*, 53, 208–214.

33. Hocini, H., Barra, A., Belec, L., Iscaki, S., Preud'homme, J. L., Pillot, J., and Bouvet, J. P. (1995), Systemic and secretory humoral immunity in the normal human vaginal tract, *Scand. J. Immunol.*, 42, 269–274.
34. Beagley, K. W., and Gockel, C. M. (2003), Regulation of innate and adaptive immunity by the female sex hormones oestradiol and progesterone, *FEMS Immunol. Med. Microbiol.*, 38, 13–22.
35. Wira, C. R., Rossoll, R. M., and Kaushic, C. (2000), Antigen-presenting cells in the female reproductive tract: Influence of estradiol on antigen presentation by vaginal cells, *Endocrinology*, 141, 2877–2885.
36. Franklin, R. D., and Kutteh, W. H. (1999), Characterization of immunoglobulins and cytokines in human cervical mucus: Influence of exogenous and endogenous hormones, *J. Reprod. Immunol.*, 42, 93–106.
37. Farage, M., and Maibach, H. (2006), Lifetime changes in the vulva and vagina, *Arch. Gynecol. Obstet.*, 273, 195–202.
38. Nilsson, K., Risberg, B., and Heimer, G. (1995), The vaginal epithelium in the postmenopause—Cytology, histology and pH as methods of assessment, *Maturitas*, 21, 51–56.
39. Vermani, K., and Garg, S. (2000), The scope and potential of vaginal drug delivery, *Pharm. Sci. Technol. Today*, 3, 359–364.
40. Bernkop-Schnurch, A., and Hornof, M. (2003), Intravaginal drug delivery systems, *Am. J. Drug Deliv.*, 1, 241–254.
41. Alexander, N. J., Baker, E., Kaptein, M., Karck, U., Miller, L., and Zampaglione, E. (2004), Why consider vaginal drug administration? *Fertil. Steril.*, 82, 1–12.
42. Hussain, A., and Ahsan, F. (2005), The vagina as a route for systemic drug delivery, *J. Controlled Release*, 103, 301–313.
43. Macht, D. I. (1918), On the absorption of drugs and poisons through the vagina, *J. Pharmacol. Exp. Ther.*, 10, 509–522.
44. Robinson, G. D. (1925), Absorption from the human vagina, *BJOG*, 32, 496–504.
45. Rock, J., Barker, R. H., and Bacon, W. B. (1947), Vaginal absorption of penicillin, *Science*, 105, 13.
46. van der Bijl, P., and van Eyk, A. D. (2003), Comparative in vitro permeability of human vaginal, small intestinal and colonic mucosa, *Int. J. Pharm.*, 261, 147–152.
47. van Eyk, A. D., and van der Bijl, P. (2004), Comparative permeability of various chemical markers through human vaginal and buccal mucosa as well as porcine buccal and mouth floor mucosa, *Arch. Oral Biol.*, 49, 387–392.
48. Villanueva, B., Casper, R. F., and Yen, S. S. (1981), Intravaginal administration of progesterone: Enhanced absorption after estrogen treatment, *Fertil. Steril.*, 35, 433–437.
49. Sassi, A. B., McCullough, K. D., Cost, M. R., Hillier, S. L., and Rohan, L. C. (2004), Permeability of tritiated water through human cervical and vaginal tissue, *J. Pharm. Sci.*, 93, 2009–2016.
50. van Eyk, A. D., and van der Bijl, P. (2005), Porcine vaginal mucosa as an in vitro permeability model for human vaginal mucosa, *Int. J. Pharm.*, 305, 105–111.
51. van der Bijl, P., and van Eyk, A. D. (2004), Human vaginal mucosa as a model of buccal mucosa for in vitro permeability studies: An overview, *Curr. Drug Deliv.*, 1, 129–135.
52. van der Bijl, P., Penkler, L., and van Eyk, A. D. (2000), Permeation of sumatriptan through human vaginal and buccal mucosa, *Headache*, 40, 137–141.
53. Corbo, D. C., Liu, J. C., and Chien, Y. W. (1989), Drug absorption through mucosal membranes: Effect of mucosal route and penetrant hydrophilicity, *Pharm. Res.*, 6, 848–852.

54. Corbo, D. C., Liu, J. C., and Chien, Y. W. (1990), Characterization of the barrier properties of mucosal membranes, *J. Pharm. Sci.*, 79, 202–206.
55. Richardson, J. L., Farraj, N. F., and Illum, L. (1992), Enhanced vaginal absorption of insulin in sheep using lysophosphatidylcholine and a bioadhesive microsphere delivery system, *Int. J. Pharm.*, 88, 319–325.
56. Okada, H., Yamazaki, I., Yashiki, T., and Mima, H. (1983), Vaginal absorption of a potent luteinizing hormone-releasing hormone analogue (leuprolide) in rats II: Mechanism of absorption enhancement with organic acids, *J. Pharm. Sci.*, 72, 75–78.
57. van der Bijl, P., van Eyk, A. D., Gareis, A. A., and Thompson, I. O. (2002), Enhancement of transmucosal permeation of cyclosporine by benzalkonium chloride, *Oral Dis.*, 8, 168–172.
58. Richardson, J. L., Thomas, N. W., and Illum, L. (1991), Recovery of the rat vaginal epithelium from the histological effects of absorptions enhancers, *Int. J. Pharm.*, 77, 75–78.
59. Richardson, J. L., Illum, L., and Thomas, N. W. (1992), Vaginal absorption of insulin in the rat: Effect of penetration enhancers on insulin uptake and mucosal histology, *Pharm. Res.*, 9, 878–883.
60. Sayani, A. P., Chun, I. K., and Chien, Y. W. (1993), Transmucosal delivery of leucine enkephalin: Stabilization in rabbit enzyme extracts and enhancement of permeation through mucosae, *J. Pharm. Sci.*, 82, 1179–1185.
61. Richardson, J. L., Minhas, P. S., Thomas, N. W., and Illum, L. (1989), Vaginal administration of gentamicin to rats. Pharmaceutical and morphological studies using absorption enhancers, *Int. J. Pharm.*, 56, 29–35.
62. Nakada, Y., Miyake, M., and Awata, N. (1993), Some factors affecting the vaginal absorption of human calcitonin in rats, *Int. J. Pharm.*, 89, 169–175.
63. van der Bijl, P., and van Eyk, A. D. (2001), Areca nut extract lowers the permeability of vaginal mucosa to reduced arecoline and arecaidine, *J. Oral Pathol. Med.*, 30, 537–541.
64. Cicinelli, E., Einer-Jensen, N., Galantino, P., Alfonso, R., and Nicoletti, R. (2004), The vascular cast of the human uterus: From anatomy to physiology, *Ann. N. Y. Acad. Sci.*, 1034, 19–26.
65. Cicinelli, E., Cignarelli, M., Sabatelli, S., Romano, F., Schonauer, L. M., Padovano, R., and Einer-Jensen, N. (1998), Plasma concentrations of progesterone are higher in the uterine artery than in the radial artery after vaginal administration of micronized progesterone in an oil-based solution to postmenopausal women, *Fertil. Steril.*, 69, 471–473.
66. Bulletti, C., de Ziegler, D., Flamigni, C., Giacomucci, E., Polli, V., Bolelli, G., and Franceschetti, F. (1997), Targeted drug delivery in gynaecology: The first uterine pass effect, *Hum. Reprod.*, 12, 1073–1079.
67. Cicinelli, E., Di Naro, E., De Ziegler, D., Matteo, M., Morgese, S., Galantino, P., Brioschi, P. A., and Schonauer, A. (2003), Placement of the vaginal 17beta-estradiol tablets in the inner or outer one third of the vagina affects the preferential delivery of 17beta-estradiol toward the uterus or periurethral areas, thereby modifying efficacy and endometrial safety, *Am. J. Obstet. Gynecol.*, 189, 55–58.
68. Thorgeirsdottir, T. O., Hilmarsson, H., Thormar, H., and Kristmundsdottir, T. (2005), Development of a virucidal cream containing the monoglyceride monocaprin, *Pharmazie*, 60, 897–899.
69. Thorgeirsdottir, T. O., Thormar, H., and Kristmundsdottir, T. (2005), The influence of formulation variables on stability and microbicidal activity of monoglyceride monocaprin, *J. Drug Deliv. Sci. Tech.*, 15, 233–236.

70. Kristmundsdottir, T., Sigurdsson, P., and Thormar, H. (2003), Effect of buffers on the properties of microbicidal hydrogels containing monoglyceride as the active ingredient, *Drug Dev. Ind. Pharm.*, 29, 121–129.

71. Garg, S., Tambwekar, K. R., Vermani, K., Garg, A., Kaul, C. L., and Zaneveld, J. D. (2001), Compendium of pharmaceutical excipients for vaginal formulations, *Pharm. Technol.*, 25, 14–24.

72. Knapczyk, J., Macura, A. B., and Pawlik, B. (1992), Simple test demonstrating the antimycotic effect of chitosan, *Int. J. Pharm.*, 80, 33–38.

73. Manson, K. H., Wyand, M. S., Miller, C., and Neurath, A. R. (2000), Effect of a cellulose acetate phthalate topical cream on vaginal transmission of simian immunodeficiency virus in rhesus monkeys, *Antimicrob. Agents Chemother.*, 44, 3199–3202.

74. Katz, D. F., and Dunmire, E. N. (1993), Cervical mucus: Problems and opportunities for drug delivery via the vagina and cervix, *Adv. Drug Deliv. Rev.*, 11, 385–401.

75. Willits, R. K., and Saltzman, W. M. (2001), Synthetic polymers alter the structure of cervical mucus, *Biomaterials*, 22, 445–452.

76. Karasulu, H. Y., Taneri, F., Sanal, E., Guneri, T., and Ertan, G. (2002), Sustained release bioadhesive effervescent ketoconazole microcapsules tabletted for vaginal delivery, *J. Microencapsul.*, 19, 357–362.

77. Knapczyk, J. (1992), Antimycotic buccal and vaginal tablets with chitosan, *Int. J. Pharm.*, 88, 9–14.

78. Petersen, E. E., and Magnani, P. (2004), Efficacy and safety of vitamin C vaginal tablets in the treatment of non-specific vaginitis. A randomised, double blind, placebo-controlled study, *Eur. J. Obstet. Gynecol. Reprod. Biol.*, 117, 70–75.

79. Digenis, G. A., Nosek, D., Mohammadi, F., Darwazeh, N. B., Anwar, H. S., and Zavos, P. M. (1999), Novel vaginal controlled-delivery systems incorporating coprecipitates of nonoxynol-9, *Pharm. Dev. Technol.*, 4, 421–430.

80. Notelovitz, M., Funk, S., Nanavati, N., and Mazzeo, M. (2002), Estradiol absorption from vaginal tablets in postmenopausal women, *Obstet. Gynecol.*, 99, 556–562.

81. Maggi, L., Mastromarino, P., Macchia, S., Brigidi, P., Pirovano, F., Matteuzzi, D., and Conte, U. (2000), Technological and biological evaluation of tablets containing different strains of lactobacilli for vaginal administration, *Eur. J. Pharm. Biopharm.*, 50, 389–395.

82. Bates, C. D., Nicoll, A. E., Mullen, A. B., Mackenzie, F., Thomson, A. J., and Norman, J. E. (2003), Serum profile of isosorbide mononitrate after vaginal administration in the third trimester, *BJOG*, 110, 64–67.

83. Prista, L. N., Alves, A. C., and Morgado, R. (1996), *Tecnologia Farmacêutica*, 4th ed., Fundação Calouste Gulbenkian, Lisboa, pp 1585–1592.

84. Ozyazici, M., Turgut, E. H., Taner, M. S., Koseoglu, K., and Ertan, G. (2003), In-vitro evaluation and vaginal absorption of metronidazole suppositories in rabbits, *J. Drug Target.*, 11, 177–185.

85. Bergren, M. S., Battle, M. M., Halstead, G. W., and Theis, D. L. (1989), Investigation of the relationship between melting-related parameters and in vitro drug release from vaginal suppositories, *J. Pharm. Biomed. Anal.*, 7, 549–561.

86. Kale, V. V., Trivedi, R. V., Wate, S. P., and Bhusari, K. P. (2005), Development and evaluation of a suppository formulation containing lactobacillus and its application in vaginal diseases, *Ann. N. Y. Acad. Sci.*, 1056, 359–365.

87. Mahaguna, V., McDermott, J. M., Zhang, F., and Ochoa, F. (2004), Investigation of product quality between extemporaneously compounded progesterone vaginal

suppositories and an approved progesterone vaginal gel, *Drug. Dev. Ind. Pharm.*, 30, 1069–1078.

88. Zaneveld, L. J., Waller, D. P., Ahmad, N., Quigg, J., Kaminski, J., Nikurs, A., and De Jonge, C. (2001), Properties of a new, long-lasting vaginal delivery system (LASRS) for contraceptive and antimicrobial agents, *J. Androl.*, 22, 481–490.
89. Ladipo, O. A., De Castro, M. P., Filho, L. C., Coutinho, E., Waller, D. P., Cone, F., and Zaneveld, L. J. (2000), A new vaginal antimicrobial contraceptive formulation: Phase I clinical pilot studies, *Contraception*, 62, 91–97.
90. Mandal, T. K. (2000), Swelling-controlled release system for the vaginal delivery of miconazole, *Eur. J Pharm. Biopharm.*, 50, 337–343.
91. Merabet, J., Thompson, D., and Saul Levinson, R. (2005), Advancing vaginal drug delivery, *Expert Opin. Drug. Deliv.*, 2, 769–777.
92. Seidman, L. S., and Skokos, C. K. (2005), An evaluation of butoconazole nitrate 2% site release vaginal cream (Gynazole-1) compared to fluconazole 150 mg tablets (Diflucan) in the time to relief of symptoms in patients with vulvovaginal candidiasis, *Infect. Dis. Obstet. Gynecol.*, 13, 197–206.
93. Levinson, R. S., Mitan, S. J., Steinmetz, J. I., Gattermeir, D. J., Schumacher, R. J., and Joffrion, J. L. (2005), An open-label, two-period, crossover study of the systemic bioavailability in healthy women of clindamycin phosphate from two vaginal cream formulations, *Clin. Ther.*, 27, 1894–1900.
94. del Palacio, A., Sanz, F., Sanchez-Alor, G., Garau, M., Calvo, M. T., Boncompte, E., Alguero, M., Pontes, C., and Gomez de la Camara, A. (2000), Double-blind randomized dose-finding study in acute vulvovaginal candidosis. Comparison of flutrimazole site-release cream (1, 2 and 4%) with placebo site-release vaginal cream, *Mycoses*, 43, 355–365.
95. Wichterle, O., and Lim, D. (1960), Hydrophilic gels in biologic use, *Nature*, 185, 117.
96. Justin-Temu, M., Damian, F., Kinget, R., and Van Den Mooter, G. (2004), Intravaginal gels as drug delivery systems, *J. Womens Health (Larchmt.)*, 13, 834–844.
97. das Neves, J., and Bahia, M. F. (2006), Gels as vaginal drug delivery systems, *Int. J. Pharm.*, 318, 1–14.
98. Milani, M., Molteni, B., and Silvani, I. (2000), Effect on vaginal pH of a polycarbophil vaginal gel compared with an acidic douche in women with suspected bacterial vaginosis: A randomized, controlled study, *Curr. Ther. Res. Clin. Exp.*, 61, 781–788.
99. Robinson, J. R., and Bologna, W. J. (1994), Vaginal and reproductive system treatments using a bioadhesive polymer, *J. Controlled Release*, 28, 87–94.
100. D'Cruz, O. J., and Uckun, F. M. (2001), Gel-microemulsions as vaginal spermicides and intravaginal drug delivery vehicles, *Contraception*, 64, 113–123.
101. Knuth, K., Amiji, M., and Robinson, J. R. (1993), Hydrogel delivery systems for vaginal and oral applications: Formulation and biological considerations, *Adv. Drug Deliv. Rev.*, 11, 137–167.
102. Venkatraman, S., Davar, N., Chester, A., and Kleiner, L. (2005), An overview of controlled release systems, in Wise, D. L., Ed., *Handbook of Pharmaceutical Controlled Release Technology*, Marcel Dekker, New York, pp 431–464.
103. Okada, H., Yamazaki, I., Ogawa, Y., Hirai, S., Yashiki, T., and Mima, H. (1982), Vaginal absorption of a potent luteinizing hormone-releasing hormone analog (leuprolide) in rats I: Absorption by various routes and absorption enhancement, *J. Pharm. Sci.*, 71, 1367–1371.
104. Rosenberg, M. J., Phillips, R. S., and Holmes, M. D. (1991), Vaginal douching. Who and why? *J. Reprod. Med.*, 36, 753–758.

105. McClelland, R. S., Lavreys, L., Hassan, W. M., MAndaliya, K., Ndinya-Achola, J. O., and Baeten, J. M. (2006), Vaginal washing and increased risk of HIV-1 acquisition among African women: A 10-year prospective study, *AIDS*, 20, 269–273.

106. Myer, L., Kuhn, L., Stein, Z. A., Wright, T. C., Jr., and Denny, L. (2005), Intravaginal practices, bacterial vaginosis, and women's susceptibility to HIV infection: Epidemiological evidence and biological mechanisms, *Lancet Infect. Dis.*, 5, 786–794.

107. Rajamanoharan, S., Low, N., Jones, S. B., and Pozniak, A. L. (1999), Bacterial vaginosis, ethnicity, and the use of genital cleaning agents: A case control study, *Sex. Transm. Dis.*, 26, 404–409.

108. Zhang, J., Thomas, A. G., and Leybovich, E. (1997), Vaginal douching and adverse health effects: A meta-analysis, *Am. J. Public Health*, 87, 1207–1211.

109. Sam, A. P. (1992), Controlled release contraceptive devices: A status report, *J. Controlled Release*, 22, 35–46.

110. Johansson, E. D., and Sitruk-Ware, R. (2004), New delivery systems in contraception: Vaginal rings, *Am. J. Obstet. Gynecol.*, 190, S54–59.

111. Timmer, C. J., and Mulders, T. M. (2000), Pharmacokinetics of etonogestrel and ethinylestradiol released from a combined contraceptive vaginal ring, *Clin. Pharmacokinet.*, 39, 233–242.

112. Sitruk-Ware, R. (2005), Vaginal delivery of contraceptives, *Expert Opin. Drug Deliv.*, 2, 729–736.

113. Acarturk, F., and Altug, N. (2001), In-vitro and in-vivo evaluation of a matrix-controlled bromocriptine mesilate-releasing vaginal ring, *J. Pharm. Pharmacol.*, 53, 1721–1726.

114. Igarashi, M., Iizuka, M., Abe, Y., and Ibuki, Y. (1998), Novel vaginal danazol ring therapy for pelvic endometriosis, in particular deeply infiltrating endometriosis, *Hum. Reprod.*, 13, 1952–1956.

115. Woolfson, A. D., Malcolm, R. K., and Gallagher, R. J. (2003), Design of a silicone reservoir intravaginal ring for the delivery of oxybutynin, *J. Controlled Release*, 91, 465–476.

116. Wyatt, T. L., Whaley, K. J., Cone, R. A., and Saltzman, W. M. (1998), Antigen-releasing polymer rings and microspheres stimulate mucosal immunity in the vagina, *J. Controlled Release*, 50, 93–102.

117. Malcolm, R. K., Woolfson, A. D., Toner, C. F., Morrow, R. J., and McCullagh, S. D. (2005), Long-term, controlled release of the HIV microbicide TMC120 from silicone elastomer vaginal rings, *J. Antimicrob. Chemother.*, 56, 954–956.

118. Dziuk, P. J., and Cook, B. (1966), Passage of steroids through silicone rubber, *Endocrinology*, 78, 208–211.

119. Mishell, D. R., Jr., Talas, M., Parlow, A. F., and Moyer, D. L. (1970), Contraception by means of a silastic vaginal ring impregnated with medroxyprogesterone acetate, *Am. J. Obstet. Gynecol.*, 107, 100–107.

120. Landgren, B. M., Johannisson, E., Masironi, B., and Diczfalusy, E. (1979), Pharmacokinetic and pharmacodynamic effects of small doses of norethisterone released from vaginal rings continuously during 90 days, *Contraception*, 19, 253–271.

121. Mishell, D. R., Jr., Lumkin, M., and Jackanicz, T. (1975), Initial clinical studies of intravaginal rings containing norethindrone and norgestrel, *Contraception*, 12, 253–260.

122. van Laarhoven, H., Veurink, J., Kruft, M. A., and Vromans, H. (2004), Influence of spinline stress on release properties of a coaxial controlled release device based on EVA polymers, *Pharm. Res.*, 21, 1811–1817.

123. van Laarhoven, J. A., Kruft, M. A., and Vromans, H. (2002), In vitro release properties of etonogestrel and ethinyl estradiol from a contraceptive vaginal ring, *Int. J. Pharm.*, 232, 163–173.
124. Vartiainen, J., Wahlstrom, T., and Nilsson, C. G. (1993), Effects and acceptability of a new 17 beta-oestradiol-releasing vaginal ring in the treatment of postmenopausal complaints, *Maturitas*, 17, 129–137.
125. Nash, H. A., Brache, V., Alvarez-Sanchez, F., Jackanicz, T. M., and Harmon, T. M. (1997), Estradiol delivery by vaginal rings: Potential for hormone replacement therapy, *Maturitas*, 26, 27–33.
126. Saleh, S. I., Khidr, S. H., Ahmed, S. M., Jackanicz, T. M., and Nash, H. A. (2003), Estradiol-progesterone interaction during the preparation of vaginal rings, *J. Pharm. Sci.*, 92, 258–265.
127. Metzger, A. P., and Matlack, J. D. (1968), Comparative swelling behavior of various thermoplastic polymers, *Polym. Eng. Sci.*, 8, 110–115.
128. Chokshi, R., and Zia, H. (2004), Hot-melt extrusion technique: A review, *Iran J. Pharm. Res.*, 3, 3–16.
129. Malcolm, K., Woolfson, D., Russell, J., Tallon, P., McAuley, L., and Craig, D. (2003), Influence of silicone elastomer solubility and diffusivity on the in vitro release of drugs from intravaginal rings, *J. Controlled Release*, 90, 217–225.
130. Lee, C. H., Bhatt, P. P., and Chien, Y. W. (1997), Effect of excipient on drug release and permeation from silicone-based barrier devices, *Int. J. Pharm.*, 43, 283–290.
131. Garg, S., Vermani, K., Garg, A., Anderson, R. A., Rencher, W. B., and Zaneveld, L. J. (2005), Development and characterization of bioadhesive vaginal films of sodium polystyrene sulfonate (PSS), a novel contraceptive antimicrobial agent, *Pharm. Res.*, 22, 584–595.
132. Yoo, J. W., Dharmala, K., and Lee, C. H. (2006), The physicodynamic properties of mucoadhesive polymeric films developed as female controlled drug delivery system, *Int. J. Pharm.*, 309, 139–145.
133. Mauck, C. K., Baker, J. M., Barr, S. P., Abercrombie, T. J., and Archer, D. F. (1997), A phase I comparative study of contraceptive vaginal films containing benzalkonium chloride and nonoxynol-9. Postcoital testing and colposcopy, *Contraception*, 56, 89–96.
134. Chien, Y. W., Oppermann, J., Nicolova, B., and Lambert, H. J. (1982), Medicated tampons: Intravaginal sustained administration of metronidazole and in vitro–in vivo relationships, *J. Pharm. Sci.*, 71, 767–771.
135. Brzezinski, A., Stern, T., Arbel, R., Rahav, G., and Benita, S. (2004), Efficacy of a novel pH-buffering tampon in preserving the acidic vaginal pH during menstruation, *Int. J. Gynaecol. Obstet.*, 85, 298–300.
136. Bernstein, G. S. (1974), Conventional methods of contraception: Condom, diaphragm, and vaginal foam, *Clin. Obstet. Gynecol.*, 17, 21–33.
137. Youssef, H., Crofton, V. A., Smith, S. C., and Siemens, A. J. (1987), A clinical trial of Neo Sampoon vaginal tablets and Emko foam in Alexandria, Egypt, *Contraception*, 35, 101–110.
138. Bullough, V. L., and Bullough, B. (1990), *Contraception: A Guide to Birth Control Methods*, Prometheus Books, Buffalo, NY.
139. Smith, C. B., Noble, V., Bensch, R., Ahlin, P. A., Jacobson, J. A., and Latham, R. H. (1982), Bacterial flora of the vagina during the menstrual cycle: Findings in users of tampons, napkins, and sea sponges, *Ann. Intern. Med.*, 96, 948–951.
140. Kafka, D., and Gold, R. B. (1983), Food and Drug Administration approves vaginal sponge, *Fam. Plann. Perspect.*, 15, 146–148.

141. Kuyoh, M. A., Toroitich-Ruto, C., Grimes, D. A., Schulz, K. F., and Gallo, M. F. (2003), Sponge versus diaphragm for contraception: A Cochrane review, *Contraception*, 67, 15–18.

142. Bowers, R., Ed. (2001), U. S. women are waiting for contraceptive sponge, *Contracept. Technol. Update*, 22, 6–7.

143. Smart, J. D. (2005), The basics and underlying mechanisms of mucoadhesion, *Adv. Drug Deliv. Rev.*, 57, 1556–1568.

144. Valenta, C. (2005), The use of mucoadhesive polymers in vaginal delivery, *Adv. Drug Deliv. Rev.*, 57, 1692–1712.

145. Hosny, E. A. (1993), Polycarbophil as a controlled release matrix, *Int. J. Pharm.*, 98, 235–238.

146. Kast, C. E., Valenta, C., Leopold, M., and Bernkop-Schnurch, A. (2002), Design and in vitro evaluation of a novel bioadhesive vaginal drug delivery system for clotrimazole, *J. Controlled Release*, 81, 347–354.

147. Valenta, C., Kast, C. E., Harich, I., and Bernkop-Schnurch, A. (2001), Development and in vitro evaluation of a mucoadhesive vaginal delivery system for progesterone, *J. Controlled Release*, 77, 323–332.

148. Peppas, N. A., Bures, P., Leobandung, W., and Ichikawa, H. (2000), Hydrogels in pharmaceutical formulations, *Eur. J. Pharm. Biopharm.*, 50, 27–46.

149. Ruel-Gariepy, E., and Leroux, J. C. (2004), In situ-forming hydrogels—Review of temperature-sensitive systems, *Eur. J. Pharm. Biopharm.*, 58, 409–426.

150. Chang, J. Y., Oh, Y. K., Kong, H. S., Kim, E. J., Jang, D. D., Nam, K. T., and Kim, C. K. (2002), Prolonged antifungal effects of clotrimazole-containing mucoadhesive thermosensitive gels on vaginitis, *J. Controlled Release*, 82, 39–50.

151. Pavelic, Z., Skalko-Basnet, N., and Schubert, R. (2001), Liposomal gels for vaginal drug delivery, *Int. J. Pharm.*, 219, 139–149.

152. Pavelic, Z., Skalko-Basnet, N., and Jalsenjak, I. (2004), Liposomal gel with chloramphenicol: Characterisation and in vitro release, *Acta Pharm.*, 54, 319–330.

153. Pavelic, Z., Skalko-Basnet, N., and Jalsenjak, I. (2005), Characterisation and in vitro evaluation of bioadhesive liposome gels for local therapy of vaginitis, *Int. J. Pharm.*, 301, 140–148.

154. Pavelic, Z., Skalko-Basnet, N., Filipovic-Grcic, J., Martinac, A., and Jalsenjak, I. (2005), Development and in vitro evaluation of a liposomal vaginal delivery system for acyclovir, *J. Controlled Release*, 106, 34–43.

155. Pavelic, Z., Skalko-Basnet, N., and Jalsenjak, I. (1999), Liposomes containing drugs for treatment of vaginal infections, *Eur. J. Pharm. Sci.*, 8, 345–351.

156. Payne, N. I., Timmins, P., Ambrose, C. V., Ward, M. D., and Ridgway, F. (1986), Proliposomes: A novel solution to an old problem, *J. Pharm. Sci.*, 75, 325–329.

157. Ning, M. Y., Guo, Y. Z., Pan, H. Z., Yu, H. M., and Gu, Z. W. (2005), Preparation and evaluation of proliposomes containing clotrimazole, *Chem. Pharm. Bull. (Tokyo)*, 53, 620–624.

158. Ning, M., Guo, Y., Pan, H., Yu, H., and Gu, Z. (2005), Niosomes with sorbitan monoester as a carrier for vaginal delivery of insulin: Studies in rats, *Drug Deliv.*, 12, 399–407.

159. Ning, M., Guo, Y., Pan, H., Chen, X., and Gu, Z. (2005), Preparation, in vitro and in vivo evaluation of liposomal/niosomal gel delivery systems for clotrimazole, *Drug Dev. Ind. Pharm.*, 31, 375–383.

160. Akagi, T., Kawamura, M., Ueno, M., Hiraishi, K., Adachi, M., Serizawa, T., Akashi, M., and Baba, M. (2003), Mucosal immunization with inactivated HIV-1-capturing

nanospheres induces a significant HIV1-specific vaginal antibody response in mice, *J. Med. Virol.*, 69, 163–172.

161. Bonucci, E., Ballanti, P., Ramires, P. A., Richardson, J. L., and Benedetti, L. M. (1995), Prevention of ovariectomy osteopenia in rats after vaginal administration of Hyaff 11 microspheres containing salmon calcitonin, *Calcif. Tissue Int.*, 56, 274–279.
162. Bilensoy, E., Rouf, M. A., Vural, I., Sen, M., and Hincal, A. A. (2006), Mucoadhesive, thermosensitive, prolonged-release vaginal gel for clotrimazole:beta-cyclodextrin complex, *AAPS PharmSciTech*, 7, E38.
163. Francois, M., Snoeckx, E., Putteman, P., Wouters, F., De Proost, E., Delaet, U., Peeters, J., and Brewster, M. E. (2003), A mucoadhesive, cyclodextrin-based vaginal cream formulation of itraconazole, *AAPS PharmSci*, 5, E5.
164. Ahmed, M. O., El-Gibaly, I., and Ahmed, S. M. (1998), Effect of cyclodextrins on the physicochemical properties and antimycotic activity of clotrimazole, *Int. J. Pharm.*, 171, 111–121.
165. Shihata, A. (2004), New FDA-approved woman-controlled, latex-free barrier contraceptive device "Fem Cap," *Int. Congr. Ser.*, 1271, 303–306.
166. Lee, C. H., Bagdon, R. E., Bhatt, P. P., and Chien, Y. W. (1997), Development of silicone-based barrier devices for controlled delivery of spermicidal agents, *J. Controlled Release*, 44, 43–53.
167. McCarron, P. A., Donnelly, R. F., Gilmore, B. F., Woolfson, A. D., McClelland, R., Zawislak, A., and Price, J. H. (2004), Phototoxicity of 5-aminolevulinic acid in the HeLa cell line as an indicative measure of photodynamic effect after topical administration to gynecological lesions of intraepithelial form, *Pharm. Res.*, 21, 1871–1879.
168. Woolfson, A. D., McCafferty, D. F., McCarron, P. A., and Price, J. H. (1995), A bioadhesive patch cervical drug delivery system for the administration of 5-fluorouracil to cervical tissue, *J. Controlled Release*, 35, 49–58.
169. Mazouni, C., Provensal, M., Menard, J. P., Heckenroth, H., Guidicelli, B., Gamerre, M., and Bretelle, F. (2006), Utilisation du dispositif vaginal Propess dans le déclenchement du travail: Efficacité et innocuité, *Gynecol. Obstet. Fertil.*, 34, 489–492.
170. Vollebregt, A., van't Hof, D. B., and Exalto, N. (2002), Prepidil compared to Propess for cervical ripening, *Eur. J. Obstet. Gynecol. Reprod. Biol.*, 104, 116–119.
171. Tedajo, G. M., Bouttier, S., Grossiord, J. L., Marty, J. P., Seiller, M., and Fourniat, J. (2002), In vitro microbicidal activity of W/O/W multiple emulsion for vaginal administration, *Int. J. Antimicrob. Agents*, 20, 50–56.
172. Kaushal, G., Trombetta, L., Ochs, R. S., and Shao, J. (2006), Delivery of TEM beta-lactamase by gene-transformed *Lactococcus lactis* subsp. *lactis* through cervical cell monolayer, *Int. J. Pharm.*, 313, 29–35.
173. Yao, X. Y., Yuan, M. M., and Li, D. J. (2006), Mucosal inoculation of *Lactobacillus* expressing hCGbeta induces an anti-hCGbeta antibody response in mice of different strains, *Methods*, 38, 124–132.
174. Vail, J. G., Cohen, J. A., and Kelly, K. L. (2004), Improving topical microbicide applicators for use in resource-poor settings, *Am. J. Public Health*, 94, 1089–1092.
175. Costa, P., and Sousa Lobo, J. M. (2001), Modeling and comparison of dissolution profiles, *Eur. J. Pharm. Sci.*, 13, 123–133.
176. Siewert, M., Dressman, J., Brown, C. K., and Shah, V. P. (2003), FIP/AAPS guidelines to dissolution/in vitro release testing of novel/special dosage forms, *AAPS PharmSciTech*, 4, E7.
177. Karasulu, H. Y., Hilmioglu, S., Metin, D. Y., and Guneri, T. (2004), Efficacy of a new ketoconazole bioadhesive vaginal tablet on *Candida albicans*, *Farmaco*, 59, 163–167.

178. Ondracek, J., Stoll, B., and Krifter, R. (1988), New basket dissolution method for vaginal suppositories, *Acta Pharm. Technol.*, 34, 169–171.
179. Chun, I. K., and Chien, Y. W. (1993), Transmucosal delivery of methionine enkephalin. I: Solution stability and kinetics of degradation in various rabbit mucosa extracts, *J. Pharm. Sci.*, 82, 373–378.
180. Bechgaard, E., Riis, K. J., and Jorgensen, L. (1994), The development of an Ussing chamber for isolated human vaginal mucosa, and the viability of the in vitro system, *Int. J. Pharm.*, 106, 237–242.
181. van der Bijl, P., van Eyk, A. D., and Thompson, I. O. (1998), Penetration of human vaginal and buccal mucosa by 4.4-kd and 12-kd fluorescein-isothiocyanate-labeled dextrans, *Oral Surg. Oral. Med. Oral Pathol. Oral Radiol. Endod.*, 85, 686–691.
182. van der Bijl, P., van Eyk, A. D., and Thompson, I. O. (1998), Permeation of 17beta-estradiol through human vaginal and buccal mucosa, *Oral Surg. Oral Med. Oral Pathol. Oral Radiol. Endod.*, 85, 393–398.
183. van der Bijl, P., van Eyk, A. D., van Wyk, C. W., and Stander, I. A. (2001), Diffusion of reduced arecoline and arecaidine through human vaginal and buccal mucosa, *J. Oral Pathol. Med.*, 30, 200–205.
184. Hsu, C. C., Park, J. Y., Ho, N. F., Higuchi, W. I., and Fox, J. L. (1983), Topical vaginal drug. delivery I: Effect of the estrous cycle on vaginal membrane permeability and diffusivity of vidarabine in mice, *J. Pharm. Sci.*, 72, 674–680.
185. Okada, H., Yashiki, T., and Mima, H. (1983), Vaginal absorption of a potent luteinizing hormone-releasing hormone analogue (leuprolide) in rats III: Effect of estrous cycle on vaginal absorption of hydrophilic model compounds, *J. Pharm. Sci.*, 72, 173–176.
186. Garg, S., Anderson, R. A., Chany, C. J., 2nd, Waller, D. P., Diao, X. H., Vermani, K., and Zaneveld, L. J. (2001), Properties of a new acid-buffering bioadhesive vaginal formulation (ACIDFORM), *Contraception*, 64, 67–75.
187. Castle, P. E., Karp, D. A., Zeitlin, L., Garcia-Moreno, E. B., Moench, T. R., Whaley, K. J., and Cone, R. A. (2002), Human monoclonal antibody stability and activity at vaginal pH, *J. Reprod. Immunol.*, 56, 61–76.
188. Owen, D. H., Peters, J. J., and Katz, D. F. (2000), Rheological properties of contraceptive gels, *Contraception*, 62, 321–326.
189. Chang, J. Y., Oh, Y. K., Choi, H. G., Kim, Y. B., and Kim, C. K. (2002), Rheological evaluation of thermosensitive and mucoadhesive vaginal gels in physiological conditions, *Int. J. Pharm.*, 241, 155–163.
190. Thorgeirsdottir, T. O., Kjoniksen, A. L., Knudsen, K. D., Kristmundsdottir, T., and Nystrom, B. (2005), Viscoelastic and structural properties of pharmaceutical hydrogels containing monocaprin, *Eur. J. Pharm. Biopharm.*, 59, 333–342.
191. Thorgeirsdottir, T. O., Thormar, H., and Kristmundsdottir, T. (2006), Viscoelastic properties of a virucidal cream containing the monoglyceride monocaprin: Effects of formulation variables: A technical note, *AAPS PharmSciTech*, 7, E44.
192. Owen, D. H., Peters, J. J., Katz, and D. F. (2001), Comparison of the rheological properties of Advantage-S and Replens, *Contraception*, 64, 393–396.
193. Madsen, F., Eberth, K., and Smart, J. D. (1998), A rheological examination of the mucoadhesive/mucus interaction: The effect of mucoadhesive type and concentration, *J. Controlled Release*, 50, 167–178.
194. Owen, D. H., Peters, J. J., Lavine, M. L., and Katz, D. F. (2003), Effect of temperature and pH on contraceptive gel viscosity, *Contraception*, 67, 57–64.
195. Owen, D. H., and Katz, D. F. (2005), A review of the physical and chemical properties of human semen and the formulation of a semen simulant, *J. Androl.*, 26, 459–469.

196. Jones, D. S., Irwin, C. R., Woolfson, A. D., Djokic, J., and Adams, V. (1999), Physicochemical characterization and preliminary in vivo efficacy of bioadhesive, semisolid formulations containing flurbiprofen for the treatment of gingivitis, *J. Pharm. Sci.*, 88, 592–598.

197. Jones, D. S., Lawlor, M. S., and Woolfson, A. D. (2002), Examination of the flow rheological and textural properties of polymer gels composed of poly(methylvinylether-*co*-maleic anhydride) and poly(vinylpyrrolidone): Rheological and mathematical interpretation of textural parameters, *J. Pharm. Sci.*, 91, 2090–2101.

198. Vermani, K., Garg, S., and Zaneveld, L. J. (2002), Assemblies for in vitro measurement of bioadhesive strength and retention characteristics in simulated vaginal environment, *Drug Dev. Ind. Pharm.*, 28, 1133–1146.

199. Tamburic, S., and Craig, D. Q. M. (1997), A comparision of different in vitro methods for measuring mucoadhesive performance, *Eur. J. Pharm. Biopharm.*, 44, 159–167.

200. El-Kamel, A., Sokar, M., Naggar, V., and Al Gamal, S. (2002), Chitosan and sodium alginate-based bioadhesive vaginal tablets, *AAPS PharmSci*, 4, E44.

201. Barnhart, K. T., Stolpen, A., Pretorius, E. S., and Malamud, D. (2001), Distribution of a spermicide containing Nonoxynol-9 in the vaginal canal and the upper female reproductive tract, *Hum. Reprod.*, 16, 1151–1154.

202. Barnhart, K. T., Pretorius, E. S., Shera, D. M., Shabbout, M., and Shaunik, A. (2006), The optimal analysis of MRI data to quantify the distribution of a microbicide, *Contraception*, 73, 82–87.

203. Kieweg, S. L., Geonnotti, A. R., and Katz, D. F. (2004), Gravity-induced coating flows of vaginal gel formulations: In vitro experimental analysis, *J. Pharm. Sci.*, 93, 2941–2952.

204. Geonnotti, A. R., Peters, J. J., and Katz, D. F. (2005), Erosion of microbicide formulation coating layers: Effects of contact and shearing with vaginal fluid or semen, *J. Pharm. Sci.*, 94, 1705–1712.

205. Ceschel, G. C., Maffei, P., Lombardi Borgia, S., Ronchi, C., and Rossi, S. (2001), Development of a mucoadhesive dosage form for vaginal administration, *Drug Dev. Ind. Pharm.*, 27, 541–547.

206. Barnhart, K., Pretorius, E. S., Stolpen, A., and Malamud, D. (2001), Distribution of topical medication in the human vagina as imaged by magnetic resonance imaging, *Fertil. Steril.*, 76, 189–195.

207. Brown, J., Hooper, G., Kenyon, C. J., Haines, S., Burt, J., Humphries, J. M., Newman, S. P., Davis, S. S., Sparrow, R. A., and Wilding, I. R. (1997), Spreading and retention of vaginal formulations in postmenopausal women as assessed by gamma scintigraphy, *Pharm. Res.*, 14, 1073–1078.

208. Chatterton, B. E., Penglis, S., Kovacs, J. C., Presnell, B., and Hunt, B. (2004), Retention and distribution of two 99mTc-DTPA labelled vaginal dosage forms, *Int. J. Pharm.*, 271, 137–143.

209. Henderson, M. H., Peters, J. J., Walmer, D. K., Couchman, G. M., and Katz, D. F. (2005), Optical instrument for measurement of vaginal coating thickness by drug delivery formulations, *Rev. Sci. Instrum.*, 76, 034302.

210. Braun, K. E., Boyer, J. D., Henderson, M. H., Katz, D. F., and Wax, A. (2006), Label-free measurement of microbicidal gel thickness using low-coherence interferometry, *J. Biomed. Opt.*, 11, 020504.1–020504.3.

211. Brache, V., Cohen, J. A., Cochon, L., and Alvarez, F. (2006), Evaluating the clinical safety of three vaginal applicators: A pilot study conducted in the Dominican Republic, *Contraception*, 73, 72–77.

212. Bagga, R., Raghuvanshi, P., Gopalan, S., Das, S. K., Baweja, R., Suri, S., Malhotra, D., Khare, S., and Talwar, G. P. (2006), A polyherbal vaginal pessary with spermicidal and antimicrobial action: Evaluation of its safety, *Trans. R. Soc. Trop. Med. Hyg.*, 100, 1164–1167.

213. D'Cruz, O. J., and Uckun, F. M. (2001), Lack of adverse effects on fertility of female CD-1 mice exposed to repetitive intravaginal gel-microemulsion formulation of a dual-function anti-HIV agent: Aryl phosphate derivative of bromo-methoxy-zidovudine (compound WHI-07), *J. Appl. Toxicol.*, 21, 317–322.

214. Krebs, F. C., Miller, S. R., Catalone, B. J., Fichorova, R., Anderson, D., Malamud, D., Howett, M. K., and Wigdahl, B. (2002), Comparative in vitro sensitivities of human immune cell lines, vaginal and cervical epithelial cell lines, and primary cells to candidate microbicides nonoxynol 9, C31G, and sodium dodecyl sulfate, *Antimicrob. Agents Chemother.*, 46, 2292–2298.

215. Dezzutti, C. S., James, V. N., Ramos, A., Sullivan, S. T., Siddig, A., Bush, T. J., Grohskopf, L. A., Paxton, L., Subbarao, S., and Hart, C. E. (2004), In vitro comparison of topical microbicides for prevention of human immunodeficiency virus type 1 transmission, *Antimicrob. Agents Chemother.*, 48, 3834–3844.

216. Ayehunie, S., Cannon, C., Lamore, S., Kubilus, J., Anderson, D. J., Pudney, J., and Klausner, M. (2006), Organotypic human vaginal-ectocervical tissue model for irritation studies of spermicides, microbicides, and feminine-care products, *Toxicol. In Vitro*, 20, 689–698.

217. Eckstein, P., Jackson, M. C., Millman, N., and Sobrero, A. J. (1969), Comparison of vaginal tolerance tests of spermicidal preparations in rabbits and monkeys, *J. Reprod. Fertil.*, 20, 85–93.

218. D'Cruz, O. J., and Uckun, F. M. (2001), Intravaginal toxicity studies of a gel-microemulsion formulation of spermicidal vanadocenes in rabbits, *Toxicol. Appl. Pharmacol.*, 170, 104–112.

219. Patton, D. L., Sweeney, Y. C., Tsai, C. C., and Hillier, S. L. (2004), *Macaca fascicularis* vs. *Macaca nemestrina* as a model for topical microbicide safety studies, *J. Med. Primatol.*, 33, 105–108.

220. Gray, J. E., Weaver, R. N., Lohrberg, S. M., and Larsen, E. R. (1984), Comparative responses of vaginal mucosa to chronic pyrimidinone-induced irritation, *Toxicol. Pathol.*, 12, 228–234.

221. D'Cruz, O. J., Erbeck, D., and Uckun, F. M. (2005), A study of the potential of the pig as a model for the vaginal irritancy of benzalkonium chloride in comparison to the nonirritant microbicide PHI-443 and the spermicide vanadocene dithiocarbamate, *Toxicol. Pathol.*, 33, 465–476.

222. D'Cruz, O. J., Waurzyniak, B., Yiv, S. H., and Uckun, F. M. (2000), Evaluation of subchronic (13-week) and reproductive toxicity potential of intravaginal gel-microemulsion formulation of a dual-function phenyl phosphate derivative of bromo-methoxy zidovudine (compound whi-07) in B(6)C(3)F(1) mice, *J. Appl. Toxicol.*, 20, 319–325.

223. Chvapil, M., Droegemueller, W., Owen, J. A., Eskelson, C. D., and Betts, K. (1980), Studies of nonoxynol-9. I. The effect on the vaginas of rabbits and rats, *Fertil. Steril.*, 33, 445–450.

224. CPMP/EMEA (1997), *Replacement of Animal Studies by In Vitro Models*, CPMP/EMEA, London.

225. Adriaens, E., and Remon, J. P. (1999), Gastropods as an evaluation tool for screening the irritating potency of absorption enhancers and drugs, *Pharm. Res.*, 16, 1240–1244.

226. Dhondt, M. M., Adriaens, E., Roey, J. V., and Remon, J. P. (2005), The evaluation of the local tolerance of vaginal formulations containing dapivirine using the Slug Mucosal

Irritation test and the rabbit vaginal irritation test, *Eur. J. Pharm. Biopharm.*, 60, 419–425.

227. Mauck, C. K., Weiner, D. H., Ballagh, S. A., Creinin, M. D., Archer, D. F., Schwartz, J. L., Pymar, H. C., Lai, J. J., Rencher, W. F., and Callahan, M. M. (2004), Single and multiple exposure tolerance study of polystyrene sulfonate gel: A phase I safety and colposcopy study, *Contraception*, 70, 77–83.
228. Mauck, C. K., Baker, J. M., Birnkrant, D. B., Rowe, P. J., and Gabelnick, H. L. (2000), The use of colposcopy in assessing vaginal irritation in research, *AIDS*, 14, 2221–2227.
229. WHO/CONRAD (2004), *Manual for the Standardization of Colposcopy for the Evaluation of Vaginal products, Update 2004*, CONRAD/WHO, Geneva.
230. van der Laak, J. A., de Bie, L. M., de Leeuw, H., de Wilde, P. C., and Hanselaar, A. G. (2002), The effect of Replens on vaginal cytology in the treatment of postmenopausal atrophy: Cytomorphology versus computerised cytometry, *J. Clin. Pathol.*, 55, 446–451.
231. Rosen, A. D., and Rosen, T. (1999), Study of condom integrity after brief exposure to over-the-counter vaginal preparations, *South Med. J.*, 92, 305–307.
232. Haineault, C., Gourde, P., Perron, S., Desormeaux, A., Piret, J., Omar, R. F., Tremblay, R. R., and Bergeron, M. G. (2003), Thermoreversible gel formulation containing sodium lauryl sulfate as a potential contraceptive device, *Biol. Reprod.*, 69, 687–694.
233. Ansbacher, R. (2004), Intravaginal medications, *Fertil Steril*, 82, 1474 (Letter).
234. Stone, A. (2002), Microbicides: A new approach to preventing HIV and other sexually transmitted infections, *Nat. Rev. Drug Discov.*, 1, 977–985.
235. Rowe, P. M. (1995), Research into topical microbicides against STDs, *Lancet*, 345, 1231.
236. Watts, C., and Vickerman, P. (2001), The impact of microbicides on HIV and STD transmission: Model projections, *AIDS*, 15(Suppl. 1), S43–44.
237. Lederman, M. M., Veazey, R. S., Offord, R., Mosier, D. E., Dufour, J., Mefford, M., Piatak, M., Jr., Lifson, J. D., Salkowitz, J. R., Rodriguez, B., Blauvelt, A., and Hartley, O. (2004), Prevention of vaginal SHIV transmission in rhesus macaques through inhibition of CCR5, *Science*, 306, 485–487.
238. Yedery, R. D., and Reddy, K. V. (2005), Antimicrobial peptides as microbicidal contraceptives: Prophecies for prophylactics—A mini review, *Eur. J. Contracept. Reprod. Health Care*, 10, 32–42.
239. Veazey, R. S., Shattock, R. J., Pope, M., Kirijan, J. C., Jones, J., Hu, Q., Ketas, T., Marx, P. A., Klasse, P. J., Burton, D. R., and Moore, J. P. (2003), Prevention of virus transmission to macaque monkeys by a vaginally applied monoclonal antibody to HIV-1 gp120, *Nat. Med.*, 9, 343–346.
240. Veazey, R. S., Klasse, P. J., Schader, S. M., Hu, Q., Ketas, T. J., Lu, M., Marx, P. A., Dufour, J., Colonno, R. J., Shattock, R. J., Springer, M. S., and Moore, J. P. (2005), Protection of macaques from vaginal SHIV challenge by vaginally delivered inhibitors of virus-cell fusion, *Nature*, 438, 99–102.
241. Neurath, A. R., Strick, N., Li, Y. Y., and Debnath, A. K. (2005), *Punica granatum* (pomegranate) juice provides an HIV-1 entry inhibitor and candidate topical microbicide, *Ann. N. Y. Acad. Sci.*, 1056, 311–327.
242. Liu, S., Lu, H., Neurath, A. R., and Jiang, S. (2005), Combination of candidate microbicides cellulose acetate 1,2-benzenedicarboxylate and UC781 has synergistic and complementary effects against human immunodeficiency virus type 1 infection, *Antimicrob. Agents Chemother.*, 49, 1830–1836.
243. Roddy, R. E., Zekeng, L., Ryan, K. A., Tamoufe, U., Weir, S. S., and Wong, E. L. (1998), A controlled trial of nonoxynol 9 film to reduce male-to-female transmission of sexually transmitted diseases, *N. Engl. J. Med.*, 339, 504–510.

244. International Working Group On Vaginal Microbicides (1996), Recommendations for the development of vaginal microbicides, *AIDS*, 10, 1–6.

245. Mauck, C., Rosenberg, Z., and Van Damme, L. (2001), Recommendations for the clinical development of topical microbicides: An update, *AIDS*, 15, 857–868.

246. Garg, S., Kandarapu, R., Vermani, K., Tambwekar, K. R., Garg, A., Waller, D. P., and Zaneveld, L. J. (2003), Development pharmaceutics of microbicide formulations. Part I: Preformulation considerations and challenges, *AIDS Patient Care STDS*, 17, 17–32.

247. Garg, S., Tambwekar, K. R., Vermani, K., Kandarapu, R., Garg, A., Waller, D. P., and Zaneveld, L. J. (2003), Development pharmaceutics of microbicide formulations. Part II: Formulation, evaluation, and challenges, *AIDS Patient Care STDS*, 17, 377–399.

248. Bax, R., Douville, K., McCormick, D., Rosenberg, M., Higgins, J., and Bowden, M. (2002), Microbicides—Evaluating multiple formulations of C31G, *Contraception*, 66, 365–368.

249. D'Cruz, O. J., Samuel, P., and Uckun, F. M. (2005), Conceival, a novel noncontraceptive vaginal vehicle for lipophilic microbicides, *AAPS PharmSciTech*, 6, E56–64.

250. Chang, T. L., Chang, C. H., Simpson, D. A., Xu, Q., Martin, P. K., Lagenaur, L. A., Schoolnik, G. K., Ho, D. D., Hillier, S. L., Holodniy, M., Lewicki, J. A., and Lee, P. P. (2003), Inhibition of HIV infectivity by a natural human isolate of *Lactobacillus jensenii* engineered to express functional two-domain CD4, *Proc. Natl. Acad. Sci. U.S.A.*, 100, 11672–11677.

251. Watson, M. C., Grimshaw, J. M., Bond, C. M., Mollison, J., and Ludbrook, A. (2002), Oral versus intravaginal imidazole and triazole anti-fungal agents for the treatment of uncomplicated vulvovaginal candidiasis (thrush): A systematic review, *BJOG*, 109, 85–95.

252. Sobel, J. D. (1997), Vaginitis, *N. Engl. J. Med.*, 337, 1896–1903.

253. SaiRam, M., Ilavazhagan, G., Sharma, S. K., Dhanraj, S. A., Suresh, B., Parida, M. M., Jana, A. M., Devendra, K., and Selvamurthy, W. (2000), Anti-microbial activity of a new vaginal contraceptive NIM-76 from neem oil (Azadirachta indica), *J. Ethnopharmacol.*, 71, 377–382.

254. Pina-Vaz, C., Goncalves Rodrigues, A., Pinto, E., Costa-de-Oliveira, S., Tavares, C., Salgueiro, L., Cavaleiro, C., Goncalves, M. J., and Martinez-de-Oliveira, J. (2004), Antifungal activity of Thymus oils and their major compounds, *J. Eur. Acad. Dermatol. Venereol.*, 18, 73–78.

255. Brown, D., Henzl, M. R., and Kaufman, R. H. (1999), Butoconazole nitrate 2% for vulvovaginal candidiasis. New, single-dose vaginal cream formulation vs. seven-day treatment with miconazole nitrate. Gynazole 1 Study Group, *J. Reprod. Med.*, 44, 933–938.

256. Faro, S., and Skokos, C. K. (2005), The efficacy and safety of a single dose of Clindesse vaginal cream versus a seven-dose regimen of Cleocin vaginal cream in patients with bacterial vaginosis, *Infect. Dis. Obstet. Gynecol.*, 13, 155–160.

257. Upmalis, D. H., Cone, F. L., Lamia, C. A., Reisman, H., Rodriguez-Gomez, G., Gilderman, L., and Bradley, L. (2000), Single-dose miconazole nitrate vaginal ovule in the treatment of vulvovaginal candidiasis: Two single-blind, controlled studies versus miconazole nitrate 100 mg cream for 7 days, *J. Womens Health Gend. Based Med.*, 9, 421–429.

258. Mikamo, H., Kawazoe, K., Izumi, K., Watanabe, K., Ueno, K., and Tamaya, T. (1997), Comparative study on vaginal or oral treatment of bacterial vaginosis, *Chemotherapy*, 43, 60–68.

259. Cunningham, F. E., Kraus, D. M., Brubaker, L., and Fischer, J. H. (1994), Pharmacokinetics of intravaginal metronidazole gel, *J. Clin. Pharmacol.*, 34, 1060–1065.

260. Borin, M. T. (1990), Systemic absorption of clindamycin following intravaginal application of clindamycin phosphate 1% cream, *J. Clin. Pharmacol.*, 30, 33–38.

261. Czeizel, A. E., Kazy, Z., and Vargha, P. (2003), A population-based case-control teratological study of vaginal econazole treatment during pregnancy, *Eur. J. Obstet. Gynecol. Reprod. Biol.*, 111, 135–140.

262. Polatti, F., Rampino, M., Magnani, P., and Mascarucci, P. (2006), Vaginal pH-lowering effect of locally applied vitamin C in subjects with high vaginal pH, *Gynecol. Endocrinol.*, 22, 230–234.

263. Amaral, E., Perdigao, A., Souza, M. H., Mauck, C., Waller, D., Zaneveld, L., and Faundes, A. (2006), Vaginal safety after use of a bioadhesive, acid-buffering, microbicidal contraceptive gel (ACIDFORM) and a 2% nonoxynol-9 product, *Contraception*, 73, 542–547.

264. Decena, D. C., Co, J. T., Manalastas, R. M., Jr., Palaypayon, E. P., Padolina, C. S., Sison, J. M., Dancel, L. A., and Lelis, M. A. (2006), Metronidazole with Lactacyd vaginal gel in bacterial vaginosis, *J. Obstet. Gynaecol. Res.*, 32, 243–251.

265. Fiorilli, A., Molteni, B., and Milani, M. (2005), Successful treatment of bacterial vaginosis with a policarbophil-carbopol acidic vaginal gel: Results from a randomised double-blind, placebo-controlled trial, *Eur. J. Obstet. Gynecol. Reprod. Biol.*, 120, 202–205.

266. Levine, H., and Watson, N. (2000), Comparison of the pharmacokinetics of crinone 8% administered vaginally versus Prometrium administered orally in postmenopausal women, *Fertil. Steril.*, 73, 516–521.

267. van den Heuvel, M. W., van Bragt, A. J., Alnabawy, A. K., and Kaptein, M. C. (2005), Comparison of ethinylestradiol pharmacokinetics in three hormonal contraceptive formulations: The vaginal ring, the transdermal patch and an oral contraceptive, *Contraception*, 72, 168–174.

268. Fraser, I. S., Lacarra, M., Mishell, D. R., Alvarez, F., Brache, V., Lahteenmaki, P., Elomaa, K., Weisberg, E., and Nash, H. A. (2000), Vaginal epithelial surface appearances in women using vaginal rings for contraception, *Contraception*, 61, 131–138.

269. Bjarnadóttir, R. I. (2003), Update on contraceptive vaginal rings, *Rev. Gyn. Pract.*, 3, 156–159.

270. Roumen, F. J., Apter, D., Mulders, T. M., and Dieben, T. O. (2001), Efficacy, tolerability and acceptability of a novel contraceptive vaginal ring releasing etonogestrel and ethinyl oestradiol, *Hum. Reprod.*, 16, 469–475.

271. Oddsson, K., Leifels-Fischer, B., de Melo, N. R., Wiel-Masson, D., Benedetto, C., Verhoeven, C. H., and Dieben, T. O. (2005), Efficacy and safety of a contraceptive vaginal ring (NuvaRing) compared with a combined oral contraceptive: A 1-year randomized trial, *Contraception*, 71, 176–182.

272. Dieben, T. O., Roumen, F. J., and Apter, D. (2002), Efficacy, cycle control, and user acceptability of a novel combined contraceptive vaginal ring, *Obstet. Gynecol.*, 100, 585–593.

273. Sivin, I., Mishell, D. R., Jr., Alvarez, F., Brache, V., Elomaa, K., Lahteenmaki, P., Massai, R., Miranda, P., Croxatto, H., Dean, C., Small, M., Nash, H., and Jackanicz, T. M. (2005), Contraceptive vaginal rings releasing Nestorone and ethinylestradiol: A 1-year dose-finding trial, *Contraception*, 71, 122–129.

274. Rad, M., Kluft, C., Menard, J., Burggraaf, J., de Kam, M. L., Meijer, P., Sivin, I., and Sitruk-Ware, R. L. (2006), Comparative effects of a contraceptive vaginal ring delivering a nonandrogenic progestin and continuous ethinyl estradiol and a combined oral contraceptive containing levonorgestrel on hemostasis variables, *Am. J. Obstet. Gynecol.*, 195, 72–77.

275. Devoto, L., Fuentes, A., Palomino, A., Espinoza, A., Kohen, P., Ranta, S., and von Hertzen, H. (2005), Pharmacokinetics and endometrial tissue levels of levonorgestrel after administration of a single 1.5-mg dose by the oral and vaginal route, *Fertil. Steril.*, 84, 46–51.

276. Mor, E., Saadat, P., Kives, S., White, E., Reid, R. L., Paulson, R. J., and Stanczyk, F. Z. (2005), Comparison of vaginal and oral administration of emergency contraception, *Fertil. Steril.*, 84, 40–45.

277. Kives, S., Hahn, P. M., White, E., Stanczyk, F. Z., and Reid, R. L. (2005), Bioavailability of the Yuzpe and levonorgestrel regimens of emergency contraception: Vaginal vs. oral administration, *Contraception*, 71, 197–201.

278. Yoo, J. W., and Lee, C. H. (2006), Drug delivery systems for hormone therapy, *J. Controlled Release*, 112, 1–14.

279. Beral, V. (2003), Breast cancer and hormone-replacement therapy in the Million Women Study, *Lancet*, 362, 419–427.

280. Cardozo, L., Bachmann, G., McClish, D., Fonda, D., and Birgerson, L. (1998), Meta-analysis of estrogen therapy in the management of urogenital atrophy in postmenopausal women: Second report of the Hormones and Urogenital Therapy Committee, *Obstet. Gynecol.*, 92, 722–727.

281. Crandall, C. (2002), Vaginal estrogen preparations: A review of safety and efficacy for vaginal atrophy. *J. Womens Health (Larchmt.)*, 11, 857–877.

282. Ballagh, S. A. (2004), Vaginal rings for menopausal symptom relief, *Drugs Aging*, 21, 757–766.

283. Henriksson, L., Stjernquist, M., Boquist, L., Cedergren, I., and Selinus, I. (1996), A one-year multicenter study of efficacy and safety of a continuous, low-dose, estradiol-releasing vaginal ring (Estring) in postmenopausal women with symptoms and signs of urogenital aging, *Am. J. Obstet. Gynecol.*, 174, 85–92.

284. Eriksen, B. (1999), A randomized, open, parallel-group study on the preventive effect of an estradiol-releasing vaginal ring (Estring) on recurrent urinary tract infections in postmenopausal women, *Am. J. Obstet. Gynecol.*, 180, 1072–1079.

285. Tavaniotou, A., Smitz, J., Bourgain, C., and Devroey, P. (2000), Comparison between different routes of progesterone administration as luteal phase support in infertility treatments, *Hum. Reprod. Update*, 6, 139–148.

286. Kleinstein, J. (2005), Efficacy and tolerability of vaginal progesterone capsules (Utrogest 200) compared with progesterone gel (Crinone 8%) for luteal phase support during assisted reproduction, *Fertil. Steril.*, 83, 1641–1649.

287. Zegers-Hochschild, F., Balmaceda, J. P., Fabres, C., Alam, V., Mackenna, A., Fernandez, E., Pacheco, I. M., Sepulveda, M. S., Chen, S., Borrero, C., and Borges, E. (2000), Prospective randomized trial to evaluate the efficacy of a vaginal ring releasing progesterone for IVF and oocyte donation, *Hum. Reprod.*, 15, 2093–2097.

288. Gupta, G. (2005), Microbicidal spermicide or spermicidal microbicide? *Eur. J. Contracept. Reprod. Health Care*, 10, 212–218.

289. Castle, P. E., Whaley, K. J., Hoen, T. E., Moench, T. R., and Cone, R. A. (1997), Contraceptive effect of sperm-agglutinating monoclonal antibodies in rabbits, *Biol. Reprod.*, 56, 153–159.

290. Raymond, E. G., Chen, P. L., and Luoto, J. (2004), Contraceptive effectiveness and safety of five nonoxynol-9 spermicides: A randomized trial, *Obstet. Gynecol.*, 103, 430–439.

291. Lee, C. H., Bagdon, R., and Chien, Y. W. (1996), Comparative in vitro spermicidal activity of chelating agents and synergistic effect with nonoxynol-9 on human sperm functionality, *J. Pharm. Sci.*, 85, 91–95.

292. Zavos, P. M., Correa, J. R., and Zarmakoupis-Zavos, P. N. (1998), Assessment of a tablet drug delivery system incorporating nonoxynol-9 coprecipitated with polyvinylpyrrolidone in preventing the onset of pregnancy in rabbits, *Fertil. Steril.*, 69, 768–773.

293. Fowler, P. T., Doncel, G. F., Bummer, P. M., and Digenis, G. A. (2003), Coprecipitation of nonoxynol-9 with polyvinylpyrrolidone to decrease vaginal irritation potential while maintaining spermicidal potency, *AAPS PharmSciTech*, 4, E30.

294. Zavos, P. M., and Correa, J. R. (1997), Vaginal delivery of new formulations of nonoxynol-9 coprecipitated with polyvinylpyrrolidone in rabbits. Comparisons between two formulation delivery systems, *Contraception*, 56, 123–127.

295. Gagne, N., Cormier, H., Omar, R. F., Desormeaux, A., Gourde, P., Tremblay, M. J., Juhasz, J., Beauchamp, D., Rioux, J. E., and Bergeron, M. G. (1999), Protective effect of a thermoreversible gel against the toxicity of nonoxynol-9, *Sex Transm. Dis.*, 26, 177–183.

296. El-Gizawy, S. A., and Aglan, N. I. (2003), Formulation and evaluation of metronidazole acid gel for vaginal contraception, *J. Pharm. Pharmacol.*, 55, 903–909.

297. Owen, D. H., Dunmire, E. N., Plenys, A. M., and Katz, D. F. (1999), Factors influencing nonoxynol-9 permeation and bioactivity in cervical mucus, *J. Controlled Release*, 60, 23–34.

298. Saxena, B. B., Singh, M., Gospin, R. M., Chu, C. C., and Ledger, W. J. (2004), Efficacy of nonhormonal vaginal contraceptives from a hydrogel delivery system, *Contraception*, 70, 213–219.

299. Saltzman, W. M., and Tena, L. B. (1991), Spermicide permeation through biocompatible polymers, *Contraception*, 43, 497–505.

300. Gauger, L. J., and Curet, L. B. (1991), Comparative efficacy of intravaginal prostaglandin E2 in the gel and suppository forms for cervical ripening, DICP *Ann. Pharmacother.*, 25, 456–460.

301. Chyu, J. K., and Strassner, H. T. (1997), Prostaglandin E2 for cervical ripening: A randomized comparison of Cervidil versus Prepidil, *Am. J. Obstet. Gynecol.*, 177, 606–611.

302. Carbonell, J. L., Rodriguez, J., Delgado, E., Sanchez, C., Vargas, F., Valera, L., Mari, J., Valero, F., Salvador, I., and Llorente, M. (2004), Vaginal misoprostol 800 microg every 12 h for second-trimester abortion, *Contraception*, 70, 55–60.

303. Carbonell, J. L., Rodriguez, J., Velazco, A., Tanda, R., Sanchez, C., Barambio, S., Chami, S., Valero, F., Mari, J., de Vargas, F., and Salvador, I. (2003), Oral and vaginal misoprostol 800 microg every 8 h for early abortion, *Contraception*, 67, 457–462.

304. Carbonell, J. L., Rodriguez, J., Aragon, S., Velazco, A., Tanda, R., Sanchez, C., Barambio, S., Chami, S., and Valero, F. (2001), Vaginal misoprostol 1000 microg for early abortion, *Contraception*, 63, 131–136.

305. Zieman, M., Fong, S. K., Benowitz, N. L., Banskter, D., and Darney, P. D. (1997), Absorption kinetics of misoprostol with oral or vaginal administration, *Obstet. Gynecol.*, 90, 88–92.

306. Richardson, J. L., and Illum, L. (1992), (D) Routes of delivery: Case studies: (8) the vaginal route of peptide and protein drug delivery, *Adv. Drug Deliv. Rev.*, 8, 341–366.

307. Richardson, J. L., Ramires, P. A., Miglietta, M. R., Rochira, M., Bacelle, L., Callegaro, L., and Benedetti, L. (1995), Novel vaginal delivery systems for calcitonin: I. Evaluation of HYAFF/calcitonin microspheres in rats, *Int. J. Pharm.*, 115, 9–15.

308. Rochira, M., Miglietta, M. R., Richardson, J. L., Ferrari, L., Beccaro, M., and Benedetti, L. (1996), Novel vaginal delivery systems for calcitonin: II. Preparation and characterization of HYAFF® microspheres containing calcitonin, *Int. J. Pharm.*, 144, 19–26.

309. Sherwood, J. K., Zeitlin, L., Whaley, K. J., Cone, R. A., and Saltzman, M. (1996), Controlled release of antibodies for long-term topical passive immunoprotection of female mice against genital herpes, *Nat. Biotechnol.*, 14, 468–471.

SECTION 6

TABLET PRODUCTION

6.1

PHARMACEUTICAL PREFORMULATION: PHYSICOCHEMICAL PROPERTIES OF EXCIPIENTS AND POWDERS AND TABLET CHARACTERIZATION

Beom-Jin Lee
Kangwon National University, Chuncheon, Korea

Contents

Pharmaceutical Manufacturing Handbook: Production and Processes, edited by Shayne Cox Gad

6.1.1 INTRODUCTION

Tablet is a major category of solid dosage forms which are widely used worldwide. Extensive information is required to prepare tablets with good quality at high standards. Based on preformulation studies, the optimal dosage forms are generally decided. When given orally, the solid dosage form tablet undergoes in vitro disintegration and dissolution followed by absorption through the gastrointestinal tract (GI). The in vivo biodistribution of drug which enters the systemic circulation then occurrs (Figure 1).

In this section, general preformulation approaches for tablet production are described (Figure 2). The physicochemical properties of drug and excipients, which are crucial factors at the beginning stages, are presented. The types and functions of excipients used for tablet formulation and drug–excipient incompatibility are also discussed. Because tablets are prepared by compression of the drug with powdered excipients, the physicochemical properties of excipients and their multiple functions under regulatory standards are very important. To prepare a drug substance for the final dosage forms, pharmaceutical excipients should be added. The guidelines for selection of excipients are also given, based on the type and function of excipient, drug–excipient compatibility, type of tablet, and manufacturing parameters. Drug–excipient incompatibility is a very important issue before and after tablet prepara-

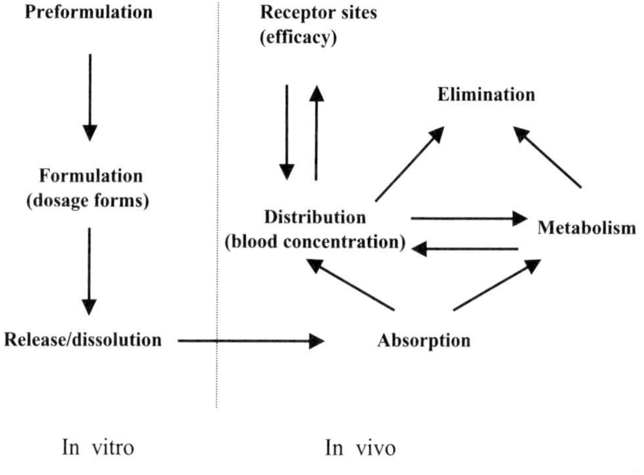

FIGURE 1 In vitro and in vivo pathways of drug which enters systemic circulation.

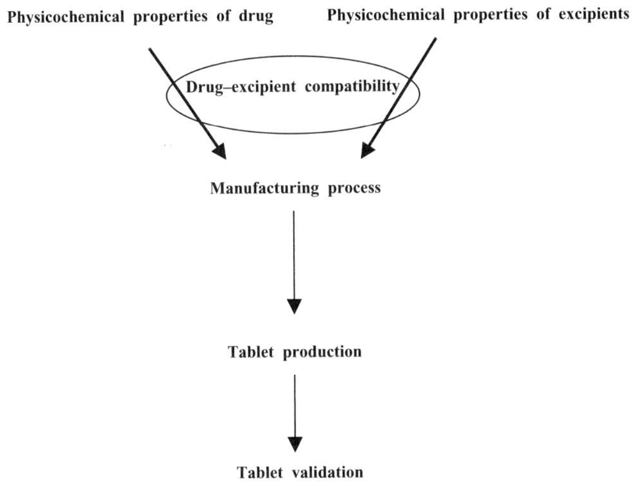

FIGURE 2 General preformulation approaches for tablet production.

tion. According to global International Conferce on Harmonisation (ICH) guidelines, we should cautiously choose the excipients in tablet formulations because the quality and sources are variable by supplier and by batch.

In general, a tablet is prepared by mixing, milling, and compression the drug–excipient mixtures using a tablet machine. Therefore, powder characteristics of drug and excipients, including particle size, flow characteristics, bulk density, hygroscopicity, mixing and milling, and compaction, are extensively discussed. If these physical properties are not fully understood, tablets with good quality at high standards are often impossible to produce.

The prepared tablet must be validated according to regulatory guidelines. In general, disintegration, dissolution, friability, hardness, and weight are characterized for quality validation of finished tablet. This biopharmaceutical preformulation information for tablet production will guide us to design the optimal tablet very efficiently in the laboratory and industrial companies.

6.1.2 SELECTION OF PHARMACEUTICAL EXCIPIENTS

6.1.2.1 Definitions and Goals

A tablet contains active ingredients as well as other substances known as excipients, which have specific functions. The types and functions of various excipients which are incorporated into tablet formulations are discussed in many textbooks [1–3].

A pharmaceutical excipient is defined as an inactive ingredient or any component other than the active ingredient added intentionally to the medicinal formulation or everything in the formulation except the active drug. Pharmaceutical excipients are also called additives, pharmaceutical ingredients, or inactive pharmaceutical ingredients. There are many reasons for selecting and adding these pharmaceutical excipients in formulations. In the preparation of various dosage forms, it is essential to combine pharmaceutical excipients with model drugs as adjuvants to prepare the

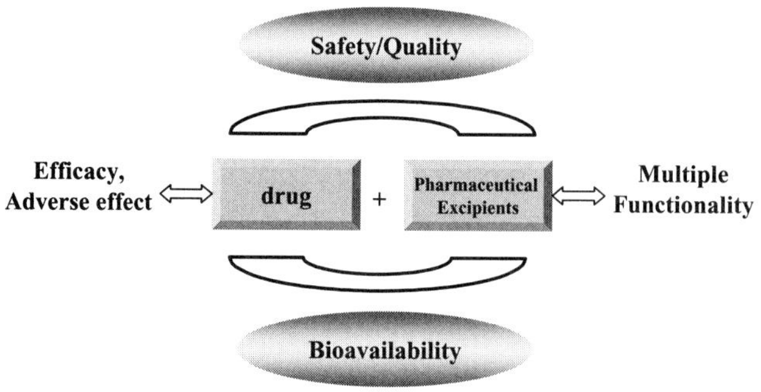

FIGURE 3 Correlation and functions of drug combined with pharmaceutical excipients in dosage form design.

solid dosage forms, mainly tablets. The pharmaceutical excipients make the drug into the final dosage forms. Physicochemical properties such as solubility, stability, metabolism, and even bioavailability of drugs can be varied by the pharmaceutical excipients [4–6]. Figure 3 shows the correlation and functions of drugs combined with pharmaceutical excipients in dosage from designs. Pharmaceutical excipients are regarded as key ingredients not only to decide optimal dosage forms but also to change the physicochemical and biological parameters of drugs. With an aid of pharmaceutical excipients, drug efficacy can be maintained. Changes of other types of dosage forms for different routes of administration are also achieved. The excipients can also function for the preparation of dosage formulation during the manufacturing processes. Patient compliance and modified releases of drugs can also be achieved if the excipients are properly applied. The colorant makes the tablet distinguishable after the coating process.

However, utilization of these pharmaceutical excipients is limited by the regulatory guidelines to be satisfied in the dosage formulations [3, 7]. In general, the regulatory guidelines require the following conditions for the use of excipients in the dosage formulations: (a) no harmful or toxicological effect and listed GRAS (generally recognized as safe), (b) good stability with no drug–excipient incompatibility and by any impurities in the excipients, (c) no interference in quality validation and analytical tests, (d) satisfaction of regulatory issues and guidelines in all countries where the product is to be marketed, (e) no instability with primary packing materials, (f) ease of accessibility, distribution, and economical cost, (g) satisfaction for environmental issues, (h) be physiologically inert, (i) be physically and chemically compatible with the active substance and the other excipients in the formulation, and (j) no unacceptable microbiological burden.

The *Handbook of Pharmaceutical Excipients* [1] contains some details of functional tests carried out on a wide range of excipients. The excipients all have pharmacopeial monographs, but it is important to understand that compliance with a monograph does not indicate equivalence between different grades or suppliers.

6.1.2.2 Types of Pharmaceutical Excipients

To prepare a drug substance into a final dosage form, pharmaceutical excipients should be added. The *Handbook of Pharmaceutical Excipients* presents more than

230 monographs of excipients used in the dosage formulations [1]. The amount and type of pharmaceutical excipients are highly dependent on the final dosage forms. In the preparation of tablets, diluents or fillers are commonly added to increase the bulk of the formulation, binders to cause the adhesion of the powdered drug and pharmaceutical substances, antiadherents or lubricants to assist in the smooth tableting process, disintegrating agents to promote tablet break-up after administration, and coating agents to improve stability, control disintegration, or enhance appearance. Thus, the pharmaceutical excipients establish the primary features and physicochemical properties of the tablet, such as the physical form, stability, dissolution, taste, and overall appearance [3]. Table 1 presents examples of pharmaceutical excipients used in tablet formulations according to principal categories.

6.1.2.3 Characteristics of Pharmaceutical Excipients

Filler (Diluent) Powder mixtures should be compacted to achieve the appropriate strength at a low compaction pressure in the tablet preparation. The compaction properties of a formulation will largely be governed by its major components. For a high-dose drug the drug itself will strongly influence the compaction, while for low-dose drugs the bulk size needs to be increased with an inactive ingredient

TABLE 1 Pharmaceutical Excipients Used in Tablet Formulations

Excipient Type	Definition	Examples
Adsorbent	Agent capable of holding other molecules onto its surface by physical or chemical (chemisorption) means	Powdered cellulose, activated charcoal
Antioxidant	Agent that inhibits oxidation and thus is used to prevent deterioration of preparations by oxidative process	Ascorbic acid, ascorbyl palmitate, butylated hydroxyanisole, butylated hydroxytoluene, hypophosphorus acid, monothioglycerol, propyl gallate, sodium ascorbate, sodium bisulfite, sodium formaldehyde, sulfoxylate, sodium metabisulfite
Colorant	Used to impart color to tablet	FD&C red no. 3, no. 20, FD&C yellow no. 6, FD&C blue no. 2, D&C green no. 5, D&C orange no. 5, D&C red no. 8, caramel, ferric oxide, red
Encapsulant	Used to form thin shells for purpose of enclosing drug substance or drug formulation for ease of administration	Gelatin, cellulose acetate phthalate
Plasticizer	Used as component of film coating solutions to enhance spread of coat over tablets, beads, and granules	Diethyl phthalate, glycerin

TABLE 1 ***Continued***

Excipient Type	Definition	Examples
Surfactant	Substance that adsorbs to surfaces or interfaces to reduce surface or interfacial tension; may be used as wetting agent, detergent, or emulsifying agents	Benzalkonium chloride, nonoxynol 10, oxtoxynol 9, polysorbate 80, sodium lauryl sulfate, sorbitan monopalmitate
Tablet antiadherent	Agent that prevents the sticking of tablet formulation ingredients to punches and dies during tablet production	Magnesium stearate, talc
Tablet binder	Substance used to cause adhesion of powder particles in tablet granulations	Acacia, alginic acid, carboxymethylcellulose sodium compressible sugar, ethylcellulose gelatin, liquid glucose, metylcellulose povidone, pregelatinized starch
Tablet diluent	Inert substance used as filler to create desired bulk, flow properties, and compression characteristics in preparation of tablets	Dibasic calcium phosphate, kaolin, lactose, mannitol, microcrystalline cellulose, powdered cellulose, precipitated calcim carbonate, sorbitol, starch
Tablet sugar and film coating excipient	Used to coat a formed tablet for purpose of protecting against drug decomposition by atmospheric oxygen or humidity, to provide desired release pattern for drug substance after administration, to mask taste or odor of drug substance, or for aesthetic purposes	Sugar: liquid glucose, sucroseFilm: hydroxyethyl cellulose, hydroxypropyl methylcellulose, methylcellulose, ethylcellulose Enteric: cellulose acetate phthalate, shellac (35% in alcohol)
Tablet direct-compression excipient	Used in direct-compression tablet formulations	Dibasic calcium phosphate
Tablet disintegrant	Used in solid dosage forms to promote disruption of solid mass into smaller particles which are more readily dispersed or dissolved	Alginic acid, carboxymethylcellulose calcium, microcrystalline cellulose, polacrilin potassium, sodium alginate, sodium starch glycollate, starch
Tablet glidant	Agent used in tablet and capsule formulations to improve flow properties of powder mixture	Colloidal silica, corn starch, talc
Tablet lubricant	Substance used in tablet formulations to reduce friction during tablet compression	Calcium stearate, magnesium stearate, mineral oil, stearic acid, zinc stearate
Tablet opaquant	Used to render tablet coating opaque; may be used alone or in combination with colorant	Titanium dioxide
Tablet-polishing agent	Used to impart an attractive sheen to coated tablets	Carnauba wax, white wax

TABLE 2 Commonly Used Tablet Diluents

Diluent	Comments
Lactose	Available as anhydrous and monohydrate; anhydrous material used for direct compression due to superior compressibility
Microcrystalline cellulose	Originally direct-compression excipient, now often included in granulations due to its excellent compressibility
Dextrose, glucose	Direct-compression diluent, often used in chewable tablets
Sucrose	Was widely used as sweetener/filler in effervescent tablets and chewable tablets; less popular nowadays due to cariogenicity
Starch and derivatives	Versatile material that can be used as diluent binder, and disintegtant
Calcium carbonate	Brittle material
Dicalcium phosphate	Excellent flow properties; brittle material
Magnesium carbonate	Direct-compression diluent

termed a diluent (or filler). High-dose formulations may also use a diluent to overcome compaction problems experienced with an active pharmaceutical substance.

There are a number of general rules for selecting a diluent. The selection of the diluent will depend on the type of processing and plasticity of materials to be used. A direct-compression formulation will require a diluent with good flow and compaction properties. If the material is extremely plastic, it is appropriated to add a diluent that compacts by brittle fracture; similarly, a brittle drug substance should be combined with a plastic filler. The solubility of the drug substance should also be considered. A soluble drug is normally formulated with an insoluble filler to optimize the disintegration and dissolution process. The hydrophilic excipients added in the formulation may also change drug solubility. Table 2 lists the more commonly used diluents in tablet formulation.

Binder (Granulating Agent) Before tableting the powder mixture via direct compression, generally powders are granulated simply by adding water or an organic solvent to form liquid bridges followed by the drying process. This granulation process can make powders of larger particle size and that are more free flowing for tablet production. The most common method of adding binders is as a solution in the granulating fluid. It is also possible add synthetic polymers such as PVP and HPMC as powders and use water as the granulating agent. When the granulate dries, the crystallization of any solids that had dissolved in the liquid will form solid bonds between the particles [8]. Inclusion of granulating agents or binders to increase granule strength is necessary. Granulating agents are usually hydrophilic polymers that have cohesive properties that both aid the granulation process and impart strength to the dried granulate.

For a granulating agent to be effective, it must form a film on the particle surface and be selected on the basis of its spreading coefficient. The spreading coefficient is defined as the difference between the *work of adhesion* of the binder and the substrate and the *work of cohesion* of the binder. Commonly used granulating agents are listed in Table 3. The binder may vary the disintegration and dissolution. Binders form hydrophilic films on the surface of the granules, which can aid in the

TABLE 3 Commonly Used Granulating (Binding) Agents

Granulating Agent	Normal Concentration (%)	Comments
Starch	5–25	Was once the most commonly used binder; starch has to be prepared as paste, which is time consuming
Pregelatinized starch	0.1–0.5	Cold-water soluble so easier to prepare than starch
Acacia	1–5	Requires preparation of past prior to use; can lead to prolonged disintegration times if used at too high a concentration
Polyvinylpyrrolidone (PVP)	2–8	Available in range of molecular weight/viscosities; soluble in water and ethanol
Hydroxypropyl methylcellulose (HPMC)	2–8	Low-viscosity grades most widely used
Methylcellulose (MC)	1–5	—

wetting of hydrophobic drugs. However, if added at too great concentrations, the films can form viscous gels on the granule surface and will retard dissolution.

Disintegrant Tablets must have sufficient strength to withstand the stresses of subsequent manufacturing operations, such as the coating, packaging, and distribution process. However, once the tablet is taken by the patient, it must break up rapidly to ensure rapid dissolution of the active ingredient in immediate-release preparations. To overcome the cohesive strength produced by the compression process and to break down the tablet into the primary particles as rapidly as possible, the disintegrants are combined with other excipients during the tableting process. Starch was the first disintegrant used in tablet manufacture. Recently, so-called superdisintegrants, including croscarmellose sodium, sodium starch glycolate, and crospovidone, display excellent disintegration activity at low concentrations and have better compression properties than starches. Traditionally, swelling and rate of swelling have been regarded as the most important characteristics of disintegrants. With the aid of these superdisintegrants, sustained-release acetaminophen tablets with biphasic patterns were successfully established to mimic the bilayered Tylenol ER tablet [9]. As a general rule, soluble drugs are formulated with insoluble fillers to maximize the effect of disintegrants. The positioning of disintegrants within the intragranular and extragranular portions of granulated formulations can affect their water uptake and disintegration time.

Commonly used disintegrants are listed in Table 4. The greater the level of disintegrant, the faster the tablet will disintegrate. The compaction properties of many disintegrants, including starch, are not satisfactory and use of high concentration could also reduce tablet strength. Disintegrants are hygroscopic materials and will absorb moisture from the atmosphere, which could negatively affect the stability of moisture-sensitive drugs if the packaging does not provide adequate protection from the environment. Disintegrant activity can be affected by mixing with hydrophobic lubricants so that care needs to be taken to optimize the manufacturing

TABLE 4 Commonly Used Disintegrants

Disintegrant	Normal Concentration (%)	Comments
Starch	5–10	Probably works by wicking; swelling minimal at body temperature
Microcrystalline cellulose		Strong wicking action; loses disintegrant action when highly compressed
Insoluble ion exchange resins		Strong wicking tendencies with some swelling action
Sodium starch glycolate[a]	2–8	Free-flowing powder that swells rapidly on contact with water
Croscarmellose sodium[a]	1–5	Swells on contact with water
Gums—agar, guar, xanthan	<5	Swell on contact with water; form viscous gels that can retard dissolution, thus limiting concentration that can be used
Alginic acid, sodium alginate	4–6	Swell like gums but form less viscous gels
Crospovidone[a]	1–5	High wicking activity

[a]Often mentioned as superdisintegrant.

process as well as the formulation. If the tablet contains a high proportion of a hydrophobic drug that has a high contact angle, a wetting agent or surfactant should be added to the formulation to modify disintegration time and subsequent dissolution of the drug from tablet. The most commonly used wetting agents are sodium lauryl sulfate and the polysorbates.

Most pharmacopeias include a disintegration test which can be applied to tablets and capsules and the detailed monograph is given in the pharmacopeias (see Section 6.1.5.1).

Lubricant The use of a lubricant is essential to increase the free flow of powders and to prevent manufacturing disorders in the tablet production. The type and amount of lubricant are cautiously selected in the formulation. The order of addition and mixing time is also considered in the tableting process.

There are three types of lubricants employed in solid dosage form manufacture. The first class of lubricant is the *glidant*. The flow properties of a powder can be enhanced by the inclusion of a glidant. These are added to overcome powder cohesiveness. The two other classes of lubricant are *antiadherent* excipients, which reduce the friction between the tablet punch faces and tablet punches, and *die wall lubricant* excipients, which reduce the friction between the tablet surface and the die wall during and after compaction to enable easy ejection of the tablet. The level of a lubricant required in a tablet is formulation dependent and can be optimized using an instrumented tableting machine.

Commonly used lubricants are listed in Table 5. Talc is traditionally one of the most commonly used glidants, having the additional benefit of being an excellent antiadherent. The level of talc that can be added to a formulation is restricted by its hydrophobic nature, too high levels resulting in decreased wetting of the tablet and a subsequent reduction in the rate of dissolution. Fumed silicon dioxides are

TABLE 5 Lubricants Commonly Used in Formulations

Lubricant	Typical Percent	Comments
		Glidants
Talc	1–5	Fine, crystalline powder; Widely used as lubricant and diluent
Fumed silicon dioxide: Aerosil, Cab-O-Sil, Syloid	0.1–0.5	Has small particle size and large surface area for good flowability; used for adsorbent, antitacking agent disintegrant, and glidant
Starch	1–10	Mainly used for binder, disintegrant, and diluent but also used for glidant
Sodium lauryl sulfate	0.2–2	Anionic surfactant, luricant and wetting agent
		Boundary Lubricants
Magnesium stearate	0.2–2	Hydrophobic, variable properties between suppliers
Calcium silicate	0.5–4	Hydrophobic
Sodium stearyl fumarate	0.5–2	Less hydrophobic than metallic stearates, partially soluble
Polyethylene glycol 4000 and 6000	2–20	Soluble, poorer lubricant activity than fatty acid ester salts
Sodium lauryl sulfate	1–3	Soluble, also acts as wetting agent
Magnesium lauryl sulfate	1–3	Acts as wetting agent
Sodium benzoate	2–5	Soluble
		Fluid Lubricants
Light mineral oil	1–3	Hydrophobic, can be applied to either formulation or tooling
Hydrogenated vegetable oil	1–5	Hydrophobic, used at higher concentrations as controlled-release agents
Stearic acid	0.25–2	Hydrophobic
Glyceryl behenate	0.5–4	Hydrophobic, also used as controlled-release agent

perhaps the most effective glidants. These are materials with very small (10-nm) spherical particles that may achieve their glidant properties. They are available in a number of grades with a range of hydrophobic and hydrophilic forms and also commercially available under diverse brand names. Starch has also been used as a glidant. The use of large amounts of starch has also aided the disintegration properties.

Die-wall lubricants can be dived into two classes, *fluid and boundary lubricants*. Fluid lubricants work by separating moving surfaces completely with a layer of lubricant. These are typically mineral oils or vegetable oils, and they may be either added to the mix or applied directly to the die wall by means of wicked punches. The oily lubricants may have a mottled appearance in the tablet due to uneven distribution, poor powder flow due to their tacky nature, and reduced tablet strength. Fluid lubricants include stearic acid, mineral oils, hydrogenated vegetable oils, glyceryl behenate, paraffins, and waxes. Boundary lubricants work by forming a thin solid film at the interface of the die and the tablet. Metallic stearates are the most widely used boundary lubricants. Such lubricants should have low shear strength and form interparticulate films.

Magnesium stearate is the most widely used lubricant. The magnesium stearate used in the pharmaceutical industry is not a pure substance but a mixture of magnesium salts of fatty acids, though predominantly magnesium stearate and magnesium palmitate. Despite its popularity, which is a reflection of its excellent lubricant properties, it has some problems associated with product consistency: its effect on tablet strength and its hydrophobicity. The U.S. Pharmacopeia (USP) requires that the stearate content should account for not less than 40% of the fatty acid content of the material, and the stearate and palmitate combined should account for not less than 90%. Within this definition, there are a range of materials to be supplied as magnesium stearate. For a given formulation, it is important that a single source of magnesium stearate be used for all batches to get product reliability.

The lubricant activity of magnesium stearate is related to its readiness to form films on the die wall surface. As a result, it has two consequences: a reduction in the ability of the powder to form strong compacts and, due to its hydrophobicity, a deleterious effect on the dissolution rate of the tablets. The hydrophobic surfaces created by magnesium stearate have been shown to reduce the rate of dissolution and bioavailability of several tablet formulations. When both lubricant and disintegrant are being added to a granulated formulation, the disintegrant should be blended with the granules prior to the addition of the lubricant to minimize the risk of forming a hydrophobic film around the disintegrant.

The third class of lubricant activity is the antiadherent. Some materials have adhesive properties and can adhere to the punch surfaces during compression. This will induce tablet disorders: sticking, with a film forming on the surface of the tablets, or picking, where solid particles from the tablet stick to the punch surface. Most die wall lubricants also have antiadherent actions, and in many formulations, the addition of a specific antiadherent will not be required separately. The antiadherent includes talc, maize starch, and microcrystalline cellulose.

Coating Materials The core compressed tablet can be used by itself, but are additional coating process of the compressed tablet can be applied for several reasons: (a) protection of the drug from the environment (moisture, air, light) for stability reasons, (b) taste masking, (c) minimizing patient/operator contact with drug substance, particularly for skin sensitizers, (d) improving product identity and appearance, (e) improving ease of swallowing, (f) improving mechanical resistance and reducing abrasion and attrition during handling, and mostly (g) modifying release properties.

There are three main methods used to coat pharmaceutical tablets: *sugar coating*, *film coating*, and *compression coating*. Sugar coating has been the most commonly used method and involves coating tablets with sucrose. A sugar-coated tablet is water based and generally starts to break up in the stomach. This is a highly skilled and multistep process that is very labour intensive. This coating process results in a 50% increase of the final tablet weight and in a significant increase in tablet size. Traditionally, sugar coating has been performed in coating pans in which the tablets are tumbled in a three-dimensional direction. The pan is supplied with a source of warm air for drying and an extraction system to remove moist air and dust. The coating solution is distributed around the tablets by their tumbling action. A dusting powder may be sprinkled onto the surface of the tablets during the drying phase to prevent the tablets from sticking together. The cycle of wetting and drying is continued until the desired amount of coating has been applied to the tablets. Typically, a sugar coating will consist of three types of coats: a sealing coat, a subcoat, and a smoothing coat. Traditionally, a seal coating of shellac dissolved in ethanol or synthetic water-resistant polymers such as cellulose acetate phthalate or polyvinylacetate phthalate is used. The subcoat is an adhesive coat on which the smoothing coat of the sharp corners of the tablet can be built. The subcoat is a mixture of a sucrose solution and an adhesive gum, such as acacia or gelatin, which rapidly distributes over the tablet surface. A dusting follows each application of solution with a subcoat powder containing materials such as calcium carbonate, calcium sulfate, acacia, talc, and kaolin. The smoothing coat consists of the majority of the tablet bulk and provides the tablet with a smooth finish. A colorant is also applied if needed. The coat consists of sucrose syrup which may contain starch or calcium carbonate. The coated tablets are usually transferred to a polishing pan and coated with a beeswax–carnauba wax mixture to provide a glossy finish to the surface.

Film coating involves the application of a polymer film to the surface of the tablet, gelatin capsules, and multiparticulate systems with a negligible increase in tablet size. The method of application of the coat differs from the sugar coating in that the coating suspension is sprayed directly onto the surface of the tablets, and drying occurs as soon as the coat hits the tablet surface. The tablet only receives a small quantity of coating solution at a time. The film coat can generally be affected by the following properties: a method of atomizing the coating suspension, the ability to heat large volumes of air (which heat the tablets and facilitate the rapid drying of the applied coat), and a method of moving the tablets that ensures all tablets are evenly sprayed.

The main methods of coating are modified conventional coating pans, side-vented pans, and fluid bed coating. The side-vented pan, now the most commonly used equipment for film coating, was designed to maximize the interaction between the tablet bed and the drying air. The mixing efficiency of the table, granules, or pellets is achieved by using appropriately designed baffles on the pan surface. Fluid-bed coating offers an alternative to pan coating and is particularly popular for coating multiparticulate systems. There are three methods by which the coating can be applied: top spraying, bottom spraying, and tangential coating. The pellets, granules, or tablets being coated are suspoended in an upward stream of air, maximizing the surface available for coating. The coating is applied by an atomizer, and this is dried by the fluidizing air.

Table 6 gives commonly used polymers for film coating of core tablet. With the exception of HPMC, the polymers are rarely used alone but are combined with other polymers to optimize the film-forming properties. A polymer for film coating will ideally meet the following criteria:

1. Solubility in the solvent selected for application: Currently the organic solvent are replaced with water as a suspension system, although certain types of film coatings may require organic solvents to be used. Commonly used solvents include alcohols (methanol, ethanol, and isopropanol), esters (ethyl acetate and ethyl lactate), ketones (acetone), and chlorinated hydrocarbons (dichloromethane and trichloroethane).

TABLE 6 Commonly Used Polymers for Film Coating of Core Tablet

Polymer	Comments
Methylcellulose (MC)	Soluble in cold water, GI fluids, and a range of organic solvents
Ethylcellulose (EC)	Soluble in organic solvents, insoluble in water and GI fluids; used alone in modified-release formulations and in combination with water-soluble cellulose for immediate-release formulations
Hydroxyethylcellulose (HEC)	Soluble in water and GI fluids
Methyl hydroxyethylcellulose (MHEC)	Soluble in water and GI fluids; has similar film-forming properties to HPMC but is less soluble in organic solvents, which limited its popularity when solvent coating was the norm
Hydroxypropyl cellulose (HPC)	Soluble in cold water, GI fluids, and polar solvents; becomes tacky when dried, so is unsuitable for use alone, often used in combination with other polymers to optimize adhesion of coat
Hydroxypropyl methylcellulose (HPMC)	Soluble in cold water, GI fluids, alcohols, and halogenated hydrocarbons; excellent film former and the most widely used polymer; can be used with lactose to improve adhesiveness
Sodium carboxymethylcellulose (NaCMC)	Soluble in water and polar solvents

2. Solubility in GI fluids: The solubility of polymers is dependent on its physicochemical nature and pH. Unless the coating is being applied for enteric coating, it should ideally be soluble across the range of pH values encountered in the GI tract.
3. Capacity to produce an elegant film even in the presence of additives such as plasticizers, pigments, and colorants.
4. Compatibility with film-coating additives and the tablet being coated.
5. Stability in the environment under normal storage conditions.
6. Freedom from undesirable taste or odor.
7. Lack of toxicity.

Enteric coating materials are also used to prevent release of the drug substance in the stomach if the drug is either an irritant to the gastric mucosa or unstable in gastric juice. Table 7 lists enteric coating polymers commonly used in tablet formulations. The choice of of enteric coating material depends on its solubility.

The third type of tablet coating is multiple-compression coating to make a bilayered, multilayered tablet (a layered tablet of two drugs) or a tablet within a tablet (a core of one drug and a shell of another). The multilayered tablet is prepared by initial compaction of a portion of the fill materials in a die followed by additional fill material and compression, depending on the number of fill materials. The layered tablet can provide some advantages. Each layer has different drug in a separate layer. The incompatible drug can be compressed simultaneously at different layers.

TABLE 7 Enteric Coating Polymers Commonly Used in Tablet Formulations

Polymer	Solubility Profile	Comments
Shellac	Above pH 7	Original enteric coating material, originally used in sugar-coated tablets; high pH required for dissolution may delay drug release; natural product which exhibits batch-to-batch variability
Cellulose acetate phthalate (CAP)	Above pH 6	High pH required for dissolution a disadvantage; forms brittle films, so must be combined with other polymers
Polyvinylacetate phthalate (PVAP)	Above pH 5	—
Hydroxypropyl methylcellulose phthalate (HPMCP)	Above pH 4.5	Optimal dissolution profile for enteric coating
Polymers of methacrylic acid and its esters	Various grades available with dissolution occurring above pH 6	—

The staged release and improved appearance from the layered tablet are also possible. In the preparation of a tablet within a tablet, a special tableting machine is required to place the intended core tablet precisely within the die, which already contains some of the coating formulation to surround the fill materials. This multilayered tablet or tablet within a tablet is also useful to get the modified release, either immediately or in a sustained manner.

Auxiliary Excipients Plasticizers are added during the tablet coating process. The film coating process involves two important stages: droplet formation and film formation. Film formation is a multistage process that involves wetting the tablet surface followed by spreading the film and eventually coalescence of the individual film particles into a continuous film. Most film-forming polymers have glass transition temperatures in excess of the temperatures reached during the coating process (typically 40–50 °C), so it is necessary to add plasticizers to the formulations which reduce the glass transition temperature. The choice of plasticizer is dependent on the type of polymer and its permanence and compatibility. Permanence is the duration of the plasticizer effect; the plasticizer should remain within the polymer film to retain its effect, so it should have a low vapor pressure and diffusion rate. Compatibility requires the plasticizer to be miscible with the polymer. Commonly used plasticizers include phthalate esters, citrate esters, triacetin, propylene glycol, polyethylene glycols (PEGs), and glycerol.

The pigments or opacifiers are also combined with the coating solution to get colored tablets. Insoluble pigments are normally preferred to soluble dyes for a number of reasons. Solid pigments produce a more opaque coat than dyes, protecting the tablet from light. The presence of insoluble particles in the suspension allows the rate of solid application to the tablet to be increased without having an adverse effect on the viscosity of the coating suspension, improving productivity.

6.1.2.4 Selection Guideline of Pharmaceutical Excipients in Tablet Formulation

In the proper selection of pharmaceutical excipients for the formulation, there are numerous factors to be considered [7]. The type of excipient is highly dependent on the model compound and its intended dosage form. The preparation method of the dosage form and proper dosage regime are also factors.

The formulation parameters of the tablet are essential. The amount and type of excipients relative to drug contents should be considered to determine the size of the dosage form. The bulk density of excipients and drug and their filling doses in the die are important factors to be considered in the tablet formulation. Uniform distribution of drug or excipient is in trouble if its contents are low in the formulation, while the use of high contents are difficult to fill in the die and the tablet size is also larger. Excipients that are potentially incompatible with the drug should be avoided. The amount and type of reactive impurities in the excipients should also be established. Batch-to-batch and supplier-to-supplier variation in impurity levels is possible. Excipients with potential adverse interaction and any unwanted incompatibility with drug should be avoided in the formulation. For an ideal formulation, it is helpful to consider these parameters as well as drug–excipient incompatibility.

The global harmonization and standardization of pharmaceutical excipients are nowadays necessary in the formulation studies [3,7]. Since all excipients in the market worldwide are supplied from many countries or companies with facilities in more than a single country, the quality of the excipients must be well documented and validated. Otherwise, the quality of the excipients should be varied batch to batch, factory to factory, and country to country. There are also hundreds of different brands and grades of excipients available, but it would be unrealistic for the formulator to expect to have a totally free choice. Most manufacturing companies select the excipients used in their factory on the basis of cost, availability, and performance. To establish the equivalency of excipients obtained from different sources, it is also necessary to perform some kind of functionality testing. Since the sources, origin, and manufacturer of the excipients are different, regulatory approval for the product is required in each country. The standards for each drug substance and excipient are contained in pharmacopeia. The four pharmacopeia with the largest international use are the United States Pharmacopeia (USP) and National Formulary (NF), British Pharmacopoeia (BP), European Pharmacopoeia (EP), and Japanese Pharmacopoeia (JP).

Unless global harmonization is established, analytical methods, testing criteria, and specification limits must be variable according to monographs of pharmacopeia from the different countries. For example, Table 8 lists the different specification limits for the viscosity of cellulose ether among three pharmacopeias. If global harmonization and standardization of pharmaceutical excipients are established, the marketing and sales of a single formulation are more facilitated worldwide. The manufacturing cost and research-and-development (R&D) cycles are also reduced. Most of all, the quality and bioequivalence of the drug products with the same formulation can be more validated since the regulatory approval of the pharmaceutical product is enhanced. The global harmonization is an ongoing effort by corporate representatives and international regulatory authorities.

TABLE 8 Specification Limits for Viscosity of Cellulose Ether

Parameter	EP	USP XXII	JP XII
Concentration of solution, %	2.0	2.0	2.0
Temperature, °C	20 ± 0.1	20 ± 0.1	20 ± 0.1
Type of viscometer	Rotating viscometer	Capillary type, Ubbelohde	Capillary type, Ubbelohde
Shear rate, S^{-1}	10	—	—
Unit	cP (Pa·s)	cP (Pa·s)	cSt
Sample preparation[a]			

[a]Very similar in all three pharmacopeia.

TABLE 9 Comparative Properties of Some Directly Compressible Fillers[a]

Filler	Compactibility	Flowability	Solubility	Disintegration	Hygroscopicity	Lubricity	Stability
Dextrose	3	2	4	2	1	2	3
Spray dried lactose	3	5	4	3	1	2	4
Fast-Flo lactose	4	4	4	4	1	2	4
Anhydrous lactose	2	3	4	4	5	2	4
Emdex (dextrates)	5	4	5	3	1	2	3
Sucrose	4	3	5	4	4	1	4
Starch	2	1	0	4	3	3	3
Starch 1500	3	2	2	4	3	2	4
Dicalcium phosphate	3	4	1	2	1	2	5
Avicel (MCC)	5	1	0	2	2	4	5

[a]Graded on a scale from 5 (good/high) to 1 (poor/low); 0 means none.

In the selection of excipients, formulation scientists should also remember that every excipient is not limited by a single function but rather can be used for many pharmaceutical applications in dosage forms. For example, cellulose and its derivatives (HPMC, EC, HPC) have been widely used as fillers, binders, suspending agents, and mainly controlled-release agents. Povidone, polymethacrylate, and cyclodextrins are also multifunctional excipients in formulations. Table 9 gives the physical properties of some directly compressible fillers. The grades of physical properties are variable among fillers used in tablet formulation. The simple and optimal formulation of drugs with multifunctional excipients can provide some advantages in tablet preparations.

6.1.3 DRUG–EXCIPIENT COMPATIBILITY

The potential for excipients to cause chemical and physical instability in drugs has been recognized for over 30 years. Drug compatibility studies have been used as an approach for accepting/rejecting excipients for use in pharmaceutical formulations [1, 10–12]. In general, drug stability can be investigated under the stress condition according to the guideline of accelerated stability testing. The factors for stress condition usually include temperature, pH, light, moisture, agitation, gravity, packaging, and method of manufacture [13, 14]. Despite this fact, approaches for excipient selection in drug formulation are often empirical. The stability issues in the development of a drug must be considered at the early and late formulation stages. Nowadays, advanced analytical instrumentation makes it possible to more rapidly identify potential excipient-induced instability to select excipients. Excipients that exhibit incompatibility with the drug are "rejected" and not included in subsequent tablet formulation studies.

6.1.3.1 Experimental Studies for Drug–Excipients Compatibility

Compatibility studies are carried out by mixing drug with one or more excipients under some type of stress condition. It has also been suggested that aqueous suspensions of the drug and excipients or drug–excipient complexes provide a better model for tablet formulations. It has been recommended that small-scale formulations using the selected excipients (this may eventually be used for the eventual formulation) be prepared using experimental processes.

Table 10 provides examples of binary and factorial design for drug–excipient compatibility studies. The two conditions of each potential parameter can be included as a binary system. All potential parameters can also be combined for factorial design. For example, water is added to the drug–excipient mixture at reasonable temperatures (50 °C or less) and then both intact drug and degradation products are measured. The high-temperature and high-humidity conditions are usually used to obtain more rapid stability assessment of drug and excipients. Generally, physicochemical properties such as drug content, color, taste, related substances, thermal analysis, and high-performance liquid chromatography (PLC) studies of evaluated for drug–excipient compatibility studies using conventional analytical techniques.

6.1.3.2 Analytical Methods for Drug–Excipient Incompatibility

The key to the early assessment of instability in formulations is the availability of analytical methods to detect low levels of degradation products, generally less than 2%. With the aid of thermal analysis and chromatographic methods [HPLC and

TABLE 10 Examples of Binary and Factorial Designs for Drug-Excipient Compatibility Studies

Binary Design	Factorial Design
Drug–excipient ratio 1:1 or 1:10	Total number of formulation = $a \times b \times c \times d \times e$
Water addition = yes/no	Type of drug–excipient (a)
Storing temperature 25 °C or 40 °C/75%	Amount of drug–excipient (b)
Storing time 1 or 4 weeks	Water addition (c)
	Storing temperature (d)
	Storing time (e)

liquid chromatography/mass spectrometry (LC/MS)], it can assign at least tentative structures for most degradation products as well as the intact drug contents [15]. Excipient compatibility screening must provide more rapid identification of excipient-mediated instability that is detected in complete formulations.

Early compatibility studies relied on color change as an indication of incompatibility. Subsequently, there were numerous reports on the use of thermal analysis techniques such as differential thermal analysis (DTA) or differential scanning calorimetry (DSC) to detect drug–excipient incompatibilities. DTA and DSC have the advantage of rapid analysis. Generally, formation of new peaks and disappearance of drug peak by the endothermic or exothermic reaction are carefully investigated. Diffuse reflectance spectroscopy (DRS) can be also used to determine the ultraviolet (UV) absorption of drug on the surface of drug–excipient mixtures.

6.1.3.3 Reaction Types and Stabilization Guidelines

Understanding the degradation chemistry of drug with excipients is essential to select proper excipients in the formulation stages [16, pp 101–151]. Drug-excipient compatibility studies are crucial to decide optimal tablet formulation and to understand the possible mechanism in many cases [10, 12, 14]. Drug instability occurs by three types of reactions: hydrolysis, oxidation, and aldehyde–amine addition. Table 11 gives reaction types of chemical and physical instability.

Hydrolysis is the most common mechanism to induce drug–excipient incompatibility. Most hydrolysis reactions are catalyzed by acids and/or bases. The microscopic pH between drug and excipients is critical. The pH degradation rate profiles are good predictors for hydrolysis reactions in solid dosage forms. The pH of optimal stability in solution is similar to the "microscopic pH" in solid dosage forms. A shift

TABLE 11 Reaction Types of Chemical and Physical Instability

Type of Instability	Order of Frequency
Chemical instability	
pH-dependent hydrolysis	12
Oxidation	
Air oxidation	5
Metal-catalyzed oxidation	3
Peroxides in excipients	2
Aldehyde–amine addition	
Aldose excipient	2
Formaldehyde from excipients	1
Michael addition with maleate salt	1
Intramolecular cyclization	1
Dimerization (Diels–Alder)	1
Racemization	1
Addition of ammonia residue from gelatin capsule	1
Esterification	1
Physical instability	
Evaporation of volatile free base	2
Reaction of methanesulfonic acid with disintegrants	1
Gelatin cross-linking by excipient impurities	1

in only one pH unit can increase or decrease the reaction rate by a factor of 10. Although solid-state compatibility studies are commonly used, solution kinetic studies are generally conducted to select excipients since it is possible to study all three of these reactions in solution. Often solution kinetic studies allow the identification of more potential degradation products and hence the development of better stability-indicating assays. The pH of optimal stability is also important for selecting the appropriate salt form of the compound and excipients in the solid formulation. Acetylsalicylic acid (ASA) is a readily hydrolyzable drug and there are literature references to its rate of hydrolysis in aqueous solution.

The moisture content of the drug and excipients plays a critical role in their incompatibility by hydrolysis. Excipients such as starch and povidone have particularly high water contents (povidone contains about 28% equilibrium moisture at 75% relative humidity), which can increase the possibility of drug contact. Magnesium trisilicate causes increased hydrolysis of aspirin in tablet form because, it is thought, of its high water content. For these reasons, some scientists recommend inclusion of water in the samples for compatibility studies. Depending on the degree of hydrolytic susceptibility, different approaches for tablet formulation can be used to minimize hydrolysis. For compounds such as ASA that are readily hydrolyzable, direct compression or dry granulation is preferable rather than wet granulation. However, drug–excipient incompatibility still occurrs. Chemical interaction between moieties of drug and excipients may lead to increased decomposition. The trans-acetylation reaction between aspirin and paracetamol and also possible direct hydrolysis of the paracetamol can occur. The amount of free salicyclic acid at 37 °C in the tablets containing paracetamol increases by the addition of talc (0.5–1%). The stearate salts should be avoided as tablet lubricants if the active component is subject to hydroxide ion–catalyzed degradation. The degradative effect of the alkali stearates is inhibited in the presence of malic, hexamic, or maleic acid.

As a general rule, in selecting excipients, it is probably best to avoid hygroscopic excipients when formulating hydrolytically labile compounds. One of the most effective ways to stabilize a pH-sensitive drug is through adjustment of the microscopic pH of the formulation. Excipients with high pH stability and buffering agents are recommended. The equilibrium moisture content (hygroscopicity) at different relative humidities for a variety of drug and excipients would be a clue to selecting optimal formulations. If dry processing and the use of nonhygroscopic excipients still result in unacceptable rates of hydrolysis, use of a dessicant and/or moisture protective packaging can further increase stability against drug hydrolysis. The manufacturing process should be conducted under low-humidity conditions and not during the hot summer season to improve drug–excipient compatibility.

Oxidation reactions are complex and it is difficult to understand the reaction mechanism. The best approach is to avoid excipients containing oxidative reactants such as peroxides and metal ions. The air oxidation or metal ion–catalyzed oxidation can be tested after storing the samples in the solutions. The impurities in excipients such as povidone or as degradation products in PEGs are organic peroxides and are typically more reactive than hydrogen peroxide. A commercially available organic peroxide such as *tert*-butyl hydroperoxide is better to evaluate the susceptibility of a compound to peroxide oxidation rather than the hydrogen peroxide. Reactive impurities such as peroxides and ionic chemicals (talc and titanium oxide) in the excipients commonly may act as catalysts for oxidation of the drug.

Therefore, it is essential to remove any oxidative peroxides and ionic chemicals. Use of free-radical scavengers such as butylated hydroxyanisole (BHA) or butylated hydroxytoluene (BHT) can stabilize the oxidation reaction of the drug with excipients via a free-radical mechanism. However, uniform distribution of the low levels of these antioxidants in solid formulations is quite difficult. The BHA is added to both lovastatin and simvastatin tablets. If an oxidation reaction is catalyzed by metal ions or excipients with high transition metal ion contents (talc), chelating agents can also be used to bind trace metal ions. However, citrate is often the agent of choice because of the potential toxicity of many chelating agents. Oxidation is less likely to occur if the oxidizable group of phenols/catechols and secondary and tertiary amines is protonated. If some oxidation reactions are catalyzed by light, the formulation can be stabilized by a light-absorbing coating.

Aldehyde–amine addition is also a typical reaction type. The potential aldehyde interaction also makes sense to screen out a compound as part of the preformulation evaluation. All aldoses such as lactose or excipients such as starch and microcrystalline cellulose that have terminal aldose groups should be avoided with the excipients having primary/secondary amines. 5-(Hydroxymethyl)-2-furaldehyde (HMF) would be a good choice since it is a degradation product of many sugars (lactose) and celluloses, at least in trace levels. The Schiff's base[–CH=N–] is formed when the sugar aldehyde (lactone) and the primary/secondary amines are mixed. The isoniazide and trace level of HMF from lactose can also readily form the Schiff's base. It has been reported that the reaction of hydrazine hydrochloride and starch to form high-molecular-weight addition products gives high-molecular-weight products. The reaction of fluoxetine hydrochloride is more rapid with spray-dried lactose monohydrate.

Many excipients are acids or bases or have acidic or basic impurities. For this reason, the reaction of amines with aldehydes requires that the amine exist in the nucleophilic-free base, rather than the protonated cationic form. For example, the reaction of fluoxetine hydrochloride with lactose was much more rapid as potassium hydroxide was added to neutralize the hydrochloride salt. Michael addition between seproxetine and maleic acid from a tablet formulation following dissociation of the hydrochloride salt occurs because of the conversion of a salt to a free acid/base.

Finally, the physical instability that can occur is the cross-linking of gelatin or its capsules. Low levels of aldehyde impurities in excipients (starch, polysorbate 80, PEG, and rayon coils) and packaging materials have been reported to cause cross-linking of gelatin. Dissolution slowing is also more pronounced for wet granulation tablets than direct-compression tablets. Dissolution slowing appears to be due to hardening of the water-soluble excipients and a reduction of disintegration time.

Once suspect excipients have been eliminated from consideration, small-scale formulations with manufacturing processes such as granulation, drying, and compression can be used to assess whether interactions between excipients or processing conditions result in any unpredicted instability. The ratio of drug and excipients is also a critical factor to consider in incompatibility studies. In addition, the lower the drug dose, the greater the possibility of degradation by low-level impurities in the excipients.

In conclusion, drug–excipient compatibility studies have a key role at the early preformulation stages to select excipients or after formulation to help identify the mechanism of any detected instability [14]. An understanding of the potential physicochemical interactions of drug with known chemical reactivities of excipients and their impurities will aid in the proper selection of excipients. Knowledge of trace impurities/additives in excipients and the consistency of these levels from batch to batch and supplies to supplier is also essential to select proper pharmaceutical excipients. Although the best solution kinetics and drug–excipient compatibility studies may be established, the possibility of unexpected instability because tablets and capsules are complex multicomponent systems must be remembered.

6.1.4 POWDER CHARACTERISTICS

Most drug and inactive excipients used in tablet formulation are in the solid state as amorphous powder or crystals of various morphological structures. There may be substantial differences in particle size, surface area, crystal morphology, wetting, and flowability as well as many physical properties of drug, excipients, and their blends [16]. Table 12 describes common micromeritic topics important to pharmaceutical preformulation.

Before their use in the solid dosage forms, it is necessary to understand and characterize the physical and chemical properties of drug, excipient, and their powder mixtures, including crystal habit, particles size, shape, flow characteristics, density, hygroscopicity, and compressibility and compaction [2, 3]. Hiestand noted that successful tableting operations require the selection of excipients that balance desirable physical, flow, and mechanical properties for tablet manufacturing [17]. The quantification of these properies using a unified approach is essential to the design and potimization of solid dosage formulations [1]. Instrumental analyses such as scanning electron microscopy (SEM), DSC, and powder X-ray diffraction (PXRD) can be very useful to characterize powder properties such as purity, polymorphism, salvation, degradation, drug–excipient compatibility, and other desirable characteristics.

6.1.4.1 Crystal Form and Habit

The morphology of a pharmaceutical solid is of importance since this property can influence the bulk powder properties. The six crystal systems are cubic, hexagonal, tetragonal, orthorhombic, monoclinic, and triclinic crystals. The observed overall

TABLE 12 Micromeritic Topics Important to Pharmaceutical Preformulation

Particle shape
Particle size distribution
Solid geometry (packing, density, porosity, void)
Surface characteristics (adsorption, area, surface energy, solubility)
Methods of determination
Chemical stability
Dynamics (flow rate, transport)
Particle separation
Processing (sieving, sedimentation, grinding, mixing, compaction)
Sampling
Drug release

shape of crystal habits comprises plate, tabular, equant, columnar, blade, and avicular as well as dendrites, treelike pattern, and spherulites, tiny crystals radiating from a center [2, 18, 19]. Crystal form (crystal habit) as well as a noncrystalline amorphous form may affect drug stability, dissolution rate, flow, mechanical properties, and ability to mix with excipients. The amorphous form of a drug has the lowest melting point and usually the fastest dissolution rate, but it is most likely to react or degrade. Mechanical properties such as flowability, miscibility, particle strength, and cohesiveness often vary among the polymorphs, crystal shapes, or habits. Cohesiveness, the surface free energy effect that results in particle aggregation, may also be different for various polymorphs and habits.

6.1.4.2 Particle Size Distribution

The determination and control of particle size distribution are often very important in pharmaceutical preformlation since the drug safety, stability, and viability of the dosage form and manufacturing process can be significantly influenced [19]. Particle shape is also important in determining particle size. The particle size of materials is readily expressed in terms of its diameter according to the definition of particle size. As the degree of asymmetry of particles increases, however, so does the difficulty of expressing size in terms of a meaningful diameter [20]. Figure 4 shows different ways of defining a diameter. The "diameter" can be simply the longest or the shortest linear dimension of the crystal. If one can calculate the area of the particle, one may obtain the diameter d_a, which is the diameter of a circle with the same area as the particle.

If all the particles in a sample are of the same size, then the powder is monodisperse. Truly, the particles have more than one size in polydisperse samples. The monodisperse particle size distribution is more desirable than the polydisperse one. Therefore, the shape and surface area of the individual particles, the size ranges based on number or weight of particles, as well as the total surface area are variable. The commonly illustrated particle forms are sphere, rod, fiber, granular, cubical, flake, condensation floc, and aggregate.

The fineness of the powder is characterized by a number (e.g., a diameter d). Particles, of course, will have different shapes so that there are different ways of defining a diameter. The technique for obtaining d_a given above has been used microscopically. More conventional is the so-called surface mean diameter, which, is the diameter of a sphere that has the same surface area as the particle. The so-called single-particle volume mean diameter is possible if there are instruments that can measure the volume of an odd-shaped particle. If the shape factor is indepen-

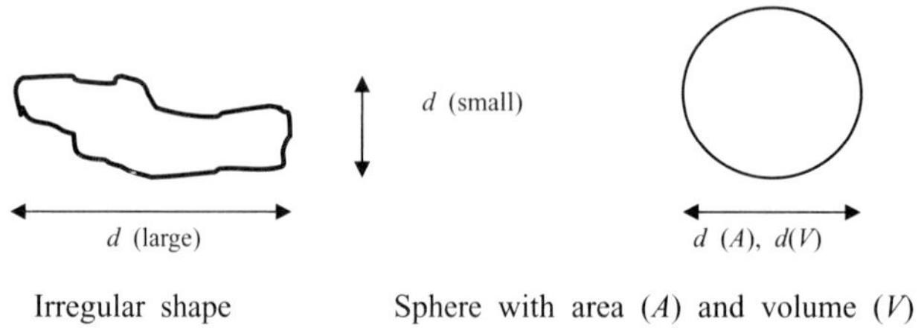

FIGURE 4 Different ways of defining a diameter.

TABLE 13 Particle Size Distribution of Powdered Cornstarch Samples

Particle Size Range (μm)	Number of Occurrences (*n*)	Percent Frequency	Cumulative Frequency (%)
25–27	1	8.3	8.3
27–29	3	25.0	33.3
29–31	4	33.3 (33.4)	66.7
31–33	3	25	91.7
33–35	1	8.3	100
Total	12	99.9 (100)	—

dent of the size of the particle, then the shape is called isometric. Examples of isometric shapes are cube, sphere, and cylinder.

Suppose a sample of 12 particles of corn starch with fairly narrow particle size distribution (25–35 μm) were measured. In this case, the distribution of particles is very narrow and approximately distributed as normal or Gaussian pattern. To convert these numbers into frequencies, it is noted that there are 12 particles in total; that is, dividing each number by 12 and multiplying this by 100 will give the percent frequency, as shown in Table 13.

If the frequency is plotted as a function of the midpoint of the diameter ranges, then a frequency histogram is obtained. It is noted that for the noncumulative curve the midpoints of the intervals are used, but for the cumulative curve the interval endpoint is used. When the number or weight of particles lying within a certain size range is plotted against the size range or mean particle size, a *frequency distribution curve* is obtained However, due to the deviation from the normal distribution of particles, a lognormal distribution of the particle size is statistically plotted against the cumulative percent frequency on a probability scale, and a linear relationship is observed (Figure 5). Probability paper is a type of paper that straightens out this type of S-shaped curve. The logarithm of the particle size is equivalent to 50% on the probability scale, that is, the 50% size is known as the *geometric mean diameter* and the slope is given by the *geometric standard deviation* σ_g, which is also the quotient of the ratio: (84% undersize or 16% oversize)/(50% size) or (50% size)/(16% undersize or 84% oversize).

Many methods are available for determining particle size in pharmaceutical practice, including microscopy, sieving, sedimentation, and determination of particle volume [19]. Sieve analysis with U.S. standard sieves is widely used to determine the particle size distribution based on powder weight. Sieves are classified according to the number of openings (Table 14) and are generally made of wire cloth woven from brass, bronze, or other suitable wire.

The USP uses descriptive terms to characterize the particle size of a given powder, which are related to the proportion of powder that is capable of passing through the openings of standardized sieves of varying dimensions in a specified time period under the mechanical sieve shaker as follows:

Coarse (or a no. 20) powder: All particles pass through a no. 20 sieve and not more than 40% through a no. 60 sieve.

Moderately coarse (or a no. 40) powder: All particles pass through a no. 40 sieve and not more than 40% through a no. 60 sieve.

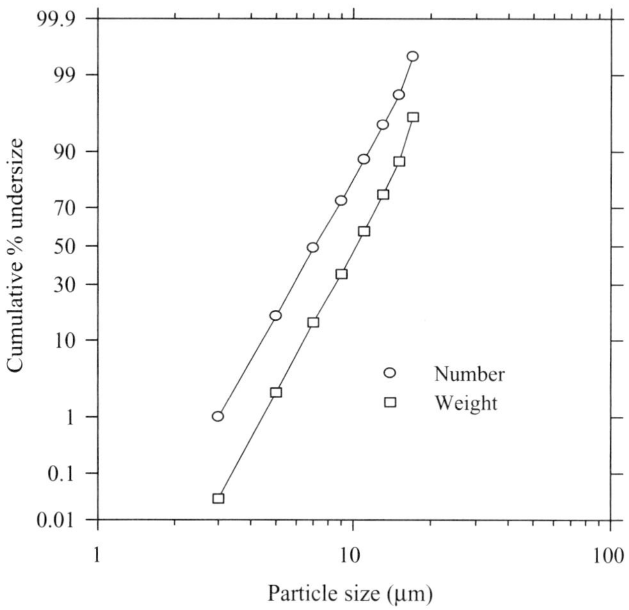

FIGURE 5 Typical lognormal distribution of particles based on weight and number.

TABLE 14 Openings of Standard Sieves

Sieve Number	Sieve Opening
2	9.5 mm
3.5	5.6 mm
4	4.75 mm
8	2.36 mm
10	2.00 mm
20	850 μm
30	600 μm
40	425 μm
50	300 μm
60	250 μm
70	212 μm
80	180 μm
100	150 μm
120	125 μm
200	75 μm
230	63 μm
270	53 μm
325	45 μm
400	38 μm

Fine (or a no. 80) powder: All particles pass through a no. 80 sieve. There is no limit as to greater fineness.

Very fine (or a no. 120) powder: All particles pass through a no. 120 sieve. There is no limit as to greater fineness.

Microscopy, in which the particles are sized through the use of a calibrated grid background or other measuring devices (range 0.2–100 μm). SEM can also readily measure the smallest particle size. The microscope allows the observer to view the actual particles, but it gives two-dimensional views. With the sedimentation rate, particle size is determined by measuring the terminal settling velocity of particles through a liquid medium in a gravitational or centrifugal environment (range 0.8–300 μm). Sedimentation rate may be calculated from the well-known Stokes equation. The sedimentation methods yield a particle size relative to the rate at which particles settle through a suspending medium, a measurement important in the development of emulsions and suspensions. With light energy diffraction (light scattering), particle size is determined by the reduction in light reaching the sensor as the particle, dispersed in a liquid or gas, passes through the sensing zone (range 0.2–500 μm). On the other hand, laser scattering utilizes a H_3–Ne laser, silicon photodiode detectors, and an ultrasonic probe for particle dispersion (range 0.02–2000 μm). The measurement of particle volume using a Coulter counter allows one to calculate an equivalent volume diameter, but no information on shape of the particles is available. Laser holography, in which a pulsed laser is fired through an aerosolized particle spray and photographed in three dimensions with a holographic camera, allows the particles to be individually imaged and sized (range 1.4–100 μm). The above methods and others may be combined to provide greater assurance of size and shape parameters. Automated particle size analyzers linked with computers are commercially available for data processing, distribution analysis, and printout.

Determination and control of particle size are often prerequisites in preformulation stages because the size distribution of excipients, drug, and their mixtures can influence safety, efficacy, stability, viability of dosage form, and manufacturing processes. Furthermore, the particle size of pharmaceuticals can affect uniform mixing, flow characteristics, formulation characteristics, dose-to-dose content uniformity, dissolution rate, and bioavailability of drug. Tablet characteristics such as porosity and flowability are highly affected by the particle size as well. The smallest particles induce electrostatic forces and aggregations while the larger particles show greater weight variations. The ideal size ranges of particles are usually 10–150 μm. Therefore, detailed information of the particle size of drug, excipients, and their blends should be required in tablet formulation as well as regulatory issues. Particle size is also important in the tableting field, since it can be very important for good homogeneity in the final tablet. The particle size should be consistent throughout the production to satisfy table formulation and regulatory demands.

6.1.4.3 Flow Characteristics

Good flow properties are a prerequisite for the successful manufacture of both tablets and powder-filled hard gelatin capsules. Proper fluidity of materials is required to be transported through the hopper of a tableting machine. The elongated particle shape and small particle size could cause high tablet weight variation, strength, unacceptable blend uniformity, and difficulty in filling containers and dies. Excipients with good flow characteristics and low cohesive powders should be more preferable in tablet production. Powder flow is affected by the numerous parameters, including purity, crystallinity, electrostatic forces, mechanical properties (brittleness, elasticity), density, size, shape, surface area, moisture content, direction and rate of shear, storage container dimension, and particle–wall interaction [19].

It is a property of all powders to resist the differential movement between particles when subjected to external stresses. A bulk powder is somewhat analogous to a non-Newtonian liquid, which exhibits plastic flow and sometimes dilatancy if the particles being influenced by attractive forces. Accordingly, powders may be free flowing or cohesive ("sticky"). The resistance is to free flow is due to the cohesive forces between particles [18]. Three principal types of interparticular forces are forces due to electrostatic changing, van der Waals forces, and forces due to moisture. Electrostatic forces are dependent on the nature of the particles, in particular their conductivity. Van der Waals forces are the most important forces for most pharmaceutical powders. These forces are inversely proportional to the square of the distance between the two particles and hence diminish rapidly as particle size and separation increase. Powders with particles below 50 μm will generally exhibit irregular or no flow due to van der Waals forces. Particle shape is also important; for example, the force between a sphere and a plane surface is about twice that between two equal-sized spheres. At low relative humidity, moisture produces a layer of adsorbed vapor on the surface of particles. Above a critical humidity, typically in the range 65–80%, it will form water liquid bridges between particles. Where a liquid bridge forms, it will give rise to an attractive force between the particles due to surface tension or capillary forces. The role of the formulator is to ensure that the flow properties of the powder are sufficient to enable its use on modern pharmaceutical equipment, powder hoppers, and flow through orifices in the tablet production.

It is important that the powder flows from the hopper to the filling station of the tablet machine at an appropriate rate and without segregation occurring. There are two types of flow that can occur from a powder hopper: *core flow* and *mass flow* [2]. Figure 6 shows the two different powder flow patterns in hoppers. When a small amount of powder is allowed to leave the hopper, there is a defined region in which downward movement takes place and the top surface begins to fall in the center. A core flow hopper is characterized by the existence of dead spaces during discharge. A mass flow hopper is one in which all the material is in motion during discharge, in particular the areas adjacent to the hopper wall. As a small amount of powder is discharged, the whole bulk of the powder will move downward. Whether core flow or mass flow is achieved is dependent on the design of the hopper (geometry and wall material) and the flow properties of the powder.

Powder flow into orifices is also important when filling dies in tablet machines and in certain types of capsule-filling machines. For a given material, the flow into or through an orifice is dependent on the particle size (Figure 7). In general, as the particle size increases, the powder flow rate also increases. However, there is practically no flow if the particle size is below 50 μm or above 1200 μm. The Carr index gives us the guidance for powder flowability. A lower Carr index of excipients is more desirable for acceptable powder flow. At the lower end of the particle size range, cohesive forces will result in poor flow. Powders with particles below 50 μm will generally exhibit irregular or no flow due to van der Waals forces. As the particle size increases, the flow rate increases until a maximum is achieved, at an orifice diameter–particle diameter ratio of 20–30. As the particle size continues to increase,

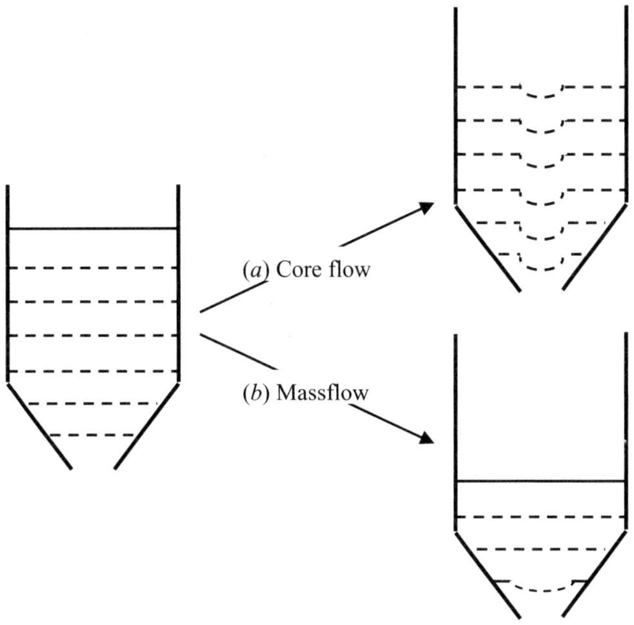

FIGURE 6 Powder flow patterns in hoppers.

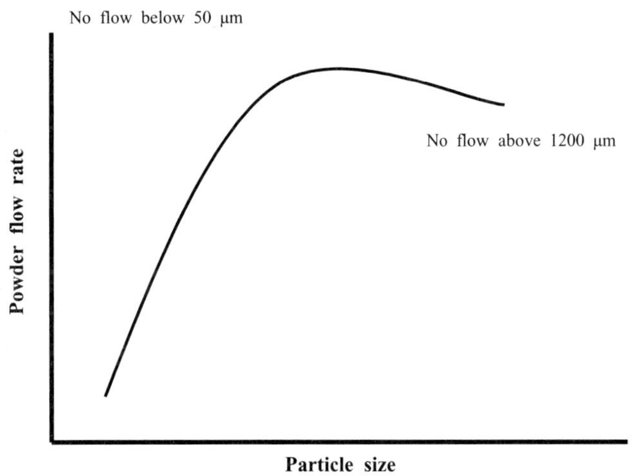

FIGURE 7 Effect of particle size on rate of powder flow through orifice.

the rate decreases due to mechanical blocking or obstruction of the orifice. Flow will stop completely when the orifice–particle ratio falls below 6 and if the size is above 1200 μm.

There are several different methods available for determining the flow properties of powders. *Shear cell methods* provide an assessment of powder flow properties as a function of consolidation load and time. There are a number of types of shear cells available, the most common being the Jenike shear cell [21].

Common indices of flowability are the *Hausner ratio* and the *Carr index* (compressibility). The increase in bulk density of a powder is related to the cohesiveness of a powder. So measurement of the bulk density of a powder is essential to define the flow characteristics. Ratios of poured-to-tapped bulk densities are expressed in two ways to give indices of flowability:

$$\text{Hausner ratio} = \frac{\text{tapped bulk density}}{\text{poured bulk density}}$$

$$\text{Carr index (compressibility)} = \frac{100 \times (\text{tapped bulkdensity} - \text{poured bulk density})}{\text{poured bulk density}}$$

The Hausner ratio varies from about 1.2 for a free-flowing powder to 1.6 for cohesive powders. Carr index classifications for flowability [2] are listed in Table 15.

Compressibility indices are a measure of the tendency for arch formation and the ease with which the arches will fail and, as such, is a useful measure of flow. A limitation of the bulk density indices for flow characteristics is that they only measure the degree of consolidation; they do not describe how rapidly consolidation occurs. *Angle of repose* is a common method used to measure powder flow with small sample quantity. If powder is poured from a funnel onto a horizontal surface, it will form a cone due to gravitational forces. The angle between the sides of the cone and the horizontal is referred to as the angle of repose. So there is a correlation between powder flow and angle of repose. The relationship between the Carr index and angle of repose is now discussed. The angle of repose is a measure of the cohesiveness of the powder, as it represents the point at which the interparticle attraction exceeds the gravitational pull on a particle. A free-following powder will form a cone with shallow sides, and hence a low angle of repose, while a cohesive powder will form a cone with steeper sides. As a rough guide, angles less than 30° are usually indicative of good flow, while powders with angles grater than 40° are likely to be problematic.

The *avalanching behavior* of powder is also a measure of flowability. If a powder is rotated in a vertical disc, the cohesion between the particles and the adhesion of the powder to the surface of the disc will lead to the powder following the direction of rotation until it reaches an unstable situation where an avalanche will occur. After the avalanche, the powder will again follow the disc prior to a further avalanche.

TABLE 15 Carr Index Classification and Powder Flowability

Carr Index (%)	Flow
5–12	Free flowing
12–16	Good
18–21	Fair
23–35	Poor
33–38	Very poor
>40	Extremely poor

Measurement of the time between avalanches and the variability in time is a measure of the flow properties of the powder.

If a powder flows poorly, the vibrator can be used, but it also causes powder segregation and stratification. The addition of glidant (occasionally lubricant) in the powder mixtures can readily increase flowability at the low concerntration. Talc or fumed silicon dioxide is an example of a glidant. If this is not sufficient to improve the flow, other means of flow improvement are necessary. There are two main factors that affect powder flow: particle size and particle shape. The more spherical a particle is, the better it flows. Small particles are very cohesive and hence do not flow well, but increasing the particle size will improve flow. With the aid of spray drying or spheronizers, particles become spherical.

In general, powder below 50 μm is not very free flowing because the cohesive forces below this size become stronger than the gravitational force, and flow through the orifice is prevented. This, of course, is a function of the size of the orifice, and flow might be possible in a larger orifice.

Particle size enlargement of the drug substance can be brought about by manipulation of the recrystallization step in the synthesis of the drug. To increase powder flow, particle size enlargement by slugging, roller compaction, and wet granulation can be used. If a powder is compressible but does not flow well, then slugging may be employed. In slugging, tablets are made of the poorly flowing substance on a high-duty, slowly operating machine into large dies. The dies are large so that the flow is sufficiently increased, but now the compression forces must be increased because the larger area dictates a larger force to attain a given pressure (the elastic limit being in stress units). In roller compacting, the powder is processed between two heavy-duty rollers, compacted between the rolls, and emerges as a compressed sheet, which is then milled. These two methods are necessary if the drug substance (e.g., aspirin) is sufficiently moisture sensitive and there are stability issues so it cannot be wet granulated. Otherwise, wet granulation is a frequently used method of particle enlargement for free-flowing powder.

6.1.4.4 Density and Bulkiness

Density When a powder is poured into a container, the volume that it occupies depends on a number of factors, such as particle size, particle shape, and surface properties. In normal circumstances, it will consist of solid particles and interparticlulate air spaces (voids or pores). The particles themselves may also contain enclosed or intraparticulate pores. If the powder bed is subjected to vibration or pressure, the particles will move relative to one another to improve their packing arrangement. Ultimately, a condition is reached where further densification is not possible without particle deformation. The density of a powder is therefore dependent on the handling conditions to which it has been subjected, and there are several definitions that can be applied either to the powder as a whole or to individual particles.

Because particles may be hard and smooth in one case and rough and spongy in another, one must express densities with great care. Density is universally defined as weight per unit volume. Three types of densities—true density, particle density, and bulk density—can be defined, depending on the volume of particles containing microscopic cracks, internal pores, and capillary spaces.

The true density is the material itself exclusive of the voids and interparticular pores larger than molecular or atomic dimension in the crystal lattice. Particle (granular) density, determined by the displacement of mercury, which does not penetrate at ordinary pressure into pores smaller than 10 μm, is the mass of the particle divided by its volume. The different terms arise from the definition of volume: *True particle density* is when the volume measured excludes both open and closed pores and is a fundamental property of a material; *apparent particle density* is when the volume measured includes intraparticulate pores; *effective particle density* is the volume "seen" by a fluid moving past the particles. *Bulk density* (powder density) is the volume in a graduated cyclinder including both the particulate volume and the pore volume. The bulk density will vary depending on the packing of the powder. Based on the definition of volume, *minimum bulk density* is when the volume of the powder is at a maximum, caused by aeration, just prior to complete breakup of the bulk. *Poured bulk density* is when the volume is measured after pouring powder into a cylinder, creating a relatively loose structure. *Tapped bulk density* is, in theory, the maximum bulk density that can be achieved without deformation of the particles.

Based on the definition of density, two new terms are defined. *Porosity* is defined as the proportion of a powder bed or compact that is occupied by pores and is a measure of the packing efficiency of a powder and *relative density* is the ratio of the measured bulk density and the true density:

$$\text{Porosity} = 1 - \frac{\text{bulk density}}{\text{true density}}$$

$$\text{Relative density} = \frac{\text{bulk density}}{\text{true density}}$$

Bulkiness The specific bulk volume, the reciprocal of bulk density, is often called *bulkiness* or *bulk*. It is an important consideration in the packaging and filling of powders for tablet production. The bulk density of calcium carbonate can vary from 0.1 to 1.3, and the lightest or bulkiest type would require a container about 13 times larger than the heaviest type. Bulkiness increases with a decrease in particle size. In a mixture of materials of different sizes, however, the smaller particles sift between the larger ones and tend to reduce the bulkiness.

To define bulkiness in detail, the porosity and density of powders should be understood. Suppose a powder, such as zinc oxide, is placed in a graduated cylinder and the total volume is noted. The volume occupied is known as the bulk volume V_b. If the powder is nonporous, that is, has no internal pores or capillary spaces, the bulk volume of the powder consists of the true volume of the solid particles plus the volume of the spaces between the particles. The volume of the spaces, known as the void volume v, is given by equation

$$v = V_b - V_p$$

where V_p is the *true volume* of the particles.

The *porosity* or *voids* (ε) of a powder is also defined as the ratio of the void volume to the bulk volume of the packing given below. Porosity is frequently expressed in percent, $\varepsilon \times 100$:

$$\varepsilon = \frac{V_b - V_p}{V_b} = 1 - \frac{V_p}{V_b}$$

6.1.4.5 Hygroscopicity

The hygroscopicity of a drug and pharmaceutical substances is a potential parameter to be considered in tablet formulation. The moisture uptake rate is quite variable depending on the type of drug and excipients as well as the environmental conditions. So, a concise definition of hygroscopicity is not possible. Powders can absorb moisture by both capillary imbibition and swelling. The instantaneous water absorption prosperties of pharmaceutical excipients correlate with total surface area while the total absorption capacity correlates with powder porosity [22].

If drug and excipients are so hygroscopic, they can readily adsorb water until they deliquesce, or begin to dissolve. Moisture adsorption is important because adsorbed water can cause incorrect weighing and degradation of drug and/or excipients. The drug, excipients, and water reaction will continue as the amount of water increases. When a solid is placed in a room, moisture will condense onto it. If this occurs simply as a limited amount of adsorbed moisture, then the substance is not hygroscopic under these conditions. These conditions exist if the water vapor pressure in the surrounding atmosphere is lower than the water vapor pressure over a saturated solution of the solid in question. However, if the water vapor pressure in the atmosphere is higher than that of the saturated solution, there will be a thermodynamic tendency for water to condense upon the solid materials (drug and excipients).

The drug and pharmaceutical excipients adsorb or lose the moisture depending on the relative humidity in the atmosphere. The nonhygroscopic materials are not affected by the moisture and are in a equilibrium state. In general, solid dosage forms such as tablets or capsules should be hydrophilic because the solid materials must dissolve after swallowing. However, solid dosage forms must also be stable against physical and chemical factors.

The moisture uptake rates (MUR) can simply be obtained by weighing the sample after a given time (six days), but in such a case it is assumed that the moisture uptake is still in the linear phase. If, for instance, the weight gain is 5 mg per 10-g sample in six days, then the MUR is $5/(10 \times 6) = 0.083$ mg/g/day. If the MUR values are plotted versus relative humidity (RH) and the curve that intercepts the x axis at 20% RH is obtained from a straight line, the compound can be stored without moisture pickup in atmospheres of less than 20% RH. In addition, the hygroscopicity of materials is indicated as follows:

$$\text{Loss of drying (LOD \%)} = \frac{\text{weight of water in sample}}{\text{total weight of wet sample}} \times 100$$

$$\text{Moisture content (MC \%)} = \frac{\text{weight of water in sample}}{\text{weight of dry sample}} \times 100$$

$$\text{Equilibrium moisture contents (EMC)} = \frac{\text{\% LOD}}{\text{\% LOD} + 100} \times 100$$

$$\text{Relative humidity (RH)} = \frac{\text{water vapor pressure in atmosphere}}{\text{saturated water vapor pressure}} \times 100$$

Depending on the hygroscopicity based on the EMC, various drug and excipients are classified in four groups (Table 16).

Figure 8 also shows an example of moisture uptake for four selected excipients as a function of relative humidity. Depending on the hygroscopicity of the exscipients, the uptake behaviors are quite variable. Excipients such as microcrystalline cellulose (MCC) and starch can pick up significant amounts of water at relatively low relative humidity. Since this water is not present as a hydrate, it is potentially free to interact with a drug.

TABLE 16 Classification of Hygroscopicity and Example Pharmaceutical Excipients

Type I: Nonhygroscopic	Type II: Slightly Hygrocopic	Type II: Moderately Hygroscopic	Type IV: Very Hygroscopic
No MC change below RH 90% or less than 20% MC changes at RH 90% after 1 week storage	No MC change below RH 80% or less than 40% MC changes at RH 80% after 1 week storage	Not more than 5% MC change below RH 60% or less than 50% MC changes at RH 70% after 1 week storage	Increase of MC at RH 40–50% or more than 70% MC changes above RH 40% after 1 week storage
	Examples		
Lactose USP	MCC	HPC	Povidone
Dicalcium phosphate	Sucrose, dextrose	HPMC	Sodium starch glycolate
Ethylcellulose	Poloxamer 188	Bentonite	Polyplasdone XL
Magnesium stearate	PEG 3350	Pregelatinized starch	Sorbitol
	CAP	Starch USP, Corn	CMC Na
		Gelatin USP	

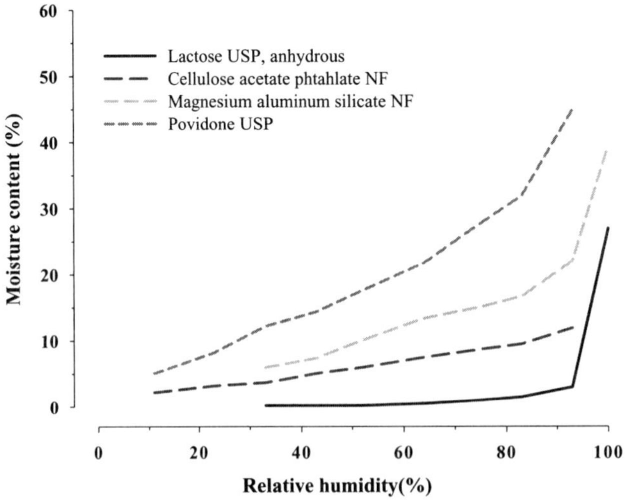

FIGURE 8 Profiles of moisture uptake for four selected excipients as function of relative humidity.

The moisture content of drug can affect the cohesiveness of particles through hydrogen bonding or by changing surface energy effects. Water also can act as a plasticizing agent and can lower the glass transition temperature of amorphous polymorphs, thus allowing a rubbery state to exist at a lower temperature. Water also can take up intra- and intergranular pore spaces (capillaries) of powders by acting as an interparticle bridge through surface adsorption mechanisms.

The hygroscopicity of materials influences the flowing characteristics of drug and excipients by forming adsorption film with an aid of water as a solvent. Dissolution of drug and excipients from solid dosage forms can also occur. The crystallinity of troglitazone-PVP K30 solid disperisions can be changed by the water content [23]. During manufacturing operations, water's ability to act as a bridge between particles can improve the compression capabilities of powder masses and affect the tensile strength and hardness of the tablet. The milling or blending is also changed. Most of all, the chemical hydrolysis of labile materials is more facilitated, resulting in low stability during storage. Adsorbed water also can act locally as a solvent, and in many cases drug–excipient incompatibility occurs by dissolving drug or excipients in granulation or blended powder mass, as observed when aspirin tablets liberate a strong odor of free salicyclic acid. Any free, unbound water in the mass can migrate throughout the material mass and act as a reagent. A waterproof packaging or container is used for many moisture-sensitive materials.

6.1.4.6 Mixing

The mixing of powders is a key step in the manufacture of virtually all solid dosage forms. In general, particle size, shape, and surface energy are important factors for blending of drug and excipients in unit operations. The perfect mixing of powder is desirable, that is, a mixture in which the probability of finding a particle of a given component is the same at all positions in the mixture, but the powder mixing has a maximum degree of randomness (Figure 9). To determine the degree of mixing obtained in a pharmaceutical operation, it is necessary to reasonably sample the mixture and determine the variation within the mix statistically [2].

Uniform mixing of powdered materials occurs if they have similar particle size distributions and particle shapes. Spherical particles mix least well while plate and fiber shapes also do not mix well because they tend to clump. The more cohesive the material, the more difficult it is to mix that material with other materials. Similarly, cohesiveness between drug and excipient or among excipients may hinder the

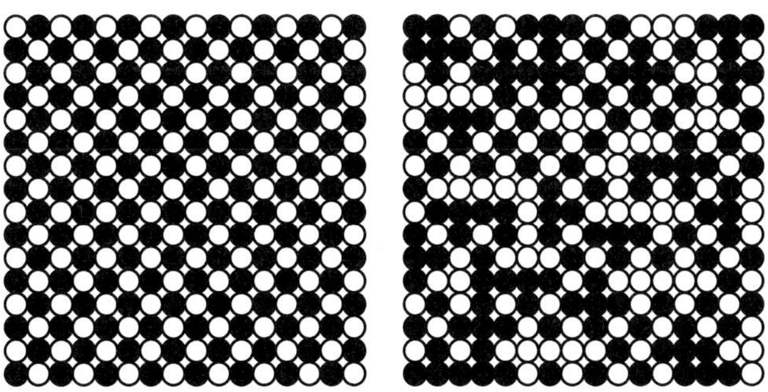

FIGURE 9 Comparison of powder mixing: perfect mixing and random mixing.

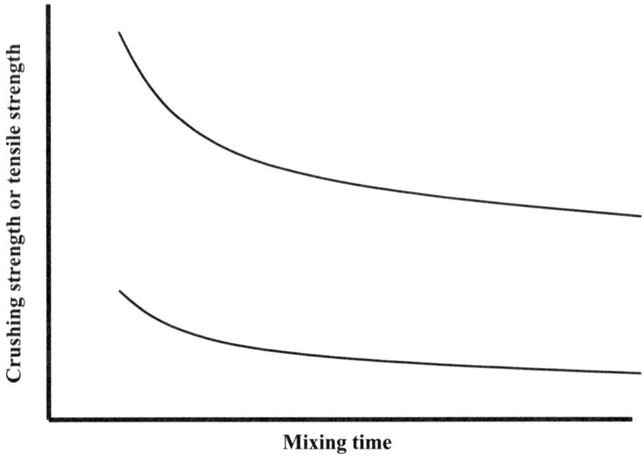

FIGURE 10 Effect of mixing time on mechanical strength of excipients.

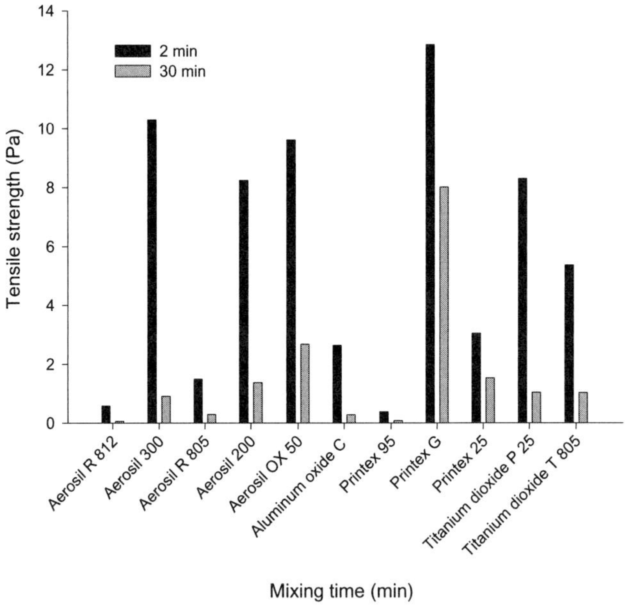

FIGURE 11 Effect of mixing time glidant tensile strength.

successful blending process. Ideally, the uniform mixing of powders should be such that the weight of sample taken is similar to the weight that the powder mix contributes to the final dosage form.

Figure 10 shows the effect of mixing time on the mechanical strength of powders. As the mixing time increases, the tensile strength of powders (or curshing strength of the tablet) gradually decreases [24, 25]. While the strength decreases with increasing mixing time for all materials tested, the effect is far more marked for materials that deform plastically. For example, the glidant tensile strength invariably decreases as the mixing time (2 min vs. 3 min) increases (Figure 11). There is no direct correlation of tensile strength with primary particle size of glidant (Figure 12). However,

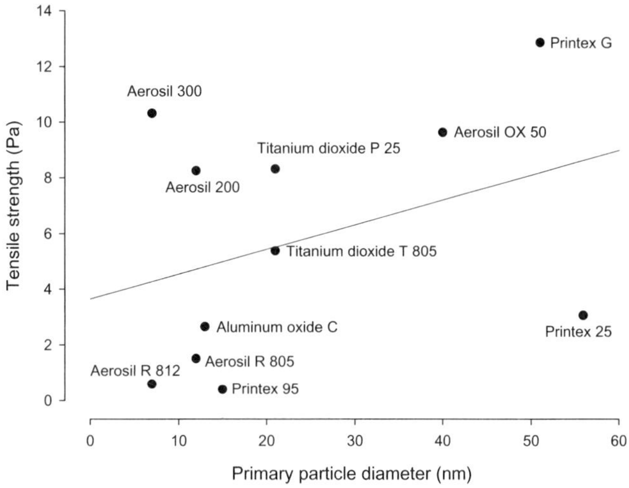

FIGURE 12 Correlation of tensile strength with primary particle size of glidant.

this correlation was further improved when some outlying glidant (Aserosil 200 and 300 and Printex 25) were excluded (R^2 values of 0.1287–0.8616).

If a powder consisting of two materials both having identical physical properties is mixed for a sufficient time, random mixing will eventually be achieved. Unfortunately, most pharmaceutical powders consist of mixtures of materials with differing physical properties, such as size, shape, density, and surface area, leading to segregation among particles, where particles of similar properties tend to collect together in part of the powder. When segregating powders are mixed, as the mixing time is extended, the powders appear to unmix. The differences in particle size are the most important for segregation in pharmaceutical powders. One exception to overcome segregation is ordered mixing rather than random mixing. When one component of a powder mix has a very small particle size (less than 5 μm) and the other is relatively large, the fine powder may coat the surface of the larger particles, and the adhesive forces will prevent segregation, known as ordered mixing. This ordered mixing makes the powders produce greater homogeneity than by random mixing. The percolation of fine particles is also a factor. If the particles sizes are quite different, the smaller particles can drop easily and move to the bottom of powder, resulting in segregation. This segregation process can occur whenever movement of particles by vibration, shaking, and pouring takes place.

6.1.4.7 Particle Size Reduction (Micronization and Milling)

Mechanical attrition, that is, high-energy ball milling of powders, is a nonequilibrium processing method that has generated the reduced particle size and the formation of physically metastable materials. It can be used to modify materials by refining the microstructure, homogenizing the composition, extending solid solubility, creating matastable crystalline phases, or producing metallic glass. High-energy ball milling is both a processing method to reduce particle size and a route to the physical synthesis of metastable materials. Early in product development, when only small amounts of drug are available, comminution (grinding/mixing) may be carried out with a mortar and pestle. For lager batches, ball milling or micronization can be used to reduce the particle size because comminution or grinding are not practical due to the length of time required [18].

There are numerous methods of particle size reduction, but their application is dependent on the intended particle size, particle distribution, cleaning convenience, operating cost, dust containment, temperature, and flexibility. The type of milling machine includes slurry, fluid energy (jet), universal, cone, and hammer. This size reduction process may reduce the risk of dissolution rate-limited bioavailability of drugs. Particle size reduction can be accomplished by using a hammer mall or a similar mill, but this process may only break up the larger crystal aggregates without significantly changing the distribution of smaller particle sizes. On the other hand, air jet mills, which impinge two streams of particles at a right angle to each other in high-velocity air streams, reduce particle size significantly within microsized ranges.

Ball milling was the most commonly used at the preformulation stage to reduce the particle size of small amounts of a compound. Ball mills reduce the size of particles through a combined process of impact and attrition. Usually they consist of a hollow cylinder that contains balls of various sizes which is rotated to initiate the grinding process. Micronized particles are typically less than 10 μm in diameter. The efficiency of the milling process is affected by rotation speed, number of balls, mill size, wet or dry milling, amount of powder, and length of time of milling. Although ball milling can effectively reduce the particle size of compounds, prolonged milling may be detrimental in terms of changes of compound crystal form from crystalline to polymorphic or amorphous form and stability.

Although ball milling on a large scale is possible, hammer milling is more preferable in the pharmaceutical industry. Powder is bled into the mill house via the hopper, and the rotating hammers impact with the powder. When this is fine enough to pass the screen, the powder will exit. The powder exiting will have a maximum particle size of that of the screen. The average particle size of the milled powder will be smaller, the smaller the feed rate, the more rapid the milling speed (rpm's), and the finer the screen. The knife has a blunt edge on one side and a knife edge on the other. Milling with the blunt edge forward gives rise to a smaller average particle size. The usual minimum particle range is about 50 μm. Large, heavy-duty hammer mills (micropulverizers) give much smaller particle size further down, typically to 20 μm. If particle sizes less than 5 μm are desired, micronizers are used.

If particle sizes in the micrometer range are required, then the attrition mill can be used. Here particles are bled into a chamber in which great turbulence has been created by two inlets of air at different pressure. The particles hit one another and are removed by centrifugal means and collected in a cyclone setup or in a bag above the cyclone. Such particles become highly electrically charged during the operation and also become very cohesive.

The milling process provides energy to the powders so that melting occurs if their melting point is sufficiently low. There is no guarantee that the original polymorph and habit will be regained upon resolidification. The milling and micronization process can also reduce the particle size of poorly soluble drugs so that the maximum surface area is exposed to enhance the solubility and dissolution properties. Although micronization of the drug offers the advantage of a small particle size and a lager surface area, it can also result in processing problems due to high dust, low density, and poor flow properties. Indeed, micronization may be counterproductive, since the micronized particles may aggregate, which may decrease the surface area. Milling and micronization process also induce the changes of crystallinity of drug into amorphous form. It has been shown that the amorphous change of the crystalline structure can be achieved by vapor condensation, supercooling of melt, rapid precipitation from solution, and mechanical applications of a crystalline mass by milling or compaction [26].

Particle size reduction of the powder has produced defects on the surface that, if enough energy is imparted, leads to amorphous regions on the surface. In turn, these regions are found to have a greater propensity to adsorb water. The dissolution rate increases as the particle size of drug powder decreases due to its greater surface area for wetting.

It is known that ball milling and other types of milling can change the morphology of a solid, for example, make a crystalline compound amorphous, increase the surface energy of a solid, and distort the crystal lattice. For example, the crystalline solid state of sorbitol exhibits a complex polymorphism made of five different forms called A-, B-, Γ-, Δ-, and E-sorbitol [27]. The structures of these polymorphs have been identified by single-crystal X-ray diffraction. The structure of the Γ form is also the most stable and the most common polymorph of sorbitol. It was reported that the Γ form of sorbitol underwent a complete gransformation toward the A form upon ball milling and also was affected by milling time. The DSC and XRPD patterns are useful in identifying these phenomena, althought the low level of amorphous character cannot be detected by techniques such as XRPD and DSC. Crystal structures by the milling process as well as compaction forces have been shown [2].

6.1.4.8 Compaction(Compressibility)

Compressibility is the property of forming a stable and intact mass when pressure is applied. The manufacture of tablets involves the process of powder compaction or compressibility, the purpose of which is to convert a loose incoherent mass of powder into a single solid object. A protocol to examine the compression properties of flowing powder should be considered by the formulation scientist when selecting the excipient, the formulation type, and the manufacturing process, for example, direct compression or granulation for the intended solid dosage form. Acetaminophen is poorly compressible whereas lactose compresses well. In general, drug and excipients, including lubricant, are blended in a tumbler mixer for a period of time and then compressed into tablets in a hydraulic press. The crushing strength is also determined to test the compressibility of tablet at room temperature.

The compression of drug powder may change the crystal structure into a polymorphic form—likewise in the milling process. The PXRD patterns and DSC of a drug can detect these phenomena. Knowledge of the crystal structure of a drug is a prerequisite for compaction.

To fully understand the compaction behavior of a material, it is necessary to be able to quantify of its elasticity, plasticity, and brittleness. A powder in a container subjected to compressive force will undergo particle rearrangement [28]. The density of the bed will increase with increasing pressure at a characteristic rate. Brittle materials will undergo fragmentation, and the fine particles formed will percolate through the bed to give secondary packing. Plastically deforming materials will distort to fill voids and may also exhibit void filling by percolation. When the limit of plastic deformation is reached, fracture occurs.

Many of the basic principles of compaction and the test methodologies are currently employed in pharmaceutical formulation. To characterize the compaction properties of a material or formulation, it must be possible to measure the relationship between the force applied to a powder bed and the volume of the powder bed. There are two principal types of compaction studies used to characterize materials: pressure–volume relationships and pressure–strength relationships. While ultimately it is the strength of a tablet that is important, the pressure–volume relationships provide information about the compaction properties of a material that allows an appropriate formulation to be developed. The instrumented tableting machine provides information for compaction that is directly relevant to production conditions. The compression profiles differ from those of rotary tableting machines used for commercial production. The profile of a single punch involves the powder bed being compressed between a moving upper punch and a stationary lower punch, while on a rotary machine, both punches move together simultaneously. A major advantage of instrumented machines is that they provide information not only on the compaction properties but also on flow and lubrication. The disadvantage of using instrumented rotary machines is the large quantity of materials required.

A large number of equations have been proposed to describe the relationship between pressure and volume reduction during the compaction process. The most widely used equation to describe the compaction of powders is the Heckel equation. Pharmaceutical powders do not produce perfect straight lines, and the type of deviation provides information about the compaction behavior of the material. A typical Heckel plot for a pharmaceutical powder is obtained showing a straight-line portion over a certain pressure range with a negative deviation at low pressures and a positive deviation at high pressures. The strength of tablets has traditionally been determined in terms of the force required to fracture a specimen across its diameter. The fracture load obtained is usually reported as a hardness value. Initially, most materials will demonstrate an increase in tensile strength proportional to the compaction pressure applied. As the compaction pressure increases, the tablet approaches zero porosity, and large increases in pressure are required to achieve small volume reductions [28]. Some materials will attain a maximum strength, and subsequent increases in pressure will produce weaker tablets. Other materials will also display an initial increase in strength proportional to the applied pressure, but the strength reaches a maximum before falling off sharply, resulting in capping or lamination in tablet production [28]. Figure 13 shows the correlation of hardness with compression pressure. Capping is the partial or complete removal of the crown of a tablet from the main body, while lamination is the separation of a tablet into two or more distinct layers. If the compressibility and flow of powders are good, it is possible to directly compress the powders into tablet. If either compressibility or flow is not satisfactory, roller compact, slugging, or we granulation can be utilized to improve compressibility.

6.1.4.9 Surface Area and Other Properties

The surface area of particles is related to particles size, as discussed previously. The surface area of powders affects the drug dissolution rate, powder flow, cohesiveness, and adsorption. Furthermore, the surface area of solid materials may also

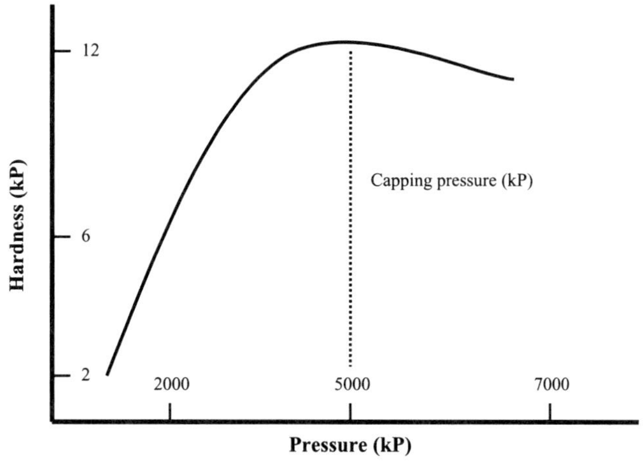

FIGURE 13 Correlation of compression pressure with hardness.

influence the physicochemical properties, adsorption, dissolution, and bioavailability of drugs.

The particle size and surface area distributions of pharmaceutical powders can be obtained by microcomputerized mercury porosimetry. Mercury porosimetry gives the volume of the pores of a powder, which is penetrated by mercury at each successive pressure; the pore volume is converted into a pore size distribution. Two other methods, adsorption and air permeability, are also available that permit direct calculation of surface area. In the adsorption method, the amount of a gas or liquid solute that is *adsorbed* onto the sample of powder to form a monolayer is a direct function of the surface area of the sample. The air permeability method depends on the fact that the rate at which a gas or liquid permeates a bed of powder is related, among other factors, to the surface area exposed to the permeant. The determination of surface area is well described by the BET (Brunauer, Emmett, and Teller) equation.

The wetting behavior of powders is an also important factor for drug dissolution. If the wetting is not satisfactory, hydrophilic excipients (lactose) and surfactant (sodium lauryl sulfate or polysorbate) are combined in the powder mixtures. The contact angle is used as an index of wetting. The lower the contact angle, the better the wetting occurs.

The general appearance of a tablet with good visual identity and overall "elegance" are essential for consumer acceptance, quality control of lot-to-lot uniformity, general tablet-to-tablet uniformity, and monitoring trouble-free manufacturing. Control of the general appearance of a tablet involves the measurement of a number of attributes, such as a tablet's size, shape, color, presence or absence of and odor, taste, surface texture, physical flaws and consistency, and legibility of any identifying markings.

6.1.5 TABLET CHARACTERIZATION

In order to formulate the optimal tablet, various properties should be considered, including drug–excipient compatibility, flowability, lubricity, appearance, dissolution, and disintegration [2]. The prepared tablet must also meet physical specification and quality standards according to the monograph of the pharmacopeia. In general, weight and its variation, content uniformity, thickness, hardness, friability disintegration, and dissolution should be considered for tablet validation [3]. These factors must be controlled during tablet production (in-process control) and are validated after the production to ensure the quality standards.

6.1.5.1 Disintegration

Immediate-release tablets should be readily disintegrated in the stomach when swallowed. This disintegration involves bursting apart the compact masses by aqueous fluid penetrating the fine pore structure of tablet. Disintegration testing is an important part of in-process control testing during production to ensure batch-to-batch uniformity, but its role in end-product testing has largely been superseded by dissolution testing because recently modified-release preparations are getting popular. It was recognized in the 1940s that tablets had to disintegrate in order for them to be bioavailable due to lack of biopharmaceutical information and primarily analytical limitations. Later, of course, in the 1950s and 1960s, the pharmaceutical scientist became aware of the importance of dissolution rates as well.

In general, for the medicinal agent in a tablet to become fully available for absorption, the tablet must first disintegrate and discharge the drug to body fluids for dissolution. The general manner in which a tablet disintegrates is as follows: (a) the tablet wets down, (b) the dissolution liquid penetrates the pore space, (c) the disintegrant absorbs water and swells, and (d) this swelling causes the tablet to break down into granules. Figure 14 shows the disintegration pathways of solid dosage forms for the dissolution and absorption of drugs. After the disintegration process, the solid dosage forms change into granules or smaller and fine particles ready for dissolution and absorption in the fluid.

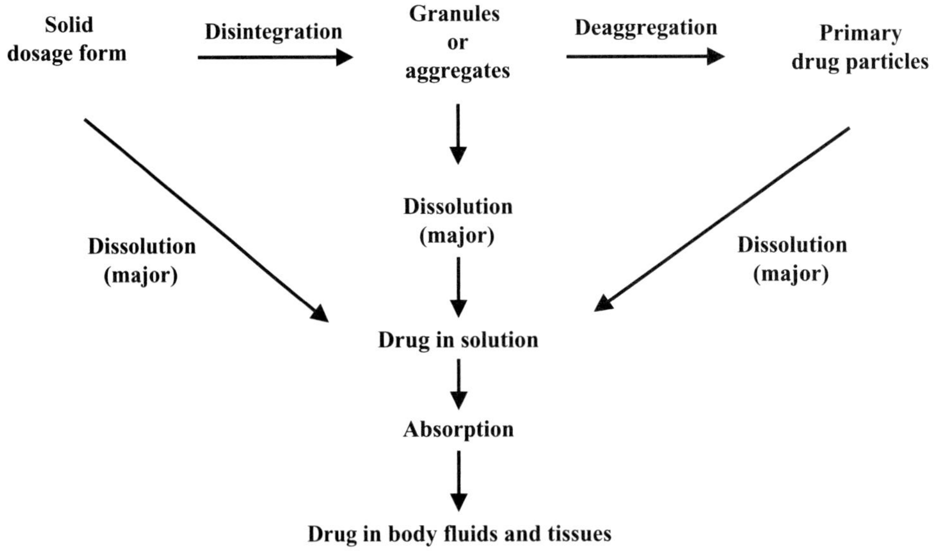

FIGURE 14 Disintegration and dissolution pathways of solid dosage forms for absorption of drug.

Tablet disintegration is the important first step in the dissolution of the drug substance contained in immediate-release tablets but dissolution is more meaningful in case of many modified release products rather than disintegration. A number of formulation and manufacturing factors can affect the disintegration and dissolution of a tablet, including the particle size of the drug substance in the formulation; the solubility and hygroscopicity of the formulation; the type and concentration of the disintegrant, binder, and lubricant used; the manufacturing method, particularly the compactness of the granulation and the compression force used in tableting; and the in-process variables which may occur. Therefore, it is vitally important for batch-to-batch consistency to establish disintegration and dissolution standards and controls for both materials and processes.

Tablet disintegration also is important for those tablets containing medicinal agents (such as antacids and antidiarrheals) that are not intended to be absorbed but rather to act locally within the GI tract. In these instances, tablet disintegration provides drug particles with an increased surface area for localized activity within the GI tract.

It is evident that there are some correlations of physical parameters with tablet disintegration time. Figure 15 shows the correlation of water penetration force, disintegration force, disintegrant contents, and compression forces of the tablet with disintegration time of the tablet. As the water penetration force increases, the tablet disintegration force also increases, resulting in shorter disintegration time [29, 30]. The amount of disintegrant in the tablet also decreases the disintegration time. The disintegration time increases as tableting pressure increases below the critical capping pressure [18]. At very low pressures the penetration of liquid into the tablet is virtually unhindered (almost like pouring water into a breaker) but the pores will be too large to allow disintegrant swelling to cause stress and the disintegration time will decrease. Once the pores are sufficiently small, penetration of the liquid into the disintegrant becomes the limiting step, and the disintegration time will increase

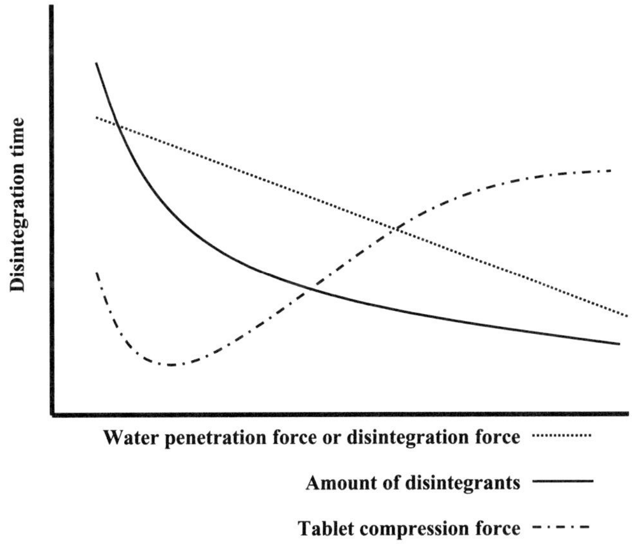

FIGURE 15 Correlation of physical parameters on tablet disintegration time.

with increasing pressure. Disintegrants and lubricants are added to wet-granulated products after the granulation has been dried. Disintegration time increases as the amount of hydrophobic lubricant increases. The mixing time for the lubricant must be kept short because otherwise the lubricant may fluidize during the mixing step and lose part of the lubricant properties that are necessary for flow in the tablet die. If the disintegration is not satisfactory, numerous types of disintegrants are added in the tablet formulations, including starch, croscarmellose sodium, sodium starch glycolate and crospovidone known as superdisintegrants. In general, the swelling rate and water uptake are the most important properties of disintegrants.

All USP tablets must pass a test for disintegration, which is conducted in vitro using a testing apparatus. The detailed monograph for disintegration testing is described in the many pharmacopeias [2, 3].

The apparatus consists of a basket-rack assembly containing dimensions held vertically upon a 10-mesh stainless steel wire screen. During testing, a tablet is placed in each of the six tubes of the basket, and the mechanical device raises and lowers the basket in the immersion fluid at a frequency of between 29 and 32 cycles per minute, the wire screen always maintained below the level of the fluid. For uncoated tablets, buccal tablets, and sublingual tablets, water maintained at about 37 °C is used as the immersion fluid unless another fluid is specified in the individual monograph. For these tests, complete disintegration is defined as that state in which any residue of the unit, except fragments of insoluble coating or capsule shell, remaining on the screen of the test apparatus is a soft mass having no palpably firm core. Buccal tablets must disintegrate within the time set forth in the individual monograph, usually 30 min, but varying from about 2 min for nitroglycerin tablets to up to 4 h. If one or more tablets fail to disintegrate, additional tests prescribed by the USP must be performed. Enteric-coated tablets are also similarly tested, except that the tablets are permitted to be tested in simulated gastric fluid for 1 h with no sign of disintegration, cracking, or softening. They are then switched to the simulated intestinal fluid for the time stated in the individual monograph during which time the tablets disintegrate completely. If 1 or 2 of the 6 tablets fails to disintegrate completely, disintegration testing is repeated on 12 additional tablets, and not less than 16 of the total 18 tablets tested must disintegrate to meet the standards.

6.1.5.2 Dissolution

Definitions Dissolution is the dynamic process by which drug is dissolved in a solvent (water) and is characterized by a rate (amount dissolved per unit time). In vitro dissolution testing of a tablet is very important for many reasons: It guides the formulation and product development process toward optimization of dosage forms for quality control and reliability. By conducting dissolution studies in the early stages of a product's development, the formulation compositions and manufacturing parameters are also tuned and monitored about the old and newly advanced tablet. The U.S. Food and Drug Administration (FDA) allows manufacturers to examine scale-up batches of 10% of the proposed size of the actual production batch or 100,000 dosage units, whichever is greater, by performing in vitro dissolution testing to assure bioequivalence from batch to batch and processing parameters. New drug applications (NDAs) submitted to the FDA contain in vitro dissolution data generally obtained from batches that have been used in pivotal clinical and/or bioavailability studies and from human studies conducted during product development. Once the specifications are established in an approved NDA, they become official (USP) specifications for all subsequent batches and bioequivalent products. The dissolution testing is also used as a tool of SUPAC (scale-up postapproval change) and variation of equipment, location, and processing factors. In addition, dissolution testing is used as a tool to examine the short- and long-term stability of dosage forms. Release mechanism and parameters which change the dissolution are also studied. Most of all, one of main goals of in vitro dissolution testing is to provide reasonable prediction and correlation with the product's in vivo bioavailability. Differentiations in the formulations and other related variables may cause deviations from in vivo bioavailability data.

A system has been developed which relates combinations of a drug's solubility (high or low) and its intestinal permeability (high or low) as a possible basis for predicting the likelihood of achieving a successful in vivo–vitro correlation (IVIVC). The four classes based on BCS are (I): high solubility and high permeability, (II): low solubility and high permeability, (III): high solubility and low permeability, and (IV): low solubility and low permeability. In class I, dissolution testing can be used as a prognostic tool to predict in vivo biovailability.

Equations Many dissolution equations are well described in the text. Most of the equation are based on the well-known Fick's law. Figure 16 shows a diagram of the concentration gradient between the matrix tablet and bulk fluid for dissolution.

From these gradient situations, the well-known Noyes–Whitney equation is given as

$$\frac{dc}{dt} = \frac{DAK}{Vh}(C_s - C_t)$$

where dc/dt = rate of drug dissolution, where $dQ/dt = V\ dc/dt$

V = volume of dissolution fluid

D = diffusion rate constant

A = surface area of dosage forms

C_s = concentration of drug in stagnant layer

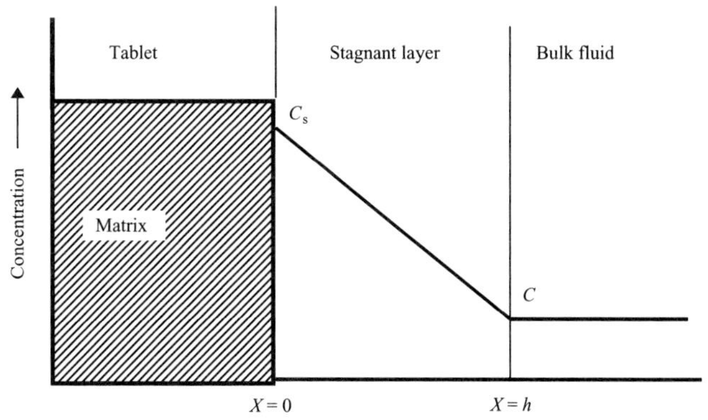

FIGURE 16 Concentration gradient from the tablet between matrix and bulk fluid for dissolution. Cs, drug solubility; C, uniform concentration; h, thickness of stagnant film; X = diffusional path length.

C_t = concentration of drug in bulk fluid at given time
K = partition coefficient
h = thickness of stagnant layer

If the bulk volume is large and the concentration of drug in the fluid is much lower than the drug solubility ($C_s >> C_t$), it is regarded as a sink condition. In this case, the equation is much simpler and the dissolution behaviors continuously occur because the chemical potential ($C_s - C_t$) approximates drug solubility (C_s).

In a matrix tablet, the following Higuchi equations are given depending on the polymeric structures of the homogenous and porous matrix [20]:

$$Q = \begin{cases} [D(2A - C_S)C_S t]^{1/2} & \text{for homogenous matrix} \\ [D\varepsilon(2A - \varepsilon C_S)C_S t]^{1/2}/\tau & \text{for porous matrix} \end{cases}$$

In the dissolution of granular powders, the Hixson–Crowell equation is also established as

$$Q_0^{1/3} - Q^{1/3} = kt \quad \text{for granular powders}$$

Testing Method In addition to formulation and manufacturing controls, the method of dissolution testing also must be controlled to minimize important variables such as paddle rotational speed, vibration, and disturbances by sampling probes. The USP includes seven apparatus designs for drug release and dissolution testing of immediate-release oral dosage forms, extended-release products, enteric-coated products, and transdermal drug delivery devices:

Apparatus I: rotating basket method, 25–150 rpm (100 rpm)
Apparatus II: rotating paddle method, 25–150 rpm (50 rpm)
Apparatus III: reciprocating cylinder method, inner tube (5–40 dips/min), outer tube (300 mL)
Apparatus IV: flow-through method: 4–16 mL/min
Apparatus V: paddle over disc
Apparatus VI: cylinder method
Apparatus VII: reciprocating holder method

Detailed guidelines for dissolution testing are described in monographs of many pharmacopeias. The USP apparatus I and USP apparatus II are used principally for tablet dissolution testing. In USP apparatus I, the dosage unit is placed inside the basket. In USP apparatus II, the dosage unit is placed on the bottom in the vessel. In each test, a volume of the dissolution medium (500–900 mL in general) is placed in the vessel and allowed to come to 37 ± 0.5 °C. Then the stirrer is rotated at the specified speed (50–200 rpm). The samples of the medium are withdrawn for analysis of the proportion of drug dissolved. The tablet or capsule must meet the stated monograph requirement for rate of dissolution. For example, "not less than 85% of the labeled amount is dissolved in 30 minutes in case of immediate release tablet." In a floating tablet, the sinker can be used in the paddle method.

Variables Affecting Dissolution In general, the dissolution profiles are highly dependent on the physicochemical properties, formulation, processing parameters, and testing conditions [2]. The physicochemical properties include particle size, surface area, crystal habit and polymorphism, solubility, molecular size, salt formation, pK_a, hydration, wetting, and surface tension. Physical factors such as viscosity, density, flocculation, and agglomeration are also considered. The formulation factors are also of importance. The amount and type of excipients and type of dosage forms play a key role in modifying dissolution behaviors. For example, Figure 17 gives the effect of lubricant and its mixing time on the dissolution rate of drugs. The presence of lubricant and its mixing time significantly changed tablet dissolution [31]. For poorly soluble drug, numerous pharmaceutical methods have been utilized to increase the dissolution rate of drug, including micronization, amorphous crystallization, spray drying, inclusion complex, microemulsion, and solid dispersion.

The processing parameters in the tablet preparation also change the dissolution, including temperature, mixing, milling, rotation speed, solvent, hardness, and surface area. The testing conditions are also important in modifying the dissolution of tablet. Therefore, the testing conditions are well defined by the regulations in the many pharmacopeias. The testing conditions include pH of the fluid, temperature, ionic strength, common ion effect, type of apparatus, rotation speed, volume size of dissolution fluid, analytical conditions, aeration, sample treatment, and mainly composition of dissolution media. Table 17 provides some of the media compositions suggested for in vitro dissolution testing of the tablet. These modified dissolution media can be used to achieve dissolution of drug under simulated in vivo conditions.

In Vitro–In Vivo Correlation In vitro dissolution testing can provide a reasonable prediction of the product's in vivo bioavailability. For a high-solubility and high-permeability drug (class I), an IVIVC may be expected if the dissolution rate is slower than the rate of gastric emptying (the rate-limiting factor). In a low-solubility and high-permeability drug, drug dissolution may be the rate-limiting step for drug absorption and an IVIVC may be expected. In a high-solubility and low-

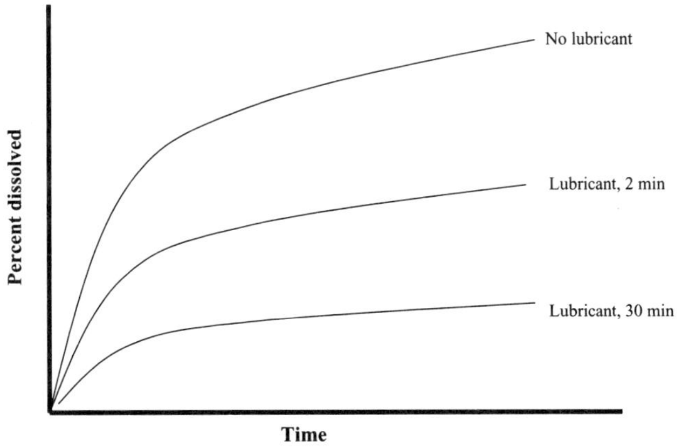

FIGURE 17 Effect of lubricant and its mixing time on dissolution rate of drugs.

TABLE 17 Suggested Media Compositions for In Vitro Dissolution Testing

Medium Type	Codes	Dissolution compositions
Water	W	Purified water
Gastric fluid (pH 1.2)	SGF	HCl–NaCl buffer
	SGF/TW	HCl–NaCl buffer/Tween 80 (1%)
	SGF/SLS	HCl–NaCl buffer/sodiun lauryl sulfate (1%)
	SGF/PEP	HCl–NaCl buffer/pepsin (0.32%)
Intestinal fluid (pH 6.8)	SIF	KH_2PO_4 buffer
	SIF/TW	KH_2PO_4 buffer/Tween 80 (1%)
	SIF/SLS	KH_2PO_4 buffer/sodiun lauryl sulfate (1%)
	SIF/Pan	KH_2PO_4 buffer/pancreatin (1%)
	SIF/fasted	0.2 *M* KH_2PO_4 buffer (3.9 g)/3 m*M* Na-taurocholate, 75 m*M* lecithin/KCl (7.7 g)/NaOH (qs)/water (qs)
Intestinal fluid (pH 5.0)	SIF/fed	Acetic acid (8.65 g)/15 m*M* Na-taurocholate/3.75 m*M*, lecithin/KCl (15.2 g)/NaOH (qs)/water (qs)

permeability drug, permeability is the rate-controlling step and only a limited IVIVC may be possible. In a drug with low solubility and low permeability, significant problems would be likely for oral drug delivery.

However, the in vivo GI condition is very complicated in terms of complex physiology and absorption process and is not simulated by the simple in vitro dissolution conditions [32]. Moreover, in vivo conditions are also complicated by food composition, type and composition of dosage formulations, relative rates of permeation, GI transit time, site of absorption, complexity of GI fluids such as pH, enzymes, bile and mucin, rate and capacity of metabolism by intestinal and hepatic enzymes, ethical difference, and patient conditions such as mood, disease, bed rest, and fasting volume of fluid given. For these reasons, in vitro dissolution is not always correlated with in vivo absorption, especially low-soluble and low-bioavailable drugs. To use dissolution testing as a prognostic tool for in vivo bioavailability, the dissolution fluid is simulated with in vivo GI condition as possible, compromising the biorelevant composition of dissolution fluid (see Table 17), gradient pH, proper stirring rate, and addition of lipid, enzyme, and surfactants.

For bioequivalence, dissolution profiles of two tablets are often compared. In this case, the difference factor and similarity factor are considered, defined by the equations

$$f_1 = \frac{\sum |R_t - T_t|}{\sum R_t} \times 100$$

$$f_2 = 50 \log \left\{ \left[1 + \frac{1}{n} \sum (R_t - T_t)^2 \right]^{-0.5} \times 100 \right\}$$

where R_t and T_t are the cumulative percents dissolved at each of the selected n time points.

The difference factor f_1 is proportional to the average difference between the two profiles while the similarity factor f_2 is inversely proportional to the average squared difference between the two profiles and measures the closeness between the two profiles. The two dissolution profiles are identical if $f_2 = 100$. An average difference of 10% at all measured time points results in a f_2 value of 50. The FDA guideline states that f_2 values of 50–100 indicate similarity between two dissolution profiles.

6.1.5.3 Weight Variation

With a tablet designed to contain a specific amount of drug in a specific amount of tablet formula, the weight of the tablet being made is routinely measured to help ensure that the tablet contains the proper amount of drug. The quantity of fill placed in the die of a tableting press determines the weight of the resulting tablet. The volume of fill is adjusted to yield tablets of desired weight and content. The depth of fill in the tablet die must be adjusted to hold a predetermined volume of powder or granulation. Each tablet weight should be calculated if the amount of drug and other excipients such as diluent, disintegrant, and binder are decided.

During production, sample tablets are periodically removed for visual inspection and automated physical measurement known as in-process control (IPC). The USP provides some guidelines for weight variation. In the test, 10 uncoated tablets are weighed individually and the average weight is calculated. The tablets are assayed and the content of active ingredient in each of the 10 tablets is calculated assuming homogeneous drug distribution.

6.1.5.4 Hardness or Breaking Strength

Tablets require a certain amount of strength, or hardness and resistance to friability, to withstand mechanical shocks of handling in manufacture, packaging, and shipping. Adequate tablet hardness and resistance to powdering and friability are necessary requisites for consumer acceptance while immediate-release tablets should readily disintegrate in the stomach as quick possible. For this reason, the relationship of harness to tablet disintegration and drug dissolution has been described.

Tablet hardness has been defined as the force required to break a tablet in a diametric compression test. To perform this test, a tablet is placed between two anvils, and the crushing strength that just causes the tablet to break is recorded. Hardness is thus sometimes termed the tablet crushing strength. It is not unusual for a tableting press to exert as little as 3000 lb and as much as 40,000 lb of force in the production of tablets. Generally, the greater the pressure applied, the harder the tablets, although the formulation composition and manufacturing process may also change tablet hardness. In general, tablets should be sufficiently hard to resist breaking during normal handling or transportation but have no problem in disintegrating and/or dissolving after swallowing.

The hardness of a tablet, like its thickness, is a function of the die fill and compression force. At a constant die fill, the hardness values increase and thickness decreases as additional compression force is applied. This relationship holds up to a maximum value for hardness and a minimum value for thickness beyond which increases in pressure cause the tablet to laminate or cap, thus destroying the integrity of the tablet. At a constant compression force (fixed distance between upper and lower punches), hardness increases with increasing die fills and decreases with lower die fills. The amount and mixing time of lubricants and excipients can affect tablet hardness. Large tablets require a greater force to cause fracture and are therefore "harder" than small tablets.

Special dedicated hardness testers or multifunctional systems are used to measure the degree of force (in kilograms, pounds, or arbitrary units) required to break a tablet. Devices to test tablet hardness include the Monsanto tester, the Strong-Cobb tester, the Pfizer tester, the Erweka tester, and the Schleuniger tester. A force of about 4 kg is considered the minimum requirement for a satisfactory tablet. Multifunctional automated equipment can determine tablet weight, hardness, thickness, and diameter. Unfortunately, these testers do not produce uniform results for the same tablet due to operator variation, lack of calibration, spring fatigue, and manufacturer variation.

6.1.5.5 Friability

Tablet hardness is not an absolute indicator of tablet strength since some formulations, when compressed into very hard tablets, tend to "cap" on attrition, losing their crown portions. Another measure of a tablet's strength is tablet friability. More powders, chips, and fragments can be produced during friability test if the tablet lacks proper strength and is manufactured in dirty processes during coating and packaging. The high friability causes lacks of elegance and consumer acceptance and even weight variation or content uniformity problems.

A tablet's durability may be determined using friability tester like Varian Friabilator testng apparatus. This apparatus determines the tablet's friability, or its tendency to crumble, by allowing it to roll and fall within the rotating apparatus. Normally, a preweighed tablet sample is placed in the friabilator, which is then operated for 100 revolutions. The tablets are then dusted and reweighed. Any loss in weight is determined. Resistance to loss of weight indicates the tablet's ability to withstand abrasion in handling, packaging, coating, and shipment. Compressed tablets that lose a maximum of not more than 0.5–1% of their weight are generally considered acceptable [3].

6.1.5.6 Content Uniformity

Tablet weight cannot be used as a potency indicator of its potency, except perhaps when the active ingredient is 90–95% of the total tablet weight. In tablets with smaller dosages, a good weight variation does not ensure good content uniformity, but a large weight variation precludes good content uniformity. The weight variation test would be a satisfactory method of determining the drug content uniformity of tablets. The content uniformity of the tablet is more important since the potency of tables is expressed on labels in terms of grams, milligrams, or micrograms. The content uniformity of tablets can be varied by three factors: (1) nonuniform distribution of the drug substance throughout the powder mixture or granulation, (2) segregation of the powder mixture or granulation during the various manufacturing processes, and (3) tablet weight variation.

By the USP method, 10 dosage units are individually assayed for their content uniformity according to the assay method described in the individual monograph.

The requirements for content uniformity are met if the amount of active ingredient in each dosage unit lies within the range of 85–115% of the label claim and the relative standard deviation is less than 6.0%. If one or more dosage units do not meet these criteria, additional tests as prescribed in the USP are required. Official compendia or other standards provide an acceptable potency range around the label potency. For highly potent, low-dose drugs such as digoxin, this range is usually not less than 90% and not more than 110% of the labeled amount. For most other larger dose drugs in tablet form, the official potency range that is permitted is not less than 95% and not more than 105% of the labeled amount.

6.1.5.7 Tablet Thickness

At a constant compressive load, tablet thickness varies with changes in die fill, particle size distribution and packing of the particle mix being compressed, and tablet weight, while with a constant die fill, thickness varies with variations in compression forces. The thickness of individual tablets may be measured with a micrometer, which permits accurate measurement and provides information on the variation between tablets. Other techniques employed in production control involve placing 5 or 10 tablets in a holding tray, where their total crown thickness may be measured with a sliding caliper scale. Tablet thickness should be controlled within ±5% variation of a standard value. Any variation in tablet thickness within a particular lot of tablets or between manufacturer's lots is not appropriate for consumer acceptance of the product. In addition, tablet thickness must be also controlled to facilitate packaging.

6.1.5.8 Tablet Shape and Size

The shape of the tablet alone can influence the choice of tableting machine used. Figure 18 shows some representative tablet shapes. The size, shape, and thickness

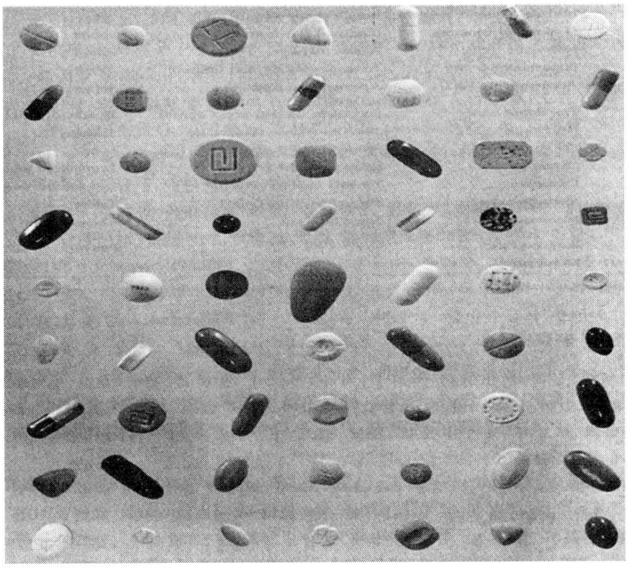

FIGURE 18 Example tablet shapes.

are very changeable. Shaped tablets requiring slotted and sophisticated punches must be run at slower speeds to avoid manufacturing disorder as compared with round tablets using conventional punches. Because of the nonuniform forces involved during compression, the more convex the tablet surface, the more likely it is to cause capping or laminating problems. The more complicated shaped tablet requires the use of a slower tableting machine or one with precompression capabilities. The size and shape of the tablet are governed by the choice of tableting machine, the best particle size for the granulation, production lot sizes, and the best type of tablet processing, packaging operation, and cost to produce the tablet.

REFERENCES

1. Kibbe, A. H. (2000), *Handbook of Pharmaceutical Excipients*, 3rd ed., American Pharmaceutical Association and Pharmaceutical Press, London.
2. Gibson, M. (2001), *Pharmaceutical Preformulation and Formulation, A Practical Guide from Candidate Drug Selection to Commercial Dosage Form*, HIS Health Group, Denver, CO.
3. Allen, L. V., Popovich, N. G., and Angel, H. C. (2005), *Pharmaceutical dosage forms and drug delivery systems*, Lippincott Williams & Wilkins, Baltimore, MD, USA.
4. Mountfield, R. J., Senepin, S., Schleimer, M., Walter, I., and Bittner, B. (2000), Potential inhibitory effects of formulation ingredients on intestinal cytochrome P450, *Int. J. Pharm.*, 211, 89–92.
5. Cornaire, G., Woodley, J., Hermann, P., Cloarec, A., Arellano, C., and Houin, G. (2004), Impact of excipients on the absorption of P-glycoprotein substrates in vitro and in vivo, *Int. J. Pharm.*, 278, 119–131.
6. Wang, S.-W., Monagle, J., McNulty, C., Putnam, D., and Chen, H. (2004), Determination of P-glycoprotein inhibition by excipients and their combinations using an integrated high-throughput process, *J. Pharm. Sci.*, 93, 2755–2767.
7. Karsa, D. R., and Stephenson, R. A. (1995), *Excipients and Delivery Systems for Pharmaceutical Formulations*, The Royal Society of Chemistry, Cambridge, pp. 1–34.
8. Farber, L., Tardos, G. I., and Michaels, J. N. (2003), Evolution and structure of drying material bridges of pharmaceutical excipients: Studies on a microscope slide, *Chem. Eng. Sci.*, 58, 4515–4525.
9. Cao, Q.-R., Choi, Y.-W., Cui, J.-H., and Lee, B.-J. (2005), Formulation, release characteristics and bioavailability of novel monolithic hydroxypropylmethylcellulose matrix tablets containing acetaminophen, *J. Controlled Release*, 108, 351–361.
10. Jackson, K., Young, D., and Pant, S. (2000), Drug-excipient interactions and their affect on absorption, *PSTT*, 3(10), 336–345.
11. Patel, H., Stalcup, A., Dansereau, R., and Sakr, A. (2003), The effect of excipients on the stability of levothyroxine sodium pentahydrate tablets, *Int. J. Pharm.*, 264, 35–43.
12. Verma, R. K., and Garg, S. (2004), Compatibility studies between isosorbide mononitrate and selected excipients used in the development of extended release formulations, *J. Pharm. Biomed. Anal.*, 35, 449–458.
13. Young, W. R. (1990), Accelerated temperature pharmaceutical product stability determinations, *Drug Dev. Ind. Pharm.*, 16(4), 551–569.
14. Waterman, K. C., and Adami, R. C. (2005), Accelerated aging: Prediction of chemical stability of pharmaceuticals, *Int. J. Pharm.*, 293, 101–125.

15. Simon, P., Veverka, M., and Okuliar, J. (2004), New screening method for the determination of stability of pharmaceuticals, *Int. J. Pharm.*, 270, 21–26.
16. Florence, A. T. and Attwood, D. (1998), *Physicochemical principles of pharmacy*, 3rd ed., Macmillan, London, pp. 5–35, 101–151.
17. Hiestand, E. N. (1989), The basis for practical applications of the tableting indices, *Pharm. Technol. Int.*, 8, 54–66.
18. Carstensen, J. T. (1993), *Pharmaceutical Principles of Solid Dosage Forms*, Technomic Publishing, Lancaster, PA.
19. Brittain, H. G. (1995), *Physical Characterization of Pharmaceutical Solids*, Marcel Dekker, New York.
20. Sinko, P. J. (2006), *Physical Pharmacy and Pharmaceutical Sciences*, Lippincott Williams and Wilkins, Baltimore, MD.
21. Carson, J. W. and Wilms, H. (2006), Development of an international standard for shear testing, *Powder Technol.*, 167, 1–9.
22. Hedenus, P., Mattsson, M. S., Niklasson, G. A., Camber, O., and Ek, R. (2000), Characterization of instantaneous water absorption properties of pharmaceutical excipients, *Int. J. Pharm.*, 141, 141–149.
23. Hasegawa, S., Hamaura, T., Furuyama, N., Kusai, A., Yonemochi, E., and Terada, K. (2005), Effects of water content in physical mixture and heating temperature on crystallinity of troglitazone-PVP K30 solid dispersions prepared by closed melting method, *Int. J. Pharm.*, 302, 103–112.
24. Mullarney, M. P., Hancock, B. C., Carlson, G. T., Ladipo D. D., and Langdon, B. A. (2003), The powder flow and compact mechanical properties of sucrose and three high-intensity sweeteners used in chewable tablets, *Int. J. Pharm.*, 257, 227–236.
25. Meyer, K., and Zimmermann, I. (2004), Effect of glidants in binary powder mixtures, *Powder Technol.*, 139, 40–54.
26. Hancock, B. C., and Zografi, G. (1997), Characteristics and significance of the amorphous state in pharmaceutical systems, *J. Pharm. Sci.*, 86(1), 1–12.
27. Willart, J., Lefebvre, J., Danède, F., Comini, S., Looten, P., and Descamps, M. (2005), Polymorphic transformation of the G-form of d-sorbitol upon milling: Structural and nanostructural analyses, *Solid State Communi.*, 135(8), 519–524.
28. Adolfsson, A., and Nystrom, C. (1996), Tablet strength, porosity, elasticity and solid state structure of tablets compressed at high loads, *Int. J. Pharm.*, 132, 95–106.
29. Caramella, C., Colombo, P., Conte, U., Ferrari, F., and Manna, A. L. (1986), Water uptake and disintegration force measurements: Towards a general understanding of disintegration mechanism, *Drug Dev. Ind. Pharm.*, 12, 1749–1766.
30. Pourkavoos, N., and Peck, G. E. (1993), The effect of swelling characteristics of superdisintegrants on the aqueous coating solution penetration into the tablet matrix during the film coating process, *Pharm. Res.*, 10(9), 1363–1371.
31. Lerk, C. F., Bolhuis, G. K., Smallenbroek, A. J., and Zuurman, K. (1982), Interaction of tablet disintegrations and magnesium stearate during mixing II. Effect on dissolution rate, *Pharm. Acta Helv*, 57, 282–286.
32. Gundert-Remy, U., and Moller, H. (1990), *Oral Controlled Release Products, Therapeutic and Biopharmaceutical Assessment*, Wissenschaftliche Verlagsgesellschaft mbH, Stuttgart, Germany, pp. 155–173.

6.2

ROLE OF PREFORMULATION IN DEVELOPMENT OF SOLID DOSAGE FORMS

OMATHANU P. PERUMAL AND SATHEESH K. PODARALLA
South Dakota State University, Brookings, South Dakota

Contents

Pharmaceutical Manufacturing Handbook: Production and Processes, edited by Shayne Cox Gad

6.2.1 INTRODUCTION

The advent of combinatorial chemistry and high-throughput screening (HTS) has exponentially increased the number of compounds synthesized and screened during the drug discovery phase. However, the overall efficiency of the drug discovery process is still exceedingly low (only 1 in 10,000 makes it to the market). Drug discovery is mostly driven by "activity screens" with little emphasis on "property screens." This is exemplified by the fact that 40% of attrition in drug discovery and development is attributed to poor biopharmaceutics and pharmacokinetic properties [1], which in turn are related to poor physicochemical properties. As a result, pharmaceutical companies have started to redesign their strategies by including property screens quite early in the discovery stage [2]. Preformulation is the study of fundamental properties and derived properties of drug substances. In other words, preformulation is the first opportunity to learn about the drug's physicochemical properties from the perspective of transforming a biologically active molecule to a "druggable" molecule. The type and extent of preformulation activities vary in a discovery and generic setting.

The main goal of a drug discovery program is to develop an orally deliverable molecule for obvious reasons of ease of manufacture and convenience of drug administration. More than 75% of the drug products in the market are oral formulations, of which more than half are solid dosage forms [3]. The "rule of five" developed by Christopher Lipinski [4] is one of the "physicochemical screens" to weed out molecules with poor physicochemical properties very early in the drug discovery process. According to Lipinski's rule, a drug will show poor oral absorption if it does not conform to any of the two physicochemical requirements listed in Table 1. The rule of five is applicable only to small molecules and it relates the chemical properties of the drug to its solubility and permeability characteristics. During the initial stages of drug discovery, the preformulation activities are mainly focused on developing a water-soluble compound for early activity studies and preclinical testing in animals. At this stage, the preformulation scientist is faced with the challenge of working with a limited quantity of compound (few milligrams) for testing a long list of physicochemical parameters. On the other hand, developing preclinical formulations can be quite a daunting task given the fact that toxicological studies require a high dose of drug (10–100 times above the effective dose) to be delivered in a small volume of the formulation. Preformulation activities increase as the molecule proceeds through the development phase. The "discovery and development phar-

TABLE 1 Lipinski Rule of Five for Orally Active Compounds

Physicochemical Parameter	Lipinski rule
Molecular weight	Not more than 500 Da
log *P*	Not more than 5
Hydrogen bond donors	Not more than 5 hydrogen bond donors expressed as the sum of OH's and NH's
Hydrogen bond acceptors	Not more than 10 expressed as the sum of OH's and NH's

maceutics" documentation forms a significant portion of the investigational new drug application (IND) application and new drug application (NDA) filed to the U.S. Food and Drug Administration (FDA). In a generic setting, preformulation studies are mainly focused on developing a formulation that is bioequivalent to the innovator's product with the main objective of filing an abbreviated new drug application (ANDA). A strong preformulation team can generate intellectual property in the form of new salts, solid-state forms, or new stabilized formulations of the drug for an innovator and/or a generic manufacturer.

In the present chapter, the discussion is mainly focused on preformulation testing for oral solid dosage forms in a drug discovery setting. The chapter address the following goals of preformulation: (i) to gain knowledge about the physicochemical characteristics of the drug, (ii) to define the physical characteristics of the drug, (iii) to understand the stability characteristics of the drug, and (iv) to determine the compatibility of excipients with the drug. In this chapter, we have grouped all the parameters under three sections and discussed in a logical sequence for the convenience of the reader.

6.2.2 PHYSICAL/BULK CHARACTERISTICS

The bulk or physical characteristics of a drug substance are mainly dictated by its solid-state properties. Purity of the drug substance is a fundamental property that is characterized at the beginning of preformulation studies. In the initial stages of drug development, the drug is usually not very pure. Nevertheless, it is essential to know the purity of the material at hand using simple measurements such as melting point. This would serve to set drug specifications during later stages of drug development. Differential scanning calorimetry (DSC) requires very little sample (1–5 mg) and is a useful tool to estimate the purity of the compound. The drug sample is heated in a crucible, where the difference in heat between the sample and a reference crucible is seen as an endotherm or exotherm in the thermogram depending on whether heat is taken up or given up, respectively, by the sample. The integrated area under the endotherm or exotherm gives a measure of the heat or enthalpy involved in this process. Melting is seen as an endothermic event and the purity of the sample will govern the peak position, shape, sharpness, and heat of fusion (ΔH_f). DSC is sensitive in detecting impurities to the extent of 0.002 mol% [5]. The DSC findings should be substantiated by a stability-indicating high-performance liquid chromatography (HPLC) assay. On the other hand, thin-layer chromatography (TLC) may be used to qualitatively detect the number of impurities in the drug sample. Impurity profiling is an important aspect of the drug development process, particularly for optimizing the synthetic process. The impurities can originate from many sources, including starting materials, intermediates, synthetic processes, or degradation reactions [6]. The regulatory guidelines stipulate that any impurity >0.05% of total daily dose (for drugs with a dose <2 g/day) or >0.15% of total daily dose (for drugs with a dose >2 g/day) should be evaluated for its safety [6]. Organoleptic properties such as color, taste, and odor are assessed qualitatively to set bulk drug specifications. If the drug has an unacceptable taste or odor, the chemistry group is advised to make a suitable salt form of the drug.

6.2.2.1 Crystallinity and Polymorphism

The majority of the drugs exist in crystalline form and are characterized by their crystal habit and crystal lattice. The crystal habit describes the external morphology of the crystal, including shape and size, while the crystal lattice describes the internal arrangement of molecules in the crystal (Figures 1*a* and *b*). Drug molecules arrange in more than one way in a crystal, and this difference in the internal arrangement of crystals is known as polymorphism. The polymorphs have the same elemental composition but differ in their physical, chemical, thermodynamic, stability, and spectroscopic properties. A crystal lattice represents the space in which molecules arrange in different ways. Organic molecules arrange in one or more of the seven possible crystal systems: triclinic, monoclinic, orthorhombic, tetragonal, rhombohedral, hexagonal, and cubic [7]. Each crystal system is characterized by its three-dimensional geometry and angles between the different crystal faces. The crystal lattice geometry is obtained using single-crystal X-ray diffractometry (XRD) and the details can be found elsewhere [7]. The difference in the crystal lattice of a drug arises as a result of the difference in packing of the molecules if the molecules are conformationally rigid (e.g., chlordiazepoxide) or due to the differences in conformation for flexible drug molecules (e.g., piroxicam). Although polymorphs differ in their internal crystal lattice, it may not be necessarily reflected in their external crystal habit (Figure 1*b*). In other words, a drug can exist in different crystal habits without any change in the internal crystal lattice (isomorphs).

Crystal habit is mainly dependent on crystal growth conditions [8]. For example, Figure 1*a* shows two different crystal habits for a given crystal lattice. A prismatic crystal habit will result if the growth is equal in all directions, while plates are formed if the growth is slow in one direction. Alternatively, needle-shaped crystals (acicular) are formed when the growth is slow in two directions. Thus, the crystal habits can

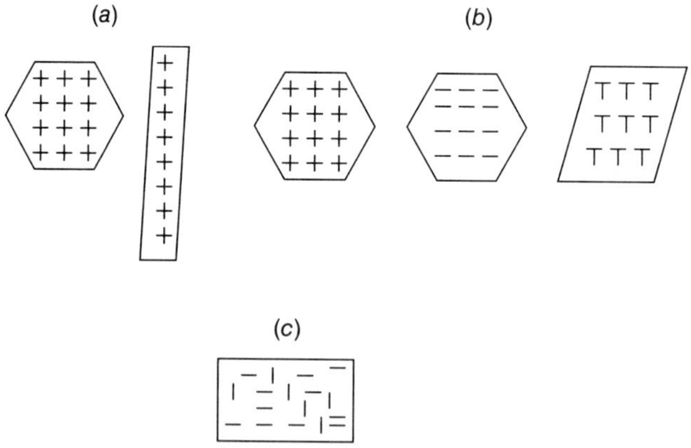

FIGURE 1 Schematic of crystal habits, polymorphs and amorphous drug forms. (*a*) Two crystal habits are shown. The internal crystal lattice is the same while the external morphology is different. (*b*) In a crystal the molecules are arranged in a regular fashion. However, the arrangement may vary depending on how the molecules orient themselves in the internal crystal lattice. The internal crystal lattice is different in all the three polymorphic forms. The polymorphs may or may not differ in their external morphology. (*c*) Random arrangement of molecules in amorphous form.

be altered without any change in the internal crystal lattice by varying the crystallization conditions. The polarity of the crystallizing solvent mainly influences the crystal habit by preferentially adsorbing to one surface of the crystal face. Similarly, surfactants or additives are added to the crystallization medium to prevent or promote the growth of a specific crystal habit [8]. Crystal habits mainly differ in physicomechanical properties such as packing, flow property, compressibility, and tablettability. Acetaminophen crystallizes as polyhedral crystals when crystallized from water and as plates when crystallized from ethanol–water (Figure 2*a*). Both these crystal habits are isomorphic [9], that is, have the same internal crystal arrangement, since their melting points and heats of fusion were similar (melting point 178 °C and ΔH_f = 177 kcal/mol). The polyhedral crystals have better flow and

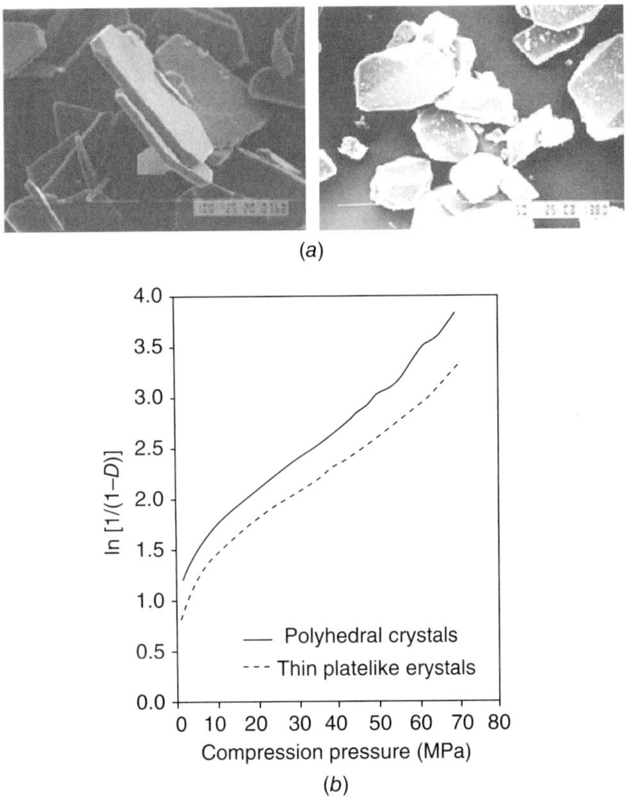

FIGURE 2 Difference in crystal habit of acetaminophen and resultant difference in compressibility (*a*) Acetaminophen crystallizes as either platy crystals or polyhedral crystals depending on the solvent of crystallization. Both crystal habits have the same internal crystal lattice since they showed the same melting point. (*b*) Difference in compression behavior of two crystal habits. The *x* axis represents the compression pressure while the *y* axis represents the densification of the drug sample on compression. This plot is known as Heckel plot. The polyhedral crystal habit shows a higher densification implying better compressibility than plate like crystals. [From Garekani, H. A., Ford, J. L., Rubinstein, M. H., and Rajabi-Sahboomi, A. R., *International Journal of Pharmaceutics*, 187, 77–89, 1999. Reproduced with permission from Elsevier.)

compression properties than platy crystals, which were brittle and fragmented during tableting (Figure 2*b*). Crystal habits are characterized using optical and electron microscopy, but their internal crystal lattice should be confirmed using DSC, XRD, and spectroscopic techniques.

Polymorphs are generated by crystallizing the drug from various solvents. The solvents are usually those that are used in the synthesis and purification of the bulk drug but may also include solvents used in drug formulations [10]. By convention, polymorphs are named based on their order of discovery, such as forms I, II or A, B or α, β. In general form I is considered the most stable and least soluble form, while form II is considered the more soluble and least stable form. The least stable and more soluble polymorphic form is usually called the metastable form. They are not "unstable" but "metastable," because the least stable form can remain stable provided the conditions are controlled to prevent its conversion to the more stable polymorphic form. Polymorphs are characterized by their solubility and stability differences with respect to temperature (Figure 3). Thermodynamically, polymorphs are classified as enantiotropic or monotropic depending on their thermal reversibility from one form to another [11]. Enantiotropic polymorphs are reversible polymorphs, where one form (form I) is more stable at higher temperature, while the other form (form II) is stable at lower temperature. They are characterized by a transition point (T_s) below the melting points of both forms (Figure 3*a*). The transition point represents the temperature at which one form converts to another. In the temperature–solubility curve, this is represented by the intersection of the solubility curves of both forms; that is, at the transition temperature, both polymorphs have the same solubility. As shown in Figure 3*a*, form II can convert to form I at a temperature above T_s, while form I can convert to form II at a temperature below T_s. On the other hand, monotropic polymorphs are not reversible but can only convert from the metastable form to the stable form. Here, T_s is higher than the melting point of both forms (Figure 3*c*). Both forms are stable in the entire temperature range below T_s.

The different polymorphs are generated based on their solubility differences in a given solvent. According to Ostwald's rule [12], the least stable or highly energetic form (form II, or metastable) will precipitate out first from a supersaturated solution followed by the stable or less energetic form (form I). Supersaturation is achieved by antisolvent addition or by altering the temperature. So, if the initial precipitate is separated rapidly, it would have predominantly the metastable form. Alternatively, the stable form can be melted and rapidly cooled to crystallize the metastable form. A stable or metastable polymorphic form is also used as a "seed" to preferentially grow and isolate the desired form during drug crystallization [13]. Several rules have been proposed to differentiate enantiotropic and monotropic polymorphs [11, 13]. A simple way to differentiate enantiotropic and monotropic polymorphs is the use of the heat–cool cycle in DSC [11]. As shown in Figure 3*b*, the enantiotropic polymorph is characterized by the appearance of solid–solid endothermic transition of form II to I followed by melting of form I. On cooling the melt of form I followed by reheating, the same thermogram is regenerated, proving the reversibility of the polymorphs. In monotropic polymorphs (Figure 3*d*), the thermogram is characterized by melting of metastable (form II) and recrystallization to form I followed by melting of form I. On cooling and reheating the sample, the transition and recrystallization peaks are not seen, indicating the irre-

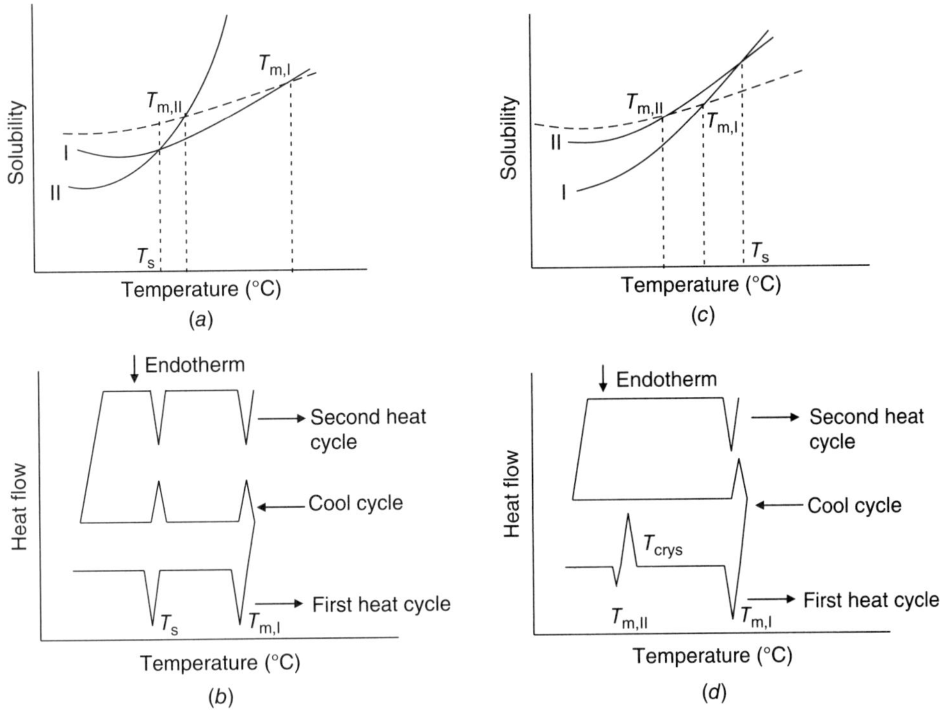

FIGURE 3 Difference between enantiotropic and monotropic polymorphs. (*a*) Solubility of enantiotropic polymorphs as function of temperature. The dotted line indicates the melting curve. Form I is less soluble below the transition temperature (*T*s), while form II is more soluble above *T*s. Form I has a higher melting point (*T*m,I) than form II (*T*m,II). Below *T*s, form I is converted to form II and above *T*s form II coverts to form I. (*b*) Thermogram generated from heat–cool–heat cycle in DSC. In the first heating cycle two endotherms are seen corresponding to conversion of form II to form I and melting of form I, respectively. On cooling both events show up as exotherms and on second heating cycle both endotherms reappear, indicating thermal reversibility of enantiotropic polymorphic pairs. (*c*). Solubility of monotropic pairs as a function of temperature. The *T*s is above melting point of both forms. Forms I and II are stable in entire temperature range and their corresponding melting points are shown. (*d*) On heating in DSC, form II melts (*T*m,II) followed by recrystallization (*T*crys) and subsequent melting of form I (*T*m,I). In cooling cycle only melting of form I is seen as an exotherm and on reheating only one endotherm corresponding to form I is seen. This is typical of monotropic polymorphs which converts from form II to stable form I and not vice versa.

versible nature of these polymorphs. The heating rate in DSC is critical for characterizing the polymorphs, as a faster heating rate may not be able to identify the transition temperature, while a lower heating rate may lead to lower resolution of peaks. Therefore, it is a usual practice to generate DSC thermograms under different heating rates during polymorph characterization [11]. Also it is important to note that the sample preparation, particle size, and crucible type can affect the quality of the thermogram [5]. XRD is also another indispensable tool in identifying polymorphs. This is based on the differential scattering of X rays when passed

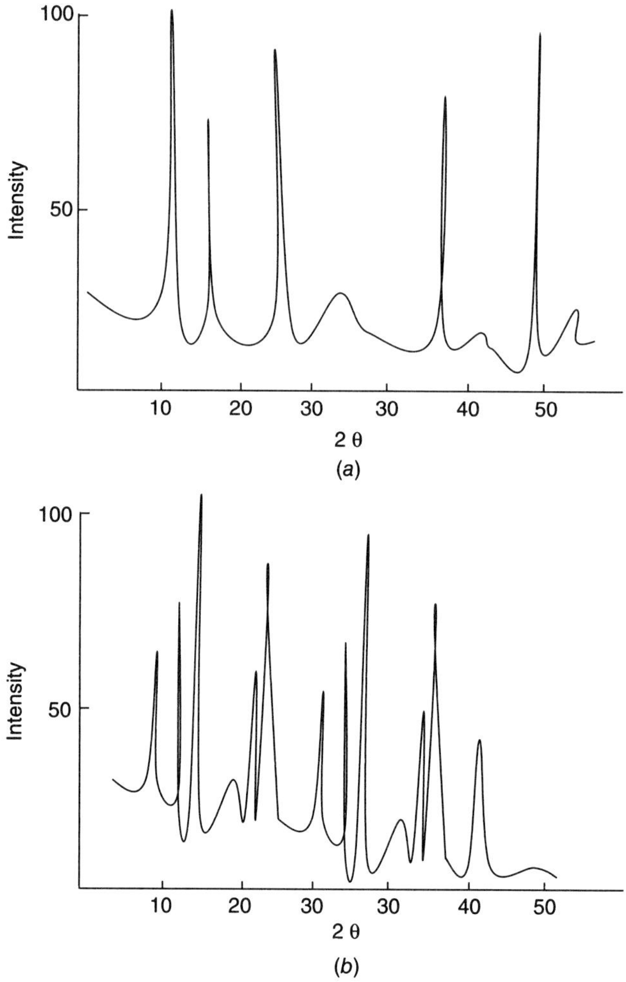

FIGURE 4 Schematic X-ray diffractograms of two polymorphic forms of a hypothetical drug. The x axis represents the detection angle and the y axis represents the intensity of the peak. As can be seen, there is a difference in the diffractograms due to the difference in the internal crystal lattice of polymorphs. The different internal arrangement in a crystal deflects the X ray at different angles.

through a powder sample. Typically, on passing through a powder sample, X rays will tend to get diffracted at various angles, and at some angle of detection, the X rays diffracted from the different planes of the crystal converge to form an amplified signal, which is detected by a photomultiplier tube. The angles at which the XRD peaks are obtained are characteristic for a polymorph (Figures 4*a* & *b*). The sample should be uniformly spread to get a good X-ray diffractogram, as an improper sample preparation may lead to variation in intensities due to the preferred orientation of a crystal in the XRD sample holder [14]. Other techniques, such as infrared (IR) spectroscopy and solid-state nuclear magnetic resonauce (NMR), are also used to characterize the polymorphs and are listed in Table 2.

TABLE 2 Techniques to Characterize Different Crystalline Forms

Technique	Applications
Thermal analysis	
Differential scanning calorimetry	Melting point, enthalpy of fusion, and crystallization; solid-state transformations
Thermogravimetric analysis	Stoichiometry of solvates and hydrates; identifying vaporization and volatilization
Hot-stage microscopy	Visualization of solid-state transformations and desolvation events
X-ray diffractometry	Identifying polymorphs; quantification of degree of crystallinity; crystal lattice geometry and solid-state transformations
Spectroscopy	
Infrared	Characterization of polymorphs based on functional groups; characterization of hydrates and solvates
Near infrared	In situ analysis of solid-state conversions; identification and quantification of polymorphs in dosage forms
Nuclear magnetic resonance	Useful to understand difference in molecular arrangement of polymorphs, hydrates, and solvates

TABLE 3 Difference in Solubilities of Polymorphs

Drug	Melting Point (°C)	Solubility Ratio[a]
Indomethacin	157, 163	1.1
Sulfathiazole	177, 202	1.7
Piroxicam	136, 154	1.3

[a]Indicates ratio of solubility of low-melting polymorphic form to solubility of high-melting polymorphic form of drug.

Polymorphism has significant implications in the solubility, bioavailability stability, processing, packaging, and storage of solid drug substances [15–17]. The metstable polymorphic form may be used to improve the solubility of drug substances. Many drugs are known to exhibit polymorphism, particularly, steroids, barbiturates, anti-inflammatory drugs, and sulfonamides, which have a high probability of exhibiting polymorphism [15]. The existing knowledge on drug polymorphism is a good starting point for a preformulation scientist to anticipate polymorphs based on the drug chemistry. In some cases, polymorphism may provide an opportunity to improve the solubility of a drug. For example, form II of chloramphenicol palmitate has a higher dissolution rate resulting in significantly higher plasma concentration than form I when administered orally [15]. However, in many cases [16] the difference in solubility may not be significant enough to cause differences in oral bioavailability (Table 3). Although the polymorphs differ in their dissolution rates, it should be realized that once the drug goes into solution, they do not differ in their properties. If a drug's absorption is limited by its poor membrane permeability, then the difference in solubility of polymorphs may not impact its bioavailability. Similarly, if the drug dissolution is rapid in comparison to the gastrointestinal (GI) transit time, then the difference in polymorph solubility will not influence its bioavailability [16].

More than the presence of metastable polymorph, it is the conversion of the metastable to the stable form during processing, storage, or use that is of great concern to the pharmaceutical scientist [17]. The unpredictability in "conditions" that result in the generation and conversion of one polymorphic form to another mainly aggravates such a situation [18]. This is exemplified by ritonavir, which is a classical case of appearance of a "new polymorphic form" after the drug was marketed [19]. Ritonavir, an anti-retroviral, drug was introduced in the market as a polymorphic form I in soft gelatin capsule in 1996. Two years later, a new polymorphic form II appeared in the formulation due to some unknown reasons causing the drug to be less soluble. The drug manufacturer withdrew the product due to failure of the batches in dissolution tests. After extensive investigation and reformulation, the drug was reintroduced in the market in 1999.

Nonetheless, in spite of their unpredictability, a good preformulation team will be able to anticipate the different polymorphs during drug development. It should not be an impossible task given the recent advancements in high-throughput generation and characterization of polymorphs [20]. From an innovator's perspective, the identification and thorough characterization of multiple polymorphs during drug discovery can extend the patent life and delay the market entry of generic manufacturers. On the other hand, it is also an opportunity for the generic manufacturers to generate new polymorphs with better solubility and stability for gaining market entry. The patent dispute on ranitidine polymorphs is a good example in this regard [21]. Two polymorphic forms of ranitidine were patented by the innovator company and the generic manufacturers had to find an appropriate method to manufacture the desired polymorph without the accompanying impurity of the other polymorph. This provided an edge for the innovator to extend the drug's market exclusivity for a little longer than they would otherwise have had. If a pure polymorph cannot be generated, the extent of polymorphic impurity should be quantified and ensured from batch to batch. The preformulation scientist closely works with the synthetic chemist in setting specifications for polymorphs.

6.2.2.2 Hydrates/Solvates

In addition to drug molecules, solvent molecules also get incorporated in the crystal lattice, resulting in altered physicochemical properties. When the solvent is water, they are known as hydrates, while if it is any other solvent, they are known as solvates. They are also known as pseudopolymorphs or solvatomorphs. Hydrates are important in this regard as one-third of all marketed drugs are hydrates [13]. Depending on how the water is arranged inside the crystals, they are classified as isolated hydrates, channel hydrates, and ion-associated hydrates [13]. In isolated hydrates, the water molecules are separated from each other by the intervening drug molecules in the crystal lattice (e.g., cephadrine dihydrate). Channel hydrates result when water molecules are linked to one another forming a channel (e.g., theophylline monohydrate). The water molecules may be present either stoichometrically or nonstoichometrically within the crystal lattice. Ion-associated hydrates are typically seen when the water is metal ion coordinated (e.g., nedocromil zinc). Nonstoichometric channel hydrates are problematic due to the presence of diffusible water, which can easily move in and out of the crystal lattices [13, 22].

Hydrates or solvates are formed by crystallizing the drug in the presence of water or solvents. The hydrate formation is dictated by water activity in a given solvent

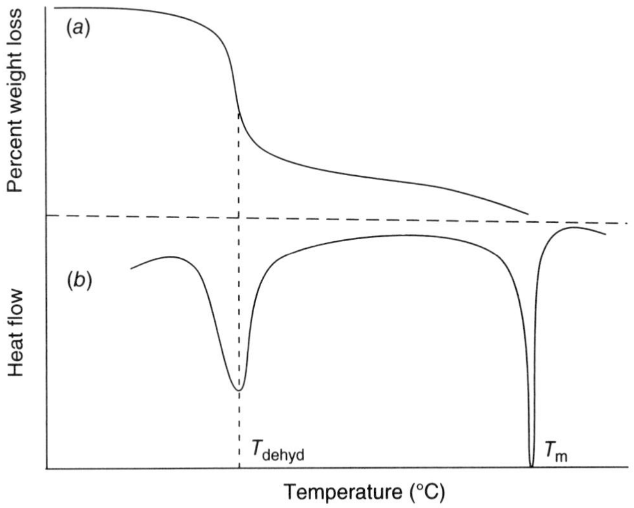

FIGURE 5 Characterization of hydrate. (*a*) TGA thermogram of monohydrate. The thermogram shows weight loss as a function of temperature. The step in the thermogram shows weight loss due to dehydration of a hydrate. (*b*) DSC thermogram showing endotherm at corresponding temperature (*T*dehyd). The second endotherm indicates the melting point of the hydrate (*T*m).

[22]. Hydrates are characterized using gravimetric methods such as thermogravimetric analysis (TGA) or by Karl Fischer titrometry [23]. In TGA, the loss of water/solvate on heating a sample is recorded as a thermogram (Figure 5). The mass change due to dehydration is seen as a step loss in the TGA thermogram. Based on the weight of the initial sample and its elemental composition, the number of water molecules can be calculated. The TGA curve in combination with a DSC thermogram helps to differentiate hydrates from other thermal transitions. In Figure 5, the endotherm in the DSC thermogram corresponds to water loss as indicated by the TGA curve. The TGA can be coupled to an IR or mass spectrometer to characterize solvates. Thermal microscopy is a useful qualitative tool to visualize the release of water from the drug crystals as a function of temperature [23].

Hydrate formation and dehydration significantly influences the processing and storage of drug products [17]. Hydrates may take up further water or dehydrate to lose water. Dehydration of hydrates leads to several possibilities [24], as shown in Figure 6. Hydrates on dehydration can form isomorphic desolvates retaining the same crystal lattice as the hydrate but without the water. Alternatively, hydrates can lose water and become anhydrous crystals. They can also lose water, forming amorphates with the loss of crystal lattice. Higher hydrates can lose water to form lower hydrates, for example, pentahydrate converting to di- or monohydrate. The hydrates also can exhibit polymorphism or on dehydration can convert to a different polymorphic form [13, 17]. Such solid-state transformations are possible during processing, such as granulation, tableting, and storage [17]. In general, hydrates are less soluble than anhydrous forms while solvates are more soluble than ansolvates in water. Ampicillin trihydrate is a classical example which shows lower solubility and lower plasma concentration than anhydrous ampicillin [15]. Preformulation studies

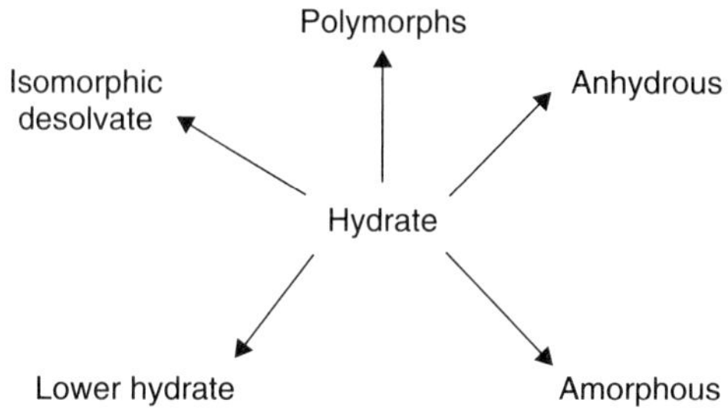

FIGURE 6 Various possibilities that arise from dehydration of hydrate. A hydrate can dehydrate reversibly into various solid-state forms. It can dehydrate to form an anhydrous form of the drug or to a lower hydrate. Hydrate can also dehydrate to form an isomorphic desolvate where the crystal lattice is retained except for the absence of water. The crystal structure may also collapse on dehydration to form an amorphous form. Hydrates on dehydration can also result in different polymorphs.

provide valuable inputs to the formulator in selecting a suitable form of the drug. For example, ampicillin is hygroscopic and hence can be used in suspension dosage forms, while ampicillin trihydrate, which is non-hygroscopic, is used in solid dosage forms.

6.2.2.3 Amorphate

Unlike a crystalline drug, an amorphous form of the drug does not have a regular crystal lattice arrangement and the molecules are arranged in random order (Figure 1*c*). Glass is a typical amorphous substance and so amorphous drugs are also known as glasses [25]. Amorphous form is prepared through milling, rapid cooling of a melt, rapid precipitation using an antisolvent, rapid dehydration of a hydrate, spray drying, or freeze drying [26]. Some of the above methods may also unintentionally produce an amorphous form during processing of the crystalline form of the drug [17]. For instance, milling during dosage form manufacture may produce an amorphous form unintentionally, as in the case of indomethacin. The amorphous form does not show a melting point but is characterized by a glass transition temperature (T_g). This temperature indicates the conversion of the amorphous form from a rigid glassy state to a more mobile rubbery state. Above T_g the amorphous form will tend to recrystallize and convert to the crystalline form, which then undergoes melting, as shown in Figure 7*a*. The T_g for an amorphous drug can vary depending on the storage conditions and thermal history of the sample and is sensitive to moisture, pressure, and temperature [26]. The T_g is seen only as a slight shift in the baseline due to a change in the specific heat capacity of the sample and is influenced by the heating rate in DSC [25]. In XRD, the amorphous form shows a shallow peak or halo, as opposed to sharp and intense peaks for a crystalline drug compound (Figure 7*b*).

The main advantage of amorphous form of the drug substance state is its significantly higher solubility than the crystalline form of the drug, primarily due to the excess surface energy [16, 27]. Therefore, conversion of a crystalline drug into an amorphous form is one of the strategies to increase drug solubility. Table 4 compares

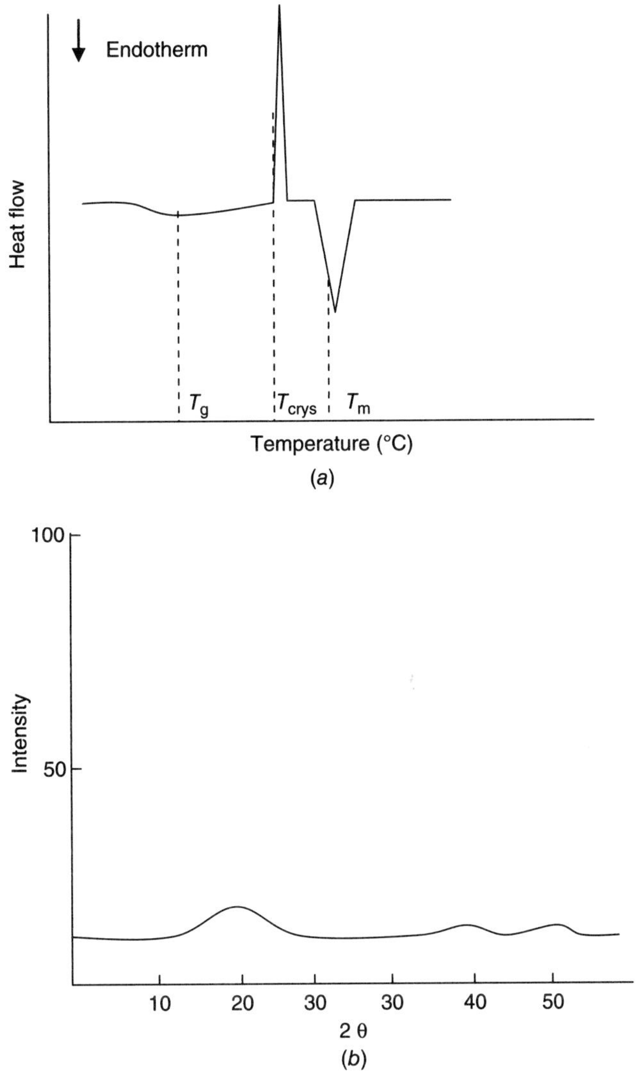

FIGURE 7 Characterization of amorphous form. (*a*) DSC thermogram of amorphous substance. Thermogram is characterized by a glass transition temperature (*T*g) above which the amorphous form is mobile and recrystallizes (*T*crys) into a crystalline form which finally melts (*T*m). (*b*) Amorphous form that does not show any peaks in XRD as it does not have regular arrangement of molecules. Shallow peaks are indicative of an amorphous drug substance.

the solubility of amorphous and crystalline forms of a few drugs. However, the biggest challenge lies in the stabilization of the amorphous form to prevent it from converting to the less soluble crystalline form during storage and use. It should be noted that they can take up moisture to convert to a crystalline form or to a hydrate, resulting in decreased drug solubility. The moisture uptake can also lead to chemical degradation [26]. The stabilization of drug amorphates is usually accomplished by increasing the T_g using polymers and thus restricting their molecular mobility and

TABLE 4 Comparative Solubilities of Amorphous and Crystalline Forms of Drugs

Drug	Solubility Ratio[a]
Carbamazepine	1.5–1.7
Griseofulvin	38–441
Glibenclamide	14

[a]Indicates ratio of solubility of amorphous form to solubility of crystalline form of drug.

chemical reactivity [26]. Particle size reduction, which is a very common processing step in dosage form manufacturing, can result in varying degrees of amorphous and crystalline forms in the drug. In such cases it is essential to quantify the degree of crystallinity using various analytical techniques such as DSC and XRD [28].

6.2.2.4 Hygroscopicity

Some solid drug substances have a tendency to absorb moisture from the atmosphere leading to physical and/or chemical instabilities. Hygroscopicity is the rate and extent of moisture adsorbed/absorbed by a solid substance. Solid drug substances may vary in their behavior to moisture and are classified as deliquescent, efflorescent, and effervescent. Deliquescent materials such as hydrochloride salts absorb moisture and become a liquid. Effervescent substances (e.g., a mixture of citric/tartaric acid and sodium bicarbonate) absorb moisture and release carbon dioxide. On the other hand, efflorescent substances such as hydrates may lose moisture depending on the relative humidity (RH). Therefore, it is important to study the moisture absorption behavior of drugs to choose the processing and storage conditions for the drug.

Hygroscopicity is measured by exposing the drug sample to various RH in a dessicator. The RH is maintained at a constant level by using salt solutions of varying concentrations (e.g., KNO_3, KCl) and the humidity is expressed with respect to the humidity of a saturated salt solution. Moisture sorption and desorption curves are generated to study the moisture uptake. The moisture absorption profile is generated by noting the increase in mass on exposure to varying RH and the desorption profile is generated by recording the change in weight with decreasing RH [23, 29]. This can be measured using a dynamic vapor sorption instrument. A typical sorption/desorption profile is shown in Figure 8 for an anhydrous and hydrate form of a hypothetical drug. As can been, for a hydrate the sorption and desorption profile is superimposable and is hence non-hygroscopic. On the other hand, the anhydrous form of the drug is hygroscopic and shows hysteresis on the sorption and desorption profile, indicating significant moisture uptake. Such profiles give useful clues to the preformulation scientist. Significant hysteresis is indicative of hydrate formation and can be used as a guide to evaluate the potential of hydrate formation. Further the profile also helps to differentiate hygroscopic and non-hygroscopic salts [29]. The profile also gives information on processing and storage conditions that can overcome solid-state transformations. For example, in Figure 8, it is seen that the anhydrous form does not take up moisture if RH is below 80% and it retains the moisture

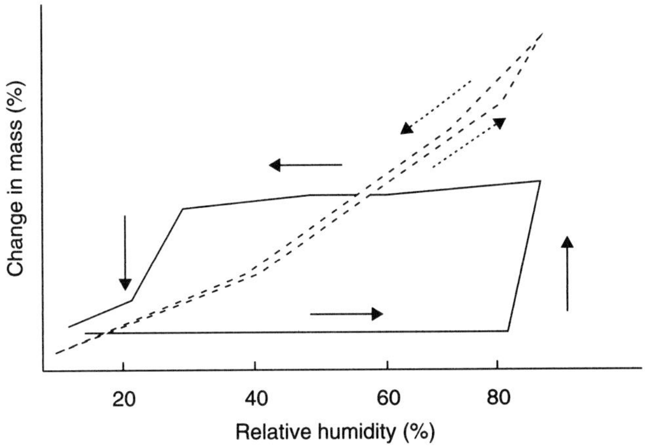

FIGURE 8 Sorption and desorption profile of hydrate and anhydrous form of hypothetical drug. The solid lines represent the profile for an anhydrous form of the drug, while the broken lines represent the profile for a hydrate form of the drug. Anhydrous form of the drug does not take up moisture until it reaches 80% RH, and on reducing the RH, it does not lose moisture until it reaches RH of 20%. The hysteresis is indicative of hygroscopicity and significant moisture uptake. The hydrate form of the drug does not show hysteresis but both the sorption and desorption curve superimpose on each other indicating that it is non-hygroscopic.

until the RH is reduced below 20%. The study can also be used to extract kinetic and temperature information on moisture uptake by drug substances. The studies conducted during preformulation testing should be representative of the anticipated processing and storage conditions of the drug.

6.2.2.5 Particle Characteristics

Drug particle characteristics such as size, shape, and surface area impact the drug's processability and product performance. Particle size is the most influential among these as the other two properties can be related to it. When sufficient drug is available, the preformulation scientist characterizes the particle size and size distribution to set specifications for formulation and future drug lots. Table 5 lists the various methods used to measure particle size. The methods differ in their principle of operation and also in the range of particle sizes they can measure [30]. Usually, the gross particle morphology is characterized using a simple optical microscope and if required is further characterized using a scanning electron microscope. Light-scattering methods are commonly used to measure particle size due to their low sample requirements and ease of measurement. The instrument readout is in the form of a graph where the particle size is plotted against the percent frequency of particles in different particular size ranges. The results are used to set particle size specifications and understand polydispersity or multimodal particle size distribution in powders. Especially if the drug is potent, a narrow size distribution is desired to ensure drug homogeneity during formulation. The surface area of a powder bed is determined using the Brunauer, Emmett, and Teller (BET) method. In this method,

TABLE 5 Methods for Particle Size Analysis

Method	Principle
Sieve	Sieve analysis utilizes a series or stack, or nest of electro brass or stainless steel sieves that have smaller mesh at the bottom followed by meshes that become progressively coarser toward the top of the series. Useful for measuring particles in size range 10–50,000 μm
Microscopy	Analysis is carried out on two-dimensional images (projected diameter) of particles which are assumed to be randomly oriented in three dimensions. Can measure particles in the size range 1–1000 μm. Electron microscopy is useful for analysis of particles in submicrometer range (0.01–100 μm). It also gives information on surface morphology and shape of the powder.
Sedimentation	Size analysis is based on sedimentation of particles as a function of their size due to gravitational pull or by using a centrifugal force. Can measure particles in the size range 0.01–100 μm.
Light scattering	This is based on the principle of light scattering of particles as a function of their hydrodynamic radius. Commonly used, as it requires a small sample size and is a rapid method for particle size measurement. It can be used to measure particles varying in size from 0.001 to 100 μm.

N_2 gas is passed through a powder bed and the surface area is calculated based on the volume of gas coming out in the absence and presence of the powder [31].

Particle size reduction is an initial step in the development of any dosage form. The data generated during preformulation testing guides the formulator in deciding about size reduction. Particles with size >100 μm generally require size reduction and particles in the size range of 10–40 μm are generally acceptable for solid dosage forms [32]. As mentioned in the earlier section, particle size reduction may lead to partial or complete amorphization of a powder, and this factor should be taken into consideration. Also worth mentioning are the other solid-state transformations that may take place during milling, such as the conversion of one polymorphic form to another or the desolvation of a hydrate [17]. It is important to maintain the particle size distribution within a narrow range to avoid powder stratification and avoid flowability issues during capsule filling or tablet compression. Particle size reduction increases drug solubility due to the enormous increase in surface area with decreasing particle size. Griseofulvin is often widely quoted in the literature in this regard, where the bioavailability of this water-insoluble drug is increased 10 times on reducing the particle size [32]. On the other hand, particle size reduction may be counterproductive for some drugs such as nitrofurantoin. Particle size reduction of nitrofurantoin causes rapid drug absorption with an associated increase in its adverse effects. In contrast, the slowly dissolving macrocrystals of nitrofurantoin do not cause adverse effects [32]. Excessive particle size reduction (<30 μm) may also lead to static charge buildup on the surface, resulting in agglomeration and reduced powder flow or reduced drug solubility. If the drug is sensitive to moisture or oxygen, then the increased surface area associated with particle size reduction may accelerate degradative reactions [32]. Therefore, it is important to study the influence of particle size on drug solubility and stability during preformualtion testing.

6.2.2.6 Powder Flow and Compressibility

The derived properties of a drug substance play a critical role in deciding about the feasibility of a solid dosage form. These include the bulk density, flow properties, and compressibility of the drug powder. Powder flow is influenced by many solid-state properties, including crystal habit, bulk density, particle size, and shape. Bulk density is an important determinant of powder flow. It is the ratio of a known weight of the powder and its bulk volume. This is determined by pouring a weighed amount of the powder into a graduated cylinder and measuring its volume. Bulk density is particularly important for high-dose drugs, where the drug would occupy a major portion of the tablet or capsule dosage form. The true density of a powder is determined using a helium densitometer [31]. The volume of helium gas that passes through an empty tube is compared with the volume of helium passing through the tube when filled with a defined weight of the powder. The difference in the volume gives the true volume of the powder, which is then used to calculate the true density of the powder. This information can be used to calculate the porosity:

$$\text{Porosity} = \frac{\text{bulk volume} - \text{true volume}}{\text{bulk volume}} \times 100 \qquad (1)$$

The porosity of a powder depends on particle size and shape; pharmaceutical powders vary in porosity from 30 to 50% [31]. Powder with varying particle size will give a porosity of less than 30%, where the small particles may occupy the pores in between the larger particles. On the other hand, powder aggregates lead to increased porosity and poor flow properties.

The powder flow is determined using the angle of repose or Carr's index (Figures 9*a* and *b*). The angle of repose is the angle that the powder makes with the horizontal surface when allowed to flow through a funnel (Figure 9*a*). This is based on the principle that powder flow is influenced by the relative influence of interparticle friction and the gravitational pull on the powder. Pharmaceutical powders have an angle of repose of 25°–40° [33], and in general, a good flowing powder will have a lower angle of repose (Figure 9*c*). The angle of repose using the funnel provides a good estimate of the influence of particle size, shape, and electrostatic interaction between the particles when the powder flows through the hopper in a tableting machine [34]. When there is a limited drug sample, Carr's index is used to estimate powder flow and compressibility. In this method, a small quantity of the powder is used to determine its bulk density and this is followed by determining the tapped density of the powder. After filling, the graduated cylinder is tapped on a hard surface (3–30 taps) until the powder consolidates and gives a constant volume (Figure 9*b*). Carr's index is calculated using the equation

$$\text{Carr's index} = \frac{\text{tapped} - \text{poured density}}{\text{tapped density}} \times 100 \qquad (2)$$

This index is a good measure of powder consolidation and compressibility for predicting the feasibility of developing a tablet dosage form. A lower Carr's index is indicative of a good flowing powder. There is a good correlation between angle of repose and Carr's index [Figure 9*c*], and depending on the quantity of the drug

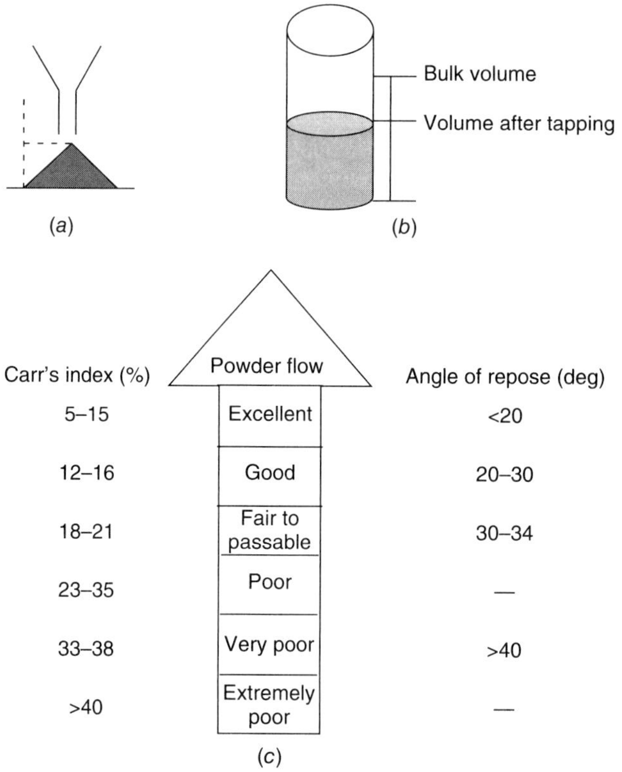

FIGURE 9 Measurement techniques for powder flow. (*a*) Angle of repose is determined by pouring a powder through a funnel and noting the angle that the powder heap makes with the horizontal surface. (*b*) Carr's index is measured by pouring a known weight of the powder into a graduated cylinder and tapping it on a hard surface until the powder is consolidated. (*c*) Carr's index is measured as percent, while angle of repose is measured in degrees, and both methods show good correlation. Powder flow is classified based on either of the measurement methods.

available for preformulation testing, either method may be used to estimate the powder flow property. Poor flowing powders may require glidants to improve their flow property. Compressibility is studied by compressing the drug in a hydraulic or IR press, and such experiments give early warnings to the formulator about capping and lamination issues in tablets. Altogether, these preformulation tests give valuable information to the formulator in deciding excipients and processes.

6.2.3 SOLUBILITY CHARACTERISTICS

Drug solubility is one of the physicochemical parameters that receives lots of attention during preformulation testing. In the initial stages, solubility studies are usually kinetically determined, where the drug is placed in contact with the solvents and then the solubility is assessed using turbidometric methods almost instantaneously.

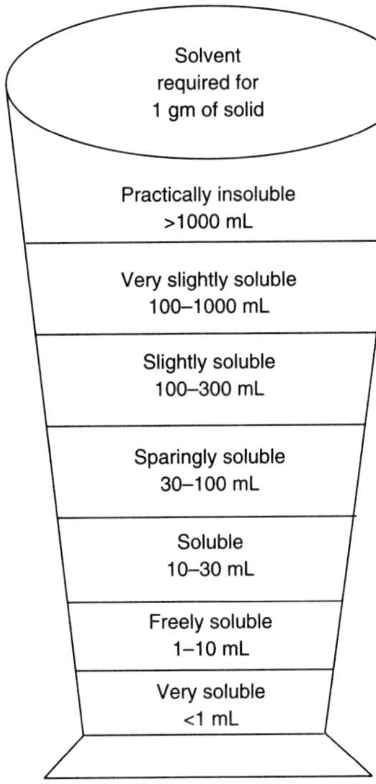

FIGURE 10 Terminologies for drug solubility. The drug solubility is qualitatively described depending on how much solvent is required to solubilize 1 g of the drug.

A high-throughput solubility screen consists of dissolving the drug in a minimal volume (few microliters) of dimethyl sulfoxide (DMSO) and then adding different volumes of water until turbidity is observed. The appearance of turbidity is considered as a rough estimate of the drug's water solubility and turbidity is measured in a 96-well plate format in a turbidometer [35]. Solubility is qualitatively described in terms of how much solvent is required to solubilize 1 g of the drug and is shown in Figure 10. Kinetic solubility measurements are only rough estimates, as they do not take into account the solid-state transitions and should be followed up with equilibrium solubility studies later when more drug is available.

Equilibrium solubility is determined by placing an excess solid drug in a few milliliters of the solvent and shaking at 37 °C for 60–72 h until equilibrium is reached. One to three samples are withdrawn, filtered, and assayed for drug content. Equilibrium solubility helps to identify polymorphic or amorphous forms of the drug, which shows an apparently higher solubility in kinetic studies. Intrinsic solubility measurements are measured for ionic compounds to determine the inherent solubility of the un-ionized form of the drug. This would mean that the solubility of a weakly acidic drug is tested in an acid medium and a weakly basic drug is tested in an alkaline medium, where the drug would remain completely un-ionized. The studies are followed by determining drug solubility at different pH and are discussed later in this section.

A drug solubility of 1 mg/mL is usually considered acceptable to avoid dissolution rate–limited absorption in vivo. If the drug solubility is <1 mg/mL, salts should be considered, if the drug is ionic and cosolvents should be considered if the drug is nonionic [33]. This brings us to a discussion of the factors that influence drug solubility: temperature, pH, crystal form, and solvents. The effect of temperature on drug solubility depends on the heat of solution (ΔH_s), which is the amount of heat given up or taken up when a drug goes into solution. The relationship between drug solubility S and temperature T is given by the van't Hoff equation, where R and C are constants:

$$\ln S = \frac{\Delta H_s}{RT} + C \qquad (3)$$

Drugs which have *−ve* ΔH_s (exothermic) generally decrease in solubility with increase in temperature (e.g., lithium salts and hydrochloride salts), while drugs which have *+ve* ΔH_s (endothermic) usually increase in solubility with increase in temperature (e.g., most organic drugs). Nonelectrolytes have a ΔH_s of 4–8 kcal/mol and show significant increase in solubility when the temperature is increased. On the other hand, salts have a ΔH_s of −2–2 kcal/mol and are therefore less sensitive to temperature [34]. Solubility studies are conducted at 25, 37, and 50 °C. In addition to providing information on the drug solubility at the body or processing temperatures, it is also useful to understand polymorphic interconversions.

With respect to selecting cosolvents, one should consider drug polarity and solvent polarity. Usually the solvents include glycerol, propylene glycol, and ethanol. Other solubilization techniques such as complexation and surfactants can also be used to enhance the solubility of the drug [36]. However, the solubilization techniques used in preclinical testing may not be same as the final formulation used in clinical studies and marketing.

For ionic substances, salt formation is the preferred strategy for drug solubilization. The salt selection is crucial during early discovery, as any change will require repeating some of the earlier preclinical studies. Salts provide wider flexibility in modulating the drug properties without changing its activity. The salt formation is used to address drug solubility, stability, and processing issues [37]. Sometimes, salt formation may be used to deliberately reduce the solubility of drug to overcome unpleasant taste or stability of the drug. For example, clindamycin pamoate is used in place of hydrochloride salt to overcome the unpleasant taste of the latter. The various factors to be considered in salt selection are discussed in the next section. In addition to improving the solubility of the drug, solubility data guides the formulator to choose a suitable granulating solvent for a tablet dosage form.

6.2.3.1 pK_a and Salt Selection

The majority of the pharmaceutical drugs are weak bases or weak acids. Among the marketed drugs, more than 75% are weak bases, 25% are weak acids, and 5% are nonionic [38]. Therefore, knowledge of pK_a is useful for enhancing drug solubility and stability. The Henderson–Hasselbalch equation is used to describe the ionization of a weak acid or base:

$$\text{Percent ionized} = \begin{cases} \dfrac{100}{1+10^{(\mathrm{p}K_\mathrm{a}-\mathrm{pH})}} & \text{for weak acid} \quad (4) \\[2ex] \dfrac{100}{1+10^{(\mathrm{pH}-\mathrm{p}K_\mathrm{a})}} & \text{for weak base} \quad (5) \end{cases}$$

The equations theoretically predict the ionization and solubility of a drug at a given pH. The $\mathrm{p}K_\mathrm{a}$, which is a characteristic property of an ionizable drug, defines the pH at which the drug is half ionized and half un-ionized. A weakly acidic drug will be predominantly in the un-ionized form two pH units below its $\mathrm{p}K_\mathrm{a}$ and predominantly ionized at two pH units above its $\mathrm{p}K_\mathrm{a}$. In a weakly basic drug, it is exactly the opposite of what is seen with an acidic drug. The $\mathrm{p}K_\mathrm{a}$ of a drug can be measured using potentiometry, solubility, conductometry, and spectroscopic techniques [35]. Potentiometry measures the change in potential when the drug is titrated with an acid or base and is suitable for drugs which have $\mathrm{p}K_\mathrm{a}$ of 3–10 [33]. If the drug is not water soluble, it is usually dissolved in a water-miscible solvent such as DMSO or methanol. In order to nullify the effect of cosolvents, measurements are made with various cosolvent concentrations and are plotted against $\mathrm{p}K_\mathrm{a}$. The intercept on the y axis gives the $\mathrm{p}K_\mathrm{a}$ at zero cosolvent concentration. Some drugs have more than one ionizable group, such as ampholytes, and are characterized by more than one $\mathrm{p}K_\mathrm{a}$ value. They are classified as ordinary or zwitterionic ampholytes [39]. In ordinary ampholytes, $\mathrm{p}K_{\mathrm{a,acidic}} > \mathrm{p}K_{\mathrm{a,basic}}$ (e.g., chlorambucil), while in zwitterionic ampholytes, $\mathrm{p}K_{\mathrm{a,acidic}} < \mathrm{p}K_{\mathrm{a,basic}}$ ampholytes (e.g., amino acids and proteins). The amino acids and proteins are characterized by their isoelectric point (pI), which is the pH at which the net charge is zero and is calculated using the formula

$$\mathrm{pI} = \frac{1}{2}(\mathrm{p}K_{\mathrm{a,acid}} + \mathrm{p}K_{\mathrm{a,base}}) \quad (6)$$

Drugs may also have more than two $\mathrm{p}K_\mathrm{a}$ values, such as polyprotic or polybasic compounds (e.g., minocycline), and such drugs exhibit a complex pH solubility profile. It is essential to know per se pH of the drug solution during preformulation studies. The pH is measured or theoretically calculated if the $\mathrm{p}K_\mathrm{a}$ and drug concentration C are known. The pH of a weak acid or the salt of a weak base and a strong acid can be calculated using the equation

$$\mathrm{pH} = \frac{(\mathrm{p}K_\mathrm{a} - \log C)}{2} \quad (7)$$

Similarly, the pH of a weak base or a salt of weak acid–strong base can be calculated using the following equation, where $\mathrm{p}K_\mathrm{w}$ is the ionization constant of water:

$$\mathrm{pH} = \frac{(\mathrm{p}K_\mathrm{w} + \mathrm{p}K_\mathrm{a} + \log C)}{2} \quad (8)$$

Solubility of a drug is directly proportional to the extent of ionization of a drug in water. Therefore, $\mathrm{p}K_\mathrm{a}$ of the drug may also be determined by noting the change in solubility of a drug as a function of pH. It is customary to check the drug solubility in the pH range usually encountered in the gastrointestinal tract (GIT; pH 1–8). The influence of pH on drug solubility can be estimated from the equations

$$S = \begin{cases} S_u[1 + 10(\text{pH} - \text{p}K_a)] & \text{for weak acid} \quad (9) \\ S_u[1 + 10(\text{p}K_a - \text{pH})] & \text{for weak base} \quad (10) \end{cases}$$

In this equation S represents the total solubility of ionized and un-ionized forms of the drug, while S_u represents the intrinsic solubility of un-ionized form of the drug. The pH solubility profile and the drug pK_a are important for salt selection. In addition, the salt selection is governed by various factors as outlined in Table 6. As a rule of thumb, a strong acid is used to form a salt with a weakly basic drug and a strong base is used to form a salt with a weakly acidic drug. The probability of salt formation can be predicted from the relative pK_a values of the drug and the counterions by using the equation [40]

$$\Delta \text{p}K_a = \text{p}K_{a,\text{drug}} - \text{p}K_{a,\text{counterion}} \quad (11)$$

The probability of salt formation is high for a weak acid if ΔpK_a is negative, that is, $\text{p}K_{a,\text{drug}} < \text{p}K_{a,\text{counterion}}$. Sodium phenytoin is an example, where phenytoin has a pK_a of 8.3 and sodium has a pK_a of 16. Similarly, a weak base will form a salt with acid counterion if ΔpK_a is positive, that is, $\text{p}K_{a,\text{drug}} > \text{p}K_{a,\text{counterion}}$. In atropine sulfate, atropine has a pK_a of 9.9 and sulfuric acid has a pK_a of –3. It is important to mention that salts do not alter the intrinsic pK_a of the drug but increase drug solubility by keeping pH on the ionization side of the drug's pK_a (Figures 11*a* and *b*). The salt formation is a futile exercise if the pK_a of a drug is <3 or >10 and other solubilization strategies have to be pursued [33].

Inorganic ions such as hydrochloride and sodium are the most frequently encountered species in pharmaceutical salts, primarily because of the ease of salt formation and their low molecular weights. They provide significant increase in drug solubility and at the same time strong acids/bases may also be hygroscopic, posing problems during processing and storage. In those cases, salts are formed using weaker organic anions or cations such as mesylate, besylate, and choline. Table 7 lists some of the commonly used inorganic and organic counterions used in pharmaceutical salts. The pH provided by the salts significantly influences the drug dissolution and subsequent drug absorption from the GIT. Though salts increase drug ionization and aqueous drug solubility, it is the un-ionized form of the drug that is absorbed through the membrane. According to the well-known pH partition hypothesis [33], a weakly

TABLE 6 Factors to Consider during Salt Selection

Relative pK_a of the drug and the counterion
Common-ion effects
Crystal habit and crystallinity
Polymorphic conversions
Hygroscopicity
Chemical stability
Manufacturability
Toxicity

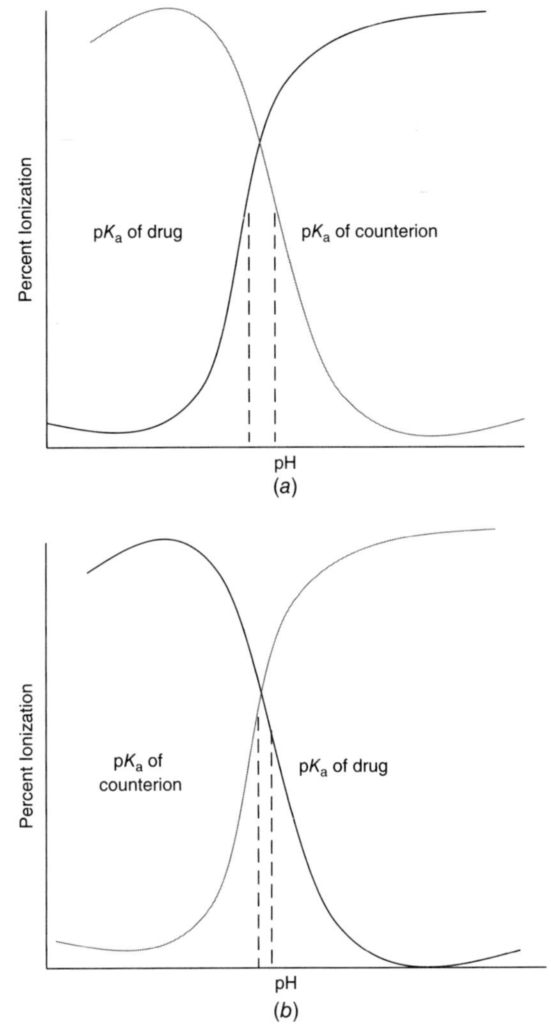

FIGURE 11 Relative p*K*a of drug and salt-forming counterion. (*a*) For a weak acid, the p*K*a of the salt-forming counterion should be higher than the drug p*K*a to keep the pH on the ionization side. (*b*) For a weak base, the salt-forming counterion should have a p*K*a less than the drug's p*K*a to keep the pH in the ionization side.

TABLE 7 Various Counterions Used to Form Drug Salts

Chemical class	Salt-Forming Counterions
Inorganic	Hydrochloride, hydrobromide, sulfate, nitrate, phosphate, sodium, potassium, calcium, and zinc
Organic	Triethanolamine, ethanolamine, lactic acid, maleic acid, citric acid, acetic acid, choline, ethanesulfonic acid, oleate, and stearate

acidic drug is primarily absorbed from the stomach (pH 1–3), where it remains in un-ionized form. A basic drug is expected to be absorbed from the intestines (pH 6.5–8), where it would be in the un-ionized form. But in many cases, the absorption of drugs cannot be satisfactorily explained by this theory. This is understandable if one considers the fact that drug dissolution from a salt is mainly influenced by the surface pH of the dissolving drug (microenvironment) rather than the bulk pH of the GI fluids. For instance, the surface pH of weakly acidic drug is 1.5 times higher than the bulk pH in the stomach, and therefore the dissolution of the salt of a weak acid will be 100 times faster than the free-acid form of the drug in stomach [40]. Similarly, it is the free base rather than its salt form which will dissolve faster in the gastric pH. So it is possible to modulate the drug solubility of a pharmaceutical salt irrespective of its location in the GIT. Considering the fact that most of the drugs are absorbed from the intestine (due to large surface area), it is desirable to select a salt that is not completely ionized or unionized at the intestinal pH to have optimal dissolution and absorption.

An important phenomenon that is often overlooked during the salt selection process is the suppression of salt ionization in GI fluids due to the common-ion effect. This is particularly important with inorganic salts, where salt ionization can be suppressed by ions such as chloride and sodium which are abundant in GI fluids. A hydrochloride salt will ionize in solution, as shown in the Equation (12), but in gastric fluid, the presence of chloride ions suppresses the drug ionization [as shown by the thicker arrow in the reverse direction in Equation (13)] to maintain an equilibrium between the ionized and un-ionized form of the drug:

$$DH^+Cl^- \rightleftarrows DH^+ + Cl^- \tag{12}$$

$$DH^+Cl^- \rightleftarrows DH^+ + Cl^- \tag{13}$$

Hence it is important to study this effect during preformulation by testing the solubility of the salt in the presence and absence of sodium chloride. Although salts do not alter the pharmacological activity of the drug, safety is an important consideration in the selection of salts. From this perspective, salts are treated as a new molecule by the FDA. The safety of the salt is evaluated with respect to its route of administration and dose of the drug [37].

6.2.3.2 Partition Coefficient

In simple terms, the partition coefficient represents the relative solubility of a drug in a hydrophobic and a hydrophilic solvent. The hydrophilic solvent is usually water or buffer (pH 7.4), while the hydrophobic phase is usually *n*-octanol. The partition coefficient ($K_{o/w}$) is defined as the ratio of concentration of the drug in the organic phase (C_o) to drug concentration in the aqueous phase (C_w):

$$K_{o/w} = \frac{C_o}{C_w} \tag{14}$$

The choice of *n*-octanol is based on its ability to mimic the lipophilicity of the biological membranes [33] and further its solubility parameter (δ = 10.24; solubility parameter is a measure of internal cohesive energy) falls within the solubility parameter range of most drugs (8–12.4). The partition coefficient is determined by dissolving the drug in one of the phases and shaking both the phases together in a flask for 30 min to achieve equilibration. Then the drug concentration is determined from one of the phases, usually the aqueous phase, and the drug concentration in the oil phase is determined by subtracting the drug concentration in the aqueous phase from the total drug that was added. This value when expressed in logarithmic form is known as log *P*. The phase volume of the two phases is 1:1 but, if the drug is less soluble in the aqueous phase, the ratio (water–octanol) is changed to 1:10 or 1:20 to have a measurable drug concentration in the aqueous phase [33]. It is important to saturate the phases with respect to the other solvent before starting the experiment to rule out the influence of solvent partitioning on drug distribution between the two phases.

Another important factor is the influence of drug ionization on the partition coefficient and this is particularly important when a buffer is used instead of water. The partition coefficient determined from Equation (14) is an apparent value rather than a true partition coefficient for ionic drugs. However, the true partition coefficient can be calculated from the apparent partition coefficient if the drug's p*K*a and the pH of the drug solution are known [39], as shown in the equations

$$\log P = \begin{cases} \log P_{\text{app}} - \log\left(\dfrac{1}{1+10^{(\text{pH}-\text{p}K_{\text{a}})}}\right) & \text{for weak acid} \quad (15) \\[2ex] \log P_{\text{app}} - \log\left(\dfrac{1}{1+10^{(\text{p}K_{\text{a}}-\text{pH})}}\right) & \text{for weak base} \quad (16) \end{cases}$$

During the initial stages of drug screening, the partition coefficient is calculated based on the chemical structure (*C* log *P*). This is done by assigning values to different fragments in the chemical structure [41]. The calculated values are only estimates, but they are useful to rank order a homologous series of compounds based on their lipophilicity for further lead optimization. Given the fact that drugs have to cross many biological membranes before reaching the site of action, the log *P* value has a significant influence on drug absorption, drug pharmacokinetics, and pharmacology. This is exemplified from the numerous structure–activity and structure–property relationships using log *P* [41]. The log *P* is important in assessing the oral absorption potential of a drug. If a drug has a low log *P* (<1), it is expected to have poor membrane permeability, while if a drug has a large log *P* (>5), it will be trapped in the lipophilic membrane. A log *P* of 1–5 is usually considered optimal for oral drug absorption [3]. For an ionic drug, the un-ionized form of the drug will be more lipophilic than the ionized form of the drug. Therefore, at any given pH in vivo, the relative proportion of ionized versus un-ionized form of the drug dictates drug dissolution and absorption through the membrane.

6.2.3.3 Drug Dissolution

Dissolution is the rate at which a solid drug goes into solution and is a critical determinant in the absorption of drugs from solid dosage forms. A drug has to go into solution before it can be absorbed. In vitro dissolution studies are a valuable tool for determining the influence of different solid-state properties on drug dissolution and vis-à-vis predict in vivo drug dissolution and absorption. It is used to screen drugs that show dissolution rate–limited absorption. The factors that influence drug dissolution rate (*dc*/*dt*) can be understood from the well-established Noyes–Whitney equation [42]

$$\frac{dC}{dt} = \frac{kSC_s}{V} \tag{17}$$

where k is the dissolution rate constant (cm/s), C_s is the saturated solubility of the drug, S is surface area of solid exposed to the solvent, and V is the volume of the dissolution medium. The dissolution rate constant is a function of the diffusion coefficient D of the drug through a stagnant aqueous layer adjacent to the dissolving surface and thickness h. Powder or particulate dissolution is carried out in a dissolution vessel in a specific volume of dissolution fluid (900 mL) which is stirred and maintained at 37 °C. Several configurations are available to study the dissolution of various dosage forms (Figures 12*a* and 12*b*). Various simulated physiological media are used to understand the in vivo behavior of drug and dosage forms [43]. The usual dissolution media include water, 0.1 N HCl, and pH 7.4 buffers. It is important to maintain sink conditions in the dissolution medium by keeping the drug concentration at 10% of saturated solubility of drug (to mimic in vivo). For poorly water soluble drugs, surfactants are often added to the dissolution medium for this purpose. During preformulation testing, particulate dissolution studies reveal the influence

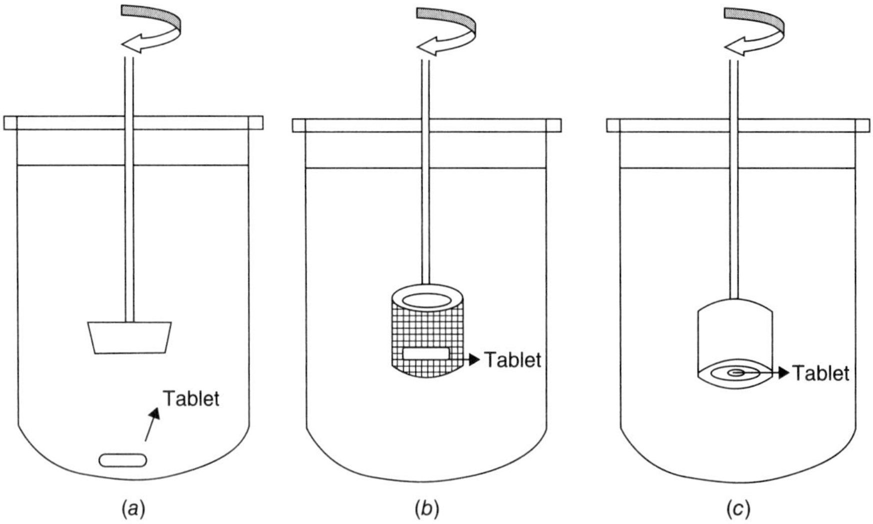

FIGURE 12 Various drug dissolution methodologies. (*a*) In the paddle method, the tablet is placed in the dissolution vessel containing dissolution medium and the paddle is rotated at defined rpm, while the dissolution vessel is maintained at body temperature. (*b*) In the basket method, the tablet is kept inside a meshed basket and rotated. (*c*) For IDR studies, the tablet is kept inside a die cavity and only one face of the tablet is exposed to the dissolution medium.

of particle size, crystal habit, and wettability of a drug substance. A formulator, on the other hand, routinely uses dissolution testing as a quality control tool in the design of dosage forms. Various models have been developed to describe the release kinetics of conventional and modified release dosage forms [44].

Since powder dissolution is influenced by changing surface area, it is not useful for delineating the effects of polymorphs, hydrates, and pharmaceutical salts. Instead, the intrinsic dissolution rate (IDR) is used. The IDR studies are conducted at a constant surface area and hence the dissolution rate observed is only a function of the intrinsic solubility of the drug. The Noyes–Whitney equation is modified for IDR, where the surface area is kept constant, and Equation (17) reduces to

$$\frac{dC}{dt} = k_1 C_s \tag{18}$$

where k_1 is the intrinsic dissolution rate constant and dC/dt is the intrinsic dissolution rate (mg·cm^2/s). For IDR studies, the drug (500 mg) is compressed in a hydraulic press (at 500 mPa) to a 13-mm disc. This disc is then loaded onto a holder which exposes only one surface of the disc to the dissolution medium (Figure 12*c*). The IDR is obtained by dividing the slope of the plot between the amount of drug dissolved and time by the area of solid exposed to the dissolution medium. The IDR predicts the influence of drug solubility on in vivo drug dissolution and absorption. A drug which has an IDR of 1 mg·cm^2/s will not generally show dissolution rate–limited absorption in vivo. However, if the IDR falls between 0.1 and 1 mg·cm^2/s, then further studies may be required to make a decision. Drugs with IDR < 0.1 mg·cm^2/s show dissolution rate–limited absorption in vivo, necessitating drug solubilization strategies [45]. With respect to pharmaceutical salts, IDR is used to understand the influence of surface pH on drug dissolution and absorption. The common-ion effect can be studied by including 0.1–0.15 *M* NaCl in the dissolution medium. Also IDR is useful to understand the difference in solubility of polymorph and amorphous forms. However, in some cases, the compression force used in making the IDR disk may by itself induce solid-state transformations [45]. DSC, IR, and XRD must be used to identify the drug's solid state before and after compression as well as at the end of the dissolution studies. A well-designed dissolution study is used as an early warning for drug molecules that would pose absorption problems in vivo.

6.2.3.4 Absorption Potential

The ultimate goal of any drug development program is to develop an orally absorbable compound. Solubility and permeability are the two most critical parameters that dictate oral absorbability of a molecule. All other parameters are directly or indirectly related to these two physicochemical properties. As can be seen from Figure 13, there are several barriers that a drug needs to overcome before reaching the systemic circulation. The oral absorption of a drug is mainly limited by drug dissolution and/or by the drug permeation across the GI membrane. Considering their importance, drug solubility and permeability are screened very early in the drug discovery process. Solubility studies are typically run in a high-throughput format using a turbidometric method as discussed earlier. Based on this primary

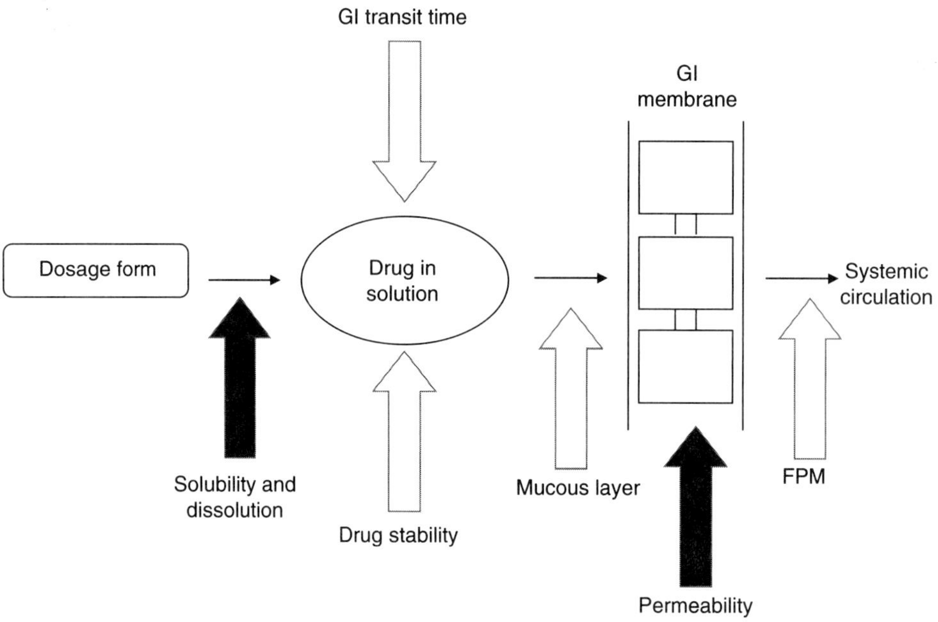

FIGURE 13 Barriers to drug absorption. Drug from the dosage form should be soluble and dissolve in GI fluids before it can be absorbed. Drug dissolution is one of the major rate-limiting steps in drug absorption. Drug absorption may be affected if the drug is unstable in GI fluids. Furthermore, drug absorption will also depend on how long the drug resides in a particular region of the GIT. The drug has to diffuse through the highly viscous mucous layer before getting absorbed through the membrane. Membrane permeability is one of major the rate-limiting steps in absorption. After absorption the drug may be subject to first-pass metabolism (FPM) in the liver before reaching the systemic circulation. Dark arrows indicate that solubility and permeability are the most influential factors.

screen, detailed solubility and dissolution studies are carried out as the compound moves through different development phases (Figure 14*a*). A usual target for drug solubility during lead selection is 1 mg/mL for avoiding dissolution rate–limited absorption [46].

Drug permeability is commonly screened in the early discovery phase using CaCo-2 cell lines, which is a human colon carcinoma cell line. Using a 96-well format, the cell lines are used as a primary screen to rank order compounds based on permeability [4]. If designed properly, the cell culture studies can be used to understand the drug absorption mechanisms [47]. To avoid permeability-limited absorption in vivo, a drug should have a permeability coefficient of 2×10^{-4}–4×10^{-4} cm/s [48]. Inputs from the drug metabolism team (based on liver microsomal studies) can give clues on the drug's susceptibility to first-pass metabolism. Following cell culture studies, a select group of compounds are studied using isolated rat intestine and segmental absorption studies to understand the drug absorption mechanism and the site of drug absorption in the GIT (Figure 15*b* and 15*a*). This is further substantiated using in situ perfusion experiments in rats [49]. The details of the studies are depicted in Figure 15*b*. Some of these compounds are studied in whole animals of which a few may make it to clinical trials in humans.

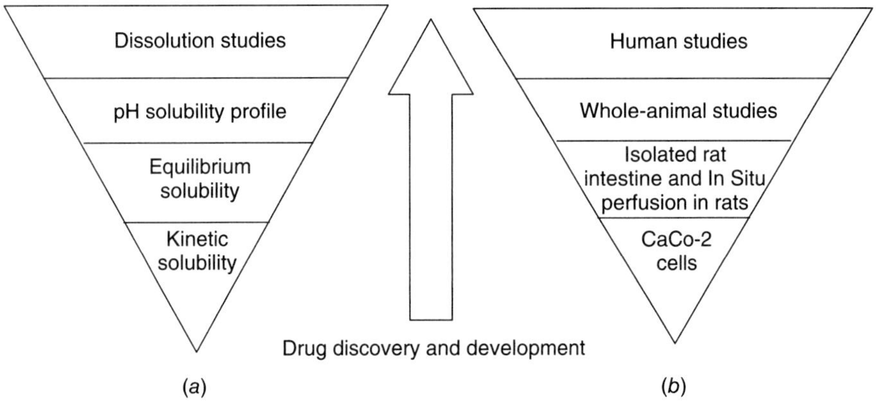

FIGURE 14 Different methods to determine solubility and permeability during various stages of drug discovery and development. (*a*) In the initial stages the drugs are screened using kinetic solubility studies, which are later followed by equilibrium solubility studies, pH solubility profile, and dissolution studies during development phase. (*b*) Drug permeability is initially screened using CaCo-2 cell lines followed by rat intestinal studies. This is followed by pharmacokinetic studies in animals and finally the potential drug molecules are tested in humans in clinical trials.

Thus, the preformulation team in coordination with other discovery teams gets useful estimates on in vivo drug absorption. In addition to drug solubility and permeability, it is also important to consider the anticipated dose while assessing the absorption potential. A useful tool to optimize the drug's physicochemical properties is the maximum absorbed dose (MAD) model [46]. The model predicts the dose that would be absorbed based on the drug's solubility (C_s; solubility in intestinal pH of 6.5), absorption rate constant (k_a; usually obtained from rat permeability studies), physiological factors such as gastric transit time (T_i; approximated as 4.5 h), and intestinal fluid volume (V_{int}; approximated as 250 mL):

$$MAD = C_s k_a V_{int} T_i \tag{19}$$

Using this model, the required solubility or permeability can be estimated for a given dose of the drug. For example, a drug with an anticipated human dose of 1 mg/kg (70 mg for a normal 70-kg subject) will require a solubility of 0.05 mg/mL provided the drug shows good permeability. Similarly, the absorption rate required to achieve the same dose for a drug with good solubility (1 mg/mL) is 0.001 min^{-1}. The MAD model is helpful in guiding development teams on optimizing drug solubility and/or permeability.

A further refinement of this model led to the evolution of the biopharmaceutics classification system (BCS), which classifies the drugs into four classes depending on their solubility and permeability (Figure 16). The BCS is applicable only to the oral route of administration, and according to this classification, a drug is considered to be highly soluble if the highest dose of the drug is soluble in a glass of water (250 mL) covering the entire pH range in GIT from 1 to 7.5, and a drug is considered to be highly permeable if the drug has more than 90% oral bioavailability [48]. The model has been developed based on the solubility and permeability characteristics

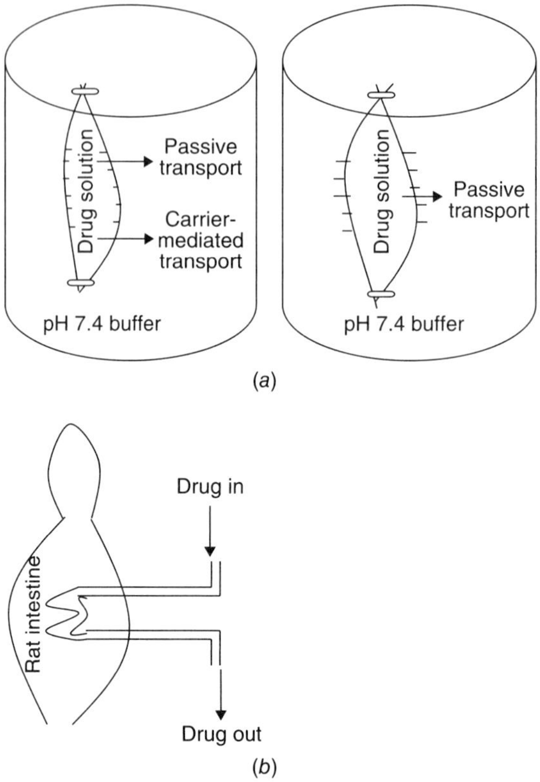

FIGURE 15 Intestinal permeability studies. (*a*) Everted rat intestine is used to study the mechanism of drug absorption. Isolated rat intestine is filled with drug solution and tied at both ends and the drug permeation into the external medium is measured (left). In another set of experiment, the rat intestine is everted with the internal mucosal membrane facing outside (right) and the serosal side facing inward. If the drug is passively absorbed, then there would not be any differences in permeability in these two experiments. If the drug is transported by carriers, then drug permeation would be seen only in the first experiment, as the carriers are present only on the mucosal side. (*b*) In situ intestinal perfusion studies are conducted in an anesthetized animal and the drug solution is pumped through a tube and drug coming out on the other side of the intestine is measured. The drug is also measured by sampling from the jugular vein in the animal. This is useful to measure the dynamic drug absorption into systemic circulation.

of marketed drugs. However, BCS is bound to become an important biopharmaceutical tool in lead optimization in drug discovery setting and also at the same time serving as a useful guide to develop new formulations in a generic setting. From a regulatory perspective, BCS provides a scientific basis to grant biowaivers. As per BCS, the dissolution test can serve as a surrogate tool for costly and time-intensive bioequivalence studies for generic drugs which are highly soluble. This requires the establishment of a good in vitro–in vivo correlation. At present, the FDA grants biowaivers to generic manufacturers of immediate-release dosage forms for drugs in class I (highly soluble and highly permeable) provided they can prove the "dissolution equivalence" of their product to that of the innovator's drug product [50].

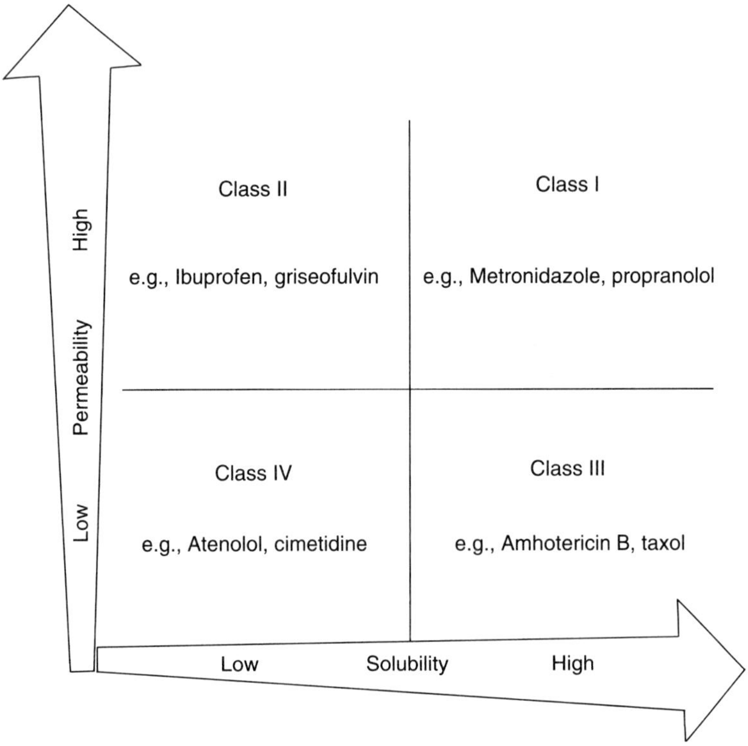

FIGURE 16 Biopharmaceutics classification system. The drugs are classified based on drug solubility and drug permeability. A drug is considered to be highly soluble if the highest dose of the drug is soluble in 250 mL of water varying in pH from 1 to 7.8 (GIT pH range). A drug is considered highly permeable if more than 90% of the drug is bioavailable by oral route. Class I drugs are highly soluble and highly permeable, class II drugs are poorly soluble but highly permeable, class III drugs are highly soluble but poorly permeable, and class IV drugs have poor solubility and permeability.

The dissolution equivalence is tested using the statistical dissolution model, termed f_2 or similarity factor, and is described by the equation

$$f_2 = 50\log\left\{\left[1+\left(\frac{1}{N}\right)\sum_{t=1-n}(R_t+T_t)^2\right]^{-0.5}\times 100\right\} \qquad (20)$$

where N is the number of dosage form units (12 units are tested), t is dissolution time points from 1 to n, R_t is the percent drug dissolved for the reference drug product, and T_t is the percent drug dissolved for the test product. Two dissolution profiles are considered similar, if $f_2 \geq 50$ and if the coefficient of variation does not exceed 20% at early dissolution time points (usually 10 min) and 10% at other time points in the pH range 1–7.5. However, if ≥85% of drug is dissolved in ≤15 min, then no comparison is required and the dissolution is based on a single time point determination [50].

The BCS paradigm can be used to develop strategies for enhancing drug solubility and/or permeability (Tables 8 and 9). Solubility enhancement may involve only

TABLE 8 Solubility Enhancement Methods

Technique	Principle
Salt formation	Increases drug solubility by keeping the pH at which the drug is ionized.
Particle size reduction	Increase in surface area increases drug solubility. Particle size is reduced to nanodimensions (nanosuspensions) for increasing drug solubility.
Change of form	Crystalline drugs are converted to amorphous forms which show higher solubility than crystalline forms of the drug.
Cosolvents	Various water miscible cosolvents are used to increase the water solubility of drug. The cosolvents are selected based on the polarity of the drug. Common cosolvents include glycerol, ethanol, and propylene glycol.
Complexation	Drug is entrapped or complexed with excipients that can mask the lipophilic groups of the drug and enhance drug solubility in water. Cyclodextrins are commonly used to entrap hydrophobic drugs in the core, while the hydrophilic groups on the periphery help to solubilize the drug.
Surfactants	Surfactants are characterized by the presence of lipophilic and hydrophilic groups and form spherical structures known as micelles in water at a certain concentration. The hydrophobic drug is entrapped in the hydrophobic core of the micelle while the hydrophilic groups on the periphery help to solubilize the drug.
Disperse systems	The hydrophobic drug is dissolved in an organic solvent and in addition may also contain an emulsifier. On contact with the intestinal fluids, the drug is emulsified (microemulsions) by bile salts and is absorbed through the intestine.

TABLE 9 Permeability Enhancement Methods

Method	Mechanism
Transcellular transport	Sorption promoters can be used to enhance the transcellular transport in the intestine, including bile salts and fatty acid esters. They tend to fluidize the lipid bilayer and enhance drug permeation across the membrane.
Paracellular transport	Enhancement is achieved by modulating the tight junctions between the cells. Chelating agents such as ethylenediamine tetraacetic acid can chelate calcium ions and transiently open the tight junctions for drug transport.
Carrier-mediated transport	Nutrient transport carriers are utilized for drug transport. Prodrugs are designed to meet the structural requirements for carrier-mediated transport.
Blocking efflux pump	P-glycoprotein is a major efflux mechanism that pumps out drug from the intestinal cells back into the intestinal fluid. Several drugs and food substances are known to inhibit p-glycoprotein and enhance drug permeation.

a physical intervention, as opposed to molecular modification for permeability enhancement. Generally [46], the permeability range of drugs varies by only 50-fold (0.001–0.5 min^{-1}) in comparison to drug solubility, which varies by six orders of magnitude (0.1 μg/mL–100 mg/mL). Hence, the formulator has greater flexibility in altering the drug's solubility in comparison to altering the drug's permeability. This

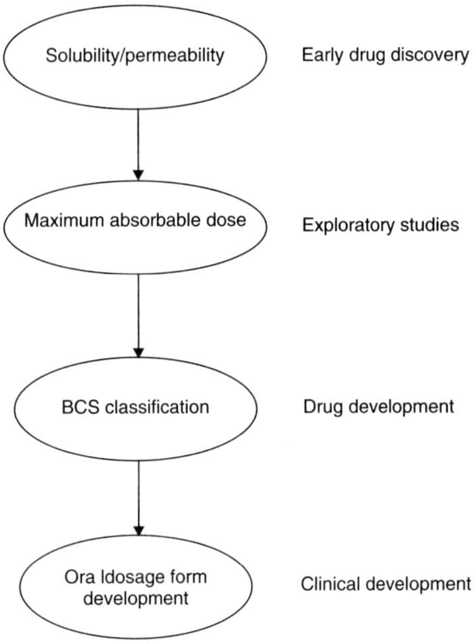

FIGURE 17 Flow chart for determining absorption potential of a drug during drug discovery and development. Solubility and permeability studies from preclinical phase are used to calculate the maximum absorbable dose and, when correlated to BCS, this can provide information on its biopharmaceutics class. This is useful to estimate if the drug absorption would be dissolution and/or permeability limited for developing appropriate drug delivery strategies.

is evident form the fact [51] that the majority of the marketed drugs are highly soluble (>55% in classes I and III). Sometimes, enhancing the permeability by altering the drug's chemical structure may be counterproductive. This because of the associated increase in molecular weight that attenuates the permeability enhancement gained with structural modification. On the other hand, optimization of drug solubility may be more fruitful if the poor permeability is overcome by increasing drug solubility to provide a high drug concentration at the absorption site. However, this may be a difficult strategy if the dose is very high. The preformulation team should use the "appropriate tools" at every stage of the drug discovery and development process to guide or alert other development teams about drug solubility and permeability issues (Figure 17).

6.2.4 STABILITY CHARACTERISTICS

Drug stability is an essential component of preformulation testing. Establishing the stability of the drug under a variety of conditions is an expensive and time-consuming process. It cannot be overemphasized that the availability of a stability-indicating assay is the key to stability studies. The preformulation scientist works closely with the analytical method development team in developing a stability-indicating assay. During the early stages, a foolproof stability-indicating assay may not be available. This is understandable considering the fact that the initial drug lots are not pure and the purity is improved as the molecule progresses to subsequent development stages. Therefore, the intention of the preformulation scientist is not to generate a thorough kinetic rate profile of the drug but to broadly define the conditions under which the drug would be stable. Only relevant stability data are generated during various phases for developing preclinical and clinical formulations [52]. The stability data are evaluated and conveyed to the formulation development team upfront if stabilization and additional packaging requirements are needed. Often the chemical structure of the drug can give clues on the drug's degradation pathway and its stability characteristics [53]. Table 10 lists some of the functional groups that are susceptible to common degradation pathways.

Stability test conditions are chosen keeping in mind the environment that the drug will encounter during drug development, processing, storage, and use (Table 11). One parameter is studied at a time while keeping all other parameters constant. Apart from classical stability studies, such as hydrolysis, oxidation, and photolysis, the preformulation scientist has to determine the stability of drugs in unconventional matrices such as animal feed used for toxicological studies [34]. The stability of a drug in the animal feed is complicated by the feed composition, including enzymes and vitamins among others. The moisture content in the feed may also vary with storage temperature. In such cases, it is appropriate to study the stability of the drug under the storage conditions encountered in the toxicological laboratory. Sensitive techniques such as liquid chromatography/mass spectrometry (LC/MS) are used to evaluate drug stability in such complex mixtures.

In general, most of the drugs undergo first-order degradation, while some drugs may follow zero-order kinetics and only a few drugs undergo second-order degradation [34]. The first- and zero-order reactions are readily differentiated by studying drug stability at two different initial drug concentrations. First-order kinetics will depend on initial drug concentration, while a zero-order reaction will be

TABLE 10 Groups Susceptible to Common Degradation Pathways

Degradation Pathway	Functional Groups
Hydrolysis	Esters, lactones, amides, lactams, oximes, imides, and malonic ureas
Oxidation	Amines, sulfides, disulfides, sulfoxides, phenol anions, thiols, nitriles, and catechols
Photolysis	Aromatic hydrocarbons, aromatic heterocyclics, aldehydes, and ketones

TABLE 11 Stability Testing Conditions

Parameters	Conditions
Temperature	5, 25, 30, 37, 40, and 60 °C
Moisture	30, 45, 60, 75, and 90% RH
pH	1, 3, 5, 7, and 9 at room and body temperature
Oxygen	Sparging with 40% oxygen or adding 100–200 ppm of hydrogen peroxide
Light	1.2 million lux hours of exposure to visible light and 200 h/m^2 exposure to UV light (360–400 nm)

independent of initial drug concentration. Accelerated stability studies are conducted to expedite the degradative reactions, where temperature is the commonly used accelerant. The influence of temperature on drug stability kinetics is described by the Arrhenius equation:

$$k = Ae^{-E_a/RT} \tag{21}$$

where k is a reaction rate constant, A is a frequency factor, E_a is activation energy, R is the gas constant, and T is absolute temperature. According to the Arrhenius equation, every 10 °C rise in temperature increases the reaction rate by two- to fivefold [31]. The usual temperatures selected for early stability studies include 5, 25, 37, 40, and 60 °C to cover the temperatures encountered in processing, use, and storage of the drug product. Using the Arrhenius equation, the rate constant from higher temperatures can be extrapolated to determine the stability at room temperature [31]. The slope of the plot of the reciprocal of temperature and the rate constant gives the activation energy. The activation energy usually varies between 15 and 60 kcal/mol with a mean value of 19.8 kcal/mol [33]. A break in the line is usually indicative of change in the activation energy due either to change in the reaction order or the mechanism of degradation at higher temperature. In such cases, it becomes imperative to conduct detailed studies to understand the drug degradation mechanism. Some of the reactions seen at higher temperature may not be representative of the reactions at room temperature. Hence, short-term high-temperature studies should be supplemented with long-term real-time stability testing at room temperature. Additionally, the drug is also exposed to moisture, oxygen, and UV light (250–360 nm). The conditions used for stress studies may vary depending on the drug type and the drug development stage [54]. The stability studies in this chapter are discussed with respect to a solid dosage form which includes solid-state stability, limited liquid state stability, and drug–excipient compatibility.

6.2.4.1 Solid-State Stability

In general, solid-state reactions are slow, complex, and at times difficult to quantify. They may manifest as either physical and/or chemical instabilities. Physical instabilities include solid–solid transformations, desolvation of hydrates, and change in color [34]. On the other hand, chemical instabilities may involve a change in drug content as a result of hydrolysis, oxidation, or light-induced degradations [32]. The influence of temperature is studied by exposing the solid drug to increasing temperatures and also exposing the drug to various relative humidities at room temperature for two to eight weeks (Table 11). If substantial change is observed at higher temperatures, the drug samples stored at 5 °C are analyzed. If no degradation is seen at higher temperature, then none can be expected at room temperature. The results from higher temperature should be carefully interpreted. For instance, a hydrate may lose moisture at higher temperatures and make a drug unstable which otherwise would be stable at room temperature. Similarly, chlortetracycline hydrochloride converts from the β form to the α form at above 65% relative humidity, in contrast to <65% relative humidity, where no transformations are observed [32].

Oxidative degradation is studied by exposing the sample to an atmosphere of 40%. The oxygen is combined with heat to accelerate the reaction. The results should be compared with samples stored in an inert atmosphere. Photolytic degradation is studied by exposing the sample to UV light (254 and 360 nm) for two to four weeks and observed for color fading and/or darkening [33]. Color fading may not always mean drug decomposition. It may be just a physical change which can be overcome by including dyes in the formulation. As indicated earlier, the solid-state reactions may be slow, sometimes generating only 1–5% of degradation products, which may be below the detection limit of HPLC [34]. A combination of qualitative and quantitative analytical tools is helpful to detect drug degradation. Nevertheless, accelerated stability studies provide an early warning of potential drug degradation and the preformulation scientist should use discretion in interpreting and sharing the results with other development teams.

6.2.4.2 Solution-State Stability

Detailed solution-state stability is of limited value if the final anticipated dosage form is a solid. However, solution-state stability studies can predict the stability of the drug in the granulating fluid and the GI fluids. Solution-state reactions are faster than the solid-state reactions. This is helpful in generating degradation products through so-called forced-degradation studies for toxicological screening and analytical method development. The studies are conducted by exposing the drug to extreme conditions such as 0.1 *N* HCl, 0.1 *N* NaOH, and water at 90 °C [34]. Forced-degradation studies are useful to qualify the safety of the degradation products if it exceeds 0.1 or 0.05% of total daily dose for drugs with <1 g or >1 g dose/day, respectively [6].

The pH rate profile is an important parameter that is studied in the solution state. In early preformulation studies, an approximate pH rate profile is generated, usually including the pH encountered in salt selection and in vivo [55]. The studies are later followed with a detailed pH profile in the whole pH range of 1–10. A typical pH rate profile is shown in Figure 18, which is useful to extract useful information on

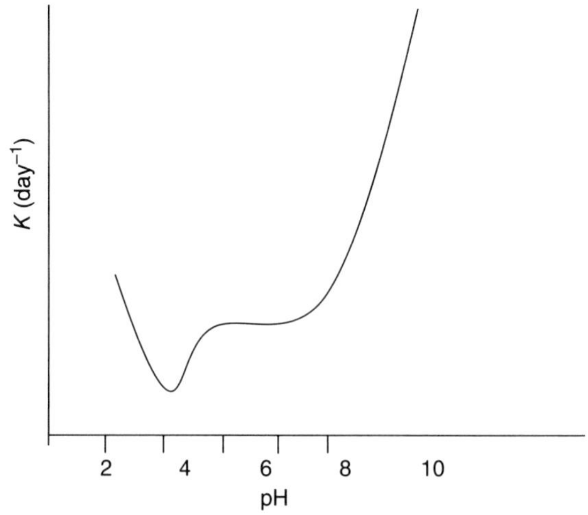

FIGURE 18 Representative pH kinetic rate profile. The drug shows a minimal degradation at around 2–4 and the degradation rate is high in alkaline pH.

drug stability. The profile is used to predict if the drug degradation is catalyzed by hydronium or hydroxyl species (specific acid–base catalysis). The minimal point in the profile is indicative of the pH at which the drug is relatively stable. However, the buffer used in the study can by itself accelerate the reaction, which is referred to as general acid–base catalysis [39]. The influence of buffer is nullified by conducting the study at one to three buffer concentrations and then plotting the rate constant against the buffer concentration. The intercept on the *y* axis gives the rate constant at zero buffer concentration.

Oxidation reactions are studied by passing oxygen in the head space of the drug solution and comparing the drug degradation with a drug solution (Table 11) filled with an inert gas in the head space. The reduced solubility of oxygen at higher temperatures may lead to an apparently reduced reaction rate in comparison to lower temperature [34]. Light sensitivity is studied by exposing the drug solution in a clear-flint bottle to UV radiation and comparing the results with the drug solution in an amber-colored container [55].

The Solution-state studies should be extrapolated to the solid state with caution. The reaction in the solution state is usually done in a dilute drug solution, in contrast to reactions in the solid state, where the saturated drug solution at the surface undergoes a multiphase reaction [33]. Moreover, the reaction order may be different in the solution and solid states. Due to the excess solvent, reactions in the solution state are usually pseudo–first order as opposed to first-order or zero-order reactions in the solid state. In spite of these limitations, solution-state studies provide clues in selecting appropriate granulating solvent and in predicting in vivo drug stability. The pH–rate profile data are also useful to predict the solid-state stability of salt forms or the stability of a drug in the presence of acidic and basic excipients [55]. Further, the pH–rate studies predict the stability of drug and its salt in the gastrointestinal pH. This is illustrated by the example of erythromycin and its salts [32]. Erythromycin is rapidly inactivated in the acidic environment of the stomach. This is overcome by using insoluble erythromycin estolate, which is stable in the gastric pH, unlike the other salts, which are easily displaced by hydrochloric acid in the stomach.

6.2.4.3 Drug–Excipient Compatibility

Excipients are the backbone of a solid dosage form, and they function as diluents, binders, disintegrants, and fillers. The excipients are in intimate contact with the drug in a solid dosage form, therefore necessating the need for drug–excipient incompatibility testing. Since the formulator has a wide selection of excipients to choose from, it would be a daunting task for the preformulation scientist to screen all possible excipients. A general practice is to select those excipients that are routinely used in in-house manufacturing of dosage forms. At least two are selected from each class of functional excipients. Excipients that are known to cause potential incompatibilities (e.g., glucose and amines) with drugs are excluded from the study [34]. The excipients are intimately mixed with the drug in the ratio that is realistic for the desired solid dosage form. For example, diluents are mixed in the ratio 20:1, while other excipients are used in the ratio 1:5 with drug [32]. The drug–excipient mixtures are then stored in a tightly sealed container (ampoule or in a bottle, where the cap is sealed with wax) in the presence and absence of 5%

w/v moisture. The mixture is stored at 40 and 25 °C and analyzed for three months. The mixture is physically observed for caking, liquefaction, and gas formation and chemically analyzed for drug degradation. A variety of techniques are used for studying drug–excipient compatibility. DSC is a useful primary screen which can rapidly detect potential drug–excipient incompatibilities [33]. For DSC studies, the drug and excipient are mixed together without moisture and are analyzed within a few hours of preparation. Moisture is avoided, as it complicates the interpretation of thermograms. If there is no interaction, then the thermogram should be the sum of thermogram of drug and excipient. The thermogram is observed for peak position, peak appearance, peak shape, appearance of new peaks, disappearance of drug peaks, peak shift, and change in enthalpy values. It is obvious that a mixture of two substances will result in depression of melting point, but if the change is significant, then it may be indicative of eutectic formation. Some of the reactions seen at high temperatures in DSC may not be reflective of room temperature. Therefore, excipients that are suspected to show incompatibilities are screened further through stability studies using TLC and HPLC (Figure 19). TLC is used to qualitatively detect any new spots with the drug–excipient mixture. The spots should be compared with the control drug sample. This is important since in the early discovery stage the drug is not pure and contains impurities. The findings from TLC should be corroborated using HPLC. The studies may be further followed-up with stability studies using multicomponent mixtures. In addition, the drug and excipient may be tested by compressing in a hydraulic press or filling in a capsule to simulate the actual dosage form [32]. A well-designed drug–excipient preformulation study can thus help the formulator in judicious choice of excipients for the final dosage form.

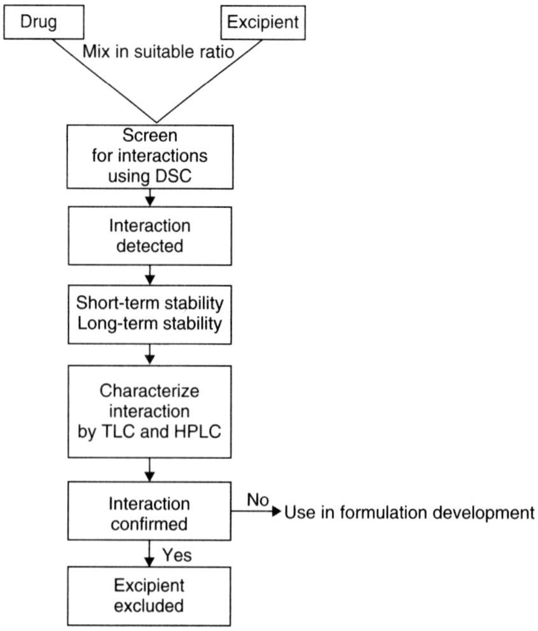

FIGURE 19 Flow chart for studying drug–excipient compatibility.

6.2.5 CONCLUSIONS

Preformulation testing has a significant role in a drug discovery and development program, as it provides valuable feedback to the various discovery and development teams in enabling druggability during lead identification and optimization. The preformulation data may mean different things to different groups in the discovery/development phase (Figure 20). For the chemistry team, the feedback from preformulation testing provides clues to optimize the chemical structure with respect to solubility, permeability, and stability. Preformulation studies give inputs to the biology group for ensuring optimal drug exposure based on solubility, permeability, and stability data, in addition to developing preclinical formulations for pharmacokinetic and toxicological studies. The analytical team gets inputs from the preformulation group on developing stability-indicating assays and setting drug specifications. Once the lead is selected and as the molecule moves to the development phase, the preformulation group provides guidance to the bulk manufacturing, formulation, and clinical evaluation teams. The bulk manufacturing team uses the data generated from the preformulation studies on salts, polymorphic purity, and particle size specifications. It is the formulation team that utilizes the maximum data from the preformulation testing to design an appropriate dosage form. The physicochemical properties are utilized for improving the drug's solubility, improving the drug's permeability, developing stabilization strategies, selecting appropriate excipients, selecting processing conditions to design, and evaluating the final dosage form. The clinical evaluation team utilizes the preformulation data along with the preclinical animal studies to understand the drug's pharmacokinetics in humans through the MAD and BCS paradigms. Therefore, a strong preformulation team in a drug discovery setting is critical for optimizing the pharmaceutical properties of the drug. This can significantly reduce the attrition rate, time, and cost of discovering a new drug. On the other hand, the preformulation team in a generic setting is valuable to optimize or further enhance the efficacy of an existing drug by designing a new drug delivery system and thus giving a new life to an old drug. As opposed

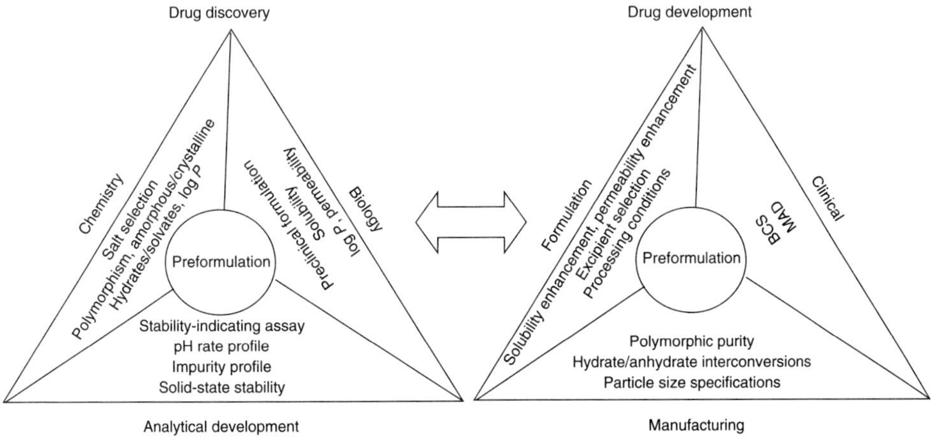

FIGURE 20 Role of preformulation in supporting other discovery and development teams.

to discovering a new drug, a new drug delivery system can be designed at one-third of the cost and money involved in a new drug discovery process [56]. The availability of high-throughput property screens and predictive models is expected to improve the efficiency and maximize the output of preformulation data in the coming years. However, the main challenge lies in intelligent use of such tools to make sense from loads of data and communicate the appropriate information to the relevant discovery/development groups for getting the drug to the marketplace in time.

REFERENCES

1. Waterbeemd, H. V., and Gifford, E. (2003), ADMET in silico modeling: Towards prediction paradise, *Nat. Rev. Drug Discov.*, 2, 192–204.
2. Kerns, E. H., and Di, L. (2003), Physicochemical profiling in drug discovery, *Drug Discov. Today*, 8, 316–323.
3. Lennarnas, H., and Abrahamsson, B. (2005), The use of biopharmaceutical classification of drugs in drug discovery and development: Current status and future extensions, *J. Pharm. Pharmacol.*, 57, 273–285.
4. Lipinski, C., Lombardo, F., Dominy B. W., and Feeney, P. J. (2001), Experimental and computational approaches to estimate solubility and permeability in drug discovery and development settings, *Adv. Drug Deliv. Rev.*, 46, 3–26.
5. Giron, D. (1998), Contribution of thermal methods and related techniques to the rational development of pharmaceuticals. Part I, *Pharm. Sci. Technol. Today*, 1, 191–199.
6. Ahuja, S. (2007), Assuring quality of drugs by monitoring impurities, *Adv. Drug Deliv. Rev.*, 59, 3–11.
7. Datta, S., and Grant, D. J. W. (2004), Crystal structures of drugs: Advances in determination, prediction and engineering, *Nat. Rev. Drug Discov.*, 42, 42–57.
8. Tiwary, A. K. (2004), Modification of crystal habit and its role in dosage form performance, *Drug Dev. Ind. Pharm.*, 27, 699–707.
9. Garekani, H. A., Ford, J. L., Rubinstein, M. H., and Rajabi-Sahboomi, A. R. (1999), Formation and compression characteristics of prismatic, polyhedral and thin plate-like crystals of paracetamol, *Int. J. Pharm.*, 187, 77–89.
10. Gu, C., Li, H., Gandhi, R. B., and Raghavan, K. (2004), Grouping solvents by statistical analysis of solvent property parameters: Implication to polymorph screening, *Int. J. Pharm.*, 283, 117–125.
11. Giron, D. (2001), Investigations of polymorphism and pseudopolymorphism in pharmaceuticals by combined thermoanalytical techniques, *J. Thermal Anal. Calor.*, 64, 37–60.
12. Threafall, T. (2003), Structural and thermodynamic aspects of Ostwald's rule, *Org. Proc. Res. Dev.*, 7, 1017–1027.
13. Vippagunata, S. R., Brittain, H. G., and Grant, D. J. W. (2001), Crystalline solids, *Adv. Drug Deliv. Rev.*, 48, 3–26.
14. Phadnis, N. V., Cavatur, R. K., and Suryanarayanan, R. (1997), Identification of drugs in pharmaceutical dosage forms by x-ray powder diffractometry, *J. Pharm. Biomed. Anal.*, 15, 929–943.
15. Haleblain, J., and McCrone, W. (1969), Pharmaceutical applications of polymorphism, *J. Pharm. Sci.*, 58, 911–929.

16. Huang, L., and Tong, W. (2004), Impact of solid state properties on developability assessment of drug candidates, *Adv. Drug Deliv. Rev.*, 56, 321–334.

17. Morris, K. R., Griesser, U. J., Eckhardt, C. J., and Stowell, J. G. (2001), Theoretical approaches to physical transformations of active pharmaceutical ingredients during manufacturing process, *Adv. Drug Deliv. Rev.*, 48, 91–114.

18. Dunitz, J. A., and Bernstein, J. (1995), Disappearing polymorphs, *Acc. Chem. Res.*, 28, 193–200.

19. Chemburkar, S. R., Bauer, J., Deming, K., Spiwek, U., Patel, K., Morris, J., Henry, R., Spanton, S., Dziki, W., Porter, W., Quick, J., Bauer, P., Donaubauer, J., Narayanan, B. A., Soldani, M., Riley, D., and McFarland, K. (2004), Dealing with the impact of ritonavir polymorphs on the late stages of bulk drug process development, *Org. Proc. Res. Dev.*, 4, 413–417.

20. Morissette, S. L., Almarsson, O., Peterson, M. L., Remenar, J. F., Read, M. J., Lemmo, A. V., Ellis, S., Cima, M. J., and Gardner, C. R. (2004), High-throughput crystallization: Polymorphs, salts, co-crystals and solvates of pharmaceutical solids, *Adv. Drug Deliv. Rev.*, 56, 275–300.

21. Davey, R. J. (2003), Pizzas, polymorphs and pills, *Chem. Commun.*, 13, 1463–1467.

22. Khankari, R. K., and Grant, D. J. W. (1995), Pharmaceutical hydrates, *Thermochim. Acta*, 248, 61–79.

23. Giron, D., Goldbrown, Ch., Mutz, M., Pfeffer, S., Piechon, Ph., and Schwab, Ph. (2002), Solid-state characterization of pharmaceutical hydrates, *J. Therm. Anal. Calor.*, 68, 453–465.

24. Byrn, S., Pfeffer, R., Ganey, M., Hoiberg, C., and Poochikian, G. (1995), Pharmaceutical solids: A strategic approach to regulatory considerations, *Pharm. Res.*, 12, 945–954.

25. Craig, D. Q. M., Royall, P. G., Kett, V. C., and Hopton, M. C. (1999), The relevance of the amorphous state to pharmaceutical dosage forms, *Int. J. Pharm.*, 179, 179–207.

26. Yu, L. (2001), Amorphous pharmaceutical solids: Preparation, characterization and stabilization, *Adv. Drug Deliv. Rev.*, 48, 27–42.

27. Hancock, B. L., and Parks, M. (2000), What is the true solubility advantage for amorphous pharmaceuticals? *Pharm. Res.*, 17, 397–404.

28. Shah, B., Kakamanu, V. K., and Bansal, A. K. (2006), Analytical techniques for quantification of amorphous/crystalline phases in pharmaceutical solids, *J. Pharm. Sci.*, 95, 1641–1665.

29. Giron, D. (2003), Characterization of salts of drug substances, *J. Thermal Anal. Calor.*, 73, 441–457.

30. Shekunov, B. Y., Chattopadhyay, P., Tong, H. H., and Chow, H. H. (2007), Particle size analysis in pharmaceutics: Principles, methods and applications, *Pharm. Res.*, 24, 203–227.

31. Sinko, P. (2006), *Martin's Physical Pharmacy and Pharmaceutical Sciences*, 5th ed., Lippincott Williams and Wilkins, Philadelphia, pp 533–560.

32. Wadke, D. A., and Jacboson, H. (1980), Preformulation testing, in Liberman, H. A., and Lachman, L., Eds., *Pharmaceutical Dosage Forms: Tablets*, Vol. 1, Marcel Dekker, New York, pp 1–59.

33. Wells, J. (2005), Pharmaceutical preformualtion, in Aulton, M. E., Ed., *Pharmaceutics, The Science of Dosage Form Design*, Churchill Livingstone, Edinburgh, pp 113–138.

34. Fiese, E. F., and Hagen, T. A. (1986), Preformulation, in Lachman, L., Liberman, H. A., and Kanig, J. A., Eds., *The Theory and Practice of Industrial Pharmacy*, 3rd ed., Lea and Febiger, Philadelphia, pp 171–196.

35. Krens, E. H. (2001), High-throughput physicochemical profiling for drug discovery, *J. Pharm. Sci.*, 90, 1838–1858.
36. Strickley, R. G. (2004), Solubilizing excipients in oral and injectable formulations, *Pharm. Res.*, 21, 201–229.
37. Bastin, R. J., Bowker, M. J., and Slates, B. J. (2000), Salt selection and optimization procedures for pharmaceutical new chemical entities, *Org. Proc. Res. Dev.*, 4, 427–435.
38. Haynes, D. A., Jones, W., and Motherwell, W. D. S. (2005), Occurrence of pharmaceutically acceptable anions and cations in the Cambridge structural databases, *J. Pharm. Sci.*, 94, 2111–2120.
39. Florence, A. T., and Atwood, D. (2005), *Physicochemical Principles of Pharmacy*, 4th ed., Pharmaceutical Press, London, pp 55–92.
40. Ando, H. Y., and Radebaugh, G. W. (2006), Property based drug design and preformulation, in Beringer, P., DerMarclerosian, A., Felton, L., Gelone, S., Gennaro, A. R., Gupta, P. K., Hoover, J. E., Popovich, N. J., Reilly Jr., W. J., and Hendrickson, R., Eds., *Remington's: The Science and Practice of Pharmacy*, Lippincott Williams and Wilkins, Philadelphia, pp 720–744.
41. Leo, A., Hansch, C., and Elkins, D. (1971), Partition coefficient and their uses, *Chem. Rev.*, 71, 524–616.
42. Dokoumetzidis, A., and Macheras, P. (2006), A century of dissolution research: From Noyes-Whitney to the biopharmaceutics classification system, *Int. J. Pharm.*, 321, 1–11.
43. Azaemi, A., Roa, W., and Lobenberg, R. (2007), Current perspectives in dissolution testing of conventional and novel dosage forms, *Int. J. Pharm.*, 328, 12–21.
44. Costa, P., and Lobo, J. M. S. (2001), Modeling and comparison of dissolution profiles, *Eur. J. Pharm. Sci.*, 13, 123–133.
45. Yu, L. X., Carlin, A. S., Amidon, G. L., and Hussain, A. S. (2004), Feasibility studies of utilizing disk intrinsic dissolution rate to classify drugs, *Int. J. Pharm.*, 270, 221–227.
46. Curatolo, W. (1998), Physical chemical properties of oral drug candidates in the discovery and exploratory development settings, *Pham. Sci. Technol. Today*, 1, 387–393.
47. Artursson, P., Palm, K., and Luthman, K. (2001), Caco-2 monolayers in experimental and theoretical predictions of drug transport, *Adv. Drug Deliv. Rev.*, 46, 27–43.
48. Amidon, G. L., Lennernas, H., Shah, V. P., and Crison, J. R. (1995), A theoretical basis for a biopharmaceutic drug classification: The correlation of in vitro drug product dissolution and in vivo bioavailability, *Pharm. Res.*, 12, 413–419.
49. Stewart, B. H., Chan, O. H., Jezyk, N., and Fleischer, D. (1997), Discrimination between candidates using models for evaluation of intestinal absorption, *Adv. Drug Deliv. Rev.*, 23, 27–45.
50. U.S. Department of Health and Human Services (DHHS) (2000, Aug.), Guidance for Industry: Waiver of in vivo bioavailability and bioequivalence studies for immediate release solid oral dosage forms based on a biopharmaceutics classification system, DHHS, Food and Drug Administration, Center for Drug Evaluation and Research, Washington, DC.
51. Yamashita, S., Yu, L. X., and Amidon, G. L. (2006), A provisional biopharmaceutical classification of the top 200 oral drug products in the United States, Great Britain, Spain and Japan, *Mol. Pharm.*, 3, 631–643.
52. Xue-Qing, C., Melissa, D. A., Christoph, G., and Gumundsson, S. O. (2006), Discovery pharmaceutics—Challenges and opportunities, *AAPS J.*, 8, E402–E408.
53. Guillory, J. K., and Poust, R. I. (2002), Chemical kinetics, in Banker, G. S., and Rhodes, C. T., Eds., *Modern pharmaceutics*, Marcel Dekker, New York, pp 139–166.

54. IFAMA (2003), ICH stability testing of new drug substances and products, International Federation of Pharmaceutical Manufacturers Associations (IFPMA), Geneva.
55. Carstensen, J. (2002), Preformulation, in Banker, G. S., and Rhodes, C. T., Eds., *Modern Pharmaceutics*, Marcel Dekker, New York, pp 167–185.
56. Pillai, O., Dhanikula, A. B., and Panchagnula, R. (2001), Drug delivery: An odyssey of 100 years, *Curr. Opin. Chem. Biol.*, 5, 439–446.

6.3

TABLET DESIGN

EDDY CASTELLANOS GIL,[1,2,3] ISIDORO CARABALLO,[2] AND BERNARD BATAILLE[3]

[1]*Center of Pharmaceutical Chemistry and University of Havana, Havana, Cuba*
[2]*University of Sevilla, Seville, Spain*
[3]*University of Montpellier 1, Montpellier, France*

Contents

Pharmaceutical Manufacturing Handbook: Production and Processes, edited by Shayne Cox Gad

6.3.1 INTRODUCTION

The compression of powders—the fourth state of matter in the words of Hans Leuenberger—is a complex task due to the variability in particle size and shape, even for particles of the same component, the unknown distribution of the particles in the die, their different ability to flow, and the forces needed to create bonds between them.

Science-based formulation is nowadays a good strategy for the development of a pharmaceutical formulation. The new regulatory environment (process analytical technology) will transform science-based formulation in the next years.

Therefore, there is a need for statistical tools able to predict the behavior of a powder mixture when it is subjected to compression forces. The purpose of this chapter is to provide a brief description of the main theoretical aspects of tablet design and formulation together with practical examples of tablet development and characterization using different techniques.

6.3.2 TABLETING CYCLE

Compressing powder or granule into a tablet is one of the simplest and oldest ways of forming a product known to humans. Nowadays, as technology has advanced, more complex machines are used with different procedures, but the basic principle, the tabletting cycle, remains the same. In tablet design many factors have to be taken into account, such as the physicolchemical properties of active compound and excipients. An important role also has to be attributed to tableting machines.

Tablet press subclasses primarily are distinguished from one another by how the powder blend is delivered to the die cavity. Tablet presses can deliver powders without mechanical assistance (gravity) (e.g., Ronchii, Manesty, Stokes, and Colton machines), with mechanical assistance (power assisted) (e.g., Ronchii, Courtoy, Kilian, Manesty, Kikusui, Fette, and Hata machines), by rotational forces (centrifugal) (e.g., the Comprima machine), and in two different locations where a tablet core is formed and subsequently an outer layer of coating material is applied (compression coating) (e.g., Kilian Manesty, and Kikusui machines).

The basic unit of any tablet press is tooling consisting of two punches and a die, called a station. The upper and lower punches come together in the die that contains the tablet formulation. Principally, two different types of machines are used, the eccentric and the rotary press. The eccentric press produces about 50–130 tablets per minute. The rotary press has a multiplicity of stations arranged on a rotating table with the dies. A few or many thousands of tablets can be produced per minute. There are numerous models of presses, manufactured by a number of companies, ranging in size, speed, and capacity.

The eccentric press is widely used early in the development stage, because the tabletting machine and the tooling are inexpensive, it can be easily instrumented, little material is needed, and setting, servicing, cleaning, and tool changeover of the machine are easy. During the manufacturing process the tablet mixture is dosed by a hopper into the die. The position of the lower punch defines the volume of the subsequent tablet mass. The compression force is given by the position of the upper punch, which defines its immersion depth into the die, and the reagent force that is built up during the densification of the material. The compressed tablet is ejected by the lower punch.

During the compression process on an eccentric press, there are other pressure ratios at the upper and lower punches. The pressure at the upper punch is usually higher than the pressure at the lower punch. A part of the pressure is lost in the material and in the resulting radial friction force against the die wall during the compression [1, 2].

Figure 1 shows tablets (300 mg) of native dextran obtained from *Leuconostoc mesenteroides* B-512F (Sigma) with a 10% water (w/w) punch in an eccentric machine (Manesty) at 15 kN with tablet side (flat-flace diameter) 10 mm. Axial displacement of water was observed according to the change in color.

On a rotary press the filling of the die and the following compression process is done at the same time at different stations. The compression is carried out in the simplest case with two rolls touching the upper and lower punches and compressing the powder mixture. In contrast to the eccentric press, the upper and lower punches exert pressure on the tablet mixture from both sides at the same time.

Tablets compressed on a rotary press generally show a more consistent hardness when the upper and lower sides of the tablets are compared, where the upper side of tablets compressed on an eccentric press is usually harder than the lower side [3].

Figure 2 illustrates the difference in compression profiles of the upper and lower punches and the punch movement with fictitious rotary and eccentric presses.

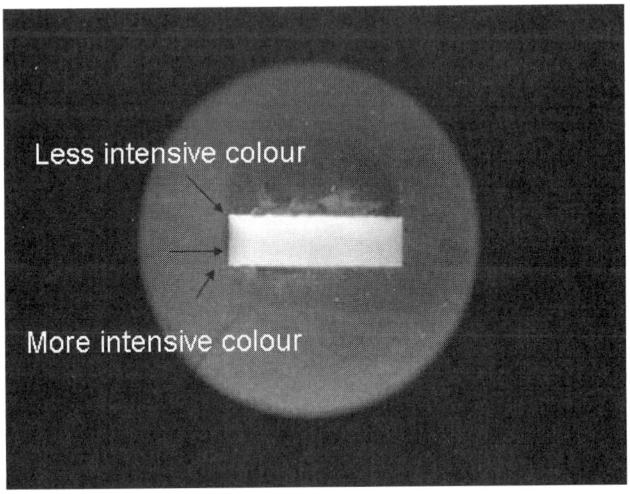

FIGURE 1 Native dextran tablet press in eccentric machine shows variability of color due to different pressure between upper and lower punches.

Another reason, to use a rotary press rather than an eccentric press for tablet compression is the dwell time is usually shorter on a rotary press (see Figure 3).

All compression mixes have an optimum compressing speed. This is why tablet press manufacturers install variable speed controls for the rotor or turret and the powder feeding mechanism. Many compression mixes are speed sensitive and will not produce satisfactory tablets at inappropriate speeds. The dwell time, where maximum pressure is applied to the mix, is relative to the peripheral speed of the turret and the diameter of the punch head flat. Any air in the compression mix must be expelled to avoid laminating or splitting of the tablet. If air is compressed within the tablet, when the pressure applied by the punch is released, the compressed air expands and breaks the tablet. Precompression or dies with a taper in the bore will

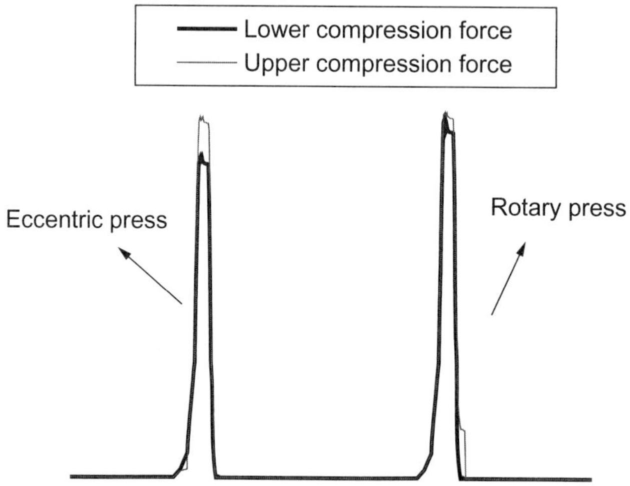

FIGURE 2 Punch movements and compression profiles of upper and lower punches of rotary and eccentric presses.

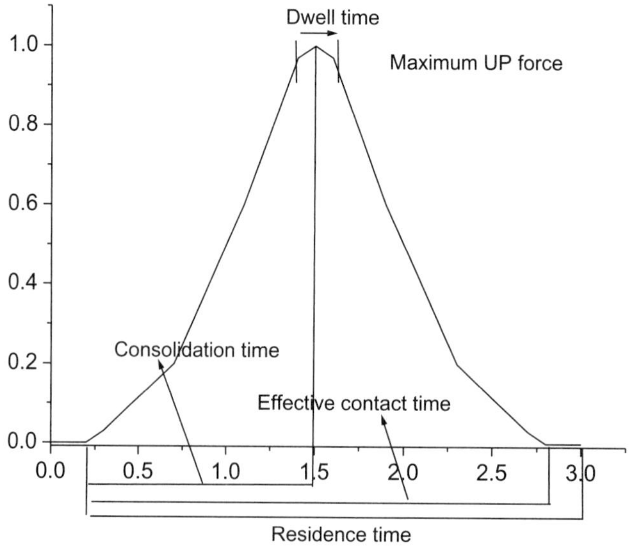

FIGURE 3 Dwell time as function of tableting machine.

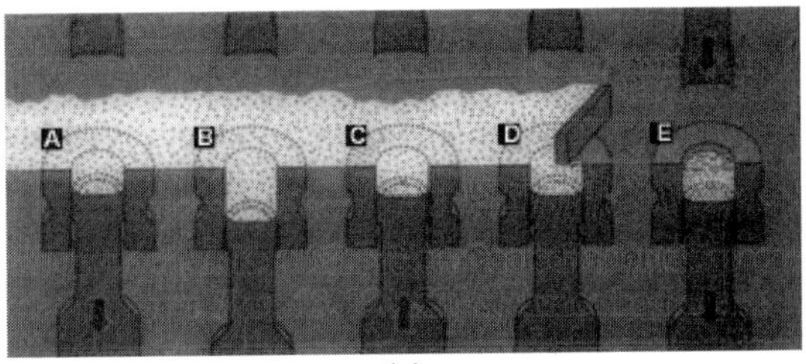

(a)

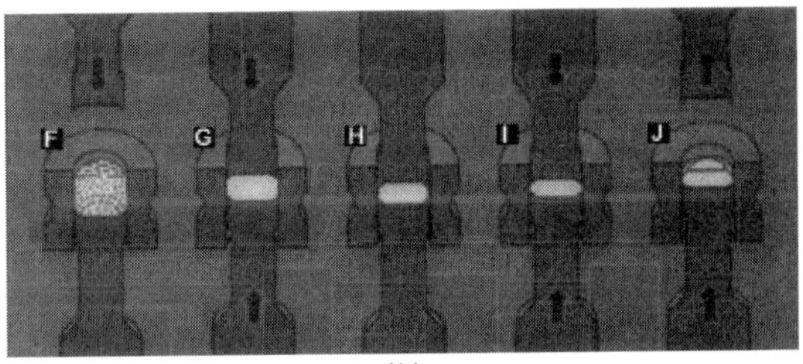

(b)

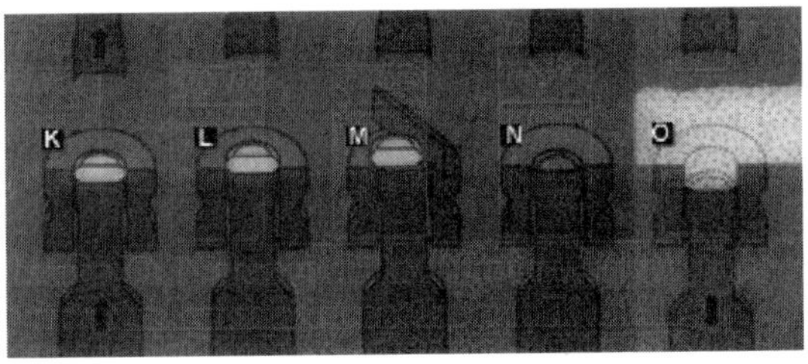

(c)

FIGURE 4 Stages of tableting process: (*a*) filling; (*b*) compression; (*c*) ejection.

often reduce this problem. Rotary presses sometimes have two pairs of compression rolls (for precompression and compression). During precompression additional compression can take place and the absolute dwell time can be prolonged. In other words, there are numerous different tablet presses with various possibilities to carry out the compression process.

A direct correlation between the results of an eccentric press and a rotary press cannot always be drawn. In addition there are many different tablet presses with different settings and possibilities. These problems can be overcome by using a compaction simulator early in the development stage. An advantage of such a simulator is its versatility, that is, all types of presses can be simulated with small amounts of solid. The problem, however, is that such a simulator is very expensive.

The tableting cycle is well explained in the literature and is broken down into three stages (Figure 4):

1. *Filling* The volume of the granule is measured (Figure 4*a*):
   A: The lower punch is allowed to descend to its lowest point.
   B: The bore of the die is filled completely with powder.
   C: The lower punch is raised to a predetermined point so that excess.
   D: The powder is leveled by passing under a blade.

E: This ensures that the bore of the die is filled with as exact volume of the material to be used, and the next stage can begin.

2. *Compression* Pressure is applied to form the granule into a solid (Figure 4*b*):

F: The upper punch is lowered into the bore of the die.

G: Precompression gives the powder an initial "punch" to remove excess air.

H: The powder is fully compressed.

I: The correct pressure is reached.

J: The upper punch is lifted out of the way ready for tablet ejection.

3. *Ejection* The tablet is ejected and the next tablet will be formed (Figure 4*c*):

K: The lower punch begins to rise in the bore of the die lifting the tablet until step L is reached.

L: Its base is level with the tap of the die.

M: The tablet is pushed aside into the take-off chute by passing under a static blade.

N: The lower punch moves to its lowest position ready for filling (**O**), similar to **A** and the entire cycle is repeated.

Guidance on compression levels for each tablet type (series 1,2, and 3) and maximum punch pressures are given in Table 1:

Series 1: Flat flace, normal concave, shallow concave tablets

Series 2: Double radius, bevel and concave tablets

Series 3: Flat beveled edge, deep concave, ball or pill

In Figure 4 the tablets progress from start to finish from left to right. On an actual machine this will be determined by the direction in which the entire turret rotates in relation to the fixed item, such as the fill hopper precompressions and compression rollers. It is important to note that this direction may vary from machine to machine, but as a general rule, British, American, and some Asian machines rotate in a clockwise direction while European machines rotate anticlockwise (see Figure 5).

6.3.3 LIMITATIONS FOR DIRECT COMPRESSION

In tablet formulation, a range of excipient materials are normally required along with the active ingredient in order to give the tablet the desired properties. For example, the reproducibility and dose homogeneity of tablets are dependent on the properties of the powder mass. The tablet should also be sufficiently strong to withstand handling but should disintegrate after intake to facilitate drug release. The choice of excipients will affect all these properties:

1. *Filler* illers are used to make tablets of sufficient size for easy handling by the patient and to facilitate production. Tablets containing a very potent active

TABLE 1 Guide Punch Pressures for Series 1, 2, and 3

Tablet Size		Pressure (kN)		
mm	in.	Series 1	Series 2	Series 3
3	$\frac{1}{8}$	5	3	2
4	$\frac{5}{32}$	10	5.6	3.7
5	$\frac{3}{16}$	15	8.8	5.8
6	$\frac{7}{32}$	22	12	8.5
7	$\frac{9}{32}$	30	17	11
8	$\frac{5}{16}$	40	22	15
9	$\frac{11}{32}$	50	28	19
10	$\frac{13}{32}$	60	35	23
11	$\frac{7}{16}$	70	42	27
12	$\frac{15}{32}$	90	50	33
13	$\frac{1}{2}$	100	59	39
14	$\frac{9}{16}$	120	67	46
15	$\frac{19}{32}$	130	78	53
16	$\frac{5}{8}$	160	90	60
17	$\frac{21}{32}$	180	102	68
18	$\frac{23}{32}$	203	114	74
19	$\frac{3}{4}$	226	127	85
20	$\frac{25}{32}$	251	141	94
25	1	393	221	147

FIGURE 5 Tableting machine (anticlockwise).

substance would be very small without excipients. A good filler will have good compactability and flow properties and acceptable taste and be non-hygroscopic and preferably chemically inert. It may also be advantageous to have a filler that fragments easily, since this counteracts the negative effects of lubricant additions to the formula [4].

2. *Binder* A material with a high bonding ability can be used as a binder to increase the mechanical strength of the tablet. A binder is usually a ductile material prone to undergo plastic (irreversible) deformation. Typically, binders are polymeric materials, often with disordered solid-state structures. Of special importance is the deformability of the peripheral parts (asperities and protrusions) of the binder particles [5].

This group of materials has the capacity of reducing interparticulate distances within the tablet, improving bond formation. If the entire bulk of the binder particles undergoes extensive plastic deformation during compression, the interparticular voids will, at least partly, be filled and tablet porosity will decrease. This increases contact area between the particles, which promotes the creation of interparticular bonds and subsequently increases tablet strength [6]. However, the effect of the binder depends on both its own properties and those of the other compounds within the tablet. A binder is often added to the granulation liquid during wet granulation to improve the cohesiveness and compactability of the powder particles, which assists the formation of agglomerates or granules. It is commonly accepted that binders added in dissolved form, during a granulation process, are more effective than those added in dry powder form during direct compression.

3. *Disintegrating Agent* A disintegrant is normally added to facilitate the rupture of bonds and subsequent disintegration of the tablets. This increases the surface area of the drug exposed to the gastrointestinal fluid; incomplete disintegration can result in incomplete absorption or a delay in the onset of action of the drug. There are several types of disintegrants, acting with different mechanisms: (a) promotion of the uptake of aqueous liquids by capillary forces, (b) swelling in contact with water, (c) release of gases when in contact with water, and (d) destruction of the binder by enzymatic action [7]. Starch is a traditional disintegrant; the concentration of starch in a conventional tablet formulation is normally up to 10% w/w. The starch particles swell moderately in contact with water, and the tablet disrupts. So-called superdisintegrants are now commonly used; since these act primarily by extensive swelling, they are effective only in small quantities [8] cross-linked sodium carboxymethyl cellulose (e.g., Ac-Di-Sol), which is effective in concentrations of 2–4%, is a commonly used superdisintegrant. Larger particles of disintegrants have been found to swell to a greater extent and with a faster rate than finer particles, resulting in more effective disintegration [9].

4. *Glidant, Antiadherent, and Lubricant* Glidants are added to increase the flowability of the powder mass, reduce interparticulate friction, and improve powder flow in the hopper shoe and die of the tableting machine. An antiadherent can be added to decrease sticking of the powder to the faces of the punches and the die walls during compaction, and a lubricant is added to decrease friction between powder and die, facilitating ejection of the tablet from the die. However, addition of lubricants (here used as a collective term and including glidants and antiadherents) can have negative effects on tablet strength, since the lubricant often reduces

the creation of interparticulate bonds [e.g., 4]. Further, lubricants can also slow the drug dissolution process by introducing hydrophobic films around drug and excipient particles [e.g., 10]. These negative effects are especially significant when long mixing times are required [11]. Therefore, the amount of lubricants should be kept relatively low and the mixing procedure kept short to avoid a homogenous distribution of lubricant throughout the powder mass. An alternative approach could then be to admix granulated qualities of lubricant [12].

5. *Flavor, Sweetener, and Colorant* Flavor and sweeteners are primarily used to improve or mask the taste of the drug, with subsequent substantial improvement in patient compliance. Coloring tablets also has aesthetic value and can improve tablet identification, especially when patients are taking a number of different tablets.

General instructions for the determination of tablet properties (e.g., hardness, disintegration, friability, dissolution profile, and stability) are contained in pharmacopeia [e.g., European Pharmacopoeia (Eur. Ph.) and U.S. Pharmacopeia (USP)].

In the manufacture of tablets it is important to define and appreciate the physical properties of the active substance, in particular particle size and flowability. The technology involved in direct compression assumes great importance in tablet formulations because it is often the least expensive, particularly in the production of generics that the active substance permits. The limiting factors are the physical properties of the active substance and its concentration in the tablets. Even substances such as ascorbic acid, which are not generally suitable for direct compression owing to the friability of the crystals, can normally be directly pressed into tablets at concentrations of 30–40%. However, this technique is not as suitable if the content of ascorbic acid is higher. This limit may be shifted upward by special direct-compression auxiliaries, for example, Ludipress (BASF).

Ludipress is derived from lactose, Kollidon 30, and Kollidon CL. It thus combines the properties of a filler, binder, disintegrant, and flow agent and also often acts as a release accelerator. By virtue of its versatility, formulations containing it are usually very simple. It can also be combined with almost all active substances with the exception of those that enter into a chemical interaction with lactose (Maillard reaction). Active substances (e.g., many analgetics) behave very differently with Ludipress when the dosage is extremely high. Acetylsalicylic acid (ASA) and metamizole can be pressed when little Ludipress has been added; ibuprofen requires a larger amount; and the fraction of Ludipress required in the tablets is too large for paracetamol (acetaminophen).

An alternative to the Ludipress grades is the outstanding dry binder Kollidon VA 64 together with excipients (e.g., calcium phosphate, microcrystalline cellulose, lactose, or starch) and a disintegrant (e.g., Kollidon CL). This combination even allows 500 mg of paracetamol to be pressed into good tablets with a weight of 700 mg.

No other dry binder has binding power and plasticity comparable to Kollidon VA 64. Plasticity, in particular, is an important parameter in direct compression. As can be seen in Figure 6 (99.5% binder + 0.5% magnesium stearate), this property of Kollidon VA 64 is not adversely affected by increasing the pressure. The beneficial

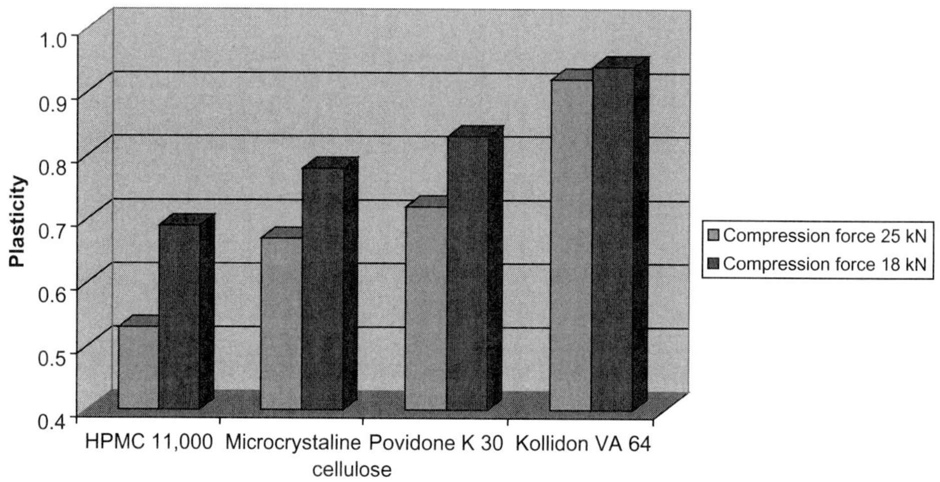

FIGURE 6 Plasticity of dry binders in tablets.

properties of Kollidon VA 64 can also be exploited for the production of concentrated active substance that is subsequently used for direct tableting. Kollidon VA 64 and Ludipress can also be combined with one another.

Acetylsalicylic Acid–Acetaminophen–Caffeine Tablet (250 mg + 250 mg + 50 mg)

Formulation

Acetaminophen powder	39.25%
Caffeine powder	7.85%
ASA powder	39.25%
Kollidon VA 64	9.4%
Kollidon CL	3.15%
Aerosil 200	0.5%
Magnesium stearate	0.6%

Manufacture by Roller Compaction The active ingredients and Kollidon VA 64 were granulated in a Gerteis Roller compactor.

Process Parameters

Force	5.0 kN/cm
Crack	2.9 mm
Looser	5.0 rpm
Dosing screw	90.5 rpm
Plug screw	108.6 rpm
Granulator	50 rpm

Tableting The granules were passed together with magnesium stearate, Aerosil 200, and Kollidon CL through an 800-μm sieve, blended for 10 min in a Turbula

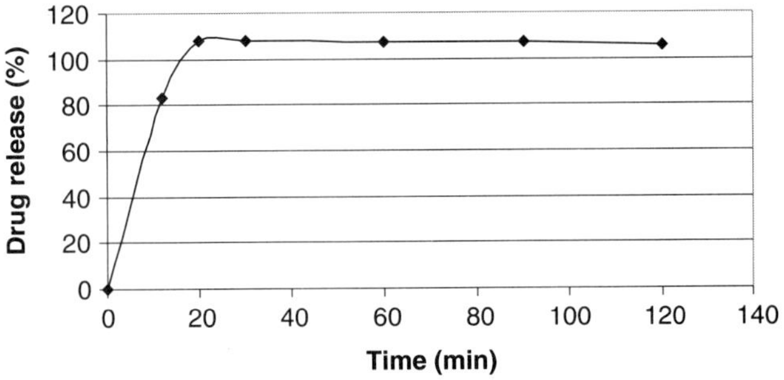

FIGURE 7 Dissolution profile for acetylsalicylic acid–acetaminophen–caffeine tablet (250 mg + 250 mg + 50 mg): paddle 50 rpm, 37 °C deionized water.

mixer, and compressed into tablets with a force of about 12 kN (tablet press Korsch PH106, 30 rpm, compression force 11.5 kN)

Tablet Properties

Weight	629.7 mg
S rel	1.4%
Diameter	12 mm
Form	Biplanar
Hardness	72 N
Friability	2.75%

Dissolution See Figure 7.

Enteric Film Coating of Tablets (Organic Solution)

Formulation of Cores

Component	Percent/Tablet (w/w)
ASA	33.4
Ludipress	49.5
Avicel PH 102	16.6
Magnesium stearate	0.5
Total	100.0

Manufacture by Direct Compression All the components were mixed for 10 min, passed through an 0.8-mm sieve, and compressed into tablets on a rotary press with a rate of 40,000 tablets/h at a compression force of 15-kN. Core shape was convex with a diameter of 9 mm and the engraving BASF. Hardness of the tablets was about 60 N.

Formulations of Coating Suspension The formulation is for 5-kg cores (diameter 9 mm; 300 mg):

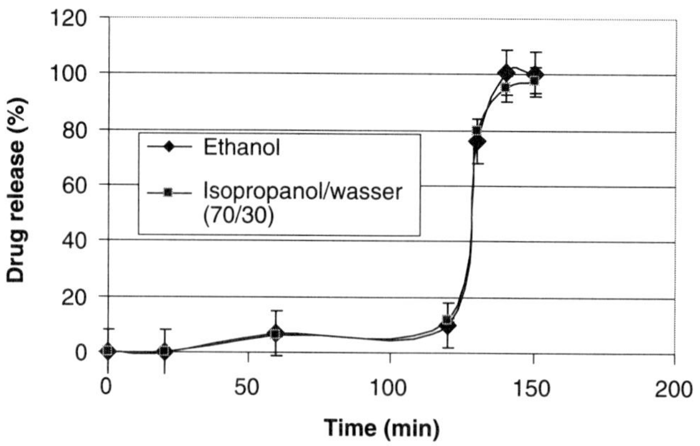

FIGURE 8 Enteric film coating of tablets, acetylsalicylic acid (organic solution): paddle 100 rpm, 37 °C; 0–2 h: 0.08 *M* HCl, 2 h+: phosphate buffer Ph 6.8.

	Parts by Weight [g]	
Component	No. 1	No. 2
Kollicoat MAE 100 P	344.52	344.52
Triethyl citrate	34.45	34.45
Ethanol	4214.60	—
Isopropanol/Wasser (70/30)	—	4214.60
Total	4593.57	4593.57
Polymer applied	10.0 mg/cm^2	
Content of polymer	7.5%	

Preparation of Spray Suspension Kollicoat MAE 100P and triethyl citrate were stirred into the solvent until complete dissolution.

Process Parameters

Coating pan	24 Accela Cota
Size of batch	5 kg
Inlet air temperature	50°C
Outlet air temperature	35–38°C
Product temperature	30–35°C
Inlet air rate	70 m
Outlet air rate	140 m
Spraying pressure	2.0 bars
Nozzle diameter	1.0 mm
Rate of spraying	30 g/min
Time of spraying	2.5 h
Preheating	3 min
Final drying	5 min

Dissolution: (See Figure 8). Dissolution was done according to USP monograph "Aspirin Delayed-Release Tablets."

Beta Carotene Tablets (15 mg)

Formulation

	Formulation 1	Formulation 2
Beta carotene dry powder 10%	160.0 g	150.0 g
Ludipress	240.0 g	—
Dicalcium phosphate, granulated with 5% Kollidon 30	—	175.0 g
Avicel PH 101	—	100.0 g
Kollidon CL	6.0 g	5.0 g
Aerosil 200	—	2.5 g
Talc	—	20.0 g
Calcium arachinate	—	2.5 g
Magnesium stearate	2.0 g	—

Manufacturing (Direct Compression) All components were mixed, passed through a 0.8-mm sieve, and pressed with a medium compression force.

Tablet Properties

	Formulation 1	Formulation 2
Weight	400 mg	502 mg
Diameter	12 mm	12 mm
Form	Biplanar	Biplanar
Hardness	59 N	57 N
Disintegration	12 min	1 min
Friability	0.1%	0%

Chemical and Physical Stability (20–25°C)

	6 Months	12 Months
Formulation 1		
Loss of beta carotene	3%	4%
Hardness	60 N	59 N
Disintegration	9 min	7 min
Friability	0.15%	0.16%
Formulation 2		
Loss of beta carotene	8%	9%

Diclofenac Na–Dispersion–Tablet (50 mg)

Formulation

Diclofenac Na	50.0 mg
Avicel PH 102	143.8 mg
Kollidon CL	50.0 mg
Aerosil 200	5.0 mg
Magnesium stearate	1.0 mg

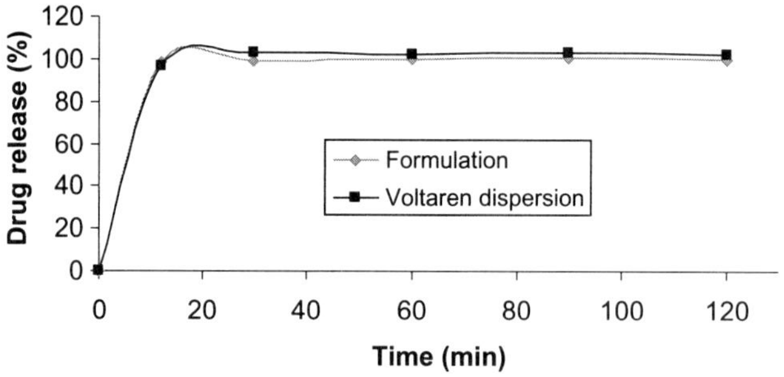

FIGURE 9 Dissolution profile for diclofenac–Na dispersion tablet (50 mg): paddle 50 rpm, 37 °C phosphate buffer, pH 7.2.

Procedure The ingredients were mixed, passed through a 0.8-mm sieve, and compressed into tablets with a force of about 10 kN. (The tablet press was Korsch PH106, 30 rpm, compression force was 11.8 kN.

Tablet Properties

Weight	248.0 mg
S rel	1.7%
Diameter	10 mm
Form	bipla Nar
Hard Ness	93 N
Friability	<0.1%

Dissolution See Figure 9.

In direct–compression formulation, there is a wide particle size distribution. Usually, the active drug is at the fine end of the range. Such a wide particle size range can easily result in significant segregation. Five primary mechanisms are responsible for most particle segregation problems [13]. Of these, only three typically occur with pharmaceutical powders: sifting, entrainment of air, and entrainment of particles in an air stream.

Sifting is a process by which smaller particles move through a matrix of larger ones. It is by far the most common method of segregation. Sifting has been found to occur with particle size ratios as low as 1.3:1 or with a sufficiently large mean particle size (the tendency to segregate by sifting decreases substantially with particle size <500 μm. Free-flowing material and interparticle motion also caused segregation by sifting.

Two techniques can be used to decrease a material's segregation tendencies: change the material or change the design of the equipment.

Lisinopril (5 mg) Reducing the ratio of excipient, lisinopril (2:1) tablets for direct compression can be obtained:

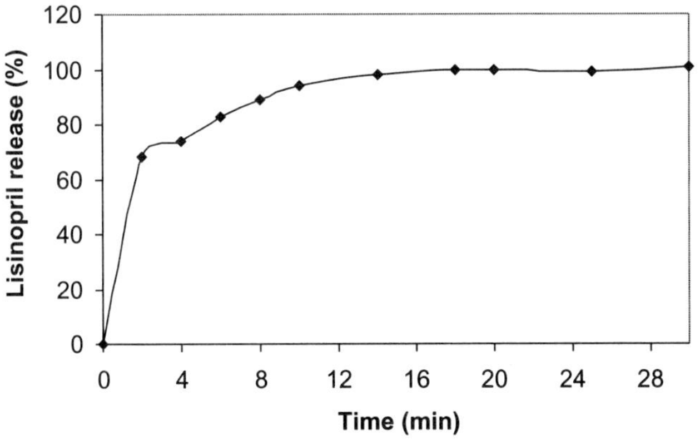

FIGURE 10 Disolution profile for lisinopril 5 mg: paddle 50 rpm, 37 °C, HCl 0.1 *N*.

Formulation	%
Lisinopril	5.0
Aerosil	0.5
Calcium phosphate dibasic	19.8
Starch 1500	2.0
Magnesium stearate	0.7
Cellulose microcrystalline (PH-250)	72.0

Lisinopril and Aerosil (<150 μm) were mixed for around 10 min. All other components were added and mixed for 15 min, passed through a 0.150-mm sieve, and pressed with 10 kN compression force.

Properties of 5-mg lisinopril tablets are as follows:

Weight	130 mg
Diameter	8 mm
Form	Biplanar
Hardness	98 N
Disintegration	2–3 min
Friability	0.05

The content uniformity of the formulation was measured at the beginning, middle, and end of the batch (50 kg) (Table 2).

Special attention should be given to the physical stability of the tablets manufactured by direct compression because some fillers/binders are known to soften or harden on storage.

6.3.4 PREVIOUS GRANULATION: BIOPHARMACEUTICAL VERSUS TECHNOLOGICAL PROPERTIES

Granulation is the process by which primary powder particles are made to adhere to form larger, multiparticle entities called granules. Pharmaceutical granules

TABLE 2 Study of Uniformity for Formulation Lisinopril 5 mg (Batch 50 kg)

Number	0–5 kg	5–10 kg	10–15 kg	15–20 kg	20–25 kg	25–30 kg	35–40 kg	40–45 kg	45–50 kg
1	4.99	5.00	5.00	4.95	5.02	5.08	5.00	4.96	5.00
2	5.00	5.06	5.00	4.92	5.00	4.99	4.83	5.11	5.17
3	5.00	4.81	4.90	5.19	4.92	5.03	4.92	5.00	5.03
4	5.05	5.00	5.00	4.99	4.82	4.96	5.00	4.93	4.99
5	4.89	4.99	4.90	5.05	5.13	5.04	4.98	4.93	4.87
6	5.02	5.06	4.85	5.10	5.13	5.00	5.18	4.81	4.93
7	5.00	5.08	5.11	5.00	5.00	4.83	4.89	5.00	5.00
8	5.00	5.00	5.03	5.03	5.00	5.06	5.00	5.18	5.00
9	4.97	4.99	5.01	5.04	5.05	4.85	5.00	5.00	4.99
10	4.91	5.00	5.01	4.98	4.97	5.00	5.00	5.11	5.05

typically have a size range between 0.2 and 4.0 mm, depending on their subsequent use. In the majority of cases this will be in the production of tablets or capsules, when granules will be made as intermediate products and have a typical size range between 0.2 and 0.5 mm, but larger granules are used as a dosage form in their own right.

Granulation normally commences after initial dry mixing of the necessary powdered ingredients so that a uniform distribution of each ingredient through the mix is achieved. After granulation the granules either will be packed or may be mixed with other excipients prior to tablet compaction.

The principal reasons for granulation are as follows:

1. To prevent segregation of the constituents of the powder mix.
2. To improve the flow properties of the mix.
3. To improve the compaction characteristics of the mix

Methods of Granulation Granulation methods can be divided into two types: *wet methods*, which use a liquid in the process, and *dry methods*, in which no liquid is used. In a suitable formulation a number of different excipients will be needed in addition to the drug. The common types used are diluents, to produce a unit dose weight of suitable size, and disintegrating agents, which are added to aid the break-up of the granule when it reaches a liquid medium (e.g., on ingestion by the patient). Adhesives in the form of a dry powder may also be added, particularly if dry granulation is employed. These ingredients will be mixed before granulation.

1. *Dry Granulation* In the dry methods of granulation the primary powder particles are aggregated under high pressure. There are two main processes: Either a large tablet (*slug*) is produced in a heavy-duty tableting press (*slugging*) or the powder is squeezed between two rollers to produce a sheet of material (*roller compaction*). In both cases these intermediate products are broken down using a suitable milling technique to produce granular material, which is usually sieved to separate the desired size fraction. This dry method may be used for drugs

that do not compress well after wet granulation or those which are sensitive to moisture.

2. *Wet Granulation* Wet granulation involves the massing of a mix of dry *primary powder particles* using a *granulating fluid*. The fluid contains a solvent which must be volatile so that it can be removed by drying and be nontoxic. Typical liquids include water, ethanol, and isopropanol, either alone or in combination. The granulation liquid may be used alone or, more usually, as a solvent containing a dissolved *adhesive* (also referred to as a *binder* or *binding agent*) which is used to ensure particle adhesion once the granule is dry. Water is commonly used for economical and ecological reasons. Its disadvantages as a solvent are that it may adversely affect drug stability, causing hydrolysis of susceptible products, and it needs a longer drying time than do organic solvents. This increases the length of the process and again may affect stability because of the extended exposure to heat.

Captopril (25 mg) + Hydrochlorothiazide (25 mg)

Formulation

	Formulation for 500 mg
Captopril	5%
Hydrochlorothiazide	5%
Lactose	65%
Carboxyethylcellulose sodium	10%
Ac-Di-Sol	3%
Starch	10%
Stearic acid	2%

Manufacturing (Wet Granulation) A mixture of all compounds (with 1.5% stearic acid) is granulated with solution 2-propanol (around 8% v/w), passed through a 0.8-mm sieve, and the rest (0.5% stearic acid) added and pressed with low compression force.

Tablet Properties (Initial Time)

Weight	500 mg
Diameter	12 mm
Form	Normal concave
Hardness	60 N
Disintegration	<4 min
Friability	<0.3%
Dissolution (captopril + hydrochlorothiazide)	
30 min	90.00%
60 min	100%

Stability of Three Batches (5 kg each) at 25 °C and 70% Relative Humidity (RH) during 12 Months

Formulation Table 3 presents a comparison of lactose monohydrate and calcium phosphate.

TABLE 3 Comparative Study of Lactose Monohydrate and Calcium Phosphate, Dibasic

	Formulation F1	Formulation F2
α-Methyldopa	275 g (78%)	275 g (78%)
Lactose monohydrate	15.5%	—
Calcium phosphate, dibasic	—	15.5%
Kollidon 30	4%	4%
Isopropanol	80 mL	80 mL
Kollidon CL	2%	2%
Magnesium stearate	0.5%	0.5%

	6 months			12 months		
Formulation	Batch 1	Batch 2	Batch 3	Batch 1	Batch 2	Batch 3
Hydrochlorothiazide (%)	98.91	99.47	100.65	99.06	100.40	102.88
Captopril (assay) (%)	101.02	100.89	100.99	100.65	100.03	100.30
Captopril disulfuric[a]	0.62	0.71	0.69	0.69	0.79	0.72
Weight	500.12	501.23	499.65	499.00	500.33	500.14
Hardness	63 N	67 N	61 N	60 N	63 N	61 N
Disintegration (min)	3	3	4	4	4	4
Friability	1.98	1.15	1.12	1.71	1.12	1.12
Dissolution						
30 min	89.45	90.54	88.77	88.00	91.00	88.05
60 min	100	100	100	100	100	100

[a]Captopril degradation product (%).

α-Methyldopa Tablet (250 mg)

Manufacturing (Wet Granulation) A mixture of α-methyldopa with lactose or calcium phosphate (for formulations **F1** or **F2**, respectively) is granulated with isopropanol solution of Kollidon 30 and passed through a sieve, the dry granules are mixed with Kollidon CL and magnesium stearate, and pressed with medium compression force.

Tablet Properties

	F1	F2
Weight	361 mg	362 mg
Diameter	11 mm	11 mm
Hardness	118 N	156 N
Disintegration	5 min	4 min
Friability	<0.1%	<0.1%
Dissolution		
10 min	45%	55%
20 min	82%	90%
30 min	90%	98%

Calcium phosphate, dibasic offers high hardness and faster dissolution profile than lactose for α-methyldopa tablets in wet granulation.

6.3.5 TABLET DESIGN FOR MATRIX SYSTEM

The advantages of controlled-release systems include maintenance of drug levels within a desired range, the need for fewer administrations, optimal use of the drug in question, and increased patient compliance. Evaluation of matrix tablets is the same as for conventional formulations but the dissolution profile and stability have to be carefully studied. Numerous methods for development of matrix tablets can be used, such as direct compression, wet granulation, pelletization, and spheronization exclusion. Nevertheless, the potential disadvantages cannot be ignored: the possible toxicity or no biocompatibility of the materials used, undesirable byproducts of degradation, any surgery required to implant or remove the system, the chance of patient discomfort from the delivery device, and the higher cost of controlled-release systems compared to traditional pharmaceutical formulations. The importance of matrix systems that they release bioactive component over an extended period of time has long been recognized in the pharmaceutical field. Matrix systems can be divided into three groups depending on the type of polymer formed:

1. *Inert Matrices* Polymers that after compression form an indigestible and insoluble porous skeleton [14] constitute the inert matrices. The main challenge in the preparation of these systems is to achieve, by means of a suitable design, total drug release from the device as well as adequate and precise drug release, guaranteeing the integrity of the matrix.

2. *Hydrophilic Matrices* Cellulose derivatives have been widely used in the formulation of hydrogel matrices for controlled drug delivery. Among them hydroxypropyl methylcellulose (HPMC) is the most extensively employed because of its ease of use, availability, and very low toxicity [15]. Drug release from these systems is controlled by the hydration of HPMC, which forms a gelatinous layer at the surface of the matrix through which the included drug diffuses.

Drug release from swellable matrix tablets is based on the glassy–rubbery transition of the polymer which occurs as a result of water penetration into the matrix. Therefore, the gel layer is physically delimited by the erosion (swollen matrix–solvent boundary) and swelling (glassy–rubbery polymer boundary) fronts.

Water-soluble drugs are released primarily by diffusion of dissolved drug molecules across the gel layer, while poorly water soluble drugs are released predominantly by erosion mechanisms.

The factors influencing the release of drugs from hydrophilic matrices include viscosity of the polymer, ratio of the polymer to drug, mixtures of polymers, compression pressure, thickness of the tablet, particle size, pH of the matrix, entrapped air in the tablet, solubility of the drug, the presence of excipients or additives, and the mode of incorporation of these substances.

3. *Lipid Matrices* These matrix tablets are formed with lipid polymers with low melting point. The drug is dissolved or solubilized in the melted lipid, such as cetyl

TABLE 4 Theophylline Formulation by Direct Compression and Wet Granulation (mg/tablet)

	Direct Compression	Wet Granulation
Granulated theophylline	264	264
90SH-4000SR HPMC (Metolose SR)	64.5	64.5
Mg stearate	1.5	1.5
Total	330	330

alcohol, ceto-stearilic alcohol, and stearic acid. Solid lipid nanoparticles are an example of an innovative lipid matrix system.

6.3.5.1 Controlled-Release Tablet by Direct Compression and Wet Granulation

Theophylline is granulated in a fluid bed with Pharmacoat 606 3% (Shin Etsu, Metolose SR) as shown in Figure 11. Table 4 gives a comparison of direct compression (DC) and wet granulation (WG) using theophylline:

Diret compression: using a twin-cell mixer, theophylline and HPMC are mixed for 10 min; then Mg stearate is added and mixed for 2 min.

Tableting conditions for DC and WG: A rotary tableting machine (KIKUSUI) is used with 12 punches (punch size 10 mm diameter, 12 mm radius, compression force 98, 147, and 196 MPa; tableting speed 20, 40, and 60 min^{-1}).

Condition for WG: Granulation machine, vertical granulator FM-VG-05; charge 300 g; binder solution ethanol–water 8:2; agitation 600 (blade)/1000 (chopper) min^{-1}; granulation time 5 min.

Powder properties for compression: Bulk density 0.35 g/mL, tapped density 0.48, average particle size 122 μm.

Theophylline tablets made by both direct compression and wet granulation have been assessed. There is almost no difference between direct compression and wet granulation methods (see Figures 12–14) under the following conditions: appropriate formulation (sufficient level of HPMC in the tablet) and precise control of the wet granulation process. Direct compression using Metolose SR is recognized as a suitable process for matrix tablets.

6.3.6 TABLET DESIGN WITH NATURAL PRODUCTS

The development and production of tablets containing a high dose of active ingredients is a complex and extensive technological challenge. Dried plant extracts are often used as therapeutically active material in the manufacture of tablets. They are

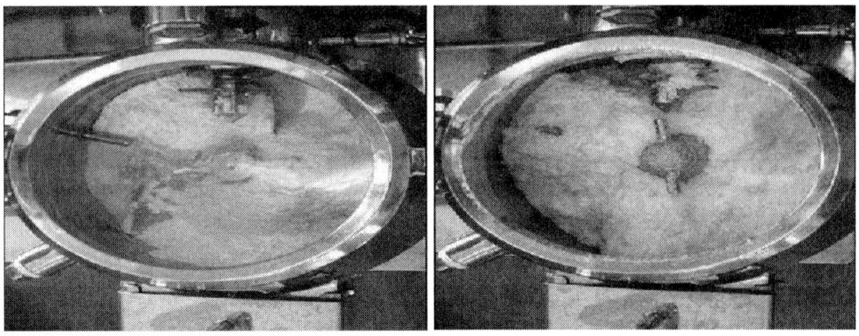

(*a*) (*b*)

FIGURE 11 Granulation process: (*a*) before 90SH-4000SR; (*b*) after 90SH-4000SR.

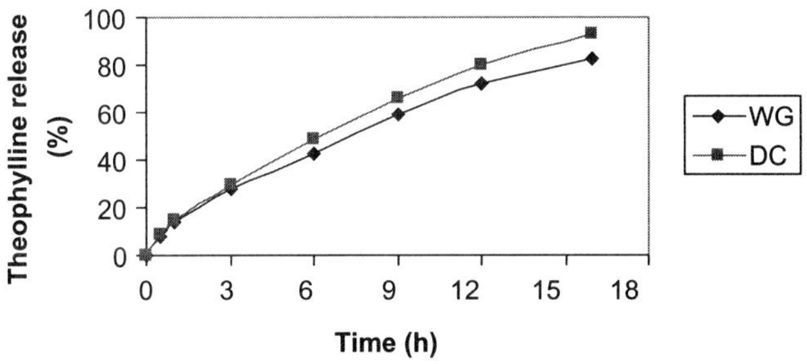

FIGURE 12 Dissolution profile for theophylline tablets (DC and WG).

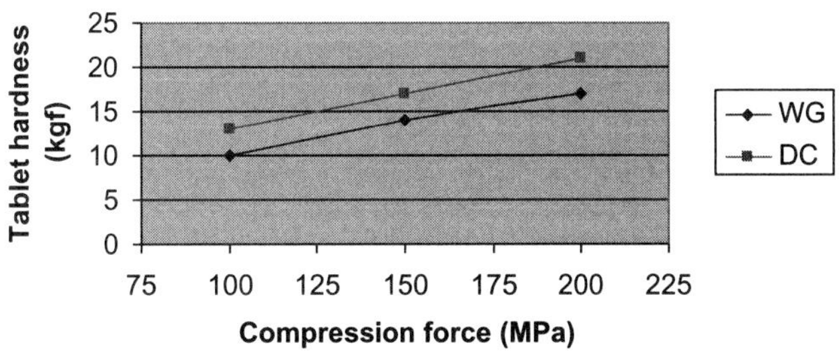

FIGURE 13 Hardness of theophylline tablets (Dc and WG).

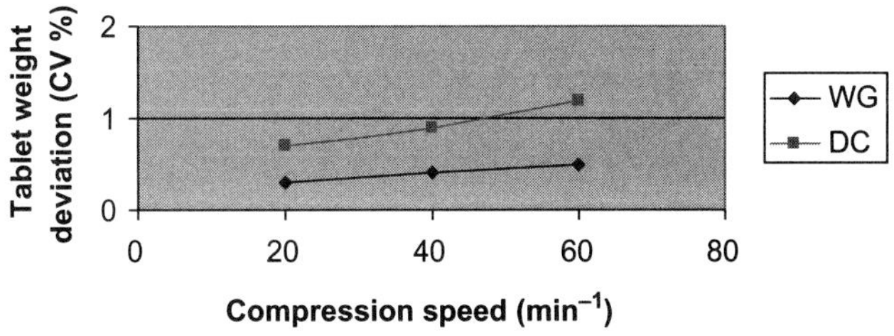

FIGURE 14 Weight deviation for theophylline tablets (DC and WG).

often very fine, poorly compressible, and very hygroscopic powders. Tablets containing a high amount of spray-dried extract show prolonged disintegration times, affecting the release of active constituents [19]. Some alternatives have been proposed to minimize these problems. Granulation is the technique most often used to improve the technological properties of these products. However, because of the products' high hygroscopicity, extracts sometimes cannot be granulated using aqueous systems. Some reports have shown that the use of lubricants during direct compression of vegetable dried extracts increased the disintegration time. According to our experience and some previous work, tablets with high amounts of some lubricant such as aerosil (up to 25% w/w) and magnesium [16] stearate incorporated into the granules had shorter disintegration time than did tablets containing the powdered mixture.

Natural products can be used as plant extracts with pharmacological activity (e.g., *Mangifera indica* L., vallerian, aloe, *Cratoxylum pruniflorum*, microporous zeolite), excipient for direct compression or granulation (chitin, chitosan, and dextran), and controlled-release systems (cellulose and native dextran). For natural products the most important factor is the standardization of the extract because properties such as the amount of active substance can be changed from batch to batch. Factors such as the origin of the extract, geographic zone, and age of the tree could affect the properties of natural extracts.

6.3.6.1 Tablet Design from Aqueous Plant Extract

A bioactive product of natural origin has been developed from folk knowledge of Asian, Latin-American, European, and U.S. ethnic medicine. We developed an extract of the *M. índica* L. (mango) stem bark, obtained by decoction of some varieties grown in tropical and subtropical climates, that is used at present as an antioxidant nutritional supplement (Vimang). The aqueous extract was dried by atomization in a spray dryer until a brown solid with 10.5% (RSD = 0.9%) water content (measured by Karl Fischer) was achieved. Tablets obtained by wet granulation (plant extract, 300 mg/unit) were used for the applications. The product is a fine brown powder that has provend to be useful in the treatment of a large population sample presenting physical stress due to age or deteriorated physiological conditions caused by chronic diseases such as cancer, diabetes, or cardiovascular disorders [17]. Recent studies have shown that treatment with the extract provided significant protection against 12-*O*-tetradecanoylphorbol-13-acetate (TPA)–induced oxidative damage and better protection when compared with other antioxidants (Vitamin C, E and beta–carotene) [18]. Furthermore, the results indicate that this extract is bioavailable for some vital target organs, including liver and brain tissues, peritoneal cell exudates, and serum. Therefore, it was concluded that it could be useful to prevent the production of reactive oxygen species (ROS) and oxidative tissue damage in vivo. All these effects are likely due to the synergic action of several compounds, such as polyphenols, terpenoids, steroids, fatty acids, and microelements, which have been reported to be present in the extract [17].

TABLE 5 Content of Mangiferin (μg/mg) in Natural Product Samples from *Mangifera indica* L. and assay Result of Pharmaceuticals

Sample	Content, μg/mg (RSD, %)	Sample	Content, μg/mg (RSD, %)
No.1 (batch 901)	254 (0.7)	No.9 (batch 0201)	125 (0.1)
No.2 (batch 903)	195 (2.0)	No.10 (batch 0202)	109 (1.5)
No.3 (batch E-1923)	187 (1.6)	No.11 (batch 0203)	116 (1.2)
No.4 (batch E-1924)	180 (1.4)	No.12 (batch 0204)	79 (2.4)*
No.5 (batch E-2032)	206 (1.6)	No.13 (batch 0205)	56 (0.5)*
No.6 (batch 0103)	149 (5.7)	No.14 (batch 0206)	55 (2.3)*
No.7 (batch 0104)	162 (0.4)	No.15 (batch 0207)	66 (1.8)*
No.8 (batch 0112)	159 (0.3)	No.16 (batch 0208)	49 (0.2)*

Pharmaceuticals from batch No.8	Amount of Vimang® (mg)	Percentage of claimed content %, (RSD, %)
Tablets (batch A)	299.91	99.97 (5.47)
Tablets (batch B)	291.36	97.12 (3.14)
Tablets (batch C)	310.51	103.51 (1.42)

*RSD = relative standard deviation.

Mangiferin (1,3,6,7-tetrahydroxyxanthone-2-*C*-β-D-glucopyranoside), a *C*-glucosylxanthone, which was first isolated from the bark, branches, and leaves of *M. indica* L., has been found to be the major component of this extract. Mangiferin is a naturally occurring chemopreventive agent in rat colon carcinogenesis [19]; exerts antidiabetic activity by increasing insulin sensitivity; shows significant inhibitory effect on bone resorption; appears to act as a potential biological response modifier with antitumor, immunomodulatory, and anti–human immunodeficiency virus (HIV) effect; is capable of providing cellular protection as an antioxidant and a radical scavenger agent; is useful as an analgesic without adverse effects; and inhibits the late event in herpes simplex virus-2 replication [20–24].

The quality control of 16 batches of Vimang active ingredient [by high-performance liquid chromatography (HPLC) and the ultraviolet (UV) method] obtained from different regions of the country (batches 1–8 from the west and batches 9–16 from the east) and its pharmaceuticals (optimum formula) were investigated and are demonstrated in Table 5. Each sample was analyzed in triplicate and the average values are listed. All assay results fell between 100 and 300 μg of mangiferin per milligram of Vimang powder, except samples 12–16, which were rejected. The differences found are probably due to the fact that the mangiferin content in the plant varies with the season of the year and the zone where it was grown. The claimed contents of this natural product required by our producers are 85–115% for tablets.

Different tablet formulations were tested, but even superdisintegrants such as Ac-Di-Sol and CMCNa up to concentrations of 5% were not sufficient to disintegrate the tablets. With the use of pH-modified product such as $NaCO_2$ or canalling as NaCl the release of extract from the tablets changed dramatically. Use of lactose

TABLE 6 Formulations by Wet Granulation

	F15	F16	F17	F18	F19	F20	F21	F22	F23	F24	F25	F26	F27	F28	F29	F30
Extract[a]	50	50	50	50	50	50	50	50	50	50	50	50	50	50	50	50
Lactose	—	—	20^a	16^a	16^a	22^a	27^a	29^a	32^a	42^a	$32^a + 10$	35^a	35^a	35^a	35^a	30^a
CMM	20^a	20^a														
NaCl	20^a	$15^a + 5$	10^a	5^a	5^a	5	5	5	5	—	—	—	5			
PEG6000	—	—	—	—	—	—	—	—	—	—	—	—	—	5	5	5
PEG400[a]	—	—	—	—	—	—	—	—	—	—	—	—	—		4	
Na_2CO_3	—	—	—	$5^a + 5$	10^a	5	5	5	5	—	—	5	—	—	—	5
CMCNa	—	—	5													
PVP[a]	5	5	5	4	4	4	4	4	4	4	4	4	4	4		4
SDS	2^a	$1^a + 1$	$1^a + 1$	1	1	2	2	2	2			2	2	2	2	2
Talc	1	1	—	1	1	2	2	2	2	4	4	4	4	4	4	4
MgEs	1	1	1	1	1											
Aerosil	1	1	$5^a + 2$	$5^a + 2$	$5^a + 2$	5^a										
Acdisol	—	—	—	5	5	5^a	5	3								
%	62%	65%	71.5%	89%	85%	88%	100%	100%	100%	46%	nF	80%	70%	71%	72%	100%
Release							IN 31	IN 34	IN 23							
In 30 Min							MIN	MIN	MIN							

Abbreviations: CMM, cellulose microcrystalline; PEG, polyethylene glycal; CMCNa, sodium carboxymethyl cellulose; PVP, Kollidon K30; SDS, sodium docecyl sulfate; MgEs, magnesium stearate; NF, no flow ability.

[a] Internal phase.

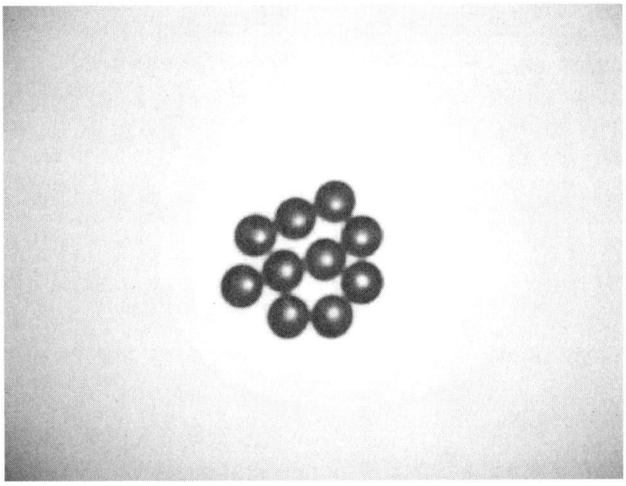

FIGURE 15 Formulation F31.

TABLE 7 Two Formulations for *mangifera indica* L. (Mango)

	Formulation F	Formulation F31
	Internal Phase	
Extract *M. indica* L.	50.0%[a]	50.0%[a]
Lactose		31.0%[a]
Povidone (PVP K30)	4.0%[a]	4.0%[a]
Starch	10.0%[a]	
CMM	28.0%[a]	
Ac-Di-Sol	5.0%[a]	
	External Phase	
SDS	—	2.0%
Na_2CO_3	—	5.0%
PEG6000	—	5.0%
Talc	2.5%	2.5%
Magnesium stearate	0.5%	0.5%

[a]Product added to internal phase.

was better than CMM, especially for flow properties. In all cases tablets always eroded and never disintegrated. All formulations studied are shown in Table 6.

F23 and **F30** are shown in Figure 15. Two formulations, **F31** (obtained from the result of **F23** and **F30**) and F, were compared for water uptake and dissolution profile, as shown in Table 7.

The properties of formulations **F31** and **F** were measured in ENSLIN equipment in order to compare the kinetics of water uptake and to study the dissolution profile, as shown in Figures 16–19. As observed in Figure 19, the dissolution of plant extract after 60 min is very slow when pH-modified (**NA_2CO_3**) and sodium dodecyl sulfate (SDS) are not used (almost not change in color for formulation **F**).

The properties of formulation F31 are as follows: diameter 12 mm, normal concave; stability studies (see Table 8) under tropical condition, 30 °C and 75% RH with desiccant (silice 1 g for plastic bottle of polyethylene no. 8 for 60 tablets).

FIGURE 16 Formulation F31 in ENSLIN equipment at initial time.

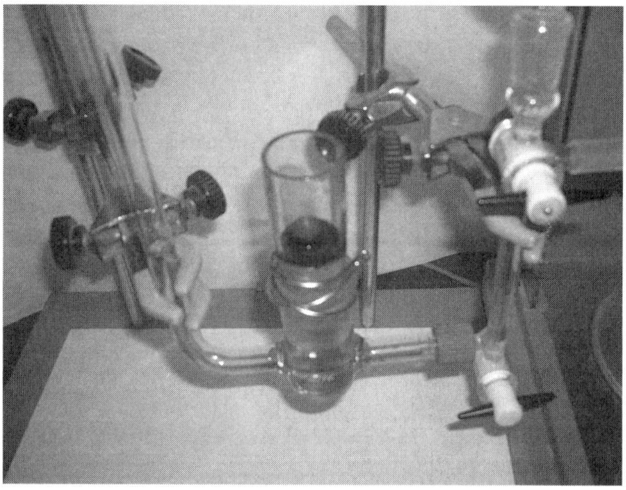

FIGURE 17 Formulation F31 in ENSLIN equipment after 60 min.

6.3.6.2 Natural Product as Vehicle for Manufactured Tablets

Chitin was evaluated as a direct-compression vehicle using powder flow properties and the physicomechanical properties of the manufactured tablets, and it was proven that this natural polymer has suitable characteristics for being used for this end.

A comparative study of chitin obtained from lobsters, starch, and carboxymethyl chitosan as disintegrating agents was conducted. The influence of the method in the preparation of tablets on the disintegrating activity of both polymers was evaluated. Chitin proved to have good characteristics as a disintegrating agent independently

FIGURE 18 Formulation F in ENSLIN equipment at initial time.

FIGURE 19 Formulation F in ENSLIN equipment after 60 min.

TABLE 8 Stability Studies for F31 under Tropical Condition

	Initial Time	3 months	6 months	9 months	12 months
Weight	600.7	601.47	601.30	602.09	601.33
Assay	99.85	100.98	100.64	99.10	101.07
Disintegration, min	27	29	30	30	30
Hardness, N	59	60	64	60	60
Friability, %	0.89	0.74	0.80	0.77	0.81
Humidity, %	5.71	5.11	5.45	5.21	5.64

TABLE 9 Formulation with Chitin, Chitosan, and Croscarmelose Sodium

	I	II	III	IV	V	VI	VII
Papaverin	19.25	18.25	17.50	18.25	17.50	18.25	17.50
Binder	10.00	10.00	10.00	10.00	10.00	10.00	10.00
Chitin	—	1.00	1.50				
Chitosan	—	—	—	—	—	1	1.50
Croscarmelose sodium	—	—	—	1.00	1.50		
Dibasic calcium phosphate	70.00	70.00	70.00	70.00	70.00	70.00	70.00
Magnesium stearate	0.75	0.75	0.75	0.75	0.75	0.75	0.75

TABLE 10 Results for Direct Compression

	I	II	III	IV	V	VI	VII
Weight, mg	348	347	345	347	348	346	347
Release at 30 min, %	<40	92.6	92.4	94.2	96.6	92.1	94.8
Disintegration, min	>60	2.30	1.40	0.28	0.22	1.30	1.10
Hardness, kgf-Erw	3.92	3.99	3.97	3.97	4.03	3.99	3.92
Friability	2.44	2.24	2.10	2.01	1.97	1.89	1.94

TABLE 11 Results for Wet Granulation

	FI	FII	FIII
Weight, mg	346.50	348.20	349.00
Release at 30 min, %	81.00	86.00	<70.00
Disintegration, min	23.70	23.00	45.30
Hardness, kgf-Erw	6.90	8.20	6.30
Friability	0.14	0.08	0.13

of the method used to make tablets (Tables 9–11). The disintegrating activity of carboxymethyl chitosan was affected by the granulation process.

Three formulas were prepared by wet granulation comparing starch, chitin, and chitosan as disintegrant (10%) formulations FI (10% starch), FII (10% chitin), and FIII (10% chitosan) (Table 11). Chitin has good properties as disintegrant. This product can be used for direct compression and wet granulation. The method for development (direct compresion or wet granulation) of tablets influences chitosan disintegrant properties.

6.3.6.3 Natural Product as Vehicle for Controlled-Release System

Dextrans are composed of chains of D-glucan ($1–20 \times 10^6$) with α-1,6 as the main-chain linkage and variable numbers of α-1,2, α-1,3, or α-1,4 branched-chain linkages. Dextran is synthesized from sucrose by dextransucrases, glucansucrases, and glucosyltransferases produced by *Leuconostoc* or *Streptococcus*. These bacteria growing in sugar juices produce dextran. High concentrations of dextran on solids (>1000 ppm) can result in severe financial losses to the sugar industry [25].

Dextran fractions obtained from enzymatic hydrolysis of native dextrans are supplied in molecular weights from 1000 to 2×10^6 Da. The molecular weight of the fraction is in most cases a key property and is defined in terms of the average molecular weight (M_w) and the number average molecular weight (M_n). The functionality of this raw material for controlling drug release is studied as a function of molecular weight. Fractions 43,000 (**F3**), 71,000 (**F2**), and 170,000 (**F1**) as native dextran 2×10^6 and 20×10^6 M_w are used [26].

Wet- and Dry-Weight Studies The method used was based on that of Tahara [15] and Jamzad [27]. The swelling and erosion of dextran polymers of differing molecular weights were examined by measuring the wet and subsequent dry weights of matrices. The experiment consisted of allowing the tablet (dextran alone) to dissolve in the medium (at the same condition described in drug release studies) for certain time periods (15, 30, 60 and 90 minutes) before being removed into a preweighed weighing boat. The excess dissolution medium was drained and blotted from around the tablet without touching it. The tablet and boat were then weighed to establish the wet weight of the tablet. The tablets were then dried to a constant weight in an oven at 105 °C. Each determination at each time point was performed in triplicate and mean values were expressed. The dissolution medium uptake per weight of dextran remaining was calculated at each time point for a particular matrix to correct for the effect of erosion and dissolution in the measurement of degree of dissolution medium uptake [Equation (1)]. Erode dextran was measured according to the equation described by Jamzad et al. [30] [Equation (2)]:

$$\begin{array}{c}\text{Water uptake per unit} \\ \text{polymer remaining (\%)}\end{array} = \frac{\text{wet weight} - \text{dry weight}}{\text{dry weight}} \times 100 \tag{1}$$

$$\text{Mass polymer loss (\%)} = \frac{\text{original weight} - \text{remaining (dry) weight}}{\text{original weight}} \times 100 \tag{2}$$

The Davidson and Peppas model [Equation (3)] was applied to these data to study the mechanism and the rate of water uptake.

$$w = K_s t^n \tag{3}$$

where w is the weight gain of the swelled matrix (water/dry polymer), K_s the kinetic constant of water penetration, t the penetration time, and n the exponent which depends on the water penetration mechanism.

Swelling and Erosion The change in wet weight, reflecting swelling, over time for the five polymer types is shown in Table 12. The higher molecular weight polymers showed the highest maximum average relative swelling, which occurred since the initial time with little erosion. In contrast, the lower molecular weight polymers (fractions **F1**, **F2**, and **F3**) exhibited minimal swelling and the erosion mechanism predominated. Consequently, tablets were dissolved very fast (100% before 45 min). These polymers and fractions show a wide range of viscosities, which cause differences in their swelling and erosion behaviors [26]. These results agree with results obtained by Sakar [C] and Walker [D] for HPMC polymer.

TABLE 12 Swelling and Erosion Properties of Dextran Tablets as Funtion of Molecular Weight

		Mass Polymer Loss (%)			
	Hardness (N)	15 min	30 min	60 min	90 min
DT, MW 2×10^6	431	11.48	14.59	27.473	36.224
DT, MW 20×10^6	482	5.29	7.33	11.835	16.153
F1	431	52.89	87.38		
F2	420	62.79	89.21		
F3	460	66.3	95.85		

	Percentage of Water Uptake			
	15 min	30 min	60 min	90 min
DT, MW 2×10^6	59.80	84.68	103.73	110.15
DT, MW 20×10^6	110.03	142.94	178.50	210.02
F1	31.61	33.14		
F2	17.61	26.89		
F3	16.82	17.65		

Note: All values are referred to applied force, 14 kN, particle size for dextran 150–200 μm, tablet weight 300 ± 11 mg, dissolution media 1000 mL distilled water, temperature 37°C, 100-rpm paddle. Abbreviations: DT, dextran; MW, molecular weight.

For native dextran a linear relationship was seen between mass polymer loss and initial dissolution time. Native dextrans also showed the highest maximum dissolution medium uptake. Here, an increase in the molecular weight of dextran resulted in an increase in water uptake (native polymer with 10 times more than fraction **F1**, 13 times more than **F2**, and 15 times more than **F3** for the first 30 min) and less erosion. Anywhere the rate of water uptake per unit weight of polymer started to decline with last initial time and in consequence for longer periods of time, nonlinear dependence could be expected. Applying the Davidson–Peppas model [Equation (3)], a value of n = 0.356 (r^2 = 0.998) for native dextran was obtained (r^2 = 0.984). An inverse relationship between erosion rate constant and molecular weight was reported by Reynolds et al. [28]. Tahara et al. [18] reported that the lower viscosity HPMC (50-cps) polymer eroded faster than the 4000-cps polymer, consistent with the current work. Thus, the higher molecular weight native dextran polymers have higher intrinsic water-holding capacity and the matrices formed from such polymers are less prone to erosion than the lower molecular weight fractions.

6.3.6.4 Mechanism of Soluble Principle Active Propranolol Hydrochloride and Lobenzarit Disodium from Dextran Tablets

Soluble drugs are considered to be released by diffusion through the matrix and poorly soluble drugs are released by erosion of the matrix. Moreover, it is considered that factors affecting swelling and erosion of these polymers may account for differences between in vitro dissolution results and subsequent in vivo performance when hydrophilic matrix tablets are compared [15].

Lobenzarit disodium (LBZ) is a drug conceived for the treatment of rheumatoid arthritis. This drug produces an improvement of immunological abnormalities and has a regulatory effect upon the antibody-producing system. Propranolol hydrochloride (PPL) is a β-adrenergic blocking agent, that is, a competitive inhibitor of the effects of catecholamines at β-adrenergic receptor sites. It is widely used in therapeutics for its antihypertensive, antiangorous, and antiarrhythmic properties. These two drugs are suitable candidates for the design of controlled-release delivery systems [25, 29]. According to their solubility in water they can be considered as high soluble (PPL) and soluble (LBZ) drugs.

A comparative study of the dissolution profile for PPL and LBZ was established as the analysis of a similarity factor defined as

$$f_2 = 50\log\left\{\left[1+\frac{1}{n}\sum_{t=1}^{n}(R_t - T_t)^2\right]^{-0.5}\times 100\right\} \tag{4}$$

In the above equation f_2 is the similarity factor, n is the number of time points, R_t is the mean percent drug dissolved of the reference formulation, and T_t is the mean percent drug dissolved of the tested formulation.

The evaluation of similarity is based on the following conditions:

- A minimum of three time points
- Twelve individual values for every time point
- Not more than one mean value of >85% dissolved
- Standard deviation of the mean that is less than 10% from the second to last time point

An f_2 value between 50 and 100 suggests that two dissolution profiles are similar [30]. In this study experimental data corresponding to 30, 60, 90, 120, 180, 240, 300, 360, 420, and 480 min were considered.

Figure 20 shows dissolution profiles for tablets of PPL or LBZ from the native dextran DTB110-1-2 matrix system, respectively 1:1 (w/w). The value for relative standard deviation (CV) was less than 5% for all points measured (n = 12).

The Higuchi and Hixson Crowell model as well as the nonlinear regression of Peppas and Peppas-Sahlin were employed to study the release data. Higuchi's slope

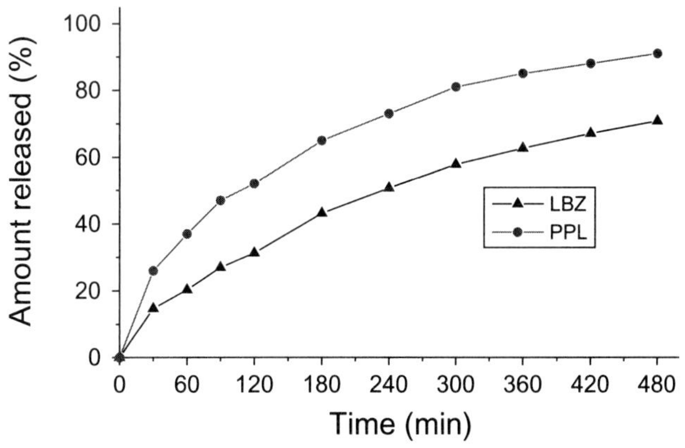

FIGURE 20 Disolution profile for PPL and LBZ dextran tablets (direct compression): paddle 100 rpm, 37 °C, deionized water.

(3.179 and 4.500% $min^{-1/2}$ for LBZ and PPL, respectively), Korsmeyer's rate constant (1.195% $min^{-0.697}$ and 4.125% $min^{-0.540}$ for LBZ and PPL, respectively), the low relaxational constant K_r (0.101% $min^{-0.898}$ for LBL and –0.040% $min^{-0.898}$ for PPL), compared with K_d values (2.941 and 6.518% $min^{-0.449}$, respectively) of Peppas-Sahlin indicated the diffusional mechanism as predominant for soluble drugs since native dextran tablets. The influence of solubility of the drug can be observed for the hydrophilic matrix (release of PPL is faster than LBZ in correspondence with its solubility in water). The value of the diffusional exponent, 0.697 (by the Korsmeyer equation), for the less soluble drug corresponds to the increment of the influence of the erosion mechanism, in agreement with other authors. This can be also observed in the Peppas y Salhin equation where the negative value obtained for K_r for the dissolution profile of PPL from dextran tablets should be interpreted in terms of a relaxation mechanism, which is insignificant compared to the diffusion process.

The dissolution profiles for LBZ:B110-1-2 and PPL:B110-1-2 tablets were also compared using similarity factor f_2. A value obtained for f_2 that is below 50 (37.48) indicates the influence of drug solubility in the dissolution profile. Furthermore, other parameters, such as dextran–drug ratio, particle size of polymer and drug, and influence of pH, have to be studied to obtain an optimum and robust formulation.

The mechanisms of drug release from dextran matrix occur in the early stage by polymer swelling, and the tablet thickness increases. Soon thereafter, polymer (and drug) dissolution starts occurring. The polymer dissolves because of chain disentanglement. Thus, there is a slow diminution of the thickness because of erosion until, finally, the tablet disappears (time > 480 min).

6.3.7 DESIGN TOOLS OF TABLET FORMULATION

Nowadays, most experimentation on tablet formulation development is still performed by changing the levels of each variable (factor) at a time, in an unsystematic way, keeping all other variables constant in order to study the effects of that specific variable on the selected response or to find the optimal conditions of a complete system. This methodology (trial and error) is based on a large number of experiments and often relies merely on the analyst's experience [31].

Statistical experimental design, also called design of experiments (DoE), is a well-established concept for planning and execution of informative experiments. DoE can be used in many applications. An important type of DoE application refers to the preparation and modification of mixtures. It involves the use of "mixture designs" for changing mixture composition and exploring how such changes will affect the properties of the mixture [32].

In the DoE approach, first process variables are "screened" to determine which are important to the outcome (excipient type, percentage, mixture time, etc.). The next step is *optimization*, when the best settings for the important variables are determined. In particular, response surface methodologies have been successfully applied in both drug discovery and development [33]. Advances in supporting software, automated synthesis instrumentation, and high-throughput analytical techniques have led to the broader adoption of this approach in pharmaceutical discovery and chemical development laboratories [34].

The benefits of using experimental design together with software to facilitate the formulation of a tablet for specific purposes, from screening to robustness testing, are well known. This technique has some advantages compared to the trial-and-error method. By applying a multivariate design for the screening experiments, many excipients are evaluated using comparatively few experiments.

The formulation work is generally based on designed experiments. Most of the experiments are fractional or full-factorial designs and are generated and evaluated in some cases with the center point replicated. The robustness of the formulation and batch-to-batch variation of the excipients and the active pharmaceutical ingredient can be evaluated with experimental designs on different occasions. Experimental design and optimization of the formulation can be performed with the use of software. Some of them have been useful in tablet design. MODDE (version 4.0 and 5.0, Umetri, Umeå, Sweden), iTAB [35], and TabletCAD are some examples.

6.3.7.1 MODDE 4.0

Tablet design for controlled-release propranolol hydrochloride was performed with the use of MODDE software [25]. A central composite design (one of the most used designs in pharmaceuticals) was applied to the optimization. This experimental design required 17 experiments ($2^k + 2k + 3$, where k is the number of variables) including three center points. Three variables and five responses (according to USP 25 tolerances for the dissolution profile for propranolol hydrochloride extended-release capsule) were involved in the experimental design. The variables and their ranges studied are summarized in Table 13. The high and low values of each variable were defined based on preliminary experiments. The critical responses were $t_{100\%}$ and $t_{30\%}$ corresponding to the time when 100 and 30% of drug contained in the tablets is delivered to the dissolution medium because this system was developed to release drug in 24 h ($t_{100\%} \sim 24$ h) and to prevent an overdose for the first minutes ($t_{30\%} > 1.5$ h). The other responses were in the amount of PPL dissolved at 4, 8, and 14 h.

Table 14 shows results obtained for every formula development according to MODDE 4.0 software. The collected experimental data were fitted by a multilinear regression (MLR) model with which several responses can be dealt with simultaneously to provide an overview of how all the factors affect all the responses. The responses of the model, R^2 and Q^2 values, were over 0.99 and 0.93 for $t_{100\%}$ and 0.98 and 0.89 for $t_{30\%}$, respectively, implying that the data fitted well with the model. Here, R^2 is the fraction of the variation of the response that can be modeled and Q^2 is the fraction of the variation of the response that can be predicted by the model. The relationship between a response y and the variables x_i, x_j, ... can be described by the polynomial:

TABLE 13 Levels of Formulation Variables (Central Composite Design)

Parameter	Low Value (−1)	Central Value (0)	High Value (+1)
Ratio DT–HPMC (w/w)	1:1	4:1	7:1
Cetyl alcohol (% w/w)	10	15	20
Ratio excipients–PPL (% w/w)	30	50	70

TABLE 14 Matrix of Central Composite Design and Results

Run Order	CEx–PPL	DT–HPMC	ce	$t_{100\%}$*	$t_{30\%}$*	t_2*	t_3*	t_4*
10	30	1:1	10	10	0.5[a]	71[a]	90[a]	—[a]
7	70	1:1	10	15	2.1[b]	44[b]	73[b]	97[a]
1	30	7:1	10	11	0.5[a]	63[a]	82[a]	—[a]
11	70	7:1	10	14	2.2[b]	40[b]	76[b]	99[a]
9	30	1:1	20	14	1.3[a]	49[b]	70[b]	99[a]
8	70	1:1	20	16	2.2[b]	46[b]	62[b]	96[a]
5	30	7:1	20	14	1.6[b]	49[b]	70[b]	99[a]
15	70	7:1	20	16	2.4[b]	43[b]	60[b]	96[a]
3	30	4:1	15	20	1.2[a]	54[b]	79[b]	94[b]
16	70	4:1	15	26	2.6[b]	39[b]	56[b]	72[b]
17	50	1:1	15	15	1.6[b]	45[b]	63[b]	98[a]
4	50	7:1	15	16	1.7[b]	43[b]	65[b]	96[a]
2	50	4:1	10	21	1.4[a]	52[b]	71[b]	93[b]
12	50	4:1	20	24	2.2[b]	45[b]	63[b]	82[b]
6	50	4:1	15	24	2.1[b]	48[b]	65[b]	85[b]
14	50	4:1	15	23.5	2.2[b]	49[b]	66[b]	86[b]
13	50	4:1	15	24	2.1[b]	48[b]	66[b]	86[b]

Note: cEx–PPL, ratio of excipients to propranolol; DT–HPMC, ratio of native dextran to hydroxypropylmethylcellulose; ce, percentage of cetyl alcohol (w/w) in the tablets. Time $t_{100\%}$ is time (hours) when 100% of PPL is dissolved in dissolution medium and $t_{30\%}$ is time (hours) when 30% of PPL is dissolved in dissolution medium; t_2 is percent of PPL dissolved at 4 h, t_3 percent of PPL dissolved at 8 h, and t_4 percent of PPL dissolved at 14 h. Values presented are the average of eighteen replicates for each batch. [a]Inside the USP range. [b]Outside USP range.

$$y = \beta_0 + \beta_i x_i + \beta_j x_j + \beta_{ij} x_i x_j + \beta_{ii} x_i^2 + \beta_{jj} x_j^2 + \cdots + E$$

where β_j are coefficients to be determined and E is the overall experimental error. Figure 21 presents the dissolution profiles of all 17 trials generated from the central composite design.

The response surface plots formed by plotting the values for $t_{30\%}$ and $t_{100\%}$ as a function of the most important variables are shown in Figure 22, where the optimum condition obtained by the model can be seen. The optimum dextran (DT)–HPMC ratio of 4:1 (w/w) gave $t_{100\%}$ equal to 24 h.

With tablet formulations composed of a matrix excipient and PPL at a ratio ranging from 40:60 to 70:30 (w/w), the values for $t_{100\%}$ were satisfactory (around 24 h). However, the respective values for $t_{30\%}$ increased according to the ratio of matrix excipient and PPL ranging from 40:60 to 70:30 (w/w), showing that early drug release was demanded and the initial dose required for pharmacological effect could not be sufficient.

Sustained-release matrix tablets with good properties were obtained with a dextran–HPMC ratio of 4:1 (w/w), with a matrix excipient–PPL ratio of 60:40 (w/w), and with a cetyl alcohol amount of 15% (w/w). The hydrophilic polymers–PPL ratio of 60:40 (w/w) is more robust for any manufactured variability than 50:50 (w/w), because the central point of the design is near the lowest desired area (Figure 22). Under the optimal conditions, the mean value of hardness was 106 ± 3 N and the friability was less than 1% (0.2%).

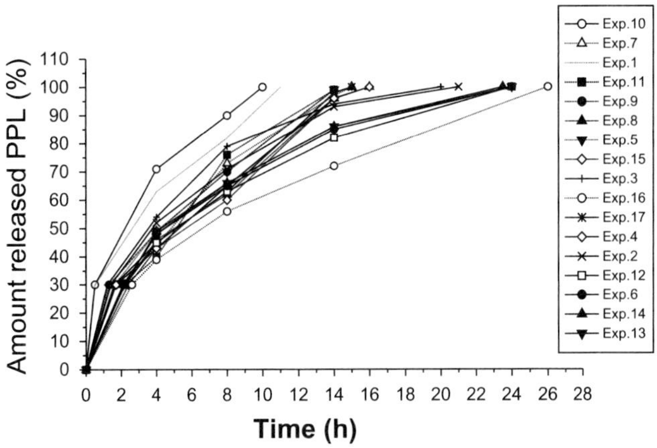

FIGURE 21 Dissolution profile generated from central composite design.

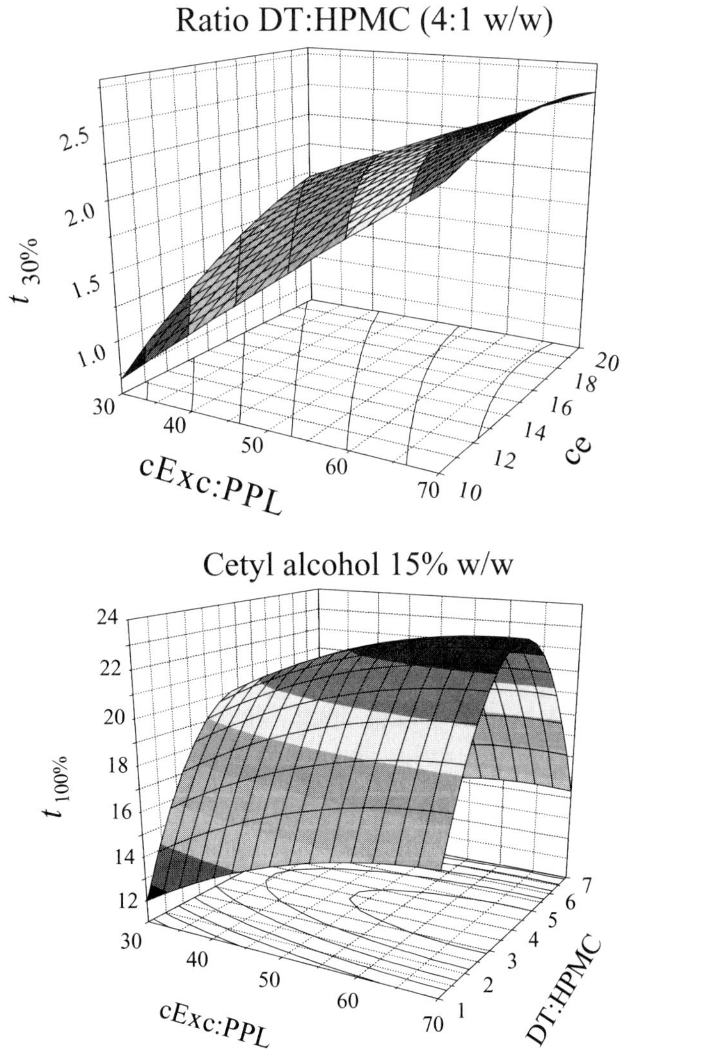

FIGURE 22 Response surface plots formed by plotting values for $t_{30\%}$ and $t_{100\%}$.

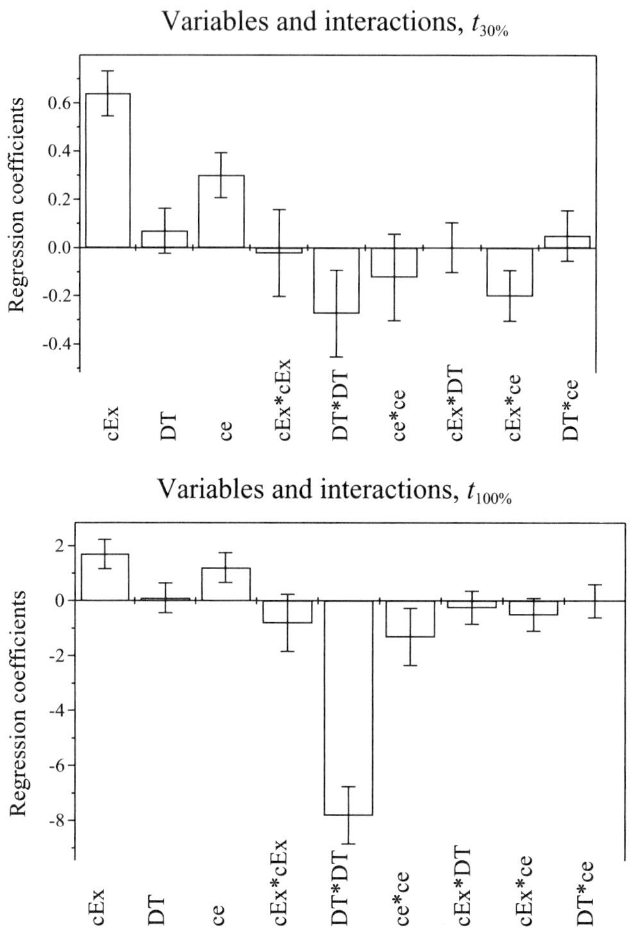

FIGURE 23 Regression coefficients.

Cetyl alcohol (ce in Figure 23) has a significant positive effect on both responses in the range studied. This may be because the hydrophobic polymer prevents the fast release of PPL for the first few hours with an increase in the diffusional path-length of the drug because the swelling of hydrophilic polymers (DT and HPMC) retards the rate of release. Interaction of cetyl alcohol and hydrophilic polymers–PPL was observed for $t_{30\%}$. If a prolonged release rate is desired during this period, the ratio of hydrophilic polymers to cetyl alcohol can be increased, resulting in a decreased interspace volume after erosion of cetyl alcohol. In contrast to other products, such as lactose [15], cetyl alcohol as hydrophobic polymer can be increased. The viscosity and texture of the gel layer and some modifications in polymer–polymer and polymer–solvent interaction are present.

6.3.7.2 iTAB

iTAB is I Holland's new tablet design aid that calculates basic tablet parameters and stress analysis for "rounds" and "shapes" in three easy steps. iTAB is Windows based, saving users from any extra set-up costs and making software upgrades easy.

Users are not required to possess any technical expertise or have any formal training, as iTAB is very user friendly and intuitive throughout. Using the drag-and-click feature, iTAB results in 2D drawings ready for production and automatically generated design reports:

Step 1 involves the selection of a specific tablet profile and modification of key parameters such as diameter, mass, volume, and surface area in real time. The iTAB safety zone ensures that unfeasible designs cannot be output from the system, meaning that only quality tablets can be produced.

Step 2 runs a real-time finite-element analysis (FEA) simulation against the design to give an indicative punch tip maximum force calculation.

Step 3 automatically produces a report summarizing key design data, for example, cup depth and surface area, that can be emailed directly to the I Holland team for further analysis, design work, and tooling production.

iTAB allows non–computer-aided design (CAD) and non-FEA users to simply and quickly design nonembossed tablets and obtain detailed information with the click of a button.

Core benefits include the avoidance of tablet manufacturing problems such as punch tip breakage from the outset. Immediate and accurate punch tip load calculations and 3D dynamic rotation of punch tip stress allow for instant decision making on design issues. iTAB is also unique in that it incorporates tooling and tablet design.

6.3.7.3 Percolation Theory

Leuenberger et al. introduced percolation theory in the pharmaceutical field in 1987 to explain the mechanical properties of compacts and the mechanisms of the formation of a tablet [36, 37]. Knowledge of the percolation thresholds of a system results in a clear improvement of the design of controlled-release dosage forms such as inert or hydrophilic matrices.

Percolation theory is a statistical theory that studies disordered or chaotic systems where the components are randomly distributed in a lattice. A cluster is defined as a group of neighboring occupied sites in the lattice, being considered an infinite or percolating cluster when it extends from one side to the rest of the sides of the lattice, that is, percolates the whole system [38].

Thus, a tablet is regarded simply as a heterogeneous binary system formed by the active principle and an excipient. As a function of their relative volume ratio, one or both components constitute a percolating cluster formed by particles of the same component that contact each other from one side to the other sides of the tablet.

In a binary pharmaceutical tablet (cylindrical lattice), the sites can be occupied by the component drug or excipient. The percolation threshold of the drug indicates at which concentration this substance dominates the drug/excipient system. The concept is very similar to the point where a component passes from being the inner to being the outer phase of an emulsion. It is not surprising that the component becoming the "outer phase" or percolating phase will have more influence on the properties of the system.

Furthermore, the concentration point at which a component is starting to percolate the system is usually related to a change in the properties of the system, which will now be more affected by this component. This is known as a critical point. Close to the critical point important changes can take place, for example, changes in the release mechanism of the active agent and modification of the tablet structure (e.g., monolith versus a desegregating device).

An important difference between particulate solids and emulsions is that in the solids two components can percolate the system at the same time, that is, two components can act as the outer phase simultaneously. In this case the system is known as a bicoherent system.

Study of Ternary Tablets Percolation theory has been developed for binary systems, however, drug delivery systems usually contain more than two components. The existence and behavior of the percolation thresholds in ternary pharmaceutical dosage forms have been studied [39] employing mixtures of three substances with very different hydrophilicity and aqueous solubility (Polyvinylpyrrolidone (PVP) cross-linked, Eudragit RS-PM, and potassium chloride).

After evaluation of the technological parameters and in vitro release behavior of the tablets, no sharp percolation thresholds were found in these ternary systems for the employed components separately. Nevertheless, a combined percolation threshold of the hydrophilic components was found, demonstrating that a multicomponent system can be reduced to a binary one using a discriminating property [39].

Matrix Systems with Different Particle Sizes Another disadvantage in the application of percolation theory to the rationalization of the pharmaceutical design was the prerequisite of an underlying regular lattice. Usually, drug delivery systems contain substances with different particle sizes. Therefore, the particles cannot be considered as each occupying one lattice site.

This problem can be initially overcome using a volume ratio instead of a lattice site ratio, expressing the percolation thresholds as critical volume fractions [36, 40–42]. Nevertheless, the influence of the particle size of the components on the percolation threshold cannot be explained using a volume fraction; that is, from this point of view, tablets with the same excipient volume are equivalent independent of their particle size. A first qualitative study of the influence of particle size on the percolation threshold [43] demonstrated that this is in clear disagreement with experimental data.

According to percolation theory, the effect of a reduction in the drug particle size should be similar to an increase in the excipient particle size in a binary system: It may be expected that the relative particle size of the component, but not its absolute particle size, will determine the properties of the system.

A quantitative study of the influence of particle size on the percolation threshold employing inert matrix tablets prepared with KCl and Eudragit RS-PM as matrix-forming material [44–46] showed that experimental data are in agreement with this hypothesis.

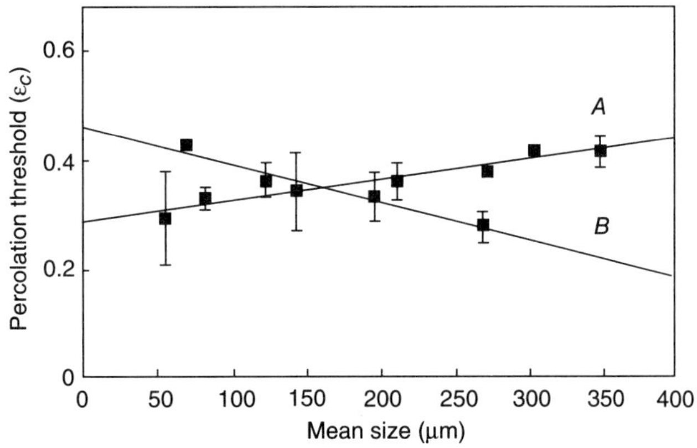

FIGURE 24 Drug percolation threshold (mean ± SE) as function of mean particle size of drug (line *A*) and excipient (line *B*) employed.

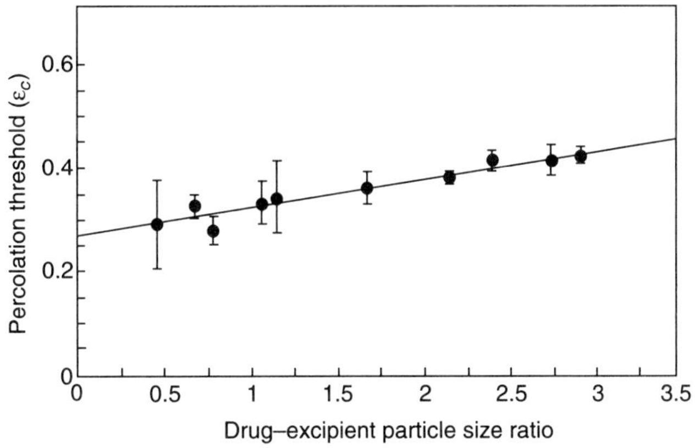

FIGURE 25 Drug percolation threshold (mean ± SE) as function of drug–excipient particle size ratio employed.

As Figure 24 shows, a linear relationship was found in this study [44] between the mean drug particle size and the corresponding drug percolation threshold (Figure 24, line *A*). Furthermore, the excipient particle size exerts a contrary effect than the drug particle size (Figure 24, line *B*); that is, the larger the excipient particle size, the lower the drug percolation threshold [46].

In addition, when the obtained drug percolation thresholds were plotted as a function of the drug–excipient particle size ratio of the matrices (see Figure 25), a linear relationship was found between the drug percolation threshold and the relative drug particle size [46]. These results are in agreement with the above exposed theoretical model based on percolation theory.

One of the advantages of the proposed model versus classical theories is its ability to explain the changes in the release behavior of the matrices by means of a change in the critical points of the system (related to the drug and excipient percolation thresholds), which can be experimentally calculated, providing a scientific basis for the optimization of these dosage forms.

Mechanical Properties The percolation approach was also employed to model the tensile strength of tablets [47, 48]. A critical tablet density was here understood as a minimal solid fraction needed to build a network of relevant contact points spanning the entire tablet. A rising tablet density led to a power law increase of the tensile strength showing an universal exponent $T_f = 2.7$.

It was shown that a power law based on percolation theory was suitable to fit the obtained tensile strength data of the binary matrix tablets studied. The best fitting was observed for a model where an initial tensile strength σ_0 was supposed [49]:

$$\sigma_t = k(\rho - \rho_c)^{2.7} + \sigma_0$$

The observed critical relative densities are understood as threshold values for the tensile strengths of the tablets. One practical consequence of these works is to avoid the manufacture of matrix tablets close to these critical densities. The formulation may not be robust in this critical range from the viewpoint of mechanical tablet stability.

6.3.7.4 Artificial Neural Networks

Artificial neural networks (ANNs) are computer programs designed to model the relationships between independent and dependent variables. They are based on the attempt to model the neural networks of the brain [50]. Functions are performed collectively and in parallel by the units, rather than there being a clear delineation of subtasks to which various units are assigned.

This methodology represents an alternative modeling technique that has been applied to pharmaceutical technology data sets, including tableting parameters [51]. The main advantage with respect to classical statistical techniques, such as response surface methodology, is that ANNs do not require the prior assumption of the nature of the relationships between input and output parameters, nor do they require the raw data to be transformed prior to model generation [51]. ANNs are capable of modeling complex, nonlinear relationships directly from the raw data.

The functional unit of ANNs is the perceptron. This is a basic unit able to generate a response as a funtion of a number of inputs received from others perceptrons. For example, the response value can be obtained as follows:

$$Y = \begin{cases} 1 & \text{if } W_0 I_0 + W_1 I_1 + W_b > 0 \\ 0 & \text{if } W_0 I_0 + W_1 I_1 + W_b \leq 0 \end{cases}$$

where I_x is the input received from perceptron x and W_x the weight assigned to this input by the perceptron. The weights can be changed to adapt the answer to the desired one using a learning algorithm.

Usually complex structures with more than 15 layers are employed, called the multilayer perceptron (MLP). Some of the commercial programs which have been used to fit tableting parameters are INForm (Intelligensys, Billingham Teesside), CAD/Chem (AI Ware, Cleveland, OH), which is no longer commercially available, and the Neural Network Toolbox of MATLAB (MathWorks, Natick, MA).

6.3.8 COATING SYSTEMS

Coating processes have come to play an important role for the protection of substances prior to application or for their sustained release. Coatings on cores usually consist of a mixture of substances. The matrix formers are responsible for the stability of the coating structure, and they also determine the coating process. Depending on the type of matrix former on binder used, three coating categories can be distinguished:

Coatings with Sucrose and Other Sugars Permit application of copious amounts of mass to the core and are widely used in the manufacture of pharmaceuticals and confectionery.

Hot Melts Add a considerable amount of mass and are applied hot and solidify while cooling on the core. They are mainly used for confectionery. The most important raw materials are fats, mostly cocoa fat, polyethylene glycol (PEG) [52], and the sugar–alcohol mixture xylitol–sorbitol.

Film Coatings Require less material, forming thin membranes which largely follow the contours of the substrate, for example, scores and engravings. The partly pH-dependent solubility and selective permeability of coatings are affected by the film formers. Such films are sometimes also used as intermediate layers in sugar coatings.

Whether or not a core is suitable for coating depends on its hardness, shape, surface, size, heat sensitivity, and tendency to interact with the coating material. Moreover, since sugar and film coating processes involve very different techniques, they place different demands on the cores to be used [53]. Tablets used as cores must be biconvex in shape to prevent them from sticking together like coins in a roll.

The ideal tablets for sugar coating will have a pronounced convex curvature and a narrow band. The consumption of coating material depends very much on the tablet shape and increases sharply if the tablets are not round. Film coating supplies coated products in which the core surface (e.g., with notches, engravings, and defects) is faithfully reproduced.The films tend to chip at sharp edges or are particularly thin in these areas. For this reason slightly curved tablet cores are preferred for film coating [54].

Film coating of pharmaceuticals is a common manufacturing stage for the following reasons: (i) to provide physical and chemical protection for the drug, (ii) to mask the taste or color of the drug, or (iii) to control the release rate or site of the drug from the tablet. When a coating composition is applied to a batch of tablets or granules (or to a batch of liquid drops or even gas bubbles), the core surfaces become covered with a polymeric film that is formed as the surfaces dries. The major component in a coating formulation is a film-forming agent which ideally is a high-molecular-weight polymer that is soluble in the proper solvent (today, most preferably in aqueous-based media). The polymer forms a gel and produces an elastic, cohesive, and adhesive film coating.

In the pharmaceutical industry, organic-solvent-based film coatings have been used for over 40 years. In the 1990s, however, interest and demands in the use of aqueous-based film coating systems rapidly increased owing to the well-documented drawbacks (unsafe, toxic, pollutive, and uneconomic) associated with organic-solvent-based coating systems. Consequently, and for the reasons mentioned above, much effort has been focused on the research and development of new aqueous-based film coating formulations. Nowadays, aqueous-soluble/dispersable polymers available on the market consist primarily of either cellulose polymers, PEGs, or acrylate copolymers. There are, however, some material-related limitations in using these polymers in aqueous-based film coatings. Consequently, application of new film formers such as chitosan, native starches, and special types of proteins for pharmaceuticals and foodstuffs has been increasingly studied.

The choice between sugars of film coatings depends not only on the desired coatings quality but also on the technical requirements, that is, on the economy of the process. The financial outlay for a selected technology often commits the manufacturer to this technology for a prolonged period of time. Film and sugar differ substantially in thickness and therefore also in the necessary mass of coating material.

The most important coating raw material is the film-forming polymer, which must be able to produce a coherent film on the substrate under the given process conditions [55, 56]. Second in importance is the solvent or dispersing system in which the polymer is applied to the surface and introduced to form a film. Other frequently used raw materials are plasticizers, glidants, fillers, and colorants. All these substances act together and influence the properties of the film.

The first decision to be made when developing a formulation concerns the desired function of the film. Depending on the requisite dissolution performance in physiological media, the film former to be tested—or several of them—is then selected from the available polymers. The economy of the coating process and the quality of the coated product depend on the correct calculation of the required amount of coating material. Empirical adjustment of new developments to existing products is therefore not recommended. The coating quantity needed for film coatings is directly related to the surface area of the core. Table 15 shows a simplified calculation of the surface area of tablets. It is based on the surface area of a cylinder circumscribing these shapes. All these data are taken from the literature. For sugar coating processes, the shape of the cores and end products have to be studied stereometrically. The sugar coating process balances irregularities, since the coating buildup does not follow the structure of the surface. Areas which require a high degree of rounding, such as high bands and very flat curvatures, attract more coating mass. Where the band height varies owing to manufacturing technique, this results in weight differ-

TABLE 15 Calculation of Surface Area (mm^2) for Different Types of Cores

	Diameter (mm)									
Height (mm)	3	4	5	6	7	8	9	10	12	14
2	33	56	70	95	120	150				
3	42	62	85	115	145	175	210	250	340	
4		75	100	130	165	200	240	280	380	485
5					185	225	270	315	415	530
6							300	345	450	570
7									505	615
8										660

ences between the individual sugar-coated tablets. This is one of the reasons why sugar-coated products always show a much wider weight distribution than film-coated tablets.

Several authors have made proposals for the most convenient size of tablet for sugar coating. A further rule has been established according to which the convex radius should be between 0.7 and 0.75 times the tablet diameter and the band height between 0.07 and 0.12 times that diameter. However, for calculation of the band height only a factor of 0.12 is advisable, since, otherwise, the minimum value of 1 mm will be fallen short of by far [53, 57–60]. For film coating the recommended convex radius is 1.5 times the tablet diameter (less curved tablets are preferred).

Numerous formulation are available for all film-forming polymers offered in the market. Thus, it is possible to dispense with many of the preliminary tests and concentrate development work on the special problems of the formulation in question. In this chapter, we will focus on SEPIFILM and Kollidon VA 64. Table 16 presents proven basic formulations reported on in the literature [61] and polymers used for the most important commercially available film formers, which also can be used as a basis for further tablet coating.

The major operational coating process parameters related to film coating are able to be measured and monitored continuously in some pan-type coaters. The inlet airflow rates influence the coating process and the subsequent quality of the coated tablets. Increasing the inlet airflow rate accelerated the drying of the tablet surface. At high inlet airflow rate, obvious film coating defects, that is, unacceptable surface roughness of the coated tablets, are observed and the loss of coating material increased.

Today advantages of aqueous film coating are well recognized and film coating technology is much developed to successfully perform these types of coatings. Process automation and monitoring of critical process parameters can be utilized to increase the overall process efficiency and predictability and to improve the homogeneity and reproducibility of the tablet batches. This will ensure high quality and safety of the final coated products, which are mandatory requirements of tablet manufacturing.

6.3.8.1 Subcoating of Tablet Cores as a Barrier to Water

As tablets are nowadays coated mostly with aqueous solutions or dispersions, it has become increasingly necessary to provide the tablet cores with a barrier layer prior to sugar or film coating. This is mainly to protect water-sensitive drugs against hydrolysis and chemical interactions, for example, between different vitamins, and to prevent the swelling of high-performance tablet desintegrants that are very sensitive even to small quantities of water. It can be especially useful when controlled-release systems with hydrophilic polymer are studied and the water contained can change the dissolution profile. Kollidon VA 64 also can be used to improve the adhesion of subsequent coatings by hydrophilization of the surface.

6.3.8.2 Kollidon VA 64

We are studying the ability of Kollidon VA 64 as a subcoating in a combined hydrophilic (dextran–HPMC)–hydrophobic (cetyl alcohol) matrix core prior to sugar

TABLE 16 Basic Formulations for Film-Coated Tablets

Formulation with Hydroxypropyl Methylcellulose (HPMC)

	A	B
Oprady	73.0%	—
Pharmacoat	—	80.0%
PEG 6000	—	8.0%
Talc	20.0%	5.0%
Pigments included, TiO_2	7.0%	7.0%
Solid content	20.0%	12.0%
Coating quantity, mg/cm^2	1–5	1–5

Formulation with Methacrylic Acid Copolymers

	A	B	C
Eudragit L100	5.0%	—	—
Eudragit S100	—	7.5%	—
Eudragit L30D-55	—	—	16.5%
PEG 6000	0.7%	1.0%	1.6%
Talc	6.0%	2.0%	4.0%
Pigments included, TiO_2	3.3%	—	—
Isopropyl alcohol	41.0%	86.5%	—
Acetone	41.0%	—	—
Water	3.0%	3.0%	77.9%
Solid content	14.0%	10.5%	22.1%
Coating quantity, mg/cm^2	2–4	2–4	3–5

Formulation with Hydroxypropyl Cellulose Acetate Succinate (HPMCAS)

Aqoat AS-MF	10.0%
Triethyl citrate	2.8%
Talc	3.0%
Sorbitan sesquioleate	0.0025%
Water	84.2%

Formulation with Ethylcellulose (EC)

	A	B
Ethylcellulose	5.0%	—
Aquacoat (30% solids)	—	30.0%
PEG 6000	—	2.0%
Glycerol triacetate	1.0%	—
Ethanol	94.0%	—
Water	—	68.0%

Formulation with Carboxymethyl Ethylcellulose (CMEC)

Duodcell	8.0%
Trisodium citrate	0.70%
Tween 80	0.04%
Glycerol monocaprylate	2.40%
Water	88.86%
Coating quantity	7 mg/cm^2

Formulation with Polyvinyl Acetate Phthalate (PVAP)

PVAP	11.0%
PEG 400	1.0%
Ethanol	66.0%
Water	22.0%

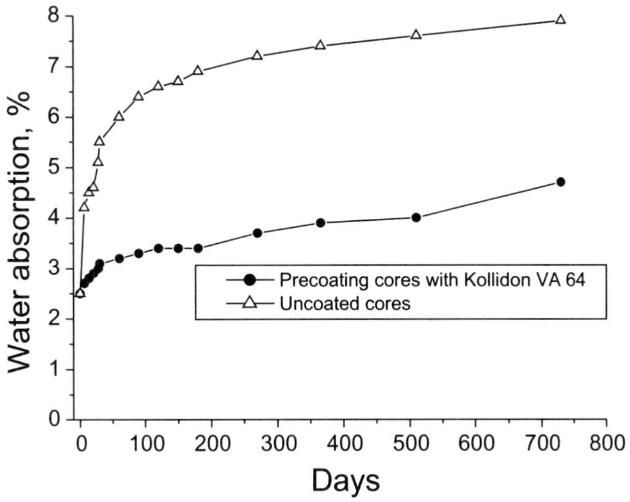

FIGURE 26 Water absorption of precoating and uncoated cores of combinated dextran–HPMC matrix tables.

coating. The copovidone (i.e., Kollidon VA 64) not only increases the mechanical properties of the tablet (less friability) but also prevents the amount of water absorbed from the air in tropical and subtropical stability conditions (25 and 75% relative humidity).

Figure 26 shows a comparative study of uncoated tablet and sugar-coated tablet after a barrier to water with copovidone 0.5 mg/cm^2 of the warm tablet cores using a 10% solution in ethanol. During two years the coating tablets (with initial humidity 2.5%) remained stable and dissolution profiles were similar to the initial time with similarity f_2 = 82 observed and friability decreased from 0.25% (uncoated cores) to 0.02% (coated tablets).

Similar results are reported in the literature when Kollidon VA 64 is compared to povidone (Kollidon K 25, K 30, and 90 F). Copovidone absorbs about three times less water than the other soluble Kollidon K 25, 30, and 90F after seven days at 25 °C up to 80% relative humidity [62]. Kollidon VA 64 is manufactured by free-radical polymetization of 6 parts of vinylpyrrolidone and 4 parts of vinyl acetate in 2-propanol. A water-soluble copolymer with a chain structure is obtained. In contrast to the soluble grades of Kollidon, the number 64 is not a *K* value but the mass ratio of the two monomers, vinylpyrrolidone and vinyl acetate. The *K* value of Kollidon VA 64 is of the same order of magnitude as that of Kollidon 30. Synonyms for Kollidon VA 64 are copovidone, copovidonum, copolyvidone, copovidon, and PVP-VAc-copolymer [Eur. Ph., Japanese Pharmaceutical Excipients, and USP National Formulary (NF)] [63].

Copovidone forms soluble films independently of the pH value, regardless of whether it is processed as a solution in water or in organic solvents. He offers better plasticity and elasticity than other povidones. On the other hand, films are also less tacky. Kollidon VA 64 usually absorbs water, and it is seldom used as the sole film-forming agent in a formulation. Normally it is better to combine it with less hygroscopic substances such as cellulose derivates [54], shellac, polyvinyl alcohol (PVA),

PEG (e.g., Macrogol 6000), or sucrose. Others plasticizers such as triethyl citrate, triacetin, or phtalates are not required. The properties of coatings can be improved with combination copovidone–cellulose derivates [64, 65]. Cellulose polymers of high viscosity, such as HPMC 2910, are used in film coating. Spray suspension at 12% HPMC K4M offers values of viscosity above 700 mPa·s and can not be normally used because 250 mPa·s is considered the limit for spraying of a coating suspension. This value of viscosity can be reduced significantly up to 250 mPa·s if 60% HPMC is substituted by Kollidon VA 64 [66] and this leads us to apply this polymer concentration and therefore to economize the spraying procedure.

6.3.8.3 SEPIFILM

SEPIFILM and SEPISPERSE Dry are ready-to-use, immediate-release, film coating compositions designed for pharmaceutical and nutritional supplement applications. Based on a unique technology, SEPIFILM/SEPISPERSE Dry are granular forms offering the following benefits:

Easy handling: **no dust**
Easy mixing: **no lumps, no foam**
Homogeneous composition: **no segregation**

Most coating compositions are based on hypromellose (HPMC) as film-forming polymer and contains microcrystalline cellulose (MCC) (see Table 17).

SEPPIC was the first company to introduce MCC in coating formulations some 20 years ago. The use of MCC allows higher solid content and enhances the adhesion of the film to the tablet core, consequently improving logo definition. Different formulas are provided by Sepifilm, such as SEPIFILM LP, SEPIFILM 003 and 752, and SEPISPERSE Dry. The last one can be provided with SEPIFILM and Kollicoat IR.

Moisture Protection of SEPIFILM LP Water vapor transmission rates were measured on free films, including titanium dioxide (Figure 27). SEPIFILM LP shows significantly lower moisture permeability compared to regular or PVA-based coating formulations. Removal of titanium dioxide (SEPIFILM LP clear) improves moisture resistance.

Like many other herbal extracts, valerian extract is very hygroscopic. Inclusion into tablets raises stability issues. Coatings reduce moisture absorption but often lead to tablet explosion or visual deterioration (black specs). Tablets formulated with 250 mg valerian extract, spray-dried lactose, and compressible starch were

TABLE 17 General Formulation for SEPIFILM Coating Film

Film-forming agent	**Hypromellose** (former HPMC)
Binder	Microcrystalline cellulose
Hydrophobic plasticizer	Stearic acid (vegetable origin)
Colors	Pigments, lakes

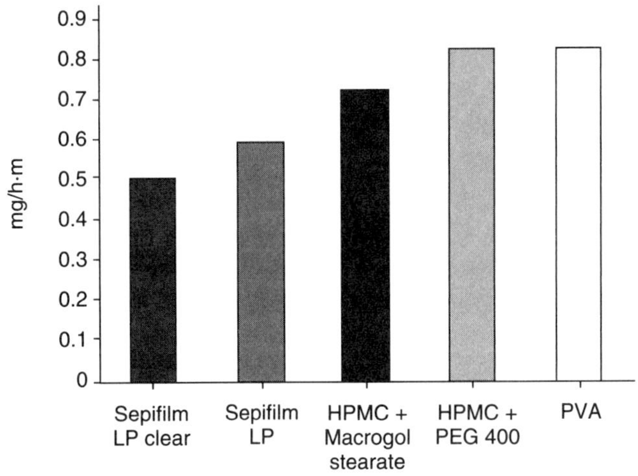

FIGURE 27 Water vapor transmission rates of free films.

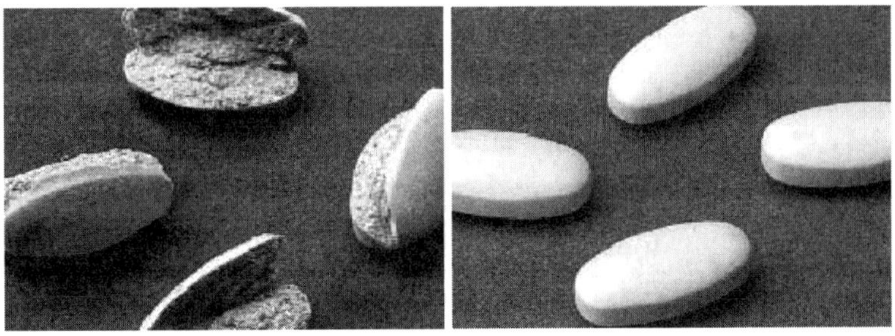

(*a*) (*b*)

FIGURE 28 Tablets coated with (*a*) conventional hypromellose and (*b*) Sepifilm LP 770 white.

stored at 40°C and 90% RH (relative humidity). Pictures were taken after one month storage (Figure 28).

Tetrazepam Tetrazepam, a well-known muscular relaxant, undergoes chemical degradation when exposed to moisture and oxygen. 3-Ketotetrazepam is one of the main degradation substances. The amount of 3-ketotetrazepam, measured by HPLC, has been monitored under 25°C/60 RH and 40°C/75% RH aging conditions.

SEPIFILM LP 770 significantly improves the stability of tetrazepam. Uncoated tablets and tablets coated with a PVA-based formulation show a strong increase in the amount of degradation substance (Figure 29).

Dissolution Profile SEPIFILM LP can efficiently improve the moisture barrier on moisture-sensitive active pharmaceutical ingredients (API) or hygroscopic cores. The breakthrough in this technology is that SEPIFILM LP does not modify the dissolution profile when compared to conventional coating (Figure 30).

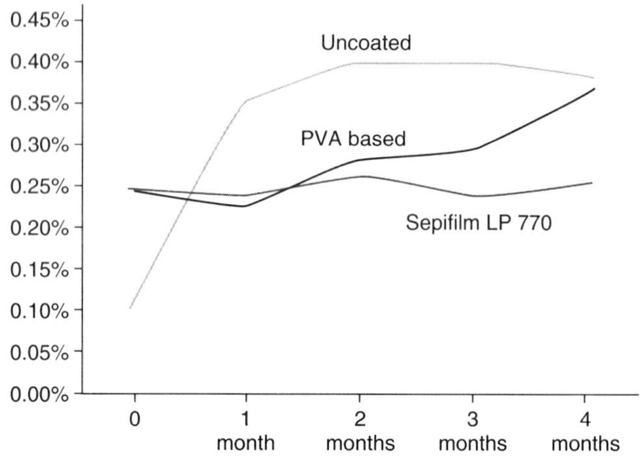

FIGURE 29 Stability of tetrazepam: uncoated tablets and different tablets coated. Dosage of 3-ketotetrazepam in 50 mg tetrazepam tablets stored at 25 °C and 60% RH leads to partial degradation of tetrazepam at T_0. Initial amount of 3-ketotetrazepam is lower in uncoated tablets: cores are likely to absorb water during the aqueous coating process.

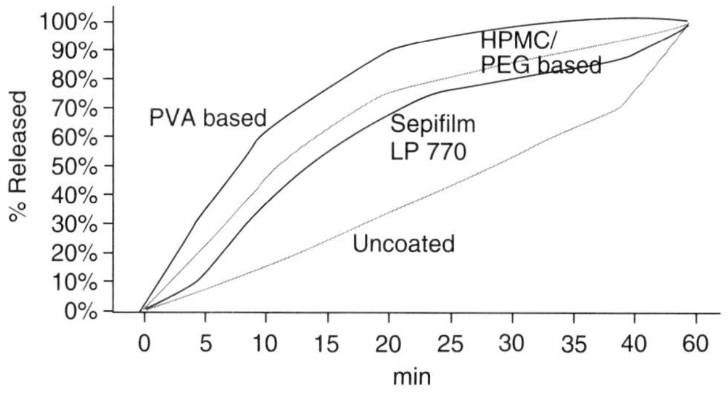

FIGURE 30 Dissolution profile of tetrazepam, 3 months storage at 40 °C and 75% RH.

SEPIFILM 003 and 752 The association of cellulose with a film coating agent was originally patented by SEPPIC. Microcrystalline cellulose is probably one of the most extensively used excipients in pharmaceutical and nutritional products. Unlike other fillers, such as lactose, cellulose is inert, vegetable derived, and accepted worldwide and its shelf life is unlimited. Figure 31 shows the advantage of cellulose microcrystalline in film coating.

Faster Film Coating Operations Cellulose is insoluble and does not increase viscosity. Dispersions with higher solid content can therefore be used and total spraying time is significantly decreased.

The maximum acceptable viscosity of a typical coating dispersion is 500 mPa·s, which corresponds to an 11% hypromellose 6 mPa·s dispersion. SEPIFILM can be dispersed up to 15% in order to reach the same viscosity (Figure 32).

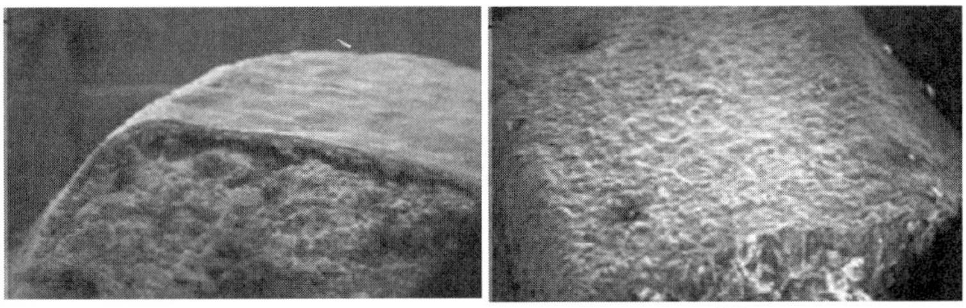

FIGURE 31 (*a*) Tablet coated with HPMC smooth film, medium discontinuity between film and core. (*b*) Tablet coated with SEPIFILM 752 white: clear edge perfectly coated, good adhesion of film to core and continuity between film and core.

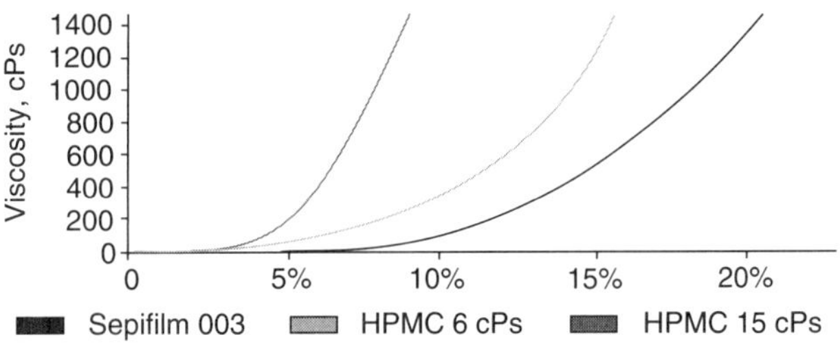

FIGURE 32 Viscosity for deferments HPMC coating dispersion.

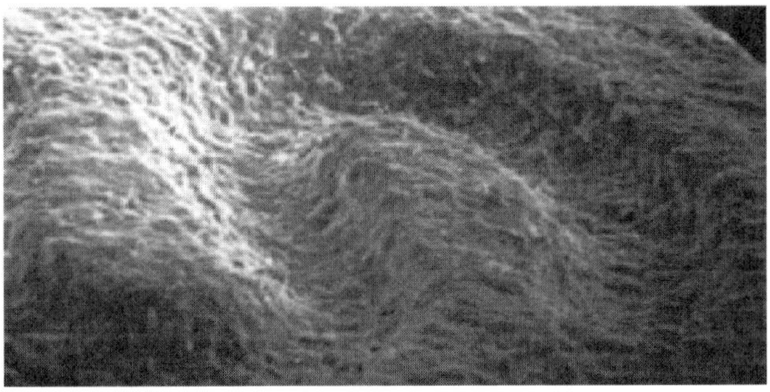

FIGURE 33 Tablet surface and edges are smoothly covered whereas no logo bridging is observed.

Enhanced Film Adhesion Comparative adhesion values were measured on 3% coated placebos using a modified crushing strength tester: Regular film formulation exhibited an adhesion value of 4 N. Addition of cellulose significantly improved film adhesion as it became impossible to remove film without disrupting tablets or film. Breakage occurred for an applied force of 6 N, which can be considered a minimum adhesion value (Figure 33).

Ingredient Segregation Avoided Powders such as hypromellose and titanium dioxide exhibit dramatically different particle size distribution and density. Such particle heterogeneity may result in constituent segregation inducing film imperfections or color deviations on tablets.

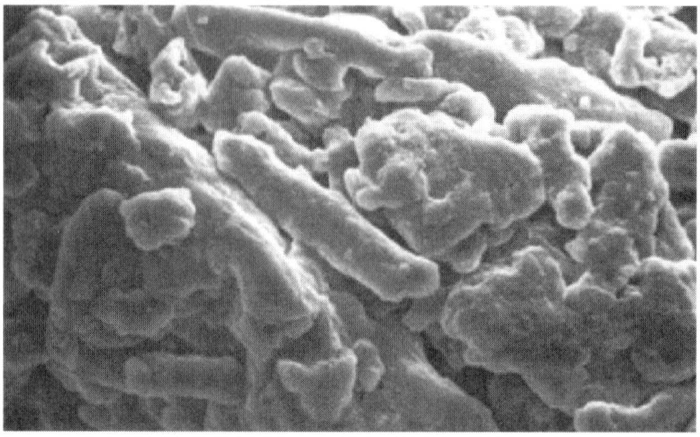

FIGURE 34 Granules of SEPIFILM 1306 green (×350). Pigments, lakes, and cellulose fibers are thoroughly coprocessed into homogeneous granules.

As seen in (Figure 34), SEPIFILM rules out this weak feature of powdered coating agents as all constituents are closely bound together. Ingredient segregation is avoided and batch-to-batch consistency is therefore guaranteed.

The PVA–PEG graft copolymer Kollicoat IR is the new instant-release, aqueous coating polymer from BASF. Due to its low viscosity and excellent mechanical properties, it permits solid content of up to 30% and leads to fine and smooth tablet film coatings.

In 2005, SEPPIC and BASF Pharma Solutions began collaboration in marketing tablet film coating systems. From this collaboration, a new range of colored SEPIFILM coating systems based on Kollicoat (a registered trademark of BASF Aktiengesellschaft) polymers will be developed to meet individual customer needs. Kollicoat IR has already been approved in Europe as a finished drug in Germany (a reference member state in a mutual recognition procedure). Common uses in pharmaceuticals depend on regulations of each country and definition of uses. Actual regulations of a specific country and/or application should be checked before use.

For pharmaceuticals, this combination is based on 88% of Kollicoat® IR White and 12% of coloring system SEPISPERSE™ Dry.

Recommended Equipment

A **propeller stirrer** is standard equipment even though a deflocculating blade is very efficient.

Deionized or distilled water at room temperature.

The **blade diameter** should be 3 times shorter than the tank's width.

The tank should be 1.5 times higher than wide. The blade should be slightly off center.

The blade should be positioned close to the bottom of the vessel.

TABLE 18 SEPIFILM LP (with or without SEPISPERSE Dry)

	Manesty XL Lab01 (A*)	Manesty Accelacota (B)	IMA-GS HT/M (C)	Driacoater 500 (C)
Solid content	LP 770 at 12%	LP 014/Dry at 12%	LP 014/Dry at 11%	LP 014/Dry at 12%
Batch size, kg	5	100	88.2	3
Spraying rate	15–27 mL/min	200–300 g/min	180–200 mL/min	7–15 g/min
Inlet air temperature, °C	60	70–75	62	55–60
Outlet air temperature, °C	47	—	—	42
Bed temperature, °C	43	40–45	36–40	—
Atomizing air/pattern pressure, bars	2	3.5–5	2.5	3
Airflow, m^3/h	440	1800	1000	270
Spraying time, min (% weight gain)	76 (1% w.g.)	60 (2% w.g.)	90 (2.2% w.g.)	70 (3% w.g.)

Mixing Procedure

SEPIFILM Product Range

1. Adjust rotation speed in order to create a vortex.
2. Quickly pour SEPIFILM into the vortex.
3. Reduce mixer speed to avoid drawing air into the liquid (risk of foam formation).
4. Increase speed again as viscosity builds up.

It is not necessary, though recommended, to keep stirring during the coating process.

Kollicoat® White + SEPISPERSE Dry

1. Adjust rotation speed to create a vortex.
2. Quickly pour SEPISPERSE Dry into the vortex.
3. Reduce mixer speed to avoid drawing air into the liquid. Stir for 15 min.
4. Pour Kollicoat IR White into the colored dispersion.
5. Stir an additional 5 min.

Some practical conditions for SEPIFILM and equipment machines are shown in Tables 18–20.

We used SEPIFILM LP in ranitidine core because this is a moisture-sensitive drug and can be a challenge to formulators because of its tendency to hydrolyze when exposed to humidity and/or high temperatures.

Cores of 300 mg that contained 50% ranitidine HCl, cellulose microcrystalline (PH-102) 27.25%, pregelatinized starch 22%, Aerosil 0.5%, and magnesium stearate

TABLE 19 SEPIFILM Formulations 050 and 752

	Driacoater 500	Glatt Coater 1000
Solid content of dispersion	SEPIFILM 050 at 15%	SEPIFILM 752 at 20%
Batch size, kg	3	80
Spraying rate, g/min/kg	7 (1 nozzle)	3.1 (3 nozzles)
Rotation speed, rpm	10	8
Inlet air temperature, °C	60	65
Outlet air temperature, °C	44	—
Bed temperature, °C	39–40	38–40
Atomizing air/pattern pressure, bars	3	3.5
Airflow, m^3/h	330	1800
Spraying time, min (3% weight gain)	85	55

TABLE 20 Kollicoat IR White + SEPISPERSE Dry

	Manesty Accelacota 24 in., 300-mg Propranolol HCl Tablets	Manesty Premier 200, Placebo Tablets
Solid content of dispersion	IR + Dry at 20%	IR + Dry at 20%
Batch size, kg	7	180
Spraying rate	5.3 g/min/kg	300–375 mL/min
Inlet air temperature, °C	60	60
Outlet air temperature, °C	38–43	45
Bed temperature, °C	39–41	—
Atomizing air/pattern pressure, bars	2/1	2.5/2.5
Airflow, m^3/h	220	2200
Spraying time, min (%weight gain)	37 (3.4% w.g.)	80 (3% w.g.)

0.5% were coated to a 3% weight gain similar to case **A*** (see Table 18). Tablet weight, diameter, thickness, hardness, and disintegration times were measured after coating. The film-coated tablets were packaged in bottles with a desiccant. Stability testing was conducted at 40 °C/75% RH for 12 months. Application of film coating (3% wg) resulted in a slight increase in tablet hardness (tablet breaking force 13–14 kp). Tablet disintegration time was not significantly affected by the film coating application (around 13 min) and 100% of the drug was released within 25 min compared to the USP limit of not less than 80% (*Q*) in 45 min. Figure 35 shows the dissolution profiles for the uncoated and coated tablets.

No significant changes were recorded for coated tablets after 12 months of storage for any property measured (Table 21). The stability of this formulation is partly due to the inclusion of pregelatinized starch in the formulation. Starch 1500 acts as a moisture scavenger and retains moisture in its complex structure of glucose polymer chains. A slight increase in tablet hardness was seen after storage (see Figure 36). No significant changes were seen in the disintegration time after storage.

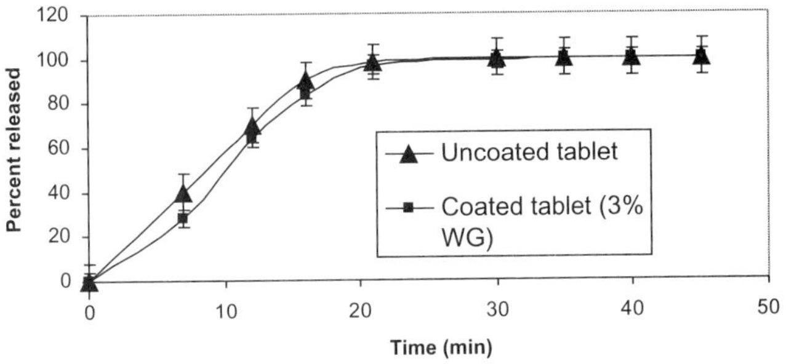

FIGURE 35 Dissolution profile of coated and uncoated ranitidine HCl tablets.

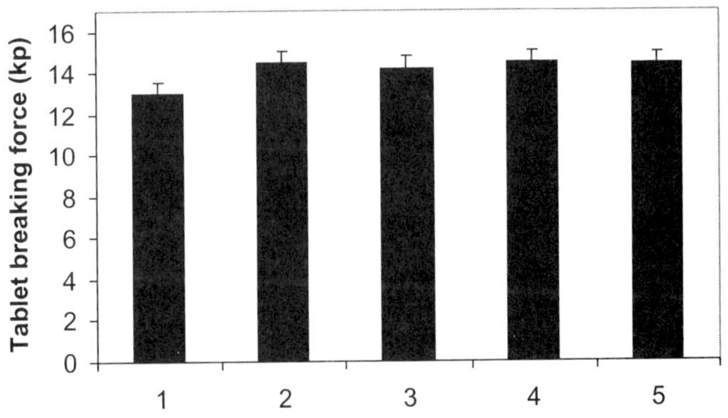

FIGURE 36 Coated tablet hardness on storage, 40 °C, 75% RH storage conditions.

TABLE 21 Stability Data Summary: Test

Test	USP Limit	Initial	1 month	3 months	6 months	12 months
Breaking force (kp)	NMT 1.00%	13.7	14.4	14.2	14.5	14.4
Friability, %		0.0	0.0	0.0	0.0	0.0
Dissolution, T85%, min	NLT 85% in 45 min	16	16	16	16	16
Assay, %	90–110	102	100	100	99	99
Impurities, %	NMT 2%	0.5	0.8	1.0	1.2	1.7

Notes: For 40 °C/75% RH storage conditions. Abbreviations: NMT, no more than; NLT, no less than.

Selected Coating Problems and Practical Solutions

(a) *Defective Coatings* Caused by poor quality of the core or inadequate coating formulations.

(b) *Chipping* Solid content is too high or it is too brittle for want of plasticizer and the film does not adhere properly to the substrate surface (too lipophilic surface or surface lacking in porosity).

Solution Revise the formulation for core and coating.

(c) *Blistering* Drying or spraying is performed at high speed, and solvent may be retained in the film. They evaporate on postdrying and may then form blisters in the film.

Solutions Lower the inlet air temperature and reduce the spray rate. Check if adhesion of the film and core is adequate.

(d) *Cracks in Film or Along Edges* Caused by too much internal stress, owing to differences in the thermal expansion of film and core or also caused by the swelling of the core during the coating operation.

(e) *Embedded Particles* Particles broken off from the core are embedded in the film during spraying.

Solutions The core lacks mechanical stability. Check the formulation of the core.

(f) *Picking* The film surface contains substances that are not molecularly dispersed and start to melt at the core bed temperature of the film coating process. These substances (e.g., PEG, stearic acid) may interfere with the film-forming polymers and produce holes in the film surface [67].

Solutions Replace these substances or lower the core bed temperature. Decrease the speed of praying.

(g) *Dull Surfaces* The film will be dull and totally devoid of gloss if a coating process does not produce the requisite smooth surface. This happens if the spray droplets start to dry before reaching the cores and are too viscous to form a smooth film.

Solutions (1) Lower the inlet air temperature and reduce the atomizing air quantity or pressure. (2) Add substances that enhance film formation, for example, plasticizer or extra solvents.

(h) *Twinning* The cores permanently stick together.

Solutions Decrease the excessive spraying. Check the tablet shape and bands. If the tablets are plane or almost plane, this continues until many of them stick together.

(i) *Bridging* the film fails to follow the contours of the tablet over break lines or engravings and settles in these without adhering to the substrate. The bridging forces in the film exceed the interfacial forces between film and core.

Solutions change the tablet surface or add plasticizer.

6.3.9 DEVELOPMENT OF PHARMACEUTICAL TABLETS USING PERCOLATION THEORY

In 1991, Bonny and Leuenberger [40] explained the changes in dissolution kinetics of a matrix controlled-release system over the whole range of drug loadings on the basis of percolation theory. For this purpose, the tablet was considered a disordered system whose particles are distributed at random. These authors derived a model for the estimation of the drug percolation thresholds from the diffusion behavior.

Knowledge of the percolation thresholds and the related critical points of the system allows a rational optimization of the matrix formulation, avoiding the trial-and-error method usually employed in the pharmaceutical industry. The ideal formulation of an inert matrix, following percolation theory, must be above the drug percolation threshold (i.e., the drug plus the initial pores percolate the system). This fact guarantees the release of the total drug dose. On the other hand, the matrix must also contain an infinite cluster of excipient (i.e., the excipient must also be above its percolation threshold). This percolating cluster of excipient avoids the disintegration of the matrix during the release process and controls the drug release [43].

This kind of system, containing percolating clusters of both drug and excipient, is called a bicoherent system. Furthermore, in order to decrease the variability in the biopharmaceutical and mechanical behavior of the matrices, due to little change in the tablet composition, it is not convenient to formulate the matrices just at the percolation threshold. In this way, knowledge of the percolation thresholds of drug and excipient supposes an important decrease in the cost of the optimization process as well as in the time to market. The percolation thresholds of different pharmaceutical powders have already been estimated, including drugs such as morphine hydrochloride [68], naltrexone hydrochloride [69], dextromethorphan hydrobromide [70], and lobenzarit disodium [71] as well as matrix-forming excipients such as hydrogenated castor oil, ethylcellulose, and acrylic polymers.

6.3.9.1 Case Study: Optimization of Inert Matrix Tablets for Controlled Release of Dextromethorphan Hydrobromide

The objective of this work was to estimate the percolation thresholds of dextromethorphan hydrobromide and Eudragit RS-PM which characterize the release behavior of these inert matrices in order to rationalize the design of these controlled-release systems.

Dextromethorphan hydrobromide is an antitussive drug with no analgesic or addictive action. Its antitussive effect is similar to codeine. The recommended oral dose for adults is 10–30 mg three to six times a day, not to exceed 120 mg daily. It is absorbed rapidly and completely when taken orally with a lag time of 15–30 min [72].

In order to estimate the percolation threshold of dextromethorphan hydrobromide, the matrices were studied from different points of view:

1. *Release Profiles and Release Kinetics* Figure 37 shows the percentage of drug released from the studied matrices. As can be appreciated, a very similar behavior was observed for matrices containing up to 50% w/w of drug. This can be attributed to the swelling process (approximately 11% v/v) undergone by the matrices during the release assay. This process makes the influence of the percolation threshold on the release profiles less evident.

Higuchi's kinetic model and Peppas' nonlinear regression ($Q = a' + b't^k$) were employed to study the release data. The results obtained are shown in Table 22. As can be seen, the exponent k underwent a change (0.4534–0.5472) between matrices containing 20 and 30% w/w of drug. Even if this is not an important change, it may be related to some changes in the matrix structure due to the drug percolation

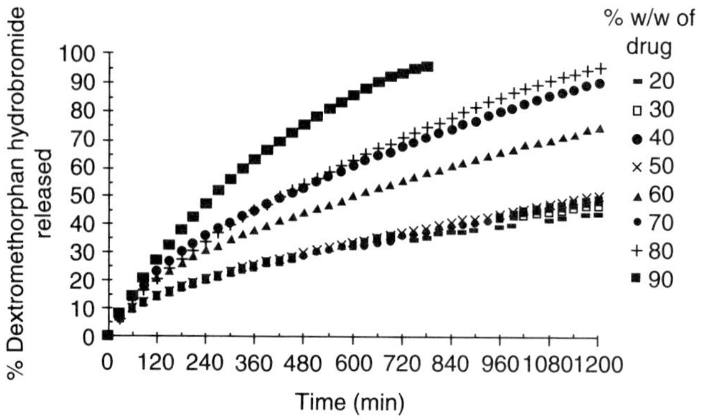

FIGURE 37 Percentage of drug released vs. time for tablets prepared with different loadings of dextromethorphan hydrobromide.

TABLE 22 Dissolution Data from Dextromethorphan–HBr/Eudragit RS-PM Matrices

Drug Load (% w/w)	n	$Q - a + b\sqrt{t}$ $b \pm$ S.E.	r	$Q = a^r + b^r t^k$ k	r
20	248	$1.43 \times 10^{-4} \pm 2.9 \times 10^{-7}$	0.999	0.45335	0.999
30	248	$2.47 \times 10^{-4} \pm 4.7 \times 10^{-7}$	0.999	0.54724	0.999
40	248	$3.38 \times 10^{-4} \pm 1.0 \times 1.0^{-6}$	0.998	0.58981	0.999
50	248	$4.43 \times 10^{-4} \pm 2.0 \times 1.0^{-6}$	0.998	0.59792	0.999
65	241	$8.60 \times 10^{-4} \pm 2.0 \times 10^{-6}$	0.999	0.54536	0.999
70	248	$1.15 \times 10^{-3} \pm 2.0 \times 10^{-6}$	0.999	0.54697	0.999
80	241	$1.45 \times 10^{-3} \pm 5.0 \times 10^{-6}$	0.999	0.57677	0.999
90	85	$2.09 \times 10^{-3} \pm 2.9 \times 10^{-5}$	0.992	0.71910	0.999

threshold. The masking effect of the swelling process on the drug percolation threshold has to be taken into account.

2. *Estimation of Drug Percolation Threshold* The drug percolation threshold was calculated using the property β described by Bonny and Leuenberger [40]. This property is defined by the equation

$$\beta = \frac{b}{\sqrt{2A - \varepsilon C_s}} \qquad (5)$$

where β is proportional to the square root of the effective diffusion coefficient D_{eff}, which is expected to obey, in the nearby of the percolation threshold, the scaling law

$$D_{eff} = kD_0(\varepsilon - \varepsilon_c)^{\mu}$$

where D_0 is the diffusion coefficient of the drug in pure solvent, k a constant, ε the total porosity of the matrix (sum of initial porosity and porosity due to the dissolution of the drug), ε_c the critical porosity or percolation threshold, and μ the critical

TABLE 23 Calculation of Tablet Property β and Related Parameters in Matrices of Dextromethorphan Hydrobromide

Drug (% w/w)	ε_0	ε	n^b	F^b	Probability[b]	A	$\beta \times 10^3$
20	0.145	0.300	248	248,833	9.9×10^{-16}	0.216	0.218
30	0.142	0.378	248	277,630	9.9×10^{-16}	0.329	0.305
40	0.129	0.453	248	88,245.6	9.9×10^{-16}	0.450	0.357
50	0.115	0.531	248	79,555.4	9.9×10^{-16}	0.579	0.413
65	0.091	0.657	241	126,817	9.9×10^{-16}	0.787	0.687
70	0.093	0.705	248	278,000	9.9×10^{-16}	0.851	0.887
80	0.083	0.799	241	97,483.4	9.9×10^{-16}	0.995	1.033
90	0.083	0.898	85	5,324.51	9.9×10^{-16}	1.133	1.395

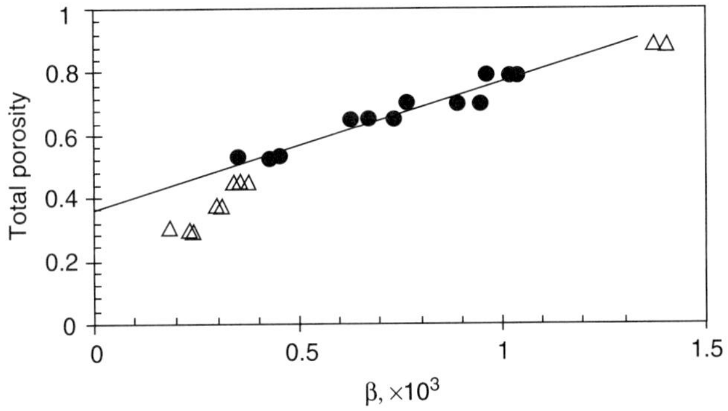

FIGURE 38 Determination of drug percolation threshold. The circles represent the values selected for the regression, according to its linear behavior. These values correspond to drug loads between 50 and 80% w/w (three tablets per lot). Each point represents one experimental datum.

exponent for conductivity. This exponent has a value of 2.0 in 3D systems. The values of β as well as the parameters involved in its calculation are shown in Table 23.

The percolation threshold of dextromethorphan hydrobromide was estimated as the intersection with the Y axis from a linear regression of the total porosity, ε, versus the property β (Figure 38). Following the method of Bonny and Leuenberger, only the β values above p_{c1} showing a linear dependence on the total porosity (circles in Figure 38) are considered in the regression. The selected β values corresponded to matrices with 50–80% w/w of drug.

The estimated critical porosity is 0.3691 ± 0.0541, considering a 95% confidence interval ($P = 0.05$). This range corresponds to a dextromethorphan hydrobromide content of between 23 and 36% w/w.

Estimation of the percolation threshold by visual methods is not very accurate, mainly due to extrapolation from 2D to 3D systems. Nevertheless, scanning electron microscopy was employed as an auxiliary technique in order to investigate the distribution of the particles of dextromethorphan hydrobromide in the matrices.

Figure 39 shows two scanning electron microscopy (SEM) micrographs corresponding to the tablet side facing the lower punch for matrices containing 20 and

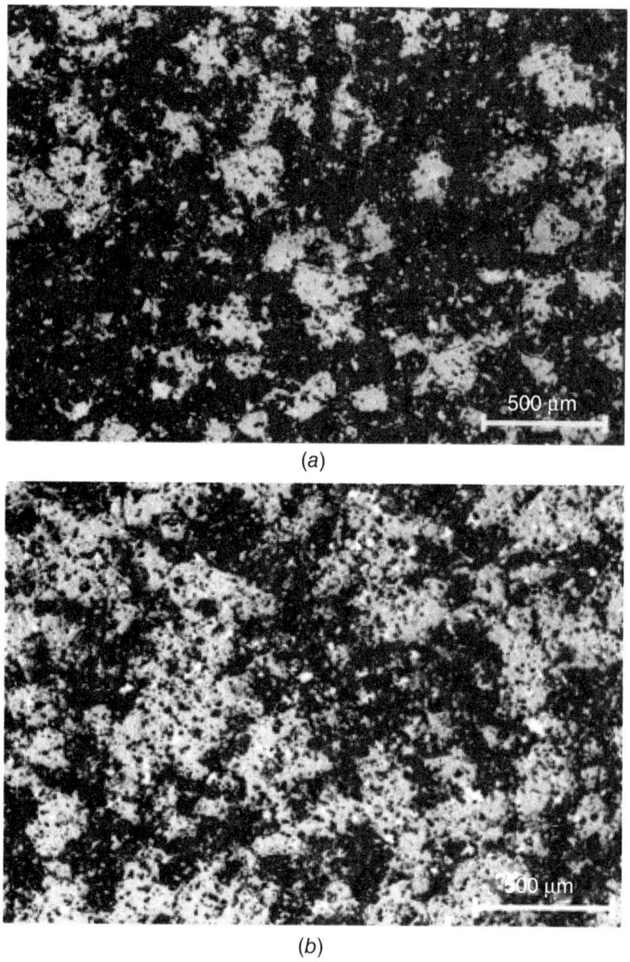

FIGURE 39 SEM micrographs corresponding to bottom side of matrices using BSE detector. The light gray particles correspond to dextromethorphan–HBr and the dark gray particles to the excipient Eudragit RS-PM. (*a*) Matrices containing 20% w/w of drug. (*b*) Matrices containing 30% w/w of drug.

30% w/w of drug using backscattering electron (BSE) detector at the same magnification. In the tablet containing 30% of drug (Figure 39*b*), an infinite drug cluster can be observed. The drug particles (light-gray particles) begin to form a connective network from the left to the right and from the top to the bottom of the micrograph. In the tablet containing 20% w/w of drug (Figure 39*a*), the particles of the drug (light-gray particles) seem to form isolated groups in the matrix.

Therefore, considering both micrographs in Figure 39, a 2D geometric phase transition can be observed. Figure 39*a* shows the drug as gray particles on a black background, whereas in Figure 39*b* there is a black-on-gray array, with black particles corresponding to Eudragit RS-PM surrounded by a gray background formed by dextromethorphan hydrobromide particles.

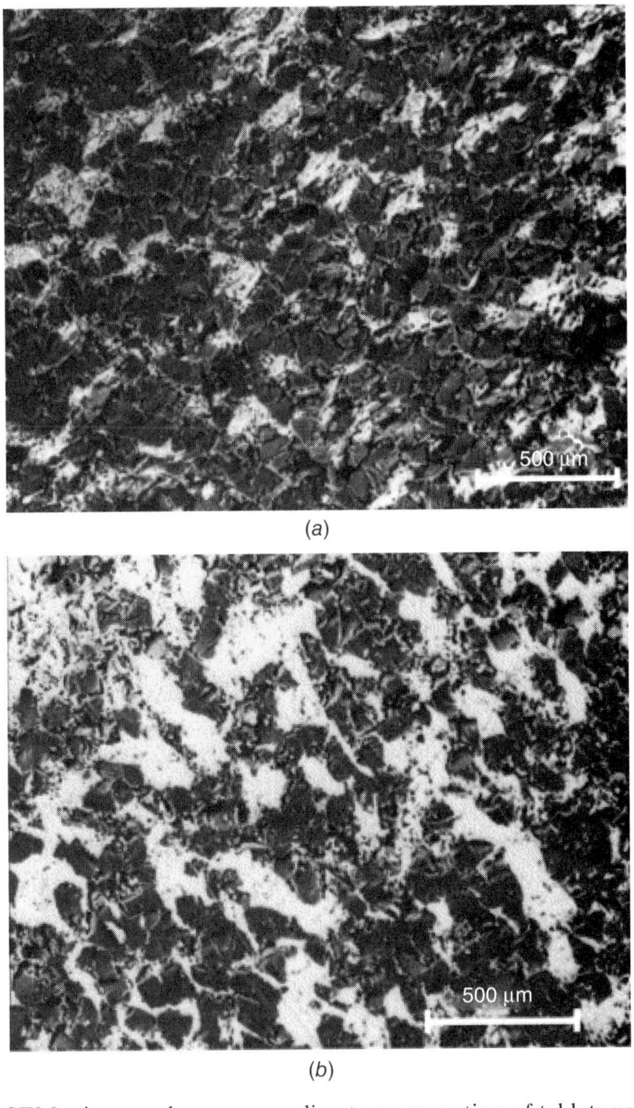

FIGURE 40 SEM micrographs corresponding to cross section of tablets using BSE detector: (*a*) matrices containing 20% w/w dextromethorphan–HBr (light gray particles); (*b*) matrices containing 30% w/w of drug.

When the cross section of these matrices (20 and 30% w/w drug loading) was observed (Figure 40), the same pattern was found, changing from gray on black (Figure 40*a*, 20% w/w of drug) to black on gray (Figure 40*b*, 30% w/w of drug).

Therefore, according to the different methods employed, the drug percolation threshold in the studied matrices is expected to be between 20 and 30% w/w of dextromethorphan hydrobromide (total porosity between 30.0 and 37.8% v/v of drug).

3. *Estimation of Excipient Percolation Threshold.* In principle, for binary pharmaceutical systems, two percolation thresholds are expected: the drug percolation

threshold p_{c1} and the excipient percolation threshold p_{c2}. The second is the point where the excipient ceases to percolate the system.

Nevertheless, in a previous study dealing with inert matrices of naltrexone–HCl [74], two different excipient percolation thresholds p_{c2} were found for the matrix-forming excipient Eudragit RS-PM: the site percolation threshold related to a change in the release kinetics and the site-bond percolation threshold derived from the mechanical properties of the tablet, where the excipient failed to maintain tablet integrity after the release assay.

An evident change in the release kinetics between tablets containing 80 and 90% w/w of drug can be observed in Table 23 (from $k \approx 0.57$ to $k \approx 0.7$ in the Peppas equation). Therefore, the site percolation threshold of the excipient can be estimated between the matrices containing 80 and 90% w/w of dextromethorphan hydrobromide (10–20% v/v of excipient). Above this threshold, a percolating cluster of excipient particles exists. These particles are able to control the drug release kinetics, but their cohesion forces can be insufficient to maintain tablet integrity after the release assay.

Formulations containing more than 65% w/w of drug were unable to maintain tablet integrity after the 20-h release assay. According to this result, the site-bond percolation threshold of the excipient ranges between 65 and 70% w/w of drug, corresponding to 29.5 and 34% v/v of excipient. Above this percolation threshold, that is, for concentrations of excipient >34% v/v, there is a percolating cluster of excipient particles bound by sufficient forces to maintain tablet integrity after drug release.

In conclusion, according to percolation theory, the studied matrices should be formulated with drug content between 30 and 65% w/w (37.8–66% v/v of total porosity). These concentrations are optimal to ensure release of the total drug dose, to have controlled release of the drug, and to avoid disintegration of the matrix. In order to increase the robustness of the formulation, the limits of this range should be avoided.

6.3.9.2 Critical Points of Hydrophilic Matrix Tablets

Recently percolation theory is starting to be applied to the study of hydrophilic matrix systems. Figure 41 shows an example of the changes observed in several release parameters employed to estimate the critical point and the related percolation threshold in hydrophilic matrices prepared using KCl as the model drug [73].

Application of the percolation theory allows explanation of the changes in the release and hydration kinetics of swellable matrix-type controlled delivery systems. According to this theory, the critical points observed in dissolution and water uptake studies can be attributed to the excipient percolation threshold. Knowledge of these thresholds is important in order to optimize the design of swellable matrix tablets. Above the excipient percolation threshold an infinite cluster of this component is formed which is able to control the hydration and release rate. Below this threshold the excipient does not percolate the system and drug release is not controlled.

Miranda et al. demonstrated experimentally the influence of the particle size of the components on the percolation threshold in hydrophilic matrices as well as the importance of the initial porosity in the formation of the gel layer (sample-spanning cluster of excipient) [74].

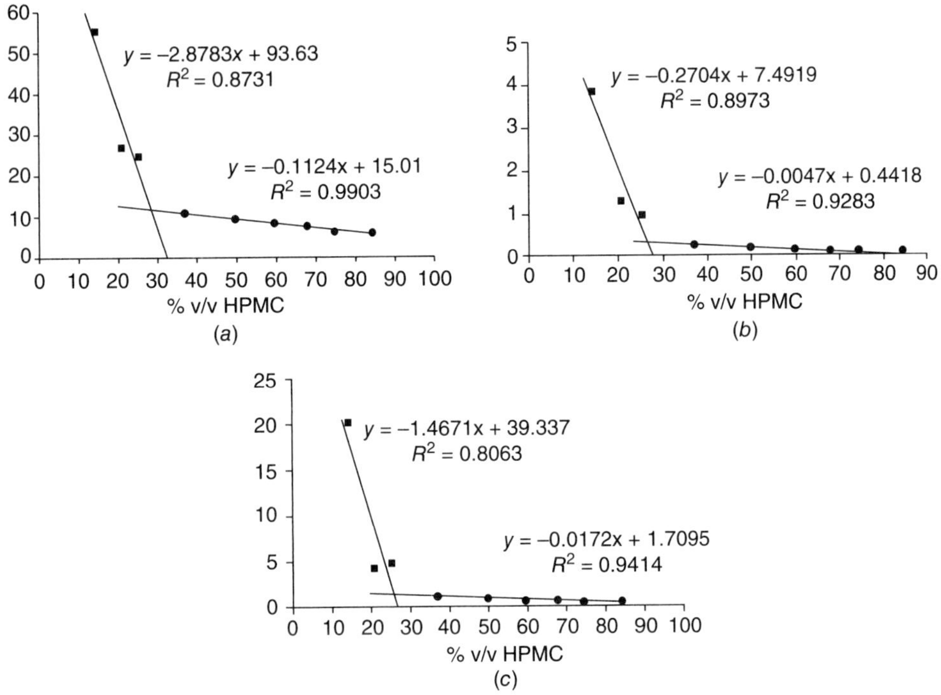

FIGURE 41 (*a*) Higuchi slope; (*b*) normalized Higuchi slope; (*c*) relaxational constant of Peppas and Sahlin versus percentage of excipient volumetric fraction for batch A (50–100 μm KCl and 150–200 μm HPMC K4M).

6.3.9.3 Case Study: Estimation of Percolation Thresholds in Acyclovir Hydrophilic Matrix Tablets

The principles of the percolation theory were applied to design controlled-release matrix tablets containing acyclovir in order to estimate the percolation threshold of the excipient in acyclovir matrix tablets and to characterize the release behavior of these hydrophilic matrices in order to rationalize the design of these controlled-release systems.

Acyclovir is a potent inhibitory of viruses of the herpes group, particularly herpes simplex virus (HSV I and II) and herpes zoster varicella virus. Unfortunately, acyclovir has a short half-life (2–3 h), and the oral dosage form must be taken five times daily, which is very inconvenient for patients [75, 76]. Consequently, the aim of this study was to develop a controlled-release formulation of acyclovir that could be taken twice daily. The materials used to prepare the tablets were acyclovir (Kern Pharma, Tarrasa, Barcelona) and hydroxypropyl methylcellulose (Methocel K4M) (Colorcon) a hydrophilic cellulose derivative as the matrix-forming material.

Binary mixtures were prepared with varying drug contents (60, 70, 80, 90, and 95%) keeping constant the drug and excipient particle size. Table 24 gives the composition of the studied batches as well as the tablet thicknesses. The mixtures were compressed on an eccentric machine (Bonals A-300) without any further excipient. Cylindrical tablets with a mean dosage of 500 mg and a diameter of 12 mm were prepared at the maximum compression force accepted by the formulations.

TABLE 24 Composition of Hydrophilic Matrices Prepared with Acyclovir/HPMC K4M (150–200 μm) and Percent HPMC plus Initial Porosity

	Percent w/w			
Batch	Acyclovir	HPMC K4M	Percent (v/v) HPMC + Initial Porosity	Tablet Thickness (mm)
A	95	5	20.76	3.01 ± 0.056
B	90	10	26.41	3.25 ± 0.052
C	80	20	37.60	3.84 ± 0.051
D	70	30	45.11	4.68 ± 0.063
E	60	40	55.82	5.81 ± 0.056

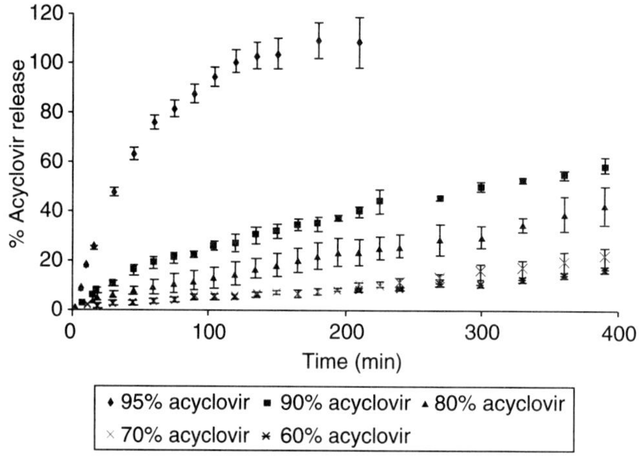

FIGURE 42 Acyclovir release from matrix tablet with total drug content of 95, 90, 80, 70, and 60 prepared with acyclovir–HPMC K4M (150–200 μm) (mean ± SD, n = 3).

The release profiles were measured with the USP 25 dissolution apparatus 2 (Turu Grau, model D-6) at 100 rpm in distilled water (900 mL) at 37 ± 0.5 °C for 12 h. Filtered samples taken at different times were determined for acyclovir content through ultraviolet absorption at λ_{max} (242 nm).

Figure 42 shows the release profiles obtained from hydrophilic matrices formulated with acyclovir and HPMC K4M 150–200 μm.

In order to study the release mechanism of acyclovir from the tablets, the fitting of the drug release data to the following kinetic equations has been studied: zero-order equation, $Q = k_0t$; Higuchi equation [77], $Q = k_Ht^{1/2}$; Korsmeyer–Peppas equation [78], $Q = kt^n$; and Peppas-Sahlin equation [79], $Q = k_d^m + k_r^{2m}$, where Q is the amount of drug remaining at time t, k_0 is the zero-order release constant, k_H is the Higuchi rate constant, k is the Korsmeyer–Peppas kinetic constant, n is the exponent indicative of the release mechanism (for matrix tablets an n value of 0.5 indicates diffusion control and an n value of 1.0 indicates erosion or relaxation control [80], intermediate values suggest that at least two processes contribute to the overall release mechanism), k_d is the diffusion rate constant, k_r is the relaxation rate constant, and m is the purely Fickian diffusion exponent for a device of any geometric shape which exhibits controlled release. In our case, the aspect ratios and exponent values (m) are shown in Table 25 [79]. The results obtained are shown in Table 26.

TABLE 25 Aspect Ratios and Exponent Values (m) for Hydrophilic Matrices Studied

Batch	Aspect Ratio	Exponent (m)
A	3.80	0.45
B	3.59	0.44
C	3.03	0.43
D	2.65	0.42
E	2.16	0.43

TABLE 26 Values of Kinetic constants Derived with Selected Equations in Range 5–70% Acyclovir Release for All Batches Studied

Batch		**A**	**B**	**C**	**D**	**E**
Acyclovir, % w/w		**95**	**90**	**80**	**70**	**60**
Zero-order equation	k_0	1.222	0.122	0.096	0.057	0.042
	r^2	0.984	0.974	0.994	0.995	0.987
	Sum of squares total	3545.9	10,352.6	6,531.4	5,149.2	1,268.0
	Sum of squares residual	53.7	278.3	51.8	24.8	16.4
Higuchi equation	k_H	12.440	3.4167	2.8518	1.7683	1.4694
	r^2	0.998	0.993	0.956	0.932	0.959
	Sum of squares total	3,545.9	10,352.6	12,367.8	5,149.2	1,268.0
	Sum of squares residual	7.0	66.9	483.6	362.4	54.6
Korsmeyer–Peppas equation	k_H	3.167	0.290	0.254	0.027	0.041
	n	0.782	0.843	0.856	1.114	1.008
	r^2	0.999	0.994	0.998	0.998	0.997
	Sum of squares total	13,220.9	26,297.6	33,309.1	13,879.4	5,673.5
	Sum of squares residual	16.7	159.5	65.5	10.1	15.2
Peppas and Sahlin equation	k_d	2.239	2.056	0.357	−0.81	−0.38
	k_r	1.615	0.161	0.205	0.202	0.127
	r^2	0.998	0.999	0.994	0.997	0.997
	Sum of squares total	49,567.0	13,220.9	28,716.1	13,879.4	5,673.5
	Sum of squares residual	42.9	23.3	165.0	43.7	15.2

Notes: k_0 (%/min), zero-order constant; k_H (%/min$^{1/2}$), Higuchi's slope; k (%/min^n), kinetic constant of Korsmeyer model; n, diffusional exponent; k_d (%/min^m), diffusional constant of Peppas and Sahlin model; k_r (%/min^{2m}), relaxational constant of Peppas and Sahlin model; m, diffusional exponent that depends on geometric shape of releasing device through its aspect ratio (see Table 25).

The analysis of the release profiles and the kinetic data indicate two different behaviors and a sudden change between them. In the first behavior, which corresponds to the matrices that release the drug at slow rates, the release was controlled by the fully hydrated gel layer. For these matrices, erosion of the hydrophilic gel structure has shown an important influence on drug release. This is indicated by the better fit of the drug release kinetics to the zero-order equation, the n value of

Korsmeyer–Peppas equation near 1, and the higher value of the relaxation constant k_r in comparison with the diffusion constant k_d in the Peppas–Sahlin equation. Taking into account the drug solubility (2.5 mg/ML), prevalence of the erosion versus swelling mechanism can be expected. After the transition point, the tablets allow the free dissolution of the drug when they are exposed to the dissolution medium due to the fact that the gel layer is not established since the first moment and, in these conditions, this structure cannot control the drug release. The Korsmeyer release rate increases from 0.290 to 3.167% $min^{-1/2}$. For these matrices, according to the Higuchi ($r^2 = 0.998$), Korsmeyer ($n = 0.782$), and Peppas–Sahlin ($k_r < k_d$) equations, drug release is governed by the diffusion process.

In hydrophilic matrices the drug threshold is less evident than the excipient threshold, which is responsible for the release control [73]. In order to estimate the percolation threshold of HPMC K4M, different kinetic parameters were studied: Higuchi rate constant, normalized Higuchi rate constant, and relaxation rate constant. The evolution of these release parameters has been studied as a function of the sum of the excipient volumetric percentage plus initial porosity. Recent studies of our research group have found the existence of a sample-spanning cluster of excipient plus pores in the hydrophilic matrix before the matrix is placed in contact with the liquid, clearly influences the release kinetics of the drug [73].

Figures 42–45 show changes in the different kinetic parameters: the Higuchi rate constant, normalized Higuchi rate constant, and relaxation rate constant. To estimate the excipient percolation threshold, these parameters were plotted versus the excipient volumetric fraction plus initial porosity.

The kinetic parameters studied show a nonlinear behavior as a function of the volumetric fraction of the excipient plus initial porosity. As an approximation for estimating the trend of the parameter, one regression line has been performed below and the other above the percolation threshold. The point of intersection between both regression lines has been taken as an estimation of the percolation threshold [73, 74].

As percolation theory predicts, the studied properties show a critical behavior as a function of the volumetric fraction of the components. A critical point has been found between 21 and 26% v/v of excipient plus initial porosity (see Table 24). This critical point can be attributed to the excipient percolation threshold.

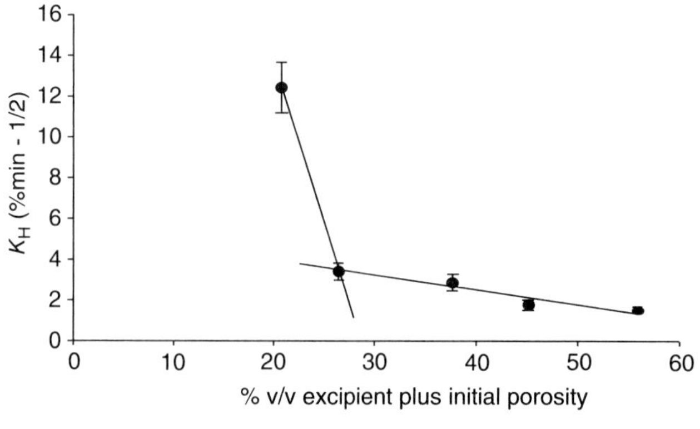

FIGURE 43 Higuchi slope (mean ± SD, $n = 3$) versus percentage of excipient volumetric fraction plus initial porosity for all batches studied.

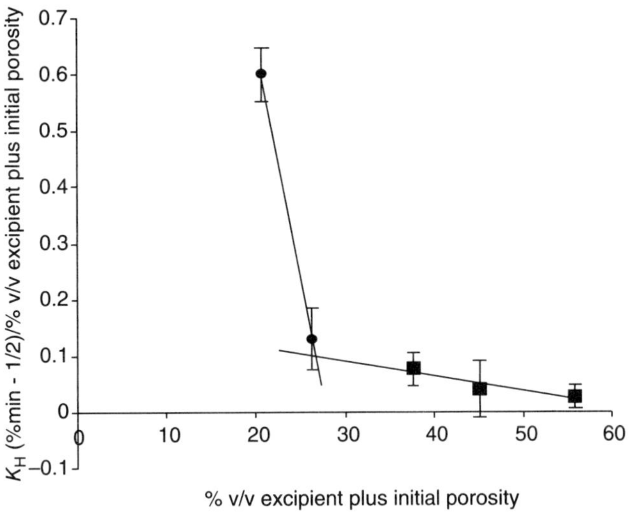

FIGURE 44 Normalized Higuchi slope (mean ± SD, $n = 3$) versus percentage of excipient volumetric fraction plus initial porosity for all batches studied.

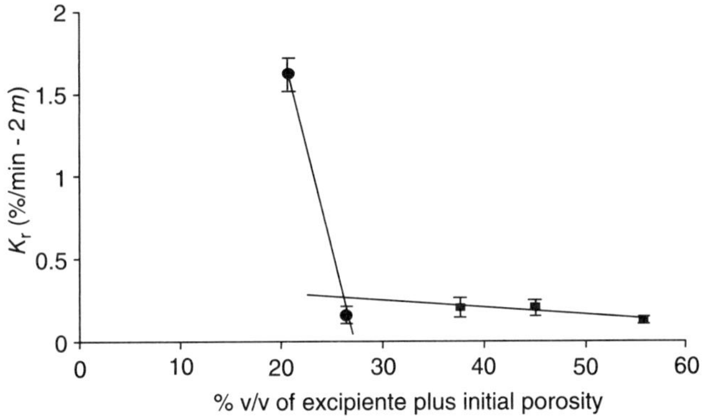

FIGURE 45 Relaxational constant of Peppas–Sahlin (mean ± SD, $n = 3$) versus percentage of excipient volumetric fraction plus initial porosity for all batches studied.

The Effective Medium Approximation (EMA), based in some assumptions, allows us to employ linear regressions as an approximation of the behavior of a disordered system outside the critical range. Based on EMA theory, two linear regressions have been performed as an approximation for estimating the percolation threshold as the point of intersection between both regression lines (see Figures 43–45). The values of the excipient percolation thresholds estimated for all the batches studied, based on the behavior of the kinetic parameters, ranged from 25.99 to 26.77%.

Therefore, the results obtained from the kinetics analysis are in agreement with the release profiles, indicating a clear change in the release rate and mechanism between matrices containing 90 and 95% w/w of drug (5–10% w/w of excipient). The existence of a critical point can be attributed to the excipient percolation threshold. From the point of view of percolation theory, this means that above 10% w/w of HPMC K4M, the existence of a network of HPMC (able to form a hydrated layer from the first moment) controls the drug release.

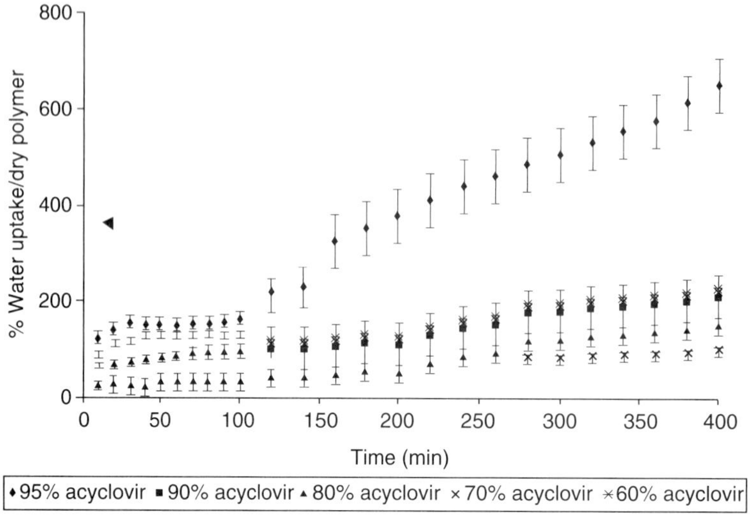

FIGURE 46 Weight gain of systems as function of swelling time for matrix tablet with total drug content of 95, 90, 80, 70, and 60% prepared with acyclovir–HPMC K4M (150–200 μm) (mean ± SE, $n = 3$).

The process of water penetration into hydrophilic matrix tablets was also studied using a modified Enslin apparatus. This apparatus contains a fritter and a system to regulate the water level. When the tablet is placed on the fritter, the water is absorbed from a reservoir which is placed on a precision balance. The amount of water uptake at each time point was read from the balance as weight loss in the reservoir. Figure 46 shows the obtained release profiles.

An increase in the rate of water uptake can be observed when the HPMC concentration decreases. A critical point was found between 90 and 95% w/w of acyclovir. This range corresponds with the critical point observed in release profile studies. The water uptake data were subjected to the Davidson and Peppas model to calculate the rate of water penetration [81]. The results show a change in the water uptake constant between the matrices containing 90–95% w/w of acyclovir, which reflects the presence of the critical point previously observed.

Knowledge of the percolation threshold of the components of the matrix formulations contributes to improve their design. First, in order to develop robust formulations, that is, to reduce variability problems when they are prepared at industrial scale, it is important to know the concentrations corresponding to the percolation thresholds. The percolation thresholds correspond to formulations showing a high variability in their properties as a function of the volume fraction of their components. Therefore, in order to design robust dosage forms, the nearby of the percolation thresholds should be avoided.

Second, the excipient percolation threshold in hydrophilic matrices represents the border between a fast release of the drug (below the threshold) and a drug release controlled by the formation of a coherent gel layer (above the excipient percolation threshold). Therefore, knowledge of this threshold will allow us to avoid the preparation of a number of unnecessary lots during the development of a pharmaceutical formulation, resulting in a reduction of the time to market.

6.3.10 ULTRASOUND-ASSISTED TABLETING (A NEW PERSPECTIVE)

The compression of a powder is a complex process that is usually affected by different kinds of problems. These problems have been widely investigated and mainly concern the volume reduction and the development of a strength between the particles of the powder sufficient to ensure tablet integrity [82]. The application of ultrasonic energy shows a great ability to reduce and even avoid these problems [83]. Ultrasound refers to mechanical waves with a frequency above 18 kHz (the approximate limit of the human ear). In an ultrasound compression machine, this vibration is obtained by means of a piezoelectric material (typically ceramics) that acts as a transducer of alternate electric energy of different frequencies in mechanical energy. An acoustic coupler, or "booster," in contact with the transducer increases the amplitude of the vibration before it is transmitted (usually in combination with mechanical pressure) to the material to be compressed.

Ultrasound-assisted powder compression has been widely employed in metallurgy as well as in the plastic and ceramic industries [84]. The first references in the pharmaceutical industry are two patents in 1993 [85] and 1994 [86]. Since then, some papers have presented experimental data in this field [45, 87–92].

Two main objectives are pursued nowadays by means of the application of ultrasound-assisted compression:

1. Increase in the drug dissolution rate due to amorphization of the drug
2. Preparation of controlled-release dosage forms with thermoplastic excipients

As a consequence of the application of ultrasonic energy, the drug can lose its crystalline structure. This will result in an increase of the dissolution rate of the active substance, which can be very adequate for slowly dissolving drugs. Nevertheless, depending on the storage conditions, the drug can recover, at least partially, its crystallinity [89, 90].

To overcome this problem, it has been proposed to use an adequate excipient, preventing the recovery of the crystallinity, leading in some cases to the preparation of solid solutions into the die of the tableting machine.

Several analytical techniques, such as infrared (IR) spectroscopy, differential scanning calorimetry, HPLC, and thin-layer chromatography (TLC), have been used to investigate possible drug degradation due to ultrasonic energy. No important permanent modification of the drug has been found, with the exception of the loss of crystallinity [89, 90].

Concerning the design of controlled-release dosage forms, using a thermoplastic excipient (e.g., copolymers of acrylic and metacrylic acid), an important decrease in the release rate has been found for tablets compressed with the assistance of ultrasonic energy in comparison with traditional tablets.

Although the effects of the ultrasonic energy on the material are not completely clarified, this slow release rate has been attributed to different phenomena:

Mechanical Pressure This pressure is exerted by the punches of the ultrasound-assisted tableting machine. This is the main compression mechanism when low ultrasonic energies are employed (below 25 J in the mixtures studied by

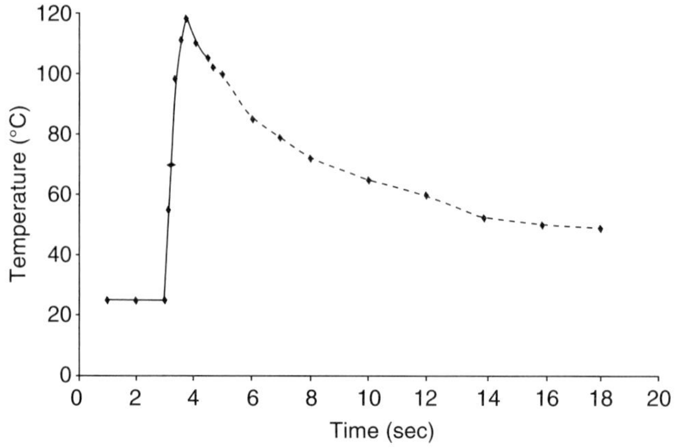

FIGURE 47 Temperature profile inside compression chamber. (Courtesy of Tecnea Srl.)

Rodriguez et al. [87, 88]) or when the materials used are not thermoplastic. In these cases the machine acts as a multiple-impact mechanical press.

Thermal Effects Due to the poor conductivity for ultrasounds (low module of elasticity and high quantity of air trapped inside) usually exhibited by the materials included in pharmaceutical formulations, a fast decay of ultrasonic energy to thermal energy is obtained. This process has been studied, monitoring the temperature inside the compression chamber by means of a thermistor. In the studied mixtures [87, 88], a fast rise in temperature was obtained in tenths of a second followed by a relatively fast decrease (see Figure 47).

The peak temperature obtained for low ultrasonic energy (25 J) is below 80 °C, whereas for high energies (125–150 J) it is above 140 °C. In mixtures of ketoprofen with acrylic polymers [90], the increase in temperature was slightly lower. In this respect it must be mentioned that a recent modification of the ultrasound-assisted tableting machine that involves the suppression of Teflon isolators in contact with the powder must result in a faster decrease in temperature inside the compression chamber. Thermal effects can cause the total or partial fusion of some components of the formulation. Nevertheless, in the assayed controlled-release formulations, the components are usually below its melting points.

Plastic Deformation Plastic deformation results from the combination of thermal and mechanical effects. The thermoplastic excipient was subjected to a temperature above its glass transition temperature (T_g) and to a high-frequency mechanical pressure that can avoid the elastic recovery of the material.

Sintering The combination of temperature, pressure, and friction effects can result in the sintering of particles, so that the limits between them are no longer evident [46, 87].

Recent studies [91–93] have shown that, for one component of the system undergoing thermoplastic deformation, the continuum percolation model can be used to predict the changes in the system with respect to a traditional pharmaceutical

FIGURE 48 SEM micrograph of matrix tablet containing potassium chloride as drug model and commercial acrylic–metacrylic copolymer. The white KCI particles are surrounded by an almost continuous dark gray mass of excipient. (Courtesy of M. Millán.)

dosage form. The continuum percolation model dispenses with the existence of a regular lattice underlying the system; therefore, the substance is not distributed into discrete lattice sites. This model deals with the volume ratio of each component and a continuum distribution function. The volume ratio is expressed as a space–occupation probability to describe the behavior of the substance [94, 95].

The continuum percolation model predicts an excipient percolation threshold around 16% v/v. This can explain the important decrease in the critical point corresponding to the excipient percolation threshold, a critical point that governs the mechanical and release properties of the matrix.

Ultrasound compaction lowers the percolation threshold of the thermoplastic excipient, resulting in a drastic reduction (about 50%) in the amount of matrix-forming excipient [98] needed to obtain the controlled-release system as well as in a better control of the drug release. The structure of the excipient inside the US-tablets does not correspond to a particulate system but to an almost continuous medium; therefore, there is no an excipient particle size inside these matrices (see Figure 48). Consequently, the percolation threshold of the active agent is higher than in traditional tablets. The insoluble excipient almost surrounds the active agent particles, slowing down the contact with the dissolution medium.

These facts can involve important advantages for the pharmaceutical industry, such as the preparation of controlled-release inert matrices containing high drug doses, with very little increase in the weight of the system. This fact is especially interesting when a high drug dose has to be included in the dosage form, as frequently occurs in controlled-release systems.

On the other hand, application of ultrasonic energy results in an increase in the temperature of the die during the compaction process. The consequences of this fact should be taken into account and cannot be neglected in the case of thermolabile drugs and/or excipients [87, 88].

Further research is needed in the area of ultrasound-assisted compression of pharmaceutical powders, including a higher number of drugs and excipients.

REFERENCES

1. Unckel, H. (1945), Processes during compression of metal powders, *Archiv fuer das Eisenhuettenwesen*, 18, 161–167.
2. Toor, H. L., and Eagleton, S. D. (1956), Plug flow and lubrication of polymer particles, *Ind. Eng. Chem.*, 48, 1825–1830.
3. Ritschel, W. A., and Bauer-Brandl, A. (2002), *Die Tablette, Handbuch der Entwicklung, Herstellung und Qualitätssicherung*, Editio Cantor Verlag Aulendorf, (Württ), Germany 505.
4. De Boer, A. H., Bolhuis, G. K., and Lerk, C. F., (1978), Bonding characteristics by scanning electron microscopy of powders mixed with magnesium stearate, *Powder Technol.*, 20, 75–82.
5. Nyström, C., Alderborn, G., Duberg, M., and Karehill, P-G. (1993), Bonding surface area and bonding mechanism—Two important factors for the understanding of powder compactability, *Drug Dev. Ind. Pharm.*, 19, 2143–2196.
6. Mattsson, S., and Nyström, C. (2000), Evaluation of strength-enhancing factors of a ductile binder in direct compression of sodium bicarbonate and calcium carbonate powders, *Eur. J. Pharm. Sci.*, 10, 53–66.
7. Rudnic, E. M., and Kottke, M. K. (1999), Tablet dosage forms, in Banker, G. S., and Rhodes, C. T., Eds., *Modern Pharmaceutics*, 3rd rev. and expanded eds., Marcel Dekker, New York, pp. 333–394.
8. Pesonen, T., Paronen, P., and Ketolainen, J. (1989), Disintegrant properties of an agglomerated cellulose powder, *Int. J. Pharm.*, 57, 139–147.
9. Rudnic, E. M., and Kottke, M. K., Tablet dosage forms. In Banker GS, Rhodes CT Eds., *Modern Pharmaceutics* 3rd eds. New York: Marcel Dekker; 1999, pp 333–394.
10. Westerberg, M., and Nystrom, C. (1991), Physicochemical aspects of drug release. XII. The effect of some carrier particle properties and lubricant admixture on drug dissolution from tableted ordered mixtures, *Int. J. Pharm.*, 69, 129–141.
11. Bolhuis, G. K., and Hölzer, A. W., Lubricant sensitivity. VIII. Effect of third components on the film formation of lubricants. In Alberborn G, Nyström C Eds. *Pharmaceutical Powder Compactation Technology*, New York: Marcel Dekker; 1996, pp 550–555.
12. Johansson, M. E. (1984), Granular magnesium stearate as a lubricant in tablet formulations, *Int. J. Pharm.*, 21, 307–315.
13. Carson, J. W. (1998 July), Overcoming particle segregation in the pharmaceutical and cosmetic industries, paper presented at Interphex USA, New York.
14. Vila-Jato, J. L., Remuñán, M. C., Seijo, B., and Torres, D. (1997), Nuevas formas de administración de medicamentos, in Vila-Jato, J. L., Ed., *Tecnología Farmacéutica II: Formas Farmacéuticas*, Editorial Síntesis, Madrid.
15. Tahara, K., Yamamoto, K., and Nishihata, T. (1995), Overall mechanism behind matrix sustained release (SR) tablets prepared with hydroxypropyl methylcellulose 2910, *J. Controlled Release*, 35, 59–66.
16. Soares, L. A. L., et al. (2005), Dry granulation and compression of spray dried plant extracts, *AAPS PharmSciTech*, 6, E359–E366.
17. Núñez Sellés, A. J., Capote, H. R., Agüero, J., Garrido, G., Delgado, R., Martinez, G., Leon, O. S., and Morales, M. (2000), New antioxidant product derived from *Mangifera indica* L, *Abstracts of Papers*, paper presented at the 220th National Meeting of the American Chemical Society, New Orleans, LA, American Chemical Society, Washington, DC.

18. Sánchez, G. M., Re, L., Giuliani, A., Núnez-Sellés, A. J., Davison, G. P., and León-Fernández, O. S. (2000), Protective effects of *Magnifera indica* L. extract, mangiferin and selected antioxidents against TPA-induced biomolecules oxidation and peritoneal macrophage activation in mice, *Pharmacol. Res.*, 42, 565–573.

19. Yoshimi, N., Matsunaga, K., Katayama, M., Yamada, Y., Kuno, T., Qiao, Z., Hara, A., Yamahara, J., and Mori, H. (2001), The inhibitory effects of mangiferin, a naturally occurring glucosylxanthone, in bowel carcinogenesis of male F344 rats, *Cancer Lett*, 163, 163–170.

20. Ichiki, H., Miura, T., Masayoshi, I., Ishihara, E., Komatsu, Y., Tanigawa, K., Ichiki, H., and Okada, M. (1998). New antidiabetic compounds, mangiferin and its glucoside, *Biol. Pharm. Bull.*, 21, 1389.

21. Li, H., Miyahara, T., Tezuka, Y., Namba, T., Nemoto, N., Tonami, S., Seto, H., Tada, T., and Kadota, S. (1998), The effect of Kampo formulae of bone resorption in vitro and in vivo. I. Active constituents of Tsu-Kan-gan. *Biol. Pharm. Bull.*, 21, 1322.

22. Guha, S., Ghosal, S., and Chattopadhyay, U. (1996), Antitumor, immunomodulatory and anti-HIV effect of mangiferin, a naturally occurring glucosylxan thone, *Chemotherapy*, 42, 443–451.

23. Ghosal, S., Rao, G., Saravanan, V., Misra, N., and Rana, D. (1996), A plausible chemical mechanism of the bioactivities of mangiferin, *Indian J. Chem.*, 35B(6), 561–566.

24. Born, M., Carrupt, P. A., Zini, R., Bree, T., Tillement, J. P., Hostettmann, K., and Testa, B. (1996), Electrochemical behaviour and antioxidant activity of some natural polyphenols, *Helv. Chim. Acta*, 79, 1147.

25. Castellanos Gil, E., Iraizoz Colarte, A., Bataille, B., Pedráz, J. L., and Heinämäki, J. (2006), Development and optimization of a novel sustained-release dextran tablet formulation for propranolol hydrochloride, *Int. J. Pharm.*, 317, 32–39.

26. Castellanos Gil, E., Bataille, B., Iraizoz Colarte, A., Delarbre, J. L., El Ghzaoui, A., and Durand, D. (2008), A sugar cane native dextran as an innovative functional excipient for the development of pharmaceutical tablets. *Eur. J. Pharm. Biopharm* 68 (2), 319–329.

27. Jamzad, S., Tutunji, L., and Fassihi, R. (2005), Analysis of macromolecular changes and drug release from hydrophilic matrix systems, *Int. J. Pharm.*, 292, 75–85.

28. Reynolds, T. D., Gehrke, S. H., Hussain, A. S., and Shenouda, L. S. (1998), Polymer erosion and drug release characterization of hydroxypropyl methylcellulose matrices. *J. Pharm. Sci.*, 87, 1115–1123.

29. Miranda, A., Millán, M., and Caraballo, I. (2006), Study of the critical points in Lobenzarit disodium hydrophilic matrices for controlled drug delivery, *Chem. Pharm. Bull.* 54, 598–602.

30. Anonymous (1999, July), Note for guidance on quality of modified release products. Oral dosage forms and transdermal dosage forms, section 1 (quality), The European Agency for the Evaluation of Medicinal Products Human Medicines Evaluation (EMEA), London.

31. Kincl, M., Turk, S., and Vrecer, F. (2005), Application of experimental design methodology (DOE) in development and optimization of drug release method, *Int. J. Pharm.*, 291, 39–49.

32. Eriksson, L., Johansson, E., and Wikstrom, C. (1998), Mixture design—Design generation, PLS analysis, and model usage, *Chemometr. Intell. Lab. Syst.*, 43, 1–24.

33. Gooding, O. W. (2004), Process optimization using combinatorial design principles: Parallel synthesis and design of experiment methods, *Curr. Opin. Chem. Biol.*, 8, 297–304.

34. Congreve, M. S., and Jamieson, C. (2002), High-throughput analytical techniques for reaction optimization, *Drug Discovery Today*, 2, 139–142.
35. I HOLLAND (2006), ITab software, available: http://www.in-pharmatechnologist.com.
36. Leuenberger, H., Rohera, B. D., and Haas, C. (1987), Percolation theory—A novel approach to solid dosage form design, *Int. J. Pharm.*, 38, 109–115.
37. Holman, L. E., and Leuenberger, H. (1988), The relationship between solid fraction and mechanical properties of compacts—The percolation theory model approach, *Int. J. Pharm.*, 46, 35–44.
38. Stauffer, D., and Aharony, A. (1992), *Introduction to Percolation Theory*, 2nd ed., Burgess Science, London.
39. Caraballo, I., Fernandez Arévalo, M., Millán, M., Rabasco, A. M., and Leuenberger, H. (1996), Study of percolation thresholds in ternary tablets, *Int. J. Pharm.* 139, 177–186.
40. Bonny, J. D., and Leuenberger, H. (1991), Matrix type controlled release systems: I. Effect of percolation on drug dissolution kinetics, *Pharm. Acta Helv.*, 66, 160–164.
41. Bonny, J. D., and Leuenberger, H. (1993), Matrix type controlled release systems. II. Percolation effects in non-swellable matrices, *Pharm. Acta Helv.*, 68, 25–33.
42. Blattner, D., Kolb, M., and Leuenberger, H. (1990), Percolation theory and compactibility of binary powder systems, *Pharm. Res.*, 7, 113–117.
43. Caraballo, I., Fernandez-Arévalo, M., Holgado, M. A., and Rabasco, A. M. (1993), Percolation theory: Application to the study of the release behaviour from inert matrix system, *Int. J. Pharm.*, 96, 175–181.
44. Caraballo, I., Millan, M., and Rabasco, A. M. (1996), Relationship between drug percolation threshold and particle size in matrix tablets, *Pharm. Res.*, 13(3), 387–390.
45. Millán, M. (1998 Apr.), Estudio del Umbral de Percolación para la optimización de matrices inertes de liberación controlada, Ph.D. thesis, University of Seville, Seville, Spain.
46. Millán, M., Caraballo, I., and Rabasco, A. M. (1998), The role of the drug/excipient particle size ratio in the percolation model for tablets, *Pharm. Res.*, 15(2), 216–220.
47. Kuentz, M. T., Leuenberger, H., and Kolb, M. (1999), Fracture in disordered media and tensile strength of microcrystalline cellulose tablets at low relative densities, *Int. J. Pharm.*, 182, 243–255.
48. Kuentz, M. T., and Leuenberger, H. (2000), A new theoretical approach to tablet strength of a binary mixture consisting of a well and a poorly compactable substance, *Eur. J. Pharm. Biopharm.*, 49, 151–159.
49. Ramírez, N., Melgoza, L. M., Kuentz, M., Sandoval, H., and Caraballo, I. (2004), Comparison of different mathematical models for the tensile strength–relative density profiles of binary tablets, *Eur. J. Pharm. Sci.*, 22(1), 19–23.
50. Bourquin, J., Schmidli, H., van Hoogevest, P., and Leuenberger, H. (1997), Basic concepts of artificial neural networks (ANN) modeling in the application to pharmaceutical development, *Pharm. Dev. Technol.*, 2, 95–109.
51. Plumb, A. P., Rowe, R. C., York, P., and Brown, M. (2005), Optimisation of the predictive ability of artificial neural network (ANN) models: A comparison of three ANN programs and four classes of training algorithm, *Eur. J. Pharm. Sci.*, 25(4–5), 395–405.
52. Gans, E. H., and Chavkin, L. (1954), The use of polyethylene glycol in tablet coating, *Pharm. Ass. Sc. (Baltin)* Ed., 43, 483–485.
53. Porter, S. C., and Bruno, C. H. (1990), Coating of pharmaceutical solid-dosage forms, in Liebermann, H. A., Lachmann, L., and Schwarz, J. B., Eds., *Pharmaceutical Dosage Forms: Tablets*, Vol. 3, 2nd eds., Marcel Dekker, New York, Chapter 2.

54. Hess, H., and Janssen, H. J. (1969), Lacquered tablets and film coated tablets, *Pharm. Acta Helv.*, 44, 581.
55. Bindschädler, C., Gurny, R., and Dolker, B. (1983), Theoretical concepts regarding the formation of films from aqueous microdispersions and application to coating, Labo-Pharma, *Probl. Tech.*, 31, 389–394.
56. Radebaugh, G. W. (1990–1996), Film coating and film-forming materials: Evaluation, in Swarbrick, J., and Boylan, J. C., Eds., *Encyclopedia of Pharmaceutical Technology*, Vol. 6, Marcel Dekker, New York.
57. Moe, E. (1945), *Overtraeking og Dragering*, Dansk Farmaceutfoeningsverl, Copenhagen.
58. Hasegava, K., Ida, T., Saika, K., and Utsumi, I. (1965), Suitable dimensions of tablets for coating, *Yakugaku Zasshi*, 85, 796.
59. Selmeczi, B., and Smogyi, J. (1996), Suitable dimensions of tablets for coating, *Pharmacie*, 21, 604.
60. Bauer, K. H., Frömming, K. H., and Führer, C. (1993), *Pharmazeutische Technologie [Pharmaceutical Technology]*, 4th ed., Thieme, Stuttgart.
61. Bauer, K. H., Lehmann, K., Osterwald, H. P., and Rothgang, G. (1998), *Coated Pharmaceutical Dosage Forms*, Medpharm Scientific Publisher, Stuttgart.
62. Bühler, V. (2003, Sept.), *Kollidon. Polyvinylpyrrolidone for the Pharmaceutical Industry*, 7th ed., BASF Pharma Ingredients, Ludwigshafen, Germany.
63. U.S. Pharmacopeial Forum (2002, May/June), Vol. 28. No. 3, 948–951.
64. Heng, P. W. S., Wan, L. S. C., and Tan, Y. T. F. (1996), Relationship between aggregation of HPMC coated spheroids and tackiness/viscosity/additives of the coating formulations, *Int. J. Pharm.*, 138, 57–66.
65. Tan, Y. T. F., Wan, L. S. C., and Heng, P. W. S. (1998), Evaluation of adhesion strength measurement for predicting aggregation propensity during fluidized bed coating, *S.T.P. Pharm. Sci.*, 8(3) 149–153.
66. Bühler, V. (2001), *Generic Drug Formulation*, 4th ed., BASF Fine Chemicals, Ludwigshafen, Germany.
67. Rowe, R. C., and Forse, S. F. (1983), Pitting—A defect on film-coated tablets, *Int. J. Pharm.*, 17(2–3), 347–349.
68. Melgoza, L. M., Caraballo, I., Alvarez-Fuentes, J., Millan, M., and Rabasco, A. M. (1998), Study of morphine hydrochloride percolation threshold in Eudragit® RS-PM matrices, *Int. J. Pharm.*, 170, 169–177.
69. Caraballo, I., Melgoza, L. M., Alvarez-Fuentes, J., Soriano, M. C., and Rabasco, A. M. (1999), Design of controlled release inert matrices of naltrexone hydrochloride based on percolation concepts, *Int. J. Pharm.*, 181, 23–30.
70. Melgoza, L. M., Rabasco, A. M., Sandoval, H., and Caraballo, I. (2001), Estimation of the percolation thresholds in dextromethorphan hydrobromide matrices, *Eur. J. Pharm. Sci.*, 12(4), 453–459.
71. Boza, A., Blanquero, R., Millán, M., and Caraballo, I. (2004), Application of a new mathematical method for the estimation of the mean surface area to calculate the percolation threshold of Lobenzarit disodium salt in controlled release matrices, *Chem. Pharm. Bull.*, 52(7), 797–801.
72. Reisine, T., and Pasternak, G. (1996), Analgésicos opioides y sus antagonistas, in Goodman, A., Hardman, J. G., Limbird, L. E., Molinoff, P. B., and Ruddon, R. W. Eds., *Las Bases Farmacológicas de la Terapéutica*, Vol. I, McGraw-Hill Interamericana, México, pp. 557–593.

73. Miranda, A., Millán, M., and Caraballo, I. (2006), Study of the critical points of HPMC hydrophilic matrices for controlled drug delivery, *Int. J. Pharm.*, 311, 75–81.
74. Miranda, A., Millán, M., and Caraballo, I. (2006), Study of the critical points in Lobenzarit disodium hydrophilic matrices for controlled drug delivery, *Chem. Pharm. Bull.*, 54, 598–602.
75. Tu, J., Wang, L., Yang, J., and Li, X. (2001), Formulation and pharmacokinetic studies of a cyclovir controlled-release capsules, *Drug Dev. Ind. Pharm.*, 27, 687–692.
76. Wangstaff, J. A., Faulds, D., and Goa, L. K. (1994), Acyclovir: A reappraisal of its antiviral activity pharmacokinetic properties and therapeutic efficacy, *Drugs*, 47(1), 153–205.
77. Higuchi, T. (1963), Mechanism of sustained-action medication. Theoretical analysis of rate of release of solid drugs dispersed in solid matrices, *J. Pharm. Sci.*, 52, 1145–1149.
78. Korsmeyer, R. W., Gurny, R., Doelker, E., Buri, P., and Peppas, N. A. (1983), Mechanisms of solute release from porous hydrophilic polymers, *Int. J. Pharm.*, 15, 25–35.
79. Peppas, N. A., and Sahlin, J. J. (1989), A simple equation for the description of solute release. 3. Coupling of diffusion and relaxation, *Int. J. Pharm.*, 57, 169–172.
80. Ford, J. L., Mitchell, K., Rowe, P., Armstrong, D. J., Elliot, P. N. C., Rostron, C., and Hogan, J. E. (1991), Mathematical modelling of drug release from hydroxypropylmethylcelluose matrices: Effect of temperature, *Int. J. Pharm.*, 71, 95–104.
81. Fuertes, I., Miranda, A., Millán, M., and Caraballo, I. (2006), Estimation of the percolation thresholds in acyclovir hydrophilic matrix tablets, *Eur. J. Pharm. Biopharm.*, 64(3), 336–342.
82. Leuenberger, H., and Rohera, B. D. (1986), Fundamentals of powder compression. II. The compression of binary powder mixtures, *Pharm. Res.*, 3, 65–74.
83. Levina, M., Rubinstein, M. H., and Rajabi-Siahboomi, A. R. (2000), Principles and application of ultrasound in pharmaceutical powder compression, *Pharm. Res.*, 17(3), 257–265.
84. Kromp, W., Trimmel, P., Prinz, F. B., and Williams, J. C. (1985), Vibratory compaction of metal powders, Mod. *Dev. Powder Metall.*, 15, 131–141.
85. Gueret, J.-L. H. (1993), Process for the compaction of a powder mixture providing an absorbent or partially friable compact product and the product obtained by this process, U.S. Patent 5,211,892.
86. Motta, G. (1994), Process for preparing controlled release pharmaceutical forms and the forms thus obtained, International Patent WO 94/14421.
87. Rodriguez, L., Cini, M., Cavallari, C., Passerini, N., Saettone, M. F., Fini, A., and Caputo, O. (1997), Physico-chemical properties of some materials compacted using an ultrasound-assisted tableting machine, in Rubinstein, M., Ed., *Proceedings 16th Pharmaceutical Technology Conference*, Athens, Greece, Vol. I, pp. 267–278.
88. Rodriguez, L., Cini, M., Cavallari, C., Passerini, N., Fabrizio Saettone, M., Fini, A., and Caputo, O. (1998), Evaluation of theophylline tablets compacted by means of novel ultrasound-assisted apparatus, *Int. J. Pharm.*, 170, 201–208.
89. Fini, A., Fernández-Hervás, M. J., Holgado, M. A., Rodriguez, L., Cavallari, C., Passerini, N., and Caputo, O. (1997), Fractal analysis of beta-cyclodextrin-indomethacin particles compacted by ultrasound, *J. Pharm. Sci.*, 86, 1303–1309.
90. Sancin, P., Caputo, O., Cavallari, C., Passerini, N., Rodriguez, L., Cini, M., and Fini, A. (1999), Effects of ultrasound-assisted compaction on ketoprogen/Eudragit S100 mixtures, *Eur. J. Pharm. Sci.*, 7, 207–213.
91. Caraballo, I., Millan, M., Fini, A., Rodriguez, L., and Cavallari, C. (2000), Percolation thresholds in ultrasound compacted tablets, *J. Controlled Release*, 69, 345–355.

92. Millán, M., and Caraballo, I. (2006), Effect of drug particle size in ultrasound compacted tablets: Continuum percolation model approach, *Int. J. Pharm.*, 310, 168–174.

93. Caraballo, I. (2001), Improvement on the release control using ultrasound assisted compression, Eurand Award 2001 "for Outstanding Novel Research in Oral Drug Delivery," organized by Eurand and the Controlled Release Society, 28th Meeting of the Controlled Release Society, San Diego, CA, June 26.

94. Efros, A. L. (1994), *Física y Geometría del desorden*, Hayka, Moscow, pp. 144–147.

95. Kuentz, M. T., and Leuenberger, H. (1998), Modified Young's modulus of microcrystalline cellulose tablets and the directed continuum percolation model, *Pharm. Dev. Technol.*, 3(1), 1–7.

FURTHER READING

A. BASF Pharma ingredients generic drug formulations, 2004, 2005.

B. Baveja, S. K., Ranga Roa, K. V., and Padmalatha Devi, K. (1987), Zero-order release hydrophilic matrix tablets of β-adrenergic blockers, *Int. J. Pharm.*, 40, 223–234.

C. Bettini, R., Colombo, P., Massimo, G., Catellani, P. L., and Vitali, T. (1994), Swelling and drug release hydrogel matrices: Polymer viscosity and matrix porosity effect, *Eur. J. Pharm. Sci.*, 2, 213–219.

D. Campos-Aldrete, M. E., and Villafuerte-Robles, L. (1997), Influence of the viscosity grade and the particle size of HPMC on metronidazole release from matrix tablet, *Eur. J. Pharm. Biopharm.*, 43, 173–178.

E. Ferrero, C., Muñoz-Ruiz, A., and Jiménez-Castellanos, M. R. (2000), Fronts movements as a useful tool for hydrophilic matrix release mechanism elucidation, *Int. J. Pharm.*, 202, 21–28.

F. U.S. Pharmacopoeia 25 (2002), National Formulary 20, U.S. Pharmacopeial Convention, Rockville, MD.

G. Vázquez, M. J., Peres-Marcos, B., Gómez-Amoza, J. L., Martínez-Pacheco, R., Souto, C., and Concheiro, A. (1992), Influence of technological variables on release of drugs from hydrophilic matrices, *Drug Dev. Ind. Pharm.*, 18, 1355–1375.

H. Velasco, M. V., Ford, J. L., Rowe, P., and Rajabi-Siahboomi, A. R. (1999), Influence of drug: hydroxypropylmethylcellulose ratio, drug and polymer particle size and compression force on the release of diclofenac sodium from HPMC tablets, *J. Controlled Release*, 57, 75–85.

I. Tu, J., Wang, L., Yang, J., Fei, H., and Li, X. (2001), Formulation and pharmacokinetic studies of acyclovir controlled-release capsules, *Drug Dev. Ind. Pharm.*, 27(7), 687–692.

6.4

TABLET PRODUCTION SYSTEMS

KATHARINA M. PICKER-FREYER
Martin-Luther-University Halle-Wittenberg, Institute of Pharmacy, Division of Pharmaceutics and Biopharmaceutics, Halle/Saale, Germany

Contents

Pharmaceutical Manufacturing Handbook: Production and Processes, edited by Shayne Cox Gad

6.4.1 INTRODUCTION

Tablet production systems can be defined as all machines which are able to produce tablets. They include tableting machines for production and research as well as tableting machine simulators, which are able to mimic the production processes of tableting machines of different size and velocity in order to facilitate scale-up.

Tablets have been produced for more than 150 years. The first tableting machine, developed by Brockedon in 1972 [1], was a manually operated single-punch machine. Currently high-speed tableting machines can produce more than a million tablets per hour.

However, the amount of drug in the early steps of development still makes the use of small tableting machines necessary. Thus, before final production a scale-up from small machines useful for the production of single tablets to high-speed machines is necessary. Since this scale-up is still based on "trial and error" [2], tableting machine simulators can be used to simulate different steps in the scale-up.

In research tableting machine simulators are often used since they allow precise and well-adjustable measurements during tableting with only a small amount of material.

The aim of this chapter is to give an overview of the different techniques used for production and research and further to show the possibilities, that instrumentation of tablet production systems gives in order to analyze the tableting process. The knowledge derived can be used for formulation development as well as to facilitate tablet production and scale-up. Applied techniques which are necessary for production in a good manufacturing practice (GMP) environment will also be discussed.

6.4.2 PHYSICS OF TABLET FORMATION

Tablets can be defined as two-phase systems which consist of a solid phase (the compressed powder) and the gaseous phase (the air). The solid phase forms a coherent network inside the tablet and a defined form with precise outer dimensions results. Here we define the *process of tablet formation* as the transformation of the powder, a noncoherent solid phase, into the compact tablet, and this process lasts until no further changes are induced by tableting. "Tableting" is only part of the process of tablet formation. It is defined as the process by which the powder is transformed into a coherent form, the tablet. In conclusion tablet formation is the result of tableting and all the changes induced by tableting [3, 4].

6.4.2.1 Tableting Process

During the tableting, the powder in the die of the tableting machine is transformed by the influence of lower and upper punches into a coherent form—"the tablet" (Figure 1). During tableting different processes occur which are responsible for the cohesion of the tablet [5, 6]. First the particles are pushed together and reoriented in the die until they have arranged in the closest packing. This process is followed by elastic and/or plastic deformation of the particles, as can be observed by confocal laser microscopy [7]. Some materials show brittle fracture. During brittle fracture new particles are produced which can again deform or fracture. Under the influence of the applied forces the particles approach each other up to bonding [8]. The bonded particle collectives continue to deform. The mechanism of deformation and bonding is dependent on material properties and process conditions. Important bonding mechanisms are van der Waals forces, surface films, liquid or solid bridges, and mechanical interlocking [9].

The nature of the resulting bonds [9, 10] and the extent of particle approach [11] determine the cohesion of the final tablet and are responsible for the compactibility of the materials [8]. Further, the formation of bonds occurs not only during the

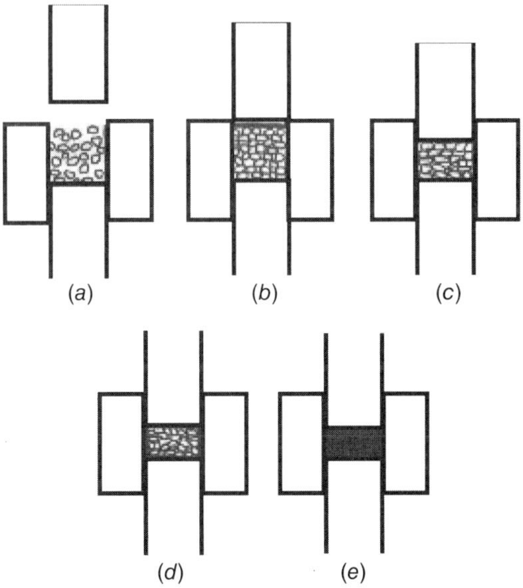

FIGURE 1 Densification of powder bed: (*a*) particles in die; (*b*) reorientation of particles; (*c*, *d*) particle deformation and fracture; (*e*) tablet in die.

compression phase but also during decompression of the tablet [12, 13]. Hence Leuenberger and Ineichen [14] describe tableting as a percolation phenomenon.

Looking closer at the tableting process it becomes clear that the applied forces during tableting are not only the result of machine movement. When punches are moving in a machine without material, no force can be measured. The force only develops when the punches come into contact with the powder bed. The materials resist pressure deformation, and while the punches are moving, a counterforce builds up which is the measured force at the punches. In conclusion punch forces are determined by the material and as a result materials can be characterized by the measured punch forces.

6.4.2.2 Final Formation of Tablet

After ejection of the tablet from the die of the tableting machine, the tableting process is finished. However, decompression continues and thus the tablet formation process is not yet finished. The remaining stress inside the tablet can resolve and produce changes in the structure of the tablet. The tablet partially releases the energy gained during tableting. The tablet often shows the occurrence of relaxation, which is called elastic recovery or relaxation of the tablet [15, 16]. This relaxation is primarily axial relaxation and to a small part radial relaxation, as Newton and Rowley [17] figured out. Hiestand [15] called this event "compression born repulsion."

Van der Voort Maarshalk [18] showed that the axial relaxation of the tablet, that is, the change in tablet height, is dependent on the stored energy and hinders the formation of bonds. The elastic recovery [19–21] can be measured by measuring tablet height at different times after tableting. Simultaneously changes in the inner structure of the tablet can occur, for example, shifts of the crystal planes [22, 23]. The fusion of amorphous regions and the recrystallization of amorphous parts are assumed [24–26]. These structural changes are induced by the tableting process and also become visible when transformation of drugs into other polymorphic forms occurs [27–29]. However, even during final formation of the tablets, bonds are formed [15, 30, 31], and thus final tablet formation can be defined as a valuable part of the tablet formation process.

6.4.3 REQUIREMENTS FOR TABLET PRODUCTION SYSTEMS

Tablet production systems should be able to produce tablets in a reasonable time and without loss of material. Usually they consist of two punches and a die, as schematically shown in Figure 1. The material is placed into the die cavity, which is closed on the lower side by the lower punch. A tablet is formed when the powder is compressed by the punches as described above. The forces evolving to produce a tablet can be up to 80 kN for pharmaceutical purposes [32]; for the production of bigger tablets they are even higher. Usually they range between 10 and 30 kN. Thus punches and dies are produced from hardened steel.

To withstand the evolving forces, the die has to be fixed tightly in the machine, and for that purpose it is accurately fit in a die table consisting of steel. The die table is further friction locked to the machine frame.

Tablet production systems can be operated manually; however most of the systems nowadays operate automatically in order to produce a sufficient number of tablets. Further requirements depend on the type of machine, the production process, the operation mode of the machine, and the production rate of the machine.

6.4.4 TABLET MANUFACTURING PROCESS

The manufacturing process of tablets principally consists of three stages: filling the die with the powder, compressing the powder, and ejecting the tablet from the die.

6.4.4.1 Filling

Die filling a tableting machine is a volumetric process. In any case the die is filled to the upper edge of the die and the powder surface flushes with the surface of the die and the die table. The filling depth is determined by the height of the die cavity, the filling volume is determined by the diameter of the die hole and the filling depth. Both the diameter of the die hole and the filling depth are used to calculate the final filling volume.

Three different techniques exist for filling the die. The simplest is to use a filling shoe which moves back and forth over the die and fills the die to its upper edge. The surface is leveled off by the filling shoe.

Another possibility is a filling shoe which does not move. In this case the dies and the accessory lower punches move below the filling shoe and the accessory upper punches move above the filling shoe. In this case the filling volume ends at the upper edge of the die; however, to level off the surface, an additional scraper is necessary. In addition, many machines use the technique of lowering the lower punch and with it the powder bed before tableting [33]. This is most helpful to avoid dusting when the upper punch moves into the die.

The newest filling technology is to fill the die by centrifugal forces [34]. The material moves through specially shaped radial channels which approach the die from the side. In this case the powder volume is determined by the positions of the lower and upper punches in the die. However, for centrifugal filling an absolute must is a free-flowing powder. This is a great disadvantage for all non-free-flowing powders.

6.4.4.2 Compression

The compression event is the central stage of tablet production. Compression not only depends on the machine but to a great deal on the material properties of a tablet formulation. The principal stages of compression have been described above. Principally, from the machine manufacturing side two different possibilities of compression exist. Either the lower punch is closing the die from the bottom side and the upper punch moves downward for compression or both upper and lower punches move simultaneously toward each other and the powder is compressed from both sides. In the first case the surface hardness of the tablet is not the same on the upper and lower sides; in the second case the surface hardness is the same on the upper and lower sides.

As already described, the tablet is not completely formed during compression [35, 36]. When the punches leave each other, the tablet relaxes during decompression, further relaxing when one punch leaves the tablet and continuing relaxation after ejection from the die. The process is called decompression as long as a force is measurable. Afterward the process is called elastic recovery or relaxation of the tablet.

6.4.4.3 Ejection

During ejection usually the lower punch moves upward to eject the tablet from the die and the upper punch has already left the die when the process of ejection starts. Only one machine is presently on the market which ejects the tablets downward at the bottom of the die [34]. After ejection from the die the tablets are collected.

6.4.5 TABLETING MACHINES

There exist tableting machines which operate in a different manner and can produce in between one single tablet and a million tablets per hour. The orientation of the particles in the machine and their rearrangement, densification, and deformation depend not only on the material but also on the tableting machine used.

The first machines for the production of tablets were simple hydraulic or manually operated presses. Later, eccentric and rotary tableting machines were developed. Today, eccentric tableting machines are only used for research, in early development, or for special applications.

The newest development for production is a machine which fills the dies by centrifugal force [34]. It is a special rotary tableting machine. Another innovation is a special machine which operates by ultrasound [37].

In the following the most important machine types and their working principles will be described.

6.4.5.1 Single-Punch Tableting Machines

Single-punch machines were the first tableting machines used at the end of the nineteenth century. The upper punch is lowered by a lever arm on the powder bed in the die and by reciprocating this procedure single tablets can be produced. Another possibility is to lower the upper punch by means of a screw. These manually operated tableting machines are no longer used.

Two other types are still in usage for special purposes. Hydraulically operated machines are used to produce tablets, for example, for Fourier transform infrared (FTIR) spectroscopy. In this case the upper punch moves hydraulically onto the powder bed and the tablet is formed in the die. The other possibility is an eccentric tableting machine, which will be described in the following.

Eccentric Tableting Machines Eccentric tableting machines are still used in research, in early development, and for material characterization. They are the machines of choice when you want to use a single-punch tableting machine. As other single-punch machines, they work with one pair of punches. In principle, they have a mobile upper punch and a lower punch which is fixed during compression. The lower punch only moves for ejection of the tablet. The densification process is unilateral.

An eccentric valve is driven by a motor and this eccentric valve is responsible for the movement of the upper punch (Figure 2). The eccentric movement determines the operation and speed of the upper punch. The upper punch is driven by the eccentric valve into the die, which is closed on the bottom side by the lower punch, and the tablet is compressed in the die.

The punch forces for compression evolve due to contact with the powder. Thus the measured punch forces at the upper and lower punches result from the movement of the upper punch and the resistance of the powder bed toward deformation. From that it becomes clear that the measured upper punch force is usually higher than the lower punch force, and thus the upper punch force is the main force used for material characterization. The tablets show different hardness on upper and lower surfaces, as do all tablets produced by single-punch machines [38].

When the upper punch is lifted from the tablet surface, the lower punch moves the tablet upward for ejection out of the die. However, as is known nowadays, the lower punch also moves slightly before due to evolving forces.

In eccentric machines, die filling is performed by a filling shoe which moves back and forth above the die. When the upper punch moves upward above the die, the die is filled, and when the upper punch starts downward movement, the powder bed is leveled off. After ejection of the tablet the filling shoe pushes the tablet down from the die table. During this filling process with a moving filling shoe, demixing of the product mixture is more easily possible than with a fixed filling shoe. This is one major disadvantage of eccentric tableting machines.

The production rate varies usually between 10 and 60 tablets/min and is determined by the number of eccentric movements. An example of an eccentric tableting machine is given in Figure 3. Other machines are described by Ritschel and Bauer-Brandl [32].

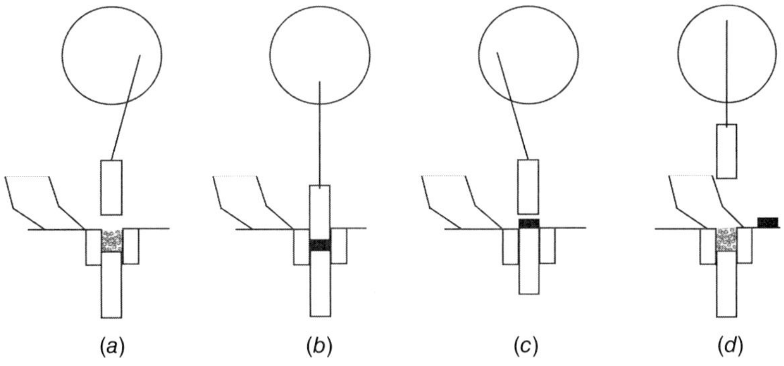

FIGURE 2 Operation of eccentric tableting machine: (*a*) filling; (*b*) compression; (*c*) ejection; (*d*) pushing from die table.

FIGURE 3 Example eccentric tableting machine. (Courtesy of Korsch XP1.)

6.4.5.2 Rotary Tableting Machines

Rotary tableting machines are commonly used for tablet production. The principle of all rotary machines is the same, with one exception, which will be discussed separately. According to Konkel and Mielck [39], the information gained with eccentric and rotary machines complement each other.

Rotary tableting machines work with a number of punch and die sets which move in a circle. The dies are fixed in a round die table and the die table circulates. Together with the dies the lower and upper punches circulate on tracks. The lower punches close the dies. The densification process is bilateral since both punches pass the compression wheels and the force is evolving on the upper as well as on the lower side of the powder bed. The produced tablets show the same hardness on the upper and lower surfaces.

There are different stages of tablet production which happen simultaneously for several tablets. The central stage of tableting occurs when the punches pass the upper and lower compression wheels and the compression wheels determine the downward movement of the upper punch and the upward movement of the lower punch: The tablet is formed by the resulting punch forces. The compression wheels can be positioned either by a flexible swing or more seldom by an eccentric valve [40].

In Figure 4 this principal stage and the other stages of tablet production are visible. Except for the compression stage the upper punch is always in an upward position. The upper punch is above the die and the lower punch closes the die, determining the filling depth; then the dies pass below the filling shoe and the powder flows into the die up to its upper edge where the powder bed is leveled off by a scraper. Whereas the upper punch is still in an upward position, the lower punch is lowered slightly on the track to keep the whole powder volume during compression and to prevent dusting [33]. Now the punches start to pass the compression wheels and the main compression event occurs. Forced by the wheels, both the upper and lower punches move toward each other, compress the tablet, and leave each other again. The upper punch lifts into an upward position and the lower punch moves upward to eject the tablet. After ejection the tablet passes a scraper and it is pushed down from the die table. An example of a rotary machine is given in Figure 5.

Additionally often rotary machines are equipped with precompression wheels (Figure 4). In this case the punches pass the precompression wheels before the main compression events starts. Precompression has the same stages of compression as the main compression, but the applied forces are lower. Precompression is deemed to be helpful to avoid, for example, dusting, capping, or lamination. After precompression only a lower main compression force is necessary.

The number of dies and complementary punches of a rotary machine can vary between one and up to hundred. The number of punches and the rotation speed

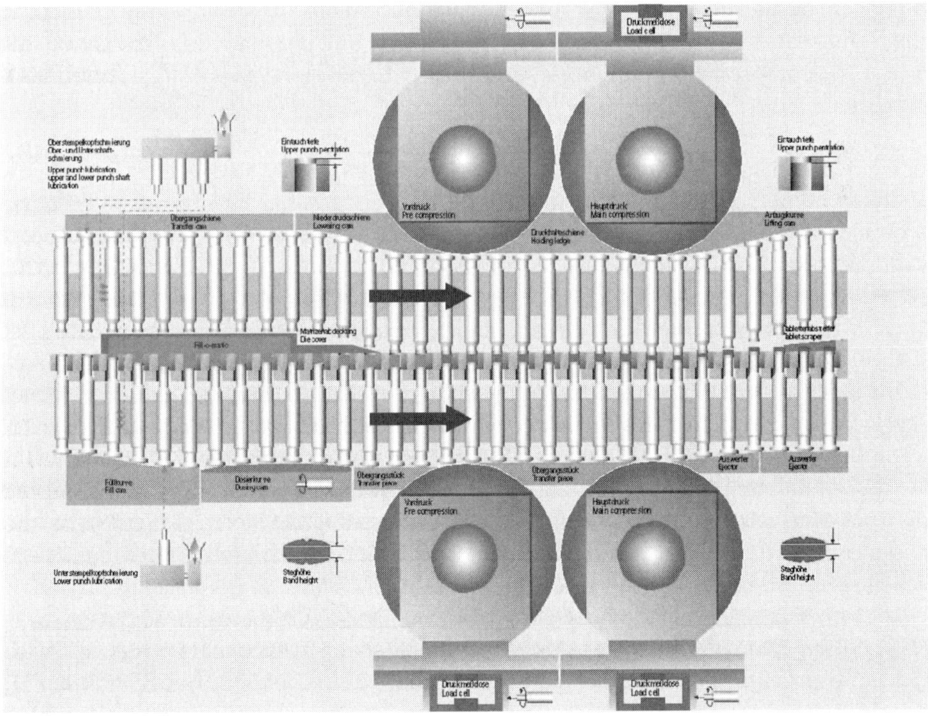

FIGURE 4 Operation of rotary tableting machine with precompression. (Courtesy of Fette.)

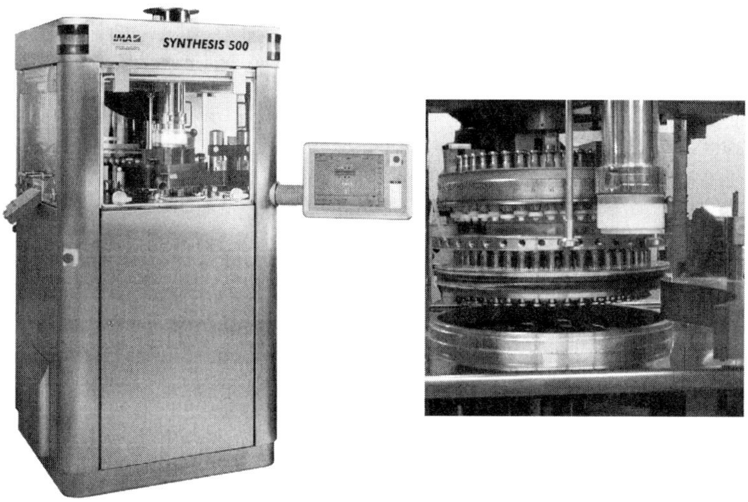

FIGURE 5 Example rotary tableting machine: left, machine view; right, detail view into compression chamber. (Courtesy of Kilian Synthesis 500.)

of the die table determine tablet production. Fast rotary machines have up to 120 rpm of the die table [32]. However, the number of tablets produced per hour not only is determined by the number of punches and the rotation speed but also is limited by the deformation properties of the tableted material. Materials need some time for deformation, and if the time during one compaction cycle is not sufficient for compression and compaction of the material, no tablets result from the process.

In order to increase the production rate of rotary tableting machines, double-sided rotary machines were build which possess two pairs of compression wheels and two filling shoes. Thus during one rotation double the number of tablets are produced compared with a one-sided rotary machine. If these machines are equipped with precompression wheels one machine contains four pairs of wheels.

Special Rotary Machines As already mentioned one special rotary machine works slightly different—called IMA Comprima (Figure 6) [34]. In this machine the material is filled by centrifugal force from the side directly into the die. The upper punch closes the die at the top and the lower punch closes the die at the bottom. However, when the given volume in the die is filled by the powder, both punches move downward until the die is completely closed. Then the compression process starts and the dies pass the compression wheels. After compression the tablet is ejected by the upper punch at the bottom of the die, contrary to all other rotary tableting machines which eject the tablet at the top of the die.

High-Speed Rotary Tableting Machines High-speed rotary machines work with the same principles as all other rotary machines. They possess a huge number of punch and die sets and often two filling stations. Another possibility is to use punch and die sets which are able to produce several tablets simultaneously. Special tooling can be used for this purpose; however this is not the subject of this chapter. As

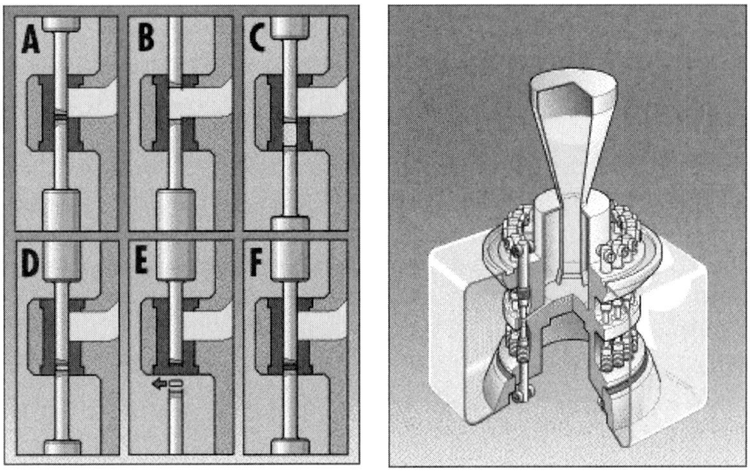

FIGURE 6 Working principle of Comprima tableting machine with centrifugal filling: left, operation mode; right, centrifugal filling. (Courtesy of IMA Comprima.)

already mentioned production speed depends to a great extent on product properties. Excellent powder flow is essential since the dies have to be filled completely. This is also most essential for the IMA Comprima, which works with centrifugal force for die filling. Often this machine cannot be used since powder flow properties are not sufficient. To improve powder flow on conventional rotary machines special filling devices have been developed (Section 6.4.11.1).

6.4.5.3 Application of Tableting Machines

In summary, the single-punch tableting machines still being used are mainly eccentric tableting machines (mostly for research), whereas rotary machines with different production output are predominantly used for production; and for rotary machines in most cases it is not machine speed that determines the production rate but material flow and compression properties.

6.4.6 TABLETING MACHINE SIMULATORS (COMPACTION SIMULATORS)

Tableting machine simulators [41–48] have been developed in order to mimic tablet production systems with a very small amount of powder. Similar to eccentric tableting machines, tableting machine simulators use one pair of punches. Working only with a single pair of punches reduces the consumption of tableting materials and facilitates instrumentation for displacement measurement.

6.4.6.1 Hydraulic Compaction Simulators

The first compaction simulators developed were hydraulic [41–45]. The hydraulic system is electronically controlled. An example is given in Figure 7. Either compression force cycles or movement of the punches was freely adjustable. This allowed

FIGURE 7 Example hydraulically working compaction simulator: left, machine view; right, detail view into compression chamber. (ESH compaction simulator, Courtesy of Huxley Bertram.)

much variation and the primary aim was to mimic the densification process of a rotary machine and the mechanical factors influencing it.

For example, theoretically, from machine geometries, the force–time profile of a rotary tableting machine [49–52] can be deduced and calculated and the data are programmed into the compaction simulator. However, the force–time profile of a tableting machine could not be calculated. Too many factors influence the measured force, for example, the tableted material, the geometries of the machine, the machine wear time, tableting speed, and tableting tools. Similarly, the displacement–time profile of a tableting machine, especially a rotary tableting machine, is very difficult to calculate. It has been shown that calculation from machine geometries is only possible to a certain extent. Mainly the mechanics of a tableting machine cannot be completely simulated [53, 54]. Thus either an approximated displacement–time profile can be used for programming the compaction simulator or approximation of real punch movement is only possible using recorded data from real tableting machines.

Thus the simulation of tableting machines needs much effort and a real simulation is almost impossible because of the hydraulic control. Further the filling process of rotary tableting machines cannot be simulated since die filling is usually processed by a filling shoe moving forth and back.

However, hydraulical compaction simulators are still used in research for basic material characterization. They show the advantage of controlling speed exactly and of using low and high punch travel speeds, between 10 and 300 mm/s. Mostly a simple displacement profile is used for characterization (e.g., a saw tooth or a sine wave profile), and the evolving forces at the lower and upper punches are measured. Further the speed of the punches can be controlled separately and both punches move freely and independently from each other. Time intervals in which the punches stand still can be freely set. Thus lots of freedom for material characterization is possible and these compaction simulators are important tools.

Another advantage of compaction simulators is that only small amounts of material are necessary to produce a tablet. One single tablet can be produced at low as well as at high speed of the punches. This is important in order to evaluate deformation properties of a formulation already in early dosage form development when only small quantities of the drug substance are available. Since nowadays the timelines for production of a new medicine are tight, this advantage of compaction simulators becomes more and more important.

6.4.6.2 Mechanical Compaction Simulators

More recently mechanical compaction simulators have been developed. The first was the linear mechanical rotary tableting machine simulator Presster (Figure 8), which was introduced in 1998 [55, 56]. It can mimic the mechanics of different rotary tableting machines and is called a linear rotary tableting machine replicator. The name Presster was combined from press and tester.

A single pair of punches moves linearly forth and back on a lower and an upper punch track. For tableting the punches pass the compression wheels which are equivalent in dimensions to those of rotary tableting machines used in practice. These compression wheels are exchangeable. Different machines are simulated by exchanging them.

The machine speed can be varied and different tableting machines are simulated by using similar dwell times between 5 and 80 ms. Special tests exhibited that rotary tableting machines can be simulated with a precision of 1–5% [57, 58]. One major disadvantage of the Presster is that it works with a moving filling shoe, and thus the filling process of rotary machine cannot be simulated.

In addition, the present model of the Presster possesses precompression wheels and thus, besides compression, precompression can be simulated. This is import

FIGURE 8 Detail of Presster. (Courtesy of MCC Corp.)

when studying the effect of precompression on the final tablet properties. However, the time between precompression and main compression is determined by the Presster geometries since the positions of the precompression and main compression wheels are fixed.

The newest development for compaction simulation is a mechanical tableting machine simulator which operates with a cam. Thus it is called Stylcam (Figure 9) [59]. The cam is positioned on the lower compression wheel and allows the simulation of different tableting machines and their dwell times due to different acceleration of the punches. It was introduced in 2005.

With the Stylcam different dwell times are obtained by adjusting the speed of the compression wheels. Precompression is simulated by compressing a tablet twice. Thus the time interval between the precompression and main compression is freely adjustable. One further advantage of the Stylcam is that it works with a fixed filling shoe, as on a conventional rotary tableting machine. However, data on the precision of this instrument are not yet available.

6.4.6.3 Application of Tableting Machine Simulators

Using mechanical compaction simulators allows us to simulate the tableting process of rotary tableting machines to a greater extent than when using hydraulical compaction simulators. Thus they will be mainly used in formulation development and scale-up.

FIGURE 9 Detail of Stylcam. (Courtesy of MedelPharm.)

However, for mechanical compaction simulators the movement of the punches is mechanically determined and, compared to hydraulic compaction simulators, not freely programmable. Thus, for basic material characterization and early formulation development, hydraulical compaction simulators can be advantageous.

6.4.7 INSTRUMENTATION OF TABLETING MACHINES

To describe the tableting process more precisely, tableting machines have been instrumented since the middle of the last century. Measured values are force, displacement, and temperature and they are always measured with dependence on time. Thus time is another variable.

6.4.7.1 Force Measurement

The first instrumentation of a tableting machine for measurement of upper punch force was performed by Brake [60]. Thus it was for the first time possible to visualize the compression process with regard to force development which results from the material stresses during tablet formation. Only shortly after that, similar measurements were published by another research group [61, 62].

Besides upper punch force, lower punch force, die wall force [63–65], ejection force [66], and tablet scraper force can be measured. Die wall force measurement will be discussed separately.

For measurement strain gages are mostly used. These strain gages consist of constantan. They are applied in eccentric tableting machines at the upper or lower punch holder and in rotary tableting machines at the machine frame or the compression roll pin. Alternatively piezoelectric crystals can be used which have to be placed inside the punches [40].

The sensitivity of force measurement is dependent on the distance between the force transducer and where the force occurs. Thus for force measurement, instrumentation of the punches is more advantageous than instrumentation of the machine frame. However, since punch and die sets have to be exchanged between different runs of the machines, instrumentation of the punch holder, the machine frame, or the roller pin is most widely spread.

The most often measured force is the upper punch force. For the eccentric machine it is the force which controls densification; for rotary tableting machines upper and lower punch forces have ideally the same values. Schmidt et al. [67] measured force with a single punch of a rotary tableting machine. Ejection force is visible as a small lower punch signal which occurs shortly after the end of one compaction cycle. It is measured by lower punch instrumentation but needs more resolution. A review of force measurement is given by Bauer-Brandl [68].

Die Wall Force Measurement During compression of the powder the forces are evolving not only at the punches but also at the die wall [63–65]. Therefore die wall force measurement complements upper and lower force measurement. Since the compression process occurs axially, these radially evolving forces are smaller than the forces at the punches. Measurement of die wall force allows, for example, for indication on die wall friction, tablet capping, and lamination. Instrumentation for die wall force measurement is difficult and different techniques have been developed [64, 69, 70]. The die can be instrumented axially or radially and strain gages or piezoelectric crystals can be used for measurement. Two main effects influence the measured signal: tablet height and tablet position. Related to this, the output signal can be nonlinear. Piezoelectric foils have been applied which possess the advantage of independence on tablet position [70].

One example most recently developed is a split die consisting of three sections (Figure 10) [64]. Integrating the sensing web in a thin middle layer isolates stress measurement to a narrow band around the tablet and gives much closer approximation to the true stress. Further die wall force measurement is linear and independent of tablet height and position as it is uncoupled from all other die wall stresses and strains. Further it is designed in the shape of a conventional die and can be mounted without modification into a die table.

6.4.7.2 Displacement Measurement

The first measurement of upper punch displacement was performed in the mid-1950s by Higuchi and co-workers [62, 71] with the aid of inductive displacement transducers. By the same instrumentation the movement of the lower punch can be visualized. Inductive transducers are mounted parallel to the punch and thus give information on punch position. Alternatively touchless measurement of displacement is possible. It is important that the transducers be positioned most closely to the punch in order to minimize the influence of machine deformation.

A measurement of displacement on a rotary tableting machine was presented in 1987 by Schmidt and Tenter [72]. Another possibility was presented by Matz and co-workes [73, 74]. Meanwhile touchless measurement systems for recording displacement were developed [73, 74]. For all measurements of displacement, correc-

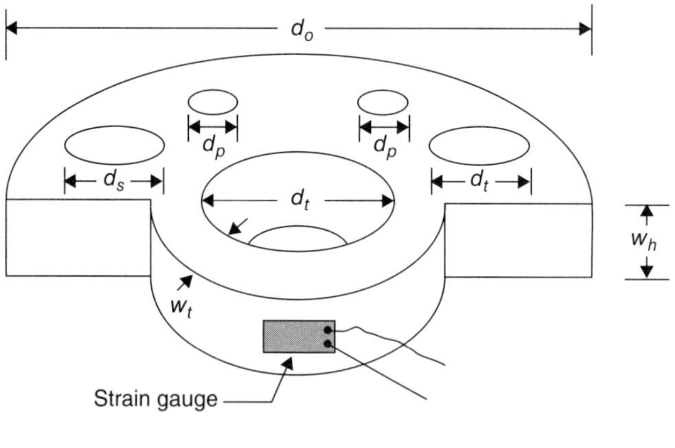

d_i: inner diameter $= \frac{3}{8}$ in.
d_o: outer diameter $= \frac{7}{8}$ in.
d_s: screw hole diameter $= \frac{5}{32}$ in.
d_p: alignment pin hole diameter $= \frac{1}{8}$ in.
w_a: web height $= \frac{1}{8}$ in.
w_t: web thickness $= \frac{1}{16}$ in., $\frac{1}{8}$ in.

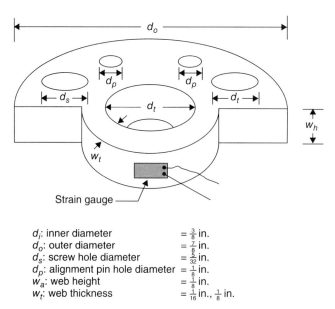

FIGURE 10 Construction details of split die. (Reproduced with permission from ref. 64.)

tions which take elastic punch and machine deformation into account are necessary. Müller and Caspar [75] showed problems which occur when machine and punch deformation are not taken into account. Krumme and co-workers gave an extensive description on this issue for eccentric tableting machines [76, 77]. From punch displacement measurements tablet height can be calculated, and this height can further be related to tablet density and porosity.

Again and again it was tried to derive displacement theoretically from machine geometry, punch geometry, and measured force [49–52]. However, until today this theoretical derivation has not been satisfying and thus experimental testing cannot be given up.

6.4.7.3 Temperature Measurement

Due to the forces evolving during tableting the compressed material can warm up. The evolving temperatures can be measured with different methods. In each case only approximated measurements are possible since either the measurement was not performed directly inside the tablet or additives were necessary, which can alter measurement.

Several methods have been applied to determine the temperature increase with thermal sensors which are installed in the punches, inside the die, or in the powder bed (Table 1) [78–87].

Most recently an analysis technique was developed which allows measurement of tablet temperature directly after ejection of the tablet on the machine (Figure 11) by an infrared sensor [85]. The temperature signal can be directly related to force and displacement measurement.

6.4.7.4 Measurement of Time

All measured variables can be determined with dependence on time. Thus the tableting process can be characterized for several variables with dependence on time.

Two time definitions are important: contact time and dwell time. Contact time can be defined as the time during which a contact of powder and punches is measurable, for example, when the force exceeds a certain limit of 100 N. Dwell time can be defined predominantly for rotary machines as the time during which the punch heads are completely under the compression wheels and thus the applied force is constant.

TABLE 1 Methods to Determine Temperature During and Shortly After Compaction

Measurement in punches [78–80]
Epoxide punches [81]
Calorimetric measurement [82, 83]
Infrared measurement [84, 85]
Measuring conductivity during tableting with conductive materials [86]
Energy calculations [78, 83, 84]
Determination of melting of materials with certain melting point [87]

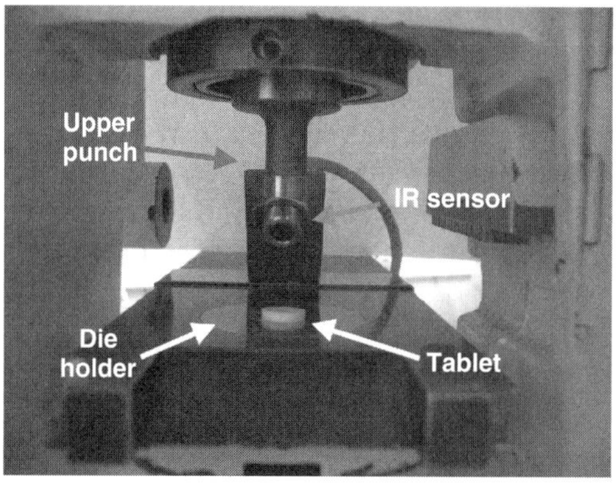

FIGURE 11 Infrared sensor unit for measuring temperature directly after tableting [85] (Martin-Luther-University Halle-Wittenberg.)

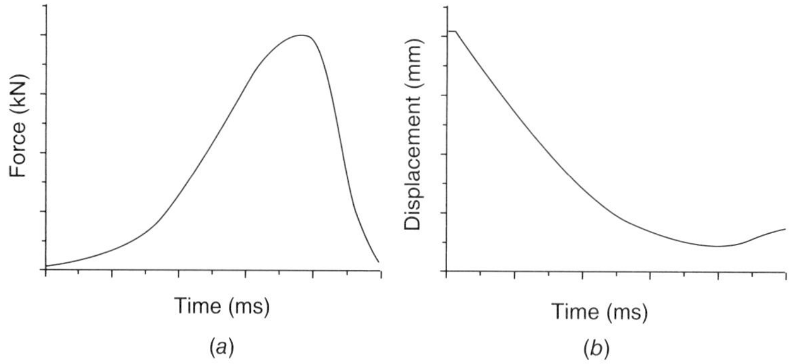

FIGURE 12 Force–time and displacement–time profile for eccentric tableting machine.

Further for description of the tableting process, the entering of the punch into the die, the compression start at the begin of contact time, and the lifting of the upper punch from the tablet are important.

6.4.8 ANALYSIS OF TABLETING PROCESS

All measured variables, namely force, time, displacement, and temperature, can be combined differently and can be analyzed afterward. From the functional relations, conclusions can be drawn about the compression and compaction behavior of the materials.

The most basic analysis is the presentation of force versus time or displacement versus time. These curves are different for eccentric and rotary tableting machines. The data given in Figures 12 and 13 are valid for the contact time of the compaction cycle of one single tablet. Due to the eccentric-driven movement of the punches, the force–time curve can be described by a sharp peak at the maximum force evolving at the punches and the displacement–time curve can be described with a sharp peak at the minimum height of the powder bed. For curves of eccentric tableting

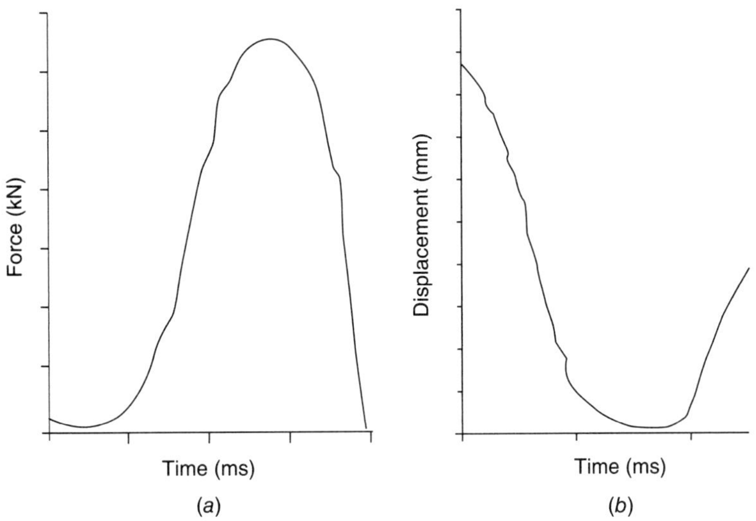

FIGURE 13 Force–time and displacement–time profile for rotary tableting machine.

machines hardly any dwell time is measurable (Figure 12). In contrast, for rotary tableting machines, the force–time curve and the displacement–time curve are flatter at the maximum peak. This is the case due to the dwell time when the punch heads move completely between the compression wheels. The dwell time is indicated in Figure 13. For rotary tableting machines with precompression wheels additionally force–time curves and displacement–time curves for precompression can be recorded. They look similar to the curves from the main compression wheels with the exception that lower forces are applied. Precompression data will not be discussed in the following since the data can be treated similar to the data of the main compression event.

Besides this presentation of force and displacement versus time, which are data directly derived from the tableting machine, other more advanced methods are possible. Extensive reviews on the methods used can be found in the literature [54, 88, 89]. In the following only the most important aspects will be discussed.

6.4.8.1 Force–Time Analysis

One method to analyze tableting data is the use of force–time or pressure–time diagrams. They are easily recordable since displacement measurement is not necessary.

Some basic parameters can be directly read from the curves. For the force values upper and lower punch forces and ejection forces should be mentioned, and for the time values contact time should be mentioned. Deduced parameters such as pressure and normalized contact time can be calculated and further statistical data are often used for characterization (Table 2). Due to the different shapes of force–time curves from eccentric tableting machines compared with those from rotary tableting machines, some parameters can only be calculated from eccentric machine data and some can only be calculated from rotary machine data.

TABLE 2 Parameters Directly Deduced from Force–Time Profiles

Maximum upper punch force (pressure)
Maximum lower punch force (pressure)
Maximum ejection force (pressure)
Contact time
Normalized contact time
Time at maximum upper compression force
Time at maximum lower compression force
Maximum upper precompression force (pressure)
Maximum lower precompression force (pressure)
Precompression contact time
Normalized precompression contact time
Time at maximum upper precompression force
Time at maximum lower precompression force

TABLE 3 Parameters Calculated for Eccentric and Rotary Tableting Machines

Source	Parameters
Chilamkurti [90]	Area under curve (AUC)
	Height of curve at maximum upper punch force
	Width of curve at half-maximum upper punch force
	Slope during compression
	Slope during decompression

Only eccentric machine data allow us to calculate the *R* value (maximum upper punch force/maximum lower punch force), which is an indication of friction. They also allow us to calculate the time difference between the maximum upper punch force and the maximum lower punch force. Only dwell time and the minimum force during the dwell time can be calculated for rotary tableting machine data. The rise time of rotary machines is defined as the time during the compression phase, and peak offset time is defined as the time difference between maximum pressure and vertical alignment of the punches. Further the inflection points during the compression and decompression phases are mostly only calculated for rotary machine data.

In addition, for force–time diagrams different methods to characterize the tableting process were developed. These methods can be divided in those applicable to force–time curves from eccentric and rotary tableting machines [90] and those applicable only to data from eccentric or rotary tableting machines (Tables 3–5).

One possibility to analyze the tableting process is to describe the areas under the curve during compression and decompression and to draw conclusions on plastic and elastic parts of deformation. Emschermann and Müller [91] applied this method to data from eccentric machines (Figure 14). Similar area comparisons were performed by the research group of Schmidt [92–94] for rotary machines (Figure 15). They tried to gain information on elasticity by calculating differences between the area under the plot in the compression phase and the area under the plot in the decompression phase. A sophisticated technique to interpret area data under one

TABLE 4 Parameters Calculated for Eccentric Tableting Machines

Source	Parameters
Emschermann [91] (Figure 14)	Area under compression curve (compression area) Area under decompression curve (decompression area) Compression area/decompression area
Pressure–time function (modified Weibull function) [39, 96–98] (Figure 16)	$p = p_{\text{max, upper punch}} \left[\frac{1-t}{\beta}\right]^{\gamma} \exp\left[-\left(1-\frac{1-t}{\beta}\right)^{\gamma}\right]$ where β = time parameter γ = asymmetry parameter
Modified Fraser–Suzuki function [99]	$f(t) = H \exp\left\{\left[\frac{-0.693}{A^2}\right] \times \left[\ln\left(\frac{\langle 1 + A \times (t_r - t)\rangle}{S}\right) \times 1.177\right]^2\right\}$ where A = asymmetry parameter t_r = time parameter S = deviation of maximum

TABLE 5 Parameters Calculated for Rotary Tableting Machines

Source	Parameters
Tenter [72]	Area center of gravity
Vogel [92–94] (Figure 15)	t_1 = compression time t_2 = time at the start of dwell time t_3 = time at the half of dwell time t_4 = time at the end of dwell time t_5 = time at the end of one compression cycle A_1 = area of densification A_4 = area of decompression A_2 = partial area A_3 = partial area A_3/A_2 = area quotient A_5 = partial area A_6 = partial area A_6/A_5 = area quotient

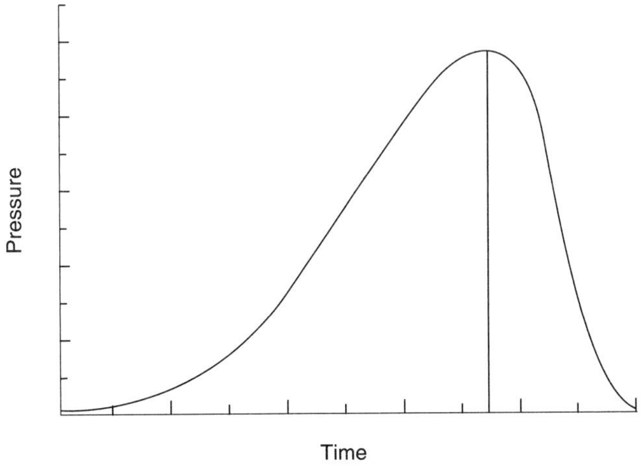

FIGURE 14 Pressure–time curve analysis [91].

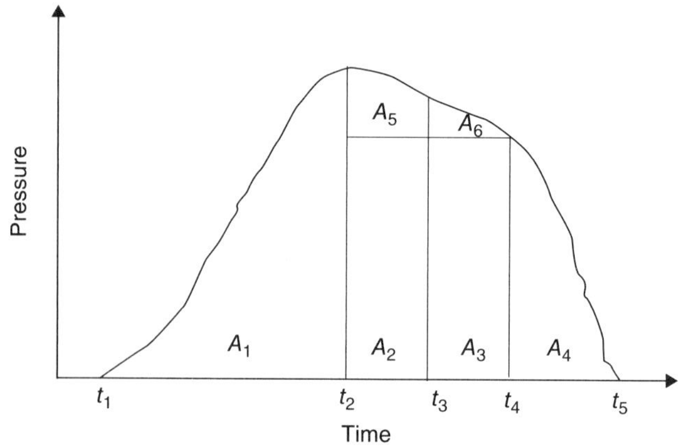

FIGURE 15 Pressure–time curve analysis [92, 93].

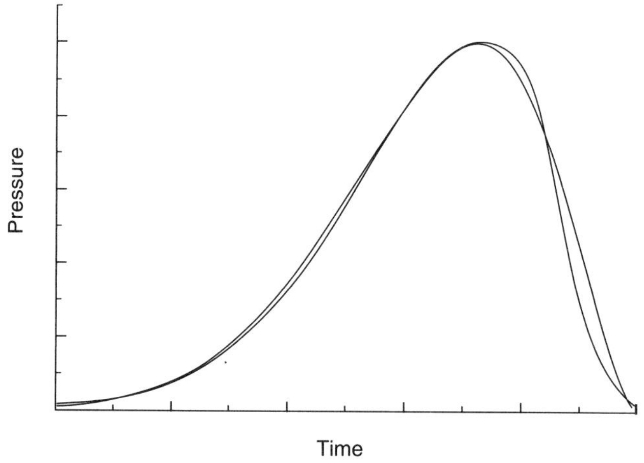

FIGURE 16 Pressure–time curve analysis with pressure–time function [39, 96–98].

compaction cycle was developed. The parameters of interest are given in Table 5. Further advances were performed by Yliruusi and co-workers [95].

Another possibility of analysis is to fit different functions to the force–time data. The research group of Mielck [39, 96–98] described the densification behavior by the pressure–time function (Figure 16). The lower the values of the parameters β and γ, the more plastically the material deforms; the higher the values, the more elastically the material deforms. The parameter γ describes the asymmetry of the curve and β the time at maximum densification. A similar function, the Fraser–Suzuki function, which originates from chemical analytics was applied to tableting data [99]. It can also be used to derive parameters that describe the deformation behavior of materials. Information on the reversible and irreversible deformation of the material can be deduced.

TABLE 6 Parameters Directly Deduced from Displacement–Time Profiles

Maximum displacement
Minimum height of tablet
Minimum volume of tablet
Maximum density of tablet
Minimum porosity of tablet
Maximum relative density of tablet
Time at maximum displacement
Time at minimum height of tablet

6.4.8.2 Displacement–Time Analysis

From displacement–time data a few parameters can be read. The most important data are given in Table 6. In addition the data allow us to calculate fast elastic recovery. The increase in tablet height from the minimum tablet height in the die up to the lifting of the upper punch from the tablet is called fast elastic recovery.

Since the travel of the punches is analyzed, the data also allow us to determine the speed of the punches at each point of the compaction cycle. Punch speed is an important parameter to compare different tableting events [100]. Maximum punch speed can be determined and used for characterization.

Only a few authors have tried to relate displacement with time. Ho and Jones [101] determined the slope of porosity over time (rise time). This slope was also used by Tsardaka [102] for analysis.

6.4.8.3 Force–Displacement Analysis

The most extensively used method to characterize the tableting process is the use of force and displacement measurements. Usually upper punch force and upper punch displacement are used. Models which relate force and displacement directly can be distinguished from those which analyze pressure and volume.

Further, some parameters can be directly read from the curves. The peak offset time of eccentric tableting machines is defined as the time difference between maximum displacement and maximum compression force [101, 103].

Relation between Force and Displacement The first information on force–displacement analysis can be found in Führer [104] and Moldenhauer et al. [105]. Force–displacement diagrams (Figure 17) are used to calculate from the areas enclosed the work or energy necessary for tableting. The force–displacement profile includes compression and decompression of the powder to the tablet. The area between compression and decompression is the area of the compaction energy, often called the energy of plastic deformation (E_P) [106]. The area E_E is the energy of elastic deformation. And the last area (E_F) to complete the triangle start of compression (D_0)–maximum force (F_{max})–displacement at maximum pressure (D_{PM}) can be interpreted as the energy of friction (E_F) [107, 108]. The sum of all three energies is the total energy E_T of the tableting machine. The energies can be displaced as an absolute value or relative to the total energy. Based on these values, Stamm and Mathis [109] developed the determination of plasticity P as

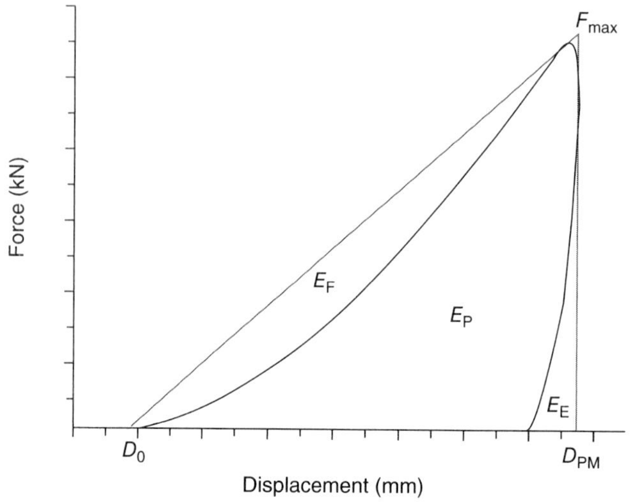

FIGURE 17 Force–displacement diagram for energy analysis [104–108].

$$P = \frac{E_P}{E_P + E_E} \quad (1)$$

Other authors tried similar attempts [110, 111] and the developed methods were regarded to be very useful [27, 112–115]. Antikainen and Yliruusi [116] more recently tried to derive further parameters from the diagrams to enable a more complete characterization. An overview on the possibilities for force–displacement analysis is given by Ragnarson [117].

Relation between Volume and Pressure The oldest method of this type of analysis is to establish a relation between the volume of the tablet and the force necessary to produce this volume [62, 71]. For exact description the height of the tablet is determined by displacement measurement and the accuracy of this measurement is extremely important. Further displacement measurement has to be corrected precisely for elastic deformation of the punches and the machine in order to use correct tablet heights.

From tablet height, volume, porosity, and relative density at different stages of densification can be deduced. These variables are plotted as a function of pressure. For analysis, for example, the equations of Heckel [118–120], Kawakita, [121, 122], Cooper and Eaton [123], Walker [124], Bal'shin [125], and Sønnergaard [126, 127] can be used. The equations are given in Table 7. A further overview of these and other equations used can be found in Celik [88].

The Heckel equation describes the densification process with first-order kinetics. A linear equation is obtained with a slope which is inversely proportional to the yield strength. The slope of the Heckel equation provides information on the plastic deformation of the powder. It has also been published that the slope of the Heckel equation can be correlated with the elastic modulus (Young's modulus).

TABLE 7 Parameters Calculated from Force–Time Profiles

Source	Parameters
Heckel [118, 119] (Figure 18)	$-\ln\varepsilon = \ln\left(\frac{1}{1-D}\right) = Kp + A$ where K = deformation parameter A = powder bed densification
Yield pressure [120]	$\text{Yield pressure} = \frac{1}{\text{Heckel slope}}$
Yield strength [118]	$\text{Yield strength} = \frac{1}{3 \times \text{Heckel slope}}$
Kawakita [121, 122]	$\frac{p}{C} = \frac{1}{ab} + \frac{1}{a}p$ where a = porosity of powder bed b = compression parameter
Walker/Balshin [124, 125]	$100V_{\text{rel}} = 100 \times \frac{V}{V_\infty} = -W \log p + C$ where W = compressibility coefficient
Sønnergard [126]	$V_{\text{rel}} = V_1 - W \log p + V_e e^{-p/p_m}$ where W = compressibility coefficient
Cooper–Eaton [123]	$V^* = \sum_{i=1}^{n} a_i V_i^* = \frac{V_0 - V}{V_0 - V_\infty} = a_1 e^{-k_1/p} + a_2 e^{-k_2/p}$ where k_1 = deformation pressure for fraction part 1 k_2 = deformation pressure for fraction part 2 a_1 = fraction part 1 of deformation a_2 = fraction part 2 of deformation
Cooper–Eaton (linearized) [123]	$\ln V^* = \ln \frac{V_0 - V}{V_0 - V_\infty} = -\frac{Q}{p} + \ln R$ where Q = extent of compressibility R = sum of fraction parts

The equation of Kawakita describes volume reduction with pressure in the form of a hyperbolic equation. Walker and Bal'shin [125] postulated a logarithmic relation between applied pressure and volume reduction, which was further modified by Sønnergard [126]. Cooper and Eaton [123] use an exponential function, which can also be linearized. Pressure thresholds for deformation mechanisms are determined. It should be noted that all of these equations and tableting models determine descriptive parameters.

The equation of Heckel is the most extensively used model and the underlying porosity–pressure plot is called a Heckel plot (Figure 18). The equation for the linear compression process follows fist-order kinetics (Table 7). Heckel

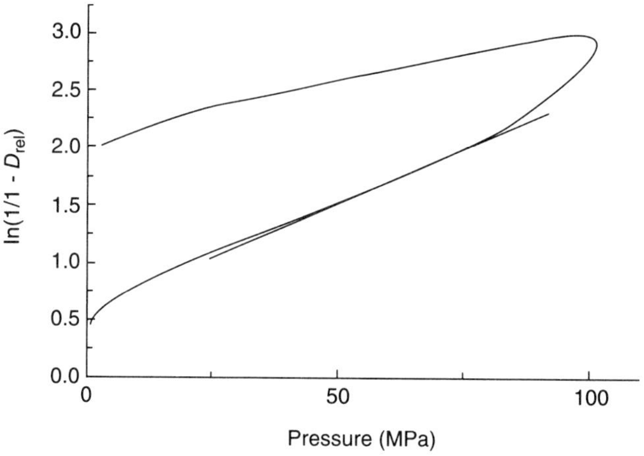

FIGURE 18 Heckel plot [118, 119].

distinguished measurements which determine the volume of the tablet without pressure (zero pressure) [119] from those measurements which determine the volume with pressure [118]. The first method allows determination of the volume after release of the elastic energy; the second method allows a higher precision repeatability since it is often difficult to determine tablet height at a defined time after ejection. Some milliseconds can cause differences [35, 36]. Years later Sun and Grant [128] tried to determine the elastic part at pressure measurements. The experiments showed that deviations in Heckel plots at high pressures are dependent on the elasticity of the material.

The equation of Heckel has been discussed again and again. One main issue of critique is that pharmaceutical powders are not purely plastically deforming materials and thus particle size and deformation mechanisms influence the derived parameters [129, 130]. Already very small errors in displacement determination or the measurement of true density can induce huge errors in the derived parameters [75–77, 129, 131, 132]. Sønnergaard [126] referred the equation of Walker and Bal'shin for his characterization of materials. He criticized further that the yield strength derived from the Heckel equation is directly dependent on the true density of the powders [127].

Despite this critique of the Heckel equation, the analysis of Heckel plots has been intensively used for the description of powder compression [128, 133–136]. Gabaude et al. [136] stated that the analysis is quite useful when defining preconditions exactly and apply correct displacement measurement.

Since the development of the equation, it has been tried to derive further information from it. Rees and Rue [129] determined the area under the Heckel plot. Duberg and Nyström [137] used the nonlinear part for characterization of particle fracture. Paronen [138] deduced elastic deformation from the appearance of the Heckel plot during decompression. Morris and Schwartz [139] analyzed different phases of the Heckel plot. Imbert et al. [134] used, in analogy to Leuenberger and Ineichen [14], percolation theory for the compression process as described by the Heckel equation. Based on the Heckel equation, Kuentz and Leuenberger [135, 140] developed a new derived equation for the pressure sensitivity of tablets.

Tsardaka and co-workers [102, 141, 142] presented the Heckel plot with dependence on time and analyzed deformation in combination with elastic recovery. Additional areas to describe plasticity were determined from two-dimensional (2D) plots [129]. Finally, the three-dimensional (3D) model [143, 144] was developed by fitting a plane to a 3D data plot on the basis of normalized time, pressure, and porosity according to Heckel.

6.4.8.4 Force–Displacement–Time Analysis

Force, displacement, and time are the three most important parameters to characterize the compaction cycles of tableting materials. Even when Hoblitzell and Rhodes postulated a linear relationship between force–time and force–displacement data [145], this could not been exactly proved until today. Thus it is important to analyze these three measured data together.

From force–displacement–time curves some parameters can be directly deduced; for example, the power during tableting and the maximum power can be determined. Another important parameter which serves as a measure for viscoelasticity is the peak offset time [101, 103]. For single-punch tableting machines it is defined as the time difference between maximum displacement and maximum force.

In addition, advanced models as those calculating viscoelasticity and the 3D model have been developed. They will be described in the following.

Viscoelasticity Models For characterization with viscoelasticity models, simulation models have been developed on the basis of Kelvin, Maxwell, and Voigt elements. These elements come from continuum mechanics and can be used to describe compression.

David and Augsburger [146] were the first to try this method of analysis. Further tests, for example, determination of complex functions based on methods of numerical mathematics, were performed by the research group of Müller [147, 148]. Although the results were helpful, for exact description the models became rather complicated and the derived parameters were complex. According to Bauer [149], these models have to be three dimensional for a reasonable description. Another similar approach was used by the research group of Rippie [12, 13]. They described the structure evolution in the tablet during decompression by the aid of vectorial 3D models and concluded that fracture and stress contribute to the final structure of the tablet.

Other research groups derived viscoelastic properties from creep experiments of the final tablet [150–154]. As Tsardaka and Rees [142] determined, stress relaxation follows a hyperbolic equation.

3D Model Most recently another technique which uses force, displacement, and time has been developed. 3D modeling is a very useful method to characterize the tableting process [3, 46, 47, 155–159]. Force is expressed as pressure, time has to be normalized, and from displacement data the porosity according to Heckel [119] is calculated. It is the only possibility to combine these variables during analysis.

To describe the tableting process the three variables were presented in a 3D plot (3D data plot) and a plane was fitted to the data twisted at $t = t_{max}$ (Figure 19*a*). From the fitting process the parameters *d* (time plasticity), *e* (pressure plasticity),

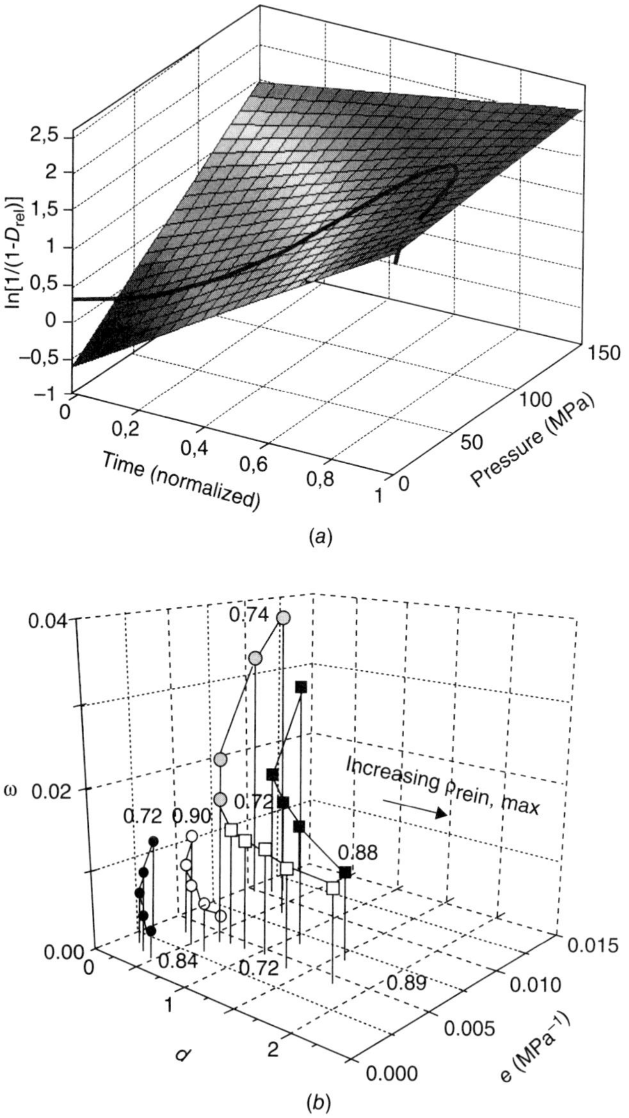

FIGURE 19 (*a*) 3D data plot with fitted plane twisted at $t = t_{max}$ and (*b*) 3D parameter plot of (●) DCPD: dicalcium phosphate dihydrate, (○) spray-dried lactose, (□) MCC: microcrystalline cellulose, (●) theophylline monohydrate, and (■) HPMC: hydroxypropyl methylcellulose for data gained with an eccentric tableting machine [47].

and ω (twisting angle, which indicates fast elastic decompression) can be derived (Table 8).

The parameters of the fitted plane (time plasticity *d*, pressure plasticity *e*, and twisting angle ω) were also exhibited in a 3D plot and this plot is called the 3D parameter plot. This plot exhibits the compression behavior of the powder. It gives a simple yet characteristic description of the tableting properties. An example is given in Figure 19*b*.

TABLE 8 Parameters Calculated from Force–Displacement–Time Profiles

Source	Parameters
3D model [143, 144, 155] (Figure 19)	$z=\ln\left(\frac{1}{1-D_{\text{rel}}}\right)=[(t-t_{\max})(d+\omega p_{\max}-p)]+(ep)+(f+dt_{\max})$ $d=\frac{\delta\ln[1/(1-D_{\text{rel}})]}{\delta t}$ $e=\frac{\delta\ln[1/(1-D_{\text{rel}})]}{\delta p}$ $f=\ln\left(\frac{1}{1-D_{\text{rel}}}\right)$ where D_{rel} = relative density t = time p = pressure $t_{\max}$ = time at maximum pressure $p_{\max}$ = maximum pressure ω = twisting angle at $t_{\max}$, indicates fast elastic decompession d = time plasticity e = pressure plasticity

Time plasticity d describes the plastic deformation of the excipient according to time [160]. It is influenced by tableting speed [157]. With increasing time plasticity d the powder deforms faster during tableting. Therefore, with increasing densification time plasticity increases. Pressure plasticity e describes the pressure-dependent increase of density. The pressure plasticity correlates with the slope of the Heckel equation [144]. With increasing pressure plasticity e the slope of the Heckel equation increases in the same direction and the necessary pressure for deformation (yield pressure [118] or yield strength [120]) decreases. The twisting angle ω is a measure of the material's elasticity and the ratio between compression and decompression. Thus, it indirectly describes fast elastic decompression during the tableting process. When ω increases, elasticity decreases. The twisting angle ω correlates with the elastic modulus [144]. In conclusion, materials which deform fast show high d values, materials which deform easily and with low pressure show high e values, and those which relax a lot show a lot elasticity and thus low ω values.

Thus the 3D model allows us to characterize the tableting process completely and to distinguish time-dependent information from pressure-dependent deformation and elasticity in one step of the analysis.

Temperature Analysis The results gained by determination of temperature during and shortly after tableting vary strongly and depend on the method used.

A temperature increase of 5 K [79, 80, 161] could be determined with conventional punches; however, with epoxide punches the measured temperature increase was as high as 30 K [81]. By calorimetric measurement an increase of 10 up to 30 K [82] was determined and by infrared measurement the increase was 10–15 K [84].

Beissenhirtz [86] measured a temperature increase of 30 K indirectly by measuring conductivity which arose during tableting with conductive materials. Energy calculations indicate a temperature increase of more than 30 K caused by tableting

[78, 84, 83]. Most recently, partial melting of drugs could be analyzed for materials whose melting temperature is as high as 94 °C [87], and the reversible transgression of a glass transition temperature of 80 °C was determined [162].

All results indicated that temperature increase depends on the material. Further temperature increase during tableting can contribute to slight changes in material structure [85].

6.4.9 ANALYSIS OF FINAL TABLET FORMATION

This characterization of the process of tablet formation has to be completed by analyzing the changes induced by tableting.

Most important is the elastic recovery of the tablets which starts during decompression and is finished dependent on the material after several days. Elastic recovery can be defined as [163].

$$\mathrm{ER}(\%) = 100 \times \frac{H_t - H_{\min}}{H_{\min}} \tag{2}$$

where $H_{\min}$ is the minimum height of the tablet under load and H_t the height of the tablet at different times t after tableting.

Elastic recovery gives information on the remaining elasticity of the materials which is only slowly released. Further it can indicate structural changes inside the materials and tablets. Structural changes induced by tableting have to be analyzed by physicochemical techniques, such as spectroscopic and thermoanalytical methods, X-ray diffraction, scanning electron microscopy, and transmission electron microscopy [35, 36, 85, 164–166]. The analyzed changes will help to better understand the process of tablet formation and identify the reasons for compactibility of materials. However, these changes are not the subject of this chapter.

6.4.10 COMPLETE DESCRIPTION OF PROCESS OF TABLET FORMATION

On the whole, the process of tablet formation can fully be described by combining the analysis of the tableting process with the final formation of the tablets. The methods which gives most detailed information of the whole process and simultaneously is a fast method is the 3D modeling technique in combination loith calculating the elastic recovery of the tablets. In addition, by combining both these methods and calculating general plasticity P from time plasticity d, pressure plasticity e, twisting angle ω, and elastic recovery ER, a more general tool for analysis of the process of tablet formation is available [3, 4].

Finally the crushing force of the tablets after relaxation gives information on the formed bonds inside the material and the compactibility. Compactibility has been described by Leuenberger [167]. For the future it can be expected that a prediction of compactibility as a result of the process of tablet formation is possible.

6.4.11 SPECIAL ACCESSORIES OF TABLETING MACHINES

In tablet production it is essential to control tablet weight and tablet homogeneity in order to ensure a uniform dosage form. For patient safety pharmacopeias demand that tablet weight is between certain limits. Tablet producers often set their more narrow specifications to ensure that they meet pharmacopeial specifications. To control tablet weight and tablet homogeneity an optimal product mixture, complete filling of the die, exact tooling, and tightly controlled machine conditions are necessary.

Another demand for patient safety and due to GMP regulations is to ensure the absence of impurities in the final dosage form, for example, residues from the previous product or residues from detergents. Thus the process of cleaning of machines has to be standardized and controlled. In addition optimized short cleaning times of the machine increase operating time for production. During the last years one innovation for tableting machines was the development of special accessories for cleaning in order to reduce standing times.

For special products (e.g., cytostatics or sterile products), it is necessary to produce tablets in a hermetically closed machine. For these products special containment solutions have been developed which allow the production in a hermetically closed machine or behind a wall. Most important is to separate the tablet production zone strongly from the mechanics of the machine.

In the following tablet filling devices, possibilities to control tablet weight and mixing homogeneity as well as advances in cleaning technology will be discussed.

6.4.11.1 Optimization of Die Filling

The basics of filling have been explained above: The filling shoe is moving back and forth, the filling shoe is fixed, or filling is centrifugally controlled. Two problems arise generally: Either the product demixes and tablet weight and content uniformity are no longer controlled or the die is not completely filled and thus tablet weight also varies.

Optimal filling of the die is determined to a great extent by the material, but the speed of the machine is also important. At low machine speeds the die is usually completely filled; at high machine speeds this becomes more difficult. Thus special filling devices using one or more paddles have been developed to improve filling. One example for a paddle feeder is given in Figure 20. Alternatively filling devices can be vibrated to improve feeding for materials with bad flow characteristics. An overview is given by Ritschel and Bauer-Brandl [32, 168].

Besides improved feeding, paddle feeders allow improved mixing uniformity since the formulation is mixed again shortly before feeding the die. This mechanical remixing is the only possibility to improve the homogeneity of the mixture. Demixing is a bigger problem for machines with a moving filling shoe than for those with a fixed filling shoe. Thus moving filling shoes are equipped with paddles as a standard. Fixed filling shoes need paddles usually only for materials with bad flowability or at high machine speeds.

FIGURE 20 Example paddle feeder. (Courtesy of Kilian.)

6.4.11.2 Tablet Weight Control

To control tablet weight different possibilities exist. The simplest method is to weigh at preset intervals (in-process control) a number of tablets manually and to adjust machine settings according to the results when necessary. For high-speed rotary machines automatically working weighing systems have been developed which determine tablet weight shortly after ejection [169, 170]. These systems can also determine tablet height and diameter, which are indirect measures for tablet uniformity. Simultaneously with tablet weight compression force drifts. A direct relation between compression force and tablet weight exists. Thus it is possible to monitor tablet uniformity also by control of compression force. For this purpose the machines are instrumented with strain gauges or piezoelectric force transducers. By control of compression, force changes in tablet weight can be directly detected.

Automatically working control systems are able to eject those tablets separately which fail the requirements (rejection mechanism), and they adjust the machine for die filling or compression force and collect only those tablets which meet the requirements. Two alternative principles for automatic tablet weight control and adjustment are possible, depending on the application and selected machine type. The principle of control of compression force is based on measurement of the final compression force under constant tablet height. This principle is used for all applications where tablet weight accuracy and constant tablet density are less critical.

The principle of control of displacement is based on measurement of tablet height variations under constant force. This principle is more accurate than the force control system. It is used for all applications where constant density of the produced tablets is critical.

Modern systems combine one of these control systems with automatic weighing of tablets. Weight control will automatically adjust the filling depth in order to keep tablet weight within specified tolerance limits.

6.4.11.3 Control of Mixing Homogeneity

The systems to control weight uniformity are not able to control uniformity of the mixture. When during filling of the die the tableted material demixes, tablet weight usually tends to vary. However, these variations can be small and not easy to detect. To monitor mixing uniformity in the final tablet, most recently spectroscopic techniques such as Raman spectroscopy and near-infrared (NIR) spectroscopy have been used [171–173]. Special online sensors have been build into the machine and they measure the spectrum for each tablet. When the mixture is homogeneous, the appearance of the spectrum will always be the same or between certain limits. If not, the production can be stopped and adjusted as far as possible. Thus a further step in quality assurance of tablets has been made. This was partially caused by the process analytical technique (PAT) initiative of the U.S. Food and Drug Administration (FDA) [174].

6.4.11.4 Cleaning

To ensure product quality, cleaning is of utmost importance. Therefore different standardized cleaning technologies have been developed [175, 176]. Detailed information may be provided by manufacturers.

Usually, the most effective way to implement cleaning is to design it into a process which has to be performed after tableting. It involves the addition of spray systems, tank cleaners, nozzles, and seals into the tableting machine in order to automate the cleaning process. The automation converts the batch processes to a continuous operation of tableting cycles and cleaning cycles. Cleaning or washing in place means an advanced wash liquid preparation system which handles all filtering, preheating, mixing, and pumping of water, detergents, and demineralized water and provides continuous monitoring and control of cleaning parameters. Another possibility for standardized cleaning is the wash-off-line procedure. In this case exchangeable compression modules are especially designed for a fast product changeover. Different techniques have been developed to exchange the modules. Either carries or arms or lifting systems are applied or additionally used. The wash-off-line procedure increases production time by cleaning the exchangeable compression modules after it is removed from the tablet press. Thus production time is increased; however, two compression modules are necessary. In this case a special separate washing system is necessary.

6.4.12 IMPORTANT FACTORS DURING MANUFACTURING PROCESS

To run a tablet production process effectively, robustly and smoothly several factors have to be kept in mind. Of utmost importance for the process are environmental humidity during tablet production and adhesion forces between machine punches and dies. These and other factors can contribute to problems during manufacturing. In the following the relevance of climatization during tableting, the necessity and methods of lubrication, and frequently occurring problems during manufacturing will be discussed.

6.4.12.1 Climatization

Relative humidity (RH) in the production room influences the water content of the materials. This has to be kept in mind during tablet production.

When the materials sorb water, they deform differently compared to the status before sorption, and compressibility changes [98, 177, 178]. Furthermore the compactibility of the materials changes and tablets with a different crushing force and friability result [179]. Even the release from the tablets can be influenced [180]. However, the influence of water content on tableting and tablet properties depends on the material; for example, hydrophilic polymers are mostly influenced by RH. Further, the influence of RH during production is most decisive when production conditions change extremely. Smaller differences up to 10% RH do not influence a robust formulation [3]; however for critical formulations even these changes are of importance. Usually the conditions at one production site do not change from day to day, but great differences have been observed between different seasons of the year. Since tablet production and the final tablet quality should be the same throughout the year, often a humidity interval between 40 and 60% RH is used for tablet production. This is a first step. However, for characterization of material properties humidity control between 40 and 60% RH is not sufficient. Material properties cannot be compared when obtained at different conditions. In this case humidity control at a certain humidity with a precision of 2–5% is absolutely necessary [3].

6.4.12.2 Lubrication

Adhesion forces between the material and punches and dies result in sticking of the tablets at the punches and dies. These adhesion forces are further influenced by RH and this has also to be kept in mind. When the adhesion forces at the punches and dies are greater than the cohesion forces between the particles inside the tablet, the tablets stick at the punches and can cap.

To overcome this problem, lubrication is the method of choice [181, 182]. Two alternatives for lubrication exist: internal lubrication and external lubrication. Internal lubrication is performed by mixing the tableted product shortly before the tableting process with a solid lubricant. Thus the lubricant is not only at the surfaces of the final tablet but also inside the tablet. As a result, internal lubrication lowers bonding, and this is especially the case for plastically deforming materials [183–186]. The most frequently used and most effective material for internal lubrication is magnesium stearate [187–189]; however, other hydrophobic or amphiphilic lubricants are also possible [190–196]. Magnesium stearate has one major disadvantage: It shows a low solubility and remains as a solid after dissolution. Thus the search for other lubricants is ongoing.

When the lubricant should not be part of the tablet formulation, for example, when bonding properties of the drug are low, external lubrication is necessary [197]. For single-tablet production the punches and dies can be manually lubricated with a fluid. In production several methods have been developed to place the fluid on the surface of punches and dies [198]. Filaments applied at the punches to lubricate the die or special caps with fluid lubricant are possible solutions. However, external lubrication also has disadvantages [199].

6.4.12.3 Occurring Problems during Manufacturing

The most frequent problems occurring during the manufacture of tablets are high tablet weight variation, capping and lamination [200], and further picking and sticking at punches and dies. Low product yield, low crushing force, and further tablet yams and chipping are other problems which have to be solved [33, 201].

High tablet weight variation can be reduced by using weight control systems. Further demixing of the tableted material has to be avoided, since demixing results in higher tablet weight variation and content uniformity can no longer be achieved. As already discussed, paddle feeders can be used to achieve mixing homogeneity for problematic products and further spectroscopic techniques can be used for control.

The problem of capping and lamination can be solved by increasing RH or adding wetting agents. Further either external lubrication may help. Picking and sticking of the tablets at punches and dies can be avoided by using lubricants as discussed above.

A low product yield is caused by loss of material during fast production processes. On rotary tableting machines this problem is solved by slightly lowering the lower punches before the compression event starts.

A low crushing force is often caused by the composition of the special formulation. If the formulation itself is not the reason for a low crushing force, compression force can easily be increased. Another possible explanation can be low humidity of the tableted product.

Finally tablets yam and chipping can occur before the tablets leave the die table of a rotary machine. Usually the lower punches or the tablet scraper of the tableting machine are not properly adjusted. Another reason can be low crushing force.

In conclusion, for smooth and perfect machine runs, product properties and machine conditions have to be tightly controlled.

6.4.13 FUTURE OF TABLET PRODUCTION SYSTEMS

In principle tablet production systems have remained the same throughout the last century. However, major improvements in instrumentation, data acquisition, and analysis techniques have been made. Nowadays more sophisticated data acquisition and analysis techniques are available which facilitate and improve interpretation of tableting data.

In order to facilitate scale-up, more sophisticated simulation systems can be thought of. They will be a real help for scale-up with small amounts of material as available in early development of formulations.

Further quality control of the tablets during tableting or shortly after has become more important. Recent trends show improvements for production in a GMP environment by isolating the production unit from the machinery. For the near future the implementation of the PAT initiative of the FDA is conceivable.

REFERENCES

1. Brockedon, W. (1843), Shaping pills, lozenges, and black lead by pressure in dies, *British Patent* 9977.
2. Leuenberger, H. (1997), Research in solid dosage forms—An obsolete topic? *Pharm. Dev. Technol.*, 2, vii–viii.

3. Picker, K. M. (2002), *New Insights in the Process of Tablet Formation—Ways to Explore Soft Tableting*, Habilitationsschrift, Martin-Luther-University Halle-Wittenberg, Halle, Germany, and also Görich und Weiershäuser Verlag, Marburg.
4. Picker, K. M. (2004), Soft tableting: A new concept to tablet pressure sensitive drugs, *Pharm. Dev. Technol.*, 9(1), 107–121.
5. Train, D. (1957), Transmission of forces through a powder mass during the process of pelleting, *Trans. Inst. Chem. Eng.*, 35, 258–266.
6. Parrott, E. L. (1990), Compression in pharmaceutical dosage forms, in Lieberman, H. A., Lachman, L., and Schwartz, J. B., Eds., *Pharmaceutical Dosage Forms, Tablets. Bd.2*, Marcel Dekker, New York, pp. 201–243.
7. Guo, H. X., Heinämäki, J., and Yliruusi, J. (1999), Characterization of particle deformation during compression measured by confocal laser scanning microscopy, *Int. J. Pharm.*, 186, 99–108.
8. Hiestand, E. N. (1997), Mechanical properties of compacts and particles that control tableting success, *J. Pharm. Sci.*, 86, 985–990.
9. Rumpf, H. (1958), Grundlagen und Methoden des Granulierens, *Chem. Ing. Tech.*, 30, 144–158.
10. Joneja, S. K., Harcum, W. W., Skinner, G. W., Barnum, P. E., and Guo, J. H. (1999), Investigating the fundamental effects of binders on pharmaceutical tablet performance, *Drug. Dev. Ind. Pharm.*, 25, 1129–1135.
11. Adolfsson, A., Gustafsson, C., and Nyström, C. (1999), Use of tablet tensile strength adjusted to surface area and mean interparticulate distance to evaluate dominating bonding mechanism, *Drug Dev. Ind. Pharm.*, 25, 753–764.
12. Rippie, E. G., and Morehead, W. T. (1994), Structure evolution of tablets during compression unloading, *J. Pharm. Sci.*, 83, 708–715.
13. Hoag, S. W., and Rippie, E. G. (1994), Thermodynamic analysis of energy dissipation by pharmaceutical tablets during stress unloading, *J. Pharm. Sci.*, 83, 903–908.
14. Leuenberger, H., and Ineichen, L. (1997), Percolation theory and physics of compression, *Eur. J. Pharm. Biopharm.*, 44, 269–272.
15. Hiestand, E. N. (1997), Principles, tenets and notions of tablet bonding and measurements of strength, *Eur. J. Pharm. Biopharm.*, 44, 229–242.
16. Armstrong, N. A., and Haines-Nutt, R. F. (1972), Elastic recovery and surface area changes in compacted powder systems, *J. Pharm. Pharmacol.*, 24S, 135P–136P.
17. Newton, J. M., and Rowley, G. (1973), The influence of tablet weight on compaction pressure/tablet density relations, *J. Pharm. Pharmacol.*, 25, 767–768.
18. van der Voort Maarschalk, K. (1997), *Tablet relaxation, origin and consequences of stress relief in tablet formation*, Ph.D. thesis, Riksuniversiteit Groningen.
19. Aulton, M. E., Travers, D. N., and White, P. J. P. (1973), Strain recovery of compacts on extended storage, *J. Pharm. Pharmacol.*, 25, 79P–86P.
20. York, P., and Baily, E. D. (1977), Dimensional changes of compacts after compression, *J. Pharm. Pharmacol.*, 29, 70–74.
21. van der Voort Maarschalk, K., Zuurman, K., Vromans, H., Bolhuis, G. K., and Lerk, C. F. (1997), Stress relaxation of compacts produced from viscoelastic materials, *Int. J. Pharm.*, 151, 27–34.
22. Führer, C. (1975), *Kristallographische Aspekte bei der Tablettenherstellung*, Fortbildungskurs Arbeitsgemeinschaft für Pharmazeutische Verfahrenstechnik, Braunschweig.
23. Hüttenrauch, R. (1978), Molekulargalenik als Grundlage moderner Arzneiformung, *Acta Pharm. Technol.*, 24(S6), 55–127.

24. Elamin, A., Sebhatu, T., and Ahlneck, C. (1995), The use of amorphous model substances to study mechanically activated materials in the solid state, *Int. J. Pharm.*, 119, 25–36.
25. Sebhatu, T., Ahlneck, C., and Alderborn, G. (1997), The effect of moisture content on the compression and bond-formation properties of amorphous lactose particles, *Int. J. Pharm.*, 146, 101–114.
26. Sebhatu, T., and Alderborn, G. (1999), Relationships between the effective interparticulate contact area and the tensile strength of tablets of amorphous and crystalline lactose of varying particle size, *Eur. J. Pharm. Sci.*, 8, 235–242.
27. Moldenhauer, H., Kala, H., Zessin, G., and Dittgen, M. (1980), Zur pharmazeutischen Technologie der Tablettierung, *Pharmazie*, 35, 714–726.
28. Chan, H. K., and Doelker, E. (1985), Polymorphic transformation of some drugs under compression, *Drug Dev. Ind. Pharm.*, 11, 315–332.
29. Pirttimäki, J., Laine, E., Ketolainen, J., and Paronen, P. (1993), Effects of grinding and compression on crystal structure of anhydrous caffeine, *Int. J. Pharm.*, 95, 93–99.
30. Hiestand, E. N. (1991), Tablet bond. I. A theoretical model, *Int. J. Pharm.*, 67, 217–229.
31. Hiestand, E. N., and Smith, D. P. (1991), Tablet bond. II. Experimental check of the model, *Int. J. Pharm.*, 67, 231–246.
32. Ritschel, W. A., and Bauer-Brandl, A. (2002), *Die Tablette—Handbuch der Entwicklung, Herstellung und Qualitätssicherung*, Editio Cantor Verlag, Aulendorf Germany.
33. Bogda, M. J. (2002), Tablet compression: Machine theory, design, and process troubleshooting, in Swarbrick, J., and Boylan, J. C., Eds., *Encyclopedia of Pharmaceutical Technology*, Marcel Dekker, New York, pp. 2669–2688.
34. Hausmann, R., Kaufmann, H.-J., and Richter, K. (1996), Neue Tablettiertechnologie mit zentrifugaler Matrizenfüllung und vollautomatischer Reinigung, *Pharm. Ind.*, 58, 842–846.
35. Picker, K. M. (2000), The automatic micrometer screw, *Eur. J. Pharm. Biopharm.*, 49(2), 171–176.
36. Picker, K. M. (2001), Time dependence of elastic recovery for characterization of tableting materials, *Pharm. Dev. Technol.*, 6(1), 61–70.
37. Caraballo, I., Millan, M., Fini, A., Rodriguez, L., and Cavallari, C. (2000), Percolation thresholds in ultrasound compacted tablets, *J. Controlled Release*, 69, 345–355.
38. Aulton, M. E. (1981), Indentation hardness profiles across the faces of some compressed tablets, *Pharm. Acta Helv.*, 56, 133–136.
39. Konkel, P., and Mielck, J. B. (1997), Associations of parameters characterizing the time course of the tabletting process on a reciprocating and on a rotary tabletting machine for high-speed compression, *Eur. J. Pharm. Biopharm.*, 44, 289–301.
40. Sucker, H., Fuchs, P., and Speiser, P. (1991), *Pharmazeutische Technologie*, Thieme Verlag, Stuttgart.
41. Bateman, S. D., Rubinstein, M. H., Rowe, R. C., Roberts, R. J., Drew, P., and Ho, A. Y. K. (1989), A comparative investigation of compression simulators, *Int. J. Pharm.*, 49, 209–212.
42. Celik, M., and Marshall, K. (1989), Use of a compaction simulator system in tabletting research, *Drug Dev. Ind. Pharm.*, 15, 759–800.
43. Celik, M., and Lordi, N. G. (1991), The pharmaceutical compaction research laboratory and information center, *Pharm. Technol.*, 15, 112–116.

44. Rubinstein, M. H., Bateman, S. D., and Thacker, H. S. (1991), Compression to constant thickness or constant force: Producing more consistent tablets, *Pharm. Technol.*, 1, 150–158.
45. Celik, M., Ong, J. T., Chowhan, Z. T., and Samuel, G. J. (1996), Compaction simulator studies of a new drug substance: Effect of particle size and shape, and its binary mixtures with microcrystalline cellulose, *Pharm. Dev. Technol.*, 1, 119–126.
46. Picker, K. M. (2000), Three-dimensional modeling to determine properties of tableting materials on rotary machines using a rotary tableting machine simulator, *Eur. J. Pharm. Biopharm.*, 50(2), 293–300.
47. Picker, K. M. (2003), The 3D model: Comparison of parameters obtained from and by simulating different tableting machines, *AAPS PharmSciTech.*, 4(3) article 35.
48. Levin, M. (1999), *Theory and practice of tablet press simulation for process scale-up*, paper presented at the Arden House Conference Harsiman, New York, USA.
49. Rippie, E. G., and Danielsson, D. W. (1981), Viscoelastic stress/strain behaviour of pharmaceutical tablets: Analysis during unloading and postcompression periods, *J. Pharm. Sci.*, 70, 476–482.
50. Oates, R. J., and Mitchell, A. G. (1989), Calculation of punch displacement and work of powder compaction on a rotary tablet press, *J. Pharm. Pharmacol.*, 41, 517–523.
51. Oates, R. J., and Mitchell, A. G. (1990), Comparison of calculated and experimentally determined displacement on a rotary tablet press using both Manesty and IPT punches, *J. Pharm. Pharmacol.*, 42, 388–396.
52. Oates, R. J., and Mitchell, A. G. (1994), A new method of estimating volume during powder compaction and the work of compaction on a rotary tablet press from measurements of applied vertical forces, *J. Pharm. Pharmacol.*, 46, 270–275.
53. Pudipeddi, M., Venkatesh, G., Faulkner, P., and Palepu, N. (1993), Correlations between compaction simulator and instrumented Betapress, *Pharm. Res.*, 10S, S165.
54. Ruegger, C. D. (1996), An investigation of the effect of compaction profiles on the tableting properties of pharmaceutical substances, Ph.D. thesis, Rutgers University, Newark, NJ.
55. Levin, M., Tsygan, L., and Dukler, S. (1998), U.S. Patent 6,106.262, International Patent Application No. PCT/US98/27421.
56. MCC Corp. (2000), *The Presster Binder*, Technical Information, MCC Corp., East Hanover, New Jersey, USA.
57. Lamey, K. (2000), *Correlations between compaction simulators and rotary tableting machines*, paper presented at the Compaction Simulator's User Meeting, Loughbourough, UK.
58. Guntermann, A. (2005), *The Presster—A tablet press simulator*, paper presented at TabletTech, Brussels.
59. MedelPharm (2005), *Technical information*, Medel Pharm., Bourg-en-Bresee, France.
60. Brake, E. F. (1951), *Development of methods for measuring pressures during tablet manufacture*, M.S. thesis, Purdue University, West Lafayette, IN.
61. Higuchi, T., Arnold, R. D., Tucker, S. J., and Busse, L. W. (1952), The physics of tablet compression. I. A preliminary report, *J. Am. Pharm. Assoc.*, 41, 93–96.
62. Higuchi, T., Nelson, E., and Busse, L. W. (1954), The physics of tablet compression. III: Design and construction of an instrumented tableting machine, *J. Am. Pharm. Assoc.*, 43, 344–348.
63. Morehead, W. T., and Rippie, E. G. (1990), Timing relationships among maxima of punch and die-wall stress and punch displacement during compaction of viscoelastic solids, *J. Pharm. Sci.*, 79, 1020–1022.

64. Yeh, C., Altaf, S. A., and Hoag, S. W. (1997), Theory of force transducer design optimization for die wall stress measurement during tablet compaction: Optimization and validation of split-web die using finite element analysis, *Pharm. Res.*, 14, 1161–1170.

65. Khossravi, D., and Morehead, W. T. (1997), Consolidation mechanisms of pharmaceutical solids: A multi-compression cycle approach, *Pharm. Res.*, 14, 1039–1045.

66. Knoechel, E. L., Sperry, C. C., Ross, H. E., and Lintner, C. J. (1967), Instrumented rotary tablet machines I, *J. Pharm. Sci.*, 56, 109–115.

67. Schmidt, P. C., Tenter, U., and Hocke, J. (1986), Presskraft-und Weg-Zeit-Charakteristik von Rundlauftablettenpressen. 1. Mitt.: Instrumentierung von Einzelstempeln zur Presskraftmessung, *Pharm. Ind.*, 48, 1546–1553.

68. Bauer-Brandl, A. (1998), Qualifizierung der Kraftmessung an Tablettenpressen, *Pharm. Ind.*, 60, 63–69.

69. Hoag, S. W., Nair, R., and Muller, F. X. (2000), Force-transducer-design optimization for the measurement of die-wall stress in a compaction simulator, *Pharm. Pharmacol. Commun.*, 6(7), 293–298.

70. Laich, T., and Kissel, T. (1995), Axial die-wall force minimum. Influences and significance for elastic behavior of single components and binary mixtures of excipients, *Pharm. Ind.* 57(2), 174–182.

71. Nelson, E., Busse, L. W., and Higuchi, T. (1955), The physics of tablet compression: VII. Determination of energy expenditure in the tablet compression process, *J. Am. Pharm. Assoc.* Sc. Ed., 44, 223.

72. Schmidt, P. C., and Tenter, U. (1987), Force and displacement characteristics of rotary tableting machines, *Pharm. Ind.*, 49, 637–642.

73. Matz, C. (1999), *Evaluation einer IR-telemetrischen Kraft/Weg-Instrumentierung für Rundlauf tabletten pressen.* Differenzierung des Verformungsverhaltens direktkomprimierbarer Tablettierhilfsstoffe, Dissertation, Universität Freiburg, Freiburg, Germany.

74. Matz, C., Bauer-Brandl, A., Rigassi, T., Schubert, R., and Becker, D. (1999), On the accuracy of a new displacement instrumentation for rotary tablet presses, *Drug Dev. Ind. Pharm.*, 25, 117–130.

75. Müller, F., and Caspar, U. (1984), Viskoelastische Phänomene während der Tablettierung, *Pharm. Ind.*, 46, 1049–1056.

76. Krumme, M. (1992), *Entwicklung rechnergestützter Verfahren zur Kompressions-und Festigkeitsanalyse von Tabletten*, Dissertation, Freie Universität, Berlin.

77. Krumme, M., Schwabe, L., and Frömming, K. H. (1998), Development of computerised procedures for the characterization of the tableting properties with eccentric machines. High precision displacement instrumentation for eccentric tablet machines, *Acta Pharm. Hung.*, 68, 322–331.

78. Rankell, A. S., and Higuchi, T. (1968), Physics of tablet compression. XV. Thermodynamic and kinetic aspects of adhesion under pressure, *J. Pharm. Sci.*, 58, 574–577.

79. DeCrosta, M. T., Schwartz, J. B., Wigent, R. J., and Marshall, K. (2000), Thermodynamic analysis of compact formation; compaction, unloading, and ejection. I. Design and development of a compaction calorimeter and mechanical and thermal energy determinations of powder compaction, *Int. J. Pharm.*, 198, 113–134.

80. DeCrosta, M. T., Schwartz, J. B., Wigent, R. J., and Marshall, K. (2001), Thermodynamic analysis of compact formation; compaction, unloading, and ejection. II. Mechanical energy (work) and thermal energy (heat) determinations of compact unloading and ejection, *Int. J. Pharm.*, 213, 45–62.

81. Bogs, H., and Lenhardt, E. (1971), Zur Kenntnis thermischer Vorgänge beim Tablettenpressen, *Pharm. Ind.*, 33, 850–854.

82. Hanus, E. J., and King, L. D. (1968), Thermodynamic effects in compression of solids, *J. Pharm. Sci.*, 57, 677–684.

83. Führer, C., and Parmentier, W. (1977), Zur Thermodynamik der Tablettierung, *Acta Pharm. Technol.*, 23, 205–213.

84. Ketolainen, J., Ilkka, J., and Paronen, P. (1993), Temperature changes during tabletting measured using infrared thermoviewer, *Int. J. Pharm.*, 92, 157–166.

85. Picker-Freyer, K. M., and Schmidt, A. G. (2004), Does temperature increase induced by tableting contribute to tablet quality? *J. Therm. Anal. Cal.*, 77, 531–539.

86. Beissenhirtz, M. (1974), Ermittlung der Verformungscharakteristik von Tablettierstoffen mit Hilfe elektrisch leitfähiger Zusätze, Dissertation, Universität Bonn, Bonn, Germany.

87. Schmidt, J. (1997), *Direktablettierung niedrigschmelzender nichtsteroidaler Antirheumatika mit mikrokristallinen Cellulosen*, Dissertation, Universität Halle-Wittenberg, Wittenberg, Germany.

88. Celik, M. (1992), Overview over compaction analysis techniques, *Drug Dev. Ind. Pharm.*, 18, 767–810.

89. Alderborn, G., and Nyström, C. (1996), *Pharmaceutical Powder Compaction Technology*, Marcel Dekker, New York.

90. Chilamkurti, R. N., Rhodes, C. T., and Schwartz, J. B. (1982), Some studies on compression properties of tablet matrices using a computerized instrumented press, *Drug Dev. Ind. Pharm.*, 8, 63–86.

91. Emschermann, B., and Müller, F. (1981), Evaluation of force measurements in tablet manufacture, *Pharm. Ind.*, 43, 191–194.

92. Vogel, P. J., and Schmidt, P. C. (1993), Force-time curves of a modern rotary tablet machine. Part 2. Influence of compression force and tableting speed of the deformation mechanisms of pharmaceutical substances, *Drug Dev. Ind. Pharm.*, 19, 1917–1930.

93. Schmidt, P. C., and Vogel, P. J. (1994), Force-time-curves of a modern rotary tablet machine. Part 1. Evaluation techniques and characterization of deformation behavior of pharmaceutical substances, *Drug Dev. Ind. Pharm.*, 20, 921–934.

94. Schmidt, P. C., and Leitritz, M. (1997), Compression force/time-profiles of microcrystalline cellulose, dicalcium phosphate dihydrate and their binary mixtures—A critical consideration of experiments and parameters, *Eur. J. Pharm. Biopharm.*, 44, 303–313.

95. Yliruusi, J. K., Merkku, P., Hellen, L., and Antikainen, O. K. (1997), A new method to evaluate the elastic behavior of tablets during compression, *Drug Dev. Ind. Pharm.*, 23, 63–68.

96. Dietrich, R., and Mielck, J. B. (1985), Eignung der Weibull-Funktion zur Charakterisierung des zeitabhängigen Verformungsverhaltens von Tablettierhilfsstoffen, *Acta Pharm. Technol.*, 31, 67–76.

97. Dietrich, R., and Mielck, J. B. (1985), Parametrisierung des zeitlichen Verlaufs der Verdichtung bei der Tablettierung mit Hilfe der modifizierten Weibull-Funktion, *Pharm. Ind.*, 47, 216–220.

98. Picker, K. M. (1995), Hydrophile Matrixtabletten: Tablettierung und Freisetzung-unter besonderer Berücksichtigung der relativen Feuchte während der Herstellung, Dissertation, Universität Hamburg, Hamburg, Germany.

99. Shlieout, G., Wiese, M., and Zessin, G. (1999), A new method to evaluate the consolidation behavior of pharmaceutical materials by using the Fraser-Suzuki function, *Drug Dev. Ind. Pharm.*, 25, 29–36.

100. Armstrong, N. A., and Palfrey, L. P. (1989), The effect of machine speed on the consolidation of four directly compressible tablet diluents, *J. Pharm. Pharmacol.*, 41, 149–151.

101. Ho, A. Y. K., and Jones, T. M. (1988), Rise time: A new index of tablet compression, *J. Pharm. Pharmacol.*, 40, 74P.

102. Tsardaka, E. D. (1990), Viscoelastic properties and compaction behavior of pharmaceutical particulate material, Ph.D. Thesis, University of Bath, Bath, England.

103. Dwivedi, S. K., Oates, R. J., and Mitchell, A. G. (1991), Peak offset times as an indication of stress relaxation during tableting on a rotary tablet press, *J. Pharm. Pharmacol.*, 43, 673–678.

104. Führer, C. (1962), Über den Druckverlauf bei der Tablettierung in Exzenterpressen, *Dtsch. Apoth. Ztg.*, 102, 827–842.

105. Moldenhauer, H., Hünerbein, B., and Kala, H. (1972), Recording of pressure-path diagrams from an eccentric press using piezoelectric measurement, *Pharmazie*, 27, 417–418.

106. Dürr, M., Hanssen, D., and Harwalik, H. (1972), Kennzahlen zur Beurteilung der Verpressbarkeit von Pulvern und Granulaten, *Pharm. Ind.*, 34, 905–911.

107. de Blaey, C. J., and Polderman, J. (1970), Compression of pharmaceuticals. I. The quantitative interpretation of force-displacement curves, *Pharm. Weekblad*, 105, 241–250.

108. de Blaey, C. J., and Polderman, J. (1971), Compression of pharmaceuticals. II. Registration and determination of force-displacement curves using a small digital computer, *Pharm. Weekblad*, 106, 57–65.

109. Stamm, A., and Mathis, C. (1976), Verpressbarkeit von festen Hilfsstoffen für die Direkttablettierung, *Acta Pharm.*Technol., 24(S1), 7–16.

110. Dürr, M. (1976), Bedeutung der Energiebilanz beim Tablettieren für die Entwicklung von Tabletten rezepturen, *Acta Pharm. Technol.*, 22, 185–194.

111. Parmentier, W. (1978), Investigation on the interpretation of the Kraft-Weg diagram, *Pharm. Ind.*, 40, 860–865.

112. Lammens, R. F., Polderman, J., and De Blaey, C. J. (1979), Evaluation of force displacement measurements during powder compaction. Part 1. Precision and accuracy of force measurements, *Int. J. Pharm. Technol. Prod. Manuf.*, 1, 26–35.

113. Lammens, R. F., Polderman, J., Armstrong, N. A., and De Blaey, C. J. (1980), Evaluation of force displacement measurements during powder compaction. Part 2. Precision and accuracy of powder height and displacement measurements, *Int. J. Pharm. Technol. Prod. Manuf.*, 1, 26–35.

114. Krycer, I., and Pope, D. G. (1982), The interpretation of powder compaction data—A critical review, *Drug Dev. Ind. Pharm.*, 8, 107–347.

115. Ragnarsson, G., and Sjogren, J. (1985), Force displacement measurements in tableting, *J. Pharm. Pharmacol.*, 37, 145–150.

116. Antikainen, O. K., and Yliruusi, J. K. (1997), New parameters derived from tablet compression curves. Part 2. Force-displacement curve, *Drug Dev. Ind. Pharm.*, 23, 81–93.

117. Ragnarson, G. (1996), Force-displacement and network measurements, in Alderborn, G., and Nyström, C., Eds., *Pharmaceutical Powder Compaction Technology*, Marcel Dekker, New York, pp. 77–97.

118. Heckel, R. W. (1961), An analysis of powder compaction phenomena, *Trans. Metall. Soc. AIME*, 221, 1001–1008.

119. Heckel, R. W. (1961), Density-pressure relationships in powder compaction, *Trans. Metall. Soc. AIME*, 221, 671–675.
120. Hersey, J. A., and Rees, J. E. (1970), Deformation of particles during briquetting, paper presented at Proc. 2nd Particle Size Analysis Conference of the Society of Analytical Chemistry, Bradford, 33.
121. Kawakita, K., and Lüdde, K. H. (1970/71), Some considerations on powder compression equations, *Powder Tech.*, 4, 61–68.
122. Lüdde, K. H., and Kawakita, K. (1966), Die Pulverkompression, *Pharmazie*, 21, 393–403.
123. Cooper, A. R., and Eaton, L. E. (1962), Compaction behaviour of several ceramic powders, *J. Am. Ceram. Soc.*, 5, 97–101.
124. Walker, E. (1923), The properties of powders. Part VII. The influence of the velocity of compression on the apparent compressibility of powders, *Trans. Faraday Soc.*, 19, 614–620.
125. Bal'shin, M. Y. (1938), Contribution to the theory of powder metallurgical processes, *Metalloprom.*, 18, 124–147.
126. Sönnergaard, J. M. (2000), Impact of particle density and initial volume on mathematical compression models, *Eur. J. Pharm. Sci.*, 11, 307–315.
127. Sönnergaard, J. M. (1999), A critical evaluation of the Heckel equation, *Int. J. Pharm.*, 193, 63–71.
128. Sun, C., and Grant, D. J. W. (2001), Influence of elastic deformation of particles on Heckel analysis, *Pharm. Dev. Technol.*, 6, 193–200.
129. Rees, J. E., and Rue, P. J. (1978), Time-dependent deformation of some direct compression excipients, *J. Pharm. Pharmacol.*, 30, 601–607.
130. York, P. (1979), A consideration of experimental variables in the analysis of powder compaction behavior, *J. Pharm. Pharmacol.*, 31, 244–246.
131. Vachon, M. G., and Chulia, D. (1999), The use of energy indices in estimating powder compaction functionality of mixtures in pharmaceutical tableting, *Int. J. Pharm.*, 177, 183–200.
132. Krumme, M., Schwabe, L., and Frömming, K. H. (2000), Development of computerised prodedures for the characterisation of the tableting properties with eccentric machines: Extended Heckel analysis, *Eur. J. Pharm. Biopharm.*, 49, 275–286.
133. Humbert-Droz, P., Gurny, R., Mordier, D., and Doelker, E. (1983), Densification behavior of drugs having bioavailability problems, *Int. J. Pharm.*, 4, 29–35.
134. Imbert, C., Tchoreloff, P., Leclerc, B., and Couarraze, G. (1997), Indices of tableting performance and application of percolation theory to powder compaction, *Eur. J. Pharm. Biopharm.*, 44, 273–282.
135. Kuentz, M., and Leuenberger, H. (2000), A new approach to tablet strength of a binary mixture consisting of a well and a poorly compactable substance, *Eur. J. Pharm. Biopharm.*, 49, 151–159.
136. Gabaude, C. M. D., Guillot, M., Gautier, J. C., Saudemon, P., and Chulia, D. (1999), Effects of true density, compacted mass, compression speed, and punch deformation on the means yield pressure, *J. Pharm. Sci.*, 88, 725–730.
137. Duberg, M., and Nyström, C. (1986), Studies on the compression of tablets. XVII. Porosity-pressure curves for the characterization of volume reduction mechanisms, *Powder Technol.*, 46, 67–75.
138. Paronen, P. (1986), Heckel-plots as indicators of elastic properties of pharmaceuticals, *Drug Dev. Ind. Pharm.*, 12, 1903–1912.
139. Morris, L. E., and Schwartz, J. B. (1995), Isolation of densification regions during powder compression, *Drug Dev. Ind. Pharm.*, 21, 427–446.

140. Kuentz, M., Leuenberger, H., and Kolb, M. (1999), Fracture in disordered media and tensile strength of microcrystalline cellulose tablets at low relative densities, *Int. J. Pharm.*, 182, 243–255.

141. Tsardaka, K. D., Rees, J. E., and Hart, J. P. (1988), Compression and recovery behaviour of compacts using extended Heckel plots, *J. Pharm. Pharmacol.*, 40, 73P.

142. Tsardaka, K. D., and Rees, J. K. (1990), Relations between viscoelastic parameters and compaction properties of two modified starches, *J. Pharm. Pharmacol.*, 42, 77P.

143. Picker, K. M. (2000), A new theoretical model to characterize the densification behavior of tableting materials, *Eur. J. Pharm. Biopharm.*, 49(3), 267–273.

144. Picker-Freger, K. M. (2007), The 3-D Model: Experimental testing of the parameters d, e, and w and validation of the analysis, *J. Pharm. Sci.*, 96(5), 1408–1417.

145. Hoblitzell, J. R., and Rhodes, C. T. (1986), Preliminary investigations on the parity of tablet compression data obtained from different instrumented presses, *Drug Dev. Ind. Pharm.*, 12, 507–525.

146. David, S. T., and Augsburger, L. L. (1977), Plastic flow during compression of directly compressible fillers and its effect on tablet strength, *J. Pharm. Sci.*, 66, 155–159.

147. Müller, F., and Caspar, U. (1984), Viskoelastische Phänomene während der Tablettierung, *Pharm. Ind.*, 46, 1049–1056.

148. Müller, F. (1996), Viscoelastic models, in Alderborn, G., and Nyström, C., Eds., *Pharmaceutical Powder Compaction Technology*, Marcel Dekker, New York, pp. 99–132.

149. Bauer, A. D. (1991), Untersuchungen zur Prozessdatengewinnung, Viskoelastizität und Struktur von Tabletten, Dissertation, Universität Bonn, Bonn, Germany.

150. Staniforth, J. N., Baichwal, A. R., and Hart, J. P. (1987), Interpretation of creep behavior of microcrystalline cellulose powders and granules during compaction, *Int. J. Pharm.*, 40, 267–269.

151. Staniforth, J. N., and Patel, C. J. (1989), Creep compliance behavior of direct compression excipients, *Powder Technol.*, 57, 83–87.

152. Tsardaka, K. D., and Rees, J. E. (1989), Plastic deformation and retarded elastic deformation of particulate solids using creep experiments, *J. Pharm. Pharmacol.*, 41, 28P.

153. Malamataris, S., and Rees, J. E. (1993), Viscoelastic properties of some pharmaceutical powders compared using creep compliance, extended Heckel analysis and tablet strength measurements, *Int. J. Pharm.*, 92, 123–135.

154. Rees, J. E., and Tsardaka, K. D. (1994), Some effects of moisture on the viscoelastic behaviour of modified starch during powder compaction, *Eur. J. Pharm. Biopharm.*, 40, 193–197.

155. Picker, K. M., and Bikane, F. (2001), An evaluation of three-dimensional modeling of compaction cycles by analyzing the densification behavior of binary and ternary mixtures, *Pharm. Dev. Technol.*, 6(3), 333–342.

156. Picker, K. M. (2004), The 3D model: Explaining densification and deformation mechanisms by using 3D parameter plots, *Drug Dev. Ind. Pharm.*, 30(4), 413–425.

157. Picker, K. M. (2003), The 3D model: Does time plasticity represent the influence of tableting speed? *AAPS PharmSciTech.*, 4(4), article 66.

158. Hauschild, K., and Picker, K. M. (2003), The 3D model: Comparison with other data analysis techniques for tableting, *AAPS PharmSci*, 5(S1), W4304.

159. Hauschild, K., and Picker-Freyer, K. M. (2004), Evaluation of a new coprocessed compound based on lactose and maize starch for tablet formulation, *AAPS PharmSci*, 6(2), article 16.

160. Picker, K. M. (2002), New insights in the process of tablet formation—Ways to explore soft tableting, Martin-Luther-University Halle-Wittenberg and Görich und Weiershäuser Verlag, Marburg, Germany.
161. Wurster, D. E., Rowlings, C. E., and Creekmore, J. R. (1995), Calorimetric analysis of powder compression: I. Design and development of a compression calorimeter, *Int. J. Pharm.*, 116, 179.
162. Picker, K. M. (2003), The relevance of glass transition temperature for the process of tablet formation, *J. Therm. Anal. Cal.*, 73(2), 597–605.
163. Armstrong, N. A., and Haines-Nutt, R. F. (1972), Elastic recovery and surface area changes in compacted powder systems, *J. Pharm. Pharmacol.*, 24, 135–136.
164. Picker-Freyer, K. M. (2005), Carrageenans: Analysis of tablet formation and properties (Part 1), *Pharm. Technol. Eur.*, 17(8), 37–44.
165. Picker-Freyer, K. M. (2005), Carrageenans: Analysis of tablet formation and properties (Part 2), *Pharm. Technol. Eur.*, 17(9), 32–44.
166. Schmidt, A. G., and Picker-Freyer, K. M. (2006), Influence of mechanical activation on the micro structure of widely used excipients for solid dosage forms, paper presented at the 5th World Meeting of Pharmaceutics, Pharmaceutical Technology and Biopharmaceutics, Geneva.
167. Leuenberger, H., Hiestand, E. N., and Sucker, H. (1981), Contribution to the theory of powder compression, *Chemie Ing. Tech.*, 53(1), 45–47.
168. Bauer-Brandl, A. (2002), Tooling for tableting, in Swarbrick, J., and Boylan, J. C., Eds., *Encyclopedia of Pharmaceutical Technology*, Marcel Dekker, New York.
169. Conte, U., Colombo, P., Caramella, C., Ferrari, F., Gazzaniga, A., Guyot, J. C., La Manna, A., and Traisnel, M. (1988), Influence of tablet weight control systems during production on biopharmaceutical properties of the tablets, *Acta Pharm. Technol.*, 34(2), 63–67.
170. Cole, G. C. (1998), *Pharmaceutical Production Facilities: Design and Applications*, Taylor and Francis, London.
171. Lyon, R. C., Lester, D. S., Lewis, E. N., Lee, E., Yu, L. X., Jefferson, E. H., and Hussain, A. S. (2002), Near-infrared spectral imaging for quality assurance of pharmaceutical products: Analysis of tablets to assess powder blend homogeneity, *AAPS PharmSciTech*, 3(3), article 17.
172. Lai, C. K., Zahari, A., Miller, B., Katstra, W. E., Cima, M. J., and Cooney, C. L. (2004), Nondestructive and on-line monitoring of tablets using light-induced fluorescence technology, *AAPS PharmSciTech*, 5(1), article 3.
173. Lee, T. H., and Lin, S. Y. (2004), Microspectroscopic FT-IR mapping system as a tool to assess blend homogeneity of drug-excipient mixtures, *Eur. J. Pharm. Sci.*, 23(2), 117–122.
174. U.S. Food and Drug Administration, Center for Drug Evaluation and Research, Office of Pharmaceutical Science, Process Analytical Technology (PAT) Initiative, assessed on 12/20/2007, available: www.fda.gov/cder/OPS/PAT.htm.
175. Oeser, W. H., and Sander, A. (1992), *Pharma-Betriebsverordnung: Grundregeln für die Herstellung von Arzneimitteln (GMP)*, Wissenschaftliche Verlagsgesellschaft Stuttgart, Germany.
176. Hyde, J. M. (1985), New developments in CIP practices, *Chem. Eng. Prog.*, 81(1), 39–41.
177. Picker, K. M., and Mielck, J. B. (1998), Effect of relative humidity during tableting on matrix formation of hydrocolloids: Densification behaviour of cellulose ethers, *Pharm. Dev. Technol.*, 3(1), 31–41.

178. Hauschild, K., and Picker-Freyer, K. M. (2006), Evaluation of tableting and tablet properties of Kollidon SR: The influence of moisture and mixtures with theophylline monohydrate, *Pharm. Technol. Dev.*, 11(1), 125–140.

179. Kachrimanis, K., Nikolakakis, I., and Malamataris, S. (2003), Tensile strength and disintegration of tableted silicified microcrystalline cellulose: Influences of interparticle bonding, *J. Pharm. Sci.*, 92(7), 1489–1501.

180. Steendam, R., Eissens, A. C. E., Frijlink, H. W., and Lerk, C. F. (2000), Plasticization of amylodextrin by moisture. Consequences for drug release from tablets, *Int. J. Pharm.*, 204(1–2), 23–33.

181. Hanssen, D., Führer, C., and Schaefer, B. (1970), Evaluation of magnesium stearate as tablet lubricating agent by electronic pressure measurements, *Pharm. Ind.*, 32(2), 97–101.

182. Bolhuis, G. K., and Lerk, C. F. (1977), Film forming of tablet lubricants during the mixing process of solids, *Acta Pharm. Technol.*, 23(1), 13–20.

183. Zuurman, K., Van der Voort Maarschalk, K., and Bolhuis, G. K. (1999), Effect of magnesium stearate on bonding and porosity expansion of tablets produced from materials with different consolidation properties, *Int. J. Pharm.*, 179(1), 107–115.

184. Van Der Voort Maarschalk, K., and Bolhuis, G. K. (1998), Improving properties of materials for direct compaction. Part II, *Pharm. Technol. Eur.*, 10(10), 28–36.

185. Shotton, E., and Lewis, C. J. (1964), Effect of lubrication on the crushing strength of tablets, *J. Pharm. Pharmacol.*, 16S, 111–120.

186. Van Der Voort Maarschalk, K., and Bolhuis, G. K. (1999), Improving properties of materials for direct compaction, *Pharm. Technol.*, 23(5), 34–46.

187. Steffens, K. J., Mueller, B. W., and List, P. H. (1982), Tribological laws and results in tablet production. 7. Studies on magnesium stearate commercial products, *Pharm. Ind.*, 44(8), 826–830.

188. Hoelzer, A. W. (1984), Batch to batch variations of commercial magnesium stearates. Chemical, physical and lubricant properties, *Labo-Pharma*, 338, 28–36.

189. Miller, T. A., and York, P. (1985), Frictional assessment of magnesium stearate and palmitate lubricant powders, *Powder Technol.*, 44(3), 219–226.

190. Lindberg, N. O. (1972), Evaluation of some tablet lubricants, *Acta Pharm. Suec.*, 9(3), 207–214.

191. Alpar, O., Deer, J. J., Hersey, J. A., and Shotton, E. (1969), Possible use of poly(tetrafluo roethylene) (Fluon) as a tablet lubricant, *J. Pharm. Pharmacol.*, 21S, 6–8.

192. Shah, N. H., Stiel, D., Weiss, M., Infeld, M. H., and Malick, A. W. (1986), Evaluation of two new tablet lubricants—Sodium stearyl fumarate and glyceryl behenate. Measurement of physical parameters (compaction, ejection and residual forces) in the tableting process and the effect of the dissolution rate, *Drug Dev. Ind. Pharm.*, 12(8–9), 1329–1346.

193. N'Diaye, A., Jannin, V., Berard, V., Andres, C., and Pourcelot, Y. (2003), Comparative study of the lubricant performance of Compritol HD5 ATO and Compritol 888 ATO: effect of polyethylene glycol behenate on lubricant capacity, *Int. J. Pharm.*, 254(2), 263–269.

194. Staniforth, J. N. (1987), Use of hydrogenated vegetable oil as a tablet lubricant, *Drug Dev. Ind. Pharm.*, 13(7), 1141–1158.

195. Dawoodbhai, S. S., Chueh, H. R., and Rhodes, C. T. (1987), Glidants and lubricant properties of several types of talcs, *Drug Dev. Ind. Pharm.*, 13(13), 2441–2467.

196. Roescheisen, G., and Schmidt, P. C. (1995), Preparation and optimization of L-leucine as lubricant for effervescent tablet formulations, *Pharm. Acta Helv.*, 70(2), 133–139.

197. Jahn, T., and Steffens, K.-J. (2005), Press chamber coating as external lubrication for high speed rotary presses: Lubricant spray rate optimization, *Drug Dev. Ind. Pharm.*, 31(10), 951–957.

198. Laich, T., and Kissel, T. (1998), Automatic adaptation of lubricant quantity by control of an external lubrication. Tests on a reciprocating and on a rotary tablet press, *Pharm. Ind.*, 60(10), 896–904.

199. Otsuka, M., Sato, M., and Matsuda, Y. (2001), Comparative evaluation of tableting compression behaviors by methods of internal and external lubricant addition: Inhibition of enzymatic activity of trypsin preparation by using external lubricant addition during the tableting compression process, *AAPS PharmSci*, 3(3), article 20.

200. Jetzer, W., and Leuenberger, H. (1984), Determination of capping tendency in pharmaceutical active ingredients and excipients, *Pharm. Acta Helv.*, 59(1), 2–7.

201. Bauer, K. H., Frömming, K. H., and Führer, C. (2002) *Lehrbuch der pharmazeutischen Technologie*, Wissenschafteiche Verlagsgesellschaft, Stuttgart, Germany.

6.5

CONTROLLED RELEASE OF DRUGS FROM TABLET COATINGS

SACIDE ALSOY ALTINKAYA
Izmir Institute of Technology, Urla-Izmir, Turkey

Contents

6.5.1 INTRODUCTION

Controlled-release technology for drug delivery applications is designed to target the drug to particular places or cells in the body, to overcome certain tissue and cellular barriers, and to control the duration and level of the drug in the body within a speficied therapeutic window. This usually implies achieving a prolonged, zero-order release rate over the desired duration of drug delivery. Controlled-release dosage forms provide sustained drug release and require less frequent drug loading than conventional forms. Thus, the toxic side effects of the drug are minimized and

Pharmaceutical Manufacturing Handbook: Production and Processes, edited by Shayne Cox Gad

patient's convenience and compliance are improved. Controlled-release systems are usually classified into four categories based on the rate-limiting step of the release process [1]: (1) diffusion-controlled systems, (2) chemically controlled systems, (3) swelling-controlled systems, and (4) magnetically controlled systems. Diffusion-controlled systems are formulated in two main geometries: reservoirs and matrices. In matrix systems, the drug is generally uniformly distributed or dissolved throughout a polymer. The release kinetics from these types of systems depend on the quantity and the diffusivity of drug in the matrix and the geometry of the system. The release rates from matrix systems usually decrease with time due to an increase in the path length for diffusion of drug and, thus, their release characteristics are not generally zero order. In the case of chemically controlled systems, the drug is either distributed through a biodegradable polymer or chemically bound to a polymer backbone chain. The drug release from the biodegradable polymer is controlled by degradation of the polymer through penetration of water or a chemical reaction [1]. The drug attached to the polymer is released by hydrolytic or enzymatic cleavage. In swelling-controlled systems, the drug is dissolved or distributed in a glassy polymer matrix. As water penetrates the dry matrix, the polymer swells and its glass transition temperature decreases below the temperature in the matrix. As a result, the glassy–rubbery transition occurs and it allows the release of the drug. The release rate of the drug out of these systems is mainly controlled by the rate and degree of swelling. Alternatively, the drug release can also be controlled magnetically by dispersing the drug and small magnetic beads within a polymer matrix and exposing them to an oscillating external magnetic field. (The drug is released as the matrix is exposed to an oscillating external magnetic field.)

Reservoir systems consist of a drug-containing core surrounded by a polymer membrane. The release rate of drug is controlled by its diffusion through the membrane [1, 2]. In addition to diffusional release, osmotic pumping mechanisms contribute to the total drug release rate if either the drug is highly soluble or an osmotic agent is added to the active core [1–4]. Reservoir systems are prepared in different geometries, such as coated tablets, beads, particles, membrane-based pouches, and microcapsules. The main advantage of these systems is their ability to maintain zero-order release rates [1, 2]. This is usually achieved by loading the powdered form of drug at a level far above the solubility of drug. Then, the concentration of drug at the internal wall of the reservoir becomes its saturation concentration and zero-order release occurs until the drug is completely dissolved. In addition, drug loading can be higher for these systems compared to other controlled-release systems; thus, the cost of formulation is minimized and the drug is released at a higher rate for a longer period of time. The major disadvantage of the reservoir system is rupture of the rate-controlling membrane if it is subject to dose dumping.

Numerous studies exist in the literature on the drug release from tablet coatings. In the majority of these studies, diffusional drug release takes place from the coatings [5–17]. Various factors such as the type [8, 10, 15], and concentration of the coating material [7, 8, 11], the type and amount of the plasticizer in the coating solution [10, 12, 15], the thickness of the coating [5], the composition of the tablet core [6, 14], the particle size of the coating material [10, 18], and the weight gain of the coating [7, 8, 11, 13, 16] were considered during the formulation of tablet coatings, and their effects on the drug release rates were investigated. The use of osmotic tablet coating systems are also described in the literature [3, 4, 17, 19–23]. Osmotic systems utilize osmotic pressure difference as a driving force, and in the simplest design they consist of an osmotic core containing drug with and without osmotic agents. Different studies have shown that the release rate of drugs from oral osmotic pump tablet coating systems is governed by various formulation variables such as solubility and osmotic pressure of core components [3, 4, 19–23], number [17] and size of the delivery orifice [3, 4, 20, 21, 23], drug loading [3, 4, 17] thickness of the coating [3, 4, 19, 21, 23], composition of the coating solution [3, 4, 17, 20, 21, 23], and weight of the coating [17, 20, 22]. The coatings prepared in most of these studies have dense structures with or without a hole drilled through the coating through which the drug is delivered. In some cases, drug delivery ports are formed by adding leachable materials to the coating [24, 25]. The main problem with these systems, in the absence of a hole, is an excessively prolonged drug release due to the low permeability of the coating. To increase permeability of the coatings, plasticizers and water-soluble additives were incorporated in the coating solution, and multilayer composite coatings [14, 15, 24, 26, 27] or multiple-compartment osmotic tablets were prepared [17, 22, 28, 29]. The permeability of the tablet coating systems were further improved by changing the structure of the coatings from dense to porous asymmetric ones [30–42]. Asymmetric membrane tablet coatings are characterized by a relatively thin, dense skin layer supported on a highly permeable, thicker, and porous sublayer. The permeabilities and the release rates of the drugs through the asymmetric-membrane capsule/tablet coatings were determined to be higher compared to conventional dense tablet coatings [30, 33, 34, 39]. The composition of the coating solution was found to have a significant effect on the structure of these type of coatings and thus on the release rate of the drug [30, 33, 35, 36, 39, 40–42]. In addition, it was reported that the drying condition had a significant impact on the structure of the asymmetric-membrane tablet coating and the release rate of the drug [41]. A review of the literature studies clearly indicates that asymmetric-type tablet coating is a new solution for developing controlled drug delivery systems, since the structure of these types of coatings can be varied easily by changing the preparation conditions without altering the coating material or significantly varying the coating thickness.

This work contains six sections in addition to the introduction. Sections 6.5.2–6.5.5 present a brief review of manufacturing methods for the application of coating materials on tablets, characterization methods used to evaluate the uniformity and defects of the tablet coatings, the techniques commonly used for manufacturing asymmetric-type membranes, and modeling approaches employed for optimization of the tablet coating formulations. The aim of the last two sections is to demonstrate the advantages of asymmetric membrane tablet coatings with respect to their drug release properties and the factors that affect the morphology of these types of tablet coatings. To achieve this goal, the in vitro release of a model compound, theophylline, from asymmetric membrane tablet coatings is determined and the morphology of the coatings is examined. In addition, the dynamics of the phase inversion is quantified in terms of ternary-phase diagrams coupled with composition paths determined from a mathematical model developed previously by our group [43]. To draw meaningful and objective conclusions from experimental data and derive an empirical expression for the release rate of drug, compositions of the coating solution are chosen using a statistical experimental design.

6.5.2 TABLET COATING METHODS

Four basic methods are commonly used for the application of coating materials on drug tablets: (1) pan coating, (2) fluidized-bed coating, (3) compression (press) coating, and (4) melt/dry coating.

Pan Coating The pan coating process is the oldest form of pharmaceutical coating for manufacturing small, coated particles or tablets. In this process, the particles are tumbled in a pan which is rotated at an angle of usually 45° to the horizontal surface at a speed between 20 and 50 rpm [44]. Coating fluid is sprayed onto the particles from a above by means of an air jet. The hot air blown through the coater evaporates the fluid and dries the film coating. Pan coating is generally preferred to coat large tablets since they are exposed to mechanical damage in other coating operations. The uniformity of the coating applied to the tablets and defects on the coating are important issues from a practical point of view. Operating conditions such as drum speed, drum solids loading, the presence/absence of baffles, air velocity, and temperature influence the movement of tablets within the moving bed, the circulation and surface exposure times spent by a tablet in the bulk of the moving bed and on the surface of the bed, respectively, and the rate of drying, which all in turn determine the uniformity of the coating in a pan coating process [45]. Different experimental tools such as light emission from a single luminous tablet, photographic and manual counting, positron emission particle tracking, magnetic resonance imaging, and near real-time video imaging techniques [45–51] were used to study the motion of tablets in the drum. Simulation of particle movement using discrete-element modeling has also been used to study the movement of tablets [52–55]. In addition, statistical experimental design was utilized to identify the critical processing variables that affect coating uniformity and loading of active agent coated on tablets [56]. Inlet airflow, pan speed, inlet air temperature, coating time, atomization pressure, and fan pressure were investigated as the process variables. Among these variables, atomization pressure, pan speed, and duration of coating were found to be critical process variables with respect to uniformity of the coating.

Fluidized-Bed Coating The basic principle of fluidized-bed coating is to suspend tablets in a moving hot-air stream in the upward direction during the coating process. The coating material is sprayed through a nozzle from the top, the side, or the bottom into the fluidized bed. The solvent in the solution is removed from the coating by the hot-air stream, which also carries the coated tablets/particles. Fluidized-bed coating provides better coating uniformity due to good solid–fluid mixing and minimizes formation of agglomerates. There are various types of fluidized-bed coating equipment, the most commonly used configuration being the Wurster column coater [44], in which a draft tube insert is placed coaxially in the bed to aid the circulation of particles. This column is not suitable for coating of large particles and tablets due to high erosion of solids associated with the higher velocity needed to circulate them [57]. Typical operating variables that affect the performance of the fluid bed coating in a Wurster column are airflow rate, air and bed temperature, spray rate, gap height between the draft tube and the air distributor plate, atomizing air pressure, humidity, and solids charge. The product coating uniformity in the Wurster column coating process is primarily determined by the variation in coating per pass and the circulation time distribution. Radioactive particle tracing, magnetic particle tracing, and controlled single- or multiple-pass coating evaluation techniques were commonly used for detecting particle circulation time and distribution [57, 58]. In addition, several models have been developed to predict the amount of material coated on the particles [59–65].

Compaction Coating In this process, the coating is compressed around a preformed core by using a special tablet press. Compaction coating is especially useful when the drug itself is unstable in the solution and precipitates from the solution in a less stable morphology. On the other hand, the large amount of coating material required limits the applicability of this technology. Another disadvantage of the process is poor adhesion at the coating–core interface which causes physical instability (i.e., friability) [66]. The Press coating process is not useful for coating relatively hard cores which provide essentially no compressibility.

Melt/Dry Coating Film coating processes require water or organic solvent(s). The use of organic solvents causes environmental pollution and excessive cost of recovery while a long time is required to remove the aqueous solvent. Both hot-melt and dry coating techniques eliminate the use of solvents; as a result the processing times become much shorter and the cost of the process is reduced. Melt coating is possible for coating materials that have a low melting temperature and acceptable thermostability. The principal stages in the film formation during dry coating are softening, melting, and curing [67–70]. The process requires larger amount of plasticizer to partially soften and dissolve the polymer.

6.5.3 CHARACTERIZATION OF TABLET COATINGS

Tablet coatings are applied to improve tablet swallowability, mask unpleasant tastes and odors, protect the tablet core against water and oxygen, which can degrade the drug in the core, and control the release rate of the drug. The rate of dissolution and bioavailability of a drug are primarily influenced by the quantity and quality of the coating applied on the tablet. Thus, the characterization of the coatings becomes essential in terms of the uniformity and integrity of the coating. A number of instrumental methods ranging from liquid chromatography [71] to various noninvasive spectroscopic probes have been introduced and evaluated as a means of monitoring the coating process. Among these methods, laser-induced breakdown spectroscopy (LIBS) has been used as a rapid technique for assessing the uniformity of the coating thickness [72]; however, the destructive nature of the method has also been reported [73]. Near-infrared spectroscopy has been employed for determining the amount of coating applied [74–76]. The main disadvantage of this method has been identified as imprecision in the calibration and validation models due to uneven distribution of coating from tablet to tablet [77]. Microscopy techniques provide direct and accurate measurement of the coating thickness but require laborious sample preparation and are therefore impractical for real-time process monitoring. Recently, Raman spectroscopy combined with multivariate data analysis has been reported as a feasible noninvasive technique to quantify tablet coating thickness and uniformity in the presence of a fluorescent ingredient in the coating formulation [77, 78]. Confocal laser scanning microscopy has been introduced as a novel technique for imaging the film–core interface and surface defects of film-coated tablets [79]. Surface roughness was also determined as an important factor in characterizing the tablet coatings using different imaging and roughness analytical techniques, including optical microscopy, scanning electron microscopy (SEM), laser profilometry, and atomic force microscopy (AFM) [80–84].

6.5.4 PREPARATION OF ASYMMETRIC MEMBRANES

Asymmetric membranes are usually produced by phase inversion techniques. In these techniques, an initially homogeneous polymer solution becomes thermodynamically unstable due to different external effects and the phase separates into polymer-lean and polymer-rich phases. The polymer-rich phase forms the matrix of the membrane, while the polymer-lean phase, rich in solvents and nonsolvents, fills the pores. Four main techniques exist to induce phase inversion and thus to prepare asymmetric porous membranes [85]: (a) thermally induced phase separation (TIPS), (b) immersion precipitation (wet casting), (c) vapor-induced phase separation (VIPS), and (d) dry (air) casting.

Thermally Induced Phase Separation In the TIPS process, an initially homogeneous solution consisting of a polymer and solvent(s) phase separates due to a decrease in the solvent quality when the temperature of the casting solution is decreased. After demixing is induced, the solvent is removed by extraction, evaporation, or freeze drying.

Immersion Precipitation (Wet Casting) A homogeneous polymer solution consisting of a polymer and solvent(s) is cast on a support and is immersed in a nonsolvent bath. During the immersion, casting solvent diffuses into the nonsolvent bath and, countercurrently, nonsolvent in the bath penetrates into the solution. The nonsolvent has a limited solubility in the polymer, and when it reaches its critical concentration in the solution, precipitation takes place. Then, the solvent and nonsolvent in the solution are extracted and film is annealed.

Vapor-Induced Phase Separation During the VIPS process, phase separation is induced by penetration of nonsolvent vapor, into the homogeneous polymer solution consisting of polymer and solvent(s). Mass transfer is usually much slower than that in the wet casting process; thus, the VIPS process has been used to obtain membranes with symmetric, cellular, and interconnected pores [86, 87].

Dry (Air) Casting In this process, the polymer dissolved in a mixture of a volatile solvent and a less volatile nonsolvent is cast on a support and exposed to an air stream. During drying of the solution, fast solvent evaporation leads to a decrease in solubility of the polymer, then phase separation takes place.

6.5.5 METHODS FOR OPTIMIZATION OF TABLET COATING FORMULATIONS

Tablet coating formulation is composed of various formulation factors and process parameters. The formulation is usually divided into modeling and optimization phases. The modeling phase consists of preparing series of experimental formulations by varying the ingredients and process conditions systematically and measuring their properties. A detailed review of the literature indicates that the response surface method (RSM) has been widely used for modeling and choosing acceptable tablet coating and other pharmaceutical formulations. The RSM includes statistical experimental designs, multiple regression analysis, and optimization algorithms to search the best formulation for a given set of constraints. Full factorial, orthogonal, Box–Behnken, central composite, pseudorandom, and Plackett–Burman designs were used to investigate the effects of tablet core formulation, coating thickness, and process parameters such as mixing time and speed in the pan coating process, inlet airflow, pan speed, inlet air temperature, coating time, atomization pressure, and fan pressure in the fluidized-bed coating process [88–92]. Artifical neural networks (ANNs) have also been investigated as an alternative method for modeling the pharmaceutical formulations [93–105]. In the RSM, the pharmaceutical responses are predicted based on the second-order polynomial equation, which is usually limited to low levels. When a nonlinear relationship between formulation factors and response variables is observed, the ANN approach was shown to give better estimations of optimal formulations [105].

6.5.6 APPLICATIONS*

Previous sections have reviewed numerous studies in the literature investigating the release mechanisms from the tablet coating systems, their advantages/disadvantages as a control release system, and the methods used to characterize and optimize their formulation. In the following sections, the in vitro performance characteristics of the asymmetric membrane tablet coatings will be illustrated using release studies of a model compound theophylline. For this purpose, first the method used for preparing asymmetric membrane tablet coatings will be explained, then the results of the dissolution studies will be discussed.

6.5.6.1 Materials and Methods

Preparation and Characterization of Tablet Coatings Tablet cores were prepared by compressing the drug without any excipient using a hydraulic press operated at 110 MPa. A stainless steel die with a diameter of 1.2 cm was used to produce 400-mg drug tablet cores. Asymmetric-membrane tablet coatings were applied

*This article was published in Biochemical Engineering Journal, 28, Sacide Alsoy Altinkaya, Hacer Yenal, In vitro drug release rates from asymmetric-membrane tablet coatings: Prediction of phase-inversion dynamics, 131–139, Copyright Elsevier (2006).

using a dip coating process (Dip Coater Nima, type D1L, serial no. 327). The tablets were dip coated in polymer solutions prepared by dissolving cellulose acetate (Aldrich) in a solution of acetone (Merck) and water. The rate of withdrawal of the tablets from the solution was adjusted to obtain similar final coating thicknesses for each coating formulation. Immediately after coating, tablets were rotated for even distribution of the viscous membrane solution, transferred into an environmental chamber (Angelantoni Industrie, Italy, Challenge series, model number CH250), and kept there for 2 h to allow for evaporation of both the solvent (acetone) and nonsolvent (water). The temperature and relative humidity of air in the environmental chamber were adjusted to 25°C and 50%, respectively. Tablets were allowed to dry further for a minimum of 24 h at room temperature prior to dissolution experiments. Faster evaporation of acetone and resulting increase in the concentration of water in the coating leads to a decrease in solubility of the cellulose acetate (CA); then phase separation takes place. Consequently, asymmetric-membrane structure forms on the tablet core. Morphology of the coatings was examined using a scanning electron microscope (Philips, XL-30SFEG). Samples were coated with gold palladium using a magnetron sputter coating instrument. The thickness of the dense skin layer, the overall porosity, and the average pore size were determined from image analysis of micrographs showing cross sections of the membranes.

Dissolution Studies The release rate of theophylline from the tablets was determined by the U.S. Pharmacopeia (USP) XXIII dissolution methodology using a dissolution tester (Caleva 10ST). According to this standard, 900 ml of dissolution medium was placed in the vessel and the temperature was maintained constant at 37°C using a constant-temperature bath. Then, the tablets were immersed in the vessel and the solution was stirred at a speed of 50 rpm. To simulate the actual dissolution environment in the body, the pH of the dissolution medium was kept at 3 during the first 3.5 h by adding 8.5 vol % phosphoric acid to 900 mL distilled water and then increased to 7.4 by adding 5.3 *M* NaOH to the dissolution medium and kept at this value until the end of the experiment. To determine the quantity of drug released from the tablets, samples were taken periodically and assayed by ultraviolet (UV) spectrophotometry (Shimadzu UV-1601) at a wavelength of 272 nm. Dissolution experiments were performed on three tablets and the release profiles were reported as the arithmetic average of the three experimental runs.

Statistical Experimental Design To determine the influence of the composition of the coating solution on the release rate of drug, experiments were statistically designed using a commercial software package called Design-Expert [106]. The system studied in this chapter consists of three components with the following compositional restrictions:

$$5 \leq \omega_1 \leq 15 \qquad 70 \leq \omega_2 \leq 90 \qquad 5 \leq \omega_3 \leq 15 \tag{1}$$

where ω_i is the weight percent of component i and 1, 2, 3 represent CA, acetone, and water, respectively. Any composition outside these limits will probably fail to

TABLE 1 Theophylline Release Rates from Asymmetric-Membrane Tablet Coatings, Results of Fitting Release Profiles to Zero-Order Kinetics, and Precipitation Times Determined from Model Predictions

Compositon (wt %)					
Cellulose Acetate	Acetone	Water	Release Rate (mg/min)	R^2	Precipitation Time[a](s)
15	80	5	0.036	0.9871	2671/2671
15	80	5	0.036	0.9864	2671/2671
5	90	5	0.45	0.9757	660/660
5	90	5	0.36	0.9864	660/660
15	70	15	0.036	0.9908	3374/3374
15	70	15	0.027	0.9925	3374/3374
5	80	15	0.36	0.9928	1000/1000
5	80	15	0.36	0.9876	1000/1000
5	85	10	0.27	0.9918	675/675
10	80	10	0.054	0.9958	1751/1751
15	75	10	0.054	0.9975	2314/2314
10	85	5	0.036	0.9887	2554/2554
12.5	77.5	10	0.036	0.9902	—
10	75	15	0.063	0.9889	1484/1484

[a]The first number corresponds to the precipitation time at the tablet–coating interface.

produce a successful asymmetric-membrane coating. In mixture experiments, the factors are the compositions of the mixture components, and the sum of the fractions of all components is equal to 1. Therefore the factor levels are mutually dependent. Thus, factorial experimental designs are not suitable for response surface modeling of mixtures since such designs require that the experimental treatment combinations be determined by independent adjustments of each component level. In addition, a standard response surface design cannot be used either due to the same constraints. Consequently using the constraint levels shown in Equation (1), a *D*-optimal design was generated by the Design-Expert software package. The 14 experimental formulations determined are shown in Table 1. The lower and upper limits on the weight fraction of each component are required to (a) obtain appropriate viscosity of the solution and coat the tablets uniformly and (b) induce phase separation, thus forming a porous membrane structure. These constraints were established based on preliminary dissolution experiments, available literature data, and the simulation results reported by Altinkaya and Ozbas [43].

Of the 14 formulations listed in Table 1, six experimental runs were required to fit the quadratic mixture model, four additional distinct runs were used to check for the lack of fit, and finally four runs were replicated to provide an estimate of pure error. Design-Expert used the vertices, the edge centers, the overall centroid, and one point located halfway between the overall centroid and one of the edge centers as candidate points. Additionally, four vertices of the design region were used as check points [106].

Determination of Phase Diagrams and Composition Paths The dynamics of the membrane formation process is predicted by combining the kinetics and thermodynamics of the system simultaneously. An appropriate thermodynamic model is necessary to construct the ternary-phase diagram and to formulate the boundary conditions of the kinetic model. Phase separation is considered to occur when a mass transfer path touches the binodal curve. In this study, a robust algorithm developed previously by our group was used to construct the phase diagram [43]. The algorithm utilizes Flory Huggins thermodynamic theory with constant interaction parameters. The composition paths were determined from the kinetic model equations which consist of coupled unsteady-state heat and mass transfer equations, film shrinkage, and complex boundary conditions. The details of both the thermodynamics and kinetic equations can be found in our previous study [43]. We have assumed that the kinetic model derived for a plane geometry can be used to predict the membrane formation process on a tablet surface. This assumption is fairly reasonable since the thickness of the coating is very small and, thus, the cylindrical geometry can be approximated as the plane geometry.

6.5.6.2 Results and Discussion

Effect of Composition of Coating Solution The effect of changing the composition of the casting solution is well documented for asymmetric membranes prepared for separation applications [107–114]. However, there are relatively few quantitative studies illustrating the relationship between the composition of the casting solution and the drug release rate from the asymmetric-membrane coated tablets/capsules [30, 33]. To investigate such a relationship, the in vitro release profiles of the model drug theophylline were measured for the 14 formulations listed in Table 1 and they are shown in the Appendix in Figures A1–A10. To find out whether the drug release from the tablet coatings provides a zero-order release kinetics, each data set was fitted to a linear equation. The quality of the fitted model is determined by the coefficient of determination R^2 and it is defined as

$$R^2 = \frac{\sigma_y - \hat{\sigma}_y}{\sigma_y} \tag{2}$$

where

$$\sigma_y = \frac{1}{n}\sum_{i=1}^{n}(Y_i - \bar{Y})^2 \tag{3}$$

$$\hat{\sigma}_y = \frac{1}{n}\sum_{i=1}^{n}(Y_i - \hat{Y}_i)^2 \tag{4}$$

denote the sample variance and the prediction error power, respectively. Additionally, n is the number of experimental data points, Y_i is the experimental observation, $\bar{Y}$ is the average of the experimental data points, and $\hat{Y}_i$ denotes the predicted value by the fitted model. The quantity R^2 lies between 0 and 1, and if the value is 1, it can be said that the fit of the model is perfect. High R^2 values listed in Table 1 for each data set indicate that there is an excellent linear relationship between the concentration of the drug and the release time; thus, all tablet coatings prepared

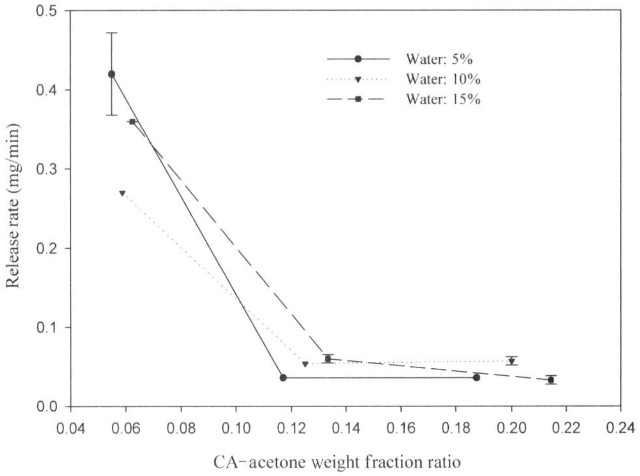

FIGURE 1 Release rate of theophylline as function of cellulose acetate(P)–acetone(S) weight fraction ratios.

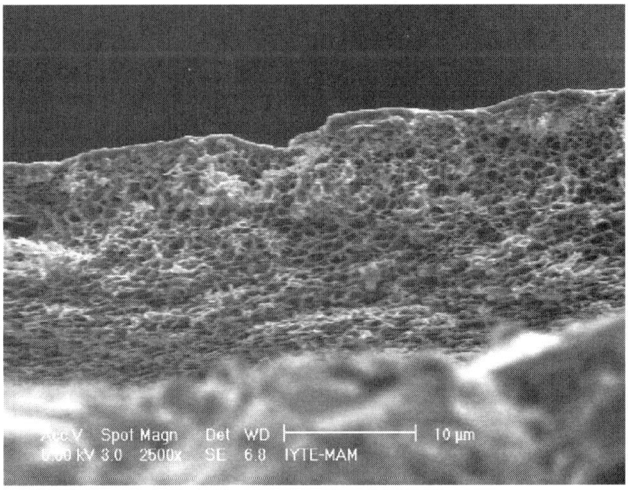

FIGURE 2 SEM of cross section of asymmetric-membrane made with 5% water and CA–acetone weight ratio of 5/90, magnification 2500×.

can provide zero-order or near-zero-order drug release. The release rates for each coating formulation were estimated from the slope of the average release profiles and they are also listed in Table 1.

The drastic change in the release rates with the composition of the coating solution is shown in Figure 1. Within the experimental composition range covered, the highest release rate was observed in the case of the lowest CA (P: polymer) and the highest acetone (S: solvent) concentrations in the casting solution [polymer (P)–solvent (S) weight fraction ratio 5/90]. This is caused by the final coating structure consisting of a very thin and dense top skin layer, and highly porous lower sublayer, as illustrated in Figure 2. At the lowest level of polymer concentration (5%), the thicknesses of the dense skin layers of the coatings are very small and similar to each other; thus, the structure of the lower sublayer becomes an important factor in determining the release rate of the drug. The porosity of the coating structure in the case of 90% acetone is so high that an increase in water concentration from 5 to 10% is not sufficient to produce a more porous final structure and, thus, the release rate of the drug decreases. A further increase of water concentration to 15% results in an opposite effect, which makes the release rate increase to a level of 0.36 mg/min due to the dominant effect of water concentration in increasing the porosity of the lower sublayer.

An increment in the CA concentration from 5 to 10% resulted in significantly lower release rates at all water concentrations because the structure of the coatings changed from porous to dense. Keeping the polymer concentration at 10% while changing the water concentration from 5 to 15% makes the release rates increase. This behavior is explained by the formation of more porous structures by adding more nonsolvent into the casting solution, which is in agreement with the observations of other groups [30, 33].

When the polymer concentration increases from 10 to 15%, no significant changes in release rates were observed. As a matter of fact, in the cases of 5% and 10% water concentrations, the release rates did not change at all. This is due to the unusual transport characteristics of asymmetric membranes which are complex functions of the properties of the different regions of the membrane. In addition to the thickness of the dense skin layer and the porosity of the lower sublayer, structural factors such as tortuosity, pore size, shape, and connectivity of the pores also strongly affect the rate of transport through the coating. The SEM pictures taken at high magnification which are shown in Figures 3*a* and 3*b* indicate that the tablet coating prepared with a P/S ratio of 10/85 has a uniform and narrow pore size distribution with regular elliptic pore shapes, while the other one (P/S: 15/80) has cylindrical pores with a wide pore size distribution, forming a connected pore network. As a result, even though the resistance of the dense skin layer of the coating made with P/S ratio of 15/80 is larger, its lower sublayer resistance is smaller due to the connected pore network. Consequently, the release rate of the theophylline becomes the same through both tablet coatings. Comparison of the scanning electron micrographs in Figures 4*a* and 4*b* indicates that the tablet coating made with the P/S ratio of 10/80 has a uniform pore size distribution, cylindrical pore shapes and high tortuosity while the tablet coating prepared with the P/S ratio of 15/75 has elliptic, irregular pore shapes and pores are isolated. Thus, the lower sublayer resistance of the coating made with the P/S ratio of 10/80 is larger due to the relatively higher tortuosity resulting in the same release rate with the coating prepared with the P/S ratio of 15/75.

At the 15% polymer concentration level, the release rate increases with the change in water concentration from 5 to 10%, which is mainly caused by the increased porosity. However, a further 5% increment in the water concentration (to 15%) made the release rate decrease back to the same level as in the case of 5% water concentration. Even though higher water concentration favors forming a more porous structure and a concomitant higher release rate, the acetone level which decreased below a critical value destroyed this mechanism.

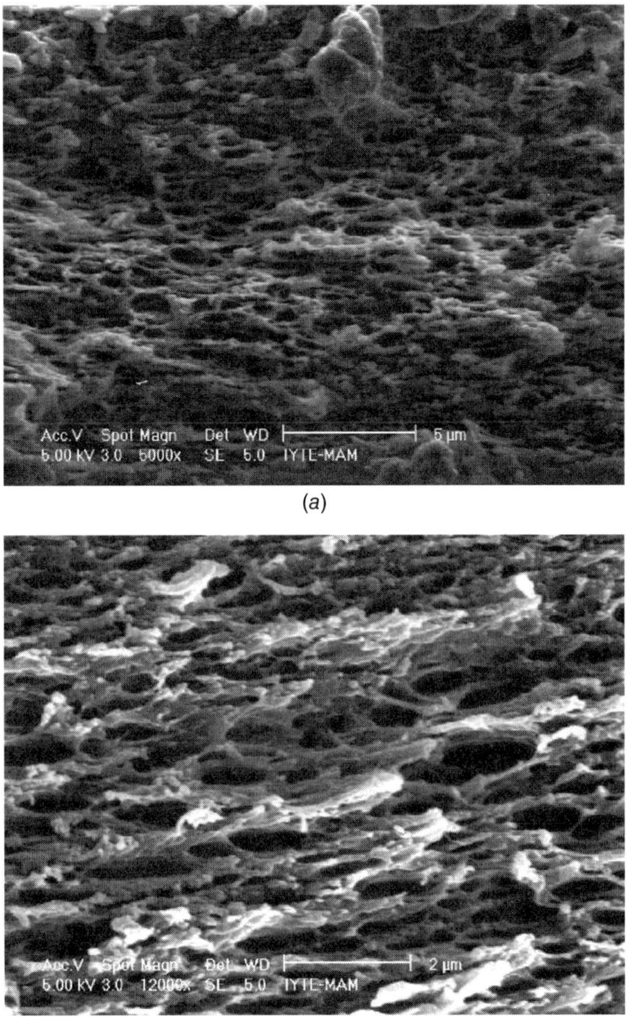

(*a*)

(*b*)

FIGURE 3 SEM of cross section of asymmetric-membrane made with 5% water and CA/acetone weight ratios of: (*a*) 10/85, magnification 5000× (*b*) 15/80, magnification 12,000×.

Results of dissolution studies along with morphological observations clearly indicate that the drug release rate is strongly influenced by the morphology of the membrane. Thus, if one wishes to control the drug release characteristics of the delivery system, a quantitative understanding of the dynamics and morphology of the phase inversion process is required. The dynamics of the phase inversion process can be quantified in terms of the ternary-phase diagram coupled with the heat and mass transfer model equations. We have obtained information about the structure of the tablet coating by plotting the composition paths on the ternary-phase diagram and the polymer concentration profile at the moment of precipitation. As an illustra-

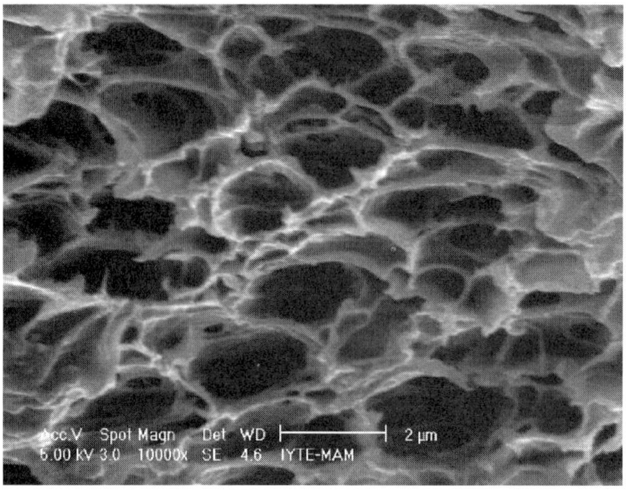

(*a*)

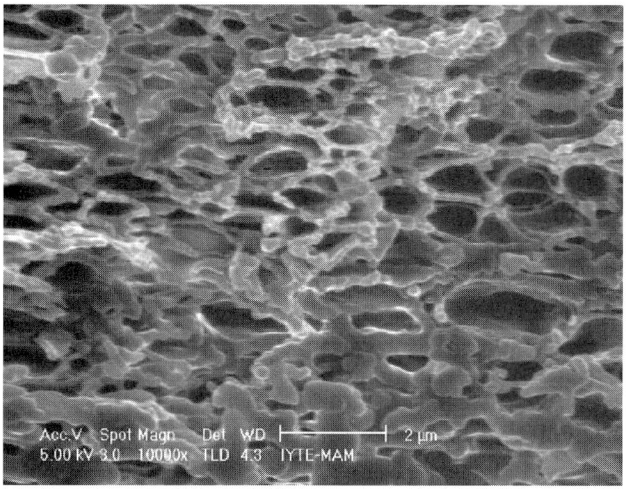

(*b*)

FIGURE 4 SEM of the cross section of the asymmetric-membrane made with 10% water and CA–acetone weight ratios of (*a*) 10/80, magnification 10,000×, and (*b*) 15/75, magnification 10,000×.

tion, in Figure 5, concentration paths in time for the tablet coating prepared with 5% CA, 90% acetone, and 5% water are shown. According to this plot, the phase separation takes place since the concentration paths in time for the drug tablet–coating and the coating–air interface cross the binodal curve at the same time. In addition, the coating–air interface enters the phase envelope at a polymer volume fraction of 0.76, while the tablet–coating interface enters with a volume fraction of 0.023. These two observations imply that the coating will be porous and asymmetric in which the top layer is more dense than the lower sublayer, which was confirmed by the SEM picture in Figure 2. The predictions, have shown that phase separation was achieved for all coating formulations, supporting the morphological observa-

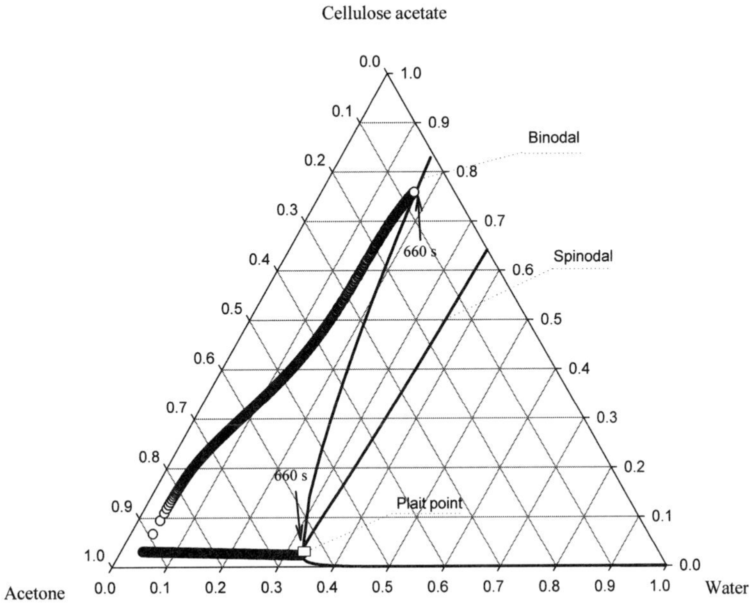

FIGURE 5 Ternary-phase diagram and concentration paths for coating made with 5% CA dissolved in 90% acetone and 5% water.

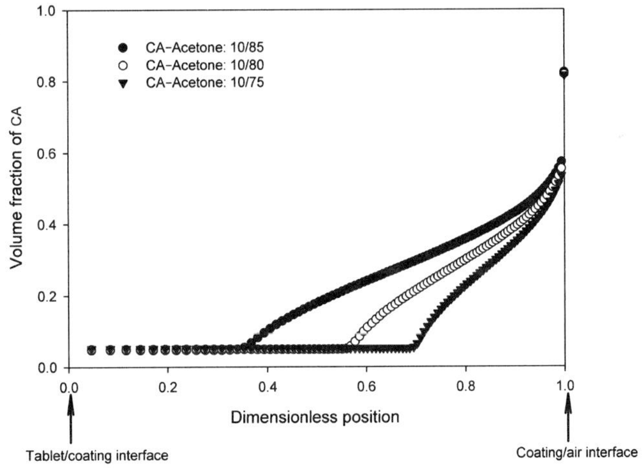

FIGURE 6 Concentration profile of CA in membrane at moment of precipitation for different CA–acetone weight fraction ratios.

tions and the precipitation times calculated for each case (listed in Table 1). Model predictions can also be used to determine a rough thickness of the high-polymer-concentration region near the coating–air interface and the pore distribution of the sublayer structure when the polymer concentration profiles at the moment of phase separation are plotted. As an illustration, such a plot is shown in Figure 6 for coatings prepared with 10% CA in the casting solution. Examination of these profiles leads to the following conclusions regarding the effect of increased water concentration: (1) The polymer concentration at the coating–air interface slightly decreases. (2) More uniform porosity distribution throughout the lower sublayer is favored. Also, porosity increases, which is in complete agreement with our release studies and the observations of other groups [30, 33].

Effect of Evaporation Condition Previous studies on more traditional applications have investigated the effect of increased air velocity, that is, forced-convection conditions for a combination of dry/wet phase inversion techniques to produce defect-free, ultrahigh flux asymmetric membranes with ultrathin skin layers [115–117]. To investigate the effect of evaporation condition on the release rate of drug, tablets were dip coated with CA solution containing 10% CA, 80% acetone, and 10% water and allowed to dry by blowing air across the surface with a blower (forced convection). As a comparison, tablets coated with the same solution were air dried under natural free-convection conditions.

As illustrated in Figure 7, the release profiles of both tablet coatings show a linear behavior only at small times, and then exponential increases in concentrations were observed. Based on this behavior, the release profiles were fitted to an empirical equation as

$$C = k_0 t \lfloor 1 + e^{b_0 C t} \rfloor \tag{5}$$

where k_0 and b_0 are fitting parameters. The accuracy of Equation (5) for correlating the release rate data in Figure 7 was confirmed by high R^2 values, very close to 1 in both cases. Due to the presence of the second term in Equation (5), the release rate

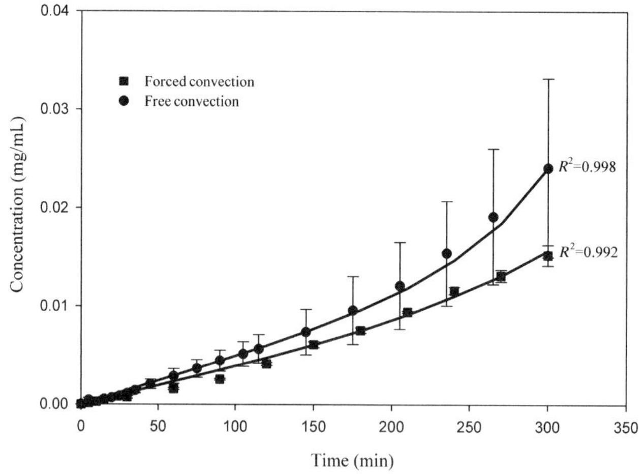

FIGURE 7 Release of theophylline from tablet coatings made with 10% CA dissolved in 80% acetone and 10% water. Coated tablets were dried under free- and forced-convection conditions. The lines correspond to prediction from Equation (5) using $k_0 = 1.92 \times 10^{-5}$ mg/(mL·min); $b_0 = 0.1179$ mL/(mg·min) for forced convection and $k_0 = 2.41 \times 10^{-5}$ mg/(mL·min); $b_0 = 0.1174$ mL/(mg·min) for free convection.

of the drug, dC/dt, is no longer constant and its dependency on the concentration of drug in the dissolution medium, C, can be expressed as

$$\frac{dC}{dt} = \frac{k_0\lfloor 1 + e^{b_0Ct}(1 + b_0Ct)\rfloor}{1 - k_0 b_0 t^2 e^{b_0Ct}} \tag{6}$$

Using Equation (6), the release rate of theophylline from the tablets dried under forced- and free-convection conditions were determined as 0.047 and 0.078 mg/min, respectively. It shoud be noted that both of these values correspond to the arithmetic average of the release rates calculated at each average concentration level shown in Figure 7. The difference in release rates can be explained by comparing the scanning electron micrographs shown in Figures 8*a* and 8*b*. It can be seen that the cross-sectional morphology of the tablet coating dried under the forced-convection condition is dense and nonporous while a porous and asymmetric structure is observed for the tablet coating dried under free-convection conditions. In the dense coating, diffusional resistance to transport of the drug occurs through the overall thickness and is larger than that in the asymmetric porous coating; hence, a lower drug release rate is observed. To understand the effect of air velocity on the formation of the coating structure, we have utilized our model predictions. The composition paths plotted in Figure 9 indicate that, when the speed of air in the drying atmosphere is significantly increased, the rate of evaporation of solvent (acetone) increases dramatically and within a short time its concentration at the surface drops to zero. This situation leads to very strong diffusional resistance within the membrane solution and, thus, slow evaporation of the nonsolvent (water). Consequently, phase separation is never achieved and the resulting membrane structure becomes dense as supported by the SEM picture in Figure 8*a*.

Effect of Nonsolvent Type A few studies in the literature have shown that various membrane morphologies can be obtained by changing the type of nonsolvent in the casting solution [118–120]. To investigate the effect of nonsolvent type, tablet coatings were prepared from a casting solution of CA in acetone as a solvent and octanol, formamide, glycerol, and hexanol as nonsolvents. The drastic change in the release rates of theophylline and percentage of dense skin layer at the surface of the tablet coating with nonsolvent type is shown in Figure 10. The decrease in the release rates is associated with the increase in the thickness of dense skin layer of the coating. The results illustrated in Figure 10 indicate that the membrane structure formation during the phase inversion process is strongly influenced by the type of nonsolvent, since each nonsolvent has different volatility, thermodynamic, and diffusion characteristics. The difference in the rates of evaporation and diffusion of nonsolvents and change in the miscibility gaps in the case of each nonsolvent lead to different mass transfer paths; consequently, the morphology of the resulting membrane structures significantly varies from porous symmetric to dense asymmetric ones.

Statistical Analysis of Experimental Design The effect of the composition of the coating solution on the release rate of drug was investigated in detail with the 14 formulations listed in Table 1. The drug release rate was chosen as an appropriate

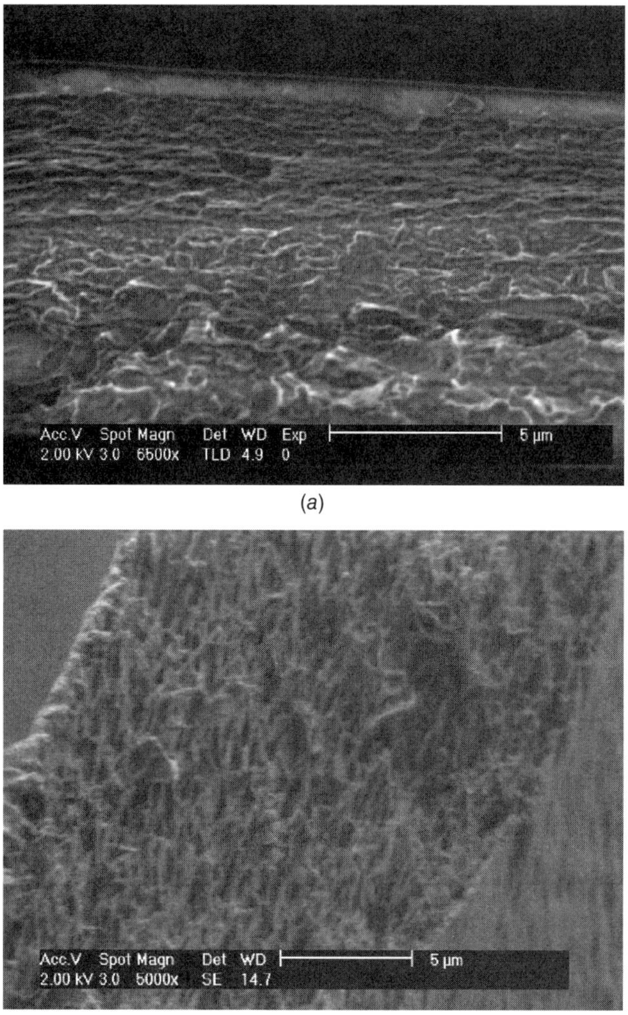

FIGURE 8 SEM of cross section of asymmetric-membrane made with 10% CA dissolved in 80% acetone and 10% water: (*a*) coating solution was dried under forced-convection condition, magnification 6500×; (*b*) coating solution was dried under free-convection condition, magnification 5000×.

response variable since zero-order release was easily achieved for all tablet coatings prepared. The data in Table 1 were best fit by the special cubic equation

$$\text{Release rate} = 70.94\omega_1 + 1.698\omega_2 + 37.55\omega_3 - 94.66\omega_1\omega_2 - 354.3\omega_1\omega_3 - 49.3\omega_2\omega_3 + 357.66\omega_1\omega_1\omega_3 \quad (7)$$

Results of the regression analysis are given in Table 2. An excellent fit of the experimental release rate data to Equation (7) was confirmed by the high R^2 value of 0.9801. In addition to the R^2 values, the significance of Equation (7) and each

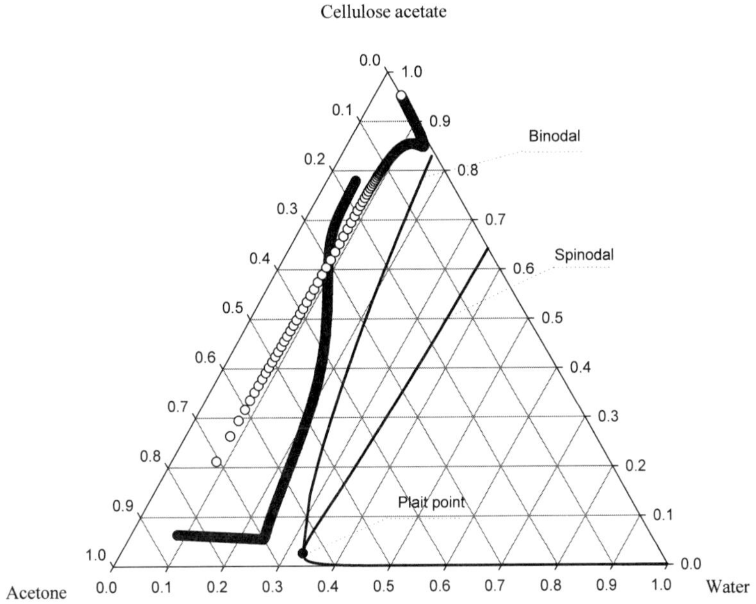

FIGURE 9 Ternary-phase diagram and concentration paths for coating made with 10% CA dissolved in 80% acetone and 10% water. Coating solution was dried under forced-convection condition.

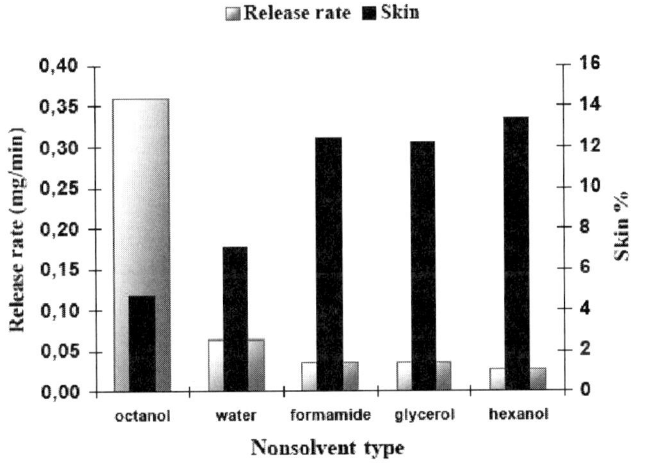

FIGURE 10 Release rates of theophylline and percentage of dense skin layer of membranes as function of nonsolvent type. Polymer: CA solvent–acetone.

term in it to the prediction of the release rate of theophylline was evaluated by the F statistic or F value. The F statistic is viewed as a ratio that expresses variance explained by the model divided by variance due to model error or experimental error and is defined as

$$F = \frac{R^2/k}{(1-R^2)/(n-k-1)} \tag{8}$$

TABLE 2 Statistical Analysis of Release Rate Data

			$\alpha = 0.05$		$\alpha = 0.01$	
Model Number	R^2	F values	F_{critic}[a]	ΔF	F_{critic}[a]	ΔF
1 (full model)	0.9801	42.21	4.21	38	8.26	33.95
2 ($\omega_1\omega_2\omega_3$)	0.955	25.13	3.87	21.26	7.19	17.94
3 ($\omega_2\omega_3$)	0.953	23.57	3.87	19.7	7.19	16.38
4 ($\omega_1\omega_3$)	0.946	20.64	3.87	16.77	7.19	13.45
5 ($\omega_1\omega_2$)	0.881	8.65	3.87	4.78	7.19	1.46

[a]Determined from statistical tables [121].

where k is the number of variables in the model. Usually the computed value of F is compared with the critical F value, $F_{k,n-k,1-\alpha}$, where α is a preselected significance level. If the value of F is substantially greater than the critical F value (F_{critic}), that is, if ΔF, the difference between F and F_{critic}, is large, then the regression equation is considered useful in predicting the response. We have assessed the contribution of each interaction term by comparing the change in ΔF and R^2 values between the full model given in Equation (7) and reduced models. The reduced models were obtained by deleting a specific interaction term in the full model; for example, model 2 includes all terms in Equation (7) except the term involving $\omega_1\omega_2\omega_3$. The results in Table 2 indicate that all binary and ternary interaction terms in the full model are needed for accurate prediction of the release rate since the largest ΔF values are calculated for the full model for both significance levels, $\alpha = 0.05$ and $\alpha = 0.01$. According to the criterion mentioned above, among all interactions, CA–acetone ($\omega_1\omega_2$) was identified as the most influential factor on the response since the largest decrease in both ΔF and R^2 values compared to those of the full model were observed when the term $\omega_1\omega_2$ was deleted from the full model. This simply implies that changing the CA–acetone ratio in the coating formulation has the most significant effect on the release rate. Specifically, increasing the ratio of CA to acetone from 5/90 to 15/80 resulted in a decrease of the release rate from 0.45 to 0.036 mg/min since the porosity of the membrane decreases and the thickness of the dense skin layer increases. The ratio of the composition of CA to water ($\omega_1\omega_3$) was also found to be an important parameter on the release rate of drug as indicated by the second largest decrease in R^2 and ΔF values compared to those of the full model. Decreasing this ratio from 15/5 to 5/15 caused an increase in the release rate by a factor of 10 since the thickness of the dense skin layer significantly decreases. Based on the decrease in ΔF and R^2 values from those of the full model given in Table 2, the relative importance of each interaction term can be ranked as follows: $\omega_1\omega_2 > \omega_1\omega_3 > \omega_2\omega_3 > \omega_1\omega_2\omega_3$.

To further illustrate the simultaneous effects of the factors on the release rate of drug, a three-dimensional response surface plot based on Equation (7) was generated, as shown in Figure 11. As can be clearly seen from this figure, the release rate can be significantly varied just by tailoring the CA–acetone and CA–water ratios without changing the coating material. In addition, Figure 11 shows that a slight maximum in release rate is observed as the ratio of composition of acetone to water increases.

In order to validate the predictive capability of the empirical expression, two formulations with compositions given in Table 3 were selected randomly from the

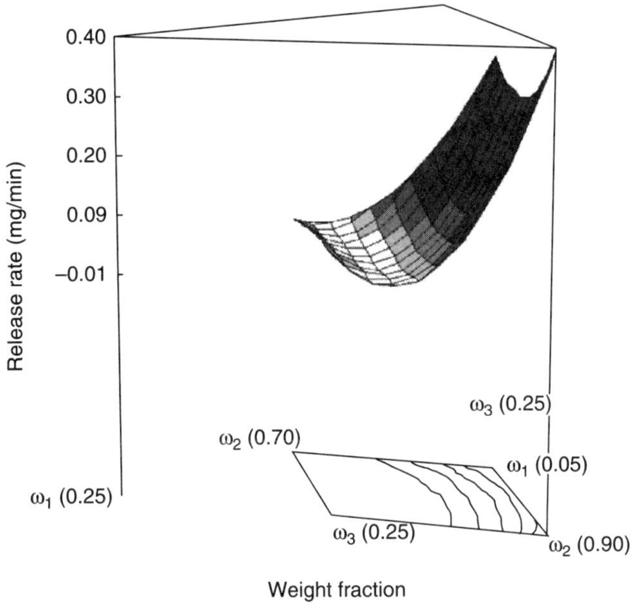

FIGURE 11 Three-dimensional response surface plot of release rate as function of composition of coating solution.

TABLE 3 Composition of Coating Solution Randomly Selected for Testing Predictive Capability of Empirical Expression

Compositon (wt %)			Experimental Release Rate (mg/min)	Predicted Release Rate (mg/min)	R^2
Cellulose Acetate	Acetone	Water			
7.5	82.5	10	0.09	0.125	0.9816
10	82.5	7.5	0.027	0.029	0.9858

experimental design region. Experimental release rates estimated from the slope of the release profiles and corresponding rates predicted from Equation (7) are also listed in Table 3. According to the results, the maximum absolute percentage difference between the experimental and predicted release rates is 3.5%. This value lies within the residuals obtained in deriving Equation (7). Based on this comparison, it is fair to conclude that the empirical expression derived in this study can be used as a tool to predict the release rate of theophylline for any composition within the experimental design region.

6.5.7 CONCLUSION

This chapter has considered the controlled release of drugs from tablet coating systems. These systems are still the preferred route of drug administration due to their advantages of ease of administration, maintaining zero-order release rates, and better patient compliance; thus, they hold the major market share in the pharmaceutical industry along with the other formulations of oral drug delivery. In tablet coated systems, diffusion of drug through the coating is usually the rate-limiting step and the desired drug release rates are achieved by properly selecting the coating material and adjusting the morphology of the coating using suitable manufacturing methods. Coating characteristics such as glass transition temperature, crystalline content, and degree of cross-linking significantly determine the release rate of the drug. For a selected coating material, the structure of the coating plays a critical role in achieving useful release rates. The morphology of the tablet coatings can be varied by incorporating plasticizer or water-soluble additives in the coating solution, by blending the polymers, or by applying multilayer composite coatings. Another approach is to apply asymmetric- and porous-type coatings. The morphological characteristics of these types of coatings, such as fraction of dense top layer and porous sublayer and size and shape of the pores, can be varied by optimizing the composition of the coating solution, evaporation conditions (temperature, relative humidity, and velocity of air), and the type of solvent or nonsolvent used in the coating solution. Asymmetric-type coatings can be used to facilitate osmotic delivery of drugs with low solubilities since high water fluxes can be achieved. These coatings allow us to control the release kinetics without altering the coating material or significantly varying the coating thickness.

Currently, considerable research efforts have been directed toward developing protein drug delivery systems due to discovery of numerous protein and peptide therapeutics. The delivery of protein drugs is usually limited to parenteral administration and frequent injections are required due to their short half-lives in the blood. In this regard, development of oral delivery systems is necessary for patient compliance and convenience. The challenge in the design of the oral delivery systems is that they should protect the incorporated drugs from chemical and enzymatic degradation until the drug reaches the delivery site. The protein drug should not be influenced by pH or bacteria and enzymes along the gastrointestinal (GI) tract and should be delivered at the desired site with a desired efficiency. To achieve site-specific delivery for protein- and peptide-based drugs, one of the strategies that has often been investigated is to coat the drug core with polymers that can respond to the stimuli of local environments such as pH and enzymes. For water-soluble protein drugs, such as insulin, an additional protective coating is usually required to isolate the drug from the surrounding water. Composite tablet coating materials which combine the enzymatic susceptibility and protective properties of polymers can be another solution for this problem. A systematic comparison study among various polymer combinations is required to select the appropriate coating materials for specific drugs. Not only the selection of the coating material but also the manufacturing technique for the preparation of tablet coatings plays a critical role and still remains one of the most challenging subjects in the controlled drug delivery area. Proteins are very sensitive to conditions that can occur during the coating process. Mechanical stresses during the preparation, exposure to a hydrophobic organic solvent, intermediate moisture level during hydration, and interaction between protein and polymer can easily inactivate the protein-based drug. Therefore, more research focused on both optimization of coating materials and manufacturing methods for encapsulating the protein and peptide drugs is necessary.

REFERENCES

1. Langer, R. S., and Peppas, N. A. (1981), Present and future applications of biomaterials in controlled drug delivery systems, *Biomaterials*, 2, 201–214.
2. Ho, W. S. W., and Sirkar, K. K. (1992), *Membrane Handbook*, Van Nostrand Reinhold, New York, pp. 915–935.
3. Liu, L., Ku, J., Khang, G., Lee, B., Rhee, J. M., and Lee, H. B. (2000), Nifedipine controlled delivery by sandwiched osmotic tablet system. *J. Controlled Release*, 68, 145–156.
4. Liu, L., Khang, G., Rhee, J. M., and Lee, H. B. (2000), Monolithic osmotic tablet system for nifedipine delivery, *J. Controlled Release*, 67, 309–322.
5. Narisawa, S., Nagata, M., Ito, T., Yoshino, H., Hirakawa, Y., and Noda, K. (1995), Drug release behavior in gastrointestinal tract of beagle dogs from multiple unit type rate-controlled or time-controlled release preparations coated with insoluble polymer-based film, *J. Controlled Release*, 33, 253–260.
6. Moussa, I. S., and Cartilier, L. H. (1997), Evaluation of cross-linked amylose press-coated tablets for sustained drug delivery, *Int. J. Pharm.*, 149, 139–149.
7. Khan, M. Z. I., Prebeg, E., and Kurjakovic, N. (1999), A pH-dependent colon targeted oral drug delivery system using methacrylic acid copolymers: I. Manipulation of drug release using Eudragit® L100-55 and Eudragit® S100 combinations, *J. Controlled Release*, 58, 215–222.
8. Krögel, I., and Bodmeier, R. (1999), Floating or pulsatile drug delivery systems based on coated effervescent cores, *Int. J. Pharm.*, 187, 175–184.
9. Macleod, G. S., Fell, J. T., and Collett, J. H. (1999), An in vitro investigation into the potential for bimodal drug release from pectin/chitosan/HPMC-coated tablets, *Int. J. Pharm.*, 188, 11–18.
10. Thoma, K., and Bechtold, K. (1999), Influence of aqueous coatings on the stability of enteric coated pellets and tablets, *Eur. J. Pharm. Biopharm.*, 47, 39–50.
11. Fukui, E., Uemura, K., and Kobayashi, M. (2000), Studies on applicability of press-coated tablets using hydroxypropylcellulose (HPC) in the outer shell for timed-release preparations, *J. Controlled Release*, 68, 215–223.
12. Crotts, G., Sheth, A., Twist, J., and Ghebre-Sellassie, I. (2001), Development of an enteric coating formulation and process for tablets primarily composed of a highly water-soluble, organic acid. *Eur. J. Pharm. Biopharm.*, 51, 71–76.
13. Ofori-Kwakye, K., and Fell, J. T. (2003), Biphasic drug release from film-coated tablets, *Int. J. Pharm.*, 250, 431–440.
14. Goto, T., Tanida, O., Yoshinaga, T., Sato, S., Ball, D. J., Wilding, I. R., Kobayashi, E., and Fujimura, A. (2004), Pharmaceutical design of a novel colon-targeted delivery system using two-layer-coated tablets of three different pharmaceutical formulations, supported by clinical evidence in humans, *J. Controlled Release*, 97, 31–42.
15. Sundy, E., and Danckwerts, M. P. (2004), A novel compression-coated doughnut-shaped tablet design for zero-order sustained release, *Eur. J. Pharm. Sci.*, 22, 477–485.
16. Tarvainen, M., Peltonen, S., Mikkonen, H., Elovaara, M., Tuunainen, M., Paronen, P., Ketolainen, J., and Sutinen, R. (2004), Aqueous starch acetate dispersion as a novel coating material for controlled release products, *J. Controlled Release*, 96, 179–191.
17. Thombre, A. G., Appel, L. E., Chidlaw, M. B., Daugherity, P. D., Dumont, F., Evans, L. A. F., and Sutton, S. C. (2004), Osmotic drug delivery using swellable-core technology, *J. Controlled Release*, 94, 75–89.

18. Kim, I. H., Park, J. H., Cheong, I. W., and Kim, J. H. (2003), Swelling and drug release behavior of tablets coated with aqueous hydroxypropyl methylcellulose phthalate (HPMCP) nanoparticles, *J. Controlled Release*, 89, 225–233.
19. Okimoto, K., Rajewski, R. A., and Stella, J. V. (1999), Release of testosterone from an osmotic pump tablet utilizing (SBE)7m-ß-cyclodextrin as both a solubilizing and an osmotic pump agent, *J. Controlled Release*, 58, 29–38.
20. Khan, M. A., Sastry, S. V., Vaithiyalingam, S. R., Agarwal, V., Nazzal, S., and Reddy, I. K. (2000), Captopril gastrointestinal therapeutic system coated with cellulose acetate pseudolatex: Evaluation of main effects of several formulation variables, *Int. J. Pharm.*, 193, 147–156.
21. Lu, E-X., Jiang, Z-Q., Zhang, Q-Z., and Jiang, X-G. (2003), A water-insoluble drug monolithic osmotic tablet system utilizing gum arabic as an osmotic, suspending and expanding agent, *J. Controlled Release*, 92, 375–382.
22. Zhang, Y., Zhang, Z., and Wu, F. (2003), A novel pulsed-release system based on swelling and osmotic pumping mechanism, *J. Controlled Release*, 89, 47–55.
23. Liu, L., and Che, B. (2006), Preparation of monolithic osmotic pump system by coating the indented core tablet, *Eur. J. Pharm. Biopharm.*, 64, 180–184.
24. Zentner, G. M., Rork, G. S., and Himmelstein, K. J. (1985), The controlled-porosity osmotic pump, *J. Controlled Release*, 1, 269–282.
25. Baker, R. W., and Brooke, J. W. (1987), Pharmaceutical drug delivery system, U.S. Patent 4,687,660.
26. Theeuwes, F., and Ayer, A. D. (1978), Osmotic device having composite walls, U.S. Patent 4,077,407.
27. Theeuwes, F. (1978), Microporous-semipermeable laminated osmotic system, U.S. Patent 4,256,108.
28. Cortese, R., and Theeuwes, F. (1982), Osmotic device with hydrogel driving member, U.S. Patent 4,327,725.
29. Swanson, D. R., Burday, B. L., Wong, P. S. L., and Theeuwes, F. (1987), Nifedipine gastrointestinal therapeutic system, *Am. J. Med.*, 83, 3–10.
30. Herbig, S. M., Cardinal, J. R., Korsmeyer, R. W., and Smith, K. L. (1995), Asymmetric membrane tablet coatings for osmotic drug delivery, *J. Controlled Release*, 35, 127–136.
31. Cardinal, J. R., Herbig, S. M., Korsmeyer, R. W., Lo, J., Smith, K. L., and Thombre, A. G. (1997), Use of asymmetric membranes in delivery devices, U.S. Patent 5,612,059.
32. Cardinal, J. R., Herbig, S. M., Korsmeyer, R. W., Lo, J., Smith, K. L., and Thombre, A. G. (1997), Asymmetric membranes in delivery devices, U.S. Patent 5,698,220.
33. Wang, D-M., Lin, F-C., Chen, L-Y., and Lai, J-Y. (1998), Application of asymmetric TPX membranes to transdermal delivery of nitroglycerin, *J. Controlled Release*, 50, 187–195.
34. Thombre, A. G., Cardinal, J. R., DeNoto, A. R., Herbig, S. M., and Smith, K. L. (1999), Asymmetric-membrane capsules for osmotic drug delivery. Part I: Development of a manufacturing process, *J. Controlled Release*, 57, 55–64.
35. Thombre, A. G., Cardinal, J. R., DeNoto, A. R., and Gibbes, D. C. (1999), Asymmetric membrane capsules for osmotic drug delivery. Part II: In vitro and in vivo drug release performance, *J. Controlled Release*, 57, 65–73.
36. Thombre, A. G., DeNoto, A. R., and Gibbes, D. C. (1999), Delivery of glipizide from asymmetric-membrane capsules using encapsulated excipients, *J. Controlled Release*, 60, 333–341.

37. Lin, Y-K., and Ho, H-O. (2003), Investigations on the drug releasing mechanism from an asymmetric-membrane-coated capsule with an in situ formed delivery orifice, *J. Controlled Release*, 89, 57–69.
38. Meier, M. M., Kanis, L. A., and Soldi, V. (2004), Characterization and drug-permeation profiles of microporous and dense cellulose acetate membranes: Influence of plasticizer and pore forming agent, *Int. J. Pharm.*, 278, 99–110.
39. Prabakaran, D., Singh, P., Jaganathan, K. S., and Vyas, S. P. (2004), Osmotically regulated asymmetric capsular systems for simultaneous sustained delivery of anti-tubercular drugs, *J. Controlled Release*, 95, 239–248.
40. Wang, C-Y., Ho, H-O., Lin, L-H., Lin, Y-K., and Ming-Thau Sheu, M-T. (2005), Asymmetric membrane capsules for delivery of poorly water-soluble drugs by osmotic effects, *Int. J. Pharm.*, 297, 89–97.
41. Altinkaya, S. A., and Yenal, H. (2006), In vitro drug release rates from asymmetric-membrane tablet coatings: Prediction of phase-inversion dynamics, *Biochem. Eng. J.*, 28, 131–139.
42. Wang, G-M., Chen, C-H., Ho, H-O., Wang, S-S., and Sheu, M-T. (2006), Novel design of osmotic chitosan capsules characterized by asymmetric membrane structure for in situ formation of delivery orifice, *Int. J. Pharm.*, 319, 71–81.
43. Altinkaya, S. A., and Ozbas, B. (2004), Modeling of asymmetric-membrane formation by dry-casting method, *J. Membr. Sci.*, 230, 71–89.
44. Mathiowitz, M. (1999), *Encyclopedia of Controlled Drug Delivery*, Wiley, New York, pp. 302–303.
45. Sandadi, S., PAndey, P., and Turton, R. (2004), In situ, near real-time acquisition of particle motion in rotating pan coating equipment using imaging techniques, *Chem. Eng. Sci.*, 59, 5807–5817.
46. Prater, D., Wilde, J., and Meakin, B. (1980), A model system for the production of aqueous tablet film coating for laboratory evaluation, *J. Pharm. Pharmacol.*, 32(Suppl.), 90.
47. Leaver, T., Shannon, H., and Rowe, R. (1985), A photometric analysis of tablet movement in a side-vented perforated drum (Accela-Cota), *J. Pharm. Pharmacol.*, 37, 17–21.
48. Nakagawa, M., Altobelli, S., Caprihan, A., Fukushima, E., and Jeong, E. (1993), Non-invasive measurements of granular flows by magnetic resonance imaging, *Exper. Fluids*, 16, 54–60.
49. Parker, D., Broadbent, C., Fowles, P., Hawkesworth, M., and McNeil, P. (1993), Positron emission particle tracking—A technique for studying flow within engineering equipment, *Nucl. Instrum. Methods Phys. Res. A*, 326, 592–607.
50. Parker, D., Dijkstra, A., Martin, T., and Seville, J. (1997), Positron emission particle tracking studies of spherical particle motion in rotating drums, *Chem. Eng. Sci.*, 52, 2011–2022.
51. Wilson, K., and Crossman, E. (1997), The influence of tablet shape and pan speed on intra-tablet film coating uniformity, *Drug Dev. Ind. Pharm.*, 23, 1239–1243.
52. Yamane, K., Sato, T., Tanaka, T., and Tsuji, Y. (1995), Computer simulation of tablet motion in coating drum, *Pharm. Res.*, 12, 1264–1268.
53. Yamane, K., Nakagawa, M., Altobelli, S., Tanaka, T., and Tsuji, Y. (1998), Steady particulate flows in a horizontal rotating cylinder, *Phys. Fluids*, 10, 1419–1427.
54. Denis, C., Hemati, M., Chulia, D., Lanne, J., Buisson, B., Daste, G., and Elbaz, F. (2003), A model of surface renewal with application to the coating of pharmaceutical tablets in rotary drums, *Powder Technol.*, 130, 174–180.

55. Pandey, P., Song, Y., Kayihan, F., and Turton, R. (2006), Simulation of particle movement in a pan coating device using discrete element modeling and its comparison with video-imaging experiments, *Powder Technol.*, 161, 79–88.
56. Rege, B. D., Gawel, J., and Kou, J. H. (2002), Identification of critical process variables for coating actives onto tablets via statistically designed experiments, *Int. J. Pharm.*, 237, 87–94.
57. Turton, R., and Cheng, X. X. (2005), The scale-up of spray coating processes for granular solids and tablets, *Powder Technol.*, 150, 78–85.
58. Xu, M., and Turton, R. (1997), A new data processing technique for noisy signals: Application to measuring particle circulation times in a draft tube equipped fluidized bed, *Powder Technol.*, 92, 111–117.
59. Sherony, D. F. (1981), A model of surface renewal with application to fluid bed coating of particles, *Chem. Eng. Sci.*, 36, 845–848.
60. Wnukowski, P., and Setterwall, F. (1989), The coating of particles in a fluidized bed (residence time distribution in a system of two coupled perfect mixers), *Chem. Eng. Sci.*, 44, 493–505.
61. Choi, M. S., and Meisen, A. (1997), Sulfur coating of urea in shallow spouted beds, *Chem. Eng. Sci.*, 52, 1073–1086.
62. Maronga, J., and Wnukowski, P. (1997), Modeling of the three-domain fluidized bed particulate coating process, *Chem. Eng. Sci.*, 52, 2915–2925.
63. Nakamura, H., Abe, E., and Yamada, N. (1998), Coating mass distributions of seed particles in a tumbling fluidized bed coater. Part II. A Monte Carlo simulation of particle coating, *Powder Technol.*, 99, 140–146.
64. Cheng, X. X., and Turton, R. (2000), The prediction of variability occurring in fluidized bed coating equipment. II: The role of nonuniform particle coverage as particle pass through the spray zone, *Pharm. Dev. Technol.*, 5, 323–332.
65. KuShaari, K., Pandey, P., Song, Y., and Turton, R. (2006), Monte Carlo simulations to determine coating uniformity in a Wurster fluidized bed coating process, *Powder Technol.*, 166, 81–90.
66. Waterman, K. C., and Fergione, M. B. (2003), Press-coating of immediate release powders onto coated controlled release tablets with adhesives, *J. Controlled Release*, 89, 387–395.
67. Leong, K. C., Lu, G. Q., and Rudolph, V. (1999), A comparative study of the fluidized-bed coating of cylindrical metal surfaces with various thermoplastic polymer powders, *J. Mater. Proc. Technol.*, 89/90, 354–360.
68. Wulf, M., Uhlmann, P., Michel, S., and Grundke, K. (2000), Surface tension studies of levelling additives in powder coatings, *Prog. Org. Coat.*, 38, 59–66.
69. Belder, E. G., Rutten, H. J. J., and Perera, D. Y. (2001), Cure characterization of powder coatings, *Prog. Org. Coat.*, 42, 142–149.
70. Pfeffer, R., Dave, R. N., Wie, D., and Ramlakhan, M. (2001), Synthesis of engineered particulates with tailored properties using dry particle coating, *Powder Technol.*, 117, 40–67.
71. McLaren, D. D., and Hollenbeck, R. G. (1987), A high performance liquid chromatographic method for the determination of the amount of hydroxypropyl methylcellulose applied to tablets during an aqueous film coating operation, *Drug Dev. Ind. Pharm.*, 13, 2179–2197.
72. Mouget, Y., Gosselin, P., Tourigny, M, and B'echard, S. (2003), Three-dimensional analyses of tablet content and film coating uniformity by laser-induced breakdown spectroscopy (LIBS), *Am. Lab.*, 35, 20–22.

73. St-Onge, L., Kwong, E., Sabsabi, M., and Vadas, E. B. (2002), Quantitative analysis of pharmaceutical products by laser-induced breakdown spectroscopy, *Spectrochim. Acta, Part B*, 57, 1131–1140.

74. Kirsch, J. D., and Drennen, J. K. (1995), Determination of film-coated tablet parameters by near-infrared spectroscopy. *J. Pharm. Biomed. Anal.*, 13, 1273–1281.

75. Kirsch, J. D., and Drennen, J. K. (1996), Near-infrared spectroscopic monitoring of the film coating process, *Pharm. Res.*, 13, 234–237.

76. Anderson, M., Josefson, M., Langkilde, F., and Wahlund, K.-G. (1999), Monitoring of a film coating process for tablets using near infrared reflectance spectrometry, *J. Pharm. Biomed. Anal.*, 20, 27–37.

77. Romero-Torres, S., Pérez-Ramos, J. D., Morris, K. R., and Grant, E. R. (2006), Raman spectroscopy for tablet coating thickness quantification and coating characterization in the presence of strong fluorescent interference, *J. Pharma. Biomed. Anal.*, 41, 811–819.

78. Romero-Torres, S., Pérez-Ramos, J. D., Morris, K. R., and Grant, E. R. (2005), Raman spectroscopic measurement of tablet-to-tablet coating variability, *J. Pharm. Biomed. Anal.*, 38, 270–274.

79. Ruotsalainen, M., Heinämäki, J., Guo, H., Laitinen, N., and Yliruusi, J. (2003), A novel technique for imaging film coating defects in the film-core interface and surface of coated tablets, *Eur. J. Pharm. Biopharm.*, 56, 381–388.

80. Podczeck, F. (1998), Measurement of surface roughness of tablets made from polyethylene glycol powders of various molecular weight, *Pharm. Pharmacol. Commun.*, 4, 179–182.

81. Riippi, M., Antikainen, O., Niskanen, T., and Yliruusi, J. (1998), The effect of compression force on surface structure, crushing strength, friability, and disintegration time of erythromycin acistrate tablets, *Eur. J. Pharm. Biopharm.*, 46, 339–345.

82. Newton, M., Petersson, J., Podczeck, F., Clarke, A., and Booth, S. (2001), The influence of formulation variables on the properties of pellets containing a self-emulsifying mixture, *J. Pharm. Sci.*, 90, 987–995.

83. Seitavuopio, P., Rantanen, J., and Yliruusi, J. (2003), Tablet surface characterisation by various imaging techniques, *Int. J. Pharm.*, 254, 281–286.

84. Seitavuopio, P., Rantanen, J., and Yliruusi, J. (2005), Use of roughness maps in visualisation of surfaces, *Eur. J. Pharm. Biopharm.*, 59, 351–358.

85. van de Witte, P., Dijkstra, P. J., Van den Berg, J. W. A., and Feijen, J. (1996), Phase separation processes in polymer solutions in relation to membrane formation, *J. Membr. Sci.*, 117, 1–31.

86. Han, M., and Bhattacharyya, D. (1995), Changes in morphology and transport characteristics of polysulfone membranes prepared by different demixing conditions, *J. Membr. Sci.*, 98, 191–200.

87. Park, H. C., Kim, Y. P., Kim, H. Y., and Kang, Y. S. (1999), Membrane formation by water vapor induced phase inversion, *J. Membr. Sci.*, 156, 169–178.

88. Tobiska, S., and Kleinebudde, P. (2001), A simple method for evaluating the mixing efficiency of a new type of pan coater, *Int. J. Pharm.*, 224, 141–149.

89. Plumb, A. P., Rowe, R. C., York, P., and Doherty, C. (2002), The effect of experimental design on the modeling of a tablet coating formulation using artificial neural networks, *Eur. J. Pharm. Sci.*, 16, 281–288.

90. Rege, B. D., Gawel, J. and Kou, H. J. (2002), Identification of critical process variables for coating actives onto tablets via statistically designed experiments, *Int. J. Pharm.*, 237, 87–94.

91. Mueller, R., and Kleinebudde, P. (in press), Influence of scale-up on the abrasion of tablets in a pan coater, *Eur. J. Pharm. Biopharm.*
92. Wang, G-M., Chen, C-H., Ho, H-O., Wang, S-S., and Sheu, M-T. (press), Novel design of osmotic chitosan capsules characterized by asymmetric membrane structure for in situ formation of delivery orifice, *Int. J. Pharm.*
93. Hussain, A. S., Shivanand, P., and Johnson, R. D. (1994), Application of neural computing in pharmaceutical product development: Computer aided formulation design, *Drug Dev. Ind. Pharm.*, 20, 1739–1752.
94. Kesavan, J. G., and Peck, G. E. (1995), Pharmaceutical formulation using neural networks. *Proc.14th Pharm. Technol. Conf. (Barcelona)*, 2, 413–431.
95. Colbourn, E. A., and Rowe, R. C. (1996), Modeling and optimization of a tablet formulation using neural networks and genetic algorithms, *Pharm. Technol. Eur.*, 8, 46–55.
96. Bourquin, J., Schmidli, H., van Hoogevest, P., and Leuenberger, H. (1997), Basic concepts of artificial neural networks (ANN) modeling in the application to pharmaceutical development, *Pharm. Dev. Technol.*, 2, 95–109.
97. Bourquin, J., Schmidli, H., van Hoogevest, P., and Leuenberger, H. (1997b), Application of artificial neural networks (ANN) in the development of solid dosage forms, *Pharm. Dev. Technol.*, 2, 111–121.
98. Ebube, N. K., McCall, T., Chen, Y., and Meyer, M. C. (1997), Relating formulation variables to in vitro dissolution using an artificial neural network, *Pharm. Dev. Technol.*, 2, 225–232.
99. Bourquin, J., Schmidli, H., van Hoogevest, P., and Leuenberger, H. (1998), Comparison of artificial neural networks (ANN) with classical modeling techniques using different experimental designs and data from a galenical study on a solid dosage form, *Eur. J. Pharm. Sci.*, 6, 287–300.
100. Bourquin, J., Schmidli, H., van Hoogevest, P., and Leuenberger, H. (1998), Advantages of artificial neural networks (ANNs) as alternative modeling technique for data sets showing non-linear relationships using data from a galenical study on a solid dosage form, *Eur. J. Pharm. Sci.*, 7, 5–16.
101. Bourquin, J., Schmidli, H., van Hoogevest, P., and Leuenberger, H. (1998), Pitfalls of artificial neural networks (ANN) modeling technique for data sets containing outlier measurements using a study on mixture properties of a direct compressed dosage form, *Eur. J. Pharm. Sci.*, 7, 17–28.
102. Chen, Y., McCall, T. W., Baichwal, A. R., and Meyer, M. C. (1999), The application of an artificial neural network and pharmacokinetic simulations in the design of controlled-release dosage forms, *J. Controlled Release*, 59, 33–41.
103. Plumb, A. P., Rowe, R. C., York, P., and Doherty, C. (2002), The effect of experimental design on the modeling of a tablet coating formulation using artificial neural networks, *Eur. J. Pharm. Sci.*, 16, 281–288.
104. Plumb, A. P., Rowe, R. C., York, P., and Doherty, C. (2003), Effect of varying optimization parameters on optimization by guided evolutionary simulated annealing (GESA) using a tablet film coat as an example formulation, *Eur. J. Pharma. Sci.*, 18, 259–266.
105. Takayama, K., Fujikawa, M., Obata, Y., and Morishita, M. (2003), Neural network based optimization of drug formulations, *Adv. Drug Deliv. Rev.*, 55, 1217–1231.
106. Montgomery, D. C. (1997), *Design and Analysis of Experiments*, 4th ed., Wiley, New York, pp. 611–616.

107. Lai, J.-Y., Lin, F.-C., Wang, C.-C., and Wang, D.-M. (1996), Effect of nonsolvent additives on the porosity and morphology of asymmetric TPX membranes, *J. Membr. Sci.*, 118, 49–61.

108. Broadhead, K. W., and Tresco, P. A. (1998), Effects of fabrication conditions on the structure and function of membranes formed from poly(acrylonitrile-vinylchloride), *J. Membr. Sci.*, 147, 235–245.

109. Won, J., Park, H. C., Kim, U. Y., Kang, S., Yoo, S. H., and Jho, J. Y. (1999), The effect of dope solution characteristics on the membrane morphology and gas transport properties: PES/-BL/NMP system, *J. Membr. Sci.*, 162, 247–255.

110. Young, T-H., Lin, D.-J., Gau, J.-J., Chuang, W.-Y., and Cheng, L.-P. (1999), Morphology of crystalline Nylon-610 membranes prepared by the immersion–precipitation process: Competition between crystallization and liquid–liquid phase separation, *Polymer*, 40, 5011–5021.

111. Chuang, W.-Y., Young, T.-H., Chiu, W.-Y., and Lin, C.-Y. (2000), The effect of polymeric additives on the structure and permeability of poly(vinyl alcohol) asymmetric-membranes, *Polymer*, 41, 5633–5641.

112. Won, J., Lee, H. J., and Kang, Y. S. (2000), The effect of dope solution characteristics on the membrane morphology and gas transport properties. Part 2: PES/BL system, *J. Membr. Sci.*, 176, 11–19.

113. Fan, S.-C., Wang, Y.-C., Li, C.-L., Lee, K.-R., Liaw, D.-J., Huang, H.-P., and Lai, J.-Y. (2002), Effect of coagulation media on membrane formation and vapor permeation performance of novel aromatic polyamide membrane, *J. Membr. Sci.*, 204, 67–79.

114. Mohamed, N. A., and Al-Dossary, A. O. H. (2003), Structure–property relationships for novel wholly aromatic polyamide-hydrazides containing various proportions of *para*-phenylene and *meta*-phenylene units. Part III: Preparation and properties of semi-permeable membranes for water desalination by reverse osmosis separation performance, *Eur. Polym. J.*, 39, 1653–1667.

115. Pinnau, I., and Koros, W. (1991), Structures and gas separation property asymmetric polysulfone membranes made by dry, wet, and dry/wet phase-inversion, *J. Appl. Polym. Sci.*, 43, 1491–1502.

116. Sharpe, I. D., Ismail, A. F., and Shilton, S. J. (1999), A study of extrusion shear and forced convection residence time in the spinning of polysulfone hollow fiber membranes for gas separation, *Sep. Purif. Technol.*, 17, 101–109.

117. Ismail, A. F., Ng, B. C., and Abdul Rahman, W. A. W. (2003), Effects of shear rate and forced convection residence time on asymmetric polysulfone membranes structure and gas separation performance, *Sep. Purif. Technol.*, 33, 255–272.

118. Matsuyama, H., Teramoto, M., and Uesaka, T. (1997), Membrane formation and structure development by dry cast process, *J. Membr. Sci.*, 135, 271–288.

119. Jansen, J. C., Macchione, M., and Drioli, E. (2005), High flux asymmetric gas separation membranes of modified poly(ether ether ketone) prepared by the dry phase inversion technique, *J. Membr. Sci.*, 255, 167–180.

120. Jansen, J. C., Buonomenna, M. G., Figoli, A., and Drioli, E. (2006), Asymmetric membranes of modified poly(ether ether ketone) with an ultra-thin skin for gas and vapour separations, *J. Membr. Sci.*, 272, 188–197.

121. Kleinbaum, D. G., Kupper, L. L., and Muller, K. E., (1987), *Applied Regression Analysis and Other Multivariable Methods*, 2nd ed., Duxbury, Belmont, pp. 658–659.

APPENDIX

The in vitro release profiles of the model drug theophylline are shown below in Figures A1 through A10.

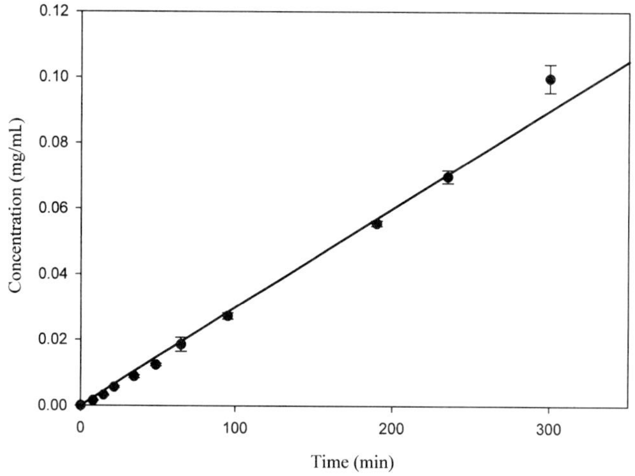

FIGURE A1 Release of theohylline from tablet coatings made with 5% CA dissolved in 90% acetone and 5% water.

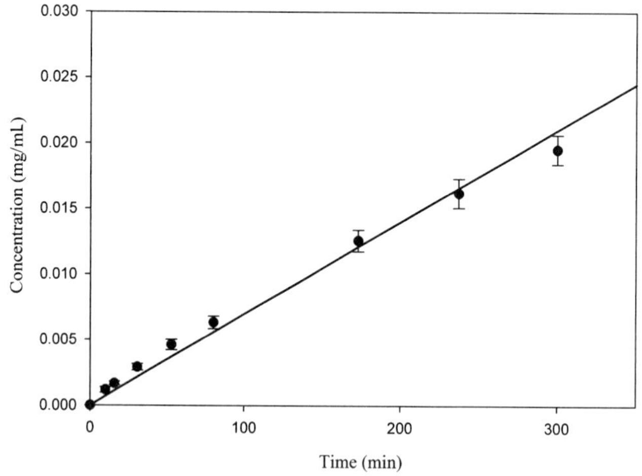

FIGURE A2 Release of theohylline from tablet coatings made with 10% CA dissolved in 85% acetone and 5% water.

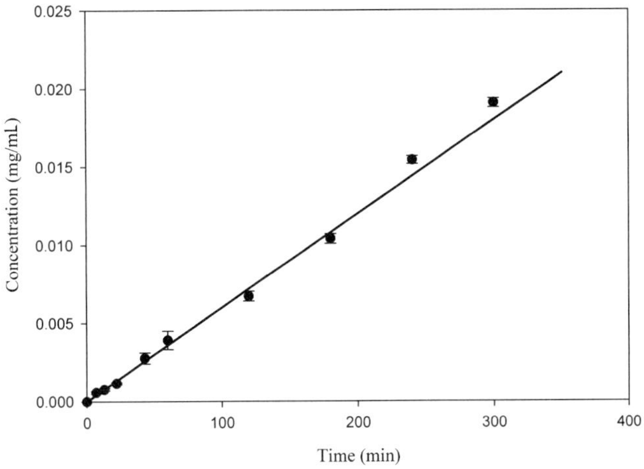

FIGURE A3 Release of theohylline from tablet coatings made with 15% CA dissolved in 80% acetone and 5% water.

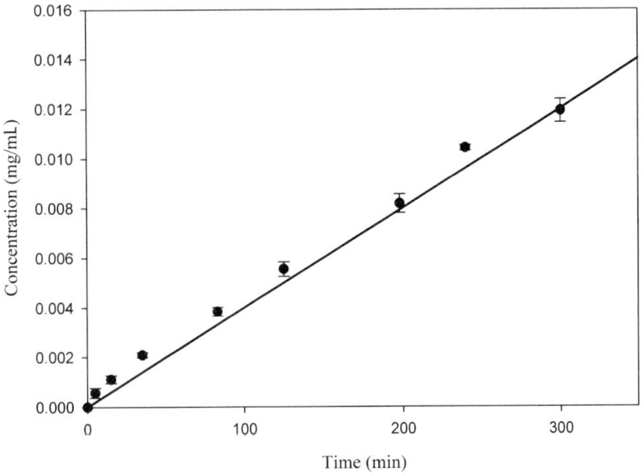

FIGURE A4 Release of theohylline from tablet coatings made with 5% CA dissolved in 85% acetone and 10% water.

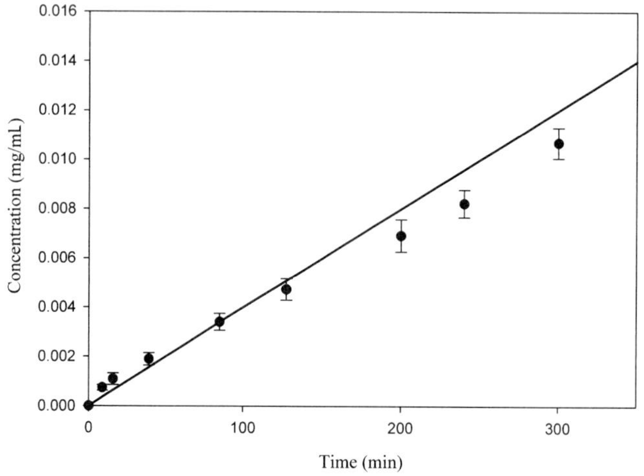

FIGURE A5 Release of theohylline from tablet coatings made with 10% CA dissolved in 80% acetone and 10% water.

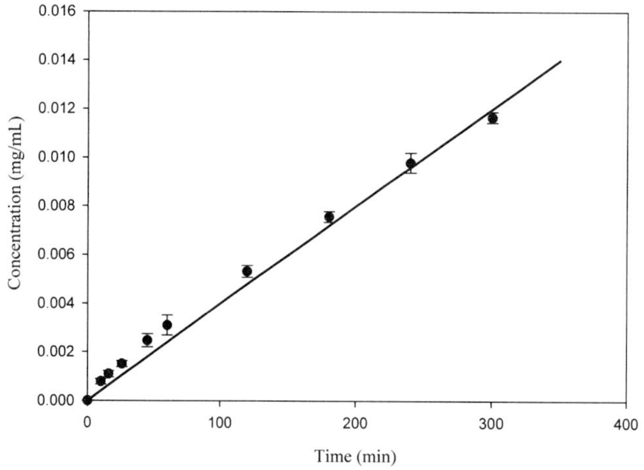

FIGURE A6 Release of theohylline from tablet coatings made with 15% CA dissolved in 75% acetone and 10% water.

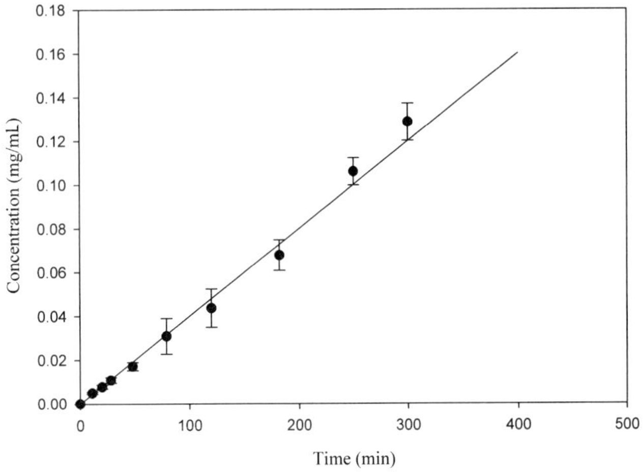

FIGURE A7 Release of theohylline from tablet coatings made with 5% CA dissolved in 80% acetone and 15% water.

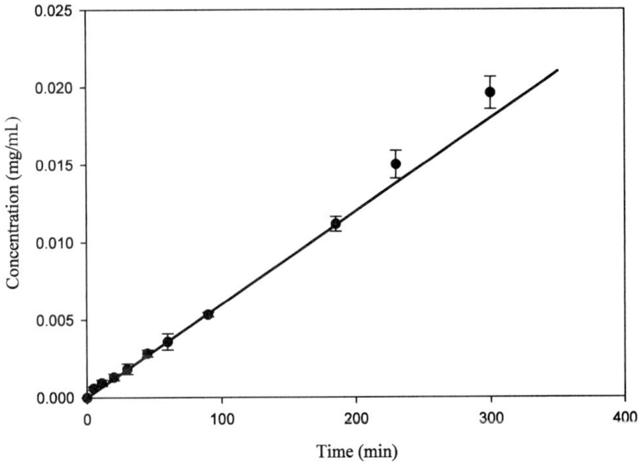

FIGURE A8 Release of theohylline from tablet coatings made with 10% CA dissolved in 75% acetone and 15% water.

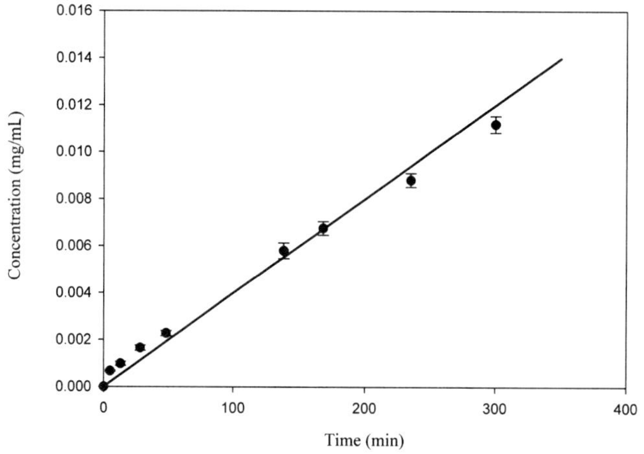

FIGURE A9 Release of theohylline from tablet coatings made with 15% CA dissolved in 70% acetone and 15% water.

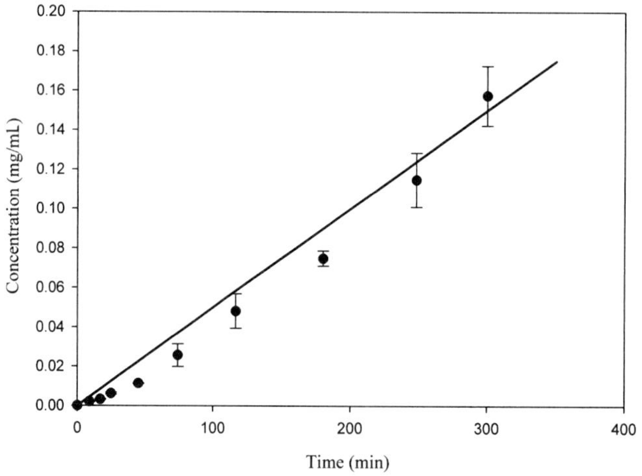

FIGURE A10 Release of theohylline from tablet coatings made with 12.5% CA dissolved in 77.5% acetone and 10% water.

6.6

TABLET COMPRESSION

Helton M. M. Santos and João J. M. S. Sousa

University of Coimbra, Coimbra, Portugal

Contents

6.6.1 INTRODUCTION

Tablets are the most important pharmaceutical dosage from and their design has always been of great interest to pharmaceutical engineering. Since tablets are made by a process of die compaction, although commonly called tablet compression, many investigations have been involved in the task to describe the mechanisms involved in this process. Nevertheless, some considerations should be taken regarding the definitions of stages involved in tablet compression. Compression is one of two

Pharmaceutical Manufacturing Handbook: Production and Processes, edited by Shayne Cox Gad

stages involved in the compaction of a two-phase system due to the application of an external force. It is defined as the reduction in the bulk volume of a material as a result of gaseous phase [1, 2]. The second stage is consolidation, which is described as an increase in the mechanical strength of the material resulting in particle-to-particle interaction [1, 2].

It is suggested that four mechanisms are basically involved in the process of compression of particles: deformation, densification, fragmentation, and attrition. The process of compression is briefly described as follows: small solid particles are filled in a die cavity and a compression force is applied to it by means of punches and then the formed monolithic dosage form is ejected. The shape of the tablet is dictated by the shape of the die while the distance between the punch tips at the point of maximum compression governs the tablet thickness, and the punch tip configuration determines the profile of the tablet. The compression cycle in a conventional rotary tablet press will be described in detail in this chapter and is illustrated in Figure 1.

The physical and mechanical properties of tablets, such as density and mechanical strength, are significantly affected by the process. Since tablet compression relies on the ability of particulates to be compacted, the need to control the critical properties of the materials with respect to readiness or ability to compact is an important issue to the formulator.

In order to compress a powder or granulation product into a tablet of specific hardness, a defined compression force must be applied. As pointed out by Shlieout et al. [3], by compressing a constant mass of powder, any variation in the applied force causes a change in the measured force. In addition, the substance itself plays an important role, that is, if it is of good compressibility, then the force needed for compression would be low. It is well known that this compressibility will depend on powder characteristics such as crystal habit and thermodynamic behavior.

The structure and strength of tablets are often discussed in terms of the relationship between the properties of the particulate material and the properties of the formed tablet. The properties of a powder that control its evolution into a tablet

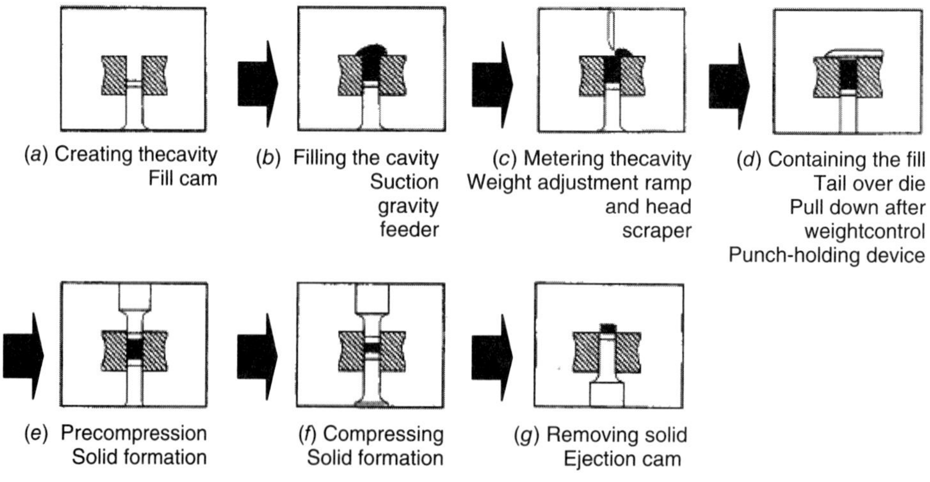

FIGURE 1 Compression cycle. (Courtesy of Thomas Engineering.)

during compression, which will also relate to the fracture toughness and the tensile strength of the tablets, are the compression mechanics of the particles and their dimensions. Generally, all materials have the ability to store some elastic strain; however, its extent will greatly vary for different materials and will depend upon the intrinsic nature of the material. There are many instances where a brittle material, or its surface, reduces significantly its cohesion or adhesion compared to that of a ductile material [4].

6.6.2 THEORY OF PARTICLE COMPACTION

Basically, the process of tablet compression starts with the rearrangement of particles within the die cavity and initial elimination of voids. As tablet formulation is a multicomponent system, its ability to form a good compact is dictated by the compressibility and compactibility characteristics of each component. Compressibility of a powder is defined as its ability to decrease in volume under pressure, and compactibility is the ability of the powdered material to be compressed into a tablet of specific tensile strength [1,2]. One emerging approach to understand the mechanism of powder consolidation and compression is known as percolation theory. In a simple way, the process of compaction can be considered a combination of site and bond percolation phenomena [5]. Percolation theory is based on the formation of clusters and the existence of a site or bond percolation phenomenon. It is possible to apply percolation theory if a system can be sufficiently well described by a lattice in which the spaces are occupied at random or all sites are already occupied and bonds between neighboring sites are formed at random.

The transitional repacking stage is driven by the particle size distribution and shape. This will determine the bulk density as the powder or granulation product is delivered into the die cavity. In this phase, the punch and particle movements occur at low pressure. The particles flow with respect to each other, with the finer particles entering the void between the larger particles, and thus the bulk density of the granulation is increased. Various techniques have been utilized to determine the degree of the two consolidation mechanisms in pharmaceutical solids (initial packing of the particles and elimination of void spaces), namely the rate dependency technique. By applying this technique, stress relaxation data based on the Maxwell model of viscoelastic behavior indicate virtually no rate dependency for elastic or brittle materials. There is also an increase in the calculated yield pressure with an increase in punch velocity for viscoplastic materials such as maize starch and polymeric materials. This is attributed to the reduction of time necessary for the plastic deformation process to occur [6]. For brittle materials such as magnesium and calcium carbonates there is no observed change in the yield pressure with increasing punch velocity [6].

When a force is applied to a material, deformation occurs. When this deformation completely disappears after cessation of the external force, further deformation occurs. Deformations that do not completely recover after release of the stress are known as plastic deformations. The force required to initiate a plastic deformation is known as the yield stress. When the particles are so closely packed that no further filling of the voids can occur, a further increase of the compressional force causes deformation at the points of contact. Both plastic and elastic deformation may occur, although one type will predominate for a given material. The ability of materials to be compressed relies on their deformation behavior. The known extreme cases are as follows. For elastic bodies, the force applied to consolidation will be fully given back (action equals reaction). This is expressed as a completely elastic deformation. For plastic bodies, the force applied will be saved as energy in the body and will express no elastic deformation at all. During tablet building, these two processes never occur alone but occur only in combination, as mentioned before.

As the external force is increased, the stresses within the particles become great enough and cracks may occur. Fragmentation furthers densification with the infiltration of the smaller fragments into the void spaces being responsible for increasing the number of particles and formation of new and clean surfaces that are potential bonding areas. The mechanisms of fragmentation and plastic deformation are not independent since both phenomena modify particle size distribution, and larger particles do not act as small particles with respect to plastic deformation [7].

The bonding of particles is governed by different mechanisms. The three most considered theories are mechanical theory, intermolecular theory, and liquid surface film theory. The first theory assumes that under pressure the individual particles experience elastic, plastic, or brittle deformation and that the edges of the particles intermesh, forming a mechanical bonding. According to Parrot [2], intermolecular theory states that under pressure the molecules at the points of true contact are close enough so that van der Waals forces interact to consolidate the particles. Liquid surface film theory relies on the presence of a thin liquid film, which may be the consequence of fusion or solution, at the surface of the particle, induced by the energy of compression. Even tough the applied force is not high, it is locally transmitted to small areas of true contact so that a very high pressure will exist at the contact surfaces. This high pressure plays an important role in the melting point and solubility of the material and proves to be essential to bonding. It follows that after releasing the pressure, solidification of the fused material would form solid bridges between the particles. An important consideration has been proposed by Zuurman et al. [8] to explain the action of some excipients during this phase. One of these excipients is magnesium stearate, which is widely used as a lubricant in order to prevent tablets from sticking to the die and punches and minimize wear of tooling. It has been proven that magnesium stearate can have an adverse effect on bonding between particles. The decrease of tablet strength is always considered to be the result of reduction of interparticle bonding due to the addition of a lubricant.

The production of tablets with the desired characteristics depends on the stresses induced by elastic rebound and the associated deformation processes during decompression and ejection. Ideally, if only elastic deformation occurred, with the sudden removal of axial pressure the particles would return to their original form, breaking any bonds that may have been under pressure.

Finally, as the lower punch rises and pushes the tablet upward, there is a continued residual die wall pressure and considerable energy may be expanded due to the die wall friction. As the tablet is removed form the die, the lateral pressure is relieved, and the tablet undergoes elastic recovery with an increase in the volume of the portion removed from the die.

The compression cycle profiles may be used to characterize the consolidation mechanisms of powders as they help to characterize the extent of pressure distribution within the powder bed as well as the formed tablet. The compression behavior of powder mixtures is usually characterized using the well-known Heckel equation [9, 10], which describes the relationship between the porosity of a compact and the applied pressure and is based on the assumption that the densification of the bulk powder in the die follows first-order kinetics:

$$\ln\frac{1}{1-\rho_r} = kP + A \qquad (1)$$

where ρ_r is the relative density of the compact at pressure σ, P is the applied pressure, and K and A are constants. The constants A and k are determined from the intercept and slope, respectively, of the extrapolated linear region of a plot of $\ln(1/1 - \rho_r)$ versus σ (compaction pressure). The Heckel constant k is related to the reciprocal of the mean yield pressure, which is the minimum pressure required to cause deformation of the material undergoing compression. The intercept obtained from the slope of the upper portion of the curve is a reflection of the densification after consolidation. A large value of k indicates the onset of plastic deformation at relatively low pressure. Thus, K appears to be a material constant. The correlation between k and the mean yield pressure P_y gives Equation (2). The constant A is related to the densification during die filling and particle rearrangement prior to bonding [11]:

$$k = \frac{1}{P_y} \qquad (2)$$

A high ρ_r value indicates that there will be a high volume reduction of the product due to particle rearrangement. The constant A has been shown to be equal to the reciprocal of the mean yield pressure required to induce plastic deformation. A larger value for A (low yield pressure) indicates the onset of plastic deformation at relatively low pressure, a sign that the material is more compressible.

The Heckel plot allows an interpretation of the mechanism of bonding. A nonlinear plot with small value for its slope (a small Heckel constant) indicates that the material undergoes fragmentation during compression. When the plot is linear, it indicates that the material undergoes plastic deformation during compression.

In addition to the Heckel approach, other techniques may be applied to the characterization of powder compression. One of these approaches was proposed by Cooper and Eaton [12]:

$$\frac{V_0 - V}{V_0 - V_s} = a_1 \exp\left(\frac{-k_1}{P}\right) + a_2 \exp\left(\frac{-k_2}{P}\right) \qquad (3)$$

where V is the volume of the compact at pressure P (m^3), V_0 is the volume of compact at zero pressure (m^3), V_s is the void-free solid material volume (m^3), a_1, a_2, k_1, and k_2 are the Cooper–Eaton constants.

The Kawakita equation [13] describes the relationship between volume reduction and applied pressure according to Equation (4), where P is the applied pressure, V_0 is the initial bulk volume, V is the volume at pressure P, a and b are the constants characteristic of the powder under compression, and C is the degree of volume reduction [Equation (5)]:

$$\frac{P}{C} = \frac{P}{a} + \frac{1}{ab} \tag{4}$$

$$C = \frac{V_0 - V}{V_0} \tag{5}$$

In the Kawakita equation the particle density is not introduced in the calculations since the model operates on the relative change in volume, which gives the same result whether the relative or the absolute volume is used. The problem in the calculation of this equation is to find the correct initial volume V_0.

6.6.3 COMPACTIBILITY

Compactibility of a powdered mixture is defined as the ability of the material to be compressed into a tablet of a specified strength without changing its composition. Investigations have demonstrated that binary mixes of identical composition could have different organizations, depending on the surface energy and particle size of the fraction used. Actually, it has been demonstrated that it is possible to control the organization of binary mixes by modifying the particle sizes of the fractions blended if they have the appropriate surface energies [6].

Generally, only powders that form hard compacts under an applied pressure without exhibiting any tendency to cap or chip can be considered as readily compactible. The compactibility of pharmaceutical powders can be characterized by its tensile strength and indentation hardness, which can be used to determine three dimensionless parameters: strain index, bonding index, and brittle fracture index.

To calculate the work of compaction during tableting, it is necessary to have accurate values of force and punch displacement. Differences in the dynamics of powder densification between eccentric and rotary machines were pointed out by Palmieri et al. [14] after compression of microcrystalline cellulose, lactose monohydrate, and dicalcium phosphate dehydrate at different compression pressures. The effect of the longer dwell time of the rotary machine press on the porosity reduction after the maximum pressure is reached is more noticeable in a ductile material such as microcrystalline cellulose. It has been shown that Heckel parameters obtained in the rotary press are in some cases different from those recovered in the eccentric machine because of the longer dwell time, machine deflection, and punch tilting occurring in the rotary press, although theoretically they could better describe the material densification in a high-speed production rotary machine.

Williams and McGinity [15] studied and compared the compaction properties of microcrystalline cellulose from six different sources using tableting indices. It was demonstrated that storage of compacts at elevated humidity conditions prior to determining the tableting indices decreased the magnitude of the tensile strength, dynamic indentation hardness, and bonding index. Based on the differences in physicomechanical properties observed for the tableting indices, the authors stated that microcrystalline cellulose products from different sources are not directly interchangeable and showed that the tensile strength, indentation hardness, bonding index, and brittle fracture index for compacts composed of microcrystalline cellulose in combination with either talc or magnesium stearate generally decreased as the amount of the lubricant was increased over the concentration range of 0–9%. Similar results were observed for admixtures of sodium sulfathiazole in combination with either talc or magnesium stearate. It was also demonstrated that the tensile strength, indentation hardness, and bonding index increased, and the brittle fracture index decreased as the percent of microcrystalline cellulose was increased in a mixture with sodium sulfathiazole.

The results of a study conducted by Muller and Augsburger [16] suggest that the pressure–volume relationship determined during powder bed compression is affected by the instantaneous punch speed profile of the displacement–time waveform for all materials studied, even though they deform by different mechanisms. It appears that the instantaneous punch speed profile of the particular displacement–time waveform is a confounding factor of Heckel analysis.

Moisute acts as a plasticizer and influences the mechanical properties of powdered materials for tablet compression. In the case of microcrystalline cellulose, at moisture levels above 5% the material exhibits significant changes consistent with a transition from the glassy state to the rubbery state [17]. The possible influence of moisture on the compaction behavior of powders was also analyzed by Gupta et al. [18]. This work evaluates the effect of variation in the ambient moisture on the compaction behavior of microcrystalline cellulose powder.

The work conducted by Gustafsson et al. [19] evaluated the particle properties and solid-state characteristics of two different brands of microcrystalline cellulose (Avicel PH101 and a brand obtained from the alga *Cladophora* sp.) and related the compaction behavior to the properties of the tablets. The difference in fibril dimension and, thereby, the fibril surface area of the two celluloses were shown to be the primary factor in determining their properties and behavior.

The compaction properties of pharmaceutical formulations can be studied experimentally using a variety of techniques, ranging from instrumented production presses to compaction simulators, and methods of analysis. The results are usually plotted as porosity–axial stress functions, which is of interest to compare different materials. However, there are some drawbacks on this type of evaluation. As mentioned by Cunningham et al. [20], this approach is deficient once it considers only the average stress along the direction of compaction, ignoring radial stress transmission and friction.

There have been some attempts to overcome the analysis of compaction problems, mostly by introducing numerical modeling approaches. The modeling approaches often used in compaction analysis are (a) phenomenological continuum models, (b) micromechanically based continuum models, and (c) discrete-element models. The parameters that should be analyzed when tableting is under development are as follows:

1. Understanding the formulation and compositional effects on the compaction process, including axial loading and unloading along with ejection
2. Determination of the stress distributions within the powder compact, including residual stresses
3. Optimization of the tablet tooling design

4. Estimation of the density distributions within a tablet that may influence dissolution or mechanical properties
5. Estimation of the compaction force necessary to obtain tablets having given properties
6. Taking into account the effect of the tablet material on the stress distribution on tooling to aid tool design
7. Assessment of the origin of defect or crack formation
8. Optimization of more complex compaction operations such as bilayer and trilayer tablets or compression-coated tablets

The demonstration of the validity of the continuum-based modelling approach to tablet compaction requires familiarity with fundamental concepts of applied mechanics. Under the theory of such a mechanism, powder compaction can be viewed as a forming event during which large irrecoverable deformation takes place as the state of the material changes from loose packing to near full density. Moreover, it is important to define the three components of the elastoplastic constitutive models which arose from the growing theory of plasticity, that is the deformation of materials such as powder within a die:

1. *Yield criterion*, which defines the transition of elastic to plastic deformation
2. *Plastic flow potential*, which dictates the relative amounts of each component of plastic flow
3. *Evolution of microstructure*, which in turn defines the resistance to further deformation

It is also known that the compression process can be described using static and dynamic models. In the case of static models, time is not considered, although it is a very important factor in the deformation process. The viscoelastic reactions are time dependent, especially for the plastic flow.

Recently Picker [21] proposed a three-dimensional (3D) model to help explaining the densification and deformation mechanism experienced by differently deforming materials during compression. According to the author, a single description of the processes during tableting is possible, and thus densification and deformation properties can be clearly distinguished with a single model. This issue has been investigated over the last years, and a comprehensive approach has been developed for the analysis of compaction using continuum mechanics principles. This approach is based on the following components:

1. Equilibrium equations (balance of forces transmitted through the material)
2. Continuity equation (conservation of mass)
3. Geometry of problem
4. Constitutive behavior of powder (stress–strain behavior)
5. Boundary conditions, including loading (e.g., displacement and velocity) and friction between tooling and powder
6. Initial conditions (e.g., initial relative density of powder)

Due to the significant nonlinearity in material properties and contact stresses, a typical powder compaction problem cannot be solved analytically without major simplifications, and thus a numerical approach is required.

The tableting properties of materials also depend on their deformation behavior. It is apparent that the tablet tensile strength is a strong function of the plastic work required for its formation but not a function of the elastic work recovered. Consequently, it is likely that strong and ductile interparticle functions, whose formation dissipated a significant plastic work, result in strong and tough compacts [4].

It should be mentioned that the material parameters do change with compaction and the use of constant material values, which is often applied, is not necessarily appropriate given the evolving microstructure of the deforming powder. The experimental characterization and accompanying analysis allow these material properties to be evaluated within the comprehensive framework of continuum mechanics, which can be useful in analyzing and predicting the effects of constitutive behavior, friction, geometry, loading schedule, and initial condition, for example, initial relative density and powder fill configuration.

Ruegger and Celik [22] investigated the effect of punch speed on the compaction properties of pharmaceutical powders with one particular objective: to separate out differences between the effect of the compression and decompression events. Tablets were prepared using an integrated compaction research system. The loading and unloading speeds were varied independently of one another. In general, when the compression speed was equal to the decompression speed, the tablet crushing strength was observed to decrease as the punch velocity increased. When the compression speed was greater than or less than the decompression speed, the results varied, depending on the material undergoing compaction. The authors also stated that the reduction of the unloading speed had a similar effect on the direct-compression ibuprofen; however, even greater improvement in the crushing strength was observed when the loading speed was reduced. As a major conclusion, it was demonstrated that the strength of tablets can be improved and some tableting problems such as capping can be minimized or prevented by modifying the rates of loading/unloading.

It is important to notice that, in the case of interacting materials, the compatibility of a binary mix will depend mostly on the compatibility of the percolating material [23]. Accordingly, several industrial applications can be made over these findings. In the development phase, it is possible to modify the formulation of interacting systems to increase the drug content without losing the compatibility of the mix, whereas in the production phase, it is possible to increase the compatibility of a poorly compatible active ingredient by sieving or preferably by milling an excipient with good compression qualities without changing the composition of the mixture.

6.6.4 TABLET COMPRESSION

The process of tablet compression is divided into three stages: filling, compression, and ejection of the tablet (see Figure 1).

During the first stage of a compression cycle the lower punch falls within the die, creating a cavity which will contain the powder or granulation product that flows from a hopper. The fill volume is determined by the depth to which the lower punch descends in the die. At this moment the particles of the powder or granulation product flow with respect to each other, thus resulting in a close packing arrangement and the physical characteristics of the material (particle size, particle size distribution, density, shape, and individual particle surface properties) associated with process parameters such as flow rate and compression rate, and the relationship between the die cavity and the particle diameter will define the number of potential bonding points between the particles. The packing characteristic of the product to be compressed is greatly affected by the shape of the particles. Since the product to be compressed comprises components of different nature, the voidage of a closely packed system is considerably changed.

When the upper punch goes down, its tip penetrates the die, confining the powder or granulation product, letting the particle rearrangement stage to continue and initiating the compression stage as the compression force is applied. As a result, forces resulting from the compression force are transmitted through the interparticulate points of contact created in the previous stage. The porosity of the powder bed is gradually decreased, the particles are forced into intimate proximity to each other, and stress is developed at the interparticulate points of contact. Once the particles have formed contacts, they will deform plastically under the applied load. Deformation of the particles will be characterized by elastic, plastic, fragmentation, or a combination of these phenomena, which will depend on the rate and magnitude of the external applied load, the duration of locally induced stress, and the physical properties of the product under compression. When the particles are in sufficiently close proximity, they are bonded. Particles bond as a result of mechanical interlocking, which is described as entanglement of the particles, phase transition at the points of contact, and intermolecular forces, namely the van der Waals force, hydrogen bonding, and ionic bonding.

After formation of the tablet by application of a compression force follows the decompression stage, where the compression force is removed and the upper punch leaves the die. Then, the formed tablet undergoes a sudden elastic expansion followed by a viscoelastic recovery during ejection when the lower punch moves upward.

6.6.5 EQUIPMENT FOR TABLET COMPRESSION

The equipment employed for tablet compression is generally categorized according to the number of compression stations and dislocation mode. Therefore, eccentric model presses have only one compression station (one die and one pair of punches, upper and lower) while rotary models have multiple compression stations (each station with one die and one pair of punches, upper and lower). The basic difference between the two types of compression equipment is that for eccentric models the compression force applied during compression is due to the upper punch whereas for rotary models it is mainly applied by the lower punch.

A rotary tablet press machine (Figure 2) comprises a housing in which the compression set and subsets (upper and lower roller assemblies) are mounted, the turret head, the upper cams, the weight control assembly and the lower cams, the hopper,

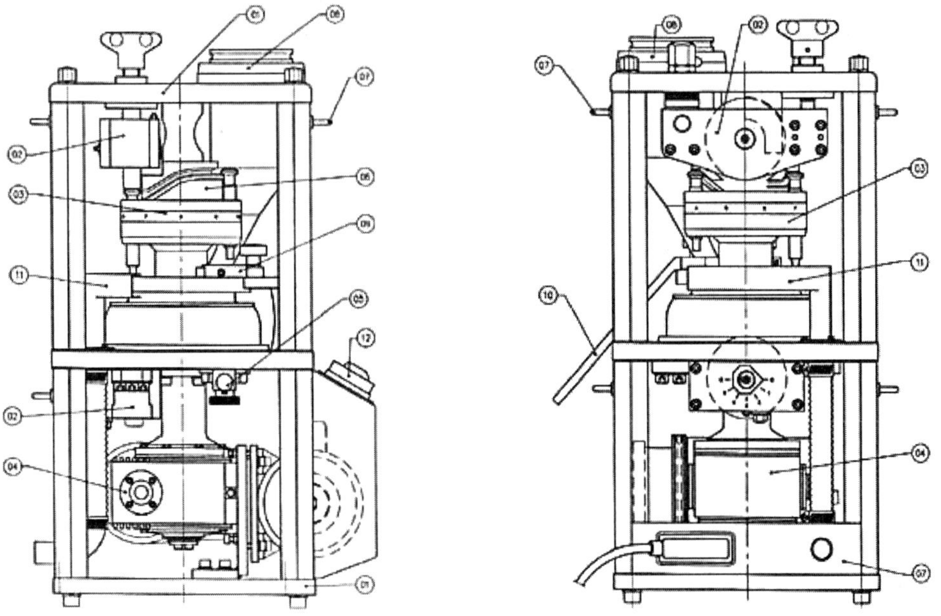

(*a*) (*b*)

FIGURE 2 Rotary tablet press machine: (*a*) left-side view; (*b*) black-side view. (1) Cabinet, (2) compression, (3) turret, (4) gear, (5) weight control assembly and lower cams, (6) plate cams, (7) guarding, (8) hopper system, (9) feed frame assembly, (10) take-off chute, (11) aspiration assembly, (12) electrical system.

the feeder assembly, the take-off chute, the aspiration assembly, the gear box and the electrical unit, and the lubrication system.

The compression zone is located on the back side of the equipment and employs a maximum load force limited by the type of tooling being used. It is of paramount importance to note that, if a load force is applied over the indicated limit, the press unit will not function properly, resulting in premature wear or possible damage to the tooling. The compression set comprises the hopper and feeder system, the die table, the upper and lower compression rollers, the upper and lower turrets, the excess-material scraper, the tablet stripper, the recirculation channel, and the aspiration system.

The hopper is usually made of stainless steel and has the shape of a funnel to contain and deliver the product to be compressed. It may be provided with a window for the observation of the product level and may also be provided with low-level sensors that signal an alarm, shut off the engine, or activate the feeding mechanism to deliver the product when it falls below this level. The feeder system usually consists of three sections (in the case of force feeders) and is ideal for press performance at high speed. The first section of a force feeder system is where the hopper is connected and is responsible for the flow of the product from the hopper to the next sections. The second section is where the die cavities are filled to their maximum capacities, and the third section is where the weight control adjustment takes place. These sections contain paddle systems which prevent packing of the product. The speed of the paddles is adjustable and should be synchronized with the die table in order to prevent tablet weight variation. Better adjustment of the paddle speed could be achieved when keeping the lowest standard deviation of the compression force. The feeder system height above the die table is usually kept between 0.05 and 0.10 mm. When the product to be compressed is of very fine particles, this height should be kept at 0.025 mm.

Presses are commonly equipped with a powder aspiration system which is connected to a vacuum source in order to remove excess powder from the die table. This assembly is essential for a high-speed press working for extended periods of time. Special attention must be taken when the powder product comprises an active ingredient of fine particle size. In this case, aspiration should be minimal in order to prevent loss of the active ingredient.

The compression subsets comprise the upper roller assembly and the lower roller assembly. The upper roller assembly is located on the roof plate of the press and utilizes an adjustment system for the regulation of the insertion depth. The lower roller assembly is located on the underside of the die table and utilizes a device for the regulation of the tablet edge thickness.

The turret head is fixed to the main shaft of the gear box. It is manufactured in two pieces (upper and lower) which guarantee the alignment between punches and dies. The gear box is mounted on the lower section under the die table and is responsible for transmitting the draft movement of the motor toward the turret head.

The upper cams are responsible for guiding the upper punches around the circumference of the turret head. It comprises the filling stage track, which guides the upper punches in an up position during its passage over the feeder system; the upper lowering cam, which guides the upper punches down in order to keep their tips covering the cavities (precompression position) and directs the upper punches to their compression stage; the upper compression roller, which guides the upper punches to their compression position; and the upper filling cam, which guides the upper punches back to the filling track.

The weight control assembly, which comprises the weight adjustment cam, is located in the lower section of the press and is regulated by an adjustment system. The lower cams are also located in the lower section of the press and comprise the preweight control (or fill cam), the weight adjustment cam, the lower lowering cam, and the ejection cam. The preweight control guides the lower punches to the full-fill position. The weight adjustment cam guides the lower punches up to the desired fill position. The lower lowering cam guides the lower punches to the precompression position. The ejection cam guides the lower punches and the formed tablets to the discharge position. It is recommended to operate the weight adjustment cam in the approximate center of the fill cam and because of this the fill cam is removable and available in different sizes having a range of approximately 10 mm with an increment range of 4 mm. The choice of the adequate fill cam for the operation of a tablet press with a particular product should be based on the density of the product. According to Figure 3, the fill cam can be adequately chosen when taking into account the density of the material to be compressed and thus the material column height in the die cavity.

Rotary tablet presses could be designed to be single, double, or triple sided. A single-sided press comprises one hopper, one set of compression rolls, and one take-off chute unit whereas double- and triple-sided presses comprise two and three each

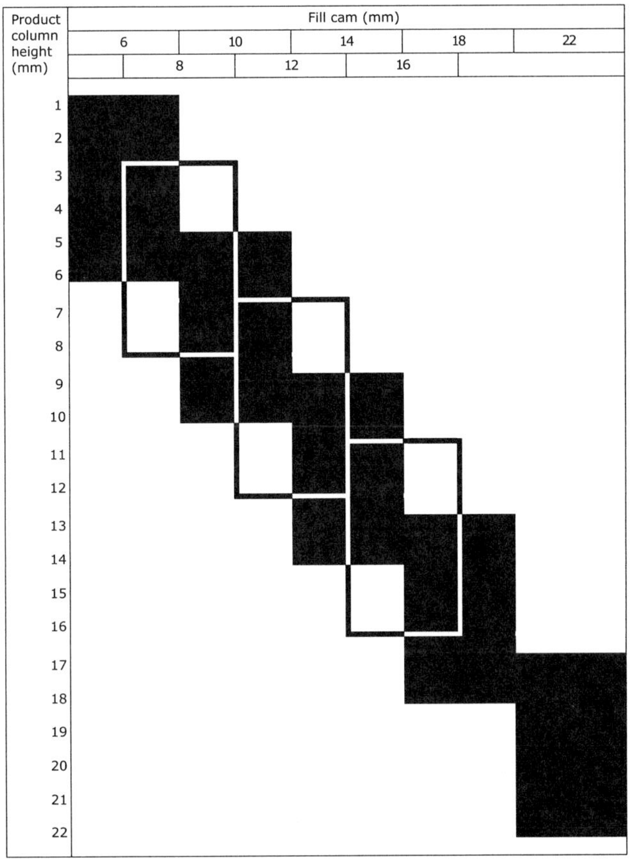

FIGURE 3 Fill cam depth as function of product column height.

of these units, respectively. Irrespective of the design of the rotary tablet press, the compression cycle is described as follows (Figures 1 and 4).

The powder or granulation product contained in the hopper flows to the feeder which spread the product through a large area over the die table in order to provide enough time to fill the die cavity. The die cavity is created when the fill cam guides the lower punches to the full-fill position and enters the feeder area. Note that the die cavity is filled with an excess of the product at this stage of the cycle. Right after, the weight adjustment ramp and head guide the lower punches to the desired fill position. The excess of the product is removed by the scraper and is pushed back by the excess product stripper when entering the recirculation channel. At this stage the lower punches are guided to the first and second lower compression rols (precompression and main compression rolls, respectively) while the upper punches are guided by the upper lowering compression roll to the precompression position and to the compression position by the main compression roll. As the upper punch penetrates the die cavity until a predefined height, the main compression roll applies the compression load over the lower punch, compressing the product in the die. Soon after compression, the upper lifting cam allows the upper punch to leave the die cavity. Simultaneously, due to the ejection cam, the lower punch is pulled,

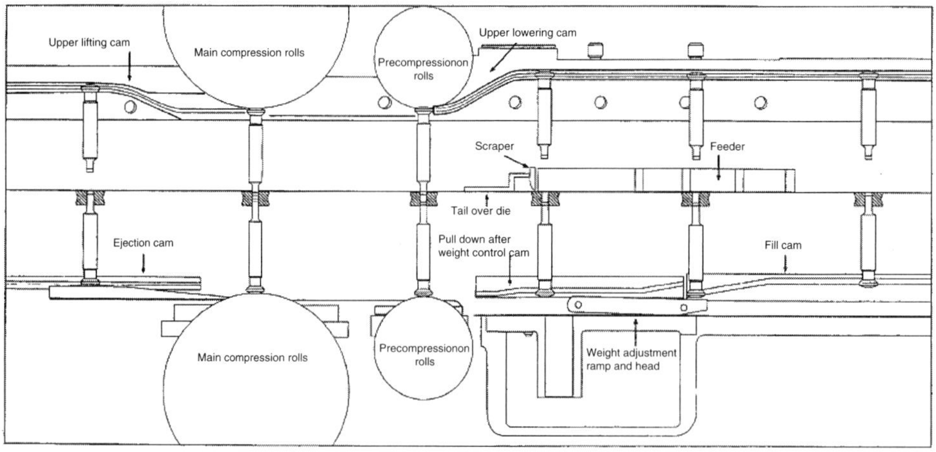

FIGURE 4 Compression cycle on rotary tablet press. (Courtesy of Thomas Engineering.)

ejecting the formed tablet to the die table. The ejected tablets are then stopped by a scraper and allowed to escape through a chute and collected. At this time the fill cam geometry makes the lower punch go down and a new compression cycle begins.

6.6.6 TABLET PRESS TOOLING

Punches and dies are essential tools in the tableting process and therefore are critical to the quality of the tablets produced. Both tools are designed for long life under normal conditions of working, but, in spite of this, they are not proof against careless handing.

It is important for those working with a tablet press to be familiar with the terminology used in the industry concerning the punches and dies. Table 1 describes the commonly used terms related to press tooling. Some of the press tooling parts can be identified in Figure 5.

When considering a tableting operation, it is important not only to select the appropriate press tooling in terms of dimensional data but also to consider the material of which tools are made. Performance of the press tooling will in part be a function of the material selected for its manufacture. Usually the material and hardness of the compression tooling are left to the manufacturer's discretion. There are various types of steel available for the manufacture of press tooling. It is important to recognize the individual characteristics of the steel regarding its composition and the percentage of each constituent element. Usually only a small amount of alloying element is added to steels (usually less than 5%) for the purpose of improving hardness and strength corrosion resistance, stability at high or low temperatures, and control of grain size. Some of these elements are as follows:

1. *Carbon* Principal hardening element. As the carbon content increases, its hardness increases. Increases the tensile strength of the steel.

TABLE 1 Tooling Terminology

Band	Area between opposing cup profiles formed by die wall
Bakelite tip relief	Undercut groove between lower punch tip straight and relief; assures sharp corner to assist in scraping product adhering to die wall
Barrel/shank	Surface controlled by turret punch guides to ensure alignment with die
Barrel-to-stem radius	Provides smooth transition from tip length to barrel
Barrel chamfer	Chamfers at ends of punch barrel, eliminates outside corners
Barrel flutes	Vertical slots machined into punch barrel to reduce bearing surface and assist in removing product in punch guides
Cup depth	Depth of cup from highest point of tip edge to lowest point of cavity
Die	Component used in conjunction with upper and lower punches; accepts product for ocmpaction and is responsible for tablet's perimeter size and configuration
Die bore	Cavity where tablet is made, shape and size determine the tablet
Die chamfer	Entry of die bore
Die groove	Groove around periphery of die to allow die to be fixed in press
Die height/ overall length	Overall height of die
Die lock	Mechanism used to lock die in position after it is installed in die table
Die outside diameter	Outside diameter of die, compatible with die pockets in press
Die taper	Gradual increase in die bore from point of compaction to mouth of bore, assists ejection
Head	End of punch which guides it through press cam track
Head/dwell flat	Flat area of head that receives full force of compression rolls at time that tablet is being formed
Inside head angle	Area of contact with lower cam and upper cam
Key	Prevents rotational movement of punches ensuring alignment to shaped and multihole dies
Keying angle	Relationship of punch key to tablet shape; position will be influenced by tablet shape, take-off angle, and turret rotation
Land	Area between edge of punch cup and outside diameter of punch tip
Neck	Relieved area between head barrel which provides clearance for die
Outside head flat angle/radius	Contact area with press cams and initial contact with pressure rolls
Overall length	Total punch length as measured from head flat to end of tip
Tip face/cup	Portion of punch tip that determines contour of tablet face including tablet embossing
Tip length	Straight portion of stem effective inside die bore
Tip relief	Portion of punch stem which is undercut or made smaller than punch tip straight; most common for lower punches in order to reduce friction from punch tip and die wall
Tooling set	Complete set of punches and dies to accommodate all stations in tablet press
Tooling station	Upper punch, lower punch, and die which accommodate one station in tablet press
Relief/undercut	Mechanical clearance between stem and die bore, sharp edge between tip straight and undercut areas acts to clean die
Stem	Area of punch opposite head which begins at end of barrel and extends to tip
Working length	Length of punch from bottom of cup to head flat; together, upper and lower working lengths control tablet thickness and weight; also known as overall length, bottom of cup (OLBC)

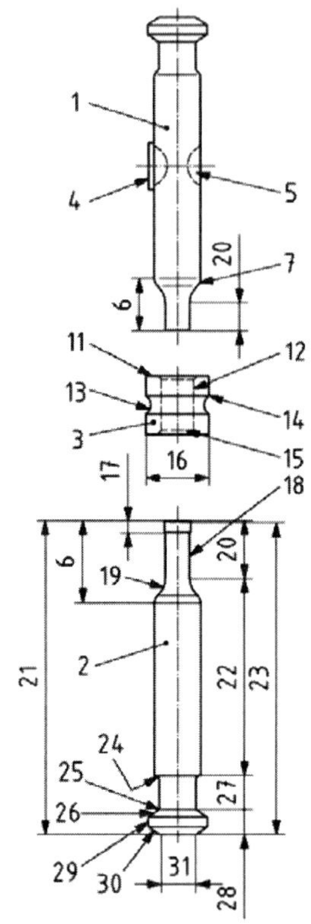

FIGURE 5 Identification of common parts of press tooling (upper and lower punches and die) according to ISO 18084, 2005: (1) upper punch, (2) lower punch, (3) die, (4) key, (5) land, (6) stem, (7) barrel-to-stem chafer, (8) cup depth, (9) tip face, (10) blended land, (11) face, (12) bore, (13) die grove, (14) protection radius or shoulder, (15) chamfer or radius, (16) outer diameter, (17) tip straight, (18) relief, (19) barrel-to-stem radius, (20) working length of tip, (21) overall length, (22) barrel, (23) working length, (24) barrel-to-neck radius, (25) neck-to-head radius, (26) inside head angle, (27) neck, (28) head, (29) head outer diameter, (30) outside head angle, (31) head flat, (32) key orientation angle, (33) upper punch face key position, (34) barrel diameter.

2. *Manganese* Increases ductility and hardenability of the steel. Also increases the rate of carbon penetration during carbonizing and imparts excellent wear resistance.
3. *Nickel* Improves the toughness and impact resistance of the steel and mildly increases its hardness.
4. *Chromium* Increases the hardness of the steel and improves its wear or abrasion resistance. It helps to limit grain size. If added in amounts greater than 5%, it can impart corrosion and wear resistances.
5. *Molybdenum* Improves hardenability and increases tensile strength of the steel.
6. *Vanadium* Produces a fine grain size and improves fatigue strength of the steel, just like molybdenum.

7. *Tungsten* Is used in tool steels to maintain hardness at elevated temperatures.
8. *Copper* Increases corrosion resistance; nevertheless its content has to be controlled, otherwise the surface quality and hot-working behavior are compromised.

The carbon steels comprise alloying elements not exceeding the defined limits of 1% carbon, 0.6% cooper, 1.65% manganese, 0.4% phosphorus, 0.6% silicon, and 0.05% sulfur. For the alloy steels, the limits exceed those for the carbon steels and may also include elements not found in carbon steels. The alloy steels have a specific designation according to the American Iron and Steel Institute (AISI). Such designation is a four-digit number where the first digit stands for the class of the alloy (e.g., 1, carbon; 2, nickel-chromium; 3, molybdenum; 4, chromium), the second digit designates the subclass of the alloy, and the last two digits designate the amount of carbon in 0.01%. The stainless steels comprise at least 10% chromium with or without the addition of any other alloying element. The tool steels are carbon steel alloys with an excess fo carbides which impart hardness and wear resistance. According to the AISI, tool steels are grouped as water hardening (W), shock resisting (S), cold-work oil hardening (O), cold-work medium-alloy air hardening (A), cold-work high-alloy high chromium (D), low alloy (L), carbon tungsten (F), low-carbon mold steels (P1–P19), other mold steels (P20–P39), chromium-based hot work (H1–H19), tungsten-based hot work (H20–H29), molybdenum-based hot work (H40–H59), high-speed tungsten based (T), and high-speed molybdenum based (M).

The appropriate steel for press tooling should be selected based upon the toughness and wear resistance required by the application, and therefore it is mandatory to have satisfactory knowledge regarding the abrasiveness, corrosiveness, and lubricity of the product intended for compression as well as the desired dimensions of the punch tip. The toughness of the steel regards its ability to resist shock and its wear resistance regards the ability to resist physical damage or erosion due to product contact.

For a clear understanding, the Thomas Engineering *Press Tooling Manual* [24] states the following:

> Punches manufactured from high carbon/high chromium steel may exhibit improved wear resistance characteristics, however under extreme compression force, the cup may crack due to the brittle nature of the steel. Steels with lower carbon and chromium levels will act conversely. While these steels may be useful in some applications, the majority will require a more moderate balance of toughness and wear resistance. Steel selection for dies is not as critical. In most cases, high wear resistance steel is preferred.
>
> The bulk of pharmaceutical tablet press punches are manufactured from S1, S5, S7, or 408 (11% chromium, 8% nickel) tool steel. The S series steels provide a good combination of shock and wear resistance and have a proven record of performance in tableting operations. At one time, 408 or 3% nickel steel was the industry standard because of its superior shock resistance toughness. The S grades however, which have only a slight loss in ductibility by comparison, offer much improved wear characteristics and have all but replaced 408 as the preferred general purpose punch steel. A2, D2 and D3 are high carbon/high chromium steel used for their excellent wear resistance. Among all the steels commonly used for press tooling, D3 has the highest wear resistance. However,

its low toughness rating typically limits its use to dies only. D2 rates slightly lower in abrasion resistance than D3 but its increased toughness makes it suitable for punch use, provided the cup design is not too fragile. A2 is a compromise between the general purpose S grades and D2 in both toughness and wear. It can be used for punches as well as dies.

Tungsten carbide, while not actually a steel, is extremely wear resistant and is commonly used to line dies. Punch tips can be manufactured from tungsten carbide; however, the cost of tooling is quite high and restricted to applications where tip fracture due to high compression forces is not likely.

Ceramic materials such as partially stabilized zirconia can also be used as die liners. Ceramics offer high wear and corrosion resistance and lower tablet ejection forces than either steel or carbide due to their low coefficient of friction.

S1, S7 and 408 provide some protection against mildly corrosive materials. More severe corrosion problems however, demand the use of stainless steel (440C) tooling. From the standpoint of wear, 440C falls between the S and D grades of tool steel. Its low toughness rating (comparable to D3) requires a strong cup design if tip fracture problems are to be avoided.

One measure of tool steel quality is the rate of inclusions. Inclusion are unwanted impurities or voids and are present to some degree in all steels. After heat treatment, inclusions give rise to localized areas of stress concentration where microscopic cracks can later develop. Remelting of the steel at the foundry will further reduce a tool steel's level of impurities; therefore improving the quality of the steel and subsequently its performance in the tooling environment. In cases where punch tip fracture is a problem, tooling suppliers may recommend a remelted or premium grade of particular steel as a means of eliminating the problem.

Concerning the configuration of compression tooling, the most commonly used are the so-called B (19 mm, or 3/4 in.) and D (21 mm, or 1 in.) tooling types. Additionally, these two types are classified into three specifications: the North American TSM (*Tableting Specification Manual*) [25], the European Union (EU) standard, and the Japan Norm (JN). The North American TSM is used in the United States and is the only standard officially supported by the governing body and published by the American Pharmacists Association. The EU standard and the JN are generally used in Europe and the Far East, respectively. In spite of the existence of these standard specifications, there are tablet press manufacturers that use their own configurations for tooling which have the disadvantage of being restricted to a specific tablet press. Figures 6 and 7 depict the three standard configurations of compression tooling. In addition, the International Organization for Standardization (ISO) standard 18084:2005 [26] comprises specifications of the main dimensions, including tolerances and characteristics of punches and dies.

Regarding the importance of compression tooling to the performance of the tablet press and the quality of the tablets, it is of paramount importance that punches and dies are handled with care. The first criterion is the identification of the tooling; that is, punches and dies should be identified according to the standard and be designated by "upper punch without key," "upper punch with key," "lower punch with key," "lower punch without key," or "die," the reference of the standard (e.g., TSM, EU, JN, ISO), and the punch or die diameter. Punches and dies should also have a marking that includes at least the manufacturer's identification, the number of the punch in the series, and/or the identification number. Upon

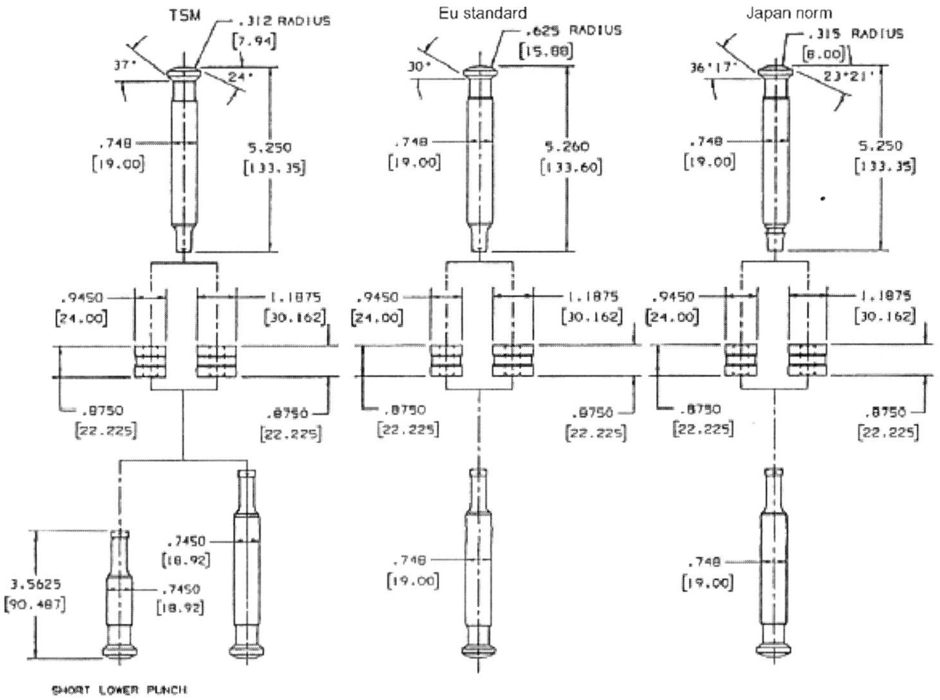

FIGURE 6 Tooling standards configurations.

receipt, after manufacturing, and prior to inspection, the punches and dies should be carefully and thoroughly cleaned and dried. Then, tooling should be lightly oiled, packed, and stored in a dry, cool place.

Damage to the punches and dies should be avoided. Therefore, they should not be transported from place to place without protective package. During transportation, installation, and removal of tooling from the tablet press, cleaning, inspection, and storage, care must be taken to avoid hitting the tips of the punches.

The visual and dimensional inspection of punches and dies should be carried out periodically. Visual inspection should be performed each time punches and dies are installed in and removed from the tablet press. Under normal conditions, slight wear is to be expected. When abnormal or excess wear is detected, the cause should be immediately investigated, inspecting the cams or components which touch the affected area of the tool. The importance of visual inspection resides on the fact that it may ensure the optimum life of the punches and dies, performance of the tablet press, and consistency and appearance of the tablets. In addition to the visual inspection, it is also recommended that dimensional inspection be performed at specific intervals throughout the life span of the punches and dies. The dimensional inspection not only ensures the consistency of hardness, weight, and thickness of the tablets, but also proves to be critical in diagnosing potential and real problem areas with regard to the tableting process and press. A typical schedule for the dimensional inspection may be as follows: 50%, 75%, 85%, 90%, and 95% of the historical or projected life cycle of the punches and dies. Therefore, the history or data base should be maintained for each set of tooling. Nevertheless, there is no general agreement on what dimensions of the punches and dies should be included in a

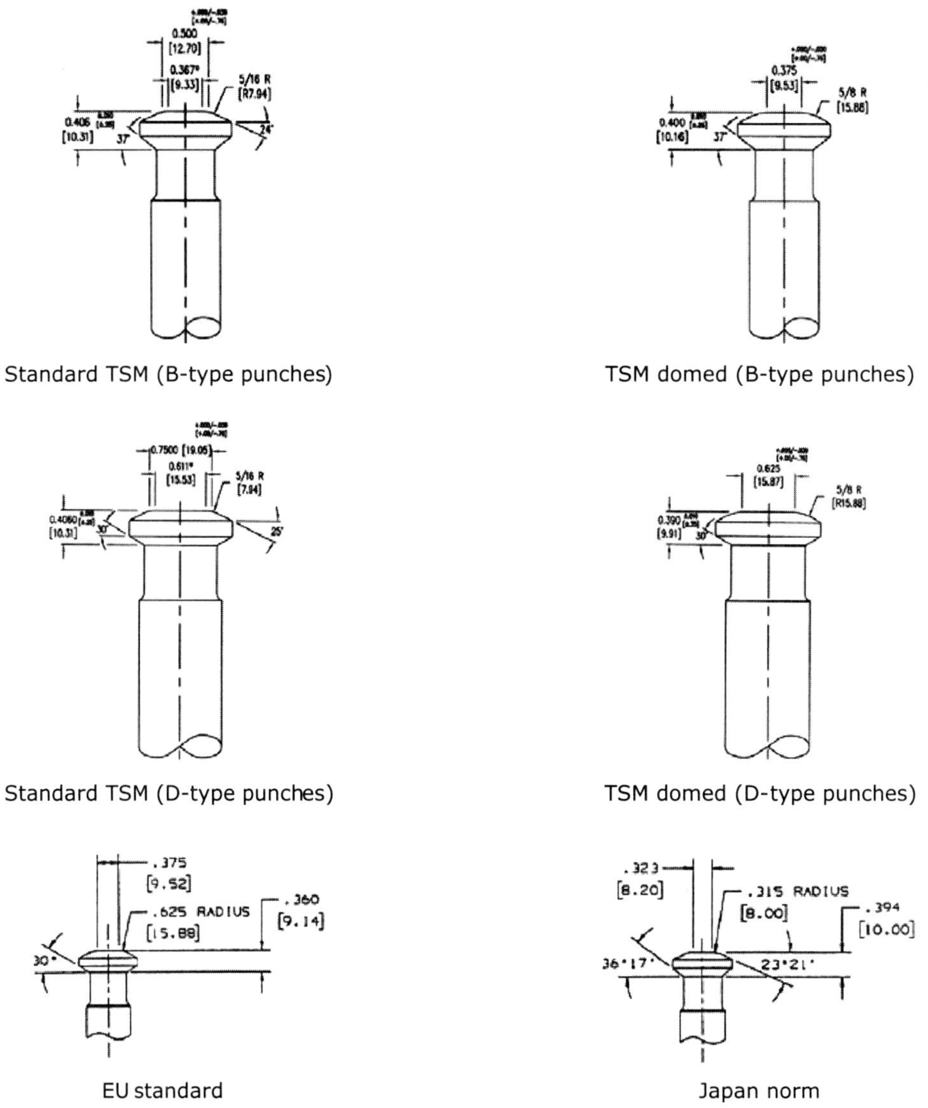

FIGURE 7 Tooling head configurations.

dimensional inspection. Some believe that a 100% inspection should be carried out while others defend that only critical dimensions (e.g., working length, cup depth, and overall length) should be inspected, believing that measuring any other dimension is either unnecessary, since it rarely if ever changes and therefore is not worthy of the time and expense of measuring, or cannot be properly measured with current equipment and is better served by a visual inspection. What is important when inspecting compression tooling is that the dimensional values are consistent within the set and tolerances and within specifications. Before proceeding with inspection, the measuring instruments should be calibrated to be certain that the dimensional values obtained are accurate and true.

6.6.7 TABLE ENGRAVING

Engraving is the most common method for tablet surface marking identification. The engraving method could be embossed (letters or symbols are raised on the tablet surface and cut into the punch tip face) or debossed (letters or symbols are cut into the tablet surface and raised on the punch tip face). For engraving on the tablet surface some specifications should be considered: stroke width, angle of engraving, radius, depth, spacing, and engraving area.

Generally, for uncoated tablet application, a stroke width between 15 and 20% of the letter height having an engraving angle of 30° is recommended. The radius should be between 50.8 μm (0.002 in.) and a value derived from dividing the stroke width by 2 times the cosine of the engraving angle. It is important to note that radii smaller than 50.8 μm or exceeding the maximum value are difficult to machine since it may decrease depth and definition of engraving. The depth is a function of the engraving angles, stroke, and radius for a given tablet size and, as a general rule, the depth should not exceed 50% of the stroke width, or no less than 88.9 μm (0.0035 in.). Spacing between letters or symbols should be a minimum of one stroke width. The available engraving area is based upon letter distortion due to the curvature or radius of the cup and thus, as a general rule, letter distortion is defined by the ratio of the outside depth of the engraving to the specified depth. Generally, distortion is present when this ratio exceeds 1:3.

When engraving is considered for fillm-coated tablets, the recommended stroke width should be the same as recommended for uncoated tablets. The recommended engraving angle should be 35°. However, engraving angles up to 40° can be used in extreme applications to allow coating solution flow. For stroke widths of 203.2 μm (0.008 in.) or less, a 30° angle is recommended to maintain minimum engraving depth. The recommended radius should be between 76.2 and 152.4 μm and a value derived by dividing the stroke width by 2 times the cosine of the engraving angle. It is important to note that radii will be determined by the flowability of the coating solution and coating process. As a general rule, stroke depth should be at least 177.8 μm. However, shallower depths can be used, provided that the film coating process is properly developed taking this factor into account. The spacing between the letters and symbols and the available engraving area considerations are the same as for uncoated tablets. The following equations can be used to determine stroke width and engraving radii:

$$S = 0.15H \leftrightarrow 0.20H \tag{6}$$

$$R_{\min} = 50.8\,\mu\text{m} \qquad R_{\max} = \frac{S}{2\cos\theta} \tag{7}$$

6.6.8 TABLET SHAPE AND PROFILE

The more popular standard geometric shapes of tablets are the round and the caplet shapes. Other tablet shapes include the oval, elliptical, square, diamond, rectangular, and polygonal. The shape of tablets plays an important role in terms of aesthetics, process (printing, film coating, packaging, and shipping), and acceptability by the consumer (identification, help with swallowing).

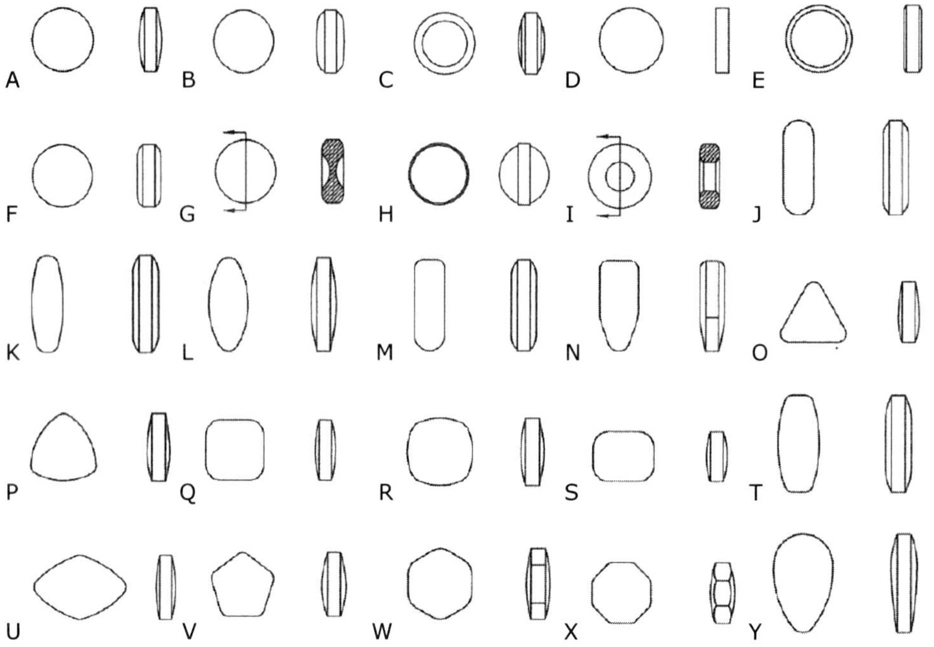

FIGURE 8 Common tablet shapes and profiles: (A) standard convex, (B) compound cup, (C) convex beveled, (D) flat faced plain, (E) flat faced bevel edged, (F) flat faced radius edged, (G) lozenge, (H) modified ball, (I) core rod with hole in center, (J) capsule, (K) modified capsule, (L) oval, (M) bullet, (N) arrow head, (O) triangle, (P) arc triangle, (Q) square, (R) pillow or arc square, (S) rectangle, (T) modified rectangle, (U) diamond, (V) pentagon, (W) hexagon, (X) octagon, (Y) almond.

In terms of design, the profile of a tablet also plays an important role in the aesthetics, packaging, orientation for printing processes, and handling. The profile of a tablet is important in the film coating process and even in helping with the swallowing. Applying a bisect score onto the tablet surface enables the tablet to be easily divided into smaller dosage amounts.

Nowadays it is common for tooling suppliers to use software to provide 2D and 3D technical drawings of tablets and tooling. Such software may provide accurate details of tablets and tooling using only tablet dimensions as input and therefore enables fast evaluation by the manufacturing department prior to ordering prototypes. Figure 8 illustrates some tablet shapes and profiles.

Flat-face, bevelled-edge tablets have many advantages due to their flatness, which provides the most compact tablet weight per volume weight, uniform hardness since the compression force is exerted evenly on the cup face, and engraving with no distortion. This tablet profile proves to be ideal for small tablets, especially when engraved, although the engraving area may be limited by the 381 μm (0.015 in.) radius on the bevel. On the other hand, compression tooling displays an inherent weakness in the punch cup design at the point where the bevel edge meets the cup flat. Attention must be paid since these types of tablets cannot be coated as they will stick together, or twin.

Shallow and standard concave tablets have the great advantage of displaying a maximum allowable area available for engraving without distortion as a result of the moderate curvature of the cup profiles and the absence of a bevel. The shallow and standard cup configurations are the strongest profiles per punch tip diameter. In addition, such profiles allow consistent distribution of the compression force over the cup face due to the slight curves involved in the cup, thus contributing to the production of tablets of uniform hardness. Nevertheless, caution must be taken concerning the cup depth since when it approaches the cup edge it may be less than the depth of the engraving. The major disadvantage of these profiles may be due to the angle of the cup profile to the tablet sidewall, which may lead to chipping at the tablet edge during film coating or handling.

Caplet-shaped tablets are easier to swallow, aesthetically pleasing, and chipping at the tablet edge generally does not occur during film coating or handling due to the angle of the cup profile to the tablet sidewall. Nevertheless, the increased curvature of the cup reduces compression force by approximately 50% compared to the shallow and standard concave profiles. Distortion of engraving may also be a problem because of the more extreme curvature. During film coating caplet tablets have the potential of sticking together, or twinning, as the tablet sidewalls are parallel. This problem could be alleviated by applying a 76.2 mm (0.003 in.) drop (15.24–20.32 cm, or 6–8 in. radius) to the sidewalls.

The concave oval profile displays a maximum allowable area for engraving and a uniform distribution of the compression force over the cup face. Such an advantage is a consequence of the mild curvature of the cup profile and absence of a bevel. Structurally, concave oval tablets are the strongest of the non-round-shaped tablets. However, due to the angle of the cup profile to the tablet sidewall, chipping at the tablet edge may occur during film coating or handling.

The compound cup profile could be used to provide round or oval tablets. This profile provides a good tablet weight per volume but simultaneously presents a weak cup edge, thus being the weakest of all cup configurations. Because of this, the maximum compression force is limited to the minor cup radius on the round shapes and the minor cup radius on the minor side for the oval shapes. In addition, the available engraving area is limited to the blending point of the two radii.

6.6.9 TABLET BISECT

Usually known as score or break line, the tablet bisect has the purpose of easily breaking the tablet in predetermined small dosages. According to the TSM, the bisect types range from the most functional (the pressure sensitive, or type G) to the least functional (partial, or type H). Each bisect type has its own characteristic, as can be seen in Figure 9. Generally, the bisect is placed on the upper punch, especially when its depth exceed 40% of the cup depth, in order to avoid problems during ejection of tablets. Nevertheless, the bisect can be placed on the lower punch either when the upper punch is supposed to contain embossed characters or printing that makes difficult the existence of the bisect or when its depth does not exceed 40% of the cup depth. When it is desired to apply a bisect to the upper tablet's surface but there is interference of engraving or printing, then a modified bisect design should be considered.

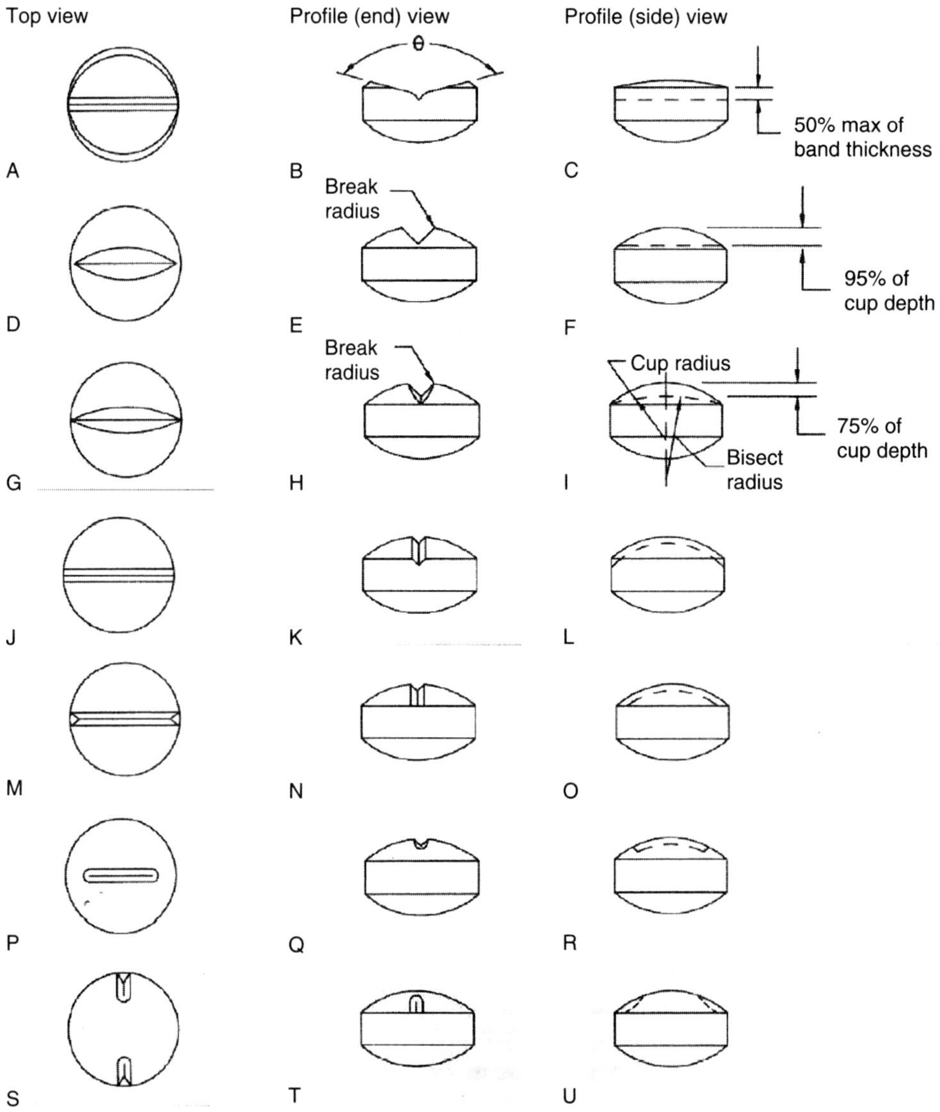

FIGURE 9 Tablet bisect for concave tooling (according to TSM): (A, B, C) pressure-sensitive (type G), (D, E, F) cut through (type D) or European style, (G, H, I) decreasing (type C), (J, K, L) standard protruding (type A), (M, N, O) standard (type E), (P, Q, R) short (type B), (S, T, U) partial (type H).

When considering applying a bisect to a tablet's surface, careful attention should be taken with respect to the tablet's cup depth, band thickness, and hardness. Considering these aspects, the specifications for the bisect size are determined taking into account the tablet's size, engraving or printing, and desired bisect design.

The TSM acknowledges two different configurations of bisect for concave tablets: protruding and cut flush. The protruding configuration follows the curvature of a radiused cup and extends past the tip edge of the punch. The cut flush configuration is the most popular bisect configuration since one may experience problems with the protruding configuration. This is explained by the fact that the protruding bisect may run into the tip edge of the lower punch if they become too close during the compression cycle of the press.

Among the bisect styles acknowledged by the TSM, the cut-through, also known as the European style, can only be applied on radiused cup designed tablets. Other styles are the standard, the short, and the partial bisects. Compared to the standard style, the cut-through style is said to have the advantage of letting patients better break the tablet into smaller subunits. On the other hand, because the cut-through is wider at the center, it decreases the available tablet surface area for engraving or printing.

6.6.10 PROBLEMS DURING TABLET MANUFACTURING

Due to either formulation or equipment, some problems can arise during the tablet compression process, such as capping and lamination, picking and sticking, mottling, double printing, weigh variation, and hardness variation. It is the early detection and accurate diagnosis of any of these flaws that can avoid tablet compression process failure and consequently improve its reliability, safety, reduce process downtime and the overall operating cost.

Often, some of the above-mentioned problems are not detected during the development of a particular tablet formulation, only appearing during scale-up as the processing speed is increased. Some of the problems experienced during tableting can be solved by shifting the formulation or alleviated by altering the tableting conditions.

6.6.10.1 Capping and Lamination

Capping and lamination are common problems that can be experienced during tableting. Capping is defined as the splitting of one or both lids of a tablet from its body. Lamination is a precursor to capping since it involves the occurrence of layers in a compact parallel to the punch face. Sometimes capping is noticed not during the process but during physical testing, such as friability and hardness.

An incipient theory proposed by Train [27] related lamination to radial elastic recovery of the compacted material during ejection. A once-accepted theory formulated that capping and lamination are the result of air entrapped in the tablet under pressure which tries to escape during ejection [28]. This theory is no longer widely accepted. Disagreement arises from the fact that some formulations cap or laminate even at low press speeds. Today, it is believed that the entrapped air may be related to capping but does not affect lamination.

A widely accepted theory for lamination presented by Long [29] and reformulated by Ritter and Sucker [30] attributes capping to the residual die wall pressure. This pressure is said to cause internal shear stresses in the tablet causing the propagation of cracks, which results in lamination or capping. The propagation of cracks can be prevented by plastic relaxation of shear stresses. Therefore, materials having sufficient plasticity may not be susceptible to lamination. Some properties of the powder mixture, such as moisture content, type and amount of the binder, and particle size, are important formulation variables that could be assessed in order to impart plasticity, thus diminishing capping and lamination tendencies.

Normally, drugs such as paracetamol, mannitol, ibuprofen, phenazone, and mefenamic acid have poor compression properties and produce tablets that are weak and frequently exhibit capping. Materials that deform elastically or exhibit time dependence are more susceptible to capping and lamination and/or strength reduction, especially as tableting rate is increased. The effect of punch velocity is most marked when transferring a material from an eccentric to a rotary press or when scaling up to larger production size tablet presses.

In addition to the possible causes of capping and lamination discussed previously, one should also consider the possibility that shape of the tooling and tooling defects are sources of capping. In such cases the problem can simply be alleviated by repairing or altering press tooling.

Usually the process of capping can be evidenced as an increase in tablet height within a few seconds after tablets are ejected from the die.

A technique generally applied to characterize and prevent the capping and lamination of a material intended to be compacted is using the brittle fracture index (BFI). The BFI was designed by Hiestand et al. [31] and measures the ability of a material to relieve stress by plastic deformation around a defect. It is obtained by applying Equation (8) and compares the tensile strength of a tablet with a hole in its center (T_0), which acts as a built-in stress concentrator defect, with the tensile strength of a similar tablet without a hole (T), both at the same relative density:

$$\mathrm{BFI} = 0.5\left[\left(\frac{T_0}{T}\right) - 1\right] \quad (8)$$

It is said that a material showing a moderate to high BFI value (>0.5) is prone to laminate and cap during the process. A low value of BFI is desirable to minimize lamination and capping during tablet production.

Indentation hardness is another measure which finds wide application in the pharmaceutical industry for the assessment of capping and lamination tendency. The indentation hardness measurement employs an indentation hardness tester and is defined as the hardness of a material determined by either the size of an indentation made by an indenting tool under a fixed load or the load necessary to produce penetration of the indenter to a predefined depth. An instrumented indentation hardness tester can be employed for that purpose since it has the ability to measure the intender penetration (H) under the applied force (F) throughout the testing cycle and is therefore capable of measuring both plastic and elastic deformation of the material under test.

Another technique for the assessment of capping and lamination tendency which has been increasingly employed in the research-and-development phase of tablet manufacturing is acoustic emission. This technique relies on the fact that an abrupt change in stress within a material to be compacted generates the release of a transient strain energy designated as acoustic emission which results in a mechanical wave that propagates within and on the surface of a structure [5, 32]. Thus, this technique can discriminate between capped and noncapped tablets based on comparing the measured level of acoustic emission energy against a decision threshold.

If it is desirable to overcome capping and lamination during the tableting process, the use of ultrasound-assisted presses could be a reliable solution. However, use this technique is still very recent since reports in the scientific literature extend only over the last decade [33].

In general, when capping and lamination are possible problems during tablet manufacture, an option could be the slower removal of force during decompression. This could be useful since capping tendency increases with increasing rates of decompression. However, better improvements could be achieved if the compression and decompression events are treated separately. By determining the effect of reducing either the loading or unloading speeds on the individual materials, it could be possible to increase crushing strength and eliminate or minimize the incidence of capping and lamination to greater extents. Thus, there is the need for a machine that is capable of customizing compaction profiles so that each formulation can be manufactured under an optimum set of conditions.

6.6.10.2 Picking and Sticking

Picking refers to adherence of powder to the punch surface. It is more problematic when the punch surfaces are engraved with logos or letters such as B, A, or O in order to produce debossed tablets. Sticking occurs when powder tends to adhere to the die leading to the development of an additional pressure to surpass friction between the formed compact and the die wall. As a result, the produced tablets show a rough surface at their edges. Furthermore, sticking can cause picking or damage the press punches by blocking the free movement of the lower punches leading to an increase of compaction pressure.

Various approaches can be used to solve picking and sticking problems during tablet manufacturing, namely optimization of press tooling, process parameters, and formulation. Generally, it is important to find the optimal combination of formulation and process parameters, particularly when market image tablets are to be produced.

In relation to formulation adjustment, an antisticking agent (talc is commonly used for this purpose) can be added to the powdered formulation in order to eliminate picking and sticking during manufacturing. Colloidal silicon dioxide may be the right choice when picking is evident since this excipient can impart smoothness to the punch surfaces. However, when adding colloidal silicon dioxide to the powder formulation, it would be necessary to add an extra lubricant in order to avoid sticking and facilitate ejection of tablets from the dies. In addition to the need for an extra excipient in the powder formulation, press tooling may need to be adjusted to improve tableting. For the production of market image tablets, logo or letters on the punches should be as big as possible. Additionally, punch tips may be plated with chrome in order to give a smooth and nonadherent surface.

When a lubricant such as stearic acid or propylene glycol or any other raw material of low melting point is present in the powder formulation, the heat generated during tableting may cause softening of these ingredients, thus leading to sticking. To overcome this problem, it may be needed to refrigerate the powder load to be tableted or to equip the press machine with a cooling unit.

6.6.10.3 Mottling

Mottling is defined as an uneven coloration of tablets or nonuniformity of color over the tablet surface. One of the possible causes of mottling may be the difference in color between the active principle and excipients, but sometimes it may be the result of degradation of the active ingredient which imparts spot zones over the surface of the tablets.

Nonetheless, when colored compressed tablets are needed for aesthetic reasons, the foremost cause of mottling is dye migration to the periphery of granules during the drying process [34]. To overcome this problem, one should consider changing the solvent used for wet granulation or the binder agent, using a low drying temperature, or decreasing the particle size of the excipient. Another way to overcome mottling was demonstrated by Zagrafi and Mattocks [35] and suggests the inclusion of an adsorbent agent such as wheat or potato starch to the formulation. The adsorbent agent is said to adsorb the dye, retarding its migration then decreasing mottling.

6.6.10.4 Weight and Hardness Variation

Weight and hardness variation are common problems experienced when tableting. Tablet weight is mainly affected by factors such as powder variation, tablet press condition and tooling, and flow of powder on the tablet press.

Inconsistent powder or granulate density and particle size distribution are common sources of weight variation during tablet compression. Problems related to the density of the powder or granulate are often associated with overfilling of the die and recirculation of the product on the tablet press. A variation of particle size distribution of the powder or granulate can be the result of segregation due to transfer or static electricity. It might also vary because the product cannot withstand the handling and mechanical stress it undergoes before reaching the tablet press.

Weight variation can arise as a result of a poorly prepared or operated tablet press. To solve this problem, one should inspect the press performance. Attention must be taken when dealing with a new die table on a load tablet press. In such a case, operation of the tablet press must regard the up-and-down motion of the punches within 76.2 μm of the setting without neglecting the conditions of the pressure rolls and cams.

Inspection of the critical dimensions of tablet press tools is recommended. At least three dimensions of the upper and lower punches should be inspected: the working length, the cup depth, and the overall length. The working length is the key factor affecting tablet weight. Therefore, the length of each punch must be correct and identical. The cup depth and the overall length are not critical with regard to controlling tablet weight. Therefore comprehensive inspection and evaluation of the press tooling are essential to minimize deviation of tablet thickness, weight, and hardness.

During the course of a compression operation it is also important to not neglect the level of the product in the hopper. Head pressure is a critical factor related to the amount of product in the hopper. The more product present in the hopper, the greater the head pressure, and vice versa. Therefore, when the head pressure varies, so does the weight of the tablets. So, in order to maintain a constant head pressure, thus reducing a potential variation of weight, compression should be conducted within a narrow range of the powder or granulate product in the hopper.

The fill cam is another factor that can have a profound effect on tablet weight. The choice of an adequate fill cam regarding some characteristics of the powder or granulate product allows the die cavity to be properly overfilled. Usually, in order to maintain consistent tablet weight during compression, it is recommended to overfill the cam by 10–30% of its volume. Basically, any tablet press part that is ultimately related to the powder product flow can have a mild or profound impact on weight control. It is important to remember that the scraper blade tends to become worn by die table rotation and powder product abrasion. Therefore, periodic inspection of its condition and replacement are recommended. Nevertheless, the scraper blade proper condition is important but also its adjustment since if it is not set up correctly, powder product may accumulate on the die table, leading to problems with weight control.

Tablet hardness variation is intimately related to weight variation and, accordingly, to the influence of compression variables such as dwell time, tablet thickness, and working length of the punches. Thus, to solve a hardness variation, consistency of the tablet weight must be checked first. If the predefined weight is achieved but hardness is out of limits, then precompression and compression forces should be adjusted while keeping tablet thickness within target limits. Although dwell time might be a source of hardness variation, adjustment of this parameter may be detrimental to the whole process since the compression rate is slowed. Occasionally, when the tablet weight target is kept within limits but hardness varies, the problem may be due to the formulation. As mentioned previously, the correct use of punches and dies is of paramount importance and periodic inspection is mandatory in order to ensure the compression process has not been compromised. So, when it becomes hard to achieve tablet hardness, it is recommended to first verify tablet weight and thickness consistency and then try to adjust the precompression. The choice to increase the tablet weight even if it is within limits or to reduce the tablet press speed is not convenient and should be used only when there are no more options.

REFERENCES

1. Çelik, M. (1994), Compaction of multiparticulate oral dosage forms, in Ghebre-Sellassier, I., Ed., *Multiparticulate Oral Drug Delivery*, Marcel Dekker, New York, pp. 181–216.
2. Parrot, E. (1990), Compression, in Lieberman, H. L., and Schwartz, J. B., Eds., *Pharmaceutical Dosage Form*, 2nd ed., Marcel Dekker, New York, pp. 201–244.
3. Shlieout, G., Wiese, M., and Zessin, G. (1999), A new method to evaluate the consolidation behavior of pharmaceutical materials by using the Fraser-Suzuki function, *Drug. Dev. Ind. Pharm.*, 25(1), 29–36.
4. Mohammed, H., Briscoe, B. J., and Pitt, K. G. (2006), A study on the coherence of compacted binary composites of microcrystalline cellulose and paracetamol, *Eur. J. Pharm. Biopharm.*, 63(1), 19–25.
5. Leuenberger, H., and Leu, R. (1992), Formation of a tablet: A site and bond percolation phenomenon, *J. Pharm. Sci.*, 81(10), 976–982.
6. Khossravi, D., and Morehead, W. T. (1997), Consolidation mechanisms of pharmaceutical solids: A multi-compression cycle approach, *Pharm. Res.*, 14(8), 1039–1045.

7. Masteau, J. C., and Thomas, G. (1999), Modelling to understand porosity and specific surface area changes during tabletting, *Powder Technol.*, 101(3), 240–248.
8. Zuurman, K., Van der Voort Maarschalk, K., and Bolhuis, G. K. (1999), Effect of magnesium stearate on bonding and porosity expansion of tablets produced from materials with different consolidation properties, *Int. J. Pharm.*, 179(1), 107–115.
9. Heckle, R. W. (1961), An analysis of powder compaction phenomena, *Trans. Metall. Soc. AIME*, 221, 1001–1008.
10. Heckle, R. W. (1961), Density-pressure relationship in powder compaction, *Trans. Metall. Soc. AIME*, 221, 671–675.
11. Shivanand, P., and Sprockel, O. L. (1992), Compaction behavior of cellulose polymers, *Powder Technol.*, (69), 177–184.
12. Cooper, A. R., and Eaton, L. E. (1962), Compaction behavior of several ceramic powders, *J. Am. Ceram. Soc.*, 45(3), 97–101.
13. Kawakita, K., and Ludde, K. H. (1970), Some considerations on powder compression equations, *Powder Technol.*, 4, 61–68.
14. Palmieri, G. F., Joiris, E., Bonacucina, G., Cespi, M., and Mercuri, A. (2005), Differences between eccentric and rotary tablet machines in the evaluation of powder densification behaviour, *Int. J. Pharm.*, 298(1), 164–175.
15. Williams, R. O.3rd, and McGinity, J. W. (1989), Compaction properties of microcrystalline cellulose and sodium sulfathiazole in combination with talc or magnesium stearate, *J. Pharm. Sci.*, 78(12), 1025–1034.
16. Muller, F. X., and Augsburger, L. L. (1994), The role of the displacement-time waveform in the determination of Heckel behaviour under dynamic conditions in a compaction simulator and a fully-instrumented rotary tablet machine, *J. Pharm. Pharmacol.*, 46(6), 468–475.
17. Amidon, G. E., and Houghton, M. E. (1995), The effect of moisture on the mechanical and powder flow properties of microcrystalline cellulose, *Pharm. Res.*, 12(6), 923–929.
18. Gupta, A., Peck, G. E., Miller, R. W., and Morris, K. R. (2005), Influence of ambient moisture on the compaction behavior of microcrystalline cellulose powder undergoing uniaxial compression and roller-compaction: A comparative study using near-infrared spectroscopy, *J. Pharm. Sci.*, 94(10), 2301–2313.
19. Gustafsson, C., Lennholm, H., Iversen, T., and Nystrom, C. (2003), Evaluation of surface and bulk characteristics of cellulose I powders in relation to compaction behavior and tablet properties, *Drug. Dev. Ind. Pharm.*, 29(10), 1095–1107.
20. Cunningham, J. C., Sinka, I. C., and Zavaliangos, A. (2004), Analysis of tablet compaction. I. Characterization of mechanical behavior of powder and powder/tooling friction, *J. Pharm. Sci.*, 93(8), 2022–2039.
21. Picker, K. M. (2004), The 3D model: Explaining densification and deformation mechanisms by using 3D parameter plots, *Drug. Dev. Ind. Pharm.*, 30(4), 413–425.
22. Ruegger, C. E., and Celik, M. (2000), The effect of compression and decompression speed on the mechanical strength of compacts, *Pharm. Dev. Technol.*, 5(4), 485–494.
23. Barra, J., Falson-Rieg, F., and Doelker, E. (1999), Influence of the organization of binary mixes on their compactibility, *Pharm. Res.*, 16(9), 1449–1455.
24. *Press Tooling Manual* (2003), Thomas Engineering Inc. Hoffman Estates, IL, USA. pp. 1–38.
25. *Tableting Specification Manual*, 7th ed., American Pharmacists Association, p. 130.
26. *International Organization for Standardization (ISO)*, 18084:2005, 1st ed, ISO, Geneva, pp. 1–13.

27. Train, D. (1956), An investigation into the compaction of powders, *J. Pharm. Pharmacol.*, 8(10), 745–761.
28. Burlinson, H. (1968), *Tablets and Tabletting*, Heinemann, London.
29. Long, W. M. (1960), Radial pressures in powder compaction, *Powder Metall.*, 6, 73–86.
30. Ritter, A., and Sucker, H. B. (1980), Studies of variables that effect tablet capping, *Pharm. Tech.*, (3), 57–65, 128.
31. Hiestand, E. N., Bane, J. M., Jr., and Strzelinski, E. P. (1971), Impact test for hardness of compressed powder compacts, *J. Pharm. Sci.*, 60(5), 758–763.
32. Joe Au, Y. H., Eissa, S., and Jones, B. E. (2004), Receiver operating characteristic analysis for the selection of threshold values for detection of capping in powder compression, *Ultrasonics*, 42(1–9), 149–153.
33. Rodriguez, L., Cini, M., Cavallari, N., Passerini, N., Saettone, M. F., Monti, D., and Caputo, O. (1995), Ultrasound-assisted compaction of pharmaceutical materials, *Farm Vestn.*, (46), 241–242.
34. Armstrong, N. A., and Palfrey, L. P. (1989), The effect of machine speed on the consolidation of four directly compressible tablet diluents, *J. Pharm. Pharmacol.*, 41(3), 149–151.
35. Zografi, G., and Mattocks, A. M. (1963), Adsorption of certified dyes by starch, *J. Pharm. Sci.*, 52(Nov.), 1103–1105.

6.7

EFFECTS OF GRINDING IN PHARMACEUTICAL TABLET PRODUCTION

GAVIN ANDREWS, DAVID JONES, HUI ZHAI, OSAMA ABU DIAK, AND GAVIN WALKER

Queen's University Belfast, Belfast, Northern Ireland

Contents

6.7.1 INTRODUCTION

The importance of size reduction in relation to pharmaceutical active agents and excipients is well known, and the aim of this chapter is to identify methods for particle size reduction, discuss how particle size and shape are characterized, and

Pharmaceutical Manufacturing Handbook: Production and Processes, edited by Shayne Cox Gad

recognize the importance of controlling particle characteristics to ensure the success of pharmaceutical powder processing and the manufacture of elegant pharmaceutical products. An initial overview of the implications of size reduction within pharmaceutics and the importance of comminution in relation to variability of active pharmaceutical ingredient (API) surface area, efficacy, and ultimately dosing regimen required to maintain optimum therapeutic effects will be addressed. This will encompass examples from a diverse range of dosage forms, including oral, parenteral, and topical systems. The effects of particle size on the essential characteristics of powders intended for compression (tablets, capsules) such as fluidity and compressibility will be addressed. The need for uniformity of size and the effects of particle size distribution on the homogeneity of mixing/blending and in essence on the uniformity of APIs within the final manufactured dosage form will be highlighted.

6.7.2 MILLING EQUIPMENT

There are many factors that must be taken into consideration in choosing milling equipment. Some of these factors are related to required product specifications such as particle size distribution, but additionally, physical and chemical properties of the material such as particle shape and moisture content must also be taken into consideration. Furthermore, other factors that are related to production requirements (mill capacity and the required production rate) must be carefully balanced to ensure the correct choice of milling equipment.

6.7.2.1 Ball Mill

A ball mill consists of a hollow cylinder mounted such that it can be rotated on its horizontal longitudinal axis (Figure 1). The length of the ball mill is slightly greater than its diameter. A ball mill reduces particle size by subjecting particles to impact and attrition forces generated by moving steel balls or pebbles (grinding medium) that typically occupy 30–50% of the total volume of the mill. It is common for a ball mill to contain balls of different diameters that aid size reduction. Generally, larger diameter balls have a higher tendency to act upon coarse feed materials

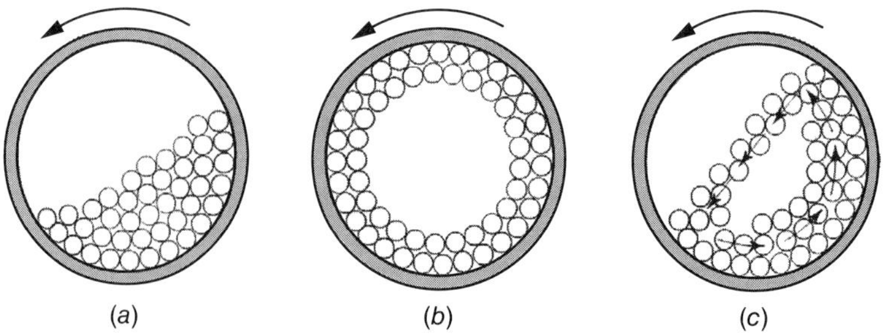

FIGURE 1 Ball mill in operation showing correct cascade action.

whereas smaller diameter balls facilitate the formation of fine product by reducing void spaces between the balls.

The most important factors governing the performance of the mill and the achievement of the desired particle size are as follows:

1. Amount of material required for subsequent testing (sample volume)
2. Speed of rotation of ball mill

A high volume of powder feed produces a cushioning effect whereas small sample volumes cause a loss of efficiency and abrasive wear of the mill parts. The amount of material to be milled in a ball mill may be expressed as a material-to-void ratio (ratio of the volume of material to that of the void in the ball charge). As the amount of material is increased, the efficiency of a ball mill is increased until the void space in the bulk volume of ball charge is filled; then, the efficiency of milling is decreased by further addition of material.

Rotational speed is the most significant factor controlling the particle size specification. The optimum speed of rotation is dependent on mill diameter. At low angular velocities the balls move with the drum until the force due to gravity exceeds the frictional force of the bed on the drum, and the balls then slide back to the base of the drum. This sequence is repeated, producing very little relative movement of balls so that size reduction is minimal. At high angular velocities the balls are thrown out onto the mill wall by centrifugal force and no size reduction occurs. At about two-thirds of the critical angular velocity where centrifuging occurs, a cascading action is produced. Balls are lifted on the rising side of the drum until their dynamic angle of repose is exceeded. At this point they fall or roll back to the base of the drum in a cascade across the diameter of the mill. By this means, the maximum size reduction occurs by impact of the particles with the balls and by attrition.

The critical speed of a ball mill is the speed at which the balls just begin to centrifuge with the mill. Thus, at the critical speed, the centrifugal force is equal to the weight of the ball. At and above the critical speed, no significant size reduction occurs. The critical speed n_c is given by the equation

$$n_c = \frac{76.6}{\sqrt{D}}$$

where D is the diameter of the mill.

A larger mill reaches its critical speed at a slower revolution rate than a smaller mill. Ball mills are operated at from 60 to 85% of the critical speed. Over this range, the output increases with the speed; however, the lower speeds are for finer grinding. An empiric rule for the optimum speed of a ball mill is

$$n = 57 - 40 \log D$$

where n is the speed in revolutions per minute and D is the inside diameter of the mill in feet.

In practice, the calculated speed should be used initially in the process and modified as required.

The use of a ball mill is advantageous in that it may be used for both wet and dry milling and additionally can be successfully employed in batch and continuous operation. Also, the installation, operation, and labor costs involved in ball milling are extremely low in comparison to other techniques, which makes this technique economically favorable.

6.7.2.2 Fluid Energy Mill

Fluid energy milling acts by particle impaction and attrition that are generated by a fluid, usually air (Figure 2). Fluid energy mills can reduce the particle size to approximately 1–20 μm. A fluid energy mill consists of a hollow toroid that has a diameter of 20–200 μm, depending on the height of the loop, which may be up to 2 m. Fluid is injected as a high-pressure jet through nozzles at the bottom of the loop with the high-velocity air, giving rise to zones of turbulence into which solid particles are fed. The high kinetic energy of the air causes the particles to impact with other particles with sufficient momentum for fracture to occur. Turbulence ensures that the high levels of particle–particle collision produce substantial size reduction by impact and attrition.

The design of fluid energy mills provides an internal classification system according to their particle size in which the finer and lighter particles are discharged and the heavier, oversized particles, under the effect of centrifugal force, are retained until reduced to a significantly smaller size.

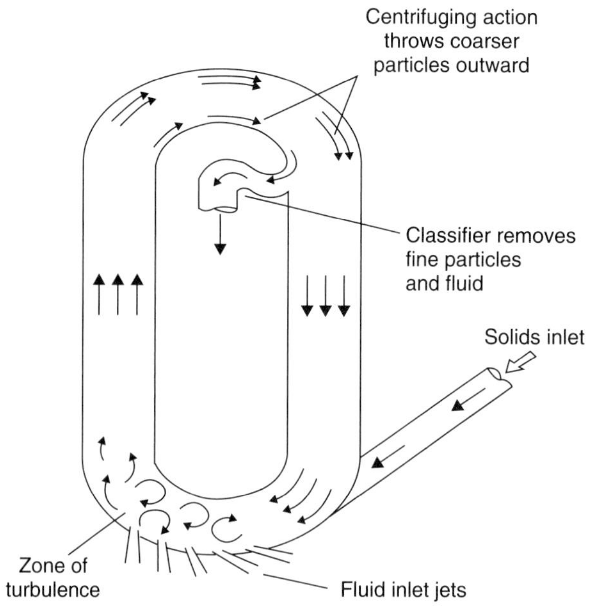

FIGURE 2 Fluid energy mill.

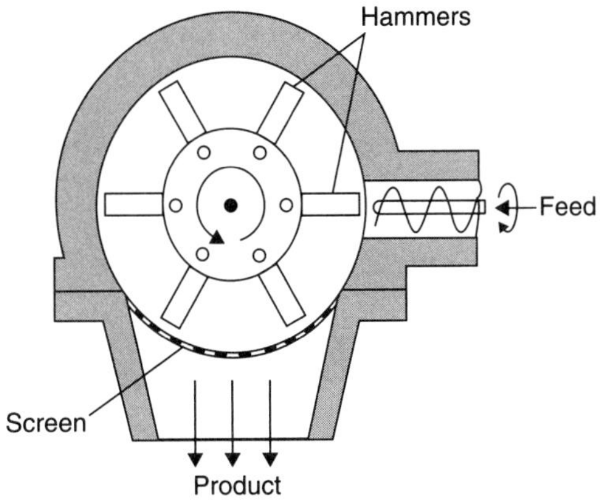

FIGURE 3 Hammer mill.

6.7.2.3 Hammer Mill

The main mechanism of size reduction produced by a hammer mill is by impaction that is generated from a series of four or more hammers hinged on a central shaft and enclosed within a rigid metal case (Figure 3). During milling the hammers swing out radially from the rotating central shaft. The angular velocity of the hammers produce strain rates up to $80\,s^{-1}$, which are so high that most particles undergo brittle fracture. As size reduction continues, the inertia of particles hitting the hammers reduces markedly and subsequent fracture is less probable, so that hammer mills tend to produce powders with narrow particle size distributions. Particle retention within the mill is achieved using a screen, which allows only sufficiently milled particles (defined particle size) to pass through. Particles passing through a given mesh can be much finer than the mesh apertures, as particles are carried around the mill by the hammers and approach the mesh tangentially. For this reason, square, rectangular, or herringbone slots are often used. According to the purpose of the operation, the hammers may be square faced or tapered to a cutting edge or have a stepped form.

The particle size achieved may be controlled variation in the speed of the hammers and additionally by careful selection of the size and type of screen. During the operation of a hammer mill the speed of rotation is critical such that below a critical impact speed the rotor turns so slowly that a blending action rather than milling is obtained. Such operating conditions result in significant rises in temperature. Moreover, at very high speeds, there is the probability of insufficient time between successive passes of the hammers for a significant mass of material to fall from the grinding zone.

The hammer mill is particularly useful in achieving particles in the approximate size range of 20–40 μm and additionally in producing a particle size distribution that is extremely narrow. The equipment offers ease of use and high levels of flexibilty (speed and screen may be rapidly changed allowing rapid variation in achievable particle size), is easy to clean, and can be operated as a closed system, thus avoiding operator exposure to potent dusts and potential explosion hazards.

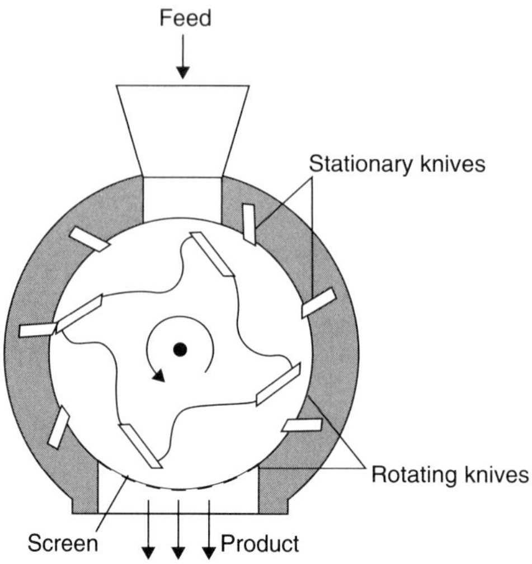

FIGURE 4 Cutter mill.

6.7.2.4 Cutting Mill

Particle size reduction using a cutting mill involves successive cutting or shearing a sample using a series of knives attached to a horizontal rotor (Figure 4). This rotary motion pushes the sample against a series of stationary knives that are attached to the mill casing. Size reduction occurs by fracture of particles between the two sets of knives, which have a clearance of approximately a few millimetres. As with a hammer mill a screen is fitted at the base of the mill casing and acts to retain material until a sufficient degree of size reduction has occurred.

6.7.3 POWDER CHARACTERIZATION TECHNIQUES

6.7.3.1 Powder Sampling

Powdered materials are used in a wide range of industries, no more so than in the pharmaceutical industry wherein powders are used for the manufacture of a wide range of dosage forms, the two most common being tablets and hard gelatin capsules. Orally administered solid dosage forms are the preferred and most patient convenient, primarily because of the ease of administration and the convenience of handling. Pharmaceutically, orally administered solid dosage forms are generally more favorable because of increased stability in comparison to their liquid counterparts (suspensions, syrups) and the increased control they offer in manipulating drug dissolution in vivo to suit end-use requirements. Solid dosage forms administered via the oral route are an intricate blend of pharmaceutical excipients (diluents, binders, disintegrants, glidants, lubricants, and flavors) and APIs. In order to successfully manufacture acceptable pharmaceutical products, these materials must be adequately mixed and/or granulated to ensure that the resultant agglomerates possess the required fluidity and compressibility and, in addition, avoid demixing during postgranulation processes. Moreover, the final characteristics of tablets or capsules such as drug dissolution rate, disintegration time, porosity, friability and hardness are significantly influenced by the properties of the powder blends used during their manufacture.

During product manufacture large volumes of powder blends are fed through production equipment/processes, and it is essential to be able to accurately determine, define, and control powder properties to ensure reproducible manufacture and product performance. Therefore the characterization of the physicochemical properties of powder blends is extremely important. It is well accepted that there are inherent difficulties in characterizing the entire mass of a bulk powder blend or process stream, so it is essential to remove and analyze discrete samples.

Sampling is a useful technique that allows an appropriate aliquot to be withdrawn from the bulk so as to collect a manageable amount of powder which is representative of the batch [3], in other words, every particle should have an equal chance of being selected [4]. However, there are many circumstances that may result in the selection of nonrepresentative samples and hence the definition of powder characteristics that are not a true estimation of the entire bulk powder. Typically, powder masses with an extremely wide particle size distribution or diverse physical properties are highly likely to be heterogeneous, which may result in high levels of variability and samples that do not represent bulk mass. Moreover, powder characteristics may change because of the attrition and segregation during transfer that can make sampling extremely difficult.

It is well accepted that two types of sampling errors are possible when removing small masses of powder from bulk [5].

1. Segregation errors, which are due to segregation within the bulk and can be minimized by suitable mixing and the use of a large number of incremental samples to form a larger test sample.
2. Statistical errors, which arise because the quantitative distribution in samples of a given magnitude is not constant but is subject to random fluctuations. Consequently, it is an example of a sampling error that cannot be prevented but can be estimated and indeed reduced by increasing the sample size.

Therefore, sampling procedures are of the greatest importance in order to reduce the effect of nonuniform size segregation and nonrandom homogeneity of a system to achieve statistically meaningful sampling results. Careful attention and faithful observance must be demonstrated and it is extremely important that sampling occurs when the powders are in motion [6] and samples are withdrawn from the whole stream for equal periods of time, rather than part of the stream for all of the time [3].

TABLE 1 Stationary Bulk Sampling

Sampling Devices	Procedure of Sampling	Application and Characteristics
Low volume powder sampler (Figure 5*a*)	In operation the sampler is inserted into the product to be sampled. At a specific sampling depth the operator pushes down on the T bar, which opens the sampling chamber. When released the spring-loaded T bar will close the sampling chamber.	Used for small quantity of sample powders. The sampler has a sampling chamber volume approximately equal to 2 mL.
Pneumatic lance sampler (Figure 5*c*)	A gentle flow of air out of the nozzle allows the probe to move through the powder bed. At the site, the air is slowly reversed to draw up a sample, which is collected against a porous plate at the end of the probe [7].	Minimizes powder disturbance and therefore is better than a sample thief, but bias still cannot be avoided [8].
Scoop sampler	A single swipe of the scoop completely across the powder bulk collects the sample. Each collection should use opposite directions.	Suitable only for materials that are homogeneous within the limits set by the quantity of material taken by the scoop. It may be used for non-free-flowing or damp materials where instrumental methods are inappropriate [9].
Thief/spear probe sampler (Figure 5*b*)	One or more cavities are stamped in a hollow cylinder enclosed by an outer rotating sleeve. The thief is inserted into sample with the cavities closed, once opened the sample fills the hole. The cavities are closed and the thief is withdrawn. It must be ensured that samples are withdrawn from different locations	Thief samplers belong to two main classes, side sampler (has one or more cavities along the probe) and end sampler (has a single cavity at the end of the probe), which are the most common used for stored non-flowing material [10].

There are a number of sampling techniques for particle sampling, which can be classified in many different ways. Here, particle sampling techniques are divided into three parts: stationary bulk sampling (Table 1 and Figure 5), flowing stream sampling (Table 2 and Figure 6), and subsampling (Table 3 and Figure 7). The sampling devices, procedures and application overview of the common used techniques in corresponding fields are shown as follow.

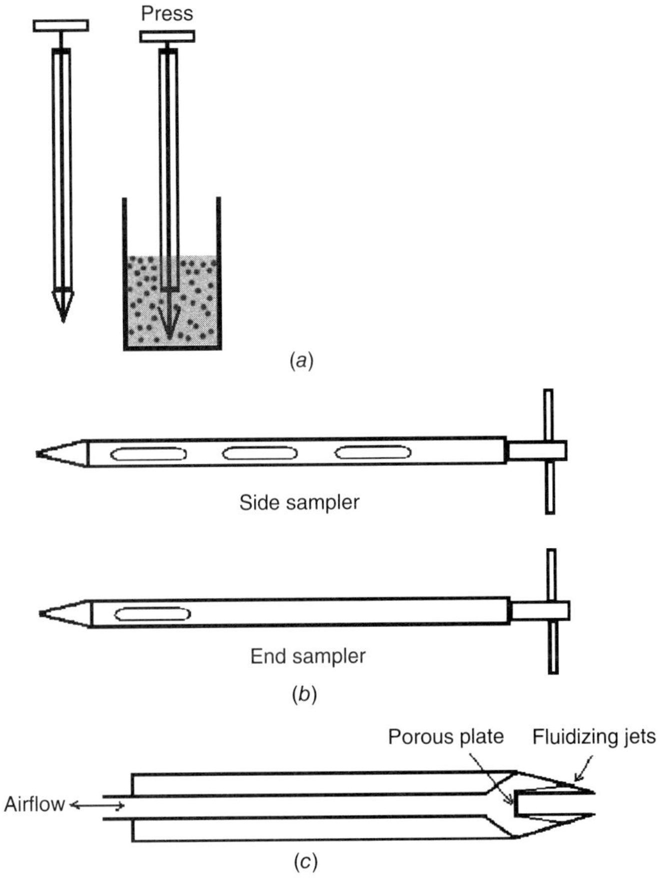

FIGURE 5 Stationary bulk sampling: (*a*) low-volume powder sampler; (*b*) thief/spear probe sampler; (*c*) pneumatic lance sampler [7].

6.7.3.3 Particle Density and Voidage

Particle density may be defined as the total mass of the particle divided by its total volume; however, depending upon the different definitions of the total volume (or the different ways to measure the particle volume), there are various definitions of particle density in existence (see Table 4).

In order to get clear understanding of the subtle differences between the definitions of various particle density types, an illustration can be formed as shown in Figure 8.

Particle Density Methods Density is defined as the ratio of mass to volume, so the density determination can be separated into two steps: measurement of mass and measurement of volume. Determining the mass of an object is rather straightforward; however, it is much more difficult to directly determine the volume of a solid. The volume of a solid object with a regular geometric shape may be calculated

TABLE 2 Flowing Stream Sampling

Sampling Devices	Procedure of Sampling	Application and Characteristics
Auger sampler (Figure 6*a*) [Line sampler for stream]	A pipe with a slot is placed inside the process stream, permitting easy capture of powder through the process stream cross section when rotated. Samples are subsequently then delivered into a separate container by gravitational forces. [6]	While this is often used for stream sampling, it is difficult to collect a representative sample when stream is heterogeneous [10].
Constant-volume sampler (Figure 6*b*) [Point sampler for stream]	Sampling occurs when the stream falls down through a pipe and a constant-volume container is inserted or withdrawn from the stream system.	Designed to extract a constant volume of homogeneous granular material for subsequent chemical analyses and is not suitable for withdrawing samples for physical analyses [11].
Diverter sampler (Figure 6*c*) [Cross-sectional sampler for stream]	The whole stream is diverted by opening a sliding cover or pivoting an external flap in the bottom of a gravity-flow chutes or pipes or screw conveyors [12]. The samples could be removed to a low-angle laser light-scattering instrument then returned to the process stream [6].	The process could be automated and highly suitable method for online particle size measurement. [7]
Full stream sampler (Figure 6*d*) [Cross-sectional sampler for stream]	Samples are withdrawn from conveyors, carried out only on the return stroke.	Extremely useful for dusty materials provided the trough extends the whole length of the stream and does not overfill [6].

mathematically; however, in most conditions, the shape of a particle is often irregular, especially in powder technology, which makes it extremely difficult to measure geometrically. Therefore, various methods have been developed to determine the volume of particles and powders. The two most in use in both laboratory and industrial settings are liquid and gas displacement methods. The different values of particle density can also be expressed in a dimensionless form, as "relative density" (or specific gravity), which is the ratio of the density of the particle to the density of water.

The discussion that follows will give an overview of the common methods used in particle density measurement.

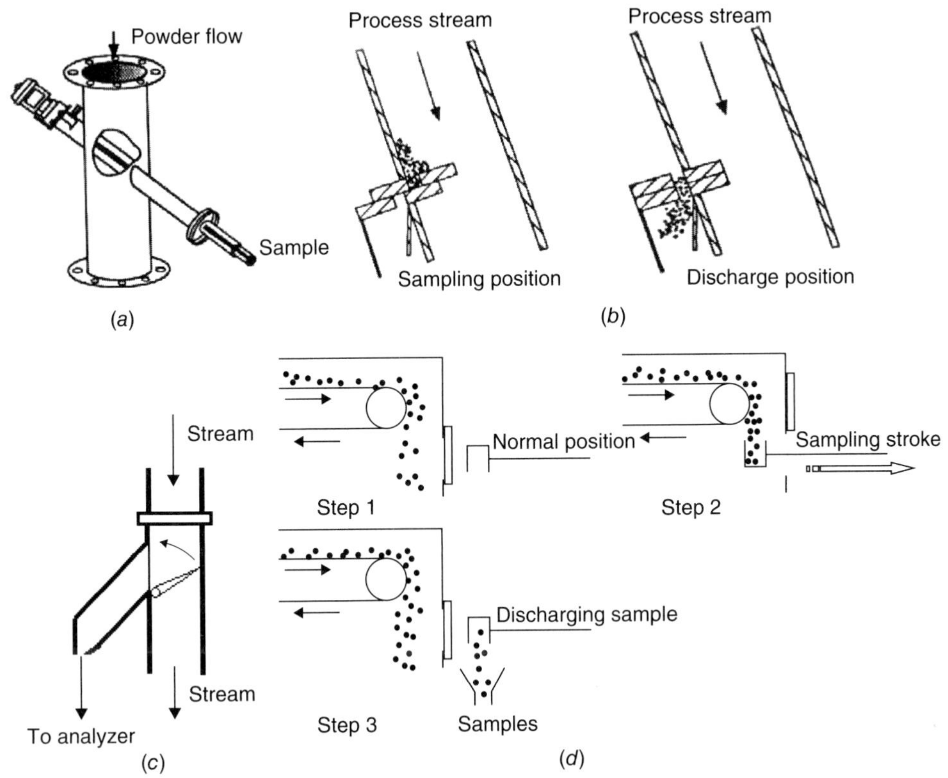

FIGURE 6 Flowing stream sampling: (*a*) auger sampler [6]; (*b*) constant-volume sampler [6]; (*c*) diverter sampler; (*d*) full-steam sampler.

Measurement of Particle Density

1. *Liquid Pycnometry Method* There are several British standards that deal with liquid pycnometry applied to specific materials [18–23]. A pycnometer bottle is weighted empty (M1), and then full of liquid (M2). Following these two initial measurements, two subsequent measurements are made: a sample of powder approximately one-third of maximum container volume (M3) and the bottle filled to capacity containing the sample and water (M4). Great care is required in the final step to ensure that the liquid is fully wetted and all the air removed. Variations in recorded weight also arise depending on how much liquid escapes when the ground glass stopper is inserted in the liquid-filled container. It is extremely important that the liquid used in this procedure does not solubilize or react with the solid particles. Moreover, the solid particles must not absorb the selected fluid.

2. *Gas Pycnometry Method* Principally this method is similar to liquid pycnometry in that volume determination is achieved by detecting the pressure or volume change associated with the displacement of a gas (rather than liquid) by a solid object. Given that this method is largely dependent upon the diffusivity of the gas, helium is often used since it has a low molecular weight and a small atomic radius, allowing high diffusivity into small pores. Sample volumes are often displayed on a

TABLE 3 Subsampling

Sampling Devices	Procedure of Sampling	Application and Characteristics
Coning and quartering (Figure 7*a*)	A cross-shaped cutter is used to separate the sample heap (which is first flattened at the top) into four equal parts. The segments are drawn apart and two opposite quadrants are combined together. This procedure is repeated at least 4 times until a small enough sample has been generated.	The first choice for non-free-flowing powders and nonflowing powders. Prone to operator bias as fine particles remain in the center of the cone and should never be used with free-flowing powders [13].
Oscillating hopper sample divider (Figure 7*c*)	Hopper (paddle) oscillates and powder falls into two collectors placed under the hopper (paddle).	Used for small quantity of samples. Sample size can be controlled by monitoring time over each collector [7].
Revolving sample splitter (Figure 7*f*)	The revolving feeder distributes the sample material equally (in time) over a number of radial chutes, assuming constant rotational speed [14].	Very easy to perform and several versions are available that are suitable for free-flowing powders, dusty powders, and cohesive powders. Handling quantities can vary from 40 L to a few grams.
Riffle/chute splitter (Figure 7*e*)	The sample is introduced to a rectangular area, divided by parallel chutes leading to two separate receptacles [14].	Well-accepted method for sample reduction that is highly suitable for free-flowing powders. Used to produce samples with a minimum volume of 5 mL.
Spinning riffler (Figure 7*d*)	a steady stream of powder is run into a rotating basket of containers [8].	Useful in subsampling large samples [15]. Suitable for free-flowing materials [13].
Table sampler (Figure 7*b*)	In a sampling table, powder flows down from the top of an inclined plane, holes and prisms splitting the powder. The powder that reaches the bottom of the plane is the sample.	Used for sample reduction with the advantages of low price and lack of moving parts.

digital counter on the testing equipment [24]; however, such volumes are easily calculated using the pressure change and the ideal gas law, $PV = nRT$. The true density of the particle can be measured using this method if the particles have no closed pores, while the apparent particle density can be measured if there are any closed pores. Additionally, if open pores are filled with wax, envelope volumes may

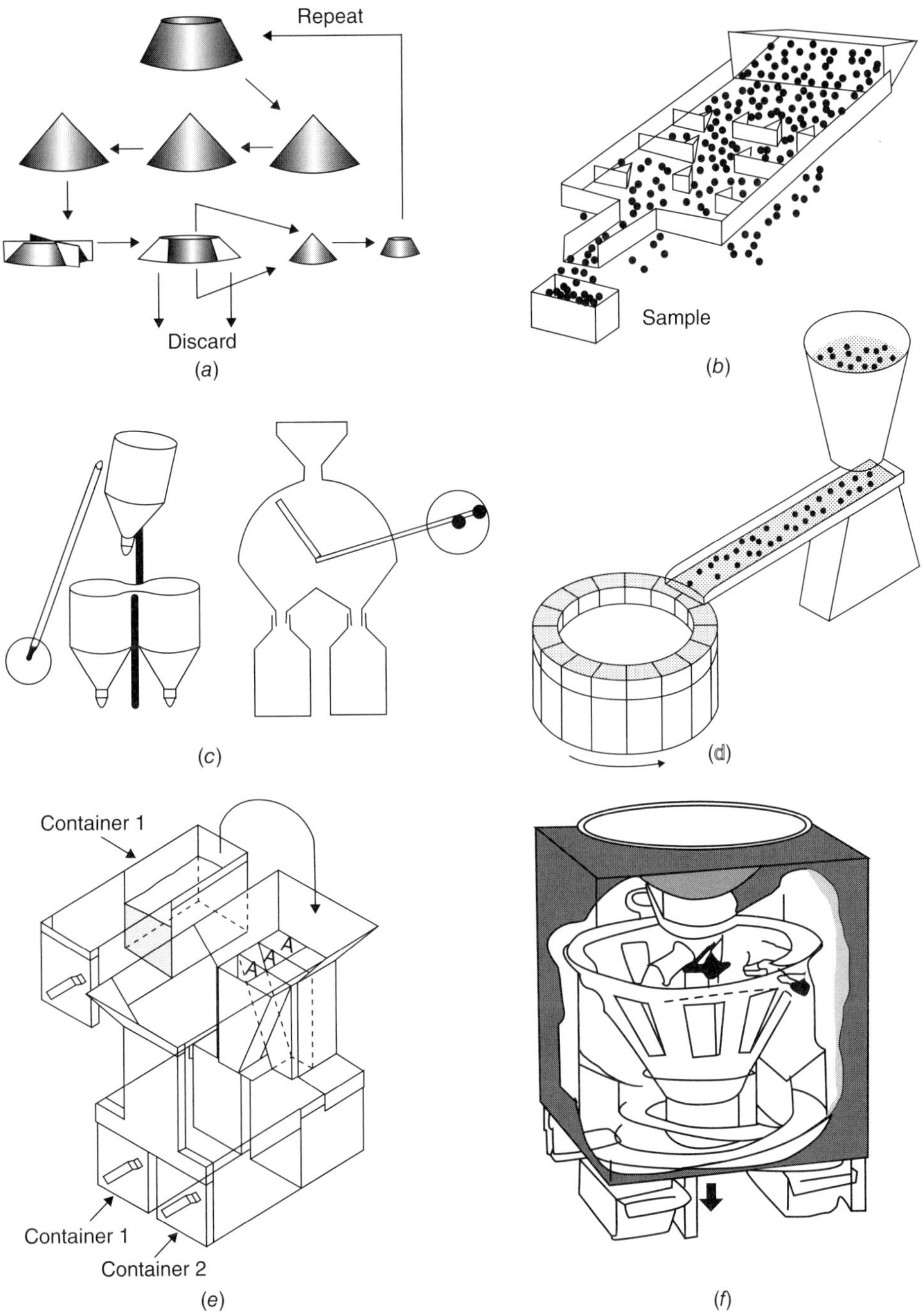

FIGURE 7 Subsampling: (*a*) coning and quartering; (*b*) table sampler [6]; (*c*) oscillating hopper (paddle) sample divider [6]; (*d*) spinning riffler (BSI); (*e*) riffle/chute splitter (BSI); (*f*) revolving sample splitter [14].

TABLE 4 Definitions of Density Terms

Density Types	Density Definitions	Volumes in Definition: Solid Material Volume	Closed-Pore Volume	Open-Pore Volume	Interparticle Void Volume
Absolute powder density	Mass of powder per unit of absolute volume, which is defined as the solid matter after exclusion of all the spaces (pores and voids) (BSI)	■	□	□	□
Apparent particle density	Mass of particles divided by its apparent particle volume, which is defined as the total volume of the particle, excluding open pores but including closed pores (BSI)	■	■	□	□
Apparent powder density	Mass of powder divided by its apparent powder volume, which is defined as the total volume of solid matter, including open pores and closed pores and interstices (BSI)	■	■	■	□
Bulk density	Mass of the particles divided by the volume they occupy, which includes the space between the particles (ASTM)	■	■	■	■
Effective particle density	Mass of a particle divided by its volume, including open pores and closed pores (BSI)	■	■	■	□
Envelope density	Ratio of the mass of a particle to the sum of the volumes of the solid in each piece and voids within each piece, which is, within close-fitting imaginary envelopes, completely surrounding each piece (ASTM)	■	■	■	□
Skeletal density	Ratio of the mass of discrete pieces of solid material to the sum of the volumes of the solid material in the pieces and closed pores within the pieces (ASTM)	■	■	□	□
Tap density	Apparent powder density obtained under stated conditions of tapping (BSI)	■	■	■	■
True density	Mass of a particle divided by its volume, excluding open pores and closed pores (BSI)	■	□	□	□

Note: BSI = British Standards Institute [16], ASTM = American Society for Testing and Material [17]. ■, included; □, excluded.

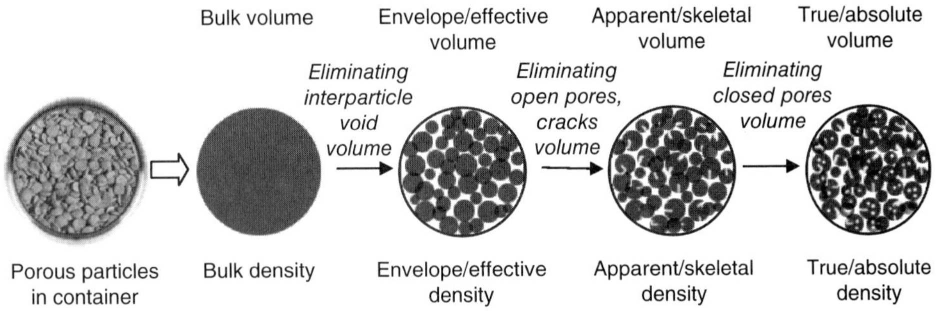

FIGURE 8 Various density types. The density value increases from bulk density to true density while the volume value decreases from bulk volume to true volume.

be determined and the difference between envelope and apparent volume can yield the open-pore volume, which is sometimes used as a measure of porosity.

3. *Hydrostatic Weighing Method* The volume of a solid sample is determined by comparing the mass of the sample in air with the mass of sample immersed in a liquid with a known density. The volume of sample may be calculated using the difference between the two measured mass values divided by the density of the liquid. This method can be used to determine the bulk or apparent volume. It is extremely important that the suspending liquid does not interact with the powder under investigation.

4. *Float–Sink or Suspension Method* This method involves placing a solid sample into a liquid with known and adjustable density. The density of liquid is incrementally adjusted until the sample begins to sink–float (ASTM C729-75 [25]), or is suspended at neutral density in the liquid (ASTM C693-93 [26]). At the point of equilibrium the density of the sample is equal to the density of the liquid.

5. *Bed Pressure Drop Method* This technique is based on making measurements of bed pressure drop as a function of gas velocity at two voidages, when gas is passed through the bed of powder in the laminar flow regime [24]. During measurement pressure changes for at least four velocities must be measured. The effective particle density ρ_p can be calculated using the equation

$$\frac{s_1}{s_2} = \left(\frac{\rho_{b1}}{\rho_{b2}}\right)\left(\frac{\rho_p - \rho_{b2}}{\rho_p - \rho_{b1}}\right)^3$$

where s is the gradient of pressure drop with gas velocity, ρ_b is the bulk density, ρ_p is the particle effective density.

6. *Sand Displacement Method* The sand displacement method is another useful way of measuring the envelope density of a particle using fine sand as the displacement media. Sand is mixed with a known amount of particles, then the density of the sample particles can be determined from the difference of the bulk density between sand alone and that with samples.

7. *Mercury Porosimetry Method* Mercury is a nonwetting liquid that must be forced to enter a pore by application of external pressure. Consequently it is an extremely useful and convenient liquid for measuring the density of powders and/or particles. This method can measure the apparent and true density of one sample by

applying different pressures. At atmospheric pressure, mercury will resist entering pores smaller than about 6 μm in diameter, but at pressures of approximately 60,000 psi (414 MPa) mercury will be forced to enter pores with diameters as small as 0.003 μm [27].

Measurement of Bulk Density Bulk density is very important in determining the size of containers used for handling, shipping, and defining storage conditions for pharmaceutical powders and granules. It is a property that is also pertinent in defining the size of hoppers and receivers for milling equipment and for sizing blending equipment in the scale-up to pilot and to commercial production [28]. The concept of bulk density is the mass of particles divided by the bulk volume, which includes not only the envelope volume of particles but also the spaces between particles, so it should not be confused with particle density [24].

The most convenient method to measure bulk density is to fill the particles into a known volume container (usually cylindrical), level the surface, and weigh the particles in the container. The bulk density is calculated by the mass of the particles divided by the volume that can be read from the scale of the measuring cylinder. In order to minimize experimental errors, the container should be ideally at least 1 L in volume, and the ratio of length and diameter should be about 2:1. Also it is recommended to leave the sample for approximately 10 min to achieve an equilibrium volume (density) value before making readings.

Given that the bulk volume associated with the particle mass is a mixture of air and solid material, the bulk density value is highly dependent on sample history prior to measurement. Calculation of the tapped density can then be achieved by tapping the bulk powder a specified number of times (to overcome cohesive forces and remove entrapped air) to determine the tapped volume of the powder. The tapped and bulk density values can be used to define the flowability and compressibility of a powder using Carr's index and the Hausner ratio.

6.7.3.4 Particle Surface Area

Surface area is one of the most important characteristics in particle technology. Particles with a different surface area will express different physical properties that will subsequently affect many applications and ultimately final dosage form properties.

Similar to particle density, there are various definitions relating to particle surface area [16]:

1. *Adsorption surface area*: the surface area calculated from an adsorption method.
2. *BET surface area*: the surface area calculated from the Brunauer, Emmett, and Teller theory of multilayer adsorption of a gas on a solid surface.
3. *Calculated surface area*: the surface area of a powder calculated from its particle size distribution.
4. *Effective permeability mass-specific surface*: the effective volume-specific surface divided by the effective solid density, determined by permeametry.

5. *Effective permeability volume-specific surface*: the effective surface area divided by the effective solid volume, determined by permeametry.
6. *Permeability surface area*: the surface area of a powder calculated from the permeability of a powder bed under stated conditions.
7. *Specific surface area* (S_w): the surface area of a unit mass of material determined under stated conditions, where S_w is usually expressed in centimeters squared per gram or meters squared per gram and can be used for quality control purposes [28].

Particle Surface Area Determination Methods From the standard definitions of particle surface area, it can be seen that various determination methods are used for surface area measurement, such as adsorption (including Langmuir's equation for monolayer adsorption and the BET equation for multilayer adsorption), particle size distribution, and permeability methods. The different methods are rarely in agreement because the value obtained depends upon the procedures used and also on the assumptions made in the theory relating the surface area to the phenomena measured. The most common methods used for measuring particle surface area are described below.

1. *Gas Adsorption Method* Gas adsorption methods measure the surface area of particles/powders through measurement of the amount of gas adsorbed onto the sample surface. The methods can measure both external and internal surfaces (including open pores in the particles) and can yield physically meaningful average particle sizes with nonporous materials [24]. The amount of gas adsorbed depends upon the nature of the solid (adsorbent) and the pressure at which adsorption takes place. The amount of gas (adsorbate) adsorbed can be found by determining the increase in weight of the solid (gravimetric method) or the amount of gas removed from the system due to adsorption by application of the gas laws (volumetric method [6]). The adsorption used in this method is physical adsorption, which is a relatively weak interaction between samples and gases and therefore can be removed by evacuation.

In this method, a graph of the number of moles of gas adsorbed per gram of solid, at constant temperature, against the equilibrium gas pressure is called an adsorption isotherm. A point must be chosen on this isotherm corresponding to the completion of the adsorbed monolayer in order to calculate S_w [29].

2. *Permeametry Method* This method is based on the fact that the flow rate of a fluid through a bed of particles depends on the pore space, the pressure drop across the bed, the fluid viscosity, dimensional factors such as the area of the bed, and specific surface area (S_w). The determination of permeability can be made either under continuous steady-state flow (constant flow rate) or under variable-flow (constant-volume) conditions.

All of the permeability methods are based on the Kozeny–Carman equation, which is used to calculate a surface area of a packed powder bed from its permeability. The Kozeny–Carman equation is expressed as [16]

$$S_k = \sqrt{\left(\frac{A\varepsilon^3 \Delta p}{K(1-\varepsilon)^2 L\eta q}\right)}$$

where S_k = effective permeability volume-specific surface of powder assuming only viscous flow occurs in determination (Kozeny–Carman term)

A = cross-sectional area of bed of powder perpendicular to direction of flow of air

ε = porosity of bed of powder

Δp = pressure difference across bed of powder

K = Kozeny constant

L = linear dimension of bed of powder parallel to direction of flow of air (commonly known as height of powder bed)

η = viscosity of air at its temperature at time of determination

q = rate of flow of incompressible fluid through bed of powder

The specific surface area calculated here only involves the walls of the pores of the bed and excludes the pores within the particles. Therefore, the surface area measured in this method can be much smaller than the total surface area measured by gas adsorption methods [24].

3. *Particle Size Distribution Method* The surface area of particles can be determined using particle size and particle shape values. The "equivalent spherical diameter" is used in this technique and many attempts to measure the surface area using this method have led to values that are significantly less than the true value (large deviations arising from inability to define particle shape due to surface irregularities and porosity). Surface area values calculated from particle size distribution methods will in effect establish the lower limit of surface area due to the implicit assumptions of sphericity or other regular geometric shapes and by ignoring the highly irregular nature of real surfaces [30].

Besides the three methods introduced above, there are many other methods of surface area determination: Any surface-dependent phenomenon can be used for such measurement [24]. Some available methods (mercury porosimetry, adsorption from solution, adsorption of dyes, chemisorption, density methods, and secondary ion mass spectroscopy) are explained in more detail elsewhere [6, 30, 31, 32].

6.7.3.5 Particle Shape

Particle behavior is a function of particle size, density, surface area, and shape. These interact in a complex manner to give the total particle behavior pattern [28]. The shape of a particle is probably the most difficult characteristic to be determined because there is such diversity in relation to particle shape. However, particle shape is a fundamental factor in powder characterization that will influence important properties such as bulk density, permeability, flowability, coatablility, particle packing arrangements, attrition, and cohesion [33–36]. Consequently it is pertinent to the successful manipulation of pharmaceutical powders that an accurate definition of particle shape is obtained prior to powder processing.

A number of methods have been proposed for particle shape analysis, including shape coefficients, shape factors, verbal descriptions, curvature signatures, moment invariants, solid shape descriptors, and mathematical functions (Fourier series expansions or fractal dimensions); these are beyond the scope of this chapter but have been adequately described in other texts [37].

In the most simplistic means of defining particle shape, measurements may be classified as either macroscopic or microscopic methods. Macroscopic methods typically determine particle shape using shape coefficients or shape factors, which are often calculated from characteristic properties of the particle such as volume, surface area, and mean particle diameter. Microscopic methods define particle texture using fractals or Fourier transforms. Additionally electron microscopy and X-ray diffraction analysis have proved useful for shape analysis of fine particles.

Particle Shape Measurement

1. *Shape Coefficients and Shape Factors* There are various types of shape factors, the majority based on statistical considerations. In essence this translates to the use of shape factors that do refer not to the shape of an individual particle but rather to the average shape of all the particles in a mass of powder. However, a method developed by Hausner [38] that uses three factors—elongation factor, bulkiness factor, and surface factor—may be used to characterize the shape of individual particles (Table 5).

2. *Determining Particle Shape by Fourier Analysis* Fourier transforms have been previously used to determine particle shape and the rollability of individual particles from the coefficients of the resulting series [39]. Moreover, fast Fourier transforms have been successfully used to determine coefficients and a particle "signature" by plotting $\ln An$ versus $\ln n$, where An is the nth Fourier coefficient and n is the frequency [29, 40, 41]. In brief, Fourier method consists of finding the centre of gravity of a particle and its perimeter, from which a polar coordinate system is set up. Amplitude spectra of a finite Fourier series in closed form are used as shape descriptors of each particle [42]. Several research papers have focused on the characterization of individual particle shape using Fourier grain analysis or morphological analysis [43–44]. The method has also been extended to the measurement of particle shapes in a blend [45] and to relate particle attrition rate in a milling operation to particle shape [46].

3. *Determining Particle Shape by Electron Microscopy* Electron microscopy has been used for the examination of fine powder dispersions and will provide information on particle shape perpendicular to the viewing direction. Standard shadowing procedures may be useful in obtaining information on shape in the third dimension. Scanning electron microscopy can give direct and valuable information on the shape of large particles [47].

4. *Determining Particle Shape by X-Ray Diffraction Broadening* The broadening of X-ray diffraction lines is primarily a measure of the departure from single-crystal perfection and regularity in a material and can therefore be used to characterize particle shape. This is the only method that gives the size of the primary crystallites, irrespective of how they are aggregated or sintered, and is of great value for determining the properties of fine powders [48, 49].

5. *Other Methods for Particle Shape Determination* Gotoh and Finney [50] proposed a mathematical method for expressing a single, three-dimensional body

TABLE 5 Shape Coefficients and Shape Factors

Coefficients and Factors	Symbols	Definitions and Equations
Volume shape coefficient	α_v	$\alpha_v = \frac{V}{d^3}$ where V = average particle volume d = mean particle diameter
Surface shape coefficient	α_s	$\alpha_s = \frac{S}{d^2}$ where S = average particle surface d = mean particle diameter
Volume–surface shape coefficient	α_{vs}	$\alpha_{vs} = \frac{\alpha_s}{\alpha_v}$ where α_v = volume shape coefficient α_s = surface shape coefficient
Shape factor	α_0	$\alpha_o = \alpha_v m \sqrt{n}$ where α_o = shape factor for equidimensional particle and thus represents part of α_v which is due to geometric shape only α_v = volume shape coefficient m = flakiness ratio, or breadth/thickness n = elongation ratio, or length/breadth
Sphericity shape factor	Ψ_w	Sphericity = (surface area of sphere having same volume as particle) / (surface area of actual particle)
Circularity shape factor		Circularity = (perimeter of particle outline)^2 / 4π(cross-sectional or projection area of particle outline)

Source: From refs. 6 and 42.

by sectioning it as an equivalent ellipsoid with the same volume, surface area, and average projected area as the original body. Moreover, wedge-shaped photodetectors to measure forward light-scattering intensity have also been explored for determination of crystal shape [51]. More recently a technique referred to as time of transition (TOT) that was first introduced in 1988 has also been used for the analysis of particle size and shape [52, 53].

6.7.4 EFFECT OF PARTICLE SIZE REDUCTION ON TABLETING PROCESSES

Particle size plays a critical role in the efficacy of a drug product. It can impact not only bioavailability but also the efficiency and success of production process and ultimately the properties of the final dosage form.

6.7.4.1 Wet Granulation Processes

The particle size of an active pharmaceutical ingredient can have significant effect on the processing behavior of a formulation, such as granule growth during wet granulation and hence the resulting granule characteristics. The particle size of the starting material can affect the strength and deformability of moist granules and hence their behavior during the wet granulation process.

The effect of particle size on granule growth is a function of several interacting factors, the balance of which largely depends on the nature of the material and the experimental conditions. Differences in granule structure and porosity, resulting from changes in starting material particle size, can also affect other characteristics (e.g., compressibility) of the granulation.

Badawy et al. [57] studied the effect of DPC 963 (a nonnucleoside reverse transcriptase inhibitor) particle size on the granule growth, porosity, and compressibility of granules manufactured by a high-shear wet granulation process. It was found that DPC 963 granule growth in the high-shear granulator and the resulting granule compressibility and porosity were sensitive to relatively small changes in drug substance particle size. Decreasing the particle size resulted in more pronounced granule growth and enhanced the porosity and compressibility of the granulation. Higher pore volume for the granulation manufactured using the active ingredient with a smaller particle size may be the reason for its higher compressibility. The high granulation porosity resulted in an increased fragmentation propensity and volume reduction behavior of the granulation that led to increased compressibility. The more porous granulation has higher tendency to densify upon application of the compression force, resulting in closer packing of the particles.

6.7.4.2 Mixing Processes

Mixing may be defined as a unit operation that aims to treat two or more components, initially in an unmixed or partially mixed state, so that each unit of the components lies as nearly as possible in contact with a unit of each of the other components [2]. Whenever a product contains more than one component, mixing will be required in the manufacturing process in order to ensure an even distribution of the active component(s).

It is well accepted that mixing solid ingredients is usually more efficient and uniform if the active ingredient and excipients are approximately the same size, which ultimately provides a greater uniformity of dose [1]. Particle size and particle size distribution are important in the powder-mixing process since they largely determine the magnitude of forces, gravitational and inertial, that can cause interparticulate movement relative to surface forces, which resist such motion. As a consequence of high interparticulate forces, as compared with the gravitational forces, powders of less than 100 μm mean particle diameter sizes are not free flowing. Powders that have high cohesive forces due to interaction of their surfaces can be expected to be more resistant to intimate mixing than those whose surfaces do not interact strongly [2].

In moving from one location to another, relative to neighboring particles, a particle must surmount a certain potential energy barrier that arises from forces resisting movement. This effect is a function of both particle size and shape and is most pronounced when high packing densities occur. Ideal mixing may be achieved when all the particles of the powder mix have similar size, shape, and density characteristics whereas segregation (demixing) may occur when powder blends are not composed of monosized near-spherical particles but contain particles that differ in size, shape, and density. Segregation is more likely to occur if the powder bed is subjected to vibration.

The main reason for segregation in powder blends is the difference in the particle size of the components of the particles contained within the blend. Due to the high diffusivity of small particles, such materials move through the voids between larger particles and so migrate to the lower regions of the powder mix. Moreover, during mixing operations, extremely fine particles have a high tendency to be forced upward by turbulent air currents as the powder blend tumbles and subsequently become isolated from the mixing process through continuous suspension above the blend. When mixing is stopped, these particles will sediment and form a layer on top of the coarser particles.

It is important to control the particle size distribution of pharmaceutical granules or powder blends because a wide size distribution can lead to a situation with a high probability of segregation. If this occurs within the hoppers of tablet machines, nonuniform products may be manufactured due principally to large weight variations. Tablet dies are filled by volume rather than weight, and consequently, the establishment of different regions within a hopper containing granules of different sizes (and hence bulk density) will contain a different mass of granules. This will lead to an unacceptable distribution of the active pharmaceutical content within the batch of finished product, even though the drug is evenly distributed by weight throughout the granules.

6.7.4.3 Flowability of Pharmaceutical Powders

Due to the relatively small particle size, irregular shape, and unique surface characteristics, many pharmaceutical powders have a high tendency to be extremely cohesive. This high level of cohesion results in "sticky" powders that have poor flowability, commonly resulting in large mass variability within the final product owing to unpredictable and variable filling of tablet dies.

Powders with different particle sizes have different flow and packing properties, which significantly alter the volume of powder expelled from manufacturing equipment during, for example, encapsulation or tablet compression. In order to avoid such problems, the particle sizes of the active pharmaceutical ingredient and other powder excipients should be defined and controlled during formulation so that problems during production are avoided. Most notably, powder flowability is of critical importance in the successful production of acceptable pharmaceutical dosage forms. High levels of flowability within pharmaceutical powders is not just important in the final stages of manufacture but is essential for many industrial processes, particularly mass transport.

Poor or uneven powder flow can result in excess entrapped air within powders, which may induce capping or lamination in specific high-speed tableting equipment. Moreover, uneven powder flow that is a direct result of the presence of excess fines within a powder blend will also promote increased particle–die wall friction, lubrication problems, and very importantly increased dust contamination hazards to operating personnel.

Although particle size is a significant factor controlling the flowability of pharmaceutical powders or granules, other factors must be considered. The presence of molecular forces between particle/granule surfaces increases the probability for cohesion and adhesion between solid particles. Cohesion may be defined as the attractive forces between like surfaces, such as component particles of a bulk solid, whereas adhesion may be defined as the attractive force between two unlike surfaces, for example, between a particle and a tablet punch. It is extremely important to appreciate that cohesive forces acting between particles in a powder bed are attributed mainly to short-range nonspecific van der Waals forces that are significantly altered as particle size and relative humidity change.

Cohesion and adhesion are phenomena that occur at the surface of a solid and hence particles with an extremely large surface area will have greater attractive forces than those with a smaller surface area. Consequently particle surface area will have a dramatic effect on the flowability of pharmaceutical powders. Typically, fine particles with very high surface-to-mass ratios will be more cohesive than larger particles, which are influenced more by gravitational forces. Particles larger than 250 μm are usually relatively free flowing, but as the size falls below 100 μm, powders become cohesive and flow problems are likely to occur. Powders having a particle size less than 10 μm are usually extremely cohesive and resist flow under gravity.

Although it has been previously stated that particles with similar particle sizes are desirable for pharmaceutical processes, a bulk powder mass with a narrow particle size distribution accompanied with dissimilar particle shapes can produce a bulk mass with inherently different flow properties, owing principally to differences in interparticle contact area.

6.7.4.4 Compression Processes

In general, the strength of a compressed powder depends on the inherent ability of the powder to reduce in volume during compression and the amount of interparticulate attraction in the final compact. The decrease in compact volume with increasing compression load is attributed normally to particle rearrangement, elastic deformation, plastic deformation, and particle fragmentation. Pharmaceutical materials normally consolidate by more than one of these mechanisms [58, 59]. Unmodified paracetamol crystals exhibit poor compressibility during compaction, resulting in weak and unacceptable tablets with a high tendency to cap [60]. Moreover the incidence of capping and lamination during production, following ejection of tablets from the die, depended on the plastic and elastic behaviors of the excipients used [61]. It has been suggested that materials undergoing plastic deformation, in contrast to elastic deformation, display enhanced bond formation and produce strong tablets.

The effect of particle size on the compression properties of paracetamol oral dosage forms has been previously reported [62]. Heckel analyses plots indicated that the predominant mechanism of compaction of paracetamol was fragmentation with larger particle fractions experiencing more fragmentation than the smaller particles. Furthermore, Heckel analysis also indicated that, for a given applied pressure, the larger particles of paracetamol produced denser compacts than the smaller particles. The results of elastic–plastic energy ratios indicated that the majority of energy involved during compaction of paracetamol was utilized as elastic energy. This suggested a massive elastic deformation of paracetamol particles under pressure, resulting in weak and capped tablets. It was found that larger particles exhibited less elastic recovery and elastic energy compared to smaller particles. This was attributed to increased fragmentation of larger particles, resulting in increased bonding between particles due to the formation of more new, fresh, and clean particle surfaces.

REFERENCES

1. Lachman, L., Lieberman, H. A., and Kanig, J. L. (1986), *The Theory and Practice of Industrial Pharmacy Textbook*, 3rd ed., Leo & Febiger, Philadelphia.
2. Aulton, M. E. (2002), *Pharmaceutics: The Science of Dosage Form Design Textbook*, 2nd ed., Churchill Livingstone, London.
3. Svarovsky, L. (1990), Characterization of powders, in Rhodes, M. J., Ed., *Principles of Powder Technology*, Wiley, Chichester, pp. 35–69.
4. Hawkins, A. E. (1990), Characterizing the single particle, in Rhodes, M. J., Ed., *Principles of Powder Technology*, Wiley, Chichester, pp. 9–34.
5. Sommer, K. (1981), Sampling error on particle analysis, *Aufbereit Tech.*, 22(2), 96–105.
6. Allen, T. (1997), *Particle Size Measurement*, 5th ed., Vol. 1, Chapman & Hall, London, pp. 1–62.
7. Venable, H. J., and Wells, J. I. (2002) Powder sampling, *Drug Dev. Ind. Pharm.*, 28(2), 107–117.
8. Kaye, B. H. (1997), *Powder Mixing*, Chapman & Hall, London, pp. 77–95.
9. British Standard, BS 3406-1 (1986), *Methods for Determination of Particle Size Distribution—Part 1: Guide to Powder Sampling*, British Standard Institution, London.
10. Bicking, C. A. (1964), Sampling, in Standen, A. Ed. *Kirk–OthmerEncyclopedia of Chemical Technology*, 2nd ed., Interscience Publishers, New York, pp. 744–762.
11. Hulley, B. J. (1970), Sampling and Sample Conditioning in On-line Fertilizer Analysis, *Chem. Eng.*, 77, 410–413.
12. Clarke, J. R. P., and Carr–Brion, K. G. (1996), Sampling systems for process analysers 2nd ed., The Bath Press, Avon, pp. 148–180.
13. Allen, T. (1964), Sampling and size analysis, *Silic. Ind.* 29(12), 509–515.
14. Petersen, L. (2004), Representative mass reduction in sampling—A critical survey of techniques and hardware, *Chemometr. Intell. Lab. Syst.*, 74, 95–114.
15. Crosby, N. T., and Patel, I. (1995), *General Principles of Good Sampling Practice*, Royal Society of Chemistry, Cambridge.
16. British Standard BS 2955 (1993), *Glossary of Terms Relating to Particle Technology*, British Standards Institution, London.
17. ASTM (1994), *Compilation of ASTM Standard Definitions*, 8th ed., American Society for Testing and Materials, Philadelphia, PA.
18. British Standard BS 812 (1995), *Testing Aggregates, Part 2: Methods of Determination of Density*, British Standards Institution, London.

19. British Standard 1016 (1980), *Methods for the Analysis and Testing of Coal and Coke, Part 13: Test Special to Coke*, British Standards Institution, London.
20. British Standard BS 1377 (1975), *Methods of Tests for Soils for Civil Engineering Purposes*, British Standards Institution, London.
21. British Standard BS 3483 (1974), *Methods for Testing Pigments for Paints, Part B8: Determination of Density Relative to Water at 4°C*, British Standards Institution, London.
22. British Standard BS 4550 (1987), *Methods for Testing Cement, Part 3.2: Density Test*, British Standards Institution, London.
23. British Standard BS 7755 (1998), *Soil Quality, Part 5: Physical Method, Section 5.3: Determination of Particle Density*, British Standards Institution, London.
24. Svarovsky, L. (1987), *Powder Testing Guide, Methods of Measuring the Physical Properties of Bulk Powders*, Published on behalf of the British Materials Handing Board by Elsevier Applied Science, London, pp. 3–33, 79–95.
25. ASTM C729-75 e1 (1995), *Standard Test Method for Density of Glass by the Sink-Float Comparator*, American Society for Testing and Materials, Philadelphia, PA.
26. ASTM C693-93 (1998), *Standard Test Method for Density of Plastics by the Density-Gradient Technique*, American Society for Testing and Materials, Philadelphia, PA.
27. Webb, P. (2001), *Volume and Density Determinations for Particle Technologists*, Micromeritics Instrument Corp, Georgia.
28. Lieberman, H., and Lachman, L. (1981), *Pharmaceutical Dosage Forms Tablets*, Vol. 2, Marcel Dekker, New York, pp. 112–150, 202–222.
29. Beddow, J. K., and Meloy, T. (1980), *Testing and Characterization of Powders and Fine Particles*, Heyden & Son, London, pp. 63–64.
30. Lowell, S., and Shields, J. E. (1984), *Powder Surface Area and Porosity*, 2nd ed., Chapman & Hall, London.
31. Beddow, J. K. (2000), *Particle Characterization in Technology*, Vol. I, CRC Press, Boca Raton, FL, 3–20.
32. Brunauer, S., Deming, L. S., Deming, W. S., and Teller, E. (1940), Adsorption of gases in multimolecular layers, *J. Am. Chem. Soc.*, 62, 1723–1732.
33. Fonner, D. E., Banker, G. S., and Swarbrick, J. (1966), Micromeritics of granular pharmaceutical solids. 1. Physical properties of particles prepared by 5 different granulation methods. *J. Pharm. Sci.*, 55, 181.
34. Ridgway, K., and Rupp, R. (1969), Effect of particle shape on particle properties, *J. Pharm. Pharmacol.*, 21, 30–39.
35. Shotton, E., and Obiorah, B. A. (1975), Effect of physical properties on compression characteristics, *J. Pharm. Sci.*, 64(7), 1213–1216.
36. Heyd, A., and Dhabbar, D., Particle shape effect on caking of coarse granulated antacid suspensions, *Drug Cosmet. Ind.* 125, 42–45.
37. Hawkins, A. E. (1993), *The Shape of Powder-Particle Outline*, Wiley, New York.
38. Hausner, H. H. (1967), *Characterization of the Powder Particle Shape in Particle Size Analysis*, Society for Analytical Chemistry, London.
39. Schwarcz, H. P., and Shane, K. C. (1969), Measurement of particle shape by Fourier analysis, *Sedimentology*, 13(3–4), 213–231.
40. Meloy, T. P. (1969), *Screening*, AIME, Washington, DC.
41. Meloy, T. P. (1977), Fast Fourier transforms applied to shape analysis of particle silhouettes to obtain morphological data, *Powder Technology*, 17, 27–35.

42. Ehrlich, R., and Full, W. E. (1984), Fourier shape analysis—a multivariate pattern recognition approach, in Beddow, J. K., Ed., *Particle Characterization in Technology*, Vol. II, Morphological Analysis, CRC Press, Boca Raton, FL.
43. Meloy, T. P., Clark, N. N., Durney, T. E., and Pitchumani, B. (1985), Measuring the particle shape mix in a powder with the cascadograph, *Chemical Engineering Science*, 40(7), pp. 1077–1084.
44. Alderliesten, M. (1991), Mean particle diameters. part II: standardization of nomenclature. *Part. Part. Syst. Charact.*, 8, 237–241.
45. Fairbridge, C., Ng, S. H., and Palmer, A. D. (1986), Fractal analysis of gas adsorption on syncrude coke, *Fuel*, 65, 1759–1762.
46. Shibata, T., and Yamaguchi, K. (1990), paper presented at the Second World Congress Particle Technology, Sept., Part 1, Kyoto, Japan.
47. Johari, O., and Bhattacharyya, S. (1969), The application of scanning electron microscopy for the characterization of powders, *Power Technol.*, 2, 335.
48. Hillard, J. E., Cohen, J. B., and Paulson, W. M. (1970), Optimum Procedures for determining ultra fine grain sizes, in Burke, J. J., Reed, N. L., and Weiss, V., Eds., *Ultrafine Grain Ceramics*, Syracuse University Press, Syracuse, New York, pp. 73.
49. Oel, H. J. (1969), Crystal growth in ceramic powders, Gray, T. J., and Frechette, V. D., Eds., *Kinetics of Reactions in Ionic Systems*, Plenum, New York, p. 249.
50. Gotoh, K., and Finney, J. L. (1975), Representation of the size and shape of a single particle, *Powder Tech.*, 12, 125–130.
51. Heffel, C., Heitzmann, D., Kramer, H., and Scarlett, B. (1995), paper presented at the 6th European Symp. Particle Size Characterization, Partec 95, Nurenberg, Germany.
52. Karasikov, N., Krauss, M., and Barazani, G. (1988), in Lloyd, P. J., Ed., *Particle Size Analysis*, Wiley, New York.
53. Manohar, B., and Sridhar, B. S. (2001), Size and shape characterization of conventionally and cryogenically ground turmeric (Curcuma domestica) particles, *Powder Technol.*, 120, 292–297.
54. Abrahamsen, A. R., and Geldart, D. (1980), Behaviour of gas-fluidized beds of fine powders, Part I. Homogeneous expansion, *Powder Technol.*, 26, 35–46.
55. Shibata, T., Tsuji, T., Uemaki, O., and Yamaguchi, K. (1994), *Am. Inst. Chem. Eng. Part* 1, 59, 95–100.
56. Realpe, A., and Velázquez, C. (2005), *Powder Technol.*, 169, 108–113.
57. Badawy, S. I., Lee, T. J., and Menning, M. M. (2000), Effect of drug substance particle size on the characteristics of granulation manufactured in a high-shear mixer, *AAPS PharmSciTech*, 1(4).
58. De Boer, A. H., Bolhuis, G. K., and Lerk, C. F. (1978), Bonding characteristics by scanning electron microscopy of powder mixed with magnesium stearate, *Powder Technol.*, 20, 75–82.
59. Duberg, M., and Nystrom, C. (1986), Studies on direct compression of tablets. 17. Porosity pressure curves for the characterization of volume reduction—Mechanisms in powder compression, *Powder Technol.*, 46, 67–75.
60. Krycer, I., Pope, D. G., and Hersey, J. A. (1982), The prediction of paracetamol capping tendencies, *J. Pharm. Pharmacol*, 34, 802–804.
61. Malamataris, S., Bin-Baie, S., and Pilpel, N. (1984), Plasto-elasticity and tableting of paracetamol, Avicel and other powders, *J. Pharm. Pharmacol.* 36, 616–617.
62. Garekani, H. A., Ford, J. L., Rubinstein, M. H., and Rajabi-Siahboomi, A. R. (2001), Effect of compression force, compression speed, and particle size on the compression properties of paracetamol, *Drug Dev. Ind. Pharm.*, 27(9), 935–942.

6.8

ORAL EXTENDED-RELEASE FORMULATIONS

ANETTE LARSSON,[1] SUSANNA ABRAHMSÉN-ALAMI,[2] AND ANNE JUPPO[3]

[1]*Chalmers University of Technology, Göteborg, Sweden*
[2]*AstraZeneca R&D Lund, Lund, Sweden*
[3]*University of Helsinki, Helsinki, Finland*

Contents

6.8.1 INTRODUCTION

6.8.1.1 Background

In order to achieve therapeutic effect, a drug needs to reach the right place in the body at the right time. For some drugs, this may be achieved by simple solutions or solid dosage forms with an instant drug release while, for others, one has to modify

Pharmaceutical Manufacturing Handbook: Production and Processes, edited by Shayne Cox Gad

the drug release. To understand the literature within the area of modified drug release, it is important to be aware of the standard terms used for dosage forms within this field. Malinowski and Marroum have summarized these terms in the book *Encyclopedia of Controlled Drug Delivery* [1]. The authors state that modified-release (MR) formulations refer to "dosage forms for which the drug release characteristics of time course and/or location are chosen to accomplish therapeutic or convenience objectives not offered by conventional dosage forms." One group of MR formulations is the delayed-release dosage form, which does not release the drug immediately after administration. One example of delayed-release formulation is the enteric-coated formulations. Another subgroup of MR formulations is the extended-release (ER) dosage forms, which are the focus of the present chapter. According to a definition from the U.S. Pharmacopeia (USP), ER formulations can be referred to as dosage forms that allow at least a twofold reduction in the dosing frequency compared to conventional dosage forms [2, 3].

The interest in oral ER formulations has dramatically increased in recent years. This can be seen in Figure 1, where the bars in the diagram correspond to the number of publications found in a search in the database SciFinder Scholar 2006 [4] that include the words *oral extended release*. This increase in publications confirms that there are many ongoing activities in this field. The expression oral extended release occurs in the database for the first time in 1954, when Yamanaka et al. utilized the slow dissolution rate of various salts of a drug (pyrimidine penicillin) to extend the period of time when the drug had a clinical effect [5]. In 1959 Robinson and Suedres made a formulation of sulfamethylthiadiazole together with hydrogenated castor oil, which was suspended in an aqueous vehicle, creating a formulation with extended drug release [6]. Later, in the late 1950s and early 1960s Sjögren and Fryklöf compressed active substances (e.g., pentobarbitone sodium and theophyl-

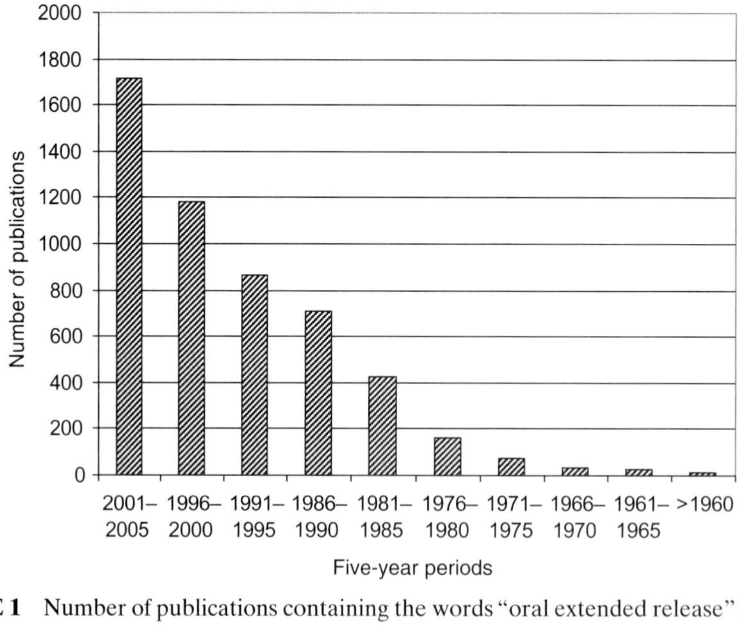

FIGURE 1 Number of publications containing the words "oral extended release" identified in the database SciFinder Scholar 2006 for the five-year period 1960–2005 [4].

line) together with polyvinyl chloride (PVC) and obtained extended drug release from these insoluble matrix tablets [7]. In an early publication they showed that increased dose loadings or addition of channeling agents increased the drug release rate. It is also interesting to note that the most popular ways to prepare oral ER formulations today were already mentioned in an early review from the early 1970s [8].

6.8.1.2 Biopharmaceutical Aspects on Oral ER Formulations

The clinical effect of low-molecular-weight substances is often related to the concentration of the drug in the blood plasma. Classical blood plasma profiles for both immediate-release (IR) and ER formulations are shown in Figure 2. It is well known that a drug only has a clinical effect when the concentration in the blood plasma is above the minimum effective concentration (MEC). If the concentration of the active substance is above the maximum safe concentration (MSC), the side effects will be unacceptable. The interval between the MEC and MSC is called the therapeutic window or therapeutic range, and the time when the concentration is above the MEC is called the "duration" of the drug. One aim of ER formulations is to increase the time the substance is above its MEC by continuous release of the drug from the formulation. Under optimal conditions the rate-limiting step in the drug absorption process of an ER formulation is its release rate, which then can be directly related to the concentration of the drug in the blood plasma. When the drug release rate from an ER formulation is constant, the blood plasma concentration will be constant under ideal conditions, whereas ER formulations with time-

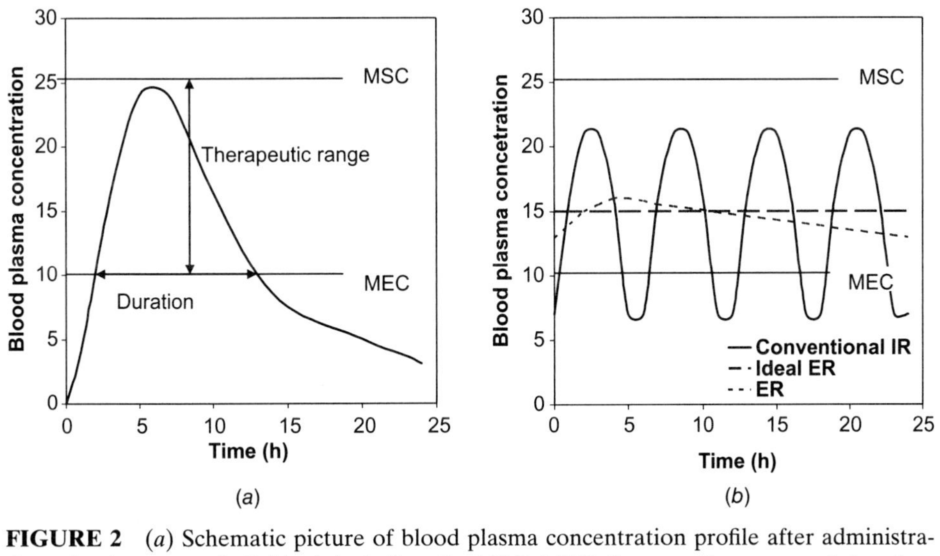

FIGURE 2 (*a*) Schematic picture of blood plasma concentration profile after administration of a drug to an individual, including the MSC, MEC, therapeutic range, and duration. (*b*) Repeated administration of IR formulation (four times daily) of a drug with short pharmacokinetic half-time and administration once daily of an ideal ER formulation with constant drug release (broken line) or ER formulation with nonconstant drug release rate (dotted line).

dependent drug release rate may give rise to time variations in the concentration of the drug in the blood plasma (Figure 2).

The ER concept might offer several advantages, such as reduction in administration frequency, reduction of side effects, less irritation in the gastrointestinal tract, and improved patient compliance. Speers and Bonnano have also mentioned some economic aspects of ER formulations, such as the possibility to patent line extensions and to reduce manufacturing costs since fewer units are required to obtain the same effect [9]. On the other hand, ER formulations may have several drawbacks, for example, large variations in effect between patients due to varying physiological factors within the patient group, limited transit time for the ER formulation, drug stability problems during the gastrointestinal passage, and more severe complications such as dose dumping.

In order to obtain a clinical effect by an orally administered drug, it is, for example, required that the drug is (i) dissolved and released from its formulation, (ii) transported over the mucosal barrier, and (iii) has passed from the lumen to the systemic blood circulation without being metabolized by, for example, the lumen or the liver. The drug dissolution rate and the rate of absorption of the dissolved active substance as well as the relation between these processes are important, in particular the dissolution process since the absorption of undissolved drug particles can be disregarded.

The ER formulations can be a single-unit, monolithic system or multiple-unit systems containing many individual units with extended release. Multiple-unit systems consist of many small pellets and are normally produced by extrusion and spheronization or coating on inert spheres [2, 10]. The composition and ER mechanism can vary for multiple-unit systems, and some examples are membrane-coated reservoir systems and polymer- or lipid-based matrix systems, where the matrices can be made of both soluble and insoluble carriers [11–16].

From a biopharmaceutical point of view, the multiple-unit systems have many advantages, for example, a more consistent gastrointestinal transit compared to larger monolithic systems [17]. The gastrointestinal transit times for monolithic and multiple-unit systems were compared in a study by Abrahamsson et al. [17]. It was found that the gastric emptying time for the small multiple units was considerably shorter than that of larger monolithic systems (on average 3.6 and 9.6 h, respectively). The transit times through the small intestine were approximately equal, whereas the transit time in the colon for the multiple units was longer compared to the monolithic system, which was explained by different influence of the motility on the different systems. Another advantage with multiple-unit compared to monolithic systems is that the effects of dose dumping become less severe [2]. A breakage and instant drug release from one pellet will have considerably lower effect than breakage of one monolithic system.

There may also be development and manufacturing advantages and disadvantages with multiple-unit systems compared to monolithic systems. The dose for multiple-unit systems may be easier to adjust since one can readily increase the number of pellets in the formulation and thus increase the dose. One can adjust the drug release profiles for multiple-unit systems by mixing pellets with different release profiles. The multiple-unit systems offer the possibility to mix pellets containing different active substances, which can be an advantage for the patients who then only need to take one formulation at a time (containing more than one active substance) instead of several formulations. Disadvantages with multiple-unit systems may also exist; for example, the time to develop the multiple-unit systems may well be longer than for monolithic systems. There are greater challenges in the scaleup procedure for the multiple-unit systems since several expensive and specialized types of equipment may be needed [18]. For film-coated formulations, additional changes in the drug release rate may be obtained upon storage as a consequence of aging of the film. However, this change can be suppressed by introducing a curing step in the production [10].

6.8.1.3 Influence of Drug Properties

Drug properties that are important to consider during development of IR tablets are metabolism, stability, permeability, and solubility [18, 19]. In the development of ER formulations, these aspects are also important, but in addition to IR formulations, they must be considered in relation to the different environments that the ER formulations meet during their passage through the gastrointestinal tract, and some of these aspects will be discussed briefly below. It can be mentioned that, based on these initial properties for drug candidates, Thrombre has constructed a feasibility assessment flow chart for ER formation development [18].

The pharmacokinetic half-life for a drug may give an indication of whether a conventional or ER formulation is to be chosen. For drugs with short biological t_{50} (less than 1–2 h), devices that continuously release the active substance are required [2, 18]. For drugs with lifetimes longer than about 10 h, ER formulations may not add any benefits compared to IR formulations. However, for drugs with half-lives between these limits, ER formulations may be a good alternative to IR formulations.

The stability of a drug in the solid state or in aqueous solution is a critical parameter when selecting an appropriate manufacturing process. A drug in an oral ER formulation reaches aqueous environments with, for example, variations in pH (1–8), ionic strength, and bile salt concentration, which requires high chemical stability of the drug [19]. Furthermore, the substance should be stable not only against chemical degradation such as hydrolysis but also against enzymatic degradation (metabolism) during the passage from the lumen to the systemic blood circulation. The drug is released from oral ER formulations along the whole gastrointestinal tract. This implies that, in contrast to IR formulations, drug permeability must be good along the whole gastrointestinal tract for drugs in ER formulations [19]. Furthermore, the solubility and the dissolution rate of the drug are extremely important to consider, since these factors will directly influence the release rate for the drugs from ER formulations. The dissolution rate can be described as the flux J of dissolved material from a drug particle and, according to the Noyes–Whitney equation, it is [20, 21]:

$$J = \frac{D}{h}(C_s - C_b) \tag{1a}$$

where D is the drug diffusion coefficient, h is the thickness of the stagnant diffusion layer around the particles, and C_s and C_b are the concentrations of the drug at the particle surface and in the surrounding bulk media, respectively. For substances in their most stable solid-state form, the concentration C_s close to the particle surface is equal to the saturated concentration. However, when a substance is in a different polymorphic state or in an amorphous state, C_s can be larger than the saturation concentration.

Many drugs are weak acids or bases with one or several pK_a values. According to the Henderson–Hasselbalch equation, the solubility of an acidic drug will depend on the pK_a and pH as

$$\mathrm{pH} = \mathrm{p}K_a + \log \frac{S_i}{S_0} \tag{2}$$

where S_i and S_0 are the concentrations of the drug's un-ionized and ionized forms, respectively. Since the pH varies along the gastrointestinal tract, the solubility and the dissolution rate [Equation (1a)] of the drug depend on the position of the drug in the gastrointestinal tract. Furthermore, for some drugs, such as indomethacin, the dissolution of the acid drug may lead to a changed microenvironmental pH within the stagnant layer and thus also influence the dissolution rate [22, 23].

6.8.1.4 Principles for Extended Drug Release

The main principles related to ER systems are as follows:

(i) Insoluble matrix formulations
(ii) Membrane-coated solid dosage forms including osmotic pump systems
(iii) Soluble hydrophilic matrix formulations

Below we will discuss each of these formulation principles in terms of basic release mechanisms and the advantages or drawbacks associated with the different formulations and manufacturing processes. However, drug release from all kinds of ER formulations starts with hydration of the formulation and water diffusion into the system. The presence of water in the formulation facilitates the start of the dissolution process of the drug, whereby the dissolved drug can be released from the formulations.

The driving forces for transport of water and drug are the differences in chemical potentials between the formulation and its surrounding. Due to the similarities in the driving forces for the dissolution of a drug and the release from an ER formulation, one can modify Equation (1a) to

$$\frac{dM_t}{dt} = JA = \frac{DA}{h}(C_s - C_b) \tag{1b}$$

where M_t is the amount of active drug that is released at the time t, D the drug diffusion coefficient, and J the flux of the drug from the formulation. The other parameters in Equation (1a) have been adjusted to fit the drug release from the formulation and therefore A becomes the surface area of the releasing system in Equation (1b) (e.g., the area of the membrane-coated tablet), h the diffusion pathlength, and C_s and C_b correspond to the concentration of the dissolved active drug at the surface of the drug particles/formulation and in the bulk solution surrounding the ER device, respectively. Depending on the exact type of ER system, some modifications in Equation (1b) may be needed to fully describe the drug release. One important factor in the equation is the difference in concentration ($C_s - C_b$), and C_b is often assumed to be zero due to release under so-called sink conditions. For active substances in their most stable solid state, the remaining concentration, C_s, is equal to the saturation concentration in that medium. However, as mentioned above, C_s can be oversaturated or depend on pH. This means that the drug release from formulations depends on the solid-state properties and pK_a of the drug as well as the pH to which the formulation is exposed [24]. Since the pH varies along the gastrointestinal tract, the drug release will be dependent on the position of the formulation in this tract. Several attempts to avoid pH-independent drug release has been made, for example, by including buffers [25–27].

6.8.2 INSOLUBLE MATRIX TABLETS

6.8.2.1 Principles of Formulation and Release Mechanisms

The history of insoluble matrix tablets goes back to the beginning of the 1960s, when Hässle and Abbott developed the Duretter and the Gradumet, respectively [2]. Since then, many ER tablets based on this principle have been developed. Looking at the homepage of the U.S. Food and Drug Administration (FDA [28]) and searching for "insoluble matrix tablets" produces more than 140 hits, which indicates that this research area is still active.

The term *insoluble matrix tablet* refers to tablets in which the drug is embedded in an inert carrier that does not dissolve in the gastrointestinal fluids. The carrier material in insoluble matrix tablets can be based on insoluble lipids or polymers, both matrix builders whose function it is to keep the matrix together during the passage through the gastrointestinal tract and thus prolong the diffusion path of the drug before it is released from the formulation. The drug can be dispersed or dissolved or both in a matrix carrier (see Figure 3) and, depending on the formulation, different mechanisms can be regarded to take place:

- Dissolved drug in the matrix diffuses through the matrix.
- Dissolved drug in the matrix diffuses through pores in the matrix.
- Dispersed drug dissolves and diffuses through the matrix.
- Dispersed drug dissolves and diffuses through pores in the matrix.

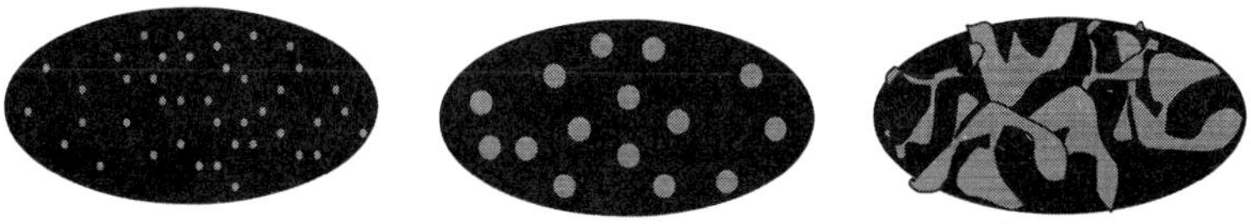

FIGURE 3 Schematic pictures of insoluble matrix systems. Left: Drug (light gray) molecularly dissolved in carrier material (black); middle: drug particles dispersed in carrier material; right: drug particles dispersed in carrier material at higher drug loading, leading to continuous network of drug.

As early as 1963, Higuchi [29] derived an expression for drug release from insoluble matrix systems. In this historical paper, Higuchi derived two equations for two different geometries, the simple planar sheet matrix system with infinite area and spherical pellets. Furthermore, two special cases were treated, one where the matrix is a homogeneous matrix without pores and another where the matrix contains pores. In the system without pores, the drug is assumed to diffuse through the homogeneous matrix. For matrix systems with pores the efficiency in transport through liquid-filled pores is greater than through the solid matrix carrier. Therefore, the main contribution to the drug release is transport in the pores. This is gained by penetration of the surrounding medium into the pores, where it dissolves the drug. The dissolved drug can diffuse through the pores and be released at the surface of the matrix. The simplest theoretical treatment of the drug dissolved in the matrix carriers assumes the following [30]:

- There is no breakage of the matrix.
- There is no dissolution of the matrix.
- There is no resistance to drug transport in the boundary layer surrounding the device.
- There is no accumulation (e.g., adsorption) of the drug in the device.
- The saturated concentration $C_{s,m}$ of the drug in the matrix is constant during the process.
- The drug-loading concentration C_0 is larger than $C_{s,m}$.
- The drug concentration around the matrix is zero, $C_b = 0$ (sink conditions).
- The diffusion constant D_m in the matrix is independent of the drug concentration.
- The partition coefficient K between the matrix material and the surrounding release medium is independent of the drug concentration.

A handy derivation of the equation describing the release from planar homogeneous matrix systems can be found in a book by Wu [30]. It is derived for a sheet with the area A and assumes that the concentration of dissolved drug inside the matrix is constant and equal to the saturation concentration $C_{s,m}$. Under these assumptions, the amount of drug, M_t, that is released at time t can be predicted as

$$M_t = A[D_m K C_{s,m} t(2C_0 - C_{s,m})]^{0.5} \tag{3}$$

When the matrix contains a drug-filled network (Figure 3, right image), water can diffuse and dissolve the drug, and this creates a pore structure. The equation describing the drug release from matrices with networks is modified to include information about the created pore structure [Equation (4)]. This can be described by the porosity ε (the volume of the pores in proportion to the total volume of the device) and the tortuosity τ of the pores (a measure of how much the diffusion path is lengthened due to lateral excursions). Also the diffusion coefficient D_m and $C_{s,m}$ in Equation (3) are replaced with D and C_s in Equation (4), corresponding to the diffusion coefficient and the solubility of the drug in the solution inside the pores, respectively:

$$M_t = A\left[\frac{D\varepsilon C_s t}{\tau}(2C_0 - \varepsilon C_s)\right]^{0.5} \tag{4}$$

The most common types of insoluble matrix tablets are those containing pores. From the equations above one can see that the drug-release depends on the solubility of the drug, the drug-loading concentration, and the diffusion coefficient, which is related to the molecular size of the drug. The area of the insoluble matrix tablet also affects the drug release and can be changed by altering the dimensions or the geometry of the tablet. The drug release from insoluble matrix tablets also depends on the porosity and pore structure of the tablet, and the drug release rate increases with increasing porosity.

A comparison of Equations (3) and (4) shows that, in both equations, the amounts of released drug are directly dependent on the area of the device, the square root of the time t, the drug-loading concentration C_0, the respective saturated drug concentrations, and the drug diffusion coefficients. In addition, the release rate (the time derivate of the amount of released material) depends on the square root of time and can be stated as

$$\frac{dM_t}{dt} \propto \frac{1}{\sqrt{t}} \tag{5}$$

As pointed out above, for ideal ER formulations, the rate-limiting step for drug absorption is the release rate from the ER formulation. Thus, since the release rate from an insoluble matrix system depends on time, the concentration of drug in the blood plasma will also be time dependent and not constant (Figure 2), which may be a therapeutic drawback. Another factor influencing the concentration of drug in the blood plasma is the gastrointestinal transit times. When the transit times of the formulations vary, the reproducibility between different administration occasions in one patient will be low, and furthermore, great variation in the patient group may be obtained. However, these conclusions are valid for all ER formulations based on matrix systems and not limited to insoluble matrix systems only.

The equations above are valid when no depletion of drug occurs inside the device. The equations for release rate will be much more complex when depletion of the drug can occur [30]. However, it has been shown that, when the amount of released material is less than approximately 60%, the release rate will depend on time as $t^{-0.5}$ [29, 30].

6.8.2.2 Manufacturing of Insoluble Matrix Tablets

Insoluble matrix tablets need a carrier, which can be a lipid- or polymer-based excipient [7, 31–36]. Some suggestions of carrier materials can be found in Table 1. The table also presents the number of hits found upon searching the FDA's homepage [37] for the number of times an excipient is registered as a component in oral extended, sustained, or controlled formulations. This list gives an indication of how often these excipients are commercially used in oral ER formulations but does not automatically tell us the exact formulation or exact mechanistic effect of the excipient. The choice of carrier material is important, and one should be aware of possible

TABLE 1 FDA Registered Oral ER Formulations Containing Commonly Used Excipients in Insoluble Matrix Formulations

Excipient	Number of Hits on FDA Homepage	Content Interval (mg)
Lipid based		
Carnauba wax	9	46–300
Stearyl alcohol	4	25–244
Glyceryl behenate	3	15–51
Castor oil	2	23
Cottonseed oil, hydrogenated	2	58–402
Cetyl alcohol	2	44–59
Paraffin	2	50–150
Stearic acid	2	26–180
Castor oil, hydrogenated	1	295–410
Vegetable oil, hydrogenated	1	228
Mineral oil	1	
Microcrystalline wax	0	
Insoluble polymer		
Ethylcellulose	9	15–309
Ammonia methacrylate copolymer	5	37–138
Polyvinyl acetate	1	46
Polyethylene	0	
Inorganic		
Calcium phosphate (dibasic)	6	33–335

Source: http://www.accessdata.fda.gov/scripts/cder/iig/index.cfm.

exposure of lipid-based formulations to erosion, which can be the result of enzymatic degradation of the lipids [38]. This will of course also influence the drug release rate.

In the compositions of insoluble matrix systems, excipients other than the carrier material are needed to obtain products with processability and that meet requirements from the pharmacopedias. Examples of categories of excipients included in insoluble matrix tablets are binders, lubricants, glidants, colorant, taste maskers, and channeling agents. As mentioned above, the drug release rate can be regulated by the porosity in the insoluble matrix system. The properties and amounts of drugs and excipients that can create pores will have a large impact on the release rate. Examples of channeling agents are sugars, salts such as sodium chloride, and polyols [2]. The pore structure also depends on other factors such as the particle sizes of the excipients and the drug, the size and porosity of the granules, and the compaction pressure.

The choice of process steps depends on the properties of the drug and the chosen excipients. For insoluble matrix tablets one often mixes the active substance with the excipients. Either this mixture can be directly compressed to matrix tablets or the powder mixture can be exposed to a granulation technology to enlarge the particle sizes. One such technique is dry granulation, that is, compaction of the mixture in, for example, a roller compactor, followed by milling to desirable granule sizes. Another granulation technique is melt granulation, where the melted granulation liquid agglomerates the particles to granules. The most common granulation method for insoluble matrix tablets is probably wet granulation [2, 7] with aqueous-based or organic granulation liquids. The wet granulated masses are dried in fluid bed driers or ovens. In order to increase the drying speed, microwaves can be used. The powder mixtures or granules are compressed in ordinary tableting machines [7]. However, it is an advantage if the compaction pressures can be carefully monitored, since this pressure may influence the porosity and thus the drug release. The final tablet can be coated to, for example, mask the taste.

Alternative production methods to the traditional compaction of powder to insoluble matrix tablets are available, some of which are based on melting technologies, but of course these methods rely on the ability of the carrier materials or additives to melt. The drug is commonly dissolved or dispersed in the melt. This melt can be filled into hard gelatin capsules [31] or it can be spray chilled by pressing it through a nozzle into a vessel containing solid carbon dioxide [39]. Hot-melt extrusion of polymer-based systems to form multiple-unit systems has been investigated. The carrier material in these cases can be, for example, Eudragit [40–42].

6.8.3 MEMBRANE-COATED ORAL EXTENDED RELEASE

6.8.3.1 Principles of Formulation and Release Mechanisms

One way to protect the drug from being directly released is to coat the system with an insoluble film. The drug is suspended or dissolved in a reservoir system which can consist of monolithic or multiple-unit systems. The MR films surrounding the reservoir will be insoluble and thus give the system extended drug release properties or they can become soluble by external trigging. The latter case is defined as a delayed-release formulation, which means that the formulation does not release directly after administration. This can be achieved, for example, by enteric coatings, where the film-forming materials are insoluble in aqueous solutions at low pH but soluble at high pH values. This results in delayed release from an enteric-coated formulation, since the pH is low in the stomach. When the units are transported into the intestine, the pH increases, the film dissolves, and the drug can be released immediately. This type of formulation will not be further discussed here, but the work of Hogan is recommended for further information on these systems [43, 44].

In ER reservoir systems, a membrane surrounds a reservoir of the drug, also called the core of the system. The membrane controls the drug release and the driving force is the difference in chemical potential over the membrane, which can be correlated with a concentration gradient over the membrane. The transport of the drug through the ER membranes can be divided into three different mechanisms [45–47]:

- Diffusion through the membrane
- Diffusion through pores and cracks in the membrane
- Osmotic transport through pores, cracks, or drilled holes

The overall drug release process for common membrane-coated systems has been shown to pass through three different time periods (Figure 4) [48]. During the initial

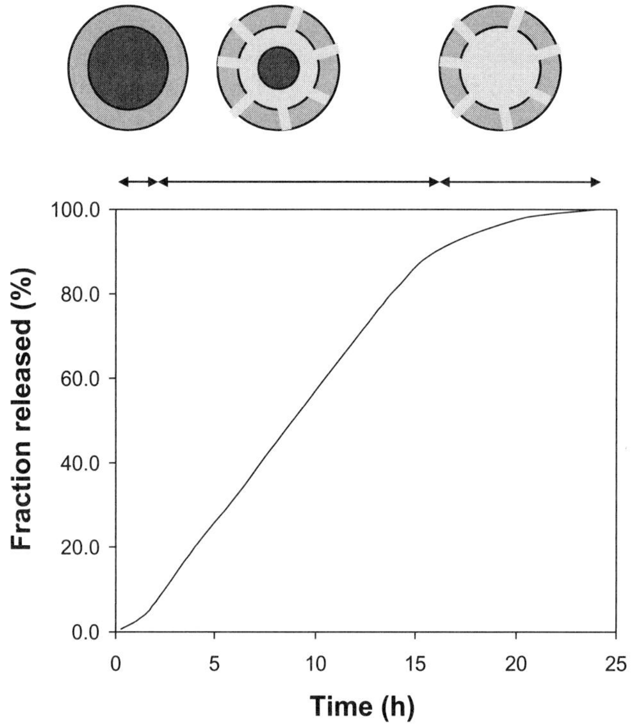

FIGURE 4 Release from membrane-coated reservoir system, where the three different stages are depicted (initial lag period, "steady-state" period, and final depletion period). A schematic picture of a formulation with the drug reservoir (dark grey) surrounded by the membrane (middle grey) is presented at the top. The dissolution medium penetrates the membrane and dissolves the drug (light gray), and pores/cracks are formed through which the drug can be released.

period, the amount of released drug will be low. The water penetrates the membrane and at the same time starts to dissolve the material in existing pores, consisting of water-soluble sugars, salt, or polymers. The water that has penetrated the membrane and reached the core starts to dissolve the active substance in the reservoir [11]. The concentration of the active drug increases continuously until its saturation concentration has been reached and pseudoequilibrium between the solid material and saturated solution inside the membrane has developed. The osmotic pressure in the reservoir will depend on the concentrations of all dissolved species inside the membrane, and an increase of the osmotic pressure may lead to membrane rupture. After an initial lag period, the number of cracks and pores becomes constant, and the osmotic pressure and the concentration of dissolved species in the reservoir and in the membrane reach their steady-state levels. When these parameters, which are the driving forces for the drug release, are constant, the drug release rate will also be constant. Therefore, a second period with time-independent drug release rate will occur.

When the solid drug material inside the membrane is completely dissolved and the concentration of the dissolved drug decreases, a final period with declining release rate will be entered. Ragnarsson et al. [49] have shown that the solid material disappears earlier as the drug solubility increases and that the third stage with time-dependent and decreasing release rate appears earlier compared to drugs with lower solubility.

The contribution of diffusion to the release process can be modeled by using Fick's first law. For diffusion of a substance through the membrane, it will turn out as (assuming sink conditions)

$$\frac{dM_t}{dt} = JA = \frac{D_m K_m A}{h}(C_s - C_b) = \frac{D_m K_m A C_s}{h} \tag{6}$$

where K_m is the partition coefficient for the drug between the membrane and solution, D_m is the diffusion coefficient in the membrane, A is the area of the membrane, h is the thickness of the membrane, and C_s and C_b are the concentrations on the inside of the membrane surface and in the bulk, respectively. The equation for diffusion through pores or cracks resembles Equation (6):

$$\frac{dM_t}{dt} = JA = \frac{D\varepsilon K A}{h\tau}(C_s - C_b) = \frac{D\varepsilon K A C_s}{h\tau} \tag{7}$$

where ε and τ are introduced to describe the porosity and tortuosity in the membrane, respectively. The parameters D_m and K_m in Equation (6) are replaced in Equation (7) by D and K, which are the diffusion coefficient in the solution inside the pores and the partition coefficient between the solution and materials surrounding the liquid-filled pores, respectively. Equations (6) and (7) depend on the concentration gradient over the membrane, and both are independent of the time.

The osmotic contribution to the drug transport is described by the so-called Kedem–Katchalsky equations (based on nonequilibrium thermodynamics) [50, 51]. A simplified version is

$$\frac{dM_t}{dt} = JA = \frac{A C_s L_p \sigma \Delta\Pi}{h} \tag{8}$$

where $\Delta\Pi$ is the osmotic pressure difference over the membrane and L_p and σ are the hydraulic permeability and the reflection, respectively.

A comparison of Equations (6)–(8) shows important similarities; they depend on the solubility of the drug, the area of the device, and the thickness of the membrane. This means that an increased solubility, larger area of the membrane, and thinner membranes will facilitate the drug release rate. This can be exemplified by a study by Ragnarsson and Johansson [11], who showed that, for different salt forms of metoprolol, an increased solubility also increased the drug release rate, which was predicted from the equations. Furthermore, Equations (6)–(8) show constant and time-independent release rates. This constant amount of released drug will be a biopharmaceutical benefit since it theoretically makes it possible to achieve a constant concentration of the drug in the blood plasma.

One special type of ER formulation based on coated reservoir systems is the so-called osmotic-controlled oral drug delivery system or osmotic pump. Pure osmotic systems have semipermeable membranes; that is, water can permeate the membrane but not other substances. These semipermeable membranes can be made of, for example, cellulose acetate [44], and for such formulations the dominating release mechanism is the osmotic pressure [Equation (8)]. The oldest formulation based on osmotic release was OROS from Alza Corporation [47]. In order to achieve drug release through the semipermeable membrane, a laser hole was drilled, but today many osmotic formulations instead use pores filled with water-soluble materials. The aspects of the formulation and different types of commercially available formulations are summarized in the review by Verma et al. [52]. Some advantages with osmotic pumps compared to other ER formulations such as hydrophilic matrix systems are (i) the time-independent drug release (often mentioned as zero-order release), (ii) the superior in vivo–in vitro correlation which facilitates further formulation development, and (iii) less variation between fasted and fed states. Potential drawbacks may be high initial development costs and lack of in-house competence. Drawbacks from an economic point of view may be the necessity to pay royalties and the need for special equipment associated with laser drilling technology.

6.8.3.2 Manufacturing of Oral Membrane-Coated Systems

The first step in the production of membrane-coated systems is to prepare the drug reservoir, the core, of the system. The process steps for producing the core depend on the size of the core. In monolithic membrane-coated systems, the core can be a filled capsule or a tablet which is produced in the traditional ways. This may include mixing of active substance and excipients, possibly granulation and drying, filling into capsules, or compaction into tablets. The production of cores for multiple-unit systems (often termed pellets) is more sophisticated and may be performed in different ways. One, and probably the most common way, is to produce cores for multiple-unit systems by extrusion and spheronizing [53–55]. An alternative methodology is to coat an inert core, e.g. glass or nonpareil beads, with the active substance and the excipients [11]. When the drug-containing core is manufactured, the process continues with the coating of the release-controlling membrane.

The composition of the core depends on the properties of the drug and the excipients, the chosen production chain, and whether the systems should be a monolithic or multiple-unit system. The compositions and production steps are reviewed by Tang et al. [56]. In general, the core will include the active substance together with the filler materials and, if necessary, solubilizers and lubricants/glidants. Classical filler materials are lactose and microcrystalline cellulose, but also other materials such as dextrose, mannitol, sorbitol, and sucrose can be used. However, it should be remembered that the dissolution of the filler material might influence the osmotic pressure. The effect of filler solubility has been investigated by Sousa et al. [54], who found a relation between the solubility of the filler materials and drug release rate. For filler materials with large water solubility, there is a great risk that the membrane will rupture due to the development of an excessive osmotic pressure, which will influence the drug release rate.

The choice of film-forming materials and film-coating techniques is critical for the drug release rate [54]. The ER membrane should remain intact during the release, which implies that it should not dissolve or erode. As film-forming material, water-insoluble substituted cellulose derivatives such as ethylcellulose have been suggested [53, 57] as well as synthetic polymers such as methylacrylates (e.g., Eudragit NE 30D, RS30D or RL30D, where NE stands for nonionic and RS/RL correspond to cationic polymers) [55]. A commercial technology platform is available under the name Eudramode, which is a platform for development of multiple-unit systems with extended drug release based on Eudragit [58, 59]. Other film-coating materials such as shellac and zein have been used, but a drawback with these naturally occurring materials is the variation in quality. To obtain a film with satisfying release properties, channeling agents such as the hydrophilic polymers hydroxypropyl methylcellulose (HPMC), hydroxypropyl cellulose (HPC), and polyethylene glycol (PEG) or other water-soluble materials such as sodium chloride or lactose may be used [2, 57]. To improve the mechanical properties of the film and thus avoid ruptures and cracks in the film, plasticizers may be added to the formulation. Examples of plasticizers are PEGs, diethyl phthalate, triacetin, mineral oils, glycerol, and chlorobutanol [60].

The film coating may be performed in different types of equipment. For coating of larger units, such as tablets or capsules, a rotating drum is often used, for example an Accela Coater, but other similar equipment may also be used [10, 53, 57, 61]. This type of equipment contains a perforated pan that rotates and thus mixes the bed of units. At the same time, a coating liquid is applied to the moving units by means of a spray gun, where the mixing of the units ensures uniform coverage of the coating. The film coating is dried by blowing a stream of hot air onto the surface of the tablets. For all types of coating processes, there are many parameters, such as the temperature and relative humidity of the inlet air, drum rotation speed, spraying rate, and droplet size of the coating liquid, that have to be adjusted in order to produce good coated films.

For coating smaller units, such as pellets, the fluid bed coating technique is used [56]. This is an attractive technique in which the starting material is placed in the coater and heated air is blown through a base plate. This leads to vigorous mixing of the pellet units. By changing the pressure of the incoming air stream, the material becomes suspended in the air. This happens when the bed starts to fluidize and the bed will then have fluidized properties similar to the properties of ordinary liquids. There are different designs of fluid bed: top spray coating, bottom spray coating (or Wurster coating), or tangential coating. They differ with regard to placement of the spray guns. In top spray coating, the liquid is sprayed from the top of the equipment and the droplets hit the particles moving in opposite direction. In the bottom spray coating, the sprayed liquid drops and particles flow in the same direction, which avoids the problem of blocking the spray guns that may occur in top spray equipment. In the more rarely tangential spray coating equipment the base plate rotates and the spray guns are spraying in a tangential direction to the spiral moving particles [10, 62].

A critical parameter for obtaining films with desirable properties is the creation of coating droplets, a process often referred to as atomization, that is, when a bulk liquid is dispersed in air to form a spray or a mist [61]. Atomization is done by letting pressurized air and bulk solution pass simultaneously through a nozzle (spray gun). The air divides the droplets into smaller units, but in spite of the name atomization, the droplets are dispersed not at atomic scale but rather in the nano- to micrometer-scale range [61].

The mechanism for film formation depends on whether the polymer is dissolved or dispersed as small latex particles in the solution. For both technologies the film formation process starts with wetting and spreading of the coating droplets on the surface of the reservoir system [10]. For the case with polymer dissolved in solution, the film formation mechanism continues with evaporation of the solvent. This leads to an increase of the polymer concentration, and at a certain limit the polymers precipitate and a coated film is formed. In the case of ER films, the most commonly dissolved polymer is ethyl cellulose and the most commonly used solvent is ethanol but other organic solvents can also be used. The use of organic solvent is problematic from the SHE (safety, health, and environmental) point of view and therefore these aspects must be considered before starting to use organic-based coating process technologies. An alternative process methodology is to disperse particles of the film-forming polymer in an aqueous solution [63]. Film formation of dispersed particles undergoes the following steps: evaporation of water, close packing of the particles, deformation of the particles, and annealing of the particles by migration of individual polymer chains between the particles to form a coherent, smooth film coating. The first steps of the process occur in the coating equipment but the last, the annealing step, may continue days after the coated product has left the coating equipment.

6.8.4 HYDROPHILIC MATRIX TABLETS

6.8.4.1 Principles of Formulation and Release Mechanisms

Several recent informative review articles on hydrophilic matrix systems have been published [e.g., 24, 64]. Hydrophilic matrix tablets are composed of an active substance, a hydrophilic polymer, release modifiers, lubricants, and glidants. The technology goes back to the mid-1960s when Lapidus and Lordi [65, 66] and Huber et al. [67] determined the drug release from hydrophilic matrix systems. The release mechanism for this type of formulation starts with dissolution of hydrophilic matrix polymers and the formation of a highly viscous polymer layer around the tablet (Figure 5) [64]. This layer is often referred to as a gel layer even though it normally contains only physical entanglements and not chemical cross-linkers, which is traditionally required for gels. However, the gel layer surrounds the inner (more or less dry) part of the tablet and this part is called the core. In traditional hydrophilic tablets, the active substance particles are embedded in the matrix carrier. The dissolution process of the active substance can start when the carrier material has dissolved in water and formed the aqueous gel layer, since without exposure to water the active substance cannot dissolve. Therefore, the "dry" core will shield the active drug from dissolution, which is one reason for the extended drug release from this type of formulation.

The release process for hydrophilic matrix tablets can be schematically described as in Figure 5. The left side of the figure shows a hydrophilic tablet undergoing dissolution, swelling, and release. An interface between the solution and gel layer, here

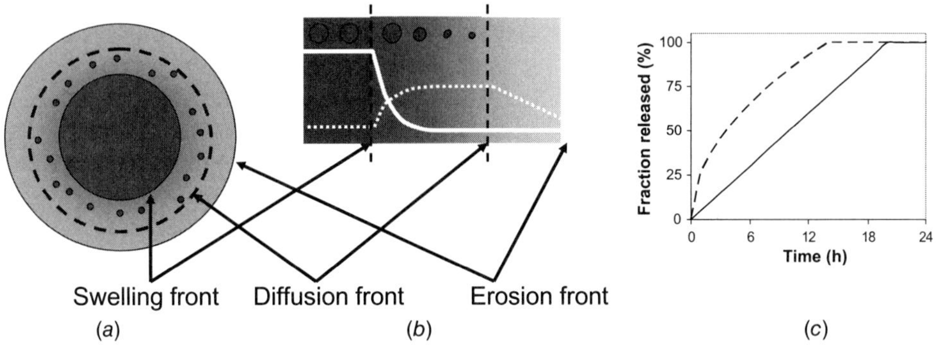

FIGURE 5 (*a*) Hydrophilic matrix system shown with core (dark gray) and drug particles (small dark gray particles). The swelling, diffusion, and erosion fronts are depicted. (*b*) Dependencies of volume fraction polymer (solid line) and dissolved drug (broken line) as function of position in matrix system together with swelling, diffusion, and erosion fronts. Top shows how the solid drug particles diminish in size. (*c*) Examples of drug release as function of time for erosion-controlled (n = 1, solid line) and diffusion-controlled systems (broken line, n = 0.5).

called the erosion front, can be identified, and the polymer chains and drug molecules are released at this front. In the gel layer, the polymer concentration will decrease (Figure 5) from a highly concentrated solution at the swelling front, the interface between the gel layer and more or less dry core, to a diluted polymer solution at the erosion front [68]. Some authors have suggested that the polymer concentration at the erosion front is related to the overlap concentration [68–70]. At the same time, the water content in the gel layer gradually increases from the center of the tablet toward the erosion front, and thus the dissolution of the active substance particles can start already in the gel layer. However, the volume fraction of dissolved drug depends on the amount of available water, which in turn is a function of the position in the gel layer [64, 71]. Assuming that the drug saturation concentration is equal in water solutions and polymer gels, the volume fraction dissolved drug will correlate with the volume fraction water available, and the volume fraction drug will gradually increase with increasing distance from the core. Far from the core, the variations in the concentration of polymer and water are less pronounced and, as long as solid particles coexcite with the saturated drug solution, the volume fraction of the drug will theoretically be almost constant. A third front has also been introduced, the diffusion front, which corresponds to the position in the gel layer where all of the active substance has dissolved. Between the swelling and dissolution front, dissolved and undissolved drug particles will coexist, but between the dissolution and erosion fronts, only dissolved drug molecules occur. In this region, the diffusion of drug out from the matrix will give rise to a decrease in the volumc fraction of thc drug.

Achieving a mechanistic understanding of drug release from hydrophilic matrix tablets is not a trivial task since the release depends on the properties of both the polymer and the drug. The swelling process is directly related to the properties of the polymers, and this is an important factor for drug release, since the polymer swelling process can, for example, transport individual drug particles through the gel layer, which has been shown by Adler et al. [72]. Macroscopically, polymer swelling can be observed as an increase in the size of the tablet. On a molecular level, the swelling depends on the dilution and transport of water into the gel layer. This transport is driven by changes in the chemical potential, and the main contribution is the increase of conformational entropy when the polymer chains are diluted [73]. The kinetics of the swelling process may vary. A faster polymer release rate compared to the swelling rate results in a movement of the erosion front toward the center of the tablet and the size of the tablet will diminish. Conversely, the tablet size increases when the polymer release is slower than the swelling process. When the swelling rate is on the same order as the polymer release rate, the position of the erosion front (i.e., tablet size) will remain constant.

The swelling front between the gel layer and amorphous (or semicrystalline) core material has traditionally been described as corresponding to a transition of the solid states of the polymers. The polymers in the core, for example in HPMC tablets, are in a glassy state, and the polymer material is transformed to a rubbery state due to the fact that water acts as a plasticizer and decreases the glass transition temperature [74]. This rubbery state can be regarded as a polymer solution, and therefore the glassy-to-rubbery state transition can be regarded as a dissolution process of the polymer, where the dissolution rate will determine the position of the swelling front. A commonly described special case for hydrophilic tablets is the so-called front synchronization, which is when the movements of the swelling front and the erosion front occur equally fast. This special case corresponds to a constant gel layer thickness.

Depending on the drug solubility and the dissolution, swelling, and release processes of the polymer, either of two different drug mechanisms can be observed: erosion- or diffusion-controlled drug release (Figure 5). One way to characterize these two mechanisms is to compare the drug and the polymer release. The erosion mechanism is characterized by equal release of the polymer and the drug, whereas when the release of the drug is faster than that of the polymer, this is called diffusion-controlled drug release [64]. The diffusion-controlled mechanism occurs when the diffusion front, the border between undissolved and dissolved drug, is displaced in the gel layer and the drug can efficiently diffuse out from the gel layer. The erosion mechanism dominates when the diffusion and erosion fronts overlap. This means that drug particles may be released from the surface of the gel layer. When this occurs, the drug particles will dissolve faster in the free solution than in the gel layer due to the fact that stirring is more efficient in the free solution, which results in a decreased thickness of the unstirred boundary layer around the particles and thus an increased dissolution rate [Equation (1a)]. Whether the release mechanism will be diffusion or erosion controlled depends on (i) the polymer release rate, which governs the position of the rate erosion front, and (ii) the drug dissolution rate, which governs the position of the diffusion front. The position of the diffusion front and the dissolution rate depend on the solubility of the drug. Lower solubility of the drug gives slower dissolution rates and hence the diffusion front can overlap more easily with the erosion front, which yields erosion-controlled drug release. On the other hand, large solubility of the drug will give diffusion-controlled drug release [75–78].

Traditionally, the drug release rate from hydrophilic matrix systems has been modeled as [79, 80]

$$Q = \int_0^t \frac{dM}{dt} dt = kt^n \tag{9}$$

where Q is the accumulated amount of released drug and k and n are constants. The values of n have been suggested to describe the drug release mechanisms. Release from a planar surface with $n = 1$ has been shown to correspond to erosion-controlled drug release and $n = 0.5$ to pure diffusion-controlled drug release. This is strictly only valid when the polymer release has $n = 1$, but this is often the case. In practice, for hydrophilic matrix systems, one often finds n to be between 0.5 and 1, indicating that both the diffusion of the drug and the polymer erosion influence drug release. For other geometric shapes, such as tablet shapes, the limits for n shift to 0.45 and 0.89 for diffusion- and erosion-controlled release, respectively [81–83].

Another popular way to describe the drug release is to characterize the influence of the relative contributions of erosion and diffusion to drug release as [84]

$$Q = at^m + bt^{2m} \tag{10}$$

where a, b, and m are constants. The first factor in Equation (10) should represent the Fickian diffusional contribution and the second term to the erosion contribution to the drug transport. One mechanistic drawback with this approach is that it treats the diffusion and erosion processes as independent of each other, which they are probably not in any practical case.

The drug release from hydrophilic matrix tablets has been found to vary with the polymer parameters, the composition of the formulation, and the process parameters. Examples of important polymer-related parameters with significant influence on drug release are the viscosity and the hydrophilicity of the polymer, where polymers of larger viscosity grades give lower release rates and longer durations of the release [85] and an increased hydrophilicity gives larger swelling and faster drug release, which was found by comparing different degrees of substitution of HPMC [86]. The drug release rate also depends on the composition of the formulation. When components with high water solubility, such as lactose, are included in the matrix, the drug release increases, which can be seen as a dilution of the gel-forming material [69, 87]. Similarly, an increased amount of a soluble active substance also increases the release rate of drugs, probably also due to the corresponding decrease of the relative amount of hydrophilic polymer [88, 89]. The size and shape of the tablet also influence drug release, and the release rate increases as the area-to-volume ratio increases [90–93]. The influence of particle size on the drug release rate has also been investigated [89, 94, 95]. The size of polymer particles seems to have low influence on drug release when there is enough gelling polymer available to quickly form a coherent gel layer. In contrast, at low amounts of hydrophilic matrix polymer, gel formation may be too slow, which makes polymer particle size important. In this case, the matrix may disintegrate before it develops a coherent gel layer [95–97]. The effect of drug particle size on the drug release rate depends on the solubility of the substance. Varying the particle size of drugs with high solubility seems to have little influence [89], whereas the release rate may depend on the particle sizes of drugs with moderate solubility [24, 89].

The effects of process parameters on drug release have been discussed in the literature. Different granulation technologies, such as dry granulation [98–100] or wet granulation [101–103] (which includes fluid bed granulation [104, 105]), have been used. Also direct compression has been used for production of hydrophilic matrix tablets [95]. The effects on the choice of production steps may be critical when the relative fraction of the polymers is low in the formulations. The effect of the compaction pressure on drug release has also been studied. Several authors [89, 88, 106, 107] have found that, while the compaction pressure has a significant effect on the tensile strength of the tablets, it has a minimal influence on drug release. This can be due to the fact that, when a coherent gel layer is formed, only the parameters governing the performance of this gel layer are important, and since parameters such as porosity do not affect the gel layer, they are of low importance for the drug release rate.

6.8.4.2 Manufacturing of Hydrophilic Matrix Tablets

The traditional way of producing hydrophilic matrix tablets resembles the production of the core for membrane-coated tablets and insoluble matrix tablets. It includes a mixing step, possibly a granulation step, a compaction step, and sometimes a coating step. However, one large difference between the production of insoluble and soluble matrix systems is notable; the latter matrix type has strong interactions with water, which complicates the production steps when water is present [108]. Therefore, for hydrophilic matrix systems with large fractions of hydrophilic polymers, traditional wet granulation with water as granulation liquid may cause problems with formation of hard lumps [108]. To avoid this problem, a new technique using foam granulation has been suggested [109, 110]. During granulation, the foams will flow on the top of already foam-wetted particles, which may lead to superior distribution of the granulation liquid. An alternative wet granulation method is to use organic solvents such as ethanol as granulation liquid [111]. However, even if the production of granules with good compaction properties can be maintained in this way, it may, as already mentioned, be an advantage to consider the SHE aspects before choosing organic solvents as granulation liquid. Another alternative to wet granulation is to use dry granulation technologies such as roller compaction [98, 99, 100]. Although this technique has several advantages compared to traditional wet granulation with water, it may result in lower tensile strengths of the tablets [112]. An alternative to granulation technologies is direct compression. This can be done by purchasing special direct-compression qualities of the hydrophilic polymers. A drawback that always arises in relation to direct compression is the difficulty to achieve content uniformity of the tablets, a problem that increases with decreasing doses of the active drug. Therefore, special care should be taken with regard to formulations with low doses of active substance when direct compaction is used [2].

Water-soluble hydrophilic matrix systems may also be extruded, both to monolithic systems and to multiple-unit systems [12, 14–16]. Polyethylene oxide (PEO) and chloropheniramine maleate have, for example, been extruded to a monolithic unit [113]. This manufacturing method proved more feasible for mixtures between low- and high-molecular-weight PEO, since systems containing only high-molecular-weight PEO proved too viscous and difficult to extrude. It was also

TABLE 2 FDA Registered Oral ER Formulations Containing Commonly Used Excipients in Hydrophilic Matrix Tablets

Hydrophilic Polymer[a]	Fraction of Excipient (%)	Number of Hits on FDA Homepage	Other Names of Excipients
Hydroxypropyl methyl cellulose [24]	10–80	12	Hypromellose, HPMC, MHPC
Hydroxypropylcellulose [120, 121]	15–35	4	HPC
Polyethylene oxide [68, 128]	5–85	4	PEO
Sodium alginate [134]	5–50	3	
Xanthan gum [125, 137–140]	10–99	3	
Hydroxyethyl cellulose [117–119]	40–97.5	3	
Carbomer [129]	20–75	1	Carbopol, polyacrylic acid
Methyl cellulose [121]	5–75	1	MC

Source: http://www.accessdata.fda.gov/scripts/cder/iig/index.cfm.

[a]Examples of references using them are given.

shown that degradation of PEO to lower molecular weight might occur due to the high temperature used in the extruder. Extrusion to small pellets based on hydrophilic polymers hydroxyethyl cellulose (HEC) and HPMC has been performed, and isopropyl alcohol instead of water was used to avoid lump formation [12]. Also Carbopol [14–16], pectin [114], and xanthan [115, 116] have been extruded and used as matrix carrier materials for extended release.

The choice of hydrophilic polymer is one determining factor for the drug release rate, as discussed above. The most common type used for hydrophilic matrix systems is HPMC (hypromellose, hydroxypropyl methyl cellulose) [24, 83] but some alternatives are HEC [117–119], HPC [120, 121], methyl cellulose [121], xanthan [122–127], PEO [68, 128], Carbopol [129–131], pectin [131–133], and alginate [134–136]; see Table 2. HPMCs are available in several approved degrees of substitution [60]. To achieve different release durations, different viscosity grades of the polymers may be used (the higher the viscosity grade, the slower the release rate). If the desired viscosity is not available, one can mix two polymer grades with different viscosity grades [68].

6.8.5 COMPARISON AND SUMMARY OF DIFFERENT TECHNOLOGIES

The main principles for oral extended drug release reviewed here are the membrane-coated reservoir systems and hydrophilic or hydrophobic matrix systems. They all have advantages and drawbacks and Table 3 summarizes some aspects of the different formulation principles.

There are a number of commercial variants of the above-mentioned principles. Examples of commercial matrix-based formulations can be found in a review by Varma et al. [137]. One of the oldest commercial hydrophilic matrix systems is

TABLE 3 Comparison between Three Drug Release Principles

Aspects	Insoluble Matrix Systems	Membrane-Coated Systems	Hydrophilic Matrix Systems
Drug solubility	Dose dependent [18]	Dose dependent [18]	Whole range
Release mechanism	Diffusion controlled	Diffusion and osmotic	Diffusion to erosion controlled
Release profile	$Q \propto t^{0.5}$	$Q \propto t^{1}$	$Q \propto t^{n}, 0.4 < n < 1$
Main release dependence (except drug properties)	Channeling components and processes parameters	Properties of membrane	Properties of polymer carrier
Composition alternatives	Many approved and functional excipients available Pore-channeling excipients may determine drug release Small changes in formulation may change release rate	Many approved and functional excipients available Release can be changed by film and core compositions Many polymers are based on natural material which may give large batch-to-batch variation	Many grades of available polymers exist Easy to regulate drug release by means of composition and polymer properties Many polymers are based on natural material, which may result in large batch-to-batch variation
Manufacturing aspects	Uses traditional production technologies and is inexpensive Release depends on process parameters	Coating process can be used to control drug release Film coating process dominates release, which may be sensitive in relation to many different process parameters	Release is mainly robust against process parameters Granulation with water is difficult
In vivo–in vitro correlation	Release depends on fasted and fed state	Good	Release depends on fasted and fed state

TIMERx, which is based on xanthan gum mixed with, for example, locus bean gum [138]. A mixture of these polymers has a special property; the different polymers interact strongly with each other and this interaction produces gels with large viscosity. Variants of osmotically controlled ER formulations, some of which are commercially available, are reviewed by Verma et al. [52].

6.8.6 OTHER ORAL ER FORMULATIONS

Here some new oral ER technologies based on principles other than those mentioned above will be presented. It is beyond the scope of this chapter to cover all systems and details thereof, and the interested reader is recommended to make use of the literature included in the references.

One promising technology is to let the active drug interact with the excipients, for example, by covalent binding between the drug and excipient. The azoaromatic cross-linkers between the drug and excipient can exemplify this. These special cross-linkers break due to bacterial degradation in the colon, but not until they are exposed to this bacterial flora [139–141]. This means that the drug is hindered from release before it reaches the colon. This is an example of site-specific delivery to the colon, an area that is reviewed by a special issue of the journal *Advanced Drug Delivery Reviews* 2005 (volume 57, number 2).

Another example of modifications of ER systems is when electrostatic interaction between charged drugs and excipients (with opposite charge) is used. This concept can be valuable for drugs with pH-dependent solubility. If, for example, the drug is a weak base, it will have a large solubility at low pH, and a major part of the drug may be rapidly released already in the stomach, which is not desirable. An introduction of polyions with opposite charge can result in electrostatic interaction between the drugs and the polyions, which can change the solubility or decrease the diffusion rate of the drug through the ER formulation. This concept is, for example, used when propranolol hydrochloride interacts with sodium carboxymethylcellulose (NaCMC) in HPMC matrices [142]. In addition, buffering of the ER systems by organic or inorganic buffers has been used to obtain pH-independent drug release [27, 143, 144]. A recent publication by Riis et al. showed that insoluble inorganic buffers such as magnesium hydroxide and magnesium oxide provided stable drug release over longer periods of time than when more soluble buffer systems were used [26].

Sophisticated systems based on responsive gels are promising alternatives in terms of oral extended release. The principle behind many of these systems is that the formulation should react and undergo some kind of transition due to a trigger signal [145]. In general, the triggers can be various factors such as temperature, light, pressure, electrosensitivity, or interactions with specific molecules such as glucose or antigens. For oral drug delivery systems, the triggers can also be a biological change in the gastrointestinal environment, such as a change in the bacteria flora, as mentioned earlier [139]. It can also be a physiochemical change, such as a pH change in the gastrointestinal tract. Hydrogels composed of copolymers of poly(acrylic acid) and covalently attached Pluronic surfactants is one such example that reacts on pH changes. The cross-linked microgels can be loaded with drug and tailored to collapse in low pH but swell and release the drug at high pH [146].

A novel method for producing ER formulations is a technology called three-dimensional printing (TheriForm technology) [147], which is similar to the one used in ordinary printers. The ink is here replaced with an active substance and carrier material. The layer-by-layer "printing" provides controlled placement of the active drug and thus of the release from the device [148].

Even if there have been advances in oral drug delivery technologies during the last 50 years, many highly sophisticated drug delivery systems have failed and have not reached the market [149]. This can be due to several reasons and in many cases the formulation is probably not the cause. However, on occasion one may be tempted to agree with Rocca and Park in their review of prospects and challenges in the oral drug delivery: "breakthrough technologies are required to generate novel dosage forms raising drug delivery to higher level" (p. 52) [149].

ACKNOWLEDGMENT

Sven Engström, Chalmers University of Technology, is acknowledged for his useful suggestions and comments.

REFERENCES

1. Malinowski, H. J., and Marroum, P. J. (1999), *Encyclopedia of Controlled Drug Delivery*, vols. 1 and 2, *Food and Drug Administration Requirements for Controlled Release Products*, John Wiley & Sons, New York, vol. 1, pp. 381–395.
2. Aulton, M. E. (2002), *Pharmaceutics: The Science of Dosage Form Design*, Churchill Livingstone, Hartcourt Publisher, Edinburgh.
3. Marroum, P. J. (1997), Bioavialability/Bioequivalence for Oral Release Producs, Controlled Release Drug Delivery Systems. Paper presented at the 5th Int. Symp. Drug Del., East Brunswick, NJ, May 15–17.
4. SCHOLAR, SciFinder (2006), American Chemical Society, Washington, DC.
5. Yamanaka, K., Yamamoto, H., Kawamura, Y., and Ito, S. (1954), Pyrimidine pencillin, U.S. Patent 2,681,339.
6. Robinson, M. J., and Svedres, E. V. (1957), Sustained-release pharmaceutical preparations, U.S. Patent 2805977.
7. Sjögren, J., and Fryklöf, L-E. (1960), Duretter—A new type of oral sustained action preparation, *Farmacevtisk Revy*, 59, 171–179.
8. Sjögren, J. (1971), Studies on a sustained release principle based on inert plastic matrix, *Acta Pharm. Suecica*, 8, 153–168.
9. Speers, M., and Bonnano, C. (1999), Economic aspects of controlled drug delivery, in (Mathiowitz, E., Ed.), *Encyclopedia of Controlled Drug Delivery*, Wiley, New York, pp. 341–347.
10. Cole, G., Hogan, J., and Aulton, M., Eds. (1995), *Pharmaceutical Coating Technology*, Taylor & Francis, London, 1995.
11. Ragnarsson, G., and Johansson, M. O. (1988), Coated drug cores in multiple unit preparations. Influence of particle size, *Drug Dev. Ind. Pharm.*, 14, 2285–2297.
12. Chatlapalli, R., and Rohera, B. D. (1998), Physical characterization of HPMC and HEC and investigation of their use as pelletization aids. *Int. J. Pharm.*, 161, 179–193.
13. Kojima, M., and Nakagami, H. (2002), Development of controlled release matrix pellets by annealing with micronized water-insoluble or enteric polymers, *J. Controlled Release*, 82, 335–343.
14. Young, C. R., Dietzsch, C., Cerea, M., Farrell, T., Fegely, K. A., Rajabi-Siahboomi, A., and McGinity, J. W. (2005), Physicochemical characterization and mechanisms of release of theophylline from melt-extruded dosage forms based on a methacrylic acid copolymer, *Int. J. Pharm.*, 301, 112–120.

15. Neau, S. H., Chow, M. Y., and Durrani, M. J. (1996), Fabrication and characterization of extruded and spheronized beads containing Carbopol 974P, NF resin, *Int. J. Pharm.*, 131, 47–55.

16. Bommareddy, G. S., Paker-Leggs, S., Saripella, K. K., and Neau, S. H. (2006), Extruded and spheronized beads containing Carbopol 974P to deliver nonelectrolytes and salts of weakly basic drugs, *Int. J. Pharm.*, 321, 62–71.

17. Abrahamsson, B., Alpsten, M., Jonsson, U. E., Lundberg, P. J., Sandberg, A., Sundgren, M., Svenheden, A., and Toelli, J. (1996), Gastro-intestinal transit of a multiple-unit formulation (metoprolol CR/ZOK) and a non-disintegrating tablet with the emphasis on colon, *Int. J. Pharm.*, 140, 229–235.

18. Thrombre, A. G. (2005), Assessment of feasibility of oral controlled release in an exploratory development setting, *Drug Discov. Today*, 10, 1159–1166.

19. Li, S., He, H., Parthiban, L. J., Yin, H., and Serajuddin, A. T. M. (2005), IV-IVC considerations in the development of immediate-release oral dosage form, *J. Pharm. Sci.*, 94, 1396–1417.

20. Noyes, A. A., and Whitney, W. R. (1897), The rate of solution of solid substances in their own solutions, *J. Am. Chem. Soc.*, 19, 930–934.

21. Dokoumetzidis, A., and Macheras, P. (2006), A century of dissolution research: From Noyes and Whitney to the Biopharmaceutics Classification System, *Int. J. Pharm.*, 321, 1–11.

22. Mooney, K. G., Mintun, M. A., Himmelstein, K. J., and Stella, V. J. (1981), Dissolution kinetics of carboxylic acids. I: Effect of pH under unbuffered conditions, *J. Pharm. Sci.*, 70, 13–22.

23. Mooney, K. G., Mintun, M. A., Himmelstein, K. J., and Stella, V. J. (1981), Dissolution kinetics of carboxylic acids. II: Effects of buffers, *J. Pharm. Sci.*, 70, 22–32.

24. Li, C. L., Martini, L. G., Ford, J. L., and Roberts, M. (2005), The use of hypromellose in oral drug delivery, *J. Pharm. Pharmacol.*, 57, 533–546.

25. Tatavarti, A. S., and Hoag, S. W. (2006), Microenvironmental pH modulation based release enhancement of a weakly basic drug from hydrophilic matrices, *J. Pharm. Sci.*, 95, 1459–1468.

26. Riis, T., Bauer-Brandl, A., Wagner, T., and Krantz, H. (2007), pH-independent drug release of an extremely poorly soluble weakly basic acidic drug from multiparticulate extended release formulations, *Eur. J. Pharm. Biopharm.*, 65, 78–84.

27. Streubel, A., Siepmann, J., Dashevsky, A., and Bodmeier, R. (2000), pH-independent release of a weakly basic drug from water-insoluble and -soluble matrix tablets, *J. Controlled Release*, 67, 101–110.

28. U.S. Food and Drug Administration (FDA) (2006), available: http://www.fda.gov.search.html.

29. Higuchi, T. (1963), Mechanism of sustained-action medication. Theoretical analysis of rate of release of solid drugs dispersed in solid matrices, *J. Pharm. Sci.*, 52, 1145–1149.

30. Wu, X. S. (1996), *Controlled Drug Delivery Systems*, Technology Publishing, Lancaster, PA.

31. Jannin, V., Pochard, E., and Chambin, O. (2006), Influence of poloxamers on the dissolution performance and stability of controlled-release formulations containing Precirol ATO 5, *Int. J. Pharm.*, 309, 6–15.

32. Sinchaipanid, N., Junyaprasert, V., and Mitrevej, A. (2004), Application of hot-melt coating for controlled release of propranolol hydrochloride pellets, *Powder Technol.*, 141, 203–209.

33. Rodriguez, L., Caputo, O., Cini, M., Cavallari, C., and Greechi, R. (1993), In vitro release of theophylline from directly-compressed matrixes containing methacrylic acid copolymers and/or dicalcium phosphate dihydrate, *Farmaco*, 48, 1597–1604.

34. Ceballos, A., Cirri, M., Maestrelli, F., Corti, G., and Mura, P. (2005), Influence of formulation and process variables on in vitro release of theophylline from directly-compressed Eudragit matrix tablets, *Farmaco*, 60, 913–918.

35. Oezyazici, M., Goekce, E. H., and Ertan, G. (2006), Release and diffusional modeling of metronidazole lipid matrices, *Eur. J. Pharm. Biopharm.*, 63, 331–339.

36. Hamdani, J., Moes, A. J., and Amighi, K. (2003), Physical and thermal characterization of Precirol and Compritol as lipophilic glycerides used for the preparation of controlled-release matrix pellets, *Int. J. Pharm.*, 260, 47–57.

37. U.S. Food and Drug Administration (FDA) (2006), available: http://www.accessdata.fda.gov/scripts/cder/iig/index.cfm.

38. Mehnert, W., and Mader, K. (2001), Solid lipid nanoparticles. Production, characterization and applications, *Adv. Drug Deliv. Rev.*, 47, 165–196.

39. Savolainen, M., Herder, J., Khoo, C., Lovqvist, K., Dahlqvist, C., Glad, H., and Juppo, A. M. (2003), Evaluation of polar lipid-hydrophilic polymer microparticles, *Int. J. Pharm.*, 262, 47–62.

40. Zhu, Y., Shah, N. H., Malick, A. W., Infeld, M. H., and McGinity, J. W. (2006), Controlled release of a poorly water-soluble drug from hot-melt extrudates containing acrylic polymers, *Drug Dev. Ind. Pharm.*, 32, 569–583.

41. Young, C. R., Dietzsch, C., and McGinity, J. W. (2005), Compression of controlled-release pellets produced by a hot-melt extrusion and spheronization process, *Pharm. Dev. Technol.*, 10, 133–139.

42. Young, C. R., Koleng, J. J., and McGinity, J. W. (2002), Production of spherical pellets by a hot-melt extrusion and spheronization process, *Int. J. Pharm.*, 242, 87–92.

43. Hogan, J. E. (1995), Modified release coatings for pharmaceutics, in Cole, G., Hogan, J., and Aulton, M., Eds., *Pharmaceutical Coating Technology*, pp. 409–438.

44. Edgar, K. J., Buchanan, C. M., Debenham, J. S., Rundquist, P. A., Seiler, B. D., Shelton, M. C., and Tindall, D. (2001), Advances in cellulose ester performance and application, *Prog. Polym. Sci.*, 26, 1605–1688.

45. Hjärtstam, J. (1998), *Ethyl Cellulose Membranes Used in Modified Release Formulations*, Chalmers University of Technology, Göteborg, Sweden.

46. Savastano, L., Leuenberger, H., and Merkle, H. P. (1995), Membrane modulated dissolution of oral drug delivery systems, *Pharm. Acta Helv.*, 70, 117–124.

47. Theeuwes, F. (1975), Elementary osmotic pumps, *J. Pharm. Sci.*, 64, 147–157.

48. Ragnarsson, G., Sandberg, A., Johansson, M. O., Lindstedt, B., and Sjoegren, J. (1992), In vitro release characteristics of a membrane-coated pellet formulation. Influence of drug solubility and particle size, *Int. J. Ph+arm.*, 79, 223–232.

49. Ragnarsson, G., Sandberg, A., Jonsson, U. E., and Sjoegren, J. (1987), Development of a new controlled release metoprolol product, *Drug Dev. Ind. Pharm.*, 13, 1495–1509.

50. Kedem, O., and Katchalsky, A. (1961), *J. Gen. Phys*, 45, 143.

51. Kedem, O., and Katchalsky, A. (1958), *Biochem. Biophys. Acta*, 27, 229.

52. Verma, R. K., Krishna, D. M., and Garg, S. (2002), Formulation aspects in the development of osmotically controlled oral drug delivery systems, *J. Controlled Release*, 79, 7–27.

53. Elchidana, P. A., and Deshpande, S. G. (1999), Microporous membrane drug delivery system for indomethacin, *J. Controlled Release*, 59, 279–285.

54. Sousa, J. J., Sousa, A., Moura, M. J., Podczeck, F., and Newton, J. M. (2002), The influence of core materials and film coating on the drug release from coated pellets, *Int. J. Pharm.*, 233, 111–122.

55. Husson, I., Leclerc, B., Spenlehauer, G., Veillard, M., and Couarraze, G. (1991), Modeling of drug release from pellets coated with an insoluble polymeric membrane, *J. Controlled Release*, 17, 163–173.

56. Tang, E. S. K., Chan, L. W., and Heng, P. W. S. (2005), Coating of multiparticulates for sustained release, *Am. J. Drug Deliv.*, 3, 17–28.

57. Verma, R. K., Kaushal, A. M., and Garg, S. (2003), Development and evaluation of extended release formulations of isosorbide mononitrate based on osmotic technology, *Int. J. Pharm.*, 263, 9–24.

58. Ravishankar, H., Patil, P., Petereit, H-U., and Renner, G. (2005), EUDRAMODE: A novel approach to sustained oral drug delivery systems, *Drug Deliv. Technol.*, 5, 48–50, 52–55.

59. Ravishankar, H., Iyer-Chavan, J., Patil, P., Samel, A., and Renner, G. (2006), Clinical studies of terbutaline controlled-release formulation prepared using EUDRAMODE, *Drug Deliv. Technol.*, 6, 50–56.

60. Kibbe, A. H. (2000), *Handbook of Pharmaceutical Excipients*, Pharmaceutical, London.

61. Aulton, M. E., and Twitchell, A. M. (1995), Solution properties and atomization in film coating [of pharmaceuticals], *Pharm. Coat. Technol.*, 64–117.

62. Glatt (2006), available: www.glatt.com.

63. Harris, M. R., and Ghebre-Sellassie, I. (1989), Aqueous polymeric coating for modified-release pellets, *Drugs Pharm. Sci.*, 36, 63–79.

64. Colombo, P., Bettini, R., Santi, P., and Peppas, N. A. (2000), Swellable matrixes for controlled drug delivery: Gel-layer behavior, mechanisms and optimal performance, *Pharm. Sci. Technol. Today*, 3, 198–204.

65. Lapidus, H., and Lordi, N. G. (1966), Some factors affecting the release of a water-soluble drug from a compressed hydrophilic matrix, *J. Pharm. Sci.*, 55, 840–843.

66. Lapidus, H., and Lordi, N. G. (1968), Drug release from compressed hydrophilic matrixes, *J. Pharm. Sci.*, 57, 1292–1301.

67. Huber, H. E., Dale, L. B., and Christenson, G. L. (1966), Utilization of hydrophilic gums for the control of drug release from tablet formulations. I. Disintegration and dissolution behavior, *J. Pharm. Sci.*, 55, 974–976.

68. Koerner, A., Larsson, A., Piculell, L., and Wittgren, B. (2005), Tuning the polymer release from hydrophilic matrix tablets by mixing short and long matrix polymers, *J. Pharm. Sci.*, 94, 759–769.

69. Ju, R. T. C., Nixon, P. R., and Patel, M. V. (1995), Drug release from hydrophilic matrixes 1. New scaling laws for predicting polymer and drug release based on the polymer disentanglement concentration and the diffusion layer, *J. Pharm. Sci.*, 84, 1455–1463.

70. Ju, R. T. C., Nixon, P. R., Patel, M. V., and Tong, D. M. (1995), Drug release from hydrophilic matrixes. 2. A mathematical model based on the polymer disentanglement concentration and the diffusion layer, *J. Pharm. Sci.*, 84, 1464–1477.

71. Bettini, R., Colombo, P., Massimo, G., Catellani, P. L., and Vitali, T. (1994), Swelling and drug release in hydrogel matrixes: Polymer viscosity and matrix porosity effects, *Eur. J. Pharm. Sci.*, 2, 213–219.

72. Adler, J., Jayan, A., and Melia, C. D. (1999), A method for quantifying differential expansion within hydrating hydrophilic matrixes by tracking embedded fluorescent microspheres, *J. Pharm. Sci.*, 88, 371–377.

73. Treloar, L. R. G. (1975), *The Physics of Rubber Elasticity*, Clarendon, Oxford.

74. Siepmann, J., Kranz, H., Bodmeier, R., and Peppas, N. A. (1999), HPMC-matrices for controlled drug delivery: A new model combining diffusion, swelling, and dissolution mechanisms and predicting the release kinetics, *Pharm. Res.*, 16, 1748–1756.

75. Colombo, P., Bettini, R., Massimo, G., Catellani, P. L., Santi, P., and Peppas, N. A. (1995), Drug diffusion front movement is important in drug release control from swellable matrix tablets, *J. Pharm. Sci.*, 84, 991–997.

76. Bettini, R., Catellani, P. L., Santi, P., Massimo, G., Peppas, N. A., and Colombo, P. (2001), Translocation of drug particles in HPMC matrix gel layer: Effect of drug solubility and influence on release rate, *J. Controlled Release*, 70, 383–391.

77. Fu, X. C., Liang, W. Q., and Ma, X. W. (2003), Relationships between the release of soluble drugs from HPMC matrices and the physicochemical properties of drugs, *Pharmazie*, 58, 221–222.

78. Tahara, K., Yamamoto, K., and Nishihata, T. (1996), Application of model-independent and model analysis for the investigation of effect of drug solubility on its release rate from hydroxypropyl methyl cellulose sustained-release tablets, *Int. J. Pharm.*, 133, 17–27.

79. Korsmeyer, R. W., Gurny, R., Doelker, E., Buri, P., and Peppas, N. A. (1983), Mechanisms of solute release from porous hydrophilic polymers, *Int. J. Pharm.*, 15, 25–35.

80. Rinaki, E., Valsami, G., and Macheras, P. (2003), The power law can describe the "entire" drug release curve from HPMC-based matrix tablets: A hypothesis, *Int. J. Pharm.*, 255, 199–207.

81. Ritger, P. L., and Peppas, N. A. (1987), A simple equation for description of solute release. II. Fickian and anomalous release from swellable devices, *J. Controlled Release*, 5, 37–42.

82. Ritger, P. L., and Peppas, N. A. (1987), A simple equation for description of solute release. I. Fickian and non-Fickian release from non-swellable devices in the form of slabs, spheres, cylinders or disks, *J. Controlled Release*, 5, 23–36.

83. Siepmann, J., and Peppas, N. A. (2001), Modeling of drug release from delivery systems based on hydroxypropyl methylcellulose (HPMC), *Adv. Drug Deliv. Rev.*, 48, 139–157.

84. Peppas, N. A., and Sahlin, J. J. (1989), A simple equation for the description of solute release. III. Coupling of diffusion and relaxation, *Int. J. Pharm.*, 57, 169–72.

85. Gao, P., and Meury, R. H. (1996), Swelling of hydroxypropyl methylcellulose matrix tablets. 1. Characterization of swelling using a novel optical imaging method, *J. Pharm. Sci.*, 85, 725–731.

86. Mitchell, K., Ford, J. L., Armstrong, D. J., Elliott, P. N. C., Hogan, J. E., and Rostron, C. (1993), The influence of substitution type on the performance of methyl cellulose and hydroxypropyl methyl cellulose in gels and matrixes, *Int. J. Pharm.*, 100, 143–154.

87. Levina, M., and Rajabi-Siahboomi, A. R. (2004), The influence of excipients on drug release from hydroxypropyl methylcellulose matrices, *J. Pharm. Sci.*, 93, 2746–2754.

88. Ford, J. L., Rubinstein, M. H., and Hogan, J. E. (1985), Propranolol hydrochloride and aminophylline release from matrix tablets containing hydroxypropyl methyl cellulose, *Int. J. Pharm.*, 24, 339–350.

89. Ford, J. L., Rubinstein, M. H., and Hogan, J. E. (1985), Formulation of sustained-release promethazine hydrochloride tablets using hydroxypropyl methyl cellulose matrixes, *Int. J. Pharm.*, 24, 327–338.

90. Reynolds, T. D., Mitchell, S. A., and Balwinski, K. M. (2002), Investigation of the effect of tablet surface area/volume on drug release from hydroxypropyl methylcellulose controlled-release matrix tablets, *Drug Dev. Ind. Pharm.*, 28, 457–466.

91. Skoug, J. W., Borin, M. T., Fleishaker, J. C., and Cooper, A. M. (1991), In vitro and in vivo evaluation of whole and half tablets of sustained-release adinazolam mesylate, *Pharm. Res.*, 8, 1482–1488.

92. Sung, K. C., Nixon, P. R., Skoug, J. W., Ju, T. R., Gao, P., Topp, E. M., and Patel, M. V. (1996), Effect of formulation variables on drug and polymer release from HPMC-based matrix tablets, *Int. J. Pharm.*, 142, 53–60.

93. Ford, J. L., Rubinstein, M. H., McCaul, F., Hogan, J. E., and Edgar, P. J. (1987), Importance of drug type, tablet shape and added diluents on drug release kinetics from hydroxypropyl methyl cellulose matrix tablets, *Int. J. Pharm.*, 40, 223–234.

94. Heng, P. W., Chan, L. W., Easterbrook, M. G., and Li, X. (2001), Investigation of the influence of mean HPMC particle size and number of polymer particles on the release of aspirin from swellable hydrophilic matrix tablets, *J. Controlled Release*, 76, 39–49.

95. Velasco, M. V., and Ford, J. L., Rowe, P., and Rajabi-Siahhoomi, A. R. (1999), Influence of drug: Hydroxypropyl methyl cellulose ratio, drug and polymer particle size and compression force on the release of diclofenac sodium from HPMC tablets, *J. Controlled Release*, 57, 75–85.

96. Dabbagh, M. A., Ford, J. L., Rubinstein, M. H., and Hogan, J. E. (1996), Effects of polymer particle size, compaction pressure and hydrophilic polymers on drug release from matrixes containing ethyl cellulose, *Int. J. Pharm.*, 140, 85–95.

97. Mitchell, K., Ford, J. L., Armstrong, D. J., Elliott, P. N. C., Hogan, J. E., and Rostron, C. (1993), The influence of the particle size of hydroxypropyl methyl cellulose K15M on its hydration and performance in matrix tablets, *Int. J. Pharm.*, 100, 175–179.

98. Sheskey, P. J., Cabelka, T. D., Robb, R. T., and Boyce, B. M. (1994), Use of roller compaction in the preparation of controlled-release hydrophilic matrix tablets containing methyl cellulose and hydroxypropyl methyl cellulose polymers, *Pharm. Technol.*, 18, 132, 134, 136, 138, 140, 142, 144, 146, 148–150.

99. Sheskey, P. J., and Hendren, J. (1999), The effects of roll compaction equipment variables, granulation technique, and HPMC polymer level on a controlled-release matrix model drug formulation, *Pharm. Technol.*, 23, 90, 92, 94, 96, 98, 100, 102, 104, 106.

100. Sheskey, P., Pacholke, K., Sackett, G., Maher, L., and Polli, J. (2000), Effect of process scale-up on robustness of tablets, tablet stability, and predicted in vivo performance, *Pharm. Technol.*, 24, 30, 32, 34, 36, 38, 40, 42, 44, 46, 48, 50, 52.

101. Timmins, P., Delargy, A. M., Howard, J. R., and Rowlands, E. A. (1991), Evaluation of the granulation of a hydrophilic matrix sustained-release tablet, *Drug Dev. Ind. Pharm.*, 17, 531–550.

102. Liu, C. H., Chen, S. C., Kao, Y. H., Kao, C. C., Sokoloski, T. D., and Sheu, M. T. (1993), Properties of hydroxypropyl methyl cellulose granules produced by water spraying, *Int. J. Pharm.*, 100, 241–248.

103. Herder, J., Adolfsson, A., and Larsson, A. (2006), Initial studies of water granulation of eight grades of hypromellose (HPMC), *Int. J. Pharm.*, 313, 57–65.

104. Dahl, T. C., and Bormeth, A. P. (1990), Naproxen controlled release matrix tablets: Fluid bed granulation feasibility, *Drug Dev. Ind. Pharm.*, 16, 581–590.

105. Nellore, R. V., Rekhi, G. S., Hussain, A. S., Tillman, L. G., and Augsburger, L. L. (1998), Development of metoprolol tartrate extended-release matrix tablet formulations for regulatory policy consideration, *J. Controlled Release*, 50, 247–256.
106. Liu, C. H., Kao, Y. H., Chen, S. C., Sokoloski, T. D., and Sheu, M. T. (1995), In-vitro and in-vivo studies of the diclofenac sodium controlled-release matrix tablets, *J. Pharm. Pharmacol.*, 47, 360–364.
107. Dahl, T. C., Calderwood, T., Bormeth, A., Trimble, K., and Piepmeier, E. (1990), Influence of physicochemical properties of hydroxypropyl methyl cellulose on naproxen release from sustained release matrix tablets, *J. Controlled Release*, 14, 1–10.
108. Shah, N. H., Railkar, A. S., Phuapradit, W., Zeng, F. W., Chen, A., Infeld, M. H., and Malick, A. W. (1996), Effect of processing techniques in controlling the release rate and mechanical strength of hydroxypropyl methyl cellulose based hydrogel matrixes, *Eur. J. Pharm. Biopharm.*, 42, 183–187.
109. Sheskey, P. J., and Keary, C. M. (2002), *Aqueous air foams*, U.S. Patent 7011702.
110. Keary, C. M., and Sheskey, P. J. (2004), Preliminary report of the discovery of a new pharmaceutical granulation process using foamed aqueous binders, *Drug Dev. Ind. Pharm.*, 30, 831–845.
111. Cao, Q-R., Choi, Y-W., Cui, J-H., and Lee, B-J. (2005), Effect of solvents on physical properties and release characteristics of monolithic hydroxypropylmethylcellulose matrix granules and tablets, *Arch. Pharm. Res.*, 28, 493–501.
112. Sheskey, P. J., and Williams, D. M. (1996), Comparison of low-shear and high-shear wet granulation techniques and the influence of percent water addition in the preparation of a controlled-release matrix tablet containing HPMC and a high-dose, highly water-soluble drug, *Pharm. Technol.*, 20, 80, 82, 84, 86, 88, 90, 92.
113. Zhang, F., and McGinity, J. W. (1999), Properties of sustained-release tablets prepared by hot-melt extrusion, *Pharm. Dev. Technol.*, 4, 241–250.
114. Urbano, A. P. A., Ribeiro, A. J., and Veiga, F. (2006), Design of pectin beads for oral protein delivery, *Chem. Ind. Chem. Eng. Q.*, 12, 24–30.
115. Fukuda, M., Peppas, N. A., and McGinity, J. W. (2006), Properties of sustained release hot-melt extruded tablets containing chitosan and xanthan gum, *Int. J. Pharm.*, 310, 90–100.
116. De Brabander, C., Vervaet, C., and Remon, J. P. (2003), Development and evaluation of sustained release mini-matrices prepared via hot melt extrusion, *J. Controlled Release*, 89, 235–247.
117. Baumgartner, S., Slamersek, V., and Kristl, J. (2003), Controlled drug delivery of hydrophilic drugs from cellulose ether matrix tablets: The influence of the drug molecule size on its release mechanism and kinetics, *Farm. Vestnik (Ljubljana, Slov.)*, 54, 359–360.
118. Genc, L., Bilac, H., and Guler, E. (1999), Studies on controlled release dimenhydrinate from matrix tablet formulations, *Pharm. Acta Helv.*, 74, 43–49.
119. Sinha Roy, D., and Rohera Bhagwan, D. (2002), Comparative evaluation of rate of hydration and matrix erosion of HEC and HPC and study of drug release from their matrices, *Eur. J. Pharm. Sci.*, 16, 193–199.
120. Vueba, M. L., Batista de Carvalho, L. A. E., Veiga, F., Sousa, J. J., and Pina, M. E. (2006), Influence of cellulose ether mixtures on ibuprofen release: MC25, HPC and HPMC K100M, *Pharm. Dev. Technol.*, 11, 213–228.
121. Vueba, M. L., Batista De Carvalho, L. A. E., Veiga, F., Sousa, J. J., and Pina, M. E. (2004), Influence of cellulose ether polymers on ketoprofen release from hydrophilic matrix tablets, *Eur. J. Pharm. Biopharm.*, 58, 51–59.

122. Talukdar, M. M., Michoel, A., Rombaut, P., and Kinget, R. (1996), Comparative study on xanthan gum and hydroxypropyl methyl cellulose as matrixes for controlled-release drug delivery I. Compaction and in vitro drug release behavior, *Int. J. Pharm.*, 129, 233–241.

123. Talukdar, M. M., and Kinget, R. (1997), Comparative study on xanthan gum and hydroxypropyl methyl cellulose as matrixes for controlled-release drug delivery. II. Drug diffusion in hydrated matrixes, *Int. J. Pharm.*, 151, 99–107.

124. Andreopoulos, A. G., and Tarantili, P. A. (2002), Study of biopolymers as carriers for controlled release, *J. Macromol. Sci. Phys.*, B41, 559–578.

125. Andreopoulos, A. G., and Tarantili, P. A. (2001), Xanthan gum as a carrier for controlled release of drugs, *J. Biomater. Appl.*, 16, 34–46.

126. Talukdar, M. M., Rombaut, P., and Kinget, R. (1998), The release mechanism of an oral controlled-release delivery system for indomethacin, *Pharm. Dev. Technol.*, 3, 1–6.

127. Talukdar, M. M., and Kinget, R. (1995), Swelling and drug release behavior of xanthan gum matrix tablets, *Int. J. Pharm.*, 120, 63–72.

128. Maggi, L., Segale, L., Torre, M. L., Ochoa Machiste, E., and Conte, U. (2002), Dissolution behaviour of hydrophilic matrix tablets containing two different polyethylene oxides (PEOs) for the controlled release of a water-soluble drug. Dimensionality study, *Biomaterials*, 23, 1113–1119.

129. Choulis, N., and Papadopoulos, H. (1975), Timed-release tablets containing quinine sulfate, *J. Pharm. Sci.*, 64, 1033–1035.

130. Huang, L-L., and Schwartz, J. B. (1995), Studies on drug release from a carbomer tablet matrix, *Drug Dev. Ind. Pharm.*, 21, 1487–1501.

131. El-Sayed, G. M., El-Said, Y., Meshali, M. M., and Schwartz, J. B. (1996), Kinetics of theophylline release from different tablet matrixes, *STP Pharma Sci.*, 6, 390–397.

132. Patel, G. N., Patel, G. C., Patel, R. B., Patel, S. S., Patel, J. K., Bharadia, P. D., and Patel, M. M. (2006), Oral colon-specific drug delivery: An overview, *Drug Deliv. Technol.*, 6, 62–71.

133. Liu, L. S., Fishman, M. L., Kost, J., and Hicks, K. B. (2003), Pectin-based systems for colon-specific drug delivery via oral route, *Biomaterials*, 24, 3333–3343.

134. Liew, C. V., Chan, L. W., Ching, A. L., and Heng, P. W. S. (2006), Evaluation of sodium alginate as drug release modifier in matrix tablets, *Int. J. Pharm.*, 309, 25–37.

135. Shilpa, A., Agrawal, S. S., and Ray, A. R. (2003), Controlled delivery of drugs from alginate matrix, *J. Macromol. Sci. Polym. Rev.*, C43, 187–221.

136. Ostberg, T., and Graffner, C. (1994), Calcium alginate matrixes for oral multiple unit administration: III. Influence of calcium concentration, amount of drug added and alginate characteristics on drug release, *Int. J. Pharm.*, 111, 271–282.

137. Varma, M. V. S., Kaushal, A. M., Garg, A., and Garg, S. (2004), Factors affecting mechanism and kinetics of drug release from matrix-based oral controlled drug delivery systems, *Am. J. Drug Deliv.*, 2, 43–57.

138. McCall, T. W., Baichwal, A. R., and Staniforth, J. N. (2003), TIMERx oral controlled-release drug delivery system, *Drugs Pharm. Sci.*, 126, 11–19.

139. Friend, D. R. (2005), New oral delivery systems for treatment of inflammatory bowel disease, *Adv. Drug Deliv. Rev.*, 57, 247–265.

140. Ghandehari, H., Kopeckova, P., and Kopecek, J. (1997), In vitro degradation of pH-sensitive hydrogels containing aromatic azo bonds, *Biomaterials*, 18, 861–872.

141. Akala, E. O., Kopeckova, P., and Kopecek, J. (1998), Novel pH-sensitive hydrogels with adjustable swelling kinetics, *Biomaterials*, 19, 1037–1047.

142. Takka, S., Rajbhandari, S., and Sakr, A. (2001), Effect of anionic polymers on the release of propranolol hydrochloride from matrix tablets, *Eur. J. Pharm. Biopharm.*, 52, 75–82.

143. Nie, S., Pan, W., Li, X., and Wu, X. (2004), The effect of citric acid added to hydroxypropyl methylcellulose (HPMC) matrix tablets on the release profile of vinpocetine, *Drug Dev. Ind. Pharm.*, 30, 627–635.

144. Varma, M. V. S., Kaushal, A. M., and Garg, S. (2005), Influence of micro-environmental pH on the gel layer behavior and release of a basic drug from various hydrophilic matrices, *J. Controlled Release*, 103, 499–510.

145. Qiu, Y., and Park, K. (2001), Environment-sensitive hydrogels for drug delivery, *Adv. Drug Deliv. Rev.*, 53, 321–339.

146. Bromberg, L. (2005), Intelligent hydrogels for the oral delivery of chemotherapeutics, *Exp. Opin. Drug Deliv.*, 2, 1003–1013.

147. Rowe, C. W., Wang, C-C., and Monkhouse, D. C. (2003), TheriForm technology, *Drugs Pharm. Sci.*, 126, 77–87.

148. Lee, K-J., Kang, A., Delfino, J. J., West, T. G., Chetty, D., Monkhouse, D. C., and Yoo, J. (2003), Evaluation of critical formulation factors in the development of a rapidly dispersing captopril oral dosage form, *Drug Dev. Ind. Pharm.*, 29, 967–979.

149. Rocca, J. G., and Park, K. (2004), Oral drug delivery: Prospects & challenges, *Drug Deliv. Technol.*, 4, 52–54, 57.

SECTION 7

ROLE OF NANOTECHNOLOGY

7.1

CYCLODEXTRIN-BASED NANOMATERIALS IN PHARMACEUTICAL FIELD

Erem Bilensoy and A. Atilla Hincal

Hacettepe University Faculty of Pharmacy, Ankara, Turkey

Contents

7.1.1 INTRODUCTION

Cyclodextrins (CDs) have a wide range of application in the pharmaceutical field due to their unique structure, which allows them to include hydrophobic molecules in their apolar cavity and to mask the physicochemical properties of the included molecule. This results in the enhancement of drug bioavailability by improving aqueous solubility and the physical and chemical stability of the drug, masking undesired side effects such as irritation, taste, or odor, and overcoming compatibility problems or interactions between drugs and excipients.

Parallel to the increasing interest and successful licensing and commercialization of nanoparticulate pharmaceutical products, CDs have also been incorporated into nanoparticulate drug delivery systems for several purposes. This can be

Pharmaceutical Manufacturing Handbook: Production and Processes, edited by Shayne Cox Gad

achieved by two approaches: (1) complexation of active ingredient with an appropriate CD and entrapment into polymeric nanoparticles to solve problems arising from the drug's physicochemical properties or (2) modification of CDs to render an amphiphilic character to these molecules, which allows CDs to self-align into nanoparticles in the form of nanospheres, nanocapsules, solid lipid nanoparticles, nanosize liposomes, and nanosize vesicles with or without the presence of surface-active agents.

In light of current research, this chapter will deal with the following issues concerning the use of CDs and derivatives as nanomaterials for drug delivery: use of CDs (natural and synthetic) derivatives in the pharmaceutical field and application of CDs in nanoparticulate drug delivery systems. A major part of the chapter will be focused on new CD derivatives, amphiphilic CDs, and the characterization, efficacy, and safety of nanoparticles prepared from amphiphilic CDs.

7.1.1.1 Cyclodextrins in Pharmaceutical Field

Natural Cyclodextrins Cyclodextrins are cyclic oligosaccharides obtained by the enzymatic degradation of starch. Major natural CDs are crystalline, homogeneous, nonhygroscopic substances which have a toruslike macroring shape built up from glucopyranose units, as seen in Figure 1 [1–3]. Cyclodextrins are named depending on the number of glucopyranose units. Major industrially produced CDs are named as follows; α-CD, possessing six units, β-CD, possessing seven units; and γ-CD, possessing eight units.

Natural CDs have been demonstrated to have a special structure; that is, glucose residues in the CD ring possess the thermodynamically favored $^4C_1$ chair conformation because all substituent groups are in equatorial position. Cyclodextrins behave like rigid compounds with two degrees of freedom: rotation at the glucosidic links C(4)–O(4) and C(1)–O(4) and rotations at the O(6) primary hydroxyl groups at the C(5)–C(6) band. As a consequence of this chair conformation, all secondary hydroxyl groups at C(2) and C(3) are located at the broader side of the CD torus in the equatorial position. Hydroxyl groups on C(2) point toward the cavity and hydroxyl groups on C(3) point outward. The primary hydroxyl groups at the C6 position are located at the narrower side of the torus. These hydroxyl groups ensure good water solubility for the natural CDs. The cavity of the torus is lined with a ring of C–H groups (C3), a ring of glucosidic oxygen atoms, and another ring of C–H groups (C5). Thus, the cavity of CDs exhibits an apolar character. This is accompanied by a high electron density and Lewis base property. The physicochemical characteristics and inclusion behavior of CDs are a direct consequence of these special binding conditions [4, 5]. Certain physicochemical characteristics of natural CDs are given in Table 1.

Cyclodextrin Derivatives Natural CDs were reported to form total or partial inclusion complexes with several drugs to improve the aqueous solubility and stability under physiological or ambient conditions, reduce or mask completely the side effects associated with the included drug, and increase compatibility of the drug with other drugs in the formulation or excipients while improving patient compliance by masking the taste or odor of the active ingredient [6–8].

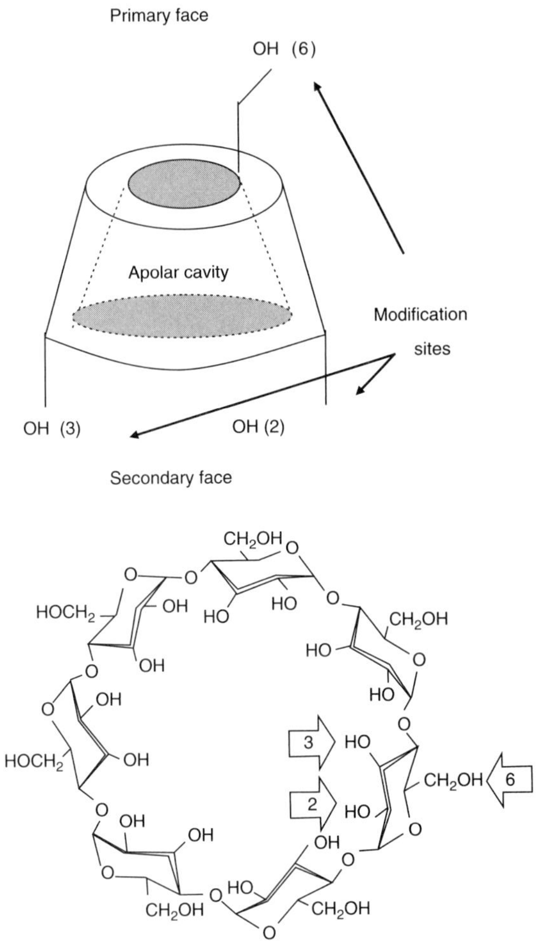

FIGURE 1 Schematic representation of natural CD structure and modification sites.

TABLE 1 Some Physicochemical Characteristics of Natural Cyclodextrins

Characteristics	α-CD	β-CD	γ-CD
Number of glucose units	6	7	8
Molecular weight, g/mol	972	1135	1297
Internal diameter, Å	4.7–5.2	6.0–6.4	7.5–8.3
External diameter, Å	14.2–15.0	15.0–15.8	17.1–17.9
Depth, Å	7.9–8.0	7.9–8.0	7.9–8.0
Solubility in water (25°C), g/L	145	18.5	232
Crystal water, w %	10.2	13.2–14.5	8.13–17.7
Approximative cavity volume in 1 mol CD, $Å^3$	174	262	472
Melting point, °C	250–260	255–265	240–245
Half-life in 1 *M* HCl at 60°C, h	6.2	5.4	3.0
Crystal forms (from water)	Hexagonal plate	Monoclinic parallelogram	Quadratic prism
Partial molar volumes in solution, mL/mol	611.4	703.8	801.2

In contrast to the advantageous nature of CDs for molecular inclusion, their surface makes it more difficult for the highly hydrophilic CD molecule to interact with lipophilic biological membranes. For this reason, natural CDs have been chemically modified to alter their water solubility, interaction with biological membranes, and drug release properties.

Two of the natural CDs are known to be parenterally unsafe due to nephrotoxic effects [9]. The etiology of the nephrotoxicity of α- and β-CD is unknown but is believed to be related to either CD uptake by kidney tubule cells resulting in disruption of intracellular function or the extraction of lipid membrane components by the CDs. The latter is suggested to be of validity since there seems to be a linear correlation between the ability of some CDs to disrupt cellular membranes and kidney nephrotoxicity [2, 6]. The ability of CDs to cause red blood cell hemolysis and membrane irritation seems also to correlate with their ability to extract lipid membrane components: cholesterol and phospholipids [10, 11].

Modification of natural CDs has been the aim of many research groups to improve safety while maintaining the ability to form inclusion complexes with various substrates. Some groups have also focused on improving the interaction between the pharmaceuticals and the CDs while others have attempted to prepare materials that can be chemically defined more precisely.

Methylated CDs are obtained by methylation of CDs on either all C2 secondary and C6 primary hydroxyl groups [dimethyl cyclodextrins (DIMEB)] or all the hydroxyl groups C2, C3, and C6 [trimethyl cyclodextrins (TRIMEB)]. Major disadvantage of methylated CDs is that their solubility decreases with increasing temperature and they are reported to be hemolytic [12]. This is a result of partial methylation of the hydroxyls of β-CD which leads to stronger drug binding but also to stronger hemolysis [12].

Hydroxypropylated CDs are statistically substituted derivatives because hydroxypropylation does not result in selective substitution as with methylation. While the reaction proceeds, the reactivity of the hydroxyl group changes, and this results in a mixture of products with various degrees of substitution. Their dissolution is endothermic so there is no decrease in solubility with increasing temperature [6, 13]. It is necessary to note that degree of substitution in hydroxypropylated CDs is inversely correlated with their inclusion capability [14, 15]. Hydroxyalkylated CDs are commercially available as tablets, ocular collyrs [16], and excipients under the trademarks Encapsin and Molecusol.

Sulfobutylether-β-cyclodextrins (SBE-β-CDs) are water soluble and parenterally safe. An advantage over hydroxypropylated-β-CDs (HP-β-CDs) is that higher sulfobutyl group substitution often results in higher drug complexation ability [17]. Inability of the SBE-β-CDs, especially the commercially available product $(SBE)_{7M}$-β-CD (Captisol), to form strong complexes with cholesterol and other membrane lipids, arising from their polyanionic nature causing Coulombic repulsions, results in a little or no membrane disruption [6, 13] and reduced hemolysis [18]. Captisol is used in parenteral and ocular systems as well as osmotic tablets and also as a freeze-drying excipient [19].

Branched CDs (mono- or di-glucosyl, maltosyl and glucopyranosyl α- and β-CDs) are more resistant to α-amylase than natural CDs. Natural β-CD and monoglucosyl-β-CD are stable in rat blood because they have no linear glycosidic bond. Branched CDs exert a lower hemolytic activity on human erythrocytes and are weaker than natural CDs. Their inclusion capability is more or less drug dependent. Steroids were reported to show a slightly higher affinity to branched CDs than to natural CDs [19].

The water solubility of acylated CDs decreases proportionally to their degree of substitution, whereas their solubility in less polar solvents such as ethanol increases. Acylated CDs of hydrophobic nature are useful for controlling the release rates of water-soluble drugs [20].

Ionizable β-CDs possess interesting solubility properties, too. Solubility in water for these derivatives depends on the pH of the solution. Carboxymethylethyl-β-cyclodextrin (CME-β-CD) is prepared as an enteric-type carrier system. The presence of a carboxymethyl group causes a pH-dependent solubility range in water, meaning that it is only slightly soluble in the low-pH region, but freely soluble in neutral and alkaline regions due to the ionization of the carboxyl group (pK_a about 3–4) [21]. Inclusion-forming capability of CME-β-CD is dependent on drug properties, including size, shape, and charge of the molecule.

Sulfated CDs are of anionic nature and are interesting from chemical and biological points of view because of their angiogenic and antiviral properties [22, 23]. Sulfated derivatives are also reported to have a heparin-like activity, resulting in increase in blood-clotting times, which limits their injectable dose [7].

Low-molecular-weight CD polymers (MW 3000–6000 Da) are soluble in water whereas high-molecular-weight CD polymers (MW > 10,000 Da) can only swell in water and form insoluble gels [24, 25]. Insoluble cross-linked bead polymers seem to be applicable as wound-healing agents for the treatment of wounds like burns or ulcers. Iodine has been complexed with such a CD polymer as an antiseptic wound healing agent [24].

7.1.2 APPLICATIONS OF CYCLODEXTRINS IN NANOPARTICLES

7.1.2.1 Incorporation of Drug–Cyclodextrin Complexes in Nanoparticulate Delivery Systems

Nanoparticles are of pharmaceutical interest due to their active and passive targeting properties and their ability to deliver poorly solube drugs and drugs with stability problems. Nanoparticles are considered more stable than liposomal delivery systems. However, a major drawback is associated with the drug-loading capacity of polymeric nanoparticles. Classical emulsion polymerization procedures result in considerably low drug-loading capacities. This results in the administration of excessive quantities of polymeric material which may impair the safety of the drug delivery system [26, 27].

For this reason, different techniques are used to improve the drug-loading properties of polymeric nanoparticles. Cyclodextrins are used for this reason to improve water solubility and sometimes the hydrolytic or photolytic stability of drugs for better loading properties [8]. Drug–CD complexes act to solubilize or stabilize active ingredients within the nanoparticles, resulting in increased drug concentration in the polymerization medium and increased hydrophobic sites in the nanosphere structure when large amounts of CDs are associated to the nanoparticles [27, 28].

The antiviral agent saguinavir was complexed to HP-β-CD to increase saquinavir loading into polyalkylcyanoacrylate nanoparticles by providing a soluble drug reservoir in the polymerization medium that is the basis of nanoparticle formation [29]. A dynamic equilibrium was observed between the complex, dissociated species of the complex, and the forming polymeric nanoparticle. During nanoparticle formation, free drug was believed to be progressively incorporated into a polymer network, driven by the drug partition coefficient between the polymer and polymerization medium. Simultaneous direct entrapment of some CD–drug complex was also suspected [28–30].

Incorporation of the steroidal drugs hydrocortisone and progesterone in complex with β-CD and HP-β-CD reduced the particle size for solid lipid nanoparticles (SLNs) below 100 nm. Steroids were demonstrated to be dispersed in the amorphous state. Compexation to CDs resulted in higher drug-loading properties for the more hydrophobic drug hydrocortisone and lower in vitro release for both drugs when they are complexed to CDs rather than their free form [31].

The in vivo behavior of nanoparticles obtained from drug–CD complexes was also evaluated. HP-β-CD addition in the polymerization medium of polyethylcyanoacrylate (PECA) nanospheres improved the subcutaneous absorption of metoclopramide in rats. PECA nanospheres with HP-β-CD provided the highest drug concentration and enhanced drug absorption compared with those with dextran or with drug solution. However, in addition to drug absorption from subcutaneous sites, HP-β-CD also enhanced the drug elimination by enhancing the drug absorption to reticuloendothelial tissues [32].

Progesterone complexed to HP-β-CD or DM-β-CD was loaded into bovine serum albumin (BSA) nanospheres. Dissolution rates of progesterone were significantly enhanced by complexation to CDs with respect to free drug. Nanospheres of 100 nm loaded with drug–CD complexes provided a pH-dependent release profile and good stability in an aqueous neutral environment [33].

In another approach, CD properties of complexation were combined with those of chitosan. Complexation with CD was believed to permit solubilization as well as protection for labile drugs while entrapment in the chitosan network was expected to facilitate absorption. Chitosan nanoparticles, including complexes of HP-β-CD with the hydrophobic model drugs triclosan and furosemide, were prepared by ionic cross-linking of chitosan with sodium tripolyphosphate (TPP) in the presence of CDs. Nanoparticles were then prepared by ionotropic gelation using the obtained drug–HP-β-CD inclusion complexes and chitosan. Cyclodextrin and TPP concentration largely affected particle size but the zeta potential remained unchanged with different parameters. On the other hand, drug entrapment increased up to 4 and 10 times by triclosan and furosemide, respectively. The release profile of nanoparticles indicated an initial burst release followed by a delayed release profile lasting up to 4 h [34].

Recently a CD–insulin complex was encapsulated in polymethacrylic acid–chitosan–polyether[polyethylene glycol (PEG)–propylene glycol] copolymer PMCP nanoparticles from the free-radical polymerization of methacrylic acid in the presence of chitosan and polyether in a medium free of solvents or surfactants. Particles had a size distribution of 500–800 nm. The HP-ß-CD inclusion complex with insulin was encapsulated into the nanoparticles, resulting in a pH-dependent release profile as seen in Figure 2. The biological activity of insulin was demonstrated with enzyme-

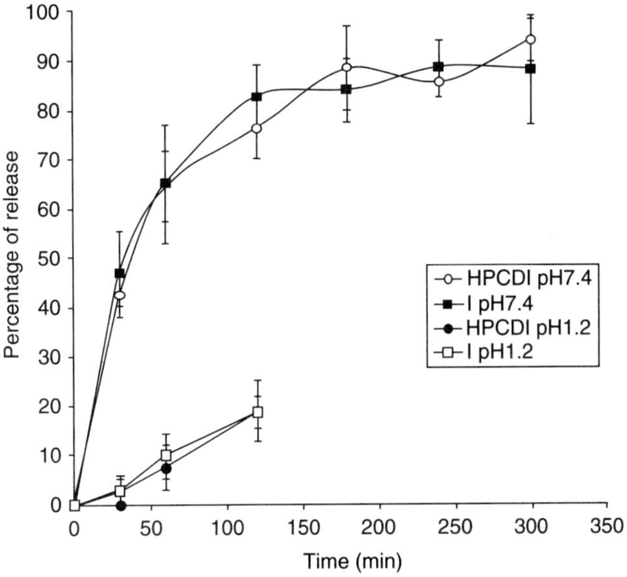

FIGURE 2 pH-dependent release profile for insulin complexed to HP-ß-CD and encapsulated in nanoparticles. (Reprinted from S. Sajeesh and C. P. Sharma, *International Journal of Pharmaceutics*, 325, 147–154, 2006, Copyright 2006, with permission from Elsevier.)

linked immunosorbent assay (ELISA). Cyclodextrin complexed to insulin encapsulated into mucoadhesive nanoparticles was believed to be a promising candidate for oral insulin delivery [35].

7.1.2.2 Preparation of Nanoparticles from Cyclodextrins

Cyclodextrins are used as excipients in the preparation of nanoparticles by three approaches:

1. Preparing nanoparticles from polymers under the presence of CDs in the medium
2. Preparing nanoparticles from polymers incorporating or conjugated to CDs
3. Preparing nanoparticles directly from amphiphilic CDs

Nanoparticles consisting copolymers of aminoethylcarbamoyl-β-cyclodextrin (AEC-β-CD) and ethylene glycol diglycidyl ether (EGDGE) were prepared by an interfacial polyaddition reaction in a mini–emulsion system. By combining these two technologies, namely, cross-linking and modification of hydroxyl groups, a novel functional nanoparticle based on β-CD was introduced as a novel material of nanobiotechnology [36].

Nanoparticles prepared fromn CDs are promising targeted delivery systems. Transferrin, an iron-binding glycoprotein ligand for tumor targeting, was conjugated to β-CD polymers and adamantane–PEG5000 through carbohydrate groups self-assembled into sub-100-nm particles in a recent study [37]. A CD-containing polycation was used for nucleic acid condensation into nanoparticles [38] and the second component, adamantane-terminated modifier for stabilizing nanoparticles to minimize interactions with plasma and to target cell surface receptors, was incorporated in this system. The particles were demonstrated to mediate transferrin-mediated delivery of nucleic acids to cultured cells [37].

Transferrin-containing CD polymer-based nanoparticles were studied as nucleic acid delivery system that can be modified for targeted delivery of small interfering ribonucleic acid (siRNA) to cancer cells. Molecular studies showed that the siRNA CD nanoparticles reduced levels of Ewing's transcript by 80% and inhibited growth of cultured Ewing's tumor cell line. It was also reported that this delivery system indicated a lack of toxicity [39].

A new tadpole-shaped polymer was synthesized via a coupling reaction of PLA onto mono[6-(2-aminoethyl)amino-6-deoxy]-β-cyclodextrin (CDenPLA). A hydrophilic head consisting of the CD group was believed to bind proteins and the PLA tail gave the amphiphilic property [40]. BSA was incorporated into nanoparticles of CDenPLA using both nanoprecipitation and double-emulsion techniques, as can be seen in Figure 3 [40]. A similar process was used to couple PLGA onto amino-β-CD and ethylenediamino-bridged bis(β-CD) to afford amphiphilic conjugates forming nanoparticles with the nanoprecipitation technique. This approach was believed to be promising for protein delivery since BSA structure was unchanged

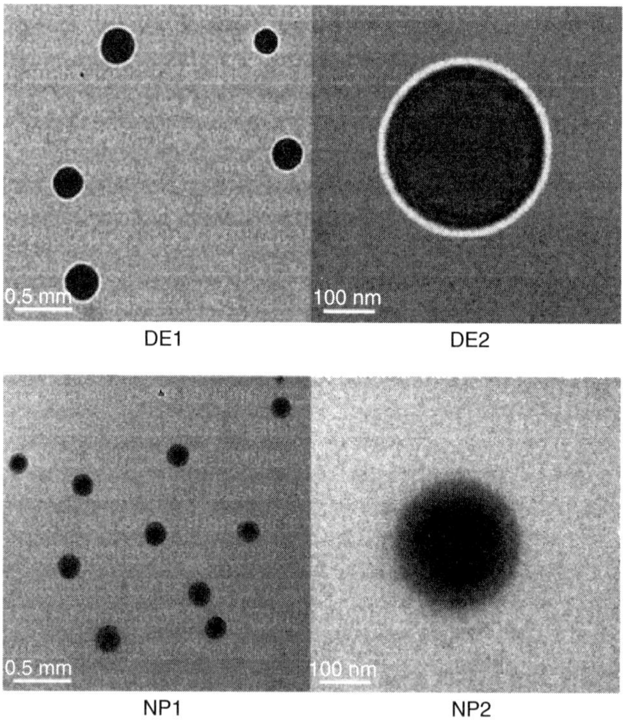

FIGURE 3 TEM photographs of BSA-loaded ßCDen47PLA nanoparticles prepared with different techniques and their magnified images. (Reprinted from H. Gao, Y. W. Yang, Y. G. Fan, and J. B., Ma, *Journal of Controlled Release*, 112, 301–311, 2006. Copyright 2006, with permission from Elsevier.)

after encapsulation into nanoparticles and during its release. Nanoparticles were reported to be stable after freeze drying [40, 41].

Nanoassemblies were formed by mixing solutions of β-CD polymer and dextran hydrophopbically modified with alkyl chains (C12) and loaded with the model hydrophobic drug benzophenone. Nanoassemblies were characterized as 110–190 nm with relatively low drug-loading values ranging between 0.3 to 1.01% w/w. Authors suggested that hydrophobic model drug and hydrophobically modified dextran compete for the apolar CD cavity; however, benzophenone does not impede the hydrophobic dextran to interact with β-CD polymer to form supramolecular assemblies at the nanoscale [42].

Another group has worked on the oligo(ethylenediamino)-β-cyclodextrin modified gold nanoparticles (OEA-CD-NP) as a vector for DNA binding. Possessing many hydrophobic cavities at the outer space, OEACD-NP was believed to have a capability of carrying biological and/or medicinal substrates into cells. Presence of the CD moieties was suggested to be the key parameter in the high affinity to DNA for the gold nanoparticles. In addition, CD moieties were demonstrated to reduce the cytotoxicity of gold nanoparticles arising from the gold clusters that impair plasma membrane functions and lead to cell death [43].

Nucleic acid delivery was also studies by Park et al. using CD-based nanoparticles prepared from β-CD-modified poly(ethylenimine) (CD-PEI). The inclusion-forming capability of β-CD was used in order to immobilize the nanoparticles on solid surfaces (adamantine-modified self-assembled monolayers). CD-PEI nanoparticles were proposed as delivery systems onto solid surfaces to attain specific and high affinity loading. The interaction is schematized in Figure 4 [44].

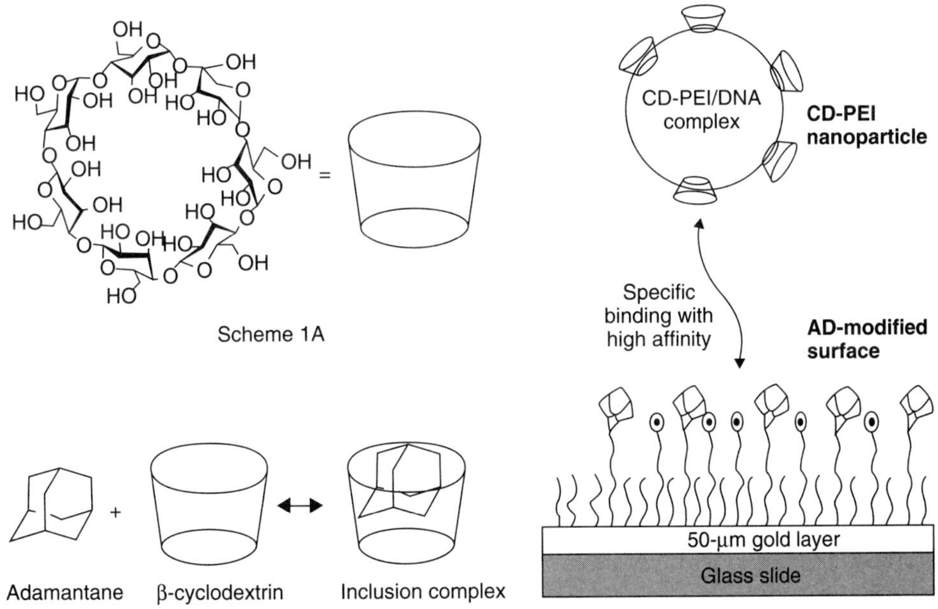

FIGURE 4 Schematic representation of β-CD–adamantane inclusion complex formation and immobilization of CD nanoparticles on adamantine-modified surface. (Reprinted with permission from ref. 44. Copyright 2006 by the American Chemical Society.)

Amphiphilic Cyclodextrins Nanoparticles have been obtained spontaneously from modified CDs of amphiphilic structure since the last decade. This approach differs from the previously discussed approaches in that amphiphilic CDs do not require the presence of another polymer or macromolecule or even surfactants.

Amphiphilic CDs have been synthesized to solve problems of natural CDs that limit their pharmaceutical applications. The main reasons for the synthesis of amphiphilic CDs are as follows:

1. Enhancement of interaction of CDs with biological membranes through a relative external hydrophobicity
2. Modification or enhancement of interaction of CDs with hydrophobic drugs arising from the high number of long aliphatic chains and by increasing the number of hydrophobic sites for possible interactions with hydrophobic molecules
3. Allowing self-assembly of CDs resulting in the spontaneous formation of nanosize carriers in the form of nanospheres and nanocapsules

The unique advantage of amphiphilic CDs is that they possess self-assembling properties that are sufficient to form nanoparticles spontaneously without the presence of a surfactant as well as the capability of including hydrophobic molecules in their cavity and within the long aliphatic chains [45–47]. Amphiphilic CDs can be classified according to their surface charge.

Nonionic Amphiphilic Cyclodextrins Nonionic amphiphilic CDs are obtained by grafting aliphatic chains of different length on the primary and/or secondary face of the CD glucopyranose unit. Different derivatives depicted in Figure 5 are named after their structure:

1. *Lollipop CDs* [48] are obtained by grafting only one aliphatic chain to 6-amino-β-CD.
2. *Cup-and-ball CDs* were synthesized by the introduction of a voluminous group such as the *tert*-butyl group which is linked to the end of the aliphatic chain in order to prevent self-inclusion of the pendant group [49, 50].

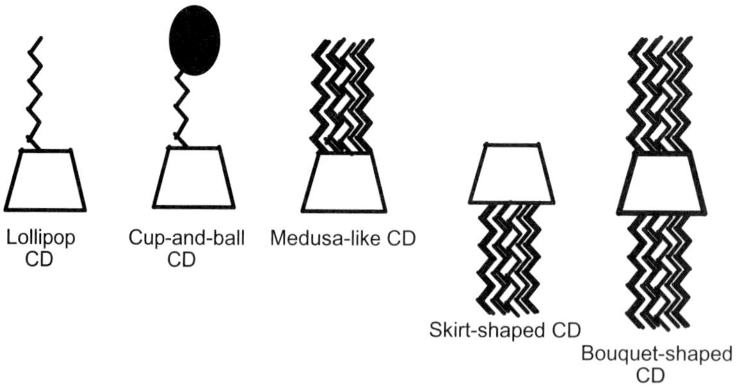

FIGURE 5 Schematic representation of some nonionic amphiphilic CDs.

3. *Medusa-like CDs* are obtained by grafting aliphatic chains with length between C10 and C16 to all the primary hydroxyls of the CD molecule [51–53].
4. *Skirt-shaped CDs* consist of β- and γ-CDs per-modified with aliphatic esters (C2–C14) on the secondary face [54–57].
5. *Bouquet-shaped CDs* result from the grafting of 14 polymethylene chains to 3-monomethylated β-cyclodextrin, meaning seven chains on each side of the CD ring molecule [58]. Per(2,6-di-*O*-alkyl) CDs where the alkyl chain may be propyl, butyl, pentyl, 3-methylbutyl, or dodecyl also take part in the bouquet family [4].
6. *Cholesteryl CDs* were recently introduced as more complicated derivatives [59]. They have been designed assuming that CD is the hydrophilic head group and cholesterol is the hydrophobic part.

Interfacial properties of nonionic amphiphilic CDs have been demonstrated by different groups [60–62]. It was found that length, structure, and bond type of the aliphatic chain play important roles upon the surface-active characteristics of amphiphilic CDs. Alignment of the amphiphilic CD molecule at the air–water interface was demonstrated to be aliphatic chains perpendicular to and a CD ring parallel to the film [62]. Inclusion-forming capability of nonionic amphiphilic CDs also has been reported with various model molecules of a different nature. It was suggested that leaving the wider side of the cavity, that is, the secondary face, unsubstituted may facilitate entrance of the drug in the cavity of the amphiphilic CD [63–65].

Cationic Amphiphilic Cyclodextrins Recently cationic amphiphilic CDs were obtained and characterized carrying an amino group as an ionic group. Heptakis(2-ω-amino-*O*-oligo(ethyleneoxide)-6-hexylthio)-β-CD, a "stealth" cationic amphiphilic CD because of the oligoethylene glycol group it carries, was synthesized [66]. The structural properties of cationic amphiphilic CDs were believed to be due to the balance between hydrophobic tails such as thioalkyl chains and hydrophilic components such as ethylene glycol oligomers. The presence of ethylene glycol chains in particular was believed to increase the colloidal stability of the supramolecular aggregates formed by cationic amphiphilic CDs. Amphiphilic alkylamino-α- and β-CDs were also reported regarding their synthesis and characterization [67].

A series of polyamino-β-CDs have been synthesized by Cryan et al. [68] and complete substitution by amine groups at the 6-position. Neutral CDs have been shown to interact with nucleic acids and nucleotides and to enhance their transfection efficiency in vivo. Cationic CDs have shown even greater ability to bind nucleotides and enhance delivery by viral vectors. The major advantage of polycationic CDs and their nanoparticles is their enhanced ability to interact with nucleic acids combined with their self-organizational properties [68].

Anionic Amphiphilic Cyclodextrins Anionic amphiphilic CDs possess a sulfate group in their structure to render an anionic nature to the molecule. An efficient regiospecific synthetic route to obtain acyl-sulfated β-CDs was introduced in which the upper rim is functionalized with sulfates and the lower rim with fatty acid esters [69]. These derivatives were able to form aggregates in aqueous medium.

Sulfated amphiphilic α-, β-, and γ-CDs were demonstrated to form 1:1 inclusion complexes with the antiviral drug acyclovir. Noncovalent interactions between acyclovir and nonsulfated amphiphilic CDs (nonionic amphiphilic CDs) appeared to take place both in the cavity of the CD and inside the hydrophobic zone generated by alkanoyl chains. However, in the case of sulfated anionic amphiphilic CDs, the interactions appear to take place only in the hydrophobic region of the alkanoyl chains [70].

Fluorine containing anionic β-CDs were first introduced by Granger et al. [71] functionalized at the 6-position by trifluoromethylthio groups. They exhibit an amphiphilic behavior at the air–water interface and are good candidates for a new class of amphiphilic carriers. Péroche et al. [72] described the synthesis of new amphiphilic perfluorohexyl- and perfluorooctyl-thio-β-CDs and their alkyl analogue, nonanethio-β-CD. The ability of these products to form nanoparticles was also investigated by photon correlation spectroscopy and imaging techniques such as scanning electron microscopy (SEM) and cryo–transmission electron microscopy (TEM).

Fluorophilic CD derivatives have been obtained as a result of combinations of CDs and a linear perfluorocarbon [73]. 2,3-Di-*O*-decafluorooctanoyl-γ-CD was obtained with a protection–deprotection synthetic method and characterized further by thin-layer chromatography (TLC), Fourier transform infrared (FTIR) spectroscopy, differential scanning calorimetry (DSC), elemental analysis, and time-of-flight mass spectrometry (OF-MS).

7.1.2.3 Safety and Efficacy of Amphiphilic Cyclodextrin Nanoparticles

Amphiphilic CDs yield nanoparticles spontaneously in the form of nanospheres or nanocapsules depending on the preparation technique. Nanoparticles have been manufactured using three different techniques. However, the nanoprecipitation technique is generally preferred since it is a simple technique resulting in unimodal distribution. The general preparation techniques for amphiphilic CD nanoparticles are as follows:

1. Nanoprecipitation [74–76]
2. Emulsion/solvent evaporation [77]
3. Detergent removal [78]

Nanocapsules are also prepared according to the same techniques. Amphiphilic CD and the oil Miglyol or benzyl benzoate are dissolved in suitable organic solvent (acetone, ethanol). The solution is poured into aqueous phase under constant stirring and the nanocapsules form spontaneously. Organic solvent is then evaporated. Resulting nanocapsules vary in size between 100 and 900 nm according to the preparation process and technological parameters [79].

Particle sizes of nanocapsules are mostly affected by the size of the oil droplet formed during the preparation along with the molar concentration and nature of amphiphilic CD. Nanospheres, on the other hand, are not significantly affected by amphiphilic CD concentration and can be formed with very high concentrations of amphiphilic CDs. The modification site of the CD (primary or secondary face) is influential for nanosphere size since modifications on the secondary face result in a larger surface area. The presence and concentration of a surfactant such as Pluronic F68 do not affect the particle size of nanospheres and nanocapsules [80]. Nano-

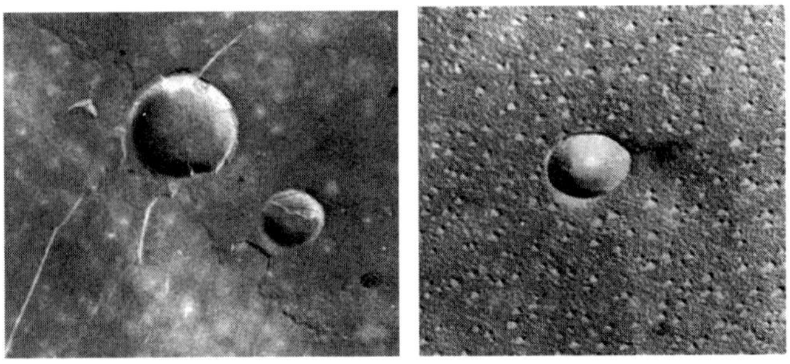

(a) (b)

FIGURE 6 Cryo-TEM images of 6-*O*-CAPRO-β-CD nanospheres (*a*) and nanocapsules (*b*). [(*a*) Reprinted from E. Memisoglu, A. Bochot, M. Şen, D. Duchene, and A. A. Hincal, *International Journal of Pharmaceutics*, 252, 143–153, 2003. Copyright 2003 with permission from Elsevier.)

spheres and nanocapsules of amphiphilic CDs were imaged with different microscopic techniques such as cryo-TEM, atomic force microscopy (AFM), and scanning transmission microscopy (STM). SEM imaging, on the other hand, results in shrinkage or disruption of the nanoparticles due to electron bombardment. Figures 6*a* and 6*b* present TEM photomicrographs after freeze fracture for 6-*O*-CAPRO-β-CD nanospheres and nanocapsules, respectively [62, 66].

Drug loading into amphiphilic nanospheres and nanocapsules is governed by the loading technique used. Amphiphilic CD nanoparticles can be loaded with the following techniques:

1. *Conventional Loading* Drug solution is added to the organic phase during preparation.
2. *Preloading* Nanoparticles are prepared directly from preformed drug–amphiphilic CD complexes.
3. *High Loading* Nanoparticles are prepared directly from preformed drug–amphiphilic CD complexes and further loaded by the addition of drug solution in the organic phase.

A high-loading technique results in two- to threefold increase in drug entrapment. Other factors influencing drug loading to amphiphilic CD nanospheres are related to drug physicochemical properties such as drug–CD association constant $k_{1:1}$, representing the affinity of the drug to the CD cavity, oil/water partition coefficient, and aqueous solubility. The affinity of the drug to the CD cavity is correlated with drug-loading capacity. Lipophilic drugs interact both with the CD cavity and the long aliphatic chains situated on either the primary or the secondary face.

Drug release properties of amphiphilic CD nanospheres are affected by various parameters, including drug lipophilicity, drug–CD association constant, and loading technique with release profiles varying from 2 to 96 h depending on the above parameters. Nanocapsules, on the other hand, exert somewhat different drug release profiles that are mostly dependent on lipophilicity and aqueous solubility of the drug. Lipophilicity of the drug is inversely correlated with the rate of release, as seen in Figure 7 [81]. Nevertheless, preparing nanoparticles directly from preformed

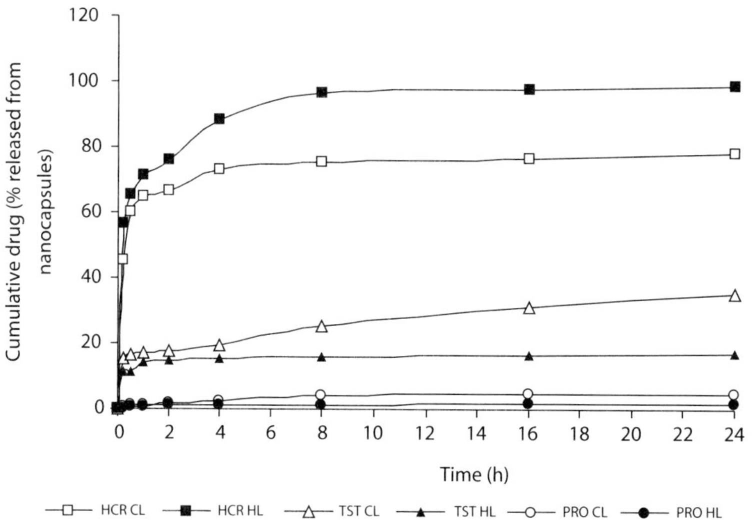

FIGURE 7 In vitro release profiles of steroids with different physicochemical properties from β-CDC6 nanocapsules (HCR HL, hydrocortisone high loaded, HCR CL, hydrocortisone conventionally loaded; TST HL, testosterone high loaded; TST CL, testosterone conventionally loaded; PRO HL, progesterone high loaded; PRO CL, progesterone conventionally loaded).

inclusion complexes helped reduce the initial burst effect observed in general for nanospheres due to their very large surface area.

Cancer Therapy Nanoparticles were first prepared with the concept of targeting colloidal carriers of nanosize to tumor tissues via the leaky vasculature in tumor regions. Since then nanoparticulate drug carriers have been associated with cancer therapy through passive and active targeting to cancer cells. Thus, amphiphilic CD nanoparticles were mainly focused on cancer therapy and its different aspects.

Tamoxifen, an antiestrogen drug used for the first-line and adjuvant therapy for metastatic breast cancer as long-term chemotherapy, has been incorporated into amphiphilic CD nanoparticles prepared using the amphiphilic CD, β-CDC6 seen in Figure 8 in order to reduce the severe side effects associated with the nonselective cytotoxicity of this drug. Tamoxifen citrate–loaded nanospheres and nanocapsules with approximately 65% entrapment efficiency liberated the drug with a controlled-release profile up to 6h when the high-loading technique is used [82]. Anticancer efficacy of tamoxifen citrate–loaded nanospheres and nanocapsules was demonstrated to be equivalent to tamoxifen citrate solution in ethanol against MCF-7 human breast cancer cells. Transcription efficiency of the tamoxifen citrate nanocapsules and nanospheres was evaluated against MELN cells in the presence of 17-β-estradiol (E2) for the inhibition of E2-mediated luciferase gene expression. It was found that transcription efficiency of tamoxifen citrate–loaded nanospheres and nanocapsules were concentration dependent [83].

Paclitaxel, an anticancer drug with bioavailability problems arising from its very low aqueous solubility, its tendency to recrystallize when diluted in aqueous media,

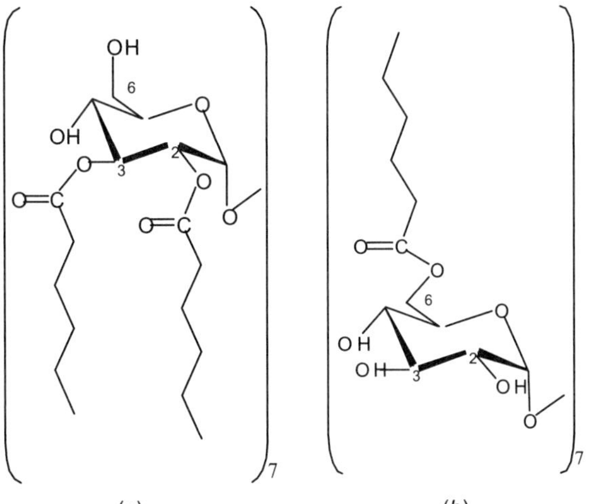

FIGURE 8 Amphiphilic β-CD derivatives modified with 6C aliphatic esters on (*a*) secondary face, β-CDC6 and (*b*) primary face, 6-*O*-CAPRO-β-CD.

and solubilizers used in its commercially available injectable formulations, has been loaded into nanoparticles prepared from amphiphilic β-CD modified on the primary face with 6C aliphatic esters, 6-*O*-CAPRO-β-CD, seen in Figure 8. Paclitaxel-loaded 6-*O*-CAPRO-β-CD nanospheres and nanocapsules were characterized with a diameter of 150 nm for nanospheres and 500 nm for nanocapsules with high entrapment efficiencies. Blank nanoparticles were proven to be physically stable in aqueous dispersion for 12 months. The in vitro release of paclitaxel from nanoparticles was completed in 24 h [84]. Amphiphilic β-CD nanoparticles were compared to the commercial vehicle Cremophor EL in terms of hemolysis and cytotoxicity. 6-*O*-CAPRO-β-CD nanospheres in particular were found to be significantly less hemolytic than paclitaxel solution in the Cremophor vehicle on human erythrocytes. Cytotoxic effects of blank nanoparticles were assessed against L929 mouse fibroblast cells and a vast difference in cytotoxicity of up to 100-fold reduction was observed for amphiphilic CD nanoparticles.

Drug-loaded nanoparticles were also evaluated for their safety and efficacy. Paclitaxel-encapsulated 6-*O*-CAPRO-β-CD nanospheres and nanocapsules were evaluated for their physical stability in a one-month period in aqueous dispersion form with repeated particle size and zeta potential measurements and AFM imaging to evaluate recrystallization in aqueous medium. Paclitaxel-loaded amphiphilic CD nanoparticles were found to be physically stable for a period of one month whereas recrystallization occurs within minutes when diluted for intravenous (IV) infusion [85]. Finally, paclitaxel-loaded amphiphilic nanoparticles were demonstrated to show similar anticancer efficacy against MCF-7 cells when compared to paclitaxel solution in a cremophor vehicle [85].

Our group is currently working on the formulation of another potent anticancer drug, camptothecin, that is clinically inactive due to its very low water solubility and poor stability under physiological pH, which causes the drug to be converted from its active lactone form to its inactive carboxylate form. Two different amphiphilic β-CD nanospheres, β-CDC6 and 6-*O*-CAPRO-β-CD, have succeeded in maintaining camptothecin in its active lactone form with considerable loading values and release profiles prolonged up to 96 h [86, 87].

Cationic amphiphilic CDs, heptakis[2-ω-amino-*O*-oligo(ethylene oxide] hexylthio-β-CD nanoparticles, have encapsulated anionic porphyrins (TPPS) by entangling these molecules within the aliphatic chains aligning both faces of the cationic amphiphilic CD. These nanoparticles were demonstrated to preserve the photodynamic properties of the entrapped photoactive agent. The photodynamic efficacy of the carrier/sensitizer nanoparticles was proven by in vitro studies on tumor HeLa cells showing significant cell death upon illumination with visible light [88].

Oxygen Delivery Amphiphilic and fluorophilic β-CD derivatives perfluoro-β-CDs were used to prepare nanocapsules with a single-step nanoprecipitation technique. Highly fluorinated materials have multiple properties, such as repellance to water and oil, unique dielectric, rheological, and optical properties, as well as exceptional chemical and biological inertness. The fluorinated chains, due to their strong hydrophobic and flurorophilic character, impart unique properties to the vesicles, including enhanced particle size stability, prolonged intravascular persistence, and increased drug encapsulation capability. Thus, 2,3-di-*O*-decafluorooctanoyl-β-CD nanoparticles were believed to be a suitable carrier for oxygen solubilization and delivery. Oxygen delivery of perfluorinated amphiphilic CD nanocapsules was compared to water and showed a prolonged delivery of oxygen. Fluorophilic nanocapsules were believed to overcome fluorocarbon emulsions as oxygen carriers due to their higher number of particles in the colloidal solution which will permit a greater rate of dissolved oxygen [73].

Oral Delivery Amphiphilic β-CD nanocapsules loaded with indomethacin have been evaluated in vivo. The nanocapsules have been applied to the rat model. It was reported that the gastrointestinal mucosa of the rat was significantly protected from the ulcerogenic effects of the active ingredient indomethacin in free form. Drug encapsulation yield in the nanocapsules were >98% and the drug content per CD unit was 7.5% w/w [89].

Cytotoxicity The cytotoxicity of nanocapsules was investigated against L929 mouse fibroblast cells and human polymorphonuclear PMNC cells with MTT assay [90]. Cell viability values of different nanocapsule and nanosphere formulations on L929 and PMNC cells indicated that nonsurfactant β-CDC6 nanocapsules were less cytotoxic than nanocapsules containing surfactants. The cytotoxicity of the nanoparticles mostly arises from surfactant presence and was concentration dependent [90].

Nanospheres of β-CDC6 prepared without surfactant and with Pluronic F68 of varying concentrations between 0.1 and 1% were found to be slightly less cytotoxic than nanocapsules to both L929 and human PMNC cells. It was concluded that cytotoxicity increased with increasing concentration of surfactant and the most suitable percentage for surfactant if required was found to be 0.1% [80].

Sterilizability Three different sterilization techniques—autoclaving, filtration, and gamma sterilization—were evaluated for amphiphilic CD nanoparticles of

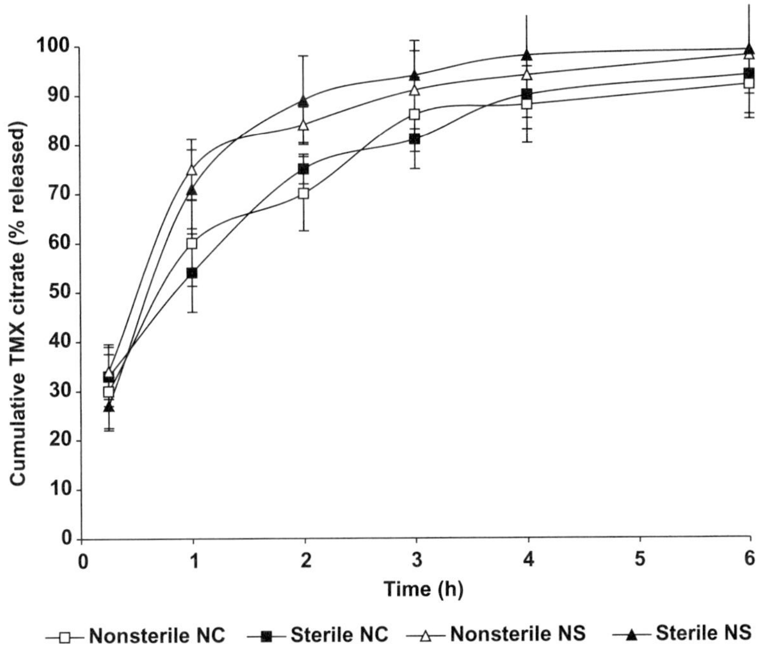

FIGURE 9 In vitro release profile of tamoxifen from ß-CDC6 nanospheres and nanocapsules before and after gamma sterilization. (Reprinted from E. Memisoglu-Bilensoy and A. A. Hincal, *International Journal of Pharmaceutics*, 311, 203–208, 2006. Copyright 2006 with permission from Elsevier.)

β-CDC6 loaded with the model drug tamoxifen [90]. It was found that filtration was not suitable for injectable amphiphilic CD nanoparticles since nanoparticle sizes were too close to filter pore sizes of 0.22 μm. Autoclaving did not affect the nanoparticle yield but caused a significant increase in particle size and aggregates. Gamma irradiation realized with a dose of 25 kGy was demonstrated to be a suitable sterilization technique since no significant change was observed for mean diameter, zeta potential, drug entrapment efficiency, and in vitro release profiles for nimodipine-loaded β-CDC6 nanospheres and nanocapsules. The in vitro release profile of sterile and nonsterile nanospheres and nanocapsules of ß-CDC6 loaded with nimodipine is seen in Figure 9 [90].

7.1.3 CONCLUSION

Cyclodextrins have been involved in nanoparticulate drug delivery systems by increasing the solubility of the drug via complex formation, forming nanoparticles in the presence of another polymer/macromolecule, forming nanoparticles by conjugation to polymers, or modification of natural CDs to render this molecule an amphiphilic character. This chapter mainly focused on the potential of amphiphilic CDs as promising carriers for anticancer drugs with bioavailability problems, oxygen delivery for the treatment of ischemia, or the safe oral administration of drugs with gastrointestinal side effects.

Many new studies are available to modify amphiphilic CDs further by giving them "stealth" properties or targeting moieties such as transferring to enable the active targeting of CD-based nanoparticles to tumor tissues. Amphiphilic CD nanocarriers now emerge as promising delivery systems for poorly soluble anticancer drugs, DNA and oligonucleotide delivery, and photodynamic and targeted tumor therapy. These systems are proven to be nonhemolytic and noncytotoxic and are capable of prolonging the release of drugs with different properties.

ACKNOWLEDGMENTS

The authors wish to thank the TUBITAK Turkish Council of Scientific and Technical Research, projects SBAG-CNRS-3 and SBAG-CD-66, and the Hacettepe University Research Fund, project 0202301005, for financial support of the amphiphilic cyclodextrin research carried out by our group at Hacettepe University, Faculty of Pharmacy, Department of Pharmaceutical Technology.

REFERENCES

1. Albers, E., and Müller, B. W. (1995), Cyclodextrin derivatives in pharmaceuticals, *CRC Crit. Rev. Ther. Drug Carrier Syst.*, 12(4), 311.
2. Loftsson, T., and Duchene, D. (2007), Cyclodextrins and their pharmaceutical applications, *Int. J. Pharm.*, 329(1–2), 1–11.
3. Uekama, K. (2004), Design and evaluation of cyclodextrin-based drug formulation, *Chem. Pharm. Bull.*, 52(8), 900–915.
4. Wenz, G. (1991), Synthesis and characterization of some lipophilic per(2,6-di-*O*-alkyl) cyclomaltooligosaccharides, *Carbohydr. Res.*, 214, 257–265.
5. Duchene, D., and Wouessidjewe, D. (1992), Industrial uses of cyclodextrins and their derivatives, *J. Coord. Chem.*, 27, 223–236.
6. Thompson, D. O. (1997), Cyclodextrins as enabling excipients: Their present and future use in pharmaceuticals, *CRC Crit. Rev. Ther. Drug Carrier Syst.*, 14(1), 1–110.
7. Rajewski, R. A., and Stella, V. J. (1996), Pharmaceutical applications of cyclodextrins 2. In vivo drug delivery, *J. Pharm. Sci.*, 85(11), 1142–1169.
8. Loftsson, T., and Brewster, M. E. (1996), Pharmaceutical applications of cyclodextrins I. Drug solubilization and stabilization, *J. Pharm. Sci.*, 85(10), 1017–1025.
9. Frank, D. W., Gray, J. E., and Weaver, R. N. (1976), Cyclodextrin nephrosis in the rat, *Am. J. Pathol.*, 83, 367–382.
10. Szejtli, J., Cserhati, T., and Szögyi, M. (1986), Interactions between cyclodextrins and cell-membrane phospholipids, *Carbohydr. Polym.*, 6, 35–49.
11. Debouzy, J. C., Fauvelle, F., Crouzy, S., Girault, L., Chapron, Y., Göschl, M., and Gadelle, A. (1997), Mechanism of α-cyclodextrin induced hemolysis 2. A study of the factors controlling the association with serine-, ethanolamine- and choline-phospholipids, *J. Pharm. Sci.*, 87(1), 59–66.
12. Uekama, K., Hirayama, F., and Irie, T. (1991), Modification of drug release by cyclodextrin derivatives, in Duchene, D., Ed., *New Trends in Cyclodextrins and Derivatives*, Editions de Santé, Paris, pp. 409–446.

13. Stella, V. J., and Rajewski, R. A. (1997), Cyclodextrins: Their future in drug formulation and delivery, *Pharm. Res.*, 14(5), 556–567.

14. Müller, B. W., and Brauns, U. (1986), Hydroxypropyl-β-cyclodextrin derivatives: Influence of average degree of substitution on complexing ability and surface activity, *J. Pharm. Sci.*, 75, 571–572.

15. Yoshida, A., Yamamoto, M., Irie, T., Hirayama, F., and Uekama, K. (1989), Some pharmaceutical properties of 3-hydroxypropyl/ and 2,3-dihydroxypropyl β-cyclodextrins and their solubilizing and stabilizing ability, *Chem. Pharm. Bull.*, 37, 1059–1063.

16. Müller, B. W. (2000), Hydroxypropyl β-cyclodextrin in drug formulation, paper presented at the CRS Workshop What's New in Cyclodextrin Delivery? Paris, July 7–8, Article 3.

17. Zia, V., Rajewski, R., Bornancini, E. R., Luna, E. A., and Stella, V. J. (1997), Effect of alkyl chain length and substitution degree on the complexation of sulfoalkyl ether β-cyclodextrins with steroids, *J. Pharm. Sci.*, 86, 220–224.

18. Stella, V. J. (2000), SBE_{7M}-β-CD or Captisol®-possible utilizations, paper presented at the CRS Workshop What's New in Cyclodextrin Delivery? Paris, July 7–8, Article 4.

19. Ma, D. Q., Rajewski, R. A., and Stella, V. J. (1999), Thermal properties and processing of (sulfobutylether)-7M-β-cyclodextrin as a freeze-drying excipient in pharmaceutical formulations, *STP Pharma Sci.*, 9(3), 261–266.

20. Komiyama, M., Yamamoto, Y., and Hirai, H. (1984), Complex formation of modified cyclodextrins with organic salts in organic solvents, *Chem. Lett.*, 1081–1084.

21. Uekama, K., Horiuchi, Y., İrie, T., and Hirayama, F. (1989), *O*-Carboxymethyl-*O*-ethylcyclomaltoheptaose as a delayed-release type drug carrier. Improvement of the oral availability of diltiazem, *Carbohydr. Res.*, 192, 323–330.

22. Uekama, K., Hirayama, F., and Irie, T. (1998), Cyclodextrin drug carrier systems, *Chem. Rev.*, 98, 2045–2076.

23. Anand, R., Nayyar, S., Pitha, J., and Merril, C. R. (1990), Sulphated sugar alphacyclodextrin sulphate, a uniquely potent anti-HIV agent, also exhibits marked synergism with AZT, and lymphoproliferative activity, Antiviral Chem. *Chemother.*, 1, 41–46.

24. Friedman, R. (1991), Cyclodextrin-containing polymers, in Duchene, D., Ed., *New Trends in Cyclodextrins and Polymers*, Editions de Santé, Paris, pp. 157–178.

25. Sebillé, B. (1987), Cyclodextrin derivatives, in Duchene, D., Ed., *Cyclodextrins and Their Industrial Uses*, Editions de Santé, Paris, pp. 351–393.

26. Memişoğlu, E., Bochot, A., Şen, M., Duchene, D., and Hincal, A. A. (2003), Non-surfactant nanospheres of progesterone inclusion complexes with amphiphilic ß-cyclodextrins, *Int. J. Pharm.*, 251, 143–153.

27. Monza de Silveira, A., Ponchel, G., Puisieux, F., and Duchene, D. (1998), Combined poly (isobutylcyanoacrylate) and cyclodextrin nanoparticles for enhancing the encapsulation of lipophilic drug, *Pharm. Res.*, 15(7), 1051–1055.

28. Duchene, D., Ponchel, G., and Wouessidjewe, D. (1999), Cyclodextrins in targeting. Application to nanoparticles, *Adv. Drug Deliv. Rev.*, 36, 29–40.

29. Boudad, H., Legrand, P., LeBas, G., Cheron, M., Duchene D., and Ponchel, G. (2001), Combined hydroxypropyl-beta-cyclodextrin and poly(alkylcyanoacrylate) nanoparticles intended for oral administration of saquinavir, *Int. J. Pharm.*, 218, 113–124.

30. Challa, R., Ahuja, A., Ali, J., and Khar, R. K. (2005), Cyclodextrins in drug delivery. An updated review, *AAPS PharmSciTech*, 6(2), E329–E357.

31. Cavalli, R., Peira, E., Caputo, O., and Gasco, M. R. (1999), Solid lipid nanoparticles as carriers of hydrocortisone and progesterone complexes with β-cyclodextrins, *Int. J. Pharm.*, 182, 59–69.

32. Radwan, M. A. (2001), Preparation and in vivo evaluation of parenteral metoclopramide-loaded poly(alkylcyanoacrylate) nanospheres in rats, *J. Microencapsul.*, 18, 467–477.

33. Luppi, B., Cerchiara,T., Bigucci, F., Caponio, D., and Zecchi, V. (2005), Bovine serum albumin nanospheres carrying progesterone inclusion complexes, *Drug Deliv.*, 12, 281–287.

34. Maestrelli, F., Garcia-Fuentes, M., Mura, P., and Alonso, M. J. (2006), A new drug nanocarrier consisting of chitosan and hydoxypropylcyclodextrin, *Eur. J. Pharm. Biopharm.*, 69(2), 79–86.

35. Sajeesh, S., and Sharma, C. P. (2006), Cyclodextrin-insulin complex encapsulated polymethacrylic acid based nanoparticles for oral insulin delivery, *Int. J. Pharm.*, 325, 147–154.

36. Eguchi, M., Da, Y. Z., Ogawa, Y, Okada, T., Yumoto, N., and Kodaka, M. (2006), Effects of conditions for preparing nanoparticles composed of aminoethylcarbamoyl-β-cyclodextrin and ethylene glycol diglycidyl ether on trap efficiency of a guest molecule, *Int. J. Pharm.*, 311, 215–222.

37. Bellocq, N. C., Pun, S. H., Jensen, G. S., and Davis, M. E. (2003), Transferrin-containing, cyclodextrin polymer-based particles for tumor-targeted gene delivery, *Bioconjug. Chem.*, 14, 1122–1132.

38. Hwang, S., Bellocq, N., and Davis, M. (2001), Effects of structure of beta-cyclodextrin-containing polymers on gene delivery, *Bioconjug. Chem.*, 12, 280–290.

39. Hede, K. (2005), Blocking cancer with RNA interference moves toward the clinic, *J. Natl. Cancer Inst.*, 97(4), 626–628.

40. Gao, H., Wang, Y. N. Fan, Y. G., and Ma, J. B. (2006), Conjugates of poly(DL-lactic acid) with ethylenediamino or diethylenetriamino bridged bis(β-cyclodextrin)s and their nanoparticles as protein delivery systems, *J. Controlled Release*, 112, 301–311.

41. Gao, H., Wang, Y. N., Fen, Y. G., and Ma, J. B. (2007), Conjugation of poly(DL-lactide-*co*-glycolide) on amino cyclodextrins and their properties as protein delivery system, *J. Biomed. Mater. Rest. A*, 80A, 111–122.

42. Daoud-Mahammed, S., Ringard-Lefebvre, C., Razzouq, N., Rosilio, V., Gillet, B., Couvreur, P., Amiel, C., and Gref, R. (2007), Spontaneous association of hydrophobized dextran and poly-β-cyclodextrin into nanoassemblies, *J. Colloid Interf. Sci.*, 307(1), 83–93.

43. Wang, H., Chen, Y., Liu, X. Y., and Liu, Y. (2007), Synthesis of oligo(ethylenediamino)-β-cyclodextrin modified gold nanoparticles as a DNA concentrator, *Mol. Pharm.*, 4(2), 189–198

44. Park, I. K., van Recum, H. A., Jiang, S., and Pun, S. H. (2006), Supramolecular assembly of cyclodextrin-based nanoparticles on solid surfaces for gene delivery, *Langmuir*, 22, 8478–8484.

45. Munoz, M., Deschenaux, R., and Coleman, A. W. (1999), Observation of microscopic patterning at the air/water interface by mixtures of amphiphilic cyclodextrins—A comparison isotherm and Brewster angle microscopy study, *J. Phys. Org. Chem.*, 12, 364–369.

46. Duchene, D., Wouessidjewe, D., and Ponchel, G. (1999), Cyclodextrins and carrier systems, *J. Controlled Release*, 62, 263–268.

47. Memisoglu-Bilensoy, E., Bochot, A., Trichard, L., Duchene, D., and Hincal, A. A. (2005), Amphiphilic cyclodextrins and microencapsulation, in Benita, S., Ed, *Microencapsulation*, 2nd rev. ed., Taylor&Francis, New York, pp. 269–295.

48. Bellanger, N., and Perly, B. (1992), NMR investigations of the conformation of new cyclodextrins-based amphiphilic transporters for hydrophobic drugs: Molecular lollipops, *J. Mol. Struct.*, 15, 215–226.

49. Dodziuk, H., Chmurski, K., Jurczak, J., Kozminski, W., Lukin, O., Sitkowski, J., and Stefaniak, J. (2000), A dynamic NMR study of self-inclusion of a pendant group in amphiphilic 6-thiophenyl-6-deoxycyclodextrins, *J Mol. Struct.*, 519, 33–36.

50. Lin, J. (1995), Synthése des cyclodextrins amphiphiles et étude de leur incorporation dans des phases phospholipidiques, Ph.D. dissertation, University of Paris, Paris.

51. Kawabata, Y., Matsumoto, M., Tanaka, M., Takahashi, H., Irinatsu, Y., Tagaki, W., Nakahara, H., and Fukuda, K. (1983), Formation and deposition of monolayers of amphiphilic β-cyclodextrin derivatives, *Chem. Lett.*, 1933–1934.

52. Djedaini, F., Coleman, A. W., and Perly, B. (1990), New cyclodextrins based media for vectorization of hydrophilic drug, mixed vesicles composed of phospholipids and lipophilic cyclodextrins, in Duchene, D., Ed., *Minutes of the 5th International Symposium on Cyclodextrins*, Editions de Santé, Paris, pp. 328–331.

53. Liu, F. Y., Kildsig, D. O., and Mitra, A. K. (1992), Complexation of 6-acyl-*O*-β-cyclodextrin derivatives with steroids, effects of chain length and substitution degree, *Drug Del. Ind. Pharm.*, 18, 1599–1612.

54. Zhang, P., Ling, C. C., Coleman, A. W., Parrot-Lopez, H., and Galons, H. (1991), Formation of amphiphilic cyclodextrins via hydrophobic esterification at the secondary hydroxyl face, *Tetrahedron Lett.*, 32, 2769–2770.

55. Zhang, P., Parrot-Lopez, H., Tchoreloff, P., Basakin, A., Ling, C. C., De Rango, C., and Coleman, A. W. (1992), Self-organizing systems based on amphiphilic cyclodextrins diesters, *J. Phys. Org. Chem.*, 5, 518–528.

56. Memişoğlu, E., Charon, D., Duchene, D., and Hıncal, A. A. (1999), Synthesis of per(2,3-di-*O*-hexanoyl)-β-cyclodextrin and characterization amphiphilic cyclodextrins nanoparticles, in Torres-Labandeira, J., and Vila-Jato, J., Eds., *Proceedings of the 9th International Symposium on Cyclodextrins*, Kluwer Academic, Dordrecht, pp. 622–624.

57. Lesieur, S., Charon, D., Lesieur, P., Ringerd-Lefebvre, C., Muguet, V., Duchene, D., and Wouessidjewe, D. (2000), Phase behaviour of fully-hydrated DMPC-amphiphilic cyclodextrin systems, *Chem. Phys. Lip.*, 10, 127–144.

58. Canceill, J., Jullien, L., Lacombe, L., and Lehn, J. M. (1992), Channel type molecular structures, Part 2, synthesis of bouquet-shaped molecules based on a β-cyclodextrin core, *Helv. Chim. Acta*, 75, 791–812.

59. Auzely-Velty, R., Djedaini-Pilard, F., Desert, S., Perly, B., and Zemb, T. (2000), Micellization of hydrophobically modified cyclodextrins 1. Miceller structures, *Langmuir*, 16, 3727–3734.

60. Tchoreloff, P., Boissonnade, M. M., Coleman, A. W., and Baszkin, A. (1995), Amphiphilic monolayers of insoluble cyclodextrins at the water/air interface. Surface pressure and surface potential studies, *Langmuir*, 11, 191–196.

61. Ringard-Lefebvre, C., Bochot, A., Memişoğlu, E., Charon, D., Duchene, D., and Baszkin, A. (2002), Effect of spread amphiphilic β-cyclodextrins on interfacial properties of the oil/water system, *Coll. Surf. B Biointerf.*, 25, 109–117.

62. Memişoğlu, E., Bochot, A., Şen, M., Charon, D., Duchene, D., and Hincal, A. A. (2002), Amphiphilic β-cyclodextrins modified on the primary face: Synthesis, characterization and evaluation of their potential as novel excipients in the preparation of nanocapsules, *J. Pharm. Sci.*, 95(1), 1214–1224.

63. Bilensoy, E., Doğan, A. L., Şen, M., and Hincal, A. A. (2007), Complexation behaviour of antiestrogen drug tamoxifen citrate with natural and modified cyclodextrins, *J. Inclus. Phenom. Macroc. Chem.*, 57, 651–655.

64. Alexandre, S., Coleman, A. W., Kasselouri, A., and Valleton, J. M. (1996), Scanning force microscopy investigation of amphiphilic cyclodextrin Langmuir-Blodgett films, *Thin Solid Films*, 284–285, 765–768.

65. Tanaka, M., Ishizuka, U., Matsumoto, M., Nakamura, T., Yabe, A., Nakanishi, H., Kawabata, Y., Takahashi, H., Tamura, S., Tagaki, W., Nakahara, H., and Fukuda, K. (1987), Host-guest complexes of an amphiphilic β-cyclodextrin and azobenzene derivatives in Langmuir-Blodgett films, *Chem. Lett.*, 1307–1310.

66. Mazzaglia, A., Angelini, N., Darcy, R., Donohue, R., Lombardo, D., Micali, N., Sciortino, M. T., Villari, V., and Scolaro, L. M. (2003), Novel heterotropic colloids of anionic porphyrins entangled in cationic amphiphilic cyclodextrins: Spectroscopic investigation and intracellular delivery, *Chem. Eur. J.*, 9, 5762–5769.

67. Matsumoto, M., Matsuzawa, Y., Noguchi, S., Sakai, H., and Abe, M. (2004), Structure of Langmuir-Blodgett films of amphiphilic cyclodextrin and water-soluble benzophenone, *Mol. Cryst. Liq. Cryst.*, 425, 197–204.

68. Cryan, S. A., Holohan, A., Donohue, R., Darcy, R., and O'Driscoll, C. M. (2004), Cell transfection with polycationic cyclodextrin vectors, *Eur. J. Pharm. Sci.*, 21, 625–633.

69. Dubes, A., Bouchu, D., Lamartine, R., and Parrot-Lopez, H. (2001), An efficient regiospecific synthetic route to multiply substituted acyl-sulphated ß-cyclodextrins, *Tetrahedron Lett.*, 42, 9147–9151.

70. Dubes, A., Degobert, G., Fessi, H., and Parrot-Lopez, H. (2003), Synthesis and characterisation of sulfated amphiphilic alpha-, beta- and gamma-cyclodextrins: Application to the complexation of acyclovir, *Carbohydr. Res.*, 338, 2185–2193.

71. Granger, C. E., Feliz, C. P., Parrot-Lopez, H., and Langlois, B. R. (2000), Fluorine containing ß-cyclodextrin: A new class of amphiphilic carriers, *Tetrahedron Lett.*, 41, 9257–9260.

72. Péroche, S., Degobert, G., Putaux, J. L., Blanchin, M. G., Fessi, H., and Parrot-Lopez, H. (2005), Synthesis and characterization of novel nanospheres made from amphiphilic perfluoroalkylthio-ß-cyclodextrins, *Eur. J. Pharm. Biopharm.*, 60, 123–131.

73. Skiba, M., Skiba-Lahiani, M., and Arnaud, P. (2002), Design of nanocapsules based on novel fluorophilic cyclodextrin derivatives and their potential role in oxygen delivery, *J. Inclus. Phenom. Macroc. Chem.*, 44, 151–154.

74. Fessi, H. C., Devissaguet, J. P., Puisieux, F., and Thies, C. (1997), Process for the preparation of dispersible colloidal systems of a substance in the form of nanoparticles, U.S. Patent 5,118,528.

75. Wouessidjewe, D., Skiba, M., Leroy-Lechat, F., Lemos-Senna, E., Puisieux, F., and Duchene, D. (1996), A new concept in drug delivery based on "skirt-shaped cyclodextrins aggregates" present state and future prospects, *STP Pharma. Sci.*, 6, 21–26.

76. Lemos-Senna, E., Wouessidjewe, D., Lesieur, S., Puisieux, F., Couarrazze, G., and Duchene, D. (1998), Evaluation of the hydrophobic drug loading characteristics in nanoprecipitated amphiphilic cyclodextrins nanospheres, *Pharm. Dev. Technol.*, 3, 1–10.

77. Lemos-Senna, E., Wouessidjewe, D., Lesieur, S., and Duchene, D. (1998), Preparation of amphiphilic cyclodextrin nanospheres using the emulsion solvent evaporation method, influence of the surfactant on preparation and hydrophobic drug loading, *Int. J. Pharm.*, 170, 119–128.

78. Lemos-Senna, E. (1998), Contribution a l'etude pharmacotechnique et physicochimique de nanospheres de cyclodextrins amphiphiles comme transporteurs de principes actifs, Ph.D. thesis, University of Paris, Paris.

79. Skiba, M., Wouessidjewe, D., Fessi, H., Devissaguet, J. P., Duchene, D., and Puisieux, F. (1992), Preparation et utilizations des nouveau systemes colloidaux dispersibles a base de cyclodextrines sous forme de nanocapsules, French Patent 92-07285.

80. Memisoglu-Bilensoy, E., Doğan, A. L., and Hincal, A. A. (2006), Cytotoxic evaluation of injectable amphiphilic cyclodextrin nanoparticles on fibroblasts and polymorphonuclear cells: Surfactant effect, *J. Pharm. Pharmacol.*, 58(5), 585–589.

81. Memişoğlu-Bilensoy, E., Şen, M., and Hincal, A. A. (2006), Effect of drug physicochemical properties on in vitro characteristics of amphiphilic cyclodextrin nanospheres and nanocapsules, *J. Microencapsul.*, 23(1), 59–68.

82. Memişoğlu-Bilensoy, E., Vural, İ., Renoir, J. M., Bochot, A., Duchene, D., and Hincal, A. A. (2005), Tamoxifen citrate loaded amphiphilic β-cyclodextrin nanoparticles: In vitro characterization and cytotoxicity, *J. Controlled Release*, 104, 489–496.

83. Vural, İ., Memişoğlu-Bilensoy, E., Renoir, J. M., Bochot, A., Duchene, D., and Hincal, A. A. (2005), Transcription efficiency of tamoxifen citrate loaded β-cyclodextrin nanoparticles, *J. Drug Deliv. Sci. Technol.*, 15(5), 339–342.

84. Bilensoy, E., Gürkaynak, O., Ertan, M., Şen, M., and Hincal, A. A., Development of nonsurfactant cyclodextrin nanoparticles loaded with anticancer drug paclitaxel, *J. Pharm. Sci.*, DOI: 10.1002/jps.2111, Published online.

85. Bilensoy, E., Gürkaynak, O., Doğan, A. L., Duman, M., and Hıncal, A. A., Safety and efficacy of amphiphilic cyclodextrin nanoparticles for paclitaxel delivery, *Int. J. Pharm.*, DOI: 10.1016/j.ijpharm.2007.06.05.

86. Çırpanlı, Y., Bilensoy, E., Çalış, S., and Hincal, A. A. (2006), Camptothecin inclusion complexes with natural and modified β-cyclodextrins, in Proceedings of the 33rd Annual Meeting and Exhibition on Controlled Release Society, Vienna, July, 22–26 pp. 988–989.

87. Çırpanlı. Y., Bilensoy, E., Çalış, S., and Hincal, A. A. (2007), Development of camptothecin loaded nanoparticles from amphiphilic β-cyclodextrin derivatives, paper presented at the Pharmaceutical Sciences World Congress PSWC, Amsterdam, April, 22–25.

88. Sortino, S., Mazzaglia, A., Scolaro, L. M., Merlo, F. M., Volveri, V., and Sciortino, M. T. (2006), Nanoparticles of cationic amphiphilic cyclodextrins entangling anionic porphyrins as carrier-sensitizer system in photodynamic cancer therapy, *Biomaterials*, 27, 4256–4265.

89. Skiba, M., Morvan, C., Duchene, D., and Puisieux, F., (1995), Evaluation of the gastrointestinal behaviour in the rat of amphiphilic β-cyclodextrin nanocapsules loaded with indomethacin, *Int. J. Pharm.*, 126, 275–279.

90. Memisoglu-Bilensoy, E., and Hincal, A. A. (2006), Sterile injectable cyclodextrin nanoparticles: Effects of gamma irradiation and autoclaving, *Int. J. Pharm.*, 311, 203–208.

7.2

NANOTECHNOLOGY IN PHARMACEUTICAL MANUFACTURING

YIGUANG JIN
Beijing Institute of Radiation Medicine, Beijing, China

Contents

7.2.1 INTRODUCTION

Nanotechnology is the ability to produce and process nanosized materials or manipulate objects within the nanoscale. The nanoscale commonly indicates the range from 1 to 100 nm [1]. However, some scientists regard the nanoscale range from 1 to 200 nm [2], even to 1000 nm [3]. Making a comparison with a human hair, it is about 80,000 nm wide. Nanotechnology is a broad, highly interdisciplinary, and still evolving field which involves the production and application of physical, chemical, and biological systems. Nanotechnology is likely to have a profound impact on our economy and society in the early twenty-first century, perhaps comparable to that of information technology or advances in cellular and molecular biology. Science and engineering research in nanotechnology promises breakthroughs in areas such

Pharmaceutical Manufacturing Handbook: Production and Processes, edited by Shayne Cox Gad

as materials, manufacturing, electronics, medicine, health care, energy, environment, biotechnology, information technology, and national security. It is widely felt that nanotechnology will lead to the next industrial revolution [4].

The idea of nanotechnology was first presented by physicist Richard Feynman. His lecture entitled "Room at the Bottom" in 1959 unveiled the possibilities available in the molecular world. Because bulk matter is built of so many atoms, there is a remarkable amount of space within which to build. Feynman's vision spawned the discipline of nanotechnology, and his dream is now coming true [5]. Along with continually increasing multidisciplinary applications of nanotechnology, many new terms with nanotechnology characteristics appear, for example, *nanomechanics* [6], *nanooptics* [7], *nanoelectronics* [8], *nanochemistry* [9], *nanomedicine* [10], *nanobiotechnology* [5, 11], *nanolithography* [12], *nanoengineering* [13], *nanofabrication* [14], and *nanomanufacturing* [15]. A very broad sense term, *nanoscience* is often used. More and more new words with *nano* as a prefix will be created to fit for the nowadays nanoworld. In fact, applications of nanotechnology in medicine and biotechnology have made great progresses in the recent two decades.

All developed countries including the United States, Japan, and Europe invest a great deal of money in nanotechnology. The National Science Foundation (NSF) of the United States is a leading agency in the national nanotechnology initiative, funding nanotechnology investments at $373 million in 2007, an increase of 8.6% from 2006 and of nearly 150% since 2001 [16]. Developing countries such as China and India also invest a lot in this increasing field so as not to stay far behind developed countries. Cancer therapy and research are hottest applied fields of bionanotechnology. In 2004, the U.S. National Cancer Institute (NCI) launched a $144 million cancer nanotechnology initiative, and the investment increased largely in the following two years [17]. At the same time, investment from public resources or companies is much higher than that from governments.

The application of nanotechnology in pharmacy has a long history, before the prevalence of the nanoconcept. It was well known 50 years ago that very small drug particles have a high solubility in solvents, resulting from the too large surface area when particle size decreased to a very small level, that is, the nanoscale, although this scale had not been mentioned yet. In 1965, Banham created liposomes (lipid vesicles) consisting of phospholipids which had a small size, typically ranging from 10 nm to several micrometers. It was soon found that liposomes were excellent drug carriers, and more importantly they had site-specific distribution capability in vivo depending on their size. It is well known that nanosized liposomes are inclined to deposit in the mononuclear phagocyte system (MPS), including liver, spleen, lung, and marrow. Therefore, nanotechnology was introduced in drug delivery very long ago. Now various nanomaterials are used to deliver drugs, and some nanosystems delivering active agents are available on the market. Undoubtedly, nanotechnology plays a key role in future pharmaceutical development and pharmacotherapy.

7.2.2 NANOMATERIALS

7.2.2.1 Types of Nanomaterials

Nanomaterial is a general term. Although nanomaterials are defined as solid or liquid materials at the nanoscale, the nanoscale range remains unclear. Many scientists regard materials that are one dimensional and 1 nm to less than 100 nm as nanomaterials. However, some scientists treat larger materials (e.g., less than 200 nm) as nanomaterials [2]. In spite of the different views, nanomaterials show unique characteristics that are different from those of bulk materials. Rapid development of nanotechnology in varied disciplines helps to create various kinds of nanomaterials. In terms of shape differences, nanomaterials can be classified as nanospheres, nanovesicles, nanoshells, nanotubes, nanohorns, nanofibers, nanowires, nanoribbons, nanorods, nanosticks, nanohelices, and so on, and they can appear in any shape imagined. In terms of state differences, nanomaterials can also be classified as nanoparticles with solid cores, nanoemulsions with liquid cores, and nanobubbles with air cores. Images of some nanomaterials of various shapes are shown in Figure 1.

Nanoparticle is the most usually used term, having a broad meaning. From a narrow sense, nanoparticles are always used to indicate all ball-like nanomaterials, and therein the term nanosphere is also used. Nanocapsules are core–shell nanoparticles, wherein trapped drugs are gathered in a core coated with a hard shell, though generally nanoparticles have uniformly dispersed drugs within the whole particle. *Nanovesicle* is not a familiar term, for example, liposomes have an inner phase and an outer phase (dispersing medium) that exist together in nanovesicles [18]. In light of drug nature, especially solubility, drugs are entrapped in an inner phase or bilayers (shells). In addition, *nanosuspension* often appears in the pharmaceutical field, meaning drug nanocrystal dispersion in liquid media [19]. Needle-shaped nanocrystals are more common than globe-shaped ones. Nanogels are newly developed based on hydrogels, being similar to nanoparticles after lyophilization [20]. Recently, a special kind of nanomaterial consisting of drugs was created for drug delivery, called self-assembled drug nanostructure (SADN), which is formed by the self-assembly of amphiphilic prodrugs in aqueous media [21, 22].

Some special nanomaterials are of great interest due to their unique properties. Dendrimers are versatile, well-defined, nanosized monodispersing macromolecules which are hyperbranched synthesized polymers constructed by repetitive monomer units. They are perfect nanoarchitectures with size from 1 nm to more than 10 nm depending on the synthesis generation. Drugs can be entrapped into the branches

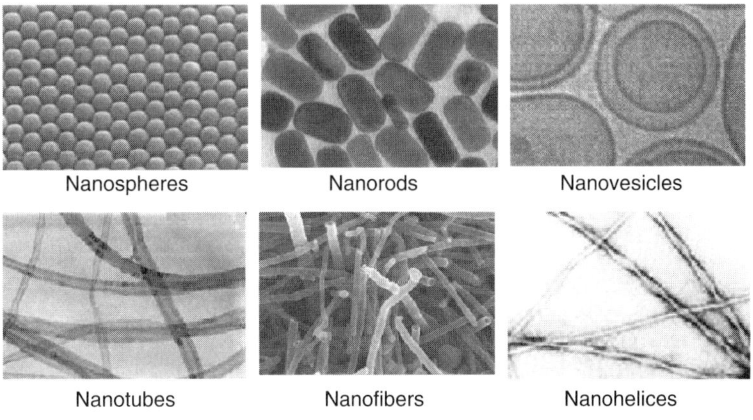

FIGURE 1 Some typical nanomaterials.

of dendrimers or conjugated with them on the high reactive surfaces [23]. The first fullerene discovered was the buckyball, also known as buckminsterfullerene. It was discovered by Smalley, Curl, and Kroto in 1985 [24], who shared a Nobel Prize in 1996 for the discovery. Buckyball is roughly spherical cages of 60 carbon atoms (C_{60}) arranged in interlocking hexagons and pentagons, like the patches on a soccer ball. Fullerenes have attracted considerable research interest, partly because of their unique structures and further because, once suitably dissolved, they display a diverse range of biological activity [25]. Quantum dots (QDs) are semiconductor nanocrystals commonly consisting of CdSe or ZnS. Besides their utilization as electronic materials, QDs have recently been applied to biomedical areas after modification. The new generations of QDs have far-reaching potential for the study of intracellular processes at the single-molecule level, high-resolution cellular imaging, long-term in vivo observation of cell trafficking, tumor targeting, and diagnostics [26].

Although many types of nanomaterials are created continually, the most important and basic issues are nanoscale effects and the subsequent particular functions. Nanomaterials with varied shapes and components provide different platforms to achieve more functions. In the area of pharmaceutical manufacturing, people focus on the drug delivery function of nanomaterials. Furthermore, the rapid development of modern medicine has led to the belief that traditional drug dosage forms such as tablets, capsules, and injections may not treat some vital diseases well, perhaps not at all. Some advanced techniques developed in other disciplines should be considered to apply to medicine. Nanomaterials can load and deliver drugs in vivo as well as display special properties such as high dispersion, adhesive property, and site-specific distribution in vivo. Modified nanomaterials further possess new functions, for example, they may be thermally sensitive, pH sensitive, magnetically sensitive, and ultrasound sensitive.

Nanotechnology has a great effect on pharmaceutical manufacturing. The unique functions of nanomaterials promise considerable benefit to pharmacotherapy over traditional drug preparations. When drug-loaded nanomaterials go through the gastrointestinal tract, high dispersion and adhesion can lead to tight contact of nanomaterials with mucous membranes, enhancing drug absorption. Nanomaterials have been applied in all routes of administration, including oral, injection (intravenous, subcutaneous, intramuscular, intra-articular cavity, and other possible injection sites), intranasal, pulmonary inhalation, conjunctiva, topical, and transdermal, possibly showing various required effects. Some of the characteristics and pharmaceutical applications of nanomaterials are given in Table 1. More applications will continue to be developed.

7.2.2.2 Manufacturing and Processing of Nanomaterials

When material dimensions reach the nanoscale, quantum mechanical and thermodynamic properties that are insignificant in bulk materials dominate, causing these nanomaterials to display new and interesting properties. The manufacturing and processing of nanomaterials may become difficult due to the unique properties. The very small size of nanomaterials produces a very large surface-to-volume ratio, that is, a great number of molecules/atoms locate on surfaces. High surface energy leads to nanomaterials easily agglomerating to diminish energy unless enough hindrance

TABLE 1 Characteristics and Applications of Some Nanomaterials in Pharmacy

Types of Nanomaterials	Characteristics	Applications in Pharmacy	References
Nanoparticles	Solid nanosized particles consisting of polymers, lipids, or inorganic materials spherically shaped most of the time, entrapped compounds dispersing in the whole particle	Loading all kinds of active agents, including drugs, vaccines, diagnostic agents, and imaging agents for good bioavailability, targeted delivery, and controlled release	27, 28
Nanocapsules	Core–shell nanoparticles with entrapped compounds gathering in the core	Loading all kinds of active agents for same aims as nanoparticles, possibly protecting entrapped agents	29
Liposomes	Lipid vesicles with entrapped compounds in inner phase or bilayers depending on physicochemical property	Loading all kinds of active agents for good bioavailability, targeted delivery, and controlled release	30, 31
Niosomes	Nonionic surfactant vesicles with similar property as liposomes	Loading all kinds of active agents for same aims as for liposomes	32
Nanoemulsions	Nanoscale emulsions	Loading drugs, as a method to prepare nanoparticles	33, 34
Polymeric micelles	Micelles consisting of amphiphilic polymers	Loading hydrophobic drugs in the core for solubilization, targeted delivery, and controlled release	35, 36
Nanogels	Nanosized hydrogels consisting of cross-linked hydrophilic polymers	Loading various compounds for controlled release or targeting	20
Dendrimers	Well-defined, nanosized, monodispersing macromolecules with hyperbranched structures	Loading all kinds of active agents for good bioavailability, targeted delivery, and controlled release	23

TABLE 1 ***Continued***

Types of Nanomaterials	Characteristics	Applications in Pharmacy	References
Carbon nanotubes (CNTs)	Nanosized tubes as if rolling up a single layer of graphite sheet (single-walled CNTs; SWNTs) or by rolling up many layers to form concentric cylinders (multiwalled CNTs; MWNTs) with diameters of ~1 nm and large length–diameter ratio	Linking a wide variety of active molecules with functionalized CNTs	37
Fullerenes	Very tiny balls consisting of 60 carbon atoms with diameter of ~0.7 nm	Water-soluble carboxylic acid C_{60} derivatives acting as antimicrobials, being linked to a variety of active molecules	25, 38, 39
Quantum dots	Tiny nanocrystals commonly consisting of semiconductor materials in the range of 2–10 nm, glowing upon ultraviolet (UV) light	Mainly as probes to track antibodies, viruses, proteins, or deoxyribonucleic acid (DNA) in vivo	26, 40
Nanosuspensions	Drug nanocrystals dispersing in aqueous media commonly stabilized by surfactants	Suitable for insoluble drugs to obtain good bioavailability and targeting	19
Self-assembled drug nanostructures	Nanostructures consisting of amphiphilic prodrugs	Suitable for hydrophilic drugs to obtain good bioavailability, targeting, and controlled release	21, 22

prevents them from agglomeration. As a result, manufacturing and processing of nanomaterials become hard issues. Anyway, many successful methods have been found to manufacture stable nanomaterials.

"Top down" and "bottom up" are two basic ways to manufacture nanomaterials. From its apparent meaning, the top-down method starts with a bulk material and then breaks it into smaller pieces using mechanical, chemical, or other forms of energy. Microchip manufacturing is the most common example of the top-down approach to produce nanomaterials. While this is an efficient approach for some industries, the process is generally labor and cost intensive. In contrast, the bottom-up method produces nanomaterials from atomic or molecular species via chemical reactions or physicochemical interactions such as self-assembly, allowing the precursor molecules/particles to grow in size. Self-assembly leads to gaining the lowest energy state of molecules and makes molecules reorient naturally to obtain ordered aggregates. Carbon nanotubes, liposomes, and the SADNs are examples of nanomaterials that are manufactured using the bottom-up approach. A deep understanding of chemical and physical properties of precursor molecules/particles is needed to design and manufacture nanomaterials using the bottom-up approach. Both top-down and bottom-up approaches can be performed in gas, liquid, supercritical fluid, solid state, or vacuum. Anyway, when bulk materials corrupt, energy is required, and certainly the obtained nanoscale materials stay at a higher energy state than their parents. Whereas in the bottom-up approach molecules self-assemble into ordered aggregates with controlled behavior. Considering the higher energy of self-assembling monomolecules dispersing in media, their aggregation should be an energy-diminishing procedure and proceed spontaneously (Figure 2).

One of the largest hurdles of nanomanufacturing is how to scale up production. In the laboratory, manufacturing nanomaterials is difficult enough as highly advanced tools and carefully clean environments are required. Therefore, scale-up manufacturing in factories becomes a great challenge, hard to achieve. The most successful mass nanomanufacturing to date has occurred with computer microprocessors where companies have been able to etch circuit boards at 65 nm or smaller. Most manufacturers are interested in the ability to control (a) particle size, (b) particle shape, (c) size distribution, (d) particle composition, and (e) degree of particle agglomeration. Neither the top-down nor bottom-up approach is superior at the moment. Each has its advantages and disadvantages. However, the bottom-up approach may have the potential to be more cost-effective in the future.

Clinical applications require that biomedical nanomaterials have good biocompatibility or biodegradability. Therefore, biodegradable polymers (synthetic or natural), small molecules such as lipids, and some bioabsorptive inorganic salts such as calcium phosphate are preferred. Other materials such as poly(ethylene glycol) (PEG) is eventually excreted from body so they can also be selected. Materials that are nonbiodegradable or not easily removed from the body, such as carbon nanotubes and quantum dots, should be carefully considered as drug carriers, although they have already been used to deliver drugs or genes. More importantly, before any nanomaterial can be used in a clinic, the acute and long-term toxicity and side effects must be estimated in detail. So a novel discipline, nanotoxicology, is of great interest [41]. In addition, problems of large-scale production of nanomaterials, for example, the uniformity and stability of products, cannot be ignored. Some nanomaterials, including liposomes, polymeric or lipid nanoparticles, nanosuspensions, and SADNs, are described in detail in the following sections. The common manufacturing methods of pharmaceutical nanomaterials are listed in Table 2, though some are only used in the laboratory.

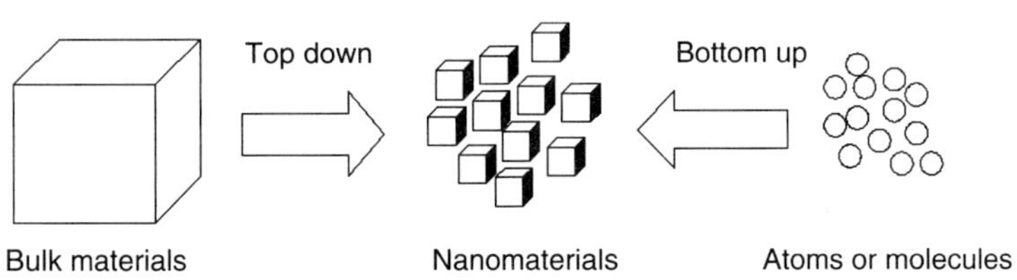

FIGURE 2 Two basic methods to manufacture nanomaterials.

TABLE 2 Manufacturing Methods of Some Nanomaterials in Pharmacy

Types of Nanomaterials	Materials Involved[a]	Manufacturing Methods	References
Nanoparticles			
Polymeric nanoparticles	Various natural polymers, e.g., albumin, gelatin, alginate, collagen, chitosan; biodegradable synthetic polymers, e.g., poly(lactic acid) (PLA), poly(lactide-*co*-glycotide) (PLGA), poly(ε-caprolactone) (PCL), poly(methyl methacrylate), and poly(alkyl cyanoacrylate); derivatives of cyclodextrin and starch; some modified polymers (e.g., PEGylated polymers) also used	Monomer polymerization, precipitation, solvent evaporation, salting out	42–44
Solid lipid nanoparticles (SLNs)	Mainly glycerides and fatty acids, surfactants also used	High-pressure homogenization, microemulsion technique, solvent evaporation	27
Inorganic nanoparticles	Calcium salts (e.g., calcium carbonate and calcium phosphate), gold	Precipitation	45–47
Nanocapsules	Various polymers, e.g., poly(*iso*-butylcyanoacrylate) (PIBCA), PLA, PLGA, PCL	Interfacial polymerization of monomers or interfacial nanodeposition of polymers	29, 48, 49
Liposomes	Phospholipids and cholesterol, phospholipid derivatives, e.g., PEG–polyethylene (PE), also added	Many methods used, mainly film hydration, reverse-phase evaporation, injection, freeze drying	50
Niosomes	Noionic surfactants, e.g., sorbitan monostearate (Span 60)	As for liposomes	32
Nanoemulsions	Oil and surfactants	High-pressure homogenization, ultrasonic emulsification, phase inversion	34

TABLE 2 *Continued*

Types of Nanomaterials	Materials Involved[a]	Manufacturing Methods	References
Polymeric micelles	Poloxamer-like block copolymers; PEG and lipophilic polymer copolymers; PEGylated lipids	Dialysis, emulsification, or film method	35, 36
Nanogels	Cross-linked hydrophilic copolymers, e.g., Pluronic–poly(ethylenimine) (PEI) and polyethylene oxide (PEO)–PEI	Covalent conjugation of polymers	20, 51, 52
Dendrimers	Dendritic macromolecules with repetitive moieties	Divergent or convergent synthesis	23, 53
Carbon nanotubes	Carbon, but only the water-soluble derivatives of CNTs used in pharmacy	CNTs formed by chemical vapor deposition (CVD) in presence of Fe catalyst, water-soluble CNT derivatives obtained by acid processing followed by conjugation with drugs	54, 55
Fullerenes (C_{60})	Carbon, but only the water-soluble derivatives of C_{60} used in pharmacy	C_{60} obtained by arc discharge method using graphite electrodes or in a benzene flame, water-soluble C_{60} derivatives obtained by acid processing followed by conjugation with drugs	38, 56
Quantum dots	Water-soluble derivatives of semiconductor materials (e.g., ZnS, PbS, CdSe, InP) used in pharmacy	QDs obtained via pyrolysis of organometallic precursors, water-soluble QD derivatives obtained by chemical reaction	57–59
Nanosuspensions	Pure drugs and stabilizers (including surfactants or polymers)	Precipitation, wet milling, homogenization	19, 60, 61
Self-assembled drug nanostructures	Polar drugs with proper conformation and lipids with long chains (e.g., glycerides, fatty acids, cholesterol)	Amphiphilic prodrugs obtained by synthesis, subsequently SADNs obtained by injection method	21, 22

[a]Organic solvents may be involved and subsequently removed.

7.2.3 NANOTECHNOLOGY FOR DRUG DELIVERY

7.2.3.1 Nanocarriers

High-throughput screening technologies in drug discovery present an efficient way to find new potential active agents. But in recent years it has become evident that the development of new drugs alone is not sufficient to ensure progress in pharmacotherapy. Poor water solubility of potential active molecules, insufficient bioavailability, fluctuating plasma levels, and high food dependency are the major and common problems. Major efforts have been spent on the development of customized drug carriers to overcome the disappointing in vivo fate of those potential drugs. For drug carriers the followings are considered: nontoxicity (acute and chronic), sufficient drug-loading capacity, possibility of drug targeting, controlled-release characteristic, chemical and physical storage stability (for both drugs and carriers), and feasibility of scaling up production with reasonable overall costs. Nanocarriers have attracted great interest because they are desirable systems to fulfill the requirements mentioned above.

Over the past decade nanocarriers as nanoparticulate pharmaceutical carriers have been shown to enhance the in vivo efficiency of many drugs both in pharmaceutical research and the clinical setting, including liposomes, micelles, nanocapsules, polymeric nanoparticles and lipid nanoparticles. They perform various therapeutically or diagnostically important functions. More importantly, many useful modifications have been made, including the increased stability and half-life of nanocarriers in the circulation, required biodistribution, passive or active targeting into the required pathological zone, responsiveness to local physiological stimuli such as pathology-associated changes in local pH and/or temperature, and ability to serve as imaging/contrast agents for various imaging modalities (gamma scintigraphy, magnetic resonance imaging, computed tomography, ultrasonography). In addition, multifunctional pharmaceutical nanocarriers have already made a promising progress [62]. Some of those pharmaceutical carriers have already found their way into clinics, while others are still under preclinical investigation. This section presents two of the most promising nanocarriers, that is, liposomes and nanoparticles, especially their manufacturing, characteristics, and applications.

Liposomes Liposomes (lipid vesicles) have a relative long history, first discovered by Banham in 1965 [63]. In the following decades, liposomes rapidly became a useful drug carrier. During the 1990s, many liposome-based drugs reached the market in the United States and Europe. The history of liposomes is the procedure of nanotechnology application to biomedicine. Phospholipids have particular structural conformation, leading to their self-assembly into bilayers with lipid chains inside and polar head groups outside during hydration. Importantly, phospholipids are the primary components of cell membranes so that liposomes have good biocompatibility without toxicity. The formation of liposomes is almost spontaneous, wherein a bottom-up procedure is involved [64]. When relatively free phospholipid molecules meet water, their polar head groups have affinity with water while lipid chains repulse water, which subsequently leads to their aggregation due to hydrophobic interaction, and then bilayers consisting of phospholipids are formed spontaneously. Closed vesicles are further formed by bilayer bending (Figure 3). Before phospho-

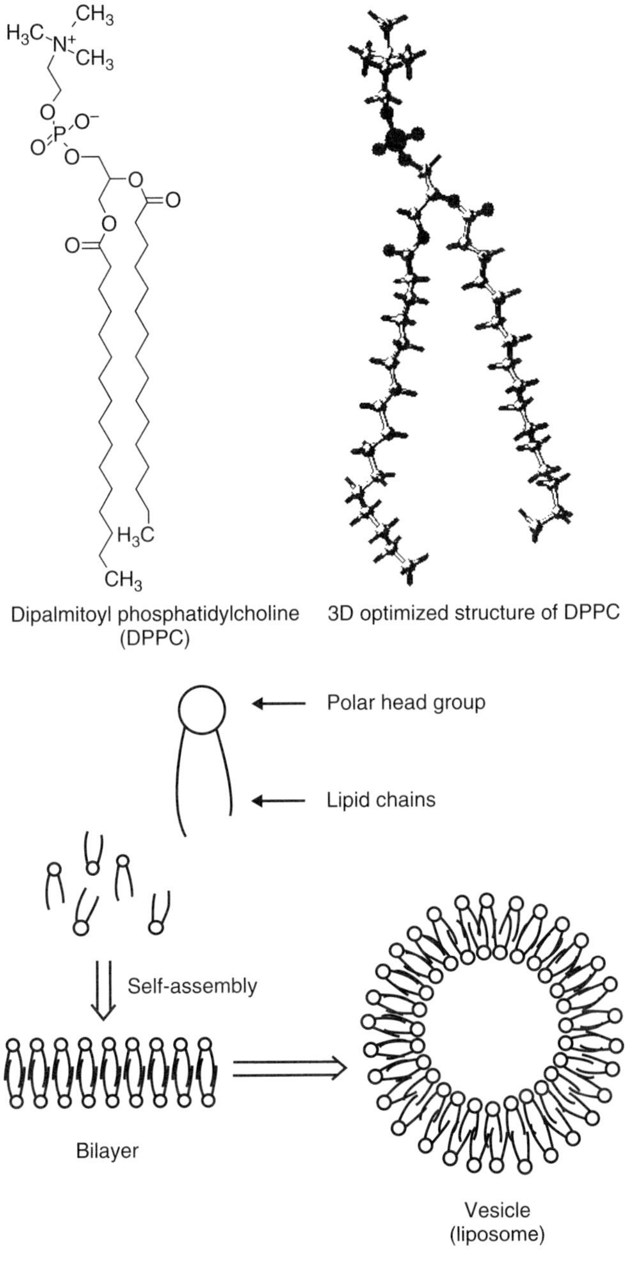

FIGURE 3 Structures of phospholipids and formation of liposomes.

lipids become "free," bulk phospholipids must be dispersed throughout, forming a thin film, dissolution or emulsification, wherein additional energy is sometimes needed. Liposomes may have a size ranging from 10 nm to more than 10 μm mainly depending on composition and manufacturing approaches. A number of reports about the preparation of liposomes can be found in the literature and a detailed description of liposomes is in Chapter 7.1 of this handbook. In this section the preparation, characteristics, and applications of nanosized liposomes are presented as well as some modified liposomes and recent progress.

Although various methods of manufacturing liposomes are reported, three types are usually involved: hydration of lipid film, interface aggregation of lipid molecules by emulsion-like process, and lipid solutions dispersing into nonsolvents by an injection-like process or controlled mixture. Practical methods are thin-film hydration [65], reverse-phase evaporation [66], ethanol injection [67], polyol dilution [68], double emulsions [69], proliposome method [70], and high-pressure homogenization [71]. Liposomes may have various morphologies related to manufacturing methods, mainly multilamellar vesicles (MLVs), large unilamellar vesicles (LUVs), and small unilamellar vesicles (SUVs). Liposomes can be further processed by sonication, detergent depletion, membrane filtration [72], and lyophilization [73] to make them finer and more uniform or stable. For example, MLVs are sonicated to SUVs.

The composition of liposomes is a key factor in their manufacturing. Phospholipids are major components of liposomes. In terms of resources, phospholipids are classified as natural, semisynthetic, and wholly synthetic phospholipids. Natural phospholipids also have different resources (e.g., soybean, egg yolk). In terms of polar head groups, phospholipids are classified as phosphatidylcholine (PC), phosphatidylethanolamine (PE), phosphatidylserine (PS), phosphatidylinositol (PI), phosphatidylglycerol (PG), and phosphatidic acid (PA), where PC and PE are the most used. Different polar head groups result in varied surface charged liposomes that then influence the stability and in vivo distribution. Because PS, PI, PG, and PA have negative charges, the liposomes containing them are negatively charged. Sometimes, other lipids such as *N,N′*-dioleoyl-*N,N′*-dimethylammonium chloride (DODAC) and stearylamine are mixed with phospholipids to prepare positively charged liposomes. Cholesterol is commonly used with phospholipids because cholesterol can make liposomal membranes stronger [50]. The mole percentage of cholesterol in the liposomal composition is commonly not more than 50%. Lecithin (an often used term in the lipid field) as a phospholipid from natural resources (e.g., soybean lecithin and egg lecithin) is often used to manufacture liposomes, which is actually a mixture composed of various kinds of phospholipids though PC dominates. The long-chain fatty acids constituting phospholipids also have many types, such as lauric (C12), myristic (C14), palmitic (C16), and stearic (C18). In general, unsaturated fatty acids occur in natural phospholipids. Dimyristoyl phosphatidylcholine (DMPC), dipalmitoyl phosphatidylcholine (DPPC), distearoyl phosphatidylcholine (DSPC), and dipalmitoyl phosphatidylethanolamine (DPPE) are the most common synthetic phospholipids. The length of the lipid chain influences the gel–liquid crystalline phase transition temperature (T_c) of phospholipids, wherein longer chained lipids lead to higher T_c. For example, DPPC has a T_c of 41°C while DSPC has a T_c of 54°C [50].

Drug entrapment is an important parameter in manufacturing liposomes which is influenced by many factors: the types, molecular weights, and physicochemical properties of drugs; the types, sizes, and compositions of liposomes; and the manufacturing methods. In addition, entrapped drugs may leak during storage. Drugs may be entrapped in one of two parts of liposomes, the inner phase or bilayers, depending on the physicochemical property of the drugs. Water-soluble drugs prefer the aqueous inner phase while lipophilic drugs prefer the hydrophobic environment of bilayers. Macromolecules such as peptides and proteins can adsorb onto bilayers, wherein electrostatic interaction can influence entrapment. Those drugs insoluble in both water and oil are hard to entrap. Rather than common manufacturing approaches, more promising methods are adopted to improve drug entrapment. Ionic gradient methods can increase the entrapment efficacy of some ionic drugs, including the pH gradient method [74], the ammonium sulfate gradient method [75, 76], the acetate gradient method [77], and the manganese ion gradient method [78]. Lyophilization of liposomes is a good preservation method which can prevent entrapped drugs from leaking, liposome precipitation and agglomeration due to gravity and thermal movement, and possible hydrolysis of phospholipids (resulting in production of toxic lyso-phospholipids). Generally, lipophilic drugs have a high entrapment efficacy, though drug loading is limited, because they insert into bilayers tightly. Therefore, an efficient method increasing entrapment efficacy is to prepare the lipophilic derivatives of hydrophilic drugs [79].

The manufacturing of nanosized liposomes can be performed using the methods mentioned above. However, the small size of nanoliposomes is difficult to achieved by methods such as film hydration. Molecular self-assembly occurs in the injection method, and then the size and morphology of obtained liposomes can be well controlled. In fact, liposomes that result from the injection method are uniform and small enough, to the nanoscale, and usually SUVs are obtained. Because of the very low toxicity of ethanol, the ethanol injection method is usually used and is described as follows to show the process of manufacturing liposomes [50]. A scale-up manufacturing process of the ethanol injection method has been established [80–82]. The obtained liposome size is mostly less than 300 nm:

(a) Handling and storage of lipids is important. Store organic solutions of phospholipids in a sealed glass container layered with argon or nitrogen below –20°C, preferably at –78°C. When transferring a portion of the material, allow it to reach room temperature before opening the bottle. Saturated phospholipids, that is, lipids composed of completely saturated fatty acids, such as DPPC, are stable as powders. However, storage of these lipids as described above is highly recommended. Unsaturated phospholipids are extremely hygroscopic as powders, which will quickly absorb moisture and become gummy upon opening the storage container. Always dissolve such lipids in a suitable solvent (preferably chloroform) and store it in a glass container at –78°C.

(b) Prepare materials such as phospholipids, cholesterol, other additives, ethanol, injector, beaker, agitation machine, and evaporation device before manufacturing. Calculate the amount of these agents according to the request of the last products. A fine-gauge needle to a 1-mL glass syringe is preferred. Dissolve lipid components (including lipophilic drugs) in ethanol. Dissolve water-soluble drugs in water or aqueous media as dispersing media.

(c) Rapidly inject the ethanol solutions into agitated aqueous media with the tip under the surface. A homogeneous and almost transparent liquid will be obtained. Repeat this process and notice that the percentage of ethanol in the last product is not more than 7.5%. Collect all liquids and remove ethanol by evaporation, dialysis, or gel filtration. The last liposomal suspensions can be further concentrated through evaporating water. Sterilize them by autoclave. They may be lyophilized when needed.

Separation of nonentrapped drugs from liposomes (or purification of liposomes) is an important process after manufacturing. The size difference between liposomes and unincorporated materials is the basis of separation. Gel chromatography, dialysis, and centrifugation are usual approaches. The drug entrapment percentage of liposomes can be obtained after separation. The whole drugs in liposomal suspensions or the entrapped drugs can be determined by dissolving liposomes with organic solvents or solubilizing liposomes with detergents to release drugs. The morphology of liposomes can be investigated by negatively stained transmission electron microscopy, cryo-electron microscopy, or freeze-fracture electron microscopy. The size distribution of liposomes is usually analysized by photon correlation spectroscopy (laser light scattering) [50].

Beyond conventional liposomes, functional liposomes are designed to achieve various therapeutic effects. Conventional liposomes manufactured by natural phospholipids or commonly used synthetic phospholipids such as PC and PE are negatively charged. However, cationic liposomes can form complexes with peptides or nucleic acids through electrostatic interaction and prefer to adsorb onto the surfaces of cell membranes, subsequently improving interaction with cells and penetrating into cytosol or phagocytosis. Therefore, cationic liposomes have become a standard transfection agent in cell manipulation [83]. Furthermore, they become primary nonviral gene delivery carriers [84].

Liposomes show site-specific distribution in the MPSs after intravenous (IV) administration due to opsonization by the complement system [85]. The diseases in MPSs can benefit from the drug targeting. But this is a bad result for diseases in other tissues. Long-circulating liposomes are then developed for targeting to non-MPS tissues. The long-circulating effect results from hydrophilic polymers coated on liposomes. For example, the half-life of the long-circulating liposomes can be extended to 20 h in rat. They are also called sterically stable liposomes or Stealth liposomes. The lipid conjugate of PEG, PEG–DSPE, is commonly used and inserts into bilayers and hinders plasma protein adsorption. The enhanced permeability and retention (EPR) effect of solid tumors makes long-circulating liposomes a very useful tool for anticancer therapy [86]. However, in recent years it was reported that in most cases PEGylated liposomes were cleared very rapidly from circulation with repeated injection. But doxorubicin PEGylated liposome is an exception. The production of anti-PEG immunoglobulin (Ig) M following injection is the major reason, and the spleen also plays a key role [87]. However, a more recent case has appeared. A modified phospholipid–methoxy(polyethylene glycol) conjugate was recently synthesized through the methylation of phosphate oxygen moiety which could prevent PEGylated liposomes from being activated by a complement system in vivo followed by achieving a true long-circulating effect [88].

Other functional liposomes are mainly stimuli-responsive liposomes. The pH-sensitive liposomes contain pH-sensitive lipids such as 1,2-dioleoyl-*sn*-3-phosphatidylethanolamine (DOPE) showing an inverted hexagonal configuration in a low-pH environment and release entrapped drugs in the low-pH environment of tumor tissues due to liposomal membrane destabilization [89]. Temperature-sensitive liposomes are prepared from special lipids such as DPPC whose phase transition temperature ($T_c = 41°C$) is proper to perform clinical anticancer therapy. When up to T_c, the fluidity of liposomal membranes increases sharply, followed by entrapped drugs releasing [90]. Some thermosensitive polymers can also be used to manufacture temperature-sensitive liposomes [91]. Magnetoliposomes load ultrafine magnetite, preferring to accumulate in the local tissue within the magnetic field [92]. Immunoliposomes load attached monoclonal antibodies to treat some severe diseases such as cancer [93].

Liposomes have been successfully applied to many drugs, diagnostic agents, imaging agents, transfection agents, vaccines, and so on. Liposomes have been tried in almost all routes of administration: oral, injection (intravenous, subcutaneous, intramuscular, intra-articular cavity, and other possible injection sites), intranasal, pulmonary inhalation, conjunctiva, topical, and transdermal. The most significant application field of liposomes is still anticancer therapy. After a long-time research for 30 years, some liposomal products have reached the market (Table 3). The major problems in manufacturing liposomes are scale-up production, efficient sterilization, and stable storage.

TABLE 3 Liposomal Drugs Approved for Clinical Application

Drug	Product Name	Composition of Liposomes and Other Major Excipients	Indication	Company
Daunorubicin	DaunoXome	DSPC, cholesterol	Kaposi's sarcoma	Gilead Sciences
Doxorubicin	Mycet	Egg PC, cholesterol	Combinational therapy of recurrent breast cancer	Zeneus
Doxorubicin	Doxil/Caelyx	MPEG–DSPE, HSPC, cholesterol, ammonium sulfate, sucrose, histidine	Refractory Kaposi's sarcoma; ovarian cancer; recurrent breast cancer	Alza/SP Europe
Amphotericin B	AmBisome (lyophilized product)	HSPC, cholesterol, DSPG, α-tocopherol, sucrose, disodium succinate	Fungal infections	Gilead Sciences
Cytarabine	DepoCyt	DOPC, DPPG, cholesterol, triolein	Lymphomatous meningitis	SkyePharma
Morphine	DepoDur	DOPC, DPPG, cholesterol, tricaprylin, triolein	Pain following major surgery	SkyePharma

MPEG = methyl PEG; HSPC = hydrogenated soy phosphatidylcholine; DSPG = disteroylphosphatidylglycerol; DOPC = dioleoylphosphatidylcholine; DPPG = dipalmitoylphosphatidylglycerol.

Nanoparticles Nanoparticles attract much attention of pharmaceutical scientists because of their controllable manufacturing, uniform preparations, and low cost. The major difference between nanoparticles and liposomes is that the former has a solid core while the latter only has inner aqueous phase and thin bilayers. In addition, in the case of liposomes, the entrapped water-soluble drugs exist only in solutions of the inner phase, while lipophilic drugs are only limited in the small space of bilayers. Therefore, the drug-loading efficiency (drug–lipid ratio) of liposomes is always limited. In the case of nanoparticles, drugs exist in the solid state, and high drug loading is possibly achieved. Unlike liposomes, nanoparticles may be composed of various materials, and biodegradable materials are preferably used. Furthermore, modified materials based on traditional natural and synthetic materials are also frequently used to manufacture nanoparticles to achieve more functions, which then highly enlarges the lists of used materials. In addition, more manufacturing methods of nanoparticles are optional than liposomes. Therefore, nanoparticles are relatively ideal nanocarriers for most drugs.

Nanoparticles can be classified as three types, polymeric nanoparticles, lipid nanoparticles, and inorganic nanoparticles, depending on the major components. Polymeric nanoparticles are exploited earlier, while lipid nanoparticles are of great interest in recent years due to very good biocompatibility. The development of polymeric nanoparticles is highly related to polymer science. Besides a great deal of natural polymers, more and more biodegradable polymers are synthesized, which allows pharmaceutical scientists to have enough optional subjects. Solid lipid nanoparticles (SLNs) composed of solid lipids have a profound advantage of no biotoxicity [94]. Inorganic nanoparticles are currently exploited only a little [45–47], and the major problems may be their poor biodegradability and relatively low drug-loading efficiency.

Polymeric Nanoparticles Polymeric materials for manufacturing nanoparticles include synthetic poly(lactic acids) (PLA), poly(lactide-*co*-glycolide) (PLGA), poly(ε-caprolactone) (PCL), poly(methyl methacrylates), and poly(alkyl cyanoacrylates); natural polymers (albumin, gelatin, alginate, collagen), and modified natural polymers (chitosan, starch). Polyesters, alone and in combination with other polymers, are most commonly used for the formulation of nanoparticles. PLGA and PLA are highly biocompatible and biodegradable. They have been used since the 1980s for numerous in vivo applications (biodegradable implants, controlled drug release). The U.S. Food and Drug Administration (FDA) has approved PLGA for human therapy [95]. More recently, formulations based on natural polymers have been developed and are on the market. For example, a wonderful natural polymer, chitosan, has permeability enhancer abilities, allowing the preparation of organic solvent free mucoadhesive particles [42].

Nanoparticles of synthetic polymers are usually manufactured by dispersion of preformed polymers. Although many methods can be used, they may be classified as monomer polymerization, nanoprecipitation, emulsion diffusion/solvent evaporation, and salting out. An appropriate method is selected mainly depending on polymer and drug natures. Polymerization of polymer monomers has been developed usually using poly(alkyl cyanoacrylate) [96, 97]. Organic solvents are usually used in polymerization. A detailed description of this method is not provided here.

The nanoprecipitation method is commonly adopted to entrap lipophilic drugs, and low polydispersity is probably achieved [42]. In general, the organic solution containing drugs and polymers is added a nonsolvent to lead to polymers precipitating together with drugs. The size of formed nanoparticles can be adjusted by the polymer and nonsolvent amounts in the organic phase. Nanoparticles can be separated from solvents and unincorporated drugs with centrifugation followed by spray drying or freeze drying when needed. The stability and drug recovery yield of nanoparticles depend on the ratio of drugs to polymers [98]. Recently, this technique has also been used to entrap hydrophilic compounds into PLGA and PLA nanoparticles [99, 100], especially peptides and proteins [101].

Another common method to manufacture polymeric nanoparticles is the emulsion diffusion or solvent evaporation technique, which is used to entrap hydrophobic or hydrophilic drugs. Generally, the polymer and hydrophobic drugs are dissolved in a partially water miscible organic phase (e.g., benzyl alcohol, propylene carbonate, and ethyl acetate). The organic solution is emulsified in aqueous media containing a suitable surfactant [i.e., anionic sodium dodecyl sulfate (SDS), nonionic poly(vinyl alcohol) (PVA) or cationic didodecyl dimethyl ammonium bromide (DMAB)] under stirring. The diffusion of the organic solvent and the counterdiffusion of water into the emulsion droplets induce polymeric nanoparticle formation. The organic solvent is evaporated. Also hydrophilic drugs could be entrapped into a water-in-oil (W/O) emulsion containing polymers and then undergo the above process. Then a water-in-oil-in-water (W/O/W) emulsion is obtained. After evaporation of total organic solvent, the drug-loaded nanoparticles can be separated. Polymer nature, polymer concentration, solvent nature, surfactant molecular mass, viscosity, phase ratio, stirring rate, temperature, and flow of water all affect nanoparticle size [102].

The salting-out method is also used. Polymers are dissolved in water-miscible organic solvents such as acetone or tetrahydrofuran (THF). The organic phase is emulsified in an aqueous phase that contains the emulsifier and salts of high concentration. Typically, the salt solution used contains 60% (w/w) magnesium chloride hexahydrate or magnesium acetate tetrahydrate with a polymer-to-salt ratio of 1:3. In contrast to the emulsion diffusion method, no diffusion of solvents occurs due to the presence of high concentrated salts. The fast addition of pure water to the O/W emulsion under mild stirring reduces the ionic strength and leads to the migration of the organic solvent to the aqueous phase, inducing nanoparticle formation. The final step is purification by cross-flow filtration or centrifugation to remove the salting-out agent. Common salting-out agents are electrolytes (sodium chloride, magnesium acetate, or magnesium chloride) or nonelectrolytes, such as sucrose. Polymer concentration and molecular weight, stirring rate and time, and the nature and concentration of surfactant and solvent are all important parameters. This method would allow avoiding the use of organic chlorinated solvents and large amounts of surfactants [102]. Furthermore, formulation of nanoparticles with natural polymers is performed by ionic gelation (chitosan), coacervation (chitosan, gelatin), and desolvation (gelatin) [102, 103]. These mild methods have the advantage of producing organic solvent-free formulations.

Additional advantages can be obtained by changing nanoparticle surface properties, for example, good stability, mucoadhesion, and long circulation time. For example, the in vivo long-circulating effect is achieved either by coating nanoparticle surfaces with hydrophilic polymers/surfactants or by incorporating biodegradable copolymers containing a hydrophilic moiety. Like long-circulating liposomes, PEG-containing polymers are frequently used to manufacture long-circulating nanoparticles. PEGylated copolymers (PLA–PEG, PLGA–PEG and PCL–PEG) are used [42, 104], and the long-circulating effect also results from the adsorption or covalent conjugation of some hydrophilic polymers with the hydrophobic surface of nanoparticles [105, 106]. Moreover, the active targeting of nanoparticles can be achieved by incorporating the conjugate of the polymer and target-directed molecule, such as (Arg-Gly-Asp) RGD, (trans-activator transcription) TAT peptides [107], and monoclonal antibody [108]. Recently, self-assembled nanoparticles have aroused great interest, consisting of amphiphilic macromolecules, such as hydrophobically modified glycol chitosan, which can also entrap drugs or peptides [109].

Before nanocarriers go into clinical applications, some issues must be considered, including drug-loading capacity, possibility of drug targeting, in vivo fate of the carrier (interaction with the biological surrounding, degradation rate, accumulation in organs), acute and chronic toxicity, scaling up of production, physical and chemical storage stability, and overall costs. A certain advantage of polymer systems is the wealth of possible chemical modifications. Possible problems of polymeric nanoparticles derive from residues of organic solvents used in the production process, polymer cytotoxicity, and the scaling up of production. Polymer hydrolysis during storage has to be taken into account and lyophilization is often required to prevent polymer degradation [94].

Solid Lipid Nanoparticles The outstanding advantage of lipid nanoparticles is perfect biocompatibility because their raw materials are the components of our body, preferring to be used or degraded by the body. Solid lipids are usually used as the major component of lipid nanoparticles—hence the name solid lipid nanoparticles. However, the used solid lipids generally become liquid at a high temperature to adapt to the preparation of SLNs. Compared with polymeric nanoparticles, the materials used for SLNs are simpler. The frequently used lipids are glycerides of various fatty acids, which also exist in the emulsions for parenteral nutrition. Large-scale production of SLNs can be achieved in a cost-effective and relatively simple way using high-pressure homogenization and microemulsion. Another useful method is solvent emulsification/evaporation. The SLN introduced in 1991 represents an alternative carrier system to traditional colloidal carriers, such as liposomes and polymeric nanoparticles. SLNs combine advantages of the traditional systems but avoid some of their major disadvantages [27]. The proposed advantages of SLNs include [94]:

- Possibility of controlled drug release and drug targeting
- Increased drug stability
- High drug payload
- Incorporation of lipophilic and hydrophilic drugs feasible
- No biotoxicity of the carrier
- Avoidance of organic solvents
- No problems with respect to large-scale production and sterilization

Solid lipids, emulsifiers, and water are generally the ingredients involved for manufacturing SLNs. The term *lipids* is used in a broader sense and includes triglycerides (e.g., stearin), partial glycerides (e.g., Imwitor), fatty acids (e.g., stearic acid), steroids (e.g., cholesterol), and waxes (e.g., cetyl palmitate). All categories of emulsifiers may be used to stabilize the lipid dispersion, and the combination of emulsifiers prevents particle agglomeration more efficiently. The choice of the emulsifier depends on the administration route and is more limited for parenteral administration.

High-pressure homogenization (HPH) has emerged as a reliable and powerful technique for the preparation of SLNs. Homogenizers of different sizes are commercially available from several manufacturers at reasonable prices. In contrast to other techniques, scaling up of HPH is out of the question in most cases. High-pressure homogenizers push a liquid with high pressure (100–2000 bars) through a narrow gap (in the range of a few micrometers). The fluid accelerates over a very short distance to very high velocity (over 1000 km/h). Very high shear stress and cavitation forces disrupt the particles down to the submicrometer range. Typical lipid contents are 5–10%, though higher lipid concentrations (up to 40%) may be used. Two general approaches of HPH, hot and cold homogenization, can be used for manufacturing SLNs (Figure 4) [27].

Microemulsions (transparently appearing with droplet size less than 100 nm) are thermodynamically stable systems, and the choice of optimal formulation containing oil, surfactant, cosurfactant, and oil–water ratio is key [110]. Generally, the solid lipid of low melting point (e.g., stearic acid) melts at a high temperature (e.g., 65–70°C), and then hot microemulsions are prepared using it. SLNs can be obtained after the hot microemulsions are rapidly cooled by injecting them into cold water (e.g., 2°C) under stirring. Emulsifiers in a formulation typically include Tween 20/60/80, lecithin, and sodium taurodeoxycholate, and coemulsifiers include alcohols and sodium monooctylphosphate. The typical volume ratios of hot microemulsions to cold water are from 1:25 to 1:50. The very low solid concentration of SLN suspensions is the disadvantage of the microemulsion method. Rapid temperature decrease in hot microemulsions is key to obtaining homogeneous and small-sized nanoparticles. A high temperature gradient can also ensure rapid lipid crystallization and prevent aggregation [94].

The solvent emulsification/evaporation method involves lipid precipitation in O/W emulsions. Solid lipids are dissolved in a water-immiscible organic solvent (e.g., cyclohexane) followed by emulsification in an aqueous medium. Upon evaporation of the solvent, the nanoparticle dispersion is formed due to lipid precipitation. Residue of organic solvents is the major problem of this method [94]. However, the microemulsion and solvent emulsification/evaporation methods can be performed conveniently in the laboratory without specific apparatuses.

During research of SLNs, some problems have continually appeared, for example, very low drug loads, drug expulsion during storage, and high water content of SLN dispersions. The α and β' crystallines of higher energy state mainly appear in conventional SLNs manufactured by hot-homogenization technique. However, these crystallines prefer to transform to the more ordered β modification of low energy state during storage. The high ordered degree improves the crystal imperfections, diminishing further to lead to drug expulsion. To solve this problem, a new kind of lipid nanoparticle was developed, called a nanostructured lipid carrier (NLC). NLCs

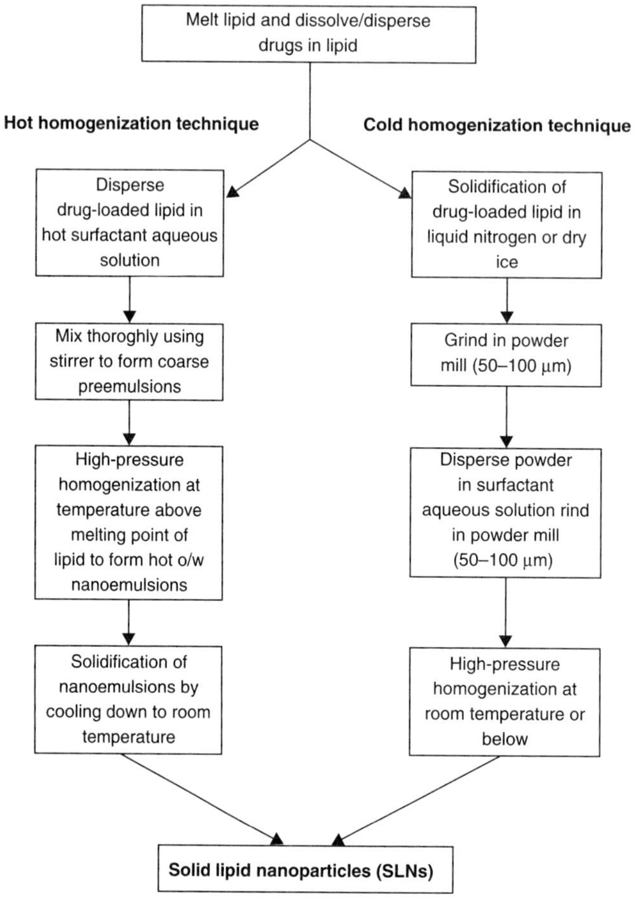

FIGURE 4 High-pressure homogenization for manufacturing SLNs.

are composed of spatially very different lipid molecules, that is, solid lipids and liquid lipids (oils). The matrix remains solid at body temperature though its melting point is lower than one of the original solid lipids. No crystallization happens in NLCs so that the drug loads can be increased and the expulsion during storage is avoided [111].

New functions can be obtained by modifications of SLNs. Incorporation of Tween 80 and Poloxamer 188 can stabilize SLNs to achieve long-circulating or crossing blood–brain barrier effects [112]. Recently, novel nanoparticles called polymer–lipid hybrid nanoparticles (PLNs) were developed [113]. They can entrap cationic anticancer agents (e.g., doxorubicin) effectively by incorporation of an anionic lipophilic polymer into lipids to treat multidrug-resistant (MDR) cancers.

In general, SLNs are used to entrap hydrophobic drugs due to their lipid nature, but a few reports show that hydrophilic drugs can also be entrapped. A hydrophilic peptide, gonadorelin, and monostearin were dissolved in acetone and ethanol at 50°C followed by pouring the resultant organic solution into an aqueous medium containing 1% PVA under agitation to obtain peptide-loaded SLNs that were subsequently separated by centrifugation. Up to 69% of gonadorelin was incorporated. The in vitro release of gonadorelin from SLNs was slow [114]. The W/O/W multiple-emulsion technique was also used to manufacture peptide-loaded SLNs. Insulin is a model peptide located in the inner water phase of the W/O/W emulsion, tripalmitin is the core of SLNs, and the surfaces are modified with PEG 2000–stearate. The insulin-loaded SLNs show good stability upon the low pH of the gastric medium and the pancreatic enzymes in intestinal medium [115].

Perspective of Nanoparticles As drug nanocarriers, nanoparticles have unique advantages: for example, high dispersing, adhesive property, targeting in vivo. Like liposomes, anticancer therapy is a major function of nanoparticles [116]. Easy modification of nanoparticles also makes them platforms to perform more functions, for example, delivering drugs across the blood–brain barrier (BBB) [117], lymphatic targeting [118], and gene delivery [119].

Abraxane is a successful paradigm of nanoparticle application. It is an albumin nanoparticle loading paclitaxel developed by American Pharmaceutical Partners (APP) and American BioScience. The outstanding advantage of Abraxane is no significant side effects, not like the traditional paclitaxel preparation with Cremophor EL (polyethoxylated castor oil) and ethanol. More nanoparticle products will reach the market in the future.

Other Nanocarriers

Nanoemulsions Lipid nanoemulsions were introduced in the 1950s as parenteral nutrition. Vegetable oils (e.g., soy oil) or middle-chain triglycerides are used, typically occupying 10–20% of the emulsion. Other ingredients include phospholipids as stabilizers and glycerol as osmolar regulation agent. In recent years this system has been further developed to load lipophilic drugs and several formulations are commercialized. Examples are etomidate (Etomidat-Lipuro), diazepam (Diazepam-Lipuro and Stesolid), propofol (Disoprivan), and dexamethasone palmitate (Lipotalon). In comparison to previous, solubilization-based formulations of these drugs, reduction of the local and systemic side effects (e.g., pain during injection) has been achieved. The possibility of controlled drug release from nanoemulsions is restricted due to the small size and the liquid state of the carrier. Most drugs show a rapid release from them. Advantages of nanoemulsions include toxicological safety and a high content of the lipid phase as well as the possibility of large-scale production by high-pressure homogenization [94].

Microemulsions Microemulsions are nanoemulsions which are optically isotropic, transparent or translucent, low-viscous, and thermodynamically stable liquid solutions, mainly containing tiny liquid droplets less than 100 nm. The manufacturing of microemulsions as a self-formed system is relatively simple. They are bicontinuous systems essentially composed of water and oil with surfactant and cosurfactant separating. A very low interfacial tension to 0 mN/m is found in microemulsions despite the large oil–water interfacial areas. A prominent example is the Sandimmun Optoral/Neoral preconcentrate for microemulsions. Now microemulsions are usually limited to dermal and peroral applications due to their high surfactant content. Because they only exist in narrow regions of phase diagrams, they are very restricted in tolerance to quantitative formulation changes [120].

Polymeric Micelles Polymeric micelles composed of amphiphilic copolymers, that is, polymers consisting of a hydrophobic block and a hydrophilic block, are gaining increasing attention. They show high stability both in vitro and in vivo and good biocompatibility, and more importantly they can solubilize a broad variety of poorly soluble drugs in their inner core. Many of these drug-loaded micelles are currently at different stages of preclinical and clinical trials. Due to their hydrophilic shell and small size, they prefer to exhibit prolonged circulation times in vivo and can accumulate in tumor tissues. Polymeric micelles are formed by block copolymers consisting of hydrophilic and hydrophobic monomer units with the length of a hydrophilic block exceeding to some extent that of a hydrophobic one. If the length of a hydrophilic block is too high, copolymers exist in water as unimers (individual molecules), while molecules with very long hydrophobic block prefer to form structures with nonmicellar morphology, such as rods and lamellae. Diblock copolymers with an A–B structure and tri- or multiblock copolymers such as poly(ethylene oxide)–poly(propylene oxide)–poly(ethylene oxide) (PEO–PPO–PEO) (A–B–A) may self-organize into micelles. The amphiphilic copolymers commonly have the critical micelle concentration (CMC) values as low as 10^{-6} *M*, which is about two orders of magnitude lower than that of such surfactants as Tween 80. As potential drug carriers, the hydrophobic core of polymeric micelles generally consists of a biodegradable polymer such as poly(β-benzyl-*l*-aspartate) (PBLA), PLA, or PCL which serves as a reservoir for an insoluble drug, protecting it from contact with the aqueous environment. The core may also consist of a water-soluble polymer [e.g., poly(aspartic acid; P(Asp)], which is rendered hydrophobic by the chemical conjugation of a hydrophobic drug or is formed through the association of two oppositely charged polyions (polyion complex micelles). Another special group of polymeric micelles is formed by lipid-core micelles, that is, micelles formed by conjugates of soluble copolymers with lipids (e.g., PEG–PE) [35, 36].

Three methods are used to manufacture polymeric micelles: dialysis, emulsification, and film methods. In the dialysis method, the drug and copolymer are dissolved together in a water-miscible solvent (e.g., ethanol) followed by dialysis against water. During the process (possibly several days), the insoluble drugs are incorporated into the formed micellar core. In the emulsification method, an O/W emulsion is first prepared using an aqueous solution of the copolymer and the drug solution in a water-insoluble volatile solvent (e.g., chloroform). The drug-loaded micelle is formed as solvent evaporation. In the film method, the copolymer solution and the drug solution are dissolved separately in miscible volatile organic solvents and are mixed followed by evaporating solvents to form a polymer/drug film. The film is hydrated in water or buffers, and then the micelle is formed by intensive shaking. If the amount of a drug exceeds the solubilization capacity of micelles, the excess drug precipitates in a crystalline form and is removed by filtration. The loading efficiency for different compounds varies from 1.5 to 50% by weight. The major driving force behind self-association of amphiphilic polymers is the decrease of free energy of the system due to removal of hydrophobic fragments from the aqueous surroundings with the formation of micelle core stabilized by hydrophilic blocks exposed to water [35, 36].

Various drugs, for example, diazepam and indomethacin, doxorubicin, anthracycline antibiotics, and polynucleotides, were effectively solubilized in polymeric micelles. Also polymeric micelles can carry various reporter (contrast) groups and become the imaging agents. Besides targeted drug delivery due to the EPR effect of tumor, specific polymeric micelles having stimuli-responsive amphiphilic block copolymers, targeting ligand molecules, or monoclonal antibody molecules are also manufactured [35, 36].

Nanogels Nanogels are colloidal stable particles made from hydrogels with nanosized hydrophilic polymeric networks. Hydrogels are the simple gels swelling strongly in aqueous media, typically composed of hydrophilic polymer components cross-linked into a network by either covalent (chemical cross-linking) or noncovalent (physical cross-linking) interactions. It is the cross-linking that provides for dimensional stability, while the high solvent content gives rise to the fluidlike transport properties. Cross-links are important to maintain the network structure of the hydrogels and prevent dissolution of the hydrophilic chains [121].

Two methods, emulsification–evaporation and the micelle/nanoparticle approach, are used to manufacture nanogels. In the former method, bis-activated PEG in dichloromethane is added dropwise to the aqueous solution of polyethylenimine (PEI) and then sonicated. The resulting white emulsion is evaporated in vacuum, producing a clear, slightly opalescent solution. This solution is stirred for less than one day at room temperature and much debris is separated by centrifugation. The nanogel suspension is obtained after dialysis against water [51]. This procedure is convenient except for using organic solvents, a vacuum evaporation step, and the obtained particles with a wide size distribution. Another method involves surface preactivated micelles or nanoparticles followed by reaction with other polymers on the surface in aqueous media. None of the organic solvents involved are of benefit. For example, a Pluronic block copolymer both ends of which are activated by 1,1′-carbonyldiimidazole is dissolved in water at a concentration above its CMC. A diluted aqueous solution of PEI is then added dropwise to the micellar solution, stirring overnight. During this procedure a covalently linked cationic polymer PEI layer is formed on the Pluronic micelles, producing nanogels with narrow size distribution. After dialysis the resulting nanogel suspension can be further lyophilized. Using this procedure, the nanogels based on Pluronic P85/PEG and F127/PEG are obtained with final yields of 55 and 70% by weight and average hydrodynamic diameters of 100 and 180 nm, respectively [20, 52].

Many drugs can be entrapped into nanogels, for example, valproic acid, nucleoside analogues, antisense oligonucleotides, adenosine triphosphate (ATP), and small interfering ribonucleic acid (siRNA). Because macromolecular drugs such as peptides and proteins need to locate in a hydrophilic environment to maintain their activity, the particular hydrophilic property of nanogels would be of benefit. Special functions such as cellular targeting, crossing the BBB, and controlled release may also be achieved by the nanogel technique. In addition, the nanogel materials should be biodegradable. In cationic nanogels, PEI and PEG are cross-linked via urethane bonds, usually considered as stable links. However, due to the presence of highly protonated PEI, hydrolysis of these bonds was significantly accelerated, and the polymer network of nanogels could rapidly degrade in aqueous solution at the physiological pH during a period of about two weeks [20].

Dendrimers Dendrimers attracted much attention after they were first investigated by Tomalia 20 years ago [122, 123], and they have become the star molecules in recent years. Dendrimers possess perfect nanoarchitectures from 1 nm to more than 10 nm, consisting of repetitive chemical moieties with tree architecture. According to repetitive folds, dendrimers with the same basic cores are divided into a series of generations. The higher generation of dendrimers represents more repetitive units. Dendrimers are hyperbranched macromolecules that can be subdivided into three architectural parts: (a) the multivalent surface, with a high number of potential reactive sites; (b) the "outer shell" just beneath the surface, having a well-defined microenvironment protected from the outside by the dendrimer surface; and (c) the core, which in higher generation dendrimers is protected from the surroundings, creating a microenvironment surrounded by dendritic branches. Therefore, the interior of dendrimers is well suited for entrapment of guest molecules. The multivalent surfaces on a higher generation dendrimer can contain a very high number of functional groups. This makes the dendritic surfaces and outer shell well suited to host–guest interactions. Dendrimers can be tailored specifically for the desired purposes, for example, as dendritic sensors, drug vehicles, or even drugs [23].

Dendrimers are synthesized through a stepwise repetitive reaction sequence, wherein a convergent or divergent approach is used. On the one hand, the most divergent dendrimer syntheses require excess monomer loading and lengthy chromatographic separations, particularly at higher generations. On the other hand, convergent synthesis strategies are generally limited to the construction of only lower generation dendrimers due to the nanoscale steric issues that are encountered when attaching the dendrons to the molecular-level core [124]. Currently much of the work on dendrimers has been based on the commercially available Starburst poly(amidoamine) (PAMAM) dendrimers that are extensively studied as drug carriers. PAMAM may be synthesized from an ammonia or ethylenediamine core (EDA) by the divergent approach, involving Michael addition followed by amidation with methyl acrylate and resulting in the production of a dendrimer family ($G = 0$–7), and half-generation dendrimers are carboxyl terminated and full-generation dendrimers are amine terminated (e.g., $G = 5.0$, 5.3 nm in size) [122, 123].

Dendrimers have been evaluated as drug nanocarriers, gene transfection agents imaging agents, and nanodrugs [124]. Also many surface-modified dendrimers have been synthesized to obtain more functions such as active targeting and gene delivery. Dendrimers may be used as drugs for antibacterial and antiviral treatment and as antitumor agents. VivaGel, a topical water-based gel based on sulfonated naphthyl-modified poly(lysine) dendrimers, has been evaluated against human immunodeficiency virus (HIV) and other sexually transmitted diseases (STDs). The cationic dendrimers prefer to destabilize cell membranes and cause cell lysis and the cytotoxicity is generation dependent with higher generation dendrimers being the most toxic. The degree of substitution as well as the type of amine functionality is important, with primary amines being more toxic than secondary or tertiary amines. Another common dendrimer, poly(propylenimine) (PPI), shows similar behavior. However, anionic dendrimers show significantly lower cytotoxicity than cationic ones. PEG or fatty acid surface-modified dendrimers can reduce the cytotoxicity of cationic dendrimers [124].

Carbon Nanotubes, Fullerenes, and Quantum Dots Carbon nanotubes and fullerenes are carbon-based nanomaterials, and quantum dots are semiconductor nanocrystals. All of them show hydrophobic property. The possibility of cytotoxicity of these materials with inorganic nature should not be ignored although low toxicity is shown [25, 40, 125, 126]. However, these seemingly good results may be partly attributed to their poor solubility in polar solvents, which subsequently makes investigation of their biological properties difficult. They hardly load any drugs unless the surface is modified hydrophilically. Because these nanomaterials are mainly produced in laboratories with the special devices, their modifications and subsequent pharmacological investigations are limited. However, a number of functional derivatives have been synthesized, and it is found that the modified products have potent and selective pharmacological effects on organs, cells, enzymes, and nucleic acids [25, 37, 57].

7.2.3.2 Nanosuspensions

Nanosuspensions of drugs are submicrometer colloidal dispersions of pure particles of drug which are stabilized by surfactants. A surprisingly large proportion of new drug candidates emerging from drug discovery programs are water insoluble, and therefore poorly bioavailable, leading to development efforts being abandoned. More than 40% of active substances during formulation development by the pharmaceutical industry are poorly water soluble. A substantial factor that prevents the development of such substances is the limited dissolution rate. Nanosuspensions are promising in addressing these so-called brickdust candidates. During the process of overcoming issues involving solubility, the additional pharmacokinetic benefits of drugs formulated in nanosuspensions come to be appreciated [19, 61].

Nanosuspensions can be used for those water-insoluble and oil-soluble compounds (high log P), although other lipidic carriers such as liposomes and emulsions can be used to formulate these compounds as well. However, nanosuspensions can be used to address other problems, such as compounds that are insoluble in both water and oil. Nanosuspensions can maintain the drug in a preferred crystalline state of size sufficiently small for pharmaceutical acceptability. For reasons of convenience to the patients, aqueous nanosuspensions can also be transformed to tablets or capsules after spray drying or freeze drying. Moreover, utilization of the dense, solid state confers an additional advantage of higher mass per volume loading. This is crucial when high dosing is required, for example, low-volume intramuscular and ophthalmic applications. Conventional approaches often attempt to solubilize insoluble drugs with the use of excessive amounts of cosolvents, but this often brings problems of toxicity. Besides, very large doses of drugs must be administered to animals when acute toxicity is investigated in preclinical research. As a result, the interference of toxic side effects caused by cosolvents cannot be ignored if using them [19].

Nanosuspensions are not nanocarriers so that what is emphasized during manufacturing is not materials but the manufacturing techniques. Only drugs and stabilizers (usually surfactants) participate in manufacturing nanosuspensions so that the process may be simple depending on drug instincts but sometimes it is not easy. The bottom-up and top-down approaches may be used in manufacturing nanosuspensions depending on drug nature and in-house devices.

Antisolvent precipitation is a bottom-up method wherein two phases are involved: the initial creation of crystal nuclei of drugs and the subsequent growth. Formation of a stable suspension with the smallest particle size requires a high nucleation rate but low growth rate. Both process rates are dependent on temperature and supersaturation. The optimum temperature for nucleation might lie below that for growth. A high-supersaturation condition is achieved by adding small amounts of a water-miscible organic solution of the drug to the nonsolvent (water) under rapid mixing, which leads to spontaneous nucleation. At high-supersaturation levels, the crystal habit or external appearance is changed to a needlelike or dendritic morphology. These crystals are easily broken, forming new smaller nuclei. Rapidly grown crystals tend to be more imperfect and often incorporate impurities and dislocations. This effect is more pronounced for flexible molecules that have many degrees of freedom [19]. The presence of stabilizing surfactants is generally necessary to assist in forming submicrometer particles, and hydrophilic groups in the surfactants lead to rapid wetting of the high-surface-area particles in aqueous media, for example, in the case of oral administration. It is well known that the unprotective surfaces of nanoparticles show a high energy that leads to particle agglomeration. Therefore, the nanoparticles must be protected by, for example, steric hindrance and electrostatic pulsion. In the case of itraconazole (ITZ) nanosuspension manufacturing, a mixture solution of ITZ and Poloxamer (P407) in THF at room temperature was mechanically injected into a P407 aqueous solution at 3°C. Magnetic stirring was utilized to enhance heat and mass transfer. Nanosuspensions containing sub-300-nm particles were obtained with drug loads as high as 86% [60].

Top-down methods are also commonly used to manufacture nanosuspensions, including wet milling and homogenization. In pearl/ball milling, the active agent, in the presence of surface stabilizer(s), is comminuted by milling media. Particle size is determined by stress intensity and the number of contact points. The drug macrosuspensions are poured into a milling container containing milling pearls from, for example, glass, zircon oxide or special polymers such as hard polystyrene derivatives. The drugs are ground to nanocrystals between the pearls. The nanosuspension-derived products, Rapamune (sirolimus tablets) and Emend (aprepitant capsules), were approved by the FDA and launched in 2000 and 2003, respectively. They are manufactured by Elan's NanoCrytal technology using a proprietary wet-milling technique. A general problem of pearl mills is potential erosion of materials from the milling pearls leading to product contamination. A polymer as substitution may minimize erosion. Scaling up with pearl mills is possible; however, there is a certain limitation in the size of the mill due to its weight. Up to about two-thirds of the mill volume are the pearls lead to heavy weight of the machinery, thus limiting the maximum batch size [127].

Homogenization can be divided into two types. One is the forcing of a suspension under pressure through a narrow-aperture valve (microfluidization). The other is high-pressure homogenization of particles in water or other media (piston gap). Microfluidization is a jet stream principle. The suspension is accelerated and passes a specially designed homogenization chamber with a high velocity. In the Z-type chamber, the suspension changes the direction of its flow a few times, leading to particle collision and shear forces, while in the Y-type chamber, the suspension stream is divided into two streams colliding frontally. Sometimes it is necessary to pass through the microfluidizer many times to minimize particle size [127].

In piston-gap homogenization, suspension contained in a cylinder passes a very thin gap with an extremely high velocity. Bubbles of water vapor are produced for compensation followed by collapsing in the valve. Cavitation-induced shock waves occur and crack the particles. Homogenization can also be utilized to further reduce the size of particles made by precipitation. Commonly dendritic crystals made by precipitation are more susceptible to rupture by the subsequent mechanical shock of homogenization. In addition, the mechanical energy supplied by the homogenizer can change the initially formed, unstable amorphous particles to a stable state through subsequent crystallization. The size of drug nanocrystals depends mainly on (a) power density of the homogenizer, (b) number of homogenization cycles, and (c) temperature. Another important determining factor for the final size of drug nanocrystals is the hardness of drugs. A relatively soft drug, paclitaxel, can reach 250 nm in size, which is less than harder drugs. The size should be homogeneous as achieved with a homogenizer to avoid physical destabilization. Stabilizers have an effect on long-term physical stability but not on maximum dispersity or the nanocrystal shape. Contamination from the production equipment is typically below 1 ppm, which is within a suitable range. Besides water, water-free media and water mixtures are used preferably due to advantages of easy evaporation or homogenization at higher temperature (with subsequent more cavitation). Oils, propylene glycol, and PEG with varieous molecular weights can be used. For PEG being solid at room temperature (e.g., PEG 1000, 6000), the obtained drug nanocrystals disperse in PEG particles at room temperature and can conveniently be put into hard capsules [19, 127].

The lyophilized drug nanosuspensions can be transferred to a final dry oral dosage form such as tablets or reconstituted prior to administration. Drug nanosuspensions can be directly used as parenteral products. A shelf life of up to three years was shown for selected nanosuspensions. Sterilization can be achieved by aseptic processing of previously sterilized components, membrane filtration for particles sufficiently small or for drugs that can withstand it, steam sterilization, or γ-irradiation.

7.2.3.3 Self-Assembled Drug Nanostructures

It is well known that liposomes are composed of amphiphilic phospholipids. The formation of liposomes is actually a procedure of molecular self-assembly. Furthermore, great amounts of amphiphilic compounds, natural or synthesized rather than phospholipid-like surfactants, can also self-assemble into ordered aggregates in aqueous media or organic solvents [128–130]. The formed aggregates are mostly nanoarchitectures with various shapes such as vesicle, rod, ribbon, fiber, tube, or helix [131–134]. They can remain relatively stable in certain environments. Many of them may become drug nanocarriers, such as liposomes, or even perform as active agents [109, 135, 136]. The research of self-assembled nanocarriers seems to go into the field of supramolecular chemistry. But these results also give us some useful information for developing new approaches of drug delivery. A novel idea may relate to why we do not try to construct a nanostructure from drugs themselves.

Twenty years ago the cardiovascular drug pindolol was conjugated with stearyl glycerol via succinyl as linker followed by forming maleate salt to obtain pindolol diglyceride. Vaizoglu and Speiser used the word "pharmacosomes" to describe the colloidal dispersions prepared from drug–lipid conjugates with or without additional surfactants [137]. Pindolol pharmacosomes (vesicle-like) were prepared from pindolol diglyceride, which showed good stability and useful pharmacokinetic parameters. Unfortunately, no more detailed research is being done about pharmacosomes, possibly because no appropriate theory supports the new dosage form and no proper drugs and lipids are selected.

The idea of manufacturing nanostructures from drugs may be resourced from liposomes, pharmacosomes, and other molecular self-assemblies in supramolecular chemistry. More importantly, this novel idea resulted from long-term efforts to work on drug delivery and to solve the disadvantages of current drug carriers. Almost all current delivery systems (usually called carriers) passively load drugs so as to always lead to low entrapment efficiency and possible drug leakage in preparation, preservation, and transportation in vivo [30, 73], and these carriers might have been destroyed in vivo before reaching target sites. In addition, lipophilic biomembranes, including cell membranes, usually prevent hydrophilic drugs from entering into target sites. If carriers cannot override cell membranes except for endocytosis/phagocytosis by cells/macrophages, the loaded hydrophilic drugs are probably released only on target surfaces, not entering. In summary, a majority of drugs could not eventually reach and get into target sites due to poor properties of carriers and drugs.

A novel technology involving prodrug, molecular self-assembly, and nanotechnology was developed to address the problems of drugs and classical carriers. The nanostructures are formed by molecular self-assembly of amphiphilic prodrugs in aqueous media generally without additional excipients. The self-assembled drug nanostructures not only possess the amphiphilic property of monomolecular drugs, benefiting to cross biomembranes, but also deliver themselves in vivo without "carriers" and then prefer to release active parent agents with a sustained rate. They may overcome some deficiencies of traditional nanocarriers such as liposomes, for example, low efficiency of drug entrapment and loading, rapid drug leakage in vitro/in vivo, and bad stability [22].

Self-assembled drug nanostructure is not a proprietary term in pharmacy currently. Herein this term is defined as the ordered nanosized self-aggregates of amphiphilic drugs in aqueous media. It is abbreviated as SADN. Another term, *self-assembled drug delivery system* (SADDS), introduced by Jin [22] obviously includes SADN. Unfortunately, most current drugs do not occupy an amphiphilic and self-assembling nature [138], so they must be modified in chemical structures before manufacturing SADNs. Then the prodrug technique is selected.

In contrast to nanosuspension technology, SADN technology is mainly applied to hydrophilic or polar drugs. These drugs are rationally modified to their amphiphilic prodrugs by lipid derivation. Molecular self-assembly in aqueous media is the key to manufacturing SADNs. According to the principles of supramolecular chemistry, the amphiphilic molecules forming self-assemblies should have proper structural conformation. The morphology of assemblies also depends on the structure of amphiphiles and the surrounding environment. Some parameters, including the optimal head area a_o, the volume v of fluid hydrophobic chain, and the maximum efficient chain length l_c, are used to describe the conformation of amphiphiles. The critical packing parameter (CPP), equal to $v/a_o l_c$, can be applied to direct self-assembly behavior. The amphiphiles prefer to form vesicles when the CPP is –1. Generally, single-chain lipids with small head group areas (e.g., SDS in a low-salt solution) are cone shaped, prone to form spherical micelles, while double-chain lipids with large head group area and fluid chains (e.g., phosphatidylcholine) are truncated-cone shaped, prone to form flexible bilayers, vesicles [139]. Therefore, the lipids used for drug covalent conjugation are rationally selected from long-chain alkyl lipids, for example, fatty acids, lipid alcohols, lipid amines, long-chain glycolipids, and cholesterol. Furthermore, too large or small polar drugs are not appropriate for preparation of self-assembling prodrugs.

Antiviral nucleoside analogues such as acyclovir, didanosine, and zidovudine were used to prepare their long-chain glyceride or cholesteryl derivatives in Jin's laboratory [140–142]. All the derivatives showed amphiphilic property and some of them self-assembled into ordered aggregates in water. Amphiphilic prodrugs were subsequently used to manufacture self-assemblies using the bottom-up approach, such as liposomes, and the self-assembly may be driven by a hydrophobic interaction, hydrogen bonding, and so on [21, 143]. The monomolecular amphiphilic prodrug is prone to incorporate into the assemblies and not to depart so that almost no drugs leak from SADNs. The whole self-assemblies are nearly composed of amphiphilic drugs, leading to high drug loading. When SADNs reach targets in vivo, the continual dissociation of aggregates and the sustained degradation of prodrugs provide controlled drug release.

Acyclovir self-assembled nanoparticles as SADNs were manufactured which showed strong targeting effect in vivo (mainly in the MPSs) and sustained release at target sites [22]. Based on this paradigm, a general process to manufacture SADNs is as follows [21, 22]:

(a) To obtain an amphiphilic prodrug with proper molecular structure, stearyl glyceride was selected to conjugate with acyclovir. Succinyl acyclovir (SACV) was synthesized and subsequently conjugated with stearyl glyceride by acylation reaction. The amphiphilic prodrug stearyl-glyceride-succinyl acyclovir (SGSA) was obtained.
(b) The injection method was used to manufacture SADNs. SGSA was dissolved in the water-miscible solvent THF. The solution containing 5 mg/mL SGSA was slowly and continually injected into vortexed water under surface via a 100-μL microsyringe. A homogeneous and slightly opalescent suspension was obtained, which was acyclovir self-assembled nanoparticles (SANs).
(c) The organic solvent was removed from the suspension through evaporation by heating, and the suspension can further be concentrated by removing water under heating until the appropriate prodrug concentration is obtained. The concentrated suspension was transferred into ampoules and sealed. It may be sterilized by autoclave.
(d) Acyclovir SANs were characterized. They were cuboidlike shaped based on transmission electron microscopy and were nanoscale with an average size of 83 nm based on dynamic light scattering. The zeta potential of −31 mV indicated the nanoparticles had negative surface charge. Hydrophobic interaction of alkyl chains improves SGSA molecules to form bilayers, and then cuboidlike nanoparticles were achieved by layer-by-layer aggregation based on inter-bilayer hydrogen bonding. The gel–liquid crystalline phase transition was about 50°C, and the mechanism of configuration changes on phase transition was analyzed [144].

(e) The in vitro and in vivo behavior of acyclovir SANs was investigated. The SANs kept the physical state stable upon centrifugation or exposure of some common additives Autoclave and bath heat for sterilization hardly influenced the state of SANs. SGSA in SANs showed good chemical stability in weak acidic or neutral buffers, although they were very sensitive to alkaline solutions and carboxylester enzymes. The SANs were rapidly removed from blood circulation after bolus IV administration to rabbits and mainly distributed in liver, spleen, and lung followed by slow elimination in these tissues.

Because nucleoside analogues are important and plentiful agents in antiviral and anticancer therapy, other polar drugs can simulate the above process to manufacture SADNs. In addition, macrophages as the reservoirs of HIV or other viruses prefer to carry viruses throughout the whole body even to the central nervous system. How to deliver drugs to macrophages has become a key issue in antiviral therapy [145]. SADNs prefer to show macrophage-specific distribution. Therefore, the antiviral SADNs show the advantages of high drug loading, controlled release, and targeting macrophage, which may provide a useful and promising way to treat increasing viral diseases. In the future SADNs will be modified to get more functions, such as long circulating effect, pH sensitivity, and use in antiviral, anticancer, and gene therapy.

7.2.4 NANOMEDICINE

Nanomedicine is a concept with broad implications. According to the definition of the European Science Foundation (ESF), the field of nanomedicine is the science and technology of diagnosing, treating, and preventing disease and traumatic injury, of relieving pain, and of preserving and improving human health using molecular tools and molecular knowledge of the human body. It is perceived as embracing five main subdisciplines that in many ways are overlapping and underpinned by the following common technical issues: (a) analytical tools, (b) nanoimaging, (c) nanomaterials and nanodevices, (d) novel therapeutics and drug delivery systems, and (e) clinical, regulatory, and toxicological issues. The ESF's scientific forward look on nanomedicine warns that nanomedicine benefits will be lost without major investment and calls for a coordinated European strategy to deliver new nanotechnology-based medical tools for diagnostics and therapeutics [146]. From a view of narrow sense, nanomedicine can be defined as the use of nanoscale or nanostructured materials in medicine that have unique medical effects according to their structure. In addition, nanostructures up to 1000 nm in size are adopted because from a technical point of view the control of materials in this size range not only results in new medical effects but also requires novel, scientifically demanding chemistry and manufacturing techniques [147].

The increasing research in nanomedicine has led to many publications, accounting for about 4% of publications on nanotechnology research (about 34,300 documents in 2004) worldwide. Also commercialization efforts in nanomedicine are increasing. About 207 companies (including 158 small- and medium-size enterprises) visibly pursue nanomedicine activities and devote either all or a significant share of their business to the development of nanomedicines. A characterizing feature of nanotechnology is its enabling function to add new functionality to existing products, making them more competitive. For example, Ambisome (Gilead), a liposomal formulation of the fungicide Fungizone (Bristol-Myers Squibb) that shows reduced kidney toxicity, had total sales of $212 million in 2004. The total sales of the 38 identified nanomedicine products from all sectors of nanomedicine are estimated to be $6.8 billion in 2004. The market is predicted to further grow to ~$12 billion by the year 2012. Currently, nanomedicine is dominated by drug delivery systems, accounting for more than 75% of the total sales. Twenty-three nanoscale drug delivery systems are available on the market, but within this group, three polymer therapeutics alone account for sales of $3.2 billion: (i) Neulasta (pegfilagrastim; recombinant methionyl human granulocyte colony stimulating factor and PEG), (ii) Pegasys (PEGylated interferon α 2a), and (iii) PEG-Intron (PEGylated interferon α 2a), all protein therapeutics to which nanoscale polymer strings of PEG have been attached to reduce immunogenicity and to prolong plasma half-life. The most widely used nanotechnology product in the field of in vitro diagnostics is colloidal gold in lateral flow assays, rapid tests for pregnancy, ovulation, HIV, and other indications. Magnetic nanoparticles are also used for cell-sorting applications in clinical diagnostics. In the field of biomaterials, the commercial status of nanotechnology-based dental restoratives is most advanced. Furthermore, nanohydroxyapatite-based products for the repair of bone defects have been successfully commercialized. Nanotechnology-based contrast agents are a market with estimated sales of about $12 million. All of the marketed contrast agents consist of superparamagnetic iron oxide nanoparticles for magnetic resonance imaging. Nanostructured electrodes are used to improve the electrode tissue contact, and nanomaterials are used to increase the biocompatibility of implant housings. Pacemakers with nanostructured (fractal) electrodes are the only active implants currently on the market that contain a nanotechnology-enabled component [147].

In spite of the great success, the safety of nanomedicine is maintained as a worrying issue. A new discipline appears to exploit the toxicological problem in nanotechnology applications, called nanotoxicology. Nanotoxicology can be defined as safety evaluation of engineered nanostructures and nanodevices. Nanomaterials could be deposited in all regions of the respiratory tract after inhalation. The small size facilitates uptake into cells and transcytosis across epithelial and endothelial cells into the blood and lymph circulation to reach potentially sensitive target sites such as bone marrow, lymph nodes, spleen, and heart. Access to the central nervous system and ganglia via translocation along axons and dendrites of neurons has also been observed. Nanomaterials could also penetrate the skin via uptake into lymphatic channels [41]. Although possible damages of those biodegradable nanomaterials for drug delivery need consideration, too much fear is needless. Usually they would be ultimately degraded nearly without any trace. However, hard or nonbiodegradable materials, including carbon nanotubes, fullerenes, quantum dots, polystyrene, and metal nanoparticles, should be thoroughly investigated about their toxic effects on our body before clinical application.

7.2.5 PERSPECTIVE

Nanotechnology has had a great effect on pharmaceutical manufacturing and strongly improves it, rapidly progressing. No one suspects the key role nanotechnology will have in future pharmaceutical research and manufacturing. The continually increasing achievements in nanotechnology will result in exciting changes in the pharmaceutical industry. Now it has gone into an era of controlling the behavior of drugs in vitro/in vivo. Although some problems such as toxicity are not addressed, the tremendous advantages that result from nanotechnology are obvious. More and more potent medicines will be manufactured and diseases such as cancer, HIV, cardiovascular diseases, and nervous system diseases may well be cured or better treated in the future by nanomedicine technology.

REFERENCES

1. Silva, G. A. (2006), Neuroscience nanotechnology: Progress, opportunities and challenges, *Nat. Rev. Neurosci.*, 7, 65–74.
2. Wang, Z. L. (2000), *Characteriaztion of Nanophase Materials*, Wiley-VCH, Weinheim, Germany, pp. 1–6.
3. Vasir, J. K., Reddy, M. K., and Labhasetwar, V. D. (2005), Nanosystems in drug targeting: Opportunities and challenges, *Curr. Nanosci.*, 1, 47–64.
4. Lane, N. F. (2004), in Bhushan, B., Ed., *Springer Handbook of Nanotechnology*, Springer-Verlag, Berlin.
5. Goodsell, D. S. (2004), *Bionanotechnology: Lessons from Nature*, Wiley-Liss, Hoboken, NJ, pp. 1–6.
6. Fritz, J., Baller, M. K., Lang, H. P., Rothuizen, H., Vettiger, P., Meyer, E., Guntherodt, H. J., Gerber, C., and Gimzewski, J. K. (2000), Translating biomolecular recognition into nanomechanics, *Science*, 288, 316–318.
7. Xu, X. H. N., Chen, J., Jeffers, R. B., and Kyriacou, S. (2002), Direct measurement of sizes and dynamics of single living membrane transporters using nanooptics, *Nano Lett.*, 2, 175–182.
8. Li, J., Papadopoulos, C., and Xu, J. (1999), Nanoelectronics—Growing Y-junction carbon nanotubes, *Nature*, 402, 253–254.
9. Sergeev, G. B. (2006), *Nanochemistry*, Elsevier, Amsterdam, The Netherlands.
10. Moghimi, S. M., Hunter, A. C., and Murray, J. C. (2005), Nanomedicine: Current status and future prospects, *FASEB J.*, 19, 311–330.
11. Whitesides, G. M. (2003), The "right" size in nanobiotechnology, *Nat. Biotechnol.*, 21, 1161–1165.
12. Lee, K. B., Park, S. J., Mirkin, C. A., Smith, J. C., and Mrksich, M. (2002), Protein nanoarrays generated by dip-pen nanolithography, *Science*, 295, 1702–1705.
13. Liu, G. Y., and Amro, N. A. (2002), Positioning protein molecules on surfaces: A nanoengineering approach to supramolecular chemistry, *Proc. Nat. Acad. Sci. U.S.A.*, 99, 5165–5170.
14. Quake, S. R., and Scherer, A. (2000), From micro- to nanofabrication with soft materials, *Science*, 290, 1536–1540.
15. Xia, Y. N., McClelland, J. J., Gupta, R., Qin, D., Zhao, X. M., Sohn, L. L., Celotta, R. J., and Whitesides, G. M. (1997), Replica molding using polymeric materials: A practical step toward nanomanufacturing, *Adv. Mater.*, 9, 147–149.

16. http://www.whitehouse.gov/omb/budget/fy2007/nsf.html 2007-1-5.

17. Service, R. F. (2005), Nanotechnology takes aim at cancer, *Science*, 310, 1132–1134.

18. Vauthey, S., Santoso, S., Gong, H., Watson, N., and Zhang, S. (2002), Molecular self-assembly of surfactant-like peptides to form nanotubes and nanovesicles, *Proc. Nat. Acad. Sci. U. S. A.*, 99, 5355–5360.

19. Rabinow, B. E. (2004), Nanosuspensions in drug delivery, *Nat. Rev. Drug Discov.*, 3, 1–12.

20. Vinogradov, S. V. (2006), Colloidal microgels in drug delivery applications, *Curr. Pharm. Des.*, 12, 4703–4712.

21. Jin, Y., Qiao, Y., Li, M., Ai, P., and Hou, X. (2005), Langmuir monolayers of the long-chain alkyl derivatives of a nucleoside analogue and the formation of self-assembled nanoparticles, *Coll. Surf. B: Biointerf.*, 42, 45–51.

22. Jin, Y., Tong, L., Ai, P., Li, M., and Hou, X. (2006), Self-assembled drug delivery systems. 1. Properties and in vitro/in vivo behavior of acyclovir self-assembled nanoparticles (SAN), *Int. J. Pharm.*, 309, 199–207.

23. Boas, U., and Heegaard, P. M. H. (2004), Dendrimers in drug research, *Chem. Soc. Rev.*, 33, 43–63.

24. Kroto, H. W., Heath, J. R., O'Brien, S.C., Curl, R. F., and Smalley, R. E. (1985), *Nature*, 318, 162–163.

25. Satoh, M., and Takayanagi, I. (2006), Pharmacological studies on fullerene (C60), a novel carbon allotrope, and its derivatives, *J. Pharm. Sci.*, 100, 513–518.

26. Michalet, X., Pinaud, F. F., Bentolila, L. A., Tsay, J. M., Doose, S., Li, J. J., Sundaresan, G., Wu, A. M., Gambhir, S. S., and Weiss, S. (2005), Quantum dots for live cells, in vivo imaging, and diagnostics, *Science*, 307, 538–544.

27. Muller, R. H., Mader, K., and Gohla, S. (2000), Solid lipid nanoparticles (SLN) for controlled drug delivery—A review of the state of the art, *Eur. J. Pharm. Biopharm.*, 50, 161–177.

28. Groneberg, D. A., Giersig, M., Welte, T., and Pison, U. (2006), Nanoparticle-based diagnosis and therapy, *Curr. Drug Targets*, 7, 643–648.

29. Couvreur, P., Barratt, G., Fattal, E., Legrand, P., and Vauthier, C. (2002), Nanocapsule technology: A review, *Crit. Rev. Ther. Drug Carrier Syst.*, 19, 99–134.

30. Barenholz, Y. (2001), Liposome application: Problems and prospects, *Curr. Opin. Coll. Interf. Sci.*, 6, 66–77.

31. Torchilin, V. P. (2005), Recent advances with liposomes as pharmaceutical carriers, *Nat. Rev. Drug Discov.*, 4, 145–160.

32. Uchegbu, I. F., and Vyas, S. P. (1998), Non-ionic surfactant based vesicles (niosomes) in drug delivery, *Int. J. Pharm.*, 172, 33–70.

33. Sadurni, N., Solans, C., Azemara, N., and Garcia-Celma, M. J. (2005), Studies on the formation of O/W nano-emulsions, by low-energy emulsification methods, suitable for pharmaceutical applications, *Eur. J. Pharm. Sci.*, 26, 438–445.

34. Solans, C., Izquierdo, P., Nolla, J., Azemara, N., and Garcia-Celma, M. J. (2005), Nano-emulsions, *Curr. Opin. Coll. Interf. Sci.*, 10, 102–110.

35. Jones, M.-C., and Leroux, J.-C. (1999), Polymeric micelles—A new generation of colloidal drug carriers, *Eur. J. Pharm. Biopharm.*, 48, 101–111.

36. Torchilin, V. P. (2007), Micellar nanocarriers: Pharmaceutical perspectives, *Pharm. Res.*, 24, 1–16.

37. Bianco, A., Kostarelos, K., and Prato, M. (2005), Applications of carbon nanotubes in drug delivery, *Curr. Opin. Chem. Biol.*, 9, 674–679.

38. Foley, S., Crowley, C., Smaihi, M., Bonfils, C., Erlanger, B. F., Seta, P., and Larroqueb, C. (2002), Cellular localisation of a water-soluble fullerene derivative, *Biochem. Biophys. Res. Commun.*, 294, 116–119.
39. Tsao, N., Luh, T.-Y., Chou, C.-K., Chang, T.-Y., Wu, J.-J., Liu, C.-C., and Lei, H.-Y. (2002), In vitro action of carboxyfullene, *J. Antimicrob. Chemother.*, 49, 641–649.
40. Ozkan, M. (2004), Quantum dots and other nanoparticles: What can they offer to drug discovery, *Drug Discov. Today*, 9, 1065–1071.
41. Oberdorster, G., Oberdorster, E., and Oberdorster, J. (2005), Nanotoxicology: An emerging discipline evolving from studies of ultrafine particles, *Environ. Health Perspect.*, 113, 823–839.
42. des Rieux, A., Fievez, V., Garinot, M., Schneider, Y.-J., and Preat, V. (2006), Nanoparticles as potential oral delivery systems of proteins and vaccines: A mechanistic approach, *J. Controlled Release*, 116, 1–27.
43. Lemos-Senna, E., Wouessidjewe, D., Lesieur, S., and Duchene, D. (1998), Preparation of amphiphilic cyclodextrin nanospheres using the emulsification solvent evaporation method. Influence of the surfactant on preparation and hydrophobic drug loading, *Int. J. Pharm.*, 170, 119–128.
44. Xiao, S., Tong, C., Liu, X., Yu, D., Liu, Q., Xue, C., Tang, D., and Zhao, L. (2006), Preparation of folate-conjugated starch nanoparticles and its application to tumor-targeted drug delivery vector, *Chin. Sci. Bull.*, 51, 1693–1697.
45. Sokolova, V. V., Radtke, I., Heumann, R., and Epple, M. (2006), Effective transfection of cells with multi-shell calcium phosphate-DNA nanoparticles, *Biomaterials*, 27, 3147–3153.
46. Ueno, Y., Futagawa, H., Takagi, Y., Ueno, A., and Mizushima, Y. (2005), Drug-incorporating calcium carbonate nanoparticles for a new delivery system, *J. Controlled Release*, 103, 93–98.
47. Souza, G., Christianson, D. R., Staquicini, F. I., Ozawa, M. G., Snyder, E. Y., Sidman, R. L., Miller, J. R., Arap, W., and Pasqualini, R. (2006), Networks of gold nanoparticles and bacteriophage as biological sensors and cell-targeting agents, *Proc. Nat. Acad. Sci. U.S.A.*, 103, 1215–1220.
48. Guinebretiere, S., Briancon, S., Lieto, J., Mayer, C., and Fessi, H. (2002), Study of the emulsion-diffusion of solvent: Preparation and characterization of nanocapsules, *Drug Dev. Res.*, 57, 18–33.
49. Hillaireau, H., Doan, T. L., Appel, M., and Couvreur, P. (2006), Hybrid polymer nanocapsules enhance in vitro delivery of azidothymidine-triphosphate to macrophages, *J. Controlled Release*, 116, 346–352.
50. New, R. R. C. (1990), *Liposomes: A Practical Approach*, Oxford University Press, Oxford, pp. 1–162.
51. Vinogradov, S., Batrakova, E., and Kabanov, A. (1999), Poly(ethylene glycol)-polyethylenimine NanoGel particles: Novel drug delivery systems for antisense oligonucleotides, *Coll. Surf. B: Biointerf.*, 16, 291–304.
52. Vinogradov, S. V., Kohli, E., and Zeman, A. D. (2005), Cross-linked polymeric nanogel formulations of 5′-triphosphates of nucleoside analogues: Role of the cellular membrane in drug release, *Mol. Pharm.*, 2, 449–461.
53. Bosman, A. W., Janssen, H. M., and Meijer, E. W. (1999), About dendrimers: Structure, physical properties, and applications, *Chem. Rev.*, 99, 1665–1688.
54. Gagner, J., Johnson, H., Watkins, E., Li, Q., Terrones, M., and Majewski, J. (2006), Carbon nanotube supported single phospholipid bilayer, *Langmuir*, 22, 10909–10911.

55. Kam, N. W. S., Liu, Z., and Dai, H. (2006), Carbon nanotubes as intracellular transporters for proteins and DNA: An investigation of the uptake mechanism and pathway, *Ang. Chem. Int. Ed.*, 45, 577–581.

56. Zakharian, T. Y., Seryshev, A., Sitharaman, B., Gilbert, B. E., Knight, V., and Wilson, L. J. (2005), A fullerene-paclitaxel chemotherapeutic: Synthesis, characterization, and study of biological activity in tissue culture, *J. Am. Chem. Soc.*, 127, 12508–12509.

57. Akerman, M. E., Chan, W. C. W., Laakkonen, P., Bhatia, S. N., and Ruoslahti, E. (2002), Nanocrystal targeting in vivo, *Proc. Nat. Acad. Sci. U.S.A.*, 99, 12617–12621.

58. Boulmedais, F., Bauchat, P., Brienne, M. J., Arnal, I., Artzner, F., Gacoin, T., Dahan, M., and Marchi-Artzner, V. (2006), Water-soluble pegylated quantum dots: From a composite hexagonal phase to isolated micelles, *Langmuir*, 22, 9797–9803.

59. Algar, W. R., and Krull, U. J. (2006), Adsorption and hybridization of oligonucleotides on mercaptoacetic acid-capped CdSe/ZnS quantum dots and quantum dot-oligonucleotide conjugates, *Langmuir*, 22, 11346–11352.

60. Matteucci, M. E., Hotze, M. A., Johnston, K. P., and Williams, III. R. O. (2006), Drug nanoparticles by antisolvent precipitation: Mixing energy versus surfactant stabilization, *Langmuir*, 22, 8951–8959.

61. Douroumis, D., and Fahr, A. (2006), Nano- and micro-particulate formulations of poorly water-soluble drugs by using a novel optimized technique, *Eur. J. Pharm. Biopharm.*, 63, 173–175.

62. Torchilin, V. P. (2006), Multifunctional nanocarriers, *Adv. Drug Deliv. Rev.*, 58, 1532–1555.

63. Banham, A. D., Standish, M. M., and Watkins, J. C. (1965), Diffusion of univalent ions across the lamellae of swollen phospholipids, *J. Mol. Biol.*, 13, 238–252.

64. Lasic, D. D., Joannic, R., Keller, B. C., Frederik, P. M., and Auvray, L. (2001), Spontaneous vesiculation, *Adv. Coll. Interf. Sci.*, 89–90, 337–349.

65. Bhalerao, S. S., and Harshal, A. R. (2003), Preparation, optimization, characterization, and stability studies of salicylic acid liposomes, *Drug Dev. Ind. Pharm.*, 29, 451–467.

66. Szoka, F., and Papahadjopoulos, D. (1978), Procedure for preparation of liposomes with large internal aqueous space and high capture by reverse-phase evaporation, *Proc. Nat. Acad. Sci. U. S. A.*, 75, 4194–4198.

67. Pons, M., Foradada, M., and Estelrich, J. (1993), Liposomes obtained by the ethanol injection method, *Int. J. Pharm.*, 95, 51–56.

68. Kikuchi, H., Yamauchi, H., and Hirota, S. A. (1994), Polyol dilution method for mass production of liposomes, *J. Liposome Res.*, 4, 71–91.

69. Wang, T., Deng, Y., Geng, Y., Gao, Z., Zou, J., and Wang, Z. (2006), Preparation of submicron unilamellar liposomes by freeze-drying double emulsions, *Biochim. Biophys. Acta*, 1758, 222–231.

70. Ning, M.-Y., Guo, Y.-Z., Pan, H.-Z., Yu, H.-M., and Gu, Z.-W. (2005), Preparation and evaluation of proliposomes containing clotrimazole, *Chem. Pharm. Bull.*, 53, 620–624.

71. Barnadas-Rodriguez, R., and Sabes, M. (2001), Factors involved in the production of liposomes with a high-pressure homogenizer, *Int. J. Pharm.*, 213, 175–186.

72. Taira, M. C., Chiaramoni, N. S., Pecuch, K. M., and Alonso-Romanowski, S. (2004), Stability of liposomal formulations in physiological conditions for oral drug delivery, *Drug Deliv.*, 11, 123–128.

73. Glavas-Dodov, M., Fredro-Kumbaradzi, E., Goracinova, K., Simonoska, M., Calis, S., Trajkovic-Jolevska, S., and Hincal, A. A. (2005), The effects of lyophilization on the stability of liposomes containing 5-FU, *Int. J. Pharm.*, 291, 79–86.

74. Liang, W., Levchenko, T. S., and Torchilin, V. P. (2004), Encapsulation of ATP into liposomes by different methods: Optimization of the procedure, *J. Microencapsul.*, 21, 251–261.
75. Haran, G., Cohen, R., Bar, L. K., and Barenholz, Y. (1993), Transmembrane ammonium sulfate gradients in liposomes produce efficient and stable entrapment of amphipathic weak bases, *Biochim. Biophys. Acta*, 1151, 201–215.
76. Zhigaltsev, I. V., Kaplun, A. P., Kucheryanu, V. G., Kryzhanovsky, G. N., Kolomeichuk, S. N., Shvets, V. I., and Yurasov, V. V. (2001), Liposomes containing dopamine entrapped in response to transmembrane ammonium sulfate gradient as carrier system for dopamine delivery into the brain of parkinsonian mice, *J. Liposome Res.*, 11, 55–71.
77. Hwang, S. H., Maitani, Y., Qi, X. R., Takayama, K., and Nagai, T. (1999), Remote loading of diclofenac, insulin and fluorescein isothiocyanate labeled insulin into liposomes by pH and acetate gradient methods, *Int. J. Pharm.*, 179, 85–95.
78. Abraham, S. A., Edwards, K., Karlsson, G., Hudon, N., Mayer, L. D., and Bally, M. B. (2004), An evaluation of transmembrane ion gradient-mediated encapsulation of topotecan within liposomes, *J. Controlled Release*, 96, 449–461.
79. Gulati, M., Grover, M., Singh, S., and Singh, M. (1998), Lipophilic drug derivatives in liposomes, *Int. J. Pharm.*, 165, 129–168.
80. Wagner, A., Vorauer-Uhl, K., and Katinger, H. (2002), Liposomes produced in a pilot scale: Production, purification and efficiency aspects, *Eur. J. Pharm. and Biopharm.*, 54, 213–219.
81. Wagner, A., Vorauer-Uhl, K., Kreismayr, G., and Katinger, H. (2002), The crossflow injection technique—An improvement of the ethanol injection method, *J. Liposome Res.*, 12, 259–270.
82. Wagner, A., Platzgummer, M., Kreismayr, G., Quendler, H., Stiegler, G., Ferko, B., Vecera, G., Vorauer-Uhl, K., and Katinger, H. (2006), GMP production of liposomes: A new industrial approach, *J. Liposome Res.*, 16, 311–319.
83. Woodle, M. C., and Scaria, P. (2001), Cationic liposomes and nucleic acids, *Curr. Opin. Coll. Interf. Sci.*, 6, 78–84.
84. Templeton, N. S. (2002), Cationic liposome-mediated gene delivery in vivo, *Biosci. Rep.*, 22, 283–295.
85. Ishida, T., Harashima, H., and Kiwada, H. (2002), Liposome clearance, *Biosci. Rep.*, 22, 197–224.
86. Winterhalter, M., Frederik, P. M., Vallner, J. J., and Lasic, D. D. (1997), Stealth® liposomes: From theory to product, *Adv. Drug Deliv. Rev.*, 24, 165–177.
87. Ishida, T., Ichihara, M., Wang, X., and Kiwada, H. (2006), Spleen plays an important role in the induction of accelerated blood clearance of PEGylated liposomes, *J. Controlled Release*, 115, 243–250.
88. Moghimi, S. M., Hamad, I., Andresen, T. L., Jorgensen, K., and Szebeni, J. (2006), Methylation of the phosphate oxygen moiety of phospholipid-methoxy(polyethylene glycol) conjugate prevents PEGylated liposome-mediated complement activation and anaphylatoxin production, *FASEB J.*, 20, 2591–2593.
89. Drummond, D. C., Zignani, M., and Leroux, J.-C. (2000), Current status of pH-sensitive liposomes in drug delivery, *Prog. Lipid Res.*, 39, 409–460.
90. Yatvin, M. B., Weinstein, J. N., Dennis, W. H., and Blumenthal, R. (1978), Liposomes and local hyperthermia: Selective delivery of methotrexate to heated tumors, *Science*, 202, 1290–1293.
91. Kono, K. (2001), Thermosensitive polymer-modified liposomes, *Adv. Drug Deliv. Rev.*, 53, 307–319.

92. Zhang, J. Q., Zhang, Z. R., Yang, H., Tan, Q. Y., Qin, S. R., and Qiu, X. L. (2005), Lyophilized paclitaxel magnetoliposomes as a potential drug delivery system for breast carcinoma via parenteral administration: In vitro and in vivo studies, *Pharm. Res.*, 22, 573–583.

93. Koning, G. A., Kamps, J. A. A. M., and Scherphof, G. L. S. (2002), Efficient intracellular delivery of 5-fluorodeoxyuridine into colon cancer cells by targeted immunoliposomes, *Cancer Detection Prevention*, 26, 299–307.

94. Mehnert, W., and Mader, K. (2001), Solid lipid nanoparticles: Production, characterization and applications, *Adv. Drug Deliv. Rev.*, 47, 165–196.

95. Shive, M. S., and Anderson, J. M. (1997), Biodegradation and biocompatibility of PLA and PLGA microspheres, *Adv. Drug Deliv. Rev.*, 28, 5–24.

96. Sommerfeld, P., Schroeder, U., and Sabel, B. A. (1997), Long-term stability of PBCA nanoparticle suspensions suggests clinical usefulness, *Int. J. Pharm.*, 155, 201–207.

97. Sommerfeld, P., Schroeder, U., and Sabel, B. A. (1998), Sterilization of unloaded polybutylcyanoacrylate nanoparticles, *Int. J. Pharm.*, 164, 113–118.

98. Chorny, M., Fishbein, I., Danenberg, H. D., and Golomb, G. (2002), Lipophilic drug loaded nanospheres prepared by nanoprecipitation: Effect of formulation variables on size, drug recovery and release kinetics, *J. Controlled Release*, 83, 389–400.

99. Govender, T., Stolnik, S., Garnett, M. C., Illum, L., and Davis, S. S. (1999), PLGA nanoparticles prepared by nanoprecipitation: Drug loading and release studies of a water soluble drug, *J. Controlled Release*, 57, 171–185.

100. Bilati, U., Allemann, E., and Doelker, E. (2005), Development of a nanoprecipitation method intended for the entrapment of hydrophilic drugs into nanoparticles, *Eur. J. Pharm. Sci.*, 24, 67–75.

101. Bilati, U., Allemann, E., and Doelker, E. (2005), Nanoprecipitation versus emulsion-based techniques for the encapsulation of proteins into biodegradable nanoparticles and process-related stability issues, *AAPS Pharm. Sci. Tech.*, 6, 74.

102. Astete, C. E., and Sabliov, C. M. (2006), Synthesis and characterization of PLGA nanoparticles, *J. Biomater. Sci. Polym. Ed.*, 17, 247–289.

103. Zwiorek, K., Kloechner, J., Wagner, E., and Coester, C. (2004), Gelatin nanoparticle as a new and simple gene delivery system, *J. Pharm. Pharm. Sci.*, 7, 22–28.

104. Cheng, J., Teply, B. A., Sherifi, I., Sung, J., Luther, G., Gu, F. X., Levy-Nissenbaum, E., Radovic-Moreno, A. F., Langer, R., and Farokhzad, O. C. (2007), Formulation of functionalized PLGA-PEG nanoparticles for in vivo targeted drug delivery, *Biomaterials*, 28, 869–876.

105. Hillery, A. M., and Florence, A. T. (1996), The effect of adsorbed poloxamer 188 and 407 surfactants on the intestinal uptake of 60-nm polystyrene particles after oral administration in the rat, *Int. J. Pharm.*, 132, 123–130.

106. Peracchia, M. T., Vauthier, C., Puisieux, F., and Couvreur, P. (1997), Development of sterically stabilized poly(isobutyl 2-cyanoacrylate) nanoparticles by chemical coupling of poly(ethylene glycol), *J. Biomed. Mater. Res.*, 34, 317–326.

107. Suk, J. S., Suh, J., Choy, K., Lai, S. K., Fu, J., and Hanes, J. (2006), Gene delivery to differentiated neurotypic cells with RGD and HIV Tat peptide functionalized polymeric nanoparticles, *Biomaterials*, 27, 5143–5150.

108. Aktas, Y., Yemisci, M., Andrieux, K., Gursoy, R. N., Alonso, M. J., FernAndez-Megia, E., Novoa-Carballal, R., Quinoa, E., Riguera, R., Sargon, M. F., Celik, H. H., Demir, A. S., Hincal, A. A., Dalkara, T., Capan, Y., and Couvreur, P. (2005), Development and brain delivery of chitosan-PEG nanoparticles: Functionalized with the monoclonal antibody OX26, *Bioconjugate Chem.*, 16, 1503–1511.

109. Park, J. H., Kwon, S., Nam, J.-O., Park, R.-W., Chung, H., Seo, S. B., Kim, I.-S., Kwon, I. C., and Jeong, S. Y. (2004), Self-assembled nanoparticles based on glycol chitosan bearing 5h-cholanic acid for RGD peptide delivery, *J. Controlled Release*, 95, 579–588.

110. Klier, J., Tucker, C. J., Kalantar, T. H., and Green, D. P. (2000), Properties and applications of microemulsions, *Adv. Mater.*, 12, 1751–1757.

111. Muller, R. H., Radtke, M., and Wissing, S. A. (2002), Solid lipid nanoparticles (SLN) and nanostructured lipid carriers (NLC) in cosmetic and dermatological preparations, *Adv. Drug Deliv. Rev.*, 54, S131-S155.

112. Goppert, T. M., and Muller, R. H. (2003), Plasma protein adsorption of Tween 80- and Poloxamer 188-stabilized solid lipid nanoparticles, *J. Drug Target.*, 11, 225–231.

113. Wong, H. L., Bendayan, R., Rauth, A. M., and Wu, X. Y. (2006), Simultaneous delivery of doxorubicin and GG918 (Elacridar) by new polymer-lipid hybrid nanoparticles (PLN) for enhanced treatment of multidrug-resistant breast cancer, *J. Controlled Release*, 116, 275–284.

114. Hu, F. Q., Hong, Y., and Yuan, H. (2004), Preparation and characterization of solid lipid nanoparticles containing peptide, *Int. J. Pharm.*, 273, 29–35.

115. Garcia-Fuentes, M., Torres, D., and Alonso, M. J. (2002), Design of lipid nanoparticles for the oral delivery of hydrophilic macromolecules, *Coll. Surf. B Biointerf.*, 27, 159–168.

116. Brannon-Peppas, L., and Blanchette, J. O. (2004), Nanoparticle and targeted systems for cancer therapy, *Adv. Drug Deliv. Rev.*, 56, 1649–1659.

117. Lockman, P. R., Mumper, R. J., Khan, M. A., and Allen, D. D. (2002), Nanoparticle technology for drug delivery across the blood-brain barrier, *Drug Dev. Ind. Pharm.*, 28, 1–13.

118. Nishioka, Y., and Yoshino, H. (2001), Lymphatic targeting with nanoparticulate system, *Adv. Drug Deliv. Rev.*, 47, 55–64.

119. Csaba, N., Caamano, P., Sanchez, A., Dominguez, F., and Alonso, M. J. (2005), PLGA: Poloxamer and PLGA:Poloxamine blend nanoparticles: New carriers for gene delivery, *Biomacromolecules*, 6, 271–278.

120. Bagwe, R. P., Kanicky, J. R., Palla, B. J., Patanjali, P. K., and Shah, D. O. (2001), Improved drug delivery using microemulsions: Rationale, recent progress, and new horizons, *Crit. Rev. Ther. Drug Carrier Syst.*, 18, 77–140.

121. Nayak, S., and Lyon, L. A. (2005), Soft nanotechnology with soft nanoparticles, *Ang. Chem. Int. Ed.*, 44, 7686–7708.

122. Tomalia, D. A., Baker, H., Dewald, J. R., Hall, M., Kallos, G., Martin, S., Roeck, J., Ryder, J., and Smith, P. (1985), A new class of polymers: Starburst-dendritic macromolecules, *Polym. J.*, 17, 117–132.

123. Tomalia, D. A., Baker, H., Dewald, J. R., Hall, M., Kallos, G., Martin, S., Roeck, J., Ryder, J., and Smith, P. (1986), Dendritic macromolecules: Synthesis of starburst dendrimers, *Macromolecules*, 19, 2466–2468.

124. Svenson, S., and Tomalia, D. A. (2005), Dendrimers in biomedical applications—Reflections on the field, *Adv. Drug Deliv. Rev.*, 57, 2106–2129.

125. Lacerda, L., Bianco, A., Prato, M., and Kostarelos, K. (2006), Carbon nanotubes as nanomedicines: From toxicology to pharmacology, *Adv. Drug Deliv. Rev.*, 58, 1460–1470.

126. Smart, S. K., Cassady, A. I., Lu, G. Q., and Martin, D. J. (2006), The biocompatibility of carbon nanotubes, *Carbon*, 44, 1034–1047.

127. Keck, C. M., and Muller, R. H. (2006), Drug nanocrystals of poorly soluble drugs produced by high pressure homogenisation, *Eur. J. Pharm. Biopharm.*, 62, 3–16.

128. Guerin, C. B. E., and Szleifer, I. (1999), Self-assembly of model nonionic amphiphilic molecules, *Langmuir*, 15, 7901–7911.

129. Choi, S. K., Vu, T. K., Jung, J. M., Kim, S. J., Jung, H. R., Chang, T., and Kim, B. H. (2005), Nucleoside-based phospholipids and their liposomes formed in water, *ChemBioChem*, 6, 432–439.

130. Snip, E., Shinkai, S., and Reinhoudt, D. N. (2001), Organogels of a nucleobase-bearing gelator and the remarkable effects of nucleoside derivatives and a porphyrin derivative on the gel stability, *Tetrahedr. Lett.*, 42, 2153–2156.

131. Engberts, J. B. F. N., and Hoekstra, D. (1995), Vesicle-forming synthetic amphiphiles, *Biochim. Biophys. Acta*, 1241, 323–340.

132. Giulieri, F., and Krafft, M. P. (2003), Tubular microstructures made from nonchiral single-chain fluorinated amphiphiles: Impact of the structure of the hydrophobic chain on the rolling-up of bilayer membrane, *J. Coll. Interf. Sci.*, 258, 335–344.

133. Giorgi, T., Lena, S., Mariani, P., Cremonini, M. A., Masiero, S., Pieraccini, S., Rabe, J. P., Samori, P., Spada, G. P., and Gottarelli, G. (2003), Supramolecular helices via self-assembly of 8-oxoguanosines, *J. Am. Chem. Soc.*, 125, 14741–14749.

134. Yanagawa, H., Ogawa, Y., Furuta, H., and Tsuno, K. (1989), Spontaneous formation of superhelical strands, *J. Am. Chem. Soc.*, 111, 4567–4570.

135. Zemel, A., Fattal, D. R., and Ben-Shaul, A. (2003), Energetics and self-assembly of amphipathic peptide pores in lipid membranes, *Biophys. J.*, 84, 2242–2255.

136. Chabaud, P., Camplo, M., Payet, D., Serin, G., Moreau, L., Barthelemy, P., and Grinstaff, M. W. (2006), Cationic nucleoside lipids for gene delivery, *Bioconjugate Chem.*, 17, 466–472.

137. Vaizoglu, M. O., and Speiser, P. P. (1986), Pharmacosomes—A novel drug delivery system, *Acta Pharm. Suec.*, 23, 163–172.

138. Schreier, S., Malheiros S. V. P., and de Paula, E. (2000), Surface active drugs: Self-association and interaction with membranes and surfactants. Physicochemical and biological aspects, *Biochim. Biophys. Acta*, 1508, 210–234.

139. Israelachvili, J. N. (1992), *Intermolecular and Surface Forces with Applications to Colloidal and Biological Systems*, Academic, London.

140. Jin, Y., Li, M., Tong, L., Wang, L., and Peng, T. (2003), Lipid derivatives of nucleoside analogues and their salts, China Patent CN1259331.

141. Jin, Y., and Ai, P. (2004), Cholesteryl derivatives of nucleoside analogues, China Patent CN1566130.

142. Jin, Y., Du, L., Xing, L., and Xin, R. (2006), Cholesteryl phosphoryl derivatives of nucleoside analogues, China Patent Application 2006101122957.

143. Jin, Y., Qiao, Y., and Hou, X. (2006), The effects of chain number and state of lipid derivatives of nucleosides on hydrogen bonding and self-assembly through the investigation of Langmuir-Blodgett films, *Appl. Surf. Sci.*, 252, 7926–7929.

144. Jin, Y. (2007), Effect of temperature on the state of the self-assembled nanoparticles prepared from an amphiphilic lipid derivative of acyclovir, *Coll. Surf. B Biointerf.*, 54, 124–125.

145. Aquaro, S., Calio, R., Balzarini, J., Bellocchi, M. C., Garaci, E., and Perno, C. F. (2002), Macrophages and HIV infection: Therapeutical approaches toward this strategic virus reservoir, *Antivir. Res.*, 55, 209–225.
146. European Science Foundation (2005), Nanomedicine, An ESF-European Medical Research Council (EMRC) forward look report, European Science Foundation, Strasbourg.
147. Wagner, V., Dullaart, A., Bock, A.-K., and Zweck, A. (2006), The emerging nanomedicine landscape, *Nat. Biotechnol.*, 24, 1211–1217.

7.3

PHARMACEUTICAL NANOSYSTEMS: MANUFACTURE, CHARACTERIZATION, AND SAFETY

D. F. Chowdhury
University of Oxford, Oxford, United Kingdom

Contents

Pharmaceutical Manufacturing Handbook: Production and Processes, edited by Shayne Cox Gad

7.3.1 DEFINITION

Nanotechnology is an enabling technology and one which is generally manifest at the primary level in the form of nanomaterials. The definition of nanotechnology therefore focuses on materials and how manipulation at the nanoscale leads to novel properties and therefore potentially new uses. The pharmaceutical industry has yet to adopt strict guidelines for what falls under the remit of nanotechnology, with numerous definitions in existence. For the purpose of this chapter, the current U.S. Food and Drug Administration (FDA) definition for nanotechnology as applied to pharmaceuticals is deemed most appropriate. The FDA describes nanotechnology as technology that includes the following [1]:

1. Research and technology development or products regulated by the FDA that are at the atomic, molecular or macromolecular levels and where at least one dimension that affects the functional behavior of the product is in the length scale range of approximately 1–100 nm
2. Creating and using structures, devices, and systems that have novel properties and functions because of their small and/or intermediate size
3. Ability to control or manipulate at the atomic scale

Nanotechnology is therefore essentially about understanding and manipulating materials at the atomic, molecular, and macromolecular level in a way that imparts properties to the material that would otherwise not exist either as individual atoms or as bulk processed macroscopic systems.

Properties that can be exploited to provide novel and unique properties to materials include surface and quantum effects, for example, van der Waals forces; electrostatic interaction; ionic, covalent and hydrogen bonding; and quantum confinement. Additionally nonconventional means of molecular assembly and atomic manipulation can lead to novel material properties. Control and exploitation of these effects can lead to new and useful changes to the thermal, magnetic, electrical, optical and mechanical, and biological and physicochemical properties of materials.

7.3.1.1 Top-Down and Bottom-Up Approaches to Nanotechnology

There are generally two approaches to nanotechnology, the top-down and bottom-up approaches. As the names suggest, the top-down approach utilizes ultraprecision machining and nanolithographic techniques among others to achieve very high definition structures with nanolevel accuracy, usually either by removing material from the surface of a larger structure until the desired structure with desired features is achieved or through deposition of material with almost atomic-scale precision and control. The bottom-up approach involves the assembly of atoms, molecules, or nanoscale components to assemble a larger structure within the nanoscale range. There are numerous methods by which this can be achieved, including conventional bulk chemical processing methods and exploitation of chemical and biological self-assembly techniques. The pharmaceutical industry is primarily involved in the application of nanomaterials rather than the discovery and development of new materials, though as this chapter will indicate, there are often areas of overlap between what is a new material and a construct of a novel material.

7.3.2 TAXONOMY OF NANOMEDICINE TECHNOLOGIES

A useful starting point would be to gauge the breadth of technologies falling under the classification of nanomedicine. Table 1 provides a means of classifying materials and processes derived from nanotechnology as relating to pharmaceuticals and medicine in general.

7.3.3 NANO–PHARMACEUTICAL SYSTEMS

Having gauged the huge scope of nanotechnology in medicine, the scope of this chapter is limited to pharmaceuticals. The term *pharmaceutical* is intended in the context of systems pertaining to drugs or dosage formulations. Nano–pharmaceutical systems generally imply products that may in their own right or in combination with another moiety bring about therapeutic benefit. They may also include engineered nanostructured systems that may act as a carrier for drugs or a delivery vehicle or a delivery system for drugs and therapeutic agents. The definition is extended to include imaging systems which may be used alone or in conjunction with therapeutic agents given the numerous nanosystems that have found application in diagnostics and imaging.

The various nanoscale architectures that can be achieved using nanotechnology include spheres (solid or hollow), tubes, porous particles, solid particles, and branched structures, and with the rapid evolution of lithographic techniques, three-dimensional objects of almost any desired shape can be achieved from both metals and polymers. Given the vast spectrum of materials of construction, size, shape, and form covered by nanosystems, a simple means of classification is needed for effective differentiation between systems. Nanosystems can be classified in a number of ways, for example, according to their elemental composition, according to size, structure, and function, or perhaps according to a structure–function relationship. It can be said that nanosystems fall into the broader category of nanostructures, which can be generalized into the following categories:

Particulate nanostructures
Capsular nanostructures
Crystalline nanostructures
Polymeric nanostructures

It can be readily appreciated that there will be an element of overlap between the broad categories above and in particular with evolving complex hybrid systems. The classification is intended as a point of reference and for ease of understanding the vast possibilities that exist with nanosystems without the need for constant reclassification as far as possible. These structures may be further differentiated according to their primary composition:

Organic
Inorganic
Organic/inorganic hybrid
Carbon based

TABLE 1 Partial Nanomedicine Technologies Taxonomy

- Raw nanomaterials
  - Nanoparticle coatings
  - Nanocrystalline materials
- Nanostructured materials
  - Cyclic peptides
  - Dendrimers
  - Detoxification agents
  - Fullerenes
  - Functional drug carriers
  - Magnetic resonance (MR) scanning (nanoparticles)
  - Nanobarcodes
  - Nanoemulsions
  - Nanofibers
  - Nanoparticles
  - Nanoshells
  - Carbon nanotubes
  - Quantum dots
- Artificial binding sites
  - Artificial antibodies
  - Artificial enzymes
  - Artificial receptors
  - Molecularly imprinted polymers
- Cell simulations and cell diagnostics
  - Cell chips
  - Cell stimulators
- DNA manipulation, sequencing, diagnostics
  - Genetic testing
  - Deoxyribonucleic acid (DNA) microarrays
  - Ultrafast DNA sequencing
  - DNA manipulation and control
- Tools and diagnostics
  - Bacterial detection systems
  - Biochips
  - Biomolecular imaging
  - Biosensors and biodetection
  - Diagnostic and defense applications
  - Endoscopic robots and microscopes
  - Fullerene-based sensors
  - Imaging (e.g., cellular)
  - Lab on chip
  - Monitoring
  - Nanosensors
  - Point-of-care diagnostics
  - Protein microarrays
  - Scanning probe microscopy
- Control of surfaces
  - Artificial surfaces, adhesive
  - Artificial surfaces, nonadhesive
  - Artificial surfaces, regulated
  - Biocompatible surfaces
  - Biofilm suppression
  - Engineered surfaces
  - Pattern surfaces (contact guidance)
  - Thin-film coatings
- Nanopores
  - Immunoisolation
  - Molecular sieves and channels
  - Nanofiltration membranes
  - Nanopores
  - Separations
- Biological research
  - Nanobiology
  - Nanoscience in life sciences
- Drug delivery
  - Drug discovery
  - Biopharmaceuticals
  - Drug delivery
  - Drug encapsulation
  - Smart drugs
- Synthetic biology and early nanodevices
  - Dynamic nanoplatform "nanosome"
  - Tecto-dendrimers
  - Artificial cells and liposomes
  - Polymeric micelles and polymersomes
- Nanorobotics
  - DNA-based devices and nanorobots
  - Diamond-based nanorobots
  - Cell repair devices
- Intracellular devices
  - Intracellular assay
  - Intracellular biocomputers
  - Intracellular sensors/reporters
  - Implant inside cells
- BioMEMS
  - Implantable materials and devices
  - Implanted bio—microelectromechanical systems (MEMSs), chips, and electrodes
  - MEMS/nanomaterial-based prosthetics
  - Sensory aids (e.g.,)
  - Microarrays
  - Microcantilever-based sensors
  - Microfluidics
  - Microneedles
  - Medical MEMS
  - MEMS surgical devices

TABLE 1 *Continued*

Molecular medicine	Biotechnology and biorobotics
Genetic therapy	Biological viral therapy
Pharmacogenomics	Virus-based hybrids
Artificial enzymes and enzyme control	Stem cells and cloning
Enzyme manipulation and control	Tissue engineering
Nanotherapeutics	Artificial organs
Antibacterial and antiviral nanoparticles	Nanobiotechnology
Fullerene-based pharmaceuticals	Biorobotics and biobots
Photodynamic therapy	
Radiopharmaceuticals	

Source: From ref. 2.

7.3.4 DESCRIPTION OF NANOSYSTEMS

There is clear evidence from Table 2 that some of the nanosystems indicated are based on conventional colloidal chemistry, and their characteristics are well established and understood. The descriptions below deal mainly with systems deemed nonconventional and cover some of the key novel properties derived from or utilized as part of their construction.

7.3.4.1 Polymeric Systems

Polymer-based systems offer numerous advantages, such as biocompatibility, biodegradability, and ability to incorporate functional groups for attachment of drugs. Drugs can be incorporated into the polymer matrix or in the cavity created by the polymeric architecture, from which the drug molecule can be released with an element of temporal control, and controlled pharmacokinetic profile with almost zero-order release achievable.

Dendrimers are large complex globular polymeric molecules [46] with well-defined chemical structure, size, and shape [47]. They consist of characteristic three-dimensional branched structures. The key components of dendrimers are their core, branches, and end groups, and precise control over these features is possible during the bottom-up synthesis process, thus allowing control over size composition, and final chemical reactivity. A more advanced form of dendrimer is the hyperbranched dendrimer, where precision control over the architectural construct is lost during the synthesis process [48]. Dendrimers are produced from monomers through an iterative sequence of reaction steps [49] using either convergent [50, 51] or divergent [52, 53] step growth polymerization [54–56] and have potential applications in gene and cancer therapy and drug delivery through complexation or encapsulation [57–59].

7.3.4.2 Quantum Dots and Quantum Confinement

Quantum dots are inorganic semiconductor nanocrystals that possess physical dimensions smaller than the exciton Bohr radius, giving rise to the unique phenom-

TABLE 2 Classification of Nanostructures According to Composition and Perceived Applications

Composition	Type of Nanostructure	Applications
Organic	Polymer micelles [3, 4]	Drug delivery
	Polymeric spheres [5]	Drug delivery
	Polymer nanoparticles [6, 7]	Drug delivery
	Polymer vesicles/containers [8–10]	Drug delivery
	Lipid nanovesicles [11]	Drug delivery
	Lipid emulsions [12]	Drug delivery
	Ring peptides [13]	Drug delivery
	Lipid nanospheres [14, 15]	Drug delivery
	Lipid nanoparticles [16, 17]	Drug delivery
	Lipid nanotubes [18]	Drug delivery
	Peptide nanoparticles [19, 20]	Drug delivery
	Nanobodies [21]	Therapeutic, diagnostic
	Dendrimers [22–24]	Drug delivery, gene delivery
Inorganic	Palladium/platinum nanoparticles [25]	Drug delivery
	Silicon nanoneedles [26]	Drug delivery
	Porous silicon [27, 28]	Drug delivery
	Gold nanoparticles [29]	Drug delivery
	Iron oxide nanoparticles [30]	Imaging
	Gold nanoshells [31]	Imaging agent, thermal ablation
	Quantum dots [32]	Imaging, cell targeting
	Metallic nanoshells [33]	Imaging, thermal ablation
	Nanocrystals [34, 35]	Drug delivery
Organic/inorganic hybrid	Nanocomposites [36]	Drug delivery
	Nanosphere–metallic particle composite [37]	Drug delivery, imaging
	Carbon nanotube clusters [38]	Drug delivery, imaging, thermal ablation
	Core–shell structures [39, 40]	Imaging, thermal ablation
Carbon based	Fullerenes [41–43]	Drug delivery, prodrug
	Carbon nanotubes [44, 45]	Drug delivery, imaging, thermal ablation

enon known as quantum confinement. Quantum confinement is the spatial confinement of charge carriers (i.e., electrons and holes) within materials. It leads to unique optical and electrical properties that are not common for bulk solids.

Quantum dots have novel and unique optical, magnetic [60], and electronic properties, exceptional imaging properties due to high color intensity, with up to 20 fluorophores, high resistance to photobleaching, and narrow spectral line widths. Their size and composition allow for tunable emission that can be excited using a single wavelength [61–64]. These properties have led to uses as fluorescent imaging probes, detection of cell signaling pathways, and cell targeting. Low depth of light penetration and relatively high background fluorescence are the key limitations of quantum dots in in vivo clinical applications. Quantum dots are generally of the size rage 2–8 nm in diameter [65] and have large molar extinction coefficients [66], thus making them very bright in vivo probes.

Applications of quantum dots include optical detection of genes and proteins in animal models and cell assays and tumor and lymph node visualization through imaging [67, 68].

7.3.4.3 Metal Nanoparticles and Surface Plasmon Resonance

Surface plasmons, also known as surface plasmon polaritons or packets of electrons, are surface electromagnetic waves that propagate parallel along a metal–dielectric interface [69, 70]. Surface plasmons exist where the complex dielectric constants of the two media are of opposite sign. The excitation of surface plasmons by light of a wavelength matching the resonant frequency of the electrons is termed *surface plasmon resonance* (SPR) for planar surfaces and *localized surface plasmon resonance* where nanometer-sized metallic structures are concerned [71, 72].

Surface plasmon effects result in useful photothermal effects [73] and have been used to enhance the surface sensitivity of various spectroscopic measurements [74], including fluorescence, Raman scattering, and second-harmonic generation.

Metallic nanostructures exhibiting SPR are composed of a dielectric core and metallic shell, for example, gold sulfide dielectric core and gold shell. By varying the core–shell thickness ratio, the surface plasmon resonance is shifted from the visible to the infrared range [75], spanning a range that is mostly transparent to human tissue, that is, has a high physiological transmissivity. Additionally, control over the particle diameter allows control over light scattering and light absorption at particle diameters below approximately 75 nm. Potential applications of the photothermal effects of engineered nanoparticles include the following:

Controlled drug delivery [76]
Analysis of controlled drug release from a matrix [77]
DNA sensor [78, 79]
Deep tissue tumor cell thermal ablation [80]
Real-time assessment of drug action [81, 82]
Immunosensor applications [83, 84]

7.3.4.4 Self-Assembled Systems

Molecular self-assembly is a synthetic technique that has been widely used to produce nano- and microstructures in a quick and efficient manner. It has become all the more crucial to the formation of nanostructures due to the control attainable over the end product and the relative ease with which nanostructures of defined structure and function can be produced using bulk manufacturing methods.

The basic principle of self-assembly is based on the simultaneous coexistence of two parallel forces [85, 86], long-range repulsive forces and short-range attractive interactions.

The types of structures attainable using molecular self-assembly are referred to as micellar structures [87] and can take on various sizes and shapes:

Direct spherical micelles [88]
Inverse spherical micelles [89, 90]

Lamellar sheets [91]
Vesicles (hollow or concentric) [92]
Body-centerd-cubic [92]
Hexagonally packed cylinders/tubes [92]
Gyroids [93]
Hollow spheres [94]

These systems have found widespread use as drug delivery vehicles, and the more advanced nanosystems are termed *smart nano-objects* due to their ability to sense local variations in physiological conditions, such as pH and temperature, and respond to the stimulus accordingly.

7.3.4.5 Nanostructures Based on Carbon

Nanotubes The structure of carbon nanotubes as observed by scanning tunneling microscopy is that of rolled grapheme sheets where endpoints of a translation vector are folded one onto another [95]. Single-walled carbon nanotubes (SWCNTs) were first reported by Iijima and Ichihashi [96] in 1993. Enormous interest in CNTs has centered around their unique properties, including high electrical conductivity, thermal conductivity, high strength and aspect ratio, ultralight weight, and excellent chemical and thermal stability.

The most common method for the production of carbon nanotubes is hydrocarbon-based chemical vapor deposition (CVD) [97] and adaptations of the CVD process [98, 99], where the nanotubes are formed by the dissolution of elemental carbon into metal nanoclusters followed by precipitation into nanotubes [100]. The CVD method is used to produce multiwalled carbon nanotubes (MWCNTs) [101] and double-walled carbon nanotubes (DWCNTs) [102] as well as SWCNTs [103]. The biomedical applications of CNTs have been made possible through surface functionalization of CNTs, which has led to drug and vaccine delivery applications [104, 105].

Fullerenes Fullerenes were first discovered in 1985 [106] and are large molecules composed exclusively of carbon atoms and manifest physically in the form of hollow spherical cagelike structures. The cages are in the region of 7–15 Å in diameter with the most common form being C_{60}, though other forms exist too, such as C_{70}, C_{76}, and C_{84}, depending on the number of carbon atoms making up the cage. Fullerenes can be produced using combustion [107] and arc discharge methods [108].

Fullerenes offer numerous points of attachment and allow precise bonding of active chemical groups in three-dimensional (3D) conformations and positional control with respect to matching conjugated fullerene compounds with a given target. Water-soluble fullerenes have shown low biological toxicity both in vitro [109] and in vivo [110]. Some of the potential applications of fullerenes in pharmaceuticals include their use in neurodegenerative and other disease conditions where oxidative stress is part of the pathogenesis due to their powerful antioxidant properties [111] and in nuclear medicine for binding of toxic metals ions, increasing therapeutic potency of radiation therapy and reducing adverse events as fullerenes do not undergo biochemical degradation within the body. Fullerene applications in photodynamic tumor therapy have also been shown [112].

7.3.5 MANUFACTURING TECHNOLOGIES

There are a diverse range of technologies being applied to the manufacture of nanosystems for pharmaceutical applications. Some of these are derived from conventional pharmaceutical technologies, such as colloidal processing, and many have been adopted from the semiconductor industry, whereby precision spatial control is achieved over the production of nanosystems and particles using fabrication techniques. To add to this, new technologies are constantly evolving through the adaptation and amalgamation of existing technologies in different fields or through pure innovation leading to completely new processes. It is outside the remit of this chapter to cover in any depth all those manufacturing technologies that may be applied to pharmaceutical manufacturing. The summary in Table 3 provides a detailed synopsis of the different types of manufacturing processes and types of technologies for each process. This is followed by a brief introduction to some of the technologies, with the omission of silicon- and carbon-based fabrication processes, which are beyond the scope of this chapter, to provide the reader with a starting point for further detailed study and investigation into those processes and technologies that may be most suited to their particular product or concept.

TABLE 3 Summary of Manufacturing Processes and Technologies for Producing Nanosystems

Manufacturing Process	Technology
Nanoscale assembly	Self-assembling micellar structures [113, 114]
	Bio-self-assembly and aggregation [115, 116]
	Nanomanipulation [117, 118]
	Soft lithography [119, 120]
	Molecular imprinting [121, 122]
	Layer-by-layer electrostatic deposition [123, 124]
	Chemical vapor deposition [125]
Nanostructuring processes for polymeric materials	Mold replication [126, 127]
	Colloidal lithography [128, 129]
	X-ray lithography [130]
	Interfacial polymerization [131, 132]
	Nanoprecipitation [133, 134]
	Emulsion solvent evaporation [135]
	Nanoimprinting [136]
	Electrospinning [137, 138]
Nanostructuring processes for silicon	Photolithographic fabrication
	X-ray lithography [130]
	Electron beam lithography [139]
	Chemical etching [140]
	Physical and chemical vapor deposition [141]
Nanostructuring processes for carbon	Electric arc discharge [142, 143]
	Laser ablation [144, 145]
	Chemical vapor deposition [146]
	Combustion [147]

7.3.5.1 Nanoscale Assembly Methods[1]

Self-Assembly through Micelle Formation Self-assembly at the nanoscale is deemed important to be able to produce commercially viable products and processes, since it offers a mode of bulk production with control over features such as size, shape, and morphology at the nanoscale. The basic principle of self-assembly is based on the simultaneous coexistence of two parallel forces:

Long-range repulsive interactions between incompatible domains
Short-range attractive interactions

If we take the example of an amphiphilic diblock copolymer, the polymer is composed of two blocks, a hydrophobic block and a hydrophilic block. When introduced to a solvent beyond a minimum concentration, the critical micelle concentration (CMC), the monomers begin to orientate such that the block that is soluble in the solvent orients itself toward the periphery, in contact with the continuous media, and the insoluble portion turns toward the core in an attempt to minimize contact with the continuous phase, thus leading to the formation of a micelle. The long-range repulsive forces arise from the relative solubilities of the blocks in the solvent, and the short-range attractive forces arise from the covalent link between the two blocks. The basic theory of micelle formation using block copolymers is outlined below since nanosystem and nano-object self-assembly is likely to be facilitated by such polymer systems, and similar principles will apply or aid toward developing self-assembling systems.

Key factors that affect micelle formation are as follows:

Equilibrium constant
Solvent type
Solvent quality
Critical micelle temperature (CMT)
Critical micelle concentration
Overall molar mass of the micelle, M_W
Micelle aggregation number, Z
Copolymer architecture
Relative block lengths
Relative geometries of copolymer blocks
Polymer composition
Core–corona interfacial tension

These key factors will influence the following micelle characteristics:

Hydrodynamic radius of micelles formed, R_H
Radius of gyration, R_G

[1]The description in this section has been summarized and adapted from J. Rodriguez-Hernandez et al., Toward "smart" nano-objects by self-assembly of block copolymers in solution, *Progress in Polymer Science*, 30 (2005), 691–724 [148].

Ratio of hydrodynamic radius, R_H, to radius of gyration, R_H
Micelle core radius, R_C
Micelle corona thickness, C

Micellar structures can be produced either by addition of the polymer solution or addition of the powdered material to the desired solvent and stirring at the optimum temperature and monomer concentration. The CMC can be determined by ultraviolet (UV) absorption or light scattering techniques such as static light scattering (SLS), dynamic light scattering (DLS), or small-angle X-ray scattering (SAXS). At the concentration at which monomers form micelles, there will be a radical drop in monomer concentration in the bulk.

The stability of micellar systems depends upon the ability to ensure the aggregated monomers do not deaggregate and that individual micelles do not coalesce to form larger aggregates. This is inevitable over a period of time, but steps can be taken to prolong the stability of the systems through various techniques, and those listed below are some of the methods used for spherical micellar systems:

Steric stabilization using emulsifiers and surfactants
Shell or core cross-linking
Viscosity-enhanced stabilization
Amine cross-linked stabilization
Thermodynamic stabilization

The types of systems that can be produced using micelle formation include spheres, shells, capsules, vesicles, clusters, and particles of various shapes and sizes, such as spheres, rods, planar structures, and layered structures. Further processing can be undertaken to add rate-controlling polymer membranes to the outer shell and to incorporate different molecules to the surface (e.g., for receptor recognition).

Biological Self-Assembly Using DNA as Construction Tool This is a technique that has been adopted to produce 2D or 3D nanosystems by utilizing the base-pairing affinity of DNA [149, 150].

Biological self-assembly using DNA can be described as a process that allows the systematic assembly of molecules with high levels of precision and accuracy without external constraints or influences. This allows the construction of nanoscale objects to the desired structure, conformation, and composition very rapidly and without the need for complex processing techniques and conditions.

DNA is a copolymer composed of a phosphate and sugar backbone and four types of bases that branch off from the backbone, A (adenine), G (guanine), C (cytosine), and T (thymidine). During DNA replication two strands of DNA come together to form a helical structure through complementary base pairing which is highly specific, whereby thymidine pairs with adenine and guanine with cytosine. When strands of DNA come together where the ends of the strands are noncomplementary, a portion of the strand extends beyond the complementary base-paring region leading to an overhang, otherwise known as a "sticky end".

The natural mechanism of DNA base pairing can be used to assemble synthetic sequences of DNA molecules by synthesizing DNA molecules such that they form stable branches [150, 151] with arms that form sticky ends that in turn can assemble to form supramolecular structures [152, 152a]. This approach may be used to produce complex 3D assemblies through sequential or layer-by-layer self-assembly and may incorporate other materials, such as particles and proteins [153–156]. The advantages of this method are as follows:

Specificity and geometry of intermolecular interactions that can be predicted
Precision control over the final structure at the nanoscale
Simple manufacturing process without external restrictions
Complex structures that can be built with defined topologies
Potential for creation of nanodevices

The types of structures that may be constructed are Branched planar/2D quadrilateral structures [157], cubes [158], octahedrons [159], and complex 2D and 3D periodic structures [160].

DNA Synthesis for Nanoconstruction Single strands of DNA, otherwise known as oligomers, are most commonly produced using a solid-support synthesis process [161, 162]. This is a cyclic process where each nucleotide is sequentially coupled to form a nucleotide chain (working from the 3′ end to the 5′ end). The 3′ end is initially covalently linked to a solid support and the nucleotide monomers are added sequentially. This is a well-established process and its key parameters and critical process steps are well documented in the literature [163, 164]. The DNA strands can be tailored according to the desired nanoconstruction scheme and target structure [165].

Nano Manipulation As the name suggests, this is quite literally a technique for physically manipulating matter at the nanoscale. Scanning probe microscopy (SPM) techniques have been most widely used to achieve this using the scanning probe tip as an implement for assembling atoms, molecules, or nanoparticles according to the desired spatial conformation [166–169].

Soft Lithography Lithography is essentially a process for printing features on a planar surface. Nanolithography tools, commonly referred to as soft lithography, allow precisely defined nanoscale features to be produced on a substrate, which can be removed from the substrate as free-standing 3D nano-objects. A number of techniques fall within the field of soft lithography, primarily for construction of micrometer-sized objects:

Replica molding
Micromolding in capillaries (MIMIC)
Microtransfer molding
Solvent-assisted microcontact molding (SAMIM)
Microcontact printing

Near-field phase shift lithography is a soft lithographic technique used to produce geometric shapes with size features at the nanoscale (approximately 40–80 nm). This involves the production of a polymer mask containing the desired pattern to be replicated on the substrate, with nanoscale features usually patterned by X-ray or electron beam exposure. The mask is then placed on the surface of the substrate and exposed to near-field light, the intensity of which leads to replication of the pattern on the mask on to the substrate. Complex geometries, shapes, and features can be produced on the substrate which can subsequently be removed to give free-standing particles or objects [170, 171].

Molecular Imprinting Molecular Imprinting is a process used to imprint or copy recognition sites from desired molecules on to polymer structures [172, 173]. The recognition sites can be produced on organic or inorganic polymers and inorganic materials such as silica and biomaterials such as proteins. A template molecule is dissolved in solvent with polymerizable monomers which undergo bond formation with the template molecule forming either noncovalent bonds through electrostatic interactions, hydrogen bonds or hydrophobic interactions, or reversible covalent bonds. The monomers are then polymerized to form a cast or semirigid polymeric structure which maintains the steric conformation of the molecule template and its recognition site upon removal of the template molecule. As a result, the molecular template affinity for molecules and analyte is mimicked by the "imprinted" polymer [174]. This has applications in chromatography and drug discovery and potential applications in targeted drug delivery.

Layer-by-Layer Electrostatic Deposition Electrostatic deposition utilizes the electrostatic bonding affinities of materials imparted by their surface charge to build highly ordered multilayered films or structures on a substrate. The process involves the successive deposition of oppositely charged polyions, exploiting the Coulombic long-range electrostatic interactions between the oppositely charged molecules, allowing formation of multilayers over a large distance. This technique can be used to build multilayer composite films on particles incorporating molecular fragments such as polymer–polymer, polymer–organic, polymer bimolecular, and polymer–mineral composition [175–177].

Chemical Vapor Deposition CVD is a crystal growth process whereby a solid material is deposited from the gas phase onto a controlled substrate using a suitable mixture of volatile precursor materials which react to produce the desired deposit on the substrate surface (Table 4). Types of films and structures that can be produced include the following:

Polycrystalline
Amorphous
Epitaxial silicon
Carbon fiber
Filaments
Carbon nanotubes
Silicon dioxide
Tungsten
Silicon nitride
Titanium nitride

TABLE 4 Chemical Vapour Deposition Methods and Their Key Features

CVD Method	Key Features
Atmospheric pressure CVD [178, 179]	Operates at atmospheric pressure
Atomic layer CVD (atomic layer epitaxy) [180, 181]	High-precision film thickness and uniformity requirements
Aerosol-assisted CVD [182, 183]	For use with involatile precursors
Direct liquid injection CVD [184, 185]	High film growth rates possible
Hot-wire CVD [186]	High growth rate, low temperature, and use of inexpensive materials such as plastics as substrate
Low-pressure CVD [187]	Improved film uniformity
Metal organic CVD [188]	Uniform and conformal deposition
Microwave plasma-assisted CVD [189]	No external heating required
Plasma-enhanced CVD [190]	Reduced substrate temperatures can be used
Rapid thermal CVD [191]	Conformal coverage over high-aspect-ratio features is possible, i.e., improved control of interfacial properties
Remote plasma-enhanced CVD [192]	Excellent conformal coverage of complex structures
	Can produce multilayer and graded layers with tailored functional group attachment
Ultrahigh vacuum CVD [193]	Reduced surface contamination

7.3.5.2 Nanostructuring Processes for Polymeric Materials

Numerous microfabrication techniques have been used to produce a wide range of implantable and oral drug delivery systems using materials ranging from silicon, glass, silicone elastomer, and plastics. Fabrication techniques have rapidly evolved to produce nanoscale objects and therapeutic systems using polymeric materials as the substrate due to their biodegradable nature. There are a number of different synthetic polymer systems that have been developed for this type of application, and the most common ones are listed below:

Poly(D-lactic acid) (PDLA)
Poly(ε-caprolactone) (PCL)
Poly(vinyl alcohol) (PVA)
Polyalkylcyanoacrylates (PACA)
Poly(L-lactide) (PLLA)
Poly(lactide-*co*-glycolide) (PLGA)
Polymethylcyanoacrylate (PMCA)

Techniques for the production of micrometer-sized features using polymers are well established and apply primarily to device construction. The techniques listed below focus primarily on attaining submicrometer, nanoscale features, and geometries using polymers such as those listed above.

Nanomold Replication A physical mold is produced that has nanoscales on the order of tens or a few hundred nanometers. To achieve such fine features with precision and repeatability, electrodeposition is used to produce the molds, otherwise referred to as a nanostamp [194]. The stamp is then use as a master stamp to duplicate the image or object by casting or embossing the polymeric material.

Colloidal Lithography Colloidal lithography is a process whereby an electrostatically self-assembled array of monodispersed colloidal nanospheres is used as a mask to construct nanoscale objects and features through deposition or etching processes. The monodisperse colloidal spheres, for example, surface-charged latex, self-organize or assemble into periodic arrays on the substrate, glass, for example, and do not aggregate due to the surface charge repulsion. This method can be used to produce 2D [195, 196] and 3D [197] nanostructures, arrays of rings, dots, honeycomb structures, pillars, and chemical patterns [198] with a high level of control over structure and conformation.

Interfacial Polymerization Interfacial polymerization is a process whereby very thin films or membranes, on the order of nanometer thickness, are produced by reacting two monomers at the interface between two immiscible solutions [199]. Nanoparticles [200] and aqueous core capsules with very thin membranes have been produced using this method for drug delivery applications.

Nanoprecipitation Nanoprecipitation is a self-assembly directed nanoparticle formation method. There are three key steps involved in this process: rapid micromixing of the solutes, the creation of a high level of supersaturation to instigate rapid nucleation and growth of precipitate, and the kinetic control and termination of growth using copolymer stabilizers. One of the drawbacks using this method is the poor incorporation of water-soluble drugs [201]. However, the main advantage associated with the production of nanospheres for drug delivery using this technique is the high degree of control attainable over particle size [202].

Emulsion Solvent Evaporation The basic concept of the emulsion solvent evaporation technique producing nanoparticles is very straightforward. The particles are formed as an emulsion of a polymer–surfactant mixture and dispersed in an organic solvent. The solvent is then evaporated to leave behind the individual emulsion droplets which form stable free nanoparticles [203]. This method is far easier and more preferable over methods such as spray drying and homogenization and operates under ambient conditions and mild emulsification conditions. The size and composition of the final particles are affected by variables such as phase ratio of the emulsion system, organic solvent composition, emulsion concentration, apparatus used, and properties of the polymer [204].

Nanoimprinting This is a lithographic technique similar to soft contact lithography discussed earlier, with the main difference being that nanoimprinting uses a hard mold to produce nanoscale features down to sub-10 nm resolution [205] by directly imprinting onto the polymer surface at high temperatures. More recently, molds produced from carbon nanotubes have been used to achieve molecular-scale resolution. Molds are generally made using electron beam lithography; however, high-definition molds are produced using molecular beam epitaxy. Some of the technical issues associated with this technique include sticking, adhesion, and material transport during imprinting [206].

Electrospinning Electrospinning is a process that uses electrostatic force to produce nanofibers from a charged polymer. An electrode is placed into a spinning polymer solution/polymer melt and the other electrode is attached to a collector plate. A high-intensity electric field is created by applying a high voltage such that the polymer solution is discharged as a jet, and during travel of this charged polymer jet toward the grounded collector plate, solvent evaporation leaves a charged polymer fiber which deposits on the collector plate [207]. These fibers have high specific surface areas and are highly flexible, and applications include the preparation of controlled drug release membranes [208].

7.3.6 CHARACTERIZATION TECHNIQUES

7.3.6.1 Nanoparticle Characterization Methods and Tools

A summary of some key properties that may be assessed as part of a characterization schedule for nanoparticles and nanostructures and a comprehensive but not exhaustive list of tools and techniques that may be used are presented in Table 5 [209–223]. The degree of characterization and method used will be determined by the intended application of the nanomaterials.

Characterization of micellar and supramolecular structures and their counterparts often require different or additional tools and techniques [224–232] and a summary is provided in Table 6 of various characterization parameters for micellar and supramolecular structures and components and analytical tools that may be applied.

7.3.6.2 Scanning Probe Technologies

Scanning probe microscopy has almost become synonymous with nanomaterial characterization [233]. This is a family of techniques that have evolved from the use of a sharp proximal probe to scan a surface in order to ascertain its properties down to atomic-scale resolution based on tip–surface interaction. There are two main SPM techniques, scanning tunneling microscopy (STM) [234, 235] and AFM [236]. Near-field scanning optical microscopy (NSOM) [237–239] also falls within the SPM family of techniques; however, this uses a subwavelength near-field light source as the scanning probe, achieving resolutions down to 50 nm, and is not discussed further here.

A host of techniques have evolved from STM and AFM, primarily involving adaptations to instrumentation depending on the material and parameter under

TABLE 5 Characterization Parameters and Tools for Nanoparticles and Nanostructures

Characterization Parameter	Analytical Tool
Composition	Liquid chromatography, e.g., high-performance liquid chromatography (HPLC), size exclusion chromatography (SLC)–HPLC
	Field flow fractionation (FFF)
	UV–visible spectrophotometry
	Refractive Index
	Inductively coupled plasma–optical emission spectrometry (ICP–OES)
	Fourier transform infrared spectrometry
	Mass spectrometry
	X-ray fluorescence
	Extended X-ray absorption fine structure (EXAFS) spectroscopy
	X-ray absorption near edge (XANES) spectroscopy
Particle diameter	Static and dynamic laser light scattering
	Scanning probe technologies
Size distribution	Static and dynamic laser light scattering
	Photon correlation spectroscopy
Surface area	BET method (Brunauer, Emmett, and Teller method)
Porosity (pore size, volume, and distribution)	Physical gas sorption
	Chemical gas sorption
	Helium picnometry
	Mercury intrusion porometry (MIP)
Core–shell thickness	Small-angle scattering of polarized neutrons (SANSPOL)
Surface structure and morphology	Small-angle neutron scattering (SANA)
	Proton nuclear magnetic resonance (^{1}H NMR) spectroscopy
	Scanning electron microscoscopy (SEM), atomic force microscopy (AFM), energy dispersive X-ray (EDXA), transmission electron microscopy (TEM), scanning probe microscopy (SPM), auger electron spectroscopy (AES), X-ray diffraction (XRD), X-ray photoelectron microscopy (APS), X-ray photoelectron spectroscopy (XPS), Vertical scanning phase shifting interferometry
Surface charge density	Zeta potential using Electrostatic light scattering (ELS)
	Zeta potential using multifrequencyelectro acoustics
	Zeta potential using phase analysis light scattering (PALS)
Shape	Electron microscopy
	Scanning probe technologies
Concentration distribution	Energy dispersive X-ray spectrometry (EDS) combined with SEM or scanning
Crystallinity, Bulk	X-ray diffraction
Crystallinity, Local	TEM/–selected area diffraction (SAD)
	Differential scanning calorimetry (DSC)
Magnetic properties	Scanning probe technologies
Electrical properties	Scanning probe technologies
Optical properties	UV–visible Spectroscopy

TABLE 6 Characterization Parameters and Analytical Tools for Micellar and Supramolecular Structures

Characterization Parameter	Analytical Tool
Critical micelle concentration	Flurimetric methods Static light scattering Dynamic Light Scattering
Aggregation number	Fluorescence correlation spectroscopy
Radius of gyration, R_G	Small-angle X-ray scattering
Hydrodynamic radius, R_H	Photon correlation spectroscopy
Core/corona size, micelle structure, overall micelle size	Small-angle X-ray scattering
Overall shape, cross section	Small-angle neutron scattering
Size, shape, and internal structure	Transmission electron microscopy Scanning probe technologies
Average molecular weight	Membrane and vapor pressure osmometry
Monitor equilibrium state, stability monitoring	Light-scattering methods
Structure elucidation, polymer architecture, polymer interactions	Nuclear magnetic resonance

investigation. It should be noted that technological advances continue unabated and new techniques are constantly being developed within the scanning probe family to cater for the characterization of new and novel materials and nanoscopic constructs. Table 7 gives a current synopsis of these techniques.

Scanning Tunneling Microscopy The scanning tunneling microscope was first described by Nobel Prize winners Binnig and Rohrer in 1982 [249] and consists of an atomically sharpened tip usually composed of tungsten, gold, or platinum–irridium. The tip is scanned within atomic distance (about 6–10 Å) of the sample under study under very high vacuum, and a bias voltage is applied between the sample and the scanning probe tip, resulting in a quantum mechanical tunneling current across the gap. The magnitude of the tunnelling current is related exponentially to the distance of separation and the local density of states (i.e., electron density in a localized region of a material) [250, 251].

The relationship between tunneling current and separation is given as:

$$I = C\rho_t\rho_s e^{d0.5}$$

where I = tunnelling current
C = constant (linear function of voltage)
ρ_t = tip electron density
ρ_s = sample electron density
$e^{d0.5}$ = separation (governed by exponential term)

The tip is scanned across the sample surface using a piezoelectric transducer in one of two modes, topographic mode or current mode. In the topographic mode a con-

TABLE 7 Summary of Scanning Probe Technologies

SPM Technique	Property Measured
Atomic force microscopy	Visualisation and measurement of surface features
Noncontact AFM [240]	Insulating substrates, atomic resolution
	Molecular systems, atomic resolution
	Biocluster and biomolecular imaging
	Imaging and spectroscopic data in liquid environments
	Nanoscale charge measurement
	Nanoscale magnetic properties
Contact AFM [241, 242]	Topographic imaging of solid substrates
	Mechanical properties
	Local adhesive properties
Piezoresponse	Characterization and domain engineering of ferroelectric materials
Lateral force	Fine structural detail
	Transitions between components on surface, e.g., polymer composites
Scanning thermal	Defects in sample based on thermal differences
Intermittent AFM (tapping mode) [243]	Biological systems: DNA/RNA analysis, protein–nucleic acid complexes, molecular crystals, biopolymers, ligand–receptor binding
Phase imaging AFM	Two phase polymer blends
	Surface contaminants
	Biological samples
Lift mode AFM techniques [244–246]	Topography
Magnetic Force	Magnetic properties/regions
Electrical Force	Electrical properties
Surface Potential	Surface potential
Scanning Capacitance	Material capacitance
Force modulation	Elasticity
Scanning tunnelling microscopy [247]	Surface imaging
	Three-dimensional profiling with vertical resolutions to 0.1 Å
	Measurement of electronic and magnetic properties
	Surface electronic state
Spin-polarized STM/STS [248]	Mapping surface magnetism at atomic scale

stant distance is maintained between the tip and sample surface using a feedback loop operated with the scanner. In the current mode, variations in current with changes in surface topography are monitored by switching off the feedback loop, thus providing a 3D image of the surface under study. The key features and limitation of STM are as follows:

Features of STM [252, 253]

Can undertake topographical imaging of surfaces with atomic-scale lateral resolution, down to 1 Å

There-dimensional profiling possible with vertical resolutions down to 0.1 Å

Wide range of materials can be analyzed

Surface electronic properties may be measured

Large field of view, from 1 Å to 100 μm

Vibrational isolation allows highly sensitive measurements to be undertaken

Ultrahigh vacuum (in the range 10^{-11} torr) minimizes sample contamination and reduces oxide layer growth, thus allowing for high sensitivity measurements

Limitations of STM

Can be difficult to differentiate between a composite of materials on the surface

Tip-induced desorption of surface molecules may occur

Ultrahigh vacuum requirements

Vibrational isolation requirements lead to increased installation costs

Low scanning speed

Atomic Force Microscopy Atomic force microscopy is a direct descendant of STM and was first described in 1986 [254]. The basic principle behind AFM is straightforward. An atomically sharp tip extending down from the end of a cantilever is scanned over the sample surface using a piezoelectric scanner. Built-in feedback mechanisms enable the tip to be maintained above the sample surface either at constant force (which allows height information to be obtained) or at constant height (to enable force information to be obtained). The detection system is usually optical whereby the upper surface of the cantilever is reflective, upon which a laser is focused which then reflects off into a dual-element photodiode, according to the motion of the cantilever as the tip is scanned across the sample surface. The tip is usually constructed from silicon or silicon nitride, and more recently carbon nanotubes have been used as very effective and highly sensitive tips.

In noncontact-mode AFM the cantilever is oscillated slightly above its resonant frequency and the tip does not make contact with the sample surface but instead oscillates just above the adsorbed fluid layer on the surface, maintaining a constant oscillation. The resonant frequency of the cantilever decreases due to van der Waals forces extending from the adsorbed fluid layer. This changes the amplitude of oscillation, the variations of which are detected using sensitive alternating current (AC) phase-sensitive devices, providing topographical information. In contact mode, AFM the tip remains in contact with the sample surface, and the feedback loop maps the vertical vibrational changes. In tapping mode, the cantilever is oscillated above the sample surface such that it intermittently contacts the sample surface. The key features and limitations of AFM are a follows:

Features of AFM [255–257]

High scan speeds

Atomic-scale resolution possible

Rough sample surfaces can be analyzed

High lateral resolutions possible

Soft samples (e.g., biological tissue) can be measured

Limitations of AFM [255–257]

Potential for image distortion due to lateral shear forces (in contact mode)
May be reduced spatial resolution due to sample scraping (in contact mode)
Tapping mode has lower scan speed compared to contact mode, though there is less susceptibility for sample damage and image distortion

7.3.7 TOXICOLOGY CONSIDERATIONS

Nanomaterials may in their own right possess novel and useful properties or as a composite of the same or different materials to form larger useful structures. Safety consideration is therefore of paramount importance since completely inert materials have the ability to exhibit toxic effects by virtue of a reduction in their size and associated increase in surface area–mass ratio, let alone materials manipulated specifically to impart novel properties.

Two obvious routes of human contact with nanoparticulates are the skin and via inhalation. Given the size of the particles, there may be a propensity for absorption into the systemic circulation. In some cases the nanosystems are engineered to achieve enhanced systemic absorption. The established methodology for toxicological assessment of new materials should be adhered to, and the discussion below is intended only to touch upon some of the immediate safety concerns that should be understood and addressed when dealing with nanomaterials.

7.3.7.1 Lung Toxicity

The safety of ultrafine particles remains to be clearly elucidated and requires the collaborative input of toxicologists (animal, cellular, molecular), epidemiologists, clinicians (pulmonary, cardiovascular, neurological), and atmospheric scientists. There are several published studies to indicate that ultrafine particles pose a higher toxicity risk compared with their larger counterparts [258–261]. Figure 1 outlines the potential effects of ultrafine particles on respiratory mucosa, the cardiovascular system, and central and peripheral nervous systems, upon inhalation.

Hohfeld et al. [262] hypothesized the toxic effects of ultrafine particles to be attributable to the following:

- High efficiency of deposition in the alveolar region due to particle size
- Large surface area
- Decreased phagocytosis leading to interaction of the particles with the epithelium, resulting in the development of conditions such as chronic diffuse interstitial fibronodular lung disease
- Dislocation from the alveolar space, leading to potential systemic effects

The hypothesis affirms the need to characterize the material's physical and chemical properties, including morphological analysis. The latter has significant ramifications on the aerodynamics of the particulate matter and hence its ultimate disposition.

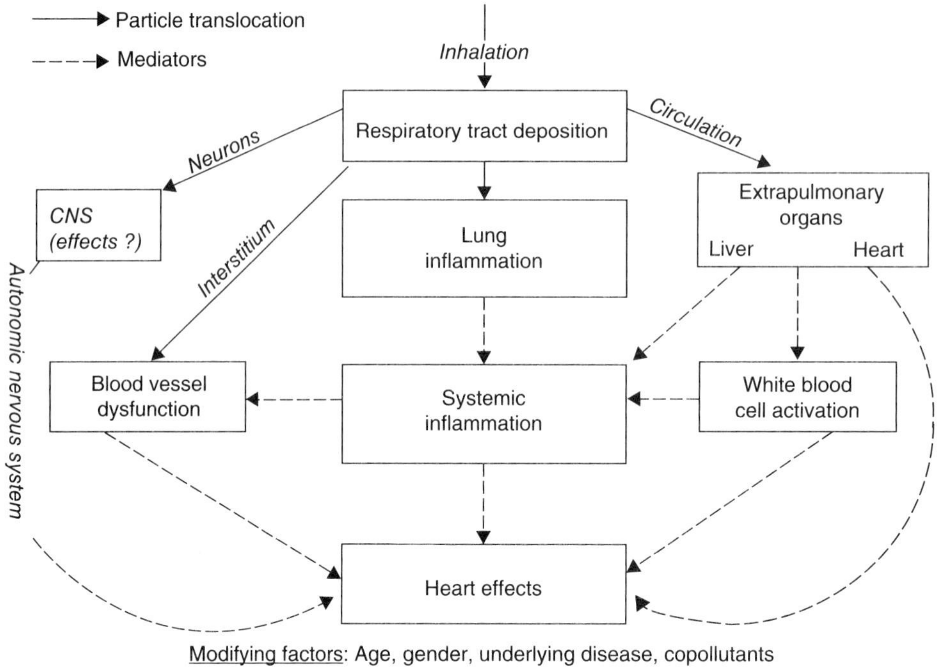

FIGURE 1 Potential mechanisms of effects of inhaled ultrafine particles. (Reproduced with permission from G. Oberdörster, Inhaled nano-sized particles: Potential effects and mechanisms, paper presented at the Symposium, Health Implications of Nanomaterials, October 2004.)

Carbon nanotubes are a class of materials finding increasingly widespread applications in drug discovery and development and may be classed as a form of ultrafine material. It has been shown that single-wall carbon nanotubes do not produce any significant respirable aerosol levels due to agglomeration resulting from the very high surface area–volume ratio and associated electrostatic interaction between the nanotubes [263]. A good deal of research has focused on developing methods for the dispersion of nanotubes for further downstream processing for conversion to useful applications. The liquid dispersions do not pose the same level of hazards posed by the dry powder material, and most of the work in the pharmaceutical industry with carbon nanotubes is focused on liquid dispersions whereby nanotubes are being functionalized and conjugated with drugs and possibly other carriers for therapeutic intent [264].

However, it has at the same time been shown that dry powder carbon nanotubes can persist in the lungs and have the potential to induce inflammatory and fibrotic reactions, evident in the form of collagen-rich granulomas in the bronchi and interstitium [265]. This emphasizes the need for caution and further work to establish the exact cause of these effects given the propensity of nanotubes to agglomerate.

7.3.7.2 Systemic Uptake

Nanoparticles, by virtue of size, have a tendency to evade phagocytosis. Uptake into the systemic circulation is thus thought to be through diffusion and via the endothelial cells, the epithelium, interstitium, and blood vessels. Translocation into the blood is thought to be through enhanced epithelial or endothelial permeability imparted by inflammatory mediators. Systemic hypercoagulation may be triggered by the inflammatory mediators in response to the diffusion of the nanoparticles through endothelium and vasculature [260].

7.3.7.3 Skin Permeation of Nanoparticles

The stratum corneum provides a formidable barrier to the entry of chemical and particulate matter into human tissue and systemic circulation. It provides a first-line defence to the ingress of foreign agents. However, there are indications that particles up to 1 μm are able to penetrate the skin's barrier and deposit in the epidermis where the antigen-presenting cells reside [266]. It follows therefore that submicrometer particles in the nanometer range have the potential to cross the stratum corneum and illicit an inflammatory response. Once again, however, the tendency of fine particles to agglomerate will to some extent inhibit penetration into the skin, in particular where the agglomerates are macroscopic. A correlation must however be drawn to establish any potential link between nanoparticle affinity for skin penetration and particle physical and morphological characteristics or indeed whether the novel and unique properties of the engineered particle in any way impart enhanced skin permeation properties and, if so, their nature and mechanisms.

REFERENCES

1. FDA and Nanotechnology Products. U.S. Food and Drug Administration (FDA), available: http://www.fda.gov/nanotechnology/faqs.html, accessed June 2006.
2. Freitas, R. A., Jr. (2005), What is nanomedicine? *Nanomed. Nanotechnol. Biol. Med.*, 1, 2–9.
3. Yang, L., et al. (2000), Physicochemical aspects of drug delivery and release from polymer-based colloids, *Curr. Opin. Coll. Interf. Sci.*, 5(1–2), 132–143.
4. Bontha, S., et al. (2006), Polymer micelles with cross-linked ionic cores for delivery of anticancer drugs, *J. Controlled Release*, 114(2), 163–174.
5. Sphurti, V., et al. (2006), Nanofibers and spheres by polymerization of cyanoacrylate monomer, *Polymer*, 47(12), 4328–4332.
6. Qian, F., et al. (2006), Preparation, characterization and enzyme inhibition of methylmethacrylate copolymer nanoparticles with different hydrophilic polymeric chains, *Eur. Polym. J.*, 44(7), 1653–1661.
7. Choi, C., et al. (2006), Thermosensitive poly(*N*-isopropylacrylamide)-*b*-poly(ε-caprolactone) nanoparticles for efficient drug delivery system, *Polymer*, 47(13), 4571–4580.
8. Gross, M., and Maskos, M. (2005), Dye loading of unimolecular, amphiphilic polymeric nanocontainers, *Polymer*, 46(10), 3329–3336.
9. Brož, P., et al. (2005), Cell targeting by a generic receptor-targeted polymer nanocontainer platform, *J. Controlled Release*, 102(2), 475–488.
10. Santoso, S. S., et al. (2002), Structures, function and applications of amphiphilic peptides, *Curr. Opin. Coll. Interf. Sci.*, 7(5–6), 262–266.
11. Heyes, J., et al. (2006), Synthesis and characterization of novel poly(ethylene glycol)-lipid conjugates suitable for use in drug delivery, *J. Controlled Release*, 112(2), 280–290.

12. Seki, J., et al. (2004), A nanometer lipid emulsion, lipid nano-sphere (LNS®), as a parenteral drug carrier for passive drug targeting, *Int. J. Pharm.*, 273(1–2), 75–83.
13. Carlson, J. C. T., and Jena, S. S. (2006), Chemically controlled self-assembly of protein nanorings, *J. Am. Chem. Soc.*, 128(23), 7630–7638.
14. Perkins, W. R., et al. (2000), Novel therapeutic nano-particles (lipocores): Trapping poorly water soluble compounds, *Int. J. Pharm.*, 200(1), 27–39.
15. Fukui, H., et al. (2003), A novel delivery system for amphotericin B with lipid nanosphere (LNS®), *Int. J. Pharm.*, 265(1–2), 37–45.
16. Müller, R. H., et al. (2006), Oral bioavailability of cyclosporine: Solid lipid nanoparticles (SLN®) versus drug nanocrystals, *Int. J. Pharm.*, 317(1), 82–89.
17. Casadei, M. A., et al. (2006), Solid lipid nanoparticles incorporated in dextran hydrogels: A new drug delivery system for oral formulations, *Int. J. Pharm*, 325(1–2), 140–146.
18. Masuda, M., and Shimizu, T. (2004), Lipid nanotubes and microtubes: experimental evidence for unsymmetrical monolayer membrane formation from unsymmetrical bolaamphiphiles, *Langmuir*, 20(14), 5969–5977.
19. Gupta, K., et al. (2006), Nanoparticle formation from poly(acrylic acid) and oppositely charged peptides, *Biophys. Chem.*, 119(3), 303–306.
20. Costantino, L., et al. (2005), Peptide-derivatized biodegradable nanoparticles able to cross the blood–brain barrier, *J. Controlled Release*, 108(1), 84–96.
21. Cortez-Retamozo, V., and Backmann, N. (2004), Efficient cancer therapy with a nanobody-based conjugate, *Cancer Res.*, 64, 2853–2857.
22. Parekh, H. S., et al. (2006), Synthesis of a library of polycationic lipid core dendrimers and their evaluation in the delivery of an oligonucleotide with hVEGF inhibition, *Bioorg. Med. Chem.*, 14(14), 4775–4780.
23. Gupta, U., et al. (2006), A review of in vitro–in vivo investigations on dendrimers: The novel nanoscopic drug carriers, *Nanomed. Nanotechnol. Biol. Med.*, 2(2), 66–73.
24. Najlah, M., et al. (2006), Synthesis, characterization and stability of dendrimer prodrugs, *Int. J. Pharm.*, 308(1–2), 175–182.
25. Koga, T., et al. (2006), Fabrication of a switchable nano-surface composed of acidic and basic block-polypeptides, *Coll. Surf. Physicochem. Eng. Aspects*, 284–285, 521–527.
26. Yin, A., and Xu, A. (2006), Fabrication of highly-ordered and densely-packed silicon nano-needle arrays for bio-sensing applications, *Mater. Res. Soc. Symp. Proc.*, 900E.
27. Vaccari, L., et al. (2006), Porous silicon as drug carrier for controlled delivery of doxorubicin anticancer agent, *Microelectron. Eng.*, 83(4–9), 1598–1601.
28. Venkatesan, N., et al. (2005), Liquid filled nanoparticles as a drug delivery tool for protein therapeutics, *Biomaterials*, 26(34), 7154–7163.
29. Xu, Z. P., et al. (2006), Inorganic nanoparticles as carriers for efficient cellular delivery, *Chem. Eng. Sci.*, 61(3), 1027–1040.
30. Neuberger, T., et al. (2005), Superparamagnetic nanoparticles for biomedical applications: Possibilities and limitations of a new drug delivery system, *J. Magnet. Magnetic Mater.*, 293(1), 483–496.
31. Kim, J-H., and Lee, T. R. (2004), Discrete thermally responsive hydrogel-coated gold nanoparticles for use as drug-delivery vehicles, *Chem. Mater*, 16, 3647–3651.
32. Gao, X., et al. (2004), *In vivo* cancer targeting and imaging with semiconductor quantum dots, *Nat. Biotechnol.*, 22, 8.
33. Wu, C., et al. (2005), Metal nanoshells as a contrast agent in near-infrared diffuse optical tomography, *Opt. Commun.*, 253(1–3), 214–221.

34. Ostrander, K. D., et al. (1999), An in-vitro assessment of a NanoCrystal™ beclomethasone dipropionate colloidal dispersion via ultrasonic nebulization, *Eur. J. Pharm. Biopharm.*, 48(3), 207–215.

35. Muller, R. H., and Keck, C. M. (2004), Challenges and solutions for the delivery of biotech drugs—A review of drug nanocrystal technology and lipid nanoparticles, *J. Biotechnol.*, 113(1–3), 151–170.

36. Cui, F., et al. (2006), Preparation and characterization of mucoadhesive polymer-coated nanoparticles, *Int. J. Pharm.*, 316(1–2), 154–161.

37. Goodson, T., et al. (2004), Optical properties and applications of dendrimer-metal nanocomposites, *Int. Rev. Phys. Chem.*, 23, 1.

38. Panchapakesan, B. (2005), Tiny technology—Tremendous therapeutic potential, in *Oncology Issues, Nanotechnology: Part 2.*

39. Lin, C-R., et al. (2006), Magnetic behavior of core–shell particles, *J. Magnet. Magnetic Mater.*, 304(1), 34–36.

40. Layre, A., et al. (2006). Novel composite core-shell nanoparticles as busulfan carriers, *J. Controlled Release*, 111(3), 271–280.

41. Foley, S., et al. (2002), Cellular localisation of a water-soluble fullerene derivative, *Biochem. Biophys. Res. Commun.*, 294(1), 116–119.

42. Friedman, S. H., et al. (2002), Optimizing the binding of fullerence inhibitors of the HIV-1 protease through predicted increases in hydrophobic desolvation, *J. Med. Chem.*, 41, 2424–2429.

43. Zhu, Z., et al. (2003), Molecular dynamics study of the connection between flap closing and binding of fullerene-based inhibitors of the HIV-1 protease, *Biochemistry*, 42, 1326–1337.

44. Smart, S. K., et al. (2006), The biocompatibility of carbon nanotubes, *Carbon*, 44(6), 1034–1047.

45. Bianco, A., et al. (2005), Applications of carbon nanotubes in drug delivery, *Curr. Opin. Chem. Biol.*, 9(6), 674–679.

46. Luan, B., and Pan, C-Y. (2006), Synthesis and characterizations of well-defined dendrimer-like copolymers with the second and third generation based on polystyrene and poly(L-lactide), *Eur. Polym. J.*, 42, 1467–1478.

47. Tomalia, D. A., et al. (1990), Starburst dendrimers: Control of size, shape, surface chemistry, topology and flexibility, *Angew. Chem., Int. Ed. Engl.*, 29, 138.

48. Petkov, V., et al. (2005), 3D structure of dendritic and hyper-branched macromolecules by X-ray diffraction, *Solid State Commun.*, 134(10), 671–675.

49. Tomalia, D. A., et al. (1985), A new class of polymers: Starburst-dendritic macromolecules, *Polym. J.*, 17, 117–132.

50. Buhleier, E., et al. (1978), Cascade and nonskid-chain-like syntheses of molecular cavity topologies, *Synthesis*, 155–158.

51. Tomalia, D. A., et al. (1991), Comb-burst dendrimer topology. New macromolecular architecture derived from dendritic grafting, *Macromolecules*, 24, 1435–1438.

52. Hawker, C. J., and Frechet, J. M. J. (1990), Preparation of polymers with controlled molecular architecture. A new convergent approach to dendritic macromolecules, *J. Am. Chem. Soc.*, 112, 7638–7647.

53. Miller, T. M., and Neenan, T. X. (1990), Convergent synthesis of monodisperse dendrimers based upon 1,3,5 trisubstituted benzenes, *Chem. Mater.*, 2, 346–349.

54. Hedrick, J. L., et al. (1998), Dendrimer-like star block and amphiphilic copolymers by combination of ring-opening and atom transfer radical polymerization, *Macromolecules*, 31, 8691–8705.

55. Hou, S. J., et al. (2003), Synthesis of water-soluble star-block and dendrimer-like copolymers based on poly(ethylene oxide) and poly(acrylic acid), *Macromolecules*, 36, 3874–3881.
56. Angot, S., et al. (2000), Amphiphilic stars and dendrimer-like architectures based on poly(ethylene oxide) and polystyrene, *Macromolecules*, 33, 5418–5426.
57. Reuter, J. D., et al. (1999), Inhibition of viral adhesion by sialic-acid-conjugated dendritic polymers, *Bioconjug. Chem.*, 10, 271–278.
58. Na, M., et al. (2006), Dendrimers as potential drug carriers. Part II. Prolonged delivery of ketoprofen by in vitro and in vivo studies, *Eur. J. Med. Chem.*, 41(5), 670–674.
59. Vandamme, Th. F., and Brobeck, L. (2005), Poly(amidoamine) dendrimers as ophthalmic vehicles for ocular delivery of pilocarpine nitrate and tropicamide, *J. Controlled Release*, 102(1), 23–38.
60. Zeng, H., et al. (2002), Exchange-coupled nanocomposite magnets via nanoparticle self-assembly, *Nature*, 420, 395–398.
61. Chan, W. C. W., et al. (2002), Luminescent quantum dots for multiplexed biological detection and imaging, *Curr. Opin. Biotechnol.*, 13, 40–46.
62. Qu, L. H., and Peng, X. G. (2002), Control of photoluminescence properties of CdSe nanocrystals in growth, *J. Am. Chem. Soc.*, 124, 2049–2055.
63. Murphy, C. J. (2002), Optical sensing with quantum dots, *Anal. Chem.*, 74, 520–526.
64. Parak, W. J., et al. (2003), Biological applications of colloidal nanocrystals, *Nanotechnology*, 14, 15–27.
65. Gao, X., et al. (2005), In vivo molecular and cellular imaging with quantum dots, *Curr. Opin. Biotechnol.*, 16, 63–72.
66. Leatherdale, C. A., et al. (2002), On the absorption cross section of CdSe nanocrystal quantum dots, *J. Phys. Chem. B*, 106, 7619–7622.
67. Gao, X. H., et al. (2004), In vivo cancer targeting and imaging with semiconductor quantum dots, *Nat. Biotechnol.*, 22, 969–976.
68. Xiao, Y., et al. (2003), Plugging into enzymes: Nanowiring of redox enzymes by a gold nanoparticle, *Science*, 299(5614), 1877–1881.
69. Liedberg, B., et al. (1995), Biosensing with surface plasmon resonance—How it all started, *Biosens. Bioelectron.*, 10, 1–4.
70. Chen, W. P., and Chen, J. M. (1981), Use of surface plasma waves for determination of the thickness and optical constants of thin metallic films, *J. Opt. Soc. Am.*, 71, 189–191.
71. Link, S., and El-Sayed, M. A. (1999), Spectral properties and relaxation dynamics of surface plasmon electronic oscillations in gold and silver nanodots and nanorods, *J. Phys. Chem.*, 103, 8410.
72. Jensen, T. R., et al. (2000), Nanosphere lithography: Tunable localized surface plasmon resonance spectra of silver nanoparticles, *J. Phys. Chem.*, 104, 10549.
73. Salmon, Z., et al. (1997), Surface plasmon resonance spectroscopy as a tool for investigating the biochemical and biophysical properties of membrane protein system. II. Application to biological system, *Biochim. Biophys. Acta*, 1331, 131–152.
74. Homola, J., et al. (2001), A novel multichannel surface plasmon resonance biosensor, *Sens. Actuators Chem.*, 76, 403–410.
75. Oldenburg, S. J. (1998), Nanoengineering of optical resonances, *Chem. Phys. Lett.*, 288, 243–247.
76. Sershen, S., et al. (2000), Temperature-sensitive polymer-nanoshell composites for photothermally modulated drug delivery, *J. Biomed. Mater. Res.*, 51(03), 293–298.

77. Muangsiri, W., and Kirsch, L. E. (2006), The protein-binding and drug release properties of macromolecular conjugates containing daptomycin and dextran, *Int. J. Pharm.*, 315(1–2), 30–43.
78. Rothenhausler, B., and Knoll, W. (1988), Surface-plasmon microscopy, *Nature*, 332, 615–617.
79. Piliarik, M., et al. (2005), A new surface plasmon resonance sensor for high-throughput screening applications, *Biosens. Bioelectron.*, 20, 2104–2110.
80. Averitt, R. D. (1997), Plasmon resonance shifts of Au-coated Au_2S nanoshells: Insight into multicomponent nanoparticle growth, *Phys. Rev. Lett.*, 78, 4217–4220.
81. McDonnell, J. M. (2001), Surface plasmon resonance: Towards an understanding of the mechanisms of biological molecular recognition, *Curr. Opin. Chem. Biol.*, 5, 572–577.
82. Baek, S. H., et al. (2004), Surface plasmon resonance imaging analysis of hexahistidine-tagged protein on the gold thin film coated with a calix crown derivative, *Biotechnol. Bioprocess. Eng.*, 9, 143–146.
83. Oh, B. K., et al. (2003), Immunosensor for detection of *Legionella pneumophila* using surface plasmon resonance, *Biosens. Bioelectron.*, 18, 605–611.
84. Oh, B. K., et al. (2004), Surface plasmon resonance immunosensor for the detection of Salmonella typhimurium, *Biosens. Bioelectron.*, 19, 1497–1504.
85. Lehn, J. M. (1993), Supramolecular chemistry, *Science*, 260, 1762–1763.
86. Tanford, C. (1974), Theory of micelle formation in aqueous solutions, *J. Phys. Chem.*, 78, 24.
87. Desai, H., et al. (2006), Micellar characteristics of diblock polyacrylate–polyethylene oxide copolymers in aqueous media, *Eur. Polym. J.*, 42(3), 593–601.
88. Yoshii, N., and Okazaki, S. (2006), A molecular dynamics study of structural stability of spherical SDS micelle as a function of its size, *Chem. Phys. Lett.*, 425(1–3), 58–61.
89. Gochman-Hecht, H., and Bianco-Peled, H. (2006), Structure modifications of AOT reverse micelles due to protein incorporation, *J. Coll. Interf. Sci.*, 297(1), 276–283.
90. Luisi, P. L., et al. (1998), Reverse micelles as hosts for proteins and small molecules, *Biochim. Biophys. Acta*, 947(1), 209–246.
91. Munch, M. R., and Gast, A. P. (1998), Block copolymers at interfaces. 1. Micelle formation, *Macromolecules*, 21, 1360–1366.
92. Siegel, D. P. (1986), Inverted micellar intermediates and the transitions between lamellar, cubic, and inverted hexagonal lipid phases. II. Implications for membrane-membrane interactions and membrane fusion, *Biophys. J.*, 49(6), 1171–1183.
93. Sokolova, I., and Kievskya, Y. (2005), 3D Design of self-assembled nanoporous colloids, *Stud. Surf. Sci. Catal.*, 56, 433–442.
94. Zhang, Y., et al. (2004), Hollow spheres from shell cross-linked, noncovalently connected micelles of carboxyl-terminated polybutadiene and poly(vinyl alcohol) in water, *Macromolecules*, 37, 1537–1543.
95. Venema, L. C., et al. (2000), Atomic structure of carbon nanotubes from scanning tunneling microscopy, *Phys. Rev. B*, 61, 2991–2996.
96. Iijima, S., and Ichihashi, T. (1993), Single-shell carbon nanotubes of 1-nm diameter, *Nature*, 363, 603–605.
97. Vinciguerra, V., et al. (2003), Growth mechanisms in chemical vapour deposited carbon nanotubes, *Nanotechnology*, 14(6), 655–660.
98. Sonoyama, N., et al. (2006), Synthesis of carbon nanotubes on carbon fibers by means of two-step thermochemical vapor deposition, *Carbon*, 44, 1754–1761.
99. Bondi, S. N., et al. (2006), Laser assisted chemical vapor deposition synthesis of carbon nanotubes and their characterization, *Carbon*, 44, 1393–1403.

100. Gavillet, J., et al. (2002), Microscopic mechanisms for the catalyst assisted growth of single-wall carbon nanotubes, *Carbon*, 40(10), 1649–1663.
101. Lee, C. J., et al. (2002), Large-scale production of aligned carbon nanotubes by the vapor phase growth method, *Chem. Phys. Lett.*, 359(1–2), 109–114.
102. Endo, M., et al. (2005), "Buckypaper" from coaxial nanotubes, *Nature*, 433, 476.
103. Hata, K., et al. (2004), Water-assisted highly efficient synthesis of impurity-free single-walled carbon nanotubes, *Science*, 306(5700), 1362–1364.
104. Pantarotto, D., et al. (2003), Synthesis, structural characterization, and immunological properties of carbon nanotubes functionalized with peptides, *J. Am. Chem. Soc.*, 125(20), 6160–6164.
105. Kam, N. W. S., and Dai, H. (2005), Carbon nanotubes as intracellular protein transporters: Generality and biological functionality, *J. Am. Chem. Soc.*, 127(16), 6021–6026.
106. Kroto, H. W., et al. (1985), C60: Buckminsterfullerene, *Nature*, 318, 162–163.
107. Goel, A., et al. (2002), Combustion synthesis of fullerenes and fullerenic nanostructures, *Carbon*, 40, 177–182.
108. Kreatschmer, W., et al. (1990), Solid C60: New form of carbon, *Nature*, 347, 354–358.
109. Baierl, T., et al. (1996), Comparison of immunological effects of fullerene C60 and raw soot from fullerene production on alveolar macrophages and macrophage like cells in vitro, *Exp. Toxicol. Pathol.*, 48, 508–511.
110. Satoh, M., et al. (1997), Inhibitory effects of a fullerene derivative, dimalonic acid C60, on nitric oxide-induced relaxation of rabbit aorta, *Eur. J. Pharmacol.*, 327, 175–181.
111. Gharbi, N., et al. (2005), Fullerene is a powerful antioxidant in vivo with no acute or subacute toxicity, *NanoLetters*, 5, 2578–2585.
112. Tabata, Y., et al. (1997), Antitumor effect of poly(ethylene glycol)-modified fullerene, *Fullerene Sci. Technol.*, 5, 989–1007.
113. Signori, F., et al. (2005), New self-assembling biocompatible–biodegradable amphiphilic block copolymers, *Polymer*, 46(23), 9642–9652.
114. Lecommandoux, S., et al. (2006), Smart hybrid magnetic self-assembled micelles and hollow capsules, *Prog. Solid State Chem.*, 34(2–4), 171–179.
115. McNally, H., et al. (2003), Self-assembly of micro- and nano-scale particles using bio-inspired events, *Appl. Surf. Sci.*, 214(1–4), 109–119.
116. Bashir, R. (2001), Invited Review: DNA-mediated artificial nanobiostructures: State of the art and future directions, *Superlatt. Microstruct.*, 29(1), 1–16.
117. Hänel, K., et al. (2006), Manipulation of organic "needles" using an STM operated under SEM control, *Surf. Sci.*, 600(12), 2411–2416.
118. Guthold, M., et al. (1999), Investigation and modification of molecular structures with the nanoManipulator, *J. Mol. Graphics Modell.*, 17(3–4), 187–197.
119. Rogers, J. A., and Nuzzo, R. G. (2005), Recent progress in soft lithography, *Mater. Today*, 8(2), 50–56.
120. Choi, D-G., et al. (2004), 2D nano/micro hybrid patterning using soft/block copolymer lithography, *Mater. Sci. Eng.*, 24(1–2), 213–216.
121. Sellergren, B., and Allender, C. J. (2005), Molecularly imprinted polymers: A bridge to advanced drug delivery, *Adv. Drug Deliv. Rev.*, 57(12), 1733–1741.
122. van Nostrum, C. F. (2005), Molecular imprinting: A new tool for drug innovation, *Drug Discov. Today Technol.*, 2(1), 119–124.
123. Caruso, F., et al. (2000), Enzyme encapsulation in layer-by-layer engineered polymer multilayer capsules, *Langmuir*, 16, 1485–1488.

124. Caruso, F., et al. (1998), Nanoengineering of inorganic and hybrid hollow spheres by colloidal templating, *Science*, 282, 1111–1114.

125. Deotare, P. B., and Kameoka, J. (2006), Fabrication of silica nanocomposite-cups using electrospraying, *Nanotechnology*, 17, 1380–1383.

126. Lu, Y., and Chen, S. C. (2004), Micro and nano-fabrication of biodegradable polymers for drug delivery, *Adv. Drug Deliv. Rev.*, 56, 1621–1633.

127. Lu, Y., et al. (2005), Shaping biodegradable polymers as nanostructures: Fabrication and applications, *Drug Discov. Today Techno.*, 2(1), 97–102.

128. Michel, R., et al. (2002), A novel approach to produce biologically relevant chemical patterns at the nanometer scale: Selective molecular assembly patterning combined with colloidal lithography, *Langmuir*, 18, 8580–8586.

129. Hanarp, P., et al. (1999), Nanostructured model biomaterial surfaces prepared by colloidal lithography, *Nanostruct. Mater.*, 12(1), 429–432.

130. Di Fabrizio, E., et al. (2004), X-ray lithography for micro- and nano-fabrications at ELETTRA for interdisciplinary applications, *J. Phys. Condens. Matter*, 16, 3517–3535.

131. Couvreur, P., et al. (2002), Nanocapsule technology: A review, *Crit. Rev. Ther. Drug Carrier Syst.*, 19(2), 99–134.

132. Watnasirichaikul, S., et al. (2000), Preparation of biodegradable insulin nanocapsules from biocompatible microemulsions, *Pharm. Res.*, 17(6), 684–689.

133. Hitt, J., et al. (2002), Nanoparticles of poorly water soluble drugs made via a continuous precipitation process, in *Proceedings, American Association of Pharmaceutical Scientists*, annual meeting, Toronto.

134. Wang, Y., et al. (2004), Polymer coating/encapsulation of nanoparticles using a supercritical anti-solvent process, *J. Supercrit. Fluids*, 28, 85–99.

135. Perez, C., et al. (2001), Poly(lactic acid)-poly(ethylene glycol) nanoparticles as new carriers for the delivery of plasmid DNA, *J. Controlled Release*, 75(1–2), 211–224.

136. Dickert, F. L. (2003), Nano- and micro-structuring of sensor materials—From molecule to cell detection, *Synthetic Metals*, 138(1–2), 65–69.

137. Zeng, J., et al. (2003), Biodegradable electrospun fibers for drug delivery, *J. Controlled Release*, 92(3), 227–231.

138. Xu, X., et al. (2005), Ultrafine medicated fibers electrospun from W/O emulsions, *J. Controlled Release*, 108(1), 33–42.

139. Nicolau, D. V., et al. (1997), Bionanostructures built on e-beam-assisted functionalized polymer surfaces, *Proc. SPIE*, 3241.

140. Sinha, P. M., et al. (2004), Nanoengineered device for drug delivery application, *Nanotechnology*, 15, 585–589.

141. Maloney, J. M., et al. (2005), In vivo biostability of CVD silicon oxide and silicon nitride films, *Mater. Res. Soc. Symp. Proc.*, 872.

142. Sugai, T., and Yoshida, H. (2003), New synthesis of high-quality double-walled carbon nanotubes by high-temperature pulsed arc discharge, *NanoLetters*, 3(6), 769–773.

143. Du, F., et al. (2006), The synthesis of single-walled carbon nanotubes with controlled length and bundle size using the electric arc method, *Carbon*, 44(7), 1327–1330.

144. Tamir, S., and Drezner, Y. (2006), New aspects on pulsed laser deposition of aligned carbon nanotubes, *Appl. Surf. Sci.*, 252(13), 4819–4823.

145. Aratono, Y., et al. (2005), Formation of fullerene(C_{60}) by laser ablation in superfluid helium at 1.5 K, *Chem. Phys. Lett.*, 408(4–6), 247–251.

146. He, C., et al. (2006), A practical method for the production of hollow carbon onion particles, *J. Alloys Compounds*, 425(1–2), 329–333.

147. Liu, T.-C., et al. (2006), Synthesis of carbon nanocapsules and carbon nanotubes by an acetylene flame method, *Carbon*, 44(10), 2045–2050.
148. Rodriguez-Hernandez, J., et al. (2005), Toward "smart" nano-objects by self-assembly of block copolymers in solution, *Prog. Polym. Sci.*, 30, 691–724.
149. Birac, J. J. (2006), Geometry based design strategy for DNA nanostructures, *J. Mol. Graphics Modell.*, 25, 470–480.
150. Samori, B., and Zuccheri, G. (2004), DNA codes for nanoscience, *Angew. Chem. Int. Ed. Engl.*, 44, 1166–1181.
151. Seeman, N. C. (1998), DNA nanotechnology: Novel DNA constructions, *Annu. Rev. Biophys. Biomol. Struct.*, 27, 225–248.
152. Seeman, N. C. (1997), DNA components for molecular architecture, *Acc. Chem. Res.*, 30, 347–391.
152a. Rothemund, P. W. K. (2006), Folding DNA to create nanoscale shapes and patterns, *Nature*, 440, 297–302.
153. Williams, K. A., et al. (2002), Nanotechnology: Carbon nanotubes with DNA recognition, *Nature*, 420, 761.
154. Niemeyer, C. M. (2004), Semi-synthetic DNA–protein conjugates: Novel tools in analytics and nanobiotechnology, *Biochem. Soc. Trans.*, 32, 51–53.
155. Asakawa, T., et al. (2006), Build-to-order nanostructures using DNA self-assembly, *Thin Solid Films*, 509(1–2), 85–93.
156. Högberg, B., et al. (2006), Study of DNA coated nanoparticles as possible programmable self-assembly building blocks, *Appl. Surf. Sci.*, 252(15), 5538–5541.
157. Seeman, N. C. (1982), Nucleic acid junctions and lattices, *J. Theor. Biol.*, 99, 237–247.
158. Chen, J., and Seeman, N. C. (1991), Synthesis from DNA of a molecule with the connectivity of a cube, *Nature*, 350, 631–633.
159. Li, X. J., et al. (1996), Antiparallel DNA double crossover molecules as components for nanoconstruction, *J. Am. Chem. Soc.*, 118, 6131–6140.
160. Fu, T. J., and Seeman, N. C. (1993), DNA double crossover structures, *Biochemistry*, 32, 3211–3220.
161. Zhang, Y., and Seeman, N. C. (1992), A solid-support methodology for the construction of geometrical objects from DNA, *J. Am. Chem. Soc.*, 114, 2656–2663.
162. Niemeyer, C. M., et al. (1999), DNA-directed immobilization: Efficient, reversible and site-selective surface binding of proteins by means of covalent DNA-streptavidin conjugates, *Anal. Biochem.*, 268, 5463.
163. Kallenbach, N. R., et al. (1983), An immobile nucleic aid junction constructed from oligonucleotides, *Nature*, 305, 829–831.
164. Ma, R. I., et al. (1986), Three-arm nucleic acid junctions are flexible, *Nucl. Acids Res.*, 14, 9745–9753.
165. Brucale, M., et al. (2006), Mastering the complexity of DNA nanostructures, *Trends Biotechnol.*, 24(5), 235–243.
166. Nakayama, Y. (2002), Scanning probe microscopy installed with nanotube probes and nanotube tweezers, *Ultramicroscopy*, 91(1–4), 49–56.
167. Lü, J. H. (2004), Nanomanipulation of extended single-DNA molecules on modified mica surfaces using the atomic force microscopy, *Coll. Surf. B Biointerf.*, 39(4), 177–180.
168. Decossas, S., et al. (2003), Nanomanipulation by atomic force microscopy of carbon nanotubes on a nanostructured surface, *Surf. Sci.*, 543(1–3), 57–62.
169. Ishii, Y., et al. (2001), Single molecule nanomanipulation of biomolecules, *Trends Biotechnol.*, 19(6), 211–216.

170. Rogers, J. A., et al. (1997), Using an elastomeric phase mask for sub-100nm photolithography in the optical near field, *Appl. Phys. Lett.*, 70, 2658–2660.
171. Rogers, J. A., et al. (1998), Generating similar to 90 nanometer features using near-field contact-mode photolithography with an elastomeric phase mask, *J. Vac. Sci. Technol.*, 16, 59–68.
172. Wulff, G., et al. (1991), Racemic resolution of free sugars with macroporous polymers prepared by molecular imprinting. Selective dependence on the arrangement of functional groups versus spatial requirements, *J. Org. Chem.*, 56(1), 395–400.
173. Sellergren, B., et al. (1988), Highly enantioselective and substrate-selective polymers obtained by molecular imprinting utilizing noncovalent interactions—NMR and chromatrographic studies on the nature of recognition, *J. Am. Chem. Soc.*, 110, 5853–5860.
174. Lubke, C., et al. (2000), Imprinted polymers prepared with stoichiometric template-monomer complexes: Efficient binding of ampicillin from aqueous solutions, *Macromolecules*, 33, 5098–5105.
175. Tskruk, V. V., et al. (1997), Self-assembled multilayer films from dendrimers, *Langmuir*, 13(8), 2171–2175.
176. Schulze, K., and Kirstein, S. (2005), Layer-by-layer deposition of TiO_2 nanoparticles, *Appl. Surf. Sci.*, 246(4), 415–419.
177. Ji, J., and Shen, J. (2005), Electrostatic self-assembly and nanomedicine. Engineering in medicine and biology society, in *IEEE-EMBS 27th Annual International Conference Proc.*, Shanghai, pp. 720–722.
178. Beers, A. M., et al. (1983), CVD silicon structures formed by amorphous and crystalline growth, *J. Crystal Growth*, 64(3), 563–557.
179. Kyung, S., et al. (2006), Deposition of carbon nanotubes by capillary-type atmospheric pressure PECVD, *Thin Solid Films*, 506–507, 268–273.
180. Goodman, C., and Pessa, M. (1986), Atomic layer epitaxy, *J. Appl. Phys.*, 60(3), 65–81.
181. Suntola, T. (1992), Atomic layer epitaxy, *Thin Solid Films*, 216(1), 84–89.
182. Xia, C., et al. (1998), Metal-organic chemical vapor deposition of Sr-Co-Fe-O films on porous substrates, *J. Mater. Res.*, 13(1), 173.
183. Choy, K. L. (2003), Chemical vapour deposition of coatings, *Prog. Mater. Sci.*, 48(2), 57–170.
184. Kaul, A. R., and Seleznev, B. V. (1993), New principle of feeding for flash evaporation MOCVD devices, *J. Phys.*, 3, 375–378.
185. Douard, A., and Maury, F. (2006), Nanocrystalline chromium-based coatings deposited by DLI-MOCVD under atmospheric pressure from Cr(CO)6, *Surf. Coat. Technol.*, 200, 6267–6271.
186. Lau, K. K. S. (2001), Hot-wire chemical vapor deposition (HWCVD) of fluorocarbon and organosilicon thin films, *Thin Solid Films*, 395(1–2), 288–291.
187. Ren, Z. F., et al. (1998), Synthesis of large arrays of well-aligned carbon nanotubes on glass, *Science*, 282(5391), 1105–1107.
188. Han, B. W., et al. (2001), Growth of *in situ* $CoSi_2$ layer by metalorganic chemical vapor deposition on Si tips and its field-emission properties, *J. Vac. Sci. Technol. Microelectron. Nanometer Struct.*, 19(2), 533–536.
189. Kuo, T.-F., et al. (2001), Microwave-assisted chemical vapor deposition process for synthesizing carbon nanotubes, *J. Vac. Sci. Technol. Microelectron. Nanometer Struct.*, 19(3), 1030–1033.
190. McSporran, N., et al. (2002), Preliminary studies of atmospheric pressure plasma enhanced CVD (AP-PECVD) of thin oxide films, *Phys. IV France*, 12, 4–17.

191. Chang, J. P., et al. (2001), Rapid thermal chemical vapor deposition of zirconium oxide for metal-oxide-semiconductor field effect transistor application, *J. Vac. Sci. Technol. B Microelectron. Nanometer Struct.*, 19(5), 1782–1787.

192. Huang, Z. P., et al. (1998), Growth of highly oriented carbon nanotubes by plasma-enhanced hot filament chemical vapor deposition, *Appl. Phys. Lett.*, 73(26), 3845–3847.

193. Huczko, A. (2002), Synthesis of aligned carbon nanotubes, *Appl. Phys. A Mater. Sci. Process.*, 74(5), 617–638.

194. Lim, C.-Y., et al. (2004), Development of an electrodeposited nanomold from compositionally modulated alloys, *J. Appl. Electrochem.*, 34, 857–866.

195. Micheletto, R., et al. (1995), A simple method for the production of a two-dimensional, ordered array of small latex particles, *Langmuir*, 11, 3333–3336.

196. Boneberg, J., et al. (1997), The formation of nano-dot and nano-ring structures in colloidal monolayer lithography, *Langmuir*, 13, 7080–7084.

197. Blanco, A., et al. (2000), Large-scale synthesis of a silicon photonic crystal with a complete three-dimensional bandgap near 1.5 micrometres, *Nature*, 405, 437–440.

198. Denis, F. A., et al. (2004), Nanoscale chemical patterns fabricated by using colloidal lithography and self-assembled monolayers, *Langmuir*, 20, 9335–9339.

199. Cadotte, J. E. (1981), U.S. Patent 4,277,344. Interfacially synthesized reverse osmosis membrane.

200. Krauel, K., et al. (2005), Using different structure types of microemulsions for the preparation of poly(alkylcyanoacrylate) nanoparticles by interfacial polymerization, *J. Controlled Release*, 106(1–2), 76–87.

201. Govender, T., et al. (1999), PLGA nanoparticles prepared by nanoprecipitation: Drug loading and release studies of a water soluble drug, *J. Controlled Release*, 57(2), 171–185.

202. Chorny, M., et al. (2002), Lipophilic drug loaded nanospheres prepared by nanoprecipitation: Effect of formulation variables on size, drug recovery and release kinetics, *J. Controlled Release*, 83(3), 389–400.

203. Desgouilles, S., et al. (2003), The design of nanoparticles obtained by solvent evaporation: A comprehensive study, *Langmuir*, 19(22), 9504–9510.

204. Zambaux, M. F. (1998), Influence of experimental parameters on the characteristics of poly(lactic acid) nanoparticles prepared by a double emulsion method, *J. Controlled Release*, 50(1–3), 31–40.

205. Cao, H., et al. (2002), Fabrication of 10nm enclosed nanofluidic channels, *Appl. Phys. Lett.*, 81(1), 174–176.

206. Scheer, H.-C., et al. (1998), Problems of the nanoimprinting technique for nanometer scale pattern definition, *J. Vac. Sci. Technol. B Microelectron. Nanometer Struct.*, 16(6), 3917–3921.

207. Kim, J-S., and Reneker, D. H. (1999), Polybenzimidazole nanofiber produced by electrospinning, *Polym. Eng. Sci.*, 39(5), 849–854.

208. Abidian, M. R., et al. (2006), Conducting-polymer nanotubes for controlled drug release, *Adv. Mater.*, 18, 405–409.

209. Arias, J. L., et al. (2006), Preparation and characterization of carbonyl iron/poly(butylcyanoacrylate) core/shell nanoparticles, *J. Collo. Interf. Sci.*, 299(2), 599–607.

210. Cui, F., et al. (2006), Preparation and characterization of mucoadhesive polymer-coated nanoparticles, *Int. J. Pharm.*, 316(1–2), 154–161.

211. Huang, C.-Y., and Lee, Y.-D. (2006), Core-shell type of nanoparticles composed of poly[(*n*-butyl cyanoacrylate)-*co*-(2-octyl cyanoacrylate)] copolymers for drug delivery

application: Synthesis, characterization and *in-vitro* degradation, *Int. J. Pharm.*, 325(1–2), 132–139.

212. Hirsjärvi, S., et al. (2006), Layer-by-layer polyelectrolyte coating of low molecular weight poly(lactic acid) nanoparticles, *Coll. Surf. B Biointerf.*, 49(1), 93–99.
213. Schubert, M. A., and Müller-Goymann, C. C. (2005), Characterisation of surface-modified solid lipid nanoparticles (SLN): Influence of lecithin and nonionic emulsifier, *Eur. J. Pharm. Biopharm.*, 61(1–2), 77–86.
214. Lo, C.-L. (2005), Preparation and characterization of intelligent core-shell nanoparticles based on poly(D,L-lactide)-*g*-poly(*N*-isopropyl acrylamide-*co*-methacrylic acid), *J. Controlled Release*, 104(3), 477–488.
215. Garcia-Fuentes, M. (2005), Design and characterization of a new drug nanocarrier made from solid–liquid lipid mixtures, *J. Coll. Interf. Sci.*, 285(2), 590–598.
216. Lochmann, D., et al. (2005), Albumin–protamine–oligonucleotide nanoparticles as a new antisense delivery system. Part 1: Physicochemical characterization, *Eur. J. Pharm. Biopharm.*, 59(3), 419–429.
217. Müller-Goymann, C. C. (2004), Physicochemical characterization of colloidal drug delivery systems such as reverse micelles, vesicles, liquid crystals and nanoparticles for topical administration, *Eur. J. Pharm. Biopharm.*, 58(2), 343–356.
218. Weyermann, J., et al. (2004), Physicochemical characterisation of cationic polybutylcyanoacrylat—Nanoparticles by fluorescence correlation spectroscopy, *Eur. J. Pharm. Biopharm.*, 58(1), 25–35.
219. Bootz, A., et al. (2004), Comparison of scanning electron microscopy, dynamic light scattering and analytical ultracentrifugation for the sizing of poly(butyl cyanoacrylate) nanoparticles, *Eur. J. Pharm. Biopharm.*, 57(2), 369–375.
220. Merroun, M., et al. (2007), Spectroscopic characterization of gold nanoparticles formed by cells and S-layer protein of *Bacillus sphaericus* JG-A12, *Mater. Sci. Eng.*, 27(1), 188–192.
221. Lu, Q. H., et al. (2006), Synthesis and characterization of composite nanoparticles comprised of gold shell and magnetic core/cores, *J. Magnet. Magnetic Mater.*, 301(1), 44–49.
222. Vdovenkova, T., et al. (2000), Silicon nanoparticles characterization by Auger electron spectroscopy, *Surf. Sci.*, 454–456, 952–956.
223. Bonini, M., et al. (2006), Synthesis and characterization of magnetic nanoparticles coated with a uniform silica shell, *Mater. Sci. Eng.*, 26(5–7), 745–750.
224. Pilon, L. N., et al. (2006), Synthesis and characterisation of new shell cross-linked micelles with amine-functional coronas, *Eur. Polym. J.*, 42(7), 1487–1498.
225. Castro, E., et al. (2006), Characterization of triblock copolymer $E_{67}S_{15}E_{67}$—Surfactant interactions, *Chem. Phys.*, 325(2–3), 492–498.
226. Tao, L., and Uhrich, K. E. (2006), Novel amphiphilic macromolecules and their in vitro characterization as stabilized micellar drug delivery systems, *J. Coll. Interf. Sci.*, 298(1), 102–110.
227. Gaucher, G., et al. (2005), Block copolymer micelles: Preparation, characterization and application in drug delivery, *J. Controlled Release*, 109(1–3), 169–188.
228. Yu, L., et al. (2006), Determination of critical micelle concentrations and aggregation numbers by fluorescence correlation spectroscopy: Aggregation of a lipopolysaccharide, *Anal. Chim. Acta*, 556(1), 216–225.
229. Kataoka, K., et al. (2001), Block copolymer micelles for drug delivery: Design, characterization and biological significance, *Adv. Drug Deliv. Rev.*, 47(1), 113–131.

230. Hagan, S. A., et al. (1996), Polylactide-poly(ethylene glycol) copolymers as drug delivery systems. 1. characterization of water dispersible micelle-forming systems, *Langmuir*, 12(9), 2153–2161.

231. Shin, I. G., et al. (1998), Methoxy poly(ethylene glycol)/epsilon-caprolactone amphiphilic block copolymeric micelle containing indomethacin. I. Preparation and characterization, *J. Controlled Release*, 51(1), 1–11.

232. Morishima, Y., et al. (1995), Characterization of unimolecular micelles of random copolymers of sodium 2-(acrylamido)-2-methylpropanesulfonate and methacrylamides bearing bulky hydrophobic substituents, *Macromolecules*, 28, 2874–2881.

233. Vansteenkiste, S. O., and Davies, M. C. (1998), Scanning probe microscopy of biomedical interfaces, *Prog. Surf. Sci.*, 57(2), 95–136.

234. Binnig, G., and Rohrer, H. (2000), Scanning tunneling microscopy, *IBM J. Res. Dev.*, 44(1–2), 279.

235. Chen, C. J., and Smith, W. F. (1994), Introduction to scanning tunneling microscopy, *Am. J. Phys.*, 62(6), 573–574.

236. Giessibl, F. J. (2005), AFM's path to atomic resolution, *Mater. Today*, 8(5), 32–41.

237. Durig, U., et al. (1986), Near-field optical-scanning microscopy, *J. Appl. Phys.*, 59(10), 3318–3327.

238. Ash, E. A., and Nichols, G. (1972), Super-resolution aperture scanning microscope, *Nature*, 237, 510.

239. Merritt, G. (1998), A compact near-field scanning optical microscope, *Ultramicroscopy*, 71(1–4), 183–189.

240. Albrecht, T. R., et al. (1991), Frequency modulation detection using high-*Q* cantilevers for enhanced force microscope sensitivity, *J. Appl. Phys.*, 69, 668–673.

241. Ohta, M., et al. (1995), Atomically resolved image of cleaved surfaces of compound semiconductors observed with an ultrahigh vacuum atomic force microscope, *J. Vac. Sci. Technol.*, 13(3), 1265–1267.

242. Binnig, G., et al. (1986), Atomic force microscope, *Phys. Rev. Lett.*, 56(9), 930–933.

243. Zhong, Q., et al. (1993), Fractured polymer/silica fiber surface studied by tapping mode atomic force microscopy, *Surf. Sci.*, 290(1–2), 688–692.

244. Stark, R. W., and Drobek, T. (1998), Determination of elastic properties of single aerogel powder particles with the AFM, *Ultramicroscopy*, 75(3), 161–169.

245. Müller, F., et al. (1997), Applications of scanning electrical force microscopy, *Microelectron. Reliabil.*, 37(10–11), 1631–1634.

246. Cappella, B., et al. (1997), Improvements in AFM imaging of the spatial variation of force–distance curves: On-line images, *Nanotechnology*, 8, 82–87.

247. Hamers, R. J. (1989), Atomic-resolution surface spectroscopy with the scanning tunneling microscope, *Annu. Rev. Phys. Chem.*, 40, 531–559.

248. Laiho, R., et al. (1997), Spin-polarized scanning tunnelling microscopy with detection of polarized luminescence emerging from a semiconductor tip, *J. Phys. Condens. Matter*, 9, 5697–5707.

249. Binnig, G., et al. (1982), Surface studies by scanning tunneling microscopy, *Phys. Rev. Lett.*, 49, 57–61.

250. Ciraci, S., and Tekman, E. (1989), Theory of transition from the tunneling regime to point contact in scanning tunneling microscopy, *Phys. Rev.*, 40, 11969–11972.

251. Doyen, G., et al. (1993), Green-function theory of scanning tunneling microscopy: Tunnel current and current density for clean metal surfaces, *Phys. Rev.*, 47, 9778–9790.

252. Fontaine, P. A., et al. (1998), Characterization of scanning tunneling microscopy and atomic force microscopy-based techniques for nanolithography on hydrogen-passivated silicon, *J. Appl. Phys.*, 84(1), 1776–1781.

253. Wiesendanger, R. (1994), Contributions of scanning probe microscopy and spectroscopy to the investigation and fabrication of nanometer-scale structures, *J. Vac. Sci. Technol. Microelectron. Nanometer Struct.*, 12(2), 515–529.

254. Binnig, G., et al. (1986), Atomic force microscopy, *Phys. Rev. Lett.*, 56, 930–933.

255. de Souza Pereira, R. (2001), Atomic force microscopy as a novel pharmacological tool, *Biochem. Pharmacol.*, 62(8), 975–983.

256. Rajagopalan, R. (2000), Atomic force and optical force microscopy: Applications to interfacial microhydrodynamics, *Coll. Surf. Physicochem. Eng. Aspects*, 174(1–2), 253–267.

257. Santos, N. C., and Castanho, M. A. R. B. (2004), An overview of the biophysical applications of atomic force microscopy, *Biophys. Chem.*, 107(2), 133–149.

258. Ferin, J., et al. (1992), Pulmonary retention of ultrafine and fine particles in rats, *Am. J. Respir. Cell Mol. Biol.*, 6, 535–542.

259. Li, X. Y., et al. (1999), Short term inflammatory responses following intratracheal instillation of fine and ultrafine carbon black in rats, *Inhal. Toxicol.*, 11, 709–731.

260. Oberdorster, G., et al. (1995), Association of particulate air pollution and acute mortality: Involvement of ultrafine particles? *Inhal. Toxicol.*, 7, 111–124.

261. Li, X. Y., et al. (1997), In vivo and in vitro pro-inflammatory effects of particulate air pollution (PM10), *Environ. Health Perspect.*, 105, 1279–1283.

262. Brüske-Hohlfeld, I., et al. (2004), Epidemiology of nanoparticles, in *Proc. of First International Symposium on Occupational Health Implications of Nanomaterials.*

263. Warheit, D. B., et al. (2004), Pulmonary bioassay toxicity study in rats with single wall carbon nanotubes, in *Proc. of First International Symposium on Occupational Health Implications of Nanomaterials.*

264. Klumpp, C., et al. (2006), Functionalized carbon nanotubes as emerging nanovectors for the delivery of therapeutics, *Biochim. Biophys. Acta*, 1758, 404–412.

265. Muller, J. (2004), Respiratory toxicity of carbon nanotubes, in *Proc. of First International Symposium on Occupational Health Implications of Nanomaterials.*

266. Tinkle, S. (2004), Dermal penetration of nanoparticles, in *Proc. of First International Symposium on Occupational Health Implications of Nanomaterials.*

SUGGESTED READING

General

Dai, L. (2004), *Intelligent Macromolecules for Smart Devices: From Materials Synthesis to Device Applications*, Springer.

Di Ventra, M., Evoy, S., and Heflin, J. R. (2004), *Introduction to Nanoscale Science and Technology*, Springer.

Lyshevski, S. E. (2005), *Nano- and Mirco-Electromechnical Systems: Fundamentals of Nano and Microengineering*, CRC Press, Boca Raton, FL.

Roco, M. C., and Bainbridge, W. S. (2003), *Converging Technologies for Improving Human Performance: Nanotechnology, Biotechnology*, Springer.

Nano–Pharmaceutical Materials and Structures

Hirsch, A., and Brettreich, M. (2004), *Fullerenes: Chemistry and Reactions*, Wiley, Hoboken, NJ.

Cao, G. (2004), *Nanostructures & Nanomaterials: Synthesis, Properties & Applications*, Imperial College Press.

Haley, M. M., and Tykwinski, R. R. (2006), *Carbon-Rich Compounds: From Molecules to Materials*, Wiley, Hoboken, NJ.

Harris, P. J. (2001), *Carbon Nanotubes and Related Structures*, Cambridge University Press.

Kumar, C. S. S. R. (2005), *Biofunctionalization of Nanomaterials*, Wiley, Hoboken, NJ.

Kumar, C. S. S. R. (2006a), *Nanomaterials for Cancer Therapy*, Wiley, Hoboken, NJ.

Kumar, C. S. S. R. (2006b), *Biological and Pharmaceutical Nanomaterials*, Wiley, Hoboken, NJ.

Lazzari, M., Liu, G., and Lecommandoux, S. (2006), *Block Copolymers in Nanoscience*, Wiley, Hoboken, NJ.

Schmid, G. (2004), *Nanoparticles: From Theory to Application*, Wiley, Hoboken, NJ.

Tadros, T. F. (2005), *Applied Surfactants: Principles and Applications*, Wiley, Hoboken, NJ.

Vogtle, F. (2000), *Dendrimers III: Design, Dimension, Function*, Springer.

Zhang, J., Zhang, J., Wang, Z., Liu, J., and Chen, S. (2002), *Self-Assembled Nanostructures*, Springer.

Manufacturing Technologies

Butt, H.-J., Graf, K., and Kappl, M. (2006), *Physics and Chemistry of Interfaces*, Wiley, Hoboken, NJ.

Champion, Y., and Fecht, H. (2004), *Nano-Architectured and Nanostructured Materials: Fabrication, Control and Properties*, Wiley, Hoboken, NJ.

Elimelech, M., Jia, X., Gregory, J., and Williams, R. (1998), *Particle Deposition & Aggregation: Measurement, Modelling and Simulation*, Elsevier, New York.

Hoch, H. C., Jelinski, L. W., and Craighead, H. G. (1996), *Nanofabrication and Biosystems*, Cambridge University Press.

Köhler, M., and Fritzsche, W. (2004), *Nanotechnology: An Introduction to Nanostructuring Techniques*, Wiley, Hoboken, NJ.

Komiyama, M., Takeuchi, T., Mukawa, T., and Asanuma, H. (2003), *Molecular Imprinting: From Fundamentals to Applications*, Wiley, Hoboken, NJ.

Kroschwitz, J. I. (1989), *Polymers: Biomaterials and Medical Applications*, Wiley, New York.

Kumar, C. S. S. R., Hormes, J., and Leuschner, C. (2005), *Nanofabrication Towards Biomedical Applications: Techniques, Tools, Applications, and Impact*, Wiley, Hoboken, NJ.

Osada, Y., and Nakagawa, T. (1992), *Membrane Science and Technology*, Marcel Dekker, New York.

Pierson, H. (1999), *Handbook of Chemical Vapor Deposition: Principles, Technologies and Applications*, William Andrew.

Ramakrishna, S., Fujihara, K., Teo, W., Lim, T., and Ma, Z. (2005), *An Introduction to Electrospinning and Nanofibers*, World Scientific.

Rotello, V. M. (2004), *Nanoparticles: Building Blocks for Nanotechnology*, Springer.

Sotomayor Torres, C. M. (2004), *Alternate Lithography: Unleashing the Potentials of Nanotechnology*, Springer.

Wise, D. L. (2000), *Handbook of Pharmaceutical Controlled Release Technology*, Marcel Dekker, New York.

Characterization Methods

Binks, B. P. (1999), *Modern Characterization Methods of Surfactant Systems*, Marcel Dekker, New York.

Brittain, H. G. (1995), *Physical Characterization of Pharmaceutical Solids*, Marcel Dekker, New York.

Brown, W., Ed. (1996), *Light Scattering: Principles and Development*, Oxford University Press.

Buyana, T. (1997), *Molecular Physics*, World Scientific.

Chung, F. H., and Smith, D. K., Eds. (1999), *Industrial Applications of X-Ray Diffraction*, Marcel Dekker, New York.

Jena, P. B., and Hoerber, J. K. H., Eds. (2006), *Force Microscopy: Applications in Biology and Medicine*, Wiley, Hoboken, NJ.

Jenkins, R., and Jenkins, J. (1995), *Quantitative X-Ray Spectrometry*, Marcel Dekker, New York.

Meyer, E., Hug, H. J., and Bennewitz, R. (2003), *Scanning Probe Microscopy*, Springer.

Pethrick, R. A., and Viney, C., Eds. (2003), *Techniques for Polymer Organisation and Morphology Characterisation*, Wiley, Hoboken, NJ.

Rosoff, M. (2002), *Nano-Surface Chemistry*, Marcel Dekker, New York.

Sharma, A., and Schulman, S. G. (1999), *Introduction to Fluorescence Spectroscopy*, Wiley, New York.

Sibilia, J. P. (1988), *A Guide to Materials Characterization and Chemical Analysis*, Wiley, New York.

Watts, J. F., and Wolstenholme, J. (2004), *An Introduction to Surface Analysis by XPS and AES*, Wiley, Hoboken, NJ.

Wyckoff, R. W. G. (1971), *Crystal Structures*, Wiley, New York.

Toxicology

Fan, A. M. (1996), *Toxicology and Risk Assessment: Principles, Methods & Applications*, Marcel Dekker, New York.

Gad, S. C. (2004), *Safety Pharmacology in Pharmaceutical Development and Approval*, CRC Press, Boca Raton, FL.

Gardner, D. E. (2005), *Toxicology of the Lung*, CRC Press, Boca Raton, FL.

Kumar, C. S. S. R. (2006), *Nanomaterials: Toxicity, Health and Environmental Issues*, Wiley, Hoboken, NJ.

Meeks, R. G. (1991), *Dermal and Ocular Toxicology: Fundamentals and Methods*, CRC Press, Boca Raton, FL.

7.4

OIL-IN-WATER NANOSIZED EMULSIONS: MEDICAL APPLICATIONS

SHUNMUGAPERUMAL TAMILVANAN*
University of Antwerp, Antwerp, Belgium

Contents

7.4.1 INTRODUCTION

It is estimated that 40% or more of bioactive substances being identified through combinatorial screening programs are poorly soluble in water [1, 2]. Consequently, the drug molecules belonging to these categories cannot be easily incorporated into aqueous-cored/based dosage forms at adequate concentrations, and thus the clinical

*Current address: Department of Pharmaceutics, Arulmigu Kalasalingam College of Pharmacy, Anand Nagar, Krishnankoil, India

Pharmaceutical Manufacturing Handbook: Production and Processes, edited by Shayne Cox Gad

efficacy of highly lipophilic drugs is being impeded. Furthermore, it is well established that the pharmaceutical industry will face more difficulties in formulating and developing novel dosage forms of new chemical entities since 50–60% of these molecules are lipophilic in nature and often exhibit hydrophobic character. Among the different innovative formulation approaches that have been suggested for enhancing lipophilic drug absorption and then clinical efficacy, lipid-based colloidal drug delivery systems such as nanosized emulsions recognized particularly for overcoming the formulation and bioavailability-related problems of such drug molecules. Other nomenclatures are also being utilized often in the medical literature to refer to nanosized emulsions, including miniemulsions [3], ultrafine emulsions [4], and submicrometer emulsions [5, 6]. The term *nanosized emulsion* [7] is preferred because in addition to giving an idea of the nanoscale size range of the dispersed droplets, it is concise and avoids misinterpretation with the term *microemulsion* (which refers to thermodynamically stable systems). Hence, nanosized emulsions are heterogenous dispersions of two immiscible liquids [oil-in-water (o/w) or water-in-oil (w/o)], and they are subjected to various instability processes such as aggregation, flocculation, coalescence, and therefore eventual phase separation according to the second law of thermodynamics. However, the physical stability of nanosized emulsions can substantially be improved with the help of suitable emulsifiers that are capable of forming a mono- or multilayer coating around the dispersed liquid droplets in such a way as to reduce interfacial tension or increase droplet–droplet repulsion. Depending on the concentrations of these three components (oil–water–emulsifier) and the efficiency of the emulsification equipment/techniques used to reduce droplet size, the final nanosized emulsion may be in the form of o/w, w/o, macroemulsion, micrometer emulsion, submicrometer emulsion, and double or multiple emulsions (o/w/o and w/o/w). Preparation know-how, potential application, and other information pertinent to w/o emulsions [8], macroemulsions [9–11], microemulsions [12, 13], and multiple emulsions [14] are thoroughly covered elsewhere. In addition, some studies have compared the performance of different emulsified systems (macroemulsions, microemulsion, multiple emulsions, and gel emulsions) prepared with similar oils and surfactants for applications such as controlled drug release [15] or drug protection [16]. Similarly the state of the art of so-called oxygen carriers or perfluorocarbon emulsions, dispersions containing submicrometer/nanosized fluoroorganic particles in water, is also thoroughly covered in the literature [17–19] and readers can refer to these complete and interesting articles.

Possible usefulness as carriers stems from the nanosized emulsion's ability to solubilize substantial amounts of hydrophilic/hydrophobic drug either at the innermost (oil or water) phase or at the o/w or w/o interfaces. While hydrophilic drugs are contained in the aqueous phase of a w/o-type emulsion or at the w/o interface of the system, hydrophobic drugs could be incorporated within the inner oil phase of an o/w-type emulsion or at the o/w interface of the system. It appears that the choice of the type of emulsion to be used therefore depends, to a large extent, upon the physicochemical properties of the drug. Between w/o and o/w types, the o/w type of nanosized emulsions would be preferred in order to successfully exploit the advantages of an emulsion carrier system. Additionally, within the o/w type, simple modifications on surface/interface structures of emulsions can be made. For instance, incorporating an emulsifier molecule alone or in a specific combination that is capable of producing either positive or negative charges over the emulsified droplets surface will lead to the formation of surface-modified emulsions. Based on these surface modifications, the o/w-type nanosized emulsions can be divided into cationic and anionic emulsions.

The o/w nanosized emulsions have many appealing properties as drug carriers. They are biocompatible, biodegradable, physically stable, and relatively easy to produce on a large scale using proven technology [20]. Due to their subcellular and submicrometer size, emulsions are expected to penetrate deep into tissues through fine capillaries and even cross the fenestration present in the epithelial lining in liver. This allows efficient delivery of therapeutic agents to target sites in the body. Not only considered as delivery carriers for lipophilic and hydrophobic drugs, nanosized emulsions can also be viewed nowadays as adjuvants to enhance the potency of deoxyribonucleic acid (DNA) vaccine. For instance, Ott et al. [21] prepared a cationic o/w emulsion based on MF59 (commercially termed Fluad), a potent squalene in water and a cationic lipid, 1,2-dioleoyl-*sn*-glycero-3-trimethylammonium propane (DOTAP). It is shown that an interaction of cationic emulsion droplets with DNA and the formed DNA-adsorbed emulsion had a higher antibody response in mice in vivo while maintaining the cellular responses equivalent to that seen with naked DNA at the same doses. Another example of o/w emulsion-based adjuvants resulting from U.S. patent literature is the Ribi adjuvant system (RAS) [22–24]. Depending on the animal species used, RAS can be classified into two types: one for use in mice, termed monophosphoryl-lipid A + trehalose dicorynomycolate emulsion (MPL + TDM emulsion), and another for use in rabbits, goats, and larger animals, called monophosphoryl-lipid A + trehalose dicorynomycolate + cell wall skeleton emulsion (MPL + TDM + CWS emulsion). Strikingly, the MPL + TDM and MPL + TDM + CWS emulsions are prepared based on 2% oil (squalene)–Tween 80–water. These adjuvants are derived from bacterial and mycobacterial cell wall components that have been prepared to reduce the undesirable side effects of toxicity and allergenicity but still provide potent stimulus to the immune system. Another example is the syntex adjuvant formulation (SAF) that contains a preformed o/w emulsion stabilized by Tween 80 and Pluronic L121 [25].

Keeping in mind the potential of o/w nanosized emulsions, the purpose of this chapter is to classify the emulsions and provide a short overview on the preparation of the new- (second- and third-) generation emulsions followed by a description of the unique property of the third-generation emulsion and examples of selected excipients used for emulsion preparation. Given a specific interest especially on the parenteral route and then on the ocular topical route, o/w nanosized emulsions for both routes share a common platform on strict criteria concerning the maximum globule size and requirement of sterility in the final emulsions. Nasal and topical routes are also covered based on published research works with nanosized emulsions. It is emphasized that the chapter focuses only on preformed nanosized emulsions (having size distribution ranging between 50 and 1000 nm with a mean droplet size of about 250 nm), which should not be confused with self-microemulsifying drug delivery systems or preformed microemulsions that are transparent, thermodynamically stable dosage forms.

7.4.2 GENERATIONS OF OIL-IN-WATER NANOSIZED EMULSIONS

In this chapter, the o/w nanosized emulsions are classified into three generations (see Figure 1) based on their development to ultimately make a better colloidal

Nanosized emulsion, NE

⇩

First-generation ne (phospholipid based)

⇩

Second-generation NE, with PEGylation on droplet surface

⇩

Third-generation NE, with PEGylation and positive charge on droplet surface

FIGURE 1 Flow chart of three generations of emulsion.

carrier for a target site within the organs/parts of the body and eye, thus allowing site-specific drug delivery and/or enhanced drug absorption.

7.4.2.1 First-Generation Emulsion

To be healthy with a quality life style is every human's desire. According to documented Indian scriptures dating back to 5000 B.C., nutritional status has always been associated with health [26]. Because nutritional depletion due to either changes in the quality or amount of dietary fat intake or abnormalities in lipid metabolism results in immunosuppression and therefore host defense impairment, favoring increased infection and mortality rates.

Traditionally depletion in dietary fats in malnourished or hypercatabolic patients is compensated through intravenous feeding using a solution containing amino acids, glucose, electrolytes, and vitamins as well as nanosized emulsions. Structurally, an o/w-type emulsion is triglyceride (TG) droplets enveloped with a stabilizing superficial layer of phospholipids [27]. Emulsions for parenteral use are complex nutrient sources composed not only of fatty acids but also substances other than TG, such as phosphatidylcholine, glycerol, and α-tocopherol in variable amounts. The emulsions also had a complex inner structure and consisted of particles with different structures, namely, oil droplets covered by an emulsifier monolayer, oil droplets covered by emulsifier oligolayers, double-emulsion droplets, and possibly small unilamellar vesicles. Commercially available nanosized emulsions used as intravenous high-calorie nutrient fluids have particle size normally around 160–400 nm in diameter and, typically, their surfaces are normally negatively charged. Larger droplets can also be detected in commercially available emulsions [28]. Lutz et al. [29] reported that the mean diameter of particles in the 20% emulsions is larger than in the 10% emulsions.

The TG in nanosized emulsions may be presented structurally in long or medium chains respectively, named LCT and MCT. The mean diameter of particles in LCT emulsions is greater than that in MCT emulsions [29]. LCT contains fatty acid chains with 14, 16, 18, 20, and 22 carbon atoms and sometimes with double bonds. The number of double bonds present defines the fatty acids in LCT as saturated, mono-unsaturad, or polyunsaturated. If the first double bond is on carbon number 3, 6, or 9 from the methyl end of the carbon chain, then the fatty acid is *n*-3, *n*-6, or *n*-9, respectively. Purified soybean or safflower oil contains LCT with a high proportion of n-6 polyunsaturated fatty acids whereas olive oil has LCT with n-9 monounsaturated fatty acids. Fish oil includes LCT with 20 or more carbon atoms where the first double bond is located between the third and fourth carbons from the methyl terminal of the fatty acids chain (omega-3 or n-3). On the other hand, MCT is derived from coconut oil and has saturated fatty acids with chains containing carbon atoms at the 6, 8, 10, or 12 positions. Both LCT and MCT either alone or MCT in combination with LCT are known for their long-term commercial acceptability in parenteral emulsions and are found in several U.S. Food and Drug Administration (FDA)–approved products. Also in Europe emulsion containing LCT/MCT enriched with fish oil became available for research. With MCT/LCT combinations in a specific ratio, nanosized emulsions appear to provide a more readily metabolizable source of energy [30]. However, LCT emulsion has been used in clinical practice for over 40 years. But for drug solubilization purpose, MCT is reported to be 100 times more soluble in water than LCT and thus to have an escalated solubilizing capability.

7.4.2.2 Second-Generation Emulsion

An easy and substantial association of lipophilic bioactive compound with the MCT or other vegetable oil–based emulsions, however, allows the emulsions to be used as vehicles/carriers for the formulation and delivery of drugs with a broad range of applications. These applications extend from enhanced solubilization or stabilization of the entrapped drug to sustained release and site-specific delivery. Hence the emulsions used for these applications are termed second-generation emulsions. Fittingly the o/w-type nanosized emulsions containing either positive or negative charge can be administered by almost all available routes, that is, topical, parenteral, oral, nasal, and even aerosolization into the lungs [31]. Despite differences in routes of administration, examples of commercially available emulsion-based formulations utilized for medical and nonmedical applications purposes are given in Table 1.

The lipid-induced enhancement in oral bioavailability of many drugs having poor water solubility is a well-known documented fact when the drugs are incorporated into emulsions [32, 33]. However, direct intravascularly or locally administered conventional first- and second-generation emulsions could be taken up rapidly by the circulating monocytes for clearance by reticuloendothelial cells (through organs such as the liver, spleen, and bone marrow) [34]. Furthermore, the extent of clearance is enhanced by the adsorption of opsonic plasma proteins onto emulsion surfaces. However, the oily hydrophobic particles of the emulsions can also be taken up by macrophages without the necessity of opsonization provided the oil phase is liberated from the emulsion through the destabilization process occurring inside the blood compartment immediately after emulsion administration intravascularly or locally. Although the core of o/w emulsions is indeed hydrophobic, the emulsion envelope is not. The exposure of the hydrophobic part to the aqueous medium will therefore destabilize the emulsion. Moreover, Sasaki et al. [35] have assumed that when the castor oil–based emulsions interact with the tears in the eye, the electrolytes in the tears elicit a physical emulsion instability resulting in some release of the oil. The electrolytes present in blood or cellular fluids can also cause a similar type of emulsion destabilization process, resulting in separation of the oil and water

TABLE 1 Selected Marketed Medical and Nonmedical Emulsions

Parenteral Fat Emulsions (o/w Type) for Nutrition		Registered Emulsions (o/w Type) Containing Drugs			
Product	Producer	Product	Drug	Producer	Application
Abbolipid/Liposyn	Abbott	Diazepam-Lipuro	Diazepam	Braun Melsungen	Intravenous
Intralipid	Pharmacia-Upjohn	Diprivan	Propofol	AstraZeneca	Intravenous
Lipofundin N or Endolipide	B. Braun	Etomidat-Lipuro	Etomidate	Braun Melsungen	Intravenous
Lipofundin MCT/LCT	—	Lipotalon (Limethason)	Dexamethasone palmitate	Merckle	Intra-arthr.
Medialipide/Vasolipid	B. Braun	Stesolid	Diazepam	Dumex	Intravenous
Medianut	B. Braun	Sandimmune	Cyclosporin A	Novartis	Oral
Lipovenos	Fresenius	Neoral	Cyclosporin A	Novartis	Oral
Ivelip/Salvilipid	Clintec/Baxter	Gengraf	Cyclosporin A	Abbott	Oral
Clinoleic	Clintec/Baxter	Norvir	Ritonavir	Abbott	Oral
Intralipos	Green Gross	Restasis	Cyclosporin A	Allergan	Ocular topical
Kabimix	Pharmacia-Upjohn	Refresh Endura	Drug free	Allergan	Ocular topical
Trivè 1000	Baxter SA	Fluad (MF59)	Adjuvant	Chiron	Parenteral

Perfluorocarbon Emulsions (Fluorocarbon-in-Water Emulsions)			Selected Topical Formulations Based on o/w or w/o Emulsions	
Product	Producer	Application	Product	Producer
Fluosol DA	Green Gross, Osaka	Blood supplement or O_2 carrier	Daivonex cream and ointment	Laboratoire Leo
Imagent	Alliance Pharmaceutical	Contrast agent to image heart	Voltaren emulgel	Ciba-Geigy
Oxygent	Alliance Pharmaceutical	Blood supplement or O_2 carrier	EMLA cream	Astra, Swedan

Promonocyte (bone marrow)

↓

Monocyte (blood)

↓

Macrophages (tissues) highly phagocytic

Connective tissue	(histiocyte)
Liver	(Kupffer cell)
Lung	(alveolar macrophage)
Spleen	(free and fixed macrophages, sinusoidal lining cell)
Lymph node	(free and fixed macrophage)
Bone marrow	(macrophages, sinusoidal lining cell)
Serous cavity	(peritoneal macrophage)
Bone tissue	(osteoclast)
Nervous system	(microglia)

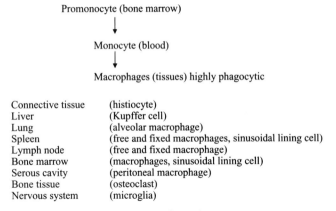

FIGURE 2 Mononuclear phagocyte system.

phases from the parenterally administered emulsions. It is thus reasonable to say that the resultant oily hydrophobic particles of the emulsions would also be taken up by macrophages independent of an opsonization process. An opsonization process is the adsorption of protein entities capable of interacting with specific plasma membrane receptors on monocytes and various subsets of tissue macrophages (see Figure 2), thus promoting particle recognition by these cells. Classical examples of opsonic molecules include various subclasses of immunoglobulins [36, 37], complement proteins such as C1q and generated C3 fragments (C3b, iC3b) [38], apolipoproteins [36, 37], von Willebrand factor, thrombospondin, fibronectin [39], and mannose-binding protein. When given by other parentral routes, for example, intraperitoneally, subcutaneously, or intramuscularly, the majority of emulsion droplets enter the lymphatic system and eventually the blood circulation where particles behave as if given intravenously. Liver, spleen, and bone marrow uptake is significantly lower. Indeed, relative to the emulsion droplet size, lymph nodes take up a much greater (over 100-fold) proportion than any other reticuloendothelial system (RES) tissue.

There is increasing interest in developing injectable emulsions that are not cleared quickly from the circulation when they are designed to reach non-RES tissues in the vascular system or extravascular sites of action or to act as circulating drug reservoirs. Earlier approaches for making long-circulating emulsions concentrate mainly on changes in the oil phase of the emulsion such as MCT versus LCT [40], use of structured lipids (SLS) having medium-chain (C_8–C_{10}) fatty acids (SLM) and short-chain (C_4) fatty acids (SLS) [41], addition of sphingomyelin [42–45] and cholesterol [46] to the emulsion, and use of hydrogenated castor oil (HCO) with at least 20 oxyethylene units (HCO20) [47–52]. Using the further established formulation approaches by which the emulsion droplet surfaces could be altered might, however, be more realistic and even more useful for a wide array of drug-targeting purposes. Steric barrier or enhanced hydrophilicity effect exerted by a polyoxyethylene (POE) chain having surfactants when added as coemulsifier into the phospholipid-stabilized first-generation emulsion allows, to some extent, the passive/inverse drug targeting to the lung, kidneys, and areas of inflammation [53, 54]. Addition of POE-based surfactants into the otherwise amphipathic phospholipid-stabilized emulsion is particularly effective against plasma protein adsorption onto emulsion surfaces because of the hydrophilicity and unique solution properties of POE-based surfactants, including minimal interfacial free energy with water, high aqueous solubility, high mobility, and large exclusion volume [54]. In addition, colloidal particles presenting hydrophilic surfaces with a low contact angle will be almost ignored by phagocytic cells [55], although emulsion particles are not supposed to be recognized as foreign by the body to some extent. Examples of POE chains containing surfactants employed so far in emulsions are Tween 80, Span 80, Brij, and Poloxamer 188 (commercially named Pluronic F-68 or Lutrol F-68). The effectiveness of these polymeric surfactant molecules to intercalate at the oil–water interface with strong bonding to the phospholipid molecules could also be checked/judged through an in vitro monolayer experiment [56].

In general, the modification of particulate carriers using amphipathic polyethylene glycol (PEG)–containing molecules results in a prolongation of their blood circulation time [57, 58]. A phosphatidylethanolamine derivative with polyethylene glycol (PEG–PE) is widely used to increase the plasma retention of particulate carriers such as liposomes [59–61], polystyrene microspheres [62], and nanospheres [63]. Therefore, similar to POE, the PEG–PE is also incorporated as a coemulsifier into emulsions (termed PEGylated emulsion) to augment its circulation half-life [64]. Liu and Liu [53] studied the biodistribution of emulsion particles coated with phosphatidylethanolamine derivatives with three different molecular weight PEGs (MW 1000, 2000, and 5000). Among them, PEG-2000 was able to prolong the circulation time of emulsion probably due to the increased hydrophilicity of the droplet surface and/or the formation of a steric barrier. However, Tirosh et al. [65] assumed, while characterizing the PEG-2000-grafted liposome by differential scanning calorimetry, densitometry, and ultrasound velocity and absorption measurements, that the steric stabilization is much more than increasing hydrophilicity. In addition, PEG-containing compounds also decrease the lipolysis of emulsion particles [47] and prevent the uptake by the mononuclear phagocytes [66].

A dipalmitoyl phosphatidylcholine (DPPC)–stabilized emulsion was prepared by Lundberg et al. [67] and the effect of addition of PEG–PE, polysorbate 80, or Pluronic F-68 on the metabolism of DPPC-stabilized emulsion was studied. Two different radioactive markers were used to investigate the fate of emulsion particles following injection into the tail vein of female BALB/c inbred mice. While ^{14}C- triolein (TO) is susceptible to the action of lipoprotein lipase (LPL), ^{3}H-cholesteryl oleate ether (CO ether) is not. Hence the removal of ^{14}C-TO represents the triglyceride metabolism, whereas the other one is a core marker to represent whole particle removal by RES organ uptake. The emulsions with DPPC as sole emulsifier were rapidly cleared from the blood with only 10–11% of CO or TO left in circulation after 1h. However, addition of PEG–PE gave a prolonged clearance rate, especially during first 3h. A further addition of cosurfactant polysorbate 80 or Pluronic F-68 resulted in a marked extension of the circulation lifetime during the first 6h. The notable effects of polysorbate 80 and Pluronic F-68 can apparently be attributed mainly to the decrease in droplet size, although an additional influence due to the increased hydrophilicity may not be ruled out.

The in vivo disposition of emulsions administered as nutrients (surface-unmodified first-generation emulsion) as well as administered as drug carriers (surface-modified second-generation emulsion) would depend on the particle properties, such as the size [68–71], zeta potential (see Sections 7.4.2.3 and 7.4.5), and compositions of phospholipids and oil phase (see the above paragraphs), which may vary among different products and the batches of each product. The size of particulate carriers such as liposomes is known to influence both the phagocytic uptake by the mononuclear phagocyte system (MPS) [68–70] and the binding of apolipoprotein (apo) to emulsions [71]. Furthermore, the particle size is a major determinant of the transfer to extravascular spaces from the blood compartment. The capillaries of the vascular system can be classified into three categories: continuous, fenestrated, and discontinuous (sinusoidal) [72]. Particulate carriers including nanosized emulsions are considered to pass through capillaries and reach extravascular cells only in organs having discontinuous capillaries such as liver, spleen, and bone marrow. In such tissues, the extravasation of particles should be regulated by their size since the largest pores in the capillary endothelium is reported to be about 100 nm [73]. In addition, tumor capillaries have unique characteristics in their structures and functions in comparison with normal tissues such as muscle [74, 75], which results in the enhanced distribution of particulate carriers to tumor tissues [76–78]. The distribution of emulsions within a tumor tissue was also regulated by the size of particulate carriers [79]. Obviously, because of the submicrometer size range (175–400 nm in diameter) of the emulsions, the more they circulate, the greater their chance of reaching respective targets. More specifically, growing solid tumors as well as regions of infection and inflammation have capillaries with increased permeability as a result of the disease process (e.g., tumor angiogenesis [74]). Pore diameters in these capillaries can range from 100 to 800 nm. Thus, drug-containing emulsion particles are small enough to extravasate from the blood into the tumor interstitial space through these pores [80]. Normal tissues, by and large, contain capillaries with tight junctions that are impermeable to emulsions and other particles of this diameter. This differential accumulation of emulsion-laden drug in tumor tissues relative to normal cells is the basis for the increased tumor specificity for the emulsion-laden drug relative to free (nonemulsion) drug. In addition, tumors lack lymphatic drainage and therefore there is low clearance of the extravasated emulsion from tumors. Thus, long-circulating lipid carriers, such as POE/PEG-coated nanosized emulsions, tend to accumulate in tumors as a result of increased microvascular permeability and defective lymphatic drainage, a process also referred to as the enhanced permeability and retention (EPR) effect [81]. Table 2 lists various formulation factors affecting the metabolism as lipoproteins, the recognition by the MPS, and the elimination from the blood circulation of both second- and third-generation nanosized emulsions after parenteral administration.

On the other hand, essential requirements of this "active" targeting approach include identification of recognition features (receptors) on the surface of the target and the corresponding molecules (ligands) that can recognize the surface. Indeed, emulsions with appropriate ligands anchored on their surface must be able to access the target, bind to its receptors, and, if needed, enter it. Furthermore, in order to bring the colloidal carrier closer to otherwise inaccessible pathological target tissues, homing devices/ligands such as antibodies and cell recognition proteins are usually linked somehow onto the particle surfaces. Various methods have been employed to couple ligands to the surface of the nanosized lipidic and polymeric carriers with reactive groups. These can be divided into covalent and noncovalent couplings. Noncovalent binding by simple physical association of targeting ligands to the

TABLE 2 Formulation Factors Affecting Metabolism as Lipoproteins, Recognition by Mononuclear Phagocyte System (MPS), and Elimination from Blood Circulation of Second- and Third-Generation Nanosized Emulsions after Parenteral Administration

	Metabolism as Lipoproteins		Recognition by MPS		Elimination from Blood Circulation	
Factor	Poor	Extensive	Poor	Extensive	Slow	Rapid
Particle size	Large	Small	Small	Large	Small	Large
Emulsifier	DPPC	EYPC	DPPC	DSPC	DPPC	EYPC
	DSPC	—	SM	—	SM	DSPC
	SM	—	—	—	—	—
Coemulsifier	Poloxamers	—	Poloxamers	—	Poloxamers	—
	HCO-60	—	HCO-60	—	HCO-60	—
	PEG–PE	—	PEG–PE	—	PEG–PE	—
	Polysorbates	—	Polysorbates	—	Polysorbates	—
	Solutol	—	Solutol	—	Solutol	—
Cationic lipid	SA/OA	—	SA/OA	—	SA/OA	—
Oil phase	LCT	MCT	—	—	LCT	MCT
	—	—	—	—	SLS	SLM

Note: DPPC, dipalmitoylphosphatidylcholine; DSPC, distearoylphosphatidylcholine; SM, sphingomyelin; EYPC, egg yolk phosphatidylcholine; HCO-60, polyoxyethylene-(60)-hydrogenated castor oil; PEG–PE, phosphotidylethanolamine derivative with polyethylene glycol; SA, stearylamine; OA, oleylamine; LCT, long-chain triglyceride; MCT, medium-chain triglyceride; SLS, structured lipid with short-chain fatty acids, C_8–C_{10}; SLM, structured lipid with medium-chain fatty acids, C_4.

nanocarrier surface has the advantage of eliminating the use of rigorous, destructive reaction agents. Common covalent coupling methods involve formation of a disulfide bond, cross-linking between two primary amines, reaction between a carboxylic acid group and primary amine, reaction between maleimide and thiol, reaction between hydrazide and aldehyde, and reaction between a primary amine and free aldehyde [82]. For antibody-conjugated second-generation anionic emulsions, the reaction of the carboxyl derivative of the coemulsifier molecule with free amine groups of the antibody and disulfide bond formation between coemulsifier derivative and reduced antibody were the two reported conjugation techniques so far [83–85]. However, by the formation of a thio-ether bond between the free maleimide reactive group already localized at the o/w interface of the emulsion oil droplets and a reduced monoclonal antibody, the antibody-tethered cationic emulsion was developed for active targeting to tumor cells [86]. It should be added that the cationic emulsion investigated for tumor-targeting purpose belongs to the third-generation emulsion category (Section 7.4.2.3).

Apart from non-RES-related disease treatment through target-specific ligand conjugation, the second-generation emulsions may also be useful for RES-related disease treatment. Certain lipoprotein or polysaccharide moiety inclusion into the emulsions would help to achieve this concept. In general, uptake of small colloidal drug carriers by the phagocytotic mononuclear cells of RES in the liver can be exploited to improve the treatment of parasitic, fungal, viral and bacterial diseases such as, for example, leishmaniasis, acquired immunodeficiency syndrome (AIDS), and hepatitis B. The approach to use emulsions as a drug carrier against microbial diseases is superior to free antimicrobial agents in terms of both distribution to the relevant intracellular sites and treating disseminated disease states effectively. As already discussed, conventional emulsion particles are capable of localizing in liver and spleen, where many pathogenic microorganisms reside.

Rensen et al. [87] demonstrated the active/selective liver targeting of an antiviral prodrug (nucleoside analogue, iododeoxyuridine) incorporated in an emulsion complexed with ligands such as recombinant apolipoprotein E (apoE) using the Wistar rat as animal model because its apoE–receptor system is comparable to that of humans [88]. Whereas the parent drug did not show any affinity for emulsion due to its hydrophilic property, derivatization with hydrophobic anchors allowed incorporation of at least 130 prodrug molecules per emulsion particle without imparting any effect on the emulsion structure and apoE association to emulsion droplets. The authors did not describe where the 130 prodrug molecules reside in the emulsion and what is the emulsion/medium partition coefficient of the prodrug. The prodrug molecules might have reasonably higher solubility in the oil or o/w interface of the emulsion possibly due to a high partition coefficient value. Plausibly, this high partition value for prodrug molecules will determine the kinetic parameter k_{off} (desorption rate of an emulsion component from the assembly), as suggested by Barenholz and Cohen [89] for liposomal technology. Furthermore, without being bound by theory, the apoE component helps to disguise the emulsion particles so that the body does not immediately recognize it as foreign but may allow the body to perceive it as native chylomicrons or very low density lipoproteins (VLDL). The small size and the approximately spherical shape allow the emulsion particles to exhibit similar physicochemical properties to native chylomicrons or VLDL (hydrolyzed by LPL) whereas the incorporated prodrug remained associated with the emulsion remnant particles following injection into the blood circulation of the rat [87]. Because the carrier particles are not recognized as foreign, the systemic circulation of the drug increases, thus increasing the likelihood of drug delivery to the target tissues (up to 700 n*M* drug concentration in liver parenchymal cells). Additionally, the clearance rate of the drug decreases, thereby reducing the likelihood of toxic effects of the drug on clearance tissues since accumulation of the drug in another part of the clearance tissues is reduced. Thus, specific organs may be targeted by using nanosized emulsion particles as described above due to target cells comprising high levels of specific receptors, for example, but not limited to apoE receptors.

To address this issue, the saccharide moieties of glycolipids and glycoproteins on the cell surface are considered to play an important role in various intercellular recognition processes. For instance, Iwamoto et al. [90] investigated the influence of coating the oil droplets in emulsion with cell-specific cholesterol bearing polysaccharide, such as mannan, amylopectin, or pullulan, on the target ability of those formulations. They observed a higher accumulation of mannan-coated emulsion in the lung in guinea pigs. Thus selective drug targeting through emulsion-bearing ligands not only leads to an improved drug effectiveness and a reduction in adverse reactions but also offers the possibility of applying highly potent drugs. Hence, the composition of the emulsion plays an important role concerning intercellular cell recognition processes and, indeed, cell recognizability is also being improved by incorporation of apoproteins or galactoproteins onto the emulsion particles to enhance their specificity [91].

Overall, although second-generation emulsion is usually used as a means of administering aqueous insoluble drugs by dissolution of the drugs within the oil phase of the emulsion, employing surface modification/PEGylation by the attachment of targeting ligands (apoE, polysaccharide, and antibody) onto the droplet surface of emulsions may be useful for both passive and active drug-targeting purposes. Thus receptor-mediated drug targeting using ligands attached to emulsions seems to hold a promising future to the achievement of cell-specific delivery of multiple classes of therapeutic cargoes, and this approach will certainly make a major contribution in treating many life-threatening diseases with a minimum of systemic side effects.

7.4.2.3 Third-Generation Emulsion

In order to increase cellular uptake, a cationization strategy is applied particularly on the surfaces of nonviral, colloidal carrier systems such as liposomes, nano- and microparticulates, and nanocapsules [92]. To make the surface of these lipidic and polymeric carrier systems a cationic property, some cationic lipids/polymers are usually added into these systems during/after preparation. But, adding only the cationic substances in phospholipids-stabilized first-generation emulsions does not help to obtain a physically stabilized emulsion for a prolonged storage period. However, using different cationic lipids as emulsifier and additional helper lipids as coemulsifier, for example, DOTAP, 1,2-dioleoyl-*sn*-glycero-3-phosphoethanolamine (DOPE), and 1-palmitoyl-2-oleoyl-*sn*-glycero-3-phosphoethanolamine-*N*-[poly(ethylene glycol)$_{2000}$] (PEG$_{2000}$PE), reports are available to prepare emulsions with positive charges on their droplet surfaces [93, 94]. Alternatively, inclusion of cation-forming substances such as lipids (stearyl or oleyl chain having primary amines) [95, 96], polymers (chitosan) [97, 98], and surfactants (cetyltrimethylammonium bromide) [99] during the preparation of second-generation emulsion allows the formation of a stabilized system with positive charges over on it. Further, the positive charge caused by stearylamine was also confirmed by a selective adsorption of thiocyanate. Its adsorption was correlated with increasing stearylamine concentration [95]. So, nanosized emulsion consisting of complex emulsifiers, that is, phospholipid–polyoxyethylene surfactant-cationized primary amine or a polymer combination, can conveniently be termed third-generation emulsion.

The extemporaneous addition of the solid drug or drug previously solubilized in another solvent or oil to the preformed first- and second-generation emulsions is not a favored approach as it might compromise the integrity of the emulsion. Since therapeutic DNA and single-stranded oligonucleotides or small interfering ribonucleic acid (siRNA) are water soluble due to their polyanionic character, the aqueous solution of these compounds need to be added directly to the preformed third-generation emulsion in order to interact electrostatically with the cationic emulsion droplets and thus associate/link superficially at the oil–water interface of the emulsion [100, 101]. During in vivo conditions when administered via parenteral, nasal, and ocular routes, the release of the DNA and oligonucleotides from the associated emulsion droplet surfaces should therefore initially be dependent solely on the affinity between the physiological anions of the biological fluid and the cationic surface of the emulsion droplets. For instance, the mono- and divalent anions containing biological fluid available in the parenteral route is plasma and in ocular topical route is tear fluid, aqueous humor, and vitreous humor. Moreover, these biofluids contain a multitude of macromolecules and nucleases. There is a possibility that endogenous negatively charged biofluid components could dissociate the DNA and oligonucleotides from cationic emulsion. It is noteworthy to conduct, during the preformulation development stages, an in vitro release study for therapeutic DNA and oligonucleotide-containing emulsion in these biological fluids, and this type of study could be considered an indicator of the strength of the interaction that occurred between DNA or oligonucleotide and the emulsion particles. However, it is interesting to see what could happen when the third-generation emulsion is applied to in vitro cell culture models in the presence of serum. The serum stability of the emulsion–DNA complex was reported [102, 103]. Further interesting investigations using third-generation emulsions in gene delivery purposes are briefly summarized in a review article [104].

7.4.2.4 Unique Property of Third-Generation Emulsion

To enhance the drug-targeting efficacy of colloidal carriers such as nanospheres and liposomes, a PEGylation/cationization strategy is traditionally made over the surface of these carriers. While surface PEGylated colloidal carriers exhibit a prolonged plasma residence time through an escaping tendency from RES uptake following parenteral administration, surface-cationized colloidal carriers can facilitate the penetration of therapeutic agents into the cell surface possibly via an endocytotic mechanism. These two facts are proved in both liposomes and nanospheres when they possess separately the cationic and PEGylatic surface modifications on them. However, a cationic emulsion colloidal carrier system developed in Simon Benita's laboratory at Hebrew University of Jerusalem, Israel, differs significantly in such a way that it holds a combination of cationic and PEGylatic surface properties on it. Benita's group have prepared a novel cationic emulsion vehicle using a combination of emulsifiers consisting of Lipoid E 80, Poloxamer 188, and stearylamine and have found the formulation suitable for parenteral use, ocular application, nasal drug delivery, and topical delivery [105].

It has been reported in an ocular pharmacokinetic study of cyclosporin A incorporated in deoxycholic acid–based anionic and stearylamine-based cationic emulsions in rabbit that, when compared to anionic emulsion, the cationic emulsion showed a significant drug reservoir effect of more than 8 h in corneal and conjunctival tissues of the rabbit eye following topical application [106]. Since cornea and conjunctiva are of anionic nature at physiological pH [107], the cationic emulsion would interact with these tissues electrostatically to implicate the observed cyclosporin A reservoir effect. This hypothesis is supported, in principle, by an ex vivo study which showed that cationic emulsion carrier exhibited better wettability properties on rabbit cornea than either saline or anionic emulsion carrier [108].

Studies [109, 110] have shown that small changes in physical properties of emulsions can influence the elimination rate of these formulations from the blood. Indeed, an organ distribution study of stearylamine-based cationic or deoxycholic acid–based anionic nanosized emulsions and Intralipid, a well-known commercial anionic emulsion, containing ^{14}C-CO was carried out following injection into the tail vein of male BALB/c mice (20–26 g) at a volume of 5 mL/kg [111, 112]. Since CO (cholesteryl oleate) is one of the most lipophilic compounds used in biopharmacy and is not prone to degradation in the body (which remains within particles even after lipolysis of emulsion), its in vivo behavior can be regarded as reflecting that of the injected nanosized emulsion in the early phase [42, 113]. Following intravenous administration of the various emulsions having ^{14}C-CO to BALB/c mice, the ^{14}C-CO was found to accumulate in organs such as lung and liver. Furthermore, it was observed that the concentration of ^{14}C-CO in the lung decreased but was again elevated over time for both the developed cationic and anionic emulsion formulations, with a concomitant decrease in the concentration of the radiolabeled compound in the liver. However, within the various emulsion distribution patterns observed in liver, a lower ^{14}C-CO concentration was observed for stearylamine-based cationic emulsion when compared to Intralipid while for deoxycholic acid–based anionic emulsion the observed concentration of ^{14}C-CO was relatively very low when compared to cationic emulsion and Intralipid. In addition, in comparison to both anionic emulsions, the stearylamine-based cationic emulsion elucidated a much longer retention time of ^{14}C-CO in the plasma, clearly indicating a long circulating half-life for cationic emulsion in the blood. Thus, the cationic nanosized emulsion can be considered a stealth long-circulating emulsion.

The above two studies clearly described the unique characteristics of third-generation emulsion in enhancing ocular drug bioavailability; on the other hand, the same emulsion has the property of circulating for a longer time in blood following parenteral administration. Excess positive charge at the oil–water interface in conjunction with the projection of highly hydrophilic POE chain (due to the presence of Poloxamer 188) toward the aqueous phase of the o/w-type nanosized emulsion is the main reason behind the emulsion attaining its unique property, which is absent in first- and second-generation emulsions. However, a better understanding of the structure of the third-generation emulsion in terms of forces involved in its formation and stabilization must ultimately be obtained in an effort to provide a clearly understood physical basis for the uniqueness in its biological efficacy following parenteral and ocular administration.

It should be added that the use of stearylamine in intravenous administered emulsion might be problematic. Stearylamine is a single-chain amphiphile having relatively high critical micellar concentration, although the concentration used in the studied emulsion is much higher than the critical micellar concentration. Therefore, due to the dilution in plasma as well as plasma lipoproteins and blood cells, there is a high probability that the emulsion will lose its stearylamine almost instantaneously. To substantiate indirectly this issue, Klang et al. [114] showed the lack of potential induced toxicity of stearylamine-based cationic emulsion in animal models in vivo and Korner et al. [115] investigated the surface properties of mixed phospholipid–stearylamine monolayers and their interaction with a nonionic surfactant (poloxamer) in vitro. Despite the presence of the stearylamine, which may be suspected of being an irritant in pure form, in the emulsifier combination, the hourly instillation of stearylamine-based cationic emulsion vehicle into rabbit eye was well tolerated without any evidence of any toxic or inflammatory response to the ocular surface during the 5 days of the study (40 single-drop instillations between 8 AM and 4 PM each day) [114]. Following 0.2-, 0.4-, and 0.6-mL single-bolus injections of the same emulsion vehicle, representing a huge single administered dose of 30 mL/kg, no animal deaths were noted over a period of 30 days, apparently indicating the absence of marked acute toxicity [114]. Furthermore, the same stearylamine-based cationic emulsion vehicle did not cause acute neurotoxicity in rats when a continuous intravenous infusion (3.3 mL) for 2 h at a rate of 27.4 μL/min was administered through the jugular vein [114]. An another study from Benita's laboratory suggests that long-term subchronic toxicity examination of the rabbit eye (healthy) following thrice-daily single-drop topical instillation of the stearylamine-based emulsion elicited an almost similar nonirritating effect to eye tissues in comparison to the thrice-daily single-drop topical instillation of the normal saline–treated control rabbit eyes (unpublished data). Thus, overall results clearly indicated that the stearylamine was strongly bound at a molecular level to the mixed interfacial film formed by Lipoid E 80 and poloxamer 188 at the oil–water interface system [115]. Such an intercalation between the emulsifiers is responsible for emulsified oil droplet stability and, in fact, prevented the stearylamine from leaking and exerting any intrinsic possible local or systemic adverse effects in model animals.

7.4.3 PREPARATION METHODS FOR DRUG-FREE/LOADED OIL-IN-WATER NANOSIZED EMULSIONS

To get a better idea of how to formulate the nanosized emulsion delivery systems suitable for parenteral, ocular, percutaneous, and nasal uses, the reader is referred to more detailed descriptions of methods of nanosized emulsion preparation [6, 116]. A hot-stage high-pressure homogenization technique or combined emulsification technique (de novo production) is frequently employed in order to prepare nanosized emulsions with desired stability even after subjection to autoclave sterilization. Therefore, the steps involved in this technique in making blank anionic and cationic emulsions were arranged in the following order:

1. Weigh the oil- and water-soluble ingredients in separate beakers.
2. Heat both oil and water phases separately to 70°C.
3. Add the oil phase to the water phase and continue the heating up to 80°C with constant stirring to form a coarse emulsion.
4. Mix at high shear to make a fine emulsion.
5. Cool the fine emulsion formed in ice bath.
6. Homogenize the fine emulsion.
7. Cool the homogenized emulsion in ice bath.
8. Filter the emulsion using a 0.5 μm membrane filter.
9. Adjust the emulsion to 7 using 0.1 *N* hydrochloric acid or 0.1 *N* sodium hydroxide solution.
10. Pass nitrogen gas into the vials containing the emulsion.
11. Sterilize the emulsion using an autoclave.

The traditional droplet size–reducing steps involved during the preparation include constant mild stirring using a magnetic stirrer when initially mixing oil and water phases, rapid Polytron mixing at high speed, and final droplet size homogenization using a two-stage homogenizer valve assembly. The initial heating is vital for the

TABLE 3 Typical Formula to Make o/w Anionic and Cationic Nanosized Emulsions

Oil Phase	Water Phase
Natural/semisynthetic oils	Poloxamer 188
Phospholipid mixture	Glycerol
Stearylamine/oleylamine[a]	Double-distilled water
Deoxycholic acid/oleic acid[b]	
Vitamin E	

[a]Necessary ingredient for cationic emulsions.
[b]Necessary ingredient for anionic emulsions.

effective solubilization of the respective oil and water phase components in their corresponding phases. Mixing the two phases with constant mild stirring and subsequently raising the temperature to 85°C are needed to form an initial coarse emulsion and to localize the surfactant molecules for better adsorption at the oil–water interface, respectively. A typical formula to make anionic and cationic nanosized emulsions is given in Table 3.

There are three different approaches to incorporate lipophilic drugs into the oil phase or at the o/w interface of the nanosized emulsions, namely, extemporaneous drug addition, de novo emulsion preparation, and an interfacial incorporation approach, which includes the recently developed SolEmul technology [117]. In principle, the lipophilic drug molecules should however be incorporated by a de novo process. Thus, the drug is initially solubilized or dispersed together with an emulsifier in suitable single-oil or oil mixture by means of slight heating. The water phase containing the osmotic agent with or without an additional emulsifier is also heated and mixed with the oil phase by means of high-speed mixers. Further homogenization takes place to obtain the needed small droplet size range of the emulsion. A terminal sterilization by filtration, steam, or autoclave then follows. The emulsion thus formed contains most of the drug molecules within its oil phase. This is a generally accepted and standard method to prepare lipophilic drug–loaded nanosized emulsions for parenteral, ocular, percutaneous, and nasal uses, as illustrated in Figure 3. This process is normally carried out under aseptic conditions and nitrogen atmosphere to prevent both contamination and potential oxidation of sensitive excipients.

7.4.4 EXCIPIENT INCLUSION: OIL-IN-WATER NANOSIZED EMULSIONS

In general, nanosized emulsions should be formulated with compatible vehicles and additives. The components of the internal and external phases of emulsion should be chosen to confer enhanced solubility and/or stability to the incorporated biologically active lipophilic drug. In addition, it should also be designed to influence biofate or therapeutic index of the incorporated drug following administration via parenteral, ocular, percutaneous, and nasal routes. In this section, general considerations concerning excipient selection and optimum concentrations are comprehen-

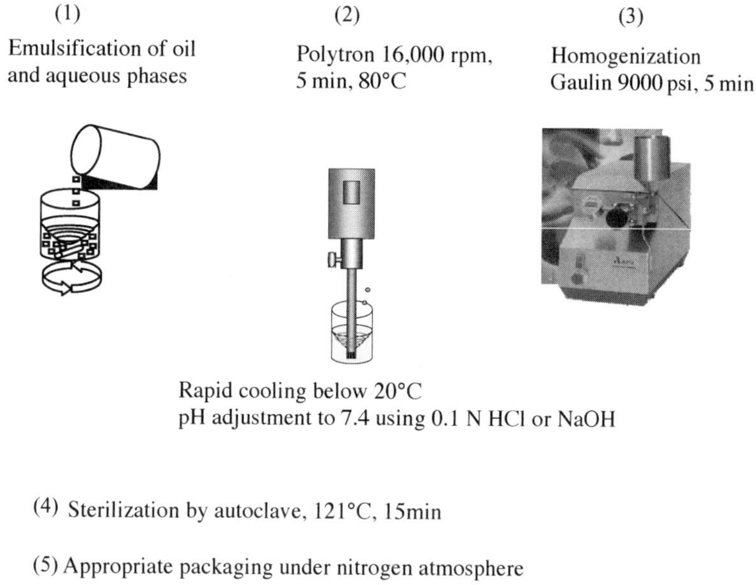

FIGURE 3 Preparation of o/w nanosized emulsion (de novo method).

sively presented mainly in their relation to the oil phase, the aqueous phase, and the emulsifiers.

Prior to the formulation design of the emulsions, data are needed concerning the drug solubility in the oil vehicle. Additionally, prerequisite information is needed about compatibility of the oil vehicle with other formulation additives and the established ocular/skin tissues–oil vehicle matching before the dosage form can be prepared. Table 4 lists the common emulsion excipients and the oils suitable for dissolving or dispersing lipophilic drugs of ocular/parenteral interests. Since oils are triglycerides, care must be taken to minimize or eliminate oxidation. Therefore, antioxidants such as α-tocopherol (0.001–0.002% w/w) should be included in a typical emulsion formulation for medical applications. The final oil-phase concentration in emulsions meant for ocular use is now widely accepted to be at or below 5% w/w taking into account that the emulsion must be kept in a low-viscosity range of between 2 and 3 centipoises, which also is the optimal viscosity for ocular preparations [118]. However, for all other medical uses, the amount of oil may be varied but generally is within 5–20% w/w. Sometimes, a mixture of oils rather than a single oil is employed since drug solubilization in the oil phase is a prerequisite to exploiting the emulsion advantages. Jumaa and Müller [98, 119] reported the effect of mixing castor oil with MCT on the viscosity of castor oil. The oil combination at the ratio of 1:1 (w/w) led to a decrease in the viscosity of castor oil and simultaneously to a decrease in the interfacial tension of the oil phase. This was related to the free fatty acids contained in castor oil, which can act as a coemulsifier resulting in lower interfacial tension and, simultaneously, in a more stable formulation in comparison with the other oil phases. In addition to the digestible oils from the family of triglycerides, including soybean oil, sesame seed oil, cottonseed oil, and safflower oil, which are routinely used for making medical emulsions, alternative biocompatible

TABLE 4 Excipients Used for Formulation of o/w Nanosized Emulsions

Oils	Emulsifiers	Cationic Lipids and Polysaccharide	Miscellaneous
Sesame oil	Cholesterol	Stearylamine	α-Tocopherol
Castor oil	Phospholipids (Lipoid)	Oleylamine	Glycerin
Soya oil	Polysorbate 80 and 20 (Tween 80 and 20)	Chitosan	Xylitol
Paraffin oil	Transcutol P		Sorbitol
Paraffin light	Cremophor RH		Thiomersal
Lanolin	Poloxamer 407		EDTA
Vaseline	Poloxamer 188		Methyl paraben
Corn oil	Miranol C_2M and MHT		Propyl paraben
Glycerin monostearate	Tyloxapol TPGS		
Medium-chain monoglycerides			
Medium-chain triglycerides			
Squalene			
Vitamin E			

Note: TPGS, α-tocopheryl-polyethylene glycol-1000-succinate; EDTA, ethylenediamine tetraacetic acid.

oils such as α-tocopherol and/or other tocols were also investigated for drug delivery purposes via o/w emulsions [120, 121]. But the emulsions formed from tocols are often considered as microemulsion systems with few exceptions being the nanosized emulsions [122, 123].

Unlike spontaneously forming thermodynamically stable microemulsion systems that require a high surfactant concentration (20% and higher), the kinetically stable nanosized emulsions can be prepared by using relatively lower surfactant concentrations. For example, a 20% o/w nanosized emulsion may only require a surfactant concentration of 5–10%. Traditionally, lecithins or phospholipids are the emulsifiers of choice to produce nanosized emulsions. However, emulsifiers of this kind are not suitable to produce submicrometer–sized emulsion droplets or to withstand the heat during steam sterilization. Therefore, additional emulsifiers preferably dissolved in the aqueous phase are usually included in the emulsion composition. A typical example of the aqueous soluble emulsifiers is nonionic surfactants (e.g., Tween 20) after taking into consideration their nonirritant nature when compared to ionic surfactants. The nonionic block copolymer of polyoxyethylene–polyoxypropylene, Pluronics F68 (Poloxamer 188), is included to stabilize the emulsion through strong steric repulsion. However, amphoteric surfactants Miranol MHT (lauroamphodiacetate and sodium tridecethsulfate) and Miranol C_2M (cocoamphodiacetate) were also used in earlier ophthalmic emulsions [124]. It should be added that commercially available cyclosporin A–loaded anionic emulsion (Restasis) contains only polysorbate 80 and carbomer 1342 at alkali pH to stabilize the drug-loaded anionic emulsion. To prepare a cationic emulsion, cationic lipids (stearyl and oleylamine) or polysaccharides (chitosan) are added to the formulation. Strikingly, if chitosan is a choice of cation producing polysaccharide emulsifier molecules, there is no need to add amphoteric or nonionic surfactants to the phospholipid or lecithin-stabilized emulsion [125]. Conversely, a cationic emulsion based on an association of poloxamer 188 and chitosan without lecithin was prepared and also showed adequate physicochemical properties regarding stability and charge effects [97, 98]. Oil-in-water emulsion compositions based on a tocopherol (or a tocopherol derivative) as the disperse phase have been described in a patent granted to Dumex [126]. The emulsion is intended for use with compounds that are sparingly soluble in water. Interestingly, the emulsifying agent used to make tocol-based emulsions are restricted to vitamin E TPGS.

Additives other than antioxidants such as preservatives (e.g., benzalkonium chloride, chlorocresol, parabens) are regularly included in emulsions to prevent microbial spoilage of multidose medical emulsions. α-Tocopherol is a good example of an antioxidant used to obtain a desirable stabilized emulsion under prolonged storage conditions. The presence of components of natural origin such as lecithin or oils with high calorific potential renders the emulsion a good medium to promote microbial growth when it is packed in multidose containers. Pharmaceutical products when distributed into multidose containers, especially for parenteral and ocular administrations, should be properly preserved against microbial contamination and proliferation during storage in normal conditions and proper use. Incorporation of preservatives in single-dose vials is also a common procedure if filtration is used as a sterilization method. Sznitowska et al. [127] studied the physicochemical compatibility between the lecithin-stabilized emulsion and 12 antimicrobial agents over two years of storage at room temperature. Preliminary physicochemical screening results indicate that addition of chlorocresol, phenol, benzyl alcohol, thiomersal, chlorhexidine gluconate, and bronopol should be avoided due to the occurrence of an unfavorable pH change followed by coalescence of lecithin-stabilized droplets of the emulsion. Furthermore, the efficacy of antimicrobial preservation was assessed using the challenge test according to the method described by the European Pharmacopoeia. Despite good physicochemical compatibility, neither parabens nor benzalkonium chloride showed satisfying antibacterial efficacy in the emulsion against the tested microorganisms and consequently did not pass the test. Therefore, higher concentrations of antimicrobial agents or their combination may be required for efficient preservation of the lecithin-stabilized emulsion probably because of unfavorable phase partitioning of the added antimicrobials within the different internal structures of the emulsion. It is interesting to note that benzalkonium chloride, a highly aqueous soluble drug, did not pass the standard challenge test even when incorporated in a cationic emulsion, particularly the third-generation category (unpublished data). This finding clearly indicates that the possible electrostatic attraction between the negatively charged lipid moieties of the mixed emulsifying film formed around the anionic emulsified oil droplets [127] and the quaternary cationic ammonium groups of the preservative is not the plausible cause for the reduced activity of the benzalkonium chloride. Thus, the possible intercalation of this surfactant in either the cationic or anionic interfacial mixed emulsifying film is likely to occur, preventing benzalkonium chloride from eliciting its adequate preservative action. Overall, it is preferable to formulate nanosized emulsions devoid of preservative agents and fill it in sterile single-dose packaging units to prevent potential contamination due to repeated use of multidose packaging. It should be pointed out that the two available ocular emulsion products (Refresh Endura and Restasis, Allergan, Irvine, CA) on the market are preservative free and packed in single-use vials only. Currently there is no commercial parenteral emulsion which contains preservatives and research concerning the problem of preservation of nanosized emulsion is very limited [128–131].

7.4.5 MEDICAL APPLICATIONS OF OIL-IN-WATER NANOSIZED EMULSIONS

It has been shown in a number of studies that the incorporation of drug in o/w nanosized emulsions significantly increased the absorption of the drug when compared with the equivalent aqueous solution administered orally [132–135]. However, the use of emulsions for oral application is limited since other attractive alternatives, such as self-emulsifying oil delivery systems, which are much less sensitive and easy to manufacture, are available [136, 137]. Thus the potential of nanosized emulsions after administration with parenteral and traditional nonparenteral topical routes such as ocular, percutaneous, and nasal is covered in this section.

7.4.5.1 Parenteral Routes

The o/w nanosized emulsion formulations of lipophilic drugs, such as propofol, etomidate, dexamethasone palmitate, and diazepam, were already developed and marketed (Table 1). Furthermore, various research groups across the world are currently undertaking projects to exploit the potential of o/w emulsions for parenteral delivery of a myriad of investigational drugs as well as other lipophilic drugs by receptor-mediated targeting to cancer cells. The important technical and clinical points to keep in mind before the use of the emulsion systems for this kind of work are given below.

It has to be clear that, once diluted and injected (or administered in ocular and other routes), the emulsion stability and fate are determined by three measurable parameters. The first is the partition coefficient of each emulsion component (including added drugs and agents) between the emulsion assembly and the medium. To some extent this partition coefficient is related to oil–water and/or octanol–water partition coefficients. For example, it was well demonstrated that per component of which $\log P$ is lower than 8, the stability upon intravenous (IV) injection is questionable [42, 138]. The other two parameters are k_{off}, a kinetic parameter which describes the desorption rate of an emulsion component from the assembly, and k_c, the rate of clearance of the emulsion from the site of administration. This approach is useful to decide if and what application a drug delivery system will have a chance to perform well [89].

Stability in plasma is an important requirement for IV emulsions as flocculated droplets may result in lung embolism. It was found that tocol-based emulsions stabilized by sodium deoxycholate/lecithins flocculated strongly when mixed with mouse, rat, and sheep plasma and serum, whereas soya oil–based emulsions with the same emulsifiers did not [123]. It was hypothesized that this effect was caused by the adsorption of plasma proteins onto the tocol droplets (opsonization). Indeed, the steric stabilization of emulsions by incorporation of emulsifiers like poloxamer 188 or PEGylated phospholipids such as $\mathrm{PEG}_{5000}\mathrm{PE}$ proved to be effective in the stabilization of tocol-based emulsions in plasma. Conversely, in vitro studies were conducted on plasma protein adsorption onto the blank second- and third-generation emulsion droplets [37, 139] to mimic the in vivo opsonization phenomenon responsible for the rapid clearance of the emulsion droplets from the blood. According to these authors, the adsorption of many protein species such as apoAs, apoCs, apoE albumin, fibrinogen, and gamma globulin onto the emulsion droplet surfaces is detectable by two-dimensional polyacrylamide gel electrophoresis.

7.4.5.2 Ocular Routes

For the eye, the method of drug delivery is important. However, when nanosized emulsion is used as a vehicle for ocular drug delivery purposes, both topical/local and intraocular routes of administration can be possible (though no data concerning intraocular drug delivery through emulsion are currently available). The o/w nanosized emulsions having both anionic and cationic charges provide a liquid-retentive carrier for ocular active agents, particularly when topically instilled into the eye. It is interesting to add here that thermodynamically stable and optically isotropic colloidal systems such as the w/o microemulsion is also designed nowadays for ocular topical. As delivery the w/o microemulsion system comprises both aqueous and oily components into its multistructure, it has the ability to incorporate considerable amounts of both hydrophilic and lipophilic drugs [140]. In fact, in comparison to ocular inserts or implants and semisolid ocular preparations, the liquid-retentive nature gives impetus to investigating further the emulsion-based ophthalmic drug delivery as it has the benefit of being comfortable to use for both ophthalmologists and patients. In addition, through topical instillation of emulsions possessing ocular active substances, the delivery of drug molecules even to the posterior portion of the eye might be of possible. In this context, the third-generation emulsion is being designed by adsorbing electrostatically the therapeutic oligonucleotides onto its surface for modulating functions of retinal pigment epithelium (RPE) cells effectively in order to treat blindness associated with age-related macular degeneration (AMD), proliferative vitreoretinopathy, retinal and choroidal neovascularization, and retinitis pigmentosa. To achieve this, it becomes necessary to know first the ocular protective mechanisms and other concomitant factors to be faced by emulsion droplets following ocular topical application. This point is further developed below.

Considerations of ocular drug delivery are not detailed in this section. Pertinent information concerning factors affecting drug permeation or retention as well as eye anatomy and physiology can be found in several reviews [141–146]. From a medical point of view, o/w nanosized emulsions for ophthalmic use aim at enhancing drug bioavailability either by providing prolonged delivery to the eye or by facilitating transcorneal/transconjunctival penetration. Drugs incorporated in o/w nanosized emulsions are lipophilic in nature, and depending on the extent of lipophilicity, either the corneal or the conjunctival/scleral route of penetration may be favored [147]. For the more lipophilic drugs the corneal route was shown to be the predominant pathway for delivering drugs to the iris, whereas the less lipophilic drugs underwent conjunctival/scleral penetration for delivery into the ciliary body [147]. Thus, transcorneal permeation has traditionally been the mechanism by which topically applied ophthalmic drugs are believed to gain access to the internal ocular structures. Relatively little attention has been given to alternate routes by which drugs may enter the eye. It was reported that drugs may be absorbed by the noncorneal route and appeared to enter certain intraocular tissues through the conjunctiva/sclera [148–150]. Indeed when compared to the cornea, drug penetration through the conjunctiva has the advantage of a larger surface area and higher permeability, at least for drugs which are not highly lipophilic. Furthermore, the lasting presence of drug molecules in the lower conjunctival cul-de-sac of the eye could result in a reservoir effect. Nevertheless, the o/w nanosized emulsions more or less physically resemble a simple aqueous-based eye drop dosage form since more than 90% of the external phase is aqueous irrespective of the formulation composition. Hence, following topical administration, nanosized emulsions would probably face almost similar ocular protective events as encountered with conventional eye drops into the eye. The o/w nanosized emulsions are likely to be destabilized by the tear fluid electrolytic and dynamic action. Because of constant eyelid movements, the basal tear flow rate (1.2 μL/min), and the reflex secretion induced by instillation (up to 400 μL/min depending on the irritating power of the topical ocular solutions [35]), topical eye drop dosage forms are known for being rapidly washed out from the eye. Therefore, the water phase of the emulsion is drained off while, probably, the oil phase of the emulsion remains in the cul-de-sac for a long period of time and functions as a drug reservoir [35]. If the volume of instilled emulsion in the eye exceeds the normal lachrymal volume of 7–10 μL, then the portion of the instilled emulsion (one or two drops, corresponding to 50–100 μL) that is not eliminated by spillage from the palpebral fissure of conjunctiva is drained quickly via the nasolacrimal system into the nasopharynx. In other words, the larger the instilled volume, the more rapidly the instilled emulsion is drained from the precorneal area. Hence the contact time of the emulsion with the absorbing surfaces (cornea and conjunctiva) is estimated to be a maximum of a few minutes, well beyond the short residence time of conventional eye drops. In order to verify the extension of the residence time of the emulsion in the conjunctival sac, Beilin et al. [151] added a fluorescent marker to the formulations. One minute after the topical instillation into eye, 39.9 ± 10.2% of the fluorescence was measured for the nanosized emulsions whereas only 6.8 ± 1.8% for regular eye drops. In addition a study was carried out in male albino rabbits to compare the corneal penetration of indomethacin from Indocollyre (a marketed hydro-PEG ocular solution) to that of negatively and positively charged emulsions [108]. By this comparison, it was intended to gain insightful mechanistic comprehension regarding the enhanced ocular penetration effect of the emulsion as a function of dosage form and surface charge. The contact angle of one droplet of the different dosage forms on the cornea was measured and found to be 70° for saline, 38° for the anionic emulsion, and 21.2° for the cationic emulsion. Respectively, the values of the calculated spreading coefficient were −47, −8.6, and −2.4 mN/m. It can clearly be deduced, owing to the marked low spreading coefficient values elicited by the emulsions, that both nanosized emulsions had better wettability properties on the cornea compared to saline. The emulsion may then prolong the residence time of the drop on the epithelial layer of the cornea, thereby enabling better drug penetration through the cornea to the internal tissues of the eye, as confirmed by animal studies [108]. It is therefore believed that drug is not released from the oil droplet and equilibrates with the tears but rather partitions directly from the oil droplets to the cell membranes on the eye surface. Therefore, it is reasonable to consider that nanosized emulsions have a real advantage since they elicit an increased ocular residence time in com-

TABLE 5 Selected List of o/w Nanosized Emulsions for Ocular Topical Drug Delivery

Emulsion Type	Drug Used	Reference
Anionic emulsion	Δ^8-THC	154
	Pilocarpine base and indomethacin	163
	Adaprolol maleate	161, 162
	Indomethacin	159, 160
	Synthetic HU-211 and pilocarpine base	124, 155, 156
	Pilocarpine base	157, 158, 164
	Cyclosporin A	168, 165–177
Cationic emulsion	Piroxicam	178
	Indomethacin	108
	Miconazole	112
	Cyclosporin A	106, 179

Note: Δ^8-THC and synthetic HU-211 are derivatives of Cannabis sativa.

parison to conventional eye drops and will significantly improve the ocular drug bioavailability [152]. This is also confirmed in numerous cited papers that are listed in Table 5.

In spite of a relatively rapid removal of conjunctivally absorbed emulsion from the eye by local circulation, direct transscleral access to some intraocular tissues cannot be excluded, especially if an electrostatic attraction does occur between the cationic oil droplets of emulsion and anionic membrane moieties of the sclera, as shown by some authors [108]. There is no doubt that the drug absorption from emulsion through the noncorneal route needs to be investigated further as it may elicit useful information on the potential of nanosized emulsions to promote drug penetration to the posterior segment through a mechanism which bypasses the anterior chamber. In addition to the above-described protective and elimination mechanisms of the eye, nanosized emulsions remaining in the precorneal area may be subject to protein binding and to metabolic degradation in the tear film. In conjunction with blood plasma, although low, tear film, aqueous humor, and vitreous humor also have varying amounts of relatively detectable proteins such as albumin, globulin, and immunoglobulins (e.g., IgA, IgG, IgM, IgE) and the enzyme lysozyme. Additional studies (at least in vitro) are necessary to understand clearly the nanosized emulsion interaction with the ocular fluid components. Although it is unlikely to happen because of the low emulsion volume remaining in the conjunctival sac, the fluid dynamics may be moderately altered by the physical and chemical properties of nanosized emulsions, which include tonicity, pH, refractive index, interfacial charge, viscosity, osmolality, and irritant ingredients. Thus, formulations of ophthalmic drug products must take into account not only the stability and compatibility of a drug in the emulsion but also the influence of the emulsion on precorneal fluid dynamics. All of the concepts exposed in this section may ultimately result in transcorneal/conjunctival absorption of 1–2% or less of the drug applied topically through the emulsions. In summary, the rate of loss of drug/emulsion from the eye can be 500–700 times greater than the rate of absorption into the eye. Thus, conventional topical delivery using emulsions cannot achieve adequate intracellular concentrations of drugs or other substances such as oligonucleotides or genes for the treatment of endophthalmitis or other sight-threatening intraocular diseases (e.g., AMD).

In order to achieve a high concentration of drug within the eye using an emulsion delivery vehicle, an approach that bypasses physiological and anatomical barriers (e.g., blood–ocular) of the eye is a more viable and attractive option. One such approach is to administer emulsion through direct intraocular injections such as periocular (subconjunctival and sub-Tenon), intracameral, intravitreal, intracapsular, or subretinal. Moreover, it is likely that intraocularly administered emulsion is able to both significantly increase drug half-life and minimize intraocular side effects that appear following intraocular injection of drug alone. In general, drugs encapsulated within emulsion are less toxic than their free counterparts. Additionally, there is a possibility of obtaining slow drug release from an intraocularly injected emulsion. Taking into account the nonphagocytic character of neural retinal cells and the ability of RPE cells to take up large molecules, including oligonucleotides, the third-generation emulsion for intravitreal or subretinal injections is more likely to be a successful approach in future. Moreover, intravitreally administered drug molecules are able to bypass the blood–ocular barrier to achieve constant therapeutic levels in the eye while minimizing systemic side effects. However, the hyalocytes, the main cellular components of the vitreous, have been classified in at least one report [153] as macrophages and thus may play a role in the uptake of intravitreally injected emulsion. It should be added that no studies are focused so far on injecting emulsion intraocularly and significant work should be devoted to generate this novel idea into a fruitful solution in ophthalmic drug delivery applications.

Over the last decade, o/w nanosized emulsions containing either anionic or cationic droplets have been recognized as interesting and promising ocular topical delivery vehicles for lipophilic drugs. Complete details are available elsewhere [117]. As an overview of this topic, important results on emulsion-based ocular topical drug delivery are covered below and are listed in Table 5.

The in vivo data obtained from studies of early formulations confirm that o/w nanosized anionic emulsions can be effective topical ophthalmic drug delivery systems [154] with a potential for sustained drug release [155]. Naveh and coworkers [156] have also noted that the intraocular pressure (IOP)–reducing effect of a single, topically administered dose of pilocarpine-loaded anionic emulsion lasted for more than 29 h in albino rabbits whereas that of the generic pilocarpine lasted only 5 h. Zurowska-Pryczkowska et al. [157] studied how nanosized emulsion as a vehicle influences the chemical stability of pilocarpine and the effect the drug has on the physical stability of nanosized emulsions. In a subsequent paper [158] from the same group on in vivo evaluation using normotensive rabbits, it was shown that the nanosized emulsion formulated with pilocarpine hydrochloride at pH 5.0 could be indicated as a preparation offering prolonged pharmacological action (miotic effect) together with satisfactory chemical stability. However, the ocular bioavailability arising from such a formulation is not significantly improved when compared to an aqueous solution. Calvo et al. [159, 160] observed an improvement in indomethacin ocular bioavailability when the drug was incorporated in an emulsion dosage form with respect to the commercial aqueous drops following topical application into rabbit eye.

In order to verify the extension of the residence time of the emulsion in the conjunctival sac, Beilin et al. [151] added a fluorescent marker to the formulations, as mentioned previously. From that observation, it is reasonable to consider that an emulsion formulation has the real advantage of increasing ocular residence time in comparison to eye drops. Anselem et al. [161] and Melamed et al. [162] prepared a nanosized emulsion containing adaprolol maleate, a novel soft β-blocking agent, and observed a delayed IOP depressant effect in human volunteers. A similar pharmacological effect was also observed in human volunteers by Aviv et al. [163] using pilocarpine base-loaded emulsion. Another randomized human trial conducted by Garty et al. [164] compared the activity of the pilocarpine base-laden nanosized emulsion instilled twice daily with a generic dosage form instilled four times a day to 40 hypertensive patients for seven days. No local side effects were observed. The IOP decreased 25% in both formulations during this time period. No significant difference was noticed between the two treatments. These results proved that the anionic emulsion extended the action of the drug and two daily administrations have the same result as four instillations of regular eye drops.

A novel nanosized anionic emulsion incorporating the immunomodulatory agent cyclosporin A was developed and the clinical efficacy was investigated for the treatment of moderate to severe dry-eye disease in humans [165–167]. The novel cyclosporin A ophthalmic dosage form represents a real breakthrough in the formulation of a complex, highly lipophilic molecule such as cyclosporin A within an o/w nanosized emulsion. Following thorough validation of this formulation through several clinical studies in various countries [165–175], an anionic o/w emulsion containing cyclosporin A 0.05% (Restasis, Allergan, Irvine, CA) was approved for the first time by the FDA, on December 23, 2002. In addition, this anionic emulsion having cyclosporin A is now available at pharmacies in the United States for the treatment of chronic dry-eye disease (available at www.restasis.com and www.dryeye.com). Furthermore, an over-the-counter (OTC) product that features an emulsion formula, Refresh Endura, is already launched in the U.S. market for eye-lubricating purposes in patients suffering from moderate to severe dry-eye syndrome.

The effect of Restasis on contact lens comfort and reducing dry-eye symptoms in patients with contact lens intolerance was evaluated in comparison to rewetting drops (carboxymethylcellulose 0.5%, Refresh Contacts) [176]. Both formulations were applied twice per day before and after lens wear. Symptoms were assessed by lens wear time, use of rewetting drops during lens wear, subjective evaluation of dryness, and completion of the ocular surface disease index questionnaire. The results of this pilot study indicate that cyclosporin 0.05% is beneficial for contact lens wearers with dry eye and reduces contact lens intolerance [176]. Furthermore, Sall et al. [177] have recently evaluated the efficacy of marketed artificial tears (Systane and Restasis) in relieving the signs and symptoms of dry eye when used as supportive therapy to a cyclosporin-based ophthalmic emulsion (i.e., Restasis + Systane vs. Restasis + Refresh). Significant differences were seen in favor of Restasis + Systane versus Restasis + Refresh for less ocular burning, stinging, grittiness, and dryness. Systane was better than Restasis + Refresh for less burning, dryness, and scratchiness. Results indicate that the choice of concomitant therapy used with Restasis has significant effects on outcome measures and both supportive therapies were compatible with Restasis [177].

When compared to either saline or anionic emulsions, the nanosized cationic emulsions were shown to enhance the ocular bioavailability of indomethacin [108], piroxicam [178], and cyclosporin A [106, 179] following one single-drop dose instillation into the rabbit eye (Figure 4). A significant drug reservoir effect was noted in the cornea and conjunctiva even for more than 8 h following the instillation [106].

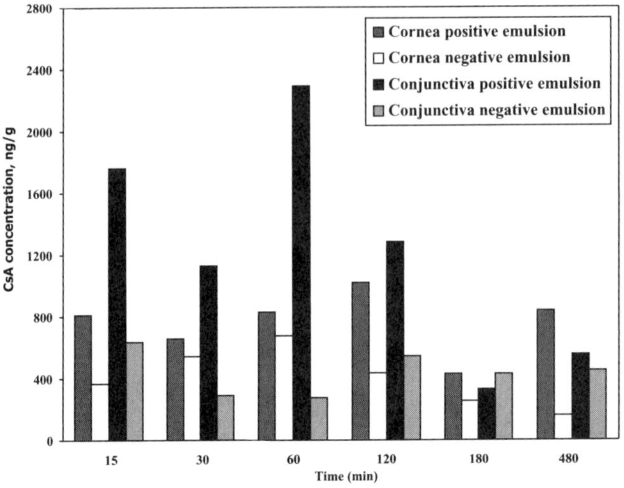

FIGURE 4 Influence of emulsion surface charges cyclosporin A (CsA) concentrations in ocular surface tissues (cornea and conjunctiva) following one single-dose (50-μL) instillation of positively (cationic) and negatively (anionic) charged CsA-loaded nanosized emulsions into albino rabbit eye.

This long residence time may help reduce evaporation of the limited volume of natural tears present in patients with dry eye. This was probably due to the adhesion of the positively charged oil droplets to the negatively charged corneal surface moieties as a result of electrostatic attraction. This hypothesis was supported by data from an ex vivo study which showed that cationic emulsion exhibited better wettability properties on albino rabbit eye cornea than either saline or anionic emulsion [108]. Associated with Poloxamer and phospholipids, a cationic primary amine, stearylamine, has been used to obtain the above-described third-generation cationic emulsions. Additionally, a cationic emulsion based on an association of Poloxamer 188 and chitosan was prepared and also showed interesting physicochemical properties on stability and charge effects [97, 98]. Moreover, the stability and ocular tolerance following topical instillation into the eye of these cationic emulsion vehicles were investigated [98, 114]. The overall studies hence stress the effectiveness of nanosized cationic emulsion, which promotes ocular drug absorption via internalization possibly through an endocytic process [112].

7.4.5.3 Nasal Route

The nasal route is still receiving great attention due to a number of advantages over parenteral and oral administration [180], particularly when first-pass metabolism makes the drug ineffective. The approach of an o/w emulsion formulation of the drug may increase absorption by solubilizing the drug in the inner phase of the emulsion and prolonging contact time between emulsion droplets and nasal mucosa.

One of the first examples for systemic delivery of peptides concerned a lipid-soluble rennin inhibitor [181]. The peptide was solubilized in the oil phase of an o/w emulsion containing membrane adjuvants such as oleic acid and mono- and diglycerides. Emulsion formulations have been proposed to simultaneously increase the solubility of the peptide and to enhance membrane permeability through interaction between the membrane and the oil components. Enhanced and prolonged in vivo nasal absorption of the rennin inhibitor was observed in emulsion formulation compared to aqueous suspension. From morphological studies, the emulsions did not provoke any significant changes to the nasal mucosa [181]. Such a formulation approach was also used for the administration of a steroidal male sex hormone testosterone [182]. The steroid was solubilized in the oil phase of the o/w emulsion and the ionic composition of the aqueous phase was modified in order to produce electrically positive, negative, and neutral droplets. Droplets with a surface charge led to better bioavailability than neutral droplets, but contrary to the above-described topical applicabilities of cationic emulsions over anionic emulsions, positively charged droplets did not provide the best results [182]. However, the emulsion approach was advantageous since it helped to overcome the solubility problem of the hydrophobic compounds.

In another study which does not involve peptide drugs, various emulsion formulations were prepared in order to modulate the partitioning of the drug between the aqueous phase and the oil phase [183]. The disappearance of a drug from the nasal cavity was determined by an in situ perfusion technique. When the drug was solubilized in the aqueous phase, the formulation did not have a significant effect on the drug disappearance rate. However, partitioning of the drug in the oil phase resulted in delaying absorption. It was suggested that oil droplets containing medium-chain triglycerides formed a pseudo–oily layer on the mucous membrane, which slowed down the drug disappearance from the nasal cavity [183]. Another interesting study reported nasal delivery of insulin formulated in both o/w and w/o emulsions [184]. As insulin partitions into the aqueous phase of the emulsion, the peptide is either incorporated within the continuous phase of the o/w emulsion or encapsulated in the aqueous droplets of the w/o emulsion. Following nasal perfusion experiments, plasma insulin concentration profiles showed enhanced insulin absorption when the peptide was formulated as an o/w emulsion compared to an aqueous solution. However, a w/o emulsion did not cause any significant increase in plasma insulin concentration. Delivery of insulin by administration of nasal drops also revealed a large dose-dependent increase in plasma insulin concentration. It also needs to be pointed out that the emulsifier mixture alone did not promote any absorption. It was suggested that insulin molecules probably were adsorbed at the surface of the oil droplets. Adhesion of the oil droplets on the mucosal membrane then induced a local increase in insulin concentration at the membrane surface. However, the number of droplets in contact with the surface had to be small enough. Otherwise, a stagnant oil layer is formed which acts as an additional barrier to the transport, as was observed with the w/o emulsion [183].

Other recent applications of emulsion formulation involve mucosal gene and vaccine delivery [185–187] and the preparation of polymeric microspheres by the w/o emulsification solvent extraction technique [188].

7.4.5.4 Topical Route

Many drugs exhibit low skin penetration, which results in poor efficacy. As opposed to common chemical skin penetration enhancers, organic solvents, which are generally associated to some degree with skin irritation, toxicity and sensitization, a solvent-free topical vehicle based on drug entrapment in o/w emulsion droplets of submicrometer size is more efficacious in terms of percutaneous absorption and possibly devoid of adverse effects. In addition, the uniqueness of the large internal hydrophobic oil core of o/w nanosized emulsion droplets allows high solubilization capacity for water-insoluble topically active medicaments and also aids in carrying water, an excellent softener, to the skin.

The concept of using anionic nanosized emulsion vehicles for enhanced percutaneous absorption of nonsteroidal anti-inflammatory drugs (NSAIDs) and diazepam was clearly proven [189, 190]. NSAIDs and diazepam in a nanosized emulsion vehicle also demonstrated noticeable systemic activity. The o/w emulsion was tested for primary irritation in humans in a 48-h trial. Low irritancy and excellent human acceptance were observed, subsequently making the further development of a nanosized emulsion vehicle very attractive.

Even though emulsion vehicles increase dermal drug delivery of lipophilic drugs in humans, one of the problems for topical drug delivery is the difficulty of applying these vehicles to the skin because of their fluidity. Rheological properties are studied in transdermal formulations and different results are given. Realdaon and Ragazzi [191] have investigated different mechanical emulsifying conditions on o/w emulsion formulations containing methyl nicotinate. The influence of these procedures on rheological properties and in vivo availability of methyl nicotinate was evaluated. Even if various viscosities were obtained, differences between batches did not compromise drug availability. On the contrary, Welin-Berger and co-workers [192, 193] concluded in their study on nanosized emulsions containing model compounds that both release and permeation rates decrease when the apparent yield stress (i.e., the macroviscosity) increases by addition of gelling polymers. Because a topical anesthetic agent will induce a pain-suppressing anesthesia, the eutectic mixture of local anesthetics (EMLA) has proven to be a useful medication for children. It is an emulsion containing a mixture of lidocaine and prilocaine. This cream gives an effective deep sedation and can be applied half an hour to 1 h prior to the procedure. Local side effects with this emulsion are very mild [194, 195]. Systemic activity of nanosized emulsions containing diazepam was compared with regular creams or ointments by Schwarz et al. [190]. Their efficiency was tested on protection against pentamethylenetetrazole, which induces convulsive effects in mice. Diazepam applied topically in emulsion creams was strongly dependent on oil droplet size. Furthermore, nanosized emulsions increased transdermal drug delivery and prolonged protective activity for up to 6 h.

Many formulations for topical emulsions are available in the scientific literature, in patents, and on the market. Progresses made in the last years in this field are concentrated on the various aspects of drug release and the influence of droplet size.

Third-generation cationic emulsions were suggested as drug carriers for topical use in the skin. It was found that α-tocopherol-loaded cationic emulsion was able to prevent oxidative damage of cultured fibroblast cells [196]. In addition, the same cationic formulation was able to protect rat skin against oxidative stress induced by ultraviolet (UV) irradiation significantly better than either the corresponding anionic emulsion or the cationic blank emulsion, as measured with a noninvasive evaluation of the lipid hydroperoxidation process of the rat skin. However, no difference was found between cationic or anionic nanosized emulsions of α-tocopherol as assessed by endogenous peroxyl radical scavenging ability. Taken together, these results suggest that the cationic emulsion allows the α-tocopherol to remain on the surface of the skin because of electrostatic interactions between the negatively charged sites of the superficial layers of the skin and the positively charged oil droplets. Although the extent of α-tocopherol incorporation into the skin is similar for both cationic and anionic emulsions, the prolonged skin surface residence time of the cationic emulsion allows an enhanced protective effect against oxidative stress. In contrast to these results, an in vitro percutaneous absorption study on hairless rat skin found that the antifungal drugs econazole and miconazole nitrate incorporated into a similar cationic emulsion formulation were more effective in terms of skin penetration than the corresponding anionic emulsion [197]. The enhanced rate of diffusion of these antifungal drugs through the skin by the cationic emulsion suggests a new approach for dermal drug penetration enhancement [197].

7.4.6 FUTURE PERSPECTIVE

Based on the performances in previous and present decades, o/w-type nanosized emulsions can conveniently be classified into three generations. First-generation emulsions are considered primarily as nutrient carriers to be administered via intravenous routes to bed-ridden patients. Second-generation emulsions start initially as drug carrier systems by solubilizing considerable amounts of lipophilic drugs at the oil phase or at the oil–water interface of the emulsion. This particular merit of emulsions is specifically exploited even commercially for both ocular and parenteral active drugs. Modifications made either in the oil phase or at the o/w interfacial film forming emulsifier molecules allow the emulsions to be able to escape from lipolysis by lipoprotein lipase, apo adsorption, and liver uptake. Such a surface-modified emulsion would prolong the circulation time in plasma and thereby an alteration in in vivo disposition of incorporated drugs following parenteral administration. Attachment of homing devices such as antibody and apoE to the surfaces of emulsions makes the selective/active delivering of emulsion-incorporated drugs to target sites such as a tumorized organ or hepatic system. Active targeting increases the affinity of the carrier system for the target site, while passive targeting minimizes the nonspecific interaction with nontargeted sites by the RES. Having together a positive charge and a steric stabilizing effect led to the development of third-generation emulsions that contain a unique property: plasma half-life prolongation and electrostatic adhesion to ocular surface tissues after topical instillation into eye. Furthermore, the third-generation emulsion shows a potential of carrying a wide range of lipophilic, amphiphilic, and polyanionic compounds, including DNA and oligonucleotides for transdermal and nasal routes. Accumulating knowledge thus suggests that constant progress in better understanding the principles and processes governing the various issues related to o/w nanosized emulsions has surely brought major improvements in the efficacy of arenteral or nonparenteral drug delivery systems.

REFERENCES

1. Lipinski, C. A, Lombardo, F., Dominy, B. W., and Feeney, P. J. (2001), Experimental and computational approaches to estimate solubility and permeability in drug discovery and development settings, *Adv. Drug Deliv. Rev.*, 46, 3–26.
2. Lipinski, C. (2002), Poor aqueous solubility—An industry wide problem in drug discovery, *Am. Pharm. Rev.*, 5, 82–85.
3. El-Aasser, M. S., and Sudol, E. D. (2004), Miniemulsions: Overview of research and applications, *JCT Res.*, 1, 21–31.
4. Nakajima, H. (1997), Microemulsions in cosmetics, in Solans, C., and Kunieda, H., Eds., *Industrial Applications of Microemulsions*, Marcel Dekker, New York, pp. 175–197.
5. Benita, S. (1998), Introduction and overview, in Benita, S., Ed., *Submicron Emulsion in Drug Targeting & Delivery*, Harwood Academic, The Netherlands, pp.1–5.
6. Klang, S. H., Benita, S. (1998), Design and evaluation of submicron emulsions as colloidal drug carriers for intravenous administration, in Benita, S., Ed., *Submicron Emulsion in Drug Targeting & Delivery*, Harwood Academic, The Netherlands, pp. 119–152.
7. Tadros, T. F., Izquierdo, P., Esquena, J., and Solans, C. (2004), Formation and stability of nanoemulsions, *Adv. Coll. Interf. Sci.*, 108–109, 303–318.
8. Solans, C., Izquierdo, P., Nolla, J., Azemar, N., and Garcia-Celma, M. J. (2005), Nanoemulsions, *Curr. Opin. Coll. Interf. Sci.*, 10, 102–110.
9. Becher, P., and Schick, M. J. (1987), Macroemulsions, in Schick, M. J., Ed., *Non-Ionic Surfactants*, Surfactant Science Series Vol. 23., Marcel Dekker, Basel, pp. 435–493.
10. Kabalnov, A. (1998), Thermodynamic and theoretical aspects of emulsions and their stability, *Curr. Opin. Coll. Interf. Sci.*, 3, 270–275.
11. Stefan, A., Palazzo, G., Ceglie, A., Panzavolta, E., and Hochkoeppler, A. (2003), Water-in-oil macroemulsions sustain long-term viability of microbial cells in organic solvents, *Biotechnol. Bioeng.*, 81, 323–328.
12. Ceglie, A., Das, K. P., and Lindman, B. (1987), Microemulsion structure in four component systems for different surfactants, *Coll. Surf.*, 28, 29–40.
13. Attwood, D. (1994), Microemulsions, in Kreuter, J., Ed., *Colloidal Drug Delivery Systems*, Marcel Dekker, New York, pp. 31–71.
14. Hino, T., Kawashima, Y., and Shimabayashi, S. (2000), Basic study for stabilization of w/o/w emulsion and its application to transcatheter arterial embolization therapy, *Adv. Drug Deliv. Rev.*, 45, 27–45.
15. Gallarate, M., Carlotti, M. E., Trotta, M., and Bovo, S. (1999), On the stability of ascorbic acid in emulsified systems for topical and cosmetic use, *Int. J. Pharm.*, 188, 233–241.
16. ErÃfnofas, I., Csoka, I., Csany, E., Orosz, K., and Makai, M. (1998), in *Proceedings of 2nd World Meeting APGI/APV*, Paris, pp. 805–806.
17. Lowe, K. C. (1999), Perfluorinated blood substitutes and artificial oxygen carriers, *Blood Rev.*, 13, 171–184.
18. Spahn, D. R. (2000), Current status of artificial oxygen carriers, *Adv. Drug Deliv. Rev.*, 40, 143–151.
19. Krafft, M. P. (2001), Fluorocarbons and fluorinated amphiphiles in drug delivery and biomedical research, *Adv. Drug Deliv. Rev.*, 47, 209–228.
20. Fukushima, S., Kishimoto, S., Takeuchi, Y., and Fukushima, M. (2000), Preparation and evaluation of o/w type emulsions containing antitumor prostaglandin, *Adv. Drug Deli Rev.*, 45, 65–75.

21. Ott, G., Singh, M., Kazzaz, J., Briones, M., Soenawan, E., Ugozzoli, M., and O'Hagan, D. T. (2002), A cationic sub-micron emulsion (MF59/DOTAP) is an effective delivery system for DNA vaccines, *J. Controlled Release*, 79, 1–5.
22. Ribi, E. E. (1984), Refined detoxified endotoxin product, U.S. Patent 4,436,727, March 13.
23. Ribi, E., Schwartzman, S. M., and Cantrell, J. L. Refined detoxified endotoxin product, (1984), U.S. Patent 4,436,728, March 13.
24. Myers, K. R., and Truchot, A. T. (1990), Modified lipopolysaccharides and process of preparation, U.S. Patent 4,912,094, March 27.
25. Allison, A. C., and Byars, N. E. (1986), An adjuvant formulation that selectively elicits the formation of antibodies of protective isotypes and of cell-mediated immunity, *J. Immunol. Methods*, 95, 157–168.
26. Chandra, R. K. (1985), Grace A. Goldsmith Award lecture. Trace element regulation of immunity and infection, *J. Am. Coll. Nutr.*, 4, 5–16.
27. Shils, M. E. (1998), Parentral nutrition, in Shils, M. E., Olson, J. A., Shike, M. E., and Ross, O., Eds., *Modern Nutr. in Health and Disease*, 9th ed., Williams & Wilkins, Baltimore, PA, pp. 415–428.
28. Koster, V. S., Kuks, P. F. M., Langer, R., and Talsma, H. (1996), Particle size in parenteral fat emulsions, what are the true limitations? *Int. J. Pharm.*, 134, 235–238.
29. Lutz, O., Meraihi, Z., Mura, J. L., Frey, A., Riess, G. H., and Bach. A. C. (1989), Fat emulsion particle size: Influence on the clearance rate and the tissue lipolytic activity, *Am. J. Clin. Nutr.*, 50, 1370–1381.
30. Rubin, M., Harell, D., Naor, N., Moser, A., Wielunsky, E., Merlob, P., and Lichtenberg, D. (1991), Lipid infusion with different triglyceride cores (long-chain vs medium-chain/long-chain triglycerides): Effect on plasma lipids and bilirubin binding in premature infants, *J. Parenteral Enteral Nutr.*, 15, 642–646.
31. Mizushima, Y., Hoshi, K., Aihara, H., and Kurachi, M. (1983), Inhibition of bronchoconstriction by aerosol of a lipid containing prostaglandin E1, *J. Pharm. Pharmacol.*, 35, 397.
32. Craig, D. Q. M., Patel, M., and Ashford, M. (2000), Administration of emulsions to the gastrointestinal tract, in Nielloud, F., and Marti-Mestres, G., Eds., *Pharmaceutical Emulsions and Suspensions*, Marcel Dekker, New York, pp. 323–360.
33. Hauss, D. J. (2002), Oral lipid-based drug delivery—a case of implementation failing to keep up with innovation? *Am. Pharm. Rev.*, 5, 22–36.
34. Sakaeda, T., and Hirano, K. (1998), Effect of composition on biological fate of oil particles after intravenous injection of O/W lipid emulsions, *J. Drug Target.*, 6, 273–284.
35. Sasaki, H., Yamamura, K., Nishida, K., Nakamura, J., and Ichikawa, M. (1996), Delivery of drugs to the eye by topical application, *Prog. Ret. Eye Res.*, 15, 583–620.
36. Harnisch, S., and Müller, R. H. (1998), Plasma protein adsorption patterns on emulsions for parenteral administration: Establishment of a protocol for two-dimensional polyacrylamide electrophoresis, *Electrophoresis*, 19, 349–354.
37. Harnisch, S., and Müller, R. H. (2000), Adsorption kinetics of plasma proteins on oil-in-water emulsions for parenteral nutrition, *Eur. J. Pharm. Biopharm.*, 49, 41–46.
38. Szebeni, J. (2001), Complement activation-related pseudoallergy caused by liposomes, micellar carriers of intravenous drugs, and radiocontrast agents, *Crit. Rev. Ther. Drug Carrier Syst.*, 18, 567–606.
39. Price, M. E., Cornelius, R. M., and Brash, J. L. (2001), Protein adsorption to polyethylene glycol modified liposomes from fibrinogen solution and from plasma, *Biochim. Biophys. Acta*, 1512, 191–205.

40. Deckelbaum, R. J., Hamilton, J. A., Moser, A., Bengtsson-Olivecrona, G., Butbul, E., Carpentier, Y. A., Gutman, A., and Olivecrona, T. (1990), Medium-chain versus long-chain triacylglycerol emulsion hydrolysis by lipoprotein lipase and hepatic lipase: Implications for the mechanisms of lipase action, *Biochemistry*, 29, 1136–1142.
41. Hedeman, H., Brøndsted, H., Müllertz, A., and Frokjaer, S. (1996), Fat emulsions based on structured lipids (1,3-specific triglycerides): An investigation of the in vivo fate, *Pharm. Res.*, 13, 725–728.
42. Takino, T., Konishi, K., Takakura, Y., and Hashida, M. (1994), Long circulating emulsion carrier systems for highly lipophilic drugs, *Biol. Pharm. Bull.*, 17, 121–125.
43. Takino, T., Nakajima, C., Takakura, Y., Sezaki, H., and Hashida, M. (1993), Controlled biodistribution of highly lipophilic drugs with various parenteral formulations, *J. Drug Target.*, 1, 117–124.
44. Redgrave, T. G., Rakic, V., Mortimer, B.-C., and Mamo, J. C. L. (1992), Effects of sphingomyelin and phosphatidylcholine acyl chains on the clearance of triacylglycerol-rich lipoproteins from plasma. Studies with lipid emulsions in rats, *Biochim. Biophys. Acta*, 1126, 65–72.
45. Arimoto, I., Matsumoto, C., Tanaka, M., Okuhira, K., Saito, H., and Handa, T. (1998), Surface composition regulates clearance from plasma and triolein lipolysis of lipid emulsions, *Lipids*, 33, 773–779.
46. Handa, T., Eguchi, Y., and Miyajima, K. (1994), Effects of cholesterol and cholesteryl oleate on lipolysis and liver uptake of triglyceride/phosphatidylcholine emulsions in rats, *Pharm. Res.*, 11, 1283–1287.
47. Kurihara, A., Shibayama, Y., Mizota, A., Yasuno, A., Ikeda, M., and Hisaoka, M. (1996), Pharmacokinetics of highly lipophilic antitumor agent palmitoyl rhizoxin incorporated in lipid emulsions in rats, *Biol. Pharm. Bull.*, 19, 252–258.
48. Kurihara, A., Shibayama, Y., Yasuno, A., Ikeda, M., and Hisaoka, M. (1996), Lipid emulsions of palmitoylrhizoxin: Effects of particle size on blood dispositions of emulsion lipid and incorporated compound in rats, *Biopharm. Drug Dispos.*, 17, 343–353.
49. Lin, S. Y., Wu, W. H., and Lui, W. Y. (1992), In vitro release, pharmacokinetic and tissue distribution studies of doxorubicin hydrochloride (Adriamycin HCl) encapsulated in lipiodolized w/o emulsions and w/o/w multiple emulsions, *Die Pharmazie*, 47, 439–443.
50. Sakaeda, T., Takahashi, K., Nishihara, Y., and Hirano, K. (1994), O/W lipid emulsions for parenteral drug delivery. I. Pharmacokinetics of the oil particles and incorporated sudan II, *Biol. Pharm. Bull.*, 17, 1490–1495.
51. Yamaguchi, T., Nishizaki, K., Itai, S., Hayashi, H., and Ohshima, H. (1995), Physicochemical characterization of parenteral lipid emulsion: Influence of cosurfactants on flocculation and coalescence, *Pharm. Res.*, 12, 1273–1278.
52. Ueda, K., Yamazaki, Y., Noto, H., Teshima, Y., Yamashita, C., Sakaeda, T., and Iwakawa, S. (2003), Effect of oxyethylene moieties in hydrogenated castor oil on the pharmacokinetics of menatetrenone incorporated in O/W lipid emulsions prepared with hydrogenated castor oil and soybean oil in rats, *J. Drug Target.*, 11, 37–43.
53. Liu, F., and Liu, D. (1995), Long-circulating emulsions (oil-in-water) as carriers for lipophilic drugs, *Pharm. Res.*, 12, 1060–1064.
54. Lee, J. H., Lee, H. B., and Andrade, J. D. (1995), Blood compatibility of polyethylene oxide surfaces, *Prog. Polym. Sci.*, 20, 1043–1079.
55. Davis, S. S., and Hansrani, P. (1985), The influence of emulsifying agents on the phagocytosis of lipid emulsions by macrophages, *Int. J. Pharm.*, 23, 69–77.
56. Levy, M. Y., Benita, S., and Baszkin, A. (1991), Interactions of a non-ionic surfactant with mixed phospholipid-oleic acid monolayers. Studies under dynamic conditions, *Coll. Surf.*, 59, 225–241.

57. Harris, J. M., Martin, N. E., and Modi, M. (2001), Pegylation: A novel process for modifying pharmacokinetics, *Clin. Pharmacokinet.*, 40, 539–551.
58. Bhadra, D., Bhadra, S., Jain, P., and Jain, N. K. (2002), Pegnology: A review of PEG-ylated systems, *Die Pharm.*, 57, 5–29.
59. Klibanov, A. L., Maruyama, K., Torchilin, V. P., and Huang, L. (1990), Amphipathic polyethyleneglycols effectively prolong the circulation time of liposomes, *FEBS Lett.*, 268, 235–237.
60. Allen, T. M., Hansen, C., Martin, F., Redemann, C., and Yau-Young, A. (1991), Liposomes containing synthetic lipid derivatives of poly(ethylene glycol) show prolonged circulation half-lives in vivo, *Biochim. Biophys. Acta*, 1066, 29–36.
61. Woodle, M. C., and Lasic, D. D. (1992), Sterically stabilized liposomes, *Biochim. Biophys. Acta*, 1113, 171–199.
62. Dunn, S. E., Brindley, A., Davis, S. S., Davies, M. C., and Illum, L. (1994), Polystyrene-poly (ethylene glycol) (PS-PEG2000) particles as model systems for site specific drug delivery. 2. The effect of PEG surface density on the in vitro cell interaction and in vivo biodistribution, *Pharm. Res.*, 11, 1016–1022.
63. Gref, R., Minamitake, Y., Peracchia, M. T., Trubetskoy, V., Torchilin, V., and Langer, R. (1994), Biodegradable long-circulating polymeric nanospheres, *Science*, 263, 1600–1603.
64. Wheeler, J. J., Wong, K. F., Ansell, S. M., Masin, D., and Bally, M. B. (1994), Polyethylene glycol modified phospholipids stabilize emulsions prepared from triacylglycerol, *J. Pharm. Sci.*, 83, 1558–1564.
65. Tirosh, O., Barenholz, Y., Katzhendler, J., and Priev, A. (1998), Hydration of polyethylene glycol-grafted liposomes, *Biophys. J.*, 74, 1371–1379.
66. Papahadjopoulos, D., Allen, T. M., Gabizon, A., Mayhew, E., Matthay, K., Huang, S. K., Lee, K.-D., Woodle, M. C., Lasic, D. D., Redemann, C., and Martin, F. J. (1991), Sterically stabilized liposomes: Improvements in pharmacokinetics and antitumor therapeutic efficacy, *Proc. Nat. Acad. Sci. U. S. A.*, 88, 11460–11464.
67. Lundberg, B. B., Mortimer, B.-C., and Redgrave, T. G. (1996), Submicron lipid emulsions containing amphipathic polyethylene glycol for use as drug-carriers with prolonged circulation time, *Int. J. Pharm.*, 134, 119–127.
68. Allen, T. M., and Everest, J. M. (1983), Effect of liposome size and drug release properties on pharmacokinetics of encapsulated drug in rats, *J. Pharmacol. Exp. Ther.*, 226, 539–544.
69. Senior, J., Crawley, J. C. W., and Gregoriadis, G. (1985), Tissue distribution of liposomes exhibiting long half-lives in the circulation after intravenous injection, *Biochim. Biophys. Acta*, 839, 1–8.
70. Liu, D., Mori, A., and Huang, L. (1991), Large liposomes containing ganglioside GM1 accumulate effectively in spleen, *Biochim. Biophys. Acta*, 1066, 159–165.
71. Connelly, P. W., and Kuksis, A. (1981), Effect of core composition and particle size of lipid emulsions on apolipoprotein transfer of plasma lipoproteins in vivo, *Biochim. Biophys. Acta*, 666, 80–89.
72. Bundgaard, M. (1980), Transport pathways in capillaries—In search of pores, *Annu. Rev. Physiol.*, 42, 325–336.
73. Wisse, E. (1970), An electron microscopic study of the fenestrated endothelial lining of rat liver sinusoids, *J. Ultrastruct. Res.*, 31, 125–150.
74. Jain, R. K. (1987), Transport of molecules across tumor vasculature, *Cancer Metastasis Rev.*, 6, 559–593.
75. Takakura, Y., and Hashida, M. (1995), Macromolecular drug carrier systems in cancer chemotherapy: Macromolecular prodrugs, *Crit. Rev. Oncol./Hematol.*, 18, 207–231.

76. Gabizon, A., Price, D. C., Huberty, J., Bresalier, R. S., and Papahadjopoulos, D. (1990), Effect of liposome composition and other factors on the targeting of liposomes to experimental tumors: Biodistribution and imaging studies, *Cancer Res.*, 50, 6371–6378.

77. Wu, N. Z., Da, D., Rudoll, T. L., Needham, D., Whorton, A. R., and Dewhirst, M. W. (1993), Increased microvascular permeability contributes to preferential accumulation of Stealth liposomes in tumor tissue, *Cancer Res.*, 53, 3765–3770.

78. Yuan, F., Dellian, M., Fukumura, D., Leunig, M., Berk, D. A., Torchilin, V. P., and Jain, R. K. (1995), Vascular permeability in a human tumor xenograft: Molecular size dependence and cutoff size, *Cancer Res.*, 55, 3752–3756.

79. Nomura, T., Yamashita, F., Takakura, Y., and Hashida, M. (1995), *Proc. Int. Symp. Control Release Bioactive Mater.*, 22, 420–421.

80. Yuan, F., Leunig, M., Huang, S. K., Berk, D. A., Papahadjopoulos, D., and Jain, R. K. (1994), Microvascular permeability and interstitial penetration of sterically stabilized (stealth) liposomes in a human tumor xenograft, *Cancer Res.*, 54, 3352–3356.

81. Gabizon, A., Shmeeda, H., Horowitz, A. T., and Zalipsky, S. (2004), Tumor cell targeting of liposome-entrapped drugs with phospholipid-anchored folic acid-PEG conjugates, *Adv. Drug Deliv. Rev.*, 56, 1177–1192.

82. Nobs, L., Buchegger, F., Gurny, R., and Allemann, E. (2004), Current methods for attaching targeting ligands to liposomes and nanoparticles, *J. Pharm. Sci.*, 93, 1980–1992.

83. Song, Y. K., Liu, D., Maruyama, K. Z., and Takizawa, T. (1996), Antibody mediated lung targeting of long-circulating emulsions, *PDA J. Pharm. Sci. Technol.*, 50, 372–377.

84. Lundberg, B. B., Griffiths, G., and Hansen, H. J. (1999), Conjugation of an anti-B-cell lymphoma monoclonal antibody, LL2, to longcirculating drug-carrier lipid emulsions, *J. Pharm. Pharmacol.*, 51, 1099–1105.

85. Lundberg, B. B., Griffiths, G., and Hansen, H. J. (2004), Cellular association and cytotoxicity of anti-CD74-targeted lipid drug-carriers in B lymphoma cells, *J. Controlled Release*, 94, 155–161.

86. Goldstein, D., Nassar, T., Lambert, G., Kadouche, J., and Benita, S. (2005), The design and evaluation of a novel targeted drug delivery system using cationic emulsion-antibody conjugates, *J. Controlled Release*, 108, 418–432.

87. Rensen, P. C. N., Van Dijk, M. C. M, Havenaar, E. C., Bijsterbosch, M. K., Kruijt, J. K., and Van Berkel, T. J. C. (1995), Selective liver targeting of antivirals by recombinant chylomicrons—A new therapeutic approach to hepatitis B, *Nat. Med.*, 1, 221–225.

88. Mahley, R. W. (1988), Apolipoprotein E: Cholesterol transport protein with expanding role in cell biology, *Science*, 240, 622–630.

89. Barenholz, Y., and Cohen, R. (1995), Rational design of amphiphile-based drug carriers and sterically stabilized carriers, *J. Liposome Res.*, 5, 905–932.

90. Iwamoto, K., Kato, T., Kawahara, M., Koyama, N., Watanabe, S., Miyake, Y., and Sunamoto, J. (1991), Polysaccharide-coated oil droplets in oil-in-water emulsions as targetable carriers for lipophilic drugs, *J. Pharm. Sci.*, 80, 219–224.

91. Grolier, P., Azais-Braesco, V., Zelmire, L., and Fessi, H. (1992), Incorporation of carotenoids in aqueous systems: Uptake by cultured rat hepatocytes, *Biochim. Biophys. Acta*, 1111, 135–138.

92. Barratt, G. (2003), Colloidal drug carriers: Achievements and perspectives, *Cell. Mol. Life Sci.*, 60, 21–37.

93. Kim, T. W., Chung, H., Kwon, I. C., Sung, H. C., and Jeong, S. Y. (2001), Optimization of lipid composition in cationic emulsion as in vitro and in vivo transfection agents, *Pharm. Res.*, 18, 54–60.

94. Kim, Y. J., Kim, T. W., Chung, H., Kwon, I. C., Sung, H. C., and Jeong, S. Y. (2001), Counterion effects on transfection activity of cationic lipid emulsion, *Biotechnol. Bioproc. Eng.*, 6, 279–283.

95. Elbaz, E., Zeevi, A., Klang, S., and Benita, S. (1993), Positively-charged submicron emulsion—A new type of colloidal drug carrier, *Int. J. Pharm.*, 96, R1–R6.

96. Guilatt, R. L., Couvreur, P., Lambert, G., Goldstein, D., Benita, S., and Dubernet, C. (2004), Extensive surface studies help to analyse zeta potential data: The case of cationic emulsions, *Chem. Phys. Lipids*, 131, 1–13.

97. Calvo, P., Remuñá-López, C., Vila-Jato, J. L., and Alonso, M. J. (1997), Development of positively charged colloidal drug carriers: Chitosan-coated polyester nanocapsules and submicro-emulsions, *Coll. Polym. Sci.*, 275, 46–53.

98. Jumaa, M., and Müller, B. W. (1999), Physicochemical properties of chitosan-lipid emulsions and their stability during the autoclaving process, *Int. J. Pharm.*, 183, 175–184.

99. Samama, J. P., Lee, K. M., and Biellmann, J. F. (1987), Enzymes and microemulsions. Activity and kinetic properties of liver alcohol dehydrogenase in ionic water-in-oil microemulsions, *Eur. J. Biochem.*, 163, 609–617.

100. Teixeira, H., Dubernet, C., Puisieux, F., Benita, S., and Couvreur, P. (1999), Submicron cationic emulsions as a new delivery system for oligonucleotides, *Pharm. Res.*, 16, 30–36.

101. Choi, B. Y., Chung, J. W., Park, J. H., Kim, K. H., Kim, Y. I., Koh, Y. H., Kwon, J. W., Lee, K. H., Choi, H. J., Kim, T. W., Kim, Y. J., Chung, H., Kwon, I. C., and Jeong, S. Y. (2002), Gene delivery to the rat liver using cationic lipid emulsion/DNA complex: Comparison between intra-arterial, intraportal and intravenous administration, *Korean J. Radiol.*, 3, 194–198.

102. Yi, S. W., Yune, Y., Kim, T. W., Chung, H., Choi, Y. W., Kwon, I. C., Lee, E. B., and Jeong, S. Y. (2000), A cationic lipid emulsion/DNA complex as a physically stable and serum-resistant gene delivery system, *Pharm. Res.*, 17, 314–320.

103. Kim, Y. J., Kim, T. W., Chung, H., Kwon, I. C., Sung, H. C., and Jeong, S. Y. (2003), The effects of serum on the stability and the transfection activity of the cationic lipid emulsion with various oils, *Int. J. Pharm.*, 252, 241–252.

104. Tamilvanan, S. (2004), Oil-in-water lipid emulsions: Implications for parenteral and ocular delivering systems, *Prog. Lipid Res.*, 43, 489–533.

105. Benita, S. (1999), Prevention of topical and ocular oxidative stress by positively charged submicron emulsion, *Biomed. Pharm.*, 53, 193–206.

106. Abdulrazik, M., Tamilvanan, S., Khoury, K., and Benita, S. (2001), Ocular delivery of cyclosporin A II. Effect of submicron emulsion's surface charge on ocular distribution of topical cyclosporin A, *STP Pharma Sci.*, 11, 427–432.

107. Rojanasakul, Y., and Robinson, J. R. (1989), Transport mechanisms of the cornea: Characterization of barrier permselectivity, *Int. J. Pharm.*, 55, 237–246.

108. Klang, S., Abdulrazik, M., and Benita, S. (2000), Influence of emulsion droplet surface charge on indomethacin ocular tissue distribution, *Pharm. Dev. Technol.*, 5, 521–532.

109. Wretlind, A. (1981), Parenteral nutrition. *Nutr. Rev.*, 39, 257–265.

110. Davis, S. S. (1982), Emulsions systems for the delivery of drugs by the parenteral route, in Bundgaard, H., Bagger Hansen, A., and Kofod, H., Eds., *Optimization of Drug Delivery*, Munksgaard, Copenhagen, pp. 333–346.

111. Klang, S. H., Parnas, M., and Benita, S. (1998), Emulsions as drug carriers—Possibilities, limitations, and future perspectives, in Muller, R. H., Benita, S., and Bohm, H. L., Eds., *Emulsions and Nanosuspensions for the Formulation of Poorly Soluble drugs*, Medpharm, Stuttgart, pp. 31–65.

112. Yang, S. C., and Benita, S. (2000), Enhanced absorption and drug targeting by positively charged submicron emulsions, *Drug Dev. Res.*, 50, 476–486.

113. Takino, T., Koreeda, N., Nomura, T., Sakaeda, T., Yamashita, F., Takakura, Y., and Hashida, M. (1998), Control of plasma cholesterol-lowering action of probucol with various lipid carrier systems, *Biol. Pharm. Bull.*, 21, 492–497.

114. Klang, S. H, Frucht-Pery, J., Hoffman, A., and Benita, S. (1994), Physicochemical characterization and acute toxicity evaluation of a positively-charged submicron emulsion vehicle, *J. Pharm. Pharmacol.*, 46, 986–993.

115. Korner, D., Benita, S., Albrecht, G., and Baszkin, A. (1994), Surface properties of mixed phospholipid–stearylamine monolayers and their interaction with a non-ionic surfactant (poloxamer), *Coll. Surf. B BioInterf.*, 3, 101–109.

116. Benita, S., and Levy, M. Y. (1993), Submicron emulsions as colloidal drug carriers for intravenous administration: Comprehensive physicochemical characterization, *J. Pharm. Sci.*, 82, 1069–1079.

117. Tamilvanan, S., and Benita, S. (2004), The potential of lipid emulsion for ocular delivery of lipophilic drugs, *Eur. J. Pharm. Biopharm.*, 58, 357–368.

118. Lee, V. H. L., and Robinson, J. R. (1986), Review: Topical ocular drug delivery: Recent developments and future challenges, *J. Ocul. Pharmacol.*, 2, 67–108.

119. Jumaa, M., and Müller, B. W. (1998), The effect of oil components and homogenization condition on the physicochemical properties and stability of parenteral fat emulsions, *Int. J. Pharm.*, 163, 81–89.

120. Constantinides, P. P., Tustian, A., and Kessler, D. R. (2004), Tocol emulsions for drug solubilization and parenteral delivery, *Adv. Drug Deliv. Rev.*, 56, 1243–1255.

121. Constantinides, P. P., Han, J., and Davis, S. S. (2006), Advances in the use of tocols as drug delivery vehicles, *Pharm. Res.*, 23, 243–255.

122. Lundberg, B. B. (1997), A submicron lipid emulsion coated with amphiphathic polyethylene glycol for parenteral administration of paclitaxel (taxol), *J. Pharm. Pharmacol.*, 49, 16–21.

123. Han, J., Davis, S. S., Papandreou, C., Melia, C. D., and Washington, C. (2004), Design and evaluation of an emulsion vehicle for paclitaxel. I. Physicochemical properties and plasma stability, *Pharm. Res.*, 21, 1573–1580.

124. Muchtar, S., and Benita, S. (1994), Emulsions as drug carriers for ophthalmic use, *Coll. Surf. A Physicochem. Eng. Aspects*, 91, 181–190.

125. Ogawa, S., Decker, E. A., and McClements, D. J. (2002), Production and characterization of o/w emulsions containing cationic droplets stabilized by lecithin-chitosan membranes, *J. Agric. Food Chem.*, 51, 2606–2812.

126. Sonne, M. R. (2001), Tocopherol compositions for delivery of biologically active agents, USPTO Patent Full-Text and Image Database No. 6,193,985, Dumex, Copenhagen, pp. 1–21.

127. Sznitowska, M., Janicki, S., Dabrowska, E. A., and Gajewska, M. (2002), Physicochemical screening of antimicrobial agents as potential preservatives for submicron emulsions, *Eur. J. Pharm. Sci.*, 15, 489–495.

128. Jumaa, M., Furkert, F. H., and Müller, B. W. (2002), A new lipid emulsion formulation with high antimicrobial efficacy using chitosan, *Eur. J. Pharm. Biopharm.*, 53, 115–123.

129. Pongcharoenkiat, N., Narsimhan, G., Lyons, R. T., and Hem, S. L. (2002), The effect of surface charge and partition coefficient on the chemical stability of solutes in o/w emulsions, *J. Pharm. Sci.*, 91, 559–570.

130. Han, J., and Washington, C. (2005), Partition of antimicrobial additives in an intravenous emulsion and their effect on emulsion physical stability, *Int. J. Pharm.*, 288, 263–271.

131. Swietlikowska, D. W., and Sznitowska, M. (2006), Partitioning of parabens between phases of submicron emulsions stabilized with egg lecithin, *Int. J. Pharm.*, 312, 174–178.

132. Myers, R. A., and Stella, V. J. (1992), Systemic bioavailability of penclomedine (NSC-338720) from oil-in-water emulsion administered intraduodenally to rats, *Int. J. Pharm.*, 78, 217–226.

133. Ilan, E., Amselem, S., Weisspapir, M., Schwartz, J., Yogev, A., Zawoznik, E., and Friedman, D. (1996), Improved oral delivery of desmopressin via a novel vehicle: Mucoadhesive submicron emulsion, *Pharm. Res.*, 13, 1083–1087.

134. Palin, K. J., Phillips, A. J., and Ning, A. (1986), The oral absorption of cefaxitin from oil and water emulsion vehicles in rats, *Int. J. Pharm.*, 33, 99–104.

135. Kimura, T., Takeda, K., Kageyn, A., Toda, M., Kurpsaki, Y., and Nkayama, T. (1989), Intestinal absorption of dolichol from emulsions and liposomes in rats, *Chem. Pharm. Bull.*, 37, 463–466.

136. Charman, S. A., Charman, W. N., Rogge, M. C., Wildon, T. D., Dutko, F. J., and Pouton, C. W. (1992), Self-emulsifying drug delivery systems: Formulation and biopharmaceutical evaluation of an investigational lipophilic compound, *Pharm. Res.*, 9, 87–93.

137. Gursoy, R. N., and Benita, S. (2004), Self-emulsifying drug delivery systems (SEDDS) for improved oral delivery of lipophilic drugs, *Biomed. Pharmacother. Dossier: Drug Deliv. Drug Efficacy*, 58, 173–182.

138. Takino, T., Nagahama, E., Sakaeda (nee Kakutani), T., Yamashita, F., Takakura, Y., and Hashida, M. (1995), Pharmacokinetic disposition analysis of lipophilic drugs injected with various lipid carriers in the single-pass rat liver perfusion system, *Int. J. Pharm.*, 114, 43–54.

139. Tamilvanan, S., Schmidt, S., Muller, R. H., and Benita, S. (2005), In vitro adsorption of plasma proteins onto the surface (charges) modified-submicron emulsions for intravenous administration, *Eur. J. Pharm. Biopharm.*, 59, 1–7.

140. Alany, R. G., Rades, T., Nicoll, J., Tucker, I. G., and Davies, N. M. (2006), W/O microemulsions for ocular delivery: Evaluation of ocular irritation and precorneal retention, *J. Controlled Release*, 111, 145–152.

141. Stjernschantz, J., and Astin, M. (1993), Anatomy and physiology of the eye: Physiological aspects of ocular drug therapy, in Edman, P., Ed., *Biopharmaceutics of Ocular Drug Delivery*, CRC Press, Boca Raton, FL, pp. 1–25.

142. Lee, V. H. L. (1993), Precorneal, corneal and postcorneal factors, in Mitra, A. K. Ed., *Ophthalmic Drug Delivery System*, Marcel Dekker, New York, pp. 59–81.

143. Järvinen, K., Järvinen, T., and Urtti, A. (1995), Ocular absorption following topical delivery, *Adv. Drug Deliv. Rev.*, 16, 3–19.

144. Prausnitz, M. R., and Noonan, J. S. (1998), Permeability of cornea, sclera and conjunctiva: A literature analysis for drug delivery to eye, *J. Pharm. Sci.*, 87, 1479–1488.

145. Washington, N., Washington, C., and Wilson, C. (2001), Ocular drug delivery, in Washington, N., Washington, C., and Wilson, C., Eds., *Physiological Pharm., Barriers to Drug Absorption*, 2nd ed., Taylor & Francis, New York, pp. 250–270.

146. Ludwig, A. (2005), The use of mucoadhesive polymers in ocular drug delivery, *Adv. Drug Deliv. Rev.*, 57, 1595–1639

147. Chien, D. S., Homsy, J. J., Gluchowski, C., and Tang-Liu, D. D. (1990), Corneal and conjunctival/scleral penetration of *p*-aminoclonidine, AGN 190342, and clonidine in rabbit eyes, *Curr. Eye Res.*, 9, 1051–1059.

148. Ahmed, I., and Patton, T. F. (1985), Importance of the noncorneal absorption route in topical ophthalmic drug delivery, *Investigative Ophthalmol. Vis. Sci.*, 26, 584–587.

149. Kaur, I. P., and Smitha, R. (2002), Penetration enhancers and ocular bioadhesives: Two new avenues for ophthalmic drug delivery, *Drug Dev. Ind. Pharm.*, 28, 353–369.

150. Dey, S., Anand, B. S., Patel, J., and Mitra, A. K. (2003), Transporters/receptors in the anterior chamber: Pathways to explore ocular drug delivery strategies, *Expert Opin. Biol. Ther.*, 3, 23–44.

151. Beilin, M., Bar-Ilan, A., and Amselem, S. (1995), Ocular retention time of submicron emulsion (SME) and the miotic response to pilocarpine delivered in SME. *Investigative Ophthalmol. Vis. Sci.*, 36, S166.

152. Aiache, J. M., el Meski, S., Beyssac, E., and Serpin, G. (1997), The formulation of drug for ocular administration, *J. Biomater. Appl.*, 11, 329–348.

153. Maurice, D. M., and Mishima, S. (1984), Ocular pharmacokinetics, in Sears, M. L., Ed., *Pharmacology. of the Eye*, Springer Verlag, New York, pp. 20–116.

154. Muchtar, S., Almog, S., Torracca, M. T., Saettone, M. F., and Benita, S. (1992), A submicron emulsion as ocular vehicle for delta-8-tetrahydrocannabinol: Effect on intraocular pressure in rabbits, *Ophthalm. Res.*, 24, 142–149.

155. Naveh, N., Weissman, C., Muchtar, S., Benita, S., and Mechoulam, R. (2000), A submicron emulsion of HU-211, a synthetic cannabinoid, reduces intraocular pressure in rabbits, *Graefe's Arch. Clin. Exp. Ophthalmol.*, 238, 334–338.

156. Naveh, N., Muchtar, S., and Benita, S. (1994), Pilocarpine incorporated into a submicron emulsion vehicle causes an unexpectedly prolonged ocular hypotensive effect in rabbits, *J. Ocul. Pharmacol.*, 10, 509–520.

157. Zurowska-Pryczkowska, K., Sznitowska, M., and Janicki, S. (1999), Studies on the effect of pilocarpine incorporation into a submicron emulsion on the stability of the drug and the vehicle, *Eur. J. Pharm. Biopharm.*, 47, 255–260.

158. Sznitowska, M., Janicki, S., Zurowska-Pryczkowska, K., and Mackiewicz, J. (2001), In vivo evaluation of submicron emulsions with pilocarpine: The effect of pH and chemical form of the drug, *J. Microencapsul.*, 18, 173–181.

159. Calvo, P., Alonso, M. J., Vila-Jato, J. L., and Robinson, J. R. (1996), Improved ocular bioavailability of indomethacin by novel ocular drug carriers, *J. Pharm. Pharmacol.*, 48, 1147–1152.

160. Calvo, P., Vila-Jato, J. L., and Alonso, M. J. (1996), Comparative in vitro evaluation of several colloidal systems, nano-particles, nanocapsules, and nanoemulsions as ocular drug carriers, *J. Pharm. Sci.*, 85, 530–536.

161. Anselem, S., Beilin, M., and Garty, N. (1993), Submicron emulsion as ocular delivery system for adaprolol maleate, a soft β-blocker, *Pharm. Res.*, 10, S205.

162. Melamed, S., Kurtz, S., Greenbaum, A., Haves, J. F., Neumann, R., and Garty, N. (1994), Adaprolol maleate in submicron emulsion, a novel soft β-blocking agent, is safe and effective in human studies, *Investigative Ophthalmol. Vis. Sci.*, 35, 1387–1390.

163. Aviv, H., Friedman, D., Bar-Ilan, A., and Vered, M. (1995), Submicron emulsions as ocular drug delivery vehicles, U.S. Patent 5,496,811, March 5, 1996.

164. Garty, N., Lusky, M., Zalish, M., Rachmiel, R., Greenbaum, A., Desatnik, H., Neumann, R., Howes, J. F., and Melamed, S. (1994), Pilocarpine in submicron emulsion formulation for treatment of ocular hypertension: A phase II clinical trial, *Investigative Ophthalmol. Vis. Sci.*, 35, 2175.

165. Ding, S., Tien, W., and Olejnik, O. (1995), Nonirritating emulsions for sensitive tissue, U.S. Patent 5,474,979 (to Allergan), December 12.

166. Ding, S., and Olejnik, O. (1997), Cyclosporin ophthalmic oil/water emulsions. Formulation and characterization, *Pharm. Res.*, 14, S41.

167. Acheampong, A. A., Shackleton, M., Tang-Liu, DD-S., Ding, S., Stern, M. E., and Decker, R. (1999), Distribution of cyclosporin A in ocular tissues after topical administration to albino rabbits and beagle dogs, *Curr. Eye Res.*, 18, 91–103.
168. Stevenson, D., Tauber, J., and Reis, B. L. (2000), Efficacy and safety of cyclosporin A ophthalmic emulsion in the treatment of moderate-to-severe dry eye disease: A dose-ranging, randomized trial. The Cyclosporin A Phase 2 Study Group, *Ophthalmology*, 107, 967–974.
169. Sall, K., Stevenson, O. D., Mundorf, T. K., and Reis, B. L. (2000), Two multicenter, randomized studies of the efficacy and safety of cyclosporin ophthalmic emulsion in moderate-to-severe dry eye disease. CsA Phase 3 Study Group, *Ophthalmology*, 107, 631–639.
170. Turner, K., Pflugfelder, S. C., Ji, Z., Feuer, W. J., Stern, M., and Reis, B. L. (2000), Interleukin-6 levels in the conjunctival epithelium of patients with dry eye disease treated with cyclosporine ophthalmic emulsion, *Cornea*, 19, 492–496.
171. Kunert, K. S., Tisdale, A. S., Stern, M. E., Smith, J. A., and Gipson, I. K. (2000), Analysis of topical cyclosporin treatment of patients with dry eye syndrome: Effect on conjunctival lymphocytes, *Arch. Ophthalmol.*, 118, 1489–1496.
172. Brignole, F., Pisella, P. J., Saint-Jean, M. De., Goldschild, M., Goguel, A., and Baudouin, C. (2001), Flow cytometric analysis of inflammatory markers in KCS: 6-month treatment with topical cyclosporin A. *Investigative Ophthalmol. Vis. Sci.*, 42, 90–95.
173. Small, D. S., Acheampong, A., Reis, B., Stern, K., Stewart, W., Berdy, G., Epstein, R., Foerster, R., Forstot, L., and Tang-Liu, D. D.-S. (2002), Blood concentrations of cyclosporin A during long-term treatment with cyclosporin A ophthalmic emulsions in patients with moderate to severe dry eye disease, *J. Ocul. Pharmacol. Ther.*, 18, 411–418.
174. Galatoire, O., Baudouin, C., Pisella, P. J., and Brignole, F. (2003), Flow cytometry in impression cytology during keratoconjunctivitis sicca: Effects of topical cyclosporin A on HLA DR expression, *J. Francais d'Ophtalmol.*, 26, 337–343.
175. Tang-Liu, D. D., and Acheampong, A. (2005), Ocular pharmacokinetics and safety of ciclosporin, a novel topical treatment for dry eye, *Clin. Pharmacokinet.*, 44, 247–261.
176. Hom, M. M. (2006), Use of cyclosporine 0.05% ophthalmic emulsion for contact lens-intolerant patients, *Eye Contact Lens: Sci. Clin. Pract.*, 32, 109–111.
177. Sall, K. N., Cohen, S. M., Christensen, M. T., and Stein, J. M. (2006), An evaluation of the efficacy of a cyclosporine-based dry eye therapy when used with marketed artificial tears as supportive therapy in dry eye, *Eye Contact Lens: Sci. Clin. Pract.*, 32, 21–26.
178. Klang, S. H., Siganos, C. S., and Benita, S. (1999), Evaluation of a positively-charged submicron emulsion of piroxicam in the rabbit corneum healing process following alkali burn, *J. Controlled Release*, 57, 19–27.
179. Tamilvanan, S., Khoury, K., Gilhar, D., and Benita, S. (2001), Ocular delivery of cyclosporin A. I. Design and characterization of cyclosporin A-loaded positively charged submicron emulsion, *STP Pharma Sci.*, 11, 421–426.
180. Hussain, A. A. (1998), Intranasal drug delivery, *Adv. Drug Deliv. Rev.*, 29, 39–49.
181. Kararli, T. T., Needham, T. E., Schoenhard, G., Baron, D. A., Schmidt, R. E., Katz, B., and Belonio, B. (1992), Enhancement of nasal delivery of a rennin inhibitor in the rat using emulsion formulations, *Pharm. Res.*, 9, 1024–1028.
182. Ko, K. Y., Needham, T. E., and Zia, H. (1998), Emulsion formulations of testosterone for nasal administration, *J. Microencapsul.*, 15, 197–205.
183. Aikawa, K., Matsumoto, K., Uda, H., Tanaka, S., Shimamura, H., Aramaki, Y., and Tsuchiya, S. (1998), Prolonged release of drug from o/w emulsion and residence in rat nasal cavity, *Pharm. Dev. Technol.*, 3, 461–469.

184. Mitra, R., Pezron, I., Chu, W. A., and Mitra, A. K. (2000), Lipid emulsions as vehicles for enhanced nasal delivery of insulin, *Int. J. Pharm.*, 205, 127–134.

185. Kim, T. W., Chung, H., Kwon, I. C., Sung, H. C., and Jeong, S. Y. (2000), In vivo gene transfer to the mouse nasal cavity mucosa using a stable cationic lipid emulsion, *Mol. Cells*, 10, 142–147.

186. Jenkins, P. (1999), Mucosal vaccine delivery, *Exp. Opin. Ther. Patients*, 9, 255–262.

187. Benita, M. B., Oudshoorn, M., Romeijn, S., Meijgaarden, K. V., Koerten, H., Meulen, H. V. D, Lambert, G., Ottenhoff, T., Benita, S., Junginger. H., and Borchard, G. (2004), Cationic submicron emulsions for pulmonary DNA immunization, *J. Controlled Release*, 100, 145–155.

188. El-Hameed, M. D. A., and Kellaway, I. W. (1997), Preparation and in vitro characterization of mucoadhesive polymeric microspheres as intranasal delivery systems, *Eur. J. Pharm. Biopharm.*, 44, 53–60.

189. Friedman, D. I., Schwarz, J. S., and Weisspapir, M. (1995), Submicron emulsion vehicle for enhanced transdermal delivery of steroidal and nonsteroidal antiinflammatory drugs, *J. Pharm. Sci.*, 84, 324–329.

190. Schwarz, J. S., Weisspapir, M. R., and Friedman, D. I. (1995), Enhanced transdermal delivery of diazepam by submicron emulsion (SME) creams, *Pharm. Res.*, 12, 687–692.

191. Realdaon, N., and Ragazzi, E. (1998), Study of drug availability from o/w emulsions obtained in different manufacturing conditions, in Proceedings of 2nd World Meeting APGI/APV, 775–776.

192. Welin-Berger, K., and Bergenstahl, B. (2000), Inhibition of Ostwald ripening in local anesthetic emulsions—By using hydrophobic excipients in the disperse phase, *Int. J. Pharm.*, 200–202, 249–260.

193. Welin-Berger, K., Neelissen, J. A., and Bergenstahl, B. (2001), The effect of rheological behaviour of a topical anaesthetic formulation on the release and permeation rates of the active compound, *Eur. J. Pharm. Sci.*, 13, 309–318.

194. Dutta, S. (1999), Use of eutectic mixture of local anesthetics in children, *Ind. J. Pediatr.*, 66, 707–715.

195. Cordoni, A., and Cordoni, L. E. (2001), Eutectic mixture of local anesthetics reduces pain during intravenous catheter insertion in the pediatric patient, *Clin. J. Pain*, 17, 115–118.

196. Ezra, R., Benita, S., Ginsberg, I., and Kohen, R. (1996), Prevention of oxidative damage in fibroblast cell cultures and rat skin by positively-charged submicron emulsion of a-tocopherol, *Eur. J. Pharm. Biopharm.*, 42, 291–298.

197. Piemi, M. P., Korner, D., Benita, S., and Marty, J. P. (1999), Positively and negatively charged submicron emulsions for enhanced topical delivery of antifungal drugs, *J. Controlled Release*, 58, 177–187.

INDEX

Pharmaceutical Manufacturing Handbook: Production and Processes, edited by Shayne Cox Gad